2011—2020 中国矿物岩石地球化学研究进展

中国矿物岩石地球化学学会 主编

科学出版社
北京

内 容 简 介

本书收集了我国矿物学、岩石学、地球化学及沉积学等学科领域知名学者2020—2021年在《矿物岩石地球化学通报》上刊登的"学科发展十年进展"系列论文，共33篇。全书较全面、系统地综述了21世纪第二个十年（2011—2020年）我国在矿物学、岩石学、地球化学、沉积学及其相关学科以及技术、模拟领域（如微束微区原位分析、高温高压实验和计算模拟等）的主要进展和创新成果，并对未来的学科发展趋势进行了展望。

本书可供从事矿物学、岩石学、地球化学、沉积学、矿床学、能源地质学、地球动力学，以及环境科学和材料科学等领域的研究人员、高等院校师生和产业部门的科技工作者参阅。

审图号：GS（2022）1935号

图书在版编目（CIP）数据

2011—2020中国矿物岩石地球化学研究进展 / 中国矿物岩石地球化学学会主编. —北京：科学出版社，2022.10
ISBN 978-7-03-073292-7

Ⅰ. ①2… Ⅱ. ①中… Ⅲ. ①矿物学–研究进展–中国–2011–2020②岩石学–研究进展–中国–2011–2020③地球化学–研究进展–中国–2011–2020 Ⅳ. ①P5

中国版本图书馆CIP数据核字（2022）第179063号

责任编辑：韦 沁 / 责任校对：何艳萍
责任印制：吴兆东 / 封面设计：北京图阅盛世

科学出版社 出版
北京东黄城根北街16号
邮政编码：100717
http://www.sciencep.com

北京中科印刷有限公司 印刷
科学出版社发行 各地新华书店经销

*

2022年10月第 一 版　开本：889×1194 1/16
2022年10月第一次印刷　印张：45 1/4
字数：1 098 000
定价：668.00元
（如有印装质量问题，我社负责调换）

前　　言

中国矿物岩石地球化学学会自成立以来，一直有一个很好的学术传统，即对学科发展进行"十年一回顾"的总结。学会邀请各专业委员会的知名学者分析、总结并撰写分支学科的十年进展、理论创新和技术突破，以记录矿物学、岩石学、地球化学及沉积学领域的历史发展足迹。

从 2020 年开始，学会秘书处先后收到 40 余篇反映过去十年研究进展的稿件，《矿物岩石地球化学通报》已从中挑选出 33 篇陆续刊出。学会学术工作委员会于 2021 年 5 月 21～23 日在南京专门组织召开了"矿物岩石地球化学十年进展论坛（2011—2020）"，对各分支学科报告进行了交流和讨论，取得了非常好的反响。

进入 21 世纪第二个十年（2011—2020），微束微区原位分析、高温高压实验和计算模拟等技术的广泛应用极大推进了我国矿物学、岩石学与地球化学的学科发展。过去十年，无论是对地球物质的组成结构和特性本身研究，还是对地球多圈层物质构成与相互作用影响地球宜居性的研究，都取得了显著的进步。这期间，作为地球系统科学的重要组成部分，我国矿物学、岩石学和地球化学在迈向深海、深地、深空研究方面，实施了一些重大科技计划，部分方向与国际并行，有些方向处于国际领跑地位。

在矿物学领域，我国在自然元素矿物大类新矿物的发现方面居于世界前列，在稀土新矿物和冲击变质超高压新矿物的发现和研究方面也取得了令人瞩目成果。在矿物晶体生长理论、高压矿物学、矿物表-界面科学、黏土矿物学等方向均取得了重要进展，提出的"纳米晶粒堆砌生长"机制，突破了基于单原子堆积的传统理论。在矿物材料研究方面，新型矿物材料不断涌现，应用领域不断拓展。矿物成分标型理论的长足进步极大促进了对成岩成矿过程的精细刻画、热液成矿规律与找矿潜力的评价、成矿类型的甄别与划分。地表多圈层交互作用产物"矿物膜"的环境属性、矿物与微生物交互作用的环境效应、纳米矿物及其环境功能、生物矿化作用及其环境效应以及人体病理性矿化作用等方面的研究则凸显了矿物的巨大环境效应。在人工合成超强、超绝热储热、超导以及仿生矿物功能材料方面，工业岩矿（宝石）资源开发、矿物基固废资源化利用等领域的研究进一步展示了矿物对人类社会发展的重要意义。

在岩石学领域，花岗岩研究向精细化深度发展，并与数值模拟、大数据、地球物理等研究方法紧密结合，在探讨花岗岩浆的源区、陆壳物质再造以及应用实验岩石学、热力学模拟和金属稳定同位素示踪岩浆演化过程等方面取得新进展。变质岩石学在超高压和超高温变质作用、相平衡模拟、变质矿物生长和定年等方面有许多新进展，确定了中国主要超高压变质带的分布特征，识别并系统研究了华北克拉通超高温变质地体，建立了基于 ACF 组分分析的变质基性岩完整相平衡关系，确定了俯冲带高压-超高压变质流体活动及其元素迁移效应。在火山与地球内部化学研究方面，通过对火山活动及其特征的研究，探索了中国大陆火山作用的成因及控制因素，揭示地球内部物质组成、物理化学条件以及相关的地球动力学机理，为防灾、减灾研究提供了基础数据和依据。在对地幔直接样品、幔源岩浆以及大陆和大洋岩石圈等地幔矿物和岩石地球化学等研究方面取得了不少进展，特别是在俯冲带结构和过程、超高压岩石部分熔融、造山带橄榄岩与壳源流体交代作用、俯冲物质再循环及壳幔相互作用等化学地球动力学研究方面都取得突破性成果。

在地球化学及其应用领域，引进了大量高性能新型微束分析仪器设备，建立了一批高规格的实验室。在元素定量和同位素比值分析方面建立了大量新技术新方法，并在地质样品前处理、参考物质研制和定值、分析仪器和关键部件研制等方面取得原创成果。同位素理论和计算研究方面达到了量子化学水平，基于精密的量子化学第一性原理计算，率先发展了超冷体系同位素分馏等同位素理论和计算方法，为多种金属稳定同位素体系提供了大量的平衡分馏系数。在示踪大陆深俯冲/折返过程中地壳部分熔融作用、汇聚板块边缘不同类型壳幔相互作用及其化学地球动力学，在精细刻画成矿作用时代、成矿物质来源和流体演化，在探测深海金属资源的富集特征、分布规律、赋存状态、成矿机制，在确定煤中元素循环、古地温、古气候、古植被类型，在建立油气成藏年代学方法、完善天然气成藏示踪同位素体系，在采用传统化探和非传统化探方法"攻深找盲"拓展找矿空间等方面取得一系列新进展。环境地球化学方面，在古气候与全球变化研究、污染物的现代地球化学过程研究，以及通过对典型火山和地热区等释放的温室气体进行了专项调查，在地球化学组成和释放通量等方面取得了一系列进展。天体地球化学方面，在太阳星云起源与演化、火星与月球等类地行星形成与宜居性研究、小行星岩浆作用和含水蚀变及后期撞击历史等众多领域取得了重要成果。金属稳定同位素是过去十年发展最为迅猛的方向之一，其应用业已拓展到了地球科学的众多研究领域，为解决这些领域内的科学问题提供了新视角、促进了新理论形成。我国学者在同位素分馏理论、分析技术方法和应用研究等方面均取得了系列创新性成果，为技术稳定同位素地球化学的迅猛发展做出了重要贡献，并已成为国际上的重要研究力量。

过去十年，国内多家单位建成了国际水平的高温高压实验平台，并加强了与原位分析测试技术的结合，在矿物、岩石、熔体和流体的热学、电学、弹性、流变、相变、扩散、元素配分等方面开展了大量前沿性研究，深化了对地球深部物质的物理化学性质以及挥发分的赋存和效应的认识，我国的实验研究队伍已成为国际高温高压实验地球科学领域的一支重要力量。计算技术快速发展，在第一性原理计算、机器学习等方面的研究水平整体处于国际第一方阵，促进了对地球深部物质的微观结构和物理性质以及元素赋存状态的认识，在矿物表面结构和表面过程研究方面取得显著进步。数学地球科学取得了跨越式发展，其中引人注目的进步是将大数据挖掘和人工智能算法引入到地学研究中。以多层卷积神经网络算法为核心的深度学习被用于岩石标本鉴定、地球化学异常识别和智能找矿。大数据与数学地球科学专业委员会于2016年正式成立，是标志着中国地学研究进入大数据与人工智能时代的里程碑事件。在沉积学方面，盆地动力学、层序地层学、源-汇系统、沉积体系和沉积相、古地理学、前寒武纪沉积学、现代沉积环境、深时地质与沉积学、生物沉积学等领域的研究取得了重要进展。深时古气候与全球变化、古环境变化及第四纪沉积等领域的研究与国际同步发展或已迈入国际研究前沿，并形成了陆相层序地层和盆地沉积充填动力学、含油气盆地沉积学等独具特色的研究方向。

现将这些陆续刊出的33篇论文汇总编辑成《2011—2020中国矿物岩石地球化学研究进展》一书，以飨读者。衷心祝愿本书的出版能起到温故而知新的作用，更好地推动我国矿物学、岩石学、地球化学及沉积学学科下个十年的持续健康发展。

本次对我国学科发展近十年进展总结工作的顺利完成，是与学会学术工作委员会以及各专业委员会和很多学者的支持是分不开的，特别是学术工作委员会主任郑永飞院士，认真筹划、组织评审，学术工作委员会的郑建平、夏群科、吴元保、倪怀玮积极协助，这些辛勤的付出将与本书一道永远留存。

中国矿物岩石地球化学学会
2021年8月

目 录

前言

矿物结构与矿物物理研究进展 …………………………… 何宏平　朱建喜　陈　锰　陶　奇
　　　　　　　　　　　　　　　　　　　　　　　　　谭大勇　梁晓亮　鲜海洋　1

环境矿物学研究进展 ………………………………………………… 鲁安怀　王长秋　李　艳　19

矿物材料研究进展 ………………………………… 吕国诚　廖立兵　李雨鑫　田林涛　刘　昊　39

新矿物的发现与研究进展 ……………………………………………………………… 蔡剑辉　53

矿物晶体结构与晶体化学研究进展 …………………………………………………… 李国武　76

应用矿物学研究进展 …………………………… 董发勤　谭道永　王进明　徐龙华
　　　　　　　　　　　　　　　　　　　李旭娟　丁文金　胡志波　黄　腾　103

成因矿物学与找矿矿物学研究进展 ………… 申俊峰　李胜荣　黄绍锋　卿　敏　张华锋　许　博　120

火山学和地球内部化学研究进展 …………………………………… 徐义刚　郭正府　刘嘉麒　136

地幔矿物学岩石学地球化学进展 …………………………………………… 张宏福　陈立辉　152

大陆俯冲带超高压变质岩部分熔融与壳幔相互作用研究进展 ……… 赵子福　陈仁旭　陈伊翔
　　　　　　　　　　　　　　　　　　　　　　　　　　　　　　戴立群　郑永飞　182

花岗岩研究进展 ……………………………………………… 徐夕生　王孝磊　赵　凯　杜德宏　208

变质岩研究进展 ……………………… 张贵宾　刘　良　魏春景　肖益林　焦淑娟　吕　增　张立飞　222

沉积学发展现状与趋势 …………………………………………………………… 王成善　林畅松　244

古地理学主要研究进展 …………… 郑秀娟　杜远生　朱筱敏　刘招君　胡　斌　吴胜和　邵龙义
　　　　　　　　　　　　　　旷红伟　罗静兰　钟大康　李　华　何登发　朱如凯　鲍志东　258

亚洲大陆边缘沉积学研究进展 …………… 石学法　乔淑卿　杨守业　李景瑞　万世明　邹建军
　　　　　　　　　　　　　　熊志方　胡利民　姚政权　董林森　王昆山　刘升发　刘焱光　282

微束分析测试技术进展 ………… 陈　意　胡兆初　贾丽辉　李金华　李秋立　李晓光　李展平
　　　　　　　　　龙　涛　唐　旭　王　建　夏小平　杨　蔚　原江燕　张　迪　李献华　302

岩矿分析测试研究进展 ……………………… 刘勇胜　屈文俊　漆　亮　袁洪林　黄　方
　　　　　　　　　　　　　　　　　　　　　　杨岳衡　胡兆初　朱振利　张　文　340

非传统稳定同位素地球化学研究进展 …………………… 韦刚健　黄　方　马金龙　邓文峰
　　　　　　　　　　　　　　　　　　　　　　　　　于慧敏　康晋霆　陈雪霏　367

团簇同位素地球化学研究进展 ……………………………………… 邓文峰　郭炀锐　韦刚健　414

实验地球科学研究进展 …… 章军锋　倪怀玮　杨晓志　毛　竹　张宝华　熊小林　侯　通　许文良　430

同位素效应理论和计算研究进展 …………………………………………………… 刘　耘　444

环境地球化学研究进展 ………………………… 冯新斌　曹晓斌　付学吾　洪　冰　关　晖
　　　　　　　　　　　　　　　　　　　李　平　王敬富　王仕禄　张　干　赵时真　463

· iii ·

章节	作者	页码
地质源温室气体释放研究概述	郑国东　赵文斌　陈　志　胥　旺　宋之光　李　琦 徐　胜　郭正府　马向贤　梁明亮　王云鹏	501
流体包裹体研究进展	倪　培　范宏瑞　潘君屹　迟　哲　崔健铭	526
矿床地球化学研究进展	钟　宏　宋谢炎　黄智龙　蓝廷广　柏中杰 陈　伟　朱经经　阳杰华　谢卓君　王新松	544
铀矿地质科技主要进展	李子颖　秦明宽　范洪海　蔡煜琦　程纪星 郭冬发　叶发旺　范　光　刘晓阳	572
天然气地球化学研究进展	刘文汇　王　星　田　辉　郑国东　王晓锋　陶　成　刘　鹏	587
深海矿产研究进展	石学法　符亚洲　李　兵　黄　牧　任向文　刘季花　于　淼　李传顺	605
煤有机地球化学研究进展	唐跃刚　王绍清　郭　鑫　李瑞青　林雨涵	621
应用地球化学研究进展	龚庆杰　夏学齐　刘宁强	646
数学地球科学跨越发展的十年	周永章　左仁广　刘　刚　袁　峰　毛先成 郭艳军　肖　凡　廖　杰　刘艳鹏	665
勘查地球化学数据处理研究进展	左仁广　王　健　熊义辉　王子烨	686
陨石学与天体化学研究进展	缪秉魁　胡　森　陈宏毅　张川统 夏志鹏　黄丽霖　薛永丽　谢兰芳	700

矿物结构与矿物物理研究进展*

何宏平[1,2,3]，朱建喜[1,2,3]，陈　锰[1,2]，陶　奇[1,2]，谭大勇[1,2]，梁晓亮[1,2]，鲜海洋[1,2]

1. 中国科学院 广州地球化学研究所，中国科学院矿物学与成矿学重点实验室，广州 510640；
2. 中国科学院　广州地球化学研究所，广东省矿物物理与材料研究开发重点实验室，广州 510640；3. 中国科学院大学，北京 100049

摘　要：2011—2020十年间，微束微区、原位分析、高温高压和计算模拟等技术的快速发展极大地推进了矿物学研究向分子、原子尺度深入。我国学者在矿物相转变与晶体生长理论、矿物物理、矿物微结构、矿物表-界面过程等领域的研究中取得了系列创新成果，主要包括：发现了高压新矿物以及矿物相变的新机制，提出了纳米颗粒附着晶化和非晶质-结晶质转化的矿物晶体生长途径与理论，揭示了矿物表面反应性的结构本质与矿物表-界面反应的微观机理，运用计算模拟方法获得了矿物原子尺度的局域结构与性质。本文从矿物晶体生长理论、高压矿物学、矿物表-界面作用和黏土矿物学等方向综述了近十年我国矿物学研究的进展，并展望了矿物学的发展方向。

关键词：矿物晶体生长　高压矿物学　矿物表面　黏土矿物

0　引　言

矿物是固体地球的基本组成单元，是地球演化过程中一系列物理、化学和生物作用的产物。矿物不仅直接参与了地球演化的整个地质过程，而且记录了地球演化过程的重要信息。因此，矿物是我们解译地质地球化学过程、认识地球的形成与演化，乃至生命起源等重要过程与重大事件的关键信息载体。

已有研究表明，地球的演化过程也是矿物多样性不断丰富的过程（Hazen et al.，2008），由原始地球的12种组成矿物逐步演化为目前已知的5 500多种矿物。另外，随着实验手段和表征技术的快速发展及其在矿物学研究中的应用，矿物多样性的内涵得到了极大丰富，从矿物结构与组成为标志的矿物种属多样性向矿物晶体生长机制、矿物表-界面过程、矿物物理化学特性等领域拓展。特别是2011—2020这十年间，随着微束微区、原位分析、高温高压和计算模拟等技术的快速发展，促进了矿物学研究从宏观-介观尺度向分子-原子水平的跨越，同时推进了矿物学与其他学科的深度交叉与融合，为解决重大地球科学问题提供了关键支撑。这十年间，我国学者在矿物学研究领域非常活跃，在矿物晶体化学、矿物表-界面过程、矿物资源利用等传统与新兴领域皆取得了长足的进展。笔者从矿物晶体生长理论、高压矿物学、矿物表-界面作用、黏土矿物学等领域综述我国学者这十年间的主要研究进展。

1　矿物晶体生长理论

晶体成核与生长理论是矿物学研究的基础。经典均相成核理论从热力学角度将成核体系的自由能变化分解成两部分：因成核导致的母相化学势变化，以及晶核与母相间的界面能变化（Becker and Döring，1935）。母相中要形成稳定的晶核必须克服一定的能垒，只有形成的晶核大于某一临界尺寸，矿物的结晶

---

* 原文"矿物结构与矿物物理研究进展综述（2011~2020年）"刊于《矿物岩石地球化学通报》2020年第39卷第4期，本文略有修改。

生长才能发生。Bai 等（2019）使用一系列固定尺寸的纳米颗粒去探测"临界冰核"，获得了"临界冰核"的尺寸。这是经典成核理论提出一百多年来首次在实验上被证实。此外，非均相成核与均相成核类似，即在均相成核基础上，考虑成核基底/杂质与晶核、母相间的相互作用。

经典晶体生长理论主要包括层生长理论（Kossel-Stranski theory）和螺旋生长（BCF）理论。层生长理论（即二维成核理论）认为晶体生长优先发生于成键数最多的生长位（通常为扭折位），当这些最佳生长位被填满时，就需要在光滑晶面上形成一个二维核，以提供新的最佳生长位。然而，实验观测却发现，在过饱和度极低的条件下，仍能观察到层生长现象。BCF 理论认为螺旋位错可为晶体生长提供永不消逝的最佳生长位，不需要二维成核。该理论在许多晶体生长过程中得到了验证，是目前最为认同的晶体生长理论。除了螺旋位错外，刃型位错、层错都可以为晶体生长提供永不消逝的最佳生长位。虽然 BCF 理论很好地解释了二维方向生长速率相同的曲线螺旋生长，但难以解释常见的直边台阶生长现象（De Yoreo et al., 2009）。由于直边螺旋的台阶边缘并不能提供 BCF 理论中需要的扭折位，因而需要探寻新的扭折位形成机制（Vekilov, 2007）。

越来越多的研究表明，晶体生长并不完全局限于经典成核理论与生长过程，且有时难以明确区分成核和生长过程。这些不遵循经典理论的晶体生长模式被称作非经典成核生长途径，包括颗粒附着晶化、非晶转化和无临界成核尺寸的二维自组装生长等。

1.1 颗粒附着晶化（crystallization of particle attachment，CPA）

研究者使用高分辨透射电子显微镜（high resolution transmission electron microscope，HRTEM）观察到一种纳米粒子定向附着的非经典生长方式。该生长方式是基于实验观测的概念（Teng, 2013），但至今还没有完善的理论体系对其进行描述（De Yoreo, 2017）。Lee Penn 和 Banfield（1998）首次报道了纳米锐钛矿在水热条件下可形成高度有序的长链，并将其归因于晶粒间共用晶体取向。随后，他们又进一步观测到了纳米晶粒间的定向拼接（Lee Penn and Banfield, 1999），并认为这种定向附着、拼接作用的驱动力为布朗运动（Banfield et al., 2000）。大量研究表明，（氢）氧化物（Wang and Xu, 2013；丁兴等，2018）、硫化物（Xian et al., 2016）、合金（Liao et al., 2012）等宏观三维晶体和高岭石、云母、海泡石（García-Romero and Suárez, 2014, 2018）等层状硅酸盐矿物中均存在这一生长方式。虽然这些研究揭示了 CPA 晶体生长模式的普遍存在，但都只观测到这种晶体生长方式的终态，对于其过程还缺乏清楚的认识。

在发现 CPA 晶体生长现象 14 年后，Science 在同一期发表了两篇利用原位液相透射电子显微镜（电镜）观察到纳米晶粒定向附着过程的文章（Li et al., 2012；Liao et al., 2012）。Pt_3Fe 合金的液相原位定向附着生长过程表现为从单体向二聚体、三聚体等多聚体的拼接过程。针铁矿纳米晶粒在液相条件下通过"跳跃-接触"方式寻找共同晶体取向。这些观测结果使人们对 CPA 的动态过程有了直观的认识。

在 CPA 晶体生长模式中，颗粒在附着聚集前后通常被认为其晶体结构和化学成分是不变的。Liu Y. F. 等（2019）发现颗粒间还可以通过化学反应而聚集晶化（即反应型附着晶化）。当碳酸钇 $[Y_2(CO_3)_3]$ 纳米颗粒悬浮液中加入电解质（$NaHCO_3$ 或 NH_4HCO_3）时，它们将在一定温度条件下反应形成微米级的 $NaY(CO_3)_2 \cdot 6H_2O$ 或 $NH_4Y(CO_3)_2 \cdot H_2O$ 片状晶体。该发现不仅丰富了颗粒附着晶化的内涵，而且为晶体工程领域提供了一种基于化学反应过程的颗粒附着晶体生长技术。

尽管目前在 CPA 基础理论方面还有待突破，但基于 CPA 途径的应用已推动了晶体合成技术的发展。Liu Z. M. 等（2019）通过"无机离子寡聚体的聚合反应"实现了厘米尺寸碳酸钙晶体材料的快速合成，而且该方法合成的碳酸钙具有很强的可塑性，可以像塑料一样按照模具形状获得各种形态的晶体，其技术核心就是利用无机离子寡聚体的附着生长。在此基础上，该团队还发明了一种基于无机离子寡聚体附着生长的牙釉质原位修复技术（Shao et al., 2019）。

值得注意的是，已知的 CPA 机制主要出现在表生作用、生物矿化或中-低温水热反应等条件较为温和的环境中（De Yoreo et al., 2015）。那么，在一些剧烈的地质作用过程中（如岩浆作用），是否也存在类似的矿物晶体生长机制？倘若存在，那颗粒附着晶化机制将是继层生长理论和螺旋生长理论之后的又一重要晶体生长理论；同时，预示我们可以通过对矿物晶体中纳米晶粒特征的详细研究获得相关地质地球化学过程的重要信息。

1.2 非晶转化途径

根据大量晶体生长早期观测证据，在多种体系中均发现了有别于经典理论的非晶转化结晶途径。在晶体生长的初期，溶液中先析出非晶的小颗粒，然后再转变成稳定的晶体结构。目前报道的非晶转变途径主要可分为两类：逆向晶体生长和直接非晶转化。

逆向晶体生长路径最早由Chen等（2007）提出。因为早期聚集体是非晶的，它们倾向于在合成溶液中形成球形颗粒。由于颗粒表面区域与溶液接触，比聚集体的内部活性更高，它的表面首先晶化，而呈现从聚集非晶颗粒表面向内部逐渐结晶的过程。这与传统晶体生长理论描述的晶体由内至外的生长有明显区别，因此被称为逆向晶体生长。当低密度非晶内核转化为高密度的晶相时，就会在晶体内部形成空洞，因此许多空心晶体就被认为是由这种生长方式形成的（Greer et al., 2012）。

直接非晶转化过程是晶体生长过程中的一种较为普遍的现象，即在晶体生长早期先形成非晶前驱体，随着生长的进行而转变成更为稳定的结晶相。碳酸钙体系是研究较为成熟的体系，在原位透射电子显微镜（transmission electron microscope, TEM）生长实验观察中，发现有球霰石、文石、方解石直接成核结晶的同时，也有非晶碳酸钙（ACC）的形成（Nielsen et al., 2014）。ACC再进一步转变成球霰石或文石结构，从而实现从不稳定的非晶态向更稳定的晶态的逐步转变。

此外，根据上述经典成核生长理论，晶体的成核需要克服一定的能垒。然而，在辉钼矿基底上，短链多肽的二维生长过程是一次生长一个行列后再组装成二维晶体，且过程不需要尺寸能垒，生长的临界成核尺寸为0（Chen J. J. et al., 2018），即无临界成核尺寸的二维自组装生长。该研究打破了经典成核理论的预言，对经典成核理论提出了挑战。

2 高压矿物学

高压矿物学是传统矿物学在压力维度的拓展，主要研究（高温）高压环境下矿物的化学组成、内部结构、外表形态、物理化学性质、形成和变化条件等方面的现象和规律及其内在联系（Qin et al., 2016）。高压矿物学为探索地球深部、陨石和陨石撞击坑，以及其他天体的物质组成及演化规律提供重要信息，是理解地球各圈层组成与性质、板块俯冲过程的物理化学变化、全地幔元素分布和分异行为、地球深部碳、氢、氧和氮循环以及矿物冲击效应等重要地质问题的关键。近十年来，随着极端环境下矿物研究工作的不断深入，以及高压技术，微观、微区和微量分析技术和同步辐射光源的长足发展，我国高压矿物学研究取得了重大进展，发现了重要的超高压矿物新相，提出了高压结构化合物压致相变新机制，揭示了碳、氢、氧和铁等重要元素组分在高温高压下的赋存形式和演变行为。

2.1 超高压矿物的发现

超高压矿物的发现和研究是认识地球深部物质结构和组成，以及地壳、地幔和地核各圈层之间物质和能量交换过程的重要窗口。自然界中已发现的矿物多达五千多种，然而超高压矿物的种类不过几十种。Xie等（2003）和Chen等（2008，2019）通过对陨石和地球陨石坑的岩石矿物冲击效应的调查和研究，发现和参与发现了12个冲击形成的矿物高压新相（Gillet et al., 2000; El Goresy et al., 2008, 2010; Gu et al., 2013, 2017; Bindi et al., 2019; Ma et al., 2019; Yang H. X. et al., 2019），其中9个已被国际矿物协会新矿物、命名和分类委员会（International Mineralogical Association Commission on New Minerals, Nomenclature and Classification, IMA CNMNC）批准并命名为新矿物。我国岫岩陨石撞击坑中的铁镁碳酸盐

[铁白云石，$Ca(Fe^{2+},Mg)(CO_3)_2$] 在撞击产生的高温高压下发生亚固态自氧化还原反应生成金刚石(Chen et al.，2018a)，同时伴随着二价铁被氧化为三价铁，形成一种后尖晶石结构、高密度的镁铁氧化物($MgFe_2^{3+}O_4$，$Pnma$)——毛河光矿(Chen et al.，2019)。铁镁碳酸盐向金刚石的转变，表明在存在碳酸盐以及压力和温度足够高的下地幔，金刚石可能是碳的一个主要载体。以三价铁为基本组分的天然镁铁氧化物——毛河光矿的发现表明，毛河光矿是下地幔中潜在的重要组成矿物之一。这些超高压新矿物的发现不仅为矿物家族增添了重要的新成员，而且深化了对矿物冲击变质效应、高压相转变机制和地幔矿物组成的认识。

2.2　高压矿物结构与相转变

高温高压实验和第一性原理计算模拟的结合是研究地球深部矿物结构、物理性质与相转变的重要手段。下地幔主要由碱土金属硅酸盐钙钛矿矿物组成，在这种环境中其他离子能否像硅酸盐一样形成稳定的钙钛矿结构，这不仅是矿物学，也是地球科学、物理学、材料科学等领域共同关注的科学问题。Xiao等(2010)继发现铬酸铅($PbCrO_3$)立方钙钛矿压致等结构相变之后，又对钒酸铅($PbVO_3$)(Zhou et al.，2012)、锗酸铅($PbGeO_3$)(Xiao et al.，2012)和硅酸锶($SrSiO_3$)(Xiao et al.，2013)等ABO_3型化合物进行了高温高压合成和压致相变行为研究。他们的研究发现，这些不同离子组合的ABO_3型化合物在下地幔温压条件下都可形成稳定的立方钙钛矿结构。当压力小于5 GPa后，不同离子组合化合物会发生立方等结构相变、立方到四方和立方到非晶化等转变，表明下地幔环境中钙钛矿结构具有较强的空间弹性和元素相容性。

矿物弹性常数的计算(吴忠庆和王文忠，2009；Wu and Wentzcovitch，2009，2011，2014；Wu Z. Q. et al.，2013；Yang and Wu，2014)与地震波观测数据的对照是反映下地幔矿物组成的重要手段。对$MgSiO_3$-钙钛矿在下地幔条件下的弹性性质计算表明，地幔橄榄岩矿物组合模型预测的地震波速能解释地震观测值(Zhang et al.，2013)。对洋中脊玄武岩(mid-ocean-ridge basalt，MORB)含铝矿物相的弹性性质计算揭示了其强烈的地震波各向异性，该矿物相的状态方程说明下地幔条件下MORB密度高于周围地幔因而具有俯冲至下地幔的能力，支持"全地幔对流"模型(Yin K. et al.，2012，2016；Zhao M. Q. et al.，2018)。而对地幔条件下碳酸盐矿物的模拟计算表明，$MgCO_3$比$CaCO_3$更稳定，因而$CaCO_3$不可能为地球深部地幔的碳汇(Zhang Z. G. et al.，2018)。第一性原理分子动力学模拟计算对地核铁矿物结构的研究发现，一定程度的Si/S替代可以提高体心立方(bcc)铁结构的应力稳定性，但仍难以说明地核铁纯粹以体心立方结构存在，推断可能是体心立方与六方最密(hcp)结构的共存相(Cui et al.，2013)。

这些发现为研究地幔、地核的矿物结构、元素的赋存形式和丰度提供了重要的实验与理论依据，是探讨地球深部物质交换过程和元素分异行为的理论基础。

2.3　高压矿物与地球深部水的赋存形式

地球深部水的赋存形式和含量不仅对地球物理场有着重要影响，而且对地球化学动力学、地幔横向不均一性、地幔对流、低速带、板块俯冲、相平衡、熔融行为、深部地质灾害(地震与火山喷发)以及成矿作用等都具有重要意义，是地球科学研究的前沿。我国学者应用大腔体压机、活塞圆筒压机和金刚石压砧等高温高压实验装置，结合原位的阻抗谱技术、超声波波速测量技术、布里渊散射、红外光谱、拉曼(Raman)光谱和X射线衍射等多种技术手段，在高压矿物的水赋存形式和含量研究中取得重要进展。研究发现，随着含水矿物(如针铁矿，FeOOH)和水分子在下地幔条件下的稳定性下降，H元素从矿物和水分子中独立出来，形成黄铁矿结构的FeO_2H_x、六方相的$(Fe,Al)OOH$以及高含氢量的二氧化硅高压相(Hu et al.，2016；Zhang L. et al.，2018；Lin et al.，2020)。而对地幔分子氢(H_2)的研究发现，在高度还原氛围下，地幔的主要组成矿物都可以溶解一定量的分子氢；分子氢的赋存形式与矿物成分和类型无关，皆以中性分子形式填充在晶格间隙中，其在矿物中的溶解度随压强增加而增大(Yang et al.，

2016）。对下地幔条件下后钙钛矿、富铝相、含水D相和δ相中的铁和氢性质的研究发现，后钙钛矿在下地幔底部温压条件下，晶格变形产生（001）滑移面，这为解释环太平洋区域D层剪切波速的传播行为提供了有力证据（Wu et al., 2017）。水对上地幔主要造岩矿物状态方程影响的实验研究也取得了重大进展（范大伟等，2018；Xu et al., 2019），发现了造岩矿物是否含水对体弹模量及体弹模量的压力导数影响显著，而对体弹模量的温度导数和热膨胀系数的影响则很小。实验测量发现，水对高温高压下石榴子石和林伍德石中的铁镁互扩散系数有明显影响（Bai et al., 2019；Zhang et al., 2019）。对俯冲带水循环、地幔转换带水化及深部俯冲洋壳含水相等的研究表明，含水玄武岩体稳定的富铝D相和富铝H相可以携带俯冲洋壳中至少2%的水；冷热板片的俯冲不仅可以造成地幔转换带（mantle transition zone，MTZ）的水化，还可造成MTZ含水量的不均一；冷的俯冲板片可能携带自由流体进入MTZ（Liu X. C. et al., 2019；Xu et al., 2019）。近十年来，俯冲带、地幔过渡带和地幔条件下高温高压水相关问题的实验研究，使人们对深部水的赋存形式、迁移行为以及含水矿物的物理化学性质有了新认识。

3 矿物表–界面作用

近十年来，我国的矿物表–界面过程研究领域十分活跃，特别是在矿物与环境物质的表–界面作用机制及环境效应研究等方面，通过与同步辐射、原位谱学、吸附模型、计算模拟等矿物学研究最新手段的结合，从原子、分子水平揭示了矿物的表–界面过程及机理，阐明了表生矿物对环境物质地球化学行为的制约机制。

3.1 类质同象对矿物表面反应性的制约机制

地壳中有许多元素本身含量很低或不能形成独立矿物，而主要以类质同象形式赋存于矿物晶体结构中，导致矿物的一系列表面物理化学性质和反应性发生显著变化。深入研究类质同象对矿物表面反应性的制约机制，有助于掌握矿物在地表系统物质循环中的作用及其机理。国内学者重点研究了金属元素在铁锰氧化物矿物结构中的赋存状态及其对表面反应性的影响。在水钠锰矿结构中，Co^{2+}主要存在于水钠锰矿层内，而Ni^{2+}、Fe^{3+}、Al^{3+}、Cu^{2+}大多吸附于八面体空位上下方，V^{5+}则主要以多聚物形式吸附于层边缘。这些置换离子会不同程度影响锰氧化物的结构和物理化学性质，改变水钠锰矿对重金属的吸附–氧化能力（Yin H. et al., 2012, 2013, 2014, 2015, 2017）。在针铁矿结构中，Al^{3+}、$Mn^{3+}\rightarrow Fe^{3+}$置换提高了晶体宽径比，增加了比表面积和吸附能力（Liu H. B. et al., 2012, 2013；Liu H. et al., 2018）。Ti^{4+}、V^{3+}、Cr^{3+}等置换作用也改变了磁铁矿表面反应性（如吸附、氧化、还原等性能）（Liang et al., 2010a, 2010b, 2015；He et al., 2015；Tan et al., 2015）。此外，类质同象对赤铁矿、菱铁矿表面反应性的制约机制也得到了较多关注（Li W. et al., 2016；Li et al., 2019；Hu et al., 2019）。

3.2 元素在矿物表面的作用机理

长期以来，宏观模拟实验是研究元素–矿物界面过程的主要手段。近年来，原位谱学、微束微区、计算模拟等技术的快速发展，成为矿物表–界面反应研究的重要手段，并可从原子、分子水平揭示元素–矿物的界面作用机制。

X射线吸收精细结构谱（X-ray absorption fine stucture，XAFS）是研究金属元素在矿物表面赋存形态的重要手段。国内学者通过与北京、上海等的先进光源合作，利用该技术揭示了典型金属元素在矿物表面赋存形态。例如，Pb^{2+}在磁铁矿表面主要形成内层络合物，吸附构型为双核三齿共角，该构型不随吸附容量变化而改变（Liang et al., 2017）。As^{5+}在铁锰氧化物表面主要通过双齿双核的键合方式结合（Liu H. et al., 2017）。在生物成因水钠锰矿中，Cu^{2+}更多地进入锰空位和吸附于锰空位上，而非生物成因水钠锰矿则倾向于在结构层边缘吸附Cu^{2+}（Li et al., 2019）。

计算模拟方法可以直观地揭示金属元素在矿物表面的吸附形态，与实验的认识相辅相成。第一性原理分子动力学模拟离子在黏土矿物表面的络合结构发现，Ni^{2+}、Cd^{2+}、Pb^{2+}、Fe^{2+}等重金属离子以单齿或双齿络合的方式与黏土矿物的端面结合（Liu et al.，2012a；Zhang C. et al.，2016，2017，2018），表面络合的Ni^{2+}离子可以水解，为其他离子提供吸附位点，形成多核团簇及促进层状硅酸盐矿物的侧向生长（Zhang et al.，2019）。

对于矿物-水界面的含氧酸根阴离子（如磷酸根、硫酸根、硒酸根和砷酸根），以及一些低分子量有机酸基团的吸附构型和机制等研究，原位红外光谱更为有效。借助该谱学技术，国内学者揭示了无机磷/有机磷在铁/铝氧化物表面的结合类型、配位形态与分子结构，阐明了磷酸根在水铁矿表面的吸附配位机制（Wang et al.，2013；Yan et al.，2014），以及磷酸根与Cd^{2+}在水铁矿表面的三元络合机制（Liu J. et al.，2018）。量子分子动力学方法还揭示了羧酸根在黏土矿物端面以内球络合的形式吸附，形成单齿络合结构；而在层间域内则以双齿络合的形式与阳离子共存（Liu X. D. et al.，2017）。

近年来，表面络合模型不断发展，这些模型不但能描述矿物表面的吸附行为和离子形态分布，还能预测环境条件改变后矿物表面和离子形态的响应特征。同时，国内学者应用多种模型预测了土壤活性矿物表面的重金属界面行为，如从分子水平揭示了重金属与土壤有机质、铁氧化物、锰氧化物的配位机制，成功构建了铁氧化物、锰氧化物以及针铁矿-腐殖酸复合物吸附重金属的电荷分布多位点络合模型（Xiong et al.，2013，2018；Zhao W. et al.，2018）。

3.3 矿物表面反应的晶面差异性

由于实验条件的限制，绝大多数关于矿物表面反应性的研究主要以矿物粉末为研究对象，所得结果通常是众多矿物颗粒中无数组晶面性质的累加或平均，这导致了对相关地球化学过程的认识和理解存在较大分歧。因此，从天然矿物的稳定晶面入手研究矿物表面反应性，可准确、深入地掌握矿物的表-界面过程（何宏平等，2019）。

以黄铁矿为例，通过对比黄铁矿典型单形晶体的表面结构与晶面反应性，发现不同晶面的氧化-还原反应存在显著差异性（Zhu et al.，2018；Xian et al.，2019b），进而影响金在黄铁矿表面的还原-沉淀反应速率（Xian et al.，2019a）。相比于立方体（100）和八面体（111）晶面，五角十二面体（210）晶面对离子态金具有最快的还原-沉淀速率，黄铁矿的各晶面之间存在协同效应（何宏平等，2019）。相关研究以矿物表面反应性为切入点，揭示了黄铁矿晶面反应性在金富集成矿过程中所起的关键作用（图1）。赤铁矿（110）晶面的Fe^{2+}比（001）晶面的Fe^{2+}具有更强的H_2O_2分解能力，这与晶面上的Fe^{2+}配位数有关。铀酰离子在赤铁矿（001）晶面为边共享双齿单核构型，而在（001）和（110）晶面则是角共享双齿双核构型（Huang et al.，2016，2017，2018）。此外，锐钛矿的（001）与（101）晶面在光催化反应中也表现出不同的作用，具体表现为光激发产生的电子与空穴分别分布在（001）与（101）晶面上，并在这两个晶面上分别参与光催化还原反应和氧化反应。这两组晶面同时暴露可有效促进光催化反应（王翔，2013；Wang X. Y. et al.，2014；Zhang et al.，2014）。

黏土矿物的基面与端面具有显著不同的结构，对表面反应性的影响体现在差异的表面质子解离能力上。第一性原理分子动力学模拟揭示了黏土矿物不同晶面和水的界面的精细结构（Liu et al.，2012b，2012c），在此基础上进行了热力学计算，定量阐释了表面质子的解离能力差别（Liu et al.，2013b，2014）。

上述工作从晶面结构和反应性的角度揭示了发生在矿物表-界面的地球化学过程，为构建更为精确的地球化学模型提供理论支撑。

3.4 光电作用对矿物表面反应性的增强效应

半导体属性是许多表生矿物的重要物理化学特性。利用表生矿物的半导体属性，通过光、电作用可

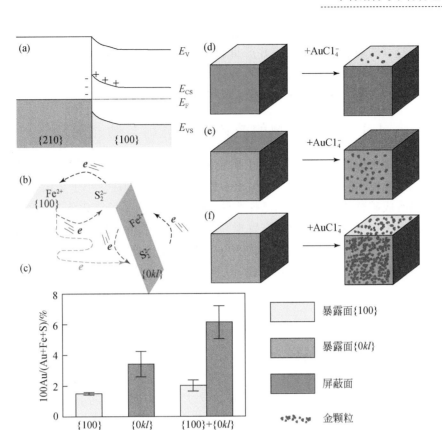

图 1 金在黄铁矿晶面的沉淀

(a) 黄铁矿 {210} 和 {100} 晶面组合的能带弯曲图；(b) 聚形黄铁矿 {100} 与 {0kl} 晶面间的电子转移路径；(c) HAuCl$_4$ 溶液中孤立 {100} 和 {0kl} 晶面及其聚形晶面沉淀的金浓度；(d) (e) (f) 孤立 {100} 和 {0kl} 晶面及其聚形晶面的金沉积行为示意图

增强其表面反应能力，这是当前环境矿物学研究的前沿热点。在太阳光照下，赤铁矿可催化 Cr^{3+}/Cr^{6+} 的氧化/还原，氧化/还原反应与吸附作用形成耦合，制约 Cr 的迁移和环境毒性（Liu J. et al., 2019）。紫外光和类质同象置换可增强磁铁矿对四溴双酚 A 等持久性有机污染物的降解能力（Zhong et al., 2012，2014）。微波诱导可增强锰氧化物的氧化能力，增大锰氧化物的电子自旋磁矩可增强其微波吸收性能，显著促进羟基自由基和超氧自由基的产生，从而提高对有机污染物的降解性能（Wang X. Y. et al., 2014；Gu et al., 2017；Lv et al., 2017）。在电化学作用下，水铁矿催化 Mn^{2+} 和 O_2 在表面进行电子传递，形成活性 Mn^{3+} 中间产物，显著促进 As^{3+} 及其他变价元素的耦合氧化（Lan et al., 2017；Wang et al., 2017）。电化学作用还使重金属离子在矿物（如铁锰氧化物、硫化矿物）界面发生转化，该机理的发现有助于研发性能优异的重金属钝化纳米矿物材料，在重金属污染水体和土壤的修复中具有较好的应用前景（Peng et al., 2016；Liu L. H. et al., 2017；Qiu et al., 2018）。上述研究不仅从分子、原子水平揭示了矿物与环境物质作用的微观机制，而且为理解矿物的环境自净化功能及其在环境修复中的应用奠定了理论基础。

4 黏土矿物学

黏土矿物是一类具有纳米结构的含水层状硅酸盐矿物，兼具环境与资源属性，是自然界中与人类活动联系最为紧密的一类矿物。一方面，它是土壤、沉积物等地球表层系统的重要矿物组分，对地球关键带的物质循环，乃至生命起源等有重要影响；另一方面，作为一类天然的纳-微米材料，黏土矿物在众多领域具有重要应用。长期以来，由于黏土矿物颗粒细小、结构复杂，研究难度大，人们对黏土矿物的晶体生长机制、不同矿物间的演化规律，以及矿物表面反应性的结构本质等问题缺乏清晰的认识，导致对相关重要地质、环境过程的理解难以深入，并严重制约了黏土矿物资源的高效利用。近十年来，我国学者利用高分辨微区、微束微区、原位分析等技术和计算模拟方法，对黏土矿物开展了深入研究，获得了一些创新认识，并在黏土矿物资源利用方面取得了长足进展。

4.1 黏土矿物的晶体生长机制与物相转变

传统的层生长和螺旋生长理论难以解释黏土矿物的生长现象。例如，同为2∶1型层状硅酸盐矿物，云母类矿物往往可以形成较大的矿物晶体，而蒙皂石族矿物的粒径则往往小于2 μm；在蒙皂石族矿物中，只有蒙脱石和皂石具有成矿意义，但两者的已探明储量差异巨大（彭杨伟和孙燕，2012）。通过黏土矿物合成、天然与合成黏土矿物微结构的对比研究发现，在蒙皂族矿物的晶化过程中，其结构中的Al^{3+}、Mg^{2+}等金属离子具有占位选择性，即Al^{3+}优先与Si^{4+}发生类质同象置换进入四面体片层。结构中，Al(IV)/Al(VI)值与Al^{3+}对Si^{4+}的置换量决定了皂石矿物片层生长与结晶度（He et al., 2014a；Tao et al., 2016；张旦等，2016；Zhang C. Q. et al., 2017）。上述认识从结构匹配性角度阐释了"为什么黏土矿物长不大"这一重要科学问题。

通过记录不同热力学条件下绿脱石生长过程的形貌特征，Decarreau等（2014）提出了"黏土矿物a-b方向的二维生长"理论。基于该理论和蒙皂石生长动力学研究，发现蒙皂石的生长实质为2∶1单元层沿a-b方向的延伸，因其层间水化阳离子阻碍了c方向的生长。此外，蒙皂石生长过程中，元素分布趋向于随机，粒径分布受2∶1单元层端面生长位点控制。该过程主要为传统的溶解再结晶，同时伴随有非传统的纳米颗粒附着生长（Zhang et al., 2020）。上述认识为进一步探索黏土矿物生长机制提供了启示。相对于传统的晶体生长研究方法，该研究采用分子探针技术记录了蒙皂石生长过程的形貌演变，并结合谱学方法揭示了不同金属离子在矿物结构中的占位，在一定程度上解决了黏土矿物的表征难题，为其他纳米矿物表征提供了范例（Zhang et al., 2020）。

一般认为，黏土矿物间的物相转变包括2∶1型→1∶1型和2∶1型之间相互转变两种方式。He等（2017）研究发现，1∶1型黏土矿物可以向2∶1型矿物转变（图2），即1∶1型蛇纹石-高岭石族黏土矿物表面可以与Si-O四面体缩合，进而转变为2∶1型膨胀性黏土矿物。高岭石等二八面体结构矿物的转变主要从矿物片层的端面开始，逐渐向片层内部延伸。对于三八面体结构的蛇纹石，由于不同蛇纹石矿物结构中的Si-O四面体片与Mg-O八面体片结合方式的差异，利蛇纹石的转变反应不仅发生在片层边缘，同时也发生在片层内部；而叶蛇纹石的转变反应仅发生在片层边缘（Ji et al., 2018）。由于1∶1型黏土矿物的形成条件与其转变为2∶1型黏土矿物的环境介质条件存在差异，导致新形成的2∶1型黏土矿物结构单元层中的两个四面体片的化学组成存在一定的差异，这很好地阐释了长期无法解决的混层黏土矿物的"极性结构"这一难题，为揭示黏土矿物的形成机制及其矿床成因提供了新的视角（He et al., 2017；Ji et al., 2018；张俊程等，2018）。

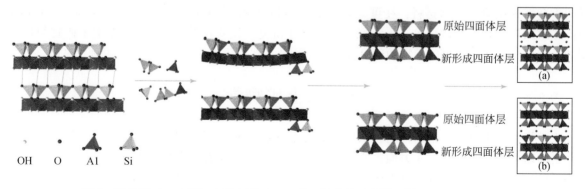

图2 高岭石（1∶1型）向蒙皂石（2∶1型）转变的示意图（据 He et al., 2017）

4.2 黏土矿物演化对重要地质过程的响应

黏土矿物的种类和晶体化学特征与其源区的古气候、古环境密切相关。例如，成土风化和成岩作用过程中，湿热的环境利于化学风化，有助于高岭石和三水铝石的形成，而干冷的环境会减缓化学风化作用，使伊利石和绿泥石得以更好地保存。蒙脱石、蛭石对水热气候条件、介质酸碱度和有机质都十分敏感，沉积地层中伊利石-蒙皂石混层矿物还与所处水体的化学特征、氧化还原条件、微生物作用等密切相关（袁慰顺，2011；周济元和崔炳芳，2015；Hong et al.，2019）。黏土矿物的矿物学特征（组成、粒度、含量、结晶度等）充分记录了成土期的气候与环境信息，并被广泛应用于海洋、湖泊等各类沉积物的物源辨识与古环境恢复、地层划分与对比、油气生成与运移、储集层性质研究等诸多领域（周济元和崔炳芳，2015）。

Fang 等（2017a）和 Hong 等（2012，2014，2015）利用高分辨透射电子显微镜结合 X 射线衍射等技术，发现了南方红土中多种过渡相混层黏土矿物，如高岭石-蒙皂石混层、伊利石-蛭石混层以及伊利石-蒙皂石-高岭石混层等。他们认为不同种类的混层矿物代表成土时期不同的环境和气候条件，根据不同矿物相含量的比值，如（Ch+I）/（K+S）与 Kübler 结晶度指数等，结合同位素定年可以对这些气候持续的时间跨度做进一步划分（Hong et al.，2013；Yin et al.，2018）。混层矿物组合特征，结合同位素分析与其他指标对比，实现对古气候环境的反演与重建（Hong et al.，2013；Wang et al.，2016；Fang et al.，2017b；Li et al.，2018；方谦等，2018）。陈漫游等（2017）研究表明，含 Mg、Al 的金属（氢）氧化物具有良好的水热反应性，它能够快速水合、吸收介质中的 CO_2，形成水滑石矿物，并可能进一步转变为 TOT 型的皂石矿物。该研究将环境中 CO_2 与黏土矿物结晶过程联系在一起，不仅提出了一种新的黏土矿物形成机制，同时也为碳循环研究提供了新思路（Tao et al.，2018，2019）。而分子模拟研究发现含 CO_3^{2-} 的水滑石矿物较之含其他阴离子的水滑石矿物的结构更为稳定，这从结构稳定性上佐证了水滑石对碳循环的作用（Chen et al.，2018b）。

黏土矿物与有机质、金属离子之间的作用一直是土壤学、环境科学、地球化学等领域重点关注的问题。在陆源碎屑物向海洋搬运与沉积过程中，陆源土壤成因的有机质-蒙脱石复合体中的有机质会被海洋成因的有机质所取代，而同为陆源的岩石成因的有机质-云母或绿泥石复合物则不发生变化，说明陆源有机质在海洋沉积物中的赋存状态是由黏土矿物这个一级因素所控制（Liu et al.，2010，2016；Blattmann et al.，2019）。近年来，页岩气的开发及其赋存机制等相关研究受到广泛关注（赵杏媛和何东博，2012）。作为页岩主要组成的黏土矿物凭借其吸附作用制约着有机质的富集与裂解、吸附态页岩气的分布（Ji et al.，2012；刘雄辉，2015；Wang et al.，2018）。经典分子动力学模拟发现，一定条件下甲烷分子与水分子形成半笼型的类水合物结构固定于蒙脱石硅氧六元环上方，这是黏土矿物赋存页岩气的结构基础（Zhou et al.，2011）。在蒙脱石向伊利石或绿泥石的转化过程中，结构层间距的减小将导致泥页岩的裂缝和空隙率的增加；层间水的析出则为页岩气的运移提供载体（Wu L. M. et al.，2012；李颖莉和蔡进功，2014；Geng et al.，2017；刘增晖，2018）。在离子吸附型稀土矿床中，黏土矿物结构的差异可导致稀土离子赋存状态的显著不同。在全风化层的稀土富集部位，稀土离子往往可以通过静电吸附和表面络合形式赋存于高岭石和埃洛石表面。而在半风化层中，由于出现了蒙皂石、伊利石、蛭石等黏土矿物，因此，稀土离子可以通过离子交换作用进入黏土矿物层间（Zhang Z. W. et al.，2016；周军明，2018；Alshameri et al.，2019；Yang M. J. et al.，2019）。另外，微生物与黏土矿物的相互作用也会显著影响离子、有机质在土壤、水体中的赋存状态和地球化学行为（Dong，2012；Sun et al.，2016；尚海丽等，2018）。

黏土矿物对地球深部地球化学和物理性质有重要影响，其中，洋壳的"蛇纹石化"（汪小妹等，2010；丁兴等，2016）和黏土矿物（层状硅酸盐矿物）对挥发分的俯冲运移机理（Yang et al.，2017；Liu W. D. et al.，2019）是地球科学的两个前沿热点。蛇纹石化可促进元素在海水与洋壳间的循环，为地球表层与深部元素提供"交换通道"（Zheng and Hermann，2014）。蛇纹石化还可能诱发洋壳扩张、断层滑移等大型地质构造运动（Wang and Karato，2013）。蛇纹石化产生的氢气、烷烃和有机酸等可为海底热液区生物提供物质和能量，因而可能对生命的起源和演化具有重要意义（黄瑞芳等，2016）。多硅白云母能够在地下300 km处稳定存在，是将氮、氢携带至地球深部的重要载体矿物。原位高温高压红外光谱揭示了多硅白云母的脱氮与脱水机制的内在联系，把俯冲带氢和氮的循环有机地联系起来，为认识地球深部的氮、水循环提供了新视角（Li et al.，2013，2019）。

此外，火星表面黏土矿物的研究近年来也受到我国学者的关注（付晓辉等，2019）。火星黏土矿物的形成机制、共生组合对火星的演化与宜居性的指示，以及火星黏土矿物的演化与有机质的可能赋存形式等正逐步成为我国矿物学研究的一个新热点。

4.3 黏土矿物资源的利用

利用表面活性剂插层技术改善黏土矿物与有机分子间的亲和性是黏土矿物资源高值利用的重要途径（He et al.，2014b）。有研究者使用绿色两性表面活性剂、有机-无机复合插层等方法，获得了多种性能优异的有机黏土（Qin et al.，2010；Zhu et al.，2011；苏琳娜，2019）。而有机嫁接方法是一种新型的黏土有机改性技术，它以可提供共价键的有机硅烷、甲醇等为有机改性剂，通过改性剂与黏土矿物表面间的缩合反应来获得有机改性黏土。目前的研究主要集中在蒙皂石族和高岭石族黏土矿物，前者具有膨胀性能，而后者的层间主要为弱的氢键作用（He et al.，2013）。蒙皂石族矿物的有机硅烷改性研究表明，嫁接产物的结构主要受矿物的电荷密度、反应溶剂的极性和嫁接量等因素制约（Su et al.，2012，2013）。对于高岭石等非膨胀性黏土矿物，一般需要采用极性小分子插层前驱体法，或高温、有机溶剂，或惰性气体保护、长时间回流等苛刻条件才能实现其表面的有机硅烷改性（Cheng et al.，2012；Yang et al.，2012；佟景贵等，2013；Tan et al.，2014；Zhang et al.，2015）。

通过黏土矿物的片层剥离可以获得纳米片层材料（Yuan et al.，2013；Jia and Song，2015；Li C. J. et al.，2016；赵子豪等，2018）。剥片后的高岭石片层会进一步发生卷曲，形成类埃洛石的管状结构材料（Yuan et al.，2013；Li et al.，2015）。带正电的黏土矿物片层与带负电的水滑石片层可以通过自组装形成一类由电性相反的片层交替排列组成的新型层状纳米材料（Li C. J. et al.，2016），该材料在储能、离子导体、催化等领域具有重要应用前景。

黏土矿物还作为吸附材料广泛用于污水处理和土壤重金属的钝化（Li et al.，2011；Wu P. X. et al.，2012；Wu Q. F. et al.，2013a；Sun et al.，2014；Zhu et al.，2016），作为催化剂或载体材料用于化学合成（An et al.，2015；王兰和曾琪静，2016），作为功能性填料用于纳米复合材料的制备（Su et al.，2012；李传常等，2017），作为缓释和靶向药物载体材料等（Hou et al.，2016；Tan et al.，2014），黏土矿物在复合材料领域展现了广泛的应用前景（Zhou and Keeling，2013）。

5 矿物学研究展望

在2011—2020年这十年间，微束微区、原位分析、高温高压和计算模拟等技术的快速发展及其在矿物学研究中的应用极大地推动了矿物学研究的深入，为相关学科的发展以及重大地质过程和事件的解译提供了关键证据。当前，"宏观更宏、微观更微，宏观微观相结合"仍将是地球科学发展的总体趋势。其中，以亚纳观-纳观尺度为特征的矿物学信息可为地球科学的宏观认识提供诊断性证据，"微观更微"在很大程度上亦有赖于矿物学研究的深入；同时，作为固体地球的基本组成单元，矿物学研究也可在地球系统内部的各种界-面过程、类地行星的形成与演化等研究中发挥关键作用（张志强，1996；Putnis and Ruiz-Agudo，2013；Putnis，2014）。例如，虽然地球的形成与演化、生命起源与进化等一直是科学家们关注的热点问题，但有关矿物演化的研究和认识并没有得到足够的重视（郭承基和王中刚，1981；蒋匡仁，1983）。事实上，地球的形成、演化与矿物的演化是密切相关的，甚至生命起源也可能与矿物演化存在内在联系。30亿年前板块构造运动的全球化启动导致了大陆地壳成分从铁镁质向长英质的过渡（Dhuime et al.，2015；Tang et al.，2016），随之发生全球大氧化事件（24亿~21亿年前）（Holland，2002；Lyons et al.，2014）。虽然许多研究认为，板块启动、地壳成分演变与大氧化事件之间存在某种联系（Kasting，2013；Lee et al.，2016），但这种联系的内在本质是什么？是否与矿物的演化有关？这是值得我们去思考的。从矿物演化的历史来看，虽然矿物多样性的突变、大气成分的变化、生物活动之间存在显著的相关性（Hazen et al.，2008），但它们之间的因果关系是什么？是大气氧含量的升高以及生物矿化导致了矿物多样性的突变，还是矿物多样性导致了大气组成的变化、甚至加快了生物进化过程？另外，虽然我们对时间尺度上的矿物演化有了一定的了解和掌握，但我们所认知的主要是在地球表层和浅部的矿物；在目前已知的五千多种矿物中，属于深部的超高压矿物种类不过几十种。那么，在空间尺度上，矿物是否也有类似于时间尺度上的矿物多样性的演化序列呢？这也是值得我们思考的问题，这无论是对认识深部物质组成与状态，还是认识地球的形成与演化都是极其重要的。

已有研究表明，地球的矿物组成是从太阳系最初的12种初始矿物（ur-minerals）演变至目前的5 562种矿物，矿物演化不仅会改变矿物晶体的组成、结构与物理化学性质，还会导致矿物表面活性及其表-界面反应性的改变，进而制约水（流体）-岩作用、元素的活化-迁移（循环）-富集（赋存）等地质地球化学过程，并进一步影响地球不同圈层间的物质与能量交换、生态系统的演化。因此，矿物演化及其表-界面过程必将是未来矿物学研究中的一个重要方向。

长期以来，矿物学研究因其所涉领域众多，所跨越的交叉领域之广，在地球科学领域恐无出其右者，这也造成了矿物学研究目标和主题不够集中、研究力量分散等问题。随着地球科学研究从学科牵引向科学问题引领的转变，矿物学研究理应做出相应的调整。因此，矿物学作为地球科学的基础和支撑学科，如何围绕深部过程与层圈作用、地球演化与宜居性等前沿科学问题，针对战略性关键金属、页岩气页岩油等国家重大需求，以及登月、火星登陆等国家战略，有效组织队伍、凝练科学目标、开展协同攻关将是近期矿物学研究的一个重要任务。

参 考 文 献

陈漫游,陶奇,何宏平,吉世超,李尚颖,张朝群. 2017. 固体核磁光谱法研究镁铝复合氧化物的水热转化机制. 岩石矿物学杂志,36(6):887-893

丁兴,刘志锋,黄瑞芳,孙卫东,陈多福. 2016. 大洋俯冲带的水岩作用:蛇纹石化. 工程研究:跨学科视野中的工程,8(3):258-268

丁兴,何俊杰,刘灼瑜. 2018. 热液条件下锐钛矿晶体生长的实验. 地球科学,43(5):1763-1772

范大伟,李博,陈伟,许金贵,匡云倩,叶之琳,周文戈,谢鸿森. 2018. 石榴子石族矿物状态方程研究进展. 高压物理学报,32(1):4-16

方谦,洪汉烈,赵璐璐,程峰,殷科,王朝文. 2018. 风化成土过程中自生矿物的气候指示意义. 地球科学,43(3):753-769

付晓辉,凌宗成,周琴,Jolliff B L,尹庆柱,王阿莲,李勃,武中臣,张江. 2019. 火星陨石和火星表面泥岩中蚀变过程对比研究. 地学前缘,doi:10.13745/j.esf.sf.2019.8.21

郭承基,王中刚. 1981. 矿物演化. 矿物学报,(1):1-9

何宏平,鲜海洋,朱建喜,谭伟,梁晓亮,陈锰. 2019. 从矿物粉晶表面反应性到矿物晶面反应性:以黄铁矿氧化行为的晶面差异性为例. 岩石学报,35(1):129-136

黄瑞芳,孙卫东,丁兴,刘金钟,彭少邦. 2016. 橄榄石和橄榄岩蛇纹石化过程中气体形成的对比研究. 中国科学:地球科学,46(1):97-108

蒋匡仁. 1983. 矿物演化周期性的探讨. 矿物学报,(4):304-312

李传常,杨立新,肖桂雨,曾铃,何忠健,陈蓉. 2017. 蛭石改型及其功能化研究进展. 硅酸盐通报,36(4):1203-1208

李颖莉,蔡进功. 2014. 泥质烃源岩中蒙脱石伊利石化对页岩气赋存的影响. 石油实验地质,36(3):352-358

刘雄辉. 2015. 页岩岩石学特征及其对吸附气量的影响. 成都:西南石油大学硕士学位论文

刘增晖. 2018. 页岩储层黏土矿物学特征及其对孔隙性影响研究. 徐州:中国矿业大学硕士学位论文

彭杨伟,孙燕. 2012. 国内外膨润土的资源特点及市场现状. 金属矿山,41(4):95-99,105

尚海丽,毕银丽,李少朋,解文武. 2018. 解钾细菌对不同黏土矿物含量土壤微生物和酶活性的影响. 安徽农业科学,46(31):124-129

苏琳娜. 2019. 黏土矿物有机硅烷改性研究进展. 矿产保护与利用,39(1):124-130

佟景贵,何宏平,陶奇. 2013. 分步升温法制备有机硅烷嫁接高岭石. 硅酸盐学报,41(11):1571-1576

汪小妹,曾志刚,欧阳荷根,殷学博,王晓媛,陈帅,张国良,武力. 2010. 大洋橄榄岩的蛇纹岩石化研究进展评述. 地球科学进展,25(6):605-616

王兰,曾琪静. 2016. 层状粘土矿物负载TiO_2光催化材料的合成与应用研究进展. 化工新型材料,44(4):11-13

王翔,李仁贵,徐倩,韩洪宪,李灿. 2013. 锐钛矿(001)与(101)晶面在光催化反应中的作用. 物理化学学报,29(07):1566-1571
吴忠庆,王文忠. 2009. 矿物高温高压下弹性的第一性原理计算研究进展. 中国科学:地球科学,46(5):582-617
袁慰顺. 2011. 膨润土的成因类型与应用方向关系探讨. 中国非金属矿工业导刊,(5):1-4
张旦,何宏平,陶奇,陈漫游,吉世超,李甜. 2016. 皂石的异相水热合成及其影响因素. 矿物学报,36(2):241-246
张俊程,李本仙,孟杰,李耀宗,王晓峰,刘晓旸,施伟光. 2018. 蒙脱石伊利石化的水热实验模拟. 世界地质,37(1):316-326
张志强. 1996. 地球科学发展的新走向. 自然杂志,18(3):168-176
赵杏媛,何东博. 2012. 黏土矿物与页岩气. 新疆石油地质,33(6):643-647
赵子豪,孙红娟,彭同江,杨梦娜. 2018. 蒙脱石的剥离分散行为及其结构、性能变化. 硅酸盐学报,46(5):739-745
周ource元,崔炳芳. 2015. 中国凹凸棒石粘土矿床成因类型探讨. 资源调查与环境,36(4):266-275
周军明,袁鹏,余亮,刘小永,张佰发,樊文枭,刘冬. 2018. 八尺风化淋积型稀土矿凝灰岩风化壳中的细粒矿物特征. 矿物学报,38(4):420-428
Alshameri A,He H P,Xin C,Zhu J X,Wei X H,Zhu R L,Wang H L. 2019. Understanding the role of natural clay minerals as effective adsorbents and alternative source of rare earth elements:adsorption operative parameters. Hydrometallurgy,185:149-161
An N,Zhou C H,Zhuang X Y,Tong D S,Yu W H. 2015. Immobilization of enzymes on clay minerals for biocatalysts and biosensors. Applied Clay Science,114:283-296
Bai G Y,Gao D,Liu Z,Zhou X,Wang J J. 2019. Probing the critical nucleus size for ice formation with graphene oxide nanosheets. Nature,576(7787):437-441
Banfield J F,Welch S A,Zhang H Z,Ebert T T,Lee Penn R. 2000. Aggregation-based crystal growth and microstructure development in natural iron oxyhydroxide biomineralization products. Science,289(5480):751-754
Becker R,Döring W. 1935. Kinetische Behandlung der Keimbildung in übersättigten Dämpfen. Annalen der Physik,416(8):719-752
Bindi L,Brenker F E,Nestola F,Koch T E,Prior D J,Lilly K,Krot A N,Bizzarro M,Xie X D. 2019. Discovery of asimowite,the Fe-analog of wadsleyite,in shock-melted silicate droplets of the Suizhou L6 and the Quebrada Chimborazo 001 CB3.0 chondrites. American Mineralogist,104(5):775-778
Blattmann T M,Liu Z,Zhang Y,Zhao Y,Haghipour N,Montluçon D B,Plötze M,Eglinton T I. 2019. Mineralogical control on the fate of continentally derived organic matter in the ocean. Science,366(6466):742-745
Chen J J,Zhu E B,Liu J,Zhang S,Lin Z Y,Duan X F,Heinz H,Huang Y,De Yoreo J J. 2018. Building two-dimensional materials one row at a time:avoiding the nucleation barrier. Science,362(6419):1135-1139
Chen M,Shu J F,Mao H K. 2008. Xieite,a new mineral of high-pressure $FeCr_2O_4$ polymorph. Chinese Science Bulletin,53(21):3341-3345
Chen M,Shu J F,Xie X D,Tan D Y,Mao H K. 2018a. Natural diamond formation by self-redox of ferromagnesian carbonate. Proceedings of the National Academy of Sciences of the United States of America,115(11):2676-2680
Chen M,Zhu R L,Lu X C,Zhu J X,He H P. 2018b. Influences of cation ratio,anion type,and water content on polytypism of layered double hydroxides. Inorganic Chemistry,57(12):7299-7313
Chen M,Shu J F,Xie X D,Tan D Y. 2019. Maohokite,a post-spinel polymorph of $MgFe_2O_4$ in shocked gneiss from the Xiuyan crater in China. Meteoritics & Planetary Science,54(3):495-502
Chen X Y,Qiao M H,Xie S H,Fan K N,Zhou W Z,He H Y. 2007. Self-construction of core-shell and hollow zeolite analcime icositetrahedra:a reversed crystal growth process via oriented aggregation of nanocrystallites and recrystallization from surface to core. Journal of the American Chemical Society,129(43):13305-13312
Cheng H F,Liu Q F,Cui X N,Zhang Q,Zhang Z L,Frost R L. 2012. Mechanism of dehydroxylation temperature decrease and high temperature phase transition of coal-bearing strata kaolinite intercalated by potassium acetate. Journal of Colloid and Interface Science,376(1):47-56
Cui H,Zhang Z G,Zhang Y G. 2013. The effect of Si and S on the stability of bcc iron with respect to tetragonal strain at the Earth's inner core conditions. Geophysical Research Letters,40(12):2958-2962
De Yoreo J J. 2017. A holistic view of nucleation and self-assembly. MRS Bulletin,42(7):525-536
De Yoreo J J,Zepeda-Ruiz L A,Friddle R W,Qiu S R,Wasylenki L E,Chernov A A,Gilmer G H,Dove P M. 2009. Rethinking classical crystal growth models through molecular scale insights:consequences of kink-limited kinetics. Crystal Growth & Design,9(12):5135-5144
De Yoreo J J,Gilbert P U P A,Sommerdijk N A J M,Lee Penn R,Whitelam S,Joester D,Zhang H Z,Rimer J D,Navrotsky A,Banfield J F,Wallace A F,Michel F M,Meldrum F C,Cölfen H,Dove P M. 2015. Crystallization by particle attachment in synthetic,biogenic,and geologic environments. Science,349(6247):aaa6760
Decarreau A,Petit S,Andrieux P,Villiéras F,Pelletier M,Razafitianamaharavo,A. 2014. Study of low-pressure argon adsorption on synthetic nontronite:implications for smectite crystal growth. Clays and Clay Minerals,62(2):102-111
Dhuime B,Wuestefeld A,Hawkesworth C J. 2015. Emergence of modern continental crust about 3 billion years ago. Nature Geoscience,8(7):552-555
Dong H L. 2012. Clay-microbe interactions and implications for environmental mitigation. Elements,8(2):113-118
El Goresy A,Dera P,Sharp T G,Prewitt C T,Chen M,Dubrovinsky L,Wopenka B,Boctor N Z,Hemley R J. 2008. Seifertite,a dense orthorhombic polymorph of silica from the Martian meteorites Shergotty and Zagami. European Journal of Mineralogy,20(4):523-528
El Goresy A,Dubrovinsky L,Gillet P,Graup G,Chen M. 2010. Akaogiite:an ultra-dense polymorph of TiO_2 with the baddeleyite-type structure,in shocked garnet gneiss from the Ries Crater,Germany. American Mineralogist,95(5-6):892-895
Fang Q,Hong H L,Chen Z Q,Yu J X,Wang C W,Yin K,Zhao L L,Liu Z,Cheng F,Gong N N,Furnes H. 2017a. Microbial proliferation coinciding with volcanism during the Permian-Triassic transition:new,direct evidence from volcanic ashes,South China. Palaeogeography, Palaeoclimatology,

Palaeoecology,474:164-186

Fang Q,Hong H L,Zhao L L,Furnes H,Lu H Y,Han W,Liu Y,Jia Z Y,Wang C W,Yin K,Algeo T J. 2017b. Tectonic uplift-influenced monsoonal changes promoted hominin occupation of the Luonan Basin:Insights from a loess-paleosol sequence,eastern Qinling Mountains,central China. Quaternary Science Reviews,169:312-329

García-Romero E,Suárez M. 2014. Sepiolite-palygorskite polysomatic series:oriented aggregation as a crystal growth mechanism in natural environments. American Mineralogist,99(8-9):1653-1661

García-Romero E,Suárez M. 2018. A structure-based argument for non-classical crystal growth in natural clay minerals. Mineralogical Magazine,82(1):171-180

Geng Y K,Jin Z K,Zhao J H,Wen X,Zhang Z P,Wang Y. 2017. Clay minerals in shales of the Lower Silurian Longmaxi Formation in the eastern Sichuan Basin,China. Clay Minerals,52(2):217-233

Gillet P,Chen M,Dubrovinsky L,El Goresy A. 2000. Natural $NaAlSi_3O_8$-hollandite in the shocked Sixiangkou meteorite. Science,287(5458):1633-1636

Greer H F,于丰娇,周午纵. 2012. 非传统晶体生长的早期观察. 中国科学:化学,42(1):74-83

Gu W L,Lv G C,Liao L B,Yang C X,Liu H,Nebendahl I,Li Z H. 2017. Fabrication of Fe-doped birnessite with tunable electron spin magnetic moments for the degradation of tetracycline under microwave irradiation. Journal of Hazardous Materials,338:428-436

Gu X P,Xie X D,Wu X B,Zhu G C,Lai J Q,Hoshino K,Huang J W. 2013. Ferrisepiolite:a new mineral from Saishitang copper skarn deposit in Xinghai County,Qinghai Province,China. European Journal of Mineralogy,25(2):177-186

Hazen R M,Papineau D,Bleeker W,Downs R T,Ferry J M,McCoy T J,Sverjensky D A,Yang H X. 2008. Mineral evolution. American Mineralogist,93(11-12):1693-1720

He H P,Tao Q,Zhu J X,Yuan P,Shen W,Yang S Q. 2013. Silylation of clay mineral surfaces. Applied Clay Science,71:15-20

He H P,Li T,Tao Q,Chen T H,Zhang D,Zhu J X,Yuan P,Zhu R L. 2014a. Aluminum ion occupancy in the structure of synthetic saponites:effect on crystallinity. American Mineralogist,99(1):109-116

He H P,Ma L Y,Zhu J X,Frost R L,Theng B K G,Bergaya F. 2014b. Synthesis of organoclays:a critical review and some unresolved issues. Applied Clay Science,100:22-28

He H P,Zhong Y H,Liang X L,Tan W,Zhu J X,Wang C Y. 2015. Natural Magnetite:an efficient catalyst for the degradation of organic contaminant. Scientific Reports,5:10139

He H P,Ji S C,Tao Q,Zhu J X,Chen T H,Liang X L,Li Z H,Dong H L. 2017. Transformation of halloysite and kaolinite into beidellite under hydrothermal condition. American Mineralogist,102(5):997-1005

Holland H D. 2002. Volcanic gases,black smokers,and the great oxidation event. Geochimica et Cosmochimica Acta,66(21):3811-3826

Hong H L,Churchman G J,Gu Y S,Yin K,Wang C W. 2012. Kaolinite-smectite mixed-layer clays in the Jiujiang red soils and their climate significance. Geoderma,173-174:75-83

Hong H L,Wang C W,Zeng K F,Gu Y S,Wu Y B,Yin K,Li Z H. 2013. Geochemical constraints on provenance of the mid-Pleistocene red earth sediments in subtropical China. Sedimentary Geology,290:97-108

Hong H L,Churchman G J,Yin K,Li R B,Li Z H. 2014. Randomly interstratified illite-vermiculite from weathering of illite in red earth sediments in Xuancheng,southeastern China. Geoderma,214-215:42-49

Hong H L,Cheng F,Yin K,Churchman G J,Wang C W. 2015. Three-component mixed-layer illite/smectite/kaolinite(I/S/K)minerals in hydromorphic soils,South China. American Mineralogist,100(8-9):1883-1891

Hong H L,Algeo T J,Fang Q,Zhao L L,Ji K P,Yin K,Wang C W,Cheng S. 2019. Facies dependence of the mineralogy and geochemistry of altered volcanic ash beds:an example from Permian-Triassic transition strata in southwestern China. Earth-Science Reviews,190:58-88

Hou D Z,Hu S,Huang Y,Gui R Y,Zhang L C,Tao Q,Zhang C M,Tian S Y,Komarneni S,Ping Q N. 2016. Preparation and in vitro study of lipid nanoparticles encapsulating drug loaded montmorillonite for ocular delivery. Applied Clay Science,119:277-283

Hu Q Y,Kim D Y,Yang W G,Yang L X,Meng Y,Zhang L,Mao H K. 2016. FeO_2 and FeOOH under deep lower-mantle conditions and Earth's oxygen-hydrogen cycles. Nature,534(7606):241-244

Hu W,Zhang Z X,Li M X,Liu H B,Zhang C A,Chen T H,Zhou Y F. 2019. Enhanced uptake capacity for uranium(VI)in aqueous solutions by activated natural siderite:performance and mechanism. Applied Geochemistry,100:96-103

Huang X P,Hou X J,Song F H,Zhao J C,Zhang L Z. 2016. Facet-dependent Cr(VI)adsorption of hematite nanocrystals. Environmental Science & Technology,50(4):1964-1972

Huang X P,Hou X J,Song F H,Zhao J C,Zhang L Z. 2017. Ascorbate induced facet dependent reductive dissolution of hematite nanocrystals. The Journal of Physical Chemistry C,121(2):1113-1121

Huang X P,Hou X J,Zhang X,Rosso K M,Zhang L Z. 2018. Facet-dependent contaminant removal properties of hematite nanocrystals and their environmental implications. Environmental Science:Nano,5(8):1790-1806

Ji L M,Zhang T W,Milliken K L,Qu J L,Zhang X L. 2012. Experimental investigation of main controls to methane adsorption in clay-rich rocks. Applied Geochemistry,27(12):2533-2545

Ji S C,Zhu J X,He H P,Tao Q,Zhu R L,Ma L Y,Chen M,Li S Y,Zhou J M. 2018. Conversion of serpentine to smectite under hydrothermal condition:implication for solid-state transformation. American Mineralogist,103(2):241-251

Jia F F, Song S X. 2015. Preparation of monolayer muscovite through exfoliation of natural muscovite. RSC Advances, 5(65):52882–52887

Kasting J F. 2013. What caused the rise of atmospheric O_2? Chemical Geology, 362:13–25

Lan S, Wang X M, Xiang Q J, Yin H, Tan W F, Qiu G H, Liu F, Zhang J, Feng X H. 2017. Mechanisms of Mn(II) catalytic oxidation on ferrihydrite surfaces and the formation of manganese(oxyhydr)oxides. Geochimica et Cosmochimica Acta, 211:79–96

Lee C T A, Yeung L Y, McKenzie N R, Yokoyama Y, Ozaki K, Lenardic A. 2016. Two-step rise of atmospheric oxygen linked to the growth of continents. Nature Geoscience, 9(6):417–424

Lee Penn R, Banfield J F. 1998. Oriented attachment and growth, twinning, polytypism, and formation of metastable phases: insights from nanocrystalline TiO_2. American Mineralogist, 83(9-10):1077–1082

Lee Penn R, Banfield J F. 1999. Morphology development and crystal growth in nanocrystalline aggregates under hydrothermal conditions: insights from Titania. Geochimica et Cosmochimica Acta, 63(10):1549–1557

Li C J, Zhang J, Lin Y Y, Chen Y, Xie X L, Wang H, Wang L J. 2016. *In situ* growth of layered double hydroxide on disordered platelets of montmorillonite. Applied Clay Science, 119:103–108

Li D S, Nielsen M H, Lee J R I, Frandsen C, Banfield J F, De Yoreo J J. 2012. Direction-specific interactions control crystal growth by oriented attachment. Science, 336(6084):1014–1018

Li M H, Sun S R, Fang X M, Wang C H, Wang Z R, Wang H L. 2018. Clay minerals and isotopes of Pleistocene lacustrine sediments from the western Qaidam Basin, NE Tibetan Plateau. Applied Clay Science, 162:382–390

Li W, Liang X L, An P F, Feng X H, Tan W F, Qiu G H, Yin H, Liu F. 2016. Mechanisms on the morphology variation of hematite crystals by Al substitution: the modification of Fe and O reticular densities. Scientific Reports, 6:35960

Li X G, Liu Q F, Cheng H F, Zhang S, Frost R L. 2015. Mechanism of kaolinite sheets curling via the intercalation and delamination process. Journal of Colloid and Interface Science, 444:74–80

Li Y, Wiedenbeck M, Shcheka S, Keppler H. 2013. Nitrogen solubility in upper mantle minerals. Earth and Planetary Science Letters, 377-378:311–323

Li Y, Liu F F, Xu X M, Liu Y W, Li Y Z, Ding H R, Chen N, Yin H, Lin H, Wang C Q, Lu A H. 2019. Influence of heavy metal sorption pathway on the structure of biogenic birnessite: insight from the band structure and photostability. Geochimica et Cosmochimica Acta, 256:116–134

Li Z H, Hong L L, Liao L B, Ackley C J, Schulz L A, MacDonald R A, Mihelich A L, Emard S M. 2011. A mechanistic study of ciprofloxacin removal by kaolinite. Colloids and Surfaces B: Biointerfaces, 88(1):339–344

Liang X L, Zhong Y H, Zhu S Y, Zhu J X, Yuan P, He H P, Zhang J. 2010a. The decolorization of Acid Orange II in non-homogeneous Fenton reaction catalyzed by natural vanadium-titanium magnetite. Journal of Hazardous Materials, 181(1-3):112–120

Liang X L, Zhu S Y, Zhong Y H, Zhu J X, Yuan P, He H P, Zhang J. 2010b. The remarkable effect of vanadium doping on the adsorption and catalytic activity of magnetite in the decolorization of methylene blue. Applied Catalysis B: Environmental, 97(1-2):151–159

Liang X L, He Z S, Tan W, Liu P, Zhu J X, Zhang J, He H P. 2015. The oxidation state and microstructural environment of transition metals (V, Co, and Ni) in magnetite: an XAFS study. Physics and Chemistry of Minerals, 42(5):373–383

Liang X L, Wei G L, Xiong J, Tan F D, He H P, Qu C C, Yin H, Zhu J X, Zhu R L, Qin Z H, Zhang J. 2017. Adsorption isotherm, mechanism, and geometry of Pb(II) on magnetites substituted with transition metals. Chemical Geology, 470:132–140

Liao H G, Cui L K, Whitelam S, Zheng H M. 2012. Real-time imaging of Pt_3Fe nanorod growth in solution. Science, 336(6084):1011–1014

Lin Y H, Hu Q Y, Meng Y, Walter M, Mao H K. 2020. Evidence for the stability of ultrahydrous stishovite in Earth's lower mantle. Proceedings of the National Academy of Sciences of the United States of America, 117(1):184–189

Liu H, Lu X C, Li J, Chen X Y, Zhu X Y, Xiang W L, Zhang R, Wang X L, Lu J J, Wang R C. 2017. Geochemical fates and unusual distribution of arsenic in natural ferromanganese duricrust. Applied Geochemistry, 76:74–87

Liu H, Lu X C, Li M, Zhang L J, Pan C, Zhang R, Li J, Xiang W L. 2018. Structural incorporation of manganese into goethite and its enhancement of Pb(II) adsorption. Environmental Science & Technology, 52(8):4719–4727

Liu H B, Chen T H, Frost R L, Chang D Y, Qing C S, Xie Q Q. 2012. Effect of aging time and Al substitution on the morphology of aluminous goethite. Journal of Colloid and Interface Science, 385(1):81–86

Liu H B, Chen T H, Chang J, Zou X H, Frost R L. 2013. The effect of hydroxyl groups and surface area of hematite derived from annealing goethite for phosphate removal. Journal of Colloid and Interface Science, 398:88–94

Liu H B, Shu D B, Sun F W, Li Q, Chen T H, Xing B B, Chen D, Qing C S. 2019. Effect of manganese substitution on the crystal structure and decomposition kinetics of siderite. Journal of Thermal Analysis and Calorimetry, 136(3):1315–1322

Liu J, Zhu R L, Liang X L, Ma L Y, Lin X J, Zhu J X, He H P, Parker S C, Molinari M. 2018. Synergistic adsorption of Cd(II) with sulfate/phosphate on ferrihydrite: an *in situ* ATR-FTIR/2D-COS study. Chemical Geology, 477:12–21

Liu J, Zhu R L, Chen Q Z, Zhou H J, Liang X L, Ma L Y, Parker S C. 2019. The significant effect of photo-catalyzed redox reactions on the immobilization of chromium by hematite. Chemical Geology, 524:228–236

Liu L H, Luo Y, Tan W F, Liu F, Suib S L, Zhang Y S, Qiu G H. 2017. Zinc removal from aqueous solution using a deionization pseudocapacitor with a high-performance nanostructured birnessite electrode. Environmental Science: Nano, 4(4):811–823

Liu W D, Yang Y, Busigny V, Xia Q K. 2019. Intimate link between ammonium loss of phengite and the deep Earth's water cycle. Earth and Planetary Science Letters, 513:95–102

Liu X C, Matsukage K N, Nishihara Y, Suzuki T, Takahashi E. 2019. Stability of the hydrous phases of Al-rich phase D and Al-rich phase H in deep subducted oceanic crust. American Mineralogist,104(1):64−72

Liu X D,Jan Meijer E,Lu X C,Wang R C. 2012a. First-principles molecular dynamics insight into Fe^{2+} complexes adsorbed on edge surfaces of clay minerals. Clays and Clay Minerals,60(4):341−347

Liu X D,Lu X C,Meijer E J,Wang R C,Zhou H Q. 2012b. Atomic-scale structures of interfaces between phyllosilicate edges and water. Geochimica et Cosmochimica Acta,81:56−68

Liu X D,Lu X C,Wang R C,Meijer E J,Zhou H Q,He H P. 2012c. Atomic scale structures of interfaces between kaolinite edges and water. Geochimica et Cosmochimica Acta,92:233−242

Liu X D,Cheng J,Sprik M,Lu X C,Wang R C. 2013a. Understanding surface acidity of gibbsite with first principles molecular dynamics simulations. Geochimica et Cosmochimica Acta,120:487−495

Liu X D,Lu X C,Sprik M,Cheng J,Meijer E J,Wang R C. 2013b. Acidity of edge surface sites of montmorillonite and kaolinite. Geochimica et Cosmochimica Acta,117:180−190

Liu X D,Cheng J,Sprik M,Lu X C,Wang R C. 2014. Surface acidity of 2:1-type dioctahedral clay minerals from first principles molecular dynamics simulations. Geochimica et Cosmochimica Acta,140:410−417

Liu X D,Lu X C,Zhang Y C,Zhang C,Wang R C. 2017. Complexation of carboxylate on smectite surfaces. Physical Chemistry Chemical Physics,19(28):18400−18406

Liu Y F,Geng H B,Qin X Y,Yang Y,Zeng Z,Chen S M,Lin Y X,Xin H X,Song C J,Zhu X G,Li D,Zhang J,Song L,Dai Z F,Kawazoe Y. 2019. Oriented attachment revisited:does a chemical reaction occur? Matter,1(3):690−704

Liu Z F,Colin C,Li X J,Zhao Y L,Tuo S T,Chen Z,Siringan F P,Liu J T,Huang C Y,You C F. 2010. Clay mineral distribution in surface sediments of the northeastern South China Sea and surrounding fluvial drainage basins:source and transport. Marine Geology,277(1-4):48−60

Liu Z F,Zhao Y L,Colin C,Stattegger K,Wiesner M G,Huh C A,Zhang Y W,Li X J,Sompongchaiyakul P,You C F,Huang C Y,Liu J T,Siringan F P,Le K P,Sathiamurthy E,Hantoro W S,Liu J G,Tuo S T,Zhao S H,Zhou S W,He Z D,Wang Y C,Bunsomboonsakul S,Li Y L. 2016. Source-to-sink transport processes of fluvial sediments in the South China Sea. Earth-Science Reviews,153:238−273

Liu Z M,Shao C Y,Jin B,Zhang Z S,Zhao Y Q,Xu X R,Tang R K. 2019. Crosslinking ionic oligomers as conformable precursors to calcium carbonate. Nature,574(7778):394−398

Lv G C,Xing X B,Liao L B,An P F,Yin H,Mei L F,Li Z H. 2017. Synthesis of birnessite with adjustable electron spin magnetic moments for the degradation of tetracycline under microwave induction. Chemical Engineering Journal,326:329−338

Lyons T W,Reinhard C T,Planavsky N J. 2014. The rise of oxygen in Earth's early ocean and atmosphere. Nature,506(7488):307−315

Ma C,Tschauner O,Bindi L,Beckett J R,Xie X D. 2019. A vacancy-rich,partially inverted spinelloid silicate,$(Mg,Fe,Si)_2(Si,\square)O_4$,as a major matrix phase in shock melt veins of the Tenham and Suizhou L6 chondrites. Meteoritics & Planetary Science,54(9):1907−1918

Nielsen M H,Aloni S,De Yoreo J J. 2014. In situ TEM imaging of $CaCO_3$ nucleation reveals coexistence of direct and indirect pathways. Science,345(6201):1158−1162

Peng Q C,Liu L H,Luo Y,Zhang Y S,Tan W F,Liu F,Sui S L,Qiu G H. 2016. Cadmium removal from aqueous solution by a deionization supercapacitor with a birnessite electrode. ACS Applied Materials & Interfaces,8(50):34405−34413

Putnis A. 2014. Why Mineral Interfaces Matter. Science,343(6178):1441−1442

Putnis C V,Ruiz-Agudo E. 2013. The mineral-water interface:Where minerals react with the environment. Elements,9(3):177−182

Qin Z H,Yuan P,Zhu J X,He H P,Liu D,Yang S Q. 2010. Influences of thermal pretreatment temperature and solvent on the organosilane modification of Al_{13}-intercalated/Al-pillared montmorillonite. Applied Clay Science,50(4):546−553

Qiu G H,Gao T Y,Hong J,Luo Y,Liu L H,Tan W F,Liu F. 2018. Mechanisms of interaction between arsenian pyrite and aqueous arsenite under anoxic and oxic conditions. Geochimica et Cosmochimica Acta,228:205−219

Shao C Y,Jin B,Mu Z,Lu H,Zhao Y Q,Wu Z F,Yan L M,Zhang Z S,Zhou Y C,Pan H H,Liu Z M,Tang R K. 2019. Repair of tooth enamel by a biomimetic mineralization frontier ensuring epitaxial growth. Science Advances,5(8):eaaw9569

Su L N,Tao Q,He H P,Zhu J X,Yuan P. 2012. Locking effect:a novel insight in the silylation of montmorillonite surfaces. Materials Chemistry and Physics,136(2-3):292−295

Su L N,Tao Q,He H P,Zhu J X,Yuan P,Zhu R L. 2013. Silylation of montmorillonite surfaces:dependence on solvent nature. Journal of Colloid and Interface Science,391:16−20

Sun S Y,Li M,Dong F Q,Wang S J,Tian L F,Mann S. 2016. Chemical signaling and functional activation in colloidosome-based protocells. Small,12(14):1920−1927

Sun Y B,Li J X,Wang X K. 2014. RETRACTED:the retention of uranium and europium onto sepiolite investigated by macroscopic,spectroscopic and modeling techniques. Geochimica et Cosmochimica Acta,140:621−643

Tan D Y,Yuan P,Annabi-Bergaya F,Liu D,He H P. 2014. High-capacity loading of 5-fluorouracil on the methoxy-modified kaolinite. Applied Clay Science,100:60−65

Tan W,Wang C Y,He H P,Xing C M,Liang X L,Dong H. 2015. Magnetite-rutile symplectite derived from ilmenite-hematite solid solution in the Xinjie Fe-Ti oxide-bearing,mafic-ultramafic layered intrusion (SW China). American Mineralogist,100(10):2348−2351

Tang M,Chen K,Rudnick R L. 2016. Archean upper crust transition from mafic to felsic marks the onset of plate tectonics. Science,351(6271):372−375

Tao Q, Fang Y, Li T, Zhang D, Chen M Y, Ji S C, He H P, Komarneni S, Zhang H B, Dong Y, Noh Y D. 2016. Silylation of saponite with 3-aminopropyl-triethoxysilane. Applied Clay Science, 132-133: 133–139

Tao Q, Chen M Y, He H P, Komarneni S. 2018. Hydrothermal transformation of mixed metal oxides and silicate anions to phyllosilicate under highly alkaline conditions. Applied Clay Science, 156: 224–230

Tao Q, Zeng Q J, Chen M Y, He H P, Komarneni S. 2019. Formation of saponite by hydrothermal alteration of metal oxides: implication for the rarity of hydrotalcite. American Mineralogist, 104(8): 1156–1164

Teng H H. 2013. How ions and molecules organize to form crystals. Elements, 9(3): 189–194

Vekilov P G. 2007. What determines the rate of growth of crystals from solution? Crystal Growth & Design, 7(12): 2796–2810

Wang C W, Hong H L, Abels H A, Li Z H, Cao K, Yin K, Song B W, Xu Y D, Ji J L, Zhang K X. 2016. Early middle Miocene tectonic uplift of the northwestern part of the Qinghai-Tibetan Plateau evidenced by geochemical and mineralogical records in the western Tarim Basin. International Journal of Earth Sciences, 105: 1021–1037

Wang D J, Karato S I. 2013. Electrical conductivity of talc aggregates at 0.5GPa: influence of dehydration. Physics and Chemistry of Minerals, 40(1): 11–17

Wang F, Zhang S T, Li C M, Liu J, He S, Zhao Y F, Yan H, Wei M, Evans D G, Duan X. 2014. Catalytic behavior of supported Ru nanoparticles on the (101) and (001) facets of anatase TiO_2. RSC Advances, 4(21): 10834–10840

Wang Q T, Lu H, Wang T L, Liu D Y, Peng P A, Zhan X, Li X Q. 2018. Pore characterization of Lower Silurian shale gas reservoirs in the Middle Yangtze region, central China. Marine and Petroleum Geology, 89: 14–26

Wang S S, Xu A W. 2013. Template-free facile solution synthesis and optical properties of ZnO mesocrystals. CrystEngComm, 15(2): 376–381

Wang X M, Li W, Harrington R, Liu F, Parise J B, Feng X H, Sparks D L. 2013. Effect of ferrihydrite crystallite size on phosphate adsorption reactivity. Environmental Science & Technology, 47(18): 10322–10331

Wang X M, Hu Y F, Tang Y D, Yang P, Feng X H, Xu W Q, Zhu M Q. 2017. Phosphate and phytate adsorption and precipitation on ferrihydrite surfaces. Environmental Science: Nano, 4(11): 2193–2204

Wang X Y, Mei L F, Xing X B, Liao L B, Lv G C, Li Z H, Wu L M. 2014. Mechanism and process of methylene blue degradation by manganese oxides under microwave irradiation. Applied Catalysis B: Environmental, 160-161: 211–216

Wu L M, Zhou C H, Keeling J, Tong D S, Yu W H. 2012. Towards an understanding of the role of clay minerals in crude oil formation, migration and accumulation. Earth-Science Reviews, 115(4): 373–386

Wu P X, Dai Y P, Long H, Zhu N W, Jai P, Wu J H, Dang Z. 2012. Characterization of organo-montmorillonites and comparison for Sr(II) removal: equilibrium and kinetic studies. Chemical Engineering Journal, 191: 288–296

Wu Q F, Li Z H, Hong H L. 2013. Adsorption of the quinolone antibiotic nalidixic acid onto montmorillonite and kaolinite. Applied Clay Science, 74: 66–73

Wu X, Lin J F, Kaercher P, Mao Z, Liu J, Wenk H R, Prakapenka V B. 2017. Seismic anisotropy of the D″ layer induced by (001) deformation of post-perovskite. Nature Communications, 8: 14669

Wu Z Q, Wentzcovitch R M. 2009. Effective semiempirical ansatz for computing anharmonic free energies. Physical Review B, 79(10): 104304

Wu Z Q, Wentzcovitch R M. 2011. Quasiharmonic thermal elasticity of crystals: an analytical approach. Physical Review B, 83(18): 184115

Wu Z Q, Wentzcovitch R M. 2014. Spin crossover in ferropericlase and velocity heterogeneities in the lower mantle. Proceedings of the National Academy of Sciences of the United States of America, 111(29): 10468–10472

Wu Z Q, Justo J F, Wentzcovitch R M. 2013. Elastic anomalies in a spin-crossover system: ferropericlase at lower mantle conditions. Physical Review Letters, 110(22): 228501

Xian H Y, Zhu J X, Tang H M, Liang X L, He H P, Xi Y F. 2016. Aggregative growth of quasi-octahedral iron pyrite mesocrystals in a polyol solution through oriented attachment. CrystEngComm, 18(46): 8823–8828

Xian H Y, He H P, Zhu J X, Du R X, Wu X, Tang H M, Tan W, Liang X L, Zhu R L, Teng H H. 2019a. Crystal habit-directed gold deposition on pyrite: surface chemical interpretation of the pyrite morphology indicative of gold enrichment. Geochimica et Cosmochimica Acta, 264: 191–204

Xian H Y, Zhu J X, Tan W, Tang H M, Liu P, Zhu R L, Liang X L, Wei J M, He H P, Teng H H. 2019b. The mechanism of defect induced hydroxylation on pyrite surfaces and implications for hydroxyl radical generation in prebiotic chemistry. Geochimica et Cosmochimica Acta, 244: 163–172

Xiao W S, Tan D Y, Xiong X L, Liu J, Xu J. 2010. Large volume collapse observed in the phase transition in cubic $PbCrO_3$ perovskite. Proceedings of the National Academy of Sciences of the United States of America, 107(32): 14026–14029

Xiao W S, Tan D Y, Zhou W, Chen M, Xiong X L, Song M S, Liu J, Ma H K, Xu J. 2012. A new cubic perovskite in $PbGeO_3$ at high pressures. American Mineralogist, 97(7): 1193–1198

Xiao W S, Tan D Y, Zhou W, Liu J, Xu J. 2013. Cubic perovskite polymorph of strontium metasilicate at high pressures. American Mineralogist, 98(11-12): 2096–2104

Xie X D, Minitti M E, Chen M, Mao H K, Wang D Q, Shu J F, Fei Y W. 2003. Tuite, γ-$Ca_3(PO_4)_2$: a new mineral from the Suizhou L6 chondrite. European Journal of Mineralogy, 15(6): 1001–1005

Xiong J, Koopal L K, Tan W F, Fang L C, Wang M X, Zhao W, Liu F, Zhang J, Weng L P. 2013. Lead binding to soil fulvic and humic acids: NICA-Donnan modeling and XAFS spectroscopy. Environmental Science & Technology, 47(20): 11634–11642

Xiong J, Weng L P, Koopal L K, Wang M X, Shi Z H, Zheng L R, Tan W F. 2018. Effect of soil fulvic and humic acids on Pb binding to the goethite/

solution interface:ligand charge distribution modeling and speciation distribution of Pb. Environmental Science & Technology,52(3):1348−1356

Xu J G,Zhang D Z,Fan D W,Dera P K,Shi F,Zhou W G. 2019. Thermoelastic properties of eclogitic garnets and omphacites:implications for deep subduction of oceanic crust and density anomalies in the upper mantle. Geophysical Research Letters,46(1):179−188

Yan Y P,Li W,Yang J,Zheng A M,Liu F,Feng X H,Sparks D L. 2014. Mechanism of Myo-inositol hexakisphosphate sorption on amorphous aluminum hydroxide:spectroscopic evidence for rapid surface precipitation. Environmental Science & Technology,48(12):6735−6742

Yang H X,Gu X P,Downs R T,Evans S H,van Nieuwenhuizen J J,Lavinsky R M,Xie X D. 2019. Meieranite,$Na_2Sr_3MgSi_6O_{17}$,a new mineral from the Wessels Mine,Kalahari Manganese Fields,South Africa. The Canadian Mineralogist,57(4):457−466

Yang M J,Liang X L,Ma L Y,Huang J,He H P,Zhu J X. 2019. Adsorption of REEs on kaolinite and halloysite:a link to the REE distribution on clays in the weathering crust of granite. Chemical Geology,525:210−217

Yang R,Wu Z Q. 2014. Elastic properties of stishovite and the $CaCl_2$-type silica at the mantle temperature and pressure:an ab initio investigation. Earth and Planetary Science Letters,404:14−21

Yang S Q,Yuan P,He H P,Qin Z H,Zhou Q,Zhu J X,Liu D. 2012. Effect of reaction temperature on grafting of γ-aminopropyl triethoxysilane (APTES) onto kaolinite. Applied Clay Science,62-63:8−14

Yang X,Keppler H,Li Y. 2016. Molecular hydrogen in mantle minerals. Geochemical Perspectives Letters,2(2):160−168

Yang Y,Busigny V,Wang Z P,Xia Q K. 2017. The fate of ammonium in phengite at high temperature. American Mineralogist,102(11):2244−2253

Yin H,Tan W F,Zheng L R,Cui H J,Qiu G H,Liu F,Feng X H. 2012. Characterization of Ni-rich hexagonal birnessite and its geochemical effects on aqueous Pb^{2+}/Zn^{2+} and As(Ⅲ). Geochimica et Cosmochimica Acta,93:47−62

Yin H,Liu F,Feng X H,Hu T D,Zheng L R,Qiu G H,Koopal L K,Tan W F. 2013. Effects of Fe doping on the structures and properties of hexagonal birnessites−comparison with Co and Ni doping. Geochimica et Cosmochimica Acta,117:1−15

Yin H,Li H,Wang Y,Ginder-Vogel M,Qiu G H,Feng X H,Zheng L R,Liu F. 2014. Effects of Co and Ni co-doping on the structure and reactivity of hexagonal birnessite. Chemical Geology,381:10−20

Yin H,Liu Y,Koopal L K,Feng X H,Chu S Q,Zhu M Q,Liu F. 2015. High co-doping promotes the transition of birnessite layer symmetry from orthogonal to hexagonal. Chemical Geology,410:12−20

Yin H,Kwon K D,Lee J Y,Shen Y,Zhao H Y,Wang X M,Liu F,Zhang J,Feng X H. 2017. Distinct effects of Al^{3+} doping on the structure and properties of hexagonal turbostratic birnessite:a comparison with Fe^{3+} doping. Geochimica et Cosmochimica Acta,208:268−284

Yin K,Zhou H Q,Huang Q,Sun Y C,Xu S J,Lu X C. 2012. First-principles study of high-pressure elasticity of CF-and CT-structure $MgAl_2O_4$. Geophysical Research Letters,39(2):L02307

Yin K,Zhou H Q,Lu X C. 2016. Thermodynamic properties of calcium ferrite-type $MgAl_2O_4$:a first principles study. Science China Earth Sciences,59(4):831−839

Yin K,Hong H L,Churchman G J,Li Z H,Fang Q. 2018. Mixed-layer illite-vermiculite as a paleoclimatic indicator in the Pleistocene red soil sediments in Jiujiang,southern China. Palaeogeography,Palaeoclimatology,Palaeoecology,510:140−151

Yuan P,Tan D Y,Annabi-Bergaya F,Yan W C,Liu D,Liu Z W. 2013. From platy kaolinite to aluminosilicate nanoroll via one-step delamination of kaolinite:effect of the temperature of intercalation. Applied Clay Science,83-84:68−76

Zhang B H,Li B W,Zhao C C,Yang X Z. 2019. Large effect of water on Fe-Mg interdiffusion in garnet. Earth and Planetary Science Letters,505:20−29

Zhang C,Liu X D,Lu X C,Meijer E J,Wang K,He M J,Wang R C. 2016. Cadmium(Ⅱ) complexes adsorbed on clay edge surfaces:insight from first principles molecular dynamics simulation. Clays and Clay Minerals,64(4):337−347

Zhang C,Liu X D,Lu X C,He M J,Meijer E J,Wang R C. 2017. Surface complexation of heavy metal cations on clay edges:insights from first principles molecular dynamics simulation of Ni(Ⅱ). Geochimica et Cosmochimica Acta,203:54−68

Zhang C,Liu X D,Lu X C,He M J. 2018. Complexation of heavy metal cations on clay edges at elevated temperatures. Chemical Geology,479:36−46

Zhang C Q,He H P,Tao Q,Ji S C,Li S Y,Ma L Y,Su X L,Zhu J X. 2017. Metal occupancy and its influence on thermal stability of synthetic saponites. Applied Clay Science,135:282−288

Zhang C Q,Petit S,He H P,Viliéras F,Razafitianamaharavo A,Baron F,Tao Q,Zhu J X. 2020. Crystal growth of smectite:a study based on the change in crystal chemistry and morphology of saponites with synthesis time. ACS Earth and Space Chemistry,4(1):14−23

Zhang L,Yuan H S,Meng Y,Mao H K. 2018. Discovery of a hexagonal ultradense hydrous phase in (Fe,Al)OOH. Proceedings of the National Academy of Sciences of the United States of America,115(12):2908−2911

Zhang S L,Liu Q F,Cheng H F,Zhang Y D,Li X G,Frost R L. 2015. Intercalation of γ-aminopropyl triethoxysilane (APTES) into kaolinite interlayer with methanol-grafted kaolinite as intermediate. Applied Clay Science,114:484−490

Zhang S T,Li C M,Yan H,Wei M,Evans D G,Duan X. 2014. Density functional theory study on the metal-support interaction between Ru cluster and Anatase $TiO_2(101)$ Surface. The Journal of Physical Chemistry C,118(7):3514−3522

Zhang Z G,Stixrude L,Brodholt J. 2013. Elastic properties of $MgSiO_3$-perovskite under lower mantle conditions and the composition of the deep Earth. Earth and Planetary Science Letters,379:1−12

Zhang Z G,Mao Z,Liu X,Zhang Y G,Brodholt J. 2018. Stability and reactions of $CaCO_3$ polymorphs in the Earth's deep mantle. Journal of Geophysical Research:Solid Earth,123(8):6491−6500

Zhang Z W,Zheng G D,Takahashi Y,Wu C Q,Zheng C F,Yao J H,Xiao C Y. 2016. Extreme enrichment of rare earth elements in hard clay rocks and its potential as a resource. Ore Geology Reviews,72:191−212

Zhao M Q, Zhou H Q, Yin K, Sun Y C, Liu X, Xu S J, Lu X C. 2018. Thermoelastic properties of aluminous phases in MORB from first-principle calculation: implications for Earth's lower mantle. Journal of Geophysical Research: Solid Earth, 123(12): 10583–10596

Zhao W, Tan W F, Wang M X, Xiong J, Liu F, Weng L P, Koopal L K. 2018. CD-MUSIC-EDL modeling of Pb^{2+} adsorption on birnessites: role of vacant and edge sites. Environmental Science & Technology, 52(18): 10522–10531

Zheng Y F, Hermann J. 2014. Geochemistry of continental subduction-zone fluids. Earth, Planets and Space, 66: 93

Zhong Y H, Liang X L, Zhong Y, Zhu J X, Zhu S Y, Yuan P, He H P, Zhang J. 2012. Heterogeneous UV/Fenton degradation of TBBPA catalyzed by titanomagnetite: catalyst characterization, performance and degradation products. Water Research, 46(15): 4633–4644

Zhong Y H, Liang X L, He Z S, Tan W, Zhu J X, Yuan P, Zhu R L, He H P. 2014. The constraints of transition metal substitutions (Ti, Cr, Mn, Co and Ni) in magnetite on its catalytic activity in heterogeneous Fenton and UV/Fenton reaction: from the perspective of hydroxyl radical generation. Applied Catalysis B: Environmental, 150-151: 612–618

Zhou C H, Keeling J. 2013. Fundamental and applied research on clay minerals: from climate and environment to nanotechnology. Applied Clay Science, 74: 3–9

Zhou Q, Lu X C, Liu X D, Zhang L H, He H P, Zhu J X, Yuan P. 2011. Hydration of methane intercalated in Na-smectites with distinct layer charge: insights from molecular simulations. Journal of Colloid and Interface Science, 355(1): 237–242

Zhou W, Tan D Y, Xiao W S, Song M S, Chen M, Xiong X L, Xu J. 2012. Structural properties of $PbVO_3$ perovskites under hydrostatic pressure conditions up to 10.6 GPa. Journal of Physics: Condensed Matter, 24(43): 435403

Zhu J X, Qing Y H, Wang T, Zhu R L, Wei J M, Tao Q, Yuan P, He H P. 2011. Preparation and characterization of zwitterionic surfactant-modified montmorillonites. Journal of Colloid and Interface Science, 360(2): 386–392

Zhu J X, Xian H Y, Lin X J, Tang H M, Du R X, Yang Y P, Zhu R L, Liang X L, Wei J M, Teng H H, He H P. 2018. Surface structure-dependent pyrite oxidation in relatively dry and moist air: implications for the reaction mechanism and sulfur evolution. Geochimica et Cosmochimica Acta, 228: 259–274

Zhu R L, Chen Q Z, Zhou Q, Xi Y F, Zhu J X, He H P. 2016. Adsorbents based on montmorillonite for contaminant removal from water: a review. Applied Clay Science, 123: 239–258

Progresses in Researches on Mineral Structure and Mineral Physics

HE Hong-ping [1,2,3], ZHU Jian-xi [1,2,3], CHEN Meng [1,2], TAO Qi [1,2], TAN Da-yong [1,2], LIANG Xiao-liang [1,2], XIAN Hai-yang [1,2]

1. CAS Key Laboratory of Mineralogy and Metallogeny, Guangzhou Institute of Geochemistry, Chinese Academy of Sciences, Guangzhou 510640; 2. Guangdong Provincial Key Laboratory of Mineral Physics and Materials, Guangzhou Institute of Geochemistry, Chinese Academy of Sciences, Guangzhou 510640; 3. University of Chinese Academy of Sciences, Beijing 100049

Abstract: During the last 10 years (2011–2020), researches on mineralogy were dramatically promoted to molecular and atomic scales (nanoscale) by the rapid developments and applications of the techniques, such as microbeam and nanobeam *in-situ* analysis, high T-P and computational simulation. In this period, Chinese mineralogists earned a series of great innovative achievements in the fields of mineral phase transition and crystal growth, mineral physics, mineral microstructure, and mineral surface-interface process science. Several new natural high-pressure minerals were discovered and the novel phase transition mechanisms involved were elucidated. Investigations on both natural and synthetic samples demonstrated that mineral crystals can grow via nanoparticle attachment crystallization and amorphous-to-crystalline transformation pathways. The principles about how mineral surface structure controlling the surface reactivity and the mineral surface/interface reaction microcosmic mechanism have been revealed. Novel insights on local structure and property of minerals at the atomic scale were obtained by using computer simulation techniques. Here, we have comprehensively reviewed the progresses in mineralogical studies in China during the last 10 years, following the lines as mineral crystal growth theory, high-pressure mineralogy, mineral surface-interface interaction science, and clay mineralogy. The challenges and directions of the future mineralogical studies have also been predicted.

Key words: mineral crystal growth; high-pressure mineralogy; mineral surface; clay mineral

环境矿物学研究进展*

鲁安怀　王长秋　李艳

北京大学 地球与空间科学学院，矿物环境功能北京市重点实验室，北京 100871

摘　要：环境矿物学聚焦矿物所禀赋的环境属性，研究矿物在记录环境、影响环境、评价环境、治理环境以及参与生物作用等方面的功能及其中蕴涵的理论方法。环境矿物学研究一方面为治理全球环境污染和生态破坏提供重要理论指导和技术支撑，同时也有助于深刻理解地球物质演化、生命起源进化及全球环境演变的宏观过程等重大理论问题。2011—2020年这十年间，我国学者在矿物日光催化及其与微生物协同作用、纳米矿物及其改性产物的环境功能、生物矿化作用等研究领域获得了新认识，提出了自然界矿物光电子能量、光电能微生物等新观点，发现了半导体矿物与微生物胞外电子传递的新机制。本文从地表多圈层交互作用产物"矿物膜"的环境属性、矿物与微生物交互作用的环境效应、纳米矿物及其环境功能、生物矿化作用及其环境效应以及人体病理性矿化作用等方面，综述了近十年我国环境矿物学研究的进展，并展望了环境矿物学的发展方向。

关键词：环境矿物学　矿物环境属性　纳米矿物　矿物与微生物　生物矿化作用

0　引　言

地球科学肩负资源和环境两大根本任务。20 世纪 90 年代在传统的资源矿物学研究基础上，国际上新型交叉学科环境矿物学应运而生。我国学者率先提出了矿物学环境属性研究方向（鲁安怀，2000），认为环境矿物学是研究天然矿物与地球各圈层之间交互作用并反映自然演变、防治生态破坏、评价环境质量、净化环境污染以及参与生物作用的科学，创立了矿物学环境属性研究体系（鲁安怀等，2015），使矿物学研究从岩石圈资源属性拓展到岩石圈与水圈、大气圈和生物圈交互作用过程的环境属性研究，丰富了现代矿物学研究内涵，整体提升了现代矿物学理论与应用研究水平。古老的地质学基础学科矿物学获得了巨大的发展机遇，现代矿物学在现代地球科学中仍占据重要基础学科地位。

环境矿物学迄今不过二十多年的发展历史，相较于有几百年历史的传统矿物学，显然很年轻。然而，惟其年轻，才更具蓬勃发展的活力。我国环境矿物学主要研究地表多圈层交互作用过程中，无机矿物发生、发展与变化过程所禀赋的生态与环境效应，特别是在无机界矿物天然自净化功能、矿物与微生物交互作用环境效应等理论研究与实际应用领域，开展了大量研究工作，取得了显著进展，甚至在某些方向领跑于国际环境矿物学发展。

本文重点回顾自 2011 年以来我国环境矿物学研究现状，主要从地表多圈层交互作用产物"矿物膜"的环境属性、矿物与微生物交互作用的环境效应、纳米矿物及其环境功能、生物矿化作用及其环境效应以及人体病理性矿化作用等方面，综述我国学者这十年间环境矿物学研究取得的重要进展。

1　地表多圈层交互作用产物"矿物膜"的环境属性

日-地系统中，暴露于太阳光下的地球表面广泛分布的天然矿物，长期受太阳光照射的响应机制一直

---

*　原文"环境矿物学研究进展（2011~2020 年）"刊于《矿物岩石地球化学通报》2020 年第 39 卷第 5 期，本文略有修改。

未被重视与理解。我国学者近年来发现太阳光辐射下地表"矿物膜"产生光电效应现象，取得矿物光电子能量研究的重大进展。在已知太阳光子和元素价电子两种基本能量形式基础上，提出矿物光电子是地表第三种能量形式的学说（鲁安怀等，2014a；Lu A. H. et al.，2019）。

1.1 地表"矿物膜"的特征与环境功能

最新研究发现，我国西北戈壁和沙漠、西南喀斯特、南方红壤等光照充足的典型景观地区，甚至植被发育地区的山崖峭壁上，暴露在阳光下的岩石表面普遍被一层厚度较薄、颜色灰黑、构造多孔的"矿物膜"所覆盖。戈壁地区的"矿物膜"，以前称为岩石漆，呈亮黑色，质地相对致密，厚数十到数百微米不等，包裹于砾石表面，与下伏基岩具有截然的分界面。红壤"矿物膜"呈红色至深褐色，厚数微米到数十微米，包裹于石英、长石颗粒及黏土矿物集合体表面，与基质矿物具有明显的分界线。喀斯特"矿物膜"呈灰黑色，覆盖于灰岩基岩表面，质地疏松多孔，灰岩表面凹陷处较厚，约为几十到上百微米。全球陆地广泛分布的多种"矿物膜"位于地壳最顶层，是地球上分布最广的自然界"太阳能薄膜"，从功能上相当于继地核、地幔和地壳之后的地球第四大圈层，构成了地球"新圈层"（鲁安怀等，2019）。岩石漆作为地表最常见的"矿物膜"，多发育于干旱、半干旱地区的砾石表面，富含水钠锰矿、赤铁矿、针铁矿、锐钛矿、金红石等半导体矿物，并富集 Mn 元素，富集程度可达地壳平均含量的 50~150 倍。Mn 在岩石漆剖面呈劈裂式分布，具显微层理结构（许晓明等，2017a；Xu et al.，2018，2019）。发育在喀斯特灰岩表面的"矿物膜"也含有以水钠锰矿为代表的锰氧化物。由于雨水冲刷，灰岩表面的"矿物膜"结构疏松多孔，膜内富含有机质组分（许晓明等，2017b）。土壤铁锰胶膜广泛发育在华中、华南等亚热带地区，富含以水钠锰矿、水羟锰矿、锰钡矿、赤铁矿、针铁矿为代表的多种铁锰氧化物矿物，具有一定的分层性，呈现外层富 Mn 内层富 Fe 的空间分布特征（Li Y. et al.，2017；Xu et al.，2018）。地表"矿物膜"的野外产出与日光照射之间存在密切联系，地表半导体矿物参与的光氧化过程在"矿物膜"的生长发育中至关重要（Xu et al.，2019）。

新近的研究显示，各种铁锰"矿物膜"均具有良好的日光响应能力，光电转换性能稳定、灵敏且长效，是天然的光电转化系统（Lu A. H. et al.，2019）。其中的水钠锰矿富集光催化性能强的 Ce 元素，可增强"矿物膜"转化太阳能效率（鲁安怀等，2019）。显然，"矿物膜"的光电效应，在矿物组合、矿物种和元素 3 个层次上，共同表现出与太阳光具有十分密切的作用关系，从内在特征上符合地表"矿物膜"是"太阳晒出来的"的外在认知，且具有潜在的产氧固碳作用（鲁安怀等，2020）。除地表"矿物膜"外，丹霞地貌的红层中也发现了类"矿物膜"结构，红层中富含以赤铁矿、锐钛矿为代表性的半导体矿物颗粒，表明地质历史时期"矿物膜"可能也具有重要的环境意义（Xiao et al.，2018；肖育雄等，2019）。地表广泛分布的半导体"矿物膜"产生的太阳光响应和光电流可能在地球圈层交互作用包括表生地球化学过程中发挥着重要作用。

1.2 矿物能带结构与光电转化效应研究方法

由于天然矿物样品成分与结构复杂，传统方法如紫外漫反射吸收谱难以得到与之禁带宽度相对应的吸收陡边。经过理论与实验验证，近年来有研究者针对天然矿物建立了紫外可见漫反射谱—同步辐射吸收和发射谱联用—第一性原理能带计算—液相光电化学性能测试—原位微区光电性能测试的系统研究方法。利用同步辐射 O 原子 K 边吸收谱和发射谱联用的方法，建立了天然金属氧化物矿物禁带宽度测定技术，并已用于测量天然铁锰氧化物矿物的禁带宽度（丁聪等，2016；Lu A. H. et al.，2019）；基于密度泛函理论，通过对晶体结构与晶格缺陷的构建与调控，从理论上分析晶体化学因素对电荷密度分布、导带与价态轨道组成、禁带宽度等半导体性质的影响（丁聪等，2015；Li et al.，2018b，2019a，2019b）；在岩石薄片上利用电子束蒸发镀膜技术蒸镀电极到薄片表面，配合电流放大器、显微镜与位移机械控制载物平台构建原位显微光电流测试装置，可有效进行微观和介观尺度下矿物界面反应机制、电子与能量转移过程、光电转换效率测量等定性与定量研究（Lu A. H. et al., 2019）。

基于微生物燃料电池（microbial fuel cell, MFC）和半导体光催化理论，新构建开发低廉高效治理污染的光燃料电池（light fuel cell, LFC）体系，能够同时实现两种污染物在微生物阳极的氧化和半导体矿物阴极的还原（Lu and Li, 2012；Lu et al., 2012a）。利用天然半导体矿物金红石构建"金红石光催化-微生物"双室体系，揭示了光电子强化微生物还原作用（丁竑瑞等，2011a）；利用天然铁氧化物、铁硫化物搭建"半导体矿物-微生物"实验体系，证明光催化增强微生物胞外电子转移能力（丁竑瑞等，2011b）；利用天然黄铁矿开展"黄铁矿-产电微生物"体系研究，明确二者之间具有良好的电子转移活性（丁竑瑞等，2012）。

1.3 矿物光催化作用与光电子能量

研究发现，地表阳光直接照射到的土壤与岩石表层区域广泛分布着金属半导体矿物。这些半导体矿物的化学成分、晶体结构、晶格缺陷等矿物学特征影响其半导体性质与日光催化能力，使其在地质时期与现代行星表面发挥多种环境效应。研究者对金红石、闪锌矿、锌铁尖晶石、赤铁矿、针铁矿及锰氧化物矿物开展了系统的半导体特性研究，结果表明天然闪锌矿与金红石中非本征金属原子的3d轨道可参与价带形成或构成禁带中分立的缺陷能级，由此可显著增强其可见光吸收和催化活性(Li et al., 2018b)；锌铁尖晶石中的反位缺陷会诱发结构中产生氧空位，降低导带底能级位置并在禁带中产生杂质能级，综合导致其光吸收范围和光催化氧化能力显著增强（Li et al., 2019a）；铁、钛氧化物半导体矿物的光催化氧化效应可促进地表Mn^{2+}的氧化与水钠锰矿的形成，氧化速率与地表矿物膜中实际聚锰速率吻合（Xu et al., 2019）；水钠锰矿中的Mn空位缺陷、重金属配位结构等精细结构特征均能不同程度地影响导、价带构成，同时引入杂质能级，影响日光照射下地表水钠锰矿的环境稳定性（Li et al., 2019b）。基于翔实的实验室研究，我国学者提出天然矿物的光电子能量是自然界中继太阳光子能量和元素价电子能量之后的第三种重要能量形式的学说，可在重大的地质历史事件（如生命起源、大氧化事件等）中发挥重要作用（鲁安怀等，2014；Lu A. H. et al., 2013, 2019）。

1.4 矿物光电子促进非光合微生物生长代谢

半导体矿物能够通过光催化作用将太阳光能转化成化学能，产生的光电子能够促进和支持非光合微生物的生长代谢。十余年来，研究者开展了半导体矿物金红石、赤铁矿、针铁矿、锰钾矿、闪锌矿、黄铁矿、磁黄铁矿和水钠锰矿等与氧化亚铁硫杆菌、粪产碱杆菌、异化金属还原菌、铜绿假单胞菌等的协同电子转移作用研究（Li et al., 2009；王鑫等，2011；丁竑瑞等，2011a, 2012；Wang et al., 2017；任桂平等，2017a, 2017c；鲁安怀等，2018）。结果表明，电极表面微生物菌体及代谢产物，能够影响半导体矿物的光子-电子的转化效率（Zeng et al., 2012）。半导体二氧化钛纳米颗粒可促进地杆菌细胞到电极的胞外电子转移，显著改善细菌的胞外电子转移能力，并通过刺激*pilA*的表达特异性诱导导电纳米线的形成（Zhou et al., 2018）。

地表"矿物膜"中的半导体矿物与微生物作用关系得到重点关注。研究者对西北戈壁岩石漆、西南喀斯特地貌、南方红壤3种生境"矿物膜"中半导体矿物与微生物作用进行了系统研究（Ren et al., 2018c, 2019；Lu A. H. et al., 2019）。在微生物群落层次，明确了天然赤铁矿间接促进红壤微生物群落胞外电子传递作用，太阳光照可显著促进化能自养型微生物：嗜酸性氧化亚铁硫杆菌（*Acidithiobacillus ferro-oxidans*）和化能异养型微生物：粪产碱杆菌（*Alcaligenes faecalis*）的生长，使红壤中微生物的群落构成发生显著改变，多样性降低（曾翠平等，2011；Lu A. H. et al., 2012）；证实了岩石漆"矿物膜"与本源微生物群落间的直接电子传递进程；构建了模拟半导体矿物光电子红壤群落演化系统，揭示了自然界"日光-半导体矿物-微生物"系统的电子传递过程（Ren et al., 2016；任桂平等，2020）。在细胞层次，构建"日光-半导体矿物-微生物"体系开展赤铁矿、水钠锰矿与铜绿假单胞菌（*Pseudomonas aeruginosa*

PAO1 电子传递研究，明确了光照下铁锰氧化物协同促进 PAO1 胞外电子传递的微观机制（Ren et al., 2017b，2018a）。进一步研究显示，一定外源电子能量可增强 A. faecalis 反硝化能力，且电子能量与反硝化能力相关（余萍等，2013）。这说明一些非光合微生物虽然不能直接利用光能合成有机物质，却可以利用半导体矿物光催化产生的光电子作为生长代谢的能量来源。基于系统深入的研究，我国学者首次证实日光下半导体矿物光电子可被非光合微生物生长代谢所利用，提出微生物光电能营养能量代谢新途径（Lu et al., 2012；鲁安怀等，2013）。

2 矿物与微生物交互作用的环境效应

在日光照射的地表半导体矿物-微生物体系中，微生物与矿物之间的协同作用可强化微生物和矿物各自拥有的某些环境功能。矿物-微生物界面电子转移过程不仅能够驱动矿物发生溶解、沉淀、转化等多种反应，而且对微生物的生长、代谢和互营协作都有重要作用，深刻影响着微生物、矿物及其与环境的共同演化，并在推动铁、碳、氮、硫等元素的地球化学循环，以及生物成矿、环境治理、新能源开发等方面都具有十分重要的意义（刘娟等，2018）。

2.1 微生物与铁氧化物界面电子传递机制

研究表明，异化铁还原菌 *Geobacter metallireducens*、*Shewanella oneidensis* 等可将铁氧化物矿物中的 $Fe(III)$ 作为终端电子受体进行胞外呼吸（Shi et al., 2016；邱轩和石良，2017）。*Hodopseudomonas palustris* TIE-1、*Sideroxydans lithotroohicus* ES-1 等金属氧化菌能以多种形态的亚铁离子作为电子的能量来源，还原氧气、二氧化碳和硝酸盐，促进自身的生长代谢，同时诱导生物成矿过程（Liu J. et al., 2012，2013）。

在无氧-微氧环境中，微生物异化铁还原、含铁矿物还原溶解等反应会释放亚铁离子，与高活性的磁铁矿、水铁矿等铁氧化物矿物反应。在地表环境氧化-还原状态发生波动时，磁铁矿可以作为天然蓄电池存储或者释放电子，调节环境的氧化-还原电势（任桂平等，2017a）。新近研究揭示了磁铁矿-亚铁离子界面电子双向流动的动态变化过程以及磁铁矿表面被腐殖酸覆盖以后，其介导微生物种间电子传递机制发生显著变化的内在原因，明晰了磁铁矿作为天然蓄电池对环境氧化-还原电势的调控机制。磁铁矿主要依靠晶格中相邻 $Fe(II)$-$Fe(III)$ 原子之间的电子跳跃传递电子，以促进不同微生物之间的电子传递；而磁铁矿表面的腐殖酸覆着层会抑制电子从微生物注入矿物晶格中，因此电子主要经由磁铁矿表面的腐殖酸进行传递，导致其促进微生物种间电子传递的效率和机制发生变化（Peng et al., 2018，2019；You et al., 2019）。此外，亚铁离子与铁氧化物之间的界面电子传递还可能催化结晶度较低的水铁矿发生晶相转化。Sheng 等（2020）利用与 $Fe(III)$ 离子具有强络合能力的二甲酚橙，提取并定量测量 $Fe(II)$ 催化水铁矿晶相转化过程中形成的中间介体活性 $Fe(III)$，揭示了亚铁催化水铁矿相变过程。将不同条件下水铁矿复杂多变的相变过程用活性 $Fe(III)$ 浓度作为唯一关键中间变量的晶相转化动力学模型进行定量描述，为进一步认识微生物介导的水铁矿相变过程提供了新的研究思路。

2.2 蛋白质与铁氧化物界面相互作用机制

蛋白质分子与矿物界面相互作用是一个复杂的动态过程，不仅与蛋白质本身的结构和性质有关，也显著受控于矿物的粒径、氧化-还原活性等性质。近年来的研究认为，嗜中性铁氧化菌 *Sideroxydans lithotroohicus* ES-1 外膜细胞色素蛋白 MtoA 可以直接氧化磁铁矿晶格中的 $Fe(II)$（Liu et al., 2012）。MtoA 与磁铁矿的界面电子传递速率与磁铁矿晶格中 $Fe(II)/Fe(III)$ 的比例，即磁铁矿的氧化-还原电位直接相关（Liu J. et al., 2013）。模式铁还原菌希瓦氏菌的跨膜电子传递链（Mtr pathway）由细胞色素蛋白 CymA、Fcc3、MtrA、MtrC、OmcA、STC 及孔蛋白 MtrB 组成（Shi et al., 2016）。外膜蛋白 OmcA 还原不同粒径赤铁矿颗粒的速率并不是随着粒径的减小而增大，而是 173 nm ≫ 15 nm > 30 nm > 55 nm（Liu et al.,

2016a)。173 nm 赤铁矿颗粒显示出更大的反应速率，与 15 nm 赤铁矿颗粒表面能更易形成结构紧密的团聚体有关（Liu et al.，2016b）。

蛋白质吸附在矿物表面，不仅造成矿物颗粒表面电荷不均一，还可能导致颗粒之间产生空间位阻力、排空作用等特殊作用力而改变矿物颗粒的团聚状态。当较低浓度的 OmcA 存在时，赤铁矿颗粒表面局部被 OmcA 覆盖。被 OmcA 覆盖的表面与暴露的矿物表面因带相反电荷而相互吸引，导致颗粒在高盐浓度下分散性增强的特殊现象（Sheng et al.，2016b）。当矿物颗粒浓度一定时，OmcA 比模式球蛋白牛血清蛋白 BSA 对赤铁矿颗粒分散状态的影响更明显（Sheng et al.，2016a）。

为达到热力学稳定状态即最小自由能，游离态蛋白质分子的多肽链会自发折叠形成一定的空间结构。与常见的有机小分子-重金属离子不同，蛋白质分子结构在矿物表面吸附过程中会发生调整，直接影响其生化活性和吸附行为。蛋白质分子在矿物表面的吸附是一个复杂的动态过程，受控于蛋白质-矿物相互作用、吸附态蛋白质空间结构变化、相邻蛋白质分子相互作用等多个因素，显著影响吸附态蛋白质的活性、矿物的表面性质及蛋白质-矿物界面电子传递等（Liu F. et al.，2019）。

2.3 厌氧微生物分解铁氧化物和硫酸盐协同作用

受电子传递速率制约，厌氧体系中微生物分解铁氧化物的速率十分缓慢。近年来的一系列研究成果，揭示了铁氧化物与硫酸盐共存体系中微生物分解矿物的反应路径、产物、电子传递方式、增强有机物厌氧转化及固碳的协同作用机制（Chen et al.，2014a）；体系中纳米磁赤铁矿-磁铁矿固相转化具有储存和释放电子的电池效应，能调控微生物代谢及其与矿物的交互作用活性（Zhou et al.，2019）；其中的铁氧化物和硫酸盐分别增强和抑制二氯酚类有毒有害物质的厌氧转化，而一定浓度的二氯酚（2~5 mg/L）对硫酸盐还原菌代谢有增强作用（Tang et al.，2018）。基于这些研究结果，研究者提出了利用铁氧化物、硫酸盐矿物的协同作用调控有机物转化、控制污染、实现低成本碳封存的新原理，可用于指导人为调控土壤、河流和湖泊沉积物、地下水、有机废物中 C、N、S、P、As 元素循环以及有机废物转化为无机碳封存，在环境污染控制领域具有一定的应用前景。例如，在高浓度硫酸盐废水中添加赤铁矿，可促使废水中的 S^{2-} 以 FeS_2 形式固定，有效加速了废水厌氧消化进程，同时降低反应器中硫化氢的浓度，从而实现甲烷的高值化产出（黄绍福等，2019）。

2.4 微生物与黏土矿物相互作用

微生物对矿物的影响，直观体现为对矿物成分、结构的改变。对胶质芽孢杆菌分解硅酸盐矿物的研究发现，微生物细胞作用于矿物晶面可导致矿物局部结构坍塌或畸变（朱云等，2011；杨晓雪等，2013）。异化铁还原菌 *Shewanella putrefaciens* CN32 和 *Shewanella oneidensis* MR-1 通过对天然蒙脱石结构中三价铁的还原促进矿物相转变（曹维政等，2011）。厌氧条件下，异化铁还原菌 *Cronobacter sakazakii* 能够还原改变针铁矿晶体结构（Wang et al.，2017）。

黏土矿物与微生物相互作用对生物还原 Cr(VI) 和降解有机污染物的影响机制研究表明：高岭石和蛭石均可显著提高铜绿假单胞菌对电子供体葡萄糖的利用率，降低反应体系中 Cr(VI) 对铜绿假单胞菌的毒害性，增强铜绿假单胞菌对 Cr(VI) 的耐受性，提高细菌的增殖速率和生物量，延长细菌的生命周期（Kang et al.，2015）；利用高岭石吸附固定和包埋法固定受污土壤筛选出的鞘胺醇单胞菌，提高了鞘胺醇单胞菌对苯酚的降解效率，且缩短降解周期，增强了鞘胺醇单胞菌对环境的适应能力（Gong et al.，2016；Ruan et al.，2018a）；蒙脱石、改性蒙脱石与细菌形成矿物-微生物复合体，不仅能够促进细菌自身生长繁殖，提高污染物的降解效率，而且与鞘胺醇单胞菌相互作用后，蒙脱石及改性蒙脱石的团聚状态明显减弱，细菌活动增强了蒙脱石结构中主要元素向溶液培养基的释放；相比八面体片层中 Al 的溶出，来自矿物四面体片层中的 Si 优先溶出且溶出量更大（Ruan et al.，2018b，2018c）。

黏土矿物的层板可以防止重金属和 DNA 酶进入层间域与 DNA 分子接触，从而保护 DNA 分子（Hou

et al., 2014a, 2014b；Wu et al., 2014a）。高岭石对于暴露于亚抑菌浓度抗生素下的大肠杆菌具有一定的缓冲和保护作用，细菌与高岭石的相互作用增加了双功能天冬氨酸激酶、对氧不敏感的烟酰胺腺嘌呤二核苷酸磷酸硝基还原酶和琥珀酸半醛脱氢酶的表达，促进了细菌的氧化还原代谢（Lai et al., 2019）；蒙脱石对金黄杆菌有保护作用，可促进细菌的生长及其对镉的耐受性，通过刺激并调节与金黄杆菌转运、代谢氧化应激镉富集及其他通路相关过程的基因表达来提高镉抗性（Wang et al., 2020）。

2.5 微生物风化蛇纹石作用

在生物风化过程中，某些真菌可以通过酸化和络合反应促进硅酸盐矿物中元素的增溶和活化。矿山尾矿土壤中筛选的显著耐受高 Mg 和 Ni 的特定菌株 *Talaromyces* sp. 与利蛇纹石的相互作用研究显示，存在真菌细胞时，由于真菌在溶液中产生有机酸，包括草酸、葡萄糖酸、甲酸和富马酸，蛇纹石的溶解会增强，Mg 的释放速率和效率显著提高；蛇纹石的溶解也与温度和矿物粒径有关，溶解速率与颗粒表面积呈非线性关系。由于 Mg 的溶出，蛇纹石表面附着一层无定形镁白炭黑。本源黄曲霉菌促进利蛇纹石溶解的研究显示，黄曲霉菌使蛇纹石中 Mg 和 Si 的释放量增加了 2 倍，而 Fe 和 Ni 的释放量增加了 10 倍以上，Fe 和 Ni 的释放主要通过连接剂促进途径进行，铁载体和草酸分别对铁和镍的增溶起作用。有研究表明，硅酸盐在生物风化过程中元素释放的速率和机制具有配体特异性，可能涉及协同效应和抑制效应。研究结果既有助于人们进一步了解真菌在根际地球化学和生态学中的关键作用，又可以提高蛇纹石等矿物的阳离子释放效率，从而达到固碳和资源回收的目的（Li Z. B. et al., 2015, 2019；Liu Y. Y. et al., 2017）。

2.6 矿物与微生物协同作用治理环境污染物

目前已发现具有半导体性质的金属氧化物和硫化物矿物大多具有可见光波段的响应与吸收，在日光作用下可产生还原性光电子和氧化性自由基，能够不同程度地降解偶氮有机物与卤代烃、还原 Cr(VI)、抑制有害微生物的活性等（Yang et al., 2011；Li et al., 2018a；Ren et al., 2018b；Li L. H. et al., 2019）。在矿物–微生物协同作用系统中，天然金红石光电子可还原 Cr(VI) 和降解偶氮染料（Li et al., 2009；丁竑瑞等，2011a）；纳米赤铁矿光电催化可降解苯酚（任桂平等，2017b）；水钠锰矿光电催化氧化能降解甲基橙（任桂平等，2017c）；人工合成的硫镉矿–硫化土杆菌，可在光的驱动下促进甲基橙（MO）的生物还原（Huang S. F. et al., 2019）；腐殖质–铁氧化物可协同促进 2,4-二氯苯氧乙酸的微生物厌氧脱氯降解（武春媛，2012）。在传统 MFC 基础上，进一步利用太阳能电池耦合转化太阳能作用，研发出两类防治污染新系统（Ding et al., 2014）：双阳极 MFC 复合系统，微生物氧化作用与半导体矿物光催化还原作用共同强化，实现 Cr(VI) 的高效阴极去除（Ren et al., 2018b）；半导体矿物（水钠锰矿）–太阳能电池复合系统，显著提高光电催化有机废水脱色性能（Ren et al., 2017a；任桂平等，2017d）。

3 纳米矿物及其环境功能

3.1 黏土矿物改型改性作用的环境功能

地表系统中广泛存在的黏土矿物因具有颗粒小、比表面积大、吸附能力强等特性，历来是广受关注的天然环境材料。下面简要回顾近十年来这方面的主要研究进展。

3.1.1 蒙脱石改性产物吸附降解污染物

蒙脱石及其热改性产物吸附性能研究表明，随着热处理温度的升高（小于 650℃），蒙脱石层间结合水甚至结构水消失，对间二硝基苯的吸附能力增强（杨青霞等，2016）。蒙脱石中引入 K^+ 可营造有利的吸附场所、增加吸附点位，从而促进并强化蒙脱石对硝基苯的吸附（江强明等，2011）。以蒙脱石为基体，烷基季铵盐和中性胺为模板剂，正硅酸乙酯（TEOS）为硅源，可合成具有更大比表面积和总孔容的多孔异质结构蒙脱石，对甲苯的平衡吸附量达到蒙脱石原土的 5.4 倍（Wang et al.，2015a，2015b，2016a，2016b）。利用聚羟基铁、谷氨酸螯合铁等对蒙脱石无机柱撑改性后，比表面积和孔容增大，并具有优异的光催化降解有机污染物性能（Huang et al.，2014，2016）。聚羟基铁-蒙脱石复合材料通过表面络合和阳离子的桥键作用增强吸附四环素（谢函芮等，2015；Wu et al.，2016）。可见光作用下，该复合矿物材料作为 Fenton 试剂可提高苯酚的降解速率（Wei et al.，2017a）。将聚羟基铁铝共同柱撑的蒙脱石负载在磁铁矿上，既能克服磁铁矿容易团聚的不足，又可提高柱撑蒙脱石的催化性能，从而快速去除水体中的苯酚（Wei et al.，2017b）。利用聚羟基铁和蛭石与二氧化锰复合，可提高二氧化锰对铊的去除率，实现同步吸附-氧化铊（Chen M. Q. et al.，2017，2019）。Fe^{2+} 在蒙脱石和高岭石表面的吸附形态与吸附量是制约构筑的矿物界面结合系统还原转化邻硝基苯酚的关键因素（梁剑滔等，2019）。利用蒙脱石和蛭石等黏土矿物作为铁、银和钯等纳米颗粒的分散载体，可有效克服纳米颗粒容易团聚的不足，增强纳米颗粒对氯酚等卤代污染物的还原能力（Wu et al.，2014b）。其中有机蒙脱石作为纳米铁颗粒的载体，形成了由纳米铁颗粒与有机蒙脱石片层堆垛而成的"卡房"结构，使得纳米铁颗粒对 Cr(Ⅵ) 的还原能力提高（Wu et al.，2012b）。蒙脱石和针铁矿等矿物独特的"模板"效应，能促进吸附在矿物层间域或表面的有机物热解形成碳纳米材料，得到具有催化性能的纳米碳-矿物复合材料，它可以高效活化过硫酸盐催化降解有机污染物（Yang S. S. et al.，2017，2020b），表明矿物衍生碳纳米材料具有优异的电催化性能和活化过硫酸盐的能力（Yang S. S. et al.，2016，2020a）。利用热还原法制备的新型零价铁-黏土矿物复合材料，可实现环境中内分泌干扰物的快速去除（Yang et al.，2018a，2018b）。

利用含有不同特征官能团的非离子型有机改性剂对蒙脱石和蛭石等黏土矿物进行有机改性，可增强黏土矿物与重金属之间的亲和力，提高黏土矿物对重金属的吸附去除能力（Wu et al.，2012b；Long et al.，2013，2014；Tran et al.，2015a，2015b）。用蒙脱石作为催化剂 BiOBr 的载体，可得到 BiOBr/蒙脱石复合材料，在可见光作用下高效降解罗丹明 B（Xu et al.，2014）。利用抗坏血酸和亚铁离子对蒙脱石改性，得到一种抗氧化型还原功能矿物材料，将重金属 Cr(Ⅵ) 还原为低毒性的 Cr(Ⅲ)，Cr(Ⅲ) 可能形成氢氧化物或被蒙脱石吸附在表面而被固定（杜巍等，2015）。两性表面活性剂的疏水碳链作用于黏土矿物能有效提高其疏水性，增强黏土矿物对有机污染物的吸附去除能力（Liu et al.，2017a，2017b）。两性表面活性剂改性的黏土矿物还可作为吸附剂处理环境中重金属-有机物复合污染问题（Liu C. M. et al.，2016，2018）。

3.1.2 黏土矿物复合材料的环境功能

不同浓度盐酸溶液热酸活化处理坡缕石，可显著提高其比表面积和总孔容，以及对苯和甲苯的动态吸附容量，且具有良好的再生性（Zhu J. X. et al.，2018）。将铜锰同时负载于坡缕石表面形成的复合材料催化活性强，对甲醛的降解有显著效果（Liu P. et al.，2018）。稀土盐类负载于高岭石和埃洛石表面制备的高岭石-碳酸氧镧、埃洛石-碳酸氧镧复合体，可大大增加对磷的吸附位点，对磷酸根的吸附量和吸附效率远高于碳酸氧镧及两种矿物（Wei et al.，2018，2019）。具特殊管状结构的埃洛石可提供更多的负载空间，对氨基三唑客体分子的负载容量比高岭石大；但氨基三唑从改性高岭石上的释放速度更慢，造成这种差异的主要机制在于片状高岭石有更高比例的层间氨基三唑，而氨基三唑在大颗粒高岭石中扩散路径更长（Tan D. Y. et al.，2015）。

采用溶胶-凝胶法制备的 TiO_2-硅藻土光催化剂，使负载在硅藻土载体上的锐钛矿相转变温度显著提高到近 900℃，光催化活性更强（Xia et al.，2014）。采用水洗净化技术对有机硅藻土提纯，并在还原气氛中炭化，硅藻土中的有机物可以转化为吸附在其表面的无定形碳，对溶液中亚甲基蓝三水合物和甲基橙的吸附能力提高。炭化硅藻土中方石英结晶相形成与石英结晶相转变温度高于非炭化硅藻土，炭化硅藻土的硅藻壳热稳定性高于非炭化硅藻壳，经 900～1 100℃ 高温炭化的硅藻土表现出较强的吸附性能（郑权男等，2013）。

3.2 纳米铁锰矿物及其环境功能

3.2.1 纳米黄铁矿环境功能

近年来的研究发现，表生自然环境中可形成天然纳米黄铁矿，如安徽铜陵地区的胶状黄铁矿。该矿区存在生物质残体包裹的磁黄铁矿，为四方硫铁矿向黄铁矿转化的中间产物，显示胶状黄铁矿的生物化学沉积成因，其中铁还原菌、硫酸盐还原菌在胶状黄铁矿–菱铁矿建造形成中发挥了重要作用（谢巧勤等，2014；Xu et al.，2017；徐亮等，2017，2019）。研究者揭示了胶状黄铁矿的纳米矿物学特征、相转变和化学反应纳米效应，发展了系列黄铁矿纳米结构化环境材料加工和应用技术（李平等，2013；Chen et al.，2014b；Yang Y. et al.，2016，2017a）。保护气氛下热处理胶状黄铁矿结构演化规律、产物的反应活性及其应用性能研究表明，随着温度升高，黄铁矿脱硫相变速度会越来越快，并且存在黄铁矿—单斜磁黄铁矿—六方磁黄铁矿—陨硫铁连续转变过程。在氮气中最佳温度下活化黄铁矿可获得主要由纳米单斜磁黄铁矿组成、具有纳米孔隙结构的功能材料。该材料回收工业、矿山含重金属废水中的铜、汞有良好效果（Chen et al.，2013，2014b；Liao et al.，2016；Yang et al.，2017b；Lu P. et al.，2019），也可用于城市生活污水、富营养化水体同步脱氮除磷深度处理等（Chen et al.，2015，2016），在治理环境污染先进矿物工程材料方面显示出巨大潜力。

3.2.2 纳米褐铁矿、针铁矿和水铁矿环境功能

褐铁矿是表生环境中广泛存在的以纳米针铁矿为主要组成的细分散混合物。对铜陵及其周边地区广泛发育的褐铁矿矿床的系统研究（Chen P. et al.，2017，2018；陈平，2019）发现，褐铁矿的矿物组成、特征元素因成因不同而呈规律性变化，针铁矿中 Al、Mn 类质同象替代以及表面吸附硅酸根、磷酸根离子影响和制约针铁矿的晶体生长、表面物理化学特性（Li M. X. et al.，2019；Zhang et al.，2019）。主要由粒径 10~90 nm 针铁矿组成的褐铁矿，针铁矿堆积还形成粒间 10~80 nm 空隙，是重要的纳米矿物资源，具有非常高的化学活性、热活性、生物活性；在空气气氛下热活化演变为不同结晶度的赤铁矿，还原气氛下演变为磁赤铁矿、磁铁矿、零价铁（Liu et al.，2013a，2013b，2015）。褐铁矿具有一定的强度，通过破碎筛分可以得到所需粒径的颗粒材料，进一步热活化可以得到具有赤铁矿、磁赤铁矿、磁铁矿、零价铁等不同物相的三级孔结构化材料（Bao et al.，2014，2017）。褐铁矿及其衍生的系列材料可用于地下水除砷、锰、铁，废水、水体深度除磷，催化氨还原烟气脱硝（Zhang et al.，2018），室温下催化净化室内空气中的甲醛，生物质气化脱焦油（Zou et al.，2016，2018），催化热化学氧化废气中 VOCs（Chen R. et al.，2017；Xiang et al.，2019）等。

利用 Fe^{2+} 对针铁矿掺杂改性，部分阳离子进入针铁矿晶格，可提高针铁矿对邻硝基苯酚和橙黄 G 的还原能力（赵丹等，2015；杜可清等，2016）。针铁矿还可作为 H_2O_2 的催化剂，实现对橙黄 G 的脱色与降解（Wu et al.，2012a）。水铁矿/二氧化钛复合材料基于光催化和 Fenton 体系可高效降解抗生素头孢噻肟（Jiang et al.，2019）。负载于水铁矿表面的 Ag/AgBr 和 Ag/AgCl 产生的光生电子既能促进水铁矿中的 Fe(Ⅲ) 还原为 Fe(Ⅱ)，又能通过减少用于还原 Fe(Ⅲ) 的 H_2O_2 消耗量，提高 H_2O_2 的有效利用率（Zhu et al.，2018a，2018b）。

传统观点认为，在河流、湖泊、海洋沉积物及水稻田等厌氧环境中，铁氧化物属于电子受体并与产甲烷菌竞争电子而抑制甲烷产生。有研究发现，纳米铁氧化物矿物在富有机物体系中能促进甲烷的产生（姚敦璠等，2013；Zhang et al.，2019），表现为添加针铁矿、赤铁矿、水铁矿等纳米铁氧化物矿物均可提高富有机物厌氧体系甲烷的产出速率、产率、产气中甲烷浓度、产气热值和底物利用率，降低有机物残余浓度及产气中二氧化碳和硫化氢浓度，促进体系中碳、硫固定。原因在于，在富营养并以较稳定有机物为主的厌氧体系中，有机物电子供体充足、铁氧化物反应动力学制约明显，即铁氧化物还原分解消耗电子流量相比产甲烷过程消耗的电子流量可以忽略不计，铁氧化物抑制产甲烷的作用无从表现（Peng et al.，2014；Tan J. et al.，2015；Yue et al.，2015a，2015b，2016；Zhu et al.，2015；Yao et al.，2016，

2017）。这些研究对推动铁氧化物纳米矿物资源在生物质可再生能源领域应用和提高生物质能源转化效率上有重要的理论和实际意义，可望应用于垃圾发电厂和垃圾填埋场产生的垃圾渗滤液、养殖粪污、餐厨垃圾等厌氧发酵沼气化工程。

3.2.3 纳米磁铁矿、磁赤铁矿和绿锈环境功能

共沉淀方法合成不同锰掺杂量磁铁矿随掺杂量从 0 增加到 1.0，氧化还原性能增强，90% 的甲醛去除率所需最低温度由 297℃ 降低至 199℃（Liu P. et al., 2016）。不同温度下热处理锰掺杂磁铁矿（掺杂量为 0.5）结果显示，400℃ 焙烧，磁赤铁矿表面 Mn^{4+} 含量和 Mn/Fe 原子比最高，甲醛催化效果最好，90% 的甲醛去除率所需最低温度为 232℃（Liang et al., 2016）。几种过渡金属离子置换的磁铁矿对 Pb(II) 的吸附研究表明，其表面羟基是主要吸附位点，Pb(II) 通过形成双核三齿内圈络合物吸附于磁铁矿表面，类质同象替代制约着磁铁矿对 Pb(II) 的吸附（Liang et al., 2017）。合成锌置换磁铁矿具有高的硝基苯还原性（Li et al., 2018c）。将水铁矿包裹在磁铁矿表面并将镧吸附于其表面，可制备一种具有高磷酸根吸附能力的易回收矿物材料（Fu et al., 2018）。

硫酸盐绿锈转化过程的吸附和共沉淀作用对 As(V) 有极强的去除能力，随 As(V) 浓度增加，绿锈氧化转化由溶解—氧化—沉淀机制向固态氧化机制过渡，产物由针铁矿和纤铁矿混合相向纤铁矿、水铁矿和高铁绿锈混合相转变，高 pH 或高空气流速或低温有利于高铁绿锈和水铁矿形成，高温有利于针铁矿形成（王小明等，2017）。

3.2.4 纳米锰氧化物环境功能

表生环境下常见的锰氧化物矿物参与土壤溶液中水溶性硫化物的氧化反应，影响硫化物的迁移、转化和归趋。缺氧环境中，水锰矿氧化水溶性硫化物主要形成单质 S，初始 pH 降低可加速 S^{2-} 的氧化反应；有氧环境中，氧气能将单质 S 进一步氧化为高价硫氧酸根离子，如 SO_4^{2-}，而水锰矿表现出良好的催化作用与化学稳定性，在溶液体系中保持较好的晶体结构和微观形貌（罗瑶等，2016）。常温碱性条件下，锰钾矿氧化 Na_2S 溶液中 S^{2-} 的主要产物是单质硫，另含少量 $S_2O_3^{2-}$、SO_3^{2-}、SO_4^{2-}；氧化速率随着温度升高、pH 降低和矿物用量增加而增大；锰氧化度越低，即 Mn(III) 含量高，锰钾矿的氧化能力越高；随着 S^{2-} 的氧化，锰钾矿首先被还原生成 $Mn(OH)_2$，后者在空气中与 O_2 作用转化成 Mn_3O_4 和 H_2O，Mn_3O_4 可进一步转化生成 MnOOH（李倩等，2011）。

生长过程中进入六方水钠锰矿中的 Ni^{2+} 大部分以 $[NiO_6]$ 八面体形式存在于水钠锰矿层内，仅有小部分存在于八面体空位上方。含 Ni 水钠锰矿层片状晶体逐渐变薄，比表面积显著增大。随着 Ni 含量增加，结构中锰氧八面体空位数减少，而层边吸附位点数基本保持不变，其对重金属离子（Pb^{2+}/Zn^{2+}）的吸附去除能力逐渐降低（殷辉等，2012）。水钠锰矿少量掺 Co 不改变其层状结构和微观形貌，但结晶度降低；随 Co 含量增加，水钠锰矿平均氧化度降低，Mn^{3+} 和 Mn^{4+} 含量分别与 Co 含量呈正相关和负相关；Co 主要以 $Co^{III}OOH$ 的形式存在于矿物结构中；掺 Co 水钠锰矿对 Pb 的吸附能力和 As 的氧化能力显著增强（殷辉等，2011）。

3.3　纳米水铝英石环境属性

水铝英石是广泛产出于火山灰风化壤、风化壳淋积层等表生风化环境的铝硅酸盐纳米矿物，是典型的环境指示矿物，其生成和演化记录了相应的地质环境特征。水铝英石及其管状多型矿物伊毛缟石形成生长过程中不同阶段产物的结构特征和形成机理研究表明，水铝英石-伊毛缟石形成初期发生的原水铝英石向原伊毛缟石的转化率，是影响地球表生风化体系中水铝英石和伊毛缟石比例的重要因素；研究提出了水铝英石-伊毛缟石的成因模式，探明了自然环境条件下原水铝英石和原伊毛缟石向水铝英石和伊毛缟石转化的机制，为理解水铝英石和伊毛缟石的广泛共生现象提供了理论依据（Du et al., 2017）。在火山喷发等热环境下水铝英石的物相结构、微区形貌和孔性等的精细变化及其机理的系统研究中，发现了水

铝英石在热作用下向莫来石转化的矿物学演化特征（Du et al.，2018）。这些认识为利用水铝英石结构反演其生成环境等提供了依据。

4 生物矿化作用及其环境效应

4.1 碳酸盐生物矿化作用的环境效应

微生物诱导的碳酸盐矿化对全球气候变化、元素地球化学循环和环境变化有着深远影响。近年来，围绕微生物在碳酸盐矿物形成过程中的作用，国内学者开展一系列深入研究，取得了显著的进展。例如，有研究者通过构建细菌原位矿化和仿生矿化相结合的特色、方法，定向研究了不同细菌组元［细菌菌体、裸菌、结合态胞外聚合物（EPS）、溶解态 EPS 以及小分子有机物］对碳酸钙变体形成及结构的影响，从分子水平揭示了不同细菌组元在生物成因碳酸钙变体形成和生长过程中的作用（Li H. et al.，2019）。大量研究显示，低温白云石的形成可能与微生物之间存在一定联系。研究者特别地选择与白云石形成环境相关的微生物作为模式菌株，开展不同细菌组元对钙镁碳酸盐形成影响的仿生矿化实验研究。结果显示，不同的细菌组元对 Ca-Mg 碳酸盐矿化有不同影响，细菌细胞在无序白云石的形成过程中发挥了决定作用；除了细菌代谢活动，无生物活性的细菌生物质也能促进无序白云石和高镁方解石的形成。这些研究不仅加深了人们对微生物诱导无序白云石及高镁方解石矿化机制的了解，也为认识自然界无代谢活性菌席中的低温白云石来源提供了实验依据，并对了解现代沉积环境和地质历史时期中无序白云石或白云石的形成具有重要指示意义（Huang Y. R. et al.，2019）。

4.2 生物矿化鸟粪石环境功能

在污水中利用微生物矿化作用回收鸟粪石（$NH_4MgPO_4 \cdot 6H_2O$），不仅能同时去除污水中的氮和磷，而且矿化产生的鸟粪石还可作为一种理想的缓释肥料，是当前水处理和环境矿物学研究的热点。研究者利用 *Shewanella oneidensis* MR-1 在仅含有机氮、磷源的模拟废水中成功矿化鸟粪石，并且首次实现了高达70% 的镁离子转化率；同时也证实了微生物能将有机氮、磷直接转化为鸟粪石，从而克服了化学沉淀鸟粪石只能回收废水中无机氮、磷的局限（Li H. et al.，2017）。在此基础上，研究者进一步利用菌株 *Shewanella oneidensis* MR-1 和海洋放线菌 *Microbacterium marinum* H207，分别在以廉价矿物方镁石和模拟海水作为镁源的情况下，实现了鸟粪石的矿化，这为鸟粪石生物矿化回收提供了更加经济有效的途径（Luo et al.，2018；Zhao et al.，2019）。这些工作不仅为实现污水不同种型氮、磷的生物去除，优化和发展鸟粪石的生物回收工艺提供了生物矿化作用指导，同时对治理水体富营养化、防止污水处理管道结垢、降低污水处理厂的氮、磷负荷等难题的解决也具有重要意义。

4.3 生物矿化及其改性蛋白石环境功能

硅藻矿化作用可形成硅藻蛋白石矿物，属于典型的生物成因矿物，具有优良的环境属性。研究表明，湖泊硅藻生物硅是地球的重要"铝汇"，揭示了"硅藻-生物硅-溶解铝"之间的独特界面反应机制，为研究硅藻驱动的 Si-C-Al 元素循环、硅藻沉积固碳作用提供了依据，对深入理解硅藻生物地球化学行为及其环境效应具有重要意义（Liu C. M. et al.，2019）。硅藻蛋白石的成岩过程记录了当时的地质过程和环境特征，其表面覆有厚几十纳米的并非源于共生黏土矿物的天然富铝铁氧化物薄膜（Yuan et al.，2019）。模拟研究（Liu D. et al.，2016）显示，硅藻蛋白石通过羟基缩合与其表面吸附铝原位形成 Si—O—Al 键，进而作为基底通过一系列界面作用形成了这层金属氧化物薄膜。研究结果"更新"了硅藻土矿物性质的认识。

研究者曾采用碱刻蚀法、气固相转晶法、表面改性+原位低温回流反应法等方法成功制备出吸附性能优异的硅藻蛋白石与沸石复合孔道材料（Yu et al., 2015a, 2015b, 2015c；Yuan et al., 2015, 2016a）。通过原位负载法将高比表面积、微孔结构丰富的水铝英石纳米颗粒均匀负载于硅藻蛋白石表面，制得具有独特多级孔道结构的、苯吸附性能良好的硅藻蛋白石–水铝英石复合材料（Deng et al., 2019）。用苯基三乙氧基硅烷对硅藻蛋白石改性，获得对苯具有强吸附作用的改性硅藻蛋白石，与水的接触角由0°增加至120°，对苯的Langmuir吸附容量比改性前提高了4.5倍（Yu et al., 2015b）。采用尿素–沉淀法将前驱体$MnCO_3$负载到硅藻蛋白石表面，经焙烧活化获得硅藻蛋白石负载型复合物，随着锰负载量的增加对甲苯的催化性能逐渐提高（Liu P. et al., 2017）。采用"硅藻蛋白石表面扩孔–离子掺杂"法制备的具多级孔结构的新型硅藻蛋白石基复合光催化材料，对亚甲基蓝的去除率最高可达99.1%（Yuan et al., 2016b）。表面负载氧化铝和氧化锰的硅藻蛋白石不仅保留了硅藻蛋白石的大孔结构，还存在微孔和介孔结构，具有较高的磷吸附容量（Song et al., 2019）。

5 人体病理性矿化作用

人体内的矿化是生命体系的一个子系统，是构成人体生命过程的若干事件之一。与人体的某些特定功能有关的功能性矿化是在人体内的特定部位发生的，并严格按特定的组成、结构和程度完成的受控过程，形成的矿物具有特殊的高级结构与组装方式（鲁安怀等，2012）。病理性矿化，也称异常矿化，属于失控过程，常常与疾病密切相关，矿化出现于不应发生的部位（即异位矿化）或在正常矿化部位矿化程度过高或过低。

软组织矿化是伴随病变出现的软组织内的矿物质沉着，是一种典型的异位病理性矿化。十余年来，乳腺、卵巢、甲状腺、心血管系统病变以及脑膜瘤等伴随的矿化得到关注与系统研究。这些发生于软组织的病理型矿化表现出许多共同特征。绝大多数的矿化产物为钙磷酸盐，少数情况如一些乳腺疾病、卵巢癌中出现钙的草酸盐矿物——草酸钙石或水草酸钙石。在钙的磷酸盐中，矿化产物主要是碳羟磷灰石（CHAP），而无定形磷酸钙（ACP）、磷酸八钙（OCP）、二羟焦磷酸钙（CPPD）和白磷钙石（WH）见于矿化初期或富镁的体液环境中，这些钙的磷酸盐矿物最终往往会转变为CHAP。矿化的归趋CHAP是有钙缺陷的B型碳羟磷灰石，即成分中的碳酸根主要取代磷酸根（Wang et al., 2011；王长秋等，2011；熊翠娥等，2011a, 2011b；Meng et al., 2015，张岩等，2017）。成分分析显示，这些矿化物的主要化学组成为Ca、P、O、C，并含Na、Mg、Zn、Sr等次要或微量元素（孟繁露等，2015）。这些钙磷酸盐矿物组成纳米多晶集合体，其中混有少量蛋白质等有机物。形态上，矿化物主要有球状和块状两种集合体形态。球状集合体有时具同心环或同心放射状构造，呈典型的砂粒体状，有时则为无圈层结构的致密球体，即类砂粒体。集合体大小不等，一般为微米级，个别可达毫米级（Li et al., 2014）。草酸钙往往具有较好的结晶，甚至形成具有良好四方晶系外形的草酸钙石单晶。

矿化物与胶原关系密切，一般依附胶原沉淀生长。在心血管系统及乳腺、卵巢、甲状腺等病变部位，矿化初始形成前驱相ACP或OCP纳米小球，这些小球会随时间推移而脱玻化或转变为更稳定的CHAP纳米晶体集合体，同时新生的ACP或OCP还会逐渐沉淀包裹在先形成的纳米小球外，形成具同心环结构的微米级矿化球。微米级矿化球内可以有数个纳米小球，纳米球体之间由ACP或OCP充填，而微米级矿化球团聚体外部也可有一层ACP或OCP。随着时间的推移，矿化沉淀量逐渐增加，分散的球状钙化集合体融合成团块状，同时，由于重结晶作用，前驱相逐渐转变为CHAP，最终形成均一的致密块状钙化（李源等，2015）。体液环境较富镁时，可先形成WH球体；随着WH的沉淀，环境的镁含量将趋于正常，进而转化为CHAP沉淀；而在缺乏足够磷的体液环境里，也会出现少见的草酸钙、甚至碳酸钙沉淀。人体其他部位的矿化集合体发生发展过程存在相似性。如颈椎黄韧带，矿化初期，首先形成柱状和粒状CPPD晶体；矿化后期，矿化集合体发生CPPD的分解和CHAP沉淀，前者为后者提供物质来源；CHAP沉积过程中，先由纳米级钙化物形成絮状矿化，后形成球状矿化（刘偌麟等，2017）。

金属元素在人体病理性矿物中的赋存状态及与外环境的离子交换关系有助于探究病理性矿化与疾病的发生发展关系。研究发现，Zn 在人体乳腺组织矿化中的含量高达 $100 \times 10^{-6} \sim 1\,000 \times 10^{-6}$ 量级；乳腺癌中 Zn/Ca 质量分数值明显高于乳腺纤维腺瘤矿化，矿化中 Zn 主要占据具有 Ca 缺陷的碳羟磷灰石的 Ca2 位点；人体乳腺组织矿化可通过对 Zn 元素的富集对人体产生重要影响（孟繁露等，2013）。

受篇幅限制，本文没有完全反映我国同行近十年来取得的环境矿物学研究成果。应该说，环境矿物学研究学科交叉性突出，在理论、方法和技术研究方面内容丰富，特别是在大气、水体和土壤等污染治理领域特色鲜明，尤其在"十三五"期间国家出台的土壤重金属防治专项研究工作中，环境矿物学研究得到充分体现，发挥着不可或缺的巨大作用。

6 结　语

环境矿物学已经成为矿物科学中一个重要的分支交叉学科。目前地球表层岩石圈与水圈、大气圈和生物圈交互作用产物中，具有环境响应的无机矿物及其形成过程，正在成为环境矿物学主要研究对象。地球关键带多个圈层交互作用中，无机矿物形成、发展与变化过程中所禀赋的生态生理效应，成为环境矿物学主要研究目标。环境矿物学又是一个多学科交叉的研究领域，矿物学、环境科学、材料科学等传统学科早已深刻地渗透其中并成为环境矿物学发展的重要基础。近年来，环境矿物学与现代生命科学的紧密融合，更从无机和有机地球系统层面上为环境矿物学发展提供了新的研究思路和手段，开辟了更为广阔的发展空间。

矿物作为反映环境变化的信息载体研究、矿物影响环境质量的本质及其防治方法研究、矿物评价环境质量的机制与方法研究、开发矿物治理环境污染与修复环境质量的重要功能研究、矿物与生物交互作用的精细过程与微观机制研究以及人体系统中矿化作用精细特征及其生理病理效应研究等，仍然是当前环境矿物学主要研究方向，可为地球环境质量管理提供矿物学方法及技术。水气土污染防治矿物环境功能研究、环境矿物材料开发应用研究、矿物半导体理论与应用研究、矿物与微生物交互作用研究、人体矿物病理与生理性研究以及矿物大数据与地球环境演化研究等是优先开展的研究方向。

发展矿物化学性质研究，深化矿物物理性质研究，揭示自然界中矿物与生物的内在联系，可成为今后矿物环境属性重点发展方向。矿物享有光电效应的特性，地表"矿物膜"具有明显的可见光光电响应，产生矿物光电子能量；矿物拥有非经典光合作用的性能，自然界无机矿物转化太阳能系统，矿物非经典光合作用具有光催化分解水产氧产氢作用以及固定二氧化碳作用，拓展自然界中有机界与无机界所共同拥有的光合作用模式；矿物具有促进生物光合作用的功能，生物光合作用中心 Mn_4CaO_5 在裂解水产氧过程中产生成分和结构类似水钠锰矿的结构中间体，探索矿物促进生物光合作用机理；均有待今后加强重点研究。地球表生系统中矿物与微生物共演化的微观机制及其环境效应研究，天然有机质和生物大分子调控矿物的形成及转化过程影响地表元素循环的机制与效应研究，地表日光照射下"矿物膜"上发生的矿物光电子传递与能量转化促进地球物质循环与环境演变以及地球生命起源与进化的微观过程与动力学机制研究等，充满着环境矿物学理论突破与技术发展的重要机遇。

参 考 文 献

曹维政,朱云,鲁安怀,李艳,王清华,张虓雷,王浩然,杨晓雪,王长秋,董海良. 2011. 两株异化铁还原菌与蒙脱石交互作用实验研究. 矿物岩石地球化学通报,30(3):311-316

陈平,陈天虎,徐亮,赵月领,周跃飞,徐晓春,谢巧勤. 2019. 铜陵叶山铁矿赤铁矿微尺度矿物学研究及地质意义. 岩石学报,35(1):177-192

丁聪,李艳,鲁安怀. 2015. 掺杂 Fe、Cd 闪锌矿电子结构的第一性原理计算. 岩石矿物学杂志,34(3):382-386

丁聪,李艳,李岩,鲁安怀. 2016. 同步辐射软 X 射线吸收谱与发射谱测定天然针铁矿能带结构. 岩石矿物学杂志,35(2):349-354

丁竑瑞,李艳,鲁安怀,王长秋. 2011a. 天然金红石可见光催化强化微生物还原作用的研究. 矿物学报,31(4):629-633

丁竑瑞,李艳,鲁安怀,王鑫,曾翠平,颜云花,王长秋. 2011b. 微生物还原铁氧化物矿物的电化学研究. 矿物岩石地球化学通报,30(3):299-303,310

丁竑瑞,李艳,鲁安怀. 2012. 双室电化学体系中产电微生物与黄铁矿单晶协同电子转移反应. 地球科学-中国地质大学学报,37(2):313-318

杜可清,吴宏海,朱慧琳,魏西鹏,管玉峰,蓝冰燕. 2016. Fe(Ⅱ)/阳离子掺杂针铁矿矿物系统对2-NP的还原转化性能研究. 岩石矿物学杂志,35(6):1075-1084

杜巍,吴宏海,魏西鹏,管玉峰,何广平,张延霖. 2015. Fe(Ⅱ)改性蒙脱石去除Cr(Ⅵ)的性能及其机理研究. 华南师范大学学报(自然科学版),47(6):63-71

黄绍福,叶捷,周顺桂. 2019. 赤铁矿抑制硫酸盐废水厌氧消化产甲烷过程中硫化氢形成与机制. 环境科学,40(4):1857-1864

江强明,张姚娜,吴宏海. 2011. 软阳离子钾改性蒙脱石矿物对硝基苯的强化吸附实验研究. 岩石矿物学杂志,30(6):1074-1080

李平,陈天虎,杨燕,谢巧勤,谢晶晶. 2013. 氮气保护下热处理胶状黄铁矿的矿物特性演化. 硅酸盐学报,41(11):1564-1570

李倩,俞颖,赵雅兰,朱丽君,冯雄汉,刘凡,邱国红. 2011. 锰钾矿氧化硫化物特性与动力学研究. 环境科学,32(7):2102-2108

梁剑滔,卢鹏澄,寇卓瑶,赵丹,吴宏海. 2019. 黏土矿物界面吸附Fe(Ⅱ)耦合对邻硝基苯酚还原转化的增强机理研究. 岩石矿物学杂志,38(6):775-781

刘娟,李晓旭,刘枫,张逸潇. 2018. 铁氧化物-微生物界面电子传递的分子机制研究进展. 矿物岩石地球化学通报,37(1):39-47

刘偌麟,姜亮,李艳,王长秋,丁竑瑞,李源,鲁安怀. 2017. 人体颈椎黄韧带中二羟焦磷酸钙和碳羟磷灰石的矿物学成因探讨. 中国科学:技术科学,47(6):646-655

鲁安怀. 2000. 矿物学研究从资源属性到环境属性的发展. 高校地质学报,6(2):245-251

鲁安怀,王长秋,李艳. 2012. 生命活动过程中无机矿化作用现象及其环境效应. 科学,64(4):14-17

鲁安怀,李艳,王鑫,丁竑瑞,曾翠平,郝瑞霞,王长秋. 2013. 半导体矿物介导非光合微生物利用光电子新途径. 微生物学通报,40(1):190-202

鲁安怀,李艳,王鑫,丁竑瑞,刘熠,王长秋. 2014a. 关键带中天然半导体矿物光电子的产生与作用. 地学前缘,21(3):256-264

鲁安怀,王鑫,李艳,丁竑瑞,王长秋,曾翠平,郝瑞霞,杨晓雪. 2014b. 矿物光电子与地球早期生命起源及演化初探. 中国科学:地球科学,44(6):1117-1123

鲁安怀,王长秋,李艳,等. 2015. 矿物学环境属性概论. 北京:科学出版社

鲁安怀,李艳,丁竑瑞,王长秋. 2018. 矿物光电子能量及矿物与微生物协同作用. 矿物岩石地球化学通报,37(3):1-15

鲁安怀,李艳,丁竑瑞,王长秋. 2019. 地表"矿物膜":地球"新圈层". 岩石学报,35(1):119-128

鲁安怀,李艳,丁竑瑞,王长秋,许晓明,刘菲菲,刘雨薇,朱莹,黎晏彰. 2020. 天然矿物光电效应:矿物非经典光合作用. 地学前缘,27,doi:10.13745/j.esf.sf.2020.5.35

罗瑶,李珊,谭文峰,刘凡,蔡家法,邱国红. 2016. 水锰矿氧化水溶性硫化物过程及其影响因素. 环境科学,37(4):1539-1545

孟繁露,王长秋,李艳,鲁安怀,梅放,柳剑英. 2013. 锌在人体病理性矿化灶中分布的地球化学讨论. 岩石矿物学杂志,32(6):789-796

孟繁露,李源,李艳,王长秋,鲁安怀,梅放. 2015. 人体乳腺癌矿化的同步辐射研究. 岩石矿物学杂志,34(6):957-962

邱轩,石良. 2017. 微生物和含铁矿物之间的电子交换. 化学学报,75(6):583-593

任桂平,孙曼仪,鲁安怀,丁竑瑞,李艳. 2017a. 天然赤铁矿促进红壤微生物胞外电子传递机制研究. 矿物岩石地球化学通报,36(1):92-97

任桂平,孙曼仪,鲁安怀,李艳,丁竑瑞. 2017b. 纳米赤铁矿电极光电催化特性及苯酚降解活性研究. 岩石矿物学杂志,36(6):825-832

任桂平,孙曼仪,鲁安怀,李艳,丁竑瑞. 2017c. 纳米水钠锰矿可见光光电化学响应与甲基橙降解活性. 矿物学报,37(4):373-379

任桂平,孙元,孙曼仪,鲁安怀,李艳,丁竑瑞. 2017d. 太阳能电池协同强化水钠锰矿光电催化染料降解研究. 岩石矿物学杂志,36(6):851-857

任桂平,鲁安怀,李艳,王长秋,丁竑瑞. 2020. 地表"矿物膜"半导体矿物光电子调控微生物群落结构演化特性研究. 地学前缘,27,doi:10.13745/j.esf.sf.2020.5.53

王长秋,赵文雯,鲁安怀,熊翠娥,梅放,柳剑英. 2011. 几种乳腺疾病矿化特征初步研究. 高校地质学报,17(1):29-38

王小明,彭晶,徐欢欢,谭文峰,刘凡,黄巧云,冯雄汉. 2017. As(V)浓度和环境因子对硫酸盐绿锈转化的影响及其机制. 化学学报,75(6):608-616

王鑫,鲁安怀,李艳,颜云花,曾翠平,丁竑瑞,王长秋. 2011. 天然闪锌矿光催化协同Acidithiobacillus ferrooxidans生长及抑制自身分解作用实验研究. 矿物学报,31(4):641-646

武春媛,周顺桂,李芳柏. 2012. 腐殖质/铁氧化物协同促进2,4-D微生物厌氧降解. 现代农药,11(5):15-19

肖育雄,黎晏彰,丁竑瑞,李艳,鲁安怀. 2019. 湖南新宁崀山丹霞红层天然半导体矿物的矿物学特征研究. 北京大学学报(自然科学版),55(5):915-923

谢函芮,吴宏海,管玉峰,何广平. 2015. 羟基铁聚阳离子插层蒙脱石对盐酸四环素的吸附特性研究. 岩石矿物学杂志,34(6):873-879

谢巧勤,陈天虎,范子良,徐晓春,周跃飞,石文兵,谢晶晶. 2014. 铜陵新桥硫铁矿床中胶状黄铁矿微尺度观察及其成因探讨. 中国科学:地球科学,44(12):2665-2674

熊翠娥,王长秋,鲁安怀,李艳,梅放,柳剑英. 2011a. 乳腺炎症及增生症病灶中钙化的矿物学研究. 岩石矿物学杂志,30(6):1014-1020

熊翠娥,王长秋,鲁安怀,李艳,梅放,柳剑英,朱梅倩,张波. 2011b. 乳腺纤维腺瘤病灶中钙化的矿物学研究. 矿物学报,31(4):713-718

徐亮,谢巧勤,陈天虎,周跃飞,徐晓春,庆承松,李平. 2017. 铜陵矿集区层状硫化物矿床成因——来自胶状黄铁矿-菱铁矿型矿石矿物学制约. 地质论评,63(6):1523-1534

徐亮,谢巧勤,周跃飞,陈平,孙少华,陈天虎. 2019. 安徽铜陵矿集区铜官山矿田胶状黄铁矿矿物学特征及其对成矿作用的制约. 岩石学报,35(12):3721-3733

许晓明,李艳,丁竑瑞,李岩,鲁安怀. 2017a. 3种典型富锰沉积物的形貌学与矿物学特征. 岩石矿物学杂志,36(6):765-778

许晓明,李艳,李岩,丁竑瑞,王浩然,鲁安怀. 2017b. 中国干旱与湿润地区岩石漆光谱学特征研究. 矿物岩石地球化学通报,36(2):299-307

杨青霞,吴宏海,杨璐瑶,李观燕,管玉峰,何广平. 2016. 蒙脱石及其热改性产物对间二硝基苯的吸附性能对比. 矿物学报,36(3):391-396

杨晓雪,王浩然,李艳,朱云,丁竑瑞,鲁安怀. 2013. 胶质芽孢杆菌3027对钙基蒙脱石的矿物结构影响. 岩石矿物学杂志,32(6):767-772
姚敦璠,陈天虎,王进,周跃飞,岳正波. 2013. 天然和水热合成针铁矿对有机物厌氧分解释放CH_4的影响. 环境科学,34(2):635-641
殷辉,冯雄汉,邱国红,谭文峰,刘凡. 2011. 掺钴水钠锰矿对铅的吸附及对砷的氧化. 环境科学,32(7):2092-2101
殷辉,谭文峰,冯雄汉,崔浩杰,邱国红,刘凡. 2012. 含Ni六方水钠锰矿的表征及其对Pb^{2+}(Zn^{2+})环境行为的影响. 土壤学报,49(3):417-427
余萍,李艳,鲁安怀,曾翠平,王鑫,丁竑瑞. 2013. 光电子作用下土壤微生物粪产碱杆菌反硝化性能研究. 岩石矿物学杂志,32(6):761-766
曾翠平,鲁安怀,李艳,吴婧,王鑫,丁竑瑞,颜云花. 2011. 红壤中微生物群落对半导体矿物日光催化作用的响应. 高校地质学报,17(1):101-106
张岩,李艳,鲁安怀,王长秋,梅放,柳剑英,杨重庆,李康. 2017. 人体几种典型病理性钙化物中碳羟磷灰石的红外光谱研究. 岩石矿物学杂志,36(6):909-915
赵丹,肖丹玲,何广平,吴宏海. 2015. Fe(Ⅱ)/针铁矿复合系统对水中橙黄G的吸附-还原脱色研究. 岩石矿物学杂志,34(1):97-102
郑权男,薛兵,徐少南,赵以辛,蒋引珊. 2013. 硅藻土中伴生有机质的炭化与吸附性能. 硅酸盐学报,41(2):230-234,239
朱云,曹维政,鲁安怀,王清华,李艳,张虓雷,王长秋. 2011. 胶质芽孢杆菌-蒙脱石相互作用实验研究. 岩石矿物学杂志,30(1):121-126
Bao T, Chen T H, Liu H B, Chen D, Qing C S, Frost R L. 2014. Preparation of magnetic porous ceramsite and its application in biological aerated filters. Journal of Water Process Engineering,4:185-195
Bao T, Chen T H, Ezzatahmadi N, Rathnayake S I, Chen D, Wille M L, Frost R. 2017. A performance evaluation of a new iron oxide-based porous ceramsite (IPC) in biological aerated filters. Environmental Technology,38(7):827-834
Chen M Q, Wu P X, Yu L F, Liu S, Ruan B, Hu H H, Zhu N W, Lin Z. 2017. FeOOH-loaded MnO_2 nano-composite:an efficient emergency material for thallium pollution incident. Journal of Environmental Management,192:31-38
Chen M Q, Wu P X, Li S S, Yang S S, Lin Z, Dang Z. 2019. The effects of interaction between vermiculite and manganese dioxide on the environmental geochemical process of thallium. Science of the Total Environment,669:903-910
Chen P, Chen T H, Xu L, Liu H B, Xie Q Q. 2017. Mn-rich limonite from the Yeshan Iron Deposit, Tongling district, China:a natural nanocomposite. Journal of Nanoscience and Nanotechnology,17(9):6931-6935
Chen P, Chen T H, Xie Q Q, Xu L, Liu H B, Zhou Y F. 2018. Mineralogy and geochemistry of limonite as a weathering product of ilvaite in the Yeshan iron Deposit, Tongling, China. Clays and Clay Minerals,66(2):190-207
Chen R, Lu J, Xiao J, Zhu C Z, Peng S C, Chen T H. 2017. α-Fe_2O_3 supported Bi_2WO_6 for photocatalytic degradation of gaseous benzene. Solid State Sciences,71:14-21
Chen T H, Yang Y, Chen D, Li P, Shi Y D, Zhu X. 2013. Structural evolution of heat-treated colloidal pyrite under inert atmosphere and its application for the removal of Cu(Ⅱ) ion from wastewater. Environmental Engineering & Management Journal,12(7):1411-1416
Chen T H, Wang J, Zhou Y F, Yue Z B, Xie Q Q, Pan M. 2014a. Synthetic effect between iron oxide and sulfate mineral on the anaerobic transformation of organic substance. Bioresource Technology,151:1-5
Chen T H, Yang Y, Li P, Liu H B, Xie J J, Xie Q Q, Zhan X M. 2014b. Performance and characterization of calcined colloidal pyrite used for copper removal from aqueous solutions in a fixed bed column. International Journal of Mineral Processing,130:82-87
Chen T H, Wang J Z, Wang J, Xie J J, Zhu C Z, Zhan X M. 2015. Phosphorus removal from aqueous solutions containing low concentration of phosphate using pyrite calcinate sorbent. International Journal of Environmental Science and Technology,12(3):885-892
Chen T H, Shi Y D, Liu H B, Chen D, Li P, Yang Y, Zhu X. 2016. A novel way to prepare pyrrhotite and its performance on removal of phosphate from aqueous solution. Desalination and Water Treatment,57(50):23864-23872
Deng L L, Du P X, Yu W B, Yuan P, Annabi-Bergaya F, Liu D, Zhou J M. 2019. Novel hierarchically porous allophane/diatomite nanocomposite for benzene adsorption. Applied Clay Science,168:155-163
Ding H R, Li Y, Lu A H, Wang X, Wang C Q. 2014. Promotion of anodic electron transfer in a microbial fuel cell combined with a silicon solar cell. Journal of Power Sources,253:177-180
Du P X, Yuan P, Thill A, Annabi-Bergaya F, Liu D, Wang S. 2017. Insights into the formation mechanism of imogolite from a full-range observation of its sol-gel growth. Applied Clay Science,150:115-124
Du P X, Yuan P, Liu D, Wang S, Song H Z, Guo H Z. 2018. Calcination-induced changes in structure, morphology, and porosity of allophane. Applied Clay Science,158:211-218
Fu H Y, Yang Y X, Zhu R L, Liu J, Usman M, Chen Q Z, He H P. 2018. Superior adsorption of phosphate by ferrihydrite-coated and lanthanum-decorated magnetite. Journal of Colloid and Interface Science,530:704-713
Gong B N, Wu P X, Huang Z J, Li Y W, Dang Z, Ruan B, Kang C X, Zhu N W. 2016. Enhanced degradation of phenol by *Sphingomonas* sp. GY2B with resistance towards suboptimal environment through adsorption on kaolinite. Chemosphere,148:388-394
Hou Y K, Wu P X, Huang Z J, Ruan B, Liu P Y, Zhu N W. 2014a. Successful intercalation of DNA into CTAB-modified clay minerals for gene protection. Journal of Materials Science,49(20):7273-7281
Hou Y K, Wu P X, Zhu N W. 2014b. The protective effect of clay minerals against damage to adsorbed DNA induced by cadmium and mercury. Chemosphere,95:206-212
Huang S F, Tang J H, Liu X, Dong G W, Zhou S G. 2019. Fast light-driven biodecolorization by a *Geobacter sulfurreducens*-CdS biohybrid. ACS Sustainable Chemistry & Engineering,7(18):15427-15433
Huang Y R, Yao Q Z, Li H, Wang F P, Zhou G T, Fu S Q. 2019. Aerobically incubated bacterial biomass-promoted formation of disordered dolomite and

implication for dolomite formation. Chemical Geology,523:19-30

Huang Z J,Wu P X,Li H L,Li W,Zhu Y J,Zhu N W. 2014. Synthesis and catalytic properties of La or Ce doped hydroxy-FeAl intercalated montmorillonite used as heterogeneous photo Fenton catalysts under sunlight irradiation. RSC Advances,4(13):6500-6507

Huang Z J,Wu P X,Gong B N,Yang S S,Li H L,Zhu Z A,Cui L H. 2016. Preservation of glutamic acid-iron chelate into montmorillonite to efficiently degrade Reactive Blue 19 in a Fenton system under sunlight irradiation at neutral pH. Applied Surface Science,370:209-217

Jiang Q,Zhu R L,Zhu Y P,Chen Q Z. 2019. Efficient degradation of cefotaxime by a UV+ferrihydrite/TiO_2+H_2O_2 process:the important role of ferrihydrite in transferring photo-generated electrons from TiO_2 to H_2O_2. Journal of Chemical Technology & Biotechnology,94(8):2512-2521

Kang C X,Wu P X,Li Y W,Ruan B,Li L P,Tran L,Zhu N W,Dang Z. 2015. Understanding the role of clay minerals in the chromium(VI) bioremoval by *Pseudomonas aeruginosa* CCTCC AB93066 under growth condition:microscopic,spectroscopic and kinetic analysis. World Journal of Microbiology and Biotechnology,31(11):1765-1779

Lai X L,Wu P X,Ruan B,Liu J,Liu Z H,Zhu N W,Dang Z. 2019. Inhibition effect of kaolinite on the development of antibiotic resistance genes in *Escherichia coli* induced by sublethal ampicillin and its molecular mechanism. Environmental Chemistry,16(5):347-359

Li H,Yao Q Z,Yu S H,Huang Y R,Chen X D,Fu S Q,Zhou G T. 2017. Bacterially mediated morphogenesis of struvite and its implication for phosphorus recovery. American Mineralogist,102(2):381-390

Li H,Yao Q Z,Wang F P,Huang Y R,Fu S Q,Zhou G T. 2019. Insights into the formation mechanism of vaterite mediated by a deep-sea bacterium *Shewanella piezotolerans* WP3. Geochimica et Cosmochimica Acta,256:35-48

Li L H,Li Y,Li Y Z,Lu A H,Ding H R,Wong P K,Sun H L,Shi J X. 2019. Natural wolframite as a novel visible-light photocatalyst towards organics degradation and bacterial inactivation. Catalysis Today,doi:10.1016/j.cattod.2019.12.013

Li M X,Liu H B,Chen T H,Wei L,Wang C,Hu W,Wang H L. 2019. The transformation of α-(Al,Fe)OOH in natural fire:effect of Al substitution amount on fixation of phosphate. Chemical Geology,524:368-382

Li Y,Lu A H,Ding H R,Jin S,Yan Y H,Wang C Q,Zeng C P,Wang X. 2009. Gr(VI) reduction at rutile-catalyzed cathod in microbial fuel cells. Electrochemistry Communications,11:1496-1499.

Li Y,Wang X,Zhu M Q,Yang C Q,Lu A H,Li K,Meng F L,Wang C Q. 2014. Mineralogical characterization of calcification in cardiovascular aortic atherosclerotic plaque:a case study. Mineralogical Magazine,78(4):775-786

Li Y,Li Y,Ding H R,Lu A H,Wang H R,Yang X X,Xu X M. 2017. Mineralogical characteristics of Fe-Mn cutans in yellow brown earth of Wuhan,China. Journal of Nanoscience and Nanotechnology,17(9):6873-6880

Li Y,Li Y Z,Yin Y D,Xia D H,Ding H R,Ding C,Wu J,Yan Y H,Liu Y,Chen N,Wong P K,Lu A H. 2018a. Facile synthesis of highly efficient ZnO/$ZnFe_2O_4$ photocatalyst using earth-abundant sphalerite and its visible light photocatalytic activity. Applied Catalysis B:Environmental,226:324-336

Li Y,Wei G L,He H P,Liang X L,Chu W,Huang D Y,Zhu J X,Tan W,Huang Q X. 2018b. Improvement of zinc substitution in the reactivity of magnetite coupled with aqueous Fe(II) towards nitrobenzene reduction. Journal of Colloid and Interface Science,517:104-112

Li Y,Xu X M,Li Y Z,Ding C,Wu J,Lu A H,Ding H R,Qin S,Wang C Q. 2018c. Absolute band structure determination on naturally occurring rutile with complex chemistry:implications for mineral photocatalysis on both Earth and Mars. Applied Surface Science,439:660-671

Li Y,Li Y Z,Xu X M,Ding C,Chen N,Ding H R,Lu A H. 2019a. Structural disorder controlled oxygen vacancy and photocatalytic activity of spinel-type minerals:a case study of $ZnFe_2O_4$. Chemical Geology,504:276-287

Li Y,Liu F F,Xu X M,Liu Y W,Li Y Z,Ding H R,Chen N,Yin H,Lin H,Wang C Q,Lu A H. 2019b. Influence of heavy metal sorption pathway on the structure of biogenic birnessite:Insight from the band structure and photostability. Geochimica et Cosmochimica Acta,256:116-134

Li Z B,Xu J,Teng H H,Liu L W,Chen J,Chen Y,Zhao L,Ji J F. 2015. Bioleaching of lizardite by magnesium- and nickel-resistant fungal isolate from serpentinite soils-implication for carbon capture and storage. Geomicrobiology Journal,32(2):181-192

Li Z B,Lu X C,Teng H H,Chen Y,Zhao L,Ji J F,Chen J,Liu L W. 2019. Specificity of low molecular weight organic acids on the release of elements from lizardite during fungal weathering. Geochimica et Cosmochimica Acta,256:20-34

Liang X L,Liu P,He H P,Wei G L,Chen T H,Tan W,Tan F D,Zhu J X,Zhu R L. 2016. The variation of cationic microstructure in Mn-doped spinel ferrite during calcination and its effect on formaldehyde catalytic oxidation. Journal of Hazardous Materials,306:305-312

Liang X L,Wei G L,Xiong J,Tan F D,He H P,Qu C C,Yin H,Zhu J X,Zhu R L,Qin Z H,Zhang J. 2017. Adsorption isotherm,mechanism,and geometry of Pb(II) on magnetites substituted with transition metals. Chemical Geology,470:132-140

Liao Y,Chen D,Zou S J,Xiong S C,Xiao X,Dang H,Chen T H,Yang S J. 2016. Recyclable naturally derived magnetic pyrrhotite for elemental mercury recovery from flue gas. Environmental Science & Technology,50(19):10562-10569

Liu C M,Wu P X,Zhu Y J,Tran L. 2016. Simultaneous adsorption of Cd^{2+} and BPA on amphoteric surfactant activated montmorillonite. Chemosphere,144:1026-1032

Liu C M,Wu P X,Tran L,Zhu N W,Dang Z. 2018. Organo-montmorillonites for efficient and rapid water remediation:sequential and simultaneous adsorption of lead and bisphenol A. Environmental Chemistry,15(5):286-295

Liu D,Yu W B,Deng L L,Yuan W W,Ma L Y,Yuan P,Du P X,He H P. 2016. Possible mechanism of structural incorporation of Al into diatomite during the deposition process I. Via a condensation reaction of hydroxyl groups. Journal of Colloid and Interface Science,461:64-68

Liu D,Yuan P,Tian Q,Liu H C,Deng L L,Song Y R,Zhou J M,Losic D,Zhou J Y,Song H Z,Guo H Z,Fan W X. 2019. Lake sedimentary biogenic silica from diatoms constitutes a significant global sink for aluminium. Nature Communications,10(1):4829

Liu F,Li X X,Sheng A X,Shang J Y,Wang Z M,Liu J. 2019. Kinetics and mechanisms of protein adsorption and conformational change on hematite particles. Environmental Science & Technology,53(17):10157-10165

Liu H B,Chen T H,Zou X H,Qing C S,Frost R L. 2013a. Thermal treatment of natural goethite:thermal transformation and physical properties. Thermochimica Acta,568:115-121

Liu H B,Chen T H,Zou X H,Xie Q Q,Qing C S,Chen D,Frost R L. 2013b. Removal of phosphorus using NZVI derived from reducing natural goethite. Chemical Engineering Journal,234:80-87

Liu H B,Chen T H,Xie Q Q,Zou X H,Chen C,Frost R L. 2015. The functionalization of limonite to prepare NZVI and its application in decomposition of p-nitrophenol. Journal of Nanoparticle Research,17(9):374

Liu J,Wang Z M,Belchik S M,Edwards M J,Liu C X,Kennedy D W,Merkley E D,Lipton M S,Butt J N,Richardson D J,Zachara J M,Fredrickson J K,Rosso K M,Shi L. 2012. Identification and characterization of M to A:a decaheme c-type cytochrome of the neutrophilic Fe(II)-oxidizing bacterium *Sideroxydans lithotrophicus* ES-1. Frontiers in Microbiology,3:37

Liu J,Pearce C I,Liu C X,Wang Z M,Shi L,Arenholz E,Rosso K M. 2013. $Fe_{3-x}Ti_xO_4$ nanoparticles as tunable probes of microbial metal oxidation. Journal of the American Chemical Society,135(24):8896-8907

Liu J,Pearce C I,Shi L,Wang Z M,Shi Z,Arenholz E,Rosso K M. 2016a. Particle size effect and the mechanism of hematite reduction by the outer membrane cytochrome OmcA of *Shewanella oneidensis* MR-1. Geochimica et Cosmochimica Acta,193:160-175

Liu J,Wang Z W,Sheng A X,Liu F,Qin F Y,Wang Z L. 2016b. *In situ* observation of hematite nanoparticle aggregates using liquid cell transmission electron microscopy. Environmental Science & Technology,50(11):5606-5613

Liu P,He H P,Wei G L,Liang X L,Qi F H,Tan F D,Tan W,Zhu J X,Zhu R L. 2016. Effect of Mn substitution on the promoted formaldehyde oxidation over spinel ferrite:catalyst characterization,performance and reaction mechanism. Applied Catalysis B:Environmental,182:476-484

Liu P,He H P,Wei G L,Liu D,Liang X L,Chen T H,Zhu J X,Zhu R L. 2017. An efficient catalyst of manganese supported on diatomite for toluene oxidation:manganese species,catalytic performance,and structure-activity relationship. Microporous and Mesoporous Materials,239:101-110

Liu P,Wei G L,Liang X L,Chen D,He H P,Chen T H,Xi Y F,Chen H L,Han D H,Zhu J X. 2018. Synergetic effect of Cu and Mn oxides supported on palygorskite for the catalytic oxidation of formaldehyde:dispersion,microstructure,and catalytic performance. Applied Clay Science,161:265-273

Liu S,Wu P X,Chen M Q,Yu L F,Kang C X,Zhu N W,Dang Z. 2017a. Amphoteric modified vermiculites as adsorbents for enhancing removal of organic pollutants:bisphenol A and tetrabromobisphenol A. Environmental Pollution,228:277-286

Liu S,Wu P X,Yu L F,Li L P,Gong B N,Zhu N W,Dang Z,Yang,C. 2017b. Preparation and characterization of organo-vermiculite based on phosphatidylcholine and adsorption of two typical antibiotics. Applied Clay Science,137:160-167

Liu Y Y,Liu C X,Nelson W C,Shi L,Xu F,Liu Y D,Yan A L,Zhong L R,Thompson C,Fredrickson J K,Zachara J M. 2017. Effect of water chemistry and hydrodynamics on nitrogen transformation activity and microbial community functional potential in hyporheic zone sediment columns. Environmental Science & Technology,51(9):4877-4886

Long H,Wu P X,Zhu N W. 2013. Evaluation of Cs^+ removal from aqueous solution by adsorption on ethylamine-modified montmorillonite. Chemical Engineering Journal,225:237-244

Long H,Wu P X,Yang L,Huang Z J,Zhu N W,Hu Z X. 2014. Efficient removal of cesium from aqueous solution with vermiculite of enhanced adsorption property through surface modification by ethylamine. Journal of Colloid and Interface Science,428:295-301

Lu A H,Li Y. 2012. Light fuel cell (LFC):a novel device for interpretation of microorganisms-involved mineral photochemical process. Geomicrobiology Journal,29(3):236-243

Lu A H,Li Y,Jin S. 2012a. Interactions between semiconducting minerals and bacteria under light. Elements,8(2):125-130

Lu A H,Li Y,Jin S,Wang X,Wu X L,Zeng C P,Li Y,Ding H R,Hao R X,Lv M,Wang C Q,Tang Y Q,Dong H L. 2012b. Growth of non-phototrophic microorganisms using solar energy through mineral photocatalysis. Nature Communications,3:768

Lu A H,Li Y,Wang X,Ding H R,Zeng C P,Yang X X,Hao R X,Wang C Q,Santosh M. 2013. Photoelectrons from minerals and microbial world:a perspective on life evolution in the early Earth. Precambrian Research,231:401-408

Lu A H,Li Y,Ding H R,Xu X M,Li Y Z,Ren G P,Liang J,Liu Y W,Hong H,Chen N,Chu S Q,Liu F F,Li Y,Wang H R,Ding C,Wang C Q,Lai Y,Liu J,Dick J,Liu K H,Hochella Jr M F. 2019. Photoelectric conversion on Earth's surface via widespread Fe- and Mn-mineral coatings. Proceedings of the National Academy of Sciences of the United States of America,116(20):9741-9746

Lu P,Chen T H,Liu H B,Li P,Peng S C,Yang Y. 2019. Green preparation of nanoporous pyrrhotite by thermal treatment of pyrite as an effective Hg(II) adsorbent:performance and mechanism. Minerals,9(2):74

Luo Y,Li H,Huang Y R,Zhao T L,Yao Q Z,Fu S Q,Zhou G T. 2018. Bacterial mineralization of struvite using MgO as magnesium source and its potential for nutrient recovery. Chemical Engineering Journal,351:195-202

Meng F L,Wang C Q,Li Y,Lu A H,Mei F,Liu J Y,Du J Y,Zhang Y. 2015. Psammoma bodies in two types of human ovarian tumours:a mineralogical study. Mineralogy and Petrology,109(3):357-365

Peng H,Pearce C I,Huang W F,Zhu Z L,N'Diaye A T,Rosso K M,Liu J. 2018. Reversible Fe(II) uptake/release by magnetite nanoparticles. Environmental Science:Nano,5(7):1545-1555

Peng H,Pearce C I,N'Diaye A T,Zhu Z L,Ni J R,Rosso K M,Liu J. 2019. Redistribution of electron equivalents between magnetite and aqueous Fe^{2+} induced by a model quinone compound AQDS. Environmental Science & Technology,53(4):1863-1873

Peng S C,Xue J,Shi C B,Wang J,Chen T H,Yue Z B. 2014. Iron-enhanced anaerobic digestion of cyanobacterial biomass from Lake Chao. Fuel,117(1):1-4

Ren G P, Ding H R, Li Y, Lu A H. 2016. Natural hematite as a low-cost and earth-abundant cathode material for performance improvement of microbial fuel cells. Catalysts, 6(10): 157

Ren G P, Sun M Y, Sun Y, Li Y, Wang C Q, Lu A H, Ding H R. 2017a. A cost-effective birnessite-silicon solar cell hybrid system with enhanced performance for dye decolorization. RSC Advances, 7(76): 47975–47982

Ren G P, Sun Y, Sun M Y, Li Y, Lu A H, Ding H R. 2017b. Visible light enhanced extracellular electron transfer between a hematite photoanode and *Pseudomonas aeruginosa*. Minerals, 7(12): 230

Ren G P, Sun Y, Ding Y, Lu A H, Li Y, Wang C Q, Ding H R. 2018a. Enhancing extracellular electron transfer between *Pseudomonas aeruginosa* PAO1 and light driven semiconducting birnessite. Bioelectrochemistry, 123: 233–240

Ren G P, Sun Y, Lu A H, Li Y, Ding H R. 2018b. Boosting electricity generation and Cr(VI) reduction based on a novel silicon solar cell coupled double-anode (photoanode/bioanode) microbial fuel cell. Journal of Power Sources, 408: 46–50

Ren G P, Yan Y C, Sun M Y, Wang X, Wu X L, Li Y, Lu A H, Ding H R. 2018c. Considerable bacterial community structure coupling with extracellular electron transfer at karst area stone in Yunnan, China. Geomicrobiology Journal, 35(5): 424–431

Ren G P, Yan Y C, Nie Y, Lu A H, Wu X L, Li Y, Wang C Q, Ding H R. 2019. Natural extracellular electron transfer between semiconducting minerals and electroactive bacterial communities occurred on the rock varnish. Frontiers in Microbiology, 10: 293

Ruan B, Wu P X, Chen M Q, Lai X L, Chen L Y, Yu L F, Gong B N, Kang C X, Dang Z, Shi Z Q, Liu Z H. 2018a. Immobilization of *Sphingomonas* sp. GY2B in polyvinyl alcohol-alginate-kaolin beads for efficient degradation of phenol against unfavorable environmental factors. Ecotoxicology and Environmental Safety, 162: 103–111

Ruan B, Wu P X, Lai X L, Wang H M, Li L P, Chen L Y, Kang C X, Zhu N W, Dang Z, Lu G N. 2018b. Effects of *Sphingomonas* sp. GY2B on the structure and physicochemical properties of stearic acid-modified montmorillonite in the biodegradation of phenanthrene. Applied Clay Science, 156: 36–44

Ruan B, Wu P X, Wang H M, Li L P, Yu L F, Chen L Y, Lai X L, Zhu N W, Dang Z, Lu G N. 2018c. Effects of interaction between montmorillonite and *Sphingomonas* sp. GY2B on the physical and chemical properties of montmorillonite in the clay-modulated biodegradation of phenanthrene. Environmental Chemistry, 15(5): 296–305

Sheng A X, Liu F, Shi L, Liu J. 2016a. Aggregation kinetics of hematite particles in the presence of outer membrane cytochrome OmcA of *Shewanella oneidenesis* MR-1. Environmental Science & Technology, 50(20): 11016–11024

Sheng A X, Liu F, Xie N, Liu J. 2016b. Impact of proteins on aggregation kinetics and adsorption ability of hematite nanoparticles in aqueous dispersions. Environmental Science & Technology, 50(5): 2228–2235

Sheng A X, Liu J, Li X X, Qafoku O, Collins R N, Jones A M, Pearce C I, Wang C M, Ni J R, Lu A H, Rosso K M. 2020. Labile Fe(III) from sorbed Fe(II) oxidation is the key intermediate in Fe(II)-catalyzed ferrihydrite transformation. Geochimica et Cosmochimica Acta, 272: 105–120

Shi L, Dong H L, Reguera G, Beyenal H, Lu A H, Liu J, Yu H Q, Fredrickson J K. 2016. Extracellular electron transfer mechanisms between microorganisms and minerals. Nature Reviews Microbiology, 14(10): 651–662

Song Y R, Yuan P, Wei Y F, Liu D, Tian Q, Zhou J M, Du P X, Deng L L, Chen F R, Wu H H. 2019. Constructing Hierarchically porous nestlike Al_2O_3-MnO_2@diatomite composite with high specific surface area for efficient phosphate removal. Industrial & Engineering Chemistry Research, 2019, 58(51): 23166–23174

Tan D Y, Yuan P, Annabi-Bergaya F, Liu D, He H P. 2015. Methoxy-modified kaolinite as a novel carrier for high-capacity loading and controlled-release of the herbicide amitrole. Scientific Reports, 5: 8870

Tan J, Wang J, Xue J, Liu S Y, Peng S C, Ma D, Chen T H, Yue Z B. 2015. Methane production and microbial community analysis in the goethite facilitated anaerobic reactors using algal biomass. Fuel, 145: 196–201

Tang T, Yue Z B, Wang J, Chen T H, Qing C S. 2018. Goethite promoted biodegradation of 2,4-dinitrophenol under nitrate reduction condition. Journal of Hazardous Materials, 343: 176–180

Tran L, Wu P X, Zhu Y J, Liu S, Zhu N W. 2015a. Comparative study of Hg(II) adsorption by thiol- and hydroxyl-containing bifunctional montmorillonite and vermiculite. Applied Surface Science, 356: 91–101

Tran L, Wu P X, Zhu Y J, Yang L, Zhu N W. 2015b. Highly enhanced adsorption for the removal of Hg(II) from aqueous solution by Mercaptoethylamine/Mercaptopropyltrimethoxysilane functionalized vermiculites. Journal of Colloid and Interface Science, 445: 348–356

Wang C Q, Yang R C, Li Y, Xiong C E, Zhao W W, Liu J Y, Zhang B, Lu A H. 2011. A study on psammoma body mineralization in meningiomas. Journal of Mineralogical and Petrological Sciences, 106(5): 229–234

Wang H M, Wu P X, Liu J, Yang S S, Ruan B, Rehman S, Liu L T, Zhu N W. 2020. The regulatory mechanism of *Chryseobacterium* sp. resistance mediated by montmorillonite upon cadmium stress. Chemosphere, 240: 124851

Wang H R, Ding H R, Li Y, Yang X X, Wang C Q, Lu A H. 2017. Mineral structure transformation of goethite in the bioreduction process by *Cronobacter sakazakii*. Journal of Nanoscience and Nanotechnology, 17(9): 7055–7060

Wang Y B, Lin X Q, Wen K, Zhu J X, He H P. 2015a. Effects of organic templates on the structural properties of porous clay heterostructures: a non-micellar template model for porous structure. Journal of Porous Materials, 22(1): 219–228

Wang Y B, Su X L, Lin X Q, Zhang P, Wen K, Zhu J X, He H P. 2015b. The non-micellar template model for porous clay heterostructures: a perspective from the layer charge of base clay. Applied Clay Science, 116-117: 102–110

Wang Y B, Su X L, Xu Z, Wen K, Zhang P, Zhu J X, He H P. 2016a. Preparation of surface-functionalized porous clay heterostructures *via* carbonization of soft-template and their adsorption performance for toluene. Applied Surface Science, 363:113-121

Wang Y B, Zhang P, Wen K, Su X L, Zhu J X, He H P. 2016b. A new insight into the compositional and structural control of porous clay heterostructures from the perspective of NMR and TEM. Microporous and Mesoporous Materials, 224:285-293

Wei X P, Wu H H, He G P, Guan Y F. 2017a. Efficient degradation of phenol using iron-montmorillonite as a Fenton catalyst: importance of visible light irradiation and intermediates. Journal of Hazardous Materials, 321:408-416

Wei X P, Wu H H, Sun F. 2017b. Magnetite/Fe-Al-montmorillonite as a Fenton catalyst with efficient degradation of phenol. Journal of Colloid and Interface Science, 504:611-619

Wei Y F, Yuan P, Song Y R, Liu D, Losic D, Tan D Y, Chen F R, Liu H C, Du P X, Zhou J M. 2018. Activating 2D nano-kaolinite using hybrid nanoparticles for enhanced phosphate capture. Chemical Communications, 54(82):11649-11652

Wei Y F, Yuan P, Liu D, Losic D, Tan D Y, Chen F R, Liu H C, Zhou J M, Du P X, Song Y R. 2019. Activation of natural halloysite nanotubes by introducing lanthanum oxycarbonate nanoparticles via co-calcination for outstanding phosphate removal. Chemical Communications, 55(14):2110-2113

Wu H H, Dou X W, Deng D Y, Guan Y F, Zhang L G, He G P. 2012. Decolourization of the azo dye Orange G in aqueous solution via a heterogeneous Fenton-like reaction catalysed by goethite. Environmental Technology, 33(13-15):1545-1552

Wu H H, Xie H R, He G P, Guan Y F, Zhang Y L. 2016. Effects of the pH and anions on the adsorption of tetracycline on iron-montmorillonite. Applied Clay Science, 119:161-169

Wu P X, Dai Y P, Long H, Zhu N W, Li P, Wu J H, Dang Z. 2012a. Characterization of organo-montmorillonites and comparison for Sr(II) removal: equilibrium and kinetic studies. Chemical Engineering Journal, 191:288-296

Wu P X, Li S Z, Ju L T, Zhu N W, Wu J H, Li P, Dang Z. 2012b. Mechanism of the reduction of hexavalent chromium by organo-montmorillonite supported iron nanoparticles. Journal of Hazardous Materials, 219-220:283-288

Wu P X, Li W, Zhu Y J, Tang Y N, Zhu N W, Guo C L. 2014a. The protective effect of layered double hydroxide against damage to DNA induced by heavy metals. Applied Clay Science, 100:76-83

Wu P X, Liu C M, Huang Z J, Wang W M. 2014b. Enhanced dechlorination performance of 2,4-dichlorophenol by vermiculite supported iron nanoparticles doped with palladium. RSC Advances, 4(49):25580-25587

Xia Y, Li F F, Jiang Y S, Xia M S, Xue B, Li Y J. 2014. Interface actions between TiO_2 and porous diatomite on the structure and photocatalytic activity of TiO_2-diatomite. Applied Surface Science, 303:290-296

Xiang Y, Zhu Y, Lu J, Zhu C Z, Zhu M Y, Xie Q Q, Chen T H. 2019. Co_3O_4/α-Fe_2O_3 catalyzed oxidative degradation of gaseous benzene: preparation, characterization and its catalytic properties. Solid State Sciences, 93:79-86

Xiao Y X, Li Y Z, Ding H R, Li Y, Lu A H. 2018. The fine characterization and potential photocatalytic effect of semiconducting metal minerals in Danxia landforms. Minerals, 8:554

Xu C Q, Wu H H, Gu F L. 2014. Efficient adsorption and photocatalytic degradation of Rhodamine B under visible light irradiation over BiOBr/montmorillonite composites. Journal of Hazardous Materials, 275:185-192

Xu L, Xie Q Q, Chen T H, Li P, Yang Y, Zhou Y F. 2017. Constraint of nanometer-sized pyrite crystals on oxidation kinetics and weathering products. Journal of Nanoscience and Nanotechnology, 17(9):6962-6966

Xu X M, Ding H R, Li Y, Lu A H, Li Y, Wang C Q. 2018. Mineralogical characteristics of Mn coatings from different weathering environments in China: clues on their formation. Mineralogy and Petrology, 112(5):671-683

Xu X M, Li Y, Li Y Z, Lu A H, Qiao R X, Liu K H, Ding H R, Wang C Q. 2019. Characteristics of desert varnish from nanometer to micrometer scale: a photo-oxidation model on its formation. Chemical Geology, 522:55-70

Yang S S, Wu P X, Chen L Y, Li L G, Huang Z J, Liu S, Li L P. 2016. A facile method to fabricate N-doped graphene-like carbon as efficient electrocatalyst using spent montmorillonite. Applied Clay Science, 132-133:731-738

Yang S S, Wu P X, Yang Q L, Zhu N W, Lu G N, Dang Z. 2017. Regeneration of iron-montmorillonite adsorbent as an efficient heterogeneous Fenton catalytic for degradation of bisphenol A: structure, performance and mechanism. Chemical Engineering Journal, 328:737-747

Yang S S, Wu P X, Liu J Q, Chen M Q, Ahmed Z, Zhu N W. 2018a. Efficient removal of bisphenol A by superoxide radical and singlet oxygen generated from peroxymonosulfate activated with Fe^0-montmorillonite. Chemical Engineering Journal, 350:484-495

Yang S S, Wu P X, Ye Q Y, Li W, Chen M Q, Zhu N W. 2018b. Efficient catalytic degradation of bisphenol A by novel Fe^0-vermiculite composite in photo-Fenton system: mechanism and effect of iron oxide shell. Chemosphere, 208:335-342

Yang S S, Duan X D, Liu J Q, Wu P X, Li C Q, Dong X B, Zhu N W, Dionysiou D D. 2020a. Efficient peroxymonosulfate activation and bisphenol A degradation derived from mineral-carbon materials: key role of double mineral-templates. Applied Catalysis B: Environmental, 267:118701

Yang S S, Huang Z Y, Wu P X, Li Y H, Dong X B, Li C Q, Zhu N W, Duan X D, Dionysiou D D. 2020b. Rapid removal of tetrabromobisphenol A by α-Fe_2O_{3-x}@Graphene@Montmorillonite catalyst with oxygen vacancies through peroxymonosulfate activation: role of halogen and α-hydroxyalkyl radicals. Applied Catalysis B: Environmental, 260:118129

Yang X G, Li Y, Lu A H, Yan Y H, Wang C Q, Wong P K. 2011. Photocatalytic reduction of carbon tetrachloride by natural sphalerite under visible light irradiation. Solar Energy Materials and Solar Cells, 95(7):1915-1921

Yang Y, Chen T H, Li P, Qing C S, Xie Q Q, Zhan X M. 2016. Immobilization of copper under an acid leach of colloidal pyrite waste rocks by a fixed-bed column. Environmental Earth Sciences, 75(3):205

Yang Y, Chen T H, Li P, Xie Q Q, Zhan X M. 2017a. Cu removal from acid mine drainage by modified pyrite: batch and column experiments. Mine Water and the Environment, 36(3): 371-378

Yang Y, Chen T H, Morrison L, Gerrity S, Collins G, Porca E, Li R H, Zhan X M. 2017b. Nanostructured pyrrhotite supports autotrophic denitrification for simultaneous nitrogen and phosphorus removal from secondary effluents. Chemical Engineering Journal, 328: 511-518

Yao D F, Wang J, Chen T H, Tan J, Wang G W. 2016. Methanogenic and carbon sequestration process facilitated by goethite and hematite in the presence of dissimilatory iron-reducing bacteria. Fresenius Environmental Bulletin, 25(6): 1883-1891

Yao D F, Zhang X, Wang G W, Chen T H, Wang J, Yue Z B, Wang Y. 2017. A novel parameter for evaluating the influence of iron oxide on the methanogenic process. Biochemical Engineering Journal, 125: 144-150

You Y S, Zheng S L, Zang H M, Liu F, Liu F H, Liu J. 2019. Stimulatory effect of magnetite on the syntrophic metabolism of *Geobacter* co-cultures: influences of surface coating. Geochimica et Cosmochimica Acta, 256: 82-96

Yu W B, Deng L L, Yuan P, Liu D, Yuan W W, Chen F R. 2015a. Preparation of hierarchically porous diatomite/MFI-type zeolite composites and their performance for benzene adsorption: the effects of desilication. Chemical Engineering Journal, 270: 450-458

Yu W B, Deng L L, Yuan P, Liu D, Yuan W W, Liu P, He H P, Li Z H, Chen F R. 2015b. Surface silylation of natural mesoporous/macroporous diatomite for adsorption of benzene. Journal of Colloid and Interface Science, 448: 545-552

Yu W B, Yuan P, Liu D, Deng L L, Yuan W W, Tao B, Cheng H F, Chen F R. 2015c. Facile preparation of hierarchically porous diatomite/MFI-type zeolite composites and their performance of benzene adsorption: the effects of NaOH etching pretreatment. Journal of Hazardous Materials, 285: 173-181

Yuan P, Liu D, Zhou J M, Tian Q, Song Y R, Wei H H, Wang S, Zhou J Y, Deng L L, Du P X. 2019. Identification of the occurrence of minor elements in the structure of diatomaceous opal using FIB and TEM-EDS. American Mineralogist, 104(9): 1323-1335

Yuan W W, Yuan P, Liu D, Yu W B, Deng L L, Chen F R. 2015. Novel hierarchically porous nanocomposites of diatomite-based ceramic monoliths coated with silicalite-1 nanoparticles for benzene adsorption. Microporous and Mesoporous Materials, 206: 184-193

Yuan W W, Yuan P, Liu D, Deng L L, Zhou J M, Yu W B, Chen F R. 2016a. A hierarchically porous diatomite/silicalite-1 composite for benzene adsorption/desorption fabricated via a facile pre-modification *in-situ* synthesis route. Chemical Engineering Journal, 294: 333-342

Yuan W W, Yuan P, Liu D, Yu W B, Laipan M, Deng L L, Chen F R. 2016b. *In-situ* hydrothermal synthesis of a novel hierarchically porous TS-1/modified-diatomite composite for methylene blue (MB) removal by the synergistic effect of adsorption and photocatalysis. Journal of Colloid and Interface Science, 462: 191-199

Yue Z B, Li Q, Li C C, Chen T H, Wang J. 2015a. Component analysis and heavy metal adsorption ability of extracellular polymeric substances (EPS) from sulfate reducing bacteria. Bioresource Technology, 194: 399-402

Yue Z B, Ma D, Wang J, Tan J, Peng S C, Chen T H. 2015b. Goethite promoted anaerobic digestion of algal biomass in continuous stirring-tank reactors. Fuel, 159: 883-886

Yue Z B, Ma D, Peng S C, Zhao X, Chen T H, Wang J. 2016. Integrated utilization of algal biomass and corn stover for biofuel production. Fuel, 168: 1-6

Zeng C P, Li Y, Lu A H, Ding H R, Wang X, Wang C Q. 2012. Electrochemical interaction of a heterotrophic bacteria *Alcaligenes faecalis* with a graphite cathode. Geomicrobiology Journal, 29(3): 244-249

Zhang C A, Chen T H, Liu H B, Chen D, Xu B, Qing C S. 2018. Low temperature SCR reaction over nano-structured Fe-Mn oxides: characterization, performance, and kinetic study. Applied Surface Science, 457: 1116-1125

Zhang X, Zhou Y F, Xie Q Q, Gao Y, Yue Z B, Chen T H. 2019. Effects of adsorbed inorganic anions on the magnetic properties of calcination-prepared porous maghemite. Physics and Chemistry of Minerals, 46(8): 751-758

Zhao T L, Li H, Huang Y R, Yao Q Z, Huang Y, Zhou G T. 2019. Microbial mineralization of struvite: salinity effect and its implication for phosphorus removal and recovery. Chemical Engineering Journal, 358: 1324-1331

Zhou S G, Tang J H, Yuan Y, Yang G Q, Xing B S. 2018. TiO_2 nanoparticle-induced nanowire formation facilitates extracellular electron transfer. Environmental Science & Technology Letters, 5(9): 564-570

Zhou Y F, Gao Y, Xie Q Q, Wang J, Yue Z B, Wei L, Yang Y, Li L, Chen T H. 2019. Reduction and transformation of nanomagnetite and nanomaghemite by a sulfate-reducing bacterium. Geochimica et Cosmochimica Acta, 256: 66-81

Zhu D, Wang J, Chen T H, Tan J, Yao D F. 2015. Comparison of hematite-facilitated anaerobic digestion of acetate and beef extract. Environmental Technology, 36(18): 2295-2299

Zhu J X, Zhang P, Wang Y B, Wen K, Su X L, Zhu R L, He H P, Xi Y F. 2018. Effect of acid activation of palygorskite on their toluene adsorption behaviors. Applied Clay Science, 159: 60-67

Zhu Y P, Zhu R L, Yan L X, Fu H Y, Xi Y F, Zhou H J, Zhu G Q, Zhu J X, He H P. 2018a. Visible-light Ag/AgBr/ferrihydrite catalyst with enhanced heterogeneous photo-Fenton reactivity via electron transfer from Ag/AgBr to ferrihydrite. Applied Catalysis B: Environmental, 239: 280-289

Zhu Y P, Zhu R L, Zhu G Q, Wang M M, Chen Y N, Zhu J X, Xi Y F, He H P. 2018b. Plasmonic Ag coated Zn/Ti-LDH with excellent photocatalytic activity. Applied Surface Science, 433: 458-467

Zou X H, Chen T H, Liu H B, Zhang P, Chen D, Zhu C Z. 2016. Catalytic cracking of toluene over hematite derived from thermally treated natural limonite. Fuel, 177: 180-189

Zou X H, Ma Z Y, Liu H B, Chen D, Wang C, Zhang P, Chen T H. 2018. Green synthesis of Ni supported hematite catalysts for syngas production from catalytic cracking of toluene as a model compound of biomass tar. Fuel, 217: 343-351

Research Progress of Environmental Mineralogy

LU An-huai, WANG Chang-qiu, LI Yan

Beijing Key Laboratory of Mineral Environmental Function, School of Earth and Space Sciences, Peking University, Beijing 100871

Abstract: Environmental mineralogy focuses on the environmental properties of minerals and studies the functions of minerals in recording environment changes, impacting on environment damage, evaluating environment quality, governing environment pollution and participating in biological interaction, as well as the theoretical methods contained therein. On the one hand, the study of environmental mineralogy provides important theoretical guidance and technical support for the treatment of global environmental pollution and ecological damage. On the other hand, it also helps to understand the major theoretical issues, such as the evolution of terrestrial materials, the origin and evolution of life and the macro process of global environmental evolution. During the decade of 2011–2020, Chinese scholars have acquired new understanding in the research fields of mineral photocatalysis and its synergism with microorganisms, environmental function of nano-minerals and their modified products, and biomineralization. They have put forward new viewpoints of photoelectric energy of natural minerals, photoelectric microorganism, and found new mechanisms of extracellular electron transfer between semiconductor minerals and microorganisms. In this paper, the research progress of environmental mineralogy in China in the recent decade has been reviewed from the aspects of environmental properties of "mineral membrane" produced by the interaction of multiple spheres on the Earth surface, environmental effects of interactions between minerals and microorganisms, nano-minerals and their environmental functions, biomineralization and its environmental effects, and pathological mineralization, and development directions of environmental mineralogy have been prospected.

Key words: environmental mineralogy; environmental properties of minerals; nanominerals; minerals and microorganisms; biomineralization

矿物材料研究进展*

吕国诚[1] 廖立兵[1] 李雨鑫[1] 田林涛[1] 刘昊[2]

1. 中国地质大学（北京）材料科学与工程学院，非金属矿物与固废资源材料化利用北京市重点实验室，北京 100083；2. 中国地质大学（北京）数理学院，北京 100083

摘 要：本文将我国矿物材料研究归纳为环境矿物材料、能源矿物材料、医药用矿物材料、农药化肥载体材料及其他电磁功能矿物材料等方向，通过归纳总结代表性成果，综述了各方向近十年的研究进展。在此基础上，结合矿物材料发展趋势和我国中长期发展规划，提出了我国矿物材料的发展方向。

关键词：矿物材料 研究进展 发展方向

0 引 言

矿物材料学是传统矿物学的延伸和拓展，与岩石学、矿物加工、材料科学与工程、化学与化工、生物医药、环境科学与工程等学科交叉，是正快速形成和发展的矿物科学与工程学科的重要组成部分（廖立兵，2011；白志民等，2013）。

矿物材料学是在全球人口快速增长和社会快速发展所导致的资源、能源过快消耗和生态环境严重恶化的背景下发展起来的。矿物材料是指以矿物为主要或重要组分的材料，有广义与狭义之分（汪灵，2006）。狭义矿物材料是指可直接利用其物理、化学性能的天然矿物岩石，或以天然矿物岩石为主要原料加工制备而成，而且组成、结构、性能和使用效能与天然矿物岩石原料存在直接继承关系的材料（廖立兵，2010）。为了更明确综述该领域近十年的研究进展，本文中的矿物材料特指"狭义矿物材料"。矿物材料不仅是冶金、化工、轻工、建材等传统产业的重要基础材料，还逐渐成为现代高新技术产业不可或缺的新型材料，在电子信息、建材、生物、食品、环境保护、生态修复、通信等现代产业领域的应用也越来越广泛。《国家创新驱动发展战略纲要》及《"十三五"国家科技创新规划》中把绿色与可持续发展作为材料产业发展的重要着力点，要求针对制约材料发展的瓶颈和薄弱环节，加快材料产业转型升级和提质增效，切实提高产业的核心竞争力和可持续发展能力。因此，矿物材料研究对于矿产资源高值和高效利用、传统产业结构优化以及促进高新技术产业发展、节能减排、生态环境保护等均具有重要意义。

近十年我国矿物材料研究发展迅速，新型矿物材料不断涌现，应用领域不断拓展，已经成为矿物科学与工程、材料科学与工程、环境科学与工程等学科活跃的前沿领域。本文综述了近十年我国矿物材料研究的主要成就，展望了未来 10~15 年我国矿物材料研究的发展方向。

---

* 原文"快速发展的我国矿物材料研究——十年进展（2011~2020 年）"刊于《矿物岩石地球化学通报》2020 年第 39 卷第 4 期，本文略有修改。

1 矿物材料研究进展

1.1 环境矿物材料

环境矿物材料是指与生态环境具有良好协调性或直接具有环境修复功能的矿物材料。环境矿物材料近十年来取得了极大的进展，其研究应用范围越来越大，除了在常见的水、气、声、土壤等领域的应用外，在荒漠治理、海上重油处理、核辐射处理等方面的应用研究还得到了加强。

1.1.1 水污染治理材料

水污染治理矿物材料是环境矿物材料最主要的部分，也是研究最活跃、成果最集中的部分。这主要是因为环境矿物材料比表面积大、离子交换能力强使其具有较强的吸附性能，可用来去除废水中的金属离子、有机污染物等。近十年，水污染治理矿物材料研究主要集中于新型材料、应用新技术研发和应用领域拓展方面。

在研发新材料方面主要是将天然矿物进行改性或复合。梁晓亮等（2012）研究了在降解不同染料的过程中类质同象置换作用对铬磁铁矿异相 Fenton 催化性能的影响，结果发现亚甲基蓝的吸附降解量随着铬置换量的增加而增加。Zhang 等（2018）通过对二维高岭土改性制备的高岭土纳米膜对刚果红显示出良好的吸附性能，具有高效的再生能力。张萍等（2018）研究了酸化对坡缕石吸附有机污染物苯的影响，结果发现酸活化可提高坡缕石在低压与高压区对苯的吸附量。张羽等（2018）、刘海波等（2016）及马文婕等（2019）通过对褐铁矿的改性，或是将沸石与 δ-MnO_2 复合，制备出高效吸附剂用于污水处理。适当改性后的蛭石可用于去除水中的氨氮（张佳萍，2012）或重金属离子（包利芳，2015）。王完牡和吴平宵（2013）将有机改性蛭石用于吸附 2,4-二氯酚。此外，将矿物与特定材料复合也可有效治理水污染。王珊等（2017）以伊利石为载体、葡萄糖为碳源，制备出纳米碳/伊利石复合材料，对溶液中的亚甲基蓝吸附效果极好。Tang（2018）研究了凹凸棒石/碳［attapulgite（APT）/C］复合材料作为可重复使用的抗生素吸附剂的吸附性能，结果表明在 300℃ 下制备的复合材料具有高吸附能力并能快速达到平衡。Xie 等（2014）及王斌等（2014）用赤泥和偏高岭石制备地聚物用于吸附水中 Pb^{2+}，该地聚物显示出良好的 Pb^{2+} 固定能力。吴向东和符勇（2019）、李辉等（2015）对膨润土和海泡石、天然沸石改性之后用于污水处理，吸附效果显著增强。

与矿物材料有关的水处理新技术近年来不断涌现。渗透反应格栅（PRB）是一种原位修复技术，是近年来研究和应用最为广泛的地下水污染修复技术（Obiri-Nyarko et al.，2014；刘菲等，2015；Faisal et al.，2018）。Wang M. S. 等（2011，2012）及 Zhang 等（2012）对蛭石、凹凸棒石静态、动态吸附去除水中低浓度氨氮和腐殖酸进行了系统的研究并用作 PRB 材料；以离子液体为绿色溶剂对天然沸石进行改性，制备了环保型 PRB 材料，并建设了示范工程，应用于云南阳宗海砷污染的治理。此外，微波降解技术应用于水处理领域也有很大进展，如在微波辅助下利用氧化锰矿物对水中抗生素（Gu et al.，2017；Lv et al.，2017）及染料（Wang et al.，2014）进行降解，效果良好。

水处理矿物材料的应用领域也在不断拓展，除用于常规的生活污水、工业废水、农业污水处理外，矿物材料还可用于海上重油的处理。邱丽娟等（2018）以石墨等为原料制备了超疏水的海绵状复合材料，该材料对水上浮油和水下重油均具有优异的吸附能力。此类复合材料在处理油脂和有机物泄漏造成的大面积污染方面有着巨大的应用前景。

1.1.2 大气污染治理材料

大气污染治理材料是环境矿物材料研究的薄弱方面，近十年来得到明显加强。一些天然矿物具有较大的比表面积和丰富的孔道结构，对气体有良好的吸附性能。沸石具有大量孔穴和孔道，对气体分子具有较强的吸附能力，常被用于吸附甲硫醇、氨气、苯气体、甲醛、甲烷等（任连海等，2014；莫奔露，

2017；张慧捷，2018）。膨润土、海泡石等具有强的气体吸附性能，常作为吸附介质用于处理 NH_3、SO_2、H_2S 等有害气体（鲁旖等，2016）。在矿物材料的改性和制备研究方面，高如琴等（2018）以硅藻土为主要原料制备出电气石修饰的硅藻土基内墙材料，对甲醛有很好的去除效果。马剑（2012）对坡缕石进行改性，用于吸附 CO_2。硅烷改性硅藻土（李铭哲等，2019）、H_2SO_4 改性海泡石（张韬和贺洋，2016）被用于吸附甲醛。综上可见，矿物材料在大气污染治理方面有很好的应用前景。

1.1.3 土壤污染治理材料

我国土壤污染问题日益严重，其中以重金属污染最为典型。矿物材料具有来源广、成本低、使用方便、对重金属污染治理效果优良等突出特点，已成为土壤污染治理的优选材料。黏土矿物是土壤的主要组分之一，通过离子交换、专性吸附或共沉淀反应可降低土壤中重金属的活性，以增强土壤的自净能力，达到钝化修复目的。海泡石与石灰石、磷肥及膨润土联合施用可促进污染土壤中镉的固定（梁学峰等，2011），坡缕石对镉的最大吸附量可达 40 mg/g，高于普通黏土矿物（Han et al.，2014）；硅藻土（Ye et al.，2015）、羟基磷灰石（Sun Y. B. et al.，2016）等也被用于重金属污染土壤的修复。此外，对天然矿物进行改性或者制备成复合材料，可进一步提高其固持重金属的性能，而成为高效的土壤污染治理材料。改性纳米沸石能显著降低土壤有效镉含量（郑荧辉等，2016）；改性海泡石可吸附固定土壤中的 As^{3+} 和 As^{5+}（张清，2015）。再有，磁性矿物材料和磁分离技术也被广泛使用。有研究用溶剂热法制备了 Fe_3O_4/膨润土复合材料（Yan et al.，2016）及磁性膨润土（李双，2016），使其具备了较高的饱和磁化强度，可作为吸附剂并能通过磁分离技术从体系中分离。

1.1.4 噪音污染治理材料

城市噪声污染是四大环境公害之一，是 21 世纪环境污染控制的主要对象，城市环境噪声污染已成为干扰人们正常生活的主要环境问题之一。全国有 3/4 以上的城市交通干线噪声平均值超过 70 dB。一些多孔矿物材料如膨胀珍珠岩板和火山岩板等具有吸声的功能，其相关研究已取得较大进展。蛭石是一种吸音性能优良的矿物，近年来常作为硬质吸声材料在建筑领域使用。习永广和彭同江（2011）将膨胀蛭石与石膏进行复合，所制备的材料可用于隔热、吸声和湿度调节。范晓瑜（2014）将蛭石与PVC复合，并对复合材料的隔声性能、力学性能、阻燃性能进行了研究，分析了材料隔声量与声压级、物料含量、面密度等之间的关系。

1.1.5 节水防渗材料

一些天然矿物由于具有较好的低渗透性和化学稳定性，以及储量丰富、价格便宜等特点，被当作防渗材料被广泛用于生活垃圾填埋场、工业危险废物填埋场、矿山尾矿处理、油槽防漏，以及地下建筑和景观工程、地铁、隧道、水利工程等领域。其中，膨润土被认为是最合适的防渗材料。刘学贵等（2010，2012）、孙志明等（2010）对改性膨润土作为垃圾填埋场的防渗材料进行了研究，结果表明聚丙烯酰胺改性膨润土作为填埋场防渗衬层的效果良好。王丫丫等（2018）以红黏土为主要原料，添加高分子羧甲基纤维素钠制备红黏土防渗材料，该材料对环境无毒无污染，原材料廉价易得，具有很大的实际应用前景。

1.1.6 荒漠治理材料

土地荒漠化、沙漠化是全世界面临的一个长期问题，是我国当前面临的最为严重的生态问题之一，也是我国生态建设的重点和难点。王爱娣（2013）以自然界分布广泛、储量丰富的天然红土和黄土为主要成分，与环境友好型高分子（CMC、PVA）和生物高分子（植物秸秆）进行复配，制备出天然黏土基固沙材料。冉飞天（2016）以聚乙烯醇、坡缕石黏土和部分中和的丙烯酸为原料制备高吸水复合材料，能够显著提高固沙试样的抗压强度。这类复合材料可以保护沙生植物的生长，使其根系免受风沙的侵蚀，同时可以为其生长提供一定的营养，有保温、保水作用，对沙漠地区植被的恢复有重要作用。

1.1.7 防辐射材料

放射性核素半衰期长，伴有放射性和化学毒性，基本上不经历生物或化学降解过程，因而在环境中可长期存留并富集。胡小强等（2016）合成了含 CsA 沸石，它对模拟核素 Cs^+ 具有较好的去除效果（去除率可达 97.95%）；李真强等（2016）采用静态实验方法研究了蒙脱石对水溶液中模拟核素 Ce^{3+} 的吸附特性，他们通过研究蒙脱石吸附 Ce^{3+} 的动力学和热力学行为来探讨其吸附机制；赖振宇（2012）将磷酸镁水泥用于固化中低放射性废物，包括用于处理中低放射性焚烧灰，结果表明磷酸镁水泥对 Cs 和 Sr 均具有较好的吸附性能，对 Sr 吸附率高达 97.72%。石墨是我国的优势矿产资源，因其具有较好的力学性能和稳定性能，而成为核科学和工程中的重要材料。例如，以熔化的氟盐做核燃料载体的第四代反应堆——熔盐堆，就是以石墨作为中子慢化剂和反射体，与燃料盐直接接触（张宝亮等，2017）。

1.1.8 保鲜防霉材料

微生物破坏商品包装致使食品腐败变质，缩短食品的货架寿命，甚至威胁人类健康和环境安全，这使得保鲜防霉材料研究受到关注。2011 年，欧洲食品安全局［EFSA Panel on Additives and Products or Substances used in Animal Feed（FEEDAP），2011］发表了一项关于层状硅酸盐特别是膨润土的研究，验证了膨润土作为食品添加剂的安全性，并证明了膨润土对牛奶中黄曲霉素的还原作用。膨润土有较强的吸附能力和黏结能力，作为防霉剂可有效防止食品含水量偏高（王凯，2014）。林宝凤等（2011）采用离子交换方法制备了壳聚糖/膨润土/锌复合物，并研究其在蒸馏水和盐水中的缓释行为及抗菌性能。段淑娥（2014）以银-组氨酸配位阳离子为前驱体，以蒙脱石为载体，制备了抗变色耐盐性的载银抗菌剂。

1.2 能源矿物材料

新兴能源的开发是世界各国关注的重点，与此相关的新材料、新技术研发是近年的研究热点。新能源矿物材料是指能实现新能源的转化和利用以及发展新能源技术所需的矿物材料，主要包括电池材料、储气材料、储热保温材料等。目前二次能源电池和太阳能电池是研究的热点，发展迅速。

1.2.1 电池材料

矿物材料可作为原料或与其他材料复合用作电池材料，其中部分矿物提纯、处理后可直接用作电池材料。在二次电池材料领域，Chen 等（2018）以天然海泡石为原料制备出一维硅纳米棒，将其作为锂离子电池的负极材料，显示出良好的循环稳定性。Fan 等（2018）将天然黑滑石进行酸处理制备出新型氟碳材料，作为锂离子电池负极，表现出优异的电化学性能。Jiang 等（2018）以天然辉钼矿为原料，经过破碎、磨矿、浮选、机械剥离和分级工艺制备出了一系列不同粒径的片状 MoS_2，获得了具有高容量的锂离子电池负极材料。Strauss 等（2000）以天然黄铁矿作为锂电池正极材料，研究了不同产地黄铁矿对电池电化学性能的影响。Zhou J. H. 等（2019）以天然黄铜矿为原料，通过简单的浮选和酸浸工艺制备得到了产率高、纯度高的微米级 $CuFeS_2$，将其作为锂离子电池负极材料，显示出了优良的倍率性能和良好的循环性能。石英是一种物理性质和化学性质均十分稳定矿产资源，属三方晶系的氧化物矿物。高纯石英砂的 SiO_2 含量高于 99.5%，采用 1~3 级天然水晶石和优质天然石类精细加工而成，是生产光纤、太阳能电池等高性能材料的主要原料。汪灵等（2013，2014）对高纯石英开展了多年研究，发明了一种以脉石英为原料加工 4N 高纯石英的方法，效果明显且用途广泛，社会、经济效益显著。Nair 等（2018）利用天然辉锑矿粉体作为真空镀膜太阳能电池的蒸发源，得到的 $Sb_2S_{0.5}Se_{2.5}$ 太阳电池的转换效率为 4.24%。

1.2.2 储气材料

化石燃料的日益消耗，使人类面临着能源短缺的严峻考验。氢能、甲烷等新型能源的有效开发和利用需要解决气体的制取、储运和应用三大问题。由于气体燃料极易着火和爆炸，其运输和储存成为其开发利用的核心。有研究显示，黏土矿物对页岩气藏的形成和开发有一定的积极意义（陈尚斌等，2011），并且具有储气性能。Liu 等（2013）研究发现在高压条件下蒙脱石、高岭石及伊利石对 CH_4 的具有良好的吸附性能；Jin 等（2014）研究了不同处理条件下的管状埃洛石的储氢能力，发现其具有良好的稳定性和高的氢吸附能力，在室温储氢方面有极大潜力；吉利明等（2012）研究了常见黏土矿物对甲烷的吸附性能,并利用扫描电子显微镜观察其微孔特征,结果发现黏土矿物吸附甲烷的能力与微孔的发育程度有关。不同矿物材料的复合,是储气材料新的研究方向。有研究以凹凸棒石为模板制得介孔凹凸棒石/碳复合材料,具有良好的储氢性能(Luo et al.,2015)。也有研究将坡缕石、微孔活性炭、沸石等进行复合用于储氢,结果表明结构可调的新型材料是未来储氢介质的发展趋势(程继鹏等,2002;刘国强,2010;杜晓明等,2010)。

1.2.3 储热保温及耐火材料

建筑能耗不断增加,建筑节能问题越来越受关注。开发绿色建筑材料尤其是外墙保温隔热材料,是实现建筑节能的重要手段。相变储热材料是建筑节能外墙保温隔热材料研究的新方向。近十年来,以矿物棉、膨胀珍珠岩、膨胀蛭石、黏土矿物(凹凸棒石、埃洛石、高岭石、蒙脱石)(胡成刚,2013)、硅藻土等为原料制备新型保温隔热材料的研究取得了较大进展,一批有应用潜力的新型矿物材料被研发出来。例如,Gao 等(2019a,2019b)利用珍珠岩尾矿制备了一种新型泡沫保温材料,其质轻且隔热性能优异。赵英良等(2016)以偏高岭土为主要原料,制备出新型保温材料。

1.2.4 发光功能矿物材料

天然萤石具有发光性能,因此萤石结构的化合物可作为很好的发光材料基质(刘涛等,2011),通过稀土离子掺杂可制备性能优异的发光材料。Yang 等(2017)采用高温固相烧结方法合成了系列萤石结构的荧光粉,并研究了 Eu^{3+} 的主要能量转移机理。除此之外,还有冰晶石结构(Yang et al.,2019a,2019b)及磷灰石结构(Ma et al.,2019;Guo et al.,2019;Zhou T. S. et al.,2019)的荧光粉被制备出,其发光性能和能量传递机理也有详细的研究。Lv 等(2018)制备了基于黏土矿物的荧光材料,可用于对液相和气相中的苯酚的检测和定量。吐尔逊·艾迪力比克(2017)用 Yb^{3+} 掺杂天然萤石,观测到了室温下 $2-Yb^{3+}$ 离子对的上转换发光;利用 Eu^{3+} 离子和 Yb^{3+} 离子共掺杂方钠石,观察到了 Eu^{3+} 离子的可见区上转换发光。

1.2.5 催化材料

一些架状、层状、链层状结构的矿物因具有复杂的孔结构和高比表面积而被广泛用作催化剂载体,因此催化用矿物材料一直是研究的热门领域。Peng 等(2017)制备出了 MoS_2/蒙脱石杂化纳米薄片,它具有较高的催化活性和稳定性,在水处理和生物医学领域具有较大的应用前景。天然沸石经酸碱改性后具有较好的催化性能,可有效提高化工产品的产量(李秉正等,2013;刘冬梅等,2015;肖欢等,2019)。光催化矿物材料一直是光催化材料研究的重要组成部分,近十年依然如此。Du 等(2015)及 Guo 等(2014)将过渡金属氧化物与黏土矿物进行复合,发现黏土矿物作为光催化剂载体能够有效固载光催化成分,有利于提高复合光催化剂的吸附性能和回收利用率。Zhu(2012)制备了 Cu_2O/海泡石复合材料,海泡石通过红移带隙提高了可见光的利用率和对污水的降解效果。Wang 等(2018)通过水热分解法制备了凹凸棒石/CdS(APT/CdS)纳米复合材料,并发现在 70 min 内对亚甲基蓝、甲基紫和刚果红的降解表现出最佳的光催化性能。李瑶等(2019)合成了 $Cu-TiO_2$/白云母复合纳米材料,用于光催化降解亚甲基蓝,其光催化性能随着掺 Cu^{2+} 量增加先提高后下降。夏悦等(2014)合成的磷掺杂纳米 TiO_2-硅藻土复合材料具有良好的光催化性能。程宏飞等(2017)采用置换插层法和煅烧法制备出 $g-C_3N_4$/高岭石复合光催化剂,发现高岭石的加入能有效避免 $g-C_3N_4$ 的团聚并显示出较好的光催化性能。此外,累托石、水滑石(艾玉明,2018)、蒙脱石(徐天缘,2017)、水钠锰矿(刘宇琪等,2019)、高岭石(杨佩和汪立今,2015)、管状埃洛石(李霞章等,2015)、坡缕石(孟双艳等,2019)等经过改性或复合后也被用于光催化领域,并且具有良好的催化效果。

1.2.6 其他能源材料

近年来,具有孔结构或纳米纤维形貌的天然矿物逐渐引起人们的注意。作为电容器电极材料,天然纳米矿物具有来源广泛、价格低廉等优势。Fan 等(2020)以天然埃洛石为模板成功合成了纳米管结构的聚苯胺作为超级电容器的电极材料,在不同电流密度下均具有较高的比电容且具有良好的电化学稳定性。

有研究利用凹凸棒石制备复合材料用于电容器和电池，可有效提升电化学性能（Zhang W. B. et al., 2014；Sun L. et al., 2016）。曹曦等（2018）及周述慧和传秀云（2014）以纳米纤维矿物蛇纹石为模板合成多孔碳并用于超级电容器中，具有良好的电化学性能。凹凸棒石（Wang et al., 2013）、硅藻土作为模板可合成具有不同结构的电学功能复合材料。

一些矿物的介电性能也受到关注，其潜在的研究和应用价值也逐渐体现出来。赵晓明和刘元军（2017）研究发现，铁氧体/碳化硅/石墨复合材料涂层厚度对介电常数实部、虚部和损耗角正切有较大影响。除天然铁氧体矿物被用于制备介电材料外，用其他矿物制备介电材料的研究也有报道。张明艳等（2016）以双酚A型环氧树脂为基体，利用碳纳米管及有机蒙脱石共同对环氧树脂进行改性，制备出了复合介电材料；秦文莉（2018）合成的黑滑石/$NiTiO_3$复合材料具有较好的电磁波吸收性能，拓展了黑滑石在电磁波吸收领域的应用。

1.3 医药用矿物材料

1.3.1 止血材料

失血是创伤致死的首要因素，目前世界各国都十分重视创伤后止血材料的研究。矿物用于医药在我国已有几千年历史，新产业、新技术的发展，推动了医药用矿物材料的研发，并逐渐成为矿物材料研究的重要方向。高岭石具有止血功能，已被应用于制备各种止血材料（干长姣等，2017），如高岭土战伤纱布、高岭土介入止血绷带、高岭土扁桃体止血海绵、高岭土止血垫等。Sun等（2017）制备了一种壳聚糖/高岭土复合多孔微球，发现其具有较好的止血效果。沸石具有离子交换、吸附、表面催化等功能，可用于制备止血剂（梅枭雄等，2012），而且无毒、无害、无过敏反应，能迅速止血，中和渗出液，并有抗炎抑菌作用，可诱导上皮再生。还有研究人员制备了石墨烯-高岭土复合海绵（Liang et al., 2018），石墨烯-蒙脱土复合海绵（Li et al., 2016），均可以快速、无风险止血。

1.3.2 药用材料

矿物药是我国的传统研究领域，近年来在抗菌抑菌、药物缓释等领域已有新的研究进展。舒展等（2018）深入研究了硅酸盐黏土矿物在抗菌抑菌领域的应用，通过控制Fe_2O_3纳米颗粒在改性高岭石纳米薄片上的分布密度，可以增强Fe_2O_3纳米颗粒的抗菌活性（Long et al., 2017）。He等（2017, 2018）通过改性或热喷涂的方法制备出埃洛石纳米管涂层，用于肿瘤细胞的捕获，实验结果显示出埃洛石纳米管在临床循环肿瘤细胞捕获中用于癌症患者的早期诊断和监测的无限潜力。在药物缓释领域，矿物材料也发挥着极大的作用。张晶宇（2013）、王知等（2014）以壳聚糖和蒙脱石为主要原料成功制备出可作为药物控制释放载体材料的复合材料，发现其在载药量、包封率及缓释方面都显示出良好的性能。胡仙超等（2014）研究了叶酸插层水滑石在模拟胃液和肠液中的释放特征。还有研究以累托石与壳聚糖为主要原料，制备出纳米复合材料并研究了纳米粒子的截留效率和释放方式（Wang X. Y. et al., 2011；Tu et al., 2015）。

1.4 农业用矿物材料

近十年来，矿物材料在农业领域的应用研究取得了很大的进展，尤其是在矿物复合肥和农药载体两方面。

1.4.1 矿物复合肥料

肥料是一种常见的改良剂，通常用于农业土壤，通过向植物提供矿物质养分来维持或提高作物的生产力。农业土壤肥料的主要营养物质是硝基、磷和钾，许多矿物中含有这类营养物质，自然成为土壤肥料的首选。例如，通过在土壤中加入沸石、膨润土等天然矿物材料，能有效改善土壤结构（燕存岳，2013），增加土壤的肥力（陈明，2014）和保水能力（徐晓敏等，2014）。郑卫红等（2016）以膨润土、硅藻土、凹凸棒土为基体制备出新型尿素包膜肥料。另外，为减少化肥的过量施用，缓释长效材料也被广泛研究。无机矿物作为缓/控释肥料的包膜材料，具有来源广泛、价格低的特点，不仅对土壤环境无污染而且还有改善土壤结构、提供某些微量元素的作用。刘爱平等（2010）以 KCl 为钾源，以新疆所产的钠基膨润土为钾肥的缓释载体制备了膨润土缓释肥。He 等（2015）采用钠基膨润土和海藻酸盐复合材料对植物拉乌尔菌 Rs-2 进行了包覆，研制出高效缓释生物肥料。还有学者对矿物的水肥保持增效机制进行了研究。刘陆涵等（2017）采用土柱淋溶实验方法，探究了土壤保水剂、腐殖酸和沸石对土壤水分保持、控氮释磷的效应。在此基础上开发外界环境响应的控释材料，有望解决我国化肥过量应用、土壤板结、污染等问题，并显著提高我国农产品的质量。

1.4.2 农药载体材料

可持续、环境友好型农药的开发是农业化学工业的重点。寻找新的活性成分和改进农药的输送系统是开发新型农药面临的主要挑战。农药制剂中作为载体剂的材料选择至关重要。黏土矿物是离子和极性化合物的优良吸附剂，具有良好的生物相容性和低毒性或无毒性特点，如高岭石、蒙脱石和海泡石等。Li 等（2012）利用羧甲基壳聚糖/膨润土复合材料开发了一种除草剂阿特拉津配方，药物释放时间延长了 50%。Jiang 等（2015）采用有机阳离子表面活性剂十二烷基三甲基氯化铵对膨润土进行改性，合成了有机膨润土制剂，该制剂成功地包封了吡虫啉的 57.58%，缓释效果显著。Tan 等（2015）研究了高岭石层间甲氧基改性对除草剂负载及释放性能的影响。

1.5 矿物材料基础研究

随着矿物材料研究的深入，新型矿物材料的研发尤需要理论指导。矿物材料基础研究，特别是成分、结构及性能关系研究近年来越来越受到关注。基础研究需要了解材料在原子分子层次上的微观信息，而目前的实验手段不能完全满足研究需求，因此理论模拟计算方法可以弥补实验研究的不足。

1.5.1 成分、结构及性能关系研究

近十年，有少数研究者对一些矿物材料的成分、结构及性能进行了较深入的研究，并以此为指导设计开发新型矿物材料。层状结构硅酸盐矿物结构单元层电荷源于硅-氧四面体和金属-氧八面体中心阳离子的不等价类质同象替代，酸、碱处理可选择性溶出四面体和八面体中的阳离子，从而改变结构单元层电荷。Lv 等（2018）通过酸改性等方法选择性溶出蒙脱石、蛭石等层状硅酸盐矿物四面体和八面体中的阳离子，结果发现酸改性可使层电荷密度增加，离子交换容量呈先增大后减小的变化趋势。他们还研究了层电荷对蒙脱石层间域和表面有机荧光分子分布形态的影响，构建了新型黏土矿物荧光材料，实现了对液相和气相中苯酚的高灵敏定性、定量检测。该课题组还通过对氧化锰矿物进行结构调控和精修，研究了氧化锰矿物的微波催化机理，揭示了电偶极矩和电子自旋磁矩对微波吸收和微波降解的影响机制，实现了在微波辐照下对污水中抗生素等有机污染物的高效降解（Gu et al., 2017; Lv et al., 2017）。沈灿等（2017）对不同产地的天然闪锌矿的矿物学和光催化性能研究发现，闪锌矿中 Fe 含量影响禁带宽度和光响应范围，Cd 含量影响光催化性能，为高附加值开发利用天然闪锌矿提供了科学依据。梁树能等（2014）对绿泥石矿物的光谱特征参量和晶体化学参数进行了分析，结果表明绿泥石矿物的吸收波长位置随绿泥石中 Fe 离子含量的增加而向长波方向移动。然而，目前矿物材料的基础研究明显薄弱，尤其缺少矿物成分、结构及性能关系的研究，严重影响新型矿物功能材料研发与应用。

1.5.2 理论模拟与计算

理论计算与实验模拟用于矿物材料的研究起步相对较晚，但近年来发展迅速。在大多数天然蒙脱石中，Na^+ 和 Ca^{2+} 作为补偿离子共同存在于层间，Zhang L. H. 等（2014）通过分子动力学模拟研究了（Na_x, Ca_y）-蒙脱石在不同含水量下的溶胀特性、水化行为和迁移率，揭示了在高 Ca^{2+}/Na^+ 的蒙脱石中，Ca^{2+} 水化复合物对 Na^+ 迁移的抑制作用。在晶体结构研究方面，周青等（2015）选取二氧化硅的两种同质多象体

作为研究模型，通过基于密度泛函理论的量子力学模拟优化各表面结构，研究了两种同质多象体表面的水化结构差异。李伟民等（2018）利用蒙特卡罗 SRIM 软件包模拟 α 粒子和 Kr^+ 离子对钙钛锆石的微观损伤机制。Zhang 等（2017）用镧掺杂纳米管黏土，通过密度泛函理论计算，揭示了金属掺杂后的几何和电子结构演化，验证了镧掺杂的原子级效应。Liu 等（2014）基于第一分子动力学的垂直能隙方法，研究了蒙脱石片层边缘上的固有酸度，发现了蒙脱石表面的主要活性位点。Lv 等（2018）、Weng 等（2018）用分子动力学模拟方法研究了蒙脱石层间离子的能量状态及其对有机物插层的影响，还研究了氧化锰矿物对抗生素的降解（Liu et al., 2019）。由此可见，理论计算与模拟可揭示矿物材料原子分子层次上的微观信息，有助于提高矿物材料的研究水平，是未来应加强的研究方向。

2 发展趋势

根据我国科技发展战略和社会对新型矿物材料的需求，在综述矿物材料领域十年进展的基础上，本文展望了未来 10~15 年我国矿物材料可能的重点研究方向。

2.1 隔热防火矿物复合材料

隔热防火材料是建筑建材领域不可或缺的材料之一，也是关系到社会公共安全和人民生命财产安全的重要材料之一。卤素阻燃材料是当前隔热防火市场上的主流产品，虽具有阻燃效率高以及用量少的特点，但因会产生具有腐蚀性以及毒性的气体，对人体和环境可造成严重危害。因此，针对目前隔热防火材料中卤系阻燃材料使用量所占比重太大的问题，制备兼具轻质高强、保温隔热、防火阻燃、耐久性强、绿色环保等性能的新型隔热防火材料是国内外隔热防火材料发展的趋势。开发隔热防火矿物新材料有望成为未来矿物材料的一个重要发展方向。

2.2 新型生物医药用矿物材料

生物医药材料是用于诊断、治疗、替代人体组织或器官或修复其功能的新型材料。近年来，随着材料科学、生命科学和医药学的不断发展，以及人们对美好生活的需求，生物医药用材料成为人类生活中不可或缺的材料之一，在我国医用材料市场中的销售额极速增加，已进入医用材料研发应用的重要机遇期，未来 10~15 年我国将成为世界生物医用材料市场最大的国家。矿物材料因其来源广泛、价格低廉、生态环保的特点而被广泛应用。因此利用矿物材料的优势，前瞻布局生物医用矿物材料的研发及推广，对推动我国矿物功能材料的研究与应用具有重要的意义。

2.3 新型农业农村用矿物材料

农业是我国经济发展的重要支柱产业之一，优质的耕地土壤对我国显得尤为重要。然而，随着我国工农业的迅猛发展，大量污染物通过各种途径进入土壤环境中造成土壤污染，严重影响了我国农业经济发展、生态环境和食品安全等。据统计，目前我国已有约 0.1 亿公顷的农业用地受到污染，其中污水灌溉污染农用地 200 多万公顷，固体废弃物污染耕地 10 多万公顷，占耕地总面积的 20% 以上，造成每年千万吨的粮食受到污染，直接经济损失至少 200 亿元。矿物材料因其资源丰富、性能优异，在土壤污染修复领域具有巨大的应用前景和价值。此外，农药的滥用与残留是农用耕地重要污染源之一，优化农药的传输系统与研究新的活性成分是开发新型农药亟待解决的问题，非金属矿物作为高效的缓释载体材料，对环境友好型农药的开发具有重要的作用，因此研发具有缓慢和持续释放或控制释放能力的农业农村用新型矿物功能材料具有重大意义。矿物还可应用于制备各类农田污染吸附降解材料、土壤改良成分调节材料、难垦土地的复垦治理材料、矿物复合肥料、农作物保鲜防霉材料、饲用霉菌毒素吸附材料和农村"厕所革命""饲养基地"污气污水治理材料及用于农村节水、防渗等的材料，为实现资源和能源消耗的优化以及生态环境的保护提供支撑材料和产品。

2.4　大气污染治理矿物材料

目前我国大气污染形势相当严峻，由山林火灾、火山喷发、燃烧石化燃料和交通尾气排放等带来的大气污染物的排放总量居高不下，以甲醛为代表的室内空气污染物也严重威胁着人们的健康。研究新型空气净化过滤材料以及异味吸附分解材料是目前空气污染治理的重要方向。很多矿物因具有良好的吸附性能、优异的氧化还原能力、成本低、来源广等特点，以及具备轻质、保温节能、环保、安全舒适和微环境调节等多种属性和功能，对研发空气污染治理与室内环境调节材料具有重大意义。

此外，针对毒性大、难处理的生物医药和化工废水、核废料、海上溢油污染以及含油废水净化等问题，研发新型环境矿物材料是重要的解决途径。

2.5　新型能源矿物材料

随着太阳能、风能、生物质能、地热能等新型能源的全面开发，新能源材料作为各种新能源技术的重要基础，在新能源系统中得到了广泛的应用。一些天然矿物具有独特的成分、结构或者形貌，并且来源广泛、价格低廉，可用作电池、电容器材料或作为载体或模板制备催化材料或电池、电容器材料。此外，矿物还可用于制备光伏用材料、储氢储气材料、相变储能材料和高导热材料等，满足新能源领域的发展需求。新能源矿物材料仍将是未来的重要发展方向。

3　结　　语

综上所述，我国矿物材料研究近十年取得了丰硕的成果，传统领域的研究不断深入，材料应用不断拓展，同时还发展了若干新的研究方向，尤其是在环境及农业领域的应用研究中取得了一系列创新成果。然而，我国矿物材料的研究仍存在局限，主要体现在两个方面，一是矿物材料基础研究，特别是矿物成分-结构-性能关系研究薄弱，新型矿物功能材料研发缺少理论指导；二是研究成果与工业生产和实际应用的需求仍存在较大差距，很多成果难以推广应用。

未来10~15年是我国经济社会发展的关键期，是我国全面建成小康社会并向中等富裕国家转变的重要过渡期。矿物材料的研究应与国家经济社会发展的需要更紧密结合，以满足国家重大战略需求。建设富强、美丽的中国，实现中国梦，需要更多性能优异的矿物材料，特别是新能源、环境及生物医药用新型矿物功能材料。因此加强矿物材料基础研究，揭示矿物成分-结构-性能关系，研发矿物功能新材料并加强成果的推广应用，成为矿物材料研究者共同的任务。矿物材料研究者任重道远。

参 考 文 献

艾玉明. 2018. 粘土矿物/TiO_2/Cu_2O光催化剂的制备及其杀藻性能的研究. 武汉:武汉工程大学硕士学位论文
白志民,马鸿文,廖立兵. 2013. 矿物材料学科建设与人才培养模式探索. 中国地质教育,22(3):21-23
包利芳. 2015. 改性蛭石处理重金属废水的研究. 成都:西华大学硕士学位论文
曹曦,传秀云,李爱军,黄杜斌. 2018. 纳米纤维矿物纤蛇纹石为模板合成多孔炭及其在超级电容器中的应用. 新型炭材料,33(3):229-236
陈明. 2014. 矿物土壤调理剂对中低产土壤改良及水稻产量的影响. 安徽农学通报,20(19):33-34,54
陈尚斌,朱炎铭,王红岩,刘洪林,魏伟,方俊华. 2011. 四川盆地南缘下志留统龙马溪组页岩气储层矿物成分特征及意义. 石油学报,32(5):775-782
程宏飞,杜贝贝,孙志明,李春全. 2017. 插层法制备g-C_3N_4/高岭石复合材料及其光学性能. 人工晶体学报,46(7):1258-1262,1266
程继鹏,张孝彬,刘芙,叶瑛,涂江平,孙诎林,陈长聘. 2002. 天然纳米矿物原位复合碳纳米管及其吸氢性能. 太阳能学报,23(6):743-747
杜晓明,李静,吴尔冬. 2010. 沸石吸附储氢研究进展. 化学进展,22(1):248-254
段淑娥. 2014. 抗变色载银抗菌剂的制备与性能研究. 见:2014年(首届)抗菌科学与技术论坛(ASTF2014)论文摘要集. 北京:全国卫生产业企业管理协会抗菌产业分会
范晓瑜. 2014. 蛭石/PVC隔声复合材料的性能研究. 杭州:浙江理工大学硕士学位论文

干长姣,甘慧,孟志云,朱晓霞,顾若兰,吴卓娜,孙文种,王东根,窦桂芳. 2017. 高岭土的止血应用研究进展. 军事医学,41(2):141-145

高如琴,谷一鸣,曹行,曹祥建,王新宁,张丽金,石泽堂. 2018. 电气石修饰硅藻土基内墙材料的制备及甲醛去除效果研究. 轻工学报,33(2):7-12

胡成刚. 2013. 建筑保温材料的应用与发展. 建材发展导向,(3):182-183

胡仙超,潘国祥,陈聪亚,曹枫,倪生良. 2014. 肠道靶向药物缓释功能叶酸插层水滑石的制备与结构表征. 矿物学报,34(1):29-34

胡小强,彭同江,孙红娟. 2016. 含CsA沸石的合成及对Cs^+的去除效果研究. 非金属矿,39(4):24-27

吉利明,邱军利,夏燕青,张同伟. 2012. 常见黏土矿物电镜扫描微孔隙特征与甲烷吸附性. 石油学报,33(2):249-256

赖振宇. 2012. 磷酸镁水泥固化中低放射性废物研究. 重庆:重庆大学博士学位论文

李秉正,黎演明,吴学众,黄日波. 2013. 酸性沸石催化葡萄糖制备5-羟甲基糠醛研究. 广西科学,20(2):158-161,164

李辉,左金龙,王军霞. 2015. 天然沸石及其改性对污水中磷的吸附. 哈尔滨商业大学学报(自然科学版),31(3):311-314

李铭哲,刘阳钰,郑水林,孙志明,贾宏伟. 2019. 改性硅藻土的甲醛吸附性能与吸附机理. 非金属矿,42(3):83-86

李双. 2016. 磁性膨润土制备及去除水中污染物的研究. 济南:济南大学硕士学位论文

李伟民,董发勤,李文周,边亮,宋功保,卢喜瑞. 2018. α粒子和Kr^+离子对钙钛锆石辐照损伤的Monte Carlo模拟. 功能材料,49(3):3001-3006

李霞章,殷禹,姚超,罗士平,左士祥,刘文杰. 2015. CeO_2-CdS/埃洛石纳米管的制备及可见光催化性能. 硅酸盐学报,43(4):482-487

李瑶,彭同江,孙红娟,肖青. 2019. Cu-TiO_2/白云母纳米复合材料的制备及结构、形貌和光催化性能. 硅酸盐学报,47(4):480-485

李真强,孙红娟,彭同江. 2016. 蒙脱石对模拟核素Ce(Ⅲ)的吸附特性和机制研究. 非金属矿,39(2):31-34

梁树он,甘甫平,闫柏琨,魏红艳,肖晨超. 2014. 绿泥石矿物成分与光谱特征关系研究. 光谱学与光谱分析,34(7):1763-1768

梁晓亮,何宏平,钟远红,袁鹏,朱建喜. 2012. 铬磁铁矿催化异相Fenton法降解不同类型染料的对比研究. 矿物学报,32(S1):150-151

梁学峰,徐应明,王林,孙国红,秦旭,孙扬. 2011. 天然黏土联合磷肥对农田土壤镉铅污染原位钝化修复效应研究. 环境科学学报,31(5):1011-1018

廖立兵. 2010. 矿物材料的定义与分类. 硅酸盐通报,29(5):1067-1071

廖立兵. 2011. 我国矿物功能材料研究的新进展. 硅酸盐学报,39(9):1523-1530

林宝凤,刘小舟,吕彦超,李术东. 2011. 壳聚糖/膨润土/锌复合物的制备与抗菌性能研究. 广西大学学报(自然科学版),36(3):429-434

刘爱平,冯启明,王维清,和雨丽. 2010. 膨润土对钾肥的吸附性能及缓释效果研究. 非金属矿,33(6):49-50,54

刘冬梅,翟玉春,马健,王海彦. 2015. Na_2CO_3处理法制备微介孔ZSM-5沸石及其催化硫醚化性能. 石油学报(石油加工),31(1):38-44

刘菲,陈亮,王广才,陈鸿汉,Gillham R W. 2015. 地下水渗流反应格栅技术发展综述. 地球科学进展,30(8):863-877

刘国强. 2010. 介孔氧化硅气凝胶和微孔活性炭的制备、织构及氢气吸附性能. 北京:北京化工大学博士学位论文

刘海波,张如玉,陈天虎,陈陈,陈冬. 2016. 褐铁矿纳米结构化相变零价铁及其除磷性能. 矿物岩石地球化学通报,35(1):64-69

刘陆涵,马妍,刘振海,彭菲,黄占斌. 2017. 三种环境材料对土壤水肥保持效应的影响研究. 农业环境科学学报,36(9):1811-1819

刘涛,吴微微,吴星艳,王亚芳. 2011. 萤石掺杂稀土发光材料的合成与应用. 科技创新导报,(22):58-59

刘学贵,刘长风,邵红,郑维涛,王恩德. 2010. 改性膨润土作为垃圾填埋场防渗材料的研究. 新型建筑材料,37(10):56-58

刘学贵,刘长风,高品一,邵红,王国胜. 2012. 聚丙烯酰胺改性膨润土防渗材料的制备及其表征. 新型建筑材料,(4):10-13

刘文琪,谢龙悦,孟繁斌,钟小梅,董发勤,刘明学. 2019. 水钠锰矿光电催化降解亚甲基蓝及其机理研究. 环境科学与技术,42(1):58-64

鲁嫡,仇丹,章凯丽. 2016. 海泡石吸附剂的应用研究进展. 宁波工程学院学报,28(1):17-22

马剑. 2012. 咪唑改性坡缕石的制备及其对CO_2的吸附. 兰州:西北师范大学硕士学位论文

马文健,陈天虎,陈冬,刘海波,程鹏,张泽鑫,陶琼,王玉珠. 2019. δ-MnO_2/沸石纳米复合材料同时去除地下水中的铁锰氨氮. 环境科学,40(10):4553-4561

梅枭雄,范承启,张继超,韩昌旭,杨霞,吴玉章. 2012. 医疗条件缺乏环境下的创伤止血药物(材料)研究现状. 中国药房,23(13):1232-1233

孟双艳,杨红菊,朱楠,杨娇,杨瑞瑞,杨志旺. 2019. BiOCl-ov/坡缕石复合可见光催化剂的制备及其对醇类的选择性氧化研究. 化学学报,77(5):461-468

莫奔露. 2017. 红辉沸石对甲醛的吸附及其在涂料中的应用. 广东经济,(16):295-296

秦文莉. 2018. 基于黑滑石的复合材料制备及其在功能材料领域的应用研究. 杭州:浙江大学博士学位论文

邱丽娟,张颖,刘帅卓,张骞,周莹. 2018. 超疏水、高强度石墨烯油水分离材料的制备及应用. 高等学校化学学报,39(12):2758-2766

冉飞天. 2016. 黏土基高分子复合材料的制备及其保水固沙性能研究. 兰州:西北师范大学硕士学位论文

任连海,郝艳,王攀. 2014. 改性沸石对餐厨垃圾释放的恶臭气体吸附研究. 环境科学与技术,37(7):137-140

沈灿,鲁安怀,谷湘平. 2017. 天然闪锌矿矿物学特征与光催化性能. 岩石矿物学杂志,36(6):807-816

舒展,张毅,谢虹忆,欧阳静,杨华明. 2018. 硅酸盐黏土矿物在抗菌方面研究进展. 材料工程,46(4):23-30

孙志明,于健,郑水林,白春华,豆中磊,孔维安,史吉刚. 2010. 离子种类及浓度对土工合成黏土垫用膨润土保水性能的影响. 硅酸盐学报,38(9):1826-1831

吐尔逊·艾迪力比克. 2017. 3-Yb^{3+}团簇合作敏化Cu^{2+}、Pb^{2+}上转换发光及团簇结构性破坏荧光猝灭研究. 长春:吉林大学博士学位论文

汪灵. 2006. 矿物材料的概念与本质. 矿物岩石,26(2):1-9

汪灵,王艳,李彩侠. 2013-01-23. 一种以脉石英为原料加工制备4N高纯石英的方法:中国,CN102887518A

汪灵,党陈萍,李彩侠,王艳,魏玉燕,夏瑾卓,潘俊良. 2014. 中国高纯石英技术现状与发展前景. 地学前缘,21(5):267-273

王爱娣. 2013. 黏土基复合固沙材料性能研究. 兰州:西北师范大学硕士学位论文

王斌,朱文凤,王林江,彭鹏,曾建民. 2014. 广西拜尔法赤泥烧胀陶粒制备及对水体中 Pb^{2+} 的吸附. 武汉理工大学学报,36(4):30-34

王凯. 2014. 畜禽饲料中防霉剂和膨润土的使用. 养殖技术顾问,(2):59

王珊,王高锋,孙文,郑水林. 2017. 富氧官能团纳米碳/伊利石复合材料的制备及吸附性能研究. 硅酸盐通报,36(7):2326-2331

王完牡,吴平霄. 2013. 有机蛭石的制备及其对2,4-二氯酚的吸附性能研究. 功能材料,44(6):835-839

王丫丫,张哲,马国富,刘瑾,张文旭,雷自强. 2018. 红土基防渗漏材料的制备及性能研究. 干旱区资源与环境,32(10):197-202

王知. 2014. α-Fe/蒙脱石磁靶向药物控制释放载体材料的制备与研究. 西安:西安科技大学硕士学位论文

吴向东,符勇. 2019. 改性膨润土和海泡石在污水处理中的实验研究. 能源与环保,41(2):115-118

习永广,彭同江. 2011. 膨胀蛭石/石膏复合保温材料的制备与表征. 复合材料学报,28(5):156-161

夏悦,李芳菲,蒋引珊,陈雪娇. 2014. 天然硅藻土负载磷掺杂 TiO_2 及其可见光催化性能. 吉林大学学报(工学版),44(2):415-420

肖欢,张维民,马静红,李瑞丰. 2019. 1,3,5-三甲苯在沸石催化剂上的催化转化. 石油学报(石油加工),35(2):369-375

徐天缘. 2017. 半导体/异相芬顿复合催化材料的构建及其光催化性能的研究. 广州:中国科学院大学(中国科学院广州地球化学研究所)博士学位论文

徐晓敏,吴淑芳,康倍铭,冯浩,杜健. 2014. 五种天然土壤改良剂的养分与保水性研究及评价. 干旱区资源与环境,28(9):85-89

燕存岳. 2013. 沸石酸化改性及其作为农用调理剂的应用研究. 长春:吉林农业大学硕士学位论文

杨佩,汪立今. 2015. 改性高岭石负载 TiO_2 光催化剂制备及分析. 非金属矿,38(5):1-4

张宝亮,戚威,夏汇浩,孙立斌,吴莘馨. 2017. 核石墨的孔结构与熔盐浸渗特性研究. 核技术,40(12):120605

张慧捷. 2018. X型沸石/活性炭吸附剂的改性对甲烷的吸附分离研究. 化工管理,(4):195-196

张佳萍. 2012. 蛭石在除氮技术中的应用研究. 环境监控与预警,4(3):45-49

张晶宇. 2013. 壳聚糖/壳聚糖-蒙脱石载药微球的制备及释放动力学研究. 哈尔滨:哈尔滨理工大学硕士学位论文

张明艳,王晨,吴淑龙,孙国辉,吴子剑. 2016. 碳纳米管/蒙脱土共掺杂环氧树脂复合材料介电性能研究. 电工技术学报,31(10):151-158

张萍,文科,王钺博,苏小丽,何宏平,朱建喜. 2018. 酸处理对坡缕石结构及其苯吸附性能的影响. 矿物学报,38(1):93-99

张清. 2015. 粘土矿物材料对水体中砷和重金属的控制研究. 南京:南京理工大学硕士学位论文

张韬,贺洋. 2016. 海泡石环境吸附材料制备研究. 非金属矿,39(4):46-47

张羽,刘海波,陈平,皮汇钰,陈天虎. 2018. 热处理褐铁矿去除水中的 Mn^{2+}. 岩石矿物学杂志,37(4):687-696

赵晓明,刘元军. 2017. 铁氧体/碳化硅/石墨三层涂层复合材料介电性能. 材料工程,45(1):33-37

赵英良,邢军,刘辉,邱景平,孙晓刚,李浩. 2016. 多孔矿物聚合材料在外墙保温方面应用的试验研究. 硅酸盐通报,35(10):3340-3344

郑卫红,潘国祥,陈伽,徐敏虹,陈海锋,曹枫,倪生良,唐培松. 2016. 非金属矿物-EC复合包膜尿素缓释肥制备与释放特征. 矿物学报,36(2):247-252

郑荧辉,熊仕娟,徐卫红,李欣忱,罗瑜,秦余丽,赵婉伊. 2016. 纳米沸石对大白菜镉吸收及土壤有效镉含量的影响. 农业环境科学学报,35(12):2353-2360

周青,魏景明,朱建喜,朱润良,唐翠华,何宏平. 2015. 二氧化硅同质多像矿物的表面结构及水化特征差异性的计算模拟. 吉林大学学报(地球科学版),45(S1):1501

周道慧,传秀云. 2014. 埃洛石为模板合成中孔炭. 无机材料学报,29(6):584-588

Chen Q Z, Zhu R L, Liu S H, Wu D C, Fu H Y, Zhu J X, He H P. 2018. Self-templating synthesis of silicon nanorods from natural sepiolite for high-performance lithium-ion battery anodes. Journal of Materials Chemistry A,6(15):6356-6362

Du Y, Tang D D, Zhang G K, Wu X Y. 2015. Facile synthesis of Ag_2O-TiO_2/sepiolite composites with enhanced visible-light photocatalytic properties. Chinese Journal of Catalysis,36(12):2219-2228

EFSA Panel on Additives and Products or Substances used in Animal Feed (FEEDAP). 2011. Scientific Opinion on the safety and efficacy of bentonite (dioctahedral montmorillonite) as feed additive for all species. EFSA Journal,9(2):2007

Faisal A A H, Sulaymon A H, Khaliefa Q M. 2018. A review of permeable reactive barrier as passive sustainable technology for groundwater remediation. International Journal of Environmental Science and Technology,15(5):1123-1138

Fan P, Liu H, Liao L B, Wang Z, Wu Y Y, Zhang Z W, Hai Y, Lv G C, Mei L F. 2018. Excellent electrochemical properties of graphene-like carbon obtained from acid-treating natural black talc as Li-ion battery anode. Electrochimica Acta,289:407-414

Fan P, Wang S N, Liu H, Liao L B, Lv G C, Mei L F. 2020. Polyaniline nanotube synthesized from natural tubular halloysite template as high performance pseudocapacitive electrode. Electrochimica Acta,331:135259

Gao H, Liu H, Liao L B, Mei L F, Lv G C, Liang L M, Zhu G D, Wang Z J, Huang D L. 2019a. Improvement of performance of foam perlite thermal insulation material by the design of a triple-hierarchical porous structure. Energy and Buildings,200:21-30

Gao H, Liu H, Liao L B, Mei L F, Shuai P F, Xi Z Y, Lv G C. 2019b. A novel inorganic thermal insulation material utilizing perlite tailings. Energy and Buildings,190:25-33

Gu W L, Lv G C, Liao L B, Yang C X, Liu H, Nebendahl I, Li Z H. 2017. Fabrication of Fe-doped birnessite with tunable electron spin magnetic moments for the degradation of tetracycline under microwave irradiation. Journal of Hazardous Materials,338:428-436

Guo Q F, Ma X X, Liao L B, Liu H K, Yang D, Liu N, Mei L F. 2019. Structure and luminescence properties of multicolor phosphor $Ba_2La_3(SiO_4)_3Cl$:Tb^{3+},Eu^{3+}. Journal of Solid State Chemistry,280:121009

Guo S, Zhang G K, Wang J Q. 2014. Photo-Fenton degradation of rhodamine B using Fe_2O_3-Kaolin as heterogeneous catalyst:characterization, process optimization and mechanism. Journal of Colloid and Interface Science,433:1-8

Han J, Xu Y M, Liang X F, Xu Y J. 2014. Sorption stability and mechanism exploration of palygorskite as immobilization agent for Cd in polluted soil. Water, Air, & Soil Pollution, 225(10):2160

He R, Liu M X, Yan S, Long Z R, Zhou C R. 2017. Large-area assembly of halloysite nanotubes for enhancing the capture of tumor cells. Journal of Materials Chemistry B, 5(9):1712-1723

He R, Liu M X, Shen Y, Liang R, Liu W, Zhou C R. 2018. Simple fabrication of rough halloysite nanotubes coatings by thermal spraying for high performance tumor cells capture. Materials Science and Engineering:C, 85:170-181

He Y H, Wu Z S, Tu L, Han Y J, Zhang G L, Li C. 2015. Encapsulation and characterization of slow-release microbial fertilizer from the composites of bentonite and alginate. Applied Clay Science, 109-110:68-75

Jiang F, Li S J, Ge P, Tang H H, Khoso S A, Zhang C Y, Yang Y, Hou H S, Hu Y H, Sun W, Ji X B. 2018. Size-tunable natural mineral-molybdenite for lithium-ion batteries toward: enhanced storage capacity and quicken ions transferring. Frontiers in Chemistry, 6:389

Jiang L, Mo J Y, Kong Z J, Qin Y P, Dai L T, Wang Y J, Ma L. 2015. Effects of organobentonites on imidacloprid release from alginate-based formulation. Applied Clay Science, 105-106:52-59

Jin J, Zhang Y, Ouyang J, Yang H M. 2014. Halloysite nanotubes as hydrogen storage materials. Physics and Chemistry of Minerals, 41(5):323-331

Li G F, Quan K C, Liang Y P, Li T Y, Yuan Q P, Tao L, Xie Q, Wang X. 2016. Graphene-montmorillonite composite sponge for safe and effective hemostasis. ACS Applied Materials & Interfaces, 8(51):35071-35080

Li J F, Yao J, Li Y M, Shao Y. 2012. Controlled release and retarded leaching of pesticides by encapsulating in carboxymethyl chitosan/bentonite composite gel. Journal of Environmental Science and Health, Part B, 47(8):795-803

Liang Y P, Xu C C, Li G F, Liu T C, Liang J F, Wang X. 2018. Graphene-kaolin composite sponge for rapid and riskless hemostasis. Colloids and Surfaces B: Biointerfaces, 169:168-175

Liu D, Yuan P, Liu H M, Li T, Tan D Y, Yuan W W, He H P. 2013. High-pressure adsorption of methane on montmorillonite, kaolinite and illite. Applied Clay Science, 85:25-30

Liu T M, Yuan G B, Lv G C, Li Y X, Liao L B, Qiu S Y, Sun C H. 2019. Synthesis of a novel catalyst MnO/CNTs for microwave-induced degradation of tetracycline. Catalysts, 9(11):911

Liu X D, Cheng J, Sprik M, Lu X C, Wang R C. 2014. Surface acidity of 2:1-type dioctahedral clay minerals from first principles molecular dynamics simulations. Geochimica et Cosmochimica Acta, 140:410-417

Long M, Zhang Y, Shu Z, Tang A D, Ouyang J, Yang H M. 2017. Fe_2O_3 nanoparticles anchored on 2D kaolinite with enhanced antibacterial activity. Chemical Communications, 53(46):6255-6258

Luo H M, Yang Y F, Sun Y X, Zhao X, Zhang J Q. 2015. Preparation of lactose-based attapulgite template carbon materials and their electrochemical performance. Journal of Solid State Electrochemistry, 19(4):1171-1180

Lv G C, Xing X B, Liao L B, An P F, Yin H, Mei L F, Li Z H. 2017. Synthesis of birnessite with adjustable electron spin magnetic moments for the degradation of tetracycline under microwave induction. Chemical Engineering Journal, 326:329-338

Lv G C, Liu S Y, Liu M, Liao L B, Wu L M, Mei L F, Li Z H, Pan C F. 2018. Detection and quantification of phenol in liquid and gas phases using a clay/dye composite. Journal of Industrial and Engineering Chemistry, 62:284-290

Ma X X, Liao L B, Guo Q F, Liu H K, Yang D, Liu N, Mei L F. 2019. Structure and luminescence properties of multicolor phosphor $Ba_2La_3(GeO_4)_3F$: Tb^{3+}, Eu^{3+}. RSC Advances, 9(61):35717-35726

Nair P K, Vázquez García G, Zamudio Medina E A, Martínez L G, Castrejón O L, Ortiz J M, Nair M T S. 2018. Antimony sulfide-selenide thin film solar cells produced from stibnite mineral. Thin Solid Films, 645:305-311

Obiri-Nyarko F, Grajales-Mesa S J, Malina G. 2014. An overview of permeable reactive barriers for in situ sustainable groundwater remediation. Chemosphere, 111:243-259

Peng K, Fu L J, Yang H M, Ouyang J, Tang A D. 2017. Hierarchical MoS_2 intercalated clay hybrid nanosheets with enhanced catalytic activity. Nano Research, 10(2):570-583

Strauss E, Ardel G, Livshits V, Burstein L, Golodnitsky D, Peled E. 2000. Lithium polymer electrolyte pyrite rechargeable battery: comparative characterization of natural pyrite from different sources as cathode material. Journal of Power Sources, 88(2):206-218

Sun L, Su T T, Xu L, Du H B. 2016. Preparation of uniform Si nanoparticles for high-performance Li-ion battery anodes. Physical Chemistry Chemical Physics, 18(3):1521-1525

Sun X, Tang Z H, Pan M, Wang Z C, Yang H Q, Liu H Q. 2017. Chitosan/kaolin composite porous microspheres with high hemostatic efficacy. Carbohydrate Polymers, 177:135-143

Sun Y B, Xu Y, Xu Y M, Wang L, Liang X F, Li Y. 2016. Reliability and stability of immobilization remediation of Cd polluted soils using sepiolite under pot and field trials. Environmental Pollution, 208:739-746

Tan D Y, Yuan P, Annabi-Bergaya F, Dong F Q, Liu D, He H P. 2015. A comparative study of tubular halloysite and platy kaolinite as carriers for the loading and release of the herbicide amitrole. Applied Clay Science, 114:190-196

Tang J, Zong L, Mu B, Kang Y R, Wang A Q. 2018. Attapulgite/carbon composites as a recyclable adsorbent for antibiotics removal. Korean Journal of Chemical Engineering, 35(8):1650-1661

Tu H, Lu Y, Wu Y, Tian J, Zhan Y F, Zeng Z Y, Deng H B, Jiang L B. 2015. Fabrication of rectorite-contained nanoparticles for drug delivery with a green and one-step synthesis method. International Journal of Pharmaceutics, 493(1-2):426-433

Wang M S, Liao L B, Zhang X L, Li Z H, Xia Z G, Cao W D. 2011. Adsorption of low-concentration ammonium onto vermiculite from Hebei Province, China. Clays and Clay Minerals, 59(5):459−465

Wang M S, Liao L B, Zhang X L, Li Z H. 2012. Adsorption of low concentration humic acid from water by palygorskite. Applied Clay Science, 67-68: 164−168

Wang X W, Mu B, An X C, Wang A Q. 2018. Insights into the relationship between the color and photocatalytic property of attapulgite/CdS nanocomposites. Applied Surface Science, 439:202−212

Wang X Y, Liu B, Wang X H, Sun R C. 2011. Amphoteric polymer-clay nanocomposites with drug-controlled release property. Current Nanoscience, 7(2):183−190

Wang X Y, Mei L F, Xing X B, Liao L B, Lv G C, Li Z H, Wu L M. 2014. Mechanism and process of methylene blue degradation by manganese oxides under microwave irradiation. Applied Catalysis B: Environmental, 160-161:211−216

Wang Y J, Liu P, Yang C, Mu B, Wang A Q. 2013. Improving capacitance performance of attapulgite/polypyrrole composites by introducing rhodamine B. Electrochimica Acta, 89:422−428

Weng J L, Liao L B, Lv G C, Li Z H, Li E W, He C, Wang S N. 2018. Probing the interactions between lucigenin and phyllosilicates with different layer structures. Dyes and Pigments, 155:135−142

Xie X L, Peng P, Zhu W F, Wang L L. 2014. Preparation and fixation ability for Pb^{2+} of geopolymeric material based on bayer process red mud. Advanced Materials Research, 1049-1050:175−179

Yan L G, Li S, Yu H Q, Shan R R, Du B, Liu T T. 2016. Facile solvothermal synthesis of Fe_3O_4/bentonite for efficient removal of heavy metals from aqueous solution. Powder Technology, 301:632−640

Yang D, Liao L B, Liu H K, Li H S, Mei L F. 2017. Structure and fluorescent properties of $Sr_{0.69}La_{0.31(1-x)}F_{2.31}:xEu^{3+}$ phosphors. Nanoscience and Nanotechnology Letters, 9(3):277−280

Yang D, Liao L B, Guo Q F, Mei L F, Liu H K, Zhou T S, Ye H. 2019a. Luminescence properties and energy transfer of $K_3LuF_6:Tb^{3+}, Eu^{3+}$ multicolor phosphors with a cryolite structure. RSC Advances, 9(8):4295−4302

Yang D, Liao L B, Guo Q F, Mei L F, Liu H K. 2019b. Crystal structure and luminescence properties of a novel cryolite-type $K_3LuF_6:Ce^{3+}$ phosphor. Journal of Solid State Chemistry, 277:32−36

Ye X X, Kang S H, Wang H M, Li H Y, Zhang Y X, Wang G Z, Zhao H J. 2015. Modified natural diatomite and its enhanced immobilization of lead, copper and cadmium in simulated contaminated soils. Journal of Hazardous Materials, 289:210−218

Zhang L H, Lu X C, Liu X D, Zhou J H, Zhou H Q. 2014. Hydration and mobility of interlayer ions of (Nax, Cay)-montmorillonite: a molecular dynamics study. The Journal of Physical Chemistry C, 118(51):29811−29821

Zhang Q, Yan Z L, Ouyang J, Zhang Y, Yang H M, Chen D L. 2018. Chemically modified kaolinite nanolayers for the removal of organic pollutants. Applied Clay Science, 157:283−290

Zhang W B, Mu B, Wang A Q, Shao S J. 2014. Attapulgite oriented carbon/polyaniline hybrid nanocomposites for electrochemical energy storage. Synthetic Metals, 192:87−92

Zhang X L, Lv G C, Liao L B, He M Q, Li Z H, Wang M S. 2012. Removal of low concentrations of ammonium and humic acid from simulated groundwater by vermiculite/palygorskite mixture. Water Environment Research, 84(8):682−688

Zhang Y, Fu L J, Shu Z, Yang H M, Tang A D, Jiang T. 2017. Substitutional doping for aluminosilicate mineral and superior water splitting performance. Nanoscale Research Letters, 12(1):456

Zhou J H, Jiang F, Li S J, Xu Z J, Sun W, Ji X B, Yang Y. 2019. $CuFeS_2$ as an anode material with an enhanced electrochemical performance for lithium-ion batteries fabricated from natural ore chalcopyrite. Journal of Solid State Electrochemistry, 23(7):1991−2000

Zhou T S, Mei L F, Zhang Y Y, Liao L B, Liu H K, Guo Q F. 2019. Color-tunable luminescence properties and energy transfer of Tb^{3+}/Sm^{3+} co-doped $Ca_9La(PO_4)_5(SiO_4)F_2$ phosphors. Optics & Laser Technology, 111:191−195

Zhu Q W, Zhang Y H, Lv F Z, Chu P K, Ye Z F, Zhou F S. 2012. Cuprous oxide created on sepiolite: preparation, characterization, and photocatalytic activity in treatment of red water from 2,4,6-trinitrotoluene manufacturing. Journal of Hazardous Materials, 217-218:11−18

Development of the Mineral Materials Research

Lv Guo-cheng[1], LIAO Li-bing[1], LI Yu-xin[1], TIAN Lin-tao[1], LIU Hao[2]

1. Beijing Key Laboratory of Materials Utilization of Nonmetallic Minerals and Solid Wastes, School of Materials Science and Technology, China University of Geosciences (Beijing), Beijing 100083;
2. School of Science, China University of Geosciences (Beijing), Beijing 100083

Abstract: In this paper, mineral materials in China are summarized into the fields of environmental mineral materials, energy mineral materials, medical mineral materials, pesticide and fertilizer carrier materials, electromagnetic functional mineral materials and other mineral materials. The research progresses in each of those fields in the past decade have been comprehensively reviewed based on the summarization of representative achievements. Then, some possible future research directions have been put forward according to the scientific development trend and the China's long-term development plan of mineral materials. This paper is hopefully to provide a reference for researchers and related departmental administrators in fields of mineral materials in China.

Key words: mineral materials; research progress; development direction

新矿物的发现与研究进展*

蔡剑辉

中国地质科学院 矿产资源研究所，自然资源部成矿作用和资源评价重点实验室，北京 100037

摘 要：从1958年我国发现第一个新矿物"香花石"开始至今，逾60年的时间里在中国发现和报道的新矿物达175种，其中142种为获得国际矿物协会新矿物、命名和分类委员会（IMA CNMNC）正式批准的有效独立矿物种。21世纪近二十年来我国新矿物研究进展尤为突出，已发现新矿物数量达60种，位列世界第六。特别是关于自然元素矿物大类新矿物的发现位于世界前列，关于稀土新矿物和冲击变质超高压新矿物的发现和研究也取得了令世界瞩目的进展。本文着重从20世纪中国新矿物种的修订、21世纪在中国发现的新矿物种的数量和晶体化学类型以及产出特征等方面对21世纪我国新矿物研究进展进行综述，以期为推进我国新矿物工作提供参考、借鉴和启迪。

关键词：新矿物 研究进展 晶体化学类型 产地与产状

0 引 言

矿物是地球和地外天体中天然产出的具有确定的化学组成和结晶学性质的固体物质，是特定地质作用的产物，蕴含着丰富的解析地球和地外天体的起源和发展、矿产资源的形成与演化、环境的变迁与修复、材料的利用和研制等诸多原始信息源。人类对自然的认识是一个不断探索、不断丰富的发展过程，对构成固体地球和地外天体中地质体的最小基本单元——矿物的发现和认识也不例外，矿物的多样性主要是由一系列自然作用过程导致的，随着地球及地外天体的演化，矿物种不断丰富，这个过程贯穿整个演化过程。因此，新矿物的发现与研究是矿物学研究中不可或缺的一个重要基础领域。发现新矿物的科学意义不仅体现在相关的创新性成果可以不断丰富矿物学研究的内容并提高矿物学的整体研究水平，进而促进地球科学的发展和矿产资源的开发利用上，更重要的是可以通过对矿物多样性的认识使人类对客观世界的认知深度及广度得到持续性的提升和拓展。发现和研究新矿物的实用价值还在于，对每一种新矿物或矿物种新的物理化学性能的认识，不仅意味着人类对地球乃至宇宙的认知又前进了一步，而且意味着可能又有一种新的固体资源和天然材料可以被利用，也意味着可能又增添了一种解析复杂自然作用过程的新途径和新依据。作为最基础的科学探索，有关新矿物的研究工作往往最能客观地反映一个国家的矿物学研究水平，同时也是反映一个国家基础科研实力的重要标志之一。

国际矿物协会新矿物和矿物命名委员会（Commission on New Minerals and Mineral Names，CNMMN）成立于1958年，2006年国际矿物协会的新矿物和矿物命名委员会（CNMMN）与矿物分类委员会（Commission on Classification of Minerals，CCM）合并为新矿物、命名和分类委员会（Commission on New Minerals, Nomenclature and Classification，CNMNC）。CNMMN及后来的CNMNC是得到国际矿物学界认可的负责指导和审订全球新矿物以及矿物命名和分类的国际学术组织，由世界各国矿物学学术组织指派的代表组成，目前CNMNC有委员37名，其中的中国代表由中国矿物岩石地球化学学会新矿物及矿物分类、命名专业委员会主任委员担任。截至2020年3月，全球已发现和认识并经IMA CNMNC批准的有效独立

---

* 原文"本世纪我国新矿物的发现与研究进展（2000~2019年）"刊于《矿物岩石地球化学通报》2021年第40卷第1期，本文略有修改。

矿物种达 5575 个。据统计，以 20 年为一周期，1800—1819 年全球发现的新矿物种为 87 种，1820—1919 年间每 20 年平均发现 185 种，1920—1939 年发现 256 种，1840—1959 年发现 342 种，1960—1979 年发现约 900 种（张建洪和彭志忠，1981）。以 1982 年我国出版的《系统矿物学》（王濮等，1982）全面收录的矿物种数计，截至 1980 年全球已发现矿物种约 2 650 种。按照 IMA CNMNC 于 2009 年 11 月公布的矿物列表，当时已批准的有效矿物种共 4 424 个，由此推算，1980—1999 年 20 年间发现的新矿物总数可达 1 170 余种。进入 21 世纪，在 2000—2019 年的 20 年间全球发现并经 IMA CNMNC 认可的新矿物种数达 1 728 个，其数量占全球已知矿物种总数的 31%。总体上看，全球发现新矿物的数量呈阶段性上升趋势。特别需要强调的是，近十年（2010—2019 年），新矿物工作的成就更为突出，全球共发现新矿物 1 126 种，亦即目前世界上已知独立矿物种总数的 1/5 还多都是近十年发现的，这充分表明，科学技术的全面飞速发展，促进了近年来国际新矿物研究的突飞猛进。一直以来，重视发现和研究新矿物都是我国矿物学研究的特色（秦善等，2016），自 1958 年黄蕴慧先生发现我国第一个新矿物"香花石"以来，中国新矿物研究开始稳步发展，并在 20 世纪 80 年代迎来蓬勃发展（王濮和李国武，2014）。截至 2019 年底，在中国发现并报道的新矿物共计 175 种，其中有 142 种为 IMA CNMNC 正式批准的有效矿物种（表 1、表 2）。与全球新矿物研究的大好局面保持同步，21 世纪（2000—2019 年）我国新矿物研究也相对活跃，也取得了非常丰硕的成果，近二十年在中国发现并经 IMA CNMNC 批准的有效矿物 61 种（包括后来撤销的 1 种），约占中国新矿物总数的 40%，占全球同期发现新矿物总数的 3.5%。近十年，我国新矿物工作再次迎来继 20 世纪 80 年代之后的又一个发展高潮，在中国发现的新矿物达 42 种之多。本文通过对 21 世纪我国新矿物研究新进展的综述，总结我国新矿物工作的成就和特点，为进一步推进我国新矿物的发现和研究工作提供参考、借鉴和启迪。

1 对 20 世纪中国新矿物种的修订

已有多位学者对在中国发现的新矿物资料进行过分阶段或分区域的整理和评述（张建洪和彭志忠，1981；李锡林，1989；周正和曹亚文，1997；张如柏，1998；张培善等，2000；Mandarino and de Fourestier，2005；徐金沙等，2012；王濮和李国武，2014；范光等，2020）。笔者在此基础上查漏补缺，对 20 世纪在中国发现的新矿物资料进行了全面而系统的梳理，获知 1958—1999 年已报道的新矿物共达 114 种之多，其中获 IMA CNMMN 批准的独立矿物 82 种，其余 32 种已发表的具矿物名称的新矿物并非国际公认的有效独立矿物种（表 1）。尽管如此，所有这些工作成果真实而全面地反映了中国新矿物工作的历程，都是值得载入史册的。新矿物研究工作由多方面的内容组成，除了发现和研究新矿物、对新矿物进行命名和分类等主要工作外，通过对已存矿物的再研究并按照新确立的矿物分类命名方案对矿物进行重新定义、重新命名等修订也是非常重要的工作内容。

表 1 在中国发现的矿物种（1958—1999 年）

序号	英文名称	中文名称	化学式
1	Aeschynite-(Nd)	钕易解石	$(Nd,Ln,Ca)(Ti,Nb)_2(O,OH)_6$
2	Anduoite	安多矿	$RuAs_2$
3	Ankangite*	安康矿*	$Ba(Ti,V^{3+},Cr)_8O_{16}$
4	Antimonselite	硒锑矿	Sb_2Se_3
5	Ashanite*	阿山矿*	$(Nb,Ta,U,Fe,Mn)_4O_8$
6	Bafertisite	钡铁钛石	$Ba(Fe^{2+},Mn)_2TiOSi_2O_7(OH,F)_2$
7	Baiyuneboite-(Ce)*	白云鄂博矿*	$NaBaCe_2(CO_3)_4F$
8	Balipholite	纤钡锂石	$LiBaMg_2Al_3(Si_2O_6)_2(OH)_8$
9	Baotite	包头矿	$Ba_4(Ti,Nb,W)_8O_{16}(SiO_3)_4Cl$
10	Caichengyunite*	蔡承云石*	$Fe_3^{2+}Al_2(SO_4)_6 \cdot 30H_2O$

续表

序号	英文名称	中文名称	化学式
11	Carboborite	水碳硼石	$Ca_2Mg[B(OH)_4]_2(CO_3)_2 \cdot 4H_2O$
12	Cebaite-(Ce)	氟碳铈钡石	$Ba_3Ce_2(CO_3)_5F_2$
13	Chaidamuite	柴达木石	$ZnFe^{3+}(SO_4)_2(OH) \cdot 4H_2O$
14	Changbaiite	长白矿	$PbNb_2O_6$
15	Changchengite	长城矿	$IrBiS$
16	Chengdeite	承德矿	Ir_3Fe
17	Chiluite	赤路矿	$Bi_3Te^{6+}Mo^{6+}O_{10.5}$
18	Chrombismite	铬铋矿	$Bi_{16}CrO_{27}$
19	Chromium	自然铬	Cr
20	Clinochalcomenite*	斜蓝硒铜矿*	$CuSeO_3 \cdot 2H_2O$
21	Clinotyrolite*	单斜铜泡石*	$Ca_2Cu_9(AsO_4)_4(SO_4)_{0.5}(OH)_9 \cdot 9H_2O$
22	D'Ansite*	氯镁芒硝(盐镁芒硝)*	$Na_{21}Mg(SO_4)_{10}Cl_3$
23	Damiaoite	大庙矿	$PtIn_2$
24	Danbaite	丹巴矿	$CuZn_2$
25	Daomanite	道马矿	$CuPtAsS_2$
26	Daqingshanite-(Ce)	大青山矿	$Sr_3CePO_4(CO_3)_3$
27	Dayingite*	大营矿*	$Cu(Pt^{3+},Co^{3+})_2S_4$
28	Diaoyudaoite	钓鱼岛石	$NaAl_{11}O_{17}$
29	Erlianite	二连石	$(Fe^{2+})_4(Fe^{3+})_2Si_6O_{15}(OH)_8$
30	Ertixiite	额尔齐斯石	$Na_2Si_4O_9$
31	Fengluanite*	丰滦矿*	$Pd_3(As,Sb)$
32	Fergusonite-(Ce)-β	褐铈铌矿-β	$(Ce,La,Nd)NbO_4$
33	Fergusonite-(Nd)-β	褐钕铌矿-β	$(Nd,Ce)NbO_4$
34	Fluorannite	氟铁云母	$K(Fe^{2+})_3(Si_3Al)O_{10}F_2$
35	Furongite	芙蓉铀矿	$Al_2(UO_2)(PO_4)_2(OH)_2 \cdot 20H_2O$
36	Gananite	赣南矿	BiF_3
37	Gaotaiite	高台矿	Ir_3Te_8
38	Guanglinite*	广林矿*	$Pd_{11}Sb_2As_2$
39	Gugiaite	顾家石	$Ca_2BeSi_2O_7$
40	Gupeiite	古北矿	Fe_3Si
41	Hexatestibiopanickelite*	六方碲锑钯镍矿*	$(Pd,Ni)(Sb,Te)$
42	Hingganite-(Y)	钇兴安石	$BeYSiO_4(OH)$
43	Hongquiite*	红旗矿*	TiO
44	Hongshiite	红石矿	$(Pt,Fe)Cu$
45	Hoshiite*	河西石*	$NiMg(CO_3)_2$
46	Hsianghualite	香花石	$Li_2Ca_3Be_3(SiO_4)_3F_2$
47	Huanghoite-(Ce)	黄河矿	$BaCe(CO_3)_2F$
48	Hunchunite	珲春矿	Au_2Pb
49	Hungchaoite	章氏硼镁石	$Mg(B_4O_7) \cdot 9H_2O)$
50	Hydroastrophyllite*	水星叶石*	$(H_3O,K,Ca)_3(Fe,Mn)_{5-65}Ti_2Si_8(O,OH)_{31}$
51	Hydrochlorborite	多水氯硼钙石	$Ca_4B_8O_{15}Cl_2 \cdot 21H_2O$
52	Hydrougrandite*	水钙榴石*	$(Ca,Mg,Fe^{2+})_3(Fe^{3+},Al)_2[(OH)_4(SiO_4)_2]$
53	Iridisite*	硫铱矿*	$(Ir,Cu,Rh,Ni,Pt)S_2$
54	Iridrhodruthenium*	铱铑钌矿*	(Ru,Rh,Ir)
55	Jianshuiite	建水矿	$(Mg,Mn)Mn_3O_7 \cdot 3H_2O$
56	Jichengite	冀承矿	$3CuIr_2S_4 \cdot (Ni,Fe)_9S_8$

续表

序号	英文名称	中文名称	化学式
57	Jinshajiangite	金沙江石	$NaKCaBa(Fe^{2+})_6Mn_2Ti_4(Si_2O_7)_4O_2(OH)_8F_2$
58	Jixianite*	蓟县矿*	$Pb(W,Fe^{3+})_2(O,OH)_7$
59	Laihunite	莱河矿	$(Fe^{3+},Fe^{2+},\square)_2SiO_4$
60	Leadamalgam	汞铅矿	$Pb_{0.7}Hg_{0.3}$
61	Liberite	锂铍石	Li_2BeSiO_4
62	Lingaitukuang	磷钙钍矿	$CaTh(PO_4)_2$
63	Lishizhenite	李时珍石	$Zn(Fe^{3+})_2(SO_4)_4 \cdot 14H_2O$
64	Liujinyinite	硫金银矿	Ag_3AuS_2
65	Luanheite	滦河矿	Ag_3Hg
66	Lunijianlaite	绿泥间蜡石	$Li_{0.7}Al_{6.2}(Si_7Al)_2(OH,O)_{10}$
67	Magbasite	硅镁钡石	$KBaMg_6AlSi_6O_2F_2$
68	Magnesiohulsite	黑硼锡镁矿	$(Mg,Fe^{2+})_2(Fe^{3+},Sn,Mg)(BO_3)O_2$
69	Magnesionigerite-2N1S	镁尼日利亚石-2N1S	$(Mg,Al,Zn)_2(Al,Sn)_6O_{11}(OH)$
70	Magnesionigerite-6N6S	镁尼日利亚石-6N6S	$(Mg,Al,Zn)_3(Al,Sn,Fe)_8O_{15}(OH)$
71	Malanite	马兰矿	$CuPt_2S_4$
72	Mayingite	马营矿	$IrBiTe$
73	Mengxianminite (of Huang et al.)*	孟宪民石（黄蕴慧等）*	$(Ca,Na)_3(Fe,Mn)_2Mg_2(Sn,Zn)_5Al_8O_{29}$
74	Mongshanite*	蒙山矿*	$(Mg,Cr,Fe^{2+})_2(Ti,Zr)_5O_{12}$
75	Muchuanite*	沐川矿*	$MoS_2 \cdot nH_2O$
76	Nanlingite	南岭石	$Na(Ca_5Li)Mg_{12}(AsO_3)_2[Fe^{2+}(AsO_3)_6]F_{14}$
77	Nanpingite	南平石	$CsAl_2(Si_3Al)O_{10}OH)_2$
78	Natrobistantite*	铋细晶石*	$(Na,Cs)Bi(Ta,Nb,Sb)_4O_{12}$
79	Omeiite	峨眉矿	$OsAs_2$
80	Orthobrannerite	斜方钛铀矿	$U^{4+}U^{6+}Ti_4O_{12}(OH)_2$
81	Parisite-(Nd)*	氟碳钙钕矿*	$Ca(Nd,Ce,La)_2(CO_3)_3F_2$
82	Pingguite	平谷矿	$Bi_6Te_2^{4+}O_{13}$
83	Qilianshanite	祁连山石	$NaHCO_3 \cdot H_3BO_3 \cdot 2H_2O$
84	Qingheiite	青河石	$Na_2MnMgAl(PO_4)_3$
85	Qitianlingite	骑田岭矿	$Fe_2^{2+}Nb_2W^{6+}O_{10}$
86	Ruarsite	硫砷钌矿	$RuAsS$
87	Shuangfengite	双峰矿	$IrTe_2$
88	Suolunite	索伦石	$Ca_2Si_2O_5(OH)_2 \cdot H_2O$
89	Taiyite*	钛钇矿*	$(Y,Ln,Ca,Th)(Ti,Nb)_2(O,OH)_6$
90	Tengchongite	腾冲铀矿	$Ca(UO_2)_6(MoO_4)_2O_5 \cdot 12H_2O$
91	Testibiopalladite*	等轴碲锑钯矿*	$PdTe(Sb,Te)$
92	Tetra-auricupride	四方铜金矿	$CuAu$
93	Tongbaite	桐柏矿	Cr_3C_2
94	Tongxinite*	铜锌矿*	Cu_2Zn
95	Trigonomagneborite*	三方硼镁石*	$MgO \cdot 3B_2O_3 \cdot 7.5H_2O$
96	Weishanite	围山矿	$(Au,Ag)_{1.2}Hg_{0.8}$
97	Xiangjiangite	湘江铀矿	$Fe^{3+}(UO_2)_4(PO_4)_2(SO_4)_2(OH) \cdot 22H_2O$
98	Xifengite	喜峰矿	Fe_5Si_3
99	Xilingolite	锡林郭勒矿	$Pb_3Bi_2S_6$
100	Ximengite	西盟石	$BiPO_4$

续表

序号	英文名称	中文名称	化学式
101	Xingzhongite	兴中矿	$Pb^{2+}Ir_2^{3+}S_4$
102	Xitieshanite	锡铁山石	$Fe^{3+}SO_4Cl \cdot 6H_2O$
103	Yanzhongite*	燕中矿*	$PdTe$
104	Yimengite	沂蒙矿	$K(Cr,Ti,Fe,Mg)_{12}O_{19}$
105	Yingjiangite	盈江铀矿	$K_2Ca(UO_2)_7(PO_4)_4(OH)_6 \cdot 6H_2O$
106	Yixunite	伊逊矿	Pt_3In
107	Yuanfuliite	袁复礼石	$Mg(Fe^{3+},Fe^{2+},Al,Ti,Mg)(BO_3)O$
108	Yuanjiangite	沅江矿	$AuSn$
109	Zabuyelite	扎布耶石	Li_2CO_3
110	Zhanghengite	张衡矿	$CuZn$
111	Zhonghuacerite-(Ce)*	中华铈矿*	$Ba_2Ce(CO_3)_3F$
112	Zincobotryogen*	锌赤铁矾*	$ZnFe^{3+}(SO_4)_2(OH) \cdot 7H_2O$
113	Zincocopiapite	锌叶绿矾	$Zn(Fe^{3+})_4(SO_4)_6(OH)_2 \cdot 7H_2O$
114	Zincovoltaite	锌绿钾铁矾	$K_2Zn_5(Fe^{3+})_{12}Al(SO_4)_{12} \cdot 18H_2O$

注："*"表示为以下3种情况之一：①从未提交 IMA CNMMN；②已提交但未获 IMA CNMMN 批准；③曾经获 IMA CNMMN 批准，但后来又被否定。按矿物英文名称字母排序。

根据 IMA CNMMN 关于确定新矿物种的原则，一个矿物种主要以其化学组成和结晶学性质为基础加以确定，新矿物种的化学组成和（或）结晶学性质与任何已存在的矿物种应有明显不同。一种新矿物及其名称被以文献的形式正式发表之前，须事先获得 CNMMN 的批准（Nickel and Mandarino，1987）。过去我国关于新矿物的发表并未严格按照 IMA CNMMN 的原则执行，因此在一定程度上造成一些关于中国新矿物的模糊认识，非常有必要予以澄清并修正。20 世纪在中国发现并报道的 114 种新矿物中只有 82 种获得 IMA CNMMN 的正式批准，有 14 种未提交 IMA CNMMN 或已提交但未获正式批准；有 15 种已发表的新矿物虽提交 IMA CNMMN 但最终被否定；有 3 种因研究数据不全当年未获 IMA CNMMN 批准，后经进一步补充研究后于 21 世纪最终获得正式批准；还有 3 个有效矿物种根据 IMA CNMNC 新颁布的相关矿物（超）族分类命名方案进行了重命名。具体情况如下。

（1）已公开发表但未提交 IMA CNMMN 批准的矿物 14 种（其中部分可能是有效矿物种）：

三方硼镁石（Trigonomagneborite），1964 年发现于青海省海西自治州大柴旦盐湖，正式发表于 1965 年（曲一华等，1965），未提交 IMA CNMMN。因为先期发现于美国加利福尼亚州 Furnace Creek 硼酸盐矿区的相同矿物种已被命名为 Mcallisterite（三方硼镁石），并且该新矿物及名称已于 1963 年先获得 IMA CNMMN 的正式批准，故 "Trigonomagneborite" 只是算 "Mcallisterite" 的同义词，而非有效矿物种名。

六方碲锑钯镍矿（Hexatestibiopanickelite），发现于中国西南地区某地（可能是四川杨柳坪），未提交 IMA CNMMN 先行发表（於祖相等，1974）。可能为有效矿物种。

水星叶石（Hydroastrophyllite），最早发现于中国四川，未提交 IMA CNMMN 批准先行发表（湖北地质学院 X 光实验室，1974）。可能为有效矿物种。

钛钇矿（Taiyite），1970 年发现于某燕山期富钇族稀土矿床的中粒白云母化花岗岩及其风化壳中，未提交 IMA CNMMN 批准即先作为新矿物种发表（齐玲仪，1974）。但钛钇矿的成分和结构均与 1879 年已发表的古老矿物 Aaeschynite-(Y)（钇易解石）相同，并非有效的新矿物种。

等轴碲锑钯矿（Testibiopalladite），发现于四川丹巴地区杨柳坪铜镍铂矿床，未提交 IMA CNMMN 批准即先行发表（中国科学院贵阳地球化学研究所，铂矿研究组，电子探针分析组，X 射线粉晶分析组，选矿组，1974）。可能为有效矿物种。

氯镁芒硝（D'Ansite），天然氯镁芒硝（盐镁芒硝）最早发现于我国某地古近纪陆相含盐岩系中，但研究资料未提交 IMA CNMMN 批准即先行发表（曲懿华等，1975）。目前 IMA CNMMN 认可的氯镁芒硝模式标本产地位于奥地利。

磷钙钍矿（Lingaitukuang），发现于新疆可可托海 3# 稀有金属花岗伟晶岩，1978 年未经 IMA CNMMN 批准先发表（王贤觉，1978）。同年产于纳米比亚 Brabant 伟晶岩中的相同矿物获得 IMA CNMMN 批准，矿物定名为 Brabantite，2007 年 Brabantite 被 IMA CNMNC 重命名为 Cheralite（Linthout，2007）。目前 Brabantite 为否定矿物种名，Lingaitukuang 算作 Brabantite（磷钙钍矿）的同义词。

斜蓝硒铜矿（Clinochalcomenite），发现于中国甘肃，未经 IMA CNMMN 批准即已先行发表（Lo et al., 1980，1984）。可能为有效矿物种。

中华铈矿［Zhonghuacerite-(Ce)］，发现于内蒙古白云鄂博矿区西矿，未提交 IMA CNMMN 批准即已先行发表（张培善和陶克捷，1981），但发表后一直被国际矿物种词汇收录（Fleischer，1983；Fleischer and Mandarino，1995），后来被 IMA CNMMN 撤销，将其划归为否定矿物种，原因是认为原样品可能相当于"黄河矿［Huanghoite-(Ce)］"或者是"库氟碳铈钡石［Kukharenkoite-(Ce)］"。

沐川矿（Muchuanite），发现于四川省沐川县某地的黑色含钼砂岩层中，未经 IMA CNMMN 批准先正式发表（张如柏等，1981）。但经进一步研究确定其为辉钼矿和胶辉钼矿的混合物，并非有效新矿物种（Fleischer et al., 1982；Fejer and Hey, 1984）。

铜锌矿（Tongxinite），1981 年发现于西藏玉龙和马拉松多斑岩铜钼矿床，1995 年发现于四川若尔盖巴西金矿床，未提交 IMA CNMMN 批准已先行发表（帅德权等，1998；罗梅和王月文，1999）。可能是一种有效新矿物种。

氟碳钙钕矿［Parisite-(Nd)］，发现于内蒙古白云鄂博矿床，未提交 IMA CNMMN 即先行发表（张培善和陶克捷，1986）。但可能为有效矿物种。

钴硅锌矿（Xingshaoite），发现于湖南省邵阳市新邵县高家坳卡林型金矿床的氧化带中，其 CoO 含量高达 25.4%，为硅锌石（Willemite，Zn_2SiO_4）的富钴新变种（郑钰纯和黄振恒，1989），而非一种有效新矿物种。

蔡承云石（Caichengyunite），发现于四川省会东县龙树村附近铅锌矿氧化带中，未提交 IMA CNMMN 即先行发表（张如柏等，2002）。但可能为有效矿物种。

（2）已公开发表并提交 IMA CNMNC 而被拒绝或否定的矿物 15 种：

河西石（Hoshiite），20 世纪 60 年代发现于甘肃某地超基性岩铜镍硫化物矿床氧化带露头中，作为新矿物先行发表（於祖相等，1964），但后经进一步研究确定为富 Ni 的菱镁矿变种，为 IMA CNMNC 认定的否定矿物种（Burke，2006）。

水钙榴石（Hydrougrandite），20 世纪 60 年代发现于内蒙古小松山橄榄岩中，作为新矿物先行发表（曹荣龙，1964），但后经进一步研究确定为富 H_2O 的钙榴石变种，为 IMA CNMNC 认定的否定矿物种。

大营矿（Dayingite），20 世纪 70 年代发现于某含铂基性–超基性岩体中（於祖相等，1974），后经进一步工作，确认其为马兰矿的富 Co 新变种（於祖相，1981），被 IMA CNMMN 否定。

丰滦矿（Fengluanite），20 世纪 70 年代发现于河北承德红石砬铂钯矿床（红区）和三道村附近含铂基性–超基性岩体（道区）（於祖相等，1974），但因化学式和晶体结构方面有误，被 IMA CNMMN 否定。后确定其为富 Sb 的等轴砷锑钯矿变种（施倪承等，1978）。

广林矿（Guanglinite），20 世纪 70 年代发现于河北承德红石砬铂钯矿床（红区）和三道村附近含铂基性–超基性岩体（道区）（於祖相等，1974），但从化学成分到结构都等同于等轴砷锑钯矿（彭志忠等，1978），被 IMA CNMMN 否定。

红旗矿（Hongquiite），20 世纪 70 年代发现于河北丰宁三道村附近含铂基性–超基性岩体（道区）（於祖相等，1974），但因化学成分与 X 射线粉晶衍射资料相矛盾（彭志忠等，1978），被 IMA CNMMN 否定。

蓟县矿（Jixianite），首次发现于天津市蓟县盘山沿河钨矿床的蓟县矿发表于 1979 年（刘建昌，1979），但由于缺乏充分的晶体结构研究，无法确定蓟县矿晶体结构中 Y 位的主导负离子，IMA CNMMN

一直将其定为存疑矿物种。按照 IMA CNMNC 新颁布的烧绿石超族矿物命名方案（Atencio et al., 2010；Christy and Atencio, 2013），确定为否定矿物种，原样品应为水空钨石族矿物的一种变种。

硫金银矿（Liujinyinite），先后发现于安徽、广东和甘肃的含金石英脉、夕卡岩型铜矿床和黄铁矿型多金属矿床中，未经 IMA CNMMN 批准先行发表（陈振玡等，1979；魏明秀，1981），但新矿物申请因资料不够充分而被 IMA CNMMN 拒绝。蔡长金（1986）经研究认为 Liujinyinite 与 1977 年 IMA CNMNC 已批准的有效矿物种 Uytenbogaardtite（Ag_3AuS_2）是同一种矿物。

蒙山矿（Mongshanite），1979 年发现于山东临沂沂蒙山金伯利岩中，未经 IMA CNMMN 批准先作为新矿物发表（周剑雄等，1984；陆琦和彭志忠，1987）。但 Mongshanite 从成分到结构均与 1982 年 IMA CNMMN 已批准的矿物 Mathiasite（Haggerty et al., 1983）相同，故 Mongshanite 只能算作 Mathiasite 的同义词。

阿山矿（Ashanite），发现于新疆伊犁自治州阿勒泰地区富蕴县可可托海伟晶岩区，首次发表于 1980 年（张如柏等，1980），一直为有争议的存疑矿物，1996 年被 IMA CNMMN 撤销，为否定矿物种（沈敢富，1998）。

铋细晶石（Natrobistantite），发现于新疆阿勒泰地区富蕴县可可托海伟晶岩区，首次发表于 1983 年（Voloshin et al., 1983），按照 IMA CNMNC 新颁布的烧绿石超族矿物命名方案（Atencio et al., 2010），确定铋细晶石为细晶石族中主导元素为零价矿物的一种变种，属于否定矿物种。

安康矿［Ankangite，$Ba(Ti, V^{3+}, Cr)_8O_{16}$］，发现于陕西省石梯重晶石矿床，1986 年经 IMA CNMMN 批准为有效矿物种（熊明等，1988）。但按照 2011 年 IMA CNMNC 批准的关于锰钡矿超族（Hollandite Supergroup）命名方案的修订建议 IMA 11-F（Williams et al., 2012a），安康矿被认定为否定矿物种，原样品相当于钡钒钛石［Mannardite，$Ba_xTi_{8-2x}V_{2x}O_{16}\cdot(2-x)H_2O$］的无水变种；

白云鄂博矿［Baiyuneboite-(Ce)］，发现于内蒙古白云鄂博稀土铁矿西矿白云岩中或东矿钠辉石型矿石中。1986 年获 IMA CNMMN 批准，1987 年正式发表（傅平秋等，1987；傅平秋和苏贤泽，1987）。但 1998 年因 Cordylite-(Ce)（氟碳钡铈石）被重新定义（沈今川和宓锦校，1992；Giester et al., 1998），而"白云鄂博矿"因与重新定义的 Cordylite-(Ce) 等同而被 IMA CNMMN 否定。Baiyuneboite-(Ce) 现算作 Cordylite-(Ce) 的同义词。

铱铑钌矿（Iridrhodruthenium），发现于我国藏北某超基性岩铬矿中，未经 IMA CNMMN 批准先发表（毛水和和周学粹，1989）。后经进一步研究确定其为富 Ir-Rh 的自然钌变种，属于否定矿物种（Jambor and Vanko, 1991）。

硫铱矿（Iridisite），发现于河北承德高寺台 Cr-PGE 矿床（高区）和武烈河砂矿中（於祖相，1995，1997）。因研究资料不足，新矿物申请被 IMA CNMMN 拒绝。

（3）早年已有报道但因数据不全当时未获 IMA CNMMN 批准，21 世纪经进一步研究终获正式批准的新矿物 3 种：

锌赤铁矾（Zincobotryogen），为 1964 年报道的在青海省锡铁山铅锌矿发现的一种新矿物（涂光炽等，1964），但当年因为需要进一步补充研究数据一直未通过 IMA CNMMN 的批准，直到 2016 年经杨主明等进一步研究并申报，才获 IMA CNMNC 正式批准（Yang et al., 2017）。

富铜泡石（Tangdanite），为 1980 年报道的一种新矿物，当时命名为"Clinotyrolite"（单斜铜泡石）（马喆生等，1980），但因未获得晶体结构数据，当时并未得到 IMA CNMMN 的批准。2012 年基于对原样品的精细研究数据，最终被 IMA CNMNC 正式批准为有效矿物种。新矿物申报最初的英文名为"Fuxiaotuite"（de Fourestier et al., 2012），后经 IMA CNMNC 建议以该矿物模式标本产地（云南东川铜矿汤丹矿）地名命名为"Tangdanite"，中文名称保留原词根"铜泡石"，命名为富铜泡石（李国武等，2014；Ma et al., 2014）。

孟宪民石（Mengxianminite），1986 年黄蕴慧等报道的"孟宪民石"产于湖南省郴州市临武县香花岭夕卡岩中，是一种含锡和铝的复杂氧化物类新矿物（Huang et al., 1986；黄蕴慧等，1988），但当时因为数据缺失并未获得 IMA CNMMN 的正式批准。2015 年饶灿等在相同产地发现相似矿物，经研究确定为自然界首次发现的含铍和锡的天然硼酸盐新矿物种，并具全新的晶体结构，命名为 Mengxianminite（孟宪民石），该矿物及其名称均获得到 IMA CNMNC 的正式批准（Rao et al., 2017）。

（4）重命名矿物 3 种：

羟硅铍钇铈矿（Yttroceberysite），1981 年发现于黑龙江省大兴安岭地区某花岗斑岩型稀有金属矿床中的一种富铈和铍的硅酸盐类新矿物（丁孝石等，1981）。1984 年发现者根据 IMA CNMMN 的建议按新矿物产地地名重新将其命名为兴安石（Hingganite）（丁孝石等，1984）。后来按照 IMA CNMMN 关于矿物命名的原则（Nickel and Mandarino，1987），重新定名为"Hingganite-(Y)"，中文译名"钇兴安石"。该矿物名称 2007 年获得 IMA CNMNC 批准。

彭志忠石-6H（Pengzhizhongite-6H）是 1989 年由陈敬中等在湖南省益阳市安化县白钨矿区发现的一种新矿物（陈敬中等，1989）。2002 年按照 IMA CNMMN 批准的尼日利亚石族矿物命名新方案（Armbruster，2002），将彭志忠石-6H（Pengzhizhongite-6H）重新命名为镁尼日利亚石-2N1S（Magnesionigerite-2N1S）。

彭志忠石-24R（Pengzhizhongite-24R）是 1989 年由陈敬中等在湖南省益阳市安化县白钨矿区发现的又一种新矿物（陈敬中等，1989）。2002 年按照 IMA CNMMN 批准的尼日利亚石族矿物命名新方案（Armbruster，2002），将彭志忠石-24R（Pengzhizhongite-24R）重新命名为镁尼日利亚石-6N6S（Magnesionigerite-6N6S）。

此外，还有产地分别为俄罗斯和加拿大的三种新矿物镁星叶石（Magnesioastrophyllite）、钡闪叶石（Barytolamprophyllite）和氟铈硅磷灰石［Fluorbritholite-(Ce)］是由中国学者主导鉴定的（彭志忠和马喆生，1963；彭志忠等，1983；Gu et al.，1994）。

2 21 世纪在中国发现的新矿物

据统计，21 世纪（2000—2019 年）以来在中国发现并经 IMA CNMNC 批准的有效独立矿物种为 61 种，详见表 2。需要指出的是，这 61 个有效矿物种中后来有一种被撤销，即碲硫银锡矿（Tellurocanfieldite），又名八家子矿（Bajiazite），是 2012 年批准的一种新矿物，后来根据其单晶结构分析结果确定其实际上是富碲的硫银锡矿变种，而非有效矿物种，故撤销批准。因此，截至 2019 年底，在中国已发现并报道的新矿物共 175 种，其中未提交、未批准或批准后被撤销的矿物共计 33 种，为 IMA CNMNC 正式批准的有效矿物种 142 种。

表 2 在中国发现的新矿物种（2000—2019 年）

序号	英文名	中文名	化学式	产地和产状	主要发现者	参考文献
1	Lingunite	玲根石	$NaAlSi_3O_8$	江苏省泰州市高岗区寺巷镇寺巷口陨石	Gillet P.（瑞士）、陈鸣、Dubrovinsky L.（德）等	Gillet et al., 2000
2	Lanmuchangite	铊明矾	$Tl^+Al(SO_4)_2 \cdot 12H_2O$	贵州省兴仁县回龙镇滥木厂铊(汞)矿床	陈代演、王冠鑫、邹振西等	陈代演等，2001
3	Hubeite	湖北石	$Ca_2Mn^{2+}Fe^{3+}Si_4O_{12}(OH) \cdot 2H_2O$	湖北省黄石市大冶铁矿	Hawthorne F. C.（加）、Cooper M. A.（加）、Grice J. D.（加）等	Hawthorne et al., 2002
4	Fluoro-nybøite	氟尼伯闪石	$NaNa_2(Mg_3Al_2)(AlSi_7O_{22})(F,OH)_2$	江苏省连云港市东海县苏鲁柯石英–榴辉岩带	Oberti R、Boiocchi M、Smith D. C.（意）	Oberti et al., 2003
5	Tuite	涂氏磷钙石	$Ca_3(PO_4)_2$	湖北省大堰坡乡随州 L6 球粒陨石	谢先德、Minitti M. E.（美）、陈鸣等	Xie et al., 2003；谢先德等，2003

续表

序号	英文名	中文名	化学式	产地和产状	主要发现者	参考文献
6	Zincospiroffite	碲锌石	$Zn_2Te_3O_8$	河北省张家口市水泉沟杂岩体东坪金矿中山沟矿床	张佩华、朱金初、赵振华等	Zhang et al., 2004
7	Maoniupingite-(Ce)	牦牛坪矿	$(Ce,Ca)_4(Fe^{3+},Ti,Fe^{2+},\square)(Ti,Fe^{3+},Fe^{2+},Nb)_4Si_4O_{22}$	四川省冕宁县牦牛坪稀土矿床	沈敢富、杨光明、徐金沙	沈敢富等, 2005
8	Luobusaite	罗布莎矿	$Fe_{0.84}Si_2$	西藏曲松县罗布莎铬铁矿区	白文吉、施倪承、方青松等	白文吉等, 2006
9	Fluoro-alumino-magnesiotaramite	氟绿闪石	$\{Na\}\{CaNa\}\{Mg_3Al_2\}(Al_2Si_6O_{22})F_2$	江苏省连云港市东海县苏鲁柯石英-榴辉岩带	Oberti R.(意)、Boiocchi M.(意)、Smith D. C.(法)等	Oberti et al., 2007
10	Ottensite	水氧硫锑钠石	$Na_3(Sb_2O_3)_3(SbS_3)\cdot 3H_2O$	贵州省黔西南州大厂锑矿田晴隆矿氧化带	Sejkora J.(捷)、Hyrsl J.(捷)	Sejkora and Hyršl, 2007
11	Chenguodaite	陈国达矿	$Ag_9FeTe_2S_4$	山东省烟台市招远埠南金矿	谷湘平、Watanabe Makoto(日)、谢先德等	谷湘平等, 2008
12	Dingdaohengite-(Ce)	丁道衡矿	$(Ce,La)_4Fe^{2+}(Ti,Fe^{2+},Mg,Fe^{3+})_2Ti_2Si_4O_{22}$	内蒙古白云鄂博稀土铁矿	徐金沙、杨光明、李国武等	Xu et al., 2008;徐金沙等, 2008;李国武等, 2005
13	Xieite	谢氏超晶石	$FeCr_2O_4$	湖北省大堰坡乡随州 L6 球粒陨石	陈鸣、束今赋、毛河光	陈鸣等, 2008
14	Zhangpeishanite	张培善石	$BaFCl$	内蒙古白云鄂博稀土铁矿东矿	Shimazaki H.(日)、Miyawaki R.(日)、Yokoyama K.(日)等	Shimazaki et al., 2008
15	Lisiguangite	李四光矿	$CuPtBiS_3$	河北省承德市滦平县三道村石榴石辉石岩体含铂钴铜硫化物脉	於祖相、陈方远、马宏伟	Yu et al., 2009; Yu and Wang, 2017
16	Qusongite	曲松矿	WC	西藏山南地区曲松县罗布莎蛇绿岩铬铁矿区	方青松、白文吉、杨经绥等	Fang et al., 2009
17	Yarlongite	雅鲁矿	$Cr_4Fe_4NiC_4$	西藏山南地区曲松县罗布莎蛇绿岩铬铁矿区	施倪承、白文吉、李国武等	施倪承等, 2009
18	Zangboite	藏布矿	$TiFeSi_2$	西藏山南地区曲松县罗布莎蛇绿岩铬铁矿区 31 矿体	李国武、方青松、施倪承等	Li et al., 2009
19	Fluorokinoshitalite	氟木下云母	$BaMg_3(Al_2Si_2O_{10})F_2$	内蒙古白云鄂博稀土铁矿东矿	Miyawaki R.(日)、Shimazaki H.(日)、Shigeoka M.(日)等	Miyawaki et al., 2011a;张培善和王中刚, 2013
20	Fluorotetraferri-phlogopite	氟铁金云母	$KMg_3(Fe^{3+}Si_3O_{10})F_2$	内蒙古白云鄂博稀土铁矿东矿	Miyawaki R.(日)、Shimazaki H.(日)、Shigeoka M.(日)等	Miyawaki et al., 2011a;张培善和王中刚, 2013
21	Yangzhumingite	杨主明云母	$KMg_{2.5}(Si_4O_{10})F_2$	内蒙古白云鄂博稀土铁矿东矿区	Miyawaki R.(日)、Shimazaki H.(日)、Shigeoka M.(日)等	Miyawaki et al., 2011b

续表

序号	英文名	中文名	化学式	产地和产状	主要发现者	参考文献
22	Ferrotaaffeite-2N'2S	铁塔菲石-2N'2S	$Be(Fe,Mg,Zn)_3Al_8O_{16}$	湖南省郴州市临武县香花岭锡多金属矿床	杨主明、丁奎首、Jeffrey de Fourestier（加）等	Yang et al.,2012a
23	Hanjiangite	汉江石	$Ba_2CaV^{3+}Al(H_2AlSi_3O_{12})(CO_3)_2F$	陕西省安康市大巴山地区石梯重晶石矿区	刘家军、李国武、毛骞等	Liu et al.,2012；刘家军等,2012
24	Hezuolinite	何作霖矿	$(Sr,REE)_4Zr(Ti,Fe^{3+})_4(Si_2O_7)_2O_8$	辽宁省凤城市赛马碱性侵入岩体	杨主明、丁奎首、Giester G.（奥）等	Yang et al.,2012b
25	Linzhiite	林芝矿	$FeSi_2$	西藏山南地区曲松县罗布莎蛇绿岩铬铁矿区31矿体	李国武、白文吉、施倪承等	Li et al.,2012；李国武等,2015
26	Naquite	那曲矿	$FeSi$	西藏山南地区曲松县罗布莎蛇绿岩铬铁矿区31矿体	施倪承、李国武、白文吉等	Shi et al.,2012；李国武等,2015
27	Tellurocanfieldite*	碲硫银锡矿*	$Ag_8Sn(S,Te)_6$	辽宁省建昌县八家子铅锌矿床	谷湘平、谢先德、鲁安怀等	Gu et al.,2012
28	Ferrisepiolite	铁海泡石	$(Fe^{3+},Fe^{2+},Mg)_4[(Si,Fe^{3+})_6O_{15}](O,OH)_2·6H_2O$	青海省兴海县赛什塘夕卡岩铜矿床	谷湘平、谢先德、吴湘滨等	Gu et al.,2013
29	Luanshiweiite-2M1	栾锂云母	$KLiAl_{1.5}\square_{0.5}(Si_{3.5}Al_{0.5})O_{10}(OH,F)_2$	河南省卢氏县官坡镇309#伟晶岩脉	范光、李国武、沈敢富等	范光等,2013
30	Titanium	自然钛	Ti	西藏山南地区曲松县罗布莎蛇绿岩铬铁矿区31矿体	方青松、白文吉、杨经绥等	Fang et al., 2013；李国武等,2015
31	Hydroxycalciopyrochlore	羟钙烧绿石	$(Ca,Na,U,\square)_2(Nb,Ti)_2O_6(OH)$	四川省冕宁县牦牛坪稀土矿床	杨光明、李国武、熊明等	Yang et al.,2014
32	Qingsongite	青松石	BN	西藏罗布莎蛇绿岩中31#铬铁矿矿床	Dobrzhinetskaya L.F.（美）、Wirth R.（德）、杨经绥等	Dobrzhinetskaya et al.,2014
33	Strontiohurlbutite	磷锶铍石	$SrBe_2(PO_4)_2$	福建省南平伟晶岩区31#伟晶岩	饶灿、王汝成、谷湘平等	Rao et al.,2014
34	Tangdanite	富铜泡石	$Ca_2Cu_9(AsO_4)_4(SO_4)_{0.5}(OH)_9·9H_2O$	云南省东川铜矿汤丹和滥泥坪矿	Jeffrey de Fourestier（加）、李国武、Glenn Poirier（加）等	Ma et al.,2014；李国武等,2014
35	Minjiangite	闽江石	$BaBe_2(PO_4)_2$	福建省南平伟晶岩区31#伟晶岩	饶灿、Hatert F.（比利时）、王汝成等	Rao et al.,2015
36	Fluorcalciopyrochlore	氟钙烧绿石	$(Ca,Na)_2(Nb,Ti)_2O_6F$	内蒙古白云鄂博稀土铁矿	李国武、杨光明、吕福德等	Li et al.,2016
37	Fluornatropyrochlore	氟钠烧绿石	$(Na,Pb,Ca,REE,U)_2Nb_2O_6F$	新疆阿克苏地区波孜果尔稀土矿床的碱性侵入花岗岩	尹京武、李国武、杨光明等	尹京武等,2016
38	Oxynatromicrolite	氧钠细晶石	$(Na,Ca,U)_2(Ta,Nb)_2O_6(O,F)$	河南省卢氏县官坡镇309#伟晶岩脉	范光、葛祥坤、李国武等	Fan et al.,2017

续表

序号	英文名	中文名	化学式	产地和产状	主要发现者	参考文献
39	Fengchengite	凤城石	$Na_{12}\square_3(Ca,Sr)_6Fe_3^{3+}Zr_3Si(Si_{25}O_{73})(H_2O,OH)_3(OH,Cl)_2$	辽宁省凤城市赛马钠质碱性正长岩	沈敢富、徐金莎、姚鹏等	沈敢富等,2017
40	Hemleyite	三方铁辉石	$(Fe_{0.48}^{2+}Mg_{0.37}Ca_{0.04}Na_{0.04}Mn_{0.03}^{2+}Al_{0.03}Cr_{0.01}^{3+})_{1.00}Si_{1.00}O_3$	湖北省大堰坡乡随州L6球粒陨石未熔主体中	Bindi L.(意)、陈鸣、谢先德等	Bindi et al.,2017
41	Hongheite	红河石	$Ca_{19}Fe^{2+}Al_4(Fe^{3+},Mg,Al)_8(\square,B)_4BSi_{18}O_{69}(O,OH)_9$	云南省个旧锡多金属矿东北马拉革夕卡岩矿床的白山冲花岗岩	徐金沙、李国武、范光等	徐金莎等,2019
42	Mengxianminite	孟宪民石	$(Ca,Na)_2Sn_2(Mg,Fe)_3Al_8[(BO_3)(BeO_4)O_6]_2$	湖南省郴州市临武县香花岭锡多金属矿床	饶灿、Hatert F.(比利时)、Dal Bo F.(比利时)等	Rao et al.,2017
43	Mianningite	冕宁铀矿	$(\square,Pb,Ca)UFe_2^{2+}(Ti,Fe^{3+})_6Ti_{12}O_{38}$	四川省冕宁县牦牛坪稀土矿附近的包子山煌斑岩破碎带	葛祥坤、范光、李国武等	Ge et al.,2017; 葛祥坤等,2018
44	Wuyanzhiite	吴延之矿	Cu_2S	湖南省衡阳市长宁县柏坊铜矿	谷湘平、Shi X、杨和雄(美)等	Gu et al.,2017
45	Zincobotryogen	锌赤铁矾	$ZnFe^{3+}(SO_4)_2(OH)\cdot 7H_2O$	青海省锡铁山铅锌矿床的氧化带	杨主明、Gerald Giester(奥地利)、毛骞等	Yang et al., 2017; 杨主明等,2017
46	Maohokite	毛河光矿	$MgFe_2O_4$	辽东半岛的岫岩陨石坑	陈鸣、束今赋、谢先德等	Chen et al.,2019
47	Shenzhuangite	沈庄硫铁镍矿	$NiFeS_2$	湖北省大堰坡乡随州L6球粒陨石	Bindi L.(意)、谢先德	Bindi and Xie,2018
48	Wumuite	乌木石	$KAl_{0.33}W_{2.67}O_9$	云南省丽江市华坪县南阳村半风化石英二长岩	李国武、薛源	Li and Xue,2018
49	Zinconigerite-2N1S	锌尼日利亚石-2N1S	$(Zn,Al,Mg)_2(Al,Sn)_6O_{11}(OH)$	湖南省郴州市临武县香花岭锡多金属矿床	饶灿、王汝成、谷湘平等	Rao et al.,2018
50	Asimowite	阿铁橄榄石	Fe_2SiO_4	湖北省大堰坡乡随州L6球粒陨石	Bindi L.(意)、Brenker F. E.(德)、Nestola F.(意)等	Bindi et al.,2019
51	Badengzhuite	巴登珠矿	TiP	西藏山南地区罗布莎铬铁矿区Cr-11矿体	熊发挥、徐向珍、Mugnaioli E.(意)等	Xiong et al.,2019a
52	Fluorluanshiweiite	氟栾锂云母	$KLiAl_{1.5}(Si_{3.5}Al_{0.5})O_{10}F_2$	河南省东部北秦岭造山带南阳山稀有金属矿床的锂铍钽花岗伟晶岩	曲凯、司马献章、李国武等	Qu et al.,2019b
53	Hiroseite	陨铁辉石	$FeSiO_3$	湖北省大堰坡乡随州L6球粒陨石	Bindi L.(意)、谢先德	Bindi et al.,2019
54	Jingsuiite	经绥矿	TiB_2	西藏山南地区的罗布莎铬铁矿矿床,以包裹体的形式产于铬铁矿矿石的刚玉中	熊发挥、徐向珍、Mugnaioli E.(意)等	Xiong et al.,2019b
55	Potassic-hastingsite	钾绿钙闪石	$\{K\}\{Ca_2\}\{Fe_4^{2+}Fe^{3+}\}(Al_2Si_6O_{22})(OH)_2$	内蒙古自治区克什克腾旗同兴镇大乃林沟	任光明、李国武、束今赋等	Ren et al.,2019

续表

序号	英文名	中文名	化学式	产地和产状	主要发现者	参考文献
56	Taipingite-(Ce)	太平石	$(Ce_7^{3+}Ca_2)_{\Sigma 9}Mg(SiO_4)_3[SiO_3(OH)]_4F_3$	河南省南阳市西峡县太平镇稀土矿床	曲凯、司马献章、范光等	Qu et al., 2019a
57	Tewite	碲钨矿	$(K_{1.5}\square_{0.5})(Te_{1.25}W_{0.25}\square_{0.5})W_5O_{19}$	云南省华坪县南阳村附近半风化状含黑云母石英二长岩	李国武、薛源、熊明等	李国武等，2018；Li et al., 2019
58	Zhiqinite	志琴矿	$TiSi_2$	西藏山南地区罗布莎铬铁矿矿区 Cr-11 矿体	熊发挥、徐向珍、Mugnaioli E.（意）等	Xiong et al., 2019c
59	Chukochenite	竺可桢石	$LiAl_5O_8$	湖南省郴州市临武县香花岭锡多金属矿床	饶灿、谷湘平、王汝成等	Rao et al., 2020
60	Lingbaoite	灵宝矿	$AgTe_3$	河南省灵宝市西南 30 km S60 含金石英脉	简伟、毛景文、Lehmann B.（德）等	Jian et al., 2020
61	Wangdaodeite	王氏钛铁矿	$FeTiO_3$	湖北省大堰坡乡随州 L6 球粒陨石的冲击熔脉内或熔脉边部	谢先德、谷湘平、杨和雄等	Xie et al., 2020

注：按新矿物公开发表时间排序；"*" 为否定矿物种。

表 2 显示，21 世纪在中国发现的新矿物中有 15 种是由外国学者主导发现和研究的，其余 46 种的第一发现人均为我国学者。从中国新矿物第一发现者的隶属机构来看，以高等院校学者最多，有 21 种新矿物由其发现，其中中国地质大学（北京）9 种、浙江大学 5 种、中南大学 4 种、中国地质大学（武汉）和南京大学及贵州工业大学各 1 种；其次是中国地质调查局直属科研机构的学者，有 15 种新矿物由其发现，其中中国地质科学院地质研究所 7 种、矿产资源研究所 1 种、成都地质调查中心 5 种、天津地质调查中心 2 种；再有就是中国科学院科研院所的学者，他们共发现 7 种新矿物，其中广州地球化学研究所 4 种、地质与地球物理研究所 3 种；此外，核工业部北京地质研究院的学者也发现了 3 种新矿物。结合周正等（1997）关于 20 世纪中国新矿物发现单位的内容来看，上述单位的新矿物发现者及其团队一直是我国新矿物研究的中坚力量。发现新矿物的确只是小概率事件，但"功夫不负有心人"，正是这些"有心人"长期坚持不懈的努力才使得我国新矿物研究的发展一直紧跟国际最新发展趋势，成就了中国在国际新矿物领域的先进地位。

2.1 在中国发现的新矿物种数量

从 21 世纪（2000—2019 年）全球新矿物产地的分布来看，尽管与发现新矿物数量最多的第一梯队俄罗斯（349 种）、美国（197 种）和意大利（160 种）相比，中国发现新矿物的数量（61 种）还有不小差距，但可与德国（72 种）、加拿大（66 种）、智利（61 种）、澳大利亚（60 种）一起列入世界新矿物第二梯队，迄今为止 21 世纪我国发现新矿物的数量已位列世界第六。

根据不同年代在中国发现新矿物种数量的统计结果（表 3），最近 10 年和 20 世纪 80 年代均为我国新矿物成果辈出的高峰时期，如果说 20 世纪 80 年代我国新矿物研究的蓬勃发展得益于新中国成立初期大规模地质勘探工作的深厚积累和改革开放后国家科技事业的空前繁荣，那么最近十年我国新矿物研究的异常活跃则有赖于世纪初期国家对地矿行业投入的加大，以及 21 世纪传统地球科学向地球系统科学的转变和现代高精尖测试技术的广泛运用。21 世纪地球科学研究已突破学科界限，地球科学各学科间以及与其他学科间的交叉融合必然加速创新性的科学发现，其中也包括新矿物的发现。近十年在中国发现的新矿物种数较之前十年（2000—2009 年）增长了一倍多，这与全球新矿物的增长趋势是一致的，2000—2009 年全球发现并经 IMA CNMNC 批准的新矿物一共 602 种，2010—2019 年增长为 1 126 种。

表3 不同年代中国发现的新矿物种数

年份	新矿物种数	年份	新矿物种数
1958—1959年	2	1990—1999年	22
1960—1969年	14	2000—2009年	19
1970—1979年	28	2010—2019年	42
1980—1989年	48	总计	175

2.2 中国发现的新矿物种的晶体化学类型

按照国际通用的矿物晶体化学分类方案，21世纪在中国已发现的61种新矿物分属于自然元素、硫化物及硫盐、卤化物、氧化物及氢氧化物、含氧盐5个矿物大类，唯有机矿物大类新矿物是空白（表4）。从各矿物大类矿物种的数量来看，最多的是含氧盐矿物大类矿物种，共计28种，包括硅酸盐类矿物21种、磷酸盐类（含砷酸盐和钒酸盐）矿物4种、硫酸盐类（含硒、碲、铬、钼、钨酸盐）矿物2种和硼酸盐类矿物1种，未发现碳酸盐类（含硝酸盐）矿物；其次为氧化物及氢氧化物大类矿物，14种均为氧化物类矿物种，未发现氢氧化物类新矿物；再次是自然元素矿物大类，共11种，其中碳化物、硅化物、氮化物及磷化物类矿物10种，单质及金属互化物类矿物1种；从次为硫化物及硫盐矿物大类，共7种，其中硫化物及其类似化合物类矿物6种，硫盐类矿物1种；最少的是卤化物矿物大类，仅有1种。

表4 在中国发现的新矿物种（2000—2019年）的晶体化学分类

矿物大类	矿物类	新矿物种	种数
自然元素矿物大类	单质及金属互化物类	自然钛	1
	碳化物、硅化物、氮化物及磷化物类	罗布莎矿、曲松矿、雅鲁矿、藏布矿、青松石、那曲矿、林芝矿、经绥矿、巴登珠矿、志琴矿	10
硫化物及硫盐矿物大类	硫化物及其类似化合物类	陈国达矿、水氧硫锑钠石、碲硫银锡矿、沈庄硫铁镍矿、吴延之矿、灵宝矿	6
	硫盐类	李四光矿	1
卤化物矿物大类	卤化物类	张培善石	1
氧化物及氢氧化物矿物大类	氧化物类	氧钠细晶石、碲锌石、谢氏超晶石、铁塔菲石-2$N'2S$、羟钙烧绿石、氟钙烧绿石、氟钠烧绿石、冕宁铀矿、碲钨矿、王氏钛铁矿、毛河光矿、乌木石、锌尼日利亚石-2$N1S$、竺可桢石	14
含氧盐矿物大类	硼酸盐类	孟宪民石	1
	硫酸盐类（包括硒、碲、铬、钼、钨酸盐）	铊明矾、锌赤铁矾	2
	磷酸盐、砷酸盐和钒酸盐类	涂氏磷钙石、磷锶铍石、闽江石、富铜泡石	4
	硅酸盐类（包括锗酸盐）	湖北石、氟尼伯闪石、牦牛坪石、玲根石、丁道衡矿、氟绿闪石、杨主明云母、汉江石、氟木下云母、氟铁金云母、铁海泡石、何作霖矿、凤城石、栾锂云母、三方铁辉石、红河石、阿铁橄榄石、钾绿钙闪石、太平石、陨铁辉石、氟栾锂云母	21
总计	—	—	61

对比21世纪全球与中国新矿物各个晶体化学类型矿物种的数目（表5），发现无论是全球还是中国新

矿物均以含氧盐大类矿物种数为最多,分别为 1 205 种和 28 种,各占全球和中国新矿物总数的 70% 和 46%。在含氧盐矿物大类中尤以硅酸盐类矿物种数最多,分别为 516 种和 21 种,分别各占全球和中国新矿物总数的 30% 和 34%,也就是说所发现新矿物种数量的约 1/3 为硅酸盐类矿物;磷酸盐(含砷酸盐和钒酸盐)类矿物次之,全球和中国分别为 391 种和 4 种,各占全球和中国新矿物总数的 23% 和 7%。硅酸盐类矿物本身就是地球上分布最广、种类最多的矿物,在发现的新矿物中占比高不足为奇,但磷酸盐类新矿物的大量发现已经成为近年来全球新矿物的"黑马",目前全球已发现的磷酸盐(含砷酸盐、钒酸盐)类矿物已近 1 000 种,是除硅酸盐类矿物外数量最多的一类含氧盐矿物,非常值得关注(蔡剑辉,2020b)。硫化物及硫盐大类矿物种数占据全球新矿物总数的第二位,氧化物及氢氧化物大类矿物占第三位,前者为 190 种,后者为 178 种。在中国新矿物总数中占据第二位的是氧化物及氢氧化物大类矿物,其次是自然元素矿物大类矿物,依次为 14 种和 11 种。从表 5 可知,我国发现的自然元素矿物大类新矿物种的数量在全球新矿物中占据相当大的比例,21 世纪全球共发现自然元素矿物大类新矿物 56 种,其中 1/5 产自中国。

表 5 新矿物(2000—2019 年)晶体化学分类对比

矿物大类	自然元素大类	硫化物及硫盐大类	卤化物大类	氧化物及氢氧化物大类	含氧盐大类					有机矿物大类	合计
					硼酸盐类	硫酸盐类	碳酸盐类	磷酸盐类	硅酸盐类		
中国新矿物种数	11	6	0	14	1	2	0	4	21	0	60
全球新矿物种数	56	190	75	178	30	221	47	391	516	24	1728

自然元素矿物大类包括单质和互化物合金,以及碳化物、硅化物、氮化物和磷化物。大量自然元素矿物大类新矿物的发现是中国新矿物工作的优势和特色。自 1958 年以来,在中国发现的自然元素矿物大类新矿物共计 27 种(表 1、表 2)。通过分析 IMA CNMNC 正式批准的中国新矿物在各矿物大类中的种数分布情况(表 6),发现其中的 16 种发现于 20 世纪八九十年代,主要以互化物合金类矿物为主(12 种),另有硅化物类矿物 2 种、碳化物和单质类矿物各 1 种。其产状较为复杂,主要发现于河北、湖南和吉林的水系砂金矿,以及河北、内蒙古、四川、西藏和新疆一些含铂族元素(platinum group element,PGE)的超基性-基性杂岩体及与其相关的铜镍、铬矿床(含矿化点)中。21 世纪共发现自然元素大类矿物 11 种,以硅化物(5 种)、碳化物(2 种)、氮化物(1 种)、磷化物(1 种)、硼化物(1 种)类矿物为主,还有 1 种单质类矿物,全部发现于西藏山南地区曲松县罗布莎蛇绿岩铬铁矿区。

表 6 中国新矿物在各矿物大类中的种数分布

发现时间	自然元素矿物大类	硫化物及硫盐大类	卤化物矿物大类	氧化物和氢氧化物矿物大类	含氧盐矿物大类					有机矿物大类	合计
					硼酸盐类	硫酸盐类	碳酸盐类	磷酸盐类	硅酸盐类		
20 世纪	16	13	1	14	5	7	5	5	16	0	82
21 世纪	11	6	1	14	1	2	0	4	21	0	60
总计	27	19	2	28	6	9	5	9	37	0	142

在矿物的晶体化学分类体系中,除矿物大类外,还包括矿物类、矿物族和矿物种,通常分 4 个层级。其中矿物族是按照矿物种晶体结构型和阳离子性质划分的,具相同晶体结构型和相似化学组成的所有矿物种构成一个矿物族。矿物族的概念在新矿物的发现与研究中具有特殊的意义。IMA CNMMN 在确定新矿物种的基本原则中指出:相对于一个已存在的矿物种的等效结构位置,一种可能的新矿物至少要有一个结构位置应当主要由一种不同的化学元素占据(Nickel and Mandarino,1987;Nickel and Grice,1999)。逆向思考的话,完全可以根据化学性质类似的不同离子占据相同结构型矿物种的等效结构位置来推导同一矿物族中可能的新矿物种,这为我们发现诸如烧绿石族、角闪石族、云母族和硅铁钛铈石族等化学组成复杂、结构位置多样的新矿物提供了很好的指导。按照 IMA CNMNC 新颁布的烧绿石超族矿物分类方案(Atencio et al., 2010),截至 2019 年底,已获得 IMA CNMNC 批准的烧绿石超族矿物已达 31 种,尚有 12

种属于推导出来的可能的新矿物种。据李国武等（2014）统计，当时被IMA CNMNC认可的烧绿石超族矿物为16种，尚有15种矿物及其名称需要完善描述并得到批准后才能成为有效的烧绿石超族矿物种。如此看来，过去的6年内，这15种烧绿石超族矿物种均已正式获批。21世纪以来中国发现了4种烧绿石超族矿物，分别是四川牦牛坪稀土矿床的羟钙烧绿石（Yang et al., 2014）、内蒙古白云鄂博稀土铁矿的氟钙烧绿石（Li et al., 2016）、新疆波孜果尔稀土矿床碱性侵入花岗岩中的氟钠烧绿石（尹京武等，2016）和河南官坡伟晶岩脉中的氧钠细晶石（Fan et al., 2017）。烧绿石超族矿物在自然界是比较稀有的，但在我国则分布广泛，种类和储量都较高（李国武等，2014）。通过一系列烧绿石超族矿物的发现，我国已基本形成了一套行之有效的研究方法，我国烧绿石超族新矿物种的发现未来可期。另外，发现角闪石超族新矿物的空间更大，因为角闪石超族的成员非常多，据不完全统计，经IMA CNMNC批准的角闪石超族矿物已逾百种，还有上百个潜在的矿物种尚待发现和研究（沈敢富等，2019）。21世纪在中国也已发现了3种角闪石超族新矿物，分别是苏鲁柯石英-榴辉岩带中发现的氟尼伯闪石和氟绿闪石（Oberti et al., 2003，2007），以及内蒙古大乃林沟火山机构中发现的钾绿钙闪石（Ren et al., 2019）。云母族同样也拥有一个庞大的矿物家族，21世纪在中国已发现5种云母族新矿物，即白云鄂博稀土铁矿床种发现的杨主明云母、氟木下云母和氟铁金云母（Miyawaki et al., 2011a, 2011b），以及河南官坡伟晶岩脉中的栾锂云母（范光等，2013）、河南南阳山稀有金属矿床的锂铯钽花岗伟晶岩中的氟栾锂云母（Qu et al., 2019b）。迄今为止，硅铁钛铈石族由14种矿物组成，21世纪在中国已发现3种硅铁钛铈石族矿物，为产于四川牦牛坪稀土矿床的牦牛坪矿（沈敢富等，2005）、内蒙古白云鄂博稀土铁矿床的丁道衡矿（Xu et al., 2008）和辽宁赛马碱性侵入岩体中的何作霖矿（Yang et al., 2012b）。可见，基于对矿物体系的系统认识，有利于根据矿物的形成规律去发现一系列新矿物种。

2.3 在中国发现的新矿物种的产地和产状特征

21世纪在中国发现的新矿物的产地主要分布在16个省、自治区，包括西藏（11种），湖北（8种），内蒙古（7种），湖南、河南（各5种），辽宁、云南（各4种），江苏、四川（各3种），此外还有福建、贵州、河北、青海（各2种），以及山东、新疆（各1种）。显然，中国新矿物的分布是比较广泛的，但相对较集中于固体矿产资源大省及地球科学研究的热点区域。

从21世纪中国新矿物的产状特征（表7）来看，64%的新矿物（38种）发现于矿区，而且以金属矿床为主，其中最引人瞩目的是在西藏雅鲁藏布江蛇绿岩带罗布莎蛇绿岩中的地幔来源豆荚状铬铁矿里发现了12种新矿物，在内蒙古白云鄂博、四川牦牛坪和河南太平镇稀土金属矿床中发现了10种新矿物，此外，在辽宁、青海、贵州、云南、湖南的铅锌、锑、铜、锡多金属等有色金属矿床中发现了10种新矿物；在河北、山东和河南的金、铂等贵金属矿床（脉）中发现4种新矿物；在贵州的稀散金属铊矿床中发现了1种新矿物。只有1种新矿物发现于非金属（重晶石）矿床中。21%的新矿物（14种）发现于各类特色岩区，如福建和河南的伟晶岩区（脉）中发现5种、辽宁和新疆的碱性岩区中发现3种、苏鲁榴辉岩带中发现2种、云南半风化石英二长岩中发现2种、云南个旧锡矿花岗岩围岩和内蒙古火山-次火山岩中各发现1种。15%的新矿物（9种）发现于陨石或陨石坑中，其中7种产于湖北随州L6球粒陨石中，其余两种分别产于江苏寺巷口陨石和辽东岫岩陨石坑。对全球和中国新矿物产状的统计分析均表明，超过一半的新矿物种都发现于矿区，大概因矿区均为地球化学异常区，而且一般都经历了十分复杂的地质过程，这为形成新矿物奠定了充分的物质基础。同时，矿区的研究程度都很高，而新矿物的发现是建立在精细研究和反复探索基础之上的。

我国疆域辽阔，地层发育齐全，岩浆构造活动频繁，地质环境复杂多变，具备了形成新矿物的良好条件。我们知道，新矿物具有与已知矿物明显不同的化学组成和（或）结晶学性质，不难推测新矿物通常形成于比较独特的地质背景。区域化学元素组成复杂多样、具有独特的地球化学异常和较极端的物理化学条件、局部物化条件变化幅度或梯度较大的地质环境均十分有利于新矿物的形成。例如，源于地球深部的幔源岩（包括基性-超基性岩、碱性岩等），往往蕴含地表或地壳浅层地质学研究难以触及的一些特殊元素组合和形成条件，因此在幔源岩区及与幔源岩相关的矿床中相对比较容易形成具有独特化学组成和全新晶体结构类型的新矿物。在辽宁赛马碱性岩和西藏与地幔岩相关的铬铁矿中发现的一系列新矿物即是最好的佐证。21世纪我国新矿物的产出环境均极具代表性，除上述与地球深部地质过程相关的矿床和岩体之外，还有地球化学背景复杂、具成因多样性和多成矿期次的稀土、稀有和稀散金属矿床与一些多金属矿床，以及具备极端温压条件的陨石和陨石坑等，这些地质背景中的新矿物的发现正是我国新矿物工作的特色与优势。

表7 中国发现的新矿物（2000—2019年）的产状特征

	新矿物产状	新矿物种	种数
矿区（38种）	黑色金属矿床：铬铁矿、铁矿床	自然钛、罗布莎矿、曲松矿、雅鲁矿、藏布矿、青松石、那曲矿、林芝矿、经绥矿、巴登珠矿、志琴矿、湖北石	12
	稀土金属矿床	太平石、氟木下云母、氟铁金云母、张培善石、杨主明云母、氟钙烧绿石、丁道衡矿、冕宁铀矿、羟钙烧绿石、牦牛坪矿	10
	有色金属矿床：铅锌、锑、铜、锡多金属矿床	碲硫银锡矿、锌赤铁矾、水氧硫锑钠石、富铜泡石、吴延之矿、铁海泡石、竺可桢石、锌尼日利亚石-2$N1S$、孟宪民石、铁塔菲石-2$N'2S$	10
	贵金属矿床：金、铂矿床	陈国达矿、碲锌石、灵宝矿、李四光矿	4
	稀散金属矿床：铊矿床	铊明矾	1
	非金属矿床：重晶石矿床	汉江石	1
岩区（14种）	伟晶岩	氧钠细晶石、栾锂云母、闽江石、磷锶铍石、氟栾锂云母	5
	碱性岩	氟钠烧绿石、何作霖矿、凤城石	3
	榴辉岩	氟绿闪石、氟尼伯闪石	2
	石英二长岩	碲钨矿、乌木矿	2
	花岗岩	红河石	1
	火山岩-次火山岩	钾绿钙闪石	1
陨石或陨石坑（9种）	陨石	玲根石、阿铁橄榄石、沈庄硫铁镍矿、陨铁辉石、三方铁辉石、涂氏磷钙石、谢氏超晶石、王氏钛铁矿	8
	陨石坑	毛河光矿	1
总计	—	—	61

由中国地质科学院地质研究所、中国地质大学（北京）与国外学者合作在西藏罗布莎蛇绿岩铬铁矿中发现的11种来自地幔深部的超高压新矿物（施倪承等，2005，2009，白文吉等，2006；Fang et al.，2009，2013；Li et al.，2009，2012；Shi et al.，2012；Dobrzhinetskaya et al.，2014；Xiong et al.，2019a，2019b，2019c），开拓了深部地幔矿物研究的新领域，加深了人类对地球深部各层圈尤其是地幔和地核物质的化学成分和晶体结构的认识，也为重新论证铬铁矿的深部成因提供了重要信息（杨经绥等，2008；李国武等，2015），这是中国新矿物领域对矿物学研究乃至地球科学研究的新贡献。

我国关于稀土金属矿床中新矿物的研究起步于1959年，截至2019年仅在内蒙古白云鄂博一个矿区就发现了16种新矿物（王濮和李国武，2014）。这些新矿物主要属于含氧盐矿物大类（主要是硅酸盐、碳酸盐和少数磷酸盐）、氧化物及氢氧化物大类和卤化物矿物大类，半数以上含有稀土元素。21世纪在白云鄂博稀土铁矿中发现的新矿物就有6种，即氟木下云母、氟铁金云母、张培善石、杨主明云母、氟钙烧绿石和丁道衡矿（徐金沙等，2008；Shimazaki et al.，2008；Miyawaki et al.，2011a，2011b；张培善和王中刚，2013；Li et al.，2016）；在四川冕宁牦牛坪稀土矿床中发现牦牛坪矿、羟钙烧绿石和冕宁铀矿3种新矿物（沈敢富等，2005；Yang et al.，2014；葛祥坤等，2018），前者为含稀土的硅酸盐类矿物，后二者属于氧化物类矿物；同时还在河南南阳太平镇稀土矿床发现含稀土元素的新矿物——太平石（Qu et al.，2019a）。稀土矿床是发现新矿物的良好场所，因为它通常都具十分复杂的地球化学背景，且具多成因和多成矿期的特点。

21世纪中国新矿物工作的另一个亮点是地外天体中一些与冲击变质有关的超高压新矿物的发现。自然界的超高压条件主要出现于巨大天体的撞击过程和地球深部环境，除了超高压地幔岩和超高压变质岩外，冲击变质成因的陨石和陨石坑是天然超高压矿物的主要产状之一。21世纪在陨石和陨石坑中发现的超高压新矿物共有8种，均为中国科学院广州地球化学研究所超高压矿物研究团队与国外学者合作发现。其中在湖北随州L6球粒陨石中发现6种：涂氏磷钙石、谢氏超晶石、王氏钛铁矿、三方铁辉石、沈庄硫铁镍矿和陨铁辉石（谢先德等，2003；陈鸣等，2008；Bindi et al., 2017；Bindi and Xie, 2018, 2019；Xie et al., 2020）。还有两种是分别发现于江苏寺巷口陨石和辽东岫岩陨石坑的玲根石和毛河光矿（Chen et al., 2019；Gillet et al., 2000）。与冲击变质作用相关的超高压新矿物的发现，不但丰富了矿物家族中数量相对较少的超高压矿物种，而且为探索天体撞击事件提供了重要信息（陈鸣，2012）。关于冲击变质及地球深部来源超高压新矿物的发现和研究表明，21世纪我国矿物学研究呈现出进一步向宇宙空间和地球深部拓展的态势。

特别值得一提的中国新矿物产地是湖南香花岭锡多金属矿区，这里是1958年中国发现的第一个新矿物"香花石"的诞生地。21世纪在香花岭矿区又发现了四种新矿物：中国科学院地质与地球物理研究所团队发现的氧化物类新矿物铁塔菲石-2N'2S（Yang et al., 2012a），浙江大学团队发现的硼酸盐类新矿物孟宪民石、氧化物类新矿物锌尼日利亚石-2N'1S和竺可桢石（Rao et al., 2017, 2018, 2020）。早在20世纪就有学者预言：湖南香花岭锡多金属矿区、东坡铅锌矿区和广西大厂锡多金属矿区的成矿地质条件和矿物共生组合均十分复杂，是形成新矿物的有利场所，可望还会有新矿物报道（李锡林，1989），如今确得实验。

除了矿区和陨石中外，特殊岩区（带）也一直是发现新矿物的重要场所。中国与全球新矿物产状的分布规律相似，在伟晶岩区（脉）和碱性岩区发现新矿物的概率高于其他岩类。21世纪浙江大学团队和核工业北京地质研究院团队分别在福建南平和河南卢氏官坡的伟晶岩区各发现两种新矿物：闽江石、磷锶铍石、栾锂云母和氧钠细晶石（范光等，2013；Rao et al., 2014, 2015；Fan et al., 2017），此外中国地质调查局天津地质调查中心团队还在河南南阳的锂铯钽花岗伟晶岩发现新矿物栾锂云母（Qu et al., 2019b）。在辽宁赛马碱性岩种发现凤城石和何作霖矿两种新矿物（Yang et al., 2012b；沈敢富等，2017），在新疆阿克苏地区波孜果尔稀土矿床的碱性侵入花岗岩中发现了氟钠烧绿石（尹京武等，2016）。前文提及的超高压变质岩是超高压矿物的储库，在我国著名的大别–苏鲁高压超高压榴辉岩带虽然至今未发现超高压新矿物，但21世纪有国外学者在该带发现了两种角闪石超族新矿物：氟尼伯闪石和氟绿闪石（Oberti et al., 2003, 2007）。比较新颖的是中国地质大学（北京）团队在攀西地区云南省华坪县境内新元古代结晶基底的半风化石英二长岩中发现了两种含钾（K）和钨（W）的氧化物类新矿物：乌木石和碲钨矿（Li and Xue, 2018；李国武等，2018；Li et al., 2019），前者是世界上首例发现的具钨青铜型结构的天然新矿物，后者是具钨青铜型结构的衍生结构，并且是首例发现的具K-Te-W-O元素组合的新矿物。具全新化学组成和晶体结构的新矿物种的发现，是产出环境具独特地质过程的标志。

相比于全球新矿物更丰富的产状类型，中国新矿物的产状类型显得略少一些。对比一下近十年全球和中国新矿物的主要产出特征，发现这期间全球发现的1 126种新矿物主要产于矿区（586种）、特殊岩区（227种）、火山喷气口（126种）、陨石（55种）、冰碛沉积物（10种）以及地层沉积建造、河流沉积物、博物馆矿石和岩石老标本、岩心和化石等（共122种）。而中国发现的42种新矿物则主要产于矿区（24种）、特殊岩区（12种）以及陨石和陨石坑（6种）（表7）。中国缺乏在火山喷气口、冰碛沉积物、地层沉积建造等一些特殊产状中发现的新矿物。我们注意到，在俄罗斯堪察加半岛托尔巴契克火山岩地区由大裂缝火山喷发形成的火山渣堆中，产出的矿物达218种之多，并且是其中119种新矿物种的模式产地，堪称闻名世界的新矿物储库（Fedotov and Markhinin, 1983；Vergasova and Filatov, 2016），仅2015年一年就发现新矿物10种，主要是硫酸盐类、磷酸盐（包括砷酸盐和钒酸盐）类和卤化物类矿物（蔡剑辉，2020a）。我国火山岩浆活动极其发育，但相关的新矿物研究却相当薄弱，今后值得重视。20世纪中国地质科学院地质研究所於祖相团队在河北滦河和潮河、湖南沅江及吉林珲春河水系砂铂或砂金矿中发现了10种新矿物，包括长城矿、滦河矿、古北矿、喜峰矿、马营矿、双峰矿、马兰矿、冀承矿、沅江矿和珲春矿（於祖相，1984，1994，1995，1996，1997；邵殿信等，1984；吴尚全等，1992；陈立昌等，1994；Yu et al.，2011），可见在中国的水系沉积物中完全有发现新矿物的空间。随着矿产资源的开发利用和地质体的自然演化，有些矿石和岩石标本在野外已经很难获取，对博物馆或其他标本收藏者保存的矿石、岩石和化石老标本及钻孔岩心样品进行再研究也是发现新矿物的一个途径，而我国这方面的新矿物工作也相对薄弱。

3 中国新矿物研究展望

新矿物的发现和研究是矿物学领域不可或缺的内容，是基础中的基础，其成果在一定程度上反映着一个国家的基础性科研实力和水平。纵观中国新矿物研究的历史，在从1958年至2019年逾60年的时间里，在中国发现并报道的新矿物达175种，其中142种获IMA CNMNC批准。虽然在数量上我国新矿物工作与国际领先水平相比还存在较大差距，但有关自然元素矿物大类、铂族元素类新矿物的发现和研究居于世界先进水平，稀土金属类新矿物和冲击变质超高压新矿物的发现和研究也取得了令世界瞩目的新进展。21世纪这20年来，中国新矿物研究成果显著，发现新矿物的种数达61种，其中60种获得IMA CNMNC的正式批准，新矿物种数量位列世界第六。西藏罗布莎蛇绿岩铬铁矿中11种超高压地幔新矿物和湖北随州球粒陨石等地外天体中8种与冲击变质相关的超高压新矿物的发现，反映出我国矿物学研究进一步向地球深部和宇宙空间拓展的趋势。

21世纪相当部分在中国发现的新矿物种是中外学者共同协作、国内不同领域不同学科学者精诚合作的成果，可以说21世纪中国新矿物工作取得突出的进展，是与这种国际、国内强强合作模式分不开的，这应该也是未来我国新矿物研究比较推行的一种工作模式。

考虑到全球新矿物具有更多样性的产状类型，中国新矿物研究可以结合中国丰富且复杂的地质背景的特点，适当拓展新矿物研究的空间和对象，以利于发现更多的新矿物。除了继续关注传统新矿物产出环境中新矿物的发现和研究外，还要加强火山岩区火山口附近、典型地层和沉积建造、冰川沉积物、老矿山老标本及钻孔样品等特殊地质背景中的新矿物工作。目前我国专门从事新矿物研究的力量非常有限，除了加强国际和国内的合作并以传帮带的形式不断扩大我国新矿物研究队伍外，还需要配合国家需求，借助国家新时期为支撑战略性新兴产业和高新技术等领域发展予以大力投入的财力、物力和人力，开展与诸如关键固体矿产资源、海洋矿产资源、新型矿物材料等领域相关的新矿物研究，以进一步增强我国新矿物研究的综合实力。

21世纪全球和中国新矿物工作均取得了突破性的进展，然而新矿物的发现和研究还有很长的路要走。有国外学者根据对地球演化大数据的分析，预测目前已知矿物的种数远未达到客观存在的上限。近年来的新矿物研究动态也表明，虽然目前发现新矿物的数量快速上升，但新矿物的发现之旅实际上将日趋艰难，因为大量稀有矿物种对应于地球历史中相应的特殊事件，往往源于偶然，现在新发现的矿物种越来越趋向微量、微观、形成条件独特且罕见。所以，新矿物的发现更需要兼具科学的前瞻性、专业的研究深度和广度、精准的分析测试技术，还要有坚忍不拔、锲而不舍的科学精神。

参 考 文 献

白文吉,施倪承,方青松,李国武,杨经绥,熊明,戎合. 2006. 新矿物:罗布莎矿. 地质学报,80(10):1487-1490
蔡长金. 1986. 金矿物及其产状. 岩石矿物学杂志,5(2):147-157
蔡剑辉. 2020a. 2016年全球发现的新矿物种. 岩石矿物学杂志,39(3):335-384
蔡剑辉. 2020b. 2017年全球发现的新矿物种. 岩石矿物学杂志,39(2):218-256
曹荣龙. 1964. 水钙榴石——一种富含铁的水柘榴石. 地质学报,44(2):219-228
陈代演,王冠鑫,邹振西,陈郁明. 2001. 新矿物——铊明矾. 矿物学报,21(3):271-277
陈敬中,杨光明,潘兆橹,施倪存,彭志忠. 1989. 新矿物——彭志忠石的发现及研究. 矿物学报,9(1):20-24
陈立昌,唐翠青,张建洪,刘振云. 1994. 沅江矿——一种金和锡的新矿物. 岩石矿物学杂志,13(3):232-238

陈鸣. 2012. 超高压矿物研究进展. 矿物岩石地球化学通报,31(5):428-432
陈鸣,束今赋,毛河光. 2008. 谢氏超晶石:一种$FeCr_2O_4$高压多形新矿物. 科学通报,53(17):2060-2063
陈振玠,郭永芬,曾骧良,许文渊,王凤阁. 1979. 硫金银矿的发现和研究. 科学通报,(18):843-848
丁孝石,白鸽,袁忠信,孙鲁仁. 1981. 羟硅铍钇铈矿——一个富铈、铍硅酸盐新矿物. 地质评论,27(5):459-466
丁孝石,白鸽,袁忠信,刘金定. 1984. 兴安石(Xinganite)的新资料. 岩石矿物及测试,3(1):46-48,45
范光,李国武,沈敢富,徐金莎,戴婕. 2013. 桼锂云母:锂云母系列的新成员. 矿物学报,33(4):713-721
范光,葛祥坤,李婷,于阿朋,王涛. 2020. 我国核地质系统发现的新矿物评述. 世界核地质科学,37(1):1-9
傅平秋,苏贤泽. 1987. 新矿物——白云鄂博矿. 矿物学报,7(4):289-297
傅平秋,孔佑华,龚国洪,邵美成,千金子. 1987. 白云鄂博矿的晶体结构. 矿物学报,7(4):299-304
葛祥坤,范光,李国武,沈敢富,陈璋如,艾钰洁. 2018. Crichtonite族矿物在我国的研究进展与冕宁铀矿的发现. 矿物学报,38(2):234-240
谷湘平,Watanabe M,谢先德,彭省临,Nakamuta Y,Ohkawa M,Hoshino K,Ohsumi K,Shibata Y. 2008. 陈国达矿($Ag_9FeTe_2S_4$):胶东地区金矿床中发现的硫碲化物新矿物. 科学通报,53(17):2064-2070
湖北地质学院X光实验室. 1974. 星叶石族矿物的晶体化学. 地质科学,(1):18-33
黄蕴慧,杜绍华,周秀仲. 1988. 香花岭岩石矿床与矿物. 北京:北京科学技术出版社,115,116
李国武,杨光明,马喆生,施倪承,熊明,沈敢富,范海福. 2005. 新矿物丁道衡矿的晶体结构. 矿物学报,25(4):313-320
李国武,马喆生,傅小土,施倪承,熊明. 2014a. 新矿物富铜泡石(Tangdanite). 高校地质学报,20(增刊):10
李国武,杨光明,熊明. 2014a. 烧绿石超族矿物分类新方案及烧绿石超族矿物. 矿物学报,34(2):153-158
李国武,施倪承,白文吉,方青松,熊明. 2015. 西藏罗布莎铬铁矿中发现的七种金属互化物新矿物. 矿物学报,35(1):13-18
李国武,薛源,谢英美. 2018. 新矿物碲钨矿的矿物学特征及成因探讨. 矿物岩石地球化学通报,37(2):186-191
李锡林. 1989. 中国科学院地球化学研究所发现的新矿物. 矿物岩石地球化学通报,8(4):265-266
刘家军,李国武,毛骞,吴胜华,柳振江,苏尚国,熊明,余晓艳. 2012. 新矿物——汉江石. 矿物学报,32(2):173-182
刘建昌. 1979. 一种钨的新矿物——蓟县矿$Pb(W,Fe^{3+})_2(O,OH)_7$. 地质学报,(1):45-49,92
陆琦,彭志忠. 1987. 蒙山矿的晶体结构. 岩石矿物学杂志,6(3):221-229
罗梅,王月文. 1999. 若尔盖巴西金矿床铜锌矿(Cu_2Zn)的发现及其地质意义. 矿物学报,19(1):20-22
马喆生,钱荣耀,彭志忠. 1980. 单斜铜泡石——云南东川发现的一种含水的铜的砷酸盐新矿物. 地质学报,54(2):134-143
毛水和,周学粹. 1989. 铱锇钌矿及其基质矿物——等轴铁铂矿. 矿物学报,9(2):136-140
彭志忠,马喆生. 1963. 星叶石的晶体结构. 科学通报,(5):67-69
彭志忠,张建洪,西门露露. 1978. 评我国近年来发现的铂族元素新矿物. 地质学报,52(4):326-336
彭志忠,张建洪,束今赋. 1983. 钡闪叶石的晶体结构. 科学通报,28(4):237-240
齐玲仪. 1974. 易解石-钇易解石族的新变种——钛钇矿. 地质学报,48(1):91-94
秦善,刘金秋,迟振卿. 2016. 矿物学发展现状及我国矿物学前景展望. 地质论评,62(4):970-978
曲一华,韩蔚田,钱自强,刘来保,闵霖生. 1965. 三方硼镁石——一种新硼酸盐矿物. 地质学报,45(3):298-305
曲懿华,韩蔚田,蔡克勤. 1975. 我国首次发现的盐镁芒硝的矿物学研究. 地质学报,49(2):180-186
邵殿信,周剑雄,张建洪,鲍大喜. 1984. 新矿物——滦河矿的研究. 矿物学报,(2):97-101
沈敢富. 1998. 阿山矿之否定. 矿物学报,18(2):230-233
沈敢富,杨光明,徐金沙. 2005. 牦牛坪矿-(Ce):一种新发现的稀土元素矿物. 沉积与特提斯地质,25(1-2):210-216
沈敢富,徐金沙,姚鹏,李国武. 2017. 凤成石:异性石族矿物N(5)位贫钠的空位类似物新种. 矿物学报,37(1-2):140-151
沈敢富,任光明,范光,曲凯,刘琰,叶青培. 2019. 角闪石超族矿物的命名:挑战与机遇——以新矿物"钾绿钙闪石"的发现为例. 世界核地质科学,36(4):193-198
沈今川,宓锦校. 1992. 对于氟碳钡铈矿(Cordylite-Ce)成分与结构的质疑. 岩石矿物学杂志,11(1):69-74
施倪承,马喆生,张乃娴,丁奎首. 1978. 等轴砷锑钯矿(丰滦矿)的晶体结构. 科学通报,(8):499-501
施倪承,李国武,熊明,马喆生. 2005. 西藏罗布莎铁族元素金属互化物矿物及其成因探讨. 岩石矿物学杂志,24(5):443-446
施倪承,白文吉,李国武,熊明,方青松,杨经绥,马喆生,戎合. 2009. 雅鲁矿:一种金属碳化物新矿物. 地质学报,83(1):25-30
帅德权,张如柏,罗梅,张经武. 1998. 天然铜锌系列中的铜锌矿(Cu_2Zn)的研究. 矿物学报,18(4):509-513
涂光炽,李锡林,谢先德,尹树森. 1964. 锌赤铁矾和锌叶绿矾——两种新的硫酸盐变种. 地质科学,5(4):313-330
王贤觉. 1978. 一个新矿物——磷钙钍矿. 科学通报,(12):743-745
王濮,李国武. 2014. 1958-2012年在中国发现的新矿物. 地学前缘,21(1):40-51
王濮,潘兆橹,翁玲宝. 1982. 系统矿物学(上册). 北京:地质出版社
魏明秀. 1981. 硫金银矿晶体结构研究的新资料. 地质科学,24(3):232-234
吴尚全,杨翼,宋群. 1992. 新的金矿物——珲春矿(Au_2Pb). 矿物学报,12(4):319-322
谢先德,Minitti M E,陈鸣,毛河光,王德强,束今赋,费英伟. 2003. 涂氏磷钙石:一种磷酸盐高压相新矿物. 地球化学,32(6):566-568
熊明,马喆生,彭志忠. 1988. 新矿物——安康矿. 科学通报,(18):1401-1404
徐金沙,杨光明,李国武,邬志亮,沈敢富. 2008. 新矿物丁道衡矿-(Ce):珀硅钛铈矿的同质多象体和硅钛铈矿亚族C(1)位上的钛类似矿物. 矿物学报,28(3):237-243
徐金沙,李国武,沈敢富. 2012. 首次在白云鄂博铁矿发现的矿物新种述评. 地质学报,86(5):842-848

徐金沙,李国武,范光,葛祥坤,祝向平,沈敢富. 2019. 红河石 $Ca_{18}(\square,Ca)_2Fe^{2+}Al_4(Fe^{3+},Mg,Al)_8(\square,B)_4BSi_{18}O_{69}(O,OH)_9$——符山石族新矿物. 地质学报,93(1):138-146

杨经绥,白文吉,方青松,戎合. 2008. 西藏罗布莎蛇绿岩铬铁矿中的超高压矿物和新矿物(综述). 地球学报,29(3):263-274

杨主明,Gerald Giester,毛蕣,马玉光,张迪,李禾. 2017. 锌赤铁矾 $ZnFe^{3+}(SO_4)_2(OH)\cdot 7H_2O$:新矿物的厘定. 中国矿物岩石地球化学学会第九次全国会员代表大会暨第16届学术年会文集. 矿物岩石地球化学通报,36(增刊):60

尹京武,李国武,杨光明,潘宝明,葛祥坤,徐海明,王军. 2016. 氟钠烧绿石——一种烧绿石超族的新矿物. 矿物学报,36(1):34-37

於祖相. 1981. 马兰矿、钴-马兰矿(大营矿)的补充与修正. 地质论评,27(1):55-57

於祖相. 1984. 古北矿、喜峰矿——燕山地区宇宙尘中的两种新矿物. 岩石矿物及测试,3(3):231-237

於祖相. 1994. 新矿物双峰矿——铱的二碲化物. 矿物学报,14(4):322-326

於祖相. 1995. 新矿物马营矿——铱的碲、铋化物. 矿物学报,15(1):5-8

於祖相. 1996. 新矿物马兰矿——自然界中首次发现三价铂、三价铱与铜的硫化物. 地质学报,70(4):309-314

於祖相. 1997. 新矿物长城矿——铱的硫铋化物. 地质学报,71(4):336-339

於祖相,傅国芬,师占义. 1964. 河西石 $NiMg(CO_3)_2$. 地质学报,44(2):213-218

於祖相,林树人,赵宝,方青松,黄其顺. 1974. 我国某地区含铂岩体中铂族元素的及伴生的新矿物初步研究. 地质学报,(2):202-218

张建洪,彭志忠. 1981. 我国发现的新矿物. 地球科学,15(2):85-98

张培善,陶克捷. 1981. 新矿物中华铈矿 $Ba_2Ce(CO_3)_3F$. 地质科学,1(2):195-196

张培善,陶克捷. 1986. 白云鄂博矿物学. 北京:科学出版社,106-107

张培善,王中刚. 2013. 白云鄂博发现三种云母族新矿物:杨主明云母、氟木下云母和氟铁金云母. 矿物学报,(增刊):282-283

张培善,陶克捷,杨主明. 2000. 我国稀土铌钽矿物学研究回顾与展望. 高校地质学报,6(2):126-131

张如柏. 1998. 中国新矿物种名录. 矿物学报,18(2):254-260

张如柏,田慧新,彭志忠,马喆生,韩凤鸣,景泽被. 1980. 新矿物——阿山矿 $(Nb,Ta,U,Fe,Mn)_4O_8$. 科学通报,(14):648-650

张如柏,龚复生,周振冬,范良明,帅德权. 1981. 沐川矿——一种新的含水的钼硫化物. 地球化学,(2):120-127

张如柏,张玉玉,洪洲,毛治华,刘发禄,习计. 2002. 蔡承云石 $(Fe^{2+}\cdot 3Al_2(SO_4)_6\cdot 30H_2O)$——一种含水的硫酸盐矿物. 中南工业大学学报(自然科学版),33(4):331-334

郑钰纯,黄振恒. 1989. 钴硅锌矿——一个富钴的硅锌矿变种. 矿物学报,9(1):33-36

中国科学院贵阳地球化学研究所,铂矿研究组,电子探针分析组,X射线粉晶分析组,选矿组. 1974. 钯和镍的碲锑化物及其他铂族金属新矿物、新变种. 地球化学,(3):169-181

周剑雄,杨国杰,张建洪. 1984. 我国金伯利岩中发现的蒙山矿. 矿物学报,(3):193-198

周正,曹亚文. 1997. 中国新矿物综述. 岩石矿物学杂志,16(1):81-90

Armbruster T. 2002. Revised nomenclature of högbomite, nigerite, and taaffeite minerals. European Journal of Mineralogy,14(2):389-395

Atencio D, Andrade M B, Christy A G, Gieré R, Kartashov P M. 2010. The pyrochlore supergroup of minerals: nomenclature. The Canadian Mineralogist,48(3):673-698

Bindi L, Xie X D. 2018. Shenzhuangite, $NiFeS_2$, the Ni-analogue of chalcopyrite from the Suizhou L6 chondrite. European Journal of Mineralogy,30(1):165-169

Bindi L, Xie X. 2019. Hiroseite, IMA 2019-019. CNMNC Newsletter No. 50, page 617. Mineralogical Magazine,83(4):615-620

Bindi L, Chen M, Xie X. 2017. Hemleyite, IMA 2016-085. CNMNC newsletter No. 35, February 2017, page 210. Mineralogical Magazine,81(1):209-213

Bindi L, Brenker F E, Nestola F, Koch T E, Prior D J, Lilly K, Krot A N, Bizzarro M, Xie X D. 2019. Discovery of asimowite, the Fe-analog of wadsleyite, in shock-melted silicate droplets of the Suizhou L6 and the Quebrada Chimborazo 001 CB3.0 chondrites. American Mineralogist,104(5):775-778

Burke E A J. 2006. A mass discreditation of GQN minerals. The Canadian Mineralogist,44(6):1557-1560

Chen M, Shu J F, Xie X D, Tan D Y. 2019. Maohokite, a post-spinel polymorph of $MgFe_2O_4$ in shocked gneiss from the Xiuyan crater in China. Eteoritics & Planetary Science,54(3):495-502

Christy A G, Atencio D. 2013. Clarification of status of species in the pyrochlore supergroup. Mineralogical Magazine,77(1):13-20

de Fourestier J, Li G, Poirier G, Chukanov N V, Ma Z. 2012. Fuxiaotuite, IMA 2011-096. CNMNC Newsletter No. 13, June 2012, page 808. Mineralogical Magazine,76(3):807-817

Dobrzhinetskaya L F, Wirth R, Yang J S, Green H W, Hutcheon I D, Weber P K, Grew E S. 2014. Qingsongite, natural cubic boron nitride: the first boron mineral from the Earth's mantle. American Mineralogist,99(4):764-772

Fan G, Ge X K, Li G W, Yu A P, Shen G F. 2017. Oxynatromicrolite, $(Na,Ca,U)_2Ta_2O_6(O,F)$, a new member of the pyrochlore supergroup from Guanpo, Henan Province, China. Mineralogical Magazine,81(4):743-751

Fang Q S, Bai W J, Yang J S, Xu X Z, Li G W, Shi N C, Xiong M, Rong H. 2009. Qusongite(WC): a new mineral. American Mineralogist,94(2-3):387-390

Fang Q S, Bai W J, Yang J S, Rong H, Shi N C, Li G W, Xiong M, Ma Z S. 2013. Titanium, Ti, a new mineral species from Luobusha, Tibet, China. Acta Geologica Sinica(English Edition),87(5):1275-1280

Fedotov S A, Markhinin Y K. 1983. The great tolbachik fissure eruption. New York: Cambridge University Press

Fejer E E, Hey M H. 1984. Thirty-third list of new mineral names. Mineralogical Magazine,48(349):569-586

Fleischer M. 1983. Glossary of mineral species 1983. Tucson,AZ:The Mineralogical Record Inc,202

Fleischer M,Mandarino J A. 1995. Glossary of Mineral Species 1995, 7th ed. Tucson,AZ:The Mineralogical Record Inc,239

Fleischer M,Chao G Y,Mandarino J A. 1982. New mineral names. American Mineralogist,67(7-8):854−860

Ge X K,Fan G,Li G W,Shen G F,Chen Z R,Ai Y J. 2017. Mianningite,($\square$,Pb,Ce,Na)(U^{4+},Mn,U^{6+})Fe_2^{3+}(Ti,Fe^{3+})$_{18}O_{38}$,a new member of the crichtonite group from Maoniuping REE deposit,Mianning County,Southwest Sichuan,China. European Journal of Mineralogy,29(2):331−338

Giester G,Ni Y X,Jarosch D,Hugihes J M,Ronsbo J G,Yang Z M,Zemann J. 1998. Cordylite-(Ce):a crystal chemical investigation of material from four localities,including type material. American Mineralogist,83(1-2):178−184

Gillet P,Chen M,Dubrovinsky L,El Goresy A. 2000. Natural $NaAlSi_3O_8$-hollandite in the shocked Sixiangkou meteorite. Science, 287 (5458): 1633−1636

Gu J,Chao G Y,Tang S. 1994. A new mineral from Mont St. Hilaire,Quebec,Canada-fluorbritholite-(Ce). Journal of Wuhan University of Technology, 9(3):9−14

Gu X,Shi X,Yang H,Lu A,Shao Y,Chen Q,Liu Z. 2017. Wuyanzhiite,IMA 2017-081. CNMNC Newsletter No. 40,December 2017,page 1581. Mineralogical Magazine,81(6):1577−1581

Gu X P,Xie X D,Lu A,Hoshino H,Huang J,Li J. 2012. Tellurocanfieldite,IMA 2012-013. CNMNC Newsletter No. 13, June 2012, page 816. Mineralogical Magazine,76(3):807−817

Gu X P,Xie X D,Wu X B,Zhu G C,Lai J Q,Hoshino K,Huang J W. 2013. Ferrisepiolite:a new mineral from Saishitang copper skarn deposit in Xinghai County,Qinghai Province,China. European Journal of Mineralogy,25(2):177−186

Haggerty S E,Smyth J R,Erlank A J,Rickard R S,Danchin R V. 1983. Lindsleyite(Ba) and mathiasite(K):two new chromium-titanates in the crichtonite series from the upper mantle. American Mineralogist,68(5-6):494−505

Hawthorne F C,Cooper M A,Grice J D,Roberts A C,Cook W R Jr,Lauf R J. 2002. Hubeite,a new mineral from the Daye mine near Huangshi,Hubei Province,China. Mineralogical Record,33(6):465−471

Huang Y H,Zhou X Z,Li G J,Du S H. 1986. Mengxianminite,a new ferro-magnesium-calcium-tin-aluminum oxide mineral. In:Abstract Program of the 14th General Meeting,International Mineralogical Association,Stanford University,USA:130

Jambor J L,Vanko D A. 1991. New mineral names. American Mineralogist,76(7-8):1434−1440

Jian W,Mao J W,Lehmann B,Li Y H,Ye H S,Cai J H,Li Z Y. 2020. Lingbaoite,$AgTe_3$,a new silver telluride from the Xiaoqinling gold district,central China. American Mineralogist,105(5):745−755

Li G W,Xue Y. 2018. Wumuite,IMA 2017-067a. CNMNC Newsletter No. 44,August 2018,page 1018. Mineralogical Magazine,82(4):1015−1021

Li G W,Fang Q S,Shi N C,Bai W J,Yang J S,Xiong M,Ma Z S,Rong H. 2009. Zangboite,$TiFeSi_2$,a new mineral species from Luobusha,Tibet,China, and its crystal structure. The Canadian Mineralogist,47(5):1265−1274

Li G W,Bai W J,Shi N C,Fang Q S,Xiong M,Yang J S,Ma Z S,Rong H. 2012. Linzhiite,$FeSi_2$,a redefined and revalidated new mineral species from Luobusha,Tibet,China. European Journal of Mineralogy,24(6):1047−1052

Li G W,Yang G M,Lu F D,Xiong M,Ge X K,Pan B M,de Fourestier J. 2016. Fluorcalciopyrochlore,a new mineral species from Bayan Obo,Inner Mongolia,P. R. China. The Canadian Mineralogist,54(5):1285−1291

Li G W,Xue Y,Xiong M. 2019. Tewite:a K-Te-W new mineral species with a modified tungsten-bronze type structure,from the Panzhihua-Xichang region,Southwest China. European Journal of Mineralogy,31(1):145−152

Linthout K. 2007. Tripartite division of the system $2REEPO_4$-$CaTh(PO_4)_2$-$2ThSiO_4$,discreditation of brabantite,and recognition of cheralite as the name for members dominated by $CaTh(PO_4)_2$. The Canadian Mineralogist,45(3):503−508

Liu J J,Li G W,Mao Q,Wu S H,Liu Z J,Su S G,Xiong M,Yu X Y. 2012. Hanjiangite,a new barium-vanadium phyllosilicate carbonate mineral from the Shiti barium deposit in the Dabashan region,China. American Mineralogist,97(2-3):281−290

Lo K D,Wei J,Zhang J Y,Gu Q F. 1980. Clinochalcomenite,a new mineral of selenite. Kexue Tongbao,25(5):427−433

Lo K D,Chen Z R,Ma Z S. 1984. The crystal structure of clinochalcomenite. Kexue Tongbao,29(3):352−355

Ma Z S,Li G W,Chukanov N V,Poirier G,Shi N C. 2014. Tangdanite,a new mineral species from the Yunnan Province,China and the discreditation of "clinotyrolite". Mineralogical Magazine,78(3):559−569

Mandarino J A,de Fourestier J. 2005. 首次在中国发现的矿物. 矿物学报,25(3):217−229

Miyawaki R,Shimazaki H,Shigeoka M,Yokoyama K,Matsubara S,Yurimoto H,Yang Z,Zhang P. 2011a. Fluorokinoshitalite and fluorotetraferriphlogopite:new species of fluoro-mica from Bayan Obo,Inner Mongolia,China. Clay Science,15(1):13−18

Miyawaki R,Shimazaki H,Shigeoka M,Yokoyama K,Matsubara S,Yurimoto H. 2011b. Yangzhumingite,$KMg_{2.5}Si_4O_{10}F_2$,a new mineral in the mica group from Bayan Obo,Inner Mongolia,China. European Journal of Mineralogy,23(3):467−473

Nickel E H,Grice J D. 1999. 国际矿物学协会新矿物及矿物命名委员会关于矿物命名的程序和原则(1997年). 岩石矿物杂志,18(3): 273−285

Nickel E H,Mandarino J A. 1987. Procedures involving the IMA Commission on New Minerals and Mineral Names and guidelines on mineral nomenclature. American Mineralogist,72(9-10):1031−1042

Oberti R,Boiocchi M,Smith D C. 2003. Fluoronyböite from Jianchang(Su-Lu,China) and nyböite from Nybö(Nordfjord,Norway):a petrological and crystal-chemical comparison of these two high-pressure amphiboles. Mineralogical Magazine,67(4):769−782

Oberti R,Boiocchi M,Smith D C,Medenbach O. 2007. Aluminotaramite, alumino-magnesiotaramite, and fluoro-alumino-magnesiotaramite:mineral data and crystal chemistry. American Mineralogist, 92(8-9):1428-1435

Qu K, Sima X, Fan G, Li G W, Shen G F, Chen H K, Liu X, Yin Q Q, Li T, Wang Y J. 2019a. Taipingite-(Ce), IMA 2018-123a. CNMNC Newsletter No. 50, page 617. Mineralogical Magazine, 83(4):615-620

Qu K, Sima X, Li G W, Fan G, Shen G F, Liu X, Xiao Z B, Hu G, Qiu L F, Wang Y J. 2019b. Fluorluanshiweiite, IMA 2019-053. CNMNC Newsletter No. 52, page 888. Mineralogical Magazine, 83(6):887-893

Rao C, Wang R C, Hatert F, Gu X P, Ottolini L, Hu H, Dong C W, Dal Bo F, Baijot M. 2014. Strontiohurlbutite, $SrBe_2(PO_4)_2$, a new mineral from Nanping No. 31 pegmatite, Fujian Province, southeastern China. American Mineralogist, 99(2-3):494-499

Rao C, Hatert F, Wang R C, Gu X P, Dal Bo F, Dong C W. 2015. Minjiangite, $BaBe_2(PO_4)_2$, a new mineral from Nanping No. 31 pegmatite, Fujian Province, southeastern China. Mineralogical Magazine, 79(5):1195-1202

Rao C, Hatert F, Dal Bo F, Wang R C, Gu X P, Baijot M. 2017. Mengxianminite, ideally $Ca_2Sn_2Mg_3Al_8[(BO_3)(BeO_4)O_6]_2$, a new borate mineral from Xianghualing skarn, Hunan Province, China, with a highly unusual chemical combination (B+Be+Sn). American Mineralogist, 102(10):2136-2141

Rao C, Wang R C, Gu X P, Xia Q K, Dong C W, Hatert F, Yu X G, Wang W Y. 2018. Zinconigerite-2N1S, IMA 2018-037. CNMNC Newsletter No. 44, August 2018, page 1020. Mineralogical Magazine, 82(4):1015-1021

Rao C, Gu X P, Wang R C, Xia Q K, Cai Y F, Dong C W, Hatert F, Hao Y T. 2020. Chukochenite, IMA 2018-132a. CNMNC Newsletter No. 54, page 361. Mineralogical Magazine, 84(2):359-365

Ren G, Li G, Shi J, Gu X, Fan G, Yu A, Liu Q, Shen G. 2019. Potassic-hastingsite, IMA 2018-160. CNMNC Newsletter No. 49, page 481. Mineralogical Magazine, 83(3):479-483

Sejkora J, Hyršl J. 2007. Ottensite: a new mineral from Qinglong, Guizhou Province, China. Mineralogical Record, 38(1):77-81

Shi N C, Bai W J, Li G W, Xiong M, Yang J S, Ma Z S, Rong H. 2012. Naquite, FeSi, a New Mineral Species from Luobusha, Tibet, western China. Acta Geologica Sinica (English Edition), 86(3):533-538

Shimazaki H, Miyawaki R, Yokoyama K, Matsubara S, Yang Z M. 2008. Zhangpeishanite, BaFCl, a new mineral in fluorite from Bayan Obo, Inner Mongolia, China. European Journal of Mineralogy, 20(6):1141-1144

Vergasova L P, Filatov S K. 2016. A study of volcanogenic exhalation mineralization. Journal of Volcanology and Seismology, 10(2):71-85

Voloshin A V, Pakhomovskii Y A, Stepanov V I and Tyusheva E N. 1983. Natrobistantite (Na, Cs) Bi (Ta, Nb, Sb)$_4$O$_{12}$-new mineral from granite pegmatites. Mineralogicheskii Zhurnal, 5(2):82-86 (in Russian with English abstract)

Williams P A, Hatert F, Pasero M, Mills S J. 2012a. New minerals and nomenclature modifications approved in 2012. Mineralogical Magazine, 76(3):807-817

Williams P A, Hatert F, Pasero M, Mills S J. 2012b. New minerals and nomenclature modifications approved in 2012. Mineralogical Magazine, 76(5):1281-1288

Xie X D, Minitti M E, Chen M, Mao H K, Wang D Q, Shu J F, Fei Y W. 2003. Tuite, γ-$Ca_3(PO_4)_2$: a new mineral from the Suizhou L6 chondrite. European Journal of Mineralogy, 15(6):1001-1005

Xie X D, Gu X P, Yang H X, Chen M, Li K. 2020. Wangdaodeite, the $LiNbO_3$-structured high-pressure polymorph of ilmenite, a new mineral from the Suizhou L6 chondrite. Meteoritics & Planetary Science, 55(1):184-192

Xiong F, Xu X, Mugnaioli E, Gemmi M, Wirth R, Grew E S, Robinson P T. 2019a. Badengzhuite, IMA 2019-076. CNMNC Newsletter No. 52, page 892. Mineralogical Magazine, 83(6):887-893

Xiong F, Xu X, Mugnaioli E, Gemmi M, Wirth R, Grew E S, Robinson P T. 2019b. Jingsuiite, IMA 2018-117b. CNMNC Newsletter No. 52, page 891. Mineralogical Magazine, 83(6):887-893

Xiong F, Xu X, Mugnaioli E, Gemmi M, Wirth R, Grew E S, Robinson P T. 2019c. Zhiqinite, IMA 2019-077. CNMNC Newsletter No. 52, page 892. Mineralogical Magazine, 83(6):887-893

Xu J S, Yang G M, Li G W, Wu Z L, Shen G F. 2008. Dingdaohengite-(Ce) from the Bayan Obo REE-Nb-Fe Mine, China: both a true polymorph of perrierite-(Ce) and a titanic analog at the C1 site of chevkinite subgroup. American Mineralogist, 93(5-6):740-744

Yang G M, Li G W, Xiong M, Pan B M, Yan C J. 2014. Hydroxycalciopyrochlore, a new mineral species from Sichuan, China. Acta Geologica Sinica (English Edition), 88(3):748-753

Yang Z M, Ding K S, de Fourestier J, Mao Q, Li H. 2012a. Ferrotaaffeite-2N'2S, a new mineral species, and the crystal structure of Fe^{2+}-rich magnesiotaaffeite-2N'2S from the Xianghualing tin-polymetallic ore field, Hunan Province, China. The Canadian Mineralogist, 50(1):21-29

Yang Z M, Giester G, Ding K S, Tillmanns E. 2012b. Hezuolinite, $(Sr, REE)_4Zr(Ti, Fe^{3+}, Fe^{2+})_2Ti_2O_8(Si_2O_7)_2$, a new mineral species of the chevkinite group from Saima alkaline complex, Liaoning Province, NE China. European Journal of Mineralogy, 24(1):189-196

Yang Z M, Giester G, Mao Q, Ma Y G, Zhang D, Li H. 2017. Zincobotryogen, $ZnFe^{3+}(SO_4)_2(OH) \cdot 7H_2O$: validation as a mineral species and new data. Mineralogy and Petrology, 111(3):363-372

Yu Z X, Wang H G. 2017. The crystal structure of lisiguangite. Acta Geologica Sinica (English Edition), 91(4):1270-1275

Yu Z X, Cheng F Y, Ma H W. 2009. Lisiguangite, $CuPtBiS_3$, a new platinum-group mineral from the Yanshan Mountains, Hebei, China. Acta Geologica Sinica (English Edition), 83(2):238-244

Yu Z X, Hao Z G, Wang H G, Yin S P, Cai J H. 2011. Jichengite $3CuIr_2S_4 \cdot (Ni, Fe)_9S_8$, a new mineral, and its crystal structure. Acta Geologica Sinica (English Edition), 85(5):1022-1027

新矿物的发现与研究进展

Zhang P H, Zhu J C, Zhao Z H, Gu X P, Lin J F. 2004. Zincospiroffite, a new tellurite mineral species from the Zhongshangou gold deposit, Hebei Province, People's Republic of China. The Canadian Mineralogist, 42(3):763-768

Research Progresses of New Minerals Discovered

CAI Jian-hui

Key Laboratory of Mineralization and Resource Evaluation, Ministry of Natural Resources, Institute of Mineral Resources Chinese Academy of Geological Sciences, Beijing 100037

Abstract: Since the discovery of the first new mineral Hsianghualite in China in 1958, 175 new minerals have been discovered and reported in China, and 142 out of them have been officially approved by IMA CNMNC as valid independent mineral species. In the past 20 years of this century, the outstanding research progresses of new minerals in China have particularly been made, as the number of new minerals discovered in China has reached to a high record of 60, with the ranking score of sixth among countries in the world. Especially, the number of discovered new species of natural element minerals ranks among the top in the world and the discovery and research of new species of rare earth minerals and shock-metamorphosed ultrahigh-pressure minerals have made significant progress in the world. This paper is focused on reviewing the revision of new mineral species discovered in China in the last century and summarizing the achievements and progresses of studies on new mineral species discovered in China in this century in terms of the quantity, crystallochemical type, and occurrence characteristics, in order to provide reference and inspiration for promoting the discovery and research of new mineral species in China in future.

Key words: new mineral species; research progress; crystallochemical type; origin and occurrence

矿物晶体结构与晶体化学研究进展*

李国武

中国地质大学(北京) 科学研究院，晶体结构实验室，北京 100083

摘　要：矿物的晶体结构与晶体化学是矿物学的重要基础研究领域之一，21 世纪以来，随着国家对地勘行业的重视，以及平面探测技术单晶衍射仪、微区、微量衍射等实验技术的应用与计算机软硬件能力的提高，我国矿物晶体结构与晶体化学研究得到飞速发展。矿物晶体结构测定与研究是伴随着新矿物的发现而发展的，新矿物的发现是矿物晶体结构研究的基石，又为新矿物提供数据支撑。目前，我国发现的各类新矿物的晶体结构大多数已被测定，精修了近 30 个新矿物的晶体结构，并发现了多种此前从未发现过的矿物新结构，对若干新矿物的晶体化学新现象有了新的认识，这些成果为新矿物的发现和申报提供了基础性数据，同时也为阐明矿物的成分结构、物理特性、成因及演化，进而为各种地质现象的解释提供科学依据。展望未来，矿物晶体结构与晶体化学研究领域机遇与挑战并存。

关键词：晶体结构　晶体化学　新矿物　研究进展

0　引　言

矿物的晶体化学分类体系是基于对矿物晶体结构的了解，而矿物的成因研究以及矿物结构与性能等都需要有矿物晶体结构的知识作为指导。因此，矿物晶体结构的测定与晶体化学研究，就成为矿物学的基本任务之一而延续至今。矿物种是根据化学成分与晶体结构共同确定的，但晶体结构方面的数据更为关键。这是由于晶体结构更能反映该物种的全貌和本质，并决定矿物的理想分子式的形式，同时对于类质同象、同质多象、多型结构中单位晶胞内各种原子的排列，通过结构精修可得到精细结构从而列出其晶体化学式。根据晶体结构提供的化学键信息可推演该矿物的硬度、解理、密度等物理性质。除个别情况外，国际矿物协会新矿物、命名和分类委员会（IMA CNMNC）对申请新矿物的基本要求是需要提供 X 射线衍射方面的数据，而且只有提供单晶 X 射线衍射结构数据才会得到数据完整性的认可。

中国对发现的新矿物的矿物晶体学基础研究一直很重视，我国学者发现的新矿物大多开展过详细的晶体学研究工作，并进行了单晶 X 射线衍射晶体结构的测定和精修，提供了完整的晶体学数据。特别是 21 世纪以来，由于新矿物及对基础矿物学研究的深入，以及实验仪器方法的进步，X 射线晶体学研究手段在光源、探测器及计算技术方面都突飞猛进式的发展。CCD 平面探测器及半导体阵列直读平面探测器单晶衍射仪的应用，使矿物结构测定的精度、速度都有了很大的提高。透射电子显微镜、同步辐射衍射、微量粉晶衍射、微区原位衍射和电子探针微区成分分析等仪器和方法的广泛使用，使矿物晶体结构与晶体化学的研究范围更为广泛，对矿物晶体结构的认识也更加深入，也由此发现了许多新矿物以及矿物新结构和有意义的晶体化学新现象，丰富了矿物晶体结构与晶体化学的研究内容，促进了传统基础矿物学的发展。笔者主要以我国 21 世纪以来与新矿物相关的矿物晶体结构与晶体化学领域为主，综述我国学者 20 年间取得的主要研究进展及新成果。

---

* 原文"我国矿物晶体结构与晶体化学研究进展及新成果（2000—2019）"刊于《矿物岩石地球化学通报》2021 年第 40 卷第 2 期，本文略有修改。

1 矿物晶体结构的新发现

矿物晶体结构测定和研究与新矿物的发现密不可分，新矿物的发现是矿物晶体结构研究的基石。目前我国发现的各类新矿物的晶体结构大多数已被测定，其中就发现了多种此前世界上从未发现过的矿物新结构，主要包括汉江石型结构、钨青铜型结构、钨青铜衍生新结构、李璞硅锰石新结构和孟宪明矿新结构。

1.1 汉江石型结构

汉江石型结构是在新矿物汉江石中发现的一种新的层状硅酸盐结构类型（李国武等，2011；Liu et al.，2012）。汉江石（Hanjiangite，IMA2009-082）的理想晶体化学式为 $Ba_2CaV_2^{3+}[(Si_3AlO_{10})(OH)_2]F(CO_3)_2$，空间群 $C2$；晶胞参数：$a = 0.5205(12)$ nm，$b = 0.9033(2)$ nm，$c = 3.2077(8)$ nm，$\beta = 93.49(8)°$，$V = 1.5054(8)$ nm³，$Z = 4$。汉江石的结构为一种全新层间域结构的层状结构（图1），是由一个2∶1型层状硅酸盐矿物的TOT层和氟碳酸盐层 $Ba_2Ca(CO_3)_2F$ 构成，即TOT型 + $Ba_2Ca(CO_3)_2F$ 层的新结构类型。八面体层（O）中充填三价离子 V^{3+}、Al^{3+}，构成二八面体型结构；四面体层（T）由Si和Al组成；Ba、Ca以氟碳酸盐形式充填于层间。$Ba_2Ca(CO_3)_2F$ 的结构与毒重石的单层结构相似，阳离子类似于方解石型排列，由于是首次发现该类型的新结构故命名为汉江石型结构。该类型的层间域也是目前发现的最宽的层状硅酸盐结构（层间宽0.9407 nm），层状硅酸盐结构型对比如图2所示。汉江石型结构的发现，不仅为层状硅酸盐矿物增添了一个新的成员，而且丰富了层状硅酸盐晶体结构与晶体化学的新内容。

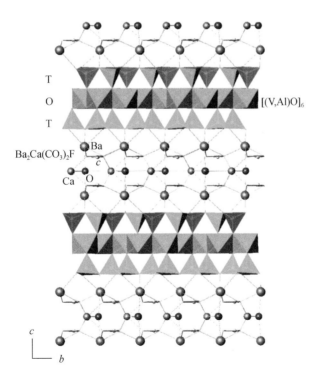

图1 汉江石晶体结构（根据结构数据绘制）

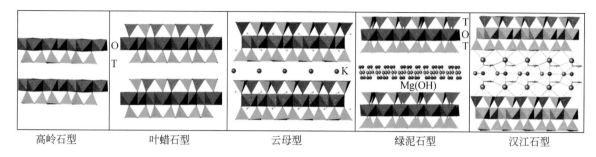

图 2　层状硅酸盐结构型对比（根据结构数据绘制）

1.2　天然钨青铜型结构

钨青铜型结构是材料中 M_xWO_3（M 为碱金属离子）氧化物的一类孔道网络结构，此前从未发现过天然的具有钨青铜型结构的矿物。乌木石（Wumuite，IMA2017-067a）是世界上首次发现的钨青铜型结构的天然矿物（Li and Xue，2018），其简化分子式为 $KAl_{0.33}W_{2.67}O_9$，六方晶系，空间群 $P6/mmm$；晶胞参数：$a = 0.729\,52(5)$ nm，$b = 0.729\,52(5)$ nm，$c = 0.377\,11(3)$ nm，$\beta = 120°$，$V = 0.173\,81(2)$ nm^3，$Z = 1$。新矿物的晶体结构具有六方钨青铜型结构[图 3（a）]，WO_6 八面体共顶角连接成六方环状孔道结构，K 分布于六方孔道中。为了平衡电价 Al^{3+} 代替 W^{4+} 是必需的，并趋向于固定比例。天然钨青铜矿物的发现为人们认知与利用自然界中新物质以及指导人工合成功能材料提供了新的依据，同时近年来发现的一系列 W、Te、K 新矿物，显示了独特的成矿地质特点，对于研究矿物成因及其地质过程有着重要的地质意义。

1.3　钨青铜衍生新结构

碲钨矿（Tewite，IMA2014-053）是一种 K-Te-W 的全新成分和全新钨青铜型衍生晶体结构的氧化物新矿物（Li et al.，2019），这也是世界上首次发现该成分及结构的新矿物，是目前唯一一种 K-Te-W 的天然矿物，此前从未发现过具有类似成分和结构的天然矿物或人工合成物。

碲钨矿的理想晶体化学式为 $(K_{1.5}\square_{0.5})_{\Sigma 2}(Te_{1.25}W_{0.25}\square_{0.5})_{\Sigma 2}W_5O_{19}$，斜方晶系，空间群 $Pban$；晶胞参数：$a = 0.725\,85(4)$ nm，$b = 2.580\,99(15)$ nm，$c = 0.381\,77(2)$ nm，$V = 0.715\,21(7)$ nm^3，$Z = 2$。碲钨矿的晶体结构为钨青铜型结构的衍生结构[图 3（b）]，WO_6 八面体共顶角连接成六方环状孔道（隧道）结构，孔道沿 c 轴延伸，大阳离子 K 充填于六方隧道中。WO_6 八面体柱间由 TeO_6 偏八面体中的弱键连接（键长 0.272 8 nm）。也可以视为 TeO_6 偏八面体连接六方钨青铜 K_xWO_3 结构中的一个六方环裂开后向 a 方向错动（1/2）a 形成的空隙之间，其中两个强键（键长为 0.195 4 nm）连接三角形状的两个 WO_6 八面体，而两个弱键（键长为 0.269 9 nm）连接断开六方环的两个 WO_6 八面体。

碲钨矿的特殊结构显示其具有独特的晶体化学特点，根据晶体结构测定（Li et al.，2019），理想的碲钨矿端元晶体化学式为 $K_2Te_2W_5O_{19}$，但该化学式的电价不平衡，正电价为 40、负电价为 38，这在天然矿物中是不能稳定存在的。在衍射实验中发现，该矿物具有非公度调制结构的衍射特征，化学分析和结构精修都显示该矿物中 K 过量和 Te 不足，其中 K 位于六方孔的中心（这与钨青铜 K_xWO_3 相同），沿孔道方向呈半有序（占位调制）分布，平均结构中指派了两个位置，但不可能全充填，K 的总占位不超过 1.5，Te 位也有占位不足，并有少量 W^{6+} 占据 Te^{4+} 的位置，总占位不超过 1.5，理想情况下 K 位和 Te 位的总正电价为 1.5×1（K^{+1}）+ 1.25×4（Te^{+4}）+ 0.25×6（W^{+6}）= 8，因此，根据晶体结构结合化学成分分析得出的实际存在的端元晶体化学式为 $(K_{1.5}\square_{0.5})_{\Sigma 2}(Te_{1.25}W_{0.25}\square_{0.5})_{\Sigma 2}W_5O_{19}$，简写为 $K_{1.5}(Te,W)_{1.5}W_5O_{19}$，化学式中空位 $\square$ 是必须存在的。由于异价离子间的类质同象置换引起的电价不平衡，由其化学成分含量及晶格空位来补充，从而使该矿物整体电荷平衡，但同时也导致了该矿物出现占位调制结构，在单晶衍

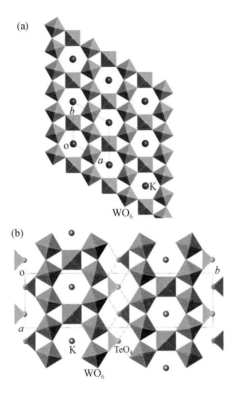

图3 乌木石和碲钨矿的晶体结构（根据结构数据绘制）

(a) 乌木石中的六方钨青铜型结构 $KAL_{0.33}W_{2.67}O_9$；(b) 碲钨矿晶体结构 $(K_{1.5}\square_{0.5})_{\Sigma 2}(Te_{1.25}W_{0.25}\square_{0.5})_{\Sigma 2}W_5O_{19}$

射中观察到弱的卫星衍射点，分析表明该调制结构为二维的非公度调制结构。调制向量 $q1 = 0.42c^*$，$q2 = 0.5c^*$，$q3 = 0.085c^*$。在同一产地发现的碲钨矿和乌木石两者的晶体结构相关，两者有成因上的联系（李国武等，2018）。

1.4 李璞硅锰石新结构

李璞硅锰石（Lipuite，IMA2014-085）发现于南非共和国北开普省卡拉哈里锰矿田，是一种硅氧四面体和磷氧四面体的复杂层状结构硅酸盐矿物（Gu et al., 2019），其理想晶体化学式为 $KNa_8Mn_5^{3+}Mg_{0.5}[Si_{12}O_{30}(OH)_4](PO_4)O_2(OH)_2·4H_2O$，斜方晶系，空间群 $Pnnm$；晶胞参数：$a = 0.908\ 0(3)$ nm，$b = 1.222\ 2(3)$ nm，$c = 1.709\ 3(5)$ nm，$V = 1.897\ 0(9)$ nm^3，$Z = 2$。其晶体结构中，SiO_4 四面体共角顶沿 [010] 构成折状层（图4），K^+、Na^+、Mn^{3+}、Mg^{2+} 和 P^{5+} 阳离子通过端氧以及氢键与 SiO_4 四面体匹配连接，形成层状。四面体层由14元环的 SiO_4 四面体组成，沿 [100] 曲折。Mn^{3+} 阳离子呈两种不同的八面体配位，该八面体在 SiO_4 四面体片之间共棱连接成五元八面体团簇。李璞硅锰石的晶体结构是一种非常独特的结构类型，其硅氧四面体片可以认为是云母中硅氧四面体片的衍生结构。粉晶X射线衍射强线 [d：nm(I/I_0)(hkl)]：0.996 5(40)(011)，0.293 8(33)(310)，0.289 5(100)(311)，0.277 7(38)(224)，0.271 3(53)(320)，0.248 3(32)(126)，0.208 6(35)(046)，0.153 4(40)(446)。

1.5 孟宪明矿新结构

孟宪明矿（Mengxianminite，IMA2015-70）是一种结构复杂的硼酸盐矿物（Rao et al., 2020），其理想晶体化学式为 $Ca_2Sn_2Mg_3Al_8[(BO_3)(BeO_4)O_6]_2$，斜方晶系，空间群 $Fdd2$；晶胞参数：$a = 6.069\ 9(4)$ nm，

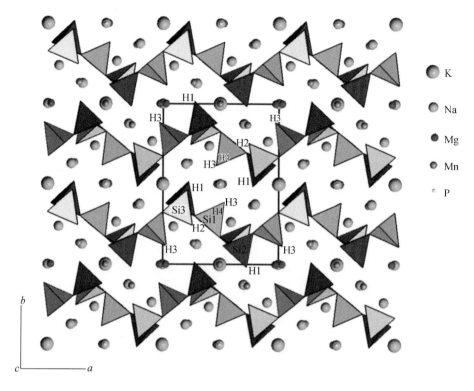

图 4　李璞硅锰石晶体结构（据 Gu et al., 2019）

$b=0.9914(1)$ nm，$c=0.5745(1)$ nm，$V=3.4574(4)$ nm^3，$Z=8$。

孟宪明矿的晶体结构较复杂，是由沿 a 轴交替叠加的 O-T1-O-T2-O′-T2 层构成（图5），可看成是两个交替的模块：A 模块（O-T1-O）相当于尖晶石结构叠加一个 O 层（AlO$_6$ 八面体层），B 模块（T2-O′-T2）为简化式为 CaSnAl(BeO$_4$)(BO$_3$) 的结构，其中 SnO$_6$ 八面体在 T2 层中呈孤岛状，并通过 BeO$_4$ 和 CaO$_{11}$ 多面体连接，AlO$_6$ 八面体共棱在 O′ 层中形成沿 {011} 或 {01$\bar{1}$} 方向的链，并在 c 方向由 BO$_3$ 三角形连接。孟宪明矿是第一种含 Sn 和 Be 的硼酸盐矿物，在香花岭夕卡岩晚期富 F 条件下结晶形成。

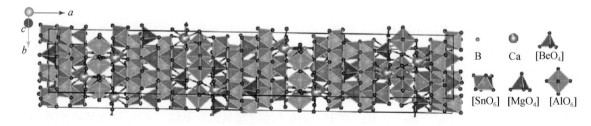

图 5　孟宪明石的晶体结构（据 Rao et al., 2020）

2　金属互化物及陨石高压新矿物

我国对于金属互化物矿物及陨石矿物研究有悠久的历史，已发现二十余种单质及金属互化物新矿物和陨石新矿物，近年来在西藏罗布莎、随州陨石及岫岩陨石坑中又发现了十余种金属互化物及高压矿物。金属互化物矿物及高压矿物的发现及其晶体结构与晶体化学的研究对于了解地球地幔矿物组成及其环境提供了重要信息。

西藏罗布莎蛇绿岩铬铁矿床是一个金属互化物矿物的宝库,迄今已发现了罗布莎矿、曲松矿、雅鲁矿、藏布矿、自然钛、那曲矿、林芝矿、巴登珠矿、志琴矿、经绥矿和青松矿等11种金属互化物新矿物。它们分别属于过渡族元素(Fe、Cr、Ni)、W及Ti的硅化物、碳化物和磷、硼、氮化物。在西藏罗布莎地区发现的地球深部矿物不是独立的个别种属,而是发现了一批能表征不同深度的地球深部矿物群体(白文吉等,2003;杨经绥等,2008)。其中金属元素(Fe、Cr、Ni、Co、Ti、W)与硅和碳形成的硅化物和碳化物特别丰富,其成因初步推断是通过地幔柱通道由深部上升至蛇绿岩后被捕掳并冷却结晶而形成的(Bai et al.,2000;施倪承等,2005a)。

西藏罗布莎蛇绿岩型铬铁矿床中地球深部矿物种群的发现加深了人类对地球深部各层圈,尤其是地幔和地核物质的化学成分和晶体结构的新认识(施倪承等,2005b;李国武等,2015),对于地球动力学及地球早期演化研究都具有重要意义,在研究地球深部矿物的物相及其相变多形的晶体结构以及深度表征方面也具有重要意义。罗布莎11种金属互化物新矿物的晶体学数据如下。

罗布莎矿(Luobusaite,IMA2005-052a)(Bai et al.,2006),其理想晶体化学式为$FeSi_2$,以Si为2计算的实验式$Fe_{0.83}Si_2$,类似于人工合成物$\beta-FeSi_2$,是林芝矿的同质多家变体。斜方晶系,空间群$Cmca$,晶胞参数:$a=0.9874$ nm,$b=0.7784$ nm,$c=0.7829$ nm,$Z=16$。用粉晶衍射数据Rietveld法精修晶体结构(李国武等,2007a),发现晶体结构中的Fe、Si原子在$b-c$面上呈互层状分布[图6(a)],Si堆积层较紧密,而Fe堆积层存在孔隙,结构精修发现结构中Fe有明显的空位缺席结构特征。最强七条粉晶衍射线[d:nm(I/I_0)(hkl)]:0.306(80)(220),0.2849(20)(221),0.2402(25)(312),0.1977(40)(313),0.1889(60)(041),0.1865(40)(114),0.1844(100)(422)。

林芝矿(Linzhiite,IMA2010-011)(Li et al.,2012),其理想晶体化学式为$FeSi_2$,与罗布莎矿互为同质多象变体。四方晶系,空间群$P4/mmm$,晶胞参数:$a=0.2725(3)$ nm,$b=0.2725(3)$ nm,$c=0.5202(10)$ nm,$V=0.3862(9)$ nm^3,$Z=1$。单晶晶体结构精修R因子为0.057(Li et al.,2012),林芝矿的晶体结构可以看作是一种堆积层状结构[图6(b)],堆积层沿ab平面展开,在该层内硅铁原子呈立方体心式堆积,立方格子顶点为Si而中心为Fe充填,由于铁的原子半径较大,将Si组成的立方最紧密堆积撑开,形成近似的最紧密堆积。最强五条粉晶衍射线为[d:nm(I/I_0)(hkl)]:0.5142(100)(001),0.2377(60)(101),0.1897(48)(110),0.1851(93)(102),0.1776(17)(200)。

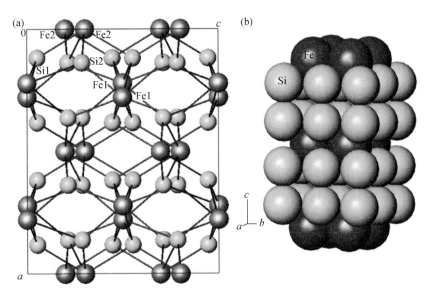

图6 罗布莎矿晶体结构图(a)和林芝矿晶体结构图(b)(根据结构数据绘制)

那曲矿(Naquite,IMA2010-010)(Shi et al.,2012),其理想晶体化学式为$FeSi$,等轴晶系,空间群

$P2_13$；粉晶衍射精修的晶胞参数是：$a=0.4486(4)$ nm，$V=0.9028(6)$ nm^3，$Z=4$，晶体结构与人工合成 FeSi 相同。粉末图中最强的衍射线为 [d：nm(I/I_0)(hkl)]：0.3174(43)(110)，0.2592(46)(111)，0.2249(25)(200)，0.2008(100)(210)，0.1831(69)(211)，0.1353(28)(311)，0.1199(38)(321)。

藏布矿（Zangboite，IMA2007-036）（Li et al.，2009），其化学成分接近理想晶体化学式 TiFeSi$_2$，是一种三元合金新矿物。斜方晶系，空间群 $Pbam$；晶胞参数：$a=0.86053$ nm，$b=0.95211$ nm，$c=0.76436$ nm，$Z=12$。单晶晶体结构精修 R 因子为 0.034（Li et al.，2009），晶体结构中 Fe、Si 构成八面体 [FeSi$_6$]，八面体共棱连接，沿 c 轴形成孔道，Ti 充填于孔道中（图7）。粉晶衍射六条强线 [d：nm(I/I_0)(hkl)]：0.38358(50)(002)，0.23010(30)(312)，0.22318(50)(312)，0.21291(100)(232)，0.20251(65)(042)，0.19155(57)(004)。

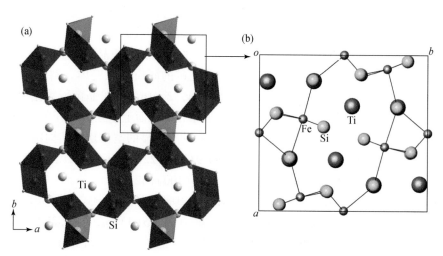

图7 藏布矿的晶体结构（根据结构数据绘制）

雅鲁矿（Yarlongite，IMA2007-035）（Shi et al.，2008），其理想晶体化学式为（Cr$_4$Fe$_4$Ni）$_9$C$_4$，成分中 C 相对稳定，而金属元素的成分变化较大；六方晶系，空间群 $P6_3mc$；晶胞参数：$a=1.8839(2)$ nm，$b=1.8839(2)$ nm，$c=0.44960(9)$ nm，$Z=6$。单晶晶体结构精修 R 因子为 0.079（Shi et al.，2008），晶体结构中 Fe、Cr、Ni 各有其不同的晶体化学位置（图8），但由于成分变化较大其占位度变化较大，Fe、Cr、Ni 总量为9，配位数12，形成带折皱的堆积层与平的堆积层的互层结构，C 的配位数为6，充填于 Fe、Cr、Ni 金属原子间构成三方柱配位多面体中，该配位多面体以共角顶或共棱方式相互连接构成了一种新型的金属碳化物结构。粉晶衍射强线 [d：nm(I/I_0)(hkl)]：0.6920(100)(110)，0.4530(35)(210)，0.3596(55)(201)，0.2493(36)(401)，0.2023(98)(421)，1998(32)(600)，0.1825(47)(601)，0.1798(54)(402)。

曲松矿（Qusongite，IMA2007-034）（Fang et al.，2009），其理想晶体化学式为 WC，六方晶系，空间群 $P\bar{6}m2$；利用粉晶衍射精修的晶胞参数：$a=0.2902(1)$ nm，$b=0.2902(1)$ nm，$c=0.2831(1)$ nm，$Z=1$。晶体结构与人工合成物的晶体结构相同，C 的配位数为6，充填于 W 金属原子间构成的三方柱配位多面体中，结构由三方柱多面体以共棱方式相互连接而成。粉晶衍射强线 [d：nm(I/I_0)(hkl)]：0.2833(44)(001)，0.2511(94)(010)，0.18778(90)(011)，0.1449(25)(110)，0.1291(36)(111)，0.1233(22)(102)，0.1149(23)(201)，0.09008(23)(021)。

自然钛（Titanium，IMA2010-044）（Fang et al.，2013），其理想晶体化学式为 Ti，成分较纯，六方晶系，空间群 $P6_3/mmc$；粉晶衍射确定的晶胞参数：$a=0.2950(2)$ nm，$c=0.4686(1)$ nm，$V=0.3532(5)$ nm^3，$Z=2$。晶体结构与化学单质中的金属 Ti 相同，粉晶衍射中最强的衍射线为 [d：nm(I/I_0)(hkl)]：0.2569(32)(010)，0.2254(100)(011)，0.1730(16)(012)，0.1478(21)(110)，0.9464(8)(121)。

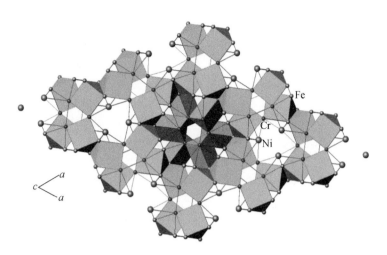

图8 雅鲁矿晶体结构（据 Shi et al., 2008）

巴登珠矿（Badengzhuite, IMA2019-076）（Xiong et al., 2019a），其理想晶体化学式为 TiP，六方晶系，空间群 $P6_3/mmc$；粉晶衍射确定的晶胞参数：$a=0.349(7)$ nm，$c=1.176(24)$ nm。晶体结构与合成物 TiP 相同，粉晶衍射中最强的衍射线为 [d: nm(I/I_0)]：0.302 2(14)，0.292 7(26)，0.268 8(31)，0.239 4(100)，0.210 7(33)，0.185 6(26)，0.174 5(44)，0.141 0(11)。

志琴矿（Zhiqinite, IMA2019-077）（Xiong et al., 2019b），硅钛互化物，其理想晶体化学式为 $TiSi_2$，斜方晶系，空间群 $Fddd$；粉晶衍射确定的晶胞参数：$a=0.810(16)$ nm，$b=0.477(9)$ nm，$c=0.852(17)$ nm。晶体结构与与合成物 $TiSi_2$ 相同，粉晶衍射中最强的衍射线为 [d: nm(I/I_0)]：3.702(9)，2.935(31)，2.265(100)，2.130(43)，2.081(78)，2.055(9)，1.810(39)，1.353(9)。

经绥矿（Jingsuiite, IMA2019-117b）（Xiong et al., 2019c），其理想晶体化学式为 TiB_2，六方晶系，空间群 $P6/mmm$；粉晶衍射确定的晶胞参数：$a=0.301(6)$ nm，$c=0.321(6)$ nm。晶体结构与合成物 TiB_2 相同，粉晶衍射中最强的衍射线为 [d: nm(I/I_0)]：0.321 8(21)，0.261 5(60)，0.202 9(100)，0.160 9(9)，0.151 0(20)，0.137 0(11)，0.121 1(12)，0.110 1(11)。

青松矿（Qingsongite, IMA2013-30），其理想晶体化学式为 BN，美国学者 Dobrzhinetskaya 等（2014）利用电子衍射及透射电子显微镜高分辨像进行了晶体学研究，得到其晶体结构为等轴晶系，空间群 $F\bar{4}3m$；晶胞参数：$a=0.361$ nm，$Z=4$。晶体结构与人工合成物 c-BN 相同；粉晶衍射中最强的衍射线为 [d: nm(I/I_0)(hkl)]：0.208 8(100)(111)，0.180 8(8)(200)，0.127 7(20)(220)，0.109 03(10)(311)，0.090 40(3)(400)，0.082 96(8)(331)。

谢氏超晶石（Xieite, IMA2007-056），陨石冲击脉体中的高压矿物（陈鸣等，2008；Chen et al., 2008），其理想晶体化学式为 $FeCr_2O_4$，斜方晶系，空间群 $Bbmm$；粉晶衍射确定的晶胞参数：$a=0.946\ 2(6)$ nm，$b=0.956\ 2(9)$ nm，$c=0.291\ 6(1)$ nm，$V=0.263\ 8(4)$ nm^3，$Z=4$。谢氏超晶石的晶体结构通过同步辐射 X 射线衍射技术在陨石薄片上进行微区原位分析获得，晶体结构与人工合成的 $CaTi_2O_4$ 氧化物等结构见图9，结构中包含共棱和角顶相连的八面体以及十二面体两种位置，Cr^{3+} 和 Al^{3+} 占据八面体位置，Mg^{2+} 和 Fe^{2+} 占据十二面体位置。X 射线衍射的特征谱线 [d: nm(I/I_0)]：0.267 5(100)，0.238 9(20)，0.208 9(10)，0.195 3(90)，0.156 6(60)，0.143 9(15)，0.142 5(15)，0.133 7(40)。矿物成因是在冲击波引起的高温高压作用下，通过固态反应从铬铁矿转变形成，是 $FeCr_2O_4$ 的高压多形矿物相。

王氏钛铁矿（Wangdaodeite, IMA2016-007），陨石冲击脉体中的高压矿物（Xie et al., 2020），其理想晶体化学式为 $FeTiO_3$，三方晶系，空间群 $R\bar{3}c$；粉晶衍射确定的晶胞参数：$a=0.513(1)$ nm，$c=1.378(1)$ nm，$V=0.314\ 06$ nm^3。王氏钛铁矿的晶体结构与人工合成物 $LiNbO_3$ 结构型的 $FeTiO_3$ 的相似，对比

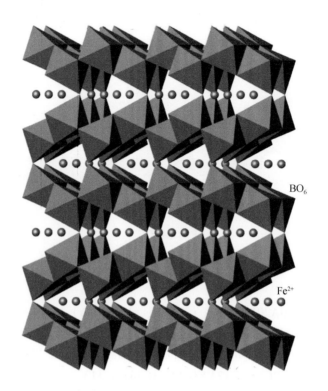

图 9 谢氏超晶石晶体结构（据陈鸣等，2008）

LiNbO$_3$ 结构型的 FeTiO$_3$ 和钛铁矿相的粉晶 X 射线衍射（X-ray diffraction，XRD）图谱和结构，LiNbO$_3$ 结构型的 FeTiO$_3$ 和钛铁矿的结构都由顶角连接的 TiO$_6$ 八面体构成，但与钛铁矿相比，LiNbO$_3$ 型 FeTiO$_3$ 的相邻八面体层是相互反转的（图10）。然而，钛铁矿相中的 TiO$_6$ 八面体共棱连接，这表明钛铁矿必须有键断开才能过渡到 LiNbO$_3$ 结构。虽然 LiNbO$_3$ 型 FeTiO$_3$ 和钛铁矿的结构非常相似，但 c/a 的轴率是区分 $R3c$ LiNbO$_3$ 型的 FeTiO$_3$（王氏钛铁矿）与 R-3 结构的钛铁矿的重要标准，一般情况下钛铁矿显示出较大的轴率（2.765），而王氏钛铁矿的轴率 c/a 为 2.694。X 射线衍射的特征谱线 [d：nm(I/I_0)]：0.374 57 (72)，0.271 5(100)，0.256 2(89)，0.223 1(57)，0.185 9(59)，0.161 9(41)，0.150 7(44)，0.147 9 (38)。

毛河光矿（Maohokite，IMA2017-047），尖晶石的同质多象变体（Chen et al.，2019），其理想晶体化学式为 MgFe$_2$O$_4$，斜方晶系，空间群 $Pnma$；同步辐射 X 射线衍射确定的晶胞参数：a = 0.890 7(1)nm，b = 0.993 7(8)nm，c = 0.298 1(1)nm，V = 0.263 8(3)nm^3。晶体结构属 CaFe$_2$O$_4$ 型超尖晶石结构，是镁铁矿的一种高压多形相。与 CaFe$_2$O$_4$(CF)、CaTi$_2$O$_4$(CT) 和 CaMn$_2$O$_4$(CM) 孔道结构有关的晶体结构被认为是超尖晶石相。粉晶衍射强线 [d：nm(I/I_0)(hkl)]：0.266 3(100)(230)，0.193 2(90)(321)，0.167 3 (20)(421)，0.153 3(50)(501)，0.143 1(12)(161)，0.115 5(15)(252)，0.113 1(12)(512)，0.108 7 (30)(823)。

毛河光矿是一种以三价铁为基本组分的镁铁氧化物高压矿物，天然毛河光矿的发现表明它是下地幔中潜在的重要矿物组成之一。其成因是岫岩陨石坑岩石中的镁铁碳酸盐矿物在陨石撞击产生的高温高压下发生了分解和重新组合而形成了毛河光矿（Chen et al.，2019）。

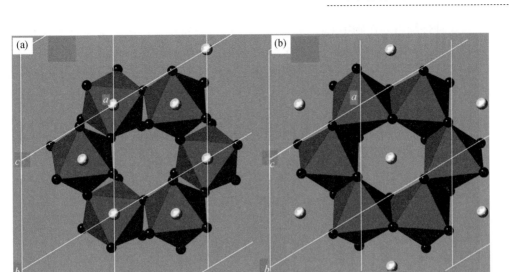

图 10　王氏钛铁矿结构图（据 Xie et al., 2020）
(a) $LiNbO_3$ 结构型的王氏钛铁矿结构；(b) 钛铁矿结构

3　若干新矿物的晶体结构、晶体化学

矿物的晶体结构是新矿物必不可缺的重要基础数据，也是新矿物晶体化学式书写的依据，21 世纪以来我国发现的新矿物有氧化物、硫化物、硅酸盐、砷酸盐、硫酸盐、硼酸盐、磷酸盐等，这些新矿物大多进行了晶体结构测定和晶体化学研究。

3.1　氧 化 物

烧绿石超族新矿物，其潜在的矿物种多达二十余种，然而目前已被国际矿物协会新矿物、命名和分类委员会（IMA CNMNC）批准认可的仅 16 种（李国武等，2014a），其中有关成分、结构、分类归属等科学问题尚未彻底解决，尚有多种潜在新矿物待确认。截至 2019 年，我国已发现 4 种在结构上具有烧绿石型结构而成分占位不同的新矿物种（表 1）：氟钙烧绿石（Fluorcalciopyrochlore，IMA2013-055）（Li et al., 2016）、氟钠烧绿石（Fluornatropyrochlore，IMA2013-056）（Yin et al., 2015）、羟钙烧绿石（Hydro-xy-calciopyrochlore，IMA2011-026）（Yang et al., 2014）和氧钠细晶石（Oxynatromicrolite，IMA2013-063）（Fan et al., 2017）。

表 1　我国发现的四种烧绿石超族新矿物的晶体学数据

名称	晶体化学式	晶系	空间群	晶胞参数/nm
氟钙烧绿石	$(Ca,Na)_2(Nb,Ti)_2O_6F$	等轴	$Fd\text{-}3m$	$a=1.04164(9)$
羟钙烧绿石	$(Ca,Na,U,\square)_2(Nb,Ti)_2O_6(OH)$	等轴	$Fd\text{-}3m$	$a=1.0381(4)$
氟钠烧绿石	$(Na,Pb,Ca,REE,U)_2Nb_2O_6F$	等轴	$Fd\text{-}3m$	$a=1.05053(10)$
氧钠细晶石	$(Na,Ca,U)_2(Ta,Nb)_2O_6(O,F)$	等轴	$Fd\text{-}3m$	$a=1.0420(6)$

烧绿石超族矿物是晶体化学通式为 $A_{2-m}B_2X_{6-w}Y_{1-n}$（$m=0\sim1.7$，$w=0\sim0.7$，$n=0\sim1.0$）成分占位非常复杂的氧化物，烧绿石超族矿物均具有烧绿石型结构（图 11），晶体结构特征为 B 位离子占据八面体中心，形成 $[BO_6]$ 八面体，八面体共棱连接，沿 [110] 方向连接呈链状，A 离子为 8 次配位分布于 $[BO_6]$ 八面体网格的空穴中 [图 11 (a)]，即占据由 O 和 OH 构成的立方体中心，形成 $[AO_8]$ 立方体，$[AO_8]$ 立方体与 $[BO_6]$ 八面体彼此共棱连接 [图 11 (b)]。通过晶体结构与晶体化学研究（李国武等，

2012),明确了这4种烧绿石新矿物中 Ca、Na、F、OH 的占位特征,即氟钙烧绿石中 A 位以 Ca 为主、Y 位以 F 为主,氟钠烧绿石中 A 位以 Na 为主、Y 位以 F 为主,羟钙烧绿石中 A 位以 Ca 为主、Y 位以 OH 为主,氧钠细晶石晶中 B 位 Ta 为主、A 位以 Na 为主、Y 位以 F 为主,晶体化学数据的获得为新矿物分类命名提供了充分的依据。

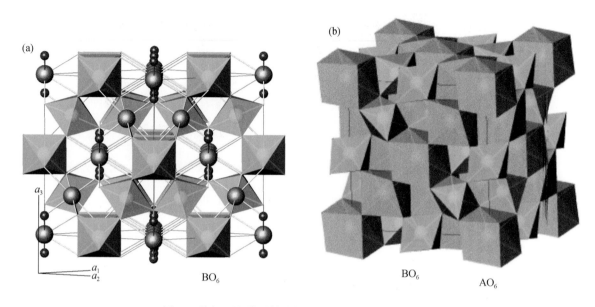

图 11　烧绿石超族晶体结构模型(根据结构数据绘制)
不同矿物种仅表现为成分不同,但结构相同

冕宁矿(Mianningite, IMA2014-072),稀土钛铁氧化物(Ge et al., 2017),其晶体化学式为($\square$, Pb, Ce, Na)(U^{4+}, Mn, U^{6+})Fe_2^{3+}(Ti, Fe^{3+})$_{18}O_{38}$,三方晶系,空间群 $R3$;晶胞参数:$a = 1.034\ 62(5)$ nm, $c = 2.083\ 7(2)$ nm, $V = 1.931\ 6(2)$ nm^3。

冕宁矿的晶体结构与其他尖钛铁矿族矿物相似,由九层紧密堆积的八面体[M(1) 和 M(3-5)]和四面体[M(2)]以及 12 配位的[M(0)]多面体组成(图12),小阳离子占据其他 19 个八面体和两个四面体的空位。Pb、Ce、La 和 Na 等大阳离子占据部分的 M0 位置,剩余 32.2% 的位置为空位。U、Mn 和 Y 位于八面体 M1 位置,M2 位置由 Fe^{3+} 四面体占据,M3—M5 位置由 Ti 和 Fe^{3+} 占据。

冕宁铀矿的晶体化学特点是其 M0 位置以空位为主,化学分析结果计算表明,在该位置上所有阳离子数为 0.678,其中 Ce、La、Nd 等三价阳离子总数为 0.217,Pb、Ba、Sr、Ca 等二价阳离子总数为 0.298,Na、K 等一价阳离子数为 0.163,剩余 0.322 为零价的空位,即 M0 位置空位数多于其他任何同价态离子数之和。根据矿物分类"价态优先"原则,即在某一结构位置,当存在不同价态的离子时,将同类价态的离子数之和统一考虑,单独作为一个组分参与矿物命名,冕宁铀矿 M0 位置零价的空位数多于其他任何同类价态离子数之和,故可以认为 M0 位置以空位为主(Ge et al., 2017)。

碲锌矿(Zincospiroffite, IMA2002-047)(Zhang et al., 2004),其理想晶体化学式为 $Zn_2Te_3O_8$,单斜晶系,空间群 $C2/c$;晶胞参数:$a = 1.272(4)$ nm, $b = 0.515(1)$ nm, $c = 1.182(3)$ nm, $\beta = 99.2(3)°$, $V = 0.764\ 57(252)$ nm^3, $Z = 4$。根据粉晶衍射特征,晶体结构与合成物 $Zn_2Te_3O_8$ 和碲锰矿等结构。粉晶衍射强线[d: nm(I)(hkl)]:0.475 8(w)(110),0.324 0(w)(31-1),0.292 8(m)(113),0.282 0(w)(20-4),0.215 5(w)(023, 511),0.198 5(w)(223),0.159 9(w)(42-5)。

锌尼日利亚石(Zinconigerite-2N1S, IMA2018-037)(Rao et al., 2018),其理想晶体化学式为 $ZnSn_2Al_{12}O_{22}(OH)_2$,三方晶系,空间群 $P3m1$;晶胞参数:$a = 0.571\ 4(1)$ nm, $c = 1.382\ 1(3)$ nm。与尼日利亚石等结构,2N1S 多型。粉晶衍射强线[d: nm(I/I_0)]:0.243 1(100),0.185 1(25),0.183 4

图12 冕宁矿的晶体结构（根据结构数据绘制）

（34），0.164 6(74)，0.154 5(81)，0.142 8(32)，0.141 7(27)。

铁塔菲石-2N'2S（Ferrotaaffeite-2N'2S，IMA2011-025）（Yang et al.，2012b），其理想晶体化学式为 BeFe$_3$Al$_8$O$_{16}$，六方晶系，空间群 $P6_3mc$；晶胞参数：a=0.569 78(8) nm，b=0.569 78(8) nm，c=1.837 3(4) nm，V=0.516 57(15) nm^3。铁塔菲石-2N'2S 晶体结构与一般塔菲石族矿物的结构类似，是基于具有3种阳离子构成的架层状结构（图13），其中 O 层是单位晶胞中由3个阳离子构成的八面体 M(1)、M(4) 层。T1'层包含一个八面体阳离子 M(2) 和一个 BeO$_4$ 四面体、一个 T(3) 四面体，T2 层为一个八面体 M(5) 和两个四面体 T(6)、T(7)。阳离子层按 O-T2-O-T1'-O-T2-O-T1'方式堆叠，其中 O-T2 构成尖晶石结构模块（S），而 O-T1'构成铁矾矿结构模块（N'）。铁塔菲石-2N'2S 晶体结构可以看作由两个尖晶石模块（S）和两个铁矾矿结构模块（N'）组合形成的 $SN'SN'$ 堆叠结构。尖晶石模块和铁矾矿模块的理想分子式分别为 T$_2$M$_4$O$_8$ 和 BeTM$_4$O$_8$，其理想的铁塔菲石-2N'2S 多型可以表示为 2×（BeT$_3$M$_8$O$_{16}$）。

铝锂矿（Chukochenite，IMA2018-132a）（Rao et al.，2020），理想晶体化学式为 LiAl$_5$O$_8$，斜方晶系，空间群 $Imma$；晶胞参数：a=0.564 2(1) nm，b=1.682 7(2) nm，c=0.801 4(1) nm，Z=6。晶体结构与人工合成 LiAl$_5$O$_8$ 同结构（图14）。粉晶衍射强线[d：nm(I/I_0)]：0.240 4(53)，0.153 8(77)，0.141 2(100)，0.125 6(52)，0.106 8(36)，0.104 0(61)，0.099 9(59)，0.094 2(35)。铝锂矿的发现，为深入了解锂的地球化学行为提供了重要信息，关键金属矿产的找矿也有重要指示意义。

3.2 硫 化 物

吴延之矿（Wuyanzhiite，IMA2017-081），辉铜矿的同质多象变体（Gu et al.，2017），其理想晶体化学式为 Cu$_2$S，四方晶系，空间群 $P4_32_12$；利用粉晶衍射精修的晶胞参数：a=0.400 08(1) nm，c=1.126 71(9) nm，晶体结构需要进一步研究。粉晶衍射强线[d：nm(I/I_0)]：0.283 3(24)，0.274 6(100)，0.230 4(97)，0.226 2(28)，0.199 8(62)，0.196 7(29)，0.188 7(26)，0.170 4(28)。

陈国达矿（Chenguodaite，IMA2004-042a）（Gu et al.，2008），其理想晶体化学式为 Ag$_9$FeTe$_2$S$_4$，斜方晶系；利用粉晶衍射精修的晶胞参数：a=1.276 9(2) nm，b=1.481 4(2) nm，c=1.623 3(1) nm，V=

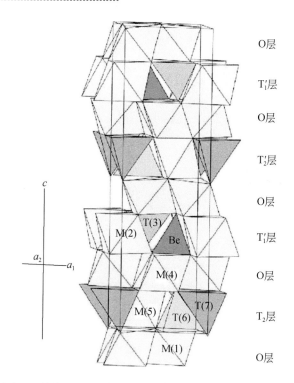

图 13　铁塔菲石-2$N'2S$ 晶体结构（据 Yang et al., 2012b）

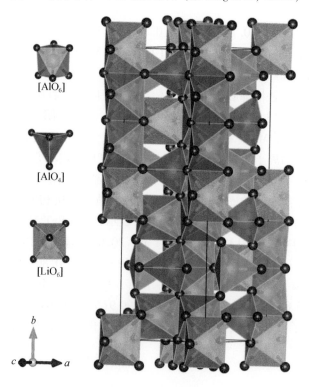

图 14　铝锂矿石晶体结构（据 Rao et al., 2020）

3.070 6 nm^3，$Z=9$。陈国达矿晶体结构尚未最终确定，但从已有的数据看不同于其他斜方结构，也不同于 $Ag_9FeTe_2S_4$ 合成相的立方结构，与 Te-Canfieldite $Ag_8SnTe_2S_4$ 相比，其成分似乎存在 $Fe^{3+}+Ag^+\rightarrow Sn^{4+}$ 和 $2Te^{2-}\rightarrow 2S^{2-}$ 的替换关系，且光学性质也非常接近。但其衍射线条有显著的差异。因此，陈国达矿与现有矿物的晶体化学的关系还有待更深入的研究（谷湘平等，2008）。利用甘多菲法粉晶衍射强线 [d：nm

(I/I_0)]: 0.674 2(69), 0.641 6(39), 0.595 1(33), 0.326 5(100), 0.298 1(24), 0.264 9(22), 0.225(24), 0.218 8(71), 0.214 2(22), 0.212 3(31), 0.204 4(23), 0.194 9(33)。

灵宝矿（Lingbaoite, IMA2018-138），金碲化物矿物（Jian et al., 2020），其理想晶体化学式为 $AgTe_3$，三方晶系，空间群 $R3m$；晶胞参数：$a=0.860(5)$ nm，$c=0.540(18)$ nm，$V=0.346(9)$ nm^3，$Z=3$。Jian 等（2020）利用电子背散射衍射（electron back scattered diffraction，EBSD）和选区电子衍射（selected area electron diffraction，SAED）研究表明，灵宝矿的晶体结构与人工合成 $AgTe_3$ 的结构相同。根据合成 AgTe 的结构特征，灵宝矿的结构可看作是 α-钋结构的有序（1∶3Ag∶Te）类似物（即简单的立方晶体结构）（图15）。银呈八面体配位，与 Te[Ag-Te：3×302.2(5)，3×308.3(5) pm]构成，每个 Te 依次被 4 个其他 Te 原子的正方形排列包围并与两个 Ag 原子一起构成 Te(Te_4Ag_2) 八面体。

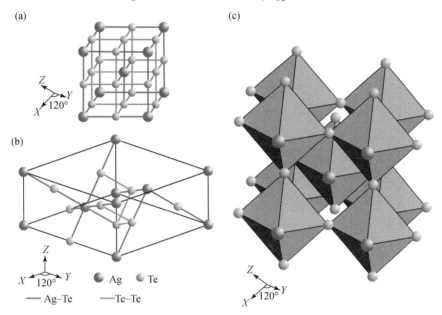

图15 灵宝矿晶体结构（据 Jian et al., 2020）

李四光石（Lisiguangite, IMA2007-003）（Yu et al., 2009），其理想晶体化学式为 $CuPtBiS_3$，斜方晶系，空间群 $P2_12_12_1$；晶胞参数：$a=0.773\ 72(15)$ nm，$b=1.284\ 4(3)$ nm，$c=0.490\ 62(10)$ nm，$V=0.487\ 57(17)$ nm^3，$Z=4$。其晶体结构与 $CuNiSbS_3$ 和 $CuNiBiS_3$ 相同，Pt^{2+} 和 Bi^{3+} 与 4 个硫共同构成一扭曲的八面体结构，4 个硫和铜构成扭曲的四面体，形成一种复杂的三维格架结构（图16）。

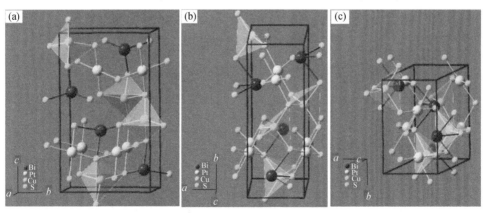

图16 李四光石晶体结构（据 Yu et al., 2009）
(a)[100]方向；(b)[010]方向；(c)[001]方向。原子颜色：Cu 绿色，Pt 白，Bi 红，S 黄色

3.3 硅 酸 盐

栾锂云母（Luanshiweiite，IMA2011-102）（范光等，2013）和氟栾锂云母（Fluorluanshiweiite，IMA2019-053）（Qu et al.，2020a）。两种新矿物是锂云母中阴离子分别为羟的端员和氟的端员，其晶体学参数见表2。

表2 栾锂云母和氟栾锂云母晶体学数据

名称	栾锂云母-2M_1	氟栾锂云母-1M
晶体化学式	$KLiAl_{1.5}(Si_{3.5}Al_{0.5})O_{10}(OH)_2$	$KLiAl_{1.5}\square_{0.5}(Si_{3.5}Al_{0.5})O_{10}F_2$
晶系	单斜晶系	单斜晶系
空间群	$C2/c$	$C2/m$
晶胞参数	$a=0.51861(7)$ nm $b=0.89817(13)$ nm $c=1.9970(3)$ nm $\beta=95.420(3)°$	$a=0.52030(5)$ nm $b=0.89894(6)$ nm $c=1.01253(9)$ nm $\beta=100.68(1)°$

两种新矿物的晶体结构与其他云母类似，TOT 型三八面体层状结构（图17），由两层四面体夹一层八面体构成，层间充填 K^+ 离子，其中栾锂云母为 $2M_1$ 多型，晶胞参数 c 是氟栾锂云母（1M 多型）的二倍。结构中 Li 在八面体中占据一个位置（等效点系 $2a$），Al 占据两个八面体位置（等效点系 $4g$），但由于电价平衡 Al 占位不足出现空位，附加阴离子位于八面体中靠硅氧骨干中六方网一侧，以 OH^- 为主时定名为栾锂云母，以 F^- 为主时为氟栾锂云母。

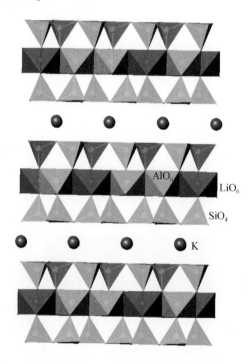

图17 栾锂云母和氟栾锂云母晶体结构模型（根据结构数据绘制）

两种新矿物结构模型相同，附加阴离子前者为 OH^-，后者为 F^-

铁海泡石（Ferrisepiolite，IMA2010-061），其理想晶体化学式为 $(Fe^{3+},Fe^{2+},Mg)_4[(Si,Fe^{3+})_6O_{15}](O,OH)_2·6H_2O$。利用粉末 X 射线衍射和透射电子显微镜进行的结构研究显示（Gu et al.，2013），该矿物与

海泡石等结构相似,斜方晶系,空间群 $Pncn$;晶胞参数:$a=1.361\ 9(8)$ nm,$b=2.695\ 9(26)$ nm,$c=0.524\ 1(7)$ nm,$V=1.924\ 08$ nm^3,$Z=4$。它是以铁(Ⅲ)为主的海泡石类似物,海泡石结构中 Fe^{3+} 和(或) Fe^{2+} 在八面体中心取代 Mg^{2+},为了保持电价平衡四面体中有少量 Fe^{3+} 代替 Si^{4+},四面体中少量 OH^- 代替 O^{2-},c 轴长有所缩小,而 a 轴有所增加。

凤成石(Fengchengite,IMA2007-018a),其理想晶体化学式为 $Na_{12}\square_3(Ca,Sr)_6Fe_3^{3+}Zr_3Si(Si_{25}O_{73})(H_2O,OH)_3(OH,Cl)$,三方晶系,空间群 $R3m$;晶胞参数:$a=1.424\ 67(5)$ nm,$c=3.003\ 3(2)$ nm,$V=5.279\ 1(5)$ nm^3;是一种异性石族矿物中 N(5)位贫 Na 的空位为主的类似物新种(沈敢富等,2017)。凤成石的晶体结构与异性石族矿物基本类似,均为环状结构,其硅氧四面体共顶角连接成三元环和九元环的独特组合环状结构(图18),[M(1)O$_6$]八面体中 M(1)以 Ca 为主组成六元环层。[M(1)O$_6$]八面体被夹在中心对称的三元和九元硅环层之间的{M(2)O$_n$}多面体层[M(2)主要为 Fe]连接形成2∶1复合层。2∶1复合层再由八面体配位的 Zr 和另一菱面体配位的 Zr 连接。对大多数异性石族矿物而言,这种开放式结构可充填五个不同的[Na$_n$]多面体。但凤成石的[Na$_n$]多面体只有3个:N(1)、N(4)和 N(5),而且 N(1)位置分裂成了 N(1a)和 N(1b)两个次级位置。一般情况下对具有中心对称、空间群为 $R\bar{3}m$ 和 c≈3.0 nm

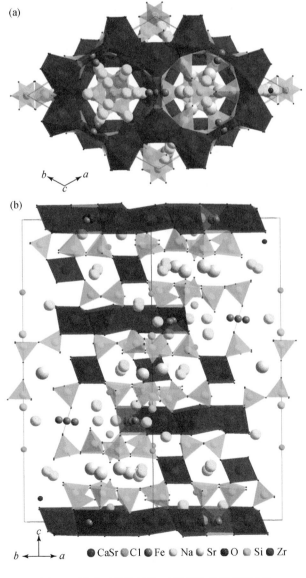

图18 凤成石晶体结构(根据结构数据绘制)

的同族成员有 3 种不同的 [Na_n] 多面体：N(1)、N(4) 和 N(5)，而且 N(1) 和 N(4) 位置各自被 6 个 Na 离子所占，而 N(5) 位置则占有 3 个 Na 离子。但凤成石的晶体化学占位与之不同，其中 N(5) 位没有被 3 个 Na 占据而是以空位为主。此外，M(1) 位也主要为 Ca 和 Sr；M(2) 位主要为 Fe^{3+}、Mn^{3+} 和 Cr^{3+} 等；M(3) 和 M(4) 为 Si；Z 位以 Zr 为主，有极少量的 Ti、Nb 置换；O10 呈无序并分裂成两个次级位置，主要被 H_2O 所占；X 位由 OH 和 Cl 共占，但 OH>Cl。

对于异性石族新矿物凤成石，其晶体化学式的确定完全依赖于该晶体结构的测定，因其在原子占位上与其他异性石族矿物不同而具有新颖性，通过晶体结构研究确定为新矿物（沈敢富等，2017）。

红河石（Hongheite，IMA2017-027），是符山石族矿物中在 X(4) 位以空位为主、Y(3) 位以 Fe^{3+} 居优和 T(2) 位为 B 的新矿物种（徐金沙等，2019）。其理想晶体化学式为 $Ca_{18}(\square,Ca)_2Fe^{2+}Al_4(Fe^{3+},Mg,Al)_8(\square,B)_4BSi_{18}O_{69}(O,OH)_9$，四方晶系，空间群 $P4/nnc$；晶胞参数：$a=1.566\,7(1)$ nm，$c=1.172\,5(2)$ nm，$Z=2$。晶体结构与符山石族矿物相似（图 19），其中 X(1-3) 和 X(4) 是 7、8 和 9 次配位的大阳离子 Ca，Y(2) 和 Y(3) 为八面体配位，为 Al、Mg、Fe^{3+} 和 Fe^{2+} 离子，Y(1) 呈四方柱状配位，主要为 Fe^{2+}、Mg、Al、Cu 和 Mn^{3+} 占据，T 有 4 次 Al、Fe 和 3 次配位 B。Z 为 SiO_4 四面体。通常 X(4) 和 Y(1) 被视为半数阳离子充填。而事实上，X(1-4) 阳离子数从 18.20 到 19.60 不等，X(1-4) 阳离子和 (Ca,Ln,Pb,Bi,Th) 的平均值为 18.9，与理想值 19 相同，即并不是所有的 X(4) 符山石族矿物都是理想的半占位。虽然红河石的晶体结构与同族矿物的晶体结构类似，但根据结构精测和化学成分计算的 Ca=18.07 apfu 显示在 X(4) 位应该是以空位（$\square$）为主，这也是与其他同族异性石矿物种的实质性区别（徐金沙等，2019）。

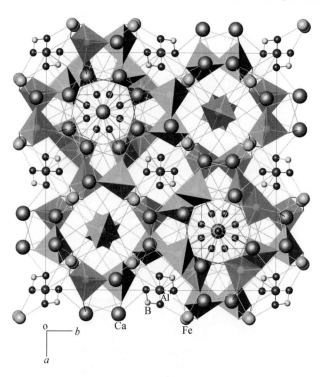

图 19　红河石晶体结构（根据结构数据绘制）

钾绿钙闪石（Potassic-hastingsite，IMA2018-160）（Ren et al.，2020），其理想晶体化学式为 $KCa_2(Fe_4^{2+}Fe^{3+})(Si_6Al_2)O_{22}(OH)_2$，单斜晶系，空间群 $C2/m$；晶胞参数：$a=0.994\,05(7)$ nm，$b=1.825\,6(2)$ nm，$c=0.535\,01(3)$ nm，$\beta=105.117(5)°$。

钾绿钙闪石作为一种具有双链结构的硅酸盐矿物，其结构与角闪石超族相似（图 20），其硅氧四面体共角顶连接成双链，[Si_4O_{11}] 双链平行于 c 轴，双链结构在 a-b 轴方向上活性氧与活性氧相对处形成 3 种较小的八面体空穴（M1，M2，M3），主要由 Fe^{2+}、Fe^{3+} 充填，并共棱连接成沿 c 轴延伸的链带。惰性氧与惰性氧相对位置形成的空隙主要由大阳离子 Ca 充填，为八次配位的畸变立方体。而在双链间较大的连续空隙，由大阳离子 K、Na 占据，且 K>(Na, □)。四面体结构中少量 Al 代替 Si，且 Si、Al 无序。

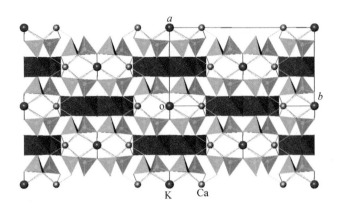

图 20　钾绿钙闪石晶体结构（根据结构数据绘制）

丁道衡矿 [Dingdaohengite-(Ce), IMA2005-014]（Xu et al., 2008），其理想晶体化学式为 $Ce_4Fe_2Ti_3Si_4O_{22}$，单斜晶系，空间群 $P2_1/a$；晶胞参数：$a = 1.34656(15)$ nm，$b = 0.57356(6)$ nm，$c = 1.10977(12)$ nm，$\beta = 100.636(2)°$，$V = 0.84239(46)$ nm^3。其晶体结构与硅钛铈矿族矿物类似（李国武等，2005），由两个单元层构成 [图 21（a）]，一层是 SiO_4 双四面体和在 B 位的 FeO_6 八面体共同组成的网层，平行（001）面排列。另一层是由 FeO_6 共顶角组成的层，结构层面平行于 a-b 平面，四面体层和八面体层交替沿 c 轴形成三维的网层，稀土大阳离子位于两个网层的空穴中。单晶衍射实验中观察到了规律的弱衍射点（李国武等，2005），发现产于白云鄂博的丁道衡矿具有赝对称空间群的超结构现象，为 $C2/m$ 赝对称的 $P2_1/a$ 空间群结构。而对四川攀枝花铁矿的硅钛铈矿的结构精测结果显示，衍射中无弱衍射点，其空间群为 $C2/m$。可见，在自然界两种结构都存在，不论是 $C2/m$ 还是 $P2_1/a$ 都可以是硅钛铈矿亚族矿物的真空间群。

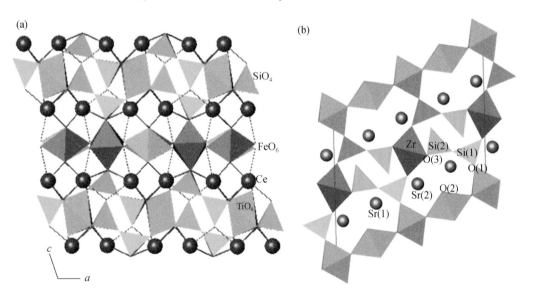

图 21　丁道衡矿晶体结构（a）和何作霖矿晶体结构（b）
（a）根据结构数据绘制；（b）据 Yang et al., 2012

何作霖矿（Hezuolinite, IMA2010-045）（Yang et al., 2012a），其理想晶体化学式为 $(Sr, R_{EE})_4Zr(Ti,$

Fe^{3+},Fe^{2+})$_2$Ti$_2$O$_8$(Si$_2$O$_7$)$_2$，单斜晶系，空间群 $C2/m$；晶胞参数：$a=1.3973(3)$ nm，$b=0.56984(11)$ nm，$c=1.1988(2)$ nm，$\beta=114.10(1)°$，$V=871.3(3)$ nm^3，$Z=2$。其晶体结构与珀硅钛铈矿相似（Yang et al.，2012a），由两个单元层构成[图21(b)]，两层交替沿 c 轴形成三维的网层，Sr、R$_{EE}$ 等大阳离子位于两个网层的空穴中。与 $P2_1/a$ 结构的 rengeite 结构比较，$C2/m$ 结构的何作霖矿结构中的 O(1)、O(2) 和 O(3) 位分别分离为 O(1) 和 O(1')、O(2) 和 O(2')、O(3) 和 O(3')。由于硅氧键角较大 [Si(1)—O(7)—Si(2)=175.8°]，硅氧四面体间连接较平直。何作霖矿是珀硅钛铈矿亚族的新成员，可被认为是硅钛铈矿–珀硅钛铈矿之间的化学过渡相。

太平石 [Taipingite-(Ce)，IMA2018-123a]（Qu et al.，2020b），其理想晶体化学式为 (Ce,La,Nd,Ca)$_9$(Mg,Fe)(SiO$_4$)$_3$[SiO$_3$(OH)]$_4$(F,OH)$_3$，三方晶系，空间群 $R3c$；晶胞参数：$a=1.07246(3)$ nm，$c=3.79528(14)$ nm。太平石的晶体结构与硅铈石型结构相似（图22），硅氧四面体 [SiO$_4$] 呈孤岛状，3 种不同的稀土阳离子与 O^{2-} 和 F$^-$、OH$^-$ 阴离子构成 8 次和 9 次配位。Mg^{2+} 离子与周围的 6 个 O^{2-} 构成略有扭曲的独立的八面体，八面体共顶角与硅氧四面体 [SiO$_4$] 连接。[SiO$_3$(OH)] 四面体在一个空间较大的特殊位置与稀土阳离子的 3 个多面体共顶连接，太平石是硅铈石族矿物中附加阴离子富 F 的类似物。其粉末 X 射线衍射数据中最强的八条线 [d：nm(I/I_0)(hkl)]：0.4518(50)(202)，0.3455(95)(122)，0.3297(85)(214)，0.3098(35)(300)，0.2941(100)(0210)，0.2683(65)(220)，0.1945(40)(238)，0.1754(40)(3018)。

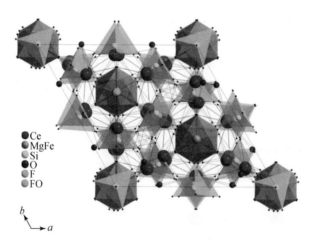

图 22 太平石晶体结构（根据结构数据绘制）

3.4 砷酸盐、硫酸盐、硼酸盐

富铜泡石（Tangdanite，IMA2011-096）（Ma et al.，2014），其理想晶体化学式为 CaCu$_9$(AsO$_4$)$_4$(SO$_4$)$_{0.5}$(OH)$_9$·9H$_2$O，单斜晶系，空间群 $C2/c$；晶胞参数：$a=5.4490(9)$ nm，$b=0.55685(9)$ nm，$c=1.04690(17)$ nm，$\beta=96.294(3)°$，$V=3.15744(90)$ nm^3，$Z=4$。富铜泡石的晶体结构与铜泡石的 2M 多型结构相似，是由 Cu、As 和 Ca 配位多面体构成的复杂板状结构（图23），可以看作铜泡石层状结构，其层的核心为铜砷酸盐亚结构构成 A 和 B 两个亚层，B 亚层为八面体 [CuO$_3$(OH)$_3$] 共棱形成的链沿 [010] 延伸，A 亚层 AsO$_4$ 四面体共顶角与 [CuO$_5$(H$_2$O)] 八面体连接，层的外侧由 Ca 八面体填充，H$_2$O 分子和 SO$_4$ 基团位于层间孔道。结构精修中 O18、O19 和 O20 各向异性参数较大，表明 O18、O19 和 O20 位置有无序现象，该位置也可能包括水分子。由于电性中和，S 分子数为 0.5。

富铜泡石的晶体结构与 2M 铜泡石不同的是成分及层间结构（李国武等，2014b），一般铜泡石的 2M 多型结构层间成分为 CO$_3^{2-}$，结构为 C—O 三角形，而富铜泡石中结构层间成分为 SO$_4^{2-}$，结构为 S—O 四面

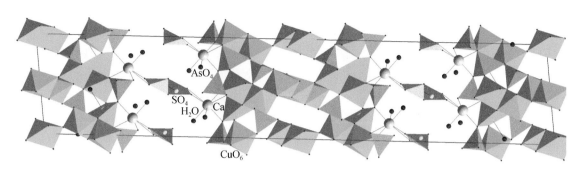

图 23　富铜泡石的晶体结构（根据结构数据绘制）
结构层间为 SO_4 四面体（黄色），但未充满

体，结构精测解决了富铜泡石的晶体结构，并发现了一种新的层间结构形式。X 射线粉晶衍射主要五条强线为 [d：nm(I/I_0)(hkl)]：0.478 2(100)(-311)，0.433 3(71)(602)，0.526 3(54)(-202)，0.394 9(47)(802)，0.297 6(46)(-15 11)。

铊明矾（Lanmuchangite，IMA2001-018）(Chen D. Y. et al., 2003)，其理想晶体化学式为 TlAl(SO_4)$_2$·12H_2O，等轴晶系，空间群 $Pa3$；晶胞参数：a=1.221 2(5) nm，V=1.821(2) nm^3，Z=4。成分及粉晶的 X 射线研究显示，铊明矾与明矾族矿物钾明矾（KAl[SO_4]$_2$·12H_2O）、钠明矾（NaAl[SO_4]$_2$·12H_2O）和铵明矾（NH_4Al[SO_4]$_2$·12H_2O）在成分上接近，仅表现为 K^+、Na^+、NH_4^+ 的位置由 Tl^+ 所代替，均为等轴晶系，空间群 $Pa3$，晶胞参数亦较接近。粉晶衍射数据与钾明矾大部分均相同，并与人工合成物 TlAl[SO_4]$_2$·12H_2O 相同，可以认为二者在结构上类似，与钾明矾等结构（陈代演等，2001）。

锌赤铁矾（Zincobotryogen，IMA2015-107）(Yang et al., 2017)，其理想晶体化学式为 ZnFe^{3+}(SO_4)$_2$(OH)·7H_2O，单斜晶系，空间群 $P2_1/n$；晶胞参数：a=1.050 4(2) nm，b=1.780 1(4) nm，c=0.712 63(14) nm，β=100.08(3)°，V=1.311 9(5) nm^3，Z=4。其晶体结构（图24）中两种八面体 [Fe(1)O_4(OH)$_2$] 和 [Fe(2)O_2(OH)$_2$(H_2O)$_2$] 共顶角连接成平行于 c 轴的链。共用顶角氧由 OH^- 离子提供，各链之间相互分离，彼此以氢键维系。两个独立的硫酸根构成四面体 S(1)O_4 和 S(2)O_4，通过侧面交替共角连接成链，并与八面体内链连接构成 [Fe^{3+}(SO_4)$_2$(OH)(H_2O)]$_2$ 结构单元。锌八面体由 [S(1)O_4] 四面体上的一个氧和 5 个水分子构成 [MO(H_2O)$_5$] 八面体，通过共享 H_2O 分子形成一条更大的链，形成一个更大的 [M^{2+}Fe^{3+}(SO_4)$_2$(OH)(H_2O)$_6$$H_2O$] 的模块，模块间由氢键连接。

闽江石（Minjiangite，IMA2013-021）(Rao et al., 2015)，其理想晶体化学式为 BaBe$_2$(PO$_4$)$_2$，六方晶系，空间群 $P6/mmm$；粉晶衍射精修的晶胞参数：a=0.503 0(8) nm，c=0.746 7(2) nm，V=0.163 96(3) nm^3。粉晶 X 射线衍射图谱与人工合成物 BaBe$_2$(PO$_4$)$_2$ 完全吻合；晶体结构与 dmisteinbergite、CaAl$_2$Si$_2$O$_8$ 类似，结构基于 Be 和 P 的四面体双层构成的六元环，平行于 c 轴堆叠形成孔道结构 [图25(a)]。在 a-b 平面上，每个环连接 6 个相邻的环形成无限的层状。沿 c 方向，四面体通过端氧共角顶连接成双层，Ba 原子位于两个双层之间的 12 配位多面体中 [图25(b)]，这种 Ba 多面体呈非常规则的六边形，12 个相同的 Be—O(2) 键为 0.297 5(2) nm。占位度精修表明，四面体位置同时被 Be 和 P 占据，占位率 0.5P+0.5Be，并且 Be、P 呈无序分布。粉晶 XRD 最强的八条线 [d：nm(I/I_0)(hkl)]：0.376 3(100)(101)，0.283 6(81)(102)，0.251 5(32)(110)，0.217 8(25)(200)，0.216 20(19)(103)，0.190 0(64)(201)，0.177 0(16)(113)，0.150 7(25)(212)。

磷锶铍石（Strontiohurlbutite，IMA2012-032）(Rao et al., 2014)，其理想晶体化学式为 SrBe$_2$(PO$_4$)$_2$，单斜晶系，空间群 $P2_1/c$；晶胞参数：a=0.799 7(3) nm，b=0.897 9(2) nm，c=0.842 0(7) nm，β=90.18(6)°，V=0.604 7(1) nm^3，Z=4。其晶体结构与磷钙铍石、CaBe$_2$(PO$_4$)$_2$、Paracelsian、BaAl$_2$Si$_2$O$_8$ 等结构。晶体结构基于由 BeO$_4$ 和 PO$_4$ 四面体共角顶组成骨架。BeO$_4$ 和 PO$_4$ 四面体分别构成四元环和八元环

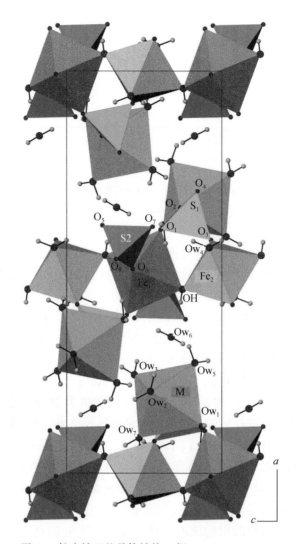

图 24 锌赤铁矾的晶体结构（据 Yang et al., 2017）

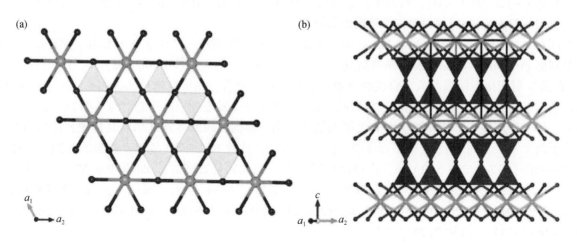

图 25 闽江石晶体结构（据 Rao et al., 2015）

（图 26），四元环由一对指向向上（U）的四面体和一对向下指向（D）的四面体组成，UUDD 型双环结构，而八元环显示为 DDUDUUDU 型结构。Sr 原子位于由八元环排列形成的通道中。在垂直于 b 轴方向上，BeO_4 和 PO_4 四面体通过共角顶连接的双轴线的链沿 a 轴延伸。Sr2 离子位于 10 配位多面体中，该 10 配位由 7 个短键 [<Sr—O>＝0.259 6(2)nm] 和三个长键 [<Sr—O>＝0.322 7(2)nm] 构成。该多面体也可以描述为一个方形棱锥和一个三方柱共面连接的组合。X 射线粉晶衍射八条强线 [d：nm (I/I_0)(hkl)]：0.355 4(100)(121)，0.335 5(51)(211)，0.307 3(38)(022)，0.254 2(67)(113)，0.223 0(42)(213)，0.221 5(87)(321)，0.204 6(54)(223)，0.171 4(32)(143)。

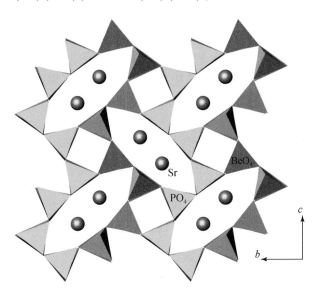

图 26　磷锶铍石晶体结构（据 Rao et al.，2014）

4　结　语

除新矿物结构外，我国学者还对硅钙锡石（Xue et al.，2017）、钡铁钛石（Li et al.，2011）、冀承矿（Yu et al.，2011）、叶绿矾（李国武等，2010）、南岭石（Yang et al.，2011）、氟铁云母（阮青锋等，2009）、草莓红绿柱石（廖尚宜等，2008）、闪叶石（李国武等，2007b）、CsNa 绿柱石（Liu et al.，2007）、镁锰磷矿（Ma et al.，2005）、铜柏矿（代明泉等，2004）、萨碳硼镁钙石（杨光明等，2004）、镁粒磷锰矿（马喆生等，2004）、富铁硅钛铈矿（杨主明等，2004）、尼日利亚石（Chen J. Z. et al.，2003）、单斜星叶石（Ma et al.，2001）、塔菲石（陈敬中等，2001）等矿物进行了晶体结构与晶体化学研究，发表了一批新数据，并发现了部分矿物中存在的占位有序-无序、超结构、调制结构以及若干矿物晶体化学新现象。这些结构精修成果对存疑矿物中的晶体结构与晶体化学有了更新的认识。此外，21 世纪相关晶体学专著《现代晶体化学》（陈敬中，2010）、《结构矿物学》（秦善，2011）的出版，对矿物晶体学中的最新成果进行了总结，为学科发展提供了更多的参考资料。

从基础矿物晶体结构与晶体化学延伸出的具体应用分支较多，虽然本文尚未涉及矿物材料晶体学、黏土矿物晶体学、高压矿物晶体学、矿物结构与性能等专门领域的矿物晶体结构与晶体化学，但是在各自领域内的矿物晶体化学研究都取得了重要的成果。就本文所关注的我国在 21 世纪基础矿物学中与新矿物相关的晶体结构与晶体化学方向取得的诸多成果，我们也可获得许多启迪。

4.1　我国新矿物晶体结构与晶体化学的特点与研究内容

（1）全新结构类型：发现了多种此前世界上从未发现过的矿物新结构，主要包括汉江石型结构、乌木石钨青铜型结构、碲钨矿钨青铜衍生新结构、李璞硅锰石新结构和孟宪明矿新结构。这些全新结构的发现是我国学者对基础矿物晶体学做出的重要新贡献。

（2）类质同象端元：新矿物的晶体结构为同族矿物的类质同象新端元，主要有雅鲁矿、氟钙烧绿石、羟钙烧绿石、氟钠烧绿石、氧钠细晶石、锌尼日利亚石、李四光石、栾锂云母、氟栾锂云母、铁海泡石、太平石、富铜泡石、铊明矾、磷锶铍石、锌赤铁矾、铁塔菲石-$2N'2S$。

（3）同质多象变体：新矿物为已知相同成分的不同结构的变体，主要有罗布莎矿、林芝矿、谢氏超晶石、王氏钛铁矿、毛河光矿。

（4）元素结构占位：由同族矿物结构中元素在某些晶体化学位置上占位优势表现出来的新矿物，主要有冕宁矿、凤成石、红河石、钾绿钙闪石、丁道衡矿、何作霖矿。

（5）人工合成物等结构：此前已经存在同结构和成分的人工合成物，首次在自然界发现的天然矿物，主要有那曲矿、藏布矿、曲松矿、自然钛、巴登珠矿、志琴矿、经绥矿、青松矿、灵宝矿、碲锌矿、闽江石、铝锂矿。

（6）尚未确定的疑难结构：少数新矿物由于样品原因，晶体结构尚未最终确定，主要有吴延之矿和陈国达矿。

（7）部分新矿物结构具有有序–无序、超结构或调制结构，如乌木石、碲钨矿、丁道衡矿。

矿物晶体结构与晶体化学是传统矿物学的基础研究领域之一，系统化、系列化研究矿物的精细结构与晶体化学仍是目前矿物晶体学研究的主要内容，这包括了研究矿物的新结构、类质同象系列矿物、发现类质同象新端元、新的同质多象变体、有序一无序、调制结构和超结构等。

4.2 基础矿物晶体学发展动力与研究方向

21世纪以来，得益于国家对地矿工作的重视和对矿物基础研究的支持，新矿物的发现及矿物晶体结构晶体化学研究亦得到了发展，同时矿物晶体结构与晶体化学研究的发展也推动了新矿物的发现，目前发现的新矿物越来越趋向于矿物包裹体、微小稀有矿物等，传统的镜下鉴定已经很难在新矿物工作中发挥主导作用，所以新矿物的发现与研究更多依赖于矿物晶体学研究。近年来，随着各种衍射方法技术（X射线衍射、同步辐射、中子衍射、电子衍射、微区衍射）的进步，特别是具有平面探测器的单晶衍射系统的普及和应用，结构分析软件功能的增强，大大提高了矿物结构分析的精度和速度。此类仪器不仅可以对单晶进行结构测定，还可以对单体颗粒细小至30 μm左右的矿物晶体获取完整的衍射数据，因此，细小包体类新矿物的发现也逐渐增加。各种谱学研究方法（红外光谱、拉曼光谱、穆斯堡尔谱、顺磁共振、核磁共振等）和高分辨率电子显微镜的应用，也极大地促进了矿物晶体化学的研究。此外，有更多学者参与到新矿物的发现工作以及材料科学发展的需求亦是促进矿物晶体学发展的动力。近年来，基于大科学装置（如同步辐射、散裂中子源等）的使用和新探测技术的应用在生物大分子领域取得了众多结构新发现和新成果，但在矿物学领域的应用极少，这方面尚有加强的空间，特别是应用于微小矿物结构研究方面，大科学装置更有优势。未来我国基础矿物晶体学的主要研究方向依然是以下几方面：

（1）新矿物晶体结构测定及晶体化学研究。新矿物种的确定需要化学成分与晶体结构两个方面的数据，因此要确定新矿物，首先要获得基本的晶体结构参数，进而研究每种矿物的晶体化学特征，对类质同象代换元素在各种结构位置中的分配进行定量精修，对于类质同象系列矿物，不再满足于仅弄清单个矿物结构，而是对全系列矿物进行系统研究，在高精度测量的基础上总结了一些新的晶体化学新现象，这不仅可进一步厘定其是否为新矿物，而且为探讨新矿物的结构、性能、成因间的关系找到内在因素。

（2）新结构类型发现与研究。新结构通常蕴藏于新矿物之中，因此寻找成分特殊、成因产状独特的新矿物是发现新结构的基石。结合现代仪器技术和结构解析方法的应用，随着对新矿物研究的广泛和深入，发现新结构依然是可能的。

（3）原有疑难矿物结构精测和再研究，深化矿物学基础研究。对已有结构分析但结构及晶体化学存疑的一些矿物利用现代测试方法重新进行结构精测和研究，这不仅是对疑难结构资料进行修正和补充，完善矿物晶体学数据以及重新认识某些矿物晶体结构与晶体化学的重要方式，而且正确而精细的晶体结构和晶体化学知识也是完善矿物晶体化学分类以及深入解释矿物成因和性质的基础，这方面也是非常有意义的矿物晶体学基础性工作。

（4）矿物中的调制结构与超结构研究。天然矿物中的有序-无序、公度（超结构）与非公度调制结构现象十分普遍。利用平面探测器X射线衍射、电子衍射、同步辐射及其他手段，研究结构中元素替换及原子分布造成的有序-无序及调制现象是现代矿物晶体学研究的内容之一。

（5）高温高压及地幔矿物晶体学研究。除进行天然地幔矿物晶体学研究外，实验室合成模拟高温高压矿物晶体学方面也有很大空间，特别是高压下的结构分析及计算机模拟结构演化等研究领域需加强。高温高压矿物结构研究可为解决地幔中矿物结构的转变提供线索，在地质学上有重要意义。

（6）进一步拓宽矿物晶体结构与晶体化学研究新领域，矿物晶体学研究不仅体现在基础研究方面，而且要在实际矿物学研究中体现其应用价值，这有较大的空间，结构与性能在矿物材料学上的研究已非常广泛，但在基础矿物学研究中还少有涉及，矿物的晶体结构决定了矿物的性质，测定矿物晶体结构的重要内容之一是加深对矿物晶体微观结构的认识，揭示其特殊物理、化学性质的内在根源。

X射线晶体学在成因矿物学方面的应用也有较大的潜力，通过X射线研究方法提供的结构标型特征反映了矿物的成因状态，可揭示其内在规律性。矿物及矿床成因的探讨通过大量矿物晶体结构资料的积累和总结，可以找出矿物结构状态的差别。类质同象系列阴离子的置换度以及晶体结构的变异等可为探讨矿物及矿床的成因以及构造环境提供重要的信息，为深入研究矿物在一定温度、压力下的演化历史有重要的作用。

4.3 机遇与挑战

本文仅概括了21世纪我国矿物晶体学发展的一个侧面，如矿物材料晶体学、黏土矿物晶体学、高压矿物晶体学等都未涉及，在一些疑难矿物结构再研究等方面也未能全部得到反映。但是窥一斑而见全豹，20年来，我国在矿物晶体结构与晶体化学基础研究方面取得了较大的进展，不仅为新矿物发现提供了新数据，而且进行了大量的晶体学系统研究工作，发现了若干新结构和新晶体化学新现象，取得了一系列令人瞩目的研究成果。21世纪有四项与新矿物和矿物晶体结构晶体化学有关的成果被选入中国地质科技十大进展：谢氏超尖晶石超高压矿物的发现（2008年）、汉江石——一种新结构类型的新矿物（2010年）、我国烧绿石超族新矿物（2013年）、7种自然界新矿物获国际认证（2019年），我国发现的碲钨矿被国际矿物学会评选为"2019年度矿物"，这些成果表明我国在新矿物及矿物晶体结构与晶体化学研究领域具有一定的学术地位。而同时，我国学者在矿物晶体结构与晶体化学领域还面临诸多挑战，如以矿物晶体学为主的研究项目极少，矿物晶体学研究较多依赖于新矿物发现，新矿物结构研究工作难度大，高水平科研队伍相对缺乏，又因为该学科规模小，对传统基础学科的矿物晶体学学科的重视程度不够，学科建设经费投入较少等都使得该领域的发展受限。此外，由于受就业面的限制，加之晶体结构分析需要相当雄厚晶体学及衍射基础的专业人才，矿物晶体学领域还存在后继人才匮乏等诸多问题。展望未来，矿物晶体学不仅要重视矿物晶体结构与晶体化学基础研究，而且要积极开拓研究新领域，在实际矿物学研究中体现矿物晶体学的应用价值，同时重视高层次人才培养，促进我国矿物晶体学学科的可持续发展。

参 考 文 献

白文吉,杨经绥,方青松,颜秉刚,史仁灯. 2003. 西藏蛇绿岩中不寻常的地幔矿物群. 中国地质,30(2):144-150
陈代演,王冠鑫,邹振西,陈郁明. 2001. 新矿物:铊明矾. 矿物学报,21(3):271-277
陈敬中. 2010. 现代晶体化学. 北京:科学出版社
陈敬中,杨建虎,王秋玲. 2001. 尼日利亚石-塔菲石-黑铝镁铁矿及相关系列矿物的晶体结构构筑原理. 矿物学报,21(1):19-26
陈鸣,束今赋,毛河光. 2008. 谢氏超晶石:一种$FeCr_2O_4$高压多形新矿物. 科学通报,53(17):2060-2063
代明泉,施倪承,马喆生,熊明,白文吉,方青松,颜秉刚,杨经绥. 2004. 桐柏矿的晶体结构测定. 矿物学报,24(1):1-6
范光,李国武,沈敢富,徐金莎,戴健. 2013. 栾锂云母:锂云母系列的新成员. 矿物学报,33(4):713-721

谷湘平,Watanabe M,谢先德,彭省临,Nakamuta Y,Ohkawa M,Hoshino K,Ohsumi K,Shibata Y. 2008. 陈国达矿($Ag_9FeTe_2S_4$):胶东地区金矿床中发现的硫碲化物新矿物. 科学通报,53(17):2064-2070

李国武,杨光明,马喆生,施倪承,熊明,沈敢富,范海福. 2005. 新矿物丁道衡矿的晶体结构. 矿物学报,25(4):313-320

李国武,施倪承,熊明,马喆生,白文吉,方青松. 2007a. 新矿物:罗布莎矿的晶体结构精修. 矿物岩石,27(5):1-5

李国武,马喆生,施倪承,熊明,沈敢富. 2007b. 赛马闪叶石的晶体结构测定. 矿物岩石地球化学通报,26(S1):252-254

李国武,熊明,丁奎首,秦克章,许英霞,方同辉. 2010. 叶绿矾矿物中的新超结构. 矿物学报,30(1):1-8

李国武,刘家军,熊明. 2011. 层状硅酸盐中发现的新结构类型. 矿物岩石地球化学通报,30(S):281

李国武,熊明,杨光明. 2012. 羟钙烧绿石及其晶体结构. 矿物学报,32(S1):26-27

李国武,杨光明,熊明. 2014a. 烧绿石超族矿物分类新方案及烧绿石超族矿物. 矿物学报,34(2):153-158

李国武,马喆生,傅小土,施倪承,熊明. 2014b. 新矿物富铜泡石(Tangdanite). 高校地质学报,20(S1):10

李国武,施倪承,白文吉,方青松,熊明. 2015. 西藏罗布莎铬铁矿中发现的七种金属互化物新矿物. 矿物学报,35(1):13-18

李国武,薛源,谢英美. 2018. 新矿物碲钨矿的矿物学特征及成因探讨. 矿物岩石地球化学通报,37(2):186-191

廖尚宜,李国武,彭明生. 2008. 草莓红绿柱石(Pezzottaite)的晶体结构测定及其意义. 矿物学报,28(4):350-356

马喆生,施倪承,叶大年. 2004. 镁粒磷锰矿Mg-Fillowite的矿物学及其晶体结构测定. 中国科学:地球科学,34(8):728-738

秦善. 2011. 结构矿物学. 北京:大学出版社

阮青锋,陆琦,邱志惠,肖平,于吉顺. 2009. 氟铁云母的晶体结构分析. 矿物岩石,29(3):1-8

沈敢富,徐金沙,姚鹏,李国武. 2017. 凤成石:异性石族矿物N(5)位贫钠的空位类似物新种. 矿物学报,37(S1):140-151

施倪承,李国武,熊明,马喆生. 2005a. 西藏罗布莎铁族元素金属互化物矿物及其成因探讨. 岩石矿物学杂志,24(5):443-446

施倪承,白文吉,李国武,方青松,马喆生,杨经绥,熊明,颜秉刚. 2005b. 地球深部金属碳化物的晶体化学. 地学前缘,12(1):29-36

徐金沙,李国武,范光,葛祥坤,祝向平,沈敢富. 2019. 红河石$Ca_{18}(\square,Ca)_2Fe^{2+}Al_4(Fe^{3+},Mg,Al)_8(\square,B)_4BSi_{18}O_{69}(O,OH)_9$:符山石族新矿物. 地质学报,93(1):138-146

杨光明,刘祥文,马喆生,施倪承. 2004. 萨碳硼镁钙石晶体结构的精确测定. 地质学报,78(2):190-194

杨经绥,白文吉,方青松,戎合. 2008. 西藏罗布莎蛇绿岩铬铁矿中的超高压矿物和新矿物(综述). 地球学报,29(3):263-274

杨主明,宋仁奎,陶克婕,张培善. 2004. 富铁硅钛铈矿的晶体化学. 中国稀土学报,22(3):398-404

Bai W J,Robinson P T,Fang Q S,Yang J S,Yan B G,Zhang Z M,Hu X F,Zhou M F,Malpas J. 2000. The PGE and base-metal alloys in the podiform chromitites of the Luobusa ophiolite,southern Tibet. The Canadian Mineralogist,38(3):585-598

Bai W J,Shi N C,Fang Q S,Li G W,Xiong M,Yang J S,Rong H. 2006. Luobusaite:a new mineral. Acta Geologica Sinica,80(5):656-659

Chen D Y,Wang G X,Zou Z X,Chen Y M. 2003. Lanmuchangite,a new thallium(Hydrous) sulphate from Lanmuchang,Guizhou Province,China. Chinese Journal of Geochemistry,22(2):185-192

Chen J Z,Wang F,Han W,Zhang R H,Peng J. 2003. Crystal chemistry research of minerals of nigerite group and its significance. Journal of China University of Geosciences,14(4):293-299

Chen M,Shu J F,Mao H K. 2008. Xieite,a new mineral of high-pressure $FeCr_2O_4$ polymorph. Chinese Science Bulletin,53(21):3341-3345

Chen M,Shu J F,Xie X D,Tan D Y. 2019. Maohokite,a post-spinel polymorph of $MgFe_2O_4$ in shocked gneiss from the Xiuyan crater in China. Meteoritics & Planetary Science,54(3):495-502

Dobrzhinetskaya L F,Wirth R,Yang J S,Green H W,Hutcheon I D,Weber P K,Grew E S. 2014. Qingsongite,natural cubic boron nitride:the first boron mineral from the Earth's mantle. American Mineralogist,99(4):764-772

Fan G,Ge X K,Li G W,Yu A P,Shen G F. 2017. Oxynatromicrolite,$(Na,Ca,U)_2Ta_2O_6(O,F)$,a new member of the pyrochlore supergroup from Guanpo,Henan Province,China. Mineralogical Magazine,81(4):743-751

Fang Q S,Bai W J,Yang J S,Xu X Z,Li G W,Shi N C,Xiong M,Rong H. 2009. Qusongite(WC):A new mineral. American Mineralogist,94(2-3):387-390

Fang Q S,Bai W J,Yang J S,Rong H,Shi N C,Li G W,Xiong M,Ma Z S. 2013. Titanium,Ti,A new mineral species from Luobusha,Tibet,China. Acta Geologica Sinica,87(5):1275-1280

Ge X K,Fan G,Li G W,Shen G F,Chen Z R,Ai Y J. 2017. Mianningite,$(\square,Pb,Ce,Na)(U^{4+},Mn,U^{6+})Fe_2^{3+}(Ti,Fe^{3+})_{18}O_{38}$,a new member of the crichtonite group from Maoniuping REE deposit,Mianning County,Southwest Sichuan,China. European Journal of Mineralogy,2(2):331-338

Gu X,Shi X,Yang H,Lu A,Shao Y,Chen Q,Liu Z. 2017. Wuyanzhiite,IMA 2017-081. CNMNC Newsletter No. 40,December 2017,page 1086. European Journal of Mineralogy,29:1083-1087

Gu X P,Watanabe M,Xie X D,Peng S L,Nakamuta Y,Ohkawa M,Hoshino K,Ohsumi K,Shibata Y. 2008. Chenguodaite($Ag_9FeTe_2S_4$):a new telluro-sulfide mineral from the gold district of East Shandong Peninsula,China. Chinese Science Bulletin,53(22):3567-3573

Gu X P,Xie X D,Wu X B,Zhu G C,Lai J Q,Hoshino K,Huang J W. 2013. Ferrisepiolite:a new mineral from Saishitang copper skarn deposit in Xinghai County,Qinghai Province,China. European Journal of Mineralogy,25(2):177-186

Gu X P,Yang H X,Xie X D,van Nieuwenhuizen J J,Downs R T,Evans S H. 2019. Lipuite,a new manganese phyllosilicate mineral from the N'Chwaning III mine,Kalahari Manganese Fields,South Africa. Mineralogical Magazine,83(5):645-654

Jian W,Mao J W,Lehmann B,Li Y H,Ye H S,Cai J H,Li Z Y. 2020. Lingbaoite,$AgTe_3$,a new silver telluride from the Xiaoqinling gold district,central China. American Mineralogist,105(5):745-755

Li G W,Xue Y. 2018. Wumuite,IMA 2017-067a. CNMNC Newsletter No. 44,August 2018,page 1018. Mineralogical Magazine,82:1015-1021

Li G W, Fang Q S, Shi N C, Bai W J, Yang J S, Ming X, Ma Z S, Rong H. 2009. Zangboite, TiFeSi$_2$, a new mineral species from luobusha, Tibet, China, and its crystal structure. The Canadian Mineralogist, 47(5): 1265-1274

Li G W, Xiong M, Shi N C, Ma Z S. 2011. A new three-imensional superstructure in bafertisite. Acta Geologica Sinica (English Edition), 85(5): 1028-1035

Li G W, Bai W J, Shi N C, Fang Q S, Xiong M, Yang J S, Ma Z S, Rong H. 2012. Linzhiite, FeSi$_2$, a redefined and revalidated new mineral species from Luobusha, Tibet, China. European Journal of Mineralogy, 24(6): 1047-1052

Li G W, Yang G M, Lu F D, Xiong M, Ge X K, Pan B M, de Fourestier J. 2016. Fluorcalciopyrochlore, a new mineral species from bayan obo, inner mongolia, P. R China. The Canadian Mineralogist, 54(5): 1285-1291

Li G W, Xue Y, Xiong M. 2019. Tewite: a K-Te-W new mineral species with a modified tungsten-bronze type structure, from the Panzhihua-Xichang region, Southwest China. European Journal of Mineralogy, 31(1): 145-152

Liu J J, Li G W, Mao Q, Wu S H, Liu Z J, Su S G, Xiong M, Yu X Y. 2012. Hanjiangite, a new barium-vanadium phyllosilicate carbonate mineral from the Shiti barium deposit in the Dabashan region, China. American Mineralogist, 97(2-3): 281-290

Liu Y, Deng J, Li G W, Shi G H. 2007. Structure refinement of Cs-rich Na-Li beryl and analysis of its Typomorphic characteristics of configurations. Acta Geologica Sinica, 81(1): 61-67

Ma Z S, Li G W, Shi N C, Zhou H Y, Ye D N. 2001. Structure refinement of astrophyllite. Science in China Series D: Earth Sciences, 44(6): 508-516

Ma Z S, Shi N C, Ye D N. 2005. Mineralogy and crystal structure determination of Mg-fillowite. Science in China Series D: Earth Sciences, 48(5): 635-646

Ma Z S, Li G W, Chukanov N V, Poirier G, Shi N C. 2014. Tangdanite, a new mineral species from the Yunnan Province, China and the discreditation of "clinotyrolite". Mineralogical Magazine, 78(3): 559-569

Qu K, Sima X Z, Li G W, Fan G, Shen G F, Liu X, Xiao Z B, Guo H, Qiu L F, Wang Y J. 2020a. Fluorluanshiweiite, KLiAl$_{1.5}$□$_{0.5}$(Si$_{3.5}$Al$_{0.5}$)O$_{10}$F$_2$, a new mineral of the mica group from the Nanyangshan LCT pegmatite deposit, North Qinling orogen, China. Minerals, 10(2): 93, doi: 10.3390/min10020093

Qu K, Sima X Z, Fan G, Li G W, Shen G F, Chen H K, Liu X, Yin Q Q, Li T, Wang Y J. 2020b. Taipingite-(Ce), (Ce$^{3+}_7$, Ca$_2$)$_{\Sigma 9}$Mg(SiO$_4$)$_3$[SiO$_3$(OH)]$_4$F$_3$, a new mineral from the Taipingzhen REE deposit, North Qinling orogen, central China. Geoscience Frontiers

Rao C, Wang R C, Hatert F, Gu X P, Ottolini L, Hu H, Dong C W, Dal Bo F, Baijot M. 2014. Strontiohurlbutite, SrBe$_2$(PO$_4$)$_2$, a new mineral from Nanping No. 31 pegmatite, Fujian Province, southeastern China. American Mineralogist, 99(2-3): 494-499

Rao C, Hatert F, Wang R C, Gu X P, Dal Bo F, Dong C W. 2015. Minjiangite, BaBe$_2$(PO$_4$)$_2$, a new mineral from Nanping No. 31 pegmatite, Fujian Province, southeastern China. Mineralogical Magazine, 79(5): 1195-1202

Rao C, Wang R, Gu X, Xia Q, Dong C, Hatert F, Yu X, Wang W. 2018. Zinconigerite-2N1S, IMA 2018-037. CNMNC Newsletter No. 44, August 2018, page 1020. Mineralogical Magazine, 82: 1015-1021

Rao C, Gu X P, Wang R C, Xia Q K, Cai Y W, Dong C W. Frédéric Hatert, Hao Y T. 2020. Chukochenite, IMA 2018-132a. CNMNC Newsletter No. 54. Mineralogical Magazine, 84: 359-360

Ren G M, Li G W, Shi J X, Gu X P, Fan G, Yu A P, Liu Q X, Shen G F. 2020. Potassic-hastingsite, KCa$_2$(Fe$^{2+}_4$Fe^{3+})(Si$_6$Al$_2$)O$_{22}$(OH)$_2$, from the Keshiketeng Banner, Inner Mongolia, China: description of the neotype and its implication. Mineralogy and Petrology, 114(5): 403-412

Shi N C, Bai W J, Li G W, Xiong M, Fang Q S, Yang J S, Ma Z S, Rong H. 2008. Yarlongite: a new metallic carbide mineral. Acta Geologica Sinica, 83(1): 52-56

Shi N C, Bai W J, Li G W, Xiong M, Yang J S, Ma Z S, Rong H. 2012. Naquite, FeSi, a new mineral species from luobusha, Tibet, western China. Acta Geologica Sinica (English Edition), 86(3): 533-538

Xie X D, Gu X P, Yang H X, Chen M, Li K. 2020. Wangdaodeite, the LiNbO$_3$-structured high-pressure polymorph of ilmenite, a new mineral from the Suizhou L6 chondrite. Meteoritics & Planetary Science, 55(1): 184-192

Xiong F, Xu X, Mugnaioli E, Gemmi M, Wirth R, Grew E, Robinson P T. 2019a. Badengzhuite, MA 2019-076. CNMNC Newsletter No. 52. Mineralogical Magazine, 83, doi: 10.1180/mgm. 2019. 73

Xiong F, Xu X, Mugnaioli E, Gemmi M, Wirth R, Grew E, Robinson P T. 2019b. Zhiqinite, IMA 2019-077. CNMNC Newsletter No. 52. Mineralogical Magazine, 83, doi: 10.1180/mgm. 2019. 73

Xiong F, Xu X, Mugnaioli E, Gemmi M, Wirth R, Grew E S, Robinson P T. 2019c. Jingsuiite, IMA 2018-117b. CNMNC Newsletter No. 52. Mineralogical Magazine, 83, doi: 10.1180/mgm. 2019. 73

Xu J S, Yang G M, Li G W, Wu Z L, Shen G F. 2008. Dingdaohengite-(Ce) from the Bayan Obo REE-Nb-Fe Mine, China: Both a true polymorph of perrierite-(Ce) and a titanic analog at the C1 site of chevkinite subgroup. American Mineralogist, 93(5-6): 740-744

Xue Y, Li G W, Yang G M. 2017. Mineralogy and crystallography of stokesite from inner Mongolia, China. The Canadian Mineralogist, 55(1): 63-74

Yang G M, Li G W, Xiong M, Pan B M, Yan C J. 2014. Hydroxycalciopyrochlore, A new mineral species from Sichuan, China. Acta Geologica Sinica (English Edition), 88(3): 748-753

Yang Z M, Giester G, Ding K S, Tillmanns E. 2011. Crystal structure of nanlingite the first mineral with a [Fe(AsO$_3$)$_6$] configuration. European Journal of Mineralogy, 23(1): 63-71

Yang Z M, Gerald G, Ding K S, Ekkehart T. 2012a. Hezuolinite, (Sr, REE)$_4$Zr(Ti, Fe^{3+}, Fe^{2+})$_2$Ti$_2$O$_8$(Si$_2$O$_7$)$_2$, a new mineral species of the chevkinite group from Saima alkaline complex, Liaoning Province, NE China. European Journal of Mineralogy, 24(1): 189-196

Yang Z M, Ding K S, de Fourestier J, Mao Q, Li H. 2012b. Ferrotaaffeite-2N'2S, a new mineral species, and the crystal structure of Fe^{2+} rich magnesiotaaffeite-2N'2S from the Xianghualing tin-polymetallic ore field, Hunan Province, China. The Canadian Mineralogist, 50(1): 21–29

Yang Z M, Giester G, Mao Q, Ma Y G, Zhang D, Li H. 2017. Zincobotryogen, $ZnFe^{3+}(SO_4)_2(OH) \cdot 7H_2O$: validation as a mineral species and new data. Mineralogy and Petrology, 111(3): 363–372

Yin J W, Li G W, Yang G M, Ge X K, Xu H M, Wang J. 2015. Fluornatropyrochlore, a new pyrochlore supergroup mineral from the boziguoer rare earth element deposit, Baicheng County, Akesu, Xinjiang, China. The Canadian Mineralogist, 53(3): 455–460

Yu Z X, Cheng F Y, Ma H W. 2009. Lisiguangite, $CuPtBiS_3$, a new platinum-group mineral from the Yanshan Mountains, Hebei, China. Acta Geologica Sinica, 83(2): 238–244

Yu Z X, Hao Z G, Wang H G, Yin S P, Cai J H. 2011. Jichengite $3CuIr_2S_4 \cdot (Ni,Fe)_9S_8$, a new mineral, and its crystal structure. Acta Geologica Sinica (English Edition), 85(5): 1022–1027

Zhang P H, Zhu J C, Zhao Z H, Gu X P, Lin J F. 2004. Zincospiroffite, a new tellurite mineral species from the zhongshangou gold deposit, Hebei Province, people's republic of China. The Canadian Mineralogist, 42(3): 763–768

Advances of Researches on the Crystal Structure and Crystal Chemistry of Minerals

LI Guo-wu

Crystal Structure Laboratory, Institute of Earth Sciences, China University of Geosciences (Beijing), Beijing 100083

Abstract: The crystal structure and crystal chemistry of minerals is one of the important basic research fields of mineralogy. Since 2000, with the attention to the geological prospecting of mineral resources by Chinese government and the improvement of experimental techniques such as single crystal diffractometer with plane detector, micro-diffraction, and computer hardware and software capability, the research discipline of crystal structure and crystal chemistry of minerals in China has been developed rapidly. The determination and research of mineral crystal structure have been developed with the discovery of new species of minerals. The discovery of new species of minerals is the cornerstone of mineral crystal structure research which provides data support for the new minerals. The crystal structures of new species of minerals discovered by Chinese scholars have mostly been determined. Especially, the crystal structures of nearly 30 kinds of the new minerals found in China have been determined and refined Some new crystal structures that were previously absent have been discovered. Some new understandings of the new crystal chemical phenomena of several new minerals have been achieved. These research results have not only been used as basic data for the discovery and declaration of new minerals, but also been used as the scientific basis for the clarification of composition and structure, physical characteristics, genesis, and evolution of minerals, and then for the interpretation of various geological phenomena. It is prospected that opportunities and challenges will be existed in the research field of mineral crystal structure and crystal chemistry in the future.

Key words: crystal structure; crystal chemistry; new species of minerals; research progress

应用矿物学研究进展*

董发勤　谭道永　王进明　徐龙华　李旭娟　丁文金　胡志波　黄　腾

西南科技大学，固体废物处理与资源化教育部重点实验室，绵阳 621010

摘　要：本文从冶金-工艺矿物学和矿物原料加工、新型工业矿物和岩石（宝石）、特种功能矿物材料、仿生矿物材料、生物矿物学、文化遗产和文物保护、矿物基固废处理与资源化利用、矿山及周边土壤污染修复8个领域综述了近十年应用矿物学研究进展，总结归纳了这些领域的相关重大进展，并提出了应用矿物学在第四次工业革命浪潮下的7个重要发展方向。

关键词：应用矿物学　工业矿物和岩石　非金属矿　矿物资源　纳米矿物　矿山污染修复

0　引　言

人类社会的发展史也是人类认识和利用矿物的历史。随着社会生产力和科技的进步，人类对矿物的认识和应用逐步深入和精细，人类社会也由石器时代、铜器时代、铁器时代、"新石器时代"等逐步发展并不断向前推进（董发勤，2015）。人类社会对矿物资源的开发利用主要体现在3个方面：一是提取矿物中的有价元素（如金属矿产矿物）；二是利用矿物中的能量（如能源矿产矿物）；三是直接利用矿物本身的物化性能（如非金属矿产矿物）。这直接衍生了应用矿物学学科。

应用矿物学是矿物科学与工程的一个分支，它是以研究矿物本身的整体应用为目的，以矿物的化学成分和晶体结构为根本，研究矿物的物化性质及其与成分结构的关系和内在机理；探索矿物性能应用和优化的方法和工艺，以便矿物在更广领域得到更充分的开发和应用。矿物应用涉及建材、冶金、化工、交通、机械、轻工等传统产业领域以及电子信息、生物医药、新能源、新材料、航空航天航海、宇宙深地深海探测等高新技术产业。矿物资源高效利用是当前地球科学与材料科学、环境科学等学科交叉的研究热点，是实现生态文明建设和社会可持续发展的重要前提，是构建人类命运共同体的重要资源构成部分，是全球共同关注的焦点。

国际矿物学协会（International Mineralogical Association，IMA）于1979年设立应用矿物学委员会（Commission for Applied Mineralogy），并于1981年举办了第一届国际应用矿物学大会（International Congress for Applied Mineralogy，ICAM），现在每两年召开一次，迄今共举办了14次，已成为应用矿物学领域全球最大规模的专业学术会议。会议旨在为从事应用矿物学的相关研究人员提供成果展示平台，加强应用矿物学领域的国际学术交流与合作，促进应用矿物学的发展。国内从事应用矿物学相关研究的科研院所和研究团队亦如雨后春笋般发展壮大，并取得了丰硕的成果、培养了大量的人才。非金属矿物资源高效利用专业委员会于2019年经中国矿物岩石地球化学学会批准成立，主要致力于非金属矿物的"政-产-学-研-用"全面发展。可以说，应用矿物学在国内外已经蓬勃发展起来了！

本文就近十年来国际ICAM大会持续活跃的八个领域进行总结，包括冶金-工艺矿物学和矿物原料加工、新型工业矿物和岩石（宝石）、特种功能矿物材料、仿生矿物材料、生物矿物学、文化遗产和文物保护、矿

---

* 原文"全球第四次工业革命前景下的应用矿物学进展（2011—2020）"刊于《矿物岩石地球化学通报》2021年第40卷第2期，本文略有修改。

物基固废处理与资源化利用、矿山及周边土壤污染修复，并就未来应用矿物学发展方向进行了展望。

1 冶金-工艺矿物学和矿物原料加工

1.1 冶金矿物学

近十年来，冶金矿物学在矿物的工艺性质、测试方法及提高冶金产品质量等方面取得了显著进展。通过原料的矿物学工艺特性及冶金性能关系研究，设计优化原料矿物对应合理的冶金预处理、冶金过程和渣相构成，相关研究集中在复杂含铜矿物、含铁矿物、稀土矿物、含钒矿物和冶金渣矿物相方面。

在湿法冶金方面，Barton等（2018）研究了内华达州某 Cu-Au(Zn) 矿的冶金矿物学特性，比较了六种浸出剂对原矿中铜的浸出效果。结果表明，砷钙铜矿和含铜的铁氧化物未能在各种浸出剂中完全溶解，是铜回收率低的主要原因。国内，张一敏团队在对石煤提钒的研究中发现，钒主要以三价类质同象形式赋存于白云母类矿物中，方解石、黄铁矿、磷灰石和赤铁矿等耗酸脉石矿物是石煤提钒工艺中的主要有害杂质；原矿中的含钒云母多呈层片状或条状与石英紧密连生或嵌布于石英脉中，导致钒浸出率不高，增加硫酸（浸出剂）浓度，能有效提高钒的浸出率并降低溶液中钒的损失（蒋谋锋等，2015）。此外，该团队开发的重-浮联合预抛尾工艺，将最终精矿 V_2O_5 品位提高至 1.01%，V_2O_5 回收率达 87.60%，较好地实现了提高酸浸给矿 V_2O_5 品位、降低酸浸作业矿量及耗酸矿物含量的目标（刘鑫等，2017）。邱廷省团队研究了某离子型稀土矿浸出前后工艺矿物学，结果表明浸出后的 Al、稀土元素（rare earth element, REE）、Fe、O 元素含量下降，K、Si 含量相对上升，Al 减少的主要原因是高岭石、褐铁矿部分的溶解，金云母在浸出后全部消失。Fe 减少的主要原因是浸出后稀土矿中的水合铁离子进入溶液中（刘庆生等，2019）。

在火法冶金方面，Tanii等（2014）研究了炼铁原料铁砂的矿物学特性对炼铁产品的影响，发现其根源在于铁砂原矿中铁氧化物所含 TiO_2 固溶体和赤铁矿含量是不同的，最终影响铁砂还原初始温度和还原过程中形成的矿物种类。牛乐乐等（2019）研究了铁矿粉矿物组成对烧结矿冶金性能的影响，通过微观矿物定量分析、微观矿相结构研究，发现赤铁矿、石英的出现能促使软化起始温度升高，赤铁矿有利于烧结矿的还原，针铁矿更易使产物中生成针状铁酸钙，高岭石分解产生的 SiO_2 形成硅酸盐相。向小平等（2020）对柳钢烧结机台车截面烧结矿的矿物学以及上、下层烧结矿的化学成分和显微结构的研究表明，各层烧结矿中的 FeO 质量分数和碱度波动很大；上层烧结矿多以赤铁矿连晶为主，铁酸钙相较少；中层烧结矿铁酸钙最多，多呈理想的针状形态；下层烧结矿中的铁酸钙生成量比中部略少。冯聪等（2016）研究了钒钛磁铁矿冶炼工艺中低、中、高含钛炉渣的物相变化情况，发现渣系热稳定性和化学稳定性先变好后变差，渣中黄长石相骤减，辉石、钙钛矿相数量增多。黄毅等（2014）对典型钢渣研究表明，转炉钢渣的主要物相是硅酸钙相、铁酸钙相、RO 相和 f-CaO 相；电炉氧化渣出现了明显的橄榄石相；精炼渣主要成分为 CaO、Al_2O_3、MgO 和 SiO_2，研究成果为钢渣的分类综合利用起到了指导作用。孙永升等（2015）研究了温度对鲕状赤铁矿石深度还原特性的影响，在 1312℃ 条件下，反应产生的 FeO 会与脉石矿物继续反应，生成铁橄榄石和尖晶石；通过控制还原温度可以调控物相转变和微观结构的破坏；还原温度越高，还原产物的物相组成越简单。

在生物冶金方面的研究主要集中在微生物品种优化、矿物中供能物质的作用与微生物组合效应方面。Nie等（2019）通过同步辐射 X 射线衍射和硫 K 边 X 射线吸收近边结构光谱等技术研究了 *Acidianus manzaensis* 菌对不同晶体结构黄铜矿的浸出作用，查明了不同晶体结构黄铜矿浸出后相态的转变规律。Liu等（2015）研究了黄铜矿在中性嗜热硫杆菌生物浸出过程中铜、铁、硫相态的 X 射线近边结构吸收光谱，对浸出后矿物表面组分进行了原位定量矿物学分析，结果查明了黄铜矿生物浸出的溶解机理。Ma等（2017）开展了铁硫氧化微生物共培养体系中微生物配比对黄铜矿浸出效果的研究，结果显示硫氧化微生物比例高的体系中拥有更好的浸出效果，这表明构建高效的共培养体系需要功能微生物特定比例的组合，也说明硫代谢在生物冶金微生物代谢功能网络中的重要性。

1.2 工艺矿物学

工艺矿物学的研究已进入微观层面，并朝着定量化和自动化方向发展。工艺矿物学参数自动测定系统（QEMSCAN）、自动矿物分析仪（MLA）、矿物表征自动定量分析系统（AMICS）和矿物模块化自动处理系统（MAPS Mineralogy）是目前工艺矿物学研究的最新手段。

QEMSCAN 测试系统的出现，是工艺矿物学领域所取得的新成就。该系统实现了解离度测定自动化，并大幅提升了解离度测定的准确性和可重现性。Nazari 等（2017）利用该系统和溶解诊断过程对难溶金进行了研究，结果发现金的难溶是因为银金矿、碲化金与其他硫化矿微细粒嵌布，金主要赋存于硫化矿物包裹体中，通过焙烧破坏包裹体可大幅提升金的浸出率。Guanira 等（2020）利用 QEMSCAN 对铜（金、银）夕卡岩型矿床尾矿进行分析，快速得到脉石矿物、粒度分布和矿物组合等信息，对该尾矿污染环境的潜在可能性进行了评价。

杨帆等（2020）利用 MLA 对煤灰的矿物学特性进行了分析，测试了含铀煤灰性质以及颗粒粒度，分析了铀矿物的解离程度及连生关系。Al-Khirbash（2020）利用 MLA 对阿曼北部山区低品位红土镍矿进行了分析，得到了矿床的粒度、矿物组成、矿物间关系度等，提出了降低冶金操作成本、提高金属回收率的方法。

AMICS 主要用于测定脉状矿石的矿物组成、粒度和嵌布特征，能够实现常规岩矿鉴定手段难以完成的矿物定量识别和鉴定。温利刚等（2018）采用 AMICS 测量了云南武定稀土矿物的种类和含量。方明山和王明燕（2018）运用 AMICS 分析了铜矿中的伴生金银的赋存状态，查明了该矿石中金银矿物的种类、金银矿物的嵌布特征和嵌布粒度，以及影响回收的最主要因素。

此外，激光消融微探针感应耦合等离子体质谱可以进行痕量检测，能够更加准确和深入地研究元素的赋存状态，也是工艺矿物学研究向微观层面迈进的推动力。二次离子质谱分析技术使矿物表面特性及其变化研究更方便可靠。这两种技术在稀贵金属的赋存状态研究方面发挥了重要作用，已成为工艺矿物学研究的重要手段。

1.3 矿物原料加工

应用矿物学在矿物原料加工方面具有突出的支撑作用，它有利于查明矿物原料加工过程中存在问题的微观症结，为矿物原料加工指标的提高开辟了有针对性的工艺路线。这在矿物超细加工、超纯净化、矿物表面改性及矿物分选等方面都有集中体现。

在矿物超细加工方面，Wang 等（2020）采用两步法制备高纯度、超细 WC-Co 复合粉体，制备过程中采用 X 射线衍射、场发射扫描电子显微镜、红外碳硫仪等矿物学研究方法，研究了所制备粉体的物相组成、形貌和碳含量，并严格测量了颗粒大小，给超细 WC-Co 复合粉体制备提供了有力指导。Li 等（2017）研究了煤在气流磨中的超细粉碎，利用偏光显微镜和扫描电子显微镜研究了原料的物相和形态特征，原生菱铁矿颗粒尺寸较大（70~100 μm），而黏土矿物颗粒非常细，呈较细的浸染状分布（约 2 μm）。要实现矿物和煤的完全解离，需要更细的粒度，从而制定了合理气流速度和相对压力参数。

高纯矿物原料加工有很大的挑战性。高纯石英具有非常重要的用途，但石英中微量杂质的剔除仍是一个难题。吴道等（2017）对青海某石英原料的分析发现，微细粒白云母是该矿中的主要杂质矿物，片径多在数十微米，紧密镶嵌在石英晶粒之间。微区成分分析表明，Al 和 K 杂质元素主要赋存于白云母中。制备高纯石英的关键是石英和白云母的解离并分离。因此通过十二胺反浮选白云母，浮选精矿进行加温酸浸，可较好地除去各种杂质，石英纯度可达 99.99%。

邵秋月等（2019）对硅灰石的表面改性及新型矿物基聚合物的合成进行研究，用扫描电子显微镜对改性前后的硅灰石的表面形貌进行对比，发现改性前硅灰石表面光滑、改性后表面有新的附着物出现。不同改性条件下，硅灰石经过浓度较高的改性剂处理，其表面的包覆层更加完整，改性效果更好。

Cabri 等（2017）对比了两种不同成矿类型铂族含镍矿体的矿物学及加工潜力差异，发现第一种矿体矿物嵌布粒度粗易于磨矿和浮选加工，而第二种只能进行湿法冶金才能回收有价金属，从而采用不同的加工方法，将该低品位矿物高效回收。Wei 等（2019）研究了黄铜矿不同解理面的反应活性差异，揭示了黄铜矿新鲜解理面与 O_2 和 H_2O 的原位非均相反应机理和矿浆中常见"难免离子"对黄铜矿浮选的影响机理。

2 新型工业矿物和岩石（宝石）

工业矿物和岩石是除金属矿产与燃料矿产以外，化学组成或物理性能可为工业利用而具有经济价值的所有非金属矿物和岩石（陈正国等，2019），其研究应用范围近年来也在逐渐拓展。如水镁石和纤维水镁石既可作为提取镁的重要原料之一，又可作为阻燃增强材料，适用于防火涂层、造纸工业、环保等领域（秦雅静和朱德山，2014）。叶蜡石不仅是制备超硬材料的必备腔体材料，也是优异的烧蚀材料，在国防、航空航天等领域具有广泛应用（杜培鑫和袁鹏，2019）。电气石作为一种含硼的环状结构硅酸盐矿物，其功能复合材料及新产品的开发也在不断深入（文圆等，2019）。王菲等（2019）开发的稀土强化电气石矿物复合功能材料，阐明了电气石矿物显微结构与内部活性位点的形成机制，拥有稀土强化电气石矿物的优选和先进加工等关键技术，形成了高性能电气石远红外辐射和自发极化的性能评价方法。玄武岩纤维作为一种新兴纤维材料，其黏结性、耐热性及抗腐蚀性等物理性能较为优越，以玄武岩纤维为添加材料可制备多种复合材料，广泛用于消防、环保、航空航天、工程塑料、建筑等不同领域（秦丽辉，2014）。何军拥和田承宇（2013）的研究表明，玄武岩纤维具有较好的水工特性，能有效提升混凝土耐酸耐碱性能、抗渗性以及抗冻性，其兼具防水特性，使其在堤坝、码头及跨海大桥等领域具有广阔的应用前景。此外，地聚合物作为硅酸盐材料，还兼具有机高聚物、玻璃和水泥等材料的特点，显示出极大的发展前景（杜保聪等，2018）。王健等（2015）针对传统注浆材料抗侵蚀性差、后期体积倒缩等问题，设计了地聚合物双液注浆材料，其凝结时间从几十秒到几十分钟可控，工程加固止水性能较传统注浆材料优势明显。

宝玉石与能源矿产、金属矿产、非金属矿产和水气矿产并列为五大类矿产。人类对宝玉石的认识历史悠久，认知度最高的是钻石、红宝石、蓝宝石和祖母绿四大珍贵宝石，地质学家和宝石学家也在不断地发掘"新成员"，如近年来对沙弗莱石（Tsavorite）、变色水铝石（Zultanite）、加斯佩石（Gaspeite）、萨索斯岛大理石（Thasos marble）等新宝石的认知和普及，非常规性宝石材料的价值也将不断体现（郭鸿旭，2017）。人工宝石目前常用的加工方法主要包括焰熔法、助溶剂法、水热法和化学沉淀法等（张志祥等，2013）。陈庆汉（2012）改进了壳熔法蓝宝石晶体生长工艺，采用壳熔定向结晶法，第一次得到了厘米级蓝宝石单晶。邢睿睿等（2019）采用溶胶-凝胶法制备合成翡翠，其宝石学特征、红外光谱及紫外-可见光谱与天然翡翠基本一致，延长晶化时间有利于晶粒生长、结构致密，获得优质的合成翡翠。

张妮和林春明（2016）考察了 X 射线衍射技术应用于单晶、多晶宝石的定名及宝石矿物的晶体结构特征，探讨了宝石矿物多型的种类。申柯娅（2011）对合成祖母绿与天然祖母绿的成分及其红外吸收光谱进行了研究，两者红外吸收光谱具有明显差异：助溶剂法合成祖母绿中没有水的吸收，而天然祖母绿中含有一定水分子吸收峰。李晓静（2016）对硅酸盐类宝石矿物、有机宝石、有机物充填宝石和合成宝石的近红外光谱进行了研究。离子注入改性是提高宝石表面性能的一种可靠方法。张海亮等（2020）通过选择不同能量和剂量的镁/钛离子注入蓝宝石，其纳米硬度、纳米划痕和红外性能等均呈现出可调节的特性。

3 特种功能矿物材料

特种功能矿物材料是指在传统矿物材料基础上，根据特定矿物原料的物理化学性质和工艺特性，经精、深加工处理或人工合成，形成具有特殊性能和多用途的功能材料。在此仅介绍超强矿物（复合）材料、超绝热/储热矿物（复合）材料和超导矿物材料近十年来的进展。

3.1 超强矿物（复合）材料

超强矿物（复合）材料包括具有超硬、超弹性等优点的高端矿物（复合）材料。如以传统超硬材料金刚石为填充料，立方氮化硼为基体，添加适量的黏结剂，可合成新型金刚石复合材料 PCBN-dia，其兼具较高的硬度和耐磨性（刘宝昌等，2018）。Gao 等（2018）将两层石墨烯相叠形成双层石墨烯，其厚度仅为 A4 纸的百万分之一，杨氏模量大于 400 GPa，硬度甚至高于金刚石。超弹性材料具备较强的记忆及可逆形变性能，常压下制备的超弹性石墨烯/羧甲基纤维素复合气凝胶（SGA/CMC），在 80% 应变下循环压缩 500 次后，仍能恢复到原来高度的 95% 左右（Xu et al.，2020）。将石墨烯进行三维交联，可实现基于单个石墨烯片层的本征弹性和片层之间的共价连接，在 4 K 超低温条件下具有与室温相同的力学性能及几乎完全可逆的超弹性行为（高达 90% 的应变）（Zhao et al.，2019）。

3.2 超绝热储热矿物（复合）材料

导热系数低于静止空气（25 mW/mK）的材料称为超绝热材料，因其具有低密度、隔热性好的特点而在军工、建筑等行业具备巨大应用潜力。以硅质矿物为原料采用 Kistler 法、焚烧法及一步法制得的纳米 SiO_2 气溶胶是制备超绝热材料的主导原料；纳米 SiO_2 气溶胶本身具有较低的导热率（0.9 mW/mK），经 400℃、2 h 热处理后大孔体积分数由 63.05% 显著降低至 24.82%，导热率则降低至 0.8 mW/mK，而热处理温度超过 800℃ 则直接破坏孔结构，即适当的高温处理可改善其隔热性能和室温下二氧化硅气凝胶的可用性（Lei et al.，2017）；采用一体冷冻干燥工艺将纳米二氧化硅与聚酰亚胺（PI）交联制备 PI/SiO_2 纳米复合气溶胶，密度及导热率分别低至 0.07 mg/cm³ 和 21.8 mW/mK，在 300℃ 下仍具有较好的绝热性能（35 mW/mK）（Fan et al.，2019）。将具有较好导热性的多孔矿物（埃洛石、硅藻土、蒙脱石等）和储能相变材料进行复合可制备矿物基储热材料，其具备导热均匀、成本低等优点，可用于太阳能光热领域（谢宝珊等，2019）。如将石蜡装载至埃洛石纳米管（HNT）中可制成硬脂酸/HNT 复合储热材料，50 次储热循环后其储热性能保持不变（Zhang et al.，2012）；还可通过对 HNT 进行表面超疏水改性，提升其负载相变材料的性能（杜培鑫等，2019）。Li 等（2011）用熔融吸附法制备二元脂肪酸/硅藻土复合相变材料，癸酸-月桂酸/硅藻土的潜热能可达 66.81 J/g，相变温度可达 16.74℃。利用十二烷基硫酸钠面活性剂扩宽蒙脱石层间距，再将正十六烷插层改性后的蒙脱石制备复合储热材料，其具备持久的储热性能，潜热值高达 126 J/g（Sarier et al.，2011）。

3.3 超导矿物材料

超导材料由于具有零电阻、完全抗磁性（迈斯纳效应）等特性，而在强电、强磁、储能和超导电子器件等领域的应用前景广阔。近年来层状过渡金属硫族化合物插层超导体（插层剂为碱金属/碱土金属离子或有机物）、插层石墨超导体、铜氧化物高温超导体（Ba-La-Cu-O 体系、Bi-Sr-Ca-Cu-O 体系、Hg-Ba-Ca-Cu-O 体系等，最高超导温度达 164 K）等陆续被发现（阮彬彬，2018；白桦，2019；邵志斌，2019），超导材料研究主要集中于铁基超导体和铋硫基超导体。铁基超导体中不导电的 LaO 层为空间层，导电的 FeAs 层为超导层，对 La 位和 Fe 位进行掺杂，可提升铁基超导体的超导温度；薛其坤等利用分子束外延

技术在 $SrTiO_3$ 衬底上制备单层 FeSe 薄膜，通过 $FeSe/SrTiO_3$ 异质结界面增强电声耦合作用，界面电荷转移以及界面应力效应共同作用下，实现了 100 K 以上的超导电性（Ge et al., 2015），这是继铜氧化物高温超导体之后的第二个常压高温超导体。Mizuguchi 等（2012）首次发现二维层状铋硫基超导体（$Bi_4O_4S_3$ 和 $LaO_{0.5}F_{0.5}BiS_2$，2012 年），BiS_2 层为其超导层。与铁基超导体类似，对铋硫基超导体的空间层掺杂、替换和改造可提升铋硫基超导体的性能。除铁基超导体和铋硫基超导体之外，曹原等开发了非传统石墨烯超导材料。他们将两层石墨烯以 1.1° 的"魔角"层叠形成"超晶格（superlattice）"材料，该材料中两层石墨烯非导电，属于莫特绝缘体（Cao et al., 2018a）；当添加少量的电子到石墨烯超晶格中，电子摆脱了起初的绝缘状态，并以零电阻的状态流动，成为超导材料（Cao et al., 2018b）。此材料在特定条件可实现超导与绝缘状态相互转换，有望成为室温超导体，为量子设备提供诸多可能性。

4 仿生矿物材料

仿生矿物材料具有精密的结构，可实现其机械性能（如韧性、减震性、比强度）、催化性能、光性能、热性能的可控设计。近十年来，仿生矿物材料的模拟合成取得了令人鼓舞的成就，除作为结构工程材料外，在催化、导热、人工肌肉、激光、探测等方面的应用也越来越广。

4.1 仿生矿物结构工程材料

生物体在自然界的漫长进化中，演化出了具有优异力学性能的生物矿物材料。例如巨骨舌鱼的鳞片具有优异的外力抵抗性；雀尾螳螂的肢部锤头具有高抗冲击性；牙齿具有超高的硬度和耐磨损性能；骨骼拥有极高的抗断裂损伤能力；贝壳珍珠层是强度高、韧性好和模量高的组合体，具有优异的综合性能等。这种在温和条件下由受限的材料组分实现的多级有序结构和性能强化，模拟制备出自然界的复杂精密结构，以获得力学性能优异的结构材料，拓展矿物材料在工程领域的应用。例如，以石墨和石墨烯为原料制备的 1D 纤维，其强度高达 140 MPa，在加入高分子改善界面性能后，纤维的强度可提升至 442 MPa（Jalili et al., 2013），甚至 652 MPa，是天然贝壳的 5 倍（Hu et al., 2013）；通过离子交联和扭曲设计，可将该类纤维的断裂伸长提升至 400%，是纯石墨烯材料的 100 倍（Cruz-Silva et al., 2014）。增强、增韧后的石墨烯纤维可用于电缆、绳索和织物等领域（Xu et al., 2013；Zhang et al., 2016）。具有仿生贝壳结构的黏土/聚合物二维复合膜也具有重要的应用价值。将蒙脱石、石墨烯、二硫化钼、二硫化钨等二维层状矿物与高分子，如 PVA、PLA 和 PDA 等，制备成具有仿生贝壳结构的薄膜，此类薄膜保有良好的机械性能、兼有导电性、气体阻隔性、耐火性、耐疲劳等特性，可广泛用于建筑、工业包装、功能涂料、柔性电子装置等领域（Liang et al., 2016；Knöller et al., 2017；Wan et al., 2017）。3D 砖墙仿生结构可显著提高砖墙的机械性能。调节仿生结构中的矿物含量，可得到韧性、强度和刚性性能优异的材料，如 Al_2O_3-氰酸酯（CE）复合材料的弯曲强度是铝板强度的 50 倍（Livanov et al., 2015；Zhao et al., 2016）。Le Ferrand 等（2015）将复合有超磁性氧化铁颗粒的 SiO_2、Al_2O_3 颗粒铝喷涂在铝片上，制备出贝壳仿生结构的 Al_2O_3 片状复合材料，其弯曲强度和韧性分别为 650 MPa 和 14 $MPa \cdot m^{1/2}$；Livanov 等（2015）将涂层中 Al_2O_3 的含量提高至 90%，断裂强度增加至 320 MPa；Bouville 等（2014）将 Al_2O_3 的含量提高至 98%，弯曲极值可达 460 MPa，是纯 Al_2O_3 的 5 倍。仿生三维（three-dimensional, 3D）矿物复合材料具有优异的力学性能且质量轻，在工业中被用于替代合金、塑料、陶瓷等工程材料。例如，Al_2O_3-CE 密度仅为 1.8 g/cm^3，强度达到钢材的 1/4，用于替代钢材（Zhao et al., 2016）。

4.2 仿生矿物功能材料

利用矿物结构和化学组成的差异，可制备出各种功能不同的仿生功能材料，广泛应用于催化、导热、绝缘、感光、探测、医药等领域。

4.2.1 仿贝壳多功能材料

张博雅等以聚乙烯醇和蒙脱石为原料,仿生构筑了贝壳结构,层叠势垒提高了绝缘体表面的电子耗散,可作为高压直流气体绝缘系统的电荷耗散涂层使用(Zhang B. Y. et al., 2019)。顾军渭等以聚多巴胺和立方氮化硼纳米片制备的具有仿贝壳结构的导热复合纸,可用于柔性电子领域(Ma et al., 2020)。柏浩等制备的仿贝壳3D网络结构的氮化硼/环氧复合物具有各向异性高导热,可用于电子行业(Han et al., 2019)。Zhang L. 等 (2019) 制备了具有仿贝壳的石墨烯/海藻酸钠薄膜制动器,该薄膜器可被水蒸气驱动,可用于微型机器人、人工肌肉和快速探测等领域。Hill 等(2017)以金纳米粒子、硅酸镁锂黏土、氧化石墨烯为原料制备了具有仿贝壳结构的复合膜,可用于催化、传感、抗菌等领域。Morits 等(2017)制备的贝壳仿生结构的霞石/聚合物复合材料,具有高的强度和韧性。Das 等(2017)以纤维素钠、蒙脱石制备了具有仿贝壳结构的涂层,该涂层可用于织物的防火。郭林等以黏土为原料制备的仿生贝壳膜,可用于防水、防腐工程中(Wu et al., 2014)。Le 等(2020)以 TiO_2、石墨烯、几丁质和液晶材料制成的仿生贝壳复合薄膜在能量储存与转换、气体探测等领域具有重要应用。

4.2.2 半球形人工视网膜结构多功能材料

将钙钛矿制成具有高密度纳米线阵列,可组成半球形人工视网膜来模拟人类视网膜上的光感受器。将这种视网膜应用到电化学眼睛中,能够执行获取图像模式的基本功能(Gu et al., 2020)。Lv 等(2019)仿生合成了以 $NiMoO_4 \cdot 2H_2O$ 纳米线排列组成的手性螺旋光子晶体,可用于偏光器、圆偏振发光、激光等领域。

4.2.3 人体纳米羟基磷灰石多功能材料

Yang 等(2012)以纳米羟基磷灰石、胶原蛋白和磷脂酰丝氨酸为原料仿生制备的多孔复合支架,可用于骨组织替代工程中;羟基磷灰石(nHA)和石墨烯(GO)增强的聚乳酸(PLA)杂化纳米复合材料(PLGA/nHA)具有很好的生物相容性,石墨烯及其衍生物可促进新骨细胞在成骨分化其表面的细胞上黏附和生长,促进线粒体的生长(Liu et al., 2016),添加 GO 的 PLGA/nHA 支架可增强小鼠成骨细胞的增殖(Liang et al., 2018)。Zhou 等(2014)开发了以磷酸钙(CaP)为核、免疫原为壳的纳米粒,用于实时生产牛痘疫苗,从而免于冷冻储存运输。Nam 等(2017)开发了贻贝激发的 CaP 聚合物纳米载体,用于细胞内传递阿霉素。以磷酸钙为核,内部封装光敏剂,外部包覆聚乙二醇的杂化纳米载体可减少光化学物质在血液中的损失,可用于光能治疗(Nomoto et al., 2016)。

5 生物矿物学

目前,自然界的生物能合成约60余种矿物,含钙矿物(磷酸钙和碳酸钙)约占整个生物矿物的50%。其中碳酸钙主要构成无脊椎动物的体内外骨骼,磷酸钙几乎完全由脊椎动物所采用;其次为非晶质氧化硅;含量较少的有铁锰氧化物、硫化物、硫酸盐、钙镁有机酸盐等。这些生物矿物除了具有保护和支撑两大基本功能外,还有很多其他特殊功能,如方解石可作为三叶虫的感光器官,作为哺乳动物内耳里则的重力感受器;文石在头足类动物的贝壳里作为浮力装置,在软体动物中作为外骨骼;磷酸钙(包括羟磷灰石、磷酸八钙和无定型)主要作为脊椎动物的牙齿和内骨骼;一水、二水草酸钙可作为植物、真菌的钙库,二氧化硅存在于硅藻的细胞壁、植物的叶子;磁铁矿在鲔鱼、鲑鱼头部里具有磁导航的作用,并存在于趋磁性细菌的细胞内;水铁矿既存在于海狸、老鼠和鱼的牙齿表面,也存在于石鳖牙齿的前驱相。

关于生物体内的矿化机理,主要开展了以下4个方面的研究:

(1)生物合成矿物。生物体内碳酸钙、磷酸钙的形成就是方解石/文石/羟基磷灰石生物矿化的结果。对于生物矿化,基质蛋白一般含有谷氨酸残基或天冬氨酸残基,这些残基可与钙离子结合,是钙离子生物矿化中重要的氨基酸。Jun 等(2015)模拟体内环境进行了钙离子生物矿化过程的体外再现,揭示了钙离子在生物体内的矿化机理。毛喆等（2016）利用两种不同的硫酸盐还原菌成功合成出具有不同比表面积、铁硫原子比、表面活性组分 FeS 相对含量的硫化亚铁矿物，并研究了其对六溴环十二烷的还原脱溴作用，结果表明两种不同类型硫酸盐还原菌合成的 FeS 矿物对六溴环十二烷均具有还原脱溴能力，并且它们的还原脱溴的能力和机制相似。

（2）生物分子调控矿物生长。赵康（2013）考察了一系列不同分级结构的生物分子（氨基酸、多肽、蛋白质）添加剂对无机矿物生长的影响，结果发现不同添加剂对方解石（104）面螺旋生长丘形貌及速率都有不同程度的影响。于洋（2017）采用纳米压痕试验对大连湾牡蛎壳中的珍珠母层的微观力学性能进行了研究，并与矿物方解石的微观力学性能进行对比，解析出生物硬组织优异性能的成因与机理。闫华晓等（2019）对 Bacillus cereus MRR2（GenBank KY810857）作用下不同钙离子浓度诱导方解石生物矿化的研究，制备出有机和无机成因方解石，结果表明生物成因方解石的结晶度和活化能均明显高于有机成因方解石。谭远（2016）以华贵栉孔扇贝和施氏獭蛤的表壳层为对象，进行了生物成因纤维状文石集合体的研究，并以其为基底进行了仿生合成试验。Huang 和 Zhang（2012）利用新鲜竹蛏和豆蛏韧带的蛋白质在不添加任何添加剂的情况下，仿生合成了与韧带中极为相似的文石纤维，其纯度几乎达到 100%。刘晓晔（2019）以河南石井剖面寒武系苗岭统为例系统探讨了由光合作用微生物组成的生物膜的钙化作用过程，研究了光合作用生物膜作用下方解石鲕粒的形成过程，丰富了鲕粒的成因类型。

（3）生物诱导结晶与溶解。Zhang J. G. 等（2017）进行了生物矿化合成碳酸钙的研究，并将其用于混凝土的自修复，取得了较好的实验效果。Liu 等（2014）通过明胶改性石墨烯模拟蛋白的作用仿生矿化羟基磷灰石。作为具有生物相容性和生物活性的材料，透明质酸也被用作模板合成碳酸羟基磷灰石：透明质酸在羟基磷灰石结晶早期可以稳定无定型磷酸钙，在晚期可以影响羟基磷灰石晶体生长中的钙空位和羟基磷灰石的形貌（Li et al.，2012）。

（4）生物模板自组装矿化。Sharma 等（2020）系统综述了人体硬组织内的生物矿化过程，特别强调了骨和牙齿矿化的机制和原理；描述了蛋白质和无机离子在矿化过程中的作用。潘海华和唐睿康（2020）以生物矿化典型无机矿物磷酸钙和碳酸钙体系为例，从生物矿物-溶液界面结构、生物分子与矿物晶面的分子识别、矿物结晶调控 3 个层面综述了生物矿化的化学调控原理，并从信息传递和转化的化学工程范式出发，分析了生物矿化中分子工程和结晶调控策略。Zhao 等（2014）采用血红蛋白为软模板，制备出三维纳米片组装成的羟基磷灰石微球。

6 文化遗产和文物保护

天然的矿物材料在文化遗产、文物及其保护领域发挥着重要作用，多数文物和文化遗产均直接或间接以天然矿物为原料进行制备、保护或维修。近十年来，有关文物成分鉴定及检测、文物加固、文物表面清洁、文物抗风化及耐腐蚀、自然岩溶景观修复保护等领域受到国内外矿物学家的重视。

6.1 文物成分鉴定及检测

以天然矿物为原料，经粉碎研磨漂洗等工艺制备的矿物颜料在秦始皇兵马俑、古希腊、古罗马雕塑中大量使用。常晶晶等（2010）以拉曼光谱作为材料的"指纹"光谱对敦煌莫高窟壁画残片中的红色颜料进行了分析和鉴别，通过对朱砂天然矿物和朱砂颜料的拉曼光谱信号的比较分析，初步发现矿物颜料中有石墨。付倩丽（2016）采用显微拉曼光谱分析对秦俑彩绘颜料样品进行了半定量分析，得出彩绘矿物颜料样品的组成配比。袁鸿等（2013）采用扫描电子显微镜和 X 射线衍射分析了荆门灌子家车马土体土样的元素成分和黏粒矿物组成，发现车马土样以富集铝铁相硅酸盐矿物为主，且含有部分含钾镁相碳酸钙成分。陈庚龄和康明大（2013）对临洮哥舒翰碑的岩层矿物分析结果表明，石碑材质为砂岩，主要矿物组成为石英、长石、绿泥石、黑云母和含钙质为主的胶结物。李欣桐等（2019）对河南淅川地处考古遗址中的绿松石制品的矿物学特征进行了分析。

6.2 文物加固

王捷等（2017）以偏高岭土、硅粉及石灰石粉为主要原料制备无机矿物聚合物，并应用于龙山石窟岩体裂隙加固，无机矿物聚合物注浆材料可提供足够的强度，且可根据表层岩体的劣化情况进行表面强度调节。李晓溪（2012）研制了纳米级 SiO_2-丙烯酸酯有机/无机复合材料，并对秦俑彩绘保护材料进行改性，提高其热稳定性、耐光老化性、渗透性，并用于酥粉陶胎的加固保护。赵林毅（2012）通过对应用于岩土质文物保护加固的两种传统材料——礓石和阿嘎土进行改性研究，并应用于内蒙古元上都遗址、南京报恩寺地宫遗址及高句丽壁画墓的保护加固，显著改善了遗址土的物理力学性能，提高了耐环境侵蚀能力。

6.3 文物表面清洁

齐扬等（2015）采用激光清洗技术对云冈石窟砂岩西 43 窟墨迹和烟熏黑垢进行清洗，获得了石灰污染物的干式和湿式清洗阈值均为 32.5 mJ。铜质文物修复中常采用机械除锈、化学镀铜、电刷镀铜和金、铜着色等表面处理工艺可形成致密保护膜，结合牢固且表面孔隙较小（牛飞，2018）。李琼芳等（2018）研究了采用生物矿化法修复寒冷地区石质文物表面，利用黄龙高寒钙华沉积区分离到两株具高产碳酸酐酶活性的嗜冷型细菌菌株，在大理石试件表面进行诱导矿化沉积，实验结果表明大理石试件表面生成了致密的方解石型碳酸钙矿化层。

6.4 文物抗风化及耐腐蚀

严绍军等（2016）对重庆合川钓鱼城古战场遗址的水质分析结果表明，风化样品中石英及方解石含量较岩心样多，长石含量不同程度降低及石膏等可溶盐对岩石的片状风化有一定作用。冯楠（2011）研究了潮湿环境下砖石类文物的风化机理，提出了保护方法。杨志法等（2013）基于风化剥落深度研究了衢州古城岩石砌块的抗风化性能，得出不同岩石抗风化能力为：黄绿色火山角砾岩>赭色细砂岩>赭色凝灰岩>灰绿色凝灰岩。Zhang G. X. 等（2019）研究了石质文化遗产在热带气候条件下的生化反应和生物腐蚀机制，发现生物膜是生物退化机制的一个重要影响因素。

6.5 自然岩溶景观修复保育

以钙华为主体的岩溶景观是大自然的瑰宝。近年来，以九寨沟、黄龙沟为代表的钙华景观出现了开裂、垮塌等退化、受损现象（Zhao et al.，2018）。董发勤团队以 Ca-C-H_2O 物质循环理论为基础，利用空-地无损探测，深入揭示了钙华天然海绵地质体构筑单元特征和生物-钙华共演化规律，以"同质同相"的学术思路，以废弃钙华为原料开发成套钙华修复保育技术并应用于震损核心遗产点抢救性恢复中，首次解决了钙华自然遗产震损及退化的科学修复保育难题，国际上首次成功应用于世界自然遗产地震损、退化钙华景观修复工程，综合效果优良，实现了遵循天然钙华演化规律快速复原钙华自然景观、保持钙华遗产的完整性、原真性目的，为世界自然遗产地钙华景观生态保护提供了新的科学认知和世界级的范例。

7 矿物基固废处理与资源化利用

近十年来，固废处理与资源化研究主要趋向于从固体废弃物的物化特征出发，结合化学、物理化学、化学工程与技术、计算机科学与技术、环境科学、材料科学与工程、数学等学科的相关理论和方法，借助飞行时间二次离子质谱仪、原子力显微镜、CT 扫描、激光剥蚀原位质谱仪、显微激光拉曼光谱仪、原位 X 射线衍射仪等现代分析测试手段，对固体废弃物的表面结构与特性、有害杂质的赋存形式及全组分在不同环境条件下的迁移机制与控制方法、结构重整、反应热力学与反应动力学等进行研究，以期建立固体废弃物中有害物质的分离、固体废弃物安全堆存和工程利用的科学体系，为有效减少固体废弃物对生态环境的影响和规模化利用奠定基础。

大宗矿物基固体废物主要包括矿山尾矿以及粉煤灰、冶炼渣、工业副产石膏和赤泥等工业废渣，其再利用技术进展主要包括两个方面：有价值组分提取及新型材料制备（董发勤等，2014）。例如，赵钰等（2019）利用微生物技术提取有色金属尾矿中的低品位铜、铅、锌等有价组分。宋鹏程等（2016）以蛇纹石尾矿为原料采用蒸氨-碳化闭路工艺，制备出纳米 SiO_2 和 MgO。孟跃辉等（2010）将铅锌矿尾矿、金尾矿、铁尾矿、铜尾矿等用于制备加气混凝土、透水砖、人工鱼礁等新型建筑材料。刘来宝等（2020）围绕铬铁渣的安全转化和再生利用研究设计并制备出具有"核-壳结构"的堇青石/贝利特复相轻集料，提高轻集料混凝土力学强度约 50%（Zhang L. H. et al., 2017），并利用铬铁渣中的镁橄榄石和尖晶石等高温耐火矿相，研究制备了铬铁渣耐火浇注料（张韶华等，2014）。

对于具有放射性的核废物，主要开展了核辐射防护材料以及核素固化处理等研究。矿物基核辐射防护或处理材料的研究主要集中在寻找高稳定性的基材和易合成的方法上。例如，杨玉山等（2013）以重晶石为主要原料开发了一种适于湿热环境的防氡涂层。易发成等以膨润土为放射性废物地质处置库缓冲材料研究了其对核素的长期阻滞效应（王哲等，2020）。核素固化包括水泥固化、玻璃固化、塑料固化、沥青固化、陶瓷固化、岩石固化及晶格固化等。放射性核素晶格固化的概念是由董发勤、周时光首先在西南科技大学和中物院设立的国家联合基金课题中提出来的（受 1978 年 Ringwood 人造岩石方法的启发），是基于天然放射性核素是源于天然矿物晶格的位置，在高放核素处理处置时回放到稳定的矿物晶格中去从而达到长期稳定且与环境安全协调的目标，西南科技大学的核环境科学与技术研究团队开发了钙钛锆石、锆石、榍石、烧绿石等系列核素晶格固化体材料。

近三年国家科技部组织重点专项全面开展固废的分类、评价、标准、安全转化、资源化增值利用研发和工程示范：研究重污染固废源头减量与生态链接技术，如石墨、多金属页岩、硼矿、中低品位磷矿资源清洁利用与重污染固废源头近零排放，典型重金属铜、锌、钒钛、铝、黄金冶炼危废全过程控制技术，难熔金属废料高效回收与清洁提取、高端铜铝废材深度净化与循环再造，复杂铅基多金属固废协同冶炼、镍钴/钨/锑冶金固废清洁提取与无害化，废弃环保催化剂金属回收、含放射性固废清洁解控与安全处置等。固废产生量大、污染重的钢铁冶炼难处理渣尘泥、精细化工园区磷硫氯、制药、造纸、制革行业典型工业污泥，城市污泥快速减量与资源化耦合利用、固废焚烧残余物稳定化无害化等。

大宗铝硅酸盐无机固废物相重构与安全转化、清洁增值利用、智能化回收与分类，如量大面广的废弃混凝土砂粉再生利用；矿业、冶金、化工、陶瓷典型行业大宗低阶固废规模化高值矿物材料化；大宗工业固废协同制备低成本胶凝材料、大掺量制备装配式预制构件等。研究集中于多产业固废的资源环境属性和生态环境影响效应，开发不同种类固废有价金属及非金属、有机资源的高效富集、定向分离与清洁提取技术；开发基于"无废城市"理念的城市多源固废综合管理系统、资源回收与废旧资源协同再利用技术，污染风险控制技术；建设城市固废资源循环经济产业园，开发多源城市固废的协同利用、协同处置技术，重点开发城市固废中金属、塑料的回收技术，水泥窑、冶金窑、厌氧处置等协同处置与资源综合利用技术；开发城市群、经济发展区域、"一带一路"重点产业群、特色产业区域固废资源回收及综合利用技术，污染综合控制技术等（董发勤等，2014）。

8　矿山及周边土壤污染修复

矿山在开发的同时也造成了严重的环境污染，给矿山及周边地区居民的食品安全、生态安全及社会和谐带来了严重隐患。因此，如何治理矿山及周边土壤污染是当前国际资源与环境研究领域的热点问题。

利用矿物的环境属性特别是大量的非金属矿物及其尾矿，经超细、改性、掺杂、复合及仿生合成等方式开发的环境矿物材料为环境污染治理和修复的提供了材料和方法。

8.1 矿山废水污染治理

矿石开采、选矿、输运、处置过程中均会产生含重金属、选矿药剂和 P-N 等酸性矿山废水。近年来，利用环境矿物材料治理矿山废水污染取得了显著成效。陈晓蕾（2018）研究发现，阳离子表面活性剂（DTAB、TTAB、CTAB）改性膨润土可在 150 min 内快速吸附选矿废水中的丁铵黑药，而且 CTAB 改性的膨润土重复利用四次对丁铵黑药的去除率仍能达到 65%。王丹等（2017）发现，$FeCl_3$ 改性膨胀石墨对苯甲羟肟酸和黄药的吸附量分别为 16.31 mg/g 和 18.87 mg/g。Liu 等（2017）和 Zhou 等（2017）构筑了灰岩与生物成因马基诺矿的复合材料，不仅可有效处理去除废水中的 As，而且能够中和酸性矿山废水。彭同江团队以沸石和蒙脱石为滞固材料有效固定了某铀矿山铀尾矿中有害元素的迁移（史鸿晋，2016；胡小强，2016；李真强，2016；刘波，2017）。此外，以矿物吸附材料作为反应介质的渗透反应墙（PRB）原位处理矿业活动导致的地下水污染也是当前的研究热点。由于零价铁在 PRB 中存在易氧化和堵塞，人们相继研究了天然或改性黄铁矿、沸石和磷灰石等矿物材料对土壤地下水的修复，取得了良好效果（刘昊等，2020）。

8.2 土壤污染修复与治理

环境矿物材料具备较强的污染固化和环境净化能力，为解决众多土壤污染问题提供了新途径，尤其是海泡石、凹凸棒石、蒙脱石、高岭石和沸石等硅酸盐矿物具有复杂的多孔道结构，能吸附重金属离子，降低重金属在土壤环境中的迁移转化能力及毒性，且天然无害、价格低廉、分布广泛，具有很大的应用潜力（Xu et al., 2017; Otunola and Ololade, 2020）。以天然黏土矿物为基础开发新型土壤修复材料是近年来土壤修复技术的研究重点之一。例如，蒙脱石和碳酸盐矿化菌的协同体系对土壤中 Sr^{2+}、Pb^{2+} 具有良好的联合滞固效应，有望为实际突发性重金属、核素污染的快速原位治理提供储备材料（许凤琴，2016）。凹凸棒石负载零价铁后对土壤中铅、镉和铬的吸附量显著提升，增加了蔬菜生长速度（Xu et al., 2019）。蒙脱石负载纳米马基诺矿后对六价铬的固定效果较好（李小飞，2019）。羟基磷灰石-凹凸棒土复合材料的比表面积远大于单独的羟基磷灰石和凹凸棒土，对重金属污染土壤钝化性能较好（徐丽莎，2019）。在大量土壤修复工程应用中，黏土基修复材料常与生物炭、石灰等修复剂复配使用，且对土壤污染修复效果好于单一修复剂（胡艳美等，2020）。

除了矿山及周边土壤环境污染外，表土荒漠化、沙漠化也是当前全球面临的一个长期问题。矿物材料在土壤荒漠化治理中起着重要的固沙保水作用。冉飞天（2016）以坡缕石黏土、聚乙烯醇和部分中和的丙烯酸为原料制备得高吸水复合材料，能够显著提高固沙试样的抗压强度，从而保护沙生植物的生长中使其根系免受风沙的侵蚀，同时为其生长提供一定的营养，有保温、保水作用，对沙漠地区植被的恢复起到促进作用。王爱娣（2013）以自然界广泛分布、储量丰富的天然红土和黄土为研究对象，与环境友好型高分子（CMC、PVA）和生物高分子（植物秸秆）进行复配制备天然黏土基固沙材料兼具工程固沙、化学固沙和生物固沙的复合效果，具有良好的应用前景。姜雄和铁生年（2014）采用溶液共混法制备了膨润土-聚丙烯酰胺复合吸水保水材料，发现膨润土与聚丙烯酰胺用量质量比为 8:2 时，复合材料吸水率、吸盐率和保水率分别为 12.28 g/g、5.9 g/g 和 42.54%，适合作为固沙材料的保水添加剂，且聚丙烯酰胺用量较少。

9 展 望

近十年，应用矿物学在矿物学和相关学科的先进理论方法技术带动下，在传统应用领域取得了丰硕的成果，在新兴领域得到大力拓展。但是，我国作为全球第二大经济体在资源供给消耗反全球化的形势下，应用矿物学如何在能源革命与面临资源危机转化为资源革命如通过资源代替、可循环、最优化方式，解决生态与环境危机中的难题；此外，以石墨烯、清洁能源、量子信息技术、人工智能、5G技术、大数据云计算、生物技术等为技术突破口的第四次工业革命（绿色工业革命）已经悄然来临，国家也早在2015年部署了全面推进实施"中国制造2025"强国战略，应用矿物学在生态环境修复、矿产资源（包括再生资源）高效绿色循环利用、清洁能源生产、先进材料、碳中和、未来技术等领域发挥重要作用。

（1）重视非金属矿物资源及其功能开发利用，重点在新型资源、新型材料、生态环境保护与修复治理等领域并扩大其空间，应将磷矿、钾盐、萤石、石墨、石英、叶蜡石、黏土、硅藻土、重晶石、硼矿、锂矿、锆英石等矿产列为我国战略性非金属矿产，并加大其新性能新功能的开发和应用，拓展新型工业矿物品种和应用领域，建立矿物资源、矿物材料产品性能和质量提升与可持续发展兼容的工业矿物发展体系。

（2）大力拓展矿物自动测量与分析技术，并配合新型遥感如矿物高光谱立体分析方法，用大数据方法挖掘综合矿物学参数的特性与规律性，建立基于云计算的三维数据表征技术模型，预测选冶矿加综合工艺指标，重点瞄准贫、细、杂、难的矿物资源品种，对低品位及结构复杂的矿物如硫化矿物、氧化物、非金属矿物、煤炭、石墨等综合运用新方法技术，形成新的应用矿物学动态数据链。

（3）加大在特定环境条件（或未知环境）下矿物与环境介质的相互作用机制与效应研究，如强辐射、强激光、强微波、强电磁、强中子、超高（低）温、超高压等独特环境下矿物的相互作用与表征（谱学响应）；特别是在常温下通过生物分子获得一些特殊结构和形貌的矿物及材料，探讨蛋白质等生物大分子利用其结构的复杂性控制生物分子在生物矿化中的作用；探讨矿物形成超材料的方法与途径，如负折射率超材料、负的电容率和磁导率超材料、微米透镜材料等。

（4）非金属矿是环境友好的生态环境材料，应大力开发低品位、尾矿和采剥形成非金属矿潜在资源与新用途。研究固废与生态环境材料、低碳绿色工艺、功能、艺术美学之间的联系，服务于无废城市，朝世间无废之理想目标迈进。固体废弃物的成分复杂、毒害组分含量高、物理化学性质差异较大，特别是其资源价值较低，材料化处理成本较高，排放量巨大，要明确部分固体废弃物的资源属性，发掘难利用固废的资源途径，特别是针对尾矿、钢渣、工业副产石膏、赤泥、粉煤灰、建筑垃圾等最难处理的几种大宗固废。大宗固废的材料化资源属性评价方法与应用技术、固废的环境危害的评价方法体系、质量标准体系，在复杂组分及结构下，固废中金属、非金属资源的全组分利用、高附加值利用的应用矿物学基础研究及其资源经济性评价的市场、技术、环境、安全等方面的政策研究需要加强，开发无机-有机固废基于物质-能量-环境自平衡耦合定向转化和系统优化新方法和危险废物毒害组分快速识别与检测新技术。

（5）氢能时代加速大型工业化应用对应用矿物学带来新的挑战和机遇。氢气冶金、氢气炼钢炼铁、氢气烧结水泥、氢气交通等新工艺带来冶金过程、热反应过程、燃烧过程等矿物相变研究热潮，带动一批相关新材料的研发，如稀土储氢材料、高效发光稀土功能、人工晶状体、碳碳复合材料、碳陶复合材料、金属基复合材料、生物医用材料、高性能钕铁硼、超固体、时间晶体、木材海绵、冷沸材料、微格金属、量子技术、光子晶体、4D打印材料、自修复材料、平皱纹材料、永湿材料、坚如岩石的涂层材料、可编程水泥、纳米点钙钛矿、遮光玻璃涂层、空调功能墙体材料、仿生塑料等。

（6）在5G通信和新一代智能互联，传感感应器件制造上应用矿物学具备广阔的应用空间。我国用于与5G相关的新材料的进口率高达86%，化工新材料产业国内保障能力只有50%；高端高温合金主要依赖进口，矿物高端原料及其精细加工如半导体材料、显示材料、新能源材料、高性能纤维、高性能膜材料、大硅片、抛光材料、偏光片、OLED发光材料、高纯溅射靶材、化学机械抛光（chemical mechanical polishing，CMP）材料、碳化硅单晶等。

（7）利用非金属矿多样性与功能或矿区矿物材料开发就地取材的场地土壤污染治理材料与技术，如铀矿区放射性核素污染控制及治理、煤化工、油田重金属污染场地防治及安全利用集成技术、煤炭产业集聚区场地污染治理、有色金属选冶渣场影响区污染修复、冶炼场地土壤-地下水协同修复、废旧电器拆解场地污染区修复技术等。

总之，我们这里不能穷尽应用矿物学在未来十年的所有发展内容，但可以预期应用矿物学和科学家在第四次产业技术革命的前景下面临的挑战和迫切任务，期待广大同行共同创造和努力。

致谢：本文撰写过程中，就相关问题与张岭、黄阳、王振等同事和陈禹衡、夏雪等研究生进行了有益探讨，特此致谢。

参 考 文 献

白桦. 2019. 层状过渡金属硫族化合物的超导电性研究. 杭州:浙江大学博士学位论文
常晶晶,张文元,徐抒平,宣旭阳,苏伯民,徐蔚青. 2010. 古代壁画中矿物颜料的拉曼光谱研究. 见:第十六届全国分子光谱学学术会议论文集. 郑州:中国光学学会,中国化学会,277-278
陈庚龄,康明大. 2012. 临洮哥舒翰纪功碑岩层矿物分析. 丝绸之路,(2):114-116
陈庆汉. 2012. 壳熔法生长蓝宝石单晶的进展. 宝石和宝石学杂志,14(3):17-21
陈晓蕾. 2018. 有机改性膨润土对选矿废水中丁铵黑药的吸附性能研究. 济南:山东大学硕士学位论文
陈正国,于海军,李朝灿,陈军元,熊军. 2019. 我国非金属矿分类探讨. 中国非金属矿工业导刊,(2):1-5
董发勤. 2015. 应用矿物学. 北京:高等教育出版社
董发勤,徐龙华,彭同江,代群威,谌书. 2014. 工业固体废物资源循环利用矿物学. 地学前缘,21(5):302-312
杜保聪,张鸣,杨鼎宜,陈霓超,周勋建,蔡雨,李书龙. 2018. 地聚合物材料研究与应用现状. 四川建材,44(12):38-41
杜培鑫,袁鹏. 2019. 叶蜡石在超硬材料等关键矿物材料领域的研究和应用. 矿产保护与利用,39(6):87-92
杜培鑫,袁鹏,庄官政. 2019. 纳米管状埃洛石的应用矿物学研究进展. 矿产保护与利用,39(6):77-86
方明山,王明燕. 2018. AMICS在铜矿伴生金银综合回收中的应用. 矿冶,27(3):104-108
冯聪,储满生,唐珏,汤雅婷,柳政根. 2016. 不同类型含钛高炉渣主要冶金性能及物相. 中南大学学报(自然科学版),47(8):2556-2562
冯楠. 2011. 潮湿环境下砖石类文物风化机理与保护方法研究. 长春:吉林大学博士学位论文
付倩丽. 2016. 古代矿物颜料拉曼光谱定量分析方法模拟实验研究. 西安:西北大学硕士学位论文
郭鸿旭. 2017. 宝石"新贵":沙弗莱石 变色水铝石 加斯佩石. 上海工艺美术,(4):62-63
何军拥,田承宇. 2013. 玄武岩纤维水工高性能混凝土的耐久性研究. 混凝土与水泥制品,(5):46-48
胡小强. 2016. 沸石矿物对铀尾矿的固化及阻滞机理研究. 绵阳:西南科技大学硕士学位论文
胡艳美,王旭军,党秀丽. 2020. 改良剂对农田土壤重金属镉修复的研究进展. 江苏农业科学,48(6):17-23
黄毅,徐国平,程慧高,蒋卓辉,杨宇. 2014. 典型钢渣的化学成分、显微形貌及物相分析. 硅酸盐通报,33(8):1902-1907
姜雄,铁生年. 2014. 膨润土/聚丙烯酰胺复合吸水保水材料的制备及性能研究. 硅酸盐通报,33(4):731-735
蒋谋锋,张一敏,包申旭,杨晓. 2015. 石英对某云母型石煤酸浸提钒的影响. 有色金属(冶炼部分),(8):34-38
李琼芳,何鑫,陈超,杨清清,吕治州. 2018. 两株嗜冷碳酸钙矿化菌对大理石表面修复效果研究. 人工晶体学报,47(1):172-178
李小飞. 2019. 天然矿物材料改性及其修复铬污染土壤效果研究. 北京:北京化工大学硕士学位论文
李晓静. 2016. 常见宝石的近红外光谱研究. 昆明:昆明理工大学硕士学位论文
李晓溪. 2012. 脆弱陶质文物加固材料的筛选及改性研究. 西安:西北大学硕士学位论文
李欣桐,先怡衡,樊静怡,张璐繁,郭靖雯,高占远,温睿. 2019. 应用扫描电镜-X射线衍射-电子探针技术研究河南淅川绿松石矿物学特征. 岩矿测试,38(4):373-381
李真强. 2016. 沸石矿物对铀尾矿堆积体的固化处理与固化体稳定性研究. 绵阳:西南科技大学硕士学位论文
刘宝昌,曹鑫,孟庆阳,朱品文,戴文昊,韩哲,赵新哲,李思奇. 2018. PCBN基体孕镶金刚石复合材料的制备与性能研究. 金刚石与磨料磨具工程,38(5):21-27
刘波. 2017. 某铀矿山尾矿中有害金属元素的迁移、阻滞及机理研究. 绵阳:西南科技大学博士学位论文
刘昊,廖立兵,吕国诚,王丽娟,梅乐夫,郭庆丰. 2020. 矿物介质材料在渗透反应格栅技术中的应用研究进展. 矿物学报,40(3):274-280
刘来宝,张礼华,唐凯靖. 2020. 高碳铬铁冶金渣资源化综合利用技术. 北京:中国建材工业出版社
刘庆生,李江霖,常晴,邱廷省. 2019. 离子型稀土矿浸出前后工艺矿物学研究. 稀有金属,43(1):92-101
刘晓晔. 2019. 光合作用生物膜形成的方解石鲕粒:以河南石井剖面寒武系苗岭统为例. 北京:中国地质大学(北京)硕士学位论文
刘鑫,张一敏,刘涛,孙坤,汪承宝. 2017. 湖北某云母型含钒石煤重-浮联合预抛尾试验. 金属矿山,(5):93-98
毛喆,李丹,钟音,朱锡芬,鲜海洋,彭平安. 2016. 硫化亚铁矿物的生物合成及其对六溴环十二烷的还原脱溴研究. 地球化学,45(6):601-613
孟跃辉,倪文,张玉燕. 2010. 我国尾矿综合利用发展现状及前景. 中国矿山工程,39(5):4-9
牛飞. 2018. 几种表面处理技术在金属文物修复中的应用. 电镀与涂饰,37(16):728-731
牛乐乐,刘征建,张建良,张宗旺,王耀祖,杜诚波. 2019. 铁矿粉矿物组成对烧结冶金性能的影响. 钢铁,54(9):27-32,38
潘海华,唐睿康. 2020. 生物矿化及仿生矿化中的信息传递和转化. 化工学报,71(1):68-80

齐扬,周伟强,陈静,叶亚云,周萍,侯静敏,吴鹏. 2015. 激光清洗云冈石窟文物表面污染物的试验研究. 安全与环境工程,22(2):32-38
秦丽辉. 2014. 玄武岩纤维布加固损伤混凝土梁力学性能研究. 哈尔滨:哈尔滨工业大学博士学位论文
秦雅静,朱德山. 2014. 我国水镁石矿资源利用现状及展望. 中国非金属矿工业导刊,(6):1-3
冉飞天. 2016. 黏土基高分子复合材料的制备及其保水固沙性能研究. 兰州:西北师范大学硕士学位论文
阮彬彬. 2018. 新型卤化物、插层结构及合金超导体探索. 北京:中国科学院大学(中国科学院物理研究所)博士学位论文
邵秋月,葛文,沈建军. 2019. 硅灰石的表面改性及新型矿物基聚合物的合成. 非金属矿,42(5):16-18
邵志斌. 2019. 常规超导体及 IV-VI 族薄膜的分子束外延生长及扫描隧道显微镜研究. 武汉:华中科技大学博士学位论文
申柯娅. 2011. 天然祖母绿与合成祖母绿的成分及红外吸收光谱研究. 岩矿测试,30(2):233-237
史鸿晋. 2016. 某铀矿山铀尾矿属性与核素铀的迁移规律及工程阻滞研究. 绵阳:西南科技大学研究硕士学位论文
宋鹏程,彭同江,孙红娟,贾蕾,张馨文,邱国华,朱琳. 2016. 纤蛇纹石石棉尾矿综合利用新进展. 中国非金属矿工业导刊,(2):14-17
孙永升,韩跃新,高鹏,王琴. 2015. 温度对鲕状赤铁矿石深度还原特性的影响. 中国矿业大学学报,44(1):132-137
谭远. 2016. 生物成因纤维状文石集合体的结构表征及其仿生制备. 南宁:广西大学硕士学位论文
王爱娣. 2013. 黏土基复合固沙材料性能研究. 兰州:西北师范大学硕士学位论文
王丹. 2017. 膨胀石墨的制备及其对捕收剂吸附研究. 昆明:昆明理工大学. 硕士学位论文
王菲,于湖生,宫怀瑞. 2019. 抗菌防螨远红外黏胶纤维的制备及性能研究. 针织工业,(10):7-9
王健,张乐文,冯啸,赵少龙,王洪波. 2015. 碱激发地聚合物双液注浆材料试验与应用研究. 岩石力学与工程学报,34(S2):4418-4425
王捷,王逢睿,申喜旺,何真. 2017. 无机矿物聚合物应用于龙山石窟岩体裂隙加固的试验研究. 新型建筑材料,44(5):58-62
王哲,易发成,刘艳,刘延. 2020. 以膨润土为基材的集成缓冲材料对铀的长期阻滞效应研究. 矿物岩石地球化学通报,39(2):200-208,169
温利刚,曾普胜,詹秀春,范晨子,孙冬阳,王广,袁继海. 2018. 矿物表征自动定量分析系统(AMICS)技术在稀土稀有矿物鉴定中的应用. 岩矿测试,37(2):121-129
文圆,黄惠宁,张国涛,黄辛辰,杨景琪,戴永刚. 2019. 电气石材料性能与应用研究进展. 陶瓷,(2):17-24
吴逍,孙红娟,彭同江,段佳琪,杨红梅,吴槿萱,喇继德. 2017. 优质石英岩作为高纯石英原料的提纯试验研究. 非金属矿,40(1):68-70,74
向小平,刘武杨,李东升,甘牧原,马承胜,石楚刚. 2020. 柳钢烧结矿的化学成分及显微结构分析. 烧结球团,45(2):21-25,30
谢宝珊,李传常,张波,赵新波,陈荐,陈中胜. 2019. 硅酸盐矿物储热特征及其复合相变材料. 硅酸盐学报,47(1):143-152
邢睿睿,陈美华,邹昱. 2019. 晶化时间对高压高温合成翡翠品质的影响. 宝石和宝石学杂志,21(1):31-39
许凤琴. 2016. 蒙脱石-碳酸盐矿化菌对 Sr^{2+}、Pb^{2+} 的联合滞固研究. 绵阳:西南科技大学硕士学位论文
徐丽莎. 2019. 羟基磷灰石/凹凸棒土复合材料制备及其对重金属污染土壤钝化性能研究. 成都:成都理工大学硕士学位论文
闫华晓,韩作振,赵辉,庄定祥. 2019. 对生物成因和非生物成因方解石的新见解. 见:中国矿物岩石地球化学学会第 17 届学术年会论文摘要集. 杭州:中国矿物岩石地球化学学会,1091
严绍军,何凯,孙鹏,窦彦,陈嘉琦. 2016. 重庆合川钓鱼城古战场遗址砂岩风化试验研究. 长江科学院院报,33(8):100-104,119
杨帆,李峰,胡南,张辉,丁德馨. 2020. 含铀煤灰工艺矿物学研究. 矿物学报,doi:10.16461/j.cnki.1000-4734.2020.40.141
杨玉山,董发勤,邓跃全,曲瑞雪. 2013. 一种适于湿热环境的防氡涂层. 原子能科学技术,47(12):2384-2388
杨志法,张中俭,周剑,李丽慧. 2013. 基于风化剥落深度的衢州古城墙小西门岩石砌块和蛎灰勾缝条长期抗风化能力研究. 工程地质学报,21(1):97-102
于洋. 2017. 生物方解石微观力学性能的纳米压痕试验研究. 长春:吉林大学硕士学位论文
袁鸿,龙永芳,赵西晨,刘雄. 2013. 荆门獾子冢搬迁车马土体分析研究. 敦煌研究,(1):82-85
张海亮,张明福,韩杰才,侯永昭,杨雯. 2020. 镁/钛离子注入对蓝宝石微结构和机械性能的影响. 人工晶体学报,49(2):195-204
张妮,林春明. 2016. X 射线衍射技术应用于宝石鉴定-合成及晶体结构研究进展. 岩矿测试,35(3):217-228
张韶华,刘来宝,谭克锋,张登科,唐凯靖. 2014. 掺铬铁渣的铝镁系浇注料的制备与性能研究. 耐火材料,48(6):436-438,442
张志祥,斯琴高娃,翁根花,李伟,哈斯. 2013. 浅谈人工宝石. 西部资源,(5):179-181
赵康. 2013. 生物分子对方解石体外仿生矿化的研究. 青岛:中国石油大学(华东)博士学位论文
赵林毅. 2012. 应用于岩土质文物保护加固的两种传统材料的改性研究. 兰州:兰州大学博士学位论文
赵钰,董颖博,林海. 2019. 有色金属矿尾矿微生物浸出技术研究进展. 金属矿山,48(11):197-203
Al-Khirbash S A. 2020. Mineralogical characterization of low-grade nickel laterites from the North Oman Mountains: using mineral liberation analyses-scanning electron microscopy-based automated quantitative mineralogy. Ore Geology Reviews, 120: 103429
Barton I, Ahn J, Lee J. 2018. Mineralogical and metallurgical study of supergene ores of the Mike Cu-Au(-Zn) deposit, Carlin trend, Nevada. Hydrometallurgy, 176: 176-191
Bouville F, Maire E, Meille S, van de Moortèle B, Stevenson A J, Deville S. 2014. Strong, tough and stiff bioinspired ceramics from brittle constituents. Nature Materials, 13(5): 508-514
Cabri L J, Wilhelmij H R, Eksteen J J. 2017. Contrasting mineralogical and processing potential of two mineralization types in the platinum group element and Ni-bearing Kapalagulu Intrusion, western Tanzania. Ore Geology Reviews, 90: 772-789
Cao Y, Fatemi V, Demir A, Fang S A, Tomarken L S, Luo Y J, Sanchez-Yamagishi D J, Watanabe K, Taniguchi T, Kaxiras E, Ashoori C R, Jarillo-Herrero P. 2018a. Correlated insulator behaviour at half-filling in magic-angle graphene superlattices. Nature, 556(7699): 80-84
Cao Y, Fatemi V, Fang S A, Watanabe K, Taniguchi T, Kaxiras E, Jarillo-Herrero P. 2018b. Unconventional superconductivity in magic-angle graphene superlattices. Nature, 556(7699): 43-50

Cruz-Silva R, Morelos-Gomez A, Kim H I, Jang H K, Tristan F, Vega-Diaz S, Rajukumar L P, Elias A L, Perea-Lopez N, Suhr J, Endo M, Terrones M. 2014. Super-stretchable graphene oxide macroscopic fibers with outstanding knotability fabricated by dry film scrolling. ACS Nano, 8(6):5959–5967

Fan W, Zhang X, Zhang Y, Zhang Y F, Liu T X. 2019. Lightweight, strong, and super-thermal insulating polyimide composite aerogels under high temperature. Composites Science and Technology, 173:47–52

Gao Y, Cao T F, Cellini F, Berger C, de Heer W A, Tosatti E, Riedo E, Bongiorno A. 2018. Ultrahard carbon film from epitaxial two-layer graphene. Nature Nanotechnology, 13(2):133–138

Ge J F, Liu Z L, Liu C H, Gao C L, Qian D, Xue Q K, Liu Y, Jia J F. 2015. Superconductivity above 100 K in single-layer FeSe films on doped $SrTiO_3$. Nature Materials, 14(3):285–289

Gu L L, Poddar S, Lin Y, Long Z H, Zhang D Q, Zhang Q P, Shu L, Qiu X, Kam M, Javey A, Fan Z Y. 2020. A biomimetic eye with a hemispherical perovskite nanowire array retina. Nature, 581(7808):278–282

Guanira K, Valente T M, Ríos C A, Castellanos O M, Salazar L, Lattanzi D, Jaime P. 2020. Methodological approach for mineralogical characterization of tailings from a Cu(Au, Ag) skarn type deposit using QEMSCAN (Quantitative Evaluation of Minerals by Scanning Electron Microscopy). Journal of Geochemical Exploration, 209:106439

Han J K, Du G L, Gao W W, Bai H. 2019. An anisotropically high thermal conductive boron nitride/epoxy composite based on nacre-mimetic 3D network. Advanced Functional Materials, 29(13):1900412

Hill E H, Hanske C, Johnson A, Yate L, Jelitto H, Schneider G A, Liz-Marzan L M. 2017. Metal nanoparticle growth within clay-polymer nacre-inspired materials for improved catalysis and plasmonic detection in complex biofluids. Langmuir, 33(35):8774–8783

Hu X Z, Xu Z, Liu Z, Gao C. 2013. Liquid crystal self-templating approach to ultrastrong and tough biomimic composites. Scientific Reports, 3:2374

Huang Z Q, Zhang G S. 2012. Biomimetic synthesis of aragonite nanorod aggregates with unusual morphologies using a novel template of natural fibrous proteins at ambient condition. Crystal Growth & Design, 12(4):1816–1822

Jalili R, Aboutalebi S H, Esrafilzadeh D, Shepherd R L, Chen J, Aminorroaya-Yamini, S, Konstantinov K, Minett A I, Razal J M, Wallace G G. 2013. Scalable one-step wet-spinning of graphene fibers and yarns from liquid crystalline dispersions of graphene oxide: towards multifunctional textiles. Advanced Functional Materials, 23(43):5345–5354

Jun J M V, Altoe M V P, Aloni S, Zuckermann R N. 2015. Peptoid nanosheets as soluble, two-dimensional templates for calcium carbonate mineralization. Chemical Communications, 51(50):10218–10221

Knöller A, Lampa C P, von Cube F, Zeng T H, Bell D C, Dresselhaus M S, Burghard Z, Bill J. 2017. Strengthening of ceramic-based artificial nacre via synergistic interactions of 1D vanadium pentoxide and 2D graphene oxide building blocks. Scientific Reports, 7:40999

Le Ferrand H, Bouville F, Niebel T P, Studart A R. 2015. Magnetically assisted slip casting of bioinspired heterogeneous composites. Nature Materials, 14(11):1172–1179

Le P T A, Vu T P, Le H T, van Phan D, Nguyen C X, Luong T D, Dang N T T, Nguyen T D. 2020. Nacre-mimicking Titania/Graphene/chitin assemblies in macroscopic layered membranes and their performance. Journal of Electronic Materials, 49(6):3791–3803

Lei Y F, Chen X H, Song H H, Hu Z J, Cao B. 2017. The influence of thermal treatment on the microstructure and thermal insulation performance of silica aerogels. Journal of Non-Crystalline Solids, 470:178–183

Li M, Wu Z S, Kao H T. 2011. Study on preparation and thermal properties of binary fatty acid/diatomite shape-stabilized phase change materials. Solar Energy Materials and Solar Cells, 95(8):2412–2416

Li Q H, Li M, Zhu P Z, Wei S C. 2012. In vitro synthesis of bioactive hydroxyapatite using sodium hyaluronate as a template. Journal of Materials Chemistry, 22(38):20257–20265

Li Z, Fu Y H, Zhou A N, Zhu C Y, Yang C, Zhang Q, 2017. Air impact pulverization-precise classification process to support ultraclean coal production. Powder Technology, 318:231–241

Liang B L, Zhao H W, Zhang Q, Fan Y Z, Yue Y H, Yin P G, Guo L. 2016. Ca^{2+} enhanced nacre-inspired montmorillonite-alginate film with superior mechanical, transparent, fire retardancy, and shape memory properties. ACS Applied Materials & Interfaces, 8(42):28816–28823

Liang C Y, Luo Y C, Yang G D, Xia D, Liu L, Zhang X M, Wang H S. 2018. Graphene oxide hybridized nHAC/PLGA scaffolds facilitate the proliferation of MC3T3-E1 cells. Nanoscale Research Letters, 13:15

Liu C, Wong H M, Yeung K W K, Tjong S C. 2016. Novel electrospun polylactic acid nanocomposite fiber mats with hybrid graphene oxide and nano-hydroxyapatite reinforcements having enhanced biocompatibility. Polymers, 8(8):287

Liu H C, Nie Z Y, Xia J L, Zhu H R, Yang Y, Zhao C H, Zheng L, Zhao Y D. 2015. Investigation of copper, iron and sulfur speciation during bioleaching of chalcopyrite by moderate thermophile *Sulfobacillus thermosulfidooxidans*. International Journal of Mineral Processing, 137:1–8

Liu H Y, Cheng J, Chen F J, Bai D C, Shao C W, Wang J, Xi P X, Zeng Z Z. 2014. Gelatin functionalized graphene oxide for mineralization of hydroxyapatite: biomimetic and in vitro evaluation. Nanoscale, 6(10):5315–5322

Liu J, Zhou L, Dong F Q, Hudson-Edwards K A. 2017. Enhancing As(V) adsorption and passivation using biologically formed nano-sized FeS coatings on limestone: implications for acid mine drainage treatment and neutralization. Chemosphere, 168:529–538

Livanov K, Jelitto H, Bar-On B, Schulte K, Schneider G A, Wagner D H. 2015. Tough alumina/polymer layered composites with high ceramic content. Journal of the American Ceramic Society, 98(4):1285–1291

Lv J W, Ding D F, Yang X K, Hou K, Miao X, Wang D W, Kou B C, Huang L, Tang Z Y. 2019. Biomimetic chiral photonic crystals. Angewandte Chemie International Edition, 58(23):7783–7787

Ma L Y,Wang X J,Feng X,Liang Y L,Xiao Y H,Hao X D,Yin H Q,Liu H W,Liu X D. 2017. Co-culture microorganisms with different initial proportions reveal the mechanism of chalcopyrite bioleaching coupling with microbial community succession. Bioresource Technology,223:121-130

Ma T B,Zhao Y S,Ruan K P,Liu X R,Zhang J L Guo Y Q,Yang X T,Kong J,Gu J W. 2020. Highly thermal conductivities, excellent mechanical robustness and flexibility, and outstanding thermal stabilities of aramid nanofiber composite papers with nacre-mimetic layered structures. ACS Applied Materials & Interfaces,12(1):1677-1686

Mizuguchi Y,Fujihisa H,Gotoh Y,Suzuki K,Usui H,Kuroki K,Demura S,Takano Y,Izawa H,Miura O. 2012. BiS_2-based layered superconductor $Bi_4O_4S_3$. Physical Review B,86(22):220510

Morits M,Verho T,Sorvari J,Liljeström V,Kostiainen M A,Gröschel A H,Ikkala O. 2017. Toughness and fracture properties in nacre-mimetic clay/polymer nanocomposites. Advanced Functional Materials,27(10):1605378

Nam H Y,Min K H,Kim D E,Choi J R,Lee H J,Lee S C. 2017. Mussel-inspired poly(l-dopa)-templated mineralization for calcium phosphate-assembled intracellular nanocarriers. Colloids and Surfaces B:Biointerfaces,157:215-222

Nazari A M,Ghahreman A,Bell S. 2017. A comparative study of gold refractoriness by the application of QEMSCAN and diagnostic leach process. International Journal of Mineral Processing,169:35-46

Nie Z Y,Zhang W W,Liu H C,Zhu H R,Zhao C H,Zhang D R,Zhu W,Ma C Y,Xia J L. 2019. Bioleaching of chalcopyrite with different crystal phases by *Acidianus manzaensis*. Transactions of Nonferrous Metals Society of China,29(3):617-624

Nomoto T,Fukushima S,Kumagai M,Miyazaki K,Inoue A,Mi P,Maeda Y,Toh K,Matsumoto Y,Morimoto Y,Kishimura A,Nishiyama N,Kataoka K. 2016. Calcium phosphate-based organic-inorganic hybrid nanocarriers with pH-responsive on/off switch for photodynamic therapy. Biomaterials Science,4(5):826-838

Otunola B O,Ololade O O. 2020. A review on the application of clay minerals as heavy metal adsorbents for remediation purposes. Environmental Technology & Innovation,18:100692

Sarier N,Onder E,Ozay S,Ozkilic Y. 2011. Preparation of phase change material-montmorillonite composites suitable for thermal energy storage. Thermochimica Acta,524(1-2):39-46

Sharma V,Srinivasan A,Nikolajeff F,Kumar S. 2020. Biomineralization process in hard tissues:the interaction complexity within protein and inorganic counterparts. Acta Biomaterialia,doi:10.1016/j.actbio.2020.04.049

Tanii H,Inazumi T,Terashima K. 2014. Mineralogical study of iron sand with different metallurgical characteristic to smelting with use of Japanese classic iron-making furnace "Tatara". ISIJ International,54(5):1044-1050

Wan S J,Zhang Q,Zhou X H,Li D C,Ji B H,Jiang L,Cheng Q F. 2017. Fatigue resistant bioinspired composite from synergistic two-dimensional nanocomponents. ACS Nano,11(7):7074-7083

Wang K F,Chou K C,Zhang G H. 2020. Preparation of high-purity and ultrafine WC-Co composite powder by a simple two-step process. Advanced Powder Technology,31(5):1940-1945

Wei Z L,Li Y B,Gao H M,Zhu Y G,Qian G J,Yao J. 2019. New insights into the surface relaxation and oxidation of chalcopyrite exposed to O_2 and H_2O:a first-principles DFT study. Applied Surface Science,492:89-98

Wu Q,Guo D,Zhang Y W,Zhao H W,Chen D Z,Nai J W,Liang J F,Li X W,Sun N,Guo L. 2014. Facile and universal superhydrophobic modification to fabricate waterborne,multifunctional nacre-mimetic films with excellent stability. ACS Applied Materials & Interfaces,6(23):20597-20602

Xu C B,Qi J,Yang W J,Chen Y,Yang C,He Y L,Wang J,Lin A J. 2019. Immobilization of heavy metals in vegetable-growing soils using nano zero-valent iron modified attapulgite clay. Science of the Total Environment,686:476-483

Xu W L,Chen S,Zhu Y N,Xiang X X,Bo Y Q,Lin Z M,Wu H,Liu H E. 2020. Preparation of hyperelastic graphene/carboxymethyl cellulose composite aerogels by ambient pressure drying and its adsorption applications. Journal of Materials Science,55(24):10543-10557

Xu Y,Liang X F,Xu Y M,Qin X,Huang Q Q,Wang L,Sun Y B. 2017. Remediation of heavy metal-polluted agricultural soils using clay minerals:a review. Pedosphere,27(2):193-204

Xu Z,Sun H Y,Zhao X L,Gao C. 2013. Ultrastrong fibers assembled from giant graphene oxide sheets. Advanced Materials,25(2):188-193

Yang C R,Wang Y J,Chen X F. 2012. Preparation and evaluation of biomimetric nano-hydroxyapatite-based composite scaffolds for bone-tissue engineering. Chinese Science Bulletin,57(21):2787-2792

Zhang B Y,Wang Q,Zhang Y X,Gao W Q,Hou Y C,Zhang G X. 2019. A self-assembled,nacre-mimetic,nano-laminar structure as a superior charge dissipation coating on insulators for HVDC gas-insulated systems. Nanoscale,11(39):8046-18051

Zhang G X,Gong C J,Gu J G,Katayama Y,Someya T,Gu J D. 2019. Biochemical reactions and mechanisms involved in the biodeterioration of stone world cultural heritage under the tropical climate conditions. International Biodeterioration & Biodegradation,143:104723

Zhang J G,Liu Y Z,Feng T,Zhou M J,Zhou L,Zhou A J,Zhu L. 2017. Immobilizing bacteria in expanded perlite for the crack self-healing in concrete. Construction and Building Materials,148:610-617

Zhang J S,Zhang X,Wan Y Z,Mei D D,Zhang B. 2012. Preparation and thermal energy properties of paraffin/halloysite nanotube composite as form-stable phase change material. Solar Energy,86(5):1142-1148

Zhang L,Zhang Y Q,Li F B,Yan S,Wang Z S,Fan L X,Zhang G Z,Li H J. 2019. Water-evaporation-powered fast actuators with multimodal motion based on robust nacre-mimetic composite film. ACS Applied Materials & Interfaces,11(13):12890-12897

Zhang L H,Zhang Y S,Liu C B,Liu L B,Tang K J. 2017. Study on microstructure and bond strength of interfacial transition zone between cement paste and high-performance lightweight aggregates prepared from ferrochromium slag. Construction and Building Materials,142:31-41

Zhang Y Y, Li Y C, Ming P, Zhang Q, Liu T X, Jiang L, Cheng Q F. 2016. Ultrastrong bioinspired graphene-based fibers via synergistic toughening. Advanced Materials, 28(14):2834-2839

Zhao B, Wang Y S, Luo Y H, Li J, Zhang X, Shen T. 2018. Landslides and dam damage resulting from the Jiuzhaigou earthquake (8 August 2017), Sichuan, China. Royal Society Open Science, 5(3):171418

Zhao H W, Yue Y H, Guo L, Wu J T, Zhang Y W, Li X D, Mao S C, Han X D. 2016. Cloning nacre's 3D interlocking skeleton in engineering composites to achieve exceptional mechanical properties. Advanced Materials, 28(25):5099-5105

Zhao K, Zhang T F, Chang H C, Yang Y, Xiao P S, Zhang H T, Li C X, Tiwary C S, Ajayan P M, Chen Y S. 2019. Super-elasticity of three-dimensionally cross-linked graphene materials all the way to deep cryogenic temperatures. Science Advances, 5(4):eaav2589

Zhao X Y, Zhu Y J, Zhao J, Lu B Q, Chen F, Qi C, Wu J. 2014. Hydroxyapatite nanosheet-assembled microspheres: hemoglobin-templated synthesis and adsorption for heavy metal ions. Journal of Colloid and Interface Science, 416:11-18

Zhou L, Dong F Q, Liu J, Hudson-Edwards K A. 2017. Coupling effect of $Fe^{3+}_{(aq)}$ and biological, nano-sized FeS-coated limestone on the removal of redox-sensitive contaminants (As, Sb and Cr): implications for *in situ* passive treatment of acid mine drainage. Applied Geochemistry, 80:102-111

Zhou W B, Moguche A O, Chiu D, Murali-Krishna K, Baneyx F. 2014. Just-in-time vaccines: biomineralized calcium phosphate core-immunogen shell nanoparticles induce long-lasting $CD8^+$ T cell responses in mice. Nanomedicine, 10(3):571-578

Reserch Progress of Applied Mineralogy

DONG Fa-qin, TAN Dao-yong, WANG Jin-ming, XU Long-hua, LI Xu-juan, DING Wen-jin, HU Zhi-bo, HUANG Teng

Key Laboratory of Solid Waste Treatment and Resource Recycle, Ministry of Education, Southwest University of Science and Technology, Mianyang 621010

Abstract: In this paper, we have comprehensively reviewed the research progresses of applied mineralogy in the last decade, mainly in aspects of metallurgical-technological mineralogy and processing of mineral raw materials, new industrial minerals and rocks (gems), specific functional mineral materials, biomimetic mineral materials, biomimetic mineralogy, cultural heritage and cultural relics preservation, treatment and resource utilization of mineral based solid waste, and remediation of polluted soils in the mine and adjacent areas. Especially, we have summarized some relevant important processes in the abovementioned fields of applied mineralogy. Seven important development directions of applied mineralogy under the global tide of Fourth Industrial Revolution has been prospected.

Key words: applied mineralogy; industrial minerals and rocks; non-metallic minerals; mineral resources; nano-minerals; remediation of pollution in the mine

成因矿物学与找矿矿物学研究进展*

申俊峰[1]　李胜荣[1]　黄绍锋[2]　卿　敏[3]　张华锋[1]　许　博[4]

1. 中国地质大学(北京)　地球科学与资源学院，北京 100083；2. 中国黄金集团香港有限公司，北京 100011；3. 中国黄金集团 资源有限公司，北京 100011；4. 中国地质大学(北京)　珠宝学院，北京 100083

摘　要：成因矿物学与找矿矿物学在我国70多年的发展中已逐渐形成一套兼具理论和应用的学科体系。近十年来，原位微区分析技术推动矿物成分标型理论及其应用取得长足进步，在成岩成矿过程精细刻画、热液成矿规律与找矿潜力评价、成矿类型甄别与划分等方面表现出巨大潜力，尤其是黄铁矿、石英、磁铁矿、绿泥石、磷灰石、石榴子石等若干典型矿物成分标型在成矿理论与找矿实践表现出独特优势。可以推测，矿物成分标型必然在未来成矿规律研究与找矿评价中体现出巨大指导价值，同时也从一个侧面证实成因矿物学与找矿矿物学学科体系在地球系统科学领域具有强大的生命力。展望未来，成因矿物学还将在环境与生命过程等其他诸多领域有很大的应用前景，特别是在矿物晶体生长、矿物相转变、矿物温压计、宇宙矿物学、深部矿物学、矿物晶体化学与综合物性、生物矿物交互作用等领域均有可能取得突破性进展。另外，采用统计学和大数据集成方法以及矿物标型定量模型，深度解析矿物学"基因"信息，发展矿物学"基因编辑"技术，将有可能为解决地球科学面临的矿产资源需求和人类健康等做出巨大贡献。

关键词：成因矿物学　找矿矿物学　成分标型　信息载体

0　引　言

在岩石圈演化过程中，矿物作为地质作用的产物展示并表征地壳的时空格局变化，是良好的记录上述变化的信息载体。也就是说，地质作用过程中的任何变化特征无不通过矿物留下深刻烙印。成因矿物学理论自诞生以来即赋予其基于矿物学现象来提取与地质作用发生发展关联的成因信息，甄别和解析地质演化规律，为深刻理解地质作用过程和成因提供科学依据的使命。因此成因矿物学思想在基础地质理论研究和地质应用领域显示出强大的生命力，特别是在地质找矿领域展现出强大的应用前景，因而衍伸出找矿矿物学分支学科。目前，我国成因矿物学与找矿矿物学理论体系伴随其70余年的实践和发展，不断积累和完善，已逐渐形成一套相对完整的学科体系（李胜荣等，2020），可以预测在未来的发展历程中必将发挥出应有的作用。

我国的成因矿物学与找矿矿物学学科框架体系在20世纪末就已形成了较为完善的以矿物发生史、矿物标型学、矿物温压计、矿物共生分析、矿物成因分类和矿物成因信息应用为主要内容的学科体系（陈光远等，1987；薛君治等，1991；李胜荣和陈光远，2001；李胜荣等，2020）。特别是历经多年发展和实践检验后逐渐在如下7个方面突显出矿物学标志在矿床学应用方面的学术应用价值（李胜荣，2013；李胜荣等，2020），如判识区域地质或地球动力学背景；表征成矿物质富集规律和成矿系统结构；限定矿床形成时限，指示物质来源、运移路径和定位富集环境；刻画矿床形成过程、形成条件和形成机理；反映矿床形成后的保存与变化，研判矿体剥蚀程度；追踪矿床深部变化趋势，扩大找矿远景；判断找矿方向、

---

* 原文"成因矿物学与找矿矿物学研究进展（2010—2020）"刊于《矿物岩石地球化学通报》2021年第40卷第3期，本文略有修改。

找矿矿种和找矿潜力等。

进入21世纪以来，矿物微区微量测试技术、同位素原位测试技术、高温高压矿物生长和矿物-流体反应实验技术的不断完善，为成因矿物学理论的发展和应用提供了新的动力，使得基于矿物学信息的成岩成矿过程理解更加深化、细化和精确化（李胜荣等，2006）。特别是采用了如激光剥蚀电感耦合等离子体质谱（laser ablation inductively coupled plasma mass spectrometry，LA-ICP-MS）和激光剥蚀多接收器电感耦合等离子体质谱（laser ablation mult-collector inductively coupled plasma mass spectrometry，LA-MC-ICP-MS）等微区微量测试技术以来，矿物晶体微量元素的赋存规律和分布特点得以清楚揭示，基于矿物晶体生长过程所留下的环带构造，采用剖面成分连续测试可以清楚地了解矿物形成过程的物化环境变化，尤其是结合原位X射线衍射（XRD）技术已能够详细解析晶体生长过程中内部结构的变化规律和重要元素赋存位置。另外，通过矿物晶体生长过程的精细剖析，可以进一步推测成矿过程的物理化学条件变化，使得各种地质作用过程能够借助矿物学信息进行清晰地定性或半定量表征与描述。这样一来，传统的全岩地球化学勘查用于寻找隐伏矿床的技术，由于强烈的测试"干扰"导致异常多解而受到极大挑战或显示出滞后性（Cooke et al.，2020），也明确预示未来采用矿物化学特征（也称矿物成分标型）是用于盲矿体勘查的重要技术工具之一。例如，利用地质体副矿物锆石、磷灰石、独居石等矿物的裂变径迹分析技术对造山带隆升历史和矿产的保存利用提供新的依据；随着矿床开采深度不断加大，对于大纵深矿化富集规律的认识可以凭借矿物学信息进行标识（又称矿物标型特征）并得以充分描述或精细解析（申俊峰等，2013）；对于大型矿集区大型-超大型矿床的成矿物质超常富集规律和强矿化作用标志，以及矿床形成后的剥蚀与保存等矿床学重要内容，均可以采用矿物定量化模型展示（李胜荣，2013）。

可以看出，成因矿物学与找矿矿物学学科体系在地质学诸多领域愈发显示出其重要的理论价值和广泛的应用价值（罗照华等，2013；蔡佳等，2013；李胜荣等，2020）。

1 矿物微区微量分析技术推动矿物成分标型理论及其应用取得长足进步

众所周知，矿物和矿物组合是标识成岩成矿特点、解析成岩成矿过程和提供成岩成矿物理化学条件信息的重要载体，也是进行矿床勘查的重要宏、微观标志。特别是矿物原位微区微量元素和同位素组成，已成为找矿研究的重要工具之一（Cooke et al.，2014，2020；Cook et al.，2016；Román et al.，2019）。赵振华和严爽（2019）认为，矿物之所以能够体现上述重要功能，主要是矿物类质同象现象记录的微量元素组合及其变化规律可以揭示成岩成矿过程，进而可以精细刻画成岩成矿地球化学规律，为矿产资源寻找提供重要信息。因此，矿物微量元素组合确定是提取找矿信息的重要途径之一。

近年来，成因矿物学非常重要的发展内容之一，是先进的微区微量分析技术方法的应用，使得矿物微量元素赋存形式及其分布特点非常容易地进行半定量或定量识别与表征，这为揭示成矿元素由流体析出的沉淀结晶过程以及对于复杂条件下矿物晶体结晶时原子（离子）排列过程的理解提供了基础信息。如扫描电子显微镜、透射电子显微镜、电子探针及其能谱/波谱学分析技术的广泛应用，特别是LA-ICP-MS和LA-MC-ICP-MS测试技术和二次离子质谱（SIMS）的普遍应用，包括一些如短波红外（short wave infrared region，SWIR）光谱、傅里叶红外光谱和热红外光谱等便携式谱学测试技术，以及阴极发光成像技术等显微光学甄别技术，原位X射线衍射技术和同步辐射测试技术的联合应用，使得充分挖掘矿物晶体化学特点，深刻揭示矿物晶体成分标型并应用于解释客观地球化学现象表现出独特优势。

目前，很多金属矿物如磁铁矿、黄铁矿、辉钼矿、锡石、白钨矿、赤铁矿等，以及脉石或副矿物如石英、方解石、绿泥石、绿帘石、石榴子石、磷灰石、锆石、金红石、独居石、榍石等已经成为矿床学和矿床地球化学的重要研究对象（赵振华和严爽，2019）。因为这些矿物的化学组成特点清楚地记录了矿床的形成过程和物理化学条件，也对成矿物质来源、矿化类型具有示踪意义，特别是一些矿物的同位素（如Sr、Pb、S、C、O等传统同位素和Fe、Cu、Mo、Ca、Mg、Zn等非传统同位素）组成能够为合理解释成岩成矿作用过程（Gregory et al., 2015）和建立正确的找矿标志（Gao et al., 2018）提供可靠依据，意味着未来会有更多矿物的化学组成可以具有标型意义，抑或一些矿物的成分标型完全可以作为直接找矿标志（张素荣等，2014；Wilkinson et al., 2017；Neal et al., 2018；赵振华和严爽，2019；Wells et al., 2020）。此外，矿床中一些矿物的微量元素组合差异恰恰是不同矿床类型的主要差别，尤其是含有变价元素（如Fe、V、Mn、Ce、Eu）的矿物（如磁铁矿、磷灰石等），其变价元素含量或比值的变化规律能够敏感地反映成矿过程中氧化还原条件和酸碱度等物化环境的变化。还有，矿物的震荡环带结构中微量元素或同位素变化特点，在有效记录晶体生长过程的同时，还可用于示踪成矿流体的来源和性质。所以，矿物及其成分标型在成矿理论研究和找矿勘查实践中已不是简单的"组成"意义，而是越来越显示出强大的"标识"功能。

我们知道，矿床矿物学研究的最大意义在于能够为找矿勘查提供指示信息。最新的研究结果（Cooke et al., 2020）显示，矿物成分标型正在找矿勘查领域挑战传统的全岩地球化学标志，已在斑岩成矿系统勘查领域显示出巨大的优势和潜力（罗照华等，2018）。例如，广泛存在于岩浆体系的锆石（Loader et al., 2017）、磷灰石（Mao et al., 2016）、磁铁矿（Dupuis and Beaudoin, 2011；Nadoll et al., 2015）和斜长石（Cao et al., 2011）等，其化学成分变化规律在判定斑岩体系是否致矿具有良好的指示效果，当数据量足够大时甚至可以编制矿化所表现出的热液蚀变系统空间结构模型。显然，这样的结构模型对于理解成矿过程和进行矿床勘查均具有不可替代的作用（Parsapoor et al., 2015）。还有，常常广泛出现于斑岩成矿系统的热液成因矿物如绿泥石、绿帘石和明矾石等，其微量元素分布特点同样能够很好地揭示斑岩成矿系统的赋矿规律，对于指示斑岩系统的成矿潜力也有独特的标识意义。勘查实践表明，基于出现于地表露头的蚀变成因绿泥石和绿帘石，采用LA-ICP-MS测试技术提取其成分标型并进行空间对比，完全可以将严重剥蚀的矿床蚀变空间模型予以充分还原。当然，结合地球物理和地球化学勘查方法，还可以对隐伏斑岩系统给出合理判断。因此，基于地表或钻孔样品的矿物化学信息，建立矿物成分标型以寻找盲矿的方法在未来找矿实践中极有可能取得重大突破。对于一些特定矿床来说甚至可以有效推定强矿化中心及其埋深、估算矿化规模等（Wilkinson et al., 2015）。Cooke等（2020）采用LA-ICP-MS和短波红外（SWIR）数据耦合方法，辅以阴极发光图像识别技术，已经将矿物化学作为新型勘查工具在斑岩型矿床勘查方面进行了有益的探索，特别是在成矿潜力区进行定位预测方面取得重要进展，正在不断应用于环太平洋多个地区的斑岩和浅成热液成矿系统的矿床勘查实践（Cooke et al., 2014；Wilkinson et al., 2015）。总之，矿物成分标型对于热液成矿系统的评价已显示出明显的优势和潜力。

2 若干典型矿物成分标型在成矿与找矿研究方面具有独特优势

2.1 黄铁矿成分标型及其指示意义

黄铁矿是金属矿床最常见的矿物之一，其标型特征很早就被人们所关注。由于微量组分具有显著标型意义（陈光远等，1987；薛建玲等，2013；李小宁等，2015），也由于90%以上的金矿床中黄铁矿是主要的载金矿物，因此其微量元素赋存特征及其分析测试技术越来越受重视（Cook et al., 2016；George et al., 2018）。

已有研究显示，黄铁矿成分标型不仅能够详细记录金成矿流体的演化过程（严育通和李胜荣，2011；Reich et al., 2013；Genna and Gaboury, 2015；Gregory et al., 2016），而且能够精确表征流体物理化学条件及其变化（Peterson and Mavrogenes, 2014；Mills et al., 2015；Román et al., 2019），因而在各类金成矿系统中常表现出强大的指示意义（Barker et al., 2009；Muntean et al., 2011；Cook et al., 2013；Reich et al., 2013；Genna and Gaboury, 2015；Belousov et al., 2016；Hazarika et al., 2017；Augustin and Gaboury, 2019）。例如，热液型金矿床中黄铁矿的As、Te和Se含量对于Au的富集显示出强烈的标型指

示意义（Reich et al., 2013; Deditius et al., 2014），可以有效指导找矿勘查。一些研究结果（Deditius et al., 2014; Morishita et al., 2018; Kusebauch et al., 2018）也证实，黄铁矿中的 Au 和 As 常出现同步异常，暗示在较为宽泛的物理化学条件下 Au 与 As 存在一致的地球化学行为，尤其是在黄铁矿结晶过程其一致性更为突出并具有指示意义（Cook and Chryssoulis, 1990; Simmons et al., 2016; Morishita et al., 2018; Kusebauch et al., 2018）。抑或是 As 的地球化学行为强烈地影响黄铁矿结晶和 Au 沉淀的物理化学条件（Reich et al., 2005; Deditius et al., 2014），所以黄铁矿中的 As 异常通常是判别黄铁矿含金性的重要参数。尽管目前还不十分清楚 As 与 Au 伴随黄铁矿结晶而沉淀的机制（Román et al., 2019），但在相对还原条件下 As 替代 S 占据黄铁矿晶格位置是毋庸置疑的。更有可能是 As 在宽泛的温度范围内会强烈影响 Au 在流体与黄铁矿之间的分配（Deditius et al., 2014），所以多种类型金矿床之黄铁矿中 As 与 Au 的含量呈显著正相关关系（Reich et al., 2005），而且在不同类型金矿床中二者还有着显著不同的差异性分配（Deditius et al., 2014）。从图 1 可以看出，低温热液矿床中的黄铁矿比斑岩型矿床黄铁矿的 Au、As 含量要相对高些，说明高温不利于 Au、As 进入黄铁矿结晶体系，只有当流体向浅部运移并降温到足够低时 Au、As 才会被分配进入黄铁矿结晶体系。同理，Te 和 Se 也可以替代 S 进入黄铁矿晶格，因此一些金矿的黄铁矿会出现富 Te 现象（Bi et al., 2011; Ciobanu et al., 2012），同时黄铁矿中的 Au、Ag 含量也显著增加。综上说明，As、Te 和 Se 等大半径离子一旦进入黄铁矿晶格并替代 S 离子，必然引起晶体结构发生严重畸变，因而需要大量的其他金属阳离子（如 Cu、Co、Au、Ag 等）同步替代黄铁矿结构中的 Fe 离子或填充由于结构畸变造成的晶格缺陷以平衡黄铁矿晶格体系，所以黄铁矿中 As、Te、Se 和 Cu、Co、Au、Ag 等常出现有规律的同步变化，也说明流体成分的变化会强烈影响金属元素选择性进入黄铁矿晶格（Tardani et al., 2017）。

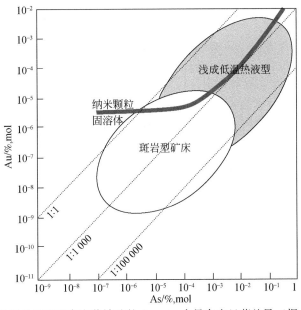

图 1　斑岩型和浅成低温热液型矿床中黄铁矿的 Au、As 含量存在显著差异（据 Deditius et al., 2014）

事实上，黄铁矿的微量元素对于指示 Pb、Zn 和 Ag 的富集规律也有特殊意义。Mukherjee 和 Large（2017）在调查澳大利亚北部中元古代麦克阿瑟盆地超大型 SEDEX 型 Zn-Pb-Ag 矿床时，对采集的 248 件黄铁矿样品进行了 14 种微量元素（Co、Ni、Cu、Zn、As、Se、Mo、Ag、Sb、Tl、Pb、Bi）含量的统计分析，结果发现矿床附近约 1 km 的沉积型黄铁矿强烈富集 Zn 和 Tl，其含量高于全球沉积型黄铁矿中该元素含量平均值的 1~2 个数量级，同时 Cu、Pb、Ag 含量也相对较高。相反，在该矿床 60 km 外的沉积型黄铁矿则 Zn 和 Tl 含量近或低于全球沉积黄铁矿平均值。说明矿床附近的黄铁矿沉淀结晶时，同时受到热液作用和吸附作用的影响，因此富集 Zn、Tl 和 Cu、Pb、Ag、Au，而远离矿床的黄铁矿结晶时仅存在吸附作用，因此只富集 Co、Ni、Cu、As、Se、Mo。

大量的实验矿物学研究（Mycroft et al.，1995）表明，Au 元素之所以能够依附于黄铁矿等硫化物沉淀富集，很大程度上归因于硫化物表面较低的氧化还原电位。也就是说溶解状态下的高价态金离子（通常以类似 $AuCl_4^-$ 等络合形式存在，金为 Au^{3+}）在靠近具有较低氧化还原电位的硫化物晶格时容易被快速还原为零价态的单质金，导致金发生沉淀。这一结果暗示黄铁矿等硫化物矿物表面富金的"还原-沉淀"机制与其形态标型特征之间存在一定内在联系。Xian 等（2019）的最新研究结果表明，黄铁矿晶格的不同取向与流体发生的"界面"反应存在显著的差异，因此黄铁矿不同面网方向是金富集的重要影响因素之一。这进一步明确，相对于立方体和八面体而言，五角十二面体的{210}结晶方向对溶液中离子态金具有最快的还原-沉淀速率，说明"界面"反应的差异性受控于不同结晶方向的原子排列形式，因而影响矿物表面的氧化-还原电位。此外他们还注意到，黄铁矿聚形晶的离子态金的界面还原-沉淀速率显著高于其单形晶，暗示聚形晶的表面电势差改变了晶体表面的电荷分布，因而促进了金的还原-沉淀反应。可见五角十二面体黄铁矿较立方体黄铁矿的载金能力强，聚形晶较单形晶载金能力强。所以金矿床中细粒五角十二面体晶形或聚形晶黄铁矿对富金矿段具有指示意义。当然，黄铁矿晶体生长过程也会受到硫逸度、温度、压力等多重因素的影响，但结晶习性对晶体表面电荷分布、表面能、氧化还原电位以及"界面"反应性和还原-沉淀速率的影响不可忽视。所以，只有多因素耦合作用于含矿流体时才能促进金离子快速沉淀，富集成矿。

总之，解析黄铁矿中微量元素的赋存特征和结构特点对于获取金矿化过程信息是非常重要的（Tanner et al.，2016；Román et al.，2019），特别是对于了解黄铁矿沉淀的物理化学条件和溶解过程的信息（George et al.，2018；Román et al.，2019）非常有帮助。因此，今后对于黄铁矿成分标型等多种微观矿物学标型特征，对成矿和找矿研究的指示意义必将受到更多的重视。

2.2 石英的成分标型及其指示意义

石英的晶体结构中常含有大量微量元素，特别是在成矿环境下晶出的石英，其赋存的微量元素种类和数量更多。一般来说，微量元素在石英晶格中多以类质同象替代 Si^{4+} 形式存在，如 Al^{3+}、Fe^{3+}、B^{3+}、Ti^{4+}、Ge^{4+}、P^{5+} 等，也有一些元素以晶格填隙的方式存在，如 Li^+、K^+、Na^+、H^+、Fe^{2+} 等，二者通常在晶格体系中具有电价补偿平衡作用。

目前，利用 LA-ICP-MS 测试技术可以在石英晶体中同时检测出 30 多种元素，常见的杂质元素有 Al、Mg、Fe、Mn、Ti 等，其次有 B、Be、Na、K、Ca、Li 等，还有少量的 Pb、Bi、Ag、Zn、Cr、Cu、Sn、Rb、Cs、Ba 等。其中，Al 的含量可高达几千 ppm（10^{-6}），其次是 Ti、Li、K、Sb、Fe、Ca、Na、P 等。由于 Al、Mg、Fe、Mn、Ti 等元素在任何成因条件下均可在石英中存在，所以被称作固定杂质元素（Rusk，2012）。

石英中的微量元素含量主要受控于形成环境和结晶速率。含矿脉石英中的 Mo、As、Pb、Bi、Ag、Zn、Cu、Y、Yb 等含量常随共生矿物而发生变化，说明形成环境对石英成分的影响较大（Acosta et al.，2020）。在岩浆和伟晶岩中，石英微量元素还受母岩结晶程度的制约，而热液成因石英主要受流体化学成分和酸碱度的影响（Jourdan et al.，2009）。

有研究表明，低温热液矿床中石英的 Al 含量通常较高，而且当 Al 含量升高时，Li、K、Na、H 等也随之升高，因此石英中 Al 的含量对其他元素含量起着重要的约束和指示作用（Thomas et al.，2011）。有学者（Wark and Watson，2006；Thomas et al.，2011；Huang and Audetat，2012）注意到，石英中 Ti 的含量明显受温度控制，而压力的影响则相对较小，因此石英中的 Ti 含量可作为估算石英形成温度的温度计。但 Ti 在石英中的含量还与结晶速度有关，结晶速度越快，进入晶格的 Ti 越多，石英 Ti 含量就越高。所以，快速冷却结晶条件下，石英的 Ti 温度计不适用。

有意思的是，石英的 Ti、Al 含量在不同成因类型的矿床中存在明显差异（图 2），因此可据此判断成矿类型。另外，由于自然金常与含 As 黄铁矿、方铅矿、黄铜矿、毒砂等共生，故与金沉淀有关的热液石英中的微量元素 Fe、Cu、Pb、Zn、As 等也明显较高，所以石英的 Fe、Cu、Pb、Zn、As 含量也可作为金沉淀的标志。

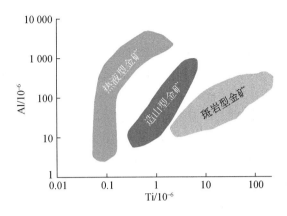

图 2　不同类型矿床中热液石英 Ti 和 Al 含量分布特征（据 Rusk，2012）

值得注意的是，中酸性岩体的石英成分也可作为判识岩浆作用与成矿作用关系的标型特征（Rusk et al.，2006）。最近，Breiter 等（2019）根据石英的微量组分对俄罗斯东特兰斯贝加里亚与 Orlovka 稀有金属矿床有关的花岗岩进行了垂直分带，并证明岩浆成因之石英对岩浆成矿作用具有明确的指示意义。他们采用 LA-ICP-MS 测定了不同空间花岗岩的石英微量组分，并结合矿物自动分析系统（TIMA）和阴极发光（cathodoluminescence，CL）图像建立了花岗岩垂直分带与矿床成因之间的关系。结果表明，侵位较深的黑云母花岗岩中石英相对富 Ti 和 Li（含量分别高达 95×10^{-6} 和 33×10^{-6}），石英以半自形–自形晶为主，不发育包裹体，CL 图像中环带结构清晰；侵位较浅的锂云母花岗岩顶部之石英则富集 Al 和 Ge（分别高达 350×10^{-6} 和 4.8×10^{-6}），晶形呈"雪球"状集合体，CL 图像不清晰，含带状分布的钠长石包裹体。由此推测深部的黑云母花岗岩中的石英经历了复杂的结晶与溶解循环作用，但是未与流体发生太多反应。而锂云母花岗岩顶部的石英则是在岩浆分异过程中与残余熔体发生了强烈的相互作用，因而成矿作用与此密切相关。因此，作为最常见的非金属矿物石英，其成分标型必然会引起广泛重视。

2.3　磁铁矿成分标型及其指示意义

磁铁矿是地壳中普遍存在于各类岩–矿石中的金属氧化物，也是重要的重砂矿物，其微量元素含量常被用作物源（Grigsby，1990；Dare et al.，2014；陈道前等，2015）和成矿类型判别的标型，也常作为找矿的直接指示标志（Dupuis and Beaudoin，2011；Nadoll et al.，2012；张聚全等，2013，2018a）。大量基于岩浆体系晶出磁铁矿的成分研究结果（Whalen and Chappel，1988；Frost and Lindsley，1991；Nadoll et al.，2014；Hensler et al.，2015）显示，磁铁矿微量成分的变化受多种因素影响，如岩浆和流体成分、温度和压力条件、冷却速率、氧逸度（f_{O_2}）、硫逸度（f_{S_2}）和硅活度（α_{SiO_2}）等，其成因意义非常重要。

LA-ICP-MS 分析结果表明，磁铁矿含有 Mg、Al、Si、P、Ca、Sc、Ti、V、Cr、Mn、Co、Ni、Cu、Zn、Ga、Ge、Y、Zr、Nb、Mo、Sn、Hf、Ta、W、Pb 等 20 多种微量元素（Deditius et al.，2018；赵振华和严爽，2019），而且这些微量元素含量在不同成因类型的磁铁矿中明显不同，特别是 Ti、V、Cr、Al、Mn、Mg、Co、Ni 等会表现出较大差异。例如，与火山作用相关的磁铁矿，其 Ti 和 V 的含量明显高于热液成因的磁铁矿；岩浆成因磁铁矿的 Ti 含量相对较高，而 Ni/Cr 值较小，但热液成因的磁铁矿则正好相反（Nadoll et al.，2014；Knipping et al.，2015），所以可以采用 Ti 和 Ni/Cr 二元图解区分岩浆型和热液型磁铁矿（Dare et al.，2014）。Dupuis 和 Beaudoin（2011）对全球 111 个不同类型矿床的磁铁矿微量元素含量的统计分析表明，不同成因矿床中磁铁矿的 Ni、Cr、Mg、Al、Si、Ca、Ti、V、Mn 存在明显差异，其中铜镍矿床具有（Ni+Cr）含量高和（Si+Mg）含量显著偏低的特点，因此可以利用（Si+Mg）与（Ni+Cr）二元成分图解清楚地识别含有铂族元素的Ni-Cu矿床。类似地，采用Ni/（Cr+Mn）与（Ti+V）二元成分图解以及（Ca+Al+Mn）与（Ti+V）二元成分图解还可以有效区分IOGC矿床、斑岩型铜矿床、BIF型铁矿床和夕卡岩型钛-钒磁铁矿床等。另外，Nadoll等（2014，2015）在综合多个斑岩和夕卡岩矿床的177件磁铁矿微量成分变化特征后发现，从高温岩浆矿床到低温热液矿床，磁铁矿中Ga、Sn、（Ti+V）和（Al+Mn）含量是逐渐降低的；而（Ti+V）和（Al+Mn）含量在BIF型矿床磁铁矿中含量是比较低的。所以，磁铁矿的（Al+Mn）与（Ti+V）成分图解也能有效区分夕卡岩型和斑岩型矿床（Nadoll et al., 2015）。

磁铁矿成分除了可以作为判别和划分矿床类型外，还可以对成矿过程的物理化学条件给出重要信息。Sun等（2011）注意到岛弧岩浆带Cu-Au矿床中的磁铁矿可以反映氧逸度的变化。因为一般岩浆体系中Fe^{3+}仅占总铁量的1/3左右，但岛弧岩浆带Cu-Au斑岩矿床中磁铁矿有2/3的晶格位置被Fe^{3+}占据，说明磁铁矿从岩浆体系晶出时大量消耗了岩浆中的Fe^{3+}，使岩浆体系氧逸度降低，致使Cu、Au在磁铁矿结晶也发生大量沉淀。同时，磁铁矿结晶还会造成氧化性岩浆中的硫酸根被还原为硫酸氢根，进而将Cu、Au等元素以硫酸氢根络合物形式萃取到流体相中，形成含矿热液。所以，在俯冲带发生部分熔融而形成的具有高初始Cu、Au含量的氧化性岩浆，极有可能形成斑岩型铜、金矿床（Sun et al., 2013；孙卫东等, 2015）。

当然，磁铁矿的微量元素含量还可以精细表征矿化过程。Sun等（2017）研究了云南北衙金多金属矿不同夕卡岩阶段磁铁矿的微量元素后发现，从早阶段到中阶段，磁铁矿的V含量有所降低，暗示热液流体的氧逸度有所增加；从中阶段到晚阶段，磁铁矿的V含量有所增加，指示热液流体的氧逸度有所降低。此外，该矿床的矿物组合显示在磁铁矿晶出的中晚阶段硫化物矿物含量明显增加，也佐证了这一规律，可见磁铁矿成分具有指示含矿热液氧逸度变化的意义。Ding等（2018）对湖南黄沙坪W-Mo-Pb-Zn多金属矿的研究也表明，矿石中磁铁矿根据微量元素特点可明显分为两组，其中一组含有较高的Mg、Al、Ti、V、Zn、Ni、Co和较低的Na、K、Ca、Si、Ge、Sn、W，暗示该组磁铁矿成因与流体作用于下伏富Zn地层有关，高含量Mg、Al、Zn主要源自流体对下伏地层的萃取，而Ti、V、Ni含量较高主要归因于较低的氧逸度；另一组磁铁矿则含有较高的Na、K、Ca、Si、Ge、Sn、W，但Mg、Al、Ti、Zn、Ni、Co含量较低，说明磁铁矿形成主要受到岩浆热液的影响。显然磁铁矿的成分标型也具有解析矿床成因意义。此外，磁铁矿的环带构造，特别是Si、Ca、Na、V、Ti和Ni等微量元素从核部到边缘呈规律性变化时，完全可以反映晶出磁铁矿的热液流体成分的变化规律（Dare et al., 2014）。

磁铁矿成分特征还可作为找矿标型应用于找矿勘查。例如，Acosta-Góngora等（2014）对加拿大Great Bear岩浆带的IOCG矿床和铁氧化物-磷灰石矿床的研究发现，成矿区域磁铁矿的Co/Ni值非常高，说明矿床的成矿后期叠加了热液矿化作用，并且由于Co比Ni更易于进入富铁流体，因此导致了高Co/Ni值磁铁矿的形成。另外，他们还发现无矿区域磁铁矿的Cr/Co值高于矿石磁铁矿，并且二者的V/Ni值也存在差异，可见磁铁矿的微量元素特征完全可以用来评价成矿潜力。

综合看来，磁铁矿的成分标型（特别是微量成分标型）具有很大的矿床成因、矿化类型和找矿勘查标识意义。特别是热液成因磁铁矿，由于流体性质直接影响磁铁矿与流体之间元素的分配，因而导致磁铁矿微量组分有规律的变化。当然，流体性质主要受控于温度、压力、酸碱度和流体组分等主要因素，因此磁铁矿的成分标型完全可以示踪热液流体源并表征流体的演化规律。

需要注意的是，磁铁矿成分标型往往与结构标型或形态标型组合出现，因而更具成因意义。丁俊和张术根（2012）在研究印尼塔里亚布铁矿床的磁铁矿时注意到，矿浆型磁铁矿晶粒较细小，Fe含量为66.08%~68.01%，晶胞参数a_0为0.839 2~0.839 9 nm；接触交代型磁铁矿多出现菱形十二面体和立方体单形，含有较多的SiO_2和Mg，a_0值偏大，为0.839 8~0.840 2 nm；热液脉型磁铁矿晶粒较粗大，Mn含量高达4.78%~6.22%，a_0值高达0.840 1~0.840 7 nm。

2.4 绿泥石成分标型及其指示意义

绿泥石可在多种地质条件下产出，并且稳定存在的温度范围较宽（赵振华和严爽，2019），其成分变化能够很好地记录形成时的物理化学条件，因而显示强烈的成岩成矿标型意义。对于热液蚀变形成的绿泥石来说，其成分标型还可以指示流体演化趋势，抑或能为判识成矿作用提供重要信息（Wilkinson et al., 2015；刘燚平等，2016）。刘燚平等（2018）通过研究山西东腰庄绿岩带型金矿后发现，矿区内有3种不同成因的绿泥石：早期区域变质成因的绿泥石（C1）、早阶段热液蚀变成因的绿泥石（C2）和成矿阶段的热液绿泥石（C3），分别表现为相对富镁、相对富铁和富集铁质，空间上表现出自围岩至矿体其Fe/(Fe+Mg)值呈逐渐升高趋势。他们同时还测得这3期绿泥石的形成温度分别为348~464℃（均值为398℃）、288~490℃（均值为380℃）和145~259℃（均值为215℃），呈明显降低趋势，且氧逸度也存在同步降低趋势。这一规律充分揭示了早期区域变质作用叠加晚期热液作用导致金富集的成矿机制，为该区进一步找矿提供了重要线索。

非常有意义的是，绿泥石作为斑岩成矿系统的前缘蚀变晕的代表性矿物被认为能够明确给出矿化信息并对深部成矿潜力有很好的指示意义（Wilkinson et al., 2015），特别是其微量元素在斑岩成矿系统中的规律性变化（Wilkinson et al., 2015, 2017；Cooke et al., 2020）。Wilkinson等（2015）认为，斑岩矿床青磐岩化带的绿泥石的Mg、Ti、V含量具有随远离矿化中心而明显降低的趋势，而其他元素则呈增高趋势。显然，可以依据绿泥石的微量元素特征进行斑岩型矿化蚀变分带，进而指导斑岩矿床的勘查（Wilkinson et al., 2015, 2017；Cooke et al., 2020）。Cooke等（2020）发现，斑岩矿床青磐岩化带绿泥石的Sr-Ca-As含量具有从中心向外逐渐增高的趋势，而Ti/Sr、Ti/Li、Mg/Ca和V/Ni值则具有逐渐降低趋势。另外，青磐岩化带外边缘绿泥石的Mn-Fe-Zn-B含量呈最高值，但向内向外则均呈显著降低趋势。另在一些斑岩型Cu、Au矿床中还发现绿泥石的$(Al^{IV}+Fe^{2+})/(Fe^{2+}+Mg^{2+})$值与Cu、Au品位呈正相关关系，由此可以指示富矿段（杨超等，2015）。

最新的研究成果（Cooke et al., 2020）表明，绿泥石和绿帘石的化学成分特征可以提供斑岩型铜-钼矿床周围及深部隐藏的地球化学信息，对隐伏斑岩矿床的矿化具有强烈的指示意义，特别是结合岩相学、蚀变岩地球化学和短波红外光谱特征其指示性更加明确，可以有效指导隐伏斑岩矿化区的圈定。此外，绿泥石的微量元素特征还可以和绿帘石、明矾石等蚀变矿物共同指示矿化中心方向，一定程度上可以预测距离矿化中心的大致距离（Huang et al., 2018；Xiao et al., 2018）。

与上述绿泥石类似，成岩-成矿过程中形成的许多暗色矿物都具有很好的矿物学指示意义，这些矿物的Mg/Fe值可以有效标识成岩成矿过程。张聚全等（2018b）对邯邢夕卡岩型铁矿化区岩浆-热液成矿系统中角闪石的成分、结构等特征的系统分析，对成矿过程中Fe质富集规律及其影响因素给出了明确合理的解释。他认为高氧逸度富铁成矿流体交代富Mg的碳酸盐围岩时消耗较少的Fe质，有利于成矿，并基于该观点建立了邯邢式铁矿的成矿模型。

2.5 磷灰石成分标型及其指示意义

磷灰石可以存在于各类岩石中，其阳离子主要是Ca和Sr、Pb、Na、REE、Ba、Mn等，磷酸根中心磷离子常被Si^{4+}、S^{6+}、As^{5+}、V^{5+}等替代，并多含有附加阴离子F^-、Cl^-和OH^-等，所以其矿物种属或亚种类型较多。由于该矿物非常容易发生类质同象替代并导致较大的晶体结构畸变且矿物相仍能够稳定存在，所以通常微量元素在该矿物与熔体和流体之间的分配系数变化较大，也因此对形成环境具有显著的示踪作用（Belousova et al., 2002；Prowatke and Klemme, 2006；赵振华和严爽，2019；Xu et al., 2020），同时可以据此进行不同类型矿化的识别。

Mao等（2016）采用电子探针和激光等离子体质谱分析技术对全球59个不同类型矿床和未矿化地质体的922件磷灰石样品进行了微量元素分析，他们发现Mg、V、Mn、Sr、Y、La、Ce、Eu、Dy、Yb、Pb、

Th、U等元素含量在不同类型矿床中的含量和组合明显不同。其中，碳酸岩型矿床的磷灰石明显富集轻稀土元素（light rare earth element，LREE），而且V、Sr、B和Nb含量较高；碱性斑岩Cu-Au矿床中的磷灰石具有V、S和Ce含量高且异常范围大的特点；钙碱性斑岩Cu-Au-Mo矿床的磷灰石Mn含量高，Eu异常变化大；IOCG铜矿床及铁氧化物-磷灰石矿床的磷灰石Mn含量较低且Eu异常变化较大；造山型Ni-Cu矿床、夕卡岩型Cu矿床、Au-Co矿床和Pb-Zn矿床中的磷灰石则杂质阳离子含量较低。这说明引起磷灰石微量组分变化的类质同象发生明显受控于不同温度条件的矿化类型，可见磷灰石成分对多种矿化类型具有明确的指示意义。

一般来说，岩浆体系中磷灰石的化学成分常伴随岩浆演化而发生明显的变化（Xu et al.，2020）。例如，在诸多斑岩型成矿系统之斑岩体中，自形斑状磷灰石晶体的S含量相对较高，且主要以S^{6+}形式存在，一般是替代P占据晶格位置（Xu et al.，2020），说明斑岩体成岩过程具有较高的氧逸度，因此是斑岩型成矿的重要识别标志（Parat et al.，2011）。智利Elteniente超大型斑岩Cu-Mo矿床、菲律宾南Mindanao Cotobato斑岩Cu矿和斐济斑岩Cu矿中，其磷灰石的SO_3含量分别达到了0.34%~0.69%、0.5%和0.31%~0.57%（赵振华和严爽，2019）。但是，日本岛弧带的多个新生代中-酸性岩体中磷灰石的SO_3含量则多不足0.1%（Imai，2004），明显低于上述斑岩型Cu-Mo或Cu矿床的S含量，暗示该地区新生代岩浆作用过程属于相对还原条件（Hattori，2018），所以日本岛弧带基本没有斑岩型Cu-Mo矿床。这些案例充分说明磷灰石中的S可以作为判别斑岩成矿与否的重要标志。实际上，成矿与不成矿斑岩的磷灰石其他组分也有明显差异。例如，张红等（2018）对于西藏玉龙地区斑岩体之磷灰石研究后发现，含矿斑岩体的磷灰石其Sr、Ba、Th、Pb和Zr含量明显高，稀土元素显著富集（除Y含量较低外），而且轻重稀土分馏明显，Ce强烈显示正异常。此外，含矿斑岩体磷灰石较不含矿斑岩其Sr/Eu值、Sr/Ce值、Sr/Y值和Th/U值均相对较低，而Ce/Pb值和Lu/Hf值则较高。此外，韩丽等（2016）在研究江西大湖塘狮尾洞钨矿区花岗岩时发现，矿区内花岗岩中的磷灰石成分变化非常大，但是蚀变岩和矿石中的磷灰石则显著富集Fe和Mn等元素，同时蚀变岩和矿石含有较多磷灰石。进一步研究认为，早期岩浆作用阶段Mn是以高价Mn^{5+}置换P^{5+}赋存于磷灰石晶格，晚期流体作用阶段则Mn呈低价Mn^{2+}置换Ca^{2+}进入磷灰石晶格。这说明流体作用较岩浆作用阶段氧逸度有所降低，加之降温使得大量Mn^{2+}和Fe^{2+}与WO_4^{2-}结合形成黑钨矿，导致了石英脉型黑钨矿的产出。所以，富Fe、Mn磷灰石具有找矿指示意义。

值得注意的是，磷灰石中的Cl元素含量也可以反映成矿体系挥发分的演化特征并对成矿作用有指示作用（Williams-Jones et al.，2012）。最近的研究表明（Xu et al.，2020），在典型的后碰撞环境斑岩系统（如伊朗、西藏和云南等）中，相比于不含矿斑岩，含矿斑岩具有明显高的S和Cl含量。同时，磷灰石中的Cl/F（>0.19），V/Y（>0.008）和Ce/Pb（>2 138）高值也说明S和Cl元素有利于成矿金属元素的迁移。结合Sr同位素数据认为，这些挥发分可能来自早期的大洋俯冲作用，在后碰撞环境下被再次激活，因而可以形成陆-陆碰撞斑岩系统。

有意思的是，阴极发光特征可以强化展示磷灰石的成分差异，对指导勘查更具实际意义。Bouzari等（2016）对比研究了斑岩型成矿系统的蚀变岩与未蚀变地质体中的磷灰石阴极发光特征，发现二者的阴极发光特征有明显差别：钾化蚀变斑岩体中磷灰石通常显示绿色，绢英岩蚀变带中的磷灰石显示灰色，而未蚀变岩石中的磷灰石则显示黄、黄-褐或褐色。对照LA-ICP-MS成分分析可知，钾化蚀变斑岩体中磷灰石Mn/Fe值低（小于1），Cl、S和Na含量也较低；绢英岩蚀变带中磷灰石Mn、Na、S、Cl、REE含量均较低；未蚀变岩石中显示黄色的磷灰石Mn含量高，Mn/Fe值也高（大于1），显示褐色的磷灰石则Mn含量低，但S和REE+Y含量高。可见，基于阴极发光特征揭示磷灰石的成分变化规律在勘查实践中进行斑岩成矿体系的蚀变分带具有显著的高效便捷性。

2.6 石榴子石成分标型及其指示意义

石榴子石因其化学组成复杂，常依据端元组分进一步划分为镁铝榴石、铁铝榴石、锰铝榴石、钙铝榴石、钙铁榴石、钙铬榴石、钙钒榴石和钙锆榴石等类质同象端元矿物种属。也由于石榴子石的岛状结构特点,在"硅氧骨干"间晶格位置能够广泛接纳不同半径和电价的多种阳离子进行晶格占位,进而造成石榴子石化学成分变化较大。以前的学者(van Westrenen et al., 2001)注意到岩浆作用和变质作用成因的石榴子石对稀土元素具有选择性接纳,因此导致稀土元素发生分异。同样地,在流体体系中晶出的石榴子石,特别是在夕卡岩化作用体系的石榴子石晶出时,会因成矿种类、热液条件和成矿过程的差异而明显不同(朱乔乔等, 2014; Deng et al., 2017)。所以,石榴子石成分的差异性应该与流体成分、结晶温度、结晶速率、晶体表面作用、共结矿物组合、水岩反应等诸多因素有关(Smith et al., 2004)。

众所周知,石榴子石多发育清晰的环带结构,能够精细记录结晶过程的体系组成和物化条件变化,对于示踪成矿流体和物质来源、揭示成矿过程的物理化学条件变化具有独特优势。尤其是 Al、Fe、Mn 等元素含量常随环带发育而有规律的变化(赵振华和严爽, 2019),所以石榴子石具有很强的成分标型意义。一些学者(D'Errico et al., 2012)还注意到夕卡岩中石榴子石环带的氧同位素组成存在规律性变化。例如,美国 Sierra Nevada 岩浆弧夕卡岩中石榴子石的 $\delta^{18}O$ 值由核部的 3.5‰降到了边缘的 -4.5‰,暗示流体演化至晚期不断有大气降水加入。因此认为随环带不同的 $\delta^{18}O$ 值变化是成矿过程中流体性质变化的反映。南非 Wesellton 金伯利岩中石榴子石由核部向边缘其 $Fe^{3+}/\Sigma Fe$ 值呈逐渐增加趋势,也说明流体演化到晚期氧逸度有所增加(Berry et al., 2013)。

总之,在夕卡岩化条件下晶出的石榴子石,其成分通常会表现出规律性变化,一般在环带核部 Al 含量较高,形成钙铝榴石系列,并富集富稀土元素(heavy rare earth element, HREE),显示弱的 Eu 异常;在环带边缘则 Fe 含量较高,形成铁铝榴石系列,并富集 LREE,显示 Eu 正异常(赵振华和严爽, 2019)。

2.7 绿帘石成分标型及其指示意义

绿帘石是斑岩系统蚀变带的常见矿物之一,常作为青磐岩化带矿物组合的重要成员,对判识斑岩体的成矿潜力具有重要意义。斑岩成矿系统的绿帘石常含有 As、Sb、Zn、Pb、Au、Ag 等成矿指示元素(赵振华和严爽, 2019),以此来区别成矿与非成矿地质事件,而且在斑岩型 Cu-Mo 和 Cu-Au 矿床中,不同空间绿帘石的 As、Sb 含量存在明显差异(Wilkinson et al., 2017),因此是鉴别成矿与否并区分矿化类型的重要标型特征。一些典型的斑岩矿床甚至可以依据绿帘石中的 As、Sb、Pb、Zn、Mn 等元素的富集程度进行矿化定位预测,因为很多情况下矿化带绿帘石中 As、Sb、Pb 含量比斑岩体本身要高出两个数量级(Cooke et al., 2014, 2020)。所以,绿帘石的微量元素含量可作为斑岩成矿系统定位成矿预测的重要工具(Cooke et al., 2020)。

2.8 锆石成分标型及其指示意义

不同地质条件下形成的锆石其成分标型显著不同。研究表明(Nardi et al., 2013),花岗岩中锆石的微量元素特征是判别岩体矿化的重要标志。热液锆石由于常含有较多的高场强元素(如 Ti、Nb、Hf 等),且稀土(特别是轻稀土)含量较高(Hoskin, 2005),因此能够提供明确的矿床成因信息(Ballard et al., 2002; 赵振华和严爽, 2019)。对于花岗岩来说,当锆石的 Th/U 值达到 1~10、Y/Ho<20、Sm/Nd>0.5,且 Nb/Y>0.8 时,通常暗示该花岗岩具有较大的成矿潜力(Nardi et al., 2013; Pizarro et al., 2020)。大量的研究(Qiu et al., 2013; Shen et al., 2015; Zhang et al., 2017; Meng et al., 2018)还表明,锆石的 Ce 异常对成矿体系的氧逸度具有指示意义,有时还可作为评价成矿规模甚至判别斑岩成矿与否的重要标志。例如,有学者(Shen et al., 2015)注意到,中亚造山带超大规模矿床的 Ce^{4+}/Ce^{3+} 值高,氧逸度也明显高;而中等规模矿床的 Ce^{4+}/Ce^{3+} 值低,氧逸度也明显低。另外,西藏玉龙斑岩铜矿区成矿岩体的 Ce^{4+}/Ce^{3+} 平均值达到了 210~334;而不成矿斑岩体的 Ce^{4+}/Ce^{3+} 平均值则仅为 93~112(Liang et al., 2006)。伊朗的 Sungun 含矿斑岩锆石 Ce^{4+}/Ce^{3+} 平均值为 296,同时存在大量石膏($CaSO_4$),说明矿床形成于强氧化环境(赵振华和严爽, 2019)。

2.9 榍石成分标型及其指示意义

榍石是岩浆岩和变质岩中常见的副矿物（王汝成等，2011），热液环境也有晶出。由于不同环境下晶出的榍石其化学成分不尽相同，因此榍石的成分变化具有成岩成矿指示意义（李华伟等，2020）。一般来说，榍石中微量和稀土元素含量较高，但成因不同其微量元素和稀土元素含量有显著差异。一些学者早就注意到（Frost et al.，2001），U、Th、Pb 和 REE 等易于替代 Ca 进入榍石晶体结构，而 Sn、Zr、Nb、Ta、Al 和 Fe^{3+} 等则易于替代 Ti 进入榍石晶体结构。当然，也常见 F、OH、Cl 等替代附加阴离子 O 的现象（Li et al.，2010）。岩浆榍石多形成于晚期岩浆结晶阶段，热液榍石主要源自热液蚀变过程。热液榍石多呈细粒（粒径一般小于 100 μm）它形产出，深褐色至褐黑色，常形成于蚀变暗色矿物的裂隙或其他含 Ti 矿物的交代边缘。通常情况下岩浆成因榍石的轻、重稀土有明显分异，并呈显著右倾的配分模式，但不具或仅为弱的 Eu 负异常；热液榍石的轻、重稀土分异不明显，Eu 异常变化范围较大。此外，岩浆成因榍石和热液成因榍石在如下方面还存在明显不同，前者一般具有高的 Th/U 值，但 Zr/Hf 和 Nb/Ta 值则通常变化范围较小；后者具有较低的 Th/U 值（多小于 1），而 Zr/Hf 和 Nb/Ta 值变化范围较大（赵振华和严爽，2019）。

从上述典型矿物黄铁矿、石英、磁铁矿、磷灰石、绿泥石、绿帘石、石榴子石、锆石和榍石的成分标型及其指示意义可以看出，近十年来矿物成分标型在成矿与找矿研究方面具有独特优势和十分重要的意义，在未来地质找矿和成矿规律研究中必将越来越表现出巨大的指导价值。这从一个侧面证实成因矿物学与找矿物学在地质与找矿理论研究和实践应用方面具有强大的生命力。

3 展　　望

纵观所有自然科学的诞生与发展历程可以发现，学科只有当其能够为解决人类重大需求提供理论或技术支撑时才能显示其强大的生命力和发展潜力。成因矿物学与找矿矿物学的诞生与发展过程也不例外，尤其是在我国地球科学的发展中，更是表现出最大限度地适应国民经济建设，不断满足我国对多种矿产资源供给的更高需求，因而在我国得以快速发展和完善。

21 世纪，应是成因矿物学与找矿矿物学又一重要发展期（李胜荣等，2020），因为该学科在我国近 70 年实践过程中，其不仅紧密结合国家需求在理论和技术方法方面不断创新，而且在地质过程和矿产资源以及环境与生命等诸多领域得以广泛的应用，意味着该学科不仅具有扎实的理论根基，而且存在巨大的拓展空间。

我们期望成因矿物学与找矿矿物学科能够与时俱进，不断学习并吸收相关学科先进的理论和方法，充分挖掘并发挥自身潜在的独特优势，进一步加强以实验矿物学为基础的矿物生长、矿物相转变、矿物温压计、宇宙矿物、矿物晶体化学与综合物性、生物矿物交互作用等研究，逐步加大成因矿物族的综合研究力度，进一步改进和完善微区微量测试技术、高温高压实验技术和计算模拟与人工智能技术，提高矿物标型信息提取的精度和信息量，以大数据和统计学思维开展矿物学定量填图，注重挖掘新的矿物标型并总结其时空规律，深度理解矿物学"基因"信息，发展矿物学"基因编辑"技术，必将在不久的将来能够为解决地球科学面临的地质过程、矿产资源、环境优化、防灾减灾、人类健康等方面做出巨大贡献。

我们也期望将成因矿物学基本理论立足我国独特的地质构造和地质环境单元，有计划并针对性地选择若干重点区域、瞄准关键矿产资源和突出环境问题开展成因矿物学研究，为发展具有我国特色的成因矿物学理论，并使其优质服务于人类健康和社会发展。

参 考 文 献

蔡佳,刘平华,刘福来,刘建辉,王舫,施建荣. 2013. 大青山–乌拉山变质杂岩带石拐地区富铝片麻岩成因矿物学与变质演化. 岩石学报,29

(2):437-461

陈道前,孙传敏,吴逍,侯兰杰,鲜海洋. 2015. 四川里伍铜矿床磁黄铁矿成因矿物学研究. 沉积与特提斯地质,35(3):75-80

陈光远,孙岱生,张立,臧维生,王健,鲁também怀. 1987a. 黄铁矿成因形态学. 现代地质,1(1):60-76

陈光远,孙岱生,殷辉安. 1987b. 成因矿物学与找矿矿物学. 重庆:重庆出版社

丁俊,张术根. 2012. 印度尼西亚塔里亚布铁矿床的磁铁矿成因矿物学特征. 中南大学学报(自然科学版),43(12):4778-4787

韩丽,黄小龙,李洁,贺鹏丽,姚军明. 2016. 江西大湖塘钨矿花岗岩的磷灰石特征及其氧逸度变化指示. 岩石学报,32(3):746-758

李华伟,董国臣,董朋生,汤家辉,王树树. 2020. 滇西北中甸弧成矿岩体中榍石化学成分特征及其成岩成矿标识. 地球科学,45(6):1999-2010

李胜荣. 2013. 成因矿物学在中国的传播与发展. 地学前缘,20(3):46-54

李胜荣,陈光远. 2001. 现代矿物学的学科体系刍议. 现代地质,15(2):157-160

李胜荣,孙丽,张华锋. 2006. 西藏曲水碰撞花岗岩的混合成因:来自成因矿物学证据. 岩石学报,22(4):884-894

李胜荣,申俊峰,董国臣,张华锋,李林,杜瑾雪,杨宗锋,李小伟. 2020. 成因矿物学:原理方法应用. 北京:科学出版社

李小宁,王翠芝,包宝同,巫伟霞. 2015. 福建紫金山金铜矿黄铁矿成因矿物学特征研究. 有色金属,67(2):36-41

刘燚平,张少颖,张华锋. 2016. 绿泥石的成因矿物学研究综述. 地球科学前沿,6(3):264-282

刘燚平,陈静,张华锋,王春亮. 2018. 山西五台山东腰庄金矿绿泥石成因矿物学研究. 矿物学报,38(5):514-529

罗照华,杨宗锋,代耕,程黎鹿,周久龙. 2013. 火成岩的晶体群与成因矿物学展望. 中国地质,40(1):176-181

罗照华,郭晶,黑慧欣,王秉璋,王涛. 2018. 东昆仑造山带家琪式斑岩型Cu-Mo矿床中花岗闪长岩的斜长石晶体群及其成矿意义. 矿物岩石地球化学通报,37(2):214-228

申俊峰,李胜荣,马广钢,刘艳,于洪军,刘海明. 2013. 玲珑金矿黄铁矿标型特征及其大纵深变化规律与找矿意义. 地学前缘,20(3):55-75

孙卫东,李贺,凌明星,丁兴,李聪颖. 2015. 磁铁矿危机与铜金热液成矿. 矿物岩石地球化学通报,34(5):895-901

王汝成,谢磊,陈骏,于阿朋,王禄斌,陆建军,朱金初. 2011. 南岭中段花岗岩中榍石对锡成矿能力的指示意义. 高校地质学报,17(3):368-380

薛建玲,李胜荣,孙文燕,张运强,张旭. 2013. 胶东邓格庄金矿黄铁矿成因矿物学特征及其找矿意义. 中国科学:地球科学,43(11):1857-1873

薛君治,白学让,陈武. 1991. 成因矿物学. 2版. 武汉:中国地质大学出版社

严育通,李胜荣. 2011. 胶东流口金矿黄铁矿成因矿物学及稳定同位素研究. 矿物岩石,31(4):58-66

杨超,唐菊兴,宋俊龙,张志,李玉彬,孙兴国,王勤,丁帅,方向,李彦波,卫鲁杰,王艺云,杨欢欢,高轲,宋扬,林彬. 2015. 西藏拿若斑岩型铜(金)矿床绿泥石特征及其地质意义. 地质学报,89(5):856-872

张红,梁华英,赵燕,凌明星,孙卫东. 2018. 藏东玉龙斑岩铜矿带磷灰石微量元素地球化学特征研究. 地球化学,47(1):14-32

张聚全,李胜荣,王吉中,白明,卢静,魏宏飞,聂潇,刘海明. 2013. 冀南邯邢地区白洞和西石门夕卡岩型铁矿磁铁矿成因矿物学研究. 地学前缘,20(3):76-87

张聚全,李胜荣,卢静. 2018a. 中酸性侵入岩的氧逸度计算. 矿物学报,38(1):1-14

张聚全,王吉中,李胜荣,申俊峰,卢静. 2018b. 南太行邯邢地区夕卡岩型铁矿成矿过程的成因矿物学解析. 矿物岩石地球化学通报,37(2):205-213

张素荣,张琳,张大可,贺福清,聂仁祥. 2014. 成因矿物学研究在重砂测量找矿中的意义. 地质通报,33(12):1956-1960

赵振华,严爽. 2019. 矿物——成矿与找矿. 岩石学报,35(1):31-68

朱乔乔,谢桂青,李伟,张帆,王建,张平,于炳飞. 2014. 湖北金山店大型夕卡岩型铁矿石榴子石原位微区分析及其地质意义. 中国地质,41(6):1944-1963

Acosta M D, Watkins J M, Reed M H, Donovan J J, DePaolo D J. 2020. Ti-in-quartz: evaluating the role of kinetics in high temperature crystal growth experiments. Geochimica et Cosmochimica Acta, 281:149-167

Acosta-Góngora P, Gleeson S A, Samson I M, Otes L, Corriveau L. 2014. Trace element geochemistry of magnetite and its relationship to Cu-Bi-Co-Au-Ag-U-W mineralization in the great Bear magmatic zone, NWT, Canada. Economic Geology, 109(7):1901-1928

Augustin J, Gaboury D. 2019. Multi-stage and multi-sourced fluid and gold in the formation of orogenic gold deposits in the world-class Mana district of Burkina Faso-Revealed by LA-ICP-MS analysis of pyrites and arsenopyrites. Ore Geology Reviews, 104:495-521

Ballard J R, Palin M J, Campbell I H. 2002. Relative oxidation states of magmas inferred from Ce(IV)/Ce(III) in zircon: application to porphyry copper deposits of northern Chile. Contributions to Mineralogy and Petrology, 144(3):347-364

Barker S L L, Hickey K A, Cline J S, Dipple G M, Kilburn M R, Vaughan J R, Longo A A. 2009. Uncloaking invisible gold: use of nano-SIMS to evaluate gold, trace elements, and sulfur isotopes in pyrite from Carlin-type gold deposits. Economic Geology, 104(7):897-904

Belousova E A, Griffin W L, O'Reilly S Y, Fisher N I. 2002. Apatite as an indicator mineral for mineral exploration: trace-element compositions and their relationship to host rock type. Journal of Geochemical Exploration, 76:45-69

Belousov I, Large R R, Meffre S, Danyushevsky L V, Steadman J, Beardsmore T. 2016. Pyrite compositions from VHMS and orogenic Au deposits in the Yilgarn Craton, western Australia: implications for gold and copper exploration. Ore Geology Reviews, 79:474-499

Berry A J, Yaxley G M, Hanger B J, Woodland A B, de Jonge M D, Howard D L, Paterson D, Kamenetsky V S. 2013. Quantitative mapping of the oxidative effects of mantle metasomatism. Geology, 41(6):683-686

Bi S J, Li J W, Zhou M F, Li Z K. 2011. Gold distribution in As-deficient pyrite and telluride mineralogy of the Yangzhaiyu gold deposit, Xiaoqinling district, southern North China Craton. Mineralium Deposita, 46(8):925-941

Bouzari F, Hart C J R, Bissig T, Barker S. 2016. Hydrothermal alteration revealed by apatite luminescence and chemistry: a potential indicator mineral for exploring covered porphyry copper deposits. Economic Geology, 111(6):1397-1410

Breiter K, Badanina E, Ďurišová J, Dosbaba M, Syritso L. 2019. Chemistry of quartz-A new insight into the origin of the Orlovka Ta-Li deposit, eastern Transbaikalia, Russia. Lithos, 348-349:105206

Cao Y, Li S R, Zhang H F, Liu X B, Li Z Z, Ao C, Yao M J. 2011. Significance of zircon trace element geochemistry, the Shihu gold deposit, western Hebei Province, North China. Journal of Rare Earths, 29(3):277-286

Ciobanu C L, Cook N J, Utsunomiya S, Kogagwa M, Green L, Gilbert S, Wade B. 2012. Gold-telluride nanoparticles revealed in arsenic-free pyrite. American Mineralogist, 97(8-9):1515-1518

Cook N J, Chryssoulis S L. 1990. Concentrations of "invisible gold" in the common sulfides. Canadian Mineralogist, 28(1):1-16

Cook N J, Ciobanu C L, Meria D, Silcock D, Wade B. 2013. Arsenopyrite-pyrite association in an orogenic gold ore: tracing mineralization history from textures and trace elements. Economic Geology, 108(6):1273-1283

Cook N J, Ciobanu C L, George L, Zhu Z Y, Wade B, Ehrig K. 2016. Trace element analysis of minerals in magmatic-hydrothermal ores by laser ablation inductively-coupled plasma mass spectrometry: approaches and opportunities. Minerals, 6(4):111

Cooke D R, Baker M, Hollings P, Sweet G, Chang Z S, Danyushevsky L, Gilbert S, Zhou T F, White N C, Gemmell J B, Inglis S. 2014. New advances in detecting the distal geochemical footprints of porphyry systems-epidote mineral chemistry as a tool for vectoring and fertility assessments. In: Kelley K D, Golden H C (eds). Building Exploration Capability for the 21st Century. Boulder, CO, USA: Society of Economic Geologists, 127-152

Cooke D R, Wilkinson J J, Baker M, Agnew P, Phillips J, Chang Z S, Chen H Y, Wilkinson C C, Inglis S, Hollings P, Zhang L J; Gemmell J B, White N C, Danyushevsky L, Martin H. 2020. Using mineral chemistry to aid exploration: a case study from the resolution porphyry cu-mo deposit, Arizona. Economic Geology, 115(4):813-840

Dare S A S, Barnes S J, Beaudoin G, Méric J, Boutroy E, Potvin-Doucet C. 2014. Trace elements in magnetite as petrogenetic indicators. Mineralium Deposita, 49(7):785-796

Deditius A P, Reich M, Kesler S E, Utsunomiya S, Chryssoulis S L, Walshe J, Ewing R C. 2014. The coupled geochemistry of Au and As in pyrite from hydrothermal ore deposits. Geochimica et Cosmochimica Acta, 140:644-670

Deditius A P, Reich M, Simon A C, Suvorova A, Knipping J, Roberts M P, Rubanov S, Dodd A, Saunders M. 2018. Nano-geochemistry of hydrothermal magnetite. Contributions to Mineralogy and Petrology, 173(6):46

Deng X D, Li J W, Luo T, Wang H Q. 2017. Dating magmatic and hydrothermal processes using andradite-rich garnet U-Pb geochronometry. Contributions to Mineralogy and Petrology, 172(9):71

D'Errico M E, Lackey J S, Surpless B E, Loewy S L, Wooden J L, Barnes J D, Strickland A, Valley J W. 2012. A detailed record of shallow hydrothermal fluid flow in the Sierra Nevada magmatic arc from low-$\delta^{18}O$ skarn garnets. Geology, 40(8):763-766

Ding T, Ma D S, Lu J J, Zhang R Q. 2018. Magnetite as an indicator of mixed sources for W-Mo-Pb-Zn mineralization in the Huangshaping polymetallic deposit, southern Hunan Province, China. Ore Geology Reviews, 95:65-78

Dupuis C, Beaudoin G. 2011. Discriminant diagrams for iron oxide trace element fingerprinting of mineral deposit types. Mineralium Deposita, 46(4):319-335

Frost B R, Lindsley D H. 1991. Occurrence of iron-titanium oxides in igneous rocks. Reviews in Mineralogy and Geochemistry, 25(1):433-468

Frost B R, Chamberlain K R, Schumacher J C. 2001. Sphene(titanite): phase relations and role as a geochronometer. Chemical Geology, 172(1-2):131-148

Gao Z F, Zhu X K, Sun J, Luo Z H, Bao C, Tang C, Ma J X. 2018. Spatial evolution of Zn-Fe-Pb isotopes of sphalerite within a single ore body: a case study from the Dongshengmiao ore deposit, Inner Mongolia, China. Mineralium Deposita, 53(1):55-65

Genna D, Gaboury D. 2015. Deciphering the hydrothermal evolution of a VMS system by LA-ICP-MS using trace elements in pyrite: an example from the Bracemac-McLeod Deposits, Abitibi, Canada, and implications for exploration. Economic Geology, 110(8):2087-2108

George L L, Biagioni C, D'Orazio M, Cook N J. 2018. Textural and trace element evolution of pyrite during greenschist facies metamorphic recrystallization in the southern Apuan Alps (Tuscany, Italy): influence on the formation of Tl-rich sulfosalt melt. Ore Geology Reviews, 102:59-105

Gregory D D, Large R R, Halpin J A, Steadman J A, Hickman A H, Ireland T R, Holden P. 2015. The chemical conditions of the Late Archean Hamersley Basin inferred from whole rock and pyrite geochemistry with $\Delta^{33}S$ and $\delta^{34}S$ isotope analyses. Geochimica et Cosmochimica Acta, 149:223-250

Gregory D D, Large R R, Bath A B, Steadman J A, Wu S, Danyushevsky L, Bull S W, Holden P, Ireland T R. 2016. Trace element content of pyrite from the Kapai Slate, St. Ives gold district, western Australia. Economic Geology, 111(6):1297-1320

Grigsby J D. 1990. Detrital magnetite as a provenance indicator. Journal of Sedimentary Research, 60(6):940-951

Hattori K. 2018. Porphyry copper potential in Japan based on magmatic oxidation state. Resource Geology, 68(2):126-137

Hazarika P, Mishra B, Pruseth K L. 2017. Trace-element geochemistry of pyrite and arsenopyrite: ore genetic implications for late Archean orogenic gold deposits in southern India. Mineralogical Magazine, 81(3):661-678

Hensler A S, Hagemann G S, Rosière A C, Angerer T, Gilbert S. 2015. Hydrothermal and metamorphic fluid-rock interaction associated with hypogene "hard" iron ore mineralisation in the Quadrilátero Ferrífero, Brazil: implications from in-situ laser ablation ICP-MS iron oxide chemistry. Ore Geology Reviews, 325-351

Hoskin P W O. 2005. Trace-element composition of hydrothermal zircon and the alteration of Hadean zircon from the Jack Hills, Australia. Geochimica et Cosmochimica Acta, 69(3):637-648

Huang J H, Chen H Y, Han J S, Deng X H, Lu W J, Zhu R L. 2018. Alteration zonation and short wavelength infrared (SWIR) characteristics of the Honghai VMS Cu-Zn deposit, eastern Tianshan, NW China. Ore Geology Reviews, 100:263-279

Huang R F and Audreas A. 2012. The titanium-in-quartz (Titani Q) themobarometer: a critical ecamination and re-calibration. Geochimica et Cosmochimica Acta,84:75-89

Imai A. 2004. Variation of Cl and SO_3 contents of microphenocrystic apatite in intermediate to silicic igneous rocks of Cenozoic Japanese island arcs: implications for porphyry Cu metallogenesis in the western Pacific island arcs. Resource Geology,54(3):357-372

Jourdan F,Bertrand H,Féraud G,Le Gall B,Watkeys M K. 2009. Lithospheric mantle evolution monitored by overlapping large igneous provinces:case study in southern Africa. Lithos,107(3-4):257-268

Knipping J L,Bilenker L D,Simon A C,Reich M,Barra R,Deditius A P,W？lle M,Heinrich C A,Holtz F,Munizaga R. 2015. Trace elements in magnetite from massive iron oxide-apatite deposits indicate a combined formation by igneous and magmatic-hydrothermal processes. Geochimica et Cosmochimica Acta,171:15-38

Kusebauch C,Oelze M,Gleeson S A. 2018. Partitioning of arsenic between hydrothermal fluid and pyrite during experimental siderite replacement. Chemical Geology,500:136-147

Li J W,Deng X D,Zhou M F,Liu Y S,Zhao X F,Guo J L. 2010. Laser ablation ICP-MS titanite U-Th-Pb dating of hydrothermal ore deposits:a case study of the Tonglushan Cu-Fe-Au skarn deposit,SE Hubei Province,China. Chemical Geology,270(1-4):56-67

Liang H Y,Campbell I H,Allen C,Sun W D,Liu C Q,Yu H X,Xie Y W,Zhang Y Q. 2006. Zircon Ce^{4+}/Ce^{3+} ratios and ages for Yulong ore-bearing porphyries in eastern Tibet. Mineralium Deposita,41(2):152-159

Loader M A,Wilkinson J J,Armstrong R N. 2017. The effect of titanite crystallisation on Eu and Ce anomalies in zircon and its implications for the assessment of porphyry Cu deposit fertility. Earth and Planetary Science Letters,472:107-119

Mao M,Rukhlov A S,Rowins S M,Spence J,Coogan L A. 2016. Apatite trace element compositions:a robust new tool for mineral exploration. Economic Geology,111(5):1187-1222

Meng X Y,Mao J W,Zhang C Q,Zhang D Y,Liu H. 2018. Melt recharge,f_{O_2}-T conditions, and metal fertility of felsic magmas:zircon trace element chemistry of Cu-Au porphyries in the Sanjiang orogenic belt,Southwest China. Mineralium Deposita,53(5):649-663

Mills S E,Tomkins A G,Weinberg R F,Fan H R. 2015. Implications of pyrite geochemistry for gold mineralisation and remobilisation in the Jiaodong gold district,Northeast China. Ore Geology Reviews,71:150-168

Morishita Y,Shimada N,Shimada K. 2018. Invisible gold in arsenian pyrite from the high-grade Hishikari gold deposit,Japan:significance of variation and distribution of Au/As ratios in pyrite. Ore Geology Reviews,95:79-93

Mukherjee I,Large R. 2017. Application of pyrite trace element chemistry to exploration for SEDEX style Zn-Pb deposits:McArthur Basin, northern Territory,Australia. Ore Geology Reviews,82:1249-1270

Muntean J L,Cline J S,Simon A C,Longo A A. 2011. Magmatic-hydrothermal origin of Nevada's Carlin-type gold deposits. Nature Geoscience,4(2):122-127

Mycroft J R,Bancroft G M,McIntyre N S,Lorimer J W. 1995. Spontaneous deposition of gold on pyrite from solutions containing Au(III) and Au(I) chlorides. Part I:a surface study. Geochimica et Cosmochimica Acta,3351-3365

Nadoll P,Mauk J L,Hayes T S,Koenig A E,Box S E. 2012. Geochemistry of magnetite from hydrothermal ore deposits and host rocks of the mesoproterozoic belt supergroup,United States. Economic Geology,107(6):1275-1292

Nadoll P,Angerer T,Mauk J L,French D,Walshe J. 2014. The chemistry of hydrothermal magnetite:a review. Ore Geology Reviews,61:1-32

Nadoll P,Mauk J L,Leveille R A,Koenig A E. 2015. Geochemistry of magnetite from porphyry Cu and skarn deposits in the southwestern United States. Mineralium Deposita,50(4):492-515

Nardi L V S,Formoso M L L,Müller I F,Fontana E,Jarvis K,Lamarão C. 2013. Zircon/rock partition coefficients of REEs,Y,Th,U,Nb,and Ta in granitic rocks:uses for provenance and mineral exploration purposes. Chemical Geology,335:1-7

Neal L C,Wilkinson J J,Mason P J,Chang Z S. 2018. Spectral characteristics of propylitic alteration minerals as a vectoring tool for porphyry copper deposits. Journal of Geochemical Exploration,184:179-198

Parat F,Holtz F,Klugel A. 2011. S-rich apatite-hosted glass inclusions in xenoliths from La Palma:constraints on the volatile partitioning in evolved alkaline magmas. Contributions to Mineralogy and Petrology,162(3):463-478

Parsapoor A,Khalili M,Tepley F,Maghami M. 2015. Mineral chemistry and isotopic composition of magmatic, re-equilibrated and hydrothermal biotites from Darreh-Zar porphyry copper deposit,Kerman (southeast of Iran). Ore Geology Reviews,66:200-218

Peterson E C,Mavrogenes J A. 2014. Linking high-grade gold mineralization to earthquake-induced fault-valve processes in the Porgera gold deposit, Papua New Guinea. Geology,42(5):383-386

Pizarro H,Campos E,Bouzari F,Rousse S,Bissig T,Gregoire M,Riquelme R. 2020. Porphyry indicator zircons (PIZs):application to exploration of porphyry copper deposits. Ore Geology Reviews,126:103771

Prowatke S,Klemme S. 2006. Trace element partitioning between apatite and silicate melts. Geochimica et Cosmochimica Acta,70(17):4513-4527

Qiu J T,Yu X Q,Santosh M,Zhang D H,Chen S Q,Li P J. 2013. Geochronology and magmatic oxygen fugacity of the Tongcun molybdenum deposit, Northwest Zhejiang,SE China. Mineralium Deposita,48(5):545-556

Reich M,Kesler S E,Utsunomiya S,Palenik C S,Chryssoulis S L,Ewing R C. 2005. Solubility of gold in arsenian pyrite. Geochimica et Cosmochimica Acta,69(11):2781-2796

Reich M,Deditius A P,Chryssoulis S,Li J W,Ma C Q,Parada M A,Barra F,Mittermayr F. 2013. Pyrite as a record of hydrothermal fluid evolution in a porphyry copper system:a SIMS/EMPA trace element study. Geochimica et Cosmochimica Acta,104:42-62

Román N, Reich M, Leisen M, Morata D, Barra F, Deditius A P. 2019. Geochemical and micro-textural fingerprints of boiling in pyrite. Geochimica et Cosmochimica Acta, 246: 60-85

Rusk B. 2012. Cathodoluminescent textures and trace elements in hydrothermal quartz. In: Götze J, Möckel R (eds). Quartz: Deposits, Mineralogy and Analytics. Berlin, Heidelberg: Springer, 307-329

Rusk B G, Reed M H, Dilles J H, Kent A J R. 2006. Intensity of quartz cathodoluminescence and trace-element content in quartz from the porphyry copper deposit at Butte, Montana. American Mineralogist, 91(8-9): 1300-1312

Shen P, Hattori K, Pan H D, Jackson S, Seitmuratova E. 2015. Oxidation condition and metal fertility of granitic magmas: zircon trace-element data from porphyry Cu deposits in the Central Asian orogenic belt. Economic Geology, 110(7): 1861-1878

Simmons S F, Brown K L, Tutolo B M. 2016. Hydrothermal transport of Ag, Au, Cu, Pb, Te, Zn, and other metals and metalloids in New Zealand geothermal systems: Spatial patterns, fluid-mineral equilibria, and implications for epithermal mineralization. Economic Geology, 111(3): 589-618

Smith M P, Henderson P, Jeffries T E R, Long J, Williams C T. 2004. The rare earth elements and uranium in garnets from the Beinn an Dubhaich Aureole, Skye, Scotland, UK: constraints on processes in a dynamic hydrothermal system. Journal of Petrology, 45(3): 457-484

Sun W D, Zhang H, Ling M X, Ding X, Chung S L, Zhou J B, Yang X Y, Fan W M. 2011. The genetic association of adakites and Cu-Au ore deposits. International Geology Review, 53(5-6): 691-703

Sun W D, Liang H Y, Ling M X, Zhan M Z, Ding X, Zhang H, Yang X Y, Li Y L, Ireland T R, Wei Q R, Fan W M. 2013. The link between reduced porphyry copper deposits and oxidized magmas. Geochimica et Cosmochimica Acta, 103: 263-275

Sun X M, Lin H, Fu Y, Li D F, Hollings P, Yang T J, Liu Z R. 2017. Trace element geochemistry of magnetite from the giant Beiya gold-polymetallic deposit in Yunnan Province, Southwest China and its implications for the ore forming processes. Ore Geology Reviews, 91: 477-490

Tanner D, Henley R W, Mavrogenes J A, Holden P. 2016. Sulfur isotope and trace element systematics of zoned pyrite crystals from the El Indio Au-Cu-Ag deposit, Chile. Contributions to Mineralogy and Petrology, 171(4): 33

Tardani D, Reich M, Deditius A P, Chryssoulis S, Sánchez-Alfaro P, Wrage J, Roberts M P. 2017. Copper-arsenic decoupling in an active geothermal system: a link between pyrite and fluid composition. Geochimica et Cosmochimica Acta, 204: 179-204

Thomas H V, Large R R, Bull S W, Maslennikov V, Berry R F, Fraser R, Froud S, Moye R. 2011. Pyrite and pyrrhotite textures and composition in sediments, laminated quartz veins, and reefs at Bendigo gold mine, Australia: insights for ore genesis. Economic Geology, 106: 1-31

van Westrenen W, Wood B J, Blundy J D. 2001. A predictive thermodynamic model of garnet-melt trace element partitioning. Contributions to Mineralogy and Petrology, 142(2): 219-234

Wark D A, Watson E B. 2006. A titanium-in-quartz geothermometer. Contributions to Mineralogy and Petrology, 152(6): 743-754

Wells T J, Meffre S, Cooke D R, Steadman J A, Hoye J L. 2020. Porphyry fertility in the Northparkes district: indicators from whole-rock geochemistry. Australian Journal of Earth Sciences, 67(5): 717-738

Whalen J B, Chappell B W. 1988. Opaque mineralogy and mafic mineral chemistry of I- and S-type granites of the Lachlan fold belt, Southeast Australia. American Mineralogist, 73(3-4): 281-296

Wilkinson J J, Chang Z S, Cooke D R, Baker M J, Wilkinson C C, Inglis S, Chen H Y, Gemmell J B. 2015. The chlorite proximitor: a new tool for detecting porphyry ore deposits. Journal of Geochemical Exploration, 152: 10-26

Wilkinson J J, Cooke D R, Baker M J, Chang Z, Wilkinson C C, Chen H, Fox N, Hollings P, White N C, Gemmell J B, Loader M A, Pacey A, Sievwright R H, Hart L A, Brugge E R. 2017. Porphyry indicator minerals and their mineral chemistry as vectoring and fertility tools. In: McClenaghan M B, Layton-Matthews D (eds). Application of Indicator Mineral Methods to Bedrock and Sediments. Toronto: Geological Survey of Canada, 67-77

Williams-Jones A E, Migdisov A A, Samson I M. 2012. Hydrothermal mobilisation of the rare earth elements—a tale of "Ceria" and "Yttria". Elements, 8(5): 355-360

Xian H Y, He H P, Zhu J X, Du R X, Wu X, Tang H M, Tan W, Liang X L, Zhu R L, Teng H H. 2019. Crystal habit-directed gold deposition on pyrite: surface chemical interpretation of the pyrite morphology indicative of gold enrichment. Geochimica et Cosmochimica Acta, 264: 191-204

Xiao B, Chen H Y, Wang Y F, Han J S, Xu C, Yang J T. 2018. Chlorite and epidote chemistry of the Yandong Cu deposit, NW China: metallogenic and exploration implications for Paleozoic porphyry Cu systems in the eastern Tianshan. Ore Geology Reviews, 100: 168-182

Xu B, Kou G Y, Etschmann B, Liu D Y, Brugger J. 2020. Spectroscopic, Raman, EMPA, micro-XRF and micro-XANES analyses of sulphur concentration and oxidation state of natural apatite crystals. Crystals, 10(11): 1032-1043

Zhang C C, Sun W D, Wang J T, Zhang L P, Sun S J, Wu K. 2017. Oxygen fugacity and porphyry mineralization: a zircon perspective of Dexing porphyry Cu deposit, China. Geochimica et Cosmochimica Acta, 206: 343-363

The Decennary New Advances on the Genetic Mineralogy and Prospecting Mineralogy

SHEN Jun-feng[1], LI Sheng-rong[1], HUANG Shao-feng[2], QING Min[3], ZHANG Hua-feng[1], XU Bo[4]

1. School of Earth Sciences and Resources, China University of Geosciences (Beijing), Beijing 100083;
2. China National Gold Group Hong Kong Limited, Beijing 100011; 3. China Gold Group Resources Co., Ltd., Beijing 100011; 4. School of Gemmology, China University of Geosciences (Beijing), Beijing 100083

Abstract: The genetic mineralogy and prospecting mineralogy have been gradually established as a set of theoretical and application subject system in more than 70 years of research and development in China. In recent decennium, the *in-situ* microanalysis technology has greatly promoted the step forward in theory and application of the mineral composition typomorph. It also has great application potentials in research fields such as the fine characterization of diagenetic and metallogenic processes, the evaluation of hydrothermal metallogenic regularity and prospective potential, and the identification and classification of metallogenic types, etc. The composition typomorphs of various typical minerals, such as pyrite, quartz, magnetite, chlorite, apatite and garnets, have shown their unique advantages in researches of metallogenic theory and prospecting practice, especially. It can be predicted that the mineral composition typomorph will play very important guiding role in the research of metallogenic regularity and the evaluation of geological prospecting in the future. At the same time, it is also proved from one side that the discipline system of genetic mineralogy and prospecting mineralogy has strong vitality in the field of earth system science. In the future, it can be prospectively expected that the development of genetic mineralogy should promote the progresses not only in researches of geological process and mineral resources but also in other fields such as environment and life science, etc. Especially, promising breakthrough could be achieved in the research fields including the mineral crystal growth, mineral phase transformation, mineral thermobarometer, cosmological mineralogy, deep mineralogy, mineral crystal chemistry and comprehensive physical properties, bio-mineral interaction, etc. Additionally, to deeply analyze mineralogical "gene" information and to develope mineralogical "gene editing" technology by using the statistic method, big data integration, and mineral typomorph quantitative models could be hopefully potential to solve problems in fields of the mineral resource demand and human health faced by the earth science.

Key words: genetic mineralogy; prospecting mineralogy; composition typomorph; information carrier

火山学和地球内部化学研究进展*

徐义刚[1] 郭正府[2] 刘嘉麒[2]

1. 中国科学院 广州地球化学研究所，同位素地球化学国家重点实验室，广州 510640；
2. 中国科学院 地质与地球物理研究所，新生代地质与环境重点实验室，北京 100029

摘　要：本文回顾并综述了 2011—2020 年十年间我国在火山学和地球内部化学领域的主要研究进展，包括：①进一步查明了我国新生代火山活动的时空分布特征，确定了火山喷发物的成因类型，建立了中国活动火山地质基础数据库；②深入研究了中国新生代火山岩成因，认为在地幔过渡带中滞留板片的脱碳和脱水作用及相关的熔融和交代作用是板内岩浆成因的主要驱动力，提出了东亚大地幔楔导致板内玄武岩成因的深源熔-流体助熔机制；通过深源包体研究为中国东部深部岩石圈的改造和演化提供了新的约束；③对峨眉山和塔里木两个二叠纪大火成岩省进行了进一步研究，揭示了两者的异同；④将我国活动火山监测与灾害防御研究推向了新的高度；⑤开展了火山喷发模拟实验与物理火山学的研究、建立了我国火山灰年代学的研究方法与测试手段，开展了火山学与地热学的交叉学科研究，进一步拓展了我国火山学研究的广度。本文还明确了学术组织在火山学学科发展中发挥的作用，并对我国火山和地球内部化学领域的未来发展进行了展望。

关键词：火山学　地球内部化学　中国

0　引　言

　　火山学和地球内部化学主要是通过对近-现代活火山的监测和研究，达到防灾、减灾的目的，通过对不同地质历史时期的火山进行研究，从而揭示地球内部物质组成和物理化学条件以及相关的地球动力学过程。在 2011—2020 年的十年间，我国学者在火山学和地球内部化学领域开展了大量的工作，取得了许多重要进展。本文旨在对这些研究进展进行全面回顾和综述，提炼近十年间我国学者在该领域提出的新观点和新认识，找出现阶段工作的不足，明确未来的发展方向，为国内同行深入了解我国在这一领域的研究现状和开展进一步的研究提供有价值的资料与参考依据，促进我国火山学和地球内部化学学科的发展。

1　进一步查明了我国新生代火山活动的时空分布特征

　　活动火山是了解现代地下深处物质物理化学性质及构造动力学的天然窗口，是新构造活动的具体表现。中国大陆新生代火山岩主要分布于大兴安岭—太行山—雪峰山一线以东地区和青藏高原周边（图 1），总面积约 19 万 km^2。其中，第四纪火山群数十处，主要包括吉林长白山天池、龙岗金龙顶子；黑龙江镜泊湖、五大连池老黑山和火烧山，科洛、二克山；内蒙古东部大兴安岭-大同火山喷发带内的鄂伦春诺敏河流域的马鞍山、达来滨呼通、河谷中 371 和 358 高地火山，柴河-阿尔山火山群的焰山、高山、十号沟盆地、小东沟和子宫山火山，锡林浩特火山群的鸽子山火山，乌兰哈达火山群的北、中、南炼丹炉火山和辉腾锡勒的玻璃敖包火山；雷琼地区的马鞍山、雷虎岭火山；广西涠洲岛；云南腾冲；新疆阿什盆地、

---

*　原文"中国火山学和地球内部化学研究进展与展望（2011～2020 年）"刊于《矿物岩石地球化学通报》2020 年第 39 卷第 4 期，本文略有修改。

青藏可可西里的火山等。近十年来，在原上新世火山台地中识别出了一批新的第四纪火山群，如在内蒙古高原南缘汉诺坝玄武岩台地中识别出乌兰哈达火山群和辉腾锡勒火山群等。对第四纪火山群基本完成了中、大比例尺的火山地质制图，进一步厘定了火山群的规模、类型及喷发历史，研究了其火山作用方式及其活动机制（白志达等，2012；樊祺诚等，2012），并发现了一些新的火山（白志达等，2012；李霓等，2014）。尤其是对大兴安岭-太行山重力梯度带以西的第四纪火山研究取得了显著进展（樊祺诚等，2015）。第四纪火山群内晚更新世以来的火山形貌保存基本完整，不同类型的火山数以千计（白志达等，2012）。对全新世以来活火山的研究也取得了突出的成果，并建立了中国活动火山地质基础数据库。

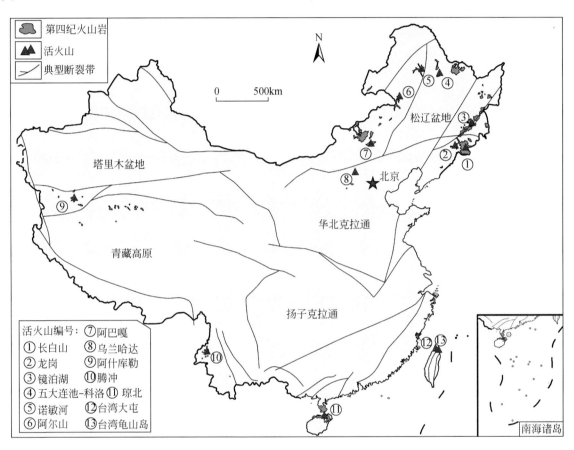

图1 中国活火山分布示意图（修改自刘嘉麒，1999；任纪舜，2003）

系统的岩性、岩相研究是识别火山喷发物成因类型的有效途径。根据火山活动的基本形式、火山产物的搬运方式、堆积机理以及火山喷发物的定位环境，结合岩相学及火山碎屑物粒度分布与形貌特征的研究，可区分出火山喷发物的成因类型，并确定其成因相及相序。中国大陆第四纪火山以中心式喷发为主，少量火山群中发育裂隙式或裂隙-中心式喷发，如乌兰哈达火山群。火山喷发物主要为熔岩和火山碎屑物（岩）。玄武质熔岩主体为溢流相，按其表面构造，可分为结壳熔岩、渣状熔岩、块状熔岩和枕状熔岩。其中，块状熔岩和枕状熔岩为近火口相熔岩流，如五大连池火山群焦得布火山和诺敏河火山群马鞍山火山的块状熔岩流；乌兰哈达火山群西南较大规模的枕状熔岩流。火山碎屑堆积物可区分为爆炸成因和非爆炸成因。爆炸成因火山碎屑物按搬运方式及堆积机理可分为崩落堆积、近源坠落堆积、火山渣降落堆积、浮岩降落堆积、火山灰降落堆积；火山碎屑流主要为浮岩流堆积，而且仅见于长白山天池火山。火山碎屑涌流主要为基底涌流堆积，广泛见于中国东部多个火山群中，构成玛珥式火山的主体。火山喷发-沉积相多见于火口湖中，以玛珥式火山区最发育。火山泥流堆积仅在长白山天池火山范围内发育。非爆炸成因碎屑物可分为熔岩自碎和淬碎两种类型，自碎碎屑物在各火山群熔岩流边缘和前缘均有出露，以角砾级碎屑为主；淬碎作用可形成集块、角砾和凝灰级不同的碎屑物，尤其是对"淬碎凝灰级"碎屑物的确定，丰富了火山灰的成因类型。侵出相与潜火山岩零星分布于各个火山群。侵出相多呈岩垄产出，如在锡林浩特及乌兰哈达火山群中均可见到（赵勇伟等，2017），潜火山岩主要以岩墙或岩栓的形式充填于火山通道及其周围区域。

在火山喷发物成因研究的基础上，识别出单成因火山和多成因火山两种类型。中国大陆第四纪火山主要为玄武质单成因火山，其喷发方式主要为夏威夷式、斯通博利式和玛珥式，其次为强斯通博利式（赵勇伟等，2010）或亚布里尼式。单成因火山大多始于中更新世，晚更新世为活动的鼎盛时期，全新世喷发活动明显减弱。这类火山的岩浆补给速率与频率都较低，喷发活动的时间短，但喷发周期间隔的时间长，每次喷发多发生于不同地点，因此形成了中国东部一系列由数十至数百座火山锥构成的单成因火山群。我国多成因火山相对较少，长白山火山群为其中的典型代表。长白山火山群火山活动主要以中酸性、酸（偏碱）性岩浆喷发为主，且经历了长期而复杂的演化过程。近十年的研究，确认了长白山地区火山活动始于29 Ma，大规模的喷发活动始于5 Ma左右，而最新的爆炸式大喷发则是距今约千年的多成因火山（Xu J. D. et al., 2012a）。长白山地区的火山喷发方式经历了从单一溢流式喷发、溢流式与爆炸式喷发交替进行到爆炸式喷发的演变过程，先后形成了三座大型的破火山（胞胎山、望天鹅和天池）及百余座熔岩渣锥。喷发产物和堆积类型多式多样、纷繁复杂。其中天池火山是一座典型的多成因复合式火山（魏海泉，2010），根据喷发方式、喷发产物类型和堆积特征等，建立了长白山天池火山的精细喷发序列与堆积相模式（Pan et al., 2017）。确定火山活动过程由早到晚依次为：溢流玄武岩（形成了区域盾状体）、溢流粗面岩（主要造锥喷发，形成了高大的火山锥体）、熔岩渣锥（长白山火山区熔岩渣锥）、空降浮岩（形成巨大的喷发柱空降浮岩堆积）、火山碎屑流（形成千年大喷发的浮岩流）、火山泥石流（千年大喷发后伴随的巨大火山泥石流）和次生堆积（天池火口内缘的垮塌堆积）。这种喷发过程与堆积相序是长白山多成因复合式火山的重要特征。

2 揭示了板内火山岩的成因新机制

地球深部过程对浅表地质起着控制作用，因此一直是地球科学领域的研究前沿。该科学问题在 Science 杂志2005年列出的125个未解问题中位列第十。火山岩及其携带的深部捕虏体是来自地球深部的使者，具有时空跨度大、分辨率高等特点，是认识地球深部的有力工具。中国东部分布有大量新生代的板内火山岩，是探讨板内玄武岩成因及地幔源区性质的理想研究对象（徐义刚和樊祺诚，2015）。近十年来新技术、新方法的兴起和使用（斑晶及其熔体包裹体的原位分析技术，名义上无水矿物中结构水的测试，非传统稳定同位素 Mg、Zn、Fe、Mo、B 等），以及与深部地球物理研究的交叉融合，使我们对板内火山岩的成因有了新的认识。

2.1 东亚大地幔楔与中国东部板内岩浆的成因

深部地球物理探测揭示西太平洋板块向东亚大陆俯冲滞留在地幔过渡带中（雷建设等，2018），这一深部结构暗示西太平洋板块俯冲对中国东部，乃至东亚大陆地质演化都产生了深远的影响。Zhao 等（2004）提出了东亚大陆边缘火山成因的大地幔楔（big mantle wedge, BMW）模式，激发了火山岩岩石地球化学研究的积极响应。Zhang 等（2009）针对苏皖新生代玄武岩类似洋岛玄武岩（ocean island basalt, OIB）的微量元素特征和橄榄石斑晶的氧同位素组成，指出其岩浆源区含有中生代太平洋俯冲板片组分。随后，岩石地球化学和同位素示踪进一步揭示了新生代岩浆源区含有再循环洋壳组分（Xu Y. G. et al., 2012；Xu Z. et al., 2012；Xu, 2014），且这些组分可能来自地幔过渡带中的古太平洋俯冲板块，引发了更加深入的系统研究，取得了一系列新认识。

2.1.1 玄武岩源区普遍存在辉石岩质的再循环组分

传统观点认为，玄武质岩浆由地幔橄榄岩（橄榄石含量大于40%的超镁铁岩）部分熔融产生。近来的实验岩石学研究则表明，干的橄榄岩部分熔融无法匹配许多板内玄武岩的地球化学组成，而辉石岩、角闪石岩、碳酸盐化的橄榄岩等可作为玄武岩的源区母岩（Pilet et al., 2008）。越来越多的研究表明，中国东部新生代玄武岩的源区普遍存在辉石岩组分。橄榄石、单斜辉石等斑晶的氧同位素组成异常（相对于正常地幔而言）以及根据单斜辉石斑晶结构水含量和全岩 Ce 含量推算的玄武岩源区的 H_2O/Ce 值异常（如相对 DMM 而言）等，进一步指示中国东部新生代玄武岩源区中普遍存在的辉石岩组分可能代表再循环洋壳或其衍生物。例如，内蒙古赤峰和华北克拉通东南缘地区新生代玄武岩中的橄榄石和（或）单斜辉石斑晶的 $\delta^{18}O$ 值都低于正常地幔值，暗示其源区存在经过变质脱水和高温水岩反应的再循环洋壳（Wang et al., 2015）。阿尔山–柴河等地区玄武岩的 H_2O/Ce 值和 Ba/Th 值之间表现为三端元混合现象，其中高 H_2O/Ce 值、中等 Ba/Th 值的端元指示再循环上洋壳和上覆沉积物组分的贡献，而低 H_2O/Ce 值、高 Ba/Th 值的端元则可能代表经历脱水的蚀变下洋壳组分（Chen et al., 2017）。这些比亏损地幔富集的再循环地壳组分贡献玄武岩的源区可以很好地解释为何中国东部新生代玄武岩普遍具有与 OIB 相似的地球化学组成（徐义刚等，2018）。值得注意的是，尽管众多研究将中国东部新生代玄武岩的成因与滞留在东亚大陆之下的地幔过渡带中的太平洋板块联系起来，目前仍无确凿的证据表明这些玄武岩源区中的再循环地壳组分一定来自滞留的太平洋板块。

2.1.2 再循环碳酸盐组分贡献玄武岩的形成

Zeng 等（2010）发现来自山东的强碱性玄武岩在原始地幔标准化的蛛网图上具有 K、Zr、Hf、Ti 的负异常，并且具有低的 SiO_2、Al_2O_3 和高的 CaO 含量，与碳酸盐化的橄榄岩低程度熔融的熔体相似，因此他们最早提出这些岩石来自碳酸盐化的橄榄岩源区。近年来众多学者对中国东部新生代玄武岩进行了 Mg 和 Zn 同位素研究，发现它们普遍具有比正常地幔偏轻的 Mg 同位素（$\delta^{26}Mg$ 为 −0.65‰ ~ −0.35‰）和偏重的 Zn 同位素组成（$\delta^{66}Zn$ 为 0.3‰ ~ 0.63‰）（Wang et al., 2018；Li et al., 2019）。由于俯冲板片携带的沉积碳酸盐具有极低的 $\delta^{26}Mg$ 值（低至−5.5‰）和高的 $\delta^{66}Zn$ 值（~0.91‰），因此这些玄武岩轻的 Mg 同位素和重的 Zn 同位素组成通常被认为继承自再循环的碳酸盐组分（Tian et al., 2016；Wang et al., 2018）。最近的 Mo 和 B 同位素研究表明，来自山东的强碱性玄武岩具有比弱碱性玄武岩具有更高的 $\delta^{98/95}Mo$（−0.31‰ ~ −0.04‰）和更低的 $\delta^{11}B$（−6.9‰ ~ −3.9‰），以及更高的 Mo 和 B 含量，也暗示它们来自碳酸盐流体交代的地幔源区（Li et al., 2016a, 2019）。Fe 同位素研究显示，中国东部新生代霞石岩具有相比于 MORB 异常高的 $\delta^{56}Fe$ 值（可达 0.29‰），这种重的 Fe 同位素要求其源区具有高的氧逸度（$Fe^{3+}/\sum Fe \geq 0.15$），而源区的高氧逸度与再循环碳酸盐的分解释氧有关。总之，目前已有大量证据支持再循环碳酸盐组分在中国东部新生代玄武岩的形成过程中扮演了十分重要的角色。然而，这些再循环碳酸盐组分是否一定来自俯冲的太平洋板块仍需更加确凿的证据。

2.1.3 东亚大地幔楔的化学不均一性和二元混合模型

Li 等（2016b, 2019）和徐义刚等（2018）发现，同中国东部玄武岩由低硅和高硅两个熔体端元混合而成，其中低硅玄武岩有高全碱、高 CaO、TFeO 和 TiO_2，高硅玄武岩具有低全碱、低 CaO、TFeO 和 TiO_2。除东北高钾岩外，无论华北和东北，还是华南均共享一个相同的低硅组分，而高硅组分在 3 个区域的表现形式有差异。如前所述，低硅组分源区为含碳酸盐的榴辉岩+橄榄岩地幔，其 $^{206}Pb/^{204}Pb$ 值较典型 HIMU 玄武岩的偏低，Nd-Hf 同位素具有（年轻的）太平洋洋壳特征，是较深（>300 km）熔融产物（徐义刚等，2018）。而高硅玄武岩的同位素组成表现出地区差异，其中华北具有 EM1 型富集组分特征，华南具有 EM2 型富集组分特点，而东北则兼有 EM1 和 EM2 两种富集组分。目前，对高硅组分的成因认识还有分歧，但多数人认为是低硅熔体与岩石圈地幔相互作用的结果。Liu 等（2016）研究发现，东北新生代钾质玄武岩的地球化学特征与玄武岩的时空分布存在明显的耦合关系。岩石圈越厚的地区产出的钾质玄武岩的 MgO 越高，K_2O/Na_2O 值和 Rb/Nb 值越低，Sr-Nd 同位素组成越亏损。晚期阶段比早期阶段喷发的玄武岩具有更高的 Rb/Nb、Ba/Nb、K/Nb 和 Ba/La 值。这些相关性表明，初始的富钾、富硅熔体由软流圈地幔中富集物质的部分熔融产生，熔体在上升过程中与亏损的岩石圈地幔发生了不同程度的反应，从而转变为钾质玄武岩。此外，Wang 等（2018）对山东新生代碱性玄武岩的研究发现，在 $\delta^{66}Zn$ 与其他元素和微量元素比值相关图上弱碱性玄武岩呈现出从强碱性玄武岩组成向类似岩石圈地幔组成过渡的趋势，据此他们指出这些玄武岩的地球化学组成的转变是由贫硅熔体与岩石圈地幔相互作用的结果。

玄武质熔体的演化过程通常发生在浅部的岩浆房内。然而，由于起源自不同富集组分（如辉石岩和碳酸盐化橄榄岩）的岩浆具有不同的组成和性质，若它们在上升过程中相遇，两者则可能发生反应，结晶出辉石和石榴子石，并产生新的岩浆组成。例如，富硅拉斑质岩浆+贫硅碱性岩浆→富硅碱性岩浆+石榴子石+单斜辉石（Zeng et al., 2017）。这种拉斑质岩浆与碱性岩浆之间的相互作用及其伴随的岩浆深部演化过程可能是造成板内玄武岩化学组成多样性的一个重要机制。由此可见，在今后的研究中还应重视大陆岩石圈地幔对中国东部新生代玄武岩地球化学组成的改造作用。

2.1.4 板内玄武岩成因新机制——大地幔楔中流体助熔机制

传统的观点笼统地认为，大陆板内玄武岩是在拉张背景下由地幔上涌产生的减压熔融所致。但中国东部新生代玄武岩中普遍富水（Xia et al., 2019），而且其源区主要为碳酸盐化地幔和再循环洋壳组分，因此流体在玄武岩的成因中发挥了重要作用。徐义刚等（2018）认为，在地幔过渡带中滞留板片的脱碳和脱水作用及相关的熔融和交代作用是大地幔楔体系中板内岩浆成因的主要驱动力（图2）。这种熔融机制在中国东部新生代玄武岩的形成中可发挥两方面的作用。一是由地幔过渡带中滞留板片的脱碳和脱水作用以及熔融作用产生的碳酸盐熔体因具有较低的密度和黏度，易于上升离开地幔过渡带进入上地幔，且在上升的过程中容易与周边地幔橄榄岩发生反应形成碳酸盐化橄榄岩，中国东部大量硅不饱和、具有轻的 Mg 和重的 Zn 同位素特征的玄武岩即被认为来源于这种地幔源区（Li et al., 2016a, 2019）。二是由于含碳酸盐的沉积物在地幔过渡带深度的低程度熔融会抽离绝大部分的 U、Th，而残留强烈富集 K、Ba、Pb 的钾锰钡矿，导致残余固相具有极低的 U/Pb 值和极高的 K/U、Ba/Th 值，这种残余物经历长期的封闭演化可形成特殊的 EM1 组分，并作为中国东北钾质玄武岩的源区组分（Wang et al., 2017）。

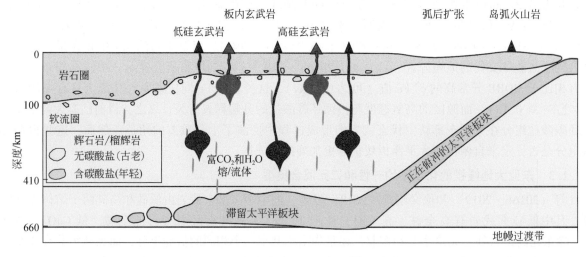

图 2 东亚大地幔楔物质组成与中国东部新生代玄武岩成因模型（据徐义刚等，2018）

2.2 利用深源包体示踪中国东部中生代深部演化

中国东部广泛分布有不同时代的幔源火山岩，其组成和时空分布规律是板块汇聚背景下板内过程与板缘过程协同作用的记录（郑建平和戴宏坤，2018；朱日祥和徐义刚，2019）。橄榄岩捕虏体是岩石圈地幔的直接样品，可以直接揭示岩石圈地幔的组成结构和演化历史。前人研究表明东部岩石圈地幔经历了强烈改造，古老难熔的克拉通型岩石圈地幔转变为新生饱满的大洋型地幔（Zhang et al., 2008），其中块体边界和内部薄弱带是地幔减薄和最终置换的优先发生区（郑建平和戴宏坤，2018），克拉通边部（Lin et al.，2019）及薄弱带地区的岩石圈地幔普遍就有饱满的组成和年轻的形成年龄（Liu et al.，2019；Wu et al.，2019），克拉通核部（如鹤壁）则保留有古老难熔的克拉通型地幔（Sun et al.，2012）。中国东部的地幔改造作用普遍认为是由（古）太平洋俯冲及后撤引起（Wu et al.，2019）。由于相对缺乏深部岩石圈捕虏体，中国西部岩石圈地幔的显著薄弱，阻碍了进一步理解周边板片俯冲对华北岩石圈地幔的改造作用。例如，在华北北缘中部地区发现的含火成碳酸岩侵入体含有金刚石等超高压矿物（Liu et al.，2016），且元素组成类似于沉积碳酸盐，Sr 同位素组成类似于古生代的海水，被解释为古亚洲洋俯冲碳酸盐底劈、上升、改造华北岩石圈的记录（Chen et al.，2017）。此外，在狼山地区发现的橄榄岩捕虏体，具有饱满的组成和弱的交代作用，且区别于典型的克拉通地幔，与中国东部新增生岩石圈地幔类似，考虑到该地区远离太平洋构造域而靠近古亚洲洋构造域，以及区域上普遍发育的与古亚洲洋俯冲相关的构造岩浆活动（Zhang et al.，2014），该地区饱满的尖晶石二辉橄榄岩可能是古亚洲洋俯冲改造华北岩石圈地幔的记录（Dai et al.，2019）。在华北南缘，橄榄岩中的锆石记录了多阶段复杂的地幔改造作用（Zheng et al.，2014），其中中生代的锆石年龄被认为是古特提斯洋向北俯冲及随后的扬子大陆深俯冲对华北岩石圈地幔的改造记录（Zhao et al.，2018），对大别造山带内地质体橄榄岩的 Mg 同位素研究证实了南侧俯冲对华北的改造作用（Shen et al.，2018）。因此，深部岩石圈地幔的改造作用是极不均一的，板块边缘和内部薄弱带是地幔改造和置换的优先发生区，岩石圈显生宙活化和深部岩石圈改造、置换作用是这些内在、外在因素共同作用的结果（郑建平等，2019）。

2.3 塔里木和峨眉山大火成岩省的成因差异

2.3.1 塔里木大火成岩省的特征与成因

塔里木大火成岩省已经成为近年研究的焦点之一，虽然该大火成岩省大部分被沙漠所覆盖，但是基于对岩心及周边出露的火山岩的研究，依然取得了许多重要进展。总体上，该大火成岩省呈现出持续时间长、岩石组合复杂、岩石富铁等特征（徐义刚等，2013；Wei et al.，2014；Xu et al.，2014b；Liu et al.，2019）。究其成因，多数学者认为其与地幔柱作用有关，相关的证据包括地壳的隆升以及存在苦橄质熔岩（Li et al.，2014），也有观点认为高的 H_2O 含量在塔里木大火成岩省岩浆形成过程中起到了重要作用（Xia et al.，2016）。早期（~290 Ma）玄武岩和晚期（~280 Ma）的侵入岩具有不同的地球化学特征，前者显示出弱的 Nb-Ta 负异常和负 ε_{Nd} 值，后者具有 Nb-Ta 正异常和正 ε_{Nd} 值，研究认为它们或分别起源于岩石圈地幔和软流圈地幔或地幔柱源区（Li et al.，2012；Xu et al.，2014b），或均起源于富集的岩石圈地幔，但随着巨量玄武岩的抽取和时间的推移，源区富集组分逐步消耗殆尽（Zhang D. Y. et al.，2018）。也有观点认为，柯坪玄武岩富集同位素组成是地幔柱来源的岩浆受地壳混染的结果（混染程度最高可达 10%；Yu et al.，2011；Zhang et al.，2012）。事实上，塔里木大火成岩省的溢流玄武岩存在多种类型，可能是岩石圈地幔和地幔热柱不同程度相互作用的产物（Yu S. Y. et al.，2017）。根据两期岩浆活动的空间分布特征，三阶段地幔柱孕育模型被提出解释该大火成岩省的深部地质过程（Xu et al.，2014a）。塔里木大火成岩省绝大多数玄武岩富铁（多数 TFeO>13%，最高可达 18.7%；李洪颜等，2013；Wei et al.，2014），可以称为铁质玄武岩（ferrobasalt）。另外，这种富铁特征在霞石岩中也有所体现，它们的形成可能与源区存在辉石岩/榴辉岩组分有关（Cheng et al.，2018；Zhang et al.，2018a）。系统的 Mg 同位素研究揭示塔里木大火成岩省主要岩石单元具有轻镁同位素异常，反映了俯冲再循环沉积型碳酸盐岩对源区的改造，但是岩石圈地幔和地幔热柱源区中碳酸盐岩的相态并不相同，前者以方解石/白云石为主，而后者为发生相变的菱镁矿/方镁石等（Cheng et al.，2017）。

2.3.2 峨眉山大火成岩省的特征与成因

与塔里木大火成岩省相比较，峨眉山大火成岩省的火成岩组合相对简单，主要为拉斑和碱性玄武岩以及少量苦橄岩和酸性岩。除云南丽江外，在云南大理、宾川和永胜等地区的熔岩底部也相继识别出苦橄岩或苦橄质熔岩（Kamenetsky et al.，2012），暗示峨眉山地幔柱的中心位于丽江-大理一带。这与峨眉山放射状岩墙指示的地幔柱中心是一致的（Li et al.，2015）。近年来，为了精确约束峨眉山地幔柱的源区特征，研究人员利用微区原位分析技术对橄榄石斑晶和其中的熔融包裹体开展了深入的研究，结果表明高钛和低钛玄武岩可能具有相似的源区，不同的 Ti 含量和 Ti/Y 值反映源区深度和部分熔融程度的差异（Zhang et al.，2019），但钛铁氧化物的分离结晶作用对 Ti 含量和 Ti/Y 值的影响也需要考虑（Hou et al.，2011）。另外，一些观点认为，该大火成岩省还可能存在辉石岩源区，推测是古俯冲事件中再循环洋壳和地幔橄榄岩相互作用的产物（Yu X. et al.，2017），尽管镁同位素并未识别出再循环沉积型碳酸盐岩的贡献（Tian et al.，2017）。有关峨眉山大火成岩省地幔柱成因的证据进一步强化，基于苦橄岩 PRIMELT2 和橄榄石-尖晶石铝温度计的计算均揭示出显著的热异常（~250℃）。地球物理解译出在地壳底部或 Moho 面之上存在着巨厚的镁铁-超镁铁层（西部 15~20 km，中东部 5~10 km），说明峨眉山大火成岩省存在大规模的底侵作用（Xu et al.，2015）。基于热化学地幔柱模型，推测峨眉山地幔柱中含有 10%~20% 高密度的再循环洋壳，因此在大规模火山喷发之前峨眉山地幔柱产生的热浮力只能导致小幅度地表隆升（~200 m），这与地质观察的结果一致（Zhu et al.，2018）。

地球物理方法是探测现代地幔柱常用的手段，但不适用于古老地幔柱的鉴别。我国地球物理学家与火山学者合作，提出了集成利用人工源地震、天然源宽频带地震探测，以及密集重力-地磁剖面测量等综合地球物理手段探测"古地幔柱"作用，在峨眉山大火成岩省研究中取得了系列成果（Chen et al.，2015；徐涛等，2015；陈赟等，2017）。他们的思路是，大火成岩省作为地球上目前已知的最大规模火山作用，不仅会在地表直观地留下大面积溢流玄武岩，而且在地球深部也会保留大规模侵入岩浆的"遗迹"，后者势必会引起岩石圈地幔和地壳的组分改造和结构变化，这些可能是深部地球物理探测可以追踪到的重要线索。通过几何结构-物性结构-动力学属性参数之间的联合约束，以及与地球化学、地质学研究成果的有机结合，Chen 等（2015）、徐涛等（2015）发现在峨眉山大火成岩省的内带，"底侵"层厚 15~20 km，底界面深度约 50~55 km，横向尺度约 150~180 km；他们还发现不同区带地壳厚度-波速比变化关系，意味着高波速比物质［组分上偏基性或（和）状态上存在熔融］在地壳增厚过程中起到了重要作用，从而揭示出与古地幔柱作用有关的大规模岩浆活动导致的地壳结构和组分变化以及深浅响应过程等。这些尝试显示了综合地球物理方法、不同学科交叉融合在重建古老重大地质事件深部过程方面的巨大潜力，对深部找矿也具有重要的指导意义。

3 深入开展了火山灾害与监测研究

目前国际上将火山监测技术分为三大类，即火山地震监测、火山形变监测和火山气体监测。中国火山监测起步较晚，始于 20 世纪 90 年代中期，经历了 20 多年的发展，目前已初步建成了由长白山、腾冲、五大连池、镜泊湖、龙岗、琼北海口等 6 个火山监测站组成的火山监测台网。中国火山监测台网整体架构分为 4 级，自下向上依次为火山观测点、火山监测站、省火山台网部和国家火山台网中心。监测数据自观测点向上通过地震专用网络上传，最终汇集到国家火山台网中心进行备份、存储和分析。目前，我国火山监测手段基本涵盖了国际常用的手段，部分达到了国际先进水平（刘国明等，2011；Xu J. D. et al.，2012）。

长白山天池火山是多成因复合火山，第四纪以来发生过多次喷发活动，其中公元 946 年左右的喷发活动是全球 2000 年以来最大规模的喷发活动之一，火山喷发指数 VEI 达到 7，此后还在 1668 年、1702 年和 1903 年发生过多次中小规模的喷发活动（Xu J. D. et al.，2012）。目前普遍认为长白山天池火山是我国最具潜在灾害性喷发危险的活火山。建站以来，长白山火山监测站观测到了 2002—2005 年的岩浆扰动事件，产出的火山监测资料对理解长白山火山现今活动状态和评估未来喷发危险起了关键作用（Xu J. D. et al.，2012；Zhang M. L. et al.，2018）。2006 年长白山火山监测站被评为科技部野外重点实验站，2019 年通过科技部评估。近十年来，长白山逐渐成为国内外火山学术界的"热点"地区，吸引了国内外大量的专家学者，在千年大喷发年代学与环境效应、火山区温室气体排放、气体地球化学、火山深部动力学等研究领域取得了重要进展（Sun et al.，2014；Zhang et al.，2015）。

腾冲火山区位于印度板块与欧亚板块碰撞带的东南缘，火山区的分布面积约 9 000 km²（赵慈平等，2006），从上新世至全新世共有 4 期活动（皇甫岗和姜朝松，2000），最近一期活动距今约 2 500～5 000 年（皇甫岗和姜朝松，2000；尹功明和李盛华，2000；Li et al.，2019）。近期的研究与监测结果表明，腾冲地区地壳内存在岩浆囊（赵慈平等，2011，2012），有再次喷发的危险（皇甫岗和姜朝松，2000；洪汉净等，2007）。自 1997 年以来的地震、GPS、水准、重力、流体地球化学等观测资料显示，在腾冲县城一带地下有岩浆侵入活动，具体表现为小震群密集活动、水平面膨胀、幔源气体释放持续增强（叶建庆等，2003；李成波等，2007；胡亚轩等，2008；赵慈平等，2011，2012）。近十年来，腾冲火山区壳内岩浆房得到进一步的确认（Hua et al.，2019；华雨淋和吕彦，2019），还发现火山区幔源气体释放存在持续增强的趋势（赵慈平等，2012），所反映的岩浆补充活动可能与火山区的显著地震事件密切相关（Lei et al.，2012；张广伟和雷建设，2015），腾冲火山区岩浆活动监测对区域 7 级强震的监测预测有重大价值。

目前我国火山大多处于休眠状态，在继续加强火山监测工作的同时，迫切需要开展系统的火山地质学、物理火山学、火山岩地球化学（包括年代学）、火山地震学、火山形变学和火山灾害学等多学科综合研究。未来十年，要有针对性地开展对每个特定火山的综合性研究，发展符合中国火山活动特点的喷发危险性预测和灾害评估技术方法，对于有潜在喷发危险的火山，加快对已有监测系统的升级改造，建成与国际接轨的火山应急预警系统平台。

4 发展了火山学领域的一些新学科

4.1 火山喷发模拟与物理火山学

火山模拟研究包括实验模拟与数字模拟，是火山地质学研究的重要补充，可弥补野外所获数据无法完整地体现火山作用过程的缺陷，在岩浆房动力学、岩墙动力学、喷发柱动力学和形态学、火山灰云扩散机制、火山碎屑流动力过程及颗粒沉降、熔岩流流变特征、岩浆-水相互作用机制研究中可发挥重要作用。实验模拟研究是使用火山喷发物或相似物模拟火山喷发现象，获取关键的流体动力学参数的特征。数值模拟则利用数学、物理和（或）化学方程描述特定的火山作用过程（Roche and Carazzo，2019）。火山模拟研究能验证火山作用过程的动力学机制，其中实验模拟研究为数字模拟研究提供关键经验参数。

近十年，国内学者开始尝试火山模拟实验与喷发动力学研究，开展了火山泥石流的流体动力学过程及其影响因素的实验室模拟研究。王雪冬（2013）利用水槽实验获得了在地形条件相同的情况下，水源流量、总水量对火山泥石流爆发量的影响规律。陈正全（2017）和陈正全等（2018）开展了火山碎屑流的流体动力学过程及其影响因素的实验室模拟研究，他们利用颗粒驱动重力流的无量纲数配比方法，在实验室中模拟了地形变化对低密度火山碎屑流的影响。实验显示火山碎屑流在通过两种障碍模型中，流体均产生"扬起"效应，翻越-通过障碍模型。流体前锋在翻越较高的横亘障碍物之后，流体特征重回"惯性-浮力平衡相"的时间越短，前锋速度衰减越快。实验流体在通过侧向部分遮挡地形之后，较窄的遮挡对流体前锋具有加速作用，并对流体产生"活化"作用，使障碍物之后的堆积物有所增加。

物理火山学是火山学的一个重要研究领域，以火山活动过程的物理变量（如温度、压力、黏度、密度、屈服强度等）及其随时空的动力变化为基础，来探讨火山不同活动过程的动力机制、时空结构及其联系。潘波等（2017）通过计算数值模拟了长白山气象站期碱流岩的熔岩流的流动速度和流动距离，提出了新疆阿什库勒火山区的熔岩流灾害区划。万园等（2011）利用 LAHARZ 数值模拟方法，对二道白河、松花江、鸭绿江及图们江 4 条火山泥石流易发河道进行了火山泥石流的物理学特征研究，结果显示该模拟所得灾害分布范围与长白山地区的历史火山泥石流分布范围吻合，说明该方法有助于其他火山区的火山泥石流灾害区划研究。

4.2 火山灰年代学

火山灰年代学是第四纪地质学与火山学相互交叉的领域，是一个新的研究方向。大规模爆炸式火山喷发会产生大量火山灰，火山灰从产生、飘散至沉降的整个过程是在短时间内完成的（数日至数年），因此从地质时间尺度来看火山灰层可以视作等时标志层。火山灰年代学就是利用火山灰等时标志层对广泛分布的地质、古气候、考古记录进行关联和定年，这包含了地层学和年代学两方面的意义（Lowe，2011）。一方面，火山灰层是火山喷发的直接证据，对重建火山喷发历史、揭示岩浆源区特征都有重要意义；另一方面，由于受到各类测年方法精度的限制，第四纪古气候与古环境记录的跨区域精确对比往往难以实现，火山灰层为这些事件的对比提供了潜在的工具（陈宣谕等，2014；刘嘉麒等，2018）。

过去，火山灰年代学家的研究对象主要是毫米至厘米级肉眼可见的火山灰层，它们主要记录了规模较大或者距离火山口较近的火山喷发，而对一些规模小或距火山口较远的喷发，肉眼可见的火山灰方法往往难以发挥作用。自 Dugmore（1989）在苏格兰泥炭中第一次提取到肉眼不可见的冰岛火山灰层之后，显微火山灰的重要性才逐渐得到关注（McLean et al.，2018）。显微火山灰层覆盖面积大，可以飘散至距离火山口数千公里以外的地区（Sun et al.，2014），因此是开展地质学和考古学时间精确对比的重要对比。

经过近十年的努力，我国学者已经掌握了显微火山灰的提取技术，并在中国科学院广州地球化学研究所建立了我国首个显微火山灰年代学实验室，可开展相关的年代学研究，并逐步开始应用到地质学和古环境等领域。显微火山灰年代学的关键是显微火山灰的提取。过去学者尝试了诸如酸化法、碱处理、烧失等各种方法，以从各类沉积物中获得纯净的火山灰（图3）。多年的实践表明，重液浮选法（Blockley et al.，2005）能较好地提取沉积物中的含量极低的显微火山灰，但缺点是较为耗时。此外还有一些学者尝试利用间接手段如 CT 扫描和 X 射线荧光（X-ray fluorescence，XRF）来识别沉积物中的火山灰层。虽然这些方法可能识别出沉积物中的火山灰层，但在进一步分离提取火山灰方面仍显不足。利用掌握的显微火山灰提取技术，我国学者已经在东亚火山灰年代学的研究领域取得了若干重要进展。Chen 等（2016）、Sun 等（2014，2015）分别报道了日本北部、我国四海龙湾、格陵兰冰芯中发现的长白山千年喷发显微火山灰（B-Tm），证明该火山灰层在东北亚地区均有广泛分布，是重要的等时标志层。同时，格陵兰冰芯年代学为限定这次火山喷发的年龄提供了重要证据，其喷发发生在公元 946—947 年（Sun et al.，2014）。Chen 等（2019）报道了日本北部首个高分辨率显微火山灰地层记录，显著扩展了区域内数层关键火山灰标志层的分布范围，并首次在日本沉积记录中识别到来自俄罗斯堪察加半岛、甚至是印度尼西亚的显微火山灰。Sun 等（2018）结合长白山近源火山灰、长白山地区湖泊沉积物火山灰、日本水月湖的火山灰记录，限定了长白山气象站期的喷发年龄距今约 8 100 年，证明了这次喷发的火山灰可以从中国东北地区扩展至日本列岛中部。

显微火山灰方法应被更广泛地应用到我国重要的地质、古气候记录中，这将会对我国火山学和第四纪地质学等相关学科的发展，尤其是火山学和第四纪地质学的交叉融合起到极大的推动作用。

4.3 与火山区相关的地热资源研究

地下岩石在特定的温度和压力条件下发生熔融、迁移、储存、分异、喷发和冷却的过程，受控于特定的岩石圈热构造背景，而同时也对岩石圈热结构必然产生深刻的影响。近十年来，我国火山学界和地热学界众多学者在中新生代火山区开展了不少相关的研究，取得了一些重要认识。

地下岩浆的产生、赋存和运移必然伴随着地球内部高温物质和能量从深部向浅部的迁移，因而往往是最具地热资源开发潜力的地区。地热流体按温度高低具有不同应用价值，高于 120℃ 的中高温流体适用于发电，而且温度越高发电效率越高；低于 100℃ 的地热流体则适用于直接采暖、温棚养殖、康乐洗浴

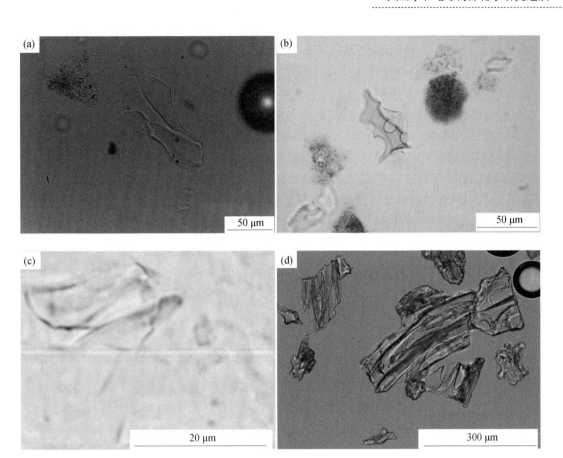

图 3 在不同地区提取到的长白山火山玻屑的镜下照片
(a) 中国四海龙湾（千年喷发火山灰）；(b) 日本久種湖（千年喷发火山灰）；
(c) 格陵兰冰芯（千年喷发火山灰）；(d) 中国圆池（气象站期火山灰）

等。目前中国大陆地区已经查明的地热资源以中低温为主（黄少鹏，2014），这是中国在地热资源直接利用方面世界领先，但在地热发电方面却远远落后于美国、印度尼西亚、菲律宾、土耳其、新西兰等国的一个重要原因，上述五国的地热发电装机容量均高于 1 000 MW，而中国地热电站在线装机容量不足 50 MW。国际上，高温地热田和绝大多数兆瓦级地热发电站均位于新生代火山活动区，但中国大陆地区目前还是例外，仅有的西藏羊八井和羊易两个兆瓦级地热发电站周边并无现代或近代火山活动，而具有新生代活动特征的火山区除腾冲以外，长白山天池-龙岗火山群、镜泊湖火山群、五大连池火山群、雷州-琼北火山群、内蒙古东部火山群等火山构造区地热显示却不那么明显，甚至还出现热泉温泊与冷泉冰洞并存的奇观。黄少鹏（2014）认为，这种奇特的冰火地质景观系广泛发育的火山熔洞和寒冷的地表温度相互作用的结果。由于空气热胀冷缩，冬季和夜晚气温低时空气密度大，熔洞洞口冷空气会沉入洞底，而洞底受地温加热的热空气则会上升溢出熔洞，只要熔洞与地表保持空气流通，即便是在像五大连池这样的年轻火山活动区的熔洞中也可常年保持冰冷的环境。冷泉冰洞的存在并不排除五大连池开发地热资源的可能。张森琦等（2017）发现五大连池北部的尾山火山区下伏岩浆囊顶界埋深在 ~6.5 km，具较好的干热岩地热地质条件。

高温岩浆在冷凝过程中要散发热量，火山体热扩散过程的数值模拟既是研究火山构造演化的重要手段，也是评估火山区地热资源的重要方法。张旗等（2014）对岩浆热场的一些基本特征进行了定性描述。段文涛等（2017）对火山传导冷却过程及其对围岩的热影响进行了定量数值模拟，认为直径为 1 000 m 的柱状侵入体的冷却时间约为 50 ka，对围岩地温场的影响时间持续不到 0.5 Ma，影响范围约为 3 km；而直径 2 000 m 的柱状侵入体的冷却时间约为 200 ka，对围岩温度场的影响可持续 1.4 Ma，影响范围可至

7 km。段文涛等（2017）模拟计算了长白山天池火山千年大喷发对于地温场的可能影响，结果发现除非有地下流体的参与，天池火山隐伏火山机构对于围岩温度分布的影响局限于约 500 m 的范围内。需要指出的是，上述数值模拟和分析都是基于单纯的热传导，而对流和辐射对火山和潜火山机构热量的扩散也可能起重要作用，特别是有地下水的参与，火山余热可能得到比热传导更加有效的运移和富集，从而在距火山口较远的地方形成有开发价值的水热系统。闫佰忠等（2017）认为环长白山地区地热异常主要分布在环天池火山口紧邻，断裂带和地热水异常区的分布有较好的对应关系。

在地热资源的勘查中，遥感技术具有经济、高效、覆盖范围广等优势，已逐渐成为地热资源探测与调查的一种新手段。彭芬等（2014）综述了国内外利用遥感技术探测火山地热的发展历史和研究现状，利用美国 Landsat-7 ETM 多波段红外遥感数据反演了内蒙古锡林郭勒盟新生代火山活动构造区的地表温度，发现研究区地表温度异常分布多与北东东与北西向线性构造的交汇处和环形构造有关，明显受控于深大断裂和火山作用。作者还圈定了 7 个地表温度异常区作为地热资源勘查的远景区。

5 学术组织在火山学学科发展中发挥积极作用

我国的火山学和地球内部化学领域的学术组织包括中国矿物岩石地球化学学会下属的火山和地球内部化学专业委员会、中国灾害防御协会下属的火山专业委员会，以及 IUGG 中国委员会下属的 IAVCEI 中国委员会。虽然我国在相关领域的人员体量较小，但上述3个学术组织在过去十年间努力工作，在发挥学术社团的功能方面取得了一些成绩。2013年、2016年和2019年分别在海南海口、内蒙古柴河和吉林长白山召开了第7、第8、第9次全国火山学术研讨会；2014年、2016年和2018年分别在北京、南京和兰州召开了第一、第二、第三届中国大地测量和地球物理学学术大会。在全国会议间隙年穿插组织了一些专题研讨会，如2011年的"柴河火山论坛"、2012年召开的"天水火山学术研讨会"、2015年在西安交通大学举行的"火山与地热学术研讨会"、2016年在腾冲主办的"灾害科普与火山研讨会"等。这些会议为我国相关领域特别是火山学领域新学科的发展起到了推动作用。另外，2016年和2017年，还成功举办了第六届国际玛珥湖学术会议和第六届国际大火成岩省学术会议，扩大了我国火山学研究的国际影响力。为系统反映我国学者在火山学和地球内部化学领域的科研进展，3个学术组织共在《岩石学报》《矿物岩石地球化学通报》和 *Lithos* 组织编辑了5个专辑（Xu et al.，2014a；徐义刚和郭正府，2018）。

6 展 望

总体上，近十年间我国火山学和地球内部化学的研究取得了较显著的进展，但是在火山学各个分支学科发展的系统性、研究特色以及分支学科的研究深度与广度等方面，与国际火山学的研究前沿相比，尚存在着明显的差距，包括火山学领域内子学科的发展不平衡、火山学与相关学科（例如，地球物理学、大气科学、环境科学、古气候学、地理科学等）的学术交流与学科交叉研究较薄弱、目前我国火山监测和减灾防灾能力与国家需求相差较远。针对这些薄弱环节，采取相关的措施并加强火山学学科建设并深化相关的研究，有望进一步缩短国内外在相关领域的研究差距。

（1）建议进一步发挥我国火山岩岩石地球化学的研究优势，提升相关研究的国际影响力。可关注的领域包括东亚大小地幔楔系统中的物质循环、大火成岩省及其环境效应、深部作用与地表过程的联系机制等。

（2）应大力扶持一些火山学领域子学科的发展，完善火山学各个分支学科发展的系统性，将火山学的基础研究与国家实际需求相结合，加强火山监测队伍和平台建设，以期在火山灾害防御、火山资源利用和科学普及等方面发挥更大的作用。

（3）加强与其他学科的交叉融合也是发展火山学的重要途径，如加强新型同位素体系在岩石成因中的应用、加强与实验岩石学的结合有助于深化对火山岩成因的认识，加强火山学与地球物理学的交叉则是深地科学的重要方向,而火山灰年代学可在古环境重建、考古学等领域发挥不可替代的作用。另外,将火成岩研究拓展到行星科学领域也是十分值得期待的。

致谢:本文是集体合作的成果,协作完成初稿的作者(按正文内容的顺序)还有:白志达、潘波、赵勇伟、樊祺诚、陈立辉、郑建平、张招崇、雷建设、许建东、赵慈平、刘国明、刘永顺、彭年、聂保锋、陈正全、陈宣谕、孙春青和黄少鹏,在此一并致谢。

参 考 文 献

白志达,谭庆伟,许桂玲,徐德斌,王妍. 2012. 内蒙东部晚第四纪火山活动与新构造. 岩石学报,28(4):1099-1107
陈宣谕,徐义刚,Menzies M. 2014. 火山灰年代学:原理与应用. 岩石学报,30(12):3491-3500
陈赟,王振华,郭希,邓阳凡,徐涛,梁晓峰,田小波,吴晶,陈林,张晰,唐国彬,徐义刚. 2017. 古地幔柱作用"遗迹"的深部地球物理探测:以峨眉山大火成岩省为例. 矿物岩石地球化学通报,36(3):394-403
陈正全. 2017. 地形对颗粒驱动重力流影响的模拟实验研究. 北京:中国地震局地质研究所博士学位论文
陈正全,许建东,魏海泉. 2018. 实验模拟侧向遮挡对低密度火山碎屑流的影响:对日本云仙岳火山1991年火山碎屑流的启示. 岩石学报,34(1):172-184
段文涛,黄少鹏,唐晓音,张炯. 2017. 利用ANSYS WORKBENCH模拟火山岩浆活动热扩散过程. 岩石学报,33(1):267-278
樊祺诚,赵勇伟,隋建立,李大明,武颖. 2012. 大兴安岭诺敏河第四纪火山岩分期:岩石学、年代学与火山地质特征. 岩石学报,28(4):1092-1098
樊祺诚,赵勇伟,陈生生,李霓,隋建立. 2015. 大兴安岭-太行山重力梯度带以西的第四纪火山活动. 矿物岩石地球化学通报,34(4):674-681
洪汉净,吴建平,王庆良,李克,赵慈平,上官志冠,杨清福,张恒荣,刘国明. 2007. 中国火山危险性等级与活动性分类. 地震地质,29(3):447-458
胡亚轩,王庆良,赵慈平,邵德晟,施发奇. 2008. 腾冲火山区形变分析. 国际地震动态,(4):42-47
华雨淋,吕彦. 2019. 腾冲火山及周边地区双差层析成像. 地球物理学报,62(8):2982-2990
皇甫岗,姜朝松. 2000. 腾冲火山研究. 昆明:云南科技出版社
黄少鹏. 2014. 中国地热能源开发的机遇与挑战. 中国能源,36(9):4-8,16
雷建设,赵大鹏,徐义刚,樊祺诚,米琦,杜沫霏,鲁明文. 2018. 长白山火山下方地幔转换带中滞留的俯冲太平洋板块存在空缺吗? 岩石学报,34(1):13-22
李成波,施行觉,刘苏苏,赵慈平,施发奇,邵德晟,姜朝松. 2007. 腾冲火山区的GPS形变特征. 地球物理学进展,22(3):765-770
李洪颜,黄小龙,李武显,曹俊,贺鹏丽,徐义刚. 2013. 塔西南其木干早二叠世玄武岩的喷发时代及地球化学特征. 岩石学报,29(10):3353-3368
李霓,魏海泉,张柳毅,赵勇伟,赵波,陈正全,陈生生,陈晓雯. 2014. 云南腾冲大六冲火山机构的发现及意义. 岩石学报,30(12):3627-3634
刘国明,孙鸿雁,郭峰. 2011. 长白山火山最新监测信息. 岩石学报,27(10):2905-2911
刘嘉麒. 1999. 中国火山. 北京:科学出版社
刘嘉麒,孙春青,游海涛. 2018. 全球火山灰年代学研究概述. 中国科学:地球科学,48(1):1-29
潘波,程滔,万园,于红梅,许建东. 2017. 基于热流变运动学模型的熔岩流灾害区划:新疆阿什库勒火山区的灾害区划. 地震地质,39(4):721-734
彭芬,黄少鹏,时庆金,程玉祥,荆勇河. 2014. 卫星热红外遥感技术在火山区地热探测中的应用:以内蒙古锡林郭勒火山区为例. 地质科学,49(3):899-914
任纪舜. 2003. 新一代中国大地构造图——中国及邻区大地构造图(1:5000000)附简要说明:从全球看中国大地构造. 地球学报,24(1):1-2
万园,许建东,林旭东,潘波. 2011. 基于数值模拟的长白山天池火山泥石流灾害展布范围分析及预测. 吉林大学学报(地球科学版),41(5):1638-1645
王雪冬. 2013. 长白山火口湖溃决引发的火山泥石流灾害危险性预测研究. 长春:吉林大学博士学位论文
魏海泉. 2010. 长白山火山岩浆柱岩浆上升作用过程. 地学前缘,17(1):11-23
徐涛,张忠杰,刘宝峰,陈赟,张明辉,田小波,徐义刚,滕吉文. 2015. 峨眉山大火成岩省地壳速度结构与古地幔柱活动遗迹:来自丽江-清镇宽角地震资料的约束. 中国科学:地球科学,45(5):561-576
徐义刚,樊祺诚. 2015. 中国东部新生代火山岩研究回顾与展望. 矿物岩石地球化学通报,34(4):682-689
徐义刚,郭正府. 2018. 火山学研究的问题与进展. 岩石学报,34(1):1-3
徐义刚,何斌,罗震宇,刘海泉. 2013. 我国大火成岩省和地幔柱研究进展与展望. 矿物岩石地球化学通报,32(1):25-39
徐义刚,李洪颜,洪路兵,马亮,马强,孙明道. 2018. 东亚大地幔楔与中国东部新生代板内玄武岩成因. 中国科学:地球科学,48(7):825-843
闫佰忠,邱淑伟,肖长来,梁秀娟. 2017. 长白山玄武岩区地热异常区遥感识别. 吉林大学学报(地球科学版),47(6):1819-1828
叶建庆,蔡绍平,刘学军,王绍晋,蔡明军. 2003. 腾冲火山地震群的活动特征. 地震地质,25(S1):128-137
尹功明,李盛华. 2000. 云南腾冲马鞍山最后一次喷发的热释光年龄. 地震研究,23(4):388-391
张广伟,雷建设. 2015. 2011年云南腾冲5.2级双震发震机理. 地球物理学报,58(4):1194-1204

张旗,金惟俊,李承东,焦守涛. 2014. 岩浆热场:它的基本特征及其与地热场的区别. 岩石学报,30(2):341-349

张淼琦,贾小丰,张杨,李胜涛,李志伟,田蒲源,明圆圆,张超. 2017. 黑龙江省五大连池尾山地区火山岩浆囊探测与干热岩地热地质条件分析. 地质学报,91(7):1506-1521

赵慈平,冉华,陈坤华. 2006. 由相对地热梯度推断的腾冲火山区现存岩浆囊. 岩石学报,22(6):1517-1528

赵慈平,冉华,陈坤华. 2011. 腾冲火山区壳内岩浆囊现今温度:来自温泉逸出气体CO_2、CH_4间碳同位素分馏的估计. 岩石学报,27(10):2883-2897

赵慈平,冉华,王云. 2012. 腾冲火山区的现代幔源氦释放:构造和岩浆活动意义. 岩石学报,28(4):1189-1204

赵勇伟,樊祺诚. 2010. 大兴安岭焰山、高山火山:一种新的火山喷发型式. 地震地质,32(1):28-37

赵勇伟,樊祺诚,李霓,龚丽文. 2017. 内蒙达里诺尔火山群晚第四纪火山地质特征与岩浆裂隙通道研究. 岩石学报,33(1):127-136

郑建平,戴宏坤. 2018. 西太平洋板片俯冲与后撤引起华北东部地幔置换并导致陆内盆-山耦合. 中国科学:地球科学,48(4):436-456

郑建平,戴宏坤,张卉. 2019. 中国东部岩石圈地幔的富化和置换过程. 矿物岩石地球化学通报,38(2):201-216

朱日祥,徐义刚. 2019. 西太平洋板块俯冲与华北克拉通破坏. 中国科学:地球科学,49(9):1346-1356

Blockley S P E, Pyne-O'Donnell S D F, Lowe J J, Matthews I P, Stone A, Pollard A M, Turney C S M, Molyneux E G. 2005. A new and less destructive laboratory procedure for the physical separation of distal glass tephra shards from sediments. Quaternary Science Reviews,24(16-17):1952-1960

Chen C F, Liu Y S, Foley S F, Ducea M N, Geng X L, Zhang W, Xu R, Hu Z C, Zhou L, Wang Z C. 2017. Carbonated sediment recycling and its contribution to lithospheric refertilization under the northern North China Craton. Chemical Geology,466:641-653

Chen X Y, Blockley S P E, Tarasov P E, Xu Y G, McLean D, Tomlinson E L, Albert P G, Liu J Q, Müller S, Wagner M, Menzies M A. 2016. Clarifying the distal to proximal tephrochronology of the Millennium (B-Tm) eruption, Changbaishan Volcano, Northeast China. Quaternary Geochronology,33:61-75

Chen X Y, McLean D, Blockley S P E, Tarasov P E, Xu Y G, Menzies M A. 2019. Developing a Holocene tephrostratigraphy for northern Japan using the sedimentary record from Lake Kushu, Rebun Island. Quaternary Science Reviews,215:272-292

Chen Y, Xu Y G, Xu T, Si S K, Liang X F, Tian X B, Deng Y F, Chen L, Wang P, Xu Y H, Lan H Q, Xiao F H, Li W, Zhang X, Yuan X H, Badal J, Teng J W. 2015. Magmatic underplating and crustal growth in the Emeishan large igneous province, SW China, revealed by a passive seismic experiment. Earth and Planetary Science Letters,432:103-114

Cheng Z G, Zhang Z C, Hou T, Santosh M, Chen L L, Ke S, Xu L J. 2017. Decoupling of Mg-C and Sr-Nd-O isotopes traces the role of recycled carbon in magnesiocarbonatites from the Tarim large igneous province. Geochimica et Cosmochimica Acta,202:159-178

Cheng Z G, Zhang Z C, Xie Q H, Hou T, Ke S. 2018. Subducted slab-plume interaction traced by magnesium isotopes in the northern margin of the Tarim large igneous province. Earth and Planetary Science Letters,489:100-110

Dai H K, Zheng J P, Xiong Q, Su Y P, Pan S K, Ping X Q, Zhou X. 2019. Fertile lithospheric mantle underlying ancient continental crust beneath the northwestern North China Craton: significant effect from the southward subduction of the Paleo-Asian Ocean. GSA Bulletin,131(1-2):3-20

Dugmore A. 1989. Icelandic volcanic ash in Scotland. Scottish Geographical Magazine,105(3):168-172

Hou T, Zhang Z C, Kusky T, Du Y S, Liu J L, Zhao Z D. 2011. A reappraisal of the high-Ti and low-Ti classification of basalts and petrogenetic linkage between basalts and mafic-ultramafic intrusions in the Emeishan large igneous province, SW China. Ore Geology Reviews,41(1):133-143

Hua Y J, Zhang S X, Li M K, Wu T F, Zou C Y, Liu L. 2019. Magma system beneath Tengchong volcanic zone inferred from local earthquake seismic tomography. Journal of Volcanology and Geothermal Research,377:1-16

Kamenetsky V S, Chung S L, Kamenetsky M B, Kuzmin D V. 2012. Picrites from the Emeishan large igneous province, SW China: a compositional continuum in primitive magmas and their respective mantle sources. Journal of Petrology,53(10):2095-2113

Lei J S, Xie F R, Mishra O P, Lu Y Z, Zhang G W, Li Y. 2012. The 2011 Yingjiang, China, earthquake: a volcano-related fluid-driven earthquake? Bulletin of the Seismological Society of America,102(1):417-425

Li D X, Yang S F, Chen H L, Cheng X G, Li K, Jin X L, Li Z L, Li Y Q, Zou S Y. 2014. Late Carboniferous crustal uplift of the Tarim Plate and its constraints on the evolution of the Early Permian Tarim large igneous province. Lithos,204:36-46

Li H B, Zhang Z C, Ernst R, Lü L S, Santosh M, Zhang D Y, Cheng Z G. 2015. Giant radiating mafic dyke swarm of the Emeishan large igneous province: identifying the mantle plume centre. Terra Nova,27(4):247-257

Li H Y, Xu Y G, Ryan J G, Huang X L, Ren Z Y, Guo H, Ning Z G. 2016a. Olivine and melt inclusion chemical constraints on the source of intracontinental basalts from the eastern North China Craton: discrimination of contributions from the subducted Pacific slab. Geochimica et Cosmochimica Acta,178:1-19

Li H Y, Zhou Z, Ryan J G, Wei G J, Xu Y G. 2016b. Boron isotopes reveal multiple metasomatic events in the mantle beneath the eastern North China Craton. Geochimica et Cosmochimica Acta,194:77-90

Li H Y, Li J, Ryan J G, Li X, Zhao R P, Ma L, Xu Y G. 2019. Molybdenum and boron isotope evidence for fluid-fluxed melting of intraplate upper mantle beneath the eastern North China Craton. Earth and Planetary Science Letters,520:105-114

Li Z L, Li Y Q, Chen H L, Santosh M, Yang S F, Xu Y G, Langmuir C H, Chen Z X, Yu X, Zou S Y. 2012. Hf isotopic characteristics of the Tarim Permian large igneous province rocks of NW China: implication for the magmatic source and evolution. Journal of Asian Earth Sciences,49:191-202

Lin A B, Zheng J P, Xiong Q, Aulbach S, Lu J G, Pan S K, Dai H K, Zhang H. 2019. A refined model for lithosphere evolution beneath the decratonized northeastern North China Craton. Contributions to Mineralogy and Petrology,174(2):15

Liu H Q, Xu Y G, Zhong Y T, Luo Z Y, Mundil R, Riley T R, Zhang L, Xie W. 2019. Crustal melting above a mantle plume: insights from the Permian Tarim large igneous province, NW China. Lithos,326-327:370-383

Liu J Q, Chen L H, Zeng G, Wang X J, Zhong Y, Yu X. 2016. Lithospheric thickness controlled compositional variations in potassic basalts of Northeast China by melt-rock interactions. Geophysical Research Letters, 43(6): 2582–2589

Lowe D J. 2011. Tephrochronology and its application: a review. Quaternary Geochronology, 6(2): 107–153

McLean D, Albert P G, Nakagawa T, Suzuki T, Staff R A, Yamada K, Kitaba I, Haraguchi T, Kitagawa J, SG14 Project Members, Smith V C. 2018. Integrating the Holocene tephrostratigraphy for East Asia using a high-resolution cryptotephra study from Lake Suigetsu (SG14 core), central Japan. Quaternary Science Reviews, 183: 36–58

Pan B, de Silva S L, Xu J D, Chen Z Q, Miggins D P, Wei H Q. 2017. The VEI-7 Millennium eruption, Changbaishan-Tianchi volcano, China/DPRK: new field, petrological, and chemical constraints on stratigraphy, volcanology, and magma dynamics. Journal of Volcanology and Geothermal Research, 343: 45–59

Pilet S, Baker M B, Stolper E M. 2008. Metasomatized lithosphere and the origin of alkaline lavas. Science, 320(5878): 916–919

Roche O, Carazzo G. 2019. The contribution of experimental volcanology to the study of the physics of eruptive processes, and related scaling issues: a review. Journal of Volcanology and Geothermal Research, 384: 103–150

Shen J, Li S G, Wang S J, Teng F Z, Li Q L, Liu Y S. 2018. Subducted Mg-rich carbonates into the deep mantle wedge. Earth and Planetary Science Letters, 503: 118–130

Sun C Q, Plunkett G, Liu J Q, Zhao H L, Sigl M, McConnell J R, Pilcher J R, Vinther B, Steffensen J P, Hall V. 2014. Ash from Changbaishan Millennium eruption recorded in Greenland ice: implications for determining the eruption's timing and impact. Geophysical Research Letters, 41(2): 694–701

Sun C Q, You H T, He H Y, Zhang L, Gao J L, Guo W F, Chen S S, Mao Q, Liu Q, Chu G Q, Liu J Q. 2015. New evidence for the presence of Changbaishan Millennium eruption ash in the Longgang volcanic field, Northeast China. Gondwana Research, 28(1): 52–60

Sun C Q, Wang L, Plunkett G, You H T, Zhu Z Y, Zhang L, Zhang B, Chu G Q, Liu J Q. 2018. Ash from the Changbaishan Qixiangzhan eruption: a new early Holocene marker horizon across East Asia. Journal of Geophysical Research: Solid Earth, 123(8): 6442–6450

Sun J, Liu C Z, Wu F Y, Yang Y H, Chu Z Y. 2012. Metasomatic origin of clinopyroxene in archean mantle xenoliths from Hebi, North China Craton: trace-element and Sr-isotope constraints. Chemical Geology, 328: 123–136

Tian H C, Yang W, Li S G, Ke S. 2017. Could sedimentary carbonates be recycled into the lower mantle? Constraints from Mg isotopic composition of Emeishan basalts. Lithos, 292-293: 250–261

Tian Y, Zhu H X, Zhao D P, Liu C, Feng X, Liu T, Ma J C. 2016. Mantle transition zone structure beneath the Changbai volcano: insight into deep slab dehydration and hot upwelling near the 410 km discontinuity. Journal of Geophysical Research: Solid Earth, 121(8): 5794–5808

Wang X C, Wilde S A, Li Q L, Yang Y N. 2015. Continental flood basalts derived from the hydrous mantle transition zone. Nature Communications, 6: 7700

Wang X J, Chen L H, Hofmann A W, Mao F G, Liu J Q, Zhong Y, Xie L W, Yang Y H. 2017. Mantle transition zone-derived EM1 component beneath NE China: geochemical evidence from Cenozoic potassic basalts. Earth and Planetary Science Letters, 465: 16–28

Wang Z Z, Liu S A, Chen L H, Li S G, Zeng G. 2018. Compositional transition in natural alkaline lavas through silica-undersaturated melt-lithosphere interaction. Geology, 46(9): 771–774

Wei X, Xu Y G, Feng Y X, Zhao J X. 2014. Plume-lithosphere interaction in the generation of the Tarim large igneous province, NW China: geochronological and geochemical constraints. American Journal of Science, 314(1): 314–356

Wu F Y, Yang J H, Xu Y G, Wilde S A, Walker R L. 2019. Destruction of the North China Craton in the Mesozoic. Annual Review of Earth and Planetary Sciences, 47: 173–195

Xia Q K, Bi Y, Li P, Tian W, Wei X, Chen H L. 2016. High water content in primitive continental flood basalts. Scientific Reports, 6: 25416

Xia Q K, Liu J, Kovács I, Hao Y T, Li P, Yang X Z, Chen H, Sheng Y M. 2019. Water in the upper mantle and deep crust of eastern China: concentration, distribution and implications. National Science Review, 6(1): 125–144

Xu J D, Liu G M, Wu J P, Ming Y H, Wang Q L, Cui D X, Shangguan Z G, Pan B, Lin X D, Liu J Q. 2012. Recent unrest of Changbaishan volcano, Northeast China: a precursor of a future eruption? Geophysical Research Letters, 39(16): L16305

Xu T, Zhang Z J, Liu B F, Chen Y, Zhang M H, Tian X B, Xu Y G, Teng J W. 2015. Crustal velocity structure in the Emeishan large igneous province and evidence of the Permian mantle plume activity. Science China: earth Sciences, 58(7): 1133–1147

Xu Y G. 2014. Recycled oceanic crust in the source of 90–40 Ma basalts in North and Northeast China: evidence, provenance and significance. Geochimica et Cosmochimica Acta, 143: 49–67

Xu Y G, Zhang H H, Qiu H N, Ge W C, Wu F Y. 2012. Oceanic crust components in continental basalts from Shuangliao, Northeast China: derived from the mantle transition zone. Chemical Geology, 328: 168–184

Xu Y G, Wang C Y, Shen S Z. 2014a. Permian large igneous provinces: characteristics, mineralization and paleo-environment effects. Lithos, 204: 1–3

Xu Y G, Wei X, Luo Z Y, Liu H Q, Cao J. 2014b. The Early Permian Tarim large igneous province: main characteristics and a plume incubation model. Lithos, 204: 20–35

Xu Z, Zhao Z F, Zheng Y F. 2012. Slab-mantle interaction for thinning of cratonic lithospheric mantle in North China: geochemical evidence from Cenozoic continental basalts in central Shandong. Lithos, 146-147: 202–217

Yu S Y, Shen N P, Song X Y, Ripley E M, Li C S, Chen L M. 2017. An integrated chemical and oxygen isotopic study of primitive olivine grains in picrites from the Emeishan large igneous province, SW China: evidence for oxygen isotope heterogeneity in mantle sources. Geochimica et Cosmochimica Acta, 215: 263–276

Yu X, Yang S F, Chen H L, Chen Z Q, Li Z L, Batt G E, Li Y Q. 2011. Permian flood basalts from the Tarim Basin, Northwest China: SHRIMP zircon U-Pb dating and geochemical characteristics. Gondwana Research, 20(2-3): 485-497

Yu X, Yang S F, Chen H L, Li Z L, Li Y Q. 2017. Petrogenetic model of the Permian Tarim large igneous province. Science China: Earth Sciences, 60(10): 1805-1816

Zeng G, Chen L H, Xu X S, Jiang S Y, Hofmann A W. 2010. Carbonated mantle sources for Cenozoic intraplate alkaline basalts in Shandong, North China. Chemical Geology, 273(1-2): 35-45

Zeng G, Chen L H, Yu X, Liu J Q, Xu X S, Erdmann S. 2017. Magma-magma interaction in the mantle beneath eastern China. Journal of Geophysical Research: solid Earth, 122(4): 2763-2779

Zhang D Y, Zhou T F, Yuan F, Jowitt S M, Fan Y, Liu S. 2012. Source, evolution and emplacement of Permian Tarim basalts: evidence from U-Pb dating, Sr-Nd-Pb-Hf isotope systematics and whole rock geochemistry of basalts from the Keping area, Xinjiang Uygur Autonomous region, Northwest China. Journal of Asian Earth Sciences, 49: 175-190

Zhang D Y, Zhang Z C, Huang H, Cheng Z G, Charlier B. 2018. Petrogenesis and metallogenesis of the Wajilitag and Puchang Fe-Ti oxide-rich intrusive complexes, northwestern Tarim large igneous province. Lithos, 304-307: 412-435

Zhang H F, Goldstein S L, Zhou X H, Sun M, Zheng J P, Cai Y. 2008. evolution of subcontinental lithospheric mantle beneath eastern China: Re-Os isotopic evidence from mantle xenoliths in Paleozoic kimberlites and Mesozoic basalts. Contributions to Mineralogy and Petrology, 155(3): 271-293

Zhang J J, Zheng Y F, Zhao Z F. 2009. Geochemical evidence for interaction between oceanic crust and lithospheric mantle in the origin of Cenozoic continental basalts in east-central China. Lithos, 110(1-4): 305-326

Zhang L, Ren Z Y, Handler M R, Wu Y D, Zhang L, Qian S P, Xia X P, Yang Q, Xu Y G. 2019. The origins of high-Ti and low-Ti magmas in large igneous provinces, insights from melt inclusion trace elements and Sr-Pb isotopes in the Emeishan large igneous province. Lithos, 344-345: 122-133

Zhang M L, Guo Z F, Sano Y, Cheng Z H, Zhang L H. 2015. Stagnant subducted Pacific slab-derived CO_2 emissions: insights into magma degassing at Changbaishan volcano, NE China. Journal of Asian Earth Sciences, 106: 49-63

Zhang M L, Guo Z F, Liu J Q, Liu G M, Zhang L H, Lei M, Zhao W B, Ma L, Sepe V, Ventura G. 2018. The intraplate Changbaishan volcanic field (China/North Korea): a review on eruptive history, magma genesis, geodynamic significance, recent dynamics and potential hazards. Earth-Science Reviews, 187: 19-52

Zhang S H, Zhao Y, Davis G A, Ye H, Wu F. 2014. Temporal and spatial variations of Mesozoic magmatism and deformation in the North China Craton: implications for lithospheric thinning and decratonization. Earth-Science Reviews, 131: 49-87

Zhao D P, Lei J S, Tang R Y. 2004. Origin of the Changbai intraplate volcanism in Northeast China: evidence from seismic tomography. Chinese Science Bulletin, 49(13): 1401-1408

Zhao Y, Zheng J P, Xiong Q, Zhang H. 2018. Destruction of the North China Craton triggered by the Triassic Yangtze continental subduction/collision: a review. Journal of Asian Earth Sciences, 164: 72-82

Zheng J P, Tang H Y, Xiong Q, Griffin W L, O'Reilly S Y, Pearson N, Zhao J H, Wu Y B, Zhang J F, Liu Y S. 2014. Linking continental deep subduction with destruction of a cratonic margin: strongly reworked North China SCLM intruded in the Triassic Sulu UHP belt. Contributions to Mineralogy and Petrology, 168(1): 1028

Zhu J, Zhang Z C, Reichow M K, Li H B, Cai W C, Pan R H. 2018. Weak vertical surface movement caused by the ascent of the Emeishan mantle anomaly. Journal of Geophysical Research: Solid Earth, 123(2): 1018-1034

Research Progress of Volcanology and Chemistry of the Earth's Interior

XU Yi-gang[1], GUO Zheng-fu[2], LIU Jia-qi[2]

1. State Key Laboratory of Isotope Geochemistry, Guangzhou Institute of Geochemistry, Chinese Academy of Sciences, Guangzhou 510640; 2. Key Laboratory of Cenozoic Geology and Environment, Institute of Geology and Geophysics, Chinese Academy of Sciences, Beijing 100029

Abstract: Major research advances in China during 2011 – 2020 on volcanology and chemistry of the Earth's interior are summarized in this paper. ①The temporal and spatial distribution of Cenozoic volcanoes in China is characterized, together with their eruptive styles. A national data base on active volcanoes is established. ②The petrogenesis of intraplate basalts has been investigated in detail. It is demonstrated that melting of the BMW to generate Cenozoic intraplate basalts is triggered by decarbonization and dehydration of the slabs stagnated in the mantle transition zone. The deep melting/fluid-assisted melting mechanism is proposed for interpreting the petrogenesis of intraplate basalts originated from the melting of the big mantle wedge in eastern Asia. The studies on mantle xenoliths also offer new constraints on the evolution and modification of the deep lithosphere underneath eastern China. ③The similarities and differences of Permian Tarim and Emeishan large igneous provinces have been elaborated. ④A new level of researches on continuous monitoring and associated disaster prevention of recent unrest volcanos in China has been achieved. ⑤Researches on the volcanic eruption simulation experiment and physical volcanology have been carried out. Several new research disciplines and analytical methods of tephrochronology have been established in China. The interdisciplinary study of volcanology and geothermal geology has been carried out. We also provide perspectives of the future study in this field.

Key words: volcanology; chemistry of the Earth's interior; China

地幔矿物学岩石学地球化学进展*

张宏福 陈立辉

西北大学 地质学系，西安 710069

摘　要：21世纪第二个十年（2010—2020年）是我国矿物岩石地球化学蓬勃发展的十年，是从克拉通破坏到造山带演化，从大陆到大洋岩石圈，从深度到广度全面进军的十年，亦是从追赶到超越的十年。这十年我国矿物岩石地球化学家在地幔矿物学、岩石学和地球化学领域取得了众多突破性进展。这些进展集中在两个方面：一是来源于地幔直接样品的，如地幔矿物学、大陆和大洋岩石圈、造山带橄榄岩和地幔中的水；二是来源于幔源岩浆的，如大陆和大洋玄武岩、基性-超基性杂岩体和火成碳酸岩。特别需要指出的是，非传统稳定同位素即金属稳定同位素新方法（如 Li、Mg、Fe 和 Ca 同位素等）的开发和广泛应用，大大提升了我国地幔岩石学和地球化学研究进程，实现了从追赶到超越的历史性跨越。

关键词：地幔　矿物学　岩石学　地球化学　金属稳定同位素　十年进展

0　引　言

我国在"十三五"国家科技创新规划中，就明确提出了要加强"三深"（深空-深海-深地）领域的战略部署。无疑，作为地球系统科学学术共同体的中国矿物岩石地球化学学会下属的地幔矿物岩石地球化学专业委员会，组织并协调所属学科专家向深地和深海领域进军是责无旁贷的。新世纪第二个十年（2010—2020年）是我国岩石地球化学从大陆到大洋、从深度到广度全面进军的十年，也是从追赶到超越蓬勃发展的十年。我国岩石地球化学家在上述领域取得了众多喜人且长足的进展。这些突破性进展集中在两个方面：一是来源于地幔直接样品的，如地幔矿物学、大陆和大洋岩石圈、造山带橄榄岩和地幔中的水；二是来源于幔源岩浆的，如大陆和大洋玄武岩、基性-超基性杂岩体和火成碳酸岩。

1　来源于地幔直接样品方面的进展

1.1　地幔矿物学

1.1.1　铬铁矿中的新矿物与蛇绿岩型金刚石的发现

铬铁矿是我国极缺矿种，对科技发展与国防建设具有战略意义。杨经绥院士团队在铬铁矿中识别出不同类型的矿物包裹体，首次发现并命名了十余种新矿物，揭示了地壳物质循环至深部地幔的过程，给人类认识地球物质组成与地球运行机制提供了实证。在地幔橄榄岩和铬铁矿中发现了大量微米-纳米级新矿物，如：经绥矿（分子式 TiB_2，空间群 $P6/mmm$）、志琴矿（分子式 $TiSi_2$，空间群 $Fddd$）、巴登珠矿（分子式 TiP，空间群 $P6_3/mmc$）、青松矿（分子式 BN，空间群 $F43m$）、罗布莎矿（分子式 $Fe_{0.83}Si_2$，空间群 $Cmca$）、林芝矿（分子式 $FeSi_2$，空间群 $P4/mmm$）、那曲矿（分子式 $FeSi$，空间群 $P2_13$）、藏布矿

---

* 原文"向'深地'和'深海'进军征途中——地幔矿物学岩石学地球化学十年进展"刊于《矿物岩石地球化学通报》2021年第40卷第4期，本文略有修改。

(分子式 $TiFeSi_2$, 空间群 Pb_am), 雅鲁矿 [分子式 $(Cr_4Fe_4Ni)_9C_4$, 空间群 $P4/mmm$], 曲松矿 (分子式 WC, 空间群 $P6m2$), 自然钛 (分子式 Ti, 空间群 $P6_3/mmc$), 这些新矿物已得到国际矿物学会的批准 (Fang et al., 2013; Yang et al., 2014; Xiong et al., 2020)。硼和磷是典型的地壳元素, 在地幔中的含量极低, 地幔岩中含硼矿物的发现改变了传统的认识, 这是壳幔循环的最直接证据。蛇绿岩型地幔橄榄岩和铬铁矿是一个重要的地幔矿物储存库, 有许多异常地幔矿物来自地幔深部强还原环境。近年来, 在全球 5 个板块古边界的 15 处蛇绿岩中找到了金刚石、方铁矿和自然铁等强还原的深地幔矿物, 表明金刚石分布于蛇绿岩层序中超基性岩层位内各个岩相中, 并在多个岩体的金刚石包裹体中发现后柯石英、钙钛矿相的 Al_2O_3 和钙钛矿, 揭示铬铁矿中保存了下地幔的物质结构 (Yang et al., 2014; Xiong F. H. et al., 2015, 2018)。对这些矿物的特征和形成条件的研究, 为我们打开了一扇了解地幔的物质组成、物理化学环境、地幔物质的运移和深部地质作用的窗口。

1.1.2 下地幔温压条件下新型铁镁氧化物的发现

北京高压科学研究中心毛河光院士团队对常见矿物磁铁矿 (Fe_3O_4)、赤铁矿 (Fe_2O_3)、针铁矿 (FeOOH) 以及地幔矿物铁方镁石 $(Mg,Fe)O$ 开展了高温高压实验研究, 结果在深下地幔温度压力环境发现地下大约 2 000~2 900 km 处的 FeOOH 通过部分脱氢作用或铁镁氧化物与俯冲含水物质发生新型高压化学反应, 形成含氢 $(Fe,Mg)O_2$ 相, 其结构与地表常见矿物黄铁矿 (FeS_2) 具有相同的立方晶体结构 (Pa-3) (Hu Q. Y. et al., 2016, 2017, 2020)。毛河光等 (Mao et al., 2017) 和刘锦等 (Liu et al., 2017b) 进一步研究发现, 在地球核幔边界极端高温高压环境下, 深部含水物质接触到炙热铁核时将产生含氢过氧化物相并释放氢气。高温高压模拟实验结果表明, 含氢 $(Fe,Mg)O_2$ 相可稳定存在于下地幔底部极端温压环境, 如 3 300 K 和 130 GPa 左右, 具有较高的结构稳定性, 长期可能产生几千米至数十千米厚度的含过氧化物相区域, 有助于核幔边界超低速区的形成。在地幔物质上升过程中, 含氢 $(Fe,Mg)O_2$ 相将通过部分脱氧和脱氢作用, 在深度约 1 000~2 000 km 处形成新型铁镁氧化物 $(Fe,Mg)_2O_3X$ 相 (X 表示该高压晶体结构中含有少量的氢和氧) (Liu J. et al., 2020)。刘锦等 (Liu J. et al., 2019, 2020) 结合同步辐射 X 射线谱学和电子能量损失谱研究, 发现在新型铁镁氧化物含氢 $(Fe,Mg)O_2$ 相和 $(Fe,Mg)_2O_3X$ 相的晶体结构中氧原子之间的强相互作用, 导致氧的价态偏移常规的负二价, 表明在深部下地幔, 氧元素将与铁元素一样价态可变, 这些发现打破了此前的传统观点, 给地球深部物质研究带来诸多全新的认识。吉林大学马琰铭教授课题组通过理论计算预测 FeO_2 晶体结构中氢可被氦取代, 表明高压结构的新特性亦对理解深部氢、氦以及其他挥发分的赋存形式和分布具有重大意义。

1.2 大陆地幔橄榄岩

以华北克拉通破坏研究为引领的大陆地幔橄榄岩研究是近十年来我国大陆岩石圈演化研究的一个显著特征。我国大陆地幔橄榄岩研究不仅揭示了稳定克拉通之所以能被破坏的本质, 而且促进了克拉通破坏研究走到了世界前列, 推动了我国固体地球科学的发展。同时, 将非传统稳定同位素地球化学研究应用于大陆岩石圈地幔研究中, 开拓了金属稳定同位素研究的新领域。

1.2.1 进一步揭示了大陆岩石圈地幔的组成特征

地幔捕虏体-捕虏晶是由寄主岩浆 (碱性玄武岩、霞石岩、黄长岩、钾镁煌斑岩和金伯利岩) 喷出地表时捕获的来自地球深部的样品, 主要包括橄榄岩、辉石岩、榴辉岩等岩石或它们的组成矿物, 它们提供了大陆岩石圈地幔组成、结构、性质和演化过程的关键信息。与年轻的、主要由饱满的二辉橄榄岩组成的大洋岩石圈地幔相比, 大陆克拉通地幔的年龄老、以高度难熔的方辉橄榄岩和二辉橄榄岩为主。近十年来, 中国学者利用不同构造环境产出的地幔捕虏体-捕虏晶重点研究了大陆岩石圈地幔的特征, 发现部分熔融 (熔体抽取) 和交代富集作用共同控制着地幔的组成, 不同来源和性质的熔体在软流圈上涌的机制驱动下与地幔橄榄岩发生相互作用, 造成不同时空范围内的地幔不均一性与壳幔解耦 (Zhang J.

et al.，2011；赵勇伟和樊祺诚，2011；Tang et al.，2011，2012，2013a，2013b；Liu et al.，2012a，2012b，2013；Pan et al.，2013，2015；Xu et al.，2013；Zhang H. F. et al.，2012，2013；Zou et al.，2014，2016，2020；Liu J. G. et al.，2015b；Wang C. Y. et al.，2016；Dai et al.，2019；Dai and Zheng，2019；Hu et al.，2019；Zhang Y. L. et al.，2019；Liu H. et al.，2019；Liu J. G. et al.，2020；Zhang et al.，2020a，2020b）。例如，古生代以来古亚洲洋板片南向俯冲、古特提斯洋板片北向俯冲，以及中新生代以来古太平洋板片西向俯冲，周边板块的多次俯冲碰撞过程中产生的熔体对华北克拉通岩石圈地幔进行了长期、多阶段的改造作用（Tang et al.，2013a；Zong and Liu，2018）。一般而言，克拉通和微陆块内部岩石圈地幔的形成时代较老（太古宙）、难熔程度较高（橄榄石 $Mg^{\#} \geqslant 91.5$）、具有富集的 Sr-Nd 同位素组成；克拉通边缘岩石圈地幔则以年轻的饱满的橄榄岩为主，具有亏损的 Sr-Nd 同位素组成，这与大陆岩石圈地幔早期的克拉通属性与后期的俯冲碰撞和熔体改造，以及软流圈-岩石圈相互作用等过程有关（Tang et al.，2013c；林阿兵等，2018）。

硅酸盐与碳酸盐熔/流体-橄榄岩反应是地幔再富集作用的两种最主要的方式，可以通过样品中的单斜辉石 Ca/Al 或 La_N/Yb_N vs. Ti/Eu 的变化趋势来判别（Zong and Liu，2018 及其参考文献）。随着微区原位分析技术的迅速发展，单斜辉石原位 Sr 同位素也被用来示踪碳酸盐熔体交代（Sun et al.，2012；Zou et al.，2014，2016，2020；Deng et al.，2017，2020；Chen R. X. et al.，2017；Wu et al.，2017；Wang C. Y. et al.，2019；Zhang et al.，2020a；Zhang L. et al.，2020）。Zong 和 Liu（2018）综合运用以上指标，系统总结了华北克拉通东部岩石圈地幔所经历的 3 类碳酸盐熔体交代作用（其中第一类交代导致单斜辉石具有异常高的 Ca/Al 值和富集的 $^{87}Sr/^{86}Sr$ 为 0.706—0.713），并且认为碳酸盐熔体交代在华北克拉通破坏和大规模金成矿过程中可能发挥了重要作用。最近，Wang Z. Z. 等（2020）对华北克拉通东部地幔捕虏体和玄武岩展开 Au 与铂族元素的对比分析，指出尽管强烈的交代作用与水化作用并不足以形成富 Au 地幔，但是被交代的地幔在减薄和熔融过程中，富挥发分的交代介质参与的确能够促进金释放，并衍生出富金流体进而形成矿床。

总之，先进的分析测试技术和非传统稳定同位素示踪方法被大量应用于岩石圈地幔演化的研究中，进一步揭示了以华北克拉通为代表的大陆岩石圈地幔的特征及其成因（Tang et al.，2007，2010，2011，2012，2014；Zhang et al.，2010；Zhao et al.，2010，2017a，2017b；Huang et al.，2011，2015；Liu S. A. et al.，2011；Su et al.，2012，2014a，2014b，2019；Xiao et al.，2013，2015，2017；Liu J. et al.，2015b；Hu Y. et al.，2016，2020；Kang et al.，2016，2017，2019，2020；Tian et al.，2016；Li H. Y. et al.，2017；Xia et al.，2017；Chen et al.，2018；Qi et al.，2019；Dai et al.，2020；汤艳杰和张宏福，2019；赵新苗，2019），不仅为地幔地球化学的研究提供了新的途径，而且开拓了金属稳定同位素研究的新领域。下面对近年来大陆岩石圈地幔再富集作用研究取得的进展进行概括。

1.2.2 大陆岩石圈地幔再富集作用

典型的克拉通岩石圈地幔被认为是原始地幔经过高度部分熔融的残余。与软流圈相比，古老的大陆岩石圈地幔高度亏损 Fe、Ca 和 Al 等玄武质组分，主要由高度难熔的方辉橄榄岩和二辉橄榄岩组成。由于古老的大陆岩石圈具有高度难熔的特点，而且密度比下伏的软流圈小，因而能够长期"漂浮"在软流圈之上，这是古老克拉通能够保持长期稳定的主要原因。然而，古老的岩石圈地幔由于受到外来熔体的改造而在组成上会变得饱满，在年龄上变得相对年轻，这种熔体改造过程被称为岩石圈地幔再富集作用（refertilization）（Tang et al.，2007；Zhang H. F. et al.，2009）。岩石圈地幔的再富集作用是通过橄榄岩-熔体反应的方式实现的，通常情况下，熔体的组成比地幔橄榄岩富 Si 和 Fe 而贫 Mg。反应的结果必然造成地幔橄榄岩向富 Si、贫 Mg 方向转化（Zhang H. F. et al.，2009）。前期大量研究表明，强烈的再富集作用不仅能够改变大陆岩石圈地幔的物质组成，而且能够改变岩石圈地幔的物理化学性质，致使古老、难熔的岩石圈地幔转变为相对年轻、饱满的岩石圈地幔，从而失去了克拉通的属性。例如，华北克拉通古老的岩石圈地幔由于遭受强烈的再富集作用，岩石圈地幔的属性发展了重大转变：从典型的克拉通型转变为大洋型，最终导致古老克拉通的破坏（Zhang H. F. et al.，2009；Tang et al.，2013b；Zheng et al.，2015）。因此，岩石圈地幔组成和性质的转变是导致克拉通破坏的关键，也是华北克拉通破坏的本质所在（张宏福，2009；朱日祥等，2020）。

再富集作用之所以能够从根本上改变华北克拉通岩石圈地幔的组成和性质，主要是因为它在华北克拉通岩石圈地幔演化的过程中广泛存在。华北北缘中生代玄武岩分布区狼山（Dai et al.，2019；Dai and Zheng，2019）和阜新（Zou et al.，2020）、新生代玄武岩分布区四子王旗（Wu et al.，2017；Zhang et al.，2020a，2020b）、阳原（Zhao et al.，2017b）、集宁（Zhang H. F. et al.，2012）、围场（Zou et al.，2016）、和汉诺坝（Chen et al.，2016；Zhao et al.，2017a）等地区的地幔橄榄岩捕虏体和（或）捕虏晶的岩石学和地球化学研究结果表明，多阶段的岩石圈再富集作用在华北北缘广泛存在，早期阶段由于熔体的多来源性造成了华北北缘中生代以来岩石地幔组成的明显不均一性，近期由于软流圈熔体的强烈改造作用，将 Sr-Nd 同位素富集的古老地幔橄榄岩转变为同位素亏损的具有大洋型地幔特征的橄榄岩。值得强调的是，阜新地区中生代玄武岩中地幔橄榄岩中单斜辉石的原位 Sr 同位素地球化学研究表明，华北克拉通岩石圈地幔在中生代（大约 100 Ma）已经发生了强烈再富集作用，岩石圈地幔的属性开始由克拉通型向大洋型转变（Zou et al.，2020）。由于之前的研究将华北克拉通岩石圈地幔属性转变发生时间界定在中生代末期（大约 67 Ma；Ying et al.，2006；Zhang J. et al.，2011）或新生代（Zhang H. F. et al.，2009 及其参考文献），这项研究将华北克拉通岩石圈地幔属性转变的时间提前至 100 Ma。

华北克拉通东部（Xiao and Zhang，2011；Tang et al.，2012；Xiao et al.，2013，2015；Deng et al.，2017，2020）和东南缘岩石圈地幔都曾经发生了强烈的再富集作用，由古生代典型的克拉通型（主量元素亏损、同位素中等程度富集、以方辉橄榄岩为主的岩石圈地幔）转变为晚中生代主量元素饱满、同位素亏损、以二辉橄榄岩为主的大洋型岩石圈地幔（张宏福，2009 及其参考文献）。华北中部新生代玄武岩中发现的同位素富集的方辉橄榄岩和同位素相对亏损的二辉橄榄岩捕虏体的研究表明，它们也是经过多期橄榄岩–熔体反应改造的产物（Tang et al.，2011，2013a，2013c，2014；Zhang H. F. et al.，2012；Chen et al.，2018）；前者可能是早期地壳熔体–古老地幔橄榄岩反应造成的结果，而后者则可能是前者与来源于软流圈的玄武质熔体在近期反应的产物。因此，中生代以来，再富集作用在华北克拉通岩石圈地幔中广泛存在，而且在克拉通的边缘和郯庐断裂带沿线地区表现得尤为突出。

通过对全球克拉通及其周边地区（包括亚洲、欧洲、非洲、美洲和澳洲）地幔橄榄岩的主、微量元素和 Sr、Nd、Re-Os 同位素组成的对比研究，揭示了大陆岩石圈地幔再富集作用的普遍性（Tang et al.，2013b）。再富集作用使古老岩石圈地幔中遭受改造的部分变得相对年轻，从而具有类似于大洋橄榄岩，即大洋型岩石圈地幔的某些特征（图1）。因此，部分具有大洋型地幔特征的二辉橄榄岩捕虏体可能是古老的岩石圈地幔遭受强烈再富集作用的结果，而不是新生的岩石圈地幔。对二者的识别将影响对克拉通岩石圈地幔演化过程的正确理解。

1.2.3 大陆岩石圈地幔定年

研究地幔捕虏体所记录的熔体亏损事件、确定大陆岩石圈地幔的年龄，对于理解岩石圈地幔的形成时代、演化历史，及其与上覆地壳之间的相互作用至关重要。Re-Os 同位素体系凭借其在地幔中强相容的优势以及对于交代作用相对不敏感的特性，被广泛应用于岩石圈地幔定年（Rudnick and Walker，2009；Luguet and Pearson，2019）。Re-Os 等时线定年方法主要通过假等时线（Al 等时线与 Fo 等时线；Liu J. G. et al.，2011，2015a，2020；Liu et al.，2012a；Liu J. Q. et al.，2017）与峰期模式年龄（T_{MA} 与 T_{RD}；Liu et al.，2012b，2013；Liu J. G. et al.，2015a，2018；Liu J. Q. et al.，2016；Zhang Y. L. et al.，2019；Tian G. C. et al.，2020）来实现。Liu S. A. 等（2011）利用橄榄岩捕虏体 Al 等时线与 Os 模式年龄限定了华北克拉通中部带不同地区岩石圈地幔年龄的差异，以山西繁峙地区为分界，太行山南部地区（鹤壁、符山）岩石圈地幔的年龄约为 2.5 Ga，与上覆地壳耦合，太行山北部和华北北缘（大同、阳原、集宁、汉诺坝）岩石圈地幔年龄约为 1.8 Ga，与上覆地壳解耦，揭示了华北克拉通北缘约 1.8 Ga 的陆–陆碰撞事件导致北

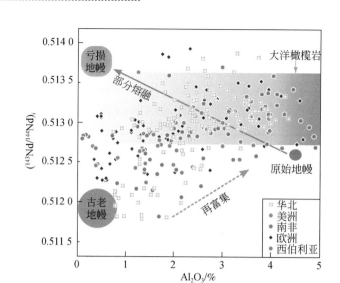

图 1 地幔橄榄岩 Al₂O₃ 与初始 Nd 同位素组成（据 Tang et al., 2013b 修改）

部原有的新太古代岩石圈地幔被改造和置换。由于岩石圈地幔中广泛存在的再富集作用会导致年轻硫化物的产生，橄榄岩全岩的 Re-Os 模式年龄可能代表橄榄岩中不同时代的硫化物的混合（表面年龄），而非岩石圈地幔真正的形成年龄，因此全岩 Re-Os 同位素年龄的解释需要慎重（Tang et al., 2013b）。应用 Re-Os 同位素体系测年，需要配合铂族元素与硫化物岩相学来评估交代作用对定年的影响（Liu S. A. et al., 2011; Liu et al., 2018; Tian G. C. et al., 2020）。

Os 模式年龄随地幔储库类型（不同球粒陨石、原始上地幔或亏损地幔）的选取而表现出较大的不确定性（~0.3 Ga）（Rudnick and Walker, 2009），而且低程度熔体抽取不会使 Re 完全耗尽。因此，Re-Os 同位素体系对于相对年轻的岩石圈地幔定年比较乏力。为了开发精准测定后太古宙岩石圈地幔形成时间的工具，Liu J. G. 等（2012，2020）和 Shu 等（2014，2019）联用橄榄岩、榴辉岩或辉石岩捕房体中的单斜辉石、斜方辉石、石榴子石等矿物的 Sm-Nd 与 Lu-Hf 同位素体系进行了尝试，发现定年之前需要先检验矿物的封闭温度与交代历史，否则只能得到无意义的或者受交代作用的年龄。Liu J. G. 等（2020）通过喀麦隆火山链 Nyos 尖晶石二辉橄榄岩的单斜辉石和斜方辉石重建的全岩 Lu-Hf 等时线得到了这些橄榄岩的年龄为（2.01±0.18）Ga，在误差范围内与该区最古老的 t_{RD} 年龄（~2.0 Ga）一致，证明了利用单矿物的 Lu-Hf 同位素体系进行地幔定年的潜力。

1.3 造山带橄榄岩

造山带橄榄岩不仅是地幔岩石地球化学，而且是造山带形成与演化过程研究的重要对象。造山带橄榄岩主要有三种类型：①阿尔卑斯型橄榄岩，即岩石圈地幔构造-热侵位就位于造山带浅部地壳的橄榄岩；②蛇绿岩型橄榄岩；③前期层状基性-超基性堆晶岩经俯冲变质形成的橄榄岩（张宏福和于红，2019）。如果按照造山带橄榄岩的原岩属性，又可分为地幔型和地壳型两类（郑建平等，2019）。由此可见，前两类橄榄岩属于地幔型造山带橄榄岩，第三类为地壳型造山带橄榄岩。

全球代表性地幔型造山带橄榄岩包括意大利东阿尔卑斯的 Ulten 橄榄岩（Rampone and Morten, 2001）、挪威西片麻岩省的 Otrøy-Fjørtoft 橄榄岩（van Roermund et al., 2002; Spengler et al., 2006; Scambelluri et al., 2008）、我国秦岭造山带的松树沟橄榄岩（Yu et al., 2016, 2017; 张宏福和于红，2019）、柴北缘带的胜利口橄榄岩（Song et al., 2007; Xiong Q. et al., 2015; Chen R. X. et al., 2017）、大别造山带的毛屋超镁铁质岩（Chen Y. et al., 2017; Malaspina et al., 2017）、苏鲁造山带的荣成、仰

口、许沟、东海-芝麻坊等地造山带橄榄岩（Zhang Z. M. et al.，2011；陈意等，2015；Su B. et al.，2017；Li et al.，2018）。

代表性的壳源型造山带橄榄岩包括：挪威西片麻岩省的 Svartberget 橄榄岩（Vrijmoed et al.，2013），我国阿尔金造山带的英格礼萨伊橄榄岩（刘良等，2002）、大别造山带的碧溪岭橄榄岩（Zheng et al.，2008）等。这些橄榄岩是早先侵位于大陆地壳的超镁铁质堆晶岩体，后伴随大陆深俯冲发生高压-超高压变质，从而形成造山带橄榄岩。这类橄榄岩的原岩不是直接地幔岩，故不赘述。

1.3.1 北秦岭松树沟橄榄岩及造山带构造演化解析

近年来，对松树沟糜棱岩化橄榄岩及其相关的高级变质岩（包括榴辉岩、退变榴辉岩、石榴斜长角闪岩）详细的岩石学和地球化学研究，不仅发现松树沟橄榄岩曾经的确是蛇绿岩的组成部分，而且还证明这些橄榄岩记录了大洋岩石圈从形成到角闪岩相变质的全过程（Yu et al.，2016，2017；张宏福和于红，2019），即 1 000~800 Ma 大洋岩石圈形成阶段，主要形成纯橄岩；800~500 Ma 洋-陆转换即陆岩石圈演化阶段，岩石圈被交代形成大量方辉橄榄岩；500~480 Ma 快速深俯冲和榴辉岩相变质阶段；460~335 Ma 角闪岩相退变质阶段，在松树沟橄榄岩中形成大量富镁的直闪石类矿物，如透闪石、阳起石和镁闪石（图2）。

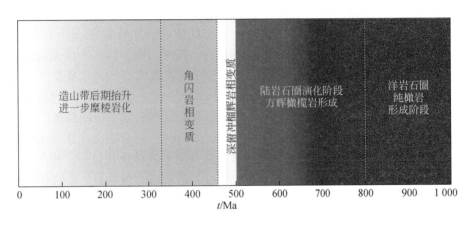

图 2　松树沟糜棱岩化橄榄岩演化历程示意图

由此可见，蛇绿岩型造山带橄榄岩能够记录造山带形成与演化的全过程，通常会经历四个形成和演化阶段：①洋岩石圈（蛇绿岩）形成阶段，形成纯橄岩；②洋-陆转换阶段，陆岩石圈演化阶段，岩石圈受交代形成方辉橄榄岩；③岩石圈深俯冲，榴辉岩相变质；④俯冲板片抬升至角闪岩相时退变质，此时在橄榄岩中形成富镁的直闪石类矿物。不同造山带中蛇绿岩型橄榄岩的区别可能只是俯冲深度和退变质程度不同而已。最后需要指出的是蛇绿岩一定要强调的是其形成时代。同时，造山带进变质作用产物经常会被后期抬升过程中退变质作用彻底改造，应引起足够重视。

1.3.2 赤城新太古代蛇绿岩的发现

上述秦岭松树沟蛇绿岩型橄榄岩记录的造山带的形成与演化过程可能具有普适性。因为我国华北中部带北部赤城蛇绿岩型橄榄岩同样记录了与秦岭松树沟橄榄岩完全类似的演化过程，唯一的差别是不同演化阶段（洋岩石圈形成、洋-陆岩石圈转化、岩石圈俯冲、俯冲岩石圈抬升）发生的时间不同而已（Zhang H. F. et al.，2016；Liu and Zhang，2019；Liu H. et al.，2019）。近年来，赤城蛇绿岩型橄榄岩的发现及其相关的高级变质岩（包括退变榴辉岩、榴闪岩、高压麻粒岩、片麻岩）详细的岩石学和地球化学研究，不仅证明赤城橄榄岩曾经的确是蛇绿岩，而且还发现这些橄榄岩记录了洋岩石圈形成到角闪岩相变质的全过程（Zhang H. F. et al.，2016；Liu and Zhang，2019；Liu H. et al.，2019），即 2 600~2 500 Ma 洋岩石圈形成阶段，主要形成纯橄岩；2 500~1 900 Ma 洋-陆转换即陆岩石圈演化阶段，岩石圈被交代形成大量方辉橄榄岩；1 900~1 850 Ma陆岩石圈深俯冲和榴辉岩相变质阶段；350~320 Ma造山带被抬升到角闪岩相，发生退变质，形成大量富镁的直闪石类矿物。需要重点指出的是赤城新太古代蛇绿岩的发现暗示现代意义的板块构造至少在新太古代晚期就已经出现（Liu and Zhang，2019；Liu H. et al.，2019）。

1.3.3 揭示了我国幔源型造山带橄榄岩普遍存在熔–岩反应

我国中央造山系的柴北缘和苏鲁–大别等地造山带橄榄岩的研究揭示这些幔源型橄榄岩普遍存在熔–岩反应，并被认为是造成这些橄榄岩组成变化的主要机制。例如，我国西部柴北缘造山带橄榄岩为含超硅石榴子石的橄榄岩杂岩体。最新研究成果揭示其来源于太古宙大陆岩石圈地幔楔（Shi et al.，2010；Xiong et al.，2011；Chen R. X. et al.，2017），并经历了Rodinia超大陆裂解时期的硅酸盐熔体交代（Xiong Q. et al.，2015）和早古生代含硅酸盐组分的超临界流体交代（Chen R. X. et al.，2017）。这些熔流体交代作用生成了金云母集合体（Xiong et al.，2014）、石榴子石和辉石内的含水矿物包裹体（Yang and Powell，2008）、锆石（Xiong et al.，2011；Chen R. X. et al.，2017）等交代矿物。

大别山造山带的毛屋橄榄岩–辉石岩杂岩体产出于三叠纪超高压岩片中，其主体为石榴子石斜方辉石岩，含少量纯橄榄岩、方辉橄榄岩、二辉石岩和单斜辉石岩。该橄榄岩–辉石岩杂岩体被认为是古生代早期时俯冲沉积物来源的富硅熔体与难熔的大陆岩石圈地幔发生熔–岩反应的产物（Chen Y. et al.，2017），随后经历了长期复杂的壳源熔流体交代过程，形成众多交代矿物，如石榴子石中含水多相矿物包裹体（Campione et al.，2017；Malaspina et al.，2017）、变质锆石、纯橄榄岩内菱镁矿、白云石和方解石等原生碳酸盐矿物，反映碳酸盐熔流体交代过程、矿物微量元素和同位素组成特征显示复杂的壳源流体交代特征（Chen Y. et al.，2017；Shen et al.，2018）。

苏鲁造山带幔源型造山带橄榄岩发育，这些橄榄岩基本来自华北陆下不同深度的岩石圈地幔楔，并经历了复杂的交代改造事件（Su B. et al.，2017；Li et al.，2018）。记录的交代作用主要表现为：硅酸盐熔体交代（日照、仰口和荣成橄榄岩）、碳酸盐熔流体交代及富水流体交代过程；相应的显性交代矿物也普遍发育，如锆石、角闪石、金云母、钛斜硅镁石、帘石、磷灰石、独居石、菱镁矿、白云石、硫化物等（Zhang Z. M. et al.，2011；Zheng et al.，2014；Su B. et al.，2017；郑建平等，2019）。这些交代矿物的出现为理解俯冲带内壳–幔相互作用提供了研究对象。

1.4 现今大洋地幔橄榄岩

大洋橄榄岩是向深海进军的重要方向之一。得益于现代海洋探测技术和地球化学分析技术的发展，我国科学家在深海橄榄岩的研究中取得重要进展。

1.4.1 发现西南印度洋深海橄榄岩与其上的洋壳存在解耦现象

大洋隆起（Oceanic Rise）长期以来被认为受到地幔柱热异常的影响而具有非常厚的洋壳，如著名的冰岛隆起。而对Marion隆起的研究则颠覆了这一传统认识。2010年中国主导的"大洋一号"21航次利用远程机器人对西南印度洋53°E地区进行了详细的样品采集，结合前期不同航次的报道，Zhou 和 Dick（2013）发现西南印度洋的Marion隆起具有非常薄的洋壳，这不同于冰岛等因受地幔柱影响而具有巨厚洋壳的大洋隆起，代表了大洋地幔的一个成分异常点。结合该地区深海橄榄岩亏损的成分特征，Zhou 和 Dick（2013）提出Marion隆起可能代表了经历古老熔融、低密度的地幔，因此可发育以53°E Dragon Bone为代表的贫岩浆活动断块；而与Dragon Bone相邻的Dragon Flag地区具有厚达10 km的洋壳，代表熔融异常区，显示超慢速扩张脊独有的熔体聚集规律（Zhou and Dick，2013；Yu and Dick，2020）。进一步的全岩主微量分析显示，Dragon Bone断块的深海橄榄岩经历了强烈的熔体交代作用，因此大量熔体在地幔中的滞留可能造成了该地区具有薄的洋壳（Chen H. et al.，2015）。而通过Dragon Bone断块深海橄榄岩和大洋玄武岩的对比研究，Gao等（2016）提出二者之间存在成分的不平衡，而且深海橄榄岩经历高程度的含水熔融，因此代表了经历古老熔融事件的再循环的地幔楔。这与近期Urann等（2020）对大西洋中脊深海橄榄岩的研究结果不谋而合，他们在大西洋中脊 16°30′N 采集到具有高难熔特征的方辉橄榄岩，单斜辉石和斜方辉石微量元素的模拟计算表明它们经历了非常高程度的含水熔融，代表了再循环的超俯冲带（supra-subduction zone，SSZ）型地幔，同时推测超难熔地幔在体积上的组成可能超过上地幔的 60%。通过原位的矿物主微量分析，Wang J. X. 等（2019）进一步提出 Dragon Bone 深海方辉橄榄岩经历了两阶段熔融历史，即早期的石榴子石相高压条件下古老的熔融和近期尖晶石相低压条件下的熔融，而后者仅产生少量熔体，因此造成了该地区的洋壳非常薄。Li 等（2019）对 Dragon Bone 和西南印度洋脊东段的深海橄榄岩样品进行 Re-Os 同位素和 HSE 含量分析，证实这些深海橄榄岩保留了古老的 Os 同位素信息（~1 Ga）。这很好地支持了 Liu C. Z. 等（2008）对北极 Gakkel 洋脊深海橄榄岩的全岩 Re-Os 同位素研究结果，即部分深海橄榄岩具有非常古老（20 亿年）的模式年龄，表明它们在进入现今洋中脊之前就曾已遭受古老的熔体事件，从而反映软流圈中存在有古老的地幔。这些古老地幔的存在，使得软流圈可能在非常小的尺度具有成分上高度的不均一性（Liu C. Z. et al., 2008a）。

1.4.2 证实大洋橄榄岩交代过程

除了先存的软流圈地幔成分和部分熔融程度差异，深海橄榄岩的成分也受到后期熔-流体交代作用的影响。对 Dragon Bone 断块和西南印度洋脊东段的深海橄榄岩中的斜方辉石进行水含量测定，得到不同的水含量变化范围，这可能是由地幔受富水熔体不同程度的交代所致，也可能是富水熔体交代作用在空间上存在差异的结果（Li W. et al., 2017）。Li W. Y. 等（2020）在大西洋中脊 Vema 转换断层处的研究对深海橄榄岩中的斜方辉石和单斜辉石进行水含量分析，进一步证实后期富水熔体对地幔的交代作用。此外，对 Gakkel 洋脊和西南印度洋中脊深海橄榄岩进行 Li 同位素分析发现，不同的矿物具有不同的 Li 同位素特征，这种差异可能由于冷却过程中 Li 元素在不同矿物之间的扩散造成（Gao et al., 2011；Liu P. P. et al., 2020），而且早期高温环境下矿物-熔体之间的 Li 元素扩散信息可能被后期低温环境下矿物-流体之间的 Li 元素扩散信息覆盖（Liu et al., P. P. 2020）。

1.5 地幔中的水

水是地幔中的重要载体。自 20 世纪 90 年代以来，以 OH 等缺陷形式存在于名义上无水矿物中的"水"开始引起人们的注意（如 Bell and Rossman, 1992）。21 世纪的头十年里，学界在分析方法开发、高温高压实验、地幔水的分布等方面开展了初步探索（周新华等，2013）。近十年来，新发展起来的利用矿物斑晶的水含量来反演岩浆水含量的方法把能够直接测量的研究对象从空间上推广到了大陆玄武岩，从时间上推广到了地球演化的早期阶段（Hamada et al., 2013；Xia et al., 2013, 2019；Liu J. et al., 2015a）。我国学者在天然样品的水含量测定、第一性原理计算、地球动力学模拟等方面取得了长足的进展（Demouchy and Bolfan-Casanova, 2016；Pesiler et al., 2017；Ni et al., 2017；Xia et al., 2019）。

1.5.1 刻画了大陆尺度的岩石圈地幔和软流圈的水含量时空分布

早期对不同构造背景下的地幔水含量的工作总体上是零散的，缺乏对同一构造地块不同圈层水含量的分布及其随时间的演化特征的系统研究。由于涉及克拉通破坏、东亚"大地幔楔"等热点问题，中国东部上地幔水的分布和演化得到了很多学者的关注，这一地区也因此成为全球上地幔水含量研究最为集中和系统的地区。继橄榄岩捕虏体的研究工作之后（Xia et al., 2010），通过玄武岩中单斜辉石斑晶反演岩浆水含量的方法，华北克拉通破坏峰期时岩石圈底部的水含量（Xia et al., 2013；Liang et al., 2019；Wang Z. Z. et al., 2020）及中国东部从东北到华南新生代软流圈地幔的水含量得以定量分析（Liu J. et al., 2015, 2017a；Chen H. et al., 2017）。这些工作揭示了随着克拉通从破坏峰期到现在，岩石圈地幔水含量逐渐降低，新生代岩石圈地幔水含量具有显著的南北差异（东北和华北块体水含量多低于 50×10^{-6}，华南块体则多高于 50×10^{-6}）；而新生代软流圈水含量具有显著的东西差异（重力梯度带以东的含量多高于 $1\,500 \times 10^{-6}$，以西地区多低于 $1\,500 \times 10^{-6}$）的特点。

1.5.2 地幔含水性对重大地质过程的影响

矿物中的水可以对黏滞度和熔融温度等性质产生与其含量不成比例的重要影响。继 Peslier 等（2010）发现南非克拉通的长期稳定性可能与岩石圈底部的低水含量（$<10\times10^{-6}$）有关后，Xia 等（2013）首次揭示了中生代华北克拉通破坏峰期的岩石圈地幔具有极高水含量（$>1\,000\times10^{-6}$），对应的流变强度与软流圈相当。这表明稳定克拉通之所以能被破坏，与其强烈水化导致的流变强度显著变低相关。华北岩石圈地幔在破坏峰期时的富水状态也得到了很多后续研究的支持（Liang et al., 2019；Wang Z. Z. et al., 2020）。在大火成岩省的成因研究方面，由于此前长期缺乏制约源区水含量的有效手段，无法综合考察高温、易融组分、高水含量等因素对其形成的作用。利用单斜辉石斑晶反演的方法，Xia 等（2016）和 Liu J. 等（2017a）确定了峨眉山和塔里木大火成岩省源区在具有高温和高比例易熔组分的同时强烈富水（$>5\,000\times10^{-6}$），这一特征适用于全球典型显生宙大火成岩省；而作为鲜明的对比，没有形成大火成岩省的海南新生代玄武岩源区虽然也具有高温和高比例易熔组分的特征，但其水含量低（$80\times10^{-6}\sim350\times10^{-6}$）（Gu et al., 2019），这就从正反两方面揭示了富水地幔源区是大火成岩省形成的必要条件。最近 Yang 和 Faccenda（2020）基于高分辨率的地球动力学数值模拟方法，认为中国东北广泛存在的新生代玄武岩以及日本海沟附近俯冲太平洋板块上的年轻点状小火山的形成，都与富水的地幔过渡带物质的上涌熔融有关。Han 等（2021）通过体波波形拟合的方法确认了中国东北地区 410 km 不连续面上方广泛存在的低速带是由太平洋板块俯冲引发的富水地幔过渡带上涌引起的含水熔体造成的。

1.6 非传统稳定同位素在大陆岩石圈地幔研究中的应用

非传统稳定同位素，亦称金属稳定同位素，是 21 世纪同位素地球化学发展最迅速的一个方向，特别是最近十年也是我国将金属稳定同位素广泛应用在大陆与大洋岩石圈演化、大陆和大洋玄武岩、地幔物质再循环、基性-超基性侵入体等地幔岩石地球化学研究的各个领域，起点高，甚至在某些方面引领了国际同位素地球化学的发展。

随着非传统金属稳定同位素分析方法的不断革新与分馏机理的不断深入，深刻地改变了大陆岩石圈地幔的研究现状，尤其是用 Li、Mg、Ca 和 Fe 同位素示踪熔体-岩石相互作用与深部碳循环等方面（Zhao et al., 2010, 2012, 2015, 2017a, 2017b；Tang et al., 2011, 2012；Su et al., 2012, 2014b, 2018；Xiao et al., 2013, 2017；Hu Y. et al., 2016；Wang Z. Z. et al., 2016；Kang et al., 2016, 2017, 2019, 2020；Chen et al., 2018；赵新苗, 2019；汤艳杰和张宏福, 2019；Dai et al., 2020）。汤艳杰和张宏福（2019）通过对比分析华北不同地区地幔橄榄岩捕虏体的 Li 同位素地球化学特征，揭示华北岩石圈地幔具有高度不均一的 Li 含量和 Li 同位素组成，反映了不同来源熔体（来自俯冲大洋板片和软流圈的熔体）对岩石圈地幔的改造作用。赵新苗（2019）系统地归纳和总结了中国东部不同类型地幔捕虏体的 Mg、Ca 和 Fe 同位素地球化学特征，探讨了地幔再富集作用（软流圈来源的熔体-橄榄岩反应）对大陆岩石圈地幔 Mg、Ca 和 Fe 同位素组成的影响。未受交代地幔捕虏体 Mg、Ca 和 Fe 同位素组成比较均一，平均值与硅酸盐地球的 Mg、Ca 和 Fe 同位素组成在误差范围内一致，并且单矿物之间 Mg-Fe 同位素分馏达到平衡，是潜在地质温度计（Huang et al., 2011；Liu J. et al., 2011）。而地幔再富集作用会导致大陆岩石圈地幔的 Mg、Ca 和 Fe 同位素组成明显不均一，说明地幔再富集作用是导致大陆岩石圈地幔 Mg、Ca 和 Fe 同位素组成变化的重要机制。这些 Li、Mg、Ca 和 Fe 同位素示踪手段，一方面进一步揭示了华北岩石圈地幔高度不均一的组成特征，以及不同来源的熔体对岩石圈地幔的改造作用，为探索大陆岩石圈地幔演化及其改造过程中熔体的性质和来源打开了新的窗口，提供了新的研究思路；另一方面，Li、Mg、Ca 和 Fe 同位素还可与岩石学、元素和同位素地球化学等传统学科相互验证以证实和证伪，这有利于正确甄别大陆岩石圈地幔演化及其改造过程熔体的性质和来源常常具有的多解性，为正确认识相关科学问题提供了必要条件。最近，Su 等（2019）发现，经受碳酸盐交代的橄榄岩捕虏体 Mg 同位素组成与正常地幔值一致，从而提出中国东部新生代玄武岩轻 Mg 同位素特征可能与强烈富集重 Mg 同位素的铬铁矿的分离结晶有关。因此在利用 Mg 同位素示踪深部碳酸盐循环时需查明铬铁矿分离作用对幔源岩浆岩 Mg 同位素组成有无影响，这为应用 Mg 同位素示踪深部碳酸盐循环提出了新的挑战。

2 来源于幔源岩浆方面的进展

2.1 中国东部新生代玄武岩

板内玄武岩的形成受深部地幔性质制约，而玄武岩的组成亦可用于揭示源区地幔的特征。与大洋板内玄武岩（如洋岛、海山玄武岩）相比，大陆板内玄武岩的形成过程受浅部岩石圈的影响较大，容易掩盖来自地幔源区的真实信息，对我们理解其成因和深部地幔性质造成了一定难度。中国东部发育大量新生代板内玄武岩，南起海南岛，北至黑龙江，断续出露长达四千余千米，是我们探讨大陆板内玄武岩成因和深部地幔性质的理想场所。近十年来，通过我国地幔研究领域学者的不懈努力，在中国东部新生代玄武岩的成因认识、深部地幔性质与动力学过程等方面取得了许多重要进展。

2.1.1 发现玄武岩的地幔源区岩性显著不均一

长期以来，人们普遍认为玄武岩是地幔橄榄岩部分熔融的产物。近来的实验岩石学研究则表明，干的橄榄岩部分熔融无法匹配许多板内玄武岩的元素组成，而辉石岩、角闪石岩、碳酸盐化橄榄岩等也可作为玄武岩的源区母岩（Dasgupta et al., 2007; Sobolev et al., 2007）。因此，地幔的岩性不均一开始被国内外学者广泛讨论，并应用到中国东部新生代玄武岩的研究中。

辉石岩被认为广泛存在于中国东部新生代玄武岩的地幔源区，是导致这些玄武岩具有比橄榄岩来源熔体明显偏低的 CaO 含量（Zeng et al., 2011; Li et al., 2016; Zhang and Guo, 2016; Zhang J. B. et al., 2017; Li P. et al., 2020）、偏高的 Fe/Mn 值（Liu Y. S. et al., 2008; Zhang J. J. et al., 2009; Wang et al., 2011; Wang et al., 2012; Xu, 2014; Li et al., 2015; Liu S. C. et al., 2016; Zhang J. B. et al., 2017）和偏高的 FC3MS（$TFeO/CaO - 3 \times MgO/SiO_2$）值（Yang and Zhou, 2013; Yang et al., 2016; Chu et al., 2017; Lei et al., 2020）的原因，在橄榄石-石英-CATS 三元相图中落在辉石岩的区域内（Zeng et al., 2017）。这些玄武岩中的橄榄石斑晶大都呈现出高 Ni 和 Fe/Mn、低 Ca 的特征（图3）（Wang et al., 2012; Hong et al., 2013; Liu J. Q. et al., 2015; Qian et al., 2015; Li et al., 2016; Zhang et al., 2016; Zhang and Guo, 2016; Zhang Y. H. et al., 2017; Yu S. Y. et al., 2018; Pang et al., 2019），同样也支持这些玄武岩起源自富辉石岩的源区。橄榄石、单斜辉石等斑晶的氧同位素异常（相对于正常地幔而言），也指示这些玄武岩源区中存在辉石岩质的再循环地壳物质（Wang et al., 2011; Xu Y. G. et al., 2012; Liu J. et al., 2015a; Liu J. G. et al., 2015a; Wang et al., 2015; Chen L. M. et al., 2017）。上述观察证实了再循环物质对中国东部新生代玄武岩地幔源区的影响与改造。但是，对于玄武岩源区辉石岩的来源目前存在多种解释，再循环的大陆地壳（Liu Y. S. et al., 2008; Chen et al., 2009; Zeng et al., 2011）、年轻的俯冲洋壳并伴随少量沉积物（Xu Z. et al., 2012; Sakuyama et al., 2013; Fan et al., 2014; Xu, 2014; Chen H. et al., 2015; Liu J. et al., 2015a; Liu J. Q. et al., 2016; Li et al., 2016; Zhang and Guo, 2016; Chen L. M. et al., 2017; Li S. G. et al., 2017; Xia et al., 2019; Yu et al., 2019）或古老的俯冲洋壳-沉积物（Kuritani et al., 2011, 2013; Li et al., 2015; Liu J. et al., 2015b; Liu F. et al., 2017; Wang X. J. et al., 2017; Zeng et al., 2017）等，均被发现存在于中国东部之下软流圈地幔之中。这一方面反映了该区地幔组成的复杂性与多样性，另一方面也表明利用玄武岩的岩石学和地球化学特征反演源区岩性还存在一定的不确定性。

2.1.2 揭示岩石圈地幔的潜在影响

近十年来的研究发现，岩石圈地幔在中国东部玄武岩的形成过程中扮演了重要角色，主要包括以下三种可能的影响方式：①熔体-岩石反应，玄武质熔体在上升过程中与岩石圈地幔发生相互作用导致其化

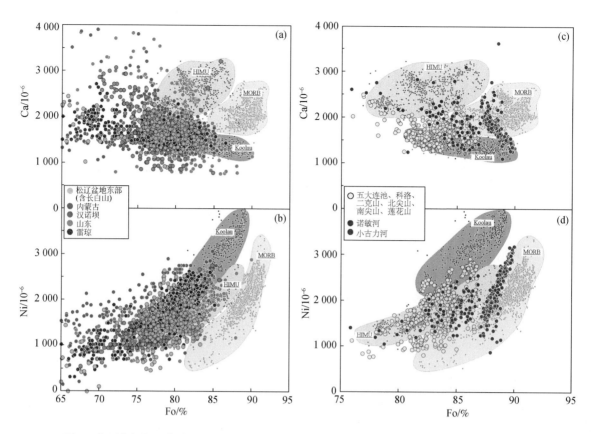

图3 中国东部新生代钠质［(a)、(b)］和钾质［(c)、(d)］玄武岩中橄榄石斑晶成分

洋中脊玄武岩（MORB）（Sobolev et al., 2007）、HIMU 型玄武岩（Weiss et al., 2016）和夏威夷 Koolau 玄武岩（Sobolev et al., 2007）结晶的橄榄石斑晶分别反映了橄榄岩、碳酸盐化橄榄岩和辉石岩来源玄武质熔体结晶形成的橄榄石斑晶成分特征。松辽盆地以东数据引自 Choi et al., 2017, 2020; Yu M. M., 2018。内蒙古数据引自 Hong et al., 2012; Wang et al., 2015; Pang et al., 2019。汉诺坝数据引自 Qian et al., 2015。山东地区数据引自 Xu Y. G., 2012; Li et al., 2016; Zhang J. B., 2017。雷琼地区数据引自 Wang et al., 2012; Liu J. Q., 2015

学组成发生变化（Liu J. Q. et al., 2016；Wang X. J. et al., 2017；Wang et al., 2018）。Liu J. Q. 等（2016）和 Liu F. 等（2017）发现东北新生代钾质玄武岩的地球化学特征与其下伏岩石圈的厚度存在耦合关系：下伏岩石圈的厚度越大，钾质玄武岩的 MgO 越高，K_2O/Na_2O 和 Rb/Nb 值越低，Sr-Nd 同位素组成越亏损。这种耦合关系表明初始的富钾富硅熔体上升过程中与亏损的岩石圈地幔发生了不同程度的反应，最终喷发的岩浆具规律性的成分变化。Wang 等（2018）发现山东新生代碱性玄武岩的 $\delta^{66}Zn$ 与 SiO_2、ε_{Nd}、Ba/Th 值等协变，且高硅玄武岩的组成呈现出从低硅玄武岩组成向类似岩石圈地幔组成过渡的趋势，据此他们提出这种组成转变是由低硅熔体与岩石圈地幔相互作用的结果。②"盖层效应"，认为岩石圈的厚度可以有效控制地幔部分熔融程度，并通过部分熔融程度的差异来影响玄武岩的化学组成，主要体现在岩石圈厚度越大的区域产出的玄武岩对应的熔融程度越低，因此具有更高的 La/Yb、La/Sm 值和更低的 SiO_2 含量等特征（Zhang L. Y. et al., 2016；Sun et al., 2017；Guo et al., 2020；Sun P. et al., 2020）。③直接源区，尽管多数学者认为中国东部新生代玄武岩起源于软流圈地幔，也有观点认为广泛分布的钠质玄武岩的源区是新生岩石圈地幔，其中富含由再循环地壳物质释放的流体-熔体交代形成的辉石岩组分（Wang et al., 2011；Xu Y. G. et al., 2012；Xu and Zheng, 2017；Xu et al., 2017）。东北钾质玄武岩因具有高钾、异常高的 Ba/Th 和 K/U 值、极低的放射成因 Pb 同位素含量（$^{206}Pb/^{204}Pb$ 低至约 16.5）显著区别于钠质玄武岩，最近的研究指出，含碳酸盐的沉积物俯冲到地幔过渡带经历少量碳酸岩熔体抽取后，富含钾锰钡矿和镁铁榴石的残留相在地幔过渡带的长期封闭演化可以很好地解释上述独特的地球化学组成（Wang X. J. et al., 2017），但也有研究者坚持钾质玄武岩起源于古老的交代岩石圈地幔这一传统观点

(Chu et al., 2013；Sun et al., 2014, 2017；Zhang L. Y. et al., 2016b)。

2.1.3 探讨东亚大地幔楔与玄武岩形成的内在联系

西太平洋板块的俯冲作用自中生代以来一直影响到东亚大陆边缘的岩浆活动。地球物理观测发现，该俯冲板片并未进入下地幔，而是平躺于地幔过渡带中被滞留（Huang and Zhao, 2006），这一平躺滞留于大洋板片之上的地幔被称为大地幔楔。尽管目前已初步建立东亚大地幔楔与中国东部新生代玄武岩之间的成因联系，但滞留板片对玄武岩形成的具体影响方式尚无定论，归纳如下：①太平洋板片的俯冲持续为地幔过渡带提供水和再循环物质（包括年轻的洋壳和沉积物），随后上述含水的板块碎片以地幔对流或者地幔柱的方式回到浅部地幔并发生部分熔融，形成中国东部的新生代玄武岩（Xu Z. et al., 2012；Xu, 2014；Chen L. et al., 2015；Wang et al., 2015；Li et al., 2016；Li S. G. et al., 2017；Chen H. et al., 2017；Xu et al., 2018）；②俯冲的太平洋板块对地幔过渡带造成扰动，导致地幔过渡带中滞留的古老再循环物质或其交代的地幔组分上涌至浅部地幔，压力降低诱发这些古老物质发生熔融并贡献玄武岩的形成（Kuritani et al., 2011, 2013；Sakuyama et al., 2013；Wang X. J. et al., 2017）；③俯冲的太平洋板块在地幔过渡带发生脱碳作用，释放的碳酸盐熔体上升并交代上地幔，导致上地幔发生碳酸盐化。CO_2 的加入导致地幔固相线显著降低，诱发地幔橄榄岩在较深压力下发生部分熔融形成玄武岩（Liu J. Q. et al., 2016；Li S. G. et al., 2017）。

2.2 大洋玄武岩

洋中脊玄武岩（MORB）和洋岛玄武岩（OIB）代表了大洋地幔在不同构造背景下从不同深度发生不同程度熔融而形成的熔体，它们携带了大量不同深度的地幔组成和演化的信息。近十年来，我国学者在利用大洋玄武岩这一地幔"探针"来制约洋脊下部地幔的组成不均一和动力学过程、南海地区地幔组成和构造演化历史、南海和西太平洋典型 OIB 的成因等方面取得了系列重要成果。

2.2.1 利用 MORB 约束洋脊下部地幔的组成不均一性

（1）东太平洋洋隆（East Pacific rise，EPR）、西南印度洋洋脊（Southwest Indian rise，SWIR）及大西洋洋中脊（Mid-Atlantic rise，MAR）MORB 样品研究。Zhang 等（2013a）对 EPR 地区不同位置［国际大洋发现计划（International Ocean Discovery Program，IODP）站位 U1367 及 U1368］的基底玄武岩进行了详细的地球化学研究，发现两地的玄武岩在 Sr-Nd-Pb-Hf 同位素组成及微量元素比值（Nb/La 和 Lu/Hf 等）上存在显著差异，并指出这种大尺度地幔组成差异是由地幔中存在的再循环洋壳引起。另一项针对 EPR 11°N 处的 MORB 研究显示该处洋脊下的地幔中含有辉石岩组分（Zhang et al., 2012b），反映了 EPR 地幔存在多尺度的组成不均一性。Yang 等（2017）发现，SWIR 50°28′E 地区存在一类以具有略富集的 Sr-Nd-Hf-Pb 同位素组成为特征的正常型 MORB（N-MORB），并揭示该富集特征反映了附近 Crozet 地幔柱的影响，地幔柱物质在向洋脊运移的途中发生了多次熔融，残余物亏损微量元素的同时保持了放射成因同位素特征，这种残余物再次熔融即可形成具富集同位素组成的 N-MORB。Wei 等（2020）发现 MAR 地区深海钻探计划（Deep Sea Drilling Program，DSDP）钻孔 559 和 561 的 E-MORB 样品不同于钻孔 561 亏损型 MORB 的同位素特征，并提出这些 E-MORB 的富集同位素特征可能分别来自洋脊外侧的地幔柱（New England 及 Azores）组分。

（2）南海地区地幔性质及构造演化历史。我国学者主导的 IODP 349 航次分别在南海东部次海盆及西南次海盆布设了 U1431 和 U1433、U1434 3 个站位，获得了南海基底玄武岩样品（Li C. F. et al., 2014）。IODP 367 航次在南海东部次海盆北缘布设了站位 U1500，取得了代表南海早期扩张形成的洋壳样品（Yu and Liu, 2020）。地球化学研究显示，站位 U1431 玄武岩以 N-MORB 为主，有少量 E-MORB，站位 U1433 和 U1434 玄武岩均为 E-MORB（Sun K. et al., 2020）；站位 U1500 为 N-MORB（Yu and Liu, 2020）。另外，站位 U1433 MORB 比站位 U1431 MORB 在 Sr-Nd-Hf 同位素组成上更亏损，$^{206}Pb/^{204}Pb$ 也较低（Zhang et al., 2018a, 2018b）。根据两个次海盆洋壳样品的 Sr-Nd-Pb-Hf 同位素组成，Zhang 等（2018a）认为，站位 U1433 MORB 较正常 MORB 更低的 $^{206}Pb/^{204}Pb$ 反映了西南次海盆洋脊下部地幔中可能有具类似下地壳组成的物质加入，而站位 U1431 MORB 更富集的 Sr-Nd-Hf 同位素组成和更高的 $^{206}Pb/^{204}Pb$ 值则与东部次海盆洋脊附近的海南热点有关。源区水含量计算显示西南次海盆样品较东部次海盆样品的源区水含量更低，Wang W. 等（2019）认为，该差异支持上述两个次海盆地幔间的组分不同的模型。站位 U1431 MORB 的橄榄石斑晶组成及其形成温度指示其源区为辉石岩质，同样支持海南热点的辉石岩质富集组分加入模型（Zhang et al., 2018b; Yang et al., 2019a）。

为了验证南海的张开是否与海南地幔柱的上涌有关，Yu M. M. 等（2018）研究了台湾南部垦丁杂岩体中年龄约 25 Ma 的 MORB 样品、年龄约 16 Ma 的站位 U1431 MORB 样品和年龄约 9 Ma 的黄岩岛海山样品。结果显示，25 Ma 样品的地幔源区为亏损 MORB 地幔的性质，而年轻样品的地幔源区则显示逐渐增强的富集作用，反映了海南地幔柱的影响。对代表南海早期扩张形成的洋壳样品（站位 U1500 玄武岩）的研究同样显示，无论是全岩还是橄榄岩斑晶的地球化学特征均与地幔柱模型不符（Yu and Liu, 2020），因此南海的张开更可能与非地幔柱过程有关，如太平洋板块的西向俯冲（Yu and Liu, 2020）。此外，Yang 等（2019b）还对站位 U1433 玄武岩中的斜长石斑晶开展了详细的矿物学工作，约束了南海洋中脊下的岩浆过程，并在此基础上探讨慢速扩张洋脊在洋壳增生中的岩浆过程。

2.2.2 约束 OIB 与地幔深部物质之间的成因联系

地幔中的碳深刻影响地幔的熔融行为（Dasgupta and Hirschmann, 2006）。高温高压实验结果显示，CO_2 参与的地幔熔融产生的硅酸盐熔体具硅不饱和特征（Dasgupta et al., 2006, 2007），与许多板内碱性火山岩的特征一致。因此，建立这些碱性火山岩与地幔碳之间的成因联系有助于揭示碳在地球深部的地球化学行为及其影响。Zhang G. L. 等（2017）在南海首次观察到天然碳酸盐化硅酸质岩浆可通过与岩石圈地幔进行连续反应而转化为碱性玄武岩，在这一过程中，碳酸岩质岩浆通过消耗岩石圈橄榄岩中的斜方辉石并形成橄榄石及单斜辉石，从而使反应熔体 SiO_2 升高、CaO 降低并发生脱 CO_2 的过程。为进一步揭示地幔中 CO_2 与碱性玄武岩的成因联系，Zhang 等（2020a, 2020b）对代表卡罗琳地幔柱晚期火山活动的波纳佩（Ponape）洋岛玄武岩进行了系统的岩石学、矿物学和地球化学研究。与南海地区极薄的岩石圈不同，卡罗琳热点下伏的中生代太平洋板块代表着巨厚的大洋岩石圈环境（Zhang et al., 2020a），有研究表明，这些火山岩具有两阶段演化过程：第一阶段类似南海板内碳酸盐化硅酸盐熔体的脱碳作用；第二阶段时，脱碳后的熔体与厚的大洋岩石圈反应达到平衡，熔体继续演化但不再与岩石圈有明显的物质交换（Zhang et al., 2020b）。

南海海盆内出露有大量扩张期后的板内火山，是探索南海深部地幔性质的理想研究对象。Yan 等（2015）通过研究南海板内玳瑁海山玄武岩的元素及 Sr-Nd-Pb 同位素组成，指出该海山与位于南海北部附近的海南地幔柱具有成因联系。为进一步验证该联系，他们分析了南海海盆内大量不同海山玄武岩的 Hf 同位素，结果显示这些玄武岩具有显著的 Hf 同位素不均一性，且 Nd-Hf 同位素之间的耦合符合亏损印度洋型地幔端元与 EM2 型富集地幔端元之间的混合，并认为该混合趋势支持南海板内岩浆作用源自海南地幔柱这一结论（Yan et al., 2019）。Zheng 等（2019）分析了南海西南次海盆南岳海山玄武岩的 Sr-Nd-Pb-Hf 同位素，结果显示它们的源区为 3 个不同组分的混合：太平洋 MORB 地幔、LCC 组分和海南地幔柱物质。Qian 等（2020）通过对 IODP 站位 U1431 板内火山碎屑中的玻璃和长石的原位分析，提出南海扩张后的板内火山岩的源区存在拆沉的大陆岩石圈，该组分可能通过大陆裂解进入地幔并引起地幔不均一性。

2.3 基性-超基性杂岩体

超基性侵入岩常与基性岩伴生，构成基性-超基性杂岩体，其产出的构造环境主要包括：与地幔柱活动有关的大火成岩省、陆内裂谷、造山晚期或造山后伸展环境。该类杂岩体因赋含铜镍硫化物、铬铁矿、铂族元素或钒钛磁铁矿等金属矿床而广受关注。

2.3.1 深入认识基性-超基性杂岩体岩石成因

有关岩浆型钒钛磁铁矿的分离结晶和岩浆不混溶成因争议已久,也是困扰岩石学家几十年的科学问题。传统观点认为,不混溶作用发生的岩浆温度为1 000℃,即只在玄武质岩浆分离结晶作用的晚期有所表现(Philpotts and Doyle,1983)。在峨眉山大火成岩省新街基性-超基性杂岩体的主要含矿层,矿物粒间发育岩浆晚期成对出现的、非反应显微结构,一种为富钛铁矿集合体,另一种为富硅集合体,表明新街岩体岩浆演化过程中确实发生过液态不混溶作用(Dong et al.,2013)。Liu 等(2014a)在对攀西白马层状岩体的研究中首次报道了橄榄石中以铁钛氧化物为主并含有少量富水矿物(角闪石和金云母)、硫化物(雌黄铁矿)和磷灰石的多相矿物包裹体。这些橄榄石中的磁铁矿包裹体在成分上与填充在橄榄石粒间的磁铁矿一致,证实这两种产状的磁铁矿具有相同的成因,进而他们提出攀枝花型钒钛磁铁矿底部上百米厚的铁矿石层是不混溶的富铁钛的岩浆结晶形成的(Liu et al.,2014b)。Hou 和 Veksler(2015)通过高温超液相线不混溶实验证实富铁和富硅熔体在 1 150~1 200℃可以稳定存在,进一步从实验岩石学角度证明了岩浆不混溶过程可以发生在岩浆演化的早期高温阶段。近年来,一些岩石显微结构、矿物熔融包裹体和高温不混溶实验岩石学方面的研究表明,尽管岩浆房尺度的液态不混溶一般很难出现,但在粒间熔体尺度上,液态不混溶作用在一些层状岩体中是普遍存在的(侯通,2017;王焰等,2017)。

同化混染作用在基性-超基性岩浆中研究得较多。玄武质岩浆在侵位过程中是否与围岩碳酸盐岩发生混染也是一个争论的热点问题,部分研究表明玄武质岩浆同化大理岩是形成岩浆矿床的一种重要机制(Tang et al.,2017a,2018)。玄武质岩浆同化混染氧化性或还原性地层将明显改变岩浆的氧逸度(柏中杰等,2019)。攀枝花层状侵入体同时代的苦橄质岩墙及攀西基性-超基性侵入体的研究表明钒钛磁铁矿矿床形成可能与玄武质岩浆混染碳酸盐岩有关(Tang et al.,2017a,2021;汤庆艳等,2020)。同样,我国近年来发现的夏日哈木含超大型铜镍矿的基性-超基性岩体的早期氧逸度升高可能与金水口群大理岩围岩的同化混染作用有关(段雪鹏等,2019)。对于造山带中的一些中小型的基性-超基性杂岩体,成矿岩浆可能具有较高的氧逸度,地壳中还原性的有机物加入有可能是导致岩浆氧逸度降低、达到硫化物饱和的重要机制(Cao et al.,2019;Wei et al.,2019)。Cao 等(2019b)利用橄榄石-尖晶石氧逸度估算出中亚造山带一系列基性-超基性质岩体的母岩浆氧逸度范围主要为FMQ+0.3 至 FMQ+2.6,落入岛弧岩浆的氧逸度范围,明显高于MORB氧逸度(FMQ-1 至 FMQ)(Wood et al.,1990)。

2.3.2 流体组成及原位微区分析技术等分析技术注入活力

流体组成的研究能够提供成矿岩浆演化过程的氧化还原条件、壳源物质类型、含硫挥发分相等的变化特征。稳定同位素和稀有气体同位素组成被广泛用于示踪幔源岩浆流体来源、岩浆演化过程以及探讨成岩动力学背景。华北克拉通大庙斜长岩杂岩体流体组成表明斜长石可能是在岩浆早期相对还原的条件下结晶的,而磁铁矿是在岩浆演化晚期相对氧化的条件下结晶的(邢长明等,2011)。峨眉山大火成岩省朱布岩浆起源于相对富H_2O、弱还原的流体环境,地壳混染和流体加入是造成朱布岩浆中硫化物熔离的关键控制因素(Tang Q. Y. et al.,2013,2017b)。攀枝花、红格和白马岩体C-O同位素表明存在来自灯影组灰岩的流体组分加入(邢长明等,2012)。C-He-Ne-Ar同位素揭示塔里木板块东北缘坡一侵入体岩浆源区较深,存在地幔柱物质作用的贡献,而坡四、坡十与坡东侵入体岩浆源区为俯冲流体交代的岩石圈地幔(Zhang M. J. et al.,2017,2019;冯鹏宇等,2018)。这一系列研究结果表明岩浆中挥发分组成对岩浆演化和成矿过程起到至关重要的作用。

近年来,矿物原位微量元素的组成为判断元素在矿物中的赋存状态,反演成矿流体的性质和演化规律,揭示岩浆房中不同矿物的生长机制和结晶历史,约束岩浆起源和演化提供了一种新的途径。如橄榄石、单斜辉石、斜方辉石、铬尖晶石、角闪石、磷灰石、磁铁矿、钛铁矿等的微量元素可以用来约束基性和超基性岩的岩浆源区组成及反演成岩成矿过程(刘飞等,2013;Xing et al.,2014;Chen L. M. et al.,2017;Wang K. Y. et al.,2019;Song et al.,2020)。夏日哈木和黄山基性-超基性杂岩体斜方辉石环带结构及微量元素结合岩相接触关系等表明环带结构是由岩浆成分变化导致,岩体由基性程度不同的岩浆多次补给形成，记录了岩浆就位与结晶过程（Mao et al.，2019；Wang K. Y. et al.，2019）。扬子板块北缘新元古代毕机沟和望江山基性–超基性杂岩体单斜辉石和斜长石微量元素组成估算表明母岩浆富集大离子亲石元素（Cs、Sr、Ba、U），亏损高场强元素（Nb、Ta、Zr、Hf），其微量元素特征与早期俯冲作用交代的地幔有关（Zhang X. Q. et al.，2020）。锆石原位的U-Pb-Hf-O同位素、橄榄石的O同位素和长石的原位Sr-Nd同位素常用来限定岩浆事件时代、地幔源区性质及演化过程（Su et al.，2011；Liu et al.，2014a，2014b；Tang et al.，2021），如峨眉山大火成岩省攀枝花、红格和太和岩体锆石Hf-O同位素研究表明地幔柱源区具有不均一性以及不同岩体岩浆演化过程存在差异。近年来磁铁矿原位微量元素分析工作迅速发展，Huang等（2014，2019）系统研究了岩浆和热液磁铁矿的成分，提出了新的分类方案，可以很好地区分岩浆和热液磁铁矿，同时对基性岩中磁铁矿床成因具有重要指示意义。

2.4 火成碳酸岩

火成碳酸岩是一类在成分上迥异于硅酸岩的幔源岩浆岩。据统计，中国已发现有27个碳酸岩岩体（Yang and Woolley，2006），对这些岩体的研究工作取得重要进展，包括如下4个方面：①碳酸岩的精确定年几年来取得突破，碳酸岩在空间上往往与各类碱性硅酸岩密切共生，过去很难对碳酸岩进行直接定年，因此通常将硅酸岩的年龄视为碳酸岩的年龄。随着测试技术的进步，对碳酸岩中各类副矿物如独居石、氟碳铈矿、铌钽矿等进行U-Pb定年成为可能（Xu et al.，2015；Ying et al.，2017）。尽管绝大多数碳酸岩与共生的碳酸岩有相同的形成时代，但也存在两者不一致的情况，如庙垭、杀熊洞等岩体中的硅酸岩为早古生代，而与之共生的碳酸岩为三叠纪。②对与稀土成矿密切相关的碳酸岩的研究不断深化，如对白云鄂博含稀土矿的白云岩和碳酸岩脉的研究表明，这些白云岩和碳酸岩脉均为幔源岩浆成因，并且经历了从铁质碳酸岩经镁质碳酸岩到钙质碳酸岩的分异演化过程（Yang K. F. et al.，2019）。对秦岭造山带晚三叠世碳酸岩的研究发现，南、北秦岭造山带碳酸岩的含矿类型存在差异，南秦岭碳酸岩以富集轻稀土为特征，而北秦岭碳酸岩以富集重稀土和钼为特征，因此推断两者经历了不同的岩浆过程。南秦岭碳酸岩源于含碳酸盐俯冲板片的部分熔融，而北秦岭碳酸岩岩浆还经历了与榴辉岩质下地壳的相互作用（Song et al.，2016）。③碳酸岩地幔源区中俯冲物质的确认，南、北秦岭晚三叠世碳酸岩的地球化学研究结果表明，其地幔源区受到源于富含碳酸盐板片熔体的交代作用（Xu et al.，2014；Song et al.，2016）；而新疆且干布拉克辉石岩–碳酸岩杂岩体的研究也表明，两者的地幔源区均经历了源自俯冲陆壳物质熔体的交代，并经历了液态不混溶分离，而与罗迪尼亚超大陆裂解有关的地幔柱是造成被交代地幔部分熔融的直接原因（Ye et al.，2013）；华北北缘新生代碳酸岩的地球化学特征显示其可能为俯冲至华北克拉通之下的古亚洲洋板片上沉积碳酸盐的部分熔融（Chen et al.，2016）。④碳酸岩地幔源区深度的限定。华北克拉通中部带北缘碳酸岩中榴辉岩包体内石榴子石以及富铁镁铁榴石的发现碳酸岩的地幔源区可以深达200~380 km（Xu et al.，2017）。

2.5 非传统稳定同位素的广泛应用

非传统稳定同位素是21世纪同位素地球化学发展最迅速的方向，特别是最近十年我国将金属稳定同位素广泛应用在大陆岩石圈演化、地幔物质再循环、基性–超基性侵入体等地幔岩石地球化学研究的各个领域，引领了国际同位素地球化学的发展。

2.5.1 地幔各种同位素储库的标定

地幔储库可大致分为富集地幔、亏损地幔和全硅酸盐地球（bulk silicate Earth，BSE）等，其中BSE最为重要，因为这是地幔演化、壳幔分异等诸多过程的起点。BSE的同位素组成是解读高温岩浆样品以及地外样品同位素数据的基准值，对稳定同位素地球化学示踪有着至关重要的意义。中国学者近年来通过研究地幔岩和幔源岩浆岩，标定了BSE多个体系的金属稳定同位素组成，如Kang等（2017）选取蒙古、西伯利亚等地的饱满橄榄岩捕虏体，对其钙同位素组成进行了精细研究，标定了BSE的$\delta^{44/40}$Ca为（0.94

±0.05)‰。在此基础上，Chen 等（2019）对意大利阿尔卑斯地区的地体橄榄岩进行了研究，发现饱满地体橄榄岩的 $\delta^{44/40}$Ca 为（0.94±0.10）‰，证明了 BSE 的 $\delta^{44/40}$Ca 约为 0.94‰。在 Cr 同位素方面，Xia 等（2017）对来自蒙古、西伯利、南非和华北克拉通的不同类型橄榄岩进行了系统的研究，标定了 BSE 的 δ^{53}Cr 为（−0.14±0.12）‰。

不相容元素在部分熔融过程会大量进入熔体，因此幔源岩浆岩（如大洋玄武岩、科马提岩、火成碳酸岩）也可用来制约 BSE 的同位素组成。Liu S. A. 等（2015）测试了橄榄岩、洋中脊玄武岩、洋岛玄武岩、弧岩浆岩的 Cu 同位素组成，并结合前人对河流、土壤、陨石的研究，估计出 BSE 的 δ^{65}Cu 为（0.06±0.2）‰。Wang Z. Z. 等（2017）测量了洋岛玄武岩及洋中脊玄武岩的 Zn 同位素，结合华北克拉通橄榄岩捕虏体以及大别-苏鲁地体橄榄岩的结果，标定 BSE 的 Zn 同位素组成为（0.20±0.05）‰。Qi 等（2019）结合饱满地幔橄榄岩和科马提岩，标定了 BSE 的 V 同位素组成为（−0.91±0.09）‰。Ba 是高度不相容元素，在橄榄岩中含量很低，且大部分赋存于粒间物质中，因此很难通过橄榄岩来标定 BSE 的 Ba 同位素组成。Li W. Y. 等（2020）研究了世界各地典型的火成碳酸岩，发现其 Ba 同位素组成均一 [$\delta^{137/134}$Ba = (0.04±0.06)‰]，且与洋中脊玄武岩的 Ba 同位素组成一致，说明 BSE 的 Ba 同位素组成可能为（0.04±0.06）‰。

2.5.2 金属稳定同位素与地幔物质循环的认识

金属稳定同位素作为新兴的地球化学工具，是示踪地壳物质再循环进入地幔的敏感指标。研究证实，在地幔部分熔融和玄武质岩浆演化过程中，Mg-Zn-Ca 同位素分馏极其有限（Teng et al., 2010；Chen et al., 2013, 2020；Wang Z. Z. et al., 2017；Huang et al., 2018a, 2018b；Zhu et al., 2018；Zhang et al., 2018b），而地表碳酸盐岩则较地幔富集轻镁、轻钙和重锌同位素（δ^{26}Mg 低至 −5.36‰，Teng, 2017；$\delta^{44/40}$Ca 低至 −1.09‰，Fantle and Tipper, 2014；δ^{66}Zn 高至 1.34‰，Pichat et al., 2003），因此 Mg-Zn-Ca 同位素是潜在的示踪地表碳酸盐深部再循环的地球化学指标。Yang 等（2012）最早发现华北克拉通小于 110 Ma 玄武岩的镁同位素组成（δ^{26}Mg 为 −0.60‰ ~ −0.03‰）轻于地幔值（δ^{26}Mg = −0.25‰±0.04‰，Teng et al., 2010），反映有富集轻镁同位素的地表碳酸盐俯冲进入地幔源区。Huang 等（2015）也发现华南新生代玄武岩同样具有轻于地幔值的镁同位素组成，结合 Yang 等（2012）的工作，他们提出了这样的认识：古太平洋板块深俯冲导致大量富集轻镁同位素的地表碳酸盐进入地幔过渡带，并在过渡带发生部分熔融，交代滞留板片上方的地幔，形成富集轻镁同位素的碳酸盐化地幔，碳酸盐化地幔发生部分熔融形成中国东部玄武岩的轻镁同位素特征。根据玄武岩的 Mg-Sr 同位素协变关系，Huang 和 Xiao（2016）发现深俯冲碳酸盐岩主要为富镁碳酸盐岩（菱镁矿占主导，白云石少量）。Li 等（2017）发现中国东部从海南到黑龙江的小于 110 Ma 的玄武岩全都具有轻镁同位素异常，说明中国东部是一个巨大的再循环碳库（Hofmann, 2017）。这一振奋人心的发现得到了后续更多玄武岩镁同位素数据的证实（Wang X. J. et al., 2017；Su et al., 2017b；Tian et al., 2018）。中国东部小于 110 Ma 玄武岩的重锌同位素组成（δ^{66}Zn 为 0.30‰ ~ 0.63‰）（Liu S. A. et al., 2016；Wang et al., 2018；Jin et al., 2020）进一步说明，通过古太平洋板块俯冲进入中国东部地幔过渡带的碳酸盐岩主要为富镁碳酸盐岩，因为富镁碳酸盐岩不仅富集重锌同位素，而且锌含量很高（Liu S. A. et al., 2016）。后续研究发现，富镁碳酸盐岩的最大俯冲深度只能到地幔过渡带（Tian et al., 2017），因为在过渡带深度压力和温度下，碳酸盐岩会发生部分熔融（Thomson et al., 2016），阻碍其进一步俯冲进入下地幔。此外，在中国东部地幔橄榄岩捕虏体（Kang et al., 2016；Chen et al., 2018）和云南腾冲火山岩（Liu et al., 2017b）中均发现了轻的钙同位素特征，也说明有富钙碳酸盐岩的深部俯冲再循环。

大量富镁碳酸盐岩的加入会将地幔中的 Fe^{2+} 氧化成 Fe^{3+}，导致中国东部地幔的 $Fe^{3+}/\Sigma Fe$ 升高，相比洋中脊地幔氧逸度更高（He Y. S. et al., 2019；Hong et al., 2020）。由于 Fe^{3+} 比 Fe^{2+} 更不相容且更富集重铁同位素（Canil et al., 1994；Dauphas et al., 2009），由富含 Fe^{3+} 的地幔部分熔融产生富 Fe 且富集重 Fe 同位素的玄武质熔体，可以解释中国东部玄武岩的高 Fe 含量和高 δ^{56}Fe 值（He Y. S. et al., 2019；He et al.,

2020a）。此外，碳酸盐会氧化地幔中的硫化物，导致硫化物分解，从而改变亲硫元素（如：铂族元素）的地球化学行为（He et al., 2020b; Cai et al., 2021）。此外，深部碳酸盐化地幔熔融会产生硅不饱和熔体，在上升过程与岩石圈地幔反应，不仅能够改变上升熔体的化学成分（Tian et al., 2016; Wang et al., 2018），而且能够诱发岩石圈地幔发生破坏（Li and Wang, 2018）。

2.5.3 非传统稳定同位素在基性-超基性杂岩体中的应用

非传统稳定同位素（Li、Mg、Ca、Fe、Cr、Ni、Cu）在探讨基性-超基性杂岩体的岩石成因和成矿过程研究中备受关注，矿物内的同位素组成还可用来制约粒间的扩散行为以及揭示缓慢冷却的岩浆过程（Chen et al., 2014; Su et al., 2017a; Zhao Y. et al., 2017, 2019; Tang et al., 2018, 2020; Bai et al., 2019; Ding et al., 2019; Tian G. C. et al., 2020）。中亚造山带峡东阿拉斯加型基性-超基性杂岩体的 Li 同位素指示岩浆分异过程中存在明显的 Li 同位素分馏，可能与岩浆的含水性、氧逸度和演化程度等因素有关（Su et al., 2017b）。同时，大型层状岩体中矿物 Li 同位素组成可以记录铬铁矿床形成过程中的矿物-粒间熔流体反应过程（Su et al., 2020）。Mg 同位素对镁铁质岩浆中碳酸盐的同化混染作用很敏感。Tang 等（2018）报道了金川基性-超基性杂岩体相对较轻的 δ^{26}Mg 值（-0.40‰ ~ -0.26‰），进而解释为岩浆演化过程中存在大理岩的混染。Cheng 等（2018）也认为塔里木大火成岩省瓦吉里塔格基性-超基性杂岩体的 Mg 同位素组成亦是碳酸盐交代作用的结果。相较于 Li 和 Mg 同位素，Fe 同位素在岩浆成矿过程中的分馏受控因素更多，不仅受到壳幔混染、岩浆演化和成矿过程的影响，而且共生矿物间也可以产生显著的 Fe 同位素分馏（Chen et al., 2014; Ding et al., 2019; Tian H. C. et al., 2020; Bai et al., 2021）。目前的研究结果总体揭示，基性-超基性层状岩体中硅酸盐矿物之间的 Fe 同位素是大致平衡的，而硅酸盐和氧化物之间的 Fe 同位素明显不平衡（Chen et al., 2014; Liu et al., 2014b; Bai et al., 2021）。对硅酸盐和氧化物之间 Fe 同位素不平衡的解释尚存在较大争议，主要包括熔-流体不混溶、亚固相元素交换以及粒间熔-流体与矿物相互反应。

近年来，开展非传统稳定同位素的联合研究已为大势所趋。新疆图拉尔根基性-超基性杂岩体 Cu-Fe 同位素研究表明，氧化还原状态的转变是造成成矿系统中 Cu-Fe 同位素显著分馏的关键因素，Cu 同位素还可用于反演和指示富集硫化物的岩浆的运动方向（Zhao et al., 2017c, 2019）。中亚造山带南缘喀拉通克和白石泉基性-超基性杂岩体黄铜矿原位 Cu-S 同位素研究表明，母岩浆来自于部分熔融程度较低的交代地幔源区（Tang et al., 2020）。北美 Stillwater 大型层状基性-超基性杂岩体 Fe-Mg 同位素与 Cr 同位素的结合揭示岩体橄榄岩带的冷却时间大概在 1 万 ~ 10 万年，为 Stillwater 岩体铬铁矿层的对流冷却机制提供了新的证据（Bai et al., 2019）。随着人们对同位素分馏机理认识的加深和单矿物原位微区分析方法的完善，结合高温高压实验模拟、岩石矿物显微结构（Mao et al., 2018; Wang K. Y. et al., 2019）和地质背景、流体组成、非传统稳定同位素和矿物原位微区成分分析将为精细刻画基性-超基性岩的形成过程、演化历史及深部动力学机制提供更多依据。

致谢：本文是中国矿物岩石地球化学学会地幔矿物岩石地球化学专委会全体同仁共同努力的结晶，是周新华研究员悉心指导的结果。各章节初稿分别由不同的撰写者提供：刘锦研究员（地幔矿物学），刘金高教授、汤艳杰研究员和赵新苗副研究员（大陆岩石圈地幔演化包括金属稳定同位素在大陆岩石圈地幔中的应用），张宏福教授和郑建平教授（造山带橄榄岩），刘传周研究员（大洋岩石圈），夏群科教授（地幔中的水），陈立辉教授和张国良教授（大陆和大洋玄武岩），汤庆艳教授和苏本勋研究员（基性-超基性杂岩体，包括非传统稳定同位素在基性-超基性杂岩体中的应用），英基丰研究员（火成碳酸岩），黄方教授（金属稳定同位素储库及其物质循环）。在此，对参与初稿撰写的各位学者表示衷心的感谢！

参 考 文 献

柏中杰, 钟宏, 朱维光. 2019. 幔源岩浆氧化还原状态及对岩浆矿床成矿的制约. 岩石学报, 35(1):204-214
陈意, 苏斌, 郭顺. 2015. 大别-苏鲁造山带橄榄岩：进展和问题. 中国科学:地球科学,45(9):1245-1269

段雪鹏,孟繁聪,范亚洲. 2019. 东昆仑夏日哈木镁铁-超镁铁岩中的钛闪石-非闪石对成矿过程的约束. 岩石学报,35(6):1819-1832

冯鹏宇,张铭杰,李立武,胡飞,孙凡婷,王亚磊,曹春辉. 2018. 新疆坡北杂岩体西端镁铁-超镁铁质岩体成因的稀有气体同位素制约. 岩石学报,34(11):3445-3454

侯通. 2017. 硅酸盐岩浆液态不混溶作用的理论基础概述. 矿物岩石地球化学通报,36(1):14-25

林阿兵,郑建平,潘少逵. 2018. 微陆块属性及过程:我国东北地区岩石圈地幔性质差异之根本. 岩石学报,34(1):143-156

刘飞,苏尚国,余晓艳,梁凤华,胡妍,牛晓露. 2013. 内蒙古文圪气镁铁-超镁铁质杂岩体中环带角闪石矿物学特征及成因. 地学前缘,20(1):206-222

刘良,孙勇,肖培喜,车自成,罗金海,陈丹玲,王焰,张安达,陈亮,王永合. 2002. 阿尔金发现超高压(>3.8 GPa)石榴二辉橄榄岩. 科学通报,47(9):657-662

汤庆艳,鲍坚,党永西,苏天宝,许仕海. 2020. 峨眉山大火成岩省白马层状侵入体的成因. 岩石学报,36(7):2163-2176

汤艳杰,张宏福. 2019. 华北克拉通岩石圈地幔的锂同位素特征与熔体改造作用. 矿物岩石地球化学通报,38(2):217-223

王焰,王坤,邢长明,魏博,董欢,曹永华. 2017. 二叠纪峨眉山地幔柱岩浆成矿作用的多样性. 矿物岩石地球化学通报,36(3):404-417

邢长明,陈伟,王焰,赵太平. 2011. 华北克拉通北缘元古宙大庙Fe-Ti-P矿床的挥发分组成和C-H-O同位素研究. 岩石学报,27(5):1500-1510

邢长明,王焰,张铭杰. 2012. 攀西地区超大型钒钛磁铁矿矿床挥发分组成及其C-H-O稳定同位素研究:对挥发分来源和矿石成因的约束. 中国科学:地球科学,42(11):1701-1715

张宏福. 2009. 橄榄岩-熔体相互作用:克拉通型岩石圈地幔能够被破坏之关键. 科学通报,54(14):2008-2026

张宏福,于红. 2019. 造山带橄榄岩岩石学与构造过程:以松树沟橄榄岩为例. 地球科学,44(4):1057-1066

赵新苗. 2019. 地幔再富集作用对岩石圈地幔Fe-Mg-Ca同位素组成的影响. 矿物岩石地球化学通报,38(4):713-724

赵勇伟,樊祺诚. 2011. 大兴安岭岩石圈地幔特征——哈拉哈河-绰尔河橄榄岩捕虏体的证据. 岩石学报,27(10):2833-2841

郑建平,熊庆,赵伊,李文博. 2019. 俯冲带橄榄岩及其记录的壳幔相互作用. 中国科学:地球科学,49(7):1037-1058

周新华,张宏福,郑建平,夏群科,刘勇胜,汤艳杰,黄方,刘传周. 2013. 新世纪十年地幔地球化学研究进展. 矿物岩石地球化学通报,32(4):379-391

朱日祥,朱光,李建威. 2020. 华北克拉通破坏. 北京:科学出版社

Bai Y, Su B X, Xiao Y, Chen C, Cui M M, He X Q, Qin L P, Charlier B. 2019. Diffusion-driven chromium isotope fractionation in ultramafic cumulate minerals: elemental and isotopic evidence from the Stillwater complex. Geochimica et Cosmochimica Acta, 263:167-181

Bai Y, Su B X, Xiao Y, Cui M M, Charlier B. 2021. Magnesium and iron isotopic evidence of inter-mineral diffusion in ultramafic cumulates of the Peridotite zone, Stillwater complex. Geochimica et Cosmochimica Acta, 292:152-169

Bell D R, Rossman G R. 1992. Water in Earth's mantle: the role of nominally anhydrous minerals. Science, 255(5050):1391-1397

Cai R H, Liu J G, Pearson D G, Li D X, Xu Y, Liu S A, Chu Z Y, Chen L H, Li S G. 2021. Oxidation of the deep big mantle wedge by recycled carbonates: constraints from highly siderophile elements and osmium isotopes. Geochimica et Cosmochimica Acta, 295:207-223

Campione M, Tumiati S, Malaspina N. 2017. Primary spinel + chlorite inclusions in mantle garnet formed at ultrahigh-pressure. Geochemical Perspectives Letters, 4:19-23

Canil D, O'Neill H S C, Pearson D G, Rudnick R L, McDonough W F, Carswell D A. 1994. Ferric iron in peridotites and mantle oxidation states. Earth and Planetary Science Letters, 123(1-3):205-220

Cao Y H, Wang C Y, Wei B. 2019. Magma oxygen fugacity of Permian to Triassic Ni-Cu sulfide-bearing mafic-ultramafic intrusions in the Central Asian orogenic belt, North China. Journal of Asian Earth Sciences, 173:250-262

Chen C F, Liu Y S, Foley S F, Ducea M N, He D T, Hu Z C, Chen W, Zong K Q. 2016. Paleo-Asian oceanic slab under the North China Craton revealed by carbonatites derived from subducted limestones. Geology, 44(12):1039-1042

Chen C F, Liu Y S, Foley S F, Ducea M N, Geng X L, Zhang W, Xu R, Hu Z C, Zhou L, Wang Z C. 2017. Carbonated sediment recycling and its contribution to lithospheric refertilization under the northern North China Craton. Chemical Geology, 466:641-653

Chen C F, Liu Y S, Feng L P, Foley S F, Zhou L, Ducea M N, Hu Z C. 2018. Calcium isotope evidence for subduction-enriched lithospheric mantle under the northern North China Craton. Geochimica et Cosmochimica Acta, 238:55-67

Chen C F, Dai W, Wang Z C, Liu Y S, Li M, Becker H, Foley S F. 2019. Calcium isotope fractionation during magmatic processes in the upper mantle. Geochimica et Cosmochimica Acta, 249:121-137

Chen C F, Ciazela J, Li W, Dai W, Wang Z C, Foley S F, Li M, Hu Z C, Liu Y S. 2020. Calcium isotopic compositions of oceanic crust at various spreading rates. Geochimica et Cosmochimica Acta, 278:272-288

Chen H, Savage P S, Teng F Z, Helz R T, Moynier F. 2013. Zinc isotope fractionation during magmatic differentiation and the isotopic composition of the bulk Earth. Earth and Planetary Science Letters, 369-370:34-42

Chen H, Xia Q K, Ingrin J, Jia Z B, Feng M. 2015. Changing recycled oceanic components in the mantle source of the Shuangliao Cenozoic basalts, NE China: new constraints from water content. Tectonophysics, 650:113-123

Chen H, Xia Q K, Ingrin J, Deloule E, Bi Y. 2017. Heterogeneous source components of intraplate basalts from NE China induced by the ongoing Pacific slab subduction. Earth and Planetary Science Letters, 459:208-220

Chen L, Chu F Y, Zhu J H, Dong Y H, Yu X, Li Z G, Tang L M. 2015. Major and trace elements of abyssal peridotites: evidence for melt refertilization beneath the ultraslow-spreading Southwest Indian Ridge (53°E segment). International Geology Review, 57(13):1715-1734

Chen L H, Zeng G, Jiang S Y, Hofmann A W, Xu X S, Pan M B. 2009. Sources of Anfengshan basalts: subducted lower crust in the Sulu UHP belt,

China. Earth and Planetary Science Letters,286(3-4):426-435

Chen L M,Song X Y,Zhu X K,Zhang X Q,Yu S Y,Yi J N. 2014. Iron isotope fractionation during crystallization and sub-solidus re-equilibration: constraints from the Baima mafic layered intrusion,SW China. Chemical Geology,380:97-109

Chen L M,Song X Y,Hu R Z,Yu S Y,He H L,Dai Z H,She Y W,Xie W. 2017. Controls on trace-element partitioning among co-crystallizing minerals: evidence from the Panzhihua layered intrusion,SW China. American Mineralogist,102(5):1006-1020

Chen R X,Li H Y,Zheng Y F,Zhang L,Gong B,Hu Z C,Yang Y H. 2017. Crust-mantle interaction in a continental subduction channel:evidence from orogenic peridotites in North Qaidam,northern Tibet. Journal of Petrology,58(2):191-226

Chen Y,Su B,Chu Z Y. 2017. Modification of an ancient subcontinental lithospheric mantle by continental subduction:insight from the Maowu garnet peridotites in the Dabie UHP belt,eastern China. Lithos,278-281:54-71

Cheng Z G,Zhang Z C,Xie Q H,Hou T,Ke S. 2018. Subducted slab-plume interaction traced by magnesium isotopes in the northern margin of the Tarim large igneous province. Earth and Planetary Science Letters,489:100-110

Choi H O,Choi S H,Schiano P,Cho M,Cluzel N,Devidal J L,Ha K. 2017. Geochemistry of olivine-hosted melt inclusions in the Baekdusan (Changbaishan) basalts:implications for recycling of oceanic crustal materials into the mantle source. Lithos,284-285:194-206

Choi H O,Choi S H,Lee Y S,Ryu J S,Lee D C,Lee S G,Sohn Y K,Liu J Q. 2020. Petrogenesis and mantle source characteristics of the late Cenozoic Baekdusan (Changbaishan) basalts,North China Craton. Gondwana Research,78:156-171

Chu Z Y,Harvey J,Liu C Z,Guo J H,Wu F Y,Tian W,Zhang Y L,Yang Y H. 2013. Source of highly potassic basalts in Northeast China:evidence from Re-Os,Sr-Nd-Hf isotopes and PGE geochemistry. Chemical Geology,357:52-66

Chu Z Y,Yan Y,Zeng G,Tian W,Li C F,Yang Y H,Guo J H. 2017. Petrogenesis of Cenozoic basalts in central-eastern China:constraints from Re-Os and PGE geochemistry. Lithos,278-281:72-83

Dai H K,Zheng J P. 2019. Mantle xenoliths and host basalts record the Paleo-Asian oceanic materials in the mantle wedge beneath northwestern North China Craton. Solid Earth Science,4(4):150-158

Dai H K,Zheng J P,Xiong Q,Su Y P,Pan S K,Ping X Q,Zhou X. 2019. Fertile lithospheric mantle underlying ancient continental crust beneath the northwestern North China Craton:significant effect from the southward subduction of the Paleo-Asian Ocean. GSA Bulletin,131(1-2):3-20

Dai W,Wang Z C,Liu Y S,Chen C F,Zong K Q,Zhou L,Zhang G L,Li M,Moynier F,Hu Z C. 2020. Calcium isotope compositions of mantle pyroxenites. Geochimica et Cosmochimica Acta,270:144-159

Dasgupta R,Hirschmann M M. 2006. Melting in the Earth's deep upper mantle caused by carbon dioxide. Nature,440(7084):659-662

Dasgupta R,Hirschmann M M,Stalker K. 2006. Immiscible transition from carbonate-rich to silicate-rich melts in the 3 GPa melting interval of eclogite + CO_2 and genesis of silica-undersaturated ocean island lavas. Journal of Petrology,47(4):647-671

Dasgupta R,Hirschmann M M,Smith N D. 2007. Partial melting experiments of peridotite + CO_2 at 3 GPa and genesis of alkalic ocean island basalts. Journal of Petrology,48(11):2093-2124

Dauphas N,Craddock P R,Asimow P D,Bennett V C,Nutman A P,Ohnenstetter D. 2009. Iron isotopes may reveal the redox conditions of mantle melting from Archean to Present. Earth and Planetary Science Letters,288(1-2):255-267

Demouchy S,Bolfan-Casanova N. 2016. Distribution and transport of hydrogen in the lithospheric mantle:a review. Lithos,240-243:402-425

Deng L X,Liu Y S,Zong K Q,Zhu L Y,Xu R,Hu Z C,Gao S. 2017. Trace element and Sr isotope records of multi-episode carbonatite metasomatism on the eastern margin of the North China Craton. Geochemistry,Geophysics,Geosystems,18(1):220-237

Deng L X,Geng X L,Liu Y S,Zong K Q,Zhu L Y,Liang Z W,Hu Z C,Zhang G D,Chen G F. 2020. Lithospheric modification by carbonatitic to alkaline melts and deep carbon cycle:insights from peridotite xenoliths of eastern China. Lithos,378-379:105789

Ding X,Ripley E M,Wang W Z,Li C H,Huang F. 2019. Iron isotope fractionation during sulfide liquid segregation and crystallization at the Lengshuiqing Ni-Cu magmatic sulfide deposit,SW China. Geochimica et Cosmochimica Acta,261:327-341

Dong H,Xing C M,Wang C Y. 2013. Textures and mineral compositions of the Xinjie layered intrusion,SW China:implications for the origin of magnetite and fractionation process of Fe-Ti-rich basaltic magmas. Geoscience Frontiers,4(5):503-515

Fan Q C,Chen S S,Zhao Y W,Zou H B,Li N,Sui J L. 2014. Petrogenesis and evolution of Quaternary basaltic rocks from the Wulanhada area,North China. Lithos,206-207:289-302

Fang Q S,Bai W J,Yang J S,Rong H,Shi N C,Li G W,Xiong M,Ma Z S. 2013. Titanium,Ti,a new mineral species from Luobusha,Tibet,China. Acta Geologica Sinica,87(5):1275-1280

Fantle M S,Tipper E T. 2014. Calcium isotopes in the global biogeochemical Ca cycle:implications for development of a Ca isotope proxy. Earth-Science Reviews,129:148-177

Gao C G,Dick H J B,Liu Y,Zhou H Y. 2016. Melt extraction and mantle source at a Southwest Indian Ridge Dragon Bone amagmatic segment on the Marion Rise. Lithos,246-247:48-60

Gao Y J,Snow J E,Casey J F,Yu J B. 2011. Cooling-induced fractionation of mantle Li isotopes from the ultraslow-spreading Gakkel Ridge. Earth and Planetary Science Letters,301(1-2):231-240

Gu X Y,Wang P Y,Kuritani T,Hanski E,Xia Q K,Wang Q Y. 2019. Low water content in the mantle source of the Hainan plume as a factor inhibiting the formation of a large igneous province. Earth and Planetary Science Letters,515:221-230

Guo P Y,Niu Y L,Sun P,Gong H M,Wang X H. 2020. Lithosphere thickness controls continental basalt compositions:an illustration using Cenozoic basalts from eastern China. Geology,48(2):128-133

Hamada M, Ushioda M, Fujii T, Takahashi E. 2013. Hydrogen concentration in plagioclase as a hygrometer of arc basaltic melts: approaches from melt inclusion analyses and hydrous melting experiments. Earth and Planetary Science Letters, 365:253-262

Han G J, Li J, Guo G R, Mooney W D, Karato S I, Yuen D A. 2021. Pervasive low-velocity layer atop the 410-km discontinuity beneath the Northwest Pacific subduction zone: implications for rheology and geodynamics. Earth and Planetary Science Letters, 554:116642

He D T, Liu Y S, Chen C F, Foley S F, Ducea M N. 2020a. Oxidation of the mantle caused by sediment recycling may contribute to the formation of iron-rich mantle melts. Science Bulletin, 65(7):519-521

He D T, Liu Y S, Moynier F, Foley S F, Chen C F. 2020b. Platinum group element mobilization in the mantle enhanced by recycled sedimentary carbonate. Earth and Planetary Science Letters, 541:116262

He Y, Chen L H, Shi J H, Zeng G, Wang X J, Xue X Q, Zhong Y, Erdmann S, Xie L W. 2019. Light Mg isotopic composition in the mantle beyond the big mantle wedge beneath eastern Asia. Journal of Geophysical Research: Solid Earth, 124(8):8043-8056

He Y S, Meng X N, Ke S, Wu H J, Zhu C W, Teng F Z, Hoefs J, Huang J, Yang W, Xu L J, Hou Z H, Ren Z Y, Li S G. 2019. A nephelinitic component with unusual δ^{56}Fe in Cenozoic basalts from eastern China and its implications for deep oxygen cycle. Earth and Planetary Science Letters, 512: 175-183

Hofmann A W. 2017. A store of subducted carbon beneath eastern China. National Science Review, 4(1):2

Hong L B, Zhang Y H, Qian S P, Liu J Q, Ren Z Y, Xu Y G. 2013. Constraints from melt inclusions and their host olivines on the petrogenesis of Oligocene-Early Miocene Xindian basalts, Chifeng area, North China Craton. Contributions to Mineralogy and Petrology, 165(2):305-326

Hong L B, Xu Y G, Zhang L, Wang Y, Ma L. 2020. Recycled carbonate-induced oxidization of the convective mantle beneath Jiaodong, eastern China. Lithos, 366-367:105544

Hou T, Veksler I V. 2015. Experimental confirmation of high-temperature silicate liquid immiscibility in multicomponent ferrobasaltic systems. American Mineralogist, 100(5-6):1304-1307

Hu J, Jiang N, Carlson R W, Guo J H, Fan W B, Huang F, Zhang S Q, Zong K Q, Li T J, Yu H M. 2019. Metasomatism of the crust-mantle boundary by melts derived from subducted sedimentary carbonates and silicates. Geochimica et Cosmochimica Acta, 260:311-328

Hu Q Y, Kim D Y, Yang W G, Yang L X, Meng Y, Zhang L, Mao H K. 2016. FeO_2 and FeOOH under deep lower-mantle conditions and Earth's oxygen-hydrogen cycles. Nature, 534(7606):241-244

Hu Q Y, Kim D Y, Liu J, Meng Y, Yang L X, Zhang D Z, Mao W L, Mao H K. 2017. Dehydrogenation of goethite in Earth's deep lower mantle. Proceedings of the National Academy of Sciences of the United States of America, 114(7):1498-1501

Hu Q Y, Liu J, Chen J H, Yan B M, Meng Y, Prakapenka B V, Mao W L, Mao H K. 2020. Mineralogy of the deep lower mantle in the presence of H_2O. National Science Review, doi:10.1093/nsr/nwaa098

Hu Y, Teng F Z, Zhang H F, Xiao Y, Su B X. 2016. Metasomatism-induced mantle magnesium isotopic heterogeneity: evidence from pyroxenites. Geochimica et Cosmochimica Acta, 185:88-111

Hu Y, Teng F Z, Plank T, Chauvel C. 2020. Potassium isotopic heterogeneity in subducting oceanic plates. Science Advances, 6(49):eabb2472

Huang F, Zhang Z F, Lundstrom C C, Zhi X C. 2011. Iron and magnesium isotopic compositions of peridotite xenoliths from eastern China. Geochimica et Cosmochimica Acta, 75(12):3318-3334

Huang J, Xiao Y L. 2016. Mg-Sr isotopes of low-δ^{26}Mg basalts tracing recycled carbonate species: implication for the initial melting depth of the carbonated mantle in eastern China. International Geology Review, 58(11):1350-1362

Huang J, Li S G, Xiao Y L, Ke S, Li W Y, Tian Y. 2015. Origin of low δ^{26}Mg Cenozoic basalts from South China Block and their geodynamic implications. Geochimica et Cosmochimica Acta, 164:298-317

Huang J, Chen S, Zhang X C, Huang F. 2018a. Effects of melt percolation on Zn isotope heterogeneity in the mantle: constraints from peridotite massifs in Ivrea-Verbano zone, Italian Alps. Journal of Geophysical Research: Solid Earth, 123(4):2706-2722

Huang J, Zhang X C, Chen S, Tang L M, Wörner G, Yu H M, Huang F. 2018b. Zinc isotopic systematics of Kamchatka-Aleutian arc magmas controlled by mantle melting. Geochimica et Cosmochimica Acta, 238:85-101

Huang J L, Zhao D P. 2006. High-resolution mantle tomography of China and surrounding regions. Journal of Geophysical Research: Solid Earth, 111(B9):B09305

Huang X W, Qi L, Meng Y M. 2014. Trace element geochemistry of magnetite from the Fe(-Cu) deposits in the Hami region, eastern Tianshan orogenic belt, NW China. Acta Geologica Sinica, 88(1):176-195

Huang X W, Sappin A A, Boutroy É, Beaudoin G, Makvandi S. 2019. Trace element composition of igneous and hydrothermal magnetite from porphyry deposits: relationship to deposit subtypes and magmatic affinity. Economic Geology, 114(5):917-952

Jin Q Z, Huang J, Liu S C, Huang F. 2020. Magnesium and zinc isotope evidence for recycled sediments and oceanic crust in the mantle sources of continental basalts from eastern China. Lithos, 370-371:105627

Kang J T, Zhu H L, Liu Y F, Liu F, Wu F, Hao Y T, Zhi X C, Zhang Z F, Huang F. 2016. Calcium isotopic composition of mantle xenoliths and minerals from eastern China. Geochimica et Cosmochimica Acta, 174:335-344

Kang J T, Ionov D A, Liu F, Zhang C L, Golovin A V, Qin L P, Zhang Z F, Huang F. 2017. Calcium isotopic fractionation in mantle peridotites by melting and metasomatism and Ca isotope composition of the Bulk Silicate Earth. Earth and Planetary Science Letters, 474:128-137

Kang J T, Ionov D A, Zhu H L, Liu F, Zhang Z F, Liu Z, Huang F. 2019. Calcium isotope sources and fractionation during melt-rock interaction in the lithospheric mantle: evidence from pyroxenites, wehrlites, and eclogites. Chemical Geology, 524:272-282

Kang J T, Zhou C, Huang J Y, Hao Y T, Liu F, Zhu H L, Zhang Z F, Huang F. 2020. Diffusion-driven Ca-Fe isotope fractionations in the upper mantle: implications for mantle cooling and melt infiltration. Geochimica et Cosmochimica Acta, 290: 41-58

Kuritani T, Ohtani E, Kimura J I. 2011. Intensive hydration of the mantle transition zone beneath China caused by ancient slab stagnation. Nature Geoscience, 4(10): 713-716

Kuritani T, Kimura J I, Ohtani E, Miyamoto H, Furuyama K. 2013. Transition zone origin of potassic basalts from Wudalianchi volcano, Northeast China. Lithos, 156-159: 1-12

Lei M, Guo Z F, Sun Y T, Zhang M L, Zhang L H, Ma L. 2020. Geochemical constraints on the origin of late Cenozoic basalts in the Mt. Changbai volcanic field, NE China: evidence for crustal recycling. International Geology Review, 62(17): 2125-2145

Li C F, Xu X, Lin J, et al. 2014. Ages and magnetic structures of the South China Sea constrained by deep tow magnetic surveys and IODP Expedition 349. Geochemistry, Geophysics, Geosystems, 15(12): 4958-4983

Li H Y, Xu Y G, Ryan J G, Huang X L, Ren Z Y, Guo H, Ning Z G. 2016. Olivine and melt inclusion chemical constraints on the source of intracontinental basalts from the eastern North China Craton: discrimination of contributions from the subducted Pacific slab. Geochimica et Cosmochimica Acta, 178: 1-19

Li H Y, Xu Y G, Ryan J G, Whattam S A. 2017. Evolution of the mantle beneath the eastern North China Craton during the Cenozoic: linking geochemical and geophysical observations. Journal of Geophysical Research: Solid Earth, 122(1): 224-246

Li H Y, Chen R X, Zheng Y F, Hu Z C, Xu L J. 2018. Crustal metasomatism at the slab-mantle interface in a continental subduction channel: geochemical evidence from orogenic peridotite in the Sulu orogen. Journal of Geophysical Research: Solid Earth, 123(3): 2174-2198

Li P, Xia Q K, Dallai L, Bonatti E, Brunelli D, Cipriani A, Ligi M. 2020. High H_2O content in pyroxenes of residual mantle peridotites at a mid Atlantic Ridge segment. Scientific Reports, 10(1): 579

Li S G, Wang Y. 2018. Formation time of the big mantle wedge beneath eastern China and a new lithospheric thinning mechanism of the North China Craton—geodynamic effects of deep recycled carbon. Science China Earth Sciences, 61(7): 853-868

Li S G, Yang W, Ke S, Meng X A, Tian H C, Xu L J, He Y S, Huang J, Wang X C, Xia Q K, Sun W D, Yang X Y, Ren Z Y, Wei H Q, Liu Y S, Meng F C, Yan J. 2017. Deep carbon cycles constrained by a large-scale mantle Mg isotope anomaly in eastern China. National Science Review, 4(1): 111-120

Li W, Soustelle V, Jin Z M, Li H M, Chen T, Tao C H. 2017. Origins of water content variations in the suboceanic upper mantle: insight from Southwest Indian Ridge abyssal peridotites. Geochemistry, Geophysics, Geosystems, 18(3): 1298-1329

Li W, Liu C Z, Tao C H, Jin Z M. 2019. Osmium isotope compositions and highly siderophile element abundances in abyssal peridotites from the Southwest Indian Ridge: implications for evolution of the oceanic upper mantle. Lithos, 346-347: 105167

Li W Y, Yu H M, Xu J, Halama R, Bell K, Nan X Y, Huang F. 2020. Barium isotopic composition of the mantle: constraints from carbonatites. Geochimica et Cosmochimica Acta, 278: 235-243

Li X Y, Zheng J P, Sun M, Pan S K, Wang W, Xia Q K. 2014. The Cenozoic lithospheric mantle beneath the interior of South China Block: constraints from mantle xenoliths in Guangxi Province. Lithos, 210-211: 14-26

Li Y Q, Ma C Q, Robinson P T, Zhou Q, Liu M L. 2015. Recycling of oceanic crust from a stagnant slab in the mantle transition zone: evidence from Cenozoic continental basalts in Zhejiang Province, SE China. Lithos, 230: 146-165

Li Y Q, Kitagawa H, Nakamura E, Ma C Q, Hu X Y, Kobayashi K, Sakaguchi C. 2020. Various ages of recycled material in the source of cenozoic basalts in SE China: implications for the role of the Hainan Plume. Journal of Petrology, 61(6): egaa060

Liang Y Y, Deng J, Liu X F, Wang Q F, Ma Y, Gao T X, Zhao L H. 2019. Water contents of Early Cretaceous mafic dikes in the Jiaodong Peninsula, eastern North China Craton: insights into an enriched lithospheric mantle source metasomatized by paleo-Pacific Plate subduction-related fluids. The Journal of Geology, 127(3): 343-362

Lin A B, Zheng J P, Xiong Q, Aulbach S, Lu J G, Pan S K, Dai H K, Zhang H. 2019. A refined model for lithosphere evolution beneath the decratonized northeastern North China Craton. Contributions to Mineralogy and Petrology, 174(2): 15

Lin A B, Zheng J P, Aulbach S, Xiong Q, Pan S K, Gerdes A. 2020. Causes and consequences of wehrlitization beneath a trans-lithospheric fault: evidence from Mesozoic basalt-borne wehrlite xenoliths from the Tan-Lu fault belt, North China Craton. Journal of Geophysical Research: Solid Earth, 125(7): e2019JB019084

Liu C Z, Snow J E, Hellebrand E, Brugmann G, von der Handt A, Büchl A, Hofmann A W. 2008. Ancient, highly heterogeneous mantle beneath Gakkel ridge, Arctic Ocean. Nature, 452(7185): 311-316

Liu C Z, Liu Z C, Wu F Y, Chu Z Y. 2012a. Mesozoic accretion of juvenile sub-continental lithospheric mantle beneath South China and its implications: geochemical and Re-Os isotopic results from Ningyuan mantle xenoliths. Chemical Geology, 291: 186-198

Liu C Z, Wu F Y, Sun J, Chu Z Y, Qiu Z L. 2012b. The Xinchang peridotite xenoliths reveal mantle replacement and accretion in southeastern China. Lithos, 150: 171-187

Liu C Z, Wu F Y, Sun J, Chu Z Y, Yu X H. 2013. Petrology, geochemistry and Re-Os isotopes of peridotite xenoliths from Maguan, Yunnan Province: implications for the Cenozoic mantle replacement in southwestern China. Lithos, 168-169: 1-14

Liu C Z, Zhang C, Liu Z C, Sun J, Chu Z Y, Qiu Z L, Wu F Y. 2017. Formation age and metasomatism of the sub-continental lithospheric mantle beneath Southeast China: Sr-Nd-Hf-Os isotopes of Mingxi mantle xenoliths. Journal of Asian Earth Sciences, 145: 591-604

Liu F, Li X, Wang G Q, Liu Y F, Zhu H L, Kang J T, Huang F, Sun W D, Xia X P, Zhang Z F. 2017. Marine carbonate component in the mantle beneath the southeastern Tibetan Plateau: evidence from magnesium and calcium isotopes. Journal of Geophysical Research: Solid Earth, 122(12): 9729-9744

Liu H, Zhang H F. 2019. Paleoproterozoic ophiolite remnants in the northern margin of the North China Craton: evidence from the Chicheng peridotite massif. Lithos, 344-345:311-323

Liu H, Zhang H F, Santosh M. 2019. Neoarchean growth and Paleoproterozoic metamorphism of an Archean ophiolite mélange in the North China Craton. Precambrian Research, 331:105377

Liu J, Xia Q K, Deloule E, Ingrin J, Chen H, Feng M. 2015a. Water content and oxygen isotopic composition of alkali basalts from the Taihang Mountains, China: recycled oceanic components in the mantle source. Journal of Petrology, 56(4):681-702

Liu J, Xia Q K, Deloule E, Chen H, Feng M. 2015b. Recycled oceanic crust and marine sediment in the source of alkali basalts in Shandong, eastern China: evidence from magma water content and oxygen isotopes. Journal of Geophysical Research: Solid Earth, 120(12):8281-8303

Liu J, Xia Q K, Kuritani T, Hanski E, Yu H R. 2017a. Mantle hydration and the role of water in the generation of large igneous provinces. Nature Communications, 8(1):1824

Liu J, Hu Q Y, Kim D Y, Wu Z Q, Wang W Z, Xiao Y M, Chow P, Meng Y, Prakapenka V B, Mao H K, Mao W L. 2017b. Hydrogen-bearing iron peroxide and the origin of ultralow-velocity zones. Nature, 551(7681):494-497

Liu J, Hu Q Y, Bi W L, Yang L X, Xiao Y M, Chow P, Meng Y, Prakapenka V B, Mao H K, Mao W L. 2019. Altered chemistry of oxygen and iron under deep Earth conditions. Nature Communications, 10(1):153

Liu J, Wang C X, Lv C J, Su X W, Liu Y J, Tang R L, Chen J H, Hu Q Y, Mao H K, Mao W L. 2020. Evidence for oxygenation of Fe-Mg oxides at mid-mantle conditions and the rise of deep oxygen. National Science Review, doi:10.1093/nsr/nwaa096

Liu J G, Rudnick R L, Walker R J, Gao S, Wu F Y, Piccoli P M, Yuan H L, Xu W L, Xu Y G. 2011. Mapping lithospheric boundaries using Os isotopes of mantle xenoliths: an example from the North China Craton. Geochimica et Cosmochimica Acta, 75(13):3881-3902

Liu J G, Carlson R W, Rudnick R L, Walker R J, Gao S, Wu F Y. 2012. Comparative Sr-Nd-Hf-Os-Pb isotope systematics of xenolithic peridotites from Yangyuan, North China Craton: additional evidence for a Paleoproterozoic age. Chemical Geology, 332-333:1-14

Liu J G, Rudnick R L, Walker R J, Xu W L, Gao S, Wu F Y. 2015a. Big insights from tiny peridotites: evidence for persistence of Precambrian lithosphere beneath the eastern North China Craton. Tectonophysics, 650:104-112

Liu J G, Scott J M, Martin C E, Pearson D G. 2015b. The longevity of Archean mantle residues in the convecting upper mantle and their role in young continent formation. Earth and Planetary Science Letters, 424:109-118

Liu J G, Riches A J V, Pearson D G, Luo Y, Kienlen B, Kjarsgaard B A, Stachel T, Armstrong J P. 2016. Age and evolution of the deep continental root beneath the central Rae Craton, northern Canada. Precambrian Research, 272:168-184

Liu J G, Brin L E, Pearson D G, Bretschneider L, Luguet A, van Acken D, Kjarsgaard B, Riches A, Mišković A. 2018. Diamondiferous Paleoproterozoic mantle roots beneath Arctic Canada: a study of mantle xenoliths from Parry Peninsula and central Victoria Island. Geochimica et Cosmochimica Acta, 239:284-311

Liu J G, Pearson D G, Shu Q, Sigurdsson H, Thomassot E, Alard O. 2020. Dating post-Archean lithospheric mantle: insights from Re-Os and Lu-Hf isotopic systematics of the Cameroon volcanic line peridotites. Geochimica et Cosmochimica Acta, 278:177-198

Liu J Q, Ren Z Y, Nichols A R L, Song M S, Qian S P, Zhang Y, Zhao P P. 2015. Petrogenesis of Late Cenozoic basalts from North Hainan Island: constraints from melt inclusions and their host olivines. Geochimica et Cosmochimica Acta, 152:89-121

Liu J Q, Chen L H, Zeng G, Wang X J, Zhong Y, Yu X. 2016. Lithospheric thickness controlled compositional variations in potassic basalts of Northeast China by melt-rock interactions. Geophysical Research Letters, 43(6):2582-2589

Liu J Q, Chen L H, Wang X J, Zhong Y, Yu X, Zeng G, Erdmann S. 2017. The role of melt-rock interaction in the formation of Quaternary high-MgO potassic basalt from the Greater Khingan Range, Northeast China. Journal of Geophysical Research: Solid Earth, 122(1):262-280

Liu P P, Zhou M F, Chen W T, Boone M, Cnudde V. 2014a. Using multiphase solid inclusions to constrain the origin of the Baima Fe-Ti-(V) oxide deposit, SW China. Journal of Petrology, 55(5):951-976

Liu P P, Zhou M F, Luais B, Cividini D, Rollion-Bard C. 2014b. Disequilibrium iron isotopic fractionation during the high-temperature magmatic differentiation of the Baima Fe-Ti oxide-bearing mafic intrusion, SW China. Earth and Planetary Science Letters, 399:21-29

Liu P P, Liang J, Dick H J B, Li X H, Chen Q, Zuo H Y, Wu J C. 2020. Enormous lithium isotopic variations of abyssal peridotites reveal fast cooling and melt/fluid-rock interactions. Journal of Geophysical Research: Solid Earth, 125(9):e2020JB020393

Liu S A, Li S G. 2019. Tracing the deep carbon cycle using metal stable isotopes: opportunities and challenges. Engineering, 5(3):448-457

Liu S A, Teng F Z, Yang W, Wu F Y. 2011. High-temperature inter-mineral magnesium isotope fractionation in mantle xenoliths from the North China Craton. Earth and Planetary Science Letters, 308(1-2):131-140

Liu S A, Huang J, Liu J G, Wörner G, Yang W, Tang Y J, Chen Y, Tang L M, Zheng J P, Li S G. 2015. Copper isotopic composition of the silicate Earth. Earth and Planetary Science Letters, 427:95-103

Liu S A, Wang Z Z, Li S G, Huang J, Yang W. 2016. Zinc isotope evidence for a large-scale carbonated mantle beneath eastern China. Earth and Planetary Science Letters, 444:169-178

Liu S C, Xia Q K, Choi S H, Deloule E, Li P, Liu J. 2016. Continuous supply of recycled Pacific oceanic materials in the source of Cenozoic basalts in SE China: the Zhejiang case. Contributions to Mineralogy and Petrology, 171(12):100

Liu Y S, Gao S, Kelemen P B, Xu W L. 2008. Recycled crust controls contrasting source compositions of Mesozoic and Cenozoic basalts in the North China Craton. Geochimica et Cosmochimica Acta, 72(9):2349-2376

Luguet A, Pearson D G. 2019. Dating mantle peridotites using Re-Os isotopes: the complex message from whole rocks, base metal sulfides, and platinum group minerals. American Mineralogist,104(2):165-189

Malaspina N,Langenhorst F,Tumiati S,Campione M,Frezzotti M L,Poli S. 2017. The redox budget of crust-derived fluid phases at the slab-mantle interface. Geochimica et Cosmochimica Acta,209:70-84

Mao H K,Hu Q Y,Yang L X,Liu J,Kim D Y,Meng Y,Zhang L,Prakapenka V B,Yang W G,Mao W L. 2017. When water meets iron at Earth's core-mantle boundary. National Science Review,4(6):870-878

Mao Y J,Barnes S J,Duan J,Qin K Z,Godel B M,Jiao J G. 2018. Morphology and particle size distribution of olivines and sulphides in the Jinchuan Ni-Cu sulphide deposit:evidence for sulphide percolation in a crystal mush. Journal of Petrology,59(9):1701-1730

Mao Y J,Barnes S J,Qin K Z,Tang D M,Martin L,Su B X,Evans N J. 2019. Rapid orthopyroxene growth induced by silica assimilation:constraints from sector-zoned orthopyroxene,olivine oxygen isotopes and trace element variations in the Huangshanxi Ni-Cu deposit,Northwest China. Contributions to Mineralogy and Petrology,174(4):33

Ni H W,Zheng Y F,Mao Z,Wang Q,Chen R X,Zhang L. 2017. Distribution,cycling and impact of water in the Earth's interior. National Science Review,4(6):879-891

Pan S K,Zheng J P,Chu L L,Griffin W L. 2013. Coexistence of the moderately refractory and fertile mantle beneath the eastern Central Asian orogenic belt. Gondwana Research,23(1):176-189

Pan S K,Zheng J P,Griffin W L,Xu Y X,Li X Y. 2015. Nature and evolution of the lithospheric mantle beneath the eastern Central Asian orogenic belt:constraints from peridotite xenoliths in the central part of the Great Xing'an Range,NE China. Lithos,238:52-63

Pang C J,Wang X C,Li C F,Wilde S A,Tian L Y. 2019. Pyroxenite-derived Cenozoic basaltic magmatism in central Inner Mongolia,eastern China:potential contributions from the subduction of the Paleo-Pacific and Paleo-Asian oceanic slabs in the mantle transition zone. Lithos,332-333:39-54

Peslier A H,Woodland A B,Bell D R,Lazarov M. 2010. Olivine water contents in the continental lithosphere and the longevity of cratons. Nature,467(7311):78-81

Peslier A H,Schönbächler M,Busemann H,Karato S I. 2017. Water in the Earth's interior:distribution and origin. Space Science Reviews,212(1-2):743-810

Philpotts A R,Doyle C D. 1983. Effect of magma oxidation state on the extent of silicate liquid immiscibility in a tholeiitic basalt. American Journal of Science,283(9):967-986

Pichat S,Douchet C,Albarède F. 2003. Zinc isotope variations in deep-sea carbonates from the eastern equatorial Pacific over the last 175 ka. Earth and Planetary Science Letters,210(1-2):167-178

Qi Y H,Wu F,Ionov D A,Puchtel I S,Carlson R W,Nicklas R W,Yu H M,Kang J T,Li C H,Huang F. 2019. Vanadium isotope composition of the Bulk Silicate Earth:constraints from peridotites and komatiites. Geochimica et Cosmochimica Acta,259:288-301

Qian S P,Ren Z Y,Zhang L,Hong L B,Liu J Q. 2015. Chemical and Pb isotope composition of olivine-hosted melt inclusions from the Hannuoba basalts,North China Craton:implications for petrogenesis and mantle source. Chemical Geology,401:111-125

Qian S P,Zhou H Y,Zhang L,Cheng R. 2020. Mantle heterogeneity beneath the South China Sea:chemical and isotopic evidence for contamination of ambient asthenospheric mantle. Lithos,354-355:105335

Rampone E,Morten L. 2001. Records of crustal metasomatism in the garnet peridotites of the Ulten zone (Upper Austroalpine,eastern Alps). Journal of Petrology,42(1):207-219

Rudnick R L,Walker R J. 2009. Interpreting ages from Re-Os isotopes in peridotites. Lithos,112(S2):1083-1095

Sakuyama T,Tian W,Kimura J I,Fukao Y,Hirahara Y,Takahashi T,Senda R,Chang Q,Miyazaki T,Obayashi M,Kawabata H,Tatsumi Y. 2013. Melting of dehydrated oceanic crust from the stagnant slab and of the hydrated mantle transition zone:constraints from Cenozoic alkaline basalts in eastern China. Chemical Geology,359:32-48

Scambelluri M,Pettke T,van Roermund H L M. 2008. Majoritic garnets monitor deep subduction fluid flow and mantle dynamics. Geology,36(1):59-62

Shen J,Qin L P,Fang Z Y,Zhang Y N,Liu J,Liu W,Wang F Y,Xiao Y,Yu H M,Wei S Q. 2018. High-temperature inter-mineral Cr isotope fractionation:a comparison of ionic model predictions and experimental investigations of mantle xenoliths from the North China Craton. Earth and Planetary Science Letters,499:278-290

Shi R D,Griffin W L,O'Reilly S Y,Zhao G C,Huang Q S,Li J,Xu J F. 2010. Evolution of the Lüliangshan garnet peridotites in the North Qaidam UHP belt,northern Tibetan Plateau:constraints from Re-Os isotopes. Lithos,117(1-4):307-321

Shu Q,Brey G P,Gerdes A,Hoefer H E. 2014. Mantle eclogites and garnet pyroxenites – the meaning of two-point isochrons,Sm-Nd and Lu-Hf closure temperatures and the cooling of the subcratonic mantle. Earth and Planetary Science Letters,389:143-154

Shu Q,Brey G P,Pearson D G,Liu J G,Gibson S A,Becker H. 2019. The evolution of the Kaapvaal Craton:a multi-isotopic perspective from lithospheric peridotites from Finsch diamond mine. Precambrian Research,331:105380

Sobolev A V,Hofmann A W,Kuzmin D V,Yaxley G M,Arndt N T,Chung S L,Danyushevsky L V,Elliott T,Frey F A,Garcia M O,Gurenko A A,Kamenetsky V S,Kerr A C,Krivolutskaya N A,Matvienkov V V,Nikogosian I K,Rocholl A,Sigurdsson I A,Sushchevskaya N M,Teklay M. 2007. The amount of recycled crust in sources of mantle-derived melts. Science,316(5823):412-417

Song S G,Su L,Niu Y L,Zhang L F,Zhang G B. 2007. Petrological and geochemical constraints on the origin of garnet peridotite in the North Qaidam ultrahigh-pressure metamorphic belt,northwestern China. Lithos,96(1-2):243-265

Song W L,Xu C,Smith M P,Kynicky J,Huang K J,Wei C W,Zhou L,Shu Q H. 2016. Origin of unusual HREE-Mo-rich carbonatites in the Qinling orogen, China. Scientific Reports, 6(1):37377

Song X Y, Wang K Y, Barnes S J, Yi J N, Chen L M, Schoneveld L E. 2020. Petrogenetic insights from chromite in ultramafic cumulates of the Xiarihamu intrusion, northern Tibet Plateau, China. American Mineralogist, 105(4):479–497

Spengler D, van Roermund H L M, Drury M R, Ottolini L, Mason P R D, Davies G R. 2006. Deep origin and hot melting of an Archaean orogenic peridotite massif in Norway. Nature, 440(7086):913–917

Su B, Chen Y, Guo S, Liu J B. 2017. Dolomite dissociation indicates ultra-deep (>150 km) subduction of a garnet-bearing dunite block (the Sulu UHP terrane). American Mineralogist, 102(11):2295–2306

Su B X, Qin K Z, Sakyi P A, Li X H, Yang Y H, Sun H, Tang D M, Liu P P, Xiao Q H, Malaviarachchi S P K. 2011. U-Pb ages and Hf-O isotopes of zircons from Late Paleozoic mafic-ultramafic units in the southern Central Asian orogenic belt: tectonic implications and evidence for an Early-Permian mantle plume. Gondwana Research, 20(2-3):516–531

Su B X, Zhang H F, Deloule E, Sakyi P A, Xiao Y, Tang Y J, Hu Y, Ying J F, Liu P P. 2012. Extremely high Li and low δ^7Li signatures in the lithospheric mantle. Chemical Geology, 292-293:149–157

Su B X, Zhang H F, Deloule E, Vigier N, Hu Y, Tang Y J, Xiao Y, Sakyi P A. 2014a. Distinguishing silicate and carbonatite mantle metasomatism by using lithium and its isotopes. Chemical Geology, 381:67–77

Su B X, Zhang H F, Deloule E, Vigier N, Sakyi P A. 2014b. Lithium elemental and isotopic variations in rock-melt interaction. Geochemistry, 74(4):705–713

Su B X, Chen C, Bai Y, Pang K N, Qin K Z, Sakyi P A. 2017a. Lithium isotopic composition of Alaskan-type intrusion and its implication. Lithos, 286-287:363–368

Su B X, Hu Y, Teng F Z, Xiao Y, Zhou X H, Sun Y, Zhou M F, Chang S C. 2017b. Magnesium isotope constraints on subduction contribution to Mesozoic and Cenozoic East Asian continental basalts. Chemical Geology, 466:116–122

Su B X, Zhou X H, Sun Y, Ying J F, Sakyi P A. 2018. Carbonatite-metasomatism signatures hidden in silicate-metasomatized mantle xenoliths from NE China. Geological Journal, 53(2):682–691

Su B X, Hu Y, Teng F Z, Xiao Y, Zhang H F, Sun Y, Bai Y, Zhu B, Zhou X H, Ying J F. 2019. Light Mg isotopes in mantle-derived lavas caused by chromite crystallization, instead of carbonatite metasomatism. Earth and Planetary Science Letters, 522:79–86

Su B X, Bai Y, Cui M M, Wang J, Xiao Y, Lenaz D, Sakyi P A, Robinson P T. 2020. Petrogenesis of the ultramafic zone of the stillwater complex in North America: constraints from mineral chemistry and stable isotopes of Li and O. Contributions to Mineralogy and Petrology, 175(7):68

Sun J, Liu C Z, Wu F Y, Yang Y H, Chu Z Y. 2012. Metasomatic origin of clinopyroxene in Archean mantle xenoliths from Hebi, North China Craton: trace-element and Sr-isotope constraints. Chemical Geology, 328:123–136

Sun K, Wu T, Liu X S, Chen X G, Li C F. 2020. Lithogeochemistry of the mid-ocean ridge basalts near the fossil ridge of the southwest sub-basin, South China Sea. Minerals, 10(5):465

Sun P, Niu Y L, Guo P Y, Duan M, Wang X H, Gong H M, Xiao Y Y. 2020. The lithospheric thickness control on the compositional variation of continental intraplate basalts: a demonstration using the cenozoic basalts and clinopyroxene megacrysts from eastern China. Journal of Geophysical Research: Solid Earth, 125(3):e2019jb019315

Sun Y, Ying J F, Zhou X H, Shao J A, Chu Z Y, Su B X. 2014. Geochemistry of ultrapotassic volcanic rocks in Xiaogulihe NE China: implications for the role of ancient subducted sediments. Lithos, 208-209:53–66

Sun Y, Teng F Z, Ying J F, Su B X, Hu Y, Fan Q C, Zhou X H. 2017. Magnesium isotopic evidence for ancient subducted oceanic crust in LOMU-like potassium-rich volcanic rocks. Journal of Geophysical Research: Solid Earth, 122(10):7562–7572

Tang D M, Qin K Z, Su B X, Mao Y J, Evans N J, Niu Y J, Kang Z. 2020. Sulfur and copper isotopic signatures of chalcopyrite at Kalatongke and Baishiquan: insights into the origin of magmatic Ni-Cu sulfide deposits. Geochimica et Cosmochimica Acta, 275:209–228

Tang Q Y, Ma Y S, Zhang M J, Li C, Zhu D, Tao Y. 2013. The origin of Ni-Cu-PGE sulfide mineralization in the margin of the Zhubu mafic-ultramafic intrusion in the Emeishan large igneous province, southwestern China. Economic Geology, 108(8):1889–1901

Tang Q Y, Li C, Tao Y, Ripley E M, Xiong F. 2017a. Association of Mg-rich olivine with magnetite as a result of brucite marble assimilation by basaltic magma in the Emeishan large igneous province, SW China. Journal of Petrology, 58(4):699–714

Tang Q Y, Zhang M J, Wang Y K, Yao Y S, Du L, Chen L M, Li Z P. 2017b. The origin of the Zhubu mafic-ultramafic intrusion of the Emeishan large igneous province, SW China: insights from volatile compositions and C-Hf-Sr-Nd isotopes. Chemical Geology, 469:47–59

Tang Q Y, Bao J, Dang Y X, Ke S, Zhao Y. 2018. Mg-Sr-Nd isotopic constraints on the genesis of the giant Jinchuan Ni-Cu-(PGE) sulfide deposit, NW China. Earth and Planetary Science Letters, 502:221–230

Tang Q Y, Li C, Ripley E M, Bao J, Su T B, Xu S H. 2021. Sr-Nd-Hf-O isotope constraints on crustal contamination and mantle source variation of three Fe-Ti-V oxide ore deposits in the Emeishan large igneous province. Geochimica et Cosmochimica Acta, 292:364–381

Tang Y J, Zhang H F, Nakamura E, Moriguti T, Kobayashi K, Ying J F. 2007. Lithium isotopic systematics of peridotite xenoliths from Hannuoba, North China Craton: implications for melt-rock interaction in the considerably thinned lithospheric mantle. Geochimica et Cosmochimica Acta, 71(17):4327–4341

Tang Y J, Zhang H F, Ying J F. 2010. A brief review of isotopically light Li—a feature of the enriched mantle. International Geology Review, 52(9):964–976

Tang Y J, Zhang H F, Nakamura E, Ying J F. 2011. Multistage melt/fluid-peridotite interactions in the refertilized lithospheric mantle beneath the North

China Craton: constraints from the Li-Sr-Nd isotopic disequilibrium between minerals of peridotite xenoliths. Contributions to Mineralogy and Petrology, 161(6): 845–861

Tang Y J, Zhang H F, Deloule E, Su B X, Ying J F, Xiao Y, Hu Y. 2012. Slab-derived lithium isotopic signatures in mantle xenoliths from northeastern North China Craton. Lithos, 149: 79–90

Tang Y J, Zhang H F, Santosh M, Ying J F. 2013a. Differential destruction of the North China Craton: a tectonic perspective. Journal of Asian Earth Sciences, 78: 71–82

Tang Y J, Zhang H F, Ying J F, Su B X. 2013b. Widespread refertilization of cratonic and circum-cratonic lithospheric mantle. Earth-Science Reviews, 118: 45–68

Tang Y J, Zhang H F, Ying J F, Su B X, Chu Z Y, Xiao Y, Zhao X M. 2013c. Highly heterogeneous lithospheric mantle beneath the central zone of the North China Craton evolved from Archean mantle through diverse melt refertilization. Gondwana Research, 23(1): 130–140

Tang Y J, Zhang H F, Deloule E, Su B X, Ying J F, Santosh M, Xiao Y. 2014. Abnormal lithium isotope composition from the ancient lithospheric mantle beneath the North China Craton. Scientific Reports, 4(1): 4274

Teng F Z. 2017. Magnesium isotope geochemistry. Reviews in Mineralogy and Geochemistry, 82(1): 219–287

Teng F Z, Li W Y, Ke S, Marty B, Dauphas N, Huang S, Wu F Y, Pourmand A. 2010. Magnesium isotopic composition of the Earth and chondrites. Geochimica et Cosmochimica Acta, 74(14): 4150–4166

Thomson A R, Walter M J, Kohn S C, Brooker R A. 2016. Slab melting as a barrier to deep carbon subduction. Nature, 529(7584): 76–79

Tian G C, Liu J G, Scott J M, Chen L H, Pearson D G, Chu Z Y, Wang Z C, Luo Y. 2020. Architecture and evolution of the lithospheric roots beneath circum-cratonic orogenic belts-the Xing'an Mongolia orogenic belt and its relationship with adjacent North China and Siberian cratonic roots. Lithos. 376-377: 105798

Tian H C, Yang W, Li S G, Ke S, Chu Z Y. 2016. Origin of low $\delta^{26}Mg$ basalts with EM-I component: evidence for interaction between enriched lithosphere and carbonated asthenosphere. Geochimica et Cosmochimica Acta, 188: 93–105

Tian H C, Yang W, Li S G, Ke S. 2017. Could sedimentary carbonates be recycled into the lower mantle: constraints from Mg isotopic composition of Emeishan basalts. Lithos, 292-293: 250–261

Tian H C, Yang W, Li S G, Ke S, Duan X Z. 2018. Low $\delta^{26}Mg$ volcanic rocks of Tengchong in southwestern China: a deep carbon cycle induced by supercritical liquids. Geochimica et Cosmochimica Acta, 240: 191–219

Tian H C, Zhang C, Teng F Z, Long Y J, Li S G, He Y S, Ke S, Chen X Y, Yang W. 2020. Diffusion-driven extreme Mg and Fe isotope fractionation in Panzhihua ilmenite: implications for the origin of mafic intrusion. Geochimica et Cosmochimica Acta, 278: 361–375

Urann B M, Dick H J B, Parnell-Turner R, Casey J F. 2020. Recycled arc mantle recovered from the Mid-Atlantic Ridge. Nature Communications, 11(1): 3887

van Roermund H L M, Carswell D A, Drury M R, Heijboer T C. 2002. Microdiamonds in a megacrystic garnet websterite pod from Bardane on the island of Fjørtoft, western Norway: evidence for diamond formation in mantle rocks during deep continental subduction. Geology, 30(11): 959–962

Vrijmoed J C, Austrheim H, John T, Hin R C, Corfu F, Davies G R. 2013. Metasomatism in the ultrahigh-pressure Svartberget garnet-peridotite (western Gneiss region, Norway): implications for the transport of crust-derived fluids within the mantle. Journal of Petrology, 54(9): 1815–1848

Wang C Y, Liu Y S, Min N, Zong K Q, Hu Z C, Gao S. 2016. Paleo-Asian oceanic subduction-related modification of the lithospheric mantle under the North China Craton: evidence from peridotite xenoliths in the Datong basalts. Lithos, 261: 109–127

Wang C Y, Liu Y S, Foley S F, Zong K Q, Hu Z C. 2019. Lithospheric transformation of the northern North China Craton by changing subduction style of the Paleo-Asian oceanic plate: constraints from peridotite and pyroxenite xenoliths in the Yangyuan basalts. Lithos, 328-329: 58–68

Wang J X, Zhou H Y, Salters V, Liu Y, Sachi-Kocher A, Dick H. 2019. Mantle melting variation and refertilization beneath the Dragon Bone amagmatic segment (53°E SWIR): major and trace element compositions of peridotites at ridge flanks. Lithos, 324-325: 325–339

Wang K Y, Song X Y, Yi J N, Barnes S J, She Y W, Zheng W Q, Schoneveld L E. 2019. Zoned orthopyroxenes in the Ni-Co sulfide ore-bearing Xiarihamu mafic-ultramafic intrusion in northern Tibetan Plateau, China: implications for multiple magma replenishments. Ore Geology Reviews, 113: 103082

Wang W, Chu F Y, Wu X C, Li Z G, Chen L, Li X H, Yan Y Z, Zhang J. 2019. Constraining mantle heterogeneity beneath the South China Sea: a new perspective on magma water content. Minerals, 9(7): 410

Wang X C, Li Z X, Li X H, Li J, Liu Y, Long W G, Zhou J B, Wang F. 2012. Temperature, pressure, and composition of the mantle source region of Late Cenozoic basalts in Hainan Island, SE Asia: a consequence of a young thermal mantle plume close to subduction zones? Journal of Petrology, 53(1): 177–233

Wang X C, Wilde S A, Li Q L, Yang Y N. 2015. Continental flood basalts derived from the hydrous mantle transition zone. Nature Communications, 6(1): 7700

Wang X J, Chen L H, Hofmann A W, Mao F G, Liu J Q, Zhong Y, Xie L W, Yang Y H. 2017. Mantle transition zone-derived EM1 component beneath NE China: geochemical evidence from Cenozoic potassic basalts. Earth and Planetary Science Letters, 465: 16–28

Wang Y, Zhao Z F, Zheng Y F, Zhang J J. 2011. Geochemical constraints on the nature of mantle source for Cenozoic continental basalts in east-central China. Lithos, 125(3-4): 940–955

Wang Z C, Cheng H, Zong K Q, Geng X L, Liu Y S, Yang J H, Wu F Y, Becker H, Foley S, Wang C Y. 2020. Metasomatized lithospheric mantle for Mesozoic giant gold deposits in the North China Craton. Geology, 48(2): 169–173

Wang Z Z, Liu S A, Ke S, Liu Y C, Li S G. 2016. Magnesium isotopic heterogeneity across the cratonic lithosphere in eastern China and its origins. Earth and Planetary Science Letters,451:77-88

Wang Z Z,Liu S A,Liu J G,Huang J,Xiao Y,Chu Z Y,Zhao X M,Tang L M. 2017. Zinc isotope fractionation during mantle melting and constraints on the Zn isotope composition of Earth's upper mantle. Geochimica et Cosmochimica Acta,198:151-167

Wang Z Z,Liu S A,Chen L H,Li S G,Zeng G. 2018. Compositional transition in natural alkaline lavas through silica-undersaturated melt-lithosphere interaction. Geology,46(9):771-774

Wang Z Z,Liu J,Xia Q K,Hao Y T,Wang Q Y. 2020. The distribution of water in the early Cretaceous lithospheric mantle of the North China Craton and implications for its destruction. Lithos,360-361:105412

Wei B,Wang C Y,Lahaye Y,Xie L H,Cao Y H. 2019. S and C isotope constraints for mantle-derived sulfur source and organic carbon-induced sulfide saturation of magmatic Ni-Cu sulfide deposits in the Central Asian orogenic belt,North China. Economic Geology,144(4):787-806

Wei X,Zhang G L,Castillo P R,Shi X F,Yan Q S,Guan Y L. 2020. New geochemical and Sr-Nd-Pb isotope evidence for FOZO and Azores plume components in the sources of DSDP Holes 559 and 561 MORBs. Chemical Geology,557:119858

Weiss Y,Class C,Goldstein S L,Hanyu T. 2016. Key new pieces of the HIMU puzzle from olivines and diamond inclusions. Nature,537(7622):666-670

Wood B J,Bryndzia L T,Johnson K E. 1990. Mantle oxidation state and its relationship to tectonic environment and fluid speciation. Science,248(4953):337-345

Wu D,Liu Y S,Chen C F,Xu R,Ducea M N,Hu Z C,Zong K Q. 2017. *In-situ* trace element and Sr isotopic compositions of mantle xenoliths constrain two-stage metasomatism beneath the northern North China Craton. Lithos,288-289:338-351

Xia J X,Qin L P,Shen J,Carlson R W,Ionov D A,Mock T D. 2017. Chromium isotope heterogeneity in the mantle. Earth and Planetary Science Letters,464:103-115

Xia Q K,Hao Y T,Li P,Deloule E,Coltorti M,Dallai L,Yang X Z,Feng M. 2010. Low water content of the Cenozoic lithospheric mantle beneath the eastern part of the North China Craton. Journal of Geophysical Research:Solid Earth,115(B7):B07207

Xia Q K,Liu J,Liu S C,Kovács I,Feng M,Dang L. 2013. High water content in Mesozoic primitive basalts of the North China Craton and implications on the destruction of cratonic mantle lithosphere. Earth and Planetary Science Letters,361:85-97

Xia Q K,Bi Y,Li P,Tian W,Wei X,Chen H L. 2016. High water content in primitive continental flood basalts. Scientific Reports,6(1):25416

Xia Q K,Liu J,Kovács I,Hao Y T,Li P,Yang X Z,Chen H,Sheng Y M. 2019. Water in the upper mantle and deep crust of eastern China:concentration,distribution and implications. National Science Review,6(1):125-144

Xiao Y,Zhang H F. 2011. Effects of melt percolation on platinum group elements and Re-Os systematics of peridotites from the Tan-Lu fault zone,eastern North China Craton. Journal of the Geological Society,London,168(5):1201-1214

Xiao Y,Teng F Z,Zhang H F,Yang W. 2013. Large magnesium isotope fractionation in peridotite xenoliths from eastern North China Craton:product of melt-rock interaction. Geochimica et Cosmochimica Acta,115:241-261

Xiao Y,Zhang H F,Deloule E,Su B X,Tang Y J,Sakyi P A,Hu Y,Ying J F. 2015. Large lithium isotopic variations in minerals from peridotite xenoliths from the eastern North China Craton. The Journal of Geology,123(1):79-94

Xiao Y,Zhang H F,Su B X,Zhu B,Chen B B,Chen C,Sakyi P A. 2017. Partial melting control of lithium concentrations and isotopes in the Cenozoic lithospheric mantle beneath Jiande area,the Cathaysia Block of SE China. Chemical Geology,466:750-761

Xing C M,Wang C Y,Li C Y. 2014. Trace element compositions of apatite from the middle zone of the Panzhihua layered intrusion,SW China:insights into the differentiation of a P- and Si-rich melt. Lithos,204:188-202

Xiong F H,Yang J S,Robinson P T,Xu X Z,Liu Z,Li Y,Li J Y,Chen S Y. 2015. Origin of podiform chromitite,a new model based on the Luobusa ophiolite,Tibet. Gondwana Research,27(2):525-542

Xiong F H,Yang J S,Dilek Y,Xu X Z,Zhang Z M. 2018. Origin and significance of diamonds and other exotic minerals in the Dingqing ophiolite peridotites,eastern Bangong-Nujiang suture zone,Tibet. Lithosphere,10(1):142-155

Xiong F H,Xu X Z,Mugnaioli E,Gemmi M,Wirth R,Grew E S,Robinson P T,Yang J S. 2020. Two new minerals,badengzhuite,TiP,and zhiqinite,$TiSi_2$,from the Cr-11 chromitite orebody,Luobusa ophiolite,Tibet,China:is this evidence for super-reduced mantle-derived fluids? European Journal of Mineralogy,32(6):557-574

Xiong Q,Zheng J P,Griffin W L,O'Reilly S Y,Zhao J H. 2011. Zircons in the Shenglikou ultrahigh-pressure garnet peridotite massif and its country rocks from the North Qaidam terrane (western China):Meso-Neoproterozoic crust-mantle coupling and early Paleozoic convergent plate-margin processes. Precambrian Research,187(1-2):33-57

Xiong Q,Zheng J P,Griffin W L,O'Reilly S Y,Pearson N J. 2014. Pyroxenite dykes in orogenic peridotite from North Qaidam (NE Tibet,China) track metasomatism and segregation in the mantle wedge. Journal of Petrology,55(12):2347-2376

Xiong Q,Griffin W L,Zheng J P,O'Reilly S Y,Pearson N J. 2015. Episodic refertilization and metasomatism of Archean mantle:evidence from an orogenic peridotite in North Qaidam (NE Tibet,China). Contributions to Mineralogy and Petrology,169(3):31

Xu C,Chakhmouradian A R,Taylor R N,Kynicky J,Li W B,Song W L,Fletcher I R. 2014. Origin of carbonatites in the South Qinling orogen:implications for crustal recycling and timing of collision between the South and North China Blocks. Geochimica et Cosmochimica Acta,143:189-206

Xu C,Kynický J,Chakhmouradian A R,Li X H,Song W L. 2015. A case example of the importance of multi-analytical approach in deciphering carbonatite petrogenesis in South Qinling orogen:Miaoya rare-metal deposit,central China. Lithos,227:107-121

Xu C,Kynický J,Tao R,Liu X,Zhang L F,Pohanka M,Song W L,Fei Y W. 2017. Recovery of an oxidized majorite inclusion from Earth's deep asthenosphere. Science Advances,3(4):e1601589

Xu W L,Zhou Q J,Pei F P,Yang D B,Gao S,Li Q L,Yang Y H. 2013. Destruction of the North China Craton:delamination or thermal/chemical erosion? Mineral chemistry and oxygen isotope insights from websterite xenoliths. Gondwana Research,23(1):119–129

Xu Y G. 2014. Recycled oceanic crust in the source of 90–40 Ma basalts in North and Northeast China:evidence,provenance and significance. Geochimica et Cosmochimica Acta,143:49–67

Xu Y G,Zhang H H,Qiu H N,Ge W C,Wu F Y. 2012. Oceanic crust components in continental basalts from Shuangliao,Northeast China:derived from the mantle transition zone. Chemical Geology,328:168–184

Xu Y G,Li H Y,Hong L B,Ma L,Ma Q,Sun M D. 2018. Generation of Cenozoic intraplate basalts in the big mantle wedge under eastern Asia. Science China Earth Sciences,61(7):869–886

Xu Z,Zheng Y F. 2017. Continental basalts record the crust-mantle interaction in oceanic subduction channel:a geochemical case study from eastern China. Journal of Asian Earth Sciences,145:233–259

Xu Z,Zhao Z F,Zheng Y F. 2012. Slab-mantle interaction for thinning of cratonic lithospheric mantle in North China:geochemical evidence from Cenozoic continental basalts in central Shandong. Lithos,146-147:202–217

Yan Q S,Castillo P,Shi X F,Wang L L,Liao L,Ren J B. 2015. Geochemistry and petrogenesis of volcanic rocks from Daimao Seamount (South China Sea) and their tectonic implications. Lithos,218-219:117–126

Yan Q S,Straub S,Shi X F. 2019. Hafnium isotopic constraints on the origin of late Miocene to Pliocene seamount basalts from the South China Sea and its tectonic implications. Journal of Asian Earth Sciences,171:162–168

Yang A Y,Zhao T P,Zhou M F,Deng X G. 2017. Isotopically enriched N-MORB:a new geochemical signature of off-axis plume-ridge interaction—a case study at 50°28′E,Southwest Indian Ridge. Journal of Geophysical Research:Solid Earth,122(1):191–213

Yang F,Huang X L,Xu Y G,He P L. 2019a. Magmatic processes associated with oceanic crustal accretion at slow-spreading ridges:evidence from plagioclase in mid-ocean ridge basalts from the South China Sea. Journal of Petrology,60(6):1135–1162

Yang F,Huang X L,Xu Y G,He P L. 2019b. Plume-ridge interaction in the South China Sea:thermometric evidence from Hole U1431E of IODP Expedition 349. Lithos,324-325:466–478

Yang J F,Faccenda M. 2020. Intraplate volcanism originating from upwelling hydrous mantle transition zone. Nature,579(7797):88–91

Yang J J,Powell R. 2008. Ultrahigh-pressure garnet peridotites from the devolatilization of sea-floor hydrated ultramafic rocks. Journal of Metamorphic Geology,26(6):695–716

Yang J S,Robinson P T,Dilek Y. 2014. Diamonds in Ophiolites:a little-known diamond occurrence. Elements,10:123–126

Yang K F,Fan H R,Pirajno F,Li X C. 2019. The Bayan Obo (China) giant REE accumulation conundrum elucidated by intense magmatic differentiation of carbonatite. Geology,47(12):1198–1202

Yang W,Teng F Z,Zhang H F,Li S G. 2012. Magnesium isotopic systematics of continental basalts from the North China craton:implications for tracing subducted carbonate in the mantle. Chemical Geology,328:185–194

Yang Z F,Zhou J H. 2013. Can we identify source lithology of basalt? Scientific Reports,3(1):1856

Yang Z F,Li J,Liang W F,Luo Z H. 2016. On the chemical markers of pyroxenite contributions in continental basalts in eastern China:implications for source lithology and the origin of basalts. Earth-Science Reviews,157:18–31

Yang Z M,Woolley A. 2006. Carbonatites in China:a review. Journal of Asian Earth Sciences,27(5):559–575

Ye H M,Li X H,Lan Z W. 2013. Geochemical and Sr-Nd-Hf-O-C isotopic constraints on the origin of the Neoproterozoic Qieganbulake ultramafic-carbonatite complex from the Tarim Block,Northwest China. Lithos,182-183:150–164

Ying J F,Zhang H F,Kita N,Morishita Y,Shimoda G. 2006. Nature and evolution of Late Cretaceous lithospheric mantle beneath the eastern North China Craton:constraints from petrology and geochemistry of peridotitic xenoliths from Jünan,Shandong Province,China. Earth and Planetary Science Letters,244(3-4):622–638

Ying Y C,Chen W,Lu J,Jiang S Y,Yang Y H. 2017. In situ U-Th-Pb ages of the Miaoya carbonatite complex in the South Qinling orogenic belt,central China. Lithos,290-291:159–171

Yu H,Zhang H F,Li X H,Zhang J,Santosh M,Yang Y H,Zhou D W. 2016. Tectonic evolution of the North Qinling orogen from subduction to collision and exhumation:evidence from zircons in metamorphic rocks of the Qinling Group. Gondwana Research,30:65–78

Yu H,Zhang H F,Santosh M. 2017. Mylonitized peridotites of Songshugou in the Qinling orogen,central China:a fragment of fossil oceanic lithosphere mantle. Gondwana Research,52:1–17

Yu M M,Yan Y,Huang C Y,Zhang X C,Tian Z X,Chen W H,Santosh M. 2018. Opening of the South China Sea and upwelling of the Hainan Plume. Geophysical Research Letters,45(6):2600–2609

Yu S Y,Xu Y G,Zhou S H,Lan J B,Chen L M,Shen N P,Zhao J X,Feng Y X. 2018. Late Cenozoic basaltic lavas from the Changbaishan-Baoqing volcanic belt,NE China:products of lithosphere-asthenosphere interaction induced by subduction of the Pacific Plate. Journal of Asian Earth Sciences,164:260–273

Yu X,Dick H J B. 2020. Plate-driven micro-hotspots and the evolution of the Dragon Flag melting anomaly,Southwest Indian Ridge. Earth and Planetary Science Letters,531:116002

Yu X,Liu Z F. 2020. Non-mantle-plume process caused the initial spreading of the South China Sea. Scientific Reports,10(1):8500

Yu X,Zeng G,Chen L H,Wang X J,Liu J Q,Xie L W,Yang T. 2019. Evidence for rutile-bearing eclogite in the mantle sources of the Cenozoic Zhejiang basalts, eastern China. Lithos, 324-325: 152–164

Zeng G, Chen L H, Xu X S, Jiang S Y, Hofmann A W. 2010. Carbonated mantle sources for Cenozoic intraplate alkaline basalts in Shandong, North China. Chemical Geology, 273(1-2): 35–45

Zeng G, Chen L H, Hofmann A W, Jiang S Y, Xu X S. 2011. Crust recycling in the sources of two parallel volcanic chains in Shandong, North China. Earth and Planetary Science Letters, 302(3-4): 359–368

Zeng G, Chen L H, Yu X, Liu J Q, Xu X S, Erdmann S. 2017. Magma-magma interaction in the mantle beneath eastern China. Journal of Geophysical Research: Solid Earth, 122(4): 2763–2779

Zhang G L, Smith-Duque C, Tang S H, Li H, Zarikian C, D'Hondt S, Inagaki F, IODP Expedition 329 Scientists. 2012a. Geochemistry of basalts from IODP site U1365: implications for magmatism and mantle source signatures of the mid-Cretaceous Osbourn Trough. Lithos, 144-145: 73–87

Zhang G L, Zong C L, Yin X B, Li H. 2012b. Geochemical constraints on a mixed pyroxenite-peridotite source for East Pacific Rise basalts. Chemical Geology, 330-331: 176–187

Zhang G L, Chen L H, Li S Z. 2013. Mantle dynamics and generation of a geochemical mantle boundary along the East Pacific Rise-Pacific/Antarctic Ridge. Earth and Planetary Science Letters, 383: 153–163

Zhang G L, Chen L H, Jackson M G, Hofmann A W. 2017. Evolution of carbonated melt to alkali basalt in the South China Sea. Nature Geoscience, 10(3): 229–235

Zhang G L, Luo Q, Zhao J, Jackson M G, Guo L S, Zhong L F. 2018a. Geochemical nature of sub-ridge mantle and opening dynamics of the South China Sea. Earth and Planetary Science Letters, 489: 145–155

Zhang G L, Sun W D, Seward G. 2018b. Mantle source and magmatic evolution of the dying spreading ridge in the South China Sea. Geochemistry, Geophysics, Geosystems, 19(11): 4385–4399

Zhang G L, Wang S, Zhang J, Zhan M J, Zhao Z H. 2020a. Evidence for the essential role of CO_2 in the volcanism of the waning Caroline mantle plume. Geochimica et Cosmochimica Acta, 290: 391–407

Zhang G L, Zhang J, Wang S, Zhao J X. 2020b. Geochemical and chronological constraints on the mantle plume origin of the Caroline Plateau. Chemical Geology, 540: 119566

Zhang H F, Goldstein S L, Zhou X H, Sun M, Cai Y. 2009. Comprehensive refertilization of lithospheric mantle beneath the North China Craton: further Os-Sr-Nd isotopic constraints. Journal of the Geological Society, London, 166(2): 249–259

Zhang H F, Deloule E, Tang Y J, Ying J F. 2010. Melt/rock interaction in remains of refertilized Archean lithospheric mantle in Jiaodong Peninsula, North China Craton: Li isotopic evidence. Contributions to Mineralogy and Petrology, 160(2): 261–277

Zhang H F, Sun Y L, Tang Y J, Xiao Y, Zhang W H, Zhao X M, Santosh M, Menzies M A. 2012. Melt-peridotite interaction in the Pre-Cambrian mantle beneath the western North China Craton: petrology, geochemistry and Sr, Nd and Re isotopes. Lithos, 149: 100–114

Zhang H F, Zhu R X, Santosh M, Ying J F, Su B X, Hu Y. 2013. Episodic widespread magma underplating beneath the North China Craton in the Phanerozoic: implications for craton destruction. Gondwana Research, 23(1): 95–107

Zhang H F, Zou D Y, Santosh M, Zhu B. 2016. Phanerozoic orogeny triggers reactivation and exhumation in the northern part of the Archean-Paleoproterozoic North China Craton. Lithos, 261: 46–54

Zhang H T, Zhang H F, Zou D Y. 2021. Comprehensive refertilization of the Archean-Paleoproterozoic lithospheric mantle beneath the northwestern North China Craton: evidence from in situ Sr isotopes of the Siziwangqi peridotites. Lithos, 380-381: 105822

Zhang J, Zhang H F, Kita N, Shimoda G, Morishita Y, Ying J F, Tang Y J. 2011. Secular evolution of the lithospheric mantle beneath the eastern North China Craton: evidence from peridotitic xenoliths from Late Cretaceous mafic rocks in the Jiaodong region, east-central China. International Geology Review, 53(2): 182–211

Zhang J B, Liu Y S, Ling W L, Gao S. 2017. Pressure-dependent compatibility of iron in garnet: insights into the origin of ferropicritic melt. Geochimica et Cosmochimica Acta, 197: 356–377

Zhang J J, Zheng Y F, Zhao Z F. 2009. Geochemical evidence for interaction between oceanic crust and lithospheric mantle in the origin of Cenozoic continental basalts in east-central China. Lithos, 110(1-4): 305–326

Zhang J R, Lv J, Li H F, Feng X L, Lu C, Redfern S A T, Liu H Y, Chen C F, Ma Y M. 2018. Rare helium-bearing compound FeO_2He stabilized at deep-earth conditions. Physical Review Letters, 121: 255703

Zhang L, Liu Y S, Wang L N, Wang C Y, Zhang G L. 2020. Multiple metasomatism of the lithospheric mantle beneath the northeastern North China Craton. Lithos, 374-375: 105719

Zhang L Y, Prelević D, Li N, Mertz-Kraus R, Buhre S. 2016. Variation of olivine composition in the volcanic rocks in the Songliao Basin, NE China: lithosphere control on the origin of the K-rich intraplate mafic lavas. Lithos, 262: 153–168

Zhang M J, Tang Q Y, Cao C H, Li W Y, Wang H, Li Z P, Yu M, Feng P Y. 2017. The Origin of Permian Pobei ultramafic complex in the northeastern Tarim Craton, western China: evidences from chemical and C-He-Ne-Ar isotopic compositions of volatiles. Chemical Geology, 469: 85–96

Zhang M J, Feng P Y, Li T, Li L W, Fu J R, Wang P, Wang Y K, Li Z P, Wang X D. 2019. The petrogenesis of the Permian Podong ultramafic intrusion in the Tarim Craton, western China: constraints from C-He-Ne-Ar isotopes. Geofluids, 2019: 6402571

Zhang M L, Guo Z F. 2016. Origin of Late Cenozoic Abaga-Dalinuoer basalts, eastern China: implications for a mixed pyroxenite-peridotite source related with deep subduction of the Pacific slab. Gondwana Research, 37: 130–151

Zhang X Q, Zhang H F, Zou H B. 2020. Rift-related Neoproterozoic tholeiitic layered mafic intrusions at northern Yangtze Block, South China: mineral chemistry evidence. Lithos,356-357:105376

Zhang Y H,Ren Z Y,Hong L B,Zhang Y,Zhang L,Qian S P,Xu Y G,Chen L L. 2017. Differential partial melting process for temporal variations of Shandong basalts revealed by melt inclusions and their host olivines. Gondwana Research,49:205-221

Zhang Y L,Liu C Z,Ge W C,Wu F Y,Chu Z Y. 2011. Ancient sub-continental lithospheric mantle (SCLM) beneath the eastern part of the Central Asian orogenic belt (CAOB):implications for crust-mantle decoupling. Lithos,126(3-4):233-247

Zhang Y L,Ge W C,Sun J,Yang H,Liu Z C,Liu J G. 2019. Age and composition of the subcontinental lithospheric mantle beneath the Xing'an-Mongolia orogenic belt:implications for the construction of microcontinents during accretionary orogenesis. Lithos,326-327:556-571

Zhang Z M,Dong X,Liou J G,Liu F,Wang W,Yui F. 2011. Metasomatism of garnet peridotite from Jiangzhuang,southern Sulu UHP belt:constraints on the interactions between crust and mantle rocks during subduction of continental lithosphere. Journal of Metamorphic Geology,29(9):917-937

Zhao X M,Zhang H F,Zhu X K,Tang S H,Tang Y J. 2010. Iron isotope variations in spinel peridotite xenoliths from North China Craton:implications for mantle metasomatism. Contributions to Mineralogy and Petrology,160(1):1-14

Zhao X M,Zhang H F,Zhu X K,Tang S H,Yan B. 2012. Iron isotope evidence for multistage melt-peridotite interactions in the lithospheric mantle of eastern China. Chemical Geology,292-293:127-139

Zhao X M,Zhang H F,Zhu X K,Zhu B,Cao H H. 2015. Effects of melt percolation on iron isotopic variation in peridotites from Yangyuan,North China Craton. Chemical Geology,401:96-110

Zhao X M,Cao H H,Mi X,Evans N J,Qi Y H,Huang F,Zhang H F. 2017a. Combined iron and magnesium isotope geochemistry of pyroxenite xenoliths from Hannuoba,North China Craton:implications for mantle metasomatism. Contributions to Mineralogy and Petrology,172(6):40

Zhao X M,Zhang Z F,Huang S C,Liu Y F,Li X,Zhang H F. 2017b. Coupled extremely light Ca and Fe isotopes in peridotites. Geochimica et Cosmochimica Acta,208:368-380

Zhao Y,Xue C J,Liu S A,Symons D T A,Zhao X B,Yang Y Q,Ke J J. 2017. Copper isotope fractionation during sulfide-magma differentiation in the Tulaergen magmatic Ni-Cu deposit,NW China. Lithos,286-287:206-215

Zhao Y,Xue C J,Liu S A,Mathur R,Zhao X B,Yang Y Q,Dai J F,Man R H,Liu X M. 2019. Redox reactions control Cu and Fe isotope fractionation in a magmatic Ni-Cu mineralization system. Geochimica et Cosmochimica Acta,249:42-58

Zheng H,Zhong L F,Kapsiotis A,Cai G Q,Wan Z F,Xia B. 2019. Post-spreading basalts from the nanyue seamount:implications for the involvement of crustal-and plume-type components in the genesis of the South China Sea Mantle. Minerals,9(6):378

Zheng J P,Sun M,Griffin W L,Zhou M F,Zhao G C,Robinson P,Tang H Y,Zhang Z H. 2008. Age and geochemistry of contrasting peridotite types in the Dabie UHP belt,eastern China:petrogenetic and geodynamic implications. Chemical Geology,247(1-2):282-304

Zheng J P,Tang H Y,Xiong Q,Griffin W L,O'Reilly S Y,Pearson N,Zhao J H,Wu Y B,Zhang J F,Liu Y S. 2014. Linking continental deep subduction with destruction of a cratonic margin:strongly reworked North China SCLM intruded in the Triassic Sulu UHP belt. Contributions to Mineralogy and Petrology,168(1):1028

Zheng J P,Lee C T A,Lu J G,Zhao J H,Wu Y B,Xia B,Li X Y,Zhang J F,Liu Y S. 2015. Refertilization-driven destabilization of subcontinental mantle and the importance of initial lithospheric thickness for the fate of continents. Earth and Planetary Science Letters,409:225-231

Zhou H Y,Dick H J B. 2013. Thin crust as evidence for depleted mantle supporting the Marion Rise. Nature,494(7436):195-200

Zhu H L,Liu F,Li X,Wang G Q,Zhang Z F,Sun W D. 2018. Calcium isotopic compositions of normal mid-ocean ridge basalts from the southern Juan de Fuca Ridge. Journal of Geophysical Research:Solid Earth,123(2):1303-1313

Zong K Q,Liu Y S. 2018. Carbonate metasomatism in the lithospheric mantle:implications for cratonic destruction in North China. Science China Earth Science,61(6):711-729

Zou D Y,Liu Y S,Hu Z C,Gao S,Zong K Q,Xu R,Deng L X,He D T,Gao C G. 2014. Pyroxenite and peridotite xenoliths from Hexigten,Inner Mongolia:insights into the Paleo-Asian Ocean subduction-related melt/fluid-peridotite interaction. Geochimica et Cosmochimica Acta,140:435-454

Zou D Y,Zhang H F,Hu Z C,Santosh M. 2016. Complex metasomatism of lithospheric mantle by asthenosphere-derived melts:evidence from peridotite xenoliths in Weichang at the northern margin of the North China Craton. Lithos,264:210-223

Zou D Y,Zhang H F,Zhang X Q,Zhang H T,Su B X. 2020. Refertilization of lithospheric mantle beneath the North China Craton in Mesozoic:evidence from in situ Sr isotopes of Fuxin peridotite. Lithos,364-365:105478

Progresses in Mineralogy, Petrology and Geochemistry of the Mantle

ZHANG Hong-fu, CHEN Li-hui

Department of Geology, Northwest University, Xi'an 710069

Abstract: The second decade of the 21st century (2010–2020) is a decade of vigorous developments in mineralogy, petrology and geochemistry in China of various fields from the craton destruction to orogenic belt evolution and from the continental to oceanic lithospheres, is a decade of comprehensive march from the depth to breadth, and is also a decade of the development from catching up to surpassing in mineralogical, petrological and geochemical researches. In the past decade, the mineralogists, petrologists and geochemists in China have made many breakthroughs in mineralogy, petrology and geochemistry of the mantle. These advances are mainly focused on two aspects including those of researches on direct samples of the mantle, such as mantle mineralogy, continental and oceanic lithospheres, peridotite of orogenic belt, and water in mantle; and those of researches on samples of magmatic rocks derived from the mantle, such as continental and oceanic basalts, basic and ultrabasic complexes, and igneous carbonatite. In particular, the development and wide application of new methods for analyzing non-traditional stable isotopes or metal stable isotopes (such as Li, Mg, Fe and Ca isotopes, etc.) have greatly promoted the research progress of mantle petrology and geochemistry in China, and achieved a historic leap from catching up to surpassing in researches on petrology and geochemistry of the mantle.

Key words: mantle; mineralogy; petrology; geochemistry; metal stable isotope; ten years' progress

大陆俯冲带超高压变质岩部分熔融与壳幔相互作用研究进展[*]

赵子福[1,2]　陈仁旭[1,2]　陈伊翔[1,2]　戴立群[1,2]　郑永飞[1,2]

1. 中国科学技术大学地球和空间科学学院，中国科学院壳幔物质与环境重点实验室，合肥 230026；
2. 中国科学院 比较行星学卓越创新中心，合肥 230026

摘　要：自 20 世纪 80 年代在大陆地壳岩石中发现柯石英和金刚石等超高压变质矿物以来，大陆深俯冲和超高压变质作用就成了固体地球科学研究的前沿和热点领域之一。经过三十余年的研究，已经在大陆地壳的俯冲深度、深俯冲岩石变质 P-T-t 轨迹、俯冲地壳岩石的折返机制、深俯冲岩石的原岩性质、大陆碰撞过程中的熔-流体活动与元素活动性、俯冲隧道内部不同类型壳幔相互作用、碰撞后岩浆岩的成因、大陆碰撞造山带成矿作用等方面取得了许多重要成果。本文重点对大陆俯冲带超高压岩石部分熔融和不同类型壳幔相互作用近十年来的研究进展进行回顾和总结，并对存在的相关科学问题和未来的研究方向进行了展望。深俯冲大陆地壳的部分熔融主要出现在两个阶段：折返的初期阶段和碰撞后阶段，前者产生了碱性熔体，后者产生了钙碱性熔体。大陆俯冲带壳幔相互作用有两种类型，涉及地幔楔与两种俯冲带流体的交代反应：一是来自深俯冲陆壳的变质脱水-熔融，二是来自先前俯冲古洋壳的变质脱水-熔融。

关键词：大陆俯冲带　部分熔融　壳幔相互作用　造山带橄榄岩　造山带岩浆岩

0　引　言

　　20 世纪 60 年代建立起来的板块构造理论是地球科学发展的一个里程碑，它极大地改变了人类对地球运作机制的认识，成为 20 世纪自然科学的重大进展之一（Zheng，2018）。按照传统的板块构造理论，大陆地壳由于其密度低，不可能俯冲到高密度的地幔中（郑永飞等，2015）。然而，20 世纪 80 年代地球科学家分别在西阿尔卑斯和挪威西部的变质表壳岩中，发现了超高压变质矿物柯石英（Chopin，1984；Smith，1984），证明大陆地壳曾俯冲到至少 80 km 的地幔深度并发生超高压变质作用，然后折返回地表。随着新的超高压指示矿物和特殊出溶结构的发现（如金刚石、α-PbO_2 结构的金红石、超硅石榴石、超硅楣石、富 Si 和 K 的单斜辉石、菱镁矿、单斜辉石中出溶斜顽辉石、斯石英假象等）（Sobolev and Shatsky，1990；Xu et al.，1992；Hwang et al.，2000；Ye et al.，2000；Zhang et al.，2002；Dobrzhinetskaya et al.，2006；Liu et al.，2007），大陆地壳的俯冲深度被不断刷新，从 80 km 到 300 km 以上。此外，长英质片麻岩锆石中柯石英包裹体的不断发现（Liu and Liou，2011），证明大规模的低密度长英质岩石曾整体俯冲到地幔深度发生超高压变质，然后又折返回浅部地壳。三十余年来，地球科学家相继在全球大陆碰撞造山带中发现了二十多个超高压变质地体（Liou et al.，2009；Zheng and Chen，2016）。大陆深俯冲和超高压变质作用研究不仅发展了板块构造理论，而且推进了大陆动力学研究。

　　通过三十余年的研究，国际学术界对大陆深俯冲和超高压变质的基本事实已经有了明确的认识，如陆壳俯冲的深度、深俯冲陆壳的变质时代、超高压岩石的全球分布位置、产出规模和 P-T-t 轨迹等（Chopin，2003；Liou et al.，2009；Zheng et al.，2009；Zheng，2012；Hermann and Rubatto，2014）。这些

---

[*] 原文刊于《矿物岩石地球化学通报》2021 年第 40 卷第 1 期，本文略有修改。

研究主要集中在超高压变质岩本身、对大陆深俯冲现象的事实认定、对大陆俯冲带熔-流体活动的探讨以及对超高压变质岩折返过程和机制的认识等方面。近年来大陆深俯冲研究的一个重要进展是将大洋俯冲隧道模型（Shreve and Cloos，1986；Cloos and Shreve，1988a，1988b）拓展到大陆俯冲带（Guillot et al.，2009；Zheng，2012；Butler et al.，2013；郑永飞等，2013；张建新，2020），这很好地解释了大陆碰撞过程中发生的构造变形、高压-超高压变质、俯冲地壳的拆离解耦、差异折返及板片-地幔相互作用等。

板块俯冲是地壳物质进入地幔并发生壳幔相互作用的主要地球动力学机制，俯冲进入地球内部的地壳物质又可以通过多种机制、不同程度地返回地壳甚至地表（Zheng and Chen，2016）。俯冲带壳幔相互作用是一个复杂的物理化学过程，涉及俯冲带变质作用、交代作用、岩浆作用等一系列地质过程。俯冲带出露的各种变质岩、造山带橄榄岩和岩浆岩分别记录了俯冲带变质作用、交代作用和岩浆作用等复杂过程，是地球深部物质循环的最终产物，为认识俯冲带壳幔相互作用的化学地球动力学机制和过程提供了直接窗口（郑永飞和陈伊翔，2019）。近年来通过对大陆俯冲带中出露的各种变质岩、造山带橄榄岩和岩浆岩的研究，在大陆俯冲带部分熔融和地壳物质循环及其壳幔相互作用方面取得了长足进展（Zhao et al.，2013；陈意等，2015；赵子福等，2015；Chen Y. X. et al.，2017；陈仁旭等，2019；郑建平等，2019a；郑永飞和陈伊翔，2019）。本文重点总结大陆俯冲带超高压岩石部分熔融和不同类型壳幔相互作用方面的研究进展，并对未来需要解决的相关科学问题和研究方向提出建议。

1 大陆俯冲带超高压岩石部分熔融

1.1 大陆俯冲带超高压岩石部分熔融的研究意义

大陆碰撞过程中高压-超高压变质岩石的部分熔融作用具有重要的研究意义。一方面，它会改变深俯冲大陆板片的流变学特征（Rosenberg and Handy，2005；Labrousse et al.，2011），影响大陆碰撞造山带的构造热演化过程（Chen Y. X. et al.，2017；Zheng and Chen，2017）；另一方面，通过地幔深度的熔体-地幔楔橄榄岩反应，导致大陆俯冲隧道内的物质迁移和壳幔相互作用（Wallis et al.，2005；Jamieson et al.，2011；Labrousse et al.，2011，2015；Zheng et al.，2011；Zheng，2012，2019；郑永飞等，2015，2016）。俯冲板片中熔体的存在会极大地降低岩石的强度，增强流动性能力（Rosenberg and Handy，2005），因此超高压变质岩石在地幔深度的部分熔融会改变碰撞动力学过程，促进甚至诱发深俯冲大陆板片从地幔深度的快速折返（Wallis et al.，2005；Labrousse et al.，2011，2015）。深俯冲陆壳部分熔融还会造成显著的壳内地球化学分异，对理解陆壳的形成和演化也具有重要意义（Chen Y. X. et al.，2017；Zheng，2019）。因此，揭示大陆碰撞造山带超高压变质岩石发生部分熔融的证据、时限、程度、温压条件和部分熔融反应机制，对认识造山带和地球壳幔系统演化均具有重要意义。

1.2 大陆俯冲带超高压岩石部分熔融的证据

陆壳具有古老、干和冷的特征，因此俯冲过程中流体诱导的岛弧岩浆作用通常不会在大陆俯冲带中发生（郑永飞等，2016），但在超高压岩石折返过程中有可能发生部分熔融（Zeng et al.，2009；Zheng et al.，2011；Xu et al.，2013；Stepanov et al.，2014；Wang et al.，2014；Zheng and Chen，2016；Chen Y. X. et al.，2017；Yu et al.，2019；Sun et al.，2020）。这种部分熔融作用在超高压变质地体中表现出不同的空间尺度。在中国苏鲁造山带东北部出露有大量的混合岩、花岗质岩脉和伟晶岩，在局部还出露有花岗岩岩体（Wallis et al.，2005；Liu F. L. et al.，2010，2012；Zhao et al.，2012，2017a；Xu et al.，2013，2016；Zhou et al.，2019）。岩石学和同位素年代学结果指示，这些长英质岩石形成于210~225 Ma，即超高压变质岩折返阶段。在苏鲁造山带的其他地方也出露有混合岩，如仰口和东海地区，但规模较小（Liu F. L. et al.，2012；Wang et al.，2014）。这种部分熔融程度的差异可能指示了超高压岩石折返过程中

经历了不同的 P-T 路径。

通过野外地质和岩石学观察及同位素定年易于识别超高压岩石发生大尺度部分熔融的现象，而识别小尺度（如手标本或薄片尺度）的部分熔融作用却很难识别。然而，这对确定造山带是否经历了大规模但小比例的部分熔融、制约造山带的构造热演化具有重要意义。前人基于高温高压部分熔融实验和低压混合岩脉体的研究发现，熔体-矿物相平衡时的二面角显著低于矿物-矿物平衡时的二面角，可见显微结构分析是揭示陆壳岩石是否发生小尺度部分熔融的有效手段（Holness，2006；Holness and Sawyer，2008；Holness et al.，2011，2012，2013）。这些认识可以应用到判别超高压岩石是否发生部分熔融的研究中，有助于识别并建立超高压岩石部分熔融的显微结构判据（Zeng et al.，2009；Chen Y. X. et al.，2013a，2013b；Liu et al.，2013）。

大别-苏鲁造山带超高压变质岩发生了广泛的低程度部分熔融，关键的显微结构证据（Zeng et al.，2009；Zheng et al.，2011；陈伊翔和郑永飞，2013；Chen Y. X. et al.，2013a，2013b，2016a，2017；Liu et al.，2013，2014；Xu et al.，2013；Wang et al.，2014）包括：①石英颗粒边界发育拉长的高度尖角状长石颗粒，或石英-石英-石英/长石构成的三联点中填充有尖角状长石颗粒；②钾长石晶面发育；③颗粒边界出现由石英+钾长石±斜长石±白云母组成的长英质显微脉体；④石英或钾长石边界发育串珠状结构。颗粒边界或三联点中尖角状长石与周围矿物的二面角很小，有时甚至连通呈网络状（图1）。这指示它们继承了熔体的显微结构，熔体-固相矿物平衡时的二面角通常小于40°，远低于固相-固相矿物结构平衡时的二面角（Holness and Sawyer，2008）。这些显微结构特征出现在榴辉岩、花岗片麻岩和钙质片麻岩等各种超高压变质岩中（Chen Y. X. et al.，2013a，2013b，2016a；Liu et al.，2013，2014；Wang et al.，2014；Xia et al.，2016），指示大别-苏鲁造山带超高压岩石发生部分熔融的规模远比之前认为的大。

深俯冲陆壳部分熔融的另一个重要证据是转熔或深熔矿物中出现长英质多相固体包裹体（图1）（Chen Y. X. et al.，2017；Gao X. Y. et al.，2017）。目前在大别-苏鲁造山带榴辉岩、副片麻岩和钙质片麻岩中已发现很多长英质多相固体包裹体，主要有石英+钠长石、石英+钠长石+钾长石、石英+钾长石、钾长石+钠长石±石英±重晶石、方解石+石英+钾长石等矿物组合（Zeng et al.，2009；Gao et al.，2012，2013，2014；Liu et al.，2013，2014，2020；Chen et al.，2014，2016a；Wang et al.，2014；Li W. C. et al.，2016）。这些包裹体的一个普遍特征是包裹体周缘寄主矿物中存在放射状裂隙，同时矿物多呈交生结构。主、微量元素特征指示它们不可能是硬玉、柯石英或钾钡铝沸石分解形成的，而很可能是结晶自先前的含水熔体或碳酸盐熔体（Zeng et al.，2009；Gao et al.，2012，2013，2014；Liu et al.，2013，2014；Chen et al.，2014；Wang et al.，2014；Li W. C. et al.，2016）。包裹体的主要成分有较大变化范围，可能与原岩组成、部分熔融温度压力条件和熔体的差异性结晶有关（Chen et al.，2014）。值得注意的是，在波西米亚造山带混合岩中发现了碳酸盐熔体和硅酸盐熔体包裹体共存的现象，反映了含碳酸盐沉积物的部分熔融（Ferrero et al.，2016）。

一些超高压变质岩中的多相固体包裹体含柯石英或金刚石，指示部分熔融发生在超高压条件下（Hwang et al.，2001，2006；Chen Y. X. et al.，2013a，2013b；Stepanov et al.，2014，2016）。波西米亚造山带 Erzgebirge 含金刚石片麻岩中的多相固体包裹体含微粒金刚石、金云母、钠云母、石英和磷灰石，指示部分熔融发生在 4~6 GPa 和 1 000℃ 条件下（Hwang et al.，2001）。在 Erzgebirge 和哈萨克斯坦的 Kokchetav 超高压片麻岩中还存在纳米尺度富 P、K 的硅酸盐玻璃质包裹体，进一步证实部分熔融发生在超高压条件下（Hwang et al.，2006）。结合高温高压均一化实验，Stepanov 等（2016）认为 Kokchetav 含金刚石超高压片麻岩中的包裹体为 4.5~6.5 GPa 和 950~1 000℃ 条件下产生的小比例熔体。

1.3　大陆俯冲带超高压岩石部分熔融的时限、温压条件与熔融机制

1.3.1　部分熔融时限和温压条件

准确限定超高压岩石部分熔融的时限对解译大陆俯冲带的构造演化至关重要，而理解部分熔融过程

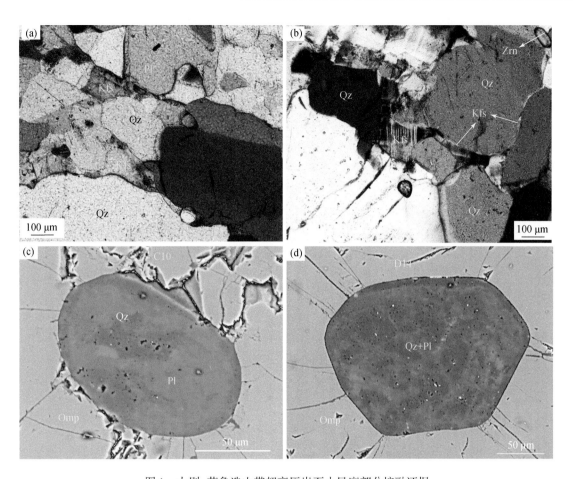

图1 大别-苏鲁造山带超高压岩石小尺度部分熔融证据
(a)(b) 超高压变质长英质岩石 (据 Chen Y. X. et al., 2013b);(c)(d) 超高压变质榴辉岩 (据 Chen et al., 2014)。
矿物符号：Pl. 斜长石；Kfs. 钾长石；Qz. 石英；Omp. 绿辉石；Zrn. 锆石

中矿物响应行为是准确定年的前提。副矿物可以通过转熔反应形成转熔矿物相，它们可以有效地溶解于含水熔体而非富水流体中，在熔体结晶时会有溶蚀或变质生长（Hermann et al., 2001; Rubatto et al., 2009; Liu et al., 2010, 2012a; Gordon et al., 2012, 2013; Chen Y. X. et al., 2013a, 2013b）。因此，这些矿物通常具有显著的主、微量元素分带。准确限定不同环带的形成机制和温压条件，可以为揭示部分熔融过程提供重要制约。由于锆石在变质和部分熔融过程中都会生长，因而被广泛用来确定变质和深熔时限（Rubatto et al., 2009; Liu F. L. et al., 2010, 2012; Chen Y. X. et al., 2013a, 2013b, 2017）。特别是结合微量元素和 Hf-O 同位素，锆石可用来制约部分熔融的压力条件和矿物组合，以及部分熔融的过程和机制。如果锆石生长于不同来源的熔体，其 Hf-O 同位素组成可能有显著差异（Chen Y. X. et al., 2013b）。

利用锆石 U-Pb 定年来确定部分熔融时限，首先要准确识别超高压变质岩是否发生部分熔融以及其中哪些锆石是通过部分熔融形成的。经历混合岩化的超高压变质岩和长英质脉体是研究部分熔融时限和过程的典型对象（Zeng et al., 2009; Liu F. L. et al., 2010, 2012; Xu et al., 2013; Wang et al., 2014; Chen R. X. et al., 2015; Li W. C. et al., 2016; Yu et al., 2019; Zhou et al., 2019）。根据锆石中矿物包裹体（特别是多相固体包裹体）、稀土元素（REE）配分型式和微量元素特征，可以准确区分超高压变质岩中的深熔锆石、残留岩浆锆石和亚固相变质生长锆石（Liu F. L. et al., 2010, 2012; Chen Y. X. et al., 2013a, 2013b）。相对于花岗片麻岩中的变质生长锆石，深熔锆石具有更陡峭的 REE 配分型式、显著升高的 U 含量和更显著的 Eu 负异常（Chen Y. X. et al., 2013a），因此，锆石的 Eu 负异常程度和 U 含量是区分长英质岩石中深熔锆石和变质锆石的有效手段。此外，矿物包裹体也可区分不同成因的锆石，石英岩深熔锆石中含有由石英±钾长石±斜长石±白云母组成的多相固体包裹体。这些包裹体显示花岗斑状结构，总体具有花岗质成分，指示先前出现的含水熔体，为锆石的深熔成因提供了直接的岩石学证据。

对深熔锆石的 U-Pb 定年可直接限定部分熔融的时限，对岩浆锆石定年给出的是相对较晚的岩浆演化时限。然而，从初始熔融到熔体结晶的时间尺度通常很短暂（Petford et al., 2000；Gordon et al., 2012）。因此，深熔熔体结晶锆石的年龄可以近似为部分熔融的时限，特别是考虑到二次离子质谱（secondary ion mass spectrometry, SIMS）和 LA-ICP-MS U-Pb 定年的误差。大量岩石学及同位素年代学研究表明，苏鲁造山带超高压岩石部分熔融的时限有多阶段特征，从折返早期的榴辉岩相–麻粒岩相阶段到晚期的角闪岩相阶段均有记录（Liu F. L. et al., 2010, 2012；Xu et al., 2013；Chen Y. X. et al., 2013a, 2013b；Wang et al., 2014；Li W. C. et al., 2016）。一些含有柯石英包裹体的深熔锆石给出的 U-Pb 年龄为（229±5）~（216±5）Ma，集中在（224±2）Ma，表明深熔作用开始于超高压岩石早期折返阶段但仍处于超高压变质条件下（图2）（Chen Y. X. et al., 2013a, 2013b）。含不同压力矿物包裹体的深熔锆石具有一致的 U-Pb 年龄（图2），表明深俯冲陆壳经历了从地幔到下地壳层位的快速折返过程。对威海混合岩的深色体、富斜长石淡色体和富钾长石脉体的对比研究揭示，折返后期冷却阶段富钾长石伟晶岩脉［(219±2) Ma］形成于富斜长石淡色体［(228±2) Ma］结晶之后，指示了较晚期的熔体分异过程（Xu et al., 2013；Song Y. R. et al., 2014）。哈萨克斯坦 Kokchetav 含金刚石片麻岩也经历了部分熔融过程（Hwang et al., 2001, 2006；Stepanov et al., 2016），对含金刚石区域的锆石 U-Pb 定年给出一致的年龄（535±9）Ma（Hermann et al., 2001），限定了 Kokchetav 造山带超高压条件下部分熔融发生的时限。

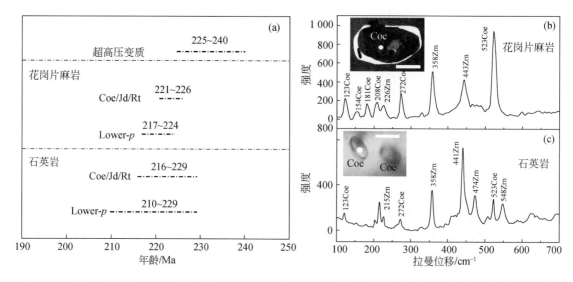

图 2　苏鲁造山带超高压岩石中深熔锆石 U-Pb 年龄和包裹体组成（据 Chen Y. X. et al., 2013a, 2013b）
(a) 深熔锆石中矿物包裹体和 U-Pb 年龄, Lower-p 指低压矿物包裹体；(b)(c) 深熔锆石中柯石英激光拉曼谱图, 标尺为 5 μm。矿物符号：Coe. 柯石英；Jd. 硬玉；Rt. 金红石；Zrn. 锆石

为了进一步制约初始熔融、熔体迁移和侵位到上地壳的时间尺度，需要高精度的同位素年龄数据。这就需要采用同位素稀释热电离质谱（isotope dilution thermal ionization mass spectrometry, ID-TIMS）U-Pb 定年方法，其精度比 SIMS 和 LA-ICP-MS 高两个数量级以上。Gordon 等（2012）对巴布亚新几内亚超高压榴辉岩相关的变形浅色体、花岗质脉体和花岗岩体进行了 ID-TIMS 锆石 U-Pb 定年，发现浅色体形成于超高压变质作用 1.1~2.6 Ma 之后，而花岗质岩脉的形成则晚于浅色体 0.5~1.0 Ma，这一结果表明深熔熔体从地幔深度初始熔融到运移到上地壳侵位的时间尺度较为短暂（~3 Ma）（Gordon et al., 2012）。

1.3.2　部分熔融机制

陆壳岩石部分熔融总体上分为两类：脱水熔融和加水熔融。对于中低压条件下的泥质岩体系，这两种方式研究的比较清楚（Clemens，2006；Brown，2013；Gao and Zeng，2014；Weinberg and Hasalová，2015；Gao L. E. et al.，2017）。然而，大陆碰撞过程中的加水熔融只有极少报道（Rubatto et al.，2009；Labrousse et al.，2011；Ganzhorn et al.，2014），主要是因为超高压变质岩在经历高压-超高压条件显著脱水后水含量很低。但超高压岩石名义上无水矿物中的结构羟基和分子水出溶可以释放显著量的自由水（Zheng et al.，1999，2003；Chen et al.，2007，2011；Zhao et al.，2007a；Zheng，2009），从而诱发超高压岩石发生加水熔融。产生的熔体从体系中带走自由水，又使体系变得相对较干，阻止了进一步的加水熔融。因此，总体上大陆碰撞过程中加水熔融不是产生大规模熔体的主要途径。

超高压岩石折返过程中的脱水熔融是主要机制，通过含水矿物，如多硅白云母和黝帘石等的分解而诱发转熔反应（Zhao et al.，2007b；Zheng Y. X. et al.，2011；Liu et al.，2013）。因此，这种部分熔融的发生取决于折返时俯冲带的热结构和超高压岩石经历的 P-T 路径。为了准确限定部分熔融机制，需要结合岩石学、地球化学和同位素定年综合分析。例如，Lang 和 Gilotti（2007）在格陵兰岛超高压变泥质岩的石榴子石中发现了多相固体包裹体，含有转熔反应的反应物和产物，即多硅白云母+石英→蓝晶石+钾长石+金红石+熔体。他们还发现包裹体的寄主石榴子石存在高钙含量斑杂状区域，与高温高压条件下石榴子石中包裹体部分熔融形成的石榴子石成分特征一致（Perchuk et al.，2005，2008）。

大别-苏鲁造山带超高压岩石为限定部分熔融机制提供了理想样品。多硅白云母脱水熔融可能是主导机制，主要证据包括：①超高压岩石折返时经历了等温甚至升温过程；②基质和深熔锆石多相包裹体中出现与钾长石共存的多硅白云母残晶（Zhao et al.，2007b；Chen Y. X. et al.，2013a，2013b）。这些观察与前人高温高压实验结果是吻合的（Vielzeuf and Holloway，1988；Hermann，2002a；Auzanneau et al.，2006）。部分熔融反应同时形成了具有较好晶形、富 Mn 的转熔石榴子石（Chen Y. X. et al.，2013a）。在苏鲁超高压石英岩中，根据石榴子石的包裹体特征、主微量元素环带和薄片中深熔锆石的 U-Pb 年龄，限定了部分熔融反应：白云母+石英+斜长石+石榴子石-I±角闪石+榍石→熔体+石榴子石-II+钾长石，并确定了熔融时限为约 214 Ma（Chen Y. X. et al.，2013b）。石英岩中深熔锆石的 $\delta^{18}O$ 值变化很大，可达 5‰，指示深熔熔体主体来自石英岩自身，而少量 $\delta^{18}O$ 值很低的熔体可能来自围岩片麻岩，暗示在折返阶段熔体的小尺度运移过程。榴辉岩俯冲折返过程中存在丰富的流体活动（Liu et al.，2013，2019；Guo S. et al.，2015，2017，2019），其部分熔融也主要以多硅白云母脱水熔融为主，黝帘石或钠云母脱水熔融可能也有少量贡献（Zeng et al.，2009；Gao et al.，2012；Liu et al.，2013，2020；Chen et al.，2014；Wang et al.，2014）。苏鲁造山带含黝帘石榴辉岩中三类多相固体包裹体具有显著不同的主量元素组成，表明它们在折返过程中通过不同的转熔反应形成（Chen et al.，2014）。由石英和斜长石组成的多相固体包裹体可能主要由黝帘石脱水熔融产生，并有大量绿辉石参与，形成了富 Na 的熔体；由石英、斜长石和少量钾长石组成的多相固体包裹体可能由黝帘石和多硅白云母共同脱水熔融形成，熔体具有变化的钾含量；由重晶石、斜长石、钾长石和绿帘石组成的多相固体包裹体可能是与熔体有关的高氧逸度富水流体与寄主矿物反应的结果。这些解释总体上与榴辉岩经历的 P-T 路径和前人对多硅白云母和黝帘石脱水熔融实验结果一致（Skjerlie and Patiño Douce，2002；Gao et al.，2012，2013）。与超高压长英质岩石不同，超高压榴辉岩只经历了低程度部分熔融，产生了结晶的多相包裹体和显微脉体，并未形成混合岩化浅色体（Zhao et al.，2007b；Gao et al.，2012，2013；Liu et al.，2013；Chen et al.，2014）。

超高压岩石脱水熔融在其他造山带也广泛存在。例如，格陵兰岛超高压变泥质岩经历了高压条件下多硅白云母+石英的脱水熔融（Lang and Gilotti，2007）；柴北缘锡铁山超高压榴辉岩中富钠长英质脉体可能经历了折返时绿辉石+黝帘石主导的脱水熔融（Chen et al.，2012）。Song 等（2014a）发现，柴北缘都兰榴辉岩中存在英云闪长质和奥长花岗质脉体及英云闪长质岩体。基于同位素定年和脉体的地球化学组成分析，他们认为这是榴辉岩在麻粒岩相条件下通过黝帘石+绿辉石脱水部分熔融形成的。值得注意的是，柴北缘榴辉岩的部分熔融形成了混合岩脉体甚至花岗质岩体，表明其经历了高程度部分熔融。这可能与其经历了高温麻粒岩相叠加有关，如都兰超高压榴辉岩部分熔融可能发生在 870～950℃和 1.9～2.0 GPa 条件下（Song et al.，2014a）。由此可见，超高压岩石的部分熔融特征与原岩属性、变质 P-T 路径密切相关。当然，俯冲带热结构的变化和折返速率也会显著影响俯冲板片的部分熔融行为（Zheng and Chen，2017；Zheng，2019）。如果超高压岩石折返较慢，受地壳放射性生热和地幔加热的影响更易引起高温麻粒岩相叠加，诱发俯冲地壳部分熔融。

1.4 大陆俯冲带超高压岩石部分熔融的地球化学效应

超高压岩石部分熔融会引起主微量元素和同位素在熔体和残留相之间的再分配，从而导致壳内地球化学分异；产生的长英质熔体可以交代上覆地幔楔橄榄岩，这对理解大陆俯冲带壳幔相互作用和地壳物质再循环具有重要意义（Zheng，2012，2019；Zheng and Hermann，2014；Chen Y. X. et al.，2017；Zhao et al.，2017a；Sun et al.，2020）。超高压岩石部分熔融形成的熔体组成受控于多种因素，如原岩成分、温压和流体条件、残留相组合等（Manning，2004；Zheng et al.，2011；Hermann et al.，2013；Zheng and Hermann，2014；Zhou et al.，2020）。

泥质岩在超高压条件下的矿物组合为石榴子石、柯石英、多硅白云母、蓝晶石、单斜辉石和金红石（Hermann and Spandler，2008；Hermann and Rubatto，2009）。因此，如果没有自由水参与，部分熔融主要是白云母脱水熔融。由于白云母富钾和其他大离子亲石元素（large ion lithophile element，LILE），产生的熔体总体上富集这些元素。其他副矿物的溶解度则控制了一些关键微量元素，如金红石（Ti、Nb、Ta）、磷灰石（P）、锆石（Zr）、独居石/褐帘石（Th、LREE）。但从天然样品中很难获取原始的超高压熔体的地球化学组成，因为超高压熔体难以保存。通过对比熔融残余相和可能的原岩组成，超高压变泥质岩来源的熔体被认为富集 LILE 和 LREE（Behn et al.，2011；Stepanov et al.，2014）。Stepanov 等（2016）通过高温高压均一化实验直接测定了 Kokchetav 变泥质岩中多相包裹体代表的熔体组成，发现超高压熔体在约 4.5 GPa 形成，具有高 LREE、Th 和 U 及较高的 LILE 含量特征，可能与独居石在熔体中的溶解有关。而 Kokchetav 变泥质岩熔融残留岩石亏损 LREE、Th、U 和 LILE，这可用超高压条件下熔体的提取来解释（Shatsky et al.，1999；Stepanov et al.，2014，2016）。深熔熔体的高 U 含量也可以解释深熔锆石的高 U 含量特征（Chen Y. X. et al.，2013a）。

花岗片麻岩体系与变泥质岩体系基本类似。苏鲁造山带超高压片麻岩部分熔融形成的混合岩浅色体具有低 REE、Th 含量特征（Xu et al.，2013；Zhou et al.，2019）。这在其他超高压变质地体也有发现，如柴北缘（Zhang et al.，2015；Yu et al.，2019）、挪威西片麻岩省（Gordon et al.，2013）。由于独居石和褐帘石是超高压片麻岩在高压-超高压条件下部分熔融时主要赋存 LREE 和 Th 的矿物相，浅色体中低的 LREE 和 Th 含量可能与其在深熔熔体中溶解极为有限有关（Hermann，2002b；Hermann and Rubatto，2009；Skora and Blundy，2012；Stepanov et al.，2014）。浅色体普遍富集 LILE（Gordon et al.，2013；Xu et al.，2013；Zhang et al.，2015），可能与部分熔融过程中富 K 和 LILE 的多硅白云母脱水分解有关（Hermann and Rubatto，2009）。

变玄武岩体系在超高压条件下的矿物组合为石榴子石、绿辉石、柯石英、多硅白云母、蓝晶石、金红石及副矿物。多硅白云母主要赋存 K 和 LILE，而其他副矿物金红石、锆石和褐帘石则分别赋存高场强元素（Ti、Nb、Ta）、Zr-Hf 和 LREE-Th（Zack et al.，2002；Rubatto and Hermann，2003）。部分熔融过程中这些矿物的行为决定了熔体的组成。对大别-苏鲁造山带代表熔体的多相固体包裹体的 LA-ICP-MS 分析发现，这些包裹体富集 LILE 和 LREE，具有低 Y 和较高的 Sr/Y 值（Gao et al.，2013，2014；Liu et al.，2013；Chen et al.，2014），这与多硅白云母脱水熔融和富 LREE 矿物溶解进入熔体有关。而榴辉岩中深熔熔体的提取则使残留相亏损 LREE 和 LILE（Zhao et al.，2007b；Wang et al.，2014；Chen et al.，2016a）。榴辉岩中混合岩浅色体、岩脉甚至岩体都表明其发生了高程度部分熔融。Chen 等（2012）发现榴辉岩中英云闪长质脉体具有高 LILE、低 LREE 和显著分异的 REE 配分模式。Song 等（2014a）发现，都兰榴辉岩的英云闪长质-奥长花岗质脉体具有高的 LILE、LREE 含量，显著分异的 REE 配分及高 Sr/Y 值特征。需要注意的是，榴辉岩部分熔融产生的熔体组成同样受控于部分熔融的温压条件、方式和副矿物的溶解。

超高压岩石的部分熔融会造成微量元素的分异和放射性成因同位素的不平衡。由于不同矿物放射性母子体比值的差异，随着时间的积累，放射性成因同位素组成会有较大差异。部分熔融过程中，如果源区矿物在放射性成因同位素组成不一致，它们以不同比例进入熔体，或由于副矿物差异性溶解进入熔体，都会造成熔体 Sr-Nd-Hf 等同位素组成的变化。这在不同的熔融体制（加水熔融 vs. 脱水熔融）下表现的更明显，在对喜马拉雅淡色花岗岩以及一些 S 型花岗岩的研究中有很好体现（Inger and Harris, 1993；Knesel and Davidson, 2002；Farina et al., 2014a, 2014b；Gao and Zeng, 2014；Gao L. E. et al., 2017）。尽管超高压岩石部分熔融的同位素行为与低压条件下地壳岩石熔融在原理上类似，但研究极少。锆石是超高压变玄武岩、变泥质岩和片麻岩中赋存 Hf 的主要矿物，具有极低的 Lu/Hf 值，随时间演化会具有极低的 $^{176}Hf/^{177}Hf$ 值。由于难熔因而残留锆石非常常见，锆石也很难与其他矿物达到 Hf 同位素均一。因此，不均一源区的部分熔融会形成 Hf 同位素组成不同的多批次熔体，并体现在结晶锆石中。目前在 S 型花岗岩的岩浆锆石中发现其 Hf 同位素组成变化很大（Farina et al., 2014b；Tang et al., 2014），对变沉积岩深熔反应的模拟解释了这种变化。苏鲁造山带混合岩浅色体、高压-超高压花岗质脉体和同折返花岗岩中的岩浆锆石相对原岩继承锆石均具有显著高的 Hf 同位素组成，指示部分熔融过程中造岩矿物提供了放射成因的 Hf（Chen Y. X. et al., 2015；Zhou et al., 2020）。总之，深熔过程中放射性成因同位素组成的变化是由于熔体-残留相同位素不平衡造成的。这种不平衡主要是不均一源区的熔融以及岩浆过程中熔体不充分混合引起的，深熔熔体的快速提取和运移使熔体倾向于保存初始的同位素不均一性。

2 大陆俯冲带壳幔相互作用的直接产物：造山带橄榄岩

2.1 造山带橄榄岩类型和识别

造山带橄榄岩与蛇绿岩不同，其原始超基性岩来自岩石圈地幔，主要经历了与大陆边缘有关的汇聚或裂解等动力学过程及其相关的物理化学性质演变，但少量蛇绿岩经历俯冲带过程叠加也可形成造山带橄榄岩（Bodinier and Godard, 2014；Chen et al., 2019；Zhang et al., 2019；张宏福和于红，2019；郑建平等，2019a）。形成于大陆碰撞过程的石榴子石橄榄岩根据岩性、地球化学和构造演化可以分为幔源型（M 型、Mg-Cr 型）和壳源型（C 型、Fe-Ti 型）（Carswell et al., 1983；Zhang et al., 2000；Zheng, 2012；郑建平等，2019a）两类。M 型橄榄岩起源于地幔楔，在大陆深俯冲过程中被俯冲-折返的陆壳刮削下来卷入俯冲隧道，与俯冲板片中的长英质和镁铁质岩石共同经历了超高压变质作用（Zheng, 2012）。C 型橄榄岩是在大陆俯冲之前，幔源岩浆侵入到俯冲陆壳发生结晶分异形成超镁铁质堆晶，随后与俯冲陆壳一起经历超高压变质作用后又折返回地表（Zhang et al., 2000；Liou et al., 2007；Zheng et al., 2008）。鉴于两者来源不同，M 型橄榄岩提供了从俯冲板片向上覆大陆岩石圈地幔楔物质传输的直接岩石学记录（Zheng, 2012；郑建平等，2019a），而 C 型橄榄岩更适合为大陆地壳俯冲和折返提供重要的温压约束。

C 型和 M 型橄榄岩可用层状结构、全岩和矿物组成来区分（Carswell et al., 1983；Zhang et al., 2000；Reverdatto et al., 2008）。层状结构可以继承自超镁铁质堆晶原岩，或通过变质分异形成，或通过流体交代橄榄岩形成（Rampone and Morten, 2001；Malaspina et al., 2006；Vrijmoed et al., 2013；陈意等，2015；Chen Y. et al., 2017），因此两种橄榄岩无法简单地通过野外观察来区别。大别造山带毛屋橄榄岩由于具有层状结构因而早期被认为是 C 型橄榄岩（Okay, 1994），近期的研究明确其为 M 型橄榄岩（Malaspina et al., 2006；Chen Y. et al., 2017）。经历了广泛地壳交代作用的 M 型橄榄岩全岩成分会发生变化，甚至会表现出与 C 型橄榄岩类似的地球化学组成（Vrijmoed et al., 2013；陈意等，2015）。因此，通过全岩组分对两类橄榄岩进行鉴别时，应充分考虑地壳交代作用对全岩组分的影响。近年来的研究发现，最亏损的橄榄岩端元（如纯橄岩）受到陆壳熔流体交代是最弱的（陈意等，2015；Chen Y. et al., 2017），对其进行地球化学分析被认为是恢复受过改造的地幔源区原岩性质的最理想的方法（Brueckner et al., 2002；

Griffin et al., 2004)。Su 等（2016a）按岩石成因将全球造山带纯橄岩分为部分熔融型橄榄岩、熔-岩反应型纯橄岩和堆晶型纯橄岩，前两类为幔源纯橄岩，它们在全岩组成、橄榄石 Fo 值及尖晶石组成上尽管存在一定重叠，但总体上可以进行区分。

壳源流体只有在硫饱和时才会显著改变橄榄岩的 Re-Os 同位素组成（Chen Y. et al., 2017）。残留橄榄岩特别是最难熔橄榄岩的 Re-Os 同位素组成可用来约束地幔亏损时间，进而为造山带橄榄岩的原岩性质提供强有力的制约。大别-苏鲁和柴北缘造山带橄榄岩的 Re-Os 同位素研究很好地识别了其中的 M 型橄榄岩（Yuan et al., 2007；Shi et al., 2010；Chen Y. et al., 2017）。铂族元素（PGE）可分为 IPGE 和 PPGE，部分熔融过程中 IPGE 为强难熔组分，保存在残留地幔中，而 PPGE 优先进入熔体（Righter et al., 2004；Brenan et al., 2005）。因此，PGE 可用于判别橄榄岩的类型（Liu Q. et al., 2012；Xie et al., 2013；Chen Y. et al., 2017；谢志鹏等, 2018）。C 型橄榄岩一般具有较低含量的 IPGE 和低的 IPGE/PPGE 值（Hattori et al., 2010；Wang et al., 2012），而 M 型橄榄岩则相反且 PPGE 显示不同程度的亏损。微区分析方法的进步能够对橄榄岩中原生矿物和次生矿物进行很好的区分，进而能够根据原位矿物地球化学组成对造山带橄榄岩的成因提供更准确的制约。Su 等（2019a）对来自大别-苏鲁、挪威西部片麻岩地体和南阿尔金造山带的造山带橄榄岩中的橄榄石进行了系统的主微量元素分析，结果发现橄榄石的 Ni/Co 值可以作为判别幔源和壳源造山带橄榄岩简单有效的矿物学指标。

2.2 造山带橄榄岩中锆石成因

锆石具有较高的物理化学稳定性和难熔性，被广泛用于包括交代作用在内的各种地质事件的年龄确定和地球化学示踪。原始地幔橄榄岩具有低的 Zr 含量和 Si 活度（Palme and O'Neill, 2007），理论上无法直接结晶出锆石（Zheng, 2012）。但在许多造山带橄榄岩、橄榄岩包体和蛇绿岩中都发现有锆石。传统认为，橄榄岩中的这些锆石是在侵位到地壳过程中受到地壳混染产生的。然而，在橄榄岩薄片中锆石颗粒和锆石中橄榄石包裹体的发现说明，锆石来自寄主橄榄岩本身而非地壳混染（Zhang et al., 2005, 2011；郑建平等, 2019b）。对于橄榄岩中锆石的成因还存在很大争议，主要机制有地幔流体交代（Grieco et al., 2001；Zheng et al., 2006）、俯冲地壳流体交代（Hermann et al., 2006）、橄榄岩侵位时花岗质岩浆注入（Belousova et al., 2015）和俯冲地壳残片（Yamamoto et al., 2013）。确定造山带橄榄岩中锆石成因对于正确解释锆石年龄及其他信息所代表的意义至关重要。

造山带橄榄岩中锆石可分为残留锆石和新生锆石。残留锆石通常保留部分锆石晶形，内部发育振荡环带（Katayama et al., 2003；Liati and Gebauer, 2009；Yang et al., 2009；Zheng et al., 2014；Li H. Y. et al., 2016），Th/U 值主体区间 0.1~1，通常具有岩浆锆石的微量元素组成。新生锆石内部均一，无明显分带，Th/U<0.01~1，化学组成各不相同，表明它们是受到不同成分流体交代作用形成的（Chen and Zheng, 2017；郑建平等, 2019b）。新生锆石中磷灰石、角闪石、氧化铀等矿物包裹体和长英质熔体包裹体的发现（Zhang et al., 2005, 2011；Hermann et al., 2006；Liati and Gebauer, 2009），指示锆石的形成可能与地壳物质的参与有关。苏鲁造山带荣成橄榄岩中新生长锆石具有负的 $\delta^{18}O$ 值和地壳特征的 Hf 同位素组成（Li H. Y. et al., 2016）。负 $\delta^{18}O$ 值是大别-苏鲁造山带深俯冲华南陆壳的典型特征（Zheng, 2012），进一步明确了橄榄岩中锆石是由地壳流体交代形成的。柴北缘绿梁山造山带橄榄岩中新生锆石也具有地壳特征的 Hf-O 同位素组成（Xiong et al., 2011；Chen R. X. et al., 2017），也证实橄榄岩中锆石是地壳流体交代成因。

造山带橄榄岩中残留锆石含有磷灰石、黑云母、角闪石、磷钇矿和硫化物等包裹体，而无橄榄石、辉石等矿物，表明残留锆石并非生长于橄榄岩中。苏鲁造山带荣成橄榄岩中残留锆石负 $\delta^{18}O$ 值的发现（Li H. Y. et al., 2016），明确指示残留锆石为地壳来源。这些残留锆石的地球化学组成与同一造山带深俯冲大陆板片超高压变质岩中的残留锆石可以类比（Zheng et al., 2014；Chen and Zheng, 2017；Chen R. X. et al., 2017；陈仁旭等, 2019）。可见造山带橄榄岩中新生锆石具有交代成因，残留锆石是俯冲地壳来源流体通过物理搬运进入橄榄岩。因此，造山带橄榄岩中交代锆石记录了地幔楔橄榄岩受到俯冲地壳来源流体交代作用的时间；残留锆石是俯冲地壳来源流体通过物理搬运进入橄榄岩中的，为交代介质来源提供矿物学制约（Chen and Zheng, 2017；陈仁旭和郑永飞, 2019）。造山带橄榄岩中锆石是记录地壳与地幔之间物质和能量交换的重要媒介，进一步为大陆俯冲带地幔楔组成演化、壳幔相互作用、流体-岩石相互作用及俯冲带地球动力学等提供重要信息（郑建平等, 2019b）。

2.3 造山带橄榄岩记录的壳幔相互作用

大陆碰撞造山带 M 型橄榄岩是大陆俯冲带壳幔相互作用最直接的记录。通过对典型大陆碰撞造山带中 M 型橄榄岩的详细研究，如挪威西片麻岩省的 Otrøy-Fjørtoft 橄榄岩（van Roermund et al., 2002；Spengler et al., 2006；Scambelluri et al., 2008, 2010；van Roermund, 2009a, 2009b；Malaspina et al., 2010；Rielli et al., 2017），意大利东阿尔卑斯的 Ulten 橄榄岩（Rampone and Morten, 2001；Tumiati et al., 2003；Scambelluri et al., 2006, 2010；Sapienza et al., 2009），大别造山带毛屋（Malaspina et al., 2006, 2009, 2015, 2017；Chen Y. et al., 2013a, 2013b, 2017）、苏鲁造山带荣成（Li H. Y et al., 2016, 2018a；Su et al., 2016b, 2017）、东海-芝麻坊（Zheng et al., 2005, 2006, 2014；Zhang et al., 2007；Malaspina et al., 2009；Ye et al., 2009）、蒋庄橄榄岩（Zhang et al., 2011；Su et al., 2019b），柴北缘造山带绿梁山橄榄岩（Song et al., 2005a, 2005b, 2007；Shi et al., 2010；Xiong et al., 2011, 2014, 2015；Chen R. X. et al., 2017），发现 M 型橄榄岩中存在诸多次生矿物，如角闪石、钛斜粒硅镁石、金云母、金红石、磷灰石、锆石、绿帘石、菱镁矿、白云石、独居石、天青石、黄铁矿、硬石膏和蓝宝石等，指示地幔楔橄榄岩受到了多期次的显性交代作用（Scambelluri et al., 2006；Zhang R. Y. et al., 2009；Zheng, 2012；陈意等, 2015；陈仁旭等, 2019；郑建平等, 2019a）。M 型橄榄岩的全岩和矿物组成还显示出多种隐性交代过程（陈意等, 2015；陈仁旭等, 2019；郑建平等, 2019a）。这两种交代作用是在机械作用（物理过程）主导下通过流体与橄榄岩间的化学反应产生的（Zheng, 2012；郑建平等, 2019a）。M 型橄榄岩全岩和矿物同位素组成进一步证实它受到了多期俯冲大陆地壳来源流体的交代作用。这些流体包括富水溶液、含水熔体、碳质流体、富含硅酸盐组分的 C-H-O 流体-超临界流体被认为都参与了交代作用，且这些熔-流体可能具有相对氧化的性质（郑建平等, 2019a）。越来越多研究在造山带橄榄岩中识别出碳酸盐熔体的交代作用（Su et al., 2017；Li et al., 2018a；郑建平等, 2019a），但还需要确定碳酸盐熔体交代作用是发生于橄榄岩进入俯冲隧道之前还是俯冲板片-地幔楔界面。

俯冲地壳来源的不同性质流体与橄榄岩相互作用会形成不同的交代岩和交代矿物（Zheng, 2012；郑建平等, 2019a）。富水溶液交代会导致地幔楔橄榄岩发生蛇纹石化和绿泥石化，富硅熔体交代会诱发贫硅矿物（橄榄石）的溶解和富硅矿物（辉石等）的结晶，硅不饱和熔体交代会消耗斜方辉石产生橄榄石（Ionov et al., 1996；Laurora et al., 2001；Xiong et al., 2014；陈意等, 2015；Li et al., 2018b；Su et al., 2019b）。从交代橄榄岩和辉石岩全岩和交代成因矿物的组成来看，进入造山带橄榄岩的流体总体上富集 Si、Ca、Al、K、Na、F、CO_2、LILE、LREE，且含有大量 HREE 和高场强元素（high field strength elements, HFSE）（Hermann et al., 2006；Scambelluri et al., 2006；陈仁旭等, 2019）。不同期次地壳交代作用涉及的流体，其性质和组成也存在很大差异。俯冲地壳来源的这些熔-流体活动可以呈弥散状扩散，可以沿物理裂隙迁移，也可以沿熔-流体通道运输（Zheng, 2012；郑建平等, 2019a）。在此过程中，这些交代流体不仅可将流体活动组分通过化学迁移带入造山带橄榄岩中，而且还将难溶的矿物机械迁移到造山带橄榄岩中（Li H. Y et al., 2016）。因此，地壳流体的交代作用不仅改变地幔楔橄榄岩的地球化学成分（Zheng et al., 2005；Zhang et al., 2007；Chen Y. et al., 2013a），而且还形成了新的交代矿物，如无水-含水硅酸盐矿物、碳酸盐矿物，以及伴生的硫酸盐矿物-硫化物、单质和氧化物等（Rampone and Morten, 2001；Scambelluri et al., 2006；Zheng et al., 2006, 2014；Xiong et al., 2011；Su et al., 2016b, 2017；Chen R. X. et al., 2017；Chen Y. et al., 2017），在地幔楔中形成各种超镁铁质交代岩（Zheng,

2012)。

对大别-苏鲁和柴北缘造山带 M 型橄榄岩中锆石的系统总结（图3）发现，俯冲大陆地壳不同部位来源的不同类型流体对橄榄岩的交代作用，在大陆碰撞期间从俯冲到峰期超高压变质再到折返过程中都有发生，且主要发生在折返阶段（Chen and Zheng，2017）。根据苏鲁芝麻坊和大别毛屋 M 型橄榄岩详细的岩石学研究推测，大别-苏鲁造山带橄榄岩曾受到先前俯冲大洋板片来源流体的交代（Ye et al.，2009；Chen Y. et al.，2013a，2013b）。大别-苏鲁造山带 M 型橄榄岩中部分新生锆石具有古生代 U-Pb 年龄（图3），早于大陆俯冲的中生代年龄，指示这些地幔楔橄榄岩受到了先前俯冲古洋壳来源流体的交代作用。Shen 等（2018）在毛屋石榴辉石岩中发现了 $\delta^{18}O$ 值为 4.3‰~12.2‰ 的古生代锆石，证实先前俯冲古洋壳来源流体对地幔楔橄榄岩的交代作用。柴北缘造山带出露有洋壳和陆壳俯冲形成的榴辉岩，其氧同位素组成存在差异（Zhang et al.，2016，2017）。绿梁山橄榄岩矿物的氧同位素组成指示其受到了洋壳和陆壳来源流体的交代作用（Chen R. X. et al.，2017）。该露头 M 型橄榄岩中除了与大陆碰撞过程相对应的交代锆石外，还存在许多约 460 Ma 的交代锆石，这与沙流河和野马滩洋壳榴辉岩的一组榴辉岩相变质年龄一致（Song et al.，2014b）。从大别-苏鲁和柴北缘这两个造山带地幔楔橄榄岩的研究实例来看，大陆俯冲带地幔楔橄榄岩经历了先前俯冲古洋壳来源的流体交代作用。因此，大陆俯冲带 M 型橄榄岩主要记录了俯冲大洋和大陆地壳与上覆大陆岩石圈地幔楔之间的相互作用，以及大陆岩石圈地幔的早期演化历史。

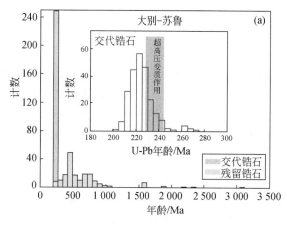
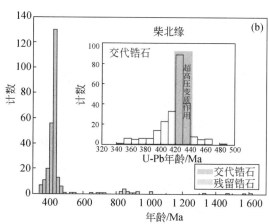

图 3　M 型橄榄岩中交代锆石和残留锆石 U-Pb 年龄

（a）大别-苏鲁造山带数据来自 Chen and Zheng，2017。（b）柴北缘造山带数据来自 Song et al.，2005b；Xiong et al.，2011；Chen R. X. et al.，2017；蔡鹏捷等，2018

2.4　大陆俯冲带壳幔相互作用及其效应

传统的板块构造理论认为，大洋俯冲带才是壳幔相互作用最为活跃的地区（Zheng and Chen，2016）。俯冲大陆地壳在折返过程中存在显著的熔-流体活动（Hermann et al.，2013；Zheng and Hermann，2014），大陆俯冲带在弧后和弧下深度可能发育有与洋壳俯冲带类似的流体活动（郑永飞等，2016）。这些熔-流体活动必然会对大陆岩石圈地幔楔和大陆俯冲隧道产生影响，主控着大陆俯冲带的物质循环。大陆俯冲隧道内壳幔相互作用在本质上类似于大洋俯冲隧道，即均发生不同块体及岩石间的机械混合、变质变形、化学反应和熔-流体活动（郑建平等，2019a；郑永飞和陈伊翔，2019）。

大陆俯冲隧道内俯冲板片会随温压的变化释放出不同性质的流体或熔体（Zheng et al.，2011；郑永飞等，2013），这些熔-流体会与地幔楔橄榄岩发生不同方式和程度的相互作用，形成不同类型的地幔交代岩（Zheng，2012；郑建平等，2019a）。根据交代介质的性质、组成及交代程度，交代岩在矿物类型、含量和成分方面是可变的，并且不同程度地富集不相容元素和放射成因同位素。这些交代岩的地壳特征主要来自俯冲地壳的交代介质，而俯冲带过程之前的镁铁质熔体提取没有明显的影响。交代岩在造山带岩石圈地幔中以大小不等的超镁铁质区域出现，造成了地幔楔的岩性和地球化学不均一。它们在地幔楔中的存在对俯冲带岩浆作用、俯冲隧道中俯冲-折返大陆地壳以及深部地幔的结构和组成具有不同的物理化学效应（Zheng，2012；Zheng and Hermann，2014；Zheng and Chen，2016）。

地幔楔中的交代岩分别在同俯冲、同折返和俯冲后阶段作为镁铁质岩浆的地幔源区（Zheng，2012；Zheng and Zhao，2017）。新形成的交代岩在受热时比正常橄榄岩更容易发生部分熔融（Zheng，2019）。但由于大陆俯冲带的冷俯冲特性，这些交代岩并不是在形成时就发生部分熔融形成岩浆岩，而是在造山带岩石圈地幔中居留几个乃至几十个百万年，受到加热后再发生部分熔融，形成同折返或碰撞后镁铁质岩浆作用（郑永飞等，2015）。然而，无论是在深俯冲大陆地壳的折返过程中，还是在碰撞后造山带岩石圈减薄过程中，地幔楔中富沃的交代岩由于加热部分熔融产生造山带岩浆活动（Zheng and Chen，2016）。低密度蛇纹石化或绿泥石化橄榄岩在大陆俯冲隧道中能引起并促进榴辉岩的折返。它们从大陆俯冲隧道中脱离，能成为富水流体的重要来源。Chen 等（2016b）对西阿尔卑斯造山带白片岩的研究表明，先前俯冲大洋板片来源的地壳流体所形成的水化橄榄岩反过来又为后期大陆俯冲隧道中的地壳岩石交代提供了富镁流体，导致了所谓的反向交代作用。

交代地幔楔橄榄岩能以碎片形式被拆离进入大陆俯冲隧道并沉入深部地幔。通过这种方式，它们将与俯冲地壳一起传输到深部地幔，从而导致地幔的不均一。地幔交代岩如碳酸盐化橄榄岩、贫硅辉石岩和角闪石岩可以在 80～160 km 的弧下（subarc）深度形成，它们在地幔楔的部分熔融就形成了常见的大洋弧玄武岩和大陆弧安山岩（Zheng，2019；Zheng et al.，2020）。如果这类交代岩形成于大于 200 km 的弧后（postarc）深度，则有可能成为一些洋岛玄武岩（Sobolev et al.，2005）和大陆玄武岩（Zhang J. J. et al.，2009；Xu et al.，2012）的地幔源区。这些洋岛型玄武岩富集 LILE、LREE 和 HFSE，如 Nb、Ta 从不亏损到富集，而 Pb 亏损，这种地球化学特征可用弧下深度俯冲洋壳的变质脱水和随后在弧后深度的部分熔融来解释（Zheng，2019）。流体活动性不相容元素如 Pb 在弧下深度明显丢失，金红石在弧后深度分解并释放 HFSE 如 Nb 和 Ta（Ringwood，1990；Zheng，2012，2019）。

3 大陆俯冲带壳幔相互作用的间接产物——造山带镁铁质岩浆岩

大陆俯冲带不仅出露大量的高压-超高压变质岩和造山带橄榄岩，还广泛发育不同类型的岩浆岩。在大陆碰撞过程中和碰撞后阶段都可以形成不同成分的岩浆岩，如大别-苏鲁造山带晚三叠世同折返和早白垩世碰撞后岩浆岩，其中的花岗岩含有大量新元古代中期和中三叠世 U-Pb 年龄的锆石继承核并具有低的锆石 $\delta^{18}O$ 值，表明它们是俯冲陆壳物质再造的产物（Zhao et al.，2017a，2017b）。除了长英质岩浆岩外，大陆碰撞造山带也出露有规模不大但分布广泛的同碰撞和碰撞后镁铁质岩浆岩（Zhao et al.，2012，2013；Guo Z. F. et al.，2015；Couzinié et al.，2016；Qi et al.，2018；Fang et al.，2020）。这些镁铁质岩浆岩为认识大陆俯冲带深部物质循环和壳幔相互作用提供了理想的天然样品，是研究和揭示造山带地幔属性与演化的直接物质记录（Zhao et al.，2013；赵子福等，2015；许文良等，2020）。在大陆碰撞造山带，俯冲陆壳和先前牵引陆壳俯冲的古洋壳物质均有可能再循环进入地幔并发生壳幔相互作用，由此产生的地幔交代岩发生部分熔融就可以形成这些同碰撞和碰撞后镁铁质岩浆岩（Zhao et al.，2013；赵子福等，2015）。

3.1 俯冲陆壳物质再循环及其衍生的镁铁质岩浆岩记录

大陆地壳俯冲-折返过程中存在显著的熔-流体活动（Zheng et al.，2011；Zheng and Hermann，2014），这些熔-流体有可能交代并改造上覆地幔楔，从而发生壳幔相互作用。大陆碰撞造山带同碰撞和碰撞后镁铁质岩浆岩与俯冲带深部壳幔相互作用密切相关，记录了深俯冲陆壳物质循环及其壳幔相互作用的相关信息。大别-苏鲁造山带是三叠纪华南陆块俯冲进入华北陆块之下形成的大陆碰撞造山带（Zheng et al.，

2003，2019；Wu and Zheng，2013），在该造山带内部和仰冲板块边缘发育有同折返或碰撞后镁铁质岩浆岩，为我们研究俯冲陆壳物质再循环及其壳幔相互作用提供了不可多得的样品（Jahn et al.，1999；Dai et al.，2011，2012；Zhao et al.，2012，2013）。这些镁铁质岩浆岩的 SiO_2 含量为 40.2%～57.3%，K_2O+Na_2O 为 0.3%～8.7%，$Mg^{\#}$ 为 42～87（Zhao et al.，2013，及其中所引参考文献）。它们具有岛弧型的微量元素分布特征，即富集 LILE、LREE 和 Pb，亏损 HFSE；富集放射成因同位素组成，即高的全岩 $(^{87}Sr/^{86}Sr)_i$ 同位素比值（0.704 0～0.711 8），低的全岩 $\varepsilon_{Nd}(t)$ 值（-21.2～-2.3）及低的锆石 $\varepsilon_{Hf}(t)$ 值（-39.7～-0.7）。它们的锆石 O 同位素比值变化较大，$\delta^{18}O$ 值为 2.0‰～7.3‰，大多不同于正常地幔锆石的 $\delta^{18}O$ 值（5.3‰±0.3‰）（Valley et al.，1998）。这些地球化学特征表明，它们来源于富沃、富集的造山带岩石圈地幔。这些镁铁质岩浆岩含有新元古代和三叠纪 U-Pb 年龄的继承锆石核（Dai et al.，2011，2015a），为俯冲华南陆壳物质再循环进入地幔源区提供了鉴定性的年代学证据（Zheng，2012）。

大别造山带碰撞后镁铁质岩浆岩中的锆石 Hf-O 同位素组成可分为 3 组（Dai et al.，2011），分别对应于俯冲华南陆壳的上、中、下地壳（Zhao et al.，2008），进一步证明不同层位的俯冲华南陆壳均再循环进入地幔。这些镁铁质岩浆岩中的辉石具有极低的 $^3He/^4He$ 值及类似于大气的 Ne、Ar 同位素组成（图4）。综合全岩地球化学数据发现，辉石的 He、Ar 同位素组成与全岩地球化学成分具有相关性（Dai L. Q. et al.，2016），表明俯冲大陆地壳不仅自身发生了再循环，其表壳岩石还携带大气稀有气体发生了再循环。通过对大别造山带碰撞后镁铁质岩浆岩中矿物（特别是单斜辉石和斜长石）主微量元素和 Sr 同位素的原位分析，进一步揭示不同地幔源区来源的镁铁质熔体间的岩浆混合（Dai et al.，2015a）。综合这些镁铁质火成岩的岩石地球化学特征，表明大陆碰撞造山带镁铁质岩浆岩的形成主要涉及以下两个过程：不同层位俯冲的华南陆壳物质在三叠纪碰撞过程中部分熔融形成的长英质熔体与上覆地幔楔橄榄岩之间的源区混合形成了富沃、富集的造山带岩石圈地幔，富集的地幔源区发生部分熔融形成了同折返和碰撞后镁铁质岩浆岩（Dai et al.，2011；Zhao et al.，2012，2013）。

大陆碰撞造山带碰撞后安山岩也间接记录了俯冲带的壳幔相互作用。Dai F. Q. 等（2016）对大别造山带碰撞后安山质火山岩进行了同位素年代学和地球化学研究，发现俯冲华南陆壳物质再循环及其壳幔相互作用的鉴定性证据：这些安山岩形成于早白垩世（124～130 Ma），含有丰富的残留锆石核，其 U-Pb 年龄为新元古代和三叠纪；具有岛弧型微量元素分布特征和富集的 Sr-Nd-Hf 同位素组成，同岩浆锆石具有变化的 $\delta^{18}O$ 值，部分同岩浆锆石和残留锆石具有低于正常地幔的 $\delta^{18}O$ 值。这些证据表明，这些安山质火山岩来源于交代富集的造山带岩石圈地幔，是三叠纪华南陆块与华北陆块碰撞过程中，深俯冲华南陆壳在弧下深度发生部分熔融所产生的长英质熔体在大陆俯冲隧道内交代上覆华北岩石圈地幔楔橄榄岩形成的。这种地幔交代岩在早白垩世发生部分熔融，形成了安山质火山岩。因此，大陆俯冲隧道内板片-地幔相互作用是形成碰撞造山带幔源岩浆岩地幔源区的关键过程，而加入地幔楔中长英质熔体的比例决定了这些幔源岩浆岩的岩石化学和地球化学成分。

青藏高原南部的喜马拉雅造山带是印度大陆向亚欧大陆俯冲碰撞的结果，是典型的新生代大陆碰撞造山带（Zheng and Wu，2018）。拉萨地体南部的冈底斯造山带属于增生造山带，对应于亚洲大陆南缘在特提斯洋壳俯冲过程中的安第斯型大陆弧。这里出露的碰撞后镁铁质岩浆岩（超钾质岩石）具有富集的放射性成因同位素组成和弧型的微量元素特征，表明其是俯冲陆壳物质再循环及其壳幔相互作用的产物（Chung et al.，2005；Guo Z. F. et al.，2015；Qi et al.，2018；Lei et al.，2019）。位于阿尔卑斯-喜马拉雅造山带西段的安纳托利亚造山带，是新特提斯洋各分支先后闭合、弧-陆和陆-陆碰撞的产物，其中出露的渐新世—中新世碰撞后镁铁质岩浆岩具有高的初始 $^{87}Sr/^{86}Sr$ 值（0.708 7～0.707 1）和低的 $\varepsilon_{Nd}(t)$ 值（-6.5～-3.5）及弧型的微量元素特征，指示其地幔源区有俯冲陆壳物质的加入（Dilek and Altunkaynak，2007，2009）。海西造山带法国中央地块碰撞后镁铁质岩浆岩具有弧型的微量元素特征、低的锆石 $\varepsilon_{Nd}(t)$ 值（-9～-2）和高的锆石 $\delta^{18}O$ 值（6.4‰～10‰），表明其地幔源区含有再循环的俯冲陆壳物质（Couzinié et al.，2016）。

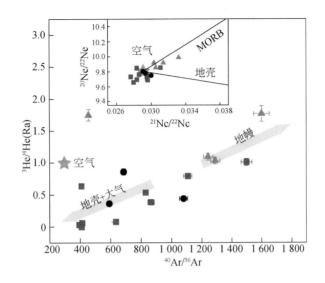

图 4　大别造山带碰撞后镁铁质岩浆岩辉石稀有气体同位素组成（据 Dai L. Q. et al., 2016）

3.2　俯冲古洋壳物质再循环及其衍生的镁铁质岩浆岩记录

低密度的大陆地壳俯冲通常被认为是受到了先前高密度俯冲洋壳的重力牵引，因此在大陆俯冲带，先前俯冲的古洋壳和随后俯冲的陆壳物质都可以再循环进入地幔（郑永飞等，2015）。在大陆碰撞造山带中寻找消失的古洋壳再循环及其壳幔相互作用的岩石学和地球化学证据，对理解大陆碰撞造山带从洋壳俯冲到大陆碰撞的构造转变、发展板块构造理论均具有重要意义。

在大洋俯冲过程中，俯冲玄武质洋壳及其上覆沉积物在弧下深度脱水形成的流体相对富集流体活动性元素如 LILE 和 LREE，亏损流体不活动性元素如 HREE 和 HFSE（Zheng, 2019）。这些流体交代上覆地幔楔形成水化的橄榄岩，该地幔源区发生部分熔融形成大洋弧玄武岩和大陆弧安山岩（Tatsumi and Eggins, 1995; Schmidt and Poli, 2014; Zheng et al., 2020）。随着洋壳的进一步俯冲，其在弧后深度（>200 km）部分熔融产生的长英质熔体在微量元素上通常表现为富集 LILE 和 LREE、不亏损 HFSE（如 Nb 和 Ta）；在放射成因同位素组成上，由玄武质洋壳部分熔融形成的熔体相对亏损，而由沉积物部分熔融产生的熔体相对富集（Zheng, 2019）。这些熔体与橄榄岩发生反应，可以形成贫橄榄石的橄榄岩、碳酸盐化橄榄岩或者不含橄榄石的辉石岩和角闪石岩（Sobolev et al., 2005; Pilet et al., 2008; Dai et al., 2017a）。这些超镁铁质的地幔交代岩可以作为一些洋岛玄武岩（Hirschmann et al., 2003; Pilet et al., 2008; Herzberg, 2011）和大陆玄武岩的地幔源区（Zhang J. J. et al., 2009; Xu et al., 2012; 郑永飞等, 2018）。因此，理解大洋俯冲过程中壳幔相互作用的化学地球动力学机制是研究镁铁质岩浆岩成因的关键。

秦岭–桐柏–红安造山带记录了华南与华北陆块从古生代到中生代时期的拼合过程，保存了一系列构造演化过程记录，包括古生代洋壳俯冲、弧地壳增生和弧–陆碰撞及早中生代陆壳俯冲和陆–陆碰撞，是典型的洋陆复合造山带（Wu and Zheng, 2013）。该造山带在形成过程中产生的同俯冲–碰撞后镁铁质岩浆岩，为研究俯冲古洋壳再循环并区分其与俯冲陆壳再循环的异同点提供了不可多得的样品。其中，桐柏造山带早古生代岛弧型镁铁质火成岩具有高的 MgO（3.3% ~ 8.4%）和低的 SiO_2（50.2% ~ 57.5%）含量，表明它们来自超镁铁质地幔源区的部分熔融（Wang et al., 2013; Zheng et al., 2019）。它们具有弧型微量元素分布特征（富集 LILE、LREE 和 Pb，亏损 HFSE）和相对亏损的放射成因同位素组成，这与典型的弧玄武岩一致，表明它们来自受俯冲洋壳来源流体（弧下深度）交代的地幔源区。模拟计算结果表明，约 1% ~ 10% 俯冲大洋地壳来源的富水溶液和 0.05% ~ 0.1% 俯冲沉积物来源的含水熔体与上覆亏损地幔橄榄岩发生反应，形成的地幔交代岩部分熔融能够解释桐柏造山带早古生代镁铁质岩浆岩的微量元素和放射性成因同位素特征（Zheng et al.，2019）。

秦岭-桐柏-红安造山带还发育有中新生代洋岛型镁铁质岩浆岩，其具有洋岛型微量元素特征及亏损的放射性成因同位素组成和变化的锆石Hf-O同位素组成（Dai et al.，2014，2015b，2017b），明显不同于正常的岛弧岩浆岩（弧型的微量元素特征和亏损的放射性成因同位素组成），也不同于大别-苏鲁造山带碰撞后镁铁质岩浆岩（弧型的微量元素特征和富集的放射性成因同位素组成）（Zhao et al.，2013）。这些特征揭示它们的地幔源区可能是由先前俯冲洋壳在弧后深度部分熔融（金红石发生分解）产生的长英质熔体与亏损的地幔楔橄榄岩反应形成的。部分镁铁质岩浆岩含有三叠纪和新元古代U-Pb年龄的残留锆石，进一步揭示俯冲的洋壳物质很有可能是靠近华南陆块边缘的古特提斯玄武质洋壳及其上覆沉积物。因此，俯冲古洋壳在弧后深度部分熔融产生的长英质熔体与上覆亏损的地幔楔橄榄岩反应形成了新生的造山带岩石圈地幔，这些地幔源区在碰撞后阶段发生部分熔融可以形成秦岭-桐柏-红安造山带中新生代洋岛型镁铁质岩浆岩。喜马拉雅造山带藏南地区出露的始新世镁铁质岩浆岩也具有洋岛型微量元素特征和相对亏损的放射性成因同位素组成 [$\varepsilon_{Nd}(t)$ = 5.1~6.1]，记录了先前俯冲新特提斯洋壳物质的再循环（Lei et al.，2019）。

此外，一些同碰撞镁铁质岩浆岩也可能记录了先前俯冲洋壳物质再循环及其壳幔相互作用。Fang等（2020）在大别-苏鲁造山带仰冲板块边缘-辽东半岛南端发现了早三叠世（244~247 Ma）基性岩脉，该时期华南-华北陆块处于大陆碰撞初始阶段。这些基性岩脉具有洋岛型微量元素特征和亏损到弱富集的Sr-Nd-Hf同位素组成。岩浆锆石Hf-O同位素组成可以分为两组，一组具有相对亏损的Hf同位素组成和高的$\delta^{18}O$值，另一组具有相对富集的Hf同位素组成和高的$\delta^{18}O$值，分别对应低温蚀变洋壳玄武岩及其上覆沉积物。同时，样品中存在早古生代年龄（422~429 Ma）和元古宙年龄（786~1059 Ma）的残留锆石，分别与古特提斯洋壳变质年龄和华南陆壳岩浆岩年龄一致，指示俯冲古特提斯洋壳及其上覆陆源沉积物参与了地幔交代作用。基于这些结果和前人对大别-苏鲁造山带构造演化的研究，古生代俯冲古特提斯洋壳及其上覆沉积物来源的熔体交代了华北大陆岩石圈地幔，产生了富化、富集的地幔源区。早三叠世，华南-华北发生陆-陆碰撞，引起俯冲古特提斯洋板片回卷，导致软流圈侧向流动，地幔楔底部受到加热发生部分熔融，产生具有洋岛型地球化学特征的同碰撞镁铁质岩浆岩。因此，这些基性脉作为华北东南缘首次发现的大陆同碰撞岩浆岩，不仅提供了大陆碰撞造山带古洋壳俯冲交代地幔楔的地球化学证据，而且记录了从洋壳俯冲到大陆碰撞过程中的构造转换。

板片俯冲过程中释放出的交代熔-流体一般可分为富水流体、硅酸盐熔体和碳酸盐熔体（Hermann et al.，2006；Zheng et al.，2011）。交代介质的成分差异直接影响上覆交代地幔楔的性质，因此有必要识别和区分交代熔-流体及其所形成的地幔交代岩性质。Dai等（2017a）对西秦岭中生代和新生代洋岛型玄武岩进行了详细的地球化学研究，认为它们的地幔源区分别受到了硅酸盐熔体和碳酸盐熔体的交代作用。新生代玄武岩具有明显低的SiO_2和Al_2O_3含量，高的CaO和MgO含量，以及高的CaO/Al_2O_3值，微量元素特征表现为富集LREE和大多数LILE、Nb和Ta，亏损K、Pb、Zr、Hf和Ti，具有高的La_N/Yb_N和低的Ti/Eu值。这些特征表明，其地幔源区经历过明显的碳酸盐熔体交代作用。另外，它们具有低的$\delta^{26}Mg$值（-0.54‰~-0.32‰），也表明它们来自于碳酸盐化的地幔源区。新生代玄武岩中碳酸盐矿物（方解石）的出现也进一步支持玄武质岩浆来源于碳酸盐化的地幔源区。与此不同，中生代玄武岩具有明显高的SiO_2和Al_2O_3含量，低的CaO和MgO含量，以及低的CaO/Al_2O_3值。它们具有洋岛型的微量元素特征，不亏损HFSE以及Pb的负异常。它们的$\delta^{26}Mg$值为-0.35‰~0.21‰，与正常地幔值（-0.25‰±0.07‰）相当。这些特征表明，中生代玄武岩可能来自于受硅酸盐熔体交代形成的地幔源区。因此，西秦岭中新生代玄武岩的地幔源区分别受到了硅酸盐熔体和碳酸盐熔体的交代作用，熔体-橄榄岩反应分别形成了角闪石岩和碳酸盐化橄榄岩的地幔源区。

大陆碰撞造山带出露的镁铁质岩浆岩记录了不同类型的壳幔相互作用（赵子福等，2015；Zheng

et al.,2020)。在大陆地壳俯冲之前，先前俯冲的古洋壳在弧下深度（80~160 km）析出的流体与上覆亏损地幔橄榄岩反应形成水化的地幔源区，这些地幔源区可以在洋壳俯冲阶段或后期发生部分熔融，形成具有岛弧型微量元素分布特征和亏损放射成因同位素组成的岛弧型镁铁质岩浆岩。赵子福等（2015）对红安-大别-苏鲁造山带早白垩世碰撞后两种类型的镁铁质岩浆岩进行了总结，并采用 SARSH（俯冲-深熔-反应-储存-加热）模型来解释其成因（Zhao et al.,2013）。第一类具有洋岛型微量元素分布特征和相对亏损的放射成因同位素组成，其地幔源区是由先前俯冲古洋壳在弧后深度发生部分熔融形成的长英质熔体与上覆亏损地幔反应形成的 [图 5（a）]；第二类显示岛弧型微量元素分布特征和相对富集的放射成因同位素组成，其地幔源区是由俯冲陆壳来源的长英质熔体与上覆古老岩石圈地幔反应形成的 [图 5（b）]。这些富化、富集的地幔交代岩储存在造山带岩石圈地幔之中保持稳定，碰撞后阶段造山带根部受

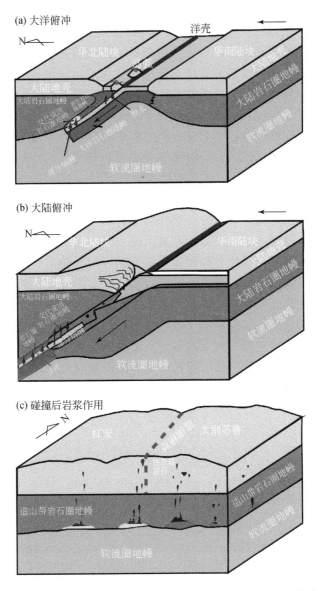

图 5　红安-大别-苏鲁造山带俯冲洋壳和陆壳物质与上覆岩石圈地幔楔相互作用
以及碰撞后镁铁质岩浆作用示意图（据赵子福等，2015）

（a）石炭纪俯冲洋壳来源的长英质熔体与上覆亏损地幔橄榄岩反应形成富集的地幔交代岩（具有洋岛型的微量元素特征和相对亏损的放射成因同位素组成）；（b）三叠纪俯冲陆壳来源的长英质熔体与上覆古老岩石圈地幔楔橄榄岩反应形成富集的地幔交代岩（具有弧型的微量元素特征和相对富集的放射成因同位素组成）；（c）早白垩世不同性质地幔交代岩部分熔融形成地球化学特征不同的碰撞后镁铁质岩浆岩

到加热引起这些地幔交代岩部分熔融,形成了两种类型的镁铁质岩浆岩[图5(c)]。因此,在大陆俯冲带存在不同类型的壳幔相互作用,造山带镁铁质岩浆岩为不同类型的板片-地幔相互作用提供了岩石学和地球化学记录,为研究俯冲带壳幔相互作用提供了一个大陆岩浆岩对象。

4 结语与展望

俯冲到地幔深度的地壳物质不可避免地在板片-地幔界面与地幔楔发生相互作用,由此形成的超镁铁质交代岩可作为造山带镁铁质火成岩的地幔源区。大洋板块俯冲过程中经历变质脱水甚至部分熔融,释放的熔流体交代上覆地幔楔并导致大规模岛弧岩浆活动。与洋壳相比,陆壳具有相对古老、干和冷的特征,因而在俯冲过程中很难发生显著的脱水和熔融过程。但是,在深俯冲陆壳折返阶段部分熔融比较普遍,产生的长英质熔体不仅可以形成同折返花岗岩,而且可以作为交代介质交代不同层位和性质的地幔。大陆碰撞造山带既经历了先前的洋壳俯冲,也经历了随后的陆壳俯冲,这些熔流体活动与板片内外的岩石反应,会记录在不同类型的变质岩和岩浆岩中。造山带橄榄岩可能保存有这些过程的直接记录,而不同阶段岩浆作用产生的镁铁质岩浆岩则间接记载了这些过程。

尽管近年来在大陆俯冲带超高压岩石部分熔融和壳幔相互作用等方面取得了重要进展,但仍有许多重要科学问题有待解决,这些问题包括:①大陆碰撞过程中超高压岩石部分熔融与造山带构造演化过程之间的联系?②原岩组成、熔融温度压力、熔融机制和副矿物等因素对熔体组成有什么影响?③岩浆过程,如熔体提取、迁移和聚集,对熔体组成及其演化有什么影响?④高压-超高压条件下不平衡部分熔融对熔体放射成因同位素组成的影响?⑤不同类型俯冲板片来源熔流体对橄榄岩交代作用发生的时空条件、过程、机制及其产物?⑥俯冲带熔流体如何改造地幔楔的物理性质?⑦俯冲板片熔流体交代橄榄岩过程中挥发性元素迁移机制及其与地幔楔氧逸度变化和造山带岩浆岩成分的联系?⑧大陆与大洋俯冲带在物质循环上的异同点?⑨大陆俯冲带壳幔相互作用发生机制和时间?⑩俯冲带熔流体活动如何影响上覆地幔楔,在不同阶段造山带中岩浆岩中的响应是什么?未来的研究需要聚焦俯冲带熔流体活动和壳幔相互作用,将俯冲带变质作用、交代作用和岩浆作用三者有机地结合起来,通过矿物学、岩石学、地球化学等多学科和现代原位微区分析技术、实验研究、数值模拟等手段的综合研究,打通宏观与微观、浅表与深部、现今地球与地质历史之间的连接,查明俯冲带变质作用、交代作用、岩浆作用,揭示包括挥发分在内的各种元素在地球内部的迁移过程、机制及其环境和资源效应,全面深入认识俯冲带壳幔相互作用的化学地球动力学机制和过程及其控制因素。

参 考 文 献

蔡鹏捷,许荣科,郑有业,陈鑫,刘嘉,俞军真. 2018. 柴北缘从大洋俯冲到陆陆碰撞:来自开屏沟造山带M型橄榄岩的证据. 地球科学,43: 2875-2892

陈仁旭,郑永飞. 2019. 造山带橄榄岩记录的大陆俯冲带多期壳幔相互作用. 地球科学,44(12):4095-4101

陈仁旭,尹壮壮,夏春鹏. 2019. 大别-苏鲁造山带橄榄岩记录的碰撞造山过程中地幔楔的地壳交代作用. 矿物岩石地球化学通报,38(3): 459-484

陈伊翔,郑永飞. 2013. 大陆碰撞过程中地壳深熔的岩石学证据:以苏鲁造山带超高压变质岩为例. 科学通报,58(22):2198-2202

陈意,苏斌,郭顺. 2015. 大别-苏鲁造山带橄榄岩:进展和问题. 中国科学:地球科学,45(9):1245-1269

谢志鹏,王建,Hattori K,薛传东,钟军伟,王泽利. 2018. 苏鲁超高压变质带胡家林超镁铁质岩成因及构造意义. 岩石学报,34(6):1539-1556

许文良,赵子福,戴立群. 2020. 碰撞后镁铁质岩浆作用:大陆造山带岩石圈地幔演化的物质记录. 中国科学:地球科学,50, doi:10.1360/SSTe-2019-026

张宏福,于红. 2019. 造山带橄榄岩岩石学与构造过程:以松树沟橄榄岩为例. 地球科学,44(4):1057-1066

张建新. 2020. 俯冲隧道研究:进展、问题及其挑战. 中国科学:地球科学,50, doi:10.1360/SSTe-2019-0312

赵子福,戴立群,郑永飞. 2015. 大陆俯冲带两类壳幔相互作用. 中国科学:地球科学,45(7):900-915

郑建平,熊庆,赵伊,李文博. 2019a. 俯冲带橄榄岩及其记录的壳幔相互作用. 中国科学:地球科学,49(7):1037-1058

郑建平,赵伊,熊庆. 2019b. 造山带橄榄岩中锆石的成因及其地质意义. 地球科学,44(4):1067-1082

郑永飞,陈伊翔. 2019. 大陆俯冲带壳幔相互作用. 地球科学,44(12):3961-3983

郑永飞,赵子福,陈伊翔. 2013. 大陆俯冲隧道过程:大陆碰撞过程中的板块界面相互作用. 科学通报,58(23):2233-2239

郑永飞,陈伊翔,戴立群,赵子福. 2015. 发展板块构造理论:从洋壳俯冲带到碰撞造山带. 中国科学:地球科学,45(6):711-735

郑永飞,陈仁旭,徐峥,张少兵. 2016. 俯冲带中的水迁移. 中国科学:地球科学,46(3):253-286

郑永飞,徐峥,赵子福,戴立群. 2018. 华北中生代镁铁质岩浆作用与克拉通减薄和破坏. 中国科学:地球科学,48(4):379-414

Auzanneau E, Vielzeuf D, Schmidt M W. 2006. Experimental evidence of decompression melting during exhumation of subducted continental crust. Contributions to Mineralogy and Petrology, 152(2):125-148

Behn M D, Kelemen P B, Hirth G, Hacker B R, Massonne H J. 2011. Diapirs as the source of the sediment signature in arc lavas. Nature Geoscience, 4(9):641-646

Belousova E A, Jiménez J M G, Graham I, Griffin W L, O'Reilly S Y, Pearson N, Martin L, Craven S, Talavera C. 2015. The enigma of crustal zircons in upper-mantle rocks: Clues from the Tumut ophiolite, Southeast Australia. Geology, 43(2):119-122

Bodinier J L, Godard M. 2014. Orogenic, ophiolitic, and abyssal peridotites. In: Holland H D, Turekian K K (eds). Treatise on Geochemistry, 2nd ed. Amsterdam: Elsevier, 103-167

Brenan J M, McDonough W F, Ash R. 2005. An experimental study of the solubility and partitioning of iridium, osmium and gold between olivine and silicate melt. Earth and Planetary Science Letters, 237(3-4):855-872

Brown M. 2013. Granite: from genesis to emplacement. GSA Bulletin, 125(7-8):1079-1113

Brueckner H K, Carswell D A, Griffin W L. 2002. Paleozoic diamonds within a Precambrian peridotite lens in UHP gneisses of the Norwegian Caledonides. Earth and Planetary Science Letters, 203(3-4):805-816

Butler J P, Beaumont C, Jamieson R A. 2013. The Alps 1: a working geodynamic model for burial and exhumation of (ultra) high-pressure rocks in Alpine-type orogens. Earth and Planetary Science Letters, 377-378:114-131

Carswell D A, Harvey M A, Al-Samman A. 1983. The petrogenesis of contrasting Fe-Ti and Mg-Cr garnet peridotite types in the high grade gneiss complex of western Norway. Bulletin de Mineralogie, 106(6):727-750

Chen D L, Liu L, Sun Y, Sun W D, Zhu X H, Liu X M, Guo C L. 2012. Felsic veins within UHP eclogite at xitieshan in North Qaidam, NW China: partial melting during exhumation. Lithos, 136-139:187-200

Chen R X, Zheng Y F. 2017. Metamorphic zirconology of continental subduction zones. Journal of Asian Earth Sciences, 145:149-176

Chen R X, Zheng Y F, Gong B, Zhao Z F, Gao T S, Chen B, Wu Y B. 2007. Origin of retrograde fluid in ultrahigh-pressure metamorphic rocks: constraints from mineral hydrogen isotope and water content changes in eclogite-gneiss transitions in the Sulu orogen. Geochimica et Cosmochimica Acta, 71(9):2299-2325

Chen R X, Zheng Y F, Gong B. 2011. Mineral hydrogen isotopes and water contents in ultrahigh-pressure metabasite and metagranite: constraints on fluid flow during continental subduction-zone metamorphism. Chemical Geology, 281(1-2):103-124

Chen R X, Ding B H, Zheng Y F, Hu Z C. 2015. Multiple episodes of anatexis in a collisional orogen: zircon evidence from migmatite in the Dabie orogen. Lithos, 212-215:247-265

Chen R X, Li H Y, Zheng Y F, Zhang L, Gong B, Hu Z C, Yang Y H. 2017. Crust-mantle interaction in a continental subduction channel: evidence from orogenic peridotites in North Qaidam, northern Tibet. Journal of Petrology, 58(2):191-226

Chen X, Schertl H P, Cambeses A, Gu P Y, Xu R K, Zheng Y Y, Jiang X J, Cai P J. 2019. From magmatic generation to UHP metamorphic overprint and subsequent exhumation: a rapid cycle of plate movement recorded by the supra-subduction zone ophiolite from the North Qaidam orogen. Lithos, 350-351:105238

Chen Y, Ye K, Wu Y W, Guo S, Su B, Liu J B. 2013a. Hydration and dehydration in the lower margin of a cold mantle wedge: implications for crust-mantle interactions and petrogeneses of arc magmas. International Geology Review, 55(12):1506-1522

Chen Y, Ye K, Guo S, Wu T F, Liu J B. 2013b. Multistage metamorphism of garnet orthopyroxenites from the Maowu mafic-ultramafic complex, Dabieshan UHP terrane, eastern China. International Geology Review, 55(10):1239-1260

Chen Y, Su B, Chu Z Y. 2017. Modification of an ancient subcontinental lithospheric mantle by continental subduction: insight from the Maowu garnet peridotites in the Dabie UHP belt, eastern China. Lithos, 278-281:54-71

Chen Y X, Zheng Y F, Hu Z C. 2013a. Synexhumation anatexis of ultrahigh-pressure metamorphic rocks: petrological evidence from granitic gneiss in the Sulu orogen. Lithos, 156-159:69-96

Chen Y X, Zheng Y F, Hu Z C. 2013b. Petrological and zircon evidence for anatexis of UHP quartzite during continental collision in the Sulu orogen. Journal of Metamorphic Geology, 31(4):389-413

Chen Y X, Zheng Y F, Gao X Y, Hu Z C. 2014. Multiphase solid inclusions in zoisite-bearing eclogite: evidence for partial melting of ultrahigh-pressure metamorphic rocks during continental collision. Lithos, 200-201:1-21

Chen Y X, Gao P, Zheng Y F. 2015. The anatectic effect on the zircon Hf isotope composition of migmatites and associated granites. Lithos, 238:174-184

Chen Y X, Tang J, Zheng Y F, Wu Y B. 2016a. Geochemical constraints on petrogenesis of marble-hosted eclogites from the Sulu orogen in China. Chemical Geology, 436:35-53

Chen Y X, Schertl H P, Zheng Y F, Huang F, Zhou K, Gong Y Z. 2016b. Mg-O isotopes trace the origin of Mg-rich fluids in the deeply subducted continental crust of western Alps. Earth and Planetary Science Letters, 456:157-167

Chen Y X, Zhou K, Gao X Y. 2017. Partial melting of ultrahigh-pressure metamorphic rocks during continental collision: evidence, time, mechanism, and effect. Journal of Asian Earth Sciences, 145:177-191

Chopin C. 1984. Coesite and pure pyrope in high-grade blueschists of the western Alps: a first record and some consequences. Contributions to Mineralogy and Petrology, 86(2): 107-118

Chopin C. 2003. Ultrahigh-pressure metamorphism: tracing continental crust into the mantle. Earth and Planetary Science Letters, 212(1): 1-14

Chung S L, Chu M F, Zhang Y Q, Xie Y W, Lo C H, Lee T Y, Lan C Y, Li X H, Zhang Q, Wang Y Z. 2005. Tibetan tectonic evolution inferred from spatial and temporal variations in post-collisional magmatism. Earth-Science Reviews, 68(3-4): 173-196

Clemens J D. 2006. Melting of the continental crust: Fluid regimes, melting reactions and source-rock fertility. In: Brown M, Rushmer T (eds). Evolution and Differentiation of the Continental Crust. Cambridge: Cambridge University Press, 296-327

Cloos M, Shreve R L. 1988a. Subduction-channel model of prism accretion, mélange formation, sediment subduction, and subduction erosion at convergent plate margins: 1. background and description. Pure and Applied Geophysics, 128(3-4): 455-500

Cloos M, Shreve R L. 1988b. Subduction-channel model of prism accretion, mélange formation, sediment subduction, and subduction erosion at convergent plate margins: 2. implications and discussion. Pure and Applied Geophysics, 128(3-4): 501-505

Couzinié S, Laurent O, Moyen J F, Zeh A, Bouilhol P, Villaros A. 2016. Post-collisional magmatism: crustal growth not identified by zircon Hf-O isotopes. Earth and Planetary Science Letters, 456: 182-195

Dai F Q, Zhao Z F, Dai L Q, Zheng Y F. 2016. Slab-mantle interaction in the petrogenesis of andesitic magmas: geochemical evidence from postcollisional intermediate volcanic rocks in the Dabie orogen, China. Journal of Petrology, 57(6): 1109-1134

Dai L Q, Zhao Z F, Zheng Y F, Li Q L, Yang Y H, Dai M N. 2011. Zircon Hf-O isotope evidence for crust-mantle interaction during continental deep subduction. Earth and Planetary Science Letters, 308(1-2): 229-244

Dai L Q, Zhao Z F, Zheng Y F, Zhang J. 2012. The nature of orogenic lithospheric mantle: geochemical constraints from postcollisional mafic-ultramafic rocks in the Dabie orogen. Chemical Geology, 334: 99-121

Dai L Q, Zhao Z F, Zheng Y F. 2014. Geochemical insights into the role of metasomatic hornblendite in generating alkali basalts. Geochemistry, Geophysics, Geosystems, 15(10): 3762-3779

Dai L Q, Zhao Z F, Zheng Y F, Zhang J. 2015a. Source and magma mixing processes in continental subduction factory: geochemical evidence from postcollisional mafic igneous rocks in the Dabie orogen. Geochemistry, Geophysics, Geosystems, 16(3): 659-680

Dai L Q, Zhao Z F, Zheng Y F. 2015b. Tectonic development from oceanic subduction to continental collision: geochemical evidence from postcollisional mafic rocks in the Hong'an-Dabie orogens. Gondwana Research, 27(3): 1236-1254

Dai L Q, Zheng Y F, He H Y, Zhao Z F. 2016. Postcollisional mafic igneous rocks record recycling of noble gases by deep subduction of the continental crust. Lithos, 252-253: 135-144

Dai L Q, Zhao Z F, Zheng Y F, An Y J, Zheng F. 2017a. Geochemical distinction between carbonate and silicate metasomatism in generating the mantle sources of Alkali basalts. Journal of Petrology, 58(5): 863-884

Dai L Q, Zheng F, Zhao Z F, Zheng Y F. 2017b. Recycling of Paleotethyan oceanic crust: geochemical record from postcollisional mafic igneous rocks in the Tongbai-Hong'an orogens. GSA Bulletin, 129(1-2): 179-192

Dilek Y, Altunkaynak S. 2007. Cenozoic crustal evolution and mantle dynamics of post-collisional magmatism in western Anatolia. International Geology Review, 49(5): 431-453

Dilek Y, Altunkaynak S. 2009. Geochemical and temporal evolution of Cenozoic magmatism in western Turkey: mantle response to collision, slab break-off, and lithospheric tearing in an orogenic belt. Geological Society, London, Special Publications, 311(1): 213-233

Dobrzhinetskaya L F, Wirth R, Green II H W. 2006. Nanometric inclusions of carbonates in Kokchetav diamonds from Kazakhstan: a new constraint for the depth of metamorphic diamond crystallization. Earth and Planetary Science Letters, 243(1-2): 85-93

Fang W, Dai L Q, Zheng Y F, Zhao Z F, Ma L T. 2020. Tectonic transition from oceanic subduction to continental collision: new geochemical evidence from early-middle Triassic mafic igneous rocks in southern Liaodong Peninsula, east-central China. GSA Bulletin, 132(7-8): 1469-1488

Farina F, Dini A, Rocchi S, Stevens G. 2014a. Extreme mineral-scale Sr isotope heterogeneity in granites by disequilibrium melting of the crust. Earth and Planetary Science Letters, 399: 103-115

Farina F, Stevens G, Gerdes A, Frei D. 2014b. Small-scale Hf isotopic variability in the Peninsula pluton (South Africa): the processes that control inheritance of source ^{176}Hf/^{177}Hf diversity in S-type granites. Contributions to Mineralogy and Petrology, 168(4): 1065

Ferrero S, Wunder B, Ziemann M A, Wälle M, O'Brien P J. 2016. Carbonatitic and granitic melts produced under conditions of primary immiscibility during anatexis in the lower crust. Earth and Planetary Science Letters, 454: 121-131

Ganzhorn A C, Labrousse L, Prouteau G, Leroy C, Vrijmoed J C, Andersen T B, Arbaret L. 2014. Structural, petrological and chemical analysis of syn-kinematic migmatites: insights from the western Gneiss Region, Norway. Journal of Metamorphic Geology, 32(6): 647-673

Gao L E, Zeng L S. 2014. Fluxed melting of metapelite and the formation of Miocene high-CaO two-mica granites in the Malashan gneiss dome, southern Tibet. Geochimica et Cosmochimica Acta, 130: 136-155

Gao L E, Zeng L S, Asimow P D. 2017. Contrasting geochemical signatures of fluid-absent versus fluid-fluxed melting of muscovite in metasedimentary sources: the Himalayan leucogranites. Geology, 45(1): 39-42

Gao X Y, Zheng Y F, Chen Y X. 2012. Dehydration melting of ultrahigh-pressure eclogite in the Dabie orogen: evidence from multiphase solid inclusions in garnet. Journal of Metamorphic Geology, 30(2): 193-212

Gao X Y, Zheng Y F, Chen Y X, Hu Z. 2013. Trace element composition of continentally subducted slab-derived melt: insight from multiphase solid inclusions in ultrahigh-pressure eclogite in the Dabie orogen. Journal of Metamorphic Geology, 31(4): 453-468

Gao X Y, Zheng Y F, Chen Y X, Hu Z. 2014. Composite carbonate and silicate multiphase solid inclusions in metamorphic garnet from ultrahigh-P eclogite in the Dabie orogen. Journal of Metamorphic Geology, 32(9):961-980

Gao X Y, Chen Y X, Zhang Q Q. 2017. Multiphase solid inclusions in ultrahigh-pressure metamorphic rocks: a snapshot of anatectic melts during continental collision. Journal of Asian Earth Sciences, 145:192-204

Gordon S M, Little T A, Hacker B R, Bowring S A, Korchinski M, Baldwin S L, Kylander-Clark A R C. 2012. Multi-stage exhumation of young UHP-HP rocks: timescales of melt crystallization in the D'Entrecasteaux Islands, southeastern Papua New Guinea. Earth and Planetary Science Letters, 351-352: 237-246

Gordon S M, Whitney D L, Teyssier C, Fossen H. 2013. U-Pb dates and trace-element geochemistry of zircon from migmatite, western Gneiss Region, Norway: significance for history of partial melting in continental subduction. Lithos, 170-171:35-53

Grieco G, Ferrario A, von Quadt A, Koeppel V, Mathez E A. 2001. The zircon-bearing chromitites of the phlogopite peridotite of Finero (Ivrea zone, southern Alps): evidence and geochronology of a metasomatized mantle slab. Journal of Petrology, 42(1):89-101

Griffin W L, Graham S, O'Reilly S Y, Pearson N J. 2004. Lithosphere evolution beneath the Kaapvaal Craton: Re-Os systematics of sulfides in mantle-derived peridotites. Chemical Geology, 208(1-4):89-118

Guillot S, Hattori K, Agard P, Schwartz S, Vidal O. 2009. Exhumation processes in oceanic and continental subduction contexts: a review. In: Lallemand S, Funiciello F (eds). Subduction Zone Geodynamics. Berlin: Springer, 175-205

Guo S, Chen Y, Ye K, Su B, Yang Y H, Zhang L M, Liu J B, Mao Q. 2015. Formation of multiple high-pressure veins in ultrahigh-pressure eclogite (Hualiangting, Dabie terrane, China): fluid source, element transfer, and closed-system metamorphic veining. Chemical Geology, 417:238-260

Guo S, Tang P, Su B, Chen Y, Ye K, Zhang L M, Gao Y J, Liu J B, Yang Y H. 2017. Unusual replacement of Fe-Ti oxides by rutile during retrogression in amphibolite-hosted veins (Dabie UHP terrane): a mineralogical record of fluid-induced oxidation processes in exhumed UHP slabs. American Mineralogist, 102(11):2268-2283

Guo S, Zhao K D, John T, Tang P, Chen Y, Su B. 2019. Metasomatic flow of metacarbonate-derived fluids carrying isotopically heavy boron in continental subduction zones: insights from tourmaline-bearing ultra-high pressure eclogites and veins (Dabie terrane, eastern China). Geochimica et Cosmochimica Acta, 253:159-200

Guo Z F, Wilson M, Zhang M L, Cheng Z H, Zhang L H. 2015. Post-collisional ultrapotassic mafic magmatism in South Tibet: products of partial melting of pyroxenite in the mantle wedge induced by roll-back and delamination of the subducted Indian continental lithosphere slab. Journal of Petrology, 56(7):1365-1405

Hattori K, Wallis S, Enami M, Mizukami T. 2010. Subduction of mantle wedge peridotites: evidence from the Higashi-akaishi ultramafic body in the Sanbagawa metamorphic belt. Island Arc, 19(1):192-207

Hermann J. 2002a. Experimental constraints on phase relations in subducted continental crust. Contributions to Mineralogy and Petrology, 143(2): 219-235

Hermann J. 2002b. Allanite: thorium and light rare earth element carrier in subducted crust. Chemical Geology, 192(3-4):289-306

Hermann J, Spandler C J. 2008. Sediment melts at sub-arc depths: an experimental study. Journal of Petrology, 49(4):717-740

Hermann J, Rubatto D. 2009. Accessory phase control on the trace element signature of sediment melts in subduction zones. Chemical Geology, 265(3-4):512-526

Hermann J, Rubatto D. 2014. Subduction of continental crust to mantle depth: geochemistry of ultrahigh-pressure rocks. Treatise on Geochemistry (Second Edition), 4:309-340

Hermann J, Rubatto D, Korsakov A, Shatsky V S. 2001. Multiple zircon growth during fast exhumation of diamondiferous, deeply subducted continental crust (Kokchetav Massif, Kazakhstan). Contributions to Mineralogy and Petrology, 141(1):66-82

Hermann J, Rubatto D, Trommsdorff V. 2006. Sub-solidus oligocene zircon formation in garnet peridotite during fast decompression and fluid infiltration (Duria, central Alps). Mineralogy and Petrology, 88(1-2):181-206

Hermann J, Zheng Y F, Rubatto D. 2013. Deep fluids in subducted continental crust. Elements, 9(4):281-288

Herzberg C. 2011. Identification of source lithology in the Hawaiian and Canary Islands: implications for origins. Journal of Petrology, 52(1):113-146

Hirschmann M M, Kogiso T, Baker M B, Stolper E M. 2003. Alkalic magmas generated by partial melting of garnet pyroxenite. Geology, 31(6):481-484

Holness M B. 2006. Melt-solid dihedral angles of common minerals in natural rocks. Journal of Petrology, 47(4):791-800

Holness M B, Sawyer E W. 2008. On the pseudomorphing of melt-filled pores during the crystallization of migmatites. Journal of Petrology, 49(7): 1343-1363

Holness M B, Cesare B, Sawyer E W. 2011. Melted rocks under the microscope: microstructures and their interpretation. Elements, 7(4):247-252

Holness M B, Humphreys M C S, Sides R, Helz R T, Tegner C. 2012. Toward an understanding of disequilibrium dihedral angles in mafic rocks. Journal of Geophysical Research: Solid Earth, 117(B6):B06207, doi:10.1029/2011jb008902

Holness M B, Namur O, Cawthorn R G. 2013. Disequilibrium dihedral angles in layered intrusions: a microstructural record of fractionation. Journal of Petrology, 54(10):2067-2093

Hwang S L, Shen P Y, Chu H T, Yui T F. 2000. Nanometer-size α-PbO_2-type TiO_2 in garnet: a thermobarometer for ultrahigh-pressure metamorphism. Science, 288(5464):321-324

Hwang S L, Shen P Y, Chu H T, Yui T F, Lin C C. 2001. Genesis of microdiamonds from melt and associated multiphase inclusions in garnet of ultrahigh-pressure gneiss from Erzgebirge, Germany. Earth and Planetary Science Letters, 188(1-2):9-15

Hwang S L, Chu H T, Yui T F, Shen P Y, Schertl H P, Liou J G, Sobolev N V. 2006. Nanometer-size P/K-rich silica glass (former melt) inclusions in microdiamond from the gneisses of Kokchetav and Erzgebirge Massifs: diversified characteristics of the formation media of metamorphic microdiamond in UHP rocks due to host-rock buffering. Earth and Planetary Science Letters, 243(1-2): 94-106

Inger S, Harris N. 1993. Geochemical constraints on leucogranite magmatism in the Langtang Valley, Nepal Himalaya. Journal of Petrology, 34(2): 345-368

Ionov D A, O'Reilly S Y, Genshaft Y S, Kopylova M G. 1996. Carbonate-bearing mantle peridotite xenoliths from Spitsbergen: phase relationships, mineral compositions and trace-element residence. Contributions to Mineralogy and Petrology, 125(4): 375-392

Jahn B M, Wu F Y, Lo C H, Tsai C H. 1999. Crust-mantle interaction induced by deep subduction of the continental crust: geochemical and Sr-Nd isotopic evidence from post-collisional mafic-ultramafic intrusions of the northern Dabie complex, central China. Chemical Geology, 157(1-2): 119-146

Jamieson R A, Unsworth M J, Harris N B W, Rosenberg C L, Schulmann K. 2011. Crustal melting and the flow of mountains. Elements, 7(4): 253-260

Katayama I, Muko A, Iizuka T, Maruyama S, Terada K, Tsutsumi Y, Sano Y, Zhang R Y, Liou J G. 2003. Dating of zircon from Ti-clinohumite bearing garnet peridotite: implication for timing of mantle metasomatism. Geology, 31(8): 713-716

Knesel K M, Davidson J P. 2002. Insights into collisional magmatism from isotopic fingerprints of melting reactions. Science, 296(5576): 2206-2208

Labrousse L, Prouteau G, Ganzhorn A C. 2011. Continental exhumation triggered by partial melting at ultrahigh pressure. Geology, 39(12): 1171-1174

Labrousse L, Duretz T, Gerya T. 2015. H_2O-fluid-saturated melting of subducted continental crust facilitates exhumation of ultrahigh-pressure rocks in continental subduction zones. Earth and Planetary Science Letters, 428: 151-161

Lang H M, Gilotti J A. 2007. Partial melting of metapelites at ultrahigh-pressure conditions, Greenland Caledonides. Journal of Metamorphic Geology, 25(2): 129-147

Laurora A, Mazzucchelli M, Rivalenti G, Vannucci R, Zanetti A, Barbieri M A, Cingolani C A. 2001. Metasomatism and melting in carbonated peridotite xenoliths from the mantle wedge: the Gobernador Gregores case (southern Patagonia). Journal of Petrology, 42(1): 69-87

Lei M, Chen J L, Tan R Y, Huang S H. 2019. Two Types of mafic rocks in southern Tibet: a mark of tectonic setting change from Neo-Tethyan oceanic crust subduction to Indian continental crust subduction. Journal of Asian Earth Sciences, 181: 103883

Li H Y, Chen R X, Zheng Y F, Hu Z C. 2016. The crust-mantle interaction in continental subduction channels: zircon evidence from orogenic peridotite in the Sulu orogen. Journal of Geophysical Research: Solid Earth, 121(2): 687-712

Li H Y, Chen R X, Zheng Y F, Hu Z C, Xu L J. 2018a. Crustal metasomatism at the slab-mantle interface in a continental subduction channel: geochemical evidence from orogenic peridotite in the Sulu orogen. Journal of Geophysical Research: Solid Earth, 123(3): 2174-2198

Li H Y, Chen R X, Zheng Y F, Hu Z C. 2018b. Water in garnet pyroxenite from the Sulu orogen: implications for crust-mantle interaction in continental subduction zone. Chemical Geology, 478: 18-38

Li W C, Chen R X, Zheng Y F, Tang H L, Hu Z C. 2016. Two episodes of partial melting in ultrahigh-pressure migmatites from deeply subducted continental crust in the Sulu orogen, China. GSA Bulletin, 128(9-10): 1521-1542

Liati A, Gebauer D. 2009. Crustal origin of zircon in a garnet peridotite: A study of U-Pb SHRIMP dating, mineral inclusions and REE geochemistry (Erzgebirge, Bohemian Massif). European Journal of Mineralogy, 21(4): 737-750

Liou J G, Zhang R Y, Ernst W G. 2007. Very high-pressure orogenic garnet peridotites. Proceedings of the National Academy of Sciences of the United States of America, 104(22): 9116-9121

Liou J G, Ernst W G, Zhang R Y, Tsujimori T, Jahn B M. 2009. Ultrahigh-pressure minerals and metamorphic terranes-the view from China. Journal of Asian Earth Sciences, 35(3-4): 199-231

Liu F L, Liou J G. 2011. Zircon as the best mineral for P-T-time history of UHP metamorphism: a review on mineral inclusions and U-Pb SHRIMP ages of zircons from the Dabie-Sulu UHP rocks. Journal of Asian Earth Sciences, 40(1): 1-39

Liu F L, Robinson P T, Gerdes A, Xue H M, Liu P H, Liou J G. 2010. Zircon U-Pb ages, REE concentrations and Hf isotope compositions of granitic leucosome and pegmatite from the north Sulu UHP terrane in China: constraints on the timing and nature of partial melting. Lithos, 117(1-4): 247-268

Liu F L, Robinson P T, Liu P H. 2012. Multiple partial melting events in the Sulu UHP terrane: zircon U-Pb dating of granitic leucosomes within amphibolite and gneiss. Journal of Metamorphic Geology, 30(8): 887-906

Liu L, Zhang J F, Green II H W, Jin Z M, Bozhilov K N. 2007. Evidence of former stishovite in metamorphosed sediments, implying subduction to >350 km. Earth and Planetary Science Letters, 263(3-4): 180-191

Liu P L, Wu Y, Liu Q, Zhang J F, Zhang L, Jin Z M. 2014. Partial melting of UHP calc-gneiss from the Dabie Mountains. Lithos, 192-195: 86-101

Liu P L, Massonne H J, Harlov D E, Jin Z M. 2019. High-pressure fluid-rock interaction and mass transfer during exhumation of deeply subducted rocks: insights from an eclogite-vein system in the ultrahigh-pressure terrane of the Dabie Shan, China. Geochemistry, Geophysics, Geosystems, 20(12): 5786-5817

Liu Q, Hou Q L, Xie L W, Li H, Ni S Q, Wu Y D. 2012. Different origins of the fractionation of platinum-group elements in Raobazhai and Bixiling mafic-ultramafic rocks from the Dabie orogen, central China. Journal of Geological Research, 2012: 631426, doi:10.1155/2012/631426

Liu Q, Hermann J, Zhang J F. 2013. Polyphase inclusions in the Shuanghe UHP eclogites formed by subsolidus transformation and incipient melting during exhumation of deeply subducted crust. Lithos, 177: 91-109

Liu Q, Hermann J, Zheng S, Zhang J F. 2020. Evidence for UHP anatexis in the Shuanghe UHP paragneiss from inclusions in clinozoisite, garnet, and zircon. Journal of Metamorphic Geology, 38(2): 129-155

Malaspina N, Hermann J, Scambelluri M, Compagnoni R. 2006. Polyphase inclusions in garnet-orthopyroxenite (Dabie Shan, China) as monitors for metasomatism and fluid-related trace element transfer in subduction zone peridotite. Earth and Planetary Science Letters, 249(3-4): 173–187

Malaspina N, Hermann J, Scambelluri M. 2009. Fluid/mineral interaction in UHP garnet peridotite. Lithos, 107(1-2): 38–52

Malaspina N, Scambelluri M, Poli S, van Roermund H L M, Langenhorst F. 2010. The oxidation state of mantle wedge majoritic garnet websterites metasomatised by C-bearing subduction fluids. Earth and Planetary Science Letters, 298(3-4): 417–426

Malaspina N, Alvaro M, Campione M, Wilhelm H, Nestola F. 2015. Dynamics of mineral crystallization from precipitated slab-derived fluid phase: first in situ synchrotron X-ray measurements. Contributions to Mineralogy and Petrology, 169(3): 26

Malaspina N, Langenhorst F, Tumiati S, Campione M, Frezzotti M L, Poli S. 2017. The redox budget of crust-derived fluid phases at the slab-mantle interface. Geochimica et Cosmochimica Acta, 209: 70–84

Manning C E. 2004. The chemistry of subduction-zone fluids. Earth and Planetary Science Letters, 223(1-2): 1–16

Okay A I. 1994. Sapphirine and Ti-clinohumite in ultra-high-pressure garnet-pyroxenite and eclogite from Dabie Shan, China. Contributions to Mineralogy and Petrology, 116(1-2): 145–155

Palme H, O'Neill H S C. 2007. Cosmochemical estimates of mantle composition. Treatise on Geochemistry, 2: 1–38

Perchuk A L, Burchard M, Maresch W V, Schertl H P. 2005. Fluid-mediated modification of garnet interiors under ultrahigh-pressure conditions. Terra Nova, 17(6): 545–553

Perchuk A L, Burchard M, Maresch W V, Schertl H P. 2008. Melting of hydrous and carbonate mineral inclusions in garnet host during ultrahigh pressure experiments. Lithos, 103(1-2): 25–45

Petford N, Cruden A R, McCaffrey K J W, Vigneresse J L. 2000. Granite magma formation, transport and emplacement in the Earth's crust. Nature, 408(6813): 669–673

Pilet S, Baker M B, Stolper E M. 2008. Metasomatized lithosphere and the origin of alkaline lavas. Science, 320(5878): 916–919

Qi Y, Gou G N, Wang Q, Wyman D A, Jiang Z Q, Li Q L, Zhang L. 2018. Cenozoic mantle composition evolution of southern Tibet indicated by Paleocene (~64 Ma) pseudoleucite phonolitic rocks in central Lhasa terrane. Lithos, 302-303: 178–188

Rampone E, Morten L. 2001. Records of crustal metasomatism in the garnet peridotites of the Ulten zone (Upper Austroalpine, eastern Alps). Journal of Petrology, 42(1): 207–219

Reverdatto V V, Selyatitskiy A Y, Carswell D A. 2008. Geochemical distinctions between "crustal" and mantle-derived peridotites/pyroxenites in high/ultrahigh pressure metamorphic complexes. Russian Geology and Geophysics, 49(2): 73–90

Rielli A, Tomkins A G, Nebel O, Brugger J, Etschmann B, Zhong R, Yaxley G M, Paterson D. 2017. Evidence of sub-arc mantle oxidation by sulphur and carbon. Geochemical Perspectives Letters, 3(2): 124–132

Righter K, Campbell A J, Humayun M, Hervig R L. 2004. Partitioning of Ru, Rh, Pd, Re, Ir, and Au between Cr-bearing spinel, olivine, pyroxene and silicate melts. Geochimica et Cosmochimica Acta, 68(4): 867–880

Ringwood A E. 1990. Slab-mantle interactions: 3. petrogenesis of intraplate magmas and structure of the upper mantle. Chemical Geology, 82: 187–207

Rosenberg C L, Handy M R. 2005. Experimental deformation of partially melted granite revisited: implications for the continental crust. Journal of Metamorphic Geology, 23(1): 19–28

Rubatto D, Hermann J. 2003. Zircon formation during fluid circulation in eclogites (Monviso, western Alps): implications for Zr and Hf budget in subduction zones. Geochimica et Cosmochimica Acta, 67(12): 2173–2187

Rubatto D, Hermann J, Berger A, Engi M. 2009. Protracted fluid-induced melting during Barrovian metamorphism in the central Alps. Contributions to Mineralogy and Petrology, 158(6): 703–722

Sapienza G T, Scambelluri M, Braga R. 2009. Dolomite-bearing orogenic garnet peridotites witness fluid-mediated carbon recycling in a mantle wedge (Ulten zone, eastern Alps, Italy). Contributions to Mineralogy and Petrology, 158(3): 401–420

Scambelluri M, Hermann J, Morten L, Rampone E. 2006. Melt-versus fluid-induced metasomatism in spinel to garnet wedge peridotites (Ulten zone, eastern Italian Alps): clues from trace element and Li abundances. Contributions to Mineralogy and Petrology, 151(4): 372–394

Scambelluri M, Pettke T, van Roermund H L M. 2008. Majoritic garnets monitor deep subduction fluid flow and mantle dynamics. Geology, 36(1): 59–62

Scambelluri M, van Roermund H L M, Pettke T. 2010. Mantle wedge peridotites: fossil reservoirs of deep subduction zone processes. Lithos, 120(1-2): 186–201

Schmidt M W, Poli S. 2014. Devolatilization during Subduction. Treatise on Geochemistry (Second Edition), 4: 669–701

Shatsky V S, Jagoutz E, Sobolev N V, Kozmenko O A, Parkhomenko V S, Troesch M. 1999. Geochemistry and age of ultrahigh pressure metamorphic rocks from the Kokchetav Massif (northern Kazakhstan). Contributions to Mineralogy and Petrology, 137(3): 185–205

Shen J, Li S G, Wang S J, Teng F Z, Li Q L, Liu Y S. 2018. Subducted Mg-rich carbonates into the deep mantle wedge. Earth and Planetary Science Letters, 503: 118–130

Shi R D, Griffin W L, O'Reilly S Y, Zhao G C, Huang Q S, Li J, Xu J F. 2010. Evolution of the Lüliangshan garnet peridotites in the North Qaidam UHP belt, northern Tibetan Plateau: constraints from Re-Os isotopes. Lithos, 117(1-4): 307–321

Shreve R L, Cloos M. 1986. Dynamics of sediment subduction, melange formation, and prism accretion. Journal of Geophysical Research, 91(B10): 10229–10245

Skjerlie K P, Patiño Douce A E. 2002. The fluid-absent partial melting of a zoisite-bearing quartz eclogite from 1.0 to 3.2 GPa: Implications for melting in thickened continental crust and for subduction-zone processes. Journal of Petrology, 43(2): 291-314

Skora S, Blundy J. 2012. Monazite solubility in hydrous silicic melts at high pressure conditions relevant to subduction zone metamorphism. Earth and Planetary Science Letters, 321-322: 104-114

Smith D C. 1984. Coesite in clinopyroxene in the Caledonides and its implications for geodynamics. Nature, 310(5979): 641-644

Sobolev A V, Hofmann A W, Sobolev S V, Nikogosian I K. 2005. An olivine-free mantle source of Hawaiian shield basalts. Nature, 434(7033): 590-597

Sobolev N V, Shatsky V S. 1990. Diamond inclusions in garnets from metamorphic rocks: a new environment for diamond formation. Nature, 343(6260): 742-746

Song S G, Zhang L F, Chen J, Liou J G, Niu Y L. 2005a. Sodic amphibole exsolutions in garnet from garnet-peridotite, North Qaidam UHPM belt, NW China: implications for ultradeep-origin and hydroxyl defects in mantle garnets. American Mineralogist, 90(5-6): 814-820

Song S G, Zhang L F, Niu Y L, Su L, Jian P, Liu D Y. 2005b. Geochronology of diamond-bearing zircons from garnet peridotite in the North Qaidam UHPM belt, northern Tibetan Plateau: a record of complex histories from oceanic lithosphere subduction to continental collision. Earth and Planetary Science Letters, 234(1-2): 99-118

Song S G, Su L, Niu Y L, Zhang L F, Zhang G B. 2007. Petrological and geochemical constraints on the origin of garnet peridotite in the North Qaidam ultrahigh-pressure metamorphic belt, northwestern China. Lithos, 96(1-2): 243-265

Song S G, Niu Y L, Su L, Wei C J, Zhang L F. 2014a. Adakitic (tonalitic-trondhjemitic) magmas resulting from eclogite decompression and dehydration melting during exhumation in response to continental collision. Geochimica et Cosmochimica Acta, 130: 42-62

Song S G, Niu Y L, Su L, Zhang C, Zhang L F. 2014b. Continental orogenesis from ocean subduction, continent collision/subduction, to orogen collapse, and orogen recycling: the example of the North Qaidam UHPM belt, NW China. Earth-Science Reviews, 129: 59-84

Song Y R, Xu H J, Zhang J F, Wang D Y, Liu E D. 2014. Syn-exhumation partial melting and melt segregation in the Sulu UHP terrane: evidences from leucosome and pegmatitic vein of migmatite. Lithos, 202-203: 55-75

Spengler D, van Roermund H L M, Drury M R, Ottolini L, Mason P R D, Davies G R. 2006. Deep origin and hot melting of an Archaean orogenic peridotite massif in Norway. Nature, 440(7086): 913-917

Stepanov A S, Hermann J, Korsakov A V, Rubatto D. 2014. Geochemistry of ultrahigh-pressure anatexis: fractionation of elements in the Kokchetav gneisses during melting at diamond-facies conditions. Contributions to Mineralogy and Petrology, 167(5): 1002

Stepanov A S, Hermann J, Rubatto D, Korsakov A V, Danyushevsky L V. 2016. Melting history of an ultrahigh-pressure paragneiss revealed by multiphase solid inclusions in garnet, Kokchetav Massif, Kazakhstan. Journal of Petrology, 57(8): 1531-1554

Su B, Chen Y, Guo S, Liu J B. 2016a. Origins of orogenic dunites: Petrology, geochemistry, and implications. Gondwana Research, 29(1): 41-59

Su B, Chen Y, Guo S, Chu Z Y, Liu J B, Gao Y J. 2016b. Carbonatitic metasomatism in orogenic dunites from Lijiatun in the Sulu UHP terrane, eastern China. Lithos, 262: 266-284

Su B, Chen Y, Guo S, Liu J B. 2017. Dolomite dissociation indicates ultra-deep (>150 km) subduction of a garnet-bearing dunite block (the Sulu UHP terrane). American Mineralogist, 102(11): 2295-2306

Su B, Chen Y, Mao Q, Zhang D, Jia L H, Guo S. 2019a. Minor elements in olivine inspect the petrogenesis of orogenic peridotites. Lithos, 344-345: 207-216

Su B, Chen Y, Guo S, Chen S, Li Y B. 2019b. Garnetite and pyroxenite in the mantle wedge formed by slab-mantle interactions at different melt/rock ratios. Journal of Geophysical Research: Solid Earth, 124(7): 6504-6522

Sun G C, Gao P, Zhao Z F, Zheng Y F. 2020. Syn-exhumation melting of the subducted continental crust: geochemical evidence from early Paleozoic granitoids in North Qaidam, northern Tibet. Lithos, 374-375: 105707

Tang M, Wang X L, Shu X J, Wang D, Yang T, Gopon P. 2014. Hafnium isotopic heterogeneity in zircons from granitic rocks: geochemical evaluation and modeling of "zircon effect" in crustal anatexis. Earth and Planetary Science Letters, 389: 188-199

Tatsumi Y, Eggins S. 1995. Subduction Zone Magmatism. Oxford: Blackwell Science

Tumiati S, Thöni M, Nimis P, Martin S, Mair V. 2003. Mantle-crust interactions during Variscan subduction in the eastern Alps (Nonsberg-Ulten zone): geochronology and new petrological constraints. Earth and Planetary Science Letters, 210(3-4): 509-526

Valley J W, Kinny P D, Schulze D J, Spicuzza M J. 1998. Zircon megacrysts from kimberlite: oxygen isotope variability among mantle melts. Contributions to Mineralogy and Petrology, 133(1-2): 1-11

van Roermund H. 2009a. Recent progress in Scandian ultrahigh-pressure metamorphism in the northernmost domain of the western Gneiss complex, SW Norway: continental subduction down to 180-200 km depth. Journal of the Geological Society, 166(4): 739-751

van Roermund H. 2009b. Mantle-wedge garnet peridotites from the northernmost ultra-high pressure domain of the wstern Gneiss region, SW Norway. European Journal of Mineralogy, 21(6): 1085-1096

van Roermund H L M, Carswell D A, Drury M R, Heijboer T C. 2002. Microdiamonds in a megacrystic garnet websterite pod from Bardane on the island of

Fjørtoft,western Norway:evidence for diamond formation in mantle rocks during deep continental subduction. Geology,30(11):959-962

Vielzeuf D,Holloway J R. 1988. Experimental determination of the fluid-absent melting relations in the pelitic system. Contributions to Mineralogy and Petrology,98(3):257-276

Vrijmoed J C,Austrheim H,John T,Hin R C,Corfu F,Davies G R. 2013. Metasomatism in the ultrahigh-pressure svartberget Garnet-peridotite (western Gneiss region,Norway):implications for the transport of crust-derived fluids within the mantle. Journal of Petrology,54(9):1815-1848

Wallis S,Tsuboi M,Suzuki K,Fanning M,Jiang L L,Tanaka T. 2005. Role of partial melting in the evolution of the Sulu (eastern China) ultrahigh-pressure terrane. Geology,33(1):129-132

Wang H,Wu Y B,Qin Z W,Zhu L Q,Liu Q,Liu X C,Gao S,Wijbrans J R,Zhou L,Gong H J,Yuan H L. 2013. Age and geochemistry of Silurian gabbroic rocks in the Tongbai orogen,central China:implications for the geodynamic evolution of the North Qinling arc-back-arc system. Lithos,179:1-15

Wang J,Hattori K,Xu W L,Yang Y Q,Xie Z P,Liu J L,Song Y. 2012. Origin of ultramafic xenoliths in high-Mg diorites from east-central China based on their oxidation state and abundance of platinum group elements. International Geology Review,54(10):1203-1218

Wang L,Kusky T M,Polat A,Wang S J,Jiang X F,Zong K Q,Wang J P,Deng H,Fu J M. 2014. Partial melting of deeply subducted eclogite from the Sulu orogen in China. Nature Communications,5(1):5604,doi:10.1038/ncomms6604

Weinberg R F,Hasalová P. 2015. Water-fluxed melting of the continental crust:a review. Lithos,212-215:158-188

Wu Y B,Zheng Y F. 2013. Tectonic evolution of a composite collision orogen:an overview on the Qinling-Tongbai-Hong'an-Dabie-Sulu orogenic belt in central China. Gondwana Research,23(4):1402-1428

Xia Q X,Wang H Z,Zhou L G,Gao X Y,Zheng Y F,van Orman J A,Xu H J,Hu Z C. 2016. Growth of metamorphic and peritectic garnets in ultrahigh-pressure metagranite during continental subduction and exhumation in the Dabie orogen. Lithos,266-267:158-181

Xie Z P,Hattori K,Wang J. 2013. Origins of ultramafic rocks in the Sulu ultrahigh-pressure terrane,eastern China. Lithos,178:158-170

Xiong Q,Zheng J P,Griffin W L,O'Reilly S Y,Zhao J H. 2011. Zircons in the Shenglikou ultrahigh-pressure garnet peridotite massif and its country rocks from the North Qaidam terrane (western China):Meso-Neoproterozoic crust-mantle coupling and early Paleozoic convergent plate-margin processes. Precambrian Research,187(1-2):33-57

Xiong Q,Zheng J P,Griffin W L,O'Reilly S Y,Pearson N J. 2014. Pyroxenite dykes in orogenic peridotite from North Qaidam (NE Tibet,China) Track metasomatism and segregation in the mantle wedge. Journal of Petrology,55(1-2):2347-2376

Xiong Q,Griffin W L,Zheng J P,O'Reilly S Y,Pearson N J. 2015. Episodic refertilization and metasomatism of Archean mantle:evidence from an orogenic peridotite in North Qaidam (NE Tibet,China). Contributions to Mineralogy and Petrology,169(3):31,doi:10.1007/s00410-015-1126-7

Xu H J,Ye K,Song Y R,Chen Y,Zhang J F,Liu Q,Guo S. 2013. Prograde metamorphism,decompressional partial melting and subsequent melt fractional crystallization in the Weihai migmatitic gneisses,Sulu UHP terrane,eastern China. Chemical Geology,341:16-37

Xu H J,Zhang J F,Wang Y F,Liu W L. 2016. Late Triassic alkaline complex in the Sulu UHP terrane:implications for post-collisional magmatism and subsequent fractional crystallization. Gondwana Research,35:390-410

Xu S T,Su W,Liu Y C,Jiang L L,Ji S Y,Okay A I,Sengör A M C. 1992. Diamond from the Dabie Shan metamorphic rocks and its implication for tectonic setting. Science,256(5053):80-82

Xu Z,Zhao Z F,Zheng Y F. 2012. Slab-mantle interaction for thinning of cratonic lithospheric mantle in North China:geochemical evidence from Cenozoic continental basalts in central Shandong. Lithos,146-147:202-217

Yamamoto S,Komiya T,Yamamoto H,Kaneko Y,Terabayashi M,Katayama I,Iizuka T,Maruyama S,Yang J S,Kon Y,Hirata T. 2013. Recycled crustal zircons from podiform chromitites in the Luobusa ophiolite,southern Tibet. Island Arc,22(1):89-103

Yang J S,Li T F,Chen S Z,Wu C L,Robinson P T,Liu D Y,Wooden J L. 2009. Genesis of garnet peridotites in the Sulu UHP belt:examples from the Chinese continental scientific drilling project-main hole,PP1 and PP3 drillholes. Tectonophysics,475(2):359-382

Ye K,Cong B L,Ye D N. 2000. The possible subduction of continental material to depths greater than 200 km. Nature,407(6805):734-736

Ye K,Song Y R,Chen Y,Xu H J,Liu J B,Sun M. 2009. Multistage metamorphism of orogenic garnet-lherzolite from Zhimafang,Sulu UHP terrane,E. China:implications for mantle wedge convection during progressive oceanic and continental subduction. Lithos,109(3-4):155-175

Yu S Y,Li S Z,Zhang J X,Peng Y B,Somerville I,Liu Y J,Wang Z Y,Li Z F,Yao Y,Li Y. 2019. Multistage anatexis during tectonic evolution from oceanic subduction to continental collision:a review of the North Qaidam UHP belt,NW China. Earth-Science Reviews,191:190-211

Yuan H L,Gao S,Rudnick R L,Jin Z M,Liu Y S,Puchtel I S,Walker R J,Yu R D. 2007. Re-Os evidence for the age and origin of peridotites from the Dabie-Sulu ultrahigh pressure metamorphic belt,China. Chemical Geology,236(3-4):323-338

Zack T,Kronz A,Foley S F,Rivers T. 2002. Trace element abundances in rutiles from eclogites and associated garnet mica schists. Chemical Geology,184(1-2):97-122

Zeng L S,Liang F H,Asimow P,Chen F Y,Chen J. 2009. Partial melting of deeply subducted continental crust and the formation of quartzofeldspathic polyphase inclusions in the Sulu UHP eclogites. Chinese Science Bulletin,54(15):2580-2594

Zhang J J, Zheng Y F, Zhao Z F. 2009. Geochemical evidence for interaction between oceanic crust and lithospheric mantle in the origin of Cenozoic continental basalts in east-central China. Lithos, 110(1-4): 305-326

Zhang L, Chen R X, Zheng Y F, Hu Z. 2015. Partial melting of deeply subducted continental crust during exhumation: insights from felsic veins and host UHP metamorphic rocks in North Qaidam, northern Tibet. Journal of Metamorphic Geology, 33(7): 671-694

Zhang L, Chen R X, Zheng Y F, Li W C, Hu Z C, Yang Y H, Tang H L. 2016. The tectonic transition from oceanic subduction to continental subduction: zirconological constraints from two types of eclogites in the North Qaidam orogen, northern Tibet. Lithos, 244: 122-139

Zhang L, Chen R X, Zheng Y F, Hu Z C, Xu L J. 2017. Whole-rock and zircon geochemical distinction between oceanic- and continental-type eclogites in the North Qaidam orogen, northern Tibet. Gondwana Research, 44: 67-88

Zhang L, Sun W W, Chen R X. 2019. Evolution of serpentinite from seafloor hydration to subduction zone metamorphism: petrology and geochemistry of serpentinite from the ultrahigh pressure North Qaidam orogen in northern Tibet. Lithos, 346-347: 105158

Zhang R Y, Liou J G, Yang J S, Yui T F. 2000. Petrochemical constraints for dual origin of garnet peridotites from the Dabie-Sulu UHP terrane, eastcentral China. Journal of Metamorphic Geology, 18(2): 149-166

Zhang R Y, Shau Y H, Liou J G, Lo C H. 2002. Discovery of clinoenstatite in garnet pyroxenites from the Dabie-Sulu ultrahigh-pressure terrane, east-central China. American Mineralogist, 87(7): 867-874

Zhang R Y, Yang J S, Wooden J L, Liou J G, Li T F. 2005. U-Pb SHRIMP geochronology of zircon in garnet peridotite from the Sulu UHP terrane, China: implications for mantle metasomatism and subduction-zone UHP metamorphism. Earth and Planetary Science Letters, 237(3-4): 729-743

Zhang R Y, Li T, Rumble D, Yui T F, Li L, Yang J S, Pan Y, Liou J G. 2007. Multiple metasomatism in Sulu ultrahigh-P garnet peridotite constrained by petrological and geochemical investigations. Journal of Metamorphic Geology, 25(2): 149-164

Zhang R Y, Liou J G, Ernst W G. 2009. The Dabie-Sulu continental collision zone: a comprehensive review. Gondwana Research, 16(1): 1-26

Zhang Z M, Dong X, Liou J G, Liu F, Wang W, Yui F. 2011. Metasomatism of garnet peridotite from Jiangzhuang, southern Sulu UHP belt: constraints on the interactions between crust and mantle rocks during subduction of continental lithosphere. Journal of Metamorphic Geology, 29(9): 917-937

Zhao Z F, Chen B, Zheng Y F, Chen R X, Wu Y B. 2007a. Mineral oxygen isotope and hydroxyl content changes in ultrahigh-pressure eclogite-gneiss contacts from Chinese Continental Scientific Drilling Project cores. Journal of Metamorphic Geology, 25(2): 165-186

Zhao Z F, Zheng Y F, Chen R X, Xia Q X, Wu Y B. 2007b. Element mobility in mafic and felsic ultrahigh-pressure metamorphic rocks during continental collision. Geochimica et Cosmochimica Acta, 71(21): 5244-5266

Zhao Z F, Zheng Y F, Wei C S, Chen F K, Liu X M, Wu F Y. 2008. Zircon U-Pb ages, Hf and O isotopes constrain the crustal architecture of the ultrahigh-pressure Dabie orogen in China. Chemical Geology, 253(3-4): 222-242

Zhao Z F, Zheng Y F, Zhang J, Dai L Q, Li Q L, Liu X M. 2012. Syn-exhumation magmatism during continental collision: evidence from alkaline intrusives of Triassic age in the Sulu orogen. Chemical Geology, 328: 70-88

Zhao Z F, Dai L Q, Zheng Y F. 2013. Postcollisional mafic igneous rocks record crust-mantle interaction during continental deep subduction. Scientific Reports, 3(1): 3413, doi: 10.1038/srep03413

Zhao Z F, Zheng Y F, Chen Y X, Sun G C. 2017a. Partial melting of subducted continental crust: geochemical evidence from synexhumation granite in the Sulu orogen. GSA Bulletin, 129(11-12): 1692-1707

Zhao Z F, Liu Z B, Chen Q. 2017b. Melting of subducted continental crust: Geochemical evidence from Mesozoic granitoids in the Dabie-Sulu orogenic belt, east-central China. Journal of Asian Earth Sciences, 145: 260-277

Zheng F, Dai L Q, Zhao Z F, Zheng Y F, Xu Z. 2019. Recycling of Paleo-oceanic crust: geochemical evidence from Early Paleozoic mafic igneous rocks in the Tongbai orogen, central China. Lithos, 328-329: 312-327

Zheng J P, Zhang R Y, Griffin W L, Liou J G, O'Reilly S Y. 2005. Heterogeneous and metasomatized mantle recorded by trace elements in minerals of the Donghai garnet peridotites, Sulu UHP terrane, China. Chemical Geology, 221(3-4): 243-259

Zheng J P, Griffin W L, O'Reilly S Y, Yang J S, Zhang R Y. 2006. A refractory mantle protolith in younger continental crust, east-central China: age and composition of zircon in the Sulu ultrahigh-pressure peridotite. Geology, 34(9): 705-708

Zheng J P, Sun M, Griffin W L, Zhou M F, Zhao G C, Robinson P, Tang H Y, Zhang Z H. 2008. Age and geochemistry of contrasting peridotite types in the Dabie UHP belt, eastern China: petrogenetic and geodynamic implications. Chemical Geology, 247(1-2): 282-304

Zheng J P, Tang H Y, Xiong Q, Griffin W L, O'Reilly S Y, Pearson N, Zhao J H, Wu Y B, Zhang J F, Liu Y S. 2014. Linking continental deep subduction with destruction of a cratonic margin: strongly reworked North China SCLM intruded in the Triassic Sulu UHP belt. Contributions to Mineralogy and Petrology, 168(1): 1028

Zheng Y F. 2009. Fluid regime in continental subduction zones: petrological insights from ultrahigh-pressure metamorphic rocks. Journal of the Geological Society, 166(4): 763-782

Zheng Y F. 2012. Metamorphic chemical geodynamics in continental subduction zones. Chemical Geology, 328: 5-48

Zheng Y F. 2018. Fifty years of plate tectonics. National Science Review, 5(2): 119

Zheng Y F. 2019. Subduction zone geochemistry. Geoscience Frontiers,10(4):1223-1254

Zheng Y F,Hermann J. 2014. Geochemistry of continental subduction-zone fluids. Earth,Planets and Space,66(1):93,doi:10.1186/1880-5981-66-93

Zheng Y F,Chen Y X. 2016. Continental versus oceanic subduction zones. National Science Review,3(4):495-519

Zheng Y F,Zhao Z F. 2017. Introduction to the structures and processes of subduction zones. Journal of Asian Earth Sciences,145:1-15

Zheng Y F,Chen R X. 2017. Regional metamorphism at extreme conditions:Implications for orogeny at convergent plate margins. Journal of Asian Earth Sciences,145:46-73

Zheng Y F,Wu F Y. 2018. The timing of continental collision between India and Asia. Science Bulletin,63(24):1649-1654

Zheng Y F,Fu B,Xiao Y L,Li Y L,Gong B. 1999. Hydrogen and oxygen isotope evidence for fluid-rock interactions in the stages of pre- and post-UHP metamorphism in the Dabie Mountains. Lithos,46(4):677-693

Zheng Y F,Fu B,Gong B,Li L. 2003. Stable isotope geochemistry of ultrahigh pressure metamorphic rocks from the Dabie-Sulu orogen in China: implications for geodynamics and fluid regime. Earth-Science Reviews,62(1-2):105-161

Zheng Y F,Chen R X,Zhao Z F. 2009. Chemical geodynamics of continental subduction-zone metamorphism:insights from studies of the Chinese Continental Scientific Drilling(CCSD)core samples. Tectonophysics,475(2):327-358

Zheng Y F,Xia Q X,Chen R X,Gao X Y. 2011. Partial melting,fluid supercriticality and element mobility in ultrahigh-pressure metamorphic rocks during continental collision. Earth-Science Reviews,107(3-4):342-374

Zheng Y F,Xu Z,Chen L,Dai L Q,Zhao Z F. 2020. Chemical geodynamics of mafic magmatism above subduction zones. Journal of Asian Earth Sciences,194:104185

Zhou K,Chen Y X,Zheng Y F,Xu L J. 2019. Migmatites record multiple episodes of crustal anatexis and geochemical differentiation in the Sulu ultrahigh-pressure metamorphic zone,eastern China. Journal of Metamorphic Geology,37(8):1099-1127

Zhou K,Che,Y X,Ma H Z,Zheng Y F,Xia X P. 2020. Geochemistry of high-pressure to ultrahigh-pressure granitic melts produced by decompressional melting of deeply subducted continental crust in the Sulu orogen,east-central China. Geochimica et Cosmochimica Acta,288:214-247

Progresses in the Study of Partial Melting of Ultrahigh-Pressure Metamorphic Rocks and Crust-Mantle Interaction in Continental Subduction Zones

ZHAO Zi-fu[1,2], CHEN Ren-xu[1,2], CHEN Yi-xiang[1,2], DAI Li-qun[1,2], ZHENG Yong-fei[1,2]

1. CAS Key Laboratory of Crust-Mantle Materials and Environments, School of Earth and Space Sciences, University of Science and Technology of China, Hefei 230026; 2. Center for Excellence in Comparative Planetology, Chinese Academy of Sciences, Hefei 230026

Abstract: Since the findings of ultrahigh-pressure (UHP) metamorphic minerals such as coesite and diamond in continental crustal rocks in the 1980s, the study of continental deep subduction and UHP metamorphism has been one of the frontiers and hotspots in solid earth science. After more than 30 years of studies, many important achievements have been made in following research aspects, including the depth of continental crust subduction, the metamorphic P-T-t path of deeply subducted rocks, the exhumation mechanism of subducted crustal rocks, the protolith nature of deeply subducted rocks, the melt/fluid activity and element mobility in the process of continental collision, the different types of crust-mantle interaction in subduction channels, the petrogenesis of post-collisional igneous rocks, and the metallogenesis in collisional orogens. In this paper, we have specially reviewed and summarized research progresses in the partial melting of UHP rocks and the different types of crust-mantle interaction in continental subduction zones in the past 10 years, and have prospected the existed relevant scientific questions and future directions of studies. It merits to be emphasized that partial melting of the deeply subducted continental crust mainly took place in two stages. The first partial melting occurred during the initial stage of exhumation, giving rise to alkaline melts. The second one happened in the post-collisional stage, yielding calc-alkaline melts. There are two types of crust-mantle interaction in continental subduction zones, involving metasomatic reaction of the mantle wedge with two kinds of subduction zone fluids. One was derived from metamorphic dehydration/melting of the deeply subducted continental crust and the other was originated from metamorphic dehydration/melting of the previously subducted oceanic crust.

Key words: continental subduction zones; partial melting; crust-mantle interaction; orogenic peridotite; orogenic igneous rocks

花岗岩研究进展*

徐夕生　王孝磊　赵　凯　杜德宏

南京大学地球科学与工程学院，内生金属矿床成矿机制研究国家重点实验室，南京 210023

摘　要：本文概述了最近十余年国内外花岗岩研究的进展，列举了一些国内外学者近十年在花岗质岩石成因和大陆演化研究方面所取得的重要成果，突出了岩浆演化精细过程研究的新成果和新进展。围绕岩体的累积生长和火山岩–侵入岩之间的联系，结合近些年花岗岩的研究前沿，具体分析了中国东南沿海花岗岩及相关岩石研究的新动向和关键科学问题，以期为我国花岗岩研究下一步工作的开展提供有益建议。

关键词：花岗岩　地壳演化　岩石成因　火山–侵入杂岩　研究进展

0　引　言

花岗岩（这里泛指"花岗质岩石"）是陆壳的重要组分，也是地球区别于其他行星的重要标志，在研究大陆的形成和演化过程中至关重要，在金属矿床成因研究和勘查中也具有十分重要的科学意义和战略价值。

对花岗岩的研究已有百余年的历史，历经了早期的"水成论"和"火成论"之争，到后来的熔融模拟实验和花岗岩成因分类研究，以及进入 21 世纪以来现代分析技术在花岗岩成因研究中的广泛应用，花岗岩岩浆作用的过程得到越来越深入的揭示，花岗岩成因理论得到不断完善，花岗岩与造山带演化、地壳增生和大陆动力学的关系得到逐步澄清，花岗岩研究显示出蓬勃发展的趋势。近十年（2011—2020 年）来，随着分析测试技术的不断进步，花岗岩研究更多地向精细化深度发展，并与数值模拟、大数据、地球物理等研究方法紧密结合，在探索地球科学的前沿重大科学问题上发挥了重要作用。由翟明国院士等召集的主题为"花岗岩：大陆形成与改造的记录"的"香山科学会议"（2015 年），旨在把花岗岩和大陆的形成演化密切联系起来，倡导花岗岩研究的革命，进而推动对大陆演化和大陆动力学研究的进程和突破，其中，花岗岩与地壳演化的联系受到重点关注，包括：岩浆形成、上升、侵位与大陆运动学、动力学的关系，大陆花岗岩的地球动力学意义，花岗岩多样性的原因（翟明国，2017）。这些问题也是近十年来国内外学者在花岗岩研究上重点关注的重要科学问题。

花岗岩的形成和演化记录了大陆地壳形成、生长和再造的重要信息。众所周知，活动大陆边缘陆壳的形成与演化涉及大陆的侧向增生和垂向增生与分异，其标志是伴随着大量花岗岩的形成，而陆内环境下的大陆再造也可以产生大量的花岗岩（Raimondo et al.，2010）。花岗质岩浆的结晶分异被认为是导致花岗岩体成分变化的重要机制和引起大陆地壳垂向成分变化的重要因素，尤其是高分异花岗岩被认为是大陆地壳高成熟度的重要岩石学标志（吴福元等，2017）。另外，花岗质岩浆的演化对大陆地壳形成演化过程中成矿元素的聚集研究（Liu et al.，2020；Wu et al.，2020）也有十分重要的意义。

本文对近十年来国内外关于花岗岩研究的现状和重要进展进行综述，并着重对中国东南沿海火山–侵入杂岩的研究进行分析，以期对未来我国花岗岩研究提供有益参考。

---

* 原文"新时期花岗岩研究的进展和趋势"刊于《矿物岩石地球化学通报》2020 年第 39 卷第 5 期，本文略有修改。

1 国内外花岗岩研究现状概述

近十年来，国内外有关花岗岩成因理论和成因过程的研究呈现精细化和高技术化的特点。如：

（1）更精细的岩浆和矿物成因过程的深入剖析。通过细致的矿物学研究来认识转熔作用与 S-型花岗岩（Díaz-Alvarado et al., 2011; Dorais and Tubrett, 2012; Bartoli et al., 2013; Rong et al., 2017）和 I-型花岗岩（Clemens et al., 2011）形成的联系，建立矿物-岩石-岩浆源区之间的耦合关系，深入揭示花岗岩的源区特征及多样性（Chappell and Wyborn, 2012; Clemens and Stevens, 2012; Hopkinson et al., 2017）。

（2）研究地壳深熔作用与花岗质熔体形成的初始过程（London et al., 2012; Wang et al., 2012, 2015; Brown, 2013; Ma et al., 2017）。利用转熔矿物中捕获的转熔熔体恢复地壳深熔初始阶段熔体的成分（包括水含量；Bartoli et al., 2013, 2014）；通过 Sr、Nd、Hf 等同位素揭示地壳物质的不平衡熔融过程（Farina et al., 2014; Tang et al., 2014; Iles et al., 2018）。

（3）通过实验岩石学（Scaillet et al., 2016; Huang et al., 2019）、热力学模拟（Nabelek et al., 2012; Lee and Morton, 2015; Zhao et al., 2017a, 2018; Ackerson et al., 2018; Chen et al., 2019）、地球物理（Liu et al., 2018; Singer et al., 2018）等方面的研究和资料综合，估算岩浆结晶的压力、温度、水含量、时间等参数，制约岩体生长的物理过程，不断深化花岗岩成因研究。

（4）利用多种同位素-微量元素分析手段来研究精细的岩浆过程。包括利用岩石学和锆石 Hf-O 同位素（Smithies et al., 2011; Foley et al., 2013; Wilcock et al., 2013）和矿物原位 Sr 同位素（Yu et al., 2018）等手段来示踪花岗岩的源区特征（Wang et al., 2011, 2013）、岩浆过程（Deering et al., 2016）、低^{18}O 事件（张少兵和郑永飞，2011; Wang W. et al., 2017）以及相关的地壳演化；利用对岩浆成分和温度敏感的锆石微量元素（Yan et al., 2020）和全岩微量元素模拟（Gelman et al., 2014; Lee and Morton, 2015）研究岩浆分异和高硅花岗岩/流纹岩的形成等。

（5）精细探讨岩体的生长和地壳增生过程。通过高精度定年（Schoene et al., 2012）和数值模型（Annen, 2011; Annen et al., 2014）来解释花岗岩体的累积生长过程和热演化史；利用锆石微区 LA-ICP-MS 和二次离子质谱（SIMS）定年厘清花岗岩时空分布框架（Ji et al., 2014; Zhu et al., 2015; Wang T. et al., 2017）；利用花岗岩 Hf-Nd 同位素填图来示踪地壳快速增生或再造过程（Pankhurst et al., 2011; Jeon et al., 2012; 王晓霞等，2014; Hou et al., 2015; Huang et al., 2020）。

（6）综合研究太古宙花岗岩与火山岩的成因联系，解译太古宙花岗岩类（主要为 TTG）的成因（Laurent et al., 2014, 2020），并以此来探究地球早期大陆形成与可能相关的板块构造启动过程（Moyen and Martin, 2012; Reimink et al., 2016; Trail et al., 2017; Moyen et al., 2019）。

（7）用非传统稳定同位素（主要是 Fe、Mg、Zn、Cu、Si 和 Li 同位素等）示踪花岗岩源区和岩浆演化过程（Telus et al., 2012; Savage et al., 2012; Foden et al., 2015; Du et al., 2017; Xu et al., 2017; Li et al., 2018）等。

需要强调的是，我国是花岗岩研究大国，随着我国地球化学分析测试手段的迅猛发展，特别是在矿物微区同位素定年和同位素组成分析测试上长足的进步，花岗岩的研究空间得到极大的拓展。近十年来，我国学者在探讨花岗岩的源区及相关的陆壳物质再循环中的地球化学行为、壳幔相互作用，通过实验岩石学、热力学模拟和非传统稳定同位素（如 Fe、Mg、Zn、Cu 和 Si 同位素等）示踪花岗岩岩浆演化过程等方面取得一系列重要成果和新的进展。

2 花岗岩研究重要进展列举

近年来，国内外在花岗质岩浆的形成、运移、储存、分异和侵位等精细过程方面的研究方兴未艾，并逐渐形成了一些重要共识，这些共识在一定程度上打破了花岗岩成因的传统认识，革新了花岗岩研究的思路。本文选择主要几方面的研究进展列举如下。

2.1 源区不均一和不平衡熔融

放射性成因同位素体系（如 Rb-Sr、Sm-Nd 和 Lu-Hf）是示踪花岗岩源区性质和岩浆过程（如同化混染和岩浆混合）的重要手段。例如，Gao 等（2017）利用华南三叠纪花岗岩 Sr-Nd 同位素特征，结合系统的全岩地球化学、锆石 U-Pb 年龄和氧同位素以及新元古代岩石空间分布的研究，指出三叠纪花岗岩主要来自新元古代地壳物质的部分熔融。然而，根据花岗岩同位素组成来准确识别出花岗岩的源区并不容易（Zhao et al., 2011），因为部分熔融产生的熔体同位素组成可能受到源区物质不均一（Wang et al., 2013）和熔融温度（Zhao et al., 2015）等多种因素影响，且花岗岩同位素组成还可能通过基性岩浆混合和（或）围岩同化混染等过程而改变。

应用同位素手段鉴别源区的一个基本假设是地壳部分熔融产生的熔体具有和源岩一致的同位素组成。近年来，对源区不平衡熔融过程的研究则对这一基本假设提出了挑战。由于地壳的成分极度不均一，各类源岩的元素和同位素组成千差万别，同一源岩中不同矿物的元素和同位素组成也不尽相同（Zeng et al., 2005），因此部分熔融所产生的不同批次的熔体有可能继承不同源区的成分特征，也有可能在同一源区的非实比部分熔融反应中由于参与熔融反应矿物比例的差异而导致同位素组成与源岩的不一致，即同位素不平衡（Tang et al., 2014；王孝磊，2017；Zhao et al., 2017b）。不平衡熔融导致放射性的同位素差异需要满足以下 3 个条件（Iles et al., 2018）：①原岩的矿物具有不同的母体–子体同位素比值（如斜长石和黑云母的 $^{87}Rb/^{87}Sr$ 分别为 0.06 和 39）；②原岩在地壳中存留足够长的时间，使得各矿物具有不同的同位素组成；③熔体产生后快速抽离源区，避免熔体与残留体达到化学平衡或不同批次熔体的混合均一化。

不平衡熔融过程较早是在野外露头尺度研究混合岩淡色体和暗色体同位素组成的差异性时提出的（Naslund, 1986；Zeng et al., 2005）。随着分析技术的提高，近年来越来越多的研究报道微观矿物尺度的同位素不平衡现象。例如，在安第斯中部火山岩捕获的地壳深熔捕房体中的熔体玻璃（McLeod et al., 2012）和意大利 Elba 岩体中的钾长石巨晶及其包裹的黑云母（Farina et al., 2014）均观测到 Sr 同位素不平衡现象。这种不平衡现象反映了随着熔融温度的升高，熔融反应由白云母脱水熔融变为黑云母脱水熔融，并进一步向角闪石脱水熔融转变的过程（Farina et al, 2014）。近年来，锆石的 Hf 同位素已经成为花岗岩研究中的常规手段，但熔融过程中 Lu-Hf 同位素体系的不平衡直到 2014 年才被 Tang 等（2014）所注意到。锆石相对其他矿物（如石榴子石、单斜辉石等）具有极低的 Lu/Hf 值，因此熔体的 Hf 同位素组成主要取决于锆石非放射性成因的 Hf 与其他矿物放射性成因 Hf 之间的竞争，而锆石中的 Hf 扩散极慢，所以熔体的 Hf 同位素实际上受控于锆石的熔融速率。全岩的酸–碱溶解实验进一步地证明了这一点，在较短的时间内全岩样品中只有部分的锆石发生了溶解，导致溶液比全岩具有更放射的 Hf 同位素组成；随着溶解时间的增加，溶液的 Hf 同位素值趋近于全岩值（Wang et al., 2018）。基于这样的考虑，Iles 等（2020）对澳大利亚拉克兰褶皱带的 S 和 I 型花岗岩进行了研究，发现 S 型花岗岩的 Hf 同位素和锆石的 Hf 同位素基本一致，而 I 型花岗岩相比其锆石具有更放射的 Hf 同位素组成（$\Delta\varepsilon_{Hf全岩-锆石} = 0.4 \sim 2.5$）。这一特征可以用变火成岩角闪石脱水引起的不平衡熔融来解释（Iles et al., 2020）。因此，我们在使用锆石或者全岩 Hf 同位素来判断花岗岩的源区时需要特别小心。

2.2 大型花岗岩岩体的累积生长

关于花岗岩岩体形成的传统认识是，与岩体规模相似的岩浆房在地壳中可以长时间存在，岩浆在其中缓慢冷却、结晶和分异（如 Bateman and Chappell, 1979）。21 世纪初，这种"大岩浆房"（Big magma tank；Glazner et al., 2004）的模式就已经受到高精度定年结果的挑战，如研究程度较高的美国内华达（Sierra Navada）岩基中的 Tuolumne 岩基，锆石 U-Pb 定年结果揭示该岩基是岩浆在大约 10 Ma 的时间范围内汇聚而成（Coleman et al., 2004）。而最近十年，岩体的累积生长模式得到越来越多的野外地质观察、高精度定年研究以及热演化数值模拟的支持，并涌现出一批研究程度高、多学科交叉研究的经典研究案例，如美国的 Tuolumne 岩基（Paterson et al., 2011）、意大利的 Adamello 岩体（Schoene et al., 2012）和智利的 Torres Del Paine 岩体（Leuthold et al., 2012；Annen et al., 2015）等。

传统的"大岩浆房"模式难以解释的一些地质观察，在岩体的累积生长模式下，可以得到合理的回答：①岩浆同位素组成在矿物尺度存在不均一性，可能是由于同位素组成不同的、不同批次的岩浆混合引起（Farina et al., 2014；Tang et al., 2014）；②对现今活火山区的地球物理研究并未探测到地壳内部高熔体含量（>50%）的岩浆房的存在（Glazner et al., 2004），这可能是因为单一批次岩浆的固结时间较短（一般小于几十至几百千年；Schoene et al., 2012；Barboni and Schoene, 2014），而高熔体分数的岩浆存留时间则更短；③大岩浆房在地壳中侵位所需要的空间问题一直困扰着地质学界（Hutton et al., 1990），而在岩体累积生长模式下，小体量、多批次岩浆的汇聚可伴随微小的构造应变进行，长期的汇聚则形成大型岩体（Karakas et al., 2017）。

2.3 晶体-熔体分离与高硅花岗岩/流纹岩的形成

晶体-熔体分离（即分离结晶）被认为是控制花岗岩成分变化的重要过程（McCarthy and Groves, 1979），但质疑的声音也一直存在（Glazner et al., 2008），主要原因是花岗质岩浆的黏度大，且缺乏确凿的岩石学证据。Bachmann 和 Bergantz（2004）估算了在合理的岩浆黏度范围内晶体-熔体分离所需要的时间尺度，结果表明通过受阻沉降、微沉降和（或）压实等过程，熔体可以有效抽离并聚集于岩体顶部，形成具有可喷发性的高硅熔体囊。这一设想得到火山岩成分变化的支持，即许多火山岩在垂向层序上往往从贫晶火山岩变化为更加富晶体的火山岩，而这种成分梯度指示了深部岩浆房的分带性（Hildreth and Wilson, 2007；Bachmann and Huber, 2016；Buret et al., 2017）。近年来，对火山-侵入杂岩的研究从全岩主、微量元素以及锆石微量元素研究角度揭示了高硅流纹岩和共生的次火山花岗斑岩成分的互补性（Deering et al., 2016，Yan et al., 2018, 2020）。

花岗岩记录矿物堆积和熔体抽离的完整过程，应当是研究晶体-熔体分离过程的重要对象。近年来，对花岗岩的显微组构研究表明，矿物晶形或晶格的定向性和韧性变形可能与岩浆在高结晶度条件下受阻沉降和（或）压实等作用有关（Beane and Wiebe, 2012；Fiedrich et al., 2017；Zhao et al., 2018）。这些组构的强弱与受阻沉降和（或）压实的强弱程度有关，进而导致不同比例的熔体发生抽离，从而引起花岗岩成分的变化（Lee and Morton, 2015；Jackson et al., 2018；Zhao et al., 2018）。结合热力学模拟的数值模拟可估算经过受阻沉降和（或）压实作用而抽离的熔体分数（Lee and Morton, 2015；Zhao et al., 2018）。例如，Zhao 等（2018）在对华南钦州湾地区早中生代旧州岩体的研究中，揭示出紫苏花岗岩（即含斜方辉石花岗岩）和不含斜方辉石花岗岩在显微组构上的差异 [图 1 (a)、(b)]，表明紫苏花岗岩经历较高程度的压实作用，导致约 40% 的粒间熔体抽离，而不含斜方辉石的花岗岩压实作用较弱，仅抽离约 20% 的粒间熔体，并据此提出带状岩浆房形成的模式 [图 1 (c)]。对于熔体抽离的时间尺度，高精度定年结果显示通过晶体-熔体产生数百立方千米高硅流纹岩所需要的时间可能为上百个千年（Deering et al., 2016），而岩浆中水含量的增加将缩短熔体抽离的时间（Hartung et al., 2019）。

2.4 "晶粥"模型与穿地壳岩浆演化

"晶粥"为晶体与熔体的混合物，这种混合物由于结晶度高（一般大于 40%~50%）而不具有可移动性或喷发性（Miller and Wark, 2008）。"晶粥"模型认为岩浆在地壳中绝大部分时间主要以这种高结晶度的"晶粥"形式储存（Cooper and Kent, 2014），这不同于"大岩浆房"模式中主要以熔体富集（结晶度小于 40%~50%）的岩浆形式存在的认识。在岩体增量汇聚的过程中，单一批次的岩浆固结的时间较短，能否维持一定量的熔体乃至形成岩浆房是由岩浆补给通量（magma flux）决定（Annen et al., 2015）。当岩浆补给的通量较小时，不同批次侵位的岩浆可能存在明显的接触界线（Torres Del Paine 岩体；

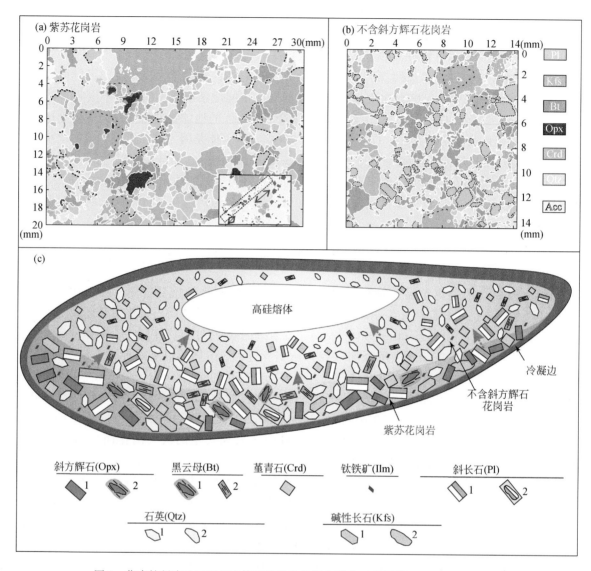

图 1 华南钦州湾地区旧州岩体岩浆演化的概念模式（底图据 Zhao et al., 2018）

该模式解释了通过晶体-熔体分离过程形成带状岩体的过程以及这一过程中矿物组合的变化。图（a）和（b）为基于冷阴极发光重绘的显微岩相图，显示了紫苏花岗岩（a）和不含斜方辉石花岗岩（b）中的矿物组合以及斜长石和碱性长石的台阶状环带（黑色和红色虚线所示）的空间分布特征。其中，紫苏花岗岩发育斜长石（蓝色所示）团、链状组构［如图（a），右下角角图显示了斜长石的空间分布型式，其中深蓝色所代表颗粒发育台阶状环带，浅蓝色颗粒不发育台阶状环带］，而在不含斜方辉石的花岗岩中斜长石多为随机分布，缺乏明显的组构特征（b）。(c) 推测的岩体相带和熔体在结晶度 40%~50%（wt.）时的迁移路径，虚线箭头指示了粒间熔体迁移的方向。图例中的数字表示大多数矿物具有两种产状或结构：自形和它形斜方辉石；自形和它形黑云母；发育或不发育台阶状环带的斜长石；石英和碱性长石堆积过程中和堆积过程后生长的晶体

Leuthold et al., 2012）；相反地，若岩浆补给的通量较大，则可在较长时间范围内维持一定量的熔体，有足够的时间进行晶体-熔体分离过程，最终形成具有可喷发性的岩浆房（Gelman et al., 2013）。

基于"晶粥"模型，Cashman 等（2017）提出在地壳垂向上广泛发育"穿地壳岩浆系统"的模式，将不同地壳水平的岩浆作用、酸性与基性岩浆作用紧密联系，整合了从幔源岩浆底侵到酸性岩浆喷出地表所涉及的岩浆产生、上升、累积、分异直至喷发等物理过程，试图建立地壳岩浆作用的统一框架。华南是国内外学者高度关注的花岗岩研究区域之一，有大量的岩石学、地球化学数据积累。Xu 等（2020）在前人发表的大量数据的基础上，将形成于浅部地壳的浙江雁荡山和福建云山破火口火山-侵入杂岩、深部地壳侵位的福建平潭-岱前山和长安山杂岩体以及中下地壳的广东麒麟辉长岩包体进行了综合研究，通

过仔细的岩相学研究、地球化学对比分析、一系列地球化学数值模拟等，阐述了中国东南沿海白垩纪火山岩和侵入岩的起源与演化过程，建立了不同地壳水平岩浆作用之间的联系及其相应的"穿地壳岩浆系统"模式，对于理解壳幔相互作用、岩浆起源和演化过程以及中国东南沿海活动大陆边缘演化具有启示意义。

2.5　金属稳定同位素约束花岗岩形成机制

金属稳定同位素研究近年来发展迅速，但对于这些同位素手段如何应用到花岗岩研究中仍然处于探索阶段。对于 Fe 同位素来说，现有数据表明，高硅花岗岩（$SiO_2 > 70\%$）相对于低硅花岗岩（$SiO_2 < 70\%$）具有更重的 Fe 同位素组成，且 $\delta^{56}Fe$ 值随着 SiO_2 含量的增加有不断升高的趋势［图 2（a）］。目前的研究认为，控制高硅花岗岩 Fe 同位素分馏的机制主要有两种：一种是含铁矿物的分离结晶（Schoenberg and Blanckenburg, 2006; Foden et al., 2015）；另一种是岩浆演化晚期的流体出溶（Poitrasson and Freydier, 2005; Heimann et al., 2008）。但高硅花岗岩缺乏明显的 Zn 同位素（流体活动性元素）分馏（Telus et al., 2012），且质量平衡计算进一步表明岩浆自身流体出溶能引起的 Fe 同位素分馏很小（<0.06‰; Du et al., 2019），这说明矿物的分离结晶才是控制大部分高硅花岗岩 Fe 同位素分馏的主要原因。从总体上看，不同类型花岗岩的 Fe 同位素组成有一定差异：其中 A 型花岗岩具有较重 Fe 同位素组成，I 型花岗岩 Fe 同位素组成偏轻，S 型花岗岩则介于两者之间［图 2（a）; Foden et al., 2015］。这是因为 A 型花岗质岩浆具有低的 H_2O 含量和氧逸度，导致富集 Fe^{3+} 的磁铁矿（富集重 Fe 同位素）饱和晚，大量分离含 Fe^{2+} 的硅酸盐矿物（富集轻 Fe 同位素）使残余熔体不断富集重 Fe 同位素；而 I 型花岗岩则具有高的 H_2O 含量和氧逸度，磁铁矿饱和早，在一定程度上抵消了 $\delta^{56}Fe$ 升高的趋势（Sossi et al., 2012; Foden et al., 2015）。

现有的数据表明，大部分的 I 型和 S 型花岗岩都具有和 MORB 相似的 Mg 同位素组成（$\delta^{26}Mg = -0.25$‰±0.06‰）（Teng et al., 2010），而一些高 SiO_2 的 I 型和大部分的 A 型花岗岩的 Mg 同位素组成则明显偏重。花岗岩中常见的含 Mg 的硅酸矿物（角闪石和黑云母）具有和全岩相似的同位素特征，分离不会引起明显的 Mg 同位素分馏（Liu et al., 2010）；磁铁矿的 Mg 同位素相比全岩偏重，分离使残余熔体的 Mg 同位素变轻（Wang et al., 2020）。这表明，矿物的分离结晶可能并不是花岗岩 Mg 同位素分馏的主要原因。化学风化会带走轻 Mg 同位素，导致残留物质富集重 Mg 同位素（Teng, 2017）。因此，高 SiO_2 的 I 型和 A 型花岗岩中高的 $\delta^{26}Mg$ 特征被认为是地表风化物质的加入引起的（Shen et al., 2009; Li et al., 2010）。但令人费解的是，起源于沉积物部分熔融的 S 型花岗岩却显示出与 MORB 相似的 Mg 同位素特征。此外，低 SiO_2 的花岗岩也可能有较多地表风化物质加入到源区之中，它们为什么没有表现出 Mg 同位素的变化呢？可见，花岗岩的 Mg 同位素分馏机制还需要更多的研究才能澄清。

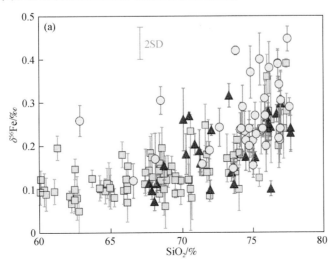

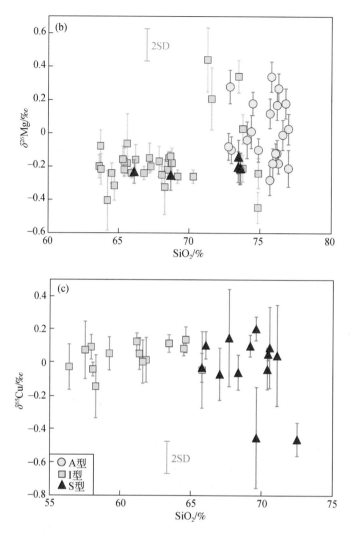

图 2 不同类型花岗岩的 Fe、Mg、Cu 同位素组成分布

Fe 同位素数据来自 Foden et al., 2015；Heimann et al., 2008；Telus et al., 2012；Poitrasson and Freydier, 2005；He et al., 2017；Du et al., 2019。Mg 同位素数据来自 Liu et al., 2010；Shen et al., 2009；Telus et al., 2012；Li et al., 2010。Cu 同位素数据来自 Li et al., 2009

对于花岗岩的 Cu 同位素来说，目前积累的数据更少，仅有 Li 等（2009）报道了拉克兰褶皱带 I 型和 S 型花岗岩的数据。这些结果表明，S 型花岗岩具有比 I 型花岗岩更大的变化，但总体上差别不大，个别变化较大的值可能是与风化地壳物质的参与或者蚀变有关。

限于篇幅，研究进展恕不一一列举。如我国学者近年来在具有埃达克质岩特征的花岗质岩石成岩和成矿研究方面也做出了具有国际影响力的重要贡献。总的来看，进入 21 世纪第二个十年，国内外花岗岩成因理论和成因过程的研究进入了新的发展阶段。以下将结合花岗岩研究前沿和笔者近年来的重点工作，简述有关中国东南沿海花岗质火山-侵入杂岩的研究现状和问题。

3 中国东南沿海花岗质火山-侵入杂岩研究及其关键科学问题

3.1 花岗质火山-侵入杂岩研究动态

有关岩浆起源、侵位、演化和火山喷发机制问题，是国际地学界火成岩研究的前沿领域之一（Petford

et al., 2000；吴福元等，2015；Keller et al., 2015；Cashman et al., 2017；Jackson et al., 2018）。近十年来，学术界高度关注火山岩-侵入岩的关系（Bachmann et al., 2007；Lipman and Bachmann, 2015；Keller et al., 2015）。2019 年于南京大学召开的第 9 届 Hutton 花岗岩会议上，大型花岗岩与酸性火山岩杂岩体的岩浆过程研究"Magmatic processes of large granitoid plutons and felsic volcanic complexes"是会议的 4 个主要议题之一。

中国东南沿海晚中生代火山-侵入杂岩带，总长约 2000 km，是环太平洋陆缘岩浆活动带的重要组成部分（谢家莹和陶奎元，1996；王德滋和周新民，2002）。该带发育多个由破火山、火山穹窿、构造洼地等组成的巨型环状火山构造，是一条呈北东向展布的巨型火山-侵入杂岩带（王德滋等，2000；Zhou et al., 2006）。

虽然近年来对几个点上的火山-侵入杂岩进行了新一轮的研究（Yan et al., 2016，2018，2020），然而，中国东南沿海火山-侵入杂岩研究急需整体推进。以往的许多研究或是偏重岩相学描述，或是偏重一般性的地球化学研究，综合性研究不够，部分杂岩体甚至尚未有可靠的年龄数据。虽然以往的认识已经注意到了火山-侵入杂岩在时、空、源上的相关性，但未能深入地约束火山岩与侵入岩在 p-t-X_{H_2O}-f_{O_2} 结晶条件上的异同，以及在岩浆物理过程上的内在联系。

3.2 花岗质火山-侵入杂岩研究中的关键科学问题

3.2.1 结晶分异-熔体抽离作用究竟发挥了多少作用？

花岗质岩浆从深部源区形成到浅部就位或喷发固结，经历了一系列复杂的过程，厘清花岗质岩浆演化机制与精细过程是花岗岩成因研究的重要挑战。其中一个关键的科学问题就是结晶分异-熔体抽离作用在花岗质岩浆形成过程中究竟发挥了多少作用？

在花岗质侵入体中熔体抽离与矿物堆晶的显微构造证据目前仍然非常缺乏（Fiedrich et al., 2017；Zhao et al., 2018），因此，当前关于晶体-熔体分离的各种机制还需深入论证（Holness et al., 2018）。吴福元等（2017）指出：单纯的重力分异并不是花岗质岩浆分离结晶作用的主导机制，流动导致的分异可能是花岗质岩浆成分变异最主要的原因。它们都将是未来花岗岩岩石学研究的前沿。中国东南沿海晚中生代火山-侵入杂岩是研究这一关键科学问题的理想对象。

3.2.2 火山-侵入杂岩的时序、跨度、脉动性和累积生长？

花岗岩体的生长和岩浆集聚过程是花岗岩研究的一个重要内容。单一批次的花岗质岩浆从形成到结晶需要多长时间？具有不同岩相的"杂岩"体形成又需要多长时间？这个问题长期受到两方面因素的制约，一是缺乏有效的手段，一般认为一个简单花岗岩体的形成可能产生于 1 Ma 以内（Petford et al., 2000），然而锆石微区 LA-ICP-MS 定年误差约 3%，这对于中生代以前的岩体来说，定年的误差已经大于了结晶的时间，从而掩盖了岩浆结晶过程的信息；二是对于岩浆中锆石结晶动力学过程还缺乏深入的了解。

二次离子质谱锆石 U-Pb 定年可降低定年误差至约 1%，化学剥蚀同位素稀释热电离质谱（chemical abrasion ID-TIMS, CA-ID-TIMS）U-Pb 定年更是能降低定年误差至 0.1% 以下（Schoene and Baxter, 2017；李献华，2018 报告），可以对较为年轻的花岗质岩体进行有效的结晶过程和岩体的生长时间约束。有了形成时间的约束，就可以研究岩浆通量（单位时间内所产生的岩浆体积）和岩体的累积生长过程。

选择合适的火山-侵入岩体进行详细解剖，将有助于破解岩浆作用过程之谜。

3.2.3 火山-侵入杂岩形成过程中幔源岩浆作用的影响

花岗岩成因研究中的另一关键科学问题是在花岗岩形成过程中幔源岩浆作用的影响。通常幔源基性岩浆会具有相对亏损的 Sr-Nd-Hf 同位素组成和接近地幔的氧同位素值，而来自地壳的酸性岩浆会具有相对富集的 Sr-Nd-Hf 同位素以及高于地幔值的氧同位素组成。不少学者认同幔源基性岩浆在一定程度上参与了花岗质岩石的形成，如 Yang 等（2007）对中国东北部早白垩世古道岭花岗岩基中的镁铁质包体和寄主花岗岩进行了精细的研究，结果显示它们具有一致的锆石 U-Pb 年龄，但 Hf 同位素组成不同，是由幔源镁铁质岩浆与壳源长英质岩浆混合形成。

但不能排除那些由富集的交代岩石圈地幔岩石熔融产出的基性岩浆与壳源酸性岩浆混合的可能性，甚至幔源组分比壳源组分的 Sr-Nd-Hf 同位素更为富集，表现为"同位素倒置"及"同位素倒置混合"（Jiang et al., 2018）。因此，花岗质岩浆形成过程中幔源岩浆作用的影响很复杂，需要精细解剖，具体分析。

在中国东南沿海地区，白垩纪幔源玄武质岩浆的底侵，已有明确的地球物理和物质成分证据（廖其林等，1988；徐夕生等，1999），同时代花岗质岩浆作用与此密切相关也已形成共识（Xu et al., 1999；Zhou and Li，2000）。但如何准确地模拟地壳熔融所需要的热，模拟并计算壳-幔组分比例，这一研究尚未全面展开。

因此，针对我国火山-侵入杂岩发育地区，开展相关的矿物学、岩石学、地球化学结合实验和数值模拟等手段的综合研究，有助于解决上述几个关键科学问题。特别是中国东南沿海发育并出露良好的晚中生代火山-侵入杂岩，已有较多的研究积累、地质背景清晰，为进一步理解结晶分异-熔体抽离作用，厘清岩浆作用的时序、跨度和脉动性，以及解析岩浆作用与幔源岩浆底侵的细节提供了优良的条件和基础。

4 研究趋势和研究意义

随着地球化学分析技术快速发展和花岗岩成岩模拟的不断深入，花岗岩成因类型与成因机制、花岗岩成因与大陆地壳演化、花岗岩形成与构造背景识别、花岗质岩浆的分异与成矿效应、花岗岩与壳幔相互作用等诸多方面取得了巨大的进展，但在花岗质岩浆起源与演进过程方面的研究还存在许多前沿科学问题。正如马昌前和李艳青（2017）指出的，只有将侵入岩与火山岩相结合，重点从侵入体形成的时间长短、岩浆相互作用的规模和频率、岩浆通量的演变、高结晶度的岩浆分异机理、侵入岩与火山岩的关系、地幔热和物质的贡献等方面入手，开展野外地质、岩石学、地球化学、同位素年代学及岩浆动力学的综合研究，才能深入认识花岗岩的成因机制，深化对大陆地壳形成和演化过程的理解。

因此，利用我国典型地区，如中国东南沿海白垩纪花岗质火山-侵入杂岩及火山机构地质出露良好的独特优势，结合精细的年代学和矿物学工作，以及大数据、数值模拟、地球物理等手段，针对关键科学问题进行解析，必将提升花岗岩研究的基础理论水平、增强我国对花岗岩成因与活动大陆边缘演化、花岗质岩浆演化与成矿效应的认知，取得开拓性的成果，具十分重要的科学意义和实际价值。

正如前文所述，花岗岩的研究还需与大陆形成和演化等紧密结合，以期在地球科学的发展中发挥更大的作用，我们认为，以下几个方面的工作值得深入开展：

（1）花岗岩类岩石的产生与大陆形成，包括地球早期花岗质岩石的特征、类型、产生的机制、演变规律，其形成与板块构造体制和大陆抬升之间的联系，以及与地球早期花岗质岩石相关的岩浆过程等。

（2）花岗岩成因与大陆地壳演化，主要包括元古宙以来花岗岩类多样性的成因与机制、花岗岩类形成的构造和岩浆动力学过程、花岗岩的源区、地壳岩浆体系与壳幔作用过程、大陆再造的机制和物理化学条件、花岗岩成因模拟，以及花岗岩形成与古环境变迁等。

（3）岩浆储库的形成和演化，包括岩浆储库（岩浆房）形成与演化的精细过程及其时间尺度，相关的岩浆动力学机制，岩浆喷发与否的控制因素，超级火山喷发的机制，可能的成矿效应、环境效应等。

（4）陆内花岗岩的形成与岩石圈演化，包括陆内岩石圈演化深部动力学过程、陆内环境下大陆再造的机制和条件、以及可能的陆内花岗岩成分多样性与成矿之间的联系等。

（5）花岗岩与地球宜居性，包括基于花岗岩形成过程的大数据分析等来研究地球内部结构、物质组成、深部动力学、壳幔相互作用等之间的联系，揭示花岗质岩浆的形成与物质循环和再造、地球深部挥发分作用和深部氧化还原条件之间的联系，探索地球宜居性的调控机制等。

致谢：感谢中国矿物岩石地球化学学会和郑永飞院士在稿件组织中给予的指导和帮助。

参 考 文 献

廖其林,王振明,王屏路,余兆康,吴宁远,刘宝诚. 1988. 福州-泉州-汕头地区地壳结构的爆炸地震研究. 地球物理学报,31(3):270-280
马昌前,李艳青. 2017. 花岗岩体的累积生长与高结晶度岩浆的分异. 岩石学报,33(5):1479-1488
王德滋,周新民. 2002. 中国东南部晚中生代花岗质火山-侵入杂岩成因与地壳演化. 北京:科学出版社,22-38
王德滋,周金城,邱检生,范洪海. 2000. 中国东南部晚中生代花岗质火山-侵入杂岩特征与成因. 高校地质学报,6(4):487-498
王晓霞,王涛,柯昌辉,杨阳,李金宝,李洪英,齐秋菊. 2014. 秦岭晚中生代花岗岩的Nd-Hf同位素填图及其对基底和钼矿的约束. 矿床地质, 33(S1):285-286
王孝磊. 2017. 花岗岩研究的若干新进展与主要科学问题. 岩石学报,33(5):1445-1458
吴福元,刘志超,刘小驰,纪伟强. 2015. 喜马拉雅淡色花岗岩. 岩石学报,31(1):1-36
吴福元,刘小驰,纪伟强,王佳敏,杨雷. 2017. 高分异花岗岩的识别与研究. 中国科学:地球科学,47(7):745-765
谢家莹,陶奎元. 1996. 中国东南大陆中生代火山地质及火山-侵入杂岩. 北京:地质出版社,1-2
徐夕生,周新民,王德滋. 1999. 壳幔作用与花岗岩成因——以中国东南沿海为例. 高校地质学报,5(3):240-250
翟明国. 2017. 花岗岩:大陆地质研究的突破口以及若干关键科学问题——"岩石学报"花岗岩专辑代序. 岩石学报,33(5):1369-1380
张少兵,郑永飞. 2011. 低$\delta^{18}O$岩浆岩的成因. 岩石学报,27(2):520-530
Ackerson M R, Mysen B O, Tailby N D, Watson E B. 2018. Low-temperature crystallization of granites and the implications for crustal magmatism. Nature,559(7712):94-97
Annen C. 2011. Implications of incremental emplacement of magma bodies for magma differentiation, thermal aureole dimensions and plutonism-volcanism relationships. Tectonophysics,500(1-4):3-10
Annen C, Paulatto M, Sparks R S J, Minshull T A, Kiddle E J. 2014. Quantification of the intrusive magma fluxes during magma chamber growth at Soufrière Hills Volcano (Montserrat, Lesser Antilles). Journal of Petrology,55(3):529-548
Annen C, Blundy J D, Leuthold J, Sparks R S J. 2015. Construction and evolution of igneous bodies: towards an integrated perspective of crustal magmatism. Lithos,230:206-221
Bachmann O, Huber C. 2016. Silicic magma reservoirs in the Earth's crust. American Mineralogist,101(11):2377-2404
Bachmann O, Bergantz G W. 2004. On the origin of crystal-poor rhyolites: extracted from batholithic crystal mushes. Journal of Petrology,45(8): 1565-1582
Bachmann O, Miller C F, de Silva S L. 2007. The volcanic-plutonic connection as a stage for understanding crustal magmatism. Journal of Volcanology and Geothermal Research,167(1-4):1-23
Barboni M, Schoene B. 2014. Short eruption window revealed by absolute crystal growth rates in a granitic magma. Nature Geoscience,7(7):524-528
Bartoli O, Cesare B, Poli S, Bodnar R J, Acosta-Vigil A, Frezzotti M L, Meli S. 2013. Recovering the composition of melt and the fluid regime at the onset of crustal anatexis and S-type granite formation. Geology,41(2):115-118
Bartoli O, Cesare B, Remusat L, Acosta-Vigil A, Poli S. 2014. The H_2O content of granite embryos. Earth and Planetary Science Letters,395:281-290
Bateman P C, Chappell B W. 1979. Crystallization, fractionation, and solidification of the Tuolumne intrusive series, Yosemite National Park, California. GSA Bulletin,90(5):465-482
Beane R, Wiebe R A. 2012. Origin of quartz clusters in Vinalhaven granite and porphyry, coastal Maine. Contributions to Mineralogy and Petrology,163 (6):1069-1082
Brown M. 2013. Granite: from genesis to emplacement. GSA Bulletin,125(7-8):1079-1113
Buret Y, Wotzlaw J F, Roozen S, Guillong M, von Quadt A, Heinrich C A. 2017. Zircon petrochronological evidence for a plutonic-volcanic connection in porphyry copper deposits. Geology,45(7):623-626
Cashman K V, Sparks R S J, Blundy J D. 2017. Vertically extensive and unstable magmatic systems: a unified view of igneous processes. Science,355 (6331):eaag3055
Chappell B W, Wyborn D. 2012. Origin of enclaves in S-type granites of the Lachlan fold belt. Lithos,154:235-247
Chen X, Lee C T A, Wang X L, Tang M. 2019. Influence of water on granite generation: modeling and perspective. Journal of Asian Earth Sciences,174: 126-134
Clemens J D, Stevens G. 2012. What controls chemical variation in granitic magmas? Lithos,134-135:317-329
Clemens J D, Stevens G, Farina F. 2011. The enigmatic sources of I-type granites: the peritectic connexion. Lithos,126(3-4):174-181
Coleman D S, Gray W, Glazner A F. 2004. Rethinking the emplacement and evolution of zoned plutons: geochronologic evidence for incremental assembly of the Tuolumne Intrusive Suite, California. Geology,32(5):433-436
Cooper K M, Kent A J R. 2014. Rapid remobilization of magmatic crystals kept in cold storage. Nature,506(7489):480-483
Deering C D, Keller B, Schoene B, Bachmann O, Beane R, Ovtcharova M. 2016. Zircon record of the plutonic-volcanic connection and protracted rhyolite melt evolution. Geology,44(4):267-270

Díaz-Alvarado J, Castro A, Fernández C, Moreno-Ventas I. 2011. Assessing bulk assimilation in cordierite-bearing granitoids from the central system batholith, Spain: experimental, geochemical and geochronological constraints. Journal of Petrology, 52(2): 223-256

Dorais M J, Tubrett M. 2012. Detecting peritectic garnet in the peraluminous Cardigan Pluton, New Hampshire. Journal of Petrology, 53(2): 299-324

Du D H, Wang X L, Yang T, Chen X, Li J Y, Li W Q. 2017. Origin of heavy Fe isotope compositions in high-silica igneous rocks: a rhyolite perspective. Geochimica et Cosmochimica Acta, 218: 58-72

Du D H, Li W Q, Wang X L, Shu X J, Yang T, Sun T. 2019. Fe isotopic fractionation during the magmatic-hydrothermal stage of granitic magmatism. Lithos, 350-351: 105265

Farina F, Dini A, Rocchi S, Stevens G. 2014. Extreme mineral-scale Sr isotope heterogeneity in granites by disequilibrium melting of the crust. Earth and Planetary Science Letters, 399: 103-115

Fiedrich A M, Bachmann O, Ulmer P, Deering C D, Kunze K, Leuthold J. 2017. Mineralogical, geochemical, and textural indicators of crystal accumulation in the Adamello Batholith (northern Italy). American Mineralogist, 102(12): 2467-2483

Foden J, Sossi P A, Wawryk C M. 2015. Fe isotopes and the contrasting petrogenesis of A-, I- and S-type granite. Lithos, 212-215: 32-44

Foley F V, Pearson N J, Rushmer T, Turner S, Adam J. 2013. Magmatic evolution and magma mixing of quaternary adakites at Solander and Little Solander Islands, New Zealand. Journal of Petrology, 54(4): 703-744

Gao P, Zheng Y F, Zhao Z F. 2017. Triassic granites in South China: a geochemical perspective on their characteristics, petrogenesis, and tectonic significance. Earth-Science Reviews, 173: 266-294

Gelman S E, Gutiérrez F J, Bachmann O. 2013. On the longevity of large upper crustal silicic magma reservoirs. Geology, 41(7): 759-762

Gelman S E, Deering C D, Bachmann O, Huber C, Gutiérrez F J. 2014. Identifying the crystal graveyards remaining after large silicic eruptions. Earth and Planetary Science Letters, 403: 299-306

Glazner A F, Bartley J M, Coleman D S, Gray W, Taylor R Z. 2004. Are plutons assembled over millions of years by amalgamation from small magma chambers? GSA Today, 14(4-5): 4-11

Glazner A F, Coleman D S, Bartley J M. 2008. The tenuous connection between high-silica rhyolites and granodiorite plutons. Geology, 36(2): 183-186

Hartung E, Weber G, Caricchi L. 2019. The role of H_2O on the extraction of melt from crystallising magmas. Earth and Planetary Science Letters, 508: 85-96

He Y S, Wu H J, Ke S, Liu S A, Wang Q. 2017. Iron isotopic compositions of adakitic and non-adakitic granitic magmas: magma compositional control and subtle residual garnet effect. Geochimica et Cosmochimica Acta, 203: 89-102

Heimann A, Beard B L, Johnson C M. 2008. The role of volatile exsolution and sub-solidus fluid/rock interactions in producing high $^{56}Fe/^{54}Fe$ ratios in siliceous igneous rocks. Geochimica et Cosmochimica Acta, 72(17): 4379-4396

Hildreth W, Wilson C J N. 2007. Compositional zoning of the bishop tuff. Journal of Petrology, 48(5): 951-999

Holness M B, Clemens J D, Vernon R H. 2018. How deceptive are microstructures in granitic rocks? Answers from integrated physical theory, phase equilibrium, and direct observations. Contributions to Mineralogy and Petrology, 173(8): 62

Hopkinson T N, Harris N B W, Warren C J, Spencer C J, Roberts N M W, Horstwood M S A, Parrish R R, EIMF. 2017. The identification and significance of pure sediment-derived granites. Earth and Planetary Science Letters, 467: 57-63

Hou Z Q, Yang Z M, Lu Y J, Kemp A, Zheng Y C, Li Q Y, Tang J X, Yang Z S, Duan L F. 2015. A genetic linkage between subduction- and collision-related porphyry Cu deposits in continental collision zones. Geology, 43(3): 247-250

Huang F F, Scaillet B, Wang R C, Erdmann S, Chen Y, Faure M, Liu H S, Xie L, Wang B, Zhu J C. 2019. Experimental constraints on intensive crystallization parameters and fractionation in A-type granites: a case study on the Qitianling Pluton, South China. Journal of Geophysical Research: Solid Earth, 124(10): 10132-10152

Huang H, Wang T, Tong Y, Qin Q, Ma X X, Yin J Y. 2020. Rejuvenation of ancient micro-continents during accretionary orogenesis: insights from the Yili Block and adjacent regions of the SW Central Asian orogenic belt. Earth-Science Reviews, 208: 103255

Hutton D H W, Dempster T J, Brown P E, Becker S D. 1990. A new mechanism of granite emplacement: intrusion in active extensional shear zones. Nature, 343(6257): 452-455

Iles K A, Hergt J M, Woodhead J D. 2018. Modelling isotopic responses to disequilibrium melting in granitic systems. Journal of Petrology, 59(1): 87-113

Iles K A, Hergt J M, Woodhead J D, Ickert R B, Williams I S. 2020. Petrogenesis of granitoids from the Lachlan fold belt, southeastern Australia: the role of disequilibrium melting. Gondwana Research, 79: 87-109

Jackson M D, Blundy J, Sparks R S J. 2018. Chemical differentiation, cold storage and remobilization of magma in the Earth's crust. Nature, 564(7736): 405-409

Jeon H, Williams I S, Chappell B W. 2012. Magma to mud to magma: rapid crustal recycling by Permian granite magmatism near the eastern Gondwana margin. Earth and Planetary Science Letters, 319-320: 104-117

Ji W Q, Wu F Y, Chung S L, Liu C Z. 2014. The Gangdese magmatic constraints on a latest Cretaceous lithospheric delamination of the Lhasa terrane, southern Tibet. Lithos, 210-211: 168-180

Jiang D S, Xu X S, Xia Y, Erdmann S. 2018. Magma mixing in a granite and related rock association: insight from its mineralogical, petrochemical, and "reversed isotope" features. Journal of Geophysical Research: Solid Earth, 123(3): 2262-2285

Karakas O, Degruyter W, Bachmann O, Dufek J. 2017. Lifetime and size of shallow magma bodies controlled by crustal-scale magmatism. Nature

Geoscience,10(6):446-450

Keller C B,Schoene B,Barboni M,Samperton K M,Husson J M. 2015. Volcanic-plutonic parity and the differentiation of the continental crust. Nature, 523(7560):301-307

Laurent O,Martin H,Moyen J F,Doucelance R. 2014. The diversity and evolution of late-Archean granitoids:evidence for the onset of "modern-style" plate tectonics between 3.0 and 2.5 Ga. Lithos,205:208-235

Laurent O,Björnsen J,Wotzlaw J F,Bretscher S,Silva M P,Moyen J F,Ulmer P,Bachmann O. 2020. Earth's earliest granitoids are crystal-rich magma reservoirs tapped by silicic eruptions. Nature Geoscience,13(2):163-169

Lee C T A,Morton D M. 2015. High silica granites:terminal porosity and crystal settling in shallow magma chambers. Earth and Planetary Science Letters,409:23-31

Leuthold J,Müntener O,Baumgartner L P,Putlitz B,Ovtcharova M,Schaltegger U. 2012. Time resolved construction of a bimodal laccolith (Torres del Paine,Patagonia). Earth and Planetary Science Letters,325-326:85-92

Li J,Huang X L,Wei G J,Liu Y,Ma J L,Han L,He P L. 2018. Lithium isotope fractionation during magmatic differentiation and hydrothermal processes in rare-metal granites. Geochimica et Cosmochimica Acta,240:64-79

Li W Q,Jackson S E,Pearson N J,Alard O,Chappell B W. 2009. The Cu isotopic signature of granites from the Lachlan fold belt,SE Australia. Chemical Geology,258(1-2):38-49

Li W Y,Teng F Z,Ke S,Rudnick R L,Gao S,Wu F Y,Chappell B W. 2010. Heterogeneous magnesium isotopic composition of the upper continental crust. Geochimica et Cosmochimica Acta,74(23):6867-6884

Lipman P W,Bachmann O. 2015. Ignimbrites to batholiths:integrating perspectives from geological,geophysical,and geochronological data. Geosphere, 11(3):705-743

Liu C,Wang R C,Wu F Y,Xie L,Liu X C,Li X K,Yang L,Li X J. 2020. Spodumene pegmatites from the Pusila pluton in the higher Himalaya,South Tibet:lithium mineralization in a highly fractionated leucogranite batholith. Lithos,358-359:105421

Liu H S,Martelet G,Wang B,Erdmann S,Chen Y,Faure M,Huang F F,Scaillet B,le-Breton N,Shu L S,Wang R C,Zhu J C. 2018. Incremental emplacement of the Late Jurassic midcrustal,lopolith-like Qitianling pluton,South China,revealed by AMS and Bouguer gravity data. Journal of Geophysical Research:Solid Earth,123(10):9249-9268

Liu S A,Teng F Z,He Y S,Ke S,Li S G. 2010. Investigation of magnesium isotope fractionation during granite differentiation:implication for Mg isotopic composition of the continental crust. Earth and Planetary Science Letters,297(3-4):646-654

London D,Morgan VI G B,Acosta-Vigil A. 2012. Experimental simulations of anatexis and assimilation involving metapelite and granitic melt. Lithos, 153:292-307

Ma L,Wang Q,Kerr A C,Yang J H,Xia X P,Ou Q,Yang Z Y,Sun P. 2017. Paleocene (c. 62 Ma) leucogranites in southern Lhasa,Tibet:products of syn-collisional crustal anatexis during slab roll-back? Journal of Petrology,58(11):2089-2114

McCarthy T C,Groves D I. 1979. The blue tier batholith,northeastern Tasmania:a cumulate-like product of fractional crystallization. Contributions to Mineralogy and Petrology,71(2):193-209

McLeod C L,Davidson J P,Nowell G M,de Silva S L. 2012. Disequilibrium melting during crustal anatexis and implications for modeling open magmatic systems. Geology,40(5):435-438

Miller C F,Wark D A. 2008. Supervolcanoes and their explosive supereruptions. Elements,4(1):11-15

Moyen J F,Martin H. 2012. Forty years of TTG research. Lithos,148:312-336

Moyen J F,Stevens G,Kisters A F M,Belcher R W,Lemirre B. 2019. TTG plutons of the barberton granitoid-greenstone terrain,South Africa. In:van Kranendonk M J,Bennett V C,Hoffmann J E (eds). Earth's Oldest Rocks,2nd ed. Amsterdam:Elsevier,615-654

Nabelek P I,Hofmeister A M,Whittington A G. 2012. The influence of temperature-dependent thermal diffusivity on the conductive cooling rates of plutons and temperature-time paths in contact aureoles. Earth and Planetary Science Letters,317-318:157-164

Naslund H R. 1986. Disequilibrium partial melting and rheomorphic layer formation in the contact aureole of the Basistoppen sill,East Greenland. Contributions to Mineralogy and Petrology,93(3):359-367

Pankhurst M J,Schaefer B F,Betts P G. 2011. Geodynamics of rapid voluminous felsic magmatism through time. Lithos,123(1-4):92-101

Paterson S R,Okaya D,Memeti V,Economos R,Miller R B. 2011. Magma addition and flux calculations of incrementally constructed magma chambers in continental margin arcs:combined field,geochronologic,and thermal modeling studies. Geosphere,7(6):1439-1468

Petford N,Cruden A R,McCaffrey K J W,Vigneresse J L. 2000. Granite magma formation,transport and emplacement in the Earth's crust. Nature,408 (6813):669-673

Poitrasson F,Freydier R. 2005. Heavy iron isotope composition of granites determined by high resolution MC-ICP-MS. Chemical Geology,222(1-2): 123-147

Raimondo T,Collins A S,Hand M,Walker-Hallam A,Smithies R H,Evins P M,Howard H M. 2010. The anatomy of a deep intracontinental orogen. Tectonics,29(4):TC4024

Reimink J R,Davies J H F L,Chacko T,Stern R A,Heaman L M,Sarkar C,Schaltegger U,Creaser R A,Pearson D G. 2016. No evidence for Hadean continental crust within Earth's oldest evolved rock unit. Nature Geoscience,9(10):777-780

Rong W,Zhang S B,Zheng Y F. 2017. Back-reaction of peritectic garnet as an explanation for the origin of mafic enclaves in S-type granite from the Jiuling batholith in South China. Journal of Petrology,58(3):569-598

Savage P S, Georg R B, Williams H M, Turner S, Halliday A N, Chappell B W. 2012. The silicon isotope composition of granites. Geochimica et Cosmochimica Acta, 92:184-202

Scaillet B, Holtz F, Pichavant M. 2016. Experimental constraints on the formation of silicic magmas. Elements, 12(2):109-114

Schoenberg R, von Blanckenburg F. 2006. Modes of planetary-scale Fe isotope fractionation. Earth and Planetary Science Letters, 252(3-4):342-359

Schoene B, Baxter E F. 2017. Petrochronology and TIMS. Reviews in Mineralogy and Geochemistry, 83(1):231-260

Schoene B, Schaltegger U, Brack P, Latkoczy C, Stracke A, Günther D. 2012. Rates of magma differentiation and emplacement in a ballooning pluton recorded by U-Pb TIMS-TEA, Adamello batholith, Italy. Earth and Planetary Science Letters, 355-356:162-173

Shen B, Jacobsen B, Lee C T A, Yin Q Z, Morton D M. 2009. The Mg isotopic systematics of granitoids in continental arcs and implications for the role of chemical weathering in crust formation. Proceedings of the National Academy of Sciences of the United States of America, 106(49):20652-20657

Singer B S, Le Mével H, Licciardi J M, Córdova L, Tikoff B, Garibaldi N, Andersen N L, Diefenbach A K, Feigl K L. 2018. Geomorphic expression of rapid Holocene silicic magma reservoir growth beneath Laguna del Maule, Chile. Science Advances, 4(6):eaat1513

Smithies R H, Howard H M, Evins P M, Kirkland C L, Kelsey D E, Hand M, Wingate M T D, Collins A S, Belousova E. 2011. High-temperature granite magmatism, crust-mantle interaction and the Mesoproterozoic intracontinental evolution of the Musgrave Province, central Australia. Journal of Petrology, 52(5):931-958

Sossi P A, Foden J D, Halverson G P. 2012. Redox-controlled iron isotope fractionation during magmatic differentiation: an example from the Red Hill intrusion, S. Tasmania. Contributions to Mineralogy and Petrology, 164(5):757-772

Tang M, Wang X L, Shu X J, Wang D, Yang T, Gopon P. 2014. Hafnium isotopic heterogeneity in zircons from granitic rocks: geochemical evaluation and modeling of "zircon effect" in crustal anatexis. Earth and Planetary Science Letters, 389:188-199

Telus M, Dauphas N, Moynier F, Tissot F L H, Teng F Z, Nabelek P I, Craddock P R, Groat L A. 2012. Iron, zinc, magnesium and uranium isotopic fractionation during continental crust differentiation: the tale from migmatites, granitoids, and pegmatites. Geochimica et Cosmochimica Acta, 97:247-265

Teng F Z. 2017. Magnesium isotope geochemistry. Reviews in Mineralogy and Geochemistry, 82(1):219-287

Teng F Z, Li W Y, Ke S, Marty B, Dauphas N, Huang S C, Wu F Y, Pourmand A. 2010. Magnesium isotopic composition of the Earth and chondrites. Geochimica et Cosmochimica Acta, 74(14):4150-4166

Trail D, Tailby N, Wang Y L, Harrison T M, Boehnke P. 2017. Aluminum in zircon as evidence for peraluminous and metaluminous melts from the Hadean to present. Geochemistry, Geophysics, Geosystems, 18(4):1580-1593

Wang D, Wang X L, Cai Y, Goldstein S L, Yang T. 2018. Do Hf isotopes in magmatic zircons represent those of their host rocks? Journal of Asian Earth Sciences, 154:202-212

Wang J M, Rubatto D, Zhang J J. 2015. Timing of partial melting and cooling across the Greater Himalayan crystalline complex (Nyalam, central Himalaya): in-sequence thrusting and its implications. Journal of Petrology, 56(9):1677-1702

Wang Q, Chung S L, Li X H, Wyman D, Li Z X, Sun W D, Qiu H N, Liu Y S, Zhu Y T. 2012. Crustal melting and flow beneath northern Tibet: evidence from mid-miocene to quaternary strongly peraluminous rhyolites in the southern Kunlun Range. Journal of Petrology, 53(12):2523-2566

Wang T, Tong Y, Zhang L, Li S, Huang H, Zhang J J, Guo L, Yang Q D, Hong D W, Donskaya T, Gladkochub D, Tserendash N. 2017. Phanerozoic granitoids in the central and eastern parts of Central Asia and their tectonic significance. Journal of Asian Earth Sciences, 145:368-392

Wang W, Cawood P A, Zhou M F, Pandit M K, Xia X P, Zhao J H. 2017. Low-δ^{18}O rhyolites from the Malani igneous suite: a positive test for South China and NW India linkage in Rodinia. Geophysical Research Letters, 44(20):10298-10305

Wang X C, Li Z X, Li X H, Li Q L, Tang G Q, Zhang Q R, Liu Y. 2011. Nonglacial origin for low-δ^{18}O Neoproterozoic magmas in the South China Block: evidence from new in-situ oxygen isotope analyses using SIMS. Geology, 39(8):735-738

Wang X L, Zhou J C, Wan Y S, Kitajima K, Wang D, Bonamici C, Qiu J S, Sun T. 2013. Magmatic evolution and crustal recycling for Neoproterozoic strongly peraluminous granitoids from southern China: Hf and O isotopes in zircon. Earth and Planetary Science Letters, 366:71-82

Wang Y, He Y S, Ke S. 2020. Mg isotope fractionation during partial melting of garnet-bearing sources: an adakite perspective. Chemical Geology, 537:119478

Wilcock J, Goff F, Minarik W G, Stix J. 2013. Magmatic recharge during the formation and resurgence of the Valles Caldera, New Mexico, USA: evidence from quartz compositional zoning and geothermometry. Journal of Petrology, 54(4):635-664

Wu F Y, Liu X C, Liu Z C, Wang R C, Xie L, Wang J M, Ji W Q, Yang L, Liu C, Khanal G P, He S X. 2020. Highly fractionated Himalayan leucogranites and associated rare-metal mineralization. Lithos, 352-353:105319

Xu L J, He Y S, Wang S J, Wu H J, Li S G. 2017. Iron isotope fractionation during crustal anatexis: constraints from migmatites from the Dabie orogen, central China. Lithos, 284-285:171-179

Xu X S, Dong C W, Li W X, Zhou X M. 1999. Late Mesozoic intrusive complexes in the coastal area of Fujian, SE China: the significance of the gabbro-diorite-granite association. Lithos, 46(2):299-315

Xu X S, Zhao K, He Z Y, Liu L, Hong W T. 2020. Cretaceous volcanic-plutonic magmatism in SE China and a genetic model. Lithos, https://doi.org/10.1016/j.lithos.2020.105728

Yan L L, He Z Y, Jahn B M, Zhao Z D. 2016. Formation of the Yandangshan volcanic-plutonic complex (SE China) by melt extraction and crystal accumulation. Lithos, 266-267:287-308

Yan L L, He Z Y, Beier C, Klemd R. 2018. Geochemical constraints on the link between volcanism and plutonism at the Yunshan caldera complex, SE China. Contributions to Mineralogy and Petrology, 173(1):4

Yan L L, He Z Y, Klemd R, Beier C, Xu X S. 2020. Tracking crystal-melt segregation and magma recharge using zircon trace element data. Chemical Geology, 542:119596

Yang J H, Wu F Y, Wilde S A, Xie L W, Yang Y H, Liu X M. 2007. Tracing magma mixing in granite genesis: in situ U-Pb dating and Hf-isotope analysis of zircons. Contributions to Mineralogy and Petrology, 153(2):177–190

Yu K Z, Liu Y S, Hu Q H, Ducea M N, Hu Z C, Zong K Q, Chen H H. 2018. Magma recharge and reactive bulk assimilation in enclave-bearing granitoids, Tonglu, South China. Journal of Petrology, 59(5):795–824

Zeng L S, Asimow P D, Saleeby J B. 2005. Coupling of anatectic reactions and dissolution of accessory phases and the Sr and Nd isotope systematics of anatectic melts from a metasedimentary source. Geochimica et Cosmochimica Acta, 69(14):3671–3682

Zhao K, Xu X S, Erdmann S. 2017a. Crystallization conditions of peraluminous charnockites: constraints from mineral thermometry and thermodynamic modelling. Contributions to Mineralogy and Petrology, 172(5):26

Zhao K, Xu X S, Erdmann S, Liu L, Xia Y. 2017b. Rapid migration of a magma source from mid- to deep-crustal levels: insights from restitic granulite enclaves and anatectic granite. GSA Bulletin, 129(11-12):1708–1725

Zhao K, Xu X S, Erdmann S. 2018. Thermodynamic modeling for an incrementally fractionated granite magma system: implications for the origin of igneous charnockite. Earth and Planetary Science Letters, 499:230–242

Zhao Z F, Zheng Y F, Wei C S, Wu F Y. 2011. Origin of postcollisional magmatic rocks in the Dabie orogen: implications for crust-mantle interaction and crustal architecture. Lithos, 126(1-2):99–114

Zhao Z F, Gao P, Zheng Y F. 2015. The source of Mesozoic granitoids in South China: integrated geochemical constraints from the Taoshan batholith in the Nanling Range. Chemical Geology, 395:11–26

Zhou X M, Li W X. 2000. Origin of Late Mesozoic igneous rocks in southeastern China: implications for lithosphere subduction and underplating of mafic magmas. Tectonophysics, 326(3-4):269–287

Zhou X M, Sun T, Shen W Z, Shu L S, Niu Y L. 2006. Petrogenesis of Mesozoic granitoids and volcanic rocks in South China: a response to tectonic evolution. Episodes, 29(1):26–33

Zhu D C, Wang Q, Zhao Z, D, Chung S L, Cawood P A, Niu Y L, Liu S A, Wu F Y, Mo X X. 2015. Magmatic record of India-Asia collision. Scientific Reports, 5:14289

Progresses of Granite Researches

XU Xi-sheng, WANG Xiao-lei, ZHAO Kai, DU De-hong

State Key Laboratory for Mineral Deposits Research, School of Earth Sciences and Engineering,
Nanjing University, Nanjing 210023

Abstract: In this paper, we have summarized the status of granite researches and listed some important achievements on the petrogenesis of granitic rocks and crustal evolution by domestic and international researchers in the recent ten years, and highlighted the new findings and advances of researches on the detailed processes of magma evolution. Focusing on the incremental growth of large granitoid plutons and the connection of felsic volcanic and plutonic rocks, and combining the cutting edge studies on granites in recent years, we have concretely analyzed the new tendencies and key scientific problems in the researches of granites and related rocks in the southeastern China, aiming to provide helpful suggestions for the future researches on granites in China.

Key words: granite; crustal evolution; petrogenesis; volcanic-plutonic complex; research progress

变质岩研究进展*

张贵宾[1] 刘 良[2] 魏春景[1] 肖益林[3] 焦淑娟[4] 吕 增[1] 张立飞[1]

1. 北京大学 地球与空间科学学院,造山带与地壳演化教育部重点实验室,北京 100871;2. 西北大学 地质学系,大陆动力学国家重点实验室,西安 710127;3. 中国科学技术大学 地球和空间科学学院,合肥 230026;4. 中国科学院 地质与地球物理研究所,北京 100029

摘 要: 变质岩是组成硅酸盐地球的三大岩石之一,对了解深部地壳的组成和地壳演化、研究地壳热结构历史记录、恢复变质岩原岩建造和指导找矿都具有重要意义。本文综述了近十年来我国变质岩学科在超高压/高温变质作用、相平衡、变质流体以及变质岩地球化学等几方面的研究进展。在已有研究基础上,近十年来我国学者又相继在柴北缘野马滩、南阿尔金、东昆仑、北秦岭及西南天山的不同类型岩石中发现了柯石英、斯石英假象(副象)以及其他一些特征的超高压变质指示矿物和结构,确定了这些超高压变质带的野外分布特征。在超高温变质作用研究方面,识别并确定了华北克拉通孔兹岩带超高温变质岩石的出露规模,准确的变质时代和时间尺度的限定以及 P-T-t 轨迹的构建,尤其是在进变质阶段的确定方面取得了一系列进展。在变质相平衡研究方面,在热力学数据库和矿物相及熔体活度模型进一步完善的基础上,开发了新的模拟计算软件 GeoPS,建立了基于 ACF 组分分析的变质基性岩完整相平衡关系;同时在深熔作用与花岗质岩石成因的定量模拟方面也有重要进展。在矿物温压计研究方面,首次建立了与斜长石无关的 GBAQ 压力计、二云母压力计和白云母 Ti 温度计。在元素地球化学方面,高场强元素(如 Ti、Nb、Ta、Zr、Hf 等)和卤族元素(如 F、Cl、Br 和 I)在变质脱水过程中的迁移和分异,以及变价元素(如 V、Fe、W 和 Mo 等)对指示氧逸度的变化研究方面都取得了一系列进展。在变质流体研究方面,对俯冲带高压-超高压流体活动的证据、流体成分的确定及流体活动时限等方面的研究也取得了明显进展。

关键词: 超高压变质作用 超高温变质作用 变质相平衡 变质流体 同位素定年

0 引 言

变质岩是组成硅酸盐地球的三大岩石之一,对了解深部地壳的组成和地壳演化、研究地壳热历史记录、恢复变质岩原岩建造和指导找矿具有重要意义。变质作用可以发生于各种地质环境,如大陆克拉通、穹窿构造核部、碰撞造山带和汇聚板块边缘等,其中俯冲带(大洋俯冲带和大陆俯冲带)是各级变质岩广泛出露的场所。变质岩学科包含了各种极端变质作用,如高压-超高压变质作用、超高温变质作用、冲击变质作用、甚低级变质作用等,以及变质熔流体活动、深俯冲及物质循环等众多研究方向。在过去的十年中,变质岩学科在诸多研究方向上都取得了重要进展,本文侧重介绍超高压-高温变质作用、相平衡、变质流体及变质岩地球化学等方面近十年的研究进展。

1 超高压变质作用研究进展

超高压变质作用是推动板块构造关于造山作用理论具有里程碑性质的突破性进展。Chopin(1984)和 Smith(1984)分别在西阿尔卑斯的石英岩和挪威的榴辉岩中发现了柯石英,这一发现使我们认识到大陆地壳可以俯冲到 90 km 以上的地下深度。之后在哈萨克斯坦的 Kokchetav 地体的壳源岩石、大别榴辉岩中

---

* 原文"中国变质岩研究近十年新进展"刊于《矿物岩石地球化学通报》2021 年第 40 卷第 6 期,本文略有修改。

金刚石包体的研究报道又将俯冲深度推进到了 120 km 以上（Sobolev and Shatsky，1990；Xu et al.，1992）。自此，柯石英和金刚石的发现和确定是厘定一个造山带存在超高压变质作用最直接的方法和途径。此外，一些特殊反应结构和矿物出溶也可以指示更大俯冲深度，如发现于德国 Erzgebirge 含金刚石变泥质片麻岩中的 $\alpha\text{-PbO}_2$ 型 TiO_2（亦称 II-type 金红石）（Hwang et al.，2000）；在哈萨克斯坦 Kokchetav 和我国的天山、大别地区报道过的碳酸盐矿物之间的反应：白云石 = 菱镁矿 + 文石（Zhang et al.，2002b；Proyer et al.，2013）；曾在多个超高压地体中被发现的超硅石榴子石出溶辉石等矿物组合（van Roermund and Drury，1998；Ye et al.，2000；Song et al.，2004，2005a）；榍石出溶石英和柯石英（Ogasawara et al.，2002），等等。这些有关超高压变质作用的进展对我们理解大陆深俯冲及折返机制提出了变革性的认识和挑战。

中国是世界上超高压变质带发育最齐全和成因类型最多的国家之一（Zheng et al.，2012a，2012b），典型的超高压变质带有世界上规模最大的大陆深俯冲形成的大别-苏鲁超高压变质带（Zheng et al.，2012a）、大陆物质俯冲深度最大的南阿尔金造山带（Liu et al.，2007，2018）、出露规模最大的洋壳深俯冲形成的西南天山超高压变质带（Zhang et al.，2019a，2019b）。我国西部发育众多不同地质时期的造山带，但由于自然条件所限，研究工作起步较晚。在 21 世纪的第一个十年中，对于西南天山、南阿尔金、柴北缘的超高压变质作用研究主要聚焦于超高压变质作用的确立。

柴北缘超高压变质带最先在副片麻岩的锆石中发现有柯石英包裹体（Yang et al.，2001；Song et al.，2003a，2003b）和在石榴橄榄岩的锆石中发现有金刚石包裹体（Song et al.，2005b），之后又在榴辉岩中发现了薄片尺度下的柯石英包体（Zhang G. B. et al.，2009；Zhang J. X. et al.，2009；Zhang et al.，2010）。都兰北带的榴辉岩中发现的 K-cymrite 包体指示其峰期压力至少大于 3 GPa（Zhang R. Y. et al.，2009），而柴北缘的石榴橄榄岩中的石榴子石出溶钠质角闪石，指示其俯冲深度大于 200 km（Song et al.，2005a）。对于南阿尔金地区，相关的超高压变质作用证据较少，但 Liu 等（2007）在非陨石撞击成因泥质岩中首次发现先存斯石英的显微出溶结构，从而超越了柯石英与金刚石记录的地壳俯冲深度，率先把陆壳俯冲-折返深度由柯石英稳定域推进到斯石英稳定域的地幔深度（>350 km）。西南天山的榴辉岩中最先报道有石英出溶及菱镁矿和文石的反应结构（Zhang et al.，2002a，2002b），之后又在榴辉岩的绿辉石中发现有柯石英出溶（Zhang et al.，2005）；直到 2008 年才见有榴辉岩的石榴子石中确凿的薄片尺度下的柯石英包裹体的报道（Lü et al.，2008，2009）。

在 21 世纪的第二个十年中，在之前研究的基础上，我国学者又在以上几个超高压变质带中陆续发现了新的超高压变质作用证据，并在东昆仑地区新发现了榴辉岩及柯石英包裹体，丰富了我国西部超高压变质带。本文仅对我国学者近十年来在我国西部超高压变质带研究方面的进展进行综述。

1.1 柴北缘超高压变质带野马滩榴辉岩中首次发现柯石英及绿辉石中的多硅白云母出溶

柴北缘超高压变质带位于青海省境内，青藏高原的东北缘，沿柴达木盆地北缘呈北西西-南东东展布。其北为祁连地体，南为柴达木地体，东接秦岭造山带，西端被阿尔金断裂所切。阿尔金断裂以西为阿尔金山主体，北临塔里木盆地，阿尔金山体主要呈北东东-南西西走向，但有研究表明阿尔金山区域内的地质体构造线走向同样为近东西走向。

1.1.1 野马滩榴辉岩中发现柯石英

野马滩榴辉岩位于柴北缘超高压变质带东端都兰地体的都兰北带，之前在泥质片麻岩的锆石中发现有柯石英包裹体（Song et al.，2003b），但榴辉岩中只见有柯石英假象和石英出溶（Song et al.，2003a；Zhang et al.，2008）。韩磊（2015）在该区域的双矿物榴辉岩薄片中发现了保存完好的柯石英包裹体，其拉曼谱图具有特征的柯石英谱峰。

1.1.2 超深俯冲绿辉石中多硅白云母的出溶证明大陆深俯冲约达 200 km

柴北缘超高压变质带中的榴辉岩和围岩片麻岩均有柯石英包裹体的报道，表明这些岩石至少曾俯冲到 80 km 的深度，但具体的俯冲深度一直缺乏有效的矿物学证据。在都兰北带野马滩榴辉岩的绿辉石中发现多硅白云母出溶结构，表明寄主绿辉石的 K_2O 含量约为 1.16%，水含量约为 10% wt.（Han et al.，2015）。实验岩石学研究表明，这一出溶现象对应的压力超过 6 GPa，推断野马滩榴辉岩的俯冲深度应大于 200 km。

1.2 东昆仑榴辉岩及围岩片岩中首次发现柯石英

东昆仑造山带位于青藏高原东北缘，东以温泉-赛什塘断裂与秦岭造山带相接，西被阿尔金断裂所截，北以柴达木地块与柴北缘和祁连造山带相隔，是我国中央造山系的重要组成部分。由于自然条件限制，东昆仑地区的高压-超高压变质作用研究起步较晚。自 2013 年起，在东昆仑造山带的夏日哈木、浪木日和温泉等地相继发现了榴辉岩（Meng et al.，2013；祁生胜等，2014；祁晓鹏等，2016），连接成了长达 500 km 的高压（超高压）榴辉岩带。这些榴辉岩呈透镜体和不规则状产于金水口岩群白沙河组的黑云斜长片麻岩中。初步的年代学研究获得榴辉岩的原岩形成时代为 934 Ma 或 486～517 Ma，变质时代为 451 Ma、432 Ma 和 412 Ma（祁晓鹏等，2016；国显正等，2017，2018）。

Bi 等（2018）在东昆仑东段的克合特地区榴辉岩的锆石中发现了柯石英。该榴辉岩呈透镜状产于变沉积岩中（泥质岩和大理岩），其主要矿物组成有石榴子石（20%～30%）、绿辉石（30%～40%）、角闪石（5%～15%）、石英（5%）、多硅白云母（5%）和金红石（2%）。绿辉石含硬玉组分（38%～43%），多具有 Cpx+Pl 后成合晶，部分完全被 Amp+Pl 替代，在部分绿辉石核部含有密集的定向石英棒状体；多硅白云母单位分子式中的 Si 值为 3.45～3.58；Grt-Omp-Phe 温压计给出的温压条件是 610～675℃ 和 2.9～3.0 GPa（Song et al.，2018）。

Bi 等（2020）在该区榴辉岩的围岩云母片岩锆石中发现了柯石英。该云母片岩的矿物组合为石榴子石（10%）、白云母（30%）、石英（40%）、斜长石（10%）和黑云母（10%），另含少量电气石、金红石和锆石。计算所得相平衡温度为（685±41）℃，压力大于 2.8 GPa，变质年龄为（426.5±0.88）Ma。

1.3 南阿尔金榴辉岩中的柯石英及斯石英假象

1.3.1 南阿尔金榴辉岩中首次发现柯石英

南阿尔金构造带是我国西部典型的高压-超高压变质岩带，其认定主要是通过一些特殊矿物出溶结构的确定，但缺乏在其他超高压变质岩带中更广泛和常见的指示矿物柯石英。

Gai 等（2017）在南阿尔金地区克其克江尕勒萨依的榴辉岩中首次发现了柯石英。该榴辉岩主要由石榴子石、绿辉石、石英及少量单斜辉石、角闪石、斜长石、金红石组成。目前发现的柯石英呈包裹体赋存于绿辉石中，其边部已退变形成多晶石英，周围具有明显的胀裂纹。拉曼光谱分析显示，柯石英包裹体核部具有 524 cm^{-1} 主峰以及 151 cm^{-1}、178 cm^{-1} 和 270 cm^{-1} 等次峰的典型柯石英拉曼谱峰值。根据柯石英包裹体的发现并结合石榴子石-单斜辉石温度计的计算结果，确定该榴辉岩的峰期变质条件为 728～880℃ 和 2.8 GPa 以上压力。

南阿尔金榴辉岩中柯石英的首次发现，为该区域经历过超高压变质作用提供了进一步的矿物学证据，也为前人关于北南阿尔金超高压变质的认识提供了进一步的证据。

1.3.2 陆壳超深俯冲到斯石英稳定域地幔深度的新证据

一种同质多象变体矿物转变为另一种变体矿物后继承保留了先期矿物晶体形态的现象称为矿物的副象，这是判断矿物之间发生过同质多象转变的重要证据。Liu 等（2018）在南阿尔金超高压榴辉岩的绿辉石和石榴子石中发现了呈长柱状或针状的多晶石英集合体，以及呈近四方形或平行四边形的单晶石英包裹体。通过斯石英与柯石英结晶习性的分析，再与高温高压实验合成斯石英与柯石英晶形的对比，确定这些特殊形态的石英多晶集合体和单晶石英包裹体的形态完全不同于以往文献中报道的呈短柱状或不规则状的柯石英假象，也不同于超高压岩石中的熔流体成因的多相固体包裹体，而应是先存四方长柱状斯石英副象不同截切面的晶体形态。据此，确定该榴辉岩的峰期压力至少应为 8~10 GPa，温度为800~1 000℃，对应斯石英稳定域的地幔深度（~300 km）。

此外，依据 SiO_2 饱和岩石体系中石榴子石超硅的最小稳定条件（≥9~10 GPa）的斯石英稳定域的高温高压实验研究结果，重新厘定出南阿尔金英格利萨依石榴辉石岩中发现石榴子石出溶单斜辉石的峰期变质压力（≥9~10 GPa），同样达到了斯石英稳定域的地幔深度（~300 km）（刘良等，2019）。

Dong 等（2018，2020）通过对南阿尔金巴什瓦克地区长英质麻粒岩和石榴子石岩开展了岩相学研究和相平衡模拟计算，得到石榴子石环带记录的峰期压力分别为 3.0~9.0 GPa 和 6.5~7.0 GPa，前者的最大压力可能进入斯石英稳定域的变质条件。他们还发现石榴子石岩中的绿辉石内有三组分别沿着主晶（001）、（100）和（401）面出溶的易变辉石，其中沿（401）面出溶的易变辉石与主晶 c 轴夹角高达约 18°，指示其可能具有高压 $C2/c$ 型结构特征，限定峰期压力在 7.0 GPa 以上。

以上这些研究表明，南阿尔金地区的榴辉岩、石榴辉石岩、长英质麻粒岩和石榴子石岩等多种岩石类型的峰期压力均达到了斯石英稳定域，指示大陆地壳俯冲到约 300 km 的地幔深度，然后再折返回地表的地质过程可能更为普遍，其岩石类型也可能具有多样性。这一认识对深入理解超深俯冲陆壳的组成、规模、形成-折返机制及其相关的壳幔相互作用提供了关键的约束条件。同时，这些研究还为在超高压变质带中识别辨认先存斯石英的研究提供了新的借鉴和研究思路。

1.4　北秦岭斜长角闪岩中首次发现柯石英

通过大量细致的岩相学观察与拉曼光谱测试，宫相宽等（2016）首次在北秦岭丹凤大寺沟斜长角闪岩的锆石中发现了柯石英、石榴子石、绿辉石和金红石等多种榴辉岩相矿物包裹体。这些柯石英具有典型的 521 cm^{-1} 主峰以及 270 cm^{-1} 和 151 cm^{-1} 次峰的拉曼光谱组合，绿辉石、石榴子石及金红石分别具有 680 cm^{-1}、910 cm^{-1} 和 611 cm^{-1} 的特征峰。柯石英、石榴子石和绿辉石显微包裹体的发现证明，北秦岭丹凤斜长角闪岩应是退变的超高压榴辉岩，其变质峰期压力应不低于 2.8 GPa。

1.5　西南天山超高压变质带多种类型岩石中发现超高压矿物证据

西南天山造山带发育世界上少有的低温超高压变质带，其岩石类型较复杂，经历了多阶段演化，记录了丰富的流体活动，但这些岩石是否共同经历了深俯冲作用还存在不同的看法（Zhang et al., 2018）。自在榴辉岩和其周围的石榴子石多硅白云母片岩中发现了柯石英之后，近些年的工作又陆续在不同的岩石类型中（如蛇纹岩、富钙岩石和变质火山碎屑岩）均发现了深俯冲的矿物学证据，并获得了它们的超高压变质条件。在超基性岩中发现了超高压指示矿物——钛粒硅镁石和钛斜硅镁石，并通过高压实验确定其稳定域，获得超高压蛇纹岩的峰期温压条件为 510~540℃、3.0~3.7 GPa（Shen T. T. et al., 2015）。在该变质带西段的木扎尔特地区的榴辉岩及其围岩片岩中均发现含柯石英岩石，确凿地证明了西南天山低温超高压变质带的空间延伸规模庞大（Lü et al., 2019a, 2019b）。通过对木扎尔特剖面变基性岩的详细岩石学研究，识别出超高压和高压两类硬柱石榴辉岩，证明西南天山古俯冲带的低温超高压属性，是地表流体向地球深部传输的重要通道，为完善地球深部水循环模型提供了天然岩石观测数据支撑（Lü et al., 2019a）。此外，西南天山还存在大量碳酸盐化变质火山碎屑岩，其中的白云石中保存了柯石英，这也是首次在低温超高压变质带中发现的白云石经历超高压变质的证据（Lü et al., 2014），从而证明了在冷俯冲带此类常见的碳酸盐矿物至少都稳定至弧岩浆源区深度，为研究深部碳循环机制提供了岩石学基础。西南天山含柯石英岩石在空间上的广泛分布以及不同岩石类型超高压条件的获得均表明，西南天山超高压变质岩构成了一个相对独立的地质单元，不支持其折返过程中的构造混杂成因。

2 超高温变质作用研究进展

超高温（ultra high temperature，UHT）变质作用是温度大于900℃、在夕线石稳定域内（0.7~1.3 GPa）的麻粒岩相变质作用（Harley，1998；Brown，2007），代表各个地质时期固相条件下最热的地壳状态。UHT鉴别性的矿物组合常发育在富Al麻粒岩中，包括假蓝宝石（Spr）+石英（Qz）、紫苏辉石（Opx）+夕线石（Sil）+石英（Qz）和含大隅石的矿物组合（Harley，2008，2021；Kelsey，2008）。全球70余处UHT麻粒岩中，变质年龄最老的约为3.1~3.0 Ga（Yu et al.，2021），最年轻的为16 Ma（Pownall et al.，2014）。UHT麻粒岩分布面积超过500 km^{2}的有15处，大都集中在前寒武纪，与超大陆的聚合过程密切相关。重建各个地质时期（造山带）深部地壳的热状态及其随时间的演化规律，是详细解读陆壳生长分异、壳幔相互作用和地球动力学过程的重要的第一手资料，直接关系到我们对板块构造启动及其转型的认识。

华北克拉通孔兹岩带呈东西向展布，长约1 000 km，自西向东可分为3个地体：贺兰山-千里山、乌拉山-大青山和集宁-凉城-丰镇地体。该岩带主要由高角闪岩相到麻粒岩相的变沉积岩（即孔兹岩系）组成，主要岩性为长英质片麻岩、夕线石榴斜长（钾长或二长）片麻岩、长石石英岩、大理岩和钙硅酸岩等，还广泛发育S型花岗岩和紫苏花岗岩，代表地壳深熔作用的结果，可能有部分幔源基性物质的贡献（金巍，1989；卢良兆等，1992；郭敬辉等，1999；刘建忠等，2000；万渝生等，2000；Xia et al.，2006；Guo J. H. et al.，2012；Yin et al.，2014；Wang et al.，2018）。区域上广泛发育辉长、苏长、闪长质岩体/岩墙，与孔兹岩系一起经历了高角闪石相-麻粒岩相变质作用。该带岩石大都经历了顺时针型P-T轨迹，峰期温压条件大都为820~880℃和0.8~1.0 GPa（Cai et al.，2014，2015；Liu P. H. et al.，2014，2017；Wu et al.，2017）。从已有的年代学资料来看，变质时代从1.81 Ga到1.96 Ga，该岩带经历了长期的或者多次幕式变质作用（Jiao et al.，2020a，2020b）。在孔兹岩带还识别出高压麻粒岩相变质岩，如贺兰山-千里山地体的变泥质岩中保存了石榴子石+蓝晶石+条纹长石的矿物组合，变质时代为1.95 Ga（Yin et al.，2009，2014）。

近十几年来，在国内外学者的共同努力下，在孔兹岩带识别出多个UHT变质岩露头（郭敬辉等，2006；Santosh et al.，2007a，2007b，2009a，2009b，2012；Liu S. J. et al.，2010，2011，2012；Jiao and Guo，2011；Jiao et al.，2011，2013a；Tsunogae et al.，2011；Guo J. H. et al.，2012；Shimizu et al.，2013；Yang et al.，2014；Li and Wei，2016，2018；Gou et al.，2018；Lobjoie et al.，2018；Liu H. et al.，2019；Wang et al.，2020），研究取得了重要进展，主要体现在：①UHT变质作用的识别及其出露规模的确定；②UHT变质作用准确地变质时代和时间尺度的限定；③UHT变质P-T-t轨迹的构建，尤其是进变质阶段的确定。

2.1 UHT变质作用的识别及其出露规模

UHT变质鉴别性的矿物组合常发育在Mg-Al质麻粒岩中，这些岩性在自然界很稀少，因此依据传统的矿物组合很难准确确定不含这些矿物组合的岩石是否经历了UHT变质作用。UHT变质作用的出露规模更加难以判定。这导致对UHT变质作用类型（接触变质和区域变质）及其加热机制存在强烈的争议（Brown，2007；Clark et al.，2011；Harley，2008，2016；Kelsey，2008）。此外，在极高的变质温度下，元素强烈的扩散作用也会导致常规的Fe-Mg交换温度计不能有效确定UHT峰期变质温度（Fitzsimons and Harley，1994；Pattison et al.，2003；吴春明，2018）。这些问题严重阻碍了人们对UHT变质岩出露规模、分布和成因的认知，需要从新的视角来识别UHT变质作用及其出露规模。

华北孔兹岩带UHT变质作用研究之初，只见有集宁地体的天皮山和大青山地体的东坡两个露头经历了UHT变质作用的报道，其中天皮山的UHT麻粒岩发育鉴别性的矿物组合（刘建忠等，2000；Guo et al., 2006; Santosh et al., 2007a, 2007b), 当时并不清楚该带广泛发育的夕线石榴片麻岩和含石榴子石的长英质片麻岩是否同样也经历了UHT变质作用, 这些岩石大都不发育UHT鉴别性的矿物组合。随后一系列采用多种不易扩散元素构成的温度计和视剖面图计算的研究工作, 不仅证明没有鉴别性矿物组合的岩石同样也经历了UHT变质作用, 还确定UHT变质作用在孔兹岩带的分布规模要大于10 000 km², (Liu et al., 2010; Jiao et al., 2011, 2013a, 2013b; Jiao and Guo, 2011; Zhang et al., 2012; Yang et al., 2014; Gou et al., 2018; Lobjoie et al., 2018; Wang et al., 2019, 2020)。有效的温度计包括二长石温度计、锆石Ti含量温度计、金红石Zr含量温度计、紫苏辉石Al含量温度计、角闪石Ti含量温度计、夕线石Fe^{3+}含量温度计等。

这些研究确定了孔兹岩带UHT变质岩"面"状分布的特征, 扩大了UHT变质岩分布规模, 明确了UHT变质作用是与造山作用密切相关的区域性变质作用, 而非局部的接触变质作用。

2.2 UHT变质作用准确的变质时代和时间尺度的限定

UHT变质时代和各变质阶段的持续时间（时间尺度）的准确限定, 直接关系到对其加热机制的判断。然而, 麻粒岩中锆石和独居石U-Pb年代学的含义一直是广为争议的科学问题 (Roberts and Finger, 1997; 吴元保和郑永飞, 2004; Kohn et al., 2015; 魏春景和朱文萍, 2016), 其根源在于不清楚这些定年副矿物是哪个变质阶段形成的 (Schaltegger et al., 1999; Rubatto et al., 2001; Kelsey and Powell, 2011; Harley and Nandakumar, 2014; Yakymchuk and Brown, 2014), 以及不确定极高的变质温度和富含熔-流体的环境对同位素体系潜在的改造作用 (Kelly et al., 2012; Taylor et al., 2014)。因此, 人们难以准确限定UHT变质作用发生的时代和时间尺度, 从而导致无法判断UHT变质作用是长期稳定的造山带自发生热作用的结果（对应"慢"的UHT变质作用）, 还是与幔源岩浆活动密切相关（对应"快"的UHT变质作用）(Harley, 2021)。

华北孔兹岩带岩石的变质时代极为复杂 (1.96~1.80 Ga), 已报道的UHT变质作用的时代也很宽泛 (1.92~1.88 Ga; Jiao et al., 2020b), 指示孔兹岩带可能经历了一个长期的演化过程。但这些变质年龄代表峰期变质还是退变质阶段? 连续长期的还是多个短期的变质作用? 仍不清楚, 从而导致对华北孔兹岩带UHT变质作用的成因有多种、甚至相互矛盾的模式。

近十年来, 该带一系列研究将锆石和独居石原位微区U-Pb同位素年代学与微量元素地球化学研究相结合, 并综合结构位置、包裹体特征、元素面扫描、与石榴子石的HREE配分模式等方法, 研究锆石和独居石在UHT变质作用中的行为, 以及UHT变质作用与熔-流体对其同位素体系的改造, 详细解析其同位素年龄的含义, 准确限定了孔兹岩带UHT变质作用的时代、期次和时间尺度 (Jiao et al., 2013b, 2020a, 2020b; Huang et al., 2019)。

这些研究揭示, UHT进变质和退变质阶段都有锆石和独居石生长, 独居石相比锆石更容易受到流体渗透作用的影响, 形成有别于原来独居石的成分环带。研究限定了孔兹岩带UHT进变质阶段的时代最大为1.94~1.90 Ga, 时间尺度不超过40 Ma, 随后经历了从1.90~1.80 Ga漫长的冷却过程, 时间尺度约为100 Ma。有研究还进一步识别出孔兹岩带存在两期高压-超高压 (HT-UHT) 变质作用: 第一期在区域上非常普遍, 峰期变质时代接近1.90 Ga; 第二期在孔兹岩带局部发生, 峰期在1.86~1.84 Ga。流体改造事件发生在1.80 Ga之后。这些研究揭示出相对快速的进变质阶段和缓慢的峰期后冷却过程, 以及冷却过程中局部地区另一期加热事件的叠加。这些研究首次识别出华北孔兹岩带经历了两期HT-UHT变质作用, 精准识别了长期造山过程中多次相对快速的加热过程。

2.3 UHT变质P-T-t轨迹的构建: 进变质阶段的确定

进变质阶段的准确限定, 是变质岩研究的一个难题, 在UHT变质岩中尤为突出。虽然UHT变质作用可能叠加了多个变质阶段, 但UHT麻粒岩通常只记录整个P-T轨迹中退变质阶段的片段 (Harley, 1998,

2008；Kelsey，2008）。由于缺少早期的矿物和成分记录，UHT 进变质阶段的限定，我们仍不清楚深部地壳如何形成区域范围内高热流的环境并发生固相条件下的 UHT 变质作用，以及 UHT 变质作用是否是部分熔融（深熔）作用之后的第二次变质作用。识别早期进变质阶段的矿物组合，确定 P-T 轨迹，对准确反演 UHT 变质作用的加热机制起到关键的约束作用。

华北孔兹岩带发育的 UHT 麻粒岩一般只记录了峰期后退变质阶段的 P-T 轨迹。尽管如此，有研究认为退变质阶段为 IBC（近等压冷却）过程（Santosh et al.，2007a，2009；Tsunogae et al.，2011），也有研究认为是 ITD（近等温减压）过程（刘建忠等，2000；Guo et al.，2006）。早期的研究认为 UHT 进变质阶段可能是一个近等温升压过程，其主要依据是假蓝宝石围绕尖晶石生长的岩相学特征（Santosh et al.，2007a，2009）。总之，UHT 变质总体的 P-T 轨迹存在很大的争议，严重制约了人们对 UHT 变质加热机制以及对孔兹岩带古元古代构造热演化过程的认识。

近十年来，华北孔兹岩带一系列研究工作将矿物学、变质岩石学与微量元素地球化学研究相结合，识别 UHT 麻粒岩中的矿物反应关系，寻找各变质阶段的矿物记录，利用石榴子石中扩散较慢的 Ca 和微量元素地球化学特征识别石榴子石的成因类型，利用相平衡模拟反演 P-T 轨迹，进而探讨 UHT 变质作用发生的构造背景及其加热机制（Jiao et al.，2013b，2015，2017，2021；Yang et al.，2014；Li and Wei，2016，2018；Gou et al.，2018；Wang et al.，2019，2020，2021；Jiao and Guo，2020）。

这些研究从不同视角共同限定了孔兹岩带 UHT 麻粒岩普遍经历了复杂的、多阶段的顺时针型 P-T 演化过程，包含峰期前减压升温阶段、峰期后近等压冷却和随后的减压升温或近等压升温阶段的 P-T 轨迹。研究发现 UHT 进变质阶段的关键性矿物学证据，识别出早期的转熔型石榴子石，形成于 UHT 进变质、压力较高的部分熔融阶段。这些研究还识别出两期 HT-UHT 变质作用，第一期在区域上普遍发育，而第二期的影响范围较小，进而提出碰撞造山作用之后存在多次岩石圈伸展阶段，HT-UHT 变质作用的成因与下伏软流圈上涌有关。这些研究深化了 UHT 变质作用在长期造山作用中的成因模式，完善了孔兹岩带在造山纪的构造热演化过程。

3　P-T 相平衡、温压计与变质反应动力学研究进展

定量计算岩石体系的 P-T-X 空间内的相平衡关系和矿物温压计是研究变质作用的最有效手段，同时变质反应动力学也得到更多关注，近年来这方面的研究进展包括如下几个方面。

3.1　热力学数据库和矿物相、熔体活度模型的完善及热力学模拟软件的开发

建立完善的内恰性热力学数据库和矿物相及熔体活度模型是进行变质相平衡研究的最重要的基础。在目前广泛使用的内恰性热力学数据库（Holland and Powell，2011）基础上，White 等（2014）对变质泥质岩中的常见铁镁矿物进行了参数化处理。对基性岩体系，Diener 和 Powell（2012）对关键的角闪石和单斜辉石的活度做了进一步完善。以往使用的熔体模型（White et al.，2007）主要适合于计算泥质岩的熔融相平衡关系，Green 等（2016）建立了适合于计算基性岩熔融相平衡关系的熔体及角闪石与单斜辉石的活度模型。

此外，国内学者在热力学模拟软件开发方面的研究，也是近十年来我国在变质地质学领域最为突出的进展之一。以往国外学者开发的几种热力学模拟软件，已得到广泛采用，但有的使用不便，有的考虑的因素还不完善。新的模拟计算软件 GeoPS（Xiang and Connolly，2021）不仅考虑的因素周全，使用起来也简便易行，其初始版已得到万余次下载和使用。

3.2　基性岩的变质相平衡关系

在以上内恰性热力学数据库和矿物相及熔体活度模型完善的基础上，基性岩变质相平衡关系研究取

得了如下 3 个方面的主要进展：

（1）基于 ACF 组分分析建立的变质基性岩完整相平衡关系。Wei 和 Duan（2019）在对洋中脊成分变质基性岩实验和相平衡模拟基础上，在 0~3.2 GPa 和 400~1 100℃条件下，做出了定性的 P-T 视剖面图；将变质基性岩复杂的相平衡关系简化为 10 个不变点，并划分出 5 个变质相和 22 个亚相组合，以及变质作用类型的四分方案。这对于理解变质基性岩的 P-T 条件、变质反应、P-T 轨迹以及变质作用的大地构造环境等有很大指导价值。

（2）基性岩高压-超高压变质相平衡关系研究。近十年来的研究发现，在俯冲造山带中折返的高压-超高压榴辉岩的压力峰期一般都位于硬柱石稳定域，而硬柱石在折返过程中通过脱水反应消失；有关不同高压-超高压榴辉岩组合中确定岩石峰期变质条件的方法也被提出（Wei and Clarke，2011；Wei and Tian，2014）。

（3）基性麻粒岩相平衡关系。麻粒岩的降温反应如果发生在地壳深部（0.7~1.0 GPa），则降温反应较缓慢，真正的峰期信息很难保存下来，传统的 Fe-Mg 交换温度计一般只能记录缺流体固相线温度或更低的温度（Li and Wei，2016）。此种情况下，可采用稀土元素温度计（Liang et al.，2013；Sun and Liang，2015）、斜长石和辉石成分进行限定（Liu and Wei，2018）或基性麻粒岩中角闪石 Ti 温度计（Liao and Wei，2019）等多种方法恢复基性麻粒岩的峰期 P-T 条件。

3.3 深熔作用与花岗质岩石成因的定量模拟研究

随着熔体活度模型的完善，很多学者试图通过变质相平衡定量计算来模拟变质沉积岩与基性岩的深熔作用过程及其与花岗质岩石的成因联系（魏春景，2016；魏春景等，2017）。有学者结合微量元素分配系数来讨论太古宙 TTG 质岩石的成因与构造环境（Johnson et al.，2017；Ge et al.，2018）。

3.4 新型矿物温度计与压力计的建立

变质泥质岩是对变质作用压力-温度条件反应最为灵敏的岩石，但由于其富铝贫钙的全岩化学特征，变质的岩石中石榴子石和斜长石往往贫钙，有时甚至不出现石榴子石或斜长石。基于此，我国学者首次建立了与斜长石无关的石榴子石-黑云母-Al_2SiO_5 矿物-石英（GBAQ）压力计（Wu，2017）、二云母压力计（Wu，2020），首次建立了简便易行的白云母 Ti 温度计（Wu and Chen，2015）。

高级变质岩石如果经历缓慢冷却过程，镁铁质矿物之间会相互交换（扩散）Fe-Mg 离子，造成 Fe-Mg 交换温度计记录的温度偏低。考虑到稀土元素扩散速率很慢，易于记录变质高峰期温度条件，我国学者建立了二辉石 REE 分配温度计（Liang et al.，2013）、石榴子石-单斜辉石 REE 分配温度计（Sun and Liang，2015）、斜长石-单斜辉石 REE 分配温度计（Sun and Liang，2017）。

3.5 变质反应动力学研究

现阶段我们对变质反应结构、变质矿物共生组合的判断，都是基于经验并辅以热力学计算来判定的。变质反应动力学（变质反应速率等方面）的研究一直缺位，严重制约了研究结果的可靠性。实际上，变质反应动力学几乎是全世界变质作用研究的盲区，不仅从业者寥若晨星，研究成果也屈指可数。目前，仅有的研究进展在于离子扩散速率方面，例如地幔岩矿物之间 REE 的扩散（Sun and Liang，2014），麻粒岩石榴子石中 Fe、Mg、Ca、Mn 离子的扩散（Zou et al.，2020）。离子扩散现象对于温度计-压力计的应用、热力学视剖面图模拟，都带来了以前意想不到的挑战。

4 变质岩元素地球化学研究进展

俯冲变质脱水过程中往往伴随有元素的迁移和分异，变质流体的性质（富水流体、含水熔体和超临界流体等）和残留矿物（如锆石、帘石、含 Ti 矿物和石榴子石等）是控制元素分异的主要因素。相比于早期对大离子亲石元素（LILE）和轻稀土元素（LREE）的研究，近十年来对高场强元素（如 Ti、Nb、Ta、Zr、Hf 等）在变质脱水过程中的迁移和分异的地球化学行为的研究取得了较大进展。Huang 等（2012）对大别山碧溪岭榴辉岩的研究发现，在水-岩相互作用较弱时，由镁铁质岩石进变质转变为榴辉岩的过程中，Nb-Ta 几乎不发生迁移和分异。相反，靠近脉体的榴辉岩其 Nb/Ta 值远高于对应的变质脉，说明局部区域强烈的水-岩相互作用能导致显著的 Nb-Ta 迁移和分异，同时超临界流体的形成是 Nb-Ta 迁移和分异能力大大提升的重要介质（Huang and Xiao, 2015）。这一观察也得到其他研究的支持（Xia et al., 2013; Zhang L. J. et al., 2016）。

在最近十年的研究中，人们进一步认识到卤族元素（如 F、Cl、Br 和 I）在变质脱水过程中对其他元素的迁移-分异能力的重要影响，如 NaCl 和 KCl 的加入可以改变流体的物理性质和溶解能力（Mantegazzi et al., 2013; Sakuma and Ichiki, 2016）；Tsay 等（2014）发现 Cl 的加入会显著增加 LREE 和 HREE 之间的分异（La/Yb = 17.4 ± 4.3），而 F 的加入只是稍微加大轻、重稀土的分异（La/Yb ≈ 4）；同时随着 Na_2CO_3 的加入，会明显的提高稀土元素在熔-流体中的溶解度（Tsay et al., 2014）。

变价元素（如 V、Fe、W 和 Mo 等）对指示氧逸度的变化具有重要意义。V 作为变质岩中的一个重要微量元素，尤其在金红石中的含量可达几百甚至上千 ppm，并且 V^{4+} 更相容于金红石晶格中。由于 Nb 和 V 取代的都是金红石中 Ti^{4+} 的八面体位置，更多的 V 进入金红石会导致 Nb 所能取代的 Ti^{4+} 位置的减少，因此它们之间是一种竞争关系。理论上更多的 V 进入金红石会显著减少 Nb 在金红石中的含量。Liu L. 等（2014）发现，大别-苏鲁造山带榴辉岩中金红石的 Nb 和 V 具有广泛的相关性，其原因可能是氧逸度变化导致了 V 分配系数的改变，从而间接影响了金红石中的 Nb 分配。在 V 易相容于金红石时，金红石中的 Nb 分配会受到抑制，反之亦然。因此，陆壳俯冲带金红石中的 V 分配行为是氧逸度的函数，而且这一效应不仅对 Nb 在金红石中的分配具有影响，对 Nb-Ta 的分异也产生一定的控制作用。

由于某些副矿物（锆石、榍石和金红石）相比于造岩矿物更加难熔，并且高场强元素具有高的价态、大的离子半径和相对不易活动的优点，因此锆石 Ti 温度计、榍石 Zr 温度计和金红石 Zr 温度计在变质岩研究中都受到广泛关注（高晓英等，2011，2013）。Gao 等（2012）根据大别造山带低温-超高压榴辉岩和中温-超高压榴辉岩的锆石 Ti 温度计研究，结合锆石 U-Pb 年龄，很好地制约了陆壳的俯冲-折返历史。

5 变质作用与变质流体活动研究进展

5.1 俯冲带流体成分的研究

变质流体根据其所含水和溶质（各种元素或其离子）比例的不同，大致可以分为富水流体、富水熔体和超临界流体（Zheng et al., 2011; Xiao et al., 2015）。近十年来对于俯冲变质过程中熔、流体活动的研究取得了长足进步，一些新的成果对理解造山带的地球动力学过程、壳幔物质再循环及其相关的弧岩浆作用和地震活动都具有重要意义。其中 H_2O 是变质流体中最关键的组分，以分子、结构羟基和流体包裹体形式存在于岩石或矿物中（Zheng and Herman, 2014）。含水矿物的赋存相与俯冲带热结构相关，如硬柱石、多硅白云母和黝帘石存在于地温梯度为 5~10℃/km 的冷或超冷俯冲带。绿泥石、绿帘石和角闪石常见于地温梯度大于 15℃/km 的暖至热俯冲带（Zheng and Herman, 2014）。深俯冲板片在进变质-折返过程中，含水矿物名义上的无水矿物的结构羟基和分子水会发生分解和减压出溶（Zheng et al., 2009, 2011, 2012b）。

CO_2 是俯冲带变质流体的重要组分，对理解深部碳循环过程具有重要意义。Ague 和 Nicolescu（2014）对希腊始新世 Cycladic 俯冲带混杂岩的地球化学、流体包裹体及 C-O 同位素的研究表明，靠近流体通道的大理岩在俯冲过程中发生了显著的溶解，造成高达 60%~90% 的全岩 CO_2 被流体带走，同时伴随着硅酸盐矿物的沉淀作用，表明俯冲带流体导致的碳酸盐矿物溶解可以释放大量的CO_2流体进入上覆的地幔楔中。Frezzotti 等（2011）在阿尔卑斯超高压变质岩的流体包裹体中发现了金刚石、Mg-方解石也说明变质流体中可以溶解一定量的 C，可能是俯冲带脱 C 的主要机制；当然，俯冲带变质脱水过程中释放的 CO_2 组分也会迁移并交代地幔橄榄岩（Berkesi et al., 2012；Frezzotti et al., 2012；Kawamoto et al., 2013）。

近年来的研究表明，硫（S）是变质流体中另一类重要组分，其富集可以显著影响岩浆上升、去气、地幔交代和成矿作用。刘海洋等（2018）在绿辉石流体包裹体中发现了含 S 还原态的子矿物，如黄铜矿和黄铁矿等，表明俯冲板片变质脱水过程释放的流体可以有效运移 S 和成矿元素。N、Cl 和 F 也是变质流体中的常见组分。Bebout 等（2013）通过对西阿尔卑斯高压-超高压变质岩的研究，指出俯冲沉积物和蚀变洋壳中 50% 以上的 N 会被俯冲板片带入 100 km 以上的地幔深度；Mukherjee 和 Sachan（2009）在喜马拉雅造山带的北部 Tso Morari 地体含柯石英的变质岩中发现了含 N_2-CH_4 共存的流体包裹体，认为含 N_2 和 CH_4 的流体来自富 NH_3 和 K 的矿物和石墨的化学分解；同样的，在 CCSD 主孔高压-超高压石英脉中也发现了 CH_4 和 N_2 共存的流体包裹体；Liu Y. C. 等（2011）在北大别经历了俯冲深度达 150 km 以上的超高压变质作用的榴辉岩中发现了含 F 达 3% 的与流体有关的磷灰石。

5.2 高压-超高压变质流体活动证据的识别

5.2.1 变质脉体

变质脉体是高压-超高压变质岩中最常见的流体活动印记之一。变质流体沿着构造剪切的薄弱地带运移、聚集和沉淀成脉。变质脉体是流体-岩石相互作用的产物，对示踪俯冲带变质流体组成和演化具有重要意义（Zheng et al., 2019），总体来说，俯冲带变质岩主要有两大类型脉体，一类是富含石英的"纯"石英脉，石英通常占脉体总体积的 98%（Li Y. C. et al., 2011；Zhang et al., 2011；Huang et al., 2012）。另一类是共生矿物组合复杂的脉体，大致有蓝晶石-黝帘石-（石榴子石-）石英脉、绿辉石-蓝晶石脉、蓝晶石-黝帘石/绿帘石-多硅白云母-石英脉和褐帘石-绿辉石-蓝晶石-石英脉等（盛英明等，2011；Spandler et al., 2011；Chen et al., 2012；Guo S. et al., 2012, 2015；Sheng et al., 2013），这类脉体可能记录的是富水流体和含水熔体完全混溶的超临界流体特征（Xia et al., 2010）。近年来在西天山高压-超高压变质带发现了大量形成于退变质阶段的高压脉体，指示比重较大的深俯冲洋壳也可以折返到地壳层位（吕增等，2013；Zhang L. J. et al., 2016；Li et al., 2017）。

5.2.2 自然样品中的超临界流体

上一个十年从实验研究角度提出俯冲变质中可能存在超临界流体，并发现相比于普通富水流体，超临界流体对于 LILE、Pb、Th、U、Sr 和 LREE 以及高场强元素具有更高的运载能力，因此对于元素迁移以及成矿等过程具有重大影响。尽管因为自然界超临界流体会随温压的降低而分溶形成富水流体和含水熔体两相，并且超临界流体会与寄主岩石-矿物或围岩发生反应而丢失超临界性质（Kawamoto et al., 2013），因此目前对于超临界流体在变质岩中的岩石学证据还存在争议（Xia et al., 2010；Zheng et al., 2011；Zheng and Hermann, 2014；Huang and Xiao, 2015；Wang et al., 2017b）。近十年来人们对于自然界特别是俯冲带超临界流体证据的识别取得了较大的进步，其中超临界流体存在的一个重要证据来自俯冲带超高压变质岩石中的多相流体-固体包裹体（高晓英和郑永飞，2013；Frezzotti and Ferrando, 2015）。Kawamoto 等（2012）的研究也表明，俯冲带形成的流体在地幔楔中与橄榄岩会发生反应形成高镁安山质的超临界流体，并随后在上升过程（~90 km）中会分离成富水流体和含水熔体，其中富水流体会交代上覆地幔楔橄榄岩并引发其部分熔融而形成岛弧岩浆岩，而含水熔体则在上升过程中与地幔橄榄岩发生反应，导致其主量元素特征发生显著改变并演化为玄武岩浆，但是其微量元素仍然继承了俯冲板片超临界流体的微量元素组成特征。大别-苏鲁造山带内石榴橄榄岩（地幔楔碎片）受到了来自俯冲板块在超高压条件下产生的超临界流体的显著交代（Zheng et al, 2011；Zheng, 2012a）。Xiao 等（2011）观察到，在手标本尺度紧密接触的超高压榴辉岩与石榴橄榄岩边界（厘米尺度），存在大量的角闪石、绿泥石等含水矿物，而且在转换区域内石榴橄榄岩受到来自俯冲板块可能的超临界流体交代相关，与主期绿辉石（CpxI）成分明显不同的次生单斜辉石（CpxII）。Ferrero 等（2016）在 Bohemian 地体西南部中低压混合岩中发现富钙的多晶包裹体，认为其是原生碳酸盐和花岗质熔体不混溶的产物；Xia 等（2010）在大别-苏鲁片麻岩的变质锆石中也发现了异常高的微量元素含量，认为是超高压变质岩石在峰期变质条件下出现的超临界流体留下的痕迹。

6 变质岩同位素地球化学研究进展

同位素在变质岩研究中有广泛的应用，如年代学研究、地质温度计、地质速率计、示踪变质岩原岩和变质流体来源（Page et al., 2010；Barnes and Sharp, 2017；Penniston-Dorland et al., 2017；Teng et al., 2017；Foster et al., 2018；He et al., 2019）。

6.1 同位素地质温度计

目前氧同位素地质温度计在变质岩研究是最成熟的，近年来随着微区分析技术的快速发展，可以精确测定变质岩中矿物包裹体的氧同位素组成，如 Quinn 等（2017）用 SIMS 测定片麻岩中石榴子石的石英包裹体的氧同位素，进而估算了峰期变质温度。但传统的氧同位素温度计只需测定 ^{18}O 和 ^{16}O 组成，而忽略了 ^{17}O，Hayles 等（2018）和 Bao 等（2016）提出的 $\Delta^{17}O$ 温度计，可以有效地反演变质流体产生的温度。

Fe-Mg 等金属稳定同位素体系也可以作为新型的地质温度计。Li W. Y. 等（2011，2016）利用大别山具有不同变质温度的榴辉岩样品"标定"获得单斜辉石-石榴子石镁同位素地质温度计的表达式：$10^3 \ln\alpha_{Cpx-Grt} = (0.99\pm0.06)\times10^6/T^2$（$T$ 为开氏温度）。Li 等（2016）根据大别山碧溪岭退变质榴辉岩中石榴子石和绿辉石的 Fe 同位素组成，拟合结果整理成矿物内的温度计形式为 $\delta^{56}Fe^{3+}_{Omp} - \delta^{56}Fe^{2+}_{Omp} = 75$ 万 ~ 84 万$/T^2$。Hu 等（2017）发现，三叠纪 GeShan 沉积碳酸盐岩经历埋藏变质作用后会重置方解石和白云石间的 Mg 同位素组成，他们根据 $\Delta^{26}Mg_{dolomite-calcite}$ 反算的变质温度约为 150~190℃，与区域地层埋藏热历史相近。

6.2 变质原岩和流体交代作用的示踪

变质岩的 Sr-Nd-C-O 同位素体系可有效的示踪变质原岩和变质流体的交代过程。以榴辉岩为例，原岩可以是经过低温蚀变的 MORB、OIB 或 IAB（Xu et al., 2016；Zhang et al., 2016b；Zhu et al., 2018；Liu et al., 2019b），高温蚀变的低 $\delta^{18}O$ 大洋辉长岩（Page et al., 2014；Zhang et al., 2016a），或经历过雪球事件的低 $\delta^{18}O$ 大陆玄武岩（Zheng et al., 2011）等。Halama 等（2011）认为变质流体交代改变了 Raspas 榴辉岩的 Sr-Nd 同位素组成，榴辉岩在镜下观察到的黝帘石也支持这结论；Wang 等（2017）认为流体交代显著提高了天山榴辉岩的 Sr 同位素组成，对 Nd 同位素组成影响较弱。

Mg 同位素具有在地表条件下分馏交代大、俯冲变质过程不发生明显分馏的特点，因此可以有效的示踪变质原岩和可能的流体交代过程（Li W. Y. et al., 2011；Wang et al., 2014；Chen et al., 2016a）。Wang 等（2012）发现，Kaalvallei 和 Bellsbank 金伯利岩携带的榴辉岩包体和低温蚀变洋壳具有相似的 Mg 同位素组成，认为榴辉岩包体的原岩为低温蚀变洋壳。值得注意的是，若俯冲板片携带的碳酸盐发生分解或溶解，会导致原岩的 Mg 含量降低、Mg 同位素组成增大（Geske et al., 2012；Wang et al., 2015），需要分离出变质岩的硅酸盐组份进行研究。此外，随着 MC-ICP-MS 测量精度的提高，Ti、Tl 和 K 等金属稳定同位素也被应用于变质原岩的示踪（Millet et al., 2016；Shu et al., 2019）。

6.3 变质过程中氧逸度变化的示踪

Cr-Fe-Zn 同位素是示踪深俯冲板片发生变质作用时氧逸度变化的有力工具。Shen J. 等（2015）发现，大别–苏鲁造山带出露的变质岩的 Cr 同位素组成差别较大，并与全岩 $Fe^{2+}/\Sigma Fe$ 呈负相关关系，反映俯冲陆壳变质脱水过程中氧逸度发生了变化。但俯冲变质过程中 Fe-Zn 的分馏也可能与含 S、Cl 和 CO_3^{2-} 的络合有关（Inglis et al., 2017）。

由于俯冲壳源物质相比于地幔具有相对氧化的特征，因此俯冲板片释放的变质流体可能具有较高的氧逸度。Li 等（2017）发现，大别山碧溪岭退变质榴辉岩的 $Fe^{3+}/\Sigma Fe$ 值较新鲜榴辉岩的 $Fe^{3+}/\Sigma Fe$ 值高，指示榴辉岩在退变质过程中经历了高氧逸度的变质流体交代反应；Malaspina 等（2017）发现，被交代的地幔橄榄岩中的多固相包裹体仍具有较高的氧逸度，认为深俯冲地壳释放的变质流体也有较高的氧逸度。

6.4 变质过程中流体行为的示踪

Li-B 同位素是示踪俯冲变质过程流体行为的有效手段。目前俯冲变质脱水过程会造成显著的 B 同位素分馏已达成了共识（Xiao et al., 2011；Yamada et al., 2019），但超高压变质过程中 Li 同位素的分馏行为还存在争议。Xiao 等（2011）认为俯冲变质脱水过程是 CCSD 榴辉岩低 Li 同位素组成的主要原因，也有研究认为俯冲变质阶段 Li 同位素分馏程度较低（Qiu et al., 2009, 2011, 2011b；Barnes et al., 2019）；Liu 等（2019a）认为高 Li 含量和低 δ^7Li 值的松多榴辉岩是俯冲板片进变质脱水和折返阶段的流体交代的综合结果。

7 变质年代学研究进展

7.1 同位素定年体系

变质岩年代学是研究区域构造变质历史（Wernert et al., 2016；Tazzo-Rangel et al., 2019）和构建变质作用 $P\text{-}T\text{-}t\text{-}X$ 轨迹（Broussolle et al., 2015；Cruz-Uribe et al., 2015；Stevens et al., 2015；Tan et al., 2017）必不可少的工具。近年来，随着 LA-ICP-MS 和 SIMS 等仪器的不断更新以及分析精度的提高，副矿物的原位定年技术取得了突破性进展，极大地丰富了变质岩年代学研究。其中，锆石因具有稳定的晶格、相对较高的体系封闭温度和环带增生明显等特点，而被广泛应用于变质岩年代学研究。Gao 等（2019）在柴达木 Yuka 超高压变质带混合岩的变质锆石中发现了 3 组年龄，指示变质原岩、峰期变质和折返阶段部分熔融的三期事件。具有核边结构的锆石也可以记录大洋俯冲向大陆俯冲的转变时间（Zhang et al., 2016b, 2017；Song et al., 2019）。Lin 等（2021）运用 LA-MC-ICP-MS 开发了锆石小束斑微区定年，剥蚀束斑可以小至 5 μm，解决了一些极窄的锆石变质边定年问题。此外，榍石（Gasser et al., 2015；Zhang et al., 2018）、磷钇矿（Elisha et al., 2019）、褐帘石（Cliff et al., 2015；Boston et al., 2017）、金红石（Gasser et al., 2015；Will et al., 2018）、钙钛矿（Shen et al., 2016）等也可作为 U-Th-Pb 定年法的对象。

7.2 变质流体年代学

流体是变质作用过程中物质迁移和能量传输的重要媒介（Zheng and Hermann, 2014；Xiao et al., 2015），变质流体的形成世代一直备受关注，赋存于变质脉体中的锆石、金红石、褐帘石和独居石等副矿物是间接限定流体活动世代的有效对象。

通过对锆石的形态学研究可以大致限定流/熔体类型：熔体中结晶的锆石通常具有相对显著的震荡生长环带，流体中结晶的锆石无环带，富水流体与超临界流体中结晶的锆石就形态学不易作区分，需结合全岩的化学组成及其他岩相学证据进行限定。大别-苏鲁地区榴辉岩和片麻岩中主要有两种类型的锆石，一为岩浆残余锆石（核），保存有较好的岩浆震荡生长环带；另一为新生的变质锆石。所以锆石既记录了早期岩浆结晶时的信息，又记录了大陆俯冲折返过程中变质作用的信息。Chen 等（2011）在苏鲁青龙山榴辉岩和片麻岩中发现了具有极低 $\delta^{18}O$ 值（可达到 $-10‰$）的锆石，其岩浆核具有 $(769±9)$ Ma 的 U-Pb 年龄，正的 $\delta^{18}O$ 值（$0.1‰\sim10.1‰$）和高 Th/U、高 $^{176}Lu/^{177}Hf$ 值；相反，新生的变质锆石则具有三叠纪的 U-Pb 年龄，负 $\delta^{18}O$ 值（$-10.00‰\sim-2.2‰$）和低 Th/U、低 $^{176}Lu/^{177}Hf$ 值。具有负 $\delta^{18}O$ 值的新生锆石与俯冲带中负 $\delta^{18}O$ 的热液蚀变岩变质脱水有关，所以超高压变质火成岩中的锆石至少记录了两期的水岩相互作用，第一期是由岩浆核所记录，表明在新元古代这些大陆岩石曾经历了高温水-岩相互作用；第二期是由变质锆石域所记录，说明大陆深俯冲带内部俯冲陆壳变质作用脱水广泛存在（Zheng et al.，2019）。

由于橄榄岩具有硅不饱和性，所以原始的橄榄岩中往往罕见锆石，但在造山带橄榄岩中却有可观的锆石量。大量研究表明，这些锆石主要是俯冲带流/熔体交代重结晶或溶解重结晶的产物（Song et al.，2005a，2005b），其形成受控于流/熔体的化学组成、来源及形成时的物理化学条件等，所以造山带橄榄岩中的锆石直接记录了俯冲带流/熔体活动。在造山带橄榄岩中也可见一些老锆石，主要为残余岩浆锆石和重结晶锆石的核部，这些锆石来自流/熔体释放端元，是变质流/熔体对原岩锆石进行物理搬运的结果。Chen 和 Zheng（2017）就大别-苏鲁造山带 M 型造山带橄榄岩进行了系统的锆石 U-Pb 年代学研究，其中新生锆石年龄为 $205\sim244$ Ma，少部分为古生代，说明 M 型造山带橄榄岩中锆石经历了多期生长，即可能在大陆深俯冲之前（对应古生代年龄）至大陆深俯冲-折返过程中经历了多期流/熔体交代。

Liu X. C. 等（2014）根据柴北缘变质带锡铁山超高压榴辉岩长英质脉和石英脉中的锆石 U-Pb 年龄和 Hf-O 同位素确定了多期熔-流体活动时间；Chen 等（2016b）在苏鲁超高压变花岗岩中识别出两期生长的榍石，分别结晶于含水熔体和富水流体。

8 结 论

（1）过去十年，在我国典型俯冲碰撞造山带中陆续发现了一些超高压变质作用的新证据，如柴北缘超高压变质带东端都兰北带的野马滩榴辉岩中首次发现了柯石英包裹体、绿辉石中的多硅白云母出溶结构；在东昆仑造山带榴辉岩及其围岩云母片岩中首次发现了柯石英，从而又确立了一条东西长约 500 km 的早古生代高压-超高压变质岩带；在南阿尔金榴辉岩中首次发现了柯石英包裹体和进一步确定了斯石英假象，发现了 $P2/c$ 型单斜辉石等，是目前折返深度最大的大陆超深俯冲岩石等。

（2）对于华北克拉通孔兹岩带，在超高温变质作用的识别及其出露规模的确定，准确的变质时代和时间尺度的限定以及 P-T-t 轨迹的构建，尤其是进变质阶段的确定方面取得了一系列突出进展。

（3）在变质相平衡研究方面，在热力学数据库和矿物相及熔体活度模型进一步完善的基础上，建立了基于 ACF 组分分析建立的变质基性岩完整相平衡关系；同时在深熔作用与花岗质岩石成因的定量模拟方面也有了重要进展。

（4）在元素地球化学方面，高场强元素（如 Ti、Nb、Ta、Zr、Hf 等）和卤族元素（如 F、Cl、Br、I）在变质脱水过程中的迁移和分异，以及变价元素（如 V、Fe、W、Mo 等）对指示氧逸度的变化取得了一系列进展。

（5）在变质流体研究方面取得了重要进展，如发现了俯冲带高压-超高压流体活动的新证据，确定了流体成分以及流体活动时限的限定方面等。

参 考 文 献

高晓英,郑永飞. 2013. 大别造山带超高压变质岩副矿物地质测温. 科学通报,58(22):2153-2158
高晓英,李姝宁,郑永飞. 2011. 超高压变质矿物中的多相固体包裹体研究进展. 岩石学报,27(2):469-489
宫相宽,陈丹玲,任云飞,刘良,高胜,杨士杰. 2016. 北秦岭含柯石英斜长角闪岩的发现及其地质意义. 科学通报,61(12):1365-1378

郭敬辉,石昕,卞爱国,许荣华,翟明国,李永刚. 1999. 桑干地区早元古代花岗岩长石Pb同位素组成和锆石U-Pb年龄:变质与地壳熔融作用及构造-热事件演化. 岩石学报,15(2):199-207

郭敬辉,陈意,彭澎,刘富,陈亮,张履桥. 2006. 内蒙古大青山假蓝宝石麻粒岩-1.8Ga的超高温(UHT)变质作用. 见:2006年全国岩石学与地球动力学研讨会论文摘要集. 南京:中国地质学会,228-231

国显正,贾群子,钱兵,弥佳茹,李金超,孔会磊,姚学钢. 2017. 东昆仑高压变质带榴辉岩和榴闪岩地球化学特征及形成动力学背景. 地球科学与环境学报,39(6):735-750

国显正,贾群子,李金超,孔会磊,姚学钢,弥佳茹,钱兵,王宇. 2018. 东昆仑高压变质带榴辉岩年代学、地球化学及其地质意义. 地球科学,43(12):4300-4318

韩磊. 2015. 柴北缘野马滩地区榴辉岩岩石学研究. 北京:北京大学博士学位论文

金巍. 1989. 华北陆台北缘(中段)早前寒武纪地质演化和其变质动力学研究. 长春:长春地质学院

刘海洋. 2018. 俯冲带变质作用与岩浆过程中的锂同位素地球化学. 合肥:中国科学技术大学博士学位论文

刘建忠,强小科,刘喜山,欧阳自远. 2000. 内蒙古大青山造山带含假蓝宝石尖晶石片麻岩的成因网格及动力学. 岩石学报,16(2):245-255

刘良,陈丹玲,章军锋,康磊,王超,杨文强,廖小莹,任云飞,盖永升. 2019. 陆壳超深俯冲到斯石英稳定域地幔深度(~300 km)的新证据. 地球科学,44(12):3998-4003

卢良兆,靳是琴,徐学纯,刘福来. 1992. 内蒙古东南部早前寒武纪孔兹岩系成因及其含矿性. 长春:吉林科学技术出版社

吕增,张立飞,陈振宇,李旭平,申婷婷. 2013. 西天山高压-超高压变质带折返过程中的流体活动证据:高压脉体和异剥钙榴岩. 科学通报,58(22):2175-2179

祁生胜,宋述光,史连昌,才加加,胡继春. 2014. 东昆仑西段夏日哈木-苏海图早古生代榴辉岩的发现及意义. 岩石学报,30(11):3345-3356

祁晓鹏,范显刚,杨杰,崔建堂,汪帮耀,范亚洲,杨高学,李真,晁文迪. 2016. 东昆仑东段浪木日上游早古生代榴辉岩的发现及其意义. 地质通报,35(11):1771-1783

盛英明,郑永飞,吴元保. 2011. 超高压岩石中变质脉的研究. 岩石学报,27(2):490-500

万渝生,耿元生,刘福来,沈其韩,刘敦一,宋彪. 2000. 华北克拉通及邻区孔兹岩系的时代及对太古宙基底组成的制约. 前寒武纪研究进展,23(4):221-237

魏春景. 2016. 麻粒岩相变质作用与花岗岩成因-II:变质泥质岩高温-超高温变质相平衡与S型花岗岩成因的定量模拟. 岩石学报,32(6):1625-1643

魏春景,朱文萍. 2016. 麻粒岩相变质作用与花岗岩成因-I:变质泥质岩/杂砂岩高温-超高温变质相平衡. 岩石学报,32(6):1611-1624

魏春景,关晓,董杰. 2017. 基性岩高温-超高温变质作用与TTG质岩成因. 岩石学报,33(5):1381-1404

吴春明. 2018. 变质地质学研究中的一些困难问题. 岩石学报,34(4):873-894

吴元保,郑永飞. 2004. 锆石成因矿物学研究及其对U-Pb年龄解释的制约. 科学通报,49(16):1589-1604

Ague J J, Nicolescu S. 2014. Carbon dioxide released from subduction zones by fluid-mediated reactions. Nature Geoscience,7(5):355-360

Bao H M, Cao X B, Hayles J A. 2016. Triple oxygen isotopes: fundamental relationships and applications. Annual Review of Earth and Planetary Sciences,44:463-492

Barnes J D, Sharp Z D. 2017. Chlorine isotope geochemistry. Reviews in Mineralogy and Geochemistry,82(1):345-378

Barnes J D, Penniston-Dorland S C, Bebout G E, Hoover W, Beaudoin G M, Agard P. 2019. Chlorine and lithium behavior in metasedimentary rocks during prograde metamorphism: a comparative study of exhumed subduction complexes (Catalina Schist and Schistes Lustrés). Lithos,336-337:40-53

Bebout G E, Fogel M L, Cartigny P. 2013. Nitrogen: highly volatile yet surprisingly compatible. Elements,9(5):333-338

Berkesi M, Guzmics T, Szabó C, Dubessy J, Bodnar R J, Hidas K, Ratter K. 2012. The role of CO_2-rich fluids in trace element transport and metasomatism in the lithospheric mantle beneath the central Pannonian Basin, Hungary, based on fluid inclusions in mantle xenoliths. Earth and Planetary Science Letters,331-332:8-20

Bi H Z, Song S G, Dong J L, Yang L M, Qi S S, Allen M B. 2018. First discovery of coesite in eclogite from East Kunlun, Northwest China. Science Bulletin,63(23):1536-1538

Bi H Z, Song S G, Yang L M, Allen M B, Qi S S, Su L. 2020. UHP metamorphism recorded by coesite-bearing metapelite in the East Kunlun orogen (NW China). Geological Magazine,157(2):160-172

Boston K R, Rubatto D, Hermann J, Engi M, Amelin Y. 2017. Geochronology of accessory allanite and monazite in the Barrovian metamorphic sequence of the central Alps, Switzerland. Lithos,286-287:502-518

Broussolle A, Štípská P, Lehmann J, Schulmann K, Hacker B R, Holder R, Kylander-Clark A R C, Hanžl P, Racek M, Hasalová P, Lexa O, Hrdličková K, Buriánek D. 2015. P-T-t-D record of crustal-scale horizontal flow and magma-assisted doming in the SW Mongolian Altai. Journal of Metamorphic Geology,33(4):359-383

Brown M. 2007. Metamorphic conditions in orogenic belts: a record of secular change. International Geology Review,49(3):193-234

Cai J, Liu F L, Liu P H, Liu C H, Wang F, Shi J R. 2014. Metamorphic P-T path and tectonic implications of pelitic granulites from the Daqingshan complex of the Khondalite Belt, North China Craton. Precambrian Research,241:161-184

Cai J, Liu F L, Liu P H, Liu C H, Wang F, Shi J R. 2015. Silica-undersaturated spinel granulites in the Daqingshan Complex of the Khondalite Belt, North China Craton: petrology and quantitative P-T-X constraints. Precambrian Research,266:119-136

Chen R X, Zheng Y F. 2017. Metamorphic zirconology of continental subduction zones. Journal of Asian Earth Sciences,145:149-176

Chen R X, Zheng Y F, Gong B. 2011. Mineral hydrogen isotopes and water contents in ultrahigh-pressure metabasite and metagranite: constraints on fluid flow during continental subduction-zone metamorphism. Chemical Geology,281(1-2):103-124

Chen R X,Zheng Y F,Hu Z C. 2012. Episodic fluid action during exhumation of deeply subducted continental crust:geochemical constraints from zoisite-quartz vein and host metabasite in the Dabie orogen. Lithos,155:146-166

Chen Y X,Schertl H P,Zheng Y F,Huang F,Zhou K,Gong Y Z. 2016a. Mg-O isotopes trace the origin of Mg-rich fluids in the deeply subducted continental crust of western Alps. Earth and Planetary Science Letters,456:157-167

Chen Y X,Zhou K,Zheng Y F,Gao X Y,Yang Y H. 2016b. Polygenetic titanite records the composition of metamorphic fluids during the exhumation of ultrahigh-pressure metagranite in the Sulu orogen. Journal of Metamorphic Geology,34(6):573-594

Chopin C. 1984. Coesite and pure pyrope in high-grade blueschists of the western Alps:a first record and some consequences. Contributions to Mineralogy and Petrology,86(2):107-118

Clark C,Fitzsimons I C W,Healy D,Harley S L. 2011. How does the continental crust get really hot? Elements,7(4):235-240

Cliff R A,Oberli F,Meier M,Droop G T R,Kelly M. 2015. syn-metamorphic folding in the Tauern Window,Austria dated by Th-Pb ages from individual allanite porphyroblasts. Journal of Metamorphic Geology,33(4):427-435

Cruz-Uribe A M,Hoisch T D,Wells M L,Vervoort J D,Mazdab F K. 2015. Linking thermodynamic modelling,Lu-Hf geochronology and trace elements in garnet:new P-T-t paths from the Sevier hinterland. Journal of Metamorphic Geology,33(7):763-781

Diener J F A,Powell R. 2012. Revised activity-composition models for clinopyroxene and amphibole. Journal of Metamorphic Geology,30(2):131-142

Dong J,Wei C J,Clarke G L,Zhang J X. 2018. Metamorphic evolution during deep subduction and exhumation of continental crust:insights from felsic granulites in South Altyn Tagh,West China. Journal of Petrology,59(10):1965-1990

Dong J,Wei C J,Chen J,Zhang J X. 2020. P-T-t path of GARNETITES in South Altyn Tagh,West China:a complete record of the ultradeep subduction and exhumation of continental crust. Journal of Geophysical Research:Solid Earth,125(2):e2019JB018881

Elisha B,Katzir Y,Kylander-Clark A R C,Golan T,Coble M A. 2019. The timing of migmatization in the northern Arabian-Nubian Shield:evidence for a juvenile sedimentary component in collision-related batholiths. Journal of Metamorphic Geology,37(5):591-610

Ferrero S,Wunder B,Ziemann M A,Wälle M,O'Brien P J. 2016. Carbonatitic and granitic melts produced under conditions of primary immiscibility during anatexis in the lower crust. Earth and Planetary Science Letters,454:121-131

Fitzsimons I C W,Harley S L. 1994. Garnet coronas in scapolite-wollastonite calc-silicates from East Antarctica:the application and limitations of activity-corrected grids. Journal of Metamorphic Geology,12(6):761-777

Foster G L,Marschall H R,Palmer M R. 2018. Boron isotope analysis of geological materials. In:Marschall H,Foster G (eds). Boron Isotopes. Cham:Springer,13-31

Frezzotti M L,Ferrando S. 2015. The chemical behavior of fluids released during deep subduction based on fluid inclusions. American Mineralogist,100(2-3):352-377

Frezzotti M L,Selverstone J,Sharp Z D,Compagnoni R. 2011. Carbonate dissolution during subduction revealed by diamond-bearing rocks from the Alps. Nature Geoscience,4(10):703-706

Frezzotti M L,Ferrando S,Tecce F,Castelli D. 2012. Water content and nature of solutes in shallow-mantle fluids from fluid inclusions. Earth and Planetary Science Letters,351-352:70-83

Gai Y S,Liu L,Wang C,Yang W Q,Kang L,Cao Y T,Liao X Y. 2017. Discovery of coesite in eclogite from Keqike Jianggalesayi:new evidence for ultrahigh-pressure metamorphism in South Altyn Tagh,northwestern China. Science Bulletin,62(15):1048-1051

Gao S B,Chen X,Xu R K,Cai P J,Lu L H,Hou W D,Guo X Z. 2019. Tracking the timing and nature of protolith,metamorphism,and partial melting of tourmaline-bearing migmatites by zircon U-Pb and Hf isotopic compositions in the Yuka terrane,North Qaidam UHP metamorphic belt. Geological Journal,54(2):1013-1036

Gao X Y,Zheng Y F,Chen Y X. 2012. Dehydration melting of ultrahigh-pressure eclogite in the Dabie orogen:evidence from multiphase solid inclusions in garnet. Journal of Metamorphic Geology,30(2):193-212

Gasser D,Jeřábek P,Faber C,Stünitz H,Menegon L,Corfu F,Erambert M,Whitehouse M J. 2015. Behaviour of geochronometers and timing of metamorphic reactions during deformation at lower crustal conditions:phase equilibrium modelling and U-Pb dating of zircon,monazite,rutile and titanite from the Kalak nappe complex,northern Norway. Journal of Metamorphic Geology,33(5):513-534

Ge R F,Zhu W B,Wilde S A,Wu H L. 2018. Remnants of Eoarchean continental crust derived from a subducted proto-arc. Science Advances,4(2):eaao3159

Geske A,Zorlu J,Richter D K,Buhl D,Niedermayr A,Immenhauser A. 2012. Impact of diagenesis and low grade metamorphosis on isotope (δ^{26}Mg,δ^{13}C,δ^{18}O and ^{87}Sr/^{86}Sr) and elemental (Ca,Mg,Mn,Fe and Sr) signatures of Triassic sabkha dolomites. Chemical Geology,332-333:45-64

Gou L L,Li Z H,Liu X M,Dong Y P,Zhao J,Zhang C L,Liu L,Long X P. 2018. Ultrahigh-temperature metamorphism in the Helanshan complex of the Khondalite Belt,North China Craton:petrology and phase equilibria of spinel-bearing pelitic granulites. Journal of Metamorphic Geology,36(9):1199-1220

Green E C R,White R W,Diener J F A,Powell R,Holland T J B,Palin R M. 2016. Activity-composition relations for the calculation of partial melting equilibria in metabasic rocks. Journal of Metamorphic Geology,34(9):845-869

Guo J H,Chen Y,Peng P,Liu F,Chen L,Zhang L Q. 2006. Sapphirine gran-ulite from Daqingshan area,Inner Mongolia - 1.85 Ga ultrahigh temperature (UHT)metamorphism. In:Proceedings of National Conference on Petrology and Geodynamics in China,Nanjing,215-218

Guo J H,Peng P,Chen Y,Jiao S J,Windley B F. 2012. UHT sapphirine granulite metamorphism at 1.93-1.92 Ga caused by gabbronorite intrusions:implications for tectonic evolution of the northern margin of the North China Craton. Precambrian Research,222-223:124–142

Guo S,Ye K,Chen Y,Liu J B,Mao Q,Ma Y G. 2012. Fluid-rock interaction and element mobilization in UHP metabasalt:constraints from an omphacite-epidote vein and host eclogites in the Dabie orogen. Lithos,136-139:145–167

Guo S,Ye K,Cheng N F,Chen Y,Su B,Liu J B. 2015. Metamorphic P-T trajectory and multi-stage fluid events of vein-bearing UHP eclogites from the Dabie terrane:insights from compositional zonations of key minerals. International Geology Review,57(9-10):1077–1102

Halama R,John T,Herms P,Hauff F,Schenk V. 2011. A stable (Li,O) and radiogenic (Sr,Nd) isotope perspective on metasomatic processes in a subducting slab. Chemical Geology,281(3-4):151–166

Han L,Zhang L F,Zhang G B. 2015. Ultra-deep subduction of Yematan eclogite in the North Qaidam UHP belt,NW China:evidence from phengite exsolution in omphacite. American Mineralogist,100(8-9):1848–1855

Harley S L. 1998. On the occurrence and characterization of ultrahigh-temperature crustal metamorphism. Geological Society, London, Special Publications,138(1):81–107

Harley S L. 2008. Refining the P-T records of UHT crustal metamorphism. Journal of Metamorphic Geology,26(2):125–154

Harley S L. 2016. A matter of time:the importance of the duration of UHT metamorphism. Journal of Mineralogical and Petrological Sciences,111(2):50–72

Harley S L. 2021. UHT metamorphism,In:Alderton D,Elias S A (eds). Encyclopedia of Geology, 2nd ed. Pittsburgh:Academic Press,522–552

Harley S L,Nandakumar V. 2014. Accessory mineral behaviour in granulite migmatites:a case study from the kerala khondalite belt,India. Journal of Petrology,55(10):1965–2002

Hayles J,Gao C H,Cao X B,Liu Y,Bao H M. 2018. Theoretical calibration of the triple oxygen isotope thermometer. Geochimica et Cosmochimica Acta,235:237–245

He M Y,Deng L,Lu H,Jin Z D. 2019. Elimination of the boron memory effect for rapid and accurate boron isotope analysis by MC-ICP-MS using NaF. Journal of Analytical Atomic Spectrometry,34(5):1026–1032

Holland T J B,Powell R. 2011. An improved and extended internally consistent thermodynamic dataset for phases of petrological interest,involving a new equation of state for solids. Journal of Metamorphic Geology,29(3):333–383

Hu Z Y,Hu W X,Wang X M,Lu Y Z,Wang L C,Liao Z W,Li W Q. 2017. Resetting of Mg isotopes between calcite and dolomite during burial metamorphism:outlook of Mg isotopes as geothermometer and seawater proxy. Geochimica et Cosmochimica Acta,208:24–40

Huang G Y,Guo J H,Jiao S J,Palin R. 2019. What drives the continental crust to be extremely hot so quickly? Journal of Geophysical Research:Solid Earth,124(11):11218–11231

Huang J,Xiao Y L. 2015. Element mobility in mafic and felsic ultrahigh-pressure metamorphic rocks from the Dabie UHP orogen,China:Insights into supercritical liquids in continental subduction zones. International Geology Review,57(9-10):1103–1129

Huang J,Xiao Y,Gao Y,Hou Z,Wu W. 2012. Nb-Ta fractionation induced by fluid-rock interaction in subduction-zones:constraints from UHP eclogite- and vein-hosted rutile from the Dabie orogen,central-eastern China. Journal of Metamorphic Geology,30(8):821–842

Hwang S L,Shen P Y,Chu H T,Yui T F. 2000. Nanometer-size α-PbO_2-Type TiO_2 in garnet:a thermobarometer for ultrahigh-pressure metamorphism. Science,288(5464):321–324

Inglis E C,Debret B,Burton K W,Millet M A,Pons M L,Dale C W,Bouilhol P,Cooper M,Nowell G M,McCoy-West A J,Williams H M. 2017. The behavior of iron and zinc stable isotopes accompanying the subduction of mafic oceanic crust:a case study from western Alpine ophiolites. Geochemistry,Geophysics,Geosystems,18(7):2562–2579

Jiao S J,Guo J H. 2011. Application of the two-feldspar geothermometer to ultrahigh-temperature (UHT) rocks in the Khondalite Belt,North China Craton and its implications. American Mineralogist,96(2-3):250–260

Jiao S J,Guo J H. 2020. Paleoproterozoic UHT metamorphism with isobaric cooling (IBC) followed by decompressionheating in the Khondalite Belt (North China Craton):new evidence from two sapphirine formation processes. Journal of Metamorphic Geology,38(4):357–378

Jiao S J,Guo J H,Mao Q,Zhao R F. 2011. Application of Zr-in-rutile thermometry:a case study from ultrahigh-temperature granulites of the Khondalite Belt,North China Craton. Contributions to Mineralogy and Petrology,162(2):379–393

Jiao S J,Guo J H,Harley S L,Peng P. 2013a. Geochronology and trace element geochemistry of zircon,monazite and garnet from the garnetite and/or associated other high-grade rocks:implications for Palaeoproterozoic tectonothermal evolution of the Khondalite Belt,North China Craton. Precambrian Research,237:78–100

Jiao S J,Guo J H,Harley S L,Windley B F. 2013b. New constraints from garnetite on the P-T path of the Khondalite Belt:implications for the tectonic evolution of the North China Craton. Journal of Petrology,54(9):1725–1758

Jiao S J,Guo J H,Wang L J,Peng P. 2015. Shortlived hightemperature prograde and retrograde metamorphism in Shaerqin sapphirine-bearing metapelites from the Daqingshan terrane,North China Craton. Precambrian Research,269:31–57

Jiao S J,Fitzsimons I C W,Guo J H. 2017. Paleoproterozoic UHT metamorphism in the Daqingshan Terrane,North China craton:new constraints from phase equilibria modeling and SIMS U-Pb zircon dating. Precambrian Research,303:208–227

Jiao S J,Fitzsimons I C W,Zi J W,Evans N J,Mcdonald B J,Guo J H. 2020a. Texturally controlled U-Th-Pb monazite geochronology reveals paleoproterozoic UHT metamorphic evolution in the Khondalite Belt,North China Craton. Journal of Petrology,61(1):egaa023

Jiao S J,Guo J H,Evans N J,Mcdonald B J,Liu P,Ouyang D J,Fitzsimons I C W. 2020b. The timing and duration of high-temperature to ultrahigh-temperature metamorphism constrained by zircon U-Pb-Hf and trace element signatures in the Khondalite Belt,North China Craton. Contributions to

Mineralogy and Petrology,175(7):66

Jiao S J,Evans N J,Guo J H,Fitzsimons I C W,Zi J W,McDonald B J. 2021. Establishing the P-T path of UHT granulites by geochemically distinguishing peritectic from retrograde garnet. American Mineralogist (in press)

Johnson T E,Brown M,Gardiner N J,Kirkland C L,Smithies R H. 2017. Earth's first stable continents did not form by subduction. Nature,543(7644):239-242

Kawamoto T,Yoshikawa M,Kumagai Y,Mirabueno M H T,Okuno M,Kobayashi T. 2013. Mantle wedge infiltrated with saline fluids from dehydration and decarbonation of subducting slab. Proceedings of the National Academy of Sciences of the United States of America,110(24):9663-9668

Kelly N M,Harley S L,Möller A. 2012. Complexity in the behavior and recrystallization of monazite during high-T metamorphism and fluid infiltration. Chemical Geology,322-323:192-208

Kelsey D E. 2008. On ultrahigh-temperature crustal metamorphism. Gondwana Research,13(1):1-29

Kelsey D E,Powell R. 2011. Progress in linking accessory mineral growth and breakdown to major mineral evolution in metamorphic rocks: a thermodynamic approach in the Na_2O-CaO-K_2O-FeO-MgO-Al_2O_3-SiO_2-H_2O-TiO_2-ZrO_2 system. Journal of Metamorphic Geology,29(1):151-166

Kohn M J,Corrie S L,Markley C. 2015. The fall and rise of metamorphic zircon. American Mineralogist,100(4):897-908

Li W Y,Teng F Z,Xiao Y L,Huang J. 2011. High-temperature inter-mineral magnesium isotope fractionation in eclogite from the Dabie orogen,China. Earth and Planetary Science Letters,304(1-2):224-230

Li W Y,Teng F Z,Xiao Y L,Gu H O,Zha X P,Huang J. 2016. Empirical calibration of the clinopyroxene-garnet magnesium isotope geothermometer and implications. Contributions to Mineralogy and Petrology,171(7):61

Li X F,Rusk B,Wang R C,Morishita Y,Watanabe Y,Chen Z Y. 2011. Rutile inclusions in quartz crystals record decreasing temperature and pressure during the exhumation of the Su-Lu UHP metamorphic belt in Donghai,East China. American Mineralogist,96(7):964-973

Li X H,Chen Y,Tchouankoue J P,Liu C Z,Li J,Ling X X,Tang G Q,Liu Y. 2017. Improving geochronological framework of the Pan-African orogeny in Cameroon:new SIMS zircon and monazite U-Pb age constraints. Precambrian Research,294:307-321

Li X W,Wei C J. 2016. Phase equilibria modelling and zircon age dating of pelitic granulites in Zhaojiayao,from the Jining Group of the Khondalite Belt,North China Craton. Journal of Metamorphic Geology,34(6):595-615

Li X W,Wei C J. 2018. Ultrahigh-temperature metamorphism in the Tuguiwula area,Khondalite Belt,North China Craton. Journal of Metamorphic Geology,36(4):489-509

Liang Y,Sun C G,Yao L J. 2013. A REE-in-two-pyroxene thermometer for mafic and ultramafic rocks. Geochimica et Cosmochimica Acta,102:246-260

Liao Y,Wei C J. 2019. Ultrahigh-temperature mafic granulite in the Huai'an complex,North China Craton:evidence from phase equilibria modelling and amphibole thermometers. Gondwana Research,76:62-76

Lin M,Zhang G B,Li N,Li H J,Wang J X. 2021. An improved in situ Zircon U-Pb dating method at high spatial resolution (≤10 μm Spot) by LA-MC-ICP-MS and its application. Geostandards and Geoanalytical Research,45(2):265-285

Liu H,Li X P,Kong F M,Santosh M,Wang H. 2019. Ultrahigh-temperature overprinting of high pressure pelitic granulites in the Huai'an Complex,North China Craton:evidence from thermodynamic modeling and isotope geochronology. Gondwana Research,72:15-33

Liu H Y,Sun H,Xiao Y L,Wang Y Y,Zeng L S,Li W Y,Guo H H,Yu H M,Pack A. 2019a. Lithium isotope systematics of the Sumdo Eclogite,Tibet:tracing fluid/rock interaction of subducted low-T altered oceanic crust. Geochimica et Cosmochimica Acta,246:385-405

Liu H Y,Xiao Y L,van den Kerkhof A,Wang Y Y,Zeng L S,Guo H H. 2019b. Metamorphism and fluid evolution of the Sumdo eclogite,Tibet:constraints from mineral chemistry,fluid inclusions and oxygen isotopes. Journal of Asian Earth Sciences,172:292-307

Liu L,Zhang J F,Green II H W,Jin Z M,Bozhilov K N. 2007. Evidence of former stishovite in metamorphosed sediments,implying subduction to >350 km. Earth and Planetary Science Letters,263(3-4):180-191

Liu L,Xiao Y L,Aulbach S,Li D Y,Hou Z H. 2014. Vanadium and niobium behavior in rutile as a function of oxygen fugacity:evidence from natural samples. Contributions to Mineralogy and Petrology,167(6):1026

Liu L,Zhang J F,Cao Y T,Green II H W,Yang W Q,Xu H J,Liao X Y,Kang L. 2018. Evidence of former stishovite in UHP eclogite from the South Altyn Tagh,western China. Earth and Planetary Science Letters,484:353-362

Liu P H,Liu F L,Liu C H,Liu J H,Wang F,Xiao L L,Cai J,Shi J R. 2014. Multiple mafic magmatic and high-grade metamorphic events revealed by zircons from meta-mafic rocks in the Daqingshan-Wulashan complex of the Khondalite Belt,North China Craton. Precambrian Research,246:334-357

Liu P H,Liu F L,Cai J,Liu C H,Liu J H,Wang F,Xiao L L,Shi J R. 2017. Spatial distribution,P-T-t paths,and tectonic significance of high-pressure mafic granulites from the Daqingshan-Wulashan complex in the Khondalite Belt,North China Craton. Precambrian Research,303:687-708

Liu S J,Li J H,Santosh M. 2010. First application of the revised Ti-in-zircon geothermometer to Paleoproterozoic ultrahigh-temperature granulites of Tuguiwula,Inner Mongolia,North China Craton. Contributions to Mineralogy and Petrology,159(2):225

Liu S J,Bai X,Li J H,Santosh M. 2011. Retrograde metamorphism of ultrahigh-temperature granulites from the Khondalite Belt in Inner Mongolia,North China Craton:evidence from aluminous orthopyroxenes. Geological Journal,46(2-3):263-275

Liu S J,Tsunogae T,Li W S,Shimizu H,Santosh M,Wan Y S,Li J H. 2012. Paleoproterozoic granulites from Heling'er:implications for regional ultrahigh-temperature metamorphism in the North China Craton. Lithos,148:54-70

Liu T,Wei C J. 2018. Metamorphic evolution of Archean ultrahigh-temperature mafic granulites from the western margin of Qian'an gneiss dome,eastern Hebei Province,North China Craton:insights into the Archean tectonic regime. Precambrian Research,318:170-187

Liu X C,Wu Y B,Gao S,Wang H,Zheng J P,Hu Z C,Zhou L,Yang S H. 2014. Record of multiple stage channelized fluid and melt activities in deeply subducted slab from zircon U-Pb age and Hf-O isotope compositions. Geochimica et Cosmochimica Acta,144:1-24

Liu Y C,Gu X F,Rolfo F,Chen Z Y. 2011. Ultrahigh-pressure metamorphism and multistage exhumation of eclogite of the Luotian dome, North Dabie complex zone (central China):evidence from mineral inclusions and decompression textures. Journal of Asian Earth Sciences,42(4):607-617

Lobjoie C,Lin W,Trap P,Goncalves P,Li Q L,Marquer D,Bruguier O,Devoir A. 2018. Ultra-high temperature metamorphism recorded in Fe-rich olivine-bearing migmatite from the Khondalite Belt, North China Craton. Journal of Metamorphic Geology,36(3):343-368

Lü Z,Zhang L F,Du J X,Bucher K. 2008. Coesite inclusions in garnet from eclogitic rocks in western Tianshan, Northwest China:convincing proof of UHP metamorphism. American Mineralogist,93(11-12):1845-1850

Lü Z,Zhang L,Du J,Bucher K. 2009. Petrology of coesite-bearing eclogite from Habutengsu Valley, western Tianshan, NW China and its tectonometamorphic implication. Journal of Metamorphic Geology,27(9):773-787

Lü Z,Zhang L F,Chen Z Y. 2014. Jadeite-and dolomite-bearing coesite eclogite from western Tianshan,NW China. European Journal of Mineralogy,26(2):245-256

Lü Z,Zhang L F,Yue J, Li X L. 2019a. Ultrahigh-pressure and high-P lawsonite eclogites in Muzhaerte, Chinese western Tianshan. Journal of Metamorphic Geology,37(5):717-743

Lü Z,Zhang L F,Yue J. 2019b. Coesite in metasediments from the Muzhaerte valley,southwestern Tianshan. Science Bulletin,64(2):78-80

Malaspina N,Langenhorst F,Tumiati S,Campione M,Frezzotti M L,Poli S. 2017. The redox budget of crust-derived fluid phases at the slab-mantle interface. Geochimica et Cosmochimica Acta,209:70-84

Mantegazzi D,Sanchez-Valle C,Driesner T. 2013. Thermodynamic properties of aqueous NaCl solutions to 1073 K and 4.5 GPa,and implications for dehydration reactions in subducting slabs. Geochimica et Cosmochimica Acta,121:263-290

Meng F C,Zhang J X,Cui M H. 2013. Discovery of Early Paleozoic eclogite from the East Kunlun, western China and its tectonic significance. Gondwana Research,23(2):825-836

Millet M A,Dauphas N,Greber N D,Burton K W,Dale C W,Debret B,Macpherson C G,Nowell G M,Williams H M. 2016. Titanium stable isotope investigation of magmatic processes on the Earth and Moon. Earth and Planetary Science Letters,449:197-205

Mukherjee B K,Sachan H K. 2009. Fluids in coesite-bearing rocks of the Tso Morari complex,NW Himalaya:evidence for entrapment during peak metamorphism and subsequent uplift. Geological Magazine,146(6):876-889

Ogasawara Y, Fukasawa K, Maruyama S. 2002. Coesite exsolution from supersilicic titanite in UHP marble from the Kokchetav Massif, northern Kazakhstan. American Mineralogist,87(4):454-461

Page F Z,Kita N T,Valley J W. 2010. Ion microprobe analysis of oxygen isotopes in garnets of complex chemistry. Chemical Geology,270(1-4):9-19

Page F Z,Essene E J,Mukasa S B,Valley J W. 2014. A garnet-zircon oxygen isotope record of subduction and exhumation fluids from the Franciscan complex,California. Journal of Petrology,55(1):103-131

Pattison D R M,Chacko T,Farquhar J,McFarlane C R M. 2003. Temperatures of granulite-facies metamorphism:constraints from experimental phase equilibria and thermobarometry corrected for retrograde exchange. Journal of Petrology,44(5):867-900

Penniston-Dorland S,Liu X M,Rudnick R L. 2017. Lithium isotope geochemistry. Reviews in Mineralogy and Geochemistry,82(1):165-217

Pownall J M,Hall R,Armstrong R A,Forster M A. 2014. Earth's youngest known ultrahigh-temperature granulites discovered on Seram, eastern Indonesia. Geology,42(4):279-282

Proyer A,Rolfo F,Zhu Y F,Castelli D,Compagnoni R. 2013. Ultrahigh-pressure metamorphism in the magnesite+aragonite stability field:evidence from two impure marbles from the Dabie-Sulu UHPM belt. Journal of Metamorphic Geology,31(1):35-48

Qiu L,Rudnick R L,McDonough W F,Merriman R J. 2009. Li and δ^7Li in mudrocks from the British Caledonides:metamorphism and source influences. Geochimica et Cosmochimica Acta,73(24):7325-7340

Qiu L,Rudnick R L,Ague J J,McDonough W F. 2011. A lithium isotopic study of sub-greenschist to greenschist facies metamorphism in an accretionary prism,New Zealand. Earth and Planetary Science Letters,301(1-2):213-221

Quinn R J,Kitajima K,Nakashima D,Spicuzza M J,Valley J W. 2017. Oxygen isotope thermometry using quartz inclusions in garnet. Journal of Metamorphic Geology,35(2):231-252

Roberts M P,Finger F. 1997. Do U-Pb zircon ages from granulites reflect peak metamorphic conditions? Geology,25(4):319-322

Rubatto D,Williams I S,Buick I S. 2001. Zircon and monazite response to prograde metamorphism in the Reynolds Range, central Australia. Contributions to Mineralogy and Petrology,140(4):458-468

Sakuma H,Ichiki M. 2016. Density and isothermal compressibility of supercritical H_2O-NaCl fluid:molecular dynamics study from 673 to 2000 K,0.2 to 2 GPa,and 0 to 22 wt% NaCl concentrations. Geofluids,16(1):89-102

Santosh M,Tsunogae T,Li J H,Liu S J. 2007a. Discovery of sapphirine-bearing Mg-Al granulites in the North China Craton:implications for Paleoproterozoic ultrahigh temperature metamorphism. Gondwana Research,11(3):263-285

Santosh M,Wilde S,Li J H. 2007b. Timing of Paleoproterozoic ultrahigh-temperature metamorphism in the North China Craton:evidence from SHRIMP U-Pb zircon geochronology. Precambrian Research,159(3-4):178-196

Santosh M,Sajeev K,Li J H,Liu S J,Itaya T. 2009a. Counterclockwise exhumation of a hot orogen:the Paleoproterozoic ultrahigh-temperature granulites in the North China Craton. Lithos,110(1-4):140-152

Santosh M,Wan Y S,Liu D Y,Dong C Y,Li J H. 2009b. Anatomy of zircons from an ultrahot orogen:the amalgamation of the North China Craton within the supercontinent columbia. The Journal of Geology,117(4):429-443

Santosh M, Liu S J, Tsunogae T, Li J H. 2012. Paleoproterozoic ultrahigh-temperature granulites in the North China Craton: implications for tectonic models on extreme crustal metamorphism. Precambrian Research, 222-223: 77-106

Schaltegger U, Fanning C M, Günther D, Maurin J C, Schulmann K, Gebauer D. 1999. Growth, annealing and recrystallization of zircon and preservation of monazite in high-grade metamorphism: conventional and in-situ U-Pb isotope, cathodoluminescence and microchemical evidence. Contributions to Mineralogy and Petrology, 134(2-3): 186-201

Shen J, Liu J, Qin L P, Wang S J, Li S G, Xia J X, Ke S, Yang J S. 2015. Chromium isotope signature during continental crust subduction recorded in metamorphic rocks. Geochemistry, Geophysics, Geosystems, 16(11): 3840-3854

Shen T T, Hermann J, Zhang L F, Lü Z, Padrón-Navarta J A, Xia B, Bader T. 2015. UHP metamorphism documented in Ti-chondrodite and Ti-clinohumitebearing serpentinized ultramafic rocks from Chinese southwestern Tianshan. Journal of Petrology, 56(7): 1425-1458

Shen T T, Wu F Y, Zhang L F, Hermann J, Li X P, Du J X. 2016. In-situ U-Pb dating and Nd isotopic analysis of perovskite from a rodingite blackwall associated with UHP serpentinite from southwestern Tianshan, China. Chemical Geology, 431: 67-82

Sheng Y M, Zheng Y F, Li S N, Hu Z. 2013. Element mobility during continental collision: insights from polymineralic metamorphic vein within UHP eclogite in the Dabie orogen. Journal of Metamorphic Geology, 31(2): 221-241

Shimizu H, Tsunogae T, Santosh M, Liu S J, Li J H. 2013. Phase equilibrium modelling of Palaeoproterozoic ultrahigh-temperature sapphirine granulite from the Inner Mongolia suture zone, North China Craton: implications for counterclockwise P-T path. Geological Journal, 48(5): 456-466

Shu Y C, Nielsen S G, Marschall H R, John T, Blusztajn J, Auro M. 2019. Closing the loop: subducted eclogites match thallium isotope compositions of ocean island basalts. Geochimica et Cosmochimica Acta, 250: 130-148

Smith D C. 1984. Coesite in clinopyroxene in the Caledonides and its implications for geodynamics. Nature, 310(5979): 641-644

Sobolev N V, Shatsky V S. 1990. Diamond inclusions in garnets from metamorphic rocks: a new environment for diamond formation. Nature, 343(6260): 742-746

Song S G, Yang J S, Liou J G, Wu C L, Shi R D, Xu Z Q. 2003a. Petrology, geochemistry and isotopic ages of eclogites from the Dulan UHPM terrane, the North Qaidam, NW China. Lithos, 70(3-4): 195-211

Song S G, Yang J S, Xu Z Q, Liou J G, Shi R D. 2003b. Metamorphic evolution of the coesite-bearing ultrahigh-pressure terrane in the North Qaidam, northern Tibet, NW China. Journal of Metamorphic Geology, 21(6): 631-644

Song S G, Zhang L F, Niu Y L. 2004. Ultra-deep origin of garnet peridotite from the North Qaidam ultrahigh-pressure belt, northern Tibetan Plateau, NW China. American Mineralogist, 89(8-9): 1330-1336

Song S G, Zhang L F, Chen J, Liou J G, Niu Y L. 2005a. Sodic amphibole exsolutions in garnet from garnet-peridotite, North Qaidam UHPM belt, NW China: implications for ultradeep-origin and hydroxyl defects in mantle garnets. American Mineralogist, 90(5-6): 814-820

Song S G, Zhang L F, Niu Y L, Su L, Jian P, Liu D Y. 2005b. Geochronology of diamond-bearing zircons from garnet peridotite in the North Qaidam UHPM belt, northern Tibetan Plateau: a record of complex histories from oceanic lithosphere subduction to continental collision. Earth and Planetary Science Letters, 234(1-2): 99-118

Song S G, Bi H Z, Qi S S, Yang L M, Allen M B, Niu Y L, Su L, Li W F. 2018. HP-UHP metamorphic belt in the East Kunlun orogen: final closure of the Proto-Tethys ocean and formation of the Pan-North-China continent. Journal of Petrology, 59(11): 2043-2060

Song S G, Niu Y L, Zhang G B, Zhang L F. 2019. Two epochs of eclogite metamorphism link "cold" oceanic subduction and "hot" continental subduction, the North Qaidam UHP belt, NW China. Geological Society, London, Special Publications, 474(1): 275-289

Spandler C, Pettke T, Rubatto D. 2011. Internal and external fluid sources for eclogite-facies veins in the monviso meta-ophiolite, western Alps: implications for fluid flow in subduction zones. Journal of Petrology, 52(6): 1207-1236

Stevens L M, Baldwin J A, Cottle J M, Kylander-Clark A R C. 2015. Phase equilibria modelling and LASS monazite petrochronology: P-T-t constraints on the evolution of the Priest River core complex, northern Idaho. Journal of Metamorphic Geology, 33(4): 385-411

Sun C G, Liang Y. 2014. An assessment of subsolidus re-equilibration on REE distribution among mantle minerals olivine, orthopyroxene, clinopyroxene, and garnet in peridotites. Chemical Geology, 372: 80-91

Sun C G, Liang Y. 2015. A REE-in-garnet-clinopyroxene thermobarometer for eclogites, granulites and garnet peridotites. Chemical Geology, 393-394: 79-92

Sun C G, Liang Y. 2017. A REE-in-plagioclase-clinopyroxene thermometer for crustal rocks. Contributions to Mineralogy and Petrology, 172: 24

Tan Z, Agard P, Gao J, John T, Li J L, Jiang T, Bayet L, Wang X S, Zhang X. 2017. P-T-time-isotopic evolution of coesite-bearing eclogites: implications for exhumation processes in SW Tianshan. Lithos, 278-281: 1-25

Taylor R J M, Clark C, Fitzsimons I C W, Santosh M, Hand M, Evans N, McDonald B. 2014. Post-peak, fluid-mediated modification of granulite facies zircon and monazite in the Trivandrum Block, southern India. Contributions to Mineralogy and Petrology, 168(2): 1044

Tazzo-Rangel M D, Weber B, González-Guzmán R, Valencia V A, Frei D, Schaaf P, Solari L A. 2019. Multiple metamorphic events in the Palaeozoic Mérida Andes basement, Venezuela: insights from U-Pb geochronology and Hf-Nd isotope systematics. International Geology Review, 61(13): 1557-1593

Teng F Z, Dauphas N, Watkins J M. 2017. Non-traditional stable isotopes. Reviews in Mineralogy & Geochemistry, 82: 1-26

Tsay A, Zajacz Z, Sanchez-Valle C. 2014. Efficient mobilization and fractionation of rare-earth elements by aqueous fluids upon slab dehydration. Earth and Planetary Science Letters, 398: 101-112

Tsunogae T, Liu S J, Santosh M, Shimizu H, Li J H. 2011. Ultrahigh-temperature metamorphism in Daqingshan, Inner Mongolia suture zone, North China

Craton. Gondwana Research,20(1):36-47

van Roermund H L M,Drury M R. 1998. Ultra-high pressure($P>6$ GPa)garnet peridotites in western Norway:exhumation of mantle rocks from >185 km depth. Terra Nova,10(6):295-301

Wang B,Tian W,Wei C J,Di Y K. 2019. Ultrahigh metamorphic temperatures over 1050℃ recorded by Fe-Ti oxides and implications for Paleoproterozoic magma-induced crustal thermal perturbation in Jining area,North China Craton. Lithos,348-349:105180

Wang B,Wei C J,Tian W,Fu B. 2020. UHT Metamorphism Peaking Above 1100℃ with Slow Cooling:insights from Pelitic Granulites in the Jining complex,North China Craton. Journal of Petrology,61(6):egaa070

Wang B,Wei C J,Tian W. 2021. Evolution of spinel-bearing ultrahigh-temperature granulite in the Jining complex,North China Craton:constrained by phase equilibria and Monte Carlo methods. Mineralogy and Petrology,115(3):283-297

Wang L J,Guo J H,Yin C Q,Peng P,Zhang J,Spencer C J,Qian J H. 2018. High-temperature S-type granitoids(charnockites)in the Jining complex,North China Craton:restite entrainment and hybridization with mafic magma. Lithos,320-321:435-453

Wang S J,Teng F Z,Williams H M,Li S G. 2012. Magnesium isotopic variations in cratonic eclogites:origins and implications. Earth and Planetary Science Letters,359-360:219-226

Wang S J,Teng F Z,Li S G,Hong J A. 2014. Magnesium isotopic systematics of mafic rocks during continental subduction. Geochimica et Cosmochimica Acta,143:34-48

Wang S J,Teng F Z,Rudnick R L,Li S G. 2015. Magnesium isotope evidence for a recycled origin of cratonic eclogites. Geology,43(12):1071-1074

Wang S J,Teng F Z,Li S G,Zhang L F,Du J X,He Y S,Niu Y L. 2017a. Tracing subduction zone fluid-rock interactions using trace element and Mg-Sr-Nd isotopes. Lithos,290-291:94-103

Wang S J,Wang L,Brown M,Piccoli P M,Johnson T E,Feng P,Deng H,Kitajima K,Huang Y. 2017b. Fluid generation and evolution during exhumation of deeply subducted UHP continental crust:petrogenesis of composite granite-quartz veins in the Sulu Belt,China. Journal of Metamorphic Geology,35(6):601-629

Wei C J,Clarke G L. 2011. Calculated phase equilibria for MORB compositions:a reappraisal of the metamorphic evolution of lawsonite eclogite. Journal of Metamorphic Geology,29(9):939-952

Wei C J,Tian Z L. 2014. Modelling of the phase relations in high-pressure and ultrahigh-pressure eclogites. Island Arc,23(4):254-262

Wei C J,Duan Z Z. 2019. Phase Relations in metabasic rocks:Constraints from the results of experiments,phase modelling and ACF analysis. Geological Society,London,Special Publications,474(1):25-45

Wernert P,Schulmann K,Chopin F,Štípská P,Bosch D,El Houicha M. 2016. Tectonometamorphic evolution of an intracontinental orogeny inferred from P-T-t-d paths of the metapelites from the Rehamna Massif(Morocco). Journal of Metamorphic Geology,34(9):917-940

White R W,Powell R,Holland T J B. 2007. Progress relating to calculation of partial melting equilibria for metapelites. Journal of Metamorphic Geology,25(5):511-527

White R W,Powell R,Holland T J B,Johnson T E,Green E C R. 2014. New mineral activity-composition relations for thermodynamic calculations in metapelitic systems. Journal of Metamorphic Geology,32(3):261-286

Will T M,Schmädicke E,Ling X X,Li X H,Li Q L. 2018. New evidence for an old idea:Geochronological constraints for a paired metamorphic belt in the central European Variscides. Lithos,302-303:278-297

Wu C M. 2017. Calibration of the garnet-biotite-Al_2SiO_5-quartz geobarometer for metapelites. Journal of Metamorphic Geology,35:983-998

Wu C M. 2020. Calibration of the biotite-muscovite geobarometer for metapelitic assemblages devoid of garnet or plagioclase. Lithos,372-373:105668

Wu C M,Chen H X. 2015. Calibration of a Ti-in-muscovite geothermometer for ilmenite-and Al_2SiO_5-bearing metapelites. Lithos,212-215:122-127

Wu J L,Zhang H F,Zhai M G,Guo J H,Li R X,Wang H Z,Zhao L,Jia X L,Wang L J,Hu B,Zhang H D. 2017. Paleoproterozoic high-pressure-high-temperature pelitic granulites from Datong in the North China Craton and their geological implications:constraints from petrology and phase equilibrium modeling. Precambrian Research,303:727-748

Xia Q X,Zheng Y F,Hu Z C. 2010. Trace elements in zircon and coexisting minerals from low-T/UHP metagranite in the Dabie orogen:implications for action of supercritical fluid during continental subduction-zone metamorphism. Lithos,114(3-4):385-412

Xia Q X,Zheng Y F,Chen Y X. 2013. Protolith control on fluid availability for zircon growth during continental subduction-zone metamorphism in the Dabie orogen. Journal of Asian Earth Sciences,67-68:93-113

Xia X P,Sun M,Zhao G C,Wu F Y,Xu P,Zhang J H,Luo Y. 2006. U-Pb and Hf isotopic study of detrital zircons from the Wulashan khondalites:constraints on the evolution of the Ordos terrane,western block of the North China Craton. Earth and Planetary Science Letters,241(3-4):581-593

Xiao Y L,Hoefs J,Hou Z H,Simon K,Zhang Z M. 2011. Fluid/rock interaction and mass transfer in continental subduction zones:constraints from trace elements and isotopes(Li,B,O,Sr,Nd,Pb)in UHP rocks from the Chinese Continental Scientific Drilling Program,Sulu,East China. Contributions to Mineralogy and Petrology,162(4):797-819

Xiao Y L,Sun H,Gu H O,Huang J,Li W Y,Liu L. 2015. Fluid/melt in continental deep subduction zones:compositions and related geochemical fractionations. Science China Earth Sciences,58(9):1457-1476

Xiang H,Connolly J A D. 2021. GeoPS:an interactive visual computing tool for thermodynamic modelling of phase equilibria. Journal of Metamorphic Geology,https://doi.org/10.1111/jmg.12626

Xu S T,Okay A I,Ji S Y,Sengör A M C,Su W,Liu Y C,Jiang L L. 1992. Diamond from the Dabie Shan metamorphic rocks and its implication for tectonic setting. Science,256:80-82

Xu X, Song S G, Allen M B, Ernst R E, Niu Y L, Su L. 2016. An 850-820 Ma LIP dismembered during breakup of the Rodinia supercontinent and destroyed by Early Paleozoic continental subduction in the northern Tibetan Plateau, NW China. Precambrian Research, 282:52-73

Yakymchuk C, Brown M. 2014. Behaviour of zircon and monazite during crustal melting. Journal of the Geological Society, 171(4):465-479

Yamada C, Tsujimori T, Chang Q, Kimura J I. 2019. Boron isotope variations of Franciscan serpentinites, northern California. Lithos, 334-335:180-189

Yang J S, Xu Z Q, Song S G, Zhang J X, Wu C L, Shi R D, Li H B, Maurice B. 2001. Discovery of coesite in the North Qaidam early Paleozoic ultrahigh pressure metamorphic belt, NW China. Earth and Planetary Science, 333:719-724

Yang Q Y, Santosh M, Tsunogae T. 2014. Ultrahigh-temperature metamorphism under isobaric heating: new evidence from the North China Craton. Journal of Asian Earth Sciences, 95:2-16

Ye K, Cong B L, Ye D N. 2000. The possible subduction of continental material to depths greater than 200 km. Nature, 407(6805):734-736

Yin C Q, Zhao G C, Sun M, Xia X P, Wei C J, Zhou X W, Leung W H. 2009. LA-ICP-MS U-Pb zircon ages of the Qianlishan complex: constrains on the evolution of the Khondalite Belt in the western block of the North China Craton. Precambrian Research, 174(1-2):78-94

Yin C Q, Zhao G C, Wei C J, Sun M, Guo J H, Zhou X W. 2014. Metamorphism and partial melting of high-pressure pelitic granulites from the Qianlishan complex: constraints on the tectonic evolution of the Khondalite Belt in the North China Craton. Precambrian Research, 242:172-186

Yu B, Santosh M, Amaldev T, Palin R M. 2021. Mesoarchean (ultra)-high temperature and high-pressure metamorphism along a microblock suture: evidence from Earth's oldest khondalites in southern India. Gondwana Research, 91:129-151

Zhang G B, Song S G, Zhang L F, Niu Y L. 2008. The subducted oceanic crust within continental-type UHP metamorphic belt in the North Qaidam, NW China: evidence from petrology, geochemistry and geochronology. Lithos, 104(1-4):99-118

Zhang G B, Zhang L F, Song S G, Niu Y L. 2009. UHP metamorphic evolution and SHRIMP geochronology of a coesite-bearing meta-ophiolitic gabbro in the North Qaidam, NW China. Journal of Asian Earth Sciences, 35(3-4):310-322

Zhang G B, Ellis D J, Christy A G, Zhang L F, Niu Y L, Song S G. 2010. UHP metamorphic evolution of coesite-bearing eclogite from the Yuka terrane, North Qaidam UHPM belt, NW China. European Journal of Mineralogy, 21(6):1287-1300

Zhang H T, Li J H, Liu S J, Li W S, Santosh M, Wang H H. 2012. Spinel + quartz-bearing ultrahigh-temperature granulites from Xumayao, Inner Mongolia suture zone, North China Craton: Petrology, phase equilibria and counterclockwise p-T path. Geoscience Frontiers, 3(5):603-611

Zhang J X, Meng F C, Li J P, Mattinson C G. 2009. Coesite in eclogite from the North Qaidam Mountains and its implications. Chinese Science Bulletin, 54(6):1105-1110

Zhang L, Chen R X, Zheng Y F, Hu Z C, Yang Y H, Xu L J. 2016a. Geochemical constraints on the protoliths of eclogites and blueschists from North Qilian, northern Tibet. Chemical Geology, 421:26-43

Zhang L, Chen R X, Zheng Y F, Li W C, Hu Z C, Yang Y H, Tang H L. 2016b. The tectonic transition from oceanic subduction to continental subduction: zirconological constraints from two types of eclogites in the North Qaidam orogen, northern Tibet. Lithos, 244:122-139

Zhang L, Chen R X, Zheng Y F, Hu Z C, Xu L J. 2017. Whole-rock and zircon geochemical distinction between oceanic- and continental-type eclogites in the North Qaidam orogen, northern Tibet. Gondwana Research, 44:67-88

Zhang L F, Ellis D J, Jiang W B. 2002a. Ultrahigh-pressure metamorphism in western Tianshan, China: Part I. evidence from inclusions of coesite pseudomorphs in garnet and from quartz exsolution lamellae in omphacite in eclogites. American Mineralogist, 87(7):853-860

Zhang L F, Ellis D J, Williams S, Jiang W B. 2002b. Ultra-high pressure metamorphism in western Tianshan, China: Part II. evidence from magnesite in eclogite. American Mineralogist, 87(7):861-866

Zhang L F, Song S G, Liou J G, Ai Y L, Li X P. 2005. Relict coesite exsolution in omphacite from western Tianshan eclogites, China. American Mineralogist, 90(1):181-186

Zhang L F, Wang Y, Zhang L J, Lü Z. 2019a. Ultrahigh pressure metamorphism and tectonic evolution of southwestern Tianshan orogenic belt, China: a comprehensive review. Acta Geologica Sinica, 93(S2):47

Zhang L F, Zhang Z M, Schertl H P, Wei C J. 2019b. HP-UHP metamorphism and tectonic evolution of orogenic belts: introduction. Geological Society, London, Special Publications, 474(1):1-4

Zhang L J, Zhang L F, Lü Z, Bader T, Chen Z Y. 2016. Nb-Ta mobility and fractionation during exhumation of UHP eclogite from southwestern Tianshan, China. Journal of Asian Earth Sciences, 122:136-157

Zhang L J, Chu X, Zhang L F, Fu B, Bader T, Du J X, Li X L. 2018. The early exhumation history of the western Tianshan UHP metamorphic belt, China: new constraints from titanite U-Pb geochronology and thermobarometry. Journal of Metamorphic Geology, 36(5):631-651

Zhang R Y, Liou J G, Iizuka Y, Yang J S. 2009. First record of K-cymrite in North Qaidam UHP eclogite, Western China. American Mineralogist, 94(2-3):222-228

Zhang Z M, Shen K, Liou J G, Dong X, Wang W, Yu F, Liu F. 2011. Fluid-rock interactions during UHP metamorphism: a review of the Dabie-Sulu orogen, east-central China. Journal of Asian Earth Sciences, 42(3):316-329

Zheng Y F. 2009. Fluid regime in continental subduction zones: petrological insights from ultrahigh-pressure metamorphic rocks. Journal of the Geological Society, 166(4):763-782

Zheng Y F. 2012. Metamorphic chemical geodynamics in continental subduction zones. Chemical Geology, 328:5-48

Zheng Y F, Hermann J. 2014. Geochemistry of continental subduction-zone fluids. Earth, Planets and Space, 66(1):93

Zheng Y F, Xia Q X, Chen R X, Gao X Y. 2011. Partial melting, fluid supercriticality and element mobility in ultrahigh-pressure metamorphic rocks during continental collision. Earth-Science Reviews, 107(3-4):342-374

Zheng Y F, Zhang L F, McClelland W C, Cuthbert S. 2012. Processes in continental collision zones: preface. Lithos, 136-139:1-9

Zheng Y F, Zhao Z F, Chen R X. 2019. Ultrahigh-pressure metamorphic rocks in the Dabie-Sulu orogenic belt: compositional inheritance and metamorphic modification. Geological Society, London, Special Publications, 474(1):89-132

Zhu J J, Zhang L F, Lü Z, Bader T. 2018. Elemental and isotopic (C, O, Sr, Nd) compositions of Late Paleozoic carbonated eclogite and marble from the SW Tianshan UHP belt, NW China: implications for deep carbon cycle. Journal of Asian Earth Sciences, 153:307-324

Zou Y, Chu X, Li Q L, Mitchell R N, Zhai M G, Zou X Y, Zhao L, Wang Y Q, Liu B. 2020. Local rapid exhumation and fast cooling in a long-lived Paleoproterozoic orogeny. Journal of Petrology, 61:egaa091

Progresses in the Study of Metamorphic Petrology

ZHANG Gui-bin[1], LIU Liang[2], WEI Chun-jing[1], XIAO Yi-lin[3], JIAO Shu-juan[4], LV Zeng[1], ZHANG Li-fei[1]

1. Key Laboratory of Orogenic Belt and Crustal Evolution, MOE, School of Earth and Space Science, Peking University, Beijing 100871; 2. Department of Geology, Northwest University, Xi'an 710127; 3. School of Earth and Space Sciences, University of Science and Technology of China, Hefei 230026; 4. Institute of Geology and Geophysics, Chinese Academy of Sciences, Beijing 100029

Abstract: Metamorphic rock is one of the three major types of rocks of the silicate earth, and thus is of great significance for understanding the composition and evolution of the crust, studying records of geological history for the thermal structure of crust, restoring the protolith nature and guiding the mineral prospecting. In this paper, we have reviewed progresses in several aspects of the researches on metamorphic rocks, including the UHP/UHT metamorphism, phase equilibrium, metamorphic fluids, and geochemistry of metamorphic rocks, in recent ten years in China. Based on the previous studies, chinese scholars had successively discovered coesite, stishovite pseudomorph and some other UHP metamorphic indicative minerals and structures in different types of rocks in several orogenic belts, such as the northern Qaidam, southern Altun, eastern Kunlun, northern Qinling, and southwestern Tianshan, and then had determined the field distribution characteristics of these UHP metamorphic belts. In the aspect of UHT metamorphism research, outcrop scale of UHT metamorphic rocks were identified in the khondalite belt of the North China Craton; and a series of progresses have been made in the determination of the precise metamorphic age, the constraint of metamorphic time scale, and the construction of P-T-t paths, especially the determination of the prograde metamorphic stages. In the aspect of metamorphic phase equilibrium research, on the basis of further improvement of thermodynamic database and models of mineral phase and melt activity, a new phase modelling software GeoPS has been developed, and the complete phase equilibrium relationship of metamorphosed basic rocks has been established based on ACF component analysis. Meanwhile, some important progresses have also been made in the aspect of quantitative simulation of the anatexis and the genesis of granitic rocks. Moreover, GBAQ and two-mica barometry and Ti-in-white mica thermometer have been quantified. In the aspect of researches on elemental geochemistry of metamorphic rocks, a series of advances have been made in the migration and differentiation of HFSE (such as Ti, Nb, Ta, Zr, and Hf) and halogen elements (such as F, Cl, Br, and I) in the process of metamorphic dehydration, as well as the change of oxygen fugacity indicated by variational elements (such as V, Fe, W, and Mo). In the aspect of researches on metamorphic fluids, some remarkable progresses have been made in the evidence identification of HP/UHP fluid activities in subduction zone, the determination of fluid composition, and the constraint of the time of fluid activity.

Key words: UHP metamorphism; UHT metamorphism; phase equilibrium of metamorphism; metamorphic fluids; isotopic dating

沉积学发展现状与趋势*

王成善　林畅松

中国地质大学（北京），北京 100083

摘　要：中国沉积学经历了从 20 世纪初至 21 世纪百余年的发展历程。进入 21 世纪以来，中国沉积学在盆地动力学、层序地层学、源-汇系统、沉积体系和沉积相、古地理学、前寒武纪沉积学、现代沉积环境、深时地质与沉积学、生物沉积学等多领域的研究中取得了重要进展。在深时古气候与全球变化、古环境变化及第四纪沉积等领域的研究与国际同步发展或已迈入国际研究前沿。同时，形成了陆相层序地层和盆地沉积充填动力学、含油气盆地沉积学等独具我国特色的研究方向，取得了一系列重要进展并围绕国家能源需求做出了重大贡献。但目前我国沉积学研究仍以跟踪国际前沿为主，而原创性的、引领性的研究较少。我国沉积学在温室陆地气候与古地貌重建、重大地质转折期的沉积和生物过程、源-汇系统、前寒武纪超大陆演化与早期地球环境等领域有望取得具有国际前沿的开拓性成果。

关键词：中国沉积学　发展历程　研究现状　未来发展趋势

0　引　言

沉积学作为研究沉积物和沉积岩及其形成过程的一门地学分支学科，经历了百余年的发展历程。沉积学从初期以研究沉积物和沉积岩的特征和沉积机理的分析为主要任务，发展到当代兼容多学科交叉的一门综合性学科。在矿产资源日趋紧张和生态环保问题日益突出的今天，沉积学在其基础理论不断完善的同时，在促进新学科诞生、矿产资源的勘查与开发、人与自然和谐发展等领域的研究，发挥着不可替代的作用。

半个多世纪以来，沉积学在沉积体系和沉积相分析、层序地层学、沉积盆地充填分析、大地构造沉积学、资源沉积学、环境沉积学等诸多方面取得了一系列重大进展。沉积学的研究可通过构造、气候等变化的沉积记录分析，揭示地球表层层圈的相互作用及其演变历史。古大陆-古地理再造、板块构造与沉积作用、盆-山关系、盆地动力学及气候-沉积响应、源-汇系统等研究，是当前旨在揭示地球表层层圈动力学过程和演变的重大课题。深时古气候、生物沉积学、前寒武纪超大陆演化与早期地球环境等成为当前人们最为关注的热点领域。层序地层学从 20 世纪 70 年代兴起至今，方兴未艾，其发展使得在盆地或全球范围内进行等时地层对比成为可能，为揭示沉积体系域、古地理、古环境等在时空上的分布和演变带来了革命性的理论和方法体系。沉积学与地球物理、地球化学、计算机技术等多学科实现大跨度的交叉渗透，促进了多个交叉学科方向的迅速发展。结合 GIS 和计算机模拟技术进行的基于大数据的活动古地理重建，是当今地球科学研究的一个前缘热点领域（王成善等，2010）。当代沉积学已成为地球科学中最重要的基础和应用基础学科之一，其发展在认识地球演化历史、促进相关学科的发展和诞生交叉学科方向，解决能源、水资源短缺及生态、环境污染、地质灾害问题，实现人类可持续发展等，有着深远的学科意义和重大的战略价值。本文旨在通过回顾中国沉积学的发展历程，剖析中国沉积学近十年来的研究进展和存在问题，展望未来的发展趋势，为中国沉积学的发展战略和研究方向提供参考和启示。

---

* 原文"中国沉积学近十年来的发展现状与趋势"刊于《矿物岩石地球化学通报》2021 年第 40 卷第 6 期，本文略有修改。

1 中国沉积学的发展历程和发展现状

20世纪初，随着中国地质学的起步，我国的沉积学也得到了相应发展。中华人民共和国成立前后，叶连俊、业治铮、吴崇筠等一批沉积学前辈和爱国学者先后从国外返回祖国，以服务和建设国家为己任，在科学研究和人才培养工作上辛勤耕耘、勇于开拓，是中国沉积学发展的主要奠基人（叶连俊，1942；刘宝珺，2001）。中华人民共和国成立后，国家对矿产资源需求的与日俱增不断推动了我国沉积学的发展。到60年代中期，我国已基本形成了一支沉积学教学和研究队伍，沉积学理论和应用得到了重视和发展，为石油、煤炭等矿产资源的勘探提供了重要的理论支持。这一时期，我国沉积学研究与国际基本上同步发展。

经历了学科停滞的十年文革"浩劫"后，中国迎来了"科学的春天"，中国沉积学也迎来了全面蓬勃发展的新时期。伴随着改革开放，我国的科学研究从跟踪国际前缘发展，到开拓前缘和创新（刘东生等，1978；王鸿祯，1985；业治铮等，1985；叶连俊，1989；吴崇筠和薛叔浩，1992）。进入21世纪，我国的沉积学及其相关领域，如陆相层序地层学、沉积大地构造、海洋沉积地质、古海洋、古气候等研究在国际上已有不俗表现或占据了一席之地（孙枢，2005）。中国沉积学研究成果的发表数量呈现飞跃式增长态势。据统计，2001年中国沉积学发表研究成果仅占全球的4.01%，至2016年达20.95%，超过了2016年中国总人口占世界总量的比例。2006—2010年，中国成为全球排名第7的沉积学热点研究区域；2011—2015年，中国和中国南海分别成为全球排名第2与第5的沉积学热点研究区域。

2018年，由王成善院士带领的中国沉积学代表参加了第20届国际沉积学大会，取得了在中国举办第21届国际沉积学大会的举办权，标志着中国沉积学在国际上的核心竞争力得到显著提升。中国具有多样化的沉积盆地类型和丰富的沉积地质记录，为中国沉积学的理论创新和发展提供了得天独厚的条件。总体上，中国的沉积学发展已迈入国际研究前缘，发展迅猛。但是，学界也普遍认识到我们的跟踪研究占多，而原始创新、国际引领性的研究少；应用研究多，而基础、基本原理研究少。无疑，中国沉积学核心实力的提升势在必行，任重道远。

2 中国沉积学当前的主要发展趋势

近十多年来，我国沉积学研究在国际热点领域，如深时古气候与全球变化、古环境变化、海平面变化及第四纪沉积等方面的研究与国际相关研究同步发展，同时在陆相盆地沉积学和层序地层学、含油气盆地沉积学、沉积盆地动力学分析、前寒武纪沉积学等多个方向方面也形成了独具特色的研究领域，并取得了一系列重要进展。

2.1 盆地形成演化和盆地动力学机制

我国学者早在20世纪70年代末就开展了沉积盆地的整体分析，在陆相伸展型盆地、陆内拗陷盆地、大型叠合盆地、边缘海盆地等领域出版了大量有特色的专著（李思田，1988；李德生，1992；田在艺和张庆春，1996）和教材（王成善和李祥辉，2003；李思田等，2004；解习农和任建业，2013；林畅松，2016）。近十多年来，依托我国油气专项、国家自然科学基金重大研究计划、973项目等的研究，盆地整体分析和盆地形成演化研究取得了一批优秀研究成果。沉积盆地的形成演化涉及盆地形成的动力学机制、演变过程以及盆地与板块构造和地幔深部过程的动力学关系。我国西部环青藏高原巨型盆山体系对大陆会聚及碰撞造山的响应关系研究为此提供了范例（金之钧和蔡立国，2007；贾承造，2009）。近年来以周缘造山带地层及沉积记录的解析为切入点，识别了洋盆、洋岛、海山等沉积地层序列，重建了洋陆作用与转化机制，揭示了西部典型叠合盆地演化的区域动力学背景；以同造山-后造山期高分辨率沉积记录分析和不整合面分析为切入点，识别了陆内构造变形-造山阶段原型盆地大区域隆拗变迁及盆地变革期构造-古地理演化，对塔里木盆地、准噶尔盆地、四川盆地等大型叠合盆地及其相邻前陆造山带形成演化取得了一系列新成果，对大型叠合盆地演化的动力学过程和形成机制做出了合理的解释（许志琴等，2008；Lin et al.，2012；李忠和彭守涛，2013；Dong et al，2016）。此外，针对我国中东部的秦岭、大别山、燕山等造山带及其周缘盆地形成演化，多年来也已取得丰硕的研究成果（张国伟等，2001）。

我国东部滨太平洋构造域中新生代陆相和边缘海盆地，位于欧亚板块、太平洋板块和澳大利亚板块的交汇处，其演化受控于多方向洋、陆板块的相互作用，其形成和演化的动力学研究自20世纪80年代以来就一直是盆地动力学研究的前缘领域（Tapponnier et al.，1990；Lüdmann and Wong，1999）。晚中生代以来东亚大陆及其陆缘裂谷构造演化的谱系蕴含着东亚大陆岩石圈伸展、薄化、破裂扩张过程的丰富信息，通过盆地沉降与伸展裂陷过程的综合研究，阐明了中国东部中新生代盆地演化的多幕裂陷和多幕反转过程，为盆地充填的区域性沉积旋回结构、幕式沉降过程，以及幕式生、排烃过程等提供了理论解释（李思田，2015；Lei and Ren，2016；Lin et al.，2018）。我国东部的大陆边缘盆地带，一直是地球系统动力学研究的前缘与热点地区。近十多年来，由我国科学家领衔的南海大洋钻探、国家自然科学基金重大计划和多项国际合作考察航次，使我国大陆边缘盆地动力学研究推向了新的高点。近年来一项重要进展是发现在南海北部陆缘深海域存在大型的拆离断裂带，控制着拆离盆地群的发育；并揭示了南海岩石圈在新生代经历了纯剪切变形伸展、拆离薄化以及地幔剥露和洋中脊扩张等构造作用，为认识深水海域盆地的形成机制和油气潜力评价提供了重要的理论依据（Yang et al.，2018；任建业等，2018）。

2.2 层序地层学

20世纪80年代发展起来的层序地层学理论不断得到完善和发展，已成为沉积地质学和盆地分析的较为完善的理论和分析方法，成了油气勘探中不可缺少的分析理论和预测技术。但经典的层序地层学理论源于被动大陆边缘海相盆地的研究，难于应用到陆相盆地的层序研究。自20世纪八九十年代以来，我国学者结合陆相盆地的特点开展了卓有成效的创新性研究，揭示了陆相盆地层序地层的形成机理，形成了独具特色的陆相湖盆层序地层分析理论和技术方法（李思田等，1999；Lin et al.，2001），并提出了基于中国盆地特色的新概念和分析方法，如构造坡折带、构造古地貌等概念和分析方法（林畅松等，2000，2009；林畅松，2019），推动了陆相层序地层学的发展，为我国含油气盆地的生-储-盖预测提供了重要的理论基础。在陆相高精度层序地层分析方法方面也取得了一系列创新性的进展（蔡希源和李思田，2003；邓宏文，2009）。由王成善教授领导组织实施的松辽盆地国际大陆科学钻探工程，获取了世界上最连续的、长达8 191 m的陆相白垩系地质记录，完整地揭示了大型陆相盆地沉积充填演化和沉积层序系列，构建了陆地白垩纪高精度层序地层框架（王成善等，2016），为陆相高精度层序地层学研究提供了极好的范例。依赖于高分辨的三维地震等地球物理技术和以找寻油气储集体为目标的高精度层序地层学和地震沉积学的研究，已形成了一整套思路和方法体系，并在识别沉积体系和储层的几何形态及定量预测储集性能等方面的应用成效显著，将促进盆地沉积充填的研究达到更高水平。同时，我国在边缘海盆地的层序地层学研究近年来也取得了显著进展。我国学者通过对地震、测井、古生物及古地磁等多类资料的综合研究，建立了南海大陆边缘新近纪以来的高精度年代-层序地层序列及其成因机制，揭示了南海北部渐新世以来的大陆边缘沉积楔的层序结构、坡折轨迹、沉积体系域的特征和演化及其构造、气候、物源变化等的响应过程，并建立了边缘海盆地的沉积层序演化模式（Wu et al.，2014；Lin et al.，2018；Xie et al.，2019）。

2.3 "源-汇"系统

"源-汇"系统研究是近十余年沉积学的一个新的研究领域。我国对地球表面大型的源-汇系统的系统性研究整体处于起步阶段，但我国学者较早就在湖相盆地中注意到了"源-汇系统"对沉积体系和储集砂体分析和预测的重要性，在断陷盆地的研究中注意到了"源、沟、扇"成因关系的分析并应用于砂岩油气藏的预测（潘元林和李思田，2004）。关于我国青藏高原新近纪大体同时性的强烈构造隆升，为新近纪盆地快速充填提供了物源条件的研究，也取得了重要进展（Coleman and Hodges，1995；张克信等，2008）。近年来，我国沉积学者在中新生代断陷盆地古隆起物源与沉积充填过程、陆内拗陷或前陆盆地中造山带物源与前陆盆地沉积充填关系、盆内古构造地貌与沉积物搬运路径等方面的研究都取得了可喜的进展（李忠和彭守涛，2013；林畅松等，2015；徐长贵等，2017）。对我国南海的陆架边缘到深海盆地的源–汇系统近年来也开展了较为广泛的研究（Jiang et al.，2015；Shao et al.，2016；Lin et al.，2018），在物源区的构造、气候条件与盆地物源供给、从内陆架河流–三角洲到陆架边缘三角洲–海底扇的源–汇系统等方面的研究有突出进展（Wang et al.，2013）。总体来看，这些研究或局限于物源体系，或较多关注其与油气成藏要素的关系，如何从这些局部地质问题扩展成一种源–汇系统及其相关普适的地质机理或地球表层层圈的动力学过程，仍需多学科队伍的联合制定长期的研究纲要。

2.4 沉积体系和沉积相

沉积学主要的研究任务之一就是研究形成沉积物（岩）的沉积环境、沉积过程及其控制因素，沉积体系和沉积相是沉积学主要的、也是基本的研究对象。沉积学的发展过程中已建立了一系列沉积体系或沉积相模式，构成了沉积学的主要理论框架和基础（Reading，1996）。沉积学仍在不断创新沉积相理论，如深水重力流沉积、生物岩沉积、事件沉积等。近十多年来，我国在陆相湖盆沉积体系和沉积模式、古老小克拉通碳酸盐岩台地、陆架边缘沉积体系和边缘海盆地沉积动力学等方面的研究，取得了令人瞩目的研究成果。

我国中新生代陆相湖盆中的扇三角洲或河流三角洲、滨岸滩坝、重力流等沉积体系研究一直是热点课题。我国沉积学者在湖泊浅水三角洲体系、重力流或异重力流沉积的研究方面取得许多重要的进展，有效指导了油气的勘探和开发（邹才能等，2008；袁选俊等，2015；朱筱敏，2016）。近年来，在非常规油气勘探需求的推动下，湖泊细粒沉积作用及其生、储的特性的研究引起了广泛的关注（邹才能和邱振，2021）。湖泊细粒沉积的生烃潜力的研究，我国学者事实上率先于国际同行并为建立特色的中国陆相石油地质理论奠定基础。湖泊细粒沉积作用和细粒沉积岩相组构及成因等深化研究，涉及湖泊的成因、沉积物源、有机质生产率以及气候等多因素的影响，目前的研究还比较薄弱，亟须开展多学科的攻关研究，以建立细粒沉积体系成因模式。

自20世纪90年代以来，陆架边缘沉积结构和沉积体系的研究成为国际上的研究热点和油气勘探新领域，我国围绕南海大陆边缘沉积过程的研究也取得了突出的成果，如通过对南海北部大陆边缘地震、测井、岩心、薄片等多类资料的综合研究，揭示了南海北部渐新世以来的大陆边缘沉积楔的整体演化过程，并建立了沉积演化模式（Wu et al.，2014；Lin et al.，2018；Xie et al.，2019）。特别是斜坡单向迁移水道、陆架边缘三角洲及富砂斜坡扇体系等沉积模式的建立，丰富了陆架边缘斜坡带沉积动力学理论，明确了沉积物供给、气候–海平面变化、构造沉降等对陆架边缘体系和大陆边缘生长的控制作用（Gong et al.，2017；Jiang et al.，2017；Lin et al.，2018；Wu et al.，2018；Zhang et al.，2019；Tian et al.，2021）。对大陆边缘坡折轨迹及其与大陆边缘生长和深水扇发育机制等的定量研究分析方面，也做了大量有益的探讨（朱筱敏等，2017；Chen et al.，2019）。

我国的碳酸盐岩研究从20世纪七八十年代就一直在追踪国际前缘研究。我国碳酸盐岩的研究伴随着碳酸盐岩油气勘探的突破，近十余年掀起了研究的热潮，在古老小克拉通背景下的碳酸盐岩沉积环境、台地演化和规模储层的发育模式方面取得了系列创新性成果，建立了复杂台地边缘礁滩、蒸发性台地、多因素叠加白云岩等规模储层的发育模式，为油气勘探提供了重要的理论指导（马永生等，2011；赵文智等，2012；Lin et al.，2012；邹才能等，2014）。但我国沉积体系和沉积相研究多偏重沉积描述和应用，似乎缺乏系统深入的机理性探讨。

2.5 古地理

古地理的重建涵盖了对地质时期的大陆轮廓、古海洋、古环境、古气候条件等的重建，也涉及对岩石圈、大气圈、生物圈和水圈历史面貌的综合研究。古地理的再造包括从整个地球表层到局部的盆地区带多尺度的古地理研究。古地理学科的发展先后经历了前板块构造"固定论"、板块构造"活动论"到当前以大数据为基础的古地理重建的发展历程。我国学者在20世纪50年代就开始了较为系统的古地理研究和编图（刘鸿允，1959）。70年代后，我国沉积学者开始了岩相古地理研究和编图（冯增昭等，1977）；刘宝珺等出版的《岩相古地理基础和工作方法》（刘宝珺和曾允孚，1985）一书标志着我国岩相古地理研究成为沉积地质学的重要分支，具有里程碑意义（陈洪德等，2017）。这一时期出版了一系列有关我国岩相古地理的研究成果，为我国煤、油气等的预测勘探提供了理论指导（关士聪，1984；李思田，1988；刘宝珺和许效松，1994）。值得指出，王鸿祯先生以全球构造"活动论"和地球演化"阶段论"相结合的构造-古地理分析方法，先后出版《中国古地理图集》（1985年）和《中国及邻区古生代生物古地理及全球古大陆再造》（1990年），推动了我国基于板块构造理论的构造-古地理研究。

近十多年来，我国的盆地沉积古地理或岩相古地理研究面向不同层次的沉积矿产勘探需求，开展了较为系统的广泛的研究和编图。注重把盆地的构造作用与沉积体系和沉积相的结合分析、并强调在等时层序地层格架中进行盆地和岩相古地理编图研究，形成了我国具有盆地特色的层序-岩相古地理或盆地构造-岩相古地理研究方向，取得了系列性显著进展。如王成善等（王成善等，1998）出版了《中国南方海相二叠系层序地层与油气勘探》，首次按层序的体系域编制了层序-岩相古地理图；马永生和陈洪德等（马永生等，2009）将构造、层序与岩相古地理有机结合，出版了《中国南方构造-层序岩相古地理图集》；林畅松（2016）注重叠合盆地关键变革期的古构造、古地貌、古地理的结合分析，出版了《叠合盆地层序地层与构造古地理》。然而，我国的古地理研究多服务于矿产勘探和开发，对有关重大科学问题的深化研究较少。最近，随着王成善等主导的深时数字地球（Deep-Time Digital Earth，DDE）计划的开展，我国正在建设大数据古地理重建平台，开展了基于GBDB数据库与GIS技术的古地理重建，这是具有国际学科前沿的开拓性工作。

2.6 前寒武纪沉积学

前寒武纪占据整个地球演化历史近90%的漫长时期，由于地球结构、构造岩浆活动、沉积作用及生命形式等方面都与显生宙显著不同，其表层的沉积作用和演化的研究就一直是沉积学家们广泛兴趣的、也是难以探知的热点领域。前寒武纪沉积学的研究多围绕以下两大问题：一是超大陆聚合-裂解与沉积盆地的响应关系及盆地中蕴藏的矿产和油气资源潜力；二是前寒武纪环境变迁的沉积记录与早期生命演化。

我国前寒武纪沉积学研究可追溯到20世纪20年代（高振西等，1934）。80年代以来，华南中新元古代的大地构造属性和沉积古地理研究是人们广泛关注的课题（许靖华和何起祥，1980；李铨和冷坚，1991；李江海和穆剑，1999）。进入21世纪，与前寒武纪超大陆演化紧密相关的沉积学研究成为热点。我国南方新元古代的地层系统很好地记录了Rodinia超大陆聚合与裂解的沉积演化过程，包括雪球地球的形成与消融过程、古海洋环境变迁等。我国沉积学家系统研究了华南新元古代沉积盆地演化及其与Rodinia超大陆聚合裂解的关系，提出了"南华裂谷"及其开启模式，再造了华南新元古代的岩相古地理（Wang and Li，2003；王剑等，2019）。围绕油气资源潜力的研究，在沉积环境、生烃潜力等方面的研究取得不少创新成果（王铁冠和韩克猷，2011；孙枢和王铁冠，2016）。我国华北、扬子、塔里木中新元古界均具有良好的油气地质条件，已在四川安岳-威远气田区获得了令人振奋的油气发现。近年来，我国学者在国际地学刊物上发表有关前寒武纪地质研究的论文显著增多，提高了我国前寒武纪研究在国际上的地位。然而，我国前寒武纪沉积学研究主要集中在区域性的地层学、年代学、盆地演化等方面，并未追踪国际热点研究，与国际先进水平相比还有待进一步探讨和提高。

2.7 现代沉积环境

20世纪60年代许多经典的沉积模式的建立，有赖于对现代沉积环境的沉积物特征和沉积过程的研究。当代沉积学的发展仍然有赖于对现代各种沉积环境及其沉积过程的观察和分析。我国处于沉积物产出最为丰富、沉积过程最为活跃的区域之一，因而成为现代沉积过程研究的天然实验室。东南亚地区河流入海沉积物通量占到了全球的80%以上（Milliman and Farnsworth, 2011），我国近海陆架宽广，为滨浅海环境和从陆到海的源汇过程和河口径流、潮汐、波浪、陆架环流的输运堆积过程提供了广阔的空间（Yang and Youn, 2007; Liu et al., 2010）。

我国学者对黄河、长江、珠江三角洲体系，江苏海岸-陆架区沙脊群，黄东海陆架泥质沉积，杭州湾河口沉积，盐沼湿地、红树林、珊瑚礁等进行过系统性的研究，取得了显著的进展。如对现代河流三角洲的研究表明，经典的以波浪、河流、潮汐为端元的三角洲分类有相当大的局限性，提出三角洲形成的沉积物临界入海通量（Wang et al., 2008）、沉积充填演化顺序、三角洲远端泥质沉积（Liu et al., 2014; Jia et al., 2018），以及低海面期的水下三角洲沉积（Gao et al., 2015）等可能是决定三角洲类型的重要因素。对江苏海岸潮滩和辐射状沙脊群的研究揭示了沉积物供给和沿岸潮差变化对沉积地貌分带性的控制机理，成了继北海潮滩之后的又一个典型范例（Liu et al., 2011; Wang et al., 2012; Shi et al., 2017; Gao, 2019）。此外，我国陆架区潮汐沙脊群占据面积超过2万km^2，在世界上独一无二（高抒, 2014）。我国学者对其形成的水动力和沉积动力过程研究多年，解析了江苏和欧洲北海两地潮流脊的沉积与地貌差异机制和沙脊群复合堆积体的形成过程。在大型河流的流域盆地内，河流与湖泊沉积也是重要的研究对象。多年来，在资源开发、环境保护、生态建设、灾害防护的国家需求下，我国学者对这些沉积体系进行了富有特色的深入研究。

我国现代沉积过程和机理研究还有待深入，尤其是重力流和漂移沉积以及珊瑚礁等方面需要更多的观测数据和模型研究。目前，本领域研究在不断深化原有研究的基础上，与多学科交叉合作以解决人类社会发展的宏观问题，如气候变化、碳循环、人类活动影响等问题，为未来"海岸带蓝图重绘"提供解决方案（Daigle et al., 2017; Mackay et al., 2017）。

2.8 深时地质与沉积学

深时通常指不能通过冰心恢复的、必须依赖岩石记录所恢复的前第四纪的地质记录（Soreghan et al., 2005）。深时气候学研究从整个地球历史的角度，通过对前第四纪沉积记录开展多种时空尺度研究，全面深入了解地球气候系统的变化以及控制这种变化的物理、化学、生物过程，着眼于并试图为未来气候预测提供依据（孙枢和王成善, 2009）。深时气候学研究是当代地球科学研究的重要组成部分。这为沉积地质学提供了难得的发展机遇（Isaacson and Montañez, 2013; Parrish and Soreghan, 2013）。

我国在深时古气候学领域研究进展迅速，涌现出大量的优秀成果。我国学者系统建立了华南新元古代年代地层（An et al., 2015; 周传明, 2016; 张启锐和兰中伍, 2016）与全球代表性剖面成冰纪底界的对比关系，提供了Marinoan冰期时代华南与澳大利亚西北部相连的新证据（Zhang et al., 2013）。在晚古生代冰室气候的研究，建立了冈瓦纳大陆冰盖的增长和消融与海平面变化的响应机制（Wang et al., 2013; 邵龙义等, 2014; Liu et al., 2017）；发现了华北在早二叠世含煤沉积的广泛发育与欧美大陆区同时期的干旱化气候存在明显差异，揭示了冰盛末期和温室气候下古温度状态和气候的微细变化（Wang and Pfefferkorn, 2013; 李守军等, 2014）。对西藏地区东特提斯洋获得的沉积地质记录研究，如伴随着海平面急剧上升的有机质碳同位素急剧负偏移、碳酸盐台地淹没和风暴作用增强、大型底栖生物灭绝等，证实了晚中生代快速增温的气候变化事件（Chen et al., 2017; Han et al., 2018）。上述成果对深入认识快速增温事件期间海洋-气候系统响应机制具有重要意义。

受全球古地理和古气候控制，东亚地区晚中生代陆相沉积主要发育在我国大陆。近年来在我国松辽盆地大陆钻探科学工程，获取了白垩纪连续的以湖泊沉积为主的陆相沉积记录。建立了磁性地层学、锆石U-Pb年代学、天文地层学、生物地层学等的晚白垩世陆相年代地层标准（Li et al., 2011; Wu et al., 2013），并为研究白垩纪大陆环境与气候演化规律，以及与生物演化更替之间的关系提供了绝无仅有的研究材料，部分成果已发表在国际学科重要期刊上（Hu et al., 2015; Wang et al., 2016），引领了国际白垩纪陆地气候研究。我国不少地史时期的地质记录可以构建跨区域性的古气候断面。认识和发挥我国地学资源优势的意义显得非常重要。通过大陆科学钻探获得保存良好、高分辨率的沉积记录，提高地质年代学的精度，着重温室地球时期深时气候模型的建立，显然是近阶段研究中的重中之重（王成善等，2017）。

2.9 生物沉积学

生物沉积学是研究有关生物参与和诱导的沉积作用过程的一门沉积学与生物学的交叉学科，其核心内容是揭示生命参与地球环境中物理和化学沉积的过程，反映现生和深时生命-环境的相互作用与协同演化，是当前人们广泛关注的交叉性热点课题（Chen et al., 2017）。这一领域的研究发展迅速。我国学者也运用多学科手段开展了我国地史时期典型实例的重点解剖和综合研究，取得了可喜的进展。

我国华南从新元古代至中生代地层保存有丰富的生物沉积学记录，对我国微生物礁和后生生物礁的研究发现了5次由微生物主导向后生生物主导的沉积体系转换期（MMT）（Chen et al., 2019）。第一次MMT发生在埃迪卡拉纪晚期，在此之前地球生态系统以微生物席为主，此后多细胞生物开始出现。第二次MMT发生在寒武纪早-中期，与著名的"寒武纪底质革命"（Bottjer et al., 2000）紧密相关。研究表明从寒武纪后生生物礁与微生物岩/礁的交替出现到奥陶纪后生生物礁的繁盛，标志着显生宙第一次由微生物主导向后生生物主导的沉积体系的转变（齐永安等，2014; Yan et al., 2017）。其他3次MMT分别发生在志留纪早期、晚泥盆世法门阶和三叠纪早-中期，分别与奥陶纪-志留纪、弗拉斯-法门阶、二叠纪-三叠纪之交的生物大灭绝相关。我国学者还发现，在华南地区，中三叠世主要的造礁后生生物Tubipytes在奥伦尼克期地层中已经大量出现（Song et al., 2011）。因此，微生物礁在早三叠世仍占主导地位。最近，在云南东部关岭组第二段碳酸盐地层中发现发育良好的叠层石，其内部发育与现代蓝细菌无异的丝状体、管状体等，表明微生物在大灭绝之后迅速繁盛，一直持续到安尼期早期（Luo et al., 2014）。这些MMT关键时期与全球气候、环境的剧烈演变存在广泛联系，因此，MMT的生物沉积记录是研究地质历史时期生物与环境协同演化的绝佳对象。

我国生物沉积学研究还处在起步阶段，正确理解（微）生物参与各种环境的沉积过程以及可能的控制因素是生物沉积学亟待解决的重要科学问题之一。特别是由微生物主导向由后生生物主导的沉积体系转折期的生物沉积特征和驱动机制的研究既是国际生物沉积学的研究热点，也是基于我国沉积记录、可望在沉积学领域取得理论创新、突破的重要方向之一。目前，微生物碳酸盐岩研究形成了国际沉积学研究的一个新热点。我国大规模微生物碳酸盐岩多发育于下古生界—前寒武系，微生物碳酸盐岩研究不断受到了广泛的重视，将会不断取得新的进展（Chen and Benton, 2012）。

3 中国沉积学未来发展展望

近十多年来，中国沉积学取得了一系列创新进展，不断地与国际沉积学研究前沿接轨，并在一些特色的领域做出了世界瞩目的研究成果。同时，我国沉积学研究围绕国家需求做出了重大的贡献。然而，正如上文所述，许多研究偏重跟踪、应用，而涉及基础性的科学问题缺乏系统性的研究，原创性成果少。无疑，中国沉积学的未来发展涉及的核心科学问题，既要能反映学科的研究前沿，又要能满足国家和人类社会文明发展的需求。

3.1 温室陆地气候与古地貌重建

全球气候是否会从两极有冰盖的冰室气候状态，进入两极无冰盖的温室气候状态，是大众与科学界共同关注的问题,人类文明的发展也迫切要求对这种变化的趋势及其环境效应有更加深入的了解。探索地质历史中,尤其是前第四纪温室条件下气候、环境的变化规律、机制及其对生物圈的影响等,意义重大。侏罗纪、白垩纪和古近纪出现过典型的温室气候,发生过多次快速增温事件(Jenkyns,2010;Godet,2013;Foster et al., 2018),与现今人类活动造成的全球变暖极其相似。因此,深入剖析这些古环境事件,包括大洋缺氧、酸化、生物灭绝更替、碳酸盐岩台地淹没、大陆风化作用和水文循环增强等的触发机制和生物环境响应过程等,是解开温室条件下地球系统运行模式的最为宝贵的钥匙。

当前深时古气候研究的重大发现大多来自海相沉积记录,有关深时陆地气候的研究仍相对滞后。但是,陆地作为人类生存繁衍的场所,深入了解其气候系统具有迫切需求,同时也是跨越海陆界线,从全球尺度预测未来气候变化的重要基础。我国学者对松辽盆地晚白垩世古土壤碳酸盐的碳、氧同位素和介形虫化石等的研究发现,东亚地区陆地气候与全球海洋气候变化具有一致性,表现为长时间尺度上的逐渐降温和短时间尺度上对快速气候变化事件的响应(Gao et al., 2015)。然而,陆地气候对全球气候变化的响应机制仍不明确。因此,对地质历史时期的古地貌重建,探讨古地貌的变化与陆地古气候的反馈机制,对于深刻理解大陆动力学和气候-环境演化十分重要,将是未来研究的重点。松辽盆地大陆钻探项目的执行,获取了白垩纪连续的湖盆沉积记录,为研究白垩纪陆地气候提供了绝无仅有的研究材料(Wang et al., 2016)。同时,我国有望从陆地植物和古土壤等记录中获取晚中生代以来连续的、精确的陆地气候参数,包括古温度、古CO_2含量、降水量等,可开展快速气候变化的陆地响应等研究,潜力巨大。而且,晚中生代陆相沉积还是我国重要的油气勘探层位。因此,在探究陆相生油与温室气候的关系、陆地生物群与陆地气候的关系方面,中国都具有明显的地域优势。

3.2 重大地质转折期的沉积过程、生物与地球化学响应

地质历史中发生过多次生物更替和环境演化事件(如微生物-后生生物沉积体系转换期、侏罗-白垩纪大洋缺氧事件、古新世-始新世之交的极热事件等),导致生态系统和全球古环境发生了重大转折。这些快速演化事件的强度大、速率快,处在地球系统演化的极端状态,是研究深时全球变化的重要窗口,其沉积记录是研究地质历史时期生物与环境协同演化的绝佳对象。当前,"以古鉴今"的思想愈发深刻地影响着有关全球变化的研究。人类对未来世界的探索和预测,需要在认识重大地质转折期的极端事件,特别是能与人类时间尺度类比的气候、环境事件的基础上开展系统工作。显然,探索重大地质转折期的沉积过程及生物与地球化学响应是解析地球深时环境演变的关键问题,也是探索未来宜居星球演化的重要参照。

我国华南、华北和西北地区广泛出露地质历史时期的微生物成因碳酸盐岩和生物礁(Yuan et al., 2011;Xiao et al., 2014;Guan et al., 2017;Chen et al., 2019),南海还发育有许多近赤道带的后生生物礁和微生物岩,这为开展"微生物-后生生物沉积体系转换期"的研究提供了得天独厚的素材。此外,我国共有10个古生代系或阶一级的金钉子剖面,这些全球生物地层标准剖面为我国沉积学者开展"微生物-后生生物沉积体系转换期"等的极端生物、环境事件的深入研究提供了良好的地质年代约束。我国西藏出露有古生代以来的连续海相沉积,尤其是早侏罗世-古近纪的浅海碳酸盐岩是全球唯一的新特提斯洋东段南缘古环境演化的记录,为开展中新生代重大地质转折期研究提供了宝贵资料,对深化全球中新生代古环境演化认识具有十分重要的意义。另外,我国新疆南部还保存有中新生代副特提斯海沉积,对开展区域海洋条件和气候-环境演化研究也意义重大。

3.3 源-汇系统

从造山带的剥蚀区形成的沉积物经搬运通道输送至深海区沉积下来的整个过程被称为"源-汇系统"(Sømme et al., 2013),其动力学过程研究是当前国际地球科学领域的前沿和热点(林畅松等,2015)。一般来说,研究地球表层系统演化历史的一个重要途径是应用沉积学理论和方法来解译沉积记录。源-汇系统的重要组成部分,如河流和三角洲体系等往往是人口稠密的地区,因而对源-汇系统的研究显然对认识

人类生存环境、全球水循环及生态系统等具有重要意义。此外，化石能源，包括石油、天然气、煤炭等都赋存于沉积盆地中，其形成、运移、储藏均与源-汇过程息息相关。随着亚洲古地理格局的演变（新特提斯洋消亡、青藏高原隆升、东亚边缘海盆地打开等），新生代以来东亚地形发生了重大转变，这对亚洲大陆的沉积物源-汇系统造成了巨大影响，也引起了全球气候的巨大变化。这些构造和气候变化信息经由源-汇系统得以全部记录在边缘海盆地沉积中，对这些沉积物的分析研究是理解东亚大陆构造及气候演化历史的关键。因此，对从造山带到边缘海盆地的源-汇系统的解剖对地球科学理论的发展和人类生存条件的改善都具有极其重要的意义和价值。

西太平洋边缘海发育有全球最为典型的大型流域沉积物源-汇系统，也是全球物质交换最为活跃的边缘海系统。特别是以珠江、红河等河流体系和海洋多尺度洋流系统相互作用构成的南海封闭型源-汇系统，和以长江为代表的世界大河体系与山溪小河体系相互作用为特征的东海开放型源-汇系统，其物源区构造和气候背景、沉积物搬运路径和过程的控制因素、盆地的洋流系统、深海沉积动力学过程以及海盆地质演化时期的沉积源-汇格局等，不仅与青藏高原和西太平洋边缘海宏观地质演化有关，还具有大型流域源-汇系统的独特性，突显了西太平洋边缘海盆地作为开展源-汇过程及环境变迁研究的理想场所的重要性。从青藏高原至亚洲边缘海盆地构成了全球最大的源-汇系统，这一复杂而又独特的源-汇系统的主体位于我国境内，我国学者长期以来在青藏高原隆升和剥蚀历史、长江和黄河的形成演化以及南海和东海演化等方面积累了大量研究成果，为开展从青藏高原至东亚边缘海盆地的源-汇系统研究奠定了重要基础，形成了我国重要的研究优势。

3.4 前寒武纪超大陆演化、早期地球环境和生命

前寒武纪发生过一系列重大地质事件，包括超大陆（Nuna、Rodinia）的聚合与裂解、大氧化事件、"雪球地球"事件、海洋化学转变和早期生命起源与演化等，这些事件是过去几十年间前寒武纪沉积学研究的焦点。越来越多的证据表明前寒武纪地球海洋和大气中氧含量的演化历史十分复杂（Jin et al., 2016；Li et al., 2017；Stolper and Keller, 2018），而超大陆裂解与聚合可能是这一时期地球氧化事件的重要诱发因素（Müller et al., 2005；Campbell and Allen, 2008）。同时，超大陆裂解还会提高海洋生产力，增强生物光合作用，以释放出更多氧气至大气中。因此，对前寒武纪沉积学的研究对揭示大陆演化、大气圈、海洋和地球早期生命之间的相互作用意义重大。

华南、华北和塔里木地区广泛出露元古宙至早寒武世沉积，我国学者已开展过大量该时期古地理学、地层学、古生物学和地质年代学等方面的基础工作（Qiao and Wang, 2014；Su, 2016）。近年来，在湖北神农架地区（Li et al., 2013）和安徽淮南地区（Tang et al., 2013）的发现使得我国在前寒武纪沉积学研究中能够获得连续的地层记录，这将为超大陆的聚合与裂解过程及相关沉积盆地的演化等科学问题的解答提供关键线索。此外，还将为生物与环境的协同演化，包括早期生命起源及海洋和大气环境背景、中元古代真核生物的起源和演化及其与环境的响应关系、微生物对白云岩和含铁建造等的形成过程的影响等的深入研究提供重要的支撑。我国前寒武纪能源和矿产资源丰富，特别是华南新元古代页岩气、华北和塔里木元古宙油气资源潜力巨大，具有极高的经济价值，前寒武纪沉积学研究也将为相关油气勘探和预测提供重要的理论基础。

近十多年来，我国沉积学发展方兴未艾，在多个研究方向和领域与国际相关研究同步发展，成果颇多，在不少领域已迈入国际前沿，并在具有中国沉积和盆地特色的一些方向和领域上取得了突出进展。但我国的沉积学研究仍然以跟踪研究占多，而原创性的、国际引领性的研究较少。我们需一如既往，不断提高我国沉积学的核心竞争力。我们相信，中国沉积学必将迎来愈发蓬勃的发展。

致谢：本文是在由中国沉积学会组织完成的中国沉积学发展战略研讨成果的基础上编写的，这项研究成果倾注了众多中国沉积学者、国际沉积学会同行以及相关学科领域专家的努力（其中不同部分的主

要执笔人有李忠、关平、邵龙义、朱筱敏、解习农、侯明才、颜佳新、陈中强、王剑、邹才能、朱如凯、陈代钊、高抒、谢树成、王璞珺、陈曦、胡修棉、刘志飞、李超等，因篇幅所限不一一列出，详见《中国沉积学发展战略》）；在本文编写过程中还得到了陈曦副教授、张曼莉博士等的帮助，在此一并表示衷心的感谢。

参 考 文 献

蔡希源，李思田. 2003. 陆相盆地高精度层序地层学——基础理论篇. 北京：地质出版社
陈洪德，侯明才，陈安清，时志强，邢凤存，黄可可，刘欣春. 2017. 中国古地理学研究进展与关键科学问题. 沉积学报，35(5)：888-901
邓宏文. 2009. 高分辨率层序地层学应用中的问题探析. 古地理学报，11(5)：471-480
冯增昭，鲍志东，吴胜和，等. 1977. 中国南方早中三叠世岩相古地理. 北京：石油工业出版社
高抒. 2014. "江苏沿海开发的资源环境生态基础"专栏前言. 南京大学学报(自然科学)，50(5)：535-537
高振西，熊永光，高平. 1934. 中国北部震旦纪地层. 中国地质学会会志，13：243-288
关士聪. 1984. 中国海陆变迁海域沉积相与油气. 北京：科学出版社
贾承造. 2009. 环青藏高原巨型盆山体系构造与塔里木盆地油气分布规律. 大地构造与成矿学，33(1)：1-9
金之钧，蔡立国. 2007. 中国海相层系油气地质理论的继承与创新. 地质学报，81(8)：1017-1024
李德生. 1992. 李德生石油地质论文集. 北京：石油工业出版社
李江海，穆剑. 1999. 我国境内格林威尔期造山带的存在及其对中元古代末期超大陆再造的制约. 地质科学，34(3)：259-272
李铨，冷坚. 1991. 神农架上前寒武系. 天津：天津科学技术出版社
李守军，田臣龙，徐凤琳，陈茹，殷天涛，赵秀丽. 2014. 山东二叠系石盒子组孢粉特征及古气候意义. 地质论评，60(4)：765-770
李思田. 1988. 断陷盆地分析与煤聚积规律. 北京：地质出版社
李思田. 2015. 沉积盆地动力学研究的进展、发展趋向与面临的挑战. 地学前缘，22(1)：1-8
李思田，王华，路凤香. 1999. 盆地动力学-基本思路与若干研究方法. 武汉：中国地质大学出版社
李思田，解习农，王华，焦养泉，任建业，庄新国，陆永潮. 2004. 沉积盆地分析基础与应用. 北京：高等教育出版社
李忠，彭守涛. 2013. 天山南北盆中-新生界碎屑锆石U-Pb年代学记录、物源体系分析与陆内盆山演化. 岩石学报，29(3)：739-755
林畅松. 2016. 沉积盆地分析原理与应用. 北京：石油工业出版社
林畅松. 2019. 盆地沉积动力学：研究现状与未来发展趋势. 石油与天然气地质，40(4)：685-700
林畅松，潘元林，肖建新，孔凡仙，刘景彦，郑和荣. 2000. "构造坡折带"——断陷盆地层序分析和油气预测的重要概念. 地球科学-中国地质大学学报，25(3)：260-266
林畅松，杨海军，刘景彦，蔡振中，彭莉，阳孝法，杨永恒. 2009. 塔里木盆地古生代中央隆起带古构造地貌及其对沉积相发育分布的制约. 中国科学D辑：地球科学，39(3)：306-316
林畅松，夏庆龙，施和生，周心怀. 2015. 地貌演化、源-汇过程与盆地分析. 地学前缘，22(1)：9-20
刘宝珺. 2001. 中国沉积学的回顾和展望. 矿物岩石，21(3)：1-7
刘宝珺，曾允孚. 1985. 岩相古地理基础和工作方法. 北京：地质出版社
刘宝珺，许效松. 1994. 中国南方岩相古地理图集. 北京：科学出版社
刘东生，安芷生，文启忠，卢演俦，韩家懋，王俊达，刁桂仪. 1978. 中国黄土的地质环境. 科学通报，23(1)：1-9
刘鸿允. 1959. 中国古地理图. 2版. 北京：科学出版社
马永生，陈洪德，王国力. 2009. 中国南方构造-层序岩相古地理图集. 北京：科学出版社
马永生，蔡勋育，赵培荣. 2011. 深层、超深层碳酸盐岩油气储层形成机理研究综述. 地学前缘，18(4)：181-192
潘元林，李思田. 2004. 大型陆相断陷盆地层序地层与隐蔽油气藏研究：以济阳坳陷为例. 北京：石油工业出版社
齐永安，王艳鹏，代明月，李姐. 2014. 豫西登封寒武系第三统张夏组凝块石灰岩及其控制因素. 微体古生物学报，31(3)：243-255
任建业，庞雄，于鹏，雷超，罗盼. 2018. 南海北部陆缘深水-超深水盆地成因机制分析. 地球物理学报，61(12)：4901-4920
邵龙义，董大啸，李明培，王海生，王东东，鲁静，郑明泉，程爱国. 2014. 华北石炭-二叠纪层序-古地理及聚煤规律. 煤炭学报，39(8)：1725-1734
孙枢. 2005. 中国沉积学的今后发展：若干思考与建议. 地学前缘，12(2)：3-10
孙枢，王成善. 2009. "深时"(Deep Time)研究与沉积学. 沉积学报，27(5)：792-810
孙枢，王铁冠. 2016. 中国东部中—新元古界地质学与油气资源. 北京：科学出版社
田在艺，张庆春. 1996. 中国含油气沉积盆地论. 北京：石油工业出版社
王成善，李祥辉. 2003. 沉积盆地分析原理与方法. 北京：高等教育出版社
王成善，陈洪德，寿建峰. 1998. 中国南方海相二叠系层序地层与油气勘探. 成都：四川科学技术出版社
王成善，郑和荣，冉波，刘本培，李祥辉，李亚林，孙红军，陈建平，胡修棉. 2010. 活动古地理重建的实践与思考——以青藏特提斯为例. 沉积学报，28(5)：849-860
王成善，冯志强，王璞珺. 2016. 白垩纪松辽盆地松科1井大陆科学钻探工程. 北京：科学出版社
王成善，王天天，陈曦，高远，张来明. 2017. 深时古气候对未来气候变化的启示. 地学前缘，24(1)：1-17

王鸿祯. 1985. 中国古地理图集. 北京：地图出版社

王剑, 江新胜, 卓皆文, 崔晓庄, 江卓斐, 魏亚楠, 蔡娟娟, 廖忠礼. 2019. 华南新元古代裂谷盆地演化与岩相古地理(附图集). 北京：科学出版社

王铁冠, 韩克猷. 2011. 论中—新元古界的原生油气资源. 石油学报, 32(1): 1-7

吴崇筠, 薛叔浩. 1992. 中国含油气盆地沉积学. 北京：石油工业出版社

解习农, 任建业. 2013. 沉积盆地分析基础. 武汉：中国地质大学出版社

徐长贵, 杜晓峰, 徐伟, 赵梦. 2017. 沉积盆地"源-汇"系统研究新进展. 石油与天然气地质, 38(1): 1-11

许靖华, 何起祥. 1980. 薄壳板块构造模式与冲撞型造山运动. 中国科学, (11): 1081-1089

许志琴, 李廷栋, 杨经绥, 嵇少丞, 王宗起, 张泽明. 2008. 大陆动力学的过去、现在和未来——理论与应用. 岩石学报, 27(7): 1433-1444

业治铮, 何起祥, 张明书, 韩春瑞, 李浩, 吴健政, 鞠连军. 1985. 西沙石岛晚更新世风成生物砂屑灰岩的沉积构造和相模式. 沉积学报, 3(1): 1-15

叶连俊. 1942. 近世沉积学之领域及其演进——纪念朱森教授. 地质论评, 7(6): 299-312

叶连俊. 1989. 中国磷块岩. 北京：科学出版社

袁选俊, 林森虎, 刘群, 姚泾利, 王岚, 郭浩, 邓秀芹, 成大伟. 2015. 湖盆细粒沉积特征与富有机质页岩分布模式——以鄂尔多斯盆地延长组长7油层组为例. 石油勘探与开发, 42(1): 34-43

张国伟, 张本仁, 袁学诚, 肖庆辉. 2001. 秦岭造山带与大陆动力学. 北京：科学出版社

张克信, 王国灿, 曹凯, 刘超, 向树元, 洪汉烈, 寇晓虎, 徐亚东, 陈奋宁, 孟艳宁, 陈锐明. 2008. 青藏高原新生代主要隆升事件：沉积响应与热年代学记录. 中国科学：地球科学, 38(12): 1575-1588

张启锐, 兰中伍. 2016. 南华系、莲沱组年龄问题的讨论. 地层学杂志, 40(3): 297-301

赵文智, 沈安江, 胡素云, 张宝民, 潘文庆, 周进高, 汪泽成. 2012. 中国碳酸盐岩储集层大型化发育的地质条件与分布特征. 石油勘探与开发, 39(1): 1-12

周传明. 2016. 扬子区新元古代前震旦纪地层对比. 地层学杂志, 40(2): 120-135

朱筱敏, 李顺利, 潘荣, 谈明轩, 陈贺贺, 王星星, 陈锋, 张梦瑜, 侯冰洁, 董艳蕾. 2016. 沉积学研究热点与进展：第32届国际沉积学会议综述. 古地理学报, 18(5): 699-716

朱筱敏, 葛家旺, 赵宏超, 袁立忠, 刘军. 2017. 陆架边缘三角洲研究进展及实例分析. 沉积学报, 35(5): 945-957

邹才能, 邱振. 2021. 中国非常规油气沉积学新进展. 沉积学报, 39(1): 1-9

邹才能, 赵文智, 张兴阳, 罗平, 王岚, 刘柳红, 薛叔浩, 袁选俊, 朱如凯, 陶士振. 2008. 大型敞流坳陷湖盆浅水三角洲与湖盆中心砂体的形成与分布. 地质学报, 82(6): 813-825

邹才能, 杜金虎, 徐春春, 汪泽成, 张宝民, 魏国齐, 王铜山, 姚根顺, 邓胜徽, 刘静江, 周慧, 徐安娜, 杨智, 姜华, 谷志东. 2014. 四川盆地震旦系—寒武系特大型气田形成分布、资源潜力及勘探发现. 石油勘探与开发, 41(3): 278-293

An Z H, Jiang G Q, Tong J N, Tian L, Ye Q, Song H Y, Song H J. 2015. Stratigraphic position of the Ediacaran Miaohe biota and its constrains on the age of the upper Doushantuo $\delta^{13}C$ anomaly in the Yangtze Gorges area, South China. Precambrian Research, 271: 243-253

Bottjer D J, Hagadorn J W, Dornbos S Q. 2000. The Cambrian substrate revolution. GSA Today, 10(9): 1-7

Campbell I H, Allen C M. 2008. Formation of supercontinents linked to increases in atmospheric oxygen. Nature Geoscience, 1(8): 554-558

Chen S, Steel R, Wang H, Zhao R, Olariu C. 2020. Clinoform growth and sediment flux into Late Cenozoic Qiongdongnan shelf margin, South China Sea. Basin Research, 32(2): 302-319

Chen X, Idakieva V, Stoykova K, Liang H M, Yao H W, Wang C S. 2017. Ammonite biostratigraphy and organic carbon isotope chemostratigraphy of the early Aptian oceanic anoxic event (OAE 1a) in the Tethyan Himalaya of southern Tibet. Palaeogeography, Palaeoclimatology, Palaeoecology, 485: 531-542

Chen Z Q, Benton M J. 2012. The timing and pattern of biotic recovery following the end-Permian mass extinction. Nature Geoscience, 5(6): 375-383

Chen Z Q, Hu X M, Montanez I P, Ogg J G. 2019. Sedimentology as a key to understanding Earth and life processes. Earth-Science Reviews, 189: 1-5

Coleman M, Hodges K. 1995. Evidence for Tibetan Plateau uplift before 14 Myr ago from a new minimum age for east-west extension. Nature, 374(6517): 49-52

Daigle H, Worthington L L, Gulick S P S, van Avendonk H J A. 2017. Rapid sedimentation and overpressure in shallow sediments of the Bering Trough, offshore southern Alaska. Journal of Geophysical Research: Solid Earth, 122(4): 2457-2477

Dong S L, Li Z, Jiang L. 2016. The early Paleozoic sedimentary-tectonic evolution of the circum-Mangar areas, Tarim Block, NW China: constraints from integrated detrital records. Tectonophysics, 682: 17-34

Foster G L, Hull P, Lunt D J, Zachos J C. 2018. Placing our current "hyperthermal" in the context of rapid climate change in our geological past. Philosophical Transactions of the Royal Society A: Mathematical, Physical and Engineering Sciences, 376(2130): 20170086

Gao S. 2019. Geomorphology and sedimentology of tidal flats. In: Perillo G M E, Wolanski E, Cahoon D R, Hopkinson C S (eds). Coastal Wetlands: An Integrated Ecosystem Approach, 2nd ed. Amsterdam: Elsevier

Gao S, Liu Y L, Yang Y, Liu P J, Zhang Y Z, Wang Y P. 2015. Evolution status of the distal mud deposit associated with the Pearl River, northern South China Sea continental shelf. Journal of Asian Earth Sciences, 114: 562-573

Godet A. 2013. Drowning unconformities: palaeoenvironmental significance and involvement of global processes. Sedimentary Geology, 293: 45-66

Gong C L, Peakall J, Wang Y M, Wells M G, Xu J. 2017. Flow processes and sedimentation in contourite channels on the northwestern South China Sea margin: a joint 3D seismic and oceanographic perspective. Marine Geology, 393: 176-193

Guan C G, Wang W, Zhou C M, Muscente A D, Wan B, Chen X, Yuan X L, Chen Z, Ouyang Q. 2017. Controls on fossil pyritization: redox conditions, sedimentary organic matter content, and *Chuaria* preservation in the Ediacaran Lantian biota. Palaeogeography, Palaeoclimatology, Palaeoecology, 474: 26-35

Han Z, Hu X M, Kemp D B, Li J. 2018. Carbonate-platform response to the Toarcian oceanic anoxic event in the southern hemisphere: implications for climatic change and biotic platform demise. Earth and Planetary Science Letters, 489: 59-71

Hu J F, Peng P A, Liu M Y, Xi D P, Song J Z, Wan X Q, Wang C S. 2015. Seawater incursion events in a Cretaceous paleo-lake revealed by specific marine biological markers. Scientific Reports, 5: 9508

Isaacson P E, Montañez I P. 2013. A "sedimentary record" of opportunities. The Sedimentary Record, 11(1): 2-9

Jenkyns H C. 2010. Geochemistry of oceanic anoxic events. Geochemistry, Geophysics, Geosystems, 11(3): Q03004

Jia J J, Gao J H, Cai T L, Li Y, Yang Y, Wang Y P, Xia X M, Li J, Wang A J, Gao S. 2018. Sediment accumulation and retention of the Changjiang (Yangtze River) subaqueous delta and its distal muds over the last century. Marine Geology, 401: 2-16

Jiang J, Shi H S, Lin C S, Zhang Z T, Wei A, Zhang B, Shu L F, Tian H X, Tao Z, Liu H Y. 2017. Sequence architecture and depositional evolution of the Late Miocene to Quaternary northeastern shelf margin of the South China Sea. Marine and Petroleum Geology, 81: 79-97

Jiang T, Cao L C, Xie X N, Wang Z F, Li X S, Zhang Y Z, Zhang D J, Sun H. 2015. Insights from heavy minerals and zircon U-Pb ages into the middle Miocene-Pliocene provenance evolution of the Yinggehai Basin, northwestern South China Sea. Sedimentary Geology, 327: 32-42

Jin C S, Li C, Algeo T J, Planavsky N J, Cui H, Yang X L, Zhao Y L, Zhang X L, Xie S C. 2016. A highly redox-heterogeneous ocean in South China during the early Cambrian (~529-514 Ma): Implications for biota-environment co-evolution. Earth and Planetary Science Letters, 441: 38-51

Lei C, Ren J Y. 2016. Hyper-extended rift systems in the Xisha Trough, northwestern South China Sea: implications for extreme crustal thinning ahead of a propagating ocean. Marine and Petroleum Geology, 77: 846-864

Li F, Yan J X, Algeo T, Wu X. 2013. Paleoceanographic conditions following the end-Permian mass extinction recorded by giant ooids (Moyang, South China). Global and Planetary Change, 105: 102-120

Li F, Yan J X, Burne R V, Chen Z Q, Algeo T J, Zhang W, Tian L, Gan Y L, Liu K, Xie S C. 2017. Paleo-seawater REE compositions and microbial signatures preserved in laminae of Lower Triassic ooids. Palaeogeography, Palaeoclimatology, Palaeoecology, 486: 96-107

Li J G, Batten D J, Zhang Y Y. 2011. Palynological record from a composite core through Late Cretaceous-Early Paleocene deposits in the Songliao Basin, Northeast China and its biostratigraphic implications. Cretaceous Research, 32(1): 1-12

Lin C S, Liu J Y, Cai S X, Zhang Y M, Lu M, Li J. 2001. Depositional architecture and developing settings of large-scale incised valley and submarine gravity flow systems in the Yinggehai and Qiongdongnan basins, South China Sea. Chinese Science Bulletin, 46(8): 690-693

Lin C S, Li H, Liu J Y. 2012. Major unconformities, tectonostratigraphic framework, and evolution of the superimposed Tarim Basin, Northwest China. Journal of Earth Science, 23(4): 395-407

Lin C S, Jiang J, Shi H S, Zhang Z T, Liu J Y, Qin C G, Li H, Ran H J, Wei A, Tian H X, Xing Z C, Yao Q Y. 2018. Sequence architecture and depositional evolution of the northern continental slope of the South China Sea: responses to tectonic processes and changes in sea level. Basin Research, 30(S1): 568-595

Liu C, Jarochowska E, Du Y S, Vachard D, Munnecke A. 2017. Stratigraphical and $\delta^{13}C$ records of Permo-Carboniferous platform carbonates, South China: responses to late Paleozoic icehouse climate and icehouse-greenhouse transition. Palaeogeography, Palaeoclimatology, Palaeoecology, 474: 113-129

Liu J, Saito Y, Kong X H, Wang H, Xiang L H, Wen C, Nakashima R. 2010. Sedimentary record of environmental evolution off the Yangtze River estuary, East China Sea, during the last ~13,000 years, with special reference to the influence of the Yellow River on the Yangtze River delta during the last 600 years. Quaternary Science Reviews, 29(17-18): 2424-2438

Liu X J, Gao S, Wang Y P. 2011. Modeling profile shape evolution for accreting tidal flats composed of mud and sand: a case study of the central Jiangsu coast, China. Continental Shelf Research, 31(16): 1750-1760

Liu Y L, Gao S, Wang Y P, Yang Y, Long J P, Zhang Y Z, Wu X D. 2014. Distal mud deposits associated with the Pearl River over the northwestern continental shelf of the South China Sea. Marine Geology, 347: 43-57

Lüdmann T, Wong H K. 1999. Neotectonic regime on the passive continental margin of the northern South China Sea. Tectonophysics, 311(1-4): 113-138

Luo M, Chen Z Q, Zhao L S, Kershaw S, Huang J Y, Wu L L, Yang H, Fang Y H, Huang Y G, Zhang Q Y, Hu S X, Zhou C Y, Wen W, Jia Z H. 2014. Early Middle Triassic stromatolites from the Luoping area, Yunnan Province, Southwest China: geobiologic features and environmental implications. Palaeogeography, Palaeoclimatology, Palaeoecology, 412: 124-140

Mackay A W, Seddon A W R, Leng M J, Heumann G, Morley D W, Piotrowska N, Rioual P, Roberts S, Swann G E A. 2017. Holocene carbon dynamics at the forest-steppe ecotone of southern Siberia. Global Change Biology, 23(5): 1942-1960

Milliman J D, Farnsworth K L. 2011. River discharge to the coastal ocean: a global synthesis. Cambridge: Cambridge University Press, 384

Müller S G, Krapež B, Barley M E, Fletcher I R. 2005. Giant iron-ore deposits of the Hamersley Province related to the breakup of Paleoproterozoic Australia: new insights from *in situ* SHRIMP dating of baddeleyite from mafic intrusions. Geology, 33(7): 577-580

Parrish J T, Soreghan G S. 2013. Sedimentary geology and the future of paleoclimate studies. The Sedimentary Record, 11(2): 4-10

Qiao X F, Wang Y B. 2014. Discussions on the lower boundary age of the Mesoproterozoic and basin tectonic evolution of the Mesoproterozoic in North China Craton. Acta Geologica Sinica, 88(9): 1623-1637

Reading H G. 1996. Sedimentary Environments: Processes, Facies and Stratigraphy, 3rd ed. Oxford: Blackwell Publishing Ltd

Shao L, Cao L C, Pang X, Jiang T, Qiao P J, Zhao M. 2016. Detrital zircon provenance of the Paleogene syn-rift sediments in the northern South China Sea. Geochemistry, Geophysics, Geosystems, 17(2): 255-269

Shi B W, Cooper J R, Pratolongo P D, Gao S, Bouma T J, Li G C, Li C Y, Yang S L, Wang Y P. 2017. Erosion and accretion on a mudflat: the importance of very shallow-water effects. Journal of Geophysical Research: Oceans, 122(12): 9476-9499

Sømme T O, Jackson C A L, Vaksdal M. 2013. Source-to-sink analysis of ancient sedimentary systems using a subsurface case study from the Møre-Trøndelag area of southern Norway: Part 1-depositional setting and fan evolution. Basin Research, 25(5): 489-511

Song H J, Wignall P B, Chen Z Q, Tong J N, Bond D P G, Lai X L, Zhao X M, Jiang H S, Yan C B, Niu Z J, Chen J, Yang H, Wang Y B. 2011. Recovery tempo and pattern of marine ecosystems after the end-Permian mass extinction. Geology, 39(8): 739-742

Soreghan G S, Bralower T J, Chandler M A, et al. 2005. Geosystems: probing Earth's deep-time climate and linked systems. Norman: University of Oklahoma Printing Service

Stolper D A, Keller C B. 2018. A record of deep-ocean dissolved O_2 from the oxidation state of iron in submarine basalts. Nature, 553(7688): 323-327

Su W B. 2016. Revision of the Mesoproterozoic chronostratigraphic subdivision both of North China and Yangtze Cratons and the relevant issues. Earth Science Frontiers, 23(6): 156-185

Tang Q, Pang K, Xiao S H, Yuan X L, Ou Z J, Wan B. 2013. Organic-walled microfossils from the early Neoproterozoic Liulaobei Formation in the Huainan region of North China and their biostratigraphic significance. Precambrian Research, 236: 157-181

Tapponnier P, Meyer B, Avouac J P, Peltzer G, Gaudemer Y, Guo S M, Xiang H F, Yin K L, Chen Z T, Cai S H, Dai H G. 1990. Active thrusting and folding in the Qilian Shan, and decoupling between upper crust and mantle in northeastern Tibet. Earth and Planetary Science Letters, 97(3-4): 382-383, 387-403

Tian H X, Lin C S, Zhang Z T, Li H, Zhang B, Zhang M L, Liu H Y, Jiang J. 2021. Depositional architecture, evolution and controlling factors of the Miocene submarine canyon system in the Pearl River Mouth Basin, northern South China Sea. Marine and Petroleum Geology, 104990

Wang H J, Yang Z S, Wang Y, Saito Y, Liu J P. 2008. Reconstruction of sediment flux from the Changjiang (Yangtze River) to the sea since the 1860s. Journal of Hydrology, 349(3-4): 318-332

Wang J, Li Z X. 2003. History of Neoproterozoic rift basins in South China: Implications for Rodinia break-up. Precambrian Research, 122(1-4): 141-158

Wang J, Pfefferkorn H W. 2013. The Carboniferous-Permian transition on the North China microcontinent-oceanic climate in the tropics. International Journal of Coal Geology, 119: 106-113

Wang J, Deng Q, Wang Z J, Qiu Y S, Duan T Z, Jiang X S, Yang Q X. 2013. New evidences for sedimentary attributes and timing of the "Macaoyuan conglomerates" on the northern margin of the Yangtze Block in southern China. Precambrian Research, 235: 58-70

Wang P J, Mattern F, Didenko N A, Zhu D F, Singer B, Sun X M. 2016. Tectonics and cycle system of the Cretaceous Songliao Basin: an inverted active continental margin basin. Earth-Science Reviews, 159: 82-102

Wang X L, Shu L S, Xing G F, Zhou J C, Tang M, Shu X J, Qi L, Hu Y H. 2012. Post-orogenic extension in the eastern part of the Jiangnan orogen: evidence from ca 800-760 Ma volcanic rocks. Precambrian Research, 222-223: 404-423

Wu H C, Zhang S H, Jiang G Q, Hinnov L, Yang T S, Li H Y, Wan X Q, Wang C S. 2013. Astrochronology of the Early Turonian-Early Campanian terrestrial succession in the Songliao Basin, northeastern China and its implication for long-period behavior of the Solar System. Palaeogeography, Palaeoclimatology, Palaeoecology, 385: 55-70

Wu H C, Zhang S H, Hinnov L A, Jiang G Q, Yang T S, Li H Y, Wan X Q, Wang C S. 2014. Cyclostratigraphy and orbital tuning of the terrestrial upper Santonian-Lower Danian in Songliao Basin, northeastern China. Earth and Planetary Science Letters, 407: 82-95

Wu W, Li Q, Yu J, Lin C S, Li D, Yang T. 2018. The central Canyon depositional patterns and filling process in east of Lingshui Depression, Qiongdongnan Basin, northern South China Sea. Geological Journal, 53(6): 3064-3081

Xiao S H, Shen B, Tang Q, Kaufman A J, Yuan X L, Li J H, Qian M P. 2014. Biostratigraphic and chemostratigraphic constraints on the age of early Neoproterozoic carbonate successions in North China. Precambrian Research, 246: 208-225

Xie X N, Ren J Y, Pang X, Lei C, Chen H. 2019. Stratigraphic architectures and associated unconformities of Pearl River Mouth Basin during rifting and lithospheric breakup of the South China Sea. Marine Geophysical Research, 40(2): 129-144

Yan Z, Liu J B, Ezaki Y, Adachi N, Du S X. 2017. Stacking patterns and growth models of multiscopic structures within Cambrian Series 3 thrombolites at the Jiulongshan section, Shandong Province, northern China. Palaeogeography, Palaeoclimatology, Palaeoecology, 474: 45-57

Yang L L, Ren J Y, McIntosh K, Pang X, Lei C, Zhao Y H. 2018. The structure and evolution of deepwater basins in the distal margin of the northern South China Sea and their implications for the formation of the continental margin. Marine and Petroleum Geology, 92: 234-254

Yang S Y, Youn J S. 2007. Geochemical compositions and provenance discrimination of the central South Yellow Sea sediments. Marine Geology, 243(1-4): 229-241

Yuan X L, Chen Z, Xiao S H, Zhou C M, Hua H. 2011. An early Ediacaran assemblage of macroscopic and morphologically differentiated eukaryotes. Nature, 470(7334): 390-393

Zhang M L, Lin C S, He M, Zhang Z T, Li H, Feng X, Tian H X, Liu H Y. 2019. Stratigraphic architecture, shelf-edge delta and constraints on the development of the Late Oligocene to Early Miocene continental margin prism, the Pearl River Mouth Basin, northern South China Sea. Marine Geology, 416: 105982

Zhang S H, Evans D A D, Li H Y, Wu H C, Jiang G Q, Dong J, Zhao Q L, Raub T D, Yang T S. 2013. Paleomagnetism of the late Cryogenian Nantuo Formation and paleogeographic implications for the South China Block. Journal of Asian Earth Sciences, 72: 164-177

Development Status and Trend of Sedimentology

WANG Cheng-shan, LIN Chang-song

China University of Geosciences (Beijing), Beijing 100083

Abstract: Sedimentology research in China has experienced over a hundred years of development from the early 20th to 21st centuries. Since the 21st century, important progresses have been made in the fields of studies on basin dynamics, sequence stratigraphy, source-sink systems, depositional systems and sedimentary facies, palaeogeography, Precambrian sedimentology, modern sedimentary environment, deep-time geology and sedimentology, and biologic sedimentology in China. The studies in the fields of deep-time paleoclimate and global climate changes, paleo-environment changes, and quaternary sedimentology are developing synchronically with international researches or have already stepped into the international research frontiers. At the same time, some research areas with unique Chinese characteristics have been formed. They include the terrestrial sequence stratigraphy, basin sedimentary filling dynamics, and petroliferous basin sedimentology, etc., in which a series of important progresses have been achieved and some great contributions to meet the national energy demands have been made. However, at present, sedimentological studies in China are still mainly concentrated on those for tracking the international research frontiers, with relatively limited original and leading researches. Sedimentological studies in China are constantly improving their core competitiveness. Especially, it is expected that some pioneering achievements at the international frontier level will be made in future in the research fields including the greenhouse terrestrial climate and paleogeomorphology reconstruction, sedimentary and biological processes during major geological transitions, source-sink systems, Precambrian supercontinent evolution and early earth environment.

Key words: Sedimentology study in China; development history; research status; future development trend

古地理学主要研究进展 *

郑秀娟[1]　杜远生[2]　朱筱敏[1]　刘招君[3]　胡　斌[4]　吴胜和[1]　邵龙义[5]　旷红伟[6]
罗静兰[7]　钟大康[1]　李　华[8]　何登发[9]　朱如凯[10]　鲍志东[1]

1. 中国石油大学（北京），北京 102249；2. 中国地质大学（武汉），武汉 430074；3. 吉林大学，长春 130061；4. 河南理工大学，焦作 454003；5. 中国矿业大学（北京），北京 100083；6. 中国地质科学院地质研究所，北京 100037；7. 西北大学，西安 710069；8. 长江大学，武汉 430100；9. 中国地质大学（北京），北京 100083；10. 中国石油勘探开发研究院，北京 100083

摘　要：本文从古地理学研究与发展的几个主要方面对中国古地理学近十年的进展进行了归纳总结，指出了存在的问题并提出未来的研究趋势和方向。认为中国古地理学在多个方面引领了古地理学的学科发展，包括：①在古地理学基础研究方面，多学科齐头并进，支撑了古地理学科的持续发展；②在古地理研究方法与成图方面，构造古地理、生物古地理中的遗迹学研究、小尺度古地理研究和大数据古地理研究有了长足进展；③在古地理学应用方面，从含油气盆地古地理、油页岩古地理及碎屑岩成岩作用多角度研究油气资源与油页岩，并在聚煤区古地理研究与煤炭资源预测及其他矿藏古地理研究与找矿方面都有新的突破。

关键词：古地理学　地震沉积学　遗迹学　小尺度古地理　大数据古地理　聚煤区古地理　油气资源

0　概　述

进入 21 世纪，古地理学成为系统研究沉积岩形成机理、恢复古地理的一门学科。全球深海钻探、板块构造学说的兴起发展，地球物理与地球化学等学科的技术进步，极大地促进了沉积古地理学的发展，是古地理学发展的直接推动力。这一时期，对河流、三角洲、深水沉积以及古地理研究方法技术等方面的研究得到加强，而同时古地理的有关研究成果在沉积矿产资源（油气资源）勘探开发中也得到较好应用。最大的特点是，与沉积学、古地理学交叉的分支学科大量出现并在理论上逐步完善，新概念的提出，新技术、新方法的开发应用以及在能源勘探开发等方面均取得了显著进展。

21 世纪的第二个十年，古地理学在基础研究、学科研究方法与成图、古地理学及矿产应用、学科交叉方面都取得了较大进步，可以说中国的古地理学在多个方面引领了古地理学学科的发展。由中国科学家主导组织创办的国际古地理学会议，吸引了来自全球五大洲的古地理学家和青年学者；在数字古地理方面，由中国科学家牵头发起的"深时数字地球国际大科学计划"吸引了多国科学家的参与。具体体现在：①在古地理学基础研究方面，地震沉积学、重力流、深水牵引流沉积、源-汇系统及元古宙臼齿碳酸盐岩研究等都取得了长足进展，支撑了古地理学科的持续发展；②在古地理研究方法与成图方面，比较突出的表现在构造古地理、生物古地理中的遗迹学研究、小尺度古地理研究和大数据古地理研究；③从古地理学的应用来看，在含油气盆地古地理研究与油气勘探、油页岩古地理研究与找矿、碎屑岩成岩作用与油气、聚煤区古地理研究与找矿及其他矿藏古地理研究与找矿方面都有新的突破，在能源与矿产的可持续发展与国家重大决策中发挥了举足轻重的作用。

在《古地理学报》得到越来越多学者重视与关注的同时，2012 年《古地理学报》（英文版）创刊，

---

* 原文"中国古地理学近十年主要进展"刊于《矿物岩石地球化学通报》2010 年第 40 卷第 1 期，本文略有修改。

为中国学者参与国际古地理学交流提供了较好的平台。自2013年开始举办的国际古地理学会议已连续召开了4届，全国古地理学及沉积学学术会议在这十年间依然是隔年召开。总体来看，古地理学已经发展成为了一个比较成熟的学科，在地质学理论研究及国家矿产能源可持续发展方面起着越来越重要的作用。

1 古地理学基础研究

1.1 地震沉积学研究进展

最近十多年来，地震沉积学研究受到高度重视，其成果主要包括3个方面：①地震沉积学理论；②地震沉积学在陆相碎屑沉积和海相沉积的砂体识别及精细表征、碳酸盐岩和混积岩刻画、地震成岩相预测等方面的应用；③地震沉积学新技术，如地球物理新方法新技术、RGB（red-green-blue）地震属性融合、储集层预测技术以及三维可视化技术雕刻地质体等，这为精细古地理成图奠定了良好基础。

1.1.1 地震沉积学理论研究进展

地震沉积学理论研究的进展主要表现在两个方面（朱筱敏等，2020），一是频率控制了地震反射同相轴等时性新认识。随着地震资料品质提升和地球物理方法技术的发展，人们发现地震同相轴等时地质意义取决于地震反射主体频率大小，即不同品质（主频）的地震数据反映的地质信息和层序界面是不同的，这一认识改变了地震地层学-层序地层学研究的前提假设。二是不同砂体组合的频段切片响应不同。一般来说，厚层砂体组合主要对应于低频地震信号，薄层砂体组合主要对应于高频地震信号。通过三维地震资料不同频段的分频处理，就可以获得相应频段的地震数据体，分频进行地震岩性学和地震地貌学（地层切片）综合研究工作，确定不同厚度砂体的时空分布特征。

1.1.2 地震沉积学应用研究进展

（1）陆相碎屑砂体地震沉积学。主要进展包括根据沉积模式和地震地貌响应，确定古物源方向和古水流流向、砂体沉积成因类型、沉积亚相和微相边界、识别刻画薄层砂体、确立沉积演化和沉积模式、有利砂体和岩性圈闭预测等，几乎涉及所有陆相沉积类型。

（2）海相碎屑岩地震沉积学。在分析和定量刻画不同层序或体系域下切谷、低位扇、低位三角洲、海侵滨岸砂体、高位障壁砂坝、潟湖和三角洲砂体以及水道/河道的弯曲特点等方面均取得明显进展（Reijenstein et al., 2011）。东南沿海多个盆地采用先进的地球物理方法，确定了不同沉积类型沉积砂体可靠的时空分布，有效地推动了油气勘探开发（陈杨等，2019；罗泉源等，2020）。

（3）砂体精细表征。主要进展包括3个方面：①井间油藏精细地质研究由模式化外推到井震结合表征，降低了开发油藏描述的井间储集层预测不确定性；②以地震沉积学与储集层构型表征的结合为例，将储集层沉积学、油藏工程分析和应用地球物理技术结合，实现了井间四级（单一沉积微相）乃至三级构型（如点坝内部侧积体）的可靠表征（刘海等，2018）；③采用地震沉积学并结合多种地球物理技术，研究不同沉积类型砂体定量参数及其构型，如确定水道体系的弯度、水道宽度、河曲带宽度、河曲拱高，以及这些参数的相互关系（Yue et al., 2019）。

（4）碳酸盐岩和混积岩研究进展。地震沉积学与地震相控非线性随机反演相结合，开辟了识别礁滩沉积体新的思路（黄捍东等，2011）。采用多频道振幅主因子分析技术，定量预测了储集层平面分布，发现厚储集层主要发育在古地貌高部位的颗粒滩沉积中。相似性方差地震属性能反映岩体内部横向变化，结合地层切片技术，可有效揭示碳酸盐岩溶蚀成岩相分布（Zeng et al., 2018）。

（5）地震成岩相预测。曾洪流等（2013）采用常规三维地震资料，预测了松辽盆地青山口组泥质胶结砂岩和方解石胶结砂岩，并预测了胶结作用（溶蚀作用）等成岩相的平面分布。在碳酸盐岩台地相的研究中，采用地震岩性学研究思路，优选反映岩体内部横向变化的相似性方差地震属性，刻画碳酸盐岩溶蚀成岩相。地震属性切片相似性方差值越大，说明岩层物性越好（Zeng et al., 2018）。另外，深层低渗-致密储集层成岩程度高，对沉积和成岩因素及其耦合对深层强非均质性储集层体系的形成与演化具有控制作用。张宪国等（2018）提出基于多尺度信息的、弹性参数叠前反演驱动的致密储集层地震成岩相预测方法。

1.1.3 地震沉积学新技术研究进展

（1）地震岩性学。中国含油气沉积盆地因常规纵波地震资料的应用受限以及地震沉积学解释效果尚难令人满意，需进行地震岩性学综合研究。根据叠后波阻抗反演 I_{REI} 地层切片分析物源方向和砂体分布，其结果与钻井资料吻合率高，地震地貌特征符合地质背景和沉积规律（刘力辉等，2013）。

（2）RGB（red-green-blue）地震属性融合。基于颜色空间的多属性RGB融合技术，可以刻画不同地质时代的隐蔽地震异常体（朱筱敏等，2019）。这是一种基于视觉的属性分析方法，是将多个地震属性通过主成分分析（principle component aralysis，PCA）技术进行降维，取前3个（或4个）主分量利用RGBA（red-green-blue-alpha）颜色融合原理获得一张融合图，并依据颜色的区域性和突变异常等视觉特征，进行地质目标和沉积厚度解释。在刻画河道砂体时可有效降低穿时效应，准确刻画河道连续性和完整性，有效识别窄小河道分布（李明等，2019）。

（3）与储集层反演预测技术紧密结合。地震相控非线性混沌反演方法是指在层序格架-地震相和井资料约束下，对不同埋深薄互层砂层进行的非线性混沌反演（Huang et al.，2016）。混沌反演能够充分利用弱地震信号，比常规反演揭示更多细节，尤其能刻画薄层砂体，反演结果与实测数据吻合较好，突出了界面和岩性体，还提高了沉积接触面和变岩性岩体的分辨率，砂岩储集层预测吻合率达到85%以上（Luo et al.，2018）。通过此方法，可以针对复杂油气储集层进行有效分析，从而增加地震沉积学应用的可信程度，使其应用效果得到进一步改善（朱筱敏等，2019）。

（4）利用三维可视化技术进行地质体雕刻。采用三维可视化软件可从任意角度展示多维度图像，研究地下沉积界面、砂体形态、叠置样式以及构造、油藏等特征。全三维地震资料解释是针对数据体的解释，它是从三维可视化显示出发，以地质体或三维研究区块为单元，采用点、线、面、体相结合的空间可视化方法进行解释，这样的解释对地质体的空间认识更直观、准确。

1.2 深水牵引流沉积研究进展

深水牵引流主要包括内波、内潮汐、等深流等，多活跃在陆坡及盆地等深水地区，长时间、持续作用可形成一系列特殊沉积体，如内潮汐沉积、等深流沉积（等深岩）等。该研究对陆坡及盆地等深水地区的古地理精细刻画具有重要意义。近十年来的主要进展主要包括3个方面。

1.2.1 内波、内潮汐沉积特征及发现

Shanmugam（2013）系统调研并整理了大西洋、太平洋、印度洋、北极及南极海洋内波发育位置、波长、波高、速度等特征，并与表面波进行对比，较为深入地研究了内波运动特征。Bourgault等（2014）利用数字模拟研究了内波对沉积的搬运，探讨了内波运动过程中对沉积物的侵蚀、搬运、沉积及再作用过程。张洪运等（2017）基于站位水文观测及海底沙波等资料，对南海北部内波及内波沉积过程的研究取得了较好成果。同时，在鄂尔多斯盆地、广西凭祥盆地、莺歌海盆地等地区的地层记录中也发现了内波、内潮汐沉积（He Y. B. et al.，2011，2012；He H. Y. et al.，2012；Gao et al.，2013，2014；杨红君等，2013；李向东和郁雅琪，2017）。李向东（2013）基于本团队成果及国内外内波、内潮汐沉积研究，对内波及内潮汐沉积进行了分类。

1.2.2 等深流沉积及发现

在南海、东非、大西洋两岸等地都发育现代等深流沉积（郑红波等，2012；李华等，2013；Rebesco et al.，2014；李华和何幼斌，2017；Chen et al.，2019）。近年来在鄂尔多斯盆地、滇中、广西、湘北等地的古代地层记录中都见有等深流相关沉积的报道（李华等，2016；李向东和郁雅棋，2017；Li et al.，2020），

在等深流沉积类型、鉴别标准、形成过程及控制因素等方面都取得了诸多认识，划分出9种类型。

1.2.3 深水交互作用沉积特征及发现

等深流、内波、内潮汐、重力流等交互作用沉积在深水环境中普遍存在，是目前深水沉积研究的主要内容之一（Gong et al.，2018；Li H. et al.，2019；Rodríguez-Tovar et al.，2019；Fonnesu et al.，2020）。He 等（2011）对宁夏香山群徐家圈组的研究，发现了内波、内潮汐与重力流交互作用沉积；李华等（2016）在对鄂尔多斯西南缘的研究中发现了等深流及重力流互层沉积；李向东等（2019）通过对鄂尔多斯盆地西缘及贺兰山地区的研究，发现了内潮汐、等深流及重力流交互作用沉积。

1.3 重力流-异重流研究进展

重力流沉积物的成因机制很多。在洪水、地震、海啸和强烈构造活动作用下，进积三角洲前缘沉积物和大陆斜坡沉积物可向前滑塌形成重力流，也可由洪水携带大量陆源碎屑物质直接进入盆地深水区形成重力流（异重流），可见重力流沉积过程复杂多变。重力流沉积物可由砂砾岩组成，也可以泥岩为主；重力流在近源和远源均可发生沉积。其成因机制研究对发育区的古地理恢复具有重要意义。

经典浊流沉积常呈扇形-朵叶状，砂质碎屑流常呈不规则舌状体，异重流常呈水道状和朵叶状组合（Zhu et al.，2016）。浊积扇-近岸水下扇主要响应于具有水道的朵叶状地貌特征，可分布于断陷湖盆深洼和陡坡带，向盆地中央延伸数千米；砂质碎屑流常呈规模较小的、面积几平方千米的舌状体，多位于三角洲前方；异重流响应于数十千米的弯曲水道和规模较小的朵叶体，也常位于三角洲的前方（Zhu et al.，2016，2018；潘树新等，2017）。

1.4 源-汇系统古地理重建

按照剥蚀-搬运-沉积三个过程进行沉积物运转过程及演化历史分析，完整的源-汇系统分析古地理重建包括沉积区古地理重建与物源区古地理重建（邵龙义等，2019）。针对大陆边缘源-汇系统，祝彦贺等（2011）通过分析南海北部被动大陆边缘盆地的地震资料，认为相对海平面变化、陆架坡折演化、沉积物供给和海洋水动力共同作用于陆架-陆坡的源-汇系统，并综合分析了源-汇系统各要素之间的沉积响应。Zhu 等（2017）基于层序地层学、地震沉积学理论及三维可视化技术（古地貌恢复），结合三维高分辨率地震资料，对渤中凹陷西斜坡东营组层序物源通道及沉积体系开展精细刻画，并以物源通道差异建立了三种不同的源-渠-汇空间耦合模式，指导区内有利储集层预测。林畅松等（2015）在新生代珠江口盆地识别出三种具有特定物源背景、搬运通道及相应沉积体系的源-汇类型，分别为河流或辫状河流平原-陆架边缘三角洲-前三角洲斜坡或陆架斜坡扇、陆架边缘上斜坡密集的直流状沟谷-滑塌-下斜坡扇、大型下切沟道-伸长状盆底扇等源-汇系统。刘强虎等（2017）基于钻井岩心、三维地震及锆石测年等资料，分析了渤海湾盆地沙垒田凸起前古近纪基岩组成及分布，并探讨了对应源-汇体系的配置关系。朱红涛等（2017）针对陆相盆地源-汇系统研究主要集中在驱动机制及地球动力学过程、深时古气候、古物源区演化恢复与古水系重建、源-汇系统要素分析和剥蚀-搬运-沉积过程耦合模式。李双应等（2019）根据大量的锆石 U/Pb 年龄和 Th/U 值测试数据，研究了大别造山带物源区与合肥盆地南缘中生代沉积耦合关系。Wang X. T. 等（2020）利用层序地层格架分析以及岩相参数分析，恢复了晚二叠世乐平期源-汇系统古地理，利用沉积通量模型估算出峨眉山大火成岩省及地幔柱的隆升历史。邵龙义等（2019）评述了深时源-汇系统古地理重建方法，指出深时尺度下沉积区古地理重建方法随着岩相古地理、构造古地理、生物古地理及层序地层学的发展已趋完善。在源-汇分析方法体系中，物源区古流域水系形态、面积范围、地貌地势等古地理要素，可通过构造要素分析、碎屑矿物分析、沉积体积回填、地貌学参数比例关系、古水力学参数比例关系和河流沉积通量模型等方法获得。

1.5 臼齿碳酸盐岩研究进展

"臼齿构造（molar-tooth structure，MTS）"是指前寒武纪一种具有肠状褶皱等特殊形态和结构的沉积构造，含有臼齿构造的碳酸盐岩称为臼齿碳酸盐岩（molar tooth carbonate rock，MTC）。迄今为止，已在全球25个地区、50多个地层剖面中发现有MTS，国外的有格陵兰、挪威、芬兰、加拿大、美国、俄罗斯西伯利亚、印度（Saha et al.，2016）、澳大利亚、西非、南非（Kuang and Hu，2014；Smith，2016），中国的有吉林南部（万隆组）、辽东大连（南关岭组、甘井子组、营城子组和兴民村组）、山东沂水-兰陵石旺庄组、栖霞蓬莱群香夼组和苏皖等地（刘老碑组、九里桥组、赵圩组、倪园组、魏集组、张渠组和望山组等），燕山地区中元古界（高于庄组），辽宁凌源中元古界雾迷山组以及豫西新元古界何家寨组、栾川群大红口组，新疆伊犁新元古界和滇中新元古界大龙口组的碳酸盐岩中都发育有丰富的MTS（或MTC），五台山滹沱群浅变质的碳酸盐岩中都相继发现了大量MTS（柳永清等，2010；Kuang and Hu，2014），而且还不断有新的MTC剖面被发现。随着对MTS微观组构研究及碳酸盐岩地球化学研究的深入，更强调用古海洋地球化学条件变化及微生物作用来解释MTS的生物-化学成因（Kuang and Hu，2014；Shen et al.，2014；Petrov，2016；Hodgskiss et al.，2018），称为生物-化学成因观（Kuang and Hu，2014）。显然，目前MTC的成因研究已经聚焦到生物作用与古海洋地球化学性质等方面，但不同研究者的认识还存在差异（Shen et al.，2016；Hodgskiss et al.，2018）。MTS的形成不是一个孤立过程，它受控于古海洋性质（海水温度、碳酸钙溶解度、海水 Fe^{2+} 和 SO_4^{2-} 含量、氧逸度、盐度、C、O、Sr、S、Mg、Mo等同位素水平），同时还与当时的大气环境（p_{CO_2}）、古纬度（对海水温度的控制）以及生物作用（平衡氧化还原水平及海水中 p_{CO_2} 与碳循环，加速 $CaCO_3$ 结晶）密切相关（旷红伟等，2011；Kuang and Hu，2014；Petrov，2016；Smith，2016；Hodgskiss et al.，2018）。因此，研究MTC形成的地球化学临界条件至关重要。全球MTC在岩石学、形态学、形成环境、微观组构、地球化学特征等方面都具有相似性，并具有特定的形成时限和时空分布范围。该研究成果对海洋碳酸盐岩及海洋古地理具有十分重要的参考价值。

2 古地理研究方法与成图

传统的古地理学分支学科包括岩相古地理学、生物古地理学、构造古地理学、第四纪古地理学及人类历史古地理学，支撑了古地理学研究的繁荣局面，但理论上突破性的成果不多。本文重点阐述构造古地理、生物古地理、小尺度古地理和大数据古地理的研究进展。

2.1 构造古地理研究进展

构造古地理一般分为活动论构造古地理和固定论构造古地理，本文重点对前者进展进行评述。活动论构造古地理是在全球板块构造模型的基础上，开展的全球或区域性的古地理重建。古地理重建是通过解释岩石记录对沉积岩相与环境的时代分布进行制图，其思想是在地球系统科学的活动论、演化论、阶段论与转换论观念下的自然延伸。整体、动态、综合分析是活动论构造古地理研究的基本方法（图1）；确定构造古地理单元的边界、属性、组成、结构与演变的"五定"原则是工作的具体步骤；搭建数据化、标准化和智能化的古地理重建平台是研究的重要途径（何登发等，2020）。

2.1.1 全球古地理重建

关于全球板块构造、古地理演化及其重建已开展了长期研究。近年来，在对不同地质历史阶段的板块构造古地理的复原研究中取得了一系列较为突出的成果（何登发等，2020）。例如，Seton等（2012）开展了400 Ma以来全球大陆、大洋盆地重建；张国伟等（2013）指出东古特提斯洋（East Paleo-Tethys）北支（勉略洋盆）于早泥盆世（420～380 Ma）开启，石炭纪—早二叠世达最大规模，晚二叠世开始俯冲

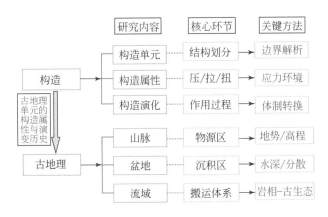

图 1 活动论构造古地理的研究内容与技术方法（据何登发等，2020）

消减，中、晚三叠世闭合，形成中国大陆主体；李江海和姜洪福（2013）系统绘制了古板块再造图、古地理环境恢复图、沉积岩相图、烃源岩分布图、主要古生物分布图等系列图件；Morra 等（2013）给出了 200 Ma 以来地表板块构造演化的镶嵌图案，再现了太平洋、非洲、欧亚等板块每隔 10 Ma 的运动学图案；Domeier 和 Torsvik（2014）建立了晚古生代（410~250 Ma）的全球板块构造模型；Domeier（2016）进行了早古生代亚皮特斯与瑞伊克洋的板块构造复原；Zhao G. C. 等（2018）建立了 750 Ma 以来突显东亚各组成陆块位置的板块构造复原系列图，这些陆块基本在 220 Ma 聚合；Zhao G. C. 等（2018）、Li S. Z. 等（2019）对全球超大陆的聚散开展了研究，基本建立了哥伦比亚、罗迪尼亚、潘吉亚 3 个超大陆的聚散模型；Young 等（2019）给出了晚古生代 410 Ma 至今板块边界闭合的全球板块构造与俯冲带的运动学模型；张光亚等（2019a，2019b）系统编制全球现今地理位置 13 个纪或世关键时间点的岩相古地理图，结合古板块恢复成果实现古构造位置下的原型盆地和岩相古地理恢复。

在全球构造模型中，将岩石圈板块运动与地幔柱耦合作用相结合是其趋势，出现了新框架下的威尔逊旋回概念（Heron，2019）。Stuar 等（2012）提出了 4D 板块的概念，主要基于高分辨率资料，可以叠加到 4~8 层数据。以前的板块复原模型不考虑板块的变形，Gurnis 等（2018）提出了刚体板块连续变形与演化的全球构造重建新方法，变形区镶嵌于三角网中，从而可以计算变形及其变化。Peace 等（2019）应用 GPlates 建立了北大西洋南部 200~0 Ma 的多期变形的板块构造模型。

2.1.2 中国古地理重建

中国古地理重建工作历史悠久，取得了一系列成果。近十年来的主要成果有：郑和荣和胡宗全（2010）出版了《中国前中生代构造-岩相古地理图集》，系统展示了海相地层的分布；肖文交等开展了中亚增生型造山带的长期研究，提出了哈萨克斯坦、蒙古山弯构造的演化模型，并指出中亚造山带是全球最大的显生宙增生型造山带，在古生代至中生代早期经历了多俯冲带、多方向的复式增生造山作用（Xiao and Santosh，2014；Xiao et al.，2015）；王剑等（2015）的"结构-成因"岩相古地理编图方法是古地理编图的一个新尝试，已应用在华南及羌塘盆地岩相古地理分析与编图中；牟传龙等（2016）研究编制了埃迪卡拉纪—志留纪中国岩相古地理图集；Zhao G. C. 等（2018）给出了东亚大陆主要地块在 750 Ma 以来的演化路径，王剑等（2019）编制了华南新元古代中期（820~635 Ma）5 个时期的岩相古地理图。

2.2　生物古地理研究进展

生物古地理研究涉及很多方面，在此着重介绍中国近十年来生物遗迹与古地理和古环境方面取得的主要进展和新认识，其中湖相沉积中遗迹组合分布模式的提出与显生宙动藻迹数据库的建立处于国际先进水平。

(1) 在显生宙海相和过渡相沉积中识别出 84 个遗迹属和 211 个遗迹种，新建 28 种遗迹组合及 26 种遗迹组构（宋慧波等，2015；Fan and Gong，2016；Zhang and Zhao，2016）；在中生代和新生代陆相冲积扇、河流与湖泊沉积中识别出44个遗迹属和107个遗迹种，新建22种遗迹组合及10种遗迹组构（Wang et al.，2016）。其中，湖相沉积中遗迹组合分布模式（图2）的提出是国际上陆相沉积环境中遗迹学研究的重要进展（Hu et al.，2014）。

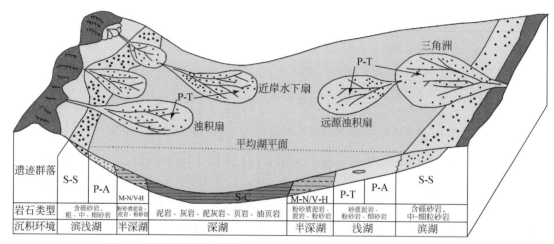

图2 湖泊沉积环境中遗迹化石组合分布模式（据 Hu et al.，2014）

S-S：*Scoyenia-Skolithos* 遗迹组合；P-A：*Palaeophycus-Arenicolites* 遗迹组合；P-T：*Planolites-Teichichnus* 遗迹组合；S-C：*Semirotundichnus-Chondrites* 遗迹组合；M-N：*Mermoides-Neonereites* 遗迹组合；V-H：*Vagorichnus-Helminthopsis* 遗迹组合

（2）在豫西寒武系张夏组碳酸盐岩中识别出六类遗迹组构（齐永安等，2012），为寒武纪沉积环境的高频交替循环变化提供了生物遗迹方面的佐证。寒武纪早期沉积中生物扰动与底质关系的研究，揭示了寒武纪底质革命对早期底栖生物的进化和生态效应的巨大影响。

（3）宋慧波等（2017）通过对豫西和晋中南地区太原组遗迹化石与古氧相的精细研究，提出了3种古氧相（富氧相、贫氧相和厌氧相）的遗迹化石响应关系及其形成的沉积背景。

（4）在华北上石炭统—下二叠统太原组灰岩中发现了大量成因与微生物相关的碳酸盐岩微形体，在动藻迹（*Zoophycos*）潜穴充填物中识别出球状、杆状、簇状、网状、瓶状和似脑球状等6种形态类型（宋慧波等，2014），并描述了由造迹者与共生微生物一起进行一系列生命活动行为所形成的复杂精美的生物成因构造。

（5）基于全球不同地区291篇论文（1821—2015年）和不同时代地层（寒武系至古近系）中180件动藻迹样品数据的分析，构建了显生宙动藻迹的数据库，从地球生物学角度系统归纳总结后，发现了显生宙动藻迹3个方面的宏演化规律及其环境背景，揭示了复杂遗迹化石动藻迹从浅阶层迁移至深阶层与从浅海迁移至半深海-深海同步的演化规律（Zhang et al.，2015）。

（6）通过对中国古生代海相和中生代陆相遗迹学与事件沉积的研究，提出了滨-浅海和湖泊环境中风暴和浊流事件沉积序列上及 P/Tr 界线附近遗迹化石的组成与分布特征（Zhao et al.，2015），并论证了 P/Tr 事件前后遗迹化石及其造迹生物生态系统的演变规律（Chen et al.，2011），为晚二叠世末生物大绝灭后的生态系统复苏过程提供了可靠的遗迹学信息。

（7）在华东、华北和西部地区白垩系中新发现大量兽脚类、蜥脚类和鸟脚类恐龙足迹群，研究认为大多数足迹都产自河流边滩、河漫滩或冲积平原和滨湖沉积环境。

（8）在华北地台中元古界和上二叠统—下三叠统碎屑岩沉积中识别多种微生物成因沉积构造，主要包括微生物席稳化波痕、微生物席平滑波痕、斑状波痕、生长脊、瘤状、刺状突起、象皮皱纹构造、皱纹构造、变余波痕、满洲藻构造和砂裂隙构造等（郑伟等，2016；Chu et al.，2017）。

（9）牛永斌等（2018）通过对塔里木盆地东河砂岩、豫西北奥陶系马家沟组和柴达木盆地石炭系碳酸盐岩中的生物扰动及遗迹组构研究，发现生物扰动作用不仅能够改善储集层的孔隙性能，而且还可使上、下储集层贯穿，增强储集层的连通性。基于对碎屑岩和碳酸盐岩储集层中的生物扰动作用和遗迹组构对储集层物性的影响分析，建立了生物扰动变化三维地质模型，并推出了生物扰动型储集层的精细表征与研究方法，从而为油气储集层预测提供遗迹学方面的证据。

2.3 小尺度古地理研究进展

古地理研究可分为多个空间尺度，包括全球尺度（整个地球范围的海陆分布）、超大尺度（如一个大洋、大洲或其内的一个较大区域，$10^6 \sim 10^7$ km²）、大尺度（如一个大盆地或小盆地群，$10^4 \sim 10^5$ km²）、中尺度（如盆地内的坳陷、凹陷或区带、$10^2 \sim 10^4$ km²）、小尺度（一个局部地理单元，如一个三角洲，一般在100 km²以内）等。沉积构型研究属于小尺度古地理与沉积学研究的范畴。近十年来，沉积构型研究从之前的以精细表征为主拓展到成因机制分析，并取得了长足进展，表现在同生逆断层控制的冲积扇、可容空间影响下的曲流河点坝、浅水缓坡背景下的三角洲指状砂坝及大陆斜坡微盆地重力流构型等方面。

2.3.1 同生断裂控制下的冲积扇构型

近年来，挤压盆地边缘同生逆断裂对冲积扇构型的控制作用取得了较大进展。在准噶尔盆地西北缘，盆地边界断裂为克-乌大断裂，此为一区域性同生逆断裂，包括大侏罗沟断裂（右行平移断裂）、花园沟断裂（逆断裂）等。在复杂的构造应力作用下，形成了正梳状、反梳状、平行状及交叉状4种断层组合样式（印森林，2014；吴胜和等，2016；印森林等，2016）。克-乌大断裂的活动强度自白碱滩至红山嘴逐步减小，导致三叠系克拉玛依组冲积扇自白碱滩向洪山嘴逐步减小（印森林等，2016），这表明同生逆断层的活动强度控制了冲积扇上游物源的供给、水动力特征的差异性，并有可能进一步导致这些扇体内部构型出现差异（吴胜和等，2016）。同生逆断裂正牵引构造发育于逆断裂上盘，引起沉积物在局限环境下的快速沉积（冯文杰，2017），形成一种特殊的冲积扇；这类正牵引构造可导致上盘区域发生一定程度的隆升，造成逆断层上盘沉积区可容空间减小、沉积物厚度减薄或地层局部缺失（印森林等，2014）。同生逆断裂断坡对冲积扇的控制断坡是冲积扇形成的必要条件，单一断坡坡度大小控制了不同相带的砂砾岩体规模，其与扇体面积、不同相带的辐向长度存在较好的指数关系；多级断坡的不同组合样式导致不同冲积扇的片流带及辫流带差异明显，控制了不同相带内部构型要素的类型及其叠置关系（印森林，2014）。

2.3.2 可容空间影响下的曲流河点坝构型

近年来，对可容空间对曲流河点坝构型的研究取得了较大的进展。可容纳空间增长速率与沉积物速率比值（A/S）控制着点坝的形态和内部构型，随着A/S值的减小，曲流河点坝从点状、单一透镜状向鳞片状发展。与经典的侧向迁移型点坝相比，顺流迁移型点坝一般发育于可容空间较小的条件下，河道两侧通常受到地貌或构造的限制（Ghinassi et al., 2016）。在沉积上，顺流迁移与旋转式顺流迁移型点坝通常伴随着逆流加积的沉积物与反向坝沉积物（Ghinassi and Ielpi, 2015；Ghinassi et al., 2016；Yan et al., 2017）。

2.3.3 浅水缓坡背景下三角洲指状沙坝构型

三角洲指状沙坝一般形成于河控较深水三角洲（如现代密西西比河三角洲），近年来在浅水缓坡背景下河控三角洲指状沙坝研究取得较大进展（Caldwell and Edmonds, 2014；Burpee et al., 2015；Tejedor et al., 2016；徐振华等，2019）。例如，对赣江浅水三角洲指状沙坝现代沉积与渤海油田明化镇组浅水三角洲指状沙坝沉积的系统研究发现，分流河道会对河口坝进行一定的改造，并形成点坝沉积（偶见心滩），但河口坝仍可成为指状沙坝的主体沉积（吴胜和等，2019；徐振华等，2019）。浅水三角洲指状沙坝的形成需满足3个条件：弱的盆地水体能量、细粒高黏沉积供给及温暖潮湿的气候。

2.3.4 大陆斜坡微盆地重力流构型

与常规深水陆坡及盆底环境相比，陆坡微盆地内的重力流构型要素受复杂地貌及构造活动的影响较为明显，因而表现出差异的空间分布样式。

（1）朵叶体构型分布样式：相比于盆底扇朵叶体，陆坡微盆地内的朵叶体规模较小，其平面展布受控于微盆地的形态规模（Zhang et al., 2016）；限制性朵叶体的面积小、厚度大，而非限制性朵叶的面积大、厚度小（Prélat et al., 2010）。

（2）水道-天然堤构型分布样式：一般当地形坡度变小时，水道弯曲度变大，加积作用及侧向迁移明显，天然堤发育程度增加；而当地形坡度变大时，水道弯曲度变小，下切侵蚀作用明显，天然堤发育程度减弱（赵晓明等，2018；Zhao X. M. et al., 2018）。

（3）块状搬运体分布样式：根据重力流块体的来源，在陆坡微盆地内部识别出外源和内源两种不同的块状搬运体。前者是由陆源碎屑物质沿大陆斜坡整体搬运至微盆地后快速堆积形成的，主要沿微盆地中心分布，底部可见因顺坡搬运产生的侵蚀擦痕（蔺鹏等，2018；张佳佳和吴胜和，2019）。后者是由微盆地边缘沉积物发生块体垮塌形成的，主要堆积在微盆地坡脚部位，在微盆地边缘斜坡部位可见明显的滑动侵蚀现象（Spychala et al., 2015；蔺鹏等，2018）。

2.4 大数据古地理研究进展

古地理学涉及的分支学科多而且综合性非常强，数字古地理重建需要综合大地学多学科的知识体系，需要收集和分析岩性、构造、古生物、地球化学等多种来源、多种类型的不同数据，而这些数据收集、挖掘、可视化等也在很大程度上依赖于计算机技术的发展。随着数据获取与分析技术的提高与古地理学知识的不断积累，可视化技术的迅猛发展及人工智能技术在固体地球科学建模中的广泛应用，将共同促使古地理重建走向标准化与智能化（Ogg et al., 2019；Zhao et al., 2019；Zahirovic et al., 2019）。

大数据驱动下的数字古地理重建思路是最近众多学者探讨的热点之一。何登发等（2020）认为，搭建数据化、标准化和智能化的古地理重建平台是古地理学研究的重要途径。张蕾等（2020）提出数字古地理重建的五个要点：①建立标准化的古地理学知识体系；②建立开放互动、动态更新的古地理数据库，并利用机器阅读技术等拓展数据来源；③建立标准化的古地理学数据质量控制体系；④利用机器学习技术建立各类型古地理重建模型，深度挖掘数据；⑤以可实时更新的智能数字地图集或多维动画形式输出成果。

古地理学知识体系繁杂，大数据收集整理需要最大限度细分研究对象（如各类沉积相、亚相、微相），最大限度罗列研究指标及其填写方法（如矿物、岩性、沉积构造、古生物等判别指标），尽可能最大范围覆盖已有的各种数据。古地理学知识体系整理过程也是对学科内容的统一化过程，学科重要概念的解释、重要分类等必须采用统一标准。一个完善的古地理学知识体系可避免对古地理数据解释的混乱，同时也是未来标准化、智能化古地理数据库与其他已有的各类专业数据库对接的重要基础，可用于指导数据平台内核和大数据分析逻辑的开发，也将成为学科专家间、数据库间以及学科专家与数据库间交流的重要纽带。

2019年2月"深时数字地球——全球古地理重建与深时大数据国际研讨会"在北京召开，会议宣布深时数字地球大科学计划正式启动，表明大数据古地理研究在全球范围展开。深时数字地球国际大科学计划，聚焦过去数十亿年的地球深时时期，涉及地球岩石圈、水圈、生物圈、大气圈等多圈层的数字耦合系统，具有指标体系复杂、内容结构封闭等特点，容易形成信息孤岛，因此需要在地质资料大数据和人工智能驱动下重新构建地球科学知识结构体系，推动大数据驱动的知识发现与创新加强。地球科学与人工智能大数据等多学科跨领域深层次的交叉渗透，围绕地球演化生命演化地球系统与环境以及能源矿产分布等核心基础科学问题，开展探索着力促进地球科学前瞻性基础研究，引领性原创成果重大突破（成秋明，2019）。

3 古地理学的应用

3.1 含油气盆地古地理研究与找油规律

不同级别、不同比例尺的岩相古地理研究，可指导石油地质早期评价、有利区评价、"甜点区"评价与目标评价，对中国油气勘探具有重要指导意义。从编图单元与范围看，一般分为大区域级、盆地级、拗陷级、区带级4个层次。近年来，全面勘探与风险勘探的需求促进了以源-汇系统思想为指导的大区域、盆地级构造-岩相古地理研究，成熟探区油气精细勘探的需求促进了区带级大比例尺沉积微相研究，页岩油气勘探的需求催生了深水细粒沉积区岩相古地理研究。在大型陆相盆地重点层系岩相古地理、中新元古界与古生界克拉通盆地构造岩相古地理、页岩层系岩相古地理、区带级大比例尺沉积微相编图等方面取得了重要进展。

3.1.1 大区域与盆地级构造-岩相古地理研究应用成果

随着大地构造学、层序地层学、盆地分析等学科的发展，大区域或盆地级岩相古地理编图逐渐形成了"以构造控盆、盆地控相、相带控源-储的指导思路，以构造运动和海（湖）平面变化或沉积基准面变化为主控因素所形成的等时层序界面及相应的地质体等为编图单元，以反映源-储分布规律为重点"的含油气盆地构造-层序岩相古地理综合编图方法，开展了相应的构造-层序岩相古地理研究和编图，有效指导了油气勘探选区、选带和部署。如扬子区中新元古界构造-岩相古地理、华北区中新元古界构造-岩相古地理（赵文智等，2019）、三大克拉通盆地台地发育模式与岩相古地理、鄂尔多斯盆地延长组长1—长10段沉积相（付金华等，2018）、准噶尔盆地中—上二叠统岩相古地理（支东明等，2019）、松辽盆地白垩系岩相古地理等研究（柳波等，2018），为烃源岩、有利储集层的评价奠定了基础，在岩性地层油气藏、海相碳酸盐岩与深层、前陆盆地、成熟探区油气勘探中发挥了重要作用。

3.1.2 页岩层系岩相古地理研究应用成果

页岩油气、致密油气形成与富集的主要地质条件包括了大面积分布的成熟优质烃源岩、物性相对较好的规模储集层、源储共生。近年来，围绕页岩层系岩相古地理与富有机质页岩分布，开展了中-上扬子区下寒武统、上奥陶统—下志留统、上二叠统页岩层系岩相古地理与富有机质页岩分布（王玉满等，2015；彭勇民等，2016）、鄂尔多斯盆地长7^1、长7^2、长7^3段沉积相与富有机质页岩分布（袁选俊等，2015；杨华等，2016；付金华等，2018）、松辽盆地青山口组、嫩江组岩相古地理与富有机质页岩分布（柳波等，2018）、准噶尔盆地二叠系岩相古地理与烃源岩分布等研究（支东明等，2019），形成了一批工业图件，指导了页岩气、页岩油选区评价与资源潜力预测。

3.1.3 区带级大比例尺沉积微相研究应用成果

为满足精细勘探的需要，开展了高精度层序地层研究，提出了陆相高精度层序地层划分与工业化应用六个步骤的研究程序和技术规范。通过沉积背景分析、层序划分对比、层序格架下的砂体等时对比、高分辨率砂体识别、大比例尺沉积微相分析、目标预测与评价，提高了层序地层学研究定量化程度和岩性地层圈闭预测精度。以扶余油层为例，将大庆长垣扶余油层纵向上划分为7个四级层序，29期河道，编制了扶余油层29期河道的大比例尺沉积微相图，分析每期河道类型及砂体分布特征。研究区整体处于基准面上升期，自下而上河道沉积作用逐渐减弱，河道厚度、河道相域分布面积逐渐减小，由多期叠置的曲流河道逐渐变为孤立的网状河道和分支河道，上部出现浅水三角洲沉积，顶部被半深湖油页岩覆盖，反映研究层段形成于可容纳空间逐渐增大的基准面上升过程，有效指导了目标评价与井位部署。

3.2 油页岩古地理研究与找矿规律

在大量勘探工作的基础上，油页岩成矿理论研究也取得丰硕成果。在前期工作和文献梳理的基础上，从古大气和古构造背景、沉积环境、地质事件等方面对油页岩古地理研究与找矿规律进行总结评述。

3.2.1 古大气背景影响

油页岩易于在相对较高的O_2和低CO_2浓度的大气背景下形成，大气组成影响油页岩的成矿期次。从盆地角度看，古气候和古构造对油页岩矿床规模起控制作用。湖盆充填状态可分为过补偿、补偿和欠补偿，而油页岩在这3种状态下均发育，只不过矿床特征有明显差别（Sun et al.，2013；刘招君等，2016）。

盆地内构造和古气候协同控制油页岩的矿床规模，补偿环境易于形成深水油页岩，而过补偿环境沉积浅水油页岩，欠补偿环境中形成的油页岩相对较少。

3.2.2 沉积环境影响

深水油页岩沉积于半深湖和深湖环境，湖底均处于贫氧-缺氧状态，生物生产力是控制油页岩品质的关键因素，有利条件的持续时间和低含氧区范围决定了油页岩矿床的规模。长期稳定的深水环境普遍发育水体分层的现象，湖底基本为缺氧-贫氧环境，这是有机质良好的保存条件。但深水环境也存在一定的水体性质差异，如松辽盆地青山口组一段沉积时期，水体基本为咸水-半咸水（Bechtel et al., 2012；孙平昌，2013），导致水体含氧层较薄，有效地减少了有机质沉降过程中氧化和呼吸等破坏作用，较大程度的保存了有机质含量。

浅水油页岩沉积于湖沼环境，富营养化和浊水藻型湖泊是沉积油页岩的前提，湖泊自身恢复和古气候调节作用导致油页岩厚度薄、非均质性强。沼泽区域水体覆盖基本可形成两种环境：一是清水草型浅湖，它具有一定的水动力条件的活水，水草改善水体营养程度，完成湖泊水体修复（Jeppesen et al., 2007），此种环境在煤层上部往往沉积含大量植物化石的粉砂岩及浅灰色泥岩，有机质丰度相对较低，如中生界中侏罗统柴达木盆地大煤沟组煤层顶底板（黄献好等，2020）；二是在沼泽之上，由于基准面抬升直接形成浊水藻型浅湖，基本为死水，湖泊发生营养化，藻类繁盛（秦伯强等，2004），沼泽聚集的大量有机质和营养化的水体共同导致水底基本为缺氧环境，极浅水促进有机质快速沉降，浊水引发相对较高的沉积速率，有机质得以快速埋藏，形成油页岩，以柴达木盆地中侏罗统石门沟组和老黑山盆地下白垩统穆棱组煤层顶底板为典型代表（宋宇，2017；Wang J. X. et al., 2020）。

3.2.3 地质事件对油页岩形成的影响

部分地质事件在油页岩成矿过程中起到促进作用，缺氧、火山活动和热液、海侵有助于提高生物生产力和形成缺氧环境，有利于油页岩的形成。

缺氧事件对中国油页岩成矿也具有重要作用。松辽盆地上白垩统年龄数据表明（He H. Y. et al., 2012；Wang et al., 2013），Cenomanian-Turonian 缺氧事件（OAE2）导致了青山口组优质油页岩的形成，嫩江组油页岩则与 Coniacian-Santonian 缺氧事件（OAE3）有关（孙平昌，2013）。羌塘地区侏罗系油页岩形成于 Toarcian 缺氧（T-OAE）背景中（伊帆等，2016）；Aptian-Albian 缺氧事件（OAE1）促进了东北地区辽西盆地群下白垩统义县组—九佛堂组油页岩沉积（万晓樵等，2017），这些缺氧事件可能还与陆上风化作用阶段性增强有关（Xu et al., 2020）。

火山灰及热液中存在大量的生命营养元素，进入水体会发生水解，成为生物生存的养料，促进了水生生物的大量繁盛，有效提高生物生产力（梁钰等，2014；邱振等，2019）。火山灰及热液对油页岩形成的影响研究，主要集中在鄂尔多斯盆地延长组（贺聪等，2017；袁伟等，2019）。

海侵会使陆相湖盆水体咸化，有利于水体分层，进而有利于藻类体的保存与生烃转化。海水在侵入过程中，向湖泊带入大量海洋浮游生物，从而促进了富有机质沉积岩的形成，水体盐度的增加易于造成水底缺氧环境，这为有机质的保存提供了良好的条件（孟志勇，2016）。众多研究表明，松辽盆地青山口组和嫩江组油页岩沉积均受到海侵作用的影响（Bechtel et al., 2012；Cao et al., 2016），而山东黄县油页岩的形成也与周期性海水进退有关（Lv et al., 2017）。

3.3 碎屑岩成岩作用研究进展

近年来，全球特别是中国在深层-超深层和非常规储集层中不断发现油气并获得突破（孙龙德等，2013；赵文智等，2014；邹才能等，2015；张功成等，2017；何登发等，2019），深层、深水、非常规储集层的成岩作用已成为不可或缺的重要研究内容，主要进展包括如下几方面。

（1）将微观尺度下的成岩作用研究，逐步向与盆地流体、盆地动力学与热动力学等宏观大尺度背景与过程相结合的流体-岩石相互作用的成岩作用系统及其时空演变机制和动态定量过程方向推进（罗静兰等，2013；孟元林等，2015；解习农等，2017）。

（2）将流体-岩石作用作为储集层成岩-成储过程研究的核心内容。随着含油气岩系中有机-无机反应机制研究的深入（Joe et al., 2014；李忠，2016；Denny et al., 2017；Zhao et al., 2017；李忠等，2018）和流体包裹体研究技术的不断改进与完善（陈红汉，2014；平宏伟等，2014；Yang et al., 2017；斯尚华等，2018；李文等，2018），人们逐渐认识到成岩作用研究中，对流体-岩石相互作用机理、形成产物及其分布特征的深入研究，有助于追踪油气运移路径，认识油藏成因和分布，同时还可获得溶蚀作用及次生孔隙形成、储集性能演化、成岩-成藏-致密化过程等重要信息。

（3）深化了热动力（地热梯度及升温速率等）与热流体对成岩作用与孔隙演化重要影响的认识。深部砂岩储集层的压实作用不仅受上覆岩石载荷、所处沉积类型控制，也受地层流体性质、盆地地温梯度场及埋藏热演化轨迹等多种因素的影响（Heap et al., 2015；Brüch et al., 2016）。由盆地热流控制的热压实效应是控制压实作用强度的重要因素，热压实效应引起的砂岩孔隙丧失在高地温场盆地中普遍存在，不同地温梯度区相同温度条件下储集层的物性可以不同（Heap et al., 2015；Nader et al., 2016；侯高峰等，2017；李弛等，2019）。在沉积条件相同情况下，随着地层温度及升温速率（地温梯度）的升高，砂岩存在压实作用增强、压实减孔速率升高的趋势；等深度条件下，地温梯度较高盆地中砂岩的孔隙度小于地温梯度低的盆地，地温梯度差异越高该差别越明显（黄志龙等，2015；张丽和陈淑慧，2017；Lei et al., 2018；罗静兰等，2019）。

（4）超压背景与超压流体对储集层孔隙演化具有重要影响。超压背景或超压孔隙流体承载了大部分的上覆地层压力，减小了岩石骨架承载的压力，可使压实作用及压实效应受到抑制，部分原生孔隙得以保存（张伙兰等，2014；Duan et al., 2018）。超压系统内水-岩反应不活跃，胶结作用变慢或停止、黏土矿物的转化受到抑制，由此减少了由于胶结作用而损失的原生孔隙（张伙兰等，2014；马勇新等，2015；Duan et al., 2018）。超压在一定程度上可使压溶作用大大减弱，从而限制自生石英的形成、减少硅质胶结物（Oye et al., 2018），并促进长石、岩屑的溶解（Xi et al., 2019）。

（5）砂岩中的硅质胶结除了在碎屑石英颗粒表面形成次生加大外，还可以结晶 c 轴平行于颗粒表面的微晶石英膜形式胶结（Worden et al., 2012）。近年来对石英次生加大的研究（Ajdukiewicz and Larese, 2012）表明，不同的温度或不同的黏土矿物类型对石英次生加大的抑制程度不一样。

（6）除了有机酸和碳酸可溶蚀砂岩中的长石、不稳定岩屑（如火山岩）和碳酸盐胶结物外，还存在碱性孔隙水或有机酸对硅质如碎屑石英颗粒和硅质胶结物的溶蚀（Zhu et al., 2015；单祥等，2018）。有机酸的溶蚀能力大于碳酸；长石或岩浆岩、变质岩岩屑溶蚀析出的物质可在垂向剖面上几米的范围内由近及远形成绿泥石、硅质及钙质胶结序列（周志恒等，2019）。

3.4 聚煤区古地理研究与煤炭资源预测

层序地层学作为一种新的年代地层格架分析方法在含煤岩系岩相古地理（邵龙义等，2014a，2014b）、聚煤规律（邵龙义等，2016；Shang et al., 2018）、成煤模式（李增学等，2001）、巨厚煤层成因（Guo et al., 2018；Wang S. et al., 2019）、煤相（鲁静等，2014）、煤系共伴生矿产资源（Lv et al., 2017）等方面均得到了广泛应用。近年来，随着"深时"研究计划在国内、国外地球科学界的逐渐形成（孙枢和王成善，2009），越来越多的学者对含煤岩系中蕴藏的"深时"古气候信息进行了深入的挖掘（邵龙义等，2011；Wang et al., 2011；Wang S. et al., 2019；Yan et al., 2019；Zhang et al., 2019）。

3.4.1 聚煤作用理论研究进展

含煤岩系沉积学的发展经历了旋回层、沉积模式及现在的层序地层学 3 个阶段（邵龙义等，2009，2017）。层序地层学与旋回地层学相结合，能够为含煤岩系对比提供等时性地层格架（邵龙义等，2017）。关于煤层的成因，有"原地堆积"和"异地堆积"之说（王东东等，2016）。煤厚的分级主要是从煤炭开采的角度来确定，一般将大于 8 m 的煤层定义为巨厚煤层。随着层序地层学引入含煤岩系中，为巨厚煤层成因研究带来了新的研究思路。此外，将可进行区域对比的等时性地层单元（层序）与传统岩相古地理研究相结合，重建中国各聚煤期等时性岩相古地理，为中国各聚煤期富煤带及聚煤中心迁移规律分析乃至煤炭资源潜力评价提供了充分的保障（邵龙义等，2014a，2014b；Shao et al.，2020）。

3.4.2 煤系共伴生矿产研究进展

不同时期的含煤岩系除赋存煤炭资源外，还发现有资源规模巨大的煤层气、煤系页岩气、致密砂岩气、天然气水合物、油页岩、煤系铀矿等能源矿产，三稀金属矿产，煤系高岭土和煤系石墨等非金属矿产。

（1）煤系气是指由整个煤系中的烃源岩母质在生物化学及物理化学煤化作用过程中演化生成的全部天然气，整个煤系中的气体都应同属一个系统（王佟等，2014）。主要包括煤层气、煤系页岩气、致密砂岩气、天然气水合物等，前三者合称为"煤系三气"（梁冰等，2016）。

（2）煤系天然气水合物首次发现于青藏高原北部祁连山木里煤田。已有研究显示，木里煤田天然气水合物的形成与煤或煤系有关，主要赋存在中侏罗统江仓组油页岩段的细粉砂岩夹层内的孔隙和裂隙中，其中的烃类气体主要来自煤层和煤系分散有机质热演化的产物，由此提出"煤型气源"天然气水合物的成因解释（王佟等，2009）。

（3）煤系铀矿包括了砂岩型铀矿和煤岩型铀矿两大类（孙升林等，2014）。砂岩型铀矿常与煤产于同一盆地中，多分布在煤层顶底板砂岩、砾岩层中，其形成与顶底板的沉积物源有关（刘池洋等，2013）。部分地区的煤层也是煤系铀矿的直接储集层，属于煤岩型铀矿。

（4）煤系伴生三稀矿产（包括稀有元素、稀土元素和稀散元素三种金属矿产）是指含煤岩系中赋存于煤层、夹矸、顶底板或煤系中的稀有、稀散和稀土金属元素矿产资源（刘东娜等，2018），是当今煤地质学研究的热点之一。

（5）煤系共伴生非金属矿产。煤系高岭土一般存在于煤层夹矸中，在煤炭开采和洗选过程中常被当作煤矸石废弃，主要由高岭石及碳质页岩等组成，其数量占原煤产量的10%～20%（孔德顺，2014）。煤系高岭土一般呈灰色或黑色，块状结构，壳状断口，隐晶质结构，蠕虫状晶体，结晶有序度高，这种高岭土与煤层具有一定的成因关系，一般厚度为0.3～0.5 m（孔德顺，2014）。煤系石墨是煤及煤系碳质页岩等在岩浆热接触变质及构造变质作用下形成的，多为隐晶质石墨，是石墨矿产的重要组成部分，也是煤系非金属矿产的一种（孙升林等，2014；曹代勇等，2017）。

3.4.3 煤系深时古气候研究进展

煤作为泥炭地的产物和重要的沉积载体，蕴含着大量的泥炭发育时期的古气候信息（邵龙义等，2009）。因此，通过研究煤中碳的聚集速率，进而分析泥炭地的碳聚集速率和净初级生产力，有助于了解地质历史时期碳循环特征，有利于为"深时"古气候的理解提供帮助。闫志明等（2016）和李雅楠等（2018）分别对不同地区进行了研究，进一步计算出泥炭地的碳聚集速率及对应的净初级生产力，从而提供了不同地区"深时"古气候信息。

3.5 其他矿藏古地理研究与找矿规律

矿产沉积学是沉积学与矿床学交叉的新兴方向，属于应用基础学科，其主要任务是应用沉积学的基本原理，探讨成矿元素的迁移-聚集机理和成矿颗粒的风化-搬运-沉积过程，从而恢复沉积矿产的古环境和形成背景，最终目的是确定矿床成因、成矿规律，建立成矿模式和找矿模型，进行成矿预测，为沉积矿产的找矿勘探提供科学依据（杜远生等，2020）。矿产沉积学发展水平明显不如能源沉积学，但随着国际大宗沉积矿产资源的需求增加以及战略紧缺矿产资源受到进一步重视（毛景文等，2019；王登红，2019；翟明国等，2019），尤其是在中国2011年开始实行的"找矿战略突破行动"的带动下，矿产沉积学也蓄势待发。最新针对矿床学研究的综述与总结在沉积矿产方面也有进一步提高与提炼的空间（李建威等，2019）。

3.5.1 中国南方铝土矿矿产沉积学研究进展

近十年来，杜远生团队对中国南方多个地区的铝土矿进行了研究，包括贵州遵义（早石炭世维宪期）、黔中清镇、修文（维宪期）、龙里（杜内期）、黔东凯里（早石炭世维宪期、早二叠世）、广西靖西–德保（中–晚二叠世之交）的铝土矿，总结了沉积型铝土矿的成矿作用和成矿规律（汪小妹等，2013；余文超等，2013，2014；杜远生等，2015；Yu et al.，2015，2016a，2019a；Weng et al.，2019），取得了很多重要认识，如认为沉积型铝土矿的含矿岩系具有类似的四层式剖面结构，尤其是遵义漏斗型铝土矿，具有白色多孔状铝土矿-暗色铝土岩或致密状铝土矿的五个旋回，表明铝土矿形成时期经历了多期次的淋滤旋回。沉积型铝土矿并非在沉积盆地内部水下形成的，而主要是在同生或准同生期陆表大气暴露条件下形成的。沉积作用仅仅提供了形成铝土矿的物质基础，淋滤作用才是铝土矿形成的真正原因，提出了沉积型铝土矿陆表淋滤成矿作用的新认识。

3.5.2 南华纪"大塘坡式"锰矿矿产沉积学研究进展

华南南华纪"大塘坡式"菱锰矿形成于新元古代古城冰期之上的大塘坡组底部，周琦、杜远生团队经过十余年的研究，对该类型锰矿的成矿年代、盆地背景、锰矿富集规律及锰矿成因有了一系列的新发现和新认识。锰矿层中的凝灰岩锆石 U-Pb 同位素年龄约为 660 Ma（余文超等，2016；Zhou et al.，2019），地垒区大塘坡组底部发育盖帽白云岩，地堑区相变为"大塘坡式"锰矿，锰质来源于深部气液流体（Yu et al.，2016b，2017）。"大塘坡式"锰矿既受北东东向地堑盆地控制，也受深部气液流体（提供锰质）的喷溢口控制，周琦等（2013，2017）划分出中心相、过渡相、边缘相，成矿理论的新认识指导了黔东锰矿的找矿重大突破，黔东锰矿新增储量 6 亿多吨。Yu 等（2019b）发现黔东南华纪大塘坡组锰矿的微生物成矿作用，其中微生物氧化固定锰离子阶段需要水体处于氧化–次氧化条件，Wang P. 等（2019）及 Ma 等（2019）也证实南华纪冰期–间冰期的气候转换引起的海洋氧化还原条件变化控制了沉积型锰矿的形成。在此基础上提出了黔东南华纪"大塘坡式"锰矿受内生作用提供锰质源，内生作用与外生作用复合成矿的新认识，以及南华纪大规模成锰作用和 Rodinia 超大陆裂解，新元古代深时古气候及大氧化事件的耦合关系的新认识。

3.5.3 贵州震旦系陡山沱组磷矿矿产沉积学研究进展

黔中地区震旦系陡山沱组磷矿是中国最重要的富磷矿矿集区，杜远生团队通过系统的含矿岩系地层学、定量古地理学和矿产沉积学研究，提出贵州震旦纪磷矿分布于黔中古陆边缘的高能无障壁海岸环境（王泽鹏等，2016；杜远生等，2018；Zhang et al.，2019），黔中震旦纪磷矿经历了初始成矿作用、簸选成矿作用和淋滤成矿作用（杜远生等，2018；张亚冠等，2019；Zhang et al.，2019），即富磷矿的"三阶段"成矿的新认识。张亚冠等（2019）根据黔中震旦纪磷矿的对比研究，认为黔中磷矿 B 矿层较 A 矿层沉积时期海洋氧化程度更高，且伴随深海充氧过程和氧化还原界面的不断变深，磷块岩沉积分布逐渐由浅水海岸延伸至深水陆棚，表明新元古代氧化事件导致的海洋氧气含量增加同时也刺激了大规模成磷作用的发生。

4 存在问题与研究方向展望

4.1 存在问题

目前，虽然古地理研究内容丰富，基础学科研究进展迅速，大量局部或单层系古地理图涌现，但最终成果的古地理图还十分欠缺，就一个盆地范围全层系古地理演化研究的系统性还远远不够，盆地完成全层系的系统古理理演化综合成图还不多。古地理学综合研究是项十分艰巨的任务，需要多学科的协同，而资金投入与项目支持力度相较于其他学科还不够，因此困难重重。同样，分层系全国范围古地理图编制研究（构造、古生物等综合应用）也存在类似问题。在构造古地理方面，全球板块造与古地理重建的研究组还不多（何登发等，2020），他们关于板块位置及古地理随时间变化的看法有一致性吗？这些研究组解释的差异在什么地方？中国典型台地区中新元古界的原生白云岩的大量存在（鲍志东等，2019），当时的沉积古地理条件，包括构造、气候、水介质等如何？这些原生白云岩广泛发育的沉积动力学还需要进一步探究。

4.2 研究方向

（1）开展多类型盆地沉积古地理学研究，恢复原型盆地的沉积面貌和古地理格局，发展构造古地理学。恢复重大构造变革期的多尺度沉积背景；说明中国不同类型沉积盆地地质背景、构造变革与沉积岩性、沉积相带的差异性、长期演化盆地的复原、造山带的古地理恢复、全球尺度的精细的活动论构造-古地理重建以及构造古地理编图等均是未来发展方向。依据中国不同类型盆地构造演化阶段特征、加强物源区母岩及汇水面积、沉积物搬运通道形成发育与定量刻画以及构造变化带对物源通道的控制，沉积区沉积类型与沉积响应特征等方面的综合研究（耦合关系研究），建立基于源-汇系统理论的湖盆沉积学理论体系；加强青藏高原、东亚、印度支那和印度板块地区再造山作用研究，了解不同级次构造活动与沉积作用之间的因果关系；加强不同构造背景、反映生物差异性沉积特征（特别是低等微生物碳酸盐岩）的碳酸盐岩台地模式研究，实现从静态碳酸盐沉积模式转变为活动构造与碳酸盐岩台地演化的动态研究，灰泥丘有可能成为继礁滩、岩溶之后又一个碳酸盐岩油气勘探的新领域；明确构造活动、生物差异性沉积与碳酸盐岩台地建造，生物作用与沉积（成岩）作用的关系。以微生物席为研究对象，研究地球早期生命演变、探索生物圈对水圈和大气圈的长时间影响。

（2）结合中国盆地类型和构造背景研究，形成具有中国区域特色的古地理学理论体系。建立多尺度层序地层格架与沉积古地理之间的关系，用源汇系统新观点探讨沉积古地理分布。加强海、陆相盆地细粒与混积沉积古地理特征和沉积动力学机理研究，形成粗粒沉积、宽缓湖盆浅水三角洲沉积、滩坝沉积、重力流等沉积古地理学理论，详细研究沉积物侵蚀、搬运、堆积过程、机制及沉积环境效应；加强事件沉积学研究，包括不同地质时期构造运动、古地震、古气候的突变、火山活动等诱因造成的正常连续沉积或不连续沉积序列，构造事件、地震事件等触发机制与事件沉积沉积作用的相互关系等；重视陆相湖盆深层细粒及有机质沉积过程、物质转化条件、作用机制以及微相划分、沉积模式研究，建立中国小克拉通盆地碳酸盐岩微地块沉积古地理模式等。

（3）多学科交叉渗透，开展综合定量古地理学研究。综合研究岩性组合、沉积构造、沉积序列、古生物、构造活动过程以及古地理地球物理响应特征，促使古地理学由定性描述向定量研究发展；由宏观古地理研究细化到高精度多信息古地理分析；现代沉积考察、水槽实验和数值模拟等将成为未来古地理学研究的重要手段，实验地质学的发展使古地理学的研究从以野外观察、描述、归纳为主，发展到归纳与演绎并重的阶段；鉴于我国陆相湖盆还存诸如砂体储集层薄、岩性-速度关系变化大、地震分辨薄层砂体难等科学和技术难题，应建立一套适合陆相盆地的地震岩性学新方法，创立各类陆相盆地的地震地貌学模式，建立不同类型陆相盆地地震沉积学研究规范；创新具有中国特色的测井沉积解释模型，发展测井沉积学研究理论、方法与技术，实现不同层次沉积类型的有效识别和岩性油气藏高效勘探；多类型（深层、深水、深海）砂体及储集层定量预测技术，是未来古地理学研究的热点和难点。

（4）构建全球性开放互动的古地理综合数据平台，全面收集、整合已有数据，最终实现基于数据平台的任意时间、任意地区的古地理重现，并为能源勘探、地灾预测、生命演化、气候变化等提供理论、技术及数据支撑，将是古地理重建新的历史使命和未来发展方向。盆地级别甚至全国、全球级别的古地理系统编图势在必行，大数据古地理研究的进一步发展，使得此项工作的可行性越来越大。

（5）创新古地理学研究方法和开展实效应用。创新研究方法开展盆地覆盖区定量古地理学研究；如何建立沉积古地理与矿产资源勘探开发之间的对应关系，以指导矿产资源的高效勘探开发；多学科交叉渗透,形成新的地质分支学科,指导沉积矿产的勘探。如何面向未来人类生存问题,科学地研究人类生存环境。随着全球人口的快速增长,伴随自然资源需求的增加,预示着地球历史研究一个新的时期(第四纪后时期)即将到来。沉积古地理学在预测自然突变(洪水、海啸及风暴……),恢复不平衡的自然系统(河谷、海滩……),控制和预测污染,矿产资源勘探和开发,城市垃圾、核废料和有毒物质处理,以及工程地质方面将发挥重要的作用。

参 考 文 献

鲍志东,季汉成,梁婷,韦明洋,史燕青,李宗峰,鲁锴,向鹏飞,张华,严睿,郭玉鑫,李卓伦,万谱,杨志波,麻晓东,刘锐,刘灿星,钟旭临,郭晓琦,蔡忠贤,张水昌. 2019. 中新元古界原生白云岩:以中国典型台地区为例. 古地理学报,21(6):869-884

曹代勇,张鹤,董业绩,吴国强,宁树正,莫佳峰,李霞. 2017. 煤系石墨矿产地质研究现状与重点方向. 地学前缘,24(5):317-327

陈红汉. 2014. 单个油包裹体显微荧光特性与热成熟度评价. 石油学报,35(3):584-590

陈杨,张建新,黄灿,焦祥燕,罗威. 2019. 莺歌海盆地黄流组轴向重力流水道充填演化特征. 东北石油大学学报,43(6):23-32,61

成秋明. 2019. 深时数字地球:全球古地理重建与深时大数据. 国际学术动态,(6):28-29

杜远生,周琦,金钟国,焦养泉. 2015. 黔务正道地区二叠系铝土矿沉积地质学. 武汉:中国地质大学出版社,27-85

杜远生,陈国勇,张亚冠,刘建中,陈庆刚,赵征. 2018. 贵州省震旦纪陡山沱组磷矿沉积地质学. 武汉:中国地质大学出版社,128-130

杜远生,余文超,张亚冠. 2020. 矿产沉积学:一个新的交叉学科方向. 古地理学报,22(4):601-619

冯文杰. 2017. 同生逆断层对冲断带渠汇体系与砂砾岩体内部构型的控制作用:以准噶尔盆地西北缘三叠系为例. 北京:中国石油大学(北京)博士学位论文

付金华,李士祥,徐黎明,牛小兵. 2018. 鄂尔多斯盆地三叠系延长组长7段古沉积环境恢复及意义. 石油勘探与开发,45(6):936-946

何登发,马永生,刘波,蔡勋育,张义杰,张健. 2019. 中国含油气盆地深层勘探的主要进展与科学问题. 地学前缘,26(1):1-12

何登发,李德生,王成善,刘少峰,陈槲俊. 2020. 活动论构造古地理的研究现状、思路与方法. 古地理学报,22(1):1-28

贺聪,吉利明,苏奥,刘颖,李剑锋,吴远东,张明震. 2017. 鄂尔多斯盆地南部延长组热水沉积作用与烃源岩发育的关系. 地学前缘,24(6):277-285

侯高峰,纪友亮,吴浩,李淋淋,王永诗,王伟. 2017. 物理模拟法定量表征碎屑岩储层物性影响因素. 地质科技情报,36(4):153-159

黄捍东,曹学虎,罗群. 2011. 地震沉积学在生物礁滩预测中的应用:以川东褶皱带建南-龙驹坝地区为例. 石油学报,32(4):629-636

黄献好,孙玉琦,王伟超,刘炳强,张少林,张语涛,邵龙义. 2020. 柴北缘西大滩地区下-中侏罗统层序-古地理及聚煤特征. 沉积学报,38(2):266-283

黄志龙,朱ígнастасения,马剑,吴红烛,张伙兰. 2015. 莺歌海盆地东方区高温高压带黄流组储层特征及高孔低渗成因. 石油与天然气地质,36(2):288-296

孔德顺. 2014. 煤系高岭土及其应用研究进展. 化工技术与开发,43(7):39-41

旷红伟,柳永清,彭楠,刘燕学,李家华. 2011. 再论白齿碳酸盐岩成因. 古地理学报,13(3):253-261

李弛,罗静兰,胡海燕,陈淑慧,王代富,柳保军,马永坤,陈亮,李晓艳. 2019. 热动力条件对白云凹陷深水区珠海组砂岩成岩演化过程的影响. 地球科学,44(2):572-587

李华,何幼斌. 2017. 等深流沉积研究进展. 沉积学报,35(2):228-240

李华,王英民,徐强,唐武,李冬. 2013. 南海北部第四系深层等深流沉积特征及类型. 古地理学报,15(5):741-750

李华,何幼斌,黄伟,刘朱睿鸷,张锦. 2016. 鄂尔多斯盆地南缘奥陶系平凉组等深流沉积. 古地理学报,18(4):631-642

李建威,赵新福,邓晓东,谭俊,胡浩,张东阳,李占轲,李欢,荣辉,杨梅珍,曹康,靳晓野,隋吉祥,祖波,昌佳,吴亚飞,文广,赵少瑞. 2019. 新中国成立以来中国矿床学研究若干重要进展. 中国科学:地球科学,49(11):1720-1771

李江海,姜洪福. 2013. 全球板块再造、岩相古地理及古环境图集. 北京:地质出版社

李明,李飞,杨宗恒,王锦西,梁锋,李正勇,梁菁,唐青松. 2019. 基于地震沉积学原理的河道砂体精细刻画:以四川盆地龙岗地区沙溪庙组致密气藏为例. 天然气勘探与开发,42(2):76-83

李双应,魏星,谢伟,程成,李敏,胡博,柴广路. 2019. 从源到汇:大别山造山带物源区与合肥盆地南缘中生代沉积耦合关系:来自碎屑锆石U-Pb年龄证据. 古地理学报,21(1):82-106

李文,何生,张柏桥,何治亮,陈曼霏,张殿伟,李天义,高键. 2018. 焦石坝背斜西缘龙马溪组页岩复合脉体中流体包裹体的古温度及古压力特征. 石油学报,39(4):402-415

李向东. 2013. 关于深水环境下内波、内潮汐沉积分类的探讨. 地质论评,59(6):1097-1109

李向东,邹雅棋. 2017. 鄂尔多斯盆地西缘桌子山地区奥陶系深水条纹条带状泥岩等深流成因分析. 古地理学报,19(6):987-997

李向东,陈海燕,陈洪达. 2019. 鄂尔多斯盆地西缘桌子山地区上奥陶统拉什仲组深水复合流沉积. 地球科学进展,34(12):1301-1315

李雅楠,邵龙义,闫志明,侯海海,唐跃,Large D J. 2018. 中侏罗世泥炭地净初级生产力及控制因素:以准噶尔盆地南缘煤田为例. 中国科学:地球科学,48(10):1324-1334

李增学,魏久传,韩美莲. 2001. 海侵事件成煤作用:一种新的聚煤模式. 地球科学进展,16(1):120-124

李忠. 2016. 盆地深层流体-岩石作用与油气形成研究前沿. 矿物岩石地球化学通报,35(5):807-816

李忠,罗威,曾冰艳,刘嘉庆,于靖波. 2018. 盆地多尺度构造驱动的流体-岩石作用及成储效应. 地球科学,43(10):3498-3510

梁冰, 石迎爽, 孙维吉, 刘强. 2016. 中国煤系"三气"成藏特征及共采可能性. 煤炭学报, 41(1): 167–173

梁钰, 侯读杰, 张金川, 杨光庆. 2014. 缺氧环境下热液活动对页岩有机质丰度的影响. 大庆石油地质与开发, 33(4): 158–165

林畅松, 夏庆龙, 施和生, 周心怀. 2015. 地貌演化、源–汇过程与盆地分析. 地学前缘, 22(1): 9–20

蔺鹏, 吴胜和, 张佳佳, 胡光义, 夏钦禹, 范洪军, 王南溆. 2018. 尼日尔三角洲盆地陆坡逆冲构造区海底扇分布规律. 石油与天然气地质, 39(5): 1073–1086

刘池洋, 毛光周, 邱欣卫, 吴柏林, 赵红格, 王建强. 2013. 有机–无机能源矿产相互作用及其共存成藏(矿). 自然杂志, 35(1): 47–55

刘东娜, 曾凡桂, 赵峰华, 王红冬, 解锡超, 邹雨. 2018. 山西省煤系伴生三稀矿产资源研究现状及找矿前景. 煤田地质与勘探, 46(4): 1–7

刘海, 林承焰, 张宪国, 栗宝鹃. 2018. 黄骅坳陷孔店地区馆陶组地震沉积特征及沉积演化模式. 中国矿业大学学报, 47(3): 549–561

刘力辉, 陈珊, 倪长宽. 2013. 叠前有色反演技术在地震岩性学研究中的应用. 石油物探, 52(2): 171–176

刘强虎, 朱筱敏, 李顺利, 徐长贵, 杜晓峰, 李慧勇, 石文龙. 2017. 沙垒田凸起西部断裂陡坡型源–汇系统. 地球科学, 42(11): 1883–1896

刘招君, 孙平昌, 柳蓉, 孟庆涛, 胡菲. 2016. 敦密断裂带盆地群油页岩特征及成矿差异分析. 吉林大学学报(地球科学版), 46(4): 1090–1099

柳波, 石佳欣, 付晓飞, 吕延防, 孙先达, 巩磊, 白云风. 2018. 陆相泥页岩层系岩相特征与页岩油富集条件: 以松辽盆地古龙凹陷白垩系青山口组一段富有机质泥页岩为例. 石油勘探与开发, 45(5): 828–838

柳永清, 旷红伟, 彭楠, 刘燕学, 江小均, 许欢. 2010. 中国元古代碳酸盐岩微亮晶构造及形成的沉积环境约束. 岩石学报, 26(7): 2122–2130

鲁静, 邵龙义, 杨敏芳, 李永红, 张正飞, 王帅, 云启成. 2014. 陆相盆地沼泽体系煤相演化、层序地层与古环境. 煤炭学报, 39(12): 2473–2481

罗静兰, 邵红梅, 杨艳芳, 李砾, 罗春燕. 2013. 松辽盆地深层火山岩储层的埋藏–烃类充注–成岩时空演化过程. 地学前缘, 20(5): 175–187

罗静兰, 何敏, 庞雄, 李驰, 柳保军, 雷川, 马永坤, 庞江. 2019. 珠江口盆地南部热演化事件与高地温梯度的成岩响应及其对油气勘探的启示. 石油学报, 40(S1): 90–104

罗泉源, 焦祥燕, 刘昆, 李安琪, 宋鹏. 2020. 乐东–陵水凹陷梅山组海底扇识别及沉积模式. 海洋地质与第四纪地质, 40(2): 90–99

马勇新, 黄银涛, 姚光庆, 成涛, 潘石坚. 2015. 莺歌海盆地DX区黄流组超压对成岩作用的影响. 地质科技情报, 34(3): 7–14

毛景文, 杨宗喜, 谢桂青, 袁顺达, 周振华. 2019. 关键矿产: 国际动向与思考. 矿床地质, 38(4): 689–698

孟元林, 吴琳, 孙洪斌, 吴晨亮, 胡安文, 张磊, 赵紫桐, 施立冬, 许丞, 李晨. 2015. 辽河西部凹陷南段异常低压背景下的成岩动力学研究与成岩相预测. 地学前缘, 22(1): 206–214

孟志勇. 2016. 四川盆地涪陵地区五峰组–龙马溪组含气页岩段纵向非均质性及其发育主控因素. 石油与天然气地质, 37(6): 838–846

牟传龙, 周恳恳, 陈小炜. 2016. 中国岩相古地理图集(埃迪卡拉纪—志留纪). 北京: 地质出版社, 1–154

牛永斌, 胡亚洲, 高文秀, 董小波, 崔胜利. 2018. 豫西北奥陶系马家沟组三段遗迹组构及沉积演化规律. 地质学报, 92(1): 15–27

潘树新, 刘化清, Zavala C, 刘彩燕, 梁苏娟, 张庆石, 白忠峰. 2017. 大型坳陷湖盆异重流成因的水道–湖底扇系统: 以松辽盆地白垩系嫩江组一段为例. 石油勘探与开发, 44(6): 860–870

彭勇民, 龙胜祥, 胡宗全, 杜伟, 顾志翔, 方屿. 2016. 四川盆地涪陵地区页岩岩石相标定方法与应用. 石油与天然气地质, 37(6): 964–970

平宏伟, 陈红汉, Thiéry R, 张晖, 李培军, 吴楠. 2014. 原油裂解对油包裹体均一温度和捕获压力的影响及其地质意义. 地球科学—中国地质大学学报, 39(5): 587–600

齐永安, 王敏, 李姐, 孙长彦, 代明月. 2012. 洛阳龙门地区中寒武统张夏组下部遗迹组构及其沉积环境. 地球科学-中国地质大学学报, 37(4): 693–706

秦伯强, 胡维平, 陈伟民. 2004. 太湖水环境演化过程与机理. 北京: 科学出版社

邱振, 卢斌, 陈振宏, 张蓉, 董大忠, 王红岩, 邱军利. 2019. 火山灰沉积与页岩有机质富集关系探讨: 以五峰组—龙马溪组含气页岩为例. 沉积学报, 37(6): 1296–1308

单祥, 郭华军, 邹志文, 李亚哲, 王力宝. 2018. 碱性环境成岩作用及其对储集层质量的影响: 以准噶尔盆地西北缘中—下二叠统碎屑岩储集层为例. 新疆石油地质, 39(1): 55–62

邵龙义, 鲁静, 汪浩, 张鹏飞. 2009. 中国含煤岩系层序地层学研究进展. 沉积学报, 27(5): 904–914

邵龙义, 汪浩, Large D J. 2011. 中国西南地区晚二叠世泥炭地净初级生产力及其控制因素. 古地理学报, 13(5): 473–480

邵龙义, 董大啸, 李明培, 王海生, 王东东, 鲁静, 郑明泉, 程爱国. 2014a. 华北石炭–二叠纪层序–古地理与聚煤规律. 煤炭学报, 39(8): 1725–1734

邵龙义, 李英娇, 靳凤仙, 高彩霞, 张超, 梁万林, 黎光明, 陈忠恕, 彭正奇, 程爱国. 2014b. 华南地区晚三叠世含煤岩系层序–古地理. 古地理学报, 16(5): 613–630

邵龙义, 张超, 闫志明, 董大啸, 高彩霞, 李英娇, 徐晓燕, 梁万林, 易同生, 徐锡惠, 黎光明, 陈忠恕, 程爱国. 2016. 华南晚二叠世层序–古地理及聚煤规律. 古地理学报, 18(6): 905–919

邵龙义, 王学天, 鲁静, 王东东, 侯海海. 2017. 再论中国含煤岩系沉积学研究进展及发展趋势. 沉积学报, 35(5): 1016–1031

邵龙义, 王学天, 李雅楠, 刘炳强. 2019. 深时源–汇系统古地理重建方法评述. 古地理学报, 21(1): 67–81

斯尚华, 陈红汉, 袁丙龙, 雷明珠, 陈杨. 2018. 利用油包裹体荧光光谱多参数划分油气充注幕次: 以塔里木盆地麦盖提斜坡巴什托构造带石炭系为例. 海相油气地质, 23(2): 25–30

宋慧波, 郭瑞睿, 王保玉, 胡斌. 2014. 华北下二叠统太原组Zoophycos潜穴中碳酸盐岩微形体的特征及其意义. 沉积学报, 32(5): 797–808

宋慧波, 王芳, 胡斌. 2015. 晋中南地区上石炭统—下二叠统太原组碳酸盐岩中遗迹组构及其沉积环境. 沉积学报, 33(6): 1126–1139

宋慧波, 毕瑜珺, 胡斌. 2017. 豫西下二叠统太原组遗迹化石与古氧相的响应特征. 古地理学报, 19(4): 653–662

宋宇. 2017. 老黑山盆地下白垩统穆棱组油页岩与煤成矿机制的精细研究. 长春: 吉林大学博士学位论文

孙龙德, 邹才能, 朱如凯, 张云辉, 张水昌, 张宝民, 朱光有, 高志勇. 2013. 中国深层油气形成、分布与潜力分析. 石油勘探与开发, 40(6):

641-649

孙平昌. 2013. 松辽盆地东南部上白垩统含油页岩系有机质富集环境动力学. 长春：吉林大学博士学位论文

孙升林, 吴国强, 曹代勇, 宁树正, 乔军伟, 朱华雄, 韩亮, 朱示乙, 苗琦, 周兢, 刘亢, 李聪聪, 陈寒勇, 蔡旭梅. 2014. 煤系矿产资源及其发展趋势. 中国煤炭地质, 26(11)：1-11

孙枢, 王成善. 2009. "深时"（Deep Time）研究与沉积学. 沉积学报, 27(5)：792-810

万晓樵, 吴怀春, 席党鹏, 刘美羽, 覃祚焕. 2017. 中国东北地区白垩纪温室时期陆相生物群与气候环境演化. 地学前缘, 24(1)：18-31

汪小妹, 焦养泉, 杜远生, 周琦, 崔滔, 计波, 雷志远, 翁申富, 金中国, 熊星. 2013. 黔北务正道地区铝土矿稀土元素地球化学特征. 地质科技情报, 32(1)：27-33

王东东, 邵龙义, 刘海燕, 邵凯, 于得明, 刘炳强. 2016. 超厚煤层成因机制研究进展. 煤炭学报, 41(6)：1487-1497

王登红. 2019. 关键矿产的研究意义、矿种厘定、资源属性、找矿进展、存在问题及主攻方向. 地质学报, 93(6)：1189-1209

王剑, 谭富文, 付修根. 2015. 沉积岩工作方法. 北京：地质出版社

王剑, 江新胜, 卓皆文, 崔晓庄, 江卓斐, 魏亚楠, 蔡娟娟, 廖忠礼. 2019. 华南新元古代裂谷盆地演化与岩相古地理. 北京：科学出版社

王佟, 刘天绩, 邵龙义, 曹代勇, 郭晋宁, 刘益芬, 文怀军, 王丹. 2009. 青海木里煤田天然气水合物特征与成因. 煤田地质与勘探, 37(6)：26-30

王佟, 王庆伟, 傅雪海. 2014. 煤系非常规天然气的系统研究及其意义. 煤田地质与勘探, 42(1)：24-27

王玉满, 董大忠, 李新景, 黄金亮, 王淑芳, 吴伟. 2015. 四川盆地及其周缘下志留统龙马溪组层序与沉积特征. 天然气工业, 35(3)：12-21

王泽鹏, 张亚冠, 杜远生, 陈国勇, 刘建中, 徐园园, 谭代卫, 李磊, 王大福, 吴文明. 2016. 黔中开阳磷矿沉积区震旦纪陡山沱期定量岩相古地理重建. 古地理学报, 18(3)：399-410

吴胜和, 冯文杰, 印森林, 喻宸, 张可. 2016. 冲积扇沉积构型研究进展. 古地理学报, 18(4)：497-512

吴胜和, 徐振华, 刘钊. 2019. 河控浅水三角洲沉积构型. 古地理学报, 21(2)：202-215

解习农, 林畅松, 李忠, 任建业, 姜涛, 姜在兴, 雷超. 2017. 中国盆地动力学研究现状及展望. 沉积学报, 35(5)：877-887

徐振华, 吴胜和, 刘钊, 赵军寿, 吴峻川, 耿红柳, 张天佑, 刘照玮. 2019. 浅水三角洲前缘指状砂坝构型特征：以渤海湾盆地渤海 BZ25 油田新近系明化镇组下段为例. 石油勘探与开发, 46(2)：322-333

闫志明, 邵龙义, 王帅, Large D J, 汪浩, 孙钦平. 2016. 早白垩世泥炭地净初级生产力及其控制因素：来自二连盆地吉尔嘎郎图凹陷6号煤的证据. 沉积学报, 34(6)：1068-1076

杨红君, 郭书生, 刘博, 卢庆治, 佟彦明. 2013. 莺歌海盆地 SE 区上中新统重力流与内波内潮汐沉积新认识. 石油实验地质, 35(6)：626-633

杨华, 牛小兵, 徐黎明, 冯胜斌, 尤源, 梁晓伟, 王芳, 张丹丹. 2016. 鄂尔多斯盆地三叠系长 7 段页岩油勘探潜力. 石油勘探与开发, 43(4)：511-520

伊帆, 朱利东, 刘显凡, 伊海生. 2016. 藏北羌塘盆地双湖地区下侏罗统油页岩的有机碳同位素异常和正构烷烃分布特征及大洋缺氧事件研究. 矿物学报, 36(3)：413-422

印森林. 2014. 同沉积逆断裂对冲积体系及其内部构型的控制作用：以准噶尔盆地西北缘三叠系为例. 北京：中国石油大学（北京）博士学位论文, 30-90

印森林, 吴胜和, 李俊飞, 冯文杰. 2014. 同生逆断层正牵引构造对高频层序地层结构及沉积充填的控制作用. 地质论评, 60(2)：310-320

印森林, 唐勇, 胡张明, 吴涛, 张磊, 张纪易. 2016. 构造活动对冲积扇及其油气成藏的控制作用：以准噶尔盆地西北缘二叠系—三叠系冲积扇为例. 新疆石油地质, 37(4)：391-400

余文超, 杜远生, 顾松竹, 崔滔, 黄兴, 喻建新, 覃永军, 雷志远, 翁申富, 曹建州. 2013. 黔北务正道地区早二叠世铝土矿多期淋滤作用及其控矿意义. 地质科技情报, 32(1)：34-39

余文超, 杜远生, 周琦, 金中国, 汪小妹, 覃永军, 崔滔. 2014. 黔北务正道地区下二叠统铝土矿层物源研究：来自碎屑锆石年代学的证据. 古地理学报, 16(1)：19-29

余文超, 杜远生, 周琦, 王萍, 袁良军, 徐源, 潘文, 谢小峰, 齐靓, 焦良轩. 2016. 黔东松桃地区大塘坡组 LA-ICP-MS 锆石 U-Pb 年龄及其地质意义. 地质论评, 62(3)：539-549

袁伟, 柳广弟, 徐黎明, 牛小兵. 2019. 鄂尔多斯盆地延长组 7 段有机质富集主控因素. 石油与天然气地质, 40(2)：326-334

袁选俊, 林森虎, 刘群, 姚泾利, 王岚, 郭浩, 邓秀芹, 成大伟. 2015. 湖盆细粒沉积特征与富有机质页岩分布模式：以鄂尔多斯盆地延长组长 7 油层组为例. 石油勘探与开发, 42(1)：34-43

曾洪流, 朱筱敏, 朱如凯, 张庆石. 2013. 砂岩成岩相地震预测：以松辽盆地齐家凹陷青山口组为例. 石油勘探与开发, 40(3)：266-274

翟明国, 吴福元, 胡瑞忠, 蒋少涌, 李文昌, 王汝成, 王登红, 齐涛, 秦克章, 温汉捷. 2019. 战略性关键金属矿产资源：现状与问题. 中国科学基金, (2)：106-111

张功成, 屈红军, 赵冲, 张凤廉, 赵钊. 2017. 全球深水油气勘探 40 年大发现及未来勘探前景. 天然气地球科学, 28(10)：1447-1477

张光亚, 童晓光, 辛仁臣, 温志新, 马锋, 黄彤飞, 王兆明, 于炳松, 李曰俊, 陈汉林, 刘小兵, 刘祚冬. 2019a. 全球岩相古地理演化与油气分布（一）. 石油勘探与开发, 46(4)：633-652

张光亚, 童晓光, 辛仁臣, 温志新, 马锋, 黄彤飞, 王兆明, 于炳松, 李曰俊, 陈汉林, 刘小兵, 刘祚冬. 2019b. 全球岩相古地理演化与油气分布（二）. 石油勘探与开发, 46(5)：848-868

张国伟, 郭安林, 王岳军, 李三忠, 董云鹏, 刘少峰, 何登发, 程顺有, 鲁如魁, 姚安平. 2013. 中国华南大陆构造与问题. 中国科学：地球科学, 43(10)：1553-1582

张洪运, 庄丽华, 阎军, 马小川. 2017. 南海北部东沙群岛西部海域的海底沙波与内波的研究进展. 海洋科学, 41(10)：149-157

张伙兰, 裴健翔, 谢金有, 于俊峰, 艾能平. 2014. 莺歌海盆地东方区黄流组一段超压储层孔隙结构特征. 中国海上油气, 26(1): 30–38

张佳佳, 吴胜和. 2019. 海底扇朵叶沉积构型研究进展. 中国海上油气, 31(5): 88–106

张蕾, 钟瀚霆, 陈安清, 赵应权, 黄可可, 李凤杰, 黄虎, 刘宇, 曹海洋, 祝圣贤, 穆财能, 侯明才, James G O. 2020. 大数据驱动下的数字古地理重建: 现状与展望. 高校地质学报, 26(1): 73–85

张丽, 陈淑慧. 2017. 珠江口盆地东部地区不同地温梯度下储层特征响应关系. 中国海上油气, 29(1): 29–38

张宪国, 张涛, 鞠传学, 林承焰, 董春梅, 林建力, 韩硕. 2018-01-09. 一种基于多尺度信息的致密储层成岩相预测方法: 中国, CN201510895321.7

张亚冠, 杜远生, 陈国勇, 刘建中, 陈庆刚, 赵征, 王泽鹏, 邓超. 2019. 富磷矿三阶段动态成矿模式: 黔中开阳式高品位磷矿成矿机制. 古地理学报, 21(2): 351–368

赵文智, 胡素云, 刘伟, 王铜山, 李永新. 2014. 再论中国陆上深层海相碳酸盐岩油气地质特征与勘探前景. 天然气工业, 34(4): 1–9

赵文智, 王晓梅, 胡素云, 张水昌, 王华建, 管树巍, 叶云涛, 任荣, 王铜山. 2019. 中国元古宇烃源岩成烃特征及勘探前景. 中国科学: 地球科学, 49(6): 939–964

赵晓明, 刘丽, 谭程鹏, 范廷恩, 胡光义, 张迎春, 张文彪, 宋来明. 2018. 海底水道体系沉积构型样式及控制因素: 以尼日尔三角洲盆地陆坡区为例. 古地理学报, 20(5): 825–840

郑和荣, 胡宗全. 2010. 中国前中生代构造-岩相古地理图集. 北京: 地质出版社

郑红波, 阎贫, 邢玉清, 王彦林. 2012. 反射地震方法研究南海北部的深水底流. 海洋学报, 34(2): 192–198

郑伟, 齐永安, 张忠慧, 邢智峰. 2016. 豫西荥阳陆相二叠纪-三叠纪之交的微生物成因构造(MISS)及其地质意义. 地球科学进展, 31(7): 737–750

支东明, 唐勇, 杨智峰, 郭旭光, 郑孟林, 万敏, 黄立良. 2019. 准噶尔盆地吉木萨尔凹陷陆相页岩油地质特征与聚集机理. 石油与天然气地质, 40(3): 524–534

周琦, 杜远生, 覃英. 2013. 古天然气渗漏沉积型锰矿床成矿系统与成矿模式: 以黔湘渝毗邻区南华纪"大塘坡式"锰矿为例. 矿床地质, 32(3): 457–466

周琦, 杜远生, 袁良军, 张遂, 杨炳南, 潘文, 余文超, 王萍, 徐源, 齐靓, 刘雨, 覃永军, 谢小峰. 2017. 古天然气渗漏沉积型锰矿床找矿模型: 以黔湘渝毗邻区南华纪"大塘坡式"锰矿为例. 地质学报, 91(10): 2285–2298

周志恒, 钟大康, 凡睿, 王爱, 唐自成, 孙海涛, 王威, 杜红权. 2019. 致密砂岩中岩屑溶蚀及其伴生胶结对孔隙发育的影响: 以川东北元坝西部须二下亚段为例. 中国矿业大学学报, 48(3): 592–603, 615

朱红涛, 徐长贵, 朱筱敏, 曾洪流, 姜在兴, 刘可禹. 2017. 陆相盆地源-汇系统要素耦合研究进展. 地球科学, 42(11): 1851–1870

朱筱敏, 董艳蕾, 曾洪流, 黄捷东, 刘强虎, 秦祎, 叶蕾. 2019. 沉积地质学发展新航程: 地震沉积学. 古地理学报, 21(2): 189–201

朱筱敏, 董艳蕾, 曾洪流, 林承焰, 张宪国. 2020. 中国地震沉积学研究现状和发展思考. 古地理学报, 22(3): 397–411

祝彦贺, 朱伟林, 徐强, 吴景富. 2011. 珠江口盆地13.8Ma陆架边缘三角洲与陆坡深水扇的"源-汇"关系. 中南大学学报(自然科学版), 42(12): 3827–3834

邹才能, 翟光明, 张光亚, 王红军, 张国生, 李建忠, 王兆明, 温志新, 马锋, 梁英波, 杨智, 李欣, 梁坤. 2015. 全球常规-非常规油气形成分布、资源潜力及趋势预测. 石油勘探与开发, 42(1): 13–25

Ajdukiewicz J M, Larese R E. 2012. How clay grain coats inhibit quartz cement and preserve porosity in deeply buried sandstones: observations and experiments. AAPG Bulletin, 96(11): 2091–2119

Bechtel A, Jia J L, Strobl S A I, Sachsenhofer R F, Liu Z J, Gratzer R, Püttmann W. 2012. Palaeoenvironmental conditions during deposition of the Upper Cretaceous oil shale sequences in the Songliao Basin (NE China): implications from geochemical analysis. Organic Geochemistry, 46: 76–95

Bourgault D, Morsilli M, Richards C, Neumeier U, Kelley D E. 2014. Sediment resuspension and nepheloid layers induced by long internal solitary waves shoaling orthogonally on uniform slopes. Continental Shelf Research, 72: 21–33

Brüch A, Maghous S, Ribeiro F L B, Dormieux L. 2016. A constitutive model for mechanical and chemo-mechanical compaction in sedimentary basins and finite element analysis. International Journal for Numerical and Analytical Methods in Geomechanics, 40(16): 2238–2270

Burpee A P, Slingerland R L, Edmonds D A, Parsons D, Best J, Cederberg J, McGuffin A, Caldwell R, Nijhuis A, Royce J. 2015. Grain-size controls on the morphology and internal geometry of river dominated deltas. Journal of Sedimentary Research, 85(6): 699–714

Caldwell R L, Edmonds D A. 2014. The effects of sediment properties on deltaic processes and morphologies: a numerical modeling study. Journal of Geophysical Research: Earth Surface, 119(5): 961–982

Cao H S, Kaufman A J, Shan X L, Cui H, Zhang G J. 2016. Sulfur isotope constraints on marine transgression in the lacustrine upper Cretaceous Songliao Basin, northeastern China. Palaeogeography, Palaeoclimatology, Palaeoecology, 451: 152–163

Chen H, Zhang W Y, Xie X N, Ren J Y. 2019. Sediment dynamics driven by contour currents and mesoscale eddies along continental slope: a case study of the northern South China Sea. Marine Geology, 409: 48–66

Chen Z Q, Tong J N, Fraiser M L. 2011. Trace fossil evidence for restoration of marine ecosystems following the end-Permian mass extinction in the Lower Yangtze region, South China. Palaeogeography, Palaeoclimatology, Palaeoecology, 299(3-4): 449–474

Chu D L, Tong J N, Bottjer D J, Song H J, Song H Y, Benton M J, Tian L, Guo W W. 2017. Microbial mats in the terrestrial Lower Triassic of North China and implications for the Permian-Triassic mass extinction. Palaeogeography, Palaeoclimatology, Palaeoecology, 474: 214–231

Denny A C, Kozdon R, Kitajima K, Valley J W. 2017. Isotopically zoned carbonate cements in Early Paleozoic sandstones of the Illinois Basin: $\delta^{18}O$ and $\delta^{13}C$ records of burial and fluid flow. Sedimentary Geology, 361: 93–110

Domeier M. 2016. A plate tectonic scenario for the Iapetus and Rheic Oceans. Gondwana Research, 36: 275–295

Domeier M, Torsvik T H. 2014. Plate tectonics in the late Paleozoic. Geoscience Frontiers, 5(3): 303–350

Duan W, Li C F, Luo C F, Chen X G, Bao X H. 2018. Effect of formation overpressure on the reservoir diagenesis and its petroleum geological significance for the DF11 Block of the Yinggehai Basin, the South China Sea. Marine and Petroleum Geology, 97: 49–65

Fan R Y, Gong Y M. 2016. Ichnological and sedimentological features of the Hongguleleng Formation (Devonian-Carboniferous transition) from the western Junggar, NW China. Palaeogeography, Palaeoclimatology, Palaeoecology, 448: 207–223

Fonnesu M, Palermo D, Galbiati M, Marchesini M, Bonamini E, Bendias D. 2020. A new world-class deep-water play-type, deposited by the syndepositional interaction of turbidity flows and bottom currents: the giant Eocene Coral Field in northern Mozambique. Marine and Petroleum Geology, 111: 179–201

Gao Z Z, He Y B, Li X D, Duan T Z, Wang Y, Liu M. 2013. Review of research in internal-wave and internal-tide deposits of China. Journal of Palaeogeography, 2(1): 56–65

Gao Z Z, He Y B, Li X D, Duan T Z, Wang Y. 2014. Reply to Shanmugam, G. Review of research in internal-wave and internal-tide deposits of China: discussion. Journal of Palaeogeography, 3(4): 351–358

Ghinassi M, Ielpi A. 2015. Stratal architecture and morphodynamics of downstream-migrating fluvial point bars (Jurassic Scalby Formation, U.K.). Journal of Sedimentary Research, 85(9): 1123–1137

Ghinassi M, Ielpi A, Aldinucci M, Fustic M. 2016. Downstream-migrating fluvial point bars in the rock record. Sedimentary Geology, 334: 66–96

Gong C L, Wang Y M, Rebesco M, Salon S, Steel R J. 2018. How do turbidity flows interact with contour currents in unidirectionally migrating deep-water channels? Geology, 46(6): 551–554

Guo B, Shao L Y, Hilton J, Wang S, Zhang L. 2018. Sequence stratigraphic interpretation of peatland evolution in thick coal seams: examples from Yimin Formation (Early Cretaceous), Hailaer Basin, China. International Journal of Coal Geology, 196: 211–231

Gurnis M, Yang T, Cannon J, Turner M, Williams S, Flament N, Müller R D. 2018. Global tectonic reconstructions with continuously deforming and evolving rigid plates. Computers & Geosciences, 116: 32–41

He H Y, Deng C L, Wang P J, Pan Y X, Zhu R X. 2012. Toward age determination of the termination of the Cretaceous normal superchron. Geochemistry, Geophysics, Geosystems, 13(2): Q02002

He Y B, Luo J X, Li X D, Gao Z Z, Wen Z. 2011. Evidence of internal-wave and internal-tide deposits in the Middle Ordovician Xujiajuan Formation of the Xiangshan Group, Ningxia, China. Geo-Marine Letter, 31(5-6): 509–523

He Y B, Luo J X, Gao Z Z, Wen Z. 2012. Reply to the discussion of He et al. (2011, Geo-Marine Letters) evidence of internal-wave and internal-tide deposits in the Middle Ordovician Xujiajuan Formation of the Xiangshan Group, Ningxia, China. Geo-Marine Letters, 32(4): 367–372

Heap M J, Brantut N, Baud P, Meredith P G. 2015. Time-dependent compaction band formation in sandstone. Journal of Geophysical Research: Solid Earth, 120(7): 4808–4830

Heron P J. 2019. Mantle plumes and mantle dynamics in the Wilson cycle. In: Wilson R W, Houseman G A, Mccaffrey K J W, Doré A G, Buiter S J H (eds). Fifty Years of the Wilson Cycle Concept in Plate Tectonics. London: Geological Society, London, Special Publications, 87–103

Hodgskiss M S W, Kunzmann M, Poirier A, Halverson G P. 2018. The role of microbial iron reduction in the formation of Proterozoic molar tooth structures. Earth and Planetary Science Letters, 482: 1–11

Hu B, Wang Y Y, Song H B, Wang Y, Liu M. 2014. The ichnofacies and ichnoassemblages in terrestrial deposits of China. Journal of Palaeogeography, 3(1): 61–73

Huang H D, Yuan S Y, Zhang Y T, Zeng J, Mu W T. 2016. Use of nonlinear chaos inversion in predicting deep thin lithologic hydrocarbon reservoirs: a case study from the Tazhong oil field of the Tarim Basin, China. Geophysics, 81(6): B221-B234

Jeppesen E, Meerhoff M, Jacobsen B A, Hansen R S, Søndergaard M, Jensen J P, Lauridsen T L, Mazzeo N, Branco C W C. 2007. Restoration of shallow lakes by nutrient control and biomanipulation: the successful strategy varies with lake size and climate. Hydrobiologia, 581(1): 269–285

Joe H S M, Kevin G T, Margaret K, David P. 2014. Compositional controls on early diagenetic pathways in fine-grained sedimentary rocks: implications for predicting unconventional reservoir attributes of mudstones. AAPG Bulletin, 98(3): 587–603

Kuang H W, Hu X F. 2014. Review of molar tooth structure research. Journal of Palaeogeography, 3(4): 359–383

Lei C, Luo J L, Pang X, Li C, Pang J, Ma Y K. 2018. Impact of temperature and geothermal gradient on sandstone reservoir quality: the baiyun sag in the Pearl River mouth basin study case (northern South China Sea). Minerals, 8(10): 452, doi: 10.3390/min8100452

Li H, van Loon A J, He Y B. 2019. Interaction between turbidity currents and a contour current: a rare example from the Ordovician of Shaanxi Province, China. Geologos, 25(1): 15–30

Li H, van Loon A J, He Y B. 2020. Cannibalism of contourites by gravity flows: explanation of the facies distribution of the Ordovician Pingliang Formation along the southern margin of the Ordos Basin, China. Canadian Journal of Earth Sciences, 57(3): 331–347

Li S Z, Li X Y, Wang G Z, Liu Y M, Wang Z C, Wang T S, Cao X Z, Guo X Y, Somerville I, Li Y, Zhou J, Dai L M, Jiang S H, Zhao H, Wang Y, Wang G, Yu S. 2019. Global Meso-Neoproterozoic plate reconstruction and formation mechanism for Precambrian basins: constraints from three cratons in China. Earth-Science Reviews, 198: 102946, doi: 10.1016/j.earscirev.2019.102946

Luo Y N, Huang H D, Yang Y D, Li Q X, Zhang S, Zhang J W. 2018. Deepwater reservoir prediction using broadband seismic-driven impedance inversion and seismic sedimentology in the South China Sea. Interpretation, 6(4): SO17-SO29

Lv D W, Wang D D, Li Z X, Liu H Y, Li Y. 2017. Depositional environment, sequence stratigraphy and sedimentary mineralization mechanism in the coal bed- and oil shale-bearing succession: a case from the Paleogene Huangxian Basin of China. Journal of Petroleum Science and Engineering, 148: 32-51

Ma Z X, Liu X T, Yu W C, Du Y S, Du Q D. 2019. Redox conditions and manganese metallogenesis in the Cryogenian Nanhua Basin: insight from the basal Datangpo Formation of South China. Palaeogeography, Palaeoclimatology, Palaeoecology, 529: 39-52

Morra G, Seton M, Quevedo L, Müller R D. 2013. Organization of the tectonic plates in the last 200 Myr. Earth and Planetary Science Letters, 373: 93-101

Nader F H, Champenois F, Barbier M, Adelinet M, Rosenberg E, Houel P, Delmas J, Swennen R. 2016. Diagenetic effects of compaction on reservoir properties: the case of early callovian "Dalle Nacrée" formation (Paris Basin, France). Journal of Geodynamics, 101: 5-29

Ogg J G, Scotese C R, Hou M C, Chen A Q, Ogg G M, Zhong H T. 2019. Global paleogeography through the Proterozoic and Phanerozoic: goals and challenges. Acta Geologica Sinica (English Edition), 93(S3): 59-60

Oye O J, Aplin A C, Jones S J, Gluyas J G, Bowen L, Orland I J, Valley J W. 2018. Vertical effective stress as a control on quartz cementation in sandstones. Marine and Petroleum Geology, 98: 640-652

Peace A L, Welford J K, Ball P J, Nirrengarten M. 2019. Deformable plate tectonic models of the southern North Atlantic. Journal of Geodynamics, 128: 11-37

Petrov P Y. 2016. Molar tooth structures and origin of peloids in proterozoic carbonate platforms (Middle Riphean of the Turukhansk Uplift, Siberia). Lithology and Mineral Resources, 51(4): 290-309, doi: 10.1134/s0024490216040064

Prélat A, Covault J A, Hodgson D M, Fildani A, Flint S A. 2010. Intrinsic controls on the range of volumes, morphologies, and dimensions of submarine lobes. Sedimentary Geology, 232(1-2): 66-76

Rebesco M, Hernández-Molina F J, van Rooij D, Wåhlin A. 2014. Contourites and associated sediments controlled by deep-water circulation processes: state-of-the-art and future Considerations. Marine Geology, 352: 111-154

Reijenstein H M, Posamentier H W, Bhattacharya J P. 2011. Seismic geomorphology and high-resolution seismic stratigraphy of inner-shelf fluvial, estuarine, deltaic, and marine sequences, Gulf of Thailand. AAPG Bulletin, 95(11): 1959-1990

Rodríguez-Tovar F J, Hernández-Molina F J, Hüneke H, Chiarella D, Llave E, Mena A, Miguez-Salas O, Dorador J, de Castro S, Stow D A V. 2019. Key evidence for distal turbiditic- and bottom-current interactions from tubular turbidite infills. Palaeogeography, Palaeoclimatology, Palaeoecology, 533: 109233, doi: 10.1016/j.palaeo.2019.109233

Saha D, Patranabis-Deb S, Collins A S. 2016. Proterozoic stratigraphy of southern Indian Cratons and global context. Stratigraphy & Timescales, 1: 1-59

Seton M, Müller D, Zahirovic S, Gaina C, Torsvik T, Shephard G, Talsma A, Gurnis M, Turner M, Maus S, Chandler M. 2012. Global continental and ocean basin reconstructions since 200 Ma. Earth-Science Reviews, 113(3-4): 212-270

Shang X X, Shao L Y, Zhang W L, Lv J G, Wang W C, Li Y H, Huang M, Lu J, Wen H J. 2018. Sequence paleogeography and coal accumulation of the Early-Middle Jurassic in central Qilian Mountain belt (Muli Basin), Qinghai Province, northwestern China. AAPG Bulletin, 102(9): 1739-1762

Shanmugam G. 2013. Modern internal waves and internal tides along oceanic pycnoclines: challenges and implications for ancient deepmarine baroclinic sands. AAPG Bulletin, 97(5): 799-843

Shao L Y, Wang X T, Wang D D, Li M P, Wang S, Li Y J, Shao K, Zhang C, Gao C X, Dong D X, Cheng A G, Lu J, Ji C W, Gao D. 2020. Sequence stratigraphy, paleogeography, and coal accumulation regularity of major coal-accumulating periods in China. International Journal of Coal Science and Technology, 7(2): 240-262

Shen B, Dong L, Xiao S, Lang X, Huang K, Peng Y, Zhou C, Ke S, Liu P. 2016. Molar tooth carbonates and benthic methane fluxes in Proterozoic oceans. Nature Communications. 7:10317. doi: 10.1038/ncomms10317

Shen Z H, Konishi H, Szlufarska I, Brown P E, Xu H F. 2014. Zcontrast imaging and ab initio study on "d" superstructure in sedimentary dolomite. American Mineralogist, 99(7): 1413-1419

Smith A G. 2016. A review of molar-tooth structures with some speculations on their origin. In: MacLean J S, Sears J W (eds). Belt Basin: Window to Mesoproterozoic Earth. Boulder: Geological Society of America, 522: 71-99

Spychala Y T, Hodgson D M, Flint S S, Mountney N P. 2015. Constraining the sedimentology and stratigraphy of submarine intraslope lobe deposits using exhumed examples from the Karoo Basin, South Africa. Sedimentary Geology, 322: 67-81

Stuart R C, Jakob S, Vidar S, Mark A S, Christian T, Are M B, Allison K T. 2012. 4D plates: on the fly visualization of multilayer geoscientific datasets in a plate tectonic environment. Computers & Geosciences, 45: 46-51

Sun P C, Sachsenhofer R F, Liu Z J, Strobl S A I, Meng Q T, Liu R, Zhen Z. 2013. Organic matter accumulation in the oil shale-and coal-bearing Huadian Basin (Eocene; NE China). International Journal of Coal Geology, 105: 1-15

Tejedor A, Longjas A, Caldwell R, Edmonds, D A, Zaliapin I, Foufoula-Georgiou E. 2016. Quantifying the signature of sediment composition on the topologic and dynamic complexity of river delta channel networks and inferences toward delta classification. Geophysical Research Letters, 43(7): 3280-3287

Wang C S, Scott R W, Wan X Q, Graham S A, Huang Y J, Wang P J, Wu H C, Dean W E, Zhang L M. 2013. Late Cretaceous climate changes recorded in eastern Asian lacustrine deposits and North American Epieric sea strata. Earth-Science Reviews, 126: 275-299

Wang C Z, Wang J, Hu B, Lu X H. 2016. Trace fossils and sedimentary environments of the upper cretaceous in the Xixia Basin, southwestern Henan Province, China. Geodinamica Acta, 28(1-2): 53-70

Wang H, Shao L Y, Large D J, Wignall P B. 2011. Constraints on carbon accumulation rate and net primary production in the Lopingian (Late Permian) tropical peatland in SW China. Palaeogeography, Palaeoclimatology, Palaeoecology, 300(1-4): 152–157

Wang J X, Sun P C, Liu Z J, Li L. 2020. Depositional environmental controls on the genesis and characteristics of oil shale: case study of the Middle Jurassic Shimengou Formation, northern Qaidam Basin, North-West China. Geological Journal, 55(6): 4585–4603

Wang P, Algeo T J, Zhou Q, Yu W C, Du Y S, Qin Y J, Xu Y, Yuan L J, Pan W. 2019. Large accumulations of ^{34}S-enriched pyrite in a low-sulfate marine basin: the Sturtian Nanhua Basin, South China. Precambrian Research, 335: 105504

Wang S, Shao L Y, Yan Z M, Shi M J, Zhang Y H. 2019. Characteristics of Early Cretaceous wildfires in peat-forming environment, NE China. Journal of Palaeogeography, 8(3): 238–250

Wang X T, Shao L Y, Eriksson K A, Yan Z M, Wang J M, Li H, Zhou R X, Lu J. 2020. Evolution of a plume-influenced source-to-sink system: an example from the coupled central Emeishan large igneous province and adjacent western Yangtze cratonic basin in the Late Permian, SW China. Earth-Science Reviews, 207: 103224

Weng S F, Yu W C, Algeo T J, Du Y S, Li P G, Lei Z Y, Zhao S. 2019. Giant bauxite deposits of South China: multistage formation linked to Late Paleozoic Ice Age (LPIA) eustatic fluctuations. Ore Geology Reviews, 104: 1–13

Worden R H, French M W, Mariani E. 2012. Amorphous silica nanofilms result in growth of misoriented microcrystalline quartz cement maintaining porosity in deeply buried sandstones. Geology, 40(2): 179–182

Xi K L, Cao Y C, Liu K Y, Wu S T, Yuan G H, Zhu R K, Kashif M, Zhao Y W. 2019. Diagenesis of tight sandstone reservoirs in the Upper Triassic Yanchang Formation, southwestern Ordos Basin, China. Marine and Petroleum Geology, 99: 548–562

Xiao W J, Santosh M. 2014. The western central Asian orogenic belt: a window to accretionary orogenesis and continental growth. Gondwana Research, 25(4): 1429–1444

Xiao W J, Windley B F, Su S, Li J L, Huang B C, Han C M, Yuan C, Sun M, Chen H L. 2015. A tale of amalgamation of three Permo-Triassic collage systems in central Asia: oroclines, sutures, and terminal accretion. Annual Review of Earth and Planetary Sciences, 43: 477–507

Xu X T, Shao L Y, Lan B, Wang S, Hilton J, Qin J Y, Hou H H, Zhao J. 2020. Continental chemical weathering during the Early Cretaceous oceanic anoxic event (OAE1b): a case study from the Fuxin fluvio-lacustrine basin, Liaoning Province, NE China. Journal of Palaeogeography, doi: 10.1186/s42501-020-00056-y

Yan N, Mountney N P, Colombera L, Dorrell R M. 2017. A 3D forward stratigraphic model of fluvial meander-bend evolution for prediction of point-bar lithofacies architecture. Computers & Geosciences, 105: 65–80

Yan Z M, Shao L Y, Large D, Wang H, Spiro B. 2019. Using geophysical logs to identify Milankovitch cycles and to calculate net primary productivity (NPP) of the Late Permian coals, western Guizhou, China. Journal of Palaeogeography, 8(1): 31–42

Yang L L, Xu T F, Liu K Y, Peng B, Yu Z C, Xu X M. 2017. Fluidrock interactions during continuous diagenesis of sandstone reservoirs and their effects on reservoir porosity. Sedimentology, 64(5): 1303–1321

Young A, Flament N, Maloney K, Williams S, Matthews K, Zahirovic S, Müller R D. 2019. Global kinematics of tectonic plates and subduction zones since the late Paleozoic Era. Geoscience Frontiers, 10(3): 989–1013

Yu W C, Du Y S, Cawood P A, Xu Y T, Yang J H. 2015. Detrital zircon evidence for the reactivation of an Early Paleozoic synorogenic basin along the North Gondwana margin in South China. Gondwana Research, 28(2): 769–780

Yu W C, Algeo T J, Du Y S, Zhang Q L, Liang Y P. 2016a. Mixed volcanogenic-lithogenic sources for Permian bauxite deposits in southwestern Youjiang Basin, South China, and their metallogenic significance. Sedimentary Geology, 341: 276–288

Yu W C, Algeo T J, Du Y S, Maynard B, Guo H, Zhou Q, Peng T P, Wang P, Yuan L J. 2016b. Genesis of Cryogenian Datangpo manganese deposit: hydrothermal influence and episodic postglacial ventilation of Nanhua Basin, South China. Palaeogeography, Palaeoclimatology, Palaeoecology, 459: 321–337

Yu W C, Algeo T J, Du Y S, Zhou Q, Wang P, Xu Y, Yuan L J, Pan W. 2017. Newly discovered Sturtian cap carbonate in the Nanhua Basin, South China. Precambrian Research, 293: 112–130

Yu W C, Algeo T J, Yan J X, Yang J H, Du Y S, Huang X, Weng S F. 2019a. Climatic and hydrologic controls on upper Paleozoic bauxite deposits in South China. Earth-Science Reviews, 189: 159–176

Yu W C, Polgári M, Gyollai I, Fintor K, Szabó M, Kovács I, Fekete J, Du Y S, Zhou Q. 2019b. Microbial metallogenesis of Cryogenian manganese ore deposits in South China. Precambrian Research, 322: 122–135

Yue D L, Li W, Wang W R, Hu G Y, Qiao H L, Hu J J, Zhang M L, Wang W F. 2019. Fused spectral-decomposition seismic attributes and forward seismic modelling to predict sand bodies in meandering fluvial reservoirs. Marine and Petroleum Geology, 99: 27–44

Zahirovic S, Salles T, Müller R D, Gurnis M, Cao W C, Braz C, Harrington L, Ibrahim Y, Garrett R, Williams S, Chen A Q, Hou M C, Ogg J G. 2019. From paleogeographic maps to evolving deeptime digital earth models. Acta Geologica Sinica, 93(S3): 73–75

Zeng H L, Zhao W Z, Xu Z H, Fu Q L, Hu S Y, Wang Z C, Li B H. 2018. Carbonate seismic sedimentology: a case study of Cambrian Longwangmiao Formation, Gaoshiti-Moxi area, Sichuan Basin, China. Petroleum Exploration and Development, 45(5): 830–839

Zhang J J, Wu S H, Fan T E, Fan H J, Jiang L, Chen C, Wu Q Y, Lin P. 2016. Research on the architecture of submarine-fan lobes in the Niger Delta Basin, offshore West Africa. Journal of Palaeogeography, 5(3): 185–204

Zhang L J, Zhao Z. 2016. Complex behavioural patterns and ethological analysis of the trace fossil *Zoophycos*: Evidence from the Lower Devonian of South China. Lethaia, 49(2): 275–284

Zhang L J, Fan R Y, Gong Y M. 2015. *Zoophycos* macroevolution since 541 Ma. Scientific Reports, 5: 14954

Zhang Y G, Pufahl P K, Du Y S, Chen G Y, Liu J Z, Chen Q G, Wang Z P, Yu W C. 2019. Economic phosphorite from the Ediacaran Doushantuo Formation, South China, and the Neoproterozoic-Cambrian phosphogenic event. Sedimentary Geology, 388: 1–19

Zhao F, He W Y, Huang C G, Wu L R, Zhang P, Wang A P. 2017. Saline fluid interaction experiment in clastic reservoir of lacustrine basin. Carbonates and Evaporites, 32(2): 167–175

Zhao G C, Wang Y J, Huang B C, Dong Y P, Li S Z, Zhang G W, Yu S. 2018. Geological reconstructions of the East Asian Blocks: from the breakup of Rodinia to the assembly of Pangea. Earth-Science Reviews, 186: 262–286

Zhao X M, Tong J N, Yao H Z, Niu Z J, Luo M, Huang Y F, Song H J. 2015. Early Triassic trace fossils from the Three Gorges area of South China: implications for the recovery of benthic ecosystems following the Permian-Triassic extinction. Palaeogeography, Palaeoclimatology, Palaeoecology, 429: 100–116

Zhao X M, Qi K, Liu L, Gong C L, McCaffrey W D. 2018. Development of a partially-avulsed submarine channel on the Niger Delta continental slope: architecture and controlling factors. Marine and Petroleum Geology, 95: 30–49

Zhao Y Q, Zhong H T, Xu S L, Hou M C, Hu X M, Zhang L, Gao Y, Zhang L M, Liu Y, Cao H Y, Mu C N, Cai P C. 2019. Quantitative expression of paleogeographic information based on big data. Acta Geologica Sinica (English Edition), 93(S3): 83–85

Zhou C M, Huyskens M H, Lang X G, Xiao S H, Yin Q Z. 2019. Calibrating the terminations of Cryogenian global glaciations. Geology, 47(3): 251–254

Zhu H H, Zhong D K, Yao J L, Sun H T, Niu X B, Liang X W, You Y, Li X. 2015. Alkaline diagenesis and its effects on reservoir porosity: a case study of Upper Triassic Chang 7 Member tight sandstone in Ordos Basin, NW China. Petroleum Exploration and Development, 42(1): 56–65

Zhu X M, Zhong D K, Yuan X J, Zhang H L, Zhu S F, Sun H T, Gao Z Y, Xian B Z. 2016. Development of sedimentary geology of petroliferous basins in China. Petroleum Exploration and Development, 43(5): 890–901

Zhu X M, Li S L, Liu Q H, Zhang Z L, Xu C G, Du X F, Li H Y, Shi W L. 2017. Source to sink studies between the Shaleitian Uplift and surrounding Sags: perspectives on the importance of hinterland relief and catchment area for sediment budget, western Bohai Bay Basin, China. Interpretation, 5(4): ST65-ST84

Zhu X M, Pan R, Li S L, Wang H B, Zhang X, Ge J W, Lu Z Y. 2018. Seismic sedimentology of sand-gravel bodies on steep slope of rift basins: a case study of Shahejie Formation, Dongying Sag, eastern China. Interpretation, 6(2): SD13-SD27

The Main Research Progresses of Palaeogeography

ZHENG Xiu-juan[1], DU Yuan-sheng[2], ZHU Xiao-min[1], LIU Zhao-jun[3], HU Bin[4], WU Sheng-he[1], SHAO Long-yi[5], KUANG Hong-wei[6], LUO Jing-lan[7], ZHONG Da-kang[1], LI Hua[8], HE Deng-fa[9], ZHU Ru-kai[10], BAO Zhi-dong[1]

1. China University of Petroleum (Beijing), Beijing 102249; 2. China University of Geosciences (Wuhan), Wuhan 430074; 3. Jilin University, Changchun 130061; 4. Henan University of Technology, Jiaozuo 454003; 5. China University of Mining and Technology (Beijing), Beijing 100083; 6. Institute of Geology, Chinese Academy of Geosciences, Beijing 100037; 7. Northwest University, Xi'an 710069; 8. Yangtze University, Wuhan 430100; 9. China University of Geosciences (Beijing), Beijing 100083; 10. Research Institute of Petroleum Exploration and Development Institute, Petro China, Beijing 100083

Abstract: In many aspects of the research and development of palaeogeography, we have discussed and summarized the progresses of China's palaeogeography in the past ten years, pointed out the current existing problems and the future research trends and directions of China's palaeogeography in this paper. It is believed that China's palaeogeography has played leading role in many aspects of the disciplinary development of palaeogeography. The main achievements are included in following aspects. ①In the aspect of basic research of palaeogeography, the joint development of many disciplines of palaeogeography has supported the sustainable development of palaeogeography. ②In the aspect of palaeogeography research methods and mapping, the researches of structural palaeogeography, bio-palaeogeography, small-scale palaeogeography, and big data palaeogeography have been developed well. ③In the aspect of application of palaeogeography, the palaeogeography of oil-bearing basins, oil shale palaeogeography, and diagenesis of clastic rocks have been applied to study oil and gas resources in oil shales, and especially some new breakthroughs in aspects of studies and explorations of the palaeogeography of coal accumulating areas and palaeogeography of other mineral deposits have been made.

Key words: palaeogeography; seismic sedimentology; ichnography; small-scale palaeogeography; big data palaeogeography; coal accumulation area palaeogeography; oil and gas resources

亚洲大陆边缘沉积学研究进展*

石学法[1,2]　乔淑卿[1,2]　杨守业[2,3]　李景瑞[2]　万世明[2,4]　邹建军[1,2]　熊志方[1,2]
胡利民[1,2]　姚政权[1,2]　董林森[1,2]　王昆山[1,2]　刘升发[1,2]　刘焱光[1,2]

1. 自然资源部 第一海洋研究所，海洋地质与成矿作用重点实验室，青岛 266061；2. 青岛海洋科学与技术试点国家实验室，海洋地质过程与环境功能实验室，青岛 266037；3. 同济大学，海洋地质国家重点实验室，上海 200092；4. 中国科学院 海洋研究所，海洋地质与环境重点实验室，青岛 266071

摘　要：近十年来，我国在亚洲大陆边缘沉积学和古海洋学研究中取得了突破性进展。在空间上，对北起拉普捷夫海、南至孟加拉湾的广大海域进行了沉积物调查取样，开展了跨纬度"源-汇"过程研究，建立了陆架第四纪高分辨率地层层序，初步揭示了构造运动、海平面变化、亚洲季风、海冰、海流以及人类活动等因素在不同时空尺度上对亚洲大陆边缘"源-汇"过程的基本控制作用。在南海通过国际大洋钻探获取的沉积记录，发现了低纬区水、碳循环直接响应地球轨道变化的证据，提出了低纬过程也能驱动全球气候变化的新认识。通过现场观测，揭示了台风、风暴潮、热带风暴等对陆架沉积和动力过程的影响，阐述了内孤立波、中尺度涡、等深流和浊流等在南海沉积物输运中的作用。对末次冰期以来暖池、黑潮、北太平洋中层水等的演化及其对沉积作用的影响研究也取得了创新性的成果。未来亚洲大陆边缘沉积学的研究应加强现代沉积过程的长期连续观测与多学科交叉研究，重视地质记录中环境演变信号的精确解译，深化数值模拟技术和海洋沉积大数据的挖掘与使用。

关键词："源-汇"过程　古海洋和古气候　海底观测　亚洲大陆边缘

0　引　言

　　大陆边缘（continental margin）长期以来一直是海洋地质学乃至地球科学研究的前沿和热点，而大陆边缘沉积也一直是海洋沉积学研究的主题。亚洲大陆边缘位于欧亚、太平洋和印度-澳大利亚三大板块的汇聚、碰撞边界，西部的青藏高原经历了强烈隆起，东部的西太平洋边缘汇集了全球约75%的边缘海，是全球独特的复合型大陆边缘（图1）。该区发育全球最具代表性河流系统（大河和山地小河）、季风系统、三角洲体系、陆架和边缘海系统，营造了世界上最大的地球表层沉积物"源-汇"过程体系。区域内以河流为纽带的"源-汇"过程系统贡献了全球约2/3的入海物质，对海洋沉积作用和生态环境都产生了巨大影响。亚洲大陆边缘发育有强大的西部边界流、暖池、北太平洋中层水和印尼贯穿流等海洋动力系统，影响着该区的沉积作用和全球气候变化（石学法等，2015）。

　　亚洲大陆边缘位于"海上丝绸之路"和"冰上丝绸之路"的核心区，它北起北冰洋的喀拉海、拉普捷夫海、东西伯利亚海、楚科奇海，东部包括太平洋的白令海、鄂霍次克海、日本海、渤-黄-东海和南海，西南部涵盖印度洋的安达曼海、孟加拉湾和阿拉伯海（图1）。这些边缘海作为陆源物质的主要沉积汇，发育建造了巨大的沉积体系，不仅直接记录了陆源物质输运的过程，也是恢复新生代亚洲大陆构造隆升、风化剥蚀、大河水系发育、季风与气候演化历史的理想载体（杨守业，2006；石学法等，2015；汪品先和翦知湣，2019）。

　　20世纪90年代，地球系统科学的兴起深刻影响和推动了海洋沉积学的发展。"从源到汇"角度研究

---

* 原文"亚洲大陆边缘沉积学研究进展（2011—2020）"刊于《矿物岩石地球化学通报》2021年第40卷第2期，本文略有修改。

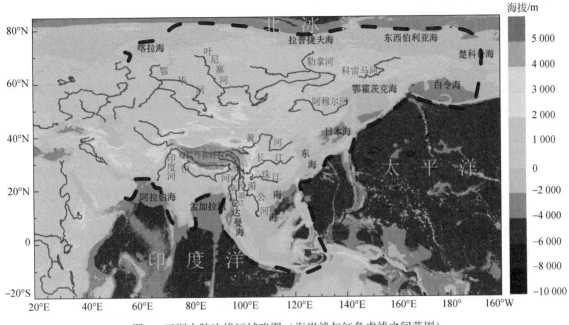

图1 亚洲大陆边缘区域略图（海岸线与红色虚线之间范围）

地表动力学逐渐成为海洋沉积学的主流和前沿方向，沉积物输运系统的概念进入海洋沉积学。所谓沉积物输运系统是一种动态系统，它将沉积物从"源"到"汇"的宿命联系在一起，并在地层记录相关的时间尺度上整合地表或其附近发生的过程及由此产生的沉积体（Allen，2017）。进入21世纪，美国启动了大陆边缘"从源到汇"计划（MARGINS，2001—2010年），欧盟发起了欧洲地层形成计划（EUROSTRATAFORM，2002—2005年），这两个研究计划影响深远，延续至今。我国从21世纪初开始关注大陆边缘沉积物"源-汇"过程研究（李铁刚等，2003；高抒，2005）。近年来，在国际上大陆边缘沉积学研究中，特别重视不同构造背景和跨气候带跨纬度的"源-汇"过程对比研究（Walsh et al.，2016）。过去十年，我国发起了区域性的亚洲大陆边缘"源-汇"过程与陆海相互作用国际合作计划，旨在选择亚洲大陆边缘的关键地区研究其"源-汇"过程及地层记录，阐明"源-汇"过程扩散系统的发育过程及机制（石学法等，2015），这是迄今为止在大陆边缘沉积学领域最为广泛的国际合作计划。与此同时，我国还开展了其他的国际合作研究，其中最引人瞩目的是国际大洋发现计划（IODP）在南海开展的卓有成效的沉积学与古海洋学研究。

本文尝试总结我国学者近十年来有关亚洲大陆边缘沉积学研究的若干主要进展。当然，国外学者在该领域也开展了大量研究，但限于篇幅仅作为背景做简要介绍。

1 调查研究概述

最近十年，我国支持了一系列有关亚洲大陆边缘沉积学调查研究项目，其中代表性的有：自然资源部"全球变化与海气相互作用"专项（2010—2020年）之"亚洲大陆边缘底质调查研究"和"'源-汇'过程与陆海相互作用"项目，国家自然科学基金委重大研究计划"南海深海过程演化（2011—2018年）"之"深海沉积过程对海盆演化的响应"和"底层海流与沉积搬运机制的变化"主题，"西太平洋地球系统多圈层相互作用"（2018—2025年）之"西太平洋流固界面跨圈层物质与能量交换过程"主题，国家自然科学基金委有关海洋沉积学重点项目，以及中国地调局系统和海洋石油产业部门的相关沉积地质学项目等。

与历史时期的研究相比，过去十年我国海洋沉积学研究的最主要特点是从全球视野开展研究，国际合作大为加强。一方面，继续与国际组织或国外科学家合作对我国管辖海域进行深入研究，这方面最具代表性的工作是南海沉积学与古海洋学研究：从 2014 年至今，IODP "决心号" 在南海实施了 4 个钻探航次，取得了近万米沉积岩心；1990—2018 年，德国 "太阳号" 和法国 "Marion Dufresne 号" 调查船在南海执行了 15 个航次，采取了 200 多个沉积柱样，使南海成为世界上沉积学调查研究程度最高的边缘海之一（汪品先和翦知湣，2019）。另一方面，通过与俄罗斯、泰国、马来西亚、印度尼西亚、缅甸、斯里兰卡、柬埔寨和孟加拉国等周边国家的合作，在我国管辖海域以外从拉普捷夫海到孟加拉湾的广大海域实施了 30 多个以海洋沉积为主题的调查航次，取得了 2 000 多站沉积物样品，使我国成为国际上对亚洲大陆边缘沉积调查研究程度最高的国家，极大地开阔了我国海洋沉积学研究范围和视野。

在研究思路和方法方面，近十年来海洋沉积学一直注重与其他学科的交叉融合，如与古海洋学和物理海洋学结合，深化了对边缘海碎屑沉积动力过程和环境响应机制的认识；与地球化学、矿物学等新指标方法结合，深化了对河流入海沉积物从源到汇过程的理解，认识了沉积旋回过程的复杂性；与海洋化学与生物地球化学交叉，揭示了亚洲大陆边缘沉积有机碳源汇过程及其主控因素。

在调查技术和观测方面，实现了大面调查、长期定点观测和不同时间尺度样品采集等手段的有机融合，形成了沉积物取样-悬浮体采集-水文要素调查-浅地层探测于一体的底质调查技术体系；建立了东海海底观测网小衢山试验站和南海北部深海沉积动力过程综合观测系统，促进了海洋沉积动力学的发展，而且应用载人深潜器直接对海底沉积过程进行观测（汪品先和翦知湣，2019；Zhang et al., 2020）；卫星与航空遥感技术、雷达技术和数值模拟方法在现代海洋沉积过程的研究中得到进一步应用，深化了对沉积物输运机制的认识。

在沉积物分析测试技术方面，先进的同位素地球化学和有机地球化学等分析方法得到广泛应用，取得了良好的效果。

通过上述科研活动，我国近十年来在亚洲大陆边缘沉积学研究方面取得了系列重要成果，在沉积物"源-汇"过程和机制、低纬气候驱动假说、沉积动力过程观测等方面实现了突破，推动了我国乃至国际上大陆边缘沉积学的发展。

2 "源-汇"过程及环境演化

亚洲大陆边缘的主要特征是地域广阔、构造复杂、气候带跨度大，其中每个海区的沉积"源-汇"过程特征、环境演化特征和控制要素都不同。各海域的主要沉积学研究进展分别阐述如下。

2.1 渤海、黄海和东海沉积

渤海、黄海与东海陆架一起组成了东中国海大陆架，是世界上最宽广的大陆架之一（秦蕴珊等，1987）。黄河和长江两条世界级大河以及山地型河流每年向中国东部海域输入巨量泥沙，在沿岸流、黑潮、潮流和风暴潮等作用下，形成了河口湾和三角洲沉积、砂质沉积、泥质沉积和混合沉积等（秦蕴珊等，1987；石学法，2012；Gao and Collins, 2014）。

2.1.1 现代三角洲和陆架沉积作用

近年来，三角洲沉积研究的热点是人类活动背景下河流入海泥沙骤减及其对三角洲沉积的影响。黄河和长江建坝之后，不仅使沉积物粒度粗化、磁性矿物含量降低和常量元素含量增加（Yang et al., 2019），河流-河口三角洲-陆架"源-汇"过程体系也发生了改变（Gao et al., 2017, 2018）。黄河"调水调沙"导致黄河入海水沙输入方式从季节性韵律转化为脉冲式输入，沉积物粒度组成由于下游河道冲刷物质的加入发生粗化，有机碳、重金属等输入方式也同步发生变化（Wang H. J. et al., 2017）。黄河现行河口水下三角洲的建造速率也明显放缓，来沙量的锐减导致整个黄河三角洲处于侵蚀状态（Wu et al., 2017）。2003 年长江三峡大坝运行之后，长江河口泥沙供给减少 70% 左右，长江水下三角洲整体呈现侵蚀、沉积物粗化的趋势（Yang et al., 2011；Dai et al., 2014），甚至东海内陆架泥质区也短时间内出现侵蚀、沉积物粗化的现象（Gao et al.，2019）。

我国东部海域潮流发育，在某些近岸海域和海峡地区，潮流流速很强。近年来潮汐沉积作用最主要的进展是发展完善了涌潮沉积、浮泥沉积和潮坪遗迹化石的识别标志与沉积过程，发现典型的涌潮沉积由平坦的侵蚀底界面、块状砂（无沉积构造）、平行层理砂以及薄层状韵律层理（砂泥间互）组成（Fan et al.，2014）。通过对东海中部陆架钻孔的研究，首次在东海陆架沉积中发现了混合事件层沉积，单一事件层中同时存在浊流沉积段（事件层下部）与碎屑流沉积段（事件层上部），表明流体由浊流向碎屑流转化，呈现浊流与碎屑流之间的过渡流态（Shan et al.，2019a，2019b）。

台风事件对河流输入物质影响的研究也备受关注。2005 年 7 月超级台风"海棠"导致高屏溪洪水，快速侵蚀的巨量泥沙以浊流甚至碎屑流的方式搬运至深海（Liu J. T. et al.，2012，2013）；而位于台湾东北部的兰阳溪是南冲绳海槽的主要物源供应者，洪水引发的异重流可将相当于 56 年来 23% 的泥沙通量在 196 h 之内输送入海（Liu J. T. et al.，2013）。建立了台风影响下陆架沉积环境演化模式，揭示了典型台风（如莫拉克台风）过程对水体-悬浮颗粒物-底质沉积物体系的影响及其控制机制（Li et al.，2012）。

2.1.2 中国东部海域"源-汇"过程

（1）大河和山地河流的输运沉积。近年来中国东部海域"源-汇"过程研究的一个重要进展是揭示出东亚大陆边缘存在两类河流"源-汇"过程体系，即大型河流（如长江）和山地河流（如台湾小河流）两类河流系统，它们处于不同的构造地貌背景，沉积物从源到汇过程和控制机制也明显不同：①两类河流流域化学风化强度的主控因素不是原岩类型，长江及我国大陆东南沿海中小河流入海物质主要受季风气候控制，而构造活跃区的台湾山地河流物质则受气候、降水和强烈的物理侵蚀等因素的控制（Bi et al.，2015）。②基于 $^{234}U/^{238}U$ 值的粉碎年龄计算显示，两类河流沉积物构造背景和滞留时间不同可能是造成其风化程度和沉积地球化学组成差异的主控因素，长江细颗粒碎屑沉积物从源到汇的时间自上游约 200 ka 增加至下游 400 ka 左右，而来自台湾岛屿河流的沉积物搬运时间只有 100 ka 左右（Li et al.，2016）；山地河流沉积物物源存在明显的空间不均一性，快速转换的沉积物搬运模式是其主要原因（Deng et al.，2019）。③ε_{Nd}、$\delta^{18}O$ 等指示全新世以来长江沉积物"源-汇"过程从自然因素控制逐渐转变为受人类活动影响，长江上游沉积物在近 2 ka 来大量进入河口地区（Bi et al.，2017）。$^{234}U/^{238}U$ 值和 ε_{Nd} 等指示了 30 ka 以来冲绳海槽南部沉积物来源逐渐从长江来源物质为主变为台湾河流来源为主（Dou et al.，2016；Li et al.，2016）。

（2）中国东部海域沉积物来源与收支平衡。渤海、黄海和东海周边入海河流物质是中国东部陆架海域沉积物的主要来源（秦蕴珊等，1987；石学法，2012）。台湾海峡现代沉积速率的空间分布显示出台湾海峡附近有 3 种不同的沉积物扩散体系，包括台湾西部浊水溪以北、以南和长江与大陆其他河流输运体系，台湾海峡每年沉积 $160×10^6$ t 沉积物，这与除长江之外这些河流入海物质量相当（Huh et al.，2011）。Qiao 等（2017）利用 413 站沉积速率、18 000 站沉积物粒度和干密度资料，划分了中国东部海域细粒级沉积区（平均粒径大于 6 Φ），首次全面阐明了中国东部海域现代沉积速率的分布特征和规律，从"源-汇"过程角度阐述了中国东部海域沉积物收支平衡，揭示出渤海、黄海和东海周边河流现在每年向海输入沉积物约为 $1\,646.6×10^6$ t，其中约 45% 沉积形成陆地三角洲，约 40%~50% 沉积在水下三角洲和陆架区域，仅有不超过 5% 的沉积物被搬运到陆架之外（图2）。Jia 等（2018）将东海内陆架划分为长江、浙江和福建泥质沉积区，进一步阐述了长江影响范围内泥质区沉积速率和沉积物收支平衡。

（3）东海内陆架泥质沉积。东海内陆架泥质沉积体是我国浅海乃至亚洲近海规模最大的全新世楔状沉积体之一，是环境和气候变化的信息载体。近年来进一步研究了东海泥质沉积物来源、沉积过程，揭示了泥质区沉积记录的全新世东亚季风演化历史和极端气候事件及其对新仙女木（Younger Dryas，YD）期、8.2 ka 和 5.5 ka 等全球气候突变事件的响应，探讨了末次冰盛期（last glacial maximum，LGM）晚期和末次冰消期长江物质、台湾入海物质及其他中小型河流对内陆架泥质区南部的贡献及输运路径（Liu et al.，2010；Gao and Collins，2014；Dong et al.，2018）。

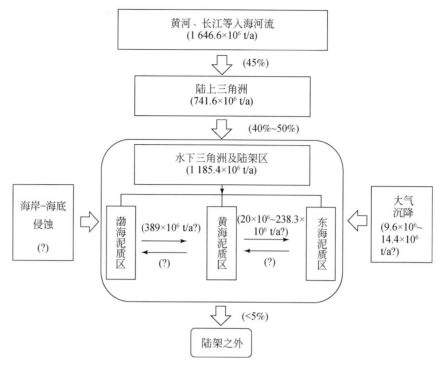

图 2 中国东部海域沉积物收支平衡（据 Qiao et al., 2017 修改）

关于东海内陆架全新世泥质沉积物来源，目前大多数研究者都认为其主要来源于长江（石学法等，2010；石学法，2012；Xu et al., 2012；Liu J. T. et al., 2018）。而 Gao 和 Collins（2014）则认为，8.0~2.0 ka 期间东海内陆架泥质沉积主要来源于经潮汐作用再次改造的已沉积地层，长江物质很少，只在最近的 2.0 ka 期间，长江物质才在浙闽沿岸流的作用下向南悬浮搬运。这个问题还需进一步研究厘清。

2.1.3 中国东部陆架第四纪年代框架与环境演化

陆架区由于沉积动力复杂多变，测年困难，因而可靠的第四纪年代框架不易建立。近十年来我国在中国东部陆架第四纪地层年代框架建立与环境演化研究方面取得了突破性进展。

（1）建立了渤海 1 Ma 以来轨道尺度上的高分辨率年代地层框架。基于渤海 BH08 钻孔岩心的精细研究建立了轨道尺度（冰期-间冰期尺度）上 1.06 Ma 以来高分辨率的年代框架（Yao et al., 2014），改变了多年来人们对渤海第四纪沉积地层年龄框架的认识。此前学术界普遍认为渤海中部地区同样深度沉积地层年龄为 200 ka（Qin et al., 1990），这为进行精确地层层序划分对比和古环境研究奠定了基础。应用这一标尺研究发现渤海中更新世以来至少发生了 10 次海侵-海退变化，且在空间上显示了海侵-海退沉积序列自海向陆逐渐减弱的趋势（Yao et al., 2014；Shi et al., 2016）。季风气候变化控制了中更新世以来渤海沉积物的供应，而海洋水动力作用对入海沉积物进行了进一步改造（Yao et al., 2020）。

（2）建立了黄海第四纪以来年代地层格架和海平面演化历史。基于对南黄海 NHH01 钻孔岩心磁性地层学研究，首次建立了南黄海 1.1 Ma 以来轨道尺度上的高分辨率年代地层格架，阐述了过去 1 Ma 以来冰期-间冰期尺度上黄海陆架区沉积环境变化及其对全球变化的响应（Liu et al., 2014）。基于南黄海 300 m 长的 CSDP-01 钻孔岩心的研究，建立了迄今黄海唯一具有较好年代控制、钻穿第四纪地层的钻孔剖面，古地磁年代学研究结果表明其底界年代为 ~3.5 Ma（Liu J. X. et al., 2016）。

NHH01 钻孔揭示黄海在过去 1 Ma 以来主要以海相沉积为主，只有在冰期极盛期如 MIS10、MIS18 和 MIS20 发现有河流相沉积（Liu et al., 2014）。而 CSDP-01 钻孔揭示了黄海在 3.5~1.66 Ma 以河流相沉积为主，在 1.66~0.83 Ma 以潮坪相和河流相交互沉积为主；而 0.83 Ma 以来，现代浅海相沉积与河流的交互沉积才主导这一区域的沉积环境变化（Liu J. et al., 2018）。这些结果说明，陆架最早海侵发生在距今约

1.7 Ma（Liu J. X. et al., 2016；Liu J. et al., 2018），且在中更新世以来的每个间冰期都发现有海侵沉积（Yao et al., 2014；Shi et al., 2016）。这说明由冰期-间冰期海平面变化导致的海侵-海退是控制中国东部陆架沉积环境变化的最主要因素。

（3）揭示了黄河至少在距今88万年前就已经贯通入海。黄河作为世界泥沙含量第一的河流，其物质输入对中国东部陆架沉积具有重要影响，因而关于黄河何时贯通入海这一问题长期受到地质学家的特别关注。通过对渤海中部BH08钻孔和南黄海NHH01钻孔的研究，获得了黄河贯通入海的直接证据，揭示了黄河至少在距今88万年前就已经入海，推进了黄河何时贯通入海这一重要科学问题的解决（Yao et al., 2017），南黄海CSDP-01钻孔沉积记录也证实了这一结论（Zhang J. et al., 2019）。

2.2 南海沉积与古环境

南海是西太平洋最大的边缘海，周边大河流和山地河流将巨量陆源物质输入海盆，东亚季风和海流系统塑造了特色的沉积作用和沉积环境，同时南海沉积物对全球气候变化具有敏感的响应（Wang et al., 2014a）。

2.2.1 南海沉积物来源与沉积作用

南海沉积物源包括陆源、火山源和生物源物质，火山物质主要分布在吕宋岛西部，生物组分分布在南沙和西沙岛屿周围，陆源物质主要来自周边的大河和山地河流（Liu J. G. et al., 2013, 2019b；Liu Z. F. et al., 2016）。

通过深水锚系长期观测和深潜技术的应用，在南海发现了深海等深流和内孤立波，实现了浊流搬运的现场观测（Zhao Y. L. et al., 2015；Zhang Y. W. et al., 2018；Jia et al., 2019）。南海等深流和内孤立波沉积的发现是我国近年来海洋沉积学研究的一个重要进展。南海等深流向西南净搬运量每年约86 Mt，具有强劲侧向搬运能力，控制了南海深海陆坡沉积搬运过程（Zhao Y. L. et al., 2015）。

台湾岸外的高屏海底峡谷锚系观测到每年平均发生6次浊流事件，其中有4次是直接由路经台湾的强台风降雨引发；通过定量测算，每年浊流沿高屏海底峡谷向下的搬运量为25.5 Mt，是陆源沉积物从近岸向深海输运的最重要方式（Zhang Y. W. et al., 2018）。基于近十年来获取的大量现场观测资料并结合沉积记录，逐步建立了以沿岸流、季风流、黑潮为基本动力，局部存在的等深流、浊流、上升流、内孤立波等特征动力类型为辅的沉积物动力输运体系（Zhao Y. L. et al., 2015；Liu Z. F. et al., 2016；Chen Z. Y. et al., 2017；Jia et al., 2019）。另外，首次观测到南海北部中尺度涡旋驱动的深海输沙过程，发现了南海锰结核、古热液口和深海冷水珊瑚林（汪品先和翦知湣，2019）。

2.2.2 沉积记录与气候演变

近年来南海研究中最具突破性的一项成果是通过沉积记录的研究，发现了低纬区水、碳循环直接响应地球轨道变化的证据，从而提出了低纬过程也能驱动全球气候变化的新认识，挑战了北极冰盖决定一切的传统观点（Wang P. X. et al., 2017；汪品先和翦知湣，2019）。过去十年中，围绕全球变化的低纬驱动机制，基于IODP钻孔沉积物地球化学、黏土矿物、有孔虫壳体元素比值和氧同位素、碳同位素等一系列数据提取出了20 ka为主的岁差周期、100 ka和400 ka为主的偏心率周期（Tian et al., 2011；Jin and Jian, 2013；Wang et al., 2014b），并提出了解释近百万年来$\delta^{13}C$发生周期由400 ka推迟至50 ka的碳储库"溶解有机碳假说"（Wang et al., 2014b），这些研究有力支持了全球变化的低纬驱动假说。

在冰期-间冰期尺度上和更短的千年乃至百年时间尺度上，南海沉积环境的演化始终与东亚季风相伴（Liu Z. F. et al., 2016）。通过分析沉积物来源与输运过程的变化、沉积记录的源区化学风化强度反演东亚季风及其控制下的海流演化历史，揭示出冰期陆海格局发生显著变化时南海北部、西部和南部具有不同的沉积物搬运模式：北部陆源物质扩散主要受物源供给和海流输送控制，西部黏土矿物组合反映了季风影响下盛行表层流的变化，南部黏土矿物组合则表明冰期和间冰期化学风化与物理侵蚀的此消彼长过程（Liu Z. F. et al., 2016）。不同物源区的相对贡献及化学风化强度参数的变化还可以揭示热带辐合带纬向移动及黑潮强度变化对物质输运的影响（Liu et al., 2011; Liu Z. F. et al., 2016）。

2.2.3 泰国湾沉积"源-汇"过程和环境演化

早在20世纪70年代，泰国及其他国家学者已开始对泰国湾沉积物组成和沉积记录等进行研究，但总体研究程度较低。基于与泰国和马来西亚的合作，开展了彭亨河、格兰丹河、湄南河、邦巴功河、拉塞河和湄干河入海泥沙特征和差异性研究，阐述了河流沉积物在巽他陆架的输运趋势和范围，其中湄南河沉积物在曼谷湾具有向东、东南方向输运的趋势（Qiao et al., 2015; 吴凯凯等, 2019）; 泰国湾南部沉积物向西北方向输运，北部物质向西南方向，都有向苏梅岛区域汇聚的趋势（张杨硕等, 2017）。黏土矿物指示泰国湾北部物质主要来源于曼谷湾北部河流，中部沉积物主要来源于南海输入物质，西南部沉积物主要来自马来半岛（Shi et al., 2015）。从碎屑矿物同样可以发现泰国湾北部和靠近马来半岛西南部主要受陆源物质影响（王昆山等, 2014b; Wang et al., 2015）。

人类活动对泰国湾沉积物中重金属、黑碳和多环芳烃有重要影响，研究发现曼谷湾沉积物中Cd和Pb存在显著的地球化学富集特征，现代Cd可能源于金属冶炼和交通污染（乔淑卿等, 2015; 郭瑜璇等, 2019）。泰国湾苏梅岛和尚塔布里省附近海域Cr、Zn、Cu和Pb含量略高，As在泰国湾北部为轻度污染（Liu et al., 2016a, 2016b）。泰国湾黑碳主要来自生物质燃烧，和世界上其他近岸区域比较，黑碳含量呈低到中等（Hu et al., 2016a）。泰国湾多环芳烃分布和组成特征与中国周边海域明显不同，反映了不同经济社会发展背景下地区之间能源结构的影响（Hu et al., 2017）。

泰国湾末次冰消期以来沉积环境演化可以划分为4个阶段：Ⅰ阶段为末次冰消期陆相沉积；Ⅱ阶段为早全新世的滨海相沉积；Ⅲ和Ⅳ阶段分别中全新世和晚全新世浅海沉积。其中，Ⅱ和Ⅳ阶段沉积物主要来自马来半岛，而Ⅲ阶段沉积物主要来自中南半岛（张杨硕等, 2017; 陈禹飞等, 2020）。

2.3 孟加拉湾和安达曼海沉积

孟加拉扇长期以来都是沉积学研究的热点区域，绝大部分工作都是基于DSDP、大洋钻探计划（Ocean Drilling Program, ODP）和IODP开展的长时间尺度的古环境和古海洋学研究（Weber et al., 2018; Weber and Reilly, 2018），而对孟加拉湾现代沉积作用和物质来源的研究却很薄弱。安达曼海位于太平洋和印度洋、东亚季风和南亚季风的交汇区，是目前世界上研究程度较低的边缘海。

2.3.1 孟加拉湾中部沉积作用和沉积模式

基于孟加拉湾中部采集的大量沉积物样品的研究，厘定了其现代沉积物来源及不同物源区的贡献。确认喜马拉雅源区是其主要物源区，主要搬运动力是浊流及其溢流体系，而印度半岛物质和缅甸物质也在不同时-空尺度上对孟加拉湾中部沉积物有所贡献，其主要输运动力是表层季风环流体系，据此建立了孟加拉湾中部现代沉积模式（Li J. R. et al., 2017, 2020; Sun et al., 2019; Liu et al., 2019a; Wang et al., 2019）。提出了孟加拉湾末次冰期以来沉积物来源"2-3-2"的分布模式，即孟加拉湾西侧沉积物来源主要为"喜马拉雅山+印度半岛"二端元，东部为"喜马拉雅山+缅甸"二端元，而中部地区则为"喜马拉雅山+印度半岛+缅甸"三端元混合模式（Li J. R. et al., 2018, 2020）。

末次冰期以来孟加拉湾沉积过程受海平面变化和印度夏季风演化的影响（Li et al., 2019），在冰期-间冰期时间尺度上，陆源输入量在末次冰期（高值）和全新世（低值）显著不同，海平面是其主要控制因素，通过控制沉积中心在孟加拉扇和陆架之间的转移从而控制深海扇沉积过程；而在更短的时间尺度上，太阳辐射量、季风降雨量及沉积记录的一致性揭示了末次冰期（尤其是18 ka以前）印度夏季风在岁差时间尺度上通过降雨量变化影响侵蚀速率控制孟加拉湾中部沉积过程，末次冰消期和早全新世陆源输入量则显示出响应于北大西洋气候变化的千年尺度变化特征。Liu等（2019a）提出的不同粒级沉积物来源差异性指示的源区变化也在千年尺度上响应于气候变化，其中细颗粒沉积物在50~45 ka、42~37 ka、31~28.5 ka、24~20 ka和14~9 ka这5个阶段主要由恒河-布拉马普特拉河系统供应。在更长时间尺度上，Chang和Zhou（2019）将释光技术应用到IODP U1444 A站位沉积物来源示踪和"源-汇"过程研究中，揭示出孟加拉湾西部 7 Ma 以来沉积物来源曾在 3.5～0.5 Ma 期间发生变化，推测其与喜马拉雅构造活动相联系的浊流活动强度变化以及海底地形的影响有关。

2.3.2 安达曼海沉积特征及夏季风沉积记录

安达曼海沉积属于大河系统（伊洛瓦底江和萨尔温江）和半岛-岛屿侵蚀输入共同控制的沉积类型，北部陆架区以伊洛瓦底江和萨尔温江入海物质为主，南部深海区则为河流沉积物与半岛-岛屿侵蚀物质的混合（曹鹏等，2015）。基于高分辨率沉积岩心和浅地层剖面资料的研究，发现沉积中心位于马塔班湾以及水深 100 m 左右的马塔班湾凹陷，目前还没有证据表明河流沉积物通过马塔班湾的深水峡谷向安达曼海深水搬运，另有部分沉积物则向西向北搬运沉积在缅甸西侧海沟俯冲带的向陆地一侧，甚至可能有大量的陆源沉积物直接注入俯冲带海沟里面（Liu J. P. et al., 2020）。

Cao 等（2015）重建了安达曼海东南部 26 ka 以来印度夏季风演化历史，通过对替代性指标周期分析提取出的 1 619 a 的控制性周期表明，百年—千年时间尺度上印度夏季风强度变化主要受太阳活动控制。安达曼海沉积记录的化学风化强度和陆源输入量变化与 20°N 太阳辐射量、降雨量和温度变化在不同的氧同位素分期具有较好的对应关系，而 18～13 ka 和 6～3 ka 的化学风化异常则与陆源侵蚀速率、温度的突变和沉积物输运入海速率的变化相关，揭示了安达曼海冰期-间冰期尺度和千年尺度上的气候驱动型化学风化机制（Liu S. F. et al., 2020）。

2.4 中高纬度边缘海和北极陆架沉积

总体看来，我国对西北太平洋中高纬度边缘海和北极陆架的研究非常薄弱，近年来主要借助中俄合作项目和中国北极科考项目开展了研究。

2.4.1 日本海沉积与古环境

日本海是西北太平洋封闭程度最高的一个边缘海，位于东亚季风下垫面，横跨温、寒两个气候带，入海河流不发育，以高海槛、发育季节性海冰、深层水和纹层沉积为显著特征（石学法等，2019）。日本科学家对日本海开展了大量的沉积学和古环境研究工作，但都集中于日本海东部靠近日本一侧（Tada et al., 2018）。近年来我国的研究主要集中在如下几个方面：

（1）系统阐述了末次间冰期以来日本海陆源碎屑沉积物"源-汇"过程。在日本海南部的郁陵海盆，其全新世细粒碎屑物质主要源自长江，冰期和冰消期沉积物则主要源自老黄河、长江的混合物，在冰消期早期存在来自朝鲜半岛物质的贡献（Zou et al., 2021）。在日本海中部，沉积物主要以粉砂和黏土质粉砂为主，碎屑物质主要源自西风携带的中亚沙漠和戈壁物质（董智等，2017）。在日本海西部，沉积物富含大量冰筏沉积和火山碎屑物质，特别是在冰消期，日本海西部海冰活动加剧（豆汝席等，2020）。总体看来，日本海陆源碎屑物质的来源、搬动和沉积过程受海平面、季风、黑潮和海冰制约（石学法等，2019）。

（2）揭示了轨道及千年尺度表层环流、深层水体通风和古生产力演化特征。在末次冰盛期，日本海沉积物发育厘米级尺度的纹层沉积，浮游有孔虫 *Neogloboquadrina pachyderma*（sinistral coiling）和 *Globigerina bulloides* 的 δ^{18}O 值均显著变轻，指示当时日本海与外界隔绝，表层海水淡化分层，深层水体通风停止；在 MIS3 期，*G. bulloides* δ^{18}O 值记录了一系列快速轻值事件，指示了在间冰阶由于东亚夏季风增强，降雨量增大，致使日本海表层生产力升高并且盐度降低，这些事件与厘米级暗色纹层沉积层相对应，指示了深层水体通风的减弱（吴永华，2016）。日本海深层水通风呈现轨道和千年尺度变化，但在空间上并不同步（石学法等，2019）。日本海南部通风加强与对马暖流入侵日本海同步，而日本海西部通风变化与海冰活动相关（豆汝席等，2020）。从日本海南部岩心孢粉记录中识别出一系列千年尺度气候事件，间冰阶喜暖植被繁盛，冰阶喜冷植被发育，显示北半球中高纬地区植被对千年尺度气候变化异常敏感（Chen J. X. et al., 2017）。

利用 TEX_{86}^L 指标与 $U_{37}^{K'}$ 指标进行了日本海古温度恢复，发现这两个指标都记录了 LGM 时期温度升高的现象，且在该段段这两个指标记录的是日本海夏季表层海水温度，这一结果合理地解释了日本海 LGM 时期的 $U_{37}^{K'}$ 温度增加的现象，解决了学术界多年来存在争议的这一问题（Wu et al., 2020）。

（3）恢复了中新世以来日本海风尘记录与古气候演化历史。发现日本海中部 IODP U1430 站位孔黏土粒级的硅酸盐组分主要源自中亚粉尘和日本列岛二端元混合，中亚粉尘供给在 11.8 Ma、8 Ma、3.5 Ma 和 1.2 Ma 显著增加，表明中亚内陆干旱化增强；其中 8 Ma 快速变干主要与青藏高原快速隆升相关，3.5 Ma 以来则主要受全球变冷控制（Shen et al., 2017）。通过对沉积岩心中风尘铁和生产力关系的研究，发现日本海 2~3 Ma 时生产力显著增加，认为这是北半球冰盖的形成强化了亚洲内陆风尘供给，提高了日本海生产力发育（Zhai et al., 2020）。

2.4.2 鄂霍次克海沉积与古环境

鄂霍次克海是西北太平洋第二大边缘海，冬季气候明显受到亚北极和北极气候影响，是现今北半球常年发育季节性海冰的最南界。国外的研究以俄罗斯科学家为主，主要关注海冰活动历史和鄂霍次克海表层水文演化历史等方面（Gorbarenko et al., 2017；Vasilenko et al., 2019）。我国以往对鄂霍次克海研究极少，近十年来才开始研究沉积学及古环境问题（石学法等，2011）。

在鄂霍次克海东南部沉积岩心中识别出陆源碎屑主要为鄂霍次克海北部和西部海冰来源，少部分由堪察加半岛南部海冰携带（王昆山等，2014c）。沉积物中粒径小于 63 μm 的组分主要来源于阿穆尔河和海冰搬运，其次为火山喷发物质，并呈现冰消期高含量的变化模式，古生产力指标也显示在冰消期鄂霍次克海生产力显著增加（Zou et al., 2015）。Zou 等（2015）提出蒙古高压的位置变化控制海冰活动，进而影响陆源物质累积变化。冰筏碎屑研究进一步发现，西伯利亚高压与阿留申低压活动中心位置和强度相对变化，是导致堪察加半岛西南部冰筏物质被输运至鄂霍次克海的主要驱动机制（Wang K. S. et al., 2017）。鄂霍次克海东南部 180 ka 以来共发生 6 次 $\delta^{13}C$ 轻值事件，其中发生于 102~90 ka 的事件降幅最大，这些轻值事件可能由表层生产力增高、中层水形成减弱、最低含氧带加强等因素共同引起（Wu et al., 2014）。

2.4.3 白令海沉积与古环境

白令海是北太平洋的半封闭边缘海，向北通过白令海峡与北冰洋相通，南部通过阿留申群岛与太平洋相连，在北太平洋和北冰洋之间起到重要的桥梁与纽带作用。

近年来的研究揭示，白令海表层沉积物来源和主要输运路径，受洋流、水团和冰川融水的强烈影响（Wang et al., 2016）。另外，白令海陆源沉积物的输入还受源区气候、海平面变化和生源物质稀释等多种因素的控制（邹建军等，2012），白令海海盆内冰筏碎屑（IRD）主要来自富碳酸盐育空河流域，其次为阿拉斯加半岛和阿留申群岛等火山岩区（陈志华等，2014）。葛淑兰等（2013）获得了白令海北部陆坡区 14 ka 以来地磁场强度和方向变化信息，相关记录可以与全新世绝对强度记录、北美和欧洲记录以及全球叠加地磁场强度曲线进行千年尺度上甚至百年尺度上的对比。

末次冰盛期白令海北部陆坡受到海冰南向扩张的影响而生产力较低，沉积物中冰筏碎屑（IRD）含量较高，HS1 时期由于受到常年冰的覆盖，沉积物 IRD 含量较低，B/A 暖期和全新世生产力较高（宋腾飞等，2018），而末次冰消期白令海北部陆坡以高生产力和底层水体缺氧为显著特征（邹建军等，2012）。在全新世高海平面时期，白令海与北太平洋、北冰洋之间的水体交换达到某种极值状态，白令海环流加强，海盆底层水含氧状况明显改善，并引发海洋生产力增加（陈志华等，2014）。黄元辉等（2013）基于硅藻种类及含量重建了近万年来白令海海冰变化历史，记录到该区末次冰消期以来包括新仙女木事件在内的 3 次冷事件及 1 次暖事件。

2.4.4 北极东西伯利亚陆架沉积

北极东西伯利亚陆架包括北冰洋的拉普捷夫海、东西伯利亚海和楚科奇海陆架，是目前北极海冰融化、快速变暖、冻土退化及植被变化最显著的地区之一，目前国际上对该区的研究程度不高，主要关注古海冰、古生产力及快速气候变化（Stein et al., 2017），我国对该区的研究更是非常薄弱。部分学者主要基于我国历次北极科考取得的样品对楚科奇陆坡海盆沉积特征、冰筏碎屑、冰期-间冰期的大西洋水团的影响、海冰形成过程及生产力和有机碳保存等方面进行了研究（梅静等，2012；Wang et al.，2013；董林森等，2014；王昆山等，2014a；章伟艳等，2015）。2016年和2018年通过两次中俄合作科考，对拉普捷夫海、东西伯利亚海和楚科奇海陆架进行了系统的沉积物调查取样。初步研究了东西伯利亚陆架现代沉积物的分布规律和来源，将研究区划分为4个沉积区：I区位于东西伯利亚海近岸河口附近，河流输入与海岸侵蚀是该区沉积物主要的贡献者；II区位于东西伯利亚海中部，沉积物来源主要为河流输入的细粒沉积物，且随着离岸距离的增加海洋自生组分开始增多；III区位于东西伯利亚海北部深水区，海底沉积物以黏土为主，细粒沉积物很可能受大西洋中层水以及波弗特环流的影响；IV区位于楚科奇海，沉积物为粉砂和砂质粉砂，主要来自于沿岸侵蚀与太平洋入流水所携带的沉积物（李秋玲，2020）。

2.5 亚洲大陆边缘风化剥蚀与沉积有机质输运

2.5.1 亚洲大陆边缘风化剥蚀与构造–气候相互作用

大陆风化和侵蚀是塑造地球表面形态、调节大气CO_2浓度和控制沉积物及溶解质从陆向海输送的关键过程，尤其硅酸盐风化与气候之间的负反馈关系被认为在稳定全球碳循环以及维系地球气候长期宜居性方面起到了关键作用（Berner and Caldeira，1997）。近年来我国学者在该领域的研究取得了突破性的进展，首次阐明了热带陆架冰期风化在碳循环中的重要性（Wan et al.，2015，2017）。

硅酸盐风化是构造时间尺度碳循环的主要碳汇，但风化过程在新生代全球变冷中的作用并不清楚。基于环喜马拉雅-青藏高原的亚洲边缘海风化记录的重建，发现中新世气候适宜期存在全球极端侵蚀和风化，与当时全球气候和大气CO_2水平耦合，而青藏高原上新世以来风化速率增加，与气候变冷趋势相反（Wan et al.，2012a），从而提出了亚洲大陆硅酸盐风化在构造稳定和活跃两种状态下对气候变化存在截然相反的反馈机制，即风化–气候之间的反馈机制在地质时期可能是动态变化的，主要受控于构造作用的活跃程度。这间接证实了高原隆升在新生代气候变冷中的驱动作用，也很好解释了Raymo等提出的构造隆升假说和BLAG的风化负反馈假说的长期争论（Wan et al.，2012a）。

轨道时间尺度碳循环研究的一个重要科学问题是冰期碳丢失之谜，即冰芯记录的大气CO_2浓度显示出明显的冰期-间冰期旋回，且冰期比间冰期的浓度低了约80×10^{-6}，那么冰期丢失的碳去了哪里？主流观点如"呼吸CO_2假说""硅渗漏假说""铁假说"等均强调海洋本身的物理、化学和生物过程影响了碳循环，而忽略了硅酸盐风化的作用，因为经典研究认为大陆风化速率在冰期相对间冰期更低，因而对冰期丢失的碳没有贡献（Foster and Vance，2006）。但上述看法忽略了冰期-间冰期时间尺度上海平面巨大变化所引起的陆架风化的反馈作用。无论是高海平面时期的南海全新世风化记录（Wan et al.，2015），还是拥有狭窄陆架和高陆源输入的阿拉伯海末次冰期以来的风化记录（Yu et al.，2020）均显示了风化与温度降水的紧密耦合。南海（Wan et al.，2017）和西菲律宾海（Xu et al.，2018）沉积记录揭示的第四纪冰期-间冰期风化演变与之前普遍认识的冰期大陆风化减弱截然相反，表明硅酸盐风化通量在冰期低海平面时增大而不是以前所认为的降低。研究表明，在全球尺度上冰期热带陆架硅酸盐风化碳消耗相当于现代全球硅酸盐风化通量的12%，可降低约9%的冰期大气CO_2，是冰期-间冰期碳循环中的一个不可忽略的重要机制（Wan et al.，2017）。

2.5.2 亚洲大陆边缘沉积有机质输运与埋藏

亚洲大陆边缘地区陆海相互作用强烈，不仅是河流入海细颗粒物的重要沉积区，也是沉积有机碳发生堆积、迁移和转化的主要区域（石学法等，2015，2016；胡利民等，2020）。

通过对我国东部陆架沉积有机碳"源-汇"过程的研究，建立了渤海、黄海沉积有机质远距离、选择性输运的概念模式，提出了大河输入和陆架沉积动力环境的控制机制（Hu et al.，2012，2013）；估算出我国东部陆架现代沉积有机碳埋藏通量约为13 Mt/a（Hu et al.，2016b），提出东部陆架泥质区是大河输入陆源沉积有机碳的重要储库，对全球范围内海洋沉积有机碳埋藏有重要贡献（石学法等，2016；Jiao et al.，2018）。

对低纬度地区有机质的来源、输运扩散机制和环境记录响应研究发现，浊流输运导致的溢流沉积是孟加拉湾有机质输运沉积的主要控制因素（李景瑞等，2017）；受控于热带季风降雨和流域较强的侵蚀作用，泰国湾入海颗粒有机碳具有显著的季节性和高强度特点（Wu et al.，2020）；在全球变暖背景下，热带河流输入和季风驱动的近岸动力环境（如层化和沿岸流等）限制了泰国湾陆源有机碳的向海扩散（Hu et al.，2017；Wu et al.，2020）。通过对比发现，泰国湾和渤海沉积黑碳具有不同源谱特征和沉积通量，揭示了区域能源结构、沉积动力条件和传输方式对黑碳空间差异性的影响机制（Hu et al.，2016a，2017）。

相比于中低纬陆架区，高纬的北极-亚北极陆架周边不仅有大河的输入，而且还发育有广袤的冻土层和季节性的海冰，使得该区域的沉积有机碳"源-汇"过程独具特色（胡利民等，2020）。通过重建百年尺度白令海陆架有机碳埋藏记录，揭示出20世纪70年代后期海冰变化对该区浮游植物群落结构演变及有机碳埋藏的制约，发现白令海陆架北部和南部近十几年的海冰变化及浮游植物群落响应具有显著的空间异质性，初步阐明了高纬度极地海域较高有机碳埋藏能力的主要影响因素（胡利民等，2015；Hu et al.，2020）。

总之，亚洲大陆边缘不同地区沉积有机质记录保存着对自然过程和人类活动影响的差异性信号。今后需要在多时间尺度上加强不同纬度陆架边缘海沉积有机质"源-汇"过程及其环境响应的对比研究。

3 古海洋与古气候的沉积记录

亚洲大陆边缘发育巨厚连续的沉积物，记录了完整的古海洋和古气候演化信息。该区最主要的海洋动力系统包括黑潮（西边界流）、北太平洋中层水、西太暖池和印尼贯穿流，它们对气候演变具有重要的影响和控制作用。

3.1 黑潮及其分支演化历史

黑潮源于北赤道流，以高温，高盐、寡营养盐为特征，是北太平洋西部风驱动向北流的暖流，对沿途的海洋环境、天气系统、渔场形成和气候变化等产生巨大影响。近年来的研究进展主要体现在以下几个方面。

末次冰盛期时黑潮主轴是否仍然流经冲绳海槽？这是古黑潮研究中一个长期存在争议的问题。国内外学者对该问题进行了长时间的研究，早期的研究者根据 LGM 低海平面时期喜暖有孔虫的低丰度，认为黑潮主流轴迁移到琉球岛弧以东，可能与台湾东部和琉球岛弧之间出现"路桥"相关（Jian et al.，1998），但近年的研究大都认为黑潮主轴在 LGM 时期仍然流经冲绳海槽（Shi et al.，2014；Zheng et al.，2016a）。

近年来有研究者发现在海洋尺度和轨道尺度上古黑潮强度/路径发生了显著变化。台湾南部珊瑚 $\delta^{18}O$ 记录表明，过去52年（1953—2004年）黑潮转移在年代际和十年际时间尺度上同步变化（Li X. H. et al.，2017）。在小冰期（1400—1850 A.D.），南黄海海表温度增加约5℃，认为系黑潮增强，受太平洋沃克环流增强驱动（Zhang Y. C. et al.，2019）。在早全新世最暖期，模拟结果显示黑潮强度明显增加（Zheng et al.，2016a）。在冰消期，多个不同指标的海表温度记录表明在 HS1 和 YD 冷期海表温度减小，并认为黑潮强度在16~15 ka 开始增加，但是在 B/A 时段 SST 模式呈现相反的模式（Zhao et al.，2015a）。Shi 等（2014）将古黑潮记录延长到88 ka，发现黑潮在全新世和 MIS5.1 期黑潮增强，在 MIS4-2 时段减弱，在 MIS3 期存在千年尺度变化规律。

黑潮对海洋沉积和生物地球化学过程有显著影响。物源分析表明，7 ka 以来黑潮带来了大量台湾源的碎屑物质到冲绳海槽，对沉积作用和沉积物分布模式产生了显著影响（Dou et al.，2016）。对冲绳海槽中部和北部沉积岩心氧化还原敏感元素指标分析发现，黑潮水团性质在千年和轨道尺度发生了显著变化，黑潮增强显著提高了深层水体通风，而其低营养盐水团也减小了表层生产力勃发和输出（Shao et al.，2016；Zou et al.，2020）。

3.2 北太平洋中层水演化

现今北太平洋中层水（North Pacific intermediate water，NPIW）主要分布在300~800 m水深，以低盐（33.8‰）为显著特征（Talley，1993）。国内对北太平洋中层水的研究程度较低。Zou等（2020）基于沉积物氧化还原敏感元素指标，发现北太平洋亚热带西部海域中层水通风模式在轨道和千年时间尺度上与北太平洋亚北极区域中层水形成演化模式一致，在冷期通风增强，暖期减弱；而在B/A时段，北太平洋亚热带涡西部中层水体显著缺氧，这与现今富氧环境形成明显差异。该研究认为缺氧扩张与该时段NPIW的形成减弱密切相关，随着全球进一步增暖，NPIW范围进一步减小，中层深度水体溶解氧损失加剧，威胁到北太平洋亚热带海洋生态系统安全。

北太平洋中层水对南海通风影响随海域和水深而发生变化。南海北部磁学记录显示，在LGM、HS1和YD时段底层环流强度显著增加，表明这与NPIW增强相关（Zheng et al.，2016b）。但沉积物氧化还原敏感元素和底栖有孔虫$\delta^{13}C$代用指标显示，LGM以来NPIW强化并没有影响南海深层水（>1 600 m）通风（Li G. et al.，2018）。南海北部深层水通风在冰消期要明显强于全新世和LGM，表明发生了北太平洋中层水强化和入侵（Wan and Jian，2014）。

3.3 西太暖池和印尼贯穿流演化

3.3.1 暖池沉积记录及其演化

西太平洋暖池（简称西太暖池）是世界上最重要的热量和水汽源区，也是全球海-气交换活跃的海区。我国科学家对西太暖池古海洋学与古气候学研究多年，在国际上形成了独特的优势。

我国研究者基于低纬海区接收的太阳辐射量主要由岁差周期控制的事实，沉积记录也表明西太暖池的气候变化并不总是受控于北半球高纬要素，提出了热带过程（如类ENSO、亚洲季风）本身在西太暖池气候变迁中起着引擎的作用（汪品先和翦知湣，2019；Dang et al.，2020）。研究还表明，在岁差周期上西太暖池生产力的变化领先全球冰体积变化2~4 ka，这很可能表明热带过程通过碳循环效应对全球气候产生影响，从而证明了热带驱动在全球气候系统中的重要性（Xiong et al.，2012）。最近Jian等（2020）提出了热带过程控制全球气候的新机制，认为西太暖池岁差驱动的海水温度变化能直接影响东西热带太平洋的带状温度梯度，从而通过改变类ENSO变化来影响高低纬地区之间的热量与水汽分配，进而影响全球气候系统。

西太暖池LGM大规模硅藻席沉积的发现为检验热带西太平洋通过碳循环控制全球气候提供了难得的素材。研究发现，LGM时大型硅藻 Ethmodiscus rex 的持续勃发最终使东菲律宾海表层由强的大气CO_2的源发展成弱的大气CO_2的汇，只占全球大洋面积约0.08%区域的大型硅藻沉积却能贡献LGM大气向大洋碳转移总量的1.3%（Xiong et al.，2013）。冰期输入增强的亚洲风尘携带丰富的营养物（硅和铁）（Wan et al.，2012b；Jiang et al.，2016），刺激了西太暖池包括大型硅藻在内的藻类的勃发及其生产力的提高，进而将更多的CO_2"泵"入海底，促进了冰期大气碳的丢失（Xiong et al.，2012，2015）。这些发现反映了西太暖池硅藻生产在全球碳循环中的重要性，凸显了热带海区硅藻勃发对冰期大气碳丢失的重要贡献。

在深部碳酸盐系统演化研究方面，发现中更新世以来西太暖池的深部$CaCO_3$埋藏呈现冰期高间冰期低的旋回特征，主要受深部$[CO_3^{2-}]$的控制（Sun et al.，2017）。基于浮游有孔虫壳体重量-深部$[CO_3^{2-}]$标定公式，定量重建了中更新世以来西太暖池深部$[CO_3^{2-}]$的演化，结果表明深部$[CO_3^{2-}]$和（或）$CaCO_3$埋藏在冰期旋回中通过$CaCO_3$补偿机制响应全球海平面变化（Qin et al.，2017）。

3.3.2 印尼贯穿流的演化

作为太平洋和印度洋之间的唯一通道，印尼贯穿流（Indonesian through flow，ITF）流经印度尼西亚海内众多海峡，调控着太平洋向印度洋的水体和热量输送，同时也可将热带气候效应放大或遥相关到高纬海区。近年来，我国学者将 LGM 以来 ITF 的演化划分为 5 个阶段：在 LGM 以及中全新世阶段，增强的东南季风（印澳季风）导致 ITF 表层暖流加强；在冰消期阶段，快速上升的海平面使 ITF 能通过龙目海峡以及翁拜海峡进入印度洋；在早全新世以及晚全新世阶段，增强的东亚季风降雨抑制了 ITF 表层流，使 ITF 温跃层流加强（Ding et al., 2013）。研究还发现，末次冰期以来 ITF 的水体性质、垂向层化和通量还受到类 ENSO 过程的调控（Xu, 2014），具体来说，类 El Niño 状态使 ITF 区域的 SST 下降、温跃层变浅（Zhang P. et al., 2018），同时，类 El Niño 状态下赤道纬向风减弱和（或）南海穿越流增强还会导致 ITF 输运量下降（Fan et al., 2018）。这显示末次冰期以来 ITF 在轨道和亚轨道时间尺度上发生了显著的变化，这些变化与海平面、东亚季风以及类 ENSO 过程密切相关。

4 结　语

近十年来，我国科学家从全球视角对亚洲大陆边缘沉积地质学开展了大量研究，取得了系列创新成果。但总体来看，我们对亚洲大陆边缘沉积的认知还很有限，今后仍需大力加强研究。特提出以下研究建议。

（1）由于海洋沉积过程受到构造、气候、海平面以及人类活动等众多因素的影响与控制，而亚洲大陆边缘所处构造位置复杂，气候梯度跨度大，地域上涉及多个国家，研究上涉及众多学科，因此亟须发起一个以亚洲大陆边缘沉积地质学为主题的大型国际合作计划开展综合对比研究。

（2）海洋沉积学研究的主要方法是"比较沉积学"，即"将今论古"的"现实主义原则"。对现代海洋沉积过程的长期连续观测是解译地层记录、认识动力学机制和进行未来演化趋势预测的基础（高抒，2011，2017）。20 世纪 90 年代前后，世界各国陆续建立了区域或者全球的海洋观测系统，包括全球海洋观测系统（GOOS）、全球实时海洋观测系统（Argo）、欧洲海洋观测数据网络（EMOD-net）、美国大洋观测计划（OOI）和加拿大海王星海底观测系统（NEPTUNE）等。近十年来，我国大力推进"海洋立体观测网"建设，并建立了区域性海底观测系统。在今后研究中需要特别加强亚洲大陆边缘现代沉积过程的长期连续观测，及时获取精细化、多层次、长时间序列的海洋环境基础数据和资料，这可深化对现代沉积过程的认知和过去环境演化的解释。

（3）数值模拟方法在海洋沉积学研究中的作用日显重要。沉积物输运系统是地球表生系统科学的重要组成部分，海洋沉积数值模型的主要目的是了解海洋沉积物的运动过程。鉴于地球表面动力学的重要性，地球表面动力学模拟系统协会（CSDMS）推动建立一个用于预测不同时空尺度内沉积物和溶解质在陆地和沉积盆地中的搬运和沉积过程的模型。目前应用的海洋沉积数值模型众多，而模拟结果不仅与建立的模型有关，还与初始条件的设立和输入参数密切相关。因此，今后海洋沉积学数值模拟研究不仅要优化现有模型和创建新模型，还应注意结合现代沉积过程的观测以获取准确参数。

（4）大数据正在成为地球科学研究的一种新思路和新方法，其实质是对海量有效数据进行挖掘、进而分析和解决问题。目前国际上与海洋沉积相关的数据库包括美国国家海洋和大气管理局（National Oceanic and Atmospheric Administration, NOAA）大数据库、全球海陆数据库（GBCO）、IODP 数据库和国家海洋科学数据中心（NMSDC）等。海洋沉积学研究涉及领域多、数据量巨大且类型多样，需通过广泛的国际合作建立亚洲大陆边缘沉积数据库，实现大数据的挖掘和应用，这是深化亚洲大陆边缘沉积学研究的基础。

致谢：本文在写作过程中，得到了于永贵、徐涛玉、单新、董江、秦秉斌、陈志华、唐正、陈金霞、陈禹飞和李秋玲等的帮助，谨致谢忱。

参 考 文 献

曹鹏,石学法,李巍然,刘升发,朱爱美,杨刚,Khokiattiwong S,Kornkanitnan N. 2015. 安达曼海东南部海域表层沉积物稀土元素特征及其物源指示意义. 海洋地质与第四纪地质,35(5):57-67

陈禹飞,乔淑卿,石学法,葛晨东,李秋玲,刘升发,张颖,王小静,Khokiattiwong S,Kornkanitnan N. 2020. 末次冰消期以来泰国湾沉积物物源变迁的元素地球化学证据. 第四纪研究,40(3):726-736

陈志华,陈毅,王汝建,黄元辉,刘欣德,王磊,邹建军. 2014. 末次冰消期以来白令海盆的冰筏碎屑事件与古海洋学演变记录. 极地研究,26(1):17-28

董林森,刘焱光,石学法,方习生,陈志华,闫仕娟,黄元辉. 2014. 西北冰洋表层沉积物黏土矿物分布特征及物质来源. 海洋学报,36(4):22-32

董智,石学法,葛晨东,邹建军,姚政权,Sergey G,王成龙,宗娴. 2017. 日本海中部 60 ka 以来的风尘沉积对西风环流演化的指示. 科学通报,62(11):1172-1184

豆汝席,邹建军,石学法,朱爱美,董智,石丰登,薛心如,Sergey G. 2020. 3 万年以来日本海西部海冰活动变化. 第四纪研究,40(3):690-703

高抒. 2005. 美国《洋陆边缘科学计划 2004》述评. 海洋地质与第四纪地质,25(1):119-123

高抒. 2011. 海洋沉积地质过程模拟:性质与问题及前景. 海洋地质与第四纪地质,31(5):1-7

高抒. 2017. 沉积记录研究的现代过程视角. 沉积学报,35(5):918-925

葛淑兰,石学法,黄元辉,陈志华,刘建兴,闫仕娟. 2013. 白令海岩心记录的冰消期 14 ka 以来地磁场强度和方向. 地球物理学报,56(9):3071-3084

郭瑜璇,乔淑卿,石学法,吴斌,袁龙,任艺君,高晶晶,朱爱美,Kornkanitnan N. 2019. 曼谷湾河口区百年来沉积物重金属变化趋势及污染来源. 海洋地质与第四纪地质,39(2):61-69

胡利民,石学法,刘焱光,白亚之,董林森,黄元辉. 2015. 白令海西部柱样沉积物中有机碳的地球化学特征与埋藏记录. 海洋地质与第四纪地质,35(3):37-47

胡利民,石学法,叶君,张钰莹. 2020. 北极东西伯利亚陆架沉积有机碳的源汇过程研究进展. 地球科学进展,35:1073-1086

黄元辉,葛淑兰,石学法,陈志华,刘焱光,王旭晨,何连花. 2013. 白令海北部陆坡 BR07 孔年龄框架重建. 海洋学报,35(6):67-74

李景瑞,刘升发,胡利民,冯秀丽,孙兴全,白亚之,石学法. 2017. 孟加拉湾中部表层沉积物有机碳分布特征及来源. 海洋科学进展,35(1):73-82

李秋玲. 2020. 北极东西伯利亚陆架沉积物特征及物源分析. 青岛:国家海洋局第一海洋研究所硕士学位论文

李铁刚,曹奇原,李安春,秦蕴珊. 2003. 从源到汇:大陆边缘的沉积作用. 地球科学进展,18(5):713-721

梅静,王汝建,陈建芳,程振波,陈志华,孙烨忱. 2012. 西北冰洋楚科奇海台 P31 孔晚第四纪的陆源沉积物记录及古海洋与古气候意义. 海洋地质与第四纪地质,32(3):77-86

乔淑卿,石学法,高晶晶,朱爱美,Kornkanitnan N,胡利民,张杨硕. 2015. 曼谷湾沉积物重金属元素的富集效应与生物有效性. 中国环境科学,35(11):3445-3451

秦蕴珊,赵一阳,陈丽蓉,赵松龄. 1987. 东海地质. 北京:科学出版社

石学法. 2012. 中国近海海洋——海洋底质. 北京:海洋出版社

石学法,刘升发,乔淑卿,刘焱光,方习生,吴永华,朱志伟. 2010. 东海闽浙沿岸泥质区沉积特征与古环境记录. 海洋地质与第四纪地质,30(4):19-30

石学法,邹建军,王昆山. 2011. 鄂霍次克海晚第四纪以来古环境演化. 海洋地质与第四纪地质,31(6):1-12

石学法,刘焱光,乔淑卿,刘升发,姚政权,邹建军,李传顺. 2015. 亚洲大陆边缘"源-汇"过程研究:沉积纪录与控制机理. 吉林大学学报(地球科学版),45(S1):12-23

石学法,胡利民,乔淑卿,白亚之. 2016. 中国东部陆架海沉积有机碳研究进展:来源、输运与埋藏. 海洋科学进展,34(3):313-327

石学法,邹建军,姚政权,豆汝席,Sergey G. 2019. 日本海末次冰期以来沉积作用和环境演化及其控制要素. 海洋地质与第四纪地质,39(3):1-11

宋腾飞,王宏雷,陈漪馨,李朝新,朱爱美,白亚之,石学法,Sergei G,Aleksandr B,刘焱光. 2018. 白令海北部陆坡 23 ka 以来古生产力变化及其对海冰扩张的响应. 海洋学报,40(5):90-106

汪品先,翦知湣. 2019. 探索南海深部的回顾与展望. 中国科学:地球科学,49(10):1590-1606

王昆山,刘焱光,董林森,陈志华. 2014a. 北冰洋西部表层沉积物重矿物特征. 极地研究,26(1):71-78

王昆山,石学法,刘升发,乔淑卿,杨刚,胡利民,Narumol K,Somkiat K. 2014b. 泰国湾西部表层沉积物重矿分布特征:对物质来源和沉积环境的指示. 第四纪研究,34(3):623-634

王昆山,石学法,吴永华,邹建军,姜晓黎. 2014c. 鄂霍次克海东南部 OS03-1 岩心重矿物分布特征及物质来源. 海洋学报,36(5):177-185

吴凯凯,刘升发,金爱民,娄章华,吴斌,李景瑞,张辉,方习生,Che Abd. Rahim Bin Mohamed,石学法. 2019. 马来半岛彭亨河和吉兰丹河沉积物稀土元素特征及其物源示踪. 海洋学报,41(7):77-91

吴永华. 2016. 日本海末次冰期以来千年尺度古海洋事件研究. 上海:同济大学博士学位论文

杨守业. 2006. 亚洲主要河流的沉积地球化学示踪研究进展. 地球科学进展,21(6):648-655

张杨硕,乔淑卿,石学法,杨刚,刘升发,杜德文,Kornkanitnan N,Khokiattiwong S,鄢全树,张海桃,曹德凯. 2017. 泰国湾底质沉积物输运

趋势. 海洋地质与第四纪地质, 37(1): 86-92

章伟艳, 于晓果, 刘焱光, 金路, 叶黎明, 许冬, 边叶萍, 张德玉, 姚旭莹, 张富元. 2015. 楚科奇海盆M04柱晚更新世以来沉积古环境记录. 海洋学报, 37(7): 85-96

邹建军, 石学法, 白亚之, 朱爱美, 陈志华, 黄元辉. 2012. 末次冰消期以来白令海古环境及古生产力演化. 地球科学(中国地质大学学报), 37(S1): 1-10

Allen P A. 2017. Sediment Routing Systems: The Fate of Sediment from Source to Sink. Cambridge: Cambridge University Press

Berner R A, Caldeira K. 1997. The need for mass balance and feedback in the geochemical carbon cycle. Geology, 25(10): 955-956

Bi L, Yang S Y, Li C, Guo Y L, Wang Q, Liu J T, Yin P. 2015. Geochemistry of river-borne clays entering the East China Sea indicates two contrasting types of weathering and sediment transport processes. Geochemistry, Geophysics, Geosystems, 16(9): 3034-3052

Bi L, Yang S Y, Zhao Y, Wang Z B, Dou Y G, Li C, Zheng H B. 2017. Provenance study of the Holocene sediments in the Changjiang (Yangtze River) Estuary and inner shelf of the East China Sea. Quaternary International, 441: 147-161

Cao P, Shi X F, Li W R, Liu S F, Yao Z Q, Hu L M, Khokiattiwong S, Kornkanitnan N. 2015. Sedimentary responses to the Indian Summer Monsoon variations recorded in the southeastern Andaman Sea slope since 26 ka. Journal of Asian Earth Sciences, 114: 512-525

Chang Z H, Zhou L P. 2019. Evidence for provenance change in deep sea sediments of the Bengal Fan: a 7 million year record from IODP U1444A. Journal of Asian Earth Sciences, 186: 104008

Chen J X, Liu Y G, Shi X F, Suk B C, Zou J J, Yao Z Q. 2017. Climate and environmental changes for the past 44 ka clarified by pollen and algae composition in the Ulleung Basin, East Sea (Japan Sea). Quaternary International, 441: 162-173

Chen Z Y, Jiang Y W, Liu J T, Gong W P. 2017. Development of upwelling on pathway and freshwater transport of Pearl River plume in northeastern South China Sea. Journal of Geophysical Research: Oceans, 122(8): 6090-6109

Dai Z J, Liu J T, Wei W, Chen J Y. 2014. Detection of the Three Gorges Dam influence on the Changjiang (Yangtze River) submerged delta. Scientific Reports, 4: 6600

Dang H W, Wu J W, Xiong Z F, Qiao P J, Li T G, Jian Z M. 2020. Orbital and sea-level changes regulate the iron-associated sediment supplies from Papua New Guinea to the equatorial Pacific. Quaternary Science Reviews, 239: 106361

Deng K, Yang S Y, Bi L, Chang Y P, Su N, Frings P, Xie X L. 2019. Small dynamic mountainous rivers in Taiwan exhibit large sedimentary geochemical and provenance heterogeneity over multispatial scales. Earth and Planetary Science Letters, 505: 96-109

Ding X, Bassinot F, Guichard F, Fang N Q. 2013. Indonesian Throughflow and monsoon activity records in the Timor Sea since the last glacial maximum. Marine Micropaleontology, 101: 115-126

Dong J, Li A C, Liu X T, Wan S M, Feng X G, Lu J, Pei W Q, Wang H L. 2018. Sea-level oscillations in the East China Sea and their implications for global seawater redistribution during 14.0-10.0 kyr BP. Palaeogeography, Palaeoclimatology, Palaeoecology, 511: 298-308

Dou Y G, Yang S Y, Shi X F, Clift P D, Liu S F, Liu J H, Li C, Bi L, Zhao Y. 2016. Provenance weathering and erosion records in southern Okinawa Trough sediments since 28 ka: geochemical and Sr-Nd-Pb isotopic evidences. Chemical Geology, 425: 93-109

Fan D D, Tu J B, Shang S, Cai G F. 2014. Characteristics of tidal-bore deposits and facies associations in the Qiantang Estuary, China. Marine Geology, 348: 1-14

Fan W J, Jian Z M, Chu Z H, Dang H W, Wang Y, Bassinot F, Han X Q, Bian Y P. 2018. Variability of the Indonesian throughflow in the Makassar Strait over the last 30 ka. Scientific Reports, 8(1): 5678

Foster G L, Vance D. 2006. Negligible glacial-interglacial variation in continental chemical weathering rates. Nature, 444(7121): 918-921

Gao J H, Jia J, Sheng H, Yu R, Li G C, Wang Y P, Yang Y, Zhao Y, Li J, Bai F, Xie W, Wang A, Zou X, Gao S. 2017. Variations in the transport, distribution, and budget of ^{210}Pb in sediment over the estuarine and inner shelf areas of the East China Sea due to Changjiang catchment changes. Journal of Geophysical Research: Earth Surface, 122(1): 235-247

Gao J H, Jia J J, Kettner A J, Xing F, Wang Y P, Li J, Bai F L, Zou X Q, Gao S. 2018. Reservoir-induced changes to fluvial fluxes and their downstream impacts on sedimentary processes: the Changjiang (Yangtze) River, China. Quaternary International, 493: 187-197

Gao J H, Shi Y, Sheng H, Kettner A J, Yang Y, Jia J J, Wang Y P, Li J, Chen Y N, Zou X Q, Gao S. 2019. Rapid response of the Changjiang (Yangtze) River and East China Sea source-to-sink conveying system to human induced catchment perturbations. Marine Geology, 414: 1-17

Gao S, Collins M B. 2014. Holocene sedimentary systems on continental shelves. Marine Geology, 352: 268-294

Gorbarenko S, Velivetskaya T, Malakhov M, Bosin A. 2017. Glacial terminations and the Last Interglacial in the Okhotsk Sea: their implication to global climatic changes. Global and Planetary Change, 152: 51-63

Hu L M, Shi X F, Guo Z G, Wang H J, Yang Z S. 2013. Sources, dispersal and preservation of sedimentary organic matter in the Yellow Sea: the importance of depositional hydrodynamic forcing. Marine Geology, 335: 52-63

Hu L M, Shi X F, Bai Y Z, Fang Y, Chen Y J, Qiao S Q, Liu S F, Yang G, Kornkanitnan N, Khokiattiwong S. 2016a. Distribution, input pathway and mass inventory of black carbon in sediments of the Gulf of Thailand, SE Asia. Estuarine, Coastal and Shelf Science, 170: 10-19

Hu L M, Shi X F, Bai Y Z, Qiao S Q, Li L, Yu Y G, Yang G, Ma D Y, Guo Z G. 2016b. Recent organic carbon sequestration in the shelf sediments of the Bohai Sea and Yellow Sea, China. Journal of Marine Systems, 155: 50-58

Hu L M, Shi X F, Qiao S Q, Lin T, Li Y Y, Bai Y Z, Wu B, Liu S F, Kornkanitnan N, Khokiattiwong S. 2017. Sources and mass inventory of sedimentary polycyclic aromatic hydrocarbons in the Gulf of Thailand: implications for pathways and energy structure in SE Asia. Science of The Total Environment, 575: 982-995

Hu L M, Liu Y G, Xiao X T, Gong, X, Zou J J, Bai Y Z, Gorbarenko S, Fahl K, Stein, R, Shi X F. 2020. Sedimentary records of bulk organic matter and lipid biomarkers in the Bering Sea: a centennial perspective of sea-ice variability and phytoplankton community. Marine Geology, 429: 106308

Huh C A, Chen W F, Hsu F H, Su C C, Chiu J K, Lin S, Liu C S, Huang B J. 2011. Modern (<100 years) sedimentation in the Taiwan Strait: rates and source-to-sink pathways elucidated from radionuclides and particle size distribution. Continental Shelf Research, 31(1): 47–63

Jia J J, Gao J H, Cai T L, Li Y, Yang Y, Wang Y P, Xia X M, Li J, Wang A J, Gao S. 2018. Sediment accumulation and retention of the Changjiang (Yangtze River) subaqueous delta and its distal muds over the last century. Marine Geology, 401: 2–16

Jia Y G, Tian Z C, Shi X F, Liu J P, Chen J X, Liu X L, Ye R J, Ren Z Y, Tian J W. 2019. Deep-sea sediment resuspension by internal solitary waves in the northern South China Sea. Scientific Reports, 9(1): 12137

Jian Z M, Saito Y, Wang P X, Li B H, Chen R H. 1998. Shifts of the Kuroshio axis over the last 20000 years. Chinese Science Bulletin, 43(12): 1053–1056

Jian Z M, Wang Y, Dang H W, Lea D W, Liu Z Y, Jin H Y, Yin Y Q. 2020. Half-precessional cycle of thermocline temperature in the western equatorial Pacific and its bihemispheric dynamics. Proceedings of the National Academy of Sciences of the United States of America, 117(13): 7044–7051

Jiang F Q, Zhou Y, Nan Q Y, Zhou Y, Zheng X F, Li T G, Li A C, Wang H L. 2016. Contribution of Asian dust and volcanic material to the western Philippine Sea over the last 220 kyr as inferred from grain size and Sr-Nd isotopes. Journal of Geophysical Research: Oceans, 121(9): 6911–6928

Jiao N Z, Liang Y T, Zhang Y Y, Liu J H, Zhang Y, Zhang R, Zhao M X, Dai M H, Zhai W D, Gao K S, Song J M, Yuan D L, Li C, Lin G H, Huang X P, Yan H Q, Hu L M, Zhang Z H, Wang L, Cao C J, Luo Y W, Luo T W, Wang N N, Dang H Y, Wang D X, Zhang S. 2018. Carbon pools and fluxes in the China Seas and adjacent oceans. Science China Earth Sciences, 61(11): 1535–1563

Jin H Y, Jian Z M. 2013. Millennial-scale climate variability during the mid-Pleistocene transition period in the northern South China Sea. Quaternary Science Reviews, 70: 15–27

Li C, Yang S Y, Zhao J X, Dosseto A, Bi L, Clark T R. 2016. The time scale of river sediment source-to-sink processes in East Asia. Chemical Geology, 446: 138–146

Li G, Rashid H, Zhong L F, Xu X, Yan W, Chen Z. 2018. Changes in deep water oxygenation of the South China Sea since the last glacial period. Geophysical Research Letters, 45(17): 9058–9066

Li J R, Liu S F, Shi X F, Feng X L, Fang X S, Cao P, Sun X Q, Ye W X, Khokiattiwong S, Kornkanitnan N. 2017. Distributions of clay minerals in surface sediments of the middle Bay of Bengal: source and transport pattern. Continental Shelf Research, 145: 59–67

Li J R, Liu S F, Shi X F, Zhang H, Fang X S, Chen M T, Cao P, Sun X Q, Ye W X, Wu K K, Khokiattiwong S, Kornkanitnan N. 2018. Clay minerals and Sr-Nd isotopic composition of the Bay of Bengal sediments: implications for sediment provenance and climate control since 40 ka. Quaternary International, 493: 50–58

Li J R, Liu S F, Shi X F, Zhang H, Fang X S, Cao P, Yang G, Xue X R, Khokiattiwong S, Kornkanitnan N. 2019. Sedimentary responses to the sea level and Indian summer monsoon changes in the central Bay of Bengal since 40 ka. Marine Geology, 415: 105947

Li J R, Liu S F, Shi X F, Chen M T, Zhang H, Zhu A M, Cui J J, Khokiattiwong S, Kornkanitnan N. 2020. Provenance of terrigenous sediments in the central Bay of Bengal and its relationship to climate changes since 25 ka. Progress in Earth and Planetary Science, 7(1): 1–16

Li L, Li Q Y, Tian J, Wang P X, Wang H, Liu Z H. 2011. A4-Ma record of thermal evolution in the tropical western Pacific and its implications on climate change. Earth and Planetary Science Letters, 309(1-2): 10–20

Li X H, Liu Y, Hsin Y C, Liu W G, Shi Z G, Chiang H W, Shen C C. 2017. Coral record of variability in the upstream Kuroshio Current during 1953–2004. Journal of Geophysical Research: Oceans, 122(8): 6936–6946

Li Y H, Wang A J, Qiao L, Fang J Y, Chen J. 2012. The impact of typhoon Morakot on the modern sedimentary environment of the mud deposition center off the Zhejiang-Fujian coast, China. Continental Shelf Research, 37: 92–100

Liu J, Zhang X H, Mei X, Zhao Q H, Guo X W, Zhao W N, Liu J X, Saito Y, Wu Z Q, Li J, Zhu X Q, Chu H X. 2018. The sedimentary succession of the last ~3.50 Myr in the western South Yellow Sea: paleoenvironmental and tectonic implications. Marine Geology, 399: 47–65

Liu J G, Xiang R, Chen M H, Chen Z, Yan W, Liu F. 2011. Influence of the Kuroshio current intrusion on depositional environment in the northern South China Sea: evidence from surface sediment records. Marine Geology, 285(1-4): 59–68

Liu J G, Xiang R, Chen Z, Chen M H, Yan W, Zhang L L, Chen H. 2013. Sources, transport and deposition of surface sediments from the South China Sea. Deep Sea Research Part I: Oceanographic Research Papers, 71: 92–102

Liu J G, He W, Cao L, Zhu Z, Xiang R, Li T G, Shi X F, Liu S F. 2019a. Staged fine-grained sediment supply from the Himalayas to the Bengal Fan in response to climate change over the past 50,000 years. Quaternary Science Reviews, 212: 164–177

Liu J G, Cao L, Yan W, Shi X F. 2019b. New archive of another significant potential sediment source in the South China Sea. Marine Geology, 410: 16–21

Liu J P, Kuehl S A, Pierce A C, Williams J, Blair N E, Harris C, Aung D W, Aye Y Y. 2020. Fate of Ayeyarwady and Thanlwin rivers sediments in the Andaman Sea and Bay of Bengal. Marine Geology, 423: 106137

Liu J T, Wang Y H, Yang R J, Hsu R T, Kao S J, Lin H L, Kuo F H. 2012. Cyclone-induced hyperpycnal turbidity currents in a submarine canyon. Journal of Geophysical Research: Oceans, 117(C4): C04033

Liu J T, Kao S J, Huh C A, Hung C C. 2013. Gravity flows associated with flood events and carbon burial: Taiwan as instructional source area. Annual Review of Marine Science, 5: 47–68

Liu J T, Hsu R T, Yang R J, Wang Y P, Wu H, Du X Q, Li A C, Chien S C, Lee J, Yang S Y, Zhu J R, Su C C, Chang Y, Huh C A. 2018. A comprehensive sediment dynamics study of a major mud belt system on the inner shelf along an energetic coast. Scientific Reports, 8(1): 4229

Liu J X, Shi X F, Liu Q S, Ge S L, Liu Y G, Yao Z Q, Zhao Q H, Jin C S, Jiang Z X, Liu S F, Qiao S Q, Li X Y, Li C S, Wang C J. 2014. Magnetostratigraphy of a greigite-bearing core from the South Yellow Sea: implications for remagnetization and sedimentation. Journal of Geophysical Research: Solid Earth, 119(10): 7425-7441

Liu J X, Liu Q S, Zhang X H, Liu J, Wu Z Q, Mei X, Shi X F, Zhao Q H. 2016. Magnetostratigraphy of a long Quaternary sediment core in the South Yellow Sea. Quaternary Science Reviews, 144: 1-15

Liu S F, Shi X F, Liu Y G, Qiao S Q, Yang G, Fang X S, Wu Y H, Li C X, Li X Y, Zhu A M, Gao J J. 2010. Records of the East Asian winter monsoon from the mud area on the inner shelf of the East China Sea since the mid-Holocene. Chinese Science Bulletin, 55(21): 2306-2314

Liu S F, Shi X F, Yang G, Khokiattiwong S, Kornkanitnan N. 2016a. Concentration distribution and assessment of heavy metals in the surface sediments of the western Gulf of Thailand. Environmental Earth Sciences, 75(4): 1-14

Liu S F, Shi X F, Yang G, Khokiattiwong S, Kornkanitnan N. 2016b. Distribution of major and trace elements in surface sediments of the western Gulf of Thailand: implications to modern sedimentation. Continental Shelf Research, 117: 81-91

Liu S F, Li J R, Zhang H, Cao P, Mi B B, Khokiattiwong S, Kornkanitnan N, Shi X F. 2020. Complex response of weathering intensity registered in the Andaman Sea sediments to the Indian Summer Monsoon over the last 40 kyr. Marine Geology, 426: 106206

Liu Z F, Zhao Y L, Colin C, Stattegger K, Wiesner M G, Huh C A, Zhang Y W, Li X J, Sompongchaiyakul P, You C F, Huang C Y, Liu J T, Siringan F P, Le K P, Sathiamurthy E, Hantoro W S, Liu J G, Tuo S T, Zhao S H, Zhou S W, He Z D, Wang Y C, Bunsomboonsakul S, Li Y L. 2016. Source-to-sink transport processes of fluvial sediments in the South China Sea. Earth-Science Review, 153: 238-273

Qiao S Q, Shi X F, Fang X S, Liu S F, Kornkanitnan N, Gao J J, Zhu A M, Hu L M, Yu Y G. 2015. Heavy metal and clay mineral analyses in the sediments of Upper Gulf of Thailand and their implications on sedimentary provenance and dispersion pattern. Journal of Asian Earth Sciences, 114: 488-496

Qiao S Q, Shi X F, Wang G Q, Zhou L, Hu B Q, Hu L M, Yang G, Liu Y G, Yao Z Q, Liu S F. 2017. Sediment accumulation and budget in the Bohai Sea, Yellow Sea and East China Sea. Marine Geology, 390: 270-281

Qin B B, Li T G, Xiong Z F, Algeo T J, Chang F M. 2017. Deepwater carbonate ion concentrations in the western tropical Pacific since 250 ka: evidence for oceanic carbon storage and global climate influence. Paleoceanography and Paleoclimatology, 32(4): 351-370

Qin Y S, Zhao Y Y, Chen L R, Zhao S L. 1990. Geology of Bohai Sea. Beijing: China Ocean Press

Shan X, Shi X F, Clift P D, Qiao S Q, Jin L N, Liu J X, Fang X S, Xu T Y, Li S L, Kandasamy S, Zhao M W, Zhu Y, Zhang H, Zhang D, Wang H W, Li Y L, Yao Z Q, Wang S, Xu J. 2019a. Carbon isotope and rare earth element composition of Late Quaternary sediment gravity flow deposits on the mid shelf of East China Sea: implications for provenance and origin of hybrid event beds. Sedimentology, 66(5): 1861-1895

Shan X, Shi X F, Qiao S Q, Jin L N, Otharán G A, Zavala C, Liu J X, Zhang Y Q, Zhang D, Xu T Y, Fu C. 2019b. The fluid mud flow deposits represent mud caps of Holocene hybrid event beds from the widest and gentlest shelf. Marine Geology, 415: 105959

Shao H B, Yang S Y, Cai F, Li C, Liang J, Li Q, Hyun S M, Kao S J, Dou Y G, Hu B Q, Dong G, Wang F. 2016. Sources and burial of organic carbon in the middle Okinawa Trough during late Quaternary paleoenvironmental change. Deep Sea Research Part I: Oceanographic Research Papers, 118: 46-56

Shen X Y, Wan S M, France-Lanord C, Clift P D, Tada R, Révillon S, Shi X F, Zhao D B, Liu Y G, Yin X B, Song Z H, Li A C. 2017. History of Asian eolian input to the Sea of Japan since 15 Ma: links to Tibetan uplift or global cooling. Earth and Planetary Science Letters, 474: 296-308

Shi X, Wu Y, Zou J, Liu Y, Ge S, Zhao M, Liu J, Zhu A, Meng X, Yao Z, Han Y. 2014. Multiproxy reconstruction for Kuroshio responses to northern hemispheric oceanic climate and the Asian Monsoon since Marine Isotope Stage 5.1 (~88 ka). Climate of the Past, 10: 1735-1750

Shi X F, Liu S F, Fang X S, Qiao S Q, Khokiattiwong S, Kornkanitnan N. 2015. Distribution of clay minerals in surface sediments of the western Gulf of Thailand: sources and transport patterns. Journal of Asian Earth Sciences, 105: 390-398

Shi X F, Yao Z Q, Liu Q S, Larrasoaña J C, Bai Y Z, Liu Y G, Liu J H, Cao P, Li X Y, Qiao S Q, Wang K S, Fang X S, Xu T Y. 2016. Sedimentary architecture of the Bohai Sea China over the last 1 Ma and implications for sea-level changes. Earth and Planetary Science Letters, 451: 10-21

Stein R, Fahl K, Schade I, Manerung A, Wassmuth S, Niessen F, Nam S I. 2017. Holocene variability in sea ice cover, primary production, and Pacific-Water inflow and climate change in the Chukchi and East Siberian Seas (Arctic Ocean). Journal of Quaternary Science, 32(3): 362-379

Sun H J, Li T G, Chang F M, Wan S M, Xiong Z F, An B Z, Sun R T. 2017. Deep-sea carbonate preservation in the western Philippine Sea over the past 1Ma. Quaternary International, 459: 101-115

Sun X Q, Liu S F, Li J R, Zhang H, Zhu A M, Cao P, Chen M T, Zhao G T, Khokiattiwong S, Kornkanitnan N, Shi X F. 2019. Major and trace element compositions of surface sediments from the lower Bengal Fan: implications for provenance discrimination and sedimentary environment. Journal of Asian Earth Sciences, 184: 104000

Tada R, Irino T, Ikehara K, Karasuda A, Sugisaki S, Xuan C, Sagawa T, Itaki T, Kubota Y, Lu S, Seki A, Murray R W, Alvarez-Zarikian C, Anderson W T Jr, Bassetti M A, Brace B J, Clemens S C, da Costa Gurgel M H, Dickens G R, Dunlea A G, Gallagher S J, Giosan L, Henderson A C G, Holbourn A E, Kinsley C W, Lee G S, Lee K E, Lofi J, Lopes C I C D, Saavedra-Pellitero M, Peterson L C, Singh R K, Toucanne S, Wan S M, Zheng H B, Ziegler M. 2018. High-resolution and high-precision correlation of dark and light layers in the Quaternary hemipelagic sediments of the Japan Sea recovered during IODP Expedition 346. Progress in Earth and Planetary Science, 5(1): 19

Talley L D. 1993. Distribution and formation of north Pacific intermediate water. Journal of Physical Oceanography, 23(3): 517-537

Tian J, Xie X, Ma W T, Jin H Y, Wang P X. 2011. X-ray fluorescence core scanning records of chemical weathering and monsoon evolution over the past 5 Myr in the southern South China Sea. Paleoceanography and Paleoclimatology, 26(4): PA4202

Vasilenko Y P, Gorbarenko S A, Bosin A A, Artemova A V, Yanchenko E A, Shi X F, Zou J J, Liu Y G, Toropova S I. 2019. Orbital-scale changes of sea ice conditions of Sea of Okhotsk during the last glaciation and the Holocene (MIS4-MIS1). Palaeogeography, Palaeoclimatology, Palaeoecology, 533: 109284

Walsh J P, Wiberg P L, Aalto R, Nittrouer C A, Kuehl S A. 2016. Source-to-sink research: Economy of the Earth's surface and its strata. Earth-Science Reviews, 153: 1–6

Wan S, Jian Z M. 2014. Deep water exchanges between the South China Sea and the Pacific since the last glacial period. Paleoceanography and Paleoclimatology, 29(12): 1162–1178

Wan S M, Clift P D, Li A C, Yu Z J, Li T G, Hu D K. 2012a. Tectonic and climatic controls on long-term silicate weathering in Asia since 5 Ma. Geophysical Research Letters, 39(15): L15611

Wan S M, Yu Z J, Clift P D, Sun H J, Li A C, Li T G. 2012b. History of Asian eolian input to the West Philippine Sea over the last one million years. Palaeogeography, Palaeoclimatology, Palaeoecology, 326-328: 152–159

Wan S M, Toucanne S, Clift P D, Zhao D B, Bayon G, Yu Z J, Cai G Q, Yin X B, Révillon S, Wang D W, Li A C, Li T G. 2015. Human impact overwhelms long-term climate control of weathering and erosion in Southwest China. Geology, 43(5): 439–442

Wan S M, Clift P D, Zhao D B, Hovius N, Munhoven G, France-Lanord C, Wang Y X, Xiong Z F, Huang J, Yu Z J, Zhang J, Ma W T, Zhang G L, Li A C, Li T G. 2017. Enhanced silicate weathering of tropical shelf sediments exposed during glacial lowstands: a sink for atmospheric CO_2. Geochimica et Cosmochimica Acta, 200: 123–144

Wang H J, Wu X, Bi N S, Li S, Yuan P, Wang A M, Syvitski J P M, Saito Y, Yang Z S, Liu S M, Nittrouer J. 2017. Impacts of the dam-orientated water-sediment regulation scheme on the lower reaches and delta of the Yellow River (Huanghe): a review. Global and Planetary Change, 157: 93–113

Wang K S, Shi X F, Qiao S Q, Kornkanitnan N, Khokiattiwong S. 2015. Distribution and composition of authigenic minerals in surface sediments of the western Gulf of Thailand. Acta Oceanologica Sinica, 34(12): 125–136

Wang K S, Shi X F, Zou J J, Kandasamy S, Gong X, Wu Y H, Yan Q S. 2017. Sediment provenance variations in the southern Okhotsk Sea over the last 180 ka: evidence from light and heavy minerals. Palaeogeography, Palaeoclimatology, Palaeoecology, 479: 61–70

Wang P X, Li Q Y, Li C F. 2014a. Geology of the China Seas. Amsterdam: Elsevier

Wang P X, Li Q Y, Tian J, Jian Z M, Liu C L, Li L, Ma W T. 2014b. Long-term cycles in the carbon reservoir of the Quaternary ocean: a perspective from the South China Sea. National Science Review, 1(1): 119–143

Wang P X, Wang B, Cheng H, Fasullo J, Guo Z T, Kiefer T, Liu Z Y. 2017. The global monsoon across time scales: mechanisms and outstanding issues. Earth-Science Reviews, 174: 84–121

Wang R, Xiao W, Maerz C, Li Q. 2013. Late quaternary paleoenvironmental changes revealed by multi-proxy records from the chukchi abyssal plain, western arctic ocean. Global and Planetary Change, 108: 100–118

Wang R, Biskaborn B K, Ramisch A, Ren J, Zhang Y Z, Gersonde R, Diekmann B. 2016. Modern modes of provenance and dispersal of terrigenous sediments in the North Pacific and Bering Sea: implications and perspectives for palaeoenvironmental reconstructions. Geo-Marine Letters, 36(4): 259–270

Wang S S, Chang L, Xue P F, Liu S F, Shi X F, Khokiattiwong S, Kornkanitnan N, Liu J X. 2019. Paleomagnetic Secular Variations During the Past 40,000 Years from the Bay of Bengal. Geochemistry, Geophysics, Geosystems, 20(6): 2559–2571

Weber M E, Reilly B T. 2018. Hemipelagic and turbiditic deposits constrain lower Bengal Fan depositional history through Pleistocene climate, monsoon, and sea level transitions. Quaternary Science Reviews, 199: 159–173

Weber M E, Lantzsch H, Dekens P, Das S K, Reilly B T, Martos Y M, Meyer-Jacob C, Agrahari S, Ekblad A, Titschack J, Holmes B, Wolfgramm P. 2018. 200,000 years of monsoonal history recorded on the lower Bengal Fan-strong response to insolation forcing. Global and Planetary Change, 166: 107–119

Wu B, Wu X D, Shi X F, Qiao S Q, Liu S F, Hu L M, Liu J H, Bai Y Z, Zhu A M, Kornkanitnan N, Khokiattiwong S. 2020. Influences of tropical monsoon climatology on the delivery and dispersal of organic carbon over the Upper Gulf of Thailand. Marine Geology, 426: 106209

Wu X, Bi N S, Xu J P, Nittrouer J A, Yang Z S, Saito Y, Wang H J. 2017. Stepwise morphological evolution of the active Yellow River (Huanghe) delta lobe (1976–2013): dominant roles of riverine discharge and sediment grain size. Geomorphology, 292: 115–127

Wu Y H, Shi X F, Zou J J, Cheng Z B, Wang K S, Ge S L, Shi F D. 2014. Benthic foraminiferal δ^{13}C minimum events in the southeastern Okhotsk Sea over the last 180 ka. Chinese Science Bulletin, 59(24): 3066–3074

Xiong Z F, Li T G, Algeo T, Nan Q Y, Zhai B, Lu B. 2012. Paleoproductivity and paleoredox conditions during late Pleistocene accumulation of laminated diatom mats in the tropical West Pacific. Chemical Geology, 334: 77–91

Xiong Z F, Li T G, Crosta X, Algeo T, Chang F M, Zhai B. 2013. Potential role of giant marine diatoms in sequestration of atmospheric CO_2 during the Last Glacial Maximum: δ^{13}C evidence from laminated *Ethmodiscus rex* mats in tropical West Pacific. Global and Planetary Change, 108: 1–14

Xiong Z F, Li T G, Algeo T J, Doering K, Frank M, Brzezinski M A, Chang F M, Opfergelt S, Crosta X, Jiang F Q, Wan S M, Zhai B. 2015. The silicon isotope composition of Ethmodiscus rex laminated diatom mats from the tropical West Pacific: implications for silicate cycling during the Last Glacial Maximum. Paleoceanography, 30: 803–823

Xiong Z F, Li T G, Chang F M, Algeo T J, Clift P D, Bretschneider L, Lu Y, Zhu X, Frank M, Sauer P E, Jiang F Q, Wan S M, Zhang X, Chen S X, Huang J. 2018. Rapid precipitation changes in the tropical West Pacific linked to North Atlantic climate forcing during the last deglaciation.

Quaternary Science Reviews, 197: 288-306

Xu J. 2014. Change of Indonesian Throughflow outflow in response to East Asian monsoon and ENSO activities since the Last Glacial. Science China Earth Sciences, 57(4): 791-801

Xu K H, Li A C, Liu J P, Milliman J D, Yang Z S, Liu C S, Kao S J, Wan S M, Xu F J. 2012. Provenance, structure, and formation of the mud wedge along inner continental shelf of the East China Sea: a synthesis of the Yangtze dispersal system. Marine Geology, 291-294: 176-191

Xu Z K, Li T G, Clift P D, Wan S M, Qiu X H, Lim D. 2018. Bathyal records of enhanced silicate erosion and weathering on the exposed Luzon shelf during glacial lowstands and their significance for atmospheric CO_2 sink. Chemical Geology, 476: 302-315

Yang S L, Milliman J D, Li P, Xu K. 2011. 50,000 dams later: erosion of the Yangtze River and its delta. Global and Planetary Change, 75(1-2): 14-20

Yang Y, Jia J J, Zhou L, Gao W H, Shi B W, Li Z H, Wang Y P, Gao S. 2019. Human-induced changes in sediment properties and amplified endmember differences: possible geological time markers in the future. Science of the Total Environment, 661: 63-74

Yao Z Q, Shi X F, Liu Q S, Liu Y G, Larrasoaña J C, Liu J X, Ge S L, Wang K S, Qiao S Q, Li X Y, Shi F D, Fang X S, Yu Y G, Yang G, Duan Z Q. 2014. Paleomagnetic and astronomical dating of sediment core BH08 from the Bohai Sea, China: implications for glacial-interglacial sedimentation. Palaeogeography, Palaeoclimatology, Palaeoecology, 393: 90-101

Yao Z Q, Shi X F, Qiao S Q, Liu Q S, Kandasamy S, Liu J X, Liu Y G, Liu J H, Fang X S, Gao J J, Dou Y G. 2017. Persistent effects of the Yellow River on the Chinese marginal seas began at least ~880 ka ago. Scientific Reports, 7(1): 2827

Yao Z Q, Shi X F, Liu Y G, Kandasamy S, Qiao S Q, Li X Y, Bai Y Z, Zhu A M. 2020. Sea-level and climate signatures recorded in orbitally-forced continental margin deposits over the last 1 Myr: new perspectives from the Bohai Sea. Palaeogeography, Palaeoclimatology, Palaeoecology, 550: 109736

Yu Z J, Colin C, Bassinot F, Wan S M, Bayon G. 2020. Climate-driven weathering shifts between highlands and floodplains. Geochemistry, Geophysics, Geosystems, 21(7): e2020GC008936

Zhai L N, Wan S M, Tada R, Zhao D B, Shi X F, Yin X B, Tan Y, Li A C. 2020. Links between iron supply from Asian dust and marine productivity in the Japan Sea since four million years ago. Geological Magazine, 157(5): 818-828

Zhang J, Wan S M, Clift P D, Huang J, Yu Z J, Zhang K D, Mei X, Liu J, Han Z Y, Nan Q Y, Zhao D B, Li A C, Chen L H, Zheng H B, Yang S Y, Li T G, Zhang X H. 2019. History of Yellow River and Yangtze River delivering sediment to the Yellow Sea since 3.5 Ma: tectonic or climate forcing? Quaternary Science Reviews, 216: 74-88

Zhang P, Xu J, Schröder J F, Holbourn A, Kuhnt W, Kochhann K G D, Ke F, Wang Z, Wu H N. 2018. Variability of the Indonesian Throughflow thermal profile over the last 25 kyr: a perspective from the southern Makassar Strait. Global and Planetary Change, 169: 214-223

Zhang Y, Fan D D, Qin R F. 2020. Estuary-shelf interactions off the Changjiang Delta during a dry-wet seasonal transition. Marine Geology, 426: 106211

Zhang Y C, Zhou X, He Y X, Jiang Y Q, Liu Y, Xie Z Q, Sun L G, Liu Z H. 2019. Persistent intensification of the Kuroshio Current during late Holocene cool intervals. Earth and Planetary Science Letters, 506: 15-22

Zhang Y W, Liu Z F, Zhao Y L, Colin C, Zhang X D, Wang M, Zhao S H, Kneller B. 2018. Long-term in situ observations on typhoon-triggered turbidity currents in the deep sea. Geology, 46(8): 675-678

Zhao J T, Li J, Cai F, Wei H L, Hu B Q, Dou Y G, Wang L B, Xiang R, Cheng H W, Dong L, Zhang C L. 2015. Sea surface temperature variation during the last deglaciation in the southern Okinawa Trough: modulation of high latitude teleconnections and the Kuroshio Current. Progress in Oceanography, 138: 238-248

Zhao Y L, Liu Z F, Zhang Y W, Li J R, Wang M, Wang W G, Xu J P. 2015. In situ observation of contour currents in the northern South China Sea: applications for deepwater sediment transport. Earth and Planetary Science Letters, 430: 477-485

Zheng X F, Li A C, Kao S J, Gong X, Frank M, Kuhn G, Cai W J, Yan H, Wan S M, Zhang H H, Jiang F Q, Hathorne E, Chen Z, Hu B Q. 2016a. Synchronicity of Kuroshio Current and climate system variability since the Last Glacial Maximum. Earth and Planetary Science Letters, 452: 247-257

Zheng X F, Kao S J, Chen Z, Menviel L, Chen H, Du Y, Wan S M, Yan H, Liu Z H, Zheng L W, Wang S H, Li D W, Zhang X. 2016b. Deepwater circulation variation in the South China Sea since the Last Glacial Maximum. Geophysical Research Letters, 43(16): 8590-8599

Zou J J, Shi X F, Zhu A M, Chen M T, Kao S J, Wu Y H, Selvaraj K, Scholz P, Bai Y Z, Wang K S, Ge S L. 2015. Evidence of sea ice-driven terrigenous detritus accumulation and deep ventilation changes in the southern Okhotsk Sea during the last 180 ka. Journal of Asian Earth Sciences, 114: 541-548

Zou J J, Shi X F, Zhu A M, Kandasamy S, Gong X, Lembke-Jene L, Chen M T, Wu Y H, Ge S L, Liu Y G, Xue X R, Lohmann G, Tiedemann R. 2020. Millennial-scale variations in sedimentary oxygenation in the western subtropical North Pacific and its links to North Atlantic climate. Climate of the Past, 16(1): 387-407

Zou J J, Shi X F, Zhu A M, He L H, Kandasamy S, Tiedemann R, Lembke-Jene L, Shi F D, Gong X, Liu Y G, Ikehara M, Yu P S. 2021. Paleoenvironmental implications of Sr and Nd isotopes variability over the past 48 ka from the southern Sea of Japan. Marine Geology, 432: 106393

Progress in Sedimentology Research of the Asian Continental Margin

SHI Xue-fa[1,2], QIAO Shu-qing[1,2], YANG Shou-ye[2,3], LI Jing-rui[2], WAN Shi-ming[2,4], ZOU Jian-jun[1,2], XIONG Zhi-fang[1,2], HU Li-min[1,2], YAO Zheng-quan[1,2], DONG Lin-sen[1,2], WANG Kun-shan[1,2], LIU Sheng-fa[1,2], LIU Yan-guang[1,2]

1. Key Laboratory of Marine Geology and Metallogeny, First Institute of Oceanography, MNR, Qingdao 266061;
2. Laboratory for Marine Geology, Pilot National Laboratory for Marine Science and Technology, Qingdao 266037;
3. State Key Laboratory of Marine Geology, Tongji University, Shanghai 200092; 4. Key Laboratory of Marine Geology and Environment, Institute of Oceanology, Chinese Academy of Sciences, Qingdao 266071

Abstract: The Asian continental margin (AM), a unique composite continental margin in the world, is located at the convergent and collision bands among the Eurasian, Pacific and Indian-Australian plates. It involves Kara Sea, Laptev Sea, East Siberian Sea, Chukchi Sea, Bering Sea, Okhotsk Sea, Japan Sea, eastern China seas (including Bohai Sea, Yellow Sea and East China Sea), South China Sea, Andaman Sea, Bay of Bengal and so on. During the past 10 years, great progresses in fields of sedimentology and paleoceanography of the AM had been made. A range of geological surveys were conducted from the Laptev Sea in North to the bay of Bengal in south, and the research on sediment source to sink of the vast area was carried out. The Quaternary high-resolution stratigraphic sequence was established, and major controlling effects of tectonic movement, sea-level change, Asian monsoon, sea ice, ocean current and human activities on the source to sink in the AM were revealed on different spatial and temporal scales. In the South China Sea, sedimentary records obtained through the International Ocean Drilling Program (IODP) had provided evidences that the water and carbon cycling in low latitude regions can directly respond to changes of the Earth's orbit and, subsequently, nurtured a new concept of low-latitude forcing climate change. Through the long-term monitoring and on-site observation, effects of typhoons, storm surges and tropical cyclone on the shelf sedimentation and dynamic processes were identified, and roles of internal solitary waves, mesoscale vortex, contour currents and turbidities in the sediment transport in the South China Sea were proposed. Innovative results had also been achieved in evolution of the western Pacific warm pool, Kuroshio, North Pacific intermediate water and in their effects on the sedimentation in those regions. In the future, more attention should be paid on long-term continuous observation of the modern depositional process, and interdisciplinary research research on identifying environmental signals in sedimentary records, numerical simulation technology, and mining and utilizing the big-data of marine sediment to advance the AM sedimentology.

Key words: source to sink; paleoceanography and paleoclimate; seafloor observation; Asian continental margin

微束分析测试技术进展*

陈 意[1] 胡兆初[2] 贾丽辉[1] 李金华[1] 李秋立[1] 李晓光[1] 李展平[3] 龙 涛[4]
唐 旭[1] 王 建[5] 夏小平[6] 杨 蔚[1] 原江燕[1] 张 迪[1] 李献华[1]

1. 中国科学院 地质与地球物理研究所,北京 100029;2. 中国地质大学(武汉),地质过程与矿产资源国家重点实验室,武汉 430074;3. 清华大学 分析中心,北京 100084;4. 中国地质科学院 地质研究所,北京 100037;5. Canadian Light Source Inc., University of Saskatchewan, Saskatoon, SK, Canada S7N 2V3;6. 中国科学院 广州地球化学研究所,广州 510640

摘 要:现代地球科学研究的重大突破在很大程度上取决于观测和分析技术的创新。新世纪以来,我国地球科学领域引进了一批高性能新型微束分析仪器设备,建立了一批高规格的实验室。本文回顾了近十年来微束分析技术与方法的主要进展及其在地球科学研究中的应用实例,包括电子探针、扫描电子显微镜、透射电子显微镜、大型离子探针、纳米离子探针、飞行时间二次离子质谱、激光剥蚀等离子体质谱、激光诱导原子探针、原子探针技术、显微红外光谱、同步辐射等,这些分析技术的进步和广泛应用极大地提高了我们对地球和行星演化历史及许多地质过程的理解。今后,应加快微束分析的新技术、新方法和新标准的开发,特别是高水平人才队伍建设,提高创新能力并在国际学术舞台上发挥重要作用。

关键词:微束分析 离子探针 电子探针 激光剥蚀 同步辐射

0 引 言

"工欲善其事,必先利其器"。现代地球科学研究的重大突破在很大程度上依赖于观测和分析技术的创新。近十年来,微束分析技术不仅取得了许多重要的技术进步和方法创新,而且在地球科学的各个学科领域都得到了广泛应用,极大地提高了我们对地球和行星历史及其许多地质过程的理解,而成为"深地、深海、深空、深时"研究中不可或缺的技术手段。

微束分析是在微米-纳米尺度上精确分析天然和人工合成样品的物相、形貌、结构、化学成分和同位素组成等。根据不同的一次束类型,微束分析技术可分为以下四类:①与电子束相关的微束分析技术,包括电子探针(electron probe microanalyzer, EPMA)、扫描电子显微镜(scanning electron microscope, SEM)和透射电子显微镜(TEM)等;②与离子束相关的微束分析技术,包括大型二次离子质谱(又称大型离子探针,LG-SIMS)、纳米二次离子质谱(又称纳米离子探针,NanoSIMS)和飞行时间二次离子质谱(time-of-flight SIMS TOF-SIMS)等;③与激光束相关的微束分析技术,包括激光剥蚀电感耦合等离子体质谱[LA(Q/HR/MC)ICP-MS]和激光诱导原子探针层析(lase induced atom probe tomography, LI-APT)等;④与微光束相关的微束分析技术,如傅里叶变换红外光谱(Fourier transform infrared spectrometer, FTIR)、同步辐射光源(synchrotron radiation, SR)等。各种微束分析技术的空间分辨率和成分分析检出范围如图1所示。

本文回顾了近十年来我国微束分析技术方法的重要进展及其在矿物、岩石、地球化学等研究中的一些代表性应用成果,并通过国际对比,展望各种技术方法的潜力和发展方向。

---

* 原文"微束分析测试技术十年(2011~2020)进展与展望"刊于《矿物岩石地球化学通报》2021年第40卷第1期,本文略有修改。

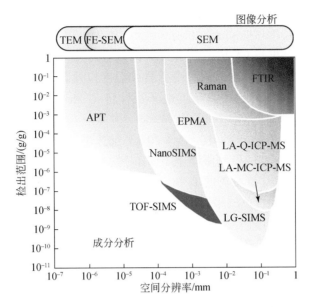

图 1　各种微束分析技术的空间分辨率和成分分析检出范围示意图（据 Li and Li，2016 修改）

TEM. 透射电子显微镜；SEM. 扫描电子显微镜；FE-SEM. 场发射扫描电子显微镜；FTIR. 傅里叶变换红外光谱；Raman. 拉曼光谱；EPMA. 电子探针；LA-Q-ICP-MS. 激光剥蚀四级杆电感耦合等离子体质谱，LA-MC-ICP-MS. 激光剥蚀多接收器电感耦合等离子体质谱；LG-SIMS. 大型二次离子质谱；NanoSIMS. 纳米二次离子质谱；TOF-SIMS. 飞行时间二次离子质谱；APT. 原子探针层析

1　电子束相关技术

1.1　电子探针

电子探针（EPMA）主要用于研究物质表面的元素组成及分布，是研究地球与行星物质组成最基础的微束分析技术。由于其具有高空间分辨率、无损、分析元素广和基体效应小等优点，而被广泛应用在岩石圈演化、矿产资源探索、环境科学、天体与行星演化等领域的研究中。由于仪器硬件设备的发展和测试软件的更新，电子探针已不再满足于仅测试样品的主量元素，开始追求"微区、微量、高精度"的测试目标，以支撑地学前沿领域研究需求（张迪等，2019）。近十年来，国内电子探针技术在矿物微量元素测试、矿物/玻璃 Fe^{3+} 分析、超轻元素分析等技术方法以及新仪器新硬件探索方面均取得了重要突破。

1.1.1　矿物微量元素分析

电子探针是矿物微量元素分析最为常用的分析手段，它的进步主要得益于电子光学系统在高束流时稳定性的提高、新型大分光晶体和高计数率波谱仪的研制以及新型分析校正软件的开发。国内电子探针在橄榄石、石英和金红石的微量元素分析方面取得了重要进展。

橄榄石是上地幔最主要的造岩矿物，除主量元素 Mg、Fe 和 Si 外，还含有丰富的 Ni、Mn、Ca、Al、Ti、Cr、Co、Zn 和 P 等微量元素，在揭示岩浆源区物质组成、温压条件和氧逸度等方面发挥着重要作用（Foley et al.，2013）。Batanova 等（2015）在 25 kV、900 nA 条件下，用能量色散 X 射线谱（X-ray energy dispersive spectrum，EDS）分析了橄榄石的主量元素 Mg、Fe 和 Si，将检测限降低至 $3\times10^{-6} \sim 9\times10^{-6}$（$1\sigma$）；用波长色散 X 射线谱（wavelength dispersion X-ray spectrum，WDS）分析微量元素，将分析误差降低至 $4\times10^{-6} \sim 14\times10^{-6}$。然而，由于 EDS 分析的主量元素误差较大，而且主量元素含量无法加入微量元素含量进行基体校正，因而降低了分析准确度。最近，Su 等（2019）进一步改进了橄榄石微量元素电子探针分析方法，他们采用"双束流全波谱"分析手段，即在单次分析过程中使用 25 kV、40 nA 和 25 kV、

900 nA 分别测试主量和微量元素，在保证分析精度的同时提高了分析效率。为有效降低检测限，Na、Cr、Ca、Mn、Ni 和 Zn 每个元素峰位测试时间为 120 s，Co、Ti 和 Al 峰位测试时间为 240 s，其检测限可降至 $5×10^{-6} \sim 16×10^{-6}$（$3\sigma$），分析结果与 LA-ICP-MS 结果在误差范围内一致。

石英中的 Ti 和 Al 被广泛用于计算岩浆岩、变质岩的温度以及估算成矿流体的酸碱度变化（Tual et al., 2018）。虽然 Donovan 等（2011）利用高精度电子探针将 Ti 和 Al 的检测限降低到 $2×10^{-6} \sim 3×10^{-6}$（3σ）和 $6×10^{-6} \sim 7×10^{-6}$（3σ），但分析时间长达 33 min，并缺乏对数据准确度及精确度的可靠评估。最近，Cui 等（2019）利用 JXA-8100 电子探针，选择 20 kV 加速电压、500 nA 束流，通过改变背景模型、采用多谱仪计数方式成功将 Ti 和 Al 的检测限降至 $10×10^{-6}$ 左右（3σ）。该方法的测试结果与石英标样参考值一致（Audétat et al., 2015）。尽管石英的 Ti 和 Al 检测限均可降至 $10×10^{-6}$ 以下，但分析束流过高，如何在保证低检测限的前提下，降低石英的束流损伤和二次荧光效应，仍是今后技术研发的难点。金红石是高场强元素和过渡元素的重要载体，其微量元素分析多采用 LA-ICP-MS 实现，但由于空间分辨率较低而难以区分环带所引起的部分微量元素含量的差异（肖益林等，2011）。此外，LA-ICP-MS 工作所用的载气 Ar 与金红石中的 Ti 生成的离子团会对 Zr 信号产生质量叠加，从而影响金红石中低含量 Zr 的测试准确度（Zhang et al., 2010；张贵宾和张立飞，2011）。因此，研究者开始重视利用电子探针测试金红石的微量元素。例如，余金杰等（2006）利用高束流（100 nA）与较长峰位测试时间（150~300 s），测试了金红石中 Cr、Nb、V、Fe 等元素，获得了 $20×10^{-6} \sim 30×10^{-6}$ 的检测限（1σ）。上述方法以人工合成的氧化物为监测标样，但缺乏对金红石标样系统的测试分析。为此，王娟等（2017）运用 CAMECA SXFive 电子探针对金红石国际标样 R10（Luvizotto et al., 2009）进行关键微量元素分析，包括 Al、Si、Fe、Cr、Zr、V、Nb、Ta，而将 Ti 和 Si 作为监测元素，分析结果显示 Zr、Nb、V、Fe、Cr 值与推荐值在误差范围内一致，大部分元素数据波动范围在 10% 以内，V、Fe 数据波动范围在 5% 以内。

1.1.2 矿物/玻璃 Fe^{3+} 分析

电子探针常规测试只能分析矿物及玻璃熔体中 FeO 含量，虽然利用电价平衡法可估算 Fe^{3+} 含量，但偏差较大。经过多年的探索及改进，Hofer 等（1994）最早提出利用 Flank Method 分析物质中的 Fe^{3+} 含量，定义不同价态 Fe 的特征 X 射线 L 能级谱线峰位（Lα 和 Lβ）的侧峰位置（flank position），对该侧峰位置和强度计数进行测量，可以准确获得矿物的 Fe^{3+} 含量。目前国内学者优化了 Flank Method，已能准确测试出玻璃、石榴子石、黑云母和角闪石中 Fe^{3+} 含量。Zhang C. 等（2018）开发了以石榴子石作为标样的准确测试硅酸盐玻璃熔体中 Fe^{2+} 含量的方法，当硅酸盐玻璃熔体 TFeO 含量的质量分数大于 5% 时，该方法对 $Fe^{2+}/\Sigma Fe$ 的分析准确度可达 ±0.1（2σ），当 TFeO 含量的质量分数小于 5% 时，其分析精度在 ±0.3 以内。李小犁等（2019）用该方法精确测定了榴辉岩中石榴子石的 Fe^{3+} 含量，结果显示其成分环带变化很好对应了榴辉岩寄主岩石的变质演化轨迹。Li X. Y. 等（2020）利用 Flank Method 开发了电子探针测试角闪石和黑云母中 $Fe^{2+}/\Sigma Fe$ 的方法，其分析误差与湿化学法相比在 ±0.1 之内（2σ）。不同矿物具有不同的 Fe Lα 和 Lβ 峰位位移差和计数强度，其基体效应、ΣFe 含量、总化学成分（矿物类型）等也会有差别，因此目前基于 Flank Method 测试其他矿物 Fe^{3+} 含量的方法亟待开发。

1.1.3 超轻元素（Be、B）分析

Be 是重要的关键金属，其特征 X 射线波长（>1.2 nm）需要大间距面网的分析晶体衍射，同时面临其他元素谱线干扰、谱峰漂移、吸收效应等问题，其精确分析一直是电子探针测试的难点。国内学者尝试利用特殊的大间距面网分析晶体（JEOL：LDEB）测试绿柱石（张文兰等，2006）中的 Be 含量，但分析误差较大，3 个分析点的标准误差为 0.75，BeO 的数据波动在 10% 左右，且数据重现性缺乏可靠的评估。吴润秋等（2020）利用岛津电子探针的 LSA300（面网间距 $2d = 30$ nm）晶体，通过选择相似的校正标样避免峰位漂移及吸收效应、运用脉冲高度分析器（PHA）过滤峰位干扰，在 12 kV、50~200 nA 条件下能够精确测试硅铍石、羟硅铍石和绿柱石中的 Be 含量。尽管 JEOL 的大间距面网晶体的 $2d$ 值（LDEB：14.5 nm；LDE3H：20 nm）远小于岛津仪器的 LSA300 晶体，但通过选择合适电压与束流提高 Be 的计数

率、改变 Be 的上背景位置避免 Si 的峰位干扰、采用 Be 金属作为校正标样，JEOL 仪器的 LDE3H 晶体仍能有效分析绿柱石中的 Be（张文兰等，2020）。该方法 25 个连续分析点的 BeO 平均值为 12.92%，与湿法化学确定的推荐值 12.23%（Liu et al.，2012）接近，标准误差为 0.52，Be 的检测限低于 $250×10^{-6}$（1σ），体现出该方法高精度的优势。未来其他含 Be 矿物分析方法的建立以及低含量 Be 的高精度分析仍是亟待突破的难题。

与 Be 类似，B 的电子探针分析同样需要大间距面网晶体，由于其特征 X 射线的计数率远高于 Be，目前电子探针已能精确测试富 B 矿物（如电气石）和硼化物的 B 含量（Meier et al.，2011）。然而，硅酸盐玻璃中微量 B 的测试却存在计数率低，Ca、Mn 和 Fe 元素的干扰，高且弯曲的背景等难题。但研究表明，在低电压（5 kV）、高束流（100 nA）、大束斑（20 μm）的测试条件下能够提高微量 B 的计数率（Cheng et al.，2019）。此外，通过利用不含 Ca、Mn 和 Fe 的富 B 玻璃作为校正标样，结合 PHA 进行峰位干扰扣除，采用指数模式模拟背景形状，Cheng 等（2019）建立了硅酸盐玻璃中微量 B（$\geq 0.2\%$ B_2O_3）的高精度分析方法，分析结果与 LA-ICP-MS 测试结果在误差范围内一致。

1.1.4 仪器硬件的发展

场发射电子探针是近年来电子探针硬件设备发展的重大突破，它实现了从传统的钨灯丝和六硼化镧（LaB_6）电子枪向场发射电子枪的转变。场发射电子探针通过一个强大的磁场吸取电子束，显著缩小电子束的尺寸并且提高束流连贯性和密度，从而提高信噪比、空间分辨率和电子枪寿命。起电子束可达亚微米级（10~80 nm）（Berger and Nissen，2014），这将有助于拓展电子探针在微米至亚微米尺度不均一复杂样品中的应用前景。然而，在低电压模式下，分析元素 L 线系或者 M 线系的特征 X 射线会面临更复杂的谱峰漂移及谱峰干扰等问题。因此，此模式下特定的分析方法亟需开发（张迪等，2019）。

近十年来，新购置的电子探针仪器多选择配备全聚焦的大晶体。全聚焦型大晶体使产生的 X 射线有相同的入射角并聚焦到探测器的一个点上，以此提高计数率。新型大晶体相对于传统的分光晶体，其计数率可提高 2~3 倍，有助于降低微量元素的检测限，提高分析精度。目前，最大的分光晶体是 Cameca 公司的"Very Large PET"晶体，其尺寸是普通 PET 晶体的 5 倍、LPET 晶体的近 2 倍，尤其适合于微量元素分析。此外，大间距面网晶体的应用也为建立超轻元素分析方法提供了必要的硬件支撑，如利用 LDEB 和 LSA300 等晶体测试硅铍石、羟硅铍石和绿柱石等矿物中的 Be 含量具有较好的结果（张文兰等，2006；吴润秋等，2020）。

1.1.5 展望

目前，随着研究的不断深入，电子探针分析方法在上述研究方向都取得了重要进展，但仍有很多难题亟待突破：①标样的开发；②二次荧光效应急需准确的校正方法；③其他含 Be 矿物的高精度分析方法有待建立；④高精度定量图像分析方法急需开发；⑤尖晶石族矿物 Fe^{3+} 的准确测定。此外，仪器硬件的发展给电子探针带来了新的机遇与挑战，如场发射电子探针低电压分析模式和软 X 射线分析谱仪新技术的开发及应用等。未来随着技术的发展，电子探针将充分发挥其"高精度、高分辨"的优势，为科研人员提供更加准确和丰富的原位微区成分信息。

1.2 扫描电子显微镜

随着现代科技的发展和学科交叉融合，科学家的研究尺度正向更宏观和更微观两极发展。扫描电子显微镜（SEM）是观测物质表面形貌的基础微束分析技术，具有成像直观、分辨率高、景深长、立体感强、样品制备简单等特点，已成为微束分析测试仪器家族中的重要成员，它显著提高了人们认知矿物组成和微观结构的能力，促进了固体地球科学、行星科学等多个学科的发展。近年来，扫描电子显微镜在低真空分辨率、附件集成、原位高温及机械台的搭建等微区微量原位观测技术的有力支撑下，已可实现在微纳尺度上观测地质样品的特性。

1.2.1 低真空分辨率提升

用常规扫描电子显微镜观察不导电或导电性能差的样品表面形貌时，由于电子聚集引起放电现象，需预先对样品喷镀导电层以消除样品表面堆积电子，实现扫描电子显微镜的高真空成像。环境扫描电子显微镜（environmental scanning electron microscope，ESEM）的低真空成像技术（高真空模式优于 6×10^{-4} Pa，低真空模式低于 2×10^2 Pa）可对未经喷镀的样品直接进行表面微观形貌和结构观察，有效地解决了水、油及非导电样品的形态结构成像问题，避免了导电层对样品表面亚微米级别信息的掩盖。近年来，ESEM 分析技术在国内外地球科学领域逐步得到了应用，例如，黏土矿物在储集层中的分布具有非均质性，遇水可能膨胀造成地层伤害，采用 ESEM 可直接观察黏土矿物的吸水膨胀行为和水敏性（Sun et al., 2019）。一些珍贵的化石胚胎样品，形态不规则，传统的扫描电子显微镜由于镀膜覆盖而难以观察其表面细节，采用 ESEM 的低真空模式，无需镀膜，通过样品台旋转和倾斜可获得不同侧面的图像，有助于恢复化石胚胎的立体形态（陈方和董熙平，2009）。黏土矿物的成分和性质对实际的油气开发过程有着非常大的影响，由于黏土矿物颗粒直径非常小，利用环境扫描电子显微镜可以有效研究其表面亚微米级形态分布和状态变化，从而有效反映成岩环境及地球化学背景（于亮等，2016）。ESEM 微观分析技术适合观察样品微观表面信息，可用于对碳酸盐岩储层进行评价和直观呈现孔隙发育情况，有助于研究溶蚀、交代过程中碳酸盐岩成岩演化过程。此外，环境扫描电子显微镜分析技术可实现微小颗粒如黄铁矿、海绿石、宇宙尘埃以及现代海洋沉积物的形貌观察。因此，ESEM 这种低真空、高空间分辨率的优势，在未来地学研究（尤其是珍贵地质样品研究）中具有广阔的应用前景。

1.2.2 原位微观实时观测技术

SEM 上可配置多种附件，如冷台、加热台、拉伸台，以及近些年研发的微操纵-微注入系统等，可进一步扩展 SEM 的功能，使 SEM 成为一个小型微观实验室，从而实现原位微观的实时观测，记录矿物生长、脱水、腐蚀、相变、形变、断裂等动态过程，研究相关的物理化学反应变化及矿物与环境的相互作用（Wang et al., 2007；Papaslioti et al., 2018；Checa et al., 2019）。

为 SEM 安装有独立温控的冷台可最大限度地维持样品在自然状态下的微观结构，实现水化物、胶凝物、液相和气相包裹体等含水样品的表面微观分析及低温气液相变研究。周广荣（2009）采用扫描电子显微镜冷台观察了干、湿状态下植物的茎、叶、花瓣、花粉粒样品的微观形态，发现表面结构差异较为明显，且获得了植物叶子清晰的气孔等干燥样品无法采集的信息。目前尚未有冷台在地学研究方面的应用，但未来可在矿床流体包裹体研究中发挥重要作用，可为研究成矿流体性质、探讨矿床成因、找矿探矿和热力学等方面提供科学的依据。

为 SEM 安装高温台，可对样品进行加热并同时采集二次电子信号（Wang et al., 2019），用于矿物晶体转变的动态实验研究。Bocker 等（2014）在高温台辅助模式下，用 SEM 实时原位观察了 1 000 ～ 1 100℃ 条件下玻璃熔体结晶为堇青石的过程。如何将 SEM 与高温实验更紧密结合，实现不同高温模式下相变观测实验研究，是今后国内地学领域 SEM 技术需面临的挑战。

1.2.3 多种成像技术集成

扫描电子显微镜通常与背散射电子（back scattered electron，BSE）探头、能量色散 X 射线谱（EDS）、电子背散射衍射（EBSD）、阴极荧光谱仪、聚焦离子束（focused ion beam，FIB）系统和共聚焦拉曼（Raman）等多种分析技术集成，使其分析能力得到多方位的扩展和提升，如除可获得样品的形貌信息外，还可对样品进行微区化学成分、晶体结构特征、阴极荧光谱和图像、矿物鉴别、背散射图像、样品切削及三维重构等分析。

元素特征 X 射线能谱仪可获得样品微区成分信息及成像区的元素分布。如采用能谱微区技术可快速、有效地解决沉积岩中的黏土矿物微粒（$<2~\mu m$）因形貌相似无法区分的问题，如伊利石、蒙脱石、绿泥石等（Li and Xiao, 2012）。最近场发射扫描电子显微镜上配备的矿物自动定量分析软件系统，如 Maps Mineralogy、Aztec Mineral、TIMA-X 和 AMICS 等，可通过内置矿物相数据库，结合能谱获得的元素组成，

直接输出矿物相的平面分布和含量图（Maruvanchery and Kim，2020）。牛津最新的 Ultim Extreme 是 Ultim Max 系列中的一款无窗椭圆形能谱，当晶体面积为 100 mm^2，加速电压小于 2 kV 时，块状样品的空间分辨率低于 10 nm，采用跑道型结构设计，工作距离短，可对 Li 元素进行检测和成像。这些软、硬件的升级，可为国内 SEM 分析技术的发展提供有力支撑。

阴极发光（CL）图像能清晰地反映矿物内部环带结构、微量元素、晶格缺陷及杂质的存在，如锆石、独居石、长石、石英和磷灰石等多种矿物受到电子激发之后，都会产生阴极荧光现象（Kayama et al.，2010；Frelinger et al.，2015）。将阴极发光探测器和高分辨率的扫描电子显微镜集成，可实现矿物的高空间分辨率阴极发光成像，因此在副矿物内部结构观测方面得到了广泛应用（陈莉等，2015）。国内已广泛应用 SEM-CL 技术开展锆石的显微结构和成因分析。除锆石外，独居石和磷灰石的 CL 成像技术也取得了显著进步，在国内地学界应用已初显成效（Li et al.，2017；Xing and Wang，2017；Gou et al.，2019）。Li 等（2017）通过对花岗岩和片麻岩中独居石的阴极发光内部结构分析，获取了岩浆和变质独居石清晰的 CL 图像，为原位高精度 U-Pb 定年提供了选点依据。Xing 和 Wang（2017）获取了辉长岩中富 F 磷灰石的高清 CL 图像，结合磷灰石微量元素特点，深入剖析了岩浆结晶和后期流体交代过程中多期次磷灰石生长历史。此外，应用石英 CL 图像可推断其内部成分的变化，在物源分析、成矿流体演化等方面发挥了重要作用（李艳青等，2011；Shu et al.，2020）。大部分矿物发光较弱，如何高效收集发光信号，实现更多矿物高清 CL 成像技术，仍是国内 SEM 分析未来的重点发展方向。

将拉曼光谱仪与 SEM 集成的 SEM-Raman 联机系统（RISE），可在单一系统内对同一点采集扫描电子显微镜与拉曼光谱数据，实现对样品表面形貌和物质分子结构关系快速直观的表征。例如，SEM-Raman 联机系统可对蛇纹石、阳起石或其他细长的闪石颗粒的形态、物相及结构进行全面分析，有效解决石棉的判定问题（Wille et al.，2019）。对于同质多相矿物，由于其化学组成相同，X 射线能谱无法区分且在背散射电子图像中衬度相差不大，采用 RISE 可有效对其进行区分和物相鉴定。此外，利用拉曼光谱对透明矿物的可穿透性，可采集样品在三维空间的拉曼数据，弥补了传统 SEM 仅关注的物质表面形貌和成分的不足。

1.2.4 展望

未来国内扫描电子显微镜分析仍有较大的提升空间，包括：①X 射线能谱仪软件的开发，如提高能谱定量分析精度，在此基础上实现对采集的能谱数据实时计算矿物晶体化学式并进行矿物相自动匹配；开发（超）轻元素 B、F、Li、Be 等元素的面分析和成分分析测定。②扫描电镜匹配的高温台、冷台及拉伸台在地学领域的应用还较少，在不同温度、压力模式下的实时观测技术亟待开发。③矿物 CL 图像内部环带的定量化分析。④SEM-Raman 联机系统在国内才刚刚起步，未来在原位微束分析领域仍有很大的技术开发空间和应用潜力。总之，扫描电子显微镜硬件和分析软件的开发和技术研究仍需不断深入，以期为地学研究提供更加直观的图像显示、精确的微区成分和结构分析。

1.3　透射电子显微镜

1.3.1　发展历史与最新进展

电子显微镜利用电子束作为照明源，突破了人类对微观世界观察认识的光学极限，被誉为"科学之眼"。与扫描电子显微镜或电子探针依靠电子束反射成像或分析不同，透射电子显微镜（TEM）是把经过加速和聚焦的电子束透过非常薄的样品而进行成像和分析，因而能获得样品内部结构信息且具有更高的空间分辨率（李金华和潘永信，2015）。

与其他显微学技术一样，追求更高空间分辨率和更强大的综合分析能力，是透射电子显微镜技术发展的核心内容。1934 年，Knoil M 和 Rusk E 采用 75 kV 的加速电压，研制了真正意义上的第一台透射电子显微镜，其空间分辨率达到 50 nm，最大放大倍数约为 12 000 倍（Rusk，1934）。此后经过 50 余年的发展，到 20 世纪八九十年代，200 kV 透射电子显微镜的空间分辨率已达 0.3 nm 左右，而超高压透射电子显微镜（500~3 000 kV 加速电压）的分辨率已提升到 0.2~0.1 nm 的原子水平，放大倍数可达 1 500 000 ［图 2（a）］（Phillipp et al.，1994）。自 20 世纪 70 年代起，随着场发射电子枪（field emission gun，FEG）的逐渐应用，促进了分析型透射电子显微镜的快速发展（Sigle，2005）。与装配钨灯丝或六硼化镧电子枪的透射电子显微镜相比，场发射源透射电子显微镜更具拓展性，通常可装配高角环形暗场（high-angle annular dark field，HAADF）探测器、扫描透射电子显微镜（scanning transmission electron microscopy，STEM）、能量色散 X 射线谱（EDS 或 EDX）、电子能量损失谱（electron energy loss spectroscopy，EELS）、能量过滤器（energy filter TEM，EFTEM）、三维重构（Tomography）和电子全息（electron holography，EH）等多个附件。采用这种高性能场发射透射电子显微镜，可以同时对样品进行微纳尺度到原子水平的"形貌结构观察、矿物相和晶体结构鉴定、化学成分探测和微磁结构观测"等综合分析［图 2（b）］（李金华和潘永信，2015）。

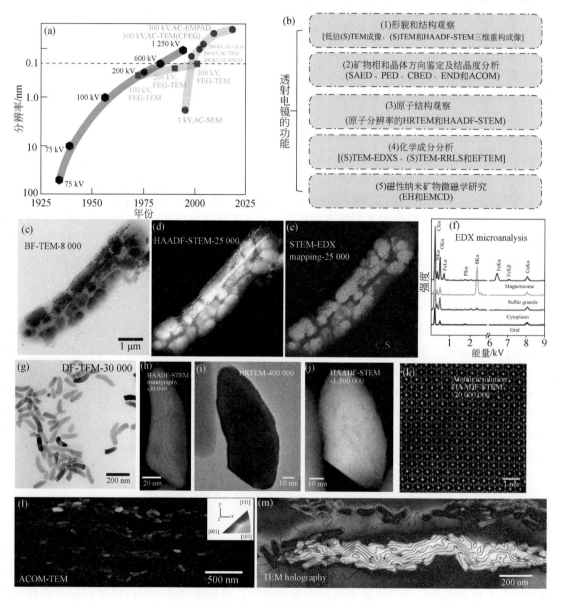

图 2　透射电子显微镜发展趋势和应用实例

（a）黑色曲线代表随加速电压升高透射电子显微镜空间分辨率的变化，蓝色曲线代表场发射透射电子显微镜（FEG-TEM）空间分辨率的变化，红色曲线代表球差矫正扫描电子显微镜（AC-SEM）和球差矫正透射电子显微镜（AC-TEM）空间分辨率的变化；（b）现代分析型透射电子显微镜的功能及涉及的相关显微学和显微谱学技术（李金华和潘永信，2015）；（c）~（m）以趋磁细菌为材料，部分展示透射电子显微镜相关技术及其应用实例（据 Li W. T. et al.，2015）

真正让透射电子显微镜观测和分析进入原子水平的要归功于电磁透镜球差矫正器的发展和应用。1992年德国3名科学家 H. Rose、K. Urban 和 M. Haider 研发了电磁透镜球差矫正器,让球差矫正透射电子显微镜（aberration-corrected TEM，AC-TEM）进入实质性应用阶段（Zach and Haider, 1995；Haider et al., 1998）。21 世纪以来,球差矫正透射电子显微镜快速发展,在不到十年时间里,其空间分辨率从 0.2 nm 提升到 0.05 nm,对样品的原位观测和综合分析能力也有了质的飞跃（Haider et al., 1998；Batson et al., 2002；Nellist et al., 2004；Kisielowski et al., 2008）。2018 年,美国阿贡国家实验室 D. Muller 领导的课题组将新一代的电子显微镜像素阵列相机（EMPAD）装配到 AC-TEM 上,采用层叠成像技术,将透射电子显微镜分辨率提升到新的世界纪录 0.039 nm（Jiang et al., 2018）。

1.3.2 在地学研究中的应用

透射电子显微镜的发展及其在生命和材料科学中的应用已有近 90 年的历史,但在地学领域的应用在近 20 年才取得明显进展。首先,聚焦离子束（focused ion beam，FIB）技术的发展和应用,让我们将扫描电子显微镜的电子束成像与镓离子束聚焦剥蚀技术相结合,可对相对复杂的地学样品进行可视条件下的精准微切割,以高效获得适用于透射电子显微镜分析的微米尺寸大小和纳米尺寸厚度样品,为透射电子显微镜在地学中的广泛应用提供了便利（Wirth, 2004；李金华和潘永信, 2015）。其次,近年来,随着地球科学与纳米科技的交叉融合,地学研究实现了从微米尺度到纳米和原子水平的跨越,即大幅度提升了我们对地球和地外广泛存在的纳米物质和结构的认知能力,从而产生了一系列新的认识和理论,支撑和促进了纳米地球科学的发展（Hochella, 2008；李金华和潘永信, 2015；Hochella et al., 2019）。例如, Keller 等 (2013) 综合应用 FIB、TEM 和 STEM 三维重构技术观测和定量研究了瑞士 Mont Terri 硬泥岩（opalinus clay）的孔隙结构特性,结果发现黏土纤维和微米级孔隙沿岩石层理面呈定向排列,孔隙之间由沿层理面定向排列的纳米孔隙而连通。这表明这些微纳空隙的膨胀与合并,可能在页岩气的大规模运移中扮演着重要角色。Rhonda 等应用其球差校正 STEM-EELS 技术,研究了原始碳质球粒陨石中纳米金刚石的原子结构,结果发现陨石中的纳米金刚石颗粒被无序、环状结构的玻璃碳包裹,从而支持了碳质球粒陨石中的纳米金刚石属于前太阳颗粒的认识（Sroud et al., 2011）。Li W. T. 等 (2015) 应用各种先进的透射电子显微镜技术 (如 HAADF-STEM-EDX、HAADF-STEM tomography、TEM Holography、HRTEM；Crystal modeling、Cs-corrected HAADF-STEM),从微纳尺度到原子水平对趋磁细菌及其磁小体的形貌结构、化学成分和磁学性质进行了系统研究 [图 2 (c) ~ (m)],发现了子弹头形磁小体的"多阶段晶体生长"规律,提出子弹头形磁小体及其 [001] 拉长是更可靠磁小体化石判据的新观点。他们与合作者已开始将这些成果应用与古老地质记录中磁小体化石的识别及其古地磁、古环境研究（Liu S. Z. et al., 2015）。最后,将透射电子显微镜与其他显微学（microscopy）或显微谱学（microspectroscopy）技术联用,发挥各自在结构鉴定和化学成分分析方面的优势,对复杂地学样品开展多尺度和多参数的综合研究,已成为近年来透射电镜技术在地学领域中应用的新趋势（Li J. H. et al., 2013；李金华和潘永信, 2015；Wacey et al., 2017；Pan et al., 2019a）。例如,Li J. H. 等 (2017) 将透射电子显微镜与荧光显微镜技术联用,首次在单细胞水平上实现了矿化微生物的种类鉴定和矿物结构的关联分析,为研究环境中未培养趋磁细菌多样性和生物矿化机制提供了高效方法。Benzerara 等将透射电子显微镜与同步辐射扫描透射 X 射线显微镜（scanning transmission X-ray microscopy，STXM）技术联用,成功地开展了太古宙—新生代不同岩石中微生物化石的识别研究（Benzerara et al., 2006；Cosmidis et al., 2013）。近年来,英国剑桥大学 R. Harrison 领导的团队将透射电子显微镜与同步辐射、原子探针等多种显微学技术联用,开展了岩石和陨石样品中纳米磁性矿物的种类鉴定、结构分析和磁性测量研究,为认识早期地球、小行星和太阳星云的磁场形成及其演化规律提供了新途径,对陨石磁学和纳米古地磁学的学科发展发挥了重要作用（Bryson et al., 2015）。

1.3.3 应用前景和建议

必须认识到,透射电子显微镜在地学领域的应用,无论其深度还是广度,都远落后于它在生命和材料等领域中的应用。首先,国内地学单位透射电子显微镜无论是仪器装备还是技术应用,都远落后于国外地学单位。20世纪90年代末，国外很多地学单位（例如，英国剑桥大学、美国弗吉尼亚理工大学、德国波茨坦地学研究中心和法国第六大学）已安装了分析型场发射透射电子显微镜，目前已进入球差矫正透射电子显微镜时代。而国内仅有少数几个地学单位在近十年才开始安装高分辨透射电子显微镜，分析型场发射透射电子显微镜仪器设备非常有限。此外，实验室的专业化程度不高，与其他显微谱学技术的平台整合和技术联用有待提高。与生命和材料科学领域相比，地学研究样品种类多、非均一性和复杂程度高，或年代久远不可重现，或来之不易稀少珍贵，这就需要搭建适合地学样品前处理和制备的配套设备并研发相关技术，更需要创建优化的方法体系，将透射电子显微镜与其他显微学技术或平台（如NanoSIMS、原子探针和同步辐射STXM等）实现有机衔接和联用，对地学样品实现多尺度多参数的综合分析，从而避免研究中出现"管中窥豹，只见一斑"的情形（李金华和潘永信，2015）。

2 离子束相关技术

2.1 大型离子探针

大型二次离子质谱（LG-SIMS；又称大型离子探针）以其高空间分辨率（亚微米到数十微米）、近无损（~1 μm剥蚀深度）、高质量分辨率（可达4万）及低检出限（<10^{-9}）等优越性能，自问世以来一直是微区地球化学分析的最强工具。近十年来，两家相关设备的主要生产商澳大利亚科学仪器公司和法国CAMECA公司均发布了新一代大型离子探针SHRIMP V和IMS 1300-HR^3。SHRIMP V主要针对轻同位素分析而特别设计，相较于之前的SHRIMP系列，主要对样品腔进行改进以提高其真空度，整个样品腔由同一块不锈钢块加工而成，并将样品台驱动马达外置；在多接收器部分则增大了单位质量的距离以便使用标准的ETP电子倍增器。IMS 1300-HR^{3}主要改进包括更换了原来的氧源系统，以亮度更大，束流更稳定的射频等离子体（radio frequency plasma, RF-plasma）氧源取代了原来的双等离子体氧源，可提高空间分辨率和分析数据重现性（Liu M. C. et al., 2018）；同时采用紫外光显微镜提高样品观察系统的光学图像分辨率（Kita et al., 2015），以更好选择样品待分析的区域，并采用了样品台Z轴自动对焦技术；在接收器部分引进了10^{12} Ω电阻的法拉第杯。中国的大型离子探针分析技术发展迅速，装备或即将装备相关大型离子探针设备的单位从十年前的两家发展到现在七家，在含U-Th矿物定年、稳定同位素及低含量挥发分等分析方法、标准物质与应用等方面均呈现出快速发展局面。

2.1.1 含U-Th矿物定年

我国目前在大型离子探针含U-Th矿物的定年技术已达到国际先进水平。十年来的进展主要体现在可定年矿物的种类越来越多、分析束斑越来越小、定年范围越来越大、定年精度越来越高。锆石一直是大型离子探针定年的首选矿物，但其他含U-Th矿物在某些情况下具备了锆石不可替代的定年价值。磷灰石、磷钇矿、独居石、金红石、斜锆石、褐帘石、钙钛矿、氟碳铈矿、榍石等含U矿物由于U含量相对低，普通Pb含量相对高，或者缺乏相应的标准物质等，长期以来在定年方法和应用研究两个方面都明显落后于锆石。近年来，中国科学院地质与地球物理研究所离子探针实验室和中国地质科学院北京离子探针中心，均建立了这些单矿物的SIMS定年方法（Li Q. L. et al., 2010, 2011a, 2011b, 2012, 2013, 2018；Ling et al., 2015, 2016; Zhang Y. B. et al., 2015; Liao et al., 2020），并针对月球和火星陨石（Zhou et al., 2013; Zhang A. C. et al., 2016; Bao et al., 2020）、沉积岩（Shi et al., 2015; Zhang Y. B. et al., 2015）、热液金矿（Chen M. H. et al., 2019）、稀土矿（Ling et al., 2016）、玉石矿（Ling et al., 2015）、钨矿（Li Q. L. et al., 2013）及低温榴辉岩（Li et al., 2011a）等其他定年方法难以准确定年的地质体系开展了年代学研究，取得了众多令人瞩目的成果。年轻地质体系由于放射性积累的子体同位素含量低，年代相对难以测定。离子探针第四纪锆石的U-Pb和U-Th定年方法也获得了发展（高钰涯等，2016），在中新世—中古新世、第四纪腾冲火山岩的年代学研究中取得较好结果（李玲等，2016; Li L. L. et al., 2017）。常规的SIMS

锆石 U-Pb 定年一般在 10～30 μm 的空间分辨尺度上进行，精度可达～1%。但对于内部结构复杂或颗粒细小的锆石，10～30 μm 空间分辨率并不能满足研究需求。中国科学院地质与地球物理研究所离子探针实验室开展了超高空间分辨率的小束斑 SIMS 定年研究，通过改进 SIMS 双等离子体源等关键部件，在延长其高功率条件下稳定工作时间的同时，将 O^- 离子束的最大束流提高了 50% 以上，在仅损失较小的分析精度（1%～2%）情况下将一次离子束斑缩小到 5 μm（Liu Y. et al., 2011）及 2～3 μm（Liu et al., 2020），从而大大提高了锆石 U-Pb 定年的空间分辨率，进一步拓展了定年研究对象。除了传统的跳峰技术外，还采用多接收器同时测试 Pb 同位素以获得更高精度的 Pb-Pb 年龄的方法被建立起来，该方法还具有不需要外部校准的显著优势（Li et al., 2009），结合传统的跳峰技术，还可以同时获得 U-Pb 年龄（Liu Y. et al., 2015），以有效限定锆石 U-Pb 体系的谐和度。

2.1.2 稳定同位素分析

我国已先后建立起锆石、磷灰石、石英、橄榄石、辉石、方解石、硫化物等单矿物以及火山玻璃的 C、Li、B、O、Si、S 同位素 SIMS 分析技术（Li X. H. et al., 2011；李献华等，2015；Yang Q. et al., 2018；Li et al., 2019；Liu Y. et al., 2019；He et al., 2020）。Tang 等（2015）研究发现，颗粒浮雕效应（样品表面不平整导致不同位置测定同位素比值不同）是影响 CAMECA IMS 1280 系列仪器高精度同位素测定的重要因素。幸运的是，可以通过二次离子峰对中参数（DTFA-X）监控和校正这一效应，通过校正后得到的石英 Si 同位素测定值可达到 0.1 per mil（2SD）超高精度（Liu Y. et al., 2019）。样品的基体效应也是影响 SIMS 同位素精确测定的不利因素，特别是橄榄石、辉石、石榴子石、独居石及玻璃等化学成分变化大的矿物，很难做到未知样品和标准物质基体完全匹配。只能研制一系列元素组成不同且尽量涵盖各个成分端元的标准物质，利用多个标准物质建立基体效应曲线来校正待测样品的仪器分馏。Su 等（2015）报道了一套 $Mg^\#$ 值为 89～94 的橄榄石和一套 $Mg^\#$ 值为 89～92 的单斜辉石和斜方辉石 Li 同位素标准物质，尽管其 $Mg^\#$ 值变化范围很小，氧同位素的基体效应难以区分（Tang et al., 2019），但这套橄榄石样品能观察到明显的 Li 同位素基体效应，并据此建立了 Li 同位素基体效应曲线（Su et al., 2015）。值得指出的是，更多的尤其是 $Mg^\#$ 值较低的相应矿物标准物质匮乏是目前 SIMS 精确测定这些矿物同位素组成亟待解决的一个瓶颈问题。而上述基体效应校正程序则需要事先额外利用 EPMA 测定待测样品的化学组成，如果待测样品化学成分不均匀，则很难保证 EPMA 测点和 SIMS 测点完全一致，从而影响了 SIMS 分析的准确度。最近，中国科学院地质与地球物理研究所离子探针实验室开发了利用单次实验同时分析 ^{16}O、^{18}O 和 ^{28}Si 同位素来"在线"获取分析点化学组成的独居石氧同位素 SIMS 分析方法（Wu et al., 2020）。这种方法在确保校正的 $\delta^{18}O$ 误差可以达到和离线 EPMA 成分校正结果同一水平（≤0.8‰，2SE）的情况下，简化了分析流程，提高了测试效率和准确度。

2.1.3 低含量挥发分等分析方法

水等挥发分也是地球和行星科学研究的前沿和热点。除了定年和稳定同位素分析，SIMS 进行水含量分析并可同时获得同位素组成的优点受到了越来越多的重视。但对水含量非常低的名义上无水矿物如橄榄石等进行水含量分析，就要求系统具有极低的背景值。大型离子探针由于腔体大，水背景值比小型离子探针要高出 1～2 个数量级。目前国际上利用大型 SIMS 对名义上无水矿物水含量分析进行了一些尝试，但获得的水背景值过高（$20×10^{-6}$～$40×10^{-6}$）（Turner et al., 2015），难以满足橄榄石等低水含量分析的要求。中国科学院广州地球化学研究所自主开发了自动化液氮加注系统以及合金超高真空制靶技术，在提高仪器自动化分析水平的同时降低了分析系统内的水背景值（Zhang W. F. et al., 2018），建立了同时测试锆石水含量和氧同位素的分析流程，氧同位素分析精度与常规氧同位素分析相当（～0.3 per mil，2SD），水测量背景值降低到 $10×10^{-6}$ 以下（Xia X. P. et al., 2019）。通过再次优化 SIMS 分析条件，进一步将分析系统的水背景值降至 $1.2×10^{-6}$ 以下（Zhang W. F. et al., 2020）。

2.1.4 标准物质

近十年来，另一个最重要的进展是标准物质的研制。SIMS 分析最大的瓶颈就是严重的基体效应，这要求样品必须有基体匹配标准物质来校正仪器分馏。SIMS 分析的标准物质必须在微米尺度上保持待测元素或同位素在仪器分析误差范围内完全均匀。目前国内外除了应用最为广泛的锆石 U-Pb 年龄和氧同位素标准物质比较充裕外，其他分析体系的标准物质都存在储备不足、质量不高的问题。即使是在国内外 SIMS 实验室广泛使用的标准物质也存在颗粒间或者不同样品批次间的不均匀现象（如磷灰石 O 同位素标准物质 Durango 和黄铁矿 S 同位素标准物质 Balmat）（Li et al., 2019; Yang et al., 2020）。即使是众多可以作为 U-Pb 年龄和氧同位素分析的锆石标准物质也仅仅只有 M257 锆石可以作为 Li 同位素分析的标准物质（Li X. H. et al., 2011）。近十年来国内相关 SIMS 微区分析的 U-Pb 年龄和 C、Li、O、S 和 Cl 等同位素以及橄榄石、锆石水含量等一系列分析标准物质从无到有开始建立，值得一提的是国家一级微区分析标准物质实现了零的突破，目前已经有 4 个: Penglai 锆石 Hf-O 同位素（GBW04482）、Oka 方解石 C-O 同位素（GBW04481）、Qinghu 锆石 U-Pb 年龄（GBW04705）及文石 Sr 同位素（GBW04704）。

2.2 纳米离子探针

纳米二次离子质谱（NanoSIMS；又称纳米离子探针）相比于传统的离子探针，具有超高空间分辨率（小至 50 nm）、高灵敏度、多接收和成像分析能力（Yang W. et al., 2015）。纳米离子探针诞生的主要驱动力来自于两个方面，一是宇宙化学研究，通过对地外样品进行同位素成像，寻找微米和次微米尺寸的太阳系外颗粒（Presolar Grain），并精确分析其同位素组成；二是生物化学研究，在单细胞或亚细胞尺度内开展元素和同位素成像，示踪细胞内发生的生物化学过程。

近十年，纳米离子探针在分析技术和科学应用两方面都获得了长足发展。通过分析方法研发、图像数据处理算法优化，无论是空间分辨率，还是分析精度，基于纳米离子探针的分析方法都有了较大的提升。在宇宙化学和生命科学两大主流研究领域中的应用，仍然是研究的前沿；特别是纳米离子探针开始被广泛应用于地球科学，为推动纳米地球科学的发展提供了重要的平台支撑。

在此仅介绍近十年来我国纳米离子探针分析技术的主要进展。这方面的进展都集中于方法研究，即针对某种特定的分析需求最优化仪器设置，从而提高分析方法的空间分辨率或者分析精度。

2.2.1 U-Pb 年代学和微量元素分析

在锆石 U-Pb 和 Pb-Pb 定年方面，十年前的空间分辨率只有 15 μm（Takahata et al., 2008），通过最优化氧离子源亮度和接收杯设置，将 Pb-Pb 定年提升至 2 μm（Yang et al., 2012）。但由于离子束亮度太高，在分析过程中形成了较深的"V"形坑，导致 U-Pb 同位素的分馏，从而带来定年偏差，为克服这一"深坑效应"，采取了 3 μm×3 μm 扫描方式，但也只能实现 5 μm 的空间分辨率。此后，通过图像定年法扫描大区域，在牺牲效率的前提下，成功克服了"深坑效应"造成的 U-Pb 分馏，进而又将锆石 U-Pb 定年的空间分辨率提升到了 2 μm（Hu et al., 2016）。

自然界中矿物的生长环带是非常普遍的现象。纳米离子探针高空间分辨率和高灵敏度的特性，使其可以在微米或亚微米区域观察矿物环带中的微量元素分布。近十年，基于纳米离子探针研发出了锆石、磷灰石、硅酸盐玻璃从 10 μm 到 1 μm 不同空间分辨率的锆石、磷灰石、硅酸盐玻璃微量元素分析方法（Hao J. L. et al., 2016; Yang et al., 2016; Zhang J. C. et al., 2016），以满足不同的研究需求，如锆石岩浆环带形成机理研究（Yang et al., 2016）、锆石–熔体分配系数研究（Hao J. L. et al., 2016）。

2.2.2 稳定同位素分析

对于硫化物的 S 同位素分析，根据空间分辨需求的不同，纳米离子探针的接收杯组合（FC 法拉第杯或 EM 电子倍增器）可以设置为 3 种模式：FC-FC-FC-EM、FC-EM-EM-EM 和 EM-EM-EM，它们分别对应不同的一次离子束，即约 5 μm、2 μm 和 1 μm 的空间分辨率，所获得的 $\delta^{33}S$ 和 $\delta^{34}S$ 的分析精度均分别为 0.3‰、0.5‰和 1‰（Zhang J. C. et al., 2014, 2017），这一系列的 S 同位素分析方法，已被大量应用于各种金属矿床的研究中。

纳米离子探针对水含量和 H 同位素的分析具有独特的优势，可以实现更高精度的分析。已建立的方法可以在 10 μm 的尺度内对硅酸盐玻璃、磷灰石开展 H_2O 含量和 H 同位素分析，H 同位素分析精度优于 40‰，MORB 玻璃（H_2O=0.258%）的 H 同位素分析精度优于 61‰，水含量分析偏差小于 6.9%（2SD）（Hu S. et al., 2015）。从而满足了对火星样品中熔体包裹体（Hu et al., 2014，2020）和磷灰石（Hu et al., 2014）的研究需求，揭示火星岩浆水和大气水的组成和演化。

基于纳米离子探针的稳定同位素分析方法还有金刚石或石墨 C 同位素分析，其空间分辨率为 3~10 μm，$\delta^{13}C$ 精度为 0.3‰~0.4‰；碳酸盐 O 同位素分析，其空间分辨率为 5 μm，$\delta^{18}O$ 精度为 0.5‰（Yang W. et al., 2015）。

2.2.3 图像算法优化和多技术联用

除了上述分析方法研发，纳米离子探针还可以通过图像处理算法的优化来提升分析效率或空间分辨率。例如，利用基于 Otsu 算法的局部动态阈值算法来对离子图像中的颗粒进行识别，避免了信号亮度不一对离子图像颗粒识别造成的影响，从而提升颗粒的识别效率（Hao et al., 2020）。另外，将低秩理论引入到纳米离子探针图像的去噪过程，不仅能对离子图像的噪声进行压制，还能很好保持原图像的边缘特性，从而提供更可靠的结构信息，实现弱信号的提取（Lin Y. et al., 2020），这为进一步的超分辨图像算法的优化提供了基础。

纳米离子探针分析技术的另一个进展是与其他显微技术的联用。与原子力显微镜（atomic force microscopy，AFM）联用可实现真正的三维元素成像（Fleming et al., 2011）；与同步辐射 X 射线荧光（XRF）联用来揭示不同空间尺度的微量元素分布（Moore et al., 2014）；与透射电子显微镜（TEM）联用可同时获取样品的纳米结构和元素信息（Xu et al., 2018b）。此外，新的射频等离子体氧离子源还被成功用于纳米离子探针，相比于传统的双等离子体（duoplasmatron）氧离子源，其空间分辨能力提升了约 4 倍。可以预期，基于新氧离子源的分析方法研究在未来会有较大的发展空间。

2.3 飞行时间二次离子质谱

二次离子质谱（SIMS），是一种检测带电粒子的质谱检测方法。当带有一定能量的一次粒子如电子、离子、中性子或光子轰击某一物体表面（常为固体表面）时，就会有粒子被发射。这些被发射出的粒子，或称为"二次"粒子，可能是电子、光子、原子、分子、原子离子或团簇离子，绝大部分发射出来的粒子是中性的，只有很小部分是带电的，能被质谱检测和分析的只有这些带电的二次离子。超过 95% 的二次粒子来源于固体最上面两层，因此 SIMS 是一种表面分析技术。飞行时间二次离子质谱（TOF-SIMS）能以极高的灵敏度（10^{-9}~10^{-6}）检测包括 H 在内的所有元素及其化合物信息。它是一种普适性的分析技术，适用于有机、无机、生物、医学、电子、地质/考古、环境等各种领域的各种固体材料的分析（图 3）。

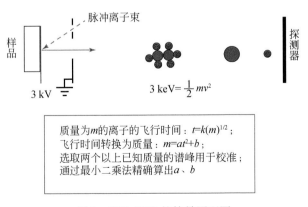

图 3　TOF-SIMS 的简易原理图

2.3.1 仪器构成

TOF-SIMS 初期的一次离子束的离子源是 Ar^+ 离子源，到 2012 年时已经有 Ar^+、O_2^+、O^-、Cs^+、Ga^+、Au（Au^+、Au_3^+）、SF_5^+、Bi（Bi^+、Bi_3^+、Bi_3^{++}、…）、C_{60}^+，GCIB [gas cluster ion beam：Ar_n^+、$(H_2O)_n^+$、…] 等，TOF-SIMS 一次离子束的发展趋势是减少脉冲宽度以提高质量分辨（ΔM）或质量分辨本领（$M/\Delta M$）；减小束斑直径以提高空间分辨率；改变离子的种类以提高二次离子，特别是高质量数分子离子的产额。不过，这三个性能指标是相互制约的，不能达到同时最佳，它们之间要取得某种平衡。1990 年以后近 30 年间，TOF-SIMS 一次离子束的束斑大小仅仅降低了约一半，最新的 Bi 离子源液态金属离子枪（liquid metal ion gun，LMIG）（Bi-LIMG）的束斑直径约 50 nm（ION-TOF GmbH，2019）。Ar、O、Cs 源是传统的 SIMS 所用的一次离子束，用于 TOF-SIMS 时只是把它脉冲化，以满足 TOF-SIMS 一次离子束的要求。为了解决 Ga-LIMG 高质量数碎片离子，特别是分子/准分子离子的二次离子产额过低的问题，2000 年前后出现了 Au-LIMG，它能提供 Au^+、Au_3^+ 的一次离子束。在这条延长线上，2003 年 Bi-LMIG 被开发出，能提供 Bi^+、Bi_3^+、Bi_3^{++} 的一次离子束，进一步提高了高质量数碎片离子，特别是分子/准分子离子的二次离子产额的同时，还减小了一次离子束束斑大小。2013 年，进一步改进 Bi 源，质材为 Bi-Mn 合金，使 Bi 源性能更加优异，并除了能给出 Bi^+、Bi_3^+ 和 Bi_3^{++} 外，还能给出 Mn^+、$BiMn^+$ 和 Bi_2Mn^{++} 等一次离子束，更适合 G-SIMS（Gentle SIMS）。一次离子束的种类对二次离子的离化率有很大的影响，使用团簇离子（如 Au_3^+、C_{60}^+、Bi_3^+ 等）比使用单原子离子（如 Ga^+、Bi^+ 等）能大大提高高质量数分子离子的离化率（ION-TOF GmbH，2012）。

现代 TOF-SIMS 均配有溅射离子枪。溅射束只用于剖蚀材料表面，实现深度剖面分析或清洁固体表面，溅射束的种类主要有 Ar、O、Cs、C_{60} 和 GCIB 等，近几年的发展主要是在尽可能实现大电流下尽量小的束斑。对溅射束主要考虑的是它和样品表面的相互作用能给测量本身带来的益处。自第一代 TOF-SIMS 出现（1982 年）以来，溅射束种类开始大幅增多，Ar^+、O_2^+、Cs^+ 主要用于纯材料或无机材料，C_{60}^+、GCIB 主要用于有机材料、高分子材料、生物材料、组织切片、细胞等，目的是在分析层面把溅射损伤对 TOF-SIMS 分析的影响降到最低，以实现对有机物、生物材料的深度剖面分析。

质量分析器是 SIMS 也包括 TOF-SIMS 的心脏部分。其是由早期的 Thomson J. J. 抛物线质量分析器，到四极杆质谱质量分析器，发展到目前广泛使用的双聚焦磁质量分析器。最早的 TOF-SIMS 质量分析器是由 William Stephens 在 1946 年提出的，现代 TOF-SIMS 以 20 世纪 80 年代初出现商品化的 TOF-SIMS 装置为起点，在 2012 年前后已发展到第五代。TOF-SIMS 质量分析器主要有反射式和静电三重聚焦式两种模式。虽然近年来这两种 TOF-SIMS 质量分析器原理变化不大，但也在不断改进中：反射式分析器增加了 EDR（extended dynamic range）（ION-TOF GmbH，2012），静电三重聚焦式分析器增加了 MS/MS（串联质谱）功能（ULVAC-PHI Inc.，2016）。

SIMS 特别是 TOF-SIMS，信息量庞大，一张质谱一般少则几百多则几万个质谱信息，在 TOF-SIMS 测试数据中要得到所需信息一般不太容易。现代 TOF-SIMS 分析技术都包含深度剖面分析，2D（面）成像分析和 3D 成像分析，因此在测试前很难确定样品需要监测哪些质谱峰。为此，TOF-SIMS 至少在 1995 年就开始引入数据再生技术，以便把遗漏的信息重新从保存的数据中再生获得。另外，TOF-SIMS 分析测试结果含有海量的信息，一张 TOF-SIMS 的质谱图所包含的质谱峰少则几百个，多则几千甚至过万个质谱峰，每个质谱峰都包含与表面成分相关的信息，如何从 TOF-SIMS 的测试数据中获取材料表面所含化学成分或感兴趣的化学成分的信息，一直是 TOF-SIMS 的重要课题（Kassenbohmer et al.，2018）。

2.3.2 数据处理及信息提取

数据的再生技术一般是指把每个周期产生每幅二维图像的像素（pixel；picture element）或 3D 图像的每个体素（voxel；voxels；corpus）对应的一张完整的质谱，每张质谱含一百万甚至几百万对数据（x，I）构成 [x 为质量数（amu）或时间（ns）或通道（channel），I 为信号强度（cnts）] 的信息全部保存在计算机里，只要按照时序读取数据，就能再现与实际的测试相同的过程。由于在每个周期产生的每幅二维

(2D）图像的像素（pixel；picture element）上，或者三维（3D）图像的每个体素（voxel；voxels；corpus）都对应有一张完整的质谱，因此，在最初测试时被遗忘或被忽视的离子、碎片离子、分子离子，在数据再生时都可以添加进去实施分析。另外，数据再生技术的一大优点在于它能在空间（面）上和在时间（深度）上，获取任意选定区域内（region of interest，ROI）和任意选定时段内的数据。

如何根据TOF-SIMS测试数据获取材料表面的化学成分或感兴趣的化学成分信息，一直是TOF-SIMS最重要的一个课题。获取材料表面化学成分的主要途径有两种，一是不断提高质量分辨本领（质量分辨率）（$M/\Delta M$），以确定尽可能高质量数离子质谱峰对应的分子式，采用融合进轨道离子阱（orbitrap）的方式，目前已达到$M/\Delta M$ = 240 000（Passarelli et al.，2017）。二是利用指纹图谱，采用多变量数学统计方法，如主成分分析（PCA）、深度机器学习等算法获取样品表面成分、特别是有机、生物成分信息（Frisz et al.，2012；Wang et al.，2017）。

2.3.3 分析技术进展与展望

自1982年第一代商品的TOF-SIMS诞生以来，至2012年已到第五代，2019年世界的两大制造厂商都发布了他们的第六代商品仪器M6（ION-TOF GmbH）以及TRIFT VI MS/MS（ULVAC-PHI Inc.）。表1是各代TOF-SIMS的发布时间和新增主要特点或主要改进点。

表1 各代TOF-SIMS的发布时间和新增主要特点或主要改进点

仪器代别	代表性仪器	出现年份	新增主要特点或主要改进点
第Ⅰ代	TOF-SIMS I	1982	一次离子束斑大小1 mm，脉冲宽度10 ns，仅用于静态SIMS
	TRIFT I	1990	脉冲化Cs离子枪，直接成像，液态金属离子枪
	7000/SAU	1989	脉冲氪准分子激光，真空紫外激光
第Ⅱ代	TOF-SIMS Ⅲ	1986	首次实现TOF-SIMS的质量分辨本领>10 000以及ppm量级的高精度分析；首次实现绝缘材料的电荷补偿
	TRIFT Ⅱ	1995	200 mm，300 mm Si晶片分析应对，In源一次离子束
	7200	1993	In源一次离子束
第Ⅲ代	TOF-SIMS Ⅲ	1993	5轴样品台，超快速无漂移的Ga枪，200 mm晶片分析应对
	TRIFT Ⅲ	2000	3D成像，Au源一次离子束，样品台移动成像技术，双源离子束
第Ⅳ代	TOF-SIMS Ⅳ	1994	首次实现双束模式的深度剖面分析，用于大面积分析的宏观扫描，SF_6^-离子源，首次实现用Au团簇离子成像分析。300 mm Si晶片分析应对
	TRIFT Ⅳ	2005	性能大幅改进的新型Au源，自动/批处理数据处理，拓扑样条数据处理（topo-strip data processing）
第Ⅴ代	TOF SIMS 5	2003	新型Bi源纳米探针，高二次离子信号线性动态范围和检测限，先进的软件，气体团簇离子源
	TRIFT Ⅴ Nano TOF	2007	5轴样品台，双束电荷补偿，Bi源一次离子束，气体团簇离子源
第Ⅵ代	Nano TOF Ⅱ	2015	新型Cs源，新型气体团簇离子源，串联质谱
	M6	2019	新型<50 nm高横向分辨率Bi离子源，质量分辨率>30 000，独特的延迟提取模式，串联质谱，多元统计分析软件包

TOF-SIMS未来的发展方向一是提高空间分辨率，提高质量分辨本领（质量分辨率），提高分子离子，特别是高质量数（>500 amu）分子的离化效率。二是发展多变量数学统计、深度机器学习等数学方法，尽可能多以及尽可能准确地获取样品表面化合物，特别是各种有机物的信息。目前TOF-SIMS最好的指标是：表征空间分辨的最小束斑直径为50 nm，质量分辨本领（质量分辨率）为12 000（^{28}Si），25 000（>200 amu）；目前最新型的一次束为Bi（Bi-Mn）液态金属离子束（LMIG），其团簇离子束Bi_3^+与初代的Ga^+离子束相比，对分子离子特别是高质量数的分子离子的离化率提高了几十倍（ION-TOF GmbH，2019）。质量范围变化不大，在20 000 amu左右。预计到2030年，一次离子束的最小束斑为20 nm左右；质量分辨本领（质量分辨率）为20 000（^{28}Si），40 000（>200 amu）；新类型的一次束将会出现，与初代的Ga^+离子束相比，对分子离子特别是高质量数的分子离子的离化率提高100倍以上。质量范围变化不大，在30 000 amu左右。与此同时，数学算法的改进，深度机器学习的完善，将进一步提高准确获取样品表面化合物，特别是各种有机物的信息的能力。另外，引入不同的基质材料，能进一步提高二次离子

的产额，特别是生物分子（>500 amu）离子的二次离子产额（Cai et al.，2017b）。与其他技术融合也是TOF-SIMS今后发展的趋势，Obi-SIMS是Obtrap与TOF-SIMS融合的例子（Passarelli et al.，2017）。

TOF-SIMS（或称静态SIMS）自其出现之日起，除了获取固体材料表面包括H在内的所有元素及其同位素的信息外，还能获取材料表面化合物的分子结构信息。所获取的分子结构信息，起初只能获取简单化合物结构，如SO_4、NO_3等酸根的信息，随后是一些简单的有机物信息。随着技术的不断发展以及研究的不断深入，2010年前后，TOF-SIMS技术开始进入到刑侦、医学、生物领域（Lanekoff et al.，2011）。特别是在目前，TOF-SIMS在生物领域，特别是对组织切片（Jung et al.，2012）、单细胞（Robinson et al.，2012；Brison et al.，2013）的分析成了热点和前沿。另外，TOF-SIMS在刑侦领域的应用研究也是个前沿课题，它在形貌和成分两个维度上对指纹实施分类鉴定，比只获得形貌信息传统的方法能得到更为丰富的信息（Cai et al.，2017a；Costa et al.，2017）。

地球化学微量元素微区原位分析方法在地球化学和宇宙化学研究中发挥着重要作用，是揭示成矿物质来源、成矿条件及矿床成因等方面有效的技术手段，也是研究月球和行星物质组成的重要方法。然而，目前的主要分析方法如激光剥蚀电感耦合等离子质谱法（LA-ICP-MS）（Azadbakht and Lentz，2020；Chen et al.，2020）对样品是半破坏的，不适用于对珍贵样品和结构、成分复杂样品的分析。二次离子磁质谱法（如IMS1280或SHRIMP II）（Giovanardi et al.，2017；Liu Y. et al.，2019）是微区原位无损分析，但是无法快速同时分析大多数元素/同位素，需要延长分析时间来完成有限种元素/同位素的分析。

TOF-SIMS是一种有效的表面分析技术，无需对样品进行化学处理，能够实现对样品表面无损伤、高灵敏度的微区原位分析。尽管该方法无法获得二次离子磁质谱仪法的分析精度，但是能够单次测量获得几乎全部同位素信息，对样品消耗量比目前常用方法低1~5个数量级，非常适用于地球特别是宇宙样品-复杂样品的微量元素微区原位分析。

近年来，TOF-SIMS也逐渐被应用于地质科学中宇宙样品、与碳酸盐岩风化-沉积有关的碳酸盐黏土型锂矿床的研究（温汉捷等，2020）、熔融包裹体分析、高铝粉煤灰（HAFA）颗粒微量元素分析（Hu et al.，2018）、矿物浮选（Chehreh et al.，2013，2014）等方面。目前商用TOF-SIMS的质量分辨率等性能无法满足地球化学微量元素全谱和准确分析的需求。但如前文所述，TOF-SIMS以其具有的对各种离子的平行检测，样品损耗小，分析速度快的优点，在对地外物质、样品量较少的（包括矿床学样品）地质学样品的分析测试中，具有广阔的应用前景。

2011年科技部启动了第一批国家重大科学仪器设备开发专项，支持基于新原理、新方法和新技术的重大科学仪器设备的开发，刘敦一研究员负责的"同位素地质学专用TOF-SIMS科学仪器"获得支持，以研制达到下列性能指标的同位素地质学专用TOF-SIMS科学仪器：质量分辨本领（质量分辨率）$R=21\,720$（FWHM）（$m/z=228$）；空间分辨率5μm；质量范围1~350 amu；质量精度优于100×10^{-6}；长期质量稳定度优于10×10^{-6}；检测限优于0.2×10^{-6}；稀土元素（NIST610）分析精度优于10%。该项目针对地质学领域，从TOF-SIMS仪器研制以及应用TOF-SIMS到地质学科学问题研究两个方面推进TOF-SIMS技术的发展（边晨光等，2015；Long et al.，2018，2020）。研究填补了该领域仪器和部分核心部件研制的空白，为自主研制TOF-SIMS奠定了理论和技术基础（图4）。

3 激光束相关技术

3.1 激光剥蚀等离子体质谱

激光剥蚀电感耦合等离子体质谱（LA-ICP-MS）可对固体样品进行微区原位元素和同位素的准确精细分析，确定化学组成在微米尺度上的分布和分配规律，完成在以往整体分析方法下难以完成的工作，因而被认为是地球化学分析方法研究领域最激动人心的重要进展之一，这为地球科学、环境科学、材料科

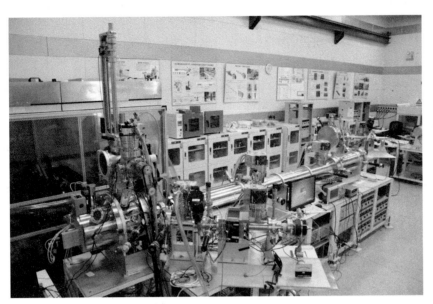

图4 研发用于微量元素分析的 TOF-SIMS

学的创新研究提供了重要支撑手段。中国是 LA-ICP-MS 技术应用的大国。据 Web of Science 数据,自2007年开始中国采用 LA-ICP-MS 技术每年发表的文章数量已超过美国成为世界第一,并且继续呈线性快速增长的趋势。2019年,中国作者发表的与 LA-ICP-MS 技术相关的论文数量占到了世界的40%。地球科学是 LA-ICP-MS 技术最重要的应用领域,以其分析为主要手段获得的地球科学研究成果非常丰硕。

3.1.1 在元素分析中的应用

LA-ICP-MS 目前已成为地质样品主微量元素分析最常用的技术之一(刘勇胜等, 2013; Li Z. et al., 2015),分析的地质样品矿物种类繁多,如硅酸盐矿物(Li Z. et al., 2015)、氧化物(Li Z. et al., 2016)、碳酸盐矿物(Chen et al., 2011)、硫化物(袁继海等, 2012; Feng et al., 2018)等。元素分馏效应是 LA-ICP-MS 分析存在的主要问题,也是当前研究的热点和难点。最新的研究表明,样品池内不同采样点处的载气流速显著影响剥蚀所产生的气溶胶颗粒尺寸大小(Luo et al., 2015),在载气流速较低区域剥蚀产生的大尺寸气溶胶颗粒在等离子体中无法完全离子化是导致元素分馏的重要因素(Luo et al., 2015)。通过降低载气流速(Luo et al., 2015)或引入氮气、水蒸气等活性基体(Luo et al., 2018a)的方式可提升等离子体温度和对基体的耐受能力,最终降低离子化过程中产生的元素分馏。飞秒激光由于与样品相互作用过程中的热效应小(Li C. Y. et al., 2016; Fu et al., 2017)、剥蚀产生的气溶胶颗粒细,可有效减小元素分馏效应和基体效应(杨文武等, 2017)。但需指出的是,飞秒激光剥蚀过程中产生的元素分馏程度与剥蚀环境紧密相关(Luo et al., 2017)。在氦气环境下,激光剥蚀直径越大获得的挥发性元素分馏效应越显著,而在小束斑剥蚀条件下则无明显分馏效应(Li Z. et al., 2015)。在氦-氩混合气剥蚀环境,飞秒激光在不同剥蚀束斑条件下均无明显的元素分馏现象(Luo et al., 2017)。在进行 LA-ICP-MS 元素分析时需要综合选择仪器参数,减小激光剥蚀、样品传输和离子化过程中的元素分馏效应,以获得准确的分析结果。

对低含量样品进行准确测定是 LA-ICP-MS 分析的发展趋势和重要方向,仪器灵敏度是制约分析结果的重要参数。目前已广泛使用氦气替代氩气作为样品载气,这可使193 nm 激光剥蚀的信号灵敏度提高2~5倍。最近的研究结果表明,该信号增敏不仅仅是因为氦气的使用提高了小尺寸气溶胶颗粒的传输效率,更主要的是因为在氦气环境下剥蚀产生的小尺寸气溶胶颗粒在 He-Ar 等离子体环境中具有更高的挥发及离子化效率(Luo et al., 2018a)。通过向等离子体中引入氮气(Hu et al., 2012a)、水蒸气或乙醇溶液(Liu S. H. et al., 2014; Luo et al., 2018a)等活性基体也可显著提升仪器的灵敏度,可观察到约2~3倍的信号增强。此外,通过改进磁质谱仪的锥接口(Hu et al., 2012a; Xu et al., 2015; Gou et al., 2018; He et al., 2018) 或提高接口真空度（Yuan et al., 2019）也可观察到显著的信号增敏。各种增敏技术的广泛应用为国内各实验室开展微量及超低含量元素分析提供了保障。

国内学者通过仪器关键硬件的研发，对提高分析信号的稳定性、增强仪器灵敏度和抑制元素分馏及基体效应等都有促进作用。如"线形"和"波形"信号匀化装置的研制显著增强了激光剥蚀信号的稳定性，很好地抑制了四极杆质谱顺序扫描时产生的光谱螺纹效应（Hu Z. C. et al., 2012b, 2015）。湿等离子体技术是目前抑制元素分馏和基体效应的有效手段，水蒸气引入装置的研发为稳定引入水蒸气提供了技术支持（Liu S. H. et al., 2014；Luo et al., 2018b）。优化剥蚀池设计也可显著降低元素分馏效应（Luo et al., 2015）及位置效应（Xie et al., 2018a）。此外，国内学者也开展了激光剥蚀系统的研发工作，相信在不久的将来，国产仪器系统的研制将为我国微区地球化学分析的发展提供新的动力。

除了对矿物样品进行微区分析外，学者们还开展了对地质全岩样品进行 LA-ICP-MS 整体分析的方法研究（Hu and Qi, 2014；Zhang and Hu, 2019）。制备激光剥蚀尺度上均匀的具有代表性的全岩样品是获取准确数据结果的前提，一般采用粉末压片法（Zhang W. et al., 2017）和熔融玻璃法（Zhu et al., 2013；Hu and Qi, 2014；Bao et al., 2016a；He et al., 2016；徐娟等，2016）来制备可直接用于 LA-ICP-MS 整体分析的全岩样品。对于粉末压片，可通过使用黏合剂直接压片或通过物理研磨或化学消解（Zhang W. et al., 2017）等方式制成超细颗粒后再压片。由于添加助熔剂制备玻璃的技术可能带来样品或仪器污染（Hu and Qi, 2014），目前国内学者主要通过直接熔融样品等方式制备熔融玻璃（Zhu et al., 2013；Bao et al., 2016a；He et al., 2016；Zhang S. Y. et al., 2019）。此外，通过使用高能红外激光直接熔融粉末样品制备熔融玻璃更是一种简便快捷、绿色环保的样品制备方法（Zhang C. X. et al., 2016）。值得一提的是，LA-ICP-MS 还被证明可直接用于液体样品中元素含量的整体分析，液体本身的高度均一性使得在微克级的进样量下，也可很好地代表样品整体的化学组成（Liao et al., 2019）。得益于激光剥蚀微量进样的特点，这种进样方法基体效应弱，与水溶剂相关（氧化物、氢氧化物）的离子干扰极低，这为某些饱受相关多原子离子干扰的元素分析提供了极大的便利，如高钡含量样品中稀土元素的测试（Liao et al., 2019）。

3.1.2 在副矿物 U-Pb 定年中的应用

近十年来，国内副矿物 U-Pb 定年测试方法也取得了快速发展并得到广泛应用。其中，锆石的分析方法尤为成熟（Xia et al., 2011；Xie et al., 2018b），在高空间分辨率条件下也可实现锆石 U-Pb 年龄的准确分析（Xie et al., 2017）。值得注意的是，目前的 LA-ICP-MS 分析技术对锆石 U-Pb 年龄测试的不确定度水平仍在 ~4%（2RSD）（Li X. H. et al., 2015），在进行锆石 U-Pb 年龄结果的解释和应用时需考虑。此外，其他含 U 副矿物的 U-Pb 年代学分析方法也日趋成熟，除了常见的榍石（Sun et al., 2012）、独居石（汪双双等，2016）、磷灰石（周红英等，2012）、磷钇矿（Liu Z. C. et al., 2011）和金红石（周红英等，2013；Xia X. P. et al., 2013）外，我国学者还率先开展了锡石（Yuan et al., 2011；Li C. Y. et al., 2016）、沥青铀矿（Zong et al., 2015）、铌铁矿（Deng et al., 2013；Che et al., 2015）、钙钛矿（Wu et al., 2013）、氟碳铈矿（Yang et al., 2014a）、石榴子石（Yang Y. H. et al., 2018）和黑钨矿（Luo et al., 2019）等副矿物的 U-Pb 年代学分析。目前，LA-ICP-MS 副矿物 U-Pb 定年分析过程中仍存在元素分馏和基体效应的问题，使用基体匹配的矿物标样校正分析是当前获得准确年龄结果的主流做法。因此，国内学者在副矿物 U-Pb 年龄分析标样的刻画上也做了大量工作，如锆石标样 Penglai（Li X. H. et al., 2010）、Qinghu（Li X. H. et al., 2013）和 SA01（Huang et al., 2020），榍石标样（Ma et al., 2019），金红石标样 RMJG（Zhang L. et al., 2020）等。除了基体匹配矿物标样的开发，国内学者对不同副矿物间 U-Pb 年龄分析时的基体效应也进行了研究（Liu Z. C. et al., 2011；Sun et al., 2012；Luo et al., 2018b）。如水蒸气辅助激光剥蚀非基体匹配副矿物 U-Pb 定年方法的建立，成功地实现了以 NIST610 玻璃为外标校正分析其他副矿物（锆石、独居石、磷钇矿、榍石和黑钨矿）的 U-Pb 年龄（Luo et al., 2018b, 2019）。整体来说，我国 LA-ICP-MS 副矿物 U-Pb 定年技术蓬勃发展，这将为探讨成岩成矿、地质演化历史等重要地质问题提供重要的技术支撑。

3.1.3 在同位素分析中的应用

激光剥蚀多接收器电感耦合等离子体质谱仪（LA-MC-ICP-MS）是最重要的微区高精度同位素分析技

术之一，已被成功应用于超过 20 个元素体系的同位素比值分析。过去十年，我国在 LA-MC-ICP-MS 新同位素分析技术开发、关键仪器部件研制、同位素分馏机理和干扰校正策略探索、固体参考物质研制等方面都获得大量原创性成果。目前，国内已建立的激光微区高精度分析方法包括放射性同位素 Sr-Nd-Hf-Os-Pb（Hu et al.，2012a；陈开运等，2013；侯可军等，2013；Yuan et al.，2013；Chen et al.，2014；Zhang et al.，2014；Xu et al.，2015，2018a；Yuan et al.，2015；Zhang W. et al.，2015，2016，2018；Bao et al.，2016a，2016b，2018；李杨等，2016；Tong et al.，2016；Zhu et al.，2016；Zhang L. et al.，2018；Yang et al.，2019）和稳定同位素 Li-B-C-Mg-S-Ca-Fe-Zr（侯可军等，2010；Xie et al.，2011；Lin et al.，2014，2017，2019；Fu et al.，2016，2017；Chen L. et al.，2017；Chen W. et al.，2017；Xie et al.，2018b；Zhang et al.，2019a，2019b）等，数据质量和空间分辨率均已位于国际前列。一批国际领先的原创技术展现了我国激光微区技术人员的创新能力和实践精神，如活性试剂辅助技术（Hu et al.，2012a；Xu et al.，2015；Fu et al.，2016，2017；Lin et al.，2019；Zhang et al.，2019b）、Split 技术（Huang et al.，2015；Dai et al.，2017；Qian and Zhang，2019）和微区同位素参考物质开发（Li X. H. et al.，2010；宋佳瑶等，2011；陈开运等，2012；李献华等，2013；Yang et al.，2014b；Bao et al.，2017；Feng et al.，2018；Huang et al.，2019，2020；Ma et al.，2019；Zhang L. et al.，2020）等。

化学辅助技术在激光微区同位素分析方法中显示了很好的应用潜力。研究表明，氮气辅助技术在激光微区 Hf、Nd、Zr 同位素分析中可以提高灵敏度 2～3 倍（Hu et al.，2012a；Xu et al.，2015；Zhang et al.，2019b），同时还在消除氧化物和氢化物干扰方面显示了极好的效果，如 Fu 等（2016）利用氮气消除 $^{32}S^1H$ 对 ^{33}S 的干扰，成功建立同时测定 $^{34}S/^{32}S$ 和 $^{33}S/^{32}S$ 的激光微区分析方法。加水辅助技术是目前解决 LA-MC-ICP-MS 基体效应的一个强大武器，Lin 等（2019）报道了向激光剥蚀池和等离子体中加入水蒸气，实现了硅酸盐玻璃对电气石样品的 Li 同位素非基体匹配校正。

由多台质谱同位素分析激光剥蚀气溶胶颗粒的 Split 技术最早由我国分析学家建立，优点是可以同时获得多种同位素或微量元素数据，提供更丰富的原位信息。近年来，该技术被继续扩展到对多同位素体系分析，如磷灰石、钙钛矿等同时分析 Rb-Sr 和 Sm-Nd（Huang et al.，2015），对斜锆石同时分析 Sm-Nd 和 Lu-Hf（Huang et al.，2015），或微量元素和同位素组成同时分析，如磷灰石稀土元素和 Nd 同位素（Qian and Zhang，2019）和硅酸盐样品的微量元素和 Pb 同位素（Dai et al.，2017）。

飞秒激光剥蚀系统（10^{-15} s）是当前最先进的激光剥蚀系统。相对于传统的纳秒（10^{-9} s）级激光器大大降低了激光剥蚀过程中产生的热分馏效应，从而拓展了该技术的应用范围并提高了分析准确度（Shaheen et al.，2012）。近年来我国将飞秒激光应用于地质样品同位素比值分析，开发了 Li（Lin et al.，2019）、S（Chen L. et al.，2017；Fu et al.，2017）、Sr（Zhang L. et al.，2018）、Pb（陈开运等，2013；Yuan et al.，2013，2015；Bao et al.，2016a，2016b；Chen et al.，2014）等分析方法。研究成果指出飞秒激光在降低激光剥蚀过程的基体效应（Fu et al.，2017；Lin et al.，2019），提高剥蚀效率（Zhang L. et al.，2018）和提高分析测试精密度（Fu et al.，2017）等方面展现出了很好效果。

固体同位素参考物质是开展激光微区同位素分析的前提。在参考物质研发方面，我国开展了大量工作，十年来推出了锆石 Hf-O 同位素和定年参考物质 Penglai（Li X. H. et al.，2010）、Qinghu（李献华等，2013）、BB（Huang et al.，2019）、SA01（Huang et al.，2020），金红石定年和 Hf 同位素参考物质 RMJG（Zhang L. et al.，2020），以及多种潜在的副矿物 Sr-Nd-Pb 同位素参考物质（Yang et al.，2014b；Tong et al.，2016；Zhang W. et al.，2016；Xu et al.，2018a；Ma et al.，2019；Yang et al.，2019）。同时在人工合成参考物质方面，机械研磨或水热合成技术制备的超细粉末压片技术（Fu et al.，2016；Zhu et al.，2016；Bao et al.，2017；Feng et al.，2018）、高温熔融玻璃（宋佳瑶等，2011；Tong et al.，2016；Chen L. et al.，2017）、人工晶体生长技术（陈开运等，2012）等已被应用于开发固体 S-Hf-Os-Pb 同位素参考物质，并取得较好的成果。

在激光微区同位素分析关键硬件和软件研制方面开展了长期研究。为解决地质样品 Pb 同位素分析过程中的 ^{204}Hg 对 ^{204}Pb 的干扰问题，Hu S. 等（2015）和 Zhang W. 等（2016）分别成功开发了镀金除汞匀化器和气体交换装置。针对激光微区稳定同位素分析中可能存在的位置效应问题，Xie 等（2018a）设计了无位置效应剥蚀池，成功用于黄铁矿的 Fe 同位素分析。Zhang W. 等（2020）开发的激光微区同位素分析专用数据处理软件 Iso-Compass，为批量化的同位素数据处理提供了便利。

3.1.4 在单个流体包裹体成分分析中的应用

LA-ICP-MS 是目前公认的单个流体包裹体成分分析最有效的方法。它结合了高空间分辨 LA 采样技术与高灵敏 ICP-MS 技术，有效地避免了传统群体萃取分析多世代流体包裹体的混合，提高了分析结果的代表性。该技术起源于 20 世纪 90 年代以苏黎世联邦理工学院（ETH）为代表的实验室，30 年来日渐成熟，相应地国内也取得了一定的进展。目前国内对该技术及相关应用的研究主要集中在中国地质大学（武汉）（马黎春等，2014；付乐兵等，2015；Li J. H. et al., 2015；Chang et al., 2018；Lin X. et al., 2020；郭伟等，2020）、中科院地球化学研究所（蓝廷广等，2017；Li C. Y. et al., 2018；Liu H. Q. et al., 2018）、南京大学（Pan et al., 2019b）。

中国地质大学（武汉）胡圣虹较早尝试采用毛细管合成流体包裹体并结合加热分析技术进行研究，结果发现加热可使流体包裹体形成均一稳定的剥蚀信号，可通过不同浓度合成的流体包裹体建立校准曲线，从而获得了绿柱石天然流体包裹体中 Li、La、Be、Cu、Pb 等元素的浓度。随后国内多个研究小组进行了石盐流体包裹体分析，如孙小虹等（2013）成功测定了单个石盐包裹体中 Ca、Sr 和 Rb 的含量，精度为 4%～20%；于倩等（2015）将合成包裹体标准校正与玻璃标准 NIST 610 结合获得了更准确的结果。近期蓝廷广等（2017）、Liu H. Q. 等（2018a）和 Pan 等（2019b）分别建立了单个流体包裹体 LA-ICP-MS 测试方法，实现了石英、黑钨矿宿主中微量元素的测定。

国内学者应用该技术研究了成矿流体来源（Jian et al., 2018）和热液流体形成和演化（孙小虹等，2013；马黎春等，2014；Li W. T. et al., 2015；Chang et al., 2018；Liu H. Q. et al., 2018；Chen P. W. et al., 2019）。Jian 等（2018）使用 UV-fs-LA-ICP-MS 冷冻流体包裹体剥蚀方法，成功分析了富 CO_2 流体包裹体，指示了矿床在未暴露和氧化的岩浆系统演化而来的两种不同的废液脉冲来源。Li W. T.（2015）利用 LA-ICP-MS 分析了辉石、石榴子石，磷灰石和石英包裹体化学成分，并系统研究了热液流体的形成和演化。Chang 等（2018）获得了藏东玉龙斑岩岩浆热液 Cu-Mo 矿床多期矿脉中单个流体包裹体的元素及元素比值信息，研究了 Cu、Mo 元素在卤水和气相流体包裹体中的分配行为，并结合 Mo/Cu、As/K、Cs/K 值变化及其他元素浓度对比，建立 Cu-Mo 矿床岩浆热液流体形成与演化模型。近期，Pan 等（2019b）通过 LA-ICP-MS 与显微红外技术相结合，重建了我国南岭地区石英脉型黑钨矿矿床精细的流体过程。

3.2 激光诱导原子探针

3.2.1 原子探针技术发展简史

所有地质过程归根结底都是发生在亚纳米尺度下的元素相互作用，因此在亚纳米尺度下进行元素和同位素的定量分析有望大幅提升对地质演化本质上的理解。原子探针层析技术就是用于亚纳米分辨本领下的元素种类和位置判别的，可提供元素周期表中所有元素定量化的三维成分和空间图形信息。该技术已有 50 多年的历史，前期仅能用于导电材料，在材料领域有广泛应用（刘文庆等，2013）。随着硬件的进步，逐渐过渡到可分析半导体材料，乃至近年来分析非导电矿物材料，于是在地学领域的应用开始显示出优势而被关注（Reddy et al., 2020）。

20 世纪末原子探针生产厂商只有法国的 CAMECA 公司和英国的 Oxford Nanoscience 公司，2002 年美国 Imago 公司投入生产的局部电极原子探针（local electrode atom probe, LEAP）大幅提高了数据采集速度并增加了分析体积。之后的几年内，Imago 公司在 LEAP 中增加了激光脉冲模块，从而拓展到了半导体材料。2007 年 Imago 公司兼并了英国的 Oxford Nanoscience 公司，整合了其具有聚焦功能的能量补偿装置，质量分辨本领提高，进一步拓宽了应用领域。2010 年，法国 CAMECA 公司收购了美国 Imago 公司，成为 LEAP 系列原子探针唯一商家。中国科学院地质与地球物理研究所最新购置了目前最新型号的 LEAP 5000

XR 原子探针（图 5 中 1）（X 表示配有激光模块，R 表示配有高质量分辨率的能量补偿装置）。

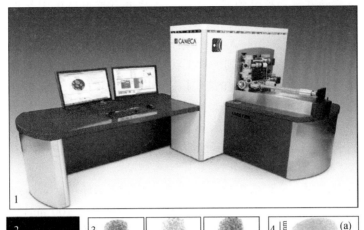
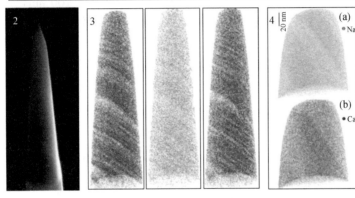

图 5 原子探针仪器、样品及研究成果示例

1. LEAP 5000XR 型原子探针；2. 针状样品图（远大于测试部位）；3. 黄铁矿中 As、Au、Cu 对应纳米尺度条带状分布图（据 Wu et al., 2019）；4. 斜长石中纳米尺度下 Na、Ca 条带分布图（据 Cao et al., 2019）

3.2.2 原子探针技术特点

与其他微束分析技术用聚焦的某种微束来探测样品不同，原子探针是将样品做成曲率半径小于 100 nm 的针状（图 5 中 2），在样品尖端接入高压使其处于待电离状态，通过脉冲电压或脉冲激光来控制逐层蒸发原子，以飞行时间质谱仪测定蒸发离子的质量/电荷值，得到质谱峰以确定元素种类，以位置敏感探头记录飞行离子在样品尖端的二维坐标，通过离子在纵向的逐层累积，确定离子的纵向坐标，进而使用计算机软件重构整个样品的三维结构信息。以 LEAP 5000XR 为例，样品离子化产率可达 50% 以上，重构的三维元素分布信息横向分辨率 0.2 nm，深度分辨率 0.1 nm，分析灵敏度接近 1×10^{-6}。

实现原子探针应用的一个重要条件是将分析样品制备成一个大高宽比、锐利的探针，针尖的尺寸控制在 100 nm 左右。非导电样品的制备需采用聚焦离子束（FIB）系统来切割和塑形样品，一般流程是在感兴趣区域表面沉淀约 1 μm 厚、2 μm 宽的铂金层，以免在减薄过程中样品表面受到离子束损伤。在保护层两侧用离子束挖出"V"形槽，将底部切开，再将一侧切开，将纳米操作手插入，用 Pt 沉积方法将机械手和样品焊接起来，再将另一侧切断，即可将样品提取出来。准备好固定样品支座，将样品用 Pt 焊接在支座上，并从长条样品上切断。在样品底部双边沉积 Pt 来增强连接效果，然后连续从外到内切除外围部分，从而形成尖锐的针尖。最后还需用离子束在低电压下抛光，以消除加工过程形成的非晶层及离子注入较多的区域。

3.2.3 原子探针当前应用

目前国内的原子探针仪器主要用于材料专业领域，地学领域仅有中国科学院地质与地球物理研究所新引进一台。然而，国内地学界已经对原子探针在地学领域的应用产生了越来越浓的兴趣，并通过国际合作开始了尝试性工作。Wu 等（2019）以西秦岭造山带大桥超大型浅成造山型金矿床为研究对象，利用纳米离子探针和原子探针等高分辨率观测和分析测试，对热液含金黄铁矿中复杂的成分环带进行了精细的显微结构和微量元素组成分析。结果显示，微米级的成分环带与流体成分的快速变化有关，而在单一成分环带的内部，微米–亚微米尺度的 Au、As、Cu 扇形环带的形成则受控于黄铁矿不同晶面的结构特征（图 5 中 3）。这些元素在微米尺度上的不规则、补丁状结构与原子尺度上的均一到岛状分布特征受黄铁矿晶格位错造成的异质外延层状–岛状生长模型的控制。此外，由 Au、As、Cu 等元素组成的纳米级韵律环带受控于黄铁矿晶体–流体界面的元素自组织配分作用，与低温条件下晶体生长界面的离子扩散速率受限有关。该研究表明，环带矿物晶格中元素固溶体在纳米尺度上的高度不均一性表现在微米以上尺度（SIMS 和 LA-ICP-MS）的成分分析过程中，将剥蚀曲线上的尖峰图谱解释为微细矿物包裹体的做法值得重新审慎思考。

斜长石晶体中常发育丰富的成分和结构环带，可用于示踪结晶环境、岩石成因和热演化过程，其环带通常表现为微米尺度（10~100 μm），对于是否存在纳米尺度（10~100 nm）环带目前尚不清楚。Cao 等（2019）对广泛发育成分环带的菲律宾黑山斑岩铜矿成矿斑岩中的同一颗斜长石斑晶开展了微米和纳米环带研究，发现这一微米和纳米尺度环带形成于不同阶段，微米尺度环带主要受控于内部晶体生长机制，纳米尺度环带形成于晶体生长后的出溶机制（图 5 中 4）。通过扩散模拟计算，两类环带分别揭示了斑岩铜矿深部岩浆房缓慢冷却过程（>0.000 5℃/a）及岩体就位时的快速冷却过程（>0.26℃/a）。

3.2.4 展望

综合来看，三维原子探针通过重建样品的元素空间含量信息，可进一步分为两类研究：可视化分析和定量分析。可视化分析包括观察样品内部晶界、相界、结构界面及位错和缺陷等不同位置形貌；定量分析包括材料内部组分确定、界面或晶界处的元素沉淀和偏聚的数量、密度和体积计算等。迄今为止，地学中的应用主要集中在副矿物的纳米级分析上，实现了元素和同位素非均匀分布的纳米级可视化，为理解和解释通过更大分析体积技术获得的地球化学数据奠定了基础。然而，由于地质样品在成分和结构上都具有极大的复杂性，样品的几何形状、晶体结构、物理特性以及多相混合情况都直接影响样品的蒸发、离子接收等，且不同数据处理方式也会出现不同的重建结果，因此在针对不同样品的测试流程和数据重构上还需加大研发力度，这将是原子探针拓展地学中应用的主要内容之一。可以预见，原子探针在不同地质背景下越来越多的矿物中的探索应用，将在基础地球科学问题研究上取得重大进展。

4 微光束相关技术

4.1 显微红外光谱

物质的分子振动能量与红外光能量相当，相应波长的红外光会被吸收，根据所得光谱图即可推知样品特定分子的结构成分信息。根据波长可将红外光谱分为近、中、远 3 个区间，由于绝大多数物质的吸收带都出现在中红外区，因此其应用最为广泛；近红外区多为倍频、合频峰；远红外区包含分子的转动光谱和某些基团的振动光谱，仅有部分金属氧化物等少数物质会产生特征吸收。近十年来，在国内外红外光谱分析技术已被广泛应用于地球科学领域，在矿物微量结构水定量分析、高温高压原位测试、微区高分辨率探测等方面取得了突出进展。

4.1.1 微量结构水定量技术

名义无水矿物中的结构水以羟基形式存在，构成了地球巨量的水储库，显著影响地球演化过程（Xia Q. K. et al., 2019）。羟基结构对中红外光的特异吸收强烈，研究者成功运用红外光谱限定了其赋存状态和含量（Hao Y. T. et al., 2016；盛英明等，2016；Wang et al., 2020）。近年来，石榴子石、辉石、橄榄石等一系列名义无水矿物的摩尔吸收系数得以标定和改进（Tian et al., 2017；Zhang B. H. et al., 2019；Li P. et al.,

2020），特定矿物吸收系数法比以往的通用标定法的精确度显著提升。目前对于多种矿物岩石中的羟基结构位置特征、原位高温下的红外吸收特征研究仍仍得进一步深入（杨燕，2017；Gu et al.，2019）。

晶格中分布的结构羟基会因晶体的各向异性而分布不均，偏振红外光谱测试方法能获取振动基团的空间取向，并且只有偏振光下的全吸光度才和水含量成正比关系，因此应用偏振红外方法可精确测量各向异性矿物的水含量（Liu Y. et al.，2014；惠鹤九等，2016）。但偏振方法操作复杂，对仪器和样品要求高。近年来，应用非偏振红外光源的近似计算方法也一直在改进，如利用多个矿物颗粒的非偏振光光谱的平均计算（Xia Q. K. et al.，2013），基线处理方法与标准峰位的参照等（Qiu et al.，2018）。总体上影响羟基定量结果准确度的最重要因素是矿物摩尔吸收系数、谱图的基线处理及积分方法选择以及偏振-非偏振红外光近似方法造成的实验计算误差（Qiu et al.，2018）。

4.1.2 高温高压原位测试技术

地球内部的矿物岩石样品赋存环境极端，无法对其物理化学性质进行直接观测，只能通过数值模拟或高温高压实验模拟参照（刘曦等，2017）。金刚石对顶砧（diamond anvil cell，DAC）高压实验装置能与红外光谱显微镜联用实现原位观测，应用于矿物相变、含水特征、流体性质等研究。目前常规的红外光谱仪仍无法进行原位高温高压条件下的测试，仅有少数研究能提供高温常压、高压室温或淬火高温的反应环境（Yang Y. et al.，2015；Liu D. et al.，2019），导致基于红外光谱的高温高压原位实验研究难以开展。因此，通过对布鲁克 Vertex 70 V 真空型红外光谱仪进行拓展，构建原位控温控压红外测试系统（苏文等，2019），可直接在高压、高/低温条件下即时获取物质的红外光谱，达到模拟地球内部环境的目的，获取矿物岩石在深地环境中的性质。

4.1.3 空间分辨率提升技术

随着研究者对时间和空间分辨率的追求，焦平面阵列探测器的发展显著提升了测试效率；计算机傅里叶变换程序已可完成针对化学反应的毫秒甚至纳秒级别的采谱时间分辨；但样品的物理空间分辨率受制于 Abbe 衍射极限（$\sim \lambda/2$），限制了化学成分不均匀样品的微区分析。红外光谱波长从 0.78~1 000 μm，其中最常用的中红外光的波长在 2.5~25 μm，实际测试中还会受物镜数值孔径及探测器类型等硬件的限制，共聚焦信号分辨率一般仅能达到 10 μm，在指纹区甚至更低一些，很难与地学样品的扫描电子显微镜图像对比而提取有用信息。近年来红外光谱通过以下几种技术显著提升了空间分辨率。

（1）同步辐射光源：同步辐射光源亮度大，通过在测试中设置极小的光圈，可达到接近 Abbe 衍射极限的空间成像分辨率，在短波长区域可达 3~5 μm（凌盛杰等，2013）。这在固体地球科学领域能带来更精细的原位微纳米尺度的矿物结构关系信息（Qian et al.，2015）。鉴于其高亮度和准直性，在某些信噪比较差的 DAC 高压测试中，甚至可以直接使用垫片透光孔作为测试光圈。但对于陨石、星尘这类珍贵且常呈微米级尺寸的样品，仍需更高的物理分辨技术。

（2）光热诱导共振技术：Dazzi 等（Dazzi et al.，2015；Dazzi and Prater，2017）基于物质吸收红外光后发生膨胀的原理开发了红外光谱与原子力显微镜（AFM）的联用技术，其原理是将可调谐的红外光源与样品作用，再直接探测样品受热后产生的尺寸膨胀，再将其转换为频率信号。这种光热诱导共振（photo thermal-induced resonance，PTIR，又称 AFM-IR）技术，突破了传统的傅里叶变换红外光谱（FTIR）及衰减全反射红外光谱（attenuated total refraction infrared，ATR-IR）的分辨极限（Jin et al.，2019），创造性地使用 AFM 探针针尖来检测样品的红外吸收，从而将空间分辨率变为 AFM 针尖的尺寸、提高至 10 nm，大大超越了红外光的衍射极限，实现了纳米尺度下的红外光谱测定。因此，对于硬度较小的样品可使用光热诱导共振技术来进行红外响应基团的高分辨率测试。

（3）散射式扫描近场显微成像技术：AFM-IR 由于特殊的采谱机理与针尖结构特征，难以应用于质硬、脆且热膨胀系数较低的无机矿物（Dazzi et al.，2015）。21 世纪以来，散射式扫描近场显微成像技术（s-SNOM）的发展带来了分辨率的革命性提升，使用合适的红外光谱光源后的近场红外技术（near field infrared，NFIR）也突破了衍射极限的制约（Chen X. Z. et al.，2019）。此技术无需使用光圈限制光束，而是用远小于波长的"点"源和"点"探测器,将足够亮度的红外光限制于光学探针之内,将其控制在离样品足够近的距离(数十纳米)之内测试,将采谱分辨率提升为钨探针的针尖尺寸,随着光源和探测针材料的更新进步,理想条件下能实现 5 nm 的空间分辨率(Mastel et al., 2018)。每种技术的发展都是相辅相成且不断完善的,包括中国上海光源在内的多个国家都建设了同步辐射光源与 s-SNOM 的整合,形成了新一代 nano-FTIR 测试平台(Freitas et al., 2018;Zhou et al., 2019)。

4.1.4 展望

红外光谱仪虽然在水含量定量分析、高温高压原位测试和空间分辨率上取得显著的进步,但仍有诸多技术难题亟待突破:①不同构造环境的名义无水矿物含水结构不一,特别是各向异性矿物的含水量随晶轴取向差异明显,尚需进一步完善其吸收模型。②原位高温高压条件下的结构水定量研究亦有待深入。目前金刚石压砧在大于 650℃ 条件下的氧化防护尚未完全解决,限制了地幔深度环境高温条件的精准获取。③目前高分辨率红外光谱技术应用于传统地学类样品的研究较为有限,对于矿物中的痕量-微量结构水,很可能由于光热膨胀效应不明显导致光热诱导共振技术难以实施,而扫描近场红外可能对其更为有效。随着纳米红外技术的不断完善,未来有望在解译矿物微纳米结构、纳米区域结构水定量测试方面实现突破。

4.2 同步辐射

4.2.1 同步辐射发展历史与最新进展

X 射线是波长(0.001~10 nm)介于紫外和 γ 射线之间的电磁波,其能量大穿透力强,很好地填补了光学与电子显微镜之间的空白,具有高空间分辨率和厚样品的成像和分析潜力,因而被广泛用于医学成像诊断和 X 射线结晶学等领域(Bunaciu et al., 2015)。但同步辐射(synchrotron radiation,SR)的出现和发展,才真正让 X 射线在自然科学研究领域的应用大放异彩(马礼敦和杨福家,2005;Willmott, 2019)。

同步辐射是速度接近光速的带电粒子在弧形轨道上运动时沿切线发生的电磁辐射,是 1947 年在美国通用电器公司的 70 MeV 电子同步加速器中被首次观察到,也因此得名。长期以来,同步辐射并不受高能物理学家欢迎,因为它消耗了加速器的能量,阻碍粒子能量的提高。然而,同步辐射具有从远红外光到硬 X 射线范围内的连续光谱,以及高亮度、高准直性、高度极化和脉冲性等优异性能,可用来开展其他光源无法实现的许多前沿科学技术研究,因而被誉为继"电光源、X 射线源和激光光源"之后的第四种"人类神光"。

从 20 世纪 60 年代中叶开始,同步辐射光源逐渐得到深入研究并投入实际应用,迄今已历经四代:第一代是在世界各国为高能物理研究建造的储存环和加速器上"寄生地"运行的兼用光源,多建于 1965—1975 年。如美国康奈尔大学 CHESS 光源,北京同步辐射装置 BSRF。第二代是同步辐射专用光源,多建于 1975—1990 年,常采用电子储存环,利用弯转磁铁产生同步辐射,但通常能量较低。如美国布鲁克海文国家实验室 NSLS 光源(800 MeV),合肥国家同步辐射实验室 NSRL 光源(800 MeV)。从 20 世纪 90 年代开始,出现了大量使用插入件的新光源——第三代同步辐射光源,其亮度比第二代提高了 100 倍,比普通实验室的 X 射线光源要亮 1 亿倍以上,使同步辐射分析的时间和空间分辨率均大幅提高,为众多自然科学领域的研究带来前所未有的新机遇。如,硬 X 射线同步辐射的成像三维空间分辨率优于 30 nm,采用软 X 射线可达到 10~30 nm,而采用相干衍射成像可进一步将分辨率提升到约 5 nm(Hitchcock, 2015)。

目前世界上第三代是同步辐射光源研究和应用的主流,代表性的有美国阿贡国家实验室的 APS 光源(7 GeV)、欧洲同步辐射装置 ESRF(6 GeV)、日本的 SPring-8 光源(8 GeV)、上海同步辐射装置 SSRF(3.5 GeV)、英国 DIAMOND 光源(3 GeV)、西班牙 ALBA 光源(3 GeV)和加拿大国家光源 CLS(2.9 GeV)等。近年来,世界各国开始建设被认为是自由电子激光(free electron laser,FEL)和衍射极限同步辐射(diffraction limited synchrotron radiation,DLSR)的第四代光源。FEL 和 DLSR 不仅能产生无与伦比的高亮度辐射,而且辐射还具有完全或高度的横向相干性,并且是快脉冲式的(Khubbutdinov et al., 2019)。代表性的 FEL 有美国的 LCLS 光源、德国的 Euro XFEL 光源等,代表性的 DLSR 有瑞典的 MAX IV 光源等。

2019年6月29日，我国在北京怀柔开始建设第四代衍射极限同步辐射"北京光源"[图6（a）]。经过五六十年的发展，同步辐射在自然科学领域应用非常广泛，已成为生命、材料、环境、物理、化学、医药学和地学等领域基础和应用研究的一种最先进的、不可替代的工具（马礼敦和杨福家，2005；Willmott，2019）。

4.2.2 同步辐射相关技术及其在地学研究中的应用

同步辐射是一种优质光源，其工作原理与其他电磁波类似，当同步辐射光与样品作用后，通过对吸收、透射、衍射、散射等信号的检测和收集来获得样品的形貌结构和化学组成，甚至磁学性质和电子结构等信息[图6（b）]（Ice et al.，2011；Willmott，2019）。近年来，随着时间和空间分辨率的提高，X射线成像（二维和三维）技术与X射线光谱学（吸收、散射、荧光等）、X射线衍射技术等结合，将同步辐射发展成为一种功能强大的多模态成像和谱学分析技术，能同时给出样品的多尺度二维和三维形貌结构、化学成分及其元素价态以及动态行为等信息（Ice et al.，2011；Wang et al.，2011；袁清习等，2019；Witte et al.，2020）。

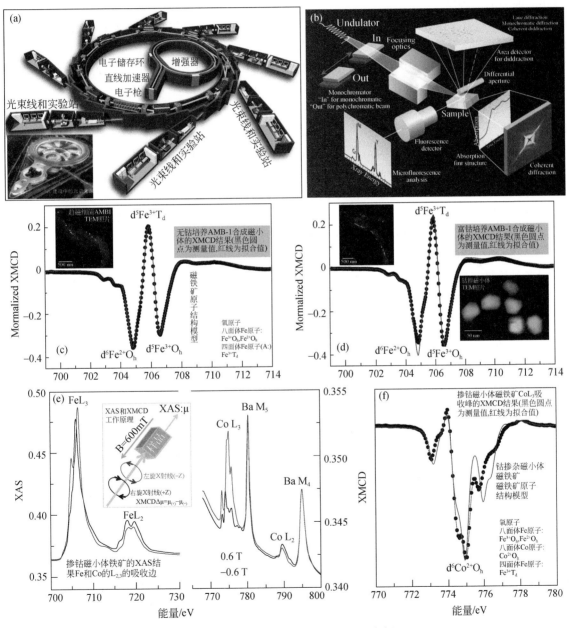

图6 同步辐射技术工作原理和应用实例

（a）常规同步辐射装置示意图，左下为正在建设中的第四代同步辐射——北京光源效果图；（b）同步辐射光的工作原理图（据Ice et al.，2011）；（c）~（f）利用同步辐射XAS和XMCD技术研究趋磁细菌合成钴掺杂磁小体磁铁矿的应用实例（据Li J. H. et al.，2016）

同步辐射技术在地学领域的应用主要有3个方面：①一些采用常规技术可以实现的研究，如激光红外和激光拉曼光谱、粉末X射线衍射（XRD）、X射线荧光（XRF）、X射线断层扫描（X-ray computed tomography，X-ray CT）、X射线散射（X-ray scattering）和穆斯堡尔谱学（Mössbauer spectroscopy）等，当采用同步辐射技术时，仅需要微量的样品，在非常短的分析时间内，就可获得高信噪比、高空间或高能量分辨率的信号（Bassett and Brown，1990；Artioli，2015；Quartieri，2015）。②有些技术是同步辐射特有的，如X射线吸收谱（X-ray absorption spectroscopy，XAS），原位高温高压XRD、XAS，以及X射线相干衍射成像（coherent X-ray diffraction imaging，CDI）技术等（Cosmidis and Benzerara，2014；Miao et al.，2015；Quartieri，2015；孙智斌等，2018）。③采用同步辐射技术，可以相对容易地实现对同一个样品在同一时刻进行两种或多种不同的分析（如XAS-XRD-XRF联用），一方面同时获得样品多种不同信息可提高效率，另一方面可有效地避免由于样品和时间造成的误差（Wogelius et al.，2011；van der Snickt et al.，2012；Edwards et al.，2018）。

由于地学样品通常复杂性和不均一程度高，有些年代久远不可再生，有些来之不易非常珍贵，同步辐射技术的高时间和高能量分辨率（相对于常规X射线仪器）、高空间分辨率（相对于光学显微镜）、无损（相对于电子显微镜）的X射线成像和光谱分析能力，使其在地学领域得到十分广泛的应用（Bassett and Brown，1990；Henderson et al.，1995；Fenter，2013；Quartieri，2015）。代表性的应用有：①利用高灵敏度的XAS及X射线磁圆二色（X-ray magnetic circular dichroism，XMCD）技术，并结合高空间分辨率的STXM、XRF和同步辐射红外光谱等技术，开展地质样品中元素分布和价态鉴定、生物矿化过程分析、古老岩石中微化石和有机质的识别鉴定、陨石和古老岩石中磁性矿物鉴定及其磁性研究［图6（c）~（f）］（Wang et al.，2011；Li J. H. et al.，2013，2016；Cosmidis and Benzerara，2014；Bryson et al.，2015；Alleon et al.，2018）。②利用同步辐射纳米成像技术，如基于CT成像（Micro-CT）和基于波带片的纳米分辨率全场透射X射线显微成像（transmission X-ray microscopy，TXM）等，对古生物化石、岩石、沉积物和陨石等样品的三维形貌观察和分析（Boller et al.，2010；Ni et al.，2013；Yin et al.，2015）。③利用原位高温高压同步辐射XRD和XAS技术，研究各种热力条件下的结构（如原子之间的亲和力、价键的强度等等），测量各种物性，以及物质之间的相互作用（如矿物之间的化学分配和化学成分等），从而在实验室模拟研究地球深部物质的物理化学性质（Saxena et al.，1995；Quartieri，2015；Liu，2016）。④利用高分辨率、高灵敏、无损伤的同步辐射X射线荧光成像和元素分析技术（有时候与其他同步辐射技术如XRD和XAS联用），研究古生物化石（如恐龙化石）、岩石和陨石等各类样品的二维/三维形貌，以及定性和定量研究样品的元素种类及其分布特征（Bergmann et al.，2010；Wogelius et al.，2011；Edwards et al.，2018；Manning et al.，2019）。

4.2.3 我国同步辐射发展现状与未来前景

经过了60多年的发展，同步辐射的理论和实验技术日臻成熟，同步辐射装置成了世界各国最具代表性的大型科研装置之一。同步辐射光源对科学技术发展的影响力受到科技界和各国政府的广泛认同和高度重视。我国的同步辐射技术的发展基本与世界同步，也拥有第一、二和三代同步辐射光源，但发展进程、研究深度和应用广度均与世界先进国家有很大的差距，具体表现在4个方面：①数量的差别。据不完全统计，目前全世界约有60多个同步辐射装置，分布在20多个国家和地区（https://lightsources.org/）。我国目前拥有同步辐射光源4个（大陆3个、台湾地区1个），而发达国家如美国拥有15个、日本拥有14个、德国拥有8个等。②我国的同步辐射光源及光束线和实验站建设相对滞后。美、德、日、英、法和意大利等发达国家从20世纪90年代起就开始大规模建设第三代中高能同步辐射光源，以日本大型同步辐射设施Spring-8为例：它始建于1991年，1997年竣工并开放设施，到2010年共建成55条光束线站，截至2009年6月，服务于日本及全世界科学、工业、企业界超过10万人次（http://www.spring8.or.jp/en/）。相比较，以我国第三代同步辐射-上海光源（SSRF）建设为例：它于2004年12月开工建设，2009年4月竣工，5月正式对用户开放，首批建成7条光束线站。2015年开始二期建设，截至2018年底，共有15条光束线站19个实验室开放运行。截至2019年4月底，接待用户约4.1万人次。目前，上海光源还在进行后续工程建设，预计到2022年，约35条光束线站和60个实验室站投入运行（http：//ssrf. sinap. cas. cn/）。③与同步辐射在物理、化学、材料、生命和环境等学科的广泛应用相比，中国地学领域对同步辐射的认知度和参与度都不高，目前仅有少数的研究者使用同步辐射装置开展研究（Li J. H. et al., 2013; Ni et al., 2013; Peng et al., 2015; Yin et al., 2015）。④近年来，随着我国对科研支持力度的不断加大，有关先进同步辐射光源的建设已被列入《国家重大科技基础设施建设中长期规划（2011—2030）》。目前，第四代高能同步辐射光源——北京光源正在建设中，预计2025年建成，届时将拥有60~70条光束线和90多个实验站（Crease, 2019）。此外，合肥和东莞等地也在积极筹备第四代同步辐射光源。未来十年，将是中国同步辐射蓬勃发展的时期，地学研究者应抓住这个百年不遇的机会，积极参与同步辐射光源、光束线和实验站的建设，让同步辐射这一大型实验装置真正成为推动中国地球科学未来发展的技术支柱。

5 结　　语

"更精、更准、更细、更多、更快"是微束分析的重要发展趋势。基体效应是微区元素和同位素定量分析的共性问题，解决这个问题的最直接方法是用基体匹配的标样进行校正。地质样品具有复杂性和多样性，对高质量微束分析标样的要求尤为突出。虽然我们在微束分析标样研发方面已经取得一些重要进展，但是高质量的标样匮乏仍然是微束分析面临的共同问题，需要加大投入进行研发。解决基体效应的另一个方法是深入研究基体效应的机理和抑制乃至消除的途径，这对于某些特定矿物的特定分析是有可能实现的。

提高空间分辨率是微束分析应用的迫切需求和技术发展一个重要方向。一方面，仪器厂商不断地努力提高仪器性能，另一方面，新技术、新方法的研发以及也能够突破现有高空间分辨率的物理极限，如在没有明显损失分析精度的情况下，LG-SIMS已经实现了2 μm尺度的精确锆石U-Pb定年；在微区图像分析方面，引入图像数据处理算法优化可以明显提高空间分辨率。

多种微束分析技术的联用是在同一个微区内高效率获取多种信息非常有效的方法，如透射电子显微镜与荧光显微镜技术联用、纳米离子探针与原子力显微镜联用等，特别是同步辐射与其他多种显微学技术联用等，这是今后一个非常值得进一步研发和应用的方向。

微束分析技术不仅在地球科学领域发挥着越来越重要的作用，而且在生物、医学、材料科学等其他领域也是不可或缺的，其中很多微束分析技术方法比在地球科学领域的应用更早、更成熟。例如，TOF-SIMS、LA-ICP-TOFMS等技术已非常成熟，并广泛用于很多学科领域的定性分析、（半）定量分析、深度剖面分析及图像分析中，但在我国地学领域中还刚刚起步。我们要虚心地向其他学科领域的微束分析专家们学习，将其先进思想、技术和方法引入地球科学领域研究中，"它山之石可以攻玉"。

过去十年里，我国地球科学领域引进了一批高性能新型微束分析仪器设备，建立了一批高规格的实验室，一批高水平微束分析研究人才也在迅速成长。然而，与"硬件"方面的投入相比，人才队伍建设和新技术新方法研发等"软件"方面的投入相对不足或滞后，这是必须重视和解决的一个重要问题。"科学引领、技术先行"，人才是关键。

参 考 文 献

边晨光, 王利, 王艳秋, 宋哲, 刘本康. 2015. 飞秒激光用作电离源的二次中性粒子质谱技术. 分析化学, 43(8): 1241-1246

陈方, 董熙平. 2009. 应用同步加速器X射线层析扫描显微（技术）揭示早寒武世化石胚胎Olivooides的内部结构. 科学通报, 54(1): 67-72

陈开运, 袁洪林, 包志安, 范超, 柳小明, 宋佳瑶. 2012. 人工合成锆石Lu-Hf同位素标样方法研究. 岩石矿物学杂志, 31(2): 279-288

陈开运, 范超, 袁洪林, 包志安, 宗春蕾, 戴梦宁, 凌雪, 杨颖. 2013. 飞秒激光剥蚀-多接收电感耦合等离子谱原位微区分析青铜中铅同位素组成——以古铜钱币为例. 光谱学与光谱分析, 33(5): 1342-1349

陈莉, 徐军, 陈晶. 2015. 扫描电子显微镜显微分析技术在地球科学中的应用. 中国科学：地球科学, 45(9): 1347-1358

付乐兵, 魏俊浩, 张道涵, 谭俊, 田宁, 赵志新. 2015. 单个流体包裹体成分 LA-ICPMS 分析与矿床学应用进展. 中南大学学报(自然科学版), 46(10): 3832-3840

高钰涯, 李秋立, 刘宇, 唐国强, 凌潇潇, 李献华. 2016. 离子探针第四纪锆石 U-Pb 和 U-Th 定年方法及应用. 第四纪研究, 36(5): 1015-1026

郭伟, 林贤, 胡圣虹. 2020. 单个流体包裹体 LA-ICP-MS 分析及应用进展. 地球科学, 45(4): 1362-1374

侯可军, 李延河, 肖应凯, 刘峰, 田有荣. 2010. LA-MC-ICP-MS 硼同位素微区原位测试技术. 科学通报, 55(22): 2207-2213

侯可军, 秦燕, 李延河, 范昌福. 2013. 磷灰石 Sr-Nd 同位素的激光剥蚀-多接收器电感耦合等离子体质谱微区分析. 岩矿测试, 32(4): 547-554

惠鹤九, 徐永江, 潘明恩. 2016. 名义上无水矿物的水含量及其地质应用. 中国科学: 地球科学, 46(5): 639-656

蓝廷广, 胡瑞忠, 范宏瑞, 毕献武, 唐燕文, 周丽, 毛伟, 陈应华. 2017. 流体包裹体及石英 LA-ICP-MS 分析方法的建立及其在矿床学中的应用. 岩石学报, 33(10): 3239-3262

李金华, 潘永信. 2015. 透射电子显微镜在地球科学研究中的应用. 中国科学: 地球科学, 45(9): 1359-1382

李玲, 夏小平, 杨晴, 李如操. 2016. 锆石 SIMS 原位微区 U-Th 不平衡定年: 以云南腾冲火山岩为例. 地球化学, 45(4): 398-406

李献华, 唐国强, 龚冰, 杨岳衡, 侯可军, 胡兆初, 李秋立, 刘宇, 李武显. 2013. Qinghu(清湖)锆石: 一个新的 U-Pb 年龄和 O, Hf 同位素微区分析工作标样. 科学通报, 58(20): 1954-1961

李献华, 刘宇, 汤艳杰, 高钰涯, 李秋立, 唐国强. 2015. 离子探针 Li 同位素微区原位分析技术与应用. 地学前缘, 22(5): 160-170

李小犁, 陶仁彪, 李清云, 朱建江, 张立飞. 2019. 石榴子石 Fe^{3+} 含量电子探针原位分析: Flank Method 方法的实例应用. 岩石学报, 35(4): 1058-1070

李艳青, 余振兵, 马昌前. 2011. 石英 SEM-CL 微结构及其在岩石学中的应用. 地球科学进展, 26(3): 325-331

李杨, 杨岳衡, 焦淑娟, 吴福元, 杨进辉, 谢烈文, 黄超. 2016. 金红石 Hf 同位素激光原位多接收等离子体质谱(LA-MC-ICP-MS)测定. 中国科学: 地球科学, 46(6): 857-869

凌盛杰, 黄郁芳, 黄蕾, 邵正中, 陈新. 2013. 同步辐射红外显微光谱学和成像技术在分析化学中的应用. 化学进展, 25(5): 821-831

刘文庆, 刘庆冬, 顾剑锋. 2013. 原子探针层析技术(APT)最新进展及应用. 金属学报, 49(9): 1025-1031

刘曦, 代立东, 邓力维, 范大伟, 刘琼, 倪怀玮, 孙樯, 巫翔, 杨晓志, 翟双猛, 张宝华, 张莉, 李和平. 2017. 近十年我国在地球内部物质高压物性实验研究方面的主要进展. 高压物理学报, 31(6): 657-681

刘勇胜, 胡兆初, 李明, 高山. 2013. LA-ICP-MS 在地质样品元素分析中的应用. 科学通报, 58(36): 3753-3769

马黎春, 汤庆峰, 张琪, 赵艳军, 孙小虹, 王鑫, 任彩霞. 2014. 蒸发岩矿物单个流体包裹体成分测定方法研究进展. 地球科学进展, 29(4): 475-481

马礼敦, 杨福家. 2005. 同步辐射应用概论. 2 版. 上海: 复旦大学出版社

盛英明, 龚冰, 李万财, 夏梅. 2016. 名义上无水矿物中微量结构水的分析方法研究进展. 中国科学: 地球科学, 46(4): 443-453

宋佳瑶, 袁洪林, 包志安, 陈开运, 贺国芬, 范超. 2011. 高温熔融研制钾长石玻璃标准物质初探. 岩矿测试, 30(4): 406-411

苏文, 刘振先, 陈菲, 高静, 李晓光. 2019. 构建红外光谱显微系统实验平台——原位模拟物质反应动力学过程. 岩石学报, 35(1): 252-260

孙小虹, 胡明月, 刘成林, 焦鹏程, 马黎春, 王鑫, 詹秀春. 2013. 激光剥蚀 ICP-MS 法测定盐类矿物单个流体包裹体的成分. 分析化学, 41(2): 235-241

孙智斌, 范家东, 江怀东. 2018. X 射线自由电子激光单颗粒成像研究. 物理, 47(8): 491-502

汪双双, 韩延兵, 李艳广, 魏小燕, 靳梦琪, 程秀花. 2016. 利用 LA-ICP-MS 在 16 μm 和 10 μm 激光束斑条件下测定独居石 U-Th-Pb 年龄. 岩矿测试, 35(4): 349-357

王娟, 陈意, 毛骞, 李秋立, 马玉光, 石永红, 宋传中. 2017. 金红石微量元素电子探针分析. 岩石学报, 33(6): 1934-1946

温汉捷, 罗重光, 杜胜江, 于文修, 顾汉念, 凌坤跃, 崔燚, 李阳, 杨季华. 2020. 碳酸盐黏土型锂资源的发现及意义. 科学通报, 65(1): 53-59

吴润秋, 饶灿, 王琪, 张迪. 2020. 关键金属铍的电子探针分析. 科学通报, 65(20): 2161-2168

肖益林, 黄建, 刘磊, 李东永. 2011. 金红石: 重要的地球化学"信息库". 岩石学报, 27(2): 398-416

徐娟, 杨守业, 胡兆初, 罗涛, 黄湘通. 2016. XRF 和 LA-ICPMS 测定硫化物熔片中的主次量元素. 光谱学与光谱分析, 36(11): 3683-3688

杨文武, 史宇宁, 商琦, 张文, 胡兆初. 2017. 飞秒激光剥蚀电感耦合等离子体质谱在地球科学中的应用进展. 光谱学与光谱分析, 37(7): 2192-2198

杨燕. 2017. 名义上无水矿物的原位高温分子光谱. 矿物岩石地球化学通报, 36(1): 48-58

于亮, 朱亚林, 闫昭圣, 吴汉宁. 2016. 环境扫描电镜在石油地质研究中的应用. 电子显微学报, 35(6): 561-566

于倩. 2015. 石盐中单个流体包裹体 LA-ICP-MS 测试方法研究. 北京: 中国地质大学(北京)硕士学位论文

余金杰, 陈振宇, 王平安, 李晓峰, 黄建平, 王辉. 2006. 苏北榴辉岩中金红石的微量元素地球化学特征. 岩石学报, 22(7): 1883-1890

袁继海, 詹秀春, 范晨子, 赵令浩, 孙冬阳, 贾泽荣, 胡明月, 蒯丽君. 2012. 玻璃标样结合硫内标归一定量技术在激光剥蚀-等离子体质谱分析硫化物矿物中的应用. 分析化学, 40(2): 201-207

袁清习, 邓彪, 关勇, 张凯, 刘宜晋. 2019. 同步辐射纳米成像技术的发展与应用. 物理, 48(4): 205-218

张迪, 陈意, 毛骞, 苏斌, 贾丽辉, 郭顺. 2019. 电子探针分析技术进展及面临的挑战. 岩石学报, 35(1): 261-274

张贵宾, 张立飞. 2011. 变质岩中金红石研究进展及存在问题. 地学前缘, 18(2): 26-32

张文兰, 王汝成, 蔡淑月. 2006. 超轻元素 Be 元素的电子探针定量分析——以绿柱石为例. 电子显微学报, 25(S1): 293-294

张文兰, 车旭东, 王汝成, 谢磊, 李晓峰, 张迪. 2020. 超轻元素铍的电子探针定量分析最佳条件探索——以绿柱石为例. 科学通报, doi: 10.1360/TB-2020-0316

周广荣. 2009. 利用扫描电镜冷台对植物样品观察初探. 电子显微学报, 28(2): 186-189

周红英, 耿建珍, 崔玉荣, 李怀坤, 李惠民. 2012. 磷灰石微区原位 LA-MC-ICP-MS U-Pb 同位素定年. 地球学报, 33(6): 857–864

周红英, 李怀坤, 崔玉荣, 耿建珍, 张健, 李惠民. 2013. 金红石 U-Pb 同位素定年技术研究. 地质学报, 87(9): 1439–1446

Alleon J, Bernard S, Le Guillou C, Beyssac O, Sugitani K, Robert F. 2018. Chemical nature of the 3.4 Ga Strelley Pool microfossils. Geochemical Perspectives Letters, 7: 37–42

Artioli G. 2015. Powder diffraction and synchrotron radiation. In: Mobilio S, Boscherini F, Meneghini C (eds). Synchrotron Radiation: Basics, Methods and Applications. Berlin, Heidelberg: Springer, 319–336

Audétat A, Garbe-Schönberg D, Kronz A, Pettke P, Rusk B, Donovan J J, Lowers H A. 2015. Characterisation of a natural quartz crystal as a reference material for microanalytical determination of Ti, Al, Li, Fe, Mn, Ga and Ge. Geostandards and Geoanalytical Research, 39(2): 171–184

Azadbakht Z, Lentz D R. 2020. High-resolution LA-ICP-MS trace-element mapping of magmatic biotite: a new approach for studying SYN-To postmagmatic evolution. The Canadian Mineralogist, 58(3): 293–311

Bao Z A, Yuan H L, Zong C L, Liu Y, Chen K Y, Zhang Y L. 2016a. Simultaneous determination of trace elements and lead isotopes in fused silicate rock powders using a boron nitride vessel and fsLA-(MC)-ICP-MS. Journal of Analytical Atomic Spectrometry, 31(4): 1012–1022

Bao Z A, Yuan W T, Yuan H L, Liu X, Chen K Y, Zong C L. 2016b. Non-matrix-matched determination of lead isotope ratios in ancient bronze artifacts by femtosecond laser ablation multi-collector inductively coupled plasma mass spectrometry. International Journal of Mass Spectrometry, 402: 12–19

Bao Z A, Chen L, Zong C L, Yuan H L, Chen K Y, Dai M N. 2017. Development of pressed sulfide powder tablets for *in situ* sulfur and lead isotope measurement using LA-MC-ICP-MS. International Journal of Mass Spectrometry, 421: 255–262

Bao Z A, Zong C L, Chen L, Lei D B, Chen K Y, Yuan H L. 2018. Determination of lead isotope ratios in Mn-Fe-rich nodules by laser ablation multi-collector inductively coupled plasma mass spectrometry. Journal of Analytical Atomic Spectrometry, 33(12): 2143–2152

Bao Z M, Shi Y R, Anderson J L, Kennedy A, Ke Z K, Gu X P, Wang P Z, Che X C, Kang Y L, Sun H Y, Wang C. 2020. Petrography and chronology of lunar meteorite Northwest Africa 6950. Science China Information Sciences, 63(4): 140902

Bassett W A, Brown Jr G E. 1990. Synchrotron radiation: applications in the earth sciences. Annual Review of Earth and Planetary Sciences, 18: 387–447

Batanova V G, Sobolev A V, Kuzmin D V. 2015. Trace element analysis of olivine: high precision analytical method for JEOL JXA-8230 electron probe microanalyser. Chemical Geology, 419: 149–157

Batson P E, Dellby N, Krivanek O L. 2002. Sub-ångstrom resolution using aberration corrected electron optics. Nature, 418(6898): 617–620

Benzerara K, Menguy N, López-García P, Yoon T H, Kazmierczak J, Tyliszczak T, Guyot F, Brown Jr G E. 2006. Nanoscale detection of organic signatures in carbonate microbialites. Proceedings of the National Academy of Sciences of the United States of America, 103(25): 9440–9445

Berger D, Nissen J. 2014. Measurement and Monte Carlo simulation of the spatial resolution i element analysis with the FEG-EPMA JEOL JXA-8530F. IOP Conference Series: Materials Science and Engineering, 55: 012002, doi: 10.1088/1757-899X/55/1/012002

Bergmann U, Morton R W, Manning P L, Sellers W I, Farrar S, Huntley K G, Wogelius R A, Larson P. 2010. Archaeopteryx feathers and bone chemistry fully revealed via synchrotron imaging. Proceedings of the National Academy of Sciences of the United States of America, 107(20): 9060–9065

Bocker C, Kouli M, Völksch G, Rüssel C. 2014. New insights into the crystallization of cordierite from a stoichiometric glass by *in situ* high-temperature SEM. Journal of Materials Science, 49(7): 2795–2801

Boller E, Cloetens P, Baruchel J, Tafforeau P, Rozenbaum O, Pourchez J. 2010. Synchrotron X-ray microtomography: a high resolution, fast and quantitative tool for rock characterization. In: Desrues J, Viggiani G, Bésuelle P (eds). Advances in X-ray Tomography for Geomaterials. London: ISTE, 125–133

Brison J, Robinson M A, Benoit D S W, Muramoto S, Stayton P S, Castner D G. 2013. TOF-SIMS 3D imaging of native and non-native species within hela cells. Analytical Chemistry, 85(22): 10869–10877

Bryson J F J, Nichols C I O, Herrero-Albillos J, Kronast F, Kasama T, Alimadadi H, van der Laan G, Nimmo F, Harrison R J. 2015. Long-lived magnetism from solidification-driven convection on the pallasite parent body. Nature, 517(7535): 472–475

Bunaciu A A, Udriștioiu E G, Aboul-Enein H Y. 2015. X-ray diffraction: instrumentation and applications. Critical Reviews in Analytical Chemistry, 45(4): 289–299

Cai L S, Sheng L F, Xia M C, Li Z P, Zhang S C, Zhang X R, Chen H Y. 2017a. Graphene oxide as a novel evenly continuous phase matrix for TOF-SIMS. Journal of the American Society for Mass Spectrometry, 28(3): 399–408

Cai L S, Xia M C, Wang Z Y, Zhao Y B, Li Z P, Zhang S C, Zhang X R. 2017b. Chemical visualization of sweat pores in fingerprints using GO-enhanced TOF-SIMS. Analytical Chemistry, 89(16): 8372–8376

Cao M J, Evans N J, Reddy S M, Fougerouse D, Hollings P, Saxey D W, McInnes B I A, Cooke D R, McDonald B J, Qin K Z. 2019. Micro- and nano-scale textural and compositional zonation in plagioclase at the Black Mountain porphyry Cu deposit: implications for magmatic processes. American Mineralogist, 104(3): 391–402

Chang J, Li J W, Audétat A. 2018. Formation and evolution of multistage magmatic-hydrothermal fluids at the Yulong porphyry Cu-Mo deposit, eastern Tibet: insights from LA-ICP-MS analysis of fluid inclusions. Geochimica et Cosmochimica Acta, 232: 181–205

Che X D, Wu F Y, Wang R C, Gerdes A, Ji W Q, Zhao Z H, Yang J H, Zhu Z Y. 2015. *In situ* U-Pb isotopic dating of columbite-tantalite by LA-ICP-MS. Ore Geology Reviews, 65: 979–989

Checa A G, Yáñez-Ávila M E, González-Segura A, Varela-Feria F, Griesshaber E, Schmahl W W. 2019. Bending and branching of calcite laths in the foliated microstructure of pectinoidean bivalves occurs at coherent crystal lattice orientation. Journal of Structural Biology, 205(3): 7–17

Chehreh Chelgani S, Hart B. 2014. TOF-SIMS studies of surface chemistry of minerals subjected to flotation separation-a review. Minerals Engineering, 57: 1-11

Chehreh Chelgani S, Hart B, Xia L. 2013. A TOF-SIMS surface chemical analytical study of rare earth element minerals from micro-flotation tests products. Minerals Engineering, 45: 32-40

Chen F C, Deng J, Wang Q F, Huizenga J M, Li G J, Gu Y W. 2020. LA-ICP-MS trace element analysis of magnetite and pyrite from the Hetaoping Fe-Zn-Pb skarn deposit in Baoshan Block, SW China: implications for ore-forming processes. Ore Geology Reviews, 117: 103309

Chen K Y, Yuan H L, Bao Z A, Zong C L, Dai M N. 2014. Precise and accurate in situ determination of lead isotope ratios in NIST, USGS, MPI-DING and CGSG glass reference materials using femtosecond laser ablation MC-ICP-MS. Geostandards and Geoanalytical Research, 38(1): 5-21

Chen L, Liu Y S, Hu Z C, Gao S, Zong K Q, Chen H H. 2011. Accurate determinations of fifty-four major and trace elements in carbonate by LA-ICP-MS using normalization strategy of bulk components as 100%. Chemical Geology, 284(3-4): 283-295

Chen L, Chen K Y, Bao Z A, Liang P, Sun T T, Yuan H L. 2017. Preparation of standards for in situ sulfur isotope measurement in sulfides using femtosecond laser ablation MC-ICP-MS. Journal of Analytical Atomic Spectrometry, 32(1): 107-116

Chen M H, Bagas L, Liao X, Zhang Z Q, Li Q L. 2019. Hydrothermal apatite SIMS Th-Pb dating: constraints on the timing of low-temperature hydrothermal Au deposits in Nibao, SW China. Lithos, 324-325: 418-428

Chen P W, Zeng Q D, Zhou T C, Wang Y B, Yu B, Chen J Q. 2019. Evolution of fluids in the Dasuji porphyry Mo deposit on the northern margin of the North China Craton: constraints from Microthermometric and LA-ICP-MS analyses of fluid inclusions. Ore Geology Reviews, 104: 26-45

Chen W, Lu J, Jiang S Y, Zhao K D, Duan D F. 2017. In situ carbon isotope analysis by laser ablation MC-ICP-MS. Analytical Chemistry, 89(24): 13415-13421

Chen X Z, Hu D B, Mescall R, You G J, Basov D N, Dai Q, Liu M K. 2019. Modern scattering-type scanning near-field optical microscopy for advanced material research. Advanced Materials, 31(24): 1804774

Cheng L N, Zhang C, Li X Y, Almeev R R, Yang X S, Holtz F. 2019. Improvement of electron probe microanalysis of boron concentration in silicate glasses. Microscopy and Microanalysis, 25(4): 874-882

Cosmidis J, Benzerara K. 2014. Soft X-ray scanning transmission micro-spectroscopy. In: Gower L, DiMasi E (eds). Handbook of Biomineralization. London: Taylor and Francis

Cosmidis J, Benzerara K, Gheerbrant E, Estève I, Bouya B, Amaghzaz M. 2013. Nanometer-scale characterization of exceptionally preserved bacterial fossils in Paleocene phosphorites from Ouled Abdoun (Morocco). Geobiology, 11(2): 139-153

Costa C, Webb R, Palitsin V, Ismail M, de Puit M, Atkinson S, Bailey M J. 2017. Rapid, secure drug testing using fingerprint development and paper spray mass spectrometry. Clinical Chemistry, 63(11): 1745-1752

Crease R P. 2019. China's next big thing. Physics World, 32(8): 21-22

Cui J Q, Yang S Y, Jiang S Y, Xie J. 2019. Improved accuracy for trace element analysis of Al and Ti in quartz by electron probe microanalysis. Microscopy and Microanalysis, 25(1): 47-57

Dai M N, Bao Z A, Chen K Y, Zong C L, Yuan H L. 2017. Simultaneous measurement of major, trace elements and Pb isotopes in silicate glasses by laser ablation quadrupole and multi-collector inductively coupled plasma mass spectrometry. Journal of Earth Science, 28(1): 92-102

Dazzi A, Prater C B. 2017. AFM-IR: technology and applications in nanoscale infrared spectroscopy and chemical imaging. Chemical Reviews, 117(7): 5146-5173

Dazzi A, Saunier J, Kjoller K, Yagoubi N. 2015. Resonance enhanced AFM-IR: a new powerful way to characterize blooming on polymers used in medical devices. International Journal of Pharmaceutics, 484(1-2): 109-114

Deng X D, Li J W, Zhao X F, Hu Z C, Hu H, Selby D, de Souza Z S. 2013. U-Pb isotope and trace element analysis of columbite-(Mn) and zircon by laser ablation ICP-MS: implications for geochronology of pegmatite and associated ore deposits. Chemical Geology, 344: 1-11

Donovan J J, Lowers H A, Rusk B G. 2011. Improved electron probe microanalysis of trace elements in quartz. American Mineralogist, 96(2-3): 274-282

Edwards N P, Webb S M, Krest C M, van Campen D, Manning P L, Wogelius R A, Bergmann U. 2018. A new synchrotron rapid-scanning X-ray fluorescence (SRS-XRF) imaging station at SSRL beamline 6-2. Journal of Synchrotron Radiation, 25(5): 1565-1573

Feng Y T, Zhang W, Hu Z C, Liu Y S, Chen K, Fu J L, Xie J Y, Shi Q H. 2018. Development of sulfide reference materials for in situ platinum group elements and S-Pb isotope analyses by LA-(MC)-ICP-MS. Journal of Analytical Atomic Spectrometry, 33(12): 2172-2183

Fenter P. 2003. Synchrotron radiation: Earth, environmental and material sciences applications: Grant S. Henderson and Don R. Baker, Editors (Short Course Series, Vol. 30, Robert Raeside, Series Editor), Mineralogical Association of Canada, Ontario, Canada, 2002, 178 pp., ISBN 0-921294-30-1 ($40). Chemical Geology, 194(4): 349-350

Fleming Y, Wirtz T, Gysin U, Glatzel T, Wegmann U, Meyer E, Maier U, Rychen J. 2011. Three dimensional imaging using secondary ion mass spectrometry and atomic force microscopy. Applied Surface Science, 258(4): 1322-1327

Foley S F, Prelevic D, Rehfeldt T, Jacob D E. 2013. Minor and trace elements in olivines as probes into early igneous and mantle melting processes. Earth and Planetary Science Letters, 363: 181-191

Freitas R O, Deneke C, Maia F C B, Medeiros H G, Moreno T, Dumas P, Petroff Y, Westfahl H. 2018. Low-aberration beamline optics for synchrotron infrared nanospectroscopy. Optics Express, 26(9): 11238-11249

Frelinger S N, Ledvina M D, Kyle J R, Zhao D G. 2015. Richard. Scanning electron microscopy cathodoluminescence of quartz: principles, techniques and applications in ore geology. Ore Geology Reviews, 65: 840-852

Frisz J F, Choi J S, Wilson R L, Harley B A C, Kraft M L. 2012. Identifying differentiation stage of individual primary hematopoietic cells from mouse bone marrow by multivariate analysis of TOF-secondary ion mass spectrometry data. Analytical Chemistry, 84(10): 4307-4313

Fu J L, Hu Z C, Zhang W, Yang L, Liu Y S, Li M, Zong K Q, Gao S, Hu S H. 2016. *In situ* sulfur isotopes ($\delta^{34}S$ and $\delta^{33}S$) analyses in sulfides and elemental sulfur using high sensitivity cones combined with the addition of nitrogen by laser ablation MC-ICP-MS. Analytica Chimica Acta, 911: 14-26

Fu J L, Hu Z C, Li J W, Yang L, Zhang W, Liu Y S, Li Q L, Zong K Q, Hu S H. 2017. Accurate determination of sulfur isotopes ($\delta^{33}S$ and $\delta^{34}S$) in sulfides and elemental sulfur by femtosecond laser ablation MC-ICP-MS with non-matrix matched calibration. Journal of Analytical Atomic Spectrometry, 32(12): 2341-2351

Giovanardi T, Girardi V A V, Correia C T, Tassinari C C G, Sato K, Cipriani A, Mazzucchelli M. 2017. New U-Pb SHRIMP-II zircon intrusion ages of the Cana Brava and Barro Alto layered complexes, central Brazil: constraints on the genesis and evolution of the Tonian Goias stratiform complex. Lithos, 282-283: 339-357

Gou L F, Jin Z D, Deng L, He M Y, Liu C Y. 2018. Effects of different cone combinations on accurate and precise determination of Li isotopic composition by MC-ICP-MS. Spectrochimica Acta Part B: Atomic Spectroscopy, 146: 1-8

Gou L L, Zi J W, Dong Y P, Liu X M, Li Z H, Xu X F, Zhang C L, Liu L, Long X P, Zhao Y H. 2019. Timing of two separate granulite-facies metamorphic events in the Helanshan Complex, North China Craton: constraints from monazite and zircon U-Pb dating of pelitic granulites. Lithos, 350-351: 105216

Gu X Y, Wang P Y, Kuritani T, Hanski E, Xia Q K, Wang Q Y. 2019. Low water content in the mantle source of the Hainan plume as a factor inhibiting the formation of a large igneous province. Earth and Planetary Science Letters, 515: 221-230

Haider M, Uhlemann S, Schwan E, Rose H, Kabius B, Urban K. 1998. Electron microscopy image enhanced. Nature, 392(6678): 768-769

Hao J L, Yang W, Luo Y, Hu S, Yin Q Z, Lin Y T. 2016. NanoSIMS measurements of trace elements at the micron scale interface between zircon and silicate glass. Journal of Analytical Atomic Spectrometry, 31(12): 2399-2409

Hao J L, Yang W, Huang W J, Xu Y C, Lin Y T, Changela H. 2020. NanoSIMS measurements of sub-micrometer particles using the local thresholding technique. Surface and Interface Analysis, 52(5): 234-239

Hao Y T, Xia Q K, Jia Z B, Zhao Q C, Li P, Feng M, Liu S C. 2016. Regional heterogeneity in the water content of the Cenozoic lithospheric mantle of eastern China. Journal of Geophysical Research, 121(2): 517-537

He M H, Xia X P, Huang X L, Ma J L, Zou J Q, Yang Q, Yang F, Zhang Y Q, Yang Y A, Wei G J. 2020. Rapid determination of the original boron isotopic composition from altered basaltic glass by *in situ* secondary ion mass spectrometry. Journal of Analytical Atomic Spectrometry, 35(2): 238-245

He T, Ni Q, Miao Q, Li M. 2018. Effects of cone combinations on the signal enhancement by nitrogen in LA-ICP-MS. Journal of Analytical Atomic Spectrometry, 33(6): 1021-1030

He Z W, Huang F, Yu H M, Xiao Y L, Wang F Y, Li Q L, Xia Y, Zhang X C. 2016. A flux-free fusion technique for rapid determination of major and trace elements in silicate rocks by LA-ICP-MS. Geostandards and Geoanalytical Research, 40(1): 5-21

Henderson C M B, Cressey G, Redfern S A T. 1995. Geological applications of synchrotron radiation. Radiation Physics and Chemistry, 45(3): 459-481

Hitchcock A P. 2015. Soft X-ray spectromicroscopy and ptychography. Journal of Electron Spectroscopy and Related Phenomena, 200: 49-63

Hochella Jr M F. 2008. Nanogeoscience: from origins to cutting-edge applications. Elements, 4(6): 373-379

Hochella Jr M F, Mogk D W, Ranville J, Allen I C, Luther G W, Marr L C, McGrail B P, Murayama M, Qafoku N P, Rosso K M, Sahai N, Schroeder P A, Vikesland P, Westerhoff P, Yang Y. 2019. Natural, incidental, and engineered nanomaterials and their impacts on the Earth system. Science, 363(6434): eaau8299

Hofer H E, Brey G P, Schulz-Dobrick B, Oberhänsli R. 1994. The determination of the oxidation state of iron by the electron microprobe. European Journal of Mineralogy, 6(3): 407-418

Hu P P, Hou X J, Zhang J B, Li S P, Wu H, Damø A J, Li H Q, Wu Q S, Xi X G. 2018. Distribution and occurrence of lithium in high-alumina-coal fly ash. International Journal of Coal Geology, 189: 27-34

Hu S, Lin Y, Zhang J, Hao J, Feng L, Xu L, Yang W, Yang J. 2014. NanoSIMS analyses of apatite and melt inclusions in the GRV 020090 Martian meteorite: hydrogen isotope evidence for recent past underground hydrothermal activity on Mars. Geochimica et Cosmochimica Acta, 140: 321-333

Hu S, Lin Y T, Zhang J C, Hao J L, Yang W, Deng L W. 2015. Measurements of water content and D/H ratio in apatite and silicate glasses using a NanoSIMS 50L. Journal of Analytical Atomic Spectrometry, 30(4): 967-978

Hu S, Lin Y T, Yang W, Wang W R Z, Zhang J C, Hao J L, Xing W F. 2016. NanoSIMS imaging method of zircon U-Pb dating. Science China Earth Sciences, 59(11): 2155-2164

Hu S, Lin Y T, Zhang J C, Hao J L, Yamaguchi A, Zhang T, Yang W, Changela H. 2020. Volatiles in the martian crust and mantle: clues from the NWA 6162 shergottite. Earth and Planetary Science Letters, 530: 115902

Hu Z C, Qi L. 2014. 15.5-Sample Digestion Methods, Volume 15: Analytical Geochemistry/Inorganic INSTR. Analysis. Treatise on Geochemistry (Second Edition), 87-109

Hu Z C, Liu Y S, Chen L, Zhou L, Li M, Zong K Q, Zhu L Y, Gao S. 2011. Contrasting matrix induced elemental fractionation in NIST SRM and rock glasses during laser ablation ICP-MS analysis at high spatial resolution. Journal of Analytical Atomic Spectrometry, 26(2): 425-430

Hu Z C, Liu Y S, Gao S, Liu W G, Zhang W, Tong X R, Lin L, Zong K Q, Li M, Chen H H, Zhou L, Yang Y. 2012a. Improved *in situ* Hf isotope ratio analysis of zircon using newly designed X skimmer cone and jet sample cone in combination with the addition of nitrogen by laser ablation multiple collector ICP-MS. Journal of Analytical Atomic Spectrometry, 27(9): 1391–1399

Hu Z C, Liu Y S, Gao S, Xiao S Q, Zhao L S, Günther D, Li M, Zhang W, Zong K Q. 2012b. A "wire" signal smoothing device for laser ablation inductively coupled plasma mass spectrometry analysis. Spectrochimica Acta Part B: Atomic Spectroscopy, 78: 50–57

Hu Z C, Zhang W, Liu Y S, Gao S, Li M, Zong K Q, Chen H H, Hu S H. 2015. "Wave" signal-smoothing and mercury-removing device for laser ablation quadrupole and multiple collector ICPMS analysis: Application to lead isotope analysis. Analytical Chemistry, 87(2): 1152–1157

Huang C, Yang Y H, Yang J H, Xie L W. 2015. In situ simultaneous measurement of Rb-Sr/Sm-Nd or Sm-Nd/Lu-Hf isotopes in natural minerals using laser ablation multi-collector ICP-MS. Journal of Analytical Atomic Spectrometry, 30(4): 994–1000

Huang C, Wang H, Yang J H, Xie L W, Yang Y H, Wu S T. 2019. Further characterization of the BB zircon via SIMS and MC-ICP-MS for Li, O, and Hf isotopic compositions. Minerals, 9(12): 774

Huang C, Wang H, Yang J H, Ramezani J, Yang C, Zhang S B, Yang Y H, Xia X P, Feng L J, Lin J, Wang T T, Ma Q, He H Y, Xie L W, Wu S T. 2020. SA01-A proposed zircon reference material for microbeam U-Pb age and Hf-O isotopic determination. Geostandards and Geoanalytical Research, 44(1): 103–123

Ice G E, Budai J D, Pang J W L. 2011. The race to X-ray microbeam and nanobeam science. Science, 334(6060): 1234–1239

ION-TOF GmbH. 2012. https://www.iontof.com/, TOF.SIMS 5 features and accessories, Extended dynamic range of up to seven orders of magnitude

ION-TOF GmbH. 2019. https://www.iontof.com/, TOF.SIMS 5 features and accessories, Wide range of ion sources (Bi_n, O_2, Ar, Xe, Cs, Arn, Ga)

Jian W, Albrecht M, Lehmann M, Mao J W, Horn I, Li Y H, Ye H S, Li Z Y, Fang G G, Xue Y S. 2018. UV-fs-LA-ICP-MS analysis of CO_2-rich fluid inclusions in a frozen state: example from the dahu Au-Mo deposit, Xiaoqinling region, central China. Geofluids, 2018: 3692180

Jiang Y, Chen Z, Han Y M, Deb P, Gao H, Xie S E, Purohit P, Tate M W, Park J, Gruner S M, Elser V, Muller D A. 2018. Electron ptychography of 2D materials to deep sub-ångström resolution. Nature, 559(7714): 343–349

Jin M Z, Belkin M A. 2019. Infrared vibrational spectroscopy of functionalized atomic force microscope probes using resonantly enhanced infrared photoexpansion nanospectroscopy. Small Methods, 3(10): 1900018

Jung S, Foston M, Kalluri U C, Tuskan G A, Ragauskas A J. 2012. 3D chemical image using TOF-SIMS revealing the biopolymer component spatial and lateral distributions in biomass. Angewandte Chemie International Edition, 51(48): 12005–12008

Kassenbohmer R, Heeger M, Dwivedi M, Körsgen M, Tyler B J, Galla H J, Arlinghaus H F. 2018. 3D molecular TOF-SIMS imaging of artificial lipid membranes using a discriminant analysis-based algorithm. Langmuir, 34(30): 8750–8757

Kayama M, Nakano S, Nishido H. 2010. Characteristics of emission centers in alkali feldspar: a new approach by using cathodoluminescence spectral deconvolution. American Mineralogist, 95(11-12): 1783–1795

Keller L M, Schuetz P, Erni R, Rossell M D, Lucas F, Gasser P, Holzer L. 2013. Characterization of multi-scale microstructural features in Opalinus Clay. Microporous and Mesoporous Materials, 170: 83–94

Khubbutdinov R, Menushenkov A P, Vartanyants I A. 2019. Coherence properties of the high-energy fourth-generation X-ray synchrotron sources. Journal of Synchrotron Radiation, 26(6): 1851–1862

Kisielowski C, Freitag B, Bischoff M, van Lin H, Lazar S, Knippels G, Tiemeijer P, van der Stam M, von Harrach S, Stekelenburg M, Haider M, Uhlemann S, Müller H, Hartel P, Kabius B, Miller D, Petrov I, Olson E A, Donchev T, Kenik E A, Lupini A R, Bentley J, Pennycook S J, Anderson I M, Minor A M, Schmid A K, Duden T, Radmilovic V, Ramasse Q M, Watanabe M, Erni R, Stach E A, Denes P, Dahmen U. 2008. Detection of single atoms and buried defects in three dimensions by aberration-corrected electron microscope with 0.5-Å information limit. Microscopy and Microanalysis, 14(5): 469–477

Kita N T, Sobol P E, Kern J R, Lord N E, Valley J W. 2015. UV-light microscope: improvements in optical imaging for a secondary ion mass spectrometer. Journal of Analytical Atomic Spectrometry, 30(5): 1207–1213

Lanekoff I, Sjövall P, Ewing A G. 2011. Relative quantification of phospholipid accumulation in the PC12 Cell plasma membrane following phospholipid incubation using TOF-SIMS imaging. Analytical Chemistry, 83(13): 5337–5343

Li C Y, Zhang R Q, Ding X, Ling M X, Fan W M, Sun W D. 2016. Dating cassiterite using laser ablation ICP-MS. Ore Geology Reviews, 72: 313–322

Li C Y, Jiang Y H, Zhao Y, Zhang C C, Ling M X, Ding X, Zhang H, Li J. 2018. Trace element analyses of fluid inclusions using laser ablation ICP-MS. Solid Earth Sciences, 3(1): 8–15

Li G C, Xiao K. 2012. Distribution characteristics of minerals and elements in chromite ore processing residue. Transactions of Tianjin University, 18(1): 52–56

Li J H, Benzerara K, Bernard S, Beyssac O. 2013. The link between biomineralization and fossilization of bacteria: insights from field and experimental studies. Chemical Geology, 359: 49–69

Li J H, Menguy N, Gatel C, Boureau V, Snoeck E, Patriarche G, Leroy E, Pan Y X. 2015. Crystal growth of bullet-shaped magnetite in magnetotactic bacteria of the Nitrospirae phylum. Journal of the Royal Society Interface, 12(103): 20141288

Li J H, Menguy N, Arrio M A, Sainctavit P, Juhin A, Wang Y Z, Chen H T, Bunau O, Otero E, Ohresser P, Pan Y X. 2016. Controlled cobalt doping in the spinel structure of magnetosome magnetite: new evidences from element-and site-specific X-ray magnetic circular dichroism analyses. Journal of The Royal Society Interface, 13(121): 20160355

Li J H, Zhang H, Menguy N, Benzerara K, Wang F X, Lin X T, Chen Z B, Pan Y X. 2017. Single-cell resolution of uncultured magnetotactic bacteria via fluorescence-coupled electron microscopy. Applied and Environmental Microbiology, 83(12): e00409–e00417

Li L L, Shi Y R, Williams I S, Anderson J L, Wu Z H, Wang S B. 2017. Geochemical and zircon isotopic evidence for extensive high level crustal contamination in Miocene to mid-Pleistocene intra-plate volcanic rocks from the Tengchong field, western Yunnan, China. Lithos, 286-287: 227–240

Li P, Xia Q K, Dallai L, Bonatti E, Brunelli D, Cipriani A, Ligi M. 2020. High H_2O content in pyroxenes of residual mantle peridotites at a mid-Atlantic Ridge segment. Scientific Reports, 10(1): 579

Li Q L, Li X H, Liu Y, Tang G Q, Yang J H, Zhu W G. 2010. Precise U-Pb and Pb-Pb dating of phanerozoic baddeleyite by SIMS with oxygen flooding technique. Journal of Analytical Atomic Spectrometry, 25(7): 1107–1113

Li Q L, Lin W, Su W, Li X H, Shi Y H, Liu Y, Tang G Q. 2011a. SIMS U-Pb rutile age of low-temperature eclogites from southwestern Chinese Tianshan, NW China. Lithos, 122(1-2): 76–86

Li Q L, Wu F Y, Li X H, Qiu Z L, Liu Y, Yang Y H, Tang G Q. 2011b. Precisely dating Paleozoic kimberlites in the North China Craton and Hf isotopic constraints on the evolution of the subcontinental lithospheric mantle. Lithos, 126(1-2): 127–134

Li Q L, Li X H, Wu F Y, Yin Q Z, Ye H M, Liu Y, Tang G Q, Zhang C L. 2012. *In-situ* SIMS U-Pb dating of phanerozoic apatite with low U and high common Pb. Gondwana Research, 21(4): 745–756

Li Q L, Li X H, Lan Z W, Guo C L, Yang Y N, Liu Y, Tang G Q. 2013. Monazite and xenotime U-Th-Pb geochronology by ion microprobe: dating highly fractionated granites at Xihuashan tungsten mine, SE China. Contributions to Mineralogy and Petrology, 166(1): 65–80

Li Q L, Liu Y, Tang G Q, Wang K Y, Ling X X, Li J. 2018. Zircon Th-Pb dating by secondary ion mass spectrometry. Journal of Analytical Atomic Spectrometry, 33(9): 1536–1544

Li R C, Xia X P, Yang S H, Chen H Y, Yang Q. 2019. Off-mount calibration and one new potential pyrrhotite reference material for sulfur isotope measurement by secondary ion mass spectrometry. Geostandards and Geoanalytical Research, 43(1): 177–187

Li R C, Xia X P, Chen H Y, Wu N P, Zhao T P, Lai C, Yang Q, Zhang Y Q. 2020. A potential new chalcopyrite reference material for secondary ion mass spectrometry sulfur isotope ratio analysis. Geostandards and Geoanalytical Research, 44(3): 485–500

Li W T, Audétat A, Zhang J. 2015. The role of evaporites in the formation of magnetite-apatite deposits along the Middle and Lower Yangtze River, China: evidence from LA-ICP-MS analysis of fluid inclusions. Ore Geology Reviews, 67: 264–278

Li X H, Li Q L. 2016. Major advances in microbeam analytical techniques and their applications in Earth Science. Science Bulletin, 61(23): 1785–1787

Li X H, Liu Y, Li Q L, Guo C H, Chamberlain K R. 2009. Precise determination of Phanerozoic zircon Pb/Pb age by multicollector SIMS without external standardization. Geochemistry, Geophysics, Geosystems, 10(4): Q04010

Li X H, Long W G, Li Q L, Liu Y, Zheng Y F, Yang Y H, Chamberlain K R, Wan D F, Guo C H, Wang X C, Tao H. 2010. Penglai zircon megacrysts: a potential new working reference material for microbeam determination of Hf-O isotopes and U-Pb age. Geostandards and Geoanalytical Research, 34(2): 117–134

Li X H, Li Q L, Liu Y, Tang G Q. 2011. Further characterization of M257 zircon standard: a working reference for SIMS analysis of Li isotopes. Journal of Analytical Atomic Spectrometry, 26(2): 352–358

Li X H, Tang G Q, Gong B, Yang Y H, Hou K J, Hu Z C, Li Q L, Liu Y, Li W X. 2013. Qinghu zircon: a working reference for microbeam analysis of U-Pb age and Hf and O isotopes. Chinese Science Bulletin, 58(36): 4647–4654

Li X H, Liu X M, Liu Y S, Su L, Sun W D, Huang H Q, Yi K. 2015. Accuracy of LA-ICPMS zircon U-Pb age determination: An inter-laboratory comparison. Science China Earth Sciences, 58(10): 1722–1730

Li X H, Chen Y, Tchouankoue J P, Liu C Z, Li J, Ling X X, Tang G Q, Liu Y. 2017. Improving geochronological framework of the Pan-African orogeny in Cameroon: new SIMS zircon and monazite U-Pb age constraints. Precambrian Research, 294: 307–321

Li X Y, Zhang C, Behrens H, Holtz F. 2020. Calculating biotite formula from electron microprobe analysis data using a machine learning method based on principal components regression. Lithos, 356-357: 105371

Li Z, Hu Z C, Liu Y S, Gao S, Li M, Zong K Q, Chen H H, Hu S H. 2015. Accurate determination of elements in silicate glass by nanosecond and femtosecond laser ablation ICP-MS at high spatial resolution. Chemical Geology, 400: 11–23

Li Z, Hu Z C, Günther D, Zong K Q, Liu Y S, Luo T, Zhang W, Gao S, Hu S H. 2016. Ablation characteristic of ilmenite using UV nanosecond and femtosecond lasers: implications for non-matrix-matched quantification. Geostandards and Geoanalytical Research, 40(4): 477–491

Liao X, Li Q L, Whitehouse M J, Yang Y H, Liu Y. 2020. Allanite U-Th-Pb geochronology by ion microprobe. Journal of Analytical Atomic Spectrometry, 35(3): 489–497

Liao X H, Hu Z C, Luo T, Zhang W, Liu Y S, Zong K Q, Zhou L, Zhang J F. 2019. Determination of major and trace elements in geological samples by laser ablation solution sampling-inductively coupled plasma mass spectrometry. Journal of Analytical Atomic Spectrometry, 34(6): 1126–1134

Lin J, Liu Y S, Tong X R, Zhu L Y, Zhang W, Hu Z C. 2017. Improved *in situ* Li isotopic ratio analysis of silicates by optimizing signal intensity, isotopic ratio stability and intensity matching using ns-LA-MC-ICP-MS. Journal of Analytical Atomic Spectrometry, 32(4): 834–842

Lin J, Liu Y S, Hu Z C, Chen W, Zhang C X, Zhao K D, Jin X Y. 2019. Accurate analysis of Li isotopes in tourmalines by LA-MC-ICP-MS under "wet" conditions with non-matrix-matched calibration. Journal of Analytical Atomic Spectrometry, 34(6): 1145–1153

Lin L, Hu Z C, Yang L, Zhang W, Liu Y S, Gao S, Hu S H. 2014. Determination of boron isotope compositions of geological materials by laser ablation MC-ICP-MS using newly designed high sensitivity skimmer and sample cones. Chemical Geology, 386: 22–30

Lin X, Guo W, Jin L L, Hu S H. 2020. Review: elemental analysis of individual fluid inclusions by laser ablation-ICP-MS. Atomic Spectroscopy, 41(1): 1–10

Lin Y, Hao J L, Miao Z Z, Zhang J H, Yang W. 2020. NanoSIMS image enhancement by reducing random noise using low-rank method. Surface and

Interface Analysis, 52(5): 240-248

Ling X X, Schmädicke E, Li Q L, Gose J, Wu R H, Wang S Q, Liu Y, Tang G Q, Li X H. 2015. Age determination of nephrite by *in-situ* SIMS U-Pb dating syngenetic titanite: a case study of the nephrite deposit from Luanchuan, Henan, China. Lithos, 220-223: 289-299

Ling X X, Li Q L, Liu Y, Yang Y H, Liu Y, Tang G Q, Li X H. 2016. *In situ* SIMS Th-Pb dating of bastnaesite: constraint on the mineralization time of the Himalayan Mianning-Dechang rare earth element deposits. Journal of Analytical Atomic Spectrometry, 31(8): 1680-1687

Liu D, Wang S, Smyth J R, Zhang J F, Wang X, Zhu X, Ye Y. 2019. *In-situ* infrared spectra for hydrous forsterite up to 1243 K: Hydration effect on thermodynamic properties. Minerals, 9(9): 512

Liu H Q, Bi X W, Lu H Z, Hu R Z, Lan T G, Wang X S, Huang M L. 2018. Nature and evolution of fluid inclusions in the Cenozoic Beiya gold deposit, SW China. Journal of Asian Earth Sciences, 161: 35-56

Liu J. 2016. High pressure X-ray diffraction techniques with synchrotron radiation. Chinese Physics B, 25(7): 076106

Liu M C, McKeegan K D, Harrison T M, Jarzebinski G, Vltava L. 2018. The Hyperion-ii radio-frequency oxygen ion source on the UCLA ims1290 ion microprobe: beam characterization and applications in geochemistry and cosmochemistry. International Journal of Mass Spectrometry, 424: 1-9

Liu S H, Hu Z C, Günther D, Ye Y H, Liu Y S, Gao S, Hu S H. 2014. Signal enhancement in laser ablation inductively coupled plasma-mass spectrometry using water and/or ethanol vapor in combination with a shielded torch. Journal of Analytical Atomic Spectrometry, 29(3): 536-544

Liu S Z, Deng C L, Xiao J L, Li J H, Paterson G A, Chang L, Yi L, Qin H F, Pan Y X, Zhu R X. 2015. Insolation driven biomagnetic response to the Holocene Warm Period in semi arid East Asia. Scientific Reports, 5: 8001

Liu Y, Li X H, Li Q L, Tang G Q, Yin Q Z. 2011. Precise U-Pb zircon dating at a scale of < 5 micron by the CAMECA 1280 SIMS using a Gaussian illumination probe. Journal of Analytical Atomic Spectrometry, 26(4): 845-851

Liu Y, Deng J, Shi G H, Sun D S. 2012. Geochemical and morphological characteristics of coarse-grained tabular beryl from the Xuebaoding W-Sn-Be deposit, Sichuan Province, western China. International Geology Review, 54(14): 1673-1684

Liu Y, Yao X, Liu Y W, Wang Y. 2014. A Fourier transform infrared spectroscopy analysis of carious dentin from transparent zone to normal zone. Caries Research, 48(4): 320-329

Liu Y, Li Q L, Tang G Q, Li X H, Yin Q Z. 2015. Towards higher precision SIMS U-Pb zircon geochronology *via* dynamic multi-collector analysis. Journal of Analytical Atomic Spectrometry, 30(4): 979-985

Liu Y, Li X H, Tang G Q, Li Q L, Liu X C, Yu H M, Huang F. 2019. Ultra-high precision silicon isotope micro-analysis using a Cameca IMS-1280 SIMS instrument by eliminating the topography effect. Journal of Analytical Atomic Spectrometry, 34(5): 906-914

Liu Y, Li X H, Li Q L, Tang G Q. 2020. Breakthrough of 2 to 3 μm scale U-Pb zircon dating using Cameca IMS-1280HR SIMS. Surface and Interface Analysis, 52(5): 214-223

Liu Z C, Wu F Y, Guo C L, Zhao Z F, Yang J H, Sun J F. 2011. *In situ* U-Pb dating of xenotime by laser ablation (LA)-ICP-MS. Chinese Science Bulletin, 56(27): 2948-2956

Long T, Clement S W J, Bao Z M, Wang P Z, Tian D, Liu D Y. 2018. High spatial resolution and high brightness ion beam probe for *in-situ* elemental and isotopic analysis. Nuclear Instruments and Methods in Physics Research Section B: Beam Interactions with Materials and Atoms, 419: 19-25

Long T, Clement S W J, Xie H Q, Liu D Y. 2020. Design, construction and performance of a TOF-SIMS for analysis of trace elements in geological materials. International Journal of Mass Spectrometry, 450: 116289

Luo T, Wang Y, Hu Z C, Günther D, Liu Y S, Gao S, Li M, Hu S H. 2015. Further investigation into ICP-induced elemental fractionation in LA-ICP-MS using a local aerosol extraction strategy. Journal of Analytical Atomic Spectrometry, 30(4): 941-949

Luo T, Ni Q, Hu Z C, Zhang W, Shi Q H, Günther D, Liu Y S, Zong K Q, Hu S H. 2017. Comparison of signal intensities and elemental fractionation in 257 nm femtosecond LA-ICP-MS using He and Ar as carrier gases. Journal of Analytical Atomic Spectrometry, 32(11): 2217-2225

Luo T, Hu Z C, Zhang W, Günther D, Liu Y S, Zong K Q, Hu S H. 2018a. Reassessment of the influence of carrier gases He and Ar on signal intensities in 193 nm excimer LA-ICP-MS analysis. Journal of Analytical Atomic Spectrometry, 33(10): 1655-1663

Luo T, Hu Z C, Zhang W, Liu Y S, Zong K Q, Zhou L, Zhang J F, Hu S H. 2018b. Water vapor-assisted "universal" nonmatrix-matched analytical method for the *in situ* U-Pb dating of zircon, monazite, titanite, and xenotime by laser ablation-inductively coupled plasma mass spectrometry. Analytical Chemistry, 90(15): 9016-9024

Luo T, Deng X D, Li J W, Hu Z C, Zhang W, Liu Y S, Zhang J F. 2019. U-Pb geochronology of wolframite by laser ablation inductively coupled plasma mass spectrometry. Journal of Analytical Atomic Spectrometry, 34(7): 1439-1446

Luvizotto G L, Zack T, Meyer H P, Ludwig T, Triebold S, Kronz A, Münker C, Stockli D F, Prowatke S, Klemme S, Jacob D E, von Eynatten H. 2009. Rutile crystals as potential trace element and isotope mineral standards for microanalysis. Chemical Geology, 261(3-4): 346-369

Ma Q, Evans N J, Ling X X, Yang J H, Wu F Y, Zhao Z D, Yang Y H. 2019. Natural titanite reference materials for *in situ* U-Pb and Sm-Nd isotopic measurements by LA-(MC)-ICP-MS. Geostandards and Geoanalytical Research, 43(3): 355-384

Manning P L, Edwards N P, Bergmann U, Anné J, Sellers W I, van Veelen A, Sokaras D, Egerton V M, Alonso-Mori R, Ignatyev K, van Dongen B E, Wakamatsu K, Ito S, Knoll F, Wogelius R A. 2019. Pheomelanin pigment remnants mapped in fossils of an extinct mammal. Nature Communications, 10(1): 2250

Maruvanchery V, Kim E. 2020. Mechanical characterization of thermally treated calcite-cemented sandstone using nanoindentation, scanning electron microscopy and automated mineralogy. International Journal of Rock Mechanics and Mining Sciences, 125: 104158

Mastel S, Govyadinov A A, Maissen C, Chuvilin A, Berger A, Hillenbrand R. 2018. Understanding the image contrast of material boundaries in ir nanoscopy reaching 5 nm spatial resolution. ACS Photonics, 5(8): 3372-3378

Meier D C, Davis J M, Vicenzi E P. 2011. An examination of kernite ($Na_2B_4O_6(OH)_2 \cdot 3H_2O$) using X-ray and electron spectroscopies: Quantitative microanalysis of a hydrated low-Z mineral. Microscopy and Microanalysis, 17(5): 718-727

Miao J W, Ishikawa T, Robinson I K, Murnane M M. 2015. Beyond crystallography: diffractive imaging using coherent x-ray light sources. Science, 348(6234): 530-535

Moore K L, Chen Y, van de Meene A M L, Hughes L, Liu W J, Geraki T, Mosselmans F, McGrath S P, Grovenor C, Zhao F J. 2014. Combined NanoSIMS and synchrotron X-ray fluorescence reveal distinct cellular and subcellular distribution patterns of trace elements in rice tissues. New Phytologist, 201(1): 104-115

Nellist P D, Chisholm M F, Dellby N, Krivanek O L, Murfitt M F, Szilagyi Z S, Lupini A R, Borisevich A, Sides Jr W H, Pennycook S J. 2004. Direct sub-angstrom imaging of a crystal lattice. Science, 305(5691): 1741

Ni X J, Gebo D L, Dagosto M, Meng J, Tafforeau P, Flynn J J, Beard K C. 2013. The oldest known primate skeleton and early haplorhine evolution. Nature, 498(7452): 60-64

Pan J Y, Ni P, Wang R C. 2019b. Comparison of fluid processes in coexisting wolframite and quartz from a giant vein-type tungsten deposit, South China: insights from detailed petrography and LA-ICP-MS analysis of fluid inclusions. American Mineralogist, 104(8): 1092-1116

Pan Y H, Hu L, Zhao T. 2019a. Applications of chemical imaging techniques in paleontology. National Science Review, 6(5): 1040-1053

Papaslioti E M, Pérez-López R, Parviainen A, Sarmiento A M, Nieto J M, Marchesi C, Delgado-Huertas A, Garrido C J. 2018. Effects of seawater mixing on the mobility of trace elements in acid phosphogypsum leachates. Marine Pollution Bulletin, 127: 695-703

Passarelli M K, Pirkl A, Moellers R, Grinfeld D, Kollmer F, Havelund R, Newman C F, Marshall P S, Arlinghaus H, Alexander M R, West A, Horning S, Niehuis E, Makarov A, Dollery C T, Gilmore I S. 2017. The 3D OrbiSIMS-label-free metabolic imaging with subcellular lateral resolution and high mass-resolving power. Nature Methods, 14(12): 1175-1183

Peng X T, Ta K W, Chen S, Zhang L J, Xu H C. 2015. Coexistence of Fe(Ⅱ)- and Mn(Ⅱ)-oxidizing bacteria govern the formation of deep sea umber deposits. Geochimica et Cosmochimica Acta, 169: 200-216

Phillipp F, Höschen R, Osaki M, Möbus G, Rühle M. 1994. New high-voltage atomic resolution microscope approaching 1 Å point resolution installed in Stuttgart. Ultramicroscopy, 56(1-3): 1-10

Qian G J, Li Y B, Gerson A R. 2015. Applications of surface analytical techniques in Earth Sciences. Surface Science Reports, 70(1): 86-133

Qian S P, Zhang L. 2019. Simultaneous in situ determination of rare earth element concentrations and Nd isotope ratio in apatite by laser ablation ICP-MS. Geochemical Journal, 53(5): 319-328

Qiu Y Y, Jiang H T, Kovács I, Xia Q K, Yang X Z. 2018. Quantitative analysis of H-species in anisotropic minerals by unpolarized infrared spectroscopy: an experimental evaluation. American Mineralogist, 103(11): 1761-1769

Quartieri S. 2015. Synchrotron radiation in the earth sciences. In: Mobilio S, Boscherini F, Meneghini C (eds). Synchrotron Radiation. Berlin, Heidelberg: Springer, 641-660

Reddy S M, Saxey D W, Rickard W D A, Fougerouse D, Montalvo S D, Verberne R, van Riessen A. 2020. Atom probe tomography: Development and application to the geosciences. Geostandards and Geoanalytical Research, 44(1): 5-50

Robinson M A, Graham D J, Castner D G. 2012. ToF-SIMS depth profiling of cells: Z-correction, 3D imaging, and sputter rate of individual NIH/3T3 fibroblasts. Analytical Chemistry, 84(11): 4880-4885

Ruska E. 1934. Über Fortschritte im Bau und in der Leistung des magnetischen Elektronenmikroskops. Zeitschrift für Physik, 87(9): 580-602

Saxena S K, Dubrovinsky L S, Häggkvist P, Cerenius Y, Shen G, Mao H K. 1995. Synchrotron X-ray study of iron at high pressure and temperature. Science, 269(5231): 1703-1704

Shaheen M E, Gagnon J E, Fryer B J. 2012. Femtosecond (fs) lasers coupled with modern ICP-MS instruments provide new and improved potential for in situ elemental and isotopic analyses in the geosciences. Chemical Geology, 330-331: 260-273

Shi Y R, Allen K, John A, Song T R, Li L L, Sun H Y. 2015. In-situ SHRIMP U-Pb dating of xenotime outgrowth on detrital zircon grains from the changzhougou formation of the ming tomb district, Beijing. Acta Geologica Sinica (English Edition), 89(1): 304-305

Shu L, Shen K, Yang R C, Song Y X, Sun Y Q, Shan W, Xiong Y X. 2020. SEM-CL study of quartz containing fluid inclusions in wangjiazhuang porphyry copper (-Molybdenum) deposit, western Shandong, China. Journal of Earth Science, 31(2): 330-341

Sigle W. 2005. Analytical transmission electron microscopy. Annual Review of Materials Research, 35: 239-314

Stroud R M, Chisholm M F, Heck P R, Alexander C M O D, Nittler L R. 2011. Supernova shock-wave-induced CO-formation of glassy carbon and nanodiamond. The Astrophysical Journal Letters, 738(2): L27

Su B, Chen Y, Mao Q, Zhang D, Jia L H, Guo S. 2019. Minor elements in olivine inspect the petrogenesis of orogenic peridotites. Lithos, 344-345: 207-216

Su B X, Gu X Y, Deloule E, Zhang H F, Li Q L, Li X H, Vigier N, Tang Y J, Tang G Q, Liu Y, Pang K N, Brewer A, Mao Q, Ma Y G. 2015. Potential orthopyroxene, clinopyroxene and olivine reference materials for in situ lithium isotope determination. Geostandards and Geoanalytical Research, 39(3): 357-369

Sun H Q, Mašín D, Najser J, Neděla V, Navrátilová E. 2019. Bentonite microstructure and saturation evolution in wetting-drying cycles evaluated using ESEM, MIP and WRC measurements. Géotechnique, 69(8): 713-726

Sun J F, Yang J H, Wu F Y, Xie L W, Yang Y H, Liu Z C, Li X H. 2012. In situ U-Pb dating of titanite by LA-ICPMS. Chinese Science Bulletin, 57

(20): 2506-2516

Takahata N, Tsutsumi Y, Sano Y. 2008. Ion microprobe U-Pb dating of zircon with a 15 micrometer spatial resolution using NanoSIMS. Gondwana Research, 14(4): 587-596

Tang G Q, Li X H, Li Q L, Liu Y, Ling X X, Yin Q Z. 2015. Deciphering the physical mechanism of the topography effect for oxygen isotope measurements using a Cameca IMS-1280 SIMS. Journal of Analytical Atomic Spectrometry, 30(4): 950-956

Tang G Q, Su B X, Li Q L, Xia X P, Jing J J, Feng L J, Martin L, Yang Q, Li X H. 2019. High-Mg# olivine, clinopyroxene and orthopyroxene reference materials for in situ oxygen isotope determination. Geostandards and Geoanalytical Research, 43(4): 585-593

Tian Z Z, Liu J, Xia Q K, Ingrin J, Hao Y T, Christophe D. 2017. Water concentration profiles in natural mantle orthopyroxenes: A geochronometer for long annealing of xenoliths within magma. Geology, 45(1): 87-90

Tong X R, Liu Y S, Hu Z C, Chen H H, Zhou L, Hu Q H, Xu R, Deng L X, Chen C H, Yang L, Gao S. 2016. Accurate determination of Sr isotopic compositions in clinopyroxene and silicate glasses by LA-MC-ICP-MS. Geostandards and Geoanalytical Research, 40(1): 85-99

Tual L, Möller C, Whitehouse M J. 2018. Tracking the prograde P-T path of Precambrian eclogite using Ti-in-quartz and Zr-in-rutile geothermobarometry. Contributions to Mineralogy and Petrology, 173(7): 56, doi: 10.1007/s00410-018-1482-1

Turner M, Ireland T, Hermann J, Holden P, Padrón-Navarta J A, Hauri E H, Turner S. 2015. Sensitive high resolution ion microprobe-stable isotope (SHRIMP-SI) analysis of water in silicate glasses and nominally anhydrous reference minerals. Journal of Analytical Atomic Spectrometry, 30(8): 1706-1722

ULVAC-PHI Inc. 2016. https://www.ulvac-phi.com/en/, Time-of-Flight SIMS / TOF-SIMS, PHI nanoTOF II™, MS/MS Option

van der Snickt G, Janssens K, Dik J, de Nolf W, Vanmeert F, Jaroszewicz J, Cotte M, Falkenberg G, van der Loeff L. 2012. Combined use of synchrotron radiation based micro-X-ray fluorescence, micro-X-ray diffraction, micro-X-ray absorption near-edge, and micro-fourier transform infrared spectroscopies for revealing an alternative degradation pathway of the pigment cadmium yellow in a painting by van Gogh. Analytical Chemistry, 84(23): 10221-10228

Wacey D, Battison L, Garwood R J, Hickman-Lewis K, Brasier M D. 2017. Advanced analytical techniques for studying the morphology and chemistry of Proterozoic microfossils. In: Brasier A T, McIlroy D, McLoughlin M (eds). Earth System Evolution and Early Life: A Celebration of the Work of Martin Brasier. Geological Society, London, Special Publication, 448(1): 81-104

Wang B, Zhu J J, Pierson E, Ramazzotti D, Batzoglou S. 2017. Visualization and analysis of single-cell RNA-seq data by kernel-based similarity learning. Nature Methods, 14(4): 414-416

Wang J, Hitchcock A P, Karunakaran C, Prange A, Franz B, Harkness T, Lu Y, Obst M, Hormes J. 2011. 3D chemical and elemental imaging by STXM spectrotomography. AIP Conference Proceedings, 1365(1): 215-218

Wang X J, Wu K, Huang W X, Zhang H F, Zheng M Y, Peng D L. 2007. Study on fracture behavior of particulate reinforced magnesium matrix composite using in situ SEM. Composites Science and Technology, 67(11-12): 2253-2260

Wang Z, Wu W W, Qian G A, Sun L J, Li X D, Correia J A F O. 2019. In-situ SEM investigation on fatigue behaviors of additive manufactured Al-Si10-Mg alloy at elevated temperature. Engineering Fracture Mechanics, 214: 149-163

Wang Z Z, Liu J, Xia Q K, Hao Y T, Wang Q Y. 2020. The distribution of water in the early Cretaceous lithospheric mantle of the North China Craton and implications for its destruction. Lithos, 360-361: 105412

Wille G, Lahondere D, Schmidt U, Duron J, Bourrat X. 2019. Coupling SEM-EDS and confocal Raman-in-SEM imaging: a new method for identification and 3D morphology of asbestos-like fibers in a mineral matrix. Journal of Hazardous Materials, 374: 447-458

Willmott P. 2019. An Introduction to Synchrotron Radiation: Techniques and Applications, 2nd ed. Hoboken: John Wiley & Sons Ltd

Wirth R. 2004. Focused ion beam (FIB): a novel technology for advanced application of micro- and nanoanalysis in geosciences and applied mineralogy. European Journal of Mineralogy, 16(6): 863-876

Witte K, Späth A, Finizio S, Donnelly C, Watts B, Sarafimov B, Odstrcil M, Guizar-Sicairos M, Holler M, Fink R H, Raabe J. 2020. From 2D STXM to 3D imaging: soft X-ray laminography of thin specimens. Nano Letters, 20(2): 1305-1314

Wogelius R A, Manning P L, Barden H E, Edwards N P, Webb S M, Sellers W I, Taylor K G, Larson P L, Dodson P, You H, Da-qing L, Bergmann U. 2011. Trace metals as biomarkers for eumelanin pigment in the fossil record. Science, 333(6049): 1622-1626

Wu F Y, Mitchell R H, Li Q L, Sun J, Liu C Z, Yang Y H. 2013. In situ U-Pb age determination and Sr-Nd isotopic analysis of perovskite from the Premier (Cullinan) kimberlite, South Africa. Chemical Geology, 353: 83-95

Wu F Y, Fougerouse D, Evans K, Reddy S M, Saxey D W, Guagliardo P, Li J W. 2019. Gold, arsenic, and copper zoning in pyrite: a record of fluid chemistry and growth kinetics. Geology, 47(7): 641-644

Wu L G, Li Q L, Liu Y, Tang G Q, Lu K, Ling X X, Li X H. 2020. Rapid and accurate SIMS microanalysis of monazite oxygen isotopes. Journal of Analytical Atomic Spectrometry, 35(8): 1607-1613

Xia Q K, Hao Y T, Liu S C, Gu X Y, Feng M. 2013. Water contents of the Cenozoic lithospheric mantle beneath the western part of the North China Craton: peridotite xenolith constraints. Gondwana Research, 23(1): 108-118

Xia Q K, Liu J, Kovács I, Hao Y T, Li P, Yang X Z, Chen H, Sheng Y M. 2019. Water in the upper mantle and deep crust of eastern China: concentration, distribution and implications. National Science Review, 6(1): 125-144

Xia X P, Sun M, Geng H Y, Sun Y L, Wang Y J, Zhao G C. 2011. Quasi-simultaneous determination of U-Pb and Hf isotope compositions of zircon by excimer laser-ablation multiple-collector ICPMS. Journal of Analytical Atomic Spectrometry, 26(9): 1868-1871

Xia X P, Ren Z Y, Wei G J, Zhang L, Sun M, Wang Y J. 2013. *In situ* rutile U-Pb dating by laser ablation-MC-ICPMS. Geochemical Journal, 47(4): 459-468

Xia X P, Cui Z X, Li W C, Zhang W F, Yang Q, Hui H J, Lai C K. 2019. Zircon water content: reference material development and simultaneous measurement of oxygen isotopes by SIMS. Journal of Analytical Atomic Spectrometry, 34(6): 1088-1097

Xie L W, Yin Q Z, Yang J H, Wu F Y, Yang Y H. 2011. High precision analysis of Mg isotopic composition in olivine by laser ablation MC-ICP-MS. Journal of Analytical Atomic Spectrometry, 26(9): 1773-1780

Xie L W, Yang J H, Yin Q Z, Yang Y H, Liu J B, Huang C. 2017. High spatial resolution *in situ* U-Pb dating using laser ablation multiple ion counting inductively coupled plasma mass spectrometry (LA-MIC-ICP-MS). Journal of Analytical Atomic Spectrometry, 32(5): 975-986

Xie L W, Xu L, Yin Q Z, Yang Y H, Huang C, Yang J H. 2018a. A novel sample cell for reducing the "*Position Effect*" in laser ablation MC-ICP-MS isotopic measurements. Journal of Analytical Atomic Spectrometry, 33(9): 1571-1578

Xie L W, Evans N J, Yang Y H, Huang C, Yang J H. 2018b. U-Th-Pb geochronology and simultaneous analysis of multiple isotope systems in geological samples by LA-MC-ICP-MS. Journal of Analytical Atomic Spectrometry, 33(10): 1600-1615

Xing C M, Wang C Y. 2017. Cathodoluminescence images and trace element compositions of fluorapatite from the Hongge layered intrusion in SW China: a record of prolonged crystallization and overprinted fluid metasomatism. American Mineralogist, 102(7): 1390-1401

Xu L, Hu Z C, Zhang W, Yang L, Liu Y S, Gao S, Luo T, Hu S H. 2015. *In situ* Nd isotope analyses in geological materials with signal enhancement and non-linear mass dependent fractionation reduction using laser ablation MC-ICP-MS. Journal of Analytical Atomic Spectrometry, 30(1): 232-244

Xu L, Yang J H, Ni Q, Yang Y H, Hu Z C, Liu Y S, Wu Y B, Luo T, Hu S H. 2018. Determination of Sm-Nd isotopic compositions in fifteen geological materials using laser ablation MC-ICP-MS and application to monazite geochronology of metasedimentary Rock in the North China Craton. Geostandards and Geoanalytical Research, 42(3): 379-394

Xu Y C, Gu L X, Li Y, Mo B, Lin Y T. 2018. Combination of focused ion beam (FIB) and microtome by ultrathin slice preparation for transmission electron microscopy (TEM) observation. Earth, Planets and Space, 70(1): 150

Yang Q, Xia X P, Zhang W F, Zhang Y Q, Xiong B Q, Xu Y G, Wang Q, Wei G J. 2018. An evaluation of precision and accuracy of SIMS oxygen isotope analysis. Solid Earth Sciences, 3(3): 81-86

Yang Q, Xia X P, Zhang L, Zhang W F, Zhang Y Q, Chen L L, Yang Y N, He M H. 2020. Oxygen isotope homogeneity assessment for apatite U-Th-Pb geochronology reference materials. Surface and Interface Analysis, 52(5): 197-213

Yang W, Lin Y T, Zhang J C, Hao J L, Shen W J, Hu S. 2012. Precise micrometer-sized Pb-Pb and U-Pb dating with NanoSIMS. Journal of Analytical Atomic Spectrometry, 27(3): 479-487

Yang W, Hu S, Zhang J C, Hao J L, Lin Y T. 2015. NanoSIMS analytical technique and its applications in earth sciences. Science China Earth Sciences, 58(10): 1758-1767

Yang W, Lin Y T, Hao J L, Zhang J C, Hu S, Ni H W. 2016. Phosphorus-controlled trace element distribution in zircon revealed by NanoSIMS. Contributions to Mineralogy and Petrology, 171(3): 28

Yang Y, Xia Q K, Zhang P P. 2015. Evolution of OH groups in diopside and feldspars with temperature. European Journal of Mineralogy, 27(2): 185-192

Yang Y H, Wu F Y, Li Y, Yang J H, Xie L W, Liu Y, Zhang Y B, Huang C. 2014a. *In situ* U-Pb dating of bastnaesite by LA-ICP-MS. Journal of Analytical Atomic Spectrometry, 29(6): 1017-1023

Yang Y H, Wu F Y, Yang J H, Chew D M, Xie L W, Chu Z Y, Zhang Y B, Huang C. 2014b. Sr and Nd isotopic compositions of apatite reference materials used in U-Th-Pb geochronology. Chemical Geology, 385: 35-55

Yang Y H, Wu F Y, Yang J H, Mitchell R H, Zhao Z F, Xie L W, Huang C, Ma Q, Yang M, Zhao H. 2018. U-Pb age determination of schorlomite garnet by laser ablation inductively coupled plasma mass spectrometry. Journal of Analytical Atomic Spectrometry, 33(2): 231-239

Yang Y H, Wu F Y, Li Q L, Rojas-Agramonte Y, Yang J H, Li Y, Ma Q, Xie L W, Huang C, Fan H R, Zhao Z F, Xu C. 2019. *In situ* U-Th-Pb dating and Sr-Nd isotope analysis of bastnäsite by LA-(MC)-ICP-MS. Geostandards and Geoanalytical Research, 43(4): 543-565

Yin Z J, Zhu M Y, Davidson E H, Bottjer D J, Zhao F C, Tafforeau P. 2015. Sponge grade body fossil with cellular resolution dating 60 Myr before the Cambrian. Proceedings of the National Academy of Sciences of the United States of America, 112(12): E1453-E1460

Yuan H L, Chen K Y, Bao Z A, Zong C L, Dai M N, Fan C, Yin C. 2013. Determination of lead isotope compositions of geological samples using femtosecond laser ablation MC-ICPMS. Chinese Science Bulletin, 58(32): 3914-3921

Yuan H L, Yin C, Liu X, Chen K Y, Bao Z A, Zong C L, Dai M N, Lai S C, Wang R, Jiang S Y. 2015. High precision *in-situ* Pb isotopic analysis of sulfide minerals by femtosecond laser ablation multi-collector inductively coupled plasma mass spectrometry. Science China Earth Sciences, 58(10): 1713-1721

Yuan H L, Bao Z A, Chen K Y, Zong C L, Chen L, Zhang T. 2019. Improving the sensitivity of a multi-collector inductively coupled plasma mass spectrometer via expansion-chamber pressure reduction. Journal of Analytical Atomic Spectrometry, 34(5): 1011-1017

Yuan S D, Peng J T, Hao S, Li H M, Geng J Z, Zhang D L. 2011. *In situ* LA-MC-ICP-MS and ID-IMS U-Pb geochronology of cassiterite in the giant Furong tin deposit, Hunan Province, South China: new constraints on the timing of tin-polymetallic mineralization. Ore Geology Reviews, 43(1): 235-242

Zach J, Haider M. 1995. Aberration correction in a low voltage SEM by a multipole corrector. Nuclear Instruments and Methods in Physics Research Section A: Accelerators, Spectrometers, Detectors and Associated Equipment, 363(1-2): 316-325

Zhang A C, Li Q L, Yurimoto H, Sakamoto N, Li X H, Hu S, Lin Y T, Wang R C. 2016. Young asteroidal fluid activity revealed by absolute age from apatite in carbonaceous chondrite. Nature Communications, 7: 12844

Zhang B H, Li B W, Zhao C C, Yang X Z. 2019. Large effect of water on Fe-Mg interdiffusion in garnet. Earth and Planetary Science Letters, 505: 20–29

Zhang C, Almeev R R, Hughes E C, Borisov A A, Wolff E P, Höfer H E, Botcharnikov R E, Koepke J. 2018. Electron microprobe technique for the determination of iron oxidation state in silicate glasses. American Mineralogist, 103(9): 1445–1454

Zhang C X, Hu Z C, Zhang W, Liu Y S, Zong K Q, Li M, Chen H H, Hu S H. 2016. Green and fast laser fusion technique for bulk silicate rock analysis by laser ablation-inductively coupled plasma mass spectrometry. Analytical Chemistry, 88(20): 10088–10094

Zhang G B, Ellis D J, Christy A G, Zhang L F, Song S S. 2010. Zr-in-rutile thermometry in HP/UHP eclogites from western China. Contributions to Mineralogy and Petrology, 160: 427–439

Zhang J C, Lin Y T, Yang W, Shen W J, Hao J L, Hu S, Cao M J. 2014. Improved precision and spatial resolution of sulfur isotope analysis using NanoSIMS. Journal of Analytical Atomic Spectrometry, 29(10): 1934–1943

Zhang J C, Lin Y T, Yang W, Hao J L, Hu S. 2016. Micro-scale (~10 μm) analyses of rare earth elements in silicate glass, zircon and apatite with NanoSIMS. International Journal of Mass Spectrometry, 406: 48–54

Zhang J C, Lin Y T, Yan J, Li J X, Yang W. 2017. Simultaneous determination of sulfur isotopes and trace elements in pyrite with a NanoSIMS 50L. Analytical Methods, 9(47): 6653–6661

Zhang L, Ren Z Y, Nichols A R L, Zhang Y H, Zhang Y, Qian S P, Liu J Q. 2014. Lead isotope analysis of melt inclusions by LA-MC-ICP-MS. Journal of Analytical Atomic Spectrometry, 29(8): 1393–1405

Zhang L, Ren Z Y, Wu Y D, Li N. 2018. Strontium isotope measurement of basaltic glasses by laser ablation multiple collector inductively coupled plasma mass spectrometry based on a linear relationship between analytical bias and Rb/Sr ratios. Rapid Communications in Mass Spectrometry, 32(2): 105–112

Zhang L, Wu J L, Tu J R, Wu D, Li N, Xia X P, Ren Z Y. 2020. RMJG rutile: a new natural reference material for microbeam U-Pb dating and Hf isotopic analysis. Geostandards and Geoanalytical Research, 44(1): 133–145

Zhang S Y, Zhang H L, Hou Z H, Ionov D A, Huang F. 2019. Rapid determination of trace element compositions in peridotites by LA-ICP-MS using an albite fusion method. Geostandards and Geoanalytical Research, 43(1): 93–111

Zhang W, Hu Z C. 2019. Recent advances in sample preparation methods for elemental and isotopic analysis of geological samples. Spectrochimica Acta Part B: Atomic Spectroscopy, 160: 105690

Zhang W, Hu Z C, Yang L, Liu Y S, Zong K Q, Xu H J, Chen H H, Gao S, Xu L. 2015. Improved inter-calibration of faraday cup and ion counting for *in situ* Pb isotope measurements using LA-MC-ICP-MS: application to the study of the origin of the fangshan pluton, North China. Geostandards and Geoanalytical Research, 39(4): 467–487

Zhang W, Hu Z C, Günther D, Liu Y S, Ling W L, Zong K Q, Chen H H, Gao S. 2016. Direct lead isotope analysis in Hg-rich sulfides by LA-MC-ICP-MS with a gas exchange device and matrix-matched calibration. Analytica Chimica Acta, 948: 9–18

Zhang W, Hu Z C, Liu Y S, Yang W W, Chen H H, Hu S H, Xiao H Y. 2017. Quantitative analysis of major and trace elements in NH_4HF_2-modified silicate rock powders by laser ablation-inductively coupled plasma mass spectrometry. Analytica Chimica Acta, 983: 149–159

Zhang W, Hu Z C, Liu Y S, Wu T, Deng X D, Guo J L, Zhao H. 2018. Improved *in situ* Sr isotopic analysis by a 257 nm femtosecond laser in combination with the addition of nitrogen for geological minerals. Chemical Geology, 479: 10–21

Zhang W, Hu Z C, Liu Y S, Feng L P, Jiang H S. 2019a. *In situ* calcium isotopic ratio determination in calcium carbonate materials and calcium phosphate materials using laser ablation-multiple collector-inductively coupled plasma mass spectrometry. Chemical Geology, 522: 16–25

Zhang W, Wang Z C, Moynier F, Inglis E, Tian S Y, Li M, Liu Y S, Hu Z C. 2019b. Determination of Zr isotopic ratios in zircons using laser-ablation multiple-collector inductively coupled-plasma mass-spectrometry. Journal of Analytical Atomic Spectrometry, 34(9): 1800–1809

Zhang W, Hu Z C, Liu Y S. 2020. Iso-compass: new freeware software for isotopic data reduction of LA-MC-ICP-MS. Journal of Analytical Atomic Spectrometry, 35(6): 1087–1096

Zhang W F, Xia X P, Zhang Y Q, Peng T P, Yang Q. 2018. A novel sample preparation method for ultra-high vacuum (UHV) secondary ion mass spectrometry (SIMS) analysis. Journal of Analytical Atomic Spectrometry, 33(9): 1559–1563

Zhang W F, Xia X P, Eiichi T, Li L, Yang Q, Zhang Y Q, Yang Y N, Liu M L, Lai C. 2020. Optimization of SIMS analytical parameters for water content measurement of olivine. Surface and Interface Analysis, 52(5): 224–233

Zhang Y B, Li Q L, Lan Z W, Wu F Y, Li X H, Yang J H, Zhai M G. 2015. Diagenetic xenotime dating to constrain the initial depositional time of the Yan-Liao Rift. Precambrian Research, 271: 20–32

Zhou Q, Herd C D K, Yin Q Z, Li X H, Wu F Y, Li Q L, Liu Y, Tang G Q, McCoy T J. 2013. Geochronology of the Martian meteorite Zagami revealed by U-Pb ion probe dating of accessory minerals. Earth and Planetary Science Letters, 374: 156–163

Zhou X J, Zhu H C, Zhong J J, Peng W W, Ji T, Lin Y C, Tang Y Z, Chen M. 2019. New status of the infrared beamlines at SSRF. Nuclear Science and Techniques, 30(12): 182

Zhu L Y, Liu Y S, Hu Z C, Hu Q H, Tong X R, Zong K Q, Chen H H, Gao S. 2013. Simultaneous determination of major and trace elements in fused volcanic rock powders using a hermetic vessel heater and LA-ICP-MS. Geostandards and Geoanalytical Research, 37(2): 207–229

Zhu L Y, Liu Y S, Ma T T, Lin J, Hu Z C, Wang C. 2016. *In situ* measurement of Os isotopic ratios in sulfides calibrated against ultra-fine particle standards using LA-MC-ICP-MS. Journal of Analytical Atomic Spectrometry, 31(7): 1414–1422

Zong K Q, Chen J Y, Hu Z C, Liu Y S, Li M, Fan H H, Meng Y N. 2015. In-situ U-Pb dating of uraninite by fs-LA-ICP-MS. Science China Earth Sciences, 58(10): 1731–1740

Progress of Microbeam Analytical Technologies

CHEN Yi[1], HU Zhao-chu[2], JIA Li-hui[1], LI Jin-hua[1], LI Qiu-li[1], LI Xiao-guang[1], LI Zhan-ping[3], LONG Tao[4], TANG Xu[1], WANG Jian[5], XIA Xiao-ping[6], YANG Wei[1], YUAN Jiang-yan[1], ZHANG Di[1], LI Xian-hua[1]

1. Institute of Geology and Geophysics, Chinese Academy of Sciences, Beijing 100029; 2. State Key Laboratory of Geological Processes and Mineral Resources, China University of Geosciences (Wuhan), Wuhan 430074; 3. Analysis Center of Tsinghua University, Beijing 100084; 4. Institute of Geology, Chinese Academy of Geological Sciences, Beijing 100037; 5. Canadian Light Source Inc., University of Saskatchewan, Saskatoon, SK, Canada S7N 2V3; 6. Guangzhou Institute of geochemistry, Chinese Academy of Sciences, Guangzhou 510640

Abstract: Major breakthroughs in the modern earth science research depend largely on the innovation of observation and analytical technologies. The Chinese geoscience community has possessed many state-of-the-art laboratories and advanced microbeam analytical instruments since the beginning of the new century. In this paper, we review the major progresses of various microbeam analytical technologies and methods, including EPMA, SEM, TEM, LG-SIMS, NanoSIMS, TOF-SIMS, LA-Q/HR/MC-ICPMS, LI-APT, FTIR, and Synchrotron Radiation, as well as their applications in earth sciences. Our understanding of the evolutionary history of the Earth and planets and many geological processes had been improved greatly by applications of microbeam analytical technologies. To improve our innovation ability and to play more important roles in the international academic arena, we need to develope more new technologies, methods and standards for microbeam analyses in the future.

Key words: microbeam analysis; SIMS; EPMA; laser ablation; synchrotron radiation

岩矿分析测试研究进展

刘勇胜[1]　屈文俊[2]　漆亮[3]　袁洪林[4]　黄方[5]　杨岳衡[6]　胡兆初[1]　朱振利[7]　张文[1]

1. 中国地质大学（武汉），地质过程与矿产资源国家重点实验室，武汉 430074；2. 国家地质实验测试中心，北京 100037；3. 中国科学院 地球化学研究所，矿床地球化学国家重点实验室，贵阳 550081；4. 西北大学 地质学系，大陆动力学国家重点实验室，西安 710069；5. 中国科学技术大学 地球和空间科学学院，合肥 230026；6. 中国科学院 地质与地球物理研究所，岩石圈演化国家重点实验室，北京 100029；7. 中国地质大学（武汉），生物地质与环境地质国家重点实验室，武汉 430074

摘　要：本文回顾并综述了 2011—2020 年间我国在岩石与矿物分析测试领域的主要研究进展，包括元素含量分析、放射性成因同位素和非传统稳定同位素分析、地质样品前处理技术、岩矿标准物质研制和定值、主流分析仪器及关键部件研发等。近十年来，我国学者在上述领域取得了大量原创性研究成果，开发出部分达到国际领先水平的岩矿分析新技术和新方法，极大地推动了我国地球科学研究。现代地球科学研究领域不断拓展，国家对自然资源开发的需求和人类保护生存环境的责任共同对岩矿分析测试工作提出新的要求，本文据此对未来岩矿分析测试领域的发展进行了展望。

关键词：元素定量分析　同位素组成分析　地质样品前处理　岩矿标准物质　仪器及关键部件

0　引　言

岩石、土壤、沉积物或矿物等地质样品的元素含量、元素形态和同位素比值等地球化学信息可以为研究地球演化、环境变化、矿床成因、矿产资源分布等重大地球科学问题提供重要信息。岩矿分析测试是获取上述地球化学资料的重要技术手段。在 2011—2020 年，我国的岩矿分析测试技术呈现出快速的多维度的发展趋势，在元素定量和同位素比值分析方面建立了大量新技术新方法，在地质样品前处理、参考物质研制和定值、分析仪器和关键部件研制等多个方面取得了众多原创性成果。本文基于我国岩矿分析测试技术 10 年的发展历程，梳理出岩矿分析测试的 7 个重要研究方向，并对各方向的进展进行了回顾和综述，同时指出这些研究方向所面临的问题和发展趋势，以期为国内同行深入了解我国这一领域的研究现状提供参考。

1　主量元素分析

硅酸盐岩石的全分析通常是指包括 SiO_2、Al_2O_3、Fe_2O_3、FeO、MgO、CaO、MnO、TiO_2、Na_2O、K_2O、P_2O_5 在内的 10 种主量元素和 H_2O^-、H_2O^+、CO_2 共 14 项组分的测定。有时还要求分析 F、Cl 和 S 等组分。经典的硅酸盐岩石分析方法（《岩石矿物分析》编委会，2011）是建立在完善的沉淀分离系统和重量法基础之上的具有较高准确度的方法。然而，复杂的分离流程和烦冗的重量法难以适应日益增长的硅酸盐岩石分析试样的需求。随着科技的发展，岩石主量元素含量的分析工作是以大型仪器为主。此外，岩石中的 H_2O^-、H_2O^+、CO_2 及 FeO 的测定，依然离不开化学分析法。在矿石样品的全分析中，除了 10 种主量元素的常规分析，往往还会有其他含量较高的元素需要关注，如天青石中的锶和钡、铅锌矿中的

---

* 原文"中国岩矿分析测试研究进展与展望（2011—2020）"刊于《矿物岩石地球化学通报》2021 年第 40 卷第 3 期，本文略有修改。

铅锌硫、铜矿石中的铜等。这些矿石样品中高含量元素的准确测定，也是近十年来研究的热点。

1.1 X射线荧光光谱分析

X射线荧光光谱（XRF）作为一种成熟的分析技术被广泛应用于硅酸盐等样品的常规分析，其主量元素测定的准确度可与化学法媲美（吉昂，2012）。用于XRF分析的常用制样方法主要有粉末压片法和熔融法。由于普通的粉末压片法受到粒度效应和矿物效应的限制，超细地质样品分析就成为研究人员的关注点（王晓红等，2010；张莉娟等，2014；李小莉等，2015；曾江萍等，2015a，2015b）。熔融法能消除样品的粒度效应和矿物效应，还可以用纯氧化物或在已有标准样品中加入氧化物来扩大元素的含量范围。李国会和李小莉（2015）对XRF分析的熔融法制样进行了系统总结，给出了氧化物、碳酸盐、硫化物、铁合金、石墨材料、铜精矿等几种典型样品熔融制备玻璃片的方法。

2011年我国建立了利用XRF分析硅酸盐样品中16种主微量元素含量的国家标准方法（中华人民共和国国土资源部，2011），该方法适用于岩石、土壤和水系沉积物等。以XRF测定为主，结合其他测定方法的化探样品分析方案（安徽省质量技术监督局，2014）已在国内广泛应用，取得了显著的经济和社会效益。XRF还被广泛应用于分析硅酸盐单矿物、有色、黑色、稀有以及非金属矿物矿石等（刘玉纯等，2019）。近年来，便携式X射线荧光光谱仪（PXRF）被用于在野外现场对岩石、矿石和化探样品进行多元素快速分析（张广玉等，2017）。

1.2 电感耦合等离子体光谱分析

电感耦合等离子体光谱（ICP-AES）是对地质样品中主量元素快速准确分析的重要手段（图1）。偏硼酸锂熔矿-超声提取-ICP-AES同时测定硅酸盐岩石样品中的10种主量元素的方法是分析硅酸岩样品效率较高的方法。该方法具有简单快速、试剂加入量少、空白低、分析结果重现性好、污染小等特点。该方法在改进后被用于测定矿石中的高含量元素（魏灵巧，2012）。采用混合酸溶解-ICP-AES测定样品中的高含量元素，有独特的优势，如工作温度低、操作简便，而且不加入任何金属离子，方法检出限低，能准确分析除SiO_2以外的其他主量元素和多种微量元素。也有研究人员直接将封闭混合酸溶后的样品，使用配备耐氢氟酸进样系统的ICP-AES测定矿石中的高含量铌、钽、钨等元素（王蕾等，2014；马生凤等，2016）。

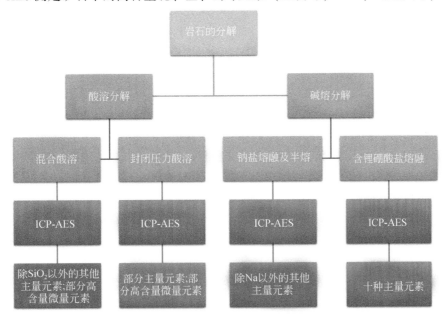

图1 ICP-AES在造岩元素分析中的应用（据李冰等，2011）

1.3 其他仪器方法

激光剥蚀电感耦合等离子体质谱（LA-ICP-MS）是固体微区分析的前沿技术。目前，LA-ICP-MS 不仅被用于需要高空间分辨的微区元素和同位素组成分析，也被用来对岩石和土壤样品进行整体组分分析，以获得精密度和准确度与电子微探针相当的主量元素含量（Liu et al.，2013；Zhu et al.，2013；陈春飞等，2014）。激光取样量较小可能会带来样品代表性不足的问题，通过将初始粉末进一步细粒化以提升样品均匀性的方法可以解决这一问题。另外，元素分析仪、高频红外分析仪等已被逐渐用于地质样品中碳-硫、碳-氢-氮或者碳-氢-氮-硫的分析。

2 微量元素分析

准确测定地质样品中的微量元素含量，对理解各种地质过程非常关键。地质样品的微量元素分析主要包括全岩/全矿物溶液 ICP-MS/ICP-AES 分析和微区原位分析。用于微量元素微区原位分析的仪器主要有电子探针、激光剥蚀电感耦合等离子体质谱及二次离子质谱等。近十年来，微量元素分析在化学前处理和仪器分析方法方面都取得了重要进展。

2.1 全岩样品溶液 ICP-MS/ICP-AES 分析

电感耦合等离子体质谱（ICP-MS）分析方法已成为全岩微量元素分析测试的主流技术，但针对不同基体的样品，往往需采取差别的前处理方法，以获得更加准确的分析结果。近十年来，化学前处理方法取得了长足进步。Zhang 等（2012b）重新评估了经典的 $HF-HNO_3$ 密闭溶样法，发现 HNO_3 的加入抑制了 HF 分解难溶硅酸盐矿物的能力，提出只使用 HF 进行第一步酸溶，在样品粉末颗粒足够细小（如 <30 μm）和样品量足够少（<100 mg）时，可避免出现氟化物沉淀。针对长英质岩石难溶矿物如锆石等分解不完全的问题，Zhang 等（2012a）采用氟化氢铵代替敞口式消解法中的氢氟酸，提高消解温度，有效分解了难溶矿物，提高了 Zr、Hf 和 HREE 等元素的回收率，且大大缩短了溶样时间；在常压消解环境下，避免了难溶氟化物 AlF_3 的形成，特别适合富铝样品如铝土矿的微量元素分析（张文等，2015）。Chen 等（2017）认为难溶氟化物沉淀可以通过加入 1 mL 逆王水和 0.5 mL 氢氟酸，在 190℃ 条件下密闭消解 2 h，便可完全溶解。张彦辉等（2018）自行设计加工了电加热增压消解装置，在同一消解罐中先微波消解（180℃，60 min）再增压消解（200~220℃，40~50 min），实现了难溶矿物（含铌钽铀矿物、锆石等）岩石样品的完全溶解。Wang 等（2015）对比了不同的溶样装置对微量元素含量测定的影响，发现密闭溶样杯消解能准确测定挥发性元素 S、Se 和 Te，但是 W 的空白较高，而高压消解仪较密闭溶样杯 Ba、Bi、Tl 的空白相对较高。

精确测定地质样品中低含量的铂族元素及 Re-Os 同位素是一项富有挑战性的工作，近年来我国在这一领域的样品消解过程取得较好的研究成果。卡洛斯管消解法是常用的铂族元素分析前处理方法，它采用密闭溶样，在加热过程中容易导致卡洛斯管爆炸。Qi 等（2010）在自行设计的分解装置中用 HNO_3 预先分解硫化物，再进行封管密闭溶解，此方法大大降低了爆炸的可能性，120 mL 的卡洛斯管可以满足 3 g 黄铁矿的溶解。一些新的消解装置被研发以提高铂族元素消解效率或者简化实验流程，如新设计的大型（120 mL）特氟龙内衬密闭溶样器结合 $HF+HNO_3$ 被用于代替卡洛斯管消解 Ni-Cu 硫化物、超基性岩和基性岩，显著提高消解效率（特别是硅酸盐矿物）并避免了卡洛斯管爆炸的风险（Qi et al.，2011），不足之处是无法分析具有挥发性的 Os。传统的卡洛斯管一次性使用，无法清洗，实验成本较高，Qi 等（2013）设计了一种新型的、可重复使用的大容量（200 mL）卡洛斯管，并配备了可拆卸的玻璃+Teflon 塞子和不锈钢密封套件，通过设定合适的溶样条件，该卡洛斯管可以用于低含量地质样品的 PGE 和 Re-Os 同位素分析。Li 等（2014a）建立了同位素稀释法 ICP-MS 和 N-TIMS 同时测定 PGE、Re 和 Os 同位素的分析流程，样品逆王水溶解后，CCl_4 萃取 Os，残余溶液经阳离子树脂（AG50W-X8），获得含 Re、PGE 和干扰元素 Mo、Zr、Hf 的溶液，再经 N-苯甲酰基苯基羟胺（BPHA）树脂将 PGE-Re 与 Mo、Zr、Hf 等基体元素分开。Wang 和 Becker（2014）实现了同一化学分析流程同时测定全岩样品中硫属元素（S、Se、Te）和高度亲铁元素的含量，并指出这些元素在岩石粉末中存在不同尺度的不均一性。另外，针对感兴趣的一些特殊元素（如 Ge、In、Te 等三稀元素）（程秀花等，2013；陈波等，2014；孙朝阳等，2016）或贵金属元素（如 Au、Ag 和 PGE 等）（杨林等，2013；李志伟等，2015），ICP-MS 准确测定还需要采用不同的溶样条件和干扰元素扣除方法。

2.2 矿物样品电子探针分析

电子探针（EPMA）常用于分析矿物的主量元素和部分微量元素含量。尽管 EPMA 具有较高的检测限，但由于其具有高空间分辨率、无损分析及低基体效应等优点而被用于测定矿物中的一些高含量微量元素。近些年，EPMA 在分析石英、橄榄石、金红石、磁铁矿等矿物的微量元素方面取得了一定进展（张迪等，2019）。

合适的空白校正方法对于电子探针微量元素分析至关重要。为了降低石英微量元素分析的检测限，Cui 等（2019）考察了石英微量元素分析时电子束束斑、电流与 X 射线信号强度的关系，发现 Al-Kα 和 Ti-Kα 的样品峰信号和背景信号强度在高电流和小束斑的情况下均发生明显变化。为了避免信号波动，对石英进行微量元素分析最好采用大束斑（如大于 20 μm）和较高电流（500 nA）。研究表明采用多点背景校正方法可以明显提高 Al 含量分析的准确度，而对 Ti 含量分析无影响，而且通过多点校正方法获得的石英样品微量元素背景信号强度可以直接用于其他未知石英样品，节约了分析时间，并减少了样品损伤（Cui et al.，2019）。

除了空白校正，针对样品选择合适的仪器分析条件，也可提高分析的准确度和精度。王娟等（2017）用 CAMECA SXFive 电子探针对金红石标样 R10 进行了 Al、Si、Ti、Fe、Cr、Zr、V、Nb 和 Ta 的分析，他们通过调整加速电压和电流、背景和峰值积分时间以及干扰谱峰处理等提高了分析精度和准确度。V 和 Fe 元素的数据波动范围在 5% 以内，而其他元素在 10% 以内，Zr、Nb、V、Fe 和 Cr 的分析结果与二次离子质谱（SIMS）和激光剥蚀电感耦合等离子体质谱（LA-ICP-MS）的推荐值在误差范围内一致。与前人电子探针分析结果相比，V、Nb 和 Fe 的测试精度有较大提高。对于磁铁矿的微量元素分析，不同的研究者也采用了不同的仪器分析条件，以达到更低的检测限。Hu 等（2014，2015）采用大电流（100 nA）和长积分时间（120~400 s）降低了磁铁矿微量元素分析的检测限，Mg、Al、Si、Ca 的检测限低于 20 μg/g，Ti、V 和 Cr 的检测限低于 30 μg/g，Mn、Co、Ni 和 Zn 的检测限低于 90 μg/g。

2.3 矿物及流体包裹体 LA-ICP-MS 原位微量元素分析

利用 LA-ICP-MS 进行矿物微区原位微量元素分析近十年也取得了重要进展。国内学者成功建立了碳酸盐矿物（Chen et al.，2011）、含水硅酸盐矿物（如角闪石、绿帘石、电气石和透闪石）（陈春飞等，2014）、石英（蓝廷广等，2017）、磷灰石（Mao et al.，2016；She et al.，2016）、磁铁矿（张德贤等，2012；Gao et al.，2013；Huang et al.，2013；孟郁苗等，2016）等矿物的 LA-ICP-MS 微量元素分析方法。对硫化物中的一些铂族元素含量也实现了 LA-ICP-MS 分析（Chen et al.，2014）。其中，对碳酸盐矿物、含水硅酸盐矿物和磁铁矿微量元素的分析（Chen et al.，2011；Huang et al.，2013；陈春飞等，2014；孟郁苗等，2016），多采用 Liu 等（2008）提出的无内标-多外标校正方法，即设定分析矿物中所有元素的氧化物含量之和为 100%。无内标-多外标校正方法的优点是无需知道内标元素的准确含量，可直接利用 LA-ICP-MS 分析矿物中所有主量和微量元素的含量，该方法省去了 EPMA 测试内标元素含量的步骤，节约了时间和成本，为了使分析结果更加准确，需分析尽可能多的元素。外部标样多为人工合成硅酸盐玻璃或者硫化物。另外一种校正方法是单内标-（单或多）外标校正方法（Norman et al.，1996）需预先用 EPMA

准确测定内标元素的含量，且需考虑外标的基体匹配问题。以磁铁矿微量元素分析为例，当以 Fe 为内标元素时，需要较大束斑（>25 μm）和 Fe 含量相对较高的硅酸盐玻璃（如 GSE-1G）作为外部校正标样才能获得准确的结果（张德贤等，2012）。除了各类型矿物，近些年也实现了透明和半透明矿物单个流体包裹体的 LA-ICP-MS 成分分析（蓝廷广等，2017）。蓝廷广等（2017）通过人工合成石英 NaCl-H_2O-Rb-Cs 和 NaCl-KCl-$CaCl_2$-H_2O-Rb-Cs 流体包裹体，使用显微测温 NaCl 等效盐度（电价平衡方法）为内标，以 NIST610 为外标，建立了针对流体包裹体成分的 LA-ICP-MS 分析方法。南京大学实验室在国内首次实现显微红外与单个包裹体 LA-ICP-MS 技术联用，目前可以对大部分透明和不透明矿物中的单个包裹体开展成分分析，其中不透明矿物包括：黑钨矿、铬铁矿、铁闪锌矿、硫砷铜矿、黄铁矿、车轮矿、辉锑矿（Pan et al.，2019）。

2.4 传统的样品制备和前处理方法与 LA-ICP-MS 的结合

传统的样品制备方法与 LA-ICP-MS 分析相结合已成为全岩样品微量元素分析的一个重要手段。全岩粉末熔融玻璃与 LA-ICP-MS 结合可以用于测定不同岩石样品的主、微量元素含量（朱律运等，2011）。针对熔融制样过程中挥发性元素 Pb 和 Zn 等容易发生丢失问题，朱律运等（2011）设计了一种双铱带高温炉，用该装置熔样可以有效抑制挥发性元素的丢失，但造成了不同程度 Cr、Ni 和 Cu 损失。Zhu 等（2013）设计了一种新的样品粉末熔融装置，由碳化硼坩埚和钼加热条构成，无需加入助熔剂便可以合成非常均一的硅酸岩玻璃。He 等（2016）采用钼纸包裹硅酸岩粉末样品置于石墨管中进行高温熔融，实现了无助熔剂的硅酸岩玻璃制备方法。岩石或矿物粉末压饼方法，也常用于制作标样，直接用于 LA-ICP-MS 的主量和微量元素分析，但是压饼法常存在样品不均一的问题。Zhang W. 等（2017）通过对岩石粉末用氟化氢铵进行预溶解，并在400℃条件下蒸干，得到非常均一的岩石粉末（<8.5 μm）后压饼，实现了硅酸盐样品 LA-ICP-MS 元素含量的准确测定。最近，Liao 等（2019）使用 LA-ICP-MS 对硅酸岩样品溶液直接进行分析，该方法改变了溶液样品传统雾化器进样的方式，有效地将与水有关的质谱干扰降低了 1~2 个数量级。而且溶液的基体效应不明显，不同的稀释倍数（80~2 000 倍）和不同浓度（2%~30% HNO_3）酸介质均获得相似的结果。与固体 LA-ICP-MS 分析相比，元素的灵敏度提高了 70~250 倍，大多数元素的检测限降低了两个数量级，且与时间分辨有关的元素分馏可以忽略。与传统雾化器进样相比，在化学前处理过程酸和超纯水的用量减少了 20~100 倍，节约了分析成本，并减少了环境污染。由于该方法很好地解决了 ICP-MS 测试时的基体效应问题，可用于直接分析具有复杂基体的样品溶液，如海水、高盐度卤水及饮料等（Liao et al.，2020）。

2.5 矿物微区面扫描技术

矿物微区面扫描技术的发展大大提高了矿物微量元素及其同位素的空间分辨率（Zhu Z. Y. et al.，2016；汪方跃等，2017；范宏瑞等，2018；周伶俐等，2019）。Zhu Z. Y. 等（2016）利用 LA-ICP-MS 对含磁铁矿、黄铁矿和菱铁矿样品进行了面扫描，发现用 Fe 作为内标元素比用 S 能够获得更加准确的结果，且磁铁矿和菱铁矿之间的基体效应可以忽略。汪方跃等（2017）建立了 LA-ICP-MS 面扫描分析技术并利用 Matlab 软件开发了一套数据处理程序，利用该技术可以在 2 h 内分析 3 mm×3 mm 区域，并同时给出单个元素或元素对比值在二维平面的分布特征。

3 放射成因同位素分析

同位素地球化学以地球和宇宙天体中同位素的形成、丰度以及自然变化过程中的分馏、演化规律为理论基础，开展地质样品或天体样品的计时、示踪和测温等研究。同位素地球化学按照应用研究可分为同位素示踪和同位素年代学两部分，按照同位素成因可分为稳定同位素地球化学和天然放射成因同位素地球化学。后者主要研究天然放射性同位素母体–子体的同位素比值及其规律和应用，如 ^{40}K-^{40}Ca-^{40}Ar、^{87}Rb-^{87}Sr、^{147}Sm-^{143}Nd、^{176}Lu-^{176}Hf、^{238}U-^{206}Pb、^{232}Th-^{208}Pb、^{187}Re-^{187}Os 和 ^{14}C 等同位素体系。下面重点对我国学者在 U-Th-Pb、Rb-Sr、Sm-Nd、Lu-Hf 和 Re-Os 五个放射性同位素体系研究中所取得的进展进行评述。

3.1 U-Th-Pb 同位素

U-Th-Pb 体系是最早使用的同位素地质年代学体系，锆石形成时富含 U 和 Th 而几乎不含 Pb，是理想的 U-Th-Pb 定年矿物。LA-ICP-MS 运用在锆石 U-Pb 年代学研究中的进展显著。Xie 等（2017）在束斑为 5.8～7.4 μm 下获得 ^{206}Pb/^{238}U 值的准确度优于 1%。Hu 等（2012）研制了国内首个解决"激光信号脉冲效应"的信号匀化装置，解决了 LA-ICP-MS 固有的激光信号脉冲效应问题，成功地建立了横向空间分辨率 10～16 μm、纵向亚微米的高精度锆石 U-Pb 定年技术。此外，其他副矿物定年也取得重要进展。Yang 等（2014）建立了氟碳铈矿、钛榴石 U-Pb 定年方法。Luo 等（2019）以锆石 91 500 为外标对黑钨矿进行校正，在剥蚀池前加入水蒸气，获得的年龄与同位素稀释热电离质谱（ID-TIMS）年龄一致。Huang 等（2020b）利用准分子激光和 MC-ICP-MS 联用，实现了轻稀土富集的副矿物如独居石、榍石、钙钛矿的 U-Pb 年龄和 Sm-Nd 同位素同时分析。SIMS 在 U-Th-Pb 年代学方面也取得进展，如氧源离子束技术实现显生宙锆石 U-Pb、Pb-Pb 精确定年方法（Li et al.，2010），模拟高斯离子束技术在小于 5 μm 束斑尺度下获得的 U-Pb 年龄的精密度和准确度为 1%～2%（Liu et al.，2011），动态多接收器法拉第杯技术准确测定显生宙锆石的 ^{207}Pb/^{206}Pb 年龄（Liu et al.，2015），并建立了氟碳铈矿和锆石 Th-Pb 定年方法（Ling et al.，2016；Li Q. L. et al.，2018）。Yang 等（2012）利用 NanoSIMS（nano-scale secondary ion mass spectrometry，纳米级二次离子质谱）分析锆石和斜锆石 U-Pb 和 Pb-Pb 年龄，锆石分析束斑 5 μm，斜锆石分析束斑 2 μm，分析精度为 2%～3%。

3.2 Rb-Sr 同位素

Rb-Sr 同位素体系广泛应用于地球科学、环境科学、考古学等领域。传统的 Rb-Sr 同位素体系常用 TIMS 和 MC-ICP-MS。全岩样品的 Rb-Sr 同位素分析方法耗时费力，对操作人员有很高的要求。另外，岩石矿物颗粒可能存在包裹体和环带，也可能形成于不同期次，这样获得的 Rb-Sr 同位素比值和年龄可能不具有地质意义，甚至误导我们的研究，因此需要发展更精细的 Rb-Sr 同位素年代学方法。在微区原位 Rb-Sr 同位素和定年方面，Huang 等（2015）利用 LA-MC-ICP-MS 气溶胶分束技术同时分析矿物中两种同位素体系，如 Rb-Sr 和 Sm-Nd 或 Sm-Nd 和 Lu-Hf 同位素。Tong 等（2016）详细调查 LA-MC-ICP-MS 分析过程中 Sr 同位素的干扰来源，通过合理的校正策略，建立了低 Sr 含量单斜辉石的 Sr 同位素分析技术。Liu W. G. 等（2020）对高 Rb/Sr 值的地质样品采用氟化物共沉淀法与阳离子交换树脂相结合的方法对 Sr 进行分离，实现了 Sr 同位素比值的高精度测定。

3.3 Sm-Nd 同位素

Sm-Nd 同位素体系是地球化学和地质年代学的重要组成部分，是探讨岩石成因和壳幔演化最重要的示踪手段之一。目前广泛采用的 Nd 同位素测定方法需要先通过化学方法使样品溶解，进行至少两次离子交换分离纯化使得 Sm 与 Nd 彻底分开，最后用 TIMS 对得到的纯净 Nd 组分进行质谱测试，从而获得研究对象的 Nd 同位素组成。对于一些 Sm 和 Nd 含量极低的超基性岩如橄榄岩，一般还需要先利用氢氧化铁共沉淀方法将稀土元素预富集。Yang 等（2011）利用 ^{149}Sm-^{150}Nd 稀释剂，用 MC-ICP-MS 测试了 USGS 系列岩石粉末标样的 Nd 同位素，长期精度优于 15×10^{-6}。Chu 等（2014）首次报道了无需预富集处理的超基性岩 Nd 同位素的测试方法。Yang 等（2019）首次报道了氟碳铈镧矿的 Sm-Nd 同位素原位分析结果。Xu 等（2015，2018）使用 Neptune 的 X 加 Jet 锥，同时在气路中通入微量 N_2，Nd 的信号可以提高 2.5～3 倍。

3.4 Lu-Hf 同位素

Lu-Hf 同位素是近年来发展最为迅速的同位素定年和地球化学示踪技术之一。锆石中含有很高含量 Hf 和较低含量的 Lu 元素，其 Hf 元素的含量可达 0.5%～2% 甚至更高，又由于其较高的封闭温度使锆石成为 Hf 同位素测定的理想矿物。锆石中由于 Lu/Hf 值较低（^{176}Lu/^{177}Hf 值通常小于 0.002），因而由 ^{176}Lu 衰变生成的 ^{176}Hf 极少。因此，锆石 Hf 同位素比值可以代表锆石形成时的 ^{176}Hf/^{177}Hf 值，从而为讨论其成因提供重要信息。Yang 等（2010）开发同一份溶液样品先后分离 Lu-Hf、Rb-Sr 和 Sm-Nd 同位素，解决了取样不同造成的同位素体系脱耦问题。Ma 等（2019b）通过加入 ^{176}Lu-^{180}Hf 混合稀释剂，采用 Ln 树脂进行化学纯化，可准确测定 Lu、Hf 含量及 Hf 同位素比值，方法空白低、灵敏度高，拓展了 Lu-Hf 同位素在地球化学、宇宙化学和环境科学等领域的应用。Yang M. 等（2020）利用 MC-ICP-MS 测定了 13 种中国地球化学岩石标准参考物质，其 Lu、Hf 含量及 Hf 同位素组成均一，表明这些标准物质可在 Lu-Hf 同位素分析中作为岩石标准参考物质。Lin 等（2020）首次采用线性回归方法对 Hf 同位素稀释剂的同位素组成进行精准标定，得到的 Hf 同位素稀释剂的测试结果在测试精度、准确度和稳定性 3 个方面均优于标准—校品间插法（standard-sample bracketing，SSB）和 C-SSBIN 校正方法。随着 Hf 稀释剂同位素组成测试精度的提高，Lu-Hf 等时线年龄的测定精度也随之提高。该方法同样可以应用到其他同位素稀释剂的准确标定中。

3.5 Re-Os 同位素

Re 和 Os 是具有亲铁性、亲铜性（亲硫性）和亲有机质的元素，在铂族矿物、硫化物、超基性岩中的含量相对较高（ng 量级），可以富集在黑色页岩中。利用 Re-Os 放射性同位素体系可以对金属硫化物、陨石、超基性岩石以及富含有机质样品进行 Re-Os 定年，这对研究岩石和金属矿床的形成时间、地幔演化、沉积地层的绝对年代、天体地球化学、岩石成因以及油气成藏时间等都有非常重要意义。

Carius 管溶样法是目前 Re-Os 同位素分析的主要样品制备方法。Qi 等（2010）采用改进的卡洛斯管原位蒸馏分离 Os，缩短了流程，提高了工作效率。Zhu L. Y. 等（2016）尝试了超细硫化物粉末压片 Os 同位素原位分析，并于 2019 年进行了改进（Zhu et al., 2019），采用全离子计数配置，对 ^{187}Os+^{187}Re 和 ^{188}Os 以及 ^{185}Re 进行静态测量，引入 Au 标准溶液交叉校准不同离子计数器的检测效率，采用标准–样品间插法（SSB）校正，当 ^{187}Re/^{188}Os 小于 0.055 时，^{187}Os/^{188}Os 的测试精度优于 0.8%。因此，LA-MC-ICP-MS 的 Re-Os 分析方法和应用是 Re-Os 同位素分析的一个重要发展方面。

4 非传统稳定同位素分析

非传统稳定同位素体系是相对传统稳定同位素体系（即 H、C、N、O 和 S 体系）而言，主要包括金属元素（如 Li、Mg、Ca、Ti、Fe、Cr、Ni、Cu、Zn、Ba、V 等）、类金属元素（如 B、Si、Ge、Sb 等）和一些非金属元素（如 Cl、Se、Br 等）的同位素体系（Teng et al., 2019）。在稳定同位素地球化学诞生后的将近 50 年内，由于传统的热电离质谱较弱的离子化效率和显著变化的仪器质量分馏，使得非传统稳定同位素测量的精确度和准确度受到极大的限制。自 20 世纪末至 21 世纪初以来，随着质谱技术的发展，人们能够利用 MC-ICP-MS 和新一代 TIMS 准确测定大量的非传统稳定同位素体系，由此诞生了非传统稳定同位素地球化学这一新兴学科。过去 20 年来，稳定同位素体系在天体化学（Zhu et al., 2001；Burkhardt et al., 2014；Bourdon et al., 2018）、地幔演化（Zhu et al., 2002；Tang et al., 2007；Li W. Q. et al., 2016）、地壳形成（Foden et al., 2015；Yang et al., 2016；Greber et al., 2017）和矿床成因（Graham et al., 2004；Wilkinson et al., 2005；Rouxel et al., 2008）等领域已经得到了广泛而深入的应用，越来越得到人们的重视，已然成为地球科学领域最具活力的学科之一，而高质量的分析技术是开展稳定同位素地球化学研究的前提和基础。

4.1 分析方法

4.1.1 质谱仪稳定性

高温地球化学过程产生的稳定同位素分馏尺度往往较小，因此对同位素数据的精度也要求特别高。为此，我们需要精心维护仪器，保持仪器状态稳定可靠。MC-ICP-MS 的稳定性比较容易受到诸多因素的干扰，如仪器参数、酸度、浓度不匹配、基体效应以及环境温度等（Yang，2009；Zhang X. C. et al.，2018）。如 Mg 同位素，图 2 揭示了仪器间温度变化对同位素测量的显著影响。对近几年测量的纯溶液国际标样（DSM-3）的 Mg 同位素组成结果的统计表明，如果温度不稳定，标准溶液的 Mg 同位素测定误差（2SD）可高达 0.08‰，这个精度不足以研究高温下的 Mg 同位素分馏。

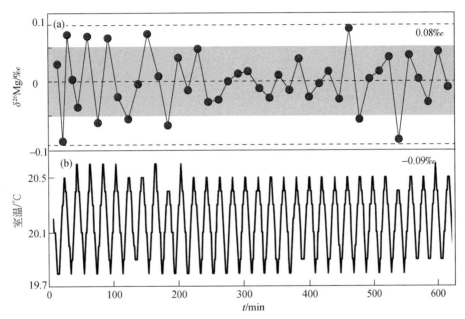

图 2 （a）振荡室温下 Mg 同位素测试数据图及（b）对应的室温记录

红色圆圈代表 DSM-3 相对于自己的 δ^{26}Mg 测量值。当室温在 10 min 内变化 0.7℃时，δ^{26}Mg 的测量值有较大的离散程度，平均 δ^{26}Mg 值为 (0.00±0.08)‰ (2SD，n = 42)。图中灰色区域代表中国科学技术大学金属稳定同位素实验室 δ^{26}Mg 测量的长期精度（2SD≤0.05‰）

4.1.2 化学提纯流程

因为 MC-ICP-MS 对溶液性质非常敏感，为获得稳定的仪器状态和进行精确的仪器质量分馏校正，自然样品在质谱测量前都必须经过高质量的化学提纯流程，以减少杂质（如基体元素、目标元素的干扰元素以及有机质等）的干扰。这些化学流程需要在超净实验室进行。此外，化学提纯过程中，为了避免同位素在离子树脂上可能产生的质量分馏，要求回收率尽可能达到 100%，或者可利用双稀释剂法校正产生的质量分馏。同样以 Mg 为例，图 3 揭示了化学提纯过程中选用不同浓度酸淋洗液时 Mg 与基体元素的分离效果。

4.1.3 质谱测量

为确保自然样品稳定同位素组成数据的准确性和精确性，在质谱测量过程中需要监测基准标样，校正基体效应和仪器测定过程中的质量歧视，严格控制数据质量。常见的同位素基准标样有：Mg，DSM-3；Si，NBS-28；Ca，SRM915a；Fe，IRMM-014；Cu，NIST976 等。除了国际基准标样，还需要对不同类型的岩石标样进行对比，例如 USGS 的橄榄岩标样（PCC-1 和 DTS-2）及玄武岩标样（BCR-2、BHVO-2 和

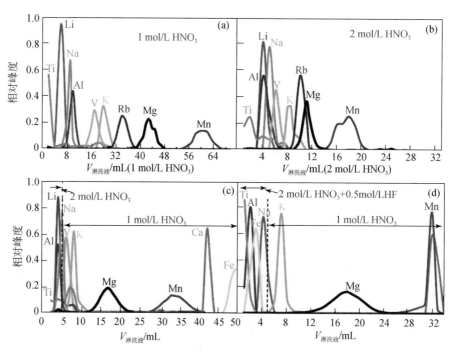

图3 不同浓度酸淋洗条件下 Mg 的化学分离效果图（据 An et al., 2014）

BIR-1）等。

样品-标样间插法和双稀释剂法是最常见的仪器质量分馏校正方法。样品-标样间插法假设标样和样品具有相同的质量分馏效应，因此可以用已知同位素组成的基准标样来校正未知样品的同位素数据（Albarède and Beard, 2004）。如果某一元素的稳定同位素数量不达到4个，那么该方法几乎是唯一的校正方法。如果某元素的稳定同位素数量达到4个，则可以利用双稀释剂法对仪器测量和化学提纯过程中产生的质量分馏进行校正（Johnson and Beard, 1999; Siebert et al., 2001）。在质谱分析时或在化学流程前，在样品中混入合适比例、已知同位素组成的双稀释剂，该方法可以一定程度上降低对回收率的要求。此外，元素添加法也可以用来校正仪器质量分馏，如利用 Cu 和 Zn 相互校正（Maréchal et al., 1999; Yang L. et al., 2018）。

按照实验室数据质量的控制规范，需要通过空白样监控化学全流程的本底，通过重复样监控分析结果的精度，以及用岩石标样监控分析结果的质量。通过测量已知成分的合成标样（即可以用单一元素标样和不同基质元素的混合得到），可以检查本实验室数据的准确性和精确性。人们通常用样品与标准的同位素比值的相对千分差（‰）来表示稳定同位素组成（即 δ 值），定义式如下：$\delta^x M = 1000 \times [(^XM/^YM)_{样品}/(^XM/^YM)_{标准}-1]$（‰）（$^XM/^YM$ 指的是元素 M 的质量数分别为 X 和 Y 的同位素的摩尔比值）。目前数据的误差一般用两倍标准偏差（2SD）表示，如果用两倍标准误差（2SE），一定要给出测量次数（n），还要考虑转换参数 Student's T（2SE = 2Student's T$\times$SD$/\sqrt{n}$）。内部测量精度可以反映仪器性能，必须小于外部测量精度。对超出数据质量控制限的测试结果要检查原因，做相应改进后重新分析。对于分馏较大的样品，最好重复测量，确保观察结果无误。

4.2 近期我国非传统稳定同位素分析方法进展

（1）Li 同位素：Lin 等（2016）在利用 MC-ICP-MS 测量 Li 同位素组成的过程中，通过使用5% NaCl 溶液冲洗显著降低了 Li 的仪器本底和记忆效应，无需严格的基体匹配便能够准确测定 Li 同位素组成。Lin 等（2017）利用 LA-MC-ICP-MS，通过优化信号强度、同位素稳定性和强度匹配，改进了硅酸盐原位 Li 同位素分析方法。Lin 等（2019）利用 MC-ICP-MS 测定了11种碳酸盐岩标准物质的 Li 同位素组成，精度可达0.4‰，拓宽了Li同位素在古气候和古环境领域的进一步应用。

（2）Mg同位素：An等（2014）针对低Mg地质样品（MgO质量分数小于1%）建立了优化的Mg化学分离流程，通过HNO_3+HF的结合有效去除了基质元素。利用MC-ICP-MS测量获得多个岩石学标样高精度的Mg同位素组成，$δ^{26}Mg$的长期测量精度好于0.05‰（2SD）。Bao等（2019）针对高K低Mg样品，在阳离子交换树脂两柱分离Mg之前加入沉淀的步骤，利用氢氧化钠置换法形成氢氧化镁沉淀，进行初步的K分离。Bao等（2020b）利用两柱分离法对生物和岩石样品中的Ca和Mg进行同时提纯，首先用DGA树脂分离出Ca，再用AG50W-X12树脂分离出Mg。

（3）Si同位素：Zhang等（2015）利用MC-ICP-MS测定稳定Si同位素组成，分析了C、N、O、H等多原子以及进样气流、Cl-基质等因素对测试的影响，通过样品标样间插法校正获得的$δ^{29}Si$和$δ^{30}Si$的长期精度分别为0.06‰~0.08‰和0.06‰~0.10‰。Yuan H. L.等（2016）采用Mg同位素作为内标校正并结合样品标样间插法进行Si同位素测试，降低了质谱及实验条件稳定性对测定结果准确度和精密度的影响，获得标样Si同位素组成在误差范围内与推荐值一致，而$δ^{30}Si$测试精度优于0.05‰（2SD）。

（4）K同位素：Li W. Q.等（2016）提出了以氘气（D_2）作为碰撞反应气体消除ArH^+对$^{41}K^+$的干扰，在IsoProbe MC-ICP-MS上实现了高精度的K同位素分析，对$^{41}K/^{39}K$值的测量达到了优于0.07‰（2SE）的内部精度和优于0.21‰（2SD）的外部重现性。使得自然界细微的低温K同位素平衡分馏也能够得以识别。

（5）Ca同位素：Feng等（2018）利用DGA树脂一柱分离的方法建立了一个简单快速分离Ca的化学流程，通过三步提纯和只用硝酸和去离子水的方式实现了节时、高回收率、低本底的化学分离。Li M.等（2018）首次在Nu Plasma 1700 MC-ICP-MS上建立高精度的Ca同位素分析方法，通过在法拉第杯（L7）旁设计一个模拟处理器（dummy bucket），用来接收$^{40}Ar^+$和$^{40}Ca^+$，以去除它们对^{42}Ca的干扰，提高Ca同位素测量的准确度。利用样品标样间插法，获得12个国际标准物质和纯Ca溶液的Ca同位素组成。$δ^{44/42}Ca$长期测量精度好于0.07‰。He D.等（2019）针对碳酸盐样品，建立了不需要化学分离流程的Ca同位素分析方法。利用MC-ICP-MS进行测量，通过有效地校正$^{40}Ar^1H_2^+$和Sr^{2+}的干扰，获得高精度的Ca同位素数据。Ca同位素的长期测量精度优于0.08‰（2SD）。

（6）Ti同位素：He等（2020）采用Ln-spec树脂结合AG50W-X12树脂的方法提纯Ti，高效节时并且保证极高的Ti回收率。利用该方法测量Ti同位素组成，发现岩石标样和石英二长岩中一组单矿物中存在重要的Ti同位素变化，表明Ti同位素是示踪地质过程的潜在工具。

（7）V同位素：Wu等（2016）优化了V的化学纯化流程和质谱分析技术。采用阳离子树脂和阴离子树脂四柱纯化，在MC-ICP-MS上利用$10^{10}Ω$放大器测量^{51}V，$10^{11}Ω$放大器测量其他同位素，V同位素长期外部精度优于0.08‰（2SD）。

（8）Cr同位素：Zhu J. M.等（2018）改进了双稀释剂MC-ICP-MS高精度测定Cr同位素的Cr纯化方法。通过三步色谱分离Cr，灵活，易操作，解决了低Cr、高基质元素样品Cr提纯困难的问题，改进后的纯化流程适用于多种地质、环境样品高精度Cr同位素分析。

（9）Fe同位素：Liu等（2014）利用AG-MP-1M阴离子树脂一柱分离Fe和Cu，有助于高效利用Fe-Cu同位素研究地质问题。He Y. S.等（2015）报道了硅酸岩、碳酸岩等常见岩浆岩及变质岩标样的高精度Fe同位素数据，并首次报道了页岩、碳酸盐及黏土等沉积岩标样的Fe同位素组成，精度优于0.03‰，有助于实验室之间的对比。Zhu C. W.等（2018）利用样品标样间插法、镍掺杂样品标样间插法、^{57}Fe-^{58}Fe双稀释剂样品间插法，分别独立校正了Fe同位素分析过程中的仪器质量分馏，获得的纯Fe溶液及标准物质测试精度都优于0.05‰，并首次给出了NIST SRM3126a的推荐值。

（10）Zn同位素：Chen等（2016）用MC-ICP-MS建立了高精度Zn同位素分析方法。发现NIST SRM 683锌块Zn同位素组成非常均一，可成为很好的Zn同位素国际标样，同时测定了17种常见岩石标样的Zn同位素组成。$δ^{66}Zn$长期测试外精度优于0.05‰。Yang等（2018b）使用双稀释剂法在3个实验室的

MC-ICP-MS 上分别测定了 7 块 NIST SRM 683 锌块共 363 个数据点的 Zn 同位素组成，结果表明 NIST SRM 683 锌块具有高度均一的 Zn 同位素组成。$\delta^{66}Zn_{JMC\text{-}Lyon}$ = 0.12‰ ± 0.04‰，非常接近全硅酸盐地球的平均锌同位素组成。推荐 NIST SRM 683 作为新一代的锌同位素国际标准物质。

（11）Rb 同位素：Zhang Z. Y. 等（2018）采用一根 Sr-Spec 树脂实现了 Rb、K、Ba、Sr 与基体元素的有效分离，并利用 MC-ICP-MS 对 Rb 同位素组成进行了精确测定。

（12）Mo 同位素：Li 等（2014b）提出采用一种新的树脂，即将 N-苯甲酰-N-苯基羟胺负载于聚甲基丙烯酸型大孔聚合物 CG-71 树脂上制成小体积的 BPHA-CG-71 树脂进行 Mo 的化学提纯。这种方法具有较高的元素分离效率和回收率。Liu 等（2016）则使用阴阳离子交换树脂柱双柱法对低 Mo 样品进行化学提纯，采用样品标样间插法和双稀释剂法分别进行仪器的质量分馏校正，获得了一系列地质标准物质的 Mo 同位素组成。NIST SRM 3134 长期外部精度优于 0.03‰（2SD），低 Mo 地质样品的精度优于 0.04‰（2SD）。Feng 等（2020b）使用 TRU 特效树脂进行快速、高效的一柱分离方法来进行 Mo 的化学提纯。

（13）Cd 同位素：Li D. D. 等（2018）利用 MC-ICP-MS 测定了 5 种 Cd 标准溶液和 7 种地质标准物质（土壤、沉积物、铁锰结核）的 Cd 同位素组成，为今后的实验室间校准提供了新的推荐值。Tan 等（2020）提出一种改进的 Cd 分离流程，加入了 ^{111}Cd-^{113}Cd 双稀释剂，采用 AGMP-1 树脂进行两柱的分离纯化，有效去除干扰元素，空白低、回收率高，可准确测定低 Cd 含量和复杂基体地质样品和环境样品。Cd 同位素准确测定拓展了在全球镉循环、壳幔演化和人、动物体内镉代谢过程等方面的应用。

（14）Ba 同位素：Nan 等（2015）采用阳离子树脂分离提纯 Ba，首次在国内利用样品-标样间插法获得高精度的 Ba 同位素数据；Zeng 等（2019）针对珊瑚等高 Ca、Sr 低 Ba 含量样品，开发了石英柱子分离提纯碳酸盐 Ba 的方法；Tian 等（2019）针对难溶的重晶石样品，建立了碳酸钠置换法-双稀释剂法测量 Ba 同位素方法。

5 地质样品前处理技术

现代主流的地质样品元素及同位素测试设备（如 ICP-MS 和 MC-ICP-MS 等质谱仪）都以溶液进样为主，通常需要将地质样品处理成均匀、澄清的溶液样品，这一过程被称为消解过程。地质样品前处理是元素和同位素分析的关键步骤，极大地影响了分析测试的数据质量、生产效率、工作成本和环境安全等（Hu and Qi，2014）。我国学者近年来在地质样品前处理方面开展了大量研究工作，取得了一系列原创性成果。

Zhang 等（2012a）和 Hu 等（2013）建立了新型氟化氢铵和氟化铵地质样品前处理方法。这是近半个世纪以来首次在国际上提出可取代氢氟酸消解硅酸盐矿物的新型无机试剂。氟化氢铵和氟化铵具有较高的沸点（238℃和250℃），可以在常压条件下快速消解各类硅酸盐矿物（1~2 h），包括锆石等。氟化氢铵和氟化铵消解法在常压下工作不易形成难溶的氟化铝沉淀，避免了难溶氟化物导致微量元素丢失的问题，确保富铝地质样品（如铝土矿、风化壳、富铝沉积岩和冰碛岩等）中微量元素的准确测试（Hu et al.，2010；Zhang et al.，2012a；Hu et al.，2013；Zhang W. et al.，2016）。氟化氢铵和氟化铵消解法具有高效、安全、低成本、灵活和工作量小等优点，是建立地质样品的绿色地球化学分析的一个重要创新，具有非常强的应用前景。目前 NH_4HF_2 消解法已经在测定富铝、富难溶副矿物大陆地壳样品中的微量元素方面显示了极好的应用效果，如 Gaschnig 等（2016）利用 NH_4HF_2 消解法测试了不同地质历史时期冰碛岩的微量元素含量，用于估算大陆上地壳的化学组成。此外，国际原子能机构近年来开展核物质野外快速检测，NH_4HF_2 消解法凭借高效、灵活和安全等优点被推荐应用于核爆炸物、核废弃物或者核常规检测等野外现场检测中（Hubley et al.，2016；Mason et al.，2017）。

近年来，卤素在行星演化、壳幔分异、矿床形成、环境演变等研究中表现出巨大的应用前景（Weis et al.，2012；Clay et al.，2017）。卤素分析的主要难点在于如何消除样品前处理过程中卤素的挥发性丢失。He 等（2018）开发了密闭快速酸消解法，将土壤、沉积物等样品置于混酸溶液（1 mL HNO_3 + 1 mL HF）中于140℃下消解 15 min，利用氨水稀释样品以避免卤素挥发，经过离心后，溴和碘被完全回收于上清液中，可以用 ICP-MS 进行准确测定。但是该方法无法消解硅酸盐矿物，He T. 等（2019）改进技术，选择氟化氢铵试剂在敞口式消解器中消解岩石粉末，高沸点的氟化氢铵在分解岩石时会释放出氨气，岩石释放的卤素与氨气结合可以形成高沸点的铵盐，从而抑制了卤素的挥发性丢失。再利用氨水稀释和离心之后，可利用 ICP-MS 准确测试上清液中的氯、溴和碘。该方法具有简单、快速、高效、经济等特点，是目前国际上效率最高的地质样品卤素分析方法。此外，Li X. L. 等（2016）利用粉末压饼技术和 XRF 直接分析地质样品中卤素含量，适用于分析一些卤素含量较高的土壤和沉积物样品。

传统溶液全岩地质样品分析技术要消耗大量的强酸、强碱试剂，排放的大量有害气体和废液将对环境造成严重污染，新型绿色环保的分析测试技术受到越来越多的关注。LA-ICP-MS 具有快速高效、无酸试剂消耗和较低多原子离子干扰等优点，成为最具潜力的"绿色"分析技术之一。但地质样品 LA-ICP-MS 整体分析长期受限于样品前处理过程，因为 LA-ICP-MS 每次分析取样量仅数百纳克，取样量太小将无法代表基体复杂的地质样品的实际化学组成。提高样品的均一性（代表性）是从根本上解决这些问题的唯一途径。Zhu 等（2013）自行研制了快速熔融设备，配合氮化硼坩埚实现火山岩快速无助熔剂熔融。Zhang C. X. 等（2016）采用高温红外激光熔融岩石粉末，实现基性火山岩到中酸性花岗闪长岩无助溶剂熔融玻璃的快速制备（图4）。Bao 等（2016）和 He 等（2016）分别选择将岩石粉末放置于氮化硼坩埚或 Mo 金属腔和石墨管组成的容器中，利用高温马弗炉可以将岩石粉末快速熔融为硅酸盐玻璃。对于熔点更高的橄榄岩（1 725～1 850℃）（图4），Zhang S. Y. 等（2019）提出利用钠长石（1 450℃）作为助熔剂，制备橄榄岩-钠长石玻璃。这些硅酸盐玻璃可以被 LA-ICP-MS 直接分析，同时报道超过40个主量元素和微量元素组成。相对于以上高温熔融玻璃制备技术，Zhang W. 等（2017）提出利用化学手段（氟化氢铵消解）实现地质样品纳米级细粒化和均匀化，结合粉末压片技术，实现对基性火山岩到中酸性花岗闪长岩10个主量元素和34个微量元素同时测定。这些新型地质样品处理技术具有高效、无试剂或弱试剂消耗、大幅降低劳动强度等优势，成为未来 LA-ICP-MS 地质样品整体分析广泛推广的重要技术支撑。

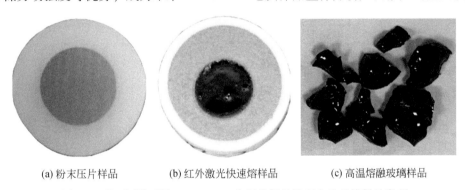

(a) 粉末压片样品　　(b) 红外激光快速熔样品　　(c) 高温熔融玻璃样品

图4　3种不同类型的 LA-ICP-MS 全岩分析前处理方法最终样品类型

6　岩矿标准物质研制和定值

近十年来，我国学者在岩矿标准物质研制方面取得了诸多可喜的成绩。一是积极参加国际实验数据的比对工作，参与国际同行的标准物质研发，作为合作者发表研究论文；二是国内同行之间的密切合作与深入交流日益加强，发挥各实验室各自的特色与专长，合作开展标准物质的研发与研究；三是开始独立自主地研发标准物质，逐步开启使用"中国标样"的新时代。我们正在逐步摆脱过度依赖欧、美、日等国标准物质的被动与不利局面，走向自立自强的喜人局面。这些岩矿标样的研究工作成果发表在 *Analytical Chemistry*、*Journal of Analytical Atomic Spectrometry*、*Chemical Geology*、*Geostandars & Geoanalytical Research*、*Analytical Methods*、*International Journal of Mass Spectrometry*、*Rapid Communications in Mass Spectrometry*、《岩石学报》、《岩矿测试》等分析地球化学权威刊物上。上述工作表明，我国的工作不但得到国际同行的认可与肯定，也充分展示出我们的实力与自信，为我国从地质大国走向地质强国迈出了关键而坚实的一步。

6.1 微区原位同位素分析标准物质研发

近十年来，我国在微区原位分析的相关标准物质研发方面做了诸多创新性的工作：①增加了锆石、独居石、钙钛矿、金红石、磷灰石、榍石、钙铁榴石、钛榴石和文石等副矿物放射性成因 Sr-Nd-Hf-U-Th-Pb 同位素标准物质数量与种类，丰富与完善了副矿物标准物质数据库，让国内外同行有了更多的选择；②矿石矿物 U-Pb 年代学标准物质的研发是近年来关键金属成矿年代学研究的热点与亮点，国际上也没有可供选择的标准物质，我国学者研发出了与成矿相关的矿石矿物 U-Pb 年代学标样（沥青铀矿、黑钨矿、锡石、铌钽矿、氟碳铈矿），为我国矿石矿物 U-Pb 成矿年代学研究提供了关键的技术支撑；③研发了稳定同位素标准物质，如锆石 Li-O 同位素、独居石 O 同位素、方解石 C-O 同位素、橄榄石和辉石系列 Li-O 同位素、磷灰石 O 同位素、黄铁矿、磁黄铁矿和黄铜矿 S 同位素、Ca-Si 同位素、Zr 同位素、锆石和橄榄石系列 H_2O 含量。此部分研究成果及参考文献请参见表 1。

表 1　天然矿物及人工合成的元素/同位素参考物质汇总

天然矿物/标准溶液/人工合成	同位素/主/微量元素	标样名称	参考文献
锆石	U-Pb-Hf-Zr-Li-O-H_2O	PL QII SA-01	Li et al., 2011, 2013；Xia et al., 2019；Huang et al., 2020a
独居石	U-Th-Pb-Nd-O	M1 M2 M3 M4 RW-1	Liu et al., 2012；Ling et al., 2017；Wu L. G. et al., 2019
钙钛矿	U-Th-Pb-Nd-Sr	AFK	Wu et al., 2013
金红石	U-Pb-Hf	JDX RMJG	Zhang L. et al., 2020
磷灰石	U-Pb/U-Th-He-O	MK-1	Wu L. et al., 2019；Yang Q. et al., 2020
榍石	U-Pb-Nd	Ontario YQ82 T3 Pakistan	Ma et al., 2019a
文石	Sr-O	GBW04704	
黑钨矿	U-Pb	MTM LB	Deng et al., 2019；Luo et al., 2019；Tang Y. W. et al., 2020
锡石	U-Pb	AY-4	Yuan et al., 2011
铌钽矿	U-Pb	Coltan 139	Che et al., 2015
氟碳铈矿	U-Th-Pb-Nd-Sr	K-9 LZ1384 MAD809	Yang et al., 2014, 2019；Ling et al., 2016
钙铁榴石	U-Pb	QC04 WS20	Yang et al., 2018a
方解石	C-O	GBW04481	Tang G. Q. et al., 2020
橄榄石和辉石	O		Su et al., 2015；Tang et al., 2019
黄铁矿、磁黄铁矿和黄铜矿	S		Bao et al., 2017；Li R. C. et al., 2019, 2020

续表

天然矿物/标准溶液/人工合成	同位素/主/微量元素	标样名称	参考文献
标准溶液	Cu	SRM3114 NWU-Cu-A NWU-Cu-B	Liu et al., 2014；Hou et al., 2016；Yuan et al., 2017
标准溶液	Ba	SRM 3104a	Nan et al., 2015；Tian et al., 2019；Zeng et al., 2019
标准溶液	Ga	SRM 994 SRM 3119a	Yuan W. et al., 2016；Zhang T. et al., 2016；Feng et al., 2019
标准溶液	Mg-Nd	GSB-Mg GSB 04-3258-2015	Li J. et al., 2017；Bao et al., 2020b
标准溶液	Mo	SRM 3134	Wen et al., 2010；Li J. et al., 2014b, 2016；Feng et al., 2020b
标准溶液	Zr	SRM 3169	Feng et al., 2020a
人工合成玻璃	同位素/主量/微量	CGSG JB GSR-2 ARM	Hu et al., 2011；包志安等，2011；宋佳瑶等，2011；Wu S. T. et al., 2019
人工合成矿物	Sr	CPX05G	Tong et al., 2016
人工合成黄铁矿与黄铜矿	Re-Os	CR-1P CO-1PF COR-1F	Zhu L. Y. et al., 2016
人工合成硫化物	PGE-Au-Pb	10th-01 10th-02 10th-03	Feng et al., 2018
人工合成白钨矿	同位素/主量/微量	CaW-1 Caw-2	Ke et al., 2020

6.2 岩石地球化学标准物质的同位素组成定值

我国的岩石地球化学标准物质的同位素组成得到了国内外同行越来越多的关注与认可。尽管我国20世纪80年代末就开始推出了系列岩石（GSR系列）、土壤（GSS系列）和水系沉积物（GSD系列）的标准物质（GBW），并对其主、微量元素组成进行了标定，但其同位素组成却极少有报道。近十年来，我国学者对这些GBW的放射性同位素、非传统同位素、Re-Os同位素和铂族元素等开展了系统的定值，例如，Li同位素（Lin et al., 2019）、B同位素（He M. Y. et al., 2011, 2013a, 2013b, 2015；Wei et al., 2014）、Mg同位素（An et al., 2014；Bao et al., 2020a）、Ca同位素（Feng et al., 2015, 2017, 2018；Zhu H. L. et al., 2016；He et al., 2017；Li M. et al., 2018；Liu F. et al., 2019；Zhang W. et al., 2019）、V同位素（Wu et al., 2016）、Cr同位素（Zhu J. M. et al., 2018；Liu C. Y. et al., 2019；Wu et al., 2020）、Fe同位素（Liu et al., 2014；He Y. S. et al., 2015；Zhu C. W. et al., 2018；Zhu et al., 2020）、Ni同位素（GBW04481）（Wu G. L. et al., 2019；Li W. H. et al., 2020）；Cu同位素（SRM3114）（Liu et al., 2014；Hou et al., 2016；Yuan et al., 2017）、Zn同位素（NIST SRM 683）（Chen et al., 2016；Yang et al., 2018b）、Ga同位素（Yuan W. et al., 2016；Zhang T. et al., 2016；Feng et al., 2019）、Rb同位素（Zhang Z. Y. et al., 2018）、Nd同位素（Ma et al., 2013）、Zr同位素（Feng et al., 2020a；Zhang et al., 2019b）、Mo同位素（Wen et al., 2010；Liu et al., 2016；Feng et al., 2020b）、Cd同位素（Li D. D. et al., 2018；Liu M. S. et al., 2020；Tan et al., 2020）、Ba同位素（NIST SRM 3104a）（Nan et al., 2015；Tian et al., 2019；Zeng et al., 2019）、W同位素（Mei et al., 2018）。这使得我们不再过度依赖美国地质调查局的USGS和日本地质调查局的GSJ岩石标准物质，同时GBW的国际使用与参考也在逐步推进，使GBW在国

际分析地球化学界占有了一席之地，而且岩石标样的同位素标定与定值也逐步向低含量的岩石样品拓展，扩展了其应用范围，充分展示了我国技术水平与能力的显著提升。

目前，广泛使用的玻璃标准是美国地质调查局的 USGS 玻璃、美国标准技术局的 NIST 玻璃和德国马普所的 MPI-DING 系列玻璃。国内地质标准玻璃物质的研究刚起步，与美国和德国玻璃标准相比，我们尽管也制备了一些玻璃标样，但其同位素定值仍然是一项长期的工作，需要进一步的研究，才能使得这些玻璃标准得到越来越多国内外同行的使用和认可（表1）。

7 岩矿分析仪器及关键部件研发

"工欲善其事，必先利其器"，地球科学的进步离不开各种现代分析仪器的快速发展。目前我国分析仪器的研发和制造整体上落后于美国、德国和日本等科技强国，我国地球化学领域很多高端分析仪器还需要进口，但近年来我国分析仪器自主创新能力不断增强，在光谱、质谱分析仪器及关键部件研发方面取得了显著进展（肖元芳等，2015；邓勃，2017；杭乐等，2019）。

7.1 光谱仪器

7.1.1 原子吸收光谱（AAS）仪器

原子吸收光谱分析的应用几乎涵盖了国民经济的各个领域，包括地质、冶金、轻工、石化、机械、环保、食品和医药等，已成为实验室必备的分析检测手段之一。相对于其他类型的分析仪器而言，我国 AAS 国产化程度较高，起步较早。近十年的进展主要体现在可靠性、稳定性、自动化程度、专用型仪器研制等方面有了进一步提高。

（1）仪器自动化和一体化程度不断提高。我国多个仪器公司在火焰石墨炉一体式设计方面取得进步，实现了由 PC 机控制软件程序切换单光束火焰与石墨炉自动切换（上海仪电科学仪器股份），自动切换单光束和双光束（安徽皖仪科技股份有限公司），首创了交直流塞曼背景同时校正技术（上海光谱仪器有限公司）。

（2）便携式、专用型原子吸收光谱研制取得进展。2010 年，北分瑞利分析仪器公司成功研发了便携式 AAS 仪器，推出属国际首创的 WFX-910 便携式原子吸收光谱仪（图5）。通过电热钨丝作为原子化器使其功耗仅为石墨炉原子化器的 6%，具有重量轻、体积小、能耗低等优势，可用于野外现场水体分析检测，获得国产原子吸收光谱仪器 BCEIA 金奖。2014 年，上海光谱仪器有限公司推出了 SP-3882 型 AAS 铅镉专用仪器，采用了小型化电热原子化器和固定波长技术，通过单光束强光光路和塞曼效应背景校正实现了铅、镉两种元素同时灵敏检测（郑国经，2017）。

图5　WFX-910 便携式原子吸收光谱仪

7.1.2 原子发射光谱仪器

原子发射光谱是研究最早、应用最为广泛的光谱分析技术，在地球科学、环境科学、材料科学、生命科学、军工生产等领域都有重要作用，主要包括火花/电弧放电光谱（Spark/Arc-AES）、电感耦合等离子体发射光谱（ICP-AES）、激光诱导击穿光谱（LIBS）以及目前尚处于实验室研究阶段的微等离子体发射光谱（micro-plasma AES）等。

Spark-AES 近年在小型台式和便携式直读仪器上有较多进展，是适用于现场快捷分析的仪器，主要类型有聚光科技 M5000 CCD 全谱火花直读光谱仪、E5000 电弧直读光谱仪、钢研纳克推出的全谱火花直读光谱仪 SparkCCD 6000 和 SparkCCD 7000 全谱火花直读光谱仪、江苏天瑞仪器的直读光谱仪 OES8000s、北分瑞利分析推出的 AES-7200 地质专用发射光谱仪，这些仪器在钢铁、地质行业等都有较好的应用。

ICP-AES 电感耦合等离子体发射光谱仪近年来国产化仪器不断涌现。聚光科技 ICP-5000 是国内率先实现商品化的 ICP 全谱型仪器。钢研纳克的 Plasma 3000 双向观测 ICP 光谱仪获得国产仪器 BCEIA 金奖。江苏天瑞仪器 ICP-3200 采用多元高斯拟合算法大大减少干扰对待定元素谱线的误判，ICP-2060T（石化专用）以突破性的油品直接进样技术，实现了汽油中硅含量的准确测定。

LIBS 激光诱导击穿光谱仪具有操作便捷，无需烦琐样品前处理，且能适应现场、高温、恶劣环境下远程分析的特点。近年来国内在 LIBS 仪器研究上取得了显著进展。例如，中国科学院的 Sun 等（2015，2018）研制出了多台套便携式 LIBS 样机，并成功实现了对钢水、选矿、有色冶金、钢铁等领域的现场分析检测。四川大学的段忆翔等（2017）发布了能量输出达到 100 mJ 以上的高能手持式 LIBS，能量强度优于国外同类产品（图6）。LIBS-拉曼组合仪器 LIBRAS 可以实现同一个样品点上原子光谱和分子光谱的分析检测。2013 年北京纳克公司推出了 LIBS-OPA100 商品仪器，这属于国际首创，它由高功率 Nd：YAG 脉冲激光器光源结合多道直读光谱仪组成，并配备大范围进行二位扫描装置，可以对元素成分及状态进行统计分布分析。此外，也有学者开发了 LIBS 在地质分析中的应用，例如对页岩剖面进行 LIBS 化学成像分析；通过 LIBS 光谱实现天然矿物的高精度分类，等等。这些都初步表明 LIBS 在地质方面的应用前景巨大。

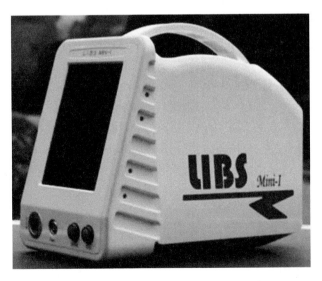

图 6　高能手提式 LIBS 仪器（据林庆宇和段忆翔，2017）

微等离子体发射光谱仪（micro-plasma AES）采用小尺寸、低功耗的微等离子体激发源，具有重量轻、功耗和气耗低、易于制造、操作简单、灵敏分析等特点，在现场原位、实时分析等方面有潜在优势，目前大多数还处于实验室研发阶段，其中国内学者也做出了许多有特色的工作。例如，中国地质大学朱振利团队（Yang et al., 2017）报道了便携式常压辉光放电原子发射（APGD-AES）重金属光谱仪样机，其整机重量小于10 kg，整机功耗约为40 W，结合氢化物发生进样可用于多种重金属元素As、Sb、Pb、Hg等的检测，检出限可达μg/L级［图7（a）］。哈尔滨工业大学的姜杰团队（Li N. et al.，2017）报道了便携式介质阻挡放电发射光谱（DBD-AES）仪器样机，其重量仅4.5 kg，由电热蒸发引入样品，可以实现Zn、Pb、Ag、Cd、Au、Cu、Mn、Fe、Cr、As等多种元素检测，检出限为0.16~11.65 μg/L［图7（b）］。中国科学院硅酸盐研究所的汪正团队（Peng et al.，2019）报道了液体阴极辉光放电发射光谱（SCGD-AES）仪器样机，可实现Cd、Hg、Pb等元素的检测，连续进样模式检出限可达7~22 μg/L。四川大学的侯贤灯团队（Li M. T. et al.，2019）报道了尖端放电发射光谱（PD-AES）仪器研制的工作，他们采用3D打印技术制造零部件（气液分离器）及固定结构，实现了快速、批量、低成本制造，通过氢化物发生进样可实现As、Bi、Ge、Hg、Pb等元素的检测，检出限可达μg/L级。

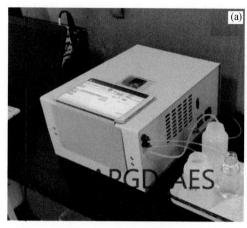

图7　(a) 常压辉光放电微等离子体发射光谱仪和 (b) 介质阻挡放电发射光谱仪

7.1.3　原子荧光仪器

原子荧光仪器是我国具有自主知识产权的特色仪器，在研发和生产上一直处于国际领先地位。近十年来，在形态分析、自动化、灯漂移校正、进样系统简化、原子化器温度控制、固体样品分析等方面取得了一定进展。目前国内很多仪器厂商都改进或推出了与高效液相色谱联用的原子荧光形态分析仪器。北京海光仪器公司在原产品基础上，进一步研发形态仪接口技术，有效解决了交叉污染，2015年推出了新一代的LC-AFS9560仪器，并荣获BCEIA金奖。其他技术，如北京瑞利分析仪器有限公司推出了首台便携式原子荧光光谱仪器PAF-1100；2012年北京吉天仪器推出了固体进样-在线捕集的DCMA-200直接进样汞镉测试仪器；北京宝德仪器有限公司采用倾斜式光学系统，显著提高了灵敏度，降低了散射光干扰，推出了BAF-400全自动四道原子荧光分析仪器；2016年金索坤推出了SK-880火焰原子荧光光谱仪，可专用于测金；2018年海光仪器公司推出的HGA-100直接测汞仪，实现了固体、液体直接分析，通过催化捕集实现不同样品的汞快速分析。

7.2　质谱仪器

7.2.1　电感耦合等离子质谱仪器

近年来，随着国家对科学仪器设备研发的重视和支持，国内仪器厂商在"国家重大科学仪器设备开发专项"的资助下，先后推出了国产化的ICP-MS。主要产品有江苏天瑞公司的ICPMS2000，钢研纳克公司的Plasma MS300，聚光科技公司的Expec7000和北京博辉公司的SOLATIONE® ICP-MS等，这些都是四级杆质谱，基本指标可以满足常规微量元素的检测。此外，北京东西分析仪器有限公司2013年收购了澳大利亚GBC公司，具备了ICP-TOFMS的生产和研发能力。需要指出的是虽然国产的ICPMS在分析性能方面与国外仪器厂商的先进型号仪器性能还有一定差距，稳定性也可能存在一定问题，但这些国产ICPMS

的研发，有望打破进口ICPMS垄断中国市场的局面。

7.2.2 二次离子质谱（SIMS）仪器

二次离子质谱（SIMS）于19世纪60年代诞生，现已广泛应用于半导体、微电子、冶金、陶瓷、地球和空间科学以及医学等领域。中国地质科学院团队研发出的首批国产飞行时间二次质谱（TOF-SIMS），可用于稀土元素分析（TOF-SIMS-REE）和稳定同位素分析（TOF-SIMS-SI），TOF-SIMS-REE空间分辨率可达5 μm，质量分辨率超过20 000，对矿物稀土元素分析精度可达10%；TOF-SIMS-SI空间分辨率可达0.5 μm，质量分辨率超过13 000（地科院地质所，2016）。

中国科学院海洋所张鑫近十年来研制了以世界首台耐高温原位拉曼光谱探针为代表的系列深海装备，构建了面向深海地球化学的原位定量探测和采样技术体系，并使用自研设备首次在南海发现裸露在海底的可燃冰，在深海热液区发现自然状态下存在超临界二氧化碳和气态水，发现全球深海沉积物中甲烷被低估20倍等成果（Zhang X. et al., 2011, 2017, 2020；Li L. F. et al., 2018, 2020）。

此外，近年来我国在质谱关键部件生产方面也有显著突破，例如，中国工程物理研究院等单位逐步攻克了高精度四级杆质量分析器的研制技术，具备了批量化制造能力；清华大学、北京大学、东华大学、中国医学科学院等单位开发了介质阻挡放电（DBDI）等系列常压开放离子源，这有利于国产仪器的研发及产业化。四级杆、离子阱、飞行时间原理的气相色谱质谱不断涌现，如北京东西电子、杭州聚光等仪器公司还研发了车载或者便携式气相色谱质谱（GC-MS）。另外，广州禾信仪器公司，基于国际领先的单颗粒气溶胶飞行时间质谱技术研发了$PM_{2.5}$在线源解析质谱，有助于解决大气颗粒物污染防治工作的开展。清浦科技等单位研制的基于常压离子源的便携式的小型化质谱，杭州聚光科技、北京普析通用仪器公司研发的液相色谱-三重四级杆质谱联用仪器，在公共卫生、食品药品等领域都具有广泛的应用前景。

8 展 望

在元素定量分析领域，现代仪器分析在主量元素分析中已占据主导地位，ICP-MS分析方法已成为微量元素分析的主流技术。未来元素分析发展热点主要在于微区和原位分析。在EPMA分析微量元素方面，监测标样开发任重道远，微量元素含量较高的硅酸盐矿物、硫化物和玻璃样品是亟待优先开发的监测标样。此外，电子探针已经可以准确测定轻元素（如Li、Be、B等）的含量，这为关键金属研究提供了技术支持。矿物原位微量元素LA-ICP-MS分析方法方面，尽管无内标-多外标校正方法在很大程度上解决了基体效应问题，但是基体匹配的外部标样对于测试一些复杂样品如富稀土元素矿物仍然重要。因此，标样的制备仍然是值得重视的方向。传统的样品制备方法（如熔融玻璃、压片法）和化学前处理方法（酸溶）与LA-ICP-MS分析相结合，融合了各自的优点，可以更加准确地分析复杂基体样品的微量元素，是未来LA-ICP-MS元素分析的重要发展趋势。

在同位素组成分析测试技术领域，传统的放射性同位素分析方法已经成熟（Sr-Nd-Hf-Pb），未来研究更多应趋向于更高的分析测试精密度和准确度，更少的样品量和更高效、环保的化学纯化流程。金属稳定同位素蓬勃发展，从样品前处理流程到仪器分析测试都有大量亟待解决的问题，例如，针对复杂的地质样品类型，化学纯化流程尚未建立可应用于所有样品类型的普适性分离纯化方法；仪器分析过程中的基体效应、质谱干扰、仪器灵敏度和稳定性、分析测试效率等问题依然需要深入研究。针对全岩同位素分析测试复杂的化学前处理流程，探索优化分离富集策略、提升生产效率、降低劳动强度等已逐渐成为新的方向，包括预先共沉淀技术、串联离子柱技术和自动化分离纯化设备研制等。微区原位同位素分析测试技术是未来同位素分析领域的重要研究方向，包括SIMS和LA-MC-ICP-MS。但是，固体同位素参考物质的缺乏是制约SIMS和LA-MC-ICP-MS开展广泛微区原位同位素分析的关键因素。尽管近年来国内学者在固体参考物质开发方面取得一定进展，但远远满足不了实际需求。LA-MC-ICP-MS在解决基体效应问题方面已经有部分研究成果，未来对非基体匹配分析的突破将有助于LA-MC-ICP-MS扩大可分析样品范围。

地质样品消解方法是开展全岩微量元素、同位素分析的基础。进一步优化酸碱消解流程，降低试剂用量、提升消解效率、克服元素丢失问题（如不完全溶解、氟化物沉淀、挥发性丢失）将继续成为地质样品消解技术革新的方向。伴随着社会的发展，绿色环保地质样品前处理技术的概念逐渐受到重视，改革现有的强酸强碱消解技术是今后地质样品前处理的发展方向。地质样品消解属于重复性的劳动，引入自动化设备、提高工作效率更符合现代高科技社会的工作模式。

近十年我国分析仪器的研发实现了规模与研发实力的显著提高，在某些方面也有显著进展，如在中低端仪器方面已能够基本满足测试和分析需求，但对高端仪器的研发还明显不足，现有国产仪器与国外进口仪器在性能指标上还有差距，特别是同位素质谱分析仪器的研发基本上还是空白，仍需要增强积累，加大投入和人才培养。

参 考 文 献

安徽省质量技术监督局. 2014. DB34/T 2127.2-2014 区域地球化学调查样品分析方法 第2部分：X 射线荧光光谱法多元素含量的测定

包志安, 袁洪林, 陈开运, 宋佳瑶, 戴梦宁, 宗春蕾. 2011. 高温熔融研制安山岩玻璃标准物质初探. 岩矿测试, 30(5): 521-527

陈波, 刘洪青, 邢应香. 2014. 电感耦合等离子体质谱法同时测定地质样品中硒硫碲. 岩矿测试, 33(2): 192-196

陈春飞, 刘先国, 胡兆初, 宋克清, 刘勇胜. 2014. LA-ICP-MS 微区原位准确分析含水硅酸盐矿物主量和微量元素. 地球科学-中国地质大学学报, 39(5): 525-536

程秀花, 唐南安, 张明祖, 黎卫亮, 王鹏, 陈陆洋. 2013. 稀有分散元素分析方法的研究进展. 理化检验：化学分册, 49(6): 757-764

邓勃. 2017. 原子吸收光谱分析在我国发展历程的回顾. 现代科学仪器, (4): 10-19

地科院地质所. 2016. 地质所完成同位素地质学专用 TOF-SIMS 科学仪器的整机装配. 地质装备, 17(2): 8-9

范宏瑞, 李兴辉, 左亚彬, 陈蕾, 刘尚, 胡芳芳, 冯凯. 2018. LA-(MC)-ICPMS 和 (Nano)SIMS 硫化物微量元素和硫同位素原位分析与矿床形成的精细过程. 岩石学报, 34(12): 3479-3496

杭乐, 徐周毅, 杭纬, 黄本立. 2019. 中国原子光谱技术及应用发展近况. 光谱学与光谱分析, 39(5): 1329-1339

吉昂. 2012. X 射线荧光光谱三十年. 岩矿测试, 31(3): 383-398

蓝廷广, 胡瑞忠, 范宏瑞, 毕献武, 唐燕文, 周丽, 毛伟, 陈应华. 2017. 流体包裹体及石英 LA-ICP-MS 分析方法的建立及其在矿床学中的应用. 岩石学报, 33(10): 3239-3262

李冰, 周剑雄, 詹秀春. 2011. 无机多元素现代仪器分析技术. 地质学报, 85(11): 1878-1916

李国会, 李小莉. 2015. X 射线荧光光谱分析熔融法制样的系统研究. 冶金分析, 35(7): 1-9

李小莉, 安树清, 徐铁民, 刘义博, 张莉娟, 曾江萍, 王娜. 2015. 超细粉末压片制样 X 射线荧光光谱测定碳酸岩样品中多种元素及 CO_2. 光谱学与光谱分析, 35(6): 1741-1745

李志伟, 高志军, 张明炜. 2015. 碱熔-电感耦合等离子体质谱法测定硫铁矿单矿物中的金、银及铂族元素. 理化检验-化学分册, 51(1): 102-104

林庆宇, 段忆翔. 2017. 激光诱导击穿光谱：从实验平台到现场仪器. 分析化学, 45(9): 1405-1414

刘玉纯, 林庆文, 马玲. 2019. X 射线荧光光谱技术在地质分析中的应用及发展动态. 化学分析计量, 28(4): 125-131

马生凤, 温宏利, 李冰, 王蕾, 朱云. 2016. 微波消解-耐氢氟酸系统电感耦合等离子体发射光谱法测定铌钽矿中的铌和钽. 岩矿测试, 35(3): 271-275

孟郁苗, 黄小文, 高剑峰, 戴智慧, 漆亮. 2016. 无内标-多外标校正激光剥蚀等离子体质谱法测定磁铁矿微量元素组成. 岩矿测试, 35(6): 585-594

宋佳瑶, 袁洪林, 包志安, 陈开运, 贺国芬, 范超. 2011. 高温熔融研制钾长石玻璃标准物质初探. 岩矿测试, 30(4): 406-411

孙朝阳, 董利明, 贺颖婷, 杨利华, 郑存江. 2016. 电感耦合等离子体质谱法测定地质样品中钪镓锗铟镉铊时的干扰及其消除方法. 理化检验-化学分册, 52(9): 1026-1030

汪方跃, 葛粲, 宁思远, 聂利青, 钟国雄, White N C. 2017. 一个新的矿物面扫描分析方法开发和地学应用. 岩石学报, 33(11): 3422-3436

王娟, 陈意, 毛骞, 李秋立, 马玉光, 石永红, 宋传中. 2017. 金红石微量元素电子探针分析. 岩石学报, 33(6): 1934-1946

王蕾, 张保科, 马生凤, 赵怀颖, 郭琳. 2014. 封闭压力酸溶-电感耦合等离子体质谱法测定钨矿石中的钨. 岩矿测试, 33(5): 661-664

王晓红, 何红蓼, 王毅民, 孙德忠, 樊兴涛, 高玉淑, 温宏利, 夏月莲. 2010. 超细样品的地质分析应用. 分析测试学报, 29(6): 578-583

魏灵巧, 付胜波, 罗磊, 黄小华, 龙安应, 帅琴. 2012. 电感耦合等离子体发射光谱法多向观测同时测定锑矿石中锑砷铜铅锌. 岩矿测试, 31(6): 967-970

肖元芳, 王小华, 杭纬. 2015. 中国原子光谱发展近况概述. 光谱学与光谱分析, 35(9): 2377-2387

《岩石矿物分析》编委会. 2011. 岩石矿物分析：第4版. 北京：地质出版社

杨林, 肖玉萍, 于珊. 2013. 低氧低分压环境下聚氨酯泡沫塑料富集-电感耦合等离子体质谱法测定地质样品中微量金. 理化检验-化学分册, 49(6): 747-748, 750

曾江萍, 李小莉, 张莉娟, 张楠, 徐铁民. 2015a. 超细粉末压片 X 射线荧光光谱法分析铬铁矿中的多种元素. 矿物学报, 35(4): 545-549

曾江萍, 张莉娟, 李小莉, 张楠, 吴良英, 王力强. 2015b. 超细粉末压片-X 射线荧光光谱法测定磷矿石中 12 种组分. 冶金分析, 35(7): 47-43

张德贤,戴塔根,胡毅. 2012. 磁铁矿中微量元素的激光剥蚀-电感耦合等离子体质谱分析方法探讨. 岩矿测试,31(1):120−126
张迪,陈意,毛骞,苏斌,贾丽辉,郭顺. 2019. 电子探针分析技术进展及面临的挑战. 岩石学报,35(1):261−274
张广玉,赵世煌,邓晃,郭跃梅,井德刚,汪艳芸. 2017. 手持式X射线荧光光谱多点测试技术在地质岩心和岩石标本预研究中的应用. 岩矿测试,36(5):501−509
张莉娟,刘义博,李小莉,徐铁民. 2014. 超细粉末压片法−X射线荧光光谱测定水系沉积物和土壤中的主量元素. 岩矿测试,33(4):517−522
张文,胡兆初,漆亮,刘勇胜,高山. 2015. 氟化氢铵消解−电感耦合等离子体质谱法测定铝土矿37个微量元素. 见:中国地球科学联合学术年会论文集. 北京
张彦辉,张良圣,常阳,范增伟,郭冬发. 2018. 增压−微波消解电感耦合等离子体质谱法测定含难溶矿物岩石样品中的微量元素. 铀矿地质, 34(2):105−111
郑国经. 2017. 原子发射光谱仪器的发展、现状及技术动向. 现代科学仪器,(4):23−36,41
中华人民共和国国土资源部. 2011. GB/T 14506.28-2010 硅酸盐岩石化学分析方法第28部分:16个主次成分量测定. 北京:中国标准出版社
周伶俐,曾庆栋,孙国涛,段晓侠,Bonnetti C, Riegler T, Long D G F, Kamber B. 2019. LA-ICPMS原位微区面扫描分析技术及其矿床学应用实例. 岩石学报,35(7):1964−1978
朱律运,刘勇胜,胡兆初,高山,王晓红,田滔. 2011. 玄武岩全岩元素含量快速、准确分析新技术:双铱带高温炉与LA-ICP-MS联用法. 地球化学,40(5):407−411
Albarède F, Beard B. 2004. Analytical methods for non-traditional isotopes. Reviews in Mineralogy and Geochemistry, 55(1):113−152
An Y J, Wu F, Xiang Y X, Nan X Y, Yu X, Yang J H, Yu H M, Xie L W, Huang F. 2014. High-precision Mg isotope analyses of low-Mg rocks by MC-ICP-MS. Chemical Geology, 390:9−21
Bao Z A, Zhang H F, Yuan H L, Liu Y, Chen K Y, Zong C L. 2016. Flux-free fusion technique using a boron nitride vessel and rapid acid digestion for determination of trace elements by ICP-MS. Journal of Analytical Atomic Spectrometry, 31(11):2261−2271
Bao Z A, Chen L, Zong C L, Yuan H L, Chen K Y, Dai M N. 2017. Development of pressed sulfide powder tablets for *in situ* sulfur and lead isotope measurement using LA-MC-ICP-MS. International Journal of Mass Spectrometry, 421:255−262
Bao Z A, Huang K J, Huang T Z, Shen B, Zong C L, Chen K Y, Yuan H L. 2019. Precise magnesium isotope analyses of high-K and low-Mg rocks by MC-ICP-MS. Journal of Analytical Atomic Spectrometry, 34(5):940−953
Bao Z A, Huang K J, Xu J, Deng L, Yang S F, Zhang P, Yuan H L. 2020a. Preparation and characterization of a new reference standard GSB-Mg for Mg isotopic analysis. Journal of Analytical Atomic Spectrometry, 35(6):1080−1086
Bao Z A, Zong C L, Chen K Y, Lv N, Yuan H L. 2020b. Chromatographic purification of Ca and Mg from biological and geological samples for isotope analysis by MC-ICP-MS. International Journal of Mass Spectrometry, 448:116268
Bourdon B, Roskosz M, Hin R C. 2018. Isotope tracers of core formation. Earth-Science Reviews, 181:61−81
Burkhardt C, Hin R C, Kleine T, Bourdon B. 2014. Evidence for Mo isotope fractionation in the solar nebula and during planetary differentiation. Earth and Planetary Science Letters, 391:201−211
Che X D, Wu F Y, Wang R C, Gerdesc A, Ji W Q, Zhao Z H, Yang J H, Zhu Z Y. 2015. *In situ* U-Pb isotopic dating of columbite-tantalite by LA-ICP-MS. Ore Geology Reviews, 65:979−989
Chen L, Liu Y S, Hu Z C, Gao S, Zong K Q, Chen H H. 2011. Accurate determinations of fifty-four major and trace elements in carbonate by LA-ICP-MS using normalization strategy of bulk components as 100%. Chemical Geology, 284(3-4):283−295
Chen L M, Song X Y, Danyushevsky L V, Wang Y S, Tian Y L, Xiao J F. 2014. A laser ablation ICP-MS study of platinum-group and chalcophile elements in base metal sulfide minerals of the Jinchuan Ni-Cu sulfide deposit, NW China. Ore Geology Reviews, 65:955−967
Chen S, Liu Y C, Hu J Y, Zhang Z F, Hou Z H, Huang F, Yu H M. 2016. Zinc isotopic compositions of NIST SRM 683 and whole-rock reference materials. Geostandards and Geoanalytical Research, 40(3):417−432
Chen S, Wang X H, Niu Y L, Sun P, Duan M, Xiao Y Y, Guo P Y, Gong H M, Wang G D, Xue Q Q. 2017. Simple and cost-effective methods for precise analysis of trace element abundances in geological materials with ICP-MS. Science Bulletin, 62(4):277−289
Chu Z Y, Guo J H, Yang Y H, Qi L, Li C F. 2014. Precise determination of Sm and Nd concentrations and Nd isotopic compositions in highly depleted ultramafic reference materials. Geostandards and Geoanalytical Research, 38(1):61−72
Clay P L, Burgess R, Busemann H, Ruzié-Hamilton L, Joachim B, Day J M D, Ballentine C J. 2017. Halogens in chondritic meteorites and terrestrial accretion. Nature, 551(7682):614−618
Cui J Q, Yang S Y, Jiang S Y, Xie J. 2019. Improved accuracy for trace element analysis of Al and Ti in quartz by electron probe microanalysis. Microscopy and Microanalysis, 25(1):47−57
Deng X D, Luo T, Li J W, Hu Z C. 2019. Direct dating of hydrothermal tungsten mineralization using *in situ* wolframite U-Pb chronology by laser ablation ICP-MS. Chemical Geology, 515:94−104
Feng L P, Zhou L, Yang L, Tong S Y, Hu Z C, Gao S. 2015. Optimization of the double spike technique using peak jump collection by a Monte Carlo method: an example for the determination of Ca isotope ratios. Journal of Analytical Atomic Spectrometry, 30(12):2403−2411
Feng L P, Zhou L, Yang L, Donald J D, Tong S Y, Liu Y S, Thomas L O, Gao S. 2017. Calcium isotopic compositions of sixteen USGS reference materials. Geostandards and Geoanalytical Research, 41(1):93−106
Feng L P, Zhou L, Yang L, Zhang W, Wang Q, Tong S Y, Hu Z C. 2018. A rapid and simple single-stage method for Ca separation from geological and biological samples for isotopic analysis by MC-ICP-MS. Journal of Analytical Atomic Spectrometry, 33(3):413−421

Feng L P, Zhou L, Liu J H, Hu Z C, Liu Y S. 2019. Determination of gallium isotopic compositions in reference materials. Geostandards and Geoanalytical Research, 43(4): 701–714

Feng L P, Hu W F, Jiao Y, Zhou L, Zhang W, Hu Z C, Liu Y S. 2020a. High-precision stable zirconium isotope ratio measurements by double spike thermal ionization mass spectrometry. Journal of Analytical Atomic Spectrometry, 35(4): 736–745

Feng L P, Zhou L, Hu W F, Zhang W, Li B C, Liu Y S, Hu Z C, Yang L. 2020b. A simple single-stage extraction method for Mo separation from geological samples for isotopic analysis by MC-ICP-MS. Journal of Analytical Atomic Spectrometry, 35(1): 145–154

Foden J, Sossi P A, Wawryk C M. 2015. Fe isotopes and the contrasting petrogenesis of A-, I- and S-type granite. Lithos, 212-215: 32–44

Gao J F, Zhou M F, Lightfoot P C, Wang C Y, Qi L, Sun M. 2013. Sulfide saturation and magma emplacement in the formation of the Permian Huangshandong Ni-Cu sulfide deposit, Xinjiang, northwestern China. Economic Geology, 108(8): 1833–1848

Gaschnig R M, Rudnick R L, McDonough W F, Kaufman A J, Valley J W, Hu Z C, Gao S, Beck M L. 2016. Compositional evolution of the upper continental crust through time, as constrained by ancient glacial diamictites. Geochimica et Cosmochimica Acta, 186: 316–343

Graham S, Pearson N, Jackson S, Griffin W, O'Reilly S Y. 2004. Tracing Cu and Fe from source to porphyry: in situ determination of Cu and Fe isotope ratios in sulfides from the Grasberg Cu-Au deposit. Chemical Geology, 207(3-4): 147–169

Greber N D, Dauphas N, Bekker A, Ptáček M P, Bindeman I N, Hofmann A. 2017. Titanium isotopic evidence for felsic crust and plate tectonics 3.5 billion years ago. Science, 357(6357): 1271–1274

He D, Zhu Z L, Zhao L Y, Belshaw N S, Zheng H T, Li X L, Hu S H. 2019. A practical method for measuring high precision calcium isotope ratios without chemical purification for calcium carbonate samples by multiple collector inductively coupled plasma mass spectrometry. Chemical Geology, 514: 105–111

He M Y, Xiao Y K, Ma Y Q, Jin Z D, Xiao J. 2011. Effective elimination of organic matter interference in boron isotopic analysis by thermal ionization mass spectrometry of coral/foraminifera: micro-sublimation technology combined with ion exchange. Rapid Communications in Mass Spectrometry, 25(6): 743–749

He M Y, Xiao Y K, Jin Z D, Liu W G, Ma Y Q, Zhang Y L, Luo C G. 2013a. Quantification of boron incorporation into synthetic calcite under controlled pH and temperature conditions using a differential solubility technique. Chemical Geology, 337-338: 67–74

He M Y, Xiao Y K, Jin Z D, Ma Y Q, Xiao J, Zhang Y L, Luo C G, Zhang F. 2013b. Accurate and precise determination of boron isotopic ratios at low concentration by positive thermal ionization mass spectrometry using static multicollection of $Cs_2BO_2^+$ ions. Analytical Chemistry, 85(15): 6248–6253

He M Y, Jin Z D, Luo C G, Deng L, Xiao J, Zhang F. 2015. Determination of boron isotope ratios in tooth enamel by inductively coupled plasma mass spectrometry (ICP-MS) after matrix separation by ion exchange chromatography. Journal of the Brazilian Chemical Society, 26(5): 949–954

He T, Xie J Y, Hu Z C, Liu T, Zhang W, Chen H H, Liu Y S, Zong K Q, Li M. 2018. A rapid acid digestion technique for the simultaneous determination of bromine and iodine in fifty-three Chinese soils and sediments by ICP-MS. Geostandards and Geoanalytical Research, 42(3): 309–318

He T, Hu Z C, Zhang W, Chen H H, Liu Y S, Wang Z C, Hu S H. 2019. Determination of Cl, Br, and I in geological materials by sector field inductively coupled plasma mass spectrometry. Analytical Chemistry, 91(13): 8109–8114

He X Y, Ma J L, Wei G J, Zhang L, Wang Z B, Wang Q S. 2020. A new procedure for titanium separation in geological samples for $^{49}Ti/^{47}Ti$ ratio measurement by MC-ICP-MS. Journal of Analytical Atomic Spectrometry, 35(1): 100–106

He Y S, Ke S, Teng F Z, Wang T T, Wu H J, Lu Y H, Li S G. 2015. High-precision iron isotope analysis of geological reference materials by high-resolution MC-ICP-MS. Geostandards and Geoanalytical Research, 39(3): 341–356

He Y S, Wang Y, Zhu C W, Huang S C, Liu S G. 2017. Mass-independent and mass-dependent Ca isotopic compositions of thirteen geological reference materials measured by thermal ionisation mass spectrometry. Geostandards and Geoanalytical Research, 41(2): 283–302

He Z W, Huang F, Yu H M, Xiao Y L, Wang F Y, Li Q L, Xia Y, Zhang X C. 2016. A flux-free fusion technique for rapid determination of major and trace elements in silicate rocks by LA-ICP-MS. Geostandards and Geoanalytical Research, 40(1): 5–21

Hou Q H, Zhou L, Gao S, Zhang T, Feng L P, Yang L. 2016. Use of Ga for mass bias correction for the accurate determination of copper isotope ratio in the NIST SRM 3114 Cu standard and geological samples by MC-ICP-MS. Journal of Analytical Atomic Spectrometry, 31(1): 280–287

Hu H, Li J W, Lentz D, Ren Z, Zhao X F, Deng X D, Hall D. 2014. Dissolution-reprecipitation process of magnetite from the Chengchao iron deposit: insights into ore genesis and implication for in-situ chemical analysis of magnetite. Ore Geology Reviews, 57: 393–405

Hu H, Lentz D, Li J W, McCarron T, Zhao X F, Hall D. 2015. Reequilibration processes in magnetite from iron skarn deposits. Economic Geology, 110(1): 1–8

Hu M Y, Fan X T, Stoll B, Kuzmin D, Liu Y, Liu Y S, Sun W D, Wang G, Zhan X C, Jochum K P. 2011. Preliminary characterisation of new reference materials for microanalysis: Chinese geological standard glasses CGSG-1, CGSG-2, CGSG-4 and CGSG-5. Geostandards and Geoanalytical Research, 35(2): 235–251

Hu Z C, Qi L. 2014. Sample digestion methods. In: Turekian H D H K (ed). Treatise on Geochemistry, 2nd ed. Oxford: Elsevier, 87–109

Hu Z C, Gao S, Liu Y S, Hu S H, Zhao L S, Li Y X, Wang Q. 2010. NH_4F assisted high pressure digestion of geological samples for multi-element analysis by ICP-MS. Journal of Analytical Atomic Spectrometry, 25(3): 408–413

Hu Z C, Liu Y S, Gao S, Xiao S Q, Zhao L S, Günther D, Li M, Zhang W, Zong K Q. 2012. A "wire" signal smoothing device for laser ablation inductively coupled plasma mass spectrometry analysis. Spectrochimica Acta Part B: Atomic Spectroscopy, 78: 50–57

Hu Z C, Zhang W, Liu Y S, Chen H H, Gaschnig R M, Zong K Q, Li M, Gao S, Hu S H. 2013. Rapid bulk rock decomposition by ammonium fluoride (NH$_4$F) in open vessels at an elevated digestion temperature. Chemical Geology, 355: 144−152

Huang C, Yang Y H, Yang J H, Xie L W. 2015. *In situ* simultaneous measurement of Rb-Sr/Sm-Nd or Sm-Nd/Lu-Hf isotopes in natural minerals using laser ablation multi-collector ICP-MS. Journal of Analytical Atomic Spectrometry, 30(4): 994−1000

Huang C, Wang H, Yang J H, Ramezani J, Yang C, Zhang S B, Yang Y H, Xia X P, Feng L J, Lin J, Wang T T, Ma Q, He H Y, Xie L W, Wu S T. 2020a. SA01-A proposed zircon reference material for microbeam U-Pb age and Hf-O isotopic determination. Geostandards and Geoanalytical Research, 44(1): 103−123

Huang C, Yang Y H, Xie L W, Wu S T, Wang H, Yang J H, Wu F Y. 2020b. *In situ* sequential U-Pb age and Sm-Nd systematics measurements of natural LREE-enriched minerals using single laser ablation multi-collector inductively coupled plasma mass spectrometry. Journal of Analytical Atomic Spectrometry, 35(3): 510−517

Huang X W, Zhou M F, Qi L, Gao J F, Wang Y W. 2013. Re-Os isotopic ages of pyrite and chemical composition of magnetite from the Cihai magmatic-hydrothermal Fe deposit, NW China. Mineralium Deposita, 48(8): 925−946

Hubley N, Brown J W N IV, Guthrie J, Robertson J D, Brockman J D. 2016. Development of ammonium bifluoride fusion method for rapid dissolution of trinitite samples and analysis by ICP-MS. Journal of Radioanalytical and Nuclear Chemistry, 307: 1777−1780

Johnson C M, Beard B L. 1999. Correction of instrumentally produced mass fractionation during isotopic analysis of Fe by thermal ionization mass spectrometry. International Journal of Mass Spectrometry, 193(1): 87−99

Ke Y Q, Zhou J Z, Yi Z Q, Sun Y J, Shao J F, You S Y, Wang W, Tang Y Z, Tu C Y. 2020. Development of REE-doped CaWO$_4$ single crystals as reference materials for *in situ* microanalysis of scheelite via LA-ICP-MS. Journal of Analytical Atomic Spectrometry, 35(5): 886−895

Li D D, Li M L, Liu W R, Qin Z Z, Liu S A. 2018. Cadmium isotope ratios of standard solutions and geological reference materials measured by MC-ICP-MS. Geostandards and Geoanalytical Research, 42(4): 593−605

Li J, Jiang X Y, Xu J F, Zhong L F, Wang X C, Wang G Q, Zhao P P. 2014a. Determination of platinum-group elements and Re-Os isotopes using ID-ICP-MS and N-TIMS from a single digestion after two-stage column separation. Geostandards and Geoanalytical Research, 38(1): 37−50

Li J, Liang X R, Zhong L F, Wang X C, Ren Z Y, Sun S L, Zhang Z F, Xu J F. 2014b. Measurement of the isotopic composition of molybdenum in geological samples by MC-ICP-MS using a novel chromatographic extraction technique. Geostandards and Geoanalytical Research, 38(3): 345−354

Li J, Zhu X K, Tang S H, Zhang K. 2016. High-precision measurement of molybdenum isotopic compositions of selected geochemical reference materials. Geostandards and Geoanalytical Research, 40(3): 405−415

Li J, Tang S H, Zhu X K, Pan C X. 2017. Production and certification of the reference material GSB 04-3258-2015 as a ^{143}Nd/^{144}Nd isotope ratio reference. Geostandards and Geoanalytical Research, 41(2): 255−262

Li L F, Zhang X, Luan Z D, Du Z F, Xi S C, Wang B, Cao L, Lian C, Yan J. 2018. *In situ* quantitative Raman detection of dissolved carbon dioxide and sulfate in deep-sea high-temperature hydrothermal vent fluids. Geochemistry, Geophysics, Geosystems, 19(6): 1809−1823

Li L F, Zhang X, Luan Z D, Du Z F, Xi S C, Wang B, Lian C, Cao L, Yan J. 2020. Hydrothermal vapor-phase fluids on the seafloor: evidence from *in situ* observations. Geophysical Research Letters, 47(10): e2019GL085778

Li M, Lei Y, Feng L P, Wang Z C, Belshaw N S, Hu Z C, Liu Y S, Zhou L, Chen H H, Chai X N. 2018. High-precision Ca isotopic measurement using a large geometry high resolution MC-ICP-MS with a dummy bucket. Journal of Analytical Atomic Spectrometry, 33(10): 1707−1719

Li M T, Li K, He L, Zheng X L, Wu X, Hou X D, Jiang X M. 2019. Point discharge microplasma optical emission spectrometer: hollow electrode for efficient volatile hydride/mercury sample introduction and 3D-printing for compact instrumentation. Analytical Chemistry, 91(11): 7001−7006

Li N, Wu Z C, Wang Y Y, Zhang J, Zhang X N, Zhang H N, Wu W H, Gao J, Jiang J. 2017. Portable dielectric barrier discharge-atomic emission spectrometer. Analytical Chemistry, 89(4): 2205−2210

Li Q L, Li X H, Liu Y, Tang G Q, Yang J H, Zhu W G. 2010. Precise U-Pb and Pb-Pb dating of Phanerozoic baddeleyite by SIMS with oxygen flooding technique. Journal of Analytical Atomic Spectrometry, 25(7): 1107−1113

Li Q L, Liu Y, Tang G Q, Wang K Y, Ling X X, Li J. 2018. Zircon Th-Pb dating by secondary ion mass spectrometry. Journal of Analytical Atomic Spectrometry, 33(9): 1536−1544

Li R C, Xia X P, Yang S H, Chen H Y, Yang Q. 2019. Off-mount calibration and one new potential pyrrhotite reference material for sulfur isotope measurement by secondary ion mass spectrometry. Geostandards and Geoanalytical Research, 43: 177−187

Li R C, Xia X P, Chen H Y, Wu N P, Zhao T P, Lai C, Yang Q. 2020. A potential new chalcopyrite reference material for secondary ion mass spectrometry sulfur isotope ratio analysis. Geostandards and Geoanalytical Research, 44(3): 485−500

Li W H, Zhu J M, Tan D C, Han G L, Zhao Z Q, Wu G L. 2020. The $\delta^{60/58}$Ni values of twenty-six selected geological reference materials. Geostandards and Geoanalytical Research, 44(3): 523−535

Li W Q, Beard B L, Li S L. 2016. Precise measurement of stable potassium isotope ratios using a single focusing collision cell multi-collector ICP-MS. Journal of Analytical Atomic Spectrometry, 31(4): 1023−1029

Li X H, Li Q L, Liu Y, Tang G Q. 2011. Further characterization of M257 zircon standard: a working reference for SIMS analysis of Li isotopes. Journal of Analytical Atomic Spectrometry, 26(2): 352−358

Li X H, Tang G Q, Gong B, Yang Y H, Hou K J, Hu Z C, Li Q L, Liu Y, Li W X. 2013. Qinghu zircon: a working reference for microbeam analysis of U-Pb age and Hf and O isotopes. Chinese Science Bulletin, 58(36): 4647−4654

Li X L, Wang Y M, Zhang Q. 2016. Determination of halogen levels in marine geological samples. Spectroscopy Letters, 49(3): 151−154

Liao X H, Hu Z C, Luo T, Zhang W, Liu Y S, Zong K Q, Zhou L, Zhang J F. 2019. Determination of major and trace elements in geological samples by laser ablation solution sampling-inductively coupled plasma mass spectrometry. Journal of Analytical Atomic Spectrometry, 34(6): 1126–1134

Liao X H, Luo T, Zhang S H, Zhang W, Zong K Q, Liu Y S, Hu Z C. 2020. Direct and rapid multi-element analysis of wine samples in their natural liquid state by laser ablation ICPMS. Journal of Analytical Atomic Spectrometry, 35(6): 1071–1079

Lin J, Liu Y S, Hu Z C, Yang L, Chen K, Chen H H, Zong K Q, Gao S. 2016. Accurate determination of lithium isotope ratios by MC-ICP-MS without strict matrix-matching by using a novel washing method. Journal of Analytical Atomic Spectrometry, 31(2): 390–397

Lin J, Liu Y S, Tong X R, Zhu L Y, Zhang W, Hu Z C. 2017. Improved in situ Li isotopic ratio analysis of silicates by optimizing signal intensity, isotopic ratio stability and intensity matching using ns-LA-MCICP-MS. Journal of Analytical Atomic Spectrometry, 32(4): 834–842

Lin J, Liu Y S, Hu Z C, Chen W, Zhang L, Chen H H. 2019. Accurate measurement of lithium isotopes in eleven carbonate reference materials by MC-ICP-MS with soft extraction mode and 10^{12} Ω resistor high gain faraday amplifiers. Geostandards and Geoanalytical Research, 43(2): 277–289

Lin R, Lin J, Zong K Q, Chen K, Tong S Y, Feng L P, Zhang W, Li M, Liu Y S, Hu Z C, Zhou L. 2020. Determination of the isotopic composition of an enriched hafnium spike by MC-ICP-MS using a regression model. Geostandards and Geoanalytical Research, 44(4): 753–762

Ling X X, Li Q L, Liu Y, Yang Y H, Liu Y, Tang G Q, Li X H. 2016. In situ SIMS Th-Pb dating of bastnaesite: constraint on the mineralization time of the Himalayan Mianning-Dechang rare earth element deposits. Journal of Analytical Atomic Spectrometry, 31(8): 1680–1687

Ling X X, Huyskens M H, Li Q L, Yin Q Z, Werner R, Liu Y, Tang G Q, Yang Y N, Li X H. 2017. Monazite RW-1: a homogenous natural reference material for SIMS U-Pb and Th-Pb isotopic analysis. Mineralogy and Petrology, 111(2): 163–172

Liu C Y, Xu L J, Liu C T, Liu J, Qin L P, Zhang Z D, Liu S A, Li S G. 2019. High-precision measurement of stable Cr isotopes in geological reference materials by a double-spike TIMS method. Geostandards and Geoanalytical Research, 43(4): 647–661

Liu F, Li X, An Y J, Li J, Zhang Z F. 2019. Calcium isotope ratio ($\delta^{44/40}$Ca) measurements of Ca-dominated minerals and rocks without column chemistry using the double-spike technique and thermal ionisation mass spectrometry. Geostandards and Geoanalytical Research, 43(3): 509–517

Liu J, Wen H J, Zhang Y X, Fan H F, Zhu C W. 2016. Precise Mo isotope ratio measurements of low-Mo (ng·g^{-1}) geological samples using MC-ICP-MS. Journal of Analytical Atomic Spectrometry, 31(6): 1287–1297

Liu M S, Zhang Q, Zhang Y, Zhang Z, Huang F, Yu H M. 2020. High-precision Cd isotope measurements of soil and rock reference materials by MC-ICP-MS with double spike correction. Geostandards and Geoanalytical Research, 44(1): 169–182

Liu S A, Li D D, Li S G, Teng F Z, Ke S, He Y S, Lu Y H. 2014. High-precision copper and iron isotope analysis of igneous rock standards by MC-ICP-MS. Journal of Analytical Atomic Spectrometry, 29(1): 122–133

Liu W G, Wei S, Zhang J, Ao C, Liu F T, Cai B, Zhou H Y, Yang J L, Li C F. 2020. An improved separation scheme for Sr through fluoride coprecipitation combined with a cation-exchange resin from geological samples with high Rb/Sr ratios for high-precision determination of Sr isotope ratios. Journal of Analytical Atomic Spectrometry, 35(5): 953–960

Liu Y, Li X H, Li Q L, Tang G Q, Yin Q Z. 2011. Precise U-Pb zircon dating at a scale of <5 micron by the CAMECA 1280 SIMS using a Gaussian illumination probe. Journal of Analytical Atomic Spectrometry, 26(4): 845–851

Liu Y, Li Q L, Tang G Q, Li X H, Yin Q Z. 2015. Towards higher precision SIMS U-Pb zircon geochronology via dynamic multi-collector analysis. Journal of Analytical Atomic Spectrometry, 30(4): 979–985

Liu Y S, Hu Z C, Gao S, Günther D, Xu J, Gao C G, Chen H H. 2008. In situ analysis of major and trace elements of anhydrous minerals by LA-ICP-MS without applying an internal standard. Chemical Geology, 257(1-2): 34–43

Liu Y S, Hu Z C, Li M, Gao S. 2013. Applications of LA-ICP-MS in the elemental analyses of geological samples. Chinese Science Bulletin, 58(32): 3863–3878

Liu Z C, Wu F Y, Yang Y H, Yang J H, Wilde S A. 2012. Neodymium isotopic compositions of the standard monazites used in U-Th-Pb geochronology. Chemical Geology, 334: 221–239

Luo T, Deng X D, Li J W, Hu Z C, Zhang W, Liu Y S, Zhang J F. 2019. U-Pb geochronology of wolframite by laser ablation inductively coupled plasma mass spectrometry. Journal of Analytical Atomic Spectrometry, 34(7): 1439–1446

Ma J L, Wei G J, Liu Y, Ren Z Y, Xu Y G, Yang Y H. 2013. Precise measurement of stable neodymium isotopes of geological materials by using MC-ICP-MS. Journal of Analytical Atomic Spectrometry, 28(12): 1926–1931

Ma Q, Evans N J, Ling X X, Yang J H, Wu F Y, Zhao Z D, Yang Y H. 2019a. Natural titanite reference materials for in situ U-Pb and Sm-Nd isotopic measurements by LA-(MC)-ICP-MS. Geostandards and Geoanalytical Research, 43(3): 355–384

Ma Q, Yang M, Zhao H, Evans N J, Chu Z Y, Xie L W, Huang C, Zhao Z D, Yang Y H. 2019b. Accurate and precise determination of Lu and Hf contents and Hf isotopic composition at the sub-nanogram level in geological samples using MC-ICP-MS. Journal of Analytical Atomic Spectrometry, 34(6): 1256–1262

Mao M, Rukhlov A S, Rowins S M, Spence J, Coogan L A. 2016. Apatite trace element compositions: a robust new tool for mineral exploration. Economic Geology, 111(5): 1187–1222

Maréchal C N, Télouk P, Albarède F. 1999. Precise analysis of copper and zinc isotopic compositions by plasma-source mass spectrometry. Chemical Geology, 156(1-4): 251–273

Mason C A, Hubley N T, Robertson J D, Wegge D L, Brockman J D. 2017. Sonication assisted dissolution of post-detonation nuclear debris using ammonium bifluoride. Radiochimica Acta, 105(12): 1059–1070

Mei Q F, Yang J H, Yang Y H. 2018. An improved extraction chromatographic purification of tungsten from a silicate matrix for high precision isotopic measurements using MC-ICPMS. journal of Analytical Atomic Spectrometry, 33(4): 569–577

Nan X Y, Wu F, Zhang Z F, Hou Z H, Huang F, Yu H M. 2015. High-precision barium isotope measurements by MC-ICP-MS. Journal of Analytical Atomic Spectrometry, 30(11): 2307–2315

Norman M D, Pearson N J, Sharma A, Griffin W L. 1996. Quantitative analysis of trace elements in geological materials by laser ablation ICPMS: instrumental operating conditions and calibration values of NIST glasses. Geostandards and Geoanalytical Research, 20(2): 247–261

Pan J Y, Ni P, Wang R C. 2019. Comparison of fluid processes in coexisting wolframite and quartz from a giant vein-type tungsten deposit, South China: insights from detailed petrography and LA-ICP-MS analysis of fluid inclusions. American Mineralogist, 104(8): 1092–1116

Peng X X, Guo X H, Ge F, Wang Z. 2019. Battery-operated portable high-throughput solution cathode glow discharge optical emission spectrometry for environmental metal detection. Journal of Analytical Atomic Spectrometry, 34(2): 394–400

Qi L, Zhou M F, Gao J F, Zhao Z. 2010. An improved Carius tube technique for determination of low concentrations of Re and Os in pyrites. Journal of Analytical Atomic Spectrometry, 25(4): 585–589

Qi L, Gao J F, Huang X W, Hu J, Zhou M F, Zhong H. 2011. An improved digestion technique for determination of platinum group elements in geological samples. Journal of Analytical Atomic Spectrometry, 26(9): 1900–1904

Qi L, Gao J F, Zhou M F, Hu J. 2013. The design of Re-usable carius tubes for the determination of rhenium, osmium and platinum-group elements in geological samples. Geostandards and Geoanalytical Research, 37(3): 345–351

Rouxel O, Shanks W C III, Bach W, Edwards K J. 2008. Integrated Fe- and S-isotope study of seafloor hydrothermal vents at East Pacific rise 9°–10° N. Chemical Geology, 252(3-4): 214–227

She Y W, Song X Y, Yu S Y, Chen L M, Zheng W Q. 2016. Apatite geochemistry of the Taihe layered intrusion, SW China: implications for the magmatic differentiation and the origin of apatite-rich Fe-Ti oxide ores. Ore Geology Reviews, 78: 151–165

Siebert C, Nägler T F, Kramers J D. 2001. Determination of molybdenum isotope fractionation by double-spike multicollector inductively coupled plasma mass spectrometry. Geochemistry, Geophysics, Geosystems, 2(7): 1032

Su B X, Gu X Y, Deloule E, Zhang H F, Li Q L, Li X H, Vigier N, Tang Y J, Tang G Q, Liu Y, Pang K N, Brewer A, Mao Q, Ma Y G. 2015. Potential orthopyroxene, clinopyroxene and olivine reference materials for in situ lithium isotope determination. Geostandards and Geoanalytical Research, 39(3): 357–369

Sun L X, Yu H B, Cong Z B, Xin Y, Li Y, Qi L F. 2015. In situ analysis of steel melt by double-pulse laser-induced breakdown spectroscopy with a Cassegrain telescope. Spectrochimica Acta Part B: Atomic Spectroscopy, 112: 40–48

Sun L X, Yu H B, Cong Z B, Lu H, Cao B, Zeng P, Dong W, Li Y. 2018. Applications of laser-induced breakdown spectroscopy in the aluminum electrolysis industry. Spectrochimica Acta Part B: Atomic Spectroscopy, 142: 29–36

Tan D C, Zhu J M, Wang X L, Han G L, Lu Z, Xu W P. 2020. High-sensitivity determination of Cd isotopes in low-Cd geological samples by double spike MC-ICP-MS. Journal of Analytical Atomic Spectrometry, 35(4): 713–727

Tang G Q, Su B X, Li Q L, Xia X P, Jing J J, Feng L J, Martin L, Yang Q, Li X H. 2019. High-Mg$^{\#}$ olivine, clinopyroxene and orthopyroxene reference materials for in situ oxygen isotope determination. Geostandards and Geoanalytical Research, 43(4): 585–593

Tang G Q, Li X H, Li Q L, Liu Y, Ling X X. 2020. A new Chinese national reference material (GBW04481) for calcite oxygen and carbon isotopic microanalysis. Surface and Interface Analysis, 52(5): 190–196

Tang Y J, Zhang H F, Nakamura E, Moriguti T, Kobayashi K, Ying J F. 2007. Lithium isotopic systematics of peridotite xenoliths from Hannuoba, North China Craton: implications for melt-rock interaction in the considerably thinned lithospheric mantle. Geochimica et Cosmochimica Acta, 71(17): 4327–4341

Tang Y W, Cui K, Zheng Z, Gao J F, Han J J, Yang J H, Liu L. 2020. LA-ICP-MS U-Pb geochronology of wolframite by combining NIST series and common lead-bearing MTM as the primary reference material: implications for metallogenesis of South China. Gondwana Research, 83: 217–231

Teng F Z, Wang S J, Frédéric M. 2019. Tracing the formation and differentiation of the Earth by non-traditional stable isotopes. Science China Earth Sciences, 62(11): 1702–1715

Tian L L, Zeng Z, Nan X Y, Yu H M, Huang F. 2019. Determining Ba isotopes of barite using the Na_2CO_3 exchange reaction and double-spike method by MC-ICP-MS. Journal of Analytical Atomic Spectrometry, 34(7): 1459–1467

Tong X R, Liu Y S, Hu Z C, Chen H H, Zhou L, Hu Q H, Xu R, Deng L X, Chen C F, Yang L, Gao S. 2016. Accurate determination of Sr isotopic compositions in clinopyroxene and silicate glasses by LA-MC-ICP-MS. Geostandards and Geoanalytical Research, 40(1): 85–99

Wang Z C, Becker H. 2014. Abundances of sulfur, selenium, tellurium, rhenium and platinum-group elements in eighteen reference materials by isotope dilution sector-field ICP-MS and negative TIMS. Geostandards Geoanalytical Research, 38(2): 189–209

Wang Z C, Becker H, Wombacher F. 2015. Mass fractions of S, Cu, Se, Mo, Ag, Cd, In, Te, Ba, Sm, W, Tl and Bi in geological reference materials and selected carbonaceous chondrites determined by isotope dilution ICP-MS. Geostandards Geoanalytical Research, 39(2): 185–208

Wei H Z, Jiang S Y, Hemming N G, Yang J H, Yang T, Wu H P, Yang T L, Yan X, Pu W. 2014. An improved procedure for separation/purification of boron from complex matrices and high-precision measurement of boron isotopes by positive thermal ionization and multicollector inductively coupled plasma mass spectrometry. Talanta, 123: 151–160

Weis P, Driesner T, Heinrich C A. 2012. Porphyry-copper ore shells form at stable pressure-temperature fronts within dynamic fluid plumes. Science, 338(6114): 1613–1616

Wen H J, Carignan J, Cloquet C, Zhu X K, Zhang Y X. 2010. Isotopic delta values of molybdenum standard reference and prepared solutions measured by MC-ICP-MS: proposition for delta zero and secondary references. Journal of Analytical Atomic Spectrometry, 25(5): 716–721

Wilkinson J J, Weiss D J, Mason T F D, Coles B J. 2005. Zinc isotope variation in hydrothermal systems: preliminary evidence from the Irish Midlands ore field. Economic Geology, 100(3): 583–590

Wu F, Qi Y H, Yu H M, Tian S Y, Hou Z H, Huang F. 2016. Vanadium isotope measurement by MC-ICP-MS. Chemical Geology, 421: 17–25

Wu F Y, Arzamastsev A A, Mitchell R H, Li Q L, Sun J, Yang Y H, Wang R C. 2013. Emplacement age and Sr-Nd isotopic compositions of the Afrikanda alkaline ultramafic complex, Kola Peninsula, Russia. Chemical Geology, 353: 210–229

Wu G L, Zhu J M, Wang X L, Han G L, Tan D C, Wang S J. 2019. A novel purification method for high precision measurement of Ni isotopes by double spike MC-ICP-MS. Journal of Analytical Atomic Spectrometry, 34(8): 1639–1651

Wu G L, Zhu J M, Wang X L, Thomas M J, Han G L. 2020. High-sensitivity measurement of Cr isotopes by double spike MC-ICP-MS at the 10 ng level. Analytical Chemistry, 92(1): 1463–1469

Wu L, Shi G H, Danišík M, Zhang Z Y, Wang Y Z, Wang F. 2019. MK-1 apatite: a new potential reference material for (U-Th)/He dating. Geostandards and Geoanalytical Research, 43(2): 301–315

Wu L G, Li X H, Ling X X, Yang Y H, Li C F, Li Y L, Mao Q, Li Q L, Putlitz B. 2019. Further characterization of the RW-1 monazite: a new working reference material for oxygen and neodymium isotopic microanalysis. Minerals, 9(10): 583

Wu S T, Wörner G, Jochum K P, Stoll B, Simon K, Kronz A. 2019. The preparation and preliminary characterisation of three synthetic andesite reference glass materials (ARM-1, ARM-2, ARM-3) for in situ microanalysis. Geostandards and Geoanalytical Research, 43(4): 567–584

Xia X P, Cui Z X, Li W C, Zhang W F, Yang Q, Hui H Q, Lai C K. 2019. Zircon water content: reference material development and simultaneous measurement of oxygen isotopes by SIMS. Journal of Analytical Atomic Spectrometry, 34(6): 1088–1097

Xie L W, Yang J H, Yin Q Z, Yang Y H, Liu J B, Huang C. 2017. High spatial resolution in situ U-Pb dating using laser ablation multiple ion counting inductively coupled plasma mass spectrometry (LA-MIC-ICP-MS). Journal of Analytical Atomic Spectrometry, 32(5): 975–986

Xu L, Hu Z C, Zhang W, Yang L, Liu Y S, Gao S, Luo T, Hu S H. 2015. In situ Nd isotope analyses in geological materials with signal enhancement and non-linear mass dependent fractionation reduction using laser ablation MC-ICP-MS. Journal of Analytical Atomic Spectrometry, 30(1): 232–244

Xu L, Yang J H, Ni Q, Yang Y H, Hu Z C, Liu Y S, Wu Y B, Luo T, Hu S H. 2018. Determination of Sm-Nd isotopic compositions in fifteen geological materials using laser ablation MC-ICP-MS and application to monazite geochronology of metasedimentary rock in the North China Craton. Geostandards and Geoanalytical Research, 42(3): 379–394

Yang C, He D, Zhu Z L, Peng H, Liu Z F, Wen G J, Bai J H, Zheng H T, Hu S H, Wang Y X. 2017. Battery-operated atomic emission analyzer for waterborne arsenic based on atmospheric pressure glow discharge excitation source. Analytical Chemistry, 89(6): 3694–3701

Yang L. 2009. Accurate and precise determination of isotopic ratios by MC-ICP-MS: A review. Mass Spectrometry Reviews, 28(6): 990–1011

Yang L, Tong S Y, Zhou L, Hu Z C, Mester Z, Meija J. 2018. A critical review on isotopic fractionation correction methods for accurate isotope amount ratio measurements by MC-ICP-MS. Journal of Analytical Atomic Spectrometry, 33(11): 1849–1861

Yang M, Yang Y H, Evans N J, Xie L W, Huang C, Wu S T, Yang J H, Wu F Y. 2020. Precise and accurate determination of Lu and Hf contents, and Hf isotopic compositions in Chinese rock reference materials by MC-ICP-MS. Geostandards and Geoanalytical Research, 44(3): 553–565

Yang Q, Xia X P, Zhang L, Zhang W F, Zhang Y Q, Chen L L, Yang Y N, He M H. 2020b. Oxygen isotope homogeneity assessment for apatite U-Th-Pb geochronology reference materials. Surface and Interface Analyses, 52(5): 197–213

Yang W, Lin Y T, Zhang J C, Hao J L, Shen W J, Hu S. 2012. Precise micrometer-sized Pb-Pb and U-Pb dating with NanoSIMS. Journal of Analytical Atomic Spectrometry, 27(3): 479–487

Yang W, Teng F Z, Li W Y, Liu S A, Ke S, Liu Y S, Zhang H F, Gao S. 2016. Magnesium isotopic composition of the deep continental crust. American Mineralogist, 101(2): 243–252

Yang Y H, Zhang H F, Chu Z Y, Xie L W, Wu F Y. 2010. Combined chemical separation of Lu, Hf, Rb, Sr, Sm and Nd from a single rock digest and precise and accurate isotope determinations of Lu-Hf, Rb-Sr and Sm-Nd isotope systems using multi-collector ICP-MS and TIMS. International Journal of Mass Spectrometry, 290(2-3): 120–126

Yang Y H, Chu Z Y, Wu F Y, Xie L W, Yang J H. 2011. Precise and accurate determination of Sm, Nd concentrations and Nd isotopic compositions in geological samples by MC-ICP-MS. Journal of Analytical Atomic Spectrometry, 26(6): 1237–1244

Yang Y H, Wu F Y, Li Y, Yang J H, Xie L W, Liu Y, Zhang Y B, Huang C. 2014. In situ U-Pb dating of bastnaesite by LA-ICP-MS. Journal of Analytical Atomic Spectrometry, 29(6): 1017–1023

Yang Y H, Wu F Y, Yang J H, Mitchell R H, Zhao Z F, Xie L W, Huang C, Ma Q, Yang M, Zhao H. 2018a. U-Pb age determination of schorlomite garnet by laser ablation inductively coupled plasma mass spectrometry. Journal of Analytical Atomic Spectrometry, 33(2): 231–239

Yang Y H, Zhang X C, Liu S A, Zhou T, Fan H F, Yu H M, Cheng W H, Huang F. 2018b. Calibrating NIST SRM 683 as a new international reference standard for Zn isotopes. Journal of Analytical Atomic Spectrometry, 33(10): 1777–1783

Yang Y H, Wu F Y, Li Q L, Rojas-Agramonte Y, Yang J H, Li Y, Ma Q, Xie L W, Huang C, Fan H R, Zhao Z F, Xu C. 2019. In situ U-Th-Pb dating and Sr-Nd isotope analysis of bastnäsite by LA-(MC)-ICP-MS. Geostandards and Geoanalytical Research, 43(4): 543–565

Yuan H L, Cheng C, Chen K Y, Bao Z A. 2016. Standard-sample bracketing calibration method combined with Mg as an internal standard for silicon isotopic compositions using multi-collector inductively coupled plasma mass spectrometry. Acta Geochimica, 35: 421–427

Yuan H L, Yuan W T, Bao Z A, Chen K Y, Huang F, Liu S A. 2017. Development of two new copper isotope standard solutions and their copper isotopic compositions. Geostandards and Geoanalytical Research, 41(1): 77–84

Yuan S D, Peng J T, Hao S, Li H M, Geng J Z, Zhang D L. 2011. In situ LA-MC-ICP-MS and ID-TIMS U-Pb geochronology of cassiterite in the giant

Furong tin deposit, Hunan Province, South China: new constraints on the timing of tin-polymetallic mineralization. Ore Geology Reviews, 43(1): 235–242

Yuan W, Chen J B, Birck J L, Yin Z Y, Yuan S L, Cai H M, Wang Z W, Huang Q, Wang Z H. 2016. Precise analysis of gallium isotopic composition by MC-ICP-MS. Analytical Chemistry, 88(19): 9606–9613

Zeng Z, Li X H, Liu Y, Huang F, Yu H M. 2019. High-precision barium isotope measurements of carbonates by MC-ICP-MS. Geostandards and Geoanalytical Research, 43(2): 291–300

Zhang A Y, Zhang J, Zhang R F, Xue Y. 2015. Determination of stable silicon isotopes using multi-collector inductively coupled plasma mass spectrometry. Chinese Journal of Analytical Chemistry, 43(9): 1353–1359

Zhang C X, Hu Z C, Zhang W, Liu Y S, Zong K Q, Li M, Chen H H, Hu S H. 2016. Green and fast laser fusion technique for bulk silicate rock analysis by laser ablation-inductively coupled plasma mass spectrometry. Analytical Chemistry, 88(20): 10088–10094

Zhang L, Wu J L, Tu J R, Wu D, Li N, Xia X P, Ren Z Y. 2020. RMJG Rutile: A new natural reference material for microbeam U-Pb dating and Hf isotopic analysis. Geostandards and Geoanalytical Research, 44(1): 133–145

Zhang S Y, Zhang H L, Hou Z H, Ionov D A, Huang F. 2019. Rapid determination of trace element compositions in peridotites by LA-ICP-MS using an albite fusion method. Geostandards and Geoanalytical Research, 43(1): 93–111

Zhang T, Zhou L, Yang L, Wang Q, Feng L P, Liu Y S. 2016. High precision measurements of gallium isotopic compositions in geological materials by MC-ICP-MS. Journal of Analytical Atomic Spectrometry, 31(8): 1673–1679

Zhang W, Hu Z C, Liu Y S, Chen H H, Gao S, Gaschnig R M. 2012a. Total rock dissolution using ammonium bifluoride (NH_4HF_2) in screw-top Teflon vials: A new development in open-vessel digestion. Analytical Chemistry, 84(24): 10686–10693

Zhang W, Hu Z C, Liu Y S, Chen L, Chen H H, Li M, Zhao L S, Hu S H, Gao S. 2012b. Reassessment of HF/HNO_3 decomposition capability in the high-pressure digestion of felsic rocks for multi-element determination by ICP-MS. Geostandards and Geoanalytical Research, 36(3): 271–289

Zhang W, Qi L, Hu Z C, Zheng C J, Liu Y S, Chen H H, Gao S, Hu S H. 2016. An investigation of digestion methods for trace elements in bauxite and their determination in ten bauxite reference materials using inductively coupled plasma-mass spectrometry. Geostandards and Geoanalytical Research, 40(2): 195–216

Zhang W, Hu Z C, Liu Y S, Yang W W, Chen H H, Hu S H, Xiao H Y. 2017. Quantitative analysis of major and trace elements in NH_4HF_2-modified silicate rock powders by laser ablation-inductively coupled plasma mass spectrometry. Analytica Chimica Acta, 983: 149–159

Zhang W, Hu Z C, Liu Y S, Feng L P, Jiang H S. 2019a. In situ calcium isotopic ratio determination in calcium carbonate materials and calcium phosphate materials using laser ablation-multiple collector-inductively coupled plasma mass spectrometry. Chemical Geology, 522: 16–25

Zhang W, Wang Z C, Moynier F, Inglis E, Tian S Y, Li M, Liu Y S, Hu Z C. 2019b. Determination of Zr isotopic ratios in zircons using laser-ablation multiple-collector inductively coupled-plasma mass-spectrometry. Journal of Analytical Atomic Spectrometry, 34(9): 1800–1809

Zhang X, Hester K C, Ussler W, Walz P M, Peltzer E T, Brewer P G. 2011. In situ Raman-based measurements of high dissolved methane concentrations in hydrate-rich ocean sediments. Geophysical Research Letters, 38(8): L08605

Zhang X, Du Z F, Zheng R, Luan Z D, Qi F J, Cheng K, Wang B, Ye W Q, Liu X R, Lian C, Chen C, Guo J J, Li Y and Yan J. 2017. Development of a new deep-sea hybrid Raman insertion probe and its application to the geochemistry of hydrothermal vent and cold seep fluids. Deep Sea Research Part I: Oceanographic Research Papers, 123: 1–12

Zhang X, Li L F, Du Z F, Hao X L, Cao L, Luan Z D, Wang B, Xi S C, Lian C, Yan J, Sun W D. 2020. Discovery of supercritical carbon dioxide in a hydrothermal system. Science Bulletin, 65(11): 958–964

Zhang X C, Zhang A Y, Zhang Z F, Huang F, Yu H M. 2018. Influence of room temperature on magnesium isotope measurements by multi-collector inductively coupled plasma mass spectrometry. Rapid Communications in Mass Spectrometry, 32(13): 1026–1030

Zhang Z Y, Ma J L, Zhang L, Liu Y, Wei G J. 2018. Rubidium purification via a single chemical column and its isotope measurement on geological standard materials by MC-ICP-MS. Journal of Analytical Atomic Spectrometry, 33(2): 322–328

Zhu C W, Lu W N, He Y S, Ke S, Wu H J, Zhang L N. 2018. Iron isotopic analyses of geological reference materials on MC-ICP-MS with instrumental mass bias corrected by three independent methods. Acta Geochimica, 37(5): 691–700

Zhu G H, Ma J L, Wei G Y, An Y J. 2020. A novel procedure for separating iron from geological materials for isotopic analysis using MC-ICP-MS. Journal of Analytical Atomic Spectrometry, 35(5): 873–877

Zhu H L, Zhang Z F, Wang G Q, Liu Y F, Liu F, Li X, Sun W D. 2016. Calcium isotopic fractionation during ion-exchange column chemistry and thermal ionisation mass spectrometry (TIMS) determination. Geostandards and Geoanalytical Research, 40(2): 185–194

Zhu J M, Wu G L, Wang X L, Han G L, Zhang L X. 2018. An improved method of Cr purification for high precision measurement of Cr isotopes by double spike MC-ICP-MS. Journal of Analytical Atomic Spectrometry, 33(5): 809–821

Zhu L Y, Liu Y S, Hu Z C, Hu Q H, Tong X R, Zong K Q, Chen H H, Gao S. 2013. Simultaneous determination of major and trace elements in fused volcanic rock powders using a hermetic vessel heater and LA-ICP-MS. Geostandards and Geoanalytical Research, 37(2): 207–229

Zhu L Y, Liu Y S, Ma T T, Lin J, Hu Z C, Wang C. 2016. In situ measurement of Os isotopic ratios in sulfides calibrated against ultra-fine particle standards using LA-MC-ICP-MS. Journal of Analytical Atomic Spectrometry, 31(7): 1414–1422

Zhu L Y, Liu Y S, Jiang S Y, Lin J. 2019. An improved in situ technique for the analysis of the Os isotope ratio in sulfides using laser ablation-multiple ion counter inductively coupled plasma mass spectrometry. Journal of Analytical Atomic Spectrometry, 34(8): 1546–1552

Zhu X K, Guo Y, O'Nions R K, Young E D, Ash R D. 2001. Isotopic homogeneity of iron in the early solar nebula. Nature, 412(6844): 311–313

Zhu X K, Guo Y, Williams R J P, O'Nions R K, Matthews A, Belshaw N S, Canters G W, de Waal E C, Weser U, Burgess B K, Salvato B. 2002. Mass fractionation processes of transition metal isotopes. Earth and Planetary Science Letters, 200(1-2): 47–62

Zhu Z Y, Cook N J, Yang T, Ciobanu C L, Zhao K D, Jiang S Y. 2016. Mapping of sulfur isotopes and trace elements in sulfides by LA-(MC)-ICP-MS: potential analytical problems, improvements and implications. Minerals, 6(4): 110

Advances of Researches on Rock and Mineral Analyses

LIU Yong-sheng[1], QU Wen-jun[2], QI Liang[3], YUAN Hong-lin[4], HUANG Fang[5], YANG Yue-heng[6], HU Zhao-chu[1], ZHU Zhen-li[7], ZHANG Wen[1]

1. State Key Laboratory of Geological Processes and Mineral Resources, China University of Geosciences (Wuhan), Wuhan 430074; 2. National Research Center for Geoanalysis, Beijing 100037; 3. State Key Laboratory of Ore Deposit Geochemistry, Institute of Geochemistry, Chinese Academy of Sciences, Guiyang 550081; 4. State Key Laboratory Continental Dynamics, Department of Geology, Northwest University, Xi'an 710069; 5. School of Earth and Space Sciences, University of Science and Technology of China, Hefei 230026; 6. State Key Laboratory of Lithospheric Evolution, Institute of Geology and Geophysics, Chinese Academy of Sciences, Beijing 100029; 7. State Key Laboratory of Biogeology and Environmental Geology, China University of Geosciences (Wuhan), Wuhan 430074

Abstract: Major advances of researches, in China during 2011–2020, on rock and mineral analyses, including the major and trace element analysis, radiogenic isotopic analysis, non-traditional stable isotopic analysis, geological sample preparation methods, development and determination of geological reference materials, research and development of mainstream analytical instruments and key components, have been reviewed and summarized in this paper. Over the past decade, domestic scholars have obtained a large number of original research results in these fields, and developed a portion of new techniques and methods for rock and mineral analyses at internationally advanced level. These excellent works have greatly promoted the development of researches on earth sciences in China. New requirements for the rock and mineral analyses are motivated due to the expanding of research fields of modern earth sciences, the China's demand for the development of natural resources, and the responsibility for protecting living environment of human beings. Therefore, we have provided perspectives for the development of researches on rock and mineral analyses in the future.

Key words: element quantitative analysis; isotopic composition analysis; geological sample preparation methods; geological reference materials; instruments and key components

非传统稳定同位素地球化学研究进展*

韦刚健[1,2] 黄 方[3] 马金龙[1,2] 邓文峰[1,2] 于慧敏[3] 康晋霆[3] 陈雪霏[1,2]

1. 中国科学院 广州地球化学研究所，同位素地球化学国家重点实验室，广州 510640；2. 中国科学院 深地科学卓越创新中心，广州 510640；3. 中国科学技术大学 地球和空间科学学院，合肥 230021

摘 要：非传统稳定同位素是过去十年国际上地球化学学科发展非常迅猛的方向，我国科研人员也以自主创新的方式参与到这一前沿研究的大潮中，研究队伍不断扩大，已成为这一学科方向的重要研究力量。本文回顾了过去十年（2010—2020 年）我国非传统稳定同位素分析技术研发方面的进展，同时对一些重要研究方向的代表性应用研究成果进行了总结。在分析技术方面，我国主要的相关实验室对众多非传统稳定同位素体系的高精度测试技术进行了探索，开拓了一些具备国际先进甚至领先水准的分析技术方法；在应用研究方面，不同的研究团队侧重于各自关注的研究领域和同位素体系，发展进程有所差别，但总体上处于蓬勃发展阶段，上升势头迅猛。有理由相信未来我国的非传统稳定同位素地球化学仍会继续快速发展，技术上会对已有的方法优化提高，并不断开拓新的体系，应用研究上将更深入探索地质、环境过程中的分馏机制，同时将在地球科学的相关研究领域得到更广泛的应用。

关键词：非传统稳定同位素 分析技术 应用研究 研究进展

0 引 言

过去十年，在前一个十年不断探索和开拓的基础上，国际上非传统稳定同位素地球化学研究在分析技术方法、分馏机理和示踪应用等方面均产出了数量巨大的成果，获得了"爆炸性"的发展。我国科学家经过前十年的学习、引进和积累（朱祥坤等，2013），在近十年里以自主创新方式参与到这一前沿研究的大潮中，使我国在这一学科领域的研究水平和能力得到了极大发展，具体体现一是研究队伍不断壮大，已成为国际上该领域的一支非常重要的研究力量；二是研究水准不断提升，在一些重要同位素体系的分析技术方法建立、分馏机理探索和各种地质过程的示踪应用等方面均取得一批具有国际先进水准乃至引领性的成果。本文尝试从分析技术方法研发、分馏机理和示踪应用等方面，对过去十年我国非传统稳定同位素地球化学研究所取得的主要成果进行总结。由于许多成果中机理探索和示踪应用难以截然分开，因此把这两个方面合并一起进行总结。

1 非传统稳定同位素地球化学研究在我国的兴起

同位素地球化学是地球科学中非常重要的分支学科，建立于放射性同位素衰变和第一性基本原理等物理化学原理基础上，是定量化的学科，可为地质过程提供年代信息、定量化示踪物质来源与演变过程，在地球科学研究的许多领域均有非常重要而且广泛的应用。经过长期的发展，同位素地球化学演化出三大支柱方向：同位素年代学、狭义的同位素地球化学（即基于放射成因同位素组成的示踪体系）和稳定同位素地球化学，其中稳定同位素占据天然存在的同位素家族的大部分。

虽然基于物理化学基本原理的理论计算对同位素地球化学学科的发展有着非常重要的作用，但对地

---

* 原文"近十年我国非传统稳定同位素地球化学研究进展"刊于《矿物岩石地球化学通报》2021 年第 40 卷第 6 期，本文略有修改。

质对象和过程的观测仍然是同位素地球化学学科最主要的研究内容。因而这一学科的发展一直都和同位素的观测技术发展密切相关。受早期高精度同位素分析技术的限制，传统的稳定同位素地球化学研究主要针对利用气体源同位素质谱测试的C-H-O-N-S五个同位素系列。20世纪90年代后期，随着同位素分析技术的发展，特别是多接收器电感耦合等离子体质谱（MC-ICP-MS）的广泛使用和技术进步，很多原先未受关注的稳定同位素组成能够被准确测定，除传统的C-H-O-N-S外的其他稳定同位素体系开始受到学界关注。相对于上述五个传统的稳定同位素体系，这类稳定同位素就统称为"non-traditional stable isotopes"，直译为中文就是"非传统稳定同位素"。这一概念较早出现于2004年5月由美国矿物学会和地球化学学会在加拿大蒙特利尔联合举办的美国地球物理学会（AGU）和加拿大地球物理学会（CGU）联合春季学术会议的会前短训班"Geochemistry of non-traditional stable isotopes"上，在此基础上 Reviews in Mineralogy and Geochemistry 出版了题为"Geochemistry of non-traditional stable isotopes"的专辑（Johnson et al., 2004）。

这个名称有明显的时代烙印，所谓的非传统是相对于20世纪末至21世纪初的分析技术而言，有些学者对这一名称持保留意见，并不断有学者提出选取一个更为贴切名称的建议。例如，这些同位素体系中绝大部分是金属元素，有学者建议使用金属稳定同位素来替代这个名称。但这些体系中还包含有一些非金属元素，如B、Si、Se等，因此也有学者建议称为"非常规稳定同位素"等。到目前为止，还难以找到一个能涵盖整个非传统稳定同位素体系而且又能体现时代特色的名称，因而这一名称还要继续使用。

早在国际上非传统稳定同位素的研究兴起之初，我国学者就已接触并开展了相关研究。早期主要是通过在国外科研机构、高校访问或者留学方式，及国内少数实验室的自主探索。21世纪初，这批访问、留学的学者陆续回国，非传统稳定同位素的研究方法和思想在国内得到迅速传播，一批重点开展非传统稳定同位素地球化学技术方法、机理和应用研究的实验室得以建立，相关研究在国内迅速兴起。例如，以中国地质科学院地质研究所、中国科学院地球化学研究所等为代表的相关机构在国内较早系统地开展Mg、Fe、Cu、Zn和Mo等同位素体系的研究，在分析技术方法和应用研究方面均取得了较高水平的成果。这一阶段的进展在中国矿物岩石地球化学学会前一个十年回顾中已有详细介绍（朱祥坤等，2013）。最近十年，在前期持续探索和发展的基础上，随着国内科研投入的不断增加，实验室的数量和规模也随之扩大，国内非传统稳定同位素地球化学的研究从传统的固体地球科学和天体化学领域拓展到地表、海洋及生命演化等领域，成了应用最广泛的地球化学研究手段之一。

与一般的同位素体系类似，非传统稳定同位素体系的研究也是按建立分析技术方法、厘清地质储库组成特征、探索地质过程中的分馏机制到应用于地质过程的示踪这样的顺序展开的。分析技术方法的建立无疑是开路先锋，同时也是过去十年我国非传统稳定同位素地球化学研究中最活跃的方向之一。本文将从分析技术和机制探索以及应用两大部分来进行回顾总结。

2 近十年我国非传统稳定同位素分析技术的进展

非传统稳定同位素组成的高精度分析基本上都是在固体源的同位素质谱仪，如多接收器电感耦合等离子体质谱仪（MC-ICP-MS）和热电离质谱仪（TIMS）上进行，其分析技术和传统的放射成因同位素体系如 $^{87}Sr/^{86}Sr$、$^{143}Nd/^{144}Nd$ 等一脉相承，均需依次进行化学纯化富集和质谱测量。但在质谱测定时 $^{87}Sr/^{86}Sr$ 和 $^{143}Nd/^{144}Nd$ 等放射成因同位素体系需选定一个固定的内部比值来校正仪器测定过程中的分馏，如此会把这些同位素比值中的质量分馏信息基本上全部抹除掉，而这些分馏信息却是非传统稳定同位素地球化学研究的核心内容。因此，非传统稳定同位素组成的质谱测量主要通过标准-样品间插法（SSB）（Zhu et al., 2000）、外标元素加入法、部分具有四个或更多稳定同位素的体系则使用双稀释剂法来进行仪器分馏校正。质谱测量的这一技术差别对非传统稳定同位素分析的化学纯化提出了更高的要求：首先要有接近100%的回收率，因为纯化过程中特别是离子交换分离过程存在极其显著的同位素分馏（Anbar et al., 2000；Zhu et al., 2002b；Ma et al., 2013a, 2013b；Zhu et al., 2016；Zhang et al., 2018），回收率不足必然导致明显的结果偏差；另外还需要尽量将非目标组分（干扰元素和基体组分）分离干净，因为使用 SSB 测试时，即便较小程度的样品和标准基体的不匹配也会导致分析结果明显的偏差。一些特别的体系可以利用质量相近的外标元素校正质谱测量过程中的分馏，如 Cu-Zn、Sr-Zr 体系，甚至一些同质异位干扰元素也可以用于质谱测量过程中的分馏校正，例如 Ca-Ti、V-Ti、Cr-Ti、Ni-Zn 和 In-Sn 体系（Shuai et al.，2020）。这些质谱测量技术虽然有所不同，但其化学处理的要求和分析方法却基本一致。

SSB 测试方法是基于待测样品的目标同位素和标准样品的相应同位素具有一致的分馏，然而样品和标准是按顺序进行测量的，并不同步进行，因此除要求样品和标准的基体匹配外，对仪器的稳定性要求也很高，仪器不可避免的短期波动往往会对测试的精度产生较大的影响。加入外标元素后，外标元素和目标元素的同位素组成是同步测试的，此时参照 Sr、Nd、Hf 等放射成因同位素组成的测定方式，利用外标元素的同位素比值作为目标同位素组成的内部标准比值进行校正，可以很大程度上消除仪器短期波动的影响。然而，外标元素和目标元素在质谱测试中如离子化、能量聚焦和质量偏转等过程中的分馏效应往往是不相同的，如果完全参照 Sr、Nd、Hf 等放射成因同位素组成的测试方式，将外标元素的同位素组成设定为一个固定的比值去校正目标元素的同位素分馏往往会得到不准确的同位素比值。

为了克服不同元素同位素分馏程度不一致而导致的外标元素校正法准确度偏离的问题，有学者发展了一种分馏系数回归校正方法。该方法基于测量过程中的经验发现，即虽然外标元素和目标元素的同位素分馏系数是不一致且随测试过程不断发生变化的，但在同一个测试序列中（仪器状态变化不是特别大时）两者之间的比值却是恒定的（Maréchal et al.，1999）。因此，样品测试前，须先通过对已知同位素比值的外标元素和目标元素标准（如参照标准）的混合溶液进行长时间的测试，获得一定范围的分馏程度的变化，通过对数线性回归［如 ln（^{65}Cu/^{63}Cu）-ln（^{66}Zn/^{64}Zn）］获取目标元素和外标元素的同位素分馏关系，再将这个关系用于单次测量中用外标元素同位素比值对目标元素的同位素比值进行分馏校正。这种方法的准确性很大程度上依赖于这个回归关系的精确建立，在保持仪器状态基本不变的情况下，往往需要非常长的测试时间（数小时甚至数十个小时）才能构建出比较精确的回归关系。一些研究者通过微调仪器测试参数（如 RF 功率），人为制造分馏程度的变化，在较短时间内（数小时）构建出外标元素和目标元素的同位素分馏关系（Malinovsky et al.，2016），进而提高分析测试的效率。不过目前这种校正方法的分析效率仍然不高，也没得到广泛应用。我国一些学者在国外实验室利用该方法对一些比较特殊的同位素如 Lu、Hf、W 和 Ir 的比值开展测试研究（Zhu et al.，2017；Tong et al.，2019；Zhang et al.，2019；He J. et al.，2020），但目前国内的实验室利用该方法开展非传统稳定同位素组成测试的尝试还比较少见。

近年来比较常见的方法是将两种校正方式相结合，在样品和标准样品中均加入合适的外标元素，然后使用 SSB 进行测试。单次测量中用外标元素的同位素比值做内标进行校正以消除仪器短期波动的影响，而外标元素的同位素比值则利用标准样品中已知的目标元素的同位素比值进行即时校正。这一方法被称为样品标准间插和内标校正结合法（combined standard sample bracketing and internal normalization，CSSBIN），可同时保证目标同位素组成测试结果的精密度和准确度，在非传统稳定同位素的测试中已得到越来越广泛的应用，必将是未来最重要的测试方法之一。

近年来，基于激光剥蚀多接收器电感耦合等离子体质谱（LA-MC-ICP-MS）和二次离子质谱（SIMS）的微区原位非传统稳定同位素分析也取得较大进展，这是一个新的分析技术体系，将单独进行总结。

2.1 碱金属和碱土金属同位素分析技术进展

碱金属中的 Li、K、Rb 和碱土金属中的 Mg、Ca、Sr、Ba 具有超出一个天然存在的稳定同位素，它们是非传统稳定同位素研究重要的组成部分，相关的研究在国内实验室均有开展。

2.1.1 Li 同位素

Li 同位素是目前应用最广的碱金属同位素体系，表述为 δ^7Li，代表样品的 ^{7}Li/^{6}Li 值相对于参照标准（L-SVEC）的千分偏差，在深部和表生地质过程中均有重要的示踪作用。早在非传统稳定同位素研究兴起之前，我国学者肖应凯就已建立了基于热电离质谱测试 Li 同位素比值的方法（肖应凯等，1982；Xiao and Beary，1989），该方法是通过测试质量数较大的含 Li 分子（如 $LiNO_3$ 或 $Li_2B_4O_7$）以减小仪器测试过程中的 Li 同位素分馏效应，但无法定量校正分馏。Li 非传统稳定同位素研究兴起后，多基于 MC-ICP-MS 采用可以进行分馏校正的测试方法，早期我国学者在这方面的研究主要借鉴国外科研团队建立的分析方法。然而 Li 同位素的分析技术远没有达到完全成熟的地步，过去十年，高精度的 Li 同位素分析方法不断得到改进，其中包括一些在国外实验室访问或工作的中国学者的工作（Gao and Casey，2012；Li W. S. et al.，2019）。最近几年，国内相关实验室也对 Li 同位素分析技术的优化开展了探索，并取得重要进展。

传统 Li 同位素分析方法的不足主要体现在化学处理过程比较费时、效率不高，对一些基体元素如 Na 等碱金属的分离效果不很理想。中国地质科学院地质研究所的团队采用双柱套接分离技术，提高了对基体元素的分离效果（Zhu Z. Y. et al.，2019），中国科学院广州地球化学研究所的团队采用 AGMP-50 树脂实现对硅酸盐和海水的 Li 同位素单柱分离，工作效率和基体元素分离效果均有明显提高（Zhu G. H. et al.，2020a，2020b）。除了对化学处理方法进行改进外，中国地质大学（武汉）的团队对利用 MC-ICP-MS 测量 Li 同位素过程中仪器记忆效应的降低进行了探索，提出了标准和样品之间 Li 含量不需要严格匹配的分析方法（Lin et al.，2016，2019a）。这些分析技术的探索使得国内实验室也可高效地产出高质量的 Li 同位素数据，标准样品的 δ^7Li 的外部重现性（2σ）基本好于±0.5‰，达到国际先进水准，为相关科学研究提供了重要技术支撑。

2.1.2 K 同位素

K 有 ^{39}K、^{40}K 和 ^{41}K 3 个天然同位素，其中 ^{39}K 和 ^{41}K 是稳定同位素，K 的稳定同位素组成表述为 $\delta^{41}K$，代表样品 $^{41}K/^{39}K$ 值相对于参照标准（NIST SRM 3141a）的千分偏差。高精度的 K 同位素测试只能依托于 MC-ICP-MS 使用 SSB 方式进行分馏校正，而 MC-ICP-MS 使用 Ar 气作为载气，Ar 的氢化物如 ArH^+ 等对 K^+ 的同位素有非常严重的质量干扰（$^{38}Ar^1H^+$ 对 $^{39}K^+$ 和 $^{40}Ar^1H^+$ 对 $^{41}K^+$），使得 K 同位素的高精度测试一直是个难题。另外，K 的纯化过程中完全将其与基体元素分离并保持接近 100% 的回收率也一直是个大的技术挑战。

近年来，K 同位素的高精度测试取得突破，在国外实验室访学或工作的我国学者高度参与了这一技术突破，并起到关键作用。降低 ArH^+ 对 K^+ 干扰主要有两种方法，一种是在等离子体激发后导入加载 H_2 或者 D_2 的碰撞室，降低 ArH^+ 能量从而减小其对 K^+ 信号的干扰，实现 K 同位素组成的高精度测试，$\delta^{41}K$ 外部重现性好于±0.10‰（2σ）（Li W. Q. et al.，2016；Wang and Jacobsen，2016）。另一种则是使用冷干等离子体激发方式，即降低 RF 发生器的功率，使用膜去溶方式去除导入仪器的样品中的溶剂（水），以大幅度降低 ArH^+ 的产率，同时使用类似高分辨率的检测方式，尽量把 ArH^+ 和 K 的信号峰稍微错开（不完全重叠），这一方法同样可以实现对 K 同位素的高精度测定，实现 $\delta^{41}K$ 外部重现性好于±0.10‰（2σ）（Hu et al.，2018；Xu et al.，2019；Chen H. et al.，2019）。

需要指出的是，以上发表的成果虽然由我国学者完成，但分析均是依托于国外实验室开展的。中国地质大学（北京）的团队通过优化 K 的化学纯化流程，以及采用冷干等离子体激发结合类似高分辨率的检测方式，首先实现了国内实验室的高精度 K 同位素分析，标样 $\delta^{41}K$ 的外部重现性好于±0.10‰（2σ）（Li X. Q. et al.，2020）。

2.1.3 Rb 同位素

Rb 有 ^{85}Rb 和 ^{87}Rb 两个天然同位素，其中 ^{87}Rb 具有放射性，半衰期较长（4.88×10^{10} a），在不涉及较长时间尺度的演化问题时可看成是稳定同位素。Rb 稳定同位素表述为 $\delta^{87}Rb$，代表样品中 $^{87}Rb/^{85}Rb$ 值相对于参照标准（NIST SRM 984）的千分偏差。Rb 同位素的研究不多，国际上也只有少数实验室开展。其测试技术的难点在于 Rb 的化学纯化处理很难将其与其他碱金属元素如 Na、K 完全分离。

到目前为止，国内只有中国科学院广州地球化学研究所的团队发表了高精度的 Rb 同位素测试方法的成果，他们的重要进展在于通过使用大体积的 Sr 特效树脂一柱顺序完全分离 Rb、K、Ba 和 Sr，从而一举突破了高精度 Rb 同位素测试的技术瓶颈，实现了 δ^{87}Rb 外部重现性好于 ±0.06‰（2σ）（Zhang Z. Y. et al., 2018）。这一化学纯化流程不仅适用于高精度 Rb 同位素组成分析，也为高精度 K 和 Ba 同位素的分析提供了更多高效的化学处理方案。

2.1.4 Mg 同位素

Mg 有 ^{24}Mg、^{25}Mg 和 ^{26}Mg 3 个稳定同位素，其组成表述为 δ^{25}Mg 或 δ^{26}Mg，分别为样品 ^{25}Mg/^{24}Mg 或者 ^{26}Mg/^{24}Mg 值相对于参照标准（DSM-3）的千分偏差。Mg 同位素是研究较早的非传统稳定同位素体系，我国不少学者在国外访学或工作期间积极参与了相关研究（Huang et al., 2009）。Mg 同位素也是在国内较早开展研究的一个非传统稳定同位素体系，在 21 世纪初的十年中，中国地质科学院地质研究所、中国科学院广州地球化学研究所等单位就已建立了高精度的 Mg 同位素测试方法（朱祥坤等，2013）。随着分析技术的逐步成熟，国内越来越多的实验室都相继建立了各自的高精度 Mg 同位素分析流程并迅速推广。

近年来，中国地质科学院地质研究所、中国科学技术大学、中国地质大学（北京）和西北大学等单位的研究团队针对一些特殊的地质样品，如高 K 低 Mg 岩石和富 REE-Nb-Fe-Mn 矿化样品等，开展了 Mg 同位素化学纯化流程和质谱测量技术的优化工作（李世珍等，2013；An and Huang, 2014；An et al., 2014；Bao et al., 2019；Gao T. et al., 2019；Gou et al., 2019a），使数据质量更稳定，δ^{25}Mg 或者 δ^{26}Mg 的外部重现性可好于 ±0.05‰（2σ），为相关研究提供了强大的技术支撑。

2.1.5 Ca 同位素

Ca 有 ^{40}Ca、^{42}Ca、^{43}Ca、^{44}Ca、^{46}Ca 和 ^{48}Ca 六个稳定同位素，因此其稳定同位素组成的表述方式比较多，根据其丰度及分析技术特点，通常表述为 $\delta^{44/40}$Ca 和 $\delta^{44/42}$Ca，分别代表样品中的 ^{44}Ca/^{40}Ca 和 ^{44}Ca/^{42}Ca 值相对于参照标准（NIST SRM 915a）的千分偏差。在一些高钾低钙样品中，^{40}K 衰变产生的 ^{40}Ca 有可能影响 Ca 的稳定同位素组成，这种情况下一般避免使用 ^{40}Ca 来表述。由于 Ca 的稳定同位素数量大于 4 个，因此可以选用双稀释剂法校正仪器测定过程中产生的 Ca 同位素分馏。使用 MC-ICP-MS 时也可选用 SSB 法进行分馏校正，但由于 MC-ICP-MS 的 Ar 载气对 ^{40}Ca 干扰非常严重，不能测量 ^{44}Ca/^{40}Ca 值而通常测量 ^{44}Ca/^{42}Ca 值。因此，测试方法的不同也会造成 Ca 稳定同位素的表述不同。

高精度 Ca 同位素分析在化学处理和质谱测量方面均存在较高的难度，虽然国际上在 20 世纪 90 年代后期就已开展了相关的研究，但分析技术瓶颈一直难以突破。近年来，我国一些实验室加强了 Ca 同位素分析技术的研发，基本上与国际上的高水准实验室同时取得了重要的技术突破。

中国科学院广州地球化学研究所的团队率先在国内建立起利用 ^{42}Ca-^{43}Ca 双稀释剂法结合 TIMS 测试技术的高精度 Ca 同位素分析方法，可以高效测试各种地质样品的 Ca 同位素组成。同时，研究人员对于其中一些关键的技术细节也进行了深入的探讨，如在稀释剂的选择（刘芳等，2016；Liu et al., 2020）、化学分离方法的建立（刘峪菲等，2015）、高钙样品免化学分离测试（Liu F. et al., 2019）、化学处理及质谱测量过程中的分馏机制和干扰校正（Zhu et al., 2016；张晨蕾等，2017）和依托双稀释剂法峰截取技术的提出（Zhu et al., 2018a）等。这些细致的技术探索使得该实验室多年持续监控的标准样品（NIST SRM 915a）$\delta^{44/40}$Ca 结果重现性达到国际领先水准 ±0.06‰（2σ）。

中国地质大学（北京）的团队则随后建立起了基于 ^{43}Ca-^{48}Ca 双稀释剂的 Triton-TIMS 分析技术并报道了多种地质标准样品的钙同位素组成（He et al., 2017）。该方法的技术特点是可以同时获取样品的 ^{44}Ca/^{40}Ca 以及 ^{44}Ca/^{42}Ca 值，因此可以对高 K 样品校正放射成因 ^{40}Ca 对 $\delta^{44/40}$Ca 值的影响。

中国地质大学（武汉）的团队在国内则率先建立起基于 MC-ICP-MS 的高精度 Ca 同位素分析方法，其标准样品的 $\delta^{44/42}$Ca 外部重现性达到 ±0.07‰（2σ）（Feng L. P. et al., 2018；Li M. et al., 2018）。目前，对于 Ca 同位素的分析技术的细节优化还在进行，而且国内也有越来越多的实验室能开展多种类型地质样品的 Ca 同位素的分析测试（Guan et al., 2020；Sun et al., 2021）。

2.1.6 稳定 Sr 同位素分析技术进展

Sr 有 ^{84}Sr、^{86}Sr、^{87}Sr 和 ^{88}Sr 4 个天然稳定同位素，其中部分 ^{87}Sr 由 ^{87}Rb 衰变产生。Sr 的稳定同位素组成

通常表述为 $\delta^{88}Sr$，代表样品中 $^{88}Sr/^{86}Sr$ 值与参照标准（NIST SRM 987）的千分偏差。传统的 Rb-Sr 年代学和示踪体系中，均假设 $^{88}Sr/^{86}Sr$ 是个定值（8.375 209），在传统的 $^{87}Sr/^{86}Sr$ 值测试过程中这个值被用来校正仪器的分馏。然而，自然界中普遍存在 $^{88}Sr/^{86}Sr$ 的分馏，稳定 Sr 同位素的研究可以为许多地质过程提供重要的示踪手段，同时也可为完善 Rb-Sr 理论体系提供重要资料。

高精度稳定 Sr 同位素可以采用双稀释剂法和 SSB 两种方式进行测试，前者测试过程相对烦琐，但稳定性相对较好，$\delta^{88}Sr$ 结果重现性一般可好于 ±0.015‰（2σ）。后者校正方法操作相对简单，但对仪器（MC-ICP-MS）的稳定性有极高要求，因此样品中往往还加入 Zr 溶液，利用 $^{92}Zr/^{90}Zr$ 或 $^{91}Zr/^{90}Zr$ 值进行内部校正以降低仪器短期波动的影响。中国科学院广州地球化学研究所的团队在国内最先开展稳定 Sr 同位素测试（韦刚健等，2015），他们利用 SSB 方式，在仪器稳定性较好的状态下 $\delta^{88}Sr$ 可以获得与双稀释法相当的外部重现性（好于 ±0.015‰）（2σ）（Ma et al., 2013b）。这一方法对拓展 Sr 同位素的示踪应用有重要价值。近年来，同济大学的团队也开展了这方面的测试，他们利用加 Zr 的 SSB 方法可达到相当的精度（Xu J. et al., 2020），并应用到表生和海洋地质过程的示踪研究中。

2.1.7 Ba 同位素分析技术进展

Ba 有 ^{130}Ba、^{132}Ba、^{134}Ba、^{135}Ba、^{136}Ba、^{137}Ba 和 ^{138}Ba 7 个稳定同位素，其稳定同位素组成通常表述为 $\delta^{137/134}Ba$ 或 $\delta^{138/134}Ba$，代表样品中 $^{137}Ba/^{134}Ba$ 或 $^{138}Ba/^{134}Ba$ 值相对于参照标准（NIST SRM 3104a）的千分偏差。Ba 的天然稳定同位素是一个较新的体系，对其研究有助于更好地了解 Ba 的地球化学特征，以及为相关的地质过程研究提供新的示踪手段。

我国学者早期曾在国外实验室开展了 Ba 在海洋过程中的示踪研究（Cao et al., 2016）。国内最先实现高精度 Ba 同位素分析的是中国科学技术大学的团队，他们利用双稀释剂（^{135}Ba-^{136}Ba）在 MC-ICP-MS 上分析各种类型地质样品的 Ba 同位素组成，标准样品的 $\delta^{137/134}Ba$ 重现性优于 ±0.05‰（2σ）（Nan et al., 2015；Tian L. L. et al., 2019；Zeng et al., 2019），达到国际先进水准。针对重晶石难以完全消解的特点，他们还建立了用 Na_2CO_3 部分提取重晶石从而获得准确的 Ba 同位素结果的分析方法（Tian L. L. et al., 2019），拓展了 Ba 同位素的示踪应用范围。An 等（2020）对各类标准样品的 Ba 同位素组成进行了测试，获得 $\delta^{138/134}Ba$ 的重现性优于 ±0.05‰（2σ）。此外，南京大学的团队利用 MC-ICP-MS 和 TIMS 也建立了高精度的 Ba 同位素分析方法（Lin et al., 2020）。这些方法的建立为推广 Ba 同位素的应用提供了技术支撑。

2.2 第三主族元素的同位素分析技术进展

第三主族元素中除 Al 只有一个天然同位素外，B、Ga、In 和 Tl 均具有两个稳定同位素。这些元素的同位素组成受到关注的程度不尽相同，在国内开展研究的程度也不太一样。

2.2.1 B 同位素分析技术进展

B 具有亲水性，也是生命必须元素，在与流体有关的地质过程中性质活泼，同时在水体特别是海洋和盐湖中富集，并积极参与生物活动。因此，B 同位素在地球科学多个领域的研究中具有极其广泛的示踪作用。早在非传统稳定同位素的研究兴起之前，B 同位素在盐湖化学、海洋科学等方面的研究就已广泛开展，在我国以肖应凯老师领导的团队等开展了大量的研究，在国际上具有较高的显示度。

B 有 ^{10}B 和 ^{11}B 两个稳定同位素，其组成表述为 $\delta^{11}B$，代表样品中 $^{11}B/^{10}B$ 值相对于参照标准（NIST SRM 951）的千分偏差。无论是化学纯化处理还是质谱测量方面，高精度 $\delta^{11}B$ 的测定均较具挑战性。由于 B 在酸介质中具有挥发性，因此对不少类型的样品进行化学处理时都存在损失而不能保证回收率，从而影响 $\delta^{11}B$ 结果的可靠性。B 的挥发性还使得一般超净实验室难以隔绝来自空气中的硼污染，无法降低实验室的硼本底，从而难以实现低硼含量样品的高精度测试。质谱测试方面，早期的 B 同位素测试往往在热电离质谱仪上进行，无论是检测负离子（BO_2^-）还是正离子（$Cs_2BO_2^+$），测试过程中的 B 同位素分馏均无法进行校正，结果的可靠性受测试过程中的主观因素影响较大。过去十年，我国 B 同位素地球化学的研究得到进一步发展，在固体地球科学和表生地质过程的研究程度逐渐深入，相应的分析技术也有较明显的发展。

在化学前处理方面，针对不同类型样品的 B 分离富集的方法不断得到优化，例如，南京大学的团队建立了能将干扰基体去除更干净的三柱分离方法（Wei et al., 2014a, 2014b），降低有机质对质谱测量干扰的化学处理方法（Wu et al., 2012），以及针对硼硅酸盐矿物的分离方法（晏雄等，2012）；中科院青海盐湖所的团队建立了系列针对黏土矿物的吸附态、硼酸盐矿物、石盐和石膏等介质的化学分离方法（马云麒等，2010；张艳灵等，2016；彭章旷等，2017；秦占杰等，2018；杨剑等，2019）。中国科学院地球环境所的团队对 B 同位素的分离技术也做了一些探索，如尝试使用微升华（micro-sublimation）方式进行生物碳酸盐样品 B 的分离和富集等（He M. Y. et al., 2011, 2015）。中国科学院广州地球化学研究所的团队则针对低硼含量的地质样品如玄武岩等，通过改造超净实验室的空气过滤系统如使用低硼本底的过滤材料等，建立了利用 AGMP 树脂在 HF 介质下的单柱分离硼技术，实现了在低本底（~2 ng）下 B 的高回收率（>99%），同时结合 MC-ICP-MS 的测试技术实现了硅酸盐样品的高精度 B 同位素测试，$\delta^{11}B$ 外部精度好于±0.3‰（2σ）（Wei et al., 2013b）。另外针对受到蚀变改造的硅酸盐样品，还可以通过化学淋滤去除蚀变组分的影响，获得准确的 $\delta^{11}B$ 结果（Li X. et al., 2019）。

2010 年代的早期，在质谱测试方面，中国科学院地球环境所的团队对在 TIMS 上利用正离子（$Cs_2BO_2^+$）的高精度 B 同位素测试方法进行了优化（贺茂勇等，2013；He et al., 2013）。不过由于 MC-ICP-MS 利用 SSB 方法可以很好校正 B 的同位素分馏，分析结果的可靠性有保证，因而逐渐成为高精度 B 同位素测试的主流方法，国内一些传统使用 TIMS 测试的团队也改用 MC-ICP-MS 进行测试（He et al., 2016）。使用 MC-ICP-MS 测试 B 同位素组成主要存在两个技术挑战，一是 B 在仪器上的记忆效应特别严重，二是标准与样品之间的基体匹配要求比较严格。针对前一个问题，中国科学院广州地球化学研究所的团队将 MC-ICP-MS 进样系统改造成了抗 HF 的组件（Teflon 的雾化器和雾室搭配蓝宝石中心管的矩管），利用低浓度 HF 淋洗，从而有效消除了记忆效应（Wei et al., 2013b）；而中国科学院地球环境研究所的团队则提出了利用 NaF 溶液清洗来消除记忆效应的解决方案（He M. Y. et al., 2019）。针对后一个问题，中国科学院广州地球化学研究所的团队通过仔细评估不同酸类型和浓度基体对 MC-ICP-MS 上高精度 B 同位素测试的影响，提出了确保测试结果可靠性的解决方案（Chen et al., 2016），使实验室实现了常规化的多介质高精度 B 同位素测试。

2.2.2 Ga 同位素分析技术进展

Ga 有 ^{69}Ga 和 ^{71}Ga 两个稳定同位素，其表述为 $\delta^{71}Ga$，代表样品中的 $^{71}Ga/^{69}Ga$ 值相对于参照标准（NIST SRM 994）的千分偏差。Ga 的同位素地球化学行为很少被探讨，主要因为 Ga 的化学纯化一直是个难题，导致高精度 Ga 同位素的测试一直比较困难。中国地质大学（武汉）的团队和中国科学院地球化学研究所的团队几乎同时在这方面取得突破，前者利用 AGMP-1M 阴离子树脂和 AG 50W-X8 阳离子树脂联用三柱离子交换实现对地质样品中的 Ga 的纯化富集（Zhang et al., 2016），后者则建立了利用 AG1-X4 阳离子树脂和 Ln-spec 特效树脂两柱离子交换的方法（Yuan et al., 2016）。Ga 同位素组成均在 MC-ICP-MS 上精确测定，标样 $\delta^{71}Ga$ 外部精度均优于±0.05‰（2σ）。利用这一领先的技术测试了系列地质标样的 Ga 同位素组成（Feng et al., 2019），并率先开拓了 Ga 同位素地球化学的研究。

2.2.3 In 同位素分析技术进展

In 有 ^{113}In 和 ^{115}In 两个天然稳定同位素，目前暂时没有看到对 In 稳定同位素组成的表述和参照标准的确立。南京大学团队成功利用 In 和 Sn 的同质异位干扰来校正质谱测量的分馏，测量出标准溶液的 $^{113}In/^{115}In$ 值为 0.044 617±0.000 013（Shuai et al., 2020）。这为开展 In 稳定同位素地球化学研究提供了技术借鉴。

2.2.4 Tl 同位素分析技术进展

Tl 有 ^{203}Tl 和 ^{205}Tl 两个天然稳定同位素，可表述为 $\varepsilon^{205}Tl$，代表样品中的 $^{205}Tl/^{203}Tl$ 值相对于参照标准

（NIST SRM 997）的万分偏差。$^{205}Tl/^{203}Tl$ 值早期被用于高精度 Pb 同位素比值质谱测量过程中的分馏校正，不过稳定 Tl 同位素也是较早开展的非传统稳定同位素体系之一（Rehkämper and Halliday，1999）。我国一些研究团队发表过对该同位素体系分析技术和应用的综述文章（贾彦龙等，2010；邱啸飞等，2014），以及依托国外实验室测试的 Tl 同位素组成开展的研究（Shu et al.，2017；Fan et al.，2020）。不过迄今还没有看到正式发表的依托国内实验室测试的稳定 Tl 同位素的结果。

2.3 第四主族元素的同位素分析技术进展

第四主族元素中的 Si、Ge 和 Sn 在非传统同位素地球化学研究中均受到关注，在我国也有一定程度的开展。

2.3.1 Si 同位素分析技术进展

Si 有 ^{28}Si、^{29}Si 和 ^{30}Si 3 个天然稳定同位素，表述为 $\delta^{29}Si$ 或 $\delta^{30}Si$，分别代表样品中 $^{29}Si/^{28}Si$ 或者 $^{30}Si/^{28}Si$ 相对于参照标准（NBS-28）的千分偏差。稳定 Si 同位素分析有两大技术体系，一是在气体源质谱上以 SiF_4 气体进样的方式测试，二是在 MC-ICP-MS 上以溶液进样方式测试。

早在非传统稳定同位素研究兴起之前，基于气体源质谱测试的稳定 Si 同位素地球化学研究已经非常普及。中国地质科学院矿床地质研究所丁悌平研究员领导的团队在 20 世纪 80 年代就已建立起高精度的 Si 同位素测试方法，标准样品 $\delta^{30}Si$ 的重现性好于 ±0.1‰（2σ）（丁悌平等，1988），并很早在矿床学等领域开展应用（蒋少涌等，1992）。这一技术体系引领了我国前期的 Si 同位素地球化学研究，在矿床学、表生地质过程、生物地球化学过程、地质古环境研究等方面都取得了不少在国际上很有显示度的成果（Ding et al.，2005，2011，2017），至今仍然发挥着重要作用。

MC-ICP-MS 技术的发展提供了另一种高效的高精度稳定 Si 同位素测试方法，我国学者早期通过在国外先进实验室的学习，引进并在国内实验室建立起该方法。华东师范大学的团队较早建立了基于 MC-ICP-MS 的稳定 Si 同位素测试技术（Zhang et al.，2014；张安余，2015），其标样的 $\delta^{30}Si$ 重现性好于 ±0.1‰（2σ），与基于气体源质谱的测试精度相当，重点应用于对河流和海洋中的溶解态 Si 以及悬浮物和颗粒物中的 Si 的研究。中国科学技术大学的团队也建立起基于 MC-ICP-MS 的高精度稳定 Si 同位素测试方法，其固体岩石标样的 $\delta^{30}Si$ 重现性达到 ±0.06‰（2σ）（Yu et al.，2018）。这些高精度的分析方法对我国 Si 同位素地球化学的应用研究起到了重要的推动作用。

2.3.2 Ge 同位素分析技术进展

Ge 有 5 个天然稳定同位素：^{70}Ge、^{72}Ge、^{73}Ge、^{74}Ge 和 ^{76}Ge，Ge 稳定同位素也有多个表述，最常见的是 $\delta^{74}Ge$，代表样品中的 $^{74}Ge/^{70}Ge$ 值与参照标准（NIST SRM 3120a）的千分偏差。

Ge 稳定同位素地球化学研究相对较少，中国科学院地球化学研究所的学者通过早期在国外相关实验室的学习（Qi et al.，2011），引进并在国内实验室建立起该分析方法。他们通过结合加 Zn 外标元素和 SSB 方法进行质谱测试过程中的分馏校正，实现了标样 $\delta^{74}Ge$ 的重现性好于 ±0.20‰（2σ）（Meng et al.，2015；Meng and Hu，2018），并在地表风化过程等方面开展示踪应用（Qi et al.，2019）。

2.3.3 Sn 同位素分析技术进展

Sn 有 10 个天然稳定同位素：^{112}Sn、^{114}Sn、^{115}Sn、^{116}Sn、^{117}Sn、^{118}Sn、^{119}Sn、^{120}Sn、^{122}Sn 和 ^{114}Sn，Sn 的稳定同位素组成也因此有很多种表述方式。已有的研究主要分两个体系：$\delta^{122/118}Sn$ 和 $\delta^{122/116}Sn$，分别代表样品中的 $^{122}Sn/^{118}Sn$ 或 $^{122}Sn/^{116}Sn$ 值相对参照标准的千分偏差，前一体系的参照标准是 Sn_IPGP，后一体系则以 NIST SRM 3161a 为参照标准。两个体系之间的结果可以相互换算。

Sn 同位素地球化学的研究目前在我国还较少开展，有学者依托国外实验室开展了稳定 Sn 同位素的测试，并据此探讨了煤矿、锡矿可能受到岩浆或热液的改造（Yao et al.，2018；Qu et al.，2020）。中国科学院广州地球化学研究所的团队尝试利用双稀释剂（^{117}Sn-^{119}Sn）法在 MC-ICP-MS 上测试 NIST SRM 3161a

标准溶液的 $\delta^{120/118}Sn$，发现有比较宽的最佳稀释范围，具备实现高精度 Sn 稳定同位素测试的潜力（Zhang L. et al., 2018）。南京大学的团队则采用添加 Sb 外部标准，结合 $^{123}Sb/^{121}Sb$ 外部校正和 SSB 方法，实现了对地质样品的高精度 Sn 同位素组成测试，标样 $\delta^{122/116}Sn$ 的重现性好于 ±0.09‰（2σ）（She et al., 2020）。这些技术将会推动我国的 Sn 稳定同位素地球化学研究。

2.4 第五和第六主族元素的同位素分析技术进展

第五和第六主族元素中除了传统的 N、O 和 S 外，只有 Sb、Se 和 Te 有超出一个的天然稳定同位素，具备开展非传统稳定同位素地球化学研究的潜力。

2.4.1 Sb 同位素分析技术进展

Sb 有 ^{121}Sb 和 ^{123}Sb 两个天然稳定同位素，可表述为 $\delta^{123}Sb$，代表样品中的 $^{123}Sb/^{121}Sb$ 值相对于参照标准（NIST SRM 3102a）的千分偏差。

Sb 稳定同位素地球化学目前受到的关注不多，国内开展的研究也较少，有学者开发了 Sb 稳定同位素的分析技术并应用于环境科学领域（Wen et al., 2018）。天津大学的团队建立了针对地质样品的高精度 Sb 同位素分析方法，并研发了利用 AG1-X4 阴离子树脂和 AG 50W-X8 阳离子树脂联用双柱分离纯化 Sb 的方法，在 MC-ICP-MS 上结合添加 Cd 外标（$^{114}Cd/^{112}Cd$）进行分馏校正建立了 C-SSBIN 测试方法。这一方法获得的标样 $\delta^{123}Sb$ 的长期重现性好于 ±0.03‰（2σ），可满足对多种地质过程中 Sb 同位素分馏示踪的要求（Liu et al., 2020）。

2.4.2 Se 同位素分析技术进展

Se 有 ^{74}Se、^{76}Se、^{77}Se、^{78}Se、^{80}Se 和 ^{82}Se 六个天然稳定同位素，其稳定同位素组成也有多种表述形式，最常见的是 $\delta^{82/76}Se$，代表样品中的 $^{82}Se/^{76}Se$ 值相对于参照标准（NIST SRM 3149）的千分偏差。

早期的高精度 Se 同位素分析存在一定的难度，但随着氢化物发生器联合 MC-ICP-MS 技术的使用，通过在 MC-ICP-MS 上测量 H_2Se^+ 的方式，使 Se 同位素组成测试的效率和精度均得到明显改善，进而推动了相关研究。

我国较早开展 Se 稳定同位素地球化学研究的是中国科学院地球化学研究所的团队，早期主要是在国外实验室开展工作，包括分析方法的建立（朱建明等，2008）以及对矿床与沉积环境的探索（Wen and Carignan, 2011; Wen et al., 2014; Zhu et al., 2014）。华东师范大学的团队率先报道了在国内实验室建立的高精度 Se 同位素组成测试方法，他们利用配备了氢化物发生器的 MC-ICP-MS，使用 SSB 的分析方式测定了海水中溶解 Se 的同位素组成，$\delta^{82/76}Se$ 的外部精度达到 ±0.16‰（2σ）（Chang et al., 2017）。中国科学院地球化学研究所的团队则使用双稀释剂法（^{74}Se-^{77}Se）在配备氢化物发生器的 MC-ICP-MS 上实现地质样品的高精度 Se 同位素测试，标样 $\delta^{82/76}Se$ 的重现性达到 ±0.10‰（2σ），为相关研究提供了技术保障（Tan et al., 2020a; Xu W. P. et al., 2020）。

2.4.3 Te 同位素分析技术进展

Te 有 ^{120}Te、^{122}Te、^{123}Te、^{124}Te、^{125}Te、^{126}Te、^{128}Te 和 ^{130}Te 八个天然稳定同位素，因此 Te 的稳定同位素有多个表述方式，通常使用的是 $\delta^{x/125}Te$（x 一般是偶数质量数的 Te 稳定同位素），代表样品中相应质量数的同位素与 ^{125}Te 值相对于参照标准（JMC Te）的千分偏差。

Te 稳定同位素是较早开展研究的非传统稳定同位素体系（Fehr et al., 2004），主要应用于天体化学和矿床学研究中 Te 的示踪（Fornadel et al., 2017; Fehr et al., 2018）。然而，到目前为止，尚未见到我国学者开展这方面研究的报道，也未见国内实验室对 Te 稳定同位素组成的分析测试的报道。

2.5 第一和第二副族元素的同位素分析技术进展

第一和第二副族元素中除 Au 只有一个天然稳定同位素外，Cu、Zn、Ag、Cd 和 Hg 都有多个天然稳定同位素，它们在非传统稳定同位素地球化学的研究中都受到不同程度的关注。

2.5.1 Cu、Zn同位素分析技术进展

Cu和Zn的同位素质量数相近而且相互间没有同质量的干扰，在MC-ICP-MS上进行同位素组成测试时可以彼此作为对方的内标来校正仪器短期波动引起的偏差，因此Cu和Zn同位素的分析技术发展是密切相关的。Cu有^{63}Cu和^{65}Cu两个天然稳定同位素，其稳定同位素表述为δ^{65}Cu，代表样品中的^{65}Cu/^{63}Cu值相对于参照标准（NIST SRM 976）的千分偏差。Zn有5个天然稳定同位素：^{64}Zn、^{66}Zn、^{67}Zn、^{68}Zn和^{70}Zn，其稳定同位素组成表述为δ^{x}Zn（x可以是66、67、68或70），代表样品中相应质量数的同位素与^{64}Zn值相对于参照标准（JMC-Lyon）的千分偏差，一般的研究重点关注δ^{66}Zn。

Cu和Zn是非常经典的非传统稳定同位素体系，我国学者较早的时候就通过国际合作在国外实验室开展这方面的研究（蒋少涌等，2001；Chen et al.，2009），而且在前一个十年，中国地质科学院地质研究所等机构的相关实验室已经建立起高精度的Cu、Zn同位素测试方法，并开展了大量的应用研究（朱祥坤等，2013），目前Cu、Zn同位素研究方法在我国许多实验室都已开展。这两个同位素的分析技术较为成熟，在过去十年，我国学者对相关的分析技术进行了进一步的探索和优化（闫斌等，2011；Zhu et al.，2015）。在MC-ICP-MS上测量时Cu只能通过SSB的方式进行分馏校正，我国学者尝试加入质量数相近的内标元素如Zn或Ga来校正仪器波动的影响（Liu et al.，2014a；Hou et al.，2016），提高了Cu同位素测试的稳定性和效率。除了SSB测试方法外，Zn同位素也可用双稀释剂校正方法来进行测试，不过Zhang L.等（2018）发现，使用^{68}Zn-^{70}Zn双稀释剂的最佳稀释范围比较窄，不容易把握测量条件因而难获得高的分析精度。而加入内标元素如Cu校正仪器的短期波动并结合SSB方法，是一种高效的获得高精度的Zn同位素结果的方法（Zhu Y. T. et al.，2019）。

除了质谱测试方法的探索外，针对一些特别的地质样品，如海水、Fe-Mn结壳和高盐度的海洋沉积物等，我国学者还发展了一些优化的化学分离纯化技术（祁昌实等，2012；冯家毅等，2013；何连花等，2016；Wang et al.，2020a），甚至提出了一些富Cu样品不需要化学分离直接测量其Cu同位素组成的方法（Bao et al.，2019；Lv et al.，2020；Zhang Y. et al.，2020），丰富了高精度Cu、Zn同位素分析技术的体系。此外，针对Cu和Zn两个同位素体系参照标准接近消耗完全的问题，我国学者还建议了新的参照物质，如以NIST SRM 683作为新的Zn同位素参照标准（Yang et al.，2018），以及潜在的Cu同位素参照标准（Yuan et al.，2017）。

2.5.2 Ag同位素分析技术进展

Ag有^{107}Ag和^{109}Ag两个天然稳定同位素，其稳定同位素通常表述为δ^{109}Ag，代表样品中的^{109}Ag/^{107}Ag值相对于参照标准（NIST SRM 978a）的千分偏差。

高精度Ag同位素组成的测试方法和其他只有两个稳定同位素的体系相似，均使用SSB方式，同时为了降低仪器短期波动的影响，测试时加入Pd内标，并利用Pd的同位素比值进行校正。这一技术方法相对比较成熟。我国开展Ag同位素地球化学研究的团队不多，中国科学院生态中心的团队报道了其主要针对基体相对简单一些的环境样品如纳米银颗粒的Ag同位素测试方法（Lu et al.，2016）；南京大学的团队报道了针对基体相对复杂的地质样品（如岩石和矿物）的高精度Ag同位素测试方法（Guo et al.，2017）。这些方法测试标准溶液δ^{109}Ag的重现性都好于±0.015‰（2σ），达到国际先进水准。

2.5.3 Cd同位素分析技术进展

Cd有8个天然稳定同位素：^{106}Cd、^{108}Cd、^{110}Cd、^{111}Cd、^{112}Cd、^{113}Cd、^{114}Cd和^{116}Cd，其稳定同位素组成有多种表述方式，常用的是$\delta^{114/110}$Cd，代表样品中的^{114}Cd/^{110}Cd值相对于参照标准（NIST SRM 3108）的千分偏差。

Cd是一种备受关注的重金属污染元素，对氧化还原条件敏感，同时广泛参与生命活动，因此利用Cd的稳定同位素对Cd的来源和迁移富集过程进行示踪，在环境科学、矿床学、沉积学和古海洋学等研究中

有较好的应用，在我国有相当多的实验室开展这方面的工作。Cd 稳定同位素的测试技术比较成熟，早期使用简单的 SSB 方法进行测试（Gao et al., 2008），标样的 $\delta^{114/110}$Cd 重现性好于±0.08‰（2σ）。不过，更多的团队使用双稀释剂法进行测试，因为使用双稀释剂（^{111}Cd-^{113}Cd）可以有比较宽的最佳稀释范围（Zhang L. et al., 2018），标样的 $\delta^{114/110}$Cd 重现性可达到±0.05‰（2σ）（Wen et al., 2015b；Li D. D. et al., 2018；Liu et al., 2020；Tan et al., 2020b）。另外，针对一些特别样品如土壤和植物，相关团队还对 Cd 的分离富集流程进行了优化（Wei et al., 2017；Lv et al., 2021）。这些方法对 Cd 稳定同位素在我国相关领域的研究应用起到了很好的推动作用。

2.5.4 Hg 同位素分析技术进展

Hg 有 7 个天然稳定同位素：^{196}Hg、^{198}Hg、^{199}Hg、^{200}Hg、^{201}Hg、^{202}Hg 和 ^{204}Hg，其稳定同位素组成通常表述为 δ^{x}Hg（x 可以为 199、200、201、202 和 204），代表样品中相应质量数同位素与 ^{198}Hg 比值相对于参照标准（NIST SRM 3133）的千分偏差。

Hg 是受到重点关注的有毒重金属元素，自然过程中普遍存在奇数和偶数 Hg 同位素非质量分馏的特别现象，因而 Hg 的稳定同位素在环境科学等领域研究中受到较大的关注。高精度的 Hg 同位素分析技术已趋成熟，针对零价 Hg 易挥发的特性，通常在 MC-ICP-MS 上配备在线的汞蒸气发生系统，将 Hg 以气体方式导入质谱，保证进样的连续性和稳定性；在质谱测量过程中，往往采用加 Tl 内标校正仪器短期波动的影响，结合 SSB 法来校正测量过程中的分馏。采用这些技术可以获得高精度的 Hg 稳定同位素结果。中国科学院地球化学研究所的团队引领了我国的 Hg 稳定同位素研究，建有成熟的高精度的测试方法（尹润生等，2010），在环境科学、生物地球化学和古海洋学等方面均已获得系列具备国际先进水准的成果（冯新斌等，2015）。

近年有关团队还针对 Hg 同位素分析的化学前处理方法进行了优化（刘锡尧等，2013；黄舒元等，2016），进一步拓展了 Hg 稳定同位素的应用范围。

2.6 第三副族元素的同位素分析技术进展

第三副族元素中的 Sc 和 Y 只有单个天然稳定同位素，镧系元素（稀土元素）中的大部分都有一个以上的天然稳定同位素，是开展稳定同位素地球化学研究的潜在对象，而锕系元素均为放射性元素，近年来发现一些长寿命的放射性同位素比值如 ^{235}U/^{238}U 也存在明显的分馏现象，因而也被纳入非传统稳定同位素研究的范畴。

2.6.1 稀土元素稳定同位素分析技术进展

稀土元素中的 La、Ce、Nd、Sm、Eu、Gd、Dy、Er、Yb 和 Lu 均有一个以上的天然稳定同位素，因此其稳定同位素组成研究非常具有潜力。然而到目前为止，针对这些体系开展的稳定同位素地球化学研究并不多，普遍存在的技术挑战是不同元素间的同质异位数的干扰比较普遍，而这些元素的化学性质又非常相似很难完全分离干净，因此大大增加了对这些体系的高精度稳定同位素组成测试的难度。目前已经开展研究的稳定同位素体系主要包括 Ce、Nd 和 Eu，国际上有个别团队开展过 Sm 的稳定同位素地球化学研究（Wakaki and Tanaka, 2016），国内尚未见这方面的报道。

Ce 有 4 个天然稳定同位素：^{136}Ce、^{138}Ce、^{140}Ce 和 ^{142}Ce，其稳定同位素组成有多种表述方法，一般表述为 $\delta^{142/140}$Ce，代表样品中的 ^{142}Ce/^{140}Ce 值相对于参照标准（JMC 304）的千分偏差。

Ce 的一个重要化学性质是在氧化条件下会变成难溶于水的正四价态而和其他稀土元素分离，这一过程中 Ce 同位素有明显的分馏，因 $\delta^{142/140}$Ce 和 Ce 异常（Ce/Ce*）一样能反映环境的氧化还原条件变化而受到学界关注（Nakada et al., 2013）。我国目前还没有正式发表的 Ce 稳定同位素的结果，仅在一些国内外学术会议中有过相关研究的报告。中国地质大学（武汉）的团队报道过 Ce 同位素的研究成果（Gao et al., 2016），不过他们是针对 La-Ce 年代学的研究，测试过程中使用了固定的 ^{136}Ce/^{142}Ce 进行分馏校正，并没有针对稳定 Ce 同位素组成开展研究。

Nd 有 7 个天然稳定同位素：^{142}Nd、^{143}Nd、^{144}Nd、^{145}Nd、^{146}Nd、^{148}Nd 和 ^{150}Nd，其中^{143}Nd 有长寿命放射性同位素^{147}Sm 衰变的贡献，一些古老的地质样品中^{142}Nd 有观察到短寿命放射性同位素^{146}Sm（半衰期约 68 Ma）的贡献。Nd 的稳定同位素组成表述为 $\delta^{x/144}$Nd（早期的表述为 $\varepsilon^{x/144}$Nd，为 $\delta^{x/144}$Nd 值的 10 倍；x 指 142、145、146、148 和 150）。在传统的 Sm-Nd 年代学研究中，这些 Nd 稳定同位素比值往往被设定为恒定值，用于^{143}Nd/^{144}Nd 测试过程中的分馏校正（使用不同的比值校正结果不尽相同，最近几十年主要使用^{146}Nd/^{144}Nd=0.7219 进行校正）。然而，地质过程中稳定同位素 Nd 同位素的分馏是比较明显的，结合稳定 Nd 同位素组成和传统的放射成因 Nd 同位素组成可以对一些地质过程有更好的示踪（Liu X. et al., 2018）。

高精度稳定 Nd 同位素的测试可以使用 SSB 法或双稀释剂法进行。中国科学院广州地球化学研究所的团队在国内最早开展地质样品稳定 Nd 同位素的分析（韦刚健等，2015），他们主要使用 SSB 测试方法，标样 $\delta^{146/144}$Nd 的外部精度好于±0.02‰（2σ），达到国际先进水准（Ma et al., 2013a）。

Eu 有^{151}Eu 和^{153}Eu 两个天然稳定同位素，但其表述方式很不一致，早期表述为 $\varepsilon^{153/151}$Eu，代表样品中的^{153}Eu/^{151}Eu 值相对于参照标准（JMC Eu）的万分偏差（Moynier et al., 2006）；近年主要表述为 $\delta^{151/153}$Eu，代表样品中的^{151}Eu/^{153}Eu 值相对于参照标准（NIST SRM 3117a）的千分偏差（de Carvalho et al., 2017；Lee and Tanaka, 2019）。

高精度 Eu 稳定同位素组成的测试通常采用加入外标元素（Sm 或者 Gd）的 SSB 方法，标样的重现性好于±0.02‰（2σ）（de Carvalho et al., 2017；Lee and Tanaka, 2019）。我国的研究团队在一些国际和国内会议中交流过 Eu 稳定同位素的测试方法（Li M. et al., 2016；朱志勇，2020），此外未见到其他正式成果发表。

2.6.2 U 稳定同位素分析技术进展

^{235}U 和^{238}U 是长寿命的放射性核素，其半衰期分别为 703.7 Ma 和 4.468 Ga，地质过程普遍存在类似稳定同位素一样的分馏，因而也被纳入非传统稳定同位素地球化学研究的范围。U 的稳定同位素组成通常表述为 δ^{238}U，代表样品中的^{238}U/^{235}U 值相对于参照标准（CRM-112a 或者 CRM-145）的千分偏差。

U 是氧化还原敏感元素，其稳定同位素也是沉积环境氧化条件的良好指标，在地球科学许多领域的研究中均有广泛的应用（Andersen et al., 2017）。一些在国外访学的中国学者也发表了不少稳定 U 同位素地球化学的研究成果（Wang X. L. et al., 2015；Chen et al., 2016；Zhang F. F. et al., 2018, 2020a, 2020b）。然而高精度的 U 稳定同位素测试需要依赖双稀释剂（^{233}U-^{236}U），这是具有放射性的人工核素，很难获取，因此国内目前还未见高精度 U 稳定同位素测试结果的报道。

2.7 第四副族元素的同位素分析技术进展

第四副族元素 Ti、Zr 和 Hf 均为高场强元素，具有难溶解难迁移等特点，在地质过程中有特别的示踪价值。同时这些元素均具有较多的稳定同位素，也是非传统稳定同位素地球化学研究中极具潜力的体系。不过真正开展稳定同位素地球化学研究的主要还是 Ti 和 Zr 体系，虽然 Hf 同位素受到的关注程度远超 Ti 和 Zr，并且在地球科学的多个领域研究中有广泛的示踪应用，但基本上都是针对 Lu-Hf 年代学体系中放射成因的 Hf 同位素组成（^{176}Hf/^{177}Hf），目前未有 Hf 的稳定同位素地球化学研究的报道。

2.7.1 Ti 稳定同位素分析技术进展

Ti 有 5 个天然稳定同位素：^{46}Ti、^{47}Ti、^{48}Ti、^{49}Ti 和 ^{50}Ti，其稳定同位素有多种表述方式，早期为 $\varepsilon^{x/46}$Ti（x 是 47、48、49 或 50），代表样品中这些同位素与^{46}Ti 的比值相对于参照标准的万分偏差。这种方式与探索天体样品中 Ti 同位素的非质量分馏的体系相似，这方面的研究在 Ti 同位素测试过程中使用固定的内部比值进行分馏校正，与一般意义上的非传统稳定同位素地球化学研究体系不同。对于基本上不存在 Ti

同位素非质量分馏的地球样品，Ti 稳定同位素一般表述为 $\delta^{x/47}Ti$（x 通常是 48、49 或 50），代表样品中这些同位素与 ^{47}Ti 的比值相对于参照标准的千分偏差。当然也有研究团队继续使用以 ^{46}Ti 为分母的表述方式。Ti 稳定同位素的参照标准也有一个发展过程，较为广泛使用的是 NIST SRM 3162a，近年来 OL-Ti 也被广泛使用。

我国学者在国外访学时，较早地开展过稳定 Ti 同位素的研究（Zhu et al.，2002a）。中国地质科学院地质研究所的团队率先在国内建立起利用 SSB 方法在 MC-ICP-MS 上测试高精度稳定 Ti 同位素的方法，标准样品 $\delta^{49/46}Ti$ 的重现性好于 ±0.06‰（2σ）（唐索寒等，2011）。近年该团队又建立起依托双稀释剂在 MC-ICP-MS 上的高精度稳定 Ti 同位素分析方法，标准样品 $\delta^{49/46}Ti$ 的重现性好于 ±0.03‰（2σ），分析精度得到进一步提高（唐索寒等，2018）。

高精度的 Ti 同位素测试的难点主要体现在化学前处理上，由于 Ti 难溶解和易水解的化学性质，介质复杂的地质样品中 Ti 的完全回收和干扰元素的完全去除一直是个挑战，同时也是 Ti 同位素分析技术改进的重点。中国科学院广州地球化学研究所的团队对 Ti 的分离纯化技术进行了探索，尝试了不同酸介质条件（如加 H_3BO_3 的 HF 介质）下的分离技术，实现了对 Ti 的高效回收和分离纯化，并利用 SSB 方法在 MC-ICP-MS 上建立起高精度的 Ti 稳定同位素测试技术，标准样品 $\delta^{49/47}Ti$ 的重现性好于 ±0.05‰（2σ）（王樵珊，2018；He X. Y. et al.，2020），为我国开展 Ti 稳定同位素地球化学的研究提供更多的技术支撑。

2.7.2　Zr 稳定同位素分析技术进展

Zr 有 5 个天然稳定同位素：^{90}Zr、^{91}Zr、^{92}Zr、^{94}Zr 和 ^{96}Zr，其中早期太阳系物质中 ^{92}Zr 有短寿命放射性核素 ^{92}Nb（半衰期约 36 Ma）的贡献。Zr 的稳定同位素通常表述为 $\delta^{94/90}Zr$，代表样品中的 $^{94}Zr/^{90}Zr$ 值相对于参照标准的千分偏差。Zr 稳定同位素是新近开始研究的体系，现有多个参照标准，包括 IPGP-Zr（Inglis et al.，2018）和 NIST SRM 3169（Feng et al.，2020a）标准溶液，以及 GJ 锆石（用于原位测试锆石的 Zr 稳定同位素）（Zhang W. et al.，2019），不同参照标准结果可以相互换算（Tian et al.，2020）。

中国地质大学（武汉）的团队在国内率先开展稳定 Zr 同位素研究，他们利用双稀释剂法（^{91}Zr-^{96}Zr）在 TIMS 上实现高精度 Zr 同位素测试，标样 $\delta^{94/90}Zr$ 的重现性好于 ±0.06‰（2σ），达到国际先进水平（Feng et al.，2020a）。除了溶液测量法外，他们还建立了基于 LA-MC-ICP-MS 的锆石原位 Zr 稳定同位素测试方法（Zhang W. et al.，2019），将在"原位分析测试技术"一节详细介绍。

2.8　第五副族元素的同位素分析技术进展

第五副族元素中的 Nb 只有一个天然稳定同位素，V 和 Ta 则有两个天然稳定同位素（分别是 ^{50}V 和 ^{51}V，^{180}Ta 和 ^{181}Ta）。这两个体系有一个相同的特点即两个稳定同位素的丰度相差悬殊：^{50}V 大约只占 0.25%，而 ^{51}V 约占 99.75%，$^{51}V/^{50}V$ 值在 400 左右；^{180}Ta 大约只占 0.012%，而 ^{181}Ta 约占 99.988%，$^{181}Ta/^{180}Ta$ 值超出 8 000。这样巨大的同位素比值在精确测量上存在很大的挑战，对质谱仪器的配置和化学纯化都有非常高的要求。到目前为止，精确测量 $^{181}Ta/^{180}Ta$ 值还比较困难（Pfeifer et al.，2017），开展 Ta 的稳定同位素地球化学研究的条件还不成熟。不过 V 同位素的高精度测试技术较早的时候已取得突破（Nielsen et al.，2011），为 V 稳定同位素地球化学的研究提供了技术支撑。

V 的稳定同位素组成可表述为 $\delta^{51}V$，代表样品中的 $^{51}V/^{50}V$ 值相对于参照标准（AA，Alfa Aesar）的千分偏差。V 是变价元素，对氧化还原条件比较敏感。由于 ^{50}V 和 ^{51}V 的丰度相差巨大，同质异位素（^{50}Ti 和 ^{50}Cr）对丰度较低的 ^{50}V 在测量时有严重干扰，因而化学分离纯化过程要求十分严格，需将 V 和其他基质元素完全分离干净，同时由于 V 同位素的自然丰度差别很大，且存在多种双原子离子的干扰，因此对质谱测试要求很高。中国科学技术大学的团队在国内率先建立起高精度的 V 同位素测试方法，他们建立了适用于处理各种岩石和沉积物样品的高效 V 分离纯化流程，相比于前人建立的化学分离方法［需将样品进行 6~8 次色谱柱化学分离流程（Prytulak et al.，2011）］，新方法显著简化了分离流程（3~4 次色谱柱化学分离流程），提高了分析效率。在 MC-ICP-MS 上利用连接不同电阻的检测器来扩大动态测量范围，结合

Aridus 膜去溶方法提高信号灵敏度，并通过 SSB 方法进行仪器质量歧视校正，可以精确测量 V 同位素组成，纯溶液标准样品 δ^{51}V 的长期重现性好于±0.1‰（2σ），这也是目前国际上最好的精度（黄方等，2016；Wu et al., 2016）。

2.9 第六和第七副族元素的同位素分析技术进展

第六副族的元素 Cr、Mo 和 W 均具有较多的天然稳定同位素（4 个及以上），在非传统稳定同位素地球化学研究中有比较高的关注度；而第七副族元素中 Mn 只有单一的天然稳定同位素，Tc 是人工核反应合成的元素，自然界中未发现，只有 Re 有两个天然同位素（其中一个 ^{187}Re 是长寿命放射性同位素），因此在非传统稳定同位素地球化学研究中较少受到关注。

2.9.1 Cr 稳定同位素分析技术进展

Cr 有 4 个天然稳定同位素 ^{50}Cr、^{52}Cr、^{53}Cr 和 ^{54}Cr，其中非常古老的样品中（如早期太阳系物质）^{53}Cr 有短寿命核素 ^{53}Mn（半衰期约 3.8 Ma）衰变的贡献，而一些天体样品中会存在 ^{54}Cr 核合成的异常，这方面的研究重点关注偏离质量相关分馏的 Cr 同位素组成变化；而对于多数地球样品则主要关注 Cr 同位素的质量相关分馏。Cr 的稳定同位素通常表述为 δ^{53}Cr，代表样品中的 ^{53}Cr/^{52}Cr 值相对于参照标准（NIST SRM 979 或 IST SRM 3112a）的千分偏差，这两个参照标准之间存在一定的系统差异，可以相互校正。

Cr 是被重点关注的重金属污染元素，也是变价元素。在自然过程中随氧化还原条件的变化 Cr 的价态也有较多的变化，而不同价态的 Cr 有不同的环境毒性，同时 Cr 同位素组成也相应发生变化，因而 Cr 稳定同位素的研究在环境科学中广受关注。中国地质大学（武汉）的团队较早之前就报道了对污染水体中的 Cr 同位素的研究（高永娟等，2009），他们利用 TIMS 测量 Cr 同位素组成，但没有采用双稀释剂法对测试过程中的分馏进行校正，而是通过严格限定测试条件，力保标准和样品的测试条件相同，并直接用标准样品的分馏系数来校正样品的分馏。这种在 TIMS 上使用类似 SSB 的方式在没有双稀释剂的情况下也不失为一种替代方法，但 TIMS 不同的单次测试条件很难严格地控制，后续基于 TIMS 更精确的 Cr 同位素测试均采用双稀释剂（^{50}Cr-^{54}Cr）方法进行分馏校正。

Cr 变价的特征使得在化学前处理过程中要严格控制氧化还原条件，从而保障 Cr 的回收率而不会产生明显的 Cr 同位素偏差。另外，在质谱测量过程中，多个元素会产生质量干扰，特别是基于 MC-ICP-MS 测量时，Ar 的氧化物（如 ^{40}Ar^{16}O$^+$ 对 ^{56}Cr）的干扰是一个很严重的问题，往往需要特别的处置方法（如采用高分辨率方式）。总体而言，高精度的 Cr 稳定同位素测试具有较高的难度。我国多个研究团队在过去十年对地质样品高精度 Cr 同位素测试方法进行了探索，在化学分离技术和质谱测量方面均做出了较重要的贡献。

针对地质样品 Cr 同位素测试的化学前处理，我国开展 Cr 稳定同位素地球化学研究的主要团队，如中国科学院地质与地球物理研究所、中国地质科学院地质研究所、中国科学技术大学和中国地质大学（北京），都各自发展了不同的 Cr 分离纯化流程，均能实现较高的 Cr 回收率（Li C. F. et al., 2017；Zhu et al., 2018；李维涵等，2019；马健雄等，2020）。质谱测试仍然有两种方案，中国科学院地质与地球物理研究所的团队使用 TIMS 结合双稀释剂测试方法，标样 δ^{53}Cr 重现性好于±0.05‰（2σ）（Li C. F. et al., 2017；Liu C. Y. et al., 2019）；其他团队则使用 MC-ICP-MS 进行测试，结合双稀释剂进行分馏校正，标样 δ^{53}Cr 重现性也均好于±0.05‰（2σ）（Zhang Q. et al., 2019；马健雄等，2020；Wu G. L. et al., 2020）。这些技术进步对我国 Cr 稳定同位素的应用研究起到了重要支撑作用。

2.9.2 Mo 稳定同位素分析技术进展

Mo 有 7 个天然稳定同位素：^{92}Mo、^{94}Mo、^{95}Mo、^{96}Mo、^{97}Mo、^{98}Mo 和 ^{100}Mo，其稳定同位素组成表述早期为 $\delta^{97/95}$Mo，而近年大多使用 $\delta^{98/95}$Mo，代表样品中相应同位素比值（^{97}Mo/^{95}Mo 或 ^{98}Mo/^{95}Mo）相对于参照标准的千分偏差。Mo 稳定同位素的参照标准有 JMC 的 Mo 标准溶液，近年主要采用 NIST SRM 3134a，两种

标准的结果可以相互换算。

Mo 是一种氧化还原敏感元素，其稳定同位素组成对沉积环境的氧化还原条件变化比较敏感，在古海洋学研究中有非常重要的示踪作用。另外，Mo 也是一种重要的金属矿产资源，Mo 同位素在示踪成矿物质来源方面也有非常重要的作用。因而，Mo 的稳定同位素体系较早就受到广泛关注。虽然在质谱测量上 Mo 同位素组成可以采用 SSB 的方式，不过使用双稀释剂（^{100}Mo-^{97}Mo）可以有非常宽的最佳稀释范围（Zhang L. et al., 2018），因此利用双稀释剂进行分馏校正在 MC-ICP-MS 上测试是目前普遍采用的方法（李津等，2011; Li. J. et al., 2016）。

由于 Mo 在多数地质样品中的含量都较低，其有效富集和分离纯化是高精度 Mo 同位素测试的难点所在。在前一个十年中，中国地质科学院地质研究所的团队已开展了 Mo 同位素的分析技术和应用研究（朱祥坤等，2013）。而在最近十年，我国主要的研究团队针对低含量地质样品的化学处理方法开展了探索，并取得重要突破。中国科学院广州地球化学研究所的团队研发了 Mo 特效树脂（BPHA 树脂），实现单柱分离地质样品的 Mo（Li et al., 2014）。该方法具有极低的全流程本底，可以准确分析如生物成因碳酸盐等 Mo 含量极低的特殊样品的 Mo 同位素组成，标准样品 $\delta^{98/95}$Mo 的重现性好于±0.06‰（2σ）（Zhao et al., 2016）。最近该团队针对一些低 Mo 含量高干扰的地质样品如淡色花岗岩采用了单-双柱配合的化学分离方法，可以实现对极低 Mo 含量（< 50 ng/g）的岩石样品的高精度 Mo 同位素测试（Fan et al., 2020a）。中国科学院地球化学研究所的团队也建立起针对低 Mo 含量样品的测试方法（Liu J. et al., 2016）；中国地质大学（武汉）的团队则利用 TRU 特效树脂也实现了 Mo 的高效单柱分离，在 MC-ICP-MS 上测量达到地质标准样品 $\delta^{98/95}$Mo 重现性为 0.03‰~0.08‰（2σ）的国际先进水准（Feng et al., 2020b）。

2.9.3　W 稳定同位素分析技术进展

W 有 5 个天然稳定同位素：^{180}W、^{182}W、^{183}W、^{184}W 和 ^{186}W，在一些非常古老的样品中（如早期太阳系物质）^{183}W 有短寿命核素 ^{183}Hf（半衰期约 8.9 Ma）衰变产生的贡献。与 Cr 同位素体系类似，W 稳定同位素研究分为两方面，一是为探索与 ^{183}Hf-^{183}W 年代学相关的偏离质量分馏的 W 稳定同位素组成变化，主要面向非常古老的地质样品和过程。其测试过程与传统的放射成因同位素组成相似，采用了固定的内部比值（^{186}W/^{184}W 或 ^{186}W/^{183}W）进行分馏校正，但无法探索地质过程中的 W 同位素分馏。另一方面则为与质量分馏相关的 W 稳定同位素探索，其重点在于了解各种地质过程中 W 稳定同位素的分馏过程与控制机制及其潜在的示踪应用，其稳定同位素组成通常表述为 δ^{186}W，代表样品中的 ^{186}W/^{184}W 值相对于参照标准（NIST SRM 3163）的千分偏差。

W 在地质样品中往往含量不高，而且化学处理过程中受到的干扰较明显，如在 HF 介质下 Zr、Ti 等容易水解，从而导致 W 的回收率不足，因而在高精度同位素组成测试时 W 的富集和纯化是比较困难的。中国科学院地质与地球物理研究所的团队对 W 的化学分离纯化流程进行了探索，建立了两种适用于地质样品的 W 富集纯化流程（Mei et al., 2018; Chu et al., 2020）。在 TIMS 上用具有较高灵敏度的负离子测试方法，精确测定地质样品的 ^{182}W/^{184}W 值。不过这些研究是针对 ^{183}Hf-^{183}W 年代学体系有关的 ^{182}W 异常，并非探索 W 同位素的质量分馏。

除了 ^{183}Hf-^{183}W 年代学体系外，国际上对于 W 同位素质量分馏的研究已有较深入开展，利用 Hf 外标加入法或双稀释剂（^{180}W-^{183}W）法在 MC-ICP-MS 上测定高精度 W 稳定同位素的方法已有报道（Breton and Quitté, 2014; Kurzweil et al., 2018），标准溶液的 δ^{186}W 重现性可达±0.07‰（2σ）的水准。不过目前尚未见国内研究团队开展这方面研究的报道。

2.9.4　Re 稳定同位素分析技术进展

Re 有 ^{185}Re 和 ^{187}Re 两个天然同位素，其中 ^{187}Re 是放射性同位素，构成的 Re-Os 同位素体系是非常重要的地质定年方法。与 Rb、U 等体系类似，在不考虑较长时间尺度变化的情况下，Re 也可以参照稳定同位素体系进行探索。Re 的稳定同位素可表述为 δ^{187}Re，代表样品中的 ^{187}Re/^{185}Re 值相对于参照标准（NIST SRM 989）的千分偏差。

Re 是氧化还原敏感元素，随着环境氧化还原条件的变化有可能发生明显的同位素分馏，因此是较有潜力的非传统稳定同位素体系。国际上较早时期就已经基于 MC-ICP-MS 建立起利用外标元素加入（W）结合 SSB 分馏校正的高精度的 Re 同位素测试方法，其标准溶液 δ^{187}Re 的重现性可达 ±0.04‰（2σ）（Miller et al., 2009; Poirier and Doucelance, 2009）。国内的研究团队也曾在一些学术会议上进行过 Re 稳定同位素的交流，但尚没有相关成果的正式报道。

2.10 第八族元素的同位素分析技术进展

第八族元素均属于重要的金属矿产资源，包括 Fe、Co 和 Ni 等社会经济发展中需求量极大的金属和比较稀少贵重的铂族元素 Ru、Rh、Pd、Os、Ir 和 Pt。这些元素中除 Co 和 Rh 只有单一的天然稳定同位素外，其他元素均有多个天然稳定同位素（Ir 有两个，其他的均有 4 个或 4 个以上），是非常具有潜力的非传统稳定同位素地球化学研究体系。

2.10.1 Fe 稳定同位素分析技术进展

Fe 有 4 个天然稳定同位素 ^{54}Fe、^{56}Fe、^{57}Fe 和 ^{58}Fe：其稳定同位素组成通常表述为 δ^{x}Fe（x 一般是 56 或 57），代表样品中相应同位素与 ^{54}Fe 的比值相对参照标准（IRMM-014）的千分偏差。

Fe 是最重要的金属之一，对社会文明和经济发展至关重要；Fe 是地球和行星的主要成分，对氧化还原条件变化敏感；Fe 也是重要的生命元素，广泛参与到各种深部和浅表地质过程及生物地球化学过程。Fe 是最早开展研究的非传统稳定同位素体系之一，Fe 同位素在地球科学的许多领域都有非常重要的示踪作用，且研究程度也比较高，因此 Fe 稳定同位素的分析技术比较成熟。Fe 同位素的质谱测量主要在 MC-ICP-MS 上进行，由于 ^{56}Fe 受到 ^{40}Ar^{16}O 的质量干扰，往往需要在高分辨情况下才能将两者区分开。质谱测量的分馏校正理论上可以采用双稀释剂法进行，不过目前通常采用 SSB 方法，或加入外标元素（Ni 或 Cu）的 SSB 方法，同样可以高效地获得高精度的 Fe 同位素结果；化学前处理主要通过离子交换树脂进行，有多种离子交换树脂及组合可供选择，近年的研究往往针对一些特殊的样品对分离纯化方式进行优化。

Fe 同位素方法较早就被我国的研究团队引进并广泛开展应用（朱祥坤等，2008b，2013），目前在各主要实验室均开展了 Fe 同位素的研究。最近十年，不同的研究团队针对不同地质样品研发了多种高精度的 Fe 同位素测试方法，对化学前处理（侯可军等，2012；孙剑等，2013；唐索寒等，2013a；Zhu G. H. et al., 2020a）、质谱测量技术（Sun et al., 2013a; He Y. S. et al., 2015; Chen et al., 2017; Gong H. M. et al., 2020）都进行了进一步的优化，分析效率得到了提高，对一些特殊的样品也获得了高精度的 Fe 同位素结果。目前常用的标准样品 δ^{56}Fe 的重现性往往能达到优于 ±0.03‰（2σ）的国际先进水准，为我国 Fe 同位素的研究提供了技术保障。

2.10.2 Ni 稳定同位素分析技术进展

Ni 有 5 个天然稳定同位素：^{58}Ni、^{60}Ni、^{61}Ni、^{62}Ni 和 ^{64}Ni，在古老样品中（如陨石）^{60}Ni 有短寿命核素 ^{60}Fe（半衰期约 2.6 Ma）衰变的贡献。与 Cr、W 等同位素体系类似，Ni 稳定同位素研究可分为两方面，一方面是陨石等研究中关注的 ^{60}Fe 衰变导致的 ^{60}Ni 异常，另一方面则关注各种地质过程中质量相关的 Ni 稳定同位素分馏，并应用于地质过程的示踪。对于后者，Ni 的稳定同位素组成通常表述为 $\delta^{x/58}$Ni（x 为 60、61、62 或 64），代表样品中相应同位素和 ^{58}Ni 的比值相对于参照标准（NIST SRM 986）的千分偏差。

Ni 是一个重要的生命元素，与产甲烷菌的活动密切相关，因而其稳定同位素在研究早期生命与环境的协同变化方面具有重要作用。我国学者在国外访学期间从事了这方面的研究工作，并且有比较重要的成果发表（Wang S. J. et al., 2019）。目前高精度 Ni 稳定同位素的测试主要利用双稀释法在 MC-ICP-MS 上测量，标准样品 $\delta^{60/58}$Ni 的重现性在 ±0.05‰（2σ）左右。不过目前还未见国内实验室测试的 Ni 稳定同位素结果的报道。

对于探索^{60}Ni异常方面的研究，中国科学院地质与地球物理研究所的团队建立了高精度的^{60}Ni/^{58}Ni值的测试方法，他们在TIMS上通过利用固定比值^{62}Ni/^{58}Ni为0.053 388 58进行分馏校正，标准溶液$\delta^{60/58}$Ni的重现性好于±0.11‰（2σ）（Li C. F. et al., 2020）。

2.10.3 铂族元素稳定同位素分析技术进展

铂族元素在多数地质样品中的含量极低，因此探索其地球化学性质有一定的难度。而这些元素的同位素体系又比较多样，包含有放射性同位素、放射成因同位素和稳定同位素。除了Rh只有单一天然同位素外，其他的5个元素均有不止一个天然稳定同位素，因而具备开展稳定同位素地球化学研究的条件。事实上，这些元素的含量虽然低，但它们在一些特别的地质过程中却有比较独特的地球化学行为，具有较高的示踪价值，因此其中一些元素的稳定同位素地球化学研究在国外的少数实验室也已开展，不过目前我国还未见有这方面的研究报道。

Ru有7个天然同位素：^{96}Ru、^{98}Ru、^{99}Ru、^{100}Ru、^{101}Ru、^{102}Ru和^{104}Ru，理论上^{98}Ru和^{99}Ru有短寿命核素^{98}Tc（半衰期约4~10 Ma）和^{99}Tc（半衰期约0.2 Ma）衰变的贡献，不过Tc是否曾经在太阳系存在过仍然缺少证据。Ru稳定同位素的研究主要关注^{100}Ru的异常，在地幔物质中存在的偏离质量分馏的^{100}Ru异常（ϵ^{100}Ru：利用固定内部比值如^{99}Ru/^{101}Ru＝0.745 075校正质量分馏）可能反映地球形成初期的物质来源（Bermingham and Walker, 2017；Fischer-Gödde et al., 2020）。而旨在探索质量分馏的Ru稳定同位素被表述为$\delta^{102/99}$Ru，代表样品中^{102}Ru/^{99}Ru值相对于参照标准（Alfa Aesar Ru标准溶液）的千分偏差，其基于MC-ICP-MS利用双稀释剂法（^{98}Ru-^{101}Ru）校正分馏的高精度测试方法也已建立，标准溶液$\delta^{102/99}$Ru的重现性可达±0.04‰（2σ）（Hopp et al., 2016）。

Pd有6个天然稳定同位素：^{102}Pd、^{104}Pd、^{105}Pd、^{106}Pd、^{108}Pd和^{110}Pd，目前Pd稳定同位素的研究主要围绕陨石样品开展，探索其在星际空间运移时受宇宙射线辐照产生的偏离质量相关分馏的情况，其测试方式采用固定的内部比值对质量相关的分馏进行校正（Ek et al., 2017）。

Os有7个天然稳定同位素：^{184}Os、^{186}Os、^{187}Os、^{188}Os、^{189}Os、^{190}Os和^{192}Os，其中^{186}Os和^{187}Os分别有长寿命放射核素^{190}Pt和^{187}Re衰变的贡献。由于Re-Os是非常重要而且应用广泛的定年体系，因此Os的同位素研究非常普遍，不过绝大部分是围绕放射成因Os同位素组成^{187}Os/^{188}Os的，这一比值测试时采用了固定的内部比值校正分馏，质量相关分馏的信息已被抹除。稳定Os同位素的研究相对少得多，其主要表述为$\delta^{190/188}$Os，代表样品中^{190}Os/^{188}Os值相对于参照标准（DROsS）的千分偏差。目前利用双稀释剂法（^{188}Os-^{190}Os）校正在MC-ICP-MS上或在TIMS上（利用负离子）均可进行高精度的稳定Os同位素组成测试，标准样品$\delta^{190/188}$Os的重现性好于±0.02‰（2σ）（Nanne et al., 2017）。

Ir有^{191}Ir和^{192}Ir两个天然稳定同位素。目前仍然未见开展Ir稳定同位素地球化学研究的报道，不过我国学者在国外访学期间曾尝试利用外标元素（Tl、Re）的回归分馏校正方法测定了高精度的^{192}Ir/^{191}Ir值（Zhu et al., 2017），有可能为Ir稳定同位素地球化学的研究提供技术支撑。

Pt有6个天然同位素：^{190}Pt、^{192}Pt、^{194}Pt、^{195}Pt、^{196}Pt和^{198}Pt，其中^{190}Pt是放射性同位素（半衰期约为10^{11} a）。Pt的稳定同位素研究分为两方面，一方面是关注其非质量分馏，测试过程中使用固定内部比值（如^{198}Pt/^{195}Pt＝0.2145）校正分馏（Hunt et al., 2017）；另一方面则关注与质量相关的分馏，Pt稳定同位素组成通常表述为δ^{198}Pt，代表样品中^{198}Pt/^{194}Pt值相对于参照标准（IRMM-010）的千分偏差。目前利用双稀释剂（^{196}Pt-^{198}Pt）在MC-ICP-MS上可进行高精度的Pt稳定同位素组成测定，标准样品的δ^{198}Pt重现性可达±0.040‰（2σ）（Creech et al., 2014）。

2.11 微区原位非传统稳定同位素分析技术进展

微区原位地球化学组成分析省去了琐碎繁杂的化学前处理流程，提高了分析测试效率，更重要的是可以提供常规化学分析不能提供的地球化学组成的微区空间分布信息，是非常重要的地球化学分析手段，在最近的十年间得到了飞速发展。在非传统稳定同位素研究兴起的初期，国际上不少研究团队就已开始探索利用微区分析手段测试高精度的非传统稳定同位素组成；随着微区原位分析技术在国内的推广应用，我国的研究团队也陆续开展了非传统稳定同位素微区分析技术的探索工作，并取得重要进展。

微区分析是直接将样品表面剥蚀的物质导入分析器中，虽然省去了化学前处理步骤，但却给相应的地球化学分析带来了许多干扰组分，这对高精度的同位素组成测试是一大挑战。目前高精度的同位素测试主要还是依赖多接收器的磁式质谱，而要克服这种高干扰的问题主要有两种方法，一种是只分析对目标元素干扰较小的简单介质样品（例如，目标元素含量高、化学组成相对均一的样品）；另一种则是配备造价较高的高分辨率磁式质谱。然而，对于非传统稳定同位素来说，微区原位测试更大的挑战则是如何可靠地校正测试过程中的同位素分馏，这是获取准确分析结果的前提。

由于没有经过化学前处理因而不具备定量添加双稀释剂或外部标准的条件，微区原位非传统稳定同位素测试的分馏校正只能通过外部标准样品来校正，即假定目标同位素在样品测试过程中的分馏与标准样品完全一致，利用标准样品的分馏系数来校正未知样品的目标同位素分馏，这与溶液法中的标准-样品间插法（SSB）的理念相同。这一方法的可靠性依赖于样品与标准之间基体（化学组成和材质）的匹配程度，因为无论是微束对样品表面的剥蚀过程，还是质谱测试过程均会产生非常显著的同位素分馏。样品与标准之间基体匹配不好，样品和标准在各自的测试过程中目标同位素的分馏程度就有显著差异，就可能得到错误的校正结果。因此，只有找到与待分析样品基体非常匹配的标准样品，才有可能用微区原位技术准确分析样品的稳定同位素组成。这也是束缚这一领域发展最大的瓶颈，目前也只有一些特别类型的样品的个别非传统稳定同位素体系的分析取得了突破。即使面临巨大困难，我国的相关研究团队还是开展了许多卓有成效的探索。

适用于微区原位高精度非传统稳定同位素测试的仪器主要包括激光剥蚀多接收器电感耦合等离子体质谱（LA-MC-ICP-MS）和二次离子质谱（SIMS）。我国配备 LA-MC-ICP-MS 的实验室比较多，而配备 SIMS 的实验室还较少，不过基于两类设备开展的相关研究却相似。

2.11.1 Li 同位素的微区原位分析技术进展

Li 同位素组成在地质过程中的变化较大，而且存在矿物内显著的扩散特征，因此利用微区原位分析探索 Li 同位素组成的变化是非常必要的。中国科学院地质与地球物理研究所的团队较早开展了利用 SIMS 测试锆石的 Li 同位素组成的尝试，实现了对低 Li 含量（<1 μg/g）锆石高精度的 Li 同位素测试，标准锆石 δ^7Li 的重现性可达±1.0‰（2σ）的国际先进水准，他们还发现了 M257 标准锆石具有非常均匀的 Li 同位素组成，可作为锆石 Li 同位素测试的校正标准（Li X. H. et al., 2011）。中国地质大学（武汉）的团队尝试在 LA-MC-ICP-MS 上测试硅酸盐矿物的 Li 同位素组成，他们利用标准岩石玻璃如 BCR-2G 作为校正标准，实现了对橄榄石斑晶的 Li 同位素组成测试（Xu et al., 2013），并且经过对测试流程的优化，使标准样品 δ^7Li 的重现性达到了±1.0‰（2σ）左右（Lin J. et al., 2017, 2019b）。

2.11.2 B 同位素的微区原位分析技术进展

B 同位素也是较早开展微区原位分析尝试的非传统稳定同位素体系，中国地质科学院地质研究所的团队较早利用 LA-MC-ICP-MS 开展了微区原位硼同位素测试，他们用 NIST 玻璃标样进行外部分馏校正，对高硼含量的矿物如电气石等的 $\delta^{11}B$ 测试可以获得好于±1.0‰（2σ）的重现性，对较低硼含量（数十 μg/g）矿物的测试也能获得好于±2.0‰（2σ）的重现性（侯可军等，2010），达到了国际先进水平。中国地质大学（武汉）的团队开展了类似的尝试，对 MC-ICP-MS 的仪器配置条件进行了探索，获得了与上述研究相当的精度和重现性（Lin et al., 2014）。该技术对于高硼含量矿物（如电气石）往往可以获得与溶液法相当的 $\delta^{11}B$ 精度［±0.5‰（2σ）］（郭海锋，2014），该方法已在国内得到广泛应用。

针对硼含量更低（<10 μg/g）的硅酸盐样品（例如火山玻璃等），中国科学院广州地球化学研究所的团队基于 SIMS，利用经溶液法标定 $\delta^{11}B$ 值的玄武岩玻璃标准样品如 BCR-2G 做外部标准校正测量过程中的分馏，实现对低硼含量火山玻璃样品的高精度硼同位素组成测定，标样 $\delta^{11}B$ 的重现性在±1.3‰（2σ）左右（He M. H. et al., 2020）。由于主要地质储库中的 $\delta^{11}B$ 差别较大，这样的微区原位 $\delta^{11}B$ 精度对于示踪

高温地质过程（如板块俯冲、岩浆演化等）中硼的来源与迁移是足够的。

2.11.3 石英 Si 同位素的微区原位分析技术进展

石英是地球上分布广泛且十分重要的含硅矿物，其化学组成比较单纯，适合于开展原位 Si 同位素组成分析。中国科学院地质与地球物理研究所的团队尝试利用 SIMS 测试石英的 Si 同位素组成，他们通过对样品表面平整程度的监控和有效分馏校正，使标样 $\delta^{30}Si$ 的重现性好于 ±0.15‰（2σ）（Liu Y. et al., 2019）。这一精度与溶液法测量的外部精度相当，可为相关的研究提供便利的 Si 稳定同位素分析技术支撑。

2.11.4 锆石 Zr 同位素的微区原位分析技术进展

锆石是微区原位分析最重要的矿物，在 U-Pb 年代学和氧同位素分析方面具有不可替代的地位，同时锆石也是最重要的含 Zr 矿物。中国地质大学（武汉）的团队尝试利用 LA-MC-ICP-MS 原位测试锆石的 Zr 同位素组成，对一些均匀的标准锆石样品其 $\delta^{94/90}Zr$ 的重现性可好于 ±0.15‰（2σ），与溶液法测量的外部精度 [通常好于 ±0.06‰（2σ）] 接近（Zhang W. et al., 2019）。该技术对拓展锆石微区原位分析技术体系及其应用范围具有重要意义。

微区原位分析技术和非传统稳定同位素地球化学研究在过去的十年均处于高速发展阶段，两者的结合也是必然趋势。随着更多研究团队的加入以及更深入细致的技术研发，相信在未来的几年，该方向的发展将会更上一个台阶，为我国相关的研究提供更强有力的技术支撑。

2.12 非传统稳定同位素分析标准样品研制进展

非传统稳定同位素主要表述为样品的同位素比值相对于某一个标准物质的千分偏差，因而标准物质在这一研究领域具有非常重要的地位。一般存在两类标准物质，一是特定同位素体系的参照标准，往往是选用该元素纯的溶液或单质（如金属）制备的纯溶液。参照标准往往是由首次报道该同位素体系测试结果的学者选定，后续的研究者沿用下来；另一类则是用于监控分析质量的标准物质，往往是一些已有元素含量的标准地质样品，不同的研究团队反复测量这些样品，逐渐确定了其相应的非传统稳定同位素组成。这些监控标准样品是不同实验室数据进行对比的基础，不管是参照标准还是监控标准，准确确定其同位素组成都是非传统稳定同位素分析技术的重要部分。

我国非传统稳定同位素的早期研究主要是借鉴国外的实验室，而对确定某个特定同位素体系的参照标准的工作开展较少，仅有少数单位开展过这方面的工作。中国地质大学（武汉）的团队建议 Zr 稳定同位素的参照标准 NIST SRM 3169（Feng et al., 2019）。目前就 Cu 和 Zn 同位素参照标准快要消耗殆尽的情况，我国学者提出了新的参照标准的建议：中国地质大学的团队建议以 NIST SRM 683 作为新的 Zn 同位素参照标准（Yang et al., 2018），西北大学的团队研制了两个潜在的 Cu 同位素参照标准的溶液（Yuan et al., 2017），中国地质科学院地质研究所的团队推出了具有作为参照标准潜力的 Fe、Cu、Zn 和 Ti 的标准溶液（唐索寒等，2013b，2016）。

针对特定地质样品监控标准的研制，中国地质科学院的团队开展了系列探索，他们早期建立了 Cu、Fe、Zn 的地质标准样品（唐索寒等，2008），近年来又研制了玄武岩的 Ti 同位素标准样品（唐索寒等，2014）、黑色页岩的 Fe 同位素标准物质（李津等，2020）。

3 近十年我国非传统稳定同位素应用研究的代表性进展

非传统稳定同位素涵括了化学性质各不相同的诸多体系，在不同的地质、环境过程中具有多种类型的分馏特征，并对地质环境因素的变化有不同程度的反映。因此，非传统稳定同位素体系除了类似经典的放射成因同位素体系可以示踪物质来源外，还能示踪各种地质过程中的物质循环和环境演变过程，因此极大地拓展了同位素地球化学的示踪范畴。在过去十年间，随着分析技术的不断提高和完善，非传统稳定同位素已被广泛应用到地球科学领域中涉及物质组成和演化的各个研究方向，以及与环境科学、生命科学等相关的边缘交叉方向，产出了极其丰富的成果。这些成果，有些已经在相关的专业委员会的进展报告中有详细的综述，有些超出了笔者的知识范围，因此，我们很难像分析技术进展那样对应用研究进行系统的梳理和详细的回顾。在此仅选取近十年来我国非传统稳定同位素应用较为丰富的典型的地球科学研究方向，包括深部（高温）地质过程、地表风化过程和气候环境演变（古海洋学）等进行回顾，以此来展示过去十年我国非传统稳定同位素地球化学应用研究的概况。

3.1 高温地质过程非传统稳定同位素代表性应用研究进展

近十年来，随着高精度质谱分析技术在国内的迅速发展，以 B、Mg、Fe、Ca、Cu、Zn、V、Ba、Si 等为代表的诸多非传统稳定同位素体系被越来越多地应用到地球科学的研究中。国内多家单位近十年来深耕非传统稳定同位素地球化学，在地质储库组成厘定、地幔过程及岩浆作用分馏机理方面取得了重要进展，并在示踪俯冲物质再循环以及矿床成因等应用方面取得了诸多成果。

3.1.1 厘定地质储库组成

Zhu 等（2001）早期的工作厘定了太阳星云初始物质的铁同位素组成；Zhang L. 等（2019）发现俯冲带地幔楔橄榄岩具有和大洋橄榄岩相似的 Fe 同位素组成；Yang 等（2009）根据橄榄岩的测试结果制约了 BSE 的 Mg 同位素组成；Liu P. P. 等（2017）研究了深海橄榄岩的 Mg 同位素组成，制约了大洋地幔的 Mg 同位素组成；An 等（2017）发现浅部上地幔和深部上地幔具有一致的 Mg-Fe 同位素组成；Kang 等（2017）以及 Chen C. F. 等（2019）分别通过捕房体橄榄岩和地体橄榄岩制约了 BSE 的 $\delta^{44/40}$Ca 约为 0.94‰；Liu S. A. 等（2015）通过测定造山带和克拉通橄榄岩、洋中脊玄武岩、洋岛玄武岩和大陆玄武岩 Cu 同位素组成，估计 BSE 的 δ^{65}Cu 为 0.06‰；Wang Z. Z. 等（2017）厘定了上地幔的 δ^{66}Zn 为 0.20‰；Liu S. A. 等（2019）根据蛇绿岩型和深海橄榄岩的组成，厘定大洋地幔的 δ^{66}Zn 为 0.19‰；Qi Y. H. 等（2019）结合饱满橄榄岩和科马提岩的测试结果，重新厘定了硅酸盐地球的 δ^{51}V 同位素组成为 -0.91‰；Li W. Y. 等（2020）对幔源火成碳酸岩进行 Ba 同位素研究，估计出上地幔平均 $\delta^{137/134}$Ba 为 0.03‰±0.06‰。

在洋壳储库方面，Chen 等（2020a）和 Zhu 等（2018b）通过研究洋中脊玄武岩制约了新鲜洋壳的钙同位素组成；Wu F. 等（2018）制约了洋中脊玄武岩的 V 同位素组成为 -0.84‰；Huang 等（2015b）厘定了年轻蚀变洋壳的 Mg 同位素组成；Huang 等（2018）发现古老的蚀变洋壳具有不均一的 Mg 同位素组成；Yu 等（2018）厘定了年轻蚀变洋壳的 Si 同位素组成；Huang 等（2016a）厘定了年轻蚀变洋壳的 Cu-Zn 同位素组成。

对于陆壳储库，Li 等（2010）制约了上地壳的镁同位素组成；Wang Y. 等（2019）制约了上地壳的钙同位素组成；Nan 等（2018）对花岗岩、黄土、河流沉积物和冰碛岩等大陆上地壳储库代表性样品进行 Ba 同位素研究，发现大陆上地壳具有高度不均一的 Ba 同位素组成，其平均 $\delta^{137/134}$Ba 为 0.00‰±0.03‰；Yang 等（2016）制约了深部大陆地壳的 Mg 同位素组成，Huang 等（2016）则限定了上陆壳的 Mg 同位素组成。

3.1.2 地幔过程与岩浆作用分馏机理

研究发现，岩浆演化过程（Fan et al., 2020b）和热液活动过程（Wang F. L. et al., 2015）均会造成 B 同位素的显著分馏。

Mg 同位素在俯冲带熔流体演化、弧岩浆作用以及多种变质过程中都会发生显著分馏（Wang et al., 2014a, 2014b, 2016, 2017; Su et al., 2015; Shen et al., 2018; Gao X. Y. et al., 2019; Chen Y. X. et al., 2020; Wang Y. et al., 2020）；此外，高温矿物间也存在显著的 Mg 同位素平衡分馏，这与温度有关，因此可用 Mg 同位素提供高温岩浆过程中的温度信息（Li W. Y. et al., 2011, 2016; Liu et al., 2010, 2011 等）。

对于 Fe 同位素，Zhu 等（2002b）首次发现高温环境下存在 Fe 同位素分馏。熔流体交代作用（Zhao et al., 2010, 2012, 2015）和俯冲作用（Sun P. et al., 2020）均可导致地幔的 Fe 同位素不均一；Mg-Fe 互扩散也可导致 Mg-Fe 同位素的动力学分馏（Wu H. J. et al., 2018）；高温矿物间的亚固相线再平衡也可导致显著的 Fe 同位素分馏（Chen et al., 2014）；受含铁矿物结晶分异或岩浆多阶段分异的影响，岩浆演化过程会造成 Fe 同位素分馏（Du et al., 2017；Li Q. W. et al., 2020）。此外，核结晶过程也可导致 Fe 同位素分馏，造成铁陨石的 Fe 同位素相对于球粒陨石偏重（Ni et al., 2020）。

对于 Ca 同位素，理论计算发现高温矿物间存在显著的平衡分馏（Feng et al., 2014；Kang et al., 2015，2019，2020；Wang W. Z. et al., 2017；Huang F. et al., 2019；Chen et al., 2020b；Dai W. et al., 2020）；此外，与部分熔融、碳酸盐熔体相关的地幔交代作用也会改变地幔的 Ca 同位素组成（Kang et al., 2017；Zhang H. M. et al., 2018；Chen et al., 2020a；Zhu H. L. et al., 2020a）；火成碳酸岩演化过程会造成显著的 Ca 同位素分馏（Sun et al., 2021）。

高温岩浆过程也会造成显著的 Cu 同位素分馏（王泽洲等，2015）。地幔交代过程中由于氧化还原作用的影响，硫化物会发生溶解再沉淀并造成明显的 Cu 同位素分馏（Liu S. A. et al., 2015；Huang et al., 2017）；地幔熔体与橄榄岩反应过程中，由于熔体性质不同，Cu 同位素会发生显著分馏（Huang et al., 2016b）；俯冲过程中，Liu S. A. 等（2015）认为俯冲带变质脱水过程中可能引起 Cu 同位素的分馏，但是 Wang Z. C. 等（2019）认为岩浆分异、脱气以及晚期热液成矿过程中 Cu 同位素的分馏有限；Zou 等（2019）对辉石岩的研究表明基性岩浆中硫化物熔体的分离结晶以及岩石圈地幔中岩浆运移的共同作用会产生明显的 Cu 同位素分馏。

对于 Zn 同位素，Huang 等（2016a）发现高温水岩反应过程中，洋壳丢失 Zn 时会发生显著的 Zn 同位素分馏。在地幔熔体与橄榄岩反应过程中，动力学扩散可以造成 Zn 同位素分馏（Huang J. et al., 2018a，2018b）。进一步的，Huang J. 等（2019）指出碳酸盐熔体交代以及熔体-橄榄岩反应过程中的动力学分馏均可导致地幔的 Zn 同位素组成不均一。Wang Z. Z. 等（2020）的研究揭示了高硅花岗岩分异过程存在显著的 Zn 同位素分馏。Xia 等（2019）的实验岩石学工作显示，含硫的金属熔体相对硅酸盐相富集轻的 Zn 同位素，而不含硫的金属熔体和硅酸盐熔体之间的分馏较小。

对于 V 同位素，Xue 等（2018）发现，L 型普通球粒陨石 V 同位素组成轻于硅酸盐地球；Wu F. 等（2018）发现 MORB 的 V 同位素组成随 SiO_2 含量的升高及 MgO 含量的降低而变重，显示岩浆演化过程中矿物的分离结晶会产生显著的 V 同位素分馏。他们的观测还显示 MORB 相对于 BSE 的 V 同位素组成系统偏重，说明地幔低比例部分熔融产生的熔体具有比源区更重的 V 同位素组成（Qi Y. H. et al., 2019）。Ding 等（2020）对美国夏威夷基拉韦厄火山熔岩的 V 同位素研究发现，Fe-Ti 氧化物的分离结晶是岩浆演化过程中驱动 V 同位素分馏的主要因素。相似的，Fe-Ti 氧化物的分离结晶也会对岩浆的 Ti 同位素组成造成显著影响（Zhao et al., 2020）。

对于 Si 同位素，依据第一性原理计算结果显示，由于不同矿物具有不同的 Si—O 键长，高温矿物间存在显著的 Si 同位素分馏（Huang et al., 2014；Wu et al., 2015；Qin et al., 2016；Li Y. H. et al., 2019），但高温岩浆过程产生的 Si 同位素分馏非常有限。因此，火成岩的 Si 同位素组成主要反映了其地幔或地壳源区的特征，如 Liu X. C. 等（2018）结合 Si-O 同位素对喜马拉雅淡色花岗岩成因进行研究，指出喜马拉雅淡色花岗岩形成于两种沉积岩熔体的混合。

Ba 同位素高温地球化学的研究刚刚起步。Deng 等（2021）对淡色花岗岩及其单矿物进行 Ba 同位素分析，发现岩浆演化过程中钾长石分离结晶可以导致花岗岩的 Ba 同位素组成明显变重。此外，Guo 等（2020）通过高温高压实验发现，流体相比熔体显著富集轻 Ba 同位素，表明岩浆流体出溶会带走更多的轻 Ba 同位素。

3.1.3 示踪壳幔物质循环

B 是亲水元素，在海水中具有较高的 $\delta^{11}B$ 组成（~40‰），与海水发生过水岩反应的蚀变洋壳玄武岩和碳酸盐等海洋沉积物往往具有较高的 $\delta^{11}B$ 值，因而 B 同位素组成是示踪洋壳物质循环的良好指标。Li H. Y. 等（2016）和 Liu H. Q. 等（2016）分别利用全岩的 B 同位素组成有效地示踪了再循环洋壳对华北克拉通地幔和吐哈盆地石炭-二叠火山岩源区的影响；将 B 同位素与另一个对流体活动敏感的 Mo 同位素相结合，能更准确地示踪俯冲板块脱流体的温压条件（金红石稳定与否）、来源（沉积物、蚀变洋壳、碳酸盐）及其对地幔熔融的影响（Li H. Y. et al., 2019；Zhang Y. Y. et al., 2019, 2020a；Fan et al., 2021）。

沉积碳酸盐岩的 Mg、Ca、Zn 同位素组成均与地幔存在显著差别，因此可作为地幔中再循环碳酸盐岩的示踪剂。已有研究发现，产自中国东部和澳大利亚的玄武岩以及部分 OIB 显示出比正常地幔更轻的 Mg 同位素组成，表明其地幔源区存在大量的再循环碳酸盐物质（Yang et al., 2012；Huang et al., 2015a；Liu D. et al., 2015；Huang and Xiao, 2016；Wang Z. Z. et al., 2016, 2017, 2018；Tian H. C. et al., 2016, 2017, 2018, 2019, 2020a；Li S. G. et al., 2017；Su et al., 2017, 2019；Sun et al., 2017；Zhao et al., 2017；Zhong et al., 2017；Wang X. J. et al., 2017, 2018；Li and Wang, 2018；Dai L. Q. et al., 2020；Jin et al., 2020）；中国东部的地幔橄榄岩及云南腾冲火山岩的异常 Ca 同位素组成揭示了碳酸盐岩再循环的影响（Kang et al., 2015；Liu et al., 2017b；Chen et al., 2018；Zhu H. L. et al., 2020b）；Liu S. A. 等（2016）系统研究了中国东部中—新生代玄武岩的 Zn 同位素组成，提出中国东部小于 110 Ma 玄武岩的高 δ^{66}Zn 值反映了再循环碳酸盐岩的影响。通过 Zn 同位素特征，Wang Z. Z. 等（2018）指出碳酸盐熔体与岩石圈地幔反应是中国东部板内碱性玄武岩成分变化的主要原因。Yang 和 Liu（2019）对峨眉山大火成岩省的 26 个苦橄岩、高钛和低钛玄武岩进行了 Zn 同位素研究，指出峨眉山苦橄岩的岩浆源区存在~15%的再循环洋壳物质。

高氧逸度的表生物质俯冲再循环将导致地幔的氧逸度升高。作为变价元素，Fe 同位素的分馏受氧逸度控制，可以示踪再循环物质引起的地幔氧逸度变化。He Y. S. 等（2019）发现，中国东部新生代玄武岩的 Fe 同位素偏重，说明碳酸盐岩再循环可导致中国东部地幔氧逸度升高。Chen Y. X. 等（2019）发现，西阿尔卑斯白片岩的 Fe 同位素偏重，揭示了俯冲带深部流体局部为还原性。Huang J. 等（2020）发现大别山超高压变质脉体的 Fe-Mg 存在异常，认为是来源于蛇纹岩的流体交代弧下地幔楔的结果。

Si 具有一定的流体活动性，因此有潜力示踪俯冲带的熔/流体活动。Yu 等（2018）对蚀变洋壳的 Si 同位素组成进行了分析，结果显示低温海水蚀变和高温热液蚀变均不会改变蚀变洋壳的 Si 同位素特征。Wang B. L. 等（2019）通过对加利福尼亚州 Franciscan 杂岩体中变质橄榄岩的 Si 同位素分析，指出俯冲带富 Si 流体的交代作用并没有显著改变地幔橄榄岩的 Si 同位素组成，质量平衡计算结果表明，俯冲带富 Si 流体的 Si 同位素组成接近于或略轻于地幔橄榄岩。Chen A. X. 等（2020）通过对缅甸硬玉岩的 Si 同位素分析，指出形成硬玉岩的俯冲流体偏重，这些富集重 Si 同位素的流体可能来源于具有重硅同位素的深海硅质岩的溶解。Li Y. H. 等（2020）测量了中国东部大别山港河和花凉亭地区高压-超高压榴辉岩-脉体的 Si 同位素，指出在变质流体演化及脉体形成过程中可以产生显著 Si 同位素分馏。

3.1.4 示踪矿床形成过程

目前 Fe 同位素已被广泛应用于矿床学研究（李志红等，2008；Wang Y. et al., 2011, 2015, 2021；孙剑等，2012a；Sun et al., 2013b；张飞飞等，2013；王跃等，2013, 2014a；齐天骄等，2017；高兆富等，2020）。在岩浆矿床中发现了氧化还原状态控制岩浆型铜镍硫化物成矿系统中 Fe 同位素分馏（Zhao Y. et al., 2019），同时岩浆不混溶过程对硫化物 Fe 同位素的影响也较大（Ding et al., 2019）。另外，研究发现攀枝花层状岩体 Fe 同位素受矿物结晶顺序控制，这对不混溶成矿模型有重要制约（Cao et al., 2019；Tian et al., 2020b）。在热液矿床中，发现角砾岩型铅锌矿床铁同位素空间变化受热液流中的瑞利分馏控制（Gao Z. F. et al., 2018）；He Z. W. 等（2020）进一步系统厘清了斑岩铜矿系统热液流体、蚀变全岩和硫化物的铁同位素变化规律，认为黄铁矿沉淀过程的铁同位素分馏效应是导致含矿流体显著 Fe 同位素变化以及共生硫化物之间不平衡 Fe 同位素分馏的主要因素。这些研究说明铁同位素可以作为有效的地球化学工具，来示踪岩浆-热液矿床的成矿物质来源和成矿过程。

Cu 同位素也已广泛应用于斑岩型铜矿床和铜镍硫化物矿床的研究（李振清等，2009；王跃等，2014b）。高氧逸度流体携带 Cu^{2+} 进入地壳形成高 δ^{65}Cu 的 Cu 富集区，为碰撞环境的斑岩铜矿提供物质来源（Zheng Y. C. et al., 2019）；另外 Cu 同位素在矿床样品中有规律的分馏现象（Wu et al., 2017；王新富和李波，2018），表明 Cu 同位素有潜力作为新的斑岩型矿床的找矿标志（陈佳等，2019）。

Zn 同位素在矿床学研究中也有广泛应用。Zhou 等（2014）利用 Zn、S、Pb 同位素研究了中国南部铅锌矿床的成因；Gao Z. F. 等（2018）对内蒙古东升庙角砾岩型 Zn-Pb 矿床的闪锌矿分析显示，Zn 同位素在热液成矿系统中的空间演化可以指示成矿流体的运移方向；Zhang H. J. 等（2019）对川滇黔地区乌斯河铅锌矿床的研究发现，瑞利分馏可导致闪锌矿的 Zn 同位素偏重。

Mg 同位素也被应用于与碳酸岩相关的矿床学研究中。孙剑等（2012b）利用 Mg 同位素探讨了白云鄂博矿床的成因；Dong 等（2016）研究了古元古代菱镁矿成矿作用的 Mg 同位素特征。

3.2 化学风化过程中非传统稳定同位素代表性示踪应用研究进展

陆壳化学风化是控制地表物质循环最重要的过程之一，其将出露地表的岩石分解成两大部分，可溶解组分被活化迁移并经河流输入海洋，而不溶解组分则存留在陆壳成为土壤和沉积物碎屑的主要组成部分。这一过程中几乎所有的非传统稳定同位素体系中均可产生显著的同位素分馏，也是造成和维持陆地和海洋之间同位素组成差异的重要原因。因此化学风化过程一直都是非传统稳定同位素地球化学研究的热点，近十年来我国的研究团队在这一方面也开展了广泛而又深入的探索，取得一系列重要研究成果。

3.2.1 碱金属和碱土金属元素的同位素体系

碱金属和碱土金属元素在化学风化过程中性质比较活泼，对化学风化过程（程度）反应灵敏，同时这些同位素体系在风化过程中的分馏显著，因而在与陆壳化学风化相关的研究中广受关注。

Li 同位素的研究起步较早，而早期我国学者对 Li 在风化过程中的应用研究多在国外实验室开展。国内研究中，Gou 等（2019b）分析了黄河流域全年的 Li 通量与 Li 同位素组成，发现黄河河水中溶解的 Li 主要来源于干旱和半干旱流域的硅酸盐和蒸发岩，在夏季风盛行季节黄土的风化是河水 Li 的主要来源，蒸发岩的贡献只在雨季有一定的增加。进而指出河水 Li 同位素的季节性变化主要受控于温度的变化，分馏梯度为 $-0.182‰/℃$。基于这一结论，作者认为与温度有关联的河水 Li 同位素变化将会直接影响到海洋的 Li 同位素变化。

K 同位素是近几年兴起的地球科学研究的热点同位素之一。表生过程中 K 同位素的分馏研究始于 2019 年，Li S. L. 等（2019）通过对国内主要大河与支流的水体溶解相与沉积物的 K 同位素研究，证实风化作用使重 K 同位素优先进入水体，而风化产物更倾向保留轻 K 同位素，这也与 Teng 等（2020）分析的广州汤塘花岗岩风化剖面的 K 同位素的结果一致。这些研究都发现，随着风化作用的增强，风化产物的 K 同位素逐渐变轻，次生矿物尤其是黏土类次生矿物如伊利石等则偏向于吸附轻 K 同位素。更为重要的是，Li S. L. 等（2019）发现河水的 K 同位素与相应流域的风化强度呈很好的负相关关系。利用这一关系，作者估算了全球河流径流的 K 同位素组成，进而结合 K 同位素的质量平衡模型，估算了现代全球 K 循环的通量，结果显示海水的 K 同位素组成对大陆风化强度极其敏感，因此可利用古海水的 K 同位素组成来推测地质时期大陆风化强度的变化。

在目前我国的非传统稳定同位素研究中，Mg 同位素无疑是应用最广泛的同位素体系之一。相比于其他非传统稳定同位素的研究，Mg 同位素在风化过程中的研究相对更为深入。Huang 等（2012）通过对海南新生代玄武岩强烈风化剖面的 Mg 同位素研究，发现在较高的 pH 条件下，重 Mg 同位素更易被高岭石吸附从而导致风化产物中的 Mg 同位素偏重；而在较低 pH 时，重 Mg 同位素又会被优先解吸附从而导致风化产物中的 Mg 同位素偏轻。风化剖面中 Mg 同位素的分馏程度与 Mg 的迁移率密切相关。在风化早期，Mg 同位素并没有显著的分馏，随着风化的加强，轻 Mg 同位素会优先被淋失，从而导致风化产物富集重 Mg 同位素。

Gao T. 等（2018）等研究了发育于水稻田剖面的 Fe-Mn 氧化物、土壤及土壤水等的 Mg 同位素组成，并对 Fe-Mn 氧化物和土壤进行了分相提取实验区分了交换态 Mg 与晶格 Mg，并分析了其中的 Mg 同位素组成，结果发现 Fe-Mn 氧化物具有比土壤更轻的 Mg 同位素组成，而 Fe-Mn 氧化物和土壤交换态都显示比结构 Mg 更轻的同位素组成，这也更精细地揭示了土壤中的 Mg 同位素的分馏机制。

Zhao T. 等（2019）对金沙江水体与悬浮物的分析发现，基于质量平衡计算结果，简单的碳酸盐与硅酸盐岩风化的混合不能很好地解释金沙江 Ca 饱和流域的 Mg 同位素组成，从而提出碳酸盐矿物在水体中的沉淀是造成该区域 Mg 同位素变化的重要机制，并指出水体 Mg 同位素的变化除了受控于流域岩体的风化外，还要考虑 Mg 在水体物理化学条件改变时发生的次生矿物沉淀的影响。Li L. B. 等（2020）分析了木孜塔格上下游水域中的冰川水、泉水、基岩和沉积物的 Mg-Sr 同位素组成，并结合上游水域的低 Mg 高 Sr 同位素组成和高 Ca/Mg 和 Ca/Na 值，他们认为主要是碳酸盐的风化控制了此流域的 Mg 同位素组成，而下游水域的高 Mg、低 Sr 同位素组成，以及低 Ca/Mg 和 Ca/Na 值信号主要是由于硅酸盐风化物质的加入所致，因此利用 Mg 同位素可以很好地区分碳酸盐与硅酸盐风化对河水 Mg 同位素的贡献。

与 Mg 同一主族的 Sr 的放射成因同位素组成被广泛用来示踪各种地质过程，而其稳定同位素组成 $\delta^{88}Sr$ 也是极好的示踪指标。Wei 等（2013b）首次将放射成因 Sr 同位素与稳定的 $\delta^{88}Sr$ 联合应用到珠江水系研究中，结果显示碳酸盐和硅酸盐风化对西江流域季节性的 Sr 含量及同位素组成的贡献明显不同。雨季 Sr 含量低，Sr 同位素值（$^{87}Sr/^{86}Sr$ 和 $\delta^{88}Sr$）高的硅酸盐风化贡献更明显，而其他季节 Sr 含量高、Sr 同位素值低的碳盐酸风化的贡献更为显著。

风化过程 Ba 同位素分馏的研究近几年才开始。Gong Y. Z. et al. 等（2019，2020）的研究表明，风化过程中矿物的溶解、沉淀都会发生 Ba 同位素分馏，而黏土矿物吸附、Fe-Mn 氧化物/氢氧化物结合都会富集轻 Ba 同位素，从而导致重 Ba 同位素更易于进入水体。这是河流比上地壳 Ba 同位素组成偏重的重要原因（Cao et al.，2016；Nan et al.，2018；Gou et al.，2020）。Gou 等（2020）的研究也表明，河流悬浮颗粒对轻 Ba 同位素的吸附是黄河河水 Ba 同位素组成变化的主要原因。

3.2.2 氧化还原敏感元素的同位素体系

陆壳的化学风化过程往往伴随显著的氧化还原条件的变化，因此对氧化还原条件敏感的元素的同位素体系往往有非常丰富的分馏特征，这也是反映环境氧化还原条件变化的重要手段，在与风化相关研究中受到广泛关注。

Mo 元素对氧化还原条件极其敏感，较早期的应用研究主要是利用 Mo 同位素制约海洋的不同氧化还原环境，同时推测陆源输入海洋的 Mo 同位素具有一致的组成，即约等于上地壳的 Mo 同位素组成。然而，更多的研究结果表明，水体输入的 Mo 普遍比陆壳的 Mo 同位素偏重，造成这一同位素分馏的原因可能来自于大陆风化过程中风化残余剖面更倾向于富集轻 Mo 同位素组成。Wang Z. B. 等（2015）对西江水域进行了一整年的水体与悬浮物的 Mo 同位素分析，结果发现珠江流域水体的 Mo 同位素组成明显高于悬浮物的、甚至高于硅酸盐基岩和碳酸盐的 Mo 同位素组成，从而推测风化过程中形成的黏土矿物、Fe-Mn 氧化物及有机质等吸附了轻 Mo 同位素从而造成水系重 Mo 同位素组成的特征。通过对比不同气候条件下的珠江、长江和黄河水系的 Mo 同位素组成，发现珠江流域的风化作用较强，水体的 Mo 同位素组成更重，黄河流域风化作用较弱，水体的 Mo 同位素组成偏轻且更接近硅酸盐等的组成，而长江流域水体的 Mo 同位素组成介于二者之间。因此可利用水体的 Mo 同位素组成来示踪不同气候条件下的风化作用差异。

对于风化残余的轻 Mo 富集机制的探索，Wang Z. B. 等（2018）对华南佛冈花岗岩风化壳进行了 Mo 同位素分析，确认了风化残余更偏向于富集轻 Mo 同位素，并通过分相提取实验发现轻 Mo 同位素主要是被 Fe-Mn 氧化物所吸附。此外，Wang Z. B. 等（2020）还对海南岛新生代的一个玄武岩风化剖面进行了 Mo 同位素分析，通过精细分相与矿物同位素组成分析确立了 Ti-Fe 氧化物是另外一个非常重要的富集轻 Mo 同位素的矿物相。Liu 等（2020）也对海南岛新生代的两个玄武岩风化剖面进行了 Mo 同位素研究，结果再一次证实了玄武岩风化过程中重 Mo 同位素会被优先迁出剖面进入水体，而轻 Mo 同位素主要被黏土矿物和 Fe-Mn 氧化物吸附，同时分相提取的实验表明有机质对重 Mo 同位素的富集也起到一定的作用。

风化过程中 Fe 是保守元素，也是氧化还原敏感的元素，其在风化过程中可有效指示氧化还原条件的变化。Song 等（2011a）系统分析了西南喀斯特地区的河水悬浮物、湖相沉积物和孔隙水等的 Fe 同位素组成，发现不同相态的Fe同位素组成的较大变化均与Fe的来源有关，煤矿Fe输入、氧化还原变化及生物作用都对Fe同位素的分馏起着重要作用。此外，Song等（2011b）对阿哈湖悬浮物的Fe同位素组成的分析发现，夏季比冬季具有更轻的Fe同位素组成，主要受黄铁矿的风化作用和土壤Fe淋滤所致，同时氧化还原条件变化导致的Fe相转换也是引起Fe同位素分馏的重要机制。Zhang R. F.等（2015）对冰川融水的Fe同位素研究表明，Fe主要是以纳米粒子或者胶体Fe的形式吸附和聚集到粒子上从而从水中清除，但这一过程中产生的Fe同位素分馏小于0.05‰，表明Fe键合环境在沉淀过程中并没有发生显著变化。Zheng X. D.等（2019）分析了红枫湖及主要支流的悬浮物的Fe同位素，并依据Fe同位素与Fe/Al的关系推测陆源输入是红枫湖Fe的主要贡献，藻类生长过程中对Fe吸附作用是引起悬浮物Fe同位素组成变化的主要原因。

除了上述风化溶解相/迁出相中Fe同位素的研究外，风化残留剖面也是Fe同位素研究的重要载体。王世霞和朱祥坤（2013）发现，玄武岩风化过程中存在比较显著的Fe同位素分馏；Liu等（2014b）分析了不同气候条件下的玄武岩风化剖面的Fe同位素组成，结果显示在热带气候下强风化的海南玄武岩剖面的Fe同位素组成的变化非常有限，Fe同位素不能很好指示氧化还原条件，而风化较弱的南卡罗来纳的风化剖面中Fe同位素组成的变化与氧化还原有着密切的关系。Li M.等（2017）基于菲律宾红壤剖面Fe同位素的研究显示，热带气候下强风化的橄榄岩剖面Fe同位素组成的变化同样非常有限。Gong等（2017）系统分析了黄土高原驿马关黄土-古土壤序列的Fe同位素，发现其Fe同位素组成均一，在0.06‰至0.12‰范围内变化，同前人基于火成岩对上地壳平均组成的估计值0.09‰±0.03‰一致，表明黄土的Fe同位素组成可以代表上地壳的平均组成。Hu等（2019）分析了亚热带地区安徽省内的两个红土剖面的Fe同位素组成，结果发现网纹层的红土具有不同的Fe同位素组成，含Fe较高的红色纹理Fe同位素偏轻，而含Fe较低的白色纹理Fe同位素偏重，指示在不同的氧化还原条件下Fe同位素的分馏具有不同的特征。Feng J. L.等（2018）发现，碳酸盐风化过程中也会出现明显的Fe同位素分馏，具有轻δ^{56}Fe的组分往往赋存于风化壳上的Fe氧化物中。Huang L. M.等（2018a，2018b）基于江西进贤和浙江慈溪稻田土剖面Fe同位素组成的研究，认为稻田土的耕作时间对剖面Fe氧化物分布和同位素组成有显著的影响，高Fe含量的层位更富集轻Fe同位素组成，还原环境下轻Fe同位素倾向于被淋失。Qi等（2020）对苏州地区稻田土剖面Fe同位素组成的分相提取实验进一步表明，在还原性层位同位素偏轻的Fe以Fe^{2+}胶体形式运移，在氧化性层位以Fe氧化物沉淀，是稻田土Fe同位素分馏的重要机理。

戚玉菡（2019）研究了热带气候下强风化的湛江玄武岩剖面中Fe和V的同位素组成，有趣的是，尽管整个剖面的V和Fe元素存在大量丢失且有显著的相关性，但V和Fe同位素组成均无显著的分馏。玄武岩发生强氧化风化的过程中，Fe在原位被氧化为Fe^{3+}后，以胶体的形式向下迁移，V吸附在Fe氧化物和黏土矿物表面或者进入Fe氧化物和黏土矿物晶格中，随着Fe的胶体发生迁移，因而两种金属元素的同位素组成均未产生显著变化。

Cu也是变价元素，也可提供氧化还原条件变化的信息。Liu等（2014b）研究发现，在南卡罗来纳的弱风化剖面中的还原层Cu同位素组成表现为恒定的值，而在氧化区Cu明显富集重同位素组成；而在风化强烈的海南玄武岩风化剖面，解吸附作用使得重Cu同位素优先向剖面下部迁移，而轻Cu同位素主要被有机物键合。两组风化剖面的Cu同位素行为的差异可能主要受到不同气候下降雨量的差异控制的元素解吸附过程和有机质生产量的影响。Lv等（2016）对茅口组黑色页岩进行了Cu-Zn同位素分析，发现风化产物Cu同位素显示非常大的变化，在强风化区域Cu丢失明显且同位素组成明显偏轻于弱风化区域，可能是Cu被运移到剖面下方并被Fe硫化物固定而引起的Cu同位素分馏所致。类似地，黑色页岩风化产物相比于原始基岩也具有偏轻的Zn同位素组成，表明风化过程中偏重的Zn会被优先释放进入水体，这与目前检测到的水体中普遍偏高的Zn同位素结果一致。Wang等（2020b）分析了扬子江及其支流的Cu同位素组成，发现从上游到中游Cu同位素明显增高，可能是颗粒态选择性吸附了轻Cu同位素所致，而从中游到下游的Cu同位素有下降的趋势，这可能与下游地区的含Cu硫化物矿床氧化性风化作用造成了轻Cu同位素组分的加入所致。另外，Cu是一种容易被生物利用的元素，植物的选择性利用能导致土壤中明显的 Cu 同位素的分馏（Li S. Z. et al., 2016, 2020）。

3.2.3 其他同位素体系

除了以上关注度比较高的同位素体系外，其他一些非传统同位素体系在陆壳化学风化过程中的研究也有开展，其中一些代表性的进展包括：

Ding 等（2011）研究了黄河及其四条支流水体中溶解 Si 及悬浮颗粒物的 Si 同位素组成。结果显示，悬浮颗粒物的 Si 同位素组成远轻于溶解态的，可能是由矿物种类差异、来自黄土高原的沉积物组成的加入以及气候和黄河流量共同作用结果，而溶解态的 Si 同位素组成主要与硅酸盐的风化、水生植物选择性利用、水合单硅酸在 Fe 氧化物上的吸附与解吸附作用，以及土壤中植硅体的溶解等多重因素共同作用有失。因此，河流 Si 同位素组成的时空变化与气候、化学风化强度和生物活动息息相关。该研究发现，相较于前人研究的其他河流，黄河的溶解态 Si 的同位素组成轻微偏重，可能与流域内较强的化学风化和生物活动相关。人类活动也可能影响世界主要河流的溶解态 Si 的同位素组成。

Zhang Y. Y. 等（2020b）对珠江水系的 Hg 同位素进行了研究，发现河流内部的环境过程对 Hg 同位素的影响非常有限。基于非质量分馏数据，作者认为虽然在部分点位发现了本地电子垃圾焚烧带来的显著负偏的 δ^{199}Hg 值，但人类活动可能对珠江全流域的溶解态 Hg 同位素没有显著的直接影响。珠江水系的 Hg 同位素组成主要受控于大气沉降和土壤的风化的输入。

Qi H. W. 等（2019）对海南新生代的玄武岩风化剖面进行了系统的 Ge 同位素研究。结果显示，风化过程中释放的 Ge 同位素相对于母岩偏重 1.38‰±0.28‰（2σ），与全球河流和海洋中偏重的同位素观测结果相符。作者认为这一变化主要受控于风化产物的吸附-解吸附作用。

与对风化剖面和风化产物的直接观测研究相比，目前非传统稳定同位素分馏在风化方面的理论计算和实验模拟还远不够。Yuan 等（2018）对 Ga 同位素在方解石与赤铁矿上的吸附进行了模拟实验，结果显示，无论是方解石还是赤铁矿都倾向于吸附轻 Ga 同位素，这一同位素分馏符合瑞利分馏模式。其主要机制是相比于 Ga 的水合物，Ga 在方解石中的配位数和Ga—O 键长均有增加，而无论是在酸性还是碱性条件下均具较轻同位素组成的四配位 Ga 更倾向于吸附在赤铁矿表面。

3.3 海洋气候环境演变研究中非传统稳定同位素代表性应用研究进展

地球宜居性的演化对生命的起源、演化和发展具有决定性作用，是地球科学研究的重要方向。在整个地球历史上，海洋均占地球表面的绝大部分，是地球生命起源和早期生命生存和发展的主要场所，因此海洋气候环境演变的研究（古海洋学）是地球宜居性演化研究的主要内容。了解与生命息息相关的关键宜居要素如温度、盐度、酸度和氧化还原程度（含氧量）的变化，以及生命应对环境变化的响应机制则是古海洋学研究的重点。非传统稳定同位素涵盖具有不同地球化学性质的体系，其中的一些体系对这些宜居要素的变化比较敏感，因而在古海洋学的研究中发挥着重要的示踪作用。近十年来，我国的研究团队在探索非传统稳定同位素体系对海洋气候环境要素的响应机制，以及对不同地质时期海洋环境变化的重建等方面均开展了大量的研究，并取得了一些重要的进展。

要了解地质时期海洋环境的变化，首先需要构建能够反映气候环境要素变化的地球化学指标（即替代指标）体系。通过对现代海洋过程进行观测，并与实测的气候环境参数相结合则是构建替代指标体系的主要研究方法。这方面的研究通常选择对环境变化敏感或者积极参与生命活动的元素，构建其与环境要素以及生命活动的定量/半定量关系。而以珊瑚、有孔虫为代表的生物成因碳酸盐持续贯穿古生代以来的整个地球历史，是记录地质时期海洋环境演变的良好载体，因而也成为这方面研究的主要研究对象。

3.3.1 B 同位素与海水 pH 的变化

生物碳酸盐的 B 同位素组成是记录钙化时海水 pH 的良好替代指标。B 同位素体系在造礁珊瑚中的研究起步较早，研究程度也较深。由于 B 在水体中的存在形式及其同位素组成均受 pH 的控制，并且海水中的硼酸根离子相较于硼酸会大比例的进入碳酸钙的晶格中，因此海洋碳酸盐的 B 同位素组成可以作为重建海水 pH 变化的重要地球化学替代指标。利用持续生长百年之久的块状滨珊瑚，通过测定其骨骼的 $\delta^{11}B$ 就可以重建数百年来海水 pH 的变化历史，评估海洋酸化的进程。Wei 等（2009）利用大堡礁滨珊瑚的 $\delta^{11}B$ 重建了南太平洋海水 pH 近 200 年的变化历史，发现自 1940 年以来海洋酸化趋势显著。Liu 等（2014）和 Wei 等（2015）分别利用海南岛沿岸的滨珊瑚进行了 $\delta^{11}B$-pH 记录重建，结果显示海南岛南部的近岸海水酸化显著，而海南岛东部近岸海水可能因受上升流影响而未呈现明显的酸化趋势。随着对珊瑚骨骼中 $\delta^{11}B$ 研究的不断深入，柯婷等（2015）发现珊瑚样品中 $\delta^{11}B$ 存在季节性周期波动，指示珊瑚 $\delta^{11}B$ 除了记录海水 pH 外还可能反映生物活动的信息。Chen X. F. 等（2019）通过 $\delta^{11}B$-B/Ca 联用方法，解析了珊瑚钙化机制的信息，发现季节时间尺度上珊瑚自身生物活动对 $\delta^{11}B$ 分馏的控制，并对滨珊瑚如何响应海洋暖化和酸化进行了探讨。

此外，珊瑚 $\delta^{11}B$-pH 记录重建也拓展到了化石样品。刘卫国等（1999）对采自南海珊瑚礁的多个中晚全新世珊瑚化石的 $\delta^{11}B$ 进行了测定，并尝试恢复了过去 7 000 年以来海水 pH 的变化趋势。Liu 等（2009）在这一工作的基础上又补充了样品，最后指出中晚全新世以来南海表层海水 pH 逐渐升高而后在现代迅速下降，并认为自然背景下冬季风的变化控制了海水 pH 的长期变化。对于更古老的珊瑚化石，受限于样品的保存程度（特别是蚀变），B 同位素的研究相对较少。马云麒等（2011）对中生代的腕足类和珊瑚化石的 B 同位素组成进行了测定，发现 $\delta^{11}B$ 比现代珊瑚（海相碳酸盐）要低 ~10‰ 之多，很可能反映海水自身 $\delta^{11}B$ 的变化。

3.3.2 碱土金属同位素对环境变化和生命活动的响应

碱土金属 Mg、Ca、Sr 和 Ba 广泛参与生命活动，同时 Ca 还是生物碳酸钙的主要组分，其同位素体系的变化可能反映特定生命活动的变化。

Ca^{2+} 和 CO_3^{2-} 是参与珊瑚钙化过程重要的两个离子。相较于 C 和 O 同位素体系，Ca 同位素在珊瑚骨骼中分馏特征和控制机理的研究程度相对薄弱。Chen 等（2016b）通过对大堡礁滨珊瑚骨骼中 $\delta^{44/40}Ca$ 的测定，发现其变化具有季节性周期，但总体上分馏程度相对较小。尽管 $\delta^{44/40}Ca$ 与温度之间存在显著相关性，但 $\delta^{44/40}Ca$ 变化对温度的敏感性同其他温度指标相比优势不大。此外，$\delta^{44/40}Ca$ 与 $\delta^{13}C$ 之间具有显著关系，指示 Ca 同位素的分馏可能仍受珊瑚自身新陈代谢的影响。

Sr 是珊瑚骨骼中含量仅次于 Ca 的金属元素，其含量可达 $7 000×10^{-6}$。尽管 Sr 在珊瑚骨骼中的变化很大程度上反映了温度对其配分的影响（因此 Sr/Ca 值是珊瑚经典的温度计），但稳定 Sr 同位素珊瑚骨骼中的分馏机制尚未查明。邓文峰等（2017）通过对现代滨珊瑚骨骼 $\delta^{88/86}Sr$ 的研究，发现其存在明显的季节性变化特征，并与太阳辐照强度有显著的对应关系，提出光照控制的珊瑚钙化过程可能是造成稳定 Sr 同位素分馏的主要原因。

Ba 是海洋中的类营养盐元素，是示踪水团混合、营养盐循环和生物生产力的重要指标。厦门大学曹知勉等分析了南海北部外陆架至陆坡真光层内相同水层的溶解态和悬浮颗粒的 $\delta^{138}Ba$ 组成，发现 150 m 以上浅水体中的溶解 Ba 和悬浮颗粒的 $\delta^{138}Ba$ 组成均在误差范围内保持不变，而后者系统轻于前者 ~0.5‰。因此，Ba 同位素在海洋真光层内发生显著分馏，较轻同位素优先从溶解态富集于颗粒态。应用瑞利和稳态分馏模型计算该过程的 Ba 同位素分馏系数分别为 –0.4‰±0.1‰ 和 –0.5‰±0.1‰，与上述差值吻合，且与大洋浅层水体和湖泊中的实测 Ba 同位素分馏系数一致。另外，海洋真光层内的颗粒 Ba 同位素信号与初级生产力之间并无明显的关联性（Cao et al., 2016）。

天津大学刘羿和中国科学技术大学于慧敏合作研究了南海 7 个不同区域浅水滨珊瑚骨骼的 Ba 同位素组成，发现除一个近岸珊瑚的 $\delta^{138/134}Ba$ 低于平均值外，其余 6 个珊瑚的 $\delta^{138/134}Ba$ 变化区间相对一致且变化幅度较小。同时，$\delta^{138/134}Ba$ 的珊瑚个体差异较小，有别于珊瑚个体中 Ba 含量的显著变化，因此珊瑚 $\delta^{138/134}Ba$ 主要记录海水 Ba 同位素组成的变化，有望成为研究过去海洋 Ba 循环的新指标（Liu Y. et al., 2019）。

对海水样品的 Mg 同位素分析发现，墨西哥湾海水的 $\delta^{26}Mg$ 和 $\delta^{25}Mg$ 在垂直和水平分布上的均一性都很好，且与夏威夷海水的 Mg 同位素值一致，综合前人发表的世界各地海水的 Mg 同位素数据，发现全球现代海水均具有均一的 Mg 同位素值，即 δ^{26}Mg 为 $-0.83‰±0.09‰$，δ^{25}Mg 为 $-0.43‰±0.06‰$（2SD，$n=90$）。由于海水容易获取，其同位素值接近于 Mg 同位素变化的中间值，可以用做 Mg 同位素分析标准物质（Ling et al., 2011）。

3.3.3 其他生源要素同位素对生命活动的响应

除碱土金属外，很多金属元素也积极参与生命活动，如 Fe 是重要的营养元素，而 Zn 和 Mo 则是某些关键生物酶的重要组分。为此，一些团队也对海洋中的 Fe、Mo 和 Zn 等同位素体系开展了探索性的研究。

Wang Z. B. 等（2019）测定了采自大堡礁和南海北部现代滨珊瑚骨骼及海水的 δ^{98}Mo，发现 Mo 同位素在珊瑚骨骼中的分馏相对较大，可达 1‰，而且分馏还受温度的控制。此外，他们还发现，δ^{98}Mo 在海水中的变化可能受生物活动（光合、呼吸作用）的影响，存在明显的昼夜变化。结合海水和珊瑚骨骼的 δ^{98}Mo 变化特征，他们认为 Mo 在进入珊瑚体内形成骨骼时的同位素分馏可能受共生体新陈代谢的影响，因此 δ^{98}Mo 有可能成为指示生物活动的替代指标。

同样，Xiao 等（2020）也发现，Zn 同位素在进入珊瑚体内形成骨骼这一过程中会产生分馏，因而珊瑚骨骼的 δ^{66}Zn 显示出季节性的周期变化，这一变化并不能完全由温度变化来解释，而有可能与珊瑚共生体新陈代谢产生的活性氧有关，指示骨骼 δ^{66}Zn 变化与珊瑚生物活动的关联。

对东海表层沉积物的 δ^{66}Zn 的研究表明，δ^{66}Zn 的变化范围从 $0.02‰±0.04‰$（2σ）到 $0.67‰±0.06‰$（2σ），接近长江口和内陆架沉积物的 δ^{66}Zn 值具有（相对较轻的 δ^{66}Zn 值，为 $0.20‰~0.31‰$），越往外陆架 δ^{66}Zn 值越偏重，达到 0.60‰，表明表层沉积物的 Zn 同位素受到了自然背景和与工业废水和污水排放有关的混合影响，因此 Zn 同位素组成可以用于示踪人类活动对海洋环境的污染（Zhang R. et al., 2018）。

对南海神户海域甲烷渗漏区钻孔岩心中自生黄铁矿的 Fe 同位素研究发现，两个岩心中自生黄铁矿的 δ^{56}Fe 值分别为 $-0.35‰~0.27‰$ 和 $-0.79‰~0.18‰$，并随埋藏深度的增加而增大，同时黄铁矿的矿化程度和 δ^{34}S 呈正相关关系。这种相关性可能是受硫酸盐化驱动的甲烷厌氧氧化作用和有机碎屑硫酸盐还原作用的控制，黄铁矿中 ^{56}Fe 和 ^{34}S 的共同富集可以作为硫酸盐化驱动的甲烷厌氧氧化作用的潜在指标（Lin Z. Y. et al., 2017）。但台西南海域甲烷渗漏区的自生黄铁矿并未显示出明显的 ^{56}Fe 的富集，两个钻孔岩心自生黄铁矿的 δ^{56}Fe 值分别为 $-0.86‰~-0.56‰$ 和 $-1.57‰~-0.25‰$，相比神户海域偏负的 δ^{56}Fe 值可能是由于异化铁还原作用释放亏损 ^{56}Fe 的二价铁到孔隙水中造成的（Lin et al., 2018）。

3.3.4 边缘海 Hg 同位素及其生物地球化学循环

近海沉积物是 Hg 生物地球化学循环中重要的汇，是陆源 Hg 迁移进入开放大洋的动态桥梁，也是深入理解陆地-海洋交界区域中 Hg 的输入和迁移的重要载体。研究人员通过对边缘海沉积物样品 Hg 同位素的研究，发现 δ^{202}Hg 在沉积物钻孔中从深层到表层逐渐增加，近海沉积物具有正的 δ^{199}Hg 值，但 1950 年以前的样品比年轻样品具有更正的 δ^{199}Hg 值。近岸沉积物样品在 1950 年以前具有负的 δ^{199}Hg 值，但年轻样品的值接近于零。在此基础上结合三元混合同位素模型，对我国四大边缘海潜在的 Hg 输入来源（工业 Hg 排放、土壤 Hg 和大气沉降 Hg）定量估算的结果表明了我国工业化和经济发展带来的工业 Hg 输入边缘海的显著增加（Yin et al., 2018）。进一步对边缘海海域近海沉积物中 Hg 的来源及其向远海区域的迁移程度研究表明，我国四大边缘海海域和河口沉积物中 Hg 的主要来源显著不同，且随着城市影响程度的降低，从近岸到近海，Hg 的主要来源从陆源输入逐渐变为大气沉降（Zhang R. et al., 2018；Meng et al., 2019；Sun X. et al., 2020）。

为深入探究海洋食物链中 Hg（尤其是甲基汞）的来源和生物地球化学转化过程，合理评估人为 Hg 排放后的环境效应，研究人员通过系统分析渤海生态系统中的全食物链（包括水生植物-大型藻类、软体动物、甲壳类和近岸-远海鱼类）中 Hg 的形态和同位素组成以及 C、O 同位素组成等，发现 Hg 在营养级传递过程并不产生同位素分馏，这为利用食物链的 Hg 同位素组成示踪水体中 Hg 的地球化学转化提供了坚实的理论基础，他们还在渤海生物中首次发现了显著偏负的偶数^{200}Hg同位素的非质量分馏现象，为未来利用偶数Hg同位素研究食物链的Hg循环提供了新的思路（Meng et al., 2020）。

对红树林生态系统Hg的迁移转化的研究表明，红树林生态系统沉积物的δ^{202}Hg和δ^{199}Hg值比周围海水更偏负，植物根系组织中Hg主要来源于沉积物并富含轻质量数的Hg同位素而具有非常负的δ^{202}Hg值，这些Hg同位素特征可能是由于多种物理和化学过程如吸附、光还原、沉淀和同化等共同导致的（Huang S. Y. et al., 2020）。

3.4 非传统稳定同位素体系在深时古海洋研究中的示踪应用进展

在整个地质历史时期，全球海洋的气候环境变化非常显著，为各种类型的生命提供了不同类型的栖息场所，同时也很大程度上控制着生命的演化进程。丰富的非传统稳定同位素体系可以准确示踪这些不同类型的环境变化，因而在古老地质时期的海洋环境演变研究中发挥重要的作用。

我国拥有非常完整的从前寒武纪到现代的海洋沉积地层，记录了丰富的地质时期海洋环境演变的历史，因而一直是深时古海洋研究的热点地区。以往的研究主要依托传统的气体稳定同位素、氧化还原敏感型元素含量和矿物学等研究方法，已经取得了非常显著的进展。非传统稳定同位素研究兴起以来，我国学者迅即将这一新方法引入深时古海洋研究中，对其中的关键科学问题进行了更深入的探索，并取得了系列重要成果。

3.4.1 示踪海洋氧化还原条件的演变

氧化还原条件（溶解氧含量）的变化是地质时期海洋最显著的环境变化之一。伴随元古宙以来大气氧含量的升高过程，海洋也从还原环境逐渐向氧化环境转变。由于海水含氧量的变化在不同区域、不同水深的进展是有差异的，因此海洋的氧化过程比大气的氧化进程更复杂，而海洋中这些氧化还原条件的变化直接影响了生命的灭绝和爆发。传统的研究中海洋氧化过程很大程度上依靠S同位素，以及对氧化还原条件敏感的元素的含量变化，而涵盖在非传统稳定同位素范畴内的氧化还原条件敏感元素的同位素体系的应用，则为相关的研究提供了更多强有力的手段。

Mo同位素是较早被广泛应用于这方面研究的体系，我国多个团队应用Mo同位素组成对我国华南新元古代—寒武纪早期海洋的氧化过程进行了重建，精确辨识了不同区域、不同水深海域的氧化过程（温汉捷等，2010；Wen et al., 2011, 2015a；Xu et al., 2012；Chen et al., 2015；Cheng et al., 2016；Luo et al., 2021）；此外，在对古生代生命大灭绝研究中，Mo同位素也可以很好地示踪海水的氧化还原条件的变化（Zhou et al., 2012）。

除Mo同位素外，其他的一些对氧化还原敏感的同位素体系，如Fe同位素（朱祥坤等，2008a；闫斌等，2010，2014；李志红和朱祥坤，2012；李志红等，2012；Fan et al., 2014, 2018a；Zhang F. F. et al., 2015；Zhu X. K. et al., 2019；Xiang et al., 2020；Wu C. et al., 2020）、Cr同位素（Huang et al., 2018c）、Se同位素（Wen et al., 2014）和Tl同位素（Fan et al., 2020）等也被用于示踪华南新元古代—寒武纪早期的海洋氧化过程。值得注意的是，目前主要依托国外实验室开展的U同位素（δ^{238}U）也被应用到我国华南地区一些重要的生命灭绝事件中海洋氧化还原条件的变化的示踪（Song et al., 2017；Zhang F. F. et al., 2018, 2020a, 2020b）。随着在国外访学人员归国和国内实验室研发工作的不断推进，将会有更多国内开发的U同位素方法应用到深时古海洋的研究中。

3.4.2 示踪地质时期海洋生产力的演变和生命活动

表层海洋生产力是海洋生命繁衍和发展的基础，同时和海洋氧化还原状态的变化息息相关。许多与生命活动密切相关的元素，如Ba等的同位素体系具有示踪古海洋生产力的潜力。近期，Wei等（2021）通过分析华南三峡地区埃迪卡拉系碳酸盐岩地层的Ba同位素组成，观察了埃迪卡拉纪这一地质历史特殊时期海洋Ba生物地球化学循环的变化，发现古代海洋Ba的生物地球化学循环可能受到海洋氧化还原状态的强烈控制，Ba同位素体系可能更适用于示踪氧化水体的生产力水平。

华南新元古代时期另一个重要的地质现象是硅质沉积岩（燧石）的大量出现，我国一些研究团队利用 Si 同位素有效分辨出这些沉积岩中 Si 的来源（Gao et al., 2020; Zhang H. J. et al., 2020），为更好地了解这一时期海洋生命大爆发提供了重要证据。

细菌是地球早期生命活动中非常重要的组成部分，产甲烷菌是其中非常重要的类型，对生命活动及温室气体的排放均具有重要意义。Ni 对产甲烷菌的生命活动具有重要作用，我国学者发现 Ni 同位素组成可以很好示踪甲烷菌的活动，指出在大氧化事件（2.3 Ga）和雪球地球（~625 Ma）两次大气升氧的关键时期，产甲烷菌活动及其对大气 CH_4 的排放均有显著变化（Wang S. J. et al., 2019; Zhao et al., 2021）。

3.4.3 示踪陆源输入、火山活动等其他影响因素

不仅仅限于对海洋环境要素的直接重建，非传统稳定同位素还能对重大环境变化和生命灭绝的驱动因素进行追踪。陆源输入和火山活动是改变海洋环境的重要因素，Wang X. 等（2018）结合 $\delta^{66}Zn$ 和放射成因 Sr 同位素，分辨出约 372 Ma 的 Frasnian-Famennian（F-F）生命大灭绝前有明显的陆源风化加强而向海洋输入增加的现象；Huang 等（2016）和 Li J. 等（2020）利用 Mg 同位素组成观察到雪球地球结束后（~625 Ma）有多次的陆壳风化加强向海洋输出明显增加的现象；而王伟中等（2020）利用 Cd 同位素也辨识出这一生命大灭绝前海洋表层生产力的显著增加；Liu S. A. 等（2017）则利用 $\delta^{66}Zn$ 分辨出 P-T 生命大灭绝事件中火山活动及陆壳风化加强向海洋输入 Zn 的过程，为更好理解 P-T 生命大灭绝的直接控制因素提供更多的证据；同样地，Zn 同位素组成也能很好地限定埃迪卡拉纪早期陆源输入的营养物质如 P 等的变化过程（Fan et al., 2018b; Yan et al., 2019）。

火山活动会向大气输入大量的 Hg，而 Hg 的同位素组成可以很好辨识其来源，可为判别地质时期重大环境变化和生命灭绝事件与火山活动的关联提供重要证据。我国学者利用 Hg 异常并结合 Hg 同位素的非质量分馏很好地构建起 P-T 生命大灭绝与西伯利亚大火成岩省活动之间的联系（Wang X. D. et al., 2018, 2019; Shen et al., 2019a, 2019b），有助于更好地了解这一地质历史上最大的生命大灭绝事件的控制机制。

3.4.4 厚层白云石成因的探索

在我国的古生代到中生代的海相沉积地层中存在着许多巨厚的白云石地层（>100 m），是关键地质时期海洋环境演变的重要研究载体，但对其成因目前仍然存在许多争论。北京大学的团队认为，引入 Mg 同位素的研究手段有可能为解决这一争论提供答案（Huang 2015; Li F. B. et al., 2016; Peng et al., 2016; Ning et al., 2019, 2020）。

4 对未来研究的展望

经过近三十年的发展，非传统稳定同位素已经渗入到涉及物质组成的地球科学各个领域，并发挥着越来越重要的作用。我国的非传统稳定同位素地球化学研究进展基本和国际同步，尤其是在过去十年更是大踏步前进，跻身国际先进行列，同时应用领域也在不断拓展。目前国内具备开展固体同位素分析测试能力的单位基本上都加入了这一研究行列，相关研究方向的科研人员对非传统同位素的了解和认可程度在不断提高。可以预见，未来十年我国非传统稳定同位素的研究将得到更快的发展，应用范围将更广泛。

从分析技术角度看，技术的进步是永无止境的。本文通过对非传统稳定同位素分析技术的全面梳理发现，虽然我国研究团队开展的非传统稳定同位素体系的技术研发已很全面，但仍然有一些体系的测试技术目前尚没有在国内实验室落地，而且现有的许多技术，在测试精度、分析效率以及适用的样品类型等方面还有非常大的提升空间，结合微区原位测试技术的非传统稳定同位素分析也才刚刚起步。这些均是未来非传统稳定同位素测试技术的重要发展方向。

未来十年，我国非传统稳定同位素发展最快的应该是在应用领域，随着分析技术的不断完善，对于各种地质过程分馏机制的探索将会成为重点，而伴随着对分馏机制的了解，大量体系将会应用到地球科

学的各个研究方向中，甚至在一些交叉学科方向，例如环境科学、生命科学、材料科学等方面均会有非常多的应用。十年后再进行回顾时，估计涉及非传统稳定同位素的研究成果将会多得无法以这样的篇幅来进行综述，很有可能是在各相关的学科方向中进行总结。因为非传统稳定同位素进展很快、成果很多，甚至在本文写作的过程中还有新的论文发表，本文难免有疏漏之处，还请广大读者谅解。

致谢：吴非、白江昊、崔灏、何妙洪、何昕悦、李洪颜、刘贲、刘熙、肖河、杨亚楠、张卓盈、朱冠虹、黄建、丁昕、程文瀚、南晓云、卫炜、宫迎增、胡霞、戚玉菡、明国栋、贺治伟、徐娟、陈振武、田兰兰、王保亮、梁文力、吕炜昕、董琳慧、方远等老师和同学在资料检索、收集及整理中给予了大量帮助，审稿专家对早期文稿提出了许多宝贵的意见和建议，对提升文章质量有较大帮助，在此一并致以衷心感谢。

参 考 文 献

陈佳, 赵云, 薛春纪, 安伟才, 郭旭东. 2019. 铜同位素在斑岩型矿床中的应用. 见：第九届全国成矿理论与找矿方法学术讨论会论文摘要集. 南京：《矿物学报》编辑部

邓文峰, 韦刚健, 马金龙, 陈雪霏, 谢露华, 曾提. 2017. 太阳辐照度控制的珊瑚骨骼稳定 Sr 同位素分馏. 见：中国矿物岩石地球化学学会第九次全国会员代表大会暨第16届学术年会文集. 西安：中国矿物岩石地球化学学会

丁悌平, 万德芳, 李金城, 蒋少涌, 宋鹤彬, 李延河, 刘志坚. 1988. 硅同位素测量方法及其地质应用. 矿床地质, 7(4)：90-96

冯家毅, 刘丛强, 赵志琦, 汪齐连, 刘文景, 灌谨, 樊宇红, 王静. 2013. 用 AGMP-1M 阴离子交换树脂直接分离 Zn 的实验方法改进. 地球与环境, 41(5)：589-595

冯新斌, 尹润生, 俞奔, 杜布云, 陈玖斌. 2015. 汞同位素地球化学概述. 地学前缘, 22(5)：124-135

高永娟, 马腾, 凌文黎, 刘存富, 李理. 2009. 铬稳定同位素分析技术及其在水污染研究中的应用. 科学通报, 54(6)：821-826

高兆富, 朱祥坤, 孙剑, 周子龙. 2020. 内蒙古炭窑口硫化物矿床 Fe、S 同位素组成及对硫化物成矿的制约. 地球学报, 41(5)：676-685

郭海锋, 夏小平, 韦刚健, 王强, 赵振华, 黄小龙, 张海祥, 袁超, 李武显. 2014. 湘南上堡花岗岩中电气石 LA-MC-ICPMS 原位微区硼同位素分析及地质意义. 地球化学, 43(1)：11-19

何连花, 刘季花, 张俊, 张辉, 高晶晶, 崔菁菁, 张颖. 2016. MC-ICP-MS 测定富钴结壳中的铜锌同位素的化学分离方法研究. 分析测试学报, 35(10)：1347-1350

贺茂勇, 马云麒, 金章东, 马海州, 张艳灵, 罗重光, 肖军, 肖应凯. 2013. 自动静态双接收高精度热电离质谱法测定硼同位素. 质谱学报, 34(2)：75-81

侯可军, 李延河, 肖应凯, 刘峰, 田有荣. 2010. LA-MC-ICP-MS 硼同位素微区原位测试技术. 科学通报, 55(22)：2207-2213

侯可军, 秦燕, 李延河. 2012. Fe 同位素的 MC-ICP-MS 测试方法. 地球学报, 33(6)：885-892

黄方, 南晓云, 吴非. 2016. V 和 Ba 同位素分析方法综述. 矿物岩石地球化学通报, 35(3)：413-421

黄舒元, 袁东星, 孙鲁闽. 2016. 环境样品的热解-溶液吸收预处理及其汞同位素组成的测定. 分析测试学报, 35(6)：704-708

贾彦龙, 肖唐付, 宁曾平, 杨菲, 姜涛. 2010. 铊同位素及环境示踪研究进展. 矿物岩石地球化学通报, 29(3)：311-316

蒋少涌, 丁悌平, 万德芳, 李延河. 1992. 辽宁弓长岭太古代条带状硅铁建造(BIF)的硅同位素组成特征. 中国科学 B 辑, (6)：626-631

蒋少涌, Woodhead J, 于富民, 潘家永, 廖启林, 吴南平. 2001. 云南金满热液脉状铜矿床 Cu 同位素组成的初步测定. 科学通报, 46(17)：1468-1471

柯婷, 韦刚健, 刘颖, 谢露华, 邓文峰, 王桂琴, 许继峰. 2015. 南海北部珊瑚高分辨率硼同位素组成及其对珊瑚礁海水 pH 变化的指示意义. 地球化学, 44(1)：1-8

李津, 朱祥坤, 唐索寒. 2011. 钼同位素比值的双稀释剂测定方法研究. 地球学报, 32(5)：601-609

李津, 马健雄, 闫斌, 唐索寒, 朱祥坤. 2020. 黑色页岩铁同位素标准物质的研制. 地球学报, 41(5)：623-629

李世珍, 房楠, 孙剑, 陈岳龙, 朱祥坤. 2013. 高 REE-Nb-Fe-Mn 样品 Mg 同位素测定的化学分离方法. 吉林大学学报(地球科学版), 43(1)：142-148

李维涵, 朱建明, 吴广亮, 谭德灿, 王静. 2019. 铬同位素纯化方法的比较研究. 矿物岩石地球化学通报, 38(5)：1024-1030

李振清, 杨志明, 朱祥坤, 侯增谦, 李世珍, 李志红, 王跃. 2009. 西藏驱龙斑岩铜矿铜同位素研究. 地质学报, 83(12)：1985-1996

李志红, 朱祥坤. 2012. 河北省宣龙式铁矿的地球化学特征及其地质意义. 岩石学报, 28(9)：2903-2911

李志红, 朱祥坤, 唐索寒. 2008. 鞍山-本溪地区条带状铁建造的铁同位素与稀土元素特征及其对成矿物质来源的指示. 岩石矿物学杂志, 27(4)：285-290

李志红, 朱祥坤, 唐索寒. 2012. 鞍山-本溪地区条带状铁矿的 Fe 同位素特征及其对成矿机理和地球早期海洋环境的制约. 岩石学报, 28(11)：3545-3558

刘芳, 祝红丽, 谭德灿, 刘峪菲, 康晋霆, 朱建明, 王桂琴, 张兆峰. 2016. 热电离质谱测定钙同位素过程中双稀释剂的选择. 质谱学报, 37(4)：310-318

刘卫国, 彭子成, 肖应凯, 王兆荣, 聂宝符, 安芷生. 1999. 南海珊瑚礁硼同位素组成及其环境意义. 地球化学, 28(6)：534-541

刘锡尧, 袁东星, 刘宝敏, 陈耀瑾, 林方方, 苏海涛, 罗苏笙. 2013. 氧化-吹扫-金柱捕集-(双柱)热脱附-电感耦合等离子体质谱测定液体和固态样品中痕量汞同位素含量. 分析科学学报, 29(1): 11-16

刘峪菲, 祝红丽, 刘芳, 王桂琴, 许继锋, 张兆峰. 2015. 钙同位素化学分离方法研究. 地球化学, 44(5): 469-476

马健雄, 李津, 陈岳龙, 朱祥坤. 2020. 地质样品中铬的化学分离及双稀释剂法铬同位素测定. 地球学报, 41(5): 630-636

马云麒, 肖应凯, 诸葛芹, 蒋生祥. 2010. 直接熔融热电离质谱法测定硼酸盐矿物中的硼同位素组成. 矿物岩石地球化学通报, 29(3): 250-255

马云麒, 肖应凯, 贺茂勇, 肖军, 沈权, 蒋生祥. 2011. 中国古生代腕足和珊瑚的硼同位素特征. 中国科学: 地球科学, 41(7): 984-999

彭章旷, 李海军, 柴小丽, 肖应凯, 张艳灵, 杨剑, 马云麒. 2017. 石盐中硼含量及其同位素的准确测定. 光谱学与光谱分析, 37(8): 2564-2568

戚玉菡. 2019. 橄榄岩、科马提岩和砖红壤的钒同位素地球化学. 博士学位论文. 合肥: 中国科学技术大学

齐天骄, 薛春纪, 朱祥坤. 2017. 新疆磁海铁矿床Fe-O-S同位素和元素地球化学示踪. 地质通报, 36(6): 1064-1076

祁昌实, 朱祥坤, 戴民汉, 唐索寒, 吴曼, 李志红, 李世珍, 李津. 2012. 海洋沉积物的铁和锌同位素测定. 地球化学, 41(3): 197-206

秦占杰, 张湘如, 彭章旷, 李庆宽, 马云麒, 樊启顺, 都永生, 王建萍, 山发寿. 2018. (硬)石膏矿物中硼的提取分离及硼同位素测定. 分析化学, 46(1): 48-54

邱啸飞, 卢山松, 谭娟娟, 杨红梅, 段瑞春. 2014. 铊同位素分析技术及其在地学中的应用. 地球科学——中国地质大学学报, 39(6): 705-715

孙剑, 朱祥坤, 陈岳龙, 房楠. 2012a. 白云鄂博地区相关地质单元的铁同位素特征及其对白云鄂博矿床成因的制约. 地质学报, 86(5): 819-828

孙剑, 房楠, 李世珍, 陈岳龙, 朱祥坤. 2012b. 白云鄂博矿床成因的Mg同位素制约. 岩石学报, 28(9): 2890-2902

孙剑, 朱祥坤, 陈岳龙. 2013. 碳酸盐矿物铁同位素测试的选择性溶解方法研究——以白云鄂博矿床赋矿白云岩为例. 岩矿测试, 32(1): 28-33

唐索寒, 朱祥坤, 李津, 闫斌. 2008. 地质样品铜、铁、锌同位素标准物质的研制. 岩石矿物学杂志, 27(4): 279-284

唐索寒, 朱祥坤, 赵新苗, 李津, 闫斌. 2011. 离子交换分离和多接收等离子体质谱法高精度测定钛同位素的组成. 分析化学, 39(12): 1830-1835

唐索寒, 闫斌, 李津. 2013a. 少量AG1-X4阴离子交换树脂分离地质标样中的铁及铁同位素测定. 地球化学, 42(1): 46-52

唐索寒, 李津, 王进辉, 潘辰旭. 2013b. 钛同位素标准溶液研制. 岩矿测试, 32(3): 377-382

唐索寒, 李津, 闫斌. 2014. 玄武岩钛同位素分析标准物质研制. 岩石矿物学杂志, 33(4): 779-784

唐索寒, 朱祥坤, 李津, 闫斌, 李世珍, 李志红, 王跃, 孙剑. 2016. 用于多接收器等离子体质谱测定的铁铜锌同位素标准溶液研制. 岩矿测试, 35(2): 127-133

唐索寒, 李津, 马健雄, 赵新苗, 朱祥坤. 2018. 地质样品中钛的化学分离及双稀释剂法钛同位素测定. 分析化学, 46(10): 1618-1627

王樵珊, 马金龙, 张乐, 韦刚健. 2018. 利用MC-ICPMS高精度测定地质样品的钛同位素组成的方法研究. 地球化学, 47(6): 604-611

王世霞, 朱祥坤. 2013. 广东湛江湖光岩地区玄武岩风化壳Fe同位素研究. 地质学报, 87(9): 1461-1468

王伟中, 张朝晖, 温汉捷, 朱传威, 张羽旭. 2020. 镉同位素在古环境重建中的应用: 以晚泥盆世弗拉期-法门期生物灭绝事件为例. 矿物岩石地球化学通报, 39(1): 87-95

王新富, 李波. 2018. 铜同位素组成在铜矿床中的变化规律. 地质科技情报, 37(3): 159-168

王跃, 朱祥坤, 程彦博, 李志红. 2013. 安徽新桥矿床矿相学与Fe同位素特征及其对矿床成因的制约. 吉林大学学报(地球科学版), 43(6): 1787-1798

王跃, 朱祥坤, 毛景文, 程彦博. 2014a. 安徽姑山矿浆型铁矿床Fe同位素初步研究. 矿床地质, 33(4): 689-696

王跃, 朱祥坤, 毛景文, 程彦博, 李志红. 2014b. 铜陵矿集区冬瓜山矿床斑岩-矽卡岩型矿床成矿作用过程中的Cu同位素地球化学行为初步研究. 地质学报, 88(12): 2413-2422

王泽洲, 刘盛遨, 李丹丹, 吕逸文, 吴松, 赵云. 2015. 铜同位素地球化学及研究新进展. 地学前缘, 22(5): 72-83

韦刚健, 马金龙, 刘颖, 徐义刚. 2015. 稳定Sr-Nd同位素体系及其传统放射成因锶钕同位素组成的影响. 地学前缘, 22(5): 136-142

温汉捷, 张羽旭, 樊海峰, 胡瑞忠. 2010. 华南下寒武统地层的Mo同位素组成特征及其古海洋环境意义. 科学通报, 55(2): 176-181

肖应凯, 王蕴慧, 曹海霞. 1982. 盐湖水中锂同位素丰度比值的质谱法测定. 质谱, (2): 40-45

闫斌, 朱祥坤, 唐索寒, 朱茂炎. 2010. 广西新元古代BIF的铁同位素特征及其地质意义. 地质学报, 84(7): 1080-1086

闫斌, 朱祥坤, 陈岳龙. 2011. 样品量的大小对铜锌同位素测定值的影响. 岩矿测试, 30(4): 400-405

闫斌, 朱祥坤, 张飞飞, 唐索寒. 2014. 峡东地区埃迪卡拉系黑色页岩的微量元素和Fe同位素特征及其古环境意义. 地质学报, 88(8): 1603-1615

晏雄, 蒋少涌, 魏海珍, 颜妍, 吴赫嫔, 濮巍. 2012. 硼硅酸盐矿物中硼的化学分离纯化与同位素测定方法. 分析化学, 40(11): 1654-1660

杨剑, 马云麒, 李兴意, 刘玉秀, 韩风清, 张艳灵, Hussain S A, 李海军. 2019. 乙二胺四乙酸二钠用于沉积物酸溶相中高精度硼同位素测定方法研究. 分析化学, 47(9): 1433-1439

尹润生, 冯新斌, Delphine F, 侍文芳, 赵志琦, 王静. 2010. 多接收电感耦合等离子体质谱法高精密度测定汞同位素组成. 分析化学, 38(7): 929-934

张安余, 张经, 张瑞峰, 薛云. 2015. 多接收电感耦合等离子体质谱仪测定稳定硅同位素. 分析化学, 43(9): 1353-1359

张晨蕾, 祝红丽, 刘峪菲, 刘芳, 张兆峰. 2017. 热电离质谱(TIMS)测定Ca同位素时Sr干扰影响的实验评价. 质谱学报, 38(5): 567-573

张飞飞, 彭乾云, 朱祥坤, 闫斌, 李津, 程龙, 斯小华. 2013. 湖北古城锰矿Fe同位素特征及其环境意义. 地质学报, 87(9): 1411-1418

张艳灵, 肖应凯, 马云麒, 刘志启, 罗重光, 贺茂勇, 刘玉秀. 2016. 三步离子交换方法用于粘土沉积物酸溶相中硼同位素测定. 分析化学, 44(5): 809-815

朱建明, Johnson T M, Clark S K, 朱祥坤. 2008. 氢化物发生-多接收杯电感耦合等离子体质谱同位素稀释法测定硒同位素. 分析化学, 36(10): 1385-1390

朱祥坤, 李志红, 唐索寒, 李延河. 2008a. 早前寒武纪硫铁矿矿床Fe同位素特征及其地质意义——以山东石河庄和河北大川为例. 岩石矿物学杂志, 27(5): 429–434

朱祥坤, 李志红, 赵新苗, 唐索寒, 何学贤, Belshaw N S. 2008b. 铁同位素的MC-ICP-MS测定方法与地质标准物质的铁同位素组成. 岩石矿物学杂志, 27(4): 263–272

朱祥坤, 王跃, 闫斌, 李津, 董爱国, 李志红, 孙剑. 2013. 非传统稳定同位素地球化学的创建与发展. 矿物岩石地球化学通报, 32(6): 651–688

朱志勇. 2020. Eu的分离提纯及其稳定同位素分析. 见: 第十二届全国同位素地质年代学与同位素地球化学学术讨论会. 武汉: 中国地质大学

An Y J, Huang F. 2014. A review of Mg isotope analytical methods by MC-ICP-MS. Journal of Earth Science, 25(5): 822–840

An Y J, Wu F, Xiang Y X, Nan X Y, Yu X, Yang J H, Yu H M, Xie L W, Huang F. 2014. High-precision Mg isotope analyses of low-Mg rocks by MC-ICP-MS. Chemical Geology, 390: 9–21

An Y J, Huang J X, Griffin W L, Liu C Z, Huang F. 2017. Isotopic composition of Mg and Fe in garnet peridotites from the Kaapvaal and Siberian Cratons. Geochimica et Cosmochimica Acta, 200: 167–185

An Y J, Li X, Zhang Z F. 2020. Barium isotopic compositions in thirty-four geological reference materials analysed by MC-ICP-MS. Geostandards and Geoanalytical Research, 44(1): 183–199

Anbar A D, Roe J E, Barling J, Nealson K H. 2000. Nonbiological fractionation of iron isotopes. Science, 288(5463): 126–128

Andersen M B, Stirling C H, Weyer S. 2017. Uranium isotope fractionation. Reviews in Mineralogy and Geochemistry, 82(1): 799–850

Bao Z A, Huang K J, Huang T Z, Shen B, Zong C L, Chen K Y, Yuan H L. 2019. Precise magnesium isotope analyses of high-K and low-Mg rocks by MC-ICP-MS. Journal of Analytical Atomic Spectrometry, 34(5): 940–953

Bermingham K R, Walker R J. 2017. The ruthenium isotopic composition of the oceanic mantle. Earth and Planetary Science Letters, 474: 466–473

Breton T, Quitté G. 2014. High-precision measurements of tungsten stable isotopes and application to earth sciences. Journal of Analytical Atomic Spectrometry, 29(12): 2284–2293

Cao Y H, Wang C Y, Huang F, Zhang Z F. 2019. Iron isotope systematics of the Panzhihua mafic layered intrusion associated with giant Fe-Ti oxide deposit in the Emeishan large igneous province, SW China. Journal of Geophysical Research: Solid Earth, 124(1): 358–375

Cao Z M, Siebert C, Hathorne E C, Dai M H, Frank M. 2016. Constraining the oceanic barium cycle with stable barium isotopes. Earth and Planetary Science Letters, 434: 1–9

Chang Y, Zhang J, Qu J Q, Xue Y. 2017. Precise selenium isotope measurement in seawater by carbon-containing hydride generation-Desolvation-MC-ICP-MS after thiol resin preconcentration. Chemical Geology, 471: 65–73

Chen A X, Li Y H, Chen Y, Yu H M, Huang F. 2020. Silicon isotope composition of subduction zone fluids as recorded by jadeitites from Myanmar. Contributions to Mineralogy and Petrology, 175(1): 6

Chen C F, Liu Y S, Feng L P, Foley S F, Zhou L, Ducea M N, Hu Z C. 2018. Calcium isotope evidence for subduction-enriched lithospheric mantle under the northern North China Craton. Geochimica et Cosmochimica Acta, 238: 55–67

Chen C F, Dai W, Wang Z C, Liu Y S, Li M, Becker H, Foley S F. 2019. Calcium isotope fractionation during magmatic processes in the upper mantle. Geochimica et Cosmochimica Acta, 249: 121–137

Chen C F, Ciazela J, Li W, Dai W, Wang Z C, Foley S F, Li M, Hu Z C, Liu Y S. 2020a. Calcium isotopic compositions of oceanic crust at various spreading rates. Geochimica et Cosmochimica Acta, 278: 272–288

Chen C F, Huang J X, Foley S F, Wang Z C, Moynier F, Liu Y S, Dai W, Li M. 2020b. Compositional and pressure controls on calcium and magnesium isotope fractionation in magmatic systems. Geochimica et Cosmochimica Acta, 290: 257–270

Chen H, Tian Z, Tuller-Ross B, Korotev R L, Wang K. 2019. High-precision potassium isotopic analysis by MC-ICP-MS: an inter-laboratory comparison and refined K atomic weight. Journal of Analytical Atomic Spectrometry, 34(1): 160–171

Chen J B, Gaillardet J, Louvat P, Huon S. 2009. Zn isotopes in the suspended load of the Seine River, France: isotopic variations and source determination. Geochimica et Cosmochimica Acta, 73(14): 4060–4076

Chen K Y, Yuan H L, Liang P, Bao Z A, Chen L. 2017. Improved nickel-corrected isotopic analysis of iron using high-resolution multi-collector inductively coupled plasma mass spectrometry. International Journal of Mass Spectrometry, 421: 196–203

Chen L M, Song X Y, Zhu X K, Zhang X Q, Yu S Y, Yi J N. 2014. Iron isotope fractionation during crystallization and sub-solidus re-equilibration: constraints from the Baima mafic layered intrusion, SW China. Chemical Geology, 380: 97–109

Chen X, Ling H F, Vance D, Shields-Zhou G A, Zhu M Y, Poulton S W, Och L M, Jiang S Y, Li D, Cremonese L, Archer C. 2015. Rise to modern levels of ocean oxygenation coincided with the Cambrian radiation of animals. Nature Communications, 6(1): 7142

Chen X F, Zhang L, Wei G J, Ma J L. 2016a. Matrix effects and mass bias caused by inorganic acids on boron isotope determination by multi-collector ICP-MS. Journal of Analytical Atomic Spectrometry, 31(12): 2410–2417

Chen X F, Deng W F, Zhu H L, Zhang Z F, Wei G J, McCulloch M T. 2016b. Assessment of coral $\delta^{44/40}$Ca as a paleoclimate proxy in the Great Barrier Reef of Australia. Chemical Geology, 435: 71–78

Chen X F, D'Olivo J P, Wei G J, McCulloch M. 2019. Anthropogenic ocean warming and acidification recorded by Sr/Ca, Li/Mg, δ^{11}B and B/Ca in Porites coral from the Kimberley region of northwestern Australia. Palaeogeography, Palaeoclimatology, Palaeoecology, 528: 50–59

Chen X M, Romaniello S J, Herrmann A D, Wasylenki L E, Anbar A D. 2016. Uranium isotope fractionation during coprecipitation with aragonite and calcite. Geochimica et Cosmochimica Acta, 188: 189–207

Chen Y X, Lu W N, He Y S, Schertl H P, Zheng Y F, Xiong J W, Zhou K. 2019. Tracking Fe mobility and Fe speciation in subduction zone fluids at the slab-mantle interface in a subduction channel: a tale of whiteschist from the western Alps. Geochimica et Cosmochimica Acta, 267: 1–16

Chen Y X, Demény A, Schertl H P, Zheng Y F, Huang F, Zhou K, Jin Q Z, Xia X P. 2020. Tracing subduction zone fluids with distinct Mg isotope compositions: insights from high-pressure metasomatic rocks (leucophyllites) from the eastern Alps. Geochimica et Cosmochimica Acta, 271: 154–178

Cheng M, Li C, Zhou L, Algeo T J, Zhang F F, Romaniello S, Jin C S, Lei L D, Feng L J, Jiang S Y. 2016. Marine Mo biogeochemistry in the context of dynamically euxinic mid-depth waters: a case study of the lower Cambrian Niutitang shales, South China. Geochimica et Cosmochimica Acta, 183: 79–93

Chu Z Y, Xu J J, Li C F, Yang Y H, Guo J H. 2020. A chromatographic method for separation of tungsten (W) from silicate samples for High-precision isotope analysis using negative thermal ionization mass spectrometry. Analytical Chemistry, 92(17): 11987–11993

Creech J B, Baker J A, Handler M R, Bizzarro M. 2014. Platinum stable isotope analysis of geological standard reference materials by double-spike MC-ICPMS. Chemical Geology, 363: 293–300

Dai L Q, Zhao K, Zhao Z F, Zheng Y F, Fang W, Zha X P, An Y J. 2020. Magnesium-carbon isotopes trace carbon recycling in continental subduction zone. Lithos, 376-377: 105774

Dai W, Wang Z C, Liu Y S, Chen C F, Zong K Q, Zhou L, Zhang G L, Li M, Moynier F, Hu Z C. 2020. Calcium isotope compositions of mantle pyroxenites. Geochimica et Cosmochimica Acta, 270: 144–159

de Carvalho G G A, Oliveira P V, Yang L. 2017. Determination of europium isotope ratiosin natural waters by MC-ICP-MS. Journal of Analytical Atomic Spectrometry, 32(5): 987–995

Deng G X, Kang J T, Nan X Y, Li Y L, Guo J H, Ding X, Huang F. 2021. Barium isotope evidence for crystal-melt separation in granitic magma reservoirs. Geochimica et Cosmochimica Acta, 292: 115–129

Ding T P, Wan D, Bai R, Zhang Z, Shen Y, Meng R. 2005. Silicon isotope abundance ratios and atomic weights of NBS-28 and other reference materials. Geochimica et Cosmochimica Acta, 69(23): 5487–5494

Ding T P, Gao J F, Tian S H, Wang H B, Li M. 2011. Silicon isotopic composition of dissolved silicon and suspended particulate matter in the Yellow River, China, with implications for the global silicon cycle. Geochimica et Cosmochimica Acta, 75(21): 6672–6689

Ding T P, Gao J F, Tian S H, Fan C F, Zhao Y, Wan D F, Zhou J X. 2017. The δ^{30}Si-peak value discovered in middle Proterozoic chert and its implication for environmental variations in the ancient ocean. Scientific Reports, 7(1): 44000

Ding X, Ripley E M, Wang W Z, Li C H, Huang F. 2019. Iron isotope fractionation during sulfide liquid segregation and crystallization at the Lengshuiqing Ni-Cu magmatic sulfide deposit, SW China. Geochimica et Cosmochimica Acta, 261: 327–341

Ding X, Helz R T, Qi Y H, Huang F. 2020. Vanadium isotope fractionation during differentiation of Kilauea Iki lava lake, Hawaii. Geochimica et Cosmochimica Acta, 289: 114–129

Dong A G, Zhu X K, Li S Z, Kendall B, Wang Y, Gao Z F. 2016. Genesis of a giant Paleoproterozoic strata-bound magnesite deposit: constraints from Mg isotopes. Precambrian Research, 281: 673–683

Du D H, Wang X L, Yang T, Chen X, Li J Y, Li W Q. 2017. Origin of heavy Fe isotope compositions in high-silica igneous rocks: a rhyolite perspective. Geochimica et Cosmochimica Acta, 218: 58–72

Ek M, Hunt A C, Schönbächler M. 2017. A new method for High-precision palladium isotope analyses of iron meteorites and other metal samples. Journal of Analytical Atomic Spectrometry, 32(3): 647–656

Fan H F, Zhu X K, Wen H J, Yan B, Li J, Feng L J. 2014. Oxygenation of Ediacaran Ocean recorded by iron isotopes. Geochimica et Cosmochimica Acta, 140: 80–94

Fan H F, Wen H J, Han T, Zhu X K, Feng L J, Chang H J. 2018a. Oceanic redox condition during the late Ediacaran (551–541 Ma), South China. Geochimica et Cosmochimica Acta, 238: 343–356

Fan H F, Wen H J, Xiao C Y, Zhou T, Cloquet C, Zhu X K. 2018b. Zinc geochemical cycling in a phosphorus-rich ocean during the Early Ediacaran. Journal of Geophysical Research: Oceans, 123(8): 5248–5260

Fan H F, Nielsen S G, Owens J D, Auro M, Shu Y C, Hardisty D S, Horner T J, Bowman C N, Young S A, Wen H J. 2020. Constraining oceanic oxygenation during the Shuram excursion in South China using thallium isotopes. Geobiology, 18(3): 348–365

Fan J J, Li J, Wang Q, Zhang L, Zhang J, Zeng X L, Ma L, Wang Z L. 2020a. High-precision molybdenum isotope analysis of low-Mo igneous rock samples by MC-ICP-MS. Chemical Geology, 545: 119648

Fan J J, Wang Q, Li J, Wei G J, Wyman D, Zhao Z H, Liu Y, Ma J L, Zhang L, Wang Z L. 2020b. Molybdenum and boron isotopic compositions of porphyry Cu mineralization-related adakitic rocks in central-eastern China: new insights into their petrogenesis and crust-mantle interaction. Journal of Geophysical Research: Solid Earth, 125(12): e2020JB020474

Fan J J, Wang Q, Li J, Wei G J, Ma J L, Ma L, Li Q W, Jiang Z Q, Zhang L, Wang Z L, Zhang L. 2021. Boron and molybdenum isotopic fractionation during crustal anatexis: constraints from the Conadong leucogranites in the Himalayan Block, South Tibet. Geochimica et Cosmochimica Acta, 297: 120–142

Fehr M A, Rehkämper M, Halliday A N. 2004. Application of MC-ICPMS to the precise determination of tellurium isotope compositions in chondrites, iron meteorites and sulfides. International Journal of Mass Spectrometry, 232(1): 83–94

Fehr M A, Hammond S J, Parkinson I J. 2018. Tellurium stable isotope fractionation in chondritic meteorites and some terrestrial samples. Geochimica et Cosmochimica Acta, 222: 17–33

Feng C Q, Qin T, Huang S C, Wu Z Q, Huang F. 2014. First-principles investigations of equilibrium calcium isotope fractionation between clinopyroxene and Ca-doped orthopyroxene. Geochimica et Cosmochimica Acta, 143: 132–142

Feng J L, Pei L L, Zhu X K, Ju J T, Gao S P. 2018. Absolute accumulation and isotope fractionation of Si and Fe during dolomite weathering and terra rossa formation. Chemical Geology, 496: 43–56

Feng L P, Zhou L, Yang L, Zhang W, Wang Q, Tong S Y, Hu Z C. 2018. A rapid and simple single-stage method for Ca separation from geological and biological samples for isotopic analysis by MC-ICP-MS. Journal of Analytical Atomic Spectrometry, 33(3): 413–421

Feng L P, Zhou L, Liu J H, Hu Z C, Liu Y S. 2019. Determination of gallium isotopic compositions in reference materials. Geostandards and Geoanalytical Research, 43(4): 701–714

Feng L P, Hu W F, Jiao Y, Zhou L, Zhang W, Hu Z C, Liu Y S. 2020a. High-precision stable zirconium isotope ratio measurements by double spike thermal ionization mass spectrometry. Journal of Analytical Atomic Spectrometry, 35(4): 736–745

Feng L P, Zhou L, Hu W F, Zhang W, Li B C, Liu Y S, Hu Z C, Yang L. 2020b. A simple single-stage extraction method for Mo separation from geological samples for isotopic analysis by MC-ICP-MS. Journal of Analytical Atomic Spectrometry, 35(1): 145–154

Fischer-Gödde M, Elfers B M, Münker C, Szilas K, Maier W D, Messling N, Morishita T, van Kranendonk M, Smithies H. 2020. Ruthenium isotope vestige of Earth's pre-late-veneer mantle preserved in Archaean rocks. Nature, 579(7798): 240–244

Fornadel A P, Spry P G, Haghnegahdar M A, Schauble E A, Jackson S E, Mills S J. 2017. Stable Te isotope fractionation in tellurium-bearing minerals from precious metal hydrothermal ore deposits. Geochimica et Cosmochimica Acta, 202: 215–230

Gao B, Liu Y, Sun K, Liang X R, Peng P A, Sheng G Y, Fu J M. 2008. Precise determination of cadmium and lead isotopic compositions in river sediments. Analytica Chimica Acta, 612(1): 114–120

Gao P, Li S J, Lash G G, He Z L, Xiao X M, Zhang D W, Hao Y Q. 2020. Silicification and Si cycling in a silica-rich ocean during the Ediacaran-Cambrian transition. Chemical Geology, 552: 119787

Gao T, Ke S, Wang S J, Li F B, Liu C S, Lei J, Liao C Z, Fei W. 2018. Contrasting Mg isotopic compositions between Fe-Mn nodules and surrounding soils: Accumulation of light Mg isotopes by Mg-depleted clay minerals and Fe oxides. Geochimica et Cosmochimica Acta, 237: 205–222

Gao T, Ke S, Li R Y, Meng X N, He Y S, Liu C S, Wang Y, Li Z J, Zhu J M. 2019. High-precision magnesium isotope analysis of geological and environmental reference materials by multiple-collector inductively coupled plasma mass spectrometry. Rapid Communications in Mass Spectrometry, 33(8): 767–777

Gao X Y, Wang L, Chen Y X, Zheng Y F, Chen R X, Huang F, Zhang Q Q, Ji M, Meng Z Y. 2019. Geochemical evidence from coesite-bearing jadeite quartzites for large-scale flow of metamorphic fluids in a continental subduction channel. Geochimica et Cosmochimica Acta, 265: 354–370

Gao Y J, Casey J F. 2012. Lithium isotope composition of ultramafic Geological Reference Materials JP-1 and DTS-2. Geostandards and Geoanalytical Research, 36(1): 75–81

Gao Y J, Ling W L, Qiu X F, Chen Z W, Lu S S, Bai X, Bai X J, Zhang J B, Yang H M, Duan R C. 2016. Decoupled Ce-Nd isotopic systematics of the neoproterozoic huangling intrusive complex and its geological significance, eastern Three Gorges, South China. Journal of Earth Science, 27(5): 864–873

Gao Z F, Zhu X K, Sun J, Luo Z H, Bao C, Tang C, Ma J X. 2018. Spatial evolution of Zn-Fe-Pb isotopes of sphalerite within a single ore body: a case study from the Dongshengmiao ore deposit, Inner Mongolia, China. Mineralium Deposita, 53(1): 55–65

Gong H M, Guo P Y, Chen S, Duan M, Sun P, Wang X H, Niu Y L. 2020. A re-assessment of nickel-doping method in iron isotope analysis on rock samples using multi-collector inductively coupled plasma mass spectrometry. Acta Geochimica, 39(3): 355–364

Gong Y Z, Xia Y, Huang F, Yu H M. 2017. Average iron isotopic compositions of the upper continental crust: constrained by loess from the Chinese Loess Plateau. Acta Geochimica, 36(2): 125–131

Gong Y Z, Zeng Z, Zhou C, Nan X Y, Yu H M, Lu Y, Li W Y, Gou W X, Cheng W H, Huang F. 2019. Barium isotope fractionation in latosol developed from strongly weathered basalt. Science of the Total Environment, 687: 1295–1304

Gong Y Z, Zeng Z, Cheng W H, Lu Y, Zhang L L, Yu H M, Huang F. 2020. Barium isotope fractionation during strong weathering of basalt in a tropical climate. Environment International, 143: 105896

Gou L F, Jin Z D, Galy A, Sun H, Deng L, Xu Y. 2019a. Effects of cone combinations on accurate and precise Mg-isotopic determination using multi-collector inductively coupled plasma mass spectrometry. Rapid Communications in Mass Spectrometry, 33(4): 351–360

Gou L F, Jin Z D, Pogge von Strandmann P A E, Li G, Qu Y X, Jun X, Deng L, Galy A. 2019b. Li isotopes in the middle Yellow River: seasonal variability, sources and fractionation. Geochimica et Cosmochimica Acta, 248: 88–108

Gou L F, Jin Z D, Galy A, Gong Y Z, Nan X Y, Jin C Y, Wang X D, Bouchez J, Cai H M, Chen J B, Yu H M, Huang F. 2020. Seasonal riverine barium isotopic variation in the middle Yellow River: sources and fractionation. Earth and Planetary Science Letters, 531: 115990

Guan Q Y, Sun Y L, Zhang Z F, Liu X M, An Y J, Liu F, Zhao S Q. 2020. Determination of $\delta^{44/40}$Ca and $\delta^{56/54}$Fe in geological materials combined with a simplified method for their separation using a single TODGA resin column. Geostandards and Geoanalytical Research, 44(4): 669–683

Guo H, Li W Y, Nan X, Huang F. 2020. Experimental evidence for light Ba isotopes favouring aqueous fluids over silicate melts. Geochemical Perspectives Letters, 16: 6–11

Guo Q, Wei H Z, Jiang S Y, Hohl S, Lin Y B, Wang Y J, Li Y C. 2017. Matrix effects originating from coexisting minerals and accurate determination of stable silver isotopes in silver deposits. Analytical Chemistry, 89(24): 13634–13641

He J, Meija J, Hou X D, Zheng C B, Mester Z, Yang L. 2020. Determination of the isotopic composition of lutetium using MC-ICPMS. Analytical and

Bioanalytical Chemistry, 412(24): 6257-6263

He M H, Xia X P, Huang X L, Ma J L, Zou J Q, Yang Q, Yang F, Zhang Y Q, Yang Y N, Wei G J. 2020. Rapid determination of the original boron isotopic composition from altered basaltic glass by $in\ situ$ secondary ion mass spectrometry. Journal of Analytical Atomic Spectrometry, 35(2): 238-245

He M Y, Xiao Y K, Ma Y Q, Jin Z D, Xiao J. 2011. Effective elimination of organic matter interference in boron isotopic analysis by thermal ionization mass spectrometry of coral/foraminifera: Micro-sublimation technology combined with ion exchange. Rapid Communications in Mass Spectrometry, 25(6): 743-749

He M Y, Xiao Y K, Jin Z D, Ma Y Q, Xiao J, Zhang Y L, Luo C G, Zhang F. 2013. Accurate and precise determination of boron isotopic ratios at low concentration by positive thermal ionization mass spectrometry using static multicollection of $Cs_2BO_2^+$ ions. Analytical Chemistry, 85(13): 6248-6253

He M Y, Jin Z D, Lu H, Ren T X. 2015. Efficient separation of boron using solid-phase extraction for boron isotope analysis by MC-ICP-MS. Analytical Methods, 7(24): 10322-10327

He M Y, Jin Z D, Lu H, Deng L, Luo C G. 2016. The different cones combination enhanced sensitivity on MC-ICP-MS: the results from boron isotope analysis. International Journal of Mass Spectrometry, 408: 33-37

He M Y, Deng L, Lu H, Jin Z D. 2019. Elimination of the boron memory effect for rapid and accurate boron isotope analysis by MC-ICP-MS using NaF. Journal of Analytical Atomic Spectrometry, 34(5): 1026-1032

He X Y, Ma J L, Wei G J, Zhang L, Wang Z B, Wang Q S. 2020. A new procedure for titanium separation in geological samples for $^{49}Ti/^{47}Ti$ ratio measurement by MC-ICP-MS. Journal of Analytical Atomic Spectrometry, 35(1): 100-106

He Y S, Ke S, Teng F Z, Wang T T, Wu H J, Lu Y H, Li S G. 2015. High-precision iron isotope analysis of Geological Reference Materials by high-resolution MC-ICP-MS. Geostandards and Geoanalytical Research, 39(3): 341-356

He Y S, Wang Y, Zhu C W, Huang S C, Li S G. 2017. Mass-independent and mass-dependent Ca isotopic compositions of thirteen geological reference materials measured by thermal ionisation mass spectrometry. Geostandards and Geoanalytical Research, 41(2): 283-302

He Y S, Meng X N, Ke S, Wu H J, Zhu C W, Teng F Z, Hoefs J, Huang J, Yang W, Xu L J, Hou Z H, Ren Z Y, Li S G. 2019. A nephelinitic component with unusual $\delta^{56}Fe$ in Cenozoic basalts from eastern China and its implications for deep oxygen cycle. Earth and Planetary Science Letters, 512: 175-183

He Z W, Zhang X C, Deng X D, Hu H, Li Y, Yu H M, Archer C, Li J W, Huang F. 2020. The behavior of Fe and S isotopes in porphyry copper systems: constraints from the Tongshankou Cu-Mo deposit, eastern China. Geochimica et Cosmochimica Acta, 270: 61-83

Hopp T, Fischer-Gödde M, Kleine T. 2016. Ruthenium stable isotope measurements by double spike MC-ICPMS. Journal of Analytical Atomic Spectrometry, 31(7): 1515-1526

Hou Q H, Zhou L, Gao S, Zhang T, Feng L P, Yang L. 2016. Use of Ga for mass bias correction for the accurate determination of copper isotope ratio in the NIST SRM 3114 Cu standard and geological samples by MC-ICPMS. Journal of Analytical Atomic Spectrometry, 31(1): 280-287

Hu X F, Zhao J L, Zhang P F, Xue Y, An B N, Huang F, Yu H M, Zhang G L, Liu X J. 2019. Fe isotopic composition of the Quaternary red clay in subtropical Southeast China: redoxic Fe mobility and its paleoenvironmental implications. Chemical Geology, 524: 356-367

Hu Y, Chen X Y, Xu Y K, Teng F Z. 2018. High-precision analysis of potassium isotopes by HR-MC-ICPMS. Chemical Geology, 493: 100-108

Huang F, Glessner J, Ianno A, Lundstrom C, Zhang Z F. 2009. Magnesium isotopic composition of igneous rock standards measured by MC-ICP-MS. Chemical Geology, 268(1-2): 15-23

Huang F, Wu Z Q, Huang S C, Wu F. 2014. First-principles calculations of equilibrium silicon isotope fractionation among mantle minerals. Geochimica et Cosmochimica Acta, 140: 509-520

Huang F, Zhou C, Wang W Z, Kang J T, Wu Z Q. 2019. First-principles calculations of equilibrium Ca isotope fractionation: implications for oldhamite formation and evolution of lunar magma ocean. Earth and Planetary Science Letters, 510: 153-160

Huang J, Xiao Y L. 2016. Mg-Sr isotopes of low-$\delta^{26}Mg$ basalts tracing recycled carbonate species: implication for the initial melting depth of the carbonated mantle in eastern China. International Geology Review, 58(11): 1350-1362

Huang J, Li S G, Xiao Y L, Ke S, Li W Y, Tian Y. 2015a. Origin of low $\delta^{26}Mg$ Cenozoic basalts from South China Block and their geodynamic implications. Geochimica et Cosmochimica Acta, 164: 298-317

Huang J, Ke S, Gao Y J, Xiao Y L, Li S G. 2015b. Magnesium isotopic compositions of altered oceanic basalts and gabbros from IODP site 1256 at the East Pacific Rise. Lithos, 231: 53-61

Huang J, Liu S A, Gao Y J, Xiao Y L, Chen S. 2016a. Copper and zinc isotope systematics of altered oceanic crust at IODP Site 1256 in the eastern equatorial Pacific. Journal of Geophysical Research: Solid Earth, 121(10): 7086-7100

Huang J, Liu S A, Wörner G, Yu H M, Xiao Y L. 2016b. Copper isotope behavior during extreme magma differentiation and degassing: a case study on Laacher See phonolite tephra (East Eifel, Germany). Contributions to Mineralogy and Petrology, 171(8): 76

Huang J, Huang F, Wang Z C, Zhang X C, Yu H M. 2017. Copper isotope fractionation during partial melting and melt percolation in the upper mantle: evidence from massif peridotites in Ivrea-Verbano zone, Italian Alps. Geochimica et Cosmochimica Acta, 211: 48-63

Huang J, Zhang X C, Chen S, Tang L M, Wörner G, Yu H M, Huang F. 2018a. Zinc isotopic systematics of Kamchatka-Aleutian arc magmas controlled by mantle melting. Geochimica et Cosmochimica Acta, 238: 85-101

Huang J, Chen S, Zhang X C, Huang F. 2018b. Effects of melt percolation on Zn isotope heterogeneity in the mantle: constraints from peridotite massifs in Ivrea-Verbano zone, Italian Alps. Journal of Geophysical Research: Solid Earth, 123(4): 2706-2722

Huang J, Liu J, Zhang Y N, Chang H J, Shen Y A, Huang F, Qin L P, 2018c. Cr isotopic composition of the Laobao cherts during the Ediacaran-Cambrian transition in South China. Chemical Geology, 482: 121-130

Huang J, Ackerman L, Zhang X C, Huang F. 2019. Mantle Zn isotopic heterogeneity caused by melt-rock reaction: evidence from Fe-rich peridotites and pyroxenites from the bohemian massif, central Europe. Journal of Geophysical Research: Solid Earth, 124(4): 3588-3604

Huang J, Guo S, Jin Q Z, Huang F. 2020. Iron and magnesium isotopic compositions of subduction-zone fluids and implications for arc volcanism. Geochimica et Cosmochimica Acta, 278: 376-391

Huang K J, Teng F Z, Wei G J, Ma J L, Bao Z Y. 2012. Adsorption- and desorption-controlled magnesium isotope fractionation during extreme weathering of basalt in Hainan Island, China. Earth and Planetary Science Letters, 359-360, 73-83

Huang K J, Teng F Z, Elsenouy A, Li W Y, Bao Z Y. 2013. Magnesium isotopic variations in loess: Origins and implications. Earth and Planetary Science Letters, 374: 60-70

Huang K J, Shen B, Lang X G, Tang W B, Peng Y, Ke S, Kaufman A J, Ma H R, Li F B. 2015. Magnesium isotopic compositions of the Mesoproterozoic dolostones: implications for Mg isotopic systematics of marine carbonates. Geochimica et Cosmochimica Acta, 164: 333-351

Huang K J, Teng F Z, Shen B, Xiao S H, Lang X G, Ma H R, Fu Y, Peng Y B. 2016. Episode of intense chemical weathering during the termination of the 635 Ma Marinoan glaciation. Proceedings of the National Academy of Sciences of the United States of America, 113(52): 14904-14909

Huang K J, Teng F Z, Plank T, Staudigel H, Hu Y, Bao Z Y. 2018. Magnesium isotopic composition of altered oceanic crust and the global Mg cycle. Geochimica et Cosmochimica Acta, 238: 357-373

Huang L M, Jia X X, Zhang G L, Thompson A, Huang F, Shao M A, Chen L M. 2018a. Variations and controls of iron oxides and isotope compositions during paddy soil evolution over a millennial time scale. Chemical Geology, 476: 340-351

Huang L M, Shao M A, Huang F, Zhang G L. 2018b. Effects of human activities on pedogenesis and iron dynamics in paddy soils developed on Quaternary red clays. CATENA, 166: 78-88

Huang S Y, Jiang R G, Song Q Y, Zhang Y B, Huang Q, Su B H, Chen Y J, Huo Y L, Lin H. 2020. Study of mercury transport and transformation in mangrove forests using stable mercury isotopes. Science of the Total Environment, 704: 135928

Hunt A C, Ek M, Schönbächler M. 2017. Separation of platinum from palladium and iridium in iron meteorites and accurate High-precision determination of platinum isotopes by multi-collector ICP-MS. Geostandards and Geoanalytical Research, 41(4): 633-647

Inglis E C, Creech J B, Deng Z B, Moynier F. 2018. High-precision zirconium stable isotope measurements of geological reference materials as measured by double-spike MC-ICPMS. Chemical Geology, 493: 544-552

Jin Q Z, Huang J, Liu S C, Huang F. 2020. Magnesium and zinc isotope evidence for recycled sediments and oceanic crust in the mantle sources of continental basalts from eastern China. Lithos, 370-371: 105627

Johnson C M, Beard B L, Albarède F. 2004. Geochemistry of non-traditional stable isotopes. Washington: Mineralogical Society of America

Kang J T, Zhu H L, Liu Y F, Liu F, Wu F, Hao Y T, Zhi X C, Zhang Z F, Huang F. 2015. Calcium isotopic composition of mantle xenoliths and minerals from eastern China. Geochimica et Cosmochimica Acta, 174: 335-344

Kang J T, Ionov D A, Liu F, Zhang C L, Golovin A V, Qin L P, Zhang Z F, Huang F. 2017. Calcium isotopic fractionation in mantle peridotites by melting and metasomatism and Ca isotope composition of the Bulk Silicate Earth. Earth and Planetary Science Letters, 474: 128-137

Kang J T, Ionov D A, Zhu H L, Liu F, Zhang Z F, Liu Z, Huang F. 2019. Calcium isotope sources and fractionation during melt-rock interaction in the lithospheric mantle: evidence from pyroxenites, wehrlites, and eclogites. Chemical Geology, 524: 272-288

Kang J T, Zhou C, Huang J Y, Hao Y T, Liu F, Zhu H L, Zhang Z F, Huang F. 2020. Diffusion-driven Ca-Fe isotope fractionations in the upper mantle: implications for mantle cooling and melt infiltration. Geochimica et Cosmochimica Acta, 290: 41-58

Kurzweil F, Münker C, Tusch J, Schoenberg R. 2018. Accurate stable tungsten isotope measurements of natural samples using a ^{180}W-^{183}W double-spike. Chemical Geology, 476: 407-417

Lee S G, Tanaka T. 2019. Determination of Eu isotopic ratio by multi-collector inductively coupled plasma mass spectrometry using a Sm internal standard. Spectrochimica Acta Part B: Atomic Spectroscopy, 156: 42-50

Li C F, Feng L J, Wang X C, Chu Z Y, Guo J H, Wilde S A. 2016. Precise measurement of Cr isotope ratios using a highly sensitive Nb_2O_5 emitter by thermal ionization mass spectrometry and an improved procedure for separating Cr from geological materials. Journal of Analytical Atomic Spectrometry, 31(12): 2375-2383

Li C F, Feng L J, Wang X C, Wilde S A, Chu Z Y, Guo J H. 2017. A low-blank two-column chromatography separation strategy based on a $KMnO_4$ oxidizing reagent for Cr isotope determination in micro-silicate samples by thermal ionization mass spectrometry. Journal of Analytical Atomic Spectrometry, 32(10): 1938-1945

Li C F, Chu Z Y, Wang X C, Feng L J, Guo J H. 2020. A highly sensitive zirconium hydrogen phosphate emitter for Ni isotope determination using thermal ionization mass spectrometry. Atomic Spectroscopy, 41(6): 249-255

Li D D, Li M L, Liu W R, Qin Z Z, Liu S A. 2018. Cadmium isotope ratios of standard solutions and geological reference materials measured by MC-ICP-MS. Geostandards and Geoanalytical Research, 42(4): 593-605

Li F B, Teng F Z, Chen J T, Huang K J, Wang S J, Lang X G, Ma H R, Peng Y B, Shen B. 2016. Constraining ribbon rock dolomitization by Mg isotopes: Implications for the 'dolomite problem'. Chemical Geology, 445: 208-220

Li H Y, Zhou Z, Ryan J G, Wei G J, Xu Y G. 2016. Boron isotopes reveal multiple metasomatic events in the mantle beneath the eastern North China Craton. Geochimica et Cosmochimica Acta, 194: 77-90

Li H Y, Li J, Ryan J G, Li X, Zhao R P, Ma L, Xu Y G. 2019. Molybdenum and boron isotope evidence for fluid-fluxed melting of intraplate upper mantle beneath the eastern North China Craton. Earth and Planetary Science Letters, 520: 105-114

Li J, Liang X R, Zhong L F, Wang X C, Ren Z Y, Sun S L, Zhang Z F, Xu J F. 2014. Measurement of the isotopic composition of molybdenum in geological samples by MC-ICP-MS using a novel chromatographic extraction technique. Geostandards and Geoanalytical Research, 38(3): 345354

Li J, Zhu X K, Tang S H, Zhang K. 2016. High-precision measurement of molybdenum isotopic compositions of selected geochemical reference materials. Geostandards and Geoanalytical Research, 40(3): 405-415

Li J, Hao C G, Wang Z H, Dong L, Wang Y W, Huang K J, Lang X G, Huang T Z, Yuan H L, Zhou C M, Shen B. 2020. Continental weathering intensity during the termination of the Marinoan snowball Earth: Mg isotope evidence from the basal Doushantuo cap carbonate in South China. Palaeogeography, Palaeoclimatology, Palaeoecology, 552: 109774

Li L B, Zhang F, Jin Z D, Xiao J, Gou L F, Xu Y. 2020. Riverine Mg isotopes response to glacial weathering within the muztag catchment of the eastern Pamir Plateau. Applied Geochemistry, 118: 104626

Li M, Chai X, Gao S, Hu Z, Liu Y, Chen H. 2016. Improvements on high-precision Eu isotope analysis by MC-ICP-MS. Goldschmidt Conference, A1759

Li M, He Y S, Kang J T, Yang X Y, He Z W, Yu H M, Huang F. 2017. Why was iron lost without significant isotope fractionation during the lateritic process in tropical environments? Geoderma, 290: 1-9

Li M, Lei Y, Feng L P, Wang Z C, Belshaw N S, Hu Z C, Liu Y S, Zhou L, Chen H H, Chai X N. 2018. High-precision Ca isotopic measurement using a large geometry high resolution MC-ICP-MS with a dummy bucket. Journal of Analytical Atomic Spectrometry, 33(10): 1707-1719

Li Q W, Zhao J H, Wang Q, Zhang Z F, An Y J, He Y T. 2020. Iron isotope fractionation in hydrous basaltic magmas in deep crustal hot zones. Geochimica et Cosmochimica Acta, 279: 29-44

Li S G, Wang Y. 2018. Formation time of the big mantle wedge beneath eastern China and a new lithospheric thinning mechanism of the North China Craton—geodynamic effects of deep recycled carbon. Science China Earth Sciences, 61(7): 853-868

Li S G, Yang W, Ke S, Meng X N, Tian H C, Xu L J, He Y S, Huang J, Wang X C, Xia Q K, Sun W D, Yang X Y, Ren Z Y, Wei H Q, Liu Y S, Meng F C, Yan J. 2017. Deep carbon cycles constrained by a large-scale mantle Mg isotope anomaly in eastern China. National Science Review, 4(1): 111-120

Li S L, Li W Q, Beard B L, Raymo M E, Wang X M, Chen Y, Chen J. 2019. K isotopes as a tracer for continental weathering and geological K cycling. Proceedings of the National Academy of Sciences of the United States of America, 116(18): 8740-8745

Li S Z, Zhu X K, Wu L H, Luo Y M. 2016. Cu isotopic compositions in *Elsholtzia splendens*: influence of soil condition and growth period on Cu isotopic fractionation in plant tissue. Chemical Geology, 444: 49-58

Li S Z, Zhu X K, Wu L H, Luo Y M. 2020. Zinc, iron, and copper isotopic fractionation in *Elsholtzia splendens* Nakai: a study of elemental uptake and (re)translocation mechanisms. Journal of Asian Earth Sciences, 192: 104227

Li W Q, Beard B L, Li S L. 2016. Precise measurement of stable potassium isotope ratios using a single focusing collision cell multi-collector ICP-MS. Journal of Analytical Atomic Spectrometry, 31(4): 1023-1029

Li W S, Liu X M, Godfrey L V. 2019. Optimisation of lithium chromatography for isotopic analysis in geological reference materials by MC-ICP-MS. Geostandards and Geoanalytical Research, 43(2): 261-276

Li W Y, Teng F Z, Ke S, Rudnick R L, Gao S, Wu F Y, Chappell B W. 2010. Heterogeneous magnesium isotopic composition of the upper continental crust. Geochimica et Cosmochimica Acta, 74(23): 6867-6884

Li W Y, Teng F Z, Xiao Y L, Huang J. 2011. High-temperature inter-mineral magnesium isotope fractionation in eclogite from the Dabie orogen, China. Earth and Planetary Science Letters, 304(1-2): 224-230

Li W Y, Teng F Z, Xiao Y L, Gu H O, Zha X P, Huang J. 2016. Empirical calibration of the clinopyroxene-garnet magnesium isotope geothermometer and implications. Contributions to Mineralogy and Petrology, 171(7): 61

Li W Y, Yu H M, Xu J, Halama R, Bell K, Nan X Y, Huang F. 2020. Barium isotopic composition of the mantle: constraints from carbonatites. Geochimica et Cosmochimica Acta, 278: 235-243

Li X, Li H Y, Ryan J G, Wei G J, Zhang L, Li N B, Huang X L, Xu Y G. 2019. High-precision measurement of B isotopes on low-boron oceanic volcanic rock samples via MC-ICPMS: evaluating acid leaching effects on boron isotope compositions, and B isotopic variability in depleted oceanic basalts. Chemical Geology, 505: 76-85

Li X H, Li Q L, Liu Y, Tang G Q. 2011. Further characterization of M257 zircon standard: a working reference for SIMS analysis of Li isotopes. Journal of Analytical Atomic Spectrometry, 26(2): 352-358

Li X Q, Han G L, Zhang Q, Miao Z. 2020. An optimal separation method for High-precision K isotope analysis by using MC-ICP-MS with a dummy bucket. Journal of Analytical Atomic Spectrometry, 35(7): 1330-1339

Li Y H, Wang W Z, Zhou C, Huang F. 2019. First-principles calculations of equilibrium silicon isotope fractionation in metamorphic silicate minerals. Solid Earth Sciences, 4(4): 142-149

Li Y H, Yu H M, Gu X F, Guo S, Huang F. 2020. Silicon isotopic fractionation during metamorphic fluid activities: constraints from eclogites and ultrahigh-pressure veins in the Dabie orogen, China. Chemical Geology, 540: 119550

Lin J, Liu Y S, Hu Z C, Yang L, Chen K, Chen H H, Zong K Q, Gao S. 2016. Accurate determination of lithium isotope ratios by MC-ICP-MS without strict matrix-matching by using a novel washing method. Journal of Analytical Atomic Spectrometry, 31(2): 390-397

Lin J, Liu Y S, Tong X R, Zhu L Y, Zhang W, Hu Z C. 2017. Improved *in situ* Li isotopic ratio analysis of silicates by optimizing signal intensity, isotopic ratio stability and intensity matching using ns-LA-MC-ICP-MS. Journal of Analytical Atomic Spectrometry, 32(4): 834–842

Lin J, Liu Y S, Hu Z C, Chen W, Zhang L, Chen H H. 2019a. Accurate measurement of lithium isotopes in eleven carbonate reference materials by MC-ICP-MS with soft extraction mode and $10^{12}\Omega$ resistor high-gain Faraday amplifiers. Geostandards and Geoanalytical Research, 43(2): 277–289

Lin J, Liu Y S, Hu Z C, Chen W, Zhang C X, Zhao K D, Jin X Y. 2019b. Accurate analysis of Li isotopes in tourmalines by LA-MC-ICP-MS under "wet" conditions with non-matrix-matched calibration. Journal of Analytical Atomic Spectrometry, 34(6): 1145–1153

Lin L, Hu Z C, Yang L, Zhang W, Liu Y S, Gao S, Hu S H. 2014. Determination of boron isotope compositions of geological materials by laser ablation MC-ICP-MS using newly designed high sensitivity skimmer and sample cones. Chemical Geology, 386: 22–30

Lin Y B, Wei H Z, Jiang S Y, Hohl S, Lei H L, Liu X, Dong G. 2020. Accurate determination of barium isotopic compositions in sequentially leached phases from carbonates by double spike-thermal ionization mass spectrometry (DS-TIMS). Analytical Chemistry, 92(3): 2417–2424

Lin Z Y, Sun X M, Lu Y, Strauss H, Xu L, Gong J L, Teichert B M A, Lu R, Lu H F, Sun W D, Peckmann J. 2017. The enrichment of heavy iron isotopes in authigenic pyrite as a possible indicator of sulfate-driven anaerobic oxidation of methane: insights from the South China Sea. Chemical Geology, 449: 15–29

Lin Z Y, Sun X M, Lu Y, Strauss H, Xu L, Chen T T, Lu H F, Peckmann J. 2018. Iron isotope constraints on diagenetic iron cycling in the Taixinan seepage area, South China Sea. Journal of Asian Earth Sciences, 168: 112–124

Ling M X, Sedaghatpour F, Teng F Z, Hays P D, Strauss J, Sun W D. 2011. Homogeneous magnesium isotopic composition of seawater: an excellent geostandard for Mg isotope analysis. Rapid Communications in Mass Spectrometry, 25(19): 2828–2836

Liu C Y, Xu L J, Liu C T, Liu J, Qin L P, Zhang Z D, Liu S A, Li S G. 2019. High-precision measurement of stable Cr isotopes in geological reference materials by a double-spike TIMS method. Geostandards and Geoanalytical Research, 43(4): 647–661

Liu D, Zhao Z D, Zhu D C, Niu Y L, Widom E, Teng F Z, DePaolo D J, Ke S, Xu J F, Wang Q, Mo X X. 2015. Identifying mantle carbonatite metasomatism through Os-Sr-Mg isotopes in Tibetan ultrapotassic rocks. Earth and Planetary Science Letters, 430: 458–469

Liu F, Zhu H L, Li X, Wang G Q, Zhang Z F. 2017a. Calcium isotopic fractionation and compositions of geochemical reference materials. Geostandards and Geoanalytical Research, 41(4): 675–688

Liu F, Li X, Wang G Q, Liu Y F, Zhu H L, Kang J T, Huang F, Sun W D, Xia X P, Zhang Z F. 2017b. Marine carbonate component in the mantle beneath the southeastern Tibetan Plateau: evidence from magnesium and calcium isotopes. Journal of Geophysical Research: Solid Earth, 122(12): 9729–9744

Liu F, Li X, An Y J, Li J, Zhang Z F. 2019. Calcium isotope ratio ($\delta^{44/40}$Ca) measurements of Ca-dominated minerals and rocks without column chemistry using the double-spike technique and thermal ionisation mass spectrometry. Geostandards and Geoanalytical Research, 43(3): 509–517

Liu F, Zhang Z F, Li X, An Y J. 2020. A practical guide to the double-spike technique for calcium isotope measurements by thermal ionization mass spectrometry (TIMS). International Journal of Mass Spectrometry, 450: 116307

Liu H Q, Xu Y G, Wei G J, Wei J X, Yang F, Chen X Y, Liu L, Wei X. 2016. B isotopes of Carboniferous-Permian volcanic rocks in the Tuha Basin mirror a transition from subduction to intraplate setting in Central Asian orogenic belt. Journal of Geophysical Research: Solid Earth, 121(11): 7946–7964

Liu J, Wen H J, Zhang Y X, Fan H F, Zhu C W. 2016. Precise Mo isotope ratio measurements of low-Mo (ng·g^{-1}) geological samples using MC-ICP-MS. Journal of Analytical Atomic Spectrometry, 31(6): 1287–1297

Liu J F, Chen J B, Zhang T, Wang Y N, Yuan W, Lang Y C, Tu C L, Liu L Z, Birck J L. 2020. Chromatographic purification of antimony for accurate isotope analysis by MC-ICP-MS. Journal of Analytical Atomic Spectrometry, 35(7): 1360–1367

Liu J H, Zhou L, Algeo T J, Wang X C, Wang Q, Wang Y, Chen M L. 2020. Molybdenum isotopic behavior during intense weathering of basalt on Hainan Island, South China. Geochimica et Cosmochimica Acta, 287: 180–204

Liu M S, Zhang Q, Zhang Y N, Zhang Z F, Huang F, Yu H M. 2020. High-precision Cd isotope measurements of soil and rock reference materials by MC-ICP-MS with double spike correction. Geostandards and Geoanalytical Research, 44(1): 169–182

Liu P P, Teng F Z, Dick H J B, Zhou M F, Chung S L. 2017. Magnesium isotopic composition of the oceanic mantle and oceanic Mg cycling. Geochimica et Cosmochimica Acta, 206: 151–165

Liu S A, Teng F Z, He Y S, Ke S, Li S G. 2010. Investigation of magnesium isotope fractionation during granite differentiation: implication for Mg isotopic composition of the continental crust. Earth and Planetary Science Letters, 297(3-4): 646–654

Liu S A, Teng F Z, Yang W, Wu F Y. 2011. High-temperature inter-mineral magnesium isotope fractionation in mantle xenoliths from the North China Craton. Earth and Planetary Science Letters, 308(1-2): 131–140

Liu S A, Li D D, Li S G, Teng F Z, Ke S, He Y S, Lu Y H. 2014a. High-precision copper and iron isotope analysis of igneous rock standards by MC-ICP-MS. Journal of Analytical Atomic Spectrometry, 29(1): 122–133

Liu S A, Teng F Z, Li S G, Wei G J, Ma J L, Li D D. 2014b. Copper and iron isotope fractionation during weathering and pedogenesis: insights from saprolite profiles. Geochimica et Cosmochimica Acta, 146: 59–75

Liu S A, Huang J, Liu J G, Wörner G, Yang W, Tang Y J, Chen Y, Tang L M, Zheng J P, Li S G. 2015. Copper isotopic composition of the silicate Earth. Earth and Planetary Science Letters, 427: 95–103

Liu S A, Wang Z Z, Li S G, Huang J, Yang W. 2016. Zinc isotope evidence for a large-scale carbonated mantle beneath eastern China. Earth and Planetary Science Letters, 444: 169–178

Liu S A, Wu H C, Shen S Z, Jiang G Q, Zhang S H, Lv Y W, Zhang H, Li S G. 2017. Zinc isotope evidence for intensive magmatism immediately before the end-Permian mass extinction. Geology, 45(4): 343–346

Liu S A, Liu P P, Lv Y W, Wang Z Z, Dai J G. 2019. Cu and Zn isotope fractionation during oceanic alteration: implications for oceanic Cu and Zn cycles. Geochimica et Cosmochimica Acta, 257: 191–205

Liu X, Wei G J, Zou J Q, Guo Y R, Ma J L, Chen X F, Liu Y, Chen J F, Li H L, Zeng T. 2018. Elemental and Sr-Nd isotope geochemistry of sinking particles in the Northern South China Sea: implications for provenance and transportation. Journal of Geophysical Research: Oceans, 123(12): 9137–9155

Liu X C, Li X H, Liu Y, Yang L, Li Q L, Wu F Y, Yu H M, Huang F. 2018. Insights into the origin of purely sediment-derived Himalayan leucogranites: Si-O isotopic constraints. Science Bulletin, 63(19): 1243–1245

Liu Y, Liu W G, Peng Z C, Xiao Y K, Wei G J, Sun W D, He J F, Liu G J, Chou C L. 2009. Instability of seawater pH in the South China Sea during the mid-late Holocene: evidence from boron isotopic composition of corals. Geochimica et Cosmochimica Acta, 73(5): 1264–1272

Liu Y, Peng Z C, Zhou R J, Song S H, Liu W G, You C F, Lin Y P, Yu K F, Wu C C, Wei G J, Xie L H, Burr G S, Shen C C. 2014. Acceleration of modern acidification in the South China Sea driven by anthropogenic CO_2. Scientific Reports, 4(1): 5148

Liu Y, Li X H, Zeng Z, Yu H M, Huang F, Felis T, Shen C C. 2019. Annually-resolved coral skeletal $\delta^{138/134}Ba$ records: a new proxy for oceanic Ba cycling. Geochimica et Cosmochimica Acta, 247: 27–39

Lu D W, Liu Q, Zhang T Y, Cai Y, Yin Y G, Jiang G B. 2016. Stable silver isotope fractionation in the natural transformation process of silver nanoparticles. Nature Nanotechnology, 11(8): 682–686

Luo J, Long X P, Bowyer F T, Mills B J W, Li J, Xiong Y J, Zhu X K, Zhang K, Poulton S W. 2021. Pulsed oxygenation events drove progressive oxygenation of the Early Mesoproterozoic ocean. Earth and Planetary Science Letters, 559: 116754

Lv N, Bao Z A, Chen L, Chen K Y, Zhang Y, Yuan H L. 2020. Accurate determination of Cu isotope compositions in Cu-bearing minerals using micro-drilling and MC-ICP-MS. International Journal of Mass Spectrometry, 457: 116414

Lv W X, Yin H M, Liu M S, Huang F, Yu H M. 2021. Effect of the dry ashing method on cadmium isotope measurements in soil and plant samples. Geostandards and Geoanalytical Research, 45(1): 245–256

Lv Y W, Liu S A, Zhu J M, Li S G. 2016. Copper and zinc isotope fractionation during deposition and weathering of highly metalliferous black shales in central China. Chemical Geology, 445: 24–35

Ma J L, Wei G J, Liu Y, Ren Z Y, Xu Y G, Yang Y H. 2013a. Precise measurement of stable neodymium isotopes of geological materials by using MC-ICP-MS. Journal of Analytical Atomic Spectrometry, 28(12): 1926–1931

Ma J L, Wei G J, Liu Y, Ren Z Y, Xu Y G, Yang Y H. 2013b. Precise measurement of stable ($\delta^{88/86}Sr$) and radiogenic ($^{87}Sr/^{86}Sr$) strontium isotope ratios in geological standard reference materials using MC-ICP-MS. Chinese Science Bulletin, 58(25): 3111–3118

Malinovsky D, Dunn P J H, Goenaga-Infante H. 2016. Calibration of Mo isotope amount ratio measurements by MC-ICPMS using normalisation to an internal standard and improved experimental design. Journal of Analytical Atomic Spectrometry, 31(10): 1978–1988

Maréchal C N, Télouk P, Albarède F. 1999. Precise analysis of copper and zinc isotopic compositions by plasma-source mass spectrometry. Chemical Geology, 156(1–4): 251–273

Mei Q F, Yang J H, Yang Y H. 2018. An improved extraction chromatographic purification of tungsten from a silicate matrix for high precision isotopic measurements using MC-ICP-MS. Journal of Analytical Atomic Spectrometry, 33(4): 569–577

Meng M, Sun R Y, Liu H W, Yu B, Yin Y G, Hu L G, Shi J B, Jiang G B. 2019. An integrated model for input and migration of mercury in Chinese coastal sediments. Environmental Science & Technology, 53(5): 2460–2471

Meng M, Sun R Y, Liu H W, Yu B, Yin Y G, Hu L G, Chen J B, Shi J B, Jiang G B. 2020. Mercury isotope variations within the marine food web of Chinese Bohai Sea: implications for mercury sources and biogeochemical cycling. Journal of Hazardous Materials, 384: 121379

Meng Y M, Hu R Z. 2018. Minireview: advances in germanium isotope analysis by multiple collector-inductively coupled plasma-mass spectrometry. Analytical Letters, 51(5): 627–647

Meng Y M, Qi H W, Hu R Z. 2015. Determination of germanium isotopic compositions of sulfides by hydride generation MC-ICP-MS and its application to the Pb-Zn deposits in SW China. Ore Geology Reviews, 65: 1095–1109

Miller C A, Peucker-Ehrenbrink B, Ball L. 2009. Precise determination of rhenium isotope composition by multi-collector inductively-coupled plasma mass spectrometry. Journal of Analytical Atomic Spectrometry, 24(8): 1069–1078

Moynier F, Bouvier A, Blichert-Toft J, Telouk P, Gasperini D, Albarède F. 2006. Europium isotopic variations in Allende CAIs and the nature of mass-dependent fractionation in the solar nebula. Geochimica et Cosmochimica Acta, 70(16): 4287–4294

Nakada R, Takahashi Y, Tanimizu M. 2013. Isotopic and speciation study on cerium during its solid-water distribution with implication for Ce stable isotope as a paleo-redox proxy. Geochimica et Cosmochimica Acta, 103: 49–62

Nan X Y, Wu F, Zhang Z F, Hou Z H, Huang F, Yu H M. 2015. High-precision barium isotope measurements by MC-ICP-MS. Journal of Analytical Atomic Spectrometry, 30(11): 2307–2315

Nan X Y, Yu H M, Rudnick R L, Gaschnig R M, Xu J, Li W Y, Zhang Q, Jin Z D, Li X H, Huang F. 2018. Barium isotopic composition of the upper continental crust. Geochimica et Cosmochimica Acta, 233: 33–49

Nanne J A M, Millet M A, Burton K W, Dale C W, Nowell G M, Williams H M. 2017. High precision osmium stable isotope measurements by double spike MC-ICP-MS and N-TIMS. Journal of Analytical Atomic Spectrometry, 32(4): 749–765

Ni P, Chabot N L, Ryan C J, Shahar A. 2020. Heavy iron isotope composition of iron meteorites explained by core crystallization. Nature Geoscience,

13(9): 611-615

Nielsen S G, Prytulak J, Halliday A N. 2011. Determination of precise and accurate $^{51}V/^{50}V$ isotope ratios by MC-ICP-MS, Part 1: chemical separation of vanadium and mass spectrometric protocols. Geostandards and Geoanalytical Research, 35(3): 293-306

Ning M, Huang K J, Lang X G, Ma H R, Yuan H L, Peng Y B, Shen B. 2019. Can crystal morphology indicate different generations of dolomites? Evidence from magnesium isotopes. Chemical Geology, 516: 1-17

Ning M, Lang X G, Huang K J, Li C, Huang T Z, Yuan H L, Xing C C, Yang R Y, Shen B. 2020. Towards understanding the origin of massive dolostones. Earth and Planetary Science Letters, 545: 116403

Peng Y, Shen B, Lang X G, Huang K J, Chen J T, Yan Z, Tang W B, Ke S, Ma H R, Li F B. 2016. Constraining dolomitization by Mg isotopes: a case study from partially dolomitized limestones of the middle Cambrian Xuzhuang Formation, North China. Geochemistry, Geophysics, Geosystems, 17(3): 1109-1129

Pfeifer M, Lloyd N S, Peters S T M, Wombacher F, Elfers B M, Schulz T, Münker C. 2017. Tantalum isotope ratio measurements and isotope abundances determined by MC-ICP-MS using amplifiers equipped with 10^{10}, 10^{12} and 10^{13} Ohm resistors. Journal of Analytical Atomic Spectrometry, 32(1): 130-143

Poirier A, Doucelance R. 2009. Effective correction of mass bias for rhenium measurements by MC-ICP-MS. Geostandards and Geoanalytical Research, 33(2): 195-204

Prytulak J, Nielsen S G, Halliday A N. 2011. Determination of precise and accurate $^{51}V/^{50}V$ isotope ratios by multi-collector ICP-MS, Part 2: isotopic composition of six reference materials plus the Allende Chondrite and verification Tests. Geostandards and Geoanalytical Research, 35(3): 307-318

Qi H W, Rouxel O, Hu R Z, Bi X W, Wen H J. 2011. Germanium isotopic systematics in Ge-rich coal from the Lincang Ge deposit, Yunnan, southwestern China. Chemical Geology, 286(3-4): 252-265

Qi H W, Hu R Z, Jiang K, Zhou T, Liu Y F, Xiong Y W. 2019. Germanium isotopes and Ge/Si fractionation under extreme tropical weathering of basalts from the Hainan Island, South China. Geochimica et Cosmochimica Acta, 253: 249-266

Qi Y H, Wu F, Ionov D A, Puchtel I S, Carlson R W, Nicklas R W, Yu H M, Kang J T, Li C H, Huang F. 2019. Vanadium isotope composition of the Bulk Silicate Earth: constraints from peridotites and komatiites. Geochimica et Cosmochimica Acta, 259: 288-301

Qi Y H, Cheng W H, Nan X Y, Yang F, Li J H Y, Li D C, Lundstrom C C, Yu H M, Zhang G L, Huang F. 2020. Iron stable isotopes in bulk soil and sequential extracted fractions trace Fe redox cycling in paddy soils. Journal of Agricultural and Food Chemistry, 68(31): 8143-8150

Qin T, Wu F, Wu Z Q, Huang F. 2016. First-principles calculations of equilibrium fractionation of O and Si isotopes in quartz, albite, anorthite, and zircon. Contributions to Mineralogy and Petrology, 171(11): 91

Qu Q Y, Liu G J, Henry M, Point D, Chmeleff J, Sun R Y, Sonke J E, Chen J B. 2020. Tin stable isotopes in magmatic-affected coal deposits: insights in the geochemical behavior of tin. Applied Geochemistry, 119: 104641

Rehkämper M, Halliday A N. 1999. The precise measurement of Tl isotopic compositions by MC-ICPMS: application to the analysis of geological materials and meteorites. Geochimica et Cosmochimica Acta, 63(6): 935-944

She J X, Wang T H, Liang H D, Muhtar M N, Li W Q, Liu X D. 2020. Sn isotope fractionation during volatilization of Sn(IV) chloride: laboratory experiments and quantum mechanical calculations. Geochimica et Cosmochimica Acta, 269: 184-202

Shen J, Li S G, Wang S J, Teng F Z, Li Q L, Liu Y S. 2018. Subducted Mg-rich carbonates into the deep mantle wedge. Earth and Planetary Science Letters, 503: 118-130

Shen J, Yu J X, Chen J B, Algeo T J, Xu G Z, Feng Q L, Shi X, Planavsky N J, Shu W C, Xie S C. 2019a. Mercury evidence of intense volcanic effects on land during the Permian-Triassic transition. Geology, 47(12): 1117-1121

Shen J, Chen J B, Algeo T J, Yuan S L, Feng Q L, Yu J X, Zhou L, O'Connell B, Planavsky N J. 2019b. Evidence for a prolonged Permian-Triassic extinction interval from global marine mercury records. Nature Communications, 10(1): 1563

Shu Y C, Nielsen S G, Zeng Z G, Shinjo R, Blusztajn J, Wang X Y, Chen S. 2017. Tracing subducted sediment inputs to the Ryukyu arc-Okinawa Trough system: evidence from thallium isotopes. Geochimica et Cosmochimica Acta, 217: 462-491

Shuai K, Li W Q, Hui H J. 2020. Isobaric spike method for absolute isotopic ratio determination by MC-ICP-MS. Analytical Chemistry, 92(7): 4820-4828

Song H Y, Song H J, Algeo T J, Tong J N, Romaniello S J, Zhu Y Y, Chu D L, Gong Y M, Anbar A D. 2017. Uranium and carbon isotopes document global-ocean redox-productivity relationships linked to cooling during the Frasnian-Famennian mass extinction. Geology, 45(10): 887-890

Song L T, Liu C Q, Wang Z L, Zhu X K, Teng Y G, Wang J S, Tang S H, Li J, Liang L L. 2011a. Iron isotope compositions of natural river and lake samples in the karst area, Guizhou Province, Southwest China. Acta Geologica Sinica-English Edition, 85(3): 712-722

Song L T, Liu C Q, Wang Z L, Zhu X K, Teng Y G, Liang L L, Tang S H, Li J. 2011b. Iron isotope fractionation during biogeochemical cycle: information from suspended particulate matter (SPM) in Aha Lake and its tributaries, Guizhou, China. Chemical Geology, 280(1-2): 170-179

Su B X, Teng F Z, Hu Y, Shi R D, Zhou M F, Zhu B, Liu F, Gong X H, Huang Q S, Xiao Y, Chen C, He Y S. 2015. Iron and magnesium isotope fractionation in oceanic lithosphere and sub-arc mantle: perspectives from ophiolites. Earth and Planetary Science Letters, 430: 523-532

Su B X, Hu Y, Teng F Z, Xiao Y, Zhou X H, Sun Y, Zhou M F, Chang S C. 2017. Magnesium isotope constraints on subduction contribution to Mesozoic and Cenozoic East Asian continental basalts. Chemical Geology, 466: 116-122

Su B X, Hu Y, Teng F Z, Xiao Y, Zhang H F, Sun Y, Bai Y, Zhu B, Zhou X H, Ying J F. 2019. Light Mg isotopes in mantle-derived lavas caused by chromite crystallization, instead of carbonatite metasomatism. Earth and Planetary Science Letters, 522: 79-86

Sun J, Zhu X K, Tang S H, Chen Y L. 2013a. Investigation of matrix effects in the MC-ICP-MS induced by Nb, W, and Cu: isotopic case studies of iron and copper. Chinese Journal of Geochemistry, 32(1): 1–6

Sun J, Zhu X K, Chen Y L, Fang N. 2013b. Iron isotopic constraints on the genesis of Bayan Obo ore deposit, Inner Mongolia, China. Precambrian Research, 235: 88–106

Sun J, Zhu X K, Belshaw N S, Chen W, Doroshkevich A G, Luo W J, Song W L, Chen B B, Cheng Z G, Li Z H, Wang Y, Kynicky J, Henderson G M. 2021. Ca isotope systematics of carbonatites: insights into carbonatite source and evolution. Geochemical Perspectives Letters, 17: 11–15

Sun P, Niu Y L, Guo P Y, Duan M, Chen S, Gong H M, Wang X H, Xiao Y Y. 2020. Large iron isotope variation in the eastern Pacific mantle as a consequence of ancient low-degree melt metasomatism. Geochimica et Cosmochimica Acta, 286: 269–288

Sun X, Yin R S, Hu L M, Guo Z G, Hurley J P, Lepak R F, Li X D. 2020. Isotopic tracing of mercury sources in estuarine-inner shelf sediments of the East China Sea. Environmental Pollution, 262: 114356

Sun Y, Teng F Z, Ying J F, Su B X, Hu Y, Fan Q C, Zhou X H. 2017. Magnesium isotopic evidence for ancient subducted oceanic crust in LOMU-like potassium-rich volcanic rocks. Journal of Geophysical Research: Solid Earth, 122(10): 7562–7572

Tan D C, Zhu J M, Wang X L, Johnson T M, Li S H, Xu W P. 2020a. Equilibrium fractionation and isotope exchange kinetics between aqueous Se(IV) and Se(VI). Geochimica et Cosmochimica Acta, 277: 21–36

Tan D C, Zhu J M, Wang X L, Han G L, Lu Z, Xu W P. 2020b. High-sensitivity determination of Cd isotopes in low-Cd geological samples by double spike MC-ICP-MS. Journal of Analytical Atomic Spectrometry, 35(4): 713–727

Teng F Z, Hu Y, Ma J L, Wei G J, Rudnick R L. 2020. Potassium isotope fractionation during continental weathering and implications for global K isotopic balance. Geochimica et Cosmochimica Acta, 278: 261–271

Tian H C, Yang W, Li S G, Ke S, Chu Z Y. 2016. Origin of low δ^{26}Mg basalts with EM-I component: evidence for interaction between enriched lithosphere and carbonated asthenosphere. Geochimica et Cosmochimica Acta, 188: 93–105

Tian H C, Yang W, Li S G, Ke S. 2017. Could sedimentary carbonates be recycled into the lower mantle? Constraints from Mg isotopic composition of Emeishan basalts. Lithos, 292-293: 250–261

Tian H C, Yang W, Li S G, Ke S, Duan X Z. 2018. Low δ^{26}Mg volcanic rocks of Tengchong in southwestern China: a deep carbon cycle induced by supercritical liquids. Geochimica et Cosmochimica Acta, 240: 191–219

Tian H C, Yang W, Li S G, Wei H Q, Yao Z S, Ke S. 2019. Approach to trace hidden paleo-weathering of basaltic crust through decoupled Mg-Sr and Nd isotopes recorded in volcanic rocks. Chemical Geology, 509: 234–248

Tian H C, Teng F Z, Hou Z Q, Tian S H, Yang W, Chen X Y, Song Y C. 2020a. Magnesium and lithium isotopic evidence for a remnant oceanic slab beneath central Tibet. Journal of Geophysical Research: Solid Earth, 125(1): e2019JB018197

Tian H C, Zhang C, Teng F Z, Long Y J, Li S G, He Y S, Ke S, Chen X Y, Yang W. 2020b. Diffusion-driven extreme Mg and Fe isotope fractionation in Panzhihua ilmenite: implications for the origin of mafic intrusion. Geochimica et Cosmochimica Acta, 278: 361–375

Tian L L, Zeng Z, Nan X Y, Yu H M, Huang F. 2019. Determining Ba isotopes of barite using the Na_2CO_3 exchange reaction and double-spike method by MC-ICP-MS. Journal of Analytical Atomic Spectrometry, 34(7): 1459–1467

Tian S Y, Inglis E C, Creech J B, Zhang W, Wang Z C, Hu Z C, Liu Y S, Moynier F. 2020. The zirconium stable isotope compositions of 22 geological reference materials, 4 zircons and 3 standard solutions. Chemical Geology, 555: 119791

Tong S Y, Meija J, Zhou L, Mester Z, Yang L. 2019. Determination of the isotopic composition of hafnium using MC-ICPMS. Metrologia, 56(4): 044008

Wakaki S, Tanaka T. 2016. Stable Sm isotopic analysis of terrestrial rock samples by double-spike thermal ionization mass spectrometry. International Journal of Mass Spectrometry, 407: 22–28

Wang B L, Li W Y, Deng G X, Huang F, Yu H M. 2019. Silicon isotope compositions of metaperidotites from the Franciscan complex of California-implications for Si isotope fractionation during subduction dehydration. Lithos, 350-351: 105228

Wang F L, Wang C Y, Zhao T P. 2015. Boron isotopic constraints on the Nb and Ta mineralization of the syenitic dikes in the ~260 Ma Emeishan large igneous province (SW China). Ore Geology Reviews, 65: 1110–1126

Wang K, Jacobsen S B. 2016. An estimate of the bulk silicate Earth potassium isotopic composition based on MC-ICPMS measurements of basalts. Geochimica et Cosmochimica Acta, 178: 223–232

Wang Q, Zhou L, Feng L P, Liu J C, Liu J H, Algeo T J, Yang L. 2020a. Use of a Cu-selective resin for Cu preconcentration from seawater prior to its isotopic analysis by MC-ICP-MS. Journal of Analytical Atomic Spectrometry, 35(11): 2732–2739

Wang Q, Zhou L, Little S H, Liu J H, Feng L P, Tong S Y. 2020b. The geochemical behavior of Cu and its isotopes in the Yangtze River. Science of the Total Environment, 728: 138428

Wang S J, Teng F Z, Li S G. 2014a. Tracing carbonate-silicate interaction during subduction using magnesium and oxygen isotopes. Nature Communications, 5(1): 5328

Wang S J, Teng F Z, Li S G, Hong J A. 2014b. Magnesium isotopic systematics of mafic rocks during continental subduction. Geochimica et Cosmochimica Acta, 143: 34–48

Wang S J, Teng F Z, Scott J M. 2016. Tracing the origin of continental HIMU-like intraplate volcanism using magnesium isotope systematics. Geochimica et Cosmochimica Acta, 185: 78–87

Wang S J, Teng F Z, Li S G, Zhang L F, Du J X, He Y S, Niu Y L. 2017. Tracing subduction zone fluid-rock interactions using trace element and Mg-

Sr-Nd isotopes. Lithos, 290-291: 94−103

Wang S J, Rudnick R L, Gaschnig R M, Wang H, Wasylenki L E. 2019. Methanogenesis sustained by sulfide weathering during the Great Oxidation Event. Nature Geoscience, 12(4): 296−300

Wang W Z, Zhou C, Qin T, Kang J T, Huang S C, Wu Z Q, Huang F. 2017. Effect of Ca content on equilibrium Ca isotope fractionation between orthopyroxene and clinopyroxene. Geochimica et Cosmochimica Acta, 219: 44−56

Wang X, Liu S A, Wang Z R, Chen D Z, Zhang L Y. 2018. Zinc and strontium isotope evidence for climate cooling and constraints on the Frasnian-Famennian (~372 Ma) mass extinction. Palaeogeography, Palaeoclimatology, Palaeoecology, 498: 68−82

Wang X D, Cawood P A, Zhao H, Zhao L S, Grasby S E, Chen Z Q, Wignall P B, Lv Z Y, Han C. 2018. Mercury anomalies across the end Permian mass extinction in South China from shallow and deep water depositional environments. Earth and Planetary Science Letters, 496: 159−167

Wang X D, Cawood P A, Zhao H, Zhao L S, Grasby S E, Chen Z Q, Zhang L. 2019. Global mercury cycle during the end-Permian mass extinction and subsequent Early Triassic recovery. Earth and Planetary Science Letters, 513: 144−155

Wang X J, Chen L H, Hofmann A W, Mao F G, Liu J Q, Zhong Y, Xie L W, Yang Y H. 2017. Mantle transition zone-derived EM1 component beneath NE China: geochemical evidence from Cenozoic potassic basalts. Earth and Planetary Science Letters, 465: 16−28

Wang X J, Chen L H, Hofmann A W, Hanyu T, Kawabata H, Zhong Y, Xie L W, Shi J H, Miyazaki T, Hirahara Y, Takahashi T, Senda R, Chang Q, Vaglarov B S, Kimura J I. 2018. Recycled ancient ghost carbonate in the Pitcairn mantle plume. Proceedings of the National Academy of Sciences of the United States of America, 115(35): 8682−8687

Wang X L, Johnson T M, Lundstrom C C. 2015. Isotope fractionation during oxidation of tetravalent uranium by dissolved oxygen. Geochimica et Cosmochimica Acta, 150: 160−170

Wang Y, Zhu X K, Mao J W, Li Z H, Cheng Y B. 2011. Iron isotope fractionation during skarn-type metallogeny: a case study of Xinqiao Cu-S-Fe-Au deposit in the Middle-Lower Yangtze valley. Ore Geology Reviews, 43(1): 194−202

Wang Y, Zhu X K, Cheng Y B. 2015. Fe isotope behaviours during sulfide-dominated skarn-type mineralisation. Journal of Asian Earth Sciences, 103: 374−392

Wang Y, He Y S, Wu H J, Zhu C W, Huang S C, Huang J. 2019. Calcium isotope fractionation during crustal melting and magma differentiation: granitoid and mineral-pair perspectives. Geochimica et Cosmochimica Acta, 259: 37−52

Wang Y, He Y S, Ke S. 2020. Mg isotope fractionation during partial melting of garnet-bearing sources: an adakite perspective. Chemical Geology, 537: 119478

Wang Y, Zhu X K, Tang C, Mao J W, Chang Z S. 2021. Discriminate between magmatic and magmatic-hydrothermal ore deposits using Fe isotopes. Ore Geology Reviews, 130: 103946

Wang Z B, Ma J L, Li J, Wei G J, Chen X F, Deng W F, Xie L H, Lu W J, Zou L. 2015. Chemical weathering controls on variations in the molybdenum isotopic composition of river water: evidence from large rivers in China. Chemical Geology, 410: 201−212

Wang Z B, Ma J L, Li J, Wei G J, Zeng T, Li L, Zhang L, Deng W F, Xie L H, Liu Z F. 2018. Fe (hydro) oxide controls Mo isotope fractionation during the weathering of granite. Geochimica et Cosmochimica Acta, 226: 1−17

Wang Z B, Li J, Wei G J, Deng W F, Chen X F, Zeng T, Wang X J, Ma J L, Zhang L, Tu X L, Wang Q, McCulloch M. 2019. Biologically controlled Mo isotope fractionation in coral reef systems. geochimica et Cosmochimica Acta, 262: 128−142

Wang Z B, Ma J L, Li J, Zeng T, Zhang Z Y, He X Y, Zhang L, Wei G J. 2020. Effect of Fe-Ti oxides on Mo isotopic variations in lateritic weathering profiles of basalt. Geochimica et Cosmochimica Acta, 286: 380−403

Wang Z C, Park J W, Wang X, Zou Z Q, Kim J, Zhang P Y, Li M. 2019. Evolution of copper isotopes in arc systems: insights from lavas and molten sulfur in Niuatahi volcano, Tonga rear arc. Geochimica et Cosmochimica Acta, 250: 18−33

Wang Z Z, Liu S A, Ke S, Liu Y C, Li S G. 2016. Magnesium isotopic heterogeneity across the cratonic lithosphere in eastern China and its origins. Earth and Planetary Science Letters, 451: 77−88

Wang Z Z, Liu S A, Liu J G, Huang J, Xiao Y, Chu Z Y, Zhao X M, Tang L M. 2017. Zinc isotope fractionation during mantle melting and constraints on the Zn isotope composition of Earth's upper mantle. Geochimica et Cosmochimica Acta, 198: 151−167

Wang Z Z, Liu S A, Chen L H, Li S G, Zeng G. 2018. Compositional transition in natural alkaline lavas through silica-undersaturated melt-lithosphere interaction. Geology, 46(9): 771−774

Wang Z Z, Liu S A, Liu Z C, Zheng Y C, Wu F Y. 2020. Extreme Mg and Zn isotope fractionation recorded in the Himalayan leucogranites. Geochimica et Cosmochimica Acta, 278: 305−321

Wei G J, McCulloch M T, Mortimer G, Deng W F, Xie L H. 2009. Evidence for ocean acidification in the Great Barrier Reef of Australia. Geochimica et Cosmochimica Acta, 73(8): 2332−2346

Wei G J, Wei J X, Liu Y, Ke T, Ren Z Y, Ma J L, Xu Y G. 2013a. Measurement on High-precision boron isotope of silicate materials by a single column purification method and MC-ICP-MS. Journal of Analytical Atomic Spectrometry, 28(4): 606−612

Wei G J, Ma J L, Liu Y, Xie L H, Lu W J, Deng W F, Ren Z Y, Zeng T, Yang Y H. 2013b. Seasonal changes in the radiogenic and stable strontium isotopic composition of Xijiang River water: implications for chemical weathering. Chemical Geology, 343: 67−75

Wei G J, Wang Z B, Ke T, Liu Y, Deng W F, Chen X F, Xu J F, Zeng T, Xie L H. 2015. Decadal variability in seawater pH in the West Pacific: evidence from coral $\delta^{11}B$ records. Journal of Geophysical Research: Oceans, 120(11): 7166−7181

Wei H Z, Jiang S Y, Hemming N G, Yang J H, Yang T, Wu H P, Yang T L, Yan X, Pu W. 2014a. An improved procedure for separation/purification of boron from complex matrices and High-precision measurement of boron isotopes by positive thermal ionization and multicollector inductively coupled plasma mass spectrometry. Talanta, 123: 151–160

Wei H Z, Jiang S Y, Yang T L, Yang J H, Yang T, Yan X, Ling B P, Liu Q, Wu H P. 2014b. Effect of metasilicate matrices on boron purification by Amberlite IRA 743 boron specific resin and isotope analysis by MC-ICP-MS. Journal of Analytical Atomic Spectrometry, 29(11): 2104–2107

Wei R F, Guo Q J, Wen H J, Peters M, Yang J X, Tian L Y, Han X K. 2017. Chromatographic separation of Cd from plants via anion-exchange resin for an isotope determination by multiple collector ICP-MS. Analytical Sciences, 33(3): 335–341

Wei W, Zeng Z, Shen J, Tian L L, Wei G Y, Ling H F, Huang F. 2021. Dramatic changes in the carbonate-hosted barium isotopic compositions in the Ediacaran Yangtze Platform. Geochimica et Cosmochimica Acta, 299: 113–129

Wen B, Zhou J W, Zhou A G, Liu C F, Li L G. 2018. A review of antimony (Sb) isotopes analytical methods and application in environmental systems. International Biodeterioration & Biodegradation, 128: 109–116

Wen H J, Carignan J. 2011. Selenium isotopes trace the source and redox processes in the black shale-hosted Se-rich deposits in China. Geochimica et Cosmochimica Acta, 75(6): 1411–1427

Wen H J, Carignan J, Zhang Y X, Fan H F, Cloquet C, Liu S R. 2011. Molybdenum isotopic records across the Precambrian-Cambrian boundary. Geology, 39(8): 775–778

Wen H J, Carignan J, Chu X L, Fan H F, Cloquet C, Huang J, Zhang Y X, Chang H J. 2014. Selenium isotopes trace anoxic and ferruginous seawater conditions in the Early Cambrian. Chemical Geology, 390: 164–172

Wen H J, Fan H F, Zhang Y X, Cloquet C, Carignan J. 2015a. Reconstruction of early Cambrian ocean chemistry from Mo isotopes. Geochimica et Cosmochimica Acta, 164: 1–16

Wen H J, Zhang Y X, Cloquet C, Zhu C W, Fan H F, Luo C G. 2015b. Tracing sources of pollution in soils from the Jinding Pb-Zn mining district in China using cadmium and lead isotopes. Applied Geochemistry, 52: 147–154

Wu C, Yang T, Shields G A, Bian X, Gao B, Ye H, Li W. 2020. Termination of Cryogenian ironstone deposition by deep ocean euxinia. Geochemical Perspectives Letters, 15: 1–5

Wu F, Qi Y H, Yu H M, Tian S Y, Hou Z H, Huang F. 2016. Vanadium isotope measurement by MC-ICP-MS. Chemical Geology, 421: 17–25

Wu F, Qi Y H, Perfit M R, Gao Y J, Langmuir C H, Wanless V D, Yu H M, Huang F. 2018. Vanadium isotope compositions of mid-ocean ridge lavas and altered oceanic crust. Earth and Planetary Science Letters, 493: 128–139

Wu G L, Zhu J M, Wang X L, Johnson T M, Han G L. 2020. High-sensitivity measurement of Cr isotopes by double spike MC-ICP-MS at the 10 ng level. Analytical Chemistry, 92(1): 1463–1469

Wu H J, He Y S, Teng F Z, Ke S, Hou Z H, Li S G. 2018. Diffusion-driven magnesium and iron isotope fractionation at a gabbro-granite boundary. Geochimica et Cosmochimica Acta, 222: 671–684

Wu H P, Jiang S Y, Wei H Z, Yan X. 2012. An experimental study of organic matters that cause isobaric ions interference for boron isotopic measurement by thermal ionization mass spectrometry. International Journal of Mass Spectrometry, 328-329: 67–77

Wu L Y, Hu R Z, Li X F, Liu S A, Tang Y W, Tang Y Y. 2017. Copper isotopic compositions of the Zijinshan high-sulfidation epithermal Cu-Au deposit, South China: implications for deposit origin. Ore Geology Reviews, 83: 191–199

Wu W, Xu Y G, Zhang Z F, Li X. 2020. Calcium isotopic composition of the lunar crust, mantle, and bulk silicate Moon: a preliminary study. Geochimica et Cosmochimica Acta, 270: 313–324

Wu Z Q, Huang F, Huang S C. 2015. Isotope fractionation induced by phase transformation: first-principles investigation for Mg_2SiO_4. Earth and Planetary Science Letters, 409: 339–347

Xia Y, Kiseeva E S, Wade J, Huang F. 2019. The effect of core segregation on the Cu and Zn isotope composition of the silicate Moon. Geochemical Perspectives Letters, 12: 12–17

Xiang L, Schoepfer S D, Zhang H, Chen Z W, Cao C Q, Shen S Z. 2020. Deep-water dissolved iron cycling and reservoir size across the Ediacaran-Cambrian transition. Chemical Geology, 541: 119575

Xiao H F, Deng W F, Wei G J, Chen J B, Zheng X Q, Shi T, Chen X F, Wang C Y, Liu X, Zeng T. 2020. A pilot study on zinc isotopic compositions in shallow-water coral skeletons. Geochemistry, Geophysics, Geosystems, 21(11): e2020GC009430

Xiao Y K, Beary E S. 1989. High-precision isotopic measurement of lithium by thermal ionization mass spectrometry. International Journal of Mass Spectrometry and Ion Processes, 94(1-2): 107–114

Xu J, Yang S Y, Yang Y H, Liu Y S, Xie X L. 2020. Precise determination of stable strontium isotopic compositions by MC-ICP-MS. Atomic Spectroscopy, 41(2): 64–73

Xu L G, Lehmann B, Mao J W, Nägler T F, Neubert N, Böttcher M E, Escher P. 2012. Mo isotope and trace element patterns of Lower Cambrian black shales in South China: multi-proxy constraints on the paleoenvironment. Chemical Geology, 318-319: 45–59

Xu R, Liu Y S, Tong X R, Hu Z C, Zong K Q, Gao S. 2013. In-situ trace elements and Li and Sr isotopes in peridotite xenoliths from Kuandian, North China Craton: insights into Pacific slab subduction-related mantle modification. Chemical Geology, 354: 107–123

Xu W P, Zhu J M, Johnson T M, Wang X L, Lin Z Q, Tan D C, Qin H B. 2020. Selenium isotope fractionation during adsorption by Fe, Mn and Al oxides. Geochimica et Cosmochimica Acta, 272: 121–136

Xu Y K, Hu Y, Chen X Y, Huang T Y, Sletten R S, Zhu D, Teng F Z. 2019. Potassium isotopic compositions of international geological reference materials. Chemical Geology, 513: 101–107

Xue Y L, Li C H, Qi Y H, Zhang C T, Miao B K, Huang F. 2018. The vanadium isotopic composition of L ordinary chondrites. Acta Geochimica, 37(4): 501-508

Yan B, Zhu X K, He X X, Tang S H. 2019. Zn isotopic evolution in early Ediacaran Ocean: a global signature. Precambrian Research, 320: 472-483

Yang C, Liu S A. 2019. Zinc isotope constraints on recycled oceanic crust in the mantle sources of the Emeishan Large Igneous province. Journal of Geophysical Research: Solid Earth, 124(12): 12537-12555

Yang W, Teng F Z, Zhang H F. 2009. Chondritic magnesium isotopic composition of the terrestrial mantle: a case study of peridotite xenoliths from the North China Craton. Earth and Planetary Science Letters, 288(3-4): 475-482

Yang W, Teng F Z, Zhang H F, Li S G. 2012. Magnesium isotopic systematics of continental basalts from the North China Craton: implications for tracing subducted carbonate in the mantle. Chemical Geology, 328: 185-194

Yang W, Teng F Z, Li W Y, Liu S A, Shan K, Liu Y S, Zhang H F, Gao S. 2016. Magnesium isotopic composition of the deep continental crust. American Mineralogist, 101(2): 243-252

Yang Y H, Zhang X C, Liu S A, Zhou T, Fan H F, Yu H M, Cheng W H, Huang F. 2018. Calibrating NIST SRM 683 as a new international reference standard for Zn isotopes. Journal of Analytical Atomic Spectrometry, 33(10): 1777-1783

Yao J M, Mathur R, Powell W, Lehmann B, Tornos F, Wilson M, Ruiz J. 2018. Sn-isotope fractionation as a record of hydrothermal redox reactions. American Mineralogist, 103(10): 1591-1598

Yin R S, Guo Z G, Hu L M, Liu W C, Hurley J P, Lepak R F, Lin T, Feng X B, Li X D. 2018. Mercury inputs to Chinese marginal seas: impact of industrialization and development of China. Journal of Geophysical Research: Oceans, 123(8): 5599-5611

Yu H M, Li Y H, Gao Y J, Huang J, Huang F. 2018. Silicon isotopic compositions of altered oceanic crust: implications for Si isotope heterogeneity in the mantle. Chemical Geology, 479: 1-9

Yuan H L, Yuan W T, Bao Z A, Chen K Y, Huang F, Liu S G. 2017. Development of two new copper isotope standard solutions and their copper isotopic compositions. Geostandards and Geoanalytical Research, 41(1): 77-84

Yuan W, Chen J B, Birck J L, Yin Z Y, Yuan S L, Cai H M, Wang Z W, Huang Q, Wang Z H. 2016. Precise analysis of gallium isotopic composition by MC-ICP-MS. Analytical Chemistry, 88(19): 9606-9613

Yuan W, Saldi G D, Chen J B, Zuccolini M V, Birck J L, Liu Y J, Schott J. 2018. Gallium isotope fractionation during Ga adsorption on calcite and goethite. Geochimica et Cosmochimica Acta, 223: 350-363

Zeng Z, Li X H, Liu Y, Huang F, Yu H M. 2019. High-precision barium isotope measurements of carbonates by MC-ICP-MS. Geostandards and Geoanalytical Research, 43(2): 291-300

Zhang A Y, Zhang J, Zhang R F, Xue Y. 2014. Modified enrichment and purification protocol for dissolved silicon-isotope determination in natural waters. Journal of Analytical Atomic Spectrometry, 29(12): 2414-2418

Zhang F F, Zhu X K, Yan B, Kendall B, Peng X, Li J, Algeo T J, Romaniello S. 2015. Oxygenation of a Cryogenian ocean (Nanhua Basin, South China) revealed by pyrite Fe isotope compositions. Earth and Planetary Science Letters, 429: 11-19

Zhang F F, Romaniello S J, Algeo T J, Lau K V, Clapham M E, Richoz S, Herrmann A D, Smith H, Horacek M, Anbar A D. 2018. Multiple episodes of extensive marine anoxia linked to global warming and continental weathering following the Latest Permian mass extinction. Science Advances, 4(4): e1602921

Zhang F F, Dahl T W, Lenton T M, Luo G M, Shen S Z, Algeo T J, Planavsky N, Liu J S, Cui Y, Qie W K, Romaniello S J, Anbar A D. 2020a. Extensive marine anoxia associated with the Late Devonian Hangenberg Crisis. Earth and Planetary Science Letters, 533: 115976

Zhang F F, Lenton T M, del Rey Á, Romaniello S J, Chen X M, Planavsky N J, Clarkson M O, Dahl T W, Lau K V, Wang W Q, Li Z H, Zhao M Y, Isson T, Algeo T J, Anbar A D. 2020b. Uranium isotopes in marine carbonates as a global ocean paleoredox proxy: a critical review. Geochimica et Cosmochimica Acta, 287: 27-49

Zhang H J, Fan H F, Xiao C Y, Wen H J, Ye L, Huang Z L, Zhou J X, Guo Q J. 2019. The mixing of multi-source fluids in the Wusihe Zn-Pb ore deposit in Sichuan Province, southwestern China. Acta Geochimica, 38(5): 642-653

Zhang H J, Fan H F, Wen H J, Zhu X K, Tian S H. 2020. Oceanic chemistry recorded by cherts during the early Cambrian Explosion, South China. Palaeogeography, Palaeoclimatology, Palaeoecology, 558: 109961

Zhang H M, Wang Y, He Y S, Teng F Z, Jacobsen S B, Helz R T, Marsh B D, Huang S C. 2018. No measurable calcium isotopic fractionation during crystallization of Kilauea Iki Lava Lake. Geochemistry, Geophysics, Geosystems, 19(9): 3128-3139

Zhang L, Li J, Xu Y G, Ren Z Y. 2018. The influence of the double spike proportion effect on stable isotope (Zn, Mo, Cd, and Sn) measurements by multicollector-inductively coupled plasma-mass spectrometry (MC-ICP-MS). Journal of Analytical Atomic Spectrometry, 33(4): 555-562

Zhang L, Sun W D, Zhang Z F, An Y J, Liu F. 2019. Iron isotopic composition of supra-subduction zone ophiolitic peridotite from northern Tibet. Geochimica et Cosmochimica Acta, 258: 274-289

Zhang Q, Liu J, Zhang Y N, Yu H M, Qin L P, Shen J. 2019. Factors affecting chromium isotope measurements using the double-spike method. Rapid Communications in Mass Spectrometry, 33(17): 1390-1400

Zhang R, Russell J, Xiao X, Zhang F, Li T G, Liu Z Y, Guan M L, Han Q, Shen L Y, Shu Y J. 2018. Historical records, distributions and sources of mercury and zinc in sediments of East China Sea: implication from stable isotopic compositions. Chemosphere, 205: 698-708

Zhang R F, John S G, Zhang J, Ren J L, Wu Y, Zhu Z Y, Liu S M, Zhu X C, Marsay C M, Wenger F. 2015. Transport and reaction of iron and iron stable isotopes in glacial meltwaters on Svalbard near Kongsfjorden: from rivers to estuary to ocean. Earth and Planetary Science Letters, 424: 201-211

Zhang R X, Meija J, Huang Y, Pei X J, Mester Z, Yang L. 2019. Determination of the isotopic composition of tungsten using MC-ICP-MS. Analytica Chimica Acta, 1089: 19-24

Zhang T, Zhou L, Yang L, Wang Q, Feng L P, Liu Y S. 2016. High precision measurements of gallium isotopic compositions in geological materials by MC-ICP-MS. Journal of Analytical Atomic Spectrometry, 31(8): 1673-1679

Zhang W, Wang Z C, Moynier F, Inglis E, Tian S Y, Li M, Liu Y S, Hu Z C. 2019. Determination of Zr isotopic ratios in zircons using laser-ablation multiple-collector inductively coupled-plasma mass-spectrometry. Journal of Analytical Atomic Spectrometry, 34(9): 1800-1809

Zhang Y, Bao Z A, Lv N, Chen K Y, Zong C L, Yuan H L. 2020. Copper isotope ratio measurements of Cu-dominated minerals without column chromatography using MC-ICP-MS. Frontiers in Chemistry, 8: 609

Zhang Y Y, Yuan C, Sun M, Chen M, Hong L B, Li J, Long X P, Li P F, Lin Z F. 2019. Recycled oceanic crust in the form of pyroxenite contributing to the Cenozoic continental basalts in Central Asia: new perspectives from olivine chemistry and whole-rock B-Mo isotopes. Contributions to Mineralogy and Petrology, 174(10): 83

Zhang Y Y, Yuan C, Sun M, Li J, Long X P, Jiang Y D, Huang Z Y. 2020a. Molybdenum and boron isotopic evidence for carbon-recycling via carbonate dissolution in subduction zones. Geochimica et Cosmochimica Acta, 278: 340-352

Zhang Y Y, Chen J B, Zheng W, Sun R Y, Yuan S L, Cai H M, Yang D A, Yuan W, Meng M, Wang Z W, Liu Y L, Liu J F. 2020b. Mercury isotope compositions in large anthropogenically impacted Pearl River, South China. Ecotoxicology and Environmental Safety, 191: 110229

Zhang Z Y, Ma J L, Zhang L, Liu Y, Wei G J. 2018. Rubidium purification via a single chemical column and its isotope measurement on geological standard materials by MC-ICP-MS. Journal of Analytical Atomic Spectrometry, 33(2): 322-328

Zhao P P, Li J, Zhang L, Wang Z B, Kong D X, Ma J L, Wei G J, Xu J F. 2016. Molybdenum mass fractions and isotopic compositions of international geological reference materials. Geostandards and Geoanalytical Research, 40(2): 217-226

Zhao T, Liu W J, Xu Z F, Sun H G, Zhou X D, Zhou L, Zhang J Y, Zhang X, Jiang H, Liu T Z. 2019. The influence of carbonate precipitation on riverine magnesium isotope signals: new constrains from Jinsha River Basin, Southeast Tibetan Plateau. Geochimica et Cosmochimica Acta, 248: 172-184

Zhao X M, Zhang H F, Zhu X K, Tang S H, Tang Y J. 2010. Iron isotope variations in spinel peridotite xenoliths from North China Craton: implications for mantle metasomatism. Contributions to Mineralogy and Petrology, 160(1): 1-14

Zhao X M, Zhang H F, Zhu X K, Tang S H, Yan B. 2012. Iron isotope evidence for multistage melt-peridotite interactions in the lithospheric mantle of eastern China. Chemical Geology, 292-293: 127-139

Zhao X M, Zhang H F, Zhu X K, Zhu B, Cao H H. 2015. Effects of melt percolation on iron isotopic variation in peridotites from Yangyuan, North China Craton. Chemical Geology, 401: 96-110

Zhao X M, Cao H H, Mi X, Evans N J, Qi Y H, Huang F, Zhang H F. 2017. Combined iron and magnesium isotope geochemistry of pyroxenite xenoliths from Hannuoba, North China Craton: implications for mantle metasomatism. Contributions to Mineralogy and Petrology, 172(6): 40

Zhao X M, Tang S H, Li J, Wang H, Helz R, Marsh B, Zhu X K, Zhang H F. 2020. Titanium isotopic fractionation during magmatic differentiation. Contributions to Mineralogy and Petrology, 175(7): 67

Zhao Y, Xue C J, Liu S A, Mathur R, Zhao X B, Yang Y Q, Dai J F, Man R H, Liu X M. 2019. Redox reactions control Cu and Fe isotope fractionation in a magmatic Ni-Cu mineralization system. Geochimica et Cosmochimica Acta, 249: 42-58

Zhao Z Q, Shen B, Zhu J M, Lang X G, Wu G L, Tan D C, Pei H X, Huang T Z, Ning M, Ma H R. 2021. Active methanogenesis during the melting of Marinoan snowball Earth. Nature Communications, 12(1): 955

Zheng X D, Teng Y G, Song L T. 2019. Iron isotopic composition of suspended particulate matter in Hongfeng Lake. Water, 11(2): 396

Zheng Y C, Liu S A, Wu C D, Griffin W L, Li Z Q, Xu B, Yang Z M, Hou Z Q, O'Reilly S Y. 2019. Cu isotopes reveal initial Cu enrichment in sources of giant porphyry deposits in a collisional setting. Geology, 47(2): 135-138

Zhong Y, Chen L H, Wang X J, Zhang G L, Xie L W, Zeng G. 2017. Magnesium isotopic variation of oceanic island basalts generated by partial melting and crustal recycling. Earth and Planetary Science Letters, 463: 127-135

Zhou J X, Huang Z L, Zhou M F, Zhu X K, Muchez P. 2014. Zinc, sulfur and lead isotopic variations in carbonate-hosted Pb-Zn sulfide deposits, Southwest China. Ore Geology Reviews, 58: 41-54

Zhou L, Wignall P B, Su J, Feng Q L, Xie S C, Zhao L S, Huang J H. 2012. U/Mo ratios and $\delta^{98/95}$Mo as local and global redox proxies during mass extinction events. Chemical Geology, 324-325: 99-107

Zhu G H, Ma J L, Wei G J, An Y J. 2020a. A novel procedure for separating iron from geological materials for isotopic analysis using MC-ICP-MS. Journal of Analytical Atomic Spectrometry, 35(5): 873-877

Zhu G H, Ma J L, Wei G J, Zhang L. 2020b. A rapid and simple method for lithium purification and isotopic analysis of geological reference materials by MC-ICP-MS. Frontiers in Chemistry, 8: 557489

Zhu H L, Zhang Z F, Wang G Q, Liu Y F, Liu F, Li X, Sun W D. 2016. Calcium isotopic fractionation during ion-exchange column chemistry and thermal ionisation mass spectrometry (TIMS) determination. Geostandards and Geoanalytical Research, 40(2): 185-194

Zhu H L, Liu F, Li X, An Y J, Wang G Q, Zhang Z F. 2018a. A "peak cut" procedure of column separation for calcium isotope measurement using the double spike technique and thermal ionization mass spectrometry (TIMS). Journal of Analytical Atomic Spectrometry, 33(4): 547-554

Zhu H L, Liu F, Li X, Wang G Q, Zhang Z F, Sun W D. 2018b. Calcium isotopic compositions of normal mid-ocean ridge basalts from the southern Juan de Fuca Ridge. Journal of Geophysical Research: Solid Earth, 123(2): 1303-1313

Zhu H L, Du L, Li X, Zhang Z F, Sun W D. 2020a. Calcium isotopic fractionation during plate subduction: constraints from back-arc basin basalts. Geochimica et Cosmochimica Acta, 270: 379-393

Zhu H L, Liu F, Li X, An Y J, Nan X Y, Du L, Huang F, Sun W D, Zhang Z F. 2020b. Significant $\delta^{44/40}$Ca variations between carbonate and clay-rich marine sediments from the Lesser Antilles forearc and implications for mantle heterogeneity. Geochimica et Cosmochimica Acta, 276: 239-257

Zhu J M, Johnson T M, Clark S K, Zhu X K, Wang X L. 2014. Selenium redox cycling during weathering of Se-rich shales: a selenium isotope study. Geochimica et Cosmochimica Acta, 126: 228-249

Zhu J M, Wu G L, Wang X L, Han G L, Zhang L X. 2018. An improved method of Cr purification for high precision measurement of Cr isotopes by double spike MC-ICP-MS. Journal of Analytical Atomic Spectrometry, 33(5): 809-821

Zhu X K, O'Nions R K, Guo Y, Belshaw N S, Rickard D. 2000. Determination of natural Cu-isotope variation by plasma-source mass spectrometry: implications for use as geochemical tracers. Chemical Geology, 163(1-4): 139-149

Zhu X K, Guo Y, O'Nions R K, Young E D, Ash R D. 2001. Isotopic homogeneity of iron in the early solar nebula. Nature, 412(6844): 311-313

Zhu X K, Makishima A, Guo Y, Belshaw N S, O'Nions R K. 2002a. High precision measurement of titanium isotope ratios by plasma source mass spectrometry. International Journal of Mass Spectrometry, 220(1): 21-29

Zhu X K, Guo Y, Williams R J P, O'Nions R K, Matthews A, Belshaw N S, Canters G W, de Waal E C, Weser U, Burgess B K, Salvato B. 2002b. Mass fractionation processes of transition metal isotopes. Earth and Planetary Science Letters, 200(1-2): 47-62

Zhu X K, Sun J, Li Z H. 2019. Iron isotopic variations of the Cryogenian banded iron formations: a new model. Precambrian Research, 331: 105359

Zhu Y T, Li M, Wang Z C, Zou Z Q, Hu Z C, Liu Y S, Zhou L, Chai X N. 2019. High-precision copper and zinc isotopic measurements in igneous rock standards using large-geometry MC-ICP-MS. Atomic Spectroscopy, 40(6): 206-214

Zhu Z H, Meija J, Zheng A R, Mester Z, Yang L. 2017. Determination of the isotopic composition of iridium using multicollector-ICPMS. Analytical Chemistry, 89(17): 9375-9382

Zhu Z Y, Jiang S Y, Yang T, Wei H Z. 2015. Improvements in Cu-Zn isotope analysis with MC-ICP-MS: a revisit of chemical purification, mass spectrometry measurement and mechanism of Cu/Zn mass bias decoupling effect. International Journal of Mass Spectrometry, 393: 34-40

Zhu Z Y, Yang T, Zhu X K. 2019. Achieving rapid analysis of Li isotopes in high-matrix and low-Li samples with MC-ICP-MS: new developments in sample preparation and mass bias behavior of Li in ICPMS. Journal of Analytical Atomic Spectrometry, 34(7): 1503-1513

Zou Z Q, Wang Z C, Li M, Becker H, Geng X L, Hu Z C, Lazarov M. 2019. Copper isotope variations during magmatic migration in the mantle: insights from mantle pyroxenites in balmuccia peridotite massif. Journal of Geophysical Research: Solid Earth, 124(11): 11130-11149

Progress of Non-traditional Stable Isotope Geochemistry

WEI Gang-jian[1,2], HUANG Fang[3], MA Jin-long[1,2], DENG Wen-feng[1,2], YU Hui-min[3], KANG Jin-ting[3], CHEN Xue-fei[1,2]

1. State Key Laboratory of Isotope Geochemistry, Guangzhou Institute of Geochemistry, Chinese Academy of Sciences, Guangzhou 510640; 2. CAS Center for Excellence in Deep Earth Science, Guangzhou 510640; 3. School of Earth and Space Sciences, University of Science and Technology of China, Hefei 230021

Abstract: As one of the most rapidly advanced branches of geochemistry, non-traditional stable isotope has been explosively developed in the world during the past decade, and Chinese researchers have also participated in a manner of independent innovation in this frontier field. This paper will systematically review the progresses in analytical technologies of non-traditional stable isotope achieved by Chinese research teams from 2010 through 2020 and will also summarize representative achievements of applying these technologies in earth sciences. It will be shown that the Chinese research teams have explored the high-precision analytical techniques in testing isotopic composition for many non-traditional stable isotope systems and have developed some internationally advanced or even leading analytical techniques. In terms of research application, different Chinese research teams have focused on different groups of non-traditional stable isotope and have applied their techniques in multiple fields of earth science study. The degree of development is different, but the overall development is in a vigorous and rapid development stage. There is reason to believe that non-traditional stable isotope geochemistry in China will continue to develop rapidly in the future, the existing methods will be optimized and improved, and new non-traditional stable isotope systems will continue to be developed. These methods will aid in studying a broader range of earth sciences and facilitate interdisciplinary researches. The mechanisms of the non-traditional stable isotope fractionation during various geological processes will be deciphered more clearly, and their applications will be more sophisticated.

Key words: non-traditional stable isotopes; analytical techniques; application studies; progress in the past decade

团簇同位素地球化学研究进展*

邓文峰 郭炀锐 韦刚健

中国科学院 广州地球化学研究所，同位素地球化学国家重点实验室，广州 510640

摘　要：团簇同位素是本世纪初发展起来的稳定同位素地球化学新的研究领域，是指自然丰度较低的重同位素相互成键（如 $^{13}C-^{18}O$ 等）形成的同位素体。其内部同位素的键合主要受控于温度，通过测定团簇同位素组成可获得温度信息。目前研究程度较深的是碳酸盐的团簇同位素温度计，其独特优势在于它不受碳酸盐形成时流体同位素组成的影响，因而在地球科学的多个研究领域被广泛应用。本文总结了近十年来我国团簇同位素地球化学在理论计算、分析方法研究、实验模拟和应用研究等方面取得的进展，建议未来除了应用研究之外，还可以利用实验模拟方法从团簇同位素的平衡理论、生物碳酸盐生命效应的影响、埋藏成岩蚀变对团簇同位素的改造等方面开展工作。另外，超高分辨率稳定同位素质谱的引进有助于推动我国 CH_4、O_2、N_2 和 N_2O 等体系的团簇同位素研究。

关键词：稳定同位素　团簇同位素　地质温度计　CO_2　CH_4

0　引　言

古温度的精确重建是古气候学研究的重点和难点，碳酸盐的氧同位素温度计是一种经典的古温度重建方法，该温度计基于 Urey（1947）发现的同位素物质的热力学性质，即平衡条件下碳酸钙从水中沉淀时，其氧同位素组成（$\delta^{18}O$）与温度呈函数关系，这也是稳定同位素地球化学的奠基石（郑永飞和陈江峰，2000）。基于与海水平衡的生物成因碳酸盐的氧同位素温度计经验方程（Epstein et al.，1951，1953），科学家利用有孔虫的氧同位素组成首次获得了更新世海水温度演化的记录（Emiliani，1955），极大地推动了古海洋学和古气候学的发展。但是，碳酸盐的 $\delta^{18}O$ 不仅受控于其沉淀时水体的温度，还受水体 $\delta^{18}O$ 的影响，因此在实际应用时还需要知道过去海水的 $\delta^{18}O$ 值的变化。在这种情况下，通常是假设海水的 $\delta^{18}O$ 在地质历史时期比较稳定，仅有约 0.5‰ 的变化（Shackleton，1967），或者利用孔隙水的 $\delta^{18}O$ 值制约历史时期的海水 $\delta^{18}O$ 值（Schrag et al.，1996，2002）。然而，海水的 $\delta^{18}O$ 值在不同时间尺度上会受到各种因素如蒸发、降水、冰盖效应等的影响（Schrag et al.，2002；Jaffrés et al.，2007；Rohling，2013）。另外，对于生物成因的碳酸盐，如生物的碳酸盐壳体或骨骼形成时的 $\delta^{18}O$ 会受到自生"生命效应（vital effects）"的影响（McConnaughey，1989；Erez，2003）。因此，碳酸盐氧同位素温度计的应用具有一定的局限性。

尽管随后建立起来的有孔虫 Mg/Ca 温度计、珊瑚 Sr/Ca 温度计等在古海水温度重建方面得到了重要的应用，但这些地球化学指标还是会受到海水相应成分和生命效应的影响（de Villiers et al.，1994，1995；Segev and Erez，2006；Hathorne et al.，2009）。因此，长期以来，找到一种不受外部环境影响的古温度重建方法一直是地球科学家梦寐以求的目标。随着科学技术的快速发展，团簇同位素地球化学（clumped isotope geochemistry）的建立使得利用独立于环境水体的地质温度计进行古温度重建成为可能。团簇同位素地球化学是美国加州理工学院的 John M. Eiler 教授及其同事在 2004—2007 年间发展起来的稳定同位素地球化学新的研究领域（Eiler and Schauble，2004；Wang et al.，2004；Eiler，2007），在过去的十几年快速发展，并成为稳定同位素地球化学的热点研究领域之一，在与古温度相关的研究领域得到了广泛应用。

---

* 原文"近十年我国团簇同位素地球化学研究进展"刊于《矿物岩石地球化学通报》2020 年第 39 卷第 5 期，本文略有修改。

我国团簇同位素地球化学的发展紧跟国际前沿，最早是由中国科学院地球化学研究所刘耘研究员在2007年引入我国，目前已在理论及应用方面取得了一批优秀的成果。本文总结了近十多年来我国团簇同位素地球化学在理论计算、分析方法研究、实验模拟和应用研究等方面取得的进展，并指出了存在的问题和未来的发展方向，期望为国内同行了解我国在这一领域的研究现状和未来开展研究工作提供参考。

1　团簇同位素地球化学

传统稳定同位素地球化学研究通常是研究天然样品的整体同位素组成（bulk isotopic compositions），即单个稀有重同位素的丰度，例如 CO_2 分子中的 $^{13}C^{16}O^{16}O$（$\delta^{13}C$）或者 $^{12}C^{18}O^{16}O$（$\delta^{18}O$）（Hoefs，2018）。但自然界中还存在浓度非常低的含有两种或者两种以上的稀有重同位素的分子（例如 $^{13}C^{18}O^{16}O$ 或 $^{13}C^{18}O^{17}O$），称为"多元取代同位素体"（multiply-substituted isotopologues）；这些同位素体具有相同的元素组成和化学结构，但具有不同的同位素组成和质量数（Eiler and Schauble，2004；Wang et al.，2004）。早期的研究认识到这种多元取代同位素体具有和相同分子的单一取代同位素体（singly substituted isotopologues）不同的独特热动力学特征（Bigeleisen and Mayer，1947；Urey 1947），因此其自然丰度可以为地质、地球化学和宇宙化学过程提供独特的制约（Wang et al.，2004）。但是，由于自然界中大部分多元取代同位素体的浓度极低（ppb-ppm级），受仪器分析的限制，直到2004年美国加州理工学院的科学家才首次实现了对 CO_2 体系多元取代同位素体组成的精准测量，并由此认识到重同位素相互键合在单相的分子中存在额外的热力学优势，即重同位素会因为温度变化而改变相互键合的程度（Eiler and Schauble，2004）。至此，一个新的稳定同位素研究领域——团簇同位素地球化学诞生并在近十多年得到迅速发展（Eiler，2007，2011）。

目前，团簇同位素地球化学领域研究程度较深的是碳酸盐团簇同位素，如利用碳酸盐晶格中离子团 $^{13}C^{18}O^{16}O^{16}O^{2-}$ 的相对含量和温度的相关性建立的团簇同位素温度计来指示碳酸盐矿物的形成温度（Ghosh et al.，2006a），其指示的温度与矿物晶格内部含有团簇同位素的化学键的排序有关，并独立于整个矿物的碳、氧同位素，理论上碳酸盐的 $\delta^{13}C$ 和 $\delta^{18}O$ 值的变化并不影响其团簇同位素组成（Affek，2012）。该温度计依赖于碳酸盐矿物单相体系的同位素体中两个重同位素的交换反应：$Ca^{12}C^{18}O^{16}O_2 + Ca^{13}C^{16}O_3 = Ca^{13}C^{18}O^{16}O_2 + Ca^{12}C^{16}O_3$。当反应达到热力学平衡时，上述反应式中含有 $^{13}C—^{18}O$ 键的碳酸根的含量与反应的平衡常数有关，而平衡常数又受控于温度，因此碳酸盐团簇同位素组成理论上可直接与矿物的形成温度建立联系（Schauble et al.，2006）。

目前还无法直接对矿物晶格中碳酸根的 $^{13}C—^{18}O$ 键含量或者团簇同位素组成——Δ_{63}（$^{13}C^{18}O^{16}O_2^{2-}$）进行测量，因此，在实际测量和应用中，一般是通过磷酸酸解碳酸盐生成 CO_2，并利用高分辨率稳定同位素比质谱测量其中质量数为47的同位素体 $^{13}C^{18}O^{16}O$ 的信号，通常以 Δ_{47} 值来表示和量化其团簇同位素组成（Eiler，2007），即

$$\Delta_{47} = [(R_{47}/R_{47}^* - 1) - (R_{46}/R_{46}^* - 1) - (R_{45}/R_{45}^* - 1)] \times 1000 \text{ (‰)}$$

式中，R_i 为质量数 i 与质量数44的实测同位素比值；R_i^* 表示在同位素体中同位素随机组合时的同位素比值（Eiler，2007）。基于量子力学和统计热力学的原理建立的温度（K）和 CO_2 团簇同位素组成 Δ_{47} 的理论关系式表达为

$$\Delta_{47} = 0.003(1000/T)^4 - 0.0438(1000/T)^3 + 0.2553(1000/T)^2 - 0.2195(1000/T) + 0.0616$$

在 270~1 000 K 范围内 Δ_{47} 的变化约 1‰（Wang et al.，2004）。国际上首个利用人工合成的无机方解石在25℃下磷酸酸解获得的 CO_2 的 Δ_{47} 和合成温度进行拟合校正获得的 Δ_{47}-T 校正方程为

$$\Delta_{47} = 0.0592 \times 10^6 \times T^{-2} - 0.02$$

适用范围为 1~50℃，在目前测量精度条件下的温度误差为 ±2℃（Ghosh et al.，2006a）。与传统碳酸盐氧同位素（$\delta^{18}O$）温度计相比（McCrea，1950；Urey et al.，1951），Δ_{47} 温度计的独特优势在于它不受碳酸盐形成时流体同位素组成的影响（Eiler，2011），因而目前被广泛应用于地球科学领域与温度相关的各种科

学问题，如古气候（Came et al., 2007；Dennis et al., 2013；Henkes et al., 2018；Meyer et al., 2019）和古高程（Ghosh et al., 2006b；Huntington et al., 2010；Zhang et al., 2018；Botsyun et al., 2019）的重建等。

2 "clumped isotope" 的中文译名

中国科学院地球化学研究所刘耘研究员团队在我国较早开展团簇同位素地球化学相关的研究。他们基于当时对 CO_2 体系中 ^{13}C-^{18}O、CH_4 体系中的 ^{13}C-D 等二元同位素体的研究，首次将"clumped isotope"翻译为"二元同位素"（唐茂等，2007；刘琪等，2009），随后，国内其他学者在发表论文时使用"耦合同位素"（马秀峰等，2012）、"双重（zhòng）同位素"（胡斌，2014）、"团簇同位素"（李平平等，2017；郭炀锐等，2019）、"簇同位素"（帅燕华等，2018）等译名。其中，"团簇同位素"这一译名最早由郑永飞院士倡导使用，2018年11月中国矿物岩石地球化学会同位素地球化学专业委员会经过讨论也建议选择该译名。"clumped isotope"本身的定义并不局限于两个稀有重同位素，随着科学技术的发展，未来有可能对含有两个以上的稀有重同位素的同位素体进行研究，另外"clump"一词本身具有"成团、成簇"的意思。因此，本文认为"团簇同位素"这一译名充分体现了专业术语翻译时遵循本意、通俗易懂、约定俗成等原则，建议国内学者统一使用这一翻译名称。目前，对"团簇同位素"这一翻译，国内研究学者已经达成共识并在发表学术研究论文时广泛应用（李平平等，2017；王晓锋等，2018；郭炀锐等，2019；徐秋晨等，2019；刘雨晨等，2020），2019年9月召开的第二届全国气体同位素技术与地球科学应用研讨会也以"团簇同位素"命名会议专题名称。

3 我国团簇同位素地球化学平台建设和分析技术研究进展

团簇同位素丰度极低，以 CO_2 体系为例，任意两个稀有重同位素（^{13}C、^{17}O 和 ^{18}O）的结合均组成团簇同位素，其中丰度最高的 $^{13}C^{18}O^{16}O$ 也仅有约 $44×10^{-6}$，因此需要灵敏度极高的同位素比质谱；多数团簇同位素比值在自然温度环境下的变化差异可能不超过 10^{-3}，在实际应用时为了满足温度分辨率，一般要求测试的比值精度至少好于 10^{-5}；此外，相同质量数的同位素体（分子）的干扰以及样品气体在纯化过程中与痕量水的交换反应干扰也是高精度团簇同位素组成分析需要解决的问题。

早期对 Δ_{47} 的测试是通过对 Finigan MAT 253 型气体稳定同位素比质谱加装额外法拉第杯使其可以接收 m/e 为 44-49 的信号来实现的（Eiler and Schauble, 2004）。中国科学院地质与地球物理研究所王旭研究员的团队在我国较早开展团簇同位素分析尝试，他们与仪器公司合作对其 Finigan MAT 253 型气体稳定同位素比质谱增配法拉第杯，并在我国首次建立了 Δ_{47} 的分析方法，内部精度可达到 0.010‰~0.014‰，外部精度可达到 0.010‰~0.020‰（Cui and Wang, 2014）。他们还首次在样品提取前处理过程中对 Porapak™ Q 吸附剂纯化 CO_2 的效率和可靠性进行了系统评估，为团簇同位素分析过程中 CO_2 气体提取方法的优化提供了非常有价值的参考（Wang et al., 2016a）。

北京大学周力平教授团队在和瑞士联邦苏黎世理工学院科研人员进行合作研究时发现，经典的双路进样系统进行 Δ_{47} 测量时通过样品和工作参考气体的多次对比和长时间积分提高测量精度。然而，该方法需要较大的样品量（~10 mg 纯碳酸盐），其中一个重要原因是传统双路进样系统对样品的利用率较低，主要原因包括：第一，要求波纹管（bellow）或毛细管中待测气体的压力至少达到 1 500~2 000 Pa，在测量结束时仍有部分气体未被利用，在使用波纹管模式时剩余气体超过 90%，使用微体积模式时也有 20%~30%；第二，压力平衡过程消耗了部分样品，在使用微体积模式时这部分气体可占 ~50%；第三，当使用双路进样方法测量工作气时，样品被切换阀（changeover valve, COV）送至废气泵缓慢排出，在使用微体积模式时，这部分气体可占到 ~10%。为此，他们依托瑞士联邦苏黎世理工学院稳定同位素实验室的 MAT 253 型同位素比质谱，建立了一种新的测量方法——"长积分双路进样（long-integration dual-inlet，LIDI)"方法，该方法改进了传统双路进样流程，避免了压力平衡过程和频繁多次的样品-工作参考气体对比以及COV切换，在相同机时内可进行更长时间积分，使得在信号的计数统计上具更低的散粒噪声（shot-noise），且能够得到与传统双路进样方法可比的精度和准确度，同时还使样品的利用率远高于传统的双路进样法，从而大大降低了团簇同位素测量所需的样品量，可以对~200 μg的有孔虫壳体进行团簇同位素测量（Hu et al.，2014；胡斌等，2015）。

中国科学院广州地球化学研究所韦刚健研究员团队也较早开展了Δ_{47}分析方法研究。他们最早在配备10个法拉第杯的Isoprime 100型的紧凑型稳定同位素比质谱（5 kV，$m/\Delta m$=110）上尝试建立分析方法，发现该型号同位素比质谱仪在性能上很难满足获得长期稳定的高精度Δ_{47}结果的需求。后来，他们利用自主搭建的CO_2提取和纯化装置结合新引进的253 Plus型大磁场稳定同位素比质谱（10 kV，$m/\Delta m$>900）成功建立了Δ_{47}的分析方法，内部精度可达到0.005‰～0.013‰，长期外部精度可达到~0.015‰（图1）。事实上，Finigan MAT 253型气体稳定同位素比质谱是Δ_{47}分析最常用的仪器，早期开展团簇同位素研究的实验室包括加州理工学院、加州大学洛杉矶分校、耶鲁大学、约翰霍普金斯大学、瑞士联邦苏黎世理工学院等的实验室都是利用该型号仪器（Eiler and Schauble，2004；Schauble et al.，2006；Huntington et al.，2009；Schmid and Bernasconi，2010；Passey and Henkes，2012），到目前为止，利用Isoprime 100型稳定同位素比质谱发表的数据主要是来自美国杜兰大学的稳定同位素实验室（Rosenheim et al.，2013；Fernandez et al.，2014；Tang et al.，2014）和法国巴黎萨克雷大学的稳定同位素实验室（Daëron et al.，2016；Peral et al.，2018）。

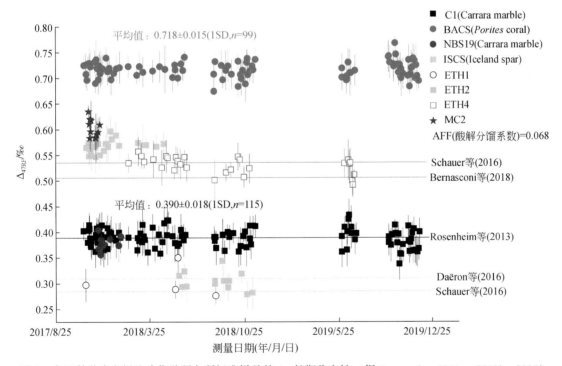

图1 中国科学院广州地球化学研究所标准样品的Δ_{47}长期稳定性（据Guo et al.，2019a，2019b，2020）

中国地质大学（武汉）是迄今为止国内第三家公开发表自己实验室产出Δ_{47}数据的单位。他们基于MAT 253型稳定同位素比质谱对一系列碳酸盐标准样品进行长期的Δ_{47}测量，发现在90℃条件下，方解石和白云石的酸解分馏系数没有统计学差异，并认为选择不同的同位素比值参数对Δ_{47}的计算会产生显著的影响，另外还建议利用碳酸盐标准物质代替绝对参考体系以提高不同实验室Δ_{47}测量结果的重现性和可比性（Chang et al.，2020a）。

近年来，随着国家对科技事业支持力度的持续加大，一些单位如中国科学院地球环境研究所、青藏高原研究所等都相继购置了新一代的253 Plus型稳定同位素比质谱开展团簇同位素研究。另外，2019年中国科学院南京地质古生物研究所引进了国内第一台商业化的用于 CO_2 体系团簇同位素分析的 IBEX 2 型气体自动提取纯化装置，并将该装置与 253 Plus 型稳定同位素比质谱联用，实现了 Δ_{47} 的自动在线分析。与以往 CO_2 的手动离线提取纯化装置相比，该自动装置可大大提高 Δ_{47} 的分析效率，减少人为干扰。

此外，上海海洋大学、中国科学院广州地球化学研究所、中国地质科学院矿产资源研究所等单位还购置了 Nu Panorama 型、天津大学等购置了 Thermo 253 Ultra 型新一代超高分辨率（$m/\Delta m$ 分别达 ~40 000 和 ~30 000）稳定同位素比质谱，利用这类超高分辨率的稳定同位素比质谱将能开展 O_2、N_2、CH_4、N_2O 等体系的团簇同位素研究（Eiler et al., 2013；Young et al., 2016），有望进一步推动我国团簇同位素地球化学相关研究的快速发展。

4 我国团簇同位素地球化学理论计算研究进展

早期的团簇同位素研究主要是基于理论计算来开展的，如基于量子力学和统计热力学等原理建立的温度与 CO_2 气体 Δ_{47} 的关系表明，在 270~1 000 K 范围内，Δ_{47} 的变化约 1‰（Wang et al., 2004）。基于对不同碳酸盐的团簇同位素温度效应的理论研究（Schauble et al., 2006），方解石、文石、白云石和菱镁矿等多种碳酸盐在 25℃ 磷酸酸解温度下和 260~1 500 K 范围内的 Δ_{47} 理论温度计被建立起来（Guo et al., 2009）。我国开展团簇同位素地球化学理论计算研究的主要是刘耘研究员团队，2007 年他们在我国首次引入团簇同位素研究时，就采用高级量子化学计算方法预测了不同温度下天然气中 CH_4 和 CO_2 气体的团簇同位素的特征，尝试利用两种团簇同位素对天然气的形成温度和迁移过程进行双重准确限定，虽然具体的应用有待于实验数据的验证，但该项研究提供了天然气成因分析的新方法（唐茂等，2007）。他们还对碳酸盐 ^{13}C-^{18}O 团簇同位素 Δ_{47}-T 关系的校正曲线进行了研究，发现磷酸酸解碳酸盐过程中同位素动力学分馏效应对 Δ_{47}-T 校正曲线具有重要影响（张思亭和刘耘，2013），对分子水平上团簇同位素动力学分馏机理的进一步研究发现，在磷酸酸解碳酸盐的过程中可能存在 H_3PO_4 催化、H_2CO_3 催化和 H_2O 催化等 3 条不同的反应路径，这些反应路径对 ^{13}C-^{18}O 团簇同位素动力学分馏均产生一定的贡献，基于此获得的新的理论 Δ_{47}-T 关系的校正曲线并不是一条孤立的直线，而是随着这 3 个反应路径贡献大小在一定的范围内变化，这可能解释了不同实验室在实验流程仅有细微差别的情况下获得的 Δ_{47}-T 校正曲线存在显著差异的原因（张思亭和刘耘，2017；Zhang et al., 2020）。

早期的团簇同位素精确计算方法在实际操作中只限于质量数很小的分子，质量数稍大的体系往往需要借助近似计算方法（Wang et al., 2004），而近似计算方法具有一定的局限性，例如不能处理含质量数相等的同位素体系，这在很大程度上限制了计算方法的应用范围；另外，该方法也只适应于单一取代同位素体 Δ_i 值接近于零的体系（Wang et al., 2004）。针对上述问题，刘耘团队对团簇同位素的计算方法进行了改进，他们将其中最关键的一个变量 Δ_i 的原始定义转换为等价的基于能量的定义，并使一些内在性质和定理凸显出来，建立了一种新的近似计算方法并成功解决了上述早期近似计算方法存在的问题，新的近似计算方法不仅有更大的应用范围，而且同精确方法的结果非常接近，并可以进一步使用校正处理使结果同精确方法几乎一样（Cao and Liu, 2012）。

团簇同位素研究总体上还处于初级阶段，为了解决不同研究团队采用的分馏表达式不一致的问题，刘耘团队重新定义了数学形式上更为严谨的团簇同位素分馏表达式，该定义具有明确的物理化学意义，即团簇同位素分馏信号代表着多元取代同位素体从单一取代同位素体获得稀有同位素的程度或能力。与加州理工学院 John Eiler 团队的定义相比，他们的定义不但与传统的地学同位素分馏定义相吻合，而且通过与反应常数建立联系计算使结果与实验学的测量结果更为契合（Liu and Liu, 2016）。另外，考虑到对团簇同位素分馏效应的预测多基于简谐近似，而简谐近似并不适用于所有同位素体系的分馏预测，尤其是含氢同位素交换的体系，他们利用超越简谐近似的量子化学方法，为 H_2O、H_2S、SO_2、NH_3 和 CH_4 等多个分子体系的团簇同位素平衡分馏信号提供了精确的团簇同位素平衡分馏参数，并在理论层面上提出了团簇同位素平衡分馏的多个共性（刘琪和刘耘，2013，2015；Liu and Liu，2016）。其研究结果显示，不同体系的团簇同位素平衡分馏信号可以与同位素取代的相对质量差以及所在原子位的键的强度建立关系，其分馏信号随着温度的升高而减小，在温度小于 500 K 的区间与 $1\,000/T$ 具有良好的线性关系。超越简谐近似的高阶能量项对于研究含氢氘交换的团簇同位素体系较为重要，其贡献值是不含氢氘交换的团簇同位素体系的 2～20 倍。他们还发现，团簇同位素体系的平衡分馏多与某些特定模式的分子振动有关（如分子伸缩振动），所以不同分子体系可以出现相似的团簇同位素平衡分馏信号，如 CH_4 和 CH_3Cl 的 ^{13}C-D 团簇同位素平衡分馏信号（Liu and Liu，2016）。此外，他们还提出了利用频率校正因子修正团簇同位素分馏温度依赖性的数学形式，这为提高团簇同位素分馏的理论预测精度提供了便利的方法（Liu and Liu，2016）。

5 我国团簇同位素地球化学应用研究进展

总体上，近十年我国团簇同位素地球化学的应用研究不算多，还处于起步阶段。这主要是受制于我国团簇同位素分析技术，虽然有不少单位引进了高分辨率的稳定同位素质谱，但目前能开展分析测试的并不多。迄今为止，国内实验室能产出团簇同位素数据并开展应用研究的只有中国科学院地质与地球物理研究所、中国科学院广州地球化学研究所和中国地质大学（武汉）3 家单位（Cui and Wang，2014；Wang et al.，2016a，2016b；Zhai et al.，2019；Guo et al.，2019a，2019b，2020；Chang et al.，2020a，2020b；Xiong et al.，2020），其他单位均是同国外单位如美国加州理工学院、加州大学洛杉矶分校、约翰霍普金斯大学以及瑞士联邦苏黎世理工学院等合作开展研究（Hu et al.，2014；Zhang et al.，2016，2018；Shuai et al.，2018a，2018b；Wu et al.，2018；Ning et al.，2019；Dong et al.，2020；郑剑锋等，2017；胡安平等，2018；徐秋晨等，2019；刘雨晨等，2020）。以下尽可能收集了近十年我国学者在团簇同位素地球化学应用方面取得的进展并进行了总结。需要说明的是，本文收集的均是已公开发表的研究成果，最近几年我国培养了很多团簇同位素地球化学研究方向的研究生，其学位论文中尚未公开或不宜公开的成果尚未包括。另外，虽然国内团簇同位素地球化学相关的应用研究比较广泛，但总体上都还处于起步阶段，大多数还只是比较零散的案例研究，研究的系统性有待完善。

5.1 古温度重建和气候变化

如前所述，与传统的 Urey 温度计基于两相或多相体系的同位素交换反应不同，碳酸盐 Δ_{47} 温度计是基于单一物相内部的同位素交换反应，由于交换反应发生在碳酸盐晶格内部，独立于外部环境的影响，在古温度重建和古气候研究中具有重要的应用价值（马秀峰等，2012；季顺川等，2013；李平平等，2017）。对于 Δ_{47} 温度计应用于珊瑚、石笋、土壤和湖泊碳酸盐来进行古温度重建和古气候研究及其需要注意的问题可以参见 Eiler（2011）的文章。最近的一项关于古温度重建应用非常有意义的研究是对波多黎各的前哥伦比亚人丢弃的双壳类外壳的 Δ_{47} 的分析结果，发现大多数贝壳被加热到 100℃ 以上但低于 200℃，这表明它们是被煮沸（标准大气压下，100℃）以外的方法（如烧烤）所烹制的（Staudigel et al.，2019），这对了解人类文明进程提供了非常重要的信息。

中国地质大学（北京）的研究人员测定了"白垩纪松辽盆地大陆科学钻探"项目岩心中古土壤碳酸盐的 Δ_{47}，并利用 Δ_{47} 温度计重建了相对连续的白垩纪-古近纪界线附近的陆相古温度记录（Zhang et al.，2018）。他们发现，在假说的小行星撞击地球之前约 30 万年，温度和大气二氧化碳浓度有明显升高，这与德干火山喷发的时间高度一致；另外，在德干火山喷发之后、小行星撞击之前，松辽盆地约 2/3 的物种发生了绝灭。这一结果可能表明，德干火山喷发导致剧烈的升温和二氧化碳浓度上升，破坏了生态系统的稳定性，造成了部分物种的绝灭（Zhang et al.，2018）。

中国石油大学（北京）的研究人员利用湖相叠层石的 Δ_{47} 重建了准噶尔盆地南缘晚新生代古温度变化

记录，结果表明，在叠层石沉积早—中期，Δ_{47}反演的碳酸盐岩生长古温度由~38℃显著降低至~32℃，反映同期准噶尔盆地南缘古气候经历了逐步趋于干冷的过程；但叠层石堆叠后期记录了再度显著升高的古温度信号（~42℃），与叠层石上部碎屑矿物组分的特征增量相吻合，有可能反映同时期大量碎屑碳酸盐岩组分的输入对Δ_{47}温度计造成了干扰，从而导致相关模型对碳酸盐岩形成古温度的过高预测（Yang et al.，2019）。

中国科学院南京地质古生物研究所的研究人员利用准噶尔盆地吉木萨尔大龙口剖面土壤碳酸盐结核Δ_{47}重建了西北晚二叠世—早三叠世温度记录，结果发现Δ_{47}值在0.58‰~0.72‰变化，对应的温度变化范围为22~55℃，虽然温度范围较大，但在整个剖面识别出两次由暖变冷又变暖的趋势，并且其碳、氧同位素值的变化均在合理范围，支持了结果的合理性；这样的结果表明，在晚二叠世—早三叠世中国西北大陆内部气候存在波动，但和生物演变、灭绝和环境的关系仍需要进一步研究（崔璨和曹长群，2018）。

5.2　生物碳酸盐团簇同位素分馏

碳酸盐Δ_{47}组成不受碳酸盐沉淀时流体组成的影响，能直接指示碳酸盐沉淀时的环境温度，具有传统地质温度计所不具备的独特优势，在古温度重建中具有重要的应用价值。作为古温度重建的重要载体，生物碳酸盐的Δ_{47}是目前团簇同位素研究领域的热点。但对于生物碳酸盐形成过程中的生物因素是否会影响Δ_{47}对温度的指示，生物碳酸盐在沉积埋藏、成岩蚀变过程中是否能保存原有的Δ_{47}组成等影响Δ_{47}温度计准确性的问题还存在广泛争议。近年来的研究发现，一些生物体如浅水珊瑚（Saenger et al.，2012）、深水珊瑚（Kimball et al.，2016；Spooner et al.，2016）、头足类动物壳体（Dennis et al.，2013）、腕足类动物壳体（Bajnai et al.，2018）和海胆骨骼（Davies and John，2019）中的碳酸盐Δ_{47}值都会受到不同程度的生命效应的影响。另外，保存完好的有孔虫壳体可以获得可靠的Δ_{47}温度，但发生严重蚀变的有孔虫壳体Δ_{47}值重建的温度则明显偏低（Leutert et al.，2019）。我国研究人员对陆生蜗牛和海洋珊瑚这两种典型生物碳酸盐的Δ_{47}进行了系统研究，获得了一系列具有国际显示度的成果（Wang et al.，2016b；Zhai et al.，2019；Guo et al.，2019b，2020；Dong et al.，2020）。

中国科学院地质与地球物理研究所、广州地球化学研究所和地球环境研究所的科研人员均对陆生蜗牛Δ_{47}开展了研究，虽然获得的结果不尽一致，但总体上体现了我国学者对蜗牛壳体Δ_{47}研究的系统性。中国科学院地质与地球物理研究所的研究团队采集了我国北方6个研究点的两种陆生蜗牛并测定其壳体Δ_{47}，结果表明 Cathaica sp. 和 Bradybaena sp. 两种蜗牛的Δ_{47}温度都和环境温度显著相关，但 Cathaica sp. 的Δ_{47}温度比 Bradybaena sp. 的Δ_{47}温度高3~5℃。具体而言，Cathaica sp. 喜欢生活在温暖潮湿的夏季，而 Bradybaena sp. 喜欢在相对冷干的冬天和（或）秋天活动。这样的结果表明陆生蜗牛壳体的Δ_{47}是季风区一种有潜力的古温度计，同时也表明季节性的气候变化对陆生蜗牛的群落组合具有重要的影响（Wang et al.，2016b）。他们进一步采集了我国从南到北68个研究点的 Cathaica sp. 和 Bradybaena sp. 两种蜗牛总共177个样本并测定其壳体Δ_{47}，结果表明壳体Δ_{47}温度和蜗牛生长季节温度具有很好的相关性，在此基础上分别推导出了两种蜗牛壳体Δ_{47}与生长季节平均温度的转化方程并验证了其可靠性，由此他们认为陆生蜗牛壳体Δ_{47}可以作为古温度计应用于古气候研究（Zhai et al.，2019）。另外，他们还发现通过蜗牛壳体Δ_{47}温度和壳体$\delta^{18}O$计算出的蜗牛体液$\delta^{18}O$组成具有重建其生长季节大气降雨$\delta^{18}O$的潜力（Zhai et al.，2019）。广州地球化学研究所的研究团队采集了我国华南到华北季风区的4种常见陆生蜗牛并测定其壳体Δ_{47}，结果表明壳体Δ_{47}温度与器测气温估计的蜗牛生长温度关系并不显著，蜗牛壳体可能因生理因素影响（如冬眠）偏向记录温暖适宜的环境温度（其生长的微环境温度）。尽管如此，利用Δ_{47}温度和壳体$\delta^{18}O$获得的蜗牛体液$\delta^{18}O$可能在一定湿度条件下能指示降雨的$\delta^{18}O$（Guo et al.，2019b）。需要指出的是上述两个研究团队关于利用Δ_{47}温度和壳体$\delta^{18}O$获得的蜗牛体液$\delta^{18}O$指示降雨$\delta^{18}O$的结论是各自独立研究获得的，两篇文章接受日期只相差3天，相互印证了各自的结论（Zhai et al.，2019；Guo et al，2019b）。中国科学院地球环境研究所的研究团队研究了洛川和渭南黄土剖面现代和末次冰期蜗牛的壳体Δ_{47}，结果表明现

代蜗牛壳体平均 Δ_{47} 温度和其生长时期的温度一致，并比末次冰期蜗牛壳体的 Δ_{47} 温度高 ~10℃；另外，他们发现 Δ_{47} 温度和壳体 $\delta^{18}O$ 获得的蜗牛体液 $\delta^{18}O$ 比蜗牛生长时期降雨的 $\delta^{18}O$ 偏正很多，这可能与强的蒸发作用有关（Dong et al., 2020）。

珊瑚骨骼碳酸盐在形成过程中通常会受到生物因素的影响，使得同位素组成无法达到热力学平衡，结果导致滨珊瑚的 Δ_{47} 值比在相同温度条件下形成的无机碳酸钙的值要偏高，这种现象是与生物因素相关的团簇同位素不平衡分馏的典型例子。然而，对不同的珊瑚个体之间以及同个珊瑚个体内部在 Δ_{47} 值不平衡分馏上是否有显著差异目前尚未明确，这将影响到珊瑚 Δ_{47} 不平衡分馏的有效校正和重建海水温度的准确性。针对以上科学问题，广州地球化学研究所的科研人员对滨珊瑚骨骼的 Δ_{47} 进行了研究，结果发现珊瑚个体内部同一时期不同生长部位的 Δ_{47} 值之间存在显著的差异而 $\delta^{18}O$ 值的差异较小（图2）。这种 Δ_{47}-$\delta^{18}O$ 的分馏关系可能指示了一种新型生物分馏现象，有别于以往研究认为的与 CO_2 扩散、羟基化、水合化反应相关的同位素非平衡分馏机制。尽管珊瑚骨骼钙化过程中同位素的分馏普遍受到其本身生物因素的影响，但总体上珊瑚 Δ_{47} 值依然与海水温度保持显著的相关性（图3），从而具有指示季节性海水温度的应用价值，这进一步巩固了滨珊瑚团簇同位素在重建海水古温度方面的应用潜能（Guo et al., 2020）。

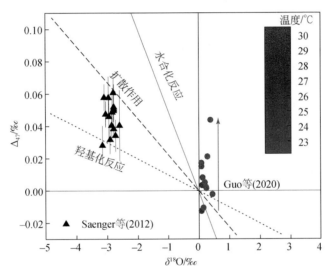

图2　滨珊瑚个体内部骨骼 Δ_{47} 和 $\delta^{18}O$ 在相同生长时期的不平衡分馏偏差（据 Guo et al., 2020）

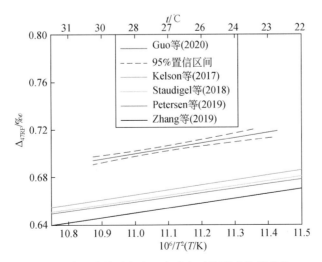

图3　滨珊瑚骨骼与人工合成和天然形成的碳酸盐的 Δ_{47} 和温度的校正关系对比（据 Guo et al., 2020）

5.3 古高程重建

Δ_{47}温度计不受碳酸盐形成时外部流体影响,可以独立获得碳酸盐的形成温度,因此,通过Δ_{47}温度计,既可直接利用碳酸盐岩的$\delta^{18}O$获得原始大气降水的$\delta^{18}O$,进而通过瑞利分馏模型或大气降水的$\delta^{18}O$与高度的经验关系重建古高度,又能利用地表高度/温度梯度计算古高度,两种方法的相互验证可以大大地提高了古高度重建的准确度(熊中玉等,2019)。由于传统氧同位素重建古高程的误差可以达到1000 m以上,近年来,利用Δ_{47}温度计进行古高程重建的误差通常小于500 m,国外研究学者对青藏高原、安第斯山脉等的隆升历史和深部动力学机制提供了重要的制约(Ghosh et al., 2006b; Huntington et al., 2015; Ingalls et al., 2018)。

中国地质大学(北京)的研究人员测量了古土壤碳酸盐Δ_{47}值,并对结果进行了季节性偏好、纬度差异和长期气候变化的校正后,利用地表高度-温度梯度重建了晚白垩世山东胶莱盆地的古高程,结果表明约80 Ma时胶莱盆地的古高程≥2.0 km,结合地层学、古地理学和古气候学证据,他们的结果支持晚白垩世东亚海岸山脉假说,即东亚大陆边缘的海岸山脉至少从约100 Ma时开始隆生并且在挤压高峰期(约90 Ma)应该大大高于2 km(Zhang et al., 2016)。他们还研究了西藏南部雅鲁藏布江缝合带始新世柳区砾岩古土壤碳酸盐结核的Δ_{47}特征,结果表明Δ_{47}温度远高于始新世合理的热带地表温度,可能受到了后期埋藏作用的改造,最大的埋藏温度约为(96.8±8.4)℃,相应的埋藏深度约为3.7~4.3 km,根据隆升过程中的Δ_{47}值及计算得到的水体$\delta^{18}O$值,推测中新世柳区砾岩的古高程约为3 km(Ning et al., 2019)。

中国地质大学(武汉)的研究人员测量了青藏高原东南缘剑川盆地始新世九子岩组下部和双河组的泥灰岩、钙质泥岩中介形虫壳体的Δ_{47},利用Δ_{47}温度计算获得了大气降水的$\delta^{18}O$值,然后通过瑞利分馏模型重建古高程,得到九子岩组古高程为(2.5±0.7) km(2σ),双河组古高程为(2.9±0.6) km(2σ),由此可以推测剑川盆地在约36 Ma之前就开始隆升,且在约36 Ma时其古高程已接近现代海拔高度(Wu et al., 2018)。中国科学院青藏高原研究所的研究人员测量了青藏高原分水岭山脉东段(横断山)贡觉盆地的古土壤钙结核样品的Δ_{47}值,并根据成岩蚀变的影响剔除了部分明显受后期改造的Δ_{47}数据,同时结合棕榈植物化石、高精度火山岩地层U-Pb年代学和气候模拟研究结果认为:藏东南地区在约54 Ma之前是低海拔(~0.7 km)的干旱炎热的沙漠,而约44 Ma以后开始抬升至与现今相似高度(3.8 km)。这种地壳的隆升可能也影响到了中国东部地区古气候和古生态环境演变(Xiong et al., 2020)。

5.4 天然气组分的团簇同位素

与CO_2的Δ_{47}相比,天然气的主要组分甲烷和乙烷的团簇同位素Δ_{18}($\Delta^{13}CH_3D$)和$\Delta^{13}C_2H_6$的分析测试方法建立得稍晚一些(Ono et al., 2014; Clog et al., 2018)。Δ_{18}温度计可以重建甲烷的形成温度(Stolper et al., 2014a, 2014b),并根据形成温度的不同可以区分热解成因和生物成因甲烷气体的来源比例(Stolper et al., 2015),正在成为天然气研究的重要手段并在天然气中烷烃气体的形成温度、形成路径以及形成机制的研究方面发挥重要作用,从而有望解决天然气的形成机理、成藏机制等科学问题(帅燕华等,2018;王晓锋等,2018)。

中国石油勘探开发研究院的科研人员通过与加州理工学院John Eiler等合作,测定了我国松辽盆地徐家围子天然气田中甲烷的Δ_{18}并重建其形成温度,结果显示甲烷的形成温度为167~213℃,低于岩浆气生成过程的温度并与当地烃源岩成熟度一致,结合$\delta^{13}C$、δD和$^3He/^4He$结果,表明松辽盆地天然气是热成因的,来源于过成熟烃源岩即高温下埋藏有机质的分解(Shuai et al., 2018a)。

5.5 海相碳酸盐岩成岩环境及热演化过程判别

海相碳酸盐岩是地球气候环境演化的记录载体,也是石油和天然气的重要储集层,对其成岩环境的准确判别对了解早期地球表层的环境演化历史、油气评价和预测及勘探具有重要的指导意义,传统研究碳酸盐岩成岩温度的方法如流体包裹体测温和$\delta^{18}O$温度计都具有一定的局限性,如找不到合适的包裹体、$\delta^{18}O$温度计会受成岩流体的影响等。碳酸盐Δ_{47}温度计可以获得准确的成岩温度,结合碳酸盐$\delta^{18}O$温度计还可以获得流体的$\delta^{18}O$组成。因此,在揭示碳酸盐岩成岩温度和成岩流体性质进而研究其成岩环境和成岩作用(Ferry et al.,2011;Meister et al.,2013;Isabel Millán et al.,2016;Lawson et al.,2018)、恢复沉积盆地热历史(Winkelstern and Lohmann,2016;Mangenot et al.,2018;Lacroix and Niemi,2019;Naylor et al.,2019)等方面具有重要的应用价值。

中国地质大学(武汉)的科研人员联合多国研究团队对我国长江三峡樟村坪地区距今5亿年前的埃迪卡拉纪陡山陀组(635~551 Ma)沉积跨度超过6 300万年白云岩地层开展了高分辨的碳酸盐团簇同位素温度研究,并结合白云石热动力学模型模拟,发现陡山陀组白云石形成温度范围为0~60℃;他们进一步结合稀土元素、流体氧同位素组成分析和显微岩相学研究,认为陡山陀组白云岩形成于一个低温且微生物活跃的海水溶液环境,这与近年来实验室和野外观察所发现的现代白云石在微生物及其有机质作用下可低温形成的机制相符,为早期地球海洋大规模白云石的成因解释提供了新的思路(Chang et al.,2020b)。

中石油杭州地质研究院的科研人员与美国加州大学洛杉矶分校合作,开展了利用Δ_{47}研究塔里木盆地、四川盆地和西沙群岛等地的碳酸盐储集层的成岩环境和成因,获得如下认识:①塔里木盆地中下寒武统颗粒白云岩形成于低温的准同生、浅埋藏环境,成岩流体为海水;细晶白云岩为深埋藏成岩环境中原岩受到了高温重结晶作用改造的产物,成岩流体为地下热卤水;孔缝中的白云石胶结物是深埋藏成岩环境富镁热卤水沉淀作用的产物(郑剑锋等,2017);②四川张村剖面茅口组斑马状白云石的围岩发生重结晶改造,Δ_{47}温度明显偏高,流体反映海水和热液卤水混合特征,鞍状白云石为典型热液作用产物;而长兴组生物礁灰岩围岩形成于海底成岩环境,块状亮晶方解石是埋藏环境下地层卤水作用形成的胶结物充填破坏孔隙(胡安平等,2018);③西沙群岛中新统白云岩形成于正常-轻微咸化海水环境(胡安平等,2018)。

中国石油大学(北京)的科研人员通过与加州大学洛杉矶分校、加州理工学院等合作,尝试利用Δ_{47}恢复塔里木盆地和四川盆地奥陶纪以来的热演化历史,获得如下结论:①Δ_{47}温度远高于可能的成岩温度,可能是受到了后期埋藏升温作用的影响,发生了团簇同位素^{13}C—^{18}O键的固态重排,塔里木顺托果勒-卡塔克地区自然演化方解石团簇同位素^{13}C—^{18}O键固态重排的"封闭温度"不高于120℃,热力学"平衡温度"不低于120℃,Δ_{47}对高温热历史(>120℃)有较好的敏感性(徐秋晨等,2019);②对于方解石来说,^{13}C—^{18}O键固态重排热演化的一阶近似模型能够较为准确地描述Δ_{47}在漫长的地质过程中的变化,而白云石Δ_{47}的热演化模型还需要进一步研究(徐秋晨等,2019);③塔里木盆地顺北、顺托和顺南地区泥晶基质的团簇同位素温度Δ_{47}平均值分别为92.34℃、124.35℃和170.27℃,通过对顺托果勒地区典型单井设置不同的热路径,进而明确最高埋藏温度的上限170~190℃,不同区域之间略有差异(刘雨晨等,2020);④二叠纪的异常高温可能是地层抬升剥蚀和岩浆活动热事件共同作用的结果,顺托果勒地区二叠纪地温梯度范围为26~46℃/km,呈现西北高东南低的趋势,地温梯度的空间变化表明可能与地幔柱活动有关(刘雨晨等,2020)。

5.6 团簇同位素实验地球化学模拟

实验地球化学模拟可以为更好地理解团簇同位素的理论和假说提供实验证据,例如,对碳酸盐团簇同位素^{13}C—^{18}O键在不同物相状态下、不同物相界面的重排已有大量模拟实验的开展(Passey and Henkes,2012;Affek,2013;Tripati et al.,2015;Brenner et al.,2018;Lloyd et al.,2018),这对深入理解地球化学过程中团簇同位素的分馏行为和控制机理,为其应用于解决实际科学问题提供了更好的制约,尤其是现今团簇同位素地球化学发展的初级阶段,团簇同位素的实验模拟显得相当重要。我国科研人员已经认识到实验模拟的重要性并已经开展了相关的研究，并取得了在国际上有显示度的成果（Shuai et al., 2018b; Guo et al., 2019a）。

中国石油勘探开发研究院的科研人员通过与加州理工学院 John Eiler 等的合作，开展了封闭体系中不含水的煤样及含水的页岩样品的热解实验，测得了产物中甲烷的 Δ_{18}，并将甲烷 Δ_{18}、热解实验温度与热力学平衡线进行比较。结果发现，热成因甲烷在特定阶段也呈现热力学不平衡状态（含量最低的 Δ_{18} 值甚至低于 0，为负值），而其他阶段甲烷簇同位素均表现为平衡特征。不平衡甲烷的生成过程对应于乙烷规模裂解阶段，他们分析认为这个阶段的乙烷裂解是甲烷产生的一个很重要的来源，而乙烷裂解来源的甲烷同位素分馏效应会明显加强，导致甲烷的簇同位素也展示动力学分馏效应而未达到平衡，这从甲烷簇同位素不平衡程度跟乙烷裂解的程度一致可以佐证（Shuai et al., 2018b）。这为研究热成因甲烷产生的化学条件提供了新的视角，为某些天然气中甲烷 Δ_{18} 记录的形成温度偏高提供了一种可能的解释（Shuai et al., 2018b）。

在碳酸盐团簇同位素的实际应用中，获得准确的 Δ_{47} 温度的前提条件是碳酸盐在形成过程中，溶解无机碳（dissolved inorganic carbon，DIC）$^{13}C-^{18}O$ 成键时达到同位素平衡。利用 DIC 氧同位素的组成特征来评估团簇同位素的平衡状态是目前普遍思路，但前人研究表明，氧同位素在 DIC 与钙化流体的交换反应速率会因其与不同质量数的碳同位素成键而有一定的动力学差异，意味着利用氧同位素评估团簇同位素平衡状态可能存在问题。而碳酸盐沉淀过程中不同相态间氧同位素交换与 $^{13}C-^{18}O$ 键达成平衡可能存在速率差异，但目前直接的实验证据还较少。中国科学院广州地球化学研究所的科研人员通过文石-方解石转变实验，获得了 $H_2O\text{-}DIC\text{-}CaCO_3$ 体系中 Δ_{47} 和 $\delta^{18}O$ 的动力学分馏特征的直接证据。他们发现，在文石转变成方解石过程中，溶解的 DIC 离子与溶液的 $\delta^{18}O$ 交换仅达到部分平衡，而 Δ_{47} 在转化的温度和反应介质条件下未发生明显的变化，基本处于完全不平衡状态，这提供了在 DIC 沉淀为方解石碳酸盐过程中 $\delta^{18}O$ 交换与 $^{13}C-^{18}O$ 键平衡存在动力学差异的直接证据（Guo et al., 2019a）。文石向方解石转化过程中 Δ_{47} 基本不变，这指示在转化过程中 $^{13}C-^{18}O$ 键基本没有发生重组，意味着 Δ_{47} 记录了文石结晶时的温度信息，这对于古气候古环境演变研究具有非常重要的意义。

6 问题与展望

团簇同位素的研究从起步到现在也不过十多年的时间，随着研究的深入，对团簇同位素研究的一些问题也凸显出来，如同位素是否平衡、生物碳酸盐生命效应的影响、埋藏成岩蚀变对团簇同位素的改造等。这些问题是整个团簇同位素研究领域都面临的问题，我国科研人员也不例外。解决这些问题一个重要的途径就是开展团簇同位素实验地球化学的研究，如 CO_2 的水合反应和去气过程是许多生物成因和无机碳酸盐形成的关键反应，相关的模拟实验不仅能验证已有理论计算结果（Guo, 2020），也有望揭示 Δ_{47} 的生物（动力学）分馏的新机制。另外，对于碳酸盐在后期改造中团簇同位素的分馏虽然破坏了原始的环境温度信息，但由于同位素的再平衡过程使得 Δ_{47} 值还可指示最后达到平衡的温度信息，这在地壳隆升、埋藏速率等方面有重要的应用价值，但前提是要明确这当中涉及的 $^{13}C-^{18}O$ 键的固态重排以及在溶液状态下 DIC 中 $^{13}C-^{18}O$ 键的重置机理。然而，目前除了方解石、文石以外其他类似矿物结构的碳酸盐，如不同有序度和 Mg 含量的白云石，其 Δ_{47} 的固态重置动力学速率还缺乏认识，而实验地球化学的模拟研究可为此提供重要的思路。再如，对于 CO_2 体系，不同团簇同位素的不平衡程度之间可能存在一定的相关性，利用 $\Delta_{48}\text{-}\Delta_{47}$ 体系在理论上可校正 Δ_{47} 的不平衡分馏，但由于缺乏 Δ_{48} 动力学分馏的实验证据，开展相关的模拟实验有望深入认识碳酸盐 $\Delta_{48}\text{-}\Delta_{47}$ 的分馏机理提供理论和实验依据，有助于利用 $\Delta_{48}\text{-}\Delta_{47}$ 体系获得更准确的温度信息。中国科学院广州地球化学研究所的团队正在开展上述这些方面的实验地球化学模拟研究。

总体而言，近十年来我国团簇同位素地球化学研究从无到有，虽然成果不多，但这些成果涉及了地球化学研究的各个方面，如理论计算、分析方法研究、实验模拟和应用研究等，并且具有较好的国际显示度，使得我国团簇同位素地球化学的研究基本与国际同行同步，其发展态势良好。随着近年来越来越多的实验室正在建立或已经建立了适用于 CO_2 体系的团簇同位素分析平台，相关的研究将进一步扩大和深入；另外，超高分辨率稳定同位素质谱的陆续引进，为我国科研人员在 CH_4、O_2、N_2 和 N_2O 等体系的团簇同位素研究领域取得突破提供了很好的条件。因此，有理由相信，未来十年我国团簇同位素地球化学将会得到快速、健康发展。

参 考 文 献

崔璨，曹长群. 2018. 土壤碳酸盐结核二元同位素反映中国西北晚二叠世—早三叠世气候波动. 见：中国古生物学会第十二次全国会员代表大会暨第29届学术年会论文摘要集(郑州)，225-226

郭炀锐，邓文峰，韦刚健. 2019. 珊瑚碳酸盐团簇同位素研究进展. 矿物岩石地球化学通报，38(4)：855-866

胡安平，沈安江，潘立银，王永生，李娴静，韦东晓. 2018. 二元同位素在碳酸盐岩储层研究中的作用. 天然气地球科学，29(1)：17-27

胡斌. 2014. 双重同位素的测量及其方法的改进. 北京：北京大学博士学位论文

胡斌，Radke J，Schlüter H J，Heine F T，周力平，Bernasconi S M. 2015. 长积分双路进样(LIDI)方法及其在Clumped同位素测量中的应用. 地质学报，89(S1)：106-109

季顺川，彭廷江，聂军胜，彭文彬. 2013. 黄土高原微生物膜类脂物和碳酸盐二元同位素重建古温度的研究进展. 海洋地质与第四纪地质，33(3)：151-158

李平平，马倩倩，邹华耀，余新亚. 2017. 团簇同位素的基本原理与地质应用. 古地理学报，19(4)：713-728

刘琪，刘耘. 2013. 采用超越简谐水平对 H_2O、H_2S、SO_2、NH_3 和 CH_4 的"clumped"同位素平衡分馏信号的理论预测. 高校地质学报，19(S1)：506

刘琪，刘耘. 2015. 有机化合物"Clumped"同位素平衡分馏信号的理论预测. 矿物学报，35(S1)：667

刘琪，唐茂，刘耘. 2009. 二元同位素(clumped isotope)方法的介绍. 矿物岩石地球化学通报，28(S1)：93

刘雨晨，邱楠生，常健，贾京坤，李慧莉，马安来. 2020. 碳酸盐团簇同位素在沉积盆地热演化中的应用——以塔里木盆地顺托果勒地区为例. 地球物理学报，63(2)：597-611

马秀峰，张兆峰，严爽，韦刚健，邓文峰，孙卫东. 2012. 耦合同位素简介. 地球环境学报，3(4)：950-959

帅燕华，张水昌，胡国艺. 2018. 天然气组分的簇同位素研究进展. 矿物岩石地球化学通报，37(4)：559-571

唐茂，赵辉，刘耘. 2007. 天然气中甲烷和 CO_2 的二元同位素特征. 矿物学报，27(3-4)：396-399

王晓锋，刘鹏，刘昌杰，孟强. 2018. 团簇同位素及特定位置同位素组成在天然气地球化学研究中的应用. 矿物岩石地球化学通报，37(4)：580-587

熊中玉，丁林，谢静. 2019. 碳酸盐耦合同位素(Δ_{47})温度计及其在古高度重建中的应用. 科学通报，64(16)：1722-1737

徐秋晨，邱楠生，刘雯，常青. 2019. 利用团簇同位素恢复沉积盆地热历史的探索. 科学通报，64(5-6)：566-578

张思亭，刘耘. 2013. 碳酸盐中 ^{13}C-^{18}O "clumped"同位素 Δ_{47} 值与温度的理论校准线. 高校地质学报，19(S1)：514-515

张思亭，刘耘. 2017. 动力学分馏对碳酸盐中 ^{13}C-^{18}O "clumped"同位素及三氧同位素关系的影响. 矿物岩石地球化学通报，36(S1)：500

郑剑锋，李晋，季汉成，黄理力，胡安平，马明璇. 2017. 二元同位素测温技术及其在白云岩储层成因研究中的应用——以塔里木盆地中下寒武统为例. 海相油气地质，22(2)：1-7

郑永飞，陈江峰. 2000. 稳定同位素地球化学. 北京：科学出版社

Affek H P. 2012. Clumped isotope paleothermometry: Principles, applications, and challenges. Paleontological Society Papers, 18: 101-114

Affek H P. 2013. Clumped isotopic equilibrium and the rate of isotope exchange between CO_2 and water. American Journal of Science, 313(4): 309-325

Bajnai D, Fiebig J, Tomašových A, Milner Garcia S, Rollion-Bard C, Raddatz J, Löffler N, Primo-Ramos C, Brand U. 2018. Assessing kinetic fractionation in brachiopod calcite using clumped isotopes. Scientific Reports, 8(1): 533

Bigeleisen J, Mayer M G. 1947. Calculation of equilibrium constants for isotopic exchange reactions. Journal of Chemical Physics, 15(5): 261-267

Botsyun S, Sepulchre P, Donnadieu Y, Risi C, Licht A, Caves Rugenstein J K. 2019. Revised paleoaltimetry data show low Tibetan Plateau elevation during the Eocene. Science, 363(6430): eaaq1436, doi: 10.1126/science.aaq1436

Brenner D C, Passey B H, Stolper D A. 2018. Influence of water on clumped-isotope bond reordering kinetics in calcite. Geochimica et Cosmochimica Acta, 224: 42-63

Came R E, Eiler J M, Veizer J, Azmy K, Brand U, Weidman C R. 2007. Coupling of surface temperatures and atmospheric CO_2 concentrations during the Palaeozoic Era. Nature, 449(7159): 198-201

Cao X B, Liu Y. 2012. Theoretical estimation of the equilibrium distribution of clumped isotopes in nature. Geochimica et Cosmochimica Acta, 77: 292-303

Chang B, Defliese W F, Li C, Huang J H, Tripati A, Algeo T J. 2020a. Effects of different constants and standards on the reproducibility of carbonate clumped isotope (Δ_{47}) measurements: insights from a long-term dataset. Rapid Communications in Mass Spectrometry, 34(8): e8678, doi: 10.1002/rcm.8678

Chang B, Li C, Liu D, Foster I, Tripati A, Lloyd M K, Maradiaga I, Luo G M, An Z H, She Z B, Xie S C, Tong J N, Huang J H, Algeo T J, Lyons T W, Immenhauser A. 2020b. Massive formation of early diagenetic dolomite in the Ediacaran ocean: constraints on the "dolomite problem".

Proceedings of the National Academy of Sciences of the United States of America, 117(25): 14005-14014

Clog M, Lawson M, Peterson B, Ferreira A A, Santos Neto E V, Eiler J M. 2018. A reconnaissance study of $^{13}C-^{13}C$ clumping in ethane from natural gas. Geochimica et Cosmochimica Acta, 223: 229-244

Cui L L, Wang X. 2014. Determination of clumped isotopes in carbonate using isotope ratio mass spectrometer: effects of extraction potential and long-term stability. International Journal of Mass Spectrometry, 372: 46-50

Daëron M, Blamart D, Peral M, Affek H P. 2016. Absolute isotopic abundance ratios and the accuracy of Δ_{47} measurements. Chemical Geology, 442: 83-96

Davies A J, John C M. 2019. The clumped ($^{13}C-^{18}O$) isotope composition of echinoid calcite: further evidence for "vital effects" in the clumped isotope proxy. Geochimica et Cosmochimica Acta, 245: 172-189

de Villiers S, Shen G T, Nelson B K. 1994. The Sr Ca-temperature relationship in coralline aragonite: influence of variability in $(SrCa)_{seawater}$ and skeletal growth parameters. Geochimica et Cosmochimica Acta, 58(1): 197-208

de Villiers S, Nelson B K, Chivas A R. 1995. Biological controls on coral Sr/Ca and $\delta^{18}O$ reconstructions of sea surface temperatures. Science, 269(5228): 1247-1249

Dennis K J, Cochran J K, Landman N H, Schrag D P. 2013. The climate of the Late Cretaceous: new insights from the application of the carbonate clumped isotope thermometer to western interior seaway macrofossil. Earth and Planetary Science Letters, 362: 51-65

Dong J B, Eiler J, An Z S, Wu N Q, Liu W G, Li X Z, Kitchen N, Lu F Y. 2020. Clumped and stable isotopes of land snail shells on the Chinese Loess Plateau and their climatic implications. Chemical Geology, 533: 119414

Eiler J M. 2007. "Clumped-isotope" geochemistry—the study of naturally-occurring, multiply-substituted isotopologues. Earth and Planetary Science Letters, 262(3-4): 309-327

Eiler J M. 2011. Paleoclimate reconstruction using carbonate clumped isotope thermometry. Quaternary Science Reviews, 30(25-26): 3575-3588

Eiler J M, Schauble E. 2004. $^{18}O^{13}C^{16}O$ in Earth's atmosphere. Geochimica et Cosmochimica Acta, 68(23): 4767-4777

Eiler J M, Clog M, Magyar P, Piasecki A, Sessions A, Stolper D, Deerberg M, Schlueter H J, Schwieters J. 2013. A high-resolution gas-source isotope ratio mass spectrometer. International Journal of Mass Spectrometry, 335: 45-56

Emiliani C. 1955. Pleistocene Temperatures. The Journal of Geology, 63(6): 538-578

Epstein S, Buchsbaum R, Lowenstam H A, Urey H C. 1951. Carbonate-water isotopic temperature scale. Geological Society of America Bulletin, 62(4): 417-426

Epstein S, Buchsbaum R, Lowenstam H A, Urey H C. 1953. Revised carbonate-water isotopic temperature scale. Geological Society of America Bulletin, 64(11): 1315-1326

Erez J. 2003. The source of ions for biomineralization in foraminifera and their implications for paleoceanographic proxies. Reviews in Mineralogy and Geochemistry, 54(1): 115-149

Fernandez A, Tang J W, Rosenheim B E. 2014. Siderite "clumped" isotope thermometry: a new paleoclimate proxy for humid continental environments. Geochimica et Cosmochimica Acta, 126: 411-421

Ferry J M, Passey B H, Vasconcelos C, Eiler J M. 2011. Formation of dolomite at 40-80℃ in the Latemar carbonate buildup, Dolomites, Italy, from clumped isotope thermometry. Geology, 39(6): 571-574

Ghosh P, Adkins J, Affek H, Balta B, Guo W F, Schauble E A, Schrag D, Eiler J M. 2006a. $^{13}C-^{18}O$ bonds in carbonate minerals: a new kind of paleothermometer. Geochimica et Cosmochimica Acta, 70(6): 1439-1456

Ghosh P, Garzione C N, Eiler J M. 2006b. Rapid uplift of the Altiplano revealed through $^{13}C-^{18}O$ bonds in paleosol carbonates. Science, 311(5760): 511-515

Guo W F. 2020. Kinetic clumped isotope fractionation in the $DIC-H_2O-CO_2$ system: patterns, controls, and implications. Geochimica et Cosmochimica Acta, 268: 230-257

Guo W F, Mosenfelder J L, Goddard III W A, Eiler J M. 2009. Isotopic fractionations associated with phosphoric acid digestion of carbonate minerals: insights from first-principles theoretical modeling and clumped isotope measurements. Geochimica et Cosmochimica Acta, 73(24): 7203-7225

Guo Y R, Deng W F, Wei G J. 2019a. Kinetic effects during the experimental transition of aragonite to calcite in aqueous solution: insights from clumped and oxygen isotope signatures. Geochimica et Cosmochimica Acta, 248: 210-230

Guo Y R, Deng W F, Wei G J, Lo L, Wang N. 2019b. Clumped isotopic signatures in land-snail shells revisited: possible palaeoenvironmental implications. Chemical Geology, 519: 83-94

Guo Y R, Deng W F, Wei G J, Chen X F, Liu X, Wang X J, Lo L, Cai G Q, Zeng T. 2020. Exploring the temperature dependence of clumped isotopes in modern *Porites* corals. Journal of Geophysical Research: Biogeosciences, 125(1): e2019JG005402

Hathorne E C, James R H, Lampitt R S. 2009. Environmental versus biomineralization controls on the intratest variation in the trace element composition of the planktonic foraminifera *G. inflata* and *G. scitula*. Paleoceanography, 24(4): PA4204

Henkes G A, Passey B H, Grossman E L, Shenton B J, Yancey T E, Pérez-Huerta A. 2018. Temperature evolution and the oxygen isotope composition of Phanerozoic oceans from carbonate clumped isotope thermometry. Earth and Planetary Science Letters, 490: 40-50

Hoefs J. 2018. Stable Isotope Geochemistry, 8th ed. Cham: Springer International Publishing

Hu B, Radke J, Schlüter H J, Heine F T, Zhou L P, Bernasconi S M. 2014. A modified procedure for gas-source isotope ratio mass spectrometry: the long-integration dual-inlet (LIDI) methodology and implications for clumped isotope measurements. Rapid Communications in Mass Spectrometry, 28

(13): 1413-1425

Huntington K W, Eiler J M, Affek H P, Guo W, Bonifacie M, Yeung L Y, Thiagarajan N, Passey B, Tripati A, Daëron M, Came R. 2009. Methods and limitations of "clumped" CO_2 isotope (Δ_{47}) analysis by gas-source isotope ratio mass spectrometry. Journal of Mass Spectrometry, 44(9): 1318-1329

Huntington K W, Wernicke B P, Eiler J M. 2010. Influence of climate change and uplift on Colorado Plateau paleotemperatures from carbonate clumped isotope thermometry. Tectonics, 29(3): TC3005, doi: 10.1029/2009TC002449

Huntington K W, Saylor J, Quade J, Hudson A M. 2015. High late Miocene-Pliocene elevation of the Zhada Basin, southwestern Tibetan Plateau, from carbonate clumped isotope thermometry. Geological Society of America Bulletin, 127(1-2): 181-199

Ingalls M, Rowley D, Olack G, Currie B, Li S Y, Schmidt J, Tremblay M, Polissar P, Shuster D L, Lin D, Colman A. 2018. Paleocene to Pliocene low-latitude, high-elevation basins of southern Tibet: implications for tectonic models of India-Asia collision, Cenozoic climate, and geochemical weathering. GSA Bulletin, 130(1-2): 307-330

Isabel Millán M, Machel H, Bernasconi S M. 2016. Constraining temperatures of formation and composition of dolomitizing fluids in the upper devonian nisku formation (alberta, Canada) with clumped isotopes. Journal of Sedimentary Research, 86(1): 107-112

Jaffrés J B D, Shields G A, Wallmann K. 2007. The oxygen isotope evolution of seawater: a critical review of a long-standing controversy and an improved geological water cycle model for the past 3.4 billion years. Earth-Science Reviews, 83(1): 83-122

Kelson J R, Huntington K W, Schauer A J, Saenger C, Lechler A R. 2017. Toward a universal carbonate clumped isotope calibration: diverse synthesis and preparatory methods suggest a single temperature relationship. Geochimica et Cosmochimica Acta, 197: 104-131

Kimball J, Eagle R, Dunbar R. 2016. Carbonate "clumped" isotope signatures in aragonitic scleractinian and calcitic gorgonian deep-sea corals. Biogeosciences, 13(23): 6487-6505

Lacroix B, Niemi N A. 2019. Investigating the effect of burial histories on the clumped isotope thermometer: an example from the Green River and Washakie Basins, Wyoming. Geochimica et Cosmochimica Acta, 247: 40-58

Lawson M, Shenton B J, Stolper D A, Eiler J M, Rasbury E T, Becker T P, Phillips-Lander C M, Buono A S, Becker S P, Pottorf R, Gray G G, Yurewicz D, Gournay J. 2018. Deciphering the diagenetic history of the El Abra Formation of eastern Mexico using reordered clumped isotope temperatures and U-Pb dating. GSA Bulletin, 130(3-4): 617-629

Leutert T J, Sexton P F, Tripati A, Piasecki A, Ho S L, Meckler A N. 2019. Sensitivity of clumped isotope temperatures in fossil benthic and planktic foraminifera to diagenetic alteration. Geochimica et Cosmochimica Acta, 257: 354-372

Liu Q, Liu Y. 2016. Clumped-isotope signatures at equilibrium of CH_4, NH_3, H_2O, H_2S and SO_2. Geochimica et Cosmochimica Acta, 175: 252-270

Lloyd M K, Ryb U, Eiler J M. 2018. Experimental calibration of clumped isotope reordering in dolomite. Geochimica et Cosmochimica Acta, 242: 1-20

Mangenot X, Gasparrini M, Rouchon V, Bonifacie M. 2018. Basin-scale thermal and fluid flow histories revealed by carbonate clumped isotopes (Δ_{47})-Middle Jurassic carbonates of the Paris Basin depocentre. Sedimentology, 65(1): 123-150

McConnaughey T. 1989. ^{13}C and ^{18}O isotopic disequilibrium in biological carbonates: I. patterns. Geochimica et Cosmochimica Acta, 53(1): 151-162

McCrea J M. 1950. On the isotopic chemistry of carbonates and a paleotemperature scale. Journal of Chemical Physics, 18(6): 849-857

Meister P, Mckenzie J A, Bernasconi S M, Brack P. 2013. Dolomite formation in the shallow seas of the Alpine Triassic. Sedimentology, 60(1): 270-291

Meyer K W, Petersen S V, Lohmann K C, Blum J D, Washburn S J, Johnson M W, Gleason J D, Kurz A Y, Winkelstern I Z. 2019. Biogenic carbonate mercury and marine temperature records reveal global influence of Late Cretaceous Deccan Traps. Nature Communications, 10(1): 5356

Naylor H N, Defliese W F, Grossman E L, Maupin C R. 2019. Investigation of the thermal history of the Delaware Basin (West Texas, USA) using carbonate clumped isotope thermometry. Basin Research, doi: 10.1111/bre.12419

Ning Z J, Zhang L M, Huntington K W, Wang C S, Dai J G, Han Z P, Passey B H, Qian X Y, Zhang J W. 2019. The burial and exhumation history of the Liuqu Conglomerate in the Yarlung Zangbo suture zone, southern Tibet: insights from clumped isotope thermometry. Journal of Asian Earth Sciences, 174: 205-217

Ono S, Wang D T, Gruen D S, Sherwood Lollar B, Zahniser M S, McManus B J, Nelson D D. 2014. Measurement of a doubly substituted methane isotopologue, $^{13}CH_3D$, by tunable infrared laser direct absorption spectroscopy. Analytical Chemistry, 86(13): 6487-6494

Passey B H, Henkes G A. 2012. Carbonate clumped isotope bond reordering and geospeedometry. Earth and Planetary Science Letters, 351-352: 223-236

Peral M, Daëron M, Blamart D, Bassinot F, Dewilde F, Smialkowski N, Isguder G, Bonnin J, Jorissen F, Kissel C, Michel E, Vázquez Riveiros N, Waelbroeck C. 2018. Updated calibration of the clumped isotope thermometer in planktonic and benthic foraminifera. Geochimica et Cosmochimica Acta, 239: 1-16

Petersen S V, Defliese W F, Saenger C, Daëron M, Huntington K W, John C M, Kelson J R, Bernasconi S M, Colman A S, Kluge T, Olack G A, Schauer A J, Bajnai D, Bonifacie M, Breitenbach S F M, Fiebig J, Fernandez A B, Henkes G A, Hodell D, Katz A, Kele S, Lohmann K C, Passey B H, Peral M Y, Petrizzo D A, Rosenheim B E, Tripati A, Venturelli R, Young E D, Winkelstern I Z. 2019. Effects of improved ^{17}O correction on interlaboratory agreement in clumped isotope calibrations, estimates of mineral-specific offsets, and temperature dependence of acid digestion fractionation. Geochemistry, Geophysics, Geosystems, 20(7): 3495-3519

Rohling E J. 2013. Paleoceanography, physical and chemical proxies | Oxygen isotope composition of seawater. In: Elias S A, Mock C J(eds). Encyclopedia of Quaternary Science, 2nd ed. Amsterdam: Elsevier, 915-922

Rosenheim B E, Tang J W, Fernandez A. 2013. Measurement of multiply substituted isotopologues ("clumped isotopes") of CO_2 using a 5 kV compact isotope ratio mass spectrometer: performance, reference frame, and carbonate paleothermometry. Rapid Communications in Mass Spectrometry, 27(16): 1847-1857

Saenger C, Affek H P, Felis T, Thiagarajan N, Lough J M, Holcomb M. 2012. Carbonate clumped isotope variability in shallow water corals: temperature dependence and growth-related vital effects. Geochimica et Cosmochimica Acta, 99: 224–242

Schauble E A, Ghosh P, Eiler J M. 2006. Preferential formation of ^{13}C-^{18}O bonds in carbonate minerals, estimated using first-principles lattice dynamics. Geochimica et Cosmochimica Acta, 70(10): 2510–2529

Schmid T W, Bernasconi S M. 2010. An automated method for "clumped-isotope" measurements on small carbonate samples. Rapid Communications in Mass Spectrometry, 24(14): 1955–1963

Schrag D P, Hampt G, Murray D W. 1996. Pore Fluid constraints on the temperature and oxygen isotopic composition of the glacial ocean. Science, 272(5270): 1930–1932

Schrag D P, Adkins J F, McIntyre K, Alexander J L, Hodell D A, Charles C D, McManus J F. 2002. The oxygen isotopic composition of seawater during the Last Glacial Maximum. Quaternary Science Reviews, 21(1-3): 331–342

Segev E, Erez J. 2006. Effect of Mg/Ca ratio in seawater on shell composition in shallow benthic foraminifera. Geochemistry, Geophysics, Geosystems, 7(2): Q02P09, doi: 10.1029/2005GC000969

Shackleton N. 1967. Oxygen isotope analyses and Pleistocene temperatures re-assessed. Nature, 215(5096): 15–17

Shuai Y H, Etiope G, Zhang S C, Douglas P M J, Huang L, Eiler J M. 2018a. Methane clumped isotopes in the Songliao Basin (China): new insights into abiotic vs. biotic hydrocarbon formation. Earth and Planetary Science Letters, 482: 213–221

Shuai Y H, Douglas P M J, Zhang S C, Stolper D A, Ellis G S, Lawson M, Lewan M D, Formolo M, Mi J K, He K, Hu G Y, Eiler J M. 2018b. Equilibrium and non-equilibrium controls on the abundances of clumped isotopologues of methane during thermogenic formation in laboratory experiments: implications for the chemistry of pyrolysis and the origins of natural gases. Geochimica et Cosmochimica Acta, 223: 159–174

Spooner P T, Guo W F, Robinson L F, Thiagarajan N, Hendry K R, Rosenheim B E, Leng M J. 2016. Clumped isotope composition of cold-water corals: a role for vital effects? Geochimica et Cosmochimica Acta, 179: 123–141

Staudigel P T, Murray S, Dunham D P, Frank T D, Fielding C R, Swart P K. 2018. Cryogenic brines as diagenetic fluids: reconstructing the diagenetic history of the Victoria Land Basin using clumped isotopes. Geochimica et Cosmochimica Acta, 224: 154–170

Staudigel P T, Swart P K, Pourmand A, Laguer-Díaz C A, Pestle W J. 2019. Boiled or roasted? Bivalve cooking methods of early Puerto Ricans elucidated using clumped isotopes. Science Advances, 5(11): eaaw5447

Stolper D A, Lawson M, Davis C L, Ferreira A A, Neto E V S, Ellis G S, Lewan M D, Martini A M, Tang Y, Schoell M, Sessions A L, Eiler J M. 2014a. Formation temperatures of thermogenic and biogenic methane. Science, 344(6191): 1500–1503

Stolper D A, Sessions A L, Ferreira A A, Santos Neto E V, Schimmelmann A, Shusta S S, Valentine D L, Eiler J M. 2014b. Combined ^{13}C-D and D-D clumping in methane: methods and preliminary results. Geochimica et Cosmochimica Acta, 126: 169–191

Stolper D A, Martini A M, Clog M, Douglas P M, Shusta S S, Valentine D L, Sessions A L, Eiler J M. 2015. Distinguishing and understanding thermogenic and biogenic sources of methane using multiply substituted isotopologues. Geochimica et Cosmochimica Acta, 161: 219–247

Tang J W, Dietzel M, Fernandez A, Tripati A K, Rosenheim B E. 2014. Evaluation of kinetic effects on clumped isotope fractionation (Δ_{47}) during inorganic calcite precipitation. Geochimica et Cosmochimica Acta, 134: 120–136

Tripati A K, Hill P S, Eagle R A, Mosenfelder J L, Tang J W, Schauble E A, Eiler J M, Zeebe R E, Uchikawa J, Coplen T B, Ries J B, Henry D. 2015. Beyond temperature: clumped isotope signatures in dissolved inorganic carbon species and the influence of solution chemistry on carbonate mineral composition. Geochimica et Cosmochimica Acta, 166: 344–371

Urey H C. 1947. The thermodynamic properties of isotopic substances. Journal of the Chemical Society, doi: 10.1039/JR9470000562

Urey H C, Lowenstam H A, Epstein S, McKinney C R. 1951. Measurement of paleotemperatures and temperatures of the upper cretaceous of England, Denmark, and the southeastern United States. GSA Bulletin, 62(4): 399–416

Wang X, Cui L L, Li Y Y, Huang X F, Zhai J X, Ding Z L. 2016a. Determination of clumped isotopes in carbonate using isotope ratio mass spectrometry: toward a systematic evaluation of a sample extraction method using a static Porapak™ Q absorbent trap. International Journal of Mass Spectrometry, 403: 8–14

Wang X, Cui L L, Zhai J X, Ding Z L. 2016b. Stable and clumped isotopes in shell carbonates of land snails *Cathaica* sp. and *Bradybaena* sp. in North China and implications for ecophysiological characteristics and paleoclimate studies. Geochemistry, Geophysics, Geosystems, 17(1): 219–231

Wang Z R, Schauble E A, Eiler J M. 2004. Equilibrium thermodynamics of multiply substituted isotopologues of molecular gases. Geochimica et Cosmochimica Acta, 68(23): 4779–4797

Winkelstern I Z, Lohmann K C. 2016. Shallow burial alteration of dolomite and limestone clumped isotope geochemistry. Geology, 44(6): 467–470

Wu J, Zhang K X, Xu Y D, Wang G C, Garzione C N, Eiler J, Leloup P H, Sorrel P, Mahéo G. 2018. Paleoelevations in the Jianchuan Basin of the southeastern Tibetan Plateau based on stable isotope and pollen grain analyses. Palaeogeography, Palaeoclimatology, Palaeoecology, 510: 93–108

Xiong Z Y, Ding L, Spicer R A, Farnsworth A, Wang X, Valdes P J, Su T, Zhang Q H, Zhang L Y, Cai F L, Wang H Q, Li Z Y, Song P P, Guo X D, Yue Y H. 2020. The early Eocene rise of the Gonjo Basin, SE Tibet: from low desert to high forest. Earth and Planetary Science Letters, 543: 116312

Yang W, Zuo R S, Wang X, Song Y, Jiang Z X, Luo Q, Zhai J X, Wang Q Y, Zhang C, Zhang Z Y. 2019. Sensitivity of lacustrine stromatolites to Cenozoic tectonic and climatic forcing in the southern Junggar Basin, NW China: new insights from mineralogical, stable and clumped isotope compositions. Palaeogeography, Palaeoclimatology, Palaeoecology, 514: 109–123

Young E D, Rumble III D, Freedman P, Mills M. 2016. A large-radius high-mass-resolution multiple-collector isotope ratio mass spectrometer for analysis of rare isotopologues of O_2, N_2, CH_4 and other gases. International Journal of Mass Spectrometry, 401: 1–10

Zhai J X, Wang X, Qin B, Cui L L, Zhang S H, Ding Z L. 2019. Clumped isotopes in land snail shells over China: towards establishing a biogenic carbonate paleothermometer. Geochimica et Cosmochimica Acta, 257: 68–79

Zhang L M, Wang C S, Cao K, Wang Q, Tan J, Gao Y. 2016. High elevation of Jiaolai Basin during the Late Cretaceous: implication for the coastal mountains along the East Asian margin. Earth and Planetary Science Letters, 456: 112–123

Zhang L M, Wang C S, Wignall P B, Kluge T, Wan X Q, Wang Q, Gao Y. 2018. Deccan volcanism caused coupled p_{CO_2} and terrestrial temperature rises, and pre-impact extinctions in northern China. Geology, 46(3): 271–274

Zhang N Z, Lin M, Snyder G T, Kakizaki Y, Yamada K, Yoshida N, Matsumoto R. 2019. Clumped isotope signatures of methane-derived authigenic carbonate presenting equilibrium values of their formation temperatures. Earth and Planetary Science Letters, 512: 207–213

Zhang S T, Liu Q, Tang M, Liu Y. 2020. Molecular-level mechanism of phosphoric acid digestion of carbonates and recalibration of the $^{13}C-^{18}O$ clumped isotope thermometer. ACS Earth and Space Chemistry, 4(3): 420–433

The Research Progress of Clumped Isotope Geochemistry

DENG Wen-feng, GUO Yang-rui, WEI Gang-jian

State Key Laboratory of Isotope Geochemistry, Guangzhou Institute of Geochemistry, Chinese Academy of Sciences, Guangzhou 510640

Abstract: Clumped isotope geochemistry concerns bonded stable isotopes in natural materials. The composition of clumped isotopes in a certain mineral is free from the effect of isotopic compositions of the surrounding fluid and is mainly controlled by its formation temperature, therefore can be used a reliable thermometer and has been widely used in Earth sciences research. This paper reviewed the research progress of theoretical calculation, analytical method, experimental simulation and application research in clumped isotope geochemistry in China during the past decade. It suggests that the future work may focus on equilibrium clumped-isotope effects, vital effects in biogenic carbonate and diagenetic reformations by using experimental simulations. In addition, clumped isotope geochemistry of other molecules, such as CH_4, O_2, N_2 and N_2O, will be promoted by the up-to-date technology of ultra-high mass resolution multiple-collector isotope ratio mass spectrometer.

Key words: stable isotope; clumped isotope; geothermometer; CO_2; CH_4

实验地球科学研究进展*

章军锋[1] 倪怀玮[2] 杨晓志[3] 毛 竹[2] 张宝华[4] 熊小林[5] 侯 通[6] 许文良[7]

1. 中国地质大学(武汉) 地球科学学院，武汉 430074；2. 中国科学技术大学 地球和空间科学学院，合肥 230026；3. 南京大学 地球科学与工程学院，南京 210023；4. 浙江大学 地球科学学院，杭州 310027；5. 中国科学院 广州地球化学研究所，同位素地球化学国家重点实验室，广州 510640；6. 中国地质大学(北京) 地球科学与资源学院，北京 100083；7. 吉林大学 地球科学学院，长春 130061

摘 要：过去十年，我国的实验地球科学快速发展，已成为国际高温高压实验领域的一支重要研究力量。代表性研究进展主要体现在地球深部重要挥发分的赋存与效应、成矿过程和机制、壳幔物质的物理化学性质和相互作用等方面。本文综述了实验地球科学 2011—2020 年间取得的主要研究进展和平台与技术建设情况，并展望了该领域未来的发展方向。

关键词：实验地球科学 高温高压 物理化学属性与地质过程

1 学科总体发展概述

实验地球科学（experimental geoscience），或称高温高压实验地球科学，是利用高温高压实验装置和技术在实验室模拟地球内部条件，开展地球内部物质（矿物、岩石、流体、熔体等）属性与过程方面的研究。早期的实验模拟研究偏重于岩石熔融和熔体结晶等内容，习惯上称为实验岩石学；高温高压实验后来被广泛应用到地球科学的多个分支学科，发展为综合性更强的实验地球科学。实验研究为解释矿物学和岩石学现象、地球化学数据以及地球物理探测资料提供了必要依据。实验结果为认识难以直接观测的地球和其他行星的内部结构、过程及其形成和演化提供了强有力的依据。根据实验研究对象和侧重点的不同，实验地球科学可分为实验岩石学、实验地球化学、实验矿床学、实验矿物学、实验矿物物理、实验流变学等分支学科。实验地球科学在精确控制温度、压力和化学组成的条件下，正演模拟地球内部的状态和过程，获取地球内部物质的物理化学参数，揭示控制地质和地球物理现象和过程的物理化学原理，剖析地球形成和演化的规律和机制。实验地球科学具有基础性、前沿性、交叉性等鲜明的学科特点。

尽管我国实验地球科学起步较晚，但通过不懈的人才培养和人才引进、加强国际合作和高水平高温高压实验平台建设，近十年（2011—2020 年）我国在实验地球科学研究上取得了长足进步，突出体现了两个特点：一是起点高，国际化程度高；二是学科跨度大，各领域发展比较均衡。目前，中国地质大学、吉林大学、中国科学技术大学、南京大学、北京大学、浙江大学和中国科学院地球化学研究所、中国科学院广州地球化学研究所、北京高压科学研究中心等单位，都初步建成了国际水平的高温高压实验平台和各具特色的研究队伍。中国的实验地球科学研究人员取得了一系列创新研究成果，在 *Nature*、*Nature Geoscience*、*Nature Communications*、*PNAS*、*NSR*、*Geology*、*EPSL*、*JGR-Solid Earth*、*GRL*、*GCA* 等权威期刊上发表了大量高水平研究论文，已成为国际高温高压实验领域的一支重要研究力量。代表性研究进展主要体现在地球深部重要挥发分的赋存与效应、成矿过程和机制、壳幔物质的物理化学性质和相互作用等方面。

---

* 原文"中国实验地球科学研究进展与展望（2011—2020）"刊于《矿物岩石地球化学通报》2021 年第 40 卷第 3 期，本文略有修改。

2 重要原创理论进展

2.1 地球深部重要挥发分的赋存与效应

氢、碳和氮等重要挥发分对地球的形成与演化有深远的影响。挥发分在地球内部的分布、储存和迁移控制着地球内部物质的物理化学性质、层圈相互作用、地幔对流与板块运动、地幔熔融以及火山作用、矿床的形成。挥发分在地球各圈层之间的循环交互也是大气圈和水圈的演化以及生物圈的形成和演化等地球宜居性的决定性因素。过去十年中，国内的实验地球科学团队在地球深部重要挥发分的赋存状态与效应的实验研究中取得了重要进展和突破。

2.1.1 地幔挥发分物质组成与效应研究

（1）分子氢（H_2）是地幔中比结构羟基更重要的水的赋存方式：南京大学杨晓志团队在地幔中 H 的赋存状态研究中取得重要进展。他们的一系列研究发现，受共存矿物、地质流体和氧逸度等的影响，上地幔中羟基（OH）形式结构水的最大储量其实很低，较前人广为接受的估计值偏低可达 20 倍以上（Yang et al., 2014, 2015），因而结构羟基所能产生的一些物理和化学效应需要重新评估。此外，他们还首次提出在深部地幔高度还原的条件下，分子氢（H_2）可以显著赋存在矿物的晶体结构中。分子氢主要以中性分子形式填充在晶格间隙中（与惰性气体分子相似），其赋存与矿物的结构和成分关系并不大，其储存能力随压力强而显著增大（Yang et al., 2016）。地幔中分子氢的赋存形式的发现，打破了前人认为地幔中的 H 主要是羟基的认识。这还意味着，早期增生和核幔分异过程中，原始大气中占主体的 H_2 可经由星云物质–岩浆海–硅酸盐矿物间的平衡直接存储在成型中的固体地球内部，这为地球上水的起源提供了新的机制。之后随着浅部地球的逐渐氧化，矿物中的 H_2 转化为 H_2O 或 OH，并最终经去气作用和俯冲作用产生了地球上现代形式的水循环。这对认识地球和类地行星上水的起源提供了全新的思路。

（2）水在上地幔和过渡带矿物中的结构特征对弹性性质的影响：浙江大学杨燕等在原位高温高压条件下对上地幔矿物中水的赋存状态进行了研究（Yang et al., 2019a, 2019b），他们发现在上地幔温压环境下，辉石和橄榄石中的氢在晶体结构中呈现无序化，并且发生占位迁移和重新组合。该研究弥补了目前在常温常压下认识地幔矿物中水的赋存方式的不足，为探索水对地幔物理化学性质的影响提供了新依据。中国地质大学（武汉）地球深部研究团队利用高温红外及高温拉曼光谱，测量了一系列含水矿物在地幔温压条件下的稳定性和结构特征（Liu et al., 2019a, 2019b; Zhu et al., 2019）。他们发现对于可以稳定存在于过渡带冷俯冲板片的重要含水矿物水镁石，氢同位素替代会对其非谐效应产生显著影响，进而影响氢同位素在地幔过渡带物相之间的平衡分馏（Zhu et al., 2019）。中国科学技术大学毛竹团队利用金刚石压砧研究了超级含水相 B 的弹性性质（Li X. Y. et al., 2016），发现含有 17%～22% 超级含水相 B 的俯冲板片在过渡带底部和下地幔顶部的堆积可以解释该区域地震波速度异常。此外，他们还利用所发展的外加热金刚石对顶砧技术在高温高压下原位测量了上地幔的主要构成矿物橄榄石的单晶弹性模量（Mao et al., 2015），推断 500 km 的深度处的亚稳态橄榄石具有低速和强横波分裂特征，这为判定深部地震是否由亚稳态橄榄石相变造成提供了重要的矿物学依据。

（3）过渡带金刚石生长的 C-H-O 流体来源：采用激光加热 X 光衍射发现，方解石在过渡带底部存在的新方解石Ⅶ相在俯冲板片温压条件下会与洋壳中的 SiO_2 发生反应，释放出的 CO_2 与水一起为超深金刚石在过渡带的生长提供所需的 C-H-O 流体（Li X. Y. et al., 2018）。

（4）下地幔深部环境存在稳定的含氢过氧化铁相（FeO_2H_x）：北京高压科学研究中心研究团队对高压下 Fe-O-H 体系相变开展了系统的金刚石压砧实验研究工作。胡清扬等（Hu Q. Y. et al., 2016, 2017）发现针铁矿（FeOOH）在下地幔 1 800 km 深度发生部分脱氢作用，形成黄铁矿立方晶体结构的含氢过氧化铁相（FeO_2H_x）；刘锦等（Liu et al., 2017）和毛河光等（Mao et al., 2017）进一步研究发现，俯冲下去的水接触到铁核时将产生 FeO_2H_x 相并释放氢气，水不仅可以与纯铁在下地幔深部高温高压环境下产生 FeO_2H_x 相，也可以与一系列地表常见铁氧化物（如 FeO 和 Fe_3O_4）反应生成 FeO_2H_x 相；刘锦等（Liu et al., 2017）的实验结果表明，FeO_2H_x 相可稳定于下地幔底部极端的温压环境，具有非常高的结构稳定性；张莉等（Zhang L. et al., 2018）发现 FeO_2H_x 相结构中的 Fe 可以被 Al 替代，形成六方晶系的含铝量为 20%～40% 的（Fe, Al）O_2H_x（HH 相）；刘锦等（Liu J. et al., 2019）研究发现 FeO_2 和含氢 FeO_2H_x 相中的铁为 +2 价，氧为 -1 价，同时氢极有可能是游离的 0 价。含氢 FeO_2 相是第一个由金刚石对顶砧实验发现的地球深部含水物质，这对于理解水在下地幔的赋存形式和分布具有重大意义。通常，深部水循环与氢循环被认为是同一个过程。然而，针铁矿在下地幔深部脱氢形成含氢 FeO_2 相，将导致地球内部氢元素循环与氧循环的解耦。同时，含氢 FeO_2 相中氧价态为 -1，而不是常规的 -2，表明在深部下地幔，氧元素将和铁元素一样价态可变。核幔边界累积 10 km 以上厚度的含氢 FeO_2 相可能是超低速区的重要潜在成因机制。这些发现打破了此前的传统观点，给地球深部物质研究带来诸多全新的认识。

2.1.2 早期地球和类地行星的挥发分物质组成与效应

（1）氮和碳在地球早期岩浆海中的赋存形式和环境效应：挥发性元素在地核、地幔及早期大气圈中的分布和丰度直接取决于氮、碳元素在岩浆海中的溶解能力和在不同圈层之间的分配行为。中国科学院广州地球化学研究所李元团队（Li et al., 2016b）发现，氮在金属相和硅酸盐熔体相之间的分配系数在 1～20，因此核幔分异过程中大量的氮可能进入到地核中。实验结果也显示在温度高达 1 800℃ 的条件下，氮同位素在金属相和硅酸盐熔体相之间仍然有巨大的分馏：地核富 ^{14}N，硅酸盐地幔富 ^{15}N。根据这些实验结果，他们推断地球氮可能主要来源于顽火辉石球粒陨石。Li 等（2015a）还对碳在岩浆海中金属相和硅酸盐熔体相中的存在形式、溶解度及其分配进行了系统研究，结果显示，碳在岩浆海硅酸盐熔体中主要以甲烷或者碳酸根的形式存在，溶解度为几十至几百 ppm；碳在金属相中的溶解度可达百分之几，具体受金属相成分控制。碳在金属相和硅酸盐熔体相之间的分配系数为 100～5 000。因此，在核、幔分异过程中，岩浆海中大量的碳进入到地核中。此外，岩浆海的去气可能主要释放 CH_4 到早期的大气中，导致最初始的地球大气圈可能相对比较富集 CH_4。Li 等（2016a）发现，如果类地行星的金属核富集硅或硫，碳将从金属核中被"驱逐"出去，形成一个富碳的行星硅酸盐幔。水星的金属核硅极度富集，这和我们今天所观测到的水星壳非常富集石墨相一致。他们据此提出在地球增生的晚期，一个类似水星的小行星冲击 44 亿年前的原始地球，导致大量的碳进入地球，这可以解释地球碳的成因和来源。

（2）地幔矿物、岩浆熔体和流体中碳、氮元素的地球化学行为：中国科学院广州地球化学研究所李元团队（Li et al., 2013）研究了氮在地幔矿物中的溶解度。他们发现，在氧化条件下，氮在上地幔矿物中的溶解度小于 $5×10^{-6}$，而在还原条件下，氮在上地幔矿物中的溶解度可达几百 ppm。基于这些实验结果，他们计算出当前地球的地幔可容纳 20～40 倍的大气氮，因此硅酸盐地幔可能是一个巨大的氮储库。在岩浆去气过程中，氮在流体相和熔体相之间的分配系数为 60～10 000，因此岩浆去气时，氮会强烈富集于流体中（Li et al., 2015b）。Li 和 Keppler（2014）指出，岩浆流体中的氮主要以氨气和（或）氮气的形式存在，早期地球的去气主要释放 NH_3 到大气中，从而导致早期地球大气圈是相对还原的。

Li 等（2017a）对碳在火星、月球以及水星玄武岩中的溶解行为的研究发现，在整个火星壳形成过程中，地幔释放到大气圈中的 CO_2 总量不足 0.1 MPa，月球玄武岩中可能含几十到几百 ppm 碳，因此 CO_2/CO 可能是月球火山喷发的重要驱动力；水星玄武岩只能携带低于 $10×10^{-6}$ 的碳释放到地表，导致玄武岩从水星幔中所携带的碳不足以成为水星玄武岩喷发的重要驱动力。此外，Li（2017）还发现在高压下存在富水流体与富碳流体（CH_4+H_2、CO_2+CH_4 或 CO_2）的不混溶现象，这挑战了地幔条件下 C-H-O 流体完全混溶的传统观点，同时他提出了一种重要的俯冲带碳循环机制。

上述成果的获得不仅推动了对地幔深部、早期地球和类地行星中挥发分地球化学行为的理解，而且为深入阐明圈层相互作用和行星演化规律提供了重要的物理化学参数和关键的原理启示。

2.2 成矿过程和机制的实验模拟

通过高温高压实验模拟矿物–熔体–流体体系的物理化学条件（压力、温度、组成、氧逸度等）、精细刻画成矿过程中元素的地球化学行为，是揭示成矿元素迁移、富集过程和成矿机制的重要途径。近十年来，我国高温高压实验在元素分配行为与成矿、中基性岩浆作用成矿效应以及热液实验地球化学研究等方面取得了重要进展。

2.2.1 元素分配行为与成矿实验

元素分配是理解地球和行星核–幔–壳分异、元素地球化学行为及成矿机制的基础，近十年来我国在该领域取得了很多重要进展。

（1）Cu、Au、Mo、Re、W等成矿金属元素在硅酸盐岩浆中的分配和迁移：由于Cu易于与贵金属形成合金使得Cu分配系数实验测定成为一个难题，中国科学院广州地球化学研究所熊小林团队通过发展含Cu贵金属样品管新技术，系统测定了Cu在硅酸盐矿物（橄榄石、斜方辉石、单斜辉石、角闪石、石榴子石、斜长石）与铁镁质–中酸性熔体之间的分配系数，同时发现Cu在硅酸盐矿物中是高度不相容的（Liu et al., 2014, 2015）。这一成果解决了Cu在硅酸盐矿物中是相容还是不相容的争论，对理解Cu在岩浆过程中的行为和成矿有重要意义，已经被收录到由美国、欧洲和日本主编的最新版 *Encyclopedia of Geochemistry*（Liu and Xiong, 2018）中，成为地球化学和成矿定量研究的重要参数，获得广泛应用。该团队最近还发现，还原条件下地幔矿物与铁镁质熔体之间Cu分配系数（0.045）是一个不受压力、温度和体系组成影响的"金钉子"分配系数（Sun et al., 2020），使用该分配系数并结合Cu的亲S行为，发展了估算地幔硫丰度的新方法，揭示了硫在上地幔的高度不均一性（$120×10^{-6} \sim 300×10^{-6}$）。中国科学院广州地球化学研究所李元团队开展了硫化物/硅酸盐熔体Au、Mo、Re分配系数研究（Li et al., 2019；Feng and Li, 2019），他们发现Mo和Re在硫化物和熔体中的分配系数有3个数量级的变化，主要受熔体FeO含量和氧逸度的控制，而且Mo和Re的分配系数呈完美的正相关。因此，可以通过研究岩浆中Mo的地球化学行为来研究Re的地球化学行为。Au在硫化物和熔体中的分配系数有4个数量级的变化，主要受硫逸度和熔体中S含量的控制。岩浆结晶分异过程中大量的Re会进入到硫化物当中，然后再重新循环到地球深部地幔。俯冲板片熔体对Au的携带能力有限，因此大量的Au会和硫一起被循环到地球深部。岛弧岩浆结晶分异过程中，即使硫化物饱和条件下，如果岩浆比较富水、富硫，氧化性适当，富Au的岩浆仍可以形成。岛弧条件下富Au岩浆的形成可能和一些大型Au矿的形成密切相连。中国科学技术大学倪怀玮团队确定了Cu、Mo、W等关键金属元素在含水硅酸盐熔体中的扩散系数，为认识岩浆热液作用过程中金属元素迁移和富集机制提供了关键制约（Ni et al., 2018；Zhang P. P. et al., 2018）。

（2）高场强元素在矿物与富溶质超临界流体间的分配和迁移：熊小林团队通过金红石/熔体和金红石/流体分配实验（Xiong et al., 2011；Chen W. et al., 2018）发现，流体中溶质含量增加能增强高场强元素（HFSE）的迁移，富溶质的超临界流体具有接近含水熔体的元素迁移能力；他们通过角闪石/熔体Nb、Ta分配实验（Li L. et al., 2017）发现H_2O是导致角闪石Nb/Ta分异的重要因素，这些成果对示踪俯冲带熔/流体性质和弧岩浆演化过程具有重要的应用价值。

（3）过渡族元素在地幔矿物与玄武质熔体间的分配：第一排过渡族元素在地幔部分熔融过程中从相容到不相容，其分配行为对理解地幔岩性不均一和地幔氧逸度有重要应用。熊小林团队的实验表明，变价元素V在地幔矿物（橄榄石、斜方辉石、单斜辉石、尖晶石）与玄武质熔体之间的分配系数不仅敏感于氧逸度，而且温度对之也有重要影响（Wang J. T. et al., 2019）；他们利用V、Ti、Sc分配系数揭示弧下地幔比大洋地幔的氧逸度高约10倍，调和了利用不同方法获得的地幔楔氧逸度的争论。

2.2.2 岩浆作用及其成矿效应的实验研究

通过实验岩石学手段，查明岩浆作用及其成矿效应，是理解岩浆系统矿床（如铁±钛±钒±磷矿床等）形成机制的关键，近十年来我国在该领域取得很多重要进展。

（1）液态不混溶作用是基鲁纳型矿床形成的关键：基鲁纳型铁矿在空间上与中酸性火山岩-次火山岩密切相关，是高品位铁矿石的主要来源之一。绝大多数矿石的主要组成矿物只有磁铁矿和磷灰石（或萤石）。对该类型富铁矿石的成因均存在岩浆成因（铁矿浆成因）和热液成因之争。由于热液蚀变广泛发育，因此厘清岩浆期该成矿系统的原生过程，即液态不混溶作用是否发生，能否直接产生铁矿浆就成为解决矿床成因的关键。以此为目标，中国地质大学（北京）侯通课题组开展了一系列的研究工作，证实了岩浆在演化过程中，当母岩浆演化形成高度氧化富水（$\Delta FMQ=3.3$，$a_{H_2O}>0.7$）的中基性岩浆时，可以发生不混溶作用形成两种成分差异很大的熔体，其中富铁相为与矿石成分类似的铁钙磷熔体（铁磷矿浆），而对于富氟体系，富铁相与结晶矿物组合形成的"晶粥"（铁氟矿浆）在冷却结晶后就可成矿。以上两个系列的实验，为全球范围内基鲁纳型矿石的成因提供了关键实验证据（Hou et al., 2017, 2018）。

（2）高温硅酸盐多组分系统液态不混溶作用的形成温度：侯通课题组开展的富铁基性岩石和酸性岩石的高温不混溶实验表明，部分样品表面不混溶作用可以在相对高温的条件下稳定存在。这比原来认定的拉斑玄武岩系统液态不混溶的温度上限（1 050℃）要高，为岩浆演化的早期（高温）阶段可以发生不混溶提供了实验证据，这在岩石学和矿床学上均具有重要意义（Hou and Veksler, 2015）。

（3）高钛富铁苦橄岩的辉石岩/榴辉岩部分熔融成因：虽然地幔柱与大规模岩浆成矿作用被认可，但地幔柱与成矿作用之间的内在联系却不甚清楚。众所周知，全球有不同时代的大火成岩省，为何仅我国峨眉山大火成岩省的攀西地区有全球最大的钒钛磁铁矿矿集区，且沿南北向分布？侯通课题组根据地球化学研究结果提出，源区存在俯冲成因的辉石岩（榴辉岩）是攀西地区钒钛磁铁矿矿床巨量富集的根本原因（Hou et al., 2011, 2013），但缺少直接的实验证据。随后，针对这个问题，他们选取高钛（$TiO_2=5.4\%$）富铁苦橄岩进行近液相线平衡实验，结果表明多组分平衡点在1 320℃和1 GPa左右，即在1 GPa以上，富铁高钛苦橄岩的液相线矿物从橄榄石转换为斜方辉石，说明高钛富铁苦橄岩在最后离开岩石圈地幔时是和无橄榄石的地幔岩石平衡的（Zhang Y. S. et al., 2018）。这为富铁高钛苦橄岩源区存在辉石岩提供了重要的实验岩石学证据。

2.2.3 热液实验地球化学

查明热液中元素的迁移和富集机制是重建热液矿床成矿过程和建立成矿模式的基础和关键。我国在热液实验领域最为显著的进展是发展了可视化、在线观测实验技术，应用新的透明高压腔和原位拉曼光谱分析技术，实现了热液条件下流体组成和结构的实时解析。

（1）流体包裹体成分的原位拉曼光谱定量分析技术：包裹体捕获温度（T）、压力（p）和流体组成（x）等可为热液实验的T-p-x条件设计提供有益借鉴。传统的包裹体成分分析技术是显微测温法，而受制于包裹体大小、流体亚稳定以及流体成分复杂等因素的制约，显微测温法存在较大的局限性。针对这些问题，南京大学王小林、中国地质大学（武汉）吕万军和中国科学院深海科学与工程研究所I-Ming Chou等研究团队应用熔融毛细硅管透明腔技术制备实验标样，建立和完善了包裹体中常见组分，如盐度、硫酸盐浓度、CO_2和CH_4密度的原位拉曼光谱分析技术，实现了单个包裹体成分的快速、无损分析（Wang X. L. et al., 2011, 2013a, 2013b; Wang W. J. et al., 2019; Qiu et al., 2020b）。目前，这些研究团队正在开发复杂的C-H-O-S-N挥发分体系的原位拉曼光谱定量分析技术（Fang et al., 2018; Qiu et al., 2020a）。这一技术被广泛应用于成岩、成矿包裹体成分的测定，同时，也为在线热液实验过程中流体成分的检测提供了技术支撑。

（2）液-液相分离是热液中元素富集的潜在机制：王小林团队针对硫酸盐-水体系高温相变开展了系统的可视化在线观察，结果发现Li、Mg、Zn、Cu等硫酸盐溶液在高温条件下都会发生液-液相分离，形成富硫酸盐和贫硫酸盐的两个不混溶液相（Wang X. L. et al., 2013a, 2016a, 2016b, 2017; Wan et al., 2015, 2017）。他们从宏观相变特征和微观离子相互作用角度揭示了液-液相分离的发生机制，即复杂的络合作用是触发液-液相分离的主要原因。另外，他们的研究还发现CO_2、CH_4等挥发分和有机组分的存在会促进液-液相分离的发生，流体减压也会降低液-液相分离的发生温度（Wan et al., 2017）。这些发现丰富了人们对含硫热液体系元素富集机制的认识。由于液–液相分离形成的高密度流体相可以作为晶体生长的前驱物，他们的实验结果也为非传统晶体生长理论研究提供了新的实验依据。

（3）热液中钨的迁移方式和成矿过程中 CO_2 的 H^+ 储库效应：王小林团队通过对 K_2WO_4-HCl-NaCl-CO_2-H_2O 体系热液条件下原位拉曼光谱的分析，揭示了复杂钨络合物稳定的温压区间。他们发现，在高盐度和较酸性的流体中，钨可能以复杂钨络合物的形式迁移，而在中性–中偏酸性流体中，复杂钨络合物的稳定性会随温度的升高而降低，在 350℃ 以上时几乎检测不到复杂钨络合物的信号（Wang C. G. et al., 2020；Wang et al., 2020a）。这些实验结果证实，复杂钨络合物的分解可能是流体中钨卸载的潜在机制，CO_2 逸出、流体混合及围岩蚀变均可降低钨络合物的稳定性。此外，他们还发现在围岩蚀变过程中，CO_2 的存在有利于维持流体的 H^+ 浓度，进而提高流体从围岩中萃取铁、锰的效率，有利于高品位矿化的发生（Wang et al., 2020b）。

2.3 壳幔物质的物理化学性质和相互作用

地球深部物质（矿物、岩石、熔体和流体）在高温高压下的物理化学性质和相互作用是认识和了解地球内部圈层结构、状态和动力学过程的重要参数。通过开展高温高压实验及相关的理论研究，系统测定和查明地球深部物质的物理化学性质和相互作用具有十分重要的意义。近十年来，我国高温高压实验在壳幔物质的物理化学性质和相互作用研究方面取得了重要进展。

2.3.1 壳幔物质的电导和热导性质

（1）岩石圈中高导异常的成因：中国地质大学（武汉）地球深部研究团队利用多面砧大压机对钠长石–石英–水体系、钠长石–石英–盐水及斜长石–盐水体系开展了系统的高温高压电导率测试，建立了 1 GPa 和 800 K 条件下全岩电导率–盐度–流体含量之间的定量关系，对我国青藏高原等地区中–下地壳中出现的高导异常的流体成因进行了定量约束（Guo et al., 2015；Li P. et al., 2018）。南京大学杨晓志研究团队发现，富氟的金云母沿特定方向的电导率可以很好地解释大陆岩石圈内出现的高导异常（Li et al., 2016c，2017b）；中国科学技术大学倪怀玮团队利用自己开发的实验方法，系统测定了不同种类熔体和流体在高温高压下的电导率（Ni et al., 2014；Guo et al., 2016a，2016b，2017，2018），查明了熔体和流体电导率随各种物理化学条件的变化规律，建立了熔体电导率定量模型，为揭示长白山天池火山等地区的岩浆房状态以及评估火山喷发前景提供了科学依据。张宝华团队通过高温高压电导率实验，指出含水斜方辉石和绿辉石（Zhang et al., 2012；Zhang et al., 2019a）以及颗粒边界碳膜模型（Zhang and Yoshino, 2017）不是上地幔高导异常的合理解释，并在国际上率先建立了剪切变形下部分熔融岩石电导率测量的新方法，提出了剪切变形下差应力诱发熔体各向异性分布是上地幔高导层形成机制的新认识（Zhang et al., 2014，2020）。

（2）与俯冲带相关的高导异常的成因：中国科学院大学的研究团队测试了蛇纹石在 3 GPa、4 GPa 下脱水前后的电导率（Wang D. J. et al., 2017），结果表明蛇纹石在脱水后电导率可以达到 1 S/m，蛇纹石在不同的地温梯度下脱水的深度与出现高导异常的深度完美耦合，因此在 40～200 km 深度出现的与俯冲带相关的高导异常很可能是由俯冲板块内蛇纹石的脱水造成的。而对于出现在上地幔中更深的、与俯冲带相关的高导异常（250～300 km）可能与多硅白云母有关。中国地质大学（武汉）地球深部研究团队测试了多硅白云母在 2.3～12 GPa 条件下多硅白云母脱水前后的电导率，结果表明多硅白云母脱水后产生的富钾流体使全岩电导率急剧增大四个数量级。通过 Cube 模型定量计算，在北菲律宾海以下 250～300 km 深度的高导异常可能是由多硅白云母分解产生的、体积含量不超过 0.9% 富钾流体造成的；而出现在阿根廷中部以下 250～300 km 深度的高导异常可能由体积含量低于 3% 的富钾流体造成（Chen S. B. et al., 2018）。中国科学院地球化学研究所李和平团队通过对绿帘石、角闪石的电导率实验研究，认为绿帘石的脱水可能对解释出现在热俯冲带内、深度为 70～120 km 的高导异常有重要作用，并提出高温高压下角闪石发生的氧化–脱氢反应，释放出氢气而非含水流体，然而脱氢后电导率的增强足以解释俯冲板片与地幔楔界面和大陆中地壳底部的高导异常（Hu H. Y. et al., 2017, 2018）。

（3）水在地球和月球熔体中的扩散系数和水扩散机制：水在熔体中的扩散系数是揭示岩浆去气动力学、进而约束岩浆源区水含量的关键参数。熔体中水的不同存在形式（H_2O 和 OH）可能具有不同的活动性，导致其对水扩散的贡献也有所不同。中国科技大学倪怀玮团队实验测定了水在地球和月球玄武质熔体中的扩散系数，发现并量化了 OH 对玄武质熔体水扩散和电导率的影响，进而揭示月幔比地幔中度贫水（Zhang L. et al., 2017, 2019）。在查明硅酸盐熔体中水扩散速率和机制的基础上，Ni 和 Zhang（2018）建立了高精度、广适用性的俯冲带钙碱性系列熔体水扩散系数定量模型，为揭示岩浆水含量演化提供了关键依据。

（4）一些常见壳幔矿物在高温高压条件下的热导率和热扩散系数：中国地质大学（武汉）地球深部研究团队在上地幔条件下（14 GPa，1 000 K）测量了绿辉石、硬玉、透辉石的热导率（Wang et al., 2014），为理解榴辉岩对俯冲板片热结构的影响提供了新认识。中国科学院地球化学研究所苗社强等利用激光闪射法（LFA427）在高温高压下测量了华北克拉通主要类型岩石和矿物的热扩散系数（Miao et al., 2014a, 2014b, 2019），为中国东部岩石圈的热结构和热演化提供了重要约束。中国科学院地球化学所张宝华课题组在多面砧大压机上建立了同时测量矿物岩石热扩散系数和热导率的瞬态平面热源法，目前已取得了花岗岩（Fu et al., 2019）、辉石岩（Ge et al., 2021）、橄榄石、长石、石榴子石等矿物岩石热物理性质测量方面的大量第一手实验数据，发现花岗岩的热物理性质主要受石英含量控制，从矿物热物理角度论证了青藏高原中上地壳存在部分熔融的合理性；同时实验揭示了水能显著降低地幔矿物（特别是橄榄石）的热导率（Zhang et al., 2019b），为精确约束地球以及类地行星内部的热结构提供了重要参数和理论基础。

这些研究成果为壳幔不同深度、不同构造背景下的高导异常的成因提供了多种可能解释机制，为深入了解地球内部物质状态和热结构提供了依据。

2.3.2 壳幔矿物和岩石的流变性质

（1）下地壳麻粒岩的流变和地震活动的相变成因：中国地质大学（武汉）地球深部研究团队深入研究了下地壳麻粒岩的高温高压变形（Wang et al., 2012），他们的实验结果表明，华北克拉通的基性麻粒岩具有较低的流变强度，比上地幔橄榄岩的强度低一个数量级，为经典的大陆岩石圈的果冻三明治强度模型中的弱下地壳和华北克拉通下地壳拆沉作用提供了实验依据。Shi 等（2018）进一步的准稳态麻粒岩变形实验研究表明，麻粒岩的榴辉岩化会导致岩石高压脆性破裂，为下地壳地震活动的成因物理机制的合理解释首次提供了实验依据。

（2）地幔过渡带的力学强度：中国地质大学（武汉）地球深部研究团队系统研究了以 Mg_2GeO_4 为相似物的地幔转换带的强度和黏度，他们的实验结果表明地幔转换带相对上地幔可能具有高的力学强度，当俯冲大洋板片中的橄榄石在进入下地幔发生相变的过程中，其力学强度由于联通相变结构可能不会发生大的变化（Zhao et al., 2012; Shi et al., 2015），这些实验结果约束了转换带的黏度变化特征，为俯冲带在转换带发生偏转的形态特征提供了一个合理的解释。

（3）后钙钛矿的塑性变形可导致下地幔 D″ 层剪切波速异常：中国地质大学（武汉）地球深部研究团队率先借助先进的同步辐射径向 X 衍射技术、激光加温技术和自动加压装置，对（$Mg_{0.75}Fe_{0.25}$）SiO_3 在对应 D″ 层温压力条件下的形变过程进行了原位表征，阐明了后钙钛矿在真实的 D″ 层高温高压环境下的主滑移面应该是（001）面（Wu et al., 2017）。此滑移属于后钙钛矿的形变产生而非布里奇曼石到后钙钛矿的相变产生。结合地球动力学模拟，指出此滑移系会造成 D″ 层的剪切波速在水平偏振方向快于垂直偏振方向，大约 3.7% 的差异。该成果的主要贡献在于揭示了环太平洋边界下 D″ 层剪切波速异常的机理。

上述研究成果为定量揭示由物质成分和热结构控制的地球不同圈层流变学性质的三维结构（横向不均一性和纵向分层性）和地震活动成因物理机制提供了重要基础数据。

2.3.3 壳幔相互作用的岩石学和地球化学过程

壳幔相互作用是重要的地球动力学过程，决定了地壳和地幔之间物质与能量的交换，它发生在俯冲带、大陆内部造山带、洋中脊等多种活动构造环境中，是全球范围岩浆活动、构造演化和成矿作用的重要环节。

（1）富硅熔体与橄榄岩反应实验揭示地幔不均一性成因和克拉通破坏机制：中国地质大学（武汉）地球深部研究研究团队系统研究了榴辉岩熔体与橄榄岩在等静压和差异应力条件下反应的高温高压实验。研究结果表明，实验产物具有与汉诺坝地区幔源复合捕虏体相似的结构，反应熔体与华北克拉通中生代高镁闪长岩具有相似的成分（Wang et al., 2010）。Zhang 等（2012）发现在等静压力条件下，无水榴辉岩部分熔融的熔体对橄榄岩的侵蚀能力较弱；在差应力存在的条件下，榴辉岩熔体对橄榄岩的侵蚀能力增强。提出华北克拉通的减薄是长期小规模拆沉作用的结果，拆沉作用导致的熔体-橄榄岩反应和纵向构造变形加快了再循环熔体对残留岩石圈地幔的侵蚀，为拆沉作用导致的华北克拉通破坏物理机制提供了实验角度的约束。吉林大学研究团队系统研究了熔体主量元素成分对壳源熔体-橄榄岩反应过程和结果的影响（Wang C. G. et al., 2013），结果发现熔体-橄榄岩反应的产物受控于反应熔体的液相线矿物组成。若反应熔体的液相线矿物为斜方辉石（±橄榄石），则反应形成方辉橄榄岩；若反应熔体的液相线矿物为橄榄石，则反应形成纯橄岩。通过实验结果与华北克拉通地幔捕虏体的对比研究，为华北克拉通岩石圈地幔受再循环陆壳熔体改造的时间和空间范围提供了实验约束。Yu 等（2014）、Wang 和 Tang（2013）开展的角闪榴辉岩和英云闪长岩熔体与地幔橄榄石反应的实验，证实了富硅熔体与橄榄岩反应由橄榄石的溶解和新矿物相的结晶主导，反应使英云闪长质熔体的 MgO 含量和 $Mg^{\#}$ 升高，说明拆沉的地壳物质部分熔融产生的熔体与地幔橄榄石反应，为克拉通内部高镁安山质岩浆的成因提供了实验依据。吉林大学研究团队系统研究了熔体中水对熔-岩反应动力学的影响（Wang C. G. et al., 2016），他们的实验结果表明，含水熔体-橄榄岩反应可以形成高孔隙度斜方辉石岩反应带，并将初始二辉橄榄岩转化为贫斜方辉石的方辉橄榄岩甚至纯橄岩（Wang C. G. et al., 2016）；他们提出水的存在提高橄榄岩的部分熔融程度，辉石优先熔融形成的部分熔体与反应熔体混合，使后者 SiO_2 过饱和，该混合熔体与橄榄岩反应形成斜方辉石岩；指出地幔岩石中斜方辉石岩的存在是含水壳源熔体运移和交代的重要标志。Wang C. G. 等（2020）发现地幔橄榄岩熔融与否在很大程度上影响榴辉岩熔体-橄榄岩反应的机制、速率和结果。当橄榄岩处于亚固相时，反应具有较低的速率，形成富含石榴石的岩性；当橄榄岩部分熔融时，反应速率明显提高，形成富含斜方辉石的岩性。提出不同类型的熔体-橄榄岩反应对岩石圈地幔造成不同的岩石学效应和不同的熔体成分变异趋势。

（2）俯冲沉积物与橄榄岩反应实验揭示大陆俯冲带后碰撞高钾岩浆的成因：中国地质大学（武汉）地球深部研究研究团队通过沉积物与橄榄岩反应的高温高压实验，发现实验熔体与东地中海地区和中国东北富钾岩石具有相似的微量地球化学特征，据此提出受到俯冲沉积物交代的地幔可能是具有富集属性岩浆的源区（Wang and Foley, 2018；Zhang Y. F. et al., 2019）。Gao 等（2019）发现，花岗质岩浆与橄榄岩反应的过程分为两个阶段：第一阶段反应消耗橄榄石和反应熔体生成斜方辉石和金云母；第二阶段使反应产物重新熔融形成富钾富硅的熔体，熔体与 Variscan 造山带后碰撞高钾岩浆岩具有相似的地球化学特征。Wang X. 等（2019）通过对比角闪石的部分熔融熔体和钾玄岩与石榴角闪岩部分熔融的熔体，发现 20% 的地幔钾玄岩的加入可以提供下地壳部分熔融所需的热，以及埃达克质岩浆中的钾和不相容元素。这些研究成果为大陆俯冲带后碰撞高钾岩浆的成因提供了重要的实验约束。

3 实验平台和技术进展

先进的高温高压实验模拟平台和技术对推动地球深部物质属性与过程的深入研究起到了至关重要的作用。过去十年，国内的实验岩石学与矿物学高温高压实验平台建设发展迅速，许多大学和科研院所先后建立了高水平的高温高压实验室，主要包括中国科学院地球化学研究所、中国地质大学（武汉）、吉林大学、中国科学院广州地球化学研究所、中国科学技术大学、南京大学、北京大学、北京高压科学研究中心、中国地震局地质所等。

中国科学院地球化学研究所拥有铰链式 600 t 和 1 400 t、Kawai 型 1 000 t、D-DIA 型 2 000 t 和 DIA 型 YJ-3000 t 多面砧压机，LPC 250-300/50 型活塞圆通压机，在地球深部物质物性研究方面具有传统优势。

中国地质大学（武汉）地球深部研究实验室拥有两台 Walker 型 1 000 t 多面砧压机、一台 5 000 t 多面砧压机、30 多套各式金刚石压腔、Quick Press 活塞圆筒压机，以及自主研发的两台 5 GPa 围压 Griggs 型高温高压流变仪和一台 3 GPa 活塞圆筒压机，在大压机和流变仪实验技术研发上具有领先优势，以地幔岩实验岩石学、壳幔矿物高压流变学实验研究为主要特色。

吉林大学拥有完整的从高温高压合成到高温高压原位物性探测的实验装备。其中大型压机包括德国 Max Voggenreiter 1 000 t 德国多面砧大腔体压机（配备自动控压和高压电学、超声测试系统）、六面顶液压机、铰链式六面顶液压机、六面顶 Walker 型压机、Cubic 型多面砧压机、Quick Press 活塞圆筒压机和高压水热装置。基于金刚石对顶砧的高压原位物性测量装置有 X 光、Raman、霍尔效应、电导率、激光加温、荧光、红外光谱、热导率等系统，可以实现从结构分析到光、电、热等输运特性以及光学和热电性质等全方位的高压原位物性探测，形成了以熔体–橄榄岩反应动力学和地球深部物质合成与物性研究为主要特色和优势。

中国科学院广州地球化学研究所拥有分别由美国 Depth of the Earth 公司生产的 Quick Press 和 Rockland Research 公司生产的两种不同类型的活塞圆筒压机和多台冷封式高压釜，最近又引进了多台 1 000 t 和 2 500 t 多面砧压机，以微量元素分配行为研究为主要特色。

中国科学技术大学高温高压岩石学实验室拥有 3 台德产 Voggenreiter LPC250 活塞圆筒压机、3 台快速淬火冷封式高压釜和 10 台各式水热金刚石压腔，在熔体和流体的物理化学性质研究方面特色鲜明。

南京大学拥有两台可自动控压的德国产 Voggenreiter LPC250 活塞圆筒压机和 5 台快速淬火冷封式高压釜，在矿物物性和挥发性元素溶解度等方向具有研究优势。

北京大学装备了六面顶压机和 Quick Press 活塞圆筒压机，在矿物晶体结构、热弹性参数和相平衡研究方面取得了一系列成果。

北京高压科学研究中心拥有 1 台德国产 Voggenreiter Kawai 型 1 500 t 多面砧压机、1 台德国产 Voggenreiter LPC250 活塞圆筒压机，以及 3 套自主研发的激光加温系统和近千套激光加温金刚石压腔，在研发金刚石压腔加热和超高压技术上具有领先优势，以深俯冲物质的物理化学属性研究为主要特色。

此外，浙江大学依托"超重力离心模拟与实验装置"国家重大科技基础设施建设，正在筹建具有自己鲜明特色的融合超重力实验装置的高温高压实验平台。中国科学院海洋研究所、自然资源部第二海洋研究所、西北大学也先后装备了 1 000 t 多面砧压机。中国地质大学（北京）正在筹建岩浆系统成岩成矿过程的高温高压实验室，目前已拥有可控氧逸度的高温管式炉，即将装备 5 台 MHC 冷封式高压釜。高温高压实验岩石学–矿物学的平台建设在国内正呈现前所未有的大发展景象。

4 学科未来发展方向和趋势

过去十年，我国实验地球科学的研究成果逐年增加，研究方向也迅速拓展，在应用大腔体压机测量矿物岩石的热学、电学、弹性、相变、扩散等物理性质，应用金刚石压砧测量高压布里渊散射和红外谱学、电学、状态方程，以及应用流变仪测量岩石和矿物流变性质等方向上具有比较优势，相关的研究方法、技术创新和体系研究成果也得到了广泛的关注和引用，为地球物理和地球化学数据解释和地球深部过程的探讨提供了实验依据。该学科目前存在以下亟需发展的问题：首先，我国还缺乏有国际影响力的稳定学术团队，尚未建成类似美国卡内基研究所、布鲁克海文国家实验室、阿贡国家实验室、德国的 BGI 和 GFZ、日本东京大学、冈山大学、爱媛大学等世界知名的研究团队；其次，在仪器设备研制和实验方法创新方面还非常薄弱，缺乏技术人才储备，缺乏同步辐射等国家大型科学装置支撑，目前主要采用国际上成熟的技术方法开展研究；最后，我国还没有形成地质天然观测、实验模拟和计算模拟协同创新的工作模式，需要国家通过资助重大科技项目或科技计划凝聚国内分散但精干的研究力量开展协作。

当前，许多国家都开展了大量的地球物理深部探测，极大促进了地球深部物质研究的发展。了解不同温度、压力、应力状态、流体活动等条件下地球深部物质物理化学性质及其变化规律，是进一步认识各种地质过程和动力学成因的先决条件，更是解释地球物理探测新发现（如各种反射层构造、流变分层、高导低速体、地震波各向异性等）和地球物理反演模型（如密度、波速、泊松比、导电率、流体逸度等）的重要基础。地球物理深部探测已经积累了大量科学问题，急需地球深部物质研究的支持，包括：大陆形成与演化过程中物质性质的变化、Moho面的组成与特殊性质、岩石圈流变状态、软流圈熔融条件、地幔物质对流特征、地球内部物质运移机理、地球内部化学动力学过程、流体运动、壳-幔和核-幔边界物质运移和相互作用、诱发地震与火山喷发的过程和机理、成矿元素的迁移、富集和沉淀等等。以科学问题为导向的高温高压实验地球科学研究，对解决深部地质、成矿作用与地球动力学所面临的这些前沿问题是十分必要和重要的。实验地球科学的未来发展方向主要包括：①地球深部（从地幔到地核）物质结构、状态和物理化学性质；②深部岩浆作用以及元素迁移和富集的机制；③地球物理探测数据的综合解释、深部过程与浅表响应、矿产资源寻找以及自然灾害机理探索的理论和实验约束；④地球早期演化过程的理论和实验约束。

当前实验地球科学在技术上主要展现出四方面的发展趋势：①实验装置实现更高更准的温度、压力以及更大的样品空间。更高的温度和压力条件一直是实验地球科学孜孜以求的目标。从水热高压釜到活塞圆筒压机、多面砧压机，再到激光加热金刚石压腔技术的出现，现在已经实现包括地球中心在内的所有温压条件的全覆盖。巨行星乃至系外行星的压力条件（千万大气压以上量级）也可以通过冲击波动态高压技术实现。实验地球科学仍在持续追求突破更高的温度压力上限，同时实现更精确的温度压力控制。更大的样品空间可以允许研究更复杂、更贴近实际的体系，也便于布设原位测量的探针。我国正在积极研发可以实现超大样品空间和超重力条件下的多面砧大压机。②高温高压实验技术与原位分析测试技术的进一步融合。从高温高压淬火到常温常压的过程中样品可能发生多种变化，造成原始信息的丢失。高温高压技术与分析测试技术相结合，开展高温高压条件下的原位实时测量已成为实验地球科学的重要发展方向。金刚石压腔与包括同步辐射在内的多项光谱分析技术的结合已经十分成熟，但受各方面条件限制，在大压机上开展原位测量仍有待进一步发展。③实验地球科学与计算地球科学相结合。计算地球科学基于统计力学和量子力学方法，利用计算机技术模拟地球物质体系，其优势在于它可以轻松实现高温高压条件，而且可以提供微观视角。随着机器学习等计算技术的发展，实验和计算两种技术相互结合，取得的结果相互比对和印证，成为地球科学研究的重要发展趋势。④实验地球科学与天然样品研究和地球物理观测的进一步结合。实验研究必须以解决地质学和地球物理学实际问题为依托。如何把高温高压实验研究与经典的天然样品岩石地球化学研究和地球物理观测更好地结合，切实解决地球科学重大问题，是实验地球科学的使命所在。

致谢：本文成文过程得到金振民院士的指导，郭新转、李元、刘锦、王春光、王小林、巫翔、杨燕、叶宇、张艳飞、高春晓等人提供了宝贵素材和资料，在此一并表示衷心感谢！

参 考 文 献

Chen S B, Guo X Z, Yoshino T, Jin Z M, Li P. 2018. Dehydration of phengite inferred by electrical conductivity measurements: implication for the high conductivity anomalies relevant to the subduction zones. Geology, 46(1): 11–14

Chen W, Xiong X L, Wang J T, Xue S, Li L, Liu X C, Ding X, Song M S. 2018. TiO_2 solubility and Nb and Ta partitioning in rutile-silica-rich super-critical fluid systems: implications for subduction zone processes. Journal of Geophysical Research: Solid Earth, 123(6): 4765–4782

Fang J, Chou I M, Chen Y. 2018. Quantitative Raman spectroscopic study of the H_2-CH_4 gaseous system. Journal of Raman Spectroscopy, 49(4): 710–720

Feng L, Li Y. 2019. Comparative partitioning of Re and Mo between sulfide phases and silicate melt and implications for the behavior of Re during magmatic processes. Earth and Planetary Science Letters, 517: 14–25

Fu H F, Zhang B H, Ge J H, Xiong Z L, Zhai S M, Shan S M, Li H P. 2019. Thermal diffusivity and thermal conductivity of granitoids at 283–988 K and 0.3–1.5 GPa. American Mineralogist, 104(11): 1533–1545

Gao M D, Xu H J, Zhang J F, Foley S F. 2019. Experimental interaction of granitic melt and peridotite at 1.5 GPa: implications for the origin of post-collisional K-rich magmatism in continental subduction zones. Lithos, 350-351: 105241

Ge J H, Zhang B H, Xiong Z L, He L F, Li H P. 2021. Thermal properties of harzburgite and dunite at 0.8-3 GPa and 300-823 K and implications for the thermal evolution of Tibet. Geoscience Frontiers, 12(2): 947-956

Guo X, Chen Q, Ni H W. 2016a. Electrical conductivity of hydrous silicate melts and aqueous fluids: measurement and applications. Science China Earth Sciences, 59(5): 889-900

Guo X, Zhang L, Behrens H, Ni H W. 2016b. Probing the status of felsic magma reservoirs: constraints from the P-T-H_2O dependences of electrical conductivity of rhyolitic melt. Earth and Planetary Science Letters, 433: 54-62

Guo X, Li B, Ni H W, Mao Z. 2017. Electrical conductivity of hydrous andesitic melts pertinent to subduction zones. Journal of Geophysical Research: Solid Earth, 122(3): 1777-1788

Guo X, Zhang L, Su X, Mao Z, Gao X Y, Yang X Z, Ni H W. 2018. Melting inside the Tibetan crust: constraint from electrical conductivity of peraluminous granitic melt. Geophysical Research Letters, 45(9): 3906-3913

Guo X Z, Yoshino T, Shimojuku A. 2015. Electrical conductivity of albite-(quartz)-water and albite-water-NaCl systems and its implication to the high conductivity anomalies in the continental crust. Earth and Planetary Science Letters, 412: 1-9

Hou T, Veksler I V. 2015. Experimental confirmation of high-temperature silicate liquid immiscibility in multicomponent ferrobasaltic systems. American Mineralogist, 100(5-6): 1304-1307

Hou T, Zhang Z C, Ye X R, Encarnacion J, Reichow M K. 2011. Noble gas isotopic systematics of Fe-Ti-V oxide ore-related mafic-ultramafic layered intrusions in the Panxi area, China: the role of recycled oceanic crust in their petrogenesis. Geochimica et Cosmochimica Acta, 75(22): 6727-6741

Hou T, Zhang Z C, Encarnacion J, Santosh M, Sun Y L. 2013. The role of recycled oceanic crust in magmatism and metallogeny: Os-Sr-Nd isotopes, U-Pb geochronology and geochemistry of picritic dykes in the Panzhihua giant Fe-Ti oxide deposit, central Emeishan large igneous province, SW China. Contributions to Mineralogy and Petrology, 165(4): 805-822

Hou T, Charlier B, Namur O, Schütte P, Schwarz-Schampera U, Zhang Z C, Holtz F. 2017. Experimental study of liquid immiscibility in the Kiruna-type Vergenoeg iron-fluorine deposit, South Africa. Geochimica et Cosmochimica Acta, 203: 303-322

Hou T, Charlier B, Holtz F, Veksler I, Zhang Z C, Thomas R, Namur O. 2018. Immiscible hydrous Fe-Ca-P melt and the origin of iron oxide-apatite ore deposits. Nature Communications, 9(1): 1415

Hu H Y, Dai L D, Li H P, Hui K S, Sun W Q. 2017. Influence of dehydration on the electrical conductivity of epidote and implications for high-conductivity anomalies in subduction zones. Journal of Geophysical Research: Solid Earth, 122(4): 2751-2762

Hu H Y, Dai L D, Li H P, Sun W Q, Li B S. 2018. Effect of dehydrogenation on the electrical conductivity of Fe-bearing amphibole: implications for high conductivity anomalies in subduction zones and continental crust. Earth and Planetary Science Letters, 498: 27-37

Hu Q Y, Kim D Y, Yang W G, Yang L X, Meng Y, Zhang L, Mao H K. 2016. FeO_2 and FeOOH under deep lower-mantle conditions and Earth's oxygen-hydrogen cycles. Nature, 534(7606): 241-244

Hu Q Y, Kim D Y, Liu J, Meng Y, Yang L X, Zhang D Z, Mao W L, Mao H K. 2017. Dehydrogenation of goethite in Earth's deep lower mantle. Proceedings of the National Academy of Sciences of the United States of America, 114(7): 1498-1501

Li L, Xiong X L, Liu X C. 2017. Nb/Ta fractionation by amphibole in hydrous basaltic systems: implications for arc magma evolution and continental crust formation. Journal of Petrology, 58(1): 3-28

Li P, Guo X Z, Chen S B, Wang C, Yang J L, Zhou X F. 2018. Electrical conductivity of the plagioclase-NaCl-water system and its implication for the high conductivity anomalies in the mid-lower crust of Tibet Plateau. Contributions to Mineralogy and Petrology, 173(2): 16

Li X Y, Mao Z, Sun N Y, Liao Y F, Zhai S M, Wang Y, Ni H W, Wang J Y, Tkachev S N, Lin J F. 2016. Elasticity of single-crystal superhydrous phase B at simultaneous high pressure-temperature conditions. Geophysical Research Letters, 43(16): 8458-8465

Li X Y, Zhang Z G, Lin J F, Ni H W, Prakapenka V B, Mao Z. 2018. New high-pressure phase of $CaCO_3$ at the topmost lower mantle: implication for the deep-mantle carbon transportation. Geophysical Research Letters, 45(3): 1355-1360

Li Y. 2017. Immiscible C-H-O fluids formed at subduction zone conditions. Geochemical Perspectives Letters, 3(2): 12-21

Li Y, Keppler H. 2014. Nitrogen speciation in mantle and crustal fluids. Geochimica et Cosmochimica Acta, 129: 13-32

Li Y, Wiedenbeck M, Shcheka S, Keppler H. 2013. Nitrogen solubility in upper mantle minerals. Earth and Planetary Science Letters, 377-378: 311-323

Li Y, Dasgupta R, Tsuno K. 2015a. The effects of sulfur, silicon, water, and oxygen fugacity on carbon solubility and partitioning in Fe-rich alloy and silicate melt systems at 3 GPa and 1600℃: implications for core-mantle differentiation and degassing of magma oceans and reduced planetary mantles. Earth and Planetary Science Letters, 415: 54-66

Li Y, Huang R F, Wiedenbeck M, Keppler H. 2015b. Nitrogen distribution between aqueous fluids and silicate melts. Earth and Planetary Science Letters, 411: 218-228

Li Y, Dasgupta R, Tsuno K, Monteleone B, Shimizu N. 2016a. Carbon and sulfur budget of the silicate Earth explained by accretion of differentiated planetary embryos. Nature Geoscience, 9(10): 781-785

Li Y, Marty B, Shcheka S, Zimmermann L, Keppler H. 2016b. Nitrogen isotope fractionation during terrestrial core-mantle separation. Geochemical Perspectives Letters, 2(2): 138-147

Li Y, Yang X Z, Yu J H, Cai Y F. 2016c. Unusually high electrical conductivity of phlogopite: the possible role of fluorine and geophysical implications. Contributions to Mineralogy and Petrology, 171(4): 37

Li Y, Dasgupta R, Tsuno K. 2017a. Carbon contents in reduced basalts at graphite saturation: implications for the degassing of Mars, Mercury, and the Moon. Journal of Geophysical Research: Planets, 122(6): 1300–1320

Li Y, Jiang H T, Yang X Z. 2017b. Fluorine follows water: Effect on electrical conductivity of silicate minerals by experimental constraints from phlogopite. Geochimica et Cosmochimica Acta, 217: 16–27

Li Y, Feng L, Kiseeva E S, Gao Z H, Guo H H, Du Z X, Wang F Y, Shi L L. 2019. An essential role for sulfur in sulfide-silicate melt partitioning of gold and magmatic gold transport at subduction settings. Earth and Planetary Science Letters, 528: 115850

Liu D, Pang Y W, Ye Y, Jin Z M, Smyth J R, Yang Y, Zhang Z M, Wang Z P. 2019a. In-situ high-temperature vibrational spectra for synthetic and natural clinohumite: implications for dense hydrous magnesium silicates in subduction zones. American Mineralogist, 104(1): 53–63

Liu D, Wang S, Smyth J R, Zhang J F, Wang X, Zhu Z, Ye Y. 2019b. In situ infrared spectra for hydrous forsterite up to 1243 K: hydration effect on thermodynamic properties. Minerals, 9(9): 512

Liu J, Hu Q Y, Kim D Y, Wu Z Q, Wang W Z, Xiao Y M, Chow P, Meng Y, Prakapenka V B, Mao H K, Mao W L. 2017. Hydrogen-bearing iron peroxide and the origin of ultralow-velocity zones. Nature, 551(7681): 494–497

Liu J, Hu Q Y, Bi W L, Yang L X, Xiao Y M, Chow P, Meng Y, Prakapenka V B, Mao H K, Mao W L. 2019. Altered chemistry of oxygen and iron under deep Earth conditions. Nature Communications, 10(1): 153

Liu X C, Xiong X L. 2018. Copper. In: White W M (ed). Encyclopedia of Geochemistry. Cham: Springer, 303–305

Liu X C, Xiong X L, Audétat A, Li Y, Song M S, Li L, Sun W D, Ding X. 2014. Partitioning of copper between olivine, orthopyroxene, clinopyroxene, spinel, garnet and silicate melts at upper mantle conditions. Geochimica et Cosmochimica Acta, 125: 1–22

Liu X C, Xiong X L, Audétat A, Li Y. 2015. Partitioning of Cu between mafic minerals, Fe-Ti oxides and intermediate to felsic melts. Geochimica et Cosmochimica Acta, 151: 86–102

Mao H K, Hu Q Y, Yang L X, Liu J, Kim D Y, Meng Y, Zhang L, Prakapenka V B, Yang W G, Mao W L. 2017. When water meets iron at Earth's core-mantle boundary. National Science Review, 4(6): 870–878

Mao Z, Fan D W, Lin J F, Yang J, Tkachev S N, Zhuravlev K, Prakapenka V B. 2015. Elasticity of single-crystal olivine at high pressures and temperatures. Earth and Planetary Science Letters, 426: 204–215

Miao S Q, Li H P, Chen G. 2014a. The temperature dependence of thermal conductivity for lherzolites from the North China Craton and the associated constraints on the thermodynamic thickness of the lithosphere. Geophysical Journal International, 197(2): 900–909

Miao S Q, Li H P, Chen G. 2014b. Temperature dependence of thermal diffusivity, specific heat capacity, and thermal conductivity for several types of rocks. Journal of Thermal Analysis and Calorimetry, 115(2): 1057–1063

Miao S Q, Zhou Y S, Li H P. 2019. Thermal diffusivity of lherzolite at high Pressures and high temperatures using pulse method. Journal of Earth Science, 30(1): 218–222

Ni H W, Zhang L. 2018. A general model of water diffusivity in calc-alkaline silicate melts and glasses. Chemical Geology, 478: 60–68

Ni H W, Chen Q, Keppler H. 2014. Electrical conductivity measurements of aqueous fluids under pressure with a hydrothermal diamond anvil cell. Review of Scientific Instruments, 85(11): 115107

Ni H W, Shi H F, Zhang L, Li W C, Guo X, Liang T. 2018. Cu diffusivity in granitic melts with application to the formation of porphyry Cu deposits. Contributions to Mineralogy and Petrology, 173(6): 50

Qiu Y, Wang X L, Liu X, Cao J, Liu Y F, Xi B B, Gao W L. 2020a. In situ Raman spectroscopic quantification of CH_4-CO_2 mixture: application to fluid inclusions hosted in quartz veins from the Longmaxi Formation shales in Sichuan Basin, southwestern China. Petroleum Science, 17(1): 23–25

Qiu Y, Yang Y X, Wang X L, Wan Y, Hu W X, Lu J J, Tao G L, Li Z, Meng F W. 2020b. In situ Raman spectroscopic quantification of aqueous sulfate: experimental calibration and application to natural fluid inclusions. Chemical Geology, 533(5): 119447

Shi F, Zhang J F, Xia G, Jin Z, Green II H W. 2015. Rheology of Mg_2GeO_4 olivine and spinel harzburgite: implications for Earth's mantle transition zone. Geophysical Research Letters, 42(7): 2212–2218

Shi F, Wang Y B, Yu T, Zhu L P, Zhang J F, Wen J G, Gasc J, Incel S, Schubnel A, Li Z Y, Chen T, Liu W L, Prakapenka V, Jin Z M. 2018. Lower-crustal earthquakes in southern Tibet are linked to eclogitization of dry metastable granulite. Nature Communications, 9(1): 3483

Sun Z X, Xiong X L, Wang J T, Liu X C, Li L, Ruan M F, Zhang L, Takahashi E. 2020. Sulfur abundance and heterogeneity in the MORB mantle estimated by copper partitioning and sulfur solubility modelling. Earth and Planetary Science Letters, 538: 116169

Wan Y, Wang X L, Hu W X, Chou I M. 2015. Raman spectroscopic observations of the ion association between Mg^{2+} and SO_4^{2-} in $MgSO_4$-Saturated droplets at temperatures of ≤380℃. Journal of Physical Chemistry A, 119(34): 9027–9036

Wan Y, Wang X L, Hu W X, Chou I M, Wang X Y, Chen Y, Xu Z M. 2017. In situ optical and Raman spectroscopic observations of the effects of pressure and fluid composition on liquid-liquid phase separation in aqueous cadmium sulfate solutions (≤400℃, 50 MPa) with geological and geochemical implications. Geochimica et Cosmochimica Acta, 211: 133–152

Wang C, Jin Z M, Gao S, Zhang J F, Zheng S. 2010. Eclogite-melt/peridotite reaction: experimental constrains on the destruction mechanism of the North China Craton. Science China Earth Science, 53(6): 797–809

Wang C, Yoneda A, Osako M, Ito E, Yoshino T, Jin Z M. 2014. Measurement of thermal conductivity of omphacite, jadeite, and diopside up to 14 GPa and 1 000 K: implication for the role of eclogite in subduction slab. Journal of Geophysical Research: Solid Earth, 119(8): 6277–6287

Wang C G, Liang Y, Xu W L, Dygert N. 2013. Effect of melt composition on basalt and peridotite interaction: laboratory dissolution experiments with

Wang C G, Liang Y, Dygert N, Xu W L. 2016. Formation of orthopyroxenite by reaction between peridotite and hydrous basaltic melt: an experimental study. Contributions to Mineralogy and Petrology, 171(8-9): 77

Wang C G, Lo Cascio M, Liang Y, Xu W L. 2020. An experimental study of peridotite dissolution in eclogite-derived melts: implications for styles of melt-rock interaction in lithospheric mantle beneath the North China Craton. Geochimica et Cosmochimica Acta, 278: 157–176

Wang D J, Liu X W, Liu T, Shen K W, Welch D O, Li B S. 2017. Constraints from the dehydration of antigorite on high-conductivity anomalies in subduction zones. Scientific Reports, 7(1): 16893

Wang J T, Xiong X L, Takahashi E, Zhang L, Li L, Liu X C. 2019. Oxidation state of arc mantle revealed by partitioning of V, Sc, and Ti between mantle minerals and basaltic melts. Journal of Geophysical Research: Solid Earth, 124(5): 4617–4638

Wang M L, Tang H F. 2013. Reaction experiments between tonalitic melt and mantle olivine and their implications for genesis of high-Mg andesites within cratons. Science China: Earth Sciences, 56(11): 1918–1925

Wang W J, Caumon M C, Tarantola A, Pironon J, Lu W J, Huang Y H. 2019. Raman spectroscopic densimeter for pure CO_2 and CO_2-H_2O-NaCl fluid systems over a wide P-T range up to 360℃ and 50 MPa. Chemical Geology, 528: 119281

Wang X, Zhang J F, Rushmer T, Adam J, Turner S, Xu W C. 2019. Adakite-like potassic magmatism and crust-mantle interaction in a postcollisional setting: an experimental study of melting beneath the Tibetan Plateau. Journal of Geophysical Research: Solid Earth, 124(12): 12782–12798

Wang X L, Chou I M, Hu W X, Burruss R C, Sun Q, Song Y C. 2011. Raman spectroscopic measurements of CO_2 density: experimental calibration with high-pressure optical cell (HPOC) and fused silica capillary capsule (FSCC) with application to fluid inclusion observations. Geochimica et Cosmochimica Acta, 75(4): 4080–4093

Wang X L, Chou I M, Hu W X, Burruss R C. 2013a. In situ observations of liquid-liquid phase separation in aqueous $MgSO_4$ solutions: geological and geochemical implications. Geochimica et Cosmochimica Acta, 103: 1–10

Wang X L, Hu W X, Chou I M. 2013b. Raman spectroscopic characterization on the OH stretching bands in NaCl-Na_2CO_3-Na_2SO_4-CO_2-H_2O systems: implications for the measurement of chloride concentrations in fluid inclusions. Journal of Geochemical Exploration, 132: 111–119

Wang X L, Wan Y, Hu W X, Chou I M, Cai S Y, Lin N, Zhu Q, Li Z. 2016a. Visual and in situ Raman spectroscopic observations of the liquid-liquid immiscibility in aqueous uranyl sulfate solutions at temperatures up to 420℃. The Journal of Supercritical Fluids, 112: 95–102

Wang X L, Wan Y, Hu W X, Chou I M, Cao J, Wang X Y, Wang M, Li Z. 2016b. In situ observations of liquid-liquid phase separation in aqueous $ZnSO_4$ solutions at temperatures up to 400℃: implications for Zn^{2+}-SO_4^{2-} association and evolution of submarine hydrothermal fluids. Geochimica et Cosmochimica Acta, 181: 126–143

Wang X L, Wang X Y, Chou I M, Hu W X, Wan Y, Li Z. 2017. Properties of lithium under hydrothermal conditions revealed by in situ Raman spectroscopic characterization of Li_2O-SO_3-$H_2O(D_2O)$ systems at temperatures up to 420℃. Chemical Geology, 451: 104–115

Wang X L, Qiu Y, Chou I M, Zhang R Q, Li G L, Zhong R C. 2020a. Effects of pH and salinity on the hydrothermal transport of tungsten: insights from in situ Raman spectroscopic characterization of K_2WO_4-NaCl-HCl-CO_2 solutions at temperatures up to 400℃. Geofluids, 2020: 2978984

Wang X L, Qiu Y, Lu J J, Chou I M, Zhang W L, Li G L, Hu W X, Li Z, Zhong R C. 2020b. In situ Raman spectroscopic investigation of the hydrothermal speciation of tungsten: implications for the ore-forming process. Chemical Geology, 532: 119299

Wang Y, Foley S F. 2018. Hybridization melting between continent-derived sediment and depleted peridotite in subduction zones. Journal of Geophysical Research: Solid Earth, 123(5): 3414–3429

Wang Y F, Zhang J F, Jin Z M, Green II H W. 2012. Mafic granulite rheology: implications for a weak continental lower crust. Earth and Planetary Science Letters, 353-354: 99–107

Wu X, Lin J F, Kaercher P, Mao Z, Liu J, Wenk H R, Prakapenka V B. 2017. Seismic anisotropy of the D" layer induced by (001) deformation of post-perovskite. Nature Communications, 8(1): 14669

Xiong X L, Keppler H, Audétat A, Ni H W, Sun W D, Li Y. 2011. Partitioning of Nb and Ta between rutile and felsic melt and the fractionation of Nb/Ta during partial melting of hydrous metabasalt. Geochimica et Cosmochimica Acta, 75(7): 1673–1692

Yang X Z. 2015. OH solubility in olivine in the peridotite-COH system under reducing conditions and implications for water storage and hydrous melting in the reducing upper mantle. Earth and Planetary Science Letters, 432: 199–209

Yang X Z, Liu D and Xia Q K. 2014. CO_2-induced small water solubility in olivine and implications for properties of the shallow mantle. Earth and Planetary Science Letters, 403: 37–47

Yang X Z, Keppler H, Li Y. 2016. Molecular hydrogen in mantle minerals. Geochemical Perspectives Letters, 2(2): 160–168

Yang Y, Ingrin J, Xia Q K, Liu W D. 2019a. Nature of hydrogen defects in clinopyroxenes from room temperature up to 1000℃: implication for the preservation of hydrogen in the upper mantle and impact on electrical conductivity. American Mineralogist, 104(1): 79–93

Yang Y, Liu W D, Qi Z M, Wang Z P, Smyth J R, Xia Q K. 2019b. Re-configuration and interaction of hydrogen sites in olivine at high temperature and high pressure. American Mineralogist, 104(6): 878–889

Yu Y, Xu W L, Wang C G. 2014. Experimental studies of melt-peridotite reactions at 1–2 GPa and 1250–1400℃ and their implications for transforming the nature of lithospheric mantle and for high-Mg signatures in adakitic rocks. Science China: Earth Sciences, 57(3): 415–427

Zhang B H, Yoshino T. 2017. Effect of graphite on the electrical conductivity of the lithospheric mantle. Geochemistry Geophysics Geosystems, 18(1): 23–40

Zhang B H, Yoshino T. 2020. Temperature-enhanced electrical conductivity anisotropy in partially molten peridotite under shear deformation. Earth and Planetary Science Letters, 530: 115922

Zhang B H, Yoshino T, Yamazaki D, Manthilake G, Katsura T. 2014. Electrical conductivity anisotropy in partially molten peridotite under shear deformation. Earth and Planetary Science Letters, 405: 98-109

Zhang B H, Ge J H, Xiong Z L, Zhai S M. 2019a. Effect of water on the thermal properties of olivine with implications for lunar internal temperature. Journal of Geophysical Research: Planets, 124(12): 3469-3481

Zhang B H, Yoshino T, Zhao C C. 2019b. The effect of water on Fe-Mg interdiffusion rates in ringwoodite and implications for the electrical conductivity in the mantle transition zone. Journal of Geophysical Research: Solid Earth, 124(3): 2510-2524

Zhang J F, Wang C, Wang Y F. 2012. Experimental constraints on the destruction mechanism of the North China Craton. Lithos, 149: 91-99

Zhang L, Guo X, Wang Q X, Ding J L, Ni H W. 2017. Diffusion of hydrous species in model basaltic melt. Geochimica et Cosmochimica Acta, 215: 377-386

Zhang L, Yuan H S, Meng Y, Mao H K. 2018. Discovery of a hexagonal ultradense hydrous phase in (Fe, Al)OOH. Proceedings of the National Academy of Sciences of the United States of America, 115(12): 2908-2911

Zhang L, Guo X, Li W C, Ding J L, Zhou D Y, Zhang L M, Ni H W. 2019. Reassessment of pre-eruptive water content of lunar volcanic glass based on new data of water diffusivity. Earth and Planetary Science Letters, 522: 40-47

Zhang P P, Zhang L, Wang Z P, Li W C, Guo X, Ni H W. 2018. Diffusion of molybdenum and tungsten in anhydrous and hydrous granitic melts. American Mineralogist, 103(12): 1966-1974

Zhang Y F, Wang C, Zhu L Y, Jin Z M, Li W. 2019. Partial melting of mixed sediment-peridotite mantle source and its implications. Journal of Geophysical Research: Solid Earth, 124(7): 6490-6503

Zhang Y S, Hou T, Veksler I V, Lesher C E, Namur O. 2018. Phase equilibria and geochemical constraints on the petrogenesis of high-Ti picrite from the Paleogene East Greenland flood basalt province. Lithos, 300-301: 20-32

Zhao S, Jin Z, Zhang J, Xu H, Xia G, Green II H W. 2012. Does subducting lithosphere weaken as it enters the lower mantle? Geophysical Research Letters, 39(10): L10311

Zhu X, Guo X Z, Smyth J R, Ye Y, Wang X, Liu D. 2019. High-temperature vibrational spectra between $Mg(OH)_2$ and $Mg(OD)_2$: anharmonic contribution to thermodynamics and D/H fractionation for brucite. Journal of Geophysical Research: Solid Earth, 124(8): 8267-8280

Research Progress of Experimental Geoscience

ZHANG Jun-feng[1], NI Huai-wei[2], YANG Xiao-zhi[3], MAO Zhu[2], ZHANG Bao-hua[4], XIONG Xiao-lin[5], HOU Tong[6], XU Wen-liang[7]

1. School of Earth Sciences, China University of Geosciences (Wuhan), Wuhan 430074; 2. School of Earth and Space Sciences, University of Science and Technology of China, Hefei 230026; 3. School of Earth Sciences and Engineering, Nanjing University, Nanjing 210023; 4. School of Earth Sciences, Zhejiang University, Hangzhou 310027; 5. State Key Laboratory of Isotope Geochemistry, Guangzhou Institute of Geochemistry, Chinese Academy of Sciences, Guangzhou 510640; 6. School of Earth Sciences and Resources, China University of Geosciences (Beijing), Beijing 100083; 7. School of Earth Sciences, Jilin University, Changchun 130061

Abstract: In the past ten years, the rapid development of experimental geoscience in China has made China an important international research force in the field of high-temperature and high-pressure experiments. The representative research progresses of experimental geoscience in China mainly included studies of occurrence and effects of important volatiles in the deep Earth, mineralization process and mechanism, and physicochemical properties and interactions of crust and mantle materials. This article reviews the major research progresses, the platform construction, and the technological advancements of experimental geoscience from 2011 to 2020 in China and looks into future directions of research in this field.

Key words: experimental geoscience; high temperature and high pressure; physicochemical properties and geological processes

同位素效应理论和计算研究进展

刘 耘[1,2,3]

1. 中国科学院 地球化学研究所，矿床地球化学国家重点实验室，贵阳 550081；2. 成都理工大学 行星科学国际研究中心，成都 610059；3. 中国科学院 比较行星学卓越创新中心，合肥 230026

摘　要：近十年来，国内稳定同位素地球化学的理论解释已普遍达到了量子化学水平。基于精密的量子化学从头或第一性原理，研究者们开始了平衡和动力学分馏系数的计算。在同位素的分析测试、野外观察、理论和计算四个方向上，国内在理论和计算方向的发展可能还算最好的，整体处于国际第一方阵的地位。国内学者率先发展了超冷体系同位素分馏、含压力效应的同位素分馏、同位素的浓度效应、团簇同位素、微小同位素异常、热梯度下同位素扩散效应、高温重复过程等方向的同位素理论和计算方法，也发展了针对固–液两相同位素分馏的可变体积的分子簇模型（variable volume cluster model，VVCM）计算新方法、针对重金属同位素固相体系的含核体积效应处理的方法，以及针对熔体中同位素扩散的动力学分馏的理论和计算方法。同时，还为大量不同的非传统（金属）同位素体系，提供了大量的平衡分馏系数，为这些同位素体系的日后深入应用奠定了较好的基础。但是，在这些成果中，真正由我国学者首先提出的原始概念、模型、体系和方法还很少，绝大多数都是对前人（主要是国外同行）提出的理论体系和新兴发展方向的修改和补充。未来我国同位素理论和计算领域应率先使用包含量子场论在内的新一代理论工具，在同位素效应的新方向、新概念的提出和新理论的建立方面，做出更多的贡献。

关键词：同位素效应　稳定同位素分馏　理论计算　量子化学　平衡分馏　超冷体系　压力效应　同位素浓度效应

0　引　言

同位素效应是指由化学和物理过程产生的同位素成分的变化，其过程也被称为同位素分馏。随着同位素测试技术的发展，同位素作为一种重要的研究工具，已被广泛应用于地球科学、行星科学、生态环境、能源等领域。同位素分馏理论研究的目的是进一步完善和发展这一工具，以便把同位素研究方法的潜力发掘出来，这是发展同位素应用最关键的环节之一。

尽管已经有少数研究者将同位素方法用于动力学和非平衡过程，但近十年来，仍然是以平衡过程的同位素分馏研究为主。本文从同位素平衡分馏理论建设、同位素动力学分馏理论建设、基本分馏参数的计算三个方面来总结国内十年的进展，最后对未来学科的发展进行了展望。

1　同位素平衡分馏的理论创新

Bigeleisen（1996）对稳定同位素平衡分馏效应做了总结，认为这个问题既包括了经典的质量依赖分馏效应，即由著名的 Bigeleisen-Mayer 公式（Bigeleisen and Mayer，1947；Urey，1947）确定的部分，还包括非谐效应、超越 Born-Oppenheimer 近似的效应、核磁效应、核场效应。对地学研究的体系而言，还必须加上同位素压力效应。

1.1　超冷体系同位素平衡分馏理论的建立

目前的稳定同位素地球化学理论体系是针对常温到高温（高压）环境建立的，这些理论并不能直接应用

---

* 原文"近十年中国同位素效应理论和计算研究进展"刊于《矿物岩石地球化学通报》2021 年第 40 卷第 6 期，本文略有修改。

于超低温环境，以下称之为"超冷体系"，笼统指约在零下100℃或更低温度体系。这是因为在超低温条件下，各种量子效应将变得十分显著，与同位素分馏相关的诸多微小的能量差别会被放大，都变得十分重要，导致无法直接忽略。另外，很多可以用于室温和高温体系的数学近似处理也将不成立，造成原有的理论推导过程无法成立。其中一个例子就是在对系统转动能量的处理，一般在常温和高温下转动能级是连续占据的，从而可以使用数学近似方法来进行处理。但在超低温下，这些转动能级仅有小部分被占据，是中断和离散的，并不能用以前的数学近似来处理。因此，原先的理论公式便无法适用于超低温体系。

遗憾的是，无论是处理各种被忽略掉的量子效应，还是重新进行同位素效应的理论推导，在超低温条件下都面临相当巨大的困难，这也是学界一直缺少相关理论工作的主要原因。超低温下的平衡实验也几乎没有，因为温度越低，越难在短时间达到平衡，人们没法做需要几万年或者几十万年才能平衡的实验。

但是，超低温环境的同位素理论又绝非鸡肋。首先，在太阳系形成初期，星云盘内部的环境是在10～15 K的超低温下，仅H、He、H_2等极少的单原子和化合物可以作为气体存在，其他所有物质都被冷凝为固体和尘埃（由于宇宙辐射，会在这些尘埃内部发生大量的基于自由基的化学反应）。随着原始太阳的形成及其产生的辐射，会导致部分星云升温，但升温的幅度，取决于距离原始太阳的距离。在这一过程中，因为原始星云盘十分巨大，绝大部分的星云物质依然处于极度低温的环境。首先，大部分气态物质在原始太阳形成后，被太阳风逐渐"吹"到外太阳系，这些被移除的气体和保留下来的固体之间，存在未知的、大概率极为夸张的同位素分馏。如果不理解这些同位素分馏过程，是无法明白内、外太阳系物质之间在同位素组成上的巨大差别的（如彗星的氢同位素与地球的巨大差异），也无法详细研究太阳系同位素分布模式和演化过程；其次，现今太阳系主要星体的表面和高空大气大多是处于超低温的环境，在这些地方依然有大量的同位素交换和分馏。随着深空探测的开展，越来越多的同位素测试数据被获得（Pinto et al., 1986；Eberhardt et al., 1987；Robert et al., 2000；Cordier et al., 2008；Robert, 2010；Nixon et al., 2012；Webster et al., 2013；Altwegg et al., 2015），中国和美国也计划在月球南极建立永久性科学基地，并且需要探测和研究在南极陨石坑永久阴影区中极低温的水冰。如何解释上述这些探测计划获得的同位素数据，将是一个严峻的挑战。

另外，天体化学研究极为依赖"同位素异常"信号来进行溯源，而忽略超低温下同位素的非质量分馏，是该研究领域一大亟待填补的缺陷。同位素异常是指在校正掉自然界或仪器分析过程产生的同位素质量依赖分馏后，某一个同位素（如^{17}O）相对于其他的同位素（如^{16}O和^{18}O）发生了富集或亏损，则称"某某同位素异常"（如"^{17}O异常"）。天体化学中假设同位素异常是由不同的核合成过程所致，其前太阳系颗粒，往往显示高达千分之几万甚至几十万的巨大异常信号；同位素异常也可以由不同浓度的放射性成因元素的母体所导致。这些都是源区自带的特殊属性，而非一般化学过程的产物。因此，天体化学大量使用同位素异常信号来进行源区的识别、陨石种类的划分，是其最重要的基本工具之一。但遗憾的是，我们已经认识到，即使在高温下，一些化学过程仍然可以发生可观测的非质量分馏（Zhang and Liu，2020）。随着温度的降低，这些非质量分馏将会快速增长至异常显著的程度。然而，目前学界近年来才刚刚开始意识到需要校正由核体积效应引发的非质量分馏，而其他几种机理引发的非质量分馏则依然未被考虑。因此，评估超低温下的同位素非质量分馏的程度，能够重新审视这些同位素异常，并修正其中核合成过程或放射性母体引发的异常份额，可以重新改进天体化学这一无比重要的研究工具。

由此可见，超低温同位素分馏理论，不仅不是一个可有可无的理论，甚至可能是未来天体化学的核心理论工具之一。基于这样的认识，Zhang Y. N. 和 Liu（2018）于2018年在 GCA 杂志上发表了题为"The theory of equilibrium isotope fractionations for gaseous molecules under super-cold conditions"的论文，这是学界首次关于超冷体系同位素分馏的理论建设工作。

Zhang Y. N. 和 Liu（2018）在经典同位素分馏理论的基础上，增加了超冷体系必须要考虑的十几种与同位素有关的量子效应，并提供了详尽的计算方法。图1显示随温度的降低，经典同位素分馏理论的误差会越来越大，无法处理超低温体系。

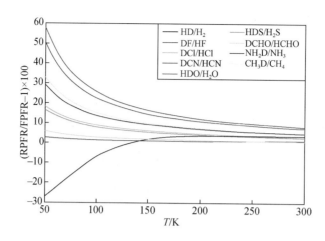

图1 超冷条件下的重要同位素体系的分馏计算结果对比（据 Zhang Y. N. and Liu，2018）
纵坐标是新建立的超冷体系的理论和经典同位素分馏理论（Bigeleisen-Mayer 公式）的计算结果差距的百分比

Zhang Y. N. 和 Liu（2018）的工作有几个缺憾：一是完全忽略了平衡分馏下的核磁效应（Magnetic isotope effect）。一些常用的同位素体系是具有核磁活性（即原子核具备非零的核自旋）的体系，包括氢同位素（H、D）、碳同位素（^{13}C）和氧同位素（^{17}O）等。但目前整个学界（包括化学和物理领域）尚无人基于精确的量子力学和统计力学方法，来建立平衡分馏的同位素磁效应的理论计算方法。前人很多针对磁效应的动力学分馏理论，如系间穿越理论（inter-system crossing），均是针对化学反应动力学方面的半经验性质的假说（Turro，1983）。因为在常温和高温下，磁效应导致的同位素平衡分馏是非常小的，Zhang Y. N. 和 Liu（2018）认为或许在超低温下也可以忽略。然而，目前尚未有超低温下同位素磁效应的数据，这个假设因而存在问题；二是他们使用了较为粗糙的方法来处理核场效应和 Born-Oppenheimer approximation（BOA）的校正。为了简化理论计算，他们将这两个效应均做了一级近似处理，其中对核场效应使用了零阶相对论效应近似（ZORA），对 BOA 使用了对角化近似的处理（DBOC），这些简化可能会导致一些较小的误差。

地球化学领域极少有人在研究同位素平衡分馏时考虑 Born-Oppenheimer approximation（BOA）的校正，而 Zhang Y. N. 和 Liu（2018）使用 DBOC 的工作提供了一个稀缺的样本，他们发现即使温度升高至室温的条件下，DBOC 校正对 H、C 同位素体系的分馏计算还是有明显的提升。换言之，如果要获得非常精确的 H、C 同位素分馏值，在室温下的分馏甚至都应该考虑 BOA 的校正。

总之，同位素核磁效应平衡分馏理论的缺失，不仅是超冷体系的缺憾，也是稳定同位素理论框架中最明显的一个缺陷。因为建立精确的核磁效应同位素分馏理论的难度巨大，学界到目前为止都是空白。

1.2 重复岩浆过程对同位素分馏放大效应的认识

同位素异常最常见的起因是因为某个同位素具有放射性同位素的母体，由于母体不断衰变，因而体系中的子体浓度逐渐累积为一个异常信号。在地幔地球化学研究中，一个被广泛接受的假设是，在极高的温度下不会发生同位素分馏，尤其是不会发生同位素非质量分馏，这是"化学地球动力学"的两大基石之一（另一是假定部分熔融是平衡分异过程）。因此，人们将经过质量歧视校正（the mass bias correction 或称 normalization）后依然存在的同位素异常，只归因于放射性衰变或核合成异常，而非其他化学和物理过程的后果。所谓的质量歧视校正指的是一种通过假定待测同位素体系中的某一同位素比值对于所有样品而言是固定不变的，利用某种事先约定的同位素分馏定律，结合质谱分析测量结果，从而确定并扣除待测样品可能发生的所有质量相关分馏（mass dependent fractionation，MDF）的校正方法。目前最为学界所广泛使用的质量歧视校正方法便是基于指数分馏定律的。该定律假定所有同位素比值发生的分馏值与同位素的质量比呈指数关系，即分馏值=（同位素质量比）$^\beta$。因此，只要计算出了 β 值，便可以对所有测量得到的同位素比值进行质量歧视校正。基于此，诸多放射性成因同位素体系，如 Sr、Nd、Pb 等被广泛用于研究地幔不均一性。采用同样的思想，同时随着质谱分析测试精度的不断提高，一些新的同位素体系（如^{182}W 等）也被用于进行类似的研究。

然而，随着核场效应或称核体积效应（nuclear volume effect，NVE）的发现（Fujii et al.，1989；Bigeleisen，1996），人们意识到对于那些大质量的同位素体系（如 W、Hg、U 等），由于不同的同位素的原子核大小略有差异，导致不同同位素之间的能量出现了差异并最终引起同位素分馏。这种由于核体积效应造成的分馏不同于传统认识的 MDF，其分馏值的大小正比于不同同位素之间的原子核电荷半径平方差（$\delta \langle r^2 \rangle$）。同时，在温度依赖性关系上，不同于 MDF 引起的分馏正比于 $1/T^2$，核体积效应引起的分馏正比于 $1/T$。这意味着即使在涉及熔融结晶过程的高温下（>1 000℃），对于那些大质量的同位素体系而言，核体积效应反而会成为造成同位素分馏的主导因素。除了温度依赖关系外，更为重要的还有一点，就是对于绝大多数同位素体系，其原子核电荷半径的平方随原子质量的变化并非线性，而多为锯齿形。这就决定了核体积效应导致的同位素分馏将是非质量相关分馏（mass independent fractionation，MIF）。Fujii 等（2006）通过理论推导发现，针对质量歧视校正最广为使用的指数分馏定律并不能完全扣除核体积效应造成的分馏，同时还结合实验结果推断：某些陨石样品中发现的 Ca、Ti、Cr、Sr 和 Ba 同位素异常信号很可能是核体积效应的结果，据此提出了同时针对质量依赖分馏和核体积效应的校正方法。

近几年来，核体积效应也逐渐被学界所认识和了解，已经有不少领域的学者意识到核体积效应造成的非质量分馏对同位素分析结果的影响，如 U-Pb 定年（Amelin et al.，2010）和同位素分析（Cook and Schönbächler，2016；Saji et al.，2016；Kruijer and Kleine，2018；Rizo et al.，2019；Tusch et al.，2019）。以^{182}W 同位素异常为例，有研究者将某些在幔源样品中发现的^{182}W 异常归因于可能的高温下核体积效应造成的非质量分馏，但却缺乏广泛的理论和实验研究支持。

基于上述研究现状，受前人提出的多阶段岩浆过程可能放大微小的同位素分馏信号的假说的启发（Moynier et al.，2013），以及采用数学建模和蒙特卡洛思想模拟复杂岩浆过程的思路（Liang and Liu，2016；Liu and Liang，2017），Zhang 和 Liu（2020）以^{17}O 和^{182}W 研究对象为例，设计了多阶段部分熔融和结晶模型（图2），在量子化学计算结果的基础上，通过随机方法模拟了多阶段岩浆过程中，研究了由核体积效应造成的同位素非质量分馏信号在经过质量歧视校正后的变化过程。

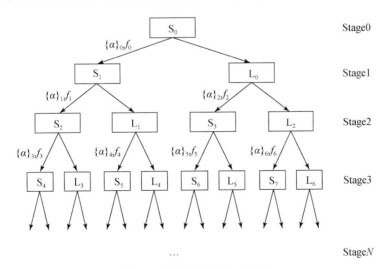

图2　多阶段熔融和结晶模型

S 代表固体相；L 代表熔体相；箭头代表一次熔融或结晶过程；$\{\alpha\}$ 代表一次熔融或结晶过程中，新生成的相与残留相之间的同位素分馏；f 代表每次熔融或结晶过程中，元素在固体相和熔体相之间的分配比例多阶段的熔融和结晶过程在梯级岩浆房系统、地幔柱系统和热管构造系统里非常常见。Zhang 和 Liu（2020）假定每次熔融和结晶过程中，两相间达到了元素和同位素平衡（如果不平衡，可以预见同位素异常信号会更大）。同时，针对具体的同位素分馏系数，该研究采用了与 Fujii 等（2006）的方法，将同位素分馏拆分为传统意义上的 MDF 以及由核体积效应造成的 MIF 两部分［公式（1）］。而在具体的模拟过程中，该研究采用随机方法确定每次熔融或结晶过程中被研究体系的元素分配系数、同位素分馏系数及其他相关的参数。就元素分配系数而言，其变化范围参考前人的实验研究；就同位素分馏系数而言，其变化范围则在前人研究结果的基础上，通过量子化学计算的方法确定核体积效应对总的同位素分馏值的贡献后才确定。

$$^{m_2/m_1}\alpha = 1 + \underbrace{A\frac{m_2 - m_1}{m_1 m_2}}_{\text{MDF}} + \underbrace{B\delta\langle r^2\rangle_{m_2/m_1}}_{\text{NVF导致的MIF}} \tag{1}$$

式中，m_1 和 m_2 代表不同同位素的质量；$\delta\langle r^2\rangle$ 代表不同同位素间原子核电荷半径的平方差；A 和 B 分别代表 MDF 和 NVE 导致的 MIF 相关的参数，具体和发生同位素分馏的两相的物理化学性质和发生分馏时的物理化学条件直接相关。

以 ^{182}W 同位素异常为例，该研究的模拟结果显示，随着模拟的进行，新生成的固体相和熔体相相对于初始的源区，呈现出更为分散的同位素组成（在统计学意义上表现为逐渐增大的标准差；图3）。

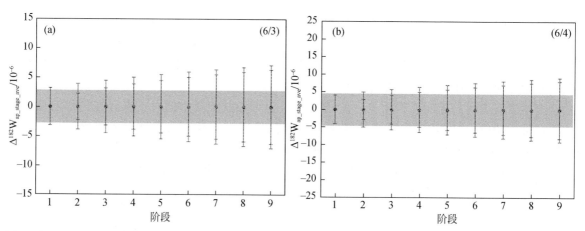

图3　每阶段新形成的固体相与熔体相的 ^{182}W 同位素异常的统计平均值及 2 倍标准差（据 Zhang and Liu，2020）
（a）基于 ^{186}W/^{183}W 进行质量歧视校正后的结果；（b）基于 ^{186}W/^{184}W 进行质量歧视校正后的结果。
蓝色代表固体相；红色代表熔体相

目前针对幔源岩石样品的 W 同位素研究表明，在部分太古宙幔源岩石样品中常会观测到正的 ^{182}W 同位素异常，而在某些年轻的洋岛玄武岩样品中则会观察到负的 ^{182}W 同位素异常。前人研究主要认为，这些在古老岩石样品中的正 ^{182}W 同位素异常是在地球形成的早期阶段，硅酸盐或硅酸盐-铁熔体分异过程中，更亲石的 ^{182}Hf 倾向于留在硅酸盐中，随后衰变形成 ^{182}W 形成的（Touboul et al.，2012）。而那些年轻玄武岩中的负 ^{182}W 异常则被认为是潜在的核-幔相互作用，使得假定严重亏损 ^{182}W 同位素的地核物质与玄武岩的地幔源区混合的结果（Mundl-Petermeier et al.，2020）。本研究通过模拟发现，正负两种 ^{182}W 同位素异常都可以分别通过硅酸盐熔体的多阶段含 W 和硫的金属相析出以及地幔源区的多阶段熔体抽离过程来解释，仅需在每次熔融或金属相析出过程中，两相间存在着约 0.01‰~0.02‰ 的 ^{186}W/^{184}W 同位素分馏（图4）。

该研究的意义是发现熔融和结晶过程中由核体积效应导致的微小同位素分馏可能会在多阶段演化模型中被放大，从而为解释地幔样品中发现的各种微小同位素异常（如 ^{182}W 和 ^{17}O 等）提供一种新思路。例如，可以解释 Cano 等（2020）发现的月球不同玄武岩样品之间的 20×10^{-6} 左右的 ^{17}O 异常的差距。

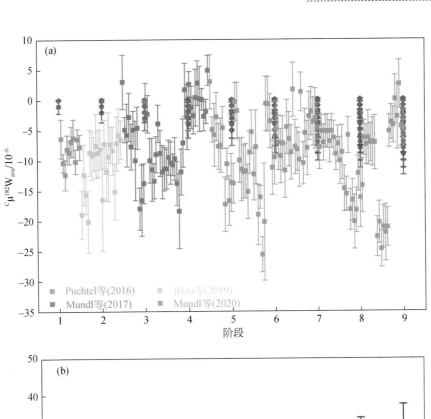

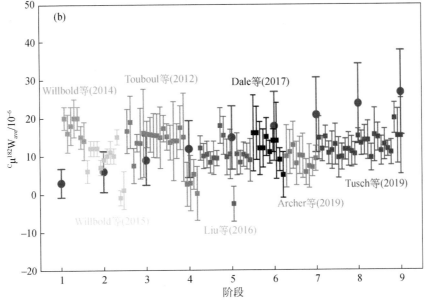

图 4 多阶段地幔源区熔体抽离过程模拟结果（a）和多阶段含 W、S 的金属相洗出过程模拟结果（b）（据 Zhang and Liu，2020）

1.3 同位素平衡分馏浓度效应的认识

随着非传统同位素研究的进行，人们已经开始测试矿物中微量元素的同位素成分。中国科学技术大学黄方和吴忠庆教授的联合课题组发现了一种普遍存在、非常重要的同位素效应，他们将其命名为"同位素浓度效应"，即两个矿物之间某个同位素的分馏值会在其他条件完全一样的情况下，因为该同位素体系的浓度不同而发生变化。这种变化主要发生在该同位素体系在矿物中浓度很低的阶段，如果浓度增加到了一个定值，则分馏又会保持基本恒定。

Feng 等（2014）于 2014 年首次发现了这个效应：斜方辉石与单斜辉石之间的 Ca 同位素分馏会在 Ca 浓度小于一个极值后，不再保持恒定，而是快速变大，Ca 浓度越低，分馏越大（图 5）。

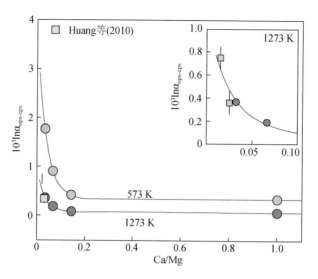

图 5　斜方辉石与单斜辉石的 Ca 同位素分馏（据 Feng et al., 2014）
在 Ca 浓度小于一个极值后，其分馏行为发生了较大变化

黄方和吴忠庆课题组连续发表多篇文章，对几个不同的非传统体系的浓度效应都给予了研究。Wang 等（2017a）发现，白云石的镁浓度对碳酸盐矿物之间的镁和钙平衡同位素分馏影响很大，镁浓度越低，方解石越富集轻镁同位素（图6）；Wang 等（2017b）还发现，斜方辉石的钙浓度在一定范围内也对斜方辉石与单斜辉石之间的钙同位素平衡分馏影响很大（图7），钙浓度越低，斜方辉石与单斜辉石之间的钙同位素平衡分馏越大，而且成分效应有两个"拐点"，即浓度在某个范围内，成分效应十分显著，而浓度高于某个值或低于某个值，浓度变化对平衡同位素分馏无显著影响；Li 等（2019a）研究了钾同位素在长石和微斜长石之间的分馏，发现同样具有显著的浓度效应（图8）。

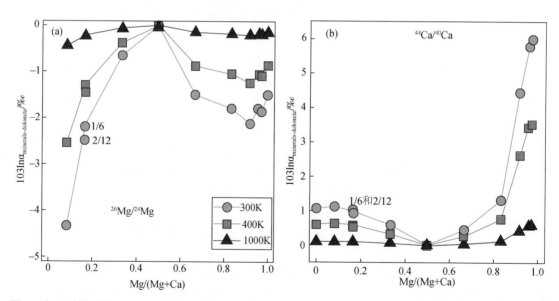

图 6　白云石的不同 Mg 和 Ca 浓度对碳酸盐矿物之间 Mg 和 Ca 同位素分馏的影响（据 Wang et al., 2017a）

众所周知，固溶体存在非线性的性质变化，同位素的浓度效应就是这种现象的表现。该工作的重要性在于，提醒学界从研究主量元素的同位素转到研究微量元素的同位素成分时，矿物-矿物之间的同位素分馏不再是一个单一的分馏值，一定还要注意它的"浓度效应"。

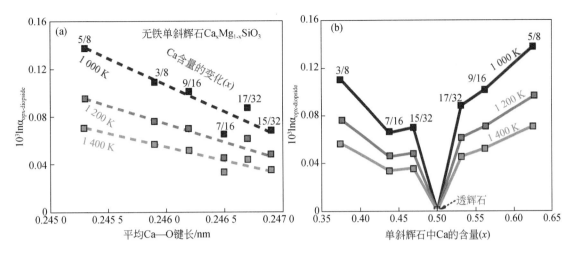

图7　斜方辉石和透辉石的Ca同位素分馏与Ca和Fe浓度的关系（据Wang et al.，2017b）

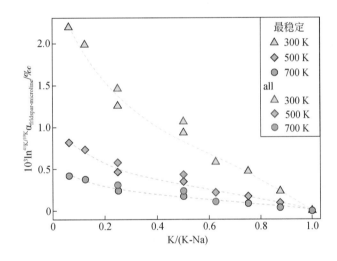

图8　长石和微斜长石之间K同位素分馏的浓度效应（据Li et al.，2019a）

1.4　含同位素压力效应计算方法的建立

Joy和Libby（1960）发表了第一篇定量估算压力效应的文章，他们预测同位素平衡分馏具有明显的压力效应。Polyakov和Kharlashina（1994）发表了同位素压力效应的重要文章，建立了一种定量计算压力效应的理论方法。他们发现，如果想要获得一定压力下的两个不同同位素体的Gibbs自由能之差，可以用在同一体积下它们的赫姆赫兹自由能之差来近似代表。又因为赫姆赫兹自由能与配分函数之间已有明确的表达式，因此可以通过计算声子频率来求得赫姆赫兹自由能之差。

黄方和吴忠庆团队是学界第一个采用量子化学计算方法来精确确定同位素压力效应的课题组。他们使用了准简谐近似的方法，获取了状态方程、赫姆赫兹自由能和不同体积的关系，然后，再用于计算含压力效应的同位素分馏。他们精确计算了Mg和Si同位素含压力效应的一些分馏系数（图9、图10）。他们建立的压力效应方法，已被广泛应用于各种涉及高压环境的同位素分馏计算中（Blanchard et al.，2017）。

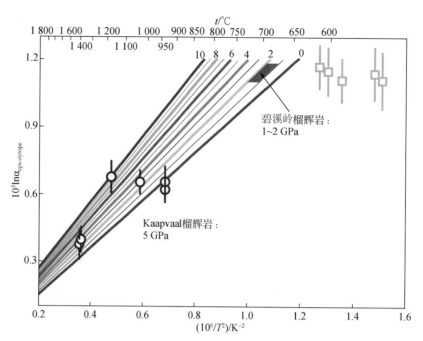

图9 单斜辉石和镁铝榴石之间包含了压力效应的 Mg 同位素的分馏（据 Huang et al., 2013）

图中 0~10 的数字是指不同压力（GPa）

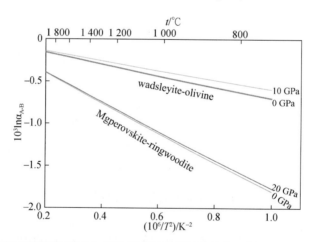

图10 不同地幔矿物之间的 Si 同位素分馏及其压力效应（据 Huang et al., 2014）

1.5 微小同位素异常计算方法的建立

该研究方向由以色列 Boaz Luz 教授在 2005 年首次提出，目前已发展成为稳定同位素地球化学的一个重要的新兴分支。微小同位素异常研究强烈依赖于不同过程、不同交换反应的高精度的三同位素关系（θ 值），但 θ 值很难通过实验获得。Cao 和 Liu（2011）首次建立了研究 θ 值的理论框架，并定义一个同位素新的变量 Kappa：

$$\theta^{E}_{a-b} = \kappa_a + (\kappa_a - \kappa_b) \frac{\ln^{18}\beta_b}{\ln^{18}\alpha_{a-b}} \tag{2}$$

这是以氧同位素为例的 θ 值的计算方法。这一工作为该领域从水循环研究拓展到固体地球科学研究奠定了基础，并被几乎所有该领域的后续文章所引用。

Cao 等（2019）又将该方法用于研究洋岛玄武岩中的 O^{17} 异常，并用于指示壳幔相互作用的程度（图11）。

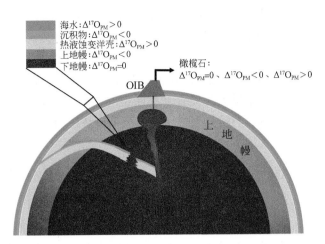

图 11 不同储库相对于地幔平均值的 ^{17}O 异常方向（据 Cao et al.，2019）

1.6 重金属同位素体系（含核体积效应）的同位素平衡分馏方法的建立

重元素需要准确计算其"量子相对论效应"，而传统上被广泛使用的基于薛定谔方程的量子化学计算方法，不能准确处理"量子相对论效应"。重金属元素的精确计算要使用四组分的 DIRAC 方程，计算量和计算难度均很大。

核场效应（或核体积效应）是一种由同位素的核电荷密度、核大小和形状差异引起的"基态电子能"不同的现象，它强烈影响到重元素的稳定同位素平衡分馏。传统理论以点电荷来代替原子核的处理，当原子较轻时，这个近似处理是合理的，但随着原子序数的增大，近似处理就越来越不合理。当原子序数较大时，不同同位素的电子能差别较大，如果不考虑电子能差别，将严重低估重同位素的分馏程度。

Bigeleisen（1996）正式将核体积效应引入同位素分馏计算的公式。Schauble（2007）首次从量子化学理论上计算了 Hg、Tl 等重金属同位素分馏系数，从此开启了核体积效应量子化学从头计算研究的大门。他的计算发现，室温时 Hg 和 Tl 体系的平衡分馏可达 2‰～3‰，远远超过传统理论方法的预测，计算结果也同实验观测相吻合。由于重金属元素的同位素体系均具有较大的核体积效应，因此建立方法才是解决这些重同位素体系问题的关键。国际上目前已有五六个课题组开展了分子体系的核体积同位素效应的研究，中国科学院地球化学研究所刘耘课题组的杨莎博士率先在国内建立了含核体积效应的重金属同位素体系的计算方法，并计算了很多 Hg、Tl、Pb、Sn 同位素体系的分馏系数（Yang and Liu，2015，2016）。

另外，由于核体积效应对温度的依赖性比质量效应弱很多，这造成在 1 000℃ 以上的体系中，几乎仅剩下核体积效应导致的同位素分馏，质量依赖分馏则几乎消失。因此研究高温体系的重金属同位素分馏、建立准确针对固相矿物的核体积计算方法是重金属同位素高温体系研究的重中之重。Schauble（2013）率先建立了固相体系的核体积效应计算方法；其后，Fang 和 Liu（2019）也建立了固相体系的核体积效应计算方法，而且他们还对现有方法的一个关键之处进行了修改：Schauble（2013）的方法是使用"接触电子密度"，而从理论上，应该使用全部起作用的电子密度，即"表观电子密度"，前者的大小一般只有后者的 90% 左右，Fang 和 Liu（2019）建立的新方法直接使用了表观电子密度，因而从理论上提高了精度。

1.7 Clumped 同位素分馏理论

Clumped 同位素方法又称"团簇同位素"方法，是研究物质中"稀-稀"同位素体的浓度，以获取温度及过程信息的一种方法。国内对 Clumped 同位素理论和计算研究较少，迄今只有中国科学院地球化学研究所刘耘课题组做了一些工作（Cao and Liu，2012；Liu and Liu，2016；Zhang et al.，2020）。

Cao 和 Liu（2012）对 Clumped 同位素的计算方法进行了基于统计力学的改进，他们将其中最关键的变量（Δ_i）的原始定义转换为等价的基于能量的定义，并使一些内在性质和定理凸显出来。

Liu 和 Liu（2016）定义了数学形式上更为严谨的 Clumped 同位素分馏表达式（Δ_i），相对于 Wang 等（2004）的定义，他们的定义不但与传统的地学同位素分馏定义相吻合，而且通过与反应常数（K）建立的联系使得计算结果与实验学的测量结果更为契合，同时也与 Ono 等（2014）的定义相呼应。这一定义具有明确的物理化学意义，即 Clumped 同位素分馏信号代表着多替换同位素体从单替换同位素体获得稀有同位素的程度或能力，具体的表达式如下：

$$\delta_i = \left(\frac{R_i}{R_i^{\infty}} - 1\right) \times 1000, \quad \Delta_{A*B*X} = \delta_{A*B*X} - \delta_{A*BX} - \delta_{AB*X} \approx \left(\frac{K_{A*B*X}}{K_{A*B*X}^{\infty}} - 1\right) \times 1000 \tag{3}$$

同时，Liu 和 Liu（2016）还针对 Clumped 同位素目前的研究趋向，利用超越简谐近似的量子化学方法为多个分子体系提供了精确的 Clumped 同位素平衡分馏参数，并在理论层面上提出了 Clumped 同位素平衡分馏的多个共性。他们的研究显示，不同体系的 Clumped 同位素平衡分馏信号可以与同位素替换的相对质量差以及所在原子位的键强建立关系。其分馏信号会随着温度的升高而减小，并在温度小于 500 K 的温度区间与 $1\,000/T$ 成良好的线性关系。超越简谐近似的高阶能量项对于研究含氢氘交换的 Clumped 同位素体系较为重要，其贡献值是不含氢氘交换的 Clumped 同位素体系的 2～20 倍。此外，他们还发现，Clumped 同位素体系的平衡分馏多与某些特定模式的分子振动有关（如分子伸缩振动），所以不同分子体系可以出现相似的 Clumped 同位素平衡分馏信号，如 CH_4 和 CH_3Cl 的 ^{13}C-D Clumped 同位素平衡分馏信号。

Zhang 等（2020）对 Clumped 同位素的一个重要体系——碳酸盐体系的磷酸酸解过程的分析测试过程的分馏进行了研究，首次指出了磷酸酸解碳酸盐矿物的路径是 3 条平行的路径，实验条件的微小改变都会影响到这 3 条路径贡献的相对比例，从而使实验结果发生上下漂移（图12）。

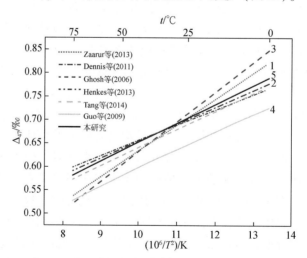

图 12　用 3 条平行反应路径生成的 Δ_{47} 与温度的关系（图中黑线）
以及与其他工作结果的对比（据 Zhang et al., 2020）
3 条路径贡献的相对比例的改变，会使图中黑线发生上下平移

1.8　VVCM 平衡分馏系数方法的建立

目前基于统计力学的 3 种固体同位素分馏计算方法（即基于爱因斯坦-德拜近似的方法、基于密度泛函微扰理论的声子计算的方法和基于分子簇模型再结合部分力常数矩阵分析的方法）都存在某些问题。例如，基于密度泛函微扰理论的声子计算的方法由于常常使用赝势，在体系涉及 H 原子或液相时，频率计算的误差较大；再如，基于分子簇模型再结合部分力常数矩阵分析的方法，在计算矿物和流体之间的

分馏时,由于流体使用的是零压下的结构,而矿物则是使用非零压下的晶体结构以及计算水平,因此也会造成误差。

刘耘课题组对基于分子簇模型再结合部分力常数矩阵分析的方法做了大幅改进,所建立的"可变体积的分子簇模型(VVCM)方法"在矿物–流体常压下的分馏计算方面、在使用高级别量子化学的理论处理方面都体现出独特的优势(Liu, 2013; Li and Liu, 2015; Gao et al., 2018)。

1.9 Position-Specific 同位素分馏方法的建立

近年来,学界逐渐开始研究一些有机物或矿物结构中特定位置的同位素成分,并用以反演物质来源、形成过程和温度等信息。Ma 等(2018, 2020)发展了一些前沿的 Position-Specific 实验技术。He 等(2020)在国内率先开展了该方向的理论计算研究,他们计算研究了几十个有机化合物官能团的 C 同位素的特定位置信息,为复杂的生物分子位置特异性同位素组成的研究奠定了理论基础。

2 对同位素动力学分馏方面的理论创新

2.1 温度梯度下同位素扩散分馏效应的理论

在有温度梯度存在的体系中,热扩散作用不仅会使元素在高温端和低温端发生明显分异,同位素也会在两端发生显著的分馏,但目前对这类过程的微观机理研究十分薄弱。Li 和 Liu(2015)基于量子力学及统计力学理论,对热梯度下同位素的分馏进行了深入研究。他们推导出了一个普适的热扩散过程的同位素分馏计算公式,以及这个公式的高温近似式:

$$\Delta_{T-T_0}{}^X M = -\frac{3}{2}\ln\frac{m^*}{m} \times \ln\frac{T}{T_0} \tag{4}$$

式中,T 和 T_0 是温度梯度两端的温度(K);m^* 和 m 是两个同位素的质量(如对 Mg 同位素,m^* 是 26,m 是 24,$^X M$ 就是 $^{26/24}$Mg)。此公式显示,在高温下同位素分馏只与温度梯度有关,而与其他化学结构因素无关,从而澄清了前人的相关争论。另外,该公式大大简化了复杂热扩散过程同位素的分馏计算,可以方便地计算出几乎所有同位素体系在高温热梯度下的分馏,因而可以方便地应用于许多地学相关方向的研究中。

2.2 岩浆不混溶过程同位素动力学分馏理论

自然界所观测到的同位素数据有许多是平衡理论解释不了的。例如,在高温下,根据平衡分馏理论,岩浆中的 Fe 同位素的分馏会很小,而许多已发表的数据却发现,实际 Fe 同位素分馏变化范围很大。这使得研究者们将注意力转移到了非平衡过程的研究上。但对于许多非平衡过程的同位素研究,理论上几乎是空白。Zhu 等(2015)对岩浆不混溶过程同位素动力学分馏理论进行了系统研究,推导出描述该过程同位素动力学分馏的公式,并建立了一个基于动力学扩散导致的同位素分馏模型。该模型可以解释目前在花岗岩体系中观测到的 Fe、Mg、Si、Li 等同位素分馏异常,并认为同位素的分馏是受岩浆演化过程的动力学效应控制的,体现在随着 SiO_2 的含量的升高,Fe、Mg 富集重同位素,而 Li 和 Si 的动力学效应由于自身原因不能显著影响其分馏的趋势。该模型的提出,将目前现存的同位素解释理论进行了新的延伸,并且该研究还基于硅酸盐熔体中矿物表面成分边界层的模型,推导出了同位素动力学分馏的计算公式。该公式有较多应用的潜力,对研究岩浆分异、A 型花岗岩的成因、岩浆去气过程的同位素分馏等方面有着重要的意义(Zhu et al., 2015)。此外,Chen 等(2018)还根据 Zhu 等(2015)的模型和公式,成功地解释了月球月海玄武岩富轻镁同位素特征的原因——是钛铁矿和周围的辉石、橄榄石发生了 Mg 离子的亚固相非平衡扩散的结果。

2.3 熔体中同位素扩散分馏

扩散是物质迁移的重要方式。理想气体（单原子或稀薄气体）向真空扩散的同位素分馏系数的大小，可直接用扩散系数的比值表示：

$$\frac{D_2}{D_1} = \left(\frac{m_1}{m_2}\right)^{1/2} \tag{5}$$

Richter 等（1999）受到上式的启发，建议将高温岩石学体系的同位素扩散系数的比值，也按照同两个同位素原子量的比值的指数形式来归纳，即

$$\frac{D_2}{D_1} = \left(\frac{m_1}{m_2}\right)^{\beta} \tag{6}$$

式中，β 为同位素动力学分馏因子。

硅酸盐熔体是地球和其他行星内部物质迁移的重要载体，了解硅酸盐熔体中金属稳定同位素的扩散系数对于利用同位素研究相关的地球化学过程和时间尺度具有重要意义。Liu 等首次利用第一性原理分子动力学的方法计算了 $MgSiO_3$ 和 Mg_2SiO_4 熔体中 Mg 同位素扩散的 β 值（Liu X. H. et al., 2018）。他们使用"假同位素"法（假设 $m_i/m_j = 1/24, 6/24, 48/24, 120/24$）得到不同温度下 Mg_2SiO_4 熔体中 Mg 同位素的 β 值：2 300 K 下为 0.184 ± 0.006，3 000 K 下为 0.245 ± 0.007，4 000 K 下为 0.257 ± 0.012；$MgSiO_3$ 熔体中 Mg 同位素的 β 值：4 000 K 下为 0.272 ± 0.005（图13）。他们发现，当温度远高于熔点时，温度变化对 β 值影响很小，但接近熔点时，由于熔体结构的变化等原因，β 值有明显改变。另外，对于这种简单体系的硅酸盐熔体，熔体成分的变化对 β 值也有微小的影响。该工作显示利用第一性原理分子动力学研究硅酸盐熔体中同位素扩散机制以及获得同位素扩散系数的巨大潜力。

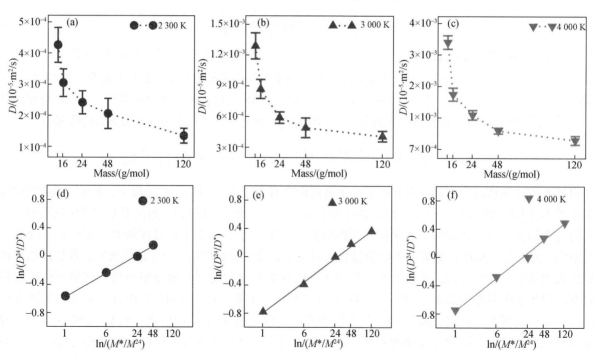

图13　2 300、3 000、4 000 K 下 Mg_2SiO_4 熔体中 Mg 同位素扩散系数与 Mg 原子质量的关系 [（a）~（c）] 及 $\ln(D^{24}/D^*)$ 与 $\ln(M^*/M^{24})$ 的线性关系 [（d）~（f）]（据 Liu X. H. et al., 2018d）
（d）（e）（f）斜率为 β 值

2.4 基于固体间隙扩散的同位素分馏

扩散作为固体中唯一一种物质传输方式，制约着矿物颗粒中元素及同位素迁移的时间及空间尺度（Chakraborty，2008），严重影响着矿物晶体中那些能反映热历史和生长历史的元素及同位素信息。同位素扩散效应是地球化学、高温岩石学、天体化学、热年代学中的重要课题。

对固体同位素扩散效应的研究，能从更微观尺度诠释同位素在一个动力学背景下的分布规律，更好地解释地质体热演化历史，以及评价矿物对同位素封闭性及保存的地质信息受扩散作用的改造程度，成为精确研究地质动力学过程的一种新型手段。

南京大学的王恺、陆现彩与国外合作者一起，首次开展了基于量子化学计算的固体间隙扩散的同位素分馏研究，他们对上地幔橄榄石中的 He 的间隙扩散进行了首次计算，结果表明在 1 Ma 的时间内，He 在上地幔可以扩散 2.7 m（Wang et al.，2015）；他们又用第一性原理量子化学计算方法计算了 He、Ne、Ar、Kr、Xe 在板块俯冲板片中的扩散和丢失的情况（Wang K. et al.，2020）。上述工作显示量子化学计算可以有效应用于与固相扩散有关的同位素分馏研究中。

3 对各同位素体系基本参数的计算进展

2011—2020 年这十年间，国内研究者通过计算获得了大量同位素平衡分馏系数，这些基本参数的获得极大地促进了相关同位素体系的应用和发展。限于篇幅，这里仅对部分工作进行简介。

中国科学技术大学的黄方和吴忠庆课题组这些年提供了大量同位素平衡分馏的计算数据。Wu Z. Q. 等（2015）计算了不同压力下的橄榄石、瓦兹利石、林伍德石、布里基曼石的 Si、Mg 和 O 同位素的平衡分馏系数；Wu F. 等（2015）计算了不同价态下一些含钒的化合物之间的钒同位素的平衡分馏系数；Qin 等（2016）计算了石英、钠长石、钙长石和锆石之间的 Si、O 同位素的平衡分馏系数；Huang 等（2019）计算了镁橄榄石、斜方辉石、钙铝榴石、易变辉石（pigeonite）、透辉石、钙长石、oldhamite 的钙同位素分馏，并以此为基础，讨论了岩浆洋固化后地幔的钙同位素分布；Wang 等（2019）计算了水溶液、各种碳酸盐、水镁石、利蛇纹石之间 Mg 同位素的分馏；Li 等（2019b）计算了 17 种矿物之间的 K 同位素平衡分馏系数；计算了不同矿物之间 Ti 同位素的平衡分馏系数。

中国科学院大学李永兵课题组计算了很多平衡分馏系数，获得了一些化合物和水溶液之间的 Cd 同位素分馏（Yang et al.，2015）、各种硫化物之间的硫同位素平衡分馏系数（Liu et al.，2014，2016，2018）、硫化物中 Ni 同位素的分馏系数（Liu et al.，2018a）及矿物之间的 Li 同位素分馏系数（Liu et al.，2018b）。

南京大学的魏海珍课题组提供了电气石类矿物和溶液之间 B 同位素的分馏系数（Li et al.，2020）。南京大学李伟强和刘显东课题组计算了大量溶液中不同物种之间的 Sn 同位素平衡分馏系数（She et al.，2020）。

中国科学院地球化学研究所刘耘课题组计算了一些平衡分馏系数。Li 和 Liu 等（2011）计算了一些 Se 同位素平衡分馏系数；He 和 Liu（2015）及 He 等（2016）提供了一些 Si 同位素平衡和动力学分馏系数；Gao 等（2018）计算了大量矿物和溶液之间 Mg 同位素平衡分馏系数；Zhang J. X. 和 Liu 等（2018）计算了 Zn 在挥发过程的一些分馏系数以及在水溶液中的分馏系数；Luo 等（2018）计算了 Mg 的硅酸盐熔体和气化物的分馏系数。

4 总结和展望

近十年来，国内同位素效应的理论和计算研究取得长足进展，从业人数快速增加。在诸如超冷体系同位素分馏理论、同位素浓度效应、同位素压力效应、热梯度过程、重复岩浆过程等方面都取得了一些原始创新性的成果。但是，我们的绝大多数工作，仍然停留在提供平衡分馏系数上，而随着计算技术的日益普及，这个性质的工作已成为越来越多的同位素观察与实验课题组的日常工作，并不需要理论和计算研究者的帮助。显然，同位素理论和计算领域的研究者需要在提出新方向、提出新概念等体现原始创新方面进一步努力。

鉴于应用同位素方法解决地学和行星科学问题的热潮还在继续之中，同位素理论和计算研究的从业者肩负的任务不仅仅是帮助解释一些具体问题，而是需要从学科建设层面进行思考，针对同位素学科发展中存在的瓶颈和痛点，从基础理论研究上给予解决，为这个领域提供更好的工具；同时，还应该提前寻找新的同位素应用方向，以不断开拓同位素的应用范围和解决问题的能力。

未来同位素理论和计算工作的任务是比较清晰的。第一，要确保这个领域基本分馏理论的建设完成。同位素核磁效应就是学科的一个瓶颈问题，这个效应已经在实验上证实是导致同位素非质量分馏的最重要的因素，而重要的同位素非质量分馏现象的持续发现，几乎构成了同位素地球化学最欣欣向荣的方面，同位素核磁效应解释理论的缺失就是整个同位素理论体系目前最大的一个缺憾。为了解决同位素核磁效应，更高精度和更高理论水平的量子理论需要及时引入同位素地球化学领域。而涉及核磁效应和超精细核能级的问题，则需要量子场论才能完美解决。因此，这个方向首要的任务是把解释理论从现在的经典量子力学升级至量子场论的理论水平。这一转变还可能会有一系列的连带收益，如会使地学领域更加明白与核有关的一系列曾经使用的实验技术，包括核磁共振和穆氏鲍尔技术等，也有可能开辟出一系列新型的同位素分析技术而非一定要使用质谱仪。第二，理论和计算的研究需要进一步引入机器学习的方法来改变研究效率。目前基于机器学习的分子动力学方法已经可以几百倍、上千倍地加速第一性原理分子动力学计算，而这与目前同位素分馏计算的趋势非常吻合，即主流方法为第一性原理的分子动力学，因此全面引入机器学习的技术已是无可阻挡的趋势。第三，通过对近十年来这一学科领域研究进展的总结，我们可以清晰地看见，绝大多数的工作还只是从同位素平衡分馏过程（包括化学反应的同位素动力学分馏）来挖掘信息，但如果从非平衡过程来挖掘则可以得到更多的过程信息。未来除了进行更多的动力学分馏研究，提供大量的动力学分馏参数，并为各种动力学过程建立理论模型之外，还需要针对各种非平衡过程进行研究。非平衡过程的同位素分馏理论几乎还没有起步，但是现在已经开始有国外研究者在尝试了，包括非平衡的相变，以及用数值模拟方法模拟一些非平衡的、非线性的、不完美的同位素分馏过程。非平衡会是未来很长一段时间内的一个重要主题。第四，在同位素地球化学领域，理论和计算研究者的工作与实验分析研究者的工作严重脱节，事实上很多异常的分馏是发生在仪器内部的，分析人员在用各种"经验方法"来校正仪器时，往往不明白其中的原因。因此，未来的理论和计算研究者应和分析人员一起，研究仪器内部的分馏行为，帮助行内提升同位素质谱仪分析的精度和可信性。

总之，国内同位素理论和计算方向不仅保持在了世界前列，还可以进一步提前进入一些完全新的领域，创新性地开始使用诸如基于量子场理论水平的工具、处理非平衡过程的物理化学和计算数学的新方法，助力学界解决各种疑难的同位素效应解释问题，并应用于地球科学和行星科学的重要科学问题研究之中。

参考文献

Altwegg K, Balsiger H, Bar-Nun A, Berthelier J J, Bieler A, Bochsler P, Briois C, Calmonte U, Combi M, de Keyser J, Eberhardt P, Fiethe B, Fuselier S, Gasc S, Gombosi T I, Hansen K C, Hässig M, Jäckel A, Kopp E, Korth A, LeRoy L, Mall U, Marty B, Mousis O, Neefs E, Owen T, Rème H, Rubin M, Sémon T, Tzou C Y, Waite H, Wurz P. 2015. 67P/Churyumov-Gerasimenko, a Jupiter family comet with a high D/H ratio. Science, 347(6220): 1261952

Amelin Y, Kaltenbach A, Iizuka T, Stirling C H, Ireland T R, Petaev M, Jacobsen S B. 2010. U-Pb chronology of the Solar system's oldest solids with variable $^{238}U/^{235}U$. Earth and Planetary Science Letters, 300(3-4): 343–350

Bigeleisen J. 1996. Nuclear size and shape effects in chemical reactions. Isotope chemistry of the heavy elements. Journal of the American Chemical Society, 118(15): 3676–3680

Bigeleisen J, Mayer M G. 1947. Calculation of equilibrium constants for isotopic exchange reactions. The Journal of Chemical Physics, 15(5): 261–267

Blanchard M, Balan E, Schauble E A. 2017. Equilibrium fractionation of non-traditional isotopes: a molecular modeling perspective. Reviews in Mineralogy and Geochemistry, 82(1): 27–63

Cano E J, Sharp Z D, Shearer C K. 2020. Distinct oxygen isotope compositions of the Earth and Moon. Nature Geoscience, 13(4): 270–274

Cao X B, Liu Y. 2011. Equilibrium mass-dependent fractionation relationships for triple oxygen isotopes. Geochimica et Cosmochimica Acta, 75(23): 7435–7445

Cao X B, Liu Y. 2012. Theoretical estimation of the equilibrium distribution of clumped isotopes in nature. Geochimica et Cosmochimica Acta, 77: 292–303

Cao X B, Bao H M, Gao C H, Liu Y, Huang F, Peng Y B, Zhang Y N. 2019. Triple oxygen isotope constraints on the origin of ocean island basalts. Acta Geochimica, 38(3): 327–334

Chakraborty S. 2008. Diffusion in solid silicates: a tool to track timescales of processes comes of age. Annual Review of Earth and Planetary Sciences, 36: 153–190

Chen L M, Teng F Z, Song X Y, Hu R Z, Yu S Y, Zhu D, Kang J. 2018. Magnesium isotopic evidence for chemical disequilibrium among cumulus minerals in layered mafic intrusion. Earth and Planetary Science Letters, 487: 74–83

Cook D L, Schönbächler M. 2016. High-precision measurement of W isotopes in Fe-Ni alloy and the effects from the nuclear field shift. Journal of Analytical Atomic Spectrometry, 31(7): 1400–1405

Cordier D, Mousis O, Lunine J I, Moudens A, Vuitton V. 2008. Photochemical enrichment of deuterium in Titan's atmosphere: new insights from Cassini-Huygens. The Astrophysical Journal, 689(1): L61–L64

Eberhardt P, Dolder U, Schulte W, Krankowsky D, Lämmerzahl P, Hoffman J H, Hodges R R, Berthelier J J, Illiano J M. 1987. The D/H ratio in water from comet P/Halley. Astronomy and Astrophysics, 187: 435–437

Fang T, Liu Y. 2019. Equilibrium thallium isotope fractionation and its constraint on Earth's late veneer. Acta Geochimica, 38(4): 459–471

Feng C Q, Qin T, Huang S C, Wu Z Q, Huang F. 2014. First-principles investigations of equilibrium calcium isotope fractionation between clinopyroxene and Ca-doped orthopyroxene. Geochimica et Cosmochimica Acta, 143: 132–142

Fujii T, Moynier F, Albarède F. 2006. Nuclear field vs. nucleosynthetic effects as cause of isotopic anomalies in the early Solar system. Earth and Planetary Science Letters, 247(1-2): 1–9

Fujii Y, Nomura M, Okamoto M, Onitsuka H, Kawakami F, Takeda K. 1989. An anomalous isotope effect of ^{235}U in U(IV)-U(VI) chemical exchange. Zeitschrift für Naturforschung A, 44(5): 395–398

Gao C H, Cao X B, Liu Q, Yang Y H, Zhang S T, He Y Y, Tang M, Liu Y. 2018. Theoretical calculation of equilibrium Mg isotope fractionations between minerals and aqueous solutions. Chemical Geology, 488: 62–75

He H T, Liu Y. 2015. Silicon isotope fractionation during the precipitation of quartz and the adsorption of $H_4SiO_{4(aq)}$ on Fe(III)-oxyhydroxide surfaces. Chinese Journal of Geochemistry, 34(4): 459–468

He H T, Zhang S T, Zhu C, Liu Y. 2016. Equilibrium and kinetic Si isotope fractionation factors and their implications for Si isotope distributions in the Earth's surface environments. Acta Geochimica, 35(1): 15–24

He Y Y, Bao H M, Liu Y. 2020. Predicting equilibrium intramolecular isotope distribution within a large organic molecule by the cutoff calculation. Geochimica et Cosmochimica Acta, 269: 292–302

Huang F, Chen L J, Wu Z Q, Wang W. 2013. First-principles calculations of equilibrium Mg isotope fractionations between garnet, clinopyroxene, orthopyroxene, and olivine: implications for Mg isotope thermometry. Earth and Planetary Science Letters, 367: 61–70

Huang F, Wu Z Q, Huang S C, Wu F. 2014. First-principles calculations of equilibrium silicon isotope fractionation among mantle minerals. Geochimica et Cosmochimica Acta, 140: 509–520

Huang F, Zhou C, Wang W Z, Kang J T, Wu Z Q. 2019. First-principles calculations of equilibrium Ca isotope fractionation: implications for oldhamite formation and evolution of lunar magma ocean. Earth and Planetary Science Letters, 510: 153–160

Joy H W, Libby W F. 1960. Size effects among isotopic molecules. The Journal of Chemical Physics, 33(4): 1276

Kruijer T S, Kleine T. 2018. No ^{182}W excess in the Ontong Java Plateau source. Chemical Geology, 485: 24–31

Li X F, Liu Y. 2011. Equilibrium Se isotope fractionation parameters: a first-principles study. Earth and Planetary Science Letters, 304(1-2): 113–120

Li X F, Liu Y. 2015. A theoretical model of isotopic fractionation by thermal diffusion and its implementation on silicate melts. Geochimica et Cosmochimica Acta, 154: 18–27

Li Y C, Chen H W, Wei H Z, Jiang S Y, Palmer M R, van de Ven T G M, Hohl S, Lu J J, Ma J. 2020. Exploration of driving mechanisms of equilibrium boron isotope fractionation in tourmaline group minerals and fluid: a density functional theory study. Chemical Geology, 536: 119466

Li Y H, Wang W Z, Huang S C, Wang K, Wu Z Q. 2019a. First-principles investigation of the concentration effect on equilibrium fractionation of K isotopes in feldspars. Geochimica et Cosmochimica Acta, 245: 374–384

Li Y H, Wang W Z, Wu Z Q, Huang S C. 2019b. First-principles investigation of equilibrium K isotope fractionation among K-bearing minerals. Geochimica et Cosmochimica Acta, 264: 30–42

Liang Y, Liu B D. 2016. Simple models for disequilibrium fractional melting and batch melting with application to REE fractionation in abyssal peridotites. Geochimica et Cosmochimica Acta, 173: 181–197

Liu B D, Liang Y. 2017. An introduction of Markov chain Monte Carlo method to geochemical inverse problems: reading melting parameters from REE abundances in abyssal peridotites. Geochimica et Cosmochimica Acta, 203: 216–234

Liu Q, Liu Y. 2016. Clumped-isotope signatures at equilibrium of CH_4, NH_3, H_2O, H_2S and SO_2. Geochimica et Cosmochimica Acta, 175: 252-270

Liu S Q, Li Y B, Tian H Q, Yang J L, Liu J M, Shi Y L. 2014. First-principles study of sulfur isotope fractionation in sulfides. European Journal of Mineralogy, 26(6): 717-725

Liu S Q, Li Y B, Gong H J, Chen C Y, Liu J M, Shi Y L. 2016. First-principles calculations of sulphur isotope fractionation in MX_2 minerals, with M = Fe, Co, Ni and X_2 = AsS, SbS. Chemical Geology, 441: 204-211

Liu S Q, Li Y B, Ju Y M, Liu J, Liu J M, Shi Y L. 2018a. Equilibrium nickel isotope fractionation in nickel sulfide minerals. Geochimica et Cosmochimica Acta, 222: 1-16

Liu S Q, Li Y B, Liu J, Ju Y W, Liu J M, Yang Z M, Shi Y L. 2018b. Equilibrium lithium isotope fractionation in Li-bearing minerals. Geochimica et Cosmochimica Acta, 235: 360-375

Liu S Q, Li Y B, Liu J, Gao T, Guo Y, Liu J M, Shi Y L. 2018c. First-principles investigation of the effect of crystal structure on sulfur isotope fractionation in sulfide polymorphs. European Journal of Mineralogy, 30(6): 1047-1061

Liu X H, Qi Y H, Zheng D Y, Zhou C, He L X, Huang F. 2018. Diffusion coefficients of Mg isotopes in $MgSiO_3$ and Mg_2SiO_4 melts calculated by first-principles molecular dynamics simulations. Geochimica et Cosmochimica Acta, 223: 364-376

Liu Y. 2013. On the test of a new volume variable cluster model method for stable isotopic fractionation of solids: equilibrium Mg isotopic fractionations between minerals and solutions. Goldschmidt 2013 Conference Abstracts 1632

Luo H Y, Bao H M, Yang Y H, Liu Y. 2018. Theoretical calculation of equilibrium Mg isotope fractionation between silicate melt and its vapor. Acta Geochimica, 37(5): 655-662

Ma R, Zhu Z Y, Wang B, Zhao Y, Yin X J, Lu F Y, Wang Y, Su J, Hocart C H, Zhou Y P. 2018. Novel position-specific $^{18}O/^{16}O$ measurement of carbohydrates. I. O-3 of glucose and confirmation of $^{18}O/^{16}O$ heterogeneity at natural abundance levels in glucose from starch in a C4 plant. Analytical Chemistry, 90(17): 10293-10301

Ma R, Zhao Y, Liu L, Zhu Z Y, Wang B, Wang Y, Yin X J, Su J, Zhou Y P. 2020. Novel Position-specific $^{18}O/^{16}O$ measurement of carbohydrates. II. the complete intramolecular $^{18}O/^{16}O$ profile of the glucose unit in a starch of C4 origin. Analytical Chemistry, 92(11): 7462-7470

Moynier F, Fujii T, Brennecka G A, Nielsen S G. 2013. Nuclear field shift in natural environments. Comptes Rendus Geoscience, 345(3): 150-159

Mundl-Petermeier A, Walker R J, Fischer R A, Lekic V, Jackson M G, Kurz M D. 2020. Anomalous ^{182}W in high $^3He/^4He$ ocean island basalts: fingerprints of Earth's core? Geochimica et Cosmochimica Acta, 271: 194-211

Nixon C A, Temelso B, Vinatier S, Teanby N A, Bézard B, Achterberg R K, Mandt K E, Sherrill C D, Irwin P G J, Jenning D E, Romani P N, Coustenis A, Flasar F M. 2012. Isotopic ratios in Titan's methane: measurements and modeling. The Astrophysical Journal, 749(2): 159

Ono S, Wang D T, Gruen D S, Lollar B S, Zahniser M S, McManus B J, Nelson D D. 2014. Measurement of a doubly substituted methane isotopologue, $^{13}CH_3D$, by tunable infrared laser direct absorption spectroscopy. Analytical Chemistry, 86(13): 6487-6494

Pinto J P, Lunine J I, Kim S J, Yung Y L. 1986. D to H ratio and the origin and evolution of Titan's atmosphere. Nature, 319(6052): 388-390

Polyakov V B, Kharlashina N N. 1994. Effect of pressure on equilibrium isotopic fractionation. Geochimica et Cosmochimica Acta, 58(21): 4739-4750

Qin T, Wu F, Wu Z Q, Huang F. 2016. First-principles calculations of equilibrium fractionation of O and Si isotopes in quartz, albite, anorthite, and zircon. Contributions to Mineralogy and Petrology, 171(11): 91

Richter F M, Liang Y, Davis A M. 1999. Isotope fractionation by diffusion in molten oxides. Geochimica et Cosmochimica Acta, 63(18): 2853-2861

Rizo H, Andrault D, Bennett N R, Humayun M, Brandon A, Vlastelic I, Moine B, Poirier A, Bouhifd M A, Murphy D T. 2019. ^{182}W evidence for core-mantle interaction in the source of mantle plumes. Geochemical Perspectives Letters, 11: 6-11

Robert F. 2010. Solar system Deuterium/Hydrogen ratio. Meteorites and the Early Solar System II, 341-351

Robert F, Gautier D, Dubrulle B. 2000. The solar system D/H ratio: observations and theories. Space Science Reviews, 92(1): 201-224

Saji N S, Wielandt D, Paton C, Bizzarro M. 2016. Ultra-high-precision Nd-isotope measurements of geological materials by MC-ICPMS. Journal of Analytical Atomic Spectrometry, 31(7): 1490-1504

Schauble E A. 2007. Role of nuclear volume in driving equilibrium stable isotope fractionation of mercury, thallium, and other very heavy elements. Geochimica et Cosmochimica Acta, 71(9): 2170-2189

Schauble E A. 2013. Modeling nuclear volume isotope effects in crystals. Proceedings of the National Academy of Sciences of the United States of America, 110(44): 17714-17719

She J X, Wang T H, Liang H D, Muhtar M N, Li W Q, Liu X D. 2020. Sn isotope fractionation during volatilization of Sn(Ⅳ) chloride: laboratory experiments and quantum mechanical calculations. Geochimica et Cosmochimica Acta, 269: 184-202

Touboul M, Puchtel I S, Walker R J. 2012. ^{182}W evidence for long-term preservation of early mantle differentiation products. Science, 335(6072): 1065-1069

Turro N J. 1983. Influence of nuclear spin on chemical reactions: Magnetic isotope and magnetic field effects (A Review). Proceedings of the National Academy of Sciences of the United States of America, 80(2): 609-621

Tusch J, Sprung P, van de Löcht J, Hoffmann J E, Boyd A J, Rosing M T, Münker C. 2019. Uniform ^{182}W isotope compositions in Eoarchean rocks from the Isua region, SW Greenland: the role of early silicate differentiation and missing late veneer. Geochimica et Cosmochimica Acta, 257: 284-310

Urey H C. 1947. The thermodynamic properties of isotopic substances. Journal of the Chemical Society (Resumed), 562-581, doi: 10.1039/jr9470000562

Wang K, Brodholt J, Lu X C. 2015. Helium diffusion in olivine based on first principles calculations. Geochimica et Cosmochimica Acta, 156: 145-153

Wang K, Lu X C, Brodholt J P. 2020. Diffusion of noble gases in subduction zone hydrous minerals. Geochimica et Cosmochimica Acta, 291: 50-61

Wang W Z, Qin T, Zhou C, Huang S C, Wu Z Q, Huang F. 2017a. Concentration effect on equilibrium fractionation of Mg-Ca isotopes in carbonate minerals: insights from first-principles calculations. Geochimica et Cosmochimica Acta, 208: 185-197

Wang W Z, Zhou C, Qin T, Kang J T, Huang S C, Wu Z Q, Huang F. 2017b. Effect of Ca content on equilibrium Ca isotope fractionation between orthopyroxene and clinopyroxene. Geochimica et Cosmochimica Acta, 219: 44-56

Wang W Z, Zhou C, Liu Y, Wu Z Q, Huang F. 2019. Equilibrium Mg isotope fractionation among aqueous Mg^{2+}, carbonates, brucite and lizardite: insights from first-principles molecular dynamics simulations. Geochimica et Cosmochimica Acta, 250: 117-129

Wang W Z, Huang S C, Huang F, Zhao X M, Wu Z Q. 2020. Equilibrium inter-mineral titanium isotope fractionation: implication for high-temperature titanium isotope geochemistry. Geochimica et Cosmochimica Acta, 269: 540-553

Wang Z R, Schauble E A, Eiler J M. 2004. Equilibrium thermodynamics of multiply substituted isotopologues of molecular gases. Geochimica et Cosmochimica Acta, 68(23): 4779-4797

Webster C R, Mahaffy P R, Flesch G J, Niles P B, Jones J H, Leshin L A, Atreya S K, Stern J C, Christensen L E, Owen T, Franz H, Pepin R O, Steele A, The MSL Science Team. 2013. Isotope ratios of H, C, and O in CO_2 and H_2O of the Martian atmosphere. Science, 341(6143): 260-263

Wu F, Qin T, Li X F, Liu Y, Huang J H, Wu Z Q, Huang F. 2015. First-principles investigation of vanadium isotope fractionation in solution and during adsorption. Earth and Planetary Science Letters, 426: 216-224

Wu Z Q, Huang F, Huang S C. 2015. Isotope fractionation induced by phase transformation: first-principles investigation for Mg_2SiO_4. Earth and Planetary Science Letters, 409: 339-347

Yang J L, Li Y B, Liu S Q, Tian H Q, Chen C Y, Liu J M, Shi Y L. 2015. Theoretical calculations of Cd isotope fractionation in hydrothermal fluids. Chemical Geology, 391: 74-82

Yang S, Liu Y. 2015. Nuclear volume effects in equilibrium stable isotope fractionations of mercury, thallium and lead. Scientific Reports, 5(1): 12626, doi: 10.1038/srep12626

Yang S, Liu Y. 2016. Nuclear field shift effects on stable isotope fractionation: a review. Acta Geochimica, 35(3): 227-239

Zhang J X, Liu Y. 2018. Zinc isotope fractionation under vaporization processes and in aqueous solutions. Acta Geochimica, 37(5): 663-675

Zhang S T, Liu Q, Tang M, Liu Y. 2020. Molecular-level mechanism of phosphoric acid digestion of carbonates and recalibration of the ^{13}C-^{18}O clumped isotope thermometer. ACS Earth and Space Chemistry, 4(3): 420-433

Zhang Y N, Liu Y. 2018. The theory of equilibrium isotope fractionations for gaseous molecules under super-cold conditions. Geochimica et Cosmochimica Acta, 238: 123-149

Zhang Y N, Liu Y. 2020. How to produce isotope anomalies in mantle by using extremely small isotope fractionations: a process-driven amplification effect? Geochimica et Cosmochimica Acta, 291: 19-49

Zhu D, Bao H M, Liu Y. 2015. Non-traditional stable isotope behaviors in immiscible silica-melts in a mafic magma chamber. Scientific Reports, 5(1): 17561

Progresses in Computational and Theoretical Isotope Effect Studies

LIU Yun[1,2,3]

1. State Key Laboratory of Ore Deposit Geochemistry, Institute of Geochemistry, Chinese Academy of Sciences, Guiyang 550081;
2. International Research Center for Planetary Science, College of Earth Sciences, Chengdu University of Technology, Chengdu 610059; 3. CAS Center for Excellence in Comparative Planetology, Hefei 230026

Abstract: During the past decade, the stable isotope geochemical theory for interpreting stable isotope fractionation has commonly reached the level of quantum chemistry in China. Based on the precise *ab initio* or first principles of quantum chemistry, researchers started to routinely undertake the equilibrium and kinetic isotope fractionation coefficient calculations. Among the four sub-fields of isotope geochemistry, i.e., laboratory analysis, field observation, computation, and theoretical studies, I am proud to say that computation and theoretical studies of isotope geochemistry in China have made important contributions over the past ten years, enjoying a status on par with the cutting-edge researches in the world. Chinese researchers have taken the lead in several new areas of the theoretical and calculation methods of isotope geochemistry, including isotope fractionations under super-cold conditions, isotope fractionation with pressure effects, concentration-dependent isotope effects, clumped isotope effects, small isotope anomaly, isotope diffusion effects under thermal gradient, and high temperature repeated-process isotope effects. They have also developed some new computational methods for stable isotope effects, including a variable volume cluster model (VVCM) method for computing equilibrium isotope fractionation between solid and liquid phases, methods considering nuclear-volume effect for heavy metal isotope in solid phase systems, and methods for the theoretical calculation of equilibrium and kinetic fractionations of the isotope diffusion in silicate melts. Meanwhile, they have provided a large number of needed equilibrium or kinetic isotope fractionation coefficients or parameters for a large number of various non-traditional or metal isotope systems which have broad applications in geological and planetary sciences. Their adventures have laid a good foundation for the future in-depth application of these isotope systems. However, Chinese researchers must recognize that most of the new concepts, models, systems, and methods in the field of theoretical and computation isotope geochemistry were originally suggested by western researchers. Most of the research efforts in China are made for the modifications and supplements to the theoretical systems and new development directions originally proposed by predecessors (mainly foreign counterparts). In the future, Chinese scholars in the field of theoritical and calculation isotope geochemistry should take the lead in using the new generation of theoretical tools including quantum field theory, the new generation of theoretical tools including quantum field theory, and make more contributions to propose novel ideas, new concepts, and new theories in the field of isotope effect research.

Key words: isotope effect; stable isotope fractionation; theoretical calculation; quantum chemistry; equilibrium isotope fractionation; super-cold condition; pressure isotope effect; isotope concentration effect

环境地球化学研究进展*

冯新斌[1,2]　曹晓斌[3]　付学吾[1,2]　洪冰[1,2]　关晖[1]　李平[1,2]　王敬富[1]
王仕禄[1]　张干[4]　赵时真[4]

1. 中国科学院 地球化学研究所，环境地球国家重点实验室，贵阳 550081；2. 中国科学院 第四纪与全球变化卓越中心，西安 710061；3. 南京大学 地球科学与工程学院，国际同位素效应研究中心，南京 210023；4. 中国科学院 广州地球化学研究所，有机地球化学国家重点实验室，广州 510640

摘　要：环境地球化学学科在中国生态文明和美丽中国建设、国际环境公约履约及过去全球变化研究中发挥着越来越大的作用。本文简要回顾了中国过去十年在环境地球化学领域的部分研究进展，介绍了汞、持久性有机污染物和其他重金属污染物长距离传输研究进展；总结了汞、镉、锑、铊等非传统稳定同位素在地表生物地球化学循环过程中的同位素分馏规律及污染来源和环境过程示踪方面的研究进展；回顾了传统稳定同位素地球化学与污染示踪及过去全球变化方面的研究进展。指出了以上研究方向还存在的问题及未来研究方向。

关键词：长距离传输　重金属　持久性有机污染物　传统稳定同位素　全球变化　非传统稳定同位素

0　引　言

20世纪以来，比利时马斯河谷事件，美国洛杉矶光化学烟雾事件、多诺拉事件，英国伦敦烟雾事件，日本水俣病事件以及神东川骨痛病等一系列震惊世界的环境污染事件引起了科学界、政府和公众对环境的高度关注，催生起共同保护家园的意识和行动。在这一背景下，环境地球化学于20世纪60年代兴起，成了研究人类赖以生存的地球环境的化学组成、化学作用、化学演化与人类相互关系的一门新的交叉学科。中国科学院地球化学研究所在国内率先开展了环境地球化学研究，20世纪60年代就开展了克山病的地球化学调查。1979年中国科学院地球化学研究所成立了环境地球化学研究室，同年中国矿物岩石地球化学学会环境地质地球化学专业委员会成立，1981年全国第一届环境地球化学与健康学术研讨会召开，1989年环境地球化学国家重点实验室组建并于1991年获科技部正式批准。半个世纪以来，环境地球化学学科面向国家和社会重大需求，面向国民经济主战场，面向世界科技前沿，积极进取，努力探索，成了一门充满活力的年轻学科。

环境地球化学与健康、环境地球化学与污染及环境地球化学与全球变化是目前环境地球学科的主要研究内容（刘丛强等，2018）。近年来，随着工业和社会的快速发展，一些全球性的环境地球化学问题逐渐凸显，受到广泛关注，如重金属环境污染、温室气体排放、持久性有机污染物污染、微塑料的环境和健康问题等，同时，全球多边应对各种环境问题的框架也逐步建立起来，已签订诸如《联合国气候变化框架公约》《京都议定书》《巴黎协定》《关于持久性有机污染物的斯德哥尔摩公约》和《有关汞的水俣公约》等多方协议。另外，有关环境地球化学的全球性学术会议（如重金属国际学术会议、全球汞污染国际学术会议、戈尔德施密特大会、国际卤代持久性有机污染物学术研讨会）也定期召开，极大地促进了学科的交流和发展。近十年，由于地球化学研究手段的进步，中国环境地球化学研究得到飞速发展，限于篇幅，本文仅就近十年中国在污染物长距离传输、非传

---

*　原文"环境地球化学研究近十年若干新进展"刊于《矿物岩石地球化学通报》2021年第40卷第2期，本文略有修改。

统稳定同位素地球化学理论与应用、传统稳定同位素发展与全球变化研究、传统稳定同位素地球地球化学示踪等研究领域的重要进展进行总结。

1 污染物的长距离传输

多类污染物通过大气可进行长距离传输，对偏远的南北极和青藏高原产生了重要影响，多类污染物已列入环境外交。近十年中国学者对污染物的长距离迁移规律进行了系统研究，这些污染物包括汞、持久性有机污染物（persistent organic pollutants，POPs）和其他污染物。

1.1 汞的长距离传输

汞（Hg）是主要通过大气进行传输的全球性污染物（Lindberg et al., 2007；Amos et al., 2014）。20 世纪七八十年代欧美地区偏远水生生态系统普遍出现的环境汞污染，就是长距离传输的大气汞通过干湿沉降输入到这些生态系统的结果（Lindqvist et al., 1991；Hammerschmidt and Fitzgerald, 2006；Harris et al., 2007）。鉴于全球汞污染的严峻形势，联合国环境规划署于 2013 年制定了一项具有法律约束力的国际汞公约——"关于汞的水俣公约"，该公约于 2017 年 8 月正式生效。可见汞的长距离传输不仅是环境汞污染的热点科学问题，也是有关国际环境外交的一个重要政治问题。

大气汞按物理化学形态可分为气态单质汞（gaseous elemental meracury，GEM）、气态氧化汞（gaseous oxidized mercury，GOM）和颗粒汞（particulate bounded mercury，PBM）。GEM 是对流层中大气汞的最主要形态，约占大气汞总量的 80% ~ 90%（Holmes et al., 2010；Lyman and Jaffe, 2012；Horowitz et al., 2017）。GEM 物理化学性质较为稳定，其大气居留时间通常为 0.25 ~ 1.0 年（Holmes et al., 2006；Holmes et al., 2010；Horowitz et al., 2017），能随大气环流进行长距离传输。近期的研究表明，全球陆地和海洋生态系统每年沉降的大气汞分别为 3 000 ~ 4 100 t 和 5 300 ~ 7 100 t，主要（约 75%）来自于人为活动的一次和二次汞排放（Selin, 2009；Holmes et al., 2010；Horowitz et al., 2017）。

由于全球人为源一次和二次汞排放的区域性差异（Agnan et al., 2016；Wang J. C. et al., 2017；Streets et al., 2019），全球大气汞具有显著的区域分布特征。有研究表明，北半球偏远地区平均大气 GEM 浓度为 1.53 ng/m³，约是赤道地区和南半球的 1.25 倍和 1.61 倍（Sprovieri et al., 2016）；全球城市地区大气 GEM 的平均浓度为 4.33 ng/m³，约是偏远地区平均浓度的 2.5 倍（Mao et al., 2016）。就北半球而言，亚洲地区大气 GEM 的浓度是最高的，较北美和欧洲地区约偏高 26% ~ 55%（Mao et al., 2016）。中国大气 GEM 的区域分布规律也非常显著：城市地区的平均值为 9.20 ng/m³，约是偏远地区的 3.2 倍；东部较之西部偏高（Fu et al., 2015a）。这些大气 GEM 区域分布差异以及全球和区域大气环流对全球大气汞的传输起着非常重要的作用。

亚洲，特别是中国，是全球大气汞污染较为严重的地区（Fu et al., 2015b；Wu et al., 2016），常被认为是全球大气汞的重要输出国。研究表明，亚洲大气汞的长距离传输是北极、北美和欧洲地区大气汞及其沉降通量的一个重要污染源。北极地区的大气 GEM 约有 43% 来自亚洲，其次是俄罗斯（27%）、北美（16%）和欧洲（14%）；大气汞沉降通量约有 33% 来自亚洲，其次是欧洲（22%）和北美（10%）（Travnikov, 2005；Durnford et al., 2010）。北美地区大气 GEM 约有 31% 来自亚洲，其次是北美（20%）和欧洲（9%）；大气汞沉降通量 33% 来自于本土，其次是亚洲（24%）和欧洲（14%）（Travnikov, 2005；Strode et al., 2008）。欧洲地区大气汞沉降通量主要来自本土汞排放（61%），但亚洲大气汞长距离传输的贡献达到了 15%。中国同时也被认为是亚洲其他国家和地区大气 GEM 的重要污染来源，如 Nguyen 等（2010）对韩国济州岛的研究结果表明，中国大气汞的输出占到了济州岛大气 GEM 的 51%，Kim 等（2009）对韩国首尔的研究表明，首尔地区 73% 的大气 GEM 污染事件与中国大气汞的输出有关，而 Chand 等（2008）对日本冲绳岛的研究也发现来自中国中东部地区的大气气团中的 GEM 浓度明显高于来自日本和韩国地区的气团。上述国外研究有较大的不确定性，可能显著高估了中国大气汞长距离传输对大气汞及其沉降通量的影响。中国学者的研究结果表明，中国大气 GEM 的年净输出量约为 645 t，这比国外的研究结果偏低了 25%（Wang et al., 2018）。此外，中国学者开展的源汇关系模型研究显示，中国人为源汞排放的长距离传输只占到韩国和日本大气 GEM 浓度和大气汞沉降通量的 5%~10% 和 5%~12%，而对欧洲和北美大气 GEM 浓度和沉降通量的贡献比例通常小于 5%（Chen et al., 2015）。

中国大气 GEM 的长距离传输主要受人为源一次、二次排放和亚洲季风的影响，因此具有较为显著的区域分布规律（Fu et al., 2015b）。有研究表明，中国东北偏远地区大气 GEM 和 GOM 主要受华北地区人为源汞排放的影响，而 PBM 则与中国东北地区冬季取暖和东北亚地区生物质的燃烧有关（Fu et al., 2012b；Liu C. et al., 2019）；东部沿海地区大气 GEM 污染主要与中国中东部地区的人为源汞排放有关（Ci et al., 2011；Yu et al., 2015；Zhang L. et al., 2017）；华南偏远地区大气 GEM 污染主要与珠三角地区大气汞排放有关（Chen et al., 2013；Liu et al., 2016）；西南偏远地区大气 GEM 污染主要来自于西南人口和工业密集地区（Fu et al., 2010；Zhang et al., 2015, 2016；刘伟明等，2016），这些研究证实了人为源一次和二次汞排放的长距离传输是中国偏远地区大气 GEM 最主要的污染来源。除受中国本土大气汞排放的影响外，中国特别是西部地区同时也受到南亚和东南亚大气汞长距离传输的影响。例如，南亚地区大气汞排放是中国青藏高原地区大气 GEM 的一个重要污染源（Fu et al., 2012a；Yin et al., 2018；Lin H. M. et al., 2019），南亚和东南亚地区大气汞排放是中国西南地区大气 GEM 和 PBM 的重要污染源（Wang et al., 2015；Zhang H. et al., 2015, 2016；Fu et al., 2019a）。南亚和东南亚每年排放的大气 GEM 约为 600 t，其中每年约有 190 t 输送到中国境内（Fu et al., 2015a；Wang et al., 2018）。

除大气汞外，河流汞的长距离传输也是海洋汞污染的一个重要来源。河流汞的传输主要以颗粒态汞为主（约占 97%），因此大部分通过河流输送的汞会沉积在近海环境中，只有少量（10%~28%）汞可以通过长距离传输进入远洋生态系统（Sunderland and Mason, 2007；Amos et al., 2014）。目前有关河流长距离传输研究还有很大争议，模型所估算的河流每年向远海海洋输送的汞为 200~1 500 t，约是海洋地区大气汞沉降量的 4%~30%（Sunderland and Mason, 2007；Selin, 2009；Mason et al., 2012；Amos et al., 2014）。相对而言，地中海由于受到周边欧洲、非洲和亚洲共同河流输送的影响，因而受河流汞长距离传输的影响相对较高，约占这一地区大气汞长距离传输沉降的 36%，其次是北太平洋、大西洋、南太平洋，其河流长距离传输约占大气汞长距离传输沉降的 29%、6% 和 5%，而在其他海洋中，河流长距离传输所占的比例通常小于 2%（Sunderland and Mason, 2007；AMAP/UN, 2019）。

1.2 持久性有机污染物的长距离传输

持久性有机污染物（POPs）是环境中最为典型的对生态环境和人体健康具潜在危害的有毒有机污染物。2001 年 5 月 17 日，包括中国在内的 127 个国家和地区联合签署了旨在减少或消除 POPs 的《斯德哥尔摩公约》，2004 年 5 月，公约正式生效。POPs 具有毒性、难以降解、可产生生物蓄积的特性，往往通过空气、水和迁徙物种作跨越国际边界的迁移，并沉积在远离其排放地点的地区，随后在那里的陆地生态系统和水域生态系统中蓄积。可见，POPs 对人体健康和生态环境的危害，以及长距离迁移和污染全球化的属性，是其成为全球环境问题、需人类社会共同应对的主要原因。大气长距离迁移是 POPs 全球扩散和生物地球化学过程的重要环节，同其大气-地表交换过程密不可分，因此对其研究包括了 POPs 的大气迁移和大气-地表交换（源区排放、汇区沉降）两类环境过程。目前主要通过固定站点连续观测、源区和受体区（偏远地区）环境中 POPs 的观测和大气-地表交换（排放或沉降）、数值模式模拟 3 条技术路线实现。由于青藏高原和亚洲季风的影响，中国 POPs 的大气长距离迁移方向和季节变化极大受控于东南季风和印度季风。因而，季风携带的 POPs 长距离大气迁移是包括中国在内的亚洲 POPs 大气传输的重要机制、区域特色。过去 10 年，中国学者在 POPs 大气迁移方面的研究工作，主要围绕青藏高原、中国东部沿海地区开展，少量工作涉及南极地区；同时在基础理论、观测技术和数值模式方面亦有所贡献。

1.2.1 低挥发性 POPs 在大气中的气–粒分配及其长距离传输意义

POPs 在大气中的气–粒分配是影响其大气环境过程的重要机制。传统基于热平衡的理论认为，POPs 在气相和颗粒相间发生热平衡配分，可由正辛醇–空气分配系数（$\lg K_{OA}$）等预测，依此，挥发性较差的 POPs，如十溴二苯醚（BDE-209）将主要赋存在颗粒物上并随之发生大气迁移。但哈尔滨工业大学李一凡等（Li Y. F. et al., 2015）通过对中国及全球观测数据的分析发现，POPs 的热平衡气–粒配分的只是真实环境中的特例，并提出以增加干湿沉降过程的稳态模型来描述 POPs 的气–粒分配行为。在该模型中，POPs 的气–粒分配系数不仅与 K_{OA} 有关，还是环境温度的函数。依该模型，对 $\lg K_{OA} > 12.5$ 的化合物，其气–粒配分熵的对数值将恒定在-1.53，意味着即使是极低挥发性的 POPs 也能以一定比例赋存于气相中，呈气态发生长距离大气迁移。由此他们认为，在偏远、低温的北极环境中观察到的 BDE-209 等的极低挥发性 POPs，依然更多地来自气态迁移（Li Y. F. et al., 2017）。该理论也被张庆华等在南极对 PCBs 的观测进一步证实（Wang P. et al., 2017）。

1.2.2 中国陆表大气 POPs 长距离迁移及其源–汇机制

中国科学院广州地球化学研究所郑苇等对中国边远山地森林土壤中 PCBs 的观测发现，PCBs 含量及其组成的空间分布在不同纬度呈轻组分"南高北低"、重组分"北高南低"的特征，不支持理想的全球蒸馏与冷凝模型（Zheng et al., 2014），而北方背景森林土壤中较高的重组分 PCBs 可能来自境外的大气传输。哈尔滨工业大学任南琪等对覆盖中国东部到西部的城市和背景土壤中 PCBs 的研究发现，PCBs 组成具有自东向西的"经向分馏"特征，表明 PCBs 存在自中国东部人口密集区向西部的扩散迁移（Ren et al., 2007）。徐玥等通过数值模拟实验发现，自 1948—2009 年，α-HCH 和 β-HCH 在亚洲土壤中的残留区逐渐向中国中–西部和喜马拉雅南坡的山地森林生态系统转移，说明山地森林生态系统是中国 POPs 的长期"汇区"，且与季风边缘高度重合，指示季风是中国 POPs 经向迁移的重要营力，并且 POPs 在季风携带下的大气迁移、降水、森林过滤和山地冷凝富集等作用，可能是造成 POPs 在森林土壤中累积成汇的重要原因（Xu et al., 2012，2013）。中国地质科学院刘咸德等（2015）观测到，典型有机氯农药在青藏高原东缘山地剖面的冷凝富集规律，证实大气 POPs 在中国中西部地形急变带的大气沉降（Chen et al., 2008）。刘昕等对具良好植被垂直地带谱的贡嘎山不同海拔高度和植被带下土壤 PCBs 的进一步分析发现，森林过滤作用（而非山地冷凝作用）是其主控机制，即温度可能并非是 POPs 在山地森林土壤中成汇的主控环境因子（Liu X. et al., 2014）。郑苇等采用稳定同位素标记示踪的大气–地表交换现场实验新方法和装置（Liu D. et al., 2013；Liu X. et al., 2013），对海南尖峰岭热带雨林凋落物中 PCBs 的挥发与渗滤通量的野外模拟实验发现，热带雨林中土壤腐殖层薄、降雨量大，PCBs 向土壤中的渗滤通量显著高于向大气的挥发通量（Zheng et al., 2015；张干等，2019）。以上在中国的数值模拟、野外观测和实验证据，均对"全球蒸馏假说"以地表温度差异为主要驱动力的观点提出了挑战。综合以上，中国陆表 POPs 的"源–汇"过程及相关机制，可能包括东部源区排放、大气经向传输、中西部山地森林过滤沉降、森林土壤残留成汇 4 个重要环节（图1），凸显山地森林生态系统在其中所扮演的关键角色，也表明中国中西部山地森林生态系统可能是中国陆表 POPs 长期归宿地和汇区。

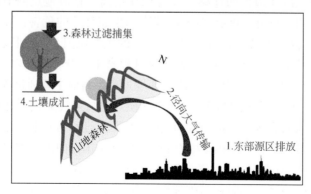

图 1 中国陆表 POPs 的"源–汇"过程"四部曲"（据张干等，2019）

1.2.3 印度季风和西风携带POPs向青藏高原的迁移沉降

中国科学院青藏高原研究所王小萍等通过开展中低纬度的冰心POPs历史记录的研究，明确了青藏高原南部和北部的POPs分别来自南亚和欧洲的输入（Wang et al.，2008，2014）。中国科学院生态中心江桂斌、王亚韡等对青藏高原不同环境介质中的POPs观测研究，亦支持此结论（Yang et al.，2010；Wu et al.，2016；Wu et al.，2017，2020）。王小萍团队基于雅鲁藏布峡谷连续3年的高频观测，进一步发现印度季风是高原南部POPs跨境传输的驱动力，潜在来源贡献函数（PSCF）模型模拟表明POPs的季风传输路径主要有二：一是自印度东海岸经孟加拉湾北上输入高原南部；二是自恒河平原随抬升气流跨越喜马拉雅山脉进入高原腹地（Sheng et al.，2013）。与高原南部不同，高原北部大气POPs的季节性波动与西风环流指数呈现较好的一致性，表明POPs自欧洲向青藏高原的输入主要是通过西风进行的（Wang X. P. et al.，2016）。近期，王小萍团队利用大气-雪冰POPs指纹特征确定了印度季风和西风对青藏高原的影响范围，分别为北纬33°和35°（化学边界），而33°~35°为季风和西风过渡区（Wang J. et al.，2019）。此外，他们还建立了适用于喜马拉雅山区的三级多介质逸度模型（Mountain-POP），得到了爬坡和山谷传输对有机氯农药（OCPs）和多氯联苯（PCBs）在青藏高原的总输入通量为2~100 t/a，山谷输送POPs的通量为高空翻越输入通量的5~10倍（Gong et al.，2019a）。

1.2.4 中国东部海岸带及邻近海区POPs的海-气平衡

由于全球范围内POPs的禁用/限用，POPs的海-气交换方向可能发生倒转，海洋由原先的大气POPs汇区，而成为大气POPs的（二次）源区，但该过程在全球海洋并非同步发生。依托中国科考航次，林田等对中国黄海、东海、南海的POPs海-气平衡动态进行了观测，发现中国海区的主要HCHs、DDTs化合物的海-气交换方向，已由沉降转为挥发，成为全球HCHs和DDTs的二次排放源（Lin et al.，2012）。多项研究发现，中国海岸带渔船防污漆中DDTs的非法使用，也可能是形成海面DDTs挥发的重要原因。在北印度洋，Huang等（2014）以α-HCH的(+/−)手性比值作为判识指标，判识出低层大气中的α-HCH有约30%来自二次挥发，说明陆地气团的POPs输送是影响其海-气交换的重要因素，而该海区的PCBs、HCB均呈现由海向大气的挥发，并存在由南亚次大陆向开放海的大气迁移趋向（Huang et al.，2014）。李军等在西印度洋海区观测到大气-海表界面上多溴联苯醚（PBDEs）的"昼挥发-夜沉降"循环现象，可能与水柱中浮游生物作用的趋光性有关，这也是继英国学者发现东大西洋海面大气多环芳烃（polycyclic aromatic hydrocarbons，PAHs）存在此现象以来，再次在寡营养海域实地验证，丰富了对POPs大气-海表交换的认识（Li et al.，2011）。

1.2.5 中国对南极和北极POPs污染的探索

作为偏远极寒地区，南极和北极POPs大气迁移研究成了全球气候变化与POPs生物地球化学研究的热点地区。张庆华等在南极乔治王岛中国长城站进行了连续3年的大气POPs观测，证实该站PCBs和PBDEs（除BDE-209外）主要受控于长距离大气迁移，而非来自长城站自身设施的污染（Wang et al.，2007）。在北极，马建民等分析了20世纪90年代以来大气POPs含量并与经模式模拟的气候变化对北极大气POPs影响的结果相比较，证实北极变暖已导致北极地表环境的POPs化合物重新释放进入大气中（Ma et al.，2011）。

1.3 其他污染物的长距离传输

除汞可以在大气中以气态形式存在并传输外，其他重金属如铅、砷、锑、镉、镍、铜、锡等（Marx and McGowan，2010）的污染物均能以气溶胶形式在环境中存留并通过大气在全球范围内长距离（高达数千千米）迁移，再由食物链进入生态系统中蓄积，最终对生物体和人类产生严重的健康危害。

首先，如何判定重金属污染物的长距离传输和沉积是一直以来备受关注的关键科学问题（Marx and McGowan，2010）。例如，先前对青藏高原地区的研究由于海拔高采样困难导致研究少，而近十年来，由于条件改善，更易于在南北极及青藏高原等偏远地区获得冰心、冰川、泥炭、雪、沉积柱、土壤剖面等样品，从而推动了这些地区的研究工作（Dong et al.，2016；Wei R. F. et al.，2019；Wei T. et al.，2019；康世昌等，2019）。相关研究也由以往集中在工业化所致大气重金属负荷的历史影响评价研究，转向对重金属浓度进行长时间和高时间分辨率的测定研究（Dong et al.，2016；McConnell et al.，2019；Zou et al.，2020），尤其是对特定历史时期如黑死病大流行（More et al.，2017）、中世纪早期的采矿和冶炼活动等对大气中重金属污染物传输迁移的影响（Gabrieli and Barbante，2014；Loveluck et al.，2018；Preunkert et al.，2019），对比研究了记录较长历史时间代表大区域甚至半球范围的大气重金属污染物变化的冰心与记录相对较短时间和局部区域信息的泥炭、沉积物记录（Preunkert et al.，2019），冰心采样区域进一步扩大由之前的格陵兰岛、帕米尔高原、南北极拓展到喜马拉雅山脉、天山山脉、西伯利亚（Liu et al.，2011；Hong et al.，2012，2015；Eichler et al.，2014；Sierra-Hernandez et al.，2018；Gabrielli et al.，2020；Zou et al.，2020）。

在众多重金属污染中，铅污染导致的长距离传输最早受到关注并得以持续研究。Marx 等（2016）将全球沉积记录中铅数据标准化，对比量化揭示了铅在全球的富集变化，人为释放的铅仍然高度富集并普遍存在于全球生态系统中，造成严重的金属污染负担，对生态系统和生物地球化学循环具有深远影响。同时很多地区缺乏铅的沉积记录，这限制了对全球污染的评估。此外，铅的变化特征和富集程度可以指示其他金属污染物的长距离传输（Marx et al.，2014）。McConnell 等（2019）认为，气候变化、大规模战争和瘟疫、矿产的开采和枯竭等是控制重金属长距离传输在年至百年际变化的驱动因素，同时结合 FLEXPART 大气传输沉降模型表明北极的铅污染来自于欧洲而非中国。近年来砷、锑、镉等元素在区域和全球的长距离传输和对生态系统的影响逐渐引起科学家和政府的重视（Wang C. M. et al.，2016；Zhang Y. L. et al.，2016）。姜珊（2012）利用湖泊和泥炭沉积记录探讨了过去 3 000 年以来北极新奥松地区重金属 As 和 Pb 污染程度较高，指出西欧和俄罗斯是该地区的潜在污染来源地；人为源的大气沉降是环南极地区的八个沉积剖面记录的重金属来源的主要因素之一。杨仲康（2019）也对南北极污染物传输过程机制做了系统分析。吴寨（2018）利用铅同位素、碳氮比和 $\delta^{13}C$ 示踪，研究了白令海、西北冰洋和中国南海沉积物中重金属来源于中国的大气传输、越南的陆源输送等。汪星星（2018）通过南极中山站到 Dome A 断面的 58 个表层雪样品，揭示了重金属污染物的来源、传输路径和沉降机制等控制因素。王超敏（2017）以东天山庙儿沟冰心为研究介质，揭示了庙儿沟地区镉在 20 世纪中后期来源于东欧及中亚地区的工业活动，而到了 21 世纪初则来源于新疆的工业活动。此外，重金属在后工业化时代对南北半球的影响并不同步（Uglietti et al.，2015），南半球相较于北半球其研究程度非常低。

大气环流、排放源、气候变化、战争和瘟疫、自然灾害等是控制重金属污染物大气传输的关键因素。除以上污染物长距离传输的历史记录，全球尤其是中国近十年建立了大量的大气干湿沉降观测站，伴随气象观测，用于监测污染物的长距离传输（Wang et al.，2011；Pan et al.，2013；王梦梦等，2017；Wei et al.，2018；林官明等，2018）。颗粒物迁移模型、主微量元素指标、富集系数、因子分析、稳定同位素、矿物组成等指标被广泛用来研究重金属污染物的长距离传输路线和排放源（Pan and Wang，2014；Dong et al.，2016；Wei et al.，2018；Yang et al.，2019），沉降通量和污染物的排放量（Gili et al.，2016；Marx，2016；王梦梦等，2017；Al Mamun et al.，2020）以及全球特别是南半球污染物羽流传输路径也是近十年关注的重点。

今后还需深入研究的问题包括：重金属污染物长距离传输的气候效应和生态效应，尤其是对海洋生态系统和人类健康的影响机制，传输过程的化学反应机制，氮−硫−臭氧等与重金属的复合污染效应，高精度污染传输模型，以及长期高分辨率的观测。

2　非传统稳定同位素的环境地球化学研究及其应用

由于 MC-ICP-MS 仪器的普及，近十年非传统稳定同位素地球化学在环境地球化学研究中得到广泛应用，中国学者也据此开展了大量的研究工作。

2.1 汞同位素地球化学

自然界中，Hg 有 7 种稳定同位素，其平均丰度分别为 0.15%（^{196}Hg）、9.97%（^{198}Hg）、16.87%（^{199}Hg）、23.10%（^{200}Hg）、13.18%（^{201}Hg）、29.86%（^{202}Hg）、6.87%（^{204}Hg）（Yin et al., 2010）。因其独特的物化性质，汞在表生环境中十分活跃，可在微生物、光和有机质等作用下发生质量相关分馏（MDF）和非质量相关分馏（MIF）（Blum et al., 2014; Feng et al., 2015）。

汞同位素质量分馏是在物理、化学和生物过程中汞同位素由于质量差异在各物相间的分配及再分配（即同位素平衡分馏和同位素动力学分馏）。汞同位素非质量分馏主要是由核体积效应（NVE）和磁同位素效应（magnetic isotope effect, MIE）引起。NVE 和 MIE 可以利用 Δ^{199}Hg/Δ^{201}Hg 的值来判断。当发生 MIE 时，Δ^{199}Hg/Δ^{201}Hg 为 1.0～1.3；当发生 NVE 时，Δ^{199}Hg/Δ^{201}Hg>1.6。MIE 主要发生在 Hg(Ⅱ) 的光还原以及 MeHg 的光降解过程；而 NVE 主要发生在 Hg^0 挥发、Hg(Ⅱ) 无光条件下的还原、Hg 和巯基之间的络合反应等过程（Blum et al., 2014; Yin et al., 2014; Kwon et al., 2020）。

近年来随着非传统稳定同位素分析技术的快速发展，汞同位素地球化学领域也取得令人瞩目的进展，特别是在分析方法、源排放特征及汞生物地球化学循环等方面。

中国科学院地球化学研究所率先利用在线流进样系统耦合多接收器电感耦合等离子体质谱，实现了高精度测量水体、大米、稻田土壤以及汞矿石等天然样品的总汞同位素，引领中国汞同位素技术快速发展，为深入认识汞的生物地球化学循环奠定基础（Yin et al., 2010）。近年来开发的主要方法包括：①大气气态总汞同位素。前人采用单根金管或多根金管并联方法富集大气总气态汞（total gaseous mercury, TGM），采样流速需要小于 2 L/min；若采集背景地区大气总气态汞（1.0～5.0 ng/m^3），长时间使用将造成金管钝化。Fu 等（2014）自制碘化和氯化活性炭预富集 TGM，采用马弗炉热解/溶液捕获获取总汞（Fu et al., 2014）。该方法可长时间和高流速采集大气样品，平均回收率大于 90%，能够满足野外大气同位素样品采集要求。②天然水体汞同位素。由于自然水体中汞含量较低（ng/L 级），测量其汞同位素组成具有较大挑战。前人利用气液分离器、离子交换树脂以及镀金吸附剂高效地吸附低汞浓度水体中的汞，进行总汞同位素分析，但是这些方法费时且样品量大。Li K. 等（2019）采用 BrCl 氧化、SnCl$_2$ 还原、吹扫至氯化活性炭富集，实现水体总汞同位素的准确测定。③大米甲基汞同位素：甲基汞是汞最为重要的有机形态，具有脂溶性和神经毒性，利用汞同位素示踪甲基汞的生成、生物富集和人群暴露具有重要意义。Li W. 等（2019）利用化学试剂提取提纯大米样品的甲基汞，发现大米甲基汞和无机汞同位素特征具有显著差异。④土壤甲基汞同位素：由于土壤甲基汞含量和甲基汞占比例较低，低甲基汞含量和无机汞干扰是极大挑战。Qin 等（2018）采用乙基化和气相色谱分离，得到纯化的甲基汞，建立稻田土壤甲基汞同位素测量方法。

中国是全球最大的汞生产、使用和排放国，燃煤、有色金属冶炼和水泥生产都是主要排汞行业（Wu et al., 2016）。中国煤炭 MDF 和 MIF 具有很大的范围（δ^{202}Hg 为 -2.36‰～-0.14‰，Δ^{199}Hg 为 -0.44‰～0.38‰），生物成因的煤与地质成因的煤具有不同的 MIF；中国煤燃烧排放汞的同位素特征为 δ^{202}Hg 为 -0.70‰，Δ^{199}Hg 为 -0.05‰（Yin et al., 2014）。中国闪锌矿矿石具有较大范围的 δ^{202}Hg 值（-1.87‰～0.70‰），Δ^{199}Hg 为 -0.24‰～0.18‰；不同类型闪锌矿的 δ^{202}Hg 值无明显差异，而不同类型闪锌矿的 Δ^{199}Hg 值却存在显著差异，可以利用汞同位素 MIF 示踪不同类型的锌矿石汞来源（Yin et al., 2016a）。万山汞矿区汞矿石的 δ^{202}Hg 平均值为 -0.74‰±0.11‰，废渣 δ^{202}Hg 平均值为 0.08‰±0.20‰，废渣相对汞矿石 δ^{202}Hg 值明显偏重，汞矿冶炼过程导致汞同位素 MDF（δ^{202}Hg）可达约 0.80‰，却不能导致明显的汞同位素 MIF（Yin et al., 2013a）。

汞是一种全球性污染物，大气汞的长距离迁移转化是全球汞生物地球化学循环的重要环节。法国南部比利牛斯山气态单质汞（GEM）的 δ^{202}Hg 值为 -0.04‰～0.52‰，有季节性变化，与 CO 浓度显著负相关，主要来源于欧洲和大西洋；大气 GEM 的 δ^{202}Hg 与气态氧化汞（GOM）浓度呈正相关，指示 GEM 氧化过程发生 MDF 且自由对流层的 GEM 易于被氧化为 GOM（Fu et al., 2016a）。中国背景区与城市区具有不同的大气汞同位素特征，大气颗粒的 δ^{202}Hg 为 $-1.45‰ \sim -0.83‰$，Δ^{199}Hg 为 $0.27‰ \sim 0.66‰$；大气 GEM 的 δ^{202}Hg 为 $-1.63‰ \sim 0.34‰$，Δ^{199}Hg 为 $-0.26‰ \sim -0.02‰$（Yu et al., 2016；Fu et al., 2018, 2019a, 2019b）。中国街道灰尘的 δ^{202}Hg 为 $-0.61‰ \pm 0.92‰$，Δ^{199}Hg 为 $-0.03‰ \pm 0.08‰$，主要来源于煤燃烧和工业活动（Sun et al., 2020）。中国背景区大气颗粒汞（PBM）具有显著偏正的 MIF（平均 Δ^{199}Hg 为 $0.27‰ \pm 0.22‰ \sim 0.66‰ \pm 0.32‰$），明显高于一次人为源（$\Delta^{199}$Hg $\approx$ 0‰）和中国城市 PBM（Δ^{199}Hg 为 $-0.02‰ \sim 0.05‰$）的 MIF，主要受二次颗粒汞来源的影响；东部和西部背景区 PBM 的 Δ^{199}Hg 值季节性变化特征相反（东部地区夏季高于冬季，西部地区冬季高于夏季），主要受长距离传输人为源和区域一次-二源混合作用的影响（Fu et al., 2019a）。

森林系统占全球陆地总面积的 31%，是全球汞生物地球化学循环最活跃的地区之一，植被-大气界面汞交换及土壤系统的迁移转化是森林生态系统最重要的生物地球化学循环过程（王训等，2017）。在温带混交林植被生长季的夜晚，GEM 的 δ^{202}Hg 值偏重而 Δ^{200}Hg 无明显变化，GEM 发生亏损而 GOM 无变化，指示森林植被对大气 GEM 有吸收作用而对 GOM 没有影响（Fu et al., 2016b）。同时森林叶片的 Δ^{199}Hg 值随生长期增长有下降趋势（$-0.16‰ \rightarrow -0.47‰ \rightarrow -0.34‰$），叶片释放的 GEM 具有明显偏正的 Δ^{199}Hg（$0.17‰ \pm 0.20‰$，高于大气 Δ^{199}Hg$_{GEM}$），说明植被-大气界面交换过程存在再释放作用（Yuan et al., 2019）。因此，森林植被-大气界面存在交互作用，植被吸收大气汞时，大气 δ^{202}Hg 正向偏移而不发生 MIF；植被再释放汞时，由于叶片 Hg（Ⅱ）光还原作用，正向 Δ^{199}Hg 的汞进入大气。Fu 等（2019b）在长白山温带森林发现 GEM 的 δ^{202}Hg 和 Δ^{199}Hg 值与植物生长指数呈显著正相关，证明植被影响大气汞同位素特征；而哀牢山亚热带森林冠层的 GEM 的 δ^{202}Hg 与人为源排放量呈显著正相关，表明植被活动和人为源释放是北半球大气汞同位素空间变化的主要驱动因子。青藏高原/亚热带山地森林的土壤与凋落物具有相似的 Δ^{199}Hg 和 Δ^{200}Hg，指示凋落物是森林土壤的主要输入汞源（Wang X. et al., 2017；Wang J. et al., 2019）；而森林土壤与降水的 Δ^{199}Hg 值有明显差异，说明湿沉降对森林土壤汞的直接贡献较低，降水主要通过影响凋落物生物量和降解过程间接影响森林土壤汞库（Wang et al., 2019a）。Wang X. 等（2020）利用汞同位素分馏建立三元混合模型，发现冰川退化区土壤也主要来源于植被演替从大气中吸收大量 GEM，可达原始冰川汞库的 10 倍，1850 年小冰期以来有 300~400 t 汞沉降累积在冰川退缩区，未来进一步的全球变暖-植被格局改变将显著影响全球汞的生物地球化学循环。

在水生生态系统中，无机汞可以转化为甲基汞并随食物链累积放大，对水生生态及人群健康造成危害，因此水生生态系统汞的地球化学循环一直是环境汞研究的重要领域。贵州红枫湖和百花湖沉积物 δ^{202}Hg 分别为 $-2.02‰ \sim -1.67‰$ 和 $-1.10‰ \sim -0.60‰$，百花湖沉积物中的汞主要来自有机化工厂排放，而汞同位素 MIF 显示红枫湖沉积物主要来自土壤侵蚀（Feng et al., 2010）。青藏高原青海湖和纳木错湖泊沉积物 δ^{202}Hg 值分别为 $-4.55‰ \sim 3.15‰$ 和 $-5.04‰ \sim -2.16‰$，表明降雨大气汞的湿沉降、植被和土壤径流输入的增加；有明显偏正的 Δ^{200}Hg（$0.05‰ \sim 0.10‰$）和 Δ^{199}Hg（$0.12‰ \sim 0.31‰$），表明降雨汞输入和水体 Hg（Ⅱ）光致还原的作用，1990 年以来两个湖泊沉积物 Δ^{199}Hg 值显著增加，表明随着冰期的缩短 Hg（Ⅱ）的光还原的速率显著增加（Yin et al., 2016b）。Liu 等（2011）利用汞同位素混合模型，定量示踪区域背景、城市源和工业源对珠江三角洲东江沉积物汞污染的贡献。中国珠江三角洲河口和南海沉积物的汞同位素特征具有显著差异，南海沉积物高 Δ^{199}Hg 值表明部分汞进入沉积物之前经历显著的 Hg^{2+} 光还原过程，而珠江三角洲河口沉积物 Δ^{199}Hg 值偏低表明光还原不是水中汞去除的主要途径，而河流输入是汞的主要来源（Yin et al., 2015）。珠三角海洋野生鱼肉的 Δ^{199}Hg 值为 $0.05‰ \pm 0.10‰ \sim 0.59‰ \pm 0.30‰$，$\Delta^{199}$Hg/$\Delta^{201}$Hg 为 ~1.26，说明水中的 MeHg 在进入食物链之前发生光降解；草食性、底栖性和肉食性鱼类的 Δ^{199}Hg 值差异显著，说明不同食性的鱼类摄食的甲基汞具有不同程度的光致去甲基化，可以利用汞同位素揭示食物链甲基汞暴露来源（Yin et al., 2016c）。

土壤是全球最大的汞库，农作物是人类基本的食物来源之一。森林、泥炭、草地和苔原生态系统表层土壤 Δ^{199}Hg 和 Δ^{200}Hg 与植被的同位素类似；其表层土壤汞的 54%±21%、12%±9% 和 34%±18% 分别来自大气 GEM、大气 GOM 和土壤母质，而农田表层土壤汞主要来自人为源（Wang et al., 2019b）。贵州雷公山背景区土壤的 δ^{202}Hg、Δ^{199}Hg 和 Δ^{201}Hg 值随海拔上升而显著下降，指示海拔上升大气汞对土壤贡献的增加，表明山地生态系统对汞具有"山地诱捕效应"（Zhang H. et al., 2013）。贵州省汞矿区、燃煤区域和锌冶炼区表层土壤 δ^{202}Hg 值分别为 −0.43‰±0.12‰、−1.59‰±0.31‰ 和 −1.32‰±0.32‰，Δ^{199}Hg 值分别为 −0.02‰Hg（Ⅱ）0.07‰、−0.09‰±0.03‰ 和 −0.12‰±0.04‰，不同地区表层土壤的 δ^{202}Hg 和 Δ^{199}Hg 的变化是由于不同汞污染源的混合作用以及自然界同位素分馏过程引起（Feng et al., 2013）。氯碱厂周边区域土壤 δ^{202}Hg 值范围较大（−2.11‰~0.72‰），而且没有明显 MIF（Δ^{199}Hg = −0.07‰±0.06‰，Δ^{200}Hg = 0.00‰±0.03‰）（Zhu W. et al., 2018）。万山汞矿区土壤水溶态汞 δ^{202}Hg（0.70‰±0.13‰）、$(NH_4)_2S_2O_3$ 提取态汞 δ^{202}Hg（1.28‰±0.25‰）相对土壤总汞 δ^{202}Hg（−0.02‰±0.16‰）显著偏正，而 Δ^{199}Hg 无显著差异，说明重同位素易富集在可迁移部分（Yin et al., 2013b）。

水稻植株的 δ^{202}Hg 为 −3.28‰~−0.96‰，Δ^{199}Hg 为 −0.29‰~0.01‰，可以利用 Δ^{199}Hg 二元混合模型来计算水稻不同部位大气和土壤汞的相对贡献（Yin et al., 2013c）。玉米地土壤的 δ^{202}Hg 值变化大（−2.38‰~0.72‰），且玉米植株、大气和土壤三者差异显著，说明玉米植株从大气和土壤吸收汞的过程发生 MDF，玉米植株地上部分 MIF 特征说明其主要来源于大气（Sun et al., 2019）。

2.2 锑、镉同位素地球化学研究

锑（Sb）在自然界中广泛存在，被巴塞尔公约列为全球优先控制的金属污染物之一。锑有两个稳定同位素：^{121}Sb 和 ^{123}Sb，其平均丰度分别为 57.2% 和 42.8%（Berglund and Wieser, 2011）。2003 年 Rouxel 等首次报道了锑同位素的分析方法及在自然界中的组成。目前锑同位素的纯化普遍采用 Biorad AG50-X8 树脂或者 Dowex AG50-X8 树脂除去基体元素，然后用巯基棉或者 Amberlite IRA 743 树脂除去 Te、Sn 等可以发生氢化物的同质异位素干扰（Asaoka et al., 2011；Tanimizu et al., 2011；Lobo et al., 2012；Resongles et al., 2015）。巯基棉的使用可使回收效率达到 96% 以上，而 Amberlite IRA 743 树脂的使用导致回收率只有 70% 多。氢化物发生−多接收器电离耦合等离子体质谱仪连用（HG-MC-ICP-MS）相较于直接气旋雾化进样灵敏度显著提高（孟郁苗等，2016；赵博等，2018），并且减少了样品用量。锑同位素测试过程中仪器质量歧视校正目前有样品-标样匹配法和 In、Cd 或 Sn 内标法 3 种。Liu J. F. 等（2020）开发了 AG1-X4 阴离子树脂和 AG50 W-X8 阳离子树脂两步过柱纯化的方法，回收率可达 98.7%。这是中国学者首次报道锑同位素的相关工作，依然存在进样浓度过高无法满足低含量样品开展锑同位素工作的需要。

已有研究表明，Sb 在氧化与还原、吸附解吸和生物作用过程中会产生同位素分馏（Rouxel et al., 2002；Resongles et al., 2015），分别可达 9ε、4.1ε 和 19ε；而蒸发与冷凝过程中发生的微小锑同位素分馏倾向于在产物中富集重锑同位素（Tanimizu et al., 2011）。氧化还原、微生物甲基化被认为是影响锑同位素分馏的关键过程，然而这些研究只是刚起步，仍需要大量精细系统的实验机理研究。

环境中不同样品的 δ^{123}Sb 值（图 2）显示，目前在海底热液硫化物样品中观测到的最大的 δ^{123}Sb 值为 −0.22~19.1ε。Rouxel 等（2002）对不同自然环境样品如海水、岩石、硫化物的锑同位素组成做了研究；Tanimizu 等（2011）对日本一个锑矿区中的辉锑矿和废水的锑同位素组成做了研究，Lobo 等（2012）对 10 个来自世界不同矿区的辉锑矿样品的锑同位素组成做了研究；Lobo 等（2013，2014）和 Degryse 等（2015）对晚青铜、希腊和古罗马时期的玻璃中的锑同位素组成进行了分析；Resongles 等（2015）对法国 OrbRiver 和 Gardon River 两条地表河流流域的地表水的锑同位素组成进行了分析。这些研究表明，不同物质端源锑同位素组成存在显著差异，如不同产地玻璃的锑同位素组成不同；不同的地球化学过程产生的同位素分馏特征不同。现有的研究表明，锑是揭示环境系统中锑的来源、迁移、转化和生物地球化学过程的有效示踪剂，可以作为古海洋系统中氧化还原条件的指示剂。

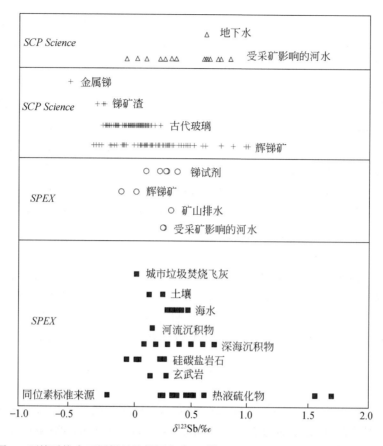

图 2 环境系统中不同样品的锑同位素（δ^{123}Sb）组成（据 Wen et al., 2015）

当前锑同位素研究中的主要问题有：①建立国际通用的同位素标准物质和基准锑溶液，以便于不同实验室测试数据对比；②高基体背景和低锑浓度的环境样品的高回收率消解纯化流程需要进一步优化和提高；③缺乏系统的理论和实验验证的锑同位素分馏机理方面的研究；④地球表层系统中各个储库的锑同位素组成的数据仍然非常缺乏。对锑同位素的研究有广阔的发展前景，将大有所为。

镉（Cd）是一个具有亲硫性、亲铜性的中等不相容元素（McDonough and Sun, 1995），它在自然界有八个稳定同位素（Pritzkow et al., 2007）：^{106}Cd（1.25%）、^{108}Cd（0.89%）、^{110}Cd（12.47%）、^{111}Cd（12.80%）、^{112}Cd（24.11%）、^{113}Cd（12.23%）、^{114}Cd（28.74%）和^{116}Cd（7.51%）。目前，在地球表生环境中观测到的镉同位素组成为 −5.05‰ ~ −0.74‰（Yang et al., 2012, 2015；Xue et al., 2013；Tan et al., 2020；Liu M. S. et al., 2020）。Cd 同位素分离提纯和仪器质量偏差校正方法近十年的发展，为获得高精度准确的 Cd 同位素数据提供了可能。Cloquet 等（2005）建立的 AG-MP-1 树脂单柱一步分离法近年来被广泛采用和改进（Wei et al., 2015；Li et al., 2018；Liu C. S. et al., 2019；Tan et al., 2020），尤其是对于超低含量样品的处理（Tan et al., 2020）使其可以将基质和干扰元素降低到合适范围，同时结合双稀释剂技术（^{111}Cd-^{113}Cd、^{110}Cd-^{1113}Cd、^{106}Cd-^{108}Cd）使其测试精度达到 0.04 ~ 0.43ε（$\delta^{114/110}$Cd）（Yang et al., 2012；Xue et al., 2013；Tan et al., 2020）。迄今国际上仍然没有统一的 Cd 同位素 "zero-delta" 标准物质，Abouchami 等（2013）对比了被 7 个实验室使用的内部标准物质，推荐美国国家标准物质中心的 NIST SRM 3108 作为国际通用的 "zero-delta" 标准物质，目前已被众多实验室采用（Liu M. S. et al., 2020；Tan et al., 2020）。

镉同位素的分馏机制主要有：①风化作用导致的同位素分馏，表现为在流体中富集镉的重同位素，而镉的轻同位素趋向于留在残留物中（Zhang Y. X. et al., 2016；Zhu C. W. et al., 2018）。在铅锌矿淋滤实验和野外土壤-沉积物实际观测中发现这一分馏（$\delta^{114/110}$Cd）可达 0.50‰（Zhang Y. X. et al., 2016）。在风化形成的次生矿物中 $\delta^{114/110}$Cd 值表现为水锌矿>大颗粒菱锌矿>小颗粒菱锌矿>硫酸铅矿（Zhu C. W. et al.,

2018）。在自然风化过程中，次生矿物的粒度和硫化物风化产物是控制 Cd 同位素分馏的主控因素（Zhu C. W. et al., 2018）。在燃煤电厂排出的飞灰的浸提实验发现了与上述一致的实验现象，浸提液中 δ^{114}Cd 达 7.1‰远远大于地表环境中观测到的镉同位素比值（Fouskas et al., 2018）。②镉同位素在蒸发–冷凝过程中的残余物会富集重同位素。在燃煤电厂的飞灰中由于冷凝作用富集镉的重同位素，而在烟气中由于蒸发作用富集镉的轻同位素（Fouskas et al., 2018）。③生物过程主要表现在海洋浮游植物优先吸收较轻的镉同位素，遵循瑞利分馏过程（Sieber et al., 2019）；而在陆地系统中土壤–小麦体系中这一过程较为复杂，小麦中镉同位素相对于土壤提取液偏负而相对于土壤偏正（Wiggenhauser et al., 2016；Imseng et al., 2019）。④无机过程中轻的 Cd 同位素优先以类质同象的方式进入方解石或者在海水体系中优先形成 CdS 沉淀，而剩余溶液富集重镉同位素（Horner et al., 2011；Guinoiseau et al., 2018）。

镉同位素被广泛应用于环境地球化学污染源示踪（Shiel et al., 2013；Zhang Y. X. et al., 2016），海洋生物地球化学循环（海水、浮游生物、铁锰结核、热液硫化物等）（Yang et al., 2015；John et al., 2017；Sieber et al., 2019），河流对海洋的输入贡献（Lambelet et al., 2013），大气粉尘对海洋的贡献（Bridgestock et al., 2017），重建古海洋环境如对二叠纪–三叠纪之交生物大灭绝事件的研究，以及碳酸盐岩中的镉同位素组成可以反演历史时期古海洋的初级生产力和营养利用率（Hohl et al., 2016；Zhang et al., 2018）。

城市化和工业化导致大量重金属被排放进入河流。Gao 等（2013）系统采集了珠江水系的北河沉积物并对其镉同位素进行了分析，发现北河沉积物中镉同位素组成范围为 $-0.35‰ \sim 0.07‰$，从而确定了北河污染的来源为：冶炼厂粉尘（$\delta^{114/110}$Cd<0）、冶炼矿渣（$\delta^{114/110}$Cd>0）、背景区土壤和开采活动（$\delta^{114/110}$Cd=0）。天然水体和沉积物中来自于酸性矿山废水的重金属污染依然是一个全球性的环境问题。Yang 等（2019）对广东大宝山流域受酸性矿山废水影响的河流水体和沉积物做了镉同位素分析，河水中溶解性镉同位素组成在 $0.21‰ \sim 1.03‰$ 范围，沉积物和水体间的镉同位素 $\Delta^{114/110}$Cd$_{river-sediment}$ 组成差异达 $1.61‰$。该研究表明河流中许多二次过程例如吸附、络合、沉淀等影响镉同位素的分馏。Wen 等（2015）应用镉同位素研究了矿区受污染土壤中镉的来源，同时 Wang P. C. 等（2019）利用镉同位素解析了农田土壤中重金属的来源。Wei R. F. 等（2019）利用镉同位素确定和量化了植物对镉的吸收和运移机制。该研究表明，随着时间的推移较轻的 Cd 同位素可以从根部转移到叶片，同时植物镉中毒越严重，Cd 同位素分馏越小。该研究表明 Cd 同位素在植物生理研究中的潜在应用。Wei 等（2015，2018）的研究也表明耐镉植物富集轻的镉同位素。张晓文（2019）明确了镉同位素在水稻各器官中的累积与转移规律及其分馏系数，同时建立了水稻果实中镉来源的同位素解析模型。

近十年镉同位素分析测试方法已非常完善，能够满足环境地球化学样品的分析。尽管已经初步了解地球表层系统中的镉同位素组成和分馏机理，但还远远不够；镉同位素在不同过程中的分馏机理认识仍然不完善，地球各储库以及人为源中 Cd 同位素组成仍然需要补充和完善。镉同位素将在海洋营养物质循环演化，示踪人为污染源，以及镉在植物–土壤–大气–水体等地球表生系统中迁移转化过程中提供强有力的技术手段。

2.3 其他新兴金属稳定同位素研究

除汞、镉和锑外，铬（Cr）、铊（Tl）、硒（Se）、镍（Ni）、钡（Ba）、银（Ag）、铜（Cu）、锌（Zn）、铁（Fe）等同位素体系在最近十年也得到了快速发展。

铬稳定同位素体系聚焦于六价铬价态转化的生物和非生物过程中的同位素分馏效应（Saad et al., 2017）和分馏系数（Jamieson-Hanes et al., 2014；Xu et al., 2015），而三价铬的氧化过程复杂需要更多的研究（Zink et al., 2010）；吸附–解吸，蒸发，溶解–稀释等过程对铬同位素的分馏系数的影响极小（Qin and Wang, 2017）。铬同位素被广泛用于地下水中示踪和评估六价铬的来源和还原程度，此外还被用于区分生物过程和非生物过程（Lu et al., 2018），以及作为古环境的指标（王相力和卫炜，2020）。

铊是一种典型的与汞、铅毒性相当的稀少元素，有两个同位素 ^{205}Tl 和 ^{203}Tl，在自然界储库中 ^{205}Tl/^{203}Tl

的变化范围可达3‰（Blusztajn et al., 2018；Shu et al., 2019；Liu J. et al., 2020）。近年来，铊同位素在研究植物、海水、黄土、人为源排放等的来源和过程方面表现出强大的潜力（Blusztajn et al., 2018；Vaněk et al., 2019；Shu et al., 2019；Liu J. et al., 2020）。

硒有6个天然稳定同位素，$^{82}Se/^{76}Se$值被用来表示硒同位素的分馏（Mitchell et al., 2012；Schilling et al., 2013；Zhu et al., 2014；Shrimpton et al., 2015；Xu et al., 2020），目前已经观测到的地球表层系统中$\delta^{82/76}Se$的变化范围是$-14.5‰\sim12‰$（Zhu et al., 2014；Stüeken, 2017；Howarth et al., 2018）。硒的生物和非生物还原是造成硒同位素分馏的主要原因，而路凯（2016）实验发现HNO_3、MnO_2氧化元素硒的过程均出现了产物极度富集重硒同位素，打破了传统认为氧化过程没有或只有极小的硒同位素分馏；Xu等（2020）填补和丰富了赤铁矿、二氧化锰和氧化铝参与吸附过程造成的硒同位素分馏的空白和认识。由于硒的不同价态的地球化学性质差异，使硒同位素特征组成成为指示地表发生的氧化还原反应过程的强有力工具（朱建明等，2015）。未来硒同位素在分馏机理、储库特征以及应用上急需开展大量研究。

钡同位素分析技术近十年才建立。Miyazaki等（2014）首次将钡同位素分析精度提高到0.03‰，之后钡同位素在水生系统中得到了广泛的研究（Horner et al., 2015；Cao et al., 2016；Bates et al., 2017；Hsieh and Henderson, 2017；Bridgestock et al., 2018；Geyman et al., 2019），表明Ba同位素是海洋碳循环过程中有效指示参数。钡还在岩石风化（Gong et al., 2019b）、沉淀（Mavromatis et al., 2020）等过程中的同位素分馏得到研究。Ba同位素的研究面临缺少统一的"zero-delta"国际同位素标样和多种钡同位素组成标准物质，限制了不同实验室间的对比。今后钡同位素在各地球储库中的组成和分馏机理研究是解决钡的环境生物地球化学问题的当务之急。

镍同位素是近十年新兴起的同位素体系。Fujii等（2011）从理论计算和实验上得出镍同位素在无机和有机结合态之间存在高达2.5‰的同位素分馏，目前在自然界观测到的镍同位素分馏可以达到3.5‰（Gueguen et al., 2013；Porter et al., 2014；Alvarez et al., 2020）。在沉积物、海水、河水、植物等介质中开展的镍同位素工作表明镍同位素可以为示踪污染来源和迁移转化过程提供有效信息（Hoefs, 2018；Wang R. M. et al., 2019）。镍同位素的研究才刚刚起步，大量的工作需要开展。

纳米银（AgNPs）在消费品中被广泛应用，并被大量排放到自然环境。银同位素技术被用于研究纳米银在氧化、沉淀等过程中的转化机理和示踪银的来源（Li W. et al., 2019）。Lu等（2016）发现纳米银在自然条件下的形成和溶解可以导致显著的银同位素分馏，分馏系数可达0.86‰；人为和自然形成的纳米银颗粒具有显著不同的银同位素组成特征。Zhang T. Y.等（2017）的研究表明纳米银颗粒的大小显著影响银同位素分馏。

铜同位素的样品前处理、纯化与分离、仪器的质量歧视校正和仪器测定等已经非常成熟，样品回收率可以达到100%，分析精度在0.04‰左右（王泽州等，2015；王倩等，2016）。近十年中国学者的研究集中在：①通过不同实验条件高岭土吸附铜过程中溶液相和吸附相铜同位素组成变化，发现黏土矿物优先吸附富集轻铜同位素，受起始铜浓度和电解质浓度影响（Li D. D. et al., 2015）；②土壤剖面风化过程中有机质倾向富集轻铜同位素而重同位素倾向向下迁移（Liu S. A. et al., 2014），同时在黑色页岩中表现出相同现象（Lv et al., 2016）；③Zeng和Han（2020）和Wang Q.等（2020）分别利用铜同位素示踪了珠江和长江中铜的来源和影响因素。

铁同位素的前处理分析测试技术已经非常成熟（秦燕等，2020；孙剑和朱祥坤，2015），但最近中国学者应用AGMP-50树脂缩短了纯化时间和减少了试剂用量，同时克服了高铜含量样品的影响（Zhu G. H. et al., 2020）。铁同位素在地球表生过程中的储库、控制过程等框架已经基本建立。中国学者的研究集中在水稻土及其剖面中铁同位素分馏特征和影响因素（Qi et al., 2020；李芳柏和李勇珠，2019）；水稻中铁的吸收转运的同位素分馏特征和机理（Liu C. S. et al., 2019）；铁同位素在风化剖面中的分馏特征及控制因素（王世霞和朱祥坤，2013；Feng et al., 2018）；大气颗粒物（Chen et al., 2020）、湖泊颗粒物（Zheng et al., 2019）以及浮游植物（Sun and Wang, 2018）吸附铁元素过程中同位素分馏特征和影响因素。

锌同位素和铜、铁同位素都是 21 世纪初就被学界广泛关注并且开展了大量的工作。中国地质科学院地质研究所研制了国产的锌同位素标准溶液（唐索寒等，2016），中国科学技术大学黄方课题组对 Nist SRM 683 锌标准物质做了系统的工作，并推荐作为锌同位素分析和研究中"zero-delta"国际同位素标样（Chen et al.，2016）。锌同位素被广泛用来示踪污染来源和影响因素：Xiao 等（2020）通过锌同位素确定了喀斯特地区人为来源锌主要是来源于远距离传输的大气降尘，Tu 等（2020）确定了台湾河流中锌的来源和王中伟（2019）确定了珠江中锌的来源及影响机制，Ma 等（2020）揭示了珠江口颗粒相和溶解相影响了河水中的锌同位素组成，Liang 等（2020）研究了湖泊生态系统中锌同位素地球化学循环，Ma 等（2019）研究了牡蛎中锌的来源；锌同位素在风化过程（Lv et al.，2020）、植物的吸收和体内的迁移转化机制（Tang et al.，2012；Deng et al.，2014）的研究起到重要作用；此外，锌同位素被用到古环境古气候揭示海洋中的地球化学循环和海洋生产力的变化（Liu S. A. et al.，2017；Lv et al.，2018；Fan et al.，2018；Yan et al.，2019）。

近十年来，非传统稳定同位素地球化学飞速发展，铬、铊、硒、镍、钡等新兴稳定同位素还有很多重要科学问题需要进一步研究，如分析方法不完善、分馏机制不清楚、同位素指纹谱缺失、示踪体系不完善等一系列重要的问题，制约了重金属同位素在示踪环境污染来源和地球化学过程中的应用前景。今后应建立大气、水体和土壤中多种金属同位素高精度的测试方法，解决复杂环境样品以及痕量金属同位素的分析测试难题；系统测定典型人为源和自然源的金属同位素组成，构建环境主要污染源的金属同位素指纹谱；开展多种金属同位素联合示踪和地球化学过程研究，加强同位素分馏系数和机理的理论计算工作等。对非传统稳定同位素的认识和理解还远远不够，许多问题仍未解决；与经过半个多世纪演变的传统稳定同位素系统相比，金属稳定同位素的研究才刚刚起步，而中国学者在其中的贡献将会越来越大。

3 传统稳定同位素与全球变化研究

近十年来，传统稳定同位素在全球变化研究中发挥了重要作用，限于篇幅，本文仅总结泥炭稳定同位素气候代用指标、团簇同位素和微小^{17}O 同位素与气候变化等方面的研究进展。

3.1 泥炭稳定同位素气候代用指标研究

21 世纪初，洪业汤等通过对中国泥炭沉积环境特点的系统调查，阐明了植物纤维素 δ^{18}O、δ^{13}C 和 ^{14}C 指示地表温度、大气降雨量及时间变化的原理，突破了国际上长期以来认为从泥炭中植物残体纤维素氧同位素组成中不可能提取出气候变化信号的认识，证明其是研究全新世气候变化的有效代用指标。之后十年间，以中国大陆地区为代表的东亚地区全新世以来亚洲季风的演化趋势及其与北半球高纬度地区气候变化关联性和对人类古文明发展影响的研究取得了引人注目的成果（Hong et al.，2000，2001，2003，2005，2009，2010；Xu et al.，2006）。

近十年来，随着研究区域的扩大和研究的持续深入，利用泥炭植物纤维素稳定同位素为主要代用指标对末次冰期以来亚洲季风、北半球西风和南半球西风演化及关联性机制等研究方面获得了一系列新的进展。

长期以来，学界都认为北半球西风是影响中国北方干旱环境，特别是西北干旱区的重要因素。随着研究工作的深入，人们逐渐注意到了西风和亚洲季风之间此消彼长的竞争关系。当前争论的焦点是，在西风和季风相互竞争的背景下，影响中国北方干旱环境降雨的主要因素是什么？

对大黑湖泥炭地多项代用指标的研究显示（Xu et al.，2019），以阿尔泰地区为代表的中国新疆北部地区，在早、晚全新世时期表现为较为湿润，而在中全新世时期较为干燥。这可能是由于该时期西风带向北移动以及夏季高温所引起的蒸发量所致。对新疆阿尔泰山脉哈拉沙子泥炭纤维素稳定同位素指标研究显示（Rao et al. 2019），在全新世阶段气候呈整体变暖的趋势，而其中在中全新世较为寒冷，早、晚全新世则较暖。这种早、晚全新世较暖，中全新世较冷的总体格局与中亚、青藏高原和华北地区的气温记录一致。对新疆柴窝堡泥炭纤维素 δ^{13}C 降雨代用记录的研究显示，自 8.5 ka B.P. 以来，该地区夏季降雨呈现出一种在波动中缓慢增加的长期趋势（Hong et al., 2014a）（图3）。这些长时间尺度降雨增加现象与同期赤道太平洋出现的类厄尔尼诺状态有关，这导致了东亚夏季风强度增强，从而给中国西北干旱区输送去更多水汽。随着工业革命后人为活动释放的大气CO_2浓度急速增加，由泥炭纤维素稳定碳同位素指示的夏季降雨也突然明显增加。该结果表明，即使在全球最干区域之一的中亚地区，在全球变暖条件下也可能在变湿。突破了既往在全球气候变暖的大格局下，"干的区域越干，湿的区域越湿"的传统观念（图3）。

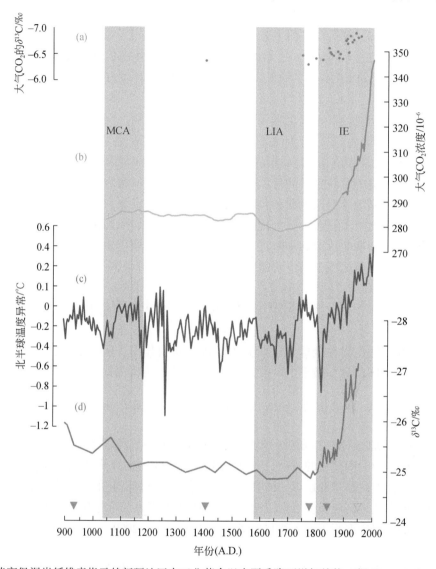

图3　柴窝堡泥炭纤维素指示的新疆地区自工业革命以来夏季降雨增加趋势（据Hong et al., 2014a）

由于缺乏足够长时间跨度的相同的气候代用记录指标，在末次冰期到全新世的转换过程中，亚洲季风在千年/百年时间尺度上的突然变化特征和中国大陆地区降雨/干湿分布格局及其机制的研究长期以来受到极大的限制。对四川越西泥炭的研究，在一定程度上弥补了这一学术空白（Hong et al., 2018）。时间跨度为33 300年的越西泥炭碳同位素序列是当前世界上所见已发表的最长的亚洲季风降雨代用记录（图4）。

研究表明，印度夏季风在末次冰期向全新世转换过程中表现出不同的变化特征。总体上，印度夏季风强度在寒冷的末次冰期较弱，而在相对较为温暖的全新世则较强。同时，对应于北半球高纬度地区千年/百年尺度气候的突然变化，印度夏季风也表现出同步的响应。这种响应在冷期和暖期都有出现。值得注意的是，这种响应关系在冷期和暖期可能表现不同。在冷期的反应似乎是一一对应的。例如，当YD和Heinrich等冷事件在北半球高纬度地区发生时，印度夏季风突然减弱；而当北半球高纬度地区发生

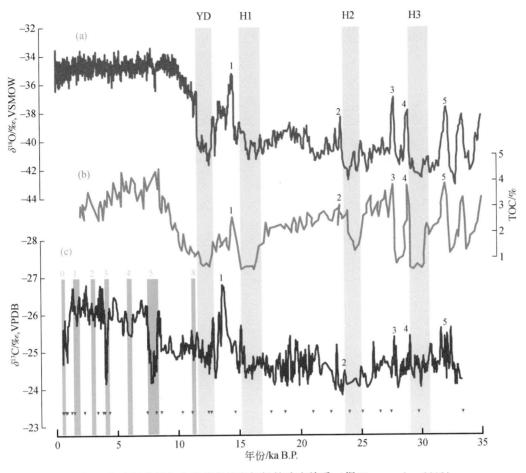

图 4 印度夏季风与北半球高纬度气候的响应关系（据 Hong et al., 2018）

Bølling/Allerød、Dansgaard-Oeschger 等突然变暖事件时，印度夏季风突然增强。而在温暖的全新世阶段，这种响应关系主要表现为印度夏季风的突然减弱，后者往往对北半球高纬度地区的突然降温事件表现出部分响应。

与此同时，受亚洲季风的影响，中国大陆地区的降雨/湿度分布模式在末次冰消期向全新世的转换过程中也表现出异常的分布模式（图5）。对应于 YD 冷期亚洲两季风的反相变化，中国大陆北方湿润而南方干旱。这主要是因为该时期东亚季风的实际增强，使季风降雨带北移，导致中国南方降雨的减少。而东亚季风减弱仅为一种表面现象。印度夏季风的实际减弱，导致了中国南方地区的干旱（Hong et al., 2014b）。

长期以来古气候学界一直存在一个未解的谜题，即为什么在一段较长的冷期之后，迅速地以一个相对较短的气候变暖期为结束，而后又再次重复变化？这种冷暖转变的过程机制是什么？这实际上涉及地球的南、北两个半球气候过程之间的相互作用及其机制问题。

结合亚洲季风与南半球西风的研究显示（Hong et al., 2019），当来自高北纬度巨量的融冰洪水倾泻入北大西洋，导致大西洋经向翻转环流的运转变慢甚至停止，致使地球系统能量重新分配，北半球变冷而南半球变暖，出现一个正相的半球热梯度或半球间温度反差，它使得地球的热带辐合带和南半球西风带的平均纬度位置向地球的南极方向移动，起源于南半球的印度洋夏季风的强度也同时变弱（图6）（Hong et al., 2019）。

对南半球西风带的活动特征的研究显示，该西风带存在一个风力最强的相对稳定的核心区域，它位于大约南纬47°附近。南半球西风带对不同相位的半球间温度反差的响应，表现为以大约南纬47°为轴，向地球南极或赤道方向摆动的过程。在 Heinrich1 事件和 YD 事件时期，当灾难性的融冰洪水对地球气候系统造成冲击时，也同时启动了地球气候系统的自我修复过程，或触发了一个南半球对北半球气候的影

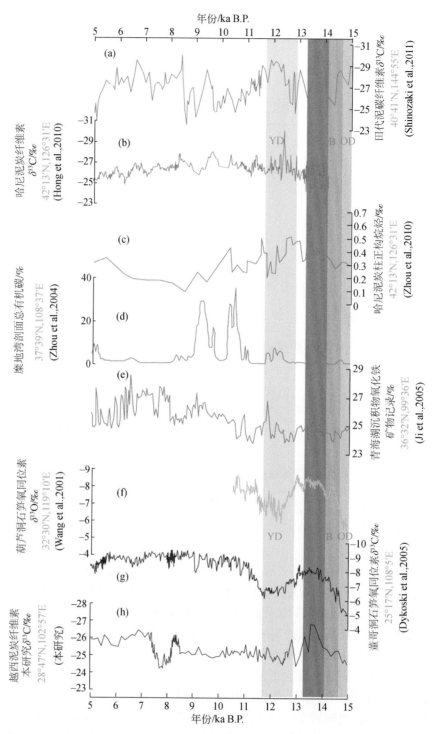

图 5 末次冰期-间冰期转换期中国大陆地区异常的降雨/干湿分布模式（据 Hong et al., 2014b）

响过程。这个自我修复过程是由于南半球西风带向南极方向的摆动触发的，它使南大洋上升涌流增强，其结果不仅把更多大洋深部的 CO_2 释放到大气环境中，导致南北半球不同的变暖，特别是增强了对南大洋深层水的拉动作用，反映了南半球西风对大西洋经向翻转环流更强的拉动影响，并最终导致大西洋经向翻转环流重新加快运转，使北半球变暖而南半球变冷（Hong et al., 2019）。该研究结果突显了地球两半球气候系统的关联特征，特别是南半球西风在其中所起的重要作用。它揭示了对于地质历史上发生的一些重大灾变，地球气候系统具有自我修复能力。

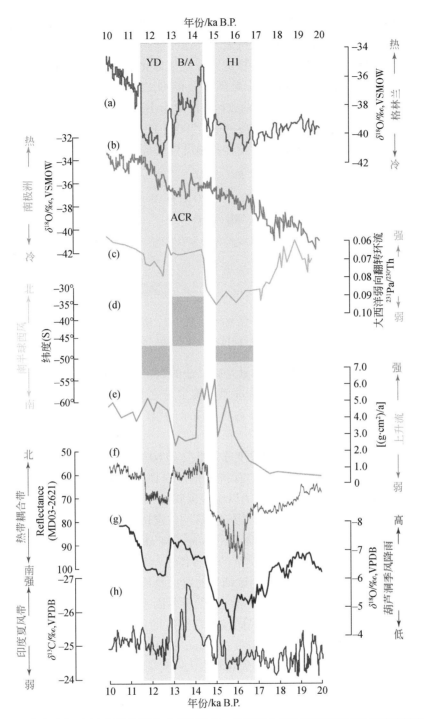

图 6　末次冰期-间冰期转换期南半球西风带、热带辐合带及印度夏季风对半球热梯度
变化响应关系（据 Hong et al., 2019）

3.2　Clumped 同位素与气候变化

气候变化是当今人类面临的严峻挑战，而温度是气候变化研究中的重要因子。自工业革命以来，人类活动导致地球大气二氧化碳浓度急剧上升，导致了全球变暖，进而引发了一系列的环境问题，如冰川冻土消融、海平面上升、极端天气频繁、降雨分布改变等（IPCC，2018）。研究古气候变化，尤其是古温度，是全面认识当今人类面临问题的重要途径。地球科学家已发展出多种指标用以研究古温度的变化，

碳酸盐团簇同位素（即 Clumped 同位素）便是其中重要的一个指标（Ghosh et al., 2006a）。下面将就团簇同位素的基本原理、碳酸盐团簇同位素的研究进展及其他团簇同位素体系做简要介绍。

团簇同位素是研究"稀-稀"同位素体的稳定同位素地球化学分支（Eiler, 2007）。以碳酸盐的团簇同位素 ^{13}C-^{18}O 为例，如果 ^{13}C 和 ^{18}O 在碳酸盐中是随机分布的，^{13}C-^{18}O 的浓度可根据 ^{13}C 和 ^{18}O 的浓度直接计算得到。但在温度较低情况下，^{13}C 和 ^{18}O 在碳酸盐中的分布不是随机的，这导致碳酸盐中 ^{13}C-^{18}O 的实际浓度偏离随机分布（Wang et al., 2004; Schauble et al., 2006; Liu J. W. et al., 2018; Liu T. B. et al., 2018）。这一偏离值依赖于温度。因此，团簇同位素可用作古温度计（图7）。

传统同位素

$$Ca^{16}O^{12}C^{16}O_2 + H_2O^{18} = Ca^{18}O^{12}C^{16}O_2 + H_2^{16}O$$

团簇同位素

$$Ca^{18}O^{12}C^{16}O_2 + Ca^{16}O^{13}C^{16}O_2 = Ca^{18}O^{13}C^{16}O_2 + Ca^{16}O^{12}C^{16}O_2$$

热 ←——————→ 冷

温度降低，同位素交换反应向右移动；温度升高，同位素交换反应向左移动

图7 传统 ^{18}O 同位素与团簇同位素温度计

目前，研究最多应用最广的团簇同位素体系是碳酸盐的 ^{13}C-^{18}O 团簇同位素体系。与传统 ^{18}O 同位素温度计相比，它的最大优势为碳酸盐单相测温，进而可确定与之并存的水的 ^{18}O 同位素组成（Eiler, 2007；图7）。如前所述，古温度是古气候研究的重要内容，而水的 ^{18}O 同位素可用于研究古气候条件下的水循环（Ghosh et al., 2006b）。

在同位素自然丰度下，碳酸盐 ^{13}C-^{18}O 的浓度非常低（如60几个ppm；Ghosh et al., 2006a），要精确测量团簇同位素以获得有效的古温度是非常具有挑战的。自团簇同位素概念在2004年提出（Eiler and Schauble, 2004）至今，研究者仍然在做各种努力以实现碳酸盐 ^{13}C-^{18}O 的高精度测量及不同实验室和仪器之间测试数据的对比。目前，已发现并解决了多种影响团簇同位素精确测量的因素，如消除压力基线效应（He et al., 2012）、建立以 CO_2 为标准的绝对参考体系（Dennis et al., 2011）、校正 ^{17}O 的影响（Daëron et al., 2016; Schauer et al., 2016）、采用以碳酸盐为标准的参考体系（Bernasconi et al., 2018）等。这些工作使碳酸盐 ^{13}C-^{18}O 的测试精度、不同实验室及仪器测试结果的对比及长时间周期的可重复性有了很大的改善。

动力学因素是影响碳酸盐 ^{13}C-^{18}O 团簇同位素应用的重要因素（Tang et al., 2014）。如果同位素动力学分馏显著的话，平衡条件下建立的 ^{13}C-^{18}O 团簇同位素与温度的关系便不再适用。近期的研究表明，地球表面形成的绝大部分碳酸盐都可能是偏离平衡的（Daëron et al., 2019），这在某种程度上动摇了碳酸盐平衡团簇同位素作为古温度计的根基。这种偏离到底有多大和是否具有内在一致性有待于进一步研究。

碳酸盐的后期改造是很难避免的问题，它决定了 ^{13}C-^{18}O 团簇同位素应用的地质时间范围（Passey and Henkes, 2012）。目前来看，500 Ma以内的碳酸盐样品还可保存初始碳酸盐 ^{13}C-^{18}O 团簇同位素的信息（Henkes et al., 2018）。但对于古老的碳酸盐样品，甄选时还是要特别小心。

团簇同位素已扩展到甲烷、氧气、和氮气体系。尽管还没有大量的研究，但已有的结果表明这些新体系的团簇同位素也可为气候变化提供重要的制约。例如，甲烷团簇同位素可制约空气中甲烷的源和汇（Haghnegahdar et al., 2017），而甲烷是很强的温室气体；氧气团簇同位素可制约对流层臭氧的含量（Yeung et al., 2019），而臭氧是重要的空气污染物及温室气体；等等。

相较于其他地球化学手段，团簇同位素在研究气候变化方面具有独特的优势。中国学者在团簇同位素理论研究方面起步较早（唐茂等，2007; Cao and Liu, 2012; Liu and Liu, 2016; Zhang et al., 2020），但在测试技术方面起步较晚，不过近些年在迎头赶上。据了解，国内已有多所实验室实现碳酸盐团簇同位素的测试（Wang X. et al., 2016; Guo and Zhou, 2019; Guo et al., 2019b; Wang N. et al., 2019; Chang et al.,

2020),并有多所实验室在建设开发研究甲烷、氧气、氮气等的团簇同位素体系。可以预见,在不久的将来,中国学者定会在团簇同位素研究领域做出重要的贡献。

3.3 微小 ^{17}O 同位素与气候变化

得益于同位素分析技术的提高,微小 ^{17}O 的同位素地球化学研究最近几年在多个方面取得了一系列重大进展,在全球变化研究领域显示出其独一无二的示踪特征。氧有3个同位素:^{16}O、^{17}O 和 ^{18}O,其中 ^{17}O 含量最低,仅占0.038%。前期研究很少关注到 ^{17}O,因为 ^{17}O 丰度低,测量困难,加之依据同位素质量分馏原理,^{17}O 同位素分馏值($\delta^{17}O$)一般认为近似为 ^{18}O 同分馏值($\delta^{18}O$)的一半($\delta^{17}O=0.52\delta^{18}O$),不具有特殊的示踪意义,因而一直未加以重视。自从 Clayton 等(1973)第一次在部分陨石中发现 $\delta^{17}O$ 与 $\delta^{18}O$ 之间的斜率不是+1/2 而是+1 以来,后来的研究发现自然界广泛存在氧同位素的非质量分馏,微小 ^{17}O 同位素在全球变化中的应用逐渐开展起来,并以其独特性显示出不可替代的潜力。三氧同位素一般用 $\Delta^{17}O$ 表示 ^{17}O 的偏离:

$$\Delta^{17}O = \delta^{17}O - C\delta^{18}O \tag{1}$$

式中,$\delta = (R_{样品}/R_{标准} - 1) \times 1000$,$R$ 为同位素比值:$^{18}O/^{16}O$ 或者 $^{17}O/^{16}O$,C 是常数。

在三氧同位素中常用 δ' 代替 δ 值:

$$\delta' = \ln(R_{样品}/R_{标准}),\ \text{即}:\delta' = \ln(\delta+1) \tag{2}$$

$$\Delta'^{17}O = \delta'^{17}O - C\delta'^{18}O \tag{3}$$

相对于 δ,使用 δ' 的优点是在 $\delta^{17}O$-$\delta^{18}O$ 图上同位素分馏呈直线,A、B 两点的直线斜率 λ 为

$$\lambda = \ln(^{17}R_A/^{17}R_B)/\ln(^{18}R_A/^{18}R_B) \tag{4}$$

α 为同位素分馏因子($\alpha = R_A/R_B$),^{17}O 与 ^{18}O 的质量分馏可以表示为

$$\theta = \ln(^{17}\alpha)/\ln(^{18}\alpha) \tag{5}$$

有关 ^{17}O 的研究初期主要集中于对不同地球化学过程对应的斜率 λ 或同位素质量分馏 θ 的实验和理论分析,进而进一步扩展它的应用。λ 和 θ 是两个不同的概念(Luz and Barkan, 2005;Cao and Liu, 2011),在 A 和 B 是同位素分馏过程的产物和底物时 λ 和 θ 具有相同的值(Angert et al., 2004)。Cao 和 Liu(2011)提出了平衡过程 θ 的计算方程。实际上,不同的研究者对上述多个参数的定义、赋值和使用方面略有差异(Miller, 2002;Luz and Barkan, 2005),Bao 等(2016)给出了系统全面的介绍和说明。

^{17}O 方法已广泛用于环境中各种含氧物质。首先是大气中的 O_2 以及水体中的溶解氧(DO),大气 O_2 的 $\Delta^{17}O$ 主要决定于平流层与对流层的交换以及光合作用产氧与呼吸作用的耗氧间的代谢平衡等过程(Bao et al., 2008;Young et al., 2014)。平流层 O_3-CO_2-O_2 之间光化学反应的非质量分馏导致 O_2 的 $\Delta^{17}O$ 为负值,Bao 等(2008)模拟的结果发现对流层 O_2 的 $\Delta^{17}O$ 信号 83% 来自平流层的光化学反应,相反,Young 等(2014)模拟的结果发现大气 O_2 的 $\Delta^{17}O$ 亏损 57% 来自 Dole 效应,平流层的贡献为 33%,10% 来自蒸发作用。导致模拟结果不同的原因之一是采用的标准不同,Young 等(2014)的模拟是以水的分馏线为基准。$\Delta^{17}O$ 以大气 O_2 为基准或者以水为基准,线的斜率 λ 不会变,但数值会有很大差异。依据大气 O_2 的 $\Delta^{17}O$ 可以恢复 CO_2 分压和初级生产力,p_{CO_2} 越高,O_2 的 $\Delta^{17}O$ 越负(Bao et al., 2008);而光合作用产生的 O_2 和呼吸作用消耗的 O_2 同位素分馏满足质量分馏,光合作用产生的氧越多,O_2 的 $\Delta^{17}O$ 越正(Luz et al., 1999),以此原理可以定量估算生产力变化(Luz and Barkan, 2000)。$\Delta^{17}O$ 在水文学研究中也显示出巨大的潜力,其核心的机理是水中 ^{17}O 和 ^{18}O 的分馏关系,汽-液平衡过程的同位素分馏 θ 值(0.526 ±0.001;0.529 ±0.001)大于水汽分子扩散过程的动力分馏 θ 值(0.511 ±0.005;0.518 5 ± 0.000 2)(Angert et al., 2004;Barkan and Luz, 2005, 2007),导致雨水的 $\Delta^{17}O$ 主要决定于(源区)蒸发过程的动力学效应,因此,$\Delta^{17}O$ 是表征相对湿度变化的一个非常好的指标(Uechi and Uemura, 2019)。蒸发过程较小的 θ 值使得降水 $\Delta^{17}O$ 偏正,蒸发后的水(湖水、叶片水、动物体内水分)$\Delta^{17}O$ 偏负

(Passey et al., 2014)。在热力学平衡条件下，沉淀的碳酸钙和文石与水体中 HCO_3^- 和 CO_3^{2-} 具有相似的 $\Delta^{17}O$（Guo and Zhou, 2019; Guo et al., 2019），湖相碳酸盐的 $\Delta^{17}O$ 与湖水蒸发线的斜率 λ 存在相关性，因而是指示湖泊系统蒸发作用的同位素指标（Passey et al., 2014; Passey and Ji, 2019），土壤碳酸盐的 $\Delta^{17}O$、石膏中晶格水的 $\Delta^{17}O$ 等也都具有指示过去水分蒸发的环境指示意义（Passey et al., 2014; Herwartz et al., 2017），与氘-盈余相比，$\Delta^{17}O$ 还有很多氘-盈余不具有的独特特性（Li S. N. et al., 2015）。此外，$\Delta^{17}O$ 还在其他领域还有很多新的应用（Bao et al., 2016），特别是在利用 CO_2-O_2-H_2O 示踪气候变化过程与水文水循环和生态系统的耦合与响应方面潜力巨大（Cao and Bao, 2013）。随着分析技术的提高，对 $\Delta^{17}O$ 同位素分馏理论的认识不断深化，$\Delta^{17}O$ 作为一种独特的手段在气候与环境变化研究中将会发挥更大的作用。

4 传统稳定同位素地球化学示踪研究

传统稳定同位素在环境地球化学过程示踪方面的应用可以追溯到 20 世纪 70 年代，近十年来中国学者在这方面的研究取得了重要进展。

4.1 氮同位素地球化学研究新进展

4.1.1 大气氮同位素地球化学

在大气环境研究中，示踪大气氮的来源是氮同位素的主要应用案例之一。Xiao 等（2012）对雨水的氮同位素组成研究充分证实了氮同位素可以有效进行氮源示踪。对大气气态 NH_3 的氮同位素研究还发现，贵阳地区雨天气态 NH_3 的 $\delta^{15}N$ 值比非雨天低，而且一旦停止降雨，$\delta^{15}N$ 值随即上升，这表明了 $^{15}NH_3$ 优先被酸雨捕获，大气氮在雨水-气态和气态-气溶胶间形态转化过程中存在氮同位素分馏，故利用氮同位素分馏进行大气氮形态转化过程的定量研究是可行的。由于大气雨水对气溶胶的捕获为物理混合过程，因此可以用稳定同位素端元混合模型进行定量化研究（Xiao et al., 2012, 2013; Liu X. Y. et al., 2017）。

对大气中不同形态氮寿命的研究是大气氮过程研究的一个重要方面，利用氮同位素来定量估算大气氮寿命则不需要考虑气象等因素，这是一种创新方法。Xiao 等（2015）最近利用雨水氮同位素组成对贵阳市大气气态 NH_3 的寿命进行了估算。研究结果表明，大气气态 $^{15}NH_3$ 优先溶于雨水中，结果大气中就会越来越富集 $^{14}NH_3$。这种现象将一直保持直至这批大气 NH_3 寿命结束。在这段时期内的后一场雨就会较前一场雨富集 ^{14}N。因此根据雨水 $\delta^{15}N$ 值的变化就可以推测大气 NH_3 的寿命（Xiao et al., 2012, 2015）。

近年来，刘学炎、宋韦等将其在氮素生物地球化学方面的研究方法和思路（Liu D. et al., 2013; Liu X. et al., 2013; Liu X. Y. et al., 2018; Hu et al., 2019）应用到大气无机氮来源和过程机制的研究中，提出了运用多源同位素质量平衡模型分析大气降水和颗粒物氮源的方案（Wang X. P. et al., 2016; Wang Y. L. et al., 2017; Zheng et al., 2018），初步评价了非化石源氮氧化物和化石源氨排放对城市氮沉降的贡献（Liu X. Y. et al., 2017; Song et al., 2019）。最近，他们发展了 $\Delta^{17}O$ 定量区分大气颗粒物硝酸根生成途径的新方法，揭示了人为污染对主要氧化过程贡献的影响（Wang Y. L. et al., 2019），并提高了大气硝酸根生成途径贡献及其同位素分馏效应的计算（Song et al., 2020）。同时，他们整合陆地环境大气主要无机硝态氮种（氮氧化物、气态硝酸、颗粒态硝酸根和降水硝酸根）的氮同位素观测，约束了自然大气环境氮氧化物转化和沉降过程的同位素效应，推动了大气硝态氮来源定量示踪方案的发展（Liu X. Y. et al., 2020）。

除了传统的大气无机氮来源和过程的研究外，肖化云、朱仁果等最近还分析了大气中氨基酸（结合态和游离态）的同位素地球化学特征，首次测定了 $PM_{2.5}$ 中结合态 Gly 的氮同位素，发现大气中结合态 Gly 转化为游离态 Gly 的过程中氮同位素效应很小，游离 Gly 和结合 Gly 的氮同位素是追踪大气中生物质燃烧源的有效工具（Zhu R. G. et al., 2019, 2020）。

4.1.2 生物氮同位素地球化学示踪大气氮沉降

近年来，越来越多的研究开始应用苔藓 $\delta^{15}N$ 指示人为成因的大气氮污染或氮沉降。朱仁果等首次使用单体氨基酸（FAA）的氮同位素组成来示踪大气氮沉降（Zhu R. G. et al., 2018）。在 FAA 代谢过程中，总的 FAA 处于同位素平衡状态，因此苔藓的 $\delta^{15}N_{TFAA}$ 值可以很好地指示大气氮源。他们还发现某些特定 FAAs 单体的 $\delta^{15}N$（如 Ala、Glu 和 Lys）和 $\delta^{15}N_{bulk}$ 一样保留了有关大气氮源的信息，这是因为这些特定 FAAs 单体的代谢途径中氮同位素分馏很小。今后的工作应包括调查不同氮沉降条件下的维管植物中游离氨基酸 $\delta^{15}N$ 值的变化性，以研究氮在不同植物组织运输过程中的动力学同位素效应。

胁迫处理同样造成了氨基酸浓度模式以及氮同位素特征的不同分布，尤其是对于某些特殊种类的氨基酸。张忠义、徐宇等研究发现，胁迫处理会使得大部分植物中氨基酸含量升高，总游离氨基酸含量升高（Xu et al., 2017）。由于特定氨基酸的生理功能的差异，这些氨基酸对逆境胁迫的响应也有明显的不同。如谷氨酸合成途径生成的 Pro 会使得其 $\delta^{15}N$ 值偏正，而精氨酸/鸟氨酸途径合成的 Pro 其 $\delta^{15}N$ 值偏负（Xu and Xiao, 2017）；由于苯丙氨酸在植物组织中苯丙酮酸途径的重要性或者其明显的抵御逆境胁迫的生理作用，逆境条件下其 $\delta^{15}N$ 往往偏正（Zhang H. et al., 2016）；Ser 对不同胁迫响应最为明显，干旱胁迫使得其含量降低而盐害胁迫使得其含量上调，结合其同位素值，Ser 上调可能是来自磷酸化途径（Zhang Y. X. et al., 2016）。说明氨基酸浓度和氮同位素值可以在一定程度上表征干旱、盐、氮供给等外部环境压力，有助于对喀斯特等生态系统的健康诊断和大气污染的研究。

最近十年，利用单体化合物（尤其是氨基酸）氮同位素分析技术追踪氮的来源、转化和归宿，已经成为生态学和生物地球化学研究中重要的手段。在氨基酸代谢过程中，以苯丙氨酸为代表的"源"氨基酸分馏极小（<0.5‰），而以谷氨酸为代表的"营养"氨基酸极度偏正（6‰~8‰）。Zhang 等（2019）解析了鄱阳湖水生生态系统中主要生物的营养级位置，结合不同食物来源的相对贡献，建立了从初级生产者到顶级捕食者的水生生物网营养结构图。表明单体氨基酸氮同位素作为一种较新的手段，能够精确和准确的确定生物体营养级位置（Zhang et al., 2019）。

饮食（食用鱼类、水产品和稻米等）是人体甲基汞暴露的主要途径。分子水平研究表明，生物体内甲基汞的代谢（摄取、储存和转运、排泄等），主要由生命基本单元——氨基酸介导。基于此，张忠义等开展了食物链水平的氨基酸代谢与汞生物富集的研究。结果初步表明，食物链水平的氨基酸特征（含量、同位素等）可以较好地表征甲基汞生物富集，证实了食物摄入是甲基汞富集的主要途径以及氨基酸代谢是主导甲基汞富集程度的主要因素（Zhang et al., 2021）。

4.1.3 水体氮同位素地球化学新进展

由于人类活动影响，水体硝酸盐（NO_3^-）污染已经成为世界范围内的环境问题。近十年来，中国主要在西南岩溶地区、东南地区、华北平原及西北干旱地区开展了 $\delta^{15}N$ 示踪水体 NO_3^- 的污染来源研究。研究表明，在城市生态系统中，地下水 NO_3^- 主要来源于生活污水以及工厂废水（刘君和陈宗宇，2009；邢萌等，2010；岳甫军等，2012）。通过 $\delta^{15}N$、$\delta^{18}O$ 同位素示踪发现，生活污水是城市生态系统主要的 NO_3^- 污染来源，同时污水中的反硝化过程也是一个重要来源（Liu et al., 2006；Yang and Toor, 2016）。在农田生态系统中，地下水 NO_3^- 可能主要来自生活污水及农田施用化肥与粪肥（贾小妨等，2010；王松等，2010；张翠云等，2010；张丽娟等，2010；Zhang X. Y. et al., 2013）。应用 $\delta^{15}N$ 对地表水 NO_3^- 来源的研究也取得了一定进展，城市生态系统地表水 NO_3^- 主要来自城市生活污水及工厂废水（邢萌等，2010；岳甫均等，2010；Ren et al., 2014）。农田生态系统地表水 NO_3^- 可能主要来自生活污水及化肥与粪肥（Li et al., 2010；邢萌等，2010；Chen et al., 2014）。而森林生态系统地表水、地下水 NO_3^- 来源主要是土壤有机氮及大气氮沉降（Jiang et al., 2009；Liu T. et al., 2013；Zhang X. Y. et al., 2013）。

当前地下水中 NH_4^+ 的运移和转化研究已成为国际上污染水文地质学的前沿课题。在国内开展了垃圾渗滤液、污灌区、污染河流等污染源中 NH_4^+ 对地下水污染的调查以及污染修复，并对其污染机制进行了模拟。吴庆乐（2015）在太湖地区水体氮的污染源和反硝化研究中，测定了河水和湖水的 $\delta^{15}\text{N-NH}_4^+$ 同位素组成，得到了河水、湖水和井水存在反硝化的证据；肖化云和刘丛强（2004）在应用氮同位素示踪贵州红枫湖河流季节性氮污染研究中，测定了废水和污染河水的 $\delta^{15}\text{N-NH}_4^+$，发现化肥厂废水的 $\delta^{15}\text{N-NH}_4^+$ 值与化肥差别很大，可以对不同类型河流氮污染源进行可靠识别；李思亮和刘丛强（2006）通过测定污水、河水和地下水的 $^{15}\text{N-NH}_4^+$ 和其余氮素的同位素组成，研究了贵阳地下水氮污染来源和转化过程。

海洋氮稳定同位素信号包含了关键的生物地球化学信息，是辨识海洋氮来源、了解海洋氮循环过程的重要途径和手段。近十几年的研究表明，固氮作用和反硝化作用在海洋氮循环中发挥着关键作用（洪义国，2013；Luo et al., 2013；Yang et al., 2017），但海洋环境的时空多变、海洋氮循环过程和物质来源复杂，未来需要结合现代观测、地质记录，综合海洋、生物、地质多个学科，考虑水文环境、地质过程和气候演变等多种因素，才能深刻理解海洋氮循环和气候、环境变化的耦合关系（Zhang R. et al., 2015；周涛等，2019）。

4.2 双碳同位素示踪体系及应用

双碳同位素（$\delta^{13}\text{C-}\Delta^{14}\text{C}$）在示踪环境中碳的来源和迁移转化过程方面具有独特优势（Chen et al., 2018）。近十年来，中国学者利用 $\delta^{13}\text{C-}\Delta^{14}\text{C}$ 在河流、湖库、海洋等水环境碳循环，以及气溶胶、干湿沉降等大气含碳物质的来源示踪等方面开展了大量研究。

河流是陆地与海洋生态系统连接的重要通道，人类活动对河流碳的来源、传输以及埋藏过程的影响越来越大。黄河每年向海洋输送大量的有机碳，大部分为千年尺度的老碳，其中颗粒有机碳（particulate organic carbon，POC）年龄远远高于溶解有机碳（dissolved organic carbon，DOC），绝大部分陆源有机碳在河口及边缘海沉积埋藏（Xue et al, 2017；Yu et al., 2019）。黄河 POC 主要来源于黄土高原的 C_3 植被、黄土-古土壤和古老岩石。生物 POC 和化石 POC 通量具有明显的季节差异，极端气候事件（暴雨）对 POC 的输出和碳循环有重要影响（Qu et al., 2020）。Wang 等（2016b）发现硅酸岩风化作用是控制黄河和长江 DIC 浓度和通量的主要过程。Liu Z. H. 等（2017）发现碳酸岩为主的珠江流域的 DIC 和 POC 的 $\Delta^{14}\text{C}$ 均偏负，有机碳主要来自水生植物光合利用碳酸岩风化产生的老碳形成的内源有机碳（AOC，表观偏"老"，实际是新近形成），并非传统认为的来自深层土壤或岩石侵蚀产生的偏老 POC（Lin B. Z. et al., 2019）。龙门山地区岷江上游的生物有机碳较老，老碳可能来自高原侵蚀，而下游生物有机碳年轻化则说明部分老碳在河流搬运过程中发生了氧化（Wang J. et al., 2019）。青藏高原河流 DOC 的 ^{14}C 年龄与世界其他河流相比明显偏老（Qu et al., 2017）。

湖泊碳循环是全球碳循环的关键组成部分，对调节营养循环和湖泊生态系统中起着重要作用。Liu J. W. 等（2018）对比中国 11 个湖泊发现湖水 DIC 的 ^{14}C 随水力停留时间增加而增加，开放湖泊通常比封闭湖泊 DIC 的 ^{14}C 值更低。Chen 等（2018）建立了湖泊不同来源有机碳的 $\Delta^{14}\text{C-}\delta^{13}\text{C}$ 指纹谱，发现抚仙湖水体 POC 的主要来源为藻类（61%），沉积物再悬浮贡献 17%，内源贡献不容忽视。陆源 POC 沉积速率比藻源 POC 快，而藻源 POC 在表层水体滞留更长时间，可长期对湖泊生态系统产生影响。青藏高原湖泊入湖河流和湖水 DOC 与 POC 的 $\Delta^{14}\text{C}$ 偏负（−569‰ ~ −49‰），浮游动物主要食物来源是 $\Delta^{14}\text{C}$ 偏负的浮游植物和沉水植物，外源有机碳和 DOC 贡献很小（Su et al., 2018）。纳木错降水 DOC 的 $\Delta^{14}\text{C}$ 年龄较南亚加德满都明显偏年轻，其中偏老的化石源 DOC 主要来自南亚的远距离传输（Li et al., 2018）。

海洋 DOC 和 DIC 是地球上最大的碳储集层之一，其生物地球化学循环对全球碳循环过程有重要作用。碳库的来源、循环时间以及吕宋海峡和南海盆地海水的快速混合交换是影响南海深部碳循环的主要因素，而黑潮和远洋水团控制着东海碳含量时空变化（Ding et al., 2019，2020）。新不列颠海沟沉积物主要源于年轻和生物成因的有机碳（organic carbon，OC）（62%），而非岩石成因的 OC（Xiao et al., 2020）。底部生物生产并非 DOC 主要来源，海底热液循环可去除 50% ~ 75% 的 DOC，细菌降解、光化学降解和对颗粒的吸附已被证明是潜在的去除机制（Lin H. T. et al., 2019；Fang et al., 2020）。微生物群落利用 DOC 作为能量来源,并促进含"极老"DOC甲烷水合物的产生(Fu et al., 2020)。Wang N. 等(2019)在海底两足动物肌肉组织中发现了核爆^{14}C信号,同时发现最深海沟的碳循环通过食物链与表层海洋紧密联系,这说明人类活动已经对全球海洋生态系统带来了极其深刻的影响。

工业革命后,全球大气CO_2浓度持续上升,全球变暖趋势不断加快,其主要原因是人为化石燃料的大量使用。目前化石燃料来源CO_2排放量的统计方法存在较大不确定性,很难获取可靠的化石源和非化石源CO_2浓度信息(Yang et al., 2005; Zhou et al., 2020)。近十年来,中国学者针对大气CO_2来源,做了大量的研究工作,认为$^{14}CO_2$是目前最适合研究化石源CO_2排放的示踪剂(Zhou et al., 2014, 2020)。牛振川等(2016)发现冬季北京市大气$\Delta^{14}CO_2$日变化十分显著($-214.2‰ \sim -82.3‰$),而$^{13}CO_2$的变化则相对小很多。受人类活动(供暖、交通等)和气象条件的影响,化石源CO_2浓度的日变化也非常显著,夜晚浓度明显高于白天,化石燃烧的贡献率在人口密集的城区可达$30\% \sim 50\%$,甚至更高水平(Wang et al., 2016a; Li et al., 2018)。

大气细颗粒物($PM_{2.5}$)与气候变化和人类健康密切相关,环境空气中细颗粒物浓度的日益增加是灰霾天气出现的主要原因,而碳质气溶胶是大气气溶胶的主要组成部分,主要由有机碳(OC)和元素碳(EC)组成。目前双碳同位素技术已应用于含碳气溶胶的源解析,有效区分化石源与生物源(Cao et al., 2017)。近十年来,国内相关学者通过有机物^{14}C特征和分子标志物等分析技术,对中国含碳气溶胶来源和时空变化进行了系统研究。Liu J. W. 等(2018)发现广州冬季2/3的总碳源于非化石来源,化石来源占1/3,雾霾天化石源对EC贡献可高达$80\% \sim 90\%$,主要来自汽车尾气排放。黄汝锦等(2019)指出近十年西安冬季雾霾中EC排放源从煤炭转变为液体化石燃料。南方背景点(海南尖峰岭)含碳气溶胶中EC和OC主要来源于生物质燃烧,这很可能与密集木材燃烧排放的有机和(或)无机物质对EC的覆盖有关(Zhang et al., 2012)。Liu D. 等(2013)发现中国东部背景点(宁波)$PM_{2.5}$中生物源对OC和EC的贡献率分别为59%和22%,夏秋季以非化石源为主,而冬春季以化石燃料排放为主。部分学者结合中国河流、边缘海沉积物数据和^{14}C同位素手段,揭示出大气沉降是向海洋输送化石源碳的重要途径,向中国边缘海输送了31%化石源碳(Yu et al., 2018)。

虽然中国学者利用双碳同位素($\delta^{13}C-\Delta^{14}C$)在示踪河流、湖泊、海洋、大气等环境中碳的来源和迁移转化过程方面已经取得了大量的研究成果,但目前的研究缺乏流域尺度长时间序列的系统性研究。建议基于双碳同位素开展如下研究:不同碳库之间的相互转化机制;生物地球化学过程对碳同位素组成的影响;外源输入碳在水生生态系统中迁移转化过程。在分析手段方面应结合其他如核磁共振(nuclear magnetic resonance, NMR)、FTICRMS(傅里叶变换离子回旋共振质谱)等先进的分析技术从化学组成和分子水平等方面精细刻画表生环境中碳的来源与循环过程。

4.3 多硫同位素相关研究

通常,多硫同位素之间存在质量分馏效应,即$\delta^{33}S \approx 0.515\delta^{34}S$和$\delta^{36}S \approx 1.90\delta^{34}S$。而一些同位素效应与电子自旋、反应中同位素对称性、核体积效应以及状态间匹配水平等因素相关,导致$\delta^{33}S$和$\delta^{36}S$偏离了质量分馏效应,称之为同位素非质量分馏,分别以$\Delta^{33}S$和$\Delta^{36}S$表示。多硫同位素对常规硫同位素地球化学的认识有重要的拓展作用,能更全面反映大气化学过程的信息。由于不同污染源的硫同位素组成可能会出现重叠效应,硫同位素指示的来源模棱两可,因此利用单一稳定硫同位素示踪大气硫酸盐来源存在局限性(Guo et al., 2010)。

在大气SO_2氧化形成硫酸根的过程中,高温环境(>500℃)可能会出现非零的$\Delta^{33}S$和$\Delta^{36}S$值(郭照冰等,2014)。此外,在平衡分馏、动力学分馏以及两者叠加的过程中,通过瑞利分馏和质量守恒效应,也会出现非零的$\Delta^{33}S$和$\Delta^{36}S$值(Guo et al., 2010)。在这个过程中,每一个SO_2氧化途径都有特定的$^{33}S, ^{34}S$和^{36}S分馏(Yang et al., 2018a)。现有的对城市地区气溶胶$\Delta^{33}S$的研究显示,$\Delta^{33}S$值的变化范围为$-0.6‰ \sim 0.5‰$(Guo et al., 2010; Han et al., 2017; Lin et al., 2018; Yang et al., 2018a, 2018b)。但

Harris 等（2013a，2013b）的研究显示，目前已知的主要 SO_2 的氧化途径（OH 自由基、H_2O_2 以及 O_2 + TMI）所产生的非质量分馏，$\Delta^{33}S$ 值的变化范围为 $-0.2‰ \sim 0.2‰$，远远小于观测到的 $\Delta^{33}S$ 值的变化范围（$-0.6‰ \sim 0.5‰$）。近十年的研究结果证明还有潜在的 SO_2 氧化途径并没有被发现，并且这些氧化途径存在明显的硫同位素非质量分馏。郭照冰等（2014）发现北京大气气溶胶存在显著的硫同位素非质量分馏效应，通过对 $\Delta^{33}S$ 与 CAPE 相关性分析，发现其形成机制不仅与平流层 SO_2 的光化学反应有关，还可能与热化学反应机制关联；Yang 等（2018a）发现，气溶胶 $\Delta^{33}S$ 在夏季和冬季较高，而春季和秋季较低，这可能是受 Criegee 臭氧化反应或 SO_2 的光化学反应的影响。

5 结　语

在过去十年，中国在环境地球化学研究方面取得长足进展，研究水平已由过去的以跟踪国际同行为主转变为部分研究方向与国际同行并行，甚至有些方向处于国际领跑地位。中国学者在污染物长距离迁移规律的研究结果，为中国参与国际环境公约谈判及履约成效评估提供了重要科技支撑。中国学者积极参与非传统稳定同位素环境地球化学这一新领域的研究，在地表重金属生物地球化学循环过程同位素分馏规律及环境示踪应用方面取得长足进展，也提高了对重金属环境地球行为认识的水平。中国学者在稳定同位素与全球变化方面，保持与国际同行同步开展团簇同位素、微小 ^{17}O 在全球变化中的应用研究，完善了利用泥碳纤维素稳定同位素示踪过去气候变化信息的研究方法。在传统稳定同位素示踪研究方面，也与国际同行同步开展了双碳同位素示踪污染来源和环境地球化学过程的研究，发展了氮同位素环境示踪技术，在大气雾霾污染来源示踪等方面发挥重要作用。应该说，环境地球化学学科无论在古气候与全球变化研究，还是污染物的现代地球化学过程研究中发挥重要作用，相信在未来环境地球化学学科在中国生态文明和美丽中国建设、国际环境公约履约及过去全球变化研究中发挥更大作用。

参 考 文 献

郭照冰，吴梦龙，刘凤玲，魏英. 2014. 北京大气气溶胶中硫氧稳定同位素组成研究. 中国科学：地球科学，44(7)：1556-1560

洪义国. 2013. 硝酸盐氮氧稳定同位素分馏过程记录的海洋氮循环研究进展. 地球科学进展，28(7)：751-764

黄汝锦，郭洁，倪海燕，曹军骥. 2019. 西安冬季元素碳气溶胶的碳同位素组成及来源变化. 矿物岩石地球化学通报，38(6)：1073-1080，1046

贾小妨，李玉中，徐春英，李巧珍，倪秀菊. 2010. 水样硝酸盐离子交换色层法氮、氧同位素分析预处理条件试验. 中国农学通报，26(16)：99-102

姜珊. 2012. 过去3000年南北极典型地区生态环境变化的沉积记录及对比. 合肥：中国科学技术大学博士学位论文

江星星. 2018. 南极中山站至 Dome A 表层雪痕量元素空间变化特征. 南京：南京大学硕士学位论文

康世昌，丛志远，王小萍，张强弓，吉振明，张玉兰，徐柏青. 2019. 大气污染物跨境传输及其对青藏高原环境影响. 科学通报，(27)：2876-2884

李芳柏，李勇珠. 2019. 稻田体系中铁的生物地球化学过程及铁同位素分馏机制研究进展. 生态环境学报，28(6)：1251-1260

李思亮，刘丛强. 2006. 贵阳地下水硝酸盐氮氧同位素特征及应用. 中国岩溶，25(2)：108-111

林官明，蔡旭晖，胡敏，李惠君. 2018. 大气气溶胶干沉降研究进展. 中国环境科学，38(9)：3211-3220

刘丛强，洪业汤，冯新斌，肖化云，陈玖斌，李社红，王仕禄，洪冰，傅平青，李思亮，徐海，骆永明，汪福顺，李向东，李高军，吴卫华，腾彦国. 2018. 环境地球化学. 见：欧阳自远，胡瑞忠，徐义刚. 中国地球化学学科发展史（上册）. 北京：科学出版社，331-392

刘君，陈宗宇. 2009. 利用稳定同位素追踪石家庄市地下水中的硝酸盐来源. 环境科学，30(6)：1602-1607

刘伟明，马明，王定勇，孙涛，魏世强. 2016. 中亚热带背景区重庆四面山大气气态总汞含量变化特征. 环境科学，37(5)：1639-1645

刘咸德，郑晓燕，王宝盛，江桂斌. 2015. 持久性有机污染物被动采样与区域大气传输，第2版. 北京：科学出版社

路凯. 2016. 元素硒的无机氧化动力学及其同位素分馏. 北京：中国科学院大学硕士学位论文

孟郁苗，胡瑞忠，高剑峰，毕献武，黄小文. 2016. 锑的地球化学行为以及锑同位素研究进展. 岩矿测试，35(4)：339-348

牛振川，周卫健，程鹏，吴书刚，卢雪峰，杜花，付云翀，熊晓虎. 2016. 北京市冬季大气化石源 CO_2 典型日变化的 ^{14}C 示踪研究. 地球环境学报，7(5)：487-493

秦燕，徐衍明，侯可军，李延河，陈蕾. 2020. 铁同位素分析测试技术研究进展. 岩矿测试，39(2)：151-161

孙剑，朱祥坤. 2015. 表生过程中铁的同位素地球化学. 地质论评，61(6)：1370-1382

唐茂，赵辉，刘耘. 2007. 天然气中甲烷和 CO_2 的二元同位素特征. 矿物学报，27(3-4)：396-399

唐索寒，朱祥坤，李津，闫斌，李世珍，李志红，王跃，孙剑. 2016. 用于多接收器等离子体质谱测定的铁铜锌同位素标准溶液研制. 岩矿测

试，35(2)：127-133

王超敏. 2017. 青藏高原及周边区域冰心放射性同位素定年及近两百年来环境记录研究. 南京：南京大学博士学位论文

王梦梦, 原梦云, 苏德纯. 2017. 中国大气重金属干湿沉降特征及时空变化规律. 中国环境科学, 37(11)：4085-4096

王世霞, 朱祥坤. 2013. 广东湛江湖光岩地区玄武岩风化壳 Fe 同位素研究. 地质学报, 87(9)：1461-1468

王松, 裴建国, 梁建宏. 2010. 利用氮氧同位素研究桂林寨底地下河硝酸盐来源. 地质灾害与环境保护, 21(4)：54-56, 81

王倩, 侯清华, 张婷, 刘金华, 陶云, 周炼. 2016. 铜同位素测定方法研究进展. 矿物岩石地球化学通报, 35(3)，497-506

王相力, 卫炜. 2020. 铬稳定同位素地球化学. 地学前缘, 27(3)：78-103

王训, 袁巍, 冯新斌. 2017. 森林生态系统汞的生物地球化学过程. 化学进展, 29(9)，970-980

王泽洲, 刘盛遨, 李丹丹, 吕逸文, 吴松, 赵云. 2015. 铜同位素地球化学及研究新进展. 地学前缘, 22(5)，72-83

王中伟. 2019. 珠江中的锌及其同位素地球化学研究. 北京：中国科学院大学博士学位论文

吴庆乐. 2015. 太湖西部南河水系入湖区域氮污染物来源及转化途径研究. 南京：南京大学硕士学位论文

吴寨. 2018. 白令海-西北冰洋和中国南海沉积物中金属的来源解析与沉降历史：基于同位素的应用. 厦门：厦门大学硕士学位论文

肖化云, 刘丛强. 2004. 氮同位素示踪贵州红枫湖河流季节性氮污染. 地球与环境, 32(1)：71-75

邢萌, 刘卫国, 胡婧. 2010. 沪河、涝河河水硝酸盐氮污染来源的氮同位素示踪. 环境科学, 31(10)：2305-2310

杨仲康. 2019. 南北极与中国近海典型地区全新世气候变化与环境事件研究. 合肥：中国科学技术大学博士学位论文

岳甫均, 李军, 刘小龙, 朱兆洲. 2010. 利用氮同位素技术探讨天津地表水氮污染. 生态学杂志, 29(7)：1403-1408

岳甫均, 李思亮, 刘丛强, 安宁, 蔡虹明. 2012. 利用反硝化细菌法测试水体硝酸盐氮同位素. 生态学杂志, 31(8)：2152-2157

张翠云, 张俊霞, 马琳娜, 张胜, 殷密英, 李政红. 2010. 硝酸盐氮氧同位素反硝化细菌法测试研究. 地球科学进展, 25(4)：360-364

张干, 李军, 田崇国. 2019. 持久性有机污染物的地球化学. 北京：科学出版社

张丽娟, 巨晓棠, 刘辰琛, 寇长林. 2010. 北方设施蔬菜种植区地下水硝酸盐来源分析：以山东省惠民县为例. 中国农业科学, 43(21)：4427-4436

张晓文. 2019. 湖南某工业区土壤及水稻重金属污染源解析. 硕士学位论文. 北京：农产品加工研究所

赵博, 朱建明, 秦海波, 谭德灿, 徐文坡. 2018. 锑同位素测试方法及其应用研究. 矿物岩石地球化学通报, 37(6)：1183-1189

周涛, 蒋壮, 耿雷. 2019. 大气氧化态活性氮循环与稳定同位素过程：问题与展望. 地球科学进展, 34(9)：922-935

朱建明, 谭德灿, 王静, 曾理. 2015. 硒同位素地球化学研究进展与应用. 地学前缘, 22(5)：102-114

Abouchami W, Galer S J G, Horner T, Rehkamper M, Wombacher F, Xue Z, Lamelet M, Gault-Ringold M, Stirling C H, Schonbachler M, Shiel A, Weis D, Holdship P F. 2013. A common reference material for cadmium isotope studies-Nist SRM 3108. Geostandards and Geoanalytical Research, 37, 5-17

Agnan Y, Le Dantec T, Moore C W, Edwards G C, Obrist D. 2016. New constraints on terrestrial surface-atmosphere fluxes of gaseous elemental mercury using a global database. Environmental Science & Technology, 50(2)：507-524

Al Mamun A, Cheng I, Zhang L M, Dabek-Zlotorzynska E, Charland J P. 2020. Overview of size distribution, concentration, and dry deposition of airborne particulate elements measured worldwide. Environmental Reviews, 28(1)：77-88

Alvarez C C, Quitté G, Schott J, Oelkers E H. 2020. Experimental determination of Ni isotope fractionation during Ni adsorption from an aqueous fluid onto calcite surfaces. Geochimica et Cosmochimica Acta, 273：26-36

AMAP/UN. 2019. Arctic monitoring and assessment programme. Oslo, Norway/UN Environment Programme, Chemicals and Health Branch, Geneva, Switzerland, viii + 426 pp including E-Annexes

Amos H M, Jacob D J, Kocman D, Horowitz H M, Zhang Y X, Dutkiewicz S, Horvat M, Corbitt E S, Krabbenhoft D P, Sunderland E M. 2014. Global biogeochemical implications of mercury discharges from rivers and sediment burial. Environmental Science & Technology, 48(16)：9514-9522

Angert A, Cappa CD, DePaolo D J. 2004. Kinetic ^{17}O effects in the hydrologic cycle：indirect evidence and implications. Geochimica et Cosmochimica Acta, 68(17)：3487-3495

Asaoka S, Takahashi Y, Araki Y, Tanimizu M. 2011. Preconcentration method of antimony using modified thiol cotton fiber for isotopic analyses of antimony in natural samples. Analytical Sciences, 27(1)：25-28

Bao H M, Lyons J R, Zhou C M. 2008. Triple oxygen isotope evidence for elevated CO_2 levels after a Neoproterozoic glaciation. Nature, 453(7194)：504-506

Bao H M, Cao X B, Hayles J A. 2016. Triple oxygen isotopes：fundamental relationships and applications. Annual Review of Earth and Planetary Sciences, 44：463-492

Barkan E, Luz B. 2005. High precision measurements of $^{17}O/^{16}O$ and $^{18}O/^{16}O$ ratios in H_2O. Rapid Communications in Mass Spectrometry, 19(24)：3737-3742

Barkan E, Luz B. 2007. Diffusivity fractionations of $H_2^{16}O/H_2^{17}O$ and $H_2^{16}O/H_2^{18}O$ in air and their implications for isotope hydrology. Rapid Communications in Mass Spectrometry, 21(18)：2999-3005

Bates S L, Hendry K R, Pryer H V, Kinsley C W, Pyle K M, Woodward E M S, Horner T J. 2017. Barium isotopes reveal role of ocean circulation on barium cycling in the Atlantic. Geochimica et Cosmochimica Acta, 204：286-299

Berglund M, Wieser M E. 2011. Isotopic compositions of the elements 2009 (IUPAC technical report). Pure and Applied Chemistry, 83(2)：397-410

Bernasconi S M, Müller I A, Bergmann K D, Breitenbach S F M, Fernandez A, Hodell D A, Jaggi M, Meckler A N, Millan I, Ziegler M. 2018. Reducing uncertainties in carbonate clumped isotope analysis through consistent carbonate-based standardization. Geochemistry, Geophysics,

Geosystems, 19(9): 2895-2914

Blum J D, Sherman L S, Johnson M W. 2014. Mercury isotopes in earth and environmental sciences. Annual Review of Earth and Planetary Sciences, 42: 249-269

Blusztajn J, Nielsen SG, Marschall H R, Shu Y C, Ostrander C M, Hanyu T. 2018. Thallium isotope systematics in volcanic rocks from St. Helena-Constraints on the origin of the HIMU reservoir. Chemical Geology, 476: 292-301

Bridgestock L, Rehkämper M, van de Flierdt T, Murphy K, Khondoker R, Baker A R, Chance R, Strekopytov S, Humphreys-Williams E, Achterberg E P. 2017. The Cd isotope composition of atmospheric aerosols from the Tropical Atlantic Ocean. Geophysical Research Letters, 44(6): 2932-2940

Bridgestock L, Hsieh Y T, Porcelli D, Homoky W B, Bryan A, Henderson G M. 2018. Controls on the barium isotope compositions of marine sediments. Earth and Planetary Science Letters, 481: 101-110

Cao F, Zhang Y L, Ren L J, Liu J W, Li J, Zhang G, Liu D, Sun Y L, Wang Z F, Shi Z B, Fu P Q. 2017. New insights into the sources and formation of carbonaceous aerosols in China: potential applications of dual-carbon isotopes. National Science Review, 4(6): 804-806

Cao X B, Bao H M. 2013. Dynamic model constraints on oxygen-17 depletion in atmospheric O_2 after a snowball Earth. Proceedings of the National Academy of Sciences of the United States of America, 110(36): 14546-14550

Cao X B, Liu Y. 2011. Equilibrium mass-dependent fractionation relationships for triple oxygen isotopes. Geochimica et Cosmochimica Acta, 75(23): 7435-7445

Cao X B, Liu Y. 2012. Theoretical estimation of the equilibrium distribution of clumped isotopes in nature. Geochimica et Cosmochimica Acta, 77: 292-303

Cao Z M, Siebert C, Hathorne E C, Dai M H, Frank M. 2016. Constraining the oceanic barium cycle with stable barium isotopes. Earth and Planetary Science Letters, 434: 1-9

Chand D, Jaffe D, Prestbo E, Swartzendruber P C, Hafner W, Weiss-Penzias P, Kato S, Takami A, Hatakeyama S, Kajii Y Z. 2008. Reactive and particulate mercury in the Asian marine boundary layer. Atmospheric Environment, 42(34): 7988-7996

Chang B, Defliese W F, Li C, Huang J H, Tripati A, Algeo T J. 2020. Effects of different constants and standards on the reproducibility of carbonate clumped isotope (Δ_{47}) measurements: insights from a long-term dataset. Rapid Communications in Mass Spectrometry, 34(8): e8678

Chen D Z, Liu W J, Liu X D, Westgate J N, Wania F. 2008. Cold-trapping of persistent organic pollutants in the mountain soils of western Sichuan, China. Environmental Science & Technology, 42(24): 9086-9091

Chen H S, Wang Z F, Li J, Tang X, Ge B Z, Wu X L, Wild O, Carmichael G R. 2015. GNAQPMS-Hg v1.0, a global nested atmospheric mercury transport model: model description, evaluation and application to trans-boundary transport of Chinese anthropogenic emissions. Geoscientific Model Development, 8(9): 2857-2876

Chen J A, Yang H Q, Zeng Y, Guo J Y, Song Y L, Ding W. 2018. Combined use of radiocarbon and stable carbon isotope to constrain the sources and cycling of particulate organic carbon in a large freshwater lake, China. Science of the Total Environment, 625: 27-38

Chen L G, Liu M, Xu Z C, Fan R F, Tao J, Chen D H, Zhang D Q, Xie D H, Sun J R. 2013. Variation trends and influencing factors of total gaseous mercury in the Pearl River Delta—a highly industrialised region in South China influenced by seasonal monsoons. Atmospheric Environment, 77: 757-766

Chen S, Liu Y C, Hu J Y, Zhang Z F, Hou Z H, Huang F, Yu H M. 2016. Zinc isotopic compositions of NIST SRM 683 and whole rock reference materials. Geostandards and Geoanalytical Research, 40(3): 417-432

Chen T Y, Li W Q, Guo B, Liu R L, Li G J, Zhao L, Ji J F. 2020. Reactive iron isotope signatures of the East Asian dust particles: implications for iron cycling in the deep North Pacific. Chemical Geology, 531: 119342

Chen Z X, Yu L, Liu W G, Lam M H W, Liu G J, Yin X B. 2014. Nitrogen and oxygen isotopic compositions of water-soluble nitrate in Taihu Lake water system, China: implication for nitrate sources and biogeochemical process. Environmental Earth Sciences, 71(1): 217-223

Ci Z J, Zhang X S, Wang Z W, Niu Z C. 2011. Atmospheric gaseous elemental mercury (GEM) over a coastal/rural site downwind of East China: temporal variation and long-range transport. Atmospheric Environment, 45(15): 2480-2487

Clayton R N, Grossman L, Mayeda T K. 1973. A component of primitive nuclear composition in carbonaceous meteorites. Science, 182(4111): 485-488

Cloquet C, Rouxel O, Carignan J, Libourel G. 2005. Natural cadmium isotopic variations in eight geological reference materials (NIST SRM 2711, BCR 176, GSS-1, GXR-1, GXR-2, GSD-12, Nod-P-1, Nod-A-1) and anthropogenic samples, measured by MC-ICP-MS. Geostandards and Geoanalytical Research, 29(1): 95-106

Daëron M, Blamart D, Peral M, Affek H P. 2016. Absolute isotopic abundance ratios and the accuracy of Δ_{47} measurements. Chemical Geology, 442: 83-96

Daëron M, Drysdale R N, Peral M, Huyghe D, Blamart D, Coplen T B, Lartaud F, Zanchetta G. 2019. Most Earth-surface calcites precipitate out of isotopic equilibrium. Nature Communications, 10: 429

Degryse P, Lobo L, Shortland A, Vanhaecke F, Blomme A, Painter J, Gimeno D, Eremin K, Greene J, Kirk S, Walton M. 2015. Isotopic investigation into the raw materials of Late Bronze Age glass making. Journal of Archaeological Science, 62: 153-160

Deng T H B, Cloquet C, Tang Y T, Sterckeman T, Echevarria G, Estrade N, Morel J L, Qiu R L. 2014. Nickel and zinc isotope fractionation in hyperaccumulating and nonaccumulating plants. Environmental Science & Technology, 48(20): 11926-11933

Dennis K J, Affek H P, Passey B H, Schrag D P, Eiler J M. 2011. Defining an absolute reference frame for "clumped" isotope studies of CO_2. Geochimica et Cosmochimica Acta, 75(22): 7117-7131

Ding L, Ge T, Wang X. 2019. Dissolved organic carbon dynamics in the East China Sea and the Northwest Pacific Ocean. Ocean Science, 15(5): 1177-1190

Ding L, Qi Y Z, Shan S, Ge T T, Luo C L, Wang X C. 2020. Radiocarbon in dissolved organic and inorganic carbon of the South China Sea. Journal of Geophysical Research: Oceans, 125(4): e2020JC016073

Dong Z W, Kang S C, Qin D H, Li Y, Wang X J, Ren J W, Li X F, Yang J, Qin X. 2016. Provenance of cryoconite deposited on the glaciers of the Tibetan Plateau: new insights from Nd-Sr isotopic composition and size distribution. Journal of Geophysical Research: Atmospheres, 121(12): 7371-7382

Durnford D, Dastoor A, Figueras-Nieto D, Ryjkov A. 2010. Long range transport of mercury to the Arctic and across Canada. Atmospheric Chemistry and Physics, 10(13): 6063-6086

Eichler A, Tobler L, Eyrikh S, Malygina N, Papina T, Schwikowski M. 2014. Ice-core based assessment of historical anthropogenic heavy metal (Cd, Cu, Sb, Zn) emissions in the soviet union. Environmental Science & Technology, 48(5): 2635-2642

Eiler J M. 2007. "Clumped-isotope" geochemistry—the study of naturally-occurring, multiply-substituted isotopologues. Earth and Planetary Science Letters, 262(3-4): 309-327

Eiler J M, Schauble E. 2004. $^{18}O^{13}C^{16}O$ in Earth's atmosphere. Geochimica et Cosmochimica Acta, 68(23): 4767-4777

Fan H F, Wen H J, Xiao C Y, Zhou T, Cloquet C, Zhu X K. 2018. Zinc geochemical cycling in a phosphorus-rich ocean during the early ediacaran. Journal of Geophysical Research: Oceans, 123(8): 5248-5260

Fang L, Lee S, Lee S A, Hahm D, Kim G, Druffel E R, Hwang J. 2020. Removal of refractory dissolved organic carbon in the Amundsen Sea, Antarctica. Scientific Reports, 10: 1213

Feng J L, Pei L L, Zhu X K, Ju J T, Gao S P. 2018. Absolute accumulation and isotope fractionation of Si and Fe during dolomite weathering and terra rossa formation. Chemical Geology, 496: 43-56

Feng X B, Foucher D, Hintelmann H, Yan H Y, He T R, Qiu G L. 2010. Tracing mercury contamination sources in sediments using mercury isotope compositions. Environmental Science & Technology, 44(9): 3363-3368

Feng X B, Yin R S, Yu B, Du B Y. 2013. Mercury isotope variations in surface soils in different contaminated areas in Guizhou Province, China. Chinese Science Bulletin, 58(2): 249-255

Fouskas F, Ma L, Engle M A, Ruppert L, Geboy N J, Costa M A. 2018. Cadmium isotope fractionation during coal combustion: insights from two U.S. coal-fired power plants. Applied Geochemistry, 96: 100-112

Fu W J, Qi Y Z, Liu Y G, Wang X C, Druffel E M, Xu X M, Ren P, Sun S W, Fan D. 2020. Production of ancient dissolved organic carbon in Arctic Ocean sediment: a pathway of carbon cycling in the extreme environment. Geophysical Research Letters, 47(5): e2020GL087119

Fu X W, Feng X, Dong Z Q, Yin R S, Wang J X, Yang Z R, Zhang H. 2010. Atmospheric gaseous elemental mercury (GEM) concentrations and mercury depositions at a high-altitude mountain peak in South China. Atmospheric Chemistry and Physics, 10(5): 2425-2437

Fu X W, Feng X, Liang P, Deliger, Zhang H, Ji J, Liu P. 2012a. Temporal trend and sources of speciated atmospheric mercury at Waliguan GAW station, northwestern China. Atmospheric Chemistry and Physics, 12: 1951-1964

Fu X W, Feng X, Shang L H, Wang S F, Zhang H. 2012b. Two years of measurements of atmospheric total gaseous mercury (TGM) at a remote site in Mt. Changbai area, northeastern China. Atmospheric Chemistry and Physics, 12(4): 4215-4226

Fu X W, Heimbürger L E, Sonke J E. 2014. Collection of atmospheric gaseous mercury for stable isotope analysis using iodine- and chlorine-impregnated activated carbon traps. Journal of Analytical Atomic Spectrometry, 29(5): 841-852

Fu X W, Zhang H, Lin C J, Feng X B, Zhou L X, Fang S X. 2015a. Correlation slopes of GEM/CO, GEM/CO_2, and GEM/CH_4 and estimated mercury emissions in China, South Asia, the Indochinese Peninsula, and Central Asia derived from observations in northwestern and southwestern China. Atmospheric Chemistry and Physics, 15(2): 1013-1028

Fu X W, Zhang H, Yu B, Wang X, Lin C J, Feng X B. 2015b. Observations of atmospheric mercury in China: a critical review. Atmospheric Chemistry and Physics, 15(16): 9455-9476

Fu X W, Marusczak N, Wang X, Gheusi F, Sonke J E. 2016a. Isotopic composition of gaseous elemental mercury in the free troposphere of the Pic du Midi observatory, France. Environmental Science & Technology, 50(11): 5641-5650

Fu X W, Zhu W, Zhang H, Sommar J, Yu B, Yang X, Wang X, Lin C J, Feng X B. 2016b. Depletion of atmospheric gaseous elemental mercury by plant uptake at Mt. Changbai, Northeast China. Atmospheric Chemistry and Physics, 16(20): 12861-12873

Fu X W, Yang X, Tang Q Y, Ming L L, Lin T, Lin C J, Li X D, Feng X B. 2018. Isotopic composition of gaseous elemental mercury in the Marine boundary layer of East China Sea. Journal of Geophysical Research: Atmospheres, 123(14): 7656-7669

Fu X W, Zhang H, Feng X B, Tan Q Y, Ming L L, Liu C, Zhang L M. 2019a. Domestic and transboundary sources of atmospheric particulate bound mercury in remote areas of China: evidence from mercury isotopes. Environmental Science & Technology, 53(4): 1947-1957

Fu X W, Zhang H, Liu C, Zhang H, Lin Cj, Feng X B. 2019b. Significant seasonal variations in isotopic composition of atmospheric total gaseous mercury at forest sites in China caused by vegetation and mercury sources. Environmental Science & Technology, 53(23): 13748-13756

Fujii T, Moynier F, Dauphas N, Abe M. 2011. Theoretical and experimental investigation of nickel isotopic fractionation in species relevant to modern and ancient oceans. Geochimica et Cosmochimica Acta, 75(2): 469-482

Gabrieli J, Barbante C. 2014. the Alps in the age of the Anthropocene: the impact of human activities on the cryosphere recorded in the Colle Gnifetti glacier. Rendiconti Lincei, 25(1): 71-83

Gabrielli P, Wegner A, Sierra-Hernández M R, Beaudon E, Davis M, Barker J, Thompson L. 2020. Early atmospheric contamination on the top of the Himalayas since the onset of the European industrial revolution. Proceedings of the National Academy of Sciences of the United States of America, 117(8): 3967−3973

Gao B, Zhou H D, Liang X R, Tu X L. 2013. Cd isotopes as a potential source tracer of metal pollution in river sediments. Environmental Pollution, 181: 340−343

Geyman B M, Ptacek J L, LaVigne M, et al. 2019. Barium in deep-sea bamboo corals: phase associations, barium stable isotopes, & prospects for paleoceanography. Earth and Planetary Science Letters, 525: 115751

Ghosh P, Adkins J, Affek H, Balta B, Guo W F, Schauble E A, Schrag D, Eller J M. 2006a. $^{13}C-^{18}O$ bonds in carbonate minerals: a new kind of paleothermometer. Geochimica et Cosmochimica Acta, 70(6): 1439−1456

Ghosh P, Garzione C N, Eiler J M. 2006b. Rapid uplift of the Altiplano revealed through $^{13}C-^{18}O$ bonds in paleosol carbonates. Science, 311(5760): 511−515

Gili S, Gaiero D M, Goldstein S L, Jr F C, Koester E, Jweda J, Vallelonga P, Kaplan M R. 2016. Provenance of dust to Antarctica: a lead isotopic perspective: Pb isotopic signature in antarctic dust. Geophysical Research Letters, 43(5): 2291−2298

Gong P, Wang X P, Pokhrel B, Wang H L, Liu X D, Liu X B, Wania F. 2019. Trans-himalayan transport of organochlorine compounds: three-year observations and model-based flux estimation. Environmental Science & Technology, 53(12): 6773−6783

Gong Y Z, Zeng Z, Zhou C, Nan X Y, Yu H M, Lu Y, Li W Y, Gou W X, Cheng W H, Huang F. 2019. Barium isotopic fractionation in latosol developed from strongly weathered basalt. Science of the Total Environment, 687: 1295−1304

Gueguen B, Rouxel O, Ponzevera E, Bekker A, Fouquet Y. 2013. Nickel isotope variations in terrestrial silicate rocks and geological reference materials measured by MC-ICP-MS. Geostandards and Geoanalytical Research, 37(3): 297−317

Guinoiseau D, Galer S J G, Abouchami W. 2018. Effect of cadmium sulphide precipitation on the partitioning of Cd isotopes: implications for the oceanic Cd cycle. Earth and Planetary Science Letters, 498: 300−308

Guo W F, Zhou C. 2019. Triple oxygen isotope fractionation in the $DIC-H_2O-CO_2$ system: a numerical framework and its implications. Geochimica et Cosmochimica Acta, 246: 541−564

Guo Y R, Deng W F, Wei G J. 2019. Kinetic effects during the experimental transition of aragonite to calcite in aqueous solution: insights from clumped and oxygen isotope signatures. Geochimica et Cosmochimica Acta, 248: 210−230

Guo Z B, Li Z Q, Farquhar J, Kaufman A J, Wu N P, Li C, Dickerson R R, Wang P C. 2010. Identification of sources and formation processes of atmospheric sulfate by sulfur isotope and scanning electron microscope measurements. Journal of Geophysical Research: Atmospheres, 115(D7): D00K07

Haghnegahdar M A, Schauble E A, Young E D. 2017. A model for $^{12}CH_2D_2$ and $^{13}CH_3D$ as complementary tracers for the budget of atmospheric CH_4. Global Biogeochemical Cycles, 31(9): 1387−1407

Hammerschmidt C R, Fitzgerald W F. 2006. Methylmercury in freshwater fish linked to atmospheric mercury deposition. Environmental Science & Technology, 40(24): 7764−7770

Han X K, Guo Q J, Strauss H, et al. 2017. Multiple sulfur isotope constraints on sources and formation processes of sulfate in Beijing $PM_{2.5}$ aerosol. Environmental Science & Technology, 51(14): 7794−7803

Harris E, Sinha B, van Pinxteren D, et al. 2013a. Enhanced role of transition metal ion catalysis during in-cloud oxidation of SO_2. Science, 340(6133): 727−730

Harris E, Sinha B, Hoppe P, Ono S. 2013b. High-precision measurements of ^{33}S and ^{34}S fractionation during SO_2 oxidation reveal causes of seasonality in SO_2 and sulfate isotopic composition. Environmental Science & Technology, 47(21): 12174−12183

Harris R C, Rudd J W M, Amyot M, Babiarz C L, Beaty K G, Blanchfield P J, Bodaly R A, Branfireun B A, Gilmour C C, Graydon J A, Heyes A, Hintelmann H, Hurley J P, Kelly C A, Krabbenhof D P, Lindberg S E, Mason R P, Paterson M J, Podemski C L, Robinson A, Sandilands K A, Southworth G R, Louis V L St, Tate M T. 2007. Whole-ecosystem study shows rapid fish-mercury response to changes in mercury deposition. Proceedings of the National Academy of Sciences of the United States of America, 104(42): 16586−16591

He B, Olack G A, Colman A S. 2012. Pressure baseline correction and high-precision CO_2 clumped-isotope (Δ_{47}) measurements in bellows and microvolume modes. Rapid Communications in Mass Spectrometry, 26(24): 2837−2853

Henkes G A, Passey B H, Grossman E L, Shenton B J, Yancey T E, Pérez-Huerta A. 2018. Temperature evolution and the oxygen isotope composition of Phanerozoic oceans from carbonate clumped isotope thermometry. Earth and Planetary Science Letters, 490: 40−50

Herwartz D, Surma J, Voigt C, Assonov S, Staubwasser M. 2017. Triple oxygen isotope systematics of structurally bonded water in gypsum. Geochimica et Cosmochimica Acta, 209: 254−266

Hoefs J. 2018. Stable Isotope Geochemistry. In: Springer Textbooks in Earth Sciences, Geography and Environment. https://doi.org/10.1007/978-3-319-78527-1

Hohl S V, Galer S J G, Gamper A, Becker H. 2016. Cadmium isotope variations in Neoproterozoic carbonates—a tracer of biologic production? Geochemical Perspective Letters, 3(1): 32−44

Holmes C D, Jacob D J, Yang X. 2006. Global lifetime of elemental mercury against oxidation by atomic bromine in the free troposphere. Geophysical Research Letters, 33(20): L20808

Holmes C D, Jacob D J, Corbitt E S, Mao J, Yang X, Talbot R, Slemr F. 2010. Global atmospheric model for mercury including oxidation by bromine atoms. Atmospheric Chemistry and Physics, 10(24): 12037–12057

Hong B, Hong Y T, Lin Q H, Shibata Y, Uchida M, Zhu Y X, Leng X T, Wang Y, Cai C C. 2010. Anti-phase oscillation of Asian monsoons during the Younger Dryas Period: evidence from peat cellulose δ^{13}C of Hani, Northeast China. Palaeogeography, Palaeoclimatology, Palaeoecology, 297(1): 214–222

Hong B, Gasse F, Uchida M, Hong Y T, Leng X T, Shibata Y, An N, Zhu Y X, Wang Y. 2014a. Increasing summer rainfall in arid eastern-Central Asia over the past 8500 years. Scitific Reports, 4: 5279

Hong B, Hong Y T, Uchida M, Shibata Y, Cai C, Peng H J, Zhu Y X, Wang Y, Yuan L G. 2014b. Abrupt variations of Indian and East Asian summer monsoons during the last deglacial stadial and interstadial. Quaternary Science Reviews, 97: 58–70

Hong B, Uchida M, Hong Y T, Peng H J, Kondo M, Ding H W. 2018. The respective characteristics of millennial-scale changes of the India summer monsoon in the Holocene and the Last Glacial. Palaeogeography, Palaeoclimatology, Palaeoecology, 496: 155–165

Hong B, Rabassa J, Uchida M, Hong Y T, Peng H J, Ding H W, Guo Q, Yao Y. 2019. Response and feedback of the Indian summer monsoon and the southern westerly winds to a temperature contrast between the hemispheres during the last glacial-interglacial transitional period. Earth-Science Reviews, 197: 102917

Hong S, Soyol-Erdene T O, Hwang H J, Hong S B, Do Hur S, Motoyama H. 2012. Evidence of global-scale As, Mo, Sb, and Tl atmospheric pollution in the Antarctic snow. Environmental Science & Technology, 46(12): 11550–11557

Hong S B, Lee K, Hur S D, Hong S, Soyol-Erdene T O, Kim S M, Chung J W, Jun S J, Kang C H. 2015. Development of melting system for measurement of trace elements and ions in ice core. Bulletin of the Korean Chemical Society, 36(4): 1069–1081

Hong Y T, Jiang H B, Liu T S, Zhou L P, Beer J, Li H D, Leng X T, Hong B, Qin X G. 2000. Response of climate to solar forcing recorded in a 6000-year δ^{18}O time-series of Chinese peat cellulose. The Holocene, 10(1): 1–7

Hong Y T, Wang Z G, Jiang H B, Lin Q H, Hong B, Zhu Y X, Wang Y, Xu L S, Leng X T, Li H D. 2001. A 6000-year record of changes in drought and precipitation in northeastern China based on a δ^{13}C time series from peat cellulose. Earth and Planetary Science Letters, 185(1-2): 111–119

Hong Y T, Hong B, Lin Q H, Zhu Y X, Shibata Y, Hirota M, Uchida M, Leng X T, Jiang H B, Xu H, Wang H, Yi L. 2003. Correlation between Indian Ocean summer monsoon and North Atlantic climate during the Holocene. Earth and Planetary Science Letters, 211(3-4): 371–380

Hong Y T, Hong B, Lin Q H, Shibata Y, Hirota M, Zhu Y X, Leng X T, Wang Y, Wang H, Yi L. 2005. Inverse phase oscillations between the East Asian and Indian Ocean summer monsoons during the last 12000 years and paleo-El Niño. Earth and Planetary Science Letters, 231(3-4): 337–346

Hong Y T, Hong B, Lin Q H, Shibata Y, Zhu Y X, Leng X T, Wang Y. 2009. Synchronous climate anomalies in the western North Pacific and North Atlantic regions during the last 14,000 years. Quaternary Science Reviews, 28(9-10): 840–849

Horner T J, Rickaby R E M, Henderson G M. 2011. Isotopic fractionation of cadmium into calcite. Earth and Planetary Science Letters, 312(1-2): 243–253

Horner T J, Kinsley C W, Nielsen S G. 2015. Barium-isotopic fractionation in seawater mediated by barite cycling and oceanic circulation. Earth and Planetary Science Letters, 430: 511–522

Horowitz H M, Jacob D J, Zhang Y X, Dibble T S, Slemr F, Amos H M, Schmidt J A, Corbitt E S, Marais E A, Sunderland E M. 2017. A new mechanism for atmospheric mercury redox chemistry: implications for the global mercury budget. Atmospheric Chemistry and Physics, 17(10): 6353–6371

Howarth S, Prytulak J, Little S H, Hammond S J, Widdowson M. 2018. Thallium concentration and thallium isotope composition of lateritic terrains. Geochimica et Cosmochimica Acta, 239: 446–462

Hsieh Y T, Henderson G M. 2017. Barium stable isotopes in the global ocean: tracer of Ba inputs and utilization. Earth and Planetary Science Letters, 473: 269–278

Hu C C, Lei Y B, Tan Y H, Sun X C, Xu H, Liu C Q, Liu X Y. 2019. Plant nitrogen and phosphorus utilization under invasive pressure in a montane ecosystem of tropical China. Journal of Ecology, 107(1): 372–386

Huang Y M, Li J, Xu Y, Xu W H, Cheng Z N, Liu J W, Wang Y, Tian C G, Luo C L, Zhang G. 2014. Polychlorinated biphenyls (PCBs) and hexachlorobenzene (HCB) in the equatorial Indian Ocean: temporal trend, continental outflow and air-water exchange. Marine Pollution Bulletin, 80(1-2): 194–199

Imseng M, Wiggenhauser M, Keller A, Müller M, Rehkämper M, Murphy K, Kreissig K, Frossard E, Wilcke W, Bigalke M. 2019. Towards an understanding of the Cd isotope fractionation during transfer from the soil to the cereal grain. Environmental Pollution, 244: 834–844

IPCC. 2018. Global Warming of 1.5℃. An IPCC special report on the impacts of global warming of 1.5℃ above pre-industrial levels and related global greenhouse gas emission pathways, in the context of strengthening the global response to the threat of climate change, sustainable development, and efforts to eradicate poverty. Geneva, Switzerland: World Meteorological Organization

Jamieson-Hanes J H, Lentz A M, Amos R T, Ptacek C J, Blowes D W. 2014. Examination of Cr(VI) treatment by zero-valent iron using *in situ*, real-time X-ray absorption spectroscopy and Cr isotope measurements. Geochimica et Cosmochimica Acta, 142: 299–313

Jiang Y J, Wu Y X, Yuan D X. 2009. Human impacts on karst groundwater contamination deduced by coupled nitrogen with strontium isotopes in the Nandong underground river system in Yunan, China. Environmental Science & Technology, 43(20): 7676–7683

John S G, Helgoe J, Townsend E. 2017. Biogeochemical cycling of Zn and Cd and their stable isotopes in the eastern tropical South Pacific. Marine Chemistry, 201: 256–262

Kim S H, Han Y J, Holsen T M, Yi S M. 2009. Characteristics of atmospheric speciated mercury concentrations [TGM, Hg(II) and Hg(p)] in Seoul, Korea. Atmospheric Environment, 43(20): 3267–3274

Kwon S Y, Blum J D, Yin R S, Tsui M T K, Yang Y H, Choi J W. 2020. Mercury stable isotopes for monitoring the effectiveness of the Minamata Convention on Mercury. Earth-Science Reviews, 203: 103111

Lambelet M, Rehkämper M, van der Flierdt T, Xue Z C, Kreissig K, Coles B, Porcelli D, Andersson P. 2013. Isotopic analysis of Cd in the mixing zone of Siberian rivers with the Arctic Ocean—new constraints on marine Cd cycling and the isotope composition of riverine Cd. Earth and Planetary Science Letters, 361: 64–73

Li C L, Chen P F, Kang S C, Yan F P, Tripathee L, Wu G J, Qu B, Sillanpää M, Yang D, Dittmar T, Stubbin A, Raymond P A. 2018. Fossil fuel combustion emission from South Asia influences precipitation dissolved organic carbon reaching the remote Tibetan Plateau: isotopic and Molecular evidence. Journal of Geophysical Research: Atmospheres, 123(11): 6248–6258

Li D D, Liu S A, Li S G. 2015. Copper isotope fractionation during adsorption onto kaolinite: experimental approach and applications. Chemical Geology, 396: 74–82

Li J, Li Q L, Gioia R, Zhang Y L, Zhang G, Li X D, Spiro B, Bhatia R S, Jones K C. 2011. PBDEs in the atmosphere over the Asian marginal seas, and the Indian and Atlantic Oceans. Atmospheric Environment, 45(37): 6622–6628

Li K, Lin C J, Yuan W, Sun G Y, Fu X W, Feng X B. 2019. An improved method for recovering and preconcentrating mercury in natural water samples for stable isotope analysis. Journal of Analytical Atomic Spectrometry, 34(11): 2303–2313

Li P, Du B Y, Maurice L, Laffont L, Lagane C, Point D, Sonke J E, Yin R S, Lin C J, Feng X B. 2017. Mercury isotope signatures of methylmercury in rice samples from the Wanshan mercury mining area, China: environmental implications. Environmental Science & Technology, 51(21): 12321–12328

Li S L, Liu C Q, Li J, Liu X L, Chetelat B, Wang B L, Wnag F S. 2010. Assessment of the sources of nitrate in the Changjiang River, China using a nitrogen and oxygen isotopic approach. Environmental Science & Technology, 44(5): 1573–1578

Li S N, Levin NE, Chesson LA. 2015. Continental scale variation in ^{17}O-excess of meteoric waters in the United States. Geochimica et Cosmochimica Acta, 164: 110–126

Li W, Gou W X, Li W Q, Zhang T Y, Yu B, Liu Q, Shi J B. 2019. Environmental applications of metal stable isotopes: silver, mercury and zinc. Environmental Pollution, 252: 1344–1356

Li Y F, Ma W L, Yang M. 2015. Prediction of gas/particle partitioning of polybrominated diphenyl ethers (PBDEs) in global air: a theoretical study. Atmospheric Chemistry and Physics, 15(4): 1669–1681

Li Y F, Qiao L N, Ren N Q, Sverko E, Mackay D, Macdonald R W. 2017. Decabrominated diphenyl ethers (BDE-209) in Chinese and global air: levels, gas/particle partitioning, and long-range transport: is long-range transport of BDE-209 really governed by the movement of particles? Environmental Science & Technology, 51(2): 1035–1042

Lin B Z, Liu Z F, Eglinton T I, Kandasamy S, Blattmann T M, Haghipour N, de Lange G J. 2019. Perspectives on provenance and alteration of suspended and sedimentary organic matter in the subtropical Pearl River system, South China. Geochimica et Cosmochimica Acta, 259: 270–287

Lin H M, Tong Y D, Yin X F, Zhang Q G, Zhang H, Zhang H R, Chen L, Kang S C, Zhang W, Schauer J, de Foy B, Bu X G, Wang X J. 2019. First measurement of atmospheric mercury species in Qomolangma Natural Nature Preserve, Tibetan Plateau, and evidence of transboundary pollutant invasion. Atmospheric Chemistry and Physics, 19(2): 1373–1391

Lin H T, Repeta D J, Xu L, Rappé M S. 2019. Dissolved organic carbon in basalt-hosted deep subseafloor fluids of the Juan de Fuca Ridge flank. Earth and Planetary Science Letters, 513: 156–165

Lin M, Kang S C, Shaheen R, Li C L, Hsu S C, Thiemens M H. 2018. Atmospheric sulfur isotopic anomalies recorded at Mt. Everest across the Anthropocene. Proceedings of the National Academy of Sciences of the United States of America, 115(27): 6964–6969

Lin T, Li J, Xu Y, Liu X, Luo C L, Cheng H R, Chen Y J, Zhang G. 2012. Organochlorine pesticides in seawater and the surrounding atmosphere of the marginal seas of China: spatial distribution, sources and air-water exchange. Science of the Total Environment, 435-436: 244–252

Lindberg S, Bullock R, Ebinghaus R, Engstrom D, Feng X B, Fitzgerald W, Pirrone N, Prestbo E, Seigneur C. 2007. A synthesis of progress and uncertainties in attributing the sources of mercury in deposition. Ambio, 36(1): 19–32

Lindqvist O, Johansson K, Bringmark L, Timm B, Aastrup M, Andersson A, Hovsenius G, Håkanson L, Iverfeldt Å, Meili M. 1991. Mercury in the Swedish environment-recent research on causes, consequences and corrective methods. Water, Air, and Soil Pollution, 55(1): R11

Liu C, Fu X W, Zhang H, Ming L L, Xu H, Zhang L M, Feng X B. 2019. Sources and outflows of atmospheric mercury at Mt. Changbai, northeastern China. Science of the Total Environment, 663: 275–284

Liu C Q, Li S L, Lang Y C, Xiao H Y. 2006. Using δ^{15}N- and δ^{18}O-values to identify nitrate sources in karst ground water, Guiyang, Southwest China. Environmental Science & Technology, 40(22): 6928–6933

Liu C S, Gao T, Liu Y H, Liu J Y, Li F B, Chen Z W, Li Y Z, Lv Y W, Song Z Y, Reinfelder J, Huang W L. 2019. Isotopic fingerprints indicate distinct strategies of Fe uptake in rice. Chemical Geology, 524: 323–328

Liu D, Li J, Zhang Y L, Xu Y, Liu X, Ding P, Shen C D, Chen Y J, Tian C G, Zhang G. 2013. The Use of Levoglucosan and radiocarbon for source apportionment of PM$_{2.5}$ carbonaceous aerosols at a background site in East China. Environmental Science & Technology, 47(18): 10454–10461

Liu E F, Zhang E L, Li K, Nath B, Li Y L, Shen J. 2013. Historical reconstruction of atmospheric lead pollution in central Yunnan Province, Southwest China: an analysis based on lacustrine sedimentary records. Environmental Science and Pollution Research, 20(12): 8739–8750

Liu J, Yin M L, Xiao T F, Zhang C S, Tsang D C W, Bao Z A, Zhou Y T, Chen Y H, Luo X W, Yuan W H, Wang J. 2020. Thallium isotopic fractionation in industrial process of pyrite smelting and environmental implications. Journal of Hazardous Materials, 384: 121378

Liu J F, Che J B, Zhang T, Wang Y N, Yuan W, Lang Y C, Tu C L, Liu L Z, Birck J L. 2020. Chromatographic purification of antimony for accurate isotope analysis by MC-ICP-MS. Journal of Analytical Atomic Spectrometry, 35(7): 1360−1367, doi: 10.1039/D0JA00136H

Liu J L, Feng X B, Yin R S, Zhu W, Li Z G. 2011. Mercury distributions and mercury isotope signatures in sediments of Dongjiang, the Pearl River Delta, China. Chemical Geology, 287(1-2): 81−89

Liu J W, Mo Y Z, Ding P, Li J, Shen C D, Zhang G. 2018. Dual carbon isotopes (^{14}C and ^{13}C) and optical properties of WSOC and HULIS-C during winter in Guangzhou, China. Science of the Total Environment, 633: 1571−1578

Liu M, Chen L G, Xie D H, Sun J R, He Q S, Cai L M, Gao Z Q, Zhang Y Q. 2016. Monsoon-driven transport of atmospheric mercury to the South China Sea from the Chinese mainland and Southeast Asia—observation of gaseous elemental mercury at a background station in South China. Environmental Science and Pollution Research, 23(21): 21631−21640

Liu M S, Zhang Q, Zhang Y N, Zhang Z F, Huang F, Yu H M. 2020. High-precision Cd isotope measurements of soil and rock reference materials by MC-ICP-MS with double spike correction. Geostandards and Geoanalytical Research, 44(1): 169−182

Liu Q, Liu Y. 2016. Clumped-isotope signatures at equilibrium of CH_4, NH_3, H_2O, H_2S and SO_2. Geochimica et Cosmochimica Acta, 175: 252−270

Liu S A, Teng F Z, Li S G, Wei G J, Ma J L, Li D D. 2014. Copper and iron isotope fractionation during weathering and pedogenesis: insights from saprolite profiles. Geochimica et Cosmochimica Acta, 146: 59−75

Liu S A, Wu H C, Shen S Z, Jiang G Q, Zhang S H, Lv Y W, Zhang H, Li S G. 2017. Zinc isotope evidence for intensive magmatism immediately before the end-Permian mass extinction. Geology, 45(4): 343−346

Liu T, Wang F, Michalski G, et al. 2013. Using ^{15}N, ^{17}O, and ^{18}O to determine nitrate sources in the Yellow River, China. Environmental Science & Technology, 47(23): 13412−13421

Liu T B, Zhou W J, Cheng P, Burr G S. 2018. A survey of the ^{14}C content of dissolved inorganic carbon in Chinese lakes. Radiocarbon, 60(2): 705−716

Liu X, Ming L L, Nizzetto L, Borgå K, Larssen T, Zheng Q, Li J, Zhang G. 2013. Critical evaluation of a new passive exchange-meter for assessing multimedia fate of persistent organic pollutants at the air-soil interface. Environmental Pollution, 181: 144−150

Liu X, Li J, Zheng Q, Bing H J, Zhang R J, Wang Y, Luo C L, Wu Y H, Pan S H, Zhang G. 2014. Forest filter effect versus cold trapping effect on the altitudinal distribution of PCBs: a case study of Mt. Gongga, eastern Tibetan Plateau. Environmental Science & Technology, 48(24): 14377−14385

Liu X Y, Koba K, Takebayashi Y, Liu C Q, Fang Y T, Yoh M. 2013a. Dual N and O isotopes of nitrate in natural plants: first insights into individual variability and organ-specific pattern. Biogeochemistry, 114(1-3): 399−411

Liu X Y, Koba K, Makabe A, Li X D, Yoh M, Liu C Q. 2013b. Ammonium first: natural mosses prefer atmospheric ammonium but vary utilization of dissolved organic nitrogen depending on habitat and nitrogen deposition. New Phytologist, 199(2): 407−419

Liu X Y, Xiao H W, Xiao H Y, Song W, Sun X C, Zheng X D, Liu C Q, Koba K. 2017. Stable isotope analyses of precipitation nitrogen sources in Guiyang, southwestern China. Environmental Pollution, 230: 486−494

Liu X Y, Koba K, Koyama L A, Hobbie S E, Weiss M S, Inagaki Y, Shaver G R, Giblin A E, Hobara S, Nadelhoffer K J, Sommerkorn M, Rastetter E B, Kling G W, Laundre J A, Yano Y, Makabe A, Yano M, Liu C Q. 2018. Nitrate is an important nitrogen source for arctic tundra plants. Proceedings of the National Academy of Sciences of the United States of America, 115(13): 3398−3403

Liu X Y, Yin Y M, Song W. 2020. Nitrogen isotope differences between major atmospheric NO_y species: implications for transformation and deposition processes. Environmental Science & Technology Letters, 7(4): 227−233

Liu Z H, Zhao M, Sun H L, Yang R, Chen B, Yang M X, Zeng Q R, Zeng H T. 2017. "Old" carbon entering the South China Sea from the carbonate-rich Pearl River Basin: coupled action of carbonate weathering and aquatic photosynthesis. Applied Geochemistry, 78: 96−104

Liang L L, Liu C, Zhu X, Ngwenya B, Wang Z, Song L, Li J. 2020. Zinc isotope characteristics in the biogeochemical cycle as revealed by analysis of suspended particulate matter (SMP) in Aha Lake and Hongfeng Lake, Guizhou, China. Journal of Earth Science, 31, 126−140

Lobo L, Devulder V, Degryse P, Vanhaecke F. 2012. Investigation of natural isotopic variation of Sb in stibnite ores via multi-collector ICP-mass spectrometry-perspectives for Sb isotopic analysis of Roman glass. Journal of Analytical Atomic Spectrometry, 27(8): 1304−1310

Lobo L, Degryse P, Shortlan A, Vanhaecke F. 2013. Isotopic analysis of antimony using multi-collector ICP-mass spectrometry for provenance determination of Roman glass. Journal of Analytical Atomic Spectrometry, 28(8): 1213−1219

Lobo L, Degryse P, Shortland A, Eremin K, Vanhaecke F. 2014. Copper and antimony isotopic analysis via multi-collector ICP-mass spectrometry for provenancing ancient glass. Journal of Analytical Atomic Spectrometry, 29(1): 58−64

Loveluck C, McCormick M, Spaulding N E, Clifford H, Handley M J, Hartman L, Hoffmann H, Korotkikh E V, Kurbatov A V, More A F, Sneed S B, Mayewski P A. 2018. Alpine ice-core evidence for the transformation of the European monetary system, AD 640−670. Antiquity, 92(366): 1571−1585

Lu D W, Liu Q, Zhang T Y, Cai Y, Yin Y G, Jiang G B. 2016. Stable silver isotope fractionation in the natural transformation process of silver nanoparticles. Nature Nanotechnology, 11(8): 682−686

Lu Y Z, Chen G J, Bai Y N, Fu L, Qin L P, Zeng R J. 2018. Chromium isotope fractionation during Cr(VI) reduction in a methane-based hollow-fiber membrane biofilm reactor. Water Research, 130: 263−270

Luo Y W, Lima I D, Karl D M, Doney S C. 2013. Data-based assessment of environmental controls on global marine nitrogen fixation. Biogeosciences Discussions, 10(4): 7367−7412

Luz B, Barkan E. 2000. Assessment of oceanic productivity with the triple-isotope composition of dissolved oxygen. Science, 288(5473): 2028−2031

Luz B, Barkan E. 2005. The isotopic ratios $^{17}O/^{16}O$ and $^{18}O/^{16}O$ in molecular oxygen and their significance in biogeochemistry. Geochimica et Cosmochimica Acta, 69(5): 1099−1110

Luz B, Barkan E, Bender M L, Thiemens M H, Boering K A. 1999. Triple-isotope composition of atmospheric oxygen as a tracer of biosphere productivity. Nature, 400(6744): 547–550

Lv Y W, Liu S A, Zhu J M, Li S G. 2016. Copper and zinc isotope fractionation during deposition and weathering of highly metalliferous black shales in central China. Chemical Geology, 445: 24–35

Lv Y W, Liu S A, Wu H C, Hohl S V, Chen S M, Li S G. 2018. Zn-Sr isotope records of the Ediacaran Doushantuo formation in South China: diagenesis assessment and implications. Geochimica et Cosmochimica Acta, 2018, 239: 330–345

Lv Y W, Liu S A, Teng F Z, Wei G J, Ma J L. 2020. Contrasting zinc isotopic fractionation in two mafic-rock weathering profiles induced by adsorption onto Fe (hydr)oxides. Chemical Geology, 539: 119504

Lyman S N, Jaffe D A. 2012. Formation and fate of oxidized mercury in the upper troposphere and lower stratosphere. Nature Geoscience, 5(2): 114–117

Ma J M, Hung H, Tian C G, Kallenborn R. 2011. Revolatilization of persistent organic pollutants in the Arctic induced by climate change. Nature Climate Change, 1(5): 255–260

Ma L, Li Y L, Wang W, Weng N Y, Evans R D, Wang W X. 2019. Zn isotope fractionation in the oyster crassostrea hongkongensis and implications for contaminant source tracking. Environmental Science & Technology, 53(11): 6402–6409

Ma L, Wang W, Xie M W, Wang W X, Evans R D. 2020. Using Zn isotopic signatures for source identification in a contaminated estuary of southern China. Environmental Science and Technology, 54(8): 5140–5149

Mao H T, Cheng I, Zhang L M. 2016. Current understanding of the driving mechanisms for spatiotemporal variations of atmospheric speciated mercury: a review. Atmospheric Chemistry and Physics, 16(20): 12897–12924

Marx S K, McGowan H A. 2010. Long-distance transport of urban and industrial metals and their incorporation into the environment: sources, transport pathways and historical trends. In: Zereini F, Wiseman C (eds). Urban Airborne Particulate Matter. Berlin, Heidelberg: Springer

Marx S K, McGowan H A, Kamber B S, Knight J M, Denholm J, Zawadzki A. 2014. Unprecedented wind erosion and perturbation of surface geochemistry marks the Anthropocene in Australia. Journal of Geophysical Research: Earth Surface, 119(1): 45–61

Marx S K, Rashid S, Stromsoe N. 2016. Global-scale patterns in anthropogenic Pb contamination reconstructed from natural archives. Environmental Pollution, 213: 283–298

Mason R P, Choi A L, Fitzgerald W F, Hammerschmidt C R, Lamborg C H, Soerensen A L, Sunderland E M. 2012. Mercury biogeochemical cycling in the ocean and policy implications. Environmental Research, 119: 101–117

Mavromatis V, van Zuilen K, Blanchard M, van Zuilen M, Dietzel M, Schott J. 2020. Experimental and theoretical modelling of kinetic and equilibrium Ba isotope fractionation during calcite and aragonite precipitation. Geochimica et Cosmochimica Acta, 269: 566–580

McConnell J R, Chellman N J, Wilson A I, Stohl A, Arienzo M M, Eckhardt S, Fritzsche D, Kipfstuhl S, Opel T, Place P F, Steffensen J P. 2019. Pervasive Arctic lead pollution suggests substantial growth in medieval silver production modulated by plague, climate, and conflict. Proceedings of the National Academy of Sciences of the United States of America, 116(30): 14910–14915

McDonough W F, Sun S S. 1995. The composition of the Earth. Chemical Geology, 120(3-4): 223–253

Miller M F. 2002. Isotopic fractionation and the quantification of ^{17}O anomalies in the oxygen three-isotope system: an appraisal and geochemical significance. Geochimica et Cosmochimica Acta, 66(11): 1881–1889

Mitchell K, Mason P R D, van Cappellen P, Johnson T M, Gill B C, Owens J D, Diaz J, Ingall E D, Reichart G J, Lyons T W. 2012. Selenium as paleo-oceanographic proxy: a first assessment. Geochimica et Cosmochimica Acta, 89: 302–317

Miyazaki T, Kimura J I, Chang Q. 2014. Analysis of stable isotope ratios of Ba by double-spike standard-sample bracketing using multiple-collector inductively coupled plasma mass spectrometry. Journal of Analytical Atomic Spectrometry, 29(3): 483–490

More A F, Spaulding N E, Bohleber P, Handley M J, Hoffmann H, Korotkikh E V, Kurbatov A V, Loveluck C P, Sneed S B, McCormick M, Mayewski P A. 2017. Next-generation ice core technology reveals true minimum natural levels of lead (Pb) in the atmosphere: insights from the Black Death. GeoHealth, 1(4): 211–219

Nguyen H T, Kim M Y, Kim K H. 2010. The influence of long-range transport on atmospheric mercury on Jeju Island, Korea. Science of the Total Environment, 408(6): 1295–1307

Pan Y P, Wang Y S. 2014. Atmospheric wet and dry deposition of trace elements at 10 sites in northern China. Atmospheric Chemistry and Physics, 15(2): 951–972

Pan Y P, Wang Y S, Sun Y, Tian S L, Cheng M T. 2013. Size-resolved aerosol trace elements at a rural mountainous site in northern China: importance of regional transport. Science of the Total Environment, 461-462: 761–771

Passey B H, Henkes G A. 2012. Carbonate clumped isotope bond reordering and geospeedometry. Earth and Planetary Science Letters, 351-352: 223–236

Passey B H, Ji H Y. 2019. Triple oxygen isotope signatures of evaporation in lake waters and carbonates: a case study from the western United States. Earth and Planetary Science Letters, 518: 1–12

Passey B H, Hu H T, Ji H Y, Montanari S, Li S N, Henkes G A, Levin N E. 2014. Triple oxygen isotopes in biogenic and sedimentary carbonates. Geochimica et Cosmochimica Acta, 141: 1–25

Porter S J, Selby D, Cameron V. 2014. Characterising the nickel isotopic composition of organic-rich marine sediments. Chemical Geology, 387: 12–21

Preunkert S, McConnell J R, Hoffmann H, Legrand M, Wilson A I, Eckhardt S, Stohl A, Chellman N J, Arienzo M M, Friedrich R. 2019. Lead and antimony in basal ice from Col Du dome (French Alps) dated with radiocarbon: a record of pollution during antiquity. Geophysical Research Letters, 46(9): 4953–4961

Pritzkow W, Wunderli S, Vogl J, Fortunato G. 2007. The isotope abundances and the atomic weight of cadmium by a metrological approach. International Journal of Mass Spectrometry, 261(1): 74-85

Qi Y H, Cheng W H, Nan X Y, Yang F, Li J, Li D C, Lundstrom C C, Yu H M, Zhang G L, Huang F. 2020. Iron stable isotopes in bulk soil and sequential extracted fractions trace Fe redox cycling in paddy soils. Journal of Agricultural and Food Chemistry, 68(31): 8143-8150

Qin C Y, Chen M, Yan H Y, Shang L H, Yao H, Li P, Feng X B. 2018. Compound specific stable isotope determination of methylmercury in contaminated soil. Science of the Total Environment, 644: 406-412

Qin L P, Wang X L. 2017. Chromium isotope geochemistry. Reviews in Mineralogy and Geochemistry, 82(1): 379-414

Qu B, Sillanpää M, Li C L, Kang S C, Stubbins A, Yan F P, Aho K S, Zhou F, Raymond P A. 2017. Aged dissolved organic carbon exported from rivers of the Tibetan Plateau. PLoS One, 12(5): e0178166

Qu Y X, Jin Z D, Wang J, Wang Y Q, Xiao J, Gou L F, Zhang F, Liu C Y, Gao Y L, Suarez M B, Xu X M. 2020. The sources and seasonal fluxes of particulate organic carbon in the Yellow River. Earth Surface Processes and Landforms, 45(9): 2004-2019, doi. 10.1002/esp.4861

Rao Z G, Huang C, Xie L H, Shi F X, Zhao Y, Cao J T, Gou X H, Chen J H, Chen F H. 2019. Long-term summer warming trend during the Holocene in Central Asia indicated by alpine peat α-cellulose $\delta^{13}C$ record. Quaternary Science Reviews, 203: 56-67

Ren N Q, Que M X, Li Y F, Liu Y, Wan X N, Xu D D, Sverko E, Ma J M. 2007. Polychlorinated biphenyls in Chinese surface soils. Environmental Science & Technology, 41(11): 3871-3876

Ren Y F, Xu Z W, Zhang X Y, Wang X K, Sun X M, Ballantine D J, Wang S Z. 2014. Nitrogen pollution and source identification of urban ecosystem surface water in Beijing. Frontiers of Environmental Science & Engineering, 8(1): 106-116

Resongles E, Freydier R, Casiot C, Viers J, Chmeleff J, Elbaz-Poulichet F. 2015. Antimony isotopic composition in river waters affected by ancient mining activity. Talanta, 144: 851-861

Rouxel O, Ludden J, Carignan J, Marin L, Fouquet Y. 2002. Natural variations of Se isotopic composition determined by hydride generation multiple collector inductively coupled plasma mass spectrometry. Geochimica et Cosmochimica Acta, 66(18): 3191-3199

Saad E M, Wang X L, Planavsky N J, Reinhard C T, Tang Y Z. 2017. Redox-independent chromium isotope fractionation induced by ligand-promoted dissolution. Nature Communications, 8: 1590

Schauble E A, Ghosh P, Eiler J M. 2006. Preferential formation of ^{13}C-^{18}O bonds in carbonate minerals, estimated using first-principles lattice dynamics. Geochimica et Cosmochimica Acta, 70(10): 2510-2529

Schauer A J, Kelson J, Saenger C, Huntington K W. 2016. Choice of ^{17}O correction affects clumped isotope (Δ_{47}) values of CO_2 measured with mass spectrometry. Rapid Communications in Mass Spectrometry, 30(24): 2607-2616

Schilling K, Johnson T M, Wilcke W. 2013. Isotope fractionation of selenium by biomethylation in microcosm incubations of soil. Chemical Geology, 352: 101-107

Selin N E. 2009. Global biogeochemical cycling of mercury: a review. Annual Review of Environment and Resources, 34: 43-63

Sheng J J, Wang X P, Gong P, Joswiak D R, Tian L D, Yao T D, Jones K C. 2013. Monsoon-driven transport of organochlorine pesticides and polychlorinated biphenyls to the Tibetan Plateau: three year atmospheric monitoring study. Environmental Science & Technology, 47(7): 3199-3208

Shiel A E, Weis D, Cossa D, Orians K J. 2013. Determining provenance of marine metal pollution in French bivalves using Cd, Zn and Pb isotopes. Geochimica et Cosmochimica Acta, 121: 155-167

Shrimpton H K, Blowes D W, Ptacek C J. 2015. Fractionation of selenium during selenate reduction by granular zerovalent iron. Environmental Science & Technology, 49(19): 11688-11696

Shu Y C, Nielsen S G, Marschall H R, John T, Blusztajn J, Auro M. 2019. Closing the loop: subducted eclogites match thallium isotope compositions of ocean island basalts. Geochimica et Cosmochimica Acta, 250: 130-148

Sieber M, Conway T M, de Souza G F, Hassler C S, Ellwood M J, Vance D. 2019. High-resolution Cd isotope systematics in multiple zones of the southern ocean from the Antarctic circumnavigation expedition. Earth and Planetary Science Letters, 527: 115799

Sierra-Hernandez M R, Gabrielli P, Beaudon E, Wegner A, Thompson L G. 2018. Atmospheric depositions of natural and anthropogenic trace elements on the Guliya ice cap (northwestern Tibetan Plateau) during the last 340 years. Atmospheric Environment, 176: 91-102

Song W, Wang Y L, Yang W, Sun X C, Tong Y D, Wang X M, Liu C Q, Bai Z P, Liu X Y. 2019. Isotopic evaluation on relative contributions of major NO_x sources to nitrate of $PM_{2.5}$ in Beijing. Environmental Pollution, 248: 183-190

Song W, Liu X Y, Wang Y L, Tong Y D, Bai Z P, Liu C Q. 2020. Nitrogen isotope differences between atmospheric nitrate and corresponding nitrogen oxides: a new constraint using oxygen isotopes. Science of the Total Environment, 701: 134515

Sprovieri F, Pirrone N, Bencardino M, D'Amore F, Carbone F, Cinnirella S, Mannarino V, Landis M, Ebinghaus R, Weigelt A, Brunke E G, Labuschagne C, Martin L, Munthe J, Wängberg I, Artaxo P, Morais F, Barbosa H D M J, Brito J, Cairns W, Barbante C, Diéguez M D, Garcia P E, Dommergue A, Angot H, Magand O, Skov H, Horvat M, Kotnik J, Read K A, Neves L M, Gawlik B M, Sena F, Mashyanov N, Obolkin V, Wip D, Feng X B, Zhang H, Fu X W, Ramachandran R, Cossa D, Knoery J, Marusczak N, Nerentorp M, Norstrom C. 2016. Atmospheric mercury concentrations observed at ground-based monitoring sites globally distributed in the framework of the GMOS network. Atmospheric Chemistry and Physics, 16(18): 11915-11935

Streets D G, Horowitz H M, Lu Z F, Levin L, Thackray C P, Sunderland E M. 2019. Global and regional trends in mercury emissions and concentrations, 2010-2015. Atmospheric Environment, 201: 417-427

Strode S A, Jaeglé L, Jaffe D A, Swartzendruber P C, Selin N E, Holmes C, Yantosca R M. 2008. Trans-Pacific transport of mercury. Journal of Geophysical Research: Atmospheres, 113(D15): D15305

Stüeken E E. 2017. Selenium isotopes as a biogeochemical proxy in deep time. Reviews in Mineralogy and Geochemistry, 82(1): 657-682

Su Y L, Hu E, Liu Z W, Jeppesen E, Middelburg J J. 2018. Assimilation of ancient organic carbon by zooplankton in Tibetan Plateau lakes is depending on watershed characteristics. Limnology and Oceanography, 63(6): 2359-2371

Sun G Y, Feng X B, Yin R S, Zhao H F, Zhang L M, Sommar J, Li Z G, Zhang H. 2019. Corn (Zea mays L.): a low methylmercury staple cereal source and an important biospheric sink of atmospheric mercury, and health risk assessment. Environment International, 131: 104971

Sun G Y, Feng X B, Yang C M, Zhang L M, Yin R S, Li Z G, Bi X Y, Wu Y J. 2020. Levels, sources, isotope signatures, and health risks of mercury in street dust across China. Journal of Hazardous Materials, 392: 122276

Sun R Y, Wang B L. 2018. Iron isotope fractionation during uptake of ferrous ion by phytoplankton. Chemical Geology, 481: 65-73

Sunderland E M, Mason R P. 2007. Human impacts on open ocean mercury concentrations. Global Biogeochemical Cycles, 21(4): Gb4022, doi 10.1029/2006gb002876

Tan D C, Zhu J M, Wang X L, Han G L, Lu Z, Xu W P. 2020. High-sensitivity determination of Cd isotopes in low-Cd geological samples by double spike MC-ICP-MS. Journal of Analytical Atomic Spectrometry, 35(4): 713-727

Tang J W, Dietzel M, Fernandez A, Tripati A K, Rosenheim B E. 2014. Evaluation of kinetic effects on clumped isotope fractionation (Δ_{47}) during inorganic calcite precipitation. Geochimica et Cosmochimica Acta, 134: 120-136

Tang Y T, Cloquet C, Stercheman T, Echevarria G, Carignan J, Qiu R L, Morel J L. 2012. Fractionation of stable zinc Iisotopes in the field-grown zinc hyperaccumulator noccaea caerulescens and the zinc-tolerant plant silene vulgaris. Environmental Science & Technology, 46, 9972-9979

Tanimizu M, Araki Y, Asaoka S, Takahashi Y. 2011. Determination of natural isotopic variation in antimony using inductively coupled plasma mass spectrometry for an uncertainty estimation of the standard atomic weight of antimony. Geochemical Journal, 45(1): 27-32

Travnikov O. 2005. Contribution of the intercontinental atmospheric transport to mercury pollution in the northern Hemisphere. Atmospheric Environment, 39(39): 7541-7548, doi: 10.1016/j.atmosenv.2005.07.066

Tu Y J, You C F, Kuo T Y. 2020. Source identification of Zn in Erren River, Taiwan: an application of Zn isotopes. Chemosphere, 248, 126044

Uechi Y, Uemura R. 2019. Dominant influence of the humidity in the moisture source region on the ^{17}O-excess in precipitation on a subtropical island. Earth and Planetary Science Letters, 513: 20-28

Uglietti C, Gabrielli P, Cooke C A, Vallelonga P, Thompson L. 2015. Widespread pollution of the South American atmosphere predates the industrial revolution by 240 y. Proceedings of the National Academy of Sciences of the United States of America, 112(8): 2349-2354

Vaněk A, Holubík O, Oborná V, Mihaljevič M, Trubač J, Ettler V P, Vokurková P, Penížek V, Zádorová T, Voegelin A. 2019. Thallium stable isotope fractionation in white mustard: implications for metal transfers and incorporation in plants. Journal of Hazardous Materials, 369: 521-527

Wang C M, Liu Y P, Zhang W B, Hong S M, Do Hur S, Lee K, Pang H X, Hou S G. 2016. High-resolution atmospheric cadmium record for AD 1776-2004 in a high-altitude ice core from the eastern Tien Shan, Central Asia. Annals of Glaciology, 57(71): 265-272

Wang J, Hilton R G, Jin Z D, Zhang F, Densmore A L, Gröcke D R, Xu X M, Li G, West A J. 2019. The isotopic composition and fluxes of particulate organic carbon exported from the eastern margin of the Tibetan Plateau. Geochimica et Cosmochimica Acta, 252: 1-15

Wang J C, Xie Z Q, Wang F Y, Kang H. 2017. Gaseous elemental mercury in the marine boundary layer and air-sea flux in the Southern Ocean in austral summer. Science of The Total Environment. 603, 510-518

Wang N, Shen C D, Sun W D, Ding P, Zhu S Y, Yi W X, Yu Z Q, Sha Z L, Mi M, He L S, Fang J S, Liu K X, Xu X M, Druffel E R M. 2019. Penetration of bomb ^{14}C into the deepest ocean trench. Geophysical Research Letters, 46(10): 5413-5419

Wang P, Li Y M, Zhang Q H, Yang Q H, Zhang L, Liu F B, Fu J J, Meng W Y, Wang D, Sun H Z, Zheng S C, Hao Y F, Liang Y, Jiang G B. 2017. Three-year monitoring of atmospheric PCBs and PBDEs at the Chinese Great Wall Station, West Antarctica: levels, chiral signature, environmental behaviors and source implication. Atmospheric Environment, 150: 407-416

Wang P C, Li Z G, Liu J L, Bi X Y, Ning Y Q, Yang S C, Yang X J. 2019. Apportionment of sources of heavy metals to agricultural soils using isotope fingerprints and multivariate statistical analyses. Environmental Pollution, 249: 208-216

Wang Q, Zhou L, Little S H, Liu J H, Feng L P, Tong S Y. 2020. The geochemical behavior of cu and its isotopes in the Yangtze River. Science of the Total Environment, 728: 138428

Wang Q Z, Zhuang G S, Li J, Huang K, Zhang R, Jiang Y L, Lin Y F, Fu J S. 2011. Mixing of dust with pollution on the transport path of Asian dust: revealed from the aerosol over Yulin, the north edge of Loess Plateau. Science of the Total Environment, 409(3): 573-581

Wang R M, Archer C, Bowie A R, Vance D. 2019. Zinc and nickel isotopes in seawater from the Indian sector of the Southern Ocean: the impact of natural iron fertilization versus Southern Ocean hydrography and biogeochemistry. Chemical Geology, 511: 452-464

Wang X, Zhang H, Lin C J, Fu X W, Zhang Y P, Feng X B. 2015. Transboundary transport and deposition of Hg emission from springtime biomass burning in the Indo-China Peninsula. Journal of Geophysical Research: Atmospheres, 120(18): 9758-9771

Wang X, Cui L L, Zhai J X, Ding Z L. 2016. Stable and clumped isotopes in shell carbonates of land snails Cathaica sp. and Bradybaena sp. in North China and implications for ecophysiological characteristics and paleoclimate studies. Geochemistry, Geophysics, Geosystems, 17(1): 219-231

Wang X, Luo J, Yin R S, Yuan W, Lin C J, Sommar J, Feng X B, Wang H M, Lin C. 2017. Using mercury isotopes to understand mercury accumulation in the montane forest floor of the eastern Tibetan Plateau. Environmental Science & Technology, 51(2): 801-809

Wang X, Lin C J, Feng X B, Yuan W, Fu X W, Zhang H, Wu Q R, Wang S X. 2018. Assessment of regional mercury deposition and emission outflow in mainland China. Journal of Geophysical Research: Atmospheres, 123(17): 9868-9890

Wang X, Yuan W, Lu Z Y, Lin C J, Yin R S, Li F, Feng X B. 2019a. Effects of precipitation on mercury accumulation on subtropical Montane forest floor: implications on climate forcing. Journal of Geophysical Research: Biogeosciences, 124(4): 959-972

Wang X, Yuan W, Lin C J, Zhang L M, Zhang H, Feng X B. 2019b. Climate and vegetation as primary drivers for global mercury storage in surface soil. Environmental Science & Technology, 53(18): 10665-10675

Wang X, Luo J, Yuan W, Lin C J, Wang F Y, Liu C, Wang G X, Feng X B. 2020. Global warming accelerates uptake of atmospheric mercury in regions experiencing glacier retreat. Proceedings of the National Academy of Sciences of the United States of America, 117(4): 2049-2055

Wang X C, Ge T T, Xu C L, Xue Y J, Luo C L. 2016a. Carbon isotopic (^{14}C and ^{13}C) characterization of fossil-fuel derived dissolved organic carbon in wet precipitation in Shandong province, China. Journal of Atmospheric Chemistry, 73(2): 207-221

Wang X C, Luo C L, Ge T T, Xu C L, Xue Y J. 2016b. Controls on the sources and cycling of dissolved inorganic carbon in the Changjiang and Huanghe Rivers estuaries, China: ^{14}C and ^{13}C studies. Limnology and Oceanography, 61(4): 1358-1374

Wang X P, Xu B Q, Kang S C, Cong Z Y, Yao T D. 2008. The historical residue trends of DDT, hexachlorocyclohexanes and polycyclic aromatic hydrocarbons in an ice core from Mt. Everest, central Himalayas, China. Atmospheric Environment, 42(27): 6699-6709

Wang X P, Halsall C, Codling G, Xie Z Y, Xu B Q, Zhao Z, Xue Y G, Ebinghaus R, Jones K C. 2014. Accumulation of perfluoroalkyl compounds in Tibetan Mountain snow: temporal patterns from 1980 to 2010. Environmental Science & Technology, 48(1): 173-181

Wang X P, Ren J, Gong P, Wang C F, Xue Y G, Yao T D, Lohmann R. 2016. Spatial distribution of the persistent organic pollutants across the Tibetan Plateau and its linkage with the climate systems: a 5-year air monitoring study. Atmospheric Chemistry and Physics, 16(11): 6901-6911

Wang X P, Chen M K, Gong P, Wang C F. 2019. Perfluorinated alkyl substances in snow as an atmospheric tracer for tracking the interactions between westerly winds and the Indian Monsoon over western China. Environment International, 124: 294-301

Wang Y L, Liu X Y, Song W, Yang W, Han B, Dou X Y, Zhao X D, Song Z L, Liu C Q, Bai Z P. 2017. Source appointment of nitrogen in $PM_{2.5}$ based on bulk $\delta^{15}N$ signatures and a Bayesian isotope mixing model. Tellus B: Chemical and Physical Meteorology, 69(1): 1299672

Wang Y L, Song W, Yang W, Sun X C, Tong Y D, Wang X M, Liu C Q, Bai Z P, Liu X Y. 2019. Influences of atmospheric pollution on the contributions of major oxidation pathways to $PM_{2.5}$ nitrate formation in Beijing. Journal of Geophysical Research: Atmospheres, 124(7): 4174-4185

Wang Z R, Schauble E A, Eiler J M. 2004. Equilibrium thermodynamics of multiply substituted isotopologues of molecular gases. Geochimica et Cosmochimica Acta, 68(23): 4779-4797

Wei R F, Guo Q J, Wen H J, Yang J X, Peters M, Zhu C W, Ma J, Zhu G X, Zhang H Z, Tian L Y, Wang C Y, Wan Y X. 2015. An analytical method for precise determination of the cadmium isotopic composition in plant samples using multiple collector inductively coupled plasma mass spectrometry. Analytical Methods, 7(6): 2479-2487

Wei R F, Guo Q J, Tian L Y, Kong J, Bai Y, Okoli C P, Wang L Y. 2019. Characteristics of cadmium accumulation and isotope fractionation in higher plants. Ecotoxicology and Environmental Safety, 174: 1-11

Wei T, Dong Z W, Kang S C, Ulbrich S. 2018. Tracing the provenance of long-range transported dust deposition in cryospheric basins of the Northeast Tibetan Plateau: REEs and trace element evidences. Atmosphere, 9(12): 461

Wei T, Dong Z W, Kang S C, Zong C L, Rostami M, Shao Y P. 2019. Atmospheric deposition and contamination of trace elements in snowpacks of mountain glaciers in the northeastern Tibetan Plateau. Science of the Total Environment, 689: 754-764

Wen H J, Zhang Y X, Cloquet C, Zhu C W, Fan H F, Luo C G. 2015. Tracing sources of pollution in soils from the Jinding Pb-Zn mining district in China using cadmium and lead isotopes. Applied Geochemistry, 52: 147-154

Wiggenhauser M, Bigalke M, Imseng M, Müller M, Keller A, Murphy K, Kreissig K, Rehkamper M, Wilcke W, Frossard E. 2016. Cadmium isotope fractionation in soil-wheat systems. Environmental Science & Technology, 50(17): 9223-9231

Wu J, Gao W, Liang Y, Fu J J, Gao Y, Wang Y W, Jiang G B. 2017. Spatiotemporal distribution and alpine behavior of short chain chlorinated paraffins in air at Shergyla Mountain and lhasa on the Tibetan Plateau of China. Environmental Science & Technology, 51(19): 11136-11144

Wu J, Gao W, Liang Y, Fu J J, Shi J B, Lu Y, Wang Y W, Jiang G B. 2020. Short- and medium-chain chlorinated paraffins in multi-environmental matrices in the Tibetan Plateau environment of China: a regional scale study. Environment International, 140: 105767

Wu Q R, Wang S X, Li G L, Liang S, Lin C J, Wang Y F, Cai S Y, Liu K Y, Hao J M. 2016. Temporal trend and spatial distribution of speciated atmospheric mercury emissions in China during 1978-2014. Environmental Science & Technology, 50(24): 13428-13435

Xiao H W, Xiao H Y, Long A M, Wang Y L. 2012. Who controls the monthly variations of NH_4^+ nitrogen isotope composition in precipitation? Atmospheric Environment, 54: 201-206

Xiao H W, Xiao H Y, Long A M, Wang Y L, Liu C Q. 2013. Chemical composition and source apportionment of rainwater at Guiyang, SW China. Journal of Atmospheric Chemistry, 70(3): 269-281

Xiao H W, Xiao H Y, Long A M, Liu C Q. 2015. $\delta^{15}N\text{-}NH_4^+$ variations of rainwater: application of the Rayleigh model. Atmospheric Research, 157: 49-55

Xiao W J, Xu Y P, Haghipour N, Montluçon, D B, Pan B B, Jia Z H, Ge H M, Yao P, Eglinton T I. 2020. Efficient sequestration of terrigenous organic carbon in the New Britain Trench. Chemical Geology, 533: 119446

Xu F, Ma T, Zhou L, Hu Z F, Shi L. 2015. Chromium isotopic fractionation during Cr(VI) reduction by *Bacillus* sp. under aerobic conditions. Chemosphere, 130: 46-51

Xu H, Hong Y T, Lin Q H, Zhu Y X, Hong B, Jiang H B. 2006. Temperature responses to quasi-100-yr solar variability during the past 6000 years based on $\delta^{18}O$ of peat cellulose in Hongyuan, eastern Qinghai-Tibet Plateau, China. Palaeogeography, Palaeoclimatology, Palaeoecology, 230(1-2):

155-164

Xu H, Zhou K E, Lan J H, Zhang G L, Zhou X Y. 2019. Arid Central Asia saw mid-Holocene drought. Geology, 47(3): 255-258

Xu W P, Zhu J M, Johnson T M, Wang X L, Lin Z Q, Tan D C, Qin H B. 2020. Selenium isotope fractionation during adsorption by Fe, Mn and Al oxides. Geochimica et Cosmochimica Acta, 272: 121-136

Xu Y, Xiao H Y. 2017. Concentrations and nitrogen isotope compositions of free amino acids in Pinus massoniana (Lamb.) needles of different ages as indicators of atmospheric nitrogen pollution. Atmospheric Environment, 164: 348-359

Xu Y, Tian C G, Ma J M, Zhang G, Li Y F, Ming L L, Li J, Chen Y J, Tang J H. 2012. Assessing environmental fate of β-HCH in asian soil and association with environmental factors. Environmental Science & Technology, 46(17): 9525-9532

Xu Y, Tian C G, Zhang G, Ming L L, Wang Y, Chen Y J, Tang J H, Li J, Luo C L. 2013. Influence of monsoon system on α-HCH fate in Asia: a model study from 1948 to 2008. Journal of Geophysical Research: Atmospheres, 118(12): 6764-6770

Xu Y, Xiao H Y, Qu L L. 2017. Nitrogen concentrations and nitrogen isotopic compositions in leaves of Cinnamomum Camphora and Pinus massoniana (Lamb.) for indicating atmospheric nitrogen deposition in Guiyang (SW China). Atmospheric Environment, 159: 1-10

Xue Y J, Zou L, Ge T T, Wang X C. 2017. Mobilization and export of millennial-aged organic carbon by the Yellow River. Limnology and Oceanography, 62(S1): S95-S111

Xue Z C, Rehkämper M, Horner T J, Abouchami W, Middag R, van de Flierdt T, de Baar H J W. 2013. Cadmium isotope variations in the Southern Ocean. Earth and Planetary Science Letters, 382: 161-172

Yan B, Zhu X K, He X X, Tang S H. 2019. Zn isotopic evolution in early Ediacaran Ocean: a global signature. Precambrian Research, 320: 472-483

Yang D A, Cartigny P, Desboeufs K, Widory D. 2018a. Seasonality in the $\Delta^{33}S$ measured in urban aerosols highlights an additional oxidation pathway for atmospheric SO_2. Atmospheric Chemistry and Physics, 19(6): 3779-3796

Yang D A, Bardoux G, Assayag N, Laskar C, Widory D, Cartigny P. 2018b. Atmospheric SO_2 oxidation by NO_2 plays no role in the mass independent sulfur isotope fractionation of urban aerosols. Atmospheric Environment, 193: 109-117

Yang F, He K, Ye B, Chen X, Cha L, Cadle S H, Chan T, Mulawa P A. 2005. One-year record of organic and elemental carbon in fine particles in downtown Beijing and Shanghai. Atmospheric Chemistry and Physics, 5(6): 1449-1457

Yang J Y T, Kao S J, Dai M H, Yan X L, Lin H L. 2017. Examining N cycling in the northern South China Sea from N isotopic signals in nitrate and particulate phases. Journal of Geophysical Research: Biogeosciences, 122(8): 2118-2136

Yang R Q, Wang Y W, Li A, Zhang Q H, Jing C Y, Wang T, Wang P, Li Y M, Jiang G B. 2010. Organochlorine pesticides and PCBs in fish from lakes of the Tibetan Plateau and the implications. Environmental Pollution, 158(6): 2310-2316

Yang S C, Lee D C, Ho T Y. 2012. The isotopic composition of Cadmium in the water column of the South China Sea. Geochimica et Cosmochimica Acta, 98: 66-77

Yang S C, Lee D C, Ho T Y. 2015. Cd isotopic composition in the suspended and sinking particles of the surface water of the South China Sea: the effects of biotic activities. Earth and Planetary Science Letters, 428: 63-72

Yang W J, Ding K B, Zhang P, Qiu H, Cloquet C, Wen H J, Morel J L, Qiu R L, Tang Y T. 2019. Cadmium stable isotope variation in a mountain area impacted by acid mine drainage. Science of the Total Environment, 646: 696-703

Yang Y Y, Toor G S. 2016. $\delta^{15}N$ and $\delta^{18}O$ reveal the sources of nitrate-nitrogen in urban residential stormwater runoff. Environmental Science & Technology, 50(6): 2881-2889

Yeung L Y, Murray L T, Martinerie P, Witrant E, Hu H T, Banerjee A, Orsi A, Chappellaz J. 2019. Isotopic constraint on the twentieth-century increase in tropospheric ozone. Nature, 570(7760): 224-227

Yin R S, Feng X B, Foucher D, Shi W F, Zhao Z Q, Wang J. 2010. High precision determination of mercury isotope ratios using online mercury vapor generation system coupled with multi-collector inductively coupled plasma-mass spectrometry. Chinese Journal of Analytical Chemistry, 38(7): 929-934

Yin R S, Feng X B, Wang J X, Li P, Liu J L, Zhang Y, Chen J B, Zheng L R, Hu T D. 2013a. Mercury speciation and mercury isotope fractionation during ore roasting process and their implication to source identification of downstream sediment in the Wanshan mercury mining area, SW China. Chemical Geology, 336: 72-79

Yin R S, Feng X B, Wang J X, Bao Z D, Yu B, Chen J B. 2013b. Mercury isotope variations between bioavailable mercury fractions and total mercury in mercury contaminated soil in Wanshan mercury mine, SW China. Chemical Geology, 336: 80-86

Yin R S, Feng X B, Meng B. 2013c. Stable mercury isotope variation in rice plants (Oryza sativa L.) from the Wanshan mercury mining district, SW China. Environmental Science & Technology, 47(5): 2238-2245

Yin R S, Feng X B, Chen J B. 2014. Mercury stable isotopic compositions in coals from major coal producing fields in China and their geochemical and environmental implications. Environmental Science & Technology, 48(10): 5565-5574

Yin R S, Feng X B, Chen B W, Zhang J J, Wang W X, Li X D. 2015. Identifying the sources and processes of mercury in subtropical estuarine and ocean sediments using Hg isotopic composition. Environmental Science & Technology, 49(3): 1347-1355

Yin R S, Feng X B, Hurley J P, Krabbenhoft D P, Lepak R F, Hu R Z, Zhang Q, Li Z G, Bi X W. 2016a. Mercury isotopes as proxies to identify sources and environmental impacts of mercury in sphalerites. Scientific Reports, 6: 18686

Yin R S, Feng X B, Hurley J P, Krabbenhoft D P, Lepak R F, Kang S C, Yang H D, Li X D. 2016b. Historical records of mercury stable isotopes in sediments of Tibetan lakes. Scientific Reports, 6: 23332

Yin R S, Feng X B, Zhang J J, Pan K, Wang W X, Li X D. 2016c. Using mercury isotopes to understand the bioaccumulation of Hg in the subtropical

Pearl River estuary, South China. Chemosphere, 147: 173-179

Yin X F, Kang S C, de Foy B, Ma Y M, Tong Y D, Zhang W, Wang X J, Zhang G S, Zhang Q G. 2018. Multi-year monitoring of atmospheric total gaseous mercury at a remote high-altitude site (Nam Co, 4730 m a. s. l.) in the inland Tibetan Plateau region. Atmospheric Chemistry and Physics, 18(14): 10557-10574

Young E D, Yeung L Y, Kohl I E. 2014. On the $\Delta^{17}O$ budget of atmospheric O_2. Geochimica et Cosmochimica Acta, 135: 102-125

Yu B, Wang X, Lin C J, Fu X W, Zhang H, Shang L H, Feng X B. 2015. Characteristics and potential sources of atmospheric mercury at a subtropical near-coastal site in East China. Journal of Geophysical Research: Atmospheres, 120(16): 8563-8574

Yu B, Fu X W, Yin R S, Zhang H, Wang X, Lin C J, Wu C S, Zhang Y P, He N N, Fu P Q, Wang Z F, Shang L H, Sommar J E, Sonke J E, Maurice L, Guinot B, Feng X B. 2016. Isotopic composition of atmospheric mercury in China: new evidence for sources and transformation processes in air and in vegetation. Environmental Science & Technology, 50(17): 9262-9269

Yu M, Guo Z G, Wang X C, Ian Eglinton T, Yuan Z N, Xing L, Zhang H L, Zhao M X. 2018. Sources and radiocarbon ages of aerosol organic carbon along the east coast of China and implications for atmospheric fossil carbon contributions to China marginal seas. Science of the Total Environment, 619-620: 957-965

Yu M, Eglinton T I, Haghipour N, Montluçon D B, Wacker L, Hou P F, Zhang H L, Zhao M X. 2019. Impacts of natural and human-induced hydrological variability on particulate organic carbon dynamics in the Yellow River. Environmental Science & Technology, 53(3): 1119-1129

Yuan W, Sommar J, Lin C J, Wang X, Li K, Liu Y, Zhang H, Lu Z Y, Wu C S, Feng X B. 2019. Stable isotope evidence shows Re-emission of elemental mercury vapor occurring after reductive loss from foliage. Environmental Science & Technology, 53(2): 651-660

Zeng J, Han G L. 2020. Preliminary copper isotope study on particulate matter in Zhujiang River, Southwest China: application for source identification. Ecotoxicology and Environmental Safety, 198: 110663

Zhang H, Yin R, Feng X B, Sommar J, Anderson C W N, Sapkota A, Fu X W, Larssen T. 2013. Atmospheric mercury inputs in montane soils increase with elevation: evidence from mercury isotope signatures. Scientific Reports, 3: 3322

Zhang H, Fu X W, Lin C J, Wang X, Feng X B. 2015. Observation and analysis of speciated atmospheric mercury in Shangri-La, Tibetan Plateau, China. Atmospheric Chemistry and Physics, 15(2): 653-665

Zhang H, Fu X W, Lin C J, Shang L H, Zhang Y P, Feng X B, Lin C. 2016. Monsoon-facilitated characteristics and transport of atmospheric mercury at a high-altitude background site in southwestern China. Atmospheric Chemistry and Physics, 16(20): 13131-13148

Zhang L, Wang L, Wang S X, Dou H Y, Li J F, Li S, Hao J M. 2017. Characteristics and sources of speciated atmospheric mercury at a coastal site in the East China Sea region. Aerosol and Air Quality Research, 17(12): 2913-2923

Zhang R, Chen M, Yang Q, Lin Y S, Mao H B, Qiu Y S, Tong J L, Lv E, Yang Z, Yang W F, Cao J P. 2015. Physical-biological coupling of N_2 fixation in the northwestern South China Sea coastal upwelling during summer. Limnology and Oceanography, 60(4): 1411-1425

Zhang S T, Liu Q, Tang M, Liu Y. 2020. Molecular-level mechanism of phosphoric acid digestion of carbonates and recalibration of the ^{13}C-^{18}O clumped isotope thermometer. ACS Earth and Space Chemistry, 4(3): 420-433

Zhang T Y, Lu D W, Zeng L X, Yin Y G, He Y J, Liu Q, Jiang G B. 2017. Role of secondary particle formation in the persistence of silver nanoparticles in humic acid containing water under light irradiation. Environmental Science & Technology, 51(24): 14164-14172

Zhang X Y, Xu Z W, Sun X M, Dong W Y, Ballantine D. 2013. Nitrate in shallow groundwater in typical agricultural and forest ecosystems in China, 2004-2010. Journal of Environmental Sciences, 25(5): 1007-1014

Zhang Y L, Perron N, Ciobanu V G, Zotter P, Minguillón M C, Wacker L, Prévôt A S H, Baltensperger U, Szidat S. 2012. On the isolation of OC and EC and the optimal strategy of radiocarbon-based source apportionment of carbonaceous aerosols. Atmospheric Chemistry and Physics, 12(22): 10841-10856

Zhang Y L, Kang S C, Chen P F, Li X F, Liu Y J, Gao T G, Guo J M, Sillanpää M. 2016. Records of anthropogenic antimony in the glacial snow from the southeastern Tibetan Plateau. Journal of Asian Earth Sciences, 131: 62-71

Zhang Y X, Wen H J, Zhu C W, Fan H F, Luo C G, Liu J, Cloquet C. 2016. Cd isotope fractionation during simulated and natural weathering. Environmental Pollution, 216: 9-17

Zhang Y X, Wen H J, Zhu C W, Fan H F, Cloquet C. 2018. Cadmium isotopic evidence for the evolution of marine primary productivity and the biological extinction event during the Permian-Triassic crisis from the Meishan section, South China. Chemical Geology, 481: 110-118

Zhang Z Y, Xiao H Y, Zheng N J, Gao X F, Zhu R G. 2016a. Compound-specific isotope analysis of amino acid labeling with stable isotope nitrogen (^{15}N) in higher plants. Chromatographia, 79(17-18): 1197-1205

Zhang Z Y, Tian J, Xiao H W, Zheng N J, Gao X F, Zhu R G, Xiao H Y. 2016b. A reliable compound-specific nitrogen isotope analysis of amino acids by GC-C-IRMS following derivatisation into N-pivaloyl-iso-propyl (NPIP) esters for high-resolution food webs estimation. Journal of Chromatography B, 1033-1034: 382-389

Zhang Z Y, Tian J, Cao Y S, Zheng N J, Zhao J J, Xiao H W, Guo W, Zhu R G, Xiao H Y. 2019. Elucidating food web structure of the Poyang Lake ecosystem using amino acid nitrogen isotopes and Bayesian mixing model. Limnology and Oceanography: Methods, 17(11): 555-564

Zhang Z Y, Wang W X, Zheng N J, Cao Y S, Xiao H W, Zhu R G, Guan H, Xiao H Y. 2021. Methylmercury biomagnification in aquatic food webs of Poyang Lake, China: insights from amino acid signatures. Journal of Hazardous Materials, 404: 123700

Zheng Q, Nizzetto L, Mulder M D, Šáňka O, Lammel G, Li J, Bing H J, Liu X, Jiang Y S, Luo C L, Zhang G. 2014. Does an analysis of polychlorinated biphenyl (PCB) distribution in mountain soils across China reveal a latitudinal fractionation paradox? Environmental Pollution, 195: 115-122

Zheng Q, Nizzetto L, Liu X, Borgå K, Starrfelt J, Li J, Jiang Y S, Liu X, Jones K C, Zhang G. 2015. Elevated mobility of persistent organic pollutants in the soil of a tropical rainforest. Environmental Science & Technology, 49(7): 4302–4309

Zheng X D, Liu X Y, Song W, Sun X C, Liu C Q. 2018. Nitrogen isotope variations of ammonium across rain events: implications for different scavenging between ammonia and particulate ammonium. Environmental Pollution, 239: 392–398

Zheng X D, Teng Y G, Song L T. 2019. Iron isotopic composition of suspended particulate matter in Hongfeng lake. Water, 11(2): 396

Zhou W J, Wu S G, Huo W W, Xiong X H, Cheng P, Lu X F, Niu Z C. 2014. Tracing fossil fuel CO_2 using $\Delta^{14}C$ in Xi'an City, China. Atmospheric Environment, 94: 538–545

Zhou W J, Niu Z C, Wu S G, Xiong X H, Hou Y Y, Wang P, Feng T, Cheng P, Du H, Lu X F, An Z S, Burr G S, Zhu Y Z. 2020. Fossil fuel CO_2 traced by radiocarbon in fifteen Chinese cities. Science of the Total Environment, 729: 138639

Zhu C W, Wen H J, Zhang Y X, Yin R S, Cloquet C. 2018. Cd isotope fractionation during sulfide mineral weathering in the Fule Zn-Pb-Cd deposit, Yunnan Province, Southwest China. Science of The Total Environment, 616–617: : 64–72

Zhu G H, Ma J L, Wei G J, An Y J. 2020. A novel procedure for separating iron from geological materials for isotopic analysis using MC-ICP-MS. Journal of Analytical Atomic Spectrometry, 35(5): 873–877

Zhu J M, Johnson T M, Clark S K, Zhu X K, Wang X L. 2014. Selenium redox cycling during weathering of Se-rich shales: a selenium isotope study. Geochimica et Cosmochimica Acta, 126: 228–249

Zhu R G, Xiao H Y, Zhang Z Y, Lai Y Y. 2018. Compound-specific $\delta^{15}N$ composition of free amino acids in moss as indicators of atmospheric nitrogen sources. Scientific Reports, 8: 14347

Zhu R G, Xiao H Y, Lv Z, Xiao H, Zhang Z Y, Zheng N J, Xiao H W. 2019. Nitrogen isotopic composition of free Gly in aerosols at a forest site. Atmospheric Environment, 222: 117179

Zhu R G, Xiao H Y, Zhu Y W, Wen Z Q, Fang X Z, Pan Y Y. 2020. Sources and transformation processes of proteinaceous matter and free amino acids in $PM_{2.5}$. Journal of Geophysical Research: Atmospheres, 125(5): e2020JD032375

Zhu W, Li Z G, Li P, Yu B, Lin C J, Sommar J, Feng X B. 2018. Re-emission of legacy mercury from soil adjacent to closed point sources of Hg emission. Environmental Pollution, 242: 718–727

Zink S, Schoenberg R, Staubwasser M. 2010. Isotopic fractionation and reaction kinetics between Cr(III) and Cr(VI) in aqueous media. Geochimica et Cosmochimica Acta, 74(20): 5729–5745

Zou X, Hou S G, Wu S Y, Zhang W B, Liu K, Yu J H, Liu Y P, Pang H X. 2020. An assessment of natural and anthropogenic trace elements in the atmospheric deposition during 1776–2004 A. D. using the Miaoergou ice core, eastern Tien Shan, China. Atmospheric Environment, 221: 117112

Progresses in Environmental Geochemistry Study

FENG Xin-bin[1,2], CAO Xiao-bin[3], FU Xue-wu[1,2], HONG Bing[1,2], GUAN Hui[1], LI Ping[1,2], WANG Jing-fu[1], WANG Shi-lu[1], ZHANG Gan[4], ZHAO Shi-zhen[4]

1. State Key Laboratory of Environmental Geochemistry, Institute of Geochemistry, Chinese Academy of Sciences, Guiyang 550081; 2. Center for Excellence in Quaternary Science and Global Change, Chinese Academy of Sciences, Xi'an 710061; 3. International Isotope Effect Research Center, College of Earth Science and Engineering, Nanjing University, Nanjing 210023; 4. State Key Laboratory of Organic Geochemistry, Guangzhou Institute of Geochemistry, Chinese Academy of Sciences, Guangzhou 510640

Abstract: In this paper, we summarized some of the research progress accomplished by Chinese scientists in the field of environmental geochemistry in the last decade. We introduced the research progress in long range transport of mercury, persistent organic pollutants (POPs) and other heavy metals in the environment. We summarized the progress in the fractionation of non-traditional stable isotopes during the biogeochemical cycling in the surface environment and application of non-traditional stable isotopes in tracing the source and environmental process of heavy metal pollution. We also introduced the progress in using traditional stable isotopes to reconstruct the climate change and to trace the sources of pollutants in the environment. Meanwhile, we suggested knowledge gaps and research needs of above research areas. We strongly believed that environmental geochemistry will play a more important role in the eco-civilization and beautiful China development plan, the implementation of international environmental conventions and the study of global change.

Key words: long range transport; heavy metals; persistent organic pollutants; traditional stable isotope; global change; non-traditional stable isotope

地质源温室气体释放研究概述*

郑国东[1] 赵文斌[2] 陈 志[3] 胥 旺[1,4] 宋之光[5] 李 琦[6] 徐 胜[7]
郭正府[2] 马向贤[1] 梁明亮[1,8] 王云鹏[5]

1. 中国科学院 西北生态环境资源研究院，甘肃省油气资源研究重点实验室，兰州 730000；2. 中国科学院 地质与地球物理研究所，北京 100029；3. 中国地震局 地震预测研究所，北京 100036；4. 成都理工大学 能源学院，成都 610059；5. 中国科学院 广州地球化学研究所，有机地球化学国家重点实验室，广州 510640；6. 中国科学院 武汉岩土力学研究所，岩土力学与工程国家重点实验室，武汉 430071；7. 天津大学 表层地球系统科学研究院，天津 300072；8. 中国地质科学院 地质力学研究所，北京 100081

摘 要：地质源温室气体是指固体地球通过各种地质作用向大气圈释放的温室气体，是固体地球与大气之间物质交换（地气交换）的重要形式，主要包括火山喷发和地热活动、断裂带构造运动、油气渗漏、天然气水合物分解、煤自燃以及岩石风化等多种地质作用过程所释放的二氧化碳、甲烷等气体。实际上，地气交换是重要的地质作用，是地球各圈层物质循环和能量交换的基本载体和重要动力学机制。全球气候变化和温室效应是人类面临的巨大挑战，地质源温室气体的类型与来源、释放机理与过程、释放通量与大气温室效应等的调查和研究已成为当今地球系统科学的热点问题和发展方向之一，对于应对全球变化和温室效应具有重要的科学意义。中国科学家积极参与和适时开展地质源温室气体调查研究，对中国大陆部分火山和地热区、地震断裂带、含油气区和泥火山等释放的温室气体进行了初步观测与调查，在地质源温室气体的地球化学组成、释放通量等方面取得了一些成果，这些工作为进一步的深入研究奠定了良好基础。

关键词：地质源温室气体 释放特征 研究现状 发展前景

0 引 言

全球环境与气候变化，是人类生存与社会发展面临的重大问题。查明各类温室气体来源及其占比、厘清大气碳收支平衡是应对全球气候变化的关键科学问题之一。近期的研究表明，大气圈温室气体浓度的增加是自然释放和人为排放共同作用的结果，其中地质源温室气体的贡献不可忽视，需要进行分类甄别和系统研究（Jenkinson et al.，1991；Etiope and Klusman，2002；郭正府等，2017）。由于地理、地貌、气候等条件的巨大差异以及地质作用的极端复杂性和历史悠久性，目前对地质源温室气体释放的认识还比较局限，地质作用向大气圈释放温室气体的机理、过程及释放通量等还很不清楚，从而导致不能准确判断地质源温室气体在大气碳收支中的占比，也无法确定自然源和人为排放量的相对比例。在此背景下制定的温室气体减排政策必然遇到研究基础不足的困惑，也会影响社会发展和经济建设的合理布局，并且阻碍"碳达峰"和"碳中和"战略的顺利实施，甚至波及国际谈判的话语权。

为了应对气候谈判，欧洲多国、美国、日本以及我国台湾都针对地质源温室气体开展了相应的碳排放专门调查，发表了大量的学术论文、研究专著和报告。我国地质源温室气体的调查研究也取得了一些积极进展，特别是在国家自然科学基金、中国地震局专项基金、中科院国际合作基金等支持下，相关研

---

* 原文"中国地质源温室气体释放近十年研究概述"刊于《矿物岩石地球化学通报》2021 年第 40 卷第 6 期，本文略有修改。

究团队对我国大陆范围的部分火山和地热区、地震断裂带、含油气盆地和泥火山等释放的温室气体进行了专项调查研究，在地质源温室气体的地球化学组成、释放通量等方面取得了一系列研究成果（Tang et al.，2007，2008；马向贤等，2012；郭正府等，2014，2015；Zheng et al.，2017；Chen et al.，2019a）。但与其他发达国家和地区相比，我国地质源温室气体的调查研究由于缺乏持续的课题和经费支持，已有的工作还比较零散，尚缺乏系统性和全面性的研究工作，全国甚至较大范围的区域性地质源温室气体释放数据几乎还是空白，亟需国家层面的有效组织和专项经费的重点支持。

1 地质源温室气体

温室气体是指大气中能够吸收地面反射的太阳辐射并重新发射辐射的一些气体，其存在和含量变化可以导致大气温度的升降，主要包括二氧化碳（CO_2）、一氧化碳（CO）等非烃气体，甲烷（CH_4）、乙烷（C_2H_6）等烃类气体，以及水蒸气等。目前，地质源温室气体的调查研究主要侧重于烃类气体（如CH_4、C_2H_6等）、CO_2和CO等非烃含碳气体。一方面，人类居住的蓝色地球拥有相对稳定的大气圈和水圈，为生命的产生和演化提供了最基本的必要条件；也正因为大气中存在温室气体，才可以保持近地表大气温度在一定范围内相对稳定，从而保证了宜居地球的生机盎然。另一方面，大气CO_2等温室气体浓度持续升高将导致全球气候变暖、气候灾害的频繁发生，自然灾害事件又威胁到人类的生存发展。很多研究结果表明，大气CO_2浓度的升高不仅是人类活动碳排放的结果，地球演化过程也一直在向大气圈释放CO_2气体（Foley and Fischer，2017）。因此，研究地质作用向大气释放CO_2等温室气体的规律和强度，准确评估自然过程对大气温室气体平衡的影响，对保护人类生存繁衍的地球环境具有重要的指导意义。

地质源温室气体释放伴随地球演化的整体过程，只是这种释放在大部分地质历史时期总体上维持着平衡状态。然而，进入工业化以来，由于化石燃料的大量使用，造成短时期内大气CO_2浓度急剧升高和严峻的温室效应问题（Brune et al.，2017；Cox et al.，2000）。在此背景下，地质作用温室气体的释放也日益受到地球科学工作者的关注和重视。地质源温室气体的调查研究就是从地质作用及其演变过程来探讨地球脱气作用与机理，包括研究现今地球脱气和大气圈温室气体的含量、来源与大气质量及平衡关系，甚至某些极端气象事件与地球释放气体的相关性等基本问题，以便为预测大气环境演变趋势和制定对策方案提供科学依据。然而，由于地质作用过程极为复杂、影响因素众多，要弄清地质源温室气体释放的过程、强度、规模等关键问题，还需要协调和组织多领域多学科的科技力量，开展广泛深入的合作研究，为我国应对气候变化以及实现"碳达峰"和"碳中和"目标提供科学依据。

地球物质的存在形态主要包括固体（态）、液体（态）、气体（态），其中根据气体化学组成又可以将气体分为烃类气体、非烃类气体和稀有气体三大类。当今全球气候异常、生态环境持续恶化，而大气二氧化碳等温室气体浓度持续升高被认为是全球气候变化的主要因素。因此，我们提出将"温室气体"单列为专门气体类型，开展针对性调查和科学研究，显然具有重要的科学意义和现实应用价值。

2 气体地球化学基本原理与研究内容

人类对空气的认识由来已久，空气与生命起源、物种进化、人类生存等都密切相关。然而，气体地球化学作为相对独立的分支学科出现的历史却比较短暂，但其发展令人鼓舞。气体地球化学，作为认识和研究地球内部地质作用机制与过程的基础学科，在研究地质作用释放气体的化学组成和同位素变化方面具有特殊的支撑作用。

2.1 气体地球化学及其研究领域

气体地球化学主要是以自然界中呈气态存在的元素及其化合物的地球化学特征、成因类型、迁移和聚集规律及其所参与的地球动力学过程、所表征的地球科学意义等为目标的科学体系。地质作用过程释放气体的现象很早就受到科学家的注意，德国、意大利、希腊等欧洲国家的科学家对地球释放气体的研究可以追溯到十八九世纪（Heinicke and Martinelli，2005），但作为一门分支学科"气体地科化学（gas geochemistry）"一词则是 1984 年在美国夏威夷召开的"火山、地震、资源勘探和地球内部的气体地球化学"国际学术会议上才正式出现，而且所研究的内容也多局限在烃类化合物的天然气和地热等资源利用方面（Heinicke and Martinelli，2005）。现代气体地球化学的研究领域已涉及地球的各个层圈，构成了气体地球化学的完整学科体系，并在资源、能源、环境、灾害等众多领域得到广泛应用。

我国天然气地球化学基础理论研究与勘探实践成绩斐然，天然气形成机理、运移与油气资源评价带二氧化碳与氢气等主要成分，也为解决我国油气资源勘探和开发生产的实际问题做出了突出贡献，创新成果不断涌现（刘文汇等，2021）。地质灾害调查研究是气体地球化学的另一个重要应用领域，尤其是地震带二氧化碳与氢气等主要成分，以及氡气和氦气等稀有气体组分的现场检测和综合研究方面取得了很多令人振奋的研究成果（陶明信等，2005；Zhou et al.，2010，2016；Han et al.，2014；周晓成等，2017），最近数年，相关研究经常与断裂带温室气体释放调查同步进行（崔月菊等，2017；Chen et al.，2019a；Sun et al.，2020a）。需要说明的是，本文将着重论述地质源温室气体释放调查研究方面的现状，而有关以上两个领域的研究进展将有专文总结，在此不再赘述。

地质源温室气体释放特征的调查研究离不开气体地球化学基础理论及其应用实践，根据地质作用释放温室气体的化学组成和相关元素的同位素特征，可以判识其来源、运移机理和释放等基础问题，进而指导地质源温室气体的类型划分和释放通量估算。通过地质源温室气体地球化学研究还可以加深对地球整体系统的了解，获得地球形成、演化的新认识，从而为解决资源、能源、环境、灾害等问题提供重要的科学依据（郑国东等，2018）。

2.2 化学组成

气体的化学组成（或组分）及其相对含量（浓度）是气体地球化学研究的基本内容，也是确定气体来源的重要标志。气体组分是指在地球系统的物理-化学条件下可以存在的各种气体分子及其所占的比例。随着科学技术的不断进步，现如今各类气体组分几乎可以实现种类确定及其相对含量的精确测定，尤其是不断创新发展的检测技术和分析仪器为地质源温室气体化学组分的高灵敏度和高精度检测分析提供了强有力的技术保障。

地球的大气系统是多种气体的混合物，根据其含量和变化可以分为三大类：恒定组分、可变组分和不定组分。恒定组分是指氮气（N_2）、氧气（O_2）和氩气（Ar）。现代大气的气体组成中，N_2 占空气体积的 78.09%、O_2 占 20.95%、Ar 占 0.93%，三者总和占空气总体积的 99.97%，其余组分是微量的氖（Ne）、氦（He）、氙（Xe）、氡（Rn）等稀有气体。可变组分是指空气中的 CO_2 和水蒸气，通常 CO_2 含量为 0.02%~0.04%，水蒸气含量小于 4%（唐孝炎等，2006）。可变组分在空气中的含量随季节、气象条件与人类活动的变化而变化。不定组分包括煤烟、尘埃、硫氧化物、氮氧化物和一氧化碳等，主要与人类活动直接相关，当这些组分达到一定浓度后就会给人类、生物造成严重的危害，因此，不定组分是大气环境科学研究的主要对象。

大气的可变组分及其相关气体正是主要的温室气体，也是地质源温室气体调查研究的直接对象。地质源温室气体主要是指含碳气体，既有烃类气体（如 CH_4、C_2H_6 等），又有非烃类气体（如 CO_2、CO 等）。火山喷发和地热活动释放的温室气体主要是 CO_2，也有 CH_4 等烃类气体；含油气盆地以及泥火山释放的温室气体主要以 CH_4、C_2H_6 等烃类气体为主（Etiope and Ciccioli，2009），并伴有 CO_2、CO 等非烃气体。不同来源的地质源温室气体，其化学组分与现今大气平均化学组成存在明显差异，因此可以利用各种气体含量之间的比值，如 O_2/N_2、$CH_4/(C_2H_6+C_4H_{10})$、CO_2/Ar 等，来区分和确认气体的主要来源和可能成因（Zheng et al.，2010；Xu et al.，2012）。

2.3 同位素组成

同位素是指具有相同原子序数（即质子数相同）但质量数不同，亦即中子数不同的一组核素，是同一种元素，但不是同一种原子，在元素周期表中占据同一个位置。气体同位素就是指气体元素不同质量数的原子，它们之间的相对关系，即同位素比值。根据其来源，可划分为放射性同位素和稳定性同位素，这也是同位素地球化学研究的主要目标。放射性同位素，半衰期是恒定的，可用来确定年龄。而稳定同位素与母源相关，因此可以作为溯源标志，即稳定同位素的示踪作用。

不同来源温室气体构成元素的同位素组成差异很大，可以用来区分气体来源和成因，揭示各种地质作用过程中相关气体的分馏与混合等问题。地质源温室气体主要包括 CH_4 等烃类气体的碳氢同位素、CO 和 CO_2 的碳同位素、稀有气体尤其是氦同位素等。当然，特定背景下的地球脱气研究，其他一些同位素的应用也经常遇到，如硫同位素、氮同位素等。

地质源温室气体在其生成-运移-释放过程中必然会经历各种地质作用过程，尤其是特定元素可以发生固体-液体-气体的相互转换，很容易导致同位素组成特征的变化，即分馏效应。例如，碳（C）元素，既是金刚石、石墨、白云石、方解石等固体矿物的组成部分，又是液态石油烃的主要元素，或者溶解在水中形成碳酸根离子，也可以成为天然气以及 CO、CO_2 等气体的主要组成，而作为生命物质构成的主要元素，其存在形式更是多种多样，C 元素在各种转化过程中的同位素分馏也很明显。因此，相关元素同位素组成特征可以提供地球各圈层物质组成和循环以及能量交换的重要信息，并利用相关同位素组成特征（比值）追索气体的来源以及迁移机理和过程，进而解析相关地质作用。

2.4 源区判识

尽管不同地质单元和地质作用类型伴随的气体化学组成和同位素比值变化很大，但随着调查领域的持续扩展和研究程度的不断深入，众多研究结果揭示，各种地质背景条件下地球脱出气体的化学组分和同位素组成均具有一定的变化范围。不少科学家也提出了一些特定地质背景条件下的地质源温室气体的划分模式和判别指标。目前被普遍采用的方法主要包括：化学组成、各种气体成分的相互比例，同位素比值，CH_4、CO_2 等气体的碳氢氧同位素，以及近些年迅速发展的团簇同位素等。当然，这些化学组成和同位素比值的应用需要考虑具体的地质背景来选择，以准确判识其来源。

地表水热系统中的气体包括了 CO_2、N_2、O_2、SO_2、H_2O 等主要组分和 H_2、He、Ne、Ar、CO 等微量组分。CO_2 的来源主要包括：大气成因、有机成因、碳酸盐岩溶解成因、变质成因及地幔来源等。其中，有机成因 CO_2 主要指地表土壤植物根系呼吸作用和土壤有机质分解形成的 CO_2；碳酸盐岩溶解成因 CO_2 主要指表层环境中土壤 CO_2 和区域碳酸盐岩溶解形成的 CO_2；变质成因 CO_2 主要包括伴随岩石变质过程的脱碳作用（decarbonation），也包括少量沉积变质岩中有机质分解和石墨的氧化等。所以，对于地球（地幔）脱气的 CO_2，就需要区分和甄别包括地表浅层大气成因 CO_2 和碳酸盐岩溶解成因 CO_2，以及除此之外的深源 CO_2，即变质成因 CO_2 和幔源成因 CO_2 等各种主要组成部分。

与 CO_2 相比较，稀有气体具有相对简单的来源，以地球原始气体 ^3He 为代表的上地幔或岩石圈地幔组分、放射性元素 U 和 Th 等衰变而来的地壳组分，以及大气组分。稀有气体具有化学惰性（不活动性）和较低的元素丰度等地球化学特征，它们在地壳流体中的变化主要取决于其溶解度、吸附性、吸着性和解吸能力等物理性质，而不涉及复杂的化学过程。大量研究结果表明，稀有气体（特别是氦）在地球各圈层具有独特的同位素组成，因此，稀有气体同位素组成（如 $^3He/^4He$、$^{20}Ne/^{22}Ne$ 等）是判识地幔和地壳来源气体最有效的同位素指标（Sano and Wakita，1985）。

随着观测数据的不断增加,各种特定地质背景端元气体组分的地球化学特征已较为清楚,各种相关端元组分之间的混合作用也已得到基本梳理,并针对壳源、幔源,有机、无机等多种来源的气体混合比例的划分等提出了一些计算模型(Kerrick and Caldeira,1998)。对于地质源温室气体,利用 CO_2 和 He 气分子组成比及其碳、氦同位素比值,可以进行来源判识和类型划分(Langmuir et al.,1978;Sano and Marty,1995),这一方法已得到普遍采纳(Sun et al.,2018;Xu et al.,2012,2013,2014;Zhang et al.,2015,2017)。各种元素及其同位素比值(如 $CO_2/^3He$、$^3He/^4He$、$^{40}Ar/^{36}Ar$、$\delta^{13}C$、$\delta^{15}N$)在气体来源判识和成因探究等研究中也应注意准确运用(Sano et al.,2017),在具体的调查研究中,区域地质背景条件的控制作用应该是解释气体来源和形成机理最重要的因素之一,而多数地球化学指标也不可能无限制地简单照搬和使用。

3 地质源温室气体野外观测与样品采集

对地质源温室气体进行的野外观测和综合研究主要包括"定性"与"定量"两个方面。前者就是要确定温室气体释放的类型,或者某种地质作用释放温室气体的特征,主要涉及气体地球化学特征研究和区域地质背景调查。后者是指特定地质单元及其地质作用释放温室气体的规模或者强度,以及释放通量。另外,还可以根据相关的地球化学指标判识不同来源气体的混合作用,计算各种来源气体的混合比例。

3.1 野外观测与测量方法

根据释放强度,地质源温室气体释放类型大致可以分为宏渗漏、小渗漏、微渗漏等。在实际调查研究中,这些类型的划分并没有严格的标准或准确的定义。通常以肉眼所见的气体释放为宏渗漏,不易觉察的为小渗漏或微渗漏,微渗漏需要借用仪器进行检测。火山地热区、大型活动断裂带、含油气沉积盆地、海底沉积物等温室气体释放的形式主要有土壤微渗漏、温泉宏渗漏、泥火山宏渗漏、海水释放等多种类型,而且宏渗漏与微渗漏经常并存。针对不同的释放形式,需要在实地考察的基础之上有所选择地制定现场观察和测量方案,优选代表性地质源温室气体释放点、线(段)、面(区),以及可对比的参照区,分别采用相应的技术方法进行现场原位调查和测量计算,以获取各研究区段温室气体释放通量的定量数据。野外测量方法主要包括:密闭气室法、气体化学与水化学法、飞行器实地测量法、遥感技术和海洋底水原位探测技术等:

(1)密闭气室法。该方法适用于土壤微渗漏释放通量的测量(图1),通过记录密闭气室内 CO_2/CH_4 浓度的累积上升与时间之间的关系,计算测量点的 CO_2/CH_4 释放通量,所用仪器为便携式 CO_2/CH_4 通量测量仪(GXH-3010E、华云 CO_2 仪、Laser One CH_4 仪、METREX 2 型等)。测量点位根据需要控制的面积进行网格化布点,实际测量过程在已测点位通量变化规律的基础上判断是否需要加密检测点位。根据公式(1)计算 CO_2/CH_4 释放通量:

$$\sum_1^i F_{microseep} = \sum_1^i \frac{(V_C/A_C) \times (C_2 - C_1)/(t_2 - t_1) \times A_S}{1000 \times 22.4} \times M \times 60 \times 24 \times 365 \times 10^{-6} (t/a) \quad (1)$$

式中,F 为各测量点 CO_2/CH_4 微观渗漏通量;V_C 为箱体体积;A_C 为箱体底面积,m^2;A_S 为观测面积;C_1、C_2 分别是时间 t_1、t_2 时箱体内的 CO_2/CH_4 浓度。

(2)气体化学法。该方法主要用于泥火山和温泉气态 CO_2 和 CH_4 等温室气体通量观测,结合气体中 CO_2/CH_4 等气体含量和溢出气体总量等资料,计算温室气体释放通量。所用的仪器包括 GL-100B 型红外线数字皂膜流量计和安捷伦便携式气相色谱等(图2),根据公式(2)计算 CO_2/CH_4 释放通量:

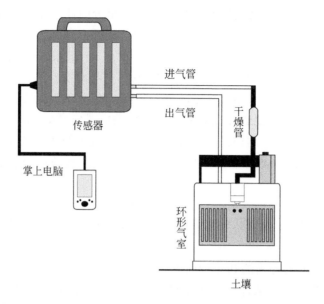

图 1 密闭气室法示意图（据郭正府等，2014）

$$E = \frac{Q_v \times T \times c}{1000 \times 22.4} \times M \times N \tag{2}$$

式中，E 为 CO_2/CH_4 的释放量；T 为时间；Q_v 为气体释放总量；c 为气体 CO_2/CH_4 浓度；N 为释放点数量；22.4 为标准状态下的气体摩尔体积；M 为 CO_2/CH_4 摩尔质量，计算时假定泥火山-温泉气体的释放通量处于恒定状态。

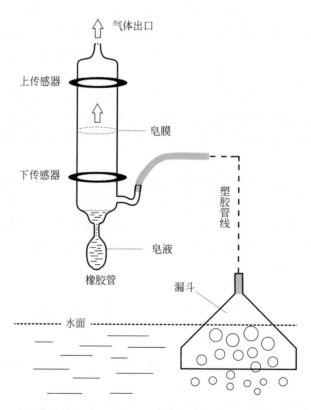

图 2 红外线数字皂膜流量计化学方法示意图（据张茂亮等，2011）

（3）水化学法。该方法可用于观测地球深部过程产生的 CO_2 溶于水形成的无机成因碳（dissolved inorganic carbon，DIC），主要为 HCO_3^- 离子。根据温泉水中逸出 CO_2 的浓度及温泉水的流量即可估算温泉水中溢出 CO_2 的通量。在计算时，需排除近地表来源的 HCO_3^{-1} 离子的贡献，公式如下：

$$C_{ex} = DIC - (Ca^{2+} + Mg^{2+} - SO_4^{2-}) \tag{3}$$

式中，C_{ex} 为温泉水中溢出的 CO_2 浓度，mol/L。该方法假定深部来源的碳在上升过程中溶于温泉水，以 DIC 形式存在，扣除来自围岩碳酸盐（方解石与白云石）中的碳，过量的碳以 CO_2 方式进入大气圈中。

（4）飞行器实地测量法。该方法用于火山喷气孔气体通量的测量，所使用的仪器为 LI-COR 非分散 CO_2 分析仪，CO_2 浓度测量范围为 $0 \sim 2\,000 \times 10^{-6}$（详见产品说明书）。

（5）遥感技术。其核心是利用相关遥感数据生成的大气 CH_4 与 CO_2 柱浓度产品反演计算近地表大气 CH_4 与 CO_2 浓度，通过历史数据对比，探讨不同时期的相关地质作用对于大气碳收支的影响，例如页岩气等油气资源开发前后对温室气体渗漏的影响，地震前后近地表 CO 释放强度变化等（Cui et al., 2013; Sun et al., 2020a）。

（6）通量塔法。该方法又称涡度协方差测量法（eddy covariance），是利用气体涡度的协方差来计算垂直方向上气体交换量，因野外测量设备形似塔状，故称通量塔法。通量塔法可以实现实时、连续全自动测量，在农田、森林、湿地生态系统的 CO_2 释放通量测量中应用较为广泛。近年来，随着传感器和计算机技术的发展，已实现多传感器（如 CO_2、CH_4、SO_2 等）同步在线监测，并应用到地质源温室气体释放量的测量中，如火山地热区（Werner et al., 2000; Anderson and Farrar, 2001; Lewicki et al., 2017）。但需要注意该项技术对工作环境的要求比较苛刻，需要测量区域下垫面尽量平坦，局地空气的温度、湿度、风力等气象条件相对稳定，因此用于地质源温室气体测量时存在较大的不确定性。

（7）海洋底水原位探测技术。主要测量海水及海底沉积物孔隙水中 CH_4 浓度。可采用 Franatech METS 灵敏性甲烷传感器与 SBE917plus CTD 联用，通过 CET 系统对 METS 进行供电和数据采集，对调查站位自近海海底到海表面海水溶解 CH_4 含量进行探测。通过对海底沉积物-水界面垂直剖面上不同位置 CH_4 进行原位测量，获得微观尺度内海底边界层 CH_4 的浓度梯度，进而估算扩散界面 CH_4 的释放通量。假定海底沉积物 CH_4 释放扩散处于恒定状态，根据 Fick 第一定律：

$$F = D \frac{dc}{dz} \tag{4}$$

式中，F 为 CH_4 通量，$mmol \cdot m^{-2} \cdot d^{-1}$；$D$ 为一定盐度和温度下海水中 CH_4 的扩散系数；dc/dz 为 CH_4 的浓度梯度 [$dc(mmol/m^3)$，$dz(m)$]。

3.2 样品采集和采样工具

由于气体的扩散性和挥发性，以及无定型性和各种气体容易相互混合等特点，地球系统的气体不仅分布极为广泛，可以说无处不在，而且能够在地球各圈层之间扩散运移，实为无孔不入。因此，包括地质源温室气体的样品采集看似简单，实际操作却很不容易，各种技术问题经常出现。对于地质源温室气体来讲，如何保证所采集的气体样品不受其他气体的污染，即保真性取样至关重要，需要采用合适的采样工具和样品容器、严格的操作程序和采集步骤、针对测试仪器进样系统而选用特定的转换接口等等，每个步骤都需要预先规划、精细操作，才能确保样品真实、数据可靠。

样品采集（物质基础）中，需要针对性地采集相关各类地质体的岩石、沉积物、气体及水体样品，用于测试分析。在火山地热区应该采集岩样和气-水样，在断裂带采集气-水样，在蛇绿岩带采集岩样和气-水样，在沉积盆地采集岩石和气-水样，在海底采集沉积物样和气-水样等。需要特别注意的是，气样采集与测试过程，一方面必须注意样品丢失问题，另一方面还要杜绝空气混染的影响。现场在线检测时，需要注意切实收集到目标气体，以及输气管线和检测仪器回路的密闭连接。气体样品的采集需要根据气体释放强度和采样现场的实际情况而定，通常采用排水法和漏斗集气法两种方法收集气体。对于冒泡频率较高的喷口，直接采用排水法在喷口液面以下收集气体；而对于气体释放强度较低或是冒泡点不容易靠近的喷口，需要采用漏斗集气法收集气体。采样过程中应做好详细记录，准确注明样品编码、采样者、日期、时间及地点等信息。

常见的气体采集装置包括不锈钢瓶、气体采样袋（铝箔袋）、不同材质的玻璃瓶、铜管等（图3）。不锈钢瓶的优势在于耐高温高压、抗腐蚀和不易燃爆，在气体样品采集和运输过程中稳定性最好，但其体积较大携带不便，可以采用体积小、重量轻、便于携带的高压铝合金双阀采样容器。气体采样袋最大的优点就是方便快捷，便于携带，操作简单，并可重复多次使用。常见的气袋有铝箔复合膜气体采样袋、聚酯（polyester）气体采样袋、PVF气体采样袋、含氟气体采样袋等。含铅玻璃瓶密闭性好，可有效防止He交换，适合各种温度温泉的采样，但价格比较昂贵，样品采集时，对排气时间需要严格的控制，排气时间足够长，确保导气管中的空气完全被排尽。无氧铜管适合于冷泉、常温温泉样品的采集，铜管在封口时两端要保留足够的空管长度，以便分析样品时与测试仪器进行连接。在样品采集过程中，要注意减压器、阀和导管都有一定的死体积，使用简单清洗操作无法有效清洗干净，从而残留气体和痕量湿气在死体积中停留并缓慢扩散进入被输送的气体中，对样品造成污染。可以采用反复增减压的清洗方法提升洗气效果。

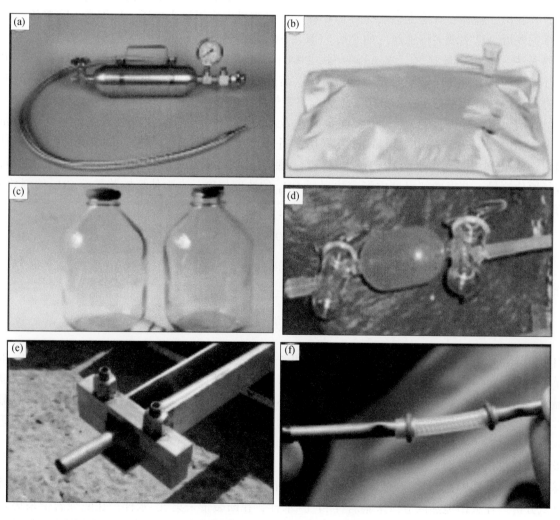

图3 常见的气体采样容器和装置

（a）双阀钢瓶：适合于高压气体收集；（b）铝箔袋：适合于空气和低压气体样品采集，需要加压取样并及时进行测试分析；（c）盐水瓶：可用于排水法气体样品的采集，但气体可溶性需要考虑；（d）含铅双耳瓶：适用于低压气体样品采集，有利于氦气、氢气等小分子气体的保存；（e）铜管：适用高低压和高温样品采集，但不适合硫化氢气体；（f）铜管+硅橡胶管：便于平行样品采集样品的运输和保存，需要注意样品容器的安全性。对于不稳定组分，尤其是氧化还原敏感性气体的变化问题，如 H_2S、CO、H_2 等，特别容易被氧化而转换成其他气体，尤其是在有氧和有水容器空间内，氧化分解速度很快。长距离样品运输和长时间存放，都需要严格的保存条件，尽可能避免样品丢失和被污染以及转化。在高原雪山冻土分布区采集流体样品需要注意样品容器的保温问题，尤其是有水存在的情况下，玻璃器皿很容易冻裂，另外，由高原低气压运送气体样品到低海拔地区的时候，由于外围气压增高可以导致样品容器内出现负压，样品的保存和转移时需要特别注意，避免相对高压的空气混入。

国际海洋调查等使用的海水地球化学勘探技术目前已经向着原位快速测试的方式方法发展。海底沉积物表层样品、沉积柱样和水柱样的采集多数都采用密闭加压技术装备以确保所采集样品不会因压力降低而改变其原始特征。沉积物水样中气体的收集可以通过减压增温等方式进行，并使用高精度分析检测仪器进行微量气体化学组成和同位素组成、沉积物烃类化合物及生物标志物等多个项目的测试鉴定，相关技术方法已经在海底天然气水合物分布区的探测方面得到实际应用（Di et al., 2014；Mau et al., 2020）。

3.3 实验室测试方法

随着分析测试技术和测试仪器的不断涌现和创新改进，几乎所有种类气体的化学成分和同位素组成目前都可以进行相对准确的测试分析，基本上可以满足包括温室气体在内的地球脱气调查研究的样品测试需求。通过对各地质体采集的水-气样进行地球化学和同位素组成的测试，结合岩石样品或近海沉积物岩石学、元素及同位素地球化学测试结果与多元混合定量计算模型等方法，探讨不同地质体释放温室气体的成因、演化及规模已经得到很好的发展。所用仪器包括：电子探针（EPMA）、多接收器电感耦合等离子体质谱（MC-ICP-MS）、3080E3 型 X 射线荧光谱仪（XRF）、MAT 271 质谱仪、Delta Plus XP 质谱仪、Agilent Macro 3000 气相色谱仪、Dionex ICS-900 型离子色谱仪以及 Noblesse 稀有气体质谱仪等。实际的仪器测试条件和操作过程需要认真记录，并在调查报告和研究论文中尽可能陈述清楚，以保证测试分析数据的可对比性。

在气体样品测试数据的分析方面，应该注意定量与定性的辩证关系，避免定量数据的无限扩大化。尤其是小量甚至微量气体，由于总量很小，气体含量很少，如氢气，但其分布广泛，单位体积样品的气体含量可能很低，但其释放规模可以很大。这种情况下，测试数据的准确性和计算过程有效数字的选择需要倍加关注。地质源温室气体的类型与释放强度与沉积、构造等地质背景条件密切相关，因此需要对区域地质背景进行全面了解和把握，进而对相关问题做出科学合理的解释。

4 中国地质源温室气体调查与研究现状

欧美日等地区以及我国台湾的地质源温室气体释放研究起步早，已有连续多年的观测积累，获取了大量的高质量观测数据，在此基础上取得了很多高水平的研究成果。相比之下，我国大陆地质源温室气体释放的调查研究工作基本上处于起步阶段，仅对个别类型的地质源温室气体进行过一些实地考察和初步研究，如长白山、腾冲、羊八井等火山地热区 CO_2 释放通量的原位测量、塔里木盆地雅克拉凝析气田和大宛齐油田 CH_4 渗漏实地检测、准噶尔盆地南缘泥火山系统油气水岩相互作用及其甲烷气体释放、汶川地震断裂带和首都圈地震断裂带体系等温室气体释放量的估算等。

4.1 火山-地热区

我国新生代火山-地热区分布广、面积大、火山成因类型丰富、所处的构造背景复杂多样，为开展火山地热区温室气体释放的观测研究提供了有利场所（Becker et al., 2008；郭正府等, 2010, 2014）。初步的研究结果显示，目前我国新生代火山地热区向大气圈释放温室气体的规模是巨大而且不可以被忽视的；

同时，处于不同构造域的火山地热区在水热活动特征、温室气体释放通量以及成因机制存在明显差异（表1）。

表1 我国新生代火山-地热区温室气体（CO_2）释放规模

编号	火山-地热区	土壤微渗漏/(t/a)	温泉/(t/a)	合计/(t/a)	构造背景
1	腾冲	4 430 000	53 000	4 480 000	特提斯构造域
2	谷露-亚东	15 000 000	—	15 000 000	
3	搭格架	—	268	268	
4	朗久	—	170	170	
5	长白山	780 000	69 000	849 000	太平洋构造域
6	五大连池	1 200 000	—	1 200 000	
合计	—	21 410 000	122 438	~2.15×10^7	—

注："—"表示无数据。数据来源：张茂亮等，2011；成智慧等，2012，2014；郭正府等，2014，2015；Zhang et al.，2015，2016，2017；赵文斌等，2018，2021；Sun et al.，2018。

青藏高原南部和腾冲火山-地热带位于特提斯构造域，主要受印度与欧亚大陆板块碰撞、俯冲的控制，以高温地热为主，深源温室气体的释放类型主要包括温泉和土壤微渗漏（郭正府等，2014，2015）。腾冲火山区每年通过土壤微渗漏、温泉等形式向大气圈释放CO_2气体约为4.5 Mt（成智慧等，2012；Zhang et al.，2016）。腾冲地区湿季和干季温室气体平均释放通量存在差别，分别为280 $g \cdot m^{-2} \cdot d^{-1}$和875 $g \cdot m^{-2} \cdot d^{-1}$（成智慧等，2014；Zhang et al.，2016），显示土壤渗透性及含水率对火山区深源温室气体释放有显著影响。青藏高原南部拉萨地块位于印度大陆俯冲的前缘，水热活动剧烈，其内部南北向谷露-亚东裂谷温室气体释放的系统研究表明，该裂谷内部地热区土壤气体释放通量为7~437 $g \cdot m^{-2} \cdot d^{-1}$，向大气圈释放$CO_2$的总通量为每年15 Mt（Zhang et al.，2017；张丽红等，2017）。此外，拉萨地块内还存在多条南北向裂谷以及大量分布在其中的温泉地热区，这些地区迄今都未开展过系统的地质源温室气体释放的调查研究（Zhang et al.，2017）。

我国东北地区新生代火山活动主要受太平洋构造域的控制，为西太平洋板块向欧亚板块俯冲的产物，以中低温地热为主（郭正府等，2014，2015；赵文斌等，2018，2021），CO_2气体释放通量相对较低（表1）。长白山火山区的温室气体释放类型主要为土壤微渗漏和温泉。其中，湖滨温泉、聚龙温泉、锦江温泉等每年通过逸出气形式释放CO_2气体约6.9万t（张茂亮等，2011）；天池火山锥体高海拔地区植被较稀疏，CO_2释放通量较接近，约为20 $g \cdot m^{-2} \cdot d^{-1}$（Zhang et al.，2015），结合其土壤微渗漏面积（110 km^2），火山锥体每年释放CO_2气体约78万t（郭正府等，2014）。而长白山天池复合成因火山锥体外围的土壤CO_2平均释放通量为41.2 $g \cdot m^{-2} \cdot d^{-1}$，要显著高于周边熔岩台地单成因火山区的平均释放通量（9.6 $g \cdot m^{-2} \cdot d^{-1}$；Sun et al.，2018）。五大连池火山区距今最近的一次喷发在约300年前（1719—1721年），与长白山火山区的显著不同之处在于，该火山区内无明显的温泉水热活动，而分布有多处水温常年低于10℃的冷泉（Xu et al.，2013），区内老黑山火山东南坡土壤微渗漏CO_2气体释放通量约11.8 $g \cdot m^{-2} \cdot d^{-1}$，全区平均释放通量为18.7 $g \cdot m^{-2} \cdot d^{-1}$（Zhao et al.，2019），初步估算，五大连池火山区通过土壤微渗漏的方式向大气圈释放CO_2的总通量约为每年1.2 Mt（表1）（赵文斌等，2021）。

火山碳观测结果表明，目前中国大陆新生代典型火山-地热区向大气圈输送的CO_2气体总通量大约为21.5 Mt/a（特提斯构造域的火山地热区贡献了其中的90%以上），释放总量相当于全球火山活动导致的温室气体（以CO_2为主）释放总量的4%左右（540 Mt/a；Burton et al.，2013）。上述部分研究结果被全球深部碳观测计划（deep carbon observatory，DCO）纳入"十年进展"专著（Orcutt et al.，2019），这是中国大陆火山地热区温室气体释放通量被国际相关组织采纳的唯一数据。

由上可知，特提斯构造域新生代火山-地热区温室气体释放通量明显高于太平洋构造域（郭正府等，

2014，2015；赵文斌等，2018）。休眠期火山区温室气体释放通量、规模与区内岩浆房的活动性、温泉水热活动的规模、土壤微渗漏面积及断裂发育等多种因素相关（郭正府等，2014，2015），此外，俯冲带火山地热区温室气体释放还与俯冲板块及其上覆板块的岩石组合、俯冲带活动性等因素密切相关（Mason et al.，2017）。

气体地球化学对比研究显示，处于太平洋构造域的长白山与五大连池火山区释放的气体以幔源为主，即具有较高的 $^3He/^4He$（Ra）值、与幔源岩浆气体一致的 $\delta^{13}C_{CO_2}$ 值等（Xu et al.，2013；Zhang et al.，2015；Zhao et al.，2019）；位于特提斯构造域的腾冲火山区气体则具有较高的 $^3He/^4He$（Ra）值和明显偏重的 $\delta^{13}C_{CO_2}$（成智慧等，2012，2014；Zhang et al.，2016）；藏南火山-地热区温泉气体具有较低的地壳成因的 $^3He/^4He$（Ra）值，其 $\delta^{13}C_{CO_2}$ 值显示出陆壳碳酸盐的明显贡献（Zhang et al.，2017）。

已有研究表明，上覆板块含有碳酸盐岩的大陆弧火山活动温室气体释放规模明显大于岛弧火山活动的温室气体释放规模（Marziano et al.，2008；Mason et al.，2017）。意大利中南部火山区（如 Vesuvio、Etna 等）陆壳碳酸盐混染作用在深源 CO_2 气体释放过程中起到关键作用（Chiodini et al.，1995；Frondini et al.，2008；Iacono-Marziano et al.，2009）。印度大陆俯冲板片与上覆欧亚大陆地壳含有大量的地壳碳酸盐岩，受俯冲再循环物质交代形成的地幔楔部分熔融产生"富碳"熔体（Zhang et al.，2017），在上升穿过陆壳的过程中与碳酸盐围岩发生热或（和）物质交换，通过陆壳同化和夕卡岩化等方式向大气圈释放巨量 CO_2 气体。

长白山、五大连池火山区距西太平洋俯冲带超过 1 000 km，而且火山气体来源可能较深（如地幔过渡带）（Zhang et al.，2015；Zhao et al.，2019；赵文斌等，2021），地表水热活动强度明显弱于特提斯构造域的腾冲火山区与藏南地热区，反映出不同构造背景下深部碳循环的成因差异（郭正府等，2014，2015；Zhang et al.，2016，2017）。与岛弧火山释放气体相对比，五大连池、辽东半岛等地区的稀有气体氦同位素比值与载气 CO_2 的碳同位素组成计算结果显示，CO_2 偏少，很可能与地球深部气体向上运移和释放过程中，由于水岩相互作用发生碳酸盐矿物沉淀有关，同时还会产生相应的碳同位素分馏效应（Xu et al.，2013，2014）。

我国大陆火山地热区温室气体观测研究起步较晚，直到 20 世纪 90 年代末才首次见有青藏高原羊八井地热区 CO_2 释放通量的报道（50 000 t/a）（Chiodini et al.，1998）。而我国大陆新生代火山区目前大多处于休眠期，这些火山-地热区由于深部上涌岩浆对围岩的持续加热烘烤，容易形成高温干热岩系统，在地表通过喷气孔、热泉、土壤微渗漏等形式向大气圈释放巨量温室气体（郭正府等，2010，2014），在开发地热资源的同时也亟需开展针对温室气体释放特征的专门调查研究。

4.2 地震断裂带

深大活动断裂带是地球内部物质和能量交换的主要地带（Kennedy et al.，1997），因其贯通了地球深部系统（岩石圈）和地表系统（土壤圈-生物圈-水圈-大气圈）（Chiodini et al.，2004，2020；陈志等，2014；Zhou et al.，2016；Chen et al.，2020），而成为地球强烈脱气的重要通道之一（Gold，1979；陶明信等，2005；Zheng et al.，2013；Capaccioni et al.，2015；Chen et al.，2018），释放规模巨大的温室气体，对全球气候和环境带来重要影响（杜乐天，2005；Zhou et al.，2016；崔月菊等，2016a，2016b）。目前，活动断裂带温室气体释放问题已得到国际同行的极大关注，以美国为代表的活动断裂带温室气体释放的专门调查和科学研究成果已逐渐涌现（Gambardella et al.，2004；Allard，2010；Lewicki et al.，2013；Capaccioni et al.，2015；Tamburello et al.，2018）。在美国圣安德烈斯断裂、日本 Yamasaki 断裂、意大利 Pernicana 断裂、中国福州断裂和呼和浩特新华广场断裂上方均观测到明显的土壤气地球化学异常（CO_2、Rn、Hg、He、H_2 和 CH_4 等）（Wakita et al.，1980；Giammanco et al.，1998；周晓成等，2007；郑国东等，2018）。实地调查发现，在美国圣安德烈斯断裂带和犹他州 Paradox 盆地北部断裂带 CO_2 的释放通量分别为每年 6.6 亿 mol（2.9 万 t）（Kulongoski et al.，2013）和 36 259 $g·m^{-2}·d^{-1}$（Jung et al.，2015）。北京西部盆岭地区断裂带CO从土壤向大气的扩散量至少为每年0.76万t，CO_2扩散量约为每年1.2万t（Sun et al., 2017）。在美国爱达荷州东南部，Sevier逆冲断裂带内的温泉群逸出CO_2的通量达每天350 t，与一些休眠火山的释放量相当（Lewicki et al., 2013）。初步调查认为，汶川M_S8.0地震后，地震破裂带CO_2的年释放总量为0.95 Mt（周晓成等，2017），其规模接近腾冲、谷露-亚东、搭格架、朗久、长白山和五大连池等中国6个新生代火山-地热区CO_2气体的年释放总量（20 Mt）的1/10（Sun et al., 2018；郭正府等，2017；赵文斌等，2018）。活动断裂带的破坏作用还是导致含油气盆地油气藏中甲烷等温室气体向大气强烈释放的重要原因之一（Zheng et al., 2013；Chen et al., 2018），而且活动断裂带也是导致火山区CO_2强烈脱气的重要因素（Sun et al., 2018, 2020b）。

众多观测表明，活动断裂带的地震活动通常会造成断裂带脱气及其地球化学特征的明显变化（King et al., 1996；Italiano et al., 2009；Chen et al., 2015）。2012年意大利艾米利亚-罗马涅地震群发生前后，意大利梅多拉地区活动断裂带土壤气CO_2、CH_4和H_2浓度出现大幅上升（Sciarra et al., 2017）。随区域地震活动强度的增大，唐山断裂带土壤气CO_2、Rn和He浓度明显上升（Han et al., 2014；Chen et al., 2019b）。2010年1月至2012年8月在全球发生的35次M_S≥7.0的地震中的12次地震的震中地区出现震前CO或O_3浓度异常升高的现象（Cui et al., 2013, 2017）。2010年4月4日墨西哥下加利福尼亚M_W7.2地震前约一个月，CO升高总量高出前两年同期平均值$2×10^{17}~4×10^{17}$ mol/cm^2，约为非地震时段的2倍（崔月菊等，2016a）。2008年5月12日的汶川地震导致短时间内龙门山断裂带向大气中至少多排放了4 740 t的CO、8 549 t的CH_4和8 715 Mt的CO_2（崔月菊等，2017），目前，汶川地表破裂带土壤气CO的释放通量（0.2~25.4 mg·m^{-2}·d^{-1}）仍明显高于背景值（5.2 mg·m^{-2}·d^{-1}），表明汶川地震后破裂区带仍持续在向大气释放温室气体（Sun et al., 2020a），某些断层泥对断层气也有一定的吸附作用（Ma et al., 2015）。

现有调查研究成果充分展示了活动断裂带温室气体释放的巨大规模，其对大气圈温室气体的贡献不可忽视。尤其是在陆地面积巨大、强震活跃的中国大陆（Zhou et al., 2017），地球深部含碳气体通过数量庞大的深大断裂带向大气圈释放更需重点关注。

4.3 沉积盆地

石油和天然气生成后就在不断地运移，并在地下适当部位聚集形成油气藏，甚至运移至地表，逸散到大气中，构成复杂的油气逸散渗漏系统（Zheng et al., 2018）。世界上大多数含油气盆地都存在各种形式、不同规模的烃类逸散和渗漏现象，主要包括宏渗漏（油苗、气苗、泥火山、沥青等）和微渗漏（肉眼不可见的烃类释放）两大类。一方面，这些烃类溢散和渗漏可以作为找油找气的直接标志（Milkov, 2005），在油气勘探和资源评价中曾经发挥过重要作用。例如，我国玉门油田和克拉玛依油田就是根据自然存在的油苗而逐渐被发现的。另一方面，油气逸散和渗漏不仅导致能源资源的极大浪费，油气逸散向大气层释放的CH_4、C_2H_6、CO_2等气体，都是重要的地质源温室气体，尤其是沉积盆地的CH_4，成为湿地之外的第二大自然源（Ciais et al, 2013）。油气逸散系统释放的C_2H_6、C_3H_8等还可以引发大气光化学污染和产生臭氧，如果释放气体中含有大量的H_2S，甚至可以导致人畜死亡（朱光有等，2004；刘文汇等，2010）。

中国陆地和近海沉积盆地广泛发育，迄今已在300多个盆地内找到石油和天然气。相关调查研究发现，我国几乎所有的含油气盆地都存在不同程度的油气渗漏现象，尤其是各种各样的油气苗，分布广泛，为寻找地下油气藏提供了直接证据，在早期的石油地质研究和勘探实践中发挥了重要作用。Zheng等（2018）综合相关资料提出了中国含油气盆地油气渗漏现状图（图4），并结合区域大地构造背景、盆地结构和发展演化，以及烃源岩类型、成熟度等资料，讨论了油气渗漏与活动构造以及盆地含油气系统之间的密切关系，结果显示：中国20个主要含油气盆地共发育932处油气渗漏点或渗漏带（710处陆上油气渗漏点和222处近海油气渗漏点位），包括81座泥火山、449处油苗、215处气苗、187处固体沥青露头。其分布也具有典型的区域性特点，我国西部地区的准噶尔盆地、柴达木盆地、民和盆地等油苗分布众多，而天然气苗则集中分布在伦坡拉盆地东部以及南方一些中小规模的含油气盆地中，与油气相关的泥火山主要在准噶尔盆地、柴达木盆地、羌塘盆地、台西南盆地等发育。这些盆地多数都经历了快速沉降，导致沉积物重力异常，为油气水等地质流体的上移扩散创造了有利条件。另外，各种油气渗漏点位的分布也明显地受到局部构造条件（Zheng et al.，2018；王国建等，2018），尤其是背斜构造和活断层的控制，渗漏油气类型则受盆地烃源岩类型、成熟度的明显影响。

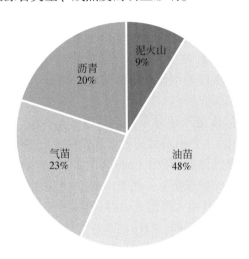

图 4　中国油气苗类型划分及其相对比重图（据 Zheng et al.，2018）

　　针对沉积盆地 CH_4 等温室气体释放的调查研究始于独山子泥火山。在石油地质界，"独山子"的知名度很高，其地名来源于境内的独山，在维吾尔语和哈萨克语中，独山子为"玛依塔克"和"玛依套"，意思是"油山"。所谓的"独山"是指位于北天山前缘以北、准噶尔盆地南缘突出地表而且存在油气苗的孤立山丘，南北长约 7 km、东西宽约 6 km，最高峰海拔 1 283 m，孤立的突兀于倾斜平原之上，西隔奎屯河与乌苏市的海烈非山相望。早在清光绪二十三年（1897 年）该地就开始了石油开采。"民国"二十五年（1936 年），从苏联引进技术装备，钻探发现了独山子油田，并在独山子（山）北坡形成石油工人聚居的矿区，即独山子矿区，现在的行政区划为克拉玛依市独山子区，因为远离克拉玛依市主城区，实属"飞地"。独山子实际上是北天山前缘一系列挤压背斜构造成山的，该背斜被命名为"独山子背斜"，周边发育的油气宏渗漏被称作"独山子泥火山"，目前主要有两处规模较大、活动较强的泥火山喷口。在独山子山北坡废弃的第一口钻井及其周围有很多油气渗出地表，人工挖坑积水后可见明显的气泡和油花，周边植被稀少而且种类单调。

　　独山子泥火山和油气田的石油地质研究历史悠久（Chen et al.，2021），成果也很丰富，独山子泥火山常被作为石油地质的教学案例、青年石油地质学工作者开始新工作的参观场所，甚至是青年学生夏令营或革命传统教育的基地。但作为地质源温室气体释放地质体的调查研究历史短暂。针对新疆泥火山源温室气体释放规模，初始的调查（Dai et al.，2012；高苑等，2012）主要是通过对气泡大小和冒泡频率的观察，对独山子泥火山喷口气体的宏观通量做了粗略估算。近些年，采纳国际上通行的密闭箱法实验性地进行了相对精确的原位测量，初步结果显示，独山子泥火山通过微渗漏每年向大气排放约 16.6 t 的甲烷温室气体（马向贤等，2014）。2014 年，采用皂膜流量计对独山子泥火山各喷口的宏渗漏进行了原位测量，并扩大了泥火山周围微渗漏的测量面积，调查结果显示，这些泥火山宏渗漏释放通量远比通过数气泡估算获得的数值高，而且，随着实测面积的扩展，微渗漏释放规模巨大，所占泥火山系统气体释放的比重也明显增加（Zheng et al.，2017）。然而，根据独山子泥火山发育区基（围）岩蚀变现象的分布范围，目前所测面积还十分局限，需要进一步扩大测量面积和提升检测频率，以获取更为准确的泥火山温室气体释放通量基础数据。

地球化学指标显示，这些逸出气体主要为热成因的煤型气（Dai et al.，2012；Chen et al.，2021）。微生物调查发现，这些泥火山环境的微生物菌群既典型特殊又丰富多彩（马小龙等，2009；Huang et al.，2016；Ren et al.，2018；Ma et al.，2021）。准噶尔盆地南缘地处内陆地区，干旱少雨，蒸发量高，而盆地之南为高耸的北天山，常年积雪为这些泥火山提供了丰富的水源，泥火山溢出的泥浆水主要为大气降水来源（Nakada et al.，2011；Li et al.，2014）。准噶尔盆地南缘中新生代地层以陆相沉积的红层为主，其中发育的泥火山和油气微渗漏导致显著的岩石去红色漂白现象，尤其是独山子背斜核部露到地表的红层蚀变现象非常典型（Schumacher，1996；Zheng et al.，2010；Xu et al.，2018）。这种类型的油气蚀变现象可以作为遥感技术寻找油气资源的典型地物标志（Fu et al.，2007）。由于压力作用，甲烷等气体和水在运移和喷出过程中，将流体输导格架的细粒黏土，甚至围岩碎屑带出地表，成为泥浆的主要成分。独山子泥火山喷出泥浆的孢粉主要来自中新世晚期沉积的塔西河组和独山子组地层（Ji et al.，2018）。这些地质地理背景使得新疆泥火山系统的研究极具特色。

除了独山子泥火山外，准噶尔盆地南缘还分布有安集海、艾奇沟、白杨沟等泥火山群，以及第四纪曾经活动过的泥火山（称为古泥火山），具有典型的岩石蚀变现象。准噶尔盆地南部地区还存在与地下深部油气藏密切相关的油气微渗漏温室气体的释放。Chen 等（2019a）通过野外观测，给出了北天山前缘四排褶皱-断裂带泥火山群 CH_4 和 CO_2 的释放情况：北天山褶皱-断裂带泥火山群 CO_2 和 CH_4 的年释放总量分别为 147.83 t 和 611.26 t，与国际上其他大型泥火山群 CO_2 和 CH_4 的释放规模相近。北天山前缘四排褶皱-断裂带泥火山区 CH_4 释放特征与 CO_2 不同，CH_4 的高通量释放点呈线状分布，其走向与北天山地区主要活动断裂带走向一致，而 CO_2 高通量释放点则呈面状分布。气体地球化学和同位素地球化学综合分析揭示，泥火山区释放的 CH_4 为深部油气藏热降解烃源气，而 CO_2 则为泥火山口喷发泥浆中有机质的微生物降解以及碳酸盐沉积物的分解所致。另外，四排褶皱-断裂带泥火山群 CH_4 和 CO_2 的释放强度随构造破坏程度由南向北减弱而降低。这些研究成果揭示，北天山四排褶皱-断裂带的活动断裂已经破坏深部地层，释放地层中赋存的有机质，是北天山泥火山形成及温室气体释放的主要控制因素。

采用静态箱法对新疆塔里木盆地不同类型的油气田（雅克拉凝析气田和大宛齐油田）进行微渗漏 CH_4 释放通量调查研究（唐俊红等，2006，2009；Tang et al.，2010，2013）结果显示，不同类型油气藏由于地质构造、油气藏控制因素等不同，微渗漏 CH_4 通量具有明显的差异。大宛齐油田埋藏浅、断层发育、封盖条件较差，其 CH_4 渗漏因子处于较高的渗漏等级，夏季 CH_4 平均通量（13.87 $mg \cdot m^{-2} \cdot d^{-1}$）远高于深层雅克拉凝析气田（1.96 $mg \cdot m^{-2} \cdot d^{-1}$）。结合大宛齐油田油气勘探资料和剖面监测结果，对比地表 CH_4 通量和地下土壤气 CH_4 浓度发现，CH_4 异常高值多分布在断层附近，并呈现沿构造带展开的趋势，由此推断断层为控制 CH_4 等烃类气体微渗漏的主要因素。新疆四大含油气盆地边缘区域和盆地内部相比较，构造背景更为复杂，气体释放特征也有明显区别，包括释放强度和稀有气体地球化学特征等方面（Xu et al.，2017）。

大宛齐油田微渗漏 CH_4 存在明显的季节性变化，夏季 76.8% 采样点表现为正通量，冬季由于积雪和冻土（>60 cm）产生了屏蔽效应，阻碍了 CH_4 微渗漏，81.4% 采样点表现为负通量，CH_4 平均通量为 -0.24 $mg \cdot m^{-2} \cdot d^{-1}$。根据 CH_4 渗漏因子方法建立浅层油气藏 CH_4 释放通量预测方程和评估模型，由此估算大宛齐油田微渗漏 CH_4 通量为 112.2 t/a，雅克拉凝析气田微渗漏 CH_4 通量在 10^2 t/a 数量级范围（Tang et al.，2017，2019）。

与国际上已经公开发表的全球陆上油气渗漏数据相比，中国可能是全世界拥有油气苗最多的国家（Zheng et al.，2018）。如此众多的油气渗漏也预示着，我国广泛发育的含油气盆地很有可能向大气释放数量可观的 CH_4、C_2H_6、CO_2 等温室气体，面对全球变化和大气碳收支问题，急需对我国含油气盆地的油气苗开展专门的系统调查和科学研究，特别是油气渗漏的现场实地测量工作。

基于油气苗及其分布特征的地表地球化学找油找气的基本原理，就是地下石油天然气可以通过流体输导格架释放到地表，一方面，天然的油气苗是地下油气藏存在的直接证据，也即常见的油气苗；另一方面，由于复杂的有机-无机相互作用可以形成特殊的地表土壤或岩石的矿物蚀变，甚至可以形成与基岩矿物组成有明显区别的蚀变矿物（Zheng et al., 2010），并且可以导致地物波谱的明显变化（Fu et al., 2007）等，成为寻找地下油气藏的直接或间接的地球化学标志。准噶尔盆地南缘的多个泥火山群都分布在一系列的背斜构造之上，尤其是独山子泥火山，直接反映了一个被第四纪新构造运动和后期风化作用破坏了的背斜油气藏。该泥火山出现在独山子背斜的核部，并被多组断层和裂隙所切割。横穿背斜的地震剖面显示，独山子背斜核部的多条断层向下穿透新近系和白垩系地层，复杂的断层裂隙系统不仅成为油气运移的通道，而且在适宜部位成为油气聚集成藏的有利场所（可形成油气藏）。伴随后来的新构造运动，部分油气藏被破坏，导致"油-气-水"地质流体沿断裂带继续向上运移，甚至逸散到地面，形成泥火山，并对流体输导格架的固体岩石产生蚀变作用。由于独山子背斜的隆起和地表剥蚀，现今地表出露的蚀变岩石可以与未受蚀变的同层岩石进行追索和对比，因此成为探讨地下油气储层地质学和地球化学特征，特别是储层物性特征及其变化的良好天然试验场。泥火山及其微渗漏系统的油-气-水-岩相互作用对甲烷等烃类物质进行改造，使得部分甲烷等烃类化合物被氧化成为 CO_2，在水中转化为 HCO_3^{1-}，遇到铁锰钙等金属阳离子后容易形成次生碳酸盐矿物沉淀，从而产生一定的"固碳作用"，并对运移的烃类物质产生改造，甚至导致油气储层的物性发生变化（Xu et al., 2018, 2020）。

4.4 近海海域温室气体释放

近海海域是海洋温室气体释放的主要场所，CH_4、CO、NO、CO_2 等气体主要来自海底沉积有机质的生物降解和海底深部沉积有机质以及油气的热裂解（Etiope et al., 2013）。中国南海北部和莺歌海海域存在显著的海底沉积物甲烷微渗漏和海底油气区甲烷宏渗漏现象。中国南海北部广泛分布的红树林、海草床、盐沼等是海洋温室气体的重要来源。通过解剖近海沉积物沉积-成岩过程中各种温室气体的生成、产出和扩散作用及其伴随的生物地球化学作用特征，估算近海沉积物以及伴生的生态系统释放温室气体的组成、通量，判断其成因来源，探讨近海海洋环境条件下各种沉积体系温室气体的产出机制、扩散模型和释放规模。

现有研究结果显示，海洋释放的地质来源甲烷气大体为 $1\sim100\ mg/m^2$，所释放的甲烷通量占全球大气甲烷通量的 $2\%\sim4\%$，但其时空变化巨大。而针对沿海植物生态系统 CH_4 通量的调查研究存在区域性差异，如 75% 以上的红树林 CH_4 通量数据点位于东半球，集中在东南亚和中国，海草床数据点集中在非洲和欧洲，几乎所有的 CH_4 通量研究都来自北半球。一般认为，CH_4 释放通量与海洋沉积有机质含量有关，沉积物中丰富的有机质可以为 CH_4 的形成提供更多的竞争性基质和（或）非竞争性基质，从而减少硫酸盐还原剂和产甲烷菌之间的竞争，红树林海草床生态通常有更高的沉积物有机质（OM）浓度。一般而言，当植物进行光合作用时会将不稳定的有机质沉积到其根部或者根际的沉积物中，从而提供了刺激 CH_4 产生的基质。在盐沼和红树林沉积物中，CH_4 产生速率与活性 OM 浓度之间正线性关系。但 OM 对 CH_4 排放的积极影响并不普遍，一些研究报告显示两者没有明显的关联性。

在开阔远洋区域，表层海水中的甲烷主要来源于海洋浮游生物和细菌新陈代谢活动，其浓度一般在 $1.81\sim3.1\ nmol/L$；在大陆架区域，海水中甲烷的浓度可达 $3.1\sim5.0\ nmol/L$；而河口和近海海域的表层海水甲烷浓度大多高于 $5.0\ nM/L$。近海虽然占海洋面积的比例较小，但其释放的甲烷气体占海洋释放量的比例却高达 75% 以上。因此，近海沉积物是研究海洋地质源温室气体释放的重要区域。我国近海地质源温室气体释放研究还比较零散，目前尚处于起步阶段。

4.5 可燃冰与煤自燃

甲烷天然气水合物（可燃冰）不仅是潜力巨大的清洁能源，水合物分解释放的 CH_4、CO_2 等也是不可忽视的温室气体，尤其是海底天然气水合物的突然释放有可能引发海洋环境，甚至全球环境的巨大变化。天然气水合物对温度和压力非常敏感，长期的气候变化可能会触发其分解，分解释放出的大量甲烷等温室气体又会导致大气温室效应的增强，而增温加剧又会促使更多的天然气水合物分解释放出更多的甲烷。这个递进循环一旦启动，就很难停下来的正反馈过程，被称为可燃冰喷射假说（clathrate gun hypothesis）（Kennett et al., 2003）。这一假说被用来解释晚第四纪（约40万年前至今）时期大气甲烷含量迅速增温的现象（Brook et al., 1996）。我国南中国海北部神狐探区发现了丰富的海底天然气水合物（陈多福等，2005），并且已经进行成功试采，海洋地球物理调查在多个区域发现似海底反射层（王家豪等，2006），深海探测也观察到烃类气体释放的海底泥火山和麻坑（Di et al., 2014, 2020），该海域的海水柱状甲烷含量的变化也与其相关（Mau et al., 2020），但其释放强度和释放通量还没有进行有效的检测和系统调查，目前的主要工作仍处于确认是否存在天然气水合物及其能否进行开采与开发利用的调查研究阶段。

煤自燃，顾名思义，就是煤不经点燃而自行着火的自然现象。煤自燃的原因，主要是煤与空气接触发生氧化反应，一方面使煤的温度升高，另一方面又使煤的燃点降低，因而易于引起煤的自燃。此外，煤层由于地质构造变动或因采掘而破碎，以及通风不良等，也是引起自燃的因素。煤的自燃不但浪费宝贵的煤炭资源，严重影响煤矿生产和安全，而且还向大气释放二氧化碳等温室气体。煤的化学成分包括有机质和无机质，其有机质的构成元素主要有碳、氢、氧、氮和硫等，此外，还有极少量的磷、氟、氯和砷等元素。碳、氢、氧是煤炭有机质的主体，占95%以上。煤中的无机物质含量很少，主要有水分和矿物质，它们的存在降低了煤的质量和利用价值。矿物质是煤炭的主要杂质，如硫化物（硫化铁和黄铁矿）、硫酸盐、碳酸盐等，其中大部分属于有害成分。煤一旦燃烧就会释放 CO_2、NO_2、SO_2 等气体，这些气体都是温室气体；如果是不完全燃烧，就会产生CO等有害气体。我国新疆、内蒙古、甘肃、青海、陕西、辽宁等省、自治区都存在煤自燃现象。许多地区的煤系地层发生自燃后，不仅导致优质煤层自身燃烧而消耗殆尽，煤层围岩以及煤矸石也被烧烤变质，被称为烧变岩，如天山北麓和准噶尔盆地西北地区广泛分布的煤自燃烧变岩（图5），主要是侏罗系煤系地层自燃后形成的砖红色或者赭红色的烧变岩。煤层围岩，尤其是煤层上覆的粉砂岩和泥岩等原有的沉积有机质也会被烘烤烧掉，形成 CO_2 等温室气体而释放。煤田地质学和煤矿开采工程都注意到对煤自燃的基础研究，但是，这个过程释放温室气体的调查研究还不多见。

图5 煤自燃烧变岩照片
（a）新疆北天山硫磺沟烧变岩；（b）烧变岩手标本；（c）准噶尔盆地火烧山烧变岩

另外，碳卫星监测到北天山和准噶尔盆地空气柱碳浓度不仅出现高值，而且四季变化不明显，推测与这些地区广泛分布的泥火山系统释放甲烷和煤系地层煤自燃释放 CO_2 等地质过程密切相关，因为这些地区经济欠发达，人烟相对较稀少，人类活动碳排放规模和强度都不大，大气碳源应该以地质自然源为主，与我国东南沿海地区的状况大不一样。

4.6 地球深部碳循环与温室气体

地质源温室气体调查研究主要侧重于以 CH_4 为主的烃类气体（也包括 C_2H_6 等）、CO_2 和 CO 等。这些气体都是地球系统碳循环的重要组成部分，并与地球系统碳循环研究密切相关，地质源温室气体的调查研究离不开对于地球系统整体碳循环问题的全面认识，也正因为如此，由美国卡内基科学理事会发起的深部地球碳观测（DCO）得到众多专家学者的热烈响应、多方支持和积极参与（Marziano et al., 2008）。我国科学家自 2009 年开始，就组织过多次的国际或国内以深部碳循环为主题的学术研讨会议，以实际行动积极响应和参与并支持相关的国际合作项目的组织和实施，并在许多方面获得显著成绩（Mao et al., 2006；Li et al., 2018）。

由地表碳俯冲和火山脱气构成的深部碳循环对于理解地质源温室气体形成机理和类型以及释放通量的长期演化具有重要意义。高精度地震层析成像显示亚洲东部的地壳–地幔结构是由滞留在地幔过渡带（深达 410～660 km）的西太平洋俯冲板片和顶部东亚大陆岩石圈，以及夹于其间的楔状上地幔构成，被称为"大地幔楔"。该大地幔楔的岩石圈华北部分曾是古老的巨厚岩石圈（200 km），被称为"华北克拉通"。其东部自 130 Ma 以来被减薄，最薄处仅有约 70 km 厚（李曙光和汪洋，2018）。该大地幔楔结构是何时形成和华北克拉通东部岩石圈是如何减薄是科学家们长期关注的两个重要地球科学问题。但以往研究忽视了大地幔楔（主要由硅酸盐组成）被俯冲板片携带的碳酸盐改造对岩石圈减薄机制的影响。近年来，我国科学家基于地表碳酸盐岩具有显著偏离地幔物质的 Mg-Zn 同位素组成特征以及中国东部中生代—新生代大陆玄武岩和环太平洋岛弧玄武岩的镁同位素研究，揭示了中国东部上地幔是一个巨大的再循环碳库（Yang et al., 2012；Liu et al., 2016，2017；Li et al., 2017）。俯冲洋壳及其携带的沉积碳酸盐在大地幔楔底部发生熔融，所产生的熔体向上渗滤交代对流上地幔，形成了碳酸盐化的橄榄岩。碳酸化橄榄岩部分熔融产生的贫硅富碳酸盐熔体自下而上作用于华北东部岩石圈地幔，可导致岩石圈底部橄榄岩碳酸盐化和部分熔融，从而使其物理性质与岩石圈下面的软流圈类似并很容易被对流软流圈地幔所置换（Li and Wang, 2018）。因此，华北克拉通岩石圈被减薄并使熔体自身由贫硅强碱性熔体向富硅弱碱性熔体转化（李曙光和汪洋，2018）。后续的铁元素同位素研究结果进一步表明，这些俯冲碳酸盐在深部地幔的还原封存及伴随的氧化性岩浆活动，可驱动氧循环，有利于促进现今地表氧化状态的形成和维持（He et al., 2019）。华北克拉通岩石圈地幔碳酸盐化交代所导致的岩石圈减薄过程很有可能为中国东部郯庐断裂带及其周边富二氧化碳气体的形成提供了物质基础和动力来源，该地区一些温泉气体的稀有气体同位素组成也显示出复杂的以壳源为主，也包含幔源物质加入的特征（Xu et al., 2012, 2013, 2014；赵文斌等，2021）。

毛河光院士率领的研究团队利用静态超高温超高压模拟实验和在线检测技术，确定了 MgO-FeO-SiO_2 系统在下地幔的温度压力条件下矿物的相关性，观察到二价铁在高温下的歧化反应，以及铁、镁的强烈分异现象，首先提出了地核存在碳化三铁，进而提出地球核幔过渡带铁–水体系释放氢和氧的概念模型（Mao et al., 2006）。相关研究还揭示，固体地球是一个巨大的碳库，地球系统超过 98% 的碳是在地球深部以多种形式存在，如碳化三铁、金刚石、石墨、碳酸盐岩等。另外，多种化学组成的地球深部地质流体携带的大量 CO_2、CH_4 等气体为地表和海洋提供了丰富的碳源，深部流体中的碳主要来源于深部碳循环过程中俯冲带和深部地幔的释放（刘全有等，2019）。深部碳循环包括俯冲带沉积碳酸盐随洋壳俯冲进入地球内部的捕获过程和以 CO_2、碳氢化合物、碳酸盐熔体等形式随洋中脊、岛弧、地幔柱等构造带岩浆活动喷发至地表及大气的释放过程，深部碳循环与地表碳循环共同构成了地球的碳循环系统。深部流体携带的大量碳，一方面为海洋、陆地生物提供了生命活动所需的碳源，促进了海洋、陆地生物的繁盛和初级生产力的提高；另一方面，为无机成因 CO_2 气藏的形成提供了重要的物质基础，成为地质源温室气体的重要来源。

4.7 CO_2 地质封存与应用

面对大气 CO_2 含量逐渐升高以及减少生产活动排放 CO_2 的迫切需求，一些科学家和工程师提出通过多种途径封存 CO_2 的概念（Holloway, 2005），并逐步得以付诸实施（Li at al., 2013；Li and Liu, 2016）。CO_2 捕集、利用和封存（CCUS）是指将 CO_2 从工业、能源利用过程或大气中分离出来，直接加以利用或

者注入地层以实现 CO_2 减排的工业过程。CCUS 是在 CO_2 捕集与封存（CCS）的基础上增加了"利用（utilization）"，这一理念是随着 CCS 技术的发展，对 CCS 技术不断深化，在中美两国的大力倡导下形成的，目前已经获得了国际上的普遍认同。CCUS 按技术流程分为捕获、输送、利用与封存等环节（蔡博峰等，2021）。中国已经投运或建设中的 CCUS 示范项目约为 40 个，捕获能力每年 300 万 t，累计封存 CO_2 超过 200 万 t。

CO_2 捕集主要分为燃烧前捕集、燃烧后捕集、富氧燃烧和直接空气捕集等。CO_2 捕集技术成熟度差异较大，目前燃烧前物理吸收法已经处于商业应用阶段（李小春，2013），燃烧后化学吸附法尚处于中试阶段，其他大部分捕获技术已处于工业示范阶段。燃烧后捕集技术是目前最成熟的捕获技术，可用于大部分火电厂的脱碳改造，国内已建有 10 万吨级的燃煤电厂 CO_2 捕获示范装置。国华锦界电厂开展的 15 万 t 碳捕集与封存示范项目已经建成，是中国规模最大的燃煤电厂燃烧后碳捕集与封存全流程示范项目。燃烧前捕集系统相对复杂，整体煤气化联合循环（IGCC）技术是典型的可进行燃烧前碳捕集的系统。国内的 IGCC 项目有华能天津 IGCC 项目以及连云港清洁能源动力系统研究设施。富氧燃烧技术是最具潜力的燃煤电厂大规模碳捕集技术之一，产生的 CO_2 浓度较高（90%~95%），更易于捕获。富氧燃烧技术发展迅速，可用于新建燃煤电厂和部分改造后的火电厂。当前第一代碳捕集技术（燃烧后捕集技术、燃烧前捕集技术、富氧燃烧技术）发展渐趋成熟，主要瓶颈为成本能耗偏高、缺乏广泛大规模示范工程经验；而第二代技术（如新型膜分离技术、新型吸收技术、新型吸附技术、增压富氧燃烧技术等）仍处于实验室研发或小试阶段，技术成熟后能耗和成本会比成熟的第一代技术降低 30% 以上，2035 年前后有望大规模推广应用（蔡博峰等，2021）。

CO_2 输送的主要方式有陆地或海底管道、船舶、铁路和公路车载等。车载运输技术比较成熟，当前我国 CCUS 示范项目均采用车载运输，至今还没有真正意义上的 CO_2 管输项目，技术成熟度较低，缺乏管输网络，但下游的油气管道规模和经验有助于我国 CO_2 管输的快速发展（陆诗建等，2019）。

CO_2 地质利用与封存是将 CO_2 注入深部地层以实现 CO_2 与大气长期隔离并（或）强化资源开采的过程（Li et al.，2018），主要技术包括强化采油、驱煤层气、强化采气、增采页岩气、溶浸采铀、封存驱水（Li et al.，2015）、增强地热系统等。我国有多种地质利用与封存场景可选择（Liu et al.，2017），理论封存总量约 2.5 万亿 t，其中封存驱水占 96%。除了溶浸采铀外，其他技术均具有较大封存容量。目前，强化采油和封存驱水在现有技术水平下可以进行大规模示范。根据国家能源产业革命路线图、中国 CCUS 科技发展路线图愿景，预计在合适的激励情景下，即 CO_2 利用补贴 250 元/t CO_2，地质封存补贴 350 元/t CO_2，2030 年我国地质利用与封存综合减排潜力可达到每年 2.4 亿 t（中值），2060 年将为每年 14 亿 t（峰值）（中国 21 世纪议程管理中心，2014）。

CO_2 化工与生物利用是指基于化学或生物技术，将 CO_2 转化为其他产品，并在此过程中具有一定减排效应的技术手段，主要包括 CO_2 化学转化制备化学品技术、CO_2 矿化利用技术、CO_2 生物利用技术三个方面。其中，CO_2 化学转化制备化学品技术主要包括制备能源化学品、精细化工品和聚合物材料技术。CO_2 矿化利用主要是通过天然矿物、工业材料和工业固废中钙、镁等碱性金属将 CO_2 进行碳酸化固定为化学性质极其稳定的碳酸盐；CO_2 生物利用技术是指以生物转化为主要特征，通过植物光合作用等，将 CO_2 用于生物质的合成，并在下游技术的辅助下实现 CO_2 资源化利用，包括微藻生物利用、CO_2 气肥利用、微生物固定 CO_2 合成苹果酸等。设施农业种植过程，通过提供二氧化碳（气肥）促进大棚蔬菜水果的优质丰产，可以获取良好的经济效益和环境效益，值得大力支持和广泛推广（中国 21 世纪议程管理中心，2014）。

整体来讲，CO_2 利用和封存都面临着长期稳定性问题。地质源温室气体的基础理论研究结果和技术方法，无疑可以提供多方面的理论借鉴和技术支撑，尤其是泥火山系统复杂的油-气-水-岩-微生物相互作用研究结果可以作为 CO_2 气体注入地层后的赋存状态及其长期稳定性、盖层密闭性、储层内 CO_2 气体的反应方向等判识和研究提供科学依据（李琦等，2019）。

5 亟需解决的相关问题和发展前景

我国地处欧亚大陆东南隅,地质背景非常复杂,第四纪构造运动异常活跃,导致我国地质源温室气体释放具有点多、线长、面广、类型复杂的特点。尽管我国地质源温室气体调查研究几乎与世界同步,但调查区域非常有限、研究工作缺乏系统性、已有观测数据比较零散,还没有形成整体数据库,急需通过多领域多单位多学科多技术的交流与合作,更新科学理念,集成和创新技术方法,尽快大规模地开展我国地质源温室气体释放的系统调查和综合研究。

5.1 加强研究队伍建设

国内从事地质源温室气体调查研究的专门队伍少而分散,目前仅有少数几个团队在进行实际的专门调查研究工作,科研人员总数20人左右,无力进行较大规模的调查研究,只能优选典型地质背景条件下的代表性地区进行有限类型的地质源温室气体现场测量,各种类型地质源温室气体释放的基础数据资料很不完善,甚至还非常欠缺。因此,需要鼓励和调动更多研究者和科研团队的参与,吸引和培养中青年专门科技人才,积极组织中国科学院、中国地震局、中国地质调查局等相关科研院所和高等院校,以及油气煤等能源产业部门的支持与合作,尽快确认和提出我国地质源温室气体主要类型、地质构造背景与区域划分;针对不同构造单元、不同构造背景实施全面调查和现场测量,创建有一定空间尺度的地质源温室气体排放清单;规范碳循环模型,预测与估算地质源温室气体释放规模,建立我国地质源温室气体释放通量数据库,综合研究地质源温室气体的碳排放与环境效应,形成地质源温室气体计量标准。

5.2 优化野外检测与样品采集

我国幅员辽阔,地质地理条件千差万别,要将高山峻岭、盆地平原、地表植被、沼泽水体等区域的监测数据进行有效对比,挑战性很大;地质源温室气体释放的定性和定量调查研究,涉及许多技术问题需要解决,如现场测试仪器精密度、检测程序和调查人员工作精细程度,样品采集运输和保存,不同实验室之间测试数据的可对比性等。目前,国际气体地球化学研究普遍面临采集样品的保真性问题,既要防止采集样品的丢失又要防止外来气体的污染,主要是空气污染,对样品容器的要求非常严格,导致甄别测试需要特殊技术和设备。因此,需要集成创新具有国际先进水平的地质源温室气体释放监测技术,进行点(释放点)-线(断裂带)-面(火山-地热区、沉积盆地)一体化的系统监测,建立时间-空间多尺度的地质源温室气体释放连续观测系统,积极发展原位测量、海底探查、空中探测、卫星遥感等检测技术。

5.3 提升综合科学研究水平

我国地质源温室气体调查研究的科研实力、人才培养等方面与国外相比差距还较大,高水平研究成果相对较少,调查范围和研究程度落后于欧美日等国。所以,时不我待,需要奋起直追,在充分发挥国内优势力量的同时加强国际合作,优化和提升我国气体地球化学测试分析水平,系统研究我国主要类型地质源温室气体的地球化学特征、成因机理与判别模式,服务于气候变化国际谈判。目前,国内外科学家已经基本达成共识,现今大气圈温室气体的浓度,尤其是CO_2浓度应是长期历史演变的结果,因而需要从历史视角,动态与静态相结合,系统而全面地评价各类温室气体的来源、演化、平衡等问题。基于我国地质源温室气体的主要类型、地质构造背景及其区域划分,综合分析各类地质源温室气体的地球化学特征、成因机制与判识方法迫在眉睫;在此基础上,创新和发展符合国际标准的监测方法和分析技术,建立我国地质源温室气体的释放通量数据库与估算模型,并对我国地质源温室气体的碳排放及其环境效应进行以我为特色但也符合国际通用准则的深入调查与综合研究,提升研究水平和国际影响力,并为国家经济建设、社会发展战略提供必要的科学依据，甚至为我国在国际气候谈判中争取主动权，确保中华民族伟大复兴事业的顺利进展提供科学支撑。

5.4 加强地质源甲烷调查研究

甲烷分子是由碳原子和氢原子构成的，热值高，燃烧时仅形成二氧化碳和水，不产生粉尘废渣等固体废弃物，因而被称为清洁能源。但是，如果甲烷释放到大气中，则会产生强烈的温室效应，等摩尔数甲烷的温室效应是 CO_2 的 21~22 倍（Houghton et al.，2001），如果考虑其光化学作用，其温室效应则更高。甲烷来源广泛，类型众多，包括沉积盆地等各种地质作用释放的甲烷在整个大气甲烷估算中占有重要份额，它既是不含放射性碳（^{14}C）甲烷源（死碳源）缺失部分的重要代表（Crutzen，1991，1995），也是甲烷重碳源的贡献者（唐俊红等，2006，2019），因此，成为大气温室气体来源探索和气候变化与应对策略制定所必须考虑的重要因素。Etiope 和 Milkov（2004）估计，全球陆地和浅海泥火山向大气释放的甲烷量大约为每年 6~9 Mt，成为湿地之后第二大最重要的大气甲烷自然源；他们还估计了世界主要泥火山周围地区的微渗漏释放甲烷的规模，大约每年 1.1~5.6 Mt。当然，这些估算都是根据有限数据计算出来的，需要开展广泛的泥火山调查研究，深入了解和全面把握泥火山系统释放甲烷的过程和规模。我国数百个含油气盆地都存在油气泄露现象，油气苗分布广泛（Zheng et al.，2018），其中准噶尔盆地、柴达木盆地、羌塘盆地等还发育一些泥火山，甲烷等烃类气体泄露特征和释放规模亟需开展系统调查和专门研究。

致谢：中国地质源温室气体调查研究一直得到刘嘉麒、李曙光、吴国雄等院士的支持，资料收集得到李营、何勇胜、陈多福、张水昌、陈践发、唐俊红、冯东、张立飞、Giovanni Martinelli、Daniele Pinti 等的热情帮助，在此一并致以衷心感谢。

参 考 文 献

陈多福,冯东,陈光谦,陈先沛,Cathles L M. 2005. 海底天然气渗漏系统演化特征及对形成水合物的影响. 沉积学报,23(2):323-328
陈志,杜建国,周晓成,崔月菊,刘雷,李营,张文来,高小其,许秋龙. 2014. 2012 年 6 月 30 日新源 M_S 6.6 地震前后北天山泥火山及温泉的水化学变化. 地震,34(3):97-107
成智慧,郭正府,张茂亮,张丽红. 2012. 腾冲新生代火山区温泉 CO_2 气体排放通量研究. 岩石学报,28(4):1217-1224
成智慧,郭正府,张茂亮,张丽红. 2014. 腾冲新生代火山区 CO_2 气体释放通量及其成因. 岩石学报,30(12):3657-3670
崔月菊,杜建国,陈杨,刘雷,刘红,易丽,孙凤霞. 2016a. 汶川 M_S 8.0 地震前后龙门山断裂带 CO 和 CH_4 排气增强. 地震研究,39(2):239-245
崔月菊,杜建国,荆凤,李新艳. 2016b. 2008 年汶川 M_S 8.0 地震前后川西含碳气体卫星高光谱特征. 地震学报,38(3):448-457
崔月菊,杜建国,李新艳,孙凤霞,邹镇宇. 2017. 汶川地震相关的断裂带碳气体排放量估算. 矿物岩石地球化学通报,36(2):222-227
杜乐天. 2005. 地球排气作用的重大意义及研究进展. 地质论评,51(2):174-180
高苑,王永莉,郑国东,孟培,吴应琴,杨辉,张虹,王有孝. 2012. 新疆准噶尔盆地独山子泥火山天然气地球化学特征. 地球学报,33(6):989-994
郭正府,李晓惠,张茂亮. 2010. 火山活动与深部碳循环的关系. 第四纪研究,30(3):497-505
郭正府,张茂亮,成智慧,张丽红,刘嘉麒. 2014. 中国大陆新生代典型火山区温室气体释放的规模及其成因. 岩石学报,30(11):3467-3480
郭正府,张茂亮,孙玉涛,成智慧,张丽红,刘嘉麒. 2015. 火山温室气体释放通量与观测的研究进展. 矿物岩石地球化学通报,34(4):690-700
郭正府,郑国东,孙玉涛,张茂亮,张丽红,成智慧. 2017. 中国大陆地质源温室气体释放. 矿物岩石地球化学通报,36(2):204-212
李琦,蔡博峰,陈帆,刘桂臻,刘兰翠. 2019. 二氧化碳地质封存的环境风险评价方法研究综述. 环境工程,37(2):13-21
李曙光,汪洋. 2018. 中国东部大地幔楔形成时代和华北克拉通岩石圈减薄新机制——深部再循环碳的地球动力学效应. 中国科学:地球科学,48(7):809-824
李小春. 2013. 二氧化碳捕集利用与封存词典. 北京:世界图书出版社
刘全有,朱东亚,孟庆强,刘佳宜,吴小奇,周冰,Fu Q,金之钧. 2019. 深部流体及有机-无机相互作用下油气形成的基本内涵. 中国科学:地球科学,49(3):499-520
刘文汇,腾格尔,高波,张中宁,张建勇,张殿伟,范明,付小东,郑伦举,刘全有. 2010. 四川盆地大中型天然气田(藏)中 H_2S 形成及富集机制. 石油勘探与开发,37(5):513-522
刘文汇,王星,田辉,郑国东,王晓锋,陶成,刘鹏. 2021. 近十年来中国天然气地球化学研究进展. 矿物岩石地球化学通报,40(3):540-555
陆诗建,赵东亚,刘建武,喻健良,李琦. 2019. 50 万 t/a CO_2 输送管道泄漏仿真模拟研究. 山东化工,48(8):229-235

马向贤,郑国东,梁收运,樊成意,王自翔,梁明亮. 2012. 地质甲烷对大气甲烷源与汇的贡献. 矿物岩石地球化学通报,31(2):139-145,183

马向贤,郑国东,郭正府,Etiope G,Fortin D,佐野有司. 2014. 准噶尔盆地南缘独山子泥火山温室气体排放通量. 科学通报,59(32):3190-3196

马小龙,王芸,杨红梅,王纯利,毛培宏,金湘,常玮,房世杰,张评浒,娄恺. 2009. 新疆泥火山细菌遗传多样性. 生态学报,29(7):3722-3728

生态环境部环境规划院. 2021. 中国二氧化碳捕集利用与封存（CCUS）报告（2021）——中国CCUS路径. 北京:生态环境部环境规划院

唐俊红,鲍征宇,向武,乔胜英,李兵. 2006. 大气甲烷碳同位素测试方法及其在雅克拉凝析气田上方大气中的应用. 环境科学,27(1):14-18

唐俊红,鲍征宇,向武. 2009. 雅克拉凝析油气田油水界面甲烷天然释放及其源示踪. 地球科学——中国地质大学学报,34(5):769-777

唐孝炎,张远航,邵敏. 2006. 大气环境化学. 北京:高等教育出版社,28

陶明信,徐永昌,史宝光,蒋忠惕,沈平,李晓斌,孙明良. 2005. 中国不同类型断裂带的地幔脱气与深部地质构造特征. 中国科学:地球科学, 35(5):441-451

王国建,汤玉平,唐俊红,李吉鹏,杨俊,李兴强. 2018. 断层对烃类微渗漏主控作用及异常分布影响的实验模拟研究. 物探与化探, 42(1):21-27

王家豪,庞雄,王存武,何敏,连世勇. 2006. 珠江口盆地白云凹陷中央底辟带的发现及识别. 地球科学——中国地质大学学报,31(2):209-213

张丽红,郭正府,郑国东,张茂亮,孙玉涛,成智慧,马向贤. 2017. 藏南新生代火山-地热区温室气体的释放通量与成因——以谷露-亚东裂谷为例. 岩石学报,33(1):250-266

张茂亮,郭正府,成智慧,张丽红,郭文峰,杨灿尧,付庆州,温心怡. 2011. 长白山火山区温泉温室气体排放通量研究. 岩石学报, 27(10):2898-2904

赵文斌,郭正府,孙玉涛,张茂亮,张丽红,雷鸣,马琳. 2018. 火山区CO_2气体释放研究进展. 矿物岩石地球化学通报,37(4):601-620

赵文斌,郭正府,刘嘉麒,张茂亮,孙玉涛,雷鸣,马琳,李菊景. 2021. 中国东北新生代火山区CO_2释放规模与成因. 岩石学报, 37(4):1255-1269

郑国东,郭正府,王云鹏,陈践发. 2018. 气体地球化学新进展——纪念著名气体地球化学专家David R. Hilton教授. 矿物岩石地球化学通报, 37(4):Ⅳ-Ⅶ

中国21世纪议程管理中心. 2014.《第三次气候变化国家评估报告》特别报告:中国二氧化碳利用技术评估报告. 北京:科学出版社

周晓成,孙凤霞,陈志,吕超甲,李静,仵柯田,杜建国. 2017. 汶川M_S 8.0地震破裂带CO_2、CH_4、Rn和Hg脱气强度. 岩石学报,33(1): 291-303

朱光有,戴金星,张水昌,李剑,金强,陈践发. 2004. 中国含硫化氢天然气的研究及勘探前景. 天然气工业,24(9):1-4

Allard P. 2010. A CO_2-rich gas trigger of explosive paroxysms at Stromboli basaltic volcano, Italy. Journal of Volcanology and Geothermal Research, 189(3-4):363-374

Anderson D E, Farrar C D. 2001. Eddy covariance measurement of CO_2 flux to the atmosphere from an area of high volcanogenic emissions, Mammoth Mountain, California. Chemical Geology,177(1-2):31-42

Becker J A, Bickle M J, Galy A, Holland T J B. 2008. Himalayan metamorphic CO_2 fluxes:quantitative constraints from hydrothermal springs. Earth and Planetary Science Letters,265(3-4):616-629

Brook E J, Sowers T, Orchardo J. 1996. Rapid variations in atmospheric methane concentration during the past 110,000 years. Science,273(5278): 1087-1091

Brune S, Williams S E, Müller R D. 2017. Potential links between continental rifting, CO_2 degassing and climate change through time. Nature Geoscience,10(12):941-946

Burton M R, Sawyer G M, Granieri D. 2013. Deep carbon emissions from volcanoes. Reviews in Mineralogy and Geochemistry,75(1):323-354

Capaccioni B, Tassi F, Cremonini S, Sciarra A, Vaselli O. 2015. Ground heating and methane oxidation processes at shallow depth in Terre Calde di Medolla (Italy):observations and conceptual model. Journal of Geophysical Research:Solid Earth,120(5):3048-3064

Chen J P, Wang X L, Sun Y G, Ni H Y, Xiang B L, Liao J D. 2021. The accumulation of natural gas and potential exploration regions in the southern margin, Junggar Basin. Frontiers in Earth Science,9:635230

Chen Z, Zhou X, Du J, Xie C Liu L, Li Y, Yi L, Liu H, Cui Y. 2015. Hydrochemical characteristics of hot spring waters in the Kangding district related to the Lushan M_S 7.0 earthquake in Sichuan,China. Natural Hazards and Earth System Sciences,15(6):1149-1156

Chen Z, Li Y, Liu Z F, Wang J, Zhou X C, Du J G. 2018. Radon emission from soil gases in the active fault zones in the capital of China and its environmental effects. Scientific Reports,8(1):16772

Chen Z, Li Y, Liu Z F, Zheng G D, Xu W, Yan W, Yi L. 2019a. CH_4 and CO_2 emissions from mud volcanoes on the southern margin of the Junggar Basin, NW China:origin, output, and relation to regional tectonics. Journal of Geophysical Research:Solid Earth,124(5):5030-5044

Chen Z, Li Y, Liu Z F, Lu C, Zhao Y X, Wang J. 2019b. Evidence of multiple sources of soil gas in the Tangshan fault zone, North China. Geofluids, 2019:1945450

Chen Z, Li Y, Martinelli G, Liu Z F, Lu C, Zhao Y X. 2020. Spatial and temporal variations of CO_2 emissions from the active fault zones in the capital area of China. Applied Geochemistry,112:104489

Chiodini G, Frondini F, Ponziani F. 1995. Deep structures and carbon dioxide degassing in central Italy. Geothermics,24(1):81-94

Chiodini G, Cioni R, Guidi M, Raco B, Marini L. 1998. Soil CO_2 flux measurements in volcanic and geothermal areas. Applied Geochemistry, 13(5):543-552

Chiodini G, Cardellini C, Amato A, Boschi E, Caliro S, Frondini F, Ventura G. 2004. Carbon dioxide Earth degassing and seismogenesis in central and southern Italy. Geophysical Research Letters,31(7):L07615

Chiodini G, Cardellini C, Di Luccio F, Selva J, Frondini F, Caliro S, Rosiello A, Beddini G, Ventura G. 2020. Correlation between tectonic CO_2 Earth degassing and seismicity is revealed by a 10-year record in the Apennines, Italy. Science Advances, 6(35): eabc2938

Ciais P, Sabine C, Bala G, Bopp L, Brovkin V, Canadell J, Chhabra A, DeFries R, Galloway J, Heimann M, Jones C, Le Quérér C, Myneni R B, Piao S L, Thornton P. 2013. Carbon and other biogeochemical cycles. In: Stocker T F, Qin D, Plattner G K, Tignor M, Allen S K, Boschung, J, Nauels A, Xia Y, Bex V, Midgley P M (eds). Climatic Change 2013: the Physical Science Basis. Contribution of Working Group I to the Fifth Assessment Report of IPCC. Cambridge: Cambridge University Press

Cox P M, Betts R A, Jones C D, Spall S A, Totterdell I J. 2000. Acceleration of global warming due to carbon-cycle feedbacks in a coupled climate model. Nature, 408(6809): 184–187

Crutzen P J. 1991. Methane's sinks and sources. Nature, 350: 380–381

Crutzen P J. 1995. On the role of CH_4 in atmospheric chemistry, sources, sinks and possible reductions in anthropogenic sources. Royal Swedish Academy of Sciences, 24(1): 52–55

Cui Y, Du J, Zhang D, Sun Y. 2013. Anomalies of total column CO and O_3 associated with great earthquakes in recent years. Natural Hazards and Earth System Science, 13(10): 2513–2519

Cui Y, Ouzounov D, Hatzopoulos N, Sun K, Zou Z, Du J. 2017. Satellite observation of CH_4 and CO anomalies associated with the Wenchuan M_S 8.0 and Lushan M_S 7.0 earthquakes in China. Chemical Geology, 469: 185–191

Dai J X, Wu X Q, Ni Y Y, Wang Z C, Zhao C Y, Wang Z Y, Liu G X. 2012. Geochemical characteristics of natural gas from mud volcanoes in the southern Junggar Basin. Earth Sciences, 55(3): 355–367

Di P F, Feng D, Chen D F. 2014. *In-situ* and on-line measurement of gas flux at a hydrocarbon seep from the northern South China Sea. Continental Shelf Research, 81: 80–87

Di P F, Feng D, Tao J, Chen D F. 2020. Using time-series videos to quantify methane bubbles flux from natural cold seeps in the South China Sea. Minerals, 10(3): 216

Etiope G, Klusman R W. 2002. Geologic emissions of methane to the atmosphere. Chemosphere, 49: 777–789

Etiope G, Milkov A V. 2004. A new estimate of global methane flux from onshore and shallow submarine mud volcanoes to the atmosphere. Environmental Geology, 46(8): 997–1002

Etiope G, Ciccioli P. 2009. Earth's degassing: a missing ethane and propane source. Science, 323(5913): 478

Etiope G, Christodoulou D, Kordella S, Marinaro G, Papatheodorou G. 2013. Offshore and onshore seepage of thermogenic gas at Katakolo Bay (western Greece). Chemical Geology, 339: 115–126

Foley S F, Fischer T P. 2017. An essential role for continental rifts and lithosphere in the deep carbon cycle. nature Geoscience, 10(12): 897–902

Frondini F, Caliro S, Cardellini C, Chiodini G, Morgantini N, Parello F. 2008. Carbon dioxide degassing from Tuscany and northern Latium (Italy). Global and Planetary Change, 61(1-2): 89–102

Fu B H, Zheng G D, Ninomiya Y, Wang C Y, Sun G Q. 2007. Mapping hydrocarbon-induced mineralogical alteration in the northern Tian Shan using ASTER multispectral data. Terra Nova, 19(4): 225–231

Gambardella B, Cardellini C, Chiodini G, Frondini F, Marini L, Ottonello G, Zuccolini M V. 2004. Fluxes of deep CO_2 in the volcanic areas of central-southern Italy. Journal of Volcanology and Geothermal Research, 136(1-2): 31–52

Giammanco S, Gurrieri S, Valenza M. 1998. Anomalous soil CO_2 degassing in relation to faults and eruptive fissures on Mount Etna (Sicily, Italy). Bulletin of Volcanology, 60(4): 252–259

Gold T. 1979. Terrestrial sources of carbon and earthquake outgassing. Journal of Petroleum Geology, 1(3): 3–19

Han X, Li Y, Du J, Zhou X, Xie C, Zhang W. 2014. Rn and CO_2 geochemistry of soil gas across the active fault zones in the capital area of China. Natural Hazards and Earth System Sciences, 14(10): 2803–2815

He Y S, Meng X N, Ke S, Wu H J, Zhu C W, Teng F Z, Hoefs J, Huang J, Yang W, Xu L J, Hou Z H, Ren Z Y, Li S G. 2019. A nephelinitic component with unusual δ^{56}Fe in Cenozoic basalts from eastern China and its implications for deep oxygen cycle. Earth and Planetary Science Letters, 512: 175–183

Heinicke J, Martinelli G. 2005. Preface: an historical overview. Annals of Geophysics, 48(1): V-VIII

Holloway S. 2005. Underground sequestration of carbon dioxide—a viable greenhouse gas mitigation option. Energy, 30(11-12): 2318–2333

Houghton J T, Ding Y, Griggs D J, Noguer M, Winden P J, Dai X. 2001. Climate Change 2001: the Scientific Basis Contribution of Working Group I to the Third Assessment Report of Intergovernmental Panel on Climate Change. Cambridge: Cambridge University Press

Huang Z L, Xu X, Han Y F, Gao Y, Wang Y L, Song C H. 2016. Comparative study of Dushanzi and Baiyanggou mud volcano microbial communities in Junggar Basin in Xinjiang, China. International Research Journal of Public and Environmental Health, 3(11): 244–256

Iacono-Marziano G, Gaillard F, Scaillet B, Pichavant M, Chiodini G. 2009. Role of non-mantle CO_2 in the dynamics of volcano degassing: the Mount Vesuvius example. geology, 37: 319–322

Italiano F, Bonfanti P, Ditta M, Petrini R, Slejko F. 2009. Helium and carbon isotopes in the dissolved gases of Friuli Region (NE Italy): geochemical evidence of CO_2 production and degassing over a seismically active area. Chemical Geology, 266(1-2): 76–85

Jenkinson D S, Adams D E, Wild A. 1991. Model estimates of CO_2 emissions from soil in response to global warming. Nature, 351(6324): 304–306

Ji L M, Zhang M Z, Ma X X, Xu W, Zheng G D. 2018. Characteristics of mixed sporopollen assemblage from sediments of Dushanzi mud volcano in southern Junggar Basin and indication to the source of mud and debris ejecta. Marine and Petroleum Geology, 89: 194–201

Jung N H, Han W S, Han K, Park E. 2015. Regional-scale advective, diffusive, and eruptive dynamics of CO_2 and brine leakage through faults and wellbores. Journal of Geophysical Research: Solid Earth, 120(5): 3003-3025

Kennedy B M, Kharaka Y K, Evans W C, Ellwood A, DePaolo D J, Thordsen J, Ambats G, Mariner R H. 1997. Mantle fluids in the San Andreas fault system, California. Science, 278(5341): 1278-1281

Kennett J P, Cannariato K G, Hendy I L, Behl R J. 2003. Methane hydrates in quaternary climate change: the clathrate gun hypothesis. Washington: American Geophysical Union, 216

Kerrick D M, Caldeira K. 1998. Metamorphic CO_2 degassing from orogenic belts. Chemical Geology, 145(3-4): 213-232

King C Y, King B S, Evans W C, Zhang W. 1996. Spatial radon anomalies on active faults in California. Applied Geochemistry, 11(4): 497-510

Kulongoski J T, Hilton D R, Barry P H, Esser B K, Hillegonds D, Belitz K. 2013. Volatile fluxes through the Big Bend section of the San Andreas Fault, California: helium and carbon-dioxide systematics. Chemical Geology, 339: 92-102

Langmuir C H, Vocke R D, Hanson J G N. 1978. A general mixing equation with applications to Icelandic basalts. Earth and Planetary Science Letters, 37: 308-392

Lewicki J L, Hilley G E, Dobeck L, McLing T L, Kennedy B M, Bill M, Marino B D V. 2013. Geologic CO_2 input into groundwater and the atmosphere, Soda Springs, ID, USA. Chemical Geology, 339: 61-70

Lewicki J L, Kelly P J, Bergfeld D, Vaughan R G, Lowenstern J B. 2017. Monitoring gas and heat emissions at Norris Geyser Basin, Yellowstone National Park, USA based on a combined eddy covariance and multi-GAS approach. Journal of Volcanology and Geothermal Research, 347: 312-326

Li N, Huang H G, Chen D F. 2014. Fluid sources and chemical processes inferred from geochemistry of pore fluids and sediments of mud volcanoes in the southern margin of the Junggar Basin, Xinjiang, northwestern China. Applied Geochemistry, 46(1): 1-9

Li Q, Liu G Z. 2016. Risk assessment of the geological storage of CO_2: a review. In: Vishal V, Singh T N (eds). Geologic Carbon Sequestration. New York: Springer, 249-284

Li Q, Liu G Z, Liu X H, Li X C. 2013. Application of a health, safety, and environmental screening and ranking framework to the Shenhua CCS project. International Journal of Greenhouse Gas Control, 17: 504-514

Li Q, Wei Y N, Liu G Z, Shi H. 2015. CO_2-EWR: a cleaner solution for coal chemical industry in China. Journal of Cleaner Production, 103: 330-337

Li Q, Song R R, Shi H, Ma J L, Liu X H, Li X C. 2018. U-tube based near-surface environmental monitoring in the Shenhua carbon dioxide capture and storage (CCS) project. Environmental Science and Pollution Research, 25(12): 12034-12052

Li S G, Wang Y. 2018. Formation time of the big mantle wedge beneath eastern China and a new lithospheric thinning mechanism of the North China Craton—geodynamic effects of deep recycled carbon. Science China Earth Sciences, 61(7): 853-868

Li S G, Yang W, Ke S, Meng X N, Tian H C, Xu L J, He Y S, Huang J, Wang X C, Xia Q K, Sun W D, Yang X Y, Ren Z Y, Wei H Q, Liu Y S, Meng F C, Yan J. 2017. Deep carbon cycles constrained by a large-scale mantle Mg isotope anomaly in eastern China. National Science Review, 4(1): 111-120

Liu H J, Were P, Li Q, Gou Y, Hou Z. 2017. Worldwide status of CCUS technologies and their development and challenges in China. Geofluids, 2017: 6126505

Liu S A, Wang Z Z, Li S G, Huang J, Yang W. 2016. Zinc isotope evidence for a large-scale carbonated mantle beneath eastern China. Earth and Planetary Science Letters, 444: 169-178

Ma K, Ma A Z, Zheng G D, Ren G, Xie F, Zhou H C, Yin J, Liang Y, Zhuang X L, Zhuang G Q. 2021. Mineralosphere microbiome leading to changed geochemical properties of sedimentary rocks from Aiqigou mud volcano, Northwest China. Microorganisms, 9(3): 560

Ma X X, Zheng G D, Liang S Y, Xu W. 2015. Geochemical characteristics of absorbed gases in fault gouge from the Daliushu dam area, NW China. Geochemical Journal, 49(4): 413-419

Mao W L, Mao H K, Meng Y, Eng P J, Hu M Y, Chow P, Cai Y Q, Shu J F, Hemley R J. 2006. X-ray-induced dissociation of H_2O and formation of an O_2-H_2 alloy at high pressure. Science, 314(5799): 636-638

Marziano G I, Gaillard F, Pichavant M. 2008. Limestone assimilation by basaltic magmas: an experimental re-assessment and application to Italian volcanoes. Contributions to Mineralogy and Petrology, 155(6): 719-738

Mason E, Edmonds M, Turchyn A V. 2017. Remobilization of crustal carbon may dominate volcanic arc emissions. Science, 357(6348): 290-294

Mau S, Tu T H, Becker M, Dos Santos Ferreira C, Chen J N, Lin L H, Wang P L, Lin S, Bohrmann G. 2020. Methane seeps and independent methane plumes in the South China Sea Offshore Taiwan. Frontiers in Marine Sciences, 7: 543

Milkov A V. 2005. Global distribution of mud volcanoes and their significance in petroleum exploration as a source of methane in the atmosphere and hydrosphere and as a geohazard. In: Martinelli G, Panahi B (eds). Mud Volcanoes, Geodynamics and Seismicity. New York: Springer, 29-34

Nakada R, Takahashi Y, Tsunogai U, Zheng G D, Shimizu H, Hattori K H. 2011. A geochemical study on mud volcanoes in the Junggar Basin, China. Applied Geochemistry, 26(7): 1065-1076

Orcutt B N, Daniel I, Dasgupta R. 2019. Deep carbon: Past to Present. London: Cambridge University Press

Ren G, Ma A Z, Zhang Y F, Deng Y, Zheng G D, Zhuang X L, Zhuang G Q, Fortin D. 2018. Electron acceptors for anaerobic oxidation of methane drive microbial community structure and diversity in mud volcanoes. Environmental Microbiology, 20(7): 2370-2385

Sano Y, Wakita H. 1985. Geographical distribution of $^3He/^4He$ ratios in Japan: implications for arc tectonics and incipient magmatism. Journal of Geophysical Research: Solid Earth, 90(B10): 8729-8741

Sano Y, Marty B. 1995. Origin of carbon in fumarolic gad from island arcs. Chemical Geology, 119(1-4): 265-274

Sano Y, Kinoshita N, Kagoshima T, Takahata N, Sakata S, Toki T, Kawagucci S, Waseda A, Lan L F, Wen H, Chen A T, Lee H, Yang T F, Zheng G D,

Tomonaga Y, Roulleau E, Pinti D L. 2017. Origin of methane-rich natural gas at the West Pacific convergent plate boundary. Scientific Reports, 7(1):15646

Schumacher D. 1996. Hydrocarbon-induced alteration of soils and sediments. In: Schumacher D, Abrams M A (eds). Hydrocarbon Migration and Its Near-surface Expression. AAPG Memoir, 66:71–89

Sciarra A, Cantucci B, Coltorti M. 2017. Learning from soil gas change and isotopic signatures during 2012 Emilia seismic sequence. Scientific Reports, 7(1):14187

Sun Y T, Zhou X C, Zheng G D, Li J, Shi H Y, Guo Z F, Du J G. 2017. Carbon monoxide degassing from seismic fault zones in the basin and range province, west of Beijing, China. Journal of Asian Earth Sciences, 149:41–48

Sun Y T, Guo Z F, Liu J Q, Du J G. 2018. CO_2 diffuse emission from maar lake: an example in Changbai volcanic field, NE China. Journal of Volcanology and Geothermal Research, 349:146–162

Sun Y T, Zhou X C, Du J G, Guo Z F. 2020a. CO diffusive emission in the co-seismic rupture zone of the Wenchuan M_S 8.0 earthquake. Geochemical Journal, 54(3):91–104

Sun Y T, Guo Z F, Fortin D. 2020b. Carbon dioxide emission from monogenetic volcanoes in the Mt. Changbai volcanic field, NE China. International Geology Review, doi:10.1080/00206814.2020.1802782

Tamburello G, Pondrelli S, Chiodini G, Rouwet D. 2018. Global-scale control of extensional tectonics on CO_2 earth degassing. Nature Communications, 9(1):4608

Tang J H, Bao Z Y, Xiang W, Gou Q H. 2007. Daily variation of natural emission of methane to the atmosphere and source identification in the Luntai fault region of the Yakela condensed oil/gas field in the Tarim Basin, Xinjiang, China. Acta Geologica Sinica, 81(5):771–778

Tang J H, Bao Z Y, Xiang W, Gou Q H. 2008. Geological emission of methane from the Yakela condensed oil/gas field in Talimu Basin, Xinjiang, China. Journal of Environmental Sciences, 20(9):1055–1062

Tang J H, Yin H Y, Wang G J, Chen Y Y. 2010. Methane microseepage from different sectors of the Yakela condensed gas field in Tarim Basin, Xinjiang, China. Applied Geochemistry, 25(8):1257–1264

Tang J H, Wang G J, Yin H Y, Li H J. 2013. Methane in soil gas and its transfer to the atmosphere in the Yakela condensed gas field in the Tarim Basin, Northwest China. Petroleum Science, 10(2):183–189

Tang J H, Xu Y, Wang G J, Etiope G, Han W, Yao Z T, Huang J G. 2017. Microseepage of methane to the atmosphere from the Dawanqi oil-gas field, Tarim Basin, China. Journal of Geophysical Research: Atmospheres, 122(8):4353–4363

Tang J H, Xu Y, Wang G J, Huang J G, Han W, Yao Z T, Zhu Z Z. 2019. Methane in soil gas and its migration to the atmosphere in the Dawanqi oilfield, Tarim Basin, China. Geofluids, 2019:1693746

Wakita H, Nakamura Y, Kita I, Fujii N, Notsu K. 1980. Hydrogen release: new Indicator of fault activity. Science, 210(4466):188–190

Werner C, Wyngaard J C, Brantley S L. 2000. Eddy-correlation measurement of hydrothermal gases. Geophysical Research Letters, 27(18):2925–2928

Xu S, Zheng G D, Xu Y C. 2012. Helium, argon and carbon isotopic compositions of spring gases in the Hainan Island, China. Acta Geologica Sinica, 86(6):1515–1523

Xu S, Zheng G D, Nakai S, Wakita S, Wang X B, Guo Z F. 2013. Hydrothermal He and CO_2 at Wudalianchi intra-plate volcano, NE China. Journal of Asian Earth Sciences, 62:526–530

Xu S, Zheng G D, Wang X B, Wang H L, Nakai S, Wakita H. 2014. Helium and carbon isotope variations in Liaodong Peninsula, NE China. Journal of Asian Earth Sciences, 90:149–156

Xu S, Zheng G D, Zheng J J, Zhou S X, Shi P L. 2017. Mantle-derived helium in foreland basins in Xinjiang, Northwest China. Tectonophysics, 694:319–331

Xu W, Zheng G D, Ma X X, Fortin D, Hilton D R, Liang S Y, Chen Z, Hu G Y. 2018. Iron speciation of mud breccia from the Dushanzi mud volcano in the Xinjiang Uygur autonomous region, NW China. Acta Geologica Sinica (English Edition), 92(6):2201–2213

Xu W, Zheng G D, Martinelli G, Ma X X, Fortin D, Fan Q H, Chen Z. 2020. Mineralogical and geochemical characteristics of hydrocarbon-bleached rocks in Baiyanggou mud volcanoes, Xinjiang, NW China. Applied Geochemistry, 116:104572

Yang D X, Li Q, Zhang L Z. 2019. Characteristics of carbon dioxide emissions from a seismically active fault. Aerosol and Air Quality Research, 19(8):1911–1919

Yang W, Teng F Z, Zhang H F, Li S G. 2012. Magnesium isotopic systematics of continental basalts from the North China Craton: implications for tracing subducted carbonate in the mantle. Chemical Geology, 328:185–194

Zhang M L, Guo Z F, Sano Y, Cheng Z H, Zhang L H. 2015. Stagnant subducted Pacific slab-derived CO_2 emissions: insights into magma degassing at Changbaishan volcano, NE China. Journal of Asian Earth Sciences, 106:49–63

Zhang M L, Guo Z F, Sano Y, Zhang L H, Sun Y T, Cheng Z H, Yang T F. 2016. Magma-derived CO_2 emissions in the Tengchong volcanic field, SE Tibet: implications for deep carbon cycle at intra-continent subduction zone. Journal of Asian Earth Sciences, 127:76–90

Zhang M L, Guo Z F, Zhang L H, Sun Y T, Cheng Z H. 2017. Geochemical constraints on origin of hydrothermal volatiles from southern Tibet and the Himalayas: understanding the degassing systems in the India-Asia continental subduction zone. Chemical Geology, 469:19–33

Zhao W B, Guo Z F, Lei M, Zhang M L, Ma L, Fortin D, Zheng G D. 2019. Volcanogenic CO_2 degassing in the Songliao continental rift system, NE China. Geofluids, 2019:8053579

Zheng G D, Fu B H, Takahashi Y, Kuno A, Matsuo M, Zhang J D. 2010. Chemical speciation of redox sensitive elements during hydrocarbon leaching in the Junggar Basin, Northwest China. Journal of Asian Earth Sciences, 39(6):713-723

Zheng G D, Xu S, Liang S Y, Shi P L, Zhao J. 2013. Gas emission from the Qingzhu River after the 2008 Wenchuan earthquake, Southwest China. Chemical Geology, 339:187-193

Zheng G D, Ma X X, Guo Z F, Hilton D R, Xu W, Liang S Y, Fan Q H, Chen W X. 2017. Gas geochemistry and methane emission from Dushanzi mud volcanoes in the southern Junggar Basin, NW China. Journal of Asian Earth Sciences, 149:184-190

Zheng G D, Xu W, Etiope G, Ma X X, Liang S Y, Fan Q H, Sajjad W, Li Y. 2018. Hydrocarbon seeps in petroliferous basins in China: a first inventory. Journal of Asian Earth Sciences, 151:269-284

Zhou X C, Du J G, Chen Z, Cheng J W, Tang Y, Yang L M, Xie C, Cui Y J, Liu L, Yi L, Yang P X, Li Y. 2010. Geochemistry of soil gas in the seismic fault zone produced by the Wenchuan M_S 8.0 earthquake, southwestern China. Geochemical Transactions, 11(1):5

Zhou X C, Chen Z, Cui Y. 2016. Environmental impact of CO_2, Rn, Hg degassing from the rupture zones produced by Wenchuan M_S 8.0 earthquake in western Sichuan, China. Environmental Geochemistry and Health, 38(5):1067-1082

Zhou X C, Liu L, Chen Z, Cui Y J, Du J G. 2017. Gas geochemistry of the hot spring in the Litang fault zone, Southeast Tibetan Plateau. Applied Geochemistry, 79:17-26

A Brief Introduction of Investigations and Studies on Geological Greenhouse Gas Emission

ZHENG Guo-dong[1], ZHAO Wen-bin[2], CHEN Zhi[3], XU Wang[1,4], SONG Zhi-guang[5], LI Qi[6], XU Sheng[7], GUO Zheng-fu[2], MA Xiang-xian[1], LIANG Ming-liang[1,8], WANG Yun-peng[5]

1. Gansu Key Laboratory for Oil and Gas Resources, Northwest Institute of Eco-Environment and Resources, Chinese Academy of Sciences, Lanzhou 730000; 2. Institute of Geology and Geophysics, Chinese Academy of Sciences, Beijing 100029; 3. Institute of Earthquake Forecasting, China Earthquake Administration, Beijing 100036; 4. College of Energy Resources, Chengdu University of Technology, Chengdu 610059; 5. State Key Laboratory of Organic Geochemistry, Guangzhou Institute of Geochemistry, Chinese Academy of Sciences, Guangzhou 510640; 6. State Key Laboratory of Geomechanics and Geotechnical Engineering, Institute of Rock and Soil Mechanics, Chinese Academy of Sciences, Wuhan 430071; 7. School of Earth System Sciences, Tianjin University, Tianjin 300072; 8. Institute of Geomechanics, Chinese Academy of Geological Sciences, Beijing 100081

Abstract: The emission of greenhouse gas of geological origins from the solid Earth into the atmosphere is one of the most important mass (and also energy) exchange processes from the interior to surface of the Earth. The greenhouse gas of geological origin mainly includes carbon dioxide and methane released through various kinds of activities including volcanic eruptions and hydrothermal activity, tectonic movements of fractures and fault zones, natural and/or anthropogenic leak of crude oils and natural gases, dissociation of methane hydrates, natural fires of coal, and weathering of carbonate rocks. In fact, the exchange of geologically originated gases (basic carriers) is an important geological process and is an important dynamic mechanism for the mass circulation and energy exchange in all circles of the earth. Global climate change and greenhouse gas effect are daunting challenges for all human beings. The investigations and researches on the types and sources, the emission mechanisms and processes, the released fluxes to the surface and the greenhouse effect on the atmosphere of greenhouse gases of geological origins have become one of the research hotspots and key development directions of the modern earth system science, with great scientific significances to deal with global climate change and to correctly understand greenhouse effect on the atmosphere. Chinese scientific community took an active part on this challenge, performed a series of key primary investigations and researches on geologically originated greenhouse gases, including those released from some volcanic areas and geothermal regions, major seismic fault zones, oil-and gas-bearing basins, and mud volcanoes within the vast and geologically varied territories of China, and obtained some achievements on the geochemical compositions and released fluxes of geologically originated greenhouse gases. These above mentioned studies have laid a good foundation for further widely in-depth studies in a near future.

Key words: geologically originated greenhouse gas; release characteristics; current study status; developing prospect

流体包裹体研究进展*

倪 培¹ 范宏瑞² 潘君屹¹ 迟 哲¹ 崔健铭¹

1. 南京大学 地球科学与工程学院，内生金属矿床成矿机制研究国家重点实验室，地质流体研究所，南京 210023；2. 中国科学院 地质与地球物理研究所，矿产资源研究重点实验室，北京 100029

摘 要：21世纪第二个十年（2011~2020年），我国流体包裹体研究与应用取得了长足进步。本文重点回顾了这一时期我国在流体包裹体研究领域的主要进展，包括流体包裹体理论、流体包裹体分析技术、矿床学和沉积成藏研究进展。流体 PVTx 性质模拟研究继续保持国际先进水平，建立了更加复杂多样的状态方程并能够较好地应用于各类天然流体体系中；同时，流体包裹体组合（fluid inclusion assemblage，FIA）概念的使用已深入人心，流体包裹体数据的获取更加科学和规范。国内学者已逐步采用或完善了一些国际前沿的流体包裹体研究新技术和新方法，如石英阴极发光技术、不透明矿物的红外显微测温技术、单个包裹体 LA-ICP-MS 分析技术、融合二氧化硅毛细管合成人工包裹体技术、金刚石对顶及相关水热实验技术、流体包裹体定年技术等，这些新方法和新技术已被应用于各类地质流体研究中，尤其是在成矿流体和成矿机制研究方面取得了大量成果。石盐包裹体和油气包裹体越来越受到环境及石油地质学家的重视，在沉积成藏尤其是在古环境和油气充注及成分演化史的研究中起着关键作用。近十年来国内学术组织发挥积极作用，定期召开学术会议，由中国学者创办的 ACROFI 系列会议已成为国际流体包裹体界的重要系列会议之一，为我国和其他亚洲国家学者提供了前沿的国际学术交流平台。本文最后展望了我国包裹体领域的发展趋势。

关键词：流体包裹体 理论研究 分析技术 矿床学研究 沉积成藏研究 进展与展望

0 引 言

流体包裹体研究是地球科学领域的一个重要分支学科，最早可以追溯到1858年英国科学家 Sorby（1858）对流体包裹体的开创性研究工作，迄今已有百余年历史。20世纪60年代以来，随着 Roedder（1960，1962，1963）对流体包裹体基础理论和分析方法的系统性完善，流体包裹体研究在地质科学领域逐渐扮演了越来越重要的角色。21世纪第二个十年（2011—2020年），流体包裹体研究取得新的进展，技术体系和实验方法得到创新、发展和完善，相关成果涉及矿床学、岩石学、宝石学、沉积成藏、行星地质、构造地质、环境地质等众多地质学领域。

20世纪60年代，流体包裹体研究在我国地学领域生根发芽，并很快成长为一门重要的分支学科。在21世纪第一个十年中，我国流体包裹体研究飞速发展（倪培等，2014），在21世纪第二个十年中更是取得长足进步，一方面表现在研究领域的扩展，从上个十年的矿床学、岩石学、沉积成藏等领域继续扩大至其他地质学新兴领域；另一方面表现在研究水平的提高，中国流体包裹体研究成果丰硕，一些研究已接近或达到国际一流水平。近十年中，中国学者创办的 ACROFI 系列学术会议共成功举办四届（2012年、2014年、2016年、2018），吸引了大量国内外学者参会交流合作，已成为国际流体包裹体学界最重要的系列会议之一。

本文对近十年来国内流体包裹体研究的主要进展进行回顾和综述，着重凝练我国学者在这一领域取得的新成果和新认识，明确未来的发展方向，期望对未来我国流体包裹体研究提供帮助。

---

* 原文"流体包裹体研究进展与展望（2011—2020）"刊于《矿物岩石地球化学通报》2021年第40卷第4期，本文略有修改。

1 流体包裹体理论研究进展

1.1 流体包裹体 PVTx 性质研究

流体包裹体主要由液相、气相和可溶解的固相部分组成，包括 H_2O 和溶解于其中的 Na^+、Mg^{2+}、K^+、Ca^{2+}、Cl^-、Br^-、CO_3^{2-}、SO_4^{2-} 等离子成分以及 H_2、O_2、N_2、Ar、CO_2、CH_4、SO_2、H_2S 等气体成分（卢焕章等，2004）。用于描述自然界流体包裹体中流体的压力、体积、温度和组成（$PVTx$）的函数关系式即为包裹体流体的状态方程（eguation of state，EOS），不断建立和完善针对自然界不同流体体系的状态方程，是流体包裹体基础理论研究的核心内容。

在 21 世纪第二个十年间，我国学者在运用状态方程进行流体性质模拟方面取得了诸多成果。刘斌（2011）选取较高压（>0.1 MPa）化学反应平衡常数进行了简单体系水溶液包裹体 pH 和 Eh 计算。对于 H_2O-$NaCl$-CO_2、H_2O-N_2 和 H_2O-H_2 的体系，有研究者采用改进的 SAFT-LJ 状态方程开展计算并扩展了这一方法在低于 573 K 的情况下对于 H_2O-$NaCl$-CO_2 体系的可使用范围，同时确认对于 H_2O-N_2 和 H_2O-H_2 体系新的分子基状态方程可用于一个更宽泛的温压范围（Sun and Dubessy，2012；Sun et al.，2015）。最近，Li 等（2020）对 H_2O-$NaCl$-CO_2 三元混合体系开展了热力学模拟和石英溶解度计算研究，系统正演了 H_2O-$NaCl$-CO_2 体系流体在 300～500℃、0.001～3 500 MPa 范围内不同 $PVTx$ 下的相行为、流体密度以及 NaCl 和 CO_2 在各相流体中的含量，构建了石英在该系统中的溶解度模型，并讨论了造山型及侵入体有关金矿床中不同类型流体包裹体和石英脉形成机制。此外，Mao 等（2011，2017）利用状态方程完成了对包括 CH_4-H_2O、CH_4-C_2H_6-C_3H_8-CO_2-N_2 在内的多个流体包裹体体系的模拟和预测，提出了依据相比例和甲烷水合物的溶解温度计算 CH_4-H_2O 包裹体成分的新模型，建立了新的亥姆霍兹自由能显式状态方程用于计算 CH_4-C_2H_6-C_3H_8-CO_2-N_2 体系的等容线。在理论计算的同时，一部分学者还采用了拉曼光谱对包裹体进行了原位的观察和分析，例如在不同温度和压力范围下氮的形态变化（Chen et al.，2019）、甲烷的形态变化（Shang et al.，2014）等。迄今中国科学家已建立了多个包裹体体系的 $PVTx$ 模型，拓宽了对流体性质的认知范围。

1.2 流体包裹体组合的应用

流体包裹体组合（FIA）代表了同时捕获的一组流体包裹体，例如矿物晶体内同一个生长带或同一条愈合裂隙中同时捕获的一系列流体包裹体可以称为一个 FIA，代表岩相学上能够划分得最细的一次流体捕获事件（Goldstein and Reynolds，1994）。FIA 是建立在详细的显微岩相学观察基础上的一种流体包裹体重要的研究方法，同时也是流体包裹体测温数据有效性的判别方法和科学的数据表达方式。FIA 概念被提出之后很快成为国际流体包裹体研究所遵循的基本准则之一，并被引入国内流体包裹体研究领域（如池国祥和卢焕章，2008）。然而，在较长的一段时间内，FIA 概念与方法并未引起国内学者的重视，21 世纪第一个十年间，国内学者发表的流体包裹体研究论文中仍鲜有提及这一概念，显微测温数据也多数缺少 FIA 的判别方法和表达方式。

近年来，越来越多的研究者认识到，使用 FIA 法则来研究流体包裹体的必要性。例如，Ni 等（2015a）采用 FIA 的概念研究了浙江璜山和平水金矿的流体包裹体特征，识别出大量水溶液包裹体和富 CO_2 水溶液包裹体构成的不混溶组合，进而判断在主矿化阶段发生的流体不混溶致使金成矿；Pan 等（2018）利用 FIA 方法对福建紫金山 Cu-Au 矿床超过 1 500 m 垂向范围内的矿体开展了流体包裹体填图工作，通过系统获取的不同空间位置 FIA 测温数据均值重建了成矿古流体的运移通道；Chi 等（2018）研究了悦洋银多金属矿床内闪锌矿和共生石英中的流体包裹体，使用 FIA 法则获取和报道了成矿过程中各阶段流体包裹体测温数据，为进一步判别成矿机制提供了更加可靠的依据；此外，Ma 等（2018）和 Zhong 等

(2013) 也分别采用 FIA 法则对胶东金亭岭金矿和霍各乞铜多金属矿床开展了测温学研究，从而对成矿流体演化过程和成矿机制提出了新的认识。这些研究实践证明，相比于传统方法，利用 FIA 法则开展流体包裹体研究具有更好的科学性，通过 FIA 方法甄别和表达包裹体数据，避免了传统方法中简单汇总大量单个包裹体数据对最终结果所产生的不可控影响。此外值得注意的是，除测温学数据应当使用 FIA 法则进行表达外，单个流体包裹体成分分析数据也同样应当基于严格的 FIA 法则进行表达（Pan et al., 2019；Zhao et al., 2020），在未来的包裹体研究中科学工作者仍需坚持和普及这一法则。

2　流体包裹体分析技术进展

流体包裹体作为古流体保存至今的唯一代表，是少数可以用来直接反演地质时期各种流体性质和过程的样本。然而，受限于它复杂的岩相学特征和微小尺寸，如何更好地对流体包裹体进行观测识别，以及如何准确获取流体包裹体中所蕴含的物理、化学信息一直是流体包裹体研究的关键内容。21 世纪的第二个十年，中国科学家在前期研究基础之上，紧跟国际前沿，革新了一部分实验分析体系，发展和完善了很多用于观测和分析流体包裹体的新技术、新方法，成果斐然。本文仅列举主要方面的成果。

2.1　阴极发光技术

受电子轰击而发光是物质的一种固有属性，扫描电子显微镜-阴极发光（SEM-CL）技术对研究矿物生长过程中的原生及次生结构具有无法替代的作用。随着该技术的不断发展，SEM-CL 已成为流体包裹体岩相学观察中不可或缺的重要环节。以流体包裹体最主要的寄主矿物石英为例，研究者通过阴极发光可以有效揭示同一石英样品中可能包含的多阶段生长结构，并据此判别包裹体在不同阶段石英中捕获的先后关系，实现对流体演化过程的精细刻画。需要注意的是，这些信息仅利用传统光学显微镜往往是极难获取的。例如，Ni 等（2017a）利用石英的阴极发光技术对桐村斑岩钼矿多阶段含矿石英网脉开展了精细的岩相学研究，发现依据宏观切穿关系划分的不同期次石英脉实际上是由至少四阶段石英溶解-沉淀过程叠加而成，而主要矿石矿物辉钼矿总是严格生长于第二期和第三期石英之间。在这一精细岩相学基础上，研究者通过不同阶段石英中原生或假次生流体包裹体精细重建了流体演化历史，并发现辉钼矿成矿与含 CO_2 流体的不混溶密切相关；然而如果不借助石英的阴极发光技术，研究者根据传统流体包裹体岩相学进行数据分析，则可能得出完全不同的结论。该研究在国内较早将 SEM-CL 技术应用于流体包裹体精细岩相学研究中，为后续同类型工作提供了参考。类似的，在对胶东金亭岭金矿的流体研究中，Ma 等（2018）进行了含矿石英脉阴极发光分析，区分出了五期石英及其中的流体包裹体，根据成矿期包裹体特征判断金的成矿与含 CO_2 流体的冷却和不混溶相关；此外，Pan 等（2019）和 Cui 等（2019）分别对瑶岗仙石英脉型黑钨矿床和小龙河石英脉型锡矿床中的石英及共生黑钨矿或锡石开展了阴极发光分析，识别出了多阶段流体活动的证据，为精细刻画黑钨矿和锡石的沉淀机制提供了重要依据。

2.2　红外显微镜技术

矿石矿物形成时捕获的流体包裹体是成矿流体的直接记录，通常比共生脉石矿物具体更好的代表性。自 20 世纪 80 年代红外显微镜诞生（Campbell et al., 1984）以来，利用不透明（或半透明）矿石矿物中的流体包裹体来直接反映成矿流体特征一直是成矿流体与成矿机制研究中的热点课题。红外显微成像技术于上个十年间引进国内，并取得了重要成果（Ni et al., 2008）。沿承上个十年的发展，我国学者于本阶段进一步推进了这一技术的完善和应用。在技术方面，我国学者较早依据红外显微镜对辉锑矿等矿石矿物中流体包裹体的测温特点评估了红外光强度对显微测温结果的影响（格西等，2011）；最近，Peng 等（2020）较为系统的定量分析了不同矿物对红外光的吸收特征以及其导致的流体包裹体测温结果偏差，发现采用最低光强进行观察有助于减少矿物对红外光能的吸收并提高测试精度，并提议了一种改进的循环测温法。在应用研究方面，Ni 等（2015c）采用红外显微测温技术对南岭地区多个典型石英脉型黑钨矿床中共生的黑钨矿和石英开展了系统的流体包裹体对比研究，揭示出形成黑钨矿和共生石英的流体存在温度和盐度的系统性差别，并依据黑钨矿中流体包裹体数据探讨了黑钨矿沉淀机制。此外，我国学者也对西华山、瑶岗仙、茅坪等石英脉型黑钨矿床开展了类似的红外显微测温研究（Wei et al., 2012；黄惠兰等，2015；Chen et al., 2018；Li W. S. et al., 2018），这些研究表明，即便是与矿石矿物密切共生的脉石矿物，其流体包裹体也不一定能够代表真实的成矿流体，因此利用红外显微技术对矿石矿物中包裹体开展直接观测是获取成矿流体信息的关键。最近，Pan 等（2019）在国内成功建立了红外显微成像和单个流体包裹体 LA-ICP-MS 联用技术，实现了对不透明矿石矿物中单个流体包裹体的主微量元素定量分析，为成矿流体和成矿机制研究增添了一项强有力的技术手段。

2.3 单个包裹体 LA-ICP-MS 分析技术

20 世纪末发展起来的单个流体包裹体 LA-ICP-MS 成分分析，是一种能够快速、原位定量测定单个流体包裹体中绝大多数主微量元素的强大分析手段（Günther et al., 1998；Heinrich et al., 2003；Pettke et al., 2012）。相比传统包裹体群体分析，该技术具有高空间分辨率、高灵敏度、高精密度、低检测限（ng/g），以及多元素同时检测等优点，从问世至今已为流体包裹体研究带来了诸多革命性进展。

然而，由于技术限制等诸多原因，单个流体包裹体的 LA-ICP-MS 成分分析在较长的一段时间内仅局限于少数西方国家实验室，国内相关技术研发进展缓慢。在 21 世纪第一个十年间，我国学者通过国际合作对这一技术进行了有益探索（Su et al., 2009），但尚未能在国内自主建立实验室，使其成为我国流体包裹体研究迈向国际先进水平的主要障碍。通过 21 世纪第二个十年间的不懈努力，目前南京大学已成功建立了单个流体（熔体）包裹体 LA-ICP-MS 分析方法，并在成矿流体和成矿机制研究中展现出了其无可比拟的巨大优势。Pan 等（2019）在精细岩相学工作基础上，采用单个流体包裹体 LA-ICP-MS 技术对瑶岗仙钨矿中黑钨矿和共生石英中的流体包裹体进行了成分分析，揭示出沉淀黑钨矿的成矿流体具有独特的元素成分特征，精细刻画了黑钨矿石英脉形成的流体演化过程，研究者还通过包裹体中不相容元素和卤族元素比值进一步判定成矿流体来自单一的稳定岩浆源区；Zhao 等（2020）对鲜花岭斑岩–浅成低温热液多金属勘查区不同空间位置的含矿石英脉中流体包裹体进行了成分分析，从元素角度证实了区内不同矿化点成矿流体均来自同一岩浆热液系统，并在流体运移过程中存在一定的大气降水混入和少量盆地卤水注入，流体包裹体中 Pb、Zn、Cu 等成矿元素浓度十分可观，预示了深部存在斑岩铜矿化的可能。此外，国内很多研究机构也在不断探索，相继建立了一些分析方法。在国内实验室自主攻坚的同时，我国学者也通过国际实验平台开展了一些研究工作。如 Shu 等（2017）利用单个流体包裹体 LA-ICP-MS 分析技术对白音诺尔铅锌矿开展研究，获取了成矿前辉石、成矿期闪锌矿及成矿后方解石中的流体包裹体元素成分，结合测温数据，揭示出成矿流体与地下水混合是铅锌成矿的主要机制，与碳酸盐岩围岩的水岩反应也促进了这一过程。总体上，虽然国内实验室建成时间尚短，相关自主研究工作仍处于起步阶段，但无论是国外已有的大量成功实践，还是近年来国内已取得的部分成果，都充分说明了单个流体包裹体成分分析在成矿流体来源示踪、金属元素迁移机制、元素配分行为研究，以及成矿机制精细解剖等诸多方面广阔的应用前景。这一新技术加深甚至改变了研究者对过往固有的成岩成矿过程的认知，支持了研究者依据全新的数据对原有模型进行补充和改进。可以预见，单个包裹体 LA-ICP-MS 成分分析技术必将在我国未来十年的流体包裹体研究中大放异彩。

2.4 金刚石压腔及相关水热实验技术

地质流体的活动范围几乎涵盖了地球的各个圈层，一些流体活动及其产物，对于一些涉及极端高压环境的流体过程和产物，利用金刚石压腔还原形成时的温压条件是必不可少的研究手段；而对于部分极难获取样品的深部地质过程，也可以通过金刚石压腔在实验室内进行模拟和反演。近十年来，国内学者利用热液金刚石压腔技术在包裹体观测和水热实验模拟方面取得了一系列重要进展。观测技术方面，在传统金刚石压腔技术基础上，Ni 等（2014）通过改进建立了对更高温压范围内流体电导率的测定方法，将压力范围提高到几个 GPa。此外，Li 等（2016）提出了一种改进的热液金刚石压腔方法（HDAC-VT），采用 3 个柱体使得受力更加均匀，同时采用 LINKAM T95 冷热台使得升降温范围更大、速率更准确稳定。在水热实验方面，Ni 和 Keppler（2012）利用热液金刚石压腔和激光拉曼光谱测定了氧化的岩浆热液中不同条件下 S 的物种，进一步证实了 S^{6+} 物种（主要是 HSO_4^- 和 H_2SO_4）可以在氧化的岩浆热液中与 SO_2 稳定共存，存在并被火山作用直接释放到大气中。针对自然界岩浆热液系统及稀土矿床中存在异常富硫酸盐流体的现象，Cui 等（2019）利用热液金刚石压腔研究了 Na_2SO_4-SiO_2-H_2O 和 Na_2SO_4-$Nd_2(SO_4)_3$-SiO_2-H_2O 体系的高温行为，模拟发现石英的存在显著改变了硫酸钠的溶解性质，并预测对于稀土而言硫酸根离子比氯离子更容易络合，真实地质环境中稀土离子可以被富含硫酸盐的流体有效运输。此外，Li 等（2018）通过 $MnWO_4$-Li_2CO_3（或 Na_2CO_3）-H_2O 体系水热实验，证实碱金属碳酸盐的存在会影响钨锰矿的溶解度，这一改变可能与碳酸盐溶解产生的高 pH 有关。

2.5 人工合成包裹体技术

实验室环境下合成的人工包裹体可以作为天然包裹体的类似物，用于验证流体包裹体研究相关的假设和校验技术方法的可行性。近十年来，我国学者利用人工合成包裹体技术开展了包括成矿元素溶解度、岩浆热液过程元素配分行为、流体包裹体形成机制等多方面工作，并引入了合成流体包裹体新方法，在流体相平衡和流体包裹体分析标定等领域取得一定进展。

利用高温高压容器在矿物裂隙愈合或硅酸盐熔体淬火过程中合成流体包裹体是对自然界流体包裹体捕获过程的最直观重现。Zhou 等（2016）对高温高压下独居石和磷钇矿在 H_2O-Na-K-Cl-F-CO_2 体系中的溶解行为进行了合成包裹体研究工作，研究表明碳酸岩流体中 Na 和 F 的富集可以提高所有稀土元素的活动性，同时在 F 或 CO_2 存在的情况下 K 的富集可能促进了轻、重稀土元素之间的分馏。另外，Zhou 等（2019）采用拉曼光谱观测了人工合成包裹体中临界点附近水的化学结构，对超临界水的物态提出了新的认识。最近，袁顺达和赵盼捞（2021）采用在硅酸盐玻璃中合成流体包裹体方法对锡在熔体-流体间的配分系数进行重新标定，一定程度上避免了利用双金属套管开展的传统溶解度实验中体系封闭性的问题。此外，人工合成流体包裹体还被应用到油气研究中，Chen 等（2015，2016）在接近实际储层温度、压力和原油含量下，在方解石中成功合成了烃类包裹体，实验发现即便很低的水含量及纯原油条件下也可形成烃类包裹体，表明水并不是矿物生长捕获烃类包裹体的必要条件。这一研究还说明在油饱和的碳酸盐储层内流体可与方解石或者灰岩发生反应，这一反应过程可能导致了次生孔隙的形成。

除上述传统合成包裹体技术外，倪培等（2011）在国内首次介绍了一种可以近似类比合成包裹体的融合二氧化硅毛细管新技术，并通过对纯 H_2O 体系、纯 CO_2 体系、H_2O-NaCl 体系和 H_2O-CO_2 体系的毛细管样品进行了显微测温和激光拉曼分析，验证了毛细管样品作为流体包裹体测试标样的可行性。随后，丁俊英等（2011）在 -120~31℃温度区间内开展了 H_2O-CO_2 体系合成包裹体的流体相平衡过程研究，首次报道了流体包裹体中 CO_2 固相和 CO_2 水溶液相的拉曼光谱实验资料。利用类似的方法，潘君屹等（2012）和 Zhu D. L. 等（2015）分别对 Na_2CO_3-H_2O 体系和 Na_2SO_4-H_2O 体系人工合成包裹体开展了拉曼定量分析，根据 CO_3^{2-} 离子和 SO_4^{2-} 离子拉曼特征峰强度与离子浓度的正相关关系，建立了适用于天然包裹体定量分析的工作曲线。

2.6 流体包裹体定年技术

大部分同位素体系定年都是针对特定矿物进行的。然而，一些金属或油气矿床由于缺少合适的定年矿物而难以获取准确的年龄，流体包裹体 $^{40}Ar/^{39}Ar$ 定年法为解决这一难题提供了新的方案（邱华宁和白秀娟，2019）。Bai 等（2013，2018）首次使用真空逐步击碎法对漂塘钨矿中矿石矿物锡石和黑钨矿开展

了流体包裹体 $^{40}Ar/^{39}Ar$ 定年,通过对比共生的富 K 长石和云母 $^{40}Ar/^{39}Ar$ 年龄,确定漂塘钨矿的成矿年龄为 160~153 Ma。类似的,Xiao 等(2019)对瑶岗仙钨矿中的黑钨矿、锡石和石英开展了流体包裹体 $^{40}Ar/^{39}Ar$ 定年,最终击碎阶段获得的气体年龄为 159~154 Ma,与前人报道的成矿年龄一致。上述研究表明,流体包裹体 $^{40}Ar/^{39}Ar$ 定年法具有相当的可靠性,并极大地拓展了金属矿床定年样品的选择范围。此外,Qiu 等(2011)利用松辽盆地油气储层中的白垩纪火山岩火成石英开展了其中富 CH_4 次生流体包裹体的 $^{40}Ar/^{39}Ar$ 定年,精确限定了松辽盆地天然气成藏年龄。

3 在矿床学研究中的进展

成矿物质的来源、迁移方式和沉淀机制是矿床学研究的核心问题,而流体包裹体研究是揭示成矿物质源-运-储过程必不可少的手段。通过包裹体的研究可以直观获知成矿流体温度、压力、化学成分等一系列信息,从而确定成矿流体的性质、成分、来源和演化,进而揭示矿床成因并建立成矿模式(倪培等,2018)。近十年来,国内学者通过流体包裹体研究,在成矿流体与成矿机制方面取得了丰硕的成果,现根据主要矿床类型分别列举如下:

3.1 浅成低温热液型矿床

浅成低温热液型矿床是一类与火山-次火山热液活动关系密切的矿床,形成深度浅(<1.5 km),成矿温度低(<320℃),是世界贵金属的主要来源之一,具有重要的经济价值。浅成低温热液矿床可进一步划分为高硫型、中硫型和低硫型,并常见不同类型相互叠加的现象,对其成因类型的详细厘定是此类矿床的重要研究内容之一。浅成低温热液矿床成矿流体演化过程通常包括流体混合、冷却、沸腾、水岩反应或多个过程的复合,金属沉淀的主导机制是研究的难点。另外,浅成低温热液矿床与斑岩型矿床常具有紧密的空间关系,两者之间的成因联系是近年来的研究热点(倪培等,2020)。

近十年来,在浅成低温热液矿床的研究中,国内学者利用流体包裹体在厘定成因类型,探索成矿机制和评价深部找矿潜力等方面取得了诸多成果。其一,详细厘定了一系列浅成低温热液矿床的成因类型。例如,对我国东南部重要的德化金矿集区开展了系统的流体包裹体研究,结合 H-O、S-Pb 同位素分析,厘定了区内诸如邱村、安村、东洋、下坂等金矿的成因类型,判断安村金矿是一个中硫型矿床,而邱村、东洋、下坂是低硫型矿床(Ni et al.,2018;Li et al.,2018a,2018b,Li S. N. et al.,2020);Fan 等(2020)对福建上山岗金矿开展了包裹体研究工作,发现其成矿流体为低盐度(1.4%~5.3% $NaCl_{eqv}$),从早期至晚期盐度明显降低,结合 H-O 同位素示踪,判断上山岗金矿为中硫型过渡至低硫型的浅成低温热液金矿床;Chi 等(2018)开展了福建悦洋银多金属矿闪锌矿和石英中的流体包裹体研究,发现闪锌矿中包裹体具有由早至晚温度和盐度明显降低的趋势,而与银矿物密切伴生的石英中发现了流体沸腾的证据,认为流体混合是铅锌沉淀的主要机制,而流体沸腾导致了银的沉淀;Zhai 等(2018)利用流体包裹体、H-O、S-Pb 同位素和数值模拟,提出围岩蚀变和大气降水混入导致了三道湾子浅成低温热液金银锑矿床的大规模成矿。另外,一些研究成功使用大比例尺流体填图重建古流体温度场。例如,Pan 等(2018)对紫金山高硫型铜金矿开展了垂向 1 500 m 范围的流体填图,揭示了古热液通道向东南方向的深部扩展的趋势,并非此前认为的罗卜岭斑岩方向,预示了紫金山深部仍有发现斑岩铜矿化的潜力;Ni 等(2019)在福建德化金矿集区进行了构造蚀变流体填图,结合土壤物化探方法,识别出该区域空间上具有蚀变类型和成矿流体温度场的系统变化,指出邱村矿床北部为该区域古热液中心,其深部具有勘探斑岩型矿床的潜力;Zhao 等(2020)利用流体填图并结合单个包裹体 LA-ICP-MS 分析技术,重建了鲜花岭中硫型浅成低温热液多金属勘查区的古流体温度场和成矿流体金属含量演化模型,预示了深部存在斑岩型铜矿化的可能。这些研究实践表明,大比例尺流体填图能够为寻找成矿热液通道和深部斑岩型矿化提供有效指示。

3.2 斑岩型矿床

斑岩型矿床是与斑岩体密切相关的一类矿床，其成矿流体主体为岩浆热液。斑岩型矿床通常具有规模大、品位低的特征，是全球铜、钼、金、银的主要来源之一，同时也是 Re、Se、Te 等关键矿产的重要来源（杨志明等，2020），其中最富经济价值的类型是斑岩钼矿和斑岩铜矿。

对斑岩钼矿成矿流体的早期认识主要来自于对美国西部典型 Climax 型矿床和分布于环太平洋的 Endako 型矿床的研究。进入 21 世纪以来，在我国的秦岭–大别地区陆续发现多个大型–特大型斑岩钼矿，如沙坪沟、汤家坪、鱼池岭等，这些矿床体现出了不同于 Climax 型和 Endako 型的构造背景、蚀变和矿化特征。近十年来，针对此类斑岩钼矿床系统开展了流体包裹体研究，完善了对其成矿流体和成矿机制的认识。例如，Chen and Wang（2011）在对大别山地区汤家坪钼矿进行的流体研究中识别出包括纯 CO_2 包裹体、CO_2-H_2O 富气相包裹体、含子晶富液相包裹体和低盐度富液相四种包裹体类型，在钼成矿阶段发现流体不混溶现象，提出不混溶是造成金属卸载的主要机制；Ni 等（2015b）在对沙坪沟钼矿的研究中详细划分了四期脉体，在辉钼矿石英脉和多金属硫化物石英脉中均发现了共生的含 CO_2 富气相包裹体和含子晶包裹体的流体沸腾组合，认为多期流体沸腾造成了钼的超常富集成矿；Li 等（2012）对东秦岭地区鱼池岭斑岩钼矿开展了系统的测温学和成分分析研究，识别出 CO_2 含量和金属元素的正相关变化关系，为 CO_2 流体不混溶作用作为一种重要的斑岩型钼矿成因机制提供了直接依据。此外，以富 CO_2 为特征的成矿流体在大别山地区的千鹅冲、姚冲、南泥湖、东沟等斑岩型钼矿也均有报道（Yang et al.，2015），这些研究表明，秦岭–大别地区的斑岩钼矿的成矿流体总体上为富 CO_2、中高温、中低盐度流体，流体不混溶或沸腾是重要的金属沉淀机制。近年来，在以铜金成矿为主的钦杭成矿带，也陆续发现了一系列斑岩钼矿，如冶岭头、桐村、十字头等。Ni 等（2017a）通过对桐村斑岩钼矿的研究，揭示出其成矿流体属富 CO_2 流体，钼成矿与含 CO_2 流体的不混溶作用密切相关；而十字头和冶岭头斑岩钼矿的研究则揭示出成矿流体为 H_2O-$NaCl$ 体系，不含 CO_2，流体沸腾促进了钼的沉淀（Ni et al.，2017b；Wang et al.，2017）。这些研究表明钦杭成矿带斑岩钼矿的成矿流体和成矿机制可能更加复杂多样。

流体沸腾在斑岩铜矿中亦十分常见，在部分矿床中被认为是主要的金属沉淀机制，如 Li 等（2017）对富家坞斑岩铜矿中多阶段脉体进行了比较研究，结果表明不同的流体包裹体组合共同记录了多阶段的流体沸腾作用；另外一些矿床的研究显示流体沸腾出现在金属沉淀之前，而沸腾后的进一步冷却直接导致了金属沉淀，如 Xiao 等（2012）分析了藏南驱龙斑岩铜矿中典型 A、B 和 D 脉中的流体包裹体，发现早期脉体中记录了流体沸腾，含铜硫化物的大量沉淀发生在进一步冷却至 320~400℃；另外，Liu 等（2016）对德兴铜厂斑岩铜矿的研究中也认为温度降低导致了辉钼矿和黄铜矿的先后沉淀。对斑岩铜矿的 He-Ar 同位素研究揭示了存在地幔组分的加入，如三江–红河成矿带铜厂斑岩铜矿硫化物 He-Ar 同位素揭示了地幔组分的存在，更多的地幔组分可能可以增大成矿的规模（Xu et al.，2014）。此外，一些研究揭示出斑岩铜矿成矿石英脉往往具有复杂的内部结构，如 Mao 等（2018）利用阴极发光成像在圆珠顶斑岩铜矿中识别出由早到晚的多期石英世代，从而对成矿流体演化进行限定。

3.3 夕卡岩型矿床

夕卡岩型矿床是指矿体产于铝硅酸盐岩（包括侵入岩、火山岩和混合岩等）和碳酸盐岩或其他钙质围岩接触带的夕卡岩及其附近的交代岩中，通过接触反应交代的方式生成的一类矿床，具有 W、Sn、Cu、Fe、Pb、Zn 等多种金属富集产出。

近年来，我国学者对不同类型夕卡岩型矿床的流体研究揭示了多样的金属富集机制。对夕卡岩型铁矿的研究揭示了流体沸腾和混合的成矿机制，并可能存在多期流体脉动；例如，Yang 等（2017）开展了马坑铁矿中磁铁矿期的透辉石和硫化物期的石英、方解石及萤石中的流体包裹体研究，结果显示磁铁矿沉淀过程中存在共生的含子晶包裹体和富气相的包裹体组合，指示了流体沸腾作用，而硫化物期包裹体具有温度、盐度明显降低的趋势，指示了流体混合作用，与 H-O 同位素证据相互印证。另外，Li 等（2019）在程潮高品位夕卡岩型铁矿中识别出三期次进变质阶段的石榴子石和辉石，其中的包裹体均具有高温（>750℃）、高盐度（>50% $NaCl_{eqv}$）及高铁含量特征，提出多期脉动的富铁流体控制了高品位铁矿的形成。对于夕卡岩型铜矿，流体沸腾可能是主导的金属沉淀机制，例如在浙江建德铜矿和云南羊拉铜矿的研究中，主成矿期石英中均发现了大量沸腾包裹体组合（Zhu J. J. et al., 2015；Chen et al., 2017）；此外，Ren 等（2020）在朗都高品位夕卡岩型铜矿中主成矿期石英中发现了含 CH_4 包裹体，推测 CH_4 的存在扩大了流体不混溶区域范围，并将 SO_4^{2-} 还原成了 S^{2-}，促进了高品位铜矿化的出现。对于夕卡岩型铅锌矿床，流体混合和水岩反应可能共同主导了金属沉淀。例如，Shu 等（2017）通过研究白音诺尔夕卡岩型铅锌矿的流体成分演化，发现具有较高金属含量的成矿前流体在高温高盐度环境中难以沉淀金属，而成矿期地下水混合导致流体温度、盐度降低，以及与碳酸盐岩围岩的水热蚀变是控制金属溶解度减小和成矿的主导因素。另外，Fang 等（2015）对东昆仑维宝夕卡岩型铅锌矿的系统测温学研究也揭示了水岩蚀变和流体混合对铅锌成矿的作用。

3.4 造山型金矿床

造山型金矿是在时间和空间上与造山作用密切相关，矿体主要以石英脉或蚀变岩赋存于脆-韧性剪切带和不同时代的变质岩中的一类矿床，是全球金的最重要来源。

近年来，我国学者通过流体包裹体研究，揭示了江南造山带中一系列赋存于韧性剪切带中的金矿属于造山型金矿。例如，在对浙江平水（深部）金矿体、璜山金矿和江西金山金矿的研究中，发现含金石英脉中发育大量富 CO_2 包裹体并出现流体不混溶现象，C-O、S-Pb 同位素进一步佐证了成矿流体来源于变质流体，符合造山型金矿的流体特征，流体包裹体压力估算也显示了其形成深度远远大于浅成低温热液金矿床的成矿极限，结合产出于韧性剪切带的地质环境，指出其属于造山型金矿（Zhao et al., 2013；Ni et al., 2015a；Xu et al., 2016）。在此基础上，Ni 等（2015a）根据平水和璜山金矿含矿脉体 400~450 Ma 的成矿年龄和相关构造背景证据，提出在江绍断裂带东段存在一期加里东期造山型金成矿作用。另外，通过系统的成矿流体演化过程的研究，揭示了流体不混溶是金的主要沉淀机制。例如，Zhao 等（2013）详细研究了金山金矿中含金石英脉形成的流体过程，发现成矿前石英脉中只有不含或含少量 CO_2 的流体包裹体，而成矿期包含了富 CO_2、含少量 CO_2 和不含 CO_2 的三类流体包裹体，共生的富 CO_2 类型和不含 CO_2 类型包裹体具有一致的均一温度，这说明在金沉淀过程中发生了 CO_2 不混溶；Chen 等（2012）对新疆南部的萨瓦亚尔顿金矿进行的研究显示，成矿期的流体包裹体的均一温度较低，但是盐度和 CO_2 含量存在极大的变化范围，指示了 CO_2 不混溶过程；另外，Zhou 等（2015）对小秦岭地区枪马金矿的研究发现，虽然 O 同位素的结果显示成矿期流体为变质流体与大气降水的混合，但是成矿期的石英内含有丰富的流体不混溶现象，推测流体不混溶仍是金的主要沉淀机制。值得注意的是，对于华北克拉通南缘一些金矿床的造山型成因仍存在争议，以熊耳山地区为例，Fan 等（2011）在祁雨沟爆破角砾岩型金矿中识别出大量沸腾包裹体组合，结合稳定同位素研究共同表明金矿化与区内花岗斑岩侵入体具有密切成因联系。此外，胶东半岛作为我国最重要的黄金产地，对其金矿成矿流体的研究备受关注（范宏瑞等，2016）。虽然这些金矿产出的大地构造背景与造山环境并不相符，可能不属于典型的造山型金矿，但其成矿流体与造山型金矿具有一定相似性。目前研究表明，胶东金矿床如玲珑、三山岛等的成矿流体组分主要为 CO_2-H_2O-$NaCl$±CH_4，温度为中低温（200~350℃），盐度小于 10% $NaCl_{eqv}$，流体不混溶是主要金属卸载机制（Wen et al., 2015, 2016；Guo et al., 2020）。另外，成矿流体可以在很大的垂向范围保持稳定，如 Hu 等（2013）对三山岛垂向 2000 m 范围内金矿化的研究显示流体温度没有明显变化。

3.5 与花岗岩有关的钨锡矿床

钨、锡属于战略性关键金属，绝大多数与花岗岩有密切成因联系。我国同时拥有世界第一的钨、锡储量，具备得天独厚的研究条件。近十年中，国内学者在与花岗岩有关钨锡矿床成矿流体与成矿机制研究方面取得了众多的新成果和新认识。

在钨矿床研究方面，Ni 等（2015c）采用红外显微镜技术，对漂塘、荡坪、大吉山、盘古山等石英脉型黑钨矿床中黑钨矿和共生石英中的原生流体包裹体开展了比较研究，发现黑钨矿和石英中的流体包裹体记录了不一致的流体过程，石英中的流体包裹体可能记录了简单冷却、流体混合与流体不混溶三种过程，而黑钨矿中的流体包裹体只记录了简单冷却这一过程，指示流体简单冷却可能是黑钨矿沉淀的重要机制。此外，在瑶岗仙钨矿和茅坪钨矿的流体研究中，Li W. S. 等（2018）和 Chen 等（2018）分别揭示出了类似的流体过程。Pan 等（2019）联合使用了红外显微镜技术、石英阴极发光技术和单个包裹体 LA-ICP-MS 分析技术，精细重建了瑶岗仙钨矿黑钨矿石英脉形成的复杂流体演化历史，揭示出形成黑钨矿和共生石英的流体存在元素成分上的显著区别，表现为黑钨矿成矿流体相比于非成矿流体显著亏损 B、As、S 元素并富集 Sr、Ca，指示黑钨矿沉淀与围岩水压破裂导致的流体沸腾和水岩反应密切相关。Peng 等（2018）研究了大湖塘钨铜矿田的大岭上矿床，对比了黑钨矿、白钨矿、磷灰石和石英中的原生流体包裹体，发现黑钨矿和白钨矿中的流体包裹体均一温度显著高于磷灰石和共生石英，显微测温数据指示了简单冷却机制，S 同位素的数据显示存在大气降水的混入，而白钨矿原位成分分析则揭示了水岩反应是重要的成矿机制，因此该矿床可能具有复杂成因。Wei 等（2019）对西华山钨矿中黄铁矿和毒砂的包裹体进行了 He-Ar 同位素分析，结果发现它们的 ^{3}He/^{4}He 和 ^{40}Ar/^{36}Ar 的值远高于地壳平均值，说明来自地幔的组分和热参与了成岩成矿。在成矿模型方面，Chen 等（2018）对茅坪钨矿缓倾斜黑钨矿石英脉进行了系统的空间采样工作，根据野外矿脉产状和不同标高石英和黑钨矿中包裹体显微测温结果，提出了茅坪钨矿"五层楼"结构和对应的流体过程模型。此外，Xie 等（2019）对湘中地区杨家山钨矿开展了成矿年代学、流体包裹体和稳定同位素综合研究，提出杨家山钨矿为一类非常特殊的加里东期石英脉型白钨矿床，流体与富钙围岩反应是造成大规模白钨矿化的主要机制。

在锡矿床研究方面，Cui 等（2019）利用阴极发光技术对小龙河石英脉型锡矿中石英及锡石的成矿期次进行了精细划分，揭示出成矿阶段存在共生的富气相和富含子晶流体包裹体沸腾组合，结合 O 同位素揭示出的大气降水混合，提出流体沸腾和流体混合是锡石的成矿机制；Liu 等（2020）对西岭锡多金属矿床不同阶段流体包裹体开展了测温学研究，结果显示成矿流体温度和盐度从早至晚不断降低，结合 H-O 同位素的变化，揭示出早期单一岩浆流体逐渐与大气降水混合的过程；在对广西大厂矿田高峰锡多金属矿床的流体包裹体研究中，赵海等（2018）发现成矿早期锡石–毒砂–磁黄铁矿阶段主要发育富含 CO_2 和 CH_4 的富气相包裹体，成矿晚期硫化物–硫盐–碳酸盐阶段石英中则仅发育气–液两相富水包裹体，从早至晚，均一温度发生明显降低而盐度未发生明显变化，推测成矿流体可能主要来源于深部夕卡岩阶段富 CO_2 气相流体的冷却收缩，流体冷却过程可能是锡石–硫化物成矿的主要控制因素。

3.6 稀土矿床

稀土矿床通常与碱性岩–碳酸岩伴生，是世界主要的轻稀土来源。近十年来，由于政策倾向和国际形势，稀土矿床受到持续研究关注，国内学者对稀土成矿的认识在不断发展。

碳酸岩流体对稀土成矿起到至关重要的作用，自 21 世纪初起开始受到极大关注（范宏瑞等，2001）。我国的原生稀土产区主要位于白云鄂博矿集区和川西冕宁–德昌成矿带（范宏瑞等，2020），近十年对此两个成矿带的包裹体研究深化了对其成矿过程的认识。白云鄂博矿床成矿过程复杂，经历多期叠加改造，对其成矿过程争议很大。研究者利用包裹体研究碳酸岩的形成过程和成矿过程方面取得了一定的成果。例如，Yang 等（2019）对白云石中包裹体进行研究，结合磷灰石的原位 Sr-Nd-O 同位素和主微量组成，提出岩浆不混溶和岩浆演化共同控制了白云鄂博中的稀土富集过程；Ni 等（2020）在紧邻碳酸岩墙的石英岩中发现了代表碳酸岩浆相关成矿流体的包裹体，Rb-Sr 等时线年龄表明碳酸岩墙侵位于加里东期俯冲背景之下，俯冲过程中释放的含稀土和 CO_2 的流体交代并改造了早期白云岩中矿化，最终形成了超大规模的稀土矿床。另外，针对川西冕宁-德昌成矿带的稀土矿床也开展了丰富的研究。对区内的牦牛坪稀土矿床的流体和熔体-流体包裹体开展的显微测温和初步成分分析结果显示，成矿流体属于K-Na-Ca-SO$_4$-H$_2$O-CO$_2$体系，认为流体不混溶和冷却对稀土成矿具有重要意义（Xie et al.，2015；Zheng and Liu，2019）。

近十年来，实验模拟在解析稀土的迁移方面也取得了较大进展。例如，Cui等（2019）发现石英的存在可以显著增大高温下硫酸盐的溶解度，促进硫酸根与稀土的络合在流体中进行迁移；Zhou等（2016）对H$_2$O-Na-K-Cl-F-CO$_2$体系中独居石和磷钇矿的溶解度研究表明碳酸岩流体中Na和F的富集可以提高所有稀土元素的活动性。

4 在沉积成藏研究中的进展

蒸发岩中的流体包裹体和油气包裹体分别是古表生环境和油气成藏过程的直接记录，是仅有的可直接反映古代沉积环境和成藏作用的流体样品。最近十年，关于蒸发岩中流体包裹体和油气包裹体分别在古环境和油气地质学领域得到了广泛应用，国内研究者对于沉积成藏方面的研究紧追国际前沿，获得了众多新的认识和成果。

4.1 古沉积环境

蒸发岩是保存古环境信息的宝库，蒸发岩中的主要矿物石盐形成于表生环境，具有较好的封闭性，石盐的内部保存的原生流体包裹体记录了原始海洋或盐湖的温度、化学组分和大气成分的信息，能够直接反映古代沉积地质环境和古气候。

Meng等（2011）研究了四川长宁2号井内震旦纪石盐中的原生流体包裹体，发现其主要的均一温度范围在（20~25±1.0）℃内，由等容线计算出了四川省震旦纪的海水最高温度在（39.4±1.0）℃，根据海水溶解氧与温度、盐度的函数关系，可以推断出埃迪卡拉期的海水中溶解氧含量已经到达了支撑更复杂的多细胞生物演化的阈值；在此基础上，Meng等（2018）继续研究了奥陶纪石盐中的原生流体包裹体，通过超微化学方法发现原生卤水包裹体流体为Na-K-Mg-Ca-Cl（富Ca）体系，填补了重大地质时期海水化学信息的空白。此外，Meng等（2014）研究了江汉盆地应城凹陷中的石盐包裹体，恢复了古水体成分，并将其与西班牙同期海相地层相比较，发现两者相差甚远，显示该凹陷的盐湖没有受到海侵的影响，属于内陆咸化盐湖。Zhang等（2016）测得老挝上白垩统农波组的石盐包裹体的均一温度为17.7~42.3℃，对比重建了沉积物海岸的古温度模型。除石盐包裹体外，Shan等（2015）采样塔里木奥陶系盆地古流体储层中方解石内的流体包裹体来还原古气候，研究发现包裹体的温度、盐度数据集中于101℃和3.5% NaCl$_{eqv}$附近，具有很高的HCO$_3^-$和变化的Cl$^-$、SO$_4^{2-}$含量，同时具有非常负的δD和δ^{18}O同位素，说明溶洞建造最有可能由表生大气水侵蚀溶解而成。在实验模拟方面，孟凡巍等（2011）在实验室成功合成了石盐包裹体，测温结果表明，形成于水体表面的漏斗晶中平行包裹体条带和在水底形成的人字晶包裹体条带具有相似的最大均一温度，因此在浅水环境下两种包裹体都可用来反映古气温。

4.2 油气成藏

油气包裹体是存在于油气储层中，被捕获、封闭于与油气过程相关的成岩自生矿物晶格缺陷或碎屑矿物成岩愈合裂隙中的流体样品。油气包裹体研究对深入认识油气成藏机理尤其是油气充注以及成分演化史具有重要的意义（张鼐，2016；杨海军等，2017；王飞宇等，2018），广泛应用于油气勘探中的盆地模拟、油气运移过程重建等方面。

在油气充注研究方面，Tao等（2014）对四川盆地侏罗纪须家河组储层中的流体包裹体的研究发现，包裹体中的气体的δ^{13}C$_{CO_2}$为-16.6‰~-9‰，指示其为有机成因，与气田中天然气的碳同位素对比，发现二者之间没有同位素分馏，因此包裹体中的 CO_2 主要来自于源岩的有机物质，少量来自无机成因气，证明了储层的油气注入发生在储层的致密化之后。Chang 等（2013）对哈拉哈塘油田中的包裹体进行了研究，根据油气样品气相质谱成分和流体包裹体均一温度的双峰式分布，确定油气储层中有两次独立的油气注入事件。在对巴什托油田油气包裹体的研究中，Cui 等（2013）发现包裹体的均一温度主要分布在 80~87℃ 和 95~100℃ 两个区间，通过重建沉积热演化史，推断两次独立的油气注入事件发生在 285~290 Ma 和 4~10 Ma 两个时间段内。此外，Song 等（2017）对比研究了塔里木盆地与奇地区多个油气储层的油气包裹体，根据等容线计算了对应的捕获温度和压力，结合气相色谱成分结果，重建了区内沉积热演化史中的两期油气运移充注过程，分别为 80~95℃、415~429 Ma 和 115~130℃、2~8 Ma。

在生烃过程研究方面，Ping 等（2020）以中国东部渤海湾盆地东濮凹陷 PS18-1 井特稠页岩油为例，在详细的脉体岩相学和油气包裹体荧光光谱及分子成分分析基础上，重建了的源岩层系内烃类流体的温压和组分演化历史，从而探讨了超压演化过程中油组分和成熟度变化，证实古超压释放导致的极性组分析出是稠油形成的一种新机制。另外，Wang 等（2020）以渤海湾东营凹陷为例，对有机质页岩中发育的平行层状纤维状方解石脉开展了流体包裹体研究，发现所有包裹体组合都确切记录了生烃过程的流体超压现象，结束了纹层状烃源岩生排烃阶段水平裂缝的垂向扩张（即纤维状晶体生长）过程中流体压力是否需要超过上覆载荷静岩压力的争论，研究发现一部分脉的扩张是通过附近岩壁的缩窄完成的，可能与 $CaCO_3$ 的溶解与再沉淀相关。

在油气运移研究方面，Li 等（2013）对大巴山地区前陆构造带烃源岩微裂隙中纤维状方解石开展了详细的油气包裹体岩相学和测温学研究，揭示出其异常超高压特征，结合 C-O 同位素反映出的有机流体与浅部流体混合特征以及区内构造演化历史，提出燕山期前陆构造作用导致大巴山褶皱隆起并伴随天然气藏破坏和改造，含烃包裹体纤维状方解石脉正是超高压构造应力驱动天然气排泄的产物。Zhuo 等（2014）在库车盆地石盐中发现了油气包裹体，通过大宛齐油田和大北气田中的油气化学性质的对比，发现油气是从蒸发岩内自下而上运移，经过了石盐层，证明存在大规模的碳氢化合物通过岩盐层的过程。陈顺勇等（2015）研究了合肥盆地石炭系—二叠系储层的流体包裹体，发现包裹体均一温度和盐度主要集中在 3 个不同的温度、盐度区间，结合构造演化、烃源岩发育背景，推测油气主要是在石炭系—二叠系内部运移。对于成藏伴生的其他事件，Lu 等（2017）系统获取了塔里木盆地顺南地区的碳酸盐岩储层中石英和大颗粒方解石中的流体包裹体温度、盐度和 Sr 同位素，识别出泥盆纪时期由构造运动引起一次热液硅化事件。

5 学术组织的作用

在 21 世纪第二个十年间，在中国矿物岩石地球化学学会矿物包裹体专业委员会的带动和引领下，我国流体包裹体研究队伍得以不断发展壮大。学会分别于 2012 年、2016 年、2018 年和 2020 年在杭州、成都、合肥、长春成功召开了第十七、十八、十九、二十届全国包裹体及地质流体学术研讨会，吸引了国内外数以千计从事包裹体研究的专家学者参与并做主题报告，接纳了大量研究生进行口头报告，为分享研究进展，每届会议都编纂了摘要论文集发布并选取部分优秀文献在增刊上发表。值得一提的是，从第十七届会议开始，每届大会主办方通过评选优秀论文，调动了广大学生的积极性；也是从第十七届会议开始，矿物包裹体专业委员会对长期以来致力于我国流体包裹体研究，并做出重要贡献的学者颁发终身成就奖。第十七届授予李兆麟教授和卢焕章教授，第十八届授予沈昆教授和刘斌教授，第十九届授予张文淮教授和魏家秀研究员，表彰他们一直以来对包裹体事业做出的杰出贡献，鼓励包裹体同行以其为榜样在包裹体科学研究中再创佳绩。除此之外，矿物包裹体专业委员会在中国矿物岩石地球化学学会每两年一次的学术年会，以及隔年间隔召开的"全国矿床会议"和"全国成矿理论与找矿方法学术讨论会"中，均设置专题进行学术讨论交流，推动流体包裹体在成岩、成矿、成藏等领域的发展。在国内会议举办如火如荼的同时，由中国学者创办的 ACORFI（亚洲流体包裹体国际会议）于第二个十年也成功召开 4

次（ACROFI IV，澳大利亚布里斯班；ACROFI V，中国西安；ACROFI VI，印度孟买；ACROFI VII，中国北京），吸引了包括阿尔及利亚、埃及、澳大利亚、英国、德国、瑞士、意大利、美国、俄罗斯、加拿大、喀麦隆、缅甸、尼泊尔、伊朗、印度以及中国等十余个国家的百余名学者和研究生参会，邀请到了包括 Andrew H. Rankin，Christoph A. Heinrich，Jean Dubessy，Robert J. Bodnar，Maria L. Frezzotti，J. D. Webster，Andrew R. Campbell 等在内的国际包裹体领域权威学者作主题报告及短期课程讲解，促进了国内外包裹体研究同行的交流，为我国和其他亚洲国家学者提供了稳定的展示平台，有力推动了国际流体包裹体研究事业的发展。为了更好地进行国内外学术交流，流体包裹体专委会于 2016 年在 Journal of Geochemical Exploration 组织 "*fluid and melt inclusions*" 专辑，2018 年在《矿物岩石地球化学通报》组织 "流体包裹体与地质流体" 专栏。

6 结语与展望

受篇幅所限，本文不能完全呈现我国包裹体领域近十年来的所有研究成果。实事求是地说，我国包裹体研究在理论、技术和应用方面内容丰富、成果众多，特别是在"十三五"期间国家战略性关键金属矿产方向研究工作中，包裹体研究工作得到充分体现，发挥了不可或缺的作用。

近年来，我国流体包裹体研究领域最重要的进展之一，就是自主建立了单个流体包裹体 LA-ICP-MS 成分分析技术。这一关键技术的突破使得我国流体包裹体研究从传统的测温学和拉曼分析进入元素组成的定量分析阶段，使从流体包裹体中获取的有效信息得到大幅提升，从而深刻影响和改变我国现有流体包裹体的研究格局。结合国际上相关研究进程来看，我国基于单个流体包裹体 LA-ICP-MS 成分分析的研究仍处于起步阶段，在分析技术和研究应用方面都有很大发展空间。在分析技术方面，可在现有四级杆电感耦合等离子质谱（Q-ICP-MS）基础上进一步开发不同类型质谱仪对单个流体包裹体的分析技术，如利用扇形磁场电感耦合等离子质谱（SF-ICP-MS）和飞行时间电感耦合等离子质谱（TOF-ICP-MS）来获取更高的测试精准度和更低的检测限（Wälle and Heinrich，2014；Harlaux et al.，2015），实现对微小（5 μm 以下）包裹体的可靠分析；也可尝试建立多接收器电感耦合等离子质谱（MC-ICP-MS）对流体包裹体中的同位素分析方法（Pettke et al.，2011）。此外，还可进一步完善流体包裹体内标获取方法来提高测试精准度（Sirbescu et al.，2013；Steele-MacInnis et al.，2016）等。在应用研究方面，应充分发挥我国矿产资源优势，着眼于关键金属矿床，尤其是我国独具优势的稀土、稀散元素矿床和稀有金属矿床开展成矿流体与成矿机制研究，填补空白。可以预见，单个流体包裹体 LA-ICP-MS 成分分析技术必将在下一个十年为我国流体包裹体研究打开全新局面，促进一大批高水平成果的产出。

本文总结的包裹体研究进展主要集中于流体包裹体领域，而作为地质流体中另一大分支的熔体包裹体研究应引起更大的重视。熔体包裹体保存有原始的岩浆成分信息，并记录了各瞬时的岩浆演化信息，在岩石成因研究中独具优势。目前，获取单个熔体包裹体的主微量成分的技术方法已较为完备，该方法提供了相比于全岩地球化学更为丰富的岩浆成分信息，揭示出传统岩石地化手段难以识别的地质过程信息，正日益成为探讨岩浆起源和演化不可或缺的重要依据（Rowe et al.，2011；Laurent et al.，2020）。精确测定熔体包裹体中同位素组成的工作方兴未艾，目前已成功应用 SIMS 和 LA-（MC）-ICP-MS 开展了熔体包裹体中 H、Li、B、C、O、S、Cl、Sr、Nd、Pb 同位素研究，在揭示地壳深循环、地幔源区不均一性、地壳岩浆混染过程等方面显示出巨大潜力（Rose-Koga et al.，2012；Hartley et al.，2013）。另外，熔体包裹体是揭示岩浆挥发分含量（如 H_2O、CO_2、F、S、Cl 等）至关重要的手段，在研究火山活动及其环境效应方面正发挥着不可替代的作用，为深入认识全球碳-水循环、气候变化和生物灭绝事件等重大关切问题提供了新的视角（Johnson et al.，2011；Capriolo et al.，2020）。尤其值得注意的是，熔体包裹体可以提供珍贵的原始岩浆中成矿元素和重要络合物含量信息，是研究成矿过程的有力手段（Grondahl and Zajacz，2017）。另外，针对成矿和贫矿系统，开展熔体和初始流体包裹体中成矿元素、重要金属络合元素含量对比研究尚处于起始阶段（Audétat，2019），需开展更多相关工作建立大数据库，进而提出判别成矿潜力的熔体-流体地球化学指标。

致谢：感谢流体包裹体领域同仁的不懈努力和对中国流体包裹体研究的巨大贡献；感谢中国矿物岩石地球化学学会和郑永飞院士在稿件组织中给予的指导和帮助；感谢张鑫、赵子豪、孟凡巍在本文撰写过程中提供的帮助。

参 考 文 献

陈顺勇,俞昊,林春明,张霞,曲长伟,张妮,倪培,丁俊英. 2015. 合肥盆地石炭系—二叠系储层流体包裹体特征与油气运移研究. 高校地质学报,21(1):131-137

池国祥,卢焕章. 2008. 流体包裹体组合对测温数据有效性的制约及数据表达方法. 岩石学报,24(9):1945-1953

丁俊英,倪培,管申进,王国光. 2011. H_2O-CO_2体系融合二氧化硅毛细管样品原位显微激光拉曼光谱研究. 地学前缘,18(5):140-146

范宏瑞,谢奕汉,王凯怡,杨学明. 2001. 碳酸岩流体及其稀土成矿作用. 地学前缘,8(4):289-295

范宏瑞,冯凯,李兴辉,胡芳芳,杨奎锋. 2016. 胶东-朝鲜半岛中生代金成矿作用. 岩石学报,32(10):3225-3238

范宏瑞,牛贺才,李晓春,杨奎锋,杨占峰,王其伟. 2020. 中国内生稀土矿床类型、成矿规律与资源展望. 科学通报,65(33):3778-3793

格西,苏文超,朱路艳,武丽艳. 2011. 红外显微镜红外光强度对测定不透明矿物中流体包裹体盐度的影响:以辉锑矿为例. 矿物学报,31(3):366-371

黄惠兰,常海亮,谭靖,李芳,张春红,周云. 2015. 共生黑钨矿与石英等多种矿物中流体包裹体的红外显微测温对比研究——以江西西华山石英脉钨矿床为例. 岩石学报,31(4):925-940

刘斌. 2011. 简单体系水溶液包裹体pH和Eh的计算. 岩石学报,27(5):1533-1542

卢焕章,范宏瑞,倪培,欧光习,沈昆,张文淮. 2004. 流体包裹体. 北京:科学出版社

孟凡巍,倪培,葛379东,王天刚,王国光,刘吉强,赵超. 2011. 实验室合成石盐包裹体的均一温度以及古气候意义. 岩石学报,27(5):1543-1547

倪培,丁俊英,Chou I M,Dubessy J. 2011. 一种新型人工"流体包裹体":融合二氧化硅毛细管技术. 地学前缘,18(5):132-139

倪培,范宏瑞,丁俊英. 2014. 流体包裹体研究进展. 矿物岩石地球化学通报,33(1):1-5

倪培,迟哲,潘君屹,王国光,陈辉,丁俊英. 2018. 热液矿床的成矿流体与成矿机制——以中国若干典型矿床为例. 矿物岩石地球化学通报,37(3):369-394

倪培,迟哲,潘君屹. 2020. 斑岩型和浅成低温热液型矿床成矿流体与找矿预测研究:以华南若干典型矿床为例. 地学前缘,27(2):60-78

潘君屹,丁俊英,倪培. 2012. $Na_2CO_3-H_2O$体系人工流体包裹体中CO_3^{2-}离子的显微拉曼光谱研究. 南京大学学报(自然科学版),48(3):328-335

邱华宁,白秀娟. 2019. 流体包裹体$^{40}Ar/^{39}Ar$定年技术与应用. 地球科学,44(3):685-697

王飞宇,冯伟平,关晶,Chao J C. 2018. 含油气盆地流体包裹体分析的关键问题和意义. 矿物岩石地球化学通报,37(3):441-450

杨海军,张宝收,肖中尧,张鼐. 2017. 塔里木盆地碳酸盐岩储集层烃包裹体研究图集. 北京:石油工业出版社

杨志明,侯增谦,周利敏,周怿惟. 2020. 中国斑岩铜矿床中的主要关键矿产. 科学通报,65(33):3653-3664

袁顺达,赵盼捞. 2021. 基于新的合成流体包裹体方法对成矿金属在熔体-流体相间分配行为的实验研究. 中国科学:地球科学,51(2):241-249

张鼐. 2016. 含油气盆地流体包裹体分析技术及应用. 北京:石油工业出版社

赵海,苏文超,沈能平,谢鹏,蔡佳丽,甘文志. 2018. 广西大厂矿田高峰锡多金属矿床流体包裹体研究. 岩石学报,34(12):3553-3566

Audétat A. 2019. The metal content of magmatic-hydrothermal fluids and its relationship to mineralization potential. Economic Geology,114(6):1033-1056

Bai X J,Wang M,Jiang Y D,Qiu H N. 2013. Direct dating of tin-tungsten mineralization of the Piaotang tungsten deposit,South China,by $^{40}Ar/^{39}Ar$ progressive crushing. Geochimica et Cosmochimica Acta,114:1-12

Bai X J,Jiang Y D,Hu R G,Gu X P,Qiu H N. 2018. Revealing mineralization and subsequent hydrothermal events:insights from $^{40}Ar/^{39}Ar$ isochron and novel gas mixing lines of hydrothermal quartzs by progressive crushing. Chemical Geology,483:332-341

Campbell A R,Hackbarth C J,Plumlee G S,Petersen U. 1984. Internal features of ore minerals seen with the infrared microscope. Economic Geology,79(6):1387-1392

Capriolo M,Marzoli A,Aradi L E,Callegaro S,Dal Corso J,Newton R J,Mills B J W,Wignall P B,Bartoli O,Baker D R,Youbi N,Remusat L,Spiess R,Szabó C. 2020. Deep CO_2 in the end-Triassic central Atlantic magmatic province. Nature Communications,11(1):1670

Chang X C,Wang T G,Li Q M,Ou G X. 2013. Charging of Ordovician reservoirs in the Halahatang depression(Tarim Basin,NW China)determined by oil geochemistry. Journal of Petroleum Geology,36(4):383-398

Chen H,Ni P,Chen R Y,Lü Z C,Ye T Z,Wang G G,Pan J Y,Pang Z S,Xue J L,Yuan H X. 2017. Constraints on the genesis of the Jiande polymetallic copper deposit in South China using fluid inclusion and O-H-Pb isotopes. Journal of the Geological Society of India,90(5):546-557

Chen H Y,Chen Y J,Baker M J. 2012. Evolution of ore-forming fluids in the Sawayaerdun gold deposit in the southwestern Chinese Tianshan metallogenic belt,Northwest China. Journal of Asian Earth Sciences,49:131-144

Chen L L,Ni P,Li W S,Ding J Y,Pan J Y,Wang G G,Yang Y L. 2018. The link between fluid evolution and vertical zonation at the Maoping tungsten deposit,southern Jiangxi,China:fluid inclusion and stable isotope evidence. Journal of Geochemical Exploration,192:18-32

Chen Q,Zhang Z G,Wang Z P,Li W C,Gao X Y,Ni H W. 2019. *In situ* Raman spectroscopic study of nitrogen speciation in aqueous fluids under pressure. Chemical Geology,506:51-57

Chen Y,Ge Y J,Zhou Z Z,Zhou Y Q. 2015. Water,is it necessary for fluid inclusions forming in calcite? Journal of Petroleum Science and Engineering, 133:103-107

Chen Y,Steele-MacInnis M,Ge Y J,Zhou Z Z,Zhou Y Q. 2016. Synthetic saline-aqueous and hydrocarbon fluid inclusions trapped in calcite at temperatures and pressures relevant to hydrocarbon basins:a reconnaissance study. Marine and Petroleum Geology,76:88-97

Chen Y J,Wang Y. 2011. Fluid inclusion study of the Tangjiaping Mo deposit,Dabie Shan,Henan Province:implications for the nature of the porphyry systems of post-collisional tectonic settings. International Geology Review,53(5-6):635-655

Chi Z,Ni P,Pan J Y,Ding J Y,Wang Y Q,Li S N,Bao T,Xue K,Wang W B. 2018. Geology,mineral paragenesis and fluid inclusion studies of the Yueyang Ag-Au-Cu deposit,South China:implications for ore genesis and exploration. Geochemistry:Exploration,Environment,Analysis,18(4): 303-318

Cui H,Zhong R C,Xie Y L,Yuan X Y,Liu W H,Brugger J,Yu C. 2020. Forming sulfate-and REE-rich fluids in the presence of quartz. Geology, 48(2):145-148

Cui J W,Wang T G,Li M J,Ou G X,Geng F,Hu J. 2013. Oil filling history of the bashituo oilfield in the markit slope,SW Tarim Basin,China. Petroleum Science,10(1):58-64

Cui X L,Wang Q F,Deng J,Wu H Y,Shu Q H. 2019. Genesis of the Xiaolonghe quartz vein type Sn deposit,SW China:insights from cathodoluminescence textures and trace elements of quartz,fluid inclusions,and oxygen isotopes. Ore Geology Reviews,111:102929

Fan H R,Hu F F,Wilde S A,Yang K F,Jin C W. 2011. The Qiyugou gold-bearing breccia pipes,Xiong'ershan region,central China:fluid-inclusion and stable-isotope evidence for an origin from magmatic fluids. International Geology Review,53(1):25-45

Fan M S,Ni P,Pan J Y,Ding J Y,Chi Z,Li W S,Zhu R Z,Li S N,Badhe K,Wang J C. 2020. Mineralogical,fluid inclusion,and stable isotopic study of the Shangshan'gang Au deposit,Southeast China:implications for ore formation and exploration. Journal of Geochemical Exploration,215:106564

Fang J,Chen H Y,Zhang L,Zheng L,Li D F,Wang C M,Shen D L. 2015. Ore genesis of the Weibao lead-zinc district,eastern Kunlun orogen,China: constrains from ore geology,fluid inclusion and isotope geochemistry. International Journal of Earth Sciences,104(5):1209-1233

Goldstein R H,Reynolds T J. 1994. Systematics of fluid inclusions in diagenetic minerals. SEPM Society for Sedimentary Geology,1-199

Grondahl C,Zajacz Z. 2017. Magmatic controls on the genesis of porphyry Cu-Mo-Au deposits:the Bingham Canyon example. Earth and Planetary Science Letters,480:53-65

Günther D,Audétat A,Frischknecht R,Heinrich C A. 1998. Quantitative analysis of major,minor and trace elements in fluid inclusions using laser ablation-inductively coupled plasma-mass spectrometry. Journal of Analytical Atomic Spectrometry,13(4):263-270

Guo L N,Deng J,Yang L Q,Wang Z L,Wang S R,Wei Y J,Chen B H. 2020. Gold deposition and resource potential of the Linglong gold deposit, Jiaodong Peninsula:geochemical comparison of ore fluids. Ore Geology Reviews,120:103434

Harlaux M,Borovinskaya O,Frick D A,Tabersky D,Gschwind S,Richard A,Günther D,Mercadier J. 2015. Capabilities of sequential and quasi-simultaneous LA-ICPMS for the multi-element analysis of small quantity of liquids (pl to nl):insights from fluid inclusion analysis. Journal of Analytical Atomic Spectrometry,30(9):1945-1969

Hartley M E,Thordarson T,Fitton J G,EIMF. 2013. Oxygen isotopes in melt inclusions and glasses from the Askja volcanic system,North Iceland. Geochimica et Cosmochimica Acta,123:55-73

Heinrich C A,Pettke T,Halter W E,Aigner-Torres M,Audétat A,Günther D,Hattendorf B,Bleiner D,Guillong M,Horn,I. 2003. Quantitative multi-element analysis of minerals,fluid and melt inclusions by laser-ablation inductively-coupled-plasma mass-spectrometry. Geochimica et Cosmochimica Acta,67(18):3473-3497

Hu F F,Fan H R,Jiang X H,Li X C,Yang K F,Mernagh T. 2013. Fluid inclusions at different depths in the Sanshandao gold deposit,Jiaodong Peninsula,China. Geofluids,13(4):528-541

Johnson E R,Kamenetsky V S,McPhie J,Wallace P J. 2011. Degassing of the H_2O-rich rhyolites of the Okataina volcanic center,Taupo volcanic zone, New Zealand. Geology,39(4):311-314

Laurent O,Björnsen J,Wotzlaw J F,Bretscher S,Silva M P,Moyen J F,Ulmer P,Bachmann O. 2020. Earth's earliest granitoids are crystal-rich magma reservoirs tapped by silicic eruptions. Nature Geoscience,13(2):163-169

Li J K,Bassett W A,Chou I M,Ding X,Li S H,Wang X Y. 2016. An improved hydrothermal diamond anvil cell. Review of Scientific Instruments, 87(5):053108

Li J K,Liu Y C,Zhao Z,Chou I M. 2018. Roles of carbonate/CO_2 in the formation of quartz-vein wolframite deposits:insight from the crystallization experiments of huebnerite in alkali-carbonate aqueous solutions in a hydrothermal diamond-anvil cell. Ore Geology Reviews,95:40-48

Li L,Ni P,Wang G G,Zhu A D,Pan J Y,Chen H,Huang B,Yuan H X,Wang Z K,Fang M H. 2017. Multi-stage fluid boiling and formation of the giant Fujiawu porphyry Cu-Mo deposit in South China. Ore Geology Reviews,81:898-911

Li N,Ulrich T,Chen Y J,Thomsen T B,Pease V,Pirajno F. 2012. Fluid evolution of the Yuchiling porphyry Mo deposit,East Qinling,China. Ore Geology Reviews,48:442-459

Li R X,Dong S W,Lehrmann D,Duan L Z. 2013. Tectonically driven organic fluid migration in the Dabashan foreland belt:evidenced by geochemistry and geothermometry of vein-filling fibrous calcite with organic inclusions. Journal of Asian Earth Sciences,75:202-212

Li S N, Ni P, Bao T, Li C Z, Xiang H L, Wang G G, Huang B, Chi Z, Dai B Z, Ding J Y. 2018a. Geology, fluid inclusion, and stable isotope systematics of the Dongyang epithermal gold deposit, Fujian Province, Southeast China: implications for ore genesis and mineral exploration. Journal of Geochemical Exploration, 195:16–30

Li S N, Ni P, Bao T, Xiang H L, Chi Z, Wang G G, Huang B, Ding J Y, Dai B Z. 2018b. Genesis of the Ancun epithermal gold deposit, Southeast China: evidence from fluid inclusion and stable isotope data. Journal of Geochemical Exploration, 195:157–177

Li S N, Ni P, Bao T, Wang G G, Chi Z, Li W S, Zhu R Z, Dai B Z, Xiang H L. 2020. Geological, fluid inclusion, and H-O-S-Pb isotopic studies of the Xiaban epithermal gold deposit, Fujian Province, Southeast China: implications for ore genesis and mineral exploration. Ore Geology Reviews, 117:103280

Li W, Xie G Q, Mao J W, Zhu Q Q, Zheng J H. 2019. Mineralogy, fluid inclusion, and stable isotope studies of the Chengchao deposit, Hubei Province, eastern China: implications for the formation of high-grade Fe skarn deposits. Economic Geology, 114(2):325–352

Li W S, Ni P, Pan J Y, Wang G G, Chen L L, Yang Y L, Ding J Y. 2018. Fluid inclusion characteristics as an indicator for tungsten mineralization in the Mesozoic Yaogangxian tungsten deposit, central Nanling district, South China. Journal of Geochemical Exploration, 192:1–17

Li X H, Klyukin Y I, Steele-MacInnis M, Fan H R, Yang K F, Zoheir B. 2020. Phase equilibria, thermodynamic properties, and solubility of quartz in saline-aqueous-carbonic fluids: application to orogenic and intrusion-related gold deposits. Geochimica et Cosmochimica Acta, 283:201–221

Liu P, Mao J W, Jian W, Mathur R. 2020. Fluid mixing leads to main-stage cassiterite precipitation at the Xiling Sn polymetallic deposit, SE China: evidence from fluid inclusions and multiple stable isotopes(H-O-S). Mineralium Deposita, 55(6):1233–1246

Liu X, Fan H R, Hu F F, Yang K F, Wen B J. 2016. Nature and evolution of the ore-forming fluids in the giant Dexing porphyry Cu-Mo-Au deposit, southeastern China. Journal of Geochemical Exploration, 171:83–95

Lu Z Y, Chen H H, Qing H R, Chi G X, Chen Q L, You D H, Yin H, Zhang S Y. 2017. Petrography, fluid inclusion and isotope studies in Ordovician carbonate reservoirs in the Shunnan area, Tarim basin, NW China: implications for the nature and timing of silicification. Sedimentary Geology, 359:29–43

Ma W D, Fan H R, Liu X, Yang K F, Hu F F, Zhao K D, Cai Y C, Hu H L. 2018. Hydrothermal fluid evolution of the Jintingling gold deposit in the Jiaodong peninsula, China: constraints from U-Pb age, CL imaging, fluid inclusion and stable isotope. Journal of Asian Earth Sciences, 160:287–303

Mao S D, Duan Z H, Zhang D H, Shi L L, Chen Y L, Li J. 2011. Thermodynamic modeling of binary CH_4-H_2O fluid inclusions. Geochimica et Cosmochimica Acta, 75(20):5892–5902

Mao S D, Lü M X, Shi Z M. 2017. Prediction of the PVTx and VLE properties of natural gases with a general Helmholtz equation of state. Part I: application to the CH_4-C_2H_6-C_3H_8-CO_2-N_2 system. Geochimica et Cosmochimica Acta, 219:74–95

Mao W, Zhong H, Zhu W G, Lin X G, Zhao X Y. 2018. Magmatic-hydrothermal evolution of the Yuanzhuding porphyry Cu-Mo deposit, South China: insights from mica and quartz geochemistry. Ore Geology Reviews, 101:765–784

Meng F W, Ni P, Schiffbauer J D, Yuan X L, Zhou C M, Wang Y G, Xia M L. 2011. Ediacaran seawater temperature: evidence from inclusions of Sinian halite. Precambrian Research, 184(1-4):63–69

Meng F W, Galamay A R, Ni P, Yang C H, Li Y P, Zhuo Q G. 2014. The major composition of a middle-late Eocene salt lake in the Yunying depression of Jianghan Basin of middle China based on analyses of fluid inclusions in halite. Journal of Asian Earth Sciences, 85:97–105

Meng F W, Zhang Y S, Galamay A R, Bukowski K, Ni P, Xing E Y, Ji L M. 2018. Ordovician seawater composition: evidence from fluid inclusions in halite. Geological Quarterly, 62(2):344–352

Ni H W, Keppler H. 2012. *In-situ* Raman spectroscopic study of sulfur speciation in oxidized magmatic-hydrothermal fluids. American Mineralogist, 97(8-9):1348–1353

Ni H W, Chen Q, Keppler H. 2014. Electrical conductivity measurements of aqueous fluids under pressure with a hydrothermal diamond anvil cell. Review of Scientific Instruments, 85(11):115107

Ni P, Zhu X, Wang R C, Shen K, Zhang Z M, Qiu J S, Huang J P. 2008. Constraining ultrahigh-pressure (UHP) metamorphism and titanium ore formation from an infrared microthermometric study of fluid inclusions in rutile from Donghai UHP eclogites, eastern China. GSA Bulletin, 120(9-10): 1296–1304

Ni P, Wang G G, Chen H, Xu Y F, Guan S J, Pan J Y, Li L. 2015a. An Early Paleozoic orogenic gold belt along the Jiang-Shao Fault, South China: evidence from fluid inclusions and Rb-Sr dating of quartz in the Huangshan and Pingshui deposits. Journal of Asian Earth Sciences, 103:87–102

Ni P, Wang G G, Yu W, Chen H, Jiang L L, Wang B H, Zhang H D, Xu Y F. 2015b. Evidence of fluid inclusions for two stages of fluid boiling in the formation of the giant Shapinggou porphyry Mo deposit, Dabie orogen, central China. Ore Geology Reviews, 65:1078–1094

Ni P, Wang X D, Wang G G, Huang J B, Pan J Y, Wang T G. 2015c. An infrared microthermometric study of fluid inclusions in coexisting quartz and wolframite from Late Mesozoic tungsten deposits in the Gannan metallogenic belt, South China. Ore Geology Reviews, 65:1062–1077

Ni P, Pan J Y, Wang G G, Chi Z, Qin H, Ding J Y, Chen H. 2017a. A CO_2-rich porphyry ore-forming fluid system constrained from a combined cathodoluminescence imaging and fluid inclusion studies of quartz veins from the Tongcun Mo deposit, South China. Ore Geology Reviews, 81:856–870

Ni P, Wang G G, Cai Y T, Zhu X T, Yuan H X, Huang B, Ding J Y, Chen H. 2017b. Genesis of the Late Jurassic Shizitou Mo deposit, South China: evidences from fluid inclusion, H-O isotope and Re-Os geochronology. Ore Geology Reviews, 81:871–883

Ni P, Pan J Y, Huang B, Wang G G, Xiang H L, Yang Y L, Li S N, Bao T. 2018. Geology, ore-forming fluid and genesis of the Qiucun gold deposit: implication for mineral exploration at Dehua prospecting region, SE China. Journal of Geochemical Exploration, 195:3–15

Ni P, Li S N, Bao T, Zheng W Y, Wang G G, Xiang H L, Chi Z, Pan J Y, Huang B, Ding J Y, Dai B Z. 2019. Mapping of fluid, alteration and soil geochemical anomaly as a guide to regional mineral exploration for the Dehua gold orefield of Fujian Province, SE China. Geochemistry: Exploration, Environment,Analysis,19(1):74–90

Ni P,Zhou J,Chi Z,Pan J Y,Li S N,Ding J Y,Han L. 2020. Carbonatite dyke and related REE mineralization in the Bayan Obo REE ore field,North China:evidence from geochemistry,C-O isotopes and Rb-Sr dating. Journal of Geochemical Exploration,215:106560

Pan J Y,Ni P,Chi Z,Yang Y L,Li S N,Bao T,Wang W B,Zeng W C,Xue K. 2018. Spatial distribution and variation of ore body,alteration and ore-forming fluid of the giant Zijinshan epithermal Cu-Au deposit,SE China:implication for mineral exploration. Geochemistry:Exploration,Environment,Analysis,18(4):279–293

Pan J Y,Ni P,Wang R C. 2019. Comparison of fluid processes in coexisting wolframite and quartz from a giant vein-type tungsten deposit,South China:insights from detailed petrography and LA-ICP-MS analysis of fluid inclusions. American Mineralogist,104(8):1092–1116

Peng H W,Fan H R,Santosh M,Hu F F,Jiang P. 2020. Infrared microthermometry of fluid inclusions in transparent to opaque minerals:challenges and new insights. Mineralium Deposita,55(7):1425–1440

Peng N J,Jiang S Y,Xiong S F,Pi D H. 2018. Fluid evolution and ore genesis of the Dalingshang deposit,Dahutang W-Cu ore field,northern Jiangxi Province,South China. Mineralium Deposita,53(8):1079–1094

Pettke T,Oberli F,Audétat A,Wiechert U,Harris C R,Heinrich C A. 2011. Quantification of transient signals in multiple collector inductively coupled plasma mass spectrometry:accurate lead isotope ratio determination by laser ablation of individual fluid inclusions. Journal of Analytical Atomic Spectrometry,26(3):475–492

Pettke T,Oberli F,Audétat A,Guillong M,Simon A C,Hanley J J,Klemm L M. 2012. Recent developments in element concentration and isotope ratio analysis of individual fluid inclusions by laser ablation single and multiple collector ICP-MS. Ore Geology Reviews,44:10–38

Ping H W,Li C Q,Chen H N,George S C,Gong S. 2020. Overpressure release:fluid inclusion evidence for a new mechanism for the formation of heavy oil. Geology,48(8):803–807

Qiu H N,Wu H Y,Yun J B,Feng Z H,Xu Y G,Mei L F,Wijbrans J R. 2011. High-precision $^{40}Ar/^{39}Ar$ age of the gas emplacement into the Songliao Basin. Geology,39(5):451–454

Ren T,Zhong H,Zhang X C. 2020. Fluid inclusion and stable isotope (C,O and S) constraints on the genesis of the high-grade Langdu Cu skarn deposit in Yunnan,SW China. Ore Geology Reviews,118:103354

Roedder E. 1960. Fluid inclusions as samples of the ore-forming fluids. Intern Geol Congr Copenhagen,16:218–229

Roedder E. 1962. Studies of fluid inclusions;Part 1,low temperature application of a dual-purpose freezing and heating stage. Economic Geology,57(7):1045–1061

Roedder E. 1963. Studies of fluid inclusions;Part 2,freezing data and their interpretation. Economic Geology,58(2):167–211

Rose-Koga E F,Koga K T,Schiano P,Le Voyer M,Shimizu N,Whitehouse M J,Clocchiatti R. 2012. Mantle source heterogeneity for South Tyrrhenian magmas revealed by Pb isotopes and halogen contents of olivine-hosted melt inclusions. Chemical Geology,334:266–279

Rowe M C,Peate D W,Peate I U. 2011. An investigation into the nature of the magmatic plumbing system at Paricutin Volcano,Mexico. Journal of Petrology,52(11):2187–2220

Shan X Q,Zhang B M,Zhang J,Zhang L P,Jia J H,Liu J J. 2015. Paleofluid restoration and its application in studies of reservoir forming:a case study of the Ordovician in Tarim Basin,NW China. Petroleum Exploration and Development,42(3):301–310

Shang L B,Chou I M,Burruss R C,Hu R Z,Bi X W. 2014. Raman spectroscopic characterization of CH_4 density over a wide range of temperature and pressure. Journal of Raman Spectroscopy,45(8):696–702

Shu Q H,Chang Z S,Hammerli J,Lai Y,Huizenga J M. 2017. Composition and evolution of fluids forming the baiyinnuo'er Zn-Pb skarn deposit,northeastern China:insights from laser ablation ICP-MS study of fluid inclusions. Economic Geology,112(6):1441–1460

Sirbescu M L C,Krukowski E G,Schmidt C,Thomas R,Samson I M,Bodnar R J. 2013. Analysis of boron in fluid inclusions by microthermometry,laser ablation ICP-MS,and Raman spectroscopy:application to the Cryo-Genie Pegmatite,San Diego County,California,USA. Chemical Geology,342:138–150

Song D F,Wang T G,Li M J,Zhang J F,Ou G X,Ni Z Y,Yang F L,Yang C Y. 2017. Geochemistry and charge history of oils from the Yuqi area of Tarim Basin,NW China. Marine and Petroleum Geology,79:81–98

Sorby H C. 1858. On the microscopical,structure of crystals,indicating the origin of minerals and rocks. Quarterly Journal of the Geological Society,14(1-2):453–500

Steele-MacInnis M,Ridley J,Lecumberri-Sanchez P,Schlegel T U,Heinrich C A. 2016. Application of low-temperature microthermometric data for interpreting multicomponent fluid inclusion compositions. Earth-Science Reviews,159:14–35

Su W C,Heinrich C A,Pettke T,Zhang X C,Hu R Z,Xia B. 2009. Sediment-hosted gold deposits in Guizhou,China:products of wall-rock sulfidation by deep crustal fluids. Economic Geology,104(1):73–93

Sun R,Dubessy J. 2012. Prediction of vapor-liquid equilibrium and $PVTx$ properties of geological fluid system with SAFT-LJ EOS including multi-polar contribution,Part II:application to H_2O-NaCl and CO_2-H_2O-NaCl system. Geochimica et Cosmochimica Acta,88:130–145

Sun R,Lai S C,Dubessy J. 2015. Calculations of vapor-liquid equilibria of the H_2O-N_2 and H_2O-H_2 systems with improved SAFT-LJ EOS. Fluid Phase Equilibria,390:23–33

Tao S Z,Zou C N,Mi J K,Gao X H,Yang C,Zhang X X,Fan J W. 2014. Geochemical comparison between gas in fluid inclusions and gas produced from the Upper Triassic Xujiahe Formation,Sichuan Basin,SW China. Organic Geochemistry,74:59–65

Wälle M,Heinrich C A. 2014. Fluid inclusion measurements by laser ablation sector-field ICP-MS. Journal of Analytical Atomic Spectrometry,29(6):

Wang G G, Ni P, Zhao C, Chen H, Yuan H X, Cai Y T, Li L, Zhu A D. 2017. A combined fluid inclusion and isotopic geochemistry study of the Zhilingtou Mo deposit, South China: implications for ore genesis and metallogenic setting. Ore Geology Reviews, 81(2): 884–897

Wang M, Chen Y, Bain W M, Song G Q, Liu K Y, Zhou Z Z, Steele-MacInnis M. 2020. Direct evidence for fluid overpressure during hydrocarbon generation and expulsion from organic-rich shales. Geology, 48(4): 374–378

Wei W F, Hu R Z, Bi X W, Peng J T, Su W C, Song S Q, Shi S H. 2012. Infrared microthermometric and stable isotopic study of fluid inclusions in wolframite at the Xihuashan tungsten deposit, Jiangxi Province, China. Mineralium Deposita, 47(6): 589–605

Wei W F, Hu R Z, Bi X W, Jiang G H, Yan B, Yin R S, Yang J H. 2019. Mantle-derived and crustal He and Ar in the ore-forming fluids of the Xihuashan granite-associated tungsten ore deposit, South China. Ore Geology Reviews, 105: 605–615

Wen B J, Fan H R, Santosh M, Hu F F, Pirajno F, Yang K F. 2015. Genesis of two different types of gold mineralization in the Linglong gold field, China: constrains from geology, fluid inclusions and stable isotope. Ore Geology Reviews, 65: 643–658

Wen B J, Fan H R, Hu F F, Liu X, Yang K F, Sun Z F, Sun Z F. 2016. Fluid evolution and ore genesis of the giant Sanshandao gold deposit, Jiaodong gold province, China: constraints from geology, fluid inclusions and H-O-S-He-Ar isotopic compositions. Journal of Geochemical Exploration, 171: 96–112

Xiao B, Qin K Z, Li G M, Li J X, Xia D X, Chen L, Zhao J X. 2012. Highly oxidized magma and fluid evolution of Miocene Qulong giant porphyry Cu-Mo deposit, southern Tibet, China. Resource Geology, 62(1): 4–18

Xiao M, Qiu H N, Jiang Y D, Cai Y, Bai X J, Zhang W F, Liu M, Qin C J. 2019. Gas release systematics of mineral-hosted fluid inclusions during stepwise crushing: implications for $^{40}Ar/^{39}Ar$ geochronology of hydrothermal fluids. Geochimica et Cosmochimica Acta, 251: 36–55

Xie G Q, Mao J W, Li W, Fu B, Zhang Z Y. 2019. Granite-related Yangjiashan tungsten deposit, southern China. Mineralium Deposita, 54(1): 67–80

Xie Y L, Li Y X, Hou Z Q, Cooke D R, Danyushevsky L, Dominy S C, Yin S P. 2015. A model for carbonatite hosted REE mineralisation—the Mianning-Dechang REE belt, western Sichuan Province, China. Ore Geology Reviews, 70: 595–612

Xu L L, Bi X W, Hu R Z, Tang Y Y, Jiang G H, Qi Y Q. 2014. Origin of the ore-forming fluids of the Tongchang porphyry Cu-Mo deposit in the Jinshajiang-Red River alkaline igneous belt, SW China: constraints from He, Ar and S isotopes. Journal of Asian Earth Sciences, 79: 884–894

Xu Y F, Ni P, Wang G G, Pan J Y, Guan S J, Chen H, Ding J Y, Li L. 2016. Geology, fluid inclusion and stable isotope study of the Huangshan orogenic gold deposit: implications for future exploration along the Jiangshan-Shaoxing fault zone, South China. Journal of Geochemical Exploration, 171: 37–54

Yang K F, Fan H R, Pirajno F, Li X C. 2019. The Bayan Obo (China) giant REE accumulation conundrum elucidated by intense magmatic differentiation of carbonatite. Geology, 47(12): 1198–1202

Yang Y F, Chen Y J, Pirajno F, Li N. 2015. Evolution of ore fluids in the Donggou giant porphyry Mo system, East Qinling, China, a new type of porphyry Mo deposit: evidence from fluid inclusion and H-O isotope systematics. Ore Geology Reviews, 65: 148–164

Yang Y L, Ni P, Pan J Y, Wang G G, Xu Y F. 2017. Constraints on the mineralization processes of the Makeng iron deposit, eastern China: fluid inclusion, H-O isotope and magnetite trace element analysis. Ore Geology Reviews, 88: 791–808

Zhai D G, William-Jones A E, Liu J J, Tombros S F, Cook N J. 2018. Mineralogical, fluid inclusion, and multiple isotope (H-O-S-Pb) constraints on the genesis of the Sandaowanzi epithermal Au-Ag-Te deposit, NE China. Economic Geology, 113(6): 1359–1382

Zhang X Y, Meng F W, Li W X, Tang Q L, Ni P. 2016. Reconstruction of Late Cretaceous coastal paleotemperature from halite deposits of the Late Cretaceous Nongbok Formation (Khorat Plateau, Laos). Palaeoworld, 25(3): 425–430

Zhao C, Ni P, Wang G G, Ding J Y, Chen H, Zhao K D, Cai Y T, Xu Y F. 2013. Geology, fluid inclusion, and isotope constraints on ore genesis of the Neoproterozoic Jinshan orogenic gold deposit, South China. Geofluids, 13(4): 506–527

Zhao Z H, Ni P, Sheng Z L, Dai B Z, Wang G G, Ding J Y, Wang B H, Zhang H D, Pan J Y, Li S N. 2020. Thermal regime reconstruction and fluid inclusion LA-ICP-MS analysis on intermediate-sulfidation epithermal Pb-Zn veins: implications for porphyry Cu deposits exploration in the Xianhualing district, Anhui, China. Ore Geology Reviews, 124: 103658

Zheng X, Liu Y. 2019. Mechanisms of element precipitation in carbonatite-related rare-earth element deposits: evidence from fluid inclusions in the Maoniuping deposit, Sichuan Province, southwestern China. Ore Geology Reviews, 107: 218–238

Zhong R C, Li W B, Chen Y J, Yue D C, Yang H F. 2013. P-T-X conditions, origin, and evolution of Cu-bearing fluids of the shear zone-hosted Huogeqi Cu-(Pb-Zn-Fe) deposit, northern China. Ore Geology Reviews, 50: 83–97

Zhou L, Mavrogenes J, Spandler C, Li H P. 2016. A synthetic fluid inclusion study of the solubility of monazite-(La) and xenotime-(Y) in H_2O-Na-K-Cl-F-CO_2 fluids at 800℃ and 0.5 GPa. Chemical Geology, 442: 121–129

Zhou L, Mernagh T P, Le Losq C. 2019. Observation of the chemical structure of water up to the critical point by raman spectroscopic analysis of fluid inclusions. Journal of Physical Chemistry B, 123(27): 5841–5847

Zhou Z J, Chen Y J, Jiang S Y, Hu C J, Qin Y, Zhao H X. 2015. Isotope and fluid inclusion geochemistry and genesis of the Qiangma gold deposit, Xiaoqinling gold field, Qinling orogen, China. Ore Geology Reviews, 66: 47–64

Zhu D L, Zhu Z A, Pan J Y, Ding J Y, Ni P. 2015. Raman micro-spectroscopic study of sulfate ion in the system Na_2SO_4-H_2O. Acta Geologica Sinica-English Edition, 89(3): 887–893

Zhu J J, Hu R Z, Richards J P, Bi X W, Zhong H. 2015. Genesis and magmatic-hydrothermal evolution of the Yangla Skarn Cu deposit, Southwest China. Economic Geology, 110(3): 631–652

Zhuo Q G, Meng F W, Song Y, Yang H J, Li Y, Ni P. 2014. Hydrocarbon migration through salt: evidence from Kelasu tectonic zone of Kuqa foreland basin in China. Carbonates and evaporites, 29(3): 291–297

Progress of Fluid Inclusion Research

NI Pei[1], FAN Hong-rui[2], PAN Jun-yi[1], CHI Zhe[1], CUI Jian-ming[1]

1. State Key Laboratory for Mineral Deposit Research, Institute of Geo-fluids, School of Earth Sciences and Engineering, Nanjing University, Nanjing 210023; 2. Key Laboratory of Mineral Resources, Institute of Geology and Geophysics, Chinese Academy of Science, Beijing 100029

Abstract: Fluid inclusion research and application had made great progress in the past decade of this century (from 2011 to 2020) in China. This paper summaries the major research advances, including: Theory progress of fluid inclusion study, Technology progress of fluid inclusion study, Advance of economic geology research, and Advance of paleo-sedimentary environment and oil & gas accumulating research. In the *PVTx* study, the research of fluid *PVTx* property simulation in China maintains the international advanced level, various state equations have been established and applied to various natural fluid systems. The concept of "Fluid inclusion assemblage" has been deeply rooted in mind of fluid inclusion researchers. Domestic scientists have adopted and improved many new advanced technologies and methods, such as cathodoluminescence imaging of quartz, infrared microthermometry of opaque minerals, LA-ICP-MS analysis of single fluid inclusion, optical fused silica capillary for synthetic fluid inclusion, hydrothermal diamond-anvil cell, and fluid inclusion dating technique, all of which have been applied in the geological research and achieved good results, especially in the study of various types of deposits. The study of petroleum inclusions and inclusions in halite is gaining more attention from environmentalists and petroleum geologists, particularly in paleo-sedimentary environment, oil charging and burial-thermal history. This paper clarifies the active role of academic organizations. The domestic academic conferences have been held regularly and smoothly. The ACROFI established by Chinese scholars has become a key international forum of the international fluid inclusion research community, providing an international academic exchange platform for researchers from China and other Asian countries. At the end, we look forward to the future development of fluid inclusion research and provide helpful suggestions for future research.

Key words: fluid inclusion; theoretical study; analytical techniques; economic geology research; sedimentary and hydrocarbon accumulation studies; progress and prospects

矿床地球化学研究进展*

钟 宏　宋谢炎　黄智龙　蓝廷广　柏中杰　陈 伟　朱经经
阳杰华　谢卓君　王新松

中国科学院 地球化学研究所，矿床地球化学国家重点实验室，贵阳 550081

摘　要：我国的矿床地球化学研究在近十年取得了众多重要进展。本文对中国岩浆型 Cu-Ni-(PGE) 硫化物和 Fe-Ti-V 矿床、斑岩型铜矿床、花岗岩型钨锡矿床、碳酸岩型稀土矿床、卡林型金矿床和密西西比河谷型（Mississippi Valley-type, MVT）铅锌矿床等的一些相关研究进展，以及原位分析技术和实验地球化学在矿床研究方面的应用进展进行了扼要论述。近十年来，造山带铜镍硫化物矿床的寻找取得突破进展，岩浆通道系统被证实对巨量钒钛磁铁矿的堆积起关键作用；碰撞型斑岩铜矿的成矿模型更趋完善，花岗岩相关钨锡矿床的成矿过程与机制获得更精细刻画，碳酸岩型稀土矿床的形成时限被精确限定；华南大规模低温成矿的时限和动力学背景研究取得重大突破，成矿物质来源和流体演化的认识更为深入；原位微区元素–同位素组成对精细刻画成矿过程发挥重要作用，实验地球化学的应用初现端倪。此外，本文还对未来需要重视的几个方面的工作提出了初步建议。

关键词：岩浆矿床　岩浆热液矿床　低温热液矿床　原位分析技术　矿床地球化学

0　引　言

矿床地球化学是矿床学与地球化学交叉融合形成的分支学科。近十多年来，成矿作用与地球动力学过程的更密切结合、成矿过程的精细化研究和先进分析测试仪器和技术（尤其是原位微区技术）的研发和应用，拓展了矿床地球化学的研究领域，创新和发展了成矿理论。我国的矿床地球化学研究在此期间取得了众多重要进展。由于矿床类型繁多、研究成果海量，本文仅总结了中国岩浆型 Cu-Ni-(PGE) 和 Fe-Ti-V 矿床、斑岩型 Cu 矿床、花岗岩相关 W-Sn 矿床、碳酸岩型 REE 矿床、卡林型 Au 矿床、密西西比河谷型（MVT）Pb-Zn 矿床的进展，同时对原位分析技术和实验地球化学在矿床研究中的一些应用进行了初步的总结。限于时间、视角和认识，相关内容难免挂一漏万，有关论述或难免偏颇，尚需有关研究者谅解和读者批评指正。

1　岩浆矿床地球化学研究进展

岩浆矿床提供了全球最重要的 Ti、V、Ni、PGE 和 Cr 资源，也是 Fe、Co 等矿产的重要来源。我国岩浆矿床的成矿作用在全球独具特色，例如，晚二叠世峨眉山地幔柱活动形成了世界上最大的 Fe-Ti-V 氧化物矿集区和若干 Cu-Ni-PGE 硫化物矿床，以及 Nb-Ta-Zr-REE 富集和矿化，同一地幔柱发生三类成矿作用的现象全球罕见；甘肃金川是全球第三大 Cu-Ni 硫化物矿床，与新元古代地幔柱/裂谷活动有关；东天山、东昆仑造山带一系列大型–超大型Cu-Ni硫化物矿床的发现而成为新的研究热点。

1.1　岩浆硫化物矿床

最近十年我国岩浆硫化物矿床的研究和找矿工作都取得了重要进展，主要体现在以下几个方面。

---

* 原文"近十年来中国矿床地球化学研究进展简述"刊于《矿物岩石地球化学通报》2021 年第 40 卷第 4 期，本文略有修改。

1.1.1 造山带找矿工作取得突破

最近十年世界新发现的最大的铜镍矿并非在大火成岩省，而是在我国青海省东昆仑造山带的夏日哈木超大型镍钴矿床。2015年该矿床的探明矿石储量为1.57亿t，镍金属储量超过108万t，Ni、Cu、Co的平均品位分别达0.65%、0.14%和0.013%，进入世界级规模镍矿床的行列（Song X. Y. et al., 2016）。2012年西澳大利亚地调局根据土壤地球化学异常在Fraser Range地区发现了Nova铜镍矿床，其矿石量达1460万t，镍金属储量达32.5万t，Cu、Ni、Co的平均品位分别达0.9%、0.08%和2.2%（Maier et al., 2016）。与大火成岩省铜镍硫化物矿床相比，造山带铜镍硫化物矿床大多铜镍品位较低、贫铂族元素，但这些发现还是点燃了各国在造山带寻找铜镍硫化物矿床的热情。鉴于造山带面积远大于大火成岩省，上述发现意义重大。

然而，关于造山带铜镍硫化物成矿规律和条件的认识却存在很大分歧。例如，我国新疆北部铜镍硫化物矿床究竟形成于俯冲阶段、碰撞或碰撞后阶段还是与地幔柱活动有关，一直存在较大争议（Qin et al., 2011；Su et al., 2011；Zhang et al., 2011；Li et al., 2012；Song et al., 2013a；Xie et al., 2014；Deng et al., 2017）。东昆仑夏日哈木矿床也存在形成于俯冲阶段还是碰撞或碰撞后阶段之争（Song X. Y. et al., 2016, 2020；Zhang Z. W. et al., 2017；Wang K. Y. et al., 2019）。扬子地块周缘元古宙的铜镍硫化物矿床（矿化），例如，四川冷水箐矿床及桂北一系列矿化岩体的矿物化学组成均显示其与俯冲过程有关，但迄今为止发现的矿床规模仍很小（Zhu et al., 2006, 2007；Zhou et al., 2017；Yao J. H. et al., 2018；Ding et al., 2019）。因此，造山带铜镍硫化物矿床成因研究还任重道远。

1.1.2 成矿构造背景的新认识

许多岩浆硫化物矿床（特别是超大型矿床）往往产于大陆岩石圈边缘。基于含矿岩体时代与地幔柱活动的耦合关系，Begg等（2010）认为，大陆岩石圈厚度大而使地幔柱到达大陆岩石圈底部时难以发生减压熔融，从而迫使地幔柱向岩石圈较薄和强度较低的大陆岩石圈边缘运移，并在较小的深度发生强烈的减压熔融、形成巨量镁铁-超镁铁岩浆，为大规模岩浆硫化物提供物质和能量条件。

造山带演化过程中，特别是碰撞阶段常伴生有大规模区域性走滑断层的形成。Lightfoot和Evans-Lamswood（2015）认为，走滑构造对铜镍硫化物成矿具有控制作用。对中国新疆黄山和黄山东岩体边缘及围岩的韧性剪切构造及岩体内部的断裂构造的分析表明，区域性剪切作用形成的次一级张性构造为幔源岩浆上升提供了通道，并为岩浆房的形成提供了空间（Wang B. et al., 2014）。宋谢炎等（2018）认为，同碰撞阶段发生大规模的区域性剪切走滑使上述俯冲洋壳的断离更加容易，剪切走滑产生的超壳断裂为幔源岩浆的上升提供了顺畅的通道，也在地壳为含矿岩体的形成创造了良好的空间。

1.1.3 揭示岩浆通道系统中硫化物熔体的运移和聚集机制

目前对岩浆通道揭露最充分的是加拿大Voisey's Bay和美国的Eagle矿床（Lightfoot and Evans-Lamswood, 2015），而更多矿床的岩浆通道，如我国甘肃金川，由于中新生代剧烈的构造作用，含矿岩体及矿体的产状被强烈改造，对矿床成因分析影响很大。Song等（2012）在对金川岩体不同矿段岩相的主、微量元素（含Cu、Ni、S、PGE等）组成对比及断裂构造分析基础上，对其原始产状进行了恢复，认为金川存在两个独立的含矿岩体，它们分别经历了不同的成岩成矿过程，并在此基础上建立了新的成岩成矿模式：东、西岩体位于岩浆通道系统的两个分枝；硫化物熔离均发生在不同的深部岩浆房；熔离的硫化物乳珠被岩浆携带到西岩体，堆积在岩体底部成矿，而东岩体的矿体更可能是有硫化物-橄榄石晶粥挤入形成（Song and Li, 2009；Chen L. M. et al., 2013）。

Mao Y. J.等（2019）和Deng Y. F.等（2017）通过全岩及矿物主、微量元素、Sr-Nd-Os同位素方法探讨了地幔源区特征、地壳同化混染、岩浆氧逸度等因素对新疆黄山矿床成矿作用的约束和贡献，认为俯冲过程对地幔源区的改造对原始岩浆地球化学特点，以及后来岩浆演化过程及硫化物熔离机制、成矿机制和矿床的各种特征都有很重要的影响。

1.1.4 贱金属硫化物中铂族元素的赋存状态新认识

岩浆硫化物矿床中铂族元素的赋存状态一直是这类矿床研究的热点问题，单矿物分析发现磁黄铁矿和镍黄铁矿中可能有很高的铂族元素含量，近年来的 LA-ICP-MS 分析也发现铂族元素的分布是不均匀的，但铂族元素究竟是以纳米颗粒还是类质同象形式分布，与哪些元素耦合却并不清楚（Chen L. M. et al., 2015）。Liang 等（2019）对峨眉山大火成岩省杨柳坪超大型铜镍铂族元素矿床中贱金属硫化物的铂族元素赋存状态进行了研究，认为铂族元素在结晶前就常常和半金属元素耦合、以配合物形式存在于硫化物熔体中。因此，尽管其含量很低，也会以液相线矿物形式最早结晶形成纳米颗粒。

1.2 岩浆氧化物矿床

钒钛磁铁矿矿床是 V 和 Ti 的主要来源，中国资源量分别占全球的48%和25%，产量占全球55%和16%（USGS, 2019）。我国钒钛磁铁矿矿床主要赋存于与地幔柱相关的大型镁铁-超镁铁层状岩体中，如峨眉山大火成岩省内带的攀枝花岩体（Zhou et al., 2005; Zhang Z. C. et al., 2014）和塔里木大火成岩省的瓦吉里塔格岩体（Cao et al., 2014）及板内伸展环境的元古宙斜长岩套中（如大庙斜长岩体；Chen W. T. et al., 2013）。此外，汇聚板块边缘的镁铁质岩体，如扬子板块北缘的毕机沟岩体也表现出一定的钒钛磁铁矿成矿潜力（Zhao et al., 2018）。

1.2.1 成矿母岩浆恢复取得突破性进展

对峨眉山和塔里木大火成岩省相关钒钛磁铁矿矿床的矿物化学研究并结合 MELTs 模拟计算发现，成矿岩体的母岩浆是由原始的铁苦橄质岩浆在地壳深部经历了橄榄石和铬铁矿分离结晶而演化来的高钛铁玄武质岩浆（Bai et al., 2012; Song et al., 2013; Cao et al., 2014; Wang C. Y. et al., 2014; Zhang D. Y. et al., 2018）。Bai 等（2014）通过单斜辉石成分结合分配系数的反演和边缘带成分估算了成矿岩体母岩浆成分，结果表明红格与攀枝花成矿岩体母岩浆分别类似于岩体附近的峨眉山玄武岩。对比发现成矿岩体母岩浆的成分差异导致了成矿岩体在含矿岩性和矿物组合的显著区别，该差异被解释为上升地幔柱岩浆对新俯冲洋壳的选择性同化。Hou 等（2011, 2013）通过稀有气体同位素及 Re-Os 同位素揭示了地幔柱-俯冲/再循环洋壳（榴辉岩）相互作用对形成铁苦橄质岩浆的贡献，并强调了源区过程对钒钛磁铁矿床的大规模成矿的控制作用。与此类似，塔里木大火成岩省相关钒钛磁铁矿矿床的铁玄武质母岩浆也被认为存在俯冲改造的岩石圈地幔的贡献（Zhang D. Y. et al., 2018; Cao J. et al., 2019）。

1.2.2 钒钛磁铁矿矿床的成因争议

钒钛磁铁矿的成矿过程存在明显的争议，目前主要的成因模式有 3 种。Howarth 等（2013）观察到钒钛磁铁矿矿层中硅酸盐矿物存在不平衡结构，提出攀枝花岩体内的主要钒钛磁铁矿矿层不是原地形成的，而是由深部岩浆房中产生的富含钛磁铁矿晶体的晶浆堆积而成；通过岩石显微结构以及运用原位微区分析技术对矿物微量元素及熔融包裹体的研究，一些学者提出镁铁-超镁铁层状岩体中赋存的钒钛磁铁矿矿床是由玄武质岩浆液态不混溶作用分离而成富铁熔体形成（Zhou et al., 2013; Liu et al., 2014; Xing et al., 2014; Wang K. et al., 2018）。类似的机制也被用来解释赋存于斜长岩套中铁钛磷灰岩（nelsonite）及相关钒钛磁铁矿矿床的成因（Chen W. T. et al., 2013; He et al., 2016）；第三种成因模式认为矿物化学垂向上的规律性变化反映磁铁矿和钛铁矿直接在玄武质岩浆演化的早期阶段结晶并发生重力分异而成矿。较高的氧逸度（Bai et al., 2012, 2016, 2019b; Ganino et al., 2013; Cao et al., 2014），富水（Howarth and Prevec, 2013; Zhang D. Y. et al., 2018），富 Fe、Ti 的母岩浆（Song et al., 2013b; She et al., 2015）成分，以及周期性岩浆的补给（Bai et al., 2012; Song et al., 2013b）是导致磁铁矿较早结晶并成矿的关键控制因素。近年来，非传统同位素（如 Fe、Mg 同位素等）被应用于钒钛磁铁矿矿床成因研究（Chen et al., 2014; Liu et al., 2014; Cao Y. H. et al., 2019）。详细的分析结果显示硅酸盐矿物与磁铁矿、钛铁矿之间存在明显的同位素不平衡，但其原因是液态不混溶还是亚固相再平衡还存在争议。

1.2.3 建立钒钛磁铁矿的岩浆通道成矿模型

研究发现,含矿层状岩体本身并不足以提供形成超大型钒钛磁铁矿矿床所需的成矿物质,说明有更多的岩浆参与了成矿过程。通过详细的矿物学和地球化学研究,Bai 等(2012)揭示了具有地球化学韵律性变化特征的赋矿层状岩体曾是多期玄武岩浆喷发的岩浆通道系统,并在此基础上建立了在岩浆通道系统中形成超大型钒钛磁铁矿矿床的成矿模式。巨量的玄武质岩浆在岩浆通道中多次卸载大量的成矿物质,是钒钛磁铁矿在一个较小岩体中超常富集的关键因素。

2 斑岩(夕卡岩)-浅成低温热液型铜矿床地球化学研究进展

作为全球主要的 Cu、Mo 及重要的 Au 等金属来源,斑岩铜矿系统(包括夕卡岩型矿床和浅成低温热液型 Au 矿床)长期是工业界和学术界重点关注的矿床类型之一(Sillitoe,2010)。就我国而言,斑岩铜矿系统供应了全国 65% 以上的 Cu、超过 1/4 的 Au 和约 90% 的 Mo(Chang et al.,2019;White et al.,2019;Yang and Cooke,2019)。

2.1 碰撞型斑岩铜矿成矿模型的创新

经典的斑岩铜矿模型认为其主要形成于与大洋俯冲有关的岛弧或陆缘弧环境(Richards,2003;Sun et al.,2013;毛景文等,2014;陈华勇和吴超,2020)。我国地处全球三大构造域即环太平洋、古亚洲洋和特提斯构造域的复合部位,增生、碰撞、陆内造山带均十分发育(Hou et al.,2007;陈衍景,2013;秦克章等,2017;侯增谦等,2020),除俯冲型斑岩铜矿外,我国还发育众多非弧环境的斑岩铜矿(Yang and Cooke,2019)。

近十年来,随着对西藏冈底斯斑岩铜矿带、玉龙斑岩铜矿带一系列非弧环境斑岩铜矿的深入研究,我国学者逐步建立了碰撞斑岩铜矿的成矿模型(Hou et al.,2015a,2017;Yang et al.,2015,2016;Wang R. et al.,2018;侯增谦等,2020)。与俯冲型斑岩铜矿一样,碰撞斑岩铜矿的岩浆源区亦具有较高的氧逸度和水含量,这些条件能保证岩浆中硫化物不会过早饱和并最终在出溶流体中富集(Richards,2009;Hou et al.,2011;Wang R. et al.,2014;Sun et al.,2017);两类斑岩铜矿的围岩蚀变和矿化特征也十分相似,但碰撞斑岩铜矿中的绢英岩化对早期钾化蚀变的叠加可能更显著,同时绢英岩化阶段可能也沉淀了大量的铜硫化物(Yang and Cooke,2019),这些新的认识丰富了斑岩铜矿的成矿理论。传统观点认为,弧环境斑岩铜矿岩浆的高氧逸度和富水特征主要来自于俯冲板片(Richards,2003;Sillitoe,2010);对碰撞斑岩铜矿而言,俯冲期形成的新生或加厚下地壳±岩石圈地幔的部分熔融可能对成矿十分关键(Li J. X. et al.,2011;Deng et al.,2015;Zhou T. F. et al.,2015;Hou et al.,2017),Hf-Nd 同位素填图也证实斑岩铜矿主要产出于较年轻的地壳分布区(Wang C. M. et al.,2016;侯增谦和王涛,2018),但目前就新生下地壳和岩石圈地幔对成矿所需挥发性组分(水、硫、氯等)和金属贡献比例尚不明确(Lu et al.,2015;Yang et al.,2015;Wang R. et al.,2018;Li W. K. et al.,2020;侯增谦等,2020)。最近,一些学者尝试通过非传统稳定同位素解决上述争议。Zheng 等(2019)发现冈底斯带成矿斑岩及黄铜矿均具有较重的 Cu 同位素组成($\delta^{65}Cu$ 为 0.18‰~1‰),据此提出具较重 Cu 同位素组成的下地壳硫化物重熔提供了冈底斯斑岩铜矿带所需的铜。值得一提的是,近年来俯冲型斑岩铜矿的成矿机理也受到了一些挑战,例如,Lee 等(2012)、Chiaradia(2014)和 Lee 和 Tang(2020)提出,下地壳富硫化物堆晶体的重熔是形成斑岩铜矿的关键,该可能与厚地壳产出更多大型斑岩铜矿这一地质事实相吻合。但就我国俯冲型斑岩铜矿而言,其初始岩浆可能更多来自于俯冲流体交代地幔楔的部分熔融(Wang et al.,2017;Yang and Cooke,2019)。此外,如前所述,传统认为硫化物过早饱和被抑制是斑岩铜矿成矿的先决条件之一,但 Du 和 Audétat(2020)通过对铜陵矿集区斑岩铜矿的研究,指出早期硫化物饱和熔离对成矿影响不大,而岩浆富水(高 Sr/Y 值)则可能对成矿更为关键;Bai 等(2020)通过铂族元素研究,发现紫金山矿集区火山岩的铂族元素较低，暗示硫化物饱和很早，但出溶流体将早期岩浆硫化物重新萃取也可能形成斑岩铜矿床。显然，有关硫化物饱和与斑岩铜矿的成因联系，仍需更多研究。

2.2 斑岩铜矿成矿过程的精细刻画

近十年来，原位元素-同位素及高精度定年分析技术的发展推动了对成矿流体来源和演化的精细刻画，同时丰富了找矿勘探手段。Mao等（2017）通过不同阶段石英微量元素的研究，深刻阐述了南岭大宝山斑岩Mo-W多金属矿床流体演化过程；Li Y.等（2017，2018）通过对驱龙斑岩铜矿各蚀变带辉钼矿Re-Os定年并基于各期次石英氧同位素周期性变化规律，阐明了成矿事件的瞬时性而热液演化的周期性特点。此外，通过激光原位测定锆石（Ce^{4+}/Ce^{3+}值、δEu等）、磷灰石（F-Cl-S含量）、榍石（Ga含量、Fe_2O_3/Al_2O_3值）、角闪石等矿物的化学成分，限定岩浆氧逸度、挥发性元素组分进而研判斑岩铜矿成矿潜力（Cao et al., 2012, 2018; Xu et al., 2012; Lu et al., 2016; Pan et al., 2016, 2018; Zhu et al., 2018；赵振华和严爽，2019），利用绿泥石、绿帘石的成分以及通过短波红外（SWIR）光谱技术识别蚀变矿物，从而约束矿床热液/矿化中心等方面过去十年也取得了诸多进展（杨志明等，2012；陈华勇和吴超，2020）。Zhu等（2018）通过研究加拿大Red Chris斑岩铜矿发现，相较于成矿前和成矿后，成矿期斑岩具有显著高的磷灰石S-Cl含量，暗示富S-Cl的基性岩浆注入可能是斑岩铜矿的成矿关键之一。杨志明等（2012）利用SWIR测量，发现西藏念村矿区东北部伊利石结晶度较大（>1.6）且Al-OH吸收峰位较小（<2 203 nm），暗示该区可能为矿化蚀变中心。另外值得指出的是，利用非传统稳定同位素如K、Mg、Fe、Cu等识别-示踪成矿流体，近十年来已有不少开创性研究，也为斑岩铜矿地球化学研究开辟了新的领域（Li Y. et al., 2018; He et al., 2020；李伟强等，2020）。李伟强等（2020）通过对德兴斑岩铜矿成矿斑岩全岩K-Mg同位素分析，发现蚀变斑岩具有显著高的K-Mg同位素比值（$\delta^{41}K$为-1.02‰~0.38‰；$\delta^{26}Mg$为-0.49‰~0.32‰），指示热液蚀变过程相对富集重的K-Mg同位素。

2.3 夕卡岩-浅成低温热液型矿床研究进展

夕卡岩型矿床是我国十分重要的Sn、Cu、Pb-Zn和Au等金属的来源（李建威等，2019；Chang et al., 2019）；浅成低温热液型金矿床提供了我国约8%的金资源（>1 460 t），其中绝大部分属于低硫型，但从金资源量上看，高硫型浅成低温热液矿床产出的金更多，如福建紫金山和台湾金瓜石矿床合计产出超过800 t的金（Chen et al., 2012; Pan et al., 2019a; White et al., 2019）。近十年来，除将上述两类矿床作为斑岩铜矿系统的重要组成部分开展研究之外，我国学者还在这两类矿床的成因机制、成矿元素共生分异及高温与低温矿床的成因联系研究方面取得了重要进展。Shu等（2013）和Zhu J. J. 等（2015）分别对我国内蒙古白音诺尔铅锌矿和云南羊拉铜矿开展了系统的地质地球化学研究，S-Pb同位素组成表明成矿物质具岩浆来源，成矿年代与矿区岩浆岩一致，从而明确了上述矿床为夕卡岩型矿床而不具同生沉积成因。Xie G. Q. 等（2015）通过对长江中下游夕卡岩型Cu-Fe和Fe矿床的研究，发现Fe矿床具有显著高的S同位素组成，暗示膏盐层的加入可能是成铁矿的关键。Zhai等（2018）通过对黑龙江三道弯子浅成低温热液型金矿床的H-O、S-Pb研究和地球化学热力学模拟，证实富Te流体可能对迁移、富集Au具有重要意义。Xie等（2019）通过榍石的U-Pb定年，发现华南曹家坝夕卡岩型钨矿床与周边低温Sb-Au矿床时代高度一致，并具显著成因联系。

3 与花岗岩相关的钨锡矿床地球化学研究进展

钨锡金属近年已成为欧美等西方发达国家高度关注的新兴战略性矿产，相关成矿作用研究及找矿勘查工作受到国内外的高度关注（Zhou et al., 2018; Mao J. et al., 2019; Lehmann, 2020；蒋少涌等，2020）。我国是钨锡资源的重要产区，其储量和产量长期居世界首位。我国钨矿床主要以石英脉型、夕卡岩、斑岩和云英岩型产出，而锡矿床主要以夕卡岩和石英脉型及云英岩型产出（Mao J. et al., 2019），分布在华南、冈底斯、三江、昆仑、秦岭、大别-苏鲁和中亚造山带等地区。

3.1 钨锡成矿时代的精确限定

近十年来，我国学者成功开发出了矿石矿物锡石和黑钨矿原位 U-Pb 同位素定年方法（Yuan S. D. et al., 2011；Deng et al., 2019；Tang et al., 2020），结合一些脉石矿物的同位素定年，如辉钼矿 Re-Os、云母 Ar-Ar 等，更精确地限定了我国钨锡矿床的形成时代（Hu et al., 2012b；Mao J. et al., 2019；Wang X. S. et al., 2019）。新元古代 Sn（W）矿床（850~790 Ma）分布在华南扬子地块西缘和南缘，早古生代 W（Sn）矿床（450~410 Ma）主要分布在北秦岭和东昆仑，晚古生代 W-Sn 矿床（310~280 Ma）主要分布在中亚造山带西部，三叠纪 W-Sn 矿床（250~210 Ma）在全国各地都有出现，从早侏罗世到白垩纪的 W-Sn 矿床（198~80 Ma）分布在中国东部，而晚白垩世到新生代 W-Sn 矿床（121~23 Ma）主要出露在冈底斯-念青唐古拉山-三江造山带（Chen X. C. et al., 2015；Yuan S. D. et al., 2018；Mao J. et al., 2019；Wang X. S. et al., 2019，阳杰华等，2017）。

3.2 钨锡富集过程与机制的精细刻画

近年来新方法的开发和原位微区技术的快速发展和不断完善，显著促进了岩浆演化 W-Sn 富集过程及 W-Sn 矿床形成与演化的精细刻画，丰富和完善了与花岗岩有关的钨锡成矿理论。传统观点认为，W-Sn 矿床与花岗岩具有密切的成因联系，但 W-Sn 在岩浆演化过程中如何迁移富集得不到有效约束。近年，随着微区技术的应用，许多学者通过研究花岗岩中稳定副矿物的矿物学和矿物化学特征（如榍石、磷灰石和石榴子石），明确揭示了岩浆演化过程存在岩浆-热液过程阶段，该阶段 W-Sn 从岩浆转移到热液流体体系并形成 W-Sn 矿床（Yang et al., 2013；Yang W. B. et al., 2014；Hulsbosch et al., 2016；Zeng et al., 2017；Zhao et al., 2019）。针对成矿流体性质研究，一些学者近年采用红外显微测温技术，建立起了不透明矿石矿物黑钨矿中流体包裹体的研究方法（Wei et al., 2012；Ni et al., 2015），直接限定成矿流体性质，弥补了以往通过研究共生脉石矿物中流体包裹体间接推断成矿流体性质带来的不足。如 Wei 等（2012）发现江西西华山钨矿床中黑钨矿流体包裹体均一温度和盐度均高于共生石英中包裹体的均一温度和盐度。针对成矿流体演化与钨锡沉淀过程研究，一些研究者对不同成矿阶段的脉石矿物（如云母、电气石等）开展原位元素和 B 同位素分析（Zhao et al., 2019；Hong et al., 2020），对矿石矿物（如锡石、黑钨矿和白钨矿）开展原位微量元素和 Sr-Hf-O 同位素等分析（Kendall-Langley et al., 2020），结合矿石矿物如黑钨矿和锡石中单个流体包裹体成分分析，确定流体演化过程有其他性质的流体加入，进一步明确了流体混合作用是钨锡沉淀的关键控制因素（Legros et al., 2016, 2018, 2019；Peng et al., 2018；Yang et al., 2019b；Zhao et al., 2019）。同时，非传统稳定同位素如 Fe、Li、Sn 等近年也逐渐应用于示踪 W-Sn 成矿流体演化与沉淀过程，取得了一些开创性研究成果，这为 W-Sn 成矿作用研究拓展了新的视角（Wawryk and Foden, 2015；Chen B. et al., 2018；Li J. et al., 2018；Yao J. M. et al., 2018）。

3.3 幔源物质参与钨锡成矿的新认识

稀有气体同位素被广泛应用于 W-Sn 矿床的成矿作用研究（Li G. L. et al., 2011；Hu et al., 2012a；翟伟等，2012；Wei W. F. et al., 2019）。传统的认识是与 W 矿床有关的花岗岩是由地壳物质重熔形成（徐克勤等，1982），但近年的稀有气体同位素研究发现，许多 W 矿床的成矿流体中存在大量幔源 He 和 Ar，如华南瑶岗仙、西华山和淘锡坑，而 W 矿床的成矿流体由花岗岩分异出来，进而推断花岗岩中必含有大量的幔源 He。因此，原被认为与 W 成矿有关的 S 型花岗岩，实际上是壳幔相互作用的产物，地幔至少提供了地壳物质重熔所需的热（Hu et al., 2012a；Wei W. F. et al., 2019）。值得一提的是，Yuan 等（2019）首次认为地幔来源的热有助于地壳物质中白云母发生熔融，同时发生 W 矿化作用；而地幔物质的直接参与可导致地壳物质中黑云母和白云母同时熔融,可产生 Sn(W)矿化作用,进一步明确了壳幔相互作用是驱使 W-Sn 共生分异的主要原因。

4 碳酸岩型稀土元素矿床地球化学研究进展

碳酸岩型稀土矿床是中国稀土的最主要来源(> 90%)(Xie et al., 2016),且具有易采、易选和易冶的特点,经济价值显著。目前为止,中国重要的碳酸岩型稀土矿床主要分布于白云鄂博、冕宁-德昌及山东微山等地区(Xie et al., 2016),但近年来在秦岭造山带也陆续发现了一系列规模不等的碳酸岩型稀土矿床,稀土氧化物(rare earth oxides,REO)总储量达 200 万 t,有望成为我国的又一重要稀土资源基地(Zhang W. et al., 2019a)。近十年来,随着分析技术的进步,该类矿床的成矿时代和背景、稀土富集机制和关键因素、流体演化过程等方面的研究也取得了一系列重要进展。

4.1 碳酸岩型稀土矿床成矿时限的精确厘定

作为世界上最大的稀土矿床,白云鄂博稀土-铌-铁矿床一直是国内外关注的焦点,但因其具有十分复杂的元素及矿物组成,同时又经历了多期地质事件的叠加改造,其成矿时代长期争论不断。随着近十年各种原位分析技术的进步,有关白云鄂博矿床的成矿时代有了一定清晰认识。例如,利用锆石 LA-ICP-MS U-Pb 及独居石 LA-MC-ICP-MS Sm-Nd 等时线定年,基本限定矿区内碳酸岩的侵位时间为 ~1.3 Ga(Yang et al., 2011;Fan et al., 2014;Yang et al., 2019b),代表了最早一期的稀土成矿/富集事件(Zhu X. K. et al., 2015;Song et al., 2018),可能与超大陆的裂解相关。然而,大多数稀土矿物的 LA-ICP-MS 和 SIMS U-Th-Pb 定年获得了一系列小于 1.3 Ga 且连续的年龄(1 000 ~ 350 Ma,峰期 ~440 Ma;Smith et al., 2015;Fan et al., 2016;Song et al., 2018),被认为记录了与古亚洲洋俯冲流体有关的热液改造或成矿事件(Yang et al., 2017)。

四川冕宁-德昌稀土矿带是我国第二大稀土成矿带,发育有牦牛坪超大型、大陆槽大型、木落寨和里庄小型矿床等一系列碳酸岩型稀土矿床(Xie et al., 2016;Liu and Hou, 2017)。稀土矿物(氟碳铈矿)的 U-Th-Pb 同位素直接定年已精确限定成矿带北部的牦牛坪、里庄和木落寨矿床主要形成于 22 ~ 27 Ma,而南部的大陆槽矿床形成于 12 Ma(Liu et al., 2015;Ling et al., 2016),均与青藏高原碰撞导致的走滑断裂相关。

秦岭造山带不同构造单元上都发现有碳酸岩型稀土矿床,包括南秦岭单元的庙垭和杀熊洞稀土-铌矿床、北秦岭单元的太平镇稀土矿床以及华北陆块南缘的黄龙铺、黄水庵和华阳川等 Mo-REE 矿床(吴昌雄等,2015;高成等,2017;李靖辉等,2017;Zhang W. et al., 2019a)。目前,利用锆石和稀土矿物(独居石和氟碳铈矿)的 LA-ICP-MS U-Pb 已基本确定该带的稀土形成于 440 ~ 410 Ma(如庙娅、杀熊洞和太平镇)和 220 ~ 200 Ma(华北南缘的 Mo-REE 矿床)两期(Zhang W. et al., 2019a),且后期事件在早期矿化上(尤其是南秦岭)可能局部有叠加。

4.2 揭示碳酸岩型稀土矿床的物质来源和成因机制

白云鄂博矿床的成因长期存在很大争议(Fan et al., 2014, 2016;Smith et al., 2015;Zhu X. K. et al., 2015;Yang et al., 2017)。尽管近十年来开展了大量的同位素地球化学和矿物学等工作(发表文章百余篇),有关稀土巨量富集原因、H8 白云岩围岩的成因、稀土与铌-铁的关系等关键问题仍未形成共识。根据现有矿物学(如岩浆矿物)研究和岩浆特点的 Sr-Nd-O 同位素组成等以及不同碳酸岩地球化学成分连续变化的特点,大多数学者倾向于认为 H8 白云岩为致矿的 ~1.3 Ga 岩浆碳酸岩体、稀土成矿与岩浆碳酸岩的演化密切相关(Wang et al., 2010;Fan et al., 2016;Yang et al., 2019b;Liu et al., 2020),而与古亚洲洋俯冲流体有关的热液事件对稀土成矿有何种贡献仍存在很大争论(Smith et al., 2015;Fan et al.，2016；Yang et al.，2017），亟待进一步研究。例如，部分研究者通过不同矿石中混合的 C-O 同位素特点认为该期热液极富稀土、通过交代 H8 围岩导致稀土成矿，代表了主成矿期（Ling et al.，2013；Yang et al.，2017）；而另一些研究者通过矿石和矿物 Nd 同位素与 1.3 Ga 火成碳酸岩相似的特点，认为该期热液贫稀土、仅对早期火成碳酸岩成因的稀土矿石进行了不同程度的改造，无明显的稀土贡献（Liu S. et al.，2018；Song et al.，2018）。

有研究者利用 Li 同位素证据证实，冕宁–德昌稀土成矿带致矿岩体的原始岩浆是受俯冲沉积物交代过的岩石圈地幔部分熔融的产物（Hou et al.，2015b；Tian et al.，2015），且原始岩浆演化过程中发生的碳酸岩与碱性岩浆不混溶作用进一步导致了稀土富集（Liu and Hou，2017）。尤其是近年来众多学者针对不同类型熔–流体包裹体开展了大量系统显微测温、SEM、激光拉曼和 LA-ICP-MS 成分等研究工作，初步确立该矿带不同矿床中成矿过程涉及流体沸腾作用而形成富稀土的高盐度流体；确定成矿流体为富 CO_2 和 SO_4^{2-} 的体系，硫酸盐对稀土的搬运和富集起重要作用；也证实温度的降低或压力释放以及浅部流体混合导致稀土矿物在晚期沉淀成矿（Xie Y. L. et al.，2015；Liu et al.，2019a；Guo and Liu，2019；Shu and Liu，2019）。然而，迄今仍有不少关键问题如流体出溶机制、稀土运移方式、硫酸盐的作用等（Cui et al.，2020b）仍需进一步研究。

对华北南缘 Mo-（REE）矿床的同位素地球化学研究表明，致矿碳酸岩的形成可能与富碳酸盐俯冲板片熔体并交代加厚榴辉岩下地壳有关（Song W. L. et al.，2016）、并在上升过程中可能有少量基底岩石的混染（Bai et al.，2019a）；稀土成矿则与熔–流体演化形成的富 LREE/HREE、Mo、Pb 和 S 流体以及伴随的矿物结晶分异过程和后期热液活化等相关（Song W. L. et al.，2016；Bai et al.，2019a）。另外，新近的元素和同位素地球化学研究表明，南秦岭单元庙垭和杀熊洞杂岩体的母岩浆可能源于被地壳物质（如碳酸盐物质）改造过的地幔源区（Xu et al.，2014；Çimen et al.，2018；Chen W. et al.，2018），并在上升过程中通过结晶分异或液态不混溶作用促使稀土元素富集成矿（Zhu et al.，2017；Su et al.，2019）。但必须提及的是，新近矿物学和 LA-ICP-MS 原位定年工作揭示，庙垭岩体中的稀土矿化在时代及矿物共生特点上却与 220～200 Ma 期热液叠加事件更为密切，可能存在后期稀土的活化和再富集（Ying et al.，2017，2020；Zhang W. et al.，2019a），抑或代表了与庙垭成岩过程无关的晚期稀土成矿事件（Çimen et al.，2018；Ying et al.，2020）。

5 低温矿床地球化学研究进展

扬子地块及邻区尤其是扬子地块西南缘矿产资源非常丰富，在面积约 50 万 km² 的范围内，卡林型金矿和锑、汞、铅、锌等低温热液矿床广泛发育，其中不乏大型–超大型矿床，构成了全球罕见的世界级多金属成矿域。区内锑矿储量占全球的 50% 以上，金矿储量约占全国的 1/5，同时还是我国铅锌矿的主要产区之一。

5.1 卡林型金矿床

全球卡林型金矿的主要产区是美国内华达和我国滇黔桂地区（右江盆地）。滇黔桂地区的卡林型金矿是我国华南低温成矿域的重要组成部分（胡瑞忠等，2016）。近十来年，由于微区–原位分析等技术的快速发展和应用，滇黔桂地区的卡林型金矿取得了重要进展。

5.1.1 成矿年代和动力学背景

诸多学者早期尝试了多种方法对滇黔桂地区的卡林型金矿进行成矿年代学研究，但由于该类型金矿固有特点（矿物颗粒细、多期次、多环带、浸染状分布）（图 1）和早期定年方法的局限，其成矿年龄一直没有得到很好的限定（胡瑞忠等，2007）。Hu 等（2017）系统总结了滇黔桂地区卡林型金矿的成矿年龄，认为该区金矿主要有两期成矿作用，分别为 200～230 Ma（相当于印支期）和 130～150 Ma（相当于燕山期）。近年来，不同研究者基于金红石和锆石 SIMS U-Pb、磷灰石 SIMS Th-Pb、磷灰石和方解石 LA-ICPMS U-Pb、锆石 U-Th/He 及绢云母 Ar-Ar 等方法，获得滇黔桂地区卡林型金矿床的成矿年龄集中在 205～239 Ma 和 129～160 Ma（Pi et al., 2017；高伟，2018；黄勇，2019；Chen M. H. et al., 2019；Zhu et al., 2020）。该区目前已发现少量 200～230 Ma 岩浆作用（Wu et al., 2019），并且继承锆石年龄分析显示深部有 130～140 Ma 和 190～220 Ma 的岩体（朱经经等，2016；高伟，2018）。上述精确定年结果很好地证实了 Hu 和 Zhou（2012）及 Hu 等（2017）提出的模式，即两期大规模低温成矿作用与其东侧华夏地块中与花岗岩浆活动有关的两期钨锡多金属成矿基本同时，扬子的低温成矿与华夏的钨锡高温成矿具有相似的成矿动力学背景（图2）（Hu and Zhou, 2012；Hu et al., 2017；胡瑞忠等，2020）。

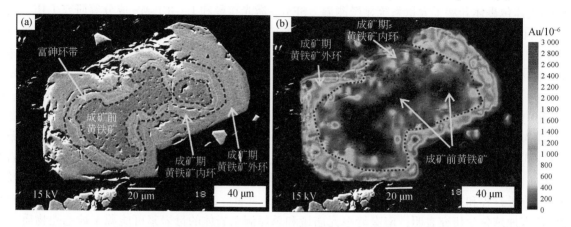

图1 锦丰金矿黄铁矿的多期次、多环带特征（据 Xie et al., 2018a）

（a）背散射图；（b）LA-ICP-MS 金面扫描图。成矿期黄铁矿通常具有内环 As 高 Au 低、外环 As 低 Au 高的特征，说明 Au 在早期 As 含量较高的成矿流体中并未大量沉淀，而在晚期 As 含量较低的成矿流体中发生大规模沉淀

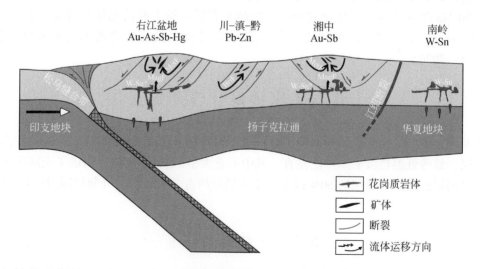

图2 扬子克拉通印支期（200～230 Ma）大规模低温成矿动力学模型（据 Hu et al., 2017；胡瑞忠等，2020）

MetF. 雨水成因流体；BriF. 卤水成因流体；MagF. 岩浆成因流体

5.1.2 成矿物质来源、流体演化和成矿作用过程

原位微区分析技术可分析不同期次、不同环带矿物的元素和同位素组成，进而能有效限定成矿物质来源和反演成矿作用过程。诸多研究者利用 EPMA、SIMS 和 LA-ICP-MS 对该区多个矿床中不同期次、不同环带黄铁矿/毒砂开展了精细的成分研究，发现成矿前与成矿期黄铁矿及矿物不同环带均具有显著的成分差异（图1）（Su et al., 2012；Hou et al., 2016；Hu et al., 2017；Xie et al., 2018b；Yan et al., 2018；

Li Z. L. et al., 2020；Wei et al., 2020），说明成矿物质不是直接来源于赋矿地层。成矿期矿物不同环带的成分记录了成矿流体从早期到晚期的演化及 Au 的沉淀过程（图 1）。此外，不少学者利用 SIMS、SHRIMP 和 LA-MC-ICP-MS 对黄铁矿/毒砂开展微区–原位硫同位素组成分析，发现成矿期黄铁矿/毒砂的硫同位素组成主要集中在 $-5‰ \sim 5‰$（Hou et al., 2016；Xie et al., 2018b；Yan et al., 2018；Li Z. L. et al., 2020；Wei et al., 2020），并与成矿晚期硫化物（如辉锑矿、雄黄、雌黄等）的硫同位组成数据相一致（Tan et al., 2015）；但不同学者对这些数据的解释不一致，主要有 3 种成矿物质来源的认识，包含岩浆热液来源（Xie et al., 2018b；Yan et al., 2018）、变质热液来源（Su et al., 2018；Li Z. L. et al., 2020；Wei et al., 2020）和盆地水萃取地层物质（Hu et al., 2017）。此外，Hg 同位素（Yin et al., 2019）和惰性气体分析（Jin et al., 2020）均显示成矿流体有岩浆流体的贡献。

5.1.3 区域成矿作用对比

Su 等（2018）系统总结和对比了右江盆地北缘和南缘的卡林型金矿，发现盆地南缘的金矿形成于印支期（232~212 Ma），富含毒砂，其 Au/As 值低（1∶100 000~1∶1 000），硫同位素值较高（δ^{34}S 为 12‰~18‰），成矿流体具有较高温度（~245℃）、低盐度（~2% NaCl$_{eqv}$）和富 CO$_2$ 等特点；而盆地北缘金矿形成于燕山期（148~134 Ma），Au/As 值较高（1∶1 000~1∶10），硫同位素组成接近零值，成矿流体为低温（~210℃）、中–低盐度（~5% NaCl$_{eqv}$）成矿流体。根据以上特征，Su 等（2018）提出印支期和燕山期造山作用释放的变质流体形成该区两期卡林型金矿。Xie 等（2018a）根据对滇黔桂地区和美国内华达卡林型金矿进行的对比研究，指出两地金矿在区域岩浆作用、蚀变特征、载金黄铁矿形貌和化学组成、成矿流体特征等方面都有差异，尤其是滇黔桂地区的金矿具有白云石化蚀变特征、流体温度–压力更高、更富集 CO$_2$，从而提出滇黔桂金矿的形成温度–压力–流体性质介于典型卡林型和造山型金矿之间（Xie et al., 2017, 2018a）。Wang 和 Grove（2018）从矿床、矿区和成矿省尺度对两地的卡林型金矿进行对比研究，发现二者具有高度相似性，从而认为滇黔桂地区的金矿应为卡林型金矿。

5.2 密西西比河谷型（MVT）铅锌矿床

扬子地台西南缘的川滇黔接壤区内成群成带分布有 400 多个大、中、小型铅锌矿床、矿点和矿化点，总面积约 17 万 km^2（柳贺昌和林文达，1999），是我国重要的 Pb、Zn、Ag 及多种分散元素生产基地之一。该矿集区是我国华南大面积低温成矿域的重要组成部分（黄智龙等，2011；Hu et al., 2017），也是我国密西西比河谷型（MVT）铅锌矿床最为集中的分布区域（Zhang et al., 2015；Hu et al., 2017）。通过对矿集区近十年的研究，在成矿理论方面取得一批新认识，成矿预测方面也取得一批新突破。

5.2.1 成矿年代学与成矿动力学背景

由于川滇黔接壤铅锌矿集区缺乏精确可靠的成矿年代学数据，其成矿动力学背景长期存在争议（黄智龙等，2011；张长青，2008；毛景文等，2012；吴越，2013；Hu et al., 2017）。不同研究者利用方解石和萤石 Sm-Nd、闪锌矿 Rb-Sr、沥青 Re-Os 和低 Re 含量硫化物 Re-Os 等时线及碳酸盐矿物原位 U-Th-Pb 等多种定年方法（文献略），获得矿集区赋存于不同时代地层的 15 个代表性矿床的成矿年龄大多集中在 190~230 Ma 范围。该年龄与扬子地块西南缘古特提洋闭合时代 205~240 Ma（钟大赉，1998）相近，为矿集区成矿动力学背景可能为印支期碰撞造山作用提供了年代学证据。Hu 等（2017）的研究表明，川滇黔接壤 Pb-Zn 矿集区成矿时代为 190~230 Ma，华南低温成矿域内的右江盆地 Au-As-Sb-Hg 矿集区（Chen M. H. et al., 2015, 2019；Pi et al., 2017；黄勇，2019）和湘中盆地 Sb-Au 矿集区（李华芹等，2008）也普遍存在该年龄段的成矿作用，并据此认为印支期奠定了华南大规模低温成矿的主体格架。

5.2.2 硫化物微量元素组成及矿床成因类型

LA-ICP-MS 技术的发展使原位准确测定硫化物的微量元素组分成为可能，因而被广泛用于矿床成因类型划分、成矿流体来源与演化、成矿预测信息提取以及伴生关键金属分布与赋存状态等研究（Cook et al., 2009；Ye et al., 2011；George et al., 2015）。不同研究者对川滇黔接壤铅锌矿集区（叶霖等，2016；Wei C. et al., 2018，2019；胡宇思等，2019；李珍立等，2019）不同铅锌矿床的硫化物原位微量元素组成研究表明，其明显不同于前人（Cook et al., 2009；Ye et al., 2011）总结的喷流沉积型、夕卡岩型和岩浆热液型铅锌矿床，而与MVT铅锌矿床总体相似；基于不同类型铅锌矿床闪锌矿的Mn-Fe、Mn-Co、Cd/Fe-Mn和In/Ge-Mn关系图，进一步确定这些矿床为MVT型铅锌矿床。

川滇黔接壤矿集区许多矿床都富含Ge、Cd等分散元素，是我国这些元素的重要生产基地之一，其分布规律、富集机理、赋存状态和综合利用长期是研究热点（温汉捷等，2019；吴越等，2019；Wei C. et al., 2019）。原位硫化物微量元素组成研究揭示，矿集区内许多铅锌矿床富含Ge、Ge富集于闪锌矿之中（吴越等，2019及其中参考文献）。进一步的研究发现，区内富集Ge的闪锌矿也相对富集Cu，且Ge与Cu之间存在明显的正相关关系（叶霖等，2016；胡宇思等，2019；李珍立等，2019；Wei C. et al., 2019）。因Cu^{2+}较Ge^{2+}更易进入闪锌矿晶格（刘英俊等，1984），两者结合的平均离子半径更接近Zn^{2+}离子半径而有利于类质同象置换，这可能是矿集区内闪锌矿不同程度富集Ge的主要原因之一（Wei C. et al., 2019）。

5.2.3 成矿物质基础和成矿流体来源与演化

（1）成矿物质来源。前人对川滇黔接壤铅锌矿集区成矿物质来源有多种认识，如赋矿地层（柳贺昌和林文达，1999）、峨眉山玄武岩（沈苏等，1988）、基底地层（Zhou et al., 2001）、多来源（黄智龙等，2004）；近年来大量Pb同位素地球化学研究成果表明，该区成矿物质主要来源于基底地层，如赋存于震旦系的天宝山（何承真等，2016；孙海瑞等，2016；Tan et al., 2019）、茂租（Wang L. J. et al., 2018；Zhang H. J. et al., 2019a）和银厂沟（Li B. et al., 2016；Tan et al., 2017），赋存于震旦–寒武系的乌斯河（Zhang H. J. et al., 2019b；Luo et al., 2020）和金沙厂（Zhou J. X. et al., 2015），赋存于寒武系的麻栗坪（Luo K. et al., 2019）和那雍枝（金中国等，2016；Zhou et al., 2018c），赋存于泥盆–石炭系的毛坪（谈树成等，2019；Xiang et al., 2020）和黔西北成矿区众多典型矿床（Li et al., 2015；Zhou et al., 2018b），赋存于石炭系的会泽矿床（Bao et al., 2017）和二叠系的富乐矿床（崔银亮等，2018；Zhou et al., 2018a；任涛等，2019）。Bao等（2017）对昆阳群不同岩石及水岩反应的淋滤液进行了成矿元素含量及Pb同位素组成研究，认为这套基底地层中富含成矿物质的岩石为铅锌矿床的主要矿源岩。

川滇黔接壤铅锌矿集区矿化与峨眉山玄武岩在空间上密切共生，很有特色。尽管玄武岩与铅锌矿床无直接成因联系，但已有研究表明峨眉山玄武岩在铅锌成矿过程中可能主要起"遮挡层"作用，为大规模流体运移、成矿元素活化–迁移–沉淀成矿营造了有利环境，还可能"保护"其下层位中的铅锌矿床免遭后期地壳全面隆升剥蚀的影响（崔银亮等，2018；李珍立等，2019；Li Z. L. et al., 2018，2020；Zhou et al., 2018a；任涛等，2019）。

（2）碳酸盐岩在成矿过程中的作用。川滇黔接壤铅锌矿集区的一个明显特征是容矿岩石为不同时代（震旦纪—二叠纪）地层中的碳酸盐岩。前人对国内外许多地区的MVT铅锌矿床进行过深入研究（文献众多，略），在成矿背景、成矿条件、成矿机制、成矿规律和成矿预测等方面都取得丰富研究成果，但有关这类铅锌矿床容矿碳酸盐岩在成矿过程中的作用，以往的研究很少深入讨论。近年来，不同研究者对矿集区典型矿床容矿碳酸盐岩进行了系统研究（Zhou et al., 2018b；Luo K. et al., 2019，2020；Xiang et al., 2020），结合水–岩相互作用C-O同位素组成模拟计算结果，揭示MVT铅锌矿床成矿过程中碳酸盐岩发挥了关键作用，主要表现为：①成矿前，碳酸盐岩与成矿流体间的水–岩相互作用，形成了强烈的白云岩化，为成矿准备了岩性、物质和空间等必要条件；②成矿期，碳酸盐矿物溶解–重结晶的循环过程，对金属大量沉淀导致成矿环境（如pH）的改变起到了缓冲作用，因而形成大量雪花状（方解石–白云石斑点）典型MVT铅锌矿床的特征矿石；③成矿后，碳酸盐矿物充填、胶结矿化场所，利于矿石保存，同时成为重要的找矿标志矿物。

（3）成矿流体来源与演化。基于川滇黔铅锌矿集区典型矿床的流体包裹体及C-H-O-S同位素组成的研究（Zhou et al., 2014, 2018a, 2018c; Luo K. et al., 2019, 2020），对成矿流体性质、来源及演化形成了较为一致的认识：成矿流体具有中-低、中-低盐度特征，成矿物质主要由基底地层提供，成矿流体中的S主要来源于赋矿地层中硫酸盐热化学还原作用（thermochemical sulfate reduction, TSR）、CO_2主要来源于碳酸盐岩地层、H_2O为盆地热卤水，流体混合作用是成矿元素沉淀成矿的重要机制。对流体混合作用成矿机制，主要证据是在矿床流体包裹体中发现流体混合现象（Zhou et al., 2014; Luo K. et al., 2019），如多相流体包裹体并存、均一温度存在两个（或多个）峰值、盐度存在两个（或多个）区间。原位S同位素组成结果显示（金中国等，2016; Zhou et al., 2018c; Luo K. et al., 2019），单颗粒硫化物中心和边缘$\delta^{34}S_{CDT}$存在很大差异，被认为是高$\delta^{34}S_{CDT}$流体与低$\delta^{34}S_{CDT}$流体混合作用的产物；刘莹莹（2017）的成矿实验也初步证实：本区富铅锌矿床可能为含Pb、Zn卤水与富还原硫流体在有利的成矿空间（如断裂带、层间破碎带、溶洞等）相互的混合结果。

5.2.4 区域成矿模型

近年来，通过对川滇黔矿集区代表性矿床的深入剖析，大都建立了矿床成因模式，如金沙厂（Zhou J. X. et al., 2015）、天宝山（何承真等，2016）、那雍枝（金中国等，2017; Zhou et al., 2018c）、银厂沟（Tan et al., 2017）、富乐（Zhou et al., 2018a）、黔西北成矿区（Zhou et al., 2018b）、乌斯河和麻栗坪（罗开，2019）、茂租（Zhang H. J. et al., 2019b）等，这些模式都具有以下特征：①矿床或矿体受构造和岩性控制，区域构造为导矿构造、层间破碎带为容矿构造，控矿岩性主要为粗晶白云岩；②成矿物质主要来源于基底地层；③成矿流体中S和CO_2具有不同来源；④矿体顶部存在"遮挡层"。

黄智龙（2020年，未发表资料）对这些成矿模式进行高度概括，建立了区域成矿模型，简述为：印支期印支地块与华南陆块后碰撞伸展作用引发大规模流体运移，大规模运移的流体淋滤基底地层中的成矿元素形成成矿流体，成矿流体沿构造通道向上运移过程中遇"遮挡层"受阻折返沿碳酸盐岩地层层间破碎带等有利空间运移，与富还原硫流体发生混合作用成矿元素沉淀成矿。

6 原位分析技术在矿床研究中的应用

最近十年，随着仪器设备的发展，从更微观尺度更高精度地示踪成矿过程，从本质上揭示成矿机理成为矿床学研究的热点及重要趋势，各种微区原位分析技术迅速发展并被广泛应用，特别是激光剥蚀电感耦合等离子质谱（LA-ICP-MS）和二次离子质谱（SIMS）分析技术，我国在微区原位定年以及元素-同位素分析方面取得诸多重要进展，极大地促进了成矿理论的发展。

6.1 矿石矿物/副矿物微区原位U-Pb定年

成矿时代的原位研究主要集中在含U矿物上，一些适用于矿床学定年的含U或低U矿物定年方法被国内学者率先开发出来并广泛应用，代表性方法有：LA-ICP-MS锡石（Yuan S. D. et al., 2011；李惠民等，2013；Li C. Y. et al., 2016；Zhang R. Q. et al., 2017）、黑钨矿（Deng et al., 2019；Luo T. et al., 2019；Tang et al., 2020；Yang et al., 2020）、石榴子石（Deng X. D. et al., 2017；Yang et al., 2018）和氟碳铈矿（Yang Y. H. et al., 2014, 2019；Ling et al., 2016；涂家润等，2017）U-Pb定年。国内近年来也建立了磷钇矿（Liu et al., 2011；Li et al., 2013）、铌钽铁矿（Che et al., 2015）以及铀矿物（晶质铀矿、沥青铀矿和铀钛矿等）（邹东风等，2011；宗克清等，2015；肖志斌等，2020a，2020b；叶丽娟等，2019）的LA-ICP-MS U-Pb定年方法，尽管这些方法国外已较早开展工作。碳酸盐LA-ICP-MS U-Pb定年是国际上最近几年发展起来的新方法（Li et al., 2014），国内已有实验室初步建立了相关方法并开展了沉积碳酸盐定年的应用研究（Shen et al., 2019；沈安江等，2019；程婷等，2020）。鉴于碳酸盐在各种热液矿床中普遍存在，其在矿床学领域具有广泛的应用前景。

6.2 同位素组成微区原位分析

矿床学研究中常用的同位素示踪体系包括 H-O-C-S 稳定同位素、Sr-Nd-Pb-Hf-Os 放射性同位素及 He-Ar 稀有气体同位素体系等，这些体系和新的同位素体系微区原位分析方法被不断开发出来，并被应用于精细揭示成矿物质来源及成矿过程研究，代表性的分析方法有 SIMS 或 LA-ICP-MS 硫化物/硫酸盐 S 同位素分析（Zhu et al., 2016；Chen et al., 2017；Fu et al., 2017；Feng et al., 2018；Yuan H. L. et al., 2018），富 Sr（如磷灰石、长石、白钨矿、钙钛矿等）和富 Nd（如独居石、榍石、氟碳铈矿等）矿物 Sr、Nd 同位素分析（Liu et al., 2012；Xu et al., 2015；Huang et al., 2015, 2020；Xu and Jiang, 2017；Li C. et al., 2018；Yang Y. H. et al., 2019），硫化物 Pb 和 Os 同位素分析（Hu et al., 2015；Yuan H. L. et al., 2015, 2018；Feng et al., 2018；Zhu et al., 2019），石英或硅酸盐 Si 同位素分析（Liu et al., 2019b），碳酸盐 Ca 同位素分析（Zhang W. et al., 2019b）以及电气石 B 和 Li 同位素分析（Hou et al., 2010；Yang and Jiang, 2012；Lin et al., 2019）等。值得一提的是，多种同位素（如 Sr-Nd、S-Pb、Sr-Pb）、同位素与 U-Pb 定年以及同位素与微量元素同时原位分析方法被越来越多地开发出来（Huang et al., 2015, 2020；Yuan H. L. et al., 2018），在综合示踪成矿过程、揭示矿床成因方面具有独特优势。此外，在仪器设备方面，一个显著的趋势是纳米离子探针以及飞秒激光被越来越多地应用于同位素微区原位分析（Zhang J. C. et al., 2014；杨蔚等，2015；袁洪林等，2015；杨文武等，2017），纳米离子探针极大地提高了空间分辨率，飞秒激光能有效地减少或消除基体效应。

6.3 单矿物微量元素组成微区原位分析

矿物微量元素可以很好地记录成矿过程及反映物质来源，随着微区原位分析技术的发展，特别是 LA-ICP-MS 方法的应用，其受到越来越多的研究和应用（刘勇胜等，2013）。近十年来，代表性分析方法包括氧化物（如磁铁矿、锡石、白钨矿、黑钨矿和石英等）（张德贤等，2012；付宇等，2013；孟郁苗等，2016；蓝廷广等，2017；Zhang Q. et al., 2018；Ke et al., 2019；Song et al., 2019）、硫化物（Ding et al., 2011；袁继海等，2011；Zhu et al., 2016；范宏瑞等，2018）、硅酸盐（Yuan J. H. et al., 2011；陈春飞等，2014；柴发达等，2018；Cao et al., 2018）、碳酸盐矿物（Chen M. H. et al., 2011；Duan et al., 2017；Yang K. F. et al., 2019）和副矿物（如锆石、磷灰石、榍石、金红石、独居石）（He et al., 2016；She et al., 2016；Song et al., 2019）等的 LA-ICP-MS 或 SIMS 微量元素分析，特别是黄铁矿和磁铁矿微量元素，被广泛应用于矿床成因分类、成矿物质来源、精细成矿过程、物理化学条件以及勘探找矿研究（范宏瑞等，2018；Huang et al., 2019；Huang and Beaudoin, 2019）。值得一提的是，微量元素 LA-ICP-MS mapping 分析方法（Zhu et al., 2016；汪方跃等，2017），结合阴极发光（CL）或背散射电子（BSE）等显微结构研究，可为成矿过程提供更精细的信息，被越来越多的研究者所应用。

6.4 单个流体包裹体组成研究

LA-ICP-MS 分析是目前单个流体包裹体组成（主、微量元素）最主要的分析手段。该方法最初是在 20 世纪 70 年代中后期尝试使用 LA-ICP-MS 分析单个流体包裹体组成建立起来的（Tsui and Holland, 1979）。随着仪器设备和技术的发展，越来越多的实验室建立了该方法，特别是近年来国内一些实验室建立了相关方法并开展了应用研究（孙小虹等，2013；蓝廷广等，2017；Li C. Y. et al., 2018；Pan et al., 2019b）。我国学者使用国内实验室建立的方法对斑岩或夕卡岩矿床（Lan et al., 2018；Liu H. Q. et al., 2018；Chen P. W. et al., 2019）、W-Sn 矿床（Pan et al., 2019b；Yang et al., 2019a）和造山型金矿床（Zhou et al., 2019）等开展了单个流体包裹体组成研究，对这些矿床成因有了新的认识。

7 实验地球化学在矿床研究中的应用

近十年来，我国实验地球化学研究在矿床学中的应用，根据实验技术或方法，主要有包括4个方面的进展：①利用活塞圆筒开展了高温高压下金属元素（Cu、Mo、Au、Re、W、Nb 和 Ta 等）及挥发分元素（S、F 和 Cl 等）在矿物、熔体和热液之间的分配行为与成矿实验，对这些元素在壳幔相互作用过程中的富集机制提供了有效限定（Xiong et al., 2011; Liu et al., 2014, 2015; Ni et al., 2018; Zhang et al., 2018; Feng and Li, 2019; Li et al., 2019; Sun et al., 2020）；②通过内加热式高压釜在精确限定氧逸度的基础上，对铁氧化物-磷灰石矿床形成过程中富铁熔体和硅酸盐熔体不混溶作用、基性岩岩浆源区以及铁钛氧化物温度-氧逸度计的使用条件等方面进行了研究（Hou et al., 2017, 2018, 2021）；③水热高压釜中金属元素在水溶液中溶解度实验法，如 Cu 在气相中的溶解迁移实验（Shang et al., 2007），W、Mo 等在水盐溶液体系及含 F 盐水溶液中溶解形式及溶解度的实验（Wang X. S. et al., 2019, 2021; Shang et al., 2020）；④原位激光拉曼观测技术，一方面以毛细石英管技术为代表，研究温度范围为 $-196 \sim 500$ ℃，压力 $100 \sim 100\,000$ kPa。该方法主要应用于水盐体系中，成矿过程氧逸度条件模拟（Shang et al., 2009）、硫酸盐热化学还原反应（TSR）观测（Yuan et al., 2013）、硫酸盐液相不混溶（Wang X. L. et al., 2013, 2016; Wan et al., 2017）、有机烃类氧化还原反应观测（Qiu et al., 2020）和成矿元素 W 在水盐体系中溶解形式研究（Wang X. L. et al., 2020）等；另一方面以热液金刚石压腔技术为代表，该技术具有极宽的研究温度和压力范围（$-196 \sim 1\,000$ ℃ 和 $100 \sim 4\times10^6$ kPa），主要应用于含碳酸盐碱性熔体中金属元素 W 熔解形式研究（Li J. K. et al., 2018）、水溶液中硫酸盐溶解控制条件及其对稀土元素溶解度的影响等研究（Cui et al., 2020b）。

8 结　　语

综上所述，我国的矿床地球化学研究立足中国大陆成矿特色，近年来取得了较多重要进展和创新成果。但是，矿床地球化学是一门理论性和实践性很强的交叉学科，且研究对象涉及极其复杂的地球各圈层起源和相互作用，我国的相关研究要总体进入国际领跑水平依然任重道远。未来的研究需要重点关注（但不限于）以下几个方面的工作。

（1）大型-超大型矿床和大型矿集区的形成往往与地球的核-幔-壳分异及壳幔相互作用有关。查明核-幔-壳分异和层圈相互作用对金属元素初始富集、有效迁移和巨量聚集形成不同类型矿床的机理仍是矿床地球化学需要解决的重大科学问题。我国矿产资源类型的极不均匀分布很可能与此密切相关。

（2）精细化、定量化描述是矿床地球化学学科发展的必然趋势。在详细的矿床地质特征和控矿因素研究基础上，依据原位微量元素-同位素组成，综合应用实验模拟和数学模拟的方法精细刻画成矿作用的地球化学和热力学过程，将为深入揭示岩浆起源-演化和流体-岩石相互作用机理、元素迁移及沉淀机制，全面理解成矿作用本质开辟新的重要途径。

（3）战略性新兴产业的蓬勃发展对新兴关键矿产（稀有、稀散、稀土、稀贵金属）需求急剧增大。目前对新兴关键矿产资源的元素行为、赋存状态、形成与分布规律、超常富集机理等基本规律的认识还很不够，制约了成矿远景区的预测和勘查突破。开展关键金属成矿背景、富集机制、控矿因素组合及勘查评价方法体系的研究是突破找矿瓶颈、保障资源安全的重要途径。

（4）深部找矿勘查是国际矿床学和勘查学界的重大前沿领域。揭示岩石圈结构和组成对矿集区分布的控制，建立隐伏矿床的地球化学异常模式和深穿透地球化学与矿物地球化学探测技术，对于突破探矿技术瓶颈、提升隐伏矿的探测能力将起到至关重要的推动作用。

（5）矿床类型和特征的多样性、区域差异性和空间不均一性要求矿床地球化学研究必须具有全球视野。我国以往的研究较少针对国外矿床和重点成矿区带。要实现成矿理论和模式的系统创新，产生具有国际影响力的原创性研究成果，系统的国际对比及全方位的国际合作势在必行。

（6）外太空蕴藏的特殊矿产资源是人类将来可能开发利用的重要资源储库。近地小行星是近期太空探索与资源开发利用最有价值的目标之一，相关资源的研究、探测和评价将是地球科学的新兴前沿研究方向。揭示行星形成演化与元素聚集机理，查明地外矿产资源的分布、储量并建立相应评价标准，是我国矿床地球化学研究亟待开拓的新领域。

致谢：本文提出的展望是与国内矿床学界诸多同事有益讨论的结果，审稿人提出了宝贵的建议，在此一并致谢。

参 考 文 献

柴发达,李全忠,闫峻,杨青亮,刘晓强,许士钊. 2018. 硅酸盐单矿物主、微量元素的 LA-ICPMS 分析及影响因素. 合肥工业大学学报(自然科学版),41(8):1093-1099,1117

陈春飞,刘先国,胡兆初,宗克清,刘勇胜. 2014. LA-ICP-MS 微区原位准确分析含水硅酸盐矿物主量和微量元素. 地球科学(中国地质大学学报),39(5):525-536

陈华勇,吴超. 2020. 俯冲带斑岩铜矿系统成矿机理与主要挑战. 中国科学:地球科学,50(7):865-886

陈衍景. 2013. 大陆碰撞成矿理论的创建及应用. 岩石学报,29(1):1-17

程婷,Zhao J X,Feng Y X,潘文庆,刘敦一. 2020. 低铀碳酸盐矿物的 LA-MC-ICPMS 微区原位 U-Pb 定年方法. 科学通报,65(2):150-154

崔银亮,周家喜,黄智龙,罗开,念红良,叶霖,李珍立. 2018. 云南富乐铅锌矿床地质、地球化学及成因. 岩石学报,34(1):194-206

范宏瑞,李兴辉,左亚彬,陈雷,刘尚,胡芳芳,冯凯. 2018. LA-(MC)-ICPMS 和(Nano)SIMS 硫化物微量元素和硫同位素原位分析与矿床形成的精细过程. 岩石学报,34(12):3479-3496

付宇,孙晓明,熊德信. 2013. 激光剥蚀-电感耦合等离子体质谱法对白钨矿中稀土元素的原位测定. 岩矿测试,32(6):875-882

高成,康清清,江宏君,郑惠,李鹏,张熊猫,宋雷,董强强,叶兴超,胡小佳. 2017. 秦岭造山带发现新型铀多金属矿:华阳川与伟晶岩脉和碳酸岩脉有关的超大型铀-铌-铅-稀土矿床. 地球化学,46(5):446-455

高伟. 2018. 桂西北卡林型金矿成矿年代学和动力学. 北京:中国科学院大学博士学位论文

何承真,肖朝益,温汉捷,周汀,朱传威,樊海峰. 2016. 四川天宝山铅锌矿床的锌-硫同位素组成及成矿物质来源. 岩石学报,32(11):3394-3406

侯增谦,王涛. 2018. 同位素填图与深部物质探测(Ⅱ):揭示地壳三维架构与区域成矿规律. 地学前缘,25(6):20-41

侯增谦,杨志明,王瑞,郑远川. 2020. 再论中国大陆斑岩 Cu-Mo-Au 矿床成矿作用. 地学前缘,27(2):20-44

胡瑞忠,彭建堂,马东升,苏文超,施春华,毕献武,涂光炽. 2007. 扬子地块西南缘大面积低温成矿时代. 矿床地质,26(6):583-596

胡瑞忠,付山岭,肖加飞. 2016. 华南大规模低温成矿的主要科学问题. 岩石学报,32(11):3239-3251

胡瑞忠,陈伟,毕献武,付山岭,尹润生,肖加飞. 2020. 扬子克拉通前寒武纪基底对中生代大面积低温成矿的制约. 地学前缘,27(2):137-150

胡宇思,叶霖,黄智龙,李珍立,韦晨,Danyushevskiy L. 2019. 滇东北麻栗坪铅锌矿床微量元素分布与赋存状态:LA-ICPMS 研究. 岩石学报,35(11):3477-3492

黄勇. 2019. 黔西南地区卡林型金矿成矿时代及成矿物质来源研究. 北京:中国科学院大学博士学位论文

黄智龙,陈进,韩润生,李文博,刘丛强,张振亮,马德云,高德荣,杨海林. 2004. 云南会泽超大型铅锌矿床地球化学及成因:兼论峨眉山玄武岩与铅锌成矿的关系. 北京:地质出版社

黄智龙,胡瑞忠,苏文超,温汉捷,刘燊,符亚洲. 2011. 西南大面积低温成矿域:研究意义、历史及新进展. 矿物学报,31(3):309-314

蒋少涌,赵葵东,姜海,苏慧敏,熊索菲,熊伊曲,徐耀明,章伟,朱律运. 2020. 中国钨锡矿床时空分布规律、地质特征与成矿机制研究进展. 科学通报,65(33):3730-3745

金中国,周家喜,黄智龙,罗开,高建国,彭松,王兵,陈兴龙. 2016. 贵州普定纳雍枝铅锌矿床成因:S 和原位 Pb 同位素证据. 岩石学报,32(11):3441-3455

金中国,周家喜,郑明泓,彭松,黄智龙,刘玲. 2017. 贵州普定五指山地区铅锌矿床成矿模式. 矿床地质,36(5):1169-1184

蓝廷广,胡瑞忠,范宏瑞,毕献武,唐燕文,周丽,毛伟,陈应华. 2017. 流体包裹体及石英 LA-ICP-MS 分析方法的建立及其在矿床学中的应用. 岩石学报,33(10):3239-3262

李华芹,王登红,陈富文,梅玉萍,蔡红. 2008. 湖南雪峰山地区铲子坪和大坪金矿成矿作用年代学研究. 地质学报,82(7):900-905

李惠民,周红英,郝爽,崔玉荣,张永清,李国占. 2013. 锡石 U-Pb 同位素定年中的年龄校正策略:兼论矿石矿物 U-Pb 同位素定年中的年龄校正问题. 矿物学报,(S1):595-596

李建威,赵新福,邓晓东,谭俊,胡浩,张东阳,李占轲,李欢,荣辉,杨梅珍,曹康,靳晓野,隋吉祥,姐波,昌佳,吴亚飞,文广,赵少瑞. 2019. 新中国成立以来中国矿床学研究若干重要进展. 中国科学:地球科学,49(11):1720-1771

李靖辉,陈化凯,张宏伟,张云海,张同林,温国栋,张盼盼. 2017. 豫西太平镇轻稀土矿床矿化特征及矿床成因. 中国地质,44(2):288-300

李伟强,赵书高,王小敏,李石磊,王国光,杨涛,金章东. 2020. 斑岩铜矿热液流体的 K-Mg 同位素示踪. 中国科学:地球科学,50(2):245-257

李珍立,叶霖,胡宇思,韦晨,黄智龙,念红良,蔡金君,Danyushevskiy L. 2019. 云南富乐铅锌矿床黄铁矿微量(稀散)元素组成及成因信息:LA-

ICPMS 研究. 岩石学报,35(11):3370-3384
刘英俊,曹励明,李兆麟,王鹤年,储同庆. 1984. 元素地球化学. 北京:科学出版社
刘莹莹. 2017. 与川-滇-黔"特富"铅锌矿床相关的实验地球化学研究. 贵阳:中国科学院地球化学研究所博士后出站报告
刘勇胜,胡兆初,李明,高山. 2013. LA-ICP-MS 在地质样品元素分析中的应用. 科学通报,58(36):3753-3769
柳贺昌,林文达. 1999. 滇东北铅锌银矿床规律研究. 昆明:云南大学出版社
罗开. 2019. 川滇黔接壤区上震旦-下寒武统地层中铅锌矿床成矿作用:以乌坡河和麻栗坪矿床为例. 北京:中国科学院大学博士学位论文
毛景文,周振华,丰成友,王义天,张长青,彭惠娟,于淼. 2012. 初论中国三叠纪大规模成矿作用及其动力学背景. 中国地质,39(6):1437-1471
毛景文,罗茂澄,谢桂青,刘军,吴胜华. 2014. 斑岩铜矿床的基本特征和研究勘查新进展. 地质学报,88(12):2153-2175
孟郁苗,黄小文,高剑峰,戴智慧,漆亮. 2016. 无内标-多外标校正激光剥蚀等离子体质谱法测定磁铁矿微量元素组成. 岩矿测试,35(6):585-594
秦克章,翟明国,李光明,赵俊兴,曾庆栋,高俊,肖文交,李继亮,孙枢. 2017. 中国陆壳演化、多块体拼合造山与特色成矿的关系. 岩石学报,33(2):305-325
任涛,周家喜,王蝶,杨光树,吕昶良. 2019. 滇东北富乐铅锌矿床微量元素和 S-Pb 同位素地球化学研究. 岩石学报,35(11):3493-3505
沈安江,胡安平,程婷,梁峰,潘文庆,俸月星,赵建新. 2019. 激光原位 U-Pb 同位素定年技术及其在碳酸盐岩成岩-孔隙演化中的应用. 石油勘探与开发,46(6):1062-1074
沈苏,金明霞,陆元法. 1988. 西昌-滇中地区主要矿产成矿规律及找矿方向. 重庆:重庆出版社
宋谢炎,邓宇峰,颉炜,陈列锰,于宋月,梁庆林. 2018. 新疆黄山-镜儿泉铜镍成矿带岩浆作用与区域走滑构造的关系. 地球科学与环境学报,40(5):505-519
孙海瑞,周家喜,黄智龙,樊海峰,叶霖,罗开,高建国. 2016. 四川会理天宝山矿床深部新发现铜矿与铅锌矿的成因关系探讨. 岩石学报,32(11):3407-3417
孙小虹,胡明月,刘成林,焦鹏程,马黎春,王鑫,詹秀春. 2013. 激光剥蚀 ICP-MS 法测定盐类矿物单个流体包裹体的成分. 分析化学,41(2):235-241
谈树成,周家喜,罗开,向震中,何小虎,张亚辉. 2019. 云南毛坪大型铅锌矿床成矿物质来源:原位 S 和 Pb 同位素制约. 岩石学报,35(11):3461-3476
涂家润,肖志斌,曲凯,李国占,周红英,李惠民,耿建珍,崔玉荣,郝爽,刘文刚. 2017. 氟碳铈矿 U-Pb 定年技术研究. 地球学报,38(6):945-951
汪方跃,葛粲,宁思远,聂利青,钟国雄,White N C. 2017. 一个新的矿物面扫描分析方法开发和地质学应用. 岩石学报,33(11):3422-3436
温汉捷,周正兵,朱传威,罗重光,王大钊,杜胜江,李晓峰,陈懋弘,李红谊. 2019. 稀散金属超常富集的主要科学问题. 岩石学报,35(11):3271-3291
吴昌雄,方鑫,鄂华. 2015. 武当地区与碱性岩有关的铌、稀土矿特征及找矿方向. 资源环境与工程,29(3):270-274,298
吴越. 2013. 川滇黔地区 MVT 铅锌矿床大规模成矿作用的时代与机制. 北京:中国地质大学(北京)博士学位论文
吴越,孔志岗,陈懋弘,张长青,曹亮,唐友军,袁鑫,张沛. 2019. 扬子板块周缘 MVT 型铅锌矿床闪锌矿微量元素组成特征与指示意义:LA-ICPMS 研究. 岩石学报,35(11):3443-3460
肖志斌,张然,叶丽娟,涂家润,耿建珍,郭虎,许雅雯,周红英,李惠民. 2020a. 沥青铀矿(GBW04420)的微区原位 U-Pb 定年分析. 地质调查与研究,43(1):1-4
肖志斌,耿建珍,涂家润,张然,叶丽娟,毕君辉,周红英. 2020b. 砂岩型铀矿微区原位 U-Pb 同位素定年技术方法研究. 岩矿测试,39(2):262-273
徐克勤,胡受奚,孙明志,叶俊. 1982. 华南两个成因系列花岗岩及其成矿特征. 矿床地质,1(2):1-14
阳杰华,刘亮,刘佳. 2017. 华南中生代大花岗岩省成岩成矿作用研究进展与展望. 矿物学报,37(6):791-800
杨蔚,胡森,张建超,郝佳龙,林杨挺. 2015. 纳米离子探针分析技术及其在地球科学中的应用. 中国科学:地球科学,45(9):1335-1346
杨文武,史光宇,商琦,张文,胡兆初. 2017. 飞秒激光剥蚀电感耦合等离子体质谱在地球科学中的应用进展. 光谱学与光谱分析,37(7):2192-2198
杨志明,侯增谦,杨竹森,曲焕春,李振清,刘云飞. 2012. 短波红外光谱技术在浅剥蚀斑岩铜矿区勘查中的应用:以西藏念村矿区为例. 矿床地质,31(4):699-717
叶丽娟,肖志斌,涂家润,耿建珍,张健,许雅雯,郭虎,崔玉荣,张永清. 2019. LA-ICPMS 与 EPMA 结合测定铀矿物微区原位 U-Pb 年龄. 地球学报,40(3):479-482
叶霖,李珍立,胡宇思,黄智龙,周家喜,樊海峰,Danyushevskiy L. 2016. 四川天宝山铅锌矿床硫化物微量元素组成:LA-ICPMS 研究. 岩石学报,32(11):3377-3393
袁洪林,殷琮,刘旭,陈开运,包志安,宗春蕾,戴梦宁,赖绍聪,王蓉,蒋少涌. 2015. 飞秒激光剥蚀多接收等离子体质谱分析硫化物中 Pb 同位素组成研究. 中国科学:地球科学,45(9):1285-1293
袁继海,詹秀春,樊兴涛,胡明月. 2011. 硫化物矿物中痕量元素的激光剥蚀-电感耦合等离子体质谱微区分析进展. 岩矿测试,30(2):121-130
翟伟,孙晓明,邬云山,孙岩岩,华仁民,叶先仁. 2012. 粤北瑶岭-梅子窝钨矿 He-Ar 同位素地球化学:对华南燕山期壳幔作用过程与成矿的制约. 科学通报,57(13):1137-1146
张长青. 2008. 中国川滇黔交界地区密西西比型(MVT)铅锌矿床成矿模型. 博士学位论文. 北京:中国地质科学院
张德贤,戴塔根,胡毅. 2012. 磁铁矿中微量元素的激光剥蚀-电感耦合等离子体质谱分析方法探讨. 岩矿测试,31(1):120-126

赵振华,严爽. 2019. 矿物-成矿与找矿. 岩石学报,35(1):31-68
钟大赉. 1998. 滇川西部古特提斯造山带. 北京:科学出版社
朱经经,钟宏,谢桂青,赵成海,胥磊落,陆刚. 2016. 右江盆地酸性脉岩继承锆石成因及地质意义. 岩石学报,32(11):3269-3280
宗克清,陈金勇,胡兆初,刘勇胜,李明,范洪海,孟艳宁. 2015. 铀矿 fs-LA-ICP-MS 原位微区 U-Pb 定年. 中国科学:地球科学,45(9):1304-1315
邹东风,李方林,张爽,黄彬,宗克清. 2011. 粤北下庄 335 矿床成矿时代的厘定:来自 LA-ICP-MS 沥青铀矿 U-Pb 年龄的制约. 矿床地质,30(5):912-922

Bai T, Chen W, Jiang S Y. 2019. Evolution of the carbonatite Mo-HREE deposits in the lesser Qinling orogen: insights from in situ geochemical investigation of calcite and sulfate. Ore Geology Reviews, 113:103069

Bai Z J, Zhong H, Naldrett A J, Zhu W G, Xu G W. 2012. Whole-rock and mineral composition constraints on the genesis of the giant Hongge Fe-Ti-V Oxide deposit in the Emeishan large igneous province, Southwest China. Economic Geology, 107(3):507-524

Bai Z J, Zhong H, Li C, Zhu W G, He D F, Qi L. 2014. Contrasting parental magma compositions for the Hongge and Panzhihua magmatic Fe-Ti-V oxide deposits, Emeishan large igneous province, SW China. Economic Geology, 109(6):1763-1785

Bai Z J, Zhong H, Li C S, Zhu W G, Hu W J. 2016. Association of cumulus apatite with compositionally unusual olivine and plagioclase in the Taihe Fe-Ti oxide ore-bearing layered mafic-ultramafic intrusion: petrogenetic significance and implications for ore genesis. American Mineralogist, 101(10):2168-2175

Bai Z J, Zhong H, Hu R Z, Zhu W G, Hu W J. 2019. Composition of the chilled marginal rocks of the Panzhihua layered intrusion, Emeishan large igneous province, SW China: implications for parental magma compositions, sulfide saturation history and Fe-Ti oxide mineralization. Journal of Petrology, 60(3):619-648

Bai Z J, Zhong H, Hu R Z, Zhu W G. 2020. Early sulfide saturation in arc volcanic rocks of Southeast China: implications for the formation of co-magmatic porphyry-epithermal Cu-Au deposits. Geochimica et Cosmochimica Acta, 280:66-84

Bao Z W, Li Q, Wang C Y. 2017. Metal source of giant Huize Zn-Pb deposit in SW China: new constraints from in situ Pb isotopic compositions of galena. Ore Geology Reviews, 91:824-836

Begg G C, Hronsky J A M, Arndt N T, Griffin W L, O'Reilly S Y, Hayward N. 2010. Lithospheric, cratonic, and geodynamic setting of Ni-Cu-PGE sulfide deposits. Economic Geology, 105:1057-1070

Cao J, Wang C Y, Xing C M, Xu Y G. 2014. Origin of the early Permian Wajilitag igneous complex and associated Fe-Ti oxide mineralization in the Tarim large igneous province, NW China. Journal of Asian Earth Sciences, 84:51-68

Cao J, Wang X, Tao J H. 2019. Petrogenesis of the Piqiang mafic-ultramafic layered intrusion and associated Fe-Ti-V oxide deposit in Tarim large igneous province, NW China. International Geology Review, 61(18):2249-2275

Cao M J, Li G M, Qin K Z, Seitmuratova E Y, Liu Y S. 2012. Major and trace element characteristics of apatites in granitoids from central Kazakhstan: implications for petrogenesis and mineralization. Resource Geology, 62(1):63-83

Cao M J, Hollings P, Cooke D R, Evans N J, McInnes B I, Qin K Z, Li G M, Sweet G, Baker M. 2018. Physicochemical processes in the magma chamber under the Black Mountain porphyry Cu-Au deposit, Philippines: insights from mineral chemistry and implications for mineralization. Economic Geology, 113(1):63-82

Cao Y H, Wang C Y, Huang F, Zhang Z F. 2019. Iron isotope systematics of the Panzhihua mafic layered intrusion associated with giant Fe-Ti oxide deposit in the Emeishan large igneous province, SW China. Journal of Geophysical Research: Solid Earth, 124(1):358-375

Chang Z S, Shu Q H, Meinert L D. 2019. Skarn deposits of China. In: Chang Z S, Goldfarb R J (eds). Mineral Deposits of China. SEG Special Publication, 22:189-234

Che X D, Wu F Y, Wang R C, Gerdes A, Ji W Q, Zhao Z H, Yang J H, Zhu Z Y. 2015. In situ U-Pb isotopic dating of columbite-tantalite by LA-ICP-MS. Ore Geology Reviews, 65:979-989

Chen B, Gu H O, Chen Y J, Sun K K, Chen W. 2018. Lithium isotope behaviour during partial melting of metapelites from the Jiangnan orogen, South China: implications for the origin of REE tetrad effect of F-rich granite and associated rare-metal mineralization. Chemical Geology, 483:372-384

Chen L, Liu Y S, Hu Z C, Gao S, Zong K Q, Chen H H. 2011. Accurate determinations of fifty-four major and trace elements in carbonate by LA-ICP-MS using normalization strategy of bulk components as 100%. Chemical Geology, 284(3-4):283-295

Chen L, Chen K Y, Bao Z A, Liang P, Sun T T, Yuan H L. 2017. Preparation of standards for in situ sulfur isotope measurement in sulfides using femtosecond laser ablation MC-ICP-MS. Journal of Analytical Atomic Spectrometry, 32:107-116

Chen L M, Song X Y, Keays R R, Qi L, Tian Y L, Wang Y S, Ba D H, Deng Y F, Xiao J F. 2013. Segregation and fractionation of magmatic Ni-Cu-PGE sulfides in the western Jinchuan intrusion, northwestern China: insights from platinum group element geochemistry. Economic Geology, 108(8):1793-1811

Chen L M, Song X Y, Zhu X K, Zhang X Q, Yu S Y, Yi J N. 2014. Iron isotope fractionation during crystallization and sub-solidus re-equilibration: constraints from the Baima mafic layered intrusion, SW China. Chemical Geology, 380:97-109

Chen L M, Song X Y, Danyushevsky L V, Wang Y S, Tian Y L, Xiao J F. 2015. A laser ablation ICP-MS study of platinum-group and chalcophile elements in base metal sulfide minerals of the Jinchuan Ni-Cu sulfide deposit, NW China. Ore Geology Reviews, 65:955-967

Chen M H, Mao J W, Bierlein F P, Norman T, Uttley P J. 2011. Structural features and metallogenesis of the Carlin-type Jinfeng (Lannigou) gold deposit, Guizhou Province, China. Ore Geology Reviews, 43(1):217-234

Chen M H, Mao J W, Li C, Zhang Z Q, Dang Y. 2015. Re-Os isochron ages for arsenopyrite from Carlin-like gold deposits in the Yunnan-Guizhou-

Guangxi "golden triangle", southwestern China. Ore Geology Reviews,64:316-327

Chen M H,Bagas L,Liao X,Zhang Z Q,Li Q L. 2019. Hydrothermal apatite SIMS Th-Pb dating:constraints on the timing of low-temperature hydrothermal Au deposits in Nibao,SW China. Lithos,324-325:418-428

Chen P W,Zeng Q D,Zhou T C,Wang Y B,Yu B,Chen J Q. 2019. Evolution of fluids in the Dasuji porphyry Mo deposit on the northern margin of the North China Craton:constraints from Microthermometric and LA-ICP-MS analyses of fluid inclusions. Ore Geology Reviews,104:26-45

Chen W,Lu J,Jiang S Y,Ying Y C,Liu Y S. 2018. Radiogenic Pb reservoir contributes to the rare earth element (REE) enrichment in South Qinling carbonatites. Chemical Geology,494:80-95

Chen W T,Zhou M F,Zhao T P. 2013. Differentiation of nelsonitic magmas in the formation of the ～1.74 Ga Damiao Fe-Ti-P ore deposit,North China. Contributions to Mineralogy and Petrology,165(6):1341-1362

Chen X C,Hu R Z,Bi X W,Zhong H,Lan J B,Zhao C H,Zhu J J. 2015. Petrogenesis of metaluminous A-type granitoids in the Tengchong-Lianghe tin belt of southwestern China:evidences from zircon U-Pb ages and Hf-O isotopes,and whole-rock Sr-Nd isotopes. Lithos,212-215:93-110

Chen Y J,Pirajno F,Wu G,Qi J P,Xiong X L. 2012. Epithermal deposits in North Xinjiang,NW China. International Journal of Earth Sciences,101(4):889-917

Chiaradia M. 2014. Copper enrichment in arc magmas controlled by overriding plate thickness. Nature Geoscience,7(1):43-46

Çimen O,Kuebler C,Monaco B,Simonetti S S,Corcoran L,Chen W,Simonetti A. 2018. Boron,carbon,oxygen and radiogenic isotope investigation of carbonatite from the Miaoya complex, central China:evidences for late-stage REE hydrothermal event and mantle source heterogeneity. Lithos,322:225-237

Cook N J,Ciobanu C L,Pring A,Skinner W,Shimizu M,Danyushevsky L,Saini-Eidukat B,Melcher F. 2009. Trace and minor elements in sphalerite:a LA-ICPMS study. Geochimica et Cosmochimica Acta,73(16):4761-4791

Cui H,Zhong R C,Wang X L,Li Z M,Ling Y F,Yu C,Chen H. 2020a. Reassessment of the zircon Raman spectroscopic pressure sensor and application to pressure determination of fused silica capillary capsule. Ore Geology Reviews,122:103540

Cui H,Zhong R C,Xie Y L,Yuan X Y,Liu W H,Brugger J,Yu C. 2020b. Forming sulfate-and REE-rich fluids in the presence of quartz. Geology,48(2):145-148

Deng J,Wang Q F,Li G J,Hou Z Q,Jiang C Z,Danyushevsky L. 2015. Geology and genesis of the giant Beiya porphyry-skarn gold deposit,northwestern Yangtze Block,China. Ore Geology Reviews,70:457-485

Deng X D,Li J W,Luo T,Wang H Q. 2017. Dating magmatic and hydrothermal processes using andradite-rich garnet U-Pb geochronometry. Contributions to Mineralogy and Petrology,172:71

Deng X D,Luo T,Li J W,Hu Z C. 2019. Direct dating of hydrothermal tungsten mineralization using in situ wolframite U-Pb chronology by laser ablation ICP-MS. Chemical Geology,515:94-104

Deng Y F,Song X Y,Hollings P,Chen L M,Zhou T F,Yuan F,Xie W,Zhang D Y,Zhao B B. 2017. Lithological and geochemical constraints on the magma conduit systems of the Huangshan Ni-Cu sulfide deposit,NW China. Mineralium Deposita,52(6):845-862

Ding L H,Yang G,Xia F,Lenehan C E,Qian G J,McFadden A,Brugger J,Zhang X H,Chen G R,Pring A. 2011. A LA-ICP-MS sulphide calibration standard based on a chalcogenide glass. Mineralogical Magazine,75(2):279-287

Ding X,Ripley E M,Wang W Z,Li C H,Huang F. 2019. Iron isotope fractionation during sulfide liquid segregation and crystallization at the Lengshuiqing Ni-Cu magmatic sulfide deposit,SW China. Geochimica et Cosmochimica Acta,261:327-341

Du J G,Audétat A. 2020. Early sulfide saturation is not detrimental to porphyry Cu-Au formation. Geology,48(5):519-524

Duan X X,Zeng Q D,Wang Y B,Zhou L L,Chen B. 2017. Genesis of the Pb-Zn deposits of the Qingchengzi ore field, eastern Liaoning, China:constraints from carbonate LA-ICPMS trace element analysis and C-O、S-Pb isotopes. Ore Geology Reviews,89:752-771

Fan H R,Hu F F,Yang K F,Pirajno F,Liu X,Wang K Y. 2014. Integrated U-Pb and Sm-Nd geochronology for a REE-rich carbonatite dyke at the giant Bayan Obo REE deposit,northern China. Ore Geology Reviews,63:510-519

Fan H R,Yang K F,Hu F F,Liu S,Wang K Y. 2016. The giant Bayan Obo REE-Nb-Fe deposit,China:controversy and ore genesis. Geoscience Frontiers,7(3):335-344

Feng L,Li Y. 2019. Comparative partitioning of Re and Mo between sulfide phases and silicate melt and implications for the behavior of Re during magmatic processes. Earth and Planetary Science Letters,517:14-25

Feng Y T,Zhang W,Hu Z C,Liu Y S,Chen K,Fu J L,Xie J Y,Shi Q H. 2018. Development of sulfide reference materials for in situ platinum group elements and S-Pb isotope analyses by LA-(MC)-ICP-MS. Journal of Analytical Atomic Spectrometry,33(12):2172-2183

Fu J L,Hu Z C,Li J W,Yang L,Zhang W,Liu Y S,Li Q L,Zong K Q,Hu S H. 2017. Accurate determination of sulfur isotopes (δ^{33}S and δ^{34}S) in sulfides and elemental sulfur by femtosecond laser ablation MC-ICP-MS with non-matrix matched calibration. Journal of Analytical Atomic Spectrometry,32(12):2341-2351

Ganino C,Harris C,Arndt N T,Prevec S A,Howarth G H. 2013. Assimilation of carbonate country rock by the parent magma of the Panzhihua Fe-Ti-V deposit (SW China):evidence from stable isotopes. Geoscience Frontiers,4(5):547-554

George L,Cook N J,Ciobanu C L,Wade B P. 2015. Trace and minor elements in galena:a reconnaissance LA-ICP-MS study. American Mineralogist,100(2):548-569

Guo D X,Liu Y. 2019. Occurrence and geochemistry of bastnäsite in carbonatite-related REE deposits,Mianning-Dechang REE belt,Sichuan Province,SW China. Ore Geology Reviews,107:266-282

He H L, Yu S Y, Song X Y, Du Z S, Dai Z H, Zhou T, Xie W. 2016. Origin of nelsonite and Fe-Ti oxides ore of the Damiao anorthosite complex, NE China: evidence from trace element geochemistry of apatite, plagioclase, magnetite and ilmenite. Ore Geology Reviews, 79:367–381

He Z W, Zhang X C, Deng X D, Hu H, Li Y, Yu H M, Archer C, Li J W, Huang F. 2020. The behavior of Fe and S isotopes in porphyry copper systems: constraints from the Tongshankou Cu-Mo deposit, eastern China. Geochimica et Cosmochimica Acta, 270:61–83

Hong W, Fox N, Cooke D R, Zhang L J, Fayek M. 2020. B- and O-isotopic compositions of tourmaline constrain late-stage magmatic volatile exsolution in Tasmanian tin-related granite systems. Mineralium Deposita, 55:63–78

Hou K J, Li Y H, Xiao Y K, Liu F, Tian Y R. 2010. In situ boron isotope measurements of natural geological materials by LA-MC-ICP-MS. Chinese Science Bulletin, 55(29):3305–3311

Hou L, Peng H J, Ding J, Zhang J R, Zhu S B, Wu S Y, Wu Y, Ouyang H G. 2016. Textures and in situ chemical and isotopic analyses of pyrite, Huijiabao trend, Youjiang Basin, China: implications for paragenesis and source of sulfur. Economic Geology, 111(2):331–353

Hou T, Zhang Z C, Ye X R, Encarnacion J, Reichow M K. 2011. Noble gas isotopic systematics of Fe-Ti-V oxide ore-related mafic-ultramafic layered intrusions in the Panxi area, China: the role of recycled oceanic crust in their petrogenesis. Geochimica et Cosmochimica Acta, 75(22):6727–6741

Hou T, Zhang Z C, Encarnacion J, Santosh M, Sun Y L. 2013. The role of recycled oceanic crust in magmatism and metallogeny: Os-Sr-Nd isotopes, U-Pb geochronology and geochemistry of picritic dykes in the Panzhihua giant Fe-Ti oxide deposit, central Emeishan large igneous province, SW China. Contributions to Mineralogy and Petrology, 165(4):805–822

Hou T, Charlier B, Namur O, Schütte P, Schwarz-Schampera U, Zhang Z, Holtz F. 2017. Experimental study of liquid immiscibility in the Kiruna-type Vergenoeg iron-fluorine deposit, South Africa. Geochimica et Cosmochimica Acta, 203:303–322

Hou T, Charlier B, Holtz F, Veksler I, Zhang Z, Thomas R, Namur O. 2018. Immiscible hydrous Fe-Ca-P melt and the origin of iron oxide-apatite ore deposits. Nature Communications, 9(1):1415

Hou T, Botcharnikov R, Moulas E, Just T, Berndt J, Koepke J, Zhang Z, Wang M, Yang Z, Holtz F. 2021. Kinetics of Fe-Ti oxide re-equilibration in magmatic systems: implications for thermo-oxybarometry. Journal of Petrology, 61(11-12), doi:10.1093/petrdogy/egaa116

Hou Z Q, Zaw K, Pan G T, Mo X X, Xu Q, Hu Y Z, Li X Z. 2007. Sanjiang Tethyan metallogenesis in SW China: tectonic setting, metallogenic epochs and deposit types. Ore Geology Reviews, 31(1-4):48–87

Hou Z Q, Yang Z M, Lu Y J, Kemp A, Zheng Y C, Li Q Y, Tang J X, Yang Z S, Duan L F. 2015a. A genetic linkage between subduction- and collision-related porphyry Cu deposits in continental collision zones. Geology, 43(3):247–250

Hou Z Q, Liu Y, Tian S H, Yang Z M, Xie Y L. 2015b. Formation of carbonatite-related giant rare-earth-element deposits by the recycling of marine sediments. Scitific Report, 5:10231

Hou Z Q, Zhou Y, Wang R, Zheng Y C, He W Y, Zhao M, Evans N J, Weinberg R F. 2017. Recycling of metal-fertilized lower continental crust: origin of non-arc Au-rich porphyry deposits at cratonic edges. Geology, 45(6):563–566

Howarth G H, Prevec S A. 2013. Hydration vs. oxidation: Modelling implications for Fe-Ti oxide crystallisation in mafic intrusions, with specific reference to the Panzhihua intrusion, SW China. Geoscience Frontiers, 4(5):555–569

Howarth G H, Prevec S A, Zhou M F. 2013. Timing of Ti-magnetite crystallisation and silicate disequilibrium in the Panzhihua mafic layered intrusion: implications for ore-forming processes. Lithos, 170-171:73–89

Hu R Z, Zhou M F. 2012. Multiple Mesozoic mineralization events in South China—an introduction to the thematic issue. Mineralium Deposita, 47(6):579–588

Hu R Z, Bi X W, Jiang G H, Chen H W, Peng J T, Qi Y Q, Wu L Y, Wei W F. 2012a. Mantle-derived noble gases in ore-forming fluids of the granite-related Yaogangxian tungsten deposit, southeastern China. Mineralium Deposita, 47(6):623–632

Hu R Z, Wei W F, Bi X W, Peng J T, Qi Y Q, Wu L Y, Chen Y W. 2012b. Molybdenite Re-Os and muscovite $^{40}Ar/^{39}Ar$ dating of the Xihuashan tungsten deposit, central Nanling district, South China. Lithos, 150:111–118

Hu R Z, Fu S L, Huang Y, Zhou M F, Fu S H, Zhao C H, Wang Y J, Bi X W, Xiao J F. 2017. The giant South China Mesozoic low-temperature metallogenic domain: reviews and a new geodynamic model. Journal of Asian Earth Sciences, 137:9–34

Hu Z C, Zhang W, Liu Y S, Gao S, Li M, Zong K Q, Chen H H, Hu S H. 2015. "Wave" signal-smoothing and mercury-removing device for laser ablation quadrupole and multiple collector ICPMS analysis: application to lead isotope analysis. Analytical Chemistry, 87(2):1152–1157

Huang C, Yang Y H, Yang J H, Xie L W. 2015. In situ simultaneous measurement of Rb-Sr/Sm-Nd or Sm-Nd/Lu-Hf isotopes in natural minerals using laser ablation multi-collector ICP-MS. Journal of Analytical Atomic Spectrometry, 30(4):994–1000

Huang C, Yang Y H, Xie L W, Wu S T, Wang H, Yang J H, Wu F Y. 2020. In situ sequential U-Pb age and Sm-Nd systematics measurements of natural LREE-enriched minerals using single laser ablation multi-collector inductively coupled plasma mass spectrometry. Journal of Analytical Atomic Spectrometry, 35(3):510–517

Huang X W, Beaudoin G. 2019. Textures and chemical compositions of magnetite from iron oxide copper-gold (IOCG) and Kiruna-type iron oxide-apatite (IOA) deposits and their implications for ore genesis and magnetite classification schemes. Economic Geology, 114(5):953–979

Huang X W, Sappin A A, Boutroy É, Beaudoin G, Makvandi S. 2019. Trace element composition of igneous and hydrothermal magnetite from porphyry deposits: relationship to deposit subtypes and magmatic affinity. Economic Geology, 114(5):917–952

Hulsbosch N, Boiron M C, Dewaele S, Muchez P. 2016. Fluid fractionation of tungsten during granite-pegmatite differentiation and the metal source of peribatholitic W quartz veins: evidence from the Karagwe-Ankole Belt (Rwanda). Geochimica et Cosmochimica Acta, 175:299–318

Jin X Y, Hofstra A H, Hunt A G, Liu J Z, Yang W, Li J W. 2020. Noble gases fingerprint the source and evolution of ore-forming fluids of Carlin-type gold deposits in the golden triangle,South China. Economic Geology,115(2):455-469

Ke Y Q,Sun Y J,Lin P J,Zhou J Z,Xu Z F,Cao C F,Yang H,Hu S H. 2019. Quantitative determination of rare earth elements in scheelite via LA-ICP-MS using REE-doped tungstate single crystals as calibration standards. Microchemical Journal,145:642-647

Kendall-Langley L A,Kemp A I S,Grigson J L,Hammerli J. 2020. U-Pb and reconnaissance Lu-Hf isotope analysis of cassiterite and columbite group minerals from Archean Li-Cs-Ta type pegmatites of western Australia. Lithos,352-353:105231

Lan T G,Hu R Z,Bi X W,Mao G J,Wen B J,Liu L,Chen Y H. 2018. Metasomatized asthenospheric mantle contributing to the generation of Cu-Mo deposits within an intracontinental setting:a case study of the ~128 Ma Wangjiazhuang Cu-Mo deposit,eastern North China Craton. Journal of Asian Earth Sciences,160:460-489

Lee C T A,Tang M. 2020. How to make porphyry copper deposits. Earth and Planetary Science Letters,529:115868

Lee C T A,Luffi P,Chin E J,Bouchet R,Dasgupta R,Morton D M,Le Roux V,Yin Q Z,Jin D. 2012. Copper systematics in arc magmas and implications for crust-mantle differentiation. Science,336(6077):64-68

Legros H,Marignac C,Mercadier J,Cuney M,Richard A,Wang R C,Charles N,Lespinasse M Y. 2016. Detailed paragenesis and Li-mica compositions as recorders of the magmatic-hydrothermal evolution of the Maoping W-Sn deposit (Jiangxi,China). Lithos,264:108-124

Legros H,Marignac C,Tabary T,Mercadier J,Richard A,Cuney M,Wang R C,Charles N,Lespinasse M Y. 2018. The ore-forming magmatic-hydrothermal system of the Piaotang W-Sn deposit (Jiangxi,China) as seen from Li-mica geochemistry. American Mineralogist,103(1):39-54

Legros H,Richard A,Tarantola A,Kouzmanov K,Mercadier J,Vennemann T,Marignac C,Cuney M,Wang R C,Charles N,Bailly L,Lespinasse M Y. 2019. Multiple fluids involved in granite-related W-Sn deposits from the world-class Jiangxi Province (China). Chemical Geology,508:92-115

Lehmann B. 2020. Formation of tin ore deposits:a reassessment. Lithos,doi:10.1016/j.lithos.2020.105756

Li B,Zhou J X,Huang Z L,Yan Z F,Bao G P,Sun H R. 2015. Geological,rare earth elemental and isotopic constraints on the origin of the Banbanqiao Zn-Pb deposit,Southwest China. Journal of Asian Earth Sciences,111:100-112

Li B,Zhou J X,Li Y S,Chen A B,Wang R X. 2016. Geology and isotope geochemistry of the Yinchanggou-Qiluogou Pb-Zn deposit,Sichuan Province,Southwest China. Acta Geologica Sinica (English Edition),90(5):1768-1779

Li C,Zhou L M,Zhao Z,Zhang Z Y,Zhao H,Li X W,Qu W J. 2018. *In-situ* Sr isotopic measurement of scheelite using fs-LA-MC-ICPMS. Journal of Asian Earth Sciences,160:38-47

Li C S,Zhang M J,Fu P E,Qian Z Z,Hu P Q,Ripley E M. 2012. The Kalatongke magmatic Ni-Cu deposits in the Central Asian orogenic belt,NW China:product of slab window magmatism? Mineralium Deposita,47(1):51-67

Li C Y,Zhang R Q,Ding X,Ling M X,Fan W M,Sun W D. 2016. Dating cassiterite using laser ablation ICP-MS. Ore Geology Reviews,72:313-322

Li C Y,Jiang Y H,Zhao Y,Zhang C C,Ling M X,Ding X,Zhang H,Li J. 2018. Trace element analyses of fluid inclusions using laser ablation ICP-MS. Solid Earth Sciences,3(1):8-15

Li G L,Hua R M,Zhang W L,Hu D Q,Wei X L,Huang X E,Xie L,Yao J M,Wang X D. 2011. He-Ar isotope composition of pyrite and wolframite in the Tieshanlong tungsten deposit,Jiangxi,China:implications for fluid evolution. Resoure Geology,61(4):356-366

Li J,Huang X L,Wei G J,Liu Y,Ma J L,Han L,He P L. 2018. Lithium isotope fractionation during magmatic differentiation and hydrothermal processes in rare-metal granites. Geochimica et Cosmochimica Acta,240:64-79

Li J K,Liu Y C,Zhao Z,Chou I M. 2018. Roles of carbonate/CO_2 in the formation of quartz-vein wolframite deposits:insight from the crystallization experiments of huebnerite in alkali-carbonate aqueous solutions in a hydrothermal diamond-anvil cell. Ore Geology Reviews,95:40-48

Li J X,Qin K Z,Li G M,Xiao B,Chen L,Zhao J X. 2011. Post-collisional ore-bearing adakitic porphyries from Gangdese porphyry copper belt,southern Tibet:melting of thickened juvenile arc lower crust. Lithos,126(3-4):265-277

Li J X,Qin K Z,Li G M,Evans N J,Huang F,Zhao J X. 2018. Iron isotope fractionation during magmatic-hydrothermal evolution:a case study from the Duolong porphyry Cu-Au deposit,Tibet. Geochimica et Cosmochimica Acta,238:1-15

Li J X,Hu R Z,Zhao C H,Zhu J J,Huang Y,Gao W,Li J W,Zhuo Y Z. 2020. Sulfur isotope and trace element compositions of pyrite determined by NanoSIMS and LA-ICP-MS:new constraints on the genesis of the Shuiyindong Carlin-like gold deposit in SW China. Mineralium Deposita,55(7):1279-1298

Li Q,Parrish R R,Horstwood M S A,McArthura J M. 2014. U-Pb dating of cements in Mesozoic Ammonites. Chemical Geology,376:76-83

Li Q L,Li X H,Lan Z W,Guo C L,Yang Y N,Liu Y,Tang G Q. 2013. Monazite and xenotime U-Th-Pb geochronology by ion microprobe:dating highly fractionated granites at Xihuashan tungsten mine,SE China. Contributions to Mineralogy and Petrology,166(1):65-80

Li W K,Yang Z M,Chiaradia M,Lai Y,Yu C,Zhang J Y. 2020. Redox state of southern Tibetan upper mantle and ultrapotassic magmas. Geology,48(7):733-736

Li Y,Selby D,Feely M,Costanzo A,Li X H. 2017. Fluid inclusion characteristics and molybdenite Re-Os geochronology of the Qulong porphyry copper-molybdenum deposit,Tibet. Mineralium Deposita,52(2):137-158

Li Y,Li X H,Selby D,Li J W. 2018. Pulsed magmatic fluid release for the formation of porphyry deposits:tracing fluid evolution in absolute time from the Tibetan Qulong Cu-Mo deposit. Geology,46(1):7-10

Li Y,Feng L,Kiseeva E S,Gao Z,Guo H,Du Z,Wang F,Shi L. 2019. An essential role for sulfur in sulfide-silicate melt partitioning of gold and magmatic gold transport at subduction settings. Earth and Planetary Science Letters,528:115850

Li Z L,Ye L,Hu Y S,Huang Z L. 2018. Geological significance of nickeliferous minerals in the Fule Pb-Zn deposit,Yunnan Province,China. Acta Geochimica,37(5):684-690

Li Z L,Ye L,Hu Y S,Huang Z L,Wei C,Wu T. 2020. Origin of the Fule Pb-Zn deposit,Yunnan Province,SW China:insight from *in situ* S isotope analysis by NanoSIMS. Geological Magazine,157(3):393–404

Liang Q L,Song X Y,Wirth R,Chen L M,Dai Z H. 2019. Implications of nano-and micrometer-size platinum-group element minerals in base metal sulfides of the Yangliuping Ni-Cu-PGE sulfide deposit,SW China. Chemical Geology,517:7–21

Lightfoot P C,Evans-Lamswood D. 2015. Structural controls on the primary distribution of mafic-ultramafic intrusions containing Ni-Cu-Co-(PGE) sulfide mineralization in the roots of large igneous provinces. Ore Geology Reviews,64:354–386

Lin J,Liu Y S,Hu Z C,Chen W,Zhang C X,Zhao K D,Jin X Y. 2019. Accurate analysis of Li isotopes in tourmalines by LA-MC-ICP-MS under "wet" conditions with non-matrix-matched calibration. Journal of Analytical Atomic Spectrometry,34(6):1145–1153

Ling M X,Liu Y L,Williams I S,Teng F Z,Yang X Y,Ding X,Wei G J,Xie L H,Deng W F,Sun W D. 2013. Formation of the world's largest REE deposit through protracted fluxing of carbonatite by subduction-derived fluids. Scientific Reports,3:1776

Ling X X,Li Q L,Liu Y,Yang Y H,Liu Y,Tang G Q,Li X H. 2016. *In situ* SIMS Th-Pb dating of bastnaesite:constraint on the mineralization time of the Himalayan Mianning-Dechang rare earth element deposits. Journal of Analytical Atomic Spectrometry,31(8):1680–1687

Liu H Q,Bi X W,Lu H Z,Hu R Z,Lan T G,Wang X S,Huang M L. 2018. Nature and evolution of fluid inclusions in the Cenozoic Beiya gold deposit,SW China. Journal of Asian Earth Sciences,161:35–56

Liu P P,Zhou M F,Luais B,Cividini D,Rollion-Bard C. 2014. Disequilibrium iron isotopic fractionation during the high-temperature magmatic differentiation of the Baima Fe-Ti oxide-bearing mafic intrusion,SW China. Earth and Planetary Science Letters,399:21–29

Liu S,Fan H R,Yang K F,Hu F F,Wang K Y,Chen F K,Yang Y H,Yang Z F,Wang Q W. 2018. Mesoproterozoic and Paleozoic hydrothermal metasomatism in the giant Bayan Obo REE-Nb-Fe deposit:constrains from trace elements and Sr-Nd isotope of fluorite and preliminary thermodynamic calculation. Precambrian Research,311:228–246

Liu S,Fan H R,Groves D I,Yang K F,Yang Z F,Wang Q W. 2020. Multiphase carbonatite-related magmatic and metasomatic processes in the genesis of the ore-hosting dolomite in the giant Bayan Obo REE-Nb-Fe deposit. Lithos,354-355:105359

Liu X,Xiong X,Audétat A,Li Y,Song M,Li L,Sun W,Ding X. 2014. Partitioning of copper between olivine,orthopyroxene,clinopyroxene,spinel,garnet and silicate melts at upper mantle conditions. Geochimica et Cosmochimica Acta,125:1–22

Liu X,Xiong X,Audétat A,Li Y. 2015. Partitioning of Cu between mafic minerals,Fe-Ti oxides and intermediate to felsic melts. Geochimica et Cosmochimica Acta,151:86–102

Liu Y,Hou Z Q. 2017. A synthesis of mineralization styles with an integrated genetic model of carbonatite-syenite-hosted REE deposits in the Cenozoic Mianning-Dechang REE metallogenic belt,the eastern Tibetan Plateau,southwestern China. Journal of Asian Earth Sciences,137:35–79

Liu Y,Hou Z Q,Tian S H,Zhang Q C,Zhu Z M,Liu J H. 2015. Zircon U-Pb ages of the Mianning-Dechang syenites,Sichuan Province,southwestern China:constraints on the giant REE mineralization belt and its regional geological setting. Ore Geology Reviews,64:554–568

Liu Y,Chakhmouradian A R,Hou Z Q,Song W L,Kynický J. 2019a. Development of REE mineralization in the giant Maoniuping deposit (Sichuan,China):insights from mineralogy,fluid inclusions,and trace-element geochemistry. Mineralium Deposita,54(5):701–718

Liu Y,Li X H,Tang G Q,Li Q L,Liu X C,Yu H M,Huang F. 2019b. Ultra-high precision silicon isotope micro-analysis using a Cameca IMS-1280 SIMS instrument by eliminating the topography effect. Journal of Analytical Atomic Spectrometry,34(5):906–914

Liu Z C,Wu F Y,Guo C L,Zhao Z F,Yang J H,Sun J F. 2011. *In situ* U-Pb dating of xenotime by laser ablation (LA)-ICP-MS. Chinese Science Bulletin,56(27):2948–2956

Liu Z C,Wu F Y,Yang Y H,Yang J H,Wilde S A. 2012. Neodymium isotopic compositions of the standard monazites used in U-Th-Pb geochronology. Chemical Geology,334:221–239

Lu Y J,Loucks R R,Fiorentini M L,Yang Z M,Hou Z Q. 2015. Fluid flux melting generated post-collisional high-Sr/Y copper-ore-forming water-rich magmas in Tibet. Geology,43(7):583–586

Lu Y J,Loucks R R,Fiorentini M,McCuaig T C,Evans N J,Yang Z M,Hou Z Q,Kirkland C L,Parra-Avila L A,Kobussen A. 2016. Zircon compositions as a pathfinder for porphyry Cu±Mo±Au deposits. SEG Special Publication,19:329–347

Luo K,Zhou J X,Huang Z L,Wang X C,Wilde S A,Zhou W,Tian L Y. 2019. New insights into the origin of early Cambrian carbonate-hosted Pb-Zn deposits in South China:a case study of the Maliping Pb-Zn deposit. Gondwana Research,70:88–103

Luo K,Zhou J X,Huang Z L,Caulfield J,Zhao J X,Feng Y X,Ouyang H G. 2020. New insights into the evolution of Mississippi valley-type hydrothermal system:a case study of the Wusihe Pb-Zn deposit,South China, using quartz *in-situ* trace elements and sulfides *in-situ* S-Pb isotopes. American Mineralogist,105:35–51

Luo T,Deng X D,Li J W,Hu Z C,Zhang W,Liu Y S,Zhang J F. 2019. U-Pb geochronology of wolframite by laser ablation inductively coupled plasma mass spectrometry. Journal of Analytical Atomic Spectrometry,34(7):1439–1446

Maier W D,Smithies R H,Spaggiari C V,Barnes S J,Kirkland C L,Yang S,Lahaye Y,Kiddie O,MacRae C. 2016. Petrogenesis and Ni-Cu sulphide potential of mafic-ultramafic rocks in the Mesoproterozoic Fraser zone within the Albany-Fraser orogen,western Australia. Precambrian Research,281:27–46

Mao J,Ouyang H,Song S,Santosh M,Yuan S,Zhou Z,Zheng W,Liu H,Liu P,Cheng Y,Chen M. 2019. Geology and metallogeny of tungsten and tin deposits in China. Society of Economic Geologists Special Publication,22:411–482

Mao W,Rusk B,Yang F,Zhang M. 2017. Physical and chemical evolution of the dabaoshan porphyry Mo deposit,South China:insights from fluid inclusions,cathodoluminescence,and trace elements in quartz. Economic Geology,112(4):889–918

Mao Y J, Barnes S J, Qin K Z, Tang D M, Martin L, Su B X, Evans N J. 2019. Rapid orthopyroxene growth induced by silica assimilation: constraints from sector-zoned orthopyroxene, olivine oxygen isotopes and trace element variations in the Huangshanxi Ni-Cu deposit, Northwest China. Contributions to Mineralogy and Petrology, 174:33

Ni H, Shi H, Zhang L, Li W C, Guo X, Liang T. 2018. Cu diffusivity in granitic melts with application to the formation of porphyry Cu deposits. Contributions to Mineralogy and Petrology, 173(6):10

Ni P, Wang X D, Wang G G, Huang J B, Pan J Y, Wang T G. 2015. An infrared microthermometric study of fluid inclusions in coexisting quartz and wolframite from Late Mesozoic tungsten deposits in the Gannan metallogenic belt, South China. Ore Geology Reviews, 65:1062–1077

Pan J Y, Ni P, Chi Z, Wang W B, Zeng W C, Xue K. 2019a. Alunite $^{40}Ar/^{39}Ar$ and zircon U-Pb constraints on the magmatic-hydrothermal history of the Zijinshan high-sulfidation epithermal Cu-Au deposit and the adjacent Luoboling porphyry Cu-Mo deposit, South China: implications for their genetic association. Economic Geology, 114:667–695

Pan J Y, Ni P, Wang R C. 2019b. Comparison of fluid processes in coexisting wolframite and quartz from a giant vein-type tungsten deposit, South China: insights from detailed petrography and LA-ICP-MS analysis of fluid inclusions. American Mineralogist, 104(8):1092–1116

Pan L C, Hu R Z, Wang X S, Bi X W, Zhu J J, Li C S. 2016. Apatite trace element and halogen compositions as petrogenetic-metallogenic indicators: examples from four granite plutons in the Sanjiang region, SW China. Lithos, 254-255:118–130

Pan L C, Hu R Z, Bi X W, Li C S, Wang X S, Zhu J J. 2018. Titanite major and trace element compositions as petrogenetic and metallogenic indicators of Mo ore deposits: examples from four granite plutons in the southern Yidun arc, SW China. American Mineralogist, 103(9):1417–1434

Peng N J, Jiang S Y, Xiong S F, Pi D H. 2018. Fluid evolution and ore genesis of the Dalingshang deposit, Dahutang W-Cu ore field, northern Jiangxi Province, South China. Mineralium Deposita, 53(8):1079–1094

Pi Q H, Hu R Z, Xiong B, Li Q L, Zhong R C. 2017. In situ SIMS U-Pb dating of hydrothermal rutile: reliable age for the Zhesang Carlin-type gold deposit in the golden triangle region, SW China. Mineralium Deposita, 52(8):1179–1190

Qin K Z, Su B X, Sakyi P A, Tang D M, Li X H, Sun H, Xiao Q H, Liu P P. 2011. SIMS zircon U-Pb geochronology and Sr-Nd isotopes of Ni-Cu-bearing mafic-ultramafic intrusions in eastern Tianshan and Beishan in correlation with flood basalts in Tarim Basin (NW China): constraints on a ca. 280 Ma mantle plume. American Journal of Science, 311(3):237–260

Qiu Y, Wang X L, Liu X, Cao J, Liu Y F, Xi B B, Gao W L. 2020. In situ Raman spectroscopic quantification of CH_4-CO_2 mixture: application to fluid inclusions hosted in quartz veins from the Longmaxi Formation shales in Sichuan Basin, southwestern China. Petroleum Science, 17(1):23–35

Richards J P. 2003. Tectono-magmatic precursors for porphyry Cu-(Mo-Au) deposit formation. Economic Geology, 98(8):1515–1533

Richards J P. 2009. Postsubduction porphyry Cu-Au and epithermal Au deposits: products of remelting of subduction-modified lithosphere. Geology, 37(3):247–250

Shang L B, Bi X W, Hu R Z, Fan W L. 2007. An experimental study on the solubility of copper bichloride in water vapor. Chinese Science Bulletin, 52(3):395–400

Shang L B, Chou I M, Lu W J, Burruss R C, Zhang Y X. 2009. Determination of diffusion coefficients of hydrogen in fused silica between 296 and 523 K by Raman spectroscopy and application of fused silica capillaries in studying redox reactions. Geochimica et Cosmochimica Acta, 73(18):5435–5443

Shang L B, Williams-Jones A, Wang X S, Timofeev A, Hu R Z, Bi X W. 2020. An experimental study of the solubility and speciation of MoO_3(s) in hydrothermal fluids at temperatures up to 350℃. Economic Geology, 115(3):661–669

She Y W, Song X Y, Yu S Y, He H L. 2015. Variations of trace element concentration of magnetite and ilmenite from the Taihe layered intrusion, Emeishan large igneous province, SW China: implications for magmatic fractionation and origin of Fe-Ti-V oxide ore deposits. Journal of Asian Earth Sciences, 113:1117–1131

She Y W, Song X Y, Yu S Y, Chen L M, Zheng W Q. 2016. Apatite geochemistry of the Taihe layered intrusion, SW China: implications for the magmatic differentiation and the origin of apatite-rich Fe-Ti oxide ores. Ore Geology Reviews, 78:151–165

Shen A J, Hu A P, Cheng T, Liang F, Pan W Q, Feng Y X, Zhao J X. 2019. Laser ablation in situ U-Pb dating and its application to diagenesis-porosity evolution of carbonate reservoirs. Petroleum Exploration and Development, 46(6):1127–1140

Shu Q, Lai Y, Sun Y, Wang C, Meng S. 2013. Ore genesis and hydrothermal evolution of the Baiyinnuo'er zinc-lead skarn deposit, Northeast China: evidence from isotopes (S, Pb) and fluid inclusions. Economic Geology, 108(4):835–860

Shu X C, Liu Y. 2019. Fluid inclusion constraints on the hydrothermal evolution of the Dalucao carbonatite-related REE deposit, Sichuan Province, China. Ore Geology Reviews, 107:41–57

Sillitoe R H. 2010. Porphyry copper systems. Economic Geology, 105(1):3–41

Smith M P, Campbell L S, Kynicky J. 2015. A review of the genesis of the world class Bayan Obo Fe-REE-Nb deposits, Inner Mongolia, China: multistage processes and outstanding questions. Ore Geology Reviews, 64:459–476

Song S W, Mao J W, Xie G Q, Chen L, Santosh M, Chen G H, Rao J F, Ouyang Y P. 2019. In situ LA-ICP-MS U-Pb geochronology and trace element analysis of hydrothermal titanite from the giant Zhuxi W (Cu) skarn deposit, South China. Mineralium Deposita, 54(4):569–590

Song W L, Xu C, Smith M P, Kynicky J, Huang K J, Wei C W, Zhou L, Shu Q H. 2016. Origin of unusual HREE-Mo-rich carbonatites in the Qinling orogen, China. Scientific Reports, 6:37377

Song W L, Xu C, Smith M P, Chakhmouradian A R, Brenna M, Kynicky J, Chen W. 2018. Genesis of the world's largest rare earth element deposit, Bayan Obo, China: protracted mineralization evolution over-1b. y. Geology, 46(4):323–326

Song X Y, Li X R. 2009. Geochemistry of the Kalatongke Ni-Cu-(PGE) sulfide deposit, NW China: implications for the formation of magmatic sulfide mineralization in a postcollisional environment. Mineralium Deposita,44(3):303-327

Song X Y,Danyushevsky L V,Keays R R,Chen L M,Tian Y L,Xiao J F. 2012. Structural,lithological,and geochemical constraints on the dynamic magma plumbing system of the Jinchuan Ni-Cu sulfide deposit,NW China. Mineralium Deposita,47(3):277-297

Song X Y,Chen L M,Deng Y F,Xie W. 2013a. Syncollisional tholeiitic magmatism induced by asthenosphere upwelling owing to slab detachment at the southern margin of the Central Asian orogenic belt. Journal of Geological Society,170(6):941-950

Song X Y,Qi H W,Hu R Z,Chen L M,Yu S Y,Zhang J F. 2013b. Formation of thick stratiform Fe-Ti oxide layers in layered intrusion and frequent replenishment of fractionated mafic magma:evidence from the Panzhihua intrusion,SW China. Geochemistry,Geophysics,Geosystems,14(3):712-732

Song X Y,Yi J N,Chen L M,She Y W,Liu C Z,Dang X Y,Yang Q A,Wu S K. 2016. The giant Xiarihamu Ni-Co sulfide deposit in the East Kunlun orogenic belt,northern Tibet Plateau,China. Economic Geology,111:29-55

Song X Y,Wang K Y,Barnes S J,Yi J N,Chen L M,Schoneveld L E. 2020. Petrogenetic insights from chromite in ultramafic cumulates of the Xiarihamu intrusion,northern Tibet Plateau,China. American Mineralogist,105(4):479-524

Su B X,Qin K Z,Sakyi P A,Li X H,Yang Y H,Sun H,Tang D M,Liu P P,Xiao Q H,Malaviarachchi S P K. 2011. U-Pb ages and Hf-O isotopes of zircons from Late Paleozoic mafic-ultramafic units in the southern Central Asian orogenic belt:tectonic implications and evidence for an Early-Permian mantle plume. Gondwana Research,20(2-3):516-531

Su J H,Zhao X F,Li X C,Hu W,Chen M,Xiong Y L. 2019. Geological and geochemical characteristics of the Miaoya syenite-carbonatite complex, central China:implications for the origin of REE-Nb-enriched carbonatite. Ore Geology Reviews,113:103101

Su W C,Zhang H T,Hu R Z,Ge X,Xia B,Chen Y Y,Zhu C. 2012. Mineralogy and geochemistry of gold-bearing arsenian pyrite from the Shuiyindong Carlin-type gold deposit,Guizhou,China:implications for gold depositional processes. Mineralium Deposita,47(6):653-662

Su W C,Dong W D,Zhang X C,Shen N P,Hu R Z,Hofstra A H,Cheng L Z,Xia Y,Yang K Y. 2018. Carlin-type gold deposits in the Dian-Qian-Gui "Golden Triangle" of Southwest China. In:Muntean J L (ed). Diversity of Carlin-Style Gold Deposits. Reviews in Economic Geology,20:157-185

Sun W D,Liang H Y,Ling M X,Zhan M Z,Ding X,Zhang H,Yang X Y,Li Y L,Ireland T R,Wei Q R,Fan W M. 2013. The link between reduced porphyry copper deposits and oxidized magmas. Geochimica et Cosmochimica Acta,103:263-275

Sun W D,Wang J T,Zhang L P,Zhang C C,Li H,Ling M X,Ding X,Li C Y,Liang H Y. 2017. The formation of porphyry copper deposits. Acta Geochimica,36(1):9-15

Sun Z,Xiong X,Wang J,Liu X,Li L,Ruan M,Zhang L,Takahashi E. 2020. Sulfur abundance and heterogeneity in the MORB mantle estimated by copper partitioning and sulfur solubility modelling. Earth and Planetary Science Letters,538:116169

Tan Q P,Xia Y,Xie Z J,Yan J,Wei D T. 2015. S,C,O,H,and Pb isotopic studies for the Shuiyindong Carlin-type gold deposit,Southwest Guizhou, China:constraints for ore genesis. Chinese Journal of Geochemistry,34(4):525-539

Tan S C,Zhou J X,Li B,Zhao J X. 2017. *In situ* Pb and bulk Sr isotope analysis of the Yinchanggou Pb-Zn deposit in Sichuan Province (SW China): constraints on the origin and evolution of hydrothermal fluids. Ore Geology Reviews,91:432-443

Tan S C,Zhou J X,Zhou M F,Ye L. 2019. *In-situ* S and Pb isotope constraints on an evolving hydrothermal system,Tianbaoshan Pb-Zn-(Cu) deposit in South China. Ore Geology Review,115:103177

Tang Y W,Cui K,Zheng Z,Gao J F,Han J J,Yang J H,Liu L. 2020. LA-ICP-MS U-Pb geochronology of wolframite by combining NIST series and common lead-bearing MTM as the primary reference material:implications for metallogenesis of South China. Gondwana Research,83:217-231

Tian S H,Hou Z Q,Su A N,Qiu L,Mo X X,Hou K J,Zhao Y,Hu W J,Yang Z S. 2015. The anomalous lithium isotopic signature of Himalayan collisional zone carbonatites in western Sichuan,SW China:enriched mantle source and petrogenesis. Geochimica et Cosmochimica Acta,159:42-60

Tsui T F,Holland H D. 1979. The analysis of fluid inclusions by laser microprobe. Economic Geology,74(7):1647-1653

USGS. 2019. National minerals information center,commodity statistics and information. https://www.usgs.gov/centers/nmic/commodity-statistics-and-information

Wan Y,Wang X L,Hu W X,Chou I M,Wang X Y,Chen Y,Xu Z M. 2017. *In situ* optical and Raman spectroscopic observations of the effects of pressure and fluid composition on liquid-liquid phase separation in aqueous cadmium sulfate solutions ($\leq 400°C$,50 MPa) with geological and geochemical implications. Geochimica et Cosmochimica Acta,211:133-152

Wang B,Cluzel D,Jahn B M,Shu L S,Chen Y,Zhai Y Z,Branquet Y,Barbanson L,Sizaret S. 2014. Late Paleozoic pre-and syn-kinematic plutons of the Kangguer-Huangshan shear zone:inference on the tectonic evolution of the eastern Chinese North Tianshan. American Journal of Science,314(1):43-79

Wang C M,Bagas L,Lu Y J,Santosh M,Du B,McCuaig T C. 2016. Terrane boundary and spatio-temporal distribution of ore deposits in the Sanjiang Tethyan orogen:insights from zircon Hf-isotopic mapping. Earth-Science Reviews,156:39-65

Wang C Y,Zhou M F,Yang S H,Qi L,Sun Y L. 2014. Geochemistry of the Abulangdang intrusion:cumulates of high-Ti picritic magmas in the Emeishan large igneous province,SW China. Chemical Geology,378-379:24-39

Wang D H,Huang F,Wang Y,He H H,Li X M,Liu X X,Sheng J F,Liang T. 2020. Regional metallogeny of Tungsten-tin-polymetallic deposits in Nanling region,South China. Ore Geology Reviews,120:103305

Wang K,Wang C Y,Ren Z Y. 2018. Apatite-hosted melt inclusions from the Panzhihua gabbroic-layered intrusion associated with a giant Fe-Ti oxide deposit in SW China:insights for magma unmixing within a crystal mush. Contributions to Mineralogy and Petrology,173(7):59

Wang K Y,Fan H R,Yang K F,Hu F F,Ma Y G. 2010. Bayan Obo carbonatites:texture evidence from polyphase intrusive and extrusive carbonatites. Acta Geologica Sinica,84(6):1365-1376

Wang K Y,Song X Y,Yi J N,Barnes S J,She Y W,Zheng W Q,Schoneveld L E. 2019. Zoned orthopyroxenes in the Ni-Co sulfide ore-bearing Xiarihamu mafic-ultramafic intrusion in northern Tibetan Plateau,China:implications for multiple magma replenishments. Ore Geology Reviews,113:103082

Wang L J,Mi M,Zhou J X,Luo K. 2018. New constraints on the origin of the Maozu carbonate-hosted epigenetic Zn-Pb deposit in NE Yunnan Province, SW China. Ore Geology Reviews,101:578-594

Wang Q F,Groves D. 2018. Carlin-style gold deposits,Youjiang Basin,China:tectono-thermal and structural analogues of the Carlin-type gold deposits, Nevada,USA. Mineralium Deposita,53(7):909-918

Wang R,Richards J P,Hou Z,Yang Z,DuFrane S A. 2014. Increased magmatic water content—the key to Oligo-Miocene porphyry Cu-Mo±Au formation in the eastern Gangdese Belt,Tibet. Economic Geology,109:1315-1339

Wang R,Tafti R,Hou Z Q,Shen Z C,Guo N,Evans N J,Jeon H,Li Q Y,Li W K. 2017. Across-arc geochemical variation in the Jurassic magmatic zone, southern Tibet:implication for continental arc-related porphyry Cu-Au mineralization. Chemical Geology,451:116-134

Wang R,Weinberg R F,Collins W J,Richards J P,Zhu D C. 2018. Origin of postcollisional magmas and formation of porphyry Cu deposits in southern Tibet. Earth-Science Reviews,181:122-143

Wang X L,Chou I M,Hu W X,Burruss R C. 2013. In situ observations of liquid-liquid phase separation in aqueous $MgSO_4$ solutions:geological and geochemical implications. Geochimica et Cosmochimica Acta,103:1-10

Wang X L,Wan Y,Hu W X,Chou I M,Cao J,Wang X Y,Wang M,Li Z. 2016. In situ observations of liquid-liquid phase separation in aqueous $ZnSO_4$ solutions at temperatures up to 400℃:implications for Zn^{2+}-SO_4^{2-} association and evolution of submarine hydrothermal fluids. Geochimica et Cosmochimica Acta,181:126-143

Wang X L,Qiu Y,Lu J J,Chou I M,Zhang W L,Li G L,Hu W X,Li Z,Zhong R C. 2020. In situ Raman spectroscopic investigation of the hydrothermal speciation of tungsten:implications for the ore-forming process. Chemical Geology,532:119299

Wang X S,Timofeev A,Williams-Jones A E,Shang L B,Bi X W. 2019. An experimental study of the solubility and speciation of tungsten in NaCl-bearing aqueous solutions at 250,300,and 350℃. Geochimica et Cosmochimica Acta,265:313-329

Wang X S,Williams-Jones A E,Hu R Z,Shang L B,Bi X W. 2021. The role of fluorine in granite-related hydrothermal tungsten ore genesis:results of experiments and modeling. Geochimica et Cosmochimica Acta,292:170-187

Wawryk C M,Foden J D. 2015. Fe-isotope fractionation in magmatic-hydrothermal mineral deposits:a case study from the Renison Sn-W deposit, Tasmania. Geochimica et Cosmochimica Acta,150:285-298

Wei C,Huang Z L,Yan Z F,Hu Y S,Ye L. 2018. Trace element contents in sphalerite from the Nayongzhi Zn-Pb deposit,northwestern Guizhou,China: insights into incorporation mechanisms,metallogenic temperature and ore genesis. Minerals,8(11):490

Wei C,Ye L,Hu Y S,Danyushevskiy L,Li Z L,Huang Z L. 2019. Distribution and occurrence of Ge and related trace elements in sphalerite from the Lehong carbonate-hosted Zn-Pb deposit,northeastern Yunnan,China:insights from SEM and LA-ICP-MS studies. Ore Geology Reviews,115:103175

Wei D T,Xia Y,Gregory D D,Steadman J A,Tan Q P,Xie Z J,Liu X J. 2020. Multistage pyrites in the Nibao disseminated gold deposit,southwestern Guizhou Province,China:insights into the origin of Au from textures,in situ trace elements,and sulfur isotope analyses. Ore Geology Reviews, 122:103446

Wei W F,Hu R Z,Bi X W,Peng J T,Su W C,Song S Q,Shi S H. 2012. Infrared microthermometric and stable isotopic study of fluid inclusions in wolframite at the Xihuashan tungsten deposit,Jiangxi Province,China. Mineralium Deposita,47(6):589-605

Wei W F,Hu R Z,Bi X W,Jiang G H,Yan B,Yin R S,Yang J H. 2019. Mantle-derived and crustal He and Ar in the ore-forming fluids of the Xihuashan granite-associated tungsten ore deposit,South China. Ore Geology Reviews,105:605-615

White N C,Zhang D Y,Hong H L,Liu L J,Sun W,Zhang M M. 2019. Epithermal gold deposits of China-an overview. SEG Special Publications,22: 235-262

Wu S Y,Hou L,Jowitt S M,Ding J,Zhang J R,Zhu S B,Zhao Z Y. 2019. Geochronology,geochemistry and petrogenesis of Late Triassic dolerites associated with the Nibao gold deposit,Youjiang Basin,southwestern China:implications for post-collisional magmatism and its relationships with Carlin-like gold mineralization. Ore Geology Reviews,111:102971

Xiang Z Z,Zhou J X,Luo K. 2020. New insights into the multi-layer metallogenesis of carbonated-hosted epigenetic Pb-Zn deposits:a case study of the Maoping Pb-Zn deposit,South China. Ore Geology Reviews,122:103538

Xie G Q,Mao J W,Zhu Q Q,Yao L,Li Y H,Li W,Zhao H J. 2015. Geochemical constraints on Cu-Fe and Fe skarn deposits in the Edong district, Middle-Lower Yangtze River metallogenic belt,China. Ore Geology Reviews,64:425-444

Xie G Q,Mao J W,Bagas L,Fu B,Zhang Z Y. 2019. Mineralogy and titanite geochronology of the Caojiaba W deposit,Xiangzhong metallogenic province, southern China:implications for a distal reduced skarn W formation. Mineralium Deposita,54(3):459-472

Xie W,Song X Y,Chen L M,Deng Y F,Zheng W Q,Wang Y S,Luan Y. 2014. Geochemistry insights on the genesis of the subduction-related Heishan magmatic Ni-Cu-(PGE) deposit,Gansu,northwestern China,at the southern margin of the Central Asian orogenic belt. Economic Geology,109(6): 1563-1583

Xie Y L,Li Y X,Hou Z Q,Cooke D R,Danyushevsky L,Dominy S C,Yin S P. 2015. A model for carbonatite hosted REE mineralisation-the Mianning-Dechang REE belt,western Sichuan Province,China. Ore Geology Reviews,70:595-612

Xie Y L,Hou Z Q,Goldfarb R J,Guo X,Wang L. 2016. Rare earth element deposits in China. In:Verplanck P L,Hitzman M W (eds). Rare Earth and Critical Elements in Ore Deposits. Society of Economic Geologists,115-136

Xie Z J,Xia Y,Cline J S,Yan B W,Wang Z P,Tan Q P,Wei D T. 2017. Comparison of the native antimony-bearing Paiting gold deposit, Guizhou

Province, China, with Carlin-type gold deposits, Nevada, USA. Mineralium Deposita, 52(1):69-84

Xie Z J, Xia Y, Cline J S, Koenig A, Wei D T, Tan Q P, Wang Z P. 2018a. Are there Carlin-type Au deposits in China? A comparison between the Guizhou China and Nevada USA deposits. In: Muntean J L (ed). Diversity of Carlin-style gold deposits. Reviews in Economic Geology, 20:187-233

Xie Z J, Xia Y, Cline J S, Pribil M J, Koenig A, Tan Q P, Wei D T, Wang Z P, Yan J. 2018b. Magmatic origin for sediment-hosted Au deposits, Guizhou Province, China: in situ chemistry and sulfur isotope composition of pyrites, Shuiyindong and Jinfeng deposits. Economic Geology, 113(7):1627-1652

Xing C M, Wang C Y, Li C Y. 2014. Trace element compositions of apatite from the middle zone of the Panzhihua layered intrusion, SW China: insights into the differentiation of a P- and Si-rich melt. Lithos, 204:188-202

Xiong X L, Keppler H, Audetat A, Ni H W, Sun, W D, Li Y A. 2011. Partitioning of Nb and Ta between rutile and felsic melt and the fractionation of Nb/Ta during partial melting of hydrous metabasalt. Geochimica Et Cosmochimica Acta, 75(7):1673-1692

Xu C, Chakhmouradian A R, Taylor R N, Kynicky J, Li W B, Song W L, Fletcher I R. 2014. Origin of carbonatites in the South Qinling orogen: implications for crustal recycling and timing of collision between the South and North China Blocks. Geochimica et Cosmochimica Acta, 143:189-206

Xu L, Hu Z C, Zhang W, Yang L, Liu Y S, Gao S, Luo T, Hu S H. 2015. In situ Nd isotope analyses in geological materials with signal enhancement and non-linear mass dependent fractionation reduction using laser ablation MC-ICP-MS. Journal of Analytical Atomic Spectrometry, 30(1):232-244

Xu L L, Bi X W, Hu R Z, Zhang X C, Su W C, Qu W J, Hu Z C, Tang Y Y. 2012. Relationships between porphyry Cu-Mo mineralization in the Jinshajiang-Red River metallogenic belt and tectonic activity: constraints from zircon U-Pb and molybdenite Re-Os geochronology. Ore Geology Reviews, 48:460-473

Xu Y M, Jiang S Y. 2017. In-situ analysis of trace elements and Sr-Pb isotopes of K-feldspars from Tongshankou Cu-Mo deposit, SE Hubei Province, China: insights into early potassic alteration of the porphyry mineralization system. Terra Nova, 29(6):343-355

Yan J, Hu R Z, Liu S, Lin Y T, Zhang J C, Fu S L. 2018. NanoSIMS element mapping and sulfur isotope analysis of Au-bearing pyrite from Lannigou Carlin-type Au deposit in SW China: new insights into the origin and evolution of Au-bearing fluids. Ore Geology Reviews, 92:29-41

Yang J H, Peng J T, Hu R Z, Bi X W, Zhao J H, Fu Y Z, Shen N P. 2013. Garnet geochemistry of tungsten-mineralized Xihuashan granites in South China. Lithos, 177:79-90

Yang J H, Kang L F, Liu L, Peng J T, Qi Y Q. 2019a. Tracing the origin of ore-forming fluids in the Piaotang tungsten deposit, South China: constraints from in-situ analyses of wolframite and individual fluid inclusion. Ore Geology Reviews, 111:102939

Yang J H, Zhang Z, Peng J T, Liu L, Leng C B. 2019b. Metal source and wolframite precipitation process at the Xihuashan tungsten deposit, South China: insights from mineralogy, fluid inclusion and stable isotope. Ore Geology Reviews, 111:102965

Yang K F, Fan H R, Santosh M, Hu F F, Wang K Y. 2011. Mesoproterozoic carbonatitic magmatism in the Bayan Obo deposit, Inner Mongolia, North China: constraints for the mechanism of super accumulation of rare earth elements. Ore Geology Reviews, 40(1):122-131

Yang K F, Fan H R, Pirajno F, Li X C. 2019. The Bayan Obo (China) giant REE accumulation conundrum elucidated by intense magmatic differentiation of carbonatite. Geology, 47:1198-1202

Yang M, Yang Y H, Wu S T, Romer R L, Che X D, Zhao Z F, Li W S, Yang J H, Wu F Y, Xie L W, Huang C, Zhang D, Zhang Y. 2020. Accurate and precise in situ U-Pb isotope dating of wolframite series minerals via LA-SF-ICP-MS. Journal of Analytical Atomic Spectrometry, 35(10):2191-2203

Yang S Y, Jiang S Y. 2012. Chemical and boron isotopic composition of tourmaline in the Xiangshan volcanic-intrusive complex, Southeast China: evidence for boron mobilization and infiltration during magmatic-hydrothermal processes. Chemical Geology, 312-313:177-189

Yang W B, Niu H C, Shan Q, Sun W D, Zhang H, Li N B, Jiang Y H, Yu X Y. 2014. Geochemistry of magmatic and hydrothermal zircon from the highly evolved Baerzhe alkaline granite: implications for Zr-REE-Nb mineralization. Mineralium Deposita, 49(4):451-470

Yang X Y, Lai X D, Pirajno F, Liu Y L, Ling M X, Sun W D. 2017. Genesis of the Bayan Obo Fe-REE-Nb formation in Inner Mongolia, North China Craton: a perspective review. Precambrian Research, 288:39-71

Yang Y H, Wu F Y, Li Y, Yang J H, Xie L W, Liu Y, Zhang Y B, Huang C. 2014. In situ U-Pb dating of bastnaesite by LA-ICP-MS. Journal of Analytical Atomic Spectrometry, 29(6):1017-1023

Yang Y H, Wu F Y, Yang J H, Mitchell R H, Zhao Z F, Xie L W, Huang C, Ma Q, Yang M, Zhao H. 2018. U-Pb age determination of schorlomite garnet by laser ablation inductively coupled plasma mass spectrometry. Journal of Analytical Atomic Spectrometry, 33(2):231-239

Yang Y H, Wu F Y, Li Q L, Rojas-Agramonte Y, Yang J H, Li Y, Ma Q, Xie L W, Huang C, Fan H R, Zhao Z F, Xu C. 2019. In situ U-Th-Pb dating and Sr-Nd isotope analysis of bastnäsite by LA-(MC)-ICP-MS. Geostandards and Geoanalytical Research, 43(4):543-565

Yang Z M, Cooke D R. 2019. Porphyry copper deposits in China. Society of Economic Geologists Special Publication, 22:133-187

Yang Z M, Lu Y J, Hou Z Q, Chang Z S. 2015. High-Mg Diorite from Qulong in southern Tibet: implications for the genesis of adakite-like intrusions and associated porphyry Cu deposits in collisional orogens. Journal of Petrology, 56(2):227-253

Yang Z M, Goldfarb R, Chang Z S. 2016. Generation of postcollisional porphyry copper deposits in southern Tibet triggered by subduction of the Indian continental plate. Society of Economic Geologists Special Publication, 19:279-300

Yao J H, Zhu W G, Li C S, Zhong H, Bai Z J, Ripley E M, Li C. 2018. Petrogenesis and ore genesis of the Lengshuiqing magmatic sulfide deposit in Southwest China: constraints from chalcophile elements (PGE, Se) and Sr-Nd-Os-S isotopes. Economic Geology, 113(3):675-698

Yao J M, Mathur R, Powell W, Lehmann B, Tornos F, Wilson M, Ruiz J. 2018. Sn-isotope fractionation as a record of hydrothermal redox reactions. American Mineralogist, 103(10):1591-1598

Ye L, Cook N J, Ciobanu C L, Liu Y P, Zhang Q, Liu T G, Gao W, Yang Y L, Danyushevsky L. 2011. Trace and minor elements in sphalerite from base metal deposits in South China: a LA-ICPMS study. Ore Geology Reviews, 39(4):188-217

Yin R S, Deng C Z, Lehmann B, Sun G Y, Lepak R F, Hurley J P, Zhao C H, Xu G W, Tan Q P, Xie Z J, Hu R Z. 2019. Magmatic-hydrothermal origin of mercury in carlin-style and epithermal gold deposits in China: evidence from mercury stable isotopes. ACS Earth and Space Chemistry, 3(8): 1631-1639

Ying Y C, Chen W, Lu J, Jiang S Y, Yang Y H. 2017. In situ U-Th-Pb ages of the Miaoya carbonatite complex in the South Qinling orogenic belt, Central China. Lithos, 290-291: 159-171

Ying Y C, Chen W, Simonetti A, Jiang S Y, Zhao K D. 2020. Significance of hydrothermal reworking for REE mineralization associated with carbonatite: constraints from in situ trace element and C-Sr isotope study of calcite and apatite from the Miaoya carbonatite complex (China). Geochimica et Cosmochimica Acta, 280: 340-359

Yuan H L, Yin C, Liu X, Chen K Y, Bao Z A, Zong C L, Dai M N, Lai S C, Wang R, Jiang S Y. 2015. High precision in-situ Pb isotopic analysis of sulfide minerals by femtosecond laser ablation multi-collector inductively coupled plasma mass spectrometry. Science China Earth Sciences, 58(10): 1713-1721

Yuan H L, Liu X, Chen L, Bao Z A, Chen K Y, Zong C L, Li X C, Qiu J W. 2018. Simultaneous measurement of sulfur and lead isotopes in sulfides using nanosecond laser ablation coupled with two multi-collector inductively coupled plasma mass spectrometers. Journal of Asian Earth Sciences, 154: 386-396

Yuan J H, Zhan X C, Sun D Y, Zhao L H, Fan C Z, Kuai L J, Hu M Y. 2011. Investigation on matrix effects in silicate minerals by laser ablation-inductively coupled plasma-mass spectrometry. Chinese Journal of Analytical Chemistry, 39(10): 1582-1587

Yuan S D, Peng J T, Hao S, Li H M, Geng J Z, Zhang D L. 2011. In situ LA-MC-ICP-MS and ID-TIMS U-Pb geochronology of cassiterite in the giant Furong tin deposit, Hunan Province, South China: new constraints on the timing of tin-polymetallic mineralization. Ore Geology Reviews, 43(1): 235-242

Yuan S D, Chou I M, Burruss R C, Wang X L, Li J K. 2013. Disproportionation and thermochemical sulfate reduction reactions in S-H_2O-CH_4 and S-D_2O-CH_4 systems from 200 to 340°C at elevated pressures. Geochimica et Cosmochimica Acta, 118: 263-275

Yuan S D, Williams-Jones A E, Mao J W, Zhao P L, Yan C, Zhang D L. 2018. The origin of the Zhangjialong tungsten deposit, South China: implications for W-Sn mineralization in large granite batholiths. Economic Geology, 113(5): 1193-1208

Yuan S D, Williams-Jones A E, Romer R L, Zhao P L, Mao J W. 2019. Protolith-related thermal controls on the decoupling of Sn and W in Sn-W metallogenic provinces: insights from the Nanling region, China. Economic Geology, 114(5): 1005-1012

Zeng L J, Niu H C, Bao Z W, Yang W B. 2017. Chemical lattice expansion of natural zircon during the magmatic-hydrothermal evolution of A-type granite. American Mineralogist, 102(3): 655-665

Zhai D G, Williams-Jones A E, Liu J J, Tombros S F, Cook N J. 2018. Mineralogical, fluid inclusion, and multiple isotope (H-O-S-Pb) constraints on the genesis of the sandaowanzi epithermal Au-Ag-Te deposit, NE China. Economic Geology, 113(6): 1359-1382

Zhang C Q, Wu Y, Hou L, Mao J W. 2015. Geodynamic setting of mineralization of Mississippi Valley-type deposits in world-class Sichuan-Yunnan-Guizhou Zn-Pb triangle, Southwest China: implications from age-dating studies in the past decade and the Sm-Nd age of Jinshachang deposit. Journal of Asian Earth Sciences, 103: 103-114

Zhang D Y, Zhang Z C, Huang H, Cheng Z G, Charlier B. 2018. Petrogenesis and metallogenesis of the Wajilitag and Puchang Fe-Ti oxide-rich intrusive complexes, northwestern Tarim large igneous province. Lithos, 304-307: 412-435

Zhang H J, Xiao C Y, Wen H J, Zhu X K, Ye L, Huang Z L, Zhou J X, Fan H F. 2019a. Homogeneous Zn isotopic compositions in the Maozu Zn-Pb ore deposit in Yunnan Province, southwestern China. Ore Geology Reviews, 109: 1-10

Zhang H J, Fan H F, Xiao C Y, Wen H J, Ye L, Huang Z L, Zhou J X, Guo Q J. 2019b. The mixing of multi-source fluids in the Wusihe Zn-Pb ore deposit in Sichuan Province, southwestern China. Acta Geochimica, 38(5): 642-653

Zhang J C, Lin Y T, Yang W, Shen W J, Hao J L, Hu S, Cao M J. 2014. Improved precision and spatial resolution of sulfur isotope analysis using NanoSIMS. Journal of Analytical Atomic Spectrometry, 29(10): 1934-1943

Zhang M J, Li C S, Fu P E, Hu P Q, Ripley E M. 2011. The Permian Huangshanxi Cu-Ni deposit in western China: intrusive-extrusive association, ore genesis, and exploration implications. Mineralium Deposita, 46(2): 153-170

Zhang P, Li Z, Wang Z, Li W C, Xuan G. 2018. Diffusion of molybdenum and tungsten in anhydrous and hydrous granitic melts. American Mineralogist, 103(12): 1966-1974

Zhang Q, Zhang R Q, Gao J F, Lu J J, Wu J W. 2018. In-situ LA-ICP-MS trace element analyses of scheelite and wolframite: constraints on the genesis of veinlet-disseminated and vein-type tungsten deposits, South China. Ore Geology Reviews, 99: 166-179

Zhang R Q, Lehmann B, Seltmann R, Sun W D, Li C Y. 2017. Cassiterite U-Pb geochronology constrains magmatic-hydrothermal evolution in complex evolved granite systems: the classic Erzgebirge tin province (Saxony and Bohemia). Geology, 45: 1095-1098

Zhang W, Chen T W, Gao J F, Chen H K, Li J H. 2019a. Two episodes of REE mineralization in the Qinling orogenic belt, central China: in-situ U-Th-Pb dating of bastnäsite and monazite. Mineralium Deposita, 54(8): 1256-1280

Zhang W, Hu Z C, Liu Y S, Feng L P, Jiang H S. 2019b. In situ calcium isotopic ratio determination in calcium carbonate materials and calcium phosphate materials using laser ablation-multiple collector-inductively coupled plasma mass spectrometry. Chemical Geology, 522: 16-25

Zhang Z C, Hou T, Santosh M, Li H M, Li J W, Zhang Z C, Song X Y, Wang M. 2014. Spatio-temporal distribution and tectonic settings of the major iron deposits in China: an overview. Ore Geology Reviews, 57: 247-263

Zhang Z W, Tang Q Y, Li C S, Wang Y L, Ripley E M. 2017. Sr-Nd-Os-S isotope and PGE geochemistry of the Xiarihamu magmatic sulfide deposit in the

Qinghai-Tibet Plateau, China. Mineralium Deposita, 52(1):51-68

Zhao H D, Zhao K D, Palmer M R, Jiang S Y. 2019. *In-situ* elemental and boron isotopic variations of tourmaline from the Sanfang granite, South China: insights into magmatic-hydrothermal evolution. Chemical Geology, 504:190-204

Zhao J H, Li Q W, Liu H, Wang W. 2018. Neoproterozoic magmatism in the western and northern margins of the Yangtze Block (South China) controlled by slab subduction and subduction-transform-edge-propagator. Earth-Science Reviews, 187:1-18

Zheng Y C, Liu S A, Wu C D, Griffin W L, Li Z Q, Xu B, Yang Z M, Hou Z Q, O'Reilly S Y. 2019. Cu isotopes reveal initial Cu enrichment in sources of giant porphyry deposits in a collisional setting. Geology, 47(2):135-138

Zhou C X, Wei C S, Guo J Y, Li C Y. 2001. The source of metals in the Qilinchang Zn-Pb deposit, northeastern Yunnan, China: Pb-Sr isotope constraints. Economic Geology, 96:583-598

Zhou J X, Huang Z L, Zhou M F, Zhu X K, Muchez P. 2014. Zinc, sulfur and lead isotopic variations in carbonate-hosted Pb-Zn sulfide Deposits, Southwest China. Ore Geology Reviews, 58:41-54

Zhou J X, Bai J H, Huang Z L, Zhu D, Yan Z F, Lv Z C. 2015. Geology, isotope geochemistry and geochronology of the Jinshachang carbonate-hosted Pb-Zn deposit, Southwest China. Journal of Asian Earth Sciences, 98:272-284

Zhou J X, Luo K, Wang X C, Wilde S A, Wu T, Huang Z L, Cui Y L, Zhao J X. 2018a. Ore genesis of the Fule Pb-Zn deposit and its relationship with the Emeishan large igneous province: evidence from mineralogy, bulk C-O-S and *in situ* S-Pb isotopes. Gondwana Research, 54:161-179

Zhou J X, Xiang Z Z, Zhou M F, Feng Y X, Luo K, Huang Z L, Wu T. 2018b. The giant Upper Yangtze Pb-Zn province in SW China: reviews, new advances and a new genetic model. Journal of Asian Earth Sciences, 154:280-315

Zhou J X, Wang X C, Wilde S A, Luo K, Huang Z L, Wu T, Jin Z G. 2018c. New insights into the metallogeny of MVT Zn-Pb deposits: a case study from the Nayongzhi in South China, using field data, fluid compositions, and *in situ* S-Pb isotopes. American Mineralogist, 103:91-108

Zhou L, Mernagh T P, Lan T G, Tang Y W, Wygrałak A. 2019. Intrusion related gold deposits in the Tanami and Kurundi-Kurinelli goldfields, northern Territory, Australia: constraints from LA-ICPMS analysis of fluid inclusions. Ore Geology Reviews, 115:103189

Zhou M F, Robinson P T, Lesher C M, Keays R R, Zhang C J, Malpas J. 2005. Geochemistry, petrogenesis and metallogenesis of the Panzhihua gabbroic layered intrusion and associated Fe-Ti-V oxide deposits, Sichuan Province, SW China. Journal of Petrology, 46(11):2253-2280

Zhou M F, Chen W T, Wang C Y, Prevec S A, Liu P P, Howarth G H. 2013. Two stages of immiscible liquid separation in the formation of Panzhihua-type Fe-Ti-V oxide deposits, SW China. Geoscience Frontiers, 4(5):481-502

Zhou M F, Gao J, Zhao Z, Zhao W W. 2018. Introduction to the special issue of Mesozoic W-Sn deposits in South China. Ore Geolog Reviews, 101:432-436

Zhou T F, Wang S W, Fan Y, Yuan F, Zhang D Y, White N C. 2015. A review of the intracontinental porphyry deposits in the Middle-Lower Yangtze River Valley metallogenic belt, eastern China. Ore Geology Reviews, 65:433-456

Zhou Y, Zhong H, Li C S, Ripley E M, Zhu W G, Bai Z J, Li C. 2017. Geochronological and geochemical constraints on sulfide mineralization in the Qingmingshan mafic intrusion in the western part of the Proterozoic Jiangnan orogenic belt along the southern margin of the Yangtze Craton. Ore Geology Reviews, 90:618-633

Zhu J, Wang L X, Peng S G, Peng L H, Wu C X, Qiu X F. 2017. U-Pb zircon age, geochemical and isotopic characteristics of the Miaoya syenite and carbonatite complex, central China. Geological Journal, 52(6):938-954

Zhu J, Zhang Z C, Santosh M, Jin Z L. 2020. Carlin-style gold province linked to the extinct Emeishan plume. Earth and Planetary Science Letters, 530:115940

Zhu J J, Hu R Z, Richards J P, Bi X W, Zhong H. 2015. Genesis and magmatic-hydrothermal evolution of the Yangla skarn Cu deposit, Southwest China. Economic Geology, 110(3):631-652

Zhu J J, Richards J P, Rees C, Creaser R A, DuFrane S A, Locock A, Petrus J A, Lang J. 2018. Elevated magmatic sulfur and chlorine contents in ore-forming magmas at the Red Chris porphyry Cu-Au deposit, northern British Columbia, Canada. Economic Geology, 113(5):1047-1075

Zhu L Y, Liu Y S, Jiang S Y, Lin J. 2019. An improved *in situ* technique for the analysis of the Os isotope ratio in sulfides using laser ablation-multiple ion counter inductively coupled plasma mass spectrometry. Journal of Analytical Atomic Spectrometry, 34(8):1546-1552

Zhu W G, Zhong H, Deng H L, Wilson A H, Liu B G, Li C Y, Qin Y. 2006. SHRIMP zircon U-Pb age, geochemistry and Nd-Sr isotopes of the Gaojiacun mafic-ultramafic intrusive complex, Southwest China. International Geology Review, 48(7):650-668

Zhu W G, Zhong H, Li X H, Liu B G, Deng H L, Qin Y. 2007. ^{40}Ar-^{39}Ar age, geochemistry and Sr-Nd-Pb isotopes of the Neoproterozoic Lengshuiqing Cu-Ni sulfide-bearing mafic-ultramafic complex, SW China. Precambrian Research, 155(1-2):98-124

Zhu X K, Sun J, Pan C X. 2015. Sm-Nd isotopic constraints on rare-earth mineralization in the Bayan Obo ore deposit, Inner Mongolia, China. Ore Geology Reviews, 64:543-553

Zhu Z Y, Cook N J, Yang T, Ciobanu C L, Zhao K D, Jiang S Y. 2016. Mapping of sulfur isotopes and trace elements in sulfides by LA-(MC)-ICP-MS: potential analytical problems, improvements and implications. Minerals, 6(4):110

Progresses in the Study of Ore Deposit Geochemistry

ZHONG Hong, SONG Xie-yan, HUANG Zhi-long, LAN Ting-guang, BAI Zhong-jie, CHEN Wei, ZHU Jing-jing, YANG Jie-hua, XIE Zhuo-jun, WANG Xin-song

State Key Laboratory of Ore Deposit Geochemistry, Institute of Geochemistry, Chinese Academy of Sciences, Guiyang 550081

Abstract: In recent decade, numerous important achievements for ore deposit geochemistry have been made in China. This paper briefly summarizes some progresses in the studies of magmatic Cu-Ni-(PGE) and Fe-Ti-V deposits, porphyry Cu deposits, granite-related W-Sn deposits, carbonatite-related REE deposits, Carlin-type Au deposits, and Mississippi Valley-type Pb-Zn deposits, and the implications of *in-situ* analytical methods and experimental geochemistry in researches of ore deposits. Notably, a breakthrough has been made in the exploration of magmatic Cu-Ni sulfide deposits in the orogenic belt, and a magma-conduit system has been confirmed to be critical for the huge accumulation of V-Ti-bearing magnetite. The metallogenic model for collision-type porphyry Cu deposits has been increasingly improved, and the ore-forming process and mechanism for granite-related W-Sn deposits has been more elaborately illustrated. The metallogenic timing for carbonatite-related REE deposits has been more precisely confined. Significant progresses in researches on timing and geodynamic background, as well as deep understandings on the sources of ore-forming materials and the evolution of ore-forming fluids for large-scale mineralizations of low-temperature hydrothermal deposits in South China have also been made. Moreover, *in-situ* micro-analyses of elemental and isotopic compositions have played important roles for elaborately illustrating the ore-forming processes, and applications of experimental geochemistry shed preliminary light on studies of ore-forming mechanisms of mineral deposits. In addition, several perspectives on future studies that need to be paid attention to have been proposed.

Key words: magmatic deposit; magmatic-hydrothermal deposit; low-temperature hydrothermal deposit; *in-situ* analytical method; ore deposit geochemistry

铀矿地质科技主要进展*

李子颖　秦明宽　范洪海　蔡煜琦　程纪星　郭冬发　叶发旺　范　光　刘晓阳

核工业北京地质研究院，北京 100029

摘　要：本文总结了近十年我国铀矿地质工作的主要进展，包括铀矿成矿理论创新、铀成矿类型和成矿区带划分、全国铀资源潜力评价、主要工业铀矿类型研究评价、相山科学深钻、零价态金属铀的发现及新矿物发现等；论述了砂岩型铀矿快速评价、热液型铀矿攻深找盲、大数据找矿、遥感高光谱、钻探工艺及分析测试等技术创新成果；概述了依据理论创新及技术方法集成创新在国内外铀矿找矿领域的重大突破；展望了铀矿地质发展方向。

关键词：铀成矿理论　铀矿勘查技术　铀矿找矿进展　展望

0　引　言

近十年来，我国铀矿地质科技工作者继承和发展了铀矿成矿理论，基于我国较复杂的地质构造背景，在砂岩铀矿成矿理论方面，突破了美国学者提出的"卷状砂岩铀成矿"和苏联学者提出的"次造山成矿"理论，提出了"叠合复成因"和"构造活动带成矿"理论；在热液铀成矿方面，创立了"热点深源铀成矿"理论，推动了我国铀矿找矿的重大突破。

在铀矿成矿规律和预测评价方面取得系列进展和突破。结合我国传统的分类和国际上铀矿的划分方案，对我国铀矿成矿类型和成矿区带进行了重新划分；针对我国发现的不同类型铀矿床，系统建立了各类型的矿床式和成矿模式，建立了铀矿资源潜力评价技术方法，完成了全国铀矿资源潜力评价以及重点成矿区带的动态评价，基本摸清了我国主要铀矿类型的铀资源潜力，提交了一批成矿远景区和找矿靶区；对我国主要工业铀矿类型砂岩型、花岗岩型、火山岩型和碳硅泥岩型铀矿进行了系统研究评价，总结了我国 60 多年来铀矿勘查工作进展和铀矿成矿规律。

在铀矿勘查技术方面，首次建立了我国热液型铀矿攻深找盲技术体系，构建了大型层间氧化带型砂岩铀矿快速评价技术方法体系，优选出适宜于砂岩铀矿成矿环境和条件的不同探测对象的技术方法，实现了地质、遥感、地球物理和地球化学及信息集成处理等多项技术的有效配置；结合计算机信息技术，基本建立了大数据找矿技术方法，向多元大数据的三维数字化-集成化-智能化-信息化预测、找矿迈进。在高光谱遥感方面，核地质系统已形成颇具特色的航空全谱带高光谱测量和应用能力以及地面-钻孔岩心成像光谱编录能力，并开发了一系列技术方法；在钻探装备及施工技术、地浸工艺孔施工技术、铀资源勘查钻探工艺等领域开展了一系列技术研发工作，成功研制了高效、耐久钻头和交流变频电动顶驱式地质岩心新型钻机以及复合式液动冲击器等；核地质分析测试新方法、新技术不断发展，总的向微区更微、精度更高，由二维向三维方向发展，包括二次离子质谱分析技术、FIB-TOF-SIMS 联用技术和基于 X-CT 的岩心三维扫描及铀矿物识别技术等。

理论创新与生产实际密切结合，实现了国内外铀矿找矿的重大突破，使我国砂岩铀矿资源占比由过去的第三位跃升为第一位，热液型铀矿深部取得重大突破，扩大了铀资源量；这些进展和成果也为海外走出去提供了理论和技术支撑，如承担了中沙双边国际合作项目——沙特铀钍资源调查评价项目，实现

---

* 原文"我国铀矿地质科技近十年的主要进展"刊于《矿物岩石地球化学通报》2021 年第 40 卷第 4 期，本文略有修改。

了铀、钍、铌、钽等找矿的重大突破。

1 铀矿成矿理论创新

我国铀矿地质工作始于1955年，经过65年的发展，已建立起完整的铀矿地质科研和勘查技术体系，在全国发现了砂岩型、花岗岩型、火山岩型和碳硅泥岩型（黑色页岩型）铀矿床350多个（张金带等，2018），为国防建设和核能发展做出了历史性贡献。

近十年来，我国铀矿地质科技工作者基于我国较复杂的地质构造背景，发展了砂岩铀矿成矿理论，提出了"叠合复成因"和"构造活动带成矿"理论（Li et al., 2008；李子颖等，2009，2019；张金带，2016）；在热液铀成矿理论方面，创立了"热点深源铀成矿"理论（李子颖等，1999，2014a；李子颖，2006）。这些成矿理论不但指导推动了我国铀矿找矿工作的重大突破，而且为世界铀矿地质科技的发展做出了贡献。

1.1 叠合复成因砂岩铀成矿

中生代以来，由于受北部西伯利亚、东部太平洋和西南部印度三大板块的夹持碰撞和俯冲作用，造就了我国及其复杂的地质构造演化特征，表现为时间上的多期多阶段，空间垂向上的叠置复合、水平上的差异非均质，且构造活动频繁、幅度大。这些特征对我国中新生代盆地砂岩型铀矿的形成具有直接的影响，反映在鄂尔多斯盆地北部超大型砂岩铀矿成矿地质特征上，如发现该区铀矿化产于侏罗系直罗组灰绿色砂岩与灰色砂岩之间的过渡带中，矿化目标层砂岩颜色均呈还原色调，矿石中铀矿物主要是铀石，此外还在矿石中发现了大量的多期次油气包裹体，这些特征不同于一般的砂岩型铀矿床，我国学者据此提出了"砂岩铀矿叠合复成因"成矿理论和模式（Li et al., 2008；李子颖等，2009，2019）。

鄂尔多斯盆地东北部砂岩型铀矿的成矿过程非常复杂，经历了构造的多期次的"动-静"耦合、潜水氧化与层间氧化成矿作用的叠加、油气-热流体的复合改造等地质成矿作用（Li et al., 2008）。砂岩型铀矿叠合复成因成矿主要体现在铀源的叠合、铀成矿流体的叠合和铀成矿作用的叠合方面（Li et al., 2008；李子颖等，2009，2019），其主要内容包括如下几个方面：

（1）铀源的叠合。主要是指成矿物质铀源来自蚀源区、含矿层和深部油气流体；铀的迁移形式除了碳酸铀酰络合物外，有机酸对铀迁移也具重要作用，铀的富集是载铀成矿流体在流经特定场所时，由于吸附和还原作用，使铀卸载、富集沉淀。影响含铀成矿流体条件改变和氧化还原作用的重要介质是还原物质，如有机质，它对铀沉淀的作用主要是还原和吸附。

（2）铀成矿流体的叠合。铀矿的形成经历了大气降水、油气流体和热流体等多种流体作用的叠合，但主体上是由氧化还原作用形成，油气作用是多期次的，它们与铀成矿作用主要表现在3个方面：一是在铀成矿作用时，油气（主要是深部上升的还原气体）形成地球化学还原障促使铀还原沉淀；二是在主要铀成矿作用后，油气的强还原作用导致原氧化砂岩再次被还原，即二次还原，使氧化色砂岩变成还原色的灰绿色砂岩；三是提供一定的铀源，该矿床在铀成矿后，还经历了较强的热流体改造作用，形成了黄铁矿、方铅矿和闪锌矿等低温矿物组合，使原先沉淀富集的铀重新组合形成铀石。

（3）铀成矿作用的叠合。鄂尔多斯砂岩型铀矿化的形成经历了原生铀的预富集、潜水氧化与层间氧化成矿作用及油气-热流体的复合改造等成矿作用，预富集阶段发生在170 Ma，古潜水氧化作用发生在135~160 Ma，古层间氧化作用阶段发生在65~125 Ma。矿床形成后，大约在8~20 Ma发生了较强烈的热改造作用，叠合复杂的成矿过程使该铀矿床具有上述独特的地质特征。

1.2 构造活动带砂岩铀成矿

根据美国、苏联的成矿理论，构造活动带难以成矿。近十多年来，我国学者通过对我国产铀盆地地质构造和铀成矿特征及成因的系统研究，创新性地构建了多种构造活动背景下的铀成矿模式，提出了构造活动区（带）砂岩型铀矿成矿理论，打破了构造活动带不能成矿的传统认识，继承发展了苏联学者提出的"次造山砂岩铀成矿作用"理论，其核心内容包括：①砂岩铀矿形成于构造动力由伸展到挤压的转换阶段，前者有利于含矿建造形成，后者有利于砂岩型铀矿成矿的改造；②构造活动是砂岩型铀矿成矿的原动力，可促进铀的活化迁移；③强构造活动区铀再迁移、再沉淀、再富集，在相对较稳定期（一般不少于 2 Ma）可形成砂岩铀矿；④后期构造活动对先前形成的砂岩型铀矿产生"改造富集"和"再造破坏"的双重作用，使砂岩型铀矿的找矿由盆缘拓展到盆中，大大提升了砂岩型铀矿的找矿前景，如鄂尔多斯盆地西部磁窑堡地区、伊犁盆地南缘中段蒙其古尔地区、松辽盆地西南钱家店地区等典型的构造活动带成矿。构造活动带铀成矿模式丰富和发展了砂岩型铀成矿理论，拓展了砂岩型铀矿的找矿空间。

1.3 热点深源热液铀成矿

热点一般被认为是地幔柱到达地表的作用形式（Wilson，1963），是在深部热动力作用下或在其影响下较长时间多期次改造深部壳幔物质于地表的综合地质作用，可起源于地壳或地幔的不同深度（李子颖等，1999，2010，2014b；李子颖，2006）。热点活动在浅部地壳或地表是以构造、岩浆、沉积、变质和成矿等地质作用的强度来体现，其变化主要取决于热点活动的强弱和发展演化阶段。它不仅控制着地质构造作用，而且也与成矿作用有着密切的关系，因此，它也是地学研究的热点。

热点铀成矿作用是在热点作用或其影响下产生的铀成矿作用，其核心内容包括：①铀是在复杂的多期次岩浆和流体作用过程中在晚期的熔体或流体中富集。铀是一种不相容元素，模拟计算表明玄武质和流纹质原岩经部分熔融和分离结晶以及流体的萃取分异可使铀富集，甚至达到矿化浓度，这也是铀矿化为什么产在多期次岩浆作用区的原因；②铀主要来自深部，成矿流体具还原性，是具复杂组成的超临界流体。这可由矿石中普遍存在的硫化物和硒化物、较高的钴、镍、铬含量、成矿流体中含较多的还原组分 CO、S、H_2 所证明；③铀主要以四价的复杂络合物或配合物形式迁移；④铀的富集沉淀主要是成矿流体在作用于近地表时，由物理化学条件的改变所致；⑤铀成矿主要形成于热点作用晚期产生的伸展构造动力学背景，控制铀矿的核心因素是热点作用与构造作用的叠合，如下庄、诸广、苗儿山、桃山、会昌、相山铀矿田等。成矿区表现于 3 个典型的异常场特征：应力形变场、交代蚀变场和元素叠置场（李子颖等，1999，2010，2014；李子颖，2006）。

热点铀成矿作用不仅从理论上可以很好地解释铀成矿区集中的地质构造作用、交代变质作用、岩浆作用和成矿作用相互之间的成生联系，而且也可解释为什么在同一构造-岩浆带有的有矿、有的无矿的现象；热液型铀矿成矿深度可达 3 km；热点作用区的构造薄弱带区（构造破碎带区、不同岩性接触界面、渗透地质体）是成矿的有利部位，而与围岩类型关系不大。这一理论大大拓展了找矿空间，对指导热液型铀矿深部找矿突破发挥了重要作用。近期，由核工业北京地质研究院李子颖团队承担的"华南热液型铀矿基地深部探测技术示范"项目在诸广地区实施的长江科钻 1 号钻孔深部取得重大找矿突破，其中在 950 m 发现厚大铀矿体，在 1 535 m 发现我国目前最深的工业铀矿化。

2 铀矿地质研究、科学深钻及成矿预测进展

2.1 铀成矿类型和成矿区带新划分

铀矿床分类是铀矿成矿作用研究和勘查工作的基础，对实际找矿和理论研究均具有重要意义，不同的学者和不同的国家采用的分类原则及方法不尽相同。国际上是以国际原子能机构的分类为代表，其分类原则以主控矿要素为基础，划分了砂岩型、不整合面型、多金属铁氧化物角砾杂岩型、古石英卵石砾岩型、花岗岩型、变质岩型、侵入岩型、火山岩型、交代岩型、表生型、碳酸盐岩型、塌陷角砾岩型、

磷块岩型、褐煤与煤岩型和黑色页岩型等15大类（IAEA，2018），为世界各国的铀矿床成矿类型的研究和类比以及促进找矿工作发挥了重要作用。

我国已发现的铀矿床产出的地质构造背景复杂，岩类众多，传统上按含矿围岩分为砂岩型、花岗岩型、火山岩型和碳硅泥岩型四大类，这在我国的实际找矿中发挥了重要作用。随着发现的矿床越来越多，特点各异，成因类型复杂，为此，在考虑传统分类的基础上，按成因大类进行了新的分类。按照我国传统的以含矿主岩为主线进行分类，再按成矿环境分亚类（张金带等，2012），将我国铀矿床主要划分为岩浆型、热液型、陆相沉积型和海相沉积型4个大类，继而分为9类21亚类（表1）。在上述9类中，花岗岩型、火山岩型、砂岩型、碳硅泥岩型（又称黑色页岩型）四大类是我国主要铀矿类型，占我国铀资源总量的90%以上。这一矿床类型的划分还兼顾了IAEA和国内铀矿的传统分类方法，对开展科学研究、找矿勘查和潜力评价及国际上的铀矿类型对比具有重要意义。

表1 中国铀矿床类型划分方案

大类	类	亚类
岩浆型	伟晶岩型	—
	碱性岩型	—
热液型	花岗岩型	岩体内带亚类
		岩体外带亚类
		岩体上覆盆地亚类
	火山岩型	火山角砾岩筒亚类
		次火山岩亚类
		密集裂隙带亚类
		层间破碎带亚型
		火山沉积碎屑岩亚类
陆相沉积型	砂岩型	层间氧化型亚类
		潜水氧化型亚类
		沉积成岩型亚类
		叠合复成因型亚类
	泥岩型	—
	煤岩型	—
海相沉积型	碳硅泥岩型（黑色页岩型）	沉积-成岩亚类
		沉积-外生改造亚类
		沉积-热液叠加亚类
		沉积-热液-淋积亚类
	磷块岩型	—

依据铀成矿区带划分方案，全国范围内共划分为4个铀成矿域、10个铀成矿省、49个铀成矿区带。4个铀成矿域是古亚洲成矿域、秦祁昆成矿域、滨太平洋成矿域和特提斯成矿域。10个铀成矿省是阿尔泰-准噶尔成矿省、天山成矿省、塔里木成矿省、祁连-秦岭成矿省、大兴安岭成矿省、吉黑（造山系）成矿省、华北陆块成矿省、扬子陆块成矿省、华东南成矿省、冈底斯-三江成矿省。

2.2 全国铀矿资源潜力评价

近十年来，针对我国发现的不同类型铀矿床，我国学者系统建立了各类型的矿床式和成矿模式，完成了全国铀矿资源潜力评价以及重点成矿区带的动态评价，基本摸清了我国主要铀矿类型的铀资源潜力，这方面取得的重要创新性成果体现在如下5个方面：

（1）结合铀矿床的含矿建造、成矿时代、成矿作用、矿化特征及其产出的地质背景，厘定了我国铀矿床式75个，完成了111个典型矿床的建模，新建立的成矿条件描述模式，具有实用性强的新特色，为全国铀矿资源潜力评价奠定了坚实基础。

（2）划分了铀矿预测类型，确定了不同类型铀矿预测工作区分布范围。在系统研究已知铀矿床的地质、物化探、遥感影像等特征基础上，提出了铀成矿预测类型划分原则，全国范围内厘定铀矿预测类型39个。分类型建立中国砂岩型、花岗岩型、火山岩型、碳硅泥岩型铀矿的成矿预测要素和预测评价模型。

（3）建立了全国铀矿资源潜力评价技术体系，确定了铀矿预测区圈定、优选的技术方法以及资源量估算方法和分类、分级原则。全国共优选出预测区342个（张金带等，2015），厘定了一批万吨级不同铀矿类型预测区，为找矿预测提供了坚实的基础。

（4）动态评估预测全国铀矿资源总量超过200万t，其中砂岩型铀矿预测资源量约占46%，其次为花岗岩型约占22%、火山岩型约占20%、碳硅泥岩型约占9%和其他类型约1%。

（5）开发了"铀矿资源潜力评价数据管理应用系统"（蔡煜琦等，2012），建立了全国铀矿资源潜力评价成果数据库。全面系统地收集了中华人民共和国成立以来全国范围铀矿相关的已有地质工作积累的资料，包括地、矿、物、化、遥和有关科研成果等，按照统一技术要求完成了地物化遥等基础类、信息类和成果类图件共3514幅，基本形成了全国铀矿资源潜力评价成果空间数据库。

2.3 铀矿地质基础研究创新

自2008年至今，为加强铀矿地质基础研究力度，形成了产、学、研联合攻关的科研模式，中国核工业地质局联合南京大学、成都理工大学、中国地质大学（武汉）、吉林大学和东华理工大学五所高校，针对铀矿地质勘查过程中的基础地质问题进行联合攻关；中国科学院地球化学研究所以胡瑞忠研究员为首的技术团队在华南铀成矿作用以及铀矿物微区原位U-Pb年代学研究领域取得了一系列创新性成果和认识。

（1）凌洪飞等（2015）提出了华南花岗岩型铀矿床形成过程。元古宙基底以泥质岩为主的富铀岩石，在印支期时通过地壳部分熔融，形成印支期强过铝的富铀花岗岩；在燕山早期，这些富铀花岗岩受到了广泛的岩浆热事件的影响而发生成矿前的面型蚀变，使含铀副矿物晶格中的铀发生活化成为活性铀，从而成为铀源体；燕山晚期华南岩石圈发生了强烈的伸展、裂解作用，诱发了地壳隆升、断陷和大规模岩浆活动，为铀成矿提供了断裂构造、热源等流体对流循环条件，以大气降水来源为主的高氧逸度流体，在花岗岩铀源体的构造裂隙系统中，经对流循环同时溶解其中的铀，成为含矿热液；当这种热液运移至扭张性、半封闭并具有还原障的断裂构造中，或遇到深源的还原性热液时，铀发生沉淀积聚富集成矿。

（2）张成江等（2015）认为，康滇地轴米易海塔地区富晶质铀矿石英脉产于新元古代受混合岩化作用影响的五马箐组黑云斜长片岩中，受韧-脆性断裂构造裂隙带控制。晶质铀矿形成于温度压力较高及深度较大的地质环境，是高温偏酸性流体在温度缓慢下降的强还原条件下结晶而成的。康滇地轴具有形成高强度铀矿化的地质背景和成矿条件，在康滇地轴混合岩地区最有找矿前景的铀矿类型应为受韧-脆性构造控制的中高温热液脉型矿化。

（3）根据构造-沉积-含铀含氧流体在时空上的耦合作用及其结果，建立了二连盆地蒸发沉积-成岩-热流体改造形成的努和廷矿床等"同盆多类型"铀矿成矿模式。同时，依据铀矿床成矿地质作用和地质特征，以及不同类型铀矿化发育的空间部位特点，指出了"同盆多类型"铀矿在盆地凹陷中发育的有利空间位置，总结了裂谷盆地不同阶段的"同盆多类型"铀矿床组合规律，为在东北亚地区类似盆地中找寻相似的铀矿床组合指明了方向（聂逢君等，2015）。

（4）当成矿期的含矿流场与沉积期的古水流体系基本一致时，铀储层砂体中层间氧化效率最高而且铀搬运通量最大，更加有利于成就大型和超大型铀矿床。在区域古构造等因素的协同影响下，同沉积期的古气候背景是制约铀储层砂体和成矿期层间氧化带发育方向和规模的极为重要的地质因素。同沉积期古气候不仅制约了铀储层砂体发育的结构和规模，同时更制约了铀储层内部和外部还原介质的类型及其空间分布规律。铀储层砂体的形态和结构制约了层间氧化带发育的方向和轨迹，而铀储层内部和外部的还原介质则控制着古层间氧化带推进的里程及前锋线位置，铀矿化作用则与氧化还原地球化学障有关（焦养泉等，2015）。

（5）依据与铀储层砂体的产出关系可划分为内部还原介质和外部还原介质，铀储层砂体的双重还原介质对铀成矿同等重要，铀储层砂体中的层间氧化作用直接与内部还原介质相关，但是当叠加有外部还原介质时，外部还原介质将通过不同的方式大大增强铀储层砂体的整体还原能力，这种组合的出现有利于稳定的层间氧化带发育，持续的铀成矿沉积期的古气候和沉积环境决定了含铀岩系还原介质的类型和丰度，以及铀储层砂体双重还原介质的组合规律，并从根本上决定了成矿期层间氧化带的发育规模（焦养泉等，2018）。

（6）在国际上首次开展了铀矿床稀有气体同位素地球化学研究（Hu et al.，2009）。基于华南铀矿床与岩石圈伸展背景下形成的断陷盆地和幔源基性脉岩伴生、铀成矿时代与基性脉岩时代一致、成矿流体中的 CO_2 和 He 具有幔源特征等事实，建立了铀矿床的地幔排气成矿模式（Hu et al.，2008，2009；胡瑞忠等，2015）。该模式认为，铀成矿需要的 CO_2，主要来自受白垩–古近纪软流圈地幔上涌、岩石圈伸展控制的地幔排气；幔源 CO_2 热流上升加入在地壳浅层断裂系统中循环的雨水成因的地下水，为浸出铀源岩石中的铀而形成成矿流体创造了条件；成矿流体中的铀在不同岩石（花岗岩、火山岩、碳硅泥岩等）的断裂中沉淀下来，分别形成了按围岩划分的 3 类铀矿床。

（7）近年来，铀矿物微区 U-Pb 定年技术已成功地应用到华南铀矿床的成矿年代学研究中，为华南铀矿床的精确定年提供了范例。如 Luo 等（2015）采用 SIMS 铀矿物 U-Pb 定年方法确定出仙石铀矿床存在 3 期铀成矿：$(135±4)$ Ma、$(113±2)$ Ma 和 $(104±2)$ Ma，且这 3 组年龄与区域上岩石圈伸展背景下形成的幔源基性岩脉的侵位年龄可一一对应；Luo 等（2017）采用沥青铀矿 SIMS 微区原位 U-Pb 定年与 U-Th-总 lPb 化学年龄相结合的方法，确定了桂北孟公界花岗岩型铀矿床的成矿年龄为 2.0 Ma，可能代表了华南新发现的一期最年轻的铀成矿事件。

2.4 中国铀矿科学深钻（CUSD1）

利用大比例尺地质调查、高精度地球物理测量、地球化学测量、放射性测量、遥感解译和三维建模等手段，系统研究了铀多金属成矿地质条件、主要控矿因素、三维地质结构、铀矿床地质特征等，遴选了科学深钻场址。

中国铀矿第一科学深钻（CUSD1）位于江西相山盆地邹家山工区东南部 2.5 km 处，创造了国内 P 口径（Φ 122 mm）绳索取心钻探深度 2 818.88 m 的记录，在成矿环境、勘查技术和铀多金属找矿方面取得一系列重大突破。

（1）重建相山盆地火山机构。基于铀矿科学深钻和高精度地球物理测量，获取了深部地质信息，揭示了火山机构和火山通道，火山主通道并不是垂直的，而是由北西向南东倾覆的，盆地存在多个岩浆–热液活动中心；相山火山盆地具有"基底–侵入岩–火山岩"三元结构特征，不是传统上认为的二元结构，构建了相山盆地三维地质模型，重塑了相山火山盆地地质结构，解决了找矿重大基础地质问题。

（2）建立标型剖面和深部勘查技术方法体系。利用科学深钻获取的岩心、深地探测数据和井中地球物理测量结果，建立了地质和物化探系列标型剖面，主要包括：岩性剖面、构造剖面、密度剖面、磁性剖面、电性剖面、放射性剖面、温度剖面、地球化学元素剖面、高光谱蚀变剖面、井中瞬变电磁测量剖面等。为相山大深度地质结构和成矿环境研究及找矿提供了标准比对和划分依据；这些标型剖面和多参数数据为深部勘查技术方法探测和标定提供了依据，提高了勘查深度和精度。

（3）铀多金属矿化发现及其成矿地质特征。首次在相山科学深钻钻孔岩心中发现了铀、铅、锌、铜、金等矿化，这些矿化在空间上具有分带性，受控于脆性-韧脆性构造带，表现为上铀-中铅锌金-下铜的多金属矿化组合（聂江涛等，2015，王健等，2016）。利用多金属矿化段中的黄铁矿进行 Rb-Sr 等时线法定年，直接获取相山多金属成矿时代为 (131.3±4.0) Ma（Guo et al.，2018），确定相山矿田多金属矿化为紧随火山-侵入岩浆作用之后的一期成矿事件。同位素年龄显示相山铀矿田主要存在两期铀矿化作用：第一期为铀-赤铁矿化阶段，成矿年龄为 (115±0.5) Ma；第二期为铀-萤石、水云母化阶段，形成年龄为 90 Ma。

铀成矿作用在相山矿田表现为热点作用诱发的高温、高压、富氟的复杂热流体体系作用产物（李子颖等，1999，2014b；李子颖，2006）。这种热系统可以产生多矿种的垂直分带（铀-多金属）、在高温碱性流体的氟碳酸稀土交代（褐帘石被氟碳钙铈矿交代）和高温富铀碱性流体条件下形成锆石、独居石、磷灰石和铀钍石新生矿物组合，在高温富氟的矿化流体条件下形成萤石-刚玉-钛铀氧化物-含钍沥青铀矿组合。

重新评价了相山盆地深部铀多金属成矿前景。该盆地具有多个岩浆-热液活动中心，不仅盆地西部和北部，而且南部均有较大的铀矿找矿前景，圈定了新的铀多金属成矿远景区。科学深钻揭露的地质条件和成矿环境表明，相山铀矿田深部具有很大的铀多金属找矿潜力，为火山岩型和花岗岩型第二找矿空间，乃至第三找矿空间铀多金属资源勘查提供了很好的示范。

2.5 零价态金属铀的发现

铀广泛分布于地球的各种地体中，由于其化学性质活泼，自然界中通常呈+3、+4、+5、+6 价态，并总以化合物状态存在，其中+4 和+6 价铀化合物稳定（李子颖等，2010）。四价态铀通常以沥青铀矿（UO_2）形成于岩浆、热液和变质作用过程中，而六价态通常以铀酰离子（UO_2^{2+}）化合物溶于水体中或在沉积、蒸发和氧化的条件下形成各种硫酸盐、碳酸盐、钒酸盐、磷酸盐等各种盐类次生铀矿物。由于铀的不稳定性和变价性，至今人们在自然界中还未发现自然金属态铀。

基于热点铀成矿作用理论认为，铀的来源具有深源性，成矿流体具还原性，铀是成矿流体进入近地表时，由于物理化学条件的改变而沉淀富集成矿的。在这一理论指导下，李子颖团队（李子颖，1999，2014b；李子颖，2006）对热液成矿流体中是否以低价铀的形式进行迁移富集成矿进行了研究。他们以我国南方典型热液铀矿床中深部原生沥青铀矿为研究对象，采用光电子能谱方法，原位分析了天然沥青铀矿中元素组成、价态并开展了与人工合成金属铀和氧化还原铀矿物的对比分析，结果发现沥青铀矿中除 U^{4+}、U^{6+}外，还有零价态，首次发现了零价态金属铀的存在，并确定了不同价态的比例（李子颖等，2015）。

这一发现为揭示热液型铀矿成矿作用机理和控矿要素提供了关键性判据，证明成矿物质铀来自深部，成矿流体具还原性，为我国铀资源的深部突破提供了理论依据。此外，根据不同价态铀的比例，可以判别矿石形成的深度，为铀成矿深度的定量预测提供依据。由于铀在自然界中是一种非常活泼的放射性元素，对不同地质作用过程是敏感的且具有时间追溯性。因此，金属铀的发现对于研究追溯地质作用过程和地球演化也具有重要意义。该研究成果是我国铀矿地质及基础地质研究领域原创性成果，突破了人们认为铀在自然界仅以价态的形式存在的惯常认识。

2.6 新矿物的发现

核地质系统是我国新矿物研究的重要力量，在 1974—1992 年间先后发现了 6 个新矿物种（范光等，2020）。近十年来，新一代矿物学工作者继续加强新矿物的研究，新发现了栾锂云母、氧钠细晶石、冕宁铀矿、羟铅烧绿石 4 个新矿物种，并获得 IMA CNMNC 的批准，续写了核地质系统在新矿物发现和研究方面的新篇章。

(1）冕宁铀矿（mianningite，IMA 2014-072）。冕宁铀矿为锶铁钛矿（crichtonite）族新矿物，其化学式为$(\square,Pb,Ce,Na)(U^{4+},Mn,U^{6+})Fe_2^{3+}(Ti_{12}Fe_6^{3+})_{18}O_{38}$，三方晶系，空间群$R$，晶胞参数：$a = 1.034\ 62(5)$ nm，$c = 2.083\ 72(2)$ nm。2013年发现于四川冕宁县牦牛坪稀土矿包子山煌斑岩破碎带中，以产地命名（Ge et al., 2017）。锶铁钛矿族矿物中的davidite（La）是澳大利亚镭山铀矿床的主要工业铀矿物，在康滇地轴发现该族的冕宁铀矿对进一步认识康滇地轴及同类铀成矿作用具有指导意义。

(2）羟铅烧绿石（hydroxyplumbopyrochlore，IMA2018-145）。羟铅烧绿石属于烧绿石超族中烧绿石族新矿物，其化学式为$(Pb_{1.5}\square_{0.5})_2Nb_2O_6(OH)$，等轴晶系，空间群$Fdm$，晶胞参数：$a = 1.055\ 78(17)$ nm，$Z = 8$。发现于沙特阿拉伯地盾的过碱性花岗质伟晶岩中（Li et al., 2020）。羟铅烧绿石富含铌、钽，甚至铀、钍，羟铅烧绿石的发现为该区铀钍铌钽的利用提供了一种工业矿物，同时为研究碱性花岗质伟晶岩的演化过程及铀钍稀有元素的富集规律提供了矿物学证据。

(3）氧钠细晶石（oxynatromicrolite，IMA 2013-063）。氧钠细晶石属于烧绿石超族中细晶石族新矿物，其化学式为$(Na,Ca,U)_2(Ta,Nb)_2O_6(O,F)$，等轴晶系，空间群为$Fdm$，晶胞参数：$a = 1.042\ 0(6)$ nm，$Z = 8$。2013年发现于豫西卢氏县官坡镇附近的官坡花岗伟晶岩密集区309花岗伟晶岩脉中（Fan et al., 2017）。氧钠细晶石是铌钽矿化的主要赋存矿物，其中含UO_2达14.6%，也是一种含铀矿物，它的发现对铌钽和铀资源的利用以及指示花岗伟晶岩的成因具有重要意义。

(4）栾锂云母（luanshiweiite，IMA2011-102）。栾锂云母为云母族新矿物，其化学式为$KLiAl_{1.5}\square_{0.5}(Si_{3.5}Al_{0.5})O_{10}(OH,F)_2$，单斜晶系，空间群$C2/c$，晶胞参数：$a = 0.518\ 61(7)$ nm，$b = 0.898\ 57(13)$ nm，$c = 1.997\ 0(3)$ nm，$V = 0.926\ 5(2)$ nm^3，$\beta = 95.420(3)°$；2M1多型。2011年发现于豫西卢氏县官坡镇附近的官坡花岗伟晶岩密集区309花岗伟晶岩脉中。该矿物原型标本来自成都理工大学栾世伟教授，以栾世伟教授的名字命名（范光等，2013）。栾锂云母的发现提供了一种新的锂元素的赋存状态，对花岗伟晶岩的成因研究具有指示意义。

3 铀矿勘查技术创新

3.1 砂岩型铀矿快速评价技术

近十年来，我国科技工作者通过对传统单项探测技术的改进和新技术的开发，优选出适宜于砂岩铀矿成矿环境和条件的不同探测对象的技术方法，实现了地质、遥感、地球物理和地球化学及信息集成处理等多项技术的有效配置，构建了新的地浸砂岩铀矿快速评价技术方法体系。该技术方法体系由区域成矿环境与远景区筛选评价技术方法子系统、成矿远景区段快速评价技术方法子系统和地浸砂岩铀资源潜力的综合识别评价技术方法子系统构成，各子系统又分别由若干技术方法有效配置而成。通过应用实现了北方重点盆地砂岩铀矿的重大突破（李子颖等，2015）。主要的技术突破和进展如下：

(1）集成了航磁、航放数据处理及弱信息提取技术。该技术可快速确定盆地基底格局、构造分区、区域放射性场（铀、钍、钾、总量）特征，圈定铀的活化区、迁出区和迁入区，区分"矿致异常"和"致矿异常"，从而为区域远景预测提供重要参数。

(2）创建了盆地古构造、古流体系统格局重塑技术，以大陆动力学、盆地动力学、区域新构造运动及叠合盆地研究的最新成果为基础，充分利用盆地构造演化、岩浆活动、同位素测年、地层学等资料，划分出盆地的构造演化阶段。在此基础上采用热年代学方法定量反演不同阶段盆山相对高差、地层剥蚀和盆地沉降幅度及不同盆段构造属性，进而重塑盆地古构造格局。以此为基础，根据古构造面貌及岩相古地理分析当时的古水动力系统格局。采用包裹体地球化学方法，对盆地沉积盖层不同层位、不同盆段（地段）开展较系统的包裹体测试研究，通过类型及期次、温度、压力、盐度、成分等特征参数，研究古流体的性质、来源、活动期次、运移途径，进而分析渗出型流体与渗入型流体的相互作用过程、古流体活动与古构造演化的时空对应性，确定古流体与砂岩铀成矿的关系。

（3）建立了含矿建造砂体识别技术。利用区域地质调查、地球物理测量、岩石露头观察、岩心编录和分析测试等资料进行盆地分析；通过盆地分析确定盆地类型、演化阶段和沉积充填特征；在构造演化格架内确定盆地不同演化阶段层序地层构成模式，建立盆地层序地层格架；在层序地层格架内确定盆地体系域和岩相分布模式，开展沉积相分析，建立相模式，预测有利沉积砂体的分布范围；根据盆地分析确定的盆地后期构造改造、埋藏史和热史等特征，确定有利沉积砂体发育的层位、规模等。

（4）发展了铀矿化深部探测技术。几乎所有的砂岩铀矿化均是深埋隐伏的铀矿，在地表没有任何矿化信息显示，且深埋地下数百米，如何探测深部成矿信息是减小找矿盲目性提高预测成功率的关键。通过铀分量化探测量、氡及其子体测量和弱信息提取技术的创新和组合应用，获得铀矿化直接和间接信息，为成功预测远景区提供了直接依据，凸显了铀矿化的空间定位。

（5）突破了开放体系 U-Pb 同位素定年方法。砂岩型铀成矿环境为开放体系，而开放体系中铀成矿时限的精确厘定是一大难题，因为非封闭体系中初始铅的扣除难以把握，铀矿物形成后因遭受后期改造而导致元素的不断带入和带出。该评价方法创新性地应用 U-Ra 平衡系数校正、古铀量恢复、微量铀矿物精确定年相结合手段，改进了开放体系 U-Pb 同位素定年方法，使所测结果最大限度地逼近其真实年龄值，为成矿过程和成矿模式研究提供了重要的依据。

（6）构建了基于 GIS 系统平台的砂岩铀矿综合预测评价系统。基于多源信息数据库和综合数字找矿模型，实现半定量和定量铀资源预测评价。

3.2 热液型铀矿攻深找盲技术

近十年来我国热液型铀矿攻深找盲技术体系首次得以建立。该体系包括深部盲矿地质评价技术体系、深部探测地球物理技术体系、深穿透地球化学技术体系及遥感影像特征纹理分形分析和亮温识别技术体系，找矿深度由 500 m 推进到 1 500 m（李子颖等，2015）。

（1）深部盲矿地质评价技术体系。包括铀成矿构造应力场恢复技术、铀成矿蚀变深部定位技术、地球化学元素组合和示踪技术、矿化体探测识别技术等，实现对岩浆活动中心识别、蚀变带矿化指示识别、成矿流体矿化识别和铀成矿构造应力场恢复等。首次将构造应力场恢复技术应用于攻深找盲铀矿预测，恢复不同时期特别是铀成矿期构造，厘定导矿构造和储矿构造，有效解决成矿空间定位问题。

（2）深部探测地球物理技术体系。通过电磁测量技术、浅层地震探测技术、高精度磁测技术、微弱信息提取技术、成矿信息地球物理直接探测技术（如氡气测量等）等试验，优化集成，创建了"电磁法+高精度磁测+高精度测氡"热液型铀矿地球物理攻深找盲技术方法组合，该方法组合集数据采集、处理、解释为一体，创新性解决了抗高压干扰技术，提出了热液型铀矿高精度磁法的连续测量和梯度测量的新方法，解决了火山岩及花岗岩地区单点磁异常真伪问题，有效探测深度达到 1 500 m，实现了对深部隐伏岩体、控矿构造及铀矿化体的有效探测。

（3）深穿透地球化学技术体系。包括放射性同位素和核素示踪、地电化学测量、分量化探等技术。创建了 3 套深穿透地球化学方法：一是首次建立铅同位素打靶法和铅同位素向量特征值法及垂向示踪组合的同位素示踪评价技术；二是以分量化探为主与 ^{210}Po 法、热释光法组合的热液型铀矿地球化学元素示踪技术；三是以地电化学勘查方法为主与土壤电导率、土壤热释汞组合的热液型铀矿物性测量找矿技术。

（4）遥感影像特征纹理分形分析和亮温识别技术体系。这是研究自然界中具有自相似形特征的图形复杂程度的一种数学方法，包括分形维数计算和多重分形谱分析两种方法，用来刻画目视解释无法区分的遥感图像的纹理特征。亮温识别方法技术是将热红外遥感传感器接收到的地表热辐射转变为图像上的亮度，并进一步利用普朗克黑体辐射模型将地物的实际辐射亮度转换为地物的温度技术，为半定量–定量解释提供了依据，首次系统建立了以遥感地质信息为主导的热液铀矿田评价标志。

3.3 大数据找矿技术方法

矿产勘查的大数据应用与数据挖掘是目前国际地质科学的重要发展方向，主要包括 3 个层面的内容：

（1）铀资源数据构建和存储管理。核地质领域开发了数字铀矿勘查系统，实现了钻孔数据采集、测井数据调用、报表输出、地质编录内容和综合柱状图计算机绘制等野外地质工作全流程的数字化，确立了铀矿勘查数字化资料检查、数据库建设等标准，建立了数据库。

（2）数据分析和各类信息提取。确立了铀资源大数据的采集、预处理、存储与管理、分析与信息挖掘以及可视化与应用 5 个方面流程；建立了分布式计算与大数据一体化等核心技术，提升了对海量的结构化、半结构化和非结构化地学大数据的存储和管理能力，开展了数据分析、信息提取等研究（蔡煜琦等，2019；叶发旺等，2019a）。

（3）机器学习与智能找矿。基本构建了局域网环境下的铀资源勘查大数据应用平台（铀矿地质云），突破了完全适合于大数据环境下的分布式计算、分布式数据挖掘、数据可视化等关键技术，在二连盆地中部基本实现了铀资源勘查大数据应用示范（刘武生等，2019），向多元大数据的三维数字化-集成化-智能化-信息化"四定"（定型、定位、定深、定量）预测技术迈进。

3.4 遥感高光谱技术

通过近十多年的发展，核地质系统已形成颇具特色的国内领先的航空全谱带高光谱测量和应用能力、地面-钻孔岩心成像光谱编录能力，并开发了一系列技术方法，在多个铀多金属成矿带中取得了明显应用效果。

（1）创建了铀矿勘查航空全谱段高光谱技术体系，突破了 CASI/SASI/TASI/MASI 多传感器航空高光谱数据获取、CASI/SASI 航空高光谱大气校正、MASI/TASI 航空中热红外高光谱温度与发射率分离、矿物填图、岩性智能识别，以及航空高光谱识别的高、中、低铝绢云母矿物成因学分析等航空高光谱数据处理与分析共性技术（叶发旺等，2013，2018，2019a），识别出了赤铁矿、褐铁矿、高铝绢云母、中铝绢云母、低铝绢云母、铁绿泥石、绿泥石、镁绿泥石、绿帘石、蒙脱石、方解石、白云石、叶蜡石、高岭石、迪开石、阳起石、蛇纹石、石膏、石英、碱性长石等 20 余种矿物，形成了国内领先的铀矿勘查航空高光谱测量技术能力。

（2）从火山岩、花岗岩等热液型铀矿地质特点出发，研发出了基于成矿有利地球化学障的航空高光谱矿物信息组合分析、铀成矿地表热液流体活动规律航空高光谱分析（Qiu et al.，2015；叶发旺等，2019b）、铀成矿环境航空高光谱蚀变矿物-岩性-构造综合分析（刘德长等，2017；叶发旺等，2019c）、以及铀成矿航空高光谱预测等具有铀矿勘查应用特色的航空高光谱信息分析与找矿预测技术，在国内地质矿产勘查领域的航空高光谱应用中具有良好的引领示范作用。

（3）利用研发的铀矿勘查航空高光谱技术在新疆雪米斯坦、甘肃龙首山、准噶尔盆地东部等重要铀成矿区带发现了十余处新的铀矿化异常，优选的找矿靶区经地面槽探查证，发现了深部控矿要素，为这些地区的铀矿勘查突破提供了重要线索；同时，还发现了十余处金、铜、铁等多金属矿化异常，优选的金铜找矿靶区十分类似紫金山金铜矿，极具找矿潜力，大大拓展了航空高光谱技术的应用效果（刘德长等，2017，2018）。

（4）建立了一套集"钻孔岩心成像光谱测量、数据处理、自动裁切、矿物识别、含量反演、三维建模"为一体的深部铀矿钻孔岩心成像光谱快速编录与分析新技术方法（Qiu et al.，2015；Zhang et al.，2019）。利用该技术方法完成我国相山铀矿第一科学深钻 2 888.8 m 的全岩心 360°扫描与信息提取分析，完成了我国数万余米的热液型和砂岩型铀矿钻孔岩心的快速扫描和矿物信息编录与分析（张川等，2019），为重要铀矿钻孔岩心资料的数字化保存和深部铀矿成矿蚀变信息分带特征与规律分析提供了重要技术支持。该技术已入选自然资源部矿产资源节约和综合利用先进适用技术目录（2019 版）。

3.5 钻探工艺新技术

近十年来,在铀矿科学深钻钻探装备及施工技术、地浸工艺孔施工技术、铀资源勘查钻探工艺等领域开展了一系列技术研发工作,取得了一系列重要成果。

(1) 研制成功高效耐久钻头。在江西相山中国铀矿科学深钻钻探过程中,使用该钻头获得岩矿心采取率达到99%以上,金刚石绳索取心钻头平均小时效率1.37 m/h,平均使用寿命115.3 m/个,最高寿命290.38 m/个。最大提钻更换钻头间隔273.61 m,最长更换钻头时间间隔427 h。创造了国内P口径(Φ 122 mm)绳索取心钻进深度纪录,标志着核工业深孔岩心钻探技术的重大突破(刘晓阳等,2013)。

(2) 研发的交流变频电动顶驱式XD-35DB型地质岩心钻机,为国内外首台,具有国际领先水平。该设备集电、液、气、信息技术为一体,大大提高了钻探装备的智能化、数字化和自动化水平(朱江龙等,2014)。

(3) 成功研制出P、H、N 3种口径的复合式液动冲击器。在节流压差式液动冲击器和射吸式液动冲击器研究的基础上(刘国经等,1985;谢文卫等,1998),采用压差原理和射吸原理相复合方式,成功研制出了P、H、N 3种口径的复合式液动冲击器。在铀矿科学深钻钻孔施工中,全孔应用研制的P口径金刚石绳索取心液动冲击器及钻进工艺技术进行施工,解决了该地区坚硬岩层钻进效率低、破碎岩层易堵卡、回次进尺长度短、金刚石钻头使用寿命短、钻孔易偏斜等钻探生产难题,使坚硬岩层钻进效率提高30%以上,钻头使用寿命提高20%以上,破碎岩层进尺长度提高100%以上,钻孔轨迹偏差控制在规范要求的范围内(叶晓平等,2018)。

(4) 成功研制出了适用于地浸砂岩铀矿卵砾石层钻进的胎体增强型孕镶金刚石钻头。在铁铜基孕镶金刚石钻头胎体中复合铸造碳化钨、碎粒硬质合金或金刚石聚晶,有效降低了孕镶金刚石取心钻头在卵砾石层钻进时金刚石换层速度,防止胎体崩裂,从而提高钻头使用寿命。胎体增强型孕镶金刚石钻头在内蒙古鄂尔多斯盆地和黑龙江三江盆地铀矿勘查项目9个钻孔中进行了钻进试验,累计试验进尺达到1 131.7 m。与普通常规钻头相比,使用寿命提高了15%以上,钻进效率提高了10%以上(刘晓阳等,2005;刘晓阳,2009)。

3.6 分析测试新技术新方法

近十年来,核地质分析测试新方法、新技术不断发展,总的向微区更微、精度更高,由二维向三维方向发展。

(1) 二次离子质谱分析技术。基于法国CAMECA公司生产的CAMECA IMS 1280HR型大尺寸高分辨二次离子质谱仪,建立了适用于复杂基体的多对象放射性同位素及稳定同位素分析技术。通过该技术,不但可以精细描绘出单颗粒矿物微米尺度三维元素同位素分布图像,还可以准确定量刻画铀矿地质研究中的形成年代、地质过程及物质来源等关键科学问题,填补了铀矿地质研究中原位微区同位素分析的空白,大大拓展了分析测试的广度和深度。该技术具有高分辨率、高效率、高精度及多维度等优势(Wu et al., 2018;He et al., 2020),对于复杂样品可以实现近乎无损的超高分辨率解析。利用该套技术,成功厘定了相山横涧铀矿床早期铀矿床年龄(He et al., 2020)及锆石氧同位素再造的现象,明确了沙特碱性伟晶岩形成中的壳幔相互作用及岩浆自氧化过程,为铀矿成矿机理研究、地球科学关键过程研究提供了强有力的技术支撑。

(2) 聚焦离子束扫描电子显微镜和飞行时间二次离子质谱联用技术。基于聚焦离子束(FIB)扫描电子显微镜和飞行时间二次离子质谱(TOF-SIMS),建立了FIB-TOF-SIMS联用技术(王涛等,2019)。该技术不但可以进行元素,尤其是超轻元素的面分析,更为重要的是还可以获得铀矿物等固体物质中元素的三维立体分布信息,解决了以往只能获得元素二维平面分布信息的技术瓶颈。该技术具有高空间分辨率(纳米级)、微区原位元素和同位素(尤其是超轻元素)分析、同时具备二维与三维分析等技术优势。利用该套联用技术，首次分析了晶质铀矿铀元素的三维空间分布，更为直观地观察到了晶质铀矿内部铀元素的分布状态，为铀矿成因、成矿机理等研究提供技术支撑。

(3) 基于X射线计算机断层扫描（X-CT）的岩心三维扫描及铀矿物识别技术。建立了X-CT三维无损成像及分析技术。相对于传统岩石矿物鉴定手段，该技术的显著优势在于可在无损状态下获得岩石样品内部三维纹理结构，结合扫描电子显微镜、激光烧蚀电感耦合等离子体质谱等岩石矿物表征技术，可有效拓展对岩石矿物内部信息的获取。通过对含铀矿物及其他常见矿物的X-CT扫描研究，初步获得了含铀矿物对X射线的衰减特征，利用该特征可实现对岩心中铀矿物颗粒有效识别，并通过三维重建获取相应矿物颗粒的物理赋存状态，为铀矿成矿理论深化研究提供技术支撑。

(4) 激光烧蚀电感耦合等离子体质谱含铀矿物分析技术。基于多接收器电感耦合等离子体质谱仪（MC-ICP-MS）和高分辨率电感耦合等离子体质谱仪（high resolution ICP-MS，HR-ICP-MS）以及准分子激光器，建立的新一代LA-ICP-MS含铀矿物分析技术，实现了含铀矿物（锆石、锡石等）、铀矿物（晶质铀矿、沥青铀矿）高空间分辨率（几十微米）微区原位U-Pb同位素定年和微量元素分析以及锆石微区原位Lu-Hf同位素的高精度测量，所获得的U-Pb同位素年龄数据可应用于揭示成岩、成矿时代（刘瑞萍等，2015，2017），微量元素则可为探讨岩石、矿床成因提供最要依据，Lu-Hf同位素可为阐明岩浆作用过程、物质来源和岩石成因提供重要的理论依据。依托该项技术，研制出了用于锡石U-Pb定年的实验室内部标准物质，获得了一项实用新型专利授权。

(5) 基于单晶衍射仪建立了单晶晶体结构解析技术。该技术通过收集10~150 μm单晶颗粒衍射数据，精确解析矿物三维结构，分析矿物中原子排列方式、原子之间键长、键角、占位率等一系列结构数据，是矿物学研究中不可缺少的技术手段。基于该技术，实验室建立了铀及含铀矿物晶体结构和晶体化学研究技术体系，为分析铀元素在复杂环境中的物理化学性质、迁移行为、赋存状态提供重要的基础数据，并发现了一种新的含铀矿物——羟铅烧绿石（Li et al.，2020），并获得了国际矿物学会新矿物命名与分类委员会的批准。

(6) 分布式实验室检测技术。通过多年的技术研发与积累，针对铀矿地质分析测试实验室的特点，建立了一套质量管理体系规范要求的分布式实验室检测技术（郭冬发等，2007）。该技术采用先进的计算机网络技术、数据库技术和标准化的管理理念，将样品的采样信息、检验流程、质量管理体系等要素有机地结合起来，构筑了一个全面、规范的信息系统，主要包括野外移动采样信息模块、实验室样品检测信息模块和实验室环境监测模块。通过该技术已将实验室管理水平、自动化水平提升到新的高度。

4　重大找矿进展

近十年来，创建和完善北方砂岩叠合复成因和构造活动带砂岩铀成矿理论，创新热液铀矿热点铀成矿理论，突破铀矿勘查系列关键技术，基本形成"天–空–地–深"三维数字勘查技术系统和铀资源"定位–定型–定深–定量"预测评价技术体系；理论技术用于实际找矿，大大拓展了找矿空间，实现了国内外铀矿找矿的重大突破。

在北方中新生代盆地伊犁、准噶尔、塔木素、鄂尔多斯、二连和松辽盆地等实现了"新类型、新层位、新区域、新深度"的突破，新类型如叠合复成因类型，新层位为侏罗系、白垩系中的6个层位，新区域指构造活动带、二次或多次再还原带地域等，新深度为找矿深度由500 m左右推进到900 m，推动落实一批大型、特大型和超大型砂岩型铀矿，形成了6个万吨级砂岩型铀矿基底，使砂岩型铀矿成为我国的第一大类型，为砂岩型铀矿基地建设提供了资源保障。

在南方相山、诸广以及苗儿山等热液型铀矿工作区，圈定了新的成矿远景区和找矿靶区，指明了找矿方向，拓展了找矿空间，扩大了已知矿床的资源储量，推动落实了一批中小型、大型和特大型铀矿床，使我国热液铀矿找矿深度由600 m左右推进到1 200 m，开辟了我国热液铀矿第二找矿空间。

在相山盆地实施的中国铀矿第一科学深钻（CUSD1）中，首次发现了上铀、中铅锌金、下铜等铀多金属矿化，铀矿化4处，铅锌矿化5段，金矿化1段，铜矿化4段，其中，2 817 m的铜矿化目前属于国内埋藏最深的铜矿化，其含量超过1%，表明第二、第三成矿空间还具有很大的铀多金属找矿潜力。

海外铀多金属找矿成果显著：在纳米比亚欢乐谷地区、尼日尔阿泽里克铀矿区等圈定了一批重要的勘查靶区，尤其是在欢乐谷地区预测的勘查靶区经钻探工程评价，落实为大型-特大型铀矿床；在近几年，完成了沙特阿拉伯铀钍资源调查评价国家项目，创下技术交叉多、完成任务快、协同联合强、取得成果多、找矿效率最高等综合成效，落实大型、超大型铀、钍、铌、钽矿床，按国际标准提交成果报告，形成了高效、快速研产结合一体化的找矿突破"沙特模式"。

5 展　望

铀资源是军民两用的战略性关键矿产，是核工业发展的物质基础。新时代，核工业的新发展对铀资源提出了新需求。"找大矿、找富矿、找经济可采矿"和"新区、新层位、新类型"的铀矿找矿新目标对铀矿地质科技创新和攻关提出了新的更高、更为紧迫的需求，铀矿地质科技工作也必将迎来一个新的发展时期。未来铀矿成矿理论的发展将不断创新完善，新一代铀矿勘查技术以大探深、智能化、绿色化、集成化为主要标志，主要重点发展方向为：

（1）大力加强铀资源重大基础前沿创新研究。主要包括砂岩型铀矿超常富集机理与板状铀成矿理论、热液型深部富大铀矿成矿机理与成矿规律、多矿种相互作用与铀成矿系统等，为铀矿勘查突破提供理论指导。

（2）大力发展砂岩型铀矿绿色智能勘查技术。主要包括绿色高效探测技术和装备、智能化勘查技术系统研发。通过高精度-高分辨-高效率砂岩铀资源勘查新技术研发、绿色高效自动化与智能化关键勘查设备研制、三维智能化勘查技术系统开发，实现三维精细化、智能化预测评价，并大幅提高铀矿勘查生产的效率。

（3）大力发展热液型铀矿绿色智能化探测技术。重点开展热液型铀矿先进勘查技术与装备研制、富大热液型铀矿四维预测与资源扩大等研究，实现更大深度富大铀矿的三维精细化、智能化预测评价。

（4）发展放射性共伴生资源高效预测评价技术。开展全国成矿区带铀钼、铀稀土、铀铍、铀铌钽以及铀磷等成矿环境、勘查技术、预测评价等研究，建立铀及共伴生多金属资源综合评价技术。

（5）持续推进铀矿科学深钻工程。铀矿科学深钻是揭示深部铀成矿地质结构与环境、探索极限成矿深度和找矿深度的"钥匙"。我国铀矿深部科学钻探工程已取得重大进展和突破，今后一段时期内，铀矿科学深钻将向纵深方向发展，在火山岩型铀矿3 000 m科学深钻的基础上，有望实施花岗岩型铀矿田铀多金属深部探测等科学钻探工程，并形成4 000 m深度的深部科学钻探技术与能力。

致谢：核工业北京地质研究院张杰林、聂江涛、刘祜、葛祥坤、刘武生、李佳丽等提供了相关素材，审稿专家提出了诸多宝贵意见，在此一并致以衷心感谢！

参 考 文 献

蔡煜琦,张文明,赵永安,田珊. 2012. 全国铀矿资源潜力评价数据管理与应用系统的研制. 铀矿地质,28(6):393-397

蔡煜琦,张金带,李子颖,郭庆银,宋继叶,范洪海,刘武生,漆富成,张明林. 2015. 中国铀矿资源特征及成矿规律概要. 地质学报,89(6):1051-1069

蔡煜琦,虞航,李晓翠,刘佳林,章看铭. 2019. 大数据时代铀矿资源预测评价的技术方法探讨. 铀矿地质,35(6):321-329

范光,李国武,沈敢富,徐金莎,戴婕. 2013. 栾锂云母：锂云母系列的新成员. 矿物学报,33(4):713-721

范光,葛祥坤,李婷,于阿朋,王涛. 2020. 我国核地质系统发现的新矿物评述. 世界核地质科学,37(1):1-9

郭冬发,武朝晖,崔建勇,欧光习,范光. 2007. 铀矿地质分析测试技术回顾与新形势下网络实验室构建. 世界核地质科学,24(1):50-62

胡瑞忠,毛景文,华仁民,范蔚茗. 2015. 华南陆块陆内成矿作用. 北京：科学出版社焦养泉,吴立群,彭云彪,荣辉,季东民,苗爱生,里宏亮. 2015. 中国北方古亚洲构造域中沉积型铀矿形成发育的沉积–构造背景综合分析. 地学前缘,22(1):189–205
焦养泉,吴立群,荣辉. 2018. 砂岩型铀矿的双重还原介质模型及其联合控矿机理:兼论大营和钱家店铀矿床. 地球科学,43(2):459–474
李军杰,刘汉彬,张佳,金贵善,张建锋,韩娟. 2016. 应用Argus多接收稀有气体质谱仪准确测量空气的Ar同位素组成. 岩矿测试,35(3):229–235
李子颖. 2006. 华南热点铀成矿作用. 铀矿地质,22(2):65–69,82
李子颖,李秀珍,林锦荣. 1999. 试论华南中新生代地幔柱构造、铀成矿作用及其找矿方向. 铀矿地质,15(1):10–17,34
李子颖,方锡珩,陈安平,欧光习,孙晔,张珂,夏毓亮,周文斌,陈法正,李满根,刘忠厚,焦养泉. 2009. 鄂尔多斯盆地东北部砂岩型铀矿叠合成矿模式. 铀矿地质,25(2):65–70,84
李子颖,黄志章,李秀珍,何建国. 2010. 南岭贵东岩浆岩与铀成矿作用. 北京:地质出版社
李子颖,张金带,秦明宽,范洪海. 2014a. 中国铀矿成矿模式. 北京:中国核工业地质局,核工业北京地质研究院
李子颖,黄志章,李秀珍,张金带,林子瑜,张玉燕. 2014b. 相山火成岩与铀成矿作用. 北京:地质出版社
李子颖,秦明宽,蔡煜琦,范洪海,程纪星,郭冬发,叶发旺,陆士立,梁春利. 2015. 铀矿地质基础研究和勘查技术研发重大进展与创新. 铀矿地质,31(S1):141–155
李子颖,方锡珩,秦明宽. 2019. 鄂尔多斯盆地北部砂岩铀成矿作用. 北京:地质出版社
凌洪飞,陈培荣,陈卫锋,孙涛,吴俊奇,吴欢. 2015. 华南中生代花岗岩与铀成矿. 见:中国地质学会2015学术年会论文摘要汇编. 西安:中国地质学会地质学报编辑部,322–327
刘德长,赵英俊,叶发旺,田丰,邱骏挺. 2017. 航空高光谱遥感区域成矿背景研究——以甘肃柳园–方山口地区为例. 遥感学报,21(1):136–148
刘德长,邱骏挺,闫柏琨,田丰. 2018. 高光谱热红外遥感技术在地质找矿中的应用. 地质论评,64(5):1190–1200
刘国经,朱万正,王有群. 1985-04-01. 射吸式冲击器:中国,CN85101970A
刘瑞萍,顾雪祥,章永梅,王佳琳,郑硌,高海军. 2015. 黑龙江东安金矿床赋矿岩浆岩锆石U-Pb年代学及岩石地球化学特征. 岩石学报,31(5):1391–1408
刘瑞萍,顾雪祥,章永梅,谢胜凯,郑硌,王佳琳,郭冬发. 2017. 黑龙江宝山夕卡岩型铜钼钨多金属矿床成岩成矿时代及其地质意义. 矿物学报,37(3):276–284
刘武生,朱鹏飞,孔维豪. 2019. 二连基地砂岩型铀矿地质数字平台和大数据找矿预测2019年度报告. 北京:核工业北京地质研究院
刘晓阳. 2009. 孕镶金刚石–针状合金复合式取心钻头的应用研究. 探矿工程(岩土钻掘工程),36(增刊):377–381
刘晓阳,杨爱军,孙建华. 2005. 混镶式硬质合金钻头在卵砾岩层中的应用. 西部探矿工程,17(12):188–189
刘晓阳,李大昌,叶雪峰. 2013. 中国铀矿第一科学深钻施工概况. 探矿工程(岩土钻掘工程),40(增刊):297–299,304
聂逢君,李满根,邓居智,严兆彬,张成勇,姜美珠,杨建新,旷文战,康世虎,申科峰. 2015. 内蒙古二连裂谷盆地"同盆多类型"铀矿床组合与找矿方向. 矿床地质,34(4):711–729
聂江涛,李子颖,王健,郭建. 2015. 江西相山矿田多金属成矿流体特征与成矿作用. 地质通报,34(2-3):535–547
王健,聂江涛,郭建,黄志章,李秀珍. 2016. 江西相山矿田深部多金属矿化特征. 地质与勘探,52(1):47–59
王涛,葛祥坤,范光,郭冬发. 2019. FIB-TOF-SIMS联用技术在矿物学研究中的应用. 铀矿地质,35(4):247–252
谢文卫,苏长寿,宋爱志. 1998. 新型高冲击功液动潜孔锤的研究. 探矿工程(岩土钻掘工程),25(6):31–32
叶发旺,王存,张川,刘洪成,武鼎. 2013. 航空高光谱遥感技术在新疆雪米斯坦铀多金属矿产勘查中的应用研究. 地质论评,59(Z1):930–931
叶发旺,孟树,张川,徐清俊,刘洪成,武鼎. 2018. 航空高光谱识别的高、中、低铝绢云母矿物成因学研究. 地质学报,92(2):395–412
叶发旺,王建刚,邱骏挺,张川. 2019a. 面向地质应用的航空高光谱CASI-SASI数据大气校正方法对比研究. 光谱学与光谱分析,39(9):2677–2685
叶发旺,张川,徐清俊,孟树,邱骏挺,王建刚. 2019b. 热液流体活动规律高光谱遥感分析示范研究——以新疆白杨河铀矿床为例. 矿床地质,38(6):1347–1364
叶发旺,孟树,张川,邱骏挺,王建刚,刘洪成,武鼎. 2019c. 甘肃龙首山芨岭铀矿床碱交代型铀矿化蚀变航空高光谱识别. 地球信息科学学报,21(2):279–292
叶晓平,李博,刘晓阳,高辉,段隆臣. 2018. 差动式双作用液动冲击器冲锤动力学方程的研究和应用. 地质与勘探,54(4):801–809
张成江,陈友良,李巨初,徐争启,姚健. 2015. 康滇地轴巨粒晶质铀矿的发现及其地质意义. 地质通报,34(12):2219–2226
张川,叶发旺,徐清俊,邱骏挺. 2019. 相山铀矿田西部深钻岩心成像光谱编录及蚀变分带特征. 国土资源遥感,31(2):231–239
张金带. 2016. 我国砂岩型铀成矿理论的创新和发展. 铀矿地质,32(6):321–332
张金带,李子颖,蔡煜琦,郭庆银,李友良,韩长青. 2012. 全国铀矿资源潜力评价工作进展与主要成果. 铀矿地质,28(6):321–326
张金带,李子颖,徐高中. 2015. 我国铀矿勘查的重大进展和突破. 北京:地质出版社
张金带,蔡煜琦,徐浩,张明林,贾立城,张字龙,黄净白,张书成,李田港,陈祖伊,仇宝聚,虞航,柯丹. 2018. 中国矿产地质志(铀矿卷). 北京:地质出版社
朱江龙,张伟,黄洪波,胡时友,刘跃进. 2014. 深孔取心钻进用高速顶驱式钻机. 探矿工程(岩土钻掘工程),41(9):114–119
Fan G, Ge X K, Li G W, Yu A P, Shen G F. 2017. Oxynatromicrolite, (Na, Ca, U)$_2$Ta$_2$O$_6$(O, F), a new member of the pyrochlore supergroup from Guanpo, Henan Province, China. Mineralogical Magazine, 81(4):743–751
Ge X K, Fan G, Li G W, Shen G F, Chen Z R, Ai Y J. 2017. Mianningite (□, Pb, Ce, Na)(U^{4+}, Mn, U^{6+})Fe$_2^{3+}$(Ti, Fe^{3+})$_{18}$O$_{38}$, a new member of the

crichtonite group from Maoniuping REE deposit, Mianning County, Southwest Sichuan, China. European Journal of Mineralogy, 29(2):331-338

Guo J, Li Z Y, Nie J T, Huang Z Z, Wang J, Lai C K. 2018. Genesis of Pb-Zn mineralization beneath the Xiangshan uranium orefield, South China: constraints from H-O-S-Pb isotopes and Rb-Sr dating. Resource Geology, 68(3):275-286

He S, Li Z Y, Guo D F, Wang Y J, Zhang C, Guo J, Fan Z W. 2020. Early mineralization age of the Hengjian uranium deposit: constraints from zircon SIMS U-Pb dating. Acta Geologica Sinica, 94(1):212-213

Hu R Z, Bi X W, Zhou M F, Peng J T, Su W C, Liu S, Qi H W. 2008. Uranium metallogenesis in South China and its relationship to crustal extension during the Cretaceous to Tertiary. Economic Geology, 103(3):583-598

Hu R Z, Burnard P G, Bi X W, Zhou M F, Peng J T, Su W C, Zhao J H. 2009. Mantle-derived gaseous components in ore-forming fluids of the Xiangshan uranium deposit, Jiangxi Province, China: evidence from He, Ar and C isotopes. Chemical Geology, 266(1-2):86-95

Li T, Li Z Y, Fan G. 2020. Hydroxyplumbopyrochlore, IMA 2018-145. CNMNC Newsletter. European Journal of Mineralogy, 53:32

Li Z Y, Fang X H, Xia Y L, Xiao X J, Sun Y, Chen A P, Jiao Y P, Zhang K. 2005, Metallogenetic conditions and exploration Criteria of Dongsheng Sandstone type uranium deposit in Inner Mongolia, China. In: Mao J W, Bierlein F B (eds). Mineral Deposit Research. Berlin: Springer-Verlag, 291-294

Li Z Y, Chen A P, Fang X H, Ou G X, Xia Y L, Sun Y. 2008. Origin and superposition metallogenic model of the sandstone-type uranium deposit in the northeastern Ordos Basin, China. Acta Geologica Sinica, 82(4):745-749

Luo J C, Hu R Z, Fayek M, Li C S, Bi X W, Abdu Y, Chen Y W. 2015. In-situ SIMS uraninite U-Pb dating and genesis of the Xianshi granite-hosted uranium deposit, South China. Ore Geology Reviews, 65:968-978

Luo J C, Hu R Z, Fayek M, Bi X W, Shi S H, Chen Y W. 2017. Newly discovered uranium mineralization at ~2.0 Ma in the Menggongjie granite-hosted uranium deposit, South China. Journal of Asian Earth Sciences, 137:241-249

Miyawaki R, Hatert F, Pasero M, Mills S J. 2020. IMA commission on new minerals, nomenclature and classification (CNMNC) newsletter 54 -. Mineralogical Magazine, 84(2):359-365

Mogan W J. 1971. Convection plumes in the lower mantle. Nature, 230:42-43

Qiu J T, Zhang C, Hu X. 2015. Integration of concentration-area fractal modeling and spectral angle mapper for ferric iron alteration mapping and uranium exploration in the Xiemisitan area, NW China. Remote Sensing, 7(10):13878-13894

Wilson J T. 1963. A possible origin of the Hawaiian Islands. Canadian Journal of Physics, 41:863-870

Wu Y, Qin M K, Guo D F, Fan G, Liu Z Y, Guo G L. 2018. The latest in-situ uraninite U-Pb age of the Guangshigou uranium deposit, northern Qinling orogen, China: constraint on the metallogenic mechanism. Acta Geologica Sinica, 92(6):2445-2447

Zhang C, Liu S F, Ye F W, Qiu J T, Zhang Z X, Wang J G. 2019. Three-dimensional modeling of alteration information with hyperspectral core imaging and application to uranium exploration in the Heyuanbei uranium deposit, Xiangshan, Jiangxi, China. Journal of Applied Remote Sensing, 13:014524

Main Progresses of Uranium Geology and Exploration Techniques

LI Zi-ying, QIN Ming-kuan, FAN Hong-hai, CAI Yu-qi, CHENG Ji-xing, GUO Dong-fa, YE Fa-wang, FAN Guang, LIU Xiao-yang

Beijing Research Institute of Uranium Geology, Beijing 100029

Abstract: The main progresses of both uranium geology and exploration techniques for the past decade in China have been summarized in this paper. They include the innovation of metallogenic theories, the classification of uranium metallogenic types and zonations, the evaluation of China's national uranium resource potential, the research and evaluation of main industrial types of uranium deposits, the Xiangshan scientific deep drilling for exploring uranium resource, and the discovery of zero-valent native uranium and new uranium minerals etc. Then, exploration techniques including technical innovation on fast evaluation and exploration of both sandstone-hosted and hydrothermal uranium deposits, hyperspectral remote sensing technology, drilling techniques, big data mining, and chemical analysis, have been described. In addition, an overview of major breakthroughs in domestic and overseas uranium prospecting using those innovative metallogenic theories and techniques has been given. Finally, the development tendency of uranium metallogeny and exploration techniques has been discussed.

Key words: uranium metallogenic theory; uranium exploration techniques; uranium exploration progress; development tendency

天然气地球化学研究进展*

刘文汇[1,5]　王　星[2,3]　田　辉[2]　郑国东[4]　王晓锋[1]　陶　成[5]　刘　鹏[1]

1. 西北大学 地质学系，大陆动力学国家重点实验室，西安 710069；2. 中国科学院 广州地球化学研究所，有机地球化学国家重点实验室，广州 510640；3. 中国科学院大学，北京 100049；4. 中国科学院 西北生态环境资源研究院，兰州 730000；5. 中国石化油气成藏重点实验室，无锡 214126

摘　要：本文简要总结了近十年来中国天然气的勘探和开发工作进展，这些进展主要包括以下几方面：①提出天然气多种来源、多元生烃机理，建立了腐泥型烃源岩生气模式，明确了腐泥型烃源岩中不同类型生气母质的生气潜力；重新认识了煤系烃源岩的生气潜力并厘定了其生气下限；②稀有气体同位素及放射性同位素定年技术不断进步，促进了我国油气成藏年代学的发展；完善了我国天然气成藏示踪体系，夯实了多源成气理论，扩大了我国天然气勘探领域；③在天然气水合物、页岩气、致密砂岩气和煤层气的生烃理论、储集物性、渗流机理、成藏过程、保存条件及开发技术等诸多方面取得了突破性进展，有效指导了非常规天然气的勘探开发；④应用甲烷团簇同位素、丙烷特位同位素、超微量气体氢同位素等分析测试手段，进一步发展了有机质生烃模拟等传统的技术方法，为研究天然气成因类型及其成藏过程提供了新的技术支持；⑤天然气地球化学理论为一系列大型气田的勘探开发提供了理论支撑，指出了微量微区分析技术是天然气地球化学发展的重要方向。

关键词：天然气　地球化学　研究进展　成藏定年　深层气成因　非常规气　大气田勘探

0　引　言

近年来，随着中浅层常规天然气年产量的下降及天然气需求量的增加，我国天然气的对外依存度从2010年之前的不足10%增至目前的40%以上（邹才能等，2020）。为满足日益增长的天然气需求，深层天然气、深水天然气和非常规天然气已经成为我国当前的勘探重点（谢玉洪，2014；魏国齐等，2017；邹才能等，2017）。"十二五"以来，在塔里木盆地、渤海湾盆地、琼东南盆地和四川盆地等地一批深层、深水及非常规天然气田先后被发现，为我国能源安全提供了有力的保障。上述重大勘探突破也是我国天然气地球化学理论不断发展和完善的结果。"十二五"以来，天然气地球化学理论不断发展和完善，有效地指导了天然气勘探开发工作（魏国齐等，2017，2018；戴金星，2019；邹才能等，2019）。然而，天然气在我国的能源结构中的占比仍远低于世界平均水平，我国天然气勘探开发仍面临着巨大的压力（魏国齐等，2018；邹才能等，2020）。这就要求我们必须加强天然气地球化学理论研究，以便更好地指导实际天然气勘探开发工作。本文综合了"十二五"以来有关天然气的科研及勘探成果，对我国天然气地球化学理论进行系统梳理，总结近年来天然气勘探开发工作的重要进展，为我国天然气的勘探开发工作提供科学依据，为保障天然气产量的稳步增长和国家能源安全提供有力支持。

1　深层天然气生烃理论及其成因判识取得重要进展

天然气具有多种来源和多元生烃的特征（刘文汇等，2017a）。近年来，随着深层、深水及非常规天然气的大规模开发，深层海相碳酸盐岩层系天然气生烃理论、腐泥型烃源岩的生气模式及煤系烃源岩的

---

* 原文"近十年来中国天然气地球化学研究进展"刊于《矿物岩石地球化学通报》2021年第40卷第3期，本文略有修改。

高演化阶段生气潜力受到了极大关注，在此基础上，一些新的用以识别不同成因来源的天然气成因判识指标被提出。

1.1　深层海相碳酸盐岩层系天然气生烃理论

碳酸盐岩是油气赋存的重要层系，全球油气产量的60%来自碳酸盐岩（王大鹏等，2016）。近年来，我国塔里木盆地、四川盆地和鄂尔多斯盆地下古生界海相碳酸盐岩层系的油气勘探取得了重大突破，表明我国下古生界海相碳酸盐岩层系油气有着良好的勘探前景，是我国现实的油气资源接替领域（金之钧，2005；赵文智等，2014；刘树根等，2016；何治亮等，2016）。

针对海相盆地深层天然气勘探的地质实际，刘文汇等（2007，2017a）提出干酪根演化过程中滞留的和运出烃源且未形成聚集油气藏的分散可溶有机质，是叠合盆地高演化阶段形成天然气的主力来源。在此基础上，利用地球化学示踪方法，证实和田河气田的气源属于分散可溶有机质的新认识，该理论的提出，明确了叠合盆地天然气的勘探方向。

另外，结合我国下古生界海相碳酸盐岩层系天然气勘探进展，特别是鄂尔多斯盆地奥陶系马家沟组盐下天然气勘探取得突破，探讨中国高热演化程度、低TOC值海相碳酸盐岩层系作为烃源岩的可能性及其生烃机理（刘文汇等，2016；Liu W. H. et al.，2017）。揭示碳酸盐岩中酸溶有机质可能是生烃物质的重要组成部分（孙敏卓等，2013；Liu et al.，2013；刘鹏等，2016）。同时，高演化碳酸盐岩烃源岩中存在规模性的有机碳向无机碳转化过程，烃源评价要避免利用残余TOC低去否定碳酸盐岩烃源曾经的生烃历程。上述认识进一步完善了我国下古生界高演化海相碳酸盐岩层系天然气勘探理论基础。

1.2　腐泥型烃源岩完整的生气模式

原油裂解气和干酪根裂解气是高成熟海相含气盆地的重要气源，前人对其生气特征进行了大量的对比研究（Wang et al.，2006；田辉等，2006；赵文智等，2006；Guo et al.，2009；Tian et al.，2009）。这些研究明确了原油裂解气和干酪根裂解气的生气潜力及主生气期，对认识高成熟海相含气盆地的天然气生成过程提供了有益的指导。而在实际地质条件下，原油和干酪根并非独立存在，烃源岩在达到排油门限以后会排出一定量的原油，这些原油以源外分散液态烃和油藏的形式存在，干酪根和滞留烃则保留在烃源岩内部（李剑等，2015，2018；赵文智等，2015）。明确干酪根和不同类型的液态烃的生气潜力与主生气期关系到深层高-过成熟地区天然气的资源潜力与勘探前景。近十年来，我国学者对腐泥型烃源岩排油效率进行了重点研究，并据此评价了腐泥型烃源岩内干酪根、源内滞留烃、源外分散液态烃和油藏液态烃的生气潜力并得到了众多认识（王东良等，2012；Gai et al.，2015；李剑等，2015，2018；赵文智等，2015）：①初步了解了腐泥型烃源岩的排油效率。指出腐泥型烃源岩的排油效率受到有机质含量、热演化程度、烃源岩厚度及岩性等因素的控制，总体为20%~80%；②明确了腐泥型烃源岩内滞留烃含量对烃源岩中有机质生气潜力的影响，对滞留烃和干酪根的生气潜力和主生气期有了更深入的认识，并对二者生气过程中的相互作用进行了研究；③建立了腐泥型烃源岩内干酪根、滞留烃、源外分散烃和油藏液态烃的生气模式，确定了各种母质的生气潜力及主生气期，并对各种母质裂解生成天然气的相对比例进行了探讨。

然而，实际地质条件下排油效率受到多种因素的影响，目前对排油效率的认识仍处于初级阶段，有必要结合实际地质条件对不同地区的排油效率进行有针对性的研究，在明确排油效率的基础上评价腐泥型烃源岩不同生气母质对天然气成藏的贡献。

1.3　煤系烃源岩高演化阶段生气潜力

自1978年我国学者首次提出"煤成气"的概念以来，我国的煤成气理论不断得到发展与完善（戴金星和龚剑明，2018）。近十年来，随着深层天然气的开发，我国学者对高过成熟阶段煤系烃源岩的生气潜力进行了研究。以往研究认为，煤系烃源岩的累计生气量不高于200 m³/t，成熟度R_o=1.2%~2.4%时生成的天然气占到总生气量的85%以上，在R_o>2.4%之后的生气量不足总气量的5%~10%（肖芝华等，2009；张水昌等，2013）。然而，近些年研究表明，煤系烃源岩在高过成熟阶段仍然具有很大的生气潜力，以前建立的煤系烃源岩生气模型可能低估了这类烃源岩的总生气量（王东良等，2012；于聪等，2013；张水昌等，2013）。我国学者通过热模拟实验发现，煤系烃源岩的最大产气量可能达到300 m³/t以上，在R_o>2.5%的高过成熟阶段，煤系烃源岩的生气量可占到总生气量的20%以上，天然气生成下限可延伸到R_o=5.0%（王东良等，2012；张水昌等，2013）。另外，在深层高温高压的地质条件下，地层水的加入还可以极大地促进煤系烃源岩的生气能力，最大促进量可达13%以上（高金亮等，2020）。我国四川盆地、鄂尔多斯盆地、塔里木盆地、松辽盆地等分布有大面积高过成熟煤系烃源岩，这些在以往勘探中被忽略的潜在产气区域有可能成为今后勘探的重点（宋岩等，2012）。

1.4 天然气成因判识指标

我国天然气成因多样、来源复杂，正确的判识天然气成因是明确天然气来源、重建天然气成藏过程的基础，对扩大天然气勘探领域有着重要意义。我国学者自20世纪90年代起便开展了大量天然气成因判识的研究工作，并利用天然气成分、碳同位素组成、轻烃、生物标志物等多种指标建立了各类天然气的判别标准和图版（戴金星，1992；刘文汇和徐永昌，1996；宋岩和徐永昌，2005；戴金星等，2008）。然而，随着我国天然气生烃理论的不断完善，勘探领域的不断扩大，这些判别指标已难以适用于高过成熟天然气的判别。为此，我国学者近年来建立了多种新的判别指标应用于干酪根裂解气和原油裂解气、油藏液态烃裂解气和源外分散液态烃裂解气、有机成因气和无机成因气等的判识。谢增业等（2016）在传统$\ln(C_1/C_2)$-$\ln(C_2/C_3)$图版的基础上，考虑天然气生成阶段建立了新的图版来判别干酪根裂解气和原油裂解气；赵文智等（2015）进一步提出利用甲基环己烷和正庚烷比值等于1来判别油藏液态烃裂解气和源外分散液态烃裂解气；李剑等（2017）则利用C_{6-7}环烷烃/(nC_6+nC_7)和甲基环己烷/nC_7两个指标建立了油藏液态烃裂解气和源外分散液态烃裂解气的判别图版并利用碳、氧同位素及惰性气体同位素等对天然气中的N_2和CO_2气体的来源进行鉴别。这些判别指标为深化天然气来源认识提供了有益指导，可进一步应用于天然气勘探。

2 油气成藏定年技术及成藏理论得到进一步发展

近年来，油气成藏定年技术得到了突破性进展，稀有气体同位素和放射性同位素体系在油气成藏绝对年龄的测定中展现出了广阔的应用前景，极大地推动了我国油气成藏年代学的发展（刘文汇等，2013a；王华建等，2013；李军杰等，2015；陶成等，2015；赛彦明等，2020）。此外，应用于天然气示踪的多种地球化学指标得以建立，成藏示踪体系得到不断完善，弥补了单一示踪指标在气源对比中的不足（宋岩等，2012；王鹏等，2015）。这些研究明确了不同来源天然气对气藏的贡献，使得多源成气、复合成藏理论得到完善（刘文汇等，2013b）。

2.1 成藏定年技术得到极大发展

随着同位素定年技术和高精度实验方法的发展，油气成藏绝对年龄的测定已成为油气领域当下研究的热点之一（陈玲等，2012；蔡长娥等，2014）。近年来，U-Pb、Sr-Nd、K-Ar、^{40}Ar-^{39}Ar及Re-Os等放射性同位素和稀有气体同位素体系在油气成藏中的应用得到了大量研究（涂湘林等，1997；张景廉等，1997；蔡李梅等，2008；刘文汇等，2013a；王华建等，2013；李军杰等，2015；李真等，2017）。我国学者曾通过测定干酪根、沥青和原油中的U-Pb和Sr-Nd同位素组成对塔里木盆地、准噶尔盆地的油气成因及成藏过程进行了探讨，有力地推动了油气成藏年代学的发展（涂湘林等，1997；张景廉等，1997）。然而这类元素在有机质中的富集程度低、提纯方法复杂，容易受到外部金属混入物的干扰，另外这些同位素体系在油气成藏后的封闭性仍存在争议，这些因素都限制了这些方法在恢复油气成藏过程中的应用（刘文汇等，2013a）。自生伊利石的K-Ar和^{40}Ar-^{39}Ar同位素体系是目前相对成熟的油气成藏定年方法，其原理是储集层中的富钾卤水环境在烃类注入后遭到破坏，储集层伊利石停止生成，因此油气成藏时间约等于或略晚于自生伊利石矿物的形成年龄（陈刚等，2012；陈玲等，2012；王华建等，2013）。该方法要求油气充注之后自生伊利石停止生长并且其中的K-Ar和^{40}Ar-^{39}Ar同位素体系保持封闭（蔡李梅等，2008；陈玲等，2012；刘文汇等，2013a）。尽管实际地质条件下能否满足上述两个条件仍存在争议，但该定年技术已成功运用于珠江口盆地、塔里木盆地、鄂尔多斯盆地及四川盆地等地的油气成藏研究中（云建兵等，2009；张有瑜和罗修泉，2011；陈刚等，2012；张有瑜等，2015）。Re和Os具有亲有机质的属性，可以在干酪根、沥青及原油等有机质中富集并长期稳定存在，因此可以通过直接测定有机质中的Re、Os元素及同位素计算Re-Os等时线年龄来确定油气成藏及演化过程（蔡长娥等，2014；沈传波等，2015；张涛等，2017；赛彦明等，2020）。陈玲等（2010）对麻江古油藏储集层沥青中的Re、Os同位素进行了分析，得到的沥青样品模式年龄为28~144 Ma（集中在85 Ma），并将这一年龄解释为古油藏遭受破坏的时间。沈传波等（2019）对四川盆地震旦-寒武系沥青样品的Re、Os元素及同位素特征进行了总结，认为四川盆地震旦-寒武系存在450 Ma和205~162 Ma两期成藏作用，原油和低熟沥青的Re-Os等时线年龄代表原油生成的时间，而高成熟焦沥青的等时线年龄则代表原油裂解生成天然气的时间。然而目前Re-Os同位素体系定年存在较大的误差，主要是因为两种元素在有机质中的富集机制和同位素体系的封闭性研究程度较低，生物降解、TSR及热裂解等次生改造对原油Re-Os同位素体系的影响仍缺乏深入的认识（沈传波等，2019）。虽然以上方法存在一定的不足，但已在油气生成和成藏等方面展现出了良好的潜力，在进一步发展和完善后应具有广阔的应用前景（刘文汇等，2013a）。

2.2 成藏示踪体系得到完善

近年来，我国天然气示踪指标的研究有了长足发展。稀有气体同位素（魏国齐等，2014；杨春等，2014；仵宗涛等，2017；王杰等，2018）、轻烃化合物（赵文智等，2015；李剑等，2017）、传统的天然气组分及碳-氢同位素（秦胜飞等，2016；周国晓等，2016；Qin et al.，2018；秦胜飞和周国晓，2018；张玉红等，2018）等天然气示踪指标不断发展与完善。这些指标已广泛应用于天然气成藏示踪研究中，如我国目前单体储量最大的安岳震旦-寒武系大气田，其气体来源存在很大争议，近年来我国学者利用天然气示踪指标对该地区的天然气来源及成藏过程进行了研究。魏国齐等（2013，2014）通过对安岳气田震旦系灯影组和寒武系龙王庙组的He、Ar、Xe等稀有气体含量及同位素分析，明确了该地区天然气为壳源成因。赵文智等（2015）和李剑等（2017）分别用轻烃指标研究了安岳气田的天然气来源，结果表明，该气田中的天然气主要来源于源外分散液态烃的热裂解。秦胜飞等近期的研究表明，水溶气的脱溶使得甲烷C、H同位素偏重（秦胜飞等，2016；周国晓等，2016；Qin et al.，2018；秦胜飞和周国晓，2018），安岳气田龙王庙组和灯影组储集层以及威远气田灯影组储集层中的天然气可能都有水溶气的贡献。

硫同位素是高硫天然气成藏示踪的重要指标，目前已广泛应用于普光气田、罗家寨气田等高硫天然气田的研究中。朱光有等（2006a）根据四川盆地不同储集层中H_2S的硫同位素特征判断出三叠系、震旦系和石炭系储集层中的H_2S是由硫酸盐热化学还原作用（TSR）形成的，而二叠系储集层中的H_2S则应来源于富硫有机质的高温裂解。刘文汇等（2017b）改进了传统的硫同位素测试方法，进一步指出四川盆地烃源岩中的含硫物质主要是由硫酸盐细菌还原作用（bacterial sulfate reduction，BSR）形成的，而储集层中的含硫物质则主要是由TSR作用形成的。然而，目前硫稳定同位素分馏机理仍存在争议，实验模拟结果与实际地质条件下硫同位素的分馏存在较大差异（朱光有等，2006b；谢增业等，2008；赵兴齐等，2011；罗厚勇等，2012）。因此，硫同位素的分馏机理及其在天然气成藏中的应用仍有待进一步研究。

此外，流体包裹体技术、无机碳、氧同位素分析技术及地层水化学特征等也被综合运用于天然气示踪研究中，使我国的成藏示踪体系得到了极大地完善（陶士振等，2012；叶素娟等，2017）。陶士振等

（2012）利用流体包裹体测温技术，烷烃气体和 CO_2 的碳同位素、稀有气体同位素组成对松辽盆地火山岩储集层中的天然气进行了示踪，结果表明该地区的天然气存在幔源无机成因气并且无机成因气主要分布在断裂附近。叶素娟等（2017）根据天然气组分和碳同位素、地层水化学特征、储集层中自生矿物碳、氧同位素组成及流体包裹体特征，提出了一系列可以指示成藏过程的地球化学示踪指标，并利用这些指标研究了川西拗陷侏罗系气藏的成藏过程。这些研究丰富了我国天然气成藏示踪体系，为进一步的天然气勘探工作提供了有益的指导。

2.3 多源成气、复合成藏理论得到完善

随着天然气成因理论及天然气成藏示踪体系的不断完善，具有多源特征的天然气藏的来源及成因类型研究取得了一定进展，开拓了天然气勘探领域。以往我国的大型气田都是以煤成气为主（表1），然而近十年来发现的储量大于 1000 亿 m^3 的 7 个大气田中有 4 个气田的天然气来源为油型气（表1），表明深层海相油型气在我国天然气勘探领域的地位越来越重要。

表1 2000年以来中国千亿方大气田特征与成因

盆地	气田	探明储量/亿 m^3	探明年份	主力气层	储集层岩性	主要气源岩	成因类型
松辽	徐深	2 719	2007	K	火山岩	下白垩统煤系地层	无机成因气和煤成气混合气
准噶尔	克拉美丽	1 053	2008	C	火山岩	石炭系煤系地层	煤成气
琼东南	陵水17-2	1 020	2014	N	砂岩	古近系陆相烃源岩	煤成气
塔里木	迪那2	1 752	2001	E	砂岩	侏罗系煤系地层	煤层气
	塔中1号	2 376	2005	C	碳酸盐岩	寒武系泥质灰岩	油型气
	大北	1 400	2010	K	砂岩	侏罗系煤系地层	煤成气
	博孜	1 100	2019	K	砂岩	侏罗系煤系地层	煤成气
	克深	548	2008	K	砂岩	侏罗系煤系地层	煤成气
	克拉2	2 840	2000	K, E	砂岩	侏罗系煤系地层	煤成气
鄂尔多斯	苏里格	16 447	2000	P_1	砂岩	石炭-二叠系煤系地层	煤成气
	大牛地	3 745	2002	C_3, P_1	砂岩	石炭-二叠系煤系地层	煤成气
	子洲	1 152	2005	P_1	砂岩	石炭-二叠系煤系地层	煤成气
	靖边	6 922	2001	O_1, O_2, P_1	碳酸盐岩	石炭-二叠系煤系地层	煤成气
	神木	3 333	2003	C_3, P_1	砂岩	石炭-二叠系煤系地层	煤成气
四川	普光	5 200	2006	T_1	碳酸盐岩	二叠系烃源岩	煤成气和油型气混合气
	广安	1 356	2006	T_3	砂岩	上三叠统煤系烃源岩	煤成气
	大天池	1 068	2006	T, P, C	碳酸盐岩	志留系和二叠系海相页岩、灰岩	油型气
	合川	2 299	2009	T_3	砂岩	上三叠统煤系烃源岩	煤成气
	元坝	2 198	2011	T_2	碳酸盐岩	三叠系煤层	煤成气
	安岳三叠系	2 082	2010	T_3	砂岩	三叠系煤层	煤成气
	安岳震旦-寒武系	10 363	2013	ϵ_1, Z_2	碳酸盐岩	下寒武统页岩	油型气
	焦石坝	6 008	2013	O-S	页岩	奥陶-志留系页岩	油型气
	长宁	4 446	2015	O-S	页岩	奥陶-志留系页岩	油型气
	威远	4 276	2015	O-S	页岩	奥陶-志留系页岩	油型气

油型气主要源于克拉通盆地内的下古生界海相地层，以台内凹陷和台缘斜坡-深水陆棚沉积为主（宋岩等，2012），有机质类型多为腐泥型或腐殖-腐泥型，成熟度通常较高，主要发育于塔里木盆地及四川盆地，如塔里木盆地的塔中1号气田及四川盆地的大天池气田、安岳震旦-寒武系气田及焦石坝、长宁和威远页岩气田等。煤成气可发育在克拉通盆地、前陆盆地和断陷盆地中，有机质类型多为腐殖型或腐泥-腐殖型，成熟度差别较大（宋岩等，2012）。其中克拉通盆地的气源岩主要为石炭-二叠系煤层，以海陆过渡相沉积为主，鄂尔多斯盆地内的诸多大气田均来源于此。前陆盆地的气源岩以侏罗系为主，沉积相多为湖沼相，另有部分滨浅湖相，主要发育于塔里木盆地，如迪那2气田和克拉2气田。断陷盆地的气源岩则以白垩系及古近系为主，以湖沼相为主，如琼东南盆地的陵水17-2气田。

3 非常规天然气地质理论及其开发技术取得突破进展

近十年是我国非常规天然气从无到有、从小到大的开拓时期，气体地球化学在非常规油气研究中具有重要作用。"十二五"以来我国非常规天然气产量增幅巨大，预计到2030年，非常规天然气年产量将占到我国天然气年产量的45%以上（徐春春等，2017）。非常规天然气勘探开发事业的发展得益于我国天然气地质理论和开发技术的不断完善。近年来，我国研究人员在天然气水合物、页岩气、致密砂岩气及煤层气等非常规天然气的地质理论和开发技术方面开展了大量的工作并取得了突破性进展，为进一步的非常规天然气勘探开发工作提供有益的指导（贾承造，2017）。

3.1 天然气水合物

天然气水合物以其能量密度高、资源量大，被认为是21世纪的理想替代能源（罗敏等，2013）。与常规天然气不同，天然气水合物以生物气为主，热成因气次之，暂未发现无机成因气的贡献（吴传芝等，2018；鲁晓兵等，2019）。目前，我国已经在南海神狐海域和祁连山冻土区取得了天然气水合物样品（赵伟等，2019）。相关研究揭示，祁连山天然气水合物以烃类气体为主，含有少量CO_2，其中甲烷含量为62.11%~99.91%，均值为98.04%，C_1/C_2值为130~11995，甲烷的$\delta^{13}C$为-62.2‰~-54.1‰，表明该地区天然气水合物以生物气为主，同时含有少量的热成因气；神狐海域存在面积广、厚度大的适合甲烷菌生存的地质环境，可为该地区的天然气水合物提供充足的气源（付少英和陆敬安，2010；黄霞和祝有海，2010）。但生物气和热成因气二者所占的比例仍然有待进一步研究。祁连山冻土带天然气水合物甲烷含量为54%~76%，乙烷含量为8%~15%，丙烷含量为4%~21%，还含有少量的丁烷、戊烷等烷烃气，显示出湿气的特征，甲烷的$\delta^{13}C$重于-50‰，均值为-47.4‰，表明该地区的天然气以热成因气为主，同时含有少量的生物气（张富贵等，2019）。

3.2 页 岩 气

北美"页岩气革命"不仅是世界油气工业的里程碑，同时也是油气地质理论的一次重大革新（贾承造，2017；邹才能等，2017）。页岩气的成因包括生物成因、热成因及混合成因。我国页岩气成熟度普遍较高，达到了干气生成阶段（肖贤明等，2013），页岩储集层中的天然气以源内滞留烃和干酪根裂解气为主，我国学者建立了不同滞留烃含量的页岩有机质生气模型，评估了液态烃对不同类型干酪根晚期生气潜力的影响（Gai et al., 2015，2018），为高-过成熟地区页岩气的来源和资源评价提供了科学依据。

四川盆地五峰-龙马溪组页岩是目前我国大规模商业化开发的页岩气层组，其中的页岩气干燥系数高，普遍达到0.99以上，是目前世界上干燥系数最高的页岩气（Dai et al., 2014）。五峰-龙马溪组页岩气的甲烷、乙烷和丙烷碳同位素都表现出完全倒转的特征（Dai et al., 2014；吴伟等，2015）。国内许多学者对天然气碳同位素倒转机理进行了研究并提出了多种成因机理，主要包括：①有机烷烃气的混合，如煤型气和油型气的混合、干酪根裂解气和原油裂解气的混合、同型不同源天然气的混合以及同源不同期天然气的混合；②扩散和吸附/解吸作用；③细菌氧化作用；④有机烷烃气与无机烷烃气的混合；⑤地层水、矿物及金属对天然气的后期改造等（盖海峰和肖贤明，2013；戴金星等，2014；秦华等，2016；陈麦雨等，2018；孙健等，2018；马中良等，2020）。然而，上述成因机理并未形成广泛认识。于聪等（2013）指出，同源不同期天然气的混合是页岩气倒转的重要原因，而孙健等（2018）则认为这种混合模型在实际条件下很难形成，一般不会导致天然气的碳同位素倒转。王亮和宁波（2020）认为，有机烷烃气的混合是四川盆地页岩气碳同位素倒转的主要原因，而马中良等（2020）则认为单纯的有机烷烃气混合并不能导致页岩气的碳同位素倒转，并指出四川盆地页岩气碳同位素倒转的原因可能是在地层抬升过程中地层水、矿物和金属等对页岩气的后期改造。尽管目前页岩气碳同位素倒转机理仍存在较大争议，但大量勘探结果表明，碳同位素组成倒转往往与页岩储集层的超压及页岩气高产有关，可用于指示页岩气富集区（盖海峰等肖贤明，2013）。陈麦雨等（2018）指出，页岩气碳同位素倒转通常发生在相对封闭的地质条件下，因此其倒转程度可以指示页岩气藏的保存条件。另外，页岩气的碳同位素组成还可指示页岩的孔隙度和渗透率，对水平井压裂位置和压裂区间的选取具有一定的参考意义（盖海峰和肖贤明，2013）。

与常规天然气相比，页岩气源储一体，表现为典型的"原地"成藏模式（宋岩等，2012）。页岩储集层以纳米孔隙为主，近年来扫描电子显微镜、压汞法、低压 N_2-CO_2 吸附、核磁共振等方法已广泛用于页岩孔隙结构的定性描述和定量表征（Tian et al.，2015；陈生蓉等，2015；张瑜等，2015；Shao et al.，2017；李志清等，2018），这些研究明确了有机质孔是纳米孔隙发育的主要载体，其中微孔（<2 nm）和介孔（2~50 nm）为占主导地位，宏孔（>50 nm）相对较少。这些孔隙决定了页岩中天然气的赋存状态，在微孔、介孔和宏孔表面，天然气烷烃分子主要以吸附态存在，而在远离介孔和宏孔表面的位置，天然气中烷烃分子则主要以游离状态存在（Krishna，2009）。明确页岩气赋存状态并确定吸附气与游离气的含量对页岩气原地气量的评价至关重要。近年来我国学者根据页岩高温高压等温吸附实验对页岩的吸附机理、吸附特征及吸附气含量进行了研究，选取三参数 SDR 模型对吸附结果进行拟合，有效地扣除了吸附相体积重复计算对页岩原地气量的高估，据此建立了地质条件下页岩原地气量模型（Pan et al.，2016；Tian et al.，2016；Zhou et al.，2018）。

北美页岩的 R_o 主值为 1.0%~2.5%，处于原油裂解或主生气期，而我国商业化开采的页岩成熟度普遍偏高，R_o 主值达到 2.5%~4.5%，在地质历史时期埋深普遍超过 7 000 m（肖贤明等，2013）。在深部高温高压条件下，页岩的孔隙结构、气体赋存状态等均会发生很大改变，从而影响页岩原地气量。近年来的研究表明，在 E_qR_o>3.5% 时，页岩孔隙度会降低，同时页岩气存在氮气含量过高的风险，明确了我国页岩成熟度范围 E_qR_o 为 2.5%~3.5% 时为页岩气勘探的有利区域，在 E_qR_o>3.5% 的区域为页岩气勘探的高风险区（Chen and Xiao，2014；Xiao et al.，2015）。

目前我国页岩气开发层位主要为四川盆地焦石坝和威远–长宁地区的五峰–龙马溪组页岩，对于时代更老、具有很大资源潜力的寒武系筇竹寺组页岩没有大规模开发（邹才能等，2015），并且我国页岩气勘探的重点集中在海相地层中，陆相和海陆过渡相页岩的资源潜力仍有待研究。此外，我国页岩气开发主要集中在 3 500 m 以浅的地层中，对于埋深大于 3500 m 的页岩有效开发仍有待开采技术的进步（邹才能等，2017）。

3.3 致密砂岩气

致密砂岩气是非常规天然气的重要组成部分，我国致密砂岩气可采资源量达 $11×10^{12}$ m^3，主要分布在鄂尔多斯盆地和四川盆地，在塔里木、吐哈、松辽和渤海湾等盆地也有发现（李建忠等，2012；魏新善等，2017b），目前致密砂岩气年产量占到我国天然气年产量的 30% 左右（魏新善等，2017b）。我国致密砂岩气地质理论经过了多阶段的发展，先后突破"深盆气"和"盆地中心气"理论，在本世纪引进了"连续型油气藏"的概念，提出了"根源气"的概念，并在致密砂岩气的成藏机理、分布及分类等方面开展了大量研究（魏新善等，2017a）。"十二五"以来，随着我国致密砂岩气勘探程度的提高，相应的地球化学理论也不断完善。早年对致密砂岩气的赋存状态研究较少，通常认为致密砂岩储集层中的天然气以游离态为主，忽略了储集层中的吸附气，近年的研究表明致密砂岩气藏中存在一定量的吸附气，这些吸附气对致密砂岩储集层的含气特征、渗流规律及储量计算等具有重要影响（魏新善等，2017b）。另外，近期研究突破了致密砂岩气"连续成藏"的概念，指出致密砂岩气藏具有多种压力系统并存，储集砂体连续而气藏不连续且呈多藏分布的特征（魏新善等，2017b），同时根据单个气藏的源储关系、储集层压力及渗流特征等将致密砂岩气藏分为"连续型"和"圈闭型"两种，并对其运聚机理、地质分布及地球化学特征等方面的差异进行了对比（郭迎春等，2013）。明确了发育在气藏不同部位的晚期裂隙对天然气成藏的影响，指出当晚期裂隙发育在气藏内部时有助于改善储集层的储集物性，形成有利于天然气运聚成藏的优势通道和储集空间，使天然气运聚成藏的动力由裂隙形成之前的分子膨胀力转变为浮力，这种情况下"连续型"气藏会转变为"圈闭型"气藏，形成有利于天然气开发的"甜点区"，而当晚期裂隙发育在构造高部位且发育程度较高时会使气藏成藏边界发生变化，甚至会使气藏完全遭到破坏（王鹏威等，2014）。

我国致密砂岩气藏通常与煤层或煤系地层紧密接触，在靠近优质烃源岩附近形成大范围、低丰度、连续性的天然气聚集（宋岩等，2012）。气藏中烷烃气含量高，普遍达到96%以上，绝大多数气田中的烷烃碳同位素组成表现出正碳序列，乙烷的$\delta^{13}C$普遍大于$-28‰$，与煤成气相同（戴金星等，2012），因为致密砂岩储集层孔渗较差，只有煤系烃源岩才能提供大规模持续不断的连续性气源。如鄂尔多斯盆地诸多致密砂岩气藏的天然气均来自石炭系本溪组、二叠系太原组和山西组3套煤系地层，四川盆地的致密砂岩气则主要来自上三叠统须家河组煤系地层（戴金星等，2012）。部分地区致密砂岩气的碳同位素序列会出现倒转，如库车凹陷大北-克深地区的深层致密砂岩气以煤成气为主，同型不同源气或者煤型气与油型气的混合是该地区致密砂岩气碳同位素序列倒转的主要原因（魏强等，2019）。另外，"连续型"和"圈闭型"致密砂岩气的地球化学特征也存在一定的差异。鄂尔多斯盆地上古生界的诸多气藏为"连续型"致密砂岩气，其天然气地球化学特征与烃源岩成熟度吻合，但甲烷和乙烷碳同位素呈现$5‰\sim10‰$的离散，这种离散性是由烃源岩在不同成熟阶段生成的天然气经过近源累积聚集形成的（郭迎春等，2016）。迪那2气田为典型的"圈闭型"致密砂岩气，其甲烷碳同位素组成偏轻，天然气成熟度低于烃源岩成熟度。随着埋深的增大，气田中甲烷/乙烷、乙烷/丙烷等值减小，CO_2含量增加且其碳同位素组成逐渐变重，郭迎春等（2016）指出，这是由长期的聚气过程和较远的运移距离导致的。

3.4 煤层气

我国煤层气资源丰富，可采资源量达$10.9\times10^{12}\ m^3$，主要分布在沁水盆地南部、鄂尔多斯盆地东缘、滇东黔西盆地北部和准噶尔盆地南部（康永尚等，2017）。煤层气包括生物气和热成因气两种来源，通常认为低煤阶煤层气以生物气为主，高煤阶煤中结构复杂的固定碳含量较高，这些碳基质难以被微生物利用生成天然气。然而近期的研究表明，高煤阶中的有机质在一定条件下同样可被微生物降解，某些微生物群落甚至可以降解无烟煤中的芳香化合物（何乔等，2013；陈林勇等，2016）。细菌成因的煤层气甲烷的碳同位素组成普遍偏轻但重于CO_2的碳同位素组成，这是由两次继承性碳同位素分馏引起的，第一次分馏发生在乙酸生成阶段，这一阶段原始煤中甲基的碳同位素组成轻于羧基的碳同位素组成的特征被保留下来。第二次分馏发生在乙酸生成CH_4和CO_2的过程中，这一阶段碳同位素组成较轻的甲基通过加氢生成CH_4，碳同位素组成较重的羧基通过去氢生成CO_2，这两个阶段的继承效应使得CH_4和CO_2的碳同位素组成呈现出明显的负相关关系，即CH_4碳同位素组成越轻，CO_2碳同位素组成越重（简阔等，2020）。

通常认为，煤层气是赋存于煤层中的自生自储式非常规天然气，而当煤层内部存在裂隙时，会形成良好的天然气运移通道，使得天然气得以解吸扩散，若煤层顶底板封闭性较强，这些天然气会在浅部再次吸附，此时煤层既具有自生自储性质，又具有他生他储的性质，呈现出多样化的气藏类型（李勇等，2020）。若构造裂隙贯穿煤层顶底板，天然气可持续运移扩散，在临近或上部砂岩中聚集成藏（欧阳永林等，2018）。

目前我国煤层气勘探开发的目的层位通常在1 000 m以内，对于占到煤层气资源60%以上的深层煤层气勘探及研究仍然不足（李辛子等，2016）。随着埋深的增大，煤层孔隙会在静岩压力作用下不断被压缩，从介孔和宏孔向微孔转变（Li et al.，2013）。煤层在深埋过程中还会受到多期构造应力的叠加改造，使得其中的裂隙系统发生很大的改变。这两种作用都会使煤层的储集物性发生很大的变化，从而影响煤层气的渗流机制及赋存状态（Krishna，2009；李松等，2016）。中浅层煤层气以吸附气为主，通常利用等温吸附实验对煤层含气量进行评价并采用两参数Langmuir方程等对实验结果进行拟合（张晓东等，2005；杨宏民等，2009），而这种方法将过剩吸附量视为绝对吸附量，忽略了游离气的密度。在中浅层压力相对较低的情况下，游离气密度相对较小，可以忽略，然而对于高温高压下的深层煤层气，游离气密度非常大，这种情况下需要区分绝对吸附量和过剩吸附量才能对吸附气含量和游离气含量进行准确评估，从而恢复煤层在高温高压下煤层的真实含气量（杨兆彪等，2011；赵龙等，2014）。因此，开展深部高温高压条件下煤层储集物性的演化机理、渗流特征及真实含气量的研究对我国深部煤层气的勘探开发具有重要的理论和实际意义。

4 天然气地球化学测试技术持续进步

近十年来，我国学者在甲烷团簇同位素、丙烷特位同位素、超微量气体组分碳-氢同位素组成、热模拟实验技术和方法上不断创新，推动了我国天然气地球化学理论的发展。

4.1 甲烷团簇同位素

团簇同位素是指含有两个或两个以上重同位素的分子的相对丰度偏离随机分布状态的程度（李平平等，2017）。在同位素平衡状态下，同位素异数体分子的相对丰度是温度的单调函数。因此，团簇同位素作为地质温度计广泛应用于古气候重建、古高程恢复、沉积盆地热史、碳酸盐岩成岩作用的研究中（李平平等，2017；徐秋晨等，2019；刘嘉庆等，2020；刘雨晨等，2020）。目前天然气领域中应用最广的是利用甲烷团簇同位素判断天然气形成温度。松辽盆地徐家围子气田天然气一直被认为是主要源于无机成因气，然而Shuai等（2018b）通过甲烷团簇同位素的Δ^{18}（$^{13}CH_3D+^{12}CH_2D_2$浓度之和）的测定计算出该地区甲烷形成的温度为167~213℃，表明该区天然气主要来源于高成熟热成因气，为该地区天然气探勘提供了新的思路。然而有些条件下甲烷团簇同位素并不能达到热力学平衡，Shuai等（2018a）的热模拟实验表明，在乙烷裂解阶段，甲烷的团簇同位素主要表现为动力学分馏效应而难以达到热力学平衡。在实际地质条件下，也有部分甲烷达不到热力学平衡，这类甲烷无法用于计算其生成温度，但却可以提供其他形成信息和历史，如区分生物成因气和酯化成因的无机气（帅燕华等，2018）。

团簇同位素测试技术是对同位素异数体分子的整体浓度的测试，主要使用高分辨率同位素质谱仪来完成。目前主要有MAT 253 Ultra（Thermo Scientific）和Panorama（Nu）两种质谱，前者可实现甲烷团簇同位素中$^{13}CH_3D$和$^{12}CH_2D_2$的总浓度（Δ^{18}）的测定；后者可分别获得甲烷的$\Delta^{13}CH_3D$和$\Delta^{12}CH_2D_2$簇同位素值（王晓锋等，2018）。然而团簇同位素的测试比较复杂，对样品的化学纯度要求极高且用样量大，这就需要复杂的前处理过程，既使得所需样品大量富集，又要在富集过程中避免同位素分馏。而且团簇同位素分析周期长、效率低，一个样品的测试时间需要8~10 h。因此，改进测试方法和提高测试效率是团簇同位素研究的重点之一。另外，目前对团簇同位素的认识仍处于初级阶段，对于影响团簇同位素分馏的诸多因素仍缺乏深入认识，在利用团簇同位素进行天然气来源和成藏过程研究中会不可避免地存在多解性，因此在利用这一方法对实际地质问题进行解释的过程中仍需谨慎。

4.2 丙烷特位同位素组成分析方法

特位同位素是指分子内不同官能团上相同原子的稳定同位素组成，其主要用于对反应过程、路径和物质来源的示踪；同时，特位同位素组成的差异反映出同位素交换反应中对于不同官能团的选择性与优先程度，以及化学键断裂和形成的断裂及结合方式（Dias et al., 2002；Hattori et al., 2011；Gilbert et al., 2012，2013）。随着测试技术的进步，逐步实现了对常见有机化合物不同位置上同位素组成的差异，常用的测试方法包括高分辨率同位素比质谱、核磁共振技术及色谱-裂解-色谱-同位素比质谱联用技术（王晓锋等，2018）。丙烷作为烷烃气体中具有特位同位素组成特征的最小分子而受到广泛的关注。Li 等（2018）对色谱-裂解-色谱-同位素比质谱联用技术进行了改善，建立了一套在线分析天然气中丙烷特位同位素的方法。该方法是将天然气样品中的丙烷经第一台色谱分离后进入高温裂解炉进行裂解，裂解碎片再次分离后进入同位素比质谱测试。该方法获取了裂解过程中的碳同位素分馏系数，对相对值进行了进一步校正。但这一方法目前仅能对丙烷碳特位同位素组成进行测试，Zhao 等（2020）利用核磁共振方法对美国 Eagle Ford 页岩气中丙烷特位碳氢同位素组成均进行了测试并对天然气来源及形成温度进行了初步的研究。但是，天然气特位同位素组成研究研究的对象是丙烷，其在天然气中含量较低。因此，利用较低含量的丙烷对天然气来源进行判识具有一定的风险。

4.3 天然气微量氢同位素组成分析方法

天然气氢同位素组成的测定一直是天然气地球化学技术中的难点，在氢含量较低的情况下准确测定氢同位素组成的难度更大。近年来，我国学者提出了新方法来尝试提高氢同位素组成测试的精确度。孟庆强等（2011）提出了一种微量氢的定量富集技术来提高氢气浓度进而达到精确测定氢同位素组成的目的。该方法根据烃类气体和氢气的物理化学差异，利用液氮分离氢气并使用定量 He 作为载气以增加系统内的压力。这种方法既不会引起氢同位素组成的变化，又不会影响同位素组成的测试。陶成等（2012）自制了氢气分离富集装置，该装置利用 5A 分子筛在 159 K 和 77 K 条件下对不同气体组分吸附性能的差异将氢气与其他气体分离并富集于冷阱中，而后在室温下释放氢气并利用 He 作为载气将氢气带入同位素质谱仪中进行 H 同位素组成的测定。该方法实现了在线连续流分析，避免了繁杂的前处理过程，极大地提高了分析效率。王希彬等（2016）利用固相萃取技术与气相色谱-同位素质谱联用实现了天然气中微量烃类化合物单分子氢同位素组成的分析，并对比了不同涂层类型对天然气微量烃类的萃取效果，优化了萃取温度和萃取时间，降低了萃取过程对氢同位素组成的影响，准确高效地测定了 C_1-C_9 化合物单体氢同位素组成。

4.4 有机质生气热模拟实验

国内主流的有机质热模拟生气实验装置主要有两种：一种为黄金管-高压釜热模拟实验装置，另一种位地层孔隙热压生烃模拟实验仪。前者已广泛用于干酪根热解生气及原油二次裂解生气研究，其优点是用量少、压力稳定，可实现干燥样品和含水样品的热模拟实验。热模拟完成后可以结合色谱对生成产物进行定量分析并利用同位素质谱仪对生成物碳同位素组成进行测定。该装置结合 Kinetics 软件和 GOR 软件可以对生烃动力学和碳同位素分馏动力学进行研究，相关成果已广泛用于我国各大含油气盆地的生烃和成藏研究（Xiao et al., 2005, 2006；Tian et al., 2007）。近年来，该装置又不断被应用于新的研究领域，Zhou 等（2014）利用该装置研究了沥青热成熟过程中拉曼光谱的演化特征，Cheng 等（2019）利用大金管对页岩热成熟过程中的水含量变化及其影响因素进行了研究，熊永强等研究了焦沥青的生成特征及其影响因素（Xiong et al., 2016；Lei et al., 2018），郭慧娟等利用大金管对样品进行了热模拟实验，研究了热成熟过程中页岩孔隙的演化特征（Ko et al., 2016；Guo et al., 2017；Liu Y. K. et al., 2017）。

后者弥补了黄金管-高压釜体系在模拟地层流体压力、生烃空间、高温高压地层水及初次排烃等方面的不足。该装置可以保留样品的原始孔隙,同时考虑到与地质条件相近的地层流体压力和上覆静岩压力。在加温加压密闭或可控生、排烃条件下对烃源岩进行热模拟实验。关德范(2018)利用该装置研究了盆地整体持续沉降阶段,生油凹陷内的烃源岩在有限孔隙空间封闭体系内的热压生烃过程,结果表明该装置能够较好地模拟含油气盆地的实际地质特征,可以真实再现烃源岩的生油过程。

5 天然气勘探取得的突破

"十二五"以来,我国天然气地球化学理论不断取得突破,为天然气勘探开发工作提供了坚实的理论基础,先后发现了一批大型、特大型天然气田。

5.1 四川盆地页岩气田

基于近年来在页岩有机质生烃、孔隙结构表征、原地气量模型及资源分布规律等方面的研究进展,我国学者总结出了页岩气富集高产的主控因素,确定了川南上奥陶统五峰组—下志留统龙马溪组页岩是我国页岩气开采的有利层位,并先后在焦石坝、长宁和威远3地实现了页岩气的商业化开采(邹才能等,2015)。五峰-龙马溪组优质页岩的厚度为20~80 m,TOC为2.0%~8.4%,页岩中硅质和钙质矿物含量高,热演化程度适中(R_o为2.0%~3.5%),孔隙度较高(大于3.0%)、裂隙发育并且具有高含气量(大于3.0 m³/t)和高地层压力(压力系数大于1.3)的特征(邹才能等,2016)。目前焦石坝、长宁和威远页岩气田探明储量超万亿方(表1),坚定了我国页岩气进一步勘探开发的信心。

5.2 安岳震旦-寒武系气田

自1964年威远气田被发现后,四川盆地古老碳酸盐岩的天然气勘探在之后的40余年里一直没有显著突破。针对这一问题,我国学者对该地区的天然气地质和地球化学特征进行了综合研究,明确了绵阳-长宁裂陷槽是川中地区的烃源灶,该裂陷槽在早寒武世达到发育高峰期,此时槽内沉积了巨厚的下寒武统筇竹寺组优质页岩。这套烃源岩生成的天然气资源量可达$135×10^{12}$ m³,是该地区的主力烃源岩(Zou et al.,2014)。两幕桐湾运动使得灯二段和灯四段顶部形成了大量的溶蚀型孔洞,这些孔洞体积大、连通性好,为优质的储集空间。烃源岩和储集层侧向交替,形成了良好的生-储-盖组合(汪泽成等,2014)。这些认识为四川盆地天然气勘探思路的转变起到了关键作用,2011年高石1井天然气勘探的突破及绵阳-长宁裂陷槽的确认使四川盆地天然气勘探的重点转向裂陷槽两侧。目前安岳震旦-寒武系大气田的探明储量已超过万亿方,成为我国已探明的单体储量最大的天然气田。

5.3 渤中19-6整装凝析油气田

渤海湾盆地是一个典型的油型盆地,其中发育沙河街组三段、沙河街组一段和东营组三段等多套腐殖-腐泥型烃源岩,晚期构造活动强烈,导致该地区天然气保存条件较差,天然气储量占比较低,长时间以来该地区的天然气勘探工作一直没有突破(薛永安和王德英,2020)。近年来的研究表明,渤中凹陷厚层腐殖-腐泥型烃源岩在深埋条件下具有高成熟阶段富气的特点,具备形成大气田的物质条件,并构建了渤中凹陷不同层系烃源岩的排烃模型,指出渤中凹陷排油量为512.5亿 m³,排气量达$311.64×10^{11}$ m³(谢玉洪等,2018;薛永安,2019)。由太古宇潜山变质花岗岩主体及上覆的古近系古新统—始新统孔店组砂砾岩组成的泛潜山储集系统发育砂砾岩孔隙带、风化壳溶蚀裂缝带和内幕裂缝带的多层次储集层结构,具有良好的储集物性(侯明才等,2019;徐长贵等,2019)。区域性稳定分布的东营组和沙河街组巨厚超压泥岩为大型凝析气田的保存提供了良好的封盖条件(施和生等,2019)。进而提出渤中19-6构造区具有"优质烃源岩深埋生气、变质岩潜山多期构造运动控储、厚层超压泥岩'被子'控制油气汇聚运移和保存"的天然气成藏模式。这一认识指导了渤中19-6构造大型凝析气田的发现，实现了渤海海域天然气勘探的领域性突破（薛永安和李慧勇，2018）。

5.4 南海陵水深水气田

琼东南盆地是我国南海重要的含油气盆地，20世纪南海西部公司在该盆地的浅水区发现了YC13-1气田并做了大量的工作，提出了崖城组海陆过渡相煤系烃源岩是该地区的主力烃源岩，储集层类型主要为古近系三角洲、滨海相储集层的认识（张迎朝等，2020）。然而，占该盆地69%的深水区的勘探工作一直停滞不前，长期没有重大发现。近年来，我国学者对该地区的天然气地质和地球化学特征开展了大量研究，结果表明天然气很可能来源于崖城组高成熟海相烃源岩，而非海陆过渡相煤系烃源岩（张迎朝等，2019a，2019b）。另外，深水区与浅水区的储集层类型不同，中央峡谷重力流沉积才是深水区有利储集体。峡谷内部充填结构具有"多期次性"，充填了相互叠置的多套有利的砂泥岩储盖组合，并依靠峡谷边界侧封和岩性尖灭形成多个构造+岩性圈闭，具有极大的勘探潜力。晚期新构造运动形成的断裂、微裂隙及底辟体系为天然气垂向运移提供了通道（谢玉洪，2014）。这些认识直接推动了中央峡谷内LS17-2大气田的发现，实现了琼东南盆地深水区油气勘探领域突破，为南海深水区油气勘探指明了方向。

6 结　　语

（1）腐泥型烃源岩和煤系烃源岩的生气机理得到了深化，定量评价了腐泥型烃源岩的排烃效率，对干酪根裂解气、滞留烃裂解气、源外分散液态烃裂解气和油藏裂解气的生气模式进行了研究并明确了各类裂解气对天然气藏的贡献；提出煤系烃源岩在$R_o>2.5\%$以后仍然有很大的生气潜力，将煤系烃源岩的生气下限拓展至$R_o=5.0\%$；建立了多种图版来区分煤成气和油型气、干酪根裂解气和原油裂解气、源外分散液态烃裂解气和油藏裂解气。

（2）稀有气体同位素和放射性同位素定年技术不断进步，极大地促进了我国油气成藏年代学的发展。稀有气体同位素、轻烃化合物及传统的天然气碳氢同位素组成等示踪指标得到了不断的发展，流体包裹体、无机矿物碳氧同位素组成及地层水地球化学特征也被应用于天然气示踪研究中，这些指标丰富了我国天然气成藏示踪体系并得到了广泛的应用。基于完善的成藏示踪体系，不同成因、不同来源的天然气对成藏的贡献得到了有效的确认，夯实了多源成气理论，扩大了我国天然气勘探领域。

（3）在非常规天然气地质理论和勘探技术上取得了突破性进展。明确了我国海域及陆上天然气水合物的气源及成藏模式，建立了多种天然气水合物开采技术；在页岩气生气机理、储集物性、原地气量评价及资源分布规律等方面取得了进展，指出$R_o>3.5\%$的地区为页岩气勘探高风险区域，寒武系筇竹寺组、陆相和海陆过渡相页岩及埋深超过3 500 m的页岩是今后页岩气勘探的重点；认识到了吸附气对致密砂岩储集层含气特征、渗流规律及储量计算的影响，创新性地提出致密砂岩储集体连续而气藏不连续，并将致密砂岩气藏划分为"连续型"和"圈闭型"两种；深入认识到生物成因气的形成机理，明确了煤层气具有多种成藏模式，既可以"自生自储"，又可以"他生他储"，指出了深层煤层气与中浅层煤层气在储集物性、天然气赋存状态、吸附性与含气量等方面的差异，指出了下一步的勘探方向。

（4）引进和创立了一系列新的分析测试技术方法，在团簇同位素、位置特异性同位素组成、超微量气体组分氢同位素组成及有机质生烃热模拟实验等方面取得了重要进展，这些新的技术方法为深化对天然气成藏过程的认识提供了新的依据。

（5）天然气地球化学理论的发展有效地指导了我国天然气勘探开发工作。页岩有机质生烃、储集物性、原地气量模型及资源分布规律等方面的研究进展促进了焦石坝、长宁和威远地区的页岩气勘探开发；绵阳-长宁裂陷槽的发现确定了川中地区的烃源灶及其资源潜力，为安岳大气田的发现奠定了基础；渤中凹陷厚层腐殖-腐泥型烃源岩生排烃模型的建立、潜山储集物性及盖层封闭性的综合研究指导了渤中19-6

整装凝析气田的勘探；琼东南盆地深水区和浅水区在烃源岩和储集层的差异研究为琼东南深水区陵水17-2气田的发现奠定了基础。

随着科学技术的不断更新，微区微量分析技术成为天然气地球化学发展的重要方向。新的分析技术同时促进了天然气地球科学的不断进步。地球化学在非常规天然气地质学、深层天然气和非生物成因天然气研究方面必将发挥更为重要的作用。

参 考 文 献

蔡长娥,邱楠生,徐少华. 2014. Re-Os同位素测年法在油气成藏年代学的研究进展. 地球科学进展,29(12):1362-1371

蔡李梅,陈红汉,李兆奇,吴悠. 2008. 油气成藏过程中的同位素测年方法评述. 沉积与特提斯地质,28(4):18-23

陈刚,徐黎明,丁超,章辉若,李书恒,胡延旭,黄得顺,李楠,李岩. 2012. 用自生伊利石定年确定鄂尔多斯盆地东北部二叠系油气成藏期次. 石油与天然气地质,33(5):713-719,729

陈林勇,王保玉,邰超,关嘉栋,赵晗,王美林,韩作颖. 2016. 无烟煤微生物成气中间代谢产物组成及其转化. 煤炭学报,41(9):2305-2311

陈玲,马昌前,凌文黎,佘振兵,陈子万. 2010. 中国南方存在印支期的油气藏——Re-Os同位素体系的制约. 地质科技情报,29(2):95-99

陈玲,张微,佘振兵. 2012. 油气成藏时间的确定方法. 新疆石油地质,33(5):550-553

陈麦雨,徐守余,张江晖,王朝,吕召宁. 2018. 页岩气碳同位素倒转成因及其意义. 海洋地质前沿,34(12):22-28

陈生蓉,帅琴,高强,田亚,徐保瑞,黄云杰. 2015. 基于扫描电镜-氮气吸脱附和压汞法的页岩孔隙结构研究. 岩矿测试,34(6):636-642

戴金星. 1992. 各类烷烃气的鉴别. 中国科学:地球科学,(2):185-193

戴金星. 2019. 中国陆上四大天然气产区. 天然气与石油,37(2):1-6

戴金星,龚剑明. 2018. 中国煤成气理论形成过程及对天然气工业发展的战略意义. 中国石油勘探,23(4):1-10

戴金星,邹才能,张水昌,李剑,倪云燕,胡国艺,罗霞,陶士振,朱光有,米敬奎,李志生,胡安平,杨春,周庆华,帅燕华,张英,马成华. 2008. 无机成因和有机成因烃烃气的鉴别. 中国科学:地球科学,38(11):1329-1341

戴金星,倪云燕,吴小奇. 2012. 中国致密砂岩气及在勘探开发上的重要意义. 石油勘探与开发,39(3):257-264

戴金星,倪云燕,胡国艺,黄士鹏,廖凤蓉,于聪,龚德瑜,吴伟. 2014. 中国致密砂岩大气田的稳定碳氢同位素组成特征. 中国科学:地球科学,44(4):563-578

付少英,陆敬安. 2010. 神狐海域天然气水合物的特征及其气源. 海洋地质动态,26(9):6-10

盖海峰,肖贤明. 2013. 页岩气碳同位素倒转:机理与应用. 煤炭学报,38(5):827-833

高金亮,倪云燕,李伟,袁懿琳. 2020. 煤系烃源岩高-过成熟阶段生气模拟实验及地质意义. 石油勘探与开发,47(4):723-729

关德范. 2018. 烃源岩生油模拟实验仪的研制与实验. 中外能源,23(5):29-36

郭迎春,庞雄奇,陈冬霞,姜福杰,汤国民. 2013. 致密砂岩气成藏研究进展及值得关注的几个问题. 石油与天然气地质,34(6):717-724

郭迎春,宋岩,庞雄奇,姜振学,付金华,杜建军. 2016. 连续型致密砂岩气近源累计聚集的特征及成因机制. 地球科学,41(3):433-440

何乔,丁晨,李贵中,陈浩,承磊,张辉. 2013. 不同成熟度煤样产甲烷潜力. 微生物学报,53(12):1307-1317

何治亮,金晓辉,沃玉进,李慧莉,白振瑞,焦存礼,张仲培. 2016. 中国海相超深层碳酸盐岩油气成藏特点及勘探领域. 中国石油勘探,21(1):3-14

侯明才,曹海洋,李慧勇,陈安清,韦阿娟,陈扬,王粤川,周雪威,叶涛. 2019. 渤海海域渤中19-6构造带深层潜山储集层特征及其控制因素. 天然气工业,39(1):33-44

黄霞,祝有海. 2010. 神狐海域水合物调查研究区天然气水合物烃类气体来源研究——与邻区LW3-1-1井烃类气体地球化学特征对比研究. 矿床地质,29(S1):1045-1046

贾承造. 2017. 论非常规油气对经典石油天然气地质学理论的突破及意义. 石油勘探与开发,44(1):1-11

简阔,傅雪海,韩作颖,许小凯,周丹,茹忠宏,王观宏,郭晨. 2020. 煤制生物气产出规律及其同位素分馏效应. 煤炭学报,45(7):2602-2609

金之钧. 2005. 中国海相碳酸盐岩层系油气勘探特殊性问题. 地学前缘,12(3):15-22

康永尚,孙良忠,张兵,顾娇杨,叶建平,姜杉钰,王金,毛得雷. 2017. 中国煤储集层渗透率主控因素和煤层气开发对策. 地质论评,63(5):1401-1418

李剑,王义凤,马卫,王东良,马成华,李志生. 2015. 深层-超深层古老烃源岩滞留烃及其裂解气资源评价. 天然气工业,35(11):9-15

李剑,李志生,王晓波,王东良,谢增业,李谨,王义凤,韩中喜,马成华,王志宏,崔会英,王蓉,郝爱胜. 2017. 多元天然气成因判识新指标及图版. 石油勘探与开发,44(4):503-512

李剑,马卫,王义凤,王东良,谢增业,李志生,马成华. 2018. 腐泥型烃源岩生排烃模拟实验与全过程生烃演化模式. 石油勘探与开发,45(3):445-454

李建忠,郭彬程,郑民,杨涛. 2012. 中国致密砂岩气主要类型、地质特征与资源潜力. 天然气地球科学,23(4):607-615

李军杰,刘汉彬,张佳,金贵善,张建锋,韩娟. 2015. K-Ar法与Ar-Ar法在油气成藏期定年的适用性. 地质学报,89(S1):87

李平平,马倩倩,邹华耀,余新亚. 2017. 团簇同位素的基本原理与地质应用. 古地理学报,19(4):713-728

李松,汤达祯,许浩,陶树. 2016. 深部煤层气储集层地质研究进展. 地学前缘,23(3):10-16

李辛子,王运海,姜昭琛,陈贞龙,王立志,吴群. 2016. 深部煤层气勘探开发进展与研究. 煤炭学报,41(1):24-31

李勇,王延斌,孟尚志,吴翔,陶传奇,许卫凯. 2020. 煤系非常规天然气合采地质基础理论进展及展望. 煤炭学报,45(4):1406-1418

李真,王选策,刘可禹,Svetlana T,杨雪梅,马行陟,孙海涛. 2017. 油气藏铼-锇同位素定年的进展与挑战. 石油学报,38(3):297-306

李志清,孙洋,胡瑞林,赵颖,彭宇. 2018. 基于核磁共振法的页岩纳米孔隙结构特征研究. 工程地质学报,26(3):758-766

刘嘉庆,李忠,颜梦珂,Swart P K,杨柳,卢朝进. 2020. 塔里木盆地塔中地区下奥陶统白云岩的成岩流体演化:来自团簇同位素的证据. 石油与天然气地质,41(1):68-82

刘鹏,王晓锋,房嬛,郑建京,李孝甫,孟强. 2016. 碳酸盐岩有机质丰度测试新方法. 沉积学报,34(1):200-206

刘树根,孙玮,李智武,邓宾,钟勇,宋金民,冉波,罗志立,韩克猷,姜磊,梁霄. 2016. 四川叠合盆地海相碳酸盐岩油气分布特征及其构造主控因素. 岩性油气藏,28(5):1-17

刘文汇,徐永昌. 1996. 天然气成因类型及判别标志. 沉积学报,14(1):110-116

刘文汇,张建勇,范明,高波,张殿伟,郑伦举. 2007. 叠合盆地天然气的重要来源—分散可溶有机质. 石油实验地质,29(1):1-6

刘文汇,王杰,陶成,胡广,卢龙飞,王萍. 2013a. 中国海相层系油气成藏年代学. 天然气地球科学,24(2):199-209

刘文汇,王晓锋,腾格尔,张殿伟,王杰,陶成,张中宁,卢龙飞. 2013b. 中国近十年天然气示踪地球化学研究进展. 矿物岩石地球化学通报,32(3):279-289

刘文汇,赵恒,刘全有,周冰,张殿伟,王杰,卢龙飞,罗厚勇,孟庆强,吴小奇. 2016. 膏盐岩层系在海相油气成藏中的潜在作用. 石油学报,37(12):1451-1462

刘文汇,腾格尔,王晓锋,黎茂稳,胡广,王杰,卢龙飞,赵恒,陈强路,罗厚勇. 2017a. 中国海相碳酸盐岩层系有机质生烃理论新解. 石油勘探与开发,44(1):155-164

刘文汇,腾格尔,张中宁,罗厚勇,张殿伟,王杰,李立武,高波,卢龙飞,赵恒. 2017b. 四川盆地高硫天然气成藏机理的同位素研究. 中国科学:地球科学,47(2):166-178

刘雨晨,邱楠生,常健,贾京坤,李慧莉,马安来. 2020. 碳酸盐团簇同位素在沉积盆地热演化中的应用——以塔里木盆地顺托果勒地区为例. 地球物理学报,63(2):597-611

鲁晓兵,张旭辉,王平康,梁前勇. 2019. 天然气水合物成藏动力学研究进展. 中国科学:物理学 力学 天文学,49(3):56-71

罗厚勇,王万春,刘文汇. 2012. TSR模拟实验研究与地质实际的异同及可能原因分析. 石油实验地质,34(2):186-192,198

罗敏,王宏斌,杨胜雄,陈多福. 2013. 南海天然气水合物研究进展. 矿物岩石地球化学通报,32(1):56-69

马中良,申宝剑,潘安阳,腾格尔,宁传祥,郑伦举. 2020. 四川盆地五峰组-龙马溪组页岩气成因与碳同位素倒转机制——来自热模拟实验的认识. 石油实验地质,42(3):428-433

孟庆强,陶成,朱东亚,金之钧,王强,郑伦举. 2011. 微量氢气定量富集方法初探. 石油实验地质,33(3):314-316

欧阳永林,田文广,孙斌,王勃,祁灵,孙钦平,杨青,董海超. 2018. 中国煤系气成藏特征及勘探对策. 天然气工业,38(3):15-23

秦华,范小军,刘明,郝景宇,梁波. 2016. 焦石坝地区龙马溪组页岩解吸气地球化学特征及地质意义. 石油学报,37(7):846-854

秦胜飞,周国晓. 2018. 气田水对甲烷氢同位素分馏作用. 天然气地球科学,29(3):311-316

秦胜飞,周国晓,李伟,侯曜华,吕芳. 2016. 四川盆地威远气田水溶气脱气成藏地球化学证据. 天然气工业,36(1):43-51

赛彦明,田辉,李杰,刘银山,张彬,刘俊杰. 2020. 含油气系统Re-Os定年及Re-Os元素和同位素体系研究新进展. 天然气地球科学,31(7):939-951

沈传波,刘泽阳,肖凡,胡迪,杜嘉祎. 2015. 石油系统Re-Os同位素体系封闭性研究进展. 地球科学进展,30(2):187-195

沈传波,葛翔,白秀娟. 2019. 四川盆地震旦-寒武系油气成藏的Re-Os年代学约束. 地球科学,44(3):713-726

施和生,王清斌,王军,刘晓健,冯冲,郝轶伟,潘文静. 2019. 渤中凹陷深层渤中19-6构造大型凝析气田的发现及勘探意义. 中国石油勘探,24(1):36-45

帅燕华,张水昌,胡国艺. 2018. 天然气组分的簇同位素研究进展. 矿物岩石地球化学通报,37(4):559-571

宋岩,徐永昌. 2005. 天然气成因类型及其鉴别. 石油勘探与开发,32(4):24-29

宋岩,赵孟军,胡国艺,朱光有. 2012. 中国天然气地球化学研究新进展及展望. 矿物岩石地球化学通报,31(6):529-542

孙健,程鹏,盖海峰. 2018. 同源不同成熟度天然气的混合对天然气碳同位素的影响. 地球化学,47(4):354-362

孙敏卓,孟仟祥,郑建京,王国仓,房嬛,王作栋. 2013. 塔里木盆地海相碳酸盐岩中有机酸盐的分析. 中南大学学报(自然科学版),44(1):216-222

陶成,刘文汇,孟庆强,杨华敏,周宇. 2012. 天然气中微痕量氢同位素的在线连续分析. 分析化学,40(3):482-486

陶成,刘文汇,腾格尔,秦建中,王杰,杨华敏,王萍. 2015. 天然气He的累积模式及定年应用初探. 地质学报,89(7):1302-1307

陶士振,戴金星,邹才能,王京红,米敬奎,汪泽成,胡素云. 2012. 松辽盆地火山岩包裹体稀有气体同位素与天然气成因成藏示踪. 岩石学报,28(3):927-938

田辉,王招明,肖中尧,李贤庆,肖贤明. 2006. 原油裂解成气动力学模拟及其意义. 科学通报,51(15):1821-1827

涂湘林,朱炳泉,张景廉,刘颖,刘菊英,施泽恩. 1997. Pb、Sr、Nd同位素体系在石油定年与成因示踪研究中的应用. 地球化学,26(2):62-72

汪泽成,姜华,王铜山,鲁卫华,谷志东,徐安娜,杨雨,徐兆辉. 2014. 四川盆地桐湾期古地貌特征及成藏意义. 石油勘探与开发,41(3):305-312

王大鹏,白国平,徐艳,陈小亮,陶家智,张明亮. 2016. 全球古生界海相碳酸盐岩大油气田特征及油气分布. 古地理学报,18(1):80-92

王东良,张英,卢双舫,国建英,李志生,莫午零,王民,王义凤. 2012. 烃源岩过成熟阶段生气潜力的实验室模拟. 沉积学报,30(6):1172-1179

王华建,张水昌,王晓梅. 2013. 如何实现油气成藏期的精确定年. 天然气地球科学,24(2):210-217

王杰,刘文汇,陶成,腾格尔,席斌斌,王萍,杨华敏. 2018. 海相油气成藏定年技术及其对元坝气田长兴组天然气成藏年代的反演. 地球科学,43(6):1817-1829

王亮,宁波. 2020. 四川盆地海相页岩气碳同位素倒转成因. 西安文理学院学报(自然科学版),23(1):81-85

王鹏,刘四兵,沈忠民,黄飞,张文凯,邹黎明. 2015. 地球化学指标示踪天然气运移机理及有效性分析——以川西坳陷侏罗系天然气为例. 天然气地球科学,26(6):1147-1155

王鹏威,陈筱,庞雄奇,姜振学,姜福杰,王迎春,郭继刚,戴琦雯,文婧. 2014. 构造裂缝对致密砂岩气成藏过程的控制作用. 天然气地球科学,25(2):185-191

王希彬,李中平,张铭杰,邢蓝田,曹ह辉,刘艳,李立武. 2016. 固相微萃取-气相色谱-同位素质谱法测定天然气中微量烃类单分子氢同位素组成. 天然气地球科学,27(6):1084-1091,1100

王晓锋,刘鹏,刘昌杰,孟强. 2018. 团簇同位素及特定位置同位素组成在天然气地球化学研究中的应用. 矿物岩石地球化学通报,37(4):580-587

魏国齐,沈平,杨威,张健,焦贵浩,谢武仁,谢增业. 2013. 四川盆地震旦系大气田形成条件与勘探远景区. 石油勘探与开发,40(2):129-138

魏国齐,王东良,王晓波,李剑,李志生,谢增业,崔会英,王志宏. 2014. 四川盆地高石梯-磨溪大气田稀有气体特征. 石油勘探与开发,41(5):533-538

魏国齐,谢增业,李剑,杨威,张水昌,张奇,刘新社,王东良,张福东,程宏岗. 2017. "十二五"中国天然气地质理论研究新进展. 天然气工业,37(8):1-13

魏国齐,李剑,杨威,谢增业,董才源,佘源琦,马卫. 2018. "十一五"以来中国天然气重大地质理论进展与勘探新发现. 天然气地球科学,29(12):1691-1705

魏强,李贤庆,梁万乐,孙可欣,谢增业,李谨,肖中尧. 2019. 库车坳陷大北-克深地区深层致密砂岩气地球化学特征及成因. 矿物岩石地球化学通报,38(2):418-427

魏新善,程国建,石晓英,赵会涛. 2017a. 致密砂岩气认知阶段讨论与启示. 西安石油大学学报(社会科学版),26(2):17-22,29

魏新善,胡爱平,赵会涛,康锐,石晓英,刘晓鹏. 2017b. 致密砂岩气地质认识新进展. 岩性油气藏,29(1):11-20

吴传芝,赵克斌,孙长青,陈银节,杨俊. 2018. 天然气水合物基本性质与主要研究方向. 非常规油气,5(4):92-99

吴伟,房忱琛,董大忠,刘丹. 2015. 页岩气地球化学异常与气源识别. 石油学报,36(11):1332-1340,1366

伍宗涛,刘兴旺,李孝甫,王晓锋,郑建京. 2017. 稀有气体同位素在四川盆地下坝气藏气源对比中的应用. 天然气地球科学,28(7):1072-1077

肖贤明,宋之光,朱炎铭,田辉,尹宏伟. 2013. 北美页岩气研究及对我国下古生界页岩气开发的启示. 煤炭学报,38(5):721-727

肖芝华,胡国艺,钟宁宁,李志生. 2009. 塔里木盆地煤系烃源岩产气率变化特征. 西南石油大学学报(自然科学版),31(1):9-13

谢玉洪. 2014. 南海北部自营深水天然气勘探重大突破及其启示. 天然气工业,34(10):1-8

谢玉洪,张功成,沈朴,刘丽芳,黄胜兵,陈少平,杨树春. 2018. 渤海湾盆地渤中凹陷大气田形成条件与勘探方向. 石油学报,39(11):1199-1210

谢增业,李志生,黄志兴,王晓波,马成华. 2008. 川东北不同含硫物质硫同位素组成及H_2S成因探讨. 地球化学,37(2):187-194

谢增业,李志生,魏国齐,李剑,王东良,王志宏,董才源. 2016. 腐泥型干酪根热降解成气潜力及裂解气判识的实验研究. 天然气地球科学,27(6):1057-1066

徐长贵,于海波,王军,刘晓健. 2019. 渤海海域渤中19-6大型凝析气田形成条件与成藏特征. 石油勘探与开发,46(1):25-38

徐春春,邹伟宏,杨跃明,段勇,沈扬,罗冰,倪超,付小东,张建勇. 2017. 中国陆上深层油气资源勘探开发现状及展望. 天然气地球科学,28(8):1139-1153

徐秋晨,邱楠生,刘雯,常青. 2019. 利用团簇同位素恢复沉积盆地热历史的探索. 科学通报,64(S1):566-578

薛永安. 2019. 渤海海域深层天然气勘探的突破与启示. 天然气工业,39(1):11-20

薛永安,李慧勇. 2018. 渤海海域深层太古界变质岩潜山大型凝析气田的发现及其地质意义. 中国海上油气,30(3):1-9

薛永安,王德英. 2020. 渤海湾油型湖盆大型天然气藏形成条件与勘探方向. 石油勘探与开发,47(2):260-271

杨春,陶士振,侯连华,米敬奎,杨帆. 2014. 松辽盆地火山岩储集层天然气藏He同位素组成累积效应. 天然气地球科学,25(1):109-115

杨宏民,任子阳,王兆丰. 2009. 煤对气体吸附特征的研究现状及应用前景展望. 煤,18(8):1-4

杨兆彪,秦勇,高弟,陈润. 2011. 超临界条件下煤层甲烷视吸附量、真实吸附量的差异及其地质意义. 天然气工业,31(4):13-16

叶素娟,朱宏权,李嵘,杨映涛,黎青. 2017. 天然气运移有机-无机地球化学示踪指标—以四川盆地川西坳陷侏罗系气藏为例. 石油勘探与开发,44(4):549-560

于聪,黄士鹏,龚德瑜,廖凤蓉,李瑾,孙庆伍. 2013. 天然气碳、氢同位素部分倒转成因—以苏里格气田为例. 石油学报,34(S1):92-101

云建兵,施和生,朱俊章,吴河勇,冯子辉,邱华宁. 2009. 砂岩储集层自生伊利石$^{40}Ar-^{39}Ar$定年技术及油气成藏年龄探讨. 地质学报,83(8):1134-1140

张富贵,孙忠军,杨志斌,周亚龙,张舜尧,曹长茂,王惠艳,唐瑞玲,庞守吉,王平康,祝有海. 2019. 中国冻土区天然气水合物的地球化学调查. 矿物岩石地球化学通报,38(6):1224-1234

张景廉,张宁,朱炳泉,涂湘林,刘菊英,刘颖,施泽恩,张平中. 1997. 克拉玛依乌尔禾沥青脉Pb-Sr-Nd同位素地球化学. 中国科学:地球科学,27(4):325-330

张水昌,胡国艺,米敬奎,帅燕华,何坤,陈建平. 2013. 三种成因天然气生成时限与生成量及其对深部油气资源预测的影响. 石油学报,34(S1):41-50

张涛,马行陟,王伦,黄家旋,赵卫卫. 2017. Re-Os同位素油气成藏定年研究进展. 石油地质与工程,31(4):30-34

张晓东,秦勇,桑树勋. 2005. 煤储集层吸附特征研究现状及展望. 中国煤田地质,17(1):16-21,29

张迎朝,李绪深,徐新德,甘军,杨希冰,梁刚,何小胡,李兴. 2019a. 琼东南盆地深水西区L25气田天然气成因、来源与成藏过程. 海相油气地

质,24(3):73-82

张迎朝,徐新德,甘军,杨希冰,朱继田,杨金海,何小胡,郭潇潇. 2019b. 琼东南盆地深水区 L18 气田上新统地层圈闭气田形成条件及成藏模式. 海洋学报,41(3):121-133

张迎朝,甘军,徐新德,梁刚,何小胡,李兴. 2020. 琼东南盆地海相烃源岩的发现与勘探意义. 煤炭技术,39(2):43-45

张有瑜,罗修泉. 2011. 英买力沥青砂岩自生伊利石 K-Ar 测年与成藏年代. 石油勘探与开发,38(2):203-210

张有瑜,陶士振,刘可禹,罗修泉. 2015. 四川盆地须家河组致密砂岩气自生伊利石年龄分布与成藏时代. 石油学报,36(11):1367-1379

张瑜,闫建萍,贾祥娟,李艳芳,邵德勇,于萍,张同伟. 2015. 四川盆地五峰组-龙马溪组富有机质泥岩孔径分布及其与页岩含气性关系. 天然气地球科学,26(9):1755-1762

张玉红,周世新,左亚彬. 2018. 碳同位素在天然气运移路径示踪中的应用研究进展. 矿物岩石地球化学通报,37(6):1198-1204

赵龙,秦勇,杨兆彪,申建,韩贝贝,张政. 2014. 煤中超临界甲烷等温吸附模型研究. 天然气地球科学,25(5):753-760

赵伟,王全胜,郑星升. 2019. 国内天然气水合物研究进展. 石化技术,26(10):165-167

赵文智,王兆云,张水昌,王红军,王云鹏. 2006. 油裂解生气是海相气源灶高效成气的重要途径. 科学通报,51(5):589-595

赵文智,胡素云,刘伟,王铜山,李永新. 2014. 再论中国陆上深层海相碳酸盐岩油气地质特征与勘探前景. 天然气工业,34(4):1-9

赵文智,王兆云,李东良,李剑,李永新,胡国义. 2015. 分散液态烃的成藏地位与意义. 石油勘探与开发,42(4):401-413

赵兴齐,陈践发,张晨,吴雪飞,刘娅昭,徐学敏. 2011. 天然气藏中硫化氢成因研究进展. 新疆石油地质,32(5):552-556

周国晓,秦胜飞,侯曜华,吕芳. 2016. 四川盆地安岳气田龙王庙组气藏天然气有水溶气贡献的迹象. 天然气地球科学,27(12):2193-2199

朱光有,张水昌,梁英波,戴金星. 2006a. 四川盆地 H_2S 的硫同位素组成及其成因探讨. 地球化学,35(4):432-442

朱光有,张水昌,梁英波,李其荣. 2006b. 四川盆地威远气田硫化氢的成因及其证据. 科学通报,51(23):2780-2788

邹才能,董大忠,王玉满,李新景,黄金亮,王淑芳,管全中,张晨晨,王红岩,刘洪林,拜文华,梁峰,吝文,赵群,刘德勋,杨智,梁萍萍,孙莎莎,邱振. 2015. 中国页岩气特征、挑战及前景(一). 石油勘探与开发,42(6):689-701

邹才能,董大忠,王玉满,李新景,黄金亮,王淑芳,管全中,张晨晨,王红岩,刘洪林,拜文华,梁峰,吝文,赵群,刘德勋,杨智,梁萍萍,孙莎莎,邱振. 2016. 中国页岩气特征、挑战及前景(二). 石油勘探与开发,43(2):166-178

邹才能,赵群,董大忠,杨智,邱振,梁峰,王南,黄勇,端安祥,张琴,胡志明. 2017. 页岩气基本特征、主要挑战与未来前景. 天然气地球科学,28(12):1781-1796

邹才能,杨智,张国生,陶士振,朱如凯,袁选俊,侯连华,董大忠,郭秋麟,宋岩,冉启全,吴松涛,白斌,王岚,王志平,杨正明,才博. 2019. 非常规油气地质学建立及实践. 地质学报,93(1):12-23

邹才能,郭建林,贾爱林,位云生,闫海军,贾成业,唐海发. 2020. 中国大气田科学开发的内涵. 天然气工业,40(3):1-12

Chen J, Xiao X M. 2014. Evolution of nanoporosity in organic-rich shales during thermal maturation. Fuel,129:173-181

Cheng P, Xiao X M, Wang X, Sun J, Wei Q. 2019. Evolution of water content in organic-rich shales with increasing maturity and its controlling factors: implications from a pyrolysis experiment on a water-saturated shale core sample. Marine and Petroleum Geology,109:291-303

Dai J X, Zou C N, Liao S M, Dong D Z, Ni Y Y, Huang J L, Wu W, Gong D Y, Huang S P, Hu G Y. 2014. Geochemistry of the extremely high thermal maturity Longmaxi shale gas, southern Sichuan Basin. Organic Geochemistry,74:3-12

Dias R F, Freeman K H, Lewan M D, Franks S G. 2002. $\delta^{13}C$ of low-molecular-weight organic acids generated by the hydrous pyrolysis of oil-prone source rocks. Geochimica et Cosmochimica Acta,66(15):2755-2769

Gai H F, Xiao X M, Cheng P, Tian H, Fu J M. 2015. Gas generation of shale organic matter with different contents of residual oil based on a pyrolysis experiment. Organic Geochemistry,78:69-78

Gai H F, Tian H, Xiao X M. 2018. Late gas generation potential for different types of shale source rocks: implications from pyrolysis experiments. International Journal of Coal Geology,193:16-29

Gilbert A, Robins R J, Remaud G S, Tcherkez G G B. 2012. Intramolecular ^{13}C pattern in hexoses from autotrophic and heterotrophic C_3 plant tissues. Proceedings of the National Academy of Sciences of the United States of America,109(44):18204-18209

Gilbert A, Yamada K, Yoshida N. 2013. Accurate method for the determination of intramolecular ^{13}C isotope composition of ethanol from aqueous solutions. Analytical Chemistry,85(14):6566-6570

Guo H J, Jia W L, Peng P A, Zeng J, He R L. 2017. Evolution of organic matter and nanometer-scale pores in an artificially matured shale undergoing two distinct types of pyrolysis: a study of the Yanchang Shale with Type II kerogen. Organic Geochemistry,105:56-66

Guo L G, Xiao X M, Tian H, Song Z G. 2009. Distinguishing gases derived from oil cracking and kerogen maturation: insights from laboratory pyrolysis experiments. Organic Geochemistry,40(10):1074-1084

Hattori R, Yamada K, Kikuchi M, Hirano S, Yoshida N. 2011. Intramolecular carbon isotope distribution of acetic acid in vinegar. Journal of Agricultural and Food Chemistry,59(17):9049-9053

Ko L T, Loucks R G, Zhang T W, Ruppel S C, Shao D Y. 2016. Pore and pore network evolution of Upper Cretaceous Boquillas (Eagle Ford-equivalent) mudrocks: results from gold tube pyrolysis experiments. AAPG Bulletin,100(11):1693-1722

Krishna R. 2009. Describing the diffusion of guest molecules inside porous structures. Journal of Physical Chemistry C,113(46):19756-19781

Lei R, Xiong Y Q, Li Y, Zhang L. 2018. Main factors influencing the formation of thermogenic solid bitumen. Organic Geochemistry,121:155-160

Li S, Tang D Z, Pan Z J, Xu H, Huang W Q. 2013. Characterization of the stress sensitivity of pores for different rank coals by nuclear magnetic resonance. Fuel,111:746-754

Li Y, Zhang L, Xiong Y Q, Gao S T, Yu Z Q, Peng P A. 2018. Determination of position-specific carbon isotope ratios of propane from natural gas.

Organic Geochemistry,119:11-21

Liu Q Y,Jin Z J,Liu W H,Lu L F,Meng Q X,Tao Y,Han P L. 2013. Presence of carboxylate salts in marine carbonate strata of the Ordos Basin and their impact on hydrocarbon generation evaluation of low TOC,high maturity source rocks. Science China Earth Sciences,56(12):2141-2149

Liu W H,Borjigin T,Wang X F,Li M W,Hu G,Wang J,Lu L F,Zhao H,Chen Q L,Luo H Y. 2017. New knowledge of hydrocarbon generating theory of organic matter in Chinese marine carbonates. Petroleum Exploration and Development,44(1):159-169

Liu Y K,Xiong Y Q,Li Y,Peng P A. 2017. Effects of oil expulsion and pressure on nanopore development in highly mature shale:evidence from a pyrolysis study of the Eocene Maoming oil shale,South China. Marine and Petroleum Geology,86:526-536

Pan L,Xiao X M,Tian H,Zhou Q,Cheng P. 2016. Geological models of gas in place of the Longmaxi shale in Southeast Chongqing,South China. Marine and Petroleum Geology,73:433-444

Qin S F,Li F,Zhou Z,Zhou G X. 2018. Geochemical characteristics of water-dissolved gases and implications on gas origin of Sinian to Cambrian reservoirs of Anyue gas field in Sichuan Basin,China. Marine and Petroleum Geology,89:83-90

Shao X H,Pang X Q,Li Q W,Wang P W,Chen D,Shen W B,Zhao Z F. 2017. Pore structure and fractal characteristics of organic-rich shales:a case study of the lower Silurian Longmaxi shales in the Sichuan Basin,SW China. Marine and Petroleum Geology,80:192-202

Shuai Y H,Douglas P M J,Zhang S C,Stolper D A,Ellis G S,Lawson M,Lewan M D,Formolo M,Mi J K,He K,Hu G Y,Eiler J M. 2018a. Equilibrium and non-equilibrium controls on the abundances of clumped isotopologues of methane during thermogenic formation in laboratory experiments: implications for the chemistry of pyrolysis and the origins of natural gases. Geochimica et Cosmochimica Acta,223:159-174

Shuai Y H,Etiope G,Zhang S C,Douglas P M J,Huang L,Eiler J M. 2018b. Methane clumped isotopes in the Songliao Basin (China):new insights into abiotic vs. biotic hydrocarbon formation. Earth and Planetary Science Letters,482:213-221

Tian H,Xiao X M,Wilkins R W T,Li X Q,Gan H J. 2007. Gas sources of the YN2 gas pool in the Tarim Basin:evidence from gas generation and methane carbon isotope fractionation kinetics of source rocks and crude oils. Marine and Petroleum Geology,24(1):29-41

Tian H,Xiao X M,Yang L G,Xiao Z Y,Guo L G,Shen J G,Lu Y H. 2009. Pyrolysis of oil at high temperatures:gas potentials,chemical and carbon isotopic signatures. Chinese Science Bulletin,54(7):1217-1224

Tian H,Pan L,Zhang T W,Xiao X M,Meng Z P,Huang B J. 2015. Pore characterization of organic-rich Lower Cambrian shales in Qiannan depression of Guizhou Province,southwestern China. Marine and Petroleum Geology,62:28-43

Tian H,Li T F,Zhang T W,Xiao X M. 2016. Characterization of methane adsorption on overmature Lower Silurian-Upper Ordovician shales in Sichuan Basin,Southwest China:experimental results and geological implications. International Journal of Coal Geology,156:36-49

Wang Y P,Zhang S C,Wang F Y,Wang Z Y,Zhao C Y,Wang H J,Liu J Z,Lu J L,Geng A S,Liu D H. 2006. Thermal cracking history by laboratory kinetic simulation of Paleozoic oil in eastern Tarim Basin,NW China,implications for the occurrence of residual oil reservoirs. Organic Geochemistry,37(12):1803-1815

Xiao X M,Zeng Q H,Tian H,Wilkins R W T,Tang Y C. 2005. Origin and accumulation model of the AK-1 natural gas pool from the Tarim Basin,China. Organic Geochemistry,36(9):1285-1298

Xiao X M,Xiong M,Tian H,Wilkins R W T,Huang B J,Tang Y C. 2006. Determination of the source area of the Ya13-1 gas pool in the Qiongdongnan Basin,South China Sea. Organic Geochemistry,37(9):990-1002

Xiao X M,Wei Q,Gai H F,Li T F,Wang M L,Pan L,Chen J,Tian H. 2015. Main controlling factors and enrichment area evaluation of shale gas of the Lower Paleozoic marine strata in South China. Petroleum Science,12(4):573-586

Xiong Y Q,Jiang W M,Wang X T,Li Y,Chen Y,Zhang L,Lei R,Peng P A. 2016. Formation and evolution of solid bitumen during oil cracking. Marine and Petroleum Geology,78:70-75

Zhao H,Liu C J,Larson T E,McGovern G P,Horita J. 2020. Bulk and position-specific isotope geochemistry of natural gases from the Late Cretaceous Eagle Ford Shale,South Texas. Marine and Petroleum Geology,122,104659

Zhou Q,Xiao X M,Pan L,Tian H. 2014. The relationship between micro-Raman spectral parameters and reflectance of solid bitumen. International Journal of Coal Geology,121:19-25

Zhou S W,Xue H Q,Ning Y,Guo W,Zhang Q. 2018. Experimental study of supercritical methane adsorption in Longmaxi shale:insights into the density of adsorbed methane. Fuel,211:140-148

Zou C N,Wei G Q,Xu C C,Du J H,Xie Z Y,Wang Z C,Hou L H,Yang C,Li J,Yang W. 2014. Geochemistry of the Sinian-Cambrian gas system in the Sichuan Basin,China. Organic Geochemistry,74:13-21

Research Progress of Natural Gas Geochemistry

LIU Wen-hui[1,5], WANG Xing[2,3], TIAN Hui[2], ZHENG Guo-dong[4],
WANG Xiao-feng[1], TAO Cheng[5], LIU Peng[1]

1. State Key Laboratory of Continental Dynamics, Department of Geology, Northwest University, Xi'an 710069; 2. State Key Laboratory of Organic Geochemistry, Guangzhou Institute of Geochemistry, Chinese Academy of Sciences, Guangzhou 510640; 3. University of Chinese Academy of Sciences, Beijing 100049; 4. Northwest Institute of Eco-Environment and Resources, Chinese Academy of Sciences, Lanzhou 730000; 5. SINOPEC Key Laboratory of Petroleum Accumulation Mechanisms, Wuxi 214126

Abstract: This paper summarized the progresses in exploration and exploitation of natural gas in China in the past decade. The major progresses lay mainly in: ①The mechanisms of multiple sources and multivariant hydrocarbon generation of natural gas were put forward and the natural gas generation pattern of sapropel source rock was set up; the hydrocarbon generation potential of different parent materials was clarified, both the gas generation potential of coal-bearing source rock and its lower limit were recognized. ②The dating technologies of rare gas isotope and radioisotope were improved continuously, which facilitated the development of hydrocarbon accumulation chronology in China and fined down the natural gas accumulation tracing system and the multi-source gas formation theory, consequently, the exploration area was expanded remarkably in China. ③Breakthroughs made in many research fields, e.g. hydrocarbon generation theory, reservoir physical property, percolation mechanism, accumulation process, preservation condition, and development technology, of natural gas hydrate, shale gas, tight sandstone gas, and coalbed methane, which had effectively guided the exploration and exploitation of unconventional natural gas. ④Several new technologies including methane cluster isotope, propane position-specific isotope and ultra-micro gas hydrogen isotope were employed to develop the traditional hydrocarbon generation simulation method of organic matter, and these new technologies provided additional supports in studying the genetic type and accumulation process of natural gas. ⑤The geochemical theory of natural gas had provided theoretical supports for the exploration and exploitation of a series of large gas fields. We believed that the microanalysis technology is becoming an important direction for the development of natural gas geochemistry.

Key words: natural gas; geochemistry; research progress; dating of hydrocarbon accumulation; the formation of deep gas; unconventional natural gas; exploration of large gas fields

深海矿产研究进展*

石学法[1,2]　符亚洲[3]　李兵[1,2]　黄牧[1,2]　任向文[1,2]　刘季花[1,2]　于淼[1,2]　李传顺[1,2]

1. 自然资源部 第一海洋研究所，自然资源部海洋地质与成矿作用重点实验室，青岛 266061；
2. 青岛海洋科学与技术试点国家实验室，海洋地质过程与环境功能实验室，青岛 266037；
3. 中国科学院 地球化学研究所，矿床地球化学国家重点实验室，贵阳 550002

摘　要：深海矿产是地球上尚未被人类充分认识和利用的最大潜在战略矿产资源，近十年我国在该领域的研究取得了重要进展。在太平洋国际海底区域申请到两块多金属结核勘探区、一块富钴结壳勘探区，在西南印度洋中脊申请到一块多金属硫化物勘探区。研究阐明了我国多金属结核和富钴结壳勘探区小尺度成矿规律，揭示了其成矿作用过程及古海洋古气候记录，探讨了关键金属元素富集机制。在西南印度洋、西北印度洋和南大西洋中脊发现了多处热液区，阐述了其成矿作用及控制因素，建立了超慢速扩洋中脊热液循环模型，探讨了拆离断层型热液成矿系统的成矿机制。在太平洋和印度洋划分了四个深海稀土成矿带，在中印度洋海盆、东南太平洋和西太平洋深海盆地发现了大面积富稀土沉积区，初步揭示了深海稀土的富集特征、分布规律、赋存状态和成矿机理。今后在继续加大深海矿产资源调查研究的同时，应聚焦深海关键金属成矿作用研究。

关键词：深海矿产　勘探合同区　成矿作用　研究进展

0　引　言

深海发育有丰富的矿产、能源和生物资源，是地球上尚未被人类充分认识和利用的最大的潜在战略资源基地。迄今已经发现的深海金属矿产主要有多金属结核、富钴结壳、多金属硫化物和深海稀土，这四类矿产资源潜力巨大，其中的 Mn、Cu、Ni、Co、Pb、Zn、REY（REE+Y）、PGE 等经济价值尤高。

20 世纪 60 年代，深海矿产资源研究进入快速发展期，世界发达国家开始进行大规模调查研究，兴起了深海"蓝色圈地运动"（许东禹，2013）。我国的深海矿产资源调查研究始于 20 世纪 70 年代末，经过四十多年的努力，迄今已遍及太平洋、印度洋和大西洋，涵盖了上述四类资源。我国作为第一批国际海底区域矿产资源勘探的"先驱投资者"，先后与国际海底管理局签订了 5 个勘探区合同，成为世界上第一个在国际海底区域拥有 3 种资源 5 个勘探区的国家，同时还获得了勘探区内相应矿产的专属勘探权。除 2001 年中国大洋协会在东太平洋 CC 区（克拉里恩-克利帕顿区，Clarion-Clipperton zone，CCZ）获得了 75 000 km² 多金属结核勘探区外，其余四个勘探区都是近十年来申请获得的：2011 年中国大洋协会在西南印度洋中脊获取了 10 000 km² 多金属硫化物勘探区，2014 年中国大洋协会在西北太平洋海山区获得了 3 000 km² 富钴结壳勘探区，2017 年中国五矿集团在东太平洋获得了 72 740 km² 多金属结核勘探区，2019 年北京先驱高技术开发公司在西太平洋获得了 74 052 km² 多金属结核勘探区。与此同时，我国对海底多金属结核、富钴结壳、多金属硫化物分布规律和成矿作用进行了深入研究，也是国际上率先开展深海稀土调查研究的国家之一。本文简要回顾了 2011—2020 年十年间我国在深海矿产资源调查与研究领域取得的主要进展和重要发现，并对今后的研究提出了若干建议。

---

* 原文"我国深海矿产研究：进展与发现（2011—2020）"刊于《矿物岩石地球化学通报》2021 年第 40 卷第 2 期，本文略有修改。

1 多金属结核和富钴结壳

多金属结核又称铁锰结核、锰结核，是人类最早发现的深海矿产，它广布于全球水深 4 000～6 500 m 的深海盆地中，主要分布在太平洋，其次是印度洋。位于东北太平洋海盆的克拉里恩-克利帕顿断裂带之间的地区（CCZ）是多金属结核经济价值最高的地区，世界上大多数国家的多金属结核专属勘探区都位于该区（许东禹，2013）。多金属结核富含多种元素，其中 Mn、Co、Ni、Cu 等元素含量高，具有潜在经济价值。多金属结核成因有水成成因、成岩成因和水成-成岩作用混合成因（许东禹，2013；Hein and Koschinsky，2014）。

富钴结壳，也称铁锰结壳、锰结壳或钴结壳（简称"结壳"），主要产在 800～3 000 m 的海山、海台、岛屿斜坡等海底高地顶部和斜坡上的坚硬岩石表面，主要是水成成因（Hein and Koschinsky，2014）。富钴结壳主要分布在太平洋，西太平洋海山区是最有经济价值的区域（称为 prime Fe-Mn crust zone，PCZ）。富钴结壳富含 Co、Ni、Te、Pt、REE 等元素，其中 Co 的含量为 0.4%～1.2%，可媲美刚果（金）成矿带内高品位钴矿床（Co 含量为 0.4%～0.5%）（杜菊民和赵学章，2010）。

多金属结核和富钴结壳金属资源潜力巨大，其中 Mn、Co、Cu、Ni、Mo、Te 可比肩陆地相应金属储量（Hein et al.，2013）。不仅如此，它们还能提供跨越数千万年的海洋和气候演变记录，在古海洋、古气候研究中有重要作用，同时对认识海洋中主要金属元素的源-汇、通量及地球化学循环过程都十分重要（Frank，2002；Hein and Koschinsky，2014；Fu，2020）。

1.1 大洋多金属结核和富钴结壳小尺度成矿规律研究及资源评价

近十年来，我国学者对大洋多金属结核和富钴结壳的调查研究主要集中在太平洋 CCZ 和 PCZ。在我国结核勘探区，从结核覆盖率分布角度查明了多金属结核小尺度分布特征，发现地形坡度对结核分布有明显影响（梁东红等，2014）。刘永刚等（2013）对多金属结核 CC 区已有调查站位进行了模型训练，分别建立了 Mn、Co、Ni、Cu 金属品位与洋壳年龄均值、沉积物厚度、地形起伏度、沉积类型等区域控矿要素之间的关系模型，得到了该区多金属结核资源定量评价有效模型。

在结壳小尺度成矿规律方面，发现尖顶海山体积小，但地形变化连续，底流及环境氧化性较强，有利于结壳的生长，结壳厚度较大，富含成矿元素和生物组分；而平顶海山山体巨大，顶部和斜坡陡崖不利于结壳生长，底流活动只发育在山顶边缘和山脊地带，后期构造活动相对频繁，结壳厚度较小（马维林等，2013，2014；杨胜雄等，2016）。

张富元等（2011，2015）建立了一种可以分别计算资源量的方法，并据此计算出太平洋海山干结壳资源量为 $5.071×10^{10}$～$1.014×10^{11}$ t，结壳分布面积为 $2.16×10^6$ km²，全球三大洋海山干结壳资源量为 $1.081×10^{11}$～$2.162×10^{11}$ t，结壳分布总面积为 $3.034×10^6$ km²。He 等（2011）通过建模研究确定，在勘探区面积为 4 856 km²，开发区域面积为 1 214 km² 条件下可满足 20 年开采，每年 100 万 t 湿结壳的年产量。

Yang 等（2020）对西太平洋深海盆地至海山的结核和结壳进行了声学定量分析，成功建立了背散射与结核、结壳和不同沉积物之间的定性和定量关系，发现高背散射强度指示海山的峰顶和侧脊上发育富钴结壳，深海盆地中发育高丰度的结核。

1.2 中国南海多金属结核和富钴结壳研究

近十年来，我国学者除了重点对 CCZ 的多金属结核和 PCZ 的富钴结壳进行研究外，还对中国南海多金属结核和结壳进行了元素富集机制、成矿物质来源和形成机制等方面的研究（张振国等，2013；Guan et al.，2017，2018，2019；Zhong et al.，2017，2019；刘兴健等，2018；殷征欣等，2019）。

南海结核和结壳主要分为三类：①南海东北坡发育以针铁矿为主的富铁结核，其 Mn/Fe 值低，微量和稀土元素含量低；②沿南海西北缘分布的表面光滑的铁锰结核和结壳，主要由钴土矿、钡镁锰矿和碳氟磷灰石（carbonate fluorapatites，CFA）组成，铁锰含量相当，微量金属富集程度中等，Ce 呈正异常；③产于南海中部盆地，由钴土矿、钡镁锰矿和水钠锰矿组成的结核和结壳，Mn/Fe 值较高（Zhong et al.，2017）。

南海多金属结核的主要矿物组成与大洋结核相似，但南海结核中的 Mn、Cu、Co、Ni 和 Zn 含量较低，而 Fe、Ti、P、Nb、Pb、Rb、Sc、Ta、Sr、Th 和 REY 等含量较高（张振国等，2013；殷征欣等，2019）；南海结核具有较快的生长速率及较高的 δCe 正异常，表明其生长在更为氧化的环境，大量陆源碎屑为其提供了丰富的成矿物质（殷征欣等，2019）。然而，南海铁锰结壳与之不同，与大洋结壳相比，南海东部海山铁锰结壳 Mn 含量较高，可能与马尼拉海沟附近的火山喷发物有关，但 Cu、Co 和 Ni 等含量较低，可能与边缘海沉积速率高或有机络合物造成水体中微量元素减少有关（刘兴健等，2018）。根据结构构造和化学成分可将南海结核中两组微层分为富 Mn-Ni-Zn-Cu-Li-Ba-Mg 微层和富 Fe-Co-Ti-Sr-Pb-REY 微层，在结核形成过程中，水成作用和成岩作用可能分别促进了两组金属元素的富集（Guan et al.，2019）。南海结壳–结核的 Fe 同位素特征表明，Fe 元素可能主要为河流输入；Os 同位素组成指示主要的成矿元素可能直接来自周围海水，而周围海水受到陆源物质输入的强烈影响，也可能有少量幔源物质的贡献（Guan et al.，2017）。结核碎屑相的 Pb、Nd 和 Sr 同位素组成特征表明，在 1.06~3.2 Ma，华南、吕宋和台湾三个陆源区向南海北部输入物质；1.06 Ma 之后，陆源组分主要是台湾岛河流和中国黄土风尘两端元的混合。南海底层水循环分别在 3.2 Ma、2.1 Ma 和 1.06 Ma 发生变化，可能是受全球变冷和吕宋弧–弧前隆起的影响（Zhong et al.，2019）。

1.3 多金属结核和富钴结壳关键金属富集机制

"关键金属"大致包括"三稀金属"及稀贵金属（铂族金属）和 W、Sn、Co、Ti、Sb、V、Ni、Cr、Mn、U 等（毛景文等，2019；王登红，2019）。多金属结核和富钴结壳主要的关键金属有 Co、Ni、Ti、Te、Pt 及 REY 等。Co 是深海铁锰矿床中最重要的关键金属元素，在结壳中影响 Co 富集的主要因素是磷酸盐化作用以及与 Fe 相矿物和碎屑矿物稀释有关的作用（任向文等，2011a）；而在水成结核中，Co 的富集主要与 Mn^{3+} 有关，而受沉积后作用影响较小（Li et al.，2020）。

1.3.1 稀土元素

稀土元素（REY）是富钴结壳和多金属结核重要的伴生有用元素（何高文等，2011；许东禹，2013）。在富钴结壳未磷酸盐化壳层中，REY 主要赋存在 $δ$-MnO_2 相中；在磷酸盐化壳层中，REY 除了赋存在 Fe、Mn 氧化物相外，还赋存在 CFA 相中（任向文等，2011a）。海水是 P 和 REY 的主要来源，当它们沉淀为 CFA 时，以高稀土元素含量和重稀土富集为特征（Jiang et al.，2020）。选择性化学提取试验显示富钴结壳新壳层中稀土元素主要赋存于铁氢氧化物相中，老壳层（磷酸盐化壳层）中稀土元素主要赋存于残渣态中，表明磷酸盐化作用对老壳层稀土元素的富集具有显著影响（高晶晶等，2015）。在水成结核中，铁氢氧化物对 REY 具有明显的清扫作用，早期形成的似叠层石构造纹层中 REY 相对亏损，表明沉积成岩过程会影响 REY 的积聚（Li et al.，2020）。

1.3.2 贵金属元素

水成结壳和结核中富集铂族元素（PGE），特别是 Pt（可达 ppm 级），具有十分重要的潜在经济价值。通过对西太平洋海山富钴结壳的铂族元素研究发现，结壳对海水中的 PGE 有选择性的吸收导致结壳的 Pt/Pd 值高且具有独特的结壳模式（任江波等，2016）。通过化学分级提取实验，发现太平洋采薇海山富钴结壳中的 PGE 富集主要是受到铁氢氧化物相和残渣态的影响，结壳中 PGE 的富集机理为铁氢氧化物胶体粒子的吸附作用（高晶晶等，2019）。

1.3.3 分散元素

与地壳和深海黏土相比，多金属结核和富钴结壳富集分散元素 Te、Tl、Se 和 Cd，而 Re、Ga、Ge 和 In 不富集；在结核圈层和结壳剖面中，Cd、Se、Te 和 Tl 的变化主要与 Mn 矿物和 Fe 矿物相对含量的变化有关，磷酸盐化壳层中分散元素的含量会受到磷酸盐化作用的影响；不同成因类型的铁锰矿床中 Cd、Se、Te 和 Tl 的含量存在显著差异，可能是由物质来源、沉积速率和矿物组成的差异所导致。在水成作用和水成–成岩混合作用下，Cd、Se、Te 和 Tl 主要通过吸附在锰氧化物或（和）铁氢氧化物进入铁锰矿床；耦合的氧化还原和共沉淀过程可以解释 Te 的高度富集（Fu and Wen，2020）。

1.4 多金属结核和富钴结壳成矿作用

1.4.1 物质来源

中太平洋海山水化学和金属元素地球化学研究表明，结壳的金属元素主要来源于海山当地的水岩反应和（或）低温热液活动，南极底层流起着输运部分成矿物质的作用（武光海和刘捷红，2012）。太平洋海山富钴结壳中生物标志物、有机碳及其稳定同位素特征表明，结壳中有机质主要来源于海洋表层水体中的低等菌藻浮游生物，以及一定量的陆源高等植物组分（李雪富等，2012；Zhao et al.，2014）。

目前对于富钴结壳稀有气体元素的来源还存在争议。李江山等（2012）认为，中太平洋富钴结壳非磷酸盐化壳层中 ^{3}He/^{4}He 与深海沉积物类似，为宇宙尘的加入引起；而 Bu 等（2014）则认为太平洋海山富钴结壳中，高 ^{3}He/^{4}He 型结壳的稀有气体主要来源于 EM 型富集地幔，低 ^{3}He/^{4}He 型结壳的稀有气体则主要来源于 HIMU 型富集地幔。

1.4.2 微生物成矿作用

微生物能够参与到结核和结壳中金属元素的迁移、转化、富集与成矿作用中。结核中的 *Pseudoalteromonas* 和 *Alteromonas* 可能对结核成矿起到诱导或控制作用（Wu et al.，2013），在深海海水–沉积物界面，微生物生态系统和结核生长之间存在着相互依赖的关系（Jiang et al.，2020）。与氧化还原反应有关的微生物能够驱动结核和结壳与海水或孔隙水之间的 Mn、Fe 及其他成矿元素的循环（Jiang et al.，2017）。

结核中自由生活和形成生物膜的细菌为锰沉积提供了基质，而结壳中颗石藻代表了作为初始锰沉积的生物种子的优势生物体（Wang et al.，2012）。盐场海芽孢杆菌能够促进释放结核中的 Fe、Mn 等元素，同时对释放出的金属离子又有富集作用，并能够诱导新矿物的形成（吕靖等，2020）。

1.4.3 富钴结壳年代学

无论是剖析富钴结壳成矿作用过程，还是利用其重建古海洋和古气候演化历史，都需要立足于准确可靠的年代学基础上。目前结壳的定年方法主要有放射性同位素测年（^{10}Be、U 系）、同位素地层学（Sr、Os）、磁性地层学、生物地层学、Co 含量经验公式和基岩年龄外推法等。近十年，我国学者主要通过以下几种方法对生长周期相对较长的结壳进行了年龄测定，获得了较为可靠的定年结果。

（1）生物地层年代学：我国学者对中、西太平洋结壳内部不同纹层的钙质超微化石进行了研究，成功确定了结壳的生长速率和生长时代。中太平洋富钴结壳的主要生长期都在晚古新世、中始新世至晚始新世、中中新世至上新世、上新世至更新世，主要的生长间断在渐新世和早始新世（武光海等，2011）；西太平洋麦哲伦海山富钴结壳中钙质超微化石可追溯的生长年龄最早可达晚白垩世（陈荣华等，2015；任向文等，2017）。

（2）宇宙成因核素 ^{129}I 年代学：^{129}I 是一种宇宙放射性同位素，其半衰期为 15.7 Ma，测年时间尺度可达 80 Ma（Fabryka-Martin et al.，1985）。新开发的 ^{129}I 年代学方法在结壳定年研究中取得成功，Dong 等（2015）应用加速器质谱（AMS）对中太平洋海山富钴结壳 CXD08-1 中 ^{129}I 进行了测试，发现 ^{129}I/^{127}I 值随着结壳深度的增加呈指数级下降；Ji 等（2015）测得中太平洋两个结壳 MP5D44 和 CXD08-1 的 ^{129}I/^{127}I 值为 $7\times10^{-14} \sim 1.27\times10^{-12}$，两个结壳的底层生长年龄分别为 54.77 Ma 和 69.69 Ma。

（3）磁性地层年代学：生长速率极其缓慢的富钴结壳能够记录地球磁极的倒转，据此可以建立高精度的磁极时间序列，获得可靠的磁性地层年代学结果（Joshima and Usui, 1998）。太平洋（PO-01）、中国南海（SCS-01、SCS-02）和印度洋（IO-01）4个水成结壳的详细磁性地层学和岩石磁学研究显示，它们的生长速率分别为 4.82 mm/Ma、4.95 mm/Ma、4.48 mm/Ma 和 11.28 mm/Ma（Yuan et al., 2017）。结壳中的纳米磁铁矿记录了结壳生长时期的地球磁场信息（Yuan et al., 2020）。

1.4.4 多金属结核和富钴结壳成矿模式

多金属结核和富钴结壳作为深海自生铁锰沉积矿产，是深海沉积动力作用的产物，其成矿作用受控于海水-孔隙水的氧逸度和pH、底流流速、碳酸盐补偿深度、沉积物来源和沉积通量、水深、地形、滑坡等多种因素，这些因素空间分布的不均一性造成了铁锰沉积矿床成因类型的多样性。本文在总结上述因素的基础上，提出了多金属结核和富钴结壳成矿模式（图1）：海山斜坡上发育水成成因的富钴结壳，相对深海铁锰沉积矿床的平均化学成分，以富Co、Ce、Mn和贫Ni、Cu为特征；在深海盆地富氧底层水活跃的区域，发育水成成因的多金属结核，具有与富钴结壳类似的成分特征；在富氧底层水活动较弱的深海盆地，发育水成-氧化成岩混合成因的多金属结核，以富Ni、Cu、Mn为特征；在富氧底层水不活跃但海表生物初级生产力高的区域，发育准厌氧成岩成因的多金属结核，以富Mn、Ni贫Co为特征。

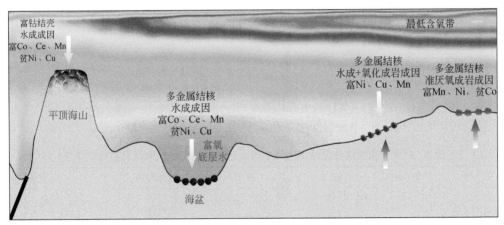

图 1　多金属结核和富钴结壳成矿模式图

1.5　结核、结壳的古海洋古气候记录

Pb、Nd、Hf、Os等放射性同位素和新兴的非传统稳定同位素是结核和结壳古海洋古气候研究的有力工具，它们能够提供风化机制和海洋环流等信息（Frank, 2002; Hein and Koschinsky, 2014; Fu, 2020）。

1.5.1　古大洋环流

富钴结壳的Nd同位素已被证明是古大洋环流变化的良好指标（Chen et al., 2011）。中北太平洋不同深度结壳表层的Nd同位素揭示了中北太平洋深水富放射成因Nd的来源：北太平洋中层水在日本岛附近形成并以平流形式将Nd传输到整个中北太平洋中层，中北太平洋深水Nd同位素在较短时间尺度上的演化并不是稳定不变的（Chen et al., 2011）。北太平洋西部马里亚纳海沟附近两块结壳Nd同位素的时间序列揭示了早中新世以来北太平洋西部深水环流的变化，发现印尼海道的关闭使太平洋和印度洋的表层和中层水交换逐渐停止（Hu et al., 2012）。另外，有研究发现，金属稳定同位素如V、Cr同位素等在重建古海洋演化方面具有潜在的应用价值（Wei et al., 2018; Wu et al., 2019），值得进一步研究。

1.5.2　古气候变化

莱恩海山链M海山结壳不完整地记录了70 Ma以来南、北半球风尘演化过程：自晚白垩世开始南半球总体风尘通量逐渐降低，至晚中新世，其Al沉积通量峰值对应着环太平洋区域的火山/热液活动期幕；

晚中新世以后，其 Al 沉积通量变化记录了北半球构造、环境变化所引起的亚洲风尘活动变化（崔迎春等，2012）。中北太平洋两块结壳的 Fe/Mn 值演化曲线与 ~1Ma 以来的深海底栖有孔虫 $\delta^{18}O$ 变化曲线能够较好地对应，表明至少 1Ma 以来结壳中 Fe 和 Mn 的含量变化与全球气候变化密切相关（胡镕等，2012）。西太平洋结壳的生长对全球冷事件具有明显响应，表明结壳生长与全球气候变化紧密相关（张海生等，2015）。

2 洋中脊热液多金属硫化物

多金属硫化物是海底高温热液活动的产物，富含 Cu、Zn、Pb、Au 和 Ag 等金属元素，据估算全球海底多金属硫化物中金属资源量约 6 亿 t（Hannington et al., 2011）。海底硫化物主要分布于大洋中脊和弧后扩张中心，至今已在世界洋底发现约 700 多处热液区（InterRidge Vents Databse Ver. 3.4. http://vents-data.interridge.org/），其中约 65% 分布在洋中脊，约 22% 分布在弧后盆地，约 12% 分布在火山弧，另外约 1% 分布在板内火山上。

2.1 调查新发现

近十年来，我国对海底硫化物调查研究主要集中在西南印度洋中脊、西北印度洋中脊（卡尔斯伯格脊）和南大西洋中脊 3 个区域，发现了多个热液区，并对其分布规律、成矿环境、成矿作用及热液循环模型开展了研究。

我国自 2005 年开始对西南印度洋中脊进行调查研究，先后发现了龙旂、玉皇、断桥、龙角、天作、天成等热液区（表 1），目前已成为国际上对西南印度洋中脊热液研究程度最高的国家（Tao et al., 2012, 2020；陶春辉等，2014）。西南印度洋中脊属于超慢速扩张洋中脊，岩浆作用弱，洋中脊构造尤其是长期活动的低角度拆离断层对热液成矿作用起着控制作用（Zhao M. H. et al., 2013；Chen et al., 2018；Tao et al., 2020）。

表 1 近十年来我国在全球三大洋中脊上新发现的海底热液区及其成矿环境

洋中脊		构造环境	热液区	基岩	控制因素	代表文献
超慢速扩张洋中脊	西南印度洋中脊	裂谷壁	龙旂、玉皇、天作、龙角	玄武岩、超基性岩及其蚀变岩类	发展期拆离断层	Tao et al., 2012；陶春辉等，2014；Liao, 2018a；Chen, 2018；Tao, 2020
		新生火山区	天成、断桥	玄武岩为主	岩浆作用为主	Yang et al., 2017；Chen et al., 2018
慢速扩张洋中脊	南大西洋中脊	新生 火山区	驹虞、德音、彤管、允臧	玄武岩为主	岩浆作用为主	Wang et al., 2016；叶俊等，2017；Li et al., 2018
			洵美	玄武岩为主	岩浆与构造双重作用	王国芝等，2019
		裂谷壁	赤狐	玄武岩、构造角砾岩	成熟拆离断层	李兵等，2019
		内角高地	采蘩	玄武岩、辉长岩及其蚀变岩类	拆离断层后期岩浆作用	Li et al., 2014, 2016
		非转换不连续带	太极	玄武岩、辉长岩及其蚀变岩类	构造作用为主	Li et al., 2018
	西北印度洋卡尔斯伯格脊	新生火山区	卧蚕 1 号、2 号	玄武岩为主	岩浆作用为主	Wang et al., 2017
			大糦	玄武岩为主	岩浆与构造双重作用	Lou et al., 2020
		裂谷壁	天休	玄武岩、超基性岩及其蚀变岩类	早期拆离断层	Han et al., 2015；周鹏等，2019

我国自 2009 年开始对南大西洋中脊热液活动开展调查研究，发现了洵美、䮝虞、太极、采蘩、德音、彤管、清扬、允臧、赤狐、凯风等热液区（表1），是目前国际上对该区调查研究程度最高的国家。南大西洋中脊属于慢速扩张洋中脊，相比于超慢速洋中脊，岩浆作用较强，构造环境表现出多样性（唐鑫等，2016；Li et al.，2018；王国芝等，2019）。

我国对西北印度洋卡尔斯伯格脊的热液活动调查始于 2012 年。卡尔斯伯格脊为慢速扩张洋中脊，与南大西洋中脊在构造和岩浆活动上表现诸多相似性，目前已在该洋中脊发现了天休、卧蚕 1 号、卧蚕 2 号和大糦等热液区（表1）（Wang et al.，2015，2017；蒋紫靖等，2017）。

2.2 成矿物质来源

热液成矿物质来源主要包括围岩、沉积物、海水及岩浆组分等，目前主流观点仍然认为热液成矿主要是高温演化的海水对基底岩石进行淋滤，成矿物质主要来自于围岩。我国学者对在各个大洋中脊采集的大量样品的研究也支持这一观点（王琰等，2012；Li et al.，2016；Zeng et al.，2017；Liao et al.，2018a）。

大部分洋中脊热液活动发育在基性岩之上，不同洋中脊热液循环体系的三维尺度差别显著。相比快速扩张脊，慢速扩张脊上热液循环系统下渗的深度更广，影响到热液循环系统对成矿物质的获取，硫化物 $\delta^{34}S$ 组成一定程度上可反映热液循环规模的相对大小（Zeng et al.，2017）。部分慢速-超慢速扩张洋中脊上发育一定量超基性岩控制的热液区，表现出较高的 Cu、Zn 和 Au 含量（Wang et al.，2014；Zhang et al.，2018）。基性岩控制的硫化物与超基性岩控制的硫化物在 Cu/Zn-(Cu+Zn) 值、Au 含量等化学元素指标存在显著差异，可作为判别成矿物质来源的辅助手段（Wang et al.，2014）。

岩浆脱气作用是否可以作为洋中脊海底热液成矿物的来源目前尚有争议，但证据在不断增加。如对南大西洋中脊德音和洵美两处热液区赋矿围岩中熔体包裹体的研究发现，在气泡部分的气泡壁上存在磁铁矿、黄铜矿、黄铁矿等金属矿物，认为岩浆脱气作用可能提供了潜在的成矿物质来源（唐鑫等，2016；王国芝等，2019）。除了岩浆脱气过程，硫化物岩浆熔离后分异也可能是潜在的成矿物质来源（Yang et al.，2014）。

2.3 成矿流体与热液羽状流

2.3.1 成矿流体

海水被认为是成矿流体最主要的组分来源，但海水是否是唯一的流体来源今尚存争论。近年来，随着调查技术的进步，我国学者对海底热液系统成矿流体的研究，逐渐由单一的流体包裹体研究转向结合海底原位喷口流体测试与分析（王晓媛等，2013；雷吉江等，2015；Ji et al.，2017；Tao et al.，2020；Zhang et al.，2020）。

西南印度洋龙旂热液区是我国目前少有的直接采取到喷口流体的热液区。Ji 等（2017）根据喷口流体的 H_2、CH_4 等资料，推测龙旂热液区基底岩石以镁铁质为主，扩张速率不是控制流体化学组成的关键因素，而基底岩石、水-岩相互作用和相分离才是主要因素。Tao 等（2020）认为龙旂热液区热液循环系统具有较长的水-岩反应区。

2.3.2 热液羽状流

海底热液喷发过程中，大约 90% 以上的金属物质会以羽状流形式扩散至海水，一部分热液羽状流中具有高于海水的 He 同位素组成，指示热液活动的存在（卢映钰等，2014）。热液羽状流对海水化学组分的改变更多地体现在其物质组成上，如在西北印度洋卡尔斯伯格脊大糦热液区附近采集的海水与颗粒物中发现了显著的 Cl、Br、Mg 负异常及 Fe、Mn 正异常（蒋紫靖等，2017）。

2.3.3 热液沉积记录

受广泛热液羽状流颗粒物沉降的影响，热液区周围沉积物中往往呈现 Zn、Cu、Fe、Mn 元素分带现象，同时还有大量热液标型矿物如黄铁矿、黄铜矿、闪锌矿、针铁矿等生成，这对识别热液活动及成矿环境具有指示意义（杨宝菊等，2019；Yang et al.，2020）。Liao 等（2018b）通过对西南印度洋典型热液区周围沉积物分析，提出热液区附近沉积物的 Cu-Cu/Fe 值可作为判别热液区成因的标志。

热液沉积物成分取决于其形成机制。杨宝菊等（2019）对南大西洋中脊多个热液沉积物样品研究后提出，热液沉积物有 3 种形成机制：①热液柱含 Fe 羟基氧化物颗粒的沉降；②非活动热液区烟囱体的风化；③低温弥散流的沉淀。

2.4 硫化物成矿年代学

近年来，我国学者建立了热液硫化物铀系同位素测年方法，并开展了硫化物成矿年代学研究（王叶剑，2012；叶俊，2017；Yang et al.，2017；Wang et al.，2019）。成矿年代学研究揭示了海底热液活动的幕式成矿特征。王叶剑（2012）应用 $^{230}Th/^{234}U$ 和 $^{210}Pb/Pb$ 法对中印度洋 Kairei 热液区的成矿年代学研究发现，该热液区至少存在 4 次成矿事件；Yang 等（2017）应用 $^{230}Th/^{238}U$ 测年方法对西南印度洋断桥热液区的成矿年龄分析发现，该区发生 4 次主要的热液事件，最早发生于（84.3±0.5）ka 前，终止于（0.737±0.023）ka 前；叶俊等（2017）应用 $^{230}Th/^{234}U$ 方法对南大西洋德音热液区硫化物测年发现，该区至少可划分出五个成矿周期，最老样品形成于 45 ka 前，并探讨了热液喷发中心的动态迁移和成矿演变历史。

2.5 新技术在热液活动研究中的应用

近年来，我国学者开始将高分辨透射电子显微镜（HRTEM）、原位显微 XRD 技术以及非传统稳定同位素（如 Zn、Fe、Cu、Hg 等）分析新技术应用到海底硫化物研究中。

HRTEM 及原位显微 XRD 技术可帮助精细刻画海底热液硫化物微观纳米结构，如 Wu 等（2016，2018）借助此类方法对 Edmond 热液区硫化物开展研究，揭示了胶状硫化物具有纤锌矿纳米片层的光学各向异性，同时发现硫化物中"隐形"金和银的出现，其分布与高密度晶格缺陷密切相关。

非传统稳定同位素可进一步揭示热液成矿作用过程。如 Zn 同位素组成的差异可指示热液成矿期次，在西南印度洋中脊玉皇热液区 SWS 区和 NES 区硫化物的 Zn 同位素组成显著不同，结合硫同位素特征和硫化物的 Zn/Cd 值，认为二者可能是不同成矿期次的产物（Liao S. L. et al.，2019）。而 Hg 同位素特征则可指示硫化物中 Hg 的来源。Zhu 等（2020）报道了西南印度洋中脊的断桥和玉皇两个热液区的硫化物 Hg 同位素组成，玉皇热液区较断桥热液区具有较高的 $\delta^{22}Hg$ 和 $\Delta^{199}Hg$，表明两个热液区具有不同的 Hg 来源：玉皇热液区 Hg 来自于岩浆和海水混合，断桥热液区仅来自于岩浆。

2.6 热液循环模型和热液成矿系统

2.6.1 超慢速扩张洋中脊热液循环模型

洋中脊热液循环系统是由热源驱动、受岩浆-构造控制、涉及水圈-岩石圈相互作用的循环体系。在对西南印度洋龙旂热液区进行地球物理、喷口流体化学、数值模拟等综合研究基础上，提出了一种超慢速扩张洋中脊热液循环模型（Yu et al.，2018；Tao et al.，2020），它具有两个特征：①超慢速扩张洋脊热液循环起源深度比其他洋中脊热液区更深，达到了 Moho 面以下；②热源可以不是在洋壳中的岩浆房，可以存在于岩石圈下部的熔体聚集带。

2.6.2 拆离断层型热液成矿系统

拆离断层是广泛发育在慢速和超慢速洋中脊上的一类特殊构造，其发育时间长，规模大，易于成为长期稳定的热液导矿构造，而沿着拆离断层发生热液-超基性岩相互作用，可为热液成矿提供丰富的物质来源。近年来我国在三大洋中脊发现了大量的拆离断层型热液系统（Li et al.，2014，2015；Chen et al.，

2020)。

西南印度洋龙旂热液区为拆离断层型热液系统，发育在发展期拆离断层的上盘（Zhao M. H. et al.，2013；Tao et al.，2020）；西北印度洋卡尔斯伯格脊天休热液区是以橄榄岩为围岩的热液系统，而天休邻区的拆离断层则处于初始发育期，此时无显著的洋底核杂岩（oceanic core complex，OCC）构造地貌形态产出（Han et al.，2015；周鹏等，2019）。

南大西洋中脊已发现的拆离断层型热液系统主要与成熟拆离断层相关，如赤狐、采蘩和凯风（表1）。其中赤狐热液区是近期新发现的高温热液区，其所处位置为拆离断层上盘构造；按照拆离断层演化阶段来看，赤狐热液区属于成熟拆离断层上盘热液成矿系统，这与已发现的西南印度洋和西北印度洋拆离断层型热液系统明显不同（李兵等，2019）。

3 深海富稀土沉积

深海富稀土沉积，也称深海稀土或富稀土泥，是指产于深海盆地中的富含稀土的沉积物，其稀土总量ΣREY一般大于700 μg/g，最高含量接近8 000 μg/g，以富含重稀土为特征[①]。深海稀土由日本科学家于2011年首次发现于太平洋（Kato et al.，2011），一经报道即引起世界各国广泛关注。深海稀土是继多金属结核、富钴结壳和多金属硫化物之后发现的第四种深海金属矿产，其资源潜力巨大。仅太平洋沉积物中，深海稀土的资源量就是已知陆地稀土资源量的1000多倍[①]（Kato et al.，2011），其成因不同于已经发现的陆地稀土矿床，属于一种新型稀土矿产。

3.1 深海稀土的调查与发现

比日本稍晚，我国于2011年开始了深海稀土研究，是国际上第二个开展深海稀土调查并取得重大发现的国家。在初步总结世界大洋地质特征和沉积物稀土元素特征的基础上，于2012年提出了多个深海稀土潜在发育区，初步划分出了四个深海稀土成矿带：西太平洋富稀土成矿带、中-东太平洋富稀土成矿带、东南太平洋富稀土成矿带和中印度洋海盆-沃顿海盆富稀土成矿带[①]（图2）。2015年我国在中印度洋海盆发现了大面积富稀土沉积区，这是国际上首次在印度洋发现大面积富稀土沉积，其后又在东南太平洋和西太平洋发现了大面积富稀土沉积。

3.2 深海稀土特征及其分布规律

深海富稀土沉积物大都表现出Ce负异常和Eu正异常，以轻稀土亏损、重稀土富集、稀土配分模式与海水相似为特征[①]（Kato et al.，2011；黄牧等，2014；张霄宇等，2019a），其ΣREY大于700 μg/g [①②]，Y元素富集程度较高，含量最高可占ΣREY的40%（黄牧，2013）。从沉积环境来看，深海富稀土沉积一般发育在低沉积速率和富氧的深海盆地中，水深超过碳酸盐补偿深度（carbonate compensation depth，CCD），表层生产力和陆源碎屑输入通量低、底层流发育（Kato et al.，2011；Sa et al.，2018）。

从沉积物类型来看，深海稀土元素主要发育于深海黏土沉积物中，并发育有少量沸石类矿物、铁锰氧化物、磷酸盐、碎屑矿物等，而钙质生物及硅质生物壳体中含量较低（刘季花，1992，2004；刘季花等，1998）。ΣREY在不同类型沉积物中呈规律性变化：从沸石黏土→远洋黏土→硅质黏土→硅质软泥，沉积物中LREE、HREY和ΣREY总体呈降低趋势，LREE与HREY分异程度增加，LREE相对更富集，Ce负异常程度降低，Eu正异常程度增加（黄牧等，2014；张霄宇等，2019b）。

从富稀土沉积分布区域来看，目前已发现西太平洋、东南太平洋、中-东太平洋、中印度洋和沃顿海

[①] 石学法，李传顺，黄牧，等．2015．国际海域资源调查与开发"十二五"课题"世界大洋海底稀土资源潜力评估"报告
[②] 石学法，杨刚，黄牧，等．2015．中国大洋34航次第五航段调查现场报告

盆尤为发育（图2），而边缘海或浅海沉积物中稀土含量较低，不会形成富稀土沉积（黄牧等，2013）。

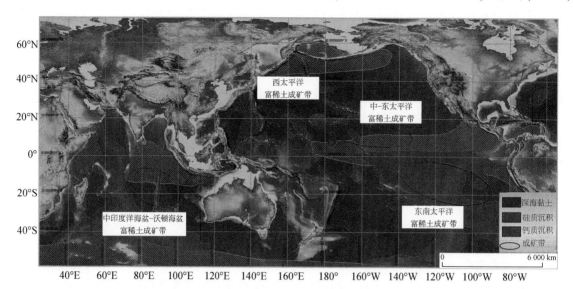

图2 太平洋和印度洋深海稀土成矿带

据石学法，李传顺，黄牧，等. 2015. 国际海域资源调查与开发"十二五"课题"世界大洋海底稀土资源潜力评估报告"修改

3.3 深海稀土来源和赋存状态

3.3.1 稀土元素来源

富稀土沉积物中稀土元素的来源主要涉及两个问题：一是沉积物中的稀土元素来源（直接来源），二是海水和孔隙水中稀土元素的来源（间接来源）。

（1）深海沉积物中稀土元素的来源 由于深海沉积物稀土元素配分型式与海水非常类似，因而长期以来认为其主要来源于海水（刘季花，1992；黄牧，2013；张霄宇等，2019a），但可能受到海水、陆源碎屑或火山源物质释放至孔隙水中稀土元素的影响（Zhao L. S. et al., 2013；于淼等，2017；Liao et al., 2019a）。近年来的研究认为沉积物-海水界面中孔隙水中的REY可能是富稀土沉积物中REY的重要来源（Deng et al., 2017）。Zhao L. S. 等（2013）认为，在沉积时间较短或者钙质生物沉积环境中，生物磷灰石的稀土元素主要来自海水；而对于沉积时间较长，以及具有陆源碎屑沉积或者火山源沉积区域，其中的稀土元素则更多地具有后期孔隙水叠加信息，反映了沉积物的物源信息。可见孔隙水也是稀土元素的重要来源。

（2）关于海水和孔隙水中稀土元素的来源。理论上进入海洋的物质都有可能为海水和孔隙水提供稀土元素。火山蚀变物质、热液喷发物质乃至进入深海的陆源碎屑物质都是深海稀土的"源"。就热液活动对富稀土沉积物的影响程度来看，在邻近洋中脊热液区的沉积物中，稀土元素的富集主要与热液活动中Fe-Mn水合（氢）氧化物的吸附作用有关（Kato et al., 2011；周天成和石学法，2019），在东南太平洋富稀土沉积区热液来源物质起到了重要的作用（周天成和石学法，2019）。

3.3.2 稀土元素赋存状态

深海沉积物稀土元素的赋存状态有两种：赋存在矿物内部发生晶格替代和被矿物表面吸附。Kato 等（2011）认为，深海稀土中稀土元素的主要载体为铁锰氧化物或氢氧化物和钙十字沸石。铁锰氧化物或氢氧化物则是通过经典的元素"清扫"机制富集稀土元素，钙十字沸石则是通过特有的晶体结构吸附稀土元素等，表面吸附是深海稀土一种重要的赋存状态。

近年来，越来越多的研究者认为，生物成因磷灰石可能是富稀土沉积中稀土元素的主要赋存矿物

（方明山等，2016；王汾连等，2016；任江波等，2017；于森等，2017；Zhang et al.，2017；Sa et al.，2018；Liao et al.，2019a）。对于稀土元素在生物磷灰石中的赋存状态也有两种看法：一种认为主要是以吸附态赋存（刘志强等，2015；周天成和石学法，2019）；另一种则认为是在其早期成岩阶段就已经进入到了晶格中，在向磷灰石等矿物中富集之前，稀土元素可能与铁锰氧化物等其他物质结合，在成岩过程中再次分配进入磷灰石等矿物中（Liao et al.，2019a；刘明等，2019）。

3.4 深海稀土大规模成矿作用的控制因素

虽然深海稀土的发现至今还不到十年，调查研究程度还较低，但研究者一般都认为深海稀土在深海中广泛发育，发生了大规模成矿作用，资源潜力巨大[①]（Kato et al.，2011）。现在要阐明深海稀土大规模成矿作用机制为时尚早，在此仅根据一些初步研究结果总结一下深海稀土成矿的主要控制因素。

控制深海稀土富集的主要因素有：构造环境、物质来源、水深、沉积速率、氧化还原环境、沉积物类型等。有利于形成大型深海稀土矿集区的条件是：构造稳定的深海盆地，水深在碳酸盐补偿深度（CCD）之下；碎屑物质输入量少，沉积速率低；南极底流（Antarctic Bottom Water，AABW）发育，水体呈氧化环境；沉积物类型主要为深海黏土。即大水深、低沉积速率和氧化环境是深海稀土大规模成矿的三大控制要素。沉积物类型实际上受水深、物质来源和沉积速率控制；低沉积速率为沉积物中磷灰石的富集和与海水的长时间接触提供了有利条件（于森等，2017），沉积速率越低，越有利于沉积物中稀土元素的富集；富氧底流则为富稀土沉积的发育提供了氧化环境，有利于重稀土微粒的封闭，以及磷酸盐、铁锰质氧化物等物质对稀土元素的吸附（邓义楠等，2018），使稀土成矿作用在海底大范围内发生。深海稀土中生物磷灰石中ΣREY可达到n万$\mu g/g$，磷酸盐组分可贡献全岩ΣREY的70%（王汾连等，2016；任江波等，2017；Liao et al.，2019a，2019b），是富稀土沉积中稀土元素可能的主要赋存矿物。

4 展　　望

近十年来，我国在深海矿产资源领域的研究取得了很大进展，目前在国际海底区域申请矿区数量位居世界第一。但总体看来，我国对深海矿产成矿规律的认知程度仍然较低，调查程度远远不够，成矿理论研究明显不足，勘查技术设备更是主要依赖进口，与西方发达国家相比还存在较大差距。我国今后特别需要重视深海勘查设备技术研发，在继续加大深海矿产调查研究力度的同时，加强深海成矿规律和成矿作用研究。

（1）加强多圈层相互作用对深海金属元素成矿的控制研究。金属元素在地球多圈层相互作用过程中的循环（壳幔相互作用与元素循环、海底-海水相互作用与元素循环、海水内部相互作用与元素循环、海底-生物圈相互作用与元素循环）从宏观和本质上控制着深海成矿作用和资源分布，通过对其研究可揭示深海金属元素超常富集的成矿背景和条件，建立深海金属富集成矿理论，实现海底成矿理论创新和指导找矿突破。

（2）聚焦深海关键金属成矿作用和分布规律研究。深海矿产最主要的特征和优势是发育战略关键金属，今后的工作中要瞄准深海关键矿产资源开展研究，揭示关键金属成矿规律，阐明深海关键金属元素赋存状态，确定深海关键金属矿床新类型，为海底关键金属矿产资源开发利用奠定科学基础。

（3）开展深海成矿作用模拟实验研究。深海成矿是海底中高压条件下，水圈、岩石圈和生物圈共同作用的结果，目前发现的深海矿产主要包括热液型矿床（多金属硫化物）和沉积型矿床（结核、结壳和稀土等）两大类。在对地质背景和地质样品开展研究的同时，需要同时进行深海成矿作用模拟实验研究，确定深海中高压条件下成矿的地球化学热力学和动力学条件，以揭示复杂的深海成矿作用过程。

[①] 石学法，李传顺，黄牧，等 . 2015. 国际海域资源调查与开发"十二五"课题"世界大洋海底稀土资源潜力评估"报告

(4) 开展海陆成矿作用对比研究。陆地成矿作用研究已经发展了比较成熟的方法学体系和理论体系，可以为现在海底成矿作用研究提供指导和借鉴。与陆地矿产研究相比，深海矿产研究属于新领域，属于当采学科。海陆对比和海陆结合必将促进深海成矿作用研究的大发展。

致谢：在本文写作过程中，叶俊、张海桃、杨宝菊等提供了有关材料；论文评审专家提出了宝贵的意见，使本文得以完善提高。在此谨致谢忱。

参 考 文 献

陈荣华,赵庆英,张海生,卢冰,Pulyaeva I A. 2015. 西太平洋富钴结壳中钙质超微化石和分子化石研究. 海洋学报,37(7):132-141
崔迎春,石学法,刘季花,马立杰. 2012. 70Ma以来风尘活动在太平洋铁锰结壳中的记录. 吉林大学学报(地球科学版),42(2):393-399
邓义楠,任江波,郭庆军,曹珺,王海峰,刘晨晖. 2018. 太平洋西部富稀土深海沉积物的地球化学特征及其指示意义. 岩石学报,34(003):733-747
杜菊民,赵学章. 2010. 刚果(金)铜-钴矿床地质特征及分布规律. 地质与勘探,46(1):165-174
方明山,石学法,肖仪武,李传顺. 2016. 太平洋深海沉积物中稀土矿物的分布特征研究. 矿冶,25(5):81-84
高晶晶,刘季花,李先国,张辉,何连花. 2015. 富钴结壳中稀土元素化学相态分析方法及其应用. 分析化学,43(12):1895-1900
高晶晶,刘季花,张辉,闫仕娟,何连花,王小静,汪虹敏. 2019. 太平洋海山富钴结壳中铂族元素赋存状态与富集机理. 海洋学报,41(8):115-124
何高文,孙晓明,杨胜雄,朱克超,宋成兵. 2011. 太平洋多金属结核和富钴结壳稀土元素地球化学对比及其地质意义. 中国地质,38(2):462-472
胡镕,陈天宇,凌洪飞. 2012. 晚第四纪中北太平洋铁锰结壳Fe/Mn变化:对古气候变化的响应. 高校地质学报,18(4):751-758
黄牧. 2013. 太平洋深海沉积物稀土元素地球化学特征及资源潜力初步研究. 青岛:国家海洋局第一海洋研究所硕士学位论文
黄牧,刘季花,石学法,朱爱美,吕华华,胡利民. 2014. 东太平洋CC区沉积物稀土元素特征及物源. 海洋科学进展,32(2):175-187
蒋紫靖,韩喜球,王叶剑,邱中炎. 2017. 印度洋卡尔斯伯格脊6°48′N附近热液羽状流水化学参数异常和颗粒物成分特征. 海洋学研究,35(4):34-43
雷吉江,初凤友,于晓果,李小虎,陶春辉,葛倩. 2015. 西南印度洋中脊热液区烃类有机质组成及其成因意义. 地学前缘,01:281-290
李兵,李传顺,石学法. 2019. 南大西洋中脊拆离断层型热液系统的调查与发现. 中国矿物岩石地球化学学会第17届学术年会论文摘要集(杭州):591
李江山,方念乔,石学法,任向文,刘季花,崔迎春. 2012. 中太平洋富钴结壳不同壳层He,Ar同位素组成. 地球科学,37(S1):93-100
李雪富,武光海,刘捷红,张宏. 2012. 中太平洋富钴结壳中生物标志物、有机碳同位素地球化学及其古海洋环境意义. 海洋学研究,30(4):29-36
梁东红,何高文,朱克超. 2014. 中国多金属结核西示范区的结核小尺度分布特征. 海洋学报,36(4):33-39
刘季花. 1992. 太平洋东部深海沉积物稀土元素地球化学. 海洋地质与第四纪地质,12(2):33-42
刘季花. 2004. 东太平洋沉积物稀土元素和Nd同位素地球化学特征及其环境指示意义. 青岛:中国科学院研究生院(海洋研究所)博士学位论文
刘季花,梁宏锋,夏宁,宋苏顷. 1998. 东太平洋深海沉积物小于2μm组分的稀土元素地球化学特征. 地球化学,27(1):49-58
刘明,孙晓霞,石学法,张文强,范德江,杨作升. 2019. 印度洋钙质软泥和硅质软泥稀土元素组成和富集机制. 海洋学报,41(1):58-71
刘兴健,唐得昊,阎贫,葛晨东. 2018. 南海东部管事海山铁锰结壳的矿物组成和地球化学特征. 海洋地质与第四纪地质,39(3):94-103
刘永刚,杜德文,曲镜如,闫仕娟,王春娟,石学法. 2013. 基于Fuzzy ARTMAP的CC区多金属结核资源定量评价. 海洋地质与第四纪地质,33(2):169-179
刘志强,吴宇坤,朱克超,李伟,朱薇,邱显扬. 2015. 太平洋中部深海黏土中稀土的赋存状态及浸出研究. 中国稀土学报,33(4):506-512
卢映钰,韩喜球,王叶剑,邱中炎. 2014. 西南印度洋49°~56°E洋脊段的热液羽状流:来自深水中的氦同位素异常证据. 海洋学报,36(6):42-49
吕靖,蓝鑫,姜明玉,曹文瑞,萨仁高娃,于心科,常凤鸣. 2020. 盐场海芽孢杆菌与大洋铁锰结核相互作用. 海洋科学,44(1):36-45
马维林,金翔龙,初凤友,李守军,杨克红. 2013. 中太平洋海山区尖顶和平顶海山结壳成矿与分布的比对研究. 海洋学报,35(2):90-112
马维林,杨克红,包更生,张恺,董如洲,初凤友. 2014. 中太平洋海山富钴结壳成矿的空间分布规律研究. 海洋学报,36(7),77-89
毛景文,杨宗喜,谢桂青,袁顺达,周振华. 2019. 关键矿产-国际动向与思考. 矿床地质,38(4):689-698
任江波,何高文,姚会强,张伙带,杨胜雄,邓希光,朱克超. 2016. 西太平洋海山富钴结壳的稀土和铂族元素特征及其意义. 地球科学:中国地质大学学报,41(10):1745-1757
任江波,何高文,朱克超,邓希光,刘纪勇,傅飘儿,姚会强,杨胜雄,孙卫东. 2017. 富稀土磷酸盐及其在深海成矿作用中的贡献. 地质学报,91(6):1312-1325
任向文,刘季花,崔迎春,石学法,尹京武. 2011a. 磷酸盐化对莱恩海山链MP2海山结壳Co富集的影响. 海洋科学进展,29(3):323-329
任向文,石学法,朱爱美,刘季花,方习生. 2011b. 麦哲伦海山群MK海山富钴结壳稀土元素的赋存相态. 吉林大学学报(地球科学版),41(3):707-714

任向文,Pulyaeva I,吕华华,石学法,曹德凯. 2017. 麦哲伦海山群 MK 海山富钴结壳钙质超微化石生物地层学研究. 地学前缘,24(1):276-296
唐鑫,杨耀民,王国芝,张海桃,范蕾,许鹏,朱志伟. 2016. 南大西洋15°S 热液区玄武岩中熔融包裹体组成及意义. 成都理工大学学报(自然科学版),43(3):362-371
陶春辉,李怀明,金肖兵,周建平,吴涛,何拥华,邓显明,顾春华,张国堙. 2014. 西南印度洋脊的海底热液活动和硫化物勘探. 科学通报,59(19):1812-1822
王登红. 2019. 关键矿产的研究意义、矿种厘定、资源属性、找矿进展、存在问题及主攻方向. 地质学报,93(6):1189-1209
王汾连,何高文,孙晓明,杨阳,赵太平. 2016. 太平洋富稀土深海沉积物中稀土元素赋存载体研究. 岩石学报,32(7):2057-2068
王国芝,石学法,雷庆,李传顺,赵甫峰,范蕾. 2019. 大西洋中脊洵美热液区赋矿围岩斑晶中熔融包裹体特征. 见:第九届全国成矿理论与找矿方法学术讨论会论文摘要集,382
王晓媛,曾志刚,陈帅,殷学博,陈镇东. 2013. 我国台湾东北部龟山岛附近海域热液流体中的稀土元素组成及其对浅海热液活动的指示. 科学通报,58(19):1874-1883
王琰,孙晓明,戴瑛知,吴仲玮. 2012. 西南印度洋中脊及东太平洋海隆海底热液硫化物硫同位素特征及其对比研究. 矿床地质,31(S1):439-440
王叶剑. 2012. 中印度洋脊 Kairei 和 Edmond 热液活动区成矿作用对比研究. 杭州:浙江大学博士学位论文
武光海,刘捷红. 2012. 海山当地物源和南极底层水对富钴结壳成矿作用的影响:来自海山周围水柱化学分析的证据. 海洋学报,34(3):92-98
武光海,Pulyaeva I A,刘捷红,李雪富. 2011. 中太平洋海山铁锰结壳生物地层学研究. 海洋学报,33(4):129-139
许东禹. 2013. 大洋矿产地质学. 北京:海洋出版社
杨宝菊,刘季花,李传顺,叶俊,李兵,石学法. 2019. 南大西洋中脊含金属沉积物稀土元素特征. 见:第九届全国成矿理论与找矿方法学术讨论会论文摘要集,387
杨胜雄,龙晓军,祁奇,冷传旭,崔尚公,郝娅楠,赵广涛. 2016. 西太平洋富钴结壳矿物学和地球化学特征:以麦哲伦海山和马尔库斯-威克海山富钴结壳为例. 中国海洋大学学报,46(2):105-116
叶俊,石学法,李兵,闫仕娟,朱志伟. 2017. 南大西洋中脊15°S 热液区成矿年代特征研究. 中国矿物岩石地球化学学会第九次全国会员代表大会暨第16届学术年会文集,1031
殷征欣,王海峰,韩金生,吕修亚,沈泽中,田静,贺惠忠,谢安远,关瑶,董超. 2019. 南海边缘海多金属结核与大洋多金属结核对比. 吉林大学学报(地球科学版),49(1):261-277
于淼,石学法,李传顺,黄牧,杨宝菊,周天成,张颖. 2017. 中印度洋海盆富稀土沉积物物质来源. 中国矿物岩石地球化学学会第17届学术年会,584
张富元,章伟艳,朱克超,张霄宇,倪建宇,赵宏樵,郑连福. 2011. 太平洋海山钴结壳资源量估算. 地球科学,36(1):1-11
张富元,章伟艳,任向文,张霄宇,朱克超. 2015. 全球三大洋海山钴结壳资源量估算. 海洋学报,37(1):88-105
张海生,胡佶,赵军,韩正兵,于培松,武光海,雷吉江,卢冰,Pulyaeva I A. 2015. 西太平洋海山富钴结壳钙质超微化石变化与 E/O 界限的地质记录. 中国科学:地球科学,45(4):508-519
张霄宇,黄牧,石学法,黄大松. 2019a. 中印度洋洋盆 GC11 岩心富稀土深海沉积的元素地球化学特征. 海洋学报,41(12):51-61
张霄宇,石学法,黄牧,滕国超,麻书畅,黄大松. 2019b. 深海富稀土沉积研究的若干问题. 中国稀土学报,5:517-529
张振国,杜远生,吴长航,方念乔,杨胜雄,刘坚,宋成兵. 2013. 南海西北陆缘大型多金属结核的生长过程及其对晚新生代古海洋环境变化的响应. 中国科学:地球科学,43(7):1168-1178
周鹏,韩喜球,王叶剑,李洪林,刘吉强,董传奇,蔡翌旸. 2019. 拆离断层对海底热液硫化物形成的制约:来自岩石学和近底观测的证据. 中国矿物岩石地球化学学会第17届学术年会论文摘要集,593-595
周天成,石学法. 2019. 东南太平洋尤潘基海盆富稀土沉积研究进展. 见:第九届全国成矿理论与找矿方法学术讨论会论文摘要集,390
Bu W R,Shi X F,Li L,Zhang M J,Glasby G P,Liu J H. 2014. Kr and Xe isotopic compositions of Fe-Mn crusts from the western and central Pacific Ocean and implications for their genesis. Acta Oceanologica Sinica,33(8):26-33
Chen J,Tao C H,Liang J,Liao S L,Dong C W,Li H M,Li W,Wang Y,Yue X H,He Y H. 2018. Newly discovered hydrothermal fields along the ultraslow-spreading Southwest Indian Ridge around 63°E. Acta Oceanologica Sinica,37(11):61-67
Chen T Y,Ling H F,Hu R. 2011. Neodymium isotopes distribution and transport in the central North Pacific deep water. Chinese Science Bulletin,56(21):2243-2250
Chen Y,Han X Q,Wang Y J,Lu J G. 2020. Precipitation of calcite veins in serpentinized harzburgite at Tianxiu hydrothermal field on Carlsberg Ridge (3.67°N),Northwest Indian Ocean:implications for fluid circulation. Journal of Earth Science,31(1):91-101
Deng Y N,Ren J B,Guo Q J,Cao J,Wang H F,Liu C H. 2017. Rare earth element geochemistry characteristics of seawater and porewater from deep sea in western Pacific. Scientific Reports,7(1):16539
Dong K J,Jiang S,He M,Lin M,Ouyang Y G,Wu S Y,Xie L B,Liu G S,Ji L H,Li Q,Wang S L. 2015. The measurement of ^{129}I in ferromanganese crusts and aerosol samples with AMS at CIAE. Nuclear Instruments and Methods in Physics Research Section B:Beam Interactions with Materials and Atoms,353:16-20
Fabryka-Martin J,Bentley H,Elmore D,Airey P L. 1985. Natural iodine-129 as an environmental tracer. Geochimica et Cosmochimica Acta,49(2):337-347
Frank M. 2002. Radiogenic isotopes:tracers of past ocean circulation and erosional input. Reviews of Geophysics,40(1):1-38
Fu Y Z. 2020. Non-traditional stable isotope geochemistry of marine ferromanganese crusts and nodules. Journal of Oceanography,76(2):71-89

Fu Y Z, Wen H J. 2020. Variabilities and enrichment mechanisms of the dispersed elements in marine Fe-Mn deposits from the Pacific Ocean. Ore Geology Reviews, 121:103470

Guan Y, Sun X M, Ren Y Z, Jiang X D. 2017. Mineralogy, geochemistry and genesis of the polymetallic crusts and nodules from the South China Sea. Ore Geology Reviews, 89:206-227

Guan Y, Ren Y Z, Sun X M, Xiao Z L, Guo Z X. 2018. Helium and argon isotopes in the Fe-Mn polymetallic crusts and nodules from the South China Sea: constraints on their genetic sources and origins. Minerals, 8(10):471

Guan Y, Ren Y Z, Sun X M, Xiao Z L, Wu Z W, Liao J L, Guo Z X, Wang Y, Huang Y. 2019. Fine scale study of major and trace elements in the Fe-Mn nodules from the South China Sea and their metallogenic constraints. Marine Geology, 416:105978

Hannington M, Jamieson J, Monecke T, Petersen S, Beaulieu S. 2011. The abundance of seafloor massive sulfide deposits. Geology, 39(12):1155-1158

He G W, Ma W L, Song C B, Yang S X, Zhu B D, Yao H Q, Jiang X X, Cheng Y S. 2011. Distribution characteristics of seamount cobalt-rich ferromanganese crusts and the determination of the size of areas for exploration and exploitation. Acta Oceanologica Sinica, 30(3):63-65

Hein J R, Koschinsky A. 2014. Deep-ocean ferromanganese crusts and nodules. Treatise on Geochemistry, 13:273-291

Hein J R, Mizell K, Koschinsky A, Conrad T A. 2013. Deep-ocean mineral deposits as a source of critical metals for high-and green-technology applications: comparison with land-based resources. Ore Geology Reviews, 51:1-14

Hu R, Chen T Y, Ling H F. 2012. Late Cenozoic history of deep water circulation in the western North Pacific: evidence from Nd isotopes of ferromanganese crusts. Chinese Science Bulletin, 57(31):4077-4086

Ji F W, Zhou H Y, Yang Q H, Gao H, Wang H, Lilley M D. 2017. Geochemistry of hydrothermal vent fluids and its implications for subsurface processes at the active Longqi hydrothermal field, Southwest Indian Ridge. Deep Sea Research Part I: Oceanographic Research Papers, 122:41-47

Ji L H, Liu G S, Chen Z G, Huang Y P, Xing N, Jiang S, He M. 2015. Measurement of ^{129}I in ferromanganese crust with AMS. Acta Oceanologica Sinica, 34(10):31-35

Jiang X D, Sun X M, Guan Y, Gong J L, Lu Y, Lu R F, Wang C. 2017. Biomineralisation of the ferromanganese crusts in the western Pacific Ocean. Journal of Asian Earth Sciences, 136:58-67

Jiang X D, Sun X M, Chou Y M, Hein J R, He G W, Fu Y, Li D F, Liao J L, Ren J B. 2020. Geochemistry and origins of carbonate fluorapatite in seamount Fe-Mn crusts from the Pacific Ocean. Marine Geology, 423:106135

Joshima M, Usui A. 1998. Magnetostratigraphy of hydrogenetic manganese crusts from northwestern Pacific seamounts. Marine Geology, 146(1-4):53-62

Kato Y, Fujinaga K, Nakamura K, Takaya Y, Kitamura K, Ohta J, Toda R, Nakashima T, Iwamori H. 2011. Deep-sea mud in the Pacific Ocean as a potential resource for rare-earth elements. Nature Geoscience, 4(8):535-539

Li B, Yang Y M, Shi X F, Ye J, Gao J J, Zhu A M, Shao M J. 2014. Characteristics of a ridge-transform inside corner intersection and associated mafic-hosted seafloor hydrothermal field (14.0°S, mid-Atlantic Ridge). Marine Geophysical Research, 35:55-68

Li B, Shi X F, Li C S, Wang J X, Pei Y L, Ye J. 2016. Lead, sulfur, and oxygen isotope systematics in hydrothermal precipitates from the 14°S hydrothermal field, South mid-Atlantic Ridge. Resource Geology, 66(3):274-285

Li B, Shi X F, Wang J X, Yan Q S, Liu C G. 2018. Tectonic environments and local geologic controls of potential hydrothermal fields along the southern mid-Atlantic Ridge (12°-14°S). Journal of Marine Systems, 181:1-13

Li D F, Fu Y, Sun X M, Wei Z Q. 2020. Critical metal enrichment mechanism of deep-sea hydrogenetic nodules: insights from mineralogy and element mobility. Ore Geology Reviews, 118:103371

Li J B, Jian H C, Chen Y J, Singh S C, Ruan A G, Qiu X L, Zhao M H, Wang X G, Niu X W, Ni J Y, Zhang J Z. 2015. Seismic observation of an extremely magmatic accretion at the ultraslow spreading Southwest Indian Ridge. Geophysical Research Letters, 42(8):2656-2663

Liao J L, Sun X M, Li D F, Sa R N, Lu Y, Lin Z Y, Xu L, Zhan R Z, Pan Y G, Xu H F. 2019a. New insights into nanostructure and geochemistry of bioapatite in REE-rich deep-sea sediments: LA-ICP-MS, TEM, and Z-contrast imaging studies. Chemical Geology, 512:58-68

Liao J L, Sun X M, Wu Z, Sa R N, Guan Y, Lu Y, Li D F, Liu Y, Deng Y N, Pan Y. 2019b. Fe-Mn (oxyhydr)oxides as an indicator of REY enrichment in deep-sea sediments from the central North Pacific. Ore Geology Reviews, 112:103044

Liao S L, Tao C H, Li H M, Barriga F J A S, Liang J, Yang W F, Yu J Y, Zhu C W. 2018a. Bulk geochemistry, sulfur isotope characteristics of the Yuhuang-1 hydrothermal field on the ultraslow-spreading Southwest Indian Ridge. Ore Geology Reviews, 96:13-27

Liao S L, Tao C H, Li H M, Zhang G Y, Liang J, Yang W F, Wang Y. 2018b. Surface sediment geochemistry and hydrothermal activity indicators in the Dragon Horn area on the Southwest Indian Ridge. Marine Geology, 398:22-34

Liao S L, Tao C H, Zhu C W, Li H M, Li X H, Liang J, Yang W F, Wang Y J. 2019. Two episodes of sulfide mineralization at the Yuhuang-1 hydrothermal field on the Southwest Indian Ridge: insight from Zn isotopes. Chemical Geology, 507:54-63

Lou Y Z, Han X Q, He Z G, Wang Y J, Qiu Z Y. 2020. Numerical modeling of hydrodynamic processes of deep-sea hydrothermal plumes: a case study on Daxi hydrothermal field, Carlsberg Ridge. Scientia Sinica Technologica, 50(2):194-208

Tao C H, Lin J, Guo S Q, Chen Y J, Wu G H, Han X Q, German C R, Yoerger D R, Zhou N, Li H M, Su X, Zhu J. 2012. First active hydrothermal vents on an ultraslow-spreading center: Southwest Indian Ridge. Geology, 40(1):47-50

Tao C H, Seyfried Jr W E, Lowell R P, Liu Y L, Liang J, Guo Z K, Ding K, Zhang H T, Liu J, Qiu L, Egorov I, Liao S L, Zhao M H, Zhou J P, Deng X M, Li H M, Wang H C, Cai W, Zhang G Y, Zhou H W, Lin J, Li W. 2020. Deep high-temperature hydrothermal circulation in a detachment faulting system on the ultra-slow spreading ridge. Nature Communications, 11(1):1300

Wang L S, Wang X F, Ye J, Ma Z B, Yang W F, Xiao J L. 2019. Separation of uranium and thorium for ^{230}Th-U dating of submarine hydrothermal sulfides. Journal of Visualized Experiments, (147), e59098, doi: 10. 3791/59098

Wang S J, Li H M, Zhai S K, Yu Z H, Shao Z Z, Cai Z W. 2016. Mineralogical characteristics of polymetallic sulfides from the deyin-1 hydrothermal field near 15°S, southern mid-Atlantic Ridge. Acta Oceanologica Sinica. 36: 22-34

Wang X H, Gan L, Wiens M, Schloßmacher U, Schröder H C, Müller W E G. 2012. Distribution of microfossils within polymetallic nodules: biogenic clusters within manganese layers. Marine Biotechnology, 14: 96-105

Wang Y J, Han X Q, Petersen S, Jin X L, Qiu Z Y, Zhu J H. 2014. Mineralogy and geochemistry of hydrothermal precipitates from Kairei hydrothermal field, central Indian Ridge. Marine Geology, 354: 69-80

Wang Y J, Han X Q, Petersen S, Frische M, Qiu Z Y, Li H M, Li H L, Wu Z C, Cui R Y. 2017. Mineralogy and trace element geochemistry of sulfide minerals from the Wocan hydrothermal field on the slow-spreading Carlsberg Ridge, Indian Ocean. Ore Geology Reviews, 84: 1-19

Wei W, Frei R, Chen T Y, Klaebe R, Liu H, Li D, Wei G Y, Ling H F. 2018. Marine ferromanganese oxide: a potentially important sink of light chromium isotopes? Chemical Geology, 495: 90-103

Wu F, Owens J D, Tang L M, Dong Y H, Huang F. 2019. Vanadium isotopic fractionation during the formation of marine ferromanganese crusts and nodules. Geochimica et Cosmochimica Acta, 265: 371-385

Wu Y H, Liao L, Wang C S, Ma W L, Meng F X, Wu M, Xu X W. 2013. A comparison of microbial communities in deep-sea polymetallic nodules and the surrounding sediments in the Pacific Ocean. Deep Sea Research Part I: Oceanographic Research Papers, 79: 40-49

Wu Z W, Sun X M, Xu H F, Konishi H, Wang Y, Wang C, Dai Y Z, Deng X G, Yu M. 2016. Occurrences and distribution of "invisible" precious metals in sulfide deposits from the Edmond hydrothermal field, central Indian Ridge. Ore Geology Reviews, 79: 105-132

Wu Z W, Sun X M, Xu H F, Konishi H, Wang Y, Lu Y, Cao K J, Wang C, Zhou H Y. 2018. Microstructural characterization and in-situ sulfur isotopic analysis of silver-bearing sphalerite from the Edmond hydrothermal field, central Indian Ridge. Ore Geology Reviews, 92: 318-347

Yang A Y, Zhou M F, Zhao T P, Deng X G, Qi L, Xu J F. 2014. Chalcophile elemental compositions of morbs from the ultraslow-spreading Southwest Indian Ridge and controls of lithospheric structure on S-saturated differentiation. Chemical Geology, 382: 1-13

Yang B, Liu J, Shi X, Zhang H, Fang X. 2020. Mineralogy and sulfur isotope characteristics of metalliferous sediments from the tangyin hydrothermal field in the southern Okinawa Trough. Ore Geology Reviews, 120: 1-14

Yang W F, Tao C H, Li H M, Liang J, Liao S L, Long J P, Ma Z B, Wang L S. 2017. ^{230}Th/^{238}U dating of hydrothermal sulfides from Duanqiao hydrothermal field, Southwest Indian Ridge. Marine Geophysical Research, 38(1-2): 71-83

Yang Y, He G W, Ma J F, Yu Z Z, Yao H Q, Deng X G, Liu F L, Wei Z Q. 2020. Acoustic quantitative analysis of ferromanganese nodules and cobalt-rich crusts distribution areas using EM122 multibeam backscatter data from deep-sea basin to seamount in western Pacific Ocean. Deep Sea Research Part I: Oceanographic Research Papers, 161: 103281

Yu Z T, Li J B, Niu X W, Rawlinson N, Ruan A G, Wang W, Hu H, Wei X D, Zhang J, Liang Y Y. 2018. Lithospheric structure and tectonic processes constrained by microearthquake activity at the central ultraslow-spreading Southwest Indian Ridge (49.2° to 50.8°E). Journal of Geophysical Research: Solid Earth, 123(8): 6247-6262

Yuan W, Zhou H Y, Zhao X X, Yang Z Y, Yang Q H, Zhu B D. 2017. Magnetic stratigraphic dating of marine hydrogenetic ferromanganese crusts. Scientific Reports, 7(1): 16748

Yuan W, Zhou H Y, Yang Z Y, Hein J R, Yang Q H. 2020. Magnetite magnetofossils record biogeochemical remanent magnetization in hydrogenetic ferromanganese crusts. Geology, 48(3): 298-302

Zeng Z G, Ma Y, Chen S, Selby D, Wang X Y, Yin X B. 2017. Sulfur and lead isotopic compositions of massive sulfides from deep-sea hydrothermal systems: implications for ore genesis and fluid circulation. Ore Geology Reviews, 87: 155-171

Zhang B S, Li Z Q, Hou Z Q, Zhang W Y, Xu B. 2018. Mineralogy and chemistry of sulfides from the Longqi and Duanqiao hydrothermal fields in the Southwest Indian Ridge. Acta Geologica Sinica, 92(5): 1798-1822

Zhang X, Li L F, Du Z F, Hao X L, Cao L, Luan Z D, Wang B, Xi S C, Lian C, Yan J, Sun W D. 2020. Discovery of supercritical carbon dioxide in a hydrothermal system. Science Bulletin, 65(11): 958-964

Zhang X Y, Tao C H, Shi X F, Li H M, Huang M, Huang D S. 2017. Geochemical characteristics of REY-rich pelagic sediments from the GC02 in central Indian Ocean Basin. Journal of Rare Earths, 35(10): 1047-1058

Zhao J, Zhang H S, Wu G H, Lu B, Pulyaeva I A, Zhang H F, Pang X H. 2014. Biomineralization of organic matter in cobalt-rich crusts from the Marcus-Wake Seamounts of the western Pacific Ocean. Acta Oceanologica Sinica, 33(12): 67-74

Zhao L S, Chen Z Q, Algeo T J, Chen J B, Chen Y L, Tong J N, Gao S, Zhou L, Hu Z C, Liu Y S. 2013. Rare-earth element patterns in conodont albid crowns: evidence for massive inputs of volcanic ash during the latest Permian biocrisis. Global and Planetary Change, 105: 135-151

Zhao M H, Qiu X L, Li J B, Sauter D, Ruan A G, Chen J, Cannat M, Singh S, Zhang J Z, Wu Z L, Niu X W. 2013. Three-dimensional seismic structure of the dragon flag oceanic core complex at the ultraslow spreading Southwest Indian Ridge (49°39′E). Geochemistry, Geophysics, Geosystems, 14(10): 4544-4563

Zhong Y, Chen Z, González F J, Hein J R, Zheng X F, Li G, Luo Y, Mo A B, Tian Y H, Wang S H. 2017. Composition and genesis of ferromanganese deposits from the northern South China Sea. Journal of Asian Earth Sciences, 138: 110-128

Zhong Y, Liu Q S, Chen Z, González F J, Hein J R, Zhang J, Zhong L F. 2019. Tectonic and paleoceanographic conditions during the formation of ferromanganese nodules from the northern South China Sea based on the high-resolution geochemistry, mineralogy and isotopes. Marine Geology, 410:

Zhu C W, Tao C H, Yin R S, Liao S L, Yang W F, Liu J, Barriga F J A S. 2020. Seawater versus mantle sources of mercury in sulfide-rich seafloor hydrothermal systems, Southwest Indian Ridge. Geochimica et Cosmochimica Acta, 281: 91–101

Research Progress of Deep-sea Minerals

SHI Xue-fa[1,2], FU Ya-zhou[3], LI Bing[1,2], HUANG Mu[1,2], REN Xiang-wen[1,2], LIU Ji-hua[1,2], YU Miao[1,2], LI Chuan-Shun[1,2]

1. Key Laboratory of Marine Geology and Metallogeny, First Institute of Oceanography, MNR, Qingdao Shandong 266061;
2. Laboratory for Marine Geology, Pilot National Laboratory for Marine Science and Technology, Qingdao Shandong 266037;
3. State Key Laboratory of Ore Deposit Geochemistry, Institute of Geochemistry, Chinese Academy of Sciences, Guiyang 550002

Abstract: Deep-sea minerals belong to potential strategic mineral resources which have been less recognized and not been utilized yet. In the last decade, China made abundant important progresses in deep sea mineral exploration and research. Two exploration areas on polymetallic nodules, one exploration area on Co-rich ferromanganese crusts in the Pacific, and one exploration area on polymetallic sulfides in the Southwest Indian Ridge were allocated by China sponsored contractors. The distribution patterns of polymetallic nodules and Co-rich ferromanganese crusts in exploration areas of small-scale were clarified. The ore-forming processes and palaeoceanographic and palaeoclimatic records were illustrated. The enrichment mechanisms of critical metals were investigated. A number of hydrothermal fields were discovered in the Southwest Indian Ridge, the Northwest Indian Ridge, and South Atlantic Ridge. In those hydrothermal fields, ore-forming conditions and mineralization mechanisms were revealed, hydrothermal recycling models for ultra-slow spreading ridge were built up, and a diversity of detachment fault-related hydrothermal systems were studied. Four metallogenetic belts of REY-rich deep-sea sediments in global deep oceans were discriminated. A few large areas covered by REY-rich deep-sea sediments were discovered respectively in the central Indian Basin, the Southeast Pacific, and the West Pacific. The preliminarily researches on enrichment feature, distribution pattern, occurrence state, and ore-forming process of the REY-rich deep-sea sediments had been carried out. The future researches on the ore-forming dynamic of critical metals in deep-sea will be prosperous.

Key words: deep-sea minerals; exploration area allocated to contractors; mineralization; research progress

煤有机地球化学研究进展*

唐跃刚　王绍清　郭　鑫　李瑞青　林雨涵

中国矿业大学（北京），北京 100083

摘　要：本文定义了煤中有机元素并充实了有机地球化学概念，综述了近十年来关于煤中有机元素现状的研究、生物标志化合物、煤分子结构和煤微观超微观研究趋势，以及煤有机地球化学在地质、工业和生态环境中的应用，提出了研究存在问题并展望了未来这一领域发展趋势。近十年来，煤有机地球化学研究主要集中于高有机硫煤硫化合物、树皮煤脂肪烃和煤多环芳烃组成。煤大分子结构研究以键合结构和嵌布结构理论为主。煤显微组分研究主要集中于类脂体、惰质体成因。特殊煤研究尤其是中国特有的树皮体研究、煤显微组分的工艺性及其热演化、燃烧与气化残渣岩石学、煤系气、煤相和层序地层学、煤基石墨烯、低碳与生态环保等领域都是当前研究热点。随着新理论与高新技术的发展，未来煤有机地球化学会出现许多新的研究领域。

关键词：煤　有机元素　地球化学　研究进展　展望

0　引　言

迄今为止，人类已认识的元素有 118 种，其中已确定命名的有 109 种。什么是有机元素？"有机"原指与生物体有关或从生物体而来的化合物，现指除 CO、CO_2、碳酸、碳酸盐、某些碳化物外的含碳原子的化合物（辞海编辑委员会，2009）。有机质来源，一般有 3 种：大气火花成因说（中国科学院地球化学研究所，1998），星云磁铁矿成因说（陈道公等，2009），热液超临界成因说（Zhang et al.，2020）。目前生命起源有五种以上假说（林巍等，2020）。有机质包括生物体和地史上由生物死亡堆积经地质作用而形成的沉积有机质（地质体有机质）。前者称为活碳，后者称为死碳。生物圈主要指前者，而有机圈则是包括了活碳的生物圈和死碳的地质体有机质。有机地球化学是研究地质体中有机质的组成、结构、起源和演化的学科（中国大百科全书总编辑委员会，1993）。一般，元素不分有机元素和无机元素，但并非所有元素都可与碳结合形成化合物，本文对有机元素的定义是：与生命有关，可直接与碳、氢（主要与碳）结合的元素。与碳弱链接（共价键或氢键）的元素，且不直接与碳连接的元素称为有机结合元素，它属无机元素，如锗与碳中氧结合。

地质体有机质（主要为沉积有机质）包括集中有机质（如煤、沥青、油页岩）和分散有机质（dispersal organic matter，DOM；如碳质泥岩、油页岩等）。有机质分析通常包括全岩鉴定（有机岩石学）、氯仿抽提分析（生物标志物研究）和干酪根研究。煤属集中有机质，碳、氢、氧、氮、硫元素为宏量元素，是主要的有机质固体可燃矿产，由地史中植物死亡堆积并经漫长地质年代的煤化作用而成；呈褐色至黑色，含碳量一般为 46%~97%，灰分低于 50%，水分低于 75%，镜质组反射率小于 10%；可分为褐煤、烟煤和无烟煤（Alpern and de Sousa，2002；辞海编辑委员会，2009；中国煤炭工业协会，2010；Dai et al.，2020b）。我们重新定义煤有机地球化学为：研究煤中与碳直接结合的元素、化合物及其衍生物的组成、成因、迁移与转化的学科。新定义更强调有机元素的分布、组成与演化和大分子结构与由有机元素组成的超微芳香晶体（超微）、显微组分（微观）和煤岩（宏观）组成、分布与演化。煤有机地球化

---

* 原文"煤有机地球化学研究进展与展望"刊于《矿物岩石地球化学通报》2021 年第 40 卷第 3 期，本文略有修改。

学的外延包括有机元素地球化学、有机分子地球化学（分子标志化合物、大分子地球化学）、显微组分地球化学和煤岩有机地球化学。研究煤有机组成已经形成较为完整的学科，有煤岩学、煤化学和煤结构学，前两者对煤中有机无机都研究。而煤有机地球化学是从地球化学观点来研究煤岩及煤化学中显微组分和大分子结构的组成、时空分布及迁变规律的学科。煤有机地球化学是有机地球化学中十分薄弱的分支（张水昌，2010；胡建芳和彭平安，2017），煤结构学衍生出煤分子工程学（曾凡桂和谢克昌，2004）。

煤是最复杂的地质体之一（Finkelman et al., 2019; Dai et al., 2020b）。要剖析由复杂有机体组成的煤，一般采用宏观、微观、超微观和分子级别4个层次研究。宏观上，即肉眼将煤划分为镜煤、亮煤、暗煤和丝炭；微观上，由各类光学显微镜将煤划分不同显微组分；超微观上，由电子显微镜、原子力显微镜和微区分析等大型物理仪器技术，揭示煤的超微结构特征、物理谱学特征和芳香晶体结构等；分子级别研究，主要集中于煤氯仿抽提的生物标志化合物和煤大分子结构构建。煤有机地球化学最早启蒙于20世纪50~60年代的煤化学和煤岩学，其代表是Teichmüller和Teichmüller（1949, 1958）用镜质组反射率来反映煤化作用程度、热成熟度和地热史（Lyons and Cross, 2005），以及H/C和O/C原子比图呈现干酪根3种类型的雏形van Krevelen（1950, 1961, 1993）；20世纪70~90年代是煤有机地球化学分析的生物标志化合物和煤成烃（傅家谟等，1990；黄第藩等，1992；傅家谟和秦匡宗，1995）及荧光分析时代；随后由于煤的清洁利用需要，煤大分子结构与模拟研究出现热潮（秦匡宗等，1998；曾凡桂和谢克昌，2004）；近十年，煤中硫（Chou, 2012; Golding et al., 2013）和树皮煤的有机地球化学、煤结构（Mathews and Chaffee, 2012）及微孔隙（Clarkson et al., 2013）与煤层气及煤基石墨烯结构、碳笼、碳纳米管研究成为热点。下面从煤中有机元素、煤生物标志化合物、煤大分子结构、煤石墨烯结构、煤微观超微观等尺度、煤孔隙结构、同位素及其方法和应用等方面，概述煤有机组成的复杂性及近十年的进展与发展趋势。

1 煤中有机元素

煤中与碳直接结合且在地史上与生命有关的元素有碳（C）、氢（H）、氧（O）、氮（N）和硫（S），也是煤质分析中元素分析的主要元素，它们构成了成煤母质中的碳水化合物（糖类，纤维素、半纤维素）、木质素、蛋白质和类脂物等有机组分（王华和严德天，2015；秦勇，2017）。煤是植物等生物转变而来的（Taylor et al., 1998; Alpernand de Sousa, 2002; Thomas, 2013, 2020），是有机成因，极个别学者提出无机成因（虞震东，2016），但无科学证据与实证。

碳构成烃类化合物的骨架，煤分子结构的主体为芳香环。煤中碳按形式可分为芳香碳和脂肪碳。芳香碳包括质子化芳碳和非质子化芳碳。非质子化芳碳包括接氧芳碳、脂取代芳碳、桥接芳碳三类（陈丽诗等，2017）。芳香碳原子占总碳原子的分数定义为芳碳率，俗称芳香度。脂肪碳中有脂环碳、烷烃中的碳、芳香环上烷基侧链中的碳和含氧官能团（如—COOH、—C=O、—OCH$_3$）中的碳等，利用^{13}C-NMR可得到各类芳碳的相对含量（陈丽诗等，2017；张锐等，2018）。

研究发现，煤中芳香碳（除羧基和羰基碳外）随煤的变质程度升高而增加，而脂肪碳则减少，含氧结构亦减少。核磁共振数据能够充分揭示煤在煤化作用期的化学变化过程：脱羧、脱羟基、脱烷基和脱氢，由脂肪链的大量断裂、含量减少和桥头碳变化的平滑趋势可看出，在中低煤化阶段是芳构化作用，缩合作用并不明显（相建华等，2016；张锐等，2018）。

煤中氢，数量上以甲基、亚甲基和羟基最多，其次是羰基、羧基和芳烃上的碳氢键。结合各官能团含量和活性大小，可认为褐煤氧化过程中的关键性官能团是羟基和甲基、亚甲基基团，其次是羧基（辛海会等，2013；安文博等，2018；郝盼云等，2020）。煤中氧是构成煤中许多有机、无机化合物以及水分等的重要元素，主要存在于煤大分子含氧官能团（如—OCH$_3$、—COOH、—OH等基）上的为有机氧；存在于氧化物、氢化物、硫酸盐矿物、碳酸盐矿物等中的为无机氧（Thomas, 2013, 2020）。氧在煤有机质中的存在形式可分为两类：即含氧官能团和醚键含氧杂环（呋喃环）。随煤的变质程度变化，含氧官能团种类和含量也随之改变；含氧官能团的数量随煤阶的升高而下降（Zhou et al., 2015；Wang K. et al., 2017）。

煤中氮，有无机化合物（Dai et al., 2012）和有机化合物（Zheng et al., 2015）。有机氮主要以吡啶型氮（N-6）、吡咯型氮（N-5）、季氮（N-Q）和氮氧化物（N-X）4种含氮官能团形式存在（吴代赦等，2006；张爱华等，2018）。

煤中有机硫主要是硫醚、硫醇、噻吩和砜等，噻吩含量随煤阶增加而相对逐渐增加（Chou, 2012；唐跃刚等，2015）。Canfield（2013）综述了煤中硫同位素的变化与显生宙硫地球化学循环之间的关系。Li和Tang（2014）在研究我国湖南辰溪高硫煤中硫同位素地球化学特征时，提出了硫"还原−再氧化−歧化模式"的成因模式，较好地解释辰溪煤中各形态硫含量及同位素组成在剖面上的变化。Li W. W.等（2015）运用X射线岩电子能谱法（X-ray photoelectron spectroscopy，XPS）研究了中国晚古生代高有机硫煤中有机硫的赋存形态，结果发现噻吩硫随碳含量的增加而增加，噻吩类和硫醇类硫化合物的相对丰度与煤级有关（Chou, 2012）。魏强等（2015）总结了煤中有机硫化合物结构的研究现状。Medunić等（2020）研究了Raša超高有机硫煤的地球化学特征并分析了其对环境的影响。

煤中元素的赋存状态分为有机态、矿物和与有机紧密结合态等3种，后者包括吸附到有机物表面、溶解在孔隙水中的和蕴藏在煤有机物或被其屏蔽的亚微米或纳米矿物的元素（Dai et al., 2020a）。过去被认为是煤中有机元素的，现在多是与有机紧密结合的元素，如煤中锗，它不与煤中碳结合，但与煤中有机官能团的氧结合（Wei et al., 2020），这也是有机与无机过渡的元素。

在我国西南部的高氟煤中检测到了有机氟化物。有机氟的成因则更偏向于成煤植物在泥炭化作用过程中，残余物与腐殖酸中的活性基因可与许多元素形成螯合物（李大华和唐跃刚，2008；Fan et al., 2016）。煤中有机氟主要为（Z)-乙基 2-(4,5-双氢-3aH 环戊烯并[d][1,3]乙二酸硫醇-2-亚基)-3,3,3-三氟丙酸酯和2-氟苯基-辛基-己二酸。煤中的有机氯有3种形式：金属离子与煤大分子上的含氧官能团通过结合形成外轨络合物；以HCl形式与煤大分子上的含氮官能团结合；直接与煤有机大分子结合（孙林兵等，2010；郭伟等，2016）。除了有机氯外，煤中常见的还有有机溴（Wei et al., 2004）。

有学者曾研究煤中镧系元素的有机亲和性和有机态稀土的赋存形式（杨建业，2010；梁虎珍等，2013）。近年来国内外也对煤中碳、氢、氧、氮、硫等元素的测定方法进行了修订（ISO 29541-2010；ASTM D 5373-16；中华人民共和国国家质量监督检验检疫总局和中国国家标准化管理委员会，2014；中华人民共和国国家质量监督检验检疫总局，2018）。

2 煤生物标志化合物

揭示煤的组成、性质、成因及演化等一般从煤岩学、煤质学以及煤氯仿抽提和干酪根等有机地球化学入手。煤氯仿抽提的沥青，主要用于研究烷烃和芳烃的生物标志化合物。在漫长的地质演化进程中，生物标志物具有很好的稳定性，基本保持了原始的碳骨架结构。因此通过煤生物标志物的组成及分布特点可很好地判别煤的有机质演化特征、生烃潜力、母质来源及沉积环境等（Ayinla et al., 2017；Patra et al., 2018），还可追溯煤形成时的古生态环境（Tewari et al., 2017）。近年来新的成果主要体现在树皮煤和超高有机硫煤的有机地球化学研究中（Zhao et al., 2014；王绍清等，2018；Wang et al., 2018a；Lin et al., 2020）。

煤生物标志化合物主要包括正构烷烃、无环的类异戊二烯烷烃、萜类及其类似的化合物、甾类和重排甾烷、卟啉化合物和芳烃化合物等。Bechtel等（2008，2018）通过研究煤系地层中的数据发现，生物标志化合物的组成在一定程度上受腐殖化作用、微生物活动以及成岩作用途径影响。因此生物标志物在泥炭及低煤级煤中的指示意义更为准确。近年来生物标志化合物在更高煤级煤中（烟煤）的研究也取得了显著进展（Fabiańska et al., 2013；Böcker et al., 2013；Schwarzbauer et al., 2013），如Naafs等（2019）详细地总结了关于泥炭和煤中一些生物地球化学指标的常规应用要点。

煤中正构烷烃的来源十分广泛，主峰碳和碳数范围的分布特征可以直接反映沉积有机质的母质来源（张珂等，2020）。藻类、细菌等低等生物的碳数范围是 C_{15}—C_{20}，而陆生高等植物由于角质层含有较多的蜡质，其主峰碳多以 C_{27}、C_{29}、C_{31} 为主，碳数范围也主要集中在 C_{27}—C_{33}（秦身钧等，2018）。不同类型的植物也会产生碳链长度不同的叶蜡正构烷烃（Diefendorf et al.，2011；Lane et al.，2017）。López-Dias 等（2013）通过 nC_{27}/nC_{31} 的值来评估灌木和树木与草本植物对泥炭形成的相对贡献。最近有研究还发现针叶树和被子植物的正构烷烃的含量和分布存在明显差异（Diefendorf et al.，2011；Lane，2017），且在美国西部裸子植物中，其正构烷烃的平均链长（average chain length，ACL）要低于相应的被子植物（Diefendorf et al.，2011）。此外，煤中正构烷烃的分布特征对有机质成熟度也有一定的指示意义（Mustapha et al.，2017）。

类异戊二烯型烷烃中最具代表性的化合物是姥鲛烷（Pr）和植烷（Ph），其含量丰富且性质稳定。Pr/Ph 及 Pr/nC_{17}（Ph/nC_{18}）通常是判别有机质沉积环境的良好指标，近年来也被广泛应用于指示煤形成时的沉积环境（Böcker et al.，2013；Mustapha et al.，2017）。Yuan 和 Zhang（2018）也认为煤中偏低的 Pr/Ph 值可能与沉积时强烈的海水影响有关。Izart 等（2012）通过 Pr/nC_{17} 和 Ph/nC_{18} 的分布图发现，澳大利亚煤主要位于Ⅲ型区域，指示成煤母质来源主要为高等植物，该结论与研究区煤的有机质类型结论一致。除了与沉积环境密切相关外，吴应琴等（2014）还发现姥鲛烷非对映异构体的异构化指数与镜质组反射率有良好线性关系，并以此提出"姥鲛烷异构化指数"（PIR），认为该值是高-过成熟的良好指标。

萜类化合物是环状的异戊二烯化合物，分为单萜、倍半萜、三萜等。Izart 等（2012）通过生物标志物中二萜烷的比值（Rdit）将源自欧洲石炭纪的煤分为三类，Rdit 是反映成煤植物类型、沼泽地下水位和气候等的有利指标。Bhattacharya 等（2017）和 Dutta 等（2017）报道了在喜马拉雅山脉东部及印度东北部的炭质页岩中收集到的上新世—更新世琥珀，其中的挥发性单萜和倍半萜成分得到了显著保护。这是由于挥发性植物代谢产物可以作为传粉引诱剂、食草动物和病原体防护剂保存，并保护植物免受非生物威胁。四环二萜烃类生物标志物曾用于指示罗汉松科和南洋杉科类植物（Lu et al.，2013）。松科对成煤植物的贡献可以通过海松烷与 $16\alpha(H)$-扁枝烷的比值来判断（Stojanović and Životić，2013；Liu B. J. et al.，2018）。五环三萜类化合物是萜类化合物中分布广泛、含量丰富且最常使用的。藿烷系列化合物中，以 $18\alpha(H)$-22,29,30-C_{27}（Ts）三降藿烷与 $17\alpha(H)$-22,29,30-C_{27}（Tm）为代表性化合物，通常依据两者比值判别沉积环境。Tm/Ts 可判别沉积环境氧化还原条件，还可用来指示有机质成熟度，但由于 Tm 和 Ts 在 R_o 为 1.4% 时几乎达到平衡状态，因此通过 Ts/(Ts+Tm) 来判别有机质成熟度更为可靠且使用更为普遍（Qi et al.，2020）。α,β-藿烷是酸催化降解产物，常用来指示泥炭形成时的环境温度和氧化还原条件（Huang et al.，2015；Inglis et al.，2018）。Hoş-Çebi 和 Korkmaz（2013）通过煤中的 $C_{31}R/C_{30}$ 值判断土耳其安纳托利亚北部大部分地区煤的沉积环境为微咸水环境，但也有少部分是沉积于淡水环境下。

地质体中的甾类化合物主要是由生物体中的甾醇转化而来。甾类化合物碳数范围分布在 C_{27}—C_{30}，碳数不同其指示的生源意义也不同。Hakimi 和 Abdullah（2014）发现，C_{29} 规则甾烷主要来源于陆源高等植物，而 C_{27}、C_{28} 规则甾烷则受低等植物影响强烈。因此，通常根据 C_{29}/C_{27} 及 C_{27}-C_{28}-C_{29} 甾烷三角图分布特征来判别沉积有机质的母质来源及沉积环境（张珂等，2020）。

煤中芳烃化合物比饱和烃更稳定，因此可用来指示更宽的成熟度范围且在有机质来源和沉积环境方面也有很好指示意义（秦身钧等，2018；Lin et al.，2020）。煤中检测出的芳烃化合物主要包括萘系列、菲系列、联苯系列、三芴系列、䓛系列以及三芳甾烷等。萘是含有两个苯环的稠环芳香烃，其化合物含量通常与陆源高等植物有关（妥进才，1996）。菲系列化合物在芳烃中含量很高，是芳烃中最常见的化合物之一，甲基菲系列化合物的相关比值是成熟度的有效指标（Fabiańska et al.，2013；Zhang and Li，2018）。吴士豪等（2020）通过计算甲基菲指数（MPI1）指出贵州松河煤矿煤的变质程度较高，但发现根据经验公式计算得出的镜质组反射率略低于实测值；三芴系列的指示意义和较高的苯并萘并噻吩系列质量分数可判断该区域煤主要形成于还原环境。Zhang 和 Li（2018）根据众多饱和烃芳烃指标判断准噶尔盆地芦草沟组富有机质页岩的成熟度达到生油窗，并发现在指示有机质成熟度时，部分芳烃指标比饱和烃指标更为灵敏。

多环芳烃（PAHs）是煤中有机碳的重要赋存形式，多环芳香化合物包括多环芳烃和杂环芳烃，并与它们的烷基和芳基衍生物相连通（Achten and Andersson, 2015）。早期认为，煤中多环芳烃是一类由 2~7 个稠合芳香环组成的有机分子，呈线性、角或簇状排列（Wang R. W. et al., 2017）。煤中多环芳烃的生成和分布很大程度上取决于煤阶、生物前驱物、沉积环境和沉积背景等因素（Laumann et al., 2011）。许多研究开始解决不同煤盆地、地质时代煤中多环芳烃的变化趋势，涉及沉积环境、煤级（Emsbo-Mattingly and Stout, 2011; Ribeiro et al., 2012）和化学类型（Wang et al., 2010）等因素。刘志华等（2010）对我国华北不同煤中多环芳烃总结发现：煤中 PAHs 的含量与镜质体含量呈负相关关系，与惰质组含量呈正相关关系。

此外，目前在煤生物标志物研究的基础上，国内外众多学者还将生物标志物与煤中碳、氮同位素结合使用（Schwarzbauer et al., 2013; Hakimi and Abdulah, 2014; Mitrović et al., 2017; Ding et al., 2018a, 2018b; Zhao et al., 2018），该方法有助于更准确地判别有机质的沉积环境等问题，是目前研究煤、烃源岩等地质体古环境变化方面的一个有效手段。

3 煤大分子结构

煤结构研究一般以镜质体作为研究对象，这是因为镜质体含量多、组成均匀、变化平稳。煤的结构包括煤化学结构（即煤的分子结构）和煤物理结构（即分子间的堆垛结构与孔隙结构）。煤的化学结构模型是煤结构的碎片特征信息和分子成键构造，能反映煤有机质的主要特征，全球范围内的研究者已提出了 130 多种煤的化学结构模型（Mathews and Chaffee, 2012）。秦匡宗等（1998）对低阶煤等的有机质物理化学结构与溶解性能进行了研究，提出煤的复合结构概念模型。曾凡桂和谢克昌（2004）构建了中国煤结构化学的理论体系与方法论，随后又提出煤分子工程及其关键问题。（Mathews and Chaffee, 2012）从分子量分布与有序化排列精度方面研究了煤的碳结构。秦志宏（2017）构建了煤的嵌布结构模型。

为探索平顶山煤中孢子体的大分子结构，Liu 等（2019）选择了等密度梯度法富集孢子体，并运用 FTIR、NMR、XPS 等手段分析了孢子体的光谱参数，在结构分析的基础上通过 Materials Studio 软件建立和优化孢子体结构模型，得到大分子结构模型的分子式为 $C_{203}H_{244}N_2O_{30}$，相对分子量为 3192.10。Wang S. Q. 等（2011, 2014, 2017b）使用现代分析方法详细研究了树皮体的化学结构特征，结果表明树皮体最明显的结构特征是脂肪族结构的富集，发现树皮体在化学结构上比镜质体具有更高的无序性，随着镜质组反射率的增加，树皮体的形态结构从纤维状向不规则的网状结构演变，定向排列似乎增强，同时 CH_2/CH_3 也随煤级的增高而降低。van Niekerk 和 Mathews（2010）根据分析数据，构建了两种二叠纪的南非煤的分子结构模型，即富含惰质组的 Highveld 煤和富含镜质体的 Waterberg 煤。蔺华林等（2013）对神东上湾煤及其岩相分离所得惰质组富集物分别进行核磁共振（^{13}C-NMR）、傅里叶变换红外光谱（FTIR）和 X 射线光电子能谱法（XPS）表征，得到煤结构单元信息，与元素分析数据相结合构建了上湾煤及其惰质组富集物的结构模型，用 ACD/CNMR predictor 软件计算了结构模型的 ^{13}C 化学位移。Wang Y. H. 等（2020）通过量子力学计算和 TG-MS 研究了黑带沟长焰煤的镜质体和惰质体的热解行为，量子力学计算表明，镜质体中的键长长且键序小，惰质体则相反；镜质体和惰质体呈现相同的断裂键顺序，但断裂键的数量有差异。

Li 等（2013）、张锐等（2018）等以低阶煤为研究对象，揭示了煤碳分子骨架中的芳香结构主要以苯环和萘环为主，也含有其他芳香环结构，且芳香层侧链数目远大于桥和环的数目，在煤碳骨架中，芳香结构在平面上的延展度和在空间上的堆垛度较低，煤中芳香结构体系小，芳核的缩聚程度低；碳原子除以 C-C 和 C-H 形式存在外，还主要以 C=O、C-O 和 O=C-O 等形式存在于这些含氧官能团中，且含氧官能团偏多。李霞等（2015, 2016a）和 Zhang（2016）对低中变质烟煤的研究表明，煤的结构演化可分为三个阶段，其镜质组反射率分别为 <0.8%、0.8%~1.3% 和 1.3%~2.0%。第一阶段以含氧官能团的脱

落和脂肪类物质的富集为主；第二阶段主要是脂肪类物质的富集和支链化程度的增加或脂环化作用与芳香化作用的协同；第三阶段镜质组反射率小于1.7%前主要为脂肪类物质的热解断裂，之后则是随脂肪类物质的断裂脱落，形成新的芳香结构体系。Yan等（2020）对中低阶煤分子结构进行了XRD、固态^{13}C核磁共振和红外光谱表征，XRD图谱显示了无处不在的无定形碳，逐渐冷凝的芳香层，以及在煤化过程中尺寸增大且石墨化的微晶；^{13}C核磁共振谱显示存在不同类型的碳，芳香组分增加而脂肪组分减少，芳香簇的尺寸和重量增大但侧链数量减少；红外光谱表明脂肪链的缩短和减少，以及随着变质程度加深芳香性的增强。秦勇（1994）和Jiang等（2019）对高煤级煤的研究发现，随着煤化程度的升高，煤的芳香层片单层距离、层片直径、平均堆砌厚度、堆砌层数都增大，但均为非线性增大。煤在石墨化过程中，发生了孔隙闭合、芳香层片增大、多个片层联结、片层边缘发生卷曲的现象。

刘振宇（2014）提出煤键合结构的认识，并用其解释煤热反应机理。随着计算机技术和三维分子模拟软件的发展，分子模拟技术已成为煤分子结构研究的重要方法。Roberts等（2015）利用HRTEM、^{13}C-NMR、XRD等研究了煤在慢加热过程中结构的变化。Pan等（2020）利用X光衍射和傅里叶变换红外光谱对受热气流作用的煤体进行了测试，结果表明在含氧环境中，随温度升高，煤大分子晶格的堆积高度（L_c）总体上升，并具有自修复能力，煤分子结构演化空间排列可逆。分子模拟是基于分子力学、分子动力学以及量子力学理论，从原子水平构建煤分子结构模型及预测煤分子结构行为的一种技术（Chen et al., 2011; Song et al., 2017a; Yang et al., 2020）。

4 煤中石墨烯结构

20世纪90年代在煤中发现了碳笼（C_{60}、C_{70}），近十年来，对煤石墨烯结构的研究是煤科学研究的热点。早在20世纪50年代，Hirsch（1954）与Cartz等（1956）就对煤中石墨烯层进行了研究，他们发现小的稠合芳族区域是形成层单元的一部分，这些层单元本身可以通过脂族或脂环族的五元环与其他类似的单元连接，形成大的桶状片。Saikia等（2009）指出煤中具有随机取向的涡轮静态结构的薄片堆叠就是石墨烯，也在分子水平上探讨了煤中石墨烯层与脂肪族碳丰度的关系，指出煤中石墨烯微晶既是多环芳烃组成的芳香片层结构。Saikia（2010）对模型化合物苯（C_6H_6）计算的原子对关联函数与模拟的一维结构函数的比较表明C_6单元是煤的主要组成。煤成熟过程以环化、去羟基化、脱烷基化、裂解、交联和芳构化/缩合为特征，最终产生仅含有杂原子的石墨，原子团之间的官能团和键的逐渐消除导致芳香片平均堆积数增加并产生更浓缩的固体残留物（Kelemen et al., 2012; Vu et al., 2013）。Manoj和Kunjomana（2014，2015）对次烟煤中石墨烯层做了系统研究，从次烟煤合成纳米尺寸的石墨烯片，其边缘具有无定形碳附加物，发现煤中的碳具有纳米晶碳的结构，L_a和L_c等结构参数随着含碳量的增加而增加，而层间距d_{002}随着含碳量、芳香性和煤阶的增加而减小。煤中多环芳香化合物包括多环芳烃和杂环芳烃，并与它们的烷基和芳基衍生物相连通（Achten and Andersson, 2015）。芳香核在结构特征上存在明显的差异，这取决于煤级。Li等（2018）在碳洋葱形成机理研究中发现，小片的石墨烯结构可以卷曲成半球状和曲面状，进而形成碳洋葱。借助于HRTEM的超高分辨率，可以直观、清晰地观察煤芳香核的结构特征，定量测量煤芳香核的结构参数，得到了无烟煤中石墨状结构存在的直接证据（郇璇，2019; Huan et al., 2020）。

5 煤有机组成的不同尺度研究

在煤的超微观研究方面，通过透射电子显微镜尤其是高分辨率透射电子显微镜，可直接在纳米尺度上检测煤中晶格条纹信息，并利用图像处理技术量化煤中晶格条纹的长度、层间距、堆垛度、取向及曲率（郭亚楠等，2013; Pan et al., 2015; 李霞等，2016b; Wang C. A. et al., 2016, 2017; Zhong等，2018; 王小令等，2020）。Oberlin（1979）首次利用透射电子显微镜（TEM）暗场技术观察到煤的微观结构特征（分层结构和堆垛结构），Bustin 等（1995）也利用此技术观察到无烟煤在应力应变作用下产生的晶格缺陷。21 世纪初，Sharma 等（2000a）首次报道了清晰的煤的 TEM 图像，这也是首次直接在电子显微镜下观察煤中存在富勒烯状结构，他们还发现煤中存在晶格条纹，证实了电子显微镜图像中边缘部分更能真实地代表煤的微观结构。Sun 等（2011）对无烟煤的高分辨透射电子显微镜研究分辨出四类芳香结构单元缺陷，直接从原子水平论证了晶格缺陷的存在。煤的微观形貌在纳米尺度范围可直观表现出来。同时，Van Niekerk 和 Mathews（2010）首先将 Sharma 等（2000b）煤的 HRTEM 图像进行二值化处理，他们通过手工提取晶格条纹，并结合图像分析算法获得了煤的晶格条纹参数，其结果与 Sharma 等（2000b）的分析结果相吻合，从而证明手工提取晶格条纹的可行性。这一将晶格条纹长度与煤中芳香环的尺寸结合，制定煤中芳香环尺寸类别的分类方法，为后来的研究提供了晶格条纹长度分类依据。由此，TEM 在煤的物理结构研究中，开始了由定性到定量的新阶段。

随着煤级增高煤中晶格条纹逐渐联结，使得煤晶格条纹变长且取向逐渐趋于一致，形成良好定向区域。在中煤级阶段，晶格条纹排列出现了 T 型堆垛排列，这可能是煤发生第二次煤化作用跃变的原因（Mathews and Sharma，2012）。Wang C. A. 等（2016，2017）利用诺丁汉大学编写的 MATLAB 代码，量化了晶格条纹的弯曲，定义并定量了曲率的表征指标（条纹弯曲度及累积角度），揭示了煤中存在非六元芳香环或晶格缺陷是导致曲率存在的原因。在对煤微晶结构深入表征的基础上，Castro-Marcano 等（2012）基于得到的多个高分辨率透射电子显微镜图像分析，结合 Fringe3D 脚本，构建了 Illinois 6 号煤的 3D 大分子物理结构模型。Mathews 和 Sharma（2012）使用 Adobe Photoshop 中的图像处理工具包（Reindeer Graphics 软件），分析了不同煤级煤的 HRTEM 二值化图像，在获得晶格条纹长度信息的基础上，得到了条纹取向，由此证明低阶煤也同样具有定向性。在低等级煤中，大多数芳族碳环是弯曲的且取向差，并有内部缺陷。高阶煤中的碳环排列紧密，晶格条纹取向进一步改善，表现为缩合环数量与直径增加（Huan et al.，2020）。Wang S. Q. 等（2017b，2018b，2020）利用原位透射电子显微镜研究了煤中芳香结构的热演化规律。此外，杨起等（1994）和姚素平等（2011b）运用原子力显微镜揭示了煤的超微结构。焦堃等（2012）和 Wang 等（2017a）分析了树皮体原子力显微镜结构与形貌特征。

显微组分（maceral）是指反射光显微镜下可辨别的单一有机成分（Taylor et al.，1998）。目前，对煤中显微组分的分类多参考国际煤与有机岩石学委员会（International Committee of Coal and Organic Petrology，ICCP）于 1998、2001、2005 和 2017 年发表的关于煤中镜质体（ICCP，1998）、惰质体（ICCP，2001）、腐殖体（Sýkorová et al.，2005）和类脂体（Pickel et al.，2017）的 1994 年分类方案。中国烟煤显微组分分类由西安煤炭勘探研究院于 1995 年提出，并于 2013 年进行了第二次修订（中华人民共和国国家质量监督检验检疫总局和中国国家标准化管理委员会，2018），其中树皮体并没有被国际煤与有机岩石学委员会等组织认可（Hower et al.，2007；唐跃刚等，2011）。近十年许多学者开展了树皮体形貌（王绍清等，2015；Wang et al.，2018b；周国庆等，2019）与结构的研究（Wang et al.，2011，2013；Mastalerz et al.，2015）。显微组分成因探讨多集中于惰质组中真菌体和粗粒体（Hower et al.，2013a，2013b）。

白向飞（2017）总结了煤岩学现状，提出干酪根的显微组分来自生物残体，由于干酪根的研究起源于煤岩学，故而其显微组分的概念及术语采用煤岩学的方式。ICCP 于 2012 年 9 月在北京举行年度学术会议，国际有机岩石学学会（The Society of Organic Petrology，TSOP）分别于 2012 年和 2018 年在中国矿业大学（北京）举办年会。

煤的宏观有机组成一直沿用镜煤、亮煤、暗煤、丝炭、烛煤和藻煤等国际岩石类型（Taylor et al.，1998），国内仍用光泽岩石类型（中华人民共和国国家质量监督检验检疫总局，2000）。2013 年中国煤炭标准委员会讨论和制定了煤体结构分类（中华人民共和国国家质量监督检验检疫总局和中国国家标准化管理委员会，2014）。全国煤炭资源潜力评价探讨了中国宏观煤岩与显微组分的地质时空分布（中国煤炭地质总局，2016），近十年来煤研究主要集中于特殊与稀缺煤（王绍清等，2015）。2010 年在国土资源部组织下，中国煤炭地质总局承担了"特殊煤稀缺煤资源调查"项目。黄文辉等（2010）、夏灵勇等（2019）和郭俊春（2018）分别分析了炼焦煤资源、高炉喷吹、活性炭、热压铸造焦、碳素制品、电石、气化等工业原料用煤资源与质量分布状况。黄文辉等（2010）报道了西北煤岩特征。2013年全国煤炭资源潜力评价工作组出版了1∶2 500 000的中国煤类分布图（中国煤炭地质总局，2016）。此外，Tang等（2020a）还对中国煤岩与煤质学的研究现状与未来发展方向进行了煤有机组成及其性质的探讨。

6 煤中孔隙结构

煤中孔隙结构研究是煤有机地球化学近十年的研究热点之一。煤作为孔隙-裂隙双重介质，其内部具有极其复杂的微观孔隙结构，煤孔隙一般可分为有机孔隙和无机孔隙，对煤的结构和性质及煤层气的储集和运移具有重要影响（Clarkson et al., 2013；Swanson et al., 2015；李祥春等，2019）。近年来，密度泛函理论（Thommes and Cychos, 2014）、核磁共振（Zhou et al., 2016）、扫描电子显微镜（Li et al., 2020）、原子力显微镜（姚素平等，2011a；Li et al., 2020）、X光微计算机断层扫描（Zhou et al., 2018）等方法手段为煤中孔隙研究带来了有力支持。

屈争辉等（2015）和侯锦绣等（2017）通过研究煤中孔隙特征及其成因，认为微孔的形成主要受控于煤的类微晶参数和芳香层片间的堆垛结构，而介孔的形成应主要受控于煤侧链的变化和煤的基本结构单元间隙。分形证据表明纳米尺度的空间有序性发生在较低的中孔-微孔范围内（Clarkson et al., 2013）。微孔呈圆形或椭圆形，而微裂缝呈弯曲不规则形状，通过圆形或椭圆形通道与其他微裂缝相连（Pan et al., 2016）。封闭孔多呈圆形、葫芦形和不规则形，孔隙壁厚，连通团呈镂空的雪片状，相互连通的孔道小且少（孟巧荣等，2011）。

Xin等（2019）用扫描电子显微镜揭示了中国低阶煤的孔隙结构类型存在明显差异。第一次煤化跃变中，煤分子的缩聚和煤基质的压实发生，导致水分、孔隙率和渗透性迅速下降，煤内部的孔隙表面和复杂性逐渐增加（Tao et al., 2018）。煤变质和物理压实后，植物组织孔隙的体积和数量减少，导致大孔显著减少。傅里叶变换红外光谱特征表明，化学结构的演化对煤的孔隙结构也有影响。芳香核比例低，官能团比例高，侧链长，煤储层空间结构"松散"，导致表面积大。随着煤成熟度的增加，侧链逐渐分解，长度减少，芳香核增多，结构变得更加"致密"（Xin et al., 2019）。Wang F.等（2014）研究了煤化作用对中高阶煤孔隙结构的影响，通过实验发现中高阶煤体的微孔、过渡孔所占的体积和比表面积比重最大。Nie等（2015）研究了不同变质程度煤样的孔隙结构，发现低阶煤中变质程度主要影响中孔结构，高阶煤随着变质程度的升高，中孔减少，微孔数量增加连通性降低。随变质程度的加深，孔隙尺寸范围而减小（Li et al., 2017），吸附能力呈先降低后升高的U形规律，这与孔径随煤级的分布有关（李祥春等，2019）。Okolo等（2015a）发现随着煤变质程度的增加，大孔对瓦斯解吸扩散的影响程度减小，微孔变得越来越重要。Chen等（2017）通过N_2吸附-解吸和CO_2吸附实验表明超微孔对煤的吸附能力有显著影响。

有学者提出显微组分是吸附孔隙度的主要控制因素，镜质体的增加和惰质体的减少导致了吸附孔隙度的不均匀分布（Li W. et al., 2015）。亮煤和半亮煤通常比半暗煤和暗煤具有相对较高的孔隙半径（Li et al., 2017）。吸附能力倾向于依赖显微组分，在某种程度上也依赖于煤化作用，这可能与煤化作用或氧化作用使得形成的镜质体，有较高的微孔隙性有关（Weishauptová and Sýkorová, 2011）。分形方法显示出分形维数受煤阶和显微组分含量的共同影响，而且分形维数随着碳氢含量的增加和灰分产率的降低而降低（Zhang et al., 2014）。唐书恒等（2013）发现在无烟煤阶段之前，腐泥煤的孔隙结构以微孔、小孔居多；至无烟煤阶段，微孔、小孔比例降低，大孔、中孔含量增加。与同煤阶腐殖煤相比，腐泥煤整体孔隙度较低，且孔径偏小，大孔不太发育。此外，埋深深度也会影响煤中孔隙特征，微孔与中孔尺寸随深度而减小，而且浅层煤具有较大的渗透性，可能是压力和煤阶随深度增加的结果（Swanson et al., 2015）。

构造煤中孔隙结构是近年来的研究热点，与原生结构煤相比，构造煤中值孔径显著偏小，构造煤平均孔容特征：微孔≈小孔>中孔的特点，原生结构煤特征：中孔>小孔≈微孔（王向浩等，2012）。宋晓夏等（2013，2014）研究表明，随着煤的变形程度增强煤中微孔比例增加，最可几孔径减小，分形维数越高，构造变形越强烈。Li 等（2014）利用扫描电子显微镜和低温氮吸附研究发现，韧性变形煤主要有亚微孔和超微孔，脆性变形煤主要有中孔和微孔，且不同尺寸的纳米孔之间会发生相互转化，特别是在韧性变形煤中。屈争辉等（2015）和 Qu 等（2017）根据煤中微孔随煤级和变形的变化规律，探讨了构造煤微孔的成因机制，分析了纳米孔的性质随煤级和变形的变化，并探讨了这些变化的原因。构造变形在增加渗透孔隙度方面影响显著（Li W. et al.，2015），构造变形可以改变煤的孔隙-微裂隙结构，从而改变煤的储层物性，总的来说，随着构造变形的增强，煤中变得越来越杂乱的微裂隙的密度逐渐变大，吸附孔隙的分形维数也增大，而渗流孔隙的分形维数减小（Song et al.，2017b；Lu et al.，2018；Li et al.，2019）。Ju 等（2018）研究了中国东部构造煤的孔隙结构特征及变形机制。Cheng 和 Pan（2020）评述了构造煤的储层性质，包括孔隙结构、吸附、扩散、渗透率和地质力学性质，并与完整煤进行了比较，研究发现，由于构造作用，构造煤的总孔容和比表面积普遍比完整煤大，但由于变质作用和构造作用的共同作用，小孔的总孔容和比表面积没有显著差异。Song 等（2020）对构造变形煤孔隙破裂研究进行了综述，并提出了构造变形煤孔隙破裂的地球动力学机制。

7 煤同位素有机地球化学

煤中常量的有机元素 C、H、O、N、S 的稳定同位素以及煤生物标志化合物、大分子结构、显微组分等同位素测定方法，都是现代煤有机地球化学研究新颖而有效的方法，是揭示煤成因、演化及有机元素循环等的有效途径。现就这 5 种有机元素同位素的研究现状阐述如下。

碳同位素，与氢同位素一起构成了煤系气是生物还是热降解成因的良好指标。近年来，碳同位素在煤及其利用中二氧化碳减排、探索古气候、植被等方面发挥了重要作用。Schwarzbauer 等（2013）根据化合物特异性同位素分析出的单个有机化合物的 $\delta^{13}C$ 值确定了 34 种晚古生代煤中脂肪族生物标志物的稳定碳同位素比率。Warwick 和 Ruppert（2016）完成了煤在燃烧过程中煤中碳同位素与二氧化碳排放之间关系的初步研究。Suto 和 Kawashima（2016）研究了来自 10 个国家的 95 种煤的 $\delta^{13}C$，结果显示测得的煤的 $\delta^{13}C$ 值变化范围很大（-27.4‰~23.7‰）。此外，新生代煤的 $\delta^{13}C$ 比古生代煤的 $\delta^{13}C$ 更低。李江涛（2018）研究得出屯兰煤的 $\delta^{13}C_{org}$ 的均值（-24.3‰）在华北地区腐殖煤的 $\delta^{13}C$ 范围，且呈逐渐正偏的总体变化趋势，表明主要是受到二叠纪华北地区温度上升的影响。Ding 等（2018b）研究发现，淮南煤 $\delta^{13}C$ 的平均值为-24.06‰，主要是由与现代气候有关的成煤植物群的 $\delta^{13}C$ 传递。Ding 等（2019）探究了煤中碳的类型对煤中碳同位素的影响机制，指出煤的 $\delta^{13}C$ 值一般为-25.1‰~-22.9‰，且随氢碳比和脂肪碳含量的降低以及芳香族碳含量的升高，$\delta^{13}C$ 会变大。

目前，氢同位素组成在煤层气（Golding et al.，2013；陶明信，2015；Wang et al.，2015）方面研究广泛。在煤研究中氢同位素主要用于煤的显微组分研究（傅家谟和秦匡宗，1995）和煤的液化（Niu et al.，2017）。

氧同位素，Solano-Acosta 等（2008）通过对印第安纳州西南部煤层中割理和裂缝填充的自生矿物进行 $\delta^{18}O$ 估值，推算出该地煤岩矿化的古温度。Warwick 和 Ruppert（2016）研究发现，煤和其他高碳燃料（泥炭和煤）大气燃烧产生 CO_2 的 $\delta^{18}O$ 为 19.03‰~27.03‰。Yang 等（2017）进行了煤氧化反应和煤热解反应产生 CO 的氧同位素（$\delta^{18}O$）测试研究，发现煤中碳氧化合物的 $\delta^{18}O$ 在氢氧反应中可逆。

煤的硫同位素的研究主要是为了查明煤中硫的来源及成煤环境（Chou，1990，2012；代世峰，2000；Dai et al.，2002；Li and Tang，2014）。煤中有机硫同位素组成与有机硫含量有一定关系：有机硫含量低于 0.8% 的低硫煤的 $\delta^{34}S_0$ 范围较窄，而有机硫含量大于 0.8% 的高硫煤的 $\delta^{34}S_0$ 分布范围宽且更负偏，这表明超高有机硫煤中的有机硫有很大一部分来源于次生硫（Chou，1990）。中国煤中硫的硫同位素组成变化大（-15‰~50‰）（Xiao and Liu，2011），分析其变化特征可模拟成煤环境，例如，内蒙古乌达矿区的煤在煤层剖面上的硫同位素组成及其变化规律显示其受到了海水的影响（代世峰等，2000；Dai et al.,

2002），湖南辰溪超高有机硫煤煤层剖面上的形态硫含量及硫同位素组成特征显示其受到了硫细菌的"还原—再氧化—歧化"影响（Li and Tang，2014）。值得注意的是，地质历史上大规模的硫同位素分馏效应不一定是硫同位素歧化效应的结果（Sim 等，2011），对显生宙硫地球化学循环的研究离不开对煤中硫同位素变化的研究（Canfield，2013）。

煤的氮主要来源于大气、生物及深部地壳（李瑾等，2013；张爱华等，2016）。氮同位素主要用于不同化石燃料使用过程中的碳氮物质与氧氮物质的演化与污染控制研究（Walters et al.，2015），煤和原油的氮的同位素以及天然气中非烃类组分 N_2 的同位素地球化学研究相对较少（张爱华等，2018）。肖化云等（2007）发现不同成煤时期和不同煤种的氮同位素组成没有显著性差异。程晨等（2018）认为煤的氮同位素组成是煤变质作用、沉积环境等多种因素耦合作用的结果。Xiao 和 Liu（2011）测得中国煤样的 $\delta^{15}N$ 集中于 $-3.0‰\sim2.0‰$，郑启明（2012）和 Zheng 等（2015）测得沁水盆地煤样的 $\delta^{15}N$ 均值为 $3.9‰$。有研究发现，泥炭的 $\delta^{15}N$ 含量变化可反映生物降解作用的氮同位素分馏效应和氮在沼泽内部的迁移作用，也反映了从古至今人类活动带来的逐渐增加的大气氮沉降影响（Esmeijer-Liu et al.，2012；Novak et al.，2014）。

8 煤有机地球化学方法

煤的有机地球化学特征复杂，其研究方法也是多种多样。传统的煤有机地球化学研究方法有溶剂抽提、质谱分析、同位素法、气相色谱、红外光谱和 X 射线光电子能谱法等。

傅里叶变换红外光谱（FTIR）可以提供有机和无机成分的分子结构的重要信息，在过去几十年中被广泛用于地质样品的化学表征（Chen et al.，2015），同时也是表征煤的分子结构与官能团特征的很好的手段（Yan et al.，2014；Jing et al.，2019）。X 射线光电子能谱法（XPS），亦称"原子指纹"，可以对煤进行结构鉴定（Xia et al.，2014）、元素定量定性分析，尤其是煤中有机硫的分析（李梅等，2013；马玲玲等，2014）。气相色谱在煤有机地球化学研究中被广泛应用，尤其是在煤热解产物分析中（Dong et al.，2012；Lievens et al.，2013；Kong et al.，2014；Yan et al.，2015）。稳定同位素法在煤有机地球化学方面的主要用途是研究煤中碳、氢、氧、氮、硫元素在煤中的分馏作用及在不同有机物中的分馏，这对于煤地质作用具有重要意义，是揭示煤成因、演化及有机元素循环等内容的有效途径（Xiao and Liu，2011）。

新兴的煤有机地球化学研究方法有拉曼光谱、高分辨透射电子显微镜、核磁共振和傅里叶变换红外质谱法等。

高分辨率透射电子显微镜（HRTEM）可对煤同一微区位置进行形貌、晶体结构和成分的全面分析。van Niekerk 和 Mathews（2010）运用 HRTEM 首次将煤的 HRTEM 晶格条纹按镜下长度做了归属分类。Huan 等（2019）运用 HRTEM 等手段表征并对比分析了煤制备石墨烯材料的结构特征。

拉曼（Raman）光谱能够有效提供分子的化学和生物结构的指纹信息，是一种简便有效的煤结构特征及其演化分析方法（Ulyanova et al.，2014）。Tselev 等（2014）、Baysal 等（2016）、He 等（2017）、Ghosh 等（2018）用 Raman 分析了原煤的结构特征，He 等（2017）还运用 FTIR 结合 Raman 研究煤结构随煤级变化特征。

核磁共振（NMR）分析可测定煤的分子结构与性质，应用最多的是 ^1H 的核磁共振和 ^{13}C 的核磁共振（Jing et al.，2019）。Wang 等（2011）利用 ^{13}C-NMR 结合 FTIR 方法，得到了树皮体最明显的结构特征是富含脂肪族结构，其脂肪族结构长，支链少。Yan 等（2014）运用 ^{13}C-NMR 研究了有机溶剂处理对褐煤化学结构和热解反应性的影响。Okolo 等（2015b）、Baysal 等（2016）利用 ^{13}C-NMR 等手段表征了原煤的分子官能团结构特征。崔馨等（2019）指出 ^{13}C-NMR 技术在了解煤分子脂肪族碳和芳香族碳结构组成等信息方面具有不可或缺的作用。

对煤进行溶剂抽提并结合仪器分析是目前十分重要和成熟的一种研究煤结构的物理化学手段。魏强等（2015）综述了煤中有机硫结构的研究方法，主要包括溶剂抽提/GC-MS、XPS/XANES、热力学方法及多手段综合分析的方法。傅里叶变换离子回旋共振质谱（Fourier transform ion cyclotron resonance mass spectrometry，FTICR-MS）的出现，使得煤、石油等沉积物里可溶有机质中极性化合物复杂质谱峰鉴定并确定质量分子精确的元素组成得以实现（胡建芳和彭平安，2017）。Tang等（2020b）运用FTICR MS对中国高有机硫煤索氏抽提产物进行了鉴定，检测到9种含硫化合物，并发现随着煤阶的增加，含氮原子的含硫化合物逐渐消失。

9 煤有机地球化学应用

近十年来，煤有机地球化学在技术和应用方面取得了长足进展。限于篇幅，本文仅从地质、工业和环境等领域阐述有机元素及其衍生物研究进展。

9.1 地质应用

近十年，煤有机地球化学地质应用大多集中于研究煤阶与煤结构的变化。Chen等（2012）应用傅里叶变换红外光谱研究了不同煤级显微组分中化学官能团的特征。Li等（2013）利用傅里叶变换红外光谱结合峰分离方法研究了不同煤阶镜质体结构的特征，发现了镜质体比原煤具有更大的脂肪烃比例，且随煤阶的增加，镜质体的含氧官能团和烷基侧链以不同的速率损失。芳构化程度随煤阶的增加而增加。在低变质阶段煤结构脱氧，脂肪结构相对增加；同时，煤结构逐步芳构化和缩合；且芳构化速度大，芳香氢含量增加（郭德勇等，2019）。在高变质阶段，煤结构芳构化，脂肪结构减少；同时，煤大分子结构不断缩合，且其缩合速度大，芳香C-C键的直接交联增加，使煤大分子中芳香氢含量减少（秦勇，1992，2017）。随煤变质程度的升高，煤中脂族结构减少，芳香结构增多，煤的结构逐渐趋向于石墨化（郎璇，2019）。煤中氮含量较少，由于煤中的氮是在泥炭化阶段固定下来，因此氮几乎全都以有机物形式存在（程晨等，2018）。煤中有机氮化物相当稳定，在成煤作用过程中氮的绝对含量几乎不发生变化，成为煤中保留的氮化物。以蛋白质形态存在的氮，仅在泥炭和褐煤中发现，烟煤中几乎没有发现（张爱华等，2016；肖剑等，2017）。煤中氮含量随煤变质程度的升高而略有上升，可能是随煤变质程度的升高，煤中含氧官能团与侧链逐渐减少、芳香化程度与分子排列规则化程度不断提高，氧含量逐渐降低，因而氮含量相对升高（Zheng et al.，2015；Ding et al.，2018a，2018b）。

此外，煤层气是近年来的煤有机地球化学研究的热点（宋党育等，2016）。在煤化阶段，煤的分子结构决定了煤的微孔隙及其演化，煤的微孔体积与芳香桥头堡碳的比例、质子化芳香碳的比例、芳香桥碳与芳香外围碳的比例（XBP）呈线性相关（Liu Y. et al.，2018）。

煤有机地球化学多被用于研究成煤植物、成煤演化和成煤环境（Böcker et al.，2013；白悦悦等，2014；Berlendis et al.，2014；秦身钧等，2018；张珂等，2020）。通过孢粉学（Mustapha et al.，2017；Eble et al.，2019）、MPI指数、DMAI指数和生物标志化合物研究（Böcker et al.，2013；Hoş-Çebi and Korkmaz，2013；Strobl et al.，2014；Oskay et al.，2019；Qi et al.，2020），可判断泥炭沼泽成煤植物的来源及微生物对成煤植物的影响，有些研究还发现微生物参与了煤有机碳的风化降解（Berlendis et al.，2014）。但秦身钧等（2018）指出，在分析变质程度较高的烟煤和无烟煤时，由于其与低成熟度煤有显著不同的有机地球化学特征，因此不能简单按传统的地质意义进行应用。煤有机地球化学还用于评价盆地煤、油、气等能源矿产成矿潜力（姚素平等，2011a；Strobl et al.，2014；Mustapha et al.，2017）、干酪根生气潜力或勘探前景（陶明信等，2014；Cheng et al.，2020），特别是有机质高演化阶段的成气潜力（孙永革等，2013；Qi et al.，2020）；判断相应煤层的生烃性能（Mardon et al.，2014；宋换新等，2015；王绍清等，2018）。近十年来不同学者对煤有机地球化学在地质应用方面的研究表明，现阶段主要研究重点和热点仍为通过煤有机地球化学特征分析煤层沉积环境并判断其生烃潜力。虽然实验室已可模拟煤显微组分产气的碳氢同位素特征，但未来同位素测定新技术的引入和发展将会有助于有机地球化学等相关领域的研究（胡建芳和彭平安，2017）。张水昌（2010）综述了我国近30年来的有机地球化学研究现状、发展方向和展望，指出煤岩组分是影响煤层生物气的主控因素。

鲁静等（2009，2014）和毛婉慧等（2011）将层序地层学、煤相学和煤岩学结合起来，探讨了煤层序、煤相和煤岩的剖面变化，丰富了成煤理论。梁虎珍等（2013）围绕镧系收缩效应对稀土-煤相互作用的影响及煤中有机态稀土的赋存形式进行了研究。侯贤旭等（2013）研究了重庆中梁山矿区主要煤层的煤岩学和煤相特征。Zhao 等（2014）、Li 和 Tang（2014）利用煤相参数分析了成煤微环境与煤中有机硫的关系。邵龙义等（2017）探讨了中国煤岩系煤相及沉积有机相的研究现状。Dai 等（2020b）全面概述了煤的前身——泥炭沉积环境的识别，厘清了许多关系。

9.2 工业应用（液化、焦化、气化与燃烧）

煤结构理论在煤的液化（Feng et al., 2013）、燃烧（Xu et al., 2018）、洗选（Xia et al., 2019）、生烃（Bulat et al., 2016）和吸附（Yu et al., 2013）等方面起到了广泛的指导作用。Singh 和 Tiwari（2020）对冈瓦纳煤中惰质体结构在炭化过程中行为的研究表明，煤在炭化时主要产生透镜状镶嵌结构，本质上是焦化，惰质体在镶嵌基质中结合良好，赋予焦炭高强度。在缺乏足够镶嵌基质的情况下，大尺寸惰质体沿其边界产生大量孔隙，并且更具反应性。Guo 等（2020a）研究了从褐煤到无烟煤等不同煤级煤的岩石学特征，并在偏光、正交偏光和辅助石膏石板条件下，系统研究了受温度影响的煤热解焦炭的岩石学特征，总结了黏结性和非黏结性煤中镜质体和惰质体在低、中、高温热解过程中的变化特征。目前煤分子结构建模与预测技术在煤的浮选中已有成功应用（Xia et al., 2019）。

Roberts 等（2015）利用核磁共振碳谱、高分辨透射电子显微镜、X 射线衍射等方法研究了煤在慢加热过程中的结构变化，结果显示不同变质程度煤，在挥发分 $V_{daf}=35\%$ 左右时，具有最高的萃取率。通过对原煤、萃取残渣和生成焦粒的对比分析发现，不同变质程度煤经过萃取后，残渣中的脂肪烃和脂环烃含量均减少，矿物质较多。肥煤和气煤氢键缔合的极性键均位于煤中大分子结构上，而焦煤和弱黏煤中的极性键大多位于小分子化合物上（张小东和张鹏，2014）。

已有研究表明，随煤中挥发分的增加，热解所得的焦油、气体和水的产率会相应增加。热解气的产率与煤化程度有关，与挥发分呈正相关。气体热值受煤种的影响并不显著（刘钦甫等，2016；张志刚，2018）。煤在低温燃烧阶段会有类似低温氧化反应，分子内部的非芳香结构侧链、活跃含氧官能团均能与氧发生反应生成 CO 和 CO_2，在高温氧化阶段，煤分子中的基团以及高能化合物会发生断裂和裂解，并与氧发生反应生成 CH_4、C_2H_6、C_3H_6 等烷烃、烯烃类气体（张玉龙，2014）。氢气是一种清洁能源，90% 的氢气是以矿物燃料为原料制备而来（姚律等，2017），其中约 1/3 是以煤为原料进行生产（王嘉琦等，2020）。煤还能光催化二氧化碳和水合成甲醇（杨瑞，2013）。

Gray 和 Devanney（1986）对焦中碳形态进行分类，该分类经过不断修正之后一直沿用至今（ASTM D5601-19）。最早的残碳研究可追溯到 20 世纪 20 年代（Heidrich et al., 2013），国际煤与有机岩石学委员会委员 Lester（2010）提出了煤燃烧炭的划分方案。而 Hower 等认为，探讨燃煤炭、气化残碳的岩石学分类，需追溯煤中显微组分的变化，即需要考虑煤地质因素，而非单纯研究生成炭的镜下形貌特征（Hower, 2012；Hower and Wagner, 2012；Suárez-Ruiz et al., 2017, Hower et al., 2017；Xing et al., 2019；Valentim, 2020；Guo et al., 2020b）。Chaves 等（2018）利用机器可视化（SVM 和深度学习）对粉煤燃烧炭进行自动化特征识别，并根据煤的反应活性将炭颗粒分为高、中、低 3 组。

石墨烯是近年来的世界材料研究热点（Ohta et al., 2006；Lee et al., 2008；Jiao et al., 2009；Novoselov et al., 2012；Cao et al., 2018a, 2018b），煤基石墨烯研究也为新材料开拓了新领域（Vijapur et al., 2013；Xu et al., 2014；Powell and Beall, 2015；Awasthi et al., 2015；Wang L. et al., 2020）。通过碳原子外层电子的 sp^n 杂化，碳元素可形成大量结构与性质完全不同的同素异形体，这些同素异形体在不同的物化条件下又可相互转化。基于该原理，人们通过不同方法，以不同碳源制造了包括石墨烯在内的多种纳米碳质材料（张亚婷，2015；郇璇，2019），已有研究通过了煤制取碳纳米管（Moothi et al.，2012）。目前，已有较多学者从煤地质学角度研究了煤级、煤中矿物质和煤中显微组分对煤制备石墨烯及其衍生物的结构与性质的影响（Tang et al.，2018；郇璇，2019；Huan et al.，2019，2020；唐跃刚等，2020）。

9.3 环境与生态应用

煤有机地球化学与环境关系的研究主要集中于煤炭利用上，尤其是燃煤排放物对环境的影响（刘惠永等，2001；李晓东等，2002；蔡昌凤和唐传罡，2012）。煤炭利用过程中还会释放较多的有机物，如多环芳烃（PAHs）、苯系物、脂环烃及直链烃等，其特殊的性质和突出的环境效益受到越来越多的关注。煤中多环芳烃的变化涉及沉积环境和化学类型（Laumann et al.，2011；Ribeiro et al.，2012）。Mathews 和 Sharma（2012）利用高分辨透射电子显微镜、激光解吸离子化质谱等方法研究了煤中分子量分布与有序化排列及多环芳烃。多环芳烃（PAHs）分布广、难降解，并具有生物积累性和三致（致癌、致畸、致突变）的慢性作用。煤基活性炭可净化污染水（Rivera-Utrilla et al.，2011；Li et al.，2012；张福凯等，2014；El-Dars et al.，2014）和烟道废气（廖继勇等，2012）。

煤层气又称煤层瓦斯，是一种形成于煤层又储集于煤层中的非常规天然气，属于强温室气体，温室效应是 CO_2 的 21 倍（Shine et al.，2005）。将煤层气加以回收利用，可直接减少甲烷排放量、间接降低 CO_2 排放量，缓解温室效应、改善空气质量并保证资源与环境安全，具有显著的环境正效益（帅官印等，2018）。低浓度瓦斯可用于提纯制液化天然气或压缩天然气（赵路正，2015）。将甲烷与空气按一定比例混合可制固体氧化物燃料电池，其转化效率远高于传统火力发电，且使用便捷、绿色环保（Kronemayer et al.，2007；PÉrillat-Merceroz et al.，2011）。以煤层气为原料制取氢气可提高热值，减少污染气体与温室气体排放。

CO_2 也是引起温室效应的主要气体，如何减少煤炭利用过程中 CO_2 的排放和提高其有效利用是国内外学界研究的重点。在众多解决方案中，CO_2 捕获和封存一直是研究热点（Martunus et al.，2012；Maheshwari et al.，2019；Ye et al.，2020）。王倩倩等（2015）通过归纳煤结构理化性质变化对煤层封存 CO_2 潜力的影响，指出煤体理化性质变化造成的环境安全与健康风险问题。煤对 CO_2 影响的另一方面在于煤中孔隙能够吸附 CO_2（翟光华等，2012）。Tang 等（2016）研究发现，煤中 CO_2 的吸附过程是整体扩散控制和表面相互作用控制相结合的过程，前者控制初始阶段，后者控制整个过程的大部分。杨瑞（2013）基于煤特殊的大分子结构和性质，探讨了煤及负载过渡金属离子煤对 CO_2 和水合成甲醇的光催化作用。煤通过光催化 CO_2 和水合成甲醇，可减少 CO_2 排放。

10 问题与展望

10.1 存 在 问 题

煤是由植物死亡堆积经地质地球化学作用转变而来的，且主要是由碳、氢、氧、氮、硫等有机元素组成演变而成的有机沉积矿产。煤有机地球化学主要研究与生命有关、与碳结合的有机元素、有机化合物、大分子结构及其显微组分在地壳中的分布及演化。目前的研究存在如下科学与工程问题：

（1）有机元素问题。煤中碳的化学结构一直是世界性难题，其结构迥异、反应性迥异。煤、石墨、金刚石的碳结构不同，性质也不同。煤中碳与全球碳循环，煤利用过程中 CO_2 的回收减排与利用，也是重大的科学与工程问题。尽管煤中氢在化石燃料中含量最低，但其赋存状态及其在煤的黏结性、煤液化等方面的机理并不十分清晰。超高有机硫煤中的硫分子结构是一世界性难题，有机硫在煤的黏结性、液化以及煤基石墨烯中的作用，一直是学界关注的科学与工程问题。

（2）不同地质时代的煤性质问题。在具有相同煤岩组成与相同镜质组反射率时，中国中生代煤与古生代煤的工艺性完全不同，煤黏结性结焦性远没古生代煤好？两个地质时代煤的分子结构有何不同？这是煤科学家和煤工程学家要努力回答的问题。

（3）树皮煤科学与工程问题。中国南方晚二叠世树皮煤的树皮体显微组分一直不被国际组织承认，它与其他类脂体显微组分的物质组成、结构与性质有何不同？世界上晚二叠世很少有煤，中国树皮煤的成因一直是世界煤科学中的难题，树皮煤的黏结性极强，大到超出现有测试基氏流动度仪器量程，这也给煤工艺工程提出树皮煤黏结机制的问题。

（4）煤有机地球化学与热工艺性问题。为什么低煤阶煤和无烟煤等非炼焦煤的液化和焦化不出现黏结性？煤有机地球化学很难解释煤有机质在煤化过程中的热工艺性质的演变问题，也很难解释煤显微组分的在煤燃烧气化中的热反应性，诸多科学与工程问题需未来年青的煤有机地球化学家去探索解决。

10.2 展望未来

中国的资源禀赋特点决定了中国仍是世界上能源消费结构以煤炭为主的大国，在煤炭的勘查、开发、加工、转化与利用以及生态环境修复等过程中，还存在许多煤科学与工程问题，世界煤研究中心理应在中国。高效与清洁的煤炭利用，是世界的要求，也是中国的要求。

（1）煤形成与碳循环。煤形成是古植物、古气候、古地理、古构造共同作用的结果，碳是主要的成煤元素，它的演化与聚煤赋煤的内外地质作用有关，碳循环在煤形成与演化过程中有必然反应。因此，煤形成与古气候关系、全球古气候及地层对比、碳循环在成煤过程中所表现的机理等，是当今煤地质学前沿课题。

（2）有机元素的利用。洁净煤技术有两个关键问题：一是最大有效地利用有用成分。煤以有机质为主，煤的利用主要是利用有机质的热值、有机转化为化工原料，近期还开发了主要利用碳元素制备碳纳米管、石墨烯等新型材料。二是煤中氢的开发与利用，这也是一大研究发展趋势，煤中氢能开发具有成本与价格优势。此外煤利用中 CO_2 的减排与利用，也是研究热点。有机元素在地质演化及煤加工利用过程中质量平衡分析，反映煤中有机元素在勘探、开发、利用及生态等生命周期的洁净煤综合指数的提出，将是煤有机地球化学中的科学与工程问题。

（3）当今生命科学的进展，将推动煤有机元素的理论研究、生命起源以及植物起源，有机元素地球化学将展示其强大优势，研究成果可为洁净煤技术与环境保护提供科学依据及工程技术保障。

（4）新技术出现。微区分辨率的提高与发展，高新科学技术仪器出现，将使煤从岩石学和地球化学两方面，从宏观肉眼→显微尺度→分子原子级别，深入探讨煤的物质组成与演化，拓展研究领域。云计算、大数据和人工智能等技术，将使以煤有机地球化学为基础的新能源和新材料在资源等方面拓展出新的研究与应用领域。

（5）深空、深地、深洋的研究。多学科的交叉与渗透、新理论的不断引入、广泛运用高新技术，使有机元素地球化学在地下深部、海洋以及太空等领域探讨有机元素赋存分布与演化中发挥重要作用。新理论、新认识及新技术将会涌现。

未来十年，煤的有机元素地球化学研究将在洁净煤利用、煤基材料利用的基础研究、煤系气的研究以及化石燃料新能源的基础研究与技术开发等领域涌现出新成果。

致谢：成文过程与中国石油大学（北京）钟宁宁和中国矿业大学（北京）任德贻、代世峰等教授进行了有益讨论，王晓帅、陈丛、杨承伟和王小令等博士研究生为本文汇集文献，部分硕士研究生查询部分文献，在此一并致谢。

参 考 文 献

安文博,王来贵,刘向峰,潘纪伟,李喜林. 2018. 基于 FTIR 和 XRD 法分析阜新长焰煤结构特征. 高分子通报,(3):67-74
白向飞. 2017. 煤岩学发展简史及其应用. 煤质技术,(Z1):1-6

白悦悦,刘招君,孙平昌,柳蓉,胡晓峰,赵汉卿,徐银波. 2014. 梅河盆地古近系梅河组下部含煤岩系有机质富集模式. 煤炭学报,39(S2):458-464

蔡昌凤,唐传罡. 2012. 焦化中水中主要有机污染物在焦煤上的竞争吸附. 煤炭学报,37(10):1753-1759

陈道公,支霞臣,杨海涛. 2009. 地球化学,第2版. 合肥:中国科学技术大学出版社

陈丽诗,王岚岚,潘铁英,周扬,张媛媛,张德祥. 2017. 固体核磁碳结构参数的修正及其在煤结构分析中的应用. 燃料化学学报,45(10):1153-1163

程晨,赵峰华,任德贻,苗雪娜. 2018. 中国煤中氮同位素组成特征初步研究. 地质学报,92(9):1959-1969

辞海编辑委员会. 2009. 辞海,第6版. 上海:上海辞书出版社,2274-2275

崔馨,严煌,赵培涛. 2019. 煤分子结构模型构建及分析方法综述. 中国矿业大学学报,48(4):704-717

代世峰. 2000. 内蒙古乌达矿区高硫煤中硫的成因. 洁净煤技术,6(1):41-45

傅家谟,秦匡宗. 1995. 干酪根地球化学. 广州:广东科技出版社

傅家谟,刘德汉,盛国英. 1990. 煤成烃地球化学. 北京:科学出版社

郭德勇,郭晓洁,刘庆军,孙向成. 2019. 烟煤级构造煤分子结构演化及动力变质作用研究. 中国矿业大学学报,48(5):1036-1044

郭俊春. 2018. 煤基活性炭制备进展及发展趋向分析. 中国石油和化工标准与质量,38(22):102-103

郭伟,秦志宏,李春生,单良. 2016. 三种煤中有机氯和溴的赋存形态及原理. 煤炭转化,39(3):11-18

郭亚楠,唐跃刚,王绍清,李薇薇,贾龙. 2013. 树皮残植煤显微组分分离及高分辨透射电镜图像分子结构. 煤炭学报,38(6):1019-1024

郝盼云,孟艳军,曾凡桂,闫涛滔,徐光波. 2020. 红外光谱定量研究不同煤阶煤的化学结构. 光谱学与光谱分析,40(3):787-792

侯锦秀,王宝俊,张玉贵,张进春. 2017. 不同煤级的微孔介孔演化特征及其成因. 煤田地质与勘探,45(5):75-81

侯贤旭,唐跃刚,宋晓夏,杨明显,郭明涛,贾龙. 2013. 重庆中梁山矿区主要煤层的煤岩学和煤相特征. 煤田地质与勘探,41(5):6-10

胡建芳,彭平安. 2017. 有机地球化学研究新进展与展望. 沉积学报,35(5):968-980

郇璇. 2019. 煤基石墨烯与煤基石墨烯量子点结构影响因素研究. 北京:中国矿业大学(北京)博士学位论文

黄第藩,华阿新,王铁冠,秦匡宗,黄晓明. 1992. 煤成油地球化学新进展. 北京:石油工业出版社

黄文辉,唐书恒,唐修义,陈萍,赵志根,万欢,敖卫华,肖秀玲,柳佳期,Finkelman B. 2010. 西北地区侏罗纪煤的煤岩学特征. 煤田地质与勘探,38(4):1-6

焦堃,姚素平,张科,胡文瑄. 2012. 树皮煤的原子力显微镜研究. 地质论评,58(4):775-782

李大华,唐跃刚. 2008. 中国西南地区煤中微量元素的分布和富集成因. 北京:地质出版社

李江涛. 2018. 西山矿区煤中有机碳同位素特征与其成熟度之间的联系. 煤矿安全,49(11):164-167

李谨,李志生,王东良,李剑,程宏岗,谢增业,王晓波,孙庆伍. 2013. 塔里木盆地含氮天然气地球化学特征及氮气来源. 石油学报,34(增刊1):102-111

李梅,杨俊和,张启锋,常海洲,孙慧. 2013. 用XPS研究新西兰高硫煤热解过程中氮、硫官能团的转变规律. 燃料化学学报,41(11):1287-1293

李霞,曾凡桂,王威,董夔,程丽媛. 2015. 低中煤级煤结构演化的FTIR表征. 煤炭学报,40(12):2900-2908

李霞,曾凡桂,王威,董夔. 2016a. 低中煤级煤结构演化的拉曼光谱表征. 煤炭学报,41(9):2298-2304

李霞,曾凡桂,司加康,王威,董夔,程丽媛. 2016b. 不同变质程度煤的高分辨率透射电镜分析. 燃料化学学报,44(3):279-286

李祥春,李忠备,张良,高佳星,聂百胜,孟洋洋. 2019. 不同煤阶煤样孔隙结构表征及其对瓦斯解吸扩散的影响. 煤炭学报,44(S1):142-156

李晓东,姚艳,严建华,徐旭,池涌,岑可法. 2002. 中国部分煤种二氯甲烷萃取液中极性和烃类有机物分布特性研究. 燃料化学学报,30(6):529-534

梁虎珍,曾凡桂,李美芬,相建华. 2013. 镧系收缩效应对稀土-煤相互作用的影响及煤中有机态稀土的赋存形式研究. 燃料化学学报,41(9):1030-1040

廖继勇,周末,李小敏. 2012. 活性炭净化技术在烧结烟气治理领域的应用. 烧结球团,37(4):61-63

林巍,李一良,王高鸿,潘永信. 2020. 天体生物学研究进展和发展趋势. 科学通报,65(5):380-391

蔺华林,李克健,章序文. 2013. 上湾煤及其惰质组富集物的结构表征与模型构建. 燃料化学学报,41(6):641-648

刘惠永,徐旭常,姚强,张爱云. 2001. 燃煤电厂飞灰碳含量与PAHs有机污染物吸附量之间相关性研究. 热能动力工程,16(4):359-362

刘钦甫,崔晓南,徐占杰,郑启明,毋应科. 2016. 煤热解气体主产物及热动力学分析. 煤田地质与勘探,44(6):27-32,37

刘振宇. 2014. 煤化学的前沿与挑战:结构与反应. 中国科学:化学,44(9):1431-1438

刘志华,刘大锰,姚艳斌. 2010. 煤中多环芳烃分布赋存规律研究. 煤炭科学技术,38(2):113-116,125

鲁静,邵龙义,鞠奇,刘天绩,文怀军,李永红,张发德,高迪. 2009. 柴北缘大煤沟矿区侏罗纪煤系层序地层及其煤岩变化特征. 煤田地质与勘探,37(4):9-14

鲁静,邵龙义,王占刚,李永红,王帅. 2014. 柴北缘侏罗纪煤层有机碳同位素组成与古气候. 中国矿业大学学报,43(4):612-618

马玲玲,秦志宏,张露,刘旭,陈航. 2014. 煤有机硫分析中XPS分峰拟合方法及参数设置. 燃料化学学报,42(3):277-283

毛婉慧,庄新国,周继兵,阮传明,雷国明. 2011. 煤相参数在煤层层序划分中的应用——以新疆准东煤田帐南西矿区为例. 煤田地质与勘探,39(1):6-10

孟巧荣,赵阳升,胡耀青,冯增朝,于艳梅. 2011. 焦煤孔隙结构形态的实验研究. 煤炭学报,36(3):487-490

秦匡宗,郭绍辉,李术元. 1998. 煤结构的新概念与煤成油机理的再认识. 科学通报,43(18):1912-1918

秦身钧,陆青锋,吴士豪,薄朋慧. 2018. 重庆中梁山晚二叠世煤有机地球化学特征. 煤炭学报,43(7):1973-1982

秦勇. 1994. 中国高煤级煤的显微岩石学特征及结构演化. 徐州:中国矿业大学出版社
秦勇. 2017. 化石能源地质学导论. 徐州:中国矿业大学出版社
秦志宏. 2017. 煤嵌布结构模型理论. 中国矿业大学学报,46(5):939-958
屈争辉,姜波,汪吉林,李明. 2015. 构造煤微孔特征及成因探讨. 煤炭学报,40(5):1093-1102
邵龙义,王学天,鲁静,王东东,侯海海. 2017. 再论中国含煤岩系沉积学研究进展及发展趋势. 沉积学报,35(5):1016-1031
帅官印,张永波,郑秀清,陈军锋,张志祥,赵雪花. 2018. 煤层气开采对地下水流场影响的数值模拟研究. 水力发电,44(11):17-20
宋党育,袁镭,白万备,宋永志,何进亚. 2016. 煤地质学研究进展与前沿. 煤田地质与勘探,44(4):1-7
宋换新,文志刚,包建平. 2015. 祁连山木里地区煤岩有机地球化学特征及生烃潜力. 天然气地球科学,26(9):1803-1813
宋晓夏,唐跃刚,李伟,王绍清,杨明显. 2013. 中梁山南矿构造煤吸附孔分形特征. 煤炭学报,38(1):134-139
宋晓夏,唐跃刚,李伟,曾凡桂,相建华. 2014. 基于小角X射线散射构造煤孔隙结构的研究. 煤炭学报,39(4):719-724
孙林兵,魏贤勇,刘晓勤,宗志敏. 2010. Illinois No.6煤中有机氮和硫赋存形态的研究. 中国矿业大学学报,39(3):437-442
孙永革,杨中威,Cramer B. 2013. 煤系有机质多阶段成气的分子碳同位素表征及其对高过成熟干酪根生气潜力评价的启示. 地球化学, 42(2):97-102
唐书恒,张静平,吴敏杰. 2013. 腐泥煤孔隙结构特征研究. 天然气地球科学,24(2):247-251
唐跃刚,郭亚楠,王绍清. 2011. 中国特殊煤种——树皮煤的研究进展. 中国科学基金,25(3):154-163
唐跃刚,贺鑫,程爱国,李薇薇,邓秀杰,魏强,李龙. 2015. 中国煤中硫含量分布特征及其沉积控制. 煤炭学报,40(9):1977-1988
唐跃刚,徐靖杰,郇璇,王绍清,陈鹏翔. 2020. 云南小发路无烟煤基石墨烯制备与谱学表征. 煤炭学报,45(2):740-748
陶明信. 2015. 中国煤层气同位素地球化学初步研究. 地质学报,89(S1):185-186
陶明信,王万春,李中平,马玉贞,李晶,李晓斌. 2014. 煤层中次生生物气的形成途径与母质综合研究. 科学通报,59(11):970-980
王华,严德天. 2015. 煤田地质学简明教程. 北京:中国地质大学出版社
王嘉琦,王秋颖,朱桐慧,朱小梅,孙冰. 2020. 甲烷重整制氢的研究现状分析. 现代化工,40(7):15-20
王倩倩,张登峰,王浩浩,顾丽莉,杨劲,杨荣,陶军. 2015. 封存过程中二氧化碳对煤体理化性质的作用规律. 化工进展,34(1):258-265
王绍清,唐跃刚,李正越,秦云虎,郭鑫,高伟程,朱士飞. 2015. 我国典型特殊煤种特性及利用研究. 洁净煤技术,21(1):32-36
王绍清,孙翎博,沙玉明. 2018. 不同聚煤区内富氢煤有机地球化学特征研究. 煤炭科学技术,46(9):233-238
王向浩,王延斌,高莎莎,洪鹏飞,张美娟. 2012. 构造煤与原生结构煤的孔隙结构及吸附性差异. 高校地质学报,18(3):528-532
王小令,李霞,曾凡桂,边洁晶. 2020. 基于HRTEM的煤中不同聚集态结构表征. 煤炭学报,45(2):749-759
魏强,唐跃刚,李薇薇,赵巧静,贺鑫,李龙,赵正福. 2015. 煤中有机硫结构研究进展. 煤炭学报,40(8):1911-1923
吴代赦,郑宝山,唐修义,王明仕,胡军,李社红,王滨滨,Finkelman R B. 2006. 中国煤中氯的含量及其分布. 地球与环境,34(1):1-6
吴士豪,薄朋慧,徐飞,王炎,陆青锋,秦身钧. 2020. 贵州松河煤矿晚二叠世煤的有机地球化学特征. 地质科技通报,39(4):141-149
吴应琴,王永莉,雷天柱,马素萍,王有孝,文启彬,夏燕青. 2014. 下古生界高过成熟烃源岩成熟度指标——姥鲛烷异构化指数研究. 质谱学报,35(4):317-323
夏灵勇,魏立勇,刘敏,田立新,桂夏辉. 2019. 稀缺炼焦煤中煤再选潜势研究. 选煤技术,(1):18-23
相建华,曾凡桂,梁虎珍,李美芬,宋晓夏,赵月圆. 2016. 不同变质程度煤的碳结构特征及其演化机制. 煤炭学报,41(6):1498-1506
肖化云,刘学炎,李友谊,刘丛强. 2007. 中国煤炭和城市苔藓的氮同位素组成特征. 矿物岩石地球化学通报,26(Z1):427
肖剑,赵云鹏,丁量,魏贤勇,宗志敏. 2017. 先锋褐煤可溶有机质中含氮化合物的组成和结构特征. 燃料化学学报,45(4):385-393
辛海会,王德明,戚绪尧,许涛,窦国兰,仲晓星. 2013. 褐煤表面官能团的分布特征及量子化学分析. 北京科技大学学报,35(2):135-139
杨建业. 2010. 煤中镧系元素有机-无机亲合性及其演变规律研究——以渭北晚古生代5#煤层为例. 中国矿业大学学报,39(3):402-407
杨起,潘治贵,汤达祯,廖立兵,马哲生,施倪承. 1994. 煤结构的STM和AFM研究. 科学通报,39(7):633-635
杨瑞. 2013. 煤光催化二氧化碳和水合成甲醇的研究. 西安:西安科技大学硕士学位论文
姚律,杨晓瑞,王倩倩,梁金花,朱建良. 2017. 甲烷催化裂解制氢及碳纳米管的研究进展. 现代化工,37(5):25-29
姚素平,胡文瑄,焦堃. 2011a. 煤、油、气共存富集的地球化学判识模式. 高校地质学报,17(2):196-205
姚素平,焦堃,张科,胡文瑄,丁海,李苗春,裴文明. 2011b. 煤纳米孔隙结构的原子力显微镜研究. 科学通报,56(22):1820-1827
虞震东. 2016. 煤的无机成因学说——兼论石油天然气和油页岩的成因. 前沿科学,10(3):33-60
曾凡桂,谢克昌. 2004. 煤结构化学的理论体系与方法论. 煤炭学报,29(4):443-447
翟光华,段利江,唐书恒,夏朝辉. 2012. 二氧化碳与煤作用机理的实验研究. 煤炭学报,37(5):788-793
张爱华,陶明信,刘朋阳,陈祥瑞,李丹丹. 2016. 煤中氮的赋存状态与含量分布研究进展. 煤田地质与勘探,44(1):9-16
张爱华,陶明信,陈祥瑞,李丹丹. 2018. 化石燃料的氮同位素地球化学研究概述. 地质学报,89(S1):231-232
张福凯,徐龙君,张丁月. 2014. 脱灰煤基活性炭吸附处理含镉废水. 环境工程学报,8(2):559-562
张珂,张绍韡,王珍珍,李进孝,李鸿豆,左贵彬,刘帮军,赵存良. 2020. 陕西省咸阳市彬长矿区胡家河4号煤有机地球化学特征. 矿物岩石地球化学通报,39(3):587-596
张锐,夏阳超,谭金龙,丁世豪,邢耀文,桂夏辉. 2018. 低阶煤分子碳结构的分析与研究. 中国煤炭,44(12):88-94,116
张水昌. 2010. 我国有机地球化学研究现状、发展方向和展望——第十二届全国有机地球化学学术会议部分总结. 石油与天然气地质,31(3):265-270,276
张小东,张鹏. 2014. 不同煤级煤分级萃取后的XRD结构特征及其演化机理. 煤炭学报,39(5):941-946
张亚婷. 2015. 煤基石墨烯的制备、修饰及应用研究. 西安:西安科技大学博士学位论文

张玉龙. 2014. 基于宏观表现与微观特性的煤低温氧化机理及其应用研究. 太原:太原理工大学博士学位论文

张志刚. 2018. 不同变质程度煤样热解气体产物逸出规律研究. 煤炭技术,37(7):293-295

赵路正. 2015. 煤矿区煤层气利用途径技术-经济-环境综合评价. 中国煤层气,12(6):42-46

郑启明. 2012. 煤层中氮及其成岩转化机理研究. 北京:中国矿业大学(北京)博士学位论文

中国大百科全书总编辑委员会. 1993. 中国大百科全书—地质学. 北京:中国大百科全书出版社

中国科学院地球化学研究所. 1998. 高等地球化学. 北京:科学出版社

中国煤炭地质总局. 2016. 中国煤炭资源赋存规律与资源评价. 北京:科学出版社

中国煤炭工业协会. 2010. GB/T 5751-2009 中国煤炭分类. 北京:中国标准出版社

中华人民共和国国家质量监督检验检疫总局. 2000. GB/T 18023-2000 烟煤的宏观煤岩类型分类. 北京:中国标准出版社

中华人民共和国国家质量监督检验检疫总局. 2018. SN/T 4764-2017 煤中碳、氢、氮、硫含量的测定元素分析仪法. 北京:中国标准出版社

中华人民共和国国家质量监督检验检疫总局,中国国家标准化管理委员会. 2014a. GB/T30050-2013 煤体结构分类. 北京:中国标准出版社

中华人民共和国国家质量监督检验检疫总局,中国国家标准化管理委员会. 2014b. GB/T30733-2014 煤中碳氢氮的测定 仪器法. 北京:中国标准出版社

中华人民共和国国家质量监督检验检疫总局,中国国家标准化管理委员会. 2018. GB/T15588-2013 烟煤显微组分分类. 北京:中国标准出版社

周国庆,姜尧发,颜跃进,王绍清. 2019. 江西乐平树皮煤的显微煤岩类型研究. 中国煤炭地质,31(7):7-11

Achten C, Andersson J T. 2015. Overview of polycyclic aromatic compounds (PAC). Polycyclic Aromatic Compounds,35(2-4):177-186

Alpern B, de Sousa M J L. 2002. Documented international enquiry on solid sedimentary fossil fuels, coal: definitions, classifications, reserves-resources, and energy potential. International Journal of Coal Geology,50(1-4):3-41

ASTM International. 2016. ASTM D5373-16 Standard test methods for determination of carbon, hydrogen and nitrogen in analysis samples of coal and carbon in analysis samples of coal and coke. Washington, DC, USA: Annual Book of ASTM Standards

ASTM International. 2019. ASTM D5601-19 Standard test method for microscopical determination of the textural components of metallurgical coke. West Conshohocken, P A, USA: Annual Book of ASTM Standards

Awasthi S, Awasthi K, Ghosh A K, Srivastava S K, Srivastava O N. 2015. Formation of single and multi-walled carbon nanotubes and graphene from Indian bituminous coal. Fuel,147:35-42

Ayinla H A, Abdullah W H, Makeen Y M, Abubakar M B, Jauro A, Yandoka B M S, Mustapha K A, Abidin N S Z. 2017. Source rock characteristics, depositional setting and hydrocarbon generation potential of Cretaceous coals and organic rich mudstones from Gombe Formation, Gongola Sub-basin, northern Benue Trough, NE Nigeria. International Journal of Coal Geology,173:212-226

Baysal M, Yürüm A, Yıldız B, Yürüm Y. 2016. Structure of some western Anatolia coals investigated by FTIR, Raman, ^{13}C solid state NMR spectroscopy and X-ray diffraction. International Journal of Coal Geology,163:166-176

Bechtel A, Püttmann W. 2018. Biomarkers: coal. In: White W M (ed). Encyclopedia of Geochemistry. Cham: Springer,123-135

Bechtel A, Gratzer R, Sachsenhofer R F, Gusterhuber J, Lücke A, Püttmann W. 2008. Biomarker and carbon isotope variation in coal and fossil wood of central Europe through the Cenozoic. Palaeogeography Palaeoclimatology Palaeoecology,262:166-175

Berlendis S, Beyssac O, Derenne S, Benzerara K, Anquetil C, Guillaumet M, Estève I, Capelle B. 2014. Comparative mineralogy, organic geochemistry and microbial diversity of the Autun black shale and Graissessac coal (France). International Journal of Coal Geology,132:147-157

Bhattacharya S, Khan M A, More S, Paruya K D, Chakraborty T, Bera S, Dutta S. 2017. Amber embalms essential oils: a rare preservation of monoterpenoids in fossil resins from eastern Himalaya. Palaios,33:218-227

Böcker J, Littke R, Hartkopf-Fröder C, Jasper K, Schwarzbauer J. 2013. Organic geochemistry of Duckmantian (Pennsylvanian) coals from the Ruhr Basin, western Germany. International Journal of Coal Geology,107:112-126

Bulat A F, Mineev S P, Prusova A A. 2016. Generating methane adsorption under relaxation of molecular structure of coal. Journal of Mining Science,52(1):70-77

Bustin R M, Rouzaud J N, Ross J V. 1995. Natural graphitization of anthracite: experimental considerations. Carbon,33(5):679-691

Canfield D E. 2013. Sulfur isotopes in coal constrain the evolution of the Phanerozoic sulfur cycle. Proceedings of the National Academy of Sciences of the United States of America,110(21):8443-8446

Cao Y, Fatemi V, Demir A, Fang S, Tomarken S L, Luo J Y, Sanchez-Yamagishi J D, Watanabe K, Taniguchi T, Kaxiras E, Ashoori R C, Jarillo-Herrero P. 2018a. Correlated insulator behaviour at half-filling in magic-angle graphene superlattices. Nature,556(7699):80-84

Cao Y, Fatemi V, Fang S, Watanabe K, Taniguchi T, Kaxiras E, Jarillo-Herrero P. 2018b. Unconventional superconductivity in magic-angle graphene superlattices. Nature,556(7699):43-50

Cartz L, Diamond R, Hirsch P B. 1956. New X-ray data on coals. Nature,177(4507):500-502

Castro-Marcano F, Lobodin V V, Rodgers R P, McKenna A M, Marshall A G, Mathews J P. 2012. A molecular model for illinois No.6 argonne premium coal: moving toward capturing the continuum structure. Fuel,95:35-49

Chaves D, Fernández-Robles L, Bernal J, Alegre E, Trujillo M. 2018. Automatic characterisation of chars from the combustion of pulverised coals using machine vision. Powder Technology,338:110-118

Chen L, Yang J L, Liu M X. 2011. Kinetic modeling of coal swelling in solvent. Industrial & Engineering Chemistry Research,50(5):2562-2568

Chen S D, Tao S, Tang, D Z, Xu H, Li S, Zhao J L, Jiang Q, Yang H X. 2017. Pore structure characterization of different rank coals using N_2 and CO_2 adsorption and its effect on CH$_4$ adsorption capacity: a case in panguan syncline, western Guizhou, China. Energy & Fuels, 31(6): 6034-6044

Chen Y Y, Mastalerz M, Schimmelmann A. 2012. Characterization of chemical functional groups in macerals across different coal ranks via micro-FTIR spectroscopy. International Journal of Coal Geology, 104: 22-33

Chen Y Y, Zou C N, Mastalerz M, Hu S Y, Gasaway C, Tao X W. 2015. Applications of micro-fourier transform infrared spectroscopy (FTIR) in the geological sciences—a review. International Journal of Molecular Sciences, 16(12): 30223-30250

Cheng X, Hou D J, Zhou X H, Liu J S, Diao H, Jiang Y H, Yu Z K. 2020. Organic geochemistry and kinetics for natural gas generation from mudstone and coal in the Xihu Sag, East China Sea Shelf Basin, China. Marine and Petroleum Geology, 118: 104405

Cheng Y P, Pan Z J. 2020. Reservoir properties of Chinese tectonic coal: a review. Fuel, 260: 116350

Chou C L. 1990. Geochemistry of sulfur in coal. In: Orr W L, White C M (eds). Geochemistry of Sulfur in Fossil Fuels. Washington, DC: American Chemical Society, 30-52

Chou C L. 2012. Sulfur in coals: a review of geochemistry and origins. International Journal of Coal Geology, 100: 1-13

Clarkson C R, Solano N, Bustin R M, Bustin A M M, Chalmers G R L, He L, Melnichenko Y B, Radliński A P, Blach T P. 2013. Pore structure characterization of North American shale gas reservoirs using USANS/SANS, gas adsorption, and mercury intrusion. Fuel, 103: 606-616

Dai S F, Ren D Y, Tang Y G, Shao L Y, Li S S. 2002. Distribution, isotopic variation and origin of sulfur in coals in the Wuda coalfield, Inner Mongolia, China. International Journal of Coal Geology, 51(4): 237-250

Dai S F, Zou J H, Jiang Y F, Ward C R, Wang X B, Li T, Xue W F, Liu S D, Tian H M, Sun X H, Zhou D. 2012. Mineralogical and geochemical compositions of the Pennsylvanian coal in the Adaohai Mine, Daqingshan coalfield, Inner Mongolia, China: modes of occurrence and origin of diaspore, gorceixite, and ammonian illite. International Journal of Coal Geology, 94: 250-270

Dai S F, Hower J C, Finkelman R B, Graham I T, French D, Ward C R, Eskenazy G, Wei Q, Zhao L. 2020a. Organic associations of non-mineral elements in coal: a review. International Journal of Coal Geology, 218: 103347

Dai S F, Bechtel A, Eble C F, Flores R M, French D, Graham I T, Hood M M, Hower J C, Korasidis V A, Moore T A, Püttmann W, Wei Q, Zhao L, O'Keefe J M K. 2020b. Recognition of peat depositional environments in coal: a review. International Journal of Coal Geology, 219: 103383

Diefendorf A F, Freeman K H, Wing S L, Graham H V, 2011. Production of n-alkyl lipids in living plants and implications for the geologic past. Geochimica Cosmochimica Acta, 75: 7472-7485

Ding D S, Liu G J, Fu B, Yuan Z J, Chen B Y. 2018a. Influence of magmatic intrusions on organic nitrogen in coal: a case study from the Zhuji mine, the Huainan coalfield, China. Fuel, 219: 88-93

Ding D S, Liu G J, Fu B, Wang W J. 2018b. New Insights into the nitrogen isotope compositions in coals from the Huainan coalfield, Anhui Province, China: influence of the distribution of nitrogen forms. Energy & Fuels, 32(9): 9380-9387

Ding D S, Liu G J, Fu B. 2019. Influence of carbon type on carbon isotopic composition of coal from the perspective of solid-state ^{13}C NMR. Fuel, 245: 174-180

Dong J, Li F, Xie K C. 2012. Study on the source of polycyclic aromatic hydrocarbons (PAHs) during coal pyrolysis by PY-GC-MS. Journal of Hazardous Materials, 243: 80-85

Dutta S, Mehrotra R C, Paul S, Tiwari S R P, Bhattacharya S, Srivastava G, Ralte V Z, Zoramthara C. 2017. Remarkable preservation of terpenoids and record of volatile signalling in plant-animal interactions from Miocene amber. Sci Rep UK, 7

Eble C F, Greb S F, Williams D A, Hower J C, O'Keefe J M K. 2019. Palynology, organic petrology and geochemistry of the Bell coal bed in western Kentucky, eastern Interior (Illinois) Basin, USA. International Journal of Coal Geology, 213: 103264

El-Dars F M S E, Ibrahim M A, Gabr A M E. 2014. Reduction of COD in water-based paint wastewater using three types of activated carbon. Desalination and Water Treatment, 52(16-18): 2975-2986

Emsbo-Mattingly S D, Stout S A. 2011. Semivolatile hydrocarbon residues of coal and coal tar. In: Stracher G B, Prakash A, Sokol E V (eds). Coal and Peat Fires: A Global Perspective. Amsterdam: Elsevier, 173-208

Esmeijer-Liu A J, Kürschner W M, Lotter A F, Verhoeven J T A, Goslar T. 2012. Stable carbon and nitrogen isotopes in a peat profile are influenced by early stage diagenesis and changes in atmospheric CO$_2$ and N deposition. Water, Air, & Soil Pollution, 223(5): 2007-2022

Fabiańska M J, Ćmiel S R, Misz-Kennan M. 2013. Biomarkers and aromatic hydrocarbons in bituminous coals of Upper Silesian Coal Basin: example from 405 coaldeam oft he Zaleskie Beds (Poland). International Journal of Coal Geology, 107: 96-111

Fan X, Jiang J, Chen L, Zhou C C, Zhu J L, Zhu T G, Wei X Y. 2016. Identification of organic fluorides and distribution of organic species in an anthracite with high content of fluorine. Fuel Processing Technology, 142: 54-58

Feng J, Li J, Li W Y. 2013. Influences of chemical structure and physical properties of coal macerals on coal liquefaction by quantum chemistry calculation. Fuel Processing Technology, 109: 19-26

Finkelman R B, Dai S F, French D. 2019. The importance of minerals in coal as the hosts of chemical elements: a review. International Journal of Coal Geology, 212: 103251

Ghosh S, Rodrigues S, Varma A K, Esterle J, Patra S, Dirghangi S S. 2018. Petrographic and Raman spectroscopic characterization of coal from Himalayan fold-thrust belts of Sikkim, India. International Journal of Coal Geology, 196: 246-259

Golding S D, Boreham C J, Esterle J S. 2013. Stable isotope geochemistry of coal bed and shale gas and related production waters: a review. International Journal of Coal Geology, 120: 24-40

Gray R J, Devanney K F. 1986. Coke carbon forms: microscopic classification and industrial applications. International Journal of Coal Geology, 6(3):

277-297

Guo X,Tang Y G,Wang Y F,Eble C F,Finkelman R B,Li P Y. 2020a. Evaluation of carbon forms and elements composition in coal gasification solid residues and their potential utilization from a view of coal geology. Waste Management,114:287-298

Guo X,Tang Y G,Eble C F,Wang Y F,Li P Y. 2020b. Study on petrographic characteristics of devolatilization char/coke related to coal rank and coal maceral. International Journal of Coal Geology,227:103504

Hakimi M H,Abdullah W H. 2014. Biological markers and carbon isotope composition of organic matter in the Upper Cretaceous coals and carbonaceous shale succession (Jiza-Qamar Basin,Yemen):origin,type and preservation. Palaeogeography,Palaeoclimatology,Palaeoecology,409:84-97

He X Q,Liu X F,Nie B S,Song D Z. 2017. FTIR and Raman spectroscopy characterization of functional groups in various rank coals. Fuel,206:555-563

Heidrich C,Feuerborn H J,Weir A. 2013. Coal combustion products:a global perspective. In:Proceedings of 2013 World of Coal Ash (WOCA) Conference. Lexington,K Y

Hirsch P B. 1954. X-ray scattering from coals. Proceedings of the Royal Society A:Mathematical,Physical and Engineering Sciences,226(1165):143-169

Hoş-Çebi F,Korkmaz S. 2013. Organic geochemistry and depositional environments of Eocene coals in northern Anatolia,Turkey. Fuel,113:481-496

Hower J C. 2012. Petrographic examination of coal-combustion fly ash. International Journal of Coal Geology,92:90-97

Hower J C,Wagner N J. 2012. Notes on the methods of the combined maceral/microlithotype determination in coal. International Journal of Coal Geology,95:47-53

Hower J C,Suárez-Ruiz I,Mastalerz M,Cook A C. 2007. The investigation of chemical structure of coal macerals via transmitted-light FT-IR microscopy by X. Sun. Spectrochimica Acta Part A:Molecular and Biomolecular Spectroscopy,67(5):1433-1437

Hower J C,O'Keefe J M K,Wagner N J,Dai S F,Wang X B,Xue W F. 2013a. An investigation of Wulantuga coal (Cretaceous,Inner Mongolia) macerals:paleopathology of faunal and fungal invasions into wood and the recognizable clues for their activity. International Journal of Coal Geology,114:44-53

Hower J C,Misz-Keenan M,O'Keefe J M K,Mastalerz M,Eble C F,Garrison T M,Johnston M N,Stucker J D. 2013b. Macrinite forms in Pennsylvanian coals. International Journal of Coal Geology,116-117:172-181

Hower J C,Groppo J G,Graham U M,Ward C R,Kostova I J,Maroto-Valer M M,Dai S F. 2017. Coal-derived unburned carbons in fly ash:a review. International Journal of Coal Geology,179:11-27

Huan X,Tang Y G,Xu J J,Lan C Y,Wang S Q. 2019. Structural characterization of graphenic material prepared from anthracites of different characteristics:a comparative analysis. Fuel Processing Technology,183:8-18

Huan X,Tang Y G,Xu J J,Xu M X. 2020. Nano-level resolution determination of aromatic nucleus in coal. Fuel,262:116532

Huang X Y,Meyers P A,Xue J T,Gong L F,Wang X X,Xie S C. 2015. Environmental factors affecting the low temperature isomerization of homohopanes in acidic peat deposits,central China. Geochimica Cosmochimica Acta,154:212-228

Inglis G N,Naafs B D A,Zeng Y,McClymont E L,Evershed R P,Pancost R D,et al. 2018. Distribution of geohopanoids in peat:implications fort he use of hopanoid-based proxies in natural archives. Geochimica Cosmochimica Acta,224:249-261

International Committee for Coal and Organic Petrology (ICCP). 1998. The new vitrinite classification (ICCP System 1994). Fuel,77(5):349-358

International Committee for Coal and Organic Petrology (ICCP). 2001. The new inertinite classification (ICCP System 1994). Fuel,80(4):459-471

Izart A,Palhol F,Gleixner G,Elie M,Blaise T,Suarez-Ruiz I,Sachsenhofer R F,Privalov V A,Panova E A. 2012. Palaeoclimate reconstruction from biomarker geochemistry and stable isotopes of n-alkanes from carboniferous and Early Permian humic coals and limnic sediments in western and eastern Europe. Organic Geochemistry,43:125-149

Izart A,Suarez-Ruiz I,Bailey J. 2015. Paleoclimate reconstruction from petrography and biomarker geochemistry from Permian humic coals in Sydney Coal Basin (Australia). International Journal of Coal Geology,138:145-157

Jiang J Y,Yang W H,Cheng Y P,Liu Z D,Zhang Q,Zhao K. 2019. Molecular structure characterization of middle-high rank coal via XRD,Raman and FTIR spectroscopy:implications for coalification. Fuel,239:559-572

Jiao L Y,Zhang L,Wang X R,Diankov G,Dai H J. 2009. Narrow graphene nanoribbons from carbon nanotubes. Nature,458(7240):877-880

Jing Z H,Rodrigues S,Strounina E,Li M R,Wood B,Underschultz J R,Esterle J S,Steel K M. 2019. Use of FTIR,XPS,NMR to characterize oxidative effects of NaClO on coal molecular structures. International Journal of Coal Geology,201:1-13

Ju Y W,Sun Y,Tan J Q,Bu H L,Han K,Li X S,Fang L Z. 2018. The composition,pore structure characterization and deformation mechanism of coal-bearing shales from tectonically altered coalfields in eastern China. Fuel,234:626-642

Kelemen S R,Sansone M,Walters C C,Kwiatek P J,Bolin T. 2012. Thermal transformations of organic and inorganic sulfur in Type II kerogen quantified by S-XANES. Geochimica et Cosmochimica Acta,83:61-78

Kong J,Zhao R F,Bai Y H,Li G L,Zhang C,Li F. 2014. Study on the formation of phenols during coal flash pyrolysis using pyrolysis-GC/MS. Fuel Processing Technology,127:41-46

Kronemayer H,Barzan D,Horiuchi M,Suganuma S,Tokutake Y,Schulz C,Bessler W G. 2007. A direct-flame solid oxide fuel cell (DFFC) operated on methane,propane,and butane. Journal of Power Sources,166(1):120-126

Lane C S. 2017. Modern n-alkane abundances and isotopic composition of vegetation in a gymnosperm-dominated ecosystem of the southeastern U. S. coastal plain. Organic Geochemistry,105:33-36

Laumann S, Micić V, Kruge M A, Achten C, Sachsenhofer R F, Schwarzbauer J, Hofmann T. 2011. Variations in concentrations and compositions of polycyclic aromatic hydrocarbons (PAHs) in coals related to the coal rank and origin. Environmental Pollution, 159(10):2690-2697

Lee C, Wei X D, Kysar J W, Hone J. 2008. Measurement of the elastic properties and intrinsic strength of monolayer graphene. Science, 321(5887):385-388

Lester E, Alvarez A, Borrego A G, Valentim B, Flores D, Clift D A, Rosenberg P, Kwiecinska B, Barranco R, Petersen H I, Mastalerz M, Milenkova K S, Panaitescu C, Marques M M, Thompson A, Watts D, Hanson S, Predeanu G, Misz M, Wu T. 2010. The procedure used to develop a coal char classification—Commission III Combustion Working Group of the International Committee for Coal and Organic Petrology. International Journal of Coal Geology, 81(4):333-342

Li F L, Jiang B, Cheng G X, Song Y, Tang Z. 2019. Structural and evolutionary characteristics of pores-microfractures and their influence on coalbed methane exploitation in high-rank brittle tectonically deformed coals of the Yangquan mining area, northeastern Qinshui Basin, China. Journal of Petroleum Science and Engineering, 174:1290-1302

Li M F, Liu W W, Zhang H X, Liang Z L, Duan P, Yan X L, Guan P F, Xu B S, Guo J J. 2018. Direct imaging of construction of carbon onions by curling few-layer graphene flakes. Physical Chemistry Chemical Physics, 20(3):2022-2027

Li W, Zhu Y M, Chen S B, Zhou Y. 2013. Research on the structural characteristics of vitrinite in different coal ranks. Fuel, 107:647-652

Li W, Liu H F, Song X X. 2015. Multifractal analysis of Hg pore size distributions of tectonically deformed coals. International Journal of Coal Geology, 144-145:138-152

Li W G, Gong X J, Li X, Zhang D Y, Gong H N. 2012. Removal of Cr(VI) from low-temperature micro-polluted surface water by tannic acid immobilized powdered activated carbon. Bioresource Technology, 113:106-113

Li W W, Tang Y G. 2014. Sulfur isotopic composition of superhigh-organic-sulfur coals from the Chenxi coalfield, southern China. International Journal of Coal Geology, 127:3-13

Li W W, Tang Y G, Zhao Q J, Wei Q. 2015. Sulfur and nitrogen in the high-sulfur coals of the Late Paleozoic from China. Fuel, 155:115-121

Li X S, Ju Y W, Hou Q L, Li Z, Wei M M, Fan J J. 2014. Characterization of coal porosity for naturally tectonically stressed coals in Huaibei coal field, China. The Scientific World Journal, 2014:560450

Li Y, Zhang C, Tang D Z, Gan Q, Niu X L, Wang K, Shen R Y. 2017. Coal pore size distributions controlled by the coalification process: an experimental study of coals from the Junggar, Ordos and Qinshui basins in China. Fuel, 206:352-363

Li Y, Yang J H, Pan Z J, Tong W S. 2020. Nanoscale pore structure and mechanical property analysis of coal: an insight combining AFM and SEM images. Fuel, 260:116352

Lievens C, Ci D H, Bai Y, Ma L G, Zhang R, Chen J Y, Gai Q Q, Long Y H, Guo X F. 2013. A study of slow pyrolysis of one low rank coal via pyrolysis-GC/MS. Fuel Processing Technology, 116:85-93

Lin Y H, Wang S Q, Sha Y M, Yang K. 2020. Organic geochemical characteristics of bark coal in Changguang area: evidence from aromatic hydrocarbons. International Journal of Coal Science & Technology, 7(2):288-298

Liu B J, Zhao C L, Ma J L, Sun Y Z, Püttmann W, 2018. The origin of pale and dark layers in Pliocene lignite deposits from Yunnan Province, Southwest China, based on coal petrological and organic geochemical analyses. International Journal of Coal Science and Technology, 195:172-188

Liu J X, Jiang Y Z, Yao W, Jiang X, Jiang X M. 2019. Molecular characterization of Henan anthracite coal. Energy Fuels, 33(7):6215-6225

Liu Y, Zhu Y M, Liu S M, Chen S B, Li W, Wang Y. 2018. Molecular structure controls on micropore evolution in coal vitrinite during coalification. International Journal of Coal Geology, 199:19-30

Lu G W, Wang J L, Wei C T, Song Y, Yan G Y, Zhang J J, Chen G H. 2018. Pore fractal model applicability and fractal characteristics of seepage and adsorption pores in middle rank tectonic deformed coals from the Huaibei coal field. Journal of Petroleum Science and Engineering, 171:808-817

Lu Y, Hautevelle Y, Michels R. 2013. Determination of the molecular signature of fossil conifers by experimental palaeochemotaxonomy-part 1: the Araucariaceae family. Biogeosciences, 10:1943-1962

Lyons P C, Cross A T. 2005. Marlies Teichmüller (1914-2000), pioneering genetic coal petrologist: some paleobotanical, palynological, and botanical influences on her research. International Journal of Coal Geology, 62(1-2):71-84

Maheshwari N, Krishna P K, Thakur I S, Srivastava S. 2019. Biological fixation of carbon dioxide and biodiesel production using microalgae isolated from sewage waste water. Environmental Science and Pollution Research, 27(22):27319-27329

Manoj B, Kunjomana A G. 2014. Systematic investigations of graphene layers in sub-bituminous coal. Russian Journal of Applied Chemistry, 87(11):1726-1733

Manoj B, Kunjomana A G. 2015. Structural characterization of graphene layers in various Indian coals by X-ray diffraction technique. IOP Conference Series: Materials Science and Engineering, 73:012096

Mardon S M, Eble C F, Hower J C, Takacs K, Mastalerz M, Bustin R M. 2014. Organic petrology, geochemistry, gas content and gas composition of Middle Pennsylvanian age coal beds in the eastern Interior (Illinois) Basin: implications for CBM development and carbon sequestration. International Journal of Coal Geology, 127:56-74

Martunus, Helwani Z, Wiheeb A D, Kim J, Othman M R. 2012. In situ carbon dioxide capture and fixation from a hot flue gas. International Journal of Greenhouse Gas Control, 6:179-188

Mastalerz M, Hower J C, Chen Y Y. 2015. Microanalysis of barkinite from Chinese coals of high volatile bituminous rank. International Journal of Coal Geology, 141-142:103-108

Mathews J P, Chaffee A L. 2012. The molecular representations of coal-a review. Fuel,96:1-14

Mathews J P, Sharma A. 2012. The structural alignment of coal and the analogous case of Argonne Upper Freeport coal. Fuel,95:19-24

Medunić G, Grigore M, Dai S F, Berti D, Hochella M F, Mastalerz M, Valentim B, Guedes A, Hower J C. 2020. Characterization of superhigh-organic-sulfur Raša coal, Istria, Croatia, and its environmental implication. International Journal of Coal Geology,217:103344

Moothi K, Iyuke S E, Meyyappan M, Rosemary F. 2012. Coal as a carbon source for carbon nanotube synthesis. Carbon,50(8):2679-2690

Mustapha K A, Abdullah W H, Konjing Z, Gee S S, Koraini A M. 2017. Organic geochemistry and palynology of coals and coal-bearing mangrove sediments of the Neogene Sandakan Formation, Northeast Sabah, Malaysia. Catena,158:30-45

Naafs B D A, Inglis G N, Blewett J, McClymont E L, Lauretano V, Xie S, Evershed R P, Pancost R D. 2019. The potential of biomarker proxies to trace climate, vegetation, and biogeochemical processes in peat: a review. Global and Planetary Change,179:57-79

Nie B S, Liu X F, Yang L L, Meng J Q, Li X C. 2015. Pore structure characterization of different rank coals using gas adsorption and scanning electron microscopy. Fuel,158:908-917

Niu B, Jin L J, Li Y, Shi Z W, Hu H Q. 2017. Isotope analysis for understanding the hydrogen transfer mechanism in direct liquefaction of Bulianta coal. Fuel,203:82-89

Novak M, Stepanova M, Jackova I, Vile M A, Wieder K, Buzek F, Adamova M, Erbanova L, Fottova D, Komarek A. 2014. Isotopic evidence for nitrogen mobility in peat bogs. Geochimica et Cosmochimica Acta,133:351-361

Novoselov K S, Fal'ko V I, Colombo L, Gellert P R, Schwab M G, Kim K. 2012. A roadmap for graphene. Nature,490(7419):192-200

Oberlin A. 1979. Application of dark-field electron microscopy to carbon study. Carbon,17(1):7-20

Ohta T, Bostwick A, Seyller T, Horn K, Rotenberg E. 2006. Controlling the electronic structure of bilayer graphene. Science,313(5789):951-954

Okolo G N, Everson R C, Neomagus H W J P, Roberts M J, Sakurovs R. 2015a. Comparing the porosity and surface areas of coal as measured by gas adsorption, mercury intrusion and SAXS techniques. Fuel,141:293-304

Okolo G N, Neomagus H W J P, Everson R C, Roberts M J, Bunt J R, Sakurovs R, Mathews J P. 2015b. Chemical-structural properties of South African bituminous coals: insights from wide angle XRD-carbon fraction analysis, ATR-FTIR, solid state ^{13}C NMR, and HRTEM techniques. Fuel,158:779-792

Oskay R G, Bechtel A, Karayiğit A I. 2019. Mineralogy, petrography and organic geochemistry of Miocene coal seams in the Kınık coalfield (Soma Basin-western Turkey): insights into depositional environment and palaeovegetation. International Journal of Coal Geology,210:103205

Pan J N, Wang S, Ju Y W, Hou Q L, Niu Q H, Wang K, Li M, Shi X H. 2015. Quantitative study of the macromolecular structures of tectonically deformed coal using high-resolution transmission electron microscopy. Journal of Natural Gas Science and Engineering,27:1852-1862

Pan J N, Wang K, Hou Q L, Niu Q H, Wang H C, Ji Z M. 2016. Micro-pores and fractures of coals analysed by field emission scanning electron microscopy and fractal theory. Fuel,164:277-285

Pan R K, Li C, Yu M G, Xiao Z J, Fu D. 2020. Evolution patterns of coal micro-structure in environments with different temperatures and oxygen conditions. Fuel,261:116425

Patra S, Dirghangi S S, Rudra A, Dutta S, Ghosh S, Varma A K, Shome D, Kalpana M S. 2018. Effects of thermal maturity on biomarker distributions in Gondwana coals from the Satpura and Damodar Valley Basins, India. International Journal of Coal Geology,196:63-81

PÉrillat-Merceroz C, Roussel P, Vannier R N, Gélin P, Rosini S, Gauthier G. 2011. Lamellar titanates: a breakthrough in the search for new solid oxide fuel cell anode materials operating on methane. Advanced Energy Materials,1(4):573-576

Pickel W, Kus J, Flores D, Kalaitzidis S, Christanis K, Cardott B J, Misz-Kennan M, Rodrigues S, Hentschel A, Hamor-Vido M, Crosdale P, Wagner N, ICCP. 2017. Classification of liptinite-ICCP system 1994. International Journal of Coal Geology,169:40-61

Powell C, Beall G W. 2015. Graphene oxide and graphene from low grade coal: synthesis, characterization and applications. Current Opinion in Colloid & Interface Science,20(5-6):362-366

Qi Y, Ju Y W, Tan J Q, Bowen L, Cai C F, Yu K, Zhu H J, Huang C, Zhang W L. 2020. Organic matter provenance and depositional environment of marine-to-continental mudstones and coals in eastern Ordos Basin, China—evidence from molecular geochemistry and petrology. International Journal of Coal Geology,217:103345

Qu Z H, Jiang B, Wang J L, Li M. 2017. Study of nanopores of tectonically deformed coal based on liquid nitrogen adsorption at low temperatures. Journal of Nanoscience and Nanotechnology,17(9):6566-6575

Ribeiro J, Silva T, Filho J G M, Flores D. 2012. Polycyclic aromatic hydrocarbons (PAHs) in burning and non-burning coal waste piles. Journal of Hazardous Materials,199-200:105-110

Rivera-Utrilla J, Sánchez-Polo M, Gómez-Serrano V, Álvarez P M, Alvim-Ferraz M C M, Dias J M. 2011. Activated carbon modifications to enhance its water treatment applications: an overview. Journal of Hazardous Materials,187(1-3):1-23

Roberts M J, Everson R C, Neomagus H W J P, van Niekerk D, Mathews J P, Branken D J. 2015. Influence of maceral composition on the structure, properties and behaviour of chars derived from South African coals. Fuel,142:9-20

Saikia B K. 2010. Inference on carbon atom arrangement in the turbostatic graphene layers in Tikak coal (India) by X-ray pair distribution function analysis. International Journal of Oil, Gas and Coal Technology,3(4):362-373

Saikia B K, Boruah R K, Gogoi P K. 2009. A X-ray diffraction analysis on graphene layers of Assam coal. Journal of Chemical Sciences,121(1):103-106

Schwarzbauer J, Littke R, Meier R, Strauss H. 2013. Stable carbon isotope ratios of aliphatic biomarkers in Late Palaeozoic coals. International Journal of

Coal Geology,107:127-140

Sharma A,Kyotani T,Tomita A. 2000a. Direct observation of layered structure of coals by a transmission electron microscope. Energy & Fuels,14(2):515-516

Sharma A,Kyotani T,Tomita A. 2000b. Direct observation of raw coals in lattice fringe mode using high-resolution transmission electron microscopy. Energy & Fuels,14(6):1219-1225

Shine K P,Fuglestvedt J S,Hailemariam K,Stuber N. 2005. Alternatives to the global warming potential for comparing climate impacts of emissions of greenhouse gases. Climatic Change,68(3):281-302

Sim M S,Bosak T,Ono S. 2011. Large sulfur isotope fractionation does not require disproportionation. Science,333(6038):74-77

Singh R,Tiwari H P. 2020. Microscopic evaluation of inertinite in the coke micro-structure:a case study from Jharia coalfield,India. International Journal of Coal Preparation and Utilization,40(1):1-11

Solano-Acosta W,Schimmelmann A,Mastalerz M,Arango I. 2008. Diagenetic mineralization in Pennsylvanian coals from Indiana,USA:$^{13}C/^{12}C$ and $^{18}O/^{16}O$ implications for cleat origin and coalbed methane generation. International Journal of Coal Geology,73(3-4):219-236

Song Y,Zhu Y M,Li W. 2017a. Macromolecule simulation and CH_4 adsorption mechanism of coal vitrinite. Applied Surface Science,396:291-302

Song Y,Jiang B,Liu J G. 2017b. Nanopore structural characteristics and their impact on methane adsorption and diffusion in low to medium tectonically deformed coals:case study in the Huaibei coal field. Energy & Fuels,31(7):6711-6723

Song Y,Jiang B,Li M,Hou C L,Xu S C. 2020. A review on pore-fractures in tectonically deformed coals. Fuel,278:118248

Stojanović K,Životić D. 2013. Comparative study of Serbian Miocene coals insights from biomarker composition. International Journal of Coal Geology,107:3-23

Strobl S A I,Sachsenhofer R F,Bechtel A,Meng Q T. 2014. Paleoenvironment of the Eocene coal seam in the Fushun Basin (NE China):implications from petrography and organic geochemistry. International Journal of Coal Geology,134-135:24-37

Suárez-Ruiz I,Valentim B,Borrego A G,Bouzinos A,Flores D,Kalaitzidis S,Malinconico M L,Marques M,Misz-Kennan M,Predeanu G,Montes J R,Rodrigues S,Siavalas G,Wagner N. 2017. Development of a petrographic classification of fly-ash components from coal combustion and co-combustion. (An ICCP Classification System,Fly-Ash Working Group-Commission III.). International Journal of Coal Geology,183:188-203

Sun Y Q,Alemany L B,Billups W E,Lu J X,Yakobson B I. 2011. Structural dislocations in anthracite. Journal of Physical Chemistry Letters,2(20):2521-2524

Suto N,Kawashima H. 2016. Global mapping of carbon isotope ratios in coal. Journal of Geochemical Exploration,167:12-19

Swansona S M,Mastalerzb M D,Englea M,Valentinea B J,Warwicka P D,Hackleya P C,Belkina H E. 2015. Pore characteristics of wilcox group coal,U.S. gulf coast region:implications for the occurrence of coalbed gas. International Journal of Coal Geology,139:80-94

Sýkorová I,Pickel W,Christanis K,Wolf M,Taylor G H,Flores D. 2005. Classification of huminite-ICCP System 1994. International Journal of Coal Geology,62(1-2):85-106

Tang X,Ripepi N,Gilliland E. 2016. Isothermal adsorption kinetics properties of carbon dioxide in crushed coal. Greenhouse Gases:Science and Technology,6(2):260-274

Tang Y G,Huan X,Lan C Y,Xu M X. 2018. Effects of coal rank and high organic sulfur on the structure and optical properties of coal-based graphene quantum dots. Acta Geologica Sinica,92(3):1218-1230

Tang Y G,Li R Q,Wang S Q. 2020a. Research progress and prospects of coal petrology and coal quality in China. International Journal of Coal Science & Technology,7(2):273-287

Tang Y G,Sun Y W,Wang X S,Yan L L,Shi Q,Ni H X,Li W W,Li X L,Finkelman R B,Pang X P. 2020b. Composition and structure of the sulfur-containing compounds in the extracts from the Chinese high-organic-sulfur coals. Energy & Fuels,34(9):10666-10675

Tao S,Chen S D,Tang D Z,Zhao X,Xu H,Li S. 2018. Material composition,pore structure and adsorption capacity of low-rank coals around the first coalification jump:a case of eastern Junggar Basin,China. Fuel,211:804-815

Taylor G H,Teichmüller M,Davis A,Diessel C F K,Littke R,Robert P. 1998. Organic Petrology. Berlin-Stuttgart:Gebrüder Borntraeger,149-150

Teichmüller M,Teichmüller R. 1949. Inkohlungsfragen im Ruhrkarbon. Zeitschrift der Deutschen Geologischen Gesellschaft,99:40-77

Teichmüller M,Teichmüller R. 1958. Inkohlungsuntersuchungen und ihre Nutzanwendung. Geologie en Mijnbouw. 20(2):41-66

Tewari A,Dutta S,Sarkar T. 2017. Biomarker signatures of Permian Gondwana coals from India and their palaeobotanical significance. Palaeogeography,Palaeoclimatology,Palaeoecology,468:414-426

Thomas L. 2013. Coal Geology. Chichester,West Sussex,England:John Wiley & Sons,Ltd

Thommes M,Cychosz K A. 2014. Physical adsorption characterization of nanoporous materials:progress and challenges. Adsorption,20(2-3):233-250

Tselev A,Ivanov I N,Lavrik N V,Belianinov A,Jesse S,Mathews J P,Mitchell G D,Kalinin S V. 2014. Mapping internal structure of coal by confocal micro-Raman spectroscopy and scanning microwave microscopy. Fuel,126:32-37

Ulyanova E V,Molchanov A N,Prokhorov I Y,Grinyov V G. 2014. Fine structure of Raman spectra in coals of different rank. International Journal of Coal Geology,121:37-43

Valentim B. 2020. Petrography of coal combustion char:a review. Fuel,277:118271

van Krevelen D W. 1950. Graphical-statistical method for the study of structure and reaction processes of coal. Fuel,29:269-284

van Krevelen D W. 1961. Coal:Typology-Physics-Chemistry-Constitution. New York:Elsevier Science Publishers Company

van Krevelen D W. 1993. Coal:Typology-Physics-Chemistry-Constitution,3rd Edition. New York:Elsevier Science Publishers Company van Niekerk D, Mathews J P. 2010. Molecular representations of Permian-aged vitrinite-rich and inertinite-rich South African coals. Fuel, 89(1):73–82

Vijapur S H, Wang D, Botte G G. 2013. Raw coal derived large area and transparent graphene films. ECS Solid State Letters, 2(7):M45–M47

Vu T T A, Horsfield B, Mahlstedt N, Schenk H J, Kelemen S R, Walters C C, Kwiatek P J, Sykes R. 2013. The structural evolution of organic matter during maturation of coals and its impact on petroleum potential and feedstock for the deep biosphere. Organic Geochemistry, 62:17–27

Walters W W, Tharp B D, Fang H, Kozak B J, Michalski M. 2015. Nitrogen isotope composition of thermally produced NO_x from various fossil-fuel combustion sources. Environmental Science & Technology, 49(19):11363–11371

Wang C A, Huddle T, Lester E H, Mathews J P. 2016. Quantifying curvature in high-resolution transmission electron microscopy lattice fringe micrographs of coals. Energy & Fuels, 30(4):2694–2704

Wang C A, Huddle T, Huang C H, Zhu W B, Wal R L V, Lester E H, Mathews J P. 2017. Improved quantification of curvature in high-resolution transmission electron microscopy lattice fringe micrographs of soots. Carbon, 117:174–181

Wang F, Cheng Y P, Lu S Q, Jin K, Zhao W. 2014. Influence of coalification on the pore characteristics of middle-high rank coal. Energy & Fuels, 28(9):5729–5736

Wang K, Du F, Wang G D. 2017. The influence of methane and CO_2 adsorption on the functional groups of coals: Insights from a Fourier transform infrared investigation. Journal of Natural Gas Science and Engineering, 45:358–367

Wang L, Zhang H, Li Y. 2020. On the difference of characterization and supercapacitive performance of graphene nanosheets from precursors of inertinite- and vitrinite-rich coal. Journal of Alloys and Compounds, 815:152502

Wang R W, Liu G J, Zhang J M, Chou C L, Liu J J. 2010. Abundances of polycyclic aromatic hydrocarbons (PAHs) in 14 Chinese and American coals and their relation to coal rank and weathering. Energy & Fuels, 24(11):6061–6066

Wang R W, Sun R Y, Liu G J, Yousaf B, Wu D, Chen J, Zhang H. 2017. A review of the biogeochemical controls on the occurrence and distribution of polycyclic aromatic compounds (PACs) in coals. Earth-Science Reviews, 171:400–418

Wang S Q, Tang Y G, Schobert H H, Guo Y N, Su Y F. 2011. FTIR and ^{13}C-NMR investigation of coal component of Late Permian coals from southern China. Energy & Fuels, 25(12):5672–5677

Wang S Q, Tang Y G, Schobert H H, Guo Y N, Gao W C, Lu X K. 2013. FTIR and simultaneous TG/MS/FTIR study of Late Permian coals from southern China. Journal of Analytical and Applied Pyrolysis, 100:75–80

Wang S Q, Cheng H F, Jiang D, Huang F, Su S, Bai H P. 2014. Raman spectroscopy of coal component of Late Permian coals from southern China. Spectrochimica Acta Part A: Molecular and Biomolecular Spectroscopy, 132:767–770

Wang S Q, Liu S M, Sun Y B, Jiang D, Zhang X M. 2017a. Investigation of coal components of Late Permian different ranks bark coal using AFM and Micro-FTIR. Fuel, 187:51–57

Wang S Q, Chen H, Ma W, Liu P H, Yang Z D. 2017b. Structural transformations of coal components upon heat treatment and explanation on their abnormal thermal behaviors. Energy Fuels, 31(11):11587–11593

Wang S Q, Tang Y G, Schobert H H, Jiang Y F, Yang Z D, Zhang X M. 2018a. Petrologic and organic geochemical characteristics of Late Permian bark coal in Mingshan coalmine, southern China. Marine and Petroleum Geology, 93:205–217

Wang S Q, Tang Y G, Chen H, Liu P H, Sha Y M. 2018b. Chemical structural transformations of different coal components at the similar coal rank by HRTEM *in situ* heating. Fuel, 218:140–147

Wang S Q, Chen H, Zhang X M. 2020. Transformation of aromatic structure of vitrinite with different coal ranks by HRTEM *in situ* heating. Fuel, 260:1–16309

Wang X F, Liu W H, Shi B G, Zhang Z N, Xu Y C, Zheng J J. 2015. Hydrogen isotope characteristics of thermogenic methane in Chinese sedimentary basins. Organic Geochemistry, 83-84:178–189

Wang Y H, Lian J, Xue Y, Liu P, Dai B, Lin H L, Han S. 2020. The pyrolysis of vitrinite and inertinite by a combination of quantum chemistry calculation and thermogravimetry-mass spectrometry. Fuel, 264:116794

Warwick P D, Ruppert L F. 2016. Carbon and oxygen isotopic composition of coal and carbon dioxide derived from laboratory coal combustion: a preliminary study. International Journal of Coal Geology, 166:128–135

Wei Q, Cui C N, Dai S F. 2020. Organic-association of Ge in the coal-hosted ore deposits: an experimental and theoretical approach. Ore Geology Reviews, 117:103291

Wei X Y, Wang X H, Zong Z M, Ni Z H, Zhang L F, Ji Y F, Xie K C, Lee C W, Liu Z X, Chu N B, Cui J Y. 2004. Identification of organochlorines and organobromines in coals. Fuel, 83(17-18):2435–2438

Weishauptová Z, Sýkorová I. 2011. Dependence of carbon dioxide sorption on the petrographic composition of bituminous coals from the Czech part of the Upper Silesian Basin, Czech Republic. Fuel, 90(1):312–323

Xia W C, Yang J G, Liang C. 2014. Investigation of changes in surface properties of bituminous coal during natural weathering processes by XPS and SEM. Applied Surface Science, 293:293–298

Xia Y C, Zhang R, Cao Y J, Xing Y W, Gui X H. 2019. Role of molecular simulation in understanding the mechanism of low-rank coal flotation: a review. Fuel, 262:116535

Xiao H Y, Liu C Q. 2011. The elemental and isotopic composition of sulfur and nitrogen in Chinese coals. Organic Geochemistry, 42(1):84–93

Xin F D, Xu H, Tang D Z, Yang J S, Chen Y P, Cao L K, Qu H X. 2019. Pore structure evolution of low-rank coal in China. International Journal of Coal Geology, 205:126–139

Xing Y W, Guo F Y, Xu M D, Gui X H, Li H S, Li G S, Xia Y C, Han H S. 2019. Separation of unburned carbon from coal fly ash: a review. Powder Technology, 353: 372-384

Xu H, Lin Q L, Zhou T H, Chen T T, Lin S P, Dong S H. 2014. Facile preparation of graphene nanosheets by pyrolysis of coal-tar pitch with the presence of aluminum. Journal of Analytical and Applied Pyrolysis, 110: 481-485

Xu J, Tang H, Su S, Liu J W, Qian K, Wang Y, Zhou Y B, Hu S, Zhang A C, Xiang J. 2018. A study of the relationships between coal structures and combustion characteristics: the insights from micro-Raman spectroscopy based on 32 kinds of Chinese coals. Applied Energy, 212: 46-56

Yan J C, Bai Z Q, Bai J, Guo Z X, Li W. 2014. Effects of organic solvent treatment on the chemical structure and pyrolysis reactivity of brown coal. Fuel, 128: 39-45

Yan J C, Lei Z P, Li Z K, Wang Z C, Ren S B, Kang S G, Wang X L, Shui H F. 2020. Molecular structure characterization of low-medium rank coals via XRD, solid state ^{13}C NMR and FTIR spectroscopy. Fuel, 268: 117038

Yan L J, Bai Y H, Zhao R F, Li F, Xie K C. 2015. Correlation between coal structure and release of the two organic compounds during pyrolysis. Fuel, 145: 12-17

Yang Y H, Pan J N, Wang K, Hou Q L. 2020. Macromolecular structural response of Wender coal under tensile stress via molecular dynamics. Fuel, 265: 116938

Yang Y L, Li Z H, Hou S S, Li J H, Si L L, Zhou Y B. 2017. Identification of primary CO in coal seam based on oxygen isotope method. Combustion Science and Technology, 189(11): 1924-1942

Ye J X, An N, Chen H, Ying Z Y, Zhang S H, Zhao J K. 2020. Performance and mechanism of carbon dioxide fixation by a newly isolated chemoautotrophic strain *Paracoccus denitrificans* PJ-1. Chemosphere, 252: 126473

Yu J L, Tahmasebi A, Han Y N, Yin F K, Li X C. 2013. A review on water in low rank coals: the existence, interaction with coal structure and effects on coal utilization. Fuel Processing Technology, 106: 9-20

Yuan Q, Zhang M. 2018. Diversities in biomarker compositions of Carboniferous-Permian humic coals in the Ordos Basin, China. Australian Journal of Earth Sciences, 65(5): 727-738

Zhang H X. 2016. Determination of aromatic structures of bituminous coal using sequential oxidation. Industrial & Engineering Chemistry Research, 55(10): 2798-2805

Zhang M M, Li Zhao. 2018. Thermal maturity of the Permian Lucaogou Formation organic-rich shale at the northern foot of Bogda Mountains, Junggar Basin (NW China): effective assessments from organic geochemistry. Fuel, 211: 278-290

Zhang S H, Tang S H, Tang D Z, Huang W H, Pan Z J. 2014. Determining fractal dimensions of coal pores by FHH model: problems and effects. Journal of Natural Gas Science and Engineering, 21: 929-939

Zhang X, Li L F, Du Z F, Hao X L, Cao L, Luan Z D, Wang B, Xi S C, Lian C, Yan J, Sun W D. 2020. Discovery of supercritical carbon dioxide in a hydrothermal system. Science Bulletin, 65(11): 958-964

Zhao B Y, Zhang Y M, Huang X Y, Qui R Y, Zhang Z Q, Meyers P A. 2018. Comparison of *n*-alkane molecular, carbon and hydrogen isotope compositions of different types of plants in the Dajiuhu peatland, central China. Organic Geochemistry, 124: 1-11

Zhao Q J, Tang Y G, Li W W, Wang S Q, Deng X J, Yu X L. 2014. Compositional characteristics of sulfur-containing compounds in high sulfur coals. Energy Exploration & Exploitation, 32(2): 301-316

Zheng Q M, Liu Q F, Huang B, Zhao W L. 2015. Isotopic composition and content of organic nitrogen in the coals of Qinshui coalfield, North China. Journal of Geochemical Exploration, 149: 120-126

Zhong Q F, Mao Q Y, Zhang L Y, Xiang J H, Xiao J, Mathews J P. 2018. Structural features of Qingdao petroleum coke from HRTEM lattice fringes: distributions of length, orientation, stacking, curvature, and a large-scale image-guided 3D atomistic representation. Carbon, 129: 790-802

Zhou G, Xu C C, Cheng W M, Zhang Q, Nie W. 2015. Effects of oxygen element and oxygen-containing functional groups on surface wettability of coal dust with various metamorphic degrees based on XPS experiment. Journal of Analytical Methods in Chemistry, 2015

Zhou H W, Zhong J C, Ren W G, Wang X Y, Yi H Y. 2018. Characterization of pore-fracture networks and their evolution at various measurement scales in coal samples using X-ray μCT and a fractal method. International Journal of Coal Geology, 189: 35-49

Zhou S D, Liu D M, Cai Y D, Yao Y B. 2016. Fractal characterization of pore-fracture in low-rank coals using a low-field NMR relaxation method. Fuel, 181: 218-226

Research Progress of the Organic Geochemistry of Coal

TANG Yue-gang, WANG Shao-qing, GUO Xin, LI Rui-qing, LIN Yu-han

China University of Mining and Technology (Beijing), Beijing 100083

Abstract: In this paper, we have defined "organic elements of the coal" and improved "organic geochemistry" concepts, reviewed the current status of researches on the organic elements in coal, the progress of researches on biomarker compounds, the trend of researches on molecular structure and microscopic and super-microscopic textures of the coal, and their applications in geology, industry and ecological environment in the past decade, pointed out current existed problems related to researches on the organic geochemistry of coal, and proposed prospects of its future development directions. In the past decade, the researches of coal organic geochemistry have mainly focused on the sulfur compounds in high-organic sulfur coal, aliphatic hydrocarbons in bark coal, and composition of coal polycyclic aromatic hydrocarbons. The main research directions of the coal macromolecular structure are of the bonding structure and embedded structure theories. The researches on microscopic components of coal are mainly focused on the genesis of liptinite and inertinite groups in coal. The current scientific research hotpots include fields of the special coal, especially the study of bark coal in China, the process ability and thermal evolution of macerals in coal, petrography of coal combustion and gasification residues, coal-measure gas, coal facies and sequence stratigraphy, coal-based graphene, and low-carbon and eco-environmental protection. With the development of new theories and high technologies, many new research fields of the organic geochemistry of coal will be emerged.

Key words: coal; organic elements; geochemistry; research progress; prospect

应用地球化学研究进展*

龚庆杰　夏学齐　刘宁强

中国地质大学（北京）地球科学与资源学院，北京 100083

摘　要：在分析应用地球化学发展历程的基础上总结了应用地球化学的研究内容，着重介绍了勘查地球化学在调查、评价、开发和修复四个阶段近十年的研究进展。调查方法可分为传统化探和非传统化探方法，传统化探方法日趋成熟、规范并得以持续推广应用，非传统化探方法在覆盖区勘查备受重视；评价主要集中在确定元素组合、圈定异常和评价异常方面，除持续应用外，其方法技术仍以对比为主导，但知识驱动技术初露端倪；在开发方面则主要体现在地浸法铀矿采选方面；修复主要是针对矿山环境的修复，由此产生了地球化学工程学这一新兴领域。

关键词：非传统化探　地球化学调查　地球化学评价　地球化学异常　地球化学修复

0　引　言

应用地球化学（applied geochemistry）这一概念诞生于 20 世纪 60 年代，以英国帝国理工学院建立应用地球化学研究组为标志，其目的是在英国自然环境研究委员会和农业研究委员会的资助下将地球化学调查应用于农业和健康（王学求和张德会，2005）。1986 年国际地球化学协会创办 *Applied Geochemistry* 杂志，标志着应用地球化学研究领域的初步形成。国际勘查地球化学家协会从 2001 年开始酝酿改名为国际应用地球化学家协会，该提议于 2003 年在爱尔兰召开的第 21 届国际勘查地球化学会议上获准通过（谢学锦，2003a；汪明启，2005），2005 年在澳大利亚召开了第一届国际应用地球化学学术会议（后称为第 22 届国际应用地球化学学术会议）。最近的会议有 2018 年在加拿大举办的第 28 届国际应用地球化学学术会议（汪明启和叶荣，2019），拟定于 2020 年 11 月举办的第 29 届国际应用地球化学学术会议因疫情可能推迟到 2021 年底在智利召开。

20 世纪最后几年，鉴于勘查地球化学向环境地球化学拓展的国内外研究动向，尤其是 1999 年全国多目标区域地球化学调查计划的实施，2001 年由於崇文、谢学锦和张本仁先生提议，中国矿物岩石地球化学学会（后文简称"本学会"）决定将元素地球化学区域地球化学专业委员会更名为应用地球化学专业委员会。首任主任委员为张德会教授，由专业委员会主办的第一届全国应用地球化学学术会议即明确了该专业委员会的宗旨：团结致力于将地球化学原理和方法应用于地球科学各个领域的科研工作者，促进和推动地球化学在地球科学及其相关领域的应用基础研究和学术交流，其主要研究领域涉及固体矿产和能源矿产的地球化学勘查、环境-生态-农业的地球化学调查与评价、地质灾害的地球化学响应等。自 2004 年至今共召开了 7 次，全国应用地球化学学术会议，第八次会议拟于 2020 年 10 月在中山大学召开。除组织学术交流外，该专业委员会还肩负总结中国应用地球化学研究进展的义务，本文针对中国 2011—2020 年的应用地球化学研究进展进行综述与展望。

1　应用地球化学概况

谢学锦（2003b）将中国勘查地球化学的发展划分为 4 个时期：1951—1957 年称为初创期，1957—

---

*　原文"2011~2020 中国应用地球化学研究进展与展望"刊于《矿物岩石地球化学通报》2020 年第 39 卷第 5 期，本文略有修改。

1966年称为孕育期，1978—1986年称为发展期，1994年至21世纪初称为新的发展高潮期。按照这一思路及"从勘查地球化学到应用地球化学"的相关论述（汪明启，2005；赵伦山和朱有光，2010；汪明启和叶荣，2019），本文将应用地球化学的发展历程划分为四个阶段：起源、初创、确立和发展，并基于这4个阶段归纳分析了应用地球化学的研究内容。

1.1 发展历程

1.1.1 起源阶段

20世纪初至20世纪70年代为应用地球化学的起源阶段，对应勘查地球化学的初创期和孕育期（谢学锦，2003b；於崇文，2005）。

在国际上的标志性事件有：① 学术专著，苏联谢尔盖耶夫于1941年出版《地球化学探矿法》一书，标志着勘查地球化学的确立；霍克斯和韦布于1962年出版了《矿产勘查的地球化学》，对以前地球化学探矿方法进行了系统总结；苏联别乌斯等于1976年出版了《环境地球化学》等。② 学术机构，如1966年国际勘查地球化学家协会成立；英国帝国理工学院于20世纪60年代建立应用地球化学研究组，标志着"应用地球化学"名词的出现。③ 1972年 *Journal of Geochemical Exploration*（《勘查地球化学》）创刊。

在国内的标志性事件有：① 科研活动，如1951年以谢学锦为首的化探小组在安徽月山首次开展勘查地球化学试验。② 研究机构，如原地质矿产部地矿司于1952年设立化探室；冶金部于1956年成立冶金化探组。③ 学科专业，如原桂林地质学校于1959年成立了金属矿地球化学勘探专业；北京地质学院于1960年建立了地球化学探矿专业。④ 出版教材，如林名章和阮天健于1961年出版了高等学校教材试用本《金属矿床地球化学探矿法》；原冶金部地质干部学习班于1973年出版《找矿地球化学概论》；原广西冶金地质学校化探教研室于1978年出版了《地球化学找矿》。⑤ 学术期刊，如创刊于1979年的《物探与化探》、《物探化探计算技术》。

这一时期的环境地球化学仍处在萌芽时期，代表性论著为苏联彼列尔曼于1955年所出版的《景观地球化学概论》，该书侧重地球化学环境与人类健康。原长春地质学院林年丰教授自60年代起就开展了地球化学环境与人类健康的研究，这是目前认为环境地球化学起源于20世纪60年代的主要依据（龙莎莎等，2005）。在学术期刊方面，1971年《环境地质与健康》创刊，1979年 *Environmental Geochemistry and Health*（《环境地球化学与健康》）创刊。

上述成果及事件标志着勘查地球化学的初创与孕育，也反映出应用地球化学源于勘查地球化学这一鲜明特征，同时环境地球化学的萌芽也为应用地球化学的起源奠定了基础。鉴于应用地球化学这一名词已经出现，但相关学术刊物、学术团体、专业教材、大科学计划等均未形成，故将该时期称为应用地球化学的起源阶段。

1.1.2 初创阶段

20世纪80年代至20世纪末为应用地球化学的初创阶段，对应勘查地球化学的发展期和环境地球化学的初创与孕育期（谢学锦，2003b；赵伦山和朱有光，2010）。

1986年 *Applied Geochemistry*（《应用地球化学》）创刊，这是国际上的标志性事件。

国内的标志性事件有：①学术团体，中国地质学会于1980年成立勘查地球化学专业委员会；中国矿物岩石地球化学学会于1981年成立元素地球化学区域地球化学矿床地球化学委员会，后经拆分，元素地球化学与区域地球化学专业委员会正式成立，此为应用地球化学专业委员会的前身。②大科学计划，如原地质矿产部于1981年正式组织实施全国区域地球化学扫面计划（或简称区域化探计划），主要为矿产资源勘查服务，调查工作集中在山区和丘陵区；1997年任天祥等在天津开始实施为环境-农业-生态服务的区域地球化学调查试点工作。③专业教材，如刘英俊和邱德同于1987年出版了《勘查地球化学》；1990年陈静生等、戎秋涛和翁焕新分别出版了《环境地球化学》；杨忠芳等于1999年出版了《现代环境地球化学》。④学术专著，如林年丰于1991年出版了《医学环境地球化学》；李惠于1998年出版了《大型、特大型金矿盲矿预测的原生叠加晕模型》；谢学锦等于1999年出版了《走向21世纪矿产勘查地球化学》文集；李家熙等于1999年出版了《中国生态环境地球化学图集》等。

这一时期的典型特征是勘查地球化学，尤其是区域地球化学调查得到充分发展；环境地球化学正式确立，其研究内容逐渐从环境地球化学与人类健康关系研究向环境地球化学与环境污染方面转变。区域地球化学调查与环境污染研究开始走向融合，从而诞生了应用地球化学这一研究领域。虽然应用地球化学这一名词及学术期刊均已出现，但相关学术团体、专业教材、大科学计划等均未形成，故称该时期为应用地球化学的初创阶段。

1.1.3 确立阶段

21世纪初十年为应用地球化学的确立阶段，体现在相应的学术团体、学术会议、专业教材的出现，以及大科学计划的实施及系列标准与规范的颁布等方面。

2003年，国际勘查地球化学家协会正式更名为国际应用地球化学家协会，是应用地球化学学科确立的标志之一。2007年，该协会在西班牙举办了第23届国际应用地球化学学术会议，并向谢学锦先生颁发了"应用地球化学家协会金奖"。

国内的标志性事件有：①学术团体与会议。中国矿物岩石地球化学学会元素地球化学与区域地球化学专业委员会于2001年更名为应用地球化学专业委员会，并于2004年在长沙召开了第一届全国应用地球化学学术会议。②研究机构。中国地质科学院地球物理地球化学勘查研究所于2004年成立了应用地球化学开放实验室。③大科学计划。全国多目标区域地球化学调查计划在1999~2001年在四川、广东、湖北三省先期试验，随后于2001年在全国逐渐铺开实施。④学术专著。伍宗华和古平等于2000年出版了《隐伏矿床的地球化学勘查》；王学求和谢学锦于2000年出版了《金的勘查地球化学理论与方法·战略与战术》；谢学锦等于2002年出版了《面向21世纪的应用地球化学——谢学锦院士从事地球化学研究五十年》；刘石年于2003年出版了《应用地球化学》；叶家瑜和江宝林于2004年出版了《区域地球化学勘查样品分析方法》；戴塔根等于2005年出版了《应用地球化学》；中国地质调查局于2006年对地质调查中关于地球化学勘查的标准汇编成册《地质调查标准汇编地球化学勘查分册》；迟清华和鄢明才于2007年出版了《应用地球化学元素丰度数据手册》；冯济舟等于2008年编辑出版了《贵州省地球化学图集》；谢学锦等于2008年编辑出版了《中国西南地区76种元素地球化学图集》；罗先熔等于2010年出版了《地电化学集成技术寻找隐伏金矿的研究及找矿预测》等。⑤专业教材。蒋敬业等于2006年出版了《应用地球化学》教材；罗先熔等于2007年出版了《勘查地球化学》教材。

上述标志性事件及成果表明，应用地球化学已在名称、期刊、教材、专著、标准与规范、学术团体、大科学计划等方面均已形成，因此称该时期为应用地球化学的确立阶段。

1.1.4 发展阶段

2010年至今为应用地球化学的发展阶段，体现在勘查地球化学和环境地球化学应用领域与技术的持续发展、深化和完善，应用地球化学的概念和内涵进一步明确。

这一阶段我国的标志性事件有：①应用地球化学的研究内容与学科体系进一步明确，如蒋敬业等于2013年修订了《应用地球化学》教材；陈岳龙等于2017年、祁士华等于2019年分别出版了《环境地球化学》教材。②围绕研究进展的全国学术交流持续跟进，如自2010年第三届全国应用地球化学学术会议起，我会应用地球化学专业委员会每两年主办一次全国应用地球化学学术会议。③修订与完善了系列标准、规范，如中华人民共和国地质矿产行业标准《区域生态地球化学评价规范》（DZ/T 0289—2015）、《土壤地球化学测量规程》（DZ/T 0145—2017）、中华人民共和国国家标准《土壤环境质量农用地土壤污染风险管控标准（试行）》（GB 15618—2018）、《土壤环境质量建设用地土壤污染风险管控标准（试行）》（GB 36600—2018）的颁布实施等。④大科学计划持续进行，如全国区域化探扫面计划和全国多目标区域地球化学调查计划持续铺开，全国矿产资源潜力评价项目取得重要进展，区域地球化学调查走向境外等。⑤系统出版了高质量的科研成果，针对传统化探方法张华等于2017年出版了《中国主要景观区区域地球化学勘查理论与方法》；针对隐伏矿勘查周奇明于 2020 年出版了《深穿透地球化学勘查技术及应用》；针对数据集成高艳芳等开发了《地球化学勘查数据一体化处理系统（Geochem Studio 3.6）》；在全国矿产资源潜力评价和多目标区域地球化学调查方面，分别从研究区、全省、大区、全国等不同层次出版了系列地球化学专著与图集。此外，奚小环（2019）系统总结了从勘查地球化学到应用地球化学的进展并指出了今后发展方向，由单目标矿产勘查到资源环境多目标调查评价的转型，从地质系统到地球系统与生态系统的升级，应用地球化学将进入大数据信息与地球系统科学时代。

1.2 研究内容

应用地球化学有广义和狭义两个层次，广义的应用地球化学是指理论地球化学（含量子地球化学、地球化学热力学、地球化学动力学等）在众多应用方面的拓展与开发，狭义的应用地球化学则是指直接服务于国家经济发展和人类社会进步的地球化学应用领域（於崇文，2005）。本文所指的即为狭义的应用地球化学。於崇文（2005）将理论地球化学和应用地球化学二者的关系表述为：理论地球化学重点在基础，应用地球化学重点在应用；基础促进应用，应用充实基础，二者互为补充、相互促进；应用基础研究则是联系二者的桥梁和纽带，应用地球化学研究固然必须充分体现其实际效果，同时也要不忘回归于应用基础研究。

结合应用地球化学的发展历史、应用领域及上述关系分析，本文给出的定义是：应用地球化学起源于勘查地球化学，兴起于环境地球化学，是利用地球化学的基本理论和方法技术为资源勘查、环境评价及社会服务的一门地球化学分支学科。它包含了勘查地球化学和环境地球化学两个比较成熟的研究领域，而在其他社会服务领域的则属于交叉或新兴研究领域。

应用地球化学基于的地球化学基本理论主要包括元素丰度、元素共生组合、元素赋存状态、元素迁移、同位素示踪和测年等理论，使用的地球化学方法技术主要包括野外采样、分析测试、数据处理、图件制作与成果表达等技术。这些地球化学的基本理论和方法技术在资源、环境及其他社会领域的应用决定了应用地球化学的研究内容。

应用地球化学在资源和环境（即勘查地球化学和环境地球化学）两个领域均可大体分为调查、评价、开发和修复四个阶段。本文针对勘查地球化学从调查、评价、开发和修复 4 个方面来分析十年来应用地球化学领域的研究进展。

2 勘查地球化学

近十年来勘查地球化学的研究进展主要集中在调查和评价两个方面，而在开发方面则主要针对采矿和选矿领域，这已超出目前勘查地球化学的研究范围；在修复方面则主要针对矿山环境修复，并由此产生了地球化学工程学这一新兴领域。

2.1 调查

勘查地球化学的调查旨在通过野外系统采样与分析测试来发现异常，侧重于样品采集与分析测试。其调查方法可分为传统化探方法和非传统化探方法（Xie and Wang，1991；王学求和谢学锦，1996）。杨少平等（2014）、王学求等（2014）针对前期勘查地球化学调查方法及进展进行了系统评述，在此仅针对 2011 年以来的主要进展从传统化探方法和非传统化探方法两方面进行综述。

2.1.1 传统化探方法

传统化探方法的采样介质为岩石、土壤和水系沉积物 3 种，分析介质中元素的全量（或总含量），主要方法有岩石地球化学测量（又称原生晕化探）、土壤地球化学测量（又称次生晕化探）和水系沉积物测量（又称分散流化探），适用于地表出露矿、半出露矿的预查、普查、详查以及针对靶区或矿区的槽探、钻探、巷道的勘查。该领域十年来的进展主要体现在化探方法技术的系统总结和持续发展两个方面。

（1）系统总结。原地质矿产部于1981年正式组织实施全国区域地球化学扫面计划，共设计约650万 km² 的陆地国土面积，在每4 km² 内采集4件样品形成一个组合样并分析其中的39种元素含量，至1990年中东部山区和丘陵区已基本完成，至2000年边远省区特殊地球化学景观区也基本完成（赵伦山和朱有光，2010）。自1999年原国土资源部启动了新一轮国土资源大调查，区域化探工作重点集中在中国北部和西部的特殊景观区，至2010年已完成勘查理论与方法的试验研究工作（张华等，2017），1:20万区域地球化学调查面积累积已完成610万 km²（向运川等，2018）。近十年来主要进展之一就是对这些成果的系统总结与规范化。

中国地质科学院地球物理地球化学勘查研究所于2013年出版了《勘查地球化学科技进展与成果（1999—2008）》，对新一轮地质大调查成果进行了系统总结。杨少平等（2014）对"十一五"以来的化探方法技术进行了总结，奚小环和李敏（2013，2017）对中国"十一五""十二五"期间的勘查地球化学工作进行了详细评述。王学求（2013）对勘查地球化学国内外80年的重大事件进行了回顾，Xie and Cheng（2014）则对中国60年来的勘查地球化学方法技术与应用进行了系统总结。张华等于2017年出版了《中国主要景观区区域地球化学勘查理论与方法》，该书是中国地质科学院地球物理地球化学勘查研究所区域化探研究组30余年几代科研工作者在区域化探理论与方法技术研究方面的集体智慧结晶。

自2006年至2013年全国矿产资源潜力评价大科学计划历时八年，随后至今进入对成果的系统整理阶段。针对化探资料应用在2010年提出了化探资料应用技术要求（向运川等，2010），从省、大区、典型区带、全国等不同层面进行了数据库建设、地球化学编图、成果综合整理等系列工作。如向运川等于2018年出版了《全国矿产资源潜力评价化探资料应用研究》，系统清理了全国截至2013年的地球化学资料，建立了包含多目标地球化学调查（分析54种元素或指标）、区域化探、化探普查、化探详查等数据的不同层次的数据库，建立了一套化探数据处理、地球化学编图、典型矿床地质地球化学建模、金属资源量定量预测的方法技术，为解决一些重大地质问题提供了地球化学依据，显示其在基础地质、资源勘查和环境评价方面的实际效用。

在1999年开始的新一轮地质大调查中，基于原区域化探组合样副样分析测试了76种元素，先期在云南、贵州、四川、重庆、广西5省（自治区、直辖市）开展试验，在此基础上谢学锦等于2008年出版了《中国西南地区76种元素地球化学图集》。随后这项工作推广至中南和华东的8个省（自治区、直辖市），基于12个省区76种元素的测试数据，程志中等于2014年出版了《中国南方地区地球化学图集》。

作为"全球地球化学基准计划"的一部分，中国于2008年开始实施了"全国地球化学基准计划"，建立了覆盖全国的地球化学基准网，至2012年完成了3 382个样品点表层和深层土壤样品的采样工作，尽管联合国教科文组织规定分析71种元素或氧化物，但中国分析测试了76种元素及5个其他指标（FeO、C_{org}、CO_2、H_2O^+、pH），编制了81个指标的地球化学基准数据和地球化学图，为资源与环境评价提供了定量标尺和长期监测的基础（王学求等，2016；Wang X. Q. et al., 2015；Wang and the CGB Sampling Team, 2015）。2016年联合国教科文组织和中国政府共建的全球尺度地球化学国际研究中心成立，旨在研究元素周期表上所有元素及其化合物在全球的含量与分布、基准与变化。

针对从区域化探分析的39种元素，到多目标地球化学调查分析的54项指标，再到全球地球化学基准计划分析的71种元素以及中国分析测试的76种元素（或81个指标），王学求（Wang and the CGB Sampling Team, 2015）、张勤等（2012）、姚文生等（2011）对中国和欧洲分析测试技术进行了系统对比，指出中国是目前唯一能够分析76种元素的国家，基于中国与国际上其他实验室对比认为，目前尚有10余种元素的分析质量有待提高，尤其是Ag的分析方法仍需突破（Yao et al., 2011）。随后，刘雪敏和王学求（2014）、王学求等（2020）对中国及全球4个大洲地球化学数据进行了系统对比研究，提出了地球化学基准与环境监测对实验室分析76种元素的建议，指出中国的分析水平总体处于国际领先水平。

在岩石地球化学调查方面，继《中国花岗岩类化学元素丰度》（史长义等，2008）出版后，为了反映花岗岩类元素丰度的空间分布规律，史长义等于2016年出版了《中国花岗岩类地球化学图集》，为开展矿产资源勘查、生态环境评价、基础地质研究等提供了基础资料。

随着传统化探方法技术的不断完善，近十年来中国地质调查局制定、修订、完善了一系列行业规范。中华人民共和国原国土资源部发布或更新了系列地质矿产行业标准，如《多目标区域地球化学调查规范（1:250 000）》（DZ/T 0258—2014）、《岩石地球化学测量技术规程》（DZ/T 0248—2014）、《地球化学普查规范（1:50 000）》（DZ/T 0011—2015）、《区域地球化学样品分析方法》（DZ/T 0279.1—2016 至 DZ/T 0279.34—2016）、《土壤地球化学测量规程》（DZ/T 0145—2017）等。

（2）持续发展。基于岩石、土壤和水系沉积物介质的传统化探工作近十年来仍在持续开展，方法技术得到了进一步规范和提高。

在区域化探扫面与方法技术研究中，西北和北部地区仍是工作持续推进的重点区域（张华等，2017）。除沙漠景观区外，羌塘高原是目前区域化探全国扫面工作最大的空白区，该区传统化探方法的技术难点在于风成沙的干扰，针对这一难题，杨少平等（2015）提出了区域化探扫面中采集 10~40 目粒级水系沉积物及采样密度为 1 点/4 km² 和 1 点/km² 的解决方案。

在化探普查方面，十年来的进展主要集中在采样技术研究和工作区调查两个方面（贾玉杰等，2013；严明书等，2016；何旺等，2019；翁望飞等，2020）。在化探详查方面，主要针对矿区开展异常查证（马生明等，2011；陈晶等，2017）。王文和刘应平（2012）报道了从 2010—2012 年对青海玉树曲麻莱县扎家同哪金矿从化探预查（1:20 万水系沉积物）、普查（1:5 万土壤）、详查（1:1 万土壤），到非传统化探地电化学测量详查，再到靶区槽探、钻探工程，最终发现了金矿体，这是从区域到靶区、传统化探与非传统化探方法相结合的一次成功试验与应用。针对传统化探从区域化探→化探普查→化探详查，陈国光等（2105）对不同阶段的目的任务与性质、野外工作方法、室内资料整理、异常分类与异常查证、分析测试等方面进行了系统梳理与分析，为传统化探从水系沉积物→土壤→岩石的扫面工作的持续发展理清了思路。

在原生晕化探方面，分带指数的计算得到发展，构造叠加晕理论持续应用，为就矿找矿、攻深找盲、危机矿山找矿提供了技术支撑（刘普凯等，2018；文雄亮等，2020）。如叶庆森（2014）对当前原生晕分带指数的计算方法进行了评述和比较。Gong 等（2016）在胶东玲珑花岗岩蚀变的模拟实验研究中，提出利用质量平衡计算中的迁移指数（或富集因子）对元素进行排序，其实验结果与李惠等（1999）基于中国 58 个金矿床得出的原生晕轴向分带规律一致，进而基于迁移指数提出了原生晕剥蚀系数的计算公式，剥蚀系数介于 0~1，其值越大越接近矿体，经玲珑金矿区穿矿剖面检验效果良好。詹胜强等（2016）在四川道孚县容须卡锂矿区首次进行了原生晕垂直分带和水平分带研究，拓宽了原生晕化探可适用的矿种。

除上述传统化探方法持续发展以外，另一大特色是"走出去"战略。由原国土资源部提出（向运川等，2015），并专门设立全球矿产资源地球化学与遥感调查工程，利用自有知识产权和中国领先的地球化学填图技术，以及开展"一带一路"国家矿产资源调查评价成为中国资源走出去的最主要方式（王学求和聂兰仕，2016），分别在哈萨克斯坦、塔吉克斯坦、吉尔吉斯斯坦、秘鲁、老挝、中蒙边界、缅甸、马达加斯加、哥伦比亚等国外地区开展了地球化学调查工作，为国外矿产资源地球化学勘查贡献了中国智慧，实现了合作共赢（李宝强等，2014；张晶等，2015；刘君安等，2017；刘汉粮等，2018；王玮等，2019）。

2.1.2 非传统化探方法

非传统化探方法的采样介质主要为气体、土壤、植物、水等，分析其中元素的全量或部分存在形式的分量，主要方法有地气测量、地电化学测量、偏提取测量、土壤微细粒测量、浅钻测量、植物测量等，主要针对隐伏矿开展化探详查。由于非传统化探主要是基于地表地球化学异常来发现深部隐伏矿床，因此许多非传统化探方法又属于穿透性地球化学测量（韩志轩等，2017）。近十年来，赵景等（2012）、刘草等（2016）、Wang 等（2016）、韩志轩等（2017）、鲁美等（2019）等对非传统化探方法的发展历史、理论支撑、技术方法、试验与应用进行了分析，尤其是周奇明等于 2020 年出版的《深穿透地球化学勘查技术及应用》是一本综合性成果的总结。本文重点综述地气测量、地电化学测量近十年来的进展。

（1）地气测量。地气测量方法最早是由瑞典科学家提出，原成都地质学院童纯菡等于 1990 年将其引入中国，王学求等于 1995 年将其进行改进，随后在中国得到广泛试验与应用。其理论支撑可表述为隐伏目标体（如矿体、断裂等）中的物质以微粒（如纳米微粒、微细粒、气溶胶等）形式随深部气体垂向或侧向迁移至近地表并被土壤吸附而富集，通过采集近地表土壤气体中的微粒，分析测试其化学组分，发现异常，评价异常与隐伏目标体的成生关系，从而达到勘探隐伏目标体的目的。其内容涉及隐伏目标体中微粒的存在形式和迁移途径、样品的采集和分析测试方法、异常的识别和评价技术等环节。

在矿石、地表土壤、地气样品等介质中微粒的特征研究方面，近十年来主要对河南唐河周庵铜镍矿床（王学求和叶荣，2011）、新疆哈密金窝子 210 金矿床（叶荣等，2012，2013）、云南昭通永胜得铜矿床（刘昶等，2011）、内蒙古克什克腾旗红山子铀矿床（王勇等，2017）开展了工作。

在地气微粒的物源示踪方面，继汪明启和高玉岩（2007）在甘肃庄浪蛟龙掌铅锌矿区进行铅同位素示踪之后，徐洋等（2014）对山东邹平王家庄斑岩铜矿区、刘雪敏等（2012）在新疆哈密金窝子金矿区和内蒙古克什克腾旗拜仁达坝-维拉斯托多金属矿区均对矿石、围岩、土壤、地气 4 种介质的铅同位素组成进行了对比研究。在稀土元素示踪方面，王勇等（2012）在新疆哈密金窝子金矿区、胡波等（2016）在广东仁化长排铀矿区、徐洋和汪明启（2018）在山东邹平王家庄铜矿区分别针对矿区矿石、围岩、土壤、地气介质的稀土元素进行研究，尽管研究者认为稀土元素对示踪地气异常样品的来源有指示意义，但证据并不完备。总的来说，针对地气样品的铅同位素和稀土元素示踪研究尚待深入，事实上基于传统岩石介质对矿床成矿物质来源的铅同位素和稀土元素示踪研究也在探索中。在从矿石到地气样品的迁移方面，唐桢等（2012，2013）和周四春等（2014）设计并进行了地气物质迁移模拟实验，他们以铅锌矿为研究对象，收集实验地气样品，发现铅锌矿找矿指示元素的地气异常，这从实验模拟层面证实了隐伏矿体可以形成地气异常，进而开展实地试验，发现矿石中的成矿及伴生元素可以作为地气法找矿的指示元素。在对铀元素的迁移方面，陈刚（2017）、李博（2018）对地质处置核废物中核素在地气作用下的迁移进行了实验模拟、计算模拟等研究，并结合实际铀矿区进行对比，结果表明埋藏于深部的核废物聚集体（或矿体）中的多种核素均能够穿越致密地质介质以地气方式发生垂向迁移。这些研究从实验模拟和实地试验两方面证实了从隐伏矿体到近地表地气样品中成矿及伴生元素迁移的可能性。

在地气样品的采集技术方面，主要有静态累积式和动态抽气式两种方法。静态累积式是将采样装置埋置在覆盖层中利用捕集剂持续（通常几十天）被动捕集覆盖层地气中的微粒，捕集剂通常为固态，如透射显微镜使用的镍网、钼网或铜网或聚氨酯泡塑（王正阳等，2013；周四春等，2014）。动态抽气式是在覆盖层打孔抽气，利用捕集剂持续主动提取覆盖层地气中的微粒，其捕集剂有液态和固态两种。为提高采样装置的灵敏度、便利性、采样一致性，刘晓辉等（2012）提出了双捕集器串联的动态抽气装置，周四春等（2012）研制了智能动态地气采样装置，吴泽民等（2019）报道了自行研制的恒流式动态地气采样器。在目前地气测量的试验和应用中，主要采用动态抽气式采样技术（王勇等，2012，2014；叶荣等，2013；王东升等，2019；吴泽民等，2019）。除了在采样装置和捕集剂的改进与选择方面外，对影响地气测量的其他因素如季节、同一天不同时段、温度、湿度等目前也进行了诸多试验（王滔和彭秀红，2013；朱剑等，2014；万卫等，2019）。不同地区的研究结果既存在一致性又存在差异，目前的研究仍处于探索阶段。在地气样品元素含量测试分析方面，目前主要采用等离子体质谱法（ICP-MS），可分析三十几种元素的含量（周四春等，2014；袁璐璐等，2014；杨吉成等，2019；韩伟等，2019；万卫等，2019）。虽然分析测试均在实验室完成，但有些元素的检出限仍不能满足要求。在针对地气样品单元素的分析测试中，赵柏宇等（2018）在江西崇仁相山铀矿区试验中，首次采用紫外可见分光光度法分析地气样品中的铀，建立起一套野外快速采样、现场快速分析的地气测量工作方案，但其检出限较高。

在地气测量实地试验和应用方面，近十年来主要围绕隐伏矿勘查、隐伏断裂勘查和地热泉勘查开展工作。对于隐伏矿勘查，近十年来涉及的金属矿种主要有铀矿（Liu et al.，2018）、金矿（王晓佳等，2016）、铜矿（赵吉海和欧阳辉，2017）、铜钼矿（卢福隆，2019）、铜镍矿（韩伟等，2019）和铅锌多金属矿（郭祥义等，2019）等。此外，对锂矿（杨吉成等，2019）和钾盐矿（王正阳等，2013）也进行了地气测量的探索性研究。

隐伏断裂地气勘探的原理是利用隐伏断裂处地气总量相对较大，断裂破碎导致地气中携带较多的微粒物质，因此地气样品中可出现元素总量或某些元素的高含量异常（周四春等，2011；廖驾等2012；李友余等，2013；葛良全等，1997）。对隐伏断裂勘查的地气方法均采用动态抽气式5%硝酸液体捕集剂法。在利用地气测量技术针对隐伏温泉勘探方面，目前仅有少量研究（刘晓辉等，2015；谢克文等，2015）。这些研究目前仅处于探索阶段，其理论、机理均尚未形成。

（2）地电化学测量。地电化学测量方法最早由苏联科学家提出，由费锡铨、徐邦梁等于1984年引入中国，潘勇飞于1989建议将相关技术进行标准化，康明和罗先熔于2003年将其进行改进，随后在中国得到了广泛试验和应用。但目前仍存在异常来源质疑、技术改进滞后、应用成果不突出等问题，尚需进一步完善理论、改进技术和扩大应用实例研究等（刘攀峰等，2018）。

地电化学测量的理论支撑可表述为：隐伏矿体周围存在离子晕或电活性物质晕（本文为继承前人文献表述将"离子晕"广义化或等同为"电活性物质晕"，并不局限于离子形态的晕），电活性物质在自然地质营力下可垂向或侧向迁移至近地表并被土壤吸附而富集，或与近地表电活性物质形成动态平衡系，在人工电场作用下通过提取近地表土壤中的电活性物质，分析测试其含量，发现异常，评价异常与隐伏矿体的成生关系，从而达到勘探隐伏矿体的目的。这一表述涉及隐伏矿体中电活性物质的存在和迁移途径、近地表土壤中电活性物质的可提取性、样品的采集与分析测试方法、异常的识别与评价技术等环节。

在矿体周围存在离子晕及在近地表土壤中可提取到离子晕的认识已得到实验、野外试验证实并被广泛接受（黄学强等，2013；孙彬彬等，2015；刘攀峰等，2017）。关于隐伏矿体周围的离子晕与近地表土壤中离子晕的关系，康明（2004）曾提出二者呈动态平衡系及外加电场下的迁移递推理论，但对地电化学测量中提取的离子晕与深部隐伏矿体周围的离子晕是否同源这一问题尚未解决（孙彬彬等，2015；刘映东等，2017），这需要从物源示踪及迁移机理两方面进行深入探讨。

关于地电化学测量的技术方法，部分学者认为包含电吸附法、电导率法、地电提取法3种，这些均是提取后生地球化学异常（李天虎等，2012；文美兰和罗先熔，2013；黄学强等，2013）。近年来倾向于认为电导率法应属于地球物理的方法，电吸附法应属于室内偏提取法的一种，只有地电提取法才属于真正意义上的（或狭义的）地电化学测量（刘攀峰等，2018；刘延斌等，2018），本文也采用这一观点。但目前较多地使用"地电化学集成技术"这一术语来包含上述地电提取法、土壤电导率法、土壤电吸附法（其至还包含土壤热释汞法或土壤吸附相态汞法）（邱炜等，2013；刘攀峰等，2016；刘延斌等，2018）。

关于地电化学测量的野外采样装置，在21世纪初的十年中桂林理工大学研制了低电压偶极地电提取装置、中国地质科学院地球物理地球化学勘查研究所研制了一种具有时间控制功能的固体载体型元素提取器，近十年来的地电化学测量则主要是基于二者的结合，即独立供电偶极子-固体泡塑载体采样装置来开展工作（孙彬彬等，2011；刘攀峰等，2017；蓝天等，2018；刘刚等，2018）。针对铀矿床的勘查，柯丹等（2016）研制了相应的地电化学野外装置，对供电装置进行了改进。而电极提取剂，目前主要采用固体泡塑载体（孙彬彬等，2011；陈亚东等，2015；刘洪军等，2017）。在地电化学提取载体样品元素含量的测试分析方面，目前主要采用等离子体质谱法（ICP-MS）、原子荧光光谱法（AFS）分析测试泡塑样品中元素的含量（聂凤莲等，2011；孙彬彬等，2016；严洪泽等，2016，2017）。

在地电化学测量方法试验和应用方面，近十年主要围绕金属矿产隐伏矿勘查开展，集中在不同景观区适宜技术参数试验（陈亚东等，2015；邱炜等，2017；张启龙等，2019）、不同景观区找矿效果试验（韦选建等，2017；杨笑笑等，2019）及不同金属矿种找矿效果的试验上。据刘攀峰等（2018）统计，截至2015年底地电化学勘查技术已在全国26个省（自治区、直辖市）130多个矿区开展了试验或应用研究，涉及的矿种包括金、银、铜、铅、锌、钨、锡、砷、锑、镍、铀等十余种。近十年来的试验和应用仍主要以金、铜、铜镍、铅锌和铀矿为主，其他矿种少有涉及。

在上述试验与应用研究中，虽然已取得显著的找矿效果，但在测试数据和技术参数方面尚待完善，不同地区测量数据缺乏可比性。尽管对不同景观区已探索出适宜的地电化学测量技术参数，但这仅是基于相对异常与矿体的空间直观对比所确定的，目前尚未提出普适的地电化学测量技术规范，这也是不同地区测量数据基本无可比性的原因之一。

（3）其他非传统化探方法。除地气测量和地电化学测量外，近十年来仍在探索改进和发展的非传统化探方法还有偏提取测量、酶提取测量、活动金属离子测量、金属活动态测量、电吸附测量、离子电导率测量、汞气测量、氡气测量、释光测量、有机烃测量、卤素测量、土壤微细粒测量、浅钻测量、生物地球化学测量等。

偏提取测量是相对于全量分析而言，即采用特殊的提取剂对一定存在形式或结合态的元素进行提取并测试其含量的技术，属于分量化探范畴（刘晓东等，2018）。偏提取测量方法可分为：① 依据提取顺序分为顺序提取和平行提取两种方法，但针对某些结合态也可采用单步提取方法（姚文生等，2012；Wang X. Q. et al., 2015；Xu et al., 2019；康欢等，2019）；② 依据提取剂成分和性质可分为酶提取测量、电吸附测量等；③ 依据提取物的成分和性质分为活动金属离子测量、金属活动态测量、热释汞测量、酸解烃测量、热释卤素测量、离子电导率测量等。

酶提取测量由美国 Clark 等（1994）提出，是利用葡萄糖酶淋滤浸出非晶质氧化膜结合的金属元素的覆盖区化探方法（周奇明等，2020），目前在中国鲜有使用。活动金属离子测量是由澳大利亚 Mann 等（1995）在第 17 届国际勘查地球化学会议上提出，这是采用一种或几种提取能力较弱的提取剂提取样品中被认为是活动态的金属离子并分析其含量的化探方法（朱炳玉等，2012），目前在中国使用较少。金属活动态测量是由王学求等（1998）提出的两阶段提取方法（Wang, 1998），第一阶段采用偏提取中的顺序提取法，将载体由弱到强依次溶解使金属释放到提取液中（不局限于金属离子，可包含非溶解态的超细粒物质等），第二阶段是对提取液的处理过程，即采用强溶剂将第一阶段释放出来的金属溶解到溶液中。金属活动态测量目前在中国使用比较广泛（赵波等，2012；叶荣等，2013；刘汉粮等，2014；申伍军等，2017；杨刚刚等，2018；付亚龙等，2019）。

电吸附测量是地电化学集成技术中的一种，是借鉴地电化学测量的原理在室内对样品进行电化学提取（柯龙跃等，2017），实质上是一种电提取的偏提取技术。相对于野外地电化学测量技术中的"提取域"原理，本方法的样品可能会有一定的局限性。离子电导率测量也是地电化学集成技术中的一种，其测量介质通常为土壤，测试参数为电导率（文美兰和欧阳菲，2013），其实质应属于地球物理勘探方法，类似土壤 pH 的测量方法。

汞气测量是针对汞的单元素测量方法，通常有壤中气汞和土壤热释汞两种方法。壤中气汞测量实质上是一种动态抽气式地气测量（李伟等，2017），只是其捕集和测试对象针对汞而已。土壤热释汞测量是将土壤样品在不同温度下加热使其释放并测试全量或某些存在形式的汞含量的方法（郑超杰等，2018），实质上是一种针对特定成分的偏提取技术。氡气测量是对地气中的氡（或其子体）进行捕集和测试以此来发现隐伏目标的勘查方法，其隐伏目标通常为铀矿体（或断裂），这实质上是具有动态瞬时式和静态累积式的一种地气测量方法（韩娟等，2013；赵丹等，2018；康欢等，2019；董昕昱等，2020）。

释光测量是指某些物质在埋藏条件下经放射性辐照后可以某种能量形式积累于矿物中，当物质经过加热或光照，其矿物中储存的能量将以发光形式释放出来，基于释光的强度与放射性辐照积累的程度进行测量的一种方法。在埋藏年龄基本相同的条件下，基于释光强度异常可以指示放射性异常，从而可达到铀矿勘查的目的（李业强等，2014；张凯等，2015）。在假设放射性辐照强度基本稳定时可用于矿物测年，若在年龄基本相同的前提下测试出异常老的年龄则同样可指示放射性异常，进而也可达到铀矿勘查的目的（Kang et al., 2020）。

有机烃测量实质上也是一种针对特定成分的偏提取技术，通常有现场抽气和酸解烃两种测量方法。现场抽气有机烃测量类似动态抽气式地气测量，其采集对象为气体、测试对象为有机烃。酸解烃测量是将土壤（或岩石等）样品在不同温度及不同分解剂作用下使其脱气并测试的方法（周奇明等，2020）。由于包裹相态的烃与成矿关系密切且受表生影响因素相对较少，故目前较多地使用酸解烃烃气测量方法（韦文定和陈金声，2013；吴二等，2014）。卤素测量是基于F、Cl、Br、I元素的测量，在实际应用中主要为F的测量。通常有两种方法，一种为传统化探法，即重点关注的是卤素的地球化学特征；另一种方法为土壤热释卤素法，即基于改变热力条件来释放和测试吸附态卤素的含量（周奇明等，2020）。该方法类似土壤热释汞的测量方法，目前在矿区应用较少（智超等，2015）。

土壤微细粒测量是近十几年发展起来的一种深穿透地球化学技术，其原理是通过分离土壤中细粒级组分以富集来自于深部矿体的活动性金属元素，从而达到识别深部隐伏矿的目的，实质上属于传统土壤测量范畴（姚文生等，2012；刘汉粮等，2014；申伍军等，2017；张必敏等，2019），只是强调采用物理分离技术以达到"偏"提取微细粒的目的。浅钻测量是介于传统化探与深穿透化探之间的适用于浅覆盖区的一种化探方法，采样是基于钻探，通常为车载空气动力反循环钻，采样对象为（浅覆盖层至）残坡积层和基岩（张必敏等，2011；Zhang et al.，2016；李小东等，2018；肖艳东等，2019），从异常形成机理上看应属于传统化探范畴，但手段应属非传统化探方法技术。

生物地球化学测量是指通过对生物器官、产物或排泄物进行元素含量分析测试并发现矿化异常的一种化探方法。最常用的是植物地球化学测量（王天刚等，2020），但一些矿区某些动物的排泄物和产物中元素含量明显高于背景区的含量，这使其也成为有效的地球化学找矿采样对象（于扬等，2019）。

综上所述，非传统化探方法是对传统化探方法的有效补充，适合在传统化探方法失效的覆盖区，所圈定的异常大多为后生异常。目前应用较广泛的是地气测量、地电化学测量、氡气测量和金属活动态测量，但大多数非传统化探方法技术目前仍处于试验阶段，期待规范化和广泛应用。非传统化探目前的状态可概括为"野外剖面对矿体，室内实验查机理；扫面类比圈靶区，工程验证添实例；技术展望在哪里，规范推广创效益"。

2.2 评 价

勘查地球化学的评价包括异常圈定和异常评价两方面，异常又可分为单元素（或单指标）异常和综合异常两类。对于综合异常的确定其关键在于综合指标应包含哪些单指标以及如何构建综合指标。不同的矿床具有不同的成矿元素组合，而特征的成矿元素组合又可指示特定的成矿信息，因此成矿指示元素通常被选为构建综合指标的指标集（向运川等，2010，2018；龚庆杰等，2015；陈晶等，2017）。在此仅从元素组合、异常圈定和异常评价3个方面进行综述。

2.2.1 元素组合

为确定找矿（或成矿）指示元素组合，通常采用的方法有多元统计法、散点图法、空间对比法等。

多元统计法主要是确定找矿指示元素组合，目前主要采用的方法有富集系数法、相关分析法、聚类分析法、因子分析法等。需要特别说明的是，这里的富集系数是指研究对象元素的含量（或研究区元素含量的平均值）与区域元素丰度（或平均值）的比值，这与元素的衬值（即元素的含量与某区域元素含量平均值的比值）相似，但与上下文中提到的富集因子（指样品中两元素的比值与参考样品中两元素比值的比值）不同。

在全国矿产资源潜力评价典型矿床地球化学建模研究中，基于从新鲜岩石→蚀变岩→矿石的研究思路中基本采用富集系数法来确定成矿（或找矿）指示元素组合（龚庆杰等，2015；向运川等，2018）。在化探普查和详查中也经常使用富集系数来确定找矿指示元素组合（申扎根，2013；张运强等，2015；何旺等，2019；翁望飞等，2020）。相关分析方法包括夹角余弦正比例相关分析、皮尔逊线性相关分析、基于排序的斯皮尔曼相关分析、基于同增长趋势的肯德尔相关分析以及基于发展态势的灰关联相关分析，目前大多使用皮尔逊线性相关分析。聚类分析法是研究物以类聚的探索性分类方法，目前在勘查地球化学领域多采用基于皮尔逊相关系数的树状层次分析图来直观表达分类结果。因子分析主要是一种降维的分析思路，基于皮尔逊相关系数的因子分析在因子命名解释环节可确定即将集成为同一因子的元素组合。

因此，在基于皮尔逊相关系数时，相关分析、聚类分析、因子分析在确定元素组合方面其实质是相同的，只是表达形式不同而已。在传统化探与非传统化探的数据处理方面这3种方法均得到了广泛应用（李天虎等，2012；张运强等，2015；刘普凯等，2018；肖高强等，2020）。

散点图法主要是为了克服多元统计法中多依赖于皮尔逊线性相关系数而忽视了非线性关系而回归原始的二维直观定性图解法。如陈晶等（2017）对河南嵩县槐树坪金矿区围岩、蚀变岩、矿石3类介质分析了Au与28种微量元素的散点图关系，进而确定了槐树坪金矿区化探找矿的指示元素组合。刘普凯等（2018）在陕西凤县庞家河金矿原生晕勘查中在双对数坐标系中研究Au与其他9种微量元素的关系以确定找矿指示元素组合等。目前的研究大多仍基于原始成分数据进行，很少采用从Aitchison空间到Euclidean空间数据转换的方法。

空间对比法是基于地球化学剖析图（或地球化学剖面剖析图）将地球化学异常区与矿床（或矿体）的地表出露区进行空间对比从而确定找矿指示元素的方法。地球化学异常剖析图是指包含有地质图的多个单元素地球化学异常图的排列放置图，地球化学剖面剖析图是指含有地质剖面图的（多个）地球化学剖面图的排列放置图。在非传统化探方法技术研究中大多采用地球化学剖面剖析图法来评定方法的可行性或确定有效指示元素组合（邱炜等，2013；王勇等，2017；杨刚刚等，2018）。在传统化探的典型矿床地球化学找矿模型研究中，通常采用地球化学异常剖析图法来确定找矿指示元素（佟依坤等，2104；龚庆杰等，2015；张晶等，2018）。

2.2.2 异常圈定

根据地球化学变量或指标性质，将地球化学异常圈定划分为单元素（或单指标）异常和综合异常圈定两大类。

（1）单元素异常圈定。单元素（或单指标）异常圈定的关键环节为确定异常下限或背景上限，目前常用的异常下限确定方法可分为定值异常下限和变值异常下限两种。前者又可分为均值-标准差法、累频法、分形法和标尺法四类，后者又可分为分区定值法和连续变化面法两类（向运川等，2018）。

就定值异常下限的确定，通常采用传统的均值-标准差法，累频法是目前绘制单元素地球化学图的常用方法，分形法是目前科研论文中较常用的方法之一（向运川等，2018）。这些方法在全国矿产资源潜力评价化探资料应用技术要求中均作为可选方法（向运川等，2010）。四分位法（即异常下限为上四分数与n倍四分位距之和，实质上类似剔除离异数据的均值-标准差法）也常用来确定异常下限（何旺等，2019），但在实际应用中并无定式，经常有将多种方法确定的异常下限进行对比后再综合确定某一值的情况（宋贺民等，2014；徐云峰等，2018）。标尺法是为便于不同研究区进行异常对比时提出的，建议选定一个基本标尺（类似行业标准或国家标准中的推荐值）作为异常下限（佟依坤等，2014；龚庆杰等，2015）。如在全国矿产资源潜力评价典型矿床地球化学建模研究中，建议将全国或成矿省等的元素含量背景值和边界品位作为标尺进而提出定值七级异常划分方案，该方法不仅可以确定异常下限，而且也可以对异常进行分级评价（向运川等，2018）。

就变值异常下限的确定，通常采用分区定值法，这是在地质地理情况复杂且面积较大的地区将其划分成一些子区，然后在各个子区内按照定值异常下限方法分别确定其异常区，从而实现在整体研究区内的异常圈定（赵禹等，2014）。连续变化面法是把异常下限（或背景值）当作一个连续变化着的地球化学面来看待，每一数据点各具有自己的异常下限。该方法主要有滑动定值法（含滑动平均剩余值法、子区中位数衬值滤波法、子区自适应衬值滤波法等）、插值背景法（含趋势面法、分形插值法等）、机器学习法、风化背景法等（袁玉涛，2015；向运川等，2018；左仁广，2019；Tian et al.，2018，2019）。

风化背景法是龚庆杰等（Gong et al.，2013）在研究胶东玲珑花岗岩风化过程时提出的一种表征元素风化行为的经验方程。该经验方程基于样品的主量元素含量可计算出微量元素的含量，并将其作为背景值，进而可采用衬度、剩余值或作为背景标尺来圈定异常（袁玉涛，2015；向运川等，2018；Gong et al.，2018；许胜超等，2019）。需要特别说明的是，风化背景法是基于样品岩性及风化程度差异而提出的经验方程，是一种真正意义上的"地球化学数据"处理方法，而上述其他处理方法则属于"数据"处理方法。

（2）综合异常圈定。综合异常圈定是指将多个指标经过某种算法而集成为综合指标，然后采用单指标确定异常的方法来对该综合指标进行异常圈定。其关键环节是如何构建综合指标。

常见的综合指标集成方法有元素含量或衬值的（加权）累加、累乘（含比值）、向量模（即平方和的开方）、回归分析值、因子得分等（向运川等，2014；陈晶等，2017；何旺等，2019；Xiang et al.，2019）。这些方法在全国矿产资源潜力评价及环境污染指数计算中经常使用（向运川等，2018）。此外，郝立波等（2016）综合考虑研究区富集与贫化元素提出投影寻踪遗传算法来构建综合指标，该指标的实质类似于灰关联系数的计算，即需要将元素划分为富集型（或效益型）和贫化型（或成本型）两类，进而将所有指标在同一样品（或方案）中的取值加权求和而形成综合指标。Liu 等（2019）综合数据驱动和知识驱动提出了成分平衡分析法来构建综合指标以及 Xiong 和 Zuo（2020）提出的混合模型等。

上述综合异常指标的集成本质上仍属于"数据"处理方法，即不是地球化学数据或元素含量数据仍可采用上述方法进行集成以构建综合指标。李睿堃等（2019）提出了一种真正意义上的"地球化学数据"找矿综合指标——金矿化地球化学基因的矿化相似度，即将样品与理想金矿石在金矿化地球化学基因上的相似度作为金矿化综合指标。

2.2.3 异常评价

异常评价是对异常的识别、评序和分类。依据异常指标的性质通常将异常评价分为单元素异常评价、综合异常评价和找矿预测区评价（向运川等，2018）。

（1）单元素异常评价。单元素异常评价首先需要进行异常识别，即综合考虑景观区、地质、矿产、特殊环境、采样域分析的偏差等因素，将异常划分为矿致异常和非矿致异常两类，然后针对矿致异常进行评序和分类。

在异常评序时通常采用的参数有异常规模（刘君安等，2017；刘汉粮等，2018）、规格化面金属量（龚鹏和马振东，2013；虞航等，2019）、七级异常中的异常分级值（Gong et al.，2018；许胜超等，2019）等。这里需要特别说明的是，龚庆杰等提出的"七级异常"方案不仅可以圈定异常，而且可以对异常进行分级评序（Gong et al.，2018；向运川等，2018），其中的"变值七级异常"方法不仅是客观的、多变量的、变值的、无需考虑数据分布形式的异常确定方法，而且也综合考虑了岩性、元素的风化行为及其边界品位的异常分级方法，是一种真正意义上的"地球化学数据"异常确定和分级方法。

异常分类是在异常评序的基础上，依据异常与矿的关系将异常划分为甲、乙、丙、丁 4 类。甲类异常是指已知矿异常或查证见矿异常，乙类异常是指推断的矿致异常，丙类异常是指性质和找矿前景不明的异常，丁类异常则是指无找矿意义和前景的异常（向运川等，2018）。

（2）综合异常评价。综合异常评价首先需要根据综合指标的含义进行异常识别，将异常分为找矿异常、岩性异常、构造异常、污染异常。对找矿综合异常的评序和分类可仿照单元素异常的评序与分类方法进行（严明书等，2016；刘攀峰等，2017；孙婷婷等，2020）。李睿堃（2019）在对河南嵩县槐树坪金矿区 1:2 万化探详查及熊耳山地区 1:20 万区域化探的数据处理中，采用基于金矿化地球化学基因的矿化相似度综合指标，不仅圈定了综合异常区而且还基于相似度值对综合异常进行了 3 级划分，这是近十年来在真正意义上的"地球化学数据"综合异常评价方面的创新进展。

（3）找矿预测区评价。找矿预测区是在单元素异常或综合异常评价的基础上圈定的具有找矿潜力的区域。依据单元素异常或综合异常的评序分类，可将找矿预测区分为 A、B、C 3 级。A 级预测区对应甲类异常或找矿潜力大的多个乙类异常，B 级预测区对应乙类异常，C 级预测区则对应丙类异常。近十年来，这一评价方法仍被广泛采用（张运强等，2015；徐云峰等，2018；肖高强等，2020），目前在评价方法上仍未出现明显的创新发展。

2.3 开　　发

勘查地球化学在开发方面主要针对采矿和选矿这两个领域，这基本超出目前勘查地球化学的研究范围，但地浸法铀矿开采与勘查地球化学的关系则比较紧密。

除上文涉及的非传统化探方法在铀矿勘查中得到广泛应用外，相应的方法技术在采矿方面，尤其是在地浸采铀方面也得到了应用（庞康，2017）。如在偏提取技术方面，同样采用 Tessier 等（1979）提出的 5 相态顺序提取技术，马强等（2012）对新疆某砂岩型铀矿矿芯提取研究发现惰性铀（即残渣态铀）是矿石中铀的重要赋存形式（约占 38%），这部分铀对地浸采铀方法来说是无效的；陈亮等（2013）对新疆伊利盆地某砂岩型铀矿石研究发现活性铀（即除残渣态以外的 4 个相态）在矿石中占绝对主导地位（约占 92%），这非常适合地浸开采。在铀矿物及其存在状态研究方面，张鑫等（2015）对新疆伊犁盆地蒙其古尔砂岩型铀矿的研究发现，铀矿物以沥青铀矿为主，含少量铀石，铀矿主要赋存于碎屑颗粒边缘和裂隙中，具明显后期吸附沉积的特征，这对地浸采铀非常有利。杨冰彬等（2020）对新疆吐哈盆地十红滩铀矿北矿带矿石研究发现，铀矿物主要为沥青铀矿、铀石和铀钛氧化物（含铀钛氧化物），且主体赋存在粒径小于 0.05 mm 的填隙物中，呈团块状、（弱）聚集状及星点状分布，这对开展地浸选矿工艺具有借鉴意义。上述这些偏提取、活动态、细粒级、后生成因的技术正是勘查地球化学在采选方面开发利用的体现。

2.4 修　　复

勘查地球化学在修复方面主要是针对矿山环境的修复，地球化学工程学这一新兴领域由此产生。地球化学工程学是由荷兰科学家 Schuiling 于 20 世纪 80 年代末提出，2002 年吴传璧在国内进行了介绍，其核心思想是把环境污染治理或修复放在天然地球化学系统中来考虑，依据地球化学基本规律充分利用废弃物及其特性实现变废为宝、以废攻废的工艺目的。由于早期的地球化学工程学强调其研究内容不仅包含围绕治理或修复的地球化学方法工艺，而且也包含地球化学调查和监测，这与中国 20 世纪 80 年代就将区域化探成果应用于环境领域以及 90 年代开展的多目标地球化学调查和 21 世纪初开展的生态地球化学评价的研究思路一致，中国的这一思路包含调查、评价（含监测与预警）、开发、修复 4 个方面，涵盖了地球化学工程学的研究内容，因此地球化学工程学这一概念并未在中国得到普及。中国学者倾向于将地球化学工程学的研究内容局限在围绕环境问题治理或修复的地球化学工艺方面（李杰等，2012），更多的学者则是倾向于直接使用污染治理、环境修复这些属于环境领域的名词。针对勘查地球化学的修复主要体现在矿山环境修复（或矿山污染治理）方面。矿山环境修复的前提是首先发现或查明环境问题，即首先是矿山环境评价环节。针对矿山环境的地球化学评价可归属到环境地球化学的评价方面（温汉辉等，2019），在此仅针对矿山环境地球化学修复（或狭义的地球化学工程学）进行综述。

矿山环境的地球化学修复针对的介质主要为土壤和地表水（或浅层地下水）（王桂芳等，2020），修复的内容主要为酸碱调节、肥力调节和重金属治理等。对矿区土壤介质的修复主要有植物修复技术法，如植物恢复对土壤 pH 与肥力的调节（赵韵美等，2014；Huang et al.，2016；赵同谦等，2017；杨鑫光等，2018）、植物对重金属的稳定与提取（毕亚凡和徐俊虎，2012），和添加改良剂修复技术法，如采用生物炭固定重金属法（徐艳等，2019），以及综合协同修复技术（肖庆超等，2015；李秀玲等，2019）等。针对矿区地表水（或浅层地下水）的修复主要有复合材料调节吸附法（肖利萍等，2014）、微生物调节吸附法（徐师等，2018）等。

除上述针对矿区土壤和地表水（或浅层地下水）的修复技术外，还可从土壤和排泄水的源头如尾矿堆或尾矿库进行控制（林海等，2019），如采用适当材料将重金属进行钝化或固定（莫斌吉等，2014；周继梅等，2018）、中和沉淀（刘强，2018）、材料吸附（黄羽飞等，2012）等治理技术。

尽管上述矿山环境治理和修复技术在思路环节已经成熟（如从源头控制治理到末端工艺技术修复），且在室内实验、现场试验、矿区复垦（或植物恢复）方面已进行了大量的研究和应用，但整体上看，基于地球化学原理的矿山修复技术或地球化学工程学目前仍处于起步探索阶段，尚未涌现出大量的成功案例以及成熟的具普适性的地球化学工艺技术。

3　总结与展望

应用地球化学起源于勘查地球化学，兴起于环境地球化学，是利用地球化学的基本理论和方法技术为资源勘查、环境评价及社会服务的一门地球化学分支学科。在资源和环境两个领域的应用大体上可划分为调查、评价、开发和修复4个方面。因其源于勘查地球化学，故将在资源领域的应用地球化学泛称为勘查地球化学。

勘查地球化学可分为传统化探方法和非传统化探方法，传统化探方法近十年来的进展主要体现在方法技术方面的系统总结、完善、规范化和应用推广方面的持续发展与"走出去"战略。随着剥蚀出露区"扫面"接近"全覆盖"，"攻深找盲"的原生晕化探将是"最后的坚守"，其应用和方法技术期待完善与创新。针对覆盖区的非传统化探近十年来得到了蓬勃发展，地气测量和地电化学测量的方法技术进一步成熟，在机理、试验、应用环节均得到了深入发展。由于覆盖区传统化探方法的"失效"，偏提取及其他非传统化探方法技术再次得到了"青睐"。

勘查地球化学的评价主要集中在确定元素组合、圈定异常、评价异常（含圈定找矿预测区）3个方面。基于"先验知识"的多元统计法及对比法仍是确定元素组合、圈定异常和评价异常的主要手段。尽管机器学习也在异常识别中得到不断尝试，但其技术仍依赖于对比思路的数据驱动，缺少真正的知识驱动。相对于各类"数据处理方法"，针对元素地球化学行为的"地球化学数据处理方法"初见成效，这将为基于知识驱动的人工智能在地球化学大数据领域的应用提供重要技术支撑。

勘查地球化学在开发方面则主要对应采矿和选矿这两个领域，这基本超出目前勘查地球化学的研究范围，但针对地浸法铀矿采选仍将发挥其技术优势。勘查地球化学在修复方面则主要是针对矿山环境的修复，由此产生了地球化学工程学这一新兴领域。虽然矿山污染治理和环境修复技术在思路环节已经成熟，但基于地球化学原理的矿山修复技术或地球化学工程学目前仍处于起步探索阶段，期待典型成功案例的涌现和广泛应用。

致谢：感谢我国应用地球化学同仁，尤其是勘查地球化学同仁，所取得的丰硕成果，这为本文的撰写提供了动力和坚实的基础。

参 考 文 献

毕亚凡,徐俊虎. 2012. 矿山重金属污染土壤的植物修复技术. 武汉工程大学学报,34(10):28-31
陈刚. 2017. 地气作用下地质处置核废物中核素迁移的实验研究. 成都:成都理工大学硕士学位论文
陈国光,马振东,吴小环,李敏,张华,湛龙,张德存,叶家瑜. 2015. 矿产地球化学勘查体系的探讨. 物探与化探,39(3):437-442
陈晶,龚庆杰,王炯辉,陈良,严桃桃,贺昕宇,李金哲. 2017. 豫西熊耳山地区槐树坪金矿岩石-土壤找矿指示元素的确定及其应用. 岩石学报,33(7):2302-2312
陈亮,谭凯旋,谢焱石,刘江,黄伟,王正庆,马强. 2013. 伊犁盆地某砂岩铀矿铀赋存形态及其地浸意义. 矿业研究与开发,33(5):118-121
陈亚东,孙彬彬,刘占元,周国华,朱晓婷. 2015. 地电化学提取有效性及提取条件试验——以半干旱草原风成砂浅覆盖景观区为例. 物探与化探,39(5):1008-1012
董昕昱,万建军,樊哲强,李仁泽,彭波. 2020. 土壤氡测量在塔拉乌苏地区砂岩型铀矿勘查中的应用. 世界核地质科学,37(1):41-46
付亚龙,常海钦,林鑫,孟刚刚,张苗苗. 2019. 金属活动态测量在冲积平原覆盖区隐伏矿的试验研究——以安徽无为龙潭头硫铁矿为例. 物探化探计算技术,41(3):401-411
葛良全,童纯菡,贺振华,李巨初,沈松平,杨凤根. 1997. 隐伏断裂上方地气异常特征及其机理研究. 成都理工学院学报,24(3):29-35
龚鹏,马振东. 2013. 矿产预测中区域化探异常的识别和评价. 地球科学——中国地质大学学报,38(S1):113-125
龚庆杰,喻劲松,韩东昱,刘宁强. 2015. 豫西牛头沟金矿地球化学找矿模型与定量预测. 北京:冶金工业出版社
郭祥义,叶荣,鲁美,韩志轩,王永康,张文慧,王可祥. 2019. 半干旱荒漠草原覆盖区地气测量方法研究——以内蒙古维拉斯托锌铜多金属矿床为例. 地质与勘探,55(3):789-800
韩娟,刘汉彬,孙晔,李军杰,韩效忠,钟芳文,金贵善. 2013. 土壤氡测量在呼斯梁-柴登壕地区砂岩型铀矿勘查中的应用. 世界核地质科学,30(1):38-43
韩伟,刘华忠,王成文,宋云涛,王乔林,孔牧. 2019. 哈密天宇铜镍矿地气测量地球化学特征及指示意义. 物探与化探,43(3):502-508

韩志轩,廖建国,张津隆,张必敏,王学求. 2017. 穿透性地球化学勘查技术综述与展望. 地球科学进展,32(8):828-838

郝立波,田密,赵新运,赵昕,张瑞森,谷雪. 2016. 基于实码加速遗传算法的投影寻踪模型在圈定水系沉积物地球化学异常中的应用——以湖南某铅锌矿床为例. 物探与化探,40(6):1151-1156

何旺,罗先熔,高文,郑超杰,周子俣,黄文斌,肖小强,王生龙. 2019. 青海省都兰县五龙沟-高地地区水系沉积物地球化学特征及找矿远景. 矿物岩石地球化学通报,38(5):1017-1023

胡波,陈刚,邱腾,谢克文. 2016. 诸广山某矿区地气稀土配分与热液成矿关系特征. 科学技术与工程,16(6):134-138

黄学强,罗先熔,刘巍,王光洪,黄蔚阁,董俊秋. 2013. 凹陷盆地铜镍多金属矿床地电化学异常特征及找矿预测. 物探与化探,37(2):199-205

黄羽飞,陈宇,刘峰彪. 2012. HDS工艺及树脂吸附法深度处理酸性矿山废水的试验研究. 有色金属(选矿部分),(6):49-52,64

贾玉杰,龚庆杰,韩东昱,刘宁强,夏旭丽,李晓蕾. 2013. 化探方法技术之取样粒度研究——以豫西牛头沟金矿1:5万化探普查为例. 地质与勘探,49(5):928-938

康欢,陈岳龙,李大鹏,徐云亮,房明亮. 2019. 二连盆地哈达图铀矿床覆盖区地球化学异常源示踪与判别. 铀矿地质,35(6):351-358

柯丹,吴国东,刘洪军,陈浩,王勇. 2016. 便携式多功能地电化学供电装置的研制. 物探与化探,40(6):1211-1216

柯龙跃,贾福聚,王峰,李星,张权. 2017. 地电提取测量法与电吸附法比较研究. 中国锰业,35(2):181-183

蓝天,罗先熔,陈晓青. 2018. 湖南国庆矿区地电化学方法寻找隐伏铜矿预测研究. 地质与勘探,54(3):563-573

李宝强,张晶,范堡程,吕鹏瑞,拉赫蒙别克. 2014. 中塔合作帕米尔地球化学调查成果与展望. 中国地质调查,1(3):22-31

李博. 2018. 地气作用下地质处置核废物中核素迁移的行为研究. 成都:成都理工大学硕士学位论文

李杰,施泽明,高琴,倪师军,张成江,李世男. 2012. 我国城市地球化学热点领域研究进展及展望. 物探与化探,36(3):429-434

李金哲. 2018. 风化过程中金含量行为的定量表征. 北京:中国地质大学(北京)博士学位论文

李金哲,龚庆杰,刘亚轩,严桃桃,李睿堃. 2018. 风化过程中硒背景值的定量表征. 现代地质,32(5):1031-1041

李睿堃. 2019. 金矿化地球化学基因:构建-检验-意义. 北京:中国地质大学(北京)博士学位论文

李天虎,罗先熔,彭桥梁,王伟,罗小平,宋忠宝,文雪琴. 2012. 甘肃金川铜镍矿床Ⅰ矿区深部边部地电化学-地电化学-地球物理多元信息成矿预测. 地质通报,31(7):1192-1200

李伟,刘翠辉,贺根文,温珍连,陈琪. 2017. 壤中汞气测量在于都营脑隐伏矿产勘查中的应用. 物探与化探,41(5):840-845

李小东,王明卫,写熹,孔广林,杜玉雕,李延月. 2018. 安徽凤阳地区浅钻地球化学方法的找矿应用研究. 地质调查与研究,41(3):217-223

李秀玲,韦岩松,辛磊,高宇星,韦诗琪,覃拥灵. 2019. 尾矿区砷污染土壤的植物、微生物协同修复. 湿法冶金,38(1):64-68,74

李业强,杨有泽,肖昶,荣耀. 2014. 土壤天然热释光法在蒙古国肯特省哈沙顿道鲁高德地区铀矿勘查中的应用. 世界核地质科学,31(3):542-546

李友余,王道永,吴德超,周琳雄. 2013. 地气测量在二郎山隧道隧址区隐伏断裂探测中的应用. 煤田地质与勘探,41(2):16-20

廖驾,吴德超,吕少辉,黄晨. 2012. 西藏加查地区某水电站坝址区隐伏断裂研究. 华南地质与矿产,28(1):64-70

林海,李真,贺银海,董颖博,李冰. 2019. 硫酸盐还原菌治理酸性矿山废水研究进展. 环境保护科学,45(5):25-31

刘草,陈远荣,鲁富云,徐建东,段炼,曾旭,洪文帅,韦永先. 2016. 基于深地勘探的化探新方法发展现状分析. 矿产与地质,30(3):446-450

刘昶,曹建劲,柯红玲. 2011. 滇东北永胜得铜矿床地气微粒特征. 化工矿产地质,33(4):201-207

刘刚,罗先熔,郑超杰,杨笑笑,蓝天,刘延斌,陈武,李武毅. 2018. 地电化学集成技术在藏南姐纳各普金多金属矿区的找矿预测研究. 矿物岩石地球化学通报,37(5):894-902

刘汉粮,王学求,张必敏,刘东盛,郭守栋. 2014. 沙泉子隐伏铜镍矿地球化学勘查方法试验. 物探化探计算技术,36(6):763-770

刘汉粮,聂兰仕,王学求,张义波,刘东盛,王玮,迟清华. 2018. 中蒙跨境阿尔泰构造带稀有元素锂区域地球化学分布. 现代地质,32(3):493-499

刘洪军,吴国东,谢胜凯. 2017. 地电化学提取铀的吸附材料制备方法. 世界核地质科学,34(1):32-35

刘君安,郭维民,徐鸣,朱云鹤,杨献忠,曾勇,姚春彦,姚仲友,李春海,张洁. 2017. 秘鲁阿雷基帕省阿蒂科地区水系沉积物地球化学特征及找矿远景预测. 地质通报,36(12):2264-2274

刘攀峰,文美兰,张佳莉. 2016. 地电化学集成技术在云南西邑铅锌矿区的找矿应用. 物探与化探,40(4):655-660

刘攀峰,文美兰,张佳莉,刘映东,王金龙. 2017. 云南西邑隐伏铅锌矿地电化学异常特征及找矿预测. 地质科技情报,36(6):197-206

刘攀峰,罗先熔,文美兰,张佳莉,郑超杰,杨龙坤,韦选建. 2018. 近三十年来我国地电化学技术研究回顾与展望. 桂林理工大学学报,38(1):47-55

刘普凯,卢飞,赵晓振. 2018. 陕西凤县庞家河金矿床18号勘探线原生晕分带特征及深部预测. 地质找矿论丛,33(1):115-123

刘强. 2018. 酸性重金属废水的联合处理试验研究. 黄金,39(2):64-66

刘晓东,董明,王庆喜. 2018. 分量化探法在连山关黄沟地区铀矿勘查中的应用. 东华理工大学学报(自然科学版),41(2):143-151

刘晓辉,周四春,童纯菡,胡波. 2012. 提高地气探测灵敏度的方法. 物探与化探,36(6):1064-1067

刘晓辉,周四春,胡波. 2015. 地气测量技术用于隐伏温泉勘探的初步研究. 矿物学报,35(S1):1110-1111

刘雪敏,王学求. 2014. 全球尺度地球化学填图计划对比研究. 地学前缘,21(2):275-285

刘雪敏,陈岳龙,王学求. 2012. 深穿透地球化学异常源同位素识别研究:以新疆金窝子金矿床、内蒙古拜仁达坝-维拉斯托多金属矿床为例. 现代地质,26(5):1104-1116

刘延斌,罗先熔,刘攀峰,郑超杰,刘刚,宋兵强,宋贵斌. 2018. 地电化学集成技术在内蒙古格鲁其堆山矿区及外围寻找隐伏铅锌矿的应用. 地质与勘探,54(5):1001-1012

刘映东,张必敏,罗先熔. 2017. 地电化学在隐伏铜镍矿勘查中的应用及异常形成机理探讨. 地质与勘探,53(4):694-703

龙莎莎,谈树成,蒋顺德. 2005. 浅析环境地球化学的研究现状. 云南地理环境研究,17(S1):81–85

卢福隆. 2019. 内蒙古阿荣旗覆盖区地气法试点研究. 北京:中国地质大学(北京)硕士学位论文

鲁美,叶荣,张必敏,王永康. 2019. 覆盖区地球化学勘查进展. 矿床地质,38(6):1408–1411

马强,冯志刚,孙静,谢二举,李小军. 2012. 新疆某地浸砂岩型铀矿中铀赋存形态的研究. 岩矿测试,31(3):501–506

马生明,朱立新,刘海良,王会强,徐甬钰. 2011. 甘肃北山辉铜山铜矿地球化学异常结构研究. 地球学报,32(4):405–412

莫斌吉,雷良奇,黄祥林,赵蛟彬,徐沛斌. 2014. 镉在硫化矿尾矿中的地球化学行为及其污染防治. 有色金属(矿山部分),66(2):34–38

聂凤莲,张蜀冀,陈雪,艾晓军. 2011. ICP-MS 法测定地电化学(泡塑)样品中痕量金. 黄金,32(12):58–61

庞康. 2017. 浅析砂岩型铀矿特征及其开采方法. 地下水,39(3):226–229

邱炜,潘彤,金永明,罗先熔. 2013. 地电化学提取技术在玉树莫海拉亨铅锌矿区的找矿应用研究. 黄金科学技术,21(2):36–39

邱炜,任华,林艳海,罗先熔,李杰. 2017. 地电化学提取技术在青藏高原干旱荒漠区的参数对比研究及找矿预测——以拉陵高里河西地区为例. 矿产勘查,8(1):117–123

申伍军,白春东,陈圆圆,张新政. 2017. 冀北蔡家营地区隐伏矿地球化学勘查方法. 物探与化探,41(2):210–218

申扎根. 2013. 水系沉积物地球化学测量在内蒙古凉城地区找矿应用. 矿产与地质,27(4):324–329

宋贺民,张辉,顾松松,王占彬,李洪杰,许鲁宁. 2014. 新疆哈拉奇地区水系沉积物地球化学特征及找矿方向. 地质通报,33(1):71–78

孙彬彬,刘占元,周国华. 2011. 固体载体型元素提取器研制. 物探与化探,35(3):375–378

孙彬彬,张学君,刘占元,周国华,张必敏,陈亚东. 2015. 地电化学异常形成机理初探. 物探与化探,39(6):1183–1187

孙彬彬,张学君,周国华,曾道明,贺灵. 2016. 地电化学泡塑载体分析质量监控研究. 物探与化探,40(3):557–560

孙婷婷,李戬,思积勇. 2020. 青海查肖玛地区1/5万水系沉积物地球化学特征及成矿预测. 中国锰业,38(1):30–35

唐桢,周四春,万志雄. 2012. 坪宝地区铅锌矿地气元素迁移规律研究. 现代矿业,(10):59–61,88

唐桢,周四春,曹勇,洪友朋,宋伟力,肖曙光,王冲. 2013. 地气物质迁移模型实验. 中国矿业,22(6):106–109,117

佟依坤,龚庆杰,韩东昱,刘宁强,徐增裕,于文龙. 2014. 化探技术之成矿指示元素组研究——以豫西牛头沟金矿为例. 地质与勘探,50(4):712–724

万卫,汪明启,高玉岩,秦欢欢,赖冬蓉. 2019. 南方红壤区隐伏矿床地气试验研究——以广西德保铜锡矿区为例. 地质找矿论丛,34(3):438–444

汪明启. 2005. 从勘查地球化学到应用地球化学——第21届国际勘查地球化学会议综述. 物探与化探,29(2):96–100

汪明启,叶荣. 2019. 面向未来资源勘查的应用地球化学——第28届 IAGS 会议综述. 物探与化探,43(4):679–691

王东升,王勇,刘洪军,王晓赛. 2019. 地气测量方法在热液型铀矿床的试验研究. 世界核地质科学,36(1):35–42

王桂芳,王翼文,张帅,杨梅金,肖慧珍,王飞龙. 2020. 硫化矿尾矿的综合利用及污染治理研究进展. 金属矿山,(2):111–116

王滔,彭秀红. 2013. 不同条件下地气异常重现性研究. 现代矿业,(7):49–51

王天刚,Fabris A,姚仲友,Hou B H,赵宇浩,赵晓丹,朱意萍. 2020. 勘查植物地球化学在我国不同地球化学景观区的应用现状及展望. 华东地质,41(1):1–7

王玮,王学求,张必敏,聂兰仕,刘汉粮,刘东盛,韩志轩,迟清华,徐善法,周建. 2019. 老挝国家尺度地球化学填图进展——以氟元素为例. 桂林理工大学学报,39(2):335–340

王文,刘应平. 2012. 青海扎家同哪金矿勘查区地球化学特征与找矿潜力分析. 地质与勘探,48(6):1206–1213

王晓佳,鲁美,王振凯,叶荣. 2016. 申家窑金矿床地气测量异常特征. 地质与勘探,52(4):667–677

王学求. 2013. 勘查地球化学80年来重大事件回顾. 中国地质,40(1):322–330

王学求,聂兰仕. 2016. "一带一路"地球化学填图实现合作共赢. 中国地质调查成果快讯,(8-9):31–33

王学求,谢学锦. 1996. 非传统金矿化探的理论与方法技术研究. 地质学报,70(1):84–95

王学求,叶荣. 2011. 纳米金属微粒发现——深穿透地球化学的微观证据. 地球学报,32(1):7–12

王学求,张德会. 2005. 应用地球化学的过去、现在与未来(序二). 地质通报,24(10-11),doi:10.3969/j.issn.1671-2552.2005.10.002

王学求,张必敏,姚文生,刘雪敏. 2014. 地球化学探测:从纳米到全球. 地学前缘,21(1):65–74

王学求,周建,徐善法,迟清华,聂兰仕,张必敏,姚文生,王玮,刘汉粮,刘东盛,韩志轩,柳青青. 2016. 全国地球化学基准网建立与土壤地球化学基准值特征. 中国地质,43(5):1469–1480

王学求,张勤,白金峰,姚文生,刘妹,刘雪敏,王玮. 2020. 地球化学基准与环境监测实验室分析指标对比与建议. 岩矿测试,39(1):1–14

王勇,叶荣,张必敏,杨榕,漆富勇. 2012. 地气测量在戈壁覆盖区210金矿的试验. 物探与化探,36(2):263–266

王勇,叶荣,张必敏,柯丹. 2014. 戈壁区210隐伏矿上方地气中 Au 微粒异常形成模式. 世界核地质科学,31(3):547–552

王勇,叶荣,刘洪军,吴国东,王东升. 2017. 热液型铀矿穿透性地球化学勘查方法的微观证据. 世界核地质科学,34(3):167–173

王正阳,曹建劲,易杰,罗松英,赖佩欣. 2013. 上升气流微粒在四川盆地钾盐卤水找矿方面的应用前景. 矿物学报,33(S2):843

韦文定,陈金声. 2013. 有机烃气综合地球化学测量在百福堂铅锌矿找矿中的应用. 物探与化探,37(3):416–421

韦选建,罗先熔,杨龙坤,刘攀峰,王艳忠,李海洋,闫伟. 2017. 吉林长白山玄武岩覆盖区地电化学找矿研究. 矿产勘查,8(3):464–472

温汉辉,韩丽杰,曲金灿,何宏伟. 2019. 广东某热液型钨矿对土壤与地下水生态环境的影响研究. 上海国土资源,40(1):91–95

文美兰,罗先熔. 2013. 金川铜镍矿床多元地学信息找矿研究. 中国地质,40(2):594–601

文美兰,欧阳菲. 2013. 广西平桂地区锡矿土壤离子电导率异常特征和离子成分及找矿预测. 地质通报,32(5):784–789

文雄亮,胡斌,柳智. 2020. 湖南临湘虎形山钨矿原生晕地球化学特征与深部找矿预测. 矿物岩石地球化学通报,https://doi.org/10.19658/j.issn.1007-2802.2020.39.024

翁望飞,王德恩,王邦民,丁勇,王拥军. 2020. 安徽省祁门—黟县地区水系沉积物地球化学特征及找矿方向. 物探与化探,44(1):1-12

吴二,陈远荣,刘巍,卢月玲,蒋慧俏,张冠清,蓝妮拉,高儇博. 2014. 烃气测量法在辽宁白云金矿找矿潜力评价中的应用. 物探与化探,38(2):248-254

吴泽民,罗齐彬,吴信民,王帅帅,符志军,肖坤,谢尚平. 2019. 地气测量在相山隐伏铀矿勘探中的试验研究. 东华理工大学学报(自然科学版),42(2):148-155

奚小环. 2019. 自然资源时期:大数据与地球系统科学——再论全面发展时期的勘查地球化学. 物探与化探,43(3):449-460

奚小环,李敏. 2013. 现代地质工作重要发展领域:"十一五"期间勘查地球化学评述. 地学前缘,20(3):161-169

奚小环,李敏. 2017. 现代勘查地球化学科学体系概论:"十二五"期间勘查成果评述. 物探与化探,41(5):779-793

向运川,牟绪赞,任天祥,等. 2010. 化探资料应用技术要求. 北京:地质出版社

向运川,龚庆杰,刘荣梅,杨万志. 2014. 区域地球化学推断地质体模型与应用——以花岗岩类侵入体为例. 岩石学报,30(9):2609-2618

向运川,元春华,陈秀法. 2015. 中国大陆周边地区主要成矿带成矿规律对比与潜力评价研究进展. 地质通报,34(4):587-598

向运川,牟绪赞,任天祥,马振东,刘荣梅,龚庆杰. 2018. 全国矿产资源潜力评价化探资料应用研究. 北京:地质出版社

肖高强,丛峰,刀艳,高晓红,李忠,张有荣. 2020. 云南凤庆鲁史-昌宁松山地区水系沉积物地球化学找矿与效果分析. 矿产勘查,11(1):183-189

肖利萍,裴格,魏芳,高小雨,丁蕊. 2014. 处理矿山废水的膨润土复合吸附剂材料筛选. 水处理技术,40(3):36-41

肖庆超,宋成怀,郝双雷. 2015. 生物炭和磷肥复合修复有色矿区重金属污染土壤的效果. 环境工程,33(S1):840-842,860

肖艳东,孟贵祥,范侥,赵玉京,王君良. 2019. 准噶尔盆地浅覆盖区浅钻化探找矿方法探索. 新疆地质,37(1):40-43

谢克文,周四春,张文宇,杨奎. 2015. 利用地气场寻找隐伏温泉带——以广东暖水村为例. 科学技术与工程,15(30):9-13

谢学锦. 2003a. 从勘查地球化学到应用地球化学. 物探与化探,27(6):412-415

谢学锦. 2003b. 2020年的勘查地球化学——从勘查地球化学到应用地球化学. 地质通报,22(11-12):863-868

徐师,张大超,吴梦,肖隆文. 2018. 硫酸盐还原菌在处理酸性矿山废水中的应用. 有色金属科学与工程,9(1):92-97

徐艳,王曙光,李娟. 2019. 生物炭对矿区重金属污染土壤养分影响及修复效果. 土地开发工程研究,4(11):33-37,53

徐洋,汪明启. 2018. 山东王家庄铜矿地气组分稀土元素特征. 内蒙古科技与经济,(12):61-64

徐洋,汪明启,高玉岩,欧阳辉. 2014. 利用铅同位素研究山东邹平王家庄铜矿地气物质来源. 物探与化探,38(1):23-27

徐云峰,王显锋,胡朝云,秦宇龙,周雪梅. 2018. 四川甲基卡地区1:5万水系沉积物地球化学特征及稀有金属找矿远景. 金属矿山,(2):121-130

许胜超,肖高强,龚庆杰,刘宁强,杨天仪,刀艳,向龙洲,李忠. 2019. 兰坪盆地区域地球化学异常特征及找矿方向. 现代地质,33(4):772-782

严洪泽,孙彬彬,徐进力,周国华,贺灵,刘银飞,王腾云. 2016. 灰化法与微波消解法处理地电化学泡塑样品的分析效果对比研究. 岩矿测试,35(3):276-283

严洪泽,陈海杰,孙彬彬,周国华,贺灵,刘银飞,王腾云. 2017. 湿法消解预处理地电化学泡塑样品有效性研究. 岩矿测试,36(5):510-518

严明书,李瑜,鲍丽然,张风雷,杨振鸿,张茂忠,杨乐超. 2016. 西藏申扎地区1:5万水系沉积物地球化学特征及找矿意义. 物探与化探,40(1):10-16

杨冰彬,乔海明,刘治国,张鑫. 2020. 吐哈盆地十红滩铀矿床北矿带铀的赋存状态研究及对地浸开采影响分析. 新疆地质,38(1):97-102

杨刚刚,李方林,张雄华. 2018. 金属活动态测量在东戈壁钼矿找矿效果研究. 新疆地质,36(2):182-188

杨吉成,周四春,刘晓辉,胡波. 2019. 卡鲁古伟晶岩锂矿的地气场特征及找矿意义. 岩石矿物学杂志,38(4):570-578

杨少平,孙跃,弓秋丽. 2014. "十一五"以来化探方法技术研究主要进展. 物探与化探,38(2):194-199

杨少平,刘华忠,孔牧,张华,刘应汉,张学君,高顺宝,郑有业. 2015. 羌塘高原典型矿区水系沉积物地球化学特征与区域化探扫面方法. 地球学报,36(3):367-376

杨笑笑,罗先熔,文美兰,欧阳菲,吕星海,尹高科,郑广明. 2019. 地电化学法在豫西崤山黄土覆盖区找矿中的应用——以洛宁县石龙山预查区为例. 物探与化探,43(2):244-256

杨鑫光,李希来,金立群,孙华方. 2018. 短期恢复下高寒矿区煤矸石山土壤变化特征研究. 草业学报,27(8):30-38

姚文生,王学求,谢学锦. 2011. 国际地球化学填图样品分析方法和数据对比. 地质通报,30(7):1111-1118

姚文生,王学求,张必敏,徐善法,申伍军,杜雪苗. 2012. 鄂尔多斯盆地砂岩型铀矿深穿透地球化学勘查方法实验. 地学前缘,19(3):167-176

叶庆森. 2014. 指示元素垂向分带序列计算方法述评. 物探化探计算技术,36(3):335-341

叶荣,张必敏,姚文生,王勇. 2012. 隐伏矿床上方纳米铜颗粒存在形式与成因. 地学前缘,19(3):120-129

叶荣,张必敏,王勇. 2013. 干旱荒漠区隐伏金矿覆盖层中金的分布与迁移:以新疆金窝子金矿田210金矿带为例. 现代地质,27(6):1265-1274

于扬,王登红,高娟琴,刘丽君,王伟,张塞. 2019. 中国三稀矿产生物找矿技术方法及其应用综述. 地质学报,93(6):1533-1542

於崇文. 2005. 地球化学的历史、发展和应用地球化学的内涵、展望(序一). 地质通报,24(10-11),doi:10.3969/j.issn.1671-2552.2005.10.001

虞航,庞雅庆,祁家明,刘佳林,陈军军,张闯. 2019. 粤北南雄地区水系沉积物地球化学特征及找矿预测. 原子能科学技术,53(3):569-576

袁璐璐,汪明启,胡佳乐. 2014. 苏尼特地球化学气体找矿研究. 煤炭技术,33(10):85-87

袁玉涛. 2015. 化探数据处理方法对比研究——以乌日尼图钨钼矿床为例. 硕士学位论文. 北京:中国地质大学(北京)

詹胜强,廖兴建,岳大斌,陈加中,周勇. 2016. 四川容须卡矿区稀有金属伟晶岩的原生晕特征. 四川地质学报,36(3):472-476

张必敏,王学求,迟清华,吕庆田. 2011. 戈壁覆盖区景观演化与Au的分散迁移. 现代地质,25(3):575-580

张必敏,王学求,叶荣,姚文生,王玮. 2019. 土壤微细粒分离测量技术在黄土覆盖区隐伏金矿勘查中的应用及异常成因探讨. 桂林理工大学学

报,39(2):301-310

张华,孔牧,杨少平,赵羽军,任天祥,孙忠军. 2017. 中国主要景观区区域地球化学勘查理论与方法. 北京:地质出版社

张金带. 2016. 我国砂岩型铀矿成矿理论的创新和发展. 铀矿地质,32(6):321-332

张晶,李宝强,孟广路,李慧英. 2015. 地球化学分析方法在吉尔吉斯斯坦库姆托尔金矿邻区找矿潜力预测中的应用. 地质通报,34(4):726-733

张晶,周军,樊会民,刘养雄,任智斌,孟广路. 2018. 西北地区典型矿床地质地球化学特征图集. 武汉:中国地质大学出版社

张凯,付锦,龚育龄,赵宁博,陈虎. 2015. 主要放射性物探方法在砂岩型铀矿勘查中的应用分析. 世界核地质科学,32(1):46-50

张启龙,杨鸿鹏,邱炜. 2019. 地电化学提取技术在青藏高原拉陵灶火中游干旱荒漠区的参数对比研究. 矿产勘查,10(7):1651-1657

张勤,白金峰,王烨. 2012. 地壳全元素配套分析方案及分析质量监控系统. 地学前缘,19(3):33-42

张鑫,聂逢君,张成勇,张虎军,董方升,董亚栋,卢亚运. 2015. 伊犁盆地蒙其古尔矿床砂岩型铀矿赋存状态研究. 科学技术与工程,15(33):18-23,47

张运强,陈海燕,张立国,陈超,刘应龙,何娇月,康璇,张金龙,彭芊芃. 2015. 冀北新杖子地区水系沉积物地球化学特征及找矿预测. 中国地质,42(6):1980-1988

赵波,龚敏,熊燃,龚鹏,曾键年,闭向阳,任利民,贾先巧,马振东. 2012. 湿润中低山景观条件下土壤金属活动态找矿试验. 物探与化探,36(6):902-906

赵柏宇,杨亚新,罗齐彬,吴泽明,王帅帅,余聪. 2018. 现场光度法测定地气中痕量铀. 西部探矿工程,30(2):121-123

赵丹,王南萍,周觅,高明山,黄志新,葛祥坤,吴儒杰. 2018. 活性炭测氡及分量化探在河北省沽源县大官厂研究区铀矿勘查中的应用研究. 铀矿地质,34(1):39-45

赵吉海,欧阳辉. 2017. 地气测量在月形铜矿中的应用研究. 西部探矿工程,29(5):161-164

赵景,邱炜,罗先熔. 2012. 两种深穿透地球化学勘查方法的研究现状及存在的主要问题. 青海大学学报(自然科学版),30(1):49-52

赵伦山,朱有光. 2010. 中国勘查地球化学60年——从勘查地球化学到应用地球化学. 见:中国地质学会地质学史专业委员会第22届学术年会论文集. 北京:中国地质学会,150-155

赵同谦,赵阳,贺玉晓,肖春艳,李鹏,王翠连. 2017. 不同植被恢复类型对矿山废弃地土壤理化性质的影响——以焦作缝山公园为例. 河南理工大学学报(自然科学版),36(4):60-67

赵禹,赵玉岩,郝立波,陆继龙. 2014. 利用快速聚类分析分区确定化探背景上限的方法. 物探化探计算技术,36(4):487-491

赵韵美,樊金拴,苏锐,郑涛,王季君. 2014. 阜新矿区不同植被恢复模式下煤矿废弃地土壤养分特征. 西北农业学报,23(8):210-216

郑超杰,罗先熔,刘攀峰,刘刚,杨笑笑,刘延斌,陈武,李武毅. 2018. 西藏隆子姐纳各普金多金属矿区常规土壤热释汞与阶梯升温式热释汞测量对比研究及找矿预测. 矿物岩石,38(1):99-110

智超,向武,曾键年,张玉成,白亮. 2015. 热释卤素法在安徽胡村铜钼矿床深部的找矿试验. 物探与化探,39(4):691-697

周继梅,岳停停,周磊,刘璟,谌书. 2018. 骨炭钝化含黄铁矿多重金属尾矿的研究. 岩石矿物学杂志,37(3):485-491

周奇明,施玉娇,赵延朋,秦国强,陆一敢,史琪. 2020. 深穿透地球化学勘查技术及应用. 北京:冶金工业出版社

周四春,刘晓辉,谷江波,吕少辉,王自运,吴丽荣. 2011. 联袂应用地气、射气与壤中α测量探测雅拉河地区隐伏断裂. 物探与化探,35(3):298-302

周四春,张保静,杨宇奇. 2012. 智能动态地气采样装置研制. 物探与化探,36(6):1059-1063

周四春,刘晓辉,童纯菡,胡波. 2014. 地气测量技术及在隐伏矿找矿中的应用研究. 地质学报,88(4):736-754

朱炳玉,顾雪祥,马华东,席蕾,李月臣,Rickleman D,潘成泽. 2012. 活动态金属离子测量(MMI)在金山金矿及外围勘查中的应用. 新疆地质,30(2):228-233

朱剑,周四春,刘俊,王兴华,张国亚. 2014. 影响地气测量结果有关因素的讨论. 现代矿业,(1):62-64,96

左仁广. 2019. 勘查地球化学数据挖掘与弱异常识别. 地学前缘,26(4):67-75

Gong Q J, Deng J, Wang C M, Wang Z L, Zhou L Z. 2013. Element behaviors due to rock weathering and its implication to geochemical anomaly recognition:a case study on Linglong biotite granite in Jiaodong peninsula, China. Journal of Geochemical Exploration,128:14-24

Gong Q J, Deng J, Jia Y J, Tong Y K, Liu N Q. 2015. Empirical equations to describe trace element behaviors due to rock weathering in China. Journal of Geochemical Exploration,152:110-117

Gong Q J, Yan T T, Li J Z, Zhang M, Liu N Q. 2016. Experimental simulation of element mass transfer and primary halo zone on water-rock interaction. Applied Geochemistry,69:1-11

Gong Q J, Li J Z, Xiang Y X, Liu R M, Wu X, Yan T T, Chen J, Li R K, Tong Y K. 2018. Determination and classification of geochemical anomalies based on backgrounds and cutoff grades of trace elements:a case study in South Nanling Range, China. Journal of Geochemical Exploration,194:44-51

Huang L, Zhang P, Hu Y G, Zhao Y. 2016. Vegetation and soil restoration in refuse dumps from open pit coal mines. Ecological Engineering,94:638-646

Kang H, Chen Y, Li D, Zhao J, Cui F, Xu Y. 2020. Deep-penetrating geochemistry for concealed sandstone-type uranium deposits:a case study of Hadatu uranium deposit in the Erenhot Basin, North China. Journal of Geochemical Exploration,211:106464

Li J Z, Gong Q J, Yan T T, Li R K, Liu N Q, Cen K. 2018. Quantitative description of geochemical backgrounds of gold due to rock weathering in Jiaodong peninsula, China. Journal of Geochemical Exploration,192:155-162

Li R K, Liu N Q, Gong Q J, Wu X, Yan T T, Li X L, Liu M X. 2019. Construction, test and application of a geochemical gold metallogene:case studies in China. Journal of Geochemical Exploration,204:1-11

Liu P F, Luo X R, Wen M L, Zhang J L, Zheng C J, Gao W, Ouyang F. 2018. Geoelectrochemical anomaly prospecting for uranium deposits in southeastern China. Applied Geochemistry, 97:226-237

Liu Y, Carranza E J M, Zhou K F, Xia Q L. 2019. Compositional balance analysis: an elegant method of geochemical pattern recognition and anomaly mapping for mineral exploration. Natural Resources Research, 28(4):1269-1283

Tessier A, Campbell P G C, Bison M. 1979. Sequential extraction procedure for the speciation particulate trace metal. Analytical Chemistry, 51:844-851

Tian M, Wang X Q, Nie L S, Zhang C S. 2018. Recognition of geochemical anomalies based on geographically weighted regression: a case study across the boundary areas of China and Mongolia. Journal of Geochemical Exploration, 190:381-389

Tian M, Wang X Q, Nie L S, Liu H L, Wang W, Yan T T. 2019. Spatial distributions and the identification of ore-related anomalies of Cu across the boundary area of China and Mongolia. Journal of Geochemical Exploration, 197:37-47

Wang M Q, Wu H, Liao Y, Fang F, Gao Y Y, Xu Y. 2015. Pilot study of partial extraction geochemistry for base metal exploration in a thick loess-covered region. Journal of Geochemical Exploration, 148:231-240

Wang X Q. 1998. Leaching of mobile forms of metals in overburden: development and application. Journal of Geochemical Exploration, 61(1-3):39-55

Wang X Q, the CGB Sampling Team. 2015. China geochemical baselines: sampling methodology. Journal of Geochemical Exploration, 148:25-39

Wang X Q, Liu X M, Han Z X, Zhou J, Xu S F, Zhang Q, Chen H J, Bo W, Xia X. 2015. Concentration and distribution of mercury in drainage catchment sediment and alluvial soil of China. Journal of Geochemical Exploration, 154:32-48

Wang X Q, Zhang B M, Lin X, Xu S F, Yao W S, Ye R. 2016. Geochemical challenges of diverse regolith-covered terrains for mineral exploration in China. Ore Geology Reviews, 73:417-431

Xiang Z L, Gu X X, Wang E Y, Wang X L, Zhang Y M, Wang Y. 2019. Delineation of deep prospecting targets by combining factor and fractal analysis in the Kekeshala skarn Cu deposit, NW China. Journal of Geochemical Exploration, 198:71-81

Xie X J, Wang X Q. 1991. Geochemical exploration for gold: a new approach to an old problem. Journal of Geochemical Exploration, 40(1-3):25-48

Xie X J, Cheng H X. 2014. Sixty years of exploration geochemistry in China. Journal of Geochemical Exploration, 139:4-8

Xiong Y H, Zuo R G. 2020. Recognizing multivariate geochemical anomalies for mineral exploration by combining deep learning and one-class support vector machine. Computers & Geosciences, 140:104484

Xu Z Q, Liang B, Geng Y, Liu T, Wang Q B. 2019. Extraction of soils above concealed lithium deposits for rare metal exploration in Jiajika area: a pilot study. Applied Geochemistry, 107:142-151

Yao W S, Xie X J, Wang X Q. 2011. Comparison of results analyzed by Chinese and European laboratories for FOREGS geochemical baselines mapping samples. Geoscience Frontiers, 2(2):247-259

Zhang B M, Wang X Q, Chi Q H, Yao W S, Liu H L, Lin X. 2016. Three-dimensional geochemical patterns of regolith over a concealed gold deposit revealed by overburden drilling in desert terrains of northwestern China. Journal of Geochemical Exploration, 164:122-135

Research Progress of Applied Geochemistry

GONG Qing-jie, XIA Xue-qi, LIU Ning-qiang

School of Earth Sciences and Resources, China University of Geosciences (Beijing), Beijing 100083

Abstract: The research progresses of applied geochemistry during the last decade in China are summarized in this paper. On the basis of the development history of applied geochemistry, the research contents are depicted, especially in the field of geochemical exploration. The research progresses of geochemistry exploration during the past decade in China can be summaried from four aspects, e.g. survey, assessment, exploitation and remediation. Traditional and non-traditional geochemical prospectings are divided according to their survey techniques, on the one hand, the traditional geochemical prospecting techniques are maturing and standarding and their applications are widening and progressing, on the another hand, the non-traditional geochemical prospecting plays an important and irreplaceable role in overburdened areas. In the aspect of the assessment, its main topics are concentrated on element association, anomaly determination and anomaly evaluation, and its applications keep progressing and widening; the comparison method is the dominated technique of the assessment, nevertheless, methods driven by geochemical knowledge become more and more important. In the aspect of the exploitation, its progresses reflect mainly on *in-situ* leaching techniques of mining sandstone hosted uranium deposits. In the aspect of the remediation, its progresses are mainly on the environment remediation of mines, which have formed a new research field of geochemical engineering.

Key words: non-traditional geochemical prospecting; geochemical survey; geochemical assessment; geochemical anomaly; geochemical remediation

数学地球科学跨越发展的十年*

周永章[1] 左仁广[2] 刘 刚[2] 袁 峰[3] 毛先成[4] 郭艳军[5] 肖 凡[1] 廖 杰[1] 刘艳鹏[1]

1. 中山大学 地球环境与地球资源研究中心，广州 510275；2. 中国地质大学（武汉），武汉 430074；3. 合肥工业大学 资源与环境工程学院，合肥 230009；4. 中南大学 地球科学与信息物理学院，长沙 410083；5. 北京大学 地球与空间科学学院，北京 100871

摘 要：近十年是科学研究从问题驱动向数据驱动转变的转折时期，科学研究的第四范式—数据密集型科学发现应势而生。这期间，大数据与人工智能算法的引入使数学地球科学实现跨越式发展，并正在改变地质学。机器学习是使计算机具有智能的根本途径。深度学习，即多层神经网络的方法，是一种实现机器学习的技术，是过去几年大数据与数学地球科学研究最重要的热点。贝叶斯网络是贝叶斯公式和图论结合的产物，可用来建立矿床地质的成因网络，进而理解矿床成因。地质大图形问题可以转化为大型的复杂网络空间问题和社区结构问题，社区分析技术可用于地震预报、地质网络分析、特殊地质现象识别、矿床预测。关联规则和推荐系统算法在地质研究中已有成功的应用实例。化探数据及其异常经常包含复杂和非线性模式，深度学习在智能识别与提取复杂地质条件下地球化学异常具有优异的能力，卷积神经网络、堆叠自编码机等是较为常用和有效的方法。非线性矿产资源预测、基于GIS和三维地质建模的三维成矿预测及相应的软件系统得到持续改进。三维虚拟仿真建模技术的应用实现了多模态、跨尺度地学虚拟现实与多维交互，地质过程数值模拟等已有创新性进展。区块链技术以及OneGeology、玻璃地球、深时数字地球等大地质科学计划，将在整合全球地质大数据、共享全球地学知识、推动数学地球科学学科发展方面起到重大的推动作用。

关键词：地质大数据 深度学习 人工智能算法 区块链 深时数字地球 矿产资源预测 数学地球科学

0 引 言

最近十年，数学地球科学的最显著发展是大数据与人工智能算法的引入。可以说，地质大数据与人工智能时代已经在这期间开启。

地质数据以指数形式增长，这是不容忽视的现实。基础地质、矿产地质、水文地质、工程地质、环境地质、灾害地质调查、勘查，产生大量的数据。各类天基、空基对地遥感观测，更产生了大量的数据。图件编绘、分析计算、模拟仿真、预测评价、管控调控，同样产生大量的数据。并且这些数据可以是结构化的，如地球化学分析和地球物理探查获得的数据；更多是非结构化的、半结构化的。

在现实面前，大数据挖掘和机器学习是地质学科跨越的必须选项。否则，就如同人用腿跟汽车、飞机、火箭赛跑，越往前走，与大数据时代的要求差距越大，最终被先进的工具所被抛弃。

尽管依托大数据的人工智能地质学还远不成熟，但已俨然成为这个时代的绚丽浪花。最近几年，国际数学地球科学协会、国际数字地球学会、中国地质学会、中国矿物岩石地球化学学会每届年会都有专题和较大篇幅的主题报告涉及地质大数据和人工智能分析。

2016年，中国矿物岩石地球化学学会大数据与数学地球科学专业委员会正式成立，这是一个里程碑的事件。大数据专委会自成立以来，坚持每年召开一次"中国大数据与数学地球科学学术讨论会"，而且

---

* 原文"数学地球科学跨越发展的十年：大数据、人工智能算法正在改变地质学"刊于《矿物岩石地球化学通报》2021年第40卷第3期，本文略有修改。

与会人数逐年增加，同时专委会还相继在《岩石学报》《地学前缘》《大地构造与成矿学》《地质通报》《矿物岩石地球化学通报》组织专辑，介绍地质大数据分析与机器学习的探索性研究成果。

2018年，周永章所著的《地球科学大数据挖掘与机器学习》出版，这是国内外相同领域首部研究型教材，它启迪一代新人，引领他们更多关注和投身地质大数据和人工智能研究（翟明国，2018；Jiao et al.，2018）。

本文试图对最近十年地质大数据与人工智能领域的研究做一粗略回顾，不可能系统和全面。作者期望读者能从中窥视到大数据与人工智能的价值，引发对地质大数据与人工智能发展的思考。

1 深度学习与人工智能地质学

1.1 机器学习、深度学习

机器学习被认为是人工智能的核心，是使计算机具有智能的根本途径。深度学习是机器学习的子集，即多层神经网络的方法，是一种实现机器学习的技术，是过去几年大数据与数学地球科学研究的最重要热点之一。

2006年，加拿大多伦多大学Geoffrey Hinton和Ruslan Salakhutdinov在Science上发表论文，开启了深度学习在学术界和工业界的浪潮（Hinton et al.，2006，2012；Lake et al.，2015；LeCun et al.，2015；Schmidhuber，2015；Karpatne et al.，2019）。此前，美国心理学家McCulloch和数学家Pitts联合提出了形式神经元的数学模型——MP模型，证明了单个神经元能执行逻辑功能，从而开创了人工神经网络研究时代。由于超大规模集成电路、脑科学、生物学、光学的迅速发展，人工神经网络的发展进入兴盛期。在分类与预测中，δ学习规则（误差校正学习算法）是使用最广泛的一种，但在人工神经网络的发展过程中，没有一种特定的学习算法适用于所有的网络结构和具体问题。

Hinton和Ruslan Salakhutdinov倡导的深度学习概念源于人工神经网络的研究，可以理解为神经网络的发展，其实质是通过构建具有很多隐层的机器学习模型和海量的训练数据，来学习更有用的特征，从而最终提升分类或预测的准确性。"深度模型"是手段，"特征学习"是目的。

卷积神经网络（图1）是深度学习中知名度最高和应用最广的一种模型，被用于图像识别和语音分析。在地质领域，徐述腾和周永章（2018）以吉林夹皮沟金矿和河北石湖金矿的黄铁矿、黄铜矿、方铅矿、闪锌矿等硫化物矿物为例，设计了有针对性的Unet卷积神经网络模型，实现了基于深度学习算法的镜下矿石矿物自动识别与分类。在Unet模型结构中一共涉及5种操作。其中紫色向右箭头为3×3卷积操作（conv3×3）和欧拉激活函数（ReLU）转换；灰色向右箭头为图像复制（copy）和截取（crop）操作；红色向下箭头表示2×2的最大池化（max pool 2×2），绿色向上箭头表示2×2的上卷积（up-conv2×2），蓝色箭头表示1×1的卷积（conv 1×1）。

深度信念网络（DBNs）由Geoffrey Hinton于2006年提出，是一种经典的深度生成式模型，通过将一系列受限玻尔兹曼机单元堆叠而进行训练。该模型在MNIST数据集上的表现超越了当时流行的SVM。张雪英等（2018）利用深度信念网络来识别地质实体。通过分析各种类型文本数据中地质实体信息的描述特点，构建地质实体信息的标注规范和语料库，设计基于深度信念网络的地质实体识别模型，解决文本数据中地质实体信息的结构化、规范化处理问题。

近几年在机器学习、深度学习领域，我国学者的代表的工作包括：韩帅等（2018）、徐述腾和周永章（2018）、焦守涛等（2018）、刘艳鹏等（2018，2020）、王怀涛等（2018）、周永章等（2018a，2018b）、王堃屹等（2019）、王语等（2020）、张野等（2020）、任秋兵等（2020）、陈进等（2020）等。从中亦折射出机器学习是当前地质大数据研究的重要热点之一。

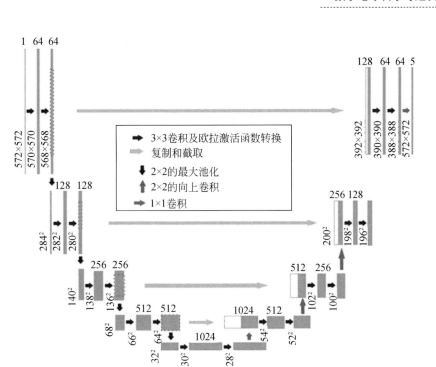

图 1 卷积神经网络模型案例

1.2 人工智能地质学

2015 年，*Science* 杂志封面发表一篇人工智能论文：3 名分别来自麻省理工学院、纽约大学和多伦多大学的研究者开发了一个"只看一眼就会写字"的计算机系统。人们只需向这个系统展示一个来自陌生文字系统的字符，它就能很快学到精髓，像人一样写出来，甚至还能写出其他类似的文字——更有甚者，它还通过了图灵测试，人们很难区分图中的字符是人类还是机器的作品。

贝叶斯原理是人工智能的最重要基础。许多人工智能系统采用的方法就是一种基于贝叶斯公式的方法——贝叶斯程序学习（bayesian program learning）。甚至有科学家认为，人类的大脑结构就是一个贝叶斯网络，贝叶斯公式是人类在没有充分或准确信息时最优的推理结构，为了提高生存效率，进化会向这个模式演进（周永章等，2017）。

科学家对自然语言处理方面的成功，开辟了一条全新的人工智能问题解决路径：原来看起来非常复杂的问题可以用贝叶斯公式转化为简单的数学问题。从实践来看它非常有效，将大量观测数据输入模型进行迭代——也就是对模型进行训练，就可以得到希望的结果。人工智能地质学还远不够成熟，但科学家在不断探索。

周永章等（2017，2018a）认为，矿床地质学家可以利用贝叶斯网络自动揭示矿床的成因机制及其背后的规律。理解矿床成因可以从理解它们的成因网络开始。贝叶斯网络是贝叶斯公式和图论结合的产物，破译矿床地质的成因网络，可以将公式本身结成贝叶斯网络。

贝叶斯网络是马尔可夫链的推广，它给复杂问题提供了一个普适性的解决框架。与马尔可夫链类似的是，贝叶斯网络中每个节点的状态值取决于其前面的有限个状态，不同的是，贝叶斯网络不受马尔可夫链的链状结构的约束，因此可以更准确地描述事件之间的相关性。为了确定各个节点之间的相关性，需要用已知数据对贝叶斯网络进行迭代和训练。贝叶斯公式的价值在于，当观测数据不充分时，它可以将专家意见和原始数据进行综合，以弥补测量中的不足。人类的认知缺陷越大，贝叶斯公式的价值就越大。

1.3 知 识 图 谱

2012 年，谷歌提出知识图谱，初衷是为了提高搜索引擎的能力，改善用户的搜索质量和搜索体验。

知识图谱以"图"的方式来描述真实世界的事物及其关系，以"实体-关系-实体"三元组的方式存储到数据库中。其中，实体是真实世界中的各种事物、存在及其概念被称为实体，关是实体与实体之间的关系，许多场景下表示为属性。从本质上讲，它是一张巨大的语义网络图，以节点表示实体或概念，边则由属性或关系构成（Pujara et al., 2013）。

知识图谱是客观世界的一种重构，与神经网络相比，它是一种可直接解释的人工智能，已逐渐成为人工智能关键技术之一，被广泛应用于智能问答（Lukovnikov et al., 2017）、智能搜索（Wang et al., 2018）和个性化推荐（Palumbo et al., 2018）等领域。

最近十年，知识图谱的构建技术一直是研究的热点之一。在信息抽取方面，Liu X. H. 等（2011）利用邻近算法（KNN）与条件随机场模型，实现了对 Twitter 文本数据中实体的识别。在知识加工方面，Wang 等（2013）利用基于主题进行层次聚类的方法得到本体结构。谷歌 Knowledge Vault 根据抽取到的结构化信息的频率对数据可信度进行评分，提高了知识图谱中知识的质量（Dong et al., 2014）。

从 2016 年开始，国家自然科学基金委与广东大数据科学研究中心联合基金持续支持城市交通、医疗、防灾、金融、管理等领域的大数据分析挖掘和智能监测、管控与预警的重大科学问题和技术问题。从已立项的项目看，其中相当一部分设有所在领域知识图谱构建的目标。在地质环境灾害领域立项的项目有："城市地质环境时空透视与大数据融合关键技术"（刘刚）、"基于地学大数据的城市地质灾害智能监测、模拟、管控与预警"（王力哲）、"基于地学大数据的城市土壤污染智能监测、模拟、管控与预警（周永章）"，等。

在"基于大数据的城市土壤污染智能监测、模拟、管控与预警"研究中，包括了异质多源时空关联的本体知识图谱的构建研究。项目从知识抽取、推理、融合、更新的角度，分别研究数据驱动的城市土壤污染知识抽取、知识推理，以及知识图谱生成与更新问题，以为全面建立起可解释的多源多层城市土壤污染知识图谱提供理论体系、框架与应用思路（图2）。知识图谱构建的重要基础是基于数据驱动方法的自动知识抽取，这需要利用深度学习模型对复杂非线性关联优异的表达能力，从数据驱动角度出发，将上阶段获得深度融合的语义表征与关联逐层分解、逐层抽象，结合多实例、多标签、多视图的建模框架，实现对地质实体提及的检测。

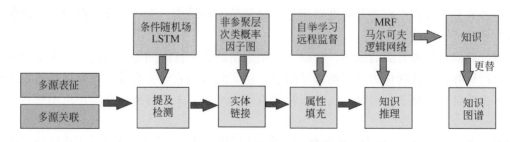

图 2　土壤污染知识图谱构建方法

2 大数据挖掘

2.1 高 维 数 据

真实的世界是一个高维空间的世界。如何快速、有效地从高维空间获得有价值的信息或发现相关目标，一直是科学家研究的目标。

高维特征集合存在以下几方面问题：大量的特征；存在许多与类别仅有微弱相关度的特征；特征相互之间存在强烈的相关度；噪声数据。解决这些问题的基本途径是降维（dimension reduction），从初始高维特征集合中选出低维特征集合，以有效地消除无关和冗余特征，改善预测精确性等学习性能，增强学习结果的易理解性。

聚类分析、主成分分析等多元统计分析方法仍然是较常用的数学降维工具（Leskovec et al., 2014）。子空间聚类是实现高维数据集聚类的有效途径，它是在高维数据空间中对传统聚类算法的一种扩展，其思想是将搜索局部化在相关维中进行。

此外，还有哈希算法等。它将任意长度的二进制值映射为较短的固定长度的二进制值（哈希值）。哈希函数可以将任意长度的输入经过变化以后得到固定长度的输出，这种单向特征和输出数据长度固定的特征使得它可以生成消息或者数据。

在最近十年间，降维技术用于化探数据的信息提取，包括基于频率分布或频率-空间分布的统计（张淼和周永章，2012；陈永良等，2014；Chen et al., 2014；Zuo et al., 2016；Chen and Wu, 2017；Grunsky and de Caritat, 2020）、分形/多重分形模型（Afzal et al., 2011；Xiao et al., 2012；Zuo and Wang, 2016；肖凡等，2017）、主成分和奇异值分解（Cheng et al., 2012；Wang et al., 2013）。左仁广（2019）分析了化探弱异常识别模式，应用局部RX方法进行多变量降维和提取弱缓地球化学异常。曹梦雪等（2018）在分析鄂尔多斯盆地北缘1∶20万地球化学土壤测量数据时，将39个元素变量分解成若干独立因子向量，将最优独立因子向量作为元素组合，利用优选后的变量和样本集合开展鄂尔多斯盆地铀资源预测。

2.2　大图形数据

形形色色的地质图凝聚着地质学家的智慧。直观上，它们是一张张大图形，是以地理空间为坐标系，包含了足够多矿床、构造及地层等地质实体的图形，本质上却是一个可视化的地质数据库。此外，这些大图形也是一张网络，连接地质实体节点的即是地质因素，有的连线是清晰的，有的则有待揭露。节点信息以及连线中隐藏的信息就是地质大图形的研究目标（周永章等，2018a）。

地质大图形问题可以转化为大型的复杂网络空间问题。复杂系统普遍具有模块结构（社区结构）特性，网络中的社区结构识别对理解整个网络的结构和功能有重要价值，可帮助分析、预测网络各元素间的交互关系。

大图形数据信息挖掘的重要思想在于，网络是描述研究系统的一种新方式，复杂网络是真实复杂系统的高度抽象。现实世界中的复杂系统都可以映射成网络，大型的网络空间则映射成为大数据空间，其中往往包含复杂的关联关系。网络研究着重于寻找关联，而这也符合地质大数据研究的核心思想：有限寻找事物间的关联关系，而非因果关系。

社区发现在梳理整个网络结构、分析各元素间的关系时发挥着重要作用，常被应用于社交网络中划分具有相同兴趣的群体（Zhai and Lin, 2018）。关于社区发现的研究最近几年受到广泛关注，例如，丁彩英等（2019）将社区检测和半监督学习框架结合，利用先验知识进行社区发现。乔少杰等（2017a）设计大规模复杂网络社区并行发现算法，提高了运行效率。针对节点隶属于多个社区的重叠社区问题，乔少杰等（2017b）和柳原和白金牛（2019）提出了面向复杂网络大数据的重叠社区检测算法。郭昆等（2019）根据密度峰值和社区归属度进行重叠社区发现。社区发现也被与数据挖掘相关技术结合，巫红霞和谢强（2019）将加权社区检测应用于高维数据特征提取中。

社区结构属于网络的中观尺度结构。除小世界效应、无标度性等复杂网络基本特征外，网络聚簇结构是复杂网络重要拓扑结构特征之一（图3）。这种结构特征隐含复杂网络中存在着社区结构，即社区内部节点之间关系相对紧密、社区之间节点关系相对稀疏。

社区发现是社区结构研究的基础和核心问题，对分析复杂网络的拓扑结构及层次结构、理解社区的形成过程、预测复杂网络的动态变化、发现复杂网络中蕴含的规律具有重要意义。

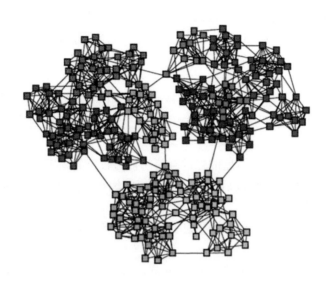

图 3　网络中的社区划分

社区发现和传统的聚类与图划分具有区别，又有联系，可以从应用背景、研究对象、研究目的等角度分析。在许多情况下，聚类和社区发现是不可以互相替代的（程学旗等，2014）。

科学家已发明多种有效的社区发现算法，如无重叠社区的图分割社区发现方法、层级聚类社区发现方法、谱聚类社区发现方法。属于图分割社区发现方法的有 KL 算法、GA 算法、谱平分方法、基于标准矩阵的谱平分法和基于电阻网络电压谱的算法。重叠社区发现算法有基于派系过滤的方法、基于局部扩张及优化的方法、基于线图、边社区的发现方法和基于图模型的统计推断方法等。由于现实中的复杂网络随着时间总是动态变化的，网络的节点和边每时每刻都可能在增加、删除或修改。发现网络中的社区结构及其演化过程，是当前复杂网络动态社区发现的热点之一。

针对地质大数据，周永章等（2018a）提出在地质复杂网络中社区结构是普遍存在的现象，使用社区分析技术识别社区结构，分析整个网络结构和功能、预测网络各元素间的交互关系具有诸多应用，例如地震预报、地质网络分析、特殊地质现象识别、矿床预测等。陈伟雄等（2019）用网络中的节点代表地质矿物构造背景，以地化元素进行社区发现，将原有构造背景划分为若干个社区，挖掘社区内部相关关系，发现对于某个社区的构造背景，其矿物主、微量元素明显区分于其他社区。当然，地质大数据多源、异构、混合性、关联关系繁杂，构成的网络结构复杂、节点和连接类型多样，使得基于地质大数据的高效计算、系统分析仍面临着巨大挑战。

基于图的拓扑结构相似度，可以开展地质文献与信息检索，Page Rank、Sim Rank 和 Page Sim 等算法都有实际应用案例。基于加权 Sim Rank 的中文查询推荐方法，主要针对很多查询之间存在隐含或间接联系，两个不能直接匹配的相关查询往往有着相同或相似的相关查询。基于它将查询映射到一个查询关系图中，图中节点表示查询，边表示查询中某种直接联系。然后，根据结构相似度算法并针对查询推荐应用问题，提出加权 Sim Rank（简称 W Sim Rank）计算图中查询（节点）间相似度。W Sim Rank 综合考虑了查询关系图的全局信息，因而能挖掘出查询间的间接关联和语义关系。

2.3　关联规则与推荐系统算法

关联规则（association rules）和推荐系统（recommender systems）算法是过去十年中数据挖掘中最活跃的研究方法之一，已引入地质领域研究。

常力恒等（2018）以全国矿产地数据库中的热液型金矿数据和潜力评价数据为研究对象，应用关联规则算法，挖掘与金矿相关的侵入岩、火山岩、变质岩建造及区域构造地质大数据的关联性，进而发现地质要素之间的共生组合规律。他们首先通过空间位置建立不同类型数据之间的联系，形成金矿属性数据库，然后基于 Apriori 算法（Agrawal and Srikant，1994）提取了大地构造环境与变质作用的频繁项集，挖掘矿产资源信息与其他信息的关联规则，发现古裂谷相、古弧盆相分别受区域动力热流变质作用和区域中高温变质作用控制明显。

刘心怡和周永章（2019）应用关联规则算法对粤西庞西垌矿床远景区 1∶5 万水系沉积物地球化学测量及异常查证数据进行分析，结果显示关联规则算法可以有效挖掘元素组合之间的关联规则数据集且符合实际，组合异常的强规则与研究区已知矿化地段的异常组合重合性较高，如当 Au、Cu、Sb 在异常值范围内时，出现 As 为异常的可能性是 95.5%。

由 Agrawal R 等首先提出的关联规则算法，旨在从大量数据的项集之中发现有趣的关联或相关性，从而达到认识事物客观规律的目的。关联规则的模式属于描述型模式，发现关联规则的算法属于无监督学习的方法。

常用的关联规则算法有 Apriori、FP-Tree、Eclat 算法和灰色关联法。其中，Apriori 算法最为经典和常用。其主要思想是找出存在于事务数据集中最大的频繁项集，利用最大频繁项目集与预先设定的最小置信度阈值生成强关联规则。置信度（support）、支持度（confidence）和提升度（lift）是 Apriori 算法的 3 个最基础概念。

目前应用的主要推荐方法包括基于关联规则推荐、基于知识推荐、基于内容推荐、协同过滤推荐、基于效用推荐和组合推荐等。推荐系统包括 3 个重要的模块：用户建模模块、推荐对象建模模块、推荐算法模块。

王堃屹等（2019）应用推荐系统算法对钦杭成矿带南段文地幅 1∶5 万地质矿产调查所获得的数据进行挖掘。研究中应用基于内容的推荐系统算法，构建能动项-因素项的效用矩阵。其中，待预测的能动项选取了中型银金矿床、小型银金矿（化）点、确证无银金矿床、未评价区域（针对银金矿床）、中型铅锌矿床、小型铅锌矿（化）点、确证无铅锌矿床、未评价区域（针对铅锌矿床），因素项选取了加里东期混合岩、燕山早期侵入岩、燕山晚期侵入岩、北东向断裂、北西向断裂以及 Au、Ag、Pb、Zn 4 种元素。通过计算效用矩阵中的已知矿床（点）与其他未评价区域之间的相似度，开展银金矿床和铅锌矿床潜在的找矿区域预测。实验结果表明，推荐系统算法能够较好地挖掘与成矿有关的信息，快速抓取出与某类矿床（点）相似的潜在成矿区域。对于银金矿床，相似度较高的区域主要分布在已知矿点周围以及 NE 向主断裂的两侧，少量分布于叠加断裂附近。对于铅锌矿床，中型铅锌矿床的结果显示出较高的区分度，高值区基本涵盖了所有已知的铅锌矿点，小型铅锌矿床的结果更加集中，除已知矿点外，还有几处高值区可作为重点的找矿靶区。

3 地球化学异常识别与提取

全球已积累了海量的岩石地球化学数据，并建立了多个相关的数据库，如 GEOROC-大陆和海洋岩石的地球化学数据库、PetDB-海底岩石学数据库、SedDB-沉积岩成分分析数据库、NAVDAT-北美火成岩成分分析数据库和 Geochron-地质年代学数据库等。

复杂地质条件下地球化学异常的识别与提取是地质大数据和人工智能算法研究应用的热点领域之一（Carranza and Laborte，2015；Aryafar and Moeini，2017）。目前在地球化学研究中应用的算法主要包括神经网络、支持向量机、随机森林、决策树、极限学习机等（Tahmasebi and Hezarkhani，2012；Izadi et al.，2013；Chen et al.，2014；Rodriguez-Galiano et al.，2014；Carranza and Laborte，2015；Harris and Grunsky，2015；Geranian et al.，2016；Chen and Wu，2017；Yu et al.，2019；Zuo et al.，2019；余晓彤等，2019；陈丽蓉，2019）。机器学习的各种算法不仅能够处理大量的与矿产预测相关的证据图层，还具有识别已知矿床与证据图层之间非线性关系的潜力（Rodriguez-Galiano et al.，2015；向杰等，2019）。

复杂化探数据及其异常经常包含复杂和非线性模式，深度学习在智能识别与提取复杂地质条件下的地球化学异常具有优异的能力。从已发表的文献可见，卷积神经网络是较为常用和有效的方法（刘艳鹏等，2018；周永章等，2018b；左仁广，2019）。这些深度学习方法将输入的数据映射到少数深层次特征中，有利于保留与复杂地质过程及其相互作用有关的深层次结构表征，最终达到提高异常识别的效果。

在传统地球化学研究中，利用Pearce图解和判别岩石的构造源区是流行的方法（Pearce et al., 1984；Verma et al., 2006；Vermeesch, 2006；赵振华，2007）。但受时代、研究区域、研究思路以及研究手段、分析技术、样本数量的限制，导致部分研究中经常出现一些困惑。大数据思维为研究岩石构造判别提供了新的思维模式：由理论驱动转变为数字驱动（张旗和周永章，2018；葛粲等，2019），成功的案例如：利用GEOROC数据库数据判别辉长岩、玄武岩、安山岩的构造环境等（杜雪亮等，2017；王金荣等，2017；韩帅等，2018；焦守涛等，2018；张旗等，2019；刘欣雨等，2019；耿厅等，2019）。

4 矿产资源预测与评价

4.1 非线性矿产资源预测

国际地科联IGCP98计划曾推出6种矿产资源定量预测评价方法，包括矿床统计预测理论及方法（Agterberg）、"三部式"资源评价法（Singer）、基于GIS的矿产资源评价（Bonham-Carter）、综合信息预测（王世称等，2000）、致矿异常预测与"三联式"预测（赵鹏大，2007）、非线性成矿预测（成秋明，2007）等。

分形、多重分形理论所提供的尺度不变性、广义自相似性及奇异性等概念和相关模型，可以较好地描述成矿过程的奇异性、成矿元素分布不均匀性及矿床空间聚散性等成矿复杂系统和矿产资源分布规律，是定量模拟、识别与提取复杂成矿异常方法之一（Cheng et al., 1994；Zhou et al., 1994；成秋明，2007；赵鹏大和夏庆霖，2009；Agterberg, 2014）。这是中国和国际数学地质界对非线性矿产资源预测研究的重大贡献。这一领域的研究始于20世纪90年代，初步成熟于21世纪初期，在最近十年仍有显著发展。

近年来，基于广义自相似性的多重分形滤波思想进一步扩展至特征值域、沃什域以及小波域，有研究者提出的混合场分解方法（如特征值N-λ分解、沃什域W-A分解以及小波域W-N分解等）（Chen and Cheng, 2016），促进了空间模式混合场广义自相似分解问题的解决。多重分形滤波方法在地球物理、地球化学以及遥感等领域被大量应用（Cheng, 2012；Zuo et al., 2016）。这些方法在不同景观（沙漠、草原等）覆盖区隐伏矿致弱缓异常识别与提取中取得了较好的应用效果（Cheng, 2012）。

4.2 基于GIS的矿产预测

基于地理信息系统（GIS）等的矿产资源预测评价研究始于20世纪80年代末（Agterberg, 2014）。最近十年，GIS预测评价仍然是矿产资源定量预测评价的主流方法，最突出的成果是全国矿集区与整装勘查区的资源潜力GIS预测评价。叶天竺、肖克炎等建立了数十个基于GIS的矿床模型，通过应用于中国1000多个已知矿床扩展到全国矿产资源潜力评估中（Bagas et al., 2017；Xiao et al., 2020）。

其他代表性研究还有：应用GIS技术对陕西潼关金矿多源信息提取，建立金成矿预测模型，圈定了多个成矿远景区（陈建平等，2014a）；对桂西-滇东南锰矿资源GIS分析与潜力评价，发展基于场模型的成矿信息提取方法和扩展的证据权法（张宝一等，2011；毛先成等，2014）；基于GIS将矿物异常特征与地质、地球化学和地球物理相结合，建立成矿预测模型，采用综合体积法进行定量预测（Cong et al., 2017）；利用GIS技术对空间数据和矿床模型进行集成，发展含矿率、体积和相似性三变量计算的资源量估算方法（Li et al., 2018）。

基于GIS的矿产资源定量评价已趋于成熟，其核心工作可概括为3个部分：①多元地学空间数据集成；②多元成矿信息提取与综合；③矿产资源潜力制图。成矿信息包括地质信息、物探信息、化探信息、遥感信息等，是解释、提取和综合其他找矿信息的基础，在美国实施的"三部式"矿产资源评价中，地质信息是圈定成矿远景区的前提。GIS空间分析技术的应用促进了地质信息提取方法的发展，利用缓冲区分析和叠置分析可以提取地层、岩浆岩、构造等成矿信息，如线性构造缓冲区、地层组合熵、岩浆岩体影响域、地质异常、线性构造异常参数等。

矿产资源GIS预测评价的优势是：可集成管理多源地学信息数据，能快速使用数据进行建模和量化分析，但对研究区域成矿规律的认识程度和预测方法的合理应用，仍然是成功利用GIS进行矿产资源预测的关键。到目前为止，GIS还不能代替地质专家在地质方面的知识素养，亦不能解决数据或数据库本身存在的问题，地质科学领域的专家依然起决定性作用，矿产预测工作离不开有经验的专家参。

4.3 基于三维地质建模的三维成矿预测

得益于三维地质建模技术的发展和数字化等信息技术的广泛运用，三维成矿预测（亦称为隐伏矿体三维预测、三维可视化成矿预测或矿产资源三维预测）理论和方法逐步获得快速发展并趋于成熟。最近十年间，基于三维地质建模的三维成矿预测是矿产勘查领域的研究热点之一。

在国外，Payne 等（2015）基于GOCAD平台，综合利用地质、地球物理及地球化学等资料构建了新西兰 Taupo 火山岩地区的三维地质模型，对矿区热液型金矿化的潜力进行预测评价。Nielsen 等（2015）对西澳大利亚 Marymia Inlier 造山型金矿，利用知识驱动方法，基于地球物理数据建立的三维预测模型，开展区域性三维成矿预测研究。Joly 等（2010）建立 Tanami 区的三维地质模型，利用证据权重方法对造山带进行了金成矿远景预测。

国内学者的代表性成果包括：基于三维可视化信息分析技术的大比例尺矿产预测方法（肖克炎等，2010），"地质信息集成–成矿信息定量提取–立体定量预测"深部矿产资源三维预测方法（毛先成等，2016），基于三维可视化技术的区域隐伏矿体三维定量预测评价方法（陈建平等，2014a），基于综合信息的"四步式"三维成矿预测方法（袁峰等，2014，2019b），对栾川钼多金属矿区开展的基于地质与重磁数据综合三维地质建模的资源预测与潜力评价方法研究（Wang et al.，2015，2017）等。

三维地质建模（three-dimensional geological modeling，3DGM）是三维成矿预测的前提和基础。它利用各种地质原始数据，采用计算机图形学与可视化技术，建立表达各种地质体几何形态、三维空间结构及相互关系的三维几何实体与属性模型。近十年来，三维地质建模方法得以飞速发展。目前，显式三维地质建模方法已趋于成熟（刘刚等，2011；孙莉等，2011；陈建平等，2014a；张明明等，2014；王琨等，2015；高乐，2016；周为喜等，2016），隐式三维地质建模方法也开始得到广泛研究和应用（袁峰等，2019b；邹艳红等，2019）。

三维成矿信息分析提取的主要方法有形态分析、缓冲区分析、距离分析与场模型分析、控制作用分析、蚀变三维建模与分析等（张明明等，2014；李楠等，2015；Li et al.，2016；Mao et al.，2016；毛先成等，2016；李晓晖等，2017）。

同时，三维地质建模软件大量涌现或持续改进，如英国 Datamine 公司的 Datamine Studio 软件，美国 Emerson Paradigm 公司的 SKUA GOCAD 地质建模软件，法国 Dassault Systemes 公司的 Surpac 软件，澳大利亚 MAPTEK 公司的 Maptek Vulcan 软件，澳大利亚 Micromine 公司的 Micromine 软件，澳大利亚 Intrepid Geophysics 公司的 GeoModeller 软件，加拿大阿波罗科技集团公司的三维建模和与分析软件 LYNX Micro Lynx，美国 DGI（Danamic Graphic）公司的三维地质建模、分析及可视化软件系统 Earth Vision，美国 Schlumberger 公司以三维地质建模为中心的一体化油藏描述软件 Petre，美国 Rockware 公司的地质分析软件 RockWare，中国三地曼公司的 3DMine 软件、迪迈公司的 Dimine 软件、中地公司的 MapGISK9、中国地质科学院矿产资源研究所的矿产资源勘探与资源评价三维软件 Minexplorer，中国地质大学的三维成矿定量预测及评价系统——GeoCube 等（肖克炎等，2010；王琨等，2015；Li et al.，2016；李晓晖等，2017）。

近十年的实践应用表明，借助三维地质建模、三维空间分析等方法、技术，三维成矿预测方法能够充分融合深部多源、多维地质及地球物理信息,实现对成矿带、矿集区、矿田、矿床多尺度的成矿有利区域三维预测和深部找矿靶区圈定(张焱,2012;余先川等,2013;陈建平等,2014a;李楠等,2015;Nielsen et al.,2015;Payne et al.,2015;Wang et al.,2015;Li et al.,2016;向杰等,2016;毛先成等,2016;柳炳利等,2016;范文遥等,2020)。

5 地质虚拟现实

虚拟现实技术(virtual reality,VR)是实现大数据可视化的重要方向。它对具有多元、异构、时空性、非线性、多尺度地质矿产勘查数据的展示要求有特别的价值。VR 提供了动态处理数据的能力,使人可以触摸数据,让大数据成为一种触觉体验,使得数据更容易理解和操纵。

近十年来,VR 技术以及应用进入高速发展时期(Goran,2016)。在水文地质领域,研究者利用虚拟现实技术沉浸感、与计算机的交互功能和实时表现功能,建立相关的地质、水文地质模型和专业模型,进而实现对含水层结构、地下水流、地下水质和环境地质问题的虚拟表达。在石油开发领域,运用三维虚拟技术创建了具有高度沉浸感的三维虚拟环境,满足企业对石油矿井等高要求、高难度职位的培训要求。交互式地形可视化技术在 2010 海地地震快速科学响应中的应用表明(Cowgill et al.,2012),基于虚拟现实的数据可视化有可能通过虚拟实地研究和大规模地形数据集的实时交互分析来改变快速的科学反应。

VR 让用户以更自然和直观方式将自己沉浸在数据中,可提高在特定时间内处理的数据量;以不同的角度查看数据,可从数据发现不同内容。目前,虚拟现实引擎日益成熟,较为通用的仿真软件包括 VRP、Quest 3D、Patchwork 3D、EON Reality 等。

多模态、跨尺度地学虚拟现实与多维交互是研究的热点之一。引入虚拟现实技术,使地质工作者以更自然和直观方式沉浸在数据中,提高处理在特定时间内数据量和动态数据的能力(郭艳军等,2019)。基于 VR、AR、MR 技术的多尺度、多模态地学数据的沉浸式虚拟现实可视化与多维交互是在传统三维地质建模的基础上,引入了无人机航拍、VR 全景技术、360°环物摄影技术、SEM 二次电子成像技术和 Unity 3D、ideaVR、VRP 等引擎进行实现。

地质环境的虚拟仿真与可视化研究领域有许多具体应用场景。Whitmeyer 等(2012)结合虚拟现实技术和成熟的谷歌地球应用来为地质实习的教学提供了辅助功能。郭艳军等(2019)应用三维虚拟仿真建模技术,建立了成矿带、地层、矿体、矿物和晶体等多分辨率的三维模型,运用光学追踪技术和图形图像渲染技术搭建了三维沉浸式交互平台,将复杂的矿产资源勘查数据系统的显示在三维沉浸式交互环境中。

6 地质过程模拟

6.1 多点地质统计学

近十年来,多点地质统计学研究得到长足的进步。它是一种随机模拟方法,是复杂地学现象随机模拟技术的一个重要分支(Caers,2011;Mariethoz and Caers,2014;Wellmann and Caumon,2018)。相应算法的提出为各向异性的复杂地质现象及过程的模拟与描述提供了技术支撑,并且已经在储层模拟、水文地质建模、多孔介质重构、遥感数据处理、图像精度改善等多个地学领域取得了很好的应用。

多点地质统计学模拟方法的重要来源之一是储层沉积相建模,它同时兼备了传统基于变差函数的两点地质统计学和基于目标的随机模拟方法的优势。在多点地质统计随机模拟中,地质变量的空间结构性被模式化在了训练图像(参考模型)中,从而克服了传统地质统计学方法不能再现目标几何形态的不足

（Allard et al., 2012；郑天成等, 2019）。此外，在多元变量对应的多个训练图像基础上，更加便于实现多源异构信息及先验模型的融合。

基于多点地质统计的随机模拟方法便于可以实现：①地质先验知识的融合。训练图像本身就是一个先验的参考模型，或者是一个包含了目标变量空间分布模式的概念模型；此外，还可将已知的地质知识作为一个先验的概率模型，去调节目标变量对应多点模式的条件概率密度分布函数，从而实现对先验知识的融合。②地质现象和过程的自动重构。多点地质统计学方法是一种既忠实于实测数据（硬数据），又便于综合使用其他间接数据（软数据）的自动模拟方法；它不但能够再现不同尺度的目标几何形态（类别变量），而且也适用于连续属性信息（连续变量）的模拟。

虽然多点地质统计学方法在地质先验知识融合和地质现象及过程的自动重构等方面具有优势，并已成功应用于地学的各个领域，但依然无法摆脱平稳性假设的束缚，在面对实际的复杂地质空间结构和属性的一体化建模及复杂地质过程动态模拟时，还存在着如下限制：①训练图像难以获取。多点地质统计学方法完全依赖于一个完整的、能够表达待模拟现象的概念模型——训练图像；但对于一个实际应用来说，难以获得一个完整的训练图像（尤其是三维训练图像），从而使得多点地质统计学方法的应用受到限制。②非平稳性的限制。实际地质勘查数据的非平稳性特征与多点地质统计学理论中平稳性假设的矛盾在一定程度上限制了其在复杂地质结构建模及地质过程模拟中的应用。因此，需要针对以上问题，引入新的模型、策略和方法，发展新型的知识驱动与数据驱动协同的多点地质统计随机模拟方法，最终实现对复杂地质空间模型的自动重构和对地质过程的动态模拟。

6.2 地球动力学数值模拟

地球动力学数值模拟一直是计算和数学地球科学的重要内容。21世纪以来，地球动力学数值模拟以大规模计算模拟结合固体地球动力学，深入研究了地球内、外核，地幔，岩石圈地幔以及地壳各个圈层的结构、变形以及相互影响作用，旨在借助模拟的方法研究和理解各种尺度上的地球动力学问题，例如，在全球尺度上探讨板块运动和地幔对流的动力学机理；在板块尺度上研究岩石圈的变形过程（如威尔逊旋回描述的从大陆张裂、海底扩张、板块俯冲到大陆碰撞）；在资源、灾害方面，计算、模拟诸如矿产资源、地震、滑坡、海啸等形成机制等。最近十年来的代表性研究包括：基于数值模拟模型分析陆内岩石圈拆沉作用的动态演化过程（Gorczyk et al., 2015），对大陆俯冲-碰撞-折返的动力学过程进行模拟推演（李忠海和石耀霖，2016），对局部褶皱的形成机理进行模拟研究（Ord and Hobbs, 2013）。此外，还被应用于矿产资源领域的相关研究（Weis et al., 2012；Ord et al., 2012）。

目前，前沿的地球动力学数值模拟往往充分考虑热-力学耦合，将过去纯粹的热模拟和力学模拟整合到一起。

热-力学耦合的地球动力学数值模拟，是通过使用各种数值离散方法，在欧拉网格节点上求解三大守恒/控制方程：质量守恒方程（方程1）、动量守恒方程（方程2）和能量守恒方程（方程3）：

$$\frac{\partial v_i}{\partial x_i}=0 \tag{1}$$

$$\frac{\partial \sigma'_{ij}}{\partial x_j}-\frac{\partial p}{\partial x_i}=-\rho g_i \tag{2}$$

$$\rho C \frac{DT}{Dt}=\frac{\partial}{\partial x_i}\left(k\frac{\partial T}{\partial x_i}\right)+H_r+H_s+H_a+H_L \tag{3}$$

式中，∂ 为偏微分符号；v_i 为速度；x_i 为空间坐标；i、j 指示空间方向（三维模型中 $i=x, y, z$；$j=x, y, z$）；σ' 为偏应力；p 为压力；ρ 为密度；g 为重力加速度；C 为热容；T 为温度；t 为时间；k 为热导率；H 为系统内部热源（进一步解释见下文）。

在热-力学耦合的动力学数值模型中，岩石的变形受控于岩石流变学性质，主要包括黏、弹、塑性。岩石的黏性变形受控于应变速率、温度和压力，这种变形往往是非线形的（即非牛顿流体变形），且往往以扩散蠕变和位错蠕变为主；在低温、高应力、大的晶体颗粒条件下（往往对应于岩石圈和浅部地幔状态），位错蠕变起主导作用；在高温、低应力、小的晶体颗粒条件下（往往对应于地幔深部状态），扩散蠕变起主导作用。当岩石的偏应力达到岩石的屈服强度时，岩石变形机制从黏性转变为塑性，这种转变通常使用德鲁克-普拉格（Drucker-Prager）屈服准则来判断。屈服强度是压力的函数，随压力的增加而增大。

基于上述，科学家模拟了大陆伸展样式之一的宽裂谷的发育和演化（Jolivet and Brun，2010）。动力学数值模拟显示，宽裂谷的形成主要受以下几个因素影响：伸展速率，地壳的岩性和流变分层，地壳厚度以及上、下地壳厚度比，岩石圈热状态。伸展速率主要影响应变集中和岩石圈的热状态。快速伸展往往可以使应变快速集中，软流圈上涌的热对流作用大于热扩散，导致岩石圈升温并弱化了岩石圈，进而促使窄裂谷的发育；相反，缓慢伸展的情况下，应变集中发生较晚，热扩散作用更加显著，导致岩石圈升温变缓，宽裂谷更容易形成（Liao and Gerya，2014）。地壳的岩性结构和上、下地壳厚度会影响地壳的流变强度结构，进而影响裂谷的伸展变形范围。对于流变强度较弱的下地壳，地壳变形往往表现为上地壳伸展集中，下地壳的均匀、扩散伸展，伴随一定的核杂岩发育（Huismans and Beaumont，2011；Brune et al.，2014）。下地壳的蠕变流动使地壳的减薄发生在很宽的范围内，导致莫霍面形态较为平坦。

科学家还模拟了地壳和岩石圈地幔优先破裂模式。结果揭示，控制地壳和岩石圈地幔优先破裂的一个重要因素是地壳和岩石圈地幔的耦合程度，即下地壳的流变强度（Huismans and Beaumont，2011；Brune et al.，2014）。当下地壳的流变强度较高时，地壳和岩石圈地幔表现为强耦合，岩石圈伸展随深度的差异性较小，应变有效集中并倾向于形成岩石圈尺度的颈缩，有利于形成地壳优先破裂模式。当下地壳的流变强度较低时，地壳和岩石圈地幔表现为弱耦合，岩石圈表现为多个流变分层，伸展变形随深度显著变化，往往表现为岩石圈地幔优先破裂。由于下地壳的流变强度控制了整个岩石圈的流变分层，导致无论是伸展环境还是挤压环境，下地壳都是影响岩石圈变形的一个重要参数。

针对克拉通岩石圈的流变分层，模拟了软弱层对伸展变形的影响作用（Liao and Gerya，2014；2015；Liu et al.，2018）。模拟结果显示，当克拉通岩石圈中存在软弱分层时（MLD层），会加速地壳变形，导致地壳优先破裂；而岩石圈地幔变形减缓。当克拉通岩石圈存在流变弱化时，岩石圈整体强度降低；岩石圈底部会发生重力不稳定性，导致部分物质剥离（Liao and Gerya，2014）。这些动力学模拟，为认识大陆岩石圈张裂提供了定量的认识。

6.3 成矿地质过程数值模拟

成矿地质过程数值模拟是以有限元和有限差分等计算方法为主要手段，配合传统矿床学等研究方法定义模拟条件，以多物理化学场控制方程来描述成矿演化过程的方法技术，已成为矿床学研究新的定量手段（Ord et al.，2012；袁峰等，2019a）。

热液矿床是最具有工业意义的矿床类型之一，其成矿作用是形变、流体流动、热传递、溶质运移及化学反应的复杂耦合过程，因此针对热液矿床成矿过程的数值模拟受到数学地球科学家的特别关注（Ord et al.，2012）。许多学者对斑岩、夕卡岩等热液矿床开展了相关研究。Weis等（2012）对斑岩铜矿床进行了成矿过程模拟，定量分析了热和流体对斑岩型矿床的规模、矿体形态及品位空间分布的影响。Zou等（2017）模拟分析了断裂构造对夕卡岩型铅锌矿成矿的控制作用。戴文强等（2019）针对安庆铜矿床，模拟分析了典型夕卡岩矿物的形成过程对后续成矿作用的影响。王语等（2020）基于成矿条件数值模拟和支持向量机算法预测了粤北凡口铅锌矿深部成矿位置。

对于深部矿产资源预测，成矿过程数值模拟方法也能够提供新的信息和支持。例如，基于成矿过程数值模拟结果圈定找矿靶区（Liu L. M. et al.，2011）、评价目标区域的成矿能力（Hu et al.，2020）、为三维成矿预测方法提供新的三维预测信息（Qin and Liu，2018；Li et al.，2018）等。

7 区块链技术应用

区块链作为一项颠覆性技术，在地质领域的应用还处于起步阶段，但可以预测，区块链与地质科学研究的结合必将成为数学地球科学和大数据研究的一个热点（周永章等，2020b）。

区块链的核心理念是去中心化、不可篡改、可追溯和匿名。区块链1.0以比特币为代表的数字货币的应用为标志。区块链2.0以智能合约的引入为标志，以计算机程序的方式来缔结和运行各种合约，可以提供更灵活的合约功能，在数字资产的确权、存证、溯源等领域已经得到了很好的应用（王继业等，2017）。区块链3.0以可编程社会为主要特征，在社会治理领域有重要的应用场景（Swan，2015；贠天一，2017），广泛应用于身份认证、公证、物流、签证、投票等领域（任明等，2018）。当应用范围扩大到整个社会时，区块链技术有可能成为"万物互联"的一种基础协议。

在大数据时代，区块链的去中心化、不可篡改、隐私保护等特性，智能合约提供的丰富交互接口，可以为区块链技术在地质领域应用提供了重要的基础。地质勘查实物、资料、数据的区块链溯源、存证管理，成为区块链地质应用的首批突破口。

地质科学是一门源远流长的学科，一代代地质工作者的心血汇成了海量的地质科学数据，建立了许多有价值的数据库。但至今，分散在科学家个人或实验室的长尾数据长期缺乏关注度，利用价值被忽视，造成科研资源的巨大浪费。在过去，科学家已经习惯了把数据孤立地存放在自己的服务器中，甚至认为这些数据是自己的隐私。数据从一开始就没有被规划进流通，数据隐私保护、数据可靠流通的激励机制的缺失，是许多科学家没有"数据上链"意识的原因。在这种情况下，中央式的信用可以满足信用问题，但成本很高，在处理一些碎片化、长尾的信用缺失场景时难以胜任。利用区块链技术，让地质数据上链是解决地质长尾数据问题的理想方案（周永章等，2020b）。

除长尾数据外，区块链技术和通证思维使打造全球地质社区成为可能。对全球地质社区的任何人，都可以使用地质通证的方式激励他们，最后地质通证的持有者形成一个全球利益共同体——全球地质社区。地质是支撑大资源大环境大生态建设的重要支柱。因而，基于区块链的大资源大环境大生态必将有宏大的发展前景。

8 大地质科学计划

8.1 OneGeology

OneGeology是2008年8月于挪威奥斯陆召开的第33届国际地质大会上发起的国际科学合作"OneGeology计划"，即"同一个地质计划"，其目标是实现全球的地质图信息共享，让所有人都可以在网络上取得全世界的动态数字化地质图，它源于2007年由全球地质调查组织在英国布莱顿发起的一项活动计划（Jackson，2007）。目前，OneGeology计划参与国达118个，它通过http://www.onegeology.org/提供全球在线地质共享数据库，提供开放的在线地质图数据和目录的访问、查询、注册共享等功能，帮助人们在网络上取得全世界的动态数字化地质图。

地质图空间数据共享通过使用WebGIS技术和OGC空间数据互操作规范实现。OneGeology应用系统是一个基于WebGIS技术的分布式应用系统，客户端Portal（门户网站）采用J2EE规范开发，服务器端以基于开放地理信息协会OGC标准的网络地图服务（web map service，WMS）和基于GeoSciML数据模型的网络要素服务（web feature service，WFS）的形式提供动态地质图数据，并允许被客户端Portal调用。WMS和WFS是OGC基于WebService技术制定的空间信息互操作标准，广泛用于分布式系统中空间数据的发布。

OneGeology 使用的地质科学标记语言 GeoSciML 数据模型（GeoScience Markup Language）是 GML 技术在地学领域的扩展，基于基础地质本体设计地质图数据编码，形成统一的接口标准，可用于地质图的网络要素服务发布。GeoSciML 定义了一种数据交换格式，具有表征地理特征的能力。GeoSciML 标记语言有12 种属性分类：地质特征（Geologic Feature）、地质单元（Geologic Unit）、地球材料（Earth Material）、地质构造（Geologic Structure）、化石（Fossil）、地质年代（Geologic Age）、钻孔和观察（Boreholes and Observations）、地质关系（Geologic Relation）、图像价值（CGI Values）、词汇（Vocabulary）、元数据（Metadata）、采集（Collection）。用户可根据数据内容选择 GeoSciML 模型的映射属性（OneGeology, 2019）。OneGeology 允许在现有空间数据或空间数据库之上增加 GeoSciML 接口，而不必更改或调整数据库结构，即可完成地质图数据共享，是一个智慧的选择。GeoSciML 数据模型可以对 Geologic Feature、Geologic Unit、Earth Material、Geologic Structure、Fossil、Geologic Age、Boreholes & Observations、Geologic Relation、CGI Values、Vocabulary、Metadata、Collection 等地质图要素进行编码（逯永光等，2011）。

与基于中间件的分布式系统比较，基于 WebService 技术的分布式系统具有很多优势。将基于 WebService 的分布式技术与 GIS 技术结合产生了基于 WebService 的 WebGIS 系统。OneGeology 选用基于 WebService 技术的网络地图服务 WMS 和网络要素服务作为实现 1∶100 万地质图空间数据共享的方法，不仅使整个 OneGeology 系统有着可扩展和跨平台的优点，而且很好地保护了服务提供者的地质图数据。1∶100 万地质图网络服务的发布分为两步进行，第一步完成网络地图服务 WMS 的发布、部署和测试；第二步完成基于 GeoSciML 数据模型的网络要素服务 WFS 的发布、部署和测试。例如，中国 1∶100 万地质图网络服务在服务器端部署完毕，可在 OneGeology 门户网站注册，以便组成一个完整的 WebGIS 分布式系统，实现中国 1∶100 万地质图数据共享。

8.2 玻璃地球

"玻璃地球"建设的倡议由澳大利亚的地学家率先提出，澳大利亚政府随即于 2001 年正式启动了"玻璃地球"计划，加拿大、法国、荷兰、英国、美国和德国等随之响应（Carr et al., 1999; Esterle and Carr, 2003）。"玻璃地球"计划的初衷是要使地下 1 000 m 变得"透明"，便于发现下一代巨型矿床；其任务是建立全国地学家联盟来填制四维"地图"，并通过高度可视化和广泛的网络服务接口传播信息。世界各国的"玻璃地球"建设计划的实施，都采取以三维区域地质填图为主导与深部探测计划相结合的方式，这是与 OneGeology 计划不同之处。

各国在"玻璃地球"建设的进展有显著差异，有些进展比较快，如荷兰、英国和德国，已建立了全国的三维地质框架模型，并融入了多源信息，甚至建立了多尺度、多分辨率的分类或分级地层框架模型。有的进展比较慢，例如美国和加拿大，虽引入了盆地分析方法和沉积格架模型，但仅仅在第四纪沉积物、水文地质和地震地质等专业领域，实现了局部三维地质框架模型构建。

英国地质调查局（British Geological Survey，BGS）在 2009—2014 发展战略中，把三维地质填图及其三维地质模型放在核心位置，并将其与深部探测密切结合起来，目前已建立了全国性分级三维地质模型（LithoFrame），开发了 GSI3D 建模软件，还逐步更新地质图及其三维模型，并结合最新的航空地球物理数据和高分辨率的地球化学数据，填制国家级的海洋地质图。LithoFrame 的核心是通过创建相互对应的不同分辨率的三维模型，形成从全国性概略模型到大比例尺详细模型的无缝过渡（Berg et al., 2011）。

中国在科技部和国家自然科学基金委的支持下设立了《深部探测技术与实验研究（2009—2012）（SinoProbe）》国家科技重大专项，完成了约 6 000 km 深地震反射剖面，研究并实验了地壳与地幔深部探测的一系列技术方法，其中包括长江中下游和南岭成矿带开展的矿集区立体探测、岩石圈三维结构与地球动力学数值模拟，并开展了深部探测数据共享平台建设和深部探测综合集成，还建立了相应的矿集区 2 000 m 以浅三维地质框架模型。并在此基础上推进实施了科技部"十三五"期间的深地探测重点研发计划项目（董树文等，2012）。

"玻璃地球"是"数字地球"在地矿领域的体现，是一种地质信息和地理信息相结合并存储于计算机网络上的、可供多用户访问和开展地质、资源、环境和地灾决策分析的三维可视化虚拟浅层地壳。"玻璃地球"是地质时空大数据的有效载体，是进行地球科学研究、协调人类社会与自然环境和谐发展的有效手段（吴冲龙和刘刚，2015）。

"玻璃地球"建设的发展趋势是：①从个别走向一般，即由个别部门的建设行为，转变为地矿行业各部门普遍参与的建设行为；②从局部走向整体，即由局部地区和局部领域，转变为全国范围的地壳整体数字化、透明化建设；③从功能走向数据，即由单纯追求建模功能，转变为全面开展以数据管理为核心的三维地质信息系统建设；④从零散走向集成，即由零散的局部三维地质建模，转为点面结合的连片集成化的多尺度三维数字地质建模；⑤从展示走向实用，即由追求单纯的三维地质体和地质结构可视化展示效果，转变为追求基于所建立的三维地质模型开展各种可视化分析、可视化仿真、可视化设计和可视化决策的能力。

8.3 化学地球

"化学地球"大科学计划启动于2016年，由联合国教科文组织全球尺度地球化学国际研究中心发起，旨在为全球自然资源与生态环境可持续发展提供系统性、长期性及权威性地球化学数据，为资源环境可持续发展提供解决方案。该计划发布时得到美国、俄罗斯、澳大利亚、南非、伊朗、土耳其等22个国家的科学家的支持（王学求，2018）。该计划将逐步建立覆盖全球地球化学基准网，获取全球地球化学基准点76个化学元素基准数据；开展地球化学填图，圈定矿产资源远景区；基于全球地球化学基准值，监测全球重金属、放射性和碳元素环境变化，分析全球性重大地质事件或污染事件的地球化学响应；建立"化学地球"大数据平台。

8.4 深时数字地球（DDE）

2018年，中国科学家倡导开展"深时数字地球"（DDE）国际大科学计划，该计划于2019年正式启动，旨在整合全球演化数据，共享全球地学知识，构建地球科学大数据平台，实现海量文件的机器阅读、海量数据的智能整合分析，助力生态环境保护研究、能源矿产勘探、气候变化研究、深时地球演化等重大科学问题和重大社会需求，成为全球地球科学研究人员的强大科研助手，推动科学技术、国民经济等领域的创新与突破。

DDE建设需要领域专家、数据专家和计算机专家的通力协作。它的提出契合了科学研究从问题驱动向数据驱动转变的关键时期。全球数据的不断激增，导致了一个前所未有的科学研究从问题驱动向数据驱动转变，科学研究的第四范式——数据密集型科学发现应势而生。第四范式基于海量数据的科研活动、过程、方法和基础设施，在海量数据和无处不在的网络上发展起来的。

DDE与更早时候开始的地球科学专业数据库建设亦有密切联系。除OneGeology外，OpenGeoscience也是比较著名的地质图数据库。它由英国地质调查局下属的英国国家地球科学数据中心（National Geoscience Data Centre，NGDC）负责管理，是世界上第一个基于整个国家地质矢量数据提供网络地图服务（WMS），目标是用户通过WMS将自己的数据与英国地质调查局（BGS）提供的地质数据集融合在一起使用。这得益于2009年英国地质调查局与美国ESRI公司的合作，它们采用ArcGIS技术推出了开放式地学信息服务。

其他著名的数据库还有美国地调所负责建立的"国家地质图数据库"（National Geologic Map Database Project，NGMDB）、全球火成岩数据库（GEOROC，http://georoc.mpch-mainz.gwdg.de/georoc/）、全球海底岩石学数据库（PetDB，https://search.earthchem.org/）、古生物学数据库（PBDB，https://paleobiodb.org/）、美国地层学数据库（Macrostrat，https://macrostrat.org/）、全球地质年代学数据库（GeoChron，http://www.geochron.org/）、沉积岩数据库（SedDB，http://www.earthchem.org/seddb）、磁学信息联盟数据库（MagIC，https://www2.earthref.org/MagIC）、中国地球物理科学数据中心（http://

geospace. geodata. cn/）等。中国地质科学院地质科学数据共享网（http：//www. geoscience. cn/）整合了中国基础地质、矿产地质、构造地质、物化探地质、水文地质、岩溶地质、岩矿测试以及环境地质数据，为国家和社会公众提供地质基础数据服务，也产生较大影响。

9 展　　望

上述对数学地球科学的简略回顾，尽管不是很全面，但足以看到，最近十年是数学地球科学跨越发展的时期，大数据、人工智能算法正在改变地质学。这种改变契合了数据密集型科学的出现，科学研究从问题驱动向数据驱动转变的转折时期。

大数据挖掘和机器学习代表了科学研究范式的变革。梳理科学发展纹理可见，人类经历过四次重要的范式变革：第一范式的核心是归纳法；第二范式的核心是以演绎法理性为主；第三范式主要针对复杂性系统进行模拟；对大数据的有效分析则成为当前第四范式的主要诉求，它形成了科学研究的第四范式。

大数据是一种思维和认知论的革命，它开启了一次重大的时代转型，因果关系不再是研究的必要前提。大数据挖掘特别适合于窥探具有高维度、全维度空间的现实世界。关联性思维作为大数据的核心思维之一，它可以从很多看似支离破碎的信息中复原一个事物的全貌，并进而能够预测或判断出尚未观察到的事物的现象。

大数据思维和大数据挖掘算法在地球资源、环境、灾害中的应用将是未来相当一个时期内数学地球科学的主要发展方向。

致谢：本文是大数据与数学地球科学专业委员会的集体成果。

参 考 文 献

曹梦雪, 路来君, 吕岩, 辛双. 2018. 鄂尔多斯盆地北缘地球化学大数据样本优选分析. 岩石学报, 34(2):363-371
常力恒, 朱月琴, 张戈一, 张旋, 胡博然. 2018. 面向矿产资源信息的空间关联性分析. 岩石学报, 34(2):314-318
陈建平, 于淼, 于萍萍, 尚北川, 郑啸, 王丽梅. 2014. 重点成矿带大中比例尺三维地质建模方法与实践. 地质学报, 88(6):1187-1195
陈进, 毛先成, 刘占坤, 邓浩. 2020. 基于随机森林算法的大尹格庄金矿床三维成矿预测. 大地构造与成矿学, 44(2):231-241
陈丽蓉. 2019. 顾及空间约束的多元地球化学异常识别自编码神经网络方法研究. 武汉: 中国地质大学博士学位论文
陈伟雄, 杨华鑫, 周泽东, 张明. 2019. 地质矿物的地球化学数据可视化分析与构造背景预测. 世界有色金属, (14):132-134
陈永良, 路来君, 李学斌. 2014. 多元地球化学异常识别的核马氏距离方法. 吉林大学学报(地球科学版), 44(1):396-408
成秋明. 2007. 成矿过程奇异性与矿产预测定量化的新理论与新方法. 地学前缘, 14(5):42-53
程学旗, 靳小龙, 王元卓, 郭嘉丰, 张铁赢, 李国杰. 2014. 大数据系统和分析技术综述. 软件学报, 25(9):1889-1908
戴文强, 李晓晖, 袁峰, 张明明, 胡训宇, 周涛发. 2019. 安庆铜矿床典型矽卡岩矿物形成过程数值模拟. 合肥工业大学学报(自然科学版), 42(3):346-354
丁彩英, 李泽鹏, 刘松华. 2019. 学习全局边函数的半监督社区检测. 太原理工大学学报, 50(2):243-250
董树文, 李廷栋, 陈宣华, 魏文博, 高锐, 吕庆田, 杨经绥, 王学求, 陈群策, 石耀霖, 黄大年, 周琦. 2012. 我国深部探测技术与实验研究进展综述. 地球物理学报, 55(12):3884-3901
杜雪亮, 张旗, 王金荣, 陈万峰, 潘振杰, 李玉琼. 2017. 全球海山玄武岩数据挖掘研究. 地质科学, 52(3):668-692
范文遥, 曹梦雪, 路来君. 2020. 基于GOCAD软件的三维地质建模可视化过程. 科学技术与工程, 20(24):9771-9778
高乐. 2016. 钦杭成矿带(南段)庞西垌地区三维地质建模及多元信息成矿预测. 广州: 中山大学博士学位论文
葛粲, 汪方跃, 顾海欧, 管怀峰, 李修钰, 袁峰. 2019. 基于卷积神经网络和火山岩大数据的构造源区判别. 地学前缘, 26(4):22-32
耿厅, 周永章, 李兴远, 王俊, 陈川, 王堃屹, 韩紫奇. 2019. 锆石微量元素对成矿岩体的判别——来自大数据思维的应用. 地质通报, 38(12):1992-1998
郭昆, 彭胜波, 张瑛瑛, 陈羽中. 2019. 基于密度峰值和社区归属度的重叠社区发现算法. 小型微型计算机系统, 40(5):1127-1136
郭艳军, 张进江, 陈斌, 崔莹, 熊文涛, 李梅, 张志诚, 秦善. 2019. 基于VR技术的多尺度地质数据3D沉浸式可视化与交互方法. 地学前缘, 26(4):146-158
韩帅, 李明超, 任秋兵, 刘承照. 2018. 基于大数据方法的玄武岩大地构造环境智能挖掘判别与分析. 岩石学报, 34(11):3207-3216
焦守涛, 周永章, 张旗, 金维浚, 刘艳鹏, 王俊. 2018. 基于GEOROC数据库的全球辉长岩大数据的大地构造环境智能判别研究. 岩石学报, 34(11):3189-3194
李楠, 肖克炎, 阴江宁, 范建福, 王琨. 2015. 表面模型缓冲区分析方法. 计算机辅助设计与图形学学报, 27(9):1625-1636

李晓晖,袁峰,马良,唐敏慧,张明明,周涛发,贾蔡,胡训宇. 2017. 三维成矿定量预测系统设计与应用实例研究. 地质科学,52(3):755-770

李忠海,石耀霖. 2016. 三维板块几何形态对大陆深俯冲动力学的制约. 地球物理学报,59(8):2806-2817

刘刚,吴冲龙,何珍文,翁正平,朱庆,张叶廷,李晓晖. 2011. 地上下一体化的三维空间数据库模型设计与应用. 地球科学(中国地质大学学报),36(2):367-374

刘心怡,周永章. 2019. 关联规则算法在粤西庞西垌地区元素异常组合研究中的应用. 地学前缘,26(4):125-130

刘欣雨,张旗,张成立. 2019. 基于大数据方法建立大洋安山岩构造环境判别图. 地质通报,38(12):1963-1970

刘艳鹏,朱立新,周永章. 2018. 卷积神经网络及其在矿床找矿预测中的应用——以安徽省兆吉口铅锌矿床为例. 岩石学报,34(11):3217-3224

刘艳鹏,朱立新,周永章. 2020. 大数据挖掘与智能预测找矿靶区实验研究——卷积神经网络模型的应用. 大地构造与成矿学,44(2):192-202

柳炳利,郭科,李程,王璐. 2016. 基于原生晕三维数据体模型的深部矿产预测. 地质学刊,40(3):403-409

柳原,白金牛. 2019. 复杂网络大数据中重叠社区自动检测仿真. 计算机仿真,36(6):393-397

逯永光,丁孝忠,李廷栋,韩坤英,剧远景,庞健峰,丁伟翠,王振洋. 2011. "OneGeology 计划"及其在中国研究新进展. 中国地质,38(3):799-808

毛先成,周尚国,张宝一,李腊梅. 2014. 锰矿 GIS 分析与评价:以桂西-滇东南地区为例. 北京:地质出版社

毛先成,张苗苗,邓浩,邹艳红,陈进. 2016. 矿区深部隐伏矿体三维可视化预测方法. 地质学刊,40(3):363-371

乔少杰,郭俊,韩楠,张小松,元昌安,唐常杰. 2017a. 大规模复杂网络社区并行发现算法. 计算机学报,40(3):687-700

乔少杰,韩楠,张凯峰,邹磊,王宏志,Gutierrez L A. 2017b. 复杂网络大数据中重叠社区检测算法. 软件学报,28(3):631-647

任明,汤红波,斯雪明,游伟. 2018. 区块链技术在政府部门的应用综述. 计算机科学,45(2):1-7

任秋兵,李明超,李玉琼,韩帅,张野,张旗. 2020. 基于全球橄榄石数据的玄武岩构造环境智能判别方法及其验证. 大地构造与成矿学,44(2):212-221

孙莉,肖克炎,唐菊兴,邹伟,李楠,孙艳. 2011. 基于 Minexplorer 探矿者软件的甲玛铜矿三维地质体建模. 成都理工大学学报(自然科学版),38(3):291-297

王怀涛,罗建民,王金荣,杨婧,宋秉田,王玉玺,王晓伟,周煜祺. 2018. 基于大数据的基性-超基性岩定量分类及成矿预测研究——以北山地区为例. 岩石学报,34(11):3195-3206

王继业,高灵超,董爱强,郭少勇,陈晖,魏欣. 2017. 基于区块链的数据安全共享网络体系研究. 计算机研究与发展,54(4):742-749

王金荣,陈万峰,张旗,焦守涛,杨婧,潘振杰,王淑华. 2017. N-MORB 和 E-MORB 数据挖掘——玄武岩判别图及洋中脊源区地幔性质的讨论. 岩石学报,33(3):993-1005

王堃屹,周永章,王俊,张奥多,余晓彤,焦守涛,刘心怡. 2019. 推荐系统算法在钦杭成矿带南段文地幅矿床预测中的应用. 地学前缘,26(4):131-137

王琨,肖克炎,李胜苗,甘曦. 2015. 基于探矿者软件(Minexplorer)的三维地质建模及储量估算——以湘西北李梅铅锌矿为例. 地质通报,34(7):1375-1385

王世称,陈永良,夏立显. 2000. 综合信息矿产预测理论与方法. 北京:科学出版社

王学求. 2018. "化学地球"国际大科学计划取得重要进展. 中国地质,45(5):858

王语,周永章,肖凡,王俊,王恺其,余晓彤. 2020. 基于成矿条件数值模拟和支持向量机算法的深部成矿预测——以粤北凡口铅锌矿为例. 大地构造与成矿学,44(2):222-230

巫红霞,谢福. 2019. 基于加权社区检测与增强人工蚁群算法的高维数据特征选择. 计算机应用与软件,36(9):285-292,301

吴冲龙,刘刚. 2015. "玻璃地球"建设的现状、问题、趋势与对策. 地质通报,34(7):1280-1287

向杰,陈建平,胡彬,胡桥,杨伟. 2016. 基于三维地质—地球物理模型的三维成矿预测——以安徽铜陵矿集区为例. 地球科学进展,31(6):603-614

向杰,陈建平,肖克炎,李诗,张志平,张烨. 2019. 基于机器学习的三维矿产定量预测——以四川拉拉铜矿为例. 地质通报,38(12):2010-2021

肖凡,陈建国,侯卫生,王正海. 2017. 钦-杭结合带南段庞西垌地区 Ag-Au 致矿地球化学异常信息识别与提取. 岩石学报,33(3):779-790

肖克炎,陈学工,李楠,邹伟,孙莉. 2010. 地质矿产勘探评价三维可视化技术及探矿者软件开发. 矿床地质,29(增刊):758-760

徐述腾,周永章. 2018. 基于深度学习的镜下矿石矿物的智能识别实验研究. 岩石学报,34(11):3244-3252

余先川,邓维科,肖克炎,邹伟. 2013. 基于三维克立格方法的可视化储量估算. 地学前缘,20(4):320-331

余晓彤,肖凡,周永章,王语,王恺其. 2019. 粤西庞西垌地区银金地球化学异常信息挖掘与提取. 地质与勘探,55(1):77-86

袁峰,李晓晖,张明明,周涛发,高道明,洪东良,刘晓明,汪启年,朱望波. 2014. 隐伏矿体三维综合信息成矿预测方法. 地质学报,88(4):630-643

袁峰,李晓晖,胡训宇,李跃,贾蔡,Alison O,张明明,戴文强,李贺. 2019a. 热液矿床成矿作用研究新途径:数值模拟. 地质科学,54(3):678-690

袁峰,张明明,李晓晖,葛粲,陆三明,李建设,周宇章,兰学毅. 2019b. 成矿预测:从二维到三维. 岩石学报,35(12):3863-3874

负天一. 2017. 区块链项目落地 3.0 时代即将开启. 中国战略新兴产业,(17):68-69

翟明国. 2018. 大数据定将改变地质——向读者推荐《地球科学大数据挖掘与机器学习》. 矿物岩石地球化学通报,37(6):1209

张宝一,毛先成,周尚国,胡超,闫芳. 2011. 基于 GIS 的证据权重法在桂西南地区优质锰矿成矿预测中的应用. 地质找矿论丛,26(4):359-366

张明明,周涛发,袁峰,李晓晖. 2014. 庐枞盆地泥河铁矿床成矿岩体三维形态学分析及找矿指标研究. 地质学报,88(4):574-583

张旗,周永章. 2018. 大数据助地质腾飞. 岩石学报,34(11):3167-3172

张旗,葛粲,焦守涛,袁峰,张明明,刘惠云. 2019. 在大数据背景下看TAS分类的不足及可能的解决方案. 地质通报,38(12):1943-1954

张雪英,叶鹏,王曙,杜咪. 2018. 基于深度信念网络的地质实体识别方法. 岩石学报,34(2):343-351

张焱. 2012. 钦杭成矿带(南段)庞西垌地区矿致地球化学异常的提取及查证研究. 广州:中山大学博士学位论文

张焱,周永章. 2012. 多重地球化学背景下地球化学弱异常增强识别与信息提取. 地球化学,41(3):278-291

张野,李明超,韩帅,任秋兵,朱月琴. 2020. 基于金矿规格单元数据的机器学习方法在成矿建模分析中的应用. 大地构造与成矿学,44(2):183-191

赵鹏大. 2007. 成矿定量预测与深部找矿. 地学前缘,14(5):1-10

赵鹏大,夏庆霖. 2009. 中国学者在数学地质学科发展中的成就与贡献. 地球科学-中国地质大学学报,34(2):225-231

赵振华. 2007. 关于岩石微量元素构造环境判别图解使用的有关问题. 大地构造与成矿学,31(1):92-103

郑天成,侯卫生,何思彤. 2019. 基于二维地质剖面的三维地质结构多点统计学模拟方法. 吉林大学学报(地球科学版),49(5):1496-1506

周为喜,陈玉华,杨永国,罗金辉. 2016. 基于角点网格的煤层三维建模与可视化研究. 煤田地质与勘探,44(5):53-57

周永章,黎培兴,王树功,肖凡,李景哲,高乐. 2017. 矿床大数据及智能矿床模型研究背景与进展. 矿物岩石地球化学通报,36(2):327-331,344

周永章,张良均,张奥多,王俊. 2018a. 地球科学大数据挖掘与机器学习. 广州:中山大学出版社,1-269

周永章,王俊,左仁广,肖凡,沈文杰,王树功. 2018b. 地质领域机器学习、深度学习及实现语言. 岩石学报,34(11):3173-3178

周永章,陈川,张旗,王功文,肖凡,沈文杰,卞静,王亚,杨威,焦守涛,刘艳鹏,韩枫. 2020a. 地质大数据分析的若干工具与应用. 大地构造与成矿学,44(2):173-182

周永章,刘楠,陈川,杨威. 2020b. 开启区块链地质应用新时代. 地质通报,39(1):1-6

邹艳红,褚慧慧,毛先成. 2019. 基于面-体布尔运算的断层切割矿体三维模型. 金属矿山,(7):153-160

左仁广. 2019. 勘查地球化学数据挖掘与弱异常识别. 地学前缘,26(4):67-75

Afzal P, Alghalandis Y F, Khakzad A, Moarefvand P, Omran N R. 2011. Delineation of mineralization zones in porphyry Cu deposits by fractal concentration-volume modeling. Journal of Geochemical Exploration,108(3):220-232

Agrawal R, Srikant R. 1994. Fast algorithms for mining association rules. In:Proceedings of the 20th VLDB Conference. Santiago,Chile,487-499

Agterberg F. 2014. Geomathematics:Theoretical Foundations,Applications and Future Developments. Switzerland:Springer

Allard D, Comunian A, Renard P. 2012. Probability aggregation methods in geoscience. Mathematical Geosciences,44(5):545-581

Aryafar A, Moeini H. 2017. Application of continuous restricted Boltzmann machine to detect multivariate anomalies from stream sediment geochemical data, Korit, East of Iran. Journal of Mining and Environment,8(4):673-682

Bagas L, Xiao K Y, Mark J, Li N. 2017. Quantitative assessment of China's mineral resources Part 1. Ore Geology Reviews,91:1081-1083

Berg R C, Mathers S J, Kessler H, Keefer D A. 2011. Synopsis of current three-dimensional geological mapping and modeling in geological survey organizations. Champaign:Illinois State Geological Survey,http://library.isgs.uiuc.edu/Pubs/pdfs/circulars/c578.pdf[2014-05-06]

Brune S, Heine C, Pérez-Gussinyé M, Sobolev S V. 2014. Rift migration explains continental margin asymmetry and crustal hyper-extension. Natural Communications,5:4014

Caers J. 2011. Modeling uncertainty in the earth sciences. Hoboken:Wiley

Carr G R, Andrew A S, Denton G J, Giblin A M, Korsch M J, Whitford D J. 1999. The "Glass Earth"—geochemical frontiers in exploration through cover. In:Exploration under Cover:Extended Abstracts. Sydney,N S W:Australian Institute of Geoscientists,33-40

Carranza E J M, Laborte A G. 2015. Random forest predictive modeling of mineral prospectivity with small number of prospects and data with missing values in Abra (Philippines). Computers & Geosciences,74(3):60-70

Chen G X, Cheng Q M. 2016. Singularity analysis based on wavelet transform of fractal measures for identifying geochemical anomaly in mineral exploration. Computers & Geosciences,87:56-66

Chen Y L, Wu W. 2017. Mapping mineral prospectivity using an extreme learning machine regression. Ore Geology Reviews,80(9):200-213

Chen Y L, Lu L J, Li X B. 2014. Application of continuous restricted Boltzmann machine to identify multivariate geochemical anomaly. Journal of Geochemical Exploration,140(3):56-63

Cheng Q M. 2012. Singularity theory and methods for mapping geochemical anomalies caused by buried sources and for predicting undiscovered mineral deposits in covered areas. Journal of Geochemical Exploration,122:55-70

Cheng Q M, Agterberg F P, Ballantyne S B. 1994. The separation of geochemical anomalies from background by fractal methods. Journal of Geochemical Exploration,51(2):109-130

Cong Y, Dong Q J, Bagas L, Xiao K Y, Wang K. 2017. Integrated GIS-based modelling for the quantitative prediction of magmatic Ti-V-Fe deposits:a case study in the Panzhihua-Xichang area of Southwest China. Ore Geology Reviews,91(2):1102-1118

Cowgill E, Bernardin T S, Oskin M E, Bowles C, Yikilmaz M B, Kreylos O, Elliott A J, Bishop S, Gold R D, Morelan A, Bawden G W, Hamann B, Kellogg L. 2012. Interactive terrain visualization enables virtual field work during rapid scientific response to the 2010 Haiti earthquake. Geosphere,8(4):787-804

Dong X, Gabrilovich E, Heitz G, Horn W, Lao N, Murphy K, Strohmann T, Sun S H, Zhang W. 2014. Knowledge vault:a web-scale approach to probabilistic knowledge fusion. In:Proceedings of the 20th ACM SIGKDD International Conference on Knowledge Discovery and Data Mining. New York:ACM,601-610

Esterle J S, Carr G R. 2003. The glass earth. Australian Institute of Geoscientists News,72:1-6

Geranian H, Tabatabaei S H, Asadi H H, Carranza E J M. 2016. Application of discriminant analysis and support vector machine in mapping gold potential areas for further drilling in the Sari-Gunay gold deposit, NW Iran. Natural Resources Research, 25(2): 145–159

Goran. 2016. Virtual reality: a brief history. http://www.useoftechnology.com/virtual-reality-history

Gorczyk W, Smithies H, Korhonen F, Howard H, de Gromard R Q. 2015. Ultra-hot Mesoproterozoic evolution of intracontinental central Australia. Geoscience Frontiers, 6(1): 23–37

Grunsky E C, de Caritat P. 2020. State-of-the-art analysis of geochemical data for mineral exploration. Geochemistry: Exploration, Environment, Analysis, 20(2): 217–232

Harris J R, Grunsky E C. 2015. Predictive lithological mapping of Canada's North using Random Forest classification applied to geophysical and geochemical data. Computers & Geosciences, 80(21): 9–25

Hinton G E, Osindero S, Teh Y W. 2006. A fast learning algorithm for deep belief nets. Neural Computation, 18(7): 1527–1554

Hinton G E, Deng L, Yu D, Dahl G E, Mohamed A R, Jaitly N, Senior A, Vanhoucke V, Nguyen P, Sainath T N, Kingsbury B. 2012. Deep neural networks for acoustic modeling in speech recognition: the shared views of four research groups. IEEE Signal Processing Magazine, 29(6): 82–97

Hu X Y, Li X H, Yuan F, Ord A, Jowitt S M, Li Y, Dai W Q, Zhou T F. 2020. Numerical modeling of ore-forming processes within the Chating Cu-Au porphyry-type deposit, China: implications for the longevity of hydrothermal systems and potential uses in mineral exploration. Ore Geology Reviews, 116: 103230

Huismans R, Beaumont C. 2011. Depth-dependent extension, two-stage breakup and cratonic underplating at rifted margins. Nature, 473(7345): 74–78

Izadi H, Sadri J, Mehran N A. 2013. Intelligent mineral identification using clustering and artificial neural networks techniques. In: Proceedings of the First Iranian Conference on Pattern Recognition and Image Analysis (PRIA). Birjand, Iran: IEEE, 1–5

Jackson I. 2007. One geology-making geological map data for the Earth accessible. Episodes, 30(1): 60–61

Jiao S T, Zhang Q, Zhou Y Z, Chen W F, Liu X Y, Gopalakrishnan G. 2018. Progress and challenges of big data research on petrology and geochemistry. Solid Earth Sciences, 3(4): 105–114

Jolivet L, Brun J P. 2010. Cenozoic geodynamic evolution of the Aegean. International Journal of Earth Sciences, 99(1): 109–138

Joly A, Porwal A, McCuaig C. 2010. 3D geophysical and geological modeling for understanding the gold mineral systems in the Tanami orogen, western Australia. In: EGU General Assembly Conference Abstracts. Vienna: EGU

Karpatne A, Ebert-Uphoff I, Ravela S, Babaie H A, Kumar V. 2019. Machine learning for the geosciences: challenges and opportunities. IEEE Transactions on Knowledge and Data Engineering, 31(8): 1544–1554

Lake B M, Salakhutdinov R, Tenenbaum J B. 2015. Human-level concept learning through probabilistic program induction. Science, 350(6266): 1332–1338

LeCun Y, Bengio Y, Hinton G. 2015. Deep learning. Nature, 521(7553): 436–444

Leskovec J, Rajaraman A, Ullman J D. 2014. Mining of Massive Datasets. Cambridge: Cambridge University Press

Li N, Bagas L, Li X H, Xiao K Y, Li Y, Ying L J, Song X L. 2016. An improved buffer analysis technique for model-based 3D mineral potential mapping and its application. Ore Geology Reviews, 76: 94–107

Li N, Song X L, Xiao K Y, Li S M, Li C B, Wang K. 2018. Part II: A demonstration of integrating multiple-scale 3D modelling into GIS-based prospectivity analysis: a case study of the Huayuan-Malichang district, China. Ore Geology Reviews, 95: 292–305

Liao J, Gerya T. 2014. Influence of lithospheric mantle stratification on craton extension: insight from two-dimensional thermo-mechanical modeling. Tectonophysics, 631: 50–64

Liao J, Gerya T. 2015. From continental rifting to seafloor spreading: insight from 3D thermo-mechanical modeling. Gondwana Research, 28(4): 1329–1343

Liu L, Morgan J P, Xu Y G, Menzies M. 2018. Craton destruction 1: cratonic keel delamination along a weak midlithospheric discontinuity layer. Journal of Geophysics Research, 123(11): 10040–10068

Liu L M, Wan C L, Zhao C B, Zhao Y L. 2011. Geodynamic constraints on orebody localization in the Anqing orefield, China: computational modeling and facilitating predictive exploration of deep deposits. Ore Geology Reviews, 43(1): 249–263

Liu X H, Zhang S D, Wei F R, Zhou M. 2011. Recognizing named entities in tweets. In: Proceedings of the 49th Annual Meeting of the Association for Computational Linguistics: Human Language Technologies-Volume 1. Stroudsburg, PA: Association for Computational Linguistics, 359–367

Lukovnikov D, Fischer A, Lehmann J, Auer S. 2017. Neural network-based question answering over knowledge graphs on word and character level. In: Proceedings of the 26th International Conference on World Wide Web. Perth, Australia: ACM, 1211–1220

Mao X C, Zhang B, Deng H, Zou Y H, Chen J. 2016. Three-dimensional morphological analysis method for geologic bodies and its parallel implementation. Computers & Geosciences, 96: 11–22

Mariethoz G, Caers J. 2014. Multiple-point Geostatistics: Stochastic Modeling with Training Images. New York: John Wiley & Sons

Nielsen S H H, Cunningham F, Hay R, Partington G, Stokes M. 2015. 3D prospectivity modelling of orogenic gold in the Marymia Inlier, western Australia. Ore Geology Reviews, 71: 578–591

Ord A, Hobbs B. 2013. Localised folding in general deformations. Tectonophysics, 587: 30–45

Ord A, Hobbs B E, Lester D R. 2012. The mechanics of hydrothermal systems: I. ore systems as chemical reactors. Ore Geology Reviews, 49: 1–44

Palumbo E, Rizzo G, Troncy R, Baralis E, Osella M, Ferro E. 2018. An empirical comparison of knowledge graph embeddings for item recommendation. DL4KGS@ ESWC: 14–20

Payne C E, Cunningham F, Peters K J, Nielsen S, Puccioni E, Wildman C, Partington G A. 2015. From 2D to 3D: prospectivity modelling in the Taupo volcanic zone, New Zealand. Ore Geology Reviews, 71: 558-577

Pearce J A, Lippard S J, Roberts S. 1984. Characteristics and tectonic significance of supra-subduction zone ophiolites. Geological Society, London, Special Publications, 16(1): 77-94

Pujara J, Miao H, Getoor L, Cohen W. 2013. Knowledge graph identification. In: International Semantic Web Conference. Berlin, Heidelberg: Springer, 542-557

Qin Y Z, Liu L M. 2018. Quantitative 3D association of geological factors and geophysical fields with mineralization and its significance for ore prediction: an example from Anqing Orefield, China. Minerals, 8(7): 300

Rodriguez-Galiano V F, Chica-Olmo M, Chica-Rivas M. 2014. Predictive modelling of gold potential with the integration of multisource information based on random forest: a case study on the Rodalquilar area, southern Spain. International Journal of Geographical Information Science, 28(7): 1336-1354

Rodriguez-Galiano V F, Sanchez-Castillo M, Chica-Olmo M, Chica-Rivas M. 2015. Machine learning predictive models for mineral prospectivity: an evaluation of neural networks, random forest, regression trees and support vector machines. Ore Geology Reviews. 71: 804-818

Schmidhuber J. 2015. Deep learning in neural networks: an overview. Neural Networks, 61: 85-117

Swan M. 2015. Blockchain: Blueprint for a New Economy. USA: O-Reilly Media Inc

Tahmasebi P, Hezarkhani A. 2012. A hybrid neural networks-fuzzy logic-genetic algorithm for grade estimation. Computers & Geosciences, 42: 18-27

Verma S P, Guevara M, Agrawal S. 2006. Discriminating four tectonic settings: five new geochemical diagrams for basic and ultrabasic volcanic rocks based on log-ratio transformation of major-element data. Journal of Earth System Science, 115(5): 485-528

Vermeesch P. 2006. Tectonic discrimination diagrams revisited. Geochemistry, Geophysics, Geosystems, 7(6): Q06017

Wang C, Yu H Z, Wan F C. 2018. Information retrieval technology based on knowledge graph. In: Proceedings of the 2018 3rd International Conference on Advances in Materials, Mechatronics and Civil Engineering (ICAMMCE 2018). Hangzhou, China: Atlantis Press, 291-296

Wang G W, Li R X, Carranza E J M, Zhang S T, Yan C H, Zhu Y Y, Qu J N, Hong D M, Song Y W, Han J W, Ma Z B, Zhang H, Yang F. 2015. 3D geological modeling for prediction of subsurface Mo targets in the Luanchuan district, China. Ore Geology Reviews, 71: 592-610

Wang G W, Ma Z B, Li R X, Song Y W, Qu J A, Zhang S T, Yan C H, Han J W. 2017. Integration of multi-source and multi-scale datasets for 3D structural modeling for subsurface exploration targeting, Luanchuan Mo-polymetallic district, China. Journal of Applied Geophysics, 139: 269-290

Wang W L, Zhao J, Cheng Q M. 2013. Fault trace-oriented singularity mapping technique to characterize anisotropic geochemical signatures in Gejiu mineral district, China. Journal of Geochemical Exploration, 134: 27-37

Weis P, Driesner T, Heinrich C A. 2012. Porphyry-copper ore shells form at stable pressure-temperature fronts within dynamic fluid plumes. Science, 338(6114): 1613-1616

Wellmann F, Caumon G. 2018. 3-D Structural geological models: Concepts, methods, and uncertainties. Advances in Geophysics, 59: 1-121

Whitmeyer S J, Bailey J E, de Paor D G, Ornduff T. 2012. Google earth and virtual visualizations in geoscience education and research. Geological Society of America, 492

Xiao F, Chen J G, Zhang Z Y, Wang C B, Wu G M, Agterberg F P. 2012. Singularity mapping and spatially weighted principal component analysis to identify geochemical anomalies associated with Ag and Pb-Zn polymetallic mineralization in Northwest Zhejiang, China. Journal of Geochemical Exploration, 122: 90-100

Xiao K Y, Li N, Jessell M. 2020. Quantitative assessment of China's mineral resources part 2. Ore Geology Reviews, 120: 103303

Yu X T, Xiao F, Zhou Y Z, Wang Y, Wang K Q. 2019. Application of hierarchical clustering, singularity mapping, and Kohonen neural network to identify Ag-Au-Pb-Zn polymetallic mineralization associated geochemical anomaly in Pangxidong district. Journal of Geochemical Exploration, 203: 87-95

Zhai J L, Lin W X. 2018. Community discovery algorithm for social networks based on parallel recommendation. IOP Conference Series: Materials Science and Engineering, 392(6): 062201

Zhou Y Z, Chown E H, Tu G Z, Guha J, Lu H Z. 1994. Geochemical migration of impurity trace elements and resultant fractal distribution patterns in source rocks. Mathematical Geosciences, 26(4): 419-435

Zou Y H, Liu Y, Dai T G, Mao X C, Lei Y B, Lai J Q, Tian H L. 2017. Finite difference modeling of metallogenic processes in the Hutouya Pb-Zn deposit, Qinghai, China: implications for hydrothermal mineralization. Ore Geology Reviews, 91: 463-476

Zuo R G, Wang J. 2016. Fractal/multifractal modeling of geochemical data: a review. Journal of Geochemical Exploration, 164: 33-41

Zuo R G, Carranza E J M, Wang J. 2016. Spatial analysis and visualization of exploration geochemical data. Earth-Science Reviews, 158: 9-18

Zuo R G, Xiong Y H, Wang J, Carranza E J M. 2019. Deep learning and its application in geochemical mapping. Earth-Science Reviews, 192: 1-14

The Great-leap-forward Development of Mathematical Geoscience During 2011−2020

ZHOU Yong-zhang[1], ZUO Ren-guang[2], LIU Gang[2], YUAN Feng[3], MAO Xian-cheng[4], GUO Yan-jun[5], XIAO Fan[1], LIAO Jie[1], LIU Yan-peng[1]

1. Center for Earth Environment & Resource, Sun Yat-sen University, Guangzhou 510275; 2. China University of Geology (Wuhan), Wuhan 430074; 3. School of Resources and Environmental Engineering, Hefei University of Technology, Hefei 230009; 4. School of Geosciences and Info-Physics, Central South University, Changsha 410083; 5. School of Earth and Space Science, Peking University, Beijing 100871

Abstract: The last decade was a historic transitional period for the transformation of scientific researches from the question-driven to data-driven. The fourth paradigm of scientific research, namely data intensive scientific discovery, has emerged in this period. The introduction of big data and artificial intelligence algorithms into the mathematical geoscience has resulted in a great-leap-forward development for mathematical geoscience in the period and is changing research patterns of geoscience. Machine learning is a fundamental way to endow computer with intelligence. Deep learning, as the multi-layer neural network algorithms, a technology for the implement of machine learning, was the most important hotspot in the study of big data and mathematical geoscience in the past few years. The Bayesian network from the combination of the Bayesian formula and the graph theory can be used to establish genetic network models of deposit geology for further understanding the mechanism of deposit formation. Big geological graph data problem can be converted to large complicated network space and community structure issues. Community detection algorithms can be applied to the earthquake prediction, geological network analysis, special geological phenomena identification and deposit prediction. The association rule and the recommendation system algorithms have been successfully applied in the geological research. Geochemical exploration data and the related anomalies generally contain complex nonlinear patterns. Deep learning is very powerful for the intelligent recognition and extraction of geochemical anomalies from various data under complicated geological conditions. Convolution neural network and stacked autoencoder are among effective methods frequently used. The methods for nonlinear mineral resources prediction, 3D metallogenic prediction based on GIS and 3D geological modeling, and corresponding software systems have been continuously improved. The 3D virtual simulation modeling technology has been applied to have realized multi-modal, trans-scale virtual reality and multi-dimensional interaction in geoscience. Innovative progresses have also been made in the numerical simulation of various geological processes. The blockchain technology together with big geoscience programs, such as the OneGeology, Glass Earth and Deep-time Digital Earth, will play very important promoting roles in fields of integrating global geological big-data systems, sharing global geoscience knowledges, and promoting the development of mathematical geoscience discipline.

Key words: geological big-data; deep learning; artificial intelligence algorithm; blockchain; Deep-time Digital Earth Program; mineral resource prediction; mathematical geoscience

勘查地球化学数据处理研究进展*

左仁广[1]　王　健[2]　熊义辉[1]　王子烨[1]

1. 中国地质大学（武汉），地质过程与矿产资源国家重点实验室，武汉 430074；
2. 成都理工大学 地球科学学院，成都 610059

摘　要：我国建立了包含海量数据的高质量的勘查地球化学数据库，为矿产勘查、环境评价和地质调查等提供了重要的数据支撑。如何高效处理勘查地球化学数据，并从中发掘和识别深层次信息一直是勘查地球化学学科研究的热点和前沿领域。本文在系统调研国内外学者过去十年发表的论著基础上，对勘查地球化学数据处理方法进行分析与对比，从勘查地球化学数据库建设、地球化学异常识别及其不确定性评价等方面概述了我国近十年来在该领域取得的主要研究进展，包括：①分形与多重分形模型由于考虑了地球化学空间模式的复杂性和尺度不变性，在全球范围内得到极大的发展和推广，我国学者引领了基于分形与多重分形的勘查地球化学数据处理；②机器学习和大数据思维开始在该领域启蒙，并迅速得到关注，正在成为研究热点和前沿领域，我国学者率先开展基于机器学习算法的勘查地球化学大数据挖掘研究；③我国学者需要进一步加强勘查地球化学数据缺失值处理以及成分数据闭合效应研究。今后该领域应进一步加强对弱缓地球化学异常识别、异常不确定性评价以及异常识别与其形成机理相结合等方面的研究。

关键词：勘查地球化学　地球化学异常　数据处理　分形　机器学习

0　引　言

勘查地球化学作为一种重要的矿产勘查方法，在找矿勘查中发挥了关键作用。不仅如此，勘查地球化学数据也为环境评价提供了重要数据源。我国区域地球化学填图计划于1978年被正式提出（Xie et al.，1997），在过去40年取得了巨大成功。截至2015年我国已累计获得1∶20万和1∶50万比例尺约700万 km^2 的水系沉积物或土壤样品的地球化学数据，覆盖70%以上的国土面积，编制完成了39种化学元素含量空间分布图，数据总量超过6 300万条。从1980年"六五计划"开始，截至2015年，采用勘查地球化学方法共发现各类矿床超过2 500处，其中大、中型规模的矿床超过70处，特别是贵金属矿产如黄金，约占全国一半的已探明储量（奚小环和李敏，2017）。

近年来，国内外对覆盖区地球化学矿产勘查越来越重视。适合不同自然地理景观的野外采样、分析方法，以及多元地球化学数据处理技术和软件研发得到飞速发展。为了满足覆盖区找矿需要，在1∶20万区域化探全国扫面计划测试的39种元素基础上，我国在未开展地球化学调查工作的厚覆盖区开展了52种元素多目标地球化学调查分析，新增与环境和农业研究等生态地球化学调查工作相关的13种元素，包括Br、C、Rb、Ti等。此外，我国在西南地区编制了76种元素的地球化学图，获得了除气体元素和人工放射性元素之外的所有元素的分布规律（程志中和谢学锦，2007），为识别和提取厚覆盖区地球化学弱缓异常信息提供了有力支撑。这些海量的地球化学数据为矿产勘查、环境评价和地质调查等提供了重要的数据支撑。如何高效处理勘查地球化学数据，并从中发掘和识别深层次信息一直是勘查地球化学学科研究的热点和前沿领域。本文在系统学习国内外学者过去10年发表的论著基础上，对该领域中的勘查地球化学数据处理这一方向进行分析与对比，概述了我国近10年来在该领域的主要研究进展。

---

* 原文"2011~2020年勘查地球化学数据处理研究进展"刊于《矿物岩石地球化学通报》2021年第40卷第1期，本文略有修改。

1 勘查地球化学数据库建设

数据库是存储、管理、输入和输出数据的软件平台。我国第一个专业的区域化探数据库（GC-81）建立于1980年，由（原）地矿部北京计算中心和物探所联合开发。为了满足日益增长和复杂的数据存储和管理需求，该中心在20世纪90年代又推出了全国区域化探数据库（REG91）。随着GIS技术的发展，中国地质科学院地球物理地球化学勘查研究所开发了基于GIS的"省级区域化探数据库信息系统（PGD2.0）"和"区域地球化学数据管理系统（GeoMDIS 2000）"，为勘查地球化学数据的存储、可视化分析和研究提供了更有效和实用的技术工具（向运川，2002）。针对全国约700万 km^2 的地球化学数据中存在的冗余、缺失、偏差和跨区拼接等问题，中国地质调查局于2001年主持研发了基于GeoExpl系统（http://www.drc.cgs.gov.cn/tech/201603/t20160309_266605.html）和基于MS SQL Server数据库的全国区域地球化学数据管理信息系统（GeoMDIS）（http://www.drc.cgs.gov.cn/tech/201603/t20160309_266561.html），并在全国范围得到推广。随后该系统不断更新和完善，补充了一系列重要数据集，建立了中国东部土壤生态地球化学基准数据库和多目标区域化探数据库等，对基础地质研究、矿产资源调查评价、生态环境可持续发展和国际交流具有重要意义（向运川等，2018）。

1.1 "化学地球"国际大科学计划

2015年9月国务院批准建立联合国教科文组织全球尺度地球化学国际研究中心。该中心主要致力于全球尺度地球化学科学研究与国际合作，建立覆盖全球的地球化学基准网和监测网，充分发挥中国在该领域的技术标准制定等方面的引领作用，为全球资源评价、环境保护和全球变化等提供基础资料，为全球资源与环境的可持续发展贡献中国力量。该国际研究中心的建立进一步确立了我国在勘查地球化学领域的核心地位。2016年我国科学家牵头启动"化学地球"国际大科学计划，建立了"化学地球"全球大数据平台。该平台基于ArcGIS Server和Web技术，旨在建立覆盖全球的地球化学基准网和关键带地球化学观测网，开展地球化学填图，估算全球资源总量，观测化学元素在岩石圈、水圈、土壤圈、大气圈和生物圈的循环，监测全球重金属、放射性元素和碳元素环境变化，从而实现化学元素在地壳的分布可视化监测和全球地球化学大数据共享（聂兰仕等，2012）。"化学地球"大科学计划的核心目标是绘制地球化学元素图谱，将元素周期表上所有化学元素的含量和空间分布绘制在地球上，持续记录全球化学元素的含量与分布、基准与变化等科学数据，通过基于互联网的"化学地球"平台，实现对全球地球化学大数据管理、展示和查询，支撑全球自然资源与环境可持续发展，为政府决策提供科学依据，为社会提供公共服务（Wang X. Q. et al., 2020）。目前，该平台已录入数据近300万条，到2021年，将建成包括10 000个基准点的地球化学基准网和100个观测点的地球化学观测网平台，并发布第一期76个化学元素基准值和基准地球化学图（王学求，2017）。同时，"化学地球"大科学计划也与"一带一路"沿线国家合作，协助蒙古、老挝、缅甸、印度尼西亚、乌兹别克斯坦、巴基斯坦等国完成了1∶100万国家尺度地球化学填图约260万 km^2，有效圈定地球化学异常700余处，为"一带一路"国家经济建设提供高质量、多元素、覆盖全国的地球化学基础数据（王学求，2018）。

1.2 多目标地球化学填图

多目标区域地球化学调查是一项前瞻性的填图工作，是我国继区域化探全国扫面计划之后勘查地球化学领域的又一重要举措，主要服务于生态地球化学评价、生态地球化学评估和生态地球化学预警等（奚小环，2007）。我国已完成700万 km^2 区域地球化学填图，为了顺应经济社会发展的需求，扩大地质工作服务领域，我国于1999年实施多目标地球化学填图项目，通过化探手段，获取了不同尺度的土壤、大气、水体、近岸海域沉积物、农作物等多介质、多种元素指标的高精度、高质量的数据，为基础地质、矿产勘查、环境、农业、生态、生物、海洋和流行病学等领域科学研究提供帮助（谢学锦，2003）。多目标地球化学填图项目将传统土壤或岩石地球化学调查的思路从地质勘查逐步延伸到大气圈、水圈乃至生态圈，其实质是对地球表层整个生态系统进行地球化学填图（刘荣梅，2013）。多目标地球化学填图项目除了需要元素分布图、成矿预测图，还需要绘制生态地球化学图、农业地球化学图和土地利用地球化学分区图等。

多目标区域地球化学工作重点部署在我国东部、中部大部分地区及西部部分经济发达和人口稠密的平原、盆地、三角洲、海岸带等（刘荣梅等，2012）。截至2019年，我国已完成1∶25万多目标区域地球化学调查面积380万 km²，获得了表层（1~20 cm）和深层（150~180 cm）土壤54种指标的土壤地球化学背景值与基准值、水地球化学和大气沉降物地球化学等数据，编制了全国地球化学分区图、土壤化学蚀变指数图、土壤环境质量分级图、土壤肥力分等图、土地质量地球化学分等图、绿色食品产地适宜性评价图、农业区划建议图等图件。多目标区域地球化学填图项目深化了地质学、土壤学、环境学、生态学、生物学、医学等领域的科学研究与应用实践（奚小环和李敏，2017）。

1.3 全球地球化学基准

地球化学基准是指系统记录地壳表层化学元素及其化合物的含量和空间分布，作为了解过去地球化学演化和预测未来地球化学变化的定量参照标尺（王学求，2012）。第一个全球地球化学基准源于1993年实施的全球地球化学基准计划（IGCP360），采用统一的地球化学填图手段，通过分析部署在全球的5 000个土壤沉积物样本来反映化学元素的空间分布特征（Darnley，1995）。虽然土壤和沉积物可以预测未来地球化学变化，但是无法反映过去的地质演化（王学求，2012）。因此，我国地球化学基准在建立的同时采用了土壤、沉积物介质和原生岩石介质。

受原国土资源部"深部探测技术与实验"专项资助，由中国地质科学院地球物理地球化学勘查研究所承担的"全国地球化学基准建立与综合研究"于2008年立项。参照全球地球化学基准网，我国地球化学基准以1∶20万图幅为标准网格单元，系统采集有代表性的岩石样品、水系沉积物和土壤样品共约2万件，经过5年攻关，建立了81项指标，涵盖大陆出露地壳76种元素的地球化学基准值，并制作了中国地球化学基准图。截至2018年，我国已完成地球化学基准观测网采样930万 km²，同时积极响应"一带一路"倡议，累计完成全球地球化学基准网建设约3 200万 km²，绘制了27个元素的地球化学基准图（王学求，2018）。中国地球化学基准的建立为衡量中国大陆化学元素演化和未来变化提供了标尺，不仅能够反映地壳元素含量的变化、分布范围和分布模式特征，还能够发现一些区域化探扫面计划尚未发现的稀土、稀有分散元素等异常区，有助于了解成矿物质时间和空间分布背景，对矿产资源预测具有重要意义。

2 数据预处理

勘查地球化学数据处理的步骤可概括为数据预处理（包括数据缺失值处理、成分数据变换、数据网格化）及地球化学异常识别与评价（包括多重分形与机器学习方法），如图1所示。

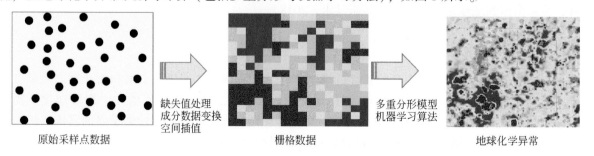

图1 勘查地球化学数据处理的一般流程

2.1 噪声及缺失值处理

勘查地球化学数据在采样、预处理及分析过程中会产生误差，由此叠加形成相应的地球化学数据噪声（裴韬和鲍征宇，1998）。另外，受仪器分析精度的影响，导致某些元素值不能被有效检测出来，使得勘查地球化学数据中某些变量会出现缺失值。地球化学噪声和数据的缺失会极大地增加地球化学异常识别和提取的不确定性（郭科等，2007）。因此，如何有效地消除噪声已成为能否正确利用地球化学信息的关键。我国学者采用的降噪处理方法有移动平均、人工免疫、卡尔曼滤波、傅里叶变换和小波变换等（郭科等，2007；陈聆等，2012；赵云华等，2013；白林等，2015）。目前，解决由于检测限造成的数据缺失问题的主要方法有：①从数据中删除低于检测限的样品；②将所有低于检测限的值设置为特定选择的低值，如检测限的一半；③通过各种方法预测低于检测限的样品值，如多元回归法等；④将低于检测限的样品值设置为区间[0，检测限]中的任意值（Reimann et al.，2008）等。我国在数据缺失值处理方面尚缺乏系统性研究。

2.2 成分数据

勘查地球化学数据是典型的成分数据，即所有元素含量的和为一定值，称为"闭合效应"。闭合效应会导致地球化学变量间产生伪相关关系，使得基于元素间相关关系的数据处理方法（如主成分分析和因子分析等）的结果具有不确定性。为了克服成分数据的闭合效应及伪相关问题，Aitchison（1982，1986）提出了 alr（additive logratio transformation）变换和 clr（centered logratio transformation）变换，Egozcue 等（2003）提出了 ilr（isometric logratio transformation）变换。通过如上成分数据变换，可以使得常规的基于欧氏空间的数据分析工具能够处理成分数据。

在2011—2020年，消除勘查地球化学数据的闭合效应逐渐成为勘查球化学数据预处理的重要步骤。Zuo 等（2013）和 Zuo（2014）利用主成分分析方法对勘查地球化学原始数据、对数变换后数据及对数比变换后数据进行对比分析，发现原始数据和基于对数变换后的数据由于受到闭合效应的影响，无法得到具有明确找矿指示意义的矿致元素组合；而经过对数比变换消除数据的闭合效应后，可得到有明确找矿指示意义的矿致元素组合。Zhao J. N. 等（2016）和 Chen 等（2018）利用因子分析对经 clr 变换后的数据进行分析，提取了与研究区矿化相关的元素组合，再利用异常识别方法（如局部奇异性分析，Cheng，2007），有效识别与提取了与矿化相关的地球化学异常；王琨等（2015）、Liu 等（2016）和刘向冲等（2017）利用对数比变换方法分别对湘西北铅锌矿区、南岭成矿区、西藏多龙矿集区水系沉积物地球化学元素进行预处理，在此基础上利用主成分分析、稳健因子分析等方法进行降维，识别的地球化学组合异常与已知矿床分布和构造特征耦合良好。除了三种对数比变换方法，余先川等（2019）通过元素值的趋势分析了地球化学元素之间的相关关系，进而提出了形态相关系数，可解决成分数据闭合效应的影响。

不仅如此，Cheng 等（2011）在利用勘查地球化学主量元素进行地质体的推断时，利用成分数据变换消除闭合效应得到了较好的结果。Zhao 等（2012）利用勘查地球化学7个主量元素和3个微量元素（Ba、Be、Li）推断东天山地区中酸性岩体，利用主成分分析和局部奇异性对 ilr 变换后的数据进行分析，其结果不仅可反映出露的中酸性岩体，还可以进一步推断出沙漠中隐伏的中酸性岩体。Wang 等（2014）利用主成分分析方法对经 ilr 变换后的数据进行分析，识别出东天山地区与矿化相关的中酸性岩体和断裂的空间分布模式，为该区矿产勘查提供了重要约束。Xiong 和 Zuo（2016a）探究了勘查地球化学数据的闭合效应对综合利用地球化学主量元素和区域地球物理数据进行中酸性岩体推断的影响，发现消除地球化学主量元素间的闭合效应后推断的中酸性岩体与已出露的中酸性岩体的空间分布范围高度吻合，其结果优于基于原始数据的推断结果。当前主量元素间的闭合效应逐渐被勘查地球化学家所关注，然而，成分数据的闭合效应对微量元素的影响尚待进一步研究。

2.3 数据网格化

数据网格化是指将空间分布不均匀的采样点数据，借助空间插值/模拟方法转换成规则分布的网格点数据的过程。数据网格化有助于压制局部噪声、估计变量在未采样位置处的取值，同时为不同变量的综合和对比提供统一的数据结构，从而便于进一步开展 GIS 空间分析与建模。常见的数据网格化技术包括最近邻插值、反距离加权插值（inverse distance weighting，IDW）、地质统计学插值/模拟、多项式回归、Delaunay 三角剖分插值、局部多项式插值等。任一网格点的元素含量值是位于该网格点附近的采样点观测值的加权平均。由规则采样点数据转换得到的网格数据中，每一网格点的元素含量值可以代表以该网格点为中心、分别以步长和扫描线间距为宽度和高度的网格单元的元素观测值。

2.3.1 网格单元大小

网格单元大小是空间插值-模拟方法中需要事先指定的一个参数。合理确定网格单元大小需要综合考虑地图比例尺、样品点的数量和分布、目标地质体的几何形态以及计算机性能等因素（Hengl，2006）。谢淑云和鲍征宇（2004）探讨了网格大小对成矿元素多重分形特征的影响，发现不同网格大小对元素分布的多重分形谱形态具有显著影响。Zuo（2012）以 IDW、描述性统计量、多重分形谱以及元素含量-面积分形模型为工具，探究了网格单元大小对地球化学元素频率及空间分布的影响。研究发现，网格单元大小对元素频率分布影响较小，但对元素空间分布模式影响较大，并可能进一步影响异常识别的效果。Lou 等（2015）和 Wang G. W. 等（2015）也开展了网格单元大小与地球化学元素分布模式之间关系的研究。

2.3.2 多重分形插值

IDW 是应用最为广泛的地球化学数据网格化方法之一，其实质是以待估点与已知样本点之间欧氏距离的倒数（或其 p 次幂）为权重的一种局部线性加权平均法。该方法形式简单，但用于模拟地球化学元素分布特征存在两方面的局限：①IDW 是一种滑动平均方法，对具有奇异性的元素空间分布可能产生平滑效应；②权重赋值过程中只考虑了样品点与待估点的几何距离，并未考虑元素含量本身的空间变异以及已知样品点的分布构型（Cheng，2008；Zuo and Wang，2016；Zuo et al.，2016）。针对第二点局限，克里金基于区域化随机变量理论，以变差函数为工具，能够获得待估点处变量取值的最优线性无偏估计，在地质、环境、水文等诸多领域获得广泛应用。克里金实质上是一种局部加权平均方法，因此，该方法同样对观测数据具有平滑作用。Cheng（1999，2008，2015）将线性插值模型与多重分形模型相结合，提出了多重分形插值方法，旨在有效刻画元素空间分布的变异性，削弱滑动加权平均和克里金产生的平滑效应。该方法基于局部邻域内元素服从分形分布，借助大尺度下的元素平均浓度来估计小尺度下的元素平均浓度。从数学形式上讲，Cheng（2008，2015）提出的多重分形插值方法将滑动加权平均得到的部分称作趋势成分，在每个局部位置处同时乘以一个表征局部奇异性特征的因子。该方法将元素含量分解为奇异成分和非奇异成分的乘积，这正是勒贝格分解定理的体现（Cheng，2008）。因此，滑动加权平均可以看作是多重分形插值方法的特例，当某局部位置处的元素分布不具有奇异性时，滑动加权平均方法等同多重分形插值（Cheng，2015）。近年来，国内外学者在开展地球化学异常识别研究中，已开始尝试运用多重分形插值方法模拟地球化学元素的复杂空间变异特征（Zuo，2011；Zuo et al.，2013；Yuan et al.，2015）。

2.3.3 序贯模拟

由于各种地质过程在多重时间和空间尺度上的相互叠加，成矿元素在表生介质中的分布往往十分复杂。但受地形、气候等自然条件的限制，以及时间、经济成本等因素的影响，地球化学观测样品往往十分有限，这些因素共同导致未采样位置处的元素含量估值具有不确定性（Costa and Koppe，1999）。当前的勘查地球化学数据处理实践通常使用空间插值（如 IDW、克里金）来推断元素的空间分布模式，插值法在表达元素分布的空间相关性和变异性方面具有优势，但难以准确有效地表征地球化学场的复杂空间格局，即插值结果存在一定的不确定性，这种不确定性将通过异常识别模型进行传递，从而对地球化学异常识别结果的可靠性产生影响（李晓晖等，2012；王健，2018；刘岳和周可法，2018）。特别是在由隐伏异常源引起的复杂地质背景下，如何在地球化学数据处理过程中有效刻画地球化学场的变异性，是提高弱缓地球化学异常识别效果、降低勘查风险的关键环节。针对这一问题，Wang 和 Zuo（2018，2019）和 Liu 等（2019a，2019b）提出以地质统计学模拟（如序贯高斯模拟、序贯指示模拟）为手段，兼顾成矿元素空间分布的结构性和变异性，通过模拟实现表征成矿元素空间分布模式的多重可能性。另外，王健（2018）从成矿元素空间分布的高阶、非线性统计特征出发，在成矿元素空间分布的多点地质统计学模拟方面进行了有益探索。目前，运用地质统计学模拟代替传统的空间插值方法，开展成矿元素空间分布不确定性建模，据此识别和评价地球化学异常，已引起国内外学者的关注。

3 地球化学异常识别与评价

3.1 成矿元素组合确定

在特定的地质环境中，某些元素往往表现出相似的地球化学行为，从而在最终形成的产物中呈现一定的共生组合关系。我国区域地球化学填图数据库中 1∶20 万水系沉积物地球化学数据共包括 39 种常量和微量元素（化合物）。选择哪些元素进行地球化学异常识别不仅要充分考虑成矿地质特征和成矿环境，而且需要分析元素的空间分布与富集特征以及与已发现矿床的空间耦合关系，这些将直接影响地球化学异常识别的效果。然而，如何选择最佳的矿致元素组合目前还没有统一的认识。常用的方法有借助多变量分析（如主成分分析、因子分析模型）（Zuo et al., 2013）以及地质背景知识（如矿床模型）（Cheng, 2007；Zhang and Zuo, 2014）来确定指示与成矿作用相关的地球化学元素组合。Cheng 等（2011）提出了空间加权主成分分析方法，基于该方法 Xiao 等（2012）识别出了与 Ag 和 Pb-Zn 矿有关的地球化学元素组合与异常。此外，基于 GIS 的空间分析方法逐渐受到地质学家的关注，并广泛应用于多元地球化学数据分析（Zuo and Wang, 2016）。如 Zuo（2018）提出利用多重分形模型和接收者操作特征曲线（receiver operating characteristic curve, ROC）寻找地球化学元素的共生组合关系，并确定最佳的矿致元素组合。该方法采用多重分形谱来定量刻画主要成岩元素、成矿元素及其伴生元素的空间分布特征，据此计算每个元素的多重分形性和多重分形谱参数，特别是不对称指数 R，以此探究地球化学元素的空间分布与富集特征。然后利用 ROC 技术，进一步分析多重分形谱左倾或不对称指数 $R>1$（代表这些元素在该区明显富集利于成矿）的主要成矿元素和伴生元素与已发现矿床的空间耦合关系，计算每个元素的 AUC（area under the curve）及 Z_{AUC} 值（Z_{AUC} 服从正态分布，95% 的置信水平下 $Z_{AUC}=1.96$），选择 AUC 值大于 0.5 且 Z_{AUC} 值大于 1.96 的元素（代表该元素与矿床具有显著的空间相关关系）作为输入变量，即与成矿作用相关的元素组合。

3.2 异常识别

3.2.1 多重分形奇异性

奇异性是指在一个狭窄的时间或空间范围内伴随巨量能量释放或物质超常富集的过程特性（Cheng, 2007, 2012；Cheng and Agterberg, 2009），它在定量表征各种非线性地质过程（如岩浆热液成矿过程）异常识别等地质变量的局部奇异结构特征有着重要的意义（Cheng, 2007）。Cheng（2007）从多重分形的角度出发，提出了基于滑动窗口的奇异性指数填图技术，用于复杂地质条件下地质异常的识别。在此基础上，Chen 等（2007）在局部奇异性分析中提出了一种迭代算法，该算法不仅能够估计由奇异性指数表征的奇异分量，还能估计由分形系数确定的非奇异分量。Zuo 等（2015）改进了奇异性分析方法，使其可应用于负值地学变量的奇异性度量（如布格重力和航磁数据）。Xiao 等（2016）提出一种基于成批滑动窗口的奇异性指数计算方法，该方法能够自动确定每个估计位置处最大窗口的最佳尺寸。Zhang 等（2016）开发了一种多参数优化算法，该算法可以产生一系列具有不同窗口大小、形状和方向的奇异性指数图。Chen 和 Cheng（2016）使用小波变换方法将奇异性指数分析模型从空间域扩展到小波域。

近年来，空间各向异性（即作用方向与范围大小的变化）在地球化学异常识别（包括奇异性分析）中的重要性愈加受到重视（Wang et al., 2013, 2018；Xiao et al., 2018）。例如，与矿化有关的地球化学异常，由于受到地层、断裂、侵入体等地质要素的控制，常具有线状或带状分布的特征。因此，空间各向异性对具有明确地质意义的异常识别至关重要。我国学者提出将奇异性分析与空间各向异性分析相结合，能够保证其所圈定的异常在地质层面上比仅利用奇异性分析所圈定的异常更加合理（Cheng, 2007；Wang et al., 2013, 2018；Zhang et al., 2016；Xiao et al., 2018）。例如，Wang 等（2018）通过定义各向异性的窗口序列，构建了一种各向异性局部奇异性指数估计方法，能够更加清晰地反映地球化学异常的局部结构特征；Xiao 等（2018）基于多重分形建模和空间加权技术，开发了一种各向异性奇异性分析方法，该方法能够表征隐含于非线性地质过程中的两个重要特性，即奇异性和空间各向异性。局部奇异性理论的普适性问题最近得到了广泛关注。目前，局部奇异性理论已被尝试用于度量洋中脊热流分布（Cheng, 2016）、火山爆发（Cheng, 2017）、大陆地壳增生（Cheng, 2018；Chen and Cheng, 2018）、地震丛聚性分布（Cheng and Sun, 2018）、大规模成矿（Zuo, 2018；Cheng, 2019）等一系列奇异地质事件当中。

3.2.2 机器学习及大数据

进行多元组合异常识别往往先借助主成分分析（或因子分析）寻求矿化元素组合，再利用局部奇异性分析等方法进行多元地球化学异常识别。然而，地质系统的复杂性以及成矿作用的多期多阶段性常导致勘查地球化学数据表现出复杂、非单一的统计分布特征。传统的多元统计方法在度量多元地球化学数据的复杂分布模式时存在一定的局限性。当前的矿产勘查重点是对现有的勘查地球化学数据进行二次开发利用，识别和提取传统方法无法识别的模式和异常，助力于覆盖区和深部找矿（左仁广，2019a）。在这种背景下，各种新的高级算法，尤其是机器学习（包括深度学习）算法不需要对数据的分布模式做出假设，可用于刻画复杂的、未知的勘查地球化学数据分布，已被引入到勘查地球化学领域，取得了可喜的成绩（Zuo et al., 2019；Zuo and Xiong, 2020）。

我国学者近年来开始并致力于基于机器学习的地球化学数据挖掘与模式识别，取得了许多创新性成果。如我国学者采用受限玻尔兹曼机（Chen et al., 2014）、核马氏距离（陈永良等，2014）、单类支持向量机（Chen and Wu, 2017）、孤立森林（Wu and Chen, 2018）、高斯混合模型（Chen and Wu, 2019）、Kohonen 神经网络（Sun et al., 2009；Yu et al., 2019）、关联规则算法（刘心怡和周永章，2019）、推荐系统算法（王堃屹等，2019）、独立成分分析（余先川等，2012；Liu et al., 2014, 2020）以及测度学习（Wang et al., 2019a, 2019b）等算法实现了地球化学异常识别。此外，我国学者也对地球化学异常识别的机器学习方法进行了对比，如李苍柏等（2020）对支持向量机、随机森林和人工神经网络算法在地球化学异常信息提取中的对比研究，构建了基于监督机器学习方法的地球化学异常信息提取流程；郑泽宇等（2018）对基于非监督的地球化学异常识别算法（孤立森林和单类支持向量机）的对比研究表明两种算法均可有效识别多元地球化学异常，但前者在数据处理耗时上表现略优。近年来，一些群体智能优化算法，如蚁群算法（Chen and An, 2016）、Bat 算法（Chen Y. L. et al., 2019），也用于地球化学异常识别和提取，识别的异常与成矿相关的地质体具有较高的空间相关性。这些研究表明群体智能优化算法是一类行之有效的地球化学异常识别方法。此外，还有一些混合的方法，如将卡尔曼滤波和盲抽取方法与支持向量机相结合，其中前者用于地球化学数据的融合及元素组合的提取，后者用于地球化学异常与背景的分离。混合方法充分吸收两种方法的优势，是地球化学数据处理的一种有益尝试（曾玉祥，2010；武飞，2014）。

深度学习作为一种具有多级非线性变换的层级机器学习算法，通过更深层次的网络模型来学习和提取深层次特征信息，达到高精度分类和预测的目的（Hinton et al., 2006; LeCun et al., 2015）。我国学者较早使用深度学习模型开展了复杂地质条件下的地球化学异常识别，如 Chen 等（2014）利用连续受限玻尔兹曼机有效识别了多元地球化学异常。在该模型的基础上，Xiong 和 Zuo（2016b）构建了基于深度自编码网络的多元地球化学异常识别模型。基于深度学习强大的特征提取能力，Xiong 和 Zuo（2020）进一步构建了深度置信网络与单类支持向量机相结合的混合模型，将深度学习模型提取的深层次特征信息作为单类支持向量机异常检测算法的输入，有效地提取了多元地球化学异常信息。Zhang 等（2019）将深度自编码网络与基于密度的聚类算法相结合，利用深度自编码网络提取深层次的地球化学特征作为密度聚类算法的输入，进行多元地球化学异常识别和提取。Wang J. 等（2020）构建了基于层次聚类的堆栈降噪自编码网络来提取地球化学数据的深层次特征信息，作为非监督孤立森林异常检测算法的输入，有效地提取了浙西北地区铅锌银多金属矿的地球化学异常。以上基于机器学习和深度学习的勘查地球化学数据处理忽略了地球化学数据的空间特性。卷积神经网络中的卷积、池化操作充分考虑了数据空间结构特征，我国学者将卷积神经网络引入地球化学数据处理当中。如 Chen L. R. 等（2019）利用卷积自编码网络提取地球化学空间结构特征，并开展了地球化学异常的识别和提取。这些研究发现深度学习模型不依赖勘查地球化学数据的分布假设，可处理复杂、非线性的空间模式，可识别传统方法无法识别的异常（Zuo and Xiong, 2018; Zuo et al., 2019）。与此同时，Zuo 和 Xiong（2018）尝试使用全部地球化学变量，将大数据思维和深度学习方法相结合，以充分考虑元素组合的复杂性和多样性，为刻画具有非线性特征的地球化学空间模式和提取隐式的异常提供了新途径，可更好的应用于地球化学异常识别。然而，大数据和深度学习模型的中间过程为"黑箱"，我们很难知道元素间的相互关系和内在联系，但这些信息具有特定的地质内涵，对矿床成因和矿产勘查具有重要的指示意义（Zuo, 2017; 左仁广, 2019a）。因此，结合勘查地球化学的特点，通过基于地质约束的机器学习和深度学习算法开展勘查地球化学大数据挖掘与集成，实现物理模型和机器学习、理论驱动与数据驱动的有机融合是未来该领域的重要研究方向。

3.2.3 经验模态分解等其他异常识别方法

成矿元素具有复杂的空间分布特征，如空间结构性、多尺度性。近年来，一系列考虑元素空间分布特征的异常识别技术被相继提出，为复杂地质条件下地球化学异常识别提供了重要的技术支撑（Zhao J. et al., 2016; Tian et al., 2018）。地理加权法以及经验模态分解法是其中的典型代表。Wang W. L. 等（2015）分别以成矿元素组合及控矿因子为目标变量和解释变量，借助地理加权回归方法识别了滇东南成矿带地球化学异常；Wang H. C. 等（2015）运用地理加权均值、变异系数等统计学方法探究了闽西南地区成矿元素的空间分布模式，为进一步的地球化学勘查提供科学依据。Xiao 等（2018）将空间加权技术与局部奇异性分析方法相结合，提出了空间加权奇异性填图方法。

地球化学背景和不同级次的异常通常是由不同期次和尺度的地质过程叠加形成的。以此为依据，经验模态分解法通过分解地球化学分布模式来提取可能指示矿化异常的组分。Xu 等（2016）运用经验模态分解法，有效识别出闽西南地区与铜多金属矿有关的地球化学异常，并指出经验模态分解是一种识别隐伏地质体的有力工具；Chen Y. Q. 等（2019）将主成分分析与二维经验模态分解相结合，有效圈定了云南腾冲锡多金属矿异常区。基于地球化学异常和背景分属于不同统计总体的认识，近十年还产生了非负矩阵分解等一系列新方法，有力提高了异常识别的效果（余先川和任雅丽，2014; Cao and Lu, 2015）。此外，Hao 等（2014）基于线性回归模型的残差置信区间识别了变换背景下的地球化学异常；Gong 等（2018）根据地球化学元素的背景和不同阈值，提出了地球化学异常的七级分类法；Zhao 等（2018）采用最大期望算法分解了勘查地球化学数据的混合分布。

3.3 不确定性评价

地球化学异常是元素富集或分散的结果，成矿作用、非矿化的地质作用、非地质作用、地球化学样品采集及加工等过程均可导致地球化学异常。因此，同一地区的地球化学异常往往具有多成因特征，这是地球化学异常具有不确定性的主要原因（Wang and Zuo，2020）。此外，地球化学异常是相对于地球化学背景定义的，在实践中需要确定区分地球化学背景及异常的阈值，不同模型或方法往往是从不同的角度出发确定分割阈值的，这进一步增加了所圈定地球化学异常的不确定性。近年来，我国学者围绕地球化学异常识别结果的不确定性开展了较为深入的研究。Wang 和 Zuo（2018，2019）和 Liu 等（2019a，2019b）指出在地球化学异常识别过程中，传统插值方法难以有效表征成矿元素复杂的空间变异特征，可能引入一定的不确定性，这种不确定性通过异常识别模型进行传递，致使所识别的地球化学异常也具有不确定性。Wang 和 Zuo（2018）和 Liu 等（2019a）提出用地质统计学模拟方法刻画成矿元素空间分布的不确定性，并结合异常识别模型表达所识别异常的可靠性，从而为地球化学勘查风险决策提供科学依据。Cheng（2007）提出的局部奇异性分析方法是当前数学地质与勘查地球化学领域的研究热点，已被诸多勘查实践证明是一种有效的弱缓地球化学异常识别方法。如何合理地确定奇异性指数的分割阈值，以圈定有价值的地球化学异常、减小其不确定性，已引起相关学者的关注。已有研究通常借助学生 t 统计量来确定合适的分割阈值；Liu 等（2017）基于奇异性指数在频率域中的分布模式，提出一种用于地球化学弱缓异常分离的奇异性指数—分位数分析方法，为客观确定奇异性指数分割阈值提供一种工具。Wang 和 Zuo（2016）提出的局部 Gap 统计量方法，不仅给出分割地球化学异常与背景的阈值，同时提供阈值所对应的置信水平，从而为表征由阈值引起的地球化学异常的不确定性提供了一种思路。

另外，地球化学异常识别模型构建以及求解过程中，也会引入不确定性。例如，局部奇异性指数的计算涉及双对数坐标系下的最小二乘估计，拟合优度的大小本身反映了奇异性指数包含一定的不确定性（Cheng，2007；Yuan et al.，2015）。局部奇异性分析方法需要设置滑动窗口的形状和大小，Zhang 等（2016）和 Xiao 等（2016）指出不同的窗口参数组合往往会导致不同的奇异性指数估计结果。通过设置一系列窗口参数，然后分析不同参数组合下获得的异常识别结果，以确定最佳参数组合，有助于降低局部奇异性分析结果的不确定性，提高异常识别的可靠性（Zhang et al.，2016；Xiao et al.，2016；Xu et al.，2020）。不确定性是客观存在的，如何有效评价并降低地球化学异常的不确定性，已成为决定复杂地质条件下地球化学勘查成败的关键环节。Wang 和 Zuo（2020）借助地理加权回归方法探究了地质和地球化学景观条件对成矿元素在表生环境下分散的制约，为进一步筛选有意义的地球化学异常提供了新的思路。

3.4 软件

我国早期的勘查地球化学数据处理与异常识别软件多嵌套在矿产资源定量预测系统中，如由成秋明院士牵头开发的 GeoDAS 矿产资源定量预测 GIS 软件系统（Cheng，2002）、由中国地质大学（武汉）数学地质与遥感地质研究所开发的 MORPAS（Mineral Ore Resources Prediction and Assessment System）（胡光道和陈建国，1998）、由中国地质科学院开发的 MRAS（Mineral Resources Assessment System）（肖克炎等，2006）、由中国地质调查局发展研究中心开发的多元地学空间数据管理与分析软件 GeoExpl（蔡煜琦，2006）和区域地球化学数据管理与分析系统 GeoMDIS（肖桂义和向运川，2001）等，都包含了勘查地球化学数据分析模块，均可较好地满足勘查地球化学数据处理的需求。近十年来，为了方便利用分形与多重分形模型处理地学数据，我国学者在 Matlab 平台及 ArcGIS 平台下开发了 GeoFrac2D 和 ArcFractal 地学数据处理软件，为基于分形与多重分形模型的勘查地球化学异常识别提供了高效、便捷的工具。GeoFrac2D 与 ArcFractal 的数据格式分别为 ASCII 文件和 ArcGIS 栅格数据，软件主要包括浓度-面积分形模型、能量谱密度-面积分形模型、局部奇异性分析方法及多重分形谱等方法（Wang and Zuo，2015；Zuo and Wang，2020）。

4 结　语

过去十年，我国学者在勘查地球化学数据处理方面做了大量探索，在勘查地球化学数据库建设、异常识别与提取等领域取得了众多进展，引领了基于分形与多重分形模型开展复杂地质条件下弱缓地球化学异常识别，在基于机器学习和深度学习模型开展勘查地球化学大数据挖掘与异常提取方面走在世界前列，但数据预处理中缺失值补全算法研发、勘查地球化学成分数据及其闭合效应影响研究还需要加强。今后还需要进一步加强弱缓地球化学异常识别、异常不确定性评价，以及将异常识别与异常形成机理相结合3个方面的研究。

传统的地球化学异常识别方法是假设元素含量服从正态或对数正态分布，没有充分考虑地球化学元素分布模式的空间结构，使得识别和提取的异常往往被光滑或被压制（de Mulder et al., 2016）。这类方法对识别浅表矿引起的"高、大、全"式的地球化学异常有效，但对于隐伏矿和深部矿引起的弱缓地球化学异常的识别却十分困难（左仁广，2019b）。当前，我国大多数地区的"高、大、全"异常已被开发殆尽，下一步主要面临弱缓地球化学异常的识别。弱缓地球化学异常相对于"高、大、全"异常而言，可能具有元素含量低、异常面积小、元素组合不全等特征。因此，还需进一步研发弱缓地球化学异常识别新方法，以有效压制噪音、提高弱缓异常的识别效果（左仁广，2019b）。

在一个研究区内，大多数算法通常会识别出多个异常，而往往只有很少一部分可能与成矿作用有关，因此需要对地球化学异常开展不确定性评价，以筛选有价值的地球化学异常，降低其不确定性。目前这方面的研究还比较薄弱，仍需加强地球化学异常识别算法研究，不仅要提取地球化学异常，而且还需要综合考虑异常所处的地质背景、构造环境及表生环境等因素，度量异常的不确定性，为后续找矿提供可靠的决策依据。

地质系统的复杂性以及成矿作用的多期多阶段性常常导致勘查地球化学数据具有复杂的空间和频率分布特性。研究表明，地球化学数据常常不满足某个单一的统计分布形式，所以传统的多元统计方法并不能很好地度量多元地球化学数据的分布情况。而机器学习不依赖数据的分布，可用于刻画复杂的、非线性的地学空间模式，可以对勘查地球化学数据进行有效处理并进行异常特征提取，从而解决传统多元统计学方法对数据分布形式依赖的局限性，并可发现常规方法发现不了的异常和模式（Zuo and Xiong, 2018）。因而，一方面需要大力推进机器学习和深度学习模型在勘查地球化学异常识别与评价中的应用，同时还需将地球化学异常形成机理（理论驱动）与机器学习（数据驱动）相结合，使其既能识别传统方法无法识别的异常，又能使所识别的异常符合地质规律。

致谢：吉林大学陈永良教授审阅了初稿，提出了很多建设性意见，在此表示衷心感谢。

参 考 文 献

白林,郭科,刘斌. 2015. 快速傅里叶变换在地球化学数据降噪中的应用. 科学技术与工程,15(26):124–127
蔡煜琦. 2006. "多元地学空间数据管理与分析系统GeoExpl"培训班在西安举办. 世界核地质科学,23(2):3168
陈聆,郭科,柳炳利,梁莉. 2012. 地球化学矿致异常非线性分析方法研究. 地球物理学进展,27(4):1701–1707
陈永良,路来君,李学斌. 2014. 多元地球化学异常识别的核马氏距离方法. 吉林大学学报(地球科学版),44(1):396–408
程志中,谢学锦. 2007. 中国西南地区76种元素地球化学填图. 物探化探计算技术,29(S1):174–179
郭科,陈聆,唐菊兴. 2007. 复杂地质地貌区地球化学异常识别非线性研究. 成都理工大学学报(自然科学版),34(6):599–604
胡光道,陈建国. 1998. 金属矿产资源评价分析系统设计. 地质科技情报,17(1):45–49
李苍柏,肖克炎,李楠,宋相龙,张帅,王凯,楚文楷,曹瑞. 2020. 支持向量机、随机森林和人工神经网络机器学习算法在地球化学异常信息提取中的对比研究. 地球学报,41(2):309–319
李晓晖,袁峰,贾蔡,张明明,周涛发. 2012. 基于地统计学插值方法的局部奇异性指数计算比较研究. 地理科学,32(2):136–142
刘荣梅. 2013. 中国多目标区域地球化学调查数据库建设研究. 北京:中国地质大学(北京)博士学位论文
刘荣梅,吴轩,向运川,耿燕婷. 2012. 中国多目标区域地球化学调查数据库建设及应用展望. 现代地质,26(5):989–995
刘向冲,王文磊,裴英茹. 2017. 西藏多龙矿集区水系沉积物地球化学数据定量分析与解释. 地质力学学报,23(5):695–706
刘心怡,周永章. 2019. 关联规则算法在粤西庞西垌地区元素异常组合研究中的应用. 地学前缘,26(4):125–130
刘岳,周可法. 2018. 西准噶尔成矿带金矿异常识别及其不确定性分析. 地球科学,43(9):3186–3199
聂兰仕,王学求,徐善法,王玮. 2012. 全球地球化学数据管理系统:"化学地球"软件研制. 地学前缘,19(3):43–48
裴韬,鲍征宇. 1998. 地球化学数据去噪方法研究. 地质地球化学,26(4):86–90
王健. 2018. 基于地质统计学模拟的地球化学异常信息提取. 武汉:中国地质大学博士学位论文

王堃屹,周永章,王俊,张奥多,余晓彤,焦守涛,刘心怡. 2019. 推荐系统算法在钦杭成矿带南段文地幅矿床预测中的应用. 地学前缘,26(4):131-137

王琨,肖克炎,丛源. 2015. 对数比变换和偏最小二乘法在地球化学组合异常提取中的应用——以湘西北铅锌矿为例. 物探与化探,39(1):141-148

王学求. 2012. 全球地球化学基准:了解过去,预测未来. 地学前缘,19(3):7-18

王学求. 2017. 透视全球资源与环境,实施"化学地球"国际大科学计划. 中国地质,44(1):201-202

王学求. 2018. "化学地球"国际大科学计划取得重要进展. 中国地质,45(5):858

武飞. 2014. 基于卡尔曼滤波与支持向量机的地球化学异常提取. 成都:成都理工大学硕士学位论文,50

奚小环. 2007. 多目标的地质大调查—21世纪勘查地球化学的战略选择. 物探与化探,31(4):283-288

奚小环,李敏. 2017. 现代勘查地球化学科学体系概论:"十二五"期间勘查成果评述. 物探与化探,41(5):779-793

向运川. 2002. 区域地球化学数据管理信息系统的实现技术. 物探与化探,26(3):209-214,217

向运川,牟绪赞,任天祥,刘荣梅,吴轩. 2018. 全国区域化探数据库. 中国地质,45(S1):32-44

肖桂义,向运川. 2001. "区域地球化学数据库管理信息系统(GeoMDIS2000)"推广班在山东举办. 物探与化探,(6):398

肖克炎,王勇毅,陈郑辉,薛群威,黄文斌,朱裕生,杨永华,张寿庭,杨毅恒,张晓华,刘锐. 2006. 中国矿产资源评价新技术与评价新模型. 北京:地质出版社

谢淑云,鲍征宇. 2004. 多重分形方法在金属成矿潜力评价中的应用. 成都理工大学学报(自然科学版),31(1):28-33

谢学锦. 2003. 从勘查地球化学到应用地球化学. 物探与化探,27(6):412-415

余先川,任雅丽. 2014. 非负矩阵分解及在地学中的应用. 地质学刊,38(2):238-244

余先川,刘立文,胡丹,王仲妮. 2012. 基于稳健有序独立成分分析(ROICA)的矿产预测. 吉林大学学报(地球科学版),42(3):872-880

余先川,张冠鹏,姚旺,杨昭颖. 2019. 形态相关系数及其在地球化学数据分析中的应用(英文). 地质学刊,43(1):103-110

曾玉祥. 2010. 盲抽取与SVM方法在地球化学异常下限提取中的应用. 硕士学位论文. 成都:成都理工大学

赵云华,李东海,柳炳利,武飞. 2013. 小波理论及其在地球化学数据处理中的应用综述. 四川理工学院学报(自然科学版),26(2):6-10

郑泽宇,赵庆英,李湜生,邱士龙. 2018. 地球化学异常识别的两种机器学习算法之比较. 世界地质,37(4):1288-1294

左仁广. 2019a. 基于深度学习的深层次矿化信息挖掘与集成. 矿物岩石地球化学通报,38(1):53-60

左仁广. 2019b. 勘查地球化学数据挖掘与弱异常识别. 地学前缘,26(4):67-75

Aitchison J. 1982. The statistical analysis of compositional data. Journal of the Royal Statistical Society:Series B (Methodological),44(2):139-160

Aitchison J. 1986. The Statistical Analysis of Compositional Data. London:Chapman and Hall

Cao M X,Lu L J. 2015. Application of the multivariate canonical trend surface method to the identification of geochemical combination anomalies. Journal of Geochemical Exploration,153:1-10

Chen G X,Cheng Q M. 2016. Singularity analysis based on wavelet transform of fractal measures for identifying geochemical anomaly in mineral exploration. Computers & Geosciences,87:56-66

Chen G X,Cheng Q M. 2018. Cyclicity and persistence of earth's evolution over time:wavelet and fractal analysis. Geophysical Research Letters,45(16):8223-8230

Chen L R,Guan Q F,Feng B,Yue H Q,Wang J Y,Zhang F. 2019. A multi-convolutional Autoencoder approach to multivariate geochemical anomaly recognition. Minerals,9(5):270

Chen X,Xu R K,Zheng Y Y,Jiang X J,Du W Y. 2018. Identifying potential Au-Pb-Ag mineralization in SE Shuangkoushan, North Qaidam,western China:combined log-ratio approach and singularity mapping. Journal of Geochemical Exploration,189:109-121

Chen Y L,An A J. 2016. Application of ant colony algorithm to geochemical anomaly detection. Journal of Geochemical Exploration,164:75-85

Chen Y L,Wu W. 2017. Mapping mineral prospectivity by using one-class support vector machine to identify multivariate geological anomalies from digital geological survey data. Australian Journal of Earth Sciences,64(5):639-651

Chen Y L,Wu W. 2019. Separation of geochemical anomalies from the sample data of unknown distribution population using Gaussian mixture model. Computers & Geosciences,125:9-18

Chen Y L,Lu L J,Li X B. 2014. Application of continuous restricted Boltzmann machine to identify multivariate geochemical anomaly. Journal of Geochemical Exploration,140:56-63

Chen Y L,Wu W,Zhao Q Y. 2019. A bat-optimized one-class support vector machine for mineral prospectivity mapping. Minerals,9(5):317

Chen Y Q,Zhang L N,Zhao B B. 2019. Identification of the anomaly component using BEMD combined with PCA from element concentrations in the Tengchongtin Belt,SW China. Geoscience Frontiers,10(4):1561-1576

Chen Z J,Cheng Q M,Chen J G,Xie S Y. 2007. A novel iterative approach for mapping local singularities from geochemical data. Nonlinear Processes in Geophysics,14(3):317-324

Cheng Q M. 1999. Multifractal interpolation. In:Lippard S J,Naess A,Sinding-Larsen R(eds). Proceedings of the 5th Annual Conference of the International Association for Mathematical Geology. Trondheim:Norwegian University of Science and Technology,245-250

Cheng Q M. 2002. GeoData analysis system(GeoDAS) for mineral exploration:user's guide and exercise manual. Material for the Training Workshop on GeoDAS Held. Toronto:York University

Cheng Q M. 2007. Mapping singularities with stream sediment geochemical data for prediction of undiscovered mineral deposits in Gejiu,Yunnan Province,China. Ore Geology Reviews,32(1-2):314-324

Cheng Q M. 2008. Modeling local scaling properties for multiscale mapping. Vadose Zone Journal,7(2):525-532

Cheng Q M. 2012. Singularity theory and methods for mapping geochemical anomalies caused by buried sources and for predicting undiscovered mineral deposits in covered areas. Journal of Geochemical Exploration,122:55-70

Cheng Q M. 2015. Multifractal interpolation method for spatial data with singularities. Journal of the Southern African Institute of Mining and Metallurgy,115(3):235-240

Cheng Q M. 2016. Fractal density and singularity analysis of heat flow over ocean ridges. Scientific Reports,6:19167

Cheng Q M. 2017. Singularity analysis of global zircon U-Pb age series and implication of continental crust evolution. Gondwana Research,51:51-63

Cheng Q M. 2018. Singularity analysis of magmatic flare-ups caused by India-Asia collisions. Journal of Geochemical Exploration,189:25-31

Cheng Q M. 2019. Integration of deep-time digital data for mapping clusters of porphyry copper mineral deposits. Acta Geologica Sinica,93(S3):8-10

Cheng Q M, Agterberg F P. 2009. Singularity analysis of ore-mineral and toxic trace elements in stream sediments. Computers & Geosciences,35(2):234-244

Cheng Q M, Sun H Y. 2018. Variation of singularity of earthquake-size distribution with respect to tectonic regime. Geoscience Frontiers,9(2):453-458

Cheng Q M,Bonham-Carter G,Wang W L,Zhang S Y,Li W C,Xia Q L. 2011. A spatially weighted principal component analysis for multi-element geochemical data for mapping locations of felsic intrusions in the Gejiu mineral district of Yunnan, China. Computers & Geosciences,37(5):662-669

Costa J F,Koppe J C. 1999. Assessing uncertainty associated with the delineation of geochemical anomalies. Natural Resources Research,8(1):59-67

Darnley A G. 1995. International geochemical mapping-a review. Journal of Geochemical Exploration,55(1-3):5-10

de Mulder E F J,Cheng Q M,Agterberg F,Goncalves M. 2016. New and game-changing developments in geochemical exploration. Episodes,39(1):70-71

Egozcue J J, Pawlowsky-Glahn V, Mateu-Figueras G, Barceló-Vidal C. 2003. Isometric logratio transformations for compositional data analysis. Mathematical Geology,35(3):279-300

Gong Q J,Li J Z,Xiang Y C,Liu R M,Wu X,Yan T T,Chen J,Li R K,Tong Y K. 2018. Determination and classification of geochemical anomalies based on backgrounds and cutoff grades of trace elements:a case study in South Nanling Range, China. Journal of Geochemical Exploration,194:44-51

Hao L B, Zhao X Y, Zhao Y Y, Lu J L, Sun L J. 2014. Determination of the geochemical background and anomalies in areas with variable lithologies. Journal of Geochemical Exploration,139:177-182

Hengl T. 2006. Finding the right pixel size. Computers & Geosciences,32(9):1283-1298

Hinton G E,Osindero S,Teh Y W. 2006. A fast learning algorithm for deep belief nets. Neural Computation,18(7):1527-1554

LeCun Y,Bengio Y,Hinton G. 2015. Deep learning. Nature,521(7553):436-444

Liu B, Guo S, Wei Y H, Zhan Z D. 2014. A fast independent component analysis algorithm for geochemical anomaly detection and its application to soil geochemistry data processing. Journal of Applied Mathematics,2014:319314

Liu B,Zhou Z L,Dai Q L,Tong W. 2020. FastICA and total variation algorithm for geochemical anomaly extraction. Earth Science Informatics,13(1):153-162

Liu Y,Cheng Q M,Zhou K F,Xia Q L,Wang X Q. 2016. Multivariate analysis for geochemical process identification using stream sediment geochemical data:a perspective from compositional data. Geochemical Journal,50(4):293-314

Liu Y,Zhou K F,Cheng Q M. 2017. A new method for geochemical anomaly separation based on the distribution patterns of singularity indices. Computers & Geosciences,105:139-147

Liu Y,Cheng Q M,Carranza E J M,Zhou K F. 2019a. Assessment of geochemical anomaly uncertainty through geostatistical simulation and singularity analysis. Natural Resources Research,28:199-212

Liu Y,Xia Q L,Carranza E J M. 2019b. Integrating sequential indicator simulation and singularity analysis to analyze uncertainty of geochemical anomaly for exploration targeting of tungsten polymetallic mineralization, Nanling Belt, South China. Journal of Geochemical Exploration,197:143-158

Lou D B, Zhang C Q, Liu H. 2015. The multifractal nature of the Ni geochemical field and implications for potential Ni mineral resources in the Huangshan-Jing'erquan area, Xinjiang, China. Journal of Geochemical Exploration,157:169-177

Reimann C,Filzmoser P,Garrett R G,Dutter R. 2008. Statistical Data Analysis Explained: Applied Environmental Statistics with R. Chichester: John Wiley & Sons

Sun X,Deng J,Gong Q J,Wang Q F,Yang L Q,Zhao Z Y. 2009. Kohonen neural network and factor analysis based approach to geochemical data pattern recognition. Journal of Geochemical Exploration,103(1):6-16

Tian M,Wang X Q,Nie L S,Zhang C S. 2018. Recognition of geochemical anomalies based on geographically weighted regression:a case study across the boundary areas of China and Mongolia. Journal of Geochemical Exploration,190:381-389

Wang G W,Li R X,Carranza E J M,Zhang S T,Yan C H,Zhu Y Y,Qu J N,Hong D M,Song Y W,Han J W,Ma Z B,Zhang H,Yang F. 2015. 3D geological modeling for prediction of subsurface Mo targets in the Luanchuan district, China. Ore Geology Reviews,71:592-610

Wang H C,Cheng Q M,Zuo R G. 2015. Quantifying the spatial characteristics of geochemical patterns via GIS-based geographically weighted statistics. Journal of Geochemical Exploration,157:110-119

Wang J,Zuo R G. 2015. A MATLAB-based program for processing geochemical data using fractal/multifractal modeling. Earth Science Informatics,8(4):937-947

Wang J,Zuo R G. 2016. An extended local gap statistic for identifying geochemical anomalies. Journal of Geochemical Exploration,164:86-93

Wang J,Zuo R G. 2018. Identification of geochemical anomalies through combined sequential Gaussian simulation and grid-based local singularity analysis. Computers & Geosciences,118:52-64

Wang J,Zuo R G. 2019. Recognizing geochemical anomalies via stochastic simulation-based local singularity analysis. Journal of Geochemical Exploration,198:29-40

Wang J,Zuo R G. 2020. Assessing geochemical anomalies using geographically weighted lasso. Applied Geochemistry,119:104668

Wang J,Zhou Y Z,Xiao F. 2020. Identification of multi-element geochemical anomalies using unsupervised machine learning algorithms: a case study from Ag-Pb-Zn deposits in northwestern Zhejiang,China. Applied Geochemistry,120:104679

Wang W L,Zhao J,Cheng Q M. 2013. Fault trace-oriented singularity mapping technique to characterize anisotropic geochemical signatures in Gejiu mineral district,China. Journal of Geochemical Exploration,134:27-37

Wang W L,Zhao J,Cheng Q M. 2014. Mapping of Fe mineralization-associated geochemical signatures using logratio transformed stream sediment geochemical data in eastern Tianshan,China. Journal of Geochemical Exploration,141:6-14

Wang W L,Zhao J,Cheng Q M,Carranza E J M. 2015. GIS-based mineral potential modeling by advanced spatial analytical methods in the southeastern Yunnan mineral district,China. Ore Geology Reviews,71:735-748

Wang W L,Cheng Q M,Zhang S Y,Zhao J. 2018. Anisotropic singularity: a novel way to characterize controlling effects of geological processes on mineralization. Journal of Geochemical Exploration,189:32-41

Wang X Q,Zhang B M,Nie L S,Wang W,Zhou J,Xu S F,Chi Q H,Liu D S,Liu H L,Han Z X,Liu Q Q,Tian M,Zhang B Y,Wu H,Li R H,Hu Q H,Yan T T,Gao Y F. 2020. Mapping Chemical Earth Program: progress and challenge. Journal of Geochemical Exploration,217:106578

Wang Z Y,Dong Y N,Zuo R G. 2019a. Mapping geochemical anomalies related to Fe-polymetallic mineralization using the maximum margin metric learning method. Ore Geology Reviews,107:258-265

Wang Z Y,Zuo R G,Dong Y N. 2019b. Mapping geochemical anomalies through integrating random forest and metric learning methods. Natural Resources Research,28(4):1285-1298

Wu W,Chen Y L. 2018. Application of isolation forest to extract multivariate anomalies from geochemical exploration data. Global Geology,21(1):36-47

Xiao F,Chen J G,Zhang Z Y,Wang C B,Wu G M,Agterberg F P. 2012. Singularity mapping and spatially weighted principal component analysis to identify geochemical anomalies associated with Ag and Pb-Zn polymetallic mineralization in Northwest Zhejiang,China. Journal of Geochemical Exploration,122:90-100

Xiao F,Chen Z J,Chen J G,Zhou Y Z. 2016. A batch sliding window method for local singularity mapping and its application for geochemical anomaly identification. Computers & Geosciences,90:189-201

Xiao F,Chen J G,Hou W S,Wang Z H,Zhou Y Z,Erten O. 2018. A spatially weighted singularity mapping method applied to identify epithermal Ag and Pb-Zn polymetallic mineralization associated geochemical anomaly in Northwest Zhejiang,China. Journal of Geochemical Exploration,189:122-137

Xie X J,Mu X Z,Ren T X. 1997. Geochemical mapping in China. Journal of Geochemical Exploration,60(1):99-113

Xiong Y H,Zuo R G. 2016a. A comparative study of two modes for mapping felsic intrusions using geoinformatics. Applied Geochemistry,75:277-283

Xiong Y H,Zuo R G. 2016b. Recognition of geochemical anomalies using a deep autoencoder network. Computers & Geosciences,86:75-82

Xiong Y H,Zuo R G. 2020. Recognizing multivariate geochemical anomalies for mineral exploration by combining deep learning and one-class support vector machine. Computers & Geosciences,140:104484

Xu G M,Cheng Q M,Zuo R G,Wang H C. 2016. Application of improved bi-dimensional empirical mode decomposition(BEMD) based on Perona-Malik to identify copper anomaly association in the southwestern Fujian (China). Journal of Geochemical Exploration,164:65-74

Xu S,Hu X Y,Carranza E J M,Wang G W. 2020. Multi-parameter analysis of local singularity mapping and its application to identify geochemical anomalies in the Xishan gold deposit,North China. Natural Resources Research,doi:10.1007/s11053-020-09669-5

Yu X T,Xiao F,Zhou Y Z,Wang Y,Wang K Q. 2019. Application of hierarchical clustering,singularity mapping,and Kohonen neural network to identify Ag-Au-Pb-Zn polymetallic mineralization associated geochemical anomaly in Pangxidong district. Journal of Geochemical Exploration,203:87-95

Yuan F,Li X H,Zhou T F,Deng Y F,Zhang D Y,Xu C,Zhang R F,Jia C,Jowitt S M. 2015. Multifractal modeling-based mapping and identification of geochemical anomalies associated with Cu and Au mineralisation in the NW Junggar area of northern Xinjiang Province,China. Journal of Geochemical Exploration,154:252-264

Zhang D J,Cheng Q M,Agterberg F,Cheng Z J. 2016. An improved solution of local window parameters setting for local singularity analysis based on Excel VBA batch processing technology. Computers & Geosciences,88:54-66

Zhang S,Xiao K Y,Carranza E J M,Yang F,Zhao Z C. 2019. Integration of auto-encoder network with density-based spatial clustering for geochemical anomaly detection for mineral exploration. Computers & Geosciences,130:43-56

Zhang Z J,Zuo R G. 2014. Sr-Nd-Pb isotope systematics of magnetite: implications for the genesis of Makeng Fe deposit,southern China. Ore Geology Reviews,57:53-60

Zhao J,Wang W L,Dong L H,Yang W Z,Cheng Q M. 2012. Application of geochemical anomaly identification methods in mapping of intermediate and felsic igneous rocks in eastern Tianshan,China. Journal of Geochemical Exploration,122:81-89

Zhao J,Wang W L,Cheng Q M,Agterberg F. 2016. Mapping of Fe mineral potential by spatially weighted principal component analysis in the eastern Tianshan mineral district,China. Journal of Geochemical Exploration,164:107-121

Zhao J N,Chen S Y,Zuo R G. 2016. Identifying geochemical anomalies associated with Au-Cu mineralization using multifractal and artificial neural network models in the Ningqiang district,Shaanxi,China. Journal of Geochemical Exploration,164:54-64

Zhao X Z, Hao L B, Lu H L, Zhao Y Y, Ma C Y, Wei Q Q. 2018. Origin of skewed frequency distribution of regional geochemical data from stream sediments and a data processing method. Journal of Geochemical Exploration, 194: 1-8

Zuo R G. 2011. Identifying geochemical anomalies associated with Cu and Pb-Zn skarn mineralization using principal component analysis and spectrum-area fractal modeling in the Gangdese Belt, Tibet (China). Journal of Geochemical Exploration, 111(1-2): 13-22

Zuo R G. 2012. Exploring the effects of cell size in geochemical mapping. Journal of Geochemical Exploration, 112: 357-367

Zuo R G. 2014. Identification of geochemical anomalies associated with mineralization in the Fanshan district, Fujian, China. Journal of Geochemical Exploration, 139: 170-176

Zuo R G. 2017. Machine learning of mineralization-related geochemical anomalies: a review of potential methods. Natural Resources Research, 26(4): 457-464

Zuo R G. 2018. Selection of an elemental association related to mineralization using spatial analysis. Journal of Geochemical Exploration, 184: 150-157

Zuo R G, Wang J. 2016. Fractal/multifractal modeling of geochemical data: a review. Journal of Geochemical Exploration, 164: 33-41

Zuo R G, Wang J L. 2020. ArcFractal: an ArcGIS add-in for processing geoscience data using fractal/multifractal models. Natural Resources Research, 29(1): 3-12

Zuo R G, Xiong Y H. 2018. Big data analytics of identifying geochemical anomalies supported by machine learning methods. Natural Resources Research, 27(1): 5-13

Zuo R G, Xiong Y H. 2020. Geodata science and geochemical mapping. Journal of Geochemical Exploration, 209: 106431

Zuo R G, Xia Q L, Wang H C. 2013. Compositional data analysis in the study of integrated geochemical anomalies associated with mineralization. Applied Geochemistry, 28: 202-211

Zuo R G, Wang J, Chen G X, Yang M G. 2015. Identification of weak anomalies: a multifractal perspective. Journal of Geochemical Exploration, 148: 12-24

Zuo R G, Carranza E J M, Wang J. 2016. Spatial analysis and visualization of exploration geochemical data. Earth-Science Reviews, 158: 9-18

Zuo R G, Xiong Y H, Wang J, Carranza E J M. 2019. Deep learning and its application in geochemical mapping. Earth-Science Reviews, 192: 1-14

Progresses of Researches on Geochemical Exploration Data Processing

ZUO Ren-guang[1], WANG Jian[2], XIONG Yi-hui[1], WANG Zi-ye[1]

1. State Key Laboratory of Geological Processes and Mineral Resources, China University of Geosciences (Wuhan), Wuhan 430074; 2. College of Earth Science, Chengdu University of Technology, Chengdu 610059

Abstract: China has established a database with large amounts of high-quality geochemical exploration data, which can provide important data support for mineral exploration, environmental assessment, and geological survey. How to efficiently process geochemical exploration data for identifying the deep-seated information has always been a hot and frontier topic in the field of exploration geochemistry. Based on systematically investigating and studying domestic and foreign scholars' publications on the geochemical exploration data processing in the past decade, in this review, we have summarized the main research progresses of the geochemical exploration data processing in China in the past decade from aspects of construction of geochemical exploration database, geochemical anomaly identification and uncertainty evaluation. In addition, this review pointed out three research directions should be further focused on geochemical exploration data processing in the future. Main progresses of researches on the geochemical exploration data processing in China in recent ten years are given below: ①The fractal and multifractal methods have been paid wide attentions globally and greatly developed and promoted due to their capabilities of characterizing the complexity and scale-invariance of geochemical spatial patterns. Chinese researchers have played leading roles in the field of exploration geochemical data processing based on fractal and multifractal models. ②The machine learning and big data thinking began to be applied to the geochemical exploration data processing and rapidly gained researcher attention in this field. They are becoming the hot topic and state-of-the-art techniques for the geochemical exploration data processing. Researchers in our country firstly explored the geochemical exploration big data using machine learning algorithms. ③More studies on the processing of missing data and closed effect of geochemical compositional data should be carried out in China. In the near future, researchers should put more efforts on studies of identifying weak geochemical anomalies, assessing the uncertainty of identified geochemical anomalies, and the integrated study of the anomaly identification and genetic mechanisms of geochemical anomalies.

Key words: geochemical exploration; geochemical anomalies; data processing; fractal; machine learning

陨石学与天体化学研究进展

缪秉魁[1,2]，胡森[3]，陈宏毅[1]，张川统[1]，夏志鹏[1]，黄丽霖[1]，薛永丽[1]，谢兰芳[1]

1. 桂林理工大学 陨石与行星物质研究中心，桂林 541006；2. 中国科学院 月球与深空探测重点实验室，北京 100101；3. 中国科学院 地质与地球物理研究所，地球与行星物理重点实验室，北京 100029

摘 要：过去十年，中国陨石学与天体化学研究迎来了一个迅速发展时期。嫦娥工程与月球采样返回，以及天问一号火星探测等一系列深空探测工程的成功实施，极大地推动了我国陨石学和天体化学的发展；南极格罗夫山陨石库的建设，为学科发展提供了充足样品；高精度同位素分析平台的建设，保障了陨石及地外返回样品的分析研究需要；此外，国内还涌现了一批优秀行星科学与天体化学青年人才。与此同时，通过对各类陨石的研究，在太阳星云起源与演化、火星与月球等类地行星形成与宜居性研究、小行星岩浆作用和含水蚀变及后期撞击历史等方面也取得了重要成果。

关键词：陨石学 天体化学 太阳星云 类地行星 小行星 南极陨石

0 引 言

天体化学是研究宇宙空间（主要是太阳系）物质的化学组成和化学演化规律的学科，是地球科学、天文学和行星科学相互渗透交互的边缘学科，同时，对了解地球起源和演化规律也具有重要意义。20世纪人类开始迈步太空以及现代科学的发展，天体化学已经成为世界前沿学科和热点领域。我国自2007年发射嫦娥一号探测器后，探月工程顺利开展，先后发射嫦娥二号、三号、四号、五号探测器，成功完成了月球背部探测和月球采样返回任务。此外，2020年7月23日成功发射了天问一号，启动了火星探测计划，标志着我国行星科学进入了新的发展阶段。天体化学为深空探测和行星科学研究的重要科学目标，陨石则是解决行星科学问题的重要窗口。反之，这些深空探测计划的顺利执行又极大地推动了我国陨石和天体化学的发展。另外，这十年间我国南极陨石库的建设也取得了重要进展。

本文总结了我国陨石学和天体化学在2011—2020年间的发展情况，侧重于以陨石为重要对象的天体化学研究内容。

1 球粒陨石研究：早期太阳星云的起源与演化

球粒陨石是早期太阳星云凝聚吸积产物，不同化学群的球粒陨石代表了不同星云区域的成分特征，因此，球粒陨石是认识和探索太阳星云形成和演化的重要窗口，也是陨石学与天体化学研究的重要对象（王道德和王桂琴，2012；林杨挺等，2014）。十年来，我国在球粒陨石方面开展了较多工作，取得一些重要进展。

1.1 前太阳系颗粒研究

前太阳系颗粒，亦称太阳系外颗粒，主要包括金刚石（C）、碳化硅（SiC）、石墨（C）、氮化碳

---

* 原文"中国陨石学与天体化学研究进展（2011—2020）"刊于《矿物岩石地球化学通报》2021年第40卷第6期，本文略有修改。

（C_3N_4）和刚玉（Al_2O_3）等，形成于早期恒星演化阶段，是认识恒星起源的重要样品对象。然而，传统的寻找前太阳系颗粒的方法为酸溶解和物理分选，这种方法既烦琐，又消耗大量珍贵的陨石样品。近年来，国内成功开发了在陨石薄片样品上原位开展同位素扫描方法，来寻找前太阳颗粒（林杨挺等，2013）。

Zhao 等（2014）借助 NanoSIMS 高分辨同位素成像技术，在 Sutter's Mill 碳质球粒陨石中搜寻前太阳颗粒。Sutter's Mill 陨石是 2012 年在美国加利福尼亚降落的一块 CM2 型碳质球粒陨石，是研究前太阳颗粒的理想样品。研究表明，该陨石的 SiC 总体丰度为 $55×10^{-6}$，与其他碳质球粒陨石相似，同时研究了不同类型前太阳颗粒的原位产状，共发现 37 个碳同位素异常颗粒和 1 个氧同位素异常颗粒。37 个碳异常颗粒中包括 31 个碳化硅颗粒和 5 个碳颗粒，氧异常颗粒为刚玉。但尚未发现前太阳系硅酸盐颗粒，可能是由于它们在行星母体中容易被水蚀变破坏。基质中 SiC 颗粒分布是不均匀的（$12×10^{-6}$ ~ $54×10^{-6}$），可能反映了细粒的基质碎屑来自不同太阳星云源区，或经历了不同的热历史（Zhao et al.，2014）。这项工作证明，NanoSIMS 原位扫描方法，不仅可以获得更丰富的信息，而且极大地提高了工作效率，为前太阳颗粒研究提供了更广阔的前景。

1.2 富铝球粒和富钙长石-橄榄石包体（POI）研究

球粒陨石的富钙铝包体（calcium-aluminium-rich inclusion，CAI）和球粒都是早期太阳星云的高温产物，它们记录了早期太阳星云演化的重要信息，对认识太阳系起源和星云演化过程具有重要意义。CAI 是由富钙铝的硅酸盐和氧化物组成的一种特殊高温凝聚集合体，是太阳早期星云最早凝聚的物质，而球粒是太阳星云高温熔融由橄榄石和低钙辉石等镁铁质硅酸盐组成的球状物。成分特征表明，CAI 和球粒没有直接联系，但是富铝球粒成分介于 CAI 和普通球粒之间。富铝球粒由钙长石和富钙辉石等富钙铝矿物组成，其 Al_2O_3 含量一般大于或等于 10%，低于 CAI，但高于普通球粒。研究表明，富铝球粒可能是 CAI 和球粒之间的过渡产物，其对了解 CAI 和球粒的成因联系具有重要意义（Zhang et al.，2014）。

CAI 形成于早期太阳星云，与蠕虫状橄榄石集合体（AOA）等难熔包体都是太阳星云降温过程的连续凝聚产物，为太阳星云凝聚理论提供了有力证据（Lin and Kimura，2003）。Dai 等（2015）和 Zhang M. M. 等（2019）研究了碳质球粒陨石中粗粒 CAI，证明粗粒 CAI 是从星云凝聚后再经历熔融结晶而形成的。球粒的形成时间可能稍晚于 CAI。CAI 的 $^{26}Al/^{27}Al$ 的初始值为 $5×10^{-5}$，而硅酸盐球粒的则约为 $1×10^{-5}$，这一差异表明凝聚成因 CAI 的形成时间要早于球粒 1~4 Ma，成分介于 CAI 和富铝球粒的 POI 也是 CAI 和球粒之间成因联系的有力证据。

氧同位素对于约束球粒陨石 CAI 和球粒在早期太阳星云形成源区具有重要意义。为了证明 CAI 和球粒之间的成因关系，国内学者对富铝球粒和 POI 开展了氧同位素研究。Zhang A. C. 等（2014，2020）和戴德求等（2020）对碳质球粒陨石的富 Al 球粒开展了岩石矿物学特征和氧同位素成分分析，结果发现富铝球粒的初始物质来源于 CAI 和镁铁质硅酸盐球粒的混合，而且富 Al 球粒中一些富 ^{16}O 的尖晶石或橄榄石颗粒可能是前期形成 CAI 在熔融结晶过程中的残留，这为 CAI 和球粒存在成因联系提供了证据。虽然大部分原始富钙铝包体具有富 ^{16}O 的同位素特征，但也有一些证据表明部分原始的超难熔富钙铝包体是从相对贫 ^{16}O 的星云环境中形成的（Zhang et al.，2015）。蒋云等（2015）对普通球粒陨石［Grove Mountains（GRV）022410（H4），GRV 052722（H3.7）和 Julesburg（L3.6）］中富铝球粒开展了氧同位素分析，获得的富铝球粒的氧同位素组成（$\delta^{18}O$ 为 -6.1‰ ~ 7.1‰，$\delta^{17}O$ 为 -4.5‰ ~ 5.1‰）与镁铁质球粒相近，远比 CAI（$\delta^{18}O = -40‰$，$\delta^{17}O = -40‰$）亏损 ^{16}O。这表明普通球粒陨石中的富铝球粒不是 CAI 与普通球粒简单混合形成的，相反，它们很有可能在多次熔融过程中与贫 ^{16}O 的星云气体储库经历了更高程度的氧同位素交换（蒋云等，2015）。

1.3 顽辉石球粒陨石研究

顽辉石球粒陨石是最还原的球粒陨石类型，对研究太阳星云的氧化还原条件具有重要意义，然而，顽辉石球粒陨石样品非常少见。GRV 13100 顽辉石球粒陨石是我国第 30 次南极科考队在南极格罗夫山发现的。该陨石的不透明矿物包括陨硫铁、陨硫铬铁矿、陨硫镁矿、陨硫钙矿、含硅铁纹石、陨磷铁矿、硅磷镍矿等金属硫化物和磷化物，总丰度达 21%，并经历了一定程度的热变质（谢兰芳等，2019）。谢兰芳等（2020）对该陨石中的硫化物开展了详细的矿物学研究，取得如下认识：①陨硫镁矿为太阳星云的直接凝聚成因，而且是在橄榄石和顽火辉石冷凝结晶之后形成的，变质温度主要为 200~300℃，个别颗粒可达 400~800℃，可能为外来吸入成因；②陨硫铁具有原生和次生两种成因，原生的陨硫铁由太阳星云直接凝聚而成，次生的陨硫铁则是在后期热变质过程中由铁镍金属经硫化作用或由陨硫镁矿分解形成；③硅磷镍矿可能是由富硅铁纹石的出溶形成。此外，由于顽辉石球粒陨石的多种同位素组成与地球的同位素组成相似，被认为是地球形成的前身物质或者类似物质。

2 原始无球粒陨石研究：行星的起源

2.1 原始无球粒陨石成因研究

原始无球粒陨石是指结构上无球粒、但化学成分上与球粒陨石相似的陨石（Weisberg et al.，2006）。这一概念最早由 Prinz 等（1983）在研究 winonaite 陨石（W 群）、橄榄古铜陨铁（lodranite 陨石，Lord 群）、富橄榄石陨石（brachinite，B 群）和 IAB 和 IIICD 铁陨石（又称 IAB 群）中的硅酸盐包体时提出的。目前原始无球粒陨石还包含 acapulcoite（A 群）和橄辉无球粒陨石（ureilite）（Weisberg et al.，2006；Collinet and Grove，2020）。此外，一些未分群的无球粒陨石，如 Tafassasset 陨石（Gardner-Vandy et al.，2012），也被认为是原始无球粒陨石。不过，对于橄辉无球粒陨石是否是原始无球粒陨石还存在争议（Krot et al.，2014）。

原始无球粒陨石的成因始终是天体化学领域所关注的问题之一。绝大部分原始无球粒陨石群的氧同位素组成不均匀，通常被认为是球粒陨石母体发生不完全熔融的产物（Rumble et al.，2008；Day et al.，2012）；然而，W 群陨石的氧同位素成分比较均一，这可能是母体较小的原因。

橄辉无球粒陨石与 B 群陨石被认为是堆晶形成的（Mittlefehldt，2008）或是熔融残余相（Weisberg et al.，2006）。当然，原始无球粒陨石存在演化程度的差异，如 A-Lord 陨石系列中，A 群陨石的演化程度不及 Lord 群陨石演化程度高，W 陨石群的不同个体也存在类似情况（Zeng et al.，2019）。原始无球粒陨石在很大程度上保留了与球粒陨石相似的初始地球化学特征，而分异的无球粒陨石的初始地球化学特征则被后期的演化所湮没。原始无球粒陨石作为球粒陨石与无球粒陨石之间的纽带，其保留了小行星母体初始分异状态（Touboul et al.，2009）。因此，原始无球粒陨石不但可以为太阳系早期小行星演化过程提供约束（如其初始球粒陨石质母体物质的性质、超变质或部分熔融的程度、变质或部分熔融的热源等），而且能为我们理解地球和月球等类地行星天体早期演化提供必要信息。

原始无球粒陨石是非常稀少的，国内非常缺乏此类陨石样品。近年来，由于南极陨石和沙漠陨石搜寻工作得到快速发展，原始无球粒陨石及其相关陨石类型实现了从无到有的突破，为我国陨石学研究提供了良机。在南极格罗夫山发现两块 W 群陨石（GRV 021663 和 GRV 022890）、一块来自 A-Lod 群母体的球粒陨石（GRV 020043）（Li et al.，2011，2018；Zeng et al.，2019）；在新疆戈壁地区，我国分别发现了一块橄辉无球粒陨石（Loulan Yizhi 034）和一块 B 群陨石（Kumtag 061）（源自国际 Meteoritical Bulletin Database）。

GRV 020043 陨石是一块 4 型球粒陨石，氧同位素和铬同位素分析表明，它属于未分群陨石，是来源于 A-Lord 系列原始无球粒陨石源区，也就是说，GRV 020043 陨石是 A-Lord 系列陨石的球粒陨石质初始物质，可能是 A-Lord 系列原始无球粒陨石的早期母体，而 A-Lord 陨石是由它热变质或部分熔融而成（Li et al., 2018）。根据对比研究，A-Lord 原始无球粒陨石母体小行星的可能结构，从核部到最外层依次为富金属 Lord 群陨石物质［如 Northwest Africa（NWA）468 和 Graves Nunataks（GRA）95209］、典型的 Lord 群陨石物质、A/Lord 过渡型、典型的 A 群陨石物质和类似 GRV020043 的球粒陨石物质（Zeng et al., 2019）。然而，有些未分群无球粒陨石（如 NWA 7325）与 A-Lord 原始无球粒陨石具有相似的氧和铬同位素组成。另外，有些原始无球粒陨石（如 Tafassasset）的成分特征与碳质球粒陨石非常相似（Gardner-Vandy et al., 2012；Kruijer et al., 2020）。这些现象说明，W 群原始无球粒陨石的成因及其与球粒陨石的关系可能是复杂的。总之，原始无球粒陨石成因还有待更多的样品和更深入的研究（Li et al., 2018）。

GRV 021663 和 GRV 022890 是两块南极 W 群原始无球粒陨石，它们的主要组成矿物为镁橄榄石、顽火辉石、铁纹石和陨硫铁等，但二者的矿物丰度差异很大，前者非常富含镁橄榄石。根据一系列的 W 群陨石（包括南极陨石 GRV 021663 和 GRV 022890）对比研究发现，W 群陨石母体仅发生了较为有限的部分熔融，冲击作用可能在母体的演化过程中扮演了重要角色，其母体小行星具有一个多层结构（Li et al., 2011；Zeng et al., 2019）。

2.2 陨石的古剩磁研究

由于磁场能够传输角动量并驱使星际增生，所以它对原始星云盘的结构和演化具有重要意义。有证据表明原始太阳行星盘存在大规模磁场，而且球粒的古地磁测量发现中盘太阳星云，即类地行星区，在太阳系形成后的 1~3 Ma 之内，其磁场强度是 5~50 μT。因此，陨石古剩磁研究对太阳系行星形成和演化具有重要意义。国内学者开展了钛辉无球粒陨石的古磁场分析，为早期类地行星区太阳星云飘散和其他大行星的形成提供时限约束（Wang et al., 2017）。他们研究的三块钛辉无球粒陨石是 D'Orbigny、Sahara 99555 和 Asuka 881371，其辉石 Pb/Pb 年龄分别是（4 563.37±0.12）Ma、（4 563.54±0.14）Ma 和（4 562.4±1.6）Ma。3 块陨石的古剩磁几乎为零（<0.3 μT），表明它们形成时（即太阳系形成之后约 3.8 Ma）类地行星区太阳星云环境没有磁场。这项研究表明，钛辉无球粒陨石磁性对于约束太阳内地行星区的消散和类木气体行星的形成具有重要意义，同时也为球粒的非星云凝聚的撞击成因提供有力证据（Wang et al., 2017）。

3 HED 族陨石：灶神星的变质历史

根据红外光谱观测、同位素研究及"黎明号"探测器数据分析，普遍认为 HED 族陨石（包括古铜钙长无球粒陨石 Howardite、钙长辉长无球粒陨石 Eucrite 和奥长古铜无球粒陨石 Diogenite）起源于 4 号小行星即灶神星（McSween et al., 2013；Mittlefehldt, 2015），因此，HED 族陨石也被称为灶神星陨石。

HED 族陨石是经过广泛熔融分异的岩浆岩，与地球岩浆岩的物质组成非常相似（Mittlefehldt, 2015）。但与地球、月球和火星的岩石组成不同的是，HED 族陨石有非常古老的岩浆结晶年龄，同位素得到其结晶年龄 4.43~4.55 Ga（Misawa et al., 2005；Mittlefehldt, 2015），是目前太阳系最古老的岩浆岩之一。灶神星的岩浆演化过程非常短暂（约 10 Ma），后期没有更年轻的岩浆活动叠加，但是在岩浆作用过程或期后一个较小的时间段内发生了热变质作用（McSween et al., 2011）。由于灶神星上缺乏地质构造运动和生物活动，而且整体挥发分含量低，其岩石应较好地保留了早期的热变质、冲击变质、流体蚀变等现象。因此，HED 族陨石是探索和研究太阳系分异型行星或小行星早期岩浆演化和变质作用的理想对象和重要窗口（陈宏毅等，2016）。

近十年来，国内开展了较多的 HED 族陨石的岩相学、地球化学以及同位素年代学等研究，并在 HED 族陨石的成因和灶神星的地质作用方面获得了一些重要认识：

（1）根据 166 块 HED 族陨石氧同位素统计分析，发现其中 146 块具有非常一致的 $\Delta^{17}O$ 值，另外 20

块的 $\Delta^{17}O$ 值出现异常。对这 20 块氧同位素异常陨石的氧同位素分析，发现其中 15 块的氧同位素异常源自内部因素（即母体成分），因此提出 HED 族陨石母体不仅仅限于灶神星，可能还有其他小行星母体（Zhang C. T. et al., 2019）。

（2）在一块苏长质 Diogenite 陨石 NWA8321 的研究中，发现了富硫化物交代橄榄石现象，证实部分 Diogenite 陨石具有部分熔融现象，对灶神星存在岩浆演化后期交代作用提供了重要依据（Zhang A. C. et al., 2020）。

（3）Eucrite 陨石 NWA 8009 是一块撞击熔融角砾岩，根据其成分环带、岩相特征及 Zr/Hf 值特征的研究发现，其中的锆石为热变质成因锆石，其 $^{207}Pb/^{206}Pb$ 年龄为（4 560±8）Ma，由此限定了灶神星的全球热变质作用为约 4.55 Ga。这一研究成果为 HED 族陨石的热变质成因机制及其热液等提供了重要的年代学依据（Liao and Hsu, 2017）。

（4）在 HED 族陨石冲击变质效应和灶神星撞击作用研究方面也取得了一些重要认识（Liao and Hsu, 2017；Pang et al., 2018；Chen et al., 2019）。Pang 等（2018）在 Eucrite NWA 8003 的冲击熔融区域中找到了柯石英、斯石英和超硅石榴子石等高温高压矿物相，并在一个富钛冲击熔融囊中发现了蓝晶石、刚玉及新矿物——灶神星矿（Vestaite），这为 HED 陨石经历的冲击变质强度（10 GPa 左右）找到了确切证据。Chen 等（2019）报道了 Eucrite NWA 2650 中长石转变为钙铝榴石+蓝晶石+二氧化硅玻璃组合、锆石转变为斜锆石等多种冲击变质现象。Liao 和 Hsu（2017）对 Eucrite NWA 8009 中磷灰石进行了 U-Pb 同位素研究，为灶神星高强度冲击变质事件提供了新的年代学数据（约 4.1~4.2Ga）。

（5）灶神星挥发分流体研究已成为国内陨石学和天体化学研究的热点。在较早期的研究中，灶神星被认为是挥发分贫瘠的小行星（Papike, 1998；McSween et al., 2011）。但近些年，国内学者通过对 HED 族陨石中次生蚀变矿物的研究，指出灶神星表面和岩浆中很可能存在挥发分-岩石相互作用过程（Zhang A. C. et al., 2013, 2020；Pang et al., 2017；Wang et al., 2019；Huang et al., 2020），主要证据包括：①发现穿插矿物和岩石的硅酸盐矿物脉。Pang 等（2017）在 Eucrite NWA 1109 陨石中观察到为三种硅酸盐脉：富铁橄榄石脉、富镁橄榄石脉和辉石脉，这些硅酸盐脉不但与撞击作用有关，还可能有流体参与。国际上认为这种脉状矿物可能是次生蚀变形成的，而且与含 H_2O 流体有关（Barrat et al., 2011；Warren et al., 2014）。②辉石、橄榄石和石英等矿物存在富硫化物的交代还原结构。Zhang 等（2013）和 Wang 等（2019）分别在 Eucrite 陨石 NWA 2339 和 NWA 1109 中发现了富铁辉石的富硫化物交代还原结构，并认为这是干的富硫蒸汽对辉石的交代作用形成的，硫蒸汽来源于撞击作用。另外，Zhang A. C. 等（2018, 2020）在 Howardite 陨石中发现了橄榄石被陨硫铁-斜方辉石交代结构，提出这种硫化物交代结构应是富硫蒸汽与橄榄石反应产物，而且硫化作用应是灶神星的内部过程。除橄榄和辉石被交代现象外，Huang 等（2020）还在单矿碎屑 Eucrite NWA 11591 中发现鳞石英的富硫化物交代结构，首次提出 HED 族陨石中鳞石英的硫化作用机制，为硫化作用增添了新的矿物类型。这些工作证实灶神星岩浆演化和变质过程中应存在挥发分流体的参与，除可能的含 H_2O 流体作用外还可能存在富硫蒸汽的硫化作用，这为重新审视贫挥发分灶神星的岩浆演化和变质历史提供了新的线索。

4 火星陨石与月球陨石：岩浆作用与行星环境

随着我国嫦娥工程的成功实施和新一轮火星探测高潮（耿言等，2018；Li et al., 2019）的到来，以及国际上月球和火星陨石的申报数量大幅度增加（据国际陨石数据库统计，2011 年 1 月至 2020 年 12 月，月球陨石从 157 块增加到 435 块，火星陨石从 105 块增加到 291 块），我国月球陨石和火星陨石的研究也产生了积极响应，并取得了重要进展。

4.1 月球陨石研究

月球是人类迈出地球走向太空的第一个跳板，因此，半个多世纪以来，月球都是国际深空探测的热点。

美国 Apollo 计划和苏联月球计划采集了 382 kg 月球样品，这些样品对月球科学研究发挥了重要作用。但由于月球采样仅仅局限于几个着陆点，而月球陨石来源是随机的，对月球成分更具有代表性，因此，月球陨石依然是认识和研究月球物质成分的重要窗口。月球陨石传统上分为高地斜长岩、月海玄武岩和混合角砾岩。由于长期的撞击和太空风化，大部分月球陨石为碎屑岩，这些碎屑岩主要具有三种端元：斜长岩、玄武岩和克里普岩（Miao et al.，2014）。近年来，国内对月球陨石进行了不少岩石学和矿物化学研究（夏志鹏等，2013；谢兰芳等，2013；姚杰等，2013；Xie et al.，2014），为了解月球表面物质成分和演化过程提供了一些基础数据。在此基础上，月球陨石研究还取得了一些有关月球成分和岩浆演化的重要认识：

（1）月球岩浆演化的认识。Xu 等（2020）研究了 10 块月球陨石的稀土元素和不相容元素成分，发现斜长石的稀土元素丰度和模式与斜长石共生的镁铁质矿物的 $Mg^{\#}$ 值没有相关性，而且斜长石的稀土元素丰度是镁铁质矿物的 40 倍，据此提出月球斜长岩不是简单的月球岩浆洋（LMO）岩浆分异产物，而是 LMO 岩浆分异和后期 KREEP 熔体或月幔部分熔融物质交代作用的结果。

（2）富尖晶石斜长岩型陨石的发现。周剑凯等（2019）发现了一块富含尖晶石的月球斜长岩的陨石 NWA 12279，这块陨石具有完整的结晶结构，主要由斜长石（70.6%）、橄榄石（11.3%）、辉石（10.0%）和镁铝尖晶石（7.0%）等组成，其中尖晶石存在两种产状：粗粒的八面体斑晶和撞击熔融再结晶的小尖晶石颗粒。NWA 12279 陨石为上月幔代表性岩石"富镁橄榄石+富铝低钙辉石+镁铝尖晶石"的矿物组合找到了直接证据（Wittmann et al.，2019），同时也印证了印度探月卫星"月船一号"M3 光谱仪在月球大型撞击坑中央山峰和环形山上发现的含镁铝尖晶石斜长岩（PSA）（Prissel et al.，2014），这对探索月球岩浆演化过程具有重要意义。

（3）极高钾（VHK）的 KREEP 岩性的发现。Lin Y. 等（2012）在月球陨石 Sayh al Uhaymir（SaU）169 撞击熔融角砾岩中发现了 K 含量异常高的 KREEP 岩屑，即极高 K（VHK）KREEP 岩性，证明 LMO 演化结晶晚期存在最后残余液相，为 LMO 晚期岩浆侵入作用的不均一性提供了证据。

（4）月球陨石冲击变质效应及锆石年龄研究。月表撞击坑形貌和月球陨石丰富的角砾结构充分证明，撞击作用在月球演化历史中具有重要地位。但有关记录月球撞击事件过程条件的研究还比较薄弱。近年来，国内在月球陨石撞击现象和撞击定年研究上取得较大进展。Zhang 等（2011）在 Dhofar 458 月球陨石中发现了锆石的多晶和多孔结构，为月球撞击事件的温度压力条件找到了有力证据。陈宏毅等（2015）在低钛玄武岩月球陨石 MIL 05035 中发现了三斜铁辉石在撞击作用下分解形成的"钙铁辉石+铁橄榄石+石英"组成的后成合晶。

为精确确定月球撞击事件的演化时间，我国学者对月球陨石和 Apollo 样品进行了同位素定年研究。Wang 等（2012）研究了月球陨石 NWA 4734 的稀土元素和斜锆石年龄，发现其中的辉石和斜长石具有与 LaPaz Icefields（LAPs）02205/02224 低钛玄武岩陨石相似的稀土元素富集规律，二者斜锆石年龄[(3 073±15)Ma] 与 LAPs [(3 039±12)Ma] 相近，提出它们来自月球上相同的或类似的陨石坑。Liu 等（2012）利用 SHRIMP-II 离子显微探针测试了月球陨石 SaU 169 高钛冲击熔融角砾岩和 Apollo 12 号月壤样品中的锆石颗粒的年龄，发现二者具有近似的结晶年龄，分别为 (3 920±13)(2σ) Ma（SaU 169）和 (3 914±7)(2σ) Ma，这是国内第一次利用阿波罗样品获得的雨海撞击事件的锆石年龄。

4.2 火星陨石研究

火星陨石是因撞击事件而随机来自火星表面的岩石，对其研究可为火星岩浆演化过程提供直接证据并限制其源区特征。国内以火星陨石为对象，开展了火星深部岩浆作用、年龄限制、流体作用和有机物探测等研究，取得了一系列成果：

（1）火星岩浆作用。通过对含粗粒橄榄石斑晶辉玻无球粒陨石 NWA 8716 的岩石学、矿物学和稀土元素地球化学研究，发现该陨石来源于轻稀土亏损的开放型火星岩浆环境，其中橄榄石斑晶是从与早期母岩浆成分相似的熔体中堆晶而成（吴蕴华等，2014）。根据 NWA 7034 火星陨石中角砾的锆石和磷灰石的原位同位素年龄分析，确定该陨石中岩屑的结晶年龄比较老，最老可达~4465 Ma，表明火星岩浆活动非常早，在太阳系早期很强烈（Hu et al.，2019）。而目前发现最年轻的火星岩石是 Zagami 陨石，它的结晶年龄仅为 180 Ma，可能代表了火星上被溅射出来的最年轻的岩浆岩（Zhou et al.，2013）。

(2) 火星壳的热液交代作用和水岩反应。NWA 7034 是首块火星表土角砾岩，主要由各种火成岩屑、矿物碎屑和冲击熔融基质组成，其全岩水含量高达 $6000×10^{-6}$，也是目前最富水的火星样品。Liu 等（2016）首次在 NWA 7034 和 NWA 7533（成对陨石）中发现了富含稀土的磷酸盐和硅酸盐，3 个磷灰石中独居石的 U-Pb 年龄分别为（1.0±0.4）Ga（2σ）、（1.1±0.5）Ga（2σ）和（2.8±0.7）Ga（2σ），记录了火星壳中热液流体和岩石相互作用的变质年龄，这指示火星在 1 Ga 或更早的时候就广泛存在热液流体。另外，Hu 等（2019）对 NWA 7034 陨石的基质中变质锆石和磷灰石进行了同位素测年，获得其锆石年龄（1634±93 Ma）和磷灰石年龄（1530±65 Ma）较年轻，而且具有富 D（δD 为 313‰~2459‰）和 ^{37}Cl（$\delta^{37}Cl$ 为 -2.2‰~+3.8‰），表明火星在 1.6 Ga 年前存在含挥发分流体参与的变质作用。

Hu 等（2019）对 NWA 7034 开展了岩矿、锆石微量元素、锆石和磷酸盐的 U-Pb 定年和磷酸盐的水含量、H 同位素和 Cl 同位素分析，发现：①火星在 15 亿年前就存在富 D 和富 P 流体的水岩作用。NWA 7034 中锆石发生蜕变质，其非晶区域明显富集 P、Y 和 U，表明有流体参与。锆石未变质区域的同位素年龄为 44 亿年，而变质非晶质区域只有 15 亿年，另外 NWA 7034 中所有磷酸盐年龄也为 15 亿年。结合磷酸盐具有富 D 和蜕变质锆石富 P 的特征，说明 15 亿年前的强烈变质事件有富 D 和富 P 流体参与，指示该流体可能来自火星浅表。②火星贫 D 和贫 ^{37}Cl 流体的水岩作用。NWA 7034 中磷灰石的水含量和 H 同位素具有明显的负相关，说明经历过两端元混合，其中一个端元富 D，另一端元贫 D，富 D 端元可能发生在 15 亿年前，与磷酸盐 U-Pb 体系被重置相关。磷灰石的剖面分析结果表明，从边部向核部水含量是降低的，而 H 和 Cl 同位素则呈升高的趋势，指示磷灰石的水从外往里扩散，且其源区来自火星内部，而非地球水的污染。

(3) 火星水与挥发分研究。火星被认为是最有可能孕育过生命的地外行星（Lin et al.，2014b），这是当前国际深空探测的热点之一。现有的火星探测和火星陨石研究表明，火星早期很可能存在海洋和大面积的水体（Ehlmann et al.，2011），为生命演化提供了基本条件。大约 30 亿年前，火星表面就已经变成现在这样极度的寒冷和干旱（Ehlmann et al.，2011）。火星剧烈的环境变迁，主要受控于内部的岩浆活动。火星内部的挥发分（水、碳、氟、硫、氯等）含量是串联火星生命、古环境和岩浆活动的重要纽带，是探索火星演化历史的钥匙。因此，近十年来，国内以陨石为对象，开展了火星挥发分和水的研究，进而讨论火星的水环境、水岩作用、火星幔挥发分等科学问题。

火星幔的水、硫和氯含量一直是火星科学的研究热点。目前普遍认为火星幔的水含量为~$20×10^{-6}$，比地球富 Cl（~$3000×10^{-6}$），而火星幔一直有贫硫和富硫的争议（Filiberto et al.，2016）。Hu 等（2014，2020）对我国南极回收的 GRV 020090 火星陨石和 NWA 6162 陨石中的岩浆包裹体和磷灰石的水含量和氢同位素进行了研究，获得有关火星水的重要认识：① 该样品岩浆包裹体的水含量和氢同位素具有非常好的对数相关性，指出这是与火星大气水交换的结果，从而推断火星大气的氢同位素组成为 6034‰±72‰，与"好奇号"的探测结果一致（Leshin et al.，2013；Webster et al.，2013）；② 这些岩浆包裹体的水含量和 D/H 值非常不均匀，二者都是从中央向外逐渐升高，说明这些水是由外部扩散进入冷却后的岩浆包裹体的。因此，这是火星大气水而不是岩浆水，这是首次发现火星存在大气降水的同位素证据；③ NWA 6162 陨石熔融包裹体的水含量和氢同位素具有非常好的对数正相关关系，指出这是火星大气水交换的结果，说明熔融包裹体中的水大部分来自火星的浅表，证明当时火星表面有水的存在（Hu et al.，2020）。

此外，Hu 等（2020）还对 NWA 6162 陨石的熔融包裹体开展了硫、氯及其与水的相关性研究，结果发现：①熔融包裹体中的 S 以纳米到微米颗粒均匀分布在熔融包裹体内，可能是金属硫化物，而且硫含量与水含量无明显相关性；②氯含量与水具有正相关关系，落在火星幔源和壳源物质的混合线上，说明熔融包裹体的氯含量也受到了水岩作用的影响；③据估算，火星幔的水、氯和硫含量分别为 $0.1×10^{-6}$~$3×10^{-6}$、$0.5×10^{-6}$~$4×10^{-6}$ 和 $0.5×10^{-6}$~$15×10^{-6}$，明显比地球地幔"干"。

（4）火星陨石中有机碳的发现。Tissint 是一块 2012 年目击降落在摩洛哥的火星陨石，也是人类能够识别火星陨石以来的第一次目击降落，极大程度避免了地球的污染。Lin Y. T. 等（2012）通过对赋存在 Tissint 火星陨石裂隙和冲击熔脉中的碳同位素分析，获得有机质的 $\delta^{13}C$ 为 $-33.1‰ \sim -12.8‰$，这比火星大气中的 CO_2 和碳酸盐要轻得多，碳质组分类似于油母岩质干酪根，推断这些碳可能来自 Tissint 母岩冲击诱发裂缝中的流体沉积，渗入火星表面岩石中的富含有机质液体的存在对火星古环境的研究具有重要意义，也有助于寻找火星上可能存在的古代生物活动痕迹。

5 陨石稀有气体研究：小行星的母体演化历史

稀有气体包括氦（He）、氖（Ne）、氩（Ar）、氪（Kr）、氙（Xe）和氡（Rn）六种元素，其不仅在地球科学中具有重要的示踪作用，也是行星物质重要的溯源途径（贺怀宇等，2010；张川统等，2018）。近十年来，国内学者积极开发陨石稀有气体测试方法，实现了微量样品激光熔融稀有气体测量，为我国快速发展的比较行星学和深空探测提供了先进的研究平台（王英等，2018），基于此陨石稀有气体研究也取得了长足进展。

5.1 月球陨石稀有气体的研究

月球表面覆盖了一层厚约 $5 \sim 12$ m 的表土层或月壤。它的形成经历了漫长的地质过程，包括太阳风植入、宇宙射线辐射、撞击及埋藏等。为解析这些地质过程和反演月球演化历史，离不开稀有气体及其宇宙射线暴露年龄研究（Herzog，2007）。

根据探月工程需求，张川统等（2018）开展了月球陨石稀有气体研究，对月球陨石稀有气体浓度、同位素比值和宇宙暴露年龄数据进行对比分析，进而讨论月球陨石演化历史。取得的主要认识有：①随着暴露程度的增大，即从非角砾岩、角砾岩到表土角砾岩，月表物质的稀有气体浓度逐渐增高；②月球陨石稀有气体成分中不存在太阳高能粒子 Ne 的组分，富重 Ne 组分是太阳风分馏作用的产物；③月球陨石存在两种宇宙暴露年龄，分别是地月转移时间的 $T_{4\pi}$ 年龄，以及指示陨石在月表受到宇宙射线辐射累积时间的 $T_{2\pi}$ 年龄，且绝大多数表土角砾岩在月球表面经历了约 $400 \sim 1\,000$ Ma 的驻留历史，而非表土角砾岩则要短得多；④月球陨石在离开月球后约 (0.4 ± 0.9) Ma 内进入地球大气层。另外，Ranjith 等（2019）分析了月球陨石 NWA 10203 不同角砾和基质的稀有气体组成，发现这些样品同时含有宇宙射线成因、放射性成因及太阳风组分的稀有气体，这与其斜长岩质表土角砾岩的岩石学研究结果相一致，另外他们计算得到的陨石角砾的 CRE 年龄约为 $66 \sim 70$ Ma。

5.2 南极陨石稀有气体研究

由于南极格罗夫山陨石富集区的发现，我国获得了大量陨石样品（Miao et al.，2018），这为开展不同类型陨石的稀有气体研究提供良机。目前已开展的南极陨石稀有气体工作主要有如下方面：

（1）普通球粒陨石。Ranjith 等（2017）和 Wang 等（2020）对南极格罗夫山 H 群和 L 群普通球粒陨石进行了稀有气体成分分析。结果表明，H 群和 L 群陨石的 K-Ar 气体保存年龄分别是 (3.67 ± 0.26) Ga 和 (459 ± 13) Ma，佐证了 H 群母体在 $3.5 \sim 4.0$ Ga 期间发生的大型撞击事件，支持了 L 群陨石母体在 470 Ma 发生的撞击破裂。3 个 H 群陨石具有不同的宇宙暴露年龄（CRE）：(3.8 ± 0.3) Ma、(5.9 ± 0.4) Ma 和 (16.6 ± 1.5) Ma，这与 H 群陨石母体发生在 4 Ma、7.2 Ma 和 $15 \sim 20$ Ma 的撞击溅射事件相一致；而两个 L 群陨石具有近似的宇宙暴露年龄，分别为 (17.4 ± 1.1) Ma 和 (17.6 ± 1.5) Ma，可能代表 L6 群陨石母体发生在 ~15 Ma 的撞击溅射事件。

（2）HED 族陨石。GRV 13001 是一块发现于南极格罗夫山的玄武质单矿碎屑角砾岩 Eucrite 陨石，Zhang 等（2021）对其进行了稀有气体研究，结果发现该陨石 CRE 年龄为 (29.9 ± 3.0) Ma，表明其很可

能在 ~30 Ma 从母体灶神星溅离。它的 K-Ar 气体保存年龄 ~3.6~4.1 Ga，小于 Eucrite 陨石（~4.55 Ga）结晶年龄，表明其 ^{40}K-^{40}Ar 体系已发生了重置，这很可能是大型撞击事件引起的。

5.3 沙漠陨石稀有气体研究

除南极外，沙漠地区也是陨石重要的富集区。近年来，我国在西北沙漠地区开展了陨石搜寻工作，找到了不少陨石，发现了我国新疆等地沙漠戈壁地区为陨石富集区。沙漠陨石的分布特征和富集规律对于了解地外物质降落通量和陨石降落历史是非常有意义的。新疆陨石猎人 2012 年在库木塔格沙漠一个陨石密集分布区发现了 24 块陨石样品。为了了解该分布区的陨石降落次数，Zeng 等（2018）对 24 块陨石开展了岩石学、矿物学、微量元素成分、密度和孔隙度等研究。通过综合对比，他们发现其中的 22 块 L5 型陨石为成对陨石，其他 2 块（Kumtag 021，L4 型和 Kumtag 032，L6 型）是不同时间降落的不同类型陨石。此外，该区域还发现了 Kumtag 003 陨石（H5 型），由此证明该区域至少存在 4 次陨石降落。为了了解陨石降落前的大小和来源历史，Zeng 等（2018）还对 9 块 L5 成对陨石和 2 块独立陨石进行了稀有气体成分分析，结果表明它们的宇宙暴露历史不同，L5 型成对陨石宇宙暴露年龄为（3.6±1.4）~（5.2±0.4）Ma，L6 型陨石宇宙暴露年龄为（4.7±0.8）Ma，L4 型陨石宇宙暴露年龄略微偏高为（5.9±0.4）Ma。根据稀有气体同位素组成，推测 3 个陨石进入地球大气层前的最小半径分别约为 76、18 和 17 cm。

6 陨石非传统同位素研究：太阳系早期演化及类地行星起源的指示

随着同位素质谱测试技术的快速发展，大量用传统的质谱仪无法精确测量的非传统同位素体系得以准确测定。过去十年，我国在陨石金属稳定同位素研究方面取得了显著进展。

6.1 非传统稳定同位素陨石储库研究

同位素储库是研究各种地质体形成与演化的前提和基础，陨石储库则是研究太阳系、行星与小行星成因和演化的同位素参照标准。近年来，国内对陨石中的 Cr（刘佳，2019）、V（Xue et al., 2018）、Ca（Wu et al., 2020）、Cu（Xia et al., 2019）、Zn（Xia et al., 2019）等同位素开展了分析研究，建立了分析流程，获得了部分陨石类型的同位素储库。

同位素测试方法和条件、样品代表性及样品量效应等，均对陨石储库同位素数据的质量有影响。使用 MC-ICP-MS 测试 Cr 同位素，容易受样品回收率和 ArN$^+$ 的干扰。与 MC-ICP-MS 相比，TIMS 测定 Cr 同位素不受上述困扰，是测试 Cr 同位素最精确的方法，秦礼萍和夏九星（2014）开发了 TIMS 测定 Cr 同位素方法，分析了球粒陨石的 Cr 同位素成分，结果表明球粒陨石的 Cr 同位素组成与地球一致，行星核幔分异不会造成 Cr 同位素的分馏或者使得地核中的 Cr 含量降低。

高精度钒（V）同位素分析技术难度大，国际上陨石 V 同位素工作也较少。Xue 等（2018）率先在国内开展 V 同位素分析和研究，他们对 11 块 L 化学群普通球粒陨石进行了系统的 V 同位素测试，测试在中分辩下进行（$\Delta M/M>5\,500$），得到的 δ^{51}V 值为 -1.76‰~-1.08‰，显示普通球粒陨石的 V 同位素组成具有一定的不均一性。

6.2 陨石同位素异常研究

同位素异常是天体化学研究的重要领域。它是同位素组成无法用质量分馏、放射性衰变和宇宙射线辐射去解释陨石和地球样品同位素组成上的差异的原因。这些异常程度非常低，但对了解和研究太阳系的演化具有重要意义，全岩同位素异常的发现对太阳系早期演化模型提出了新的挑战。与同位素质量分馏相比，同位素异常不易受后期各种地质过程的影响，这使得金属元素的同位素异常成为示踪天体物质来源、反演太阳系形成时的星际环境的最直接证据（秦礼萍，2015）。由于不受各种地质作用的影响，同位素异常可以直接用来反演太阳系形成时的星际环境、示踪来自不同核合成源区的混合过程，以及指示天体样品物质来源和成因的联系（Qin and Carlson，2016）。随着高精度质谱仪的发展，行星尺度上的多个体系的同位素异常被发现并证实，这些同位素异常颠覆了人们对于早期太阳系同位素组成在行星尺度上均一的认识，对太阳系早期演化模型提供了新的制约（Qin and Carlson，2016）。产生核合成异常的载体是陨石中的前太阳系颗粒，它们通常携带了极端的同位素核合成异常组分（Zinner，1998）。

秦礼萍（2015）对陨石金属同位素异常进行了系统总结。目前，国际上陨石中已经发现金属同位素异常主要有（秦礼萍，2015，及其引用的相关文献）：①很多金属元素在不同类型陨石中存在同位素异常，主要发现于球粒陨石 CAI 中和一些陨石全岩成分；②^{48}Ca 和 ^{50}Ti 异常主要发现于富黑铝钙石型 CAI 中（ε^{50}Ti：-600~2 500；ε^{48}Ca：-700~1 000）；③许多陨石类型都发现 ^{54}Cr 同位素异常（ε^{54}Cr），另外，它们的 ^{54}Cr/^{52}Cr 值有非常系统的变化；④Mo 同位素体系也是目前为止观测到的行星尺度上异常程度最大的金属同位素体系之一，几乎所有陨石全岩样品均显示 ε^{92}Mo 正异常，以 CM 中的异常程度最高；⑤陨石中还存在 ^{62}Ni、^{100}Ru、^{84}Sr 等其他同位素异常。

不同类型陨石的 ^{54}Cr/^{52}Cr 值存在系统变化，说明在大规模时空上原始太阳星云盘存在同位素不均一性（Qin et al.，2010），所以 Cr 同位素异常的发现是探索上述同位素不均一性的关键。为此，Qin 等（2011a，2011b）对碳质球粒陨石和普通球粒陨石开展了同位素异常研究，他们在 C1 型碳质球粒陨石 Orgueil 中发现了 ^{54}Cr（δ^{54}Cr/^{52}Cr>1 000‰）同位素载体——纳米级的含铬氧化物。据分析，估计观察到的异常只是下限，真正的异常可能达 50 000‰，这些颗粒极有可能来自两类超新星（SN Ⅱ）的内层（O/Ne 层）（Qin et al.，2011a）。因此，提出这些极端富 ^{54}Cr 的氧化物可能是随超新星的爆发加入到初始太阳星云的，因此保留了同位素分布上的不均一性（Qin et al.，2011b）。

6.3 宇宙射线对金属同位素体系的影响及校正

陨石在进入地球大气前长期暴露在宇宙射线的辐射下，样品中的核素会与宇宙射线中的高能粒子发生相互作用，产生宇宙成因核素。大量的宇宙成因核素会显著改变天体样品的同位素组成，进而影响样品宇宙射线辐射前的同位素信息。因此，有效评估和校正宇宙射线对同位素的影响是天体化学中一个十分重要的领域（刘佳，2019）。

宇宙成因 Cr 同位素的生成是由散裂反应所主导（Leya et al.，2004）。Liu 等（2019）系统分析了 25 块来自 9 个化学群的铁陨石样品的 Cr 同位素组成，他发现这些铁陨石的 Cr 同位素组成变化极大，ε^{53}Cr 的变化范围为（-0.04±0.44）~（268.29±0.14），ε^{54}Cr 的变化范围为（0.28±0.72）~（1 053.78±0.72）。ε^{53}Cr 和 ε^{54}Cr 之间存在很好的正相关关系，其最佳拟合线为 ε^{54}Cr =（3.90±0.03）×ε^{53}Cr，指示铁陨石中的 Cr 是铁陨石自生 Cr 与宇宙成因 Cr 两端元混合的结果。通过监测铁陨石中宇宙成因 ε^{54}Cr 的变化和模拟参数，计算了宇宙辐射效应对样品中 ε^{53}Cr 的影响，给出了铁陨石中 Cr 同位素组成与控制因素之间的函数关系。

月球表面长期暴露在宇宙射线辐射影响之下，月球样品普遍具有较高的辐射年龄，其原始同位素组成也很容易被改变。秦礼萍等（2017）和刘佳（2019）系统研究了三块阿波罗样品和 13 块月球陨石样品的 Cr 同位素组成。实验结果表明月球样品的 Cr 同位素组成存在显著的变化（ε^{53}Cr 约有 0.4ε 的变化，ε^{54}Cr 约有 1.0ε 的变化），大部分月球样品的 ε^{53}Cr 和 ε^{54}Cr 具有正相关关系（斜率约为 2.7），该相关关系也是由于宇宙射线辐射引起的。然而，月球样品中 ε^{53}Cr-ε^{54}Cr 的斜率低于铁陨石，可能反映了陨石组成的差异（石陨石与铁陨石）或辐射模型的差异。四块月球冲击熔融角砾岩陨石的 Cr 同位素组成偏离正相关关系线，可能是由普通球粒陨石（OC）或顽火辉石球粒陨石（EC）碰撞体的混染引起的。为了校正月球样品中 Cr 同位素受宇宙射线辐射的影响，我们用化学方法分离了阿波罗样品中具有较低 Fe/Cr 值的尖晶石相（尖晶石、铬铁矿等）和较高 Fe/Cr 值的非尖晶石相（辉石、橄榄石等），并分析了各相的 Cr 同位素组成。据对比分析，获得了月球受辐射前的 Cr 同位素组成，ε^{53}Cr 为（0.01±0.03）、ε^{54}Cr 为（0±

0.05)。校正后的结果对月碰撞体的组成及成月大碰撞过程的制约是有意义的。

6.4 金属稳定同位素对月球演化过程的制约

作为重要的示踪手段,金属稳定同位素在制约月球岩浆演化、核幔分异和月球大碰撞过程研究中发挥着重要作用。因此,国内利用 Ca、Cu、Zn 等金属稳定同位素对月球重要科学问题开展了研究。

我国学者研究了月球 Ca 同位素,为月球岩浆演化模式提供了重要约束。Huang 等(2019)根据同位素分馏系数计算模拟月球岩浆洋冷却结晶过程中的钙同位素演化。结果表明,月球斜长岩壳的 $\delta^{44/40}$Ca 比平均月球低 0.09‰~0.11‰,斜长岩月壳和月幔间的 Ca 同位素组成差异可达 0.26‰~0.33‰,证明固体月球可能存在显著的 Ca 同位素不均一。Wu 等(2020)基于两块月球玄武岩、三块长石角砾岩及长石单矿物的 Ca 同位素研究首次精确地限定了月壳、月幔和全硅酸盐月球的 Ca 同位素组成。这一研究结果表明,月球整体具有和地球、火星类似的 Ca 同位素组成,这进一步为地-月同源学说及内太阳系 Ca 同位素的均一提供了重要信息。

铜(Cu)和锌(Zn)同位素组成有潜力来示踪月球大碰撞的挥发事件和月球核幔分异过程。Xia 等(2019)采用高温高压实验岩石学的方法,精确测定了液体金属、硫化物和硅酸盐熔体之间的 Cu 和 Zn 同位素平衡分馏系数。实验测定的铜和锌同位素分馏表明,含碳、铁熔体的同位素比硅酸盐熔体重,硫化物熔体最轻。这说明,在逐渐冷却的月球岩浆海洋的硫化物分离过程中,铜强烈分离,进入到硫化物中(100<D_{Cu}硫化物/熔体<200),而使月球硅酸盐亏损 Cu,使 Cu 同位素变得更重。而月球硅酸盐 Zn 同位素表现出重的现象,则可以解释为 Zn 挥发性强,是在月球表面蒸发所致。这为月核的成分和核幔分异过程提供约束。

7 南极陨石研究进展

1969 年日本南极考察队在南极 Yamato 地区发现了 9 块不同类型的陨石之后,人们开始意识到南极存在陨石富集机制(Yoshida et al., 1971)。在此后半个多世纪,在南极发现 50 多个陨石富集区,收集到 5 万多块陨石(缪秉魁,2015)。自 1998 年至今,我国共组织了 7 次格罗夫山考察,成功发现了我国首个南极陨石富集区,收集了 12665 块陨石,使我国成为仅次于美日的南极陨石大国(表 1)(缪秉魁等,2012;Xia et al., 2016)。缪秉魁等(2012)对十年前南极陨石的研究做了较为系统的综述。本节将总结 2010—2020 年间我国在南极陨石考察与陨石库建设方面的工作进展。

表 1 我国南极格罗夫山陨石收集信息数据表

考察次序	南极科考队次	年份	参与人数	收集数量/块	总重量/kg	平均重量/g	分布区域
第 1 次	第 15 次队	1998—1999	4	4	0.53	132.5	阵风悬崖北段、中段
第 2 次	第 16 次队	1999—2000	10	28	0.59	21.1	阵风悬崖南段
第 3 次	第 19 次队	2002—2003	9	4 448	43.47	9.77	阵风悬崖全段,1 号和 4 号冰碛带,天问碎石带
第 4 次	第 22 次队	2005—2006	12	5 354	62.10	11.6	阵风悬崖全段、戴维角峰和撒哈洛夫岭东侧
第 5 次	第 26 次队	2009—2010	10	1 618	17.14	10.6	1、2、3、4 号碎石带,梅森峰南侧,撒哈洛夫岭,天问碎石带
第 6 次	第 30 次队	2013—2014	10	583	2.18	3.74	阵风悬崖中段,1、2、4 号碎石带,天问碎石带
第 7 次	第 32 次队	2015—2016	10	630	1.71	2.71	阵风悬崖北段、中段,1、2、4 号碎石带,撒哈洛夫岭东侧,天问碎石带
合计	7		65	12 665	127.72	10.1	

近十年中,我国在南极开展了两次陨石考察(2013—2014 年、2015—2016 年),分别收集到 583 块和

630 块陨石（表1）（缪秉魁，2015；夏志鹏等，2018；陈宏毅等，2020）。在 2018—2019 年，我国第 35 次南极科考队尝试性地在南极中山站所属的拉斯曼丘陵布洛克内斯半岛开展陨石考察，结果没有发现陨石，该区不具备陨石富集特征（夏志鹏和缪秉魁，2019）。在南极微陨石收集方面，我国共开展了两次采集：① 2015—2016 年，在格罗夫山地区进行南极陨石考察时，通过融化少量的表层雪和冰，收集了其中的残余物；② 2018—2019 年，在拉斯曼丘陵采集了约 60 kg 的湖底沉积物和少量冰雪残余物（夏志鹏等，2018；夏志鹏和缪秉魁，2019）。然而，至今未见与这些沉积物和残余物相关的研究报道，尚不清楚残余物中微陨石的数量及特征。

南极陨石的大量发现表明南极冰盖存在陨石富集机制。Cassidy 等（1992）提出了陨石富集的冰流模式，即：含有陨石的冰盖在向海洋流动的过程中被山脉阻挡，冰流慢慢抬升，冰在气温和下降风的作用下产生消融，冰盖中的陨石逐步暴露并富集起来（Cassidy et al.，1992）。格罗夫山地区有 64 座冰原岛峰和大面积出露的蓝冰区，地势上东南高西北低，加上强烈的下降风作用等条件，为南极陨石富集提供了优越的条件。然而，格罗夫山地区发现的大部分陨石位于阵风悬崖冰流的下方，被认为是冰流的"之"字形回流或强风搬运的结果（缪秉魁等，2012；陈宏毅等，2020）。此外，2015—2016 年，格罗夫山考察队对蓝冰消融率和冰流速度开始进行监测，计划采集五年的数据，并建立格罗夫山陨石富集规律的详细模型（陈宏毅等，2020）。

由于大量陨石的收集，陨石分类也成为陨石研究中最基础性的工作，为深入开展综合研究打下基础。2008 年前，国内九家单位共同对 2 433 块南极陨石开展了分类研究，发现了火星陨石、HED 族陨石等一批特殊陨石，建立了我国陨石分类研究的工作程序、分类标准，制定了《中国南极陨石的管理、申请及使用规定》《中国南极陨石样品收集与保存实施细则》等一系列标准。2012 年起，在科技部国家自然科技资源平台建设项目的支持下，继续开展了 3 060 块陨石分类工作（表2）。截至 2020 年 12 月，已经分类的南极陨石共 5 493 块，其中提交国际陨石命名的有 4 017 块，暂未提交命名的陨石可在中国极地标本资源共享平台中检索（http://birds.chinare.org.cn/index/）。综合 2008 年前的分类结果，我国共发现非普通球粒陨石 80 块，类型有火星陨石、HED 族陨石、原始无球粒陨石、铁陨石、中铁陨石、橄榄陨铁、碳质球粒陨石等（表3）。

表 2 历次南极格罗夫山陨石分类情况

年份	数量/块	参加单位
2000	4	北京大学、中科院地质与地球物理研究所
2001—2002	28	中科院广州地化所、中科院地质与地球物理研究所、中科院国家天文台、南京大学
2003—2004	51	
2006	600	中科院地质与地球物理研究所、中科院广州地球化学研究所、桂林理工大学、中科院国家天文台、中科院地球化学研究所、中科院紫金山天文台、南京大学和北京天文馆
2007	800	
2008	950	
2012	60	
2013	90	
2014	150	
2015	200	
2016	250	桂林理工大学
2017	250	
2018	600	
2019	860	
2020	600	
合计	5 493	

表 3 已分类南极格罗夫山陨石类型

陨石类型		数量/块	百分比/%
普通球粒陨石	3 型	80	2.33
	合计	5 413	98.50
碳质球粒陨石	CK	1	0.03
	CM	8	0.23
	CO	1	0.03
	CR	4	0.12
	CV	7	0.20
	合计	21	0.61
顽辉石球粒陨石	EH	1	0.03
	合计	2	0.06
火星陨石		2	0.06
橄辉无球粒陨石		11	0.32
HED 族陨石	Eucrites	3	0.09
原始无球粒陨石	Acapulcoite	1	0.03
	Winonaite	1	0.03
铁陨石		4	0.12
橄榄陨铁		1	0.03
中铁陨石		12	0.35
总计	—	5 493	100.00

8 展 望

由于国家空间科学战略的实施,最近十年是我国快速迈步太空和行星科学大发展的时代,我国嫦娥五号采样返回为标志性成果。在此形势下,陨石学与天体化学作为行星科学的重要专业基础,也得到快速发展并取得了许多重要成果,其中包括不同类型的陨石成因、球粒陨石以及早期太阳星云、月球陨石以及月球演化特征、火星陨石的水、稳定金属同位素的分析技术等领域。然而,陨石学与天体化学发展方面也存在一些不足或问题。针对这些问题和国家需求,下文提出一些建议和展望:

(1) 在科学技术发展日新月异的今天,迈向太空已成趋势,空间科学已经成为世界前沿科学领域,陨石学与天体化学已成为重要的基础学科。为了应对国家空间科学战略需求,国内各大高校和中国科学院也大力发展空间科学和行星科学,陨石学与天体化学的专业需求非常强劲,但是,我国目前天体化学专业力量还是比较薄弱,人才队伍相对日美发达国家还是非常小,因此,建议我国应该尽快加大天体化学人才培养,在部分具备基本条件的高校开设天体化学本科专业,为我国未来空间科学发展做好人才储备。

(2) 我国在世纪之交发现了南极格罗夫山陨石富集区,收集了上万块陨石,目前已经开展完成了5 493 块陨石的分类研究工作,为南极陨石库提供了基础数据。然而,当前仍有超过 7 000 余块南极格罗夫山陨石尚未开展任何基础分类工作,这些样品中极有可能蕴藏有丰富的稀有特殊陨石如火星陨石和月球陨石等,故而严重制约了我国陨石学与天体化学的发展。此外,与美国和日本南极陨石搜寻相比,我国南极陨石质量偏小,特殊类型陨石占比也偏低,且仅有的格罗夫山陨石富集区的陨石资源已接近枯竭。因此,建议我国未来继续开展南极陨石搜寻考察,发掘出新的南极陨石富集区,同时加快南极陨石分类

（3）深空探测是国家重大战略工程，我国已成功实施了嫦娥探月计划以及天问一号火星探测计划，并采集了月球样品。与之相比，火星陨石与月球陨石易获取、成本低，且代表了更广泛区域的样品，对它们开展研究不仅可以更好地了解火星与月球演化历史，积累必要的基础数据、实验技能和人才队伍，同时也为我国未来开展各类深空探测工程提供了不可或缺的知识背景。因此，建议未来加强火星陨石和月球陨石研究工作，迎合国家战略需求。

（4）类地行星及小行星形成演化是我国行星科学与天体化学下一步的研究重点。通过研究反映类地行星初始构成物质的球粒陨石和记载了类地行星后期演化历史的分异型无球粒陨石，目前已在类地行星形成和演化领域取得了大量成果。然而，对于记录了类地行星初始构成物质向原始行星转变历史的原始无球粒陨石而言，在成因、母体结构等方面仍存在严重争议与不足。因此，建议我国未来加强对原始无球粒陨石的研究力度，特别是对于南极格罗夫山收集到的原始无球粒陨石。

致谢：感谢中国科学院月球与深空探测重点实验室开放基金资助项目（LDSE201907）和广西科技基地及人才专项（桂科AD1850007）的支持。

参 考 文 献

陈宏毅,缪秉魁,谢兰芳,夏志鹏. 2015. 南极月球陨石 MIL05035 矿物学、岩石学及演化历史. 岩石学报,31(4):1171-1182
陈宏毅,缪秉魁,谢兰芳,黄丽霖. 2016. HED族陨石:分异型小行星物质组成和演化. 矿物岩石地球化学通报,35(5):1037-1052
陈宏毅,缪秉魁,夏志鹏,谢兰芳,赵斯哲. 2020. 南极格罗夫山陨石收集、研究进展和富集机制. 极地研究,32(4):417-434
戴德求,包海梅,刘爽,尹锋. 2020. Kainsaz(CO3)陨石中两个富Al球粒的氧同位素组成特征与形成演化. 岩石学报,36(6):1850-1856
耿言,周继时,李莎,付中梁,孟林智,刘建军,王海鹏. 2018. 我国首次火星探测任务. 深空探测学报,5(5):399-405
贺怀宇,王英,邓成龙,朱日祥. 2010. 月壤中的稀有气体. 地球化学,39(2):123-130
蒋云,徐伟彪,Guan Y B,王英. 2015. 普通球粒陨石中富铝球粒的成因:离子探针氧同位素证据. 中国科学:地球科学,45(9):1324-1334
林杨挺,缪秉魁,徐琳,胡森,冯璐,赵旭晁,杨晶. 2013. 陨石学与天体化学(2001—2010)研究进展. 矿物岩石地球化学通报,32(1):40-55
林杨挺,赵旭晁,徐于晨,胡森,杨晶. 2014. 球粒陨石与太阳星云演化. 见:第十一届"月球•行星•科学与探测"学术研讨会论文集. 贵阳:中国空间科学学会,101
刘佳. 2019. 陨石中铬同位素的宇宙射线辐射效应. 合肥:中国科学技术大学博士学位论文
缪秉魁. 2015. 格罗夫山陨石考察现状及其发展设想. 矿物岩石地球化学通报,34(6):1081-1089
缪秉魁,林杨挺,王道德,欧阳自远. 2012. 我国南极陨石收集进展(2000—2010). 矿物岩石地球化学通报,31(6):565-574
秦礼萍. 2015. 金属元素同位素异常. 矿物岩石地球化学通报,34(4):731-743
秦礼萍,夏九星. 2014. 陨石中Cr稳定同位素组成. 2014年中国地球科学联合学术年会——专题38:非传统稳定同位素的理论、分析和应用论文集. 北京:中国地球物理学会,3
秦礼萍,刘佳,何永胜. 2017. 月球陨石的铬同位素组成. 见:中国矿物岩石地球化学学会第九次全国会员代表大会暨第16届学术年会文集. 西安:中国矿物岩石地球化学学会,1159
王道德,王桂琴. 2012. 陨石学及天体化学研究某些新进展. 矿物学报,32(3):321-340
王英,贺怀宇,张川统,苏菲,Ranjith P M,马严,郑德文. 2018. 微量陨石激光熔样稀有气体测定方法. 岩石学报,34(11):3455-3466
吴蕴华,邢巍凡,徐伟彪. 2014. 球粒陨石难熔包体中贵金属的矿物岩石学分析:对太阳星云演化的启示. 天文学报,55(2):105-115
夏志鹏,缪秉魁. 2019. 南极拉斯曼丘陵布洛克内斯半岛陨石与微陨石收集. 见:中国矿物岩石地球化学学会第17届学术年会论文摘要集. 杭州:中国矿物岩石地球化学学会,1241-1242
夏志鹏,缪秉魁,陈宏毅,姚杰,谢兰芳. 2013. 南极月球陨石 EET 96008 矿物学和岩石学特征. 极地研究,25(4):352-361
夏志鹏,缪秉魁,张川统,黄丽霖. 2018. 极地微陨石的收集、研究与设想. 极地研究,30(3):314-328
谢兰芳,缪秉魁,陈宏毅,夏志鹏,姚杰. 2013. 一块新发现月球陨石 MIL090036 的岩相学和矿物学. 极地研究,25(4):342-351
谢兰芳,陈宏毅,缪秉魁,夏志鹏,邵慧敏. 2019. GRV 13100:一块在南极新发现的顽火辉石球粒陨石. 极地研究,31(2):168-178
谢兰芳,陈宏毅,缪秉魁. 2020. GRV 13100 顽火辉石球粒陨石中金属硫化物特征与成因. 矿物学报,40(1):83-91
姚杰,缪秉魁,陈宏毅,谢兰芳,夏志鹏. 2013. 南极 MIL090070 月球陨石的岩石矿物学特征. 极地研究,25(4):329-341
张川统,贺怀宇,缪秉魁. 2018. 月球陨石稀有气体和宇宙暴露年龄研究进展与展望. 矿物岩石地球化学通报,37(4):588-600,638
周剑凯,陈宏毅,谢兰芳,缪秉魁,仲艳. 2019. 一块新发现的月球陨石 NWA 12279 的岩石矿物学、源区和冲击变质作用. 岩石矿物学杂志,38(4):521-534
Barrat J A, Yamaguchi A, Bunch T E, Bohn M, Bollinger C, Ceuleneer G. 2011. Possible fluid-rock interactions on differentiated asteroids recorded in eucritic meteorites. Geochimica et Cosmochimica Acta,75(13):3839-3852

Cassidy W, Harvey R, Schutt J, Delisle G, Yanai K. 1992. The meteorite collection sites of Antarctica. Meteoritics & Planetary Science, 27(5): 490–525

Chen D L, Zhang A C, Pang R L, Chen J N, Li Y. 2019. Shock-induced phase transformation of anorthitic plagioclase in the eucrite meteorite Northwest Africa 2650. Meteoritics & Planetary Science, 54(7): 1548–1562

Collinet M, Grove T L. 2020. Formation of primitive achondrites by partial melting of alkali-undepleted planetesimals in the inner solar system. Geochimica et Cosmochimica Acta, 277: 358–376

Dai D, Zhou C, Chen X. 2015. Ca-, Al-rich inclusions in two new carbonaceous chondrites from Grove Mountains, Antarctica. Earth, Moon, and Planets, 115(1): 101–114

Day J M D, Walker R J, Ash R D, Liu Y, Rumble III D, Irving A J, Goodrich C A, Tait K, McDonough W F, Taylor L A. 2012. Origin of felsic achondrites Graves Nunataks 06128 and 06129, and ultramafic brachinites and brachinite-like achondrites by partial melting of volatile-rich primitive parent bodies. Geochimica et Cosmochimica Acta, 81: 94–128

Ehlmann B L, Mustard J F, Murchie S L, Bibring J P, Meunier A, Fraeman A A, Langevin Y. 2011. Subsurface water and clay mineral formation during the early history of Mars. Nature, 479(7371): 53–60

Filiberto J, Baratoux D, Beaty D, Breuer D, Farcy B J, Grott M, Jones J H, Kiefer W S, Mane P, McCubbin F M, Schwenzer S P. 2016. A review of volatiles in the Martian interior. Meteoritics & Planetary Science, 51(11): 1935–1958

Gardner-Vandy K G, Lauretta D S, Greenwood R C, McCoy T J, Killgore M, Franchi I A. 2012. The Tafassasset primitive achondrite: insights into initial stages of planetary differentiation. Geochimica et Cosmochimica Acta, 85: 142–159

Herzog G F. 2007. 1.13-cosmic-ray exposure ages of meteorites. Treatise on Geochemistry, 1: 1–36

Hu S, Lin Y, Zhang J, Hao J, Feng L, Xu L, Yang W, Yang J. 2014. NanoSIMS analyses of apatite and melt inclusions in the GRV 020090 martian meteorite: hydrogen isotope evidence for recent past underground hydrothermal activity on mars. Geochimica et Cosmochimica Acta, 140: 321–333

Hu S, Lin Y T, Zhang J C, Hao J L, Xing W F, Zhang T, Yang W, Changela H. 2019. Ancient geologic events on Mars revealed by zircons and apatites from the Martian regolith breccia NWA 7034. Meteoritics & Planetary Science, 54(4): 850–879

Hu S, Lin Y T, Zhang J C, Hao J L, Yamaguchi A, Zhang T, Yang W, Changela H. 2020. Volatiles in the martian crust and mantle: clues from the NWA 6162 shergottite. Earth and Planetary Science Letters, 530: 115902

Huang F, Zhou C, Wang W Z, Kang J T, Wu Z Q. 2019. First-principles calculations of equilibrium Ca isotope fractionation: Implications for oldhamite formation and evolution of lunar magma ocean. Earth and Planetary Science Letters, 510: 153–160

Huang L L, Miao B K, Chen G Z, Shao H M, Ouyang Z Y. 2020. The sulfurization recorded in tridymite in the monomict eucrite Northwest Africa 11591. Meteoritics & Planetary Science, 55(7): 1441–1457

Krot A N, Keil K, Scott E R D, Goodrich C A, Weisberg M K. 2014. 1.1-classification of meteorites and their genetic relationships. Treatise on Geochemistry, 1: 1–63

Kruijer T S, Kleine T, Borg L E. 2020. The great isotopic dichotomy of the early Solar system. Nature Astronomy, 4(1): 32–40

Leshin L A, Mahaffy P R, Webster C R, Cabane M, Coll P, Conrad P G, Archer Jr P D, Atreya S K, Brunner A E, Buch A, Eigenbrode J L, Flesch G J, Franz H B, Freissinet C, Glavin D P, McAdam A C, Miller K E, Ming D W, Morris R V, Navarro-González R, Niles P B, Owen T, Pepin R O, Squyres S, Steele A, Stern J C, Summons R E, Sumner D Y, Sutter B, Szopa C, Teinturier S, Trainer M G, Wray J J, Grotzinger J P, MSL Science Team. 2013. Volatile, isotope, and organic analysis of martian fines with the Mars Curiosity rover. Science, 341(6153): 1238937

Leya I, Begemann F, Weber H W, Wieler R, Michel R. 2004. Simulation of the interaction of galactic cosmic ray protons with meteoroids: On the production of ^{3}H and light noble gas isotopes in isotropically irradiated thick gabbro and iron targets. Meteoritics & Planetary Science, 39(3): 367–386

Li C L, Liu D W, Liu B, Ren X, Liu J J, He Z P, Zuo W, Zeng X G, Xu R, Tan X, Zhang X X, Chen W L, Shu R, Wen W B, Su Y, Zhang H B, Ouyang Z Y. 2019. Chang'E-4 initial spectroscopic identification of lunar far-side mantle-derived materials. Nature, 569(7756): 378–393

Li S J, Wang S J, Bao H M, Miao B K, Liu S, Coulson I M, Li X Y, Li Y. 2011. The antarctic achondrite, Grove Mountains 021663: an olivine-rich winonaite. Meteoritics & Planetary Science, 46(9): 1329–1344

Li S J, Yin Q Z, Bao H M, Sanborn M E, Irving A, Ziegler K, Agee C, Marti K, Miao B K, Li X Y, Li Y, Wang S J. 2018. Evidence for a multilayered internal structure of the chondritic acapulcoite-lodranite parent asteroid. Geochimica et Cosmochimica Acta, 242: 82–101

Liao S Y, Hsu W. 2017. The petrology and chronology of NWA 8009 impact melt breccia: implication for early thermal and impact histories of Vesta. Geochimica et Cosmochimica Acta, 204: 159–178

Lin Y, Kimura M. 2003. Ca-Al-rich inclusions from the Ningqiang meteorite: continuous assemblages of nebular condensates and genetic link to Type B inclusions. Geochimica et Cosmochimica Acta, 67(12): 2251–2267

Lin Y, Shen W, Liu Y, Xu L, Hofmann B A, Mao Q, Tang G Q, Wu F, Li X H. 2012. Very high-K KREEP-rich clasts in the impact melt breccia of the lunar meteorite SaU 169: new constraints on the last residue of the Lunar Magma Ocean. Geochimica et Cosmochimica Acta, 85: 19–40

Lin Y T, Hu S, Feng L, Zhang J C, Hao J L, Xu L. 2012. Petrography and shock metamorphism of the tissint olivine-phyric shergottite. In: 75th Annual Meteoritical Society Meeting. Cairns, Australia: Meteoritics & Planetary Science Supplement, 5131

Lin Y T, El Goresy A, Hu S, Zhang J C, Gillet P, Xu Y C, Hao J L, Miyahara M, Ouyang Z Y, Ohtani E, Xu L, Yang W, Feng L, Zhao X C, Yang J, Ozawa S. 2014a. NanoSIMS analysis of organic carbon from the Tissint Martian meteorite: evidence for the past existence of subsurface organic-bearing fluids on Mars. Meteoritics & Planetary Science, 49(12): 2201–2218

Lin Y T, El Goresy A, Hu S, Zhang J C, Gillet P, Xu Y C, Hao J L, Miyahara M, Ouyang Z Y, Ohtani E, Xu L, Yang W, Feng L, Zhao X C, Yang J, Ozawa S. 2014b. NanoSIMS analysis of organic carbon from the Tissint Martian meteorite: evidence for the past existence of subsurface organic-bearing fluids on Mars. Meteoritics & Planetary Science,49(12):2201-2218

Liu D Y,Jolliff B L,Zeigler R A,Korotev R L,Wan Y S,Xie H Q,Zhang Y H,Dong C Y,Wang W. 2012. Comparative zircon U-Pb geochronology of impact melt breccias from Apollo 12 and lunar meteorite SaU 169,and implications for the age of the Imbrium impact. Earth and Planetary Science Letters,319-320:277-286

Liu J,Qin L P,Xia J X,Carlson R W,Leya I,Dauphas N,He Y S. 2019. Cosmogenic effects on chromium isotopes in meteorites. Geochimica et Cosmochimica Acta,251:73-86

Liu Y,Ma C,Beckett J R,Chen Y,Guan Y B. 2016. Rare-earth-element minerals in martian breccia meteorites NWA 7034 and 7533:implications for fluid-rock interaction in the martian crust. Earth and Planetary Science Letters,451:251-262

McSween Jr H Y,Mittlefehldt D W,Beck A W,Mayne R G,McCoy T J. 2011. HED meteorites and their relationship to the geology of vesta and the dawn mission. Space Science Reviews,163(1):141-174

McSween Jr H Y,Binzel R P,de Sanctis M C,Ammannito E,Prettyman T H,Beck A W,Reddy V,Le Corre L,Gaffey M J,McCord T B,Raymond C A,Russell C T,the Dawn Science Team. 2013. Dawn:the Vesta-HED connection:and the geologic context for eucrites,diogenites,and howardites. Meteoritics & Planetary Science,48(11):2090-2104

Miao B K,Chen H Y,Xia Z P,Yao J,Xie L F,Ni W J,Zhang C T. 2014. Lunar meteorites:witnesses of the composition and evolution of the Moon. Advances in Polar Science,25(2):61-74

Miao B K,Xia Z P,Zhang C T,Ou R L,Sun Y L. 2018. Progress of Antarctic meteorite survey and research in China. Advances in Polar Science,29(2):61-77

Misawa K,Yamaguchi A,Kaiden H. 2005. U-Pb and ^{207}Pb-^{206}Pb ages of zircons from basaltic eucrites:implications for early basaltic volcanism on the eucrite parent body. Geochimica et Cosmochimica Acta,69(24):5847-5861

Mittlefehldt D W. 2008. Appendix:Meteorites-a brief tutorial. Reviews in Mineralogy and Geochemistry,68(1):571-590

Mittlefehldt D W. 2015. Asteroid(4)Vesta:I. the howardite-eucrite-diogenite(HED)clan of meteorites. Geochemistry,75(2):155-183

Pang R L,Zhang A C,Wang R C. 2017. Complex origins of silicate veinlets in HED meteorites:a case study of Northwest Africa 1109. Meteoritics & Planetary Science,52(10):2113-2131

Pang R L,Harries D,Pollok K,Zhang A C,Langenhorst F. 2018. Vestaite,(Ti^{4+}Fe^{2+})Ti$_3^{4+}$O$_9$,a new mineral in the shocked eucrite Northwest Africa 8003. American Mineralogist,103(9):1502-1511

Papike J J. 1998. Comparative planetary mineralogy:chemistry of melt-derived pyroxene,feldspar,and olivine. In:29th Annual Lunar and Planetary Science Conference. Houston

Prinz M,Nehru C E,Delaney J S,Weisberg M. 1983. Silicates in IAB and IIICD irons,winonaites,lodranites an brachina:a primitive and modified-primitive group. In:Fourteenth Lunar and Planetary Science Conference. Houston,616-617

Prissel T C,Parman S W,Jackson C R M,Rutherford M J,Hess P C,Head J W,Cheek L,Dhingra D,Pieters C M. 2014. Pink Moon:the petrogenesis of pink spinel anorthosites and implications concerning Mg-suite magmatism. Earth and Planetary Science Letters,403:144-156

Qin L P,Carlson R W. 2016. Nucleosynthetic isotope anomalies and their cosmochemical significance. Geochemical Journal,50(1):43-65

Qin L P,Alexander C M O,Carlson R W,Horan M F,Yokoyama T. 2010. Contributors to chromium isotope variation of meteorites. Geochimica et Cosmochimica Acta,74(3):1122-1145

Qin L P,Carlson R W,Alexander C M O. 2011a. Correlated nucleosynthetic isotopic variability in Cr,Sr,Ba,Sm,Nd and Hf in Murchison and QUE 97008. Geochimica et Cosmochimica Acta,75(24):7806-7828

Qin L P,Nittler L R,Alexander C M O,Wang J,Stadermann F J,Carlson R W. 2011b. Extreme ^{54}Cr-rich nano-oxides in the CI chondrite Orgueil-implication for a late supernova injection into the solar system. Geochimica et Cosmochimica Acta,75(2):629-644

Ranjith P M,He H Y,Miao B K,Su F,Zhang C T,Xia Z P,Xie L F,Zhu R X. 2017. Petrographic shock indicators and noble gas signatures in a H and an L chondrite from Antarctica. Planetary and Space Science,146:20-29

Ranjith P M,He H,Smith T M,Su F,Lin Y,Zhu R. 2019. Noble gas components in the lunar meteorite Northwest Africa 10203. In:82nd Annual Meeting of The Meteoritical Society 2019. Sapporo,6175

Rumble III D,Irving A J,Bunch T E,Wittke J H,Kuehner S M. 2008. Oxygen isotopic and petrological diversity among Brachinites NWA 4872,NWA 4874,NWA 4882 and NWA 4969:How many ancient parent bodies? In:Lunar and Planetary Science Conference XXXIX(2008). Houston,1974

Touboul M,Kleine T,Bourdon B,van Orman J A,Maden C,Zipfel J. 2009. Hf-W thermochronometry:II. accretion and thermal history of the acapulcoite-lodranite parent body. Earth and Planetary Science Letters,284(1-2):168-178

Wang H P,Weiss B P,Bai X N,Downey B G,Wang J,Wang J J,Suavet C,Fu R R,Zucolotto M E. 2017. Lifetime of the solar nebula constrained by meteorite paleomagnetism. Science,335(6325):623-627

Wang S Z,Zhang A C,Pang R L,Li Y,Chen J N. 2019. Possible records of space weathering on Vesta:case study in a brecciated eucrite Northwest Africa 1109. Meteoritics & Planetary Science,54(4):836-849

Wang Y,Hsu W,Guan Y B,Li X H,Li Q L,Liu Y,Tang G Q. 2012. Petrogenesis of the Northwest Africa 4734 basaltic lunar meteorite. Geochimica et Cosmochimica Acta,92:329-344

Wang Y,He H Y,Leya I,Ranjith P M,Su F,Stephenson P C,Zhang C T,Zheng D W. 2020. The noble gases in five ordinary chondrites from Grove Mountains in Antarctica. Planetary and Space Science,192:105045

Warren P H,Rubin A E,Isa J,Gessler N,Ahn I,Choi B G. 2014. Northwest Africa 5738:multistage fluid-driven secondary alteration in an extraordinarily evolved eucrite. Geochimica et Cosmochimica Acta,141:199-227

Webster C R,Mahaffy P R,Flesch G J,Niles P B,Jones J H,Leshin L A,Atreya S K,Stern J C,Christensen L E,Owen T,Franz H,Pepin R O,Steele A,the MSL Science Team. 2013. Isotope ratios of H,C,and O in CO_2 and H_2O of the martian atmosphere. Science,341(6143):260-263

Weisberg M K,McCoy T J,Krot A N. 2006. Systematics and evaluation of meteorite classification. In:McSween H Y,McSween Jr H Y,Binzel R P (eds). Meteorites and the Early Solar System II. Tucson:University of Arizona Press,19-52

Wittmann A,Korotev R L,Jolliff B L,Carpenter P K. 2019. Spinel assemblages in lunar meteorites Graves Nunataks 06157 and Dhofar 1528:implications for impact melting and equilibration in the Moon's upper mantle. Meteoritics & Planetary Science,54(2):379-394

Wu W,Xu Y G,Zhang Z F,Li X. 2020. Calcium isotopic composition of the lunar crust,mantle,and bulk silicate Moon:a preliminary study. Geochimica et Cosmochimica Acta,270:313-324

Xia Y,Kiseeva E S,Wade J,Huang F. 2019. The effect of core segregation on the Cu and Zn isotope composition of the silicate Moon. Geochemical Perspectives Letters,12:12-17

Xia Z P,Zhang J,Miao B K,Ou R L,Xie L F,Yang R,Jing Y. 2016. Meteorite classification for building the Chinese Antarctic meteorite depository—introduction of the classification of 500 Grove Mountains meteorites. Advances in Polar Science,27(1):56-63

Xie L F,Chen H Y,Miao B K,Xia Z Q,Yao J. 2014. Petrography and mineralogy of new lunar meteorite MIL 090036. Advances in Polar Science,25(1):17-25

Xu X Q,Hui H J,Chen W,Huang S C,Neal C R,Xu X S. 2020. Formation of lunar highlands anorthosites. Earth and Planetary Science Letters,536:116138

Xue Y L,Li C H,Qi Y H,Zhang C T,Miao B K,Huang F. 2018. The vanadium isotopic composition of L ordinary chondrites. Acta Geochimica,37(4):501-508

Yoshida M,Ando H,Omoto K,Naruse R,Ageta Y. 1971. Discovery of meteorites near Yamato Mountains,East Antarctica. Antarctic Record,39(39):62-65

Zeng X J,Li S J,Leya I,Wang S J,Smith T,Li Y,Wang P. 2018. The Kumtag 016 L5 strewn field,Xinjiang Province,China. Meteoritics & Planetary Science,53(6):1113-1130

Zeng X J,Shang Y L,Li S J,Li X Y,Wang S J,Li Y. 2019. The layered structure model for winonaite parent asteroid implicated by textural and mineralogical diversity. Earth,Planets and Space,71(1):38

Zhang A C,Hsu W B,Li X H,Ming H L,Li Q L,Liu Y,Tang G Q. 2011. Impact melting of lunar meteorite Dhofar 458:evidence from polycrystalline texture and decomposition of zircon. Meteoritics & Planetary Science,46(1):103-115

Zhang A C,Wang R C,Hsu W B,Bartoschewitz R. 2013. Record of S-rich vapors on asteroid 4 Vesta:sulfurization in the Northwest Africa 2339 eucrite. Geochimica et Cosmochimica Acta,109:1-13

Zhang A C,Itoh S,Sakamoto N,Wang R C,Yurimoto H. 2014. Origins of Al-rich chondrules:clues from a compound Al-rich chondrule in the Dar al Gani 978 carbonaceous chondrite. Geochimica et Cosmochimica Acta,130:78-92

Zhang A C,Ma C,Sakamoto N,Wang R C,Hsu W B,Yurimoto H. 2015. Mineralogical anatomy and implications of a Ti-Sc-rich ultrarefractory inclusion from Sayh al Uhaymir 290 CH3 chondrite. Geochimica et Cosmochimica Acta,163:27-39

Zhang A C,Bu Y F,Pang R L,Sakamoto N,Yurimoto H,Chen L H,Gao J F,Du D H,Wang X L,Wang R C. 2018. Origin and implications of troilite-orthopyroxene intergrowths in the brecciated diogenite Northwest Africa 7183. Geochimica et Cosmochimica Acta,220:125-145

Zhang A C,Kawasaki N,Bao H M,Liu J,Qin L P,Kuroda M,Gao J F,Chen L H,He Y,Sakamoto N,Yurimoto H. 2020. Evidence of metasomatism in the interior of Vesta. Nature Communications,11(1):1289

Zhang C T,Miao B K,He H Y. 2019. Oxygen isotopes in HED meteorites and their constraints on parent asteroids. Planetary and Space Science,168:83-94

Zhang C T,Miao B K,He H Y,Chen H Y,Ranjith P M,Xie Q L. 2021. A petrologic and noble gas isotopic study of new basaltic eucrite grove mountains 13001 from Antarctica. Minerals,11(3):279

Zhang M M,Lin Y T,Leya I,Tang G Q,Liu Y. 2019. Textural and compositional evidence for *in situ* crystallization of palisade bodies in coarse-grained Ca-Al-rich inclusions. Meteoritics & Planetary Science,54(5):1009-1023

Zhang M M,Lin Y T,Tang G Q,Liu Y,Leya I. 2020. Origin of Al-rich chondrules in CV chondrites:incorporation of diverse refractory components into the ferromagnesian chondrule-forming region. Geochimica et Cosmochimica Acta,272:198-217

Zhao X C,Lin Y T,Yin Q Z,Zhang J C,Hao J L,Zolensky M,Jenniskens P. 2014. Presolar grains in the CM2 chondrite Sutter's Mill. Meteoritics & Planetary Science,49(11):2038-2046

Zhou Q,Herd C D K,Yin Q Z,Li X H,Wu F Y,Li Q L,Liu Y,Tang G Q,McCoy T J. 2013. Geochronology of the Martian meteorite Zagami revealed by U-Pb ion probe dating of accessory minerals. Earth and Planetary Science Letters,374:156-163

Zinner E. 1998. Stellar nucleosynthesis and the isotopic composition of presolar grains from primitive meteorites. Annual Review of Earth and Planetary Sciences,26:147-188

Progresses of Researches on Meteoritics and Cosmochemistry

MIAO Bing-kui[1,2], HU Sen[3], CHEN Hong-yi[1], ZHANG Chuan-tong[1], XIA Zhi-peng[1], HUANG Li-lin[1], XUE Yong-li[1], XIE Lan-fang[1]

1. Institution of Meteorites and Planetary Materials Research, Guilin University of Technology, Guilin 541006;
2. Key Laboratory of Lunar and Deep Space Exploration, Chinese Academy of Sciences, Beijing 100101; 3. Key Laboratory of Earth and Planetary Physics, Institute of Geology and Geophysics, Chinese Academy of Sciences, Beijing 100029

Abstract: In the past decade, the researches on meteoritics and cosmochemistry in China have been developed rapidly. The successful implementation of a series of deep-space exploration projects, such as the Chang'e lunar exploration project (especially the lunar surface sampling and the return of collected lunar samples), as well as the Tianwen-1 Mars exploration project, has greatly promoted the development of meteoritics and cosmochemistry in China. The construction of database of the Grove Mountains' meteorites in Antarctica has provided sufficient samples for researchers to develop the discipline of meteoritics and cosmochemistry. The construction of high-precision isotopic analysis platform has ensured the smooth measurements and researches of the meteorites and returned extraterrestrial samples. Especially, a number of young outstanding astrochemical talents are emerged in China. At the same time, through the researches of various kinds of meteorites, many important achievements have been made in many fields, such as the origin and evolution of solar nebulae, the formation and habitability of terrestrial planets (Mars, Moon, etc.), and the magmatism, water-bearing alteration, and late impact history of asteroids.

Key words: meteoritics; cosmochemistry; solar nebula; terrestrial planets; asteroids; Antarctic meteorites